ASTRONOMY AND ASTROPHYSICS ABSTRACTS

A Publication of the Astronomisches Rechen-Institut Heidelberg
Member of the Abstracting Board of the International
Council of Scientific Unions

Volume 14
Literature 1975, Part 2

Edited by S. Böhme U. Esser W. Fricke
U. Güntzel-Lingner I. Heinrich F. Henn D. Krahn
L. Schmadel H. Scholl G. Zech

Published for
Astronomisches Rechen-Institut
by
Springer-Verlag Berlin Heidelberg New York 1976

Astronomisches Rechen-Institut
Heidelberg
Director: Prof. Dr. W. Fricke

Astronomy and Astrophysics Abstracts
Editor-in-Chief: F. Henn

Astronomy and Astrophysics Abstracts
is prepared under the auspices
of the International Astronomical Union

ISBN 3-540-07784-7 Springer-Verlag Berlin Heidelberg New York
ISBN 0-387-07784-7 Springer-Verlag New York Heidelberg Berlin

Library of Congress Catalog Card Number 72-104650.
Printed in Germany.
Offset printing: Julius Beltz, Hemsbach/Bergstr.
Bookbinding: Konrad Triltsch, Graphischer Betrieb, Würzburg.

Preface

Astronomy and Astrophysics Abstracts, which has appeared in semi-annual volumes since 1969, is devoted to the recording, summarizing and indexing of astronomical publications throughout the world. It is prepared under the auspices of the International Astronomical Union (according to a resolution adopted at the 14th General Assembly in 1970).

Astronomy and Astrophysics Abstracts aims to present a comprehensive documentation of literature in all fields of astronomy and astrophysics. Every effort will be made to ensure that the average time interval between the date of receipt of the original literature and publication of the abstracts will not exceed eight months. This time interval is near to that achieved by monthly abstracting journals, compared to which our system of accumulating abstracts for about six months offers the advantage of greater convenience for the user.

Volume 14 contains literature published in 1975 and received before March 15, 1976; some older literature which was received late and which is not recorded in earlier volumes is also included.

We acknowledge with thanks contributions to this volume by Dr. J. Bouška, who surveyed journals and publications in the Czech language and supplied us with abstracts in English, and by the Commonwealth Scientific and Industrial Research Organization (C.S.I.R.O.), Sydney, for providing titles and abstracts of papers on radio astronomy. We want to acknowledge valuable contributions to this volume by Zentralstelle für Atomkernenergie-Dokumentation, Leopoldshafen, which supported our abstracting service by sending us retrospective literature searches.

It is a pleasure to express our warmest thanks again to Miss Helga Ballmann, Mrs Monika Betz, Mrs Karola Gudé, Miss Lore Kiefert, and Mrs Ingrid Wolf, who typed the text of this volume on IBM 72 Composers and compiled the pages from abstract slips in a perfect form for offset reproduction, to Mrs Elisabeth Feigenbutz for punching material for the author index.

Heidelberg, April 1976

Siegfried Böhme
Ute Esser
Walter Fricke
Ulrich Güntzel-Lingner
Inge Heinrich
Frieda Henn
Dietlinde Krahn
Lutz Schmadel
Hans Scholl
Gert Zech

Contents

Preface
Introduction 1
Abbreviations 3

Periodicals, Proceedings, Books, Activities
001 Periodicals 5
002 Bibliographical Publications 16
003 Books 17
004 History of Astronomy, Chronology 24
005 Biography 29
006 Personal Notes 31
007 Obituaries 32
008 Observatories, Institutes 33
009 Notes on Observatories, Planetaria, and Exhibitions 49
010 Societies, Associations, Organizations 50
011 Reports on Colloquia, Congresses, Meetings, Symposia, and Expeditions . . 54
012 Proceedings of Colloquia, Congresses, Meetings, and Symposia 57
013 Reports on Astronomy in Various Countries and Particular Fields, International Cooperation 61
014 Teaching in Astronomy 62
015 Miscellanea 64

Applied Mathematics, Physics
021 Mathematics, Computing 66
022 Physical Papers Related to Astronomy and Astrophysics 68

Astronomical Instruments and Techniques
031 Astronomical Optics (77), Methods of Observation and Reduction (82), Data Processing, Automation (88) 77
032 Astronomical Instruments (90), Space Instrumentation (93) 90
033 Radio Telescopes and Equipment 97
034 Astronomical Accessories (Spectrometers, Photometers, etc.) 101
035 Clocks and Frequency Standards 110
036 Photographic Auxiliaries 112

Positional Astronomy, Celestial Mechanics

041 Positional Astronomy, Astrometry, Star Catalogues and Atlases 114
042 Celestial Mechanics, Figure of Celestial Bodies 120
043 Astronomical Constants 129
044 Time, Rotation of the Earth 130
045 Latitude Determination, Polar Motion 133
046 Astronomical Geodesy, Satellite Geodesy, Navigation 136
047 Ephemerides, Almanacs, Calendars 140

Space Research

051 Extraterrestrial Research, Spaceflight Related to Astronomy and Astrophysics 142
052 Astrodynamics and Navigation of Space Vehicles 144
053 Lunar and Planetary Probes and Satellites 148
054 Artificial Earth Satellites 150
055 Observations of Earth Satellites, Lunar and Planetary Probes 152

Theoretical Astrophysics

061 General Theoretical Problems of Astrophysics, Gravitational Instability, Neutrino Astronomy, Infrared, X-Ray, Gamma-Ray Astronomy, Abundances and Origin of Elements 154
062 Hydrodynamics, Magnetohydrodynamics, Plasma 161
063 Radiative Transfer, Scattering 166
064 Stellar Atmospheres, Stellar Envelopes, Mass Loss 171
065 Star Formation, Stellar Structure and Evolution, Neutron Stars 178
066 Relativistic Astrophysics (without Cosmology), Background Radiation, Gravitation Theory 193

Sun

071 Solar Photosphere, Spectrum 205
072 Sunspots, Faculae, Solar Activity Cycles 211
073 Solar Chromosphere, Flares, Prominences 217
074 Solar Corona, Solar Wind 229
075 Solar Patrol 240
076 Solar UV, X Rays, Gamma Radiation 242
077 Solar Radio Radiation 248
078 Solar Cosmic Radiation 254
079 Solar Eclipses 257
080 Solar Atmosphere, Figure, Internal Constitution, Neutrinos, Magnetic Fields, Rotation, Miscellanea 259

Earth

081 Figure, Composition, and Gravity of the Earth 265
082 The Earth's Atmosphere Including Refraction, Scintillation, Extinction, Airglow, Site Testing 268
083 Ionosphere 275
084 Aurorae (279), Geomagnetic Field (281), Radiation Belts (288) 279
085 Solar-Terrestrial Relations 289

Planetary System

091 Physics of the Planetary System (Planetary Atmospheres, Figure, Interior, Magnetic Fields, Rotation, etc.) 291
092 Mercury 294
093 Venus 296
094 Moon: Dynamics (300), Global Properties (303), Local Properties (311) . . 300
095 Lunar Eclipses 316
096 Lunar Occultations 317
097 Mars (318), Mars Satellites (322) 318
098 Minor Planets 323
099 Jupiter (328), Jupiter Satellites (334) 328
100 Saturn (337), Saturn Satellites (339) 337
101 Uranus, Neptune, Pluto, Transplutonian Planets 340
102 Comets (Origin, Structure, Atmospheres, Dynamics) 342
103 Comets: Listed Objects 344
104 Meteors, Meteor Streams 353
105 Meteorites, Meteorite Craters 359
106 Interplanetary Matter, Interplanetary Magnetic Field, Zodiacal Light 367
107 Cosmogony of the Planetary System 373

Stars

111 Stellar Parallaxes 375
112 Proper Motions, Radial Velocities, Space Motions 376
113 Stellar Magnitudes, Colors, Photometry 378
114 Stellar Spectra, Temperatures, Spectroscopy (384), Spectra of Individual Stars (392) 384
115 Stellar Luminosities, Masses, Diameters, HR-Diagrams and Others 400
116 Stellar Magnetic Field, Figure, Rotation 402
117 Binary and Multiple Stars, Planetary Companions, Theory 404
118 Visual Double and Multiple Stars 410
119 Spectroscopic Binaries 412
120 Variable Stars: Catalogues, Ephemerides, Miscellanea 414
121 Eclipsing Variables 415
122 Intrinsic Variables, Flare Stars, Pulsation Theory 424
123 Variable Stars: Lists of Observations, Individual Observations 439
124 Novae 442
125 Supernovae, Supernova Remnants 448
126 Low-luminosity Stars, Subdwarfs, White Dwarfs 454

Interstellar Matter, Gaseous Nebulae, Planetary Nebulae

131 Interstellar Matter, Polarization of Starlight (457),
H I, H II Regions (474) 457
132 Emission Nebulae, Reflection Nebulae 480
133 Planetary Nebulae 485
134 Crab Nebula 488

Radio Sources, Quasars, Pulsars, Infrared, X-ray, Gamma-Ray Sources, Cosmic Radiation

141 Radio Sources, Quasars (490), Pulsars (503), Infrared Sources (508) . . . 490
142 X-Ray, Gamma-Ray Sources 512
143 Cosmic Radiation 529

Stellar Systems

151 Kinematics and Dynamics of Stellar Systems 536
152 Stellar Associations 544
153 Galactic Clusters 545
154 Globular Clusters 549
155 Structure and Evolution of the Galaxy 553
156 Galactic Magnetic Field 559
157 Galactic Radio Radiation 560
158 Single and Multiple Galaxies (561), Peculiar Objects (578) 561
159 Magellanic Clouds 579
160 Clusters of Galaxies 581
161 Intergalactic Matter 585
162 Structure and Evolution of the Universe, Cosmology 587

Author Index 595
Subject Index 697

Introduction

Astronomical bibliographies

Astronomy and Astrophysics Abstracts begins documentation and abstracting as from the year 1969. For information on astronomical literature before this date consultation of one of the following bibliographies is suggested:
(1) J. J. de Lalande, Bibliographie Astronomique, Paris 1803 (this work covers the time from 480 B. C. to the year 1803, VIII + 966 pages).
(2) J. C. Houzeau, A. Lancaster, Bibliographie générale de l'astronomie, Volume I (in two parts), Bruxelles 1882, 1887, Volume II, Bruxelles 1889. The complete title of Volume II is "Bibliographie générale de l'astronomie ou catalogue méthodique des ouvrages, des mémoires et des observations astronomiques, publiés depuis l'origine de l'imprimerie jusqu'en 1880". A new edition of these volumes was prepared by D. W. Dewhirst in 1964.
(3) Bibliography of Astronomy, 1881 - 1898. The literature of this period was recorded on standard slips by the Observatoire Royal de Belgique. From the material (some 52,000 items) a microfilm version was produced by University Microfilms Limited, Tylers Green, High Wycombe, Buckinghamshire, England, in 1970.
(4) Astronomischer Jahresbericht, 1899 gegründet von Walter Wislicenus, herausgegeben vom Astronomischen Rechen-Institut in Heidelberg (formerly in Berlin), Verlag W. de Gruyter, Berlin. For the period from 1899 to 1968 sixty-eight volumes were published, each of which, in general, covers the literature of one year.
(5) Bulletin Signalétique – Section 120: Astronomie, Physique Spatiale, Géophysique. Published by Centre de Documentation du Centre National de la Recherche Scientifique, Paris. This publication is a continuation of "Bibliographie Mensuelle de l'Astronomie" founded in 1933 by the Société Astronomique de France. The publication is continued.
(6) Referativnyj Zhurnal. Founded in 1953 and published by Vsesoyuznyj Institut Nauchnoj i Tekhnicheskoj Informatsii, Akademiya Nauk, Moskva. The publication is continued.

Concept of Astronomy and Astrophysics Abstracts

This abstracting service aims to present a comprehensive documentation of the literature in all fields of astronomy and astrophysics. It appears in semi-annual volumes, two of which cover the literature of a calendar year. The half-yearly period of issue is regarded as an optimal period of time for summarizing papers into subject categories and for the presentation of abstracts as quickly as possible after the publication of the original literature. The time limits at which the documentation begins and ends for a volume are not sharply defined, except in the sense that all literature will be covered which was received by the editors within these limits.

Vol. 14 is devoted to the recording, summarizing and indexing of astronomical publications of the year 1975 received from August 16, 1975 to March 15, 1976; it also records a number of papers issued before 1975 but received within the period of time.

The main characteristics of the concept of Astronomy and Astrophysics Abstracts may be summarized briefly.

(1) Titles of papers are given in the language of their authors whenever possible. If they are not in English but supplied with English translations they will be given in English. Abstracts are presented in English, French or German. Titles of papers in Russian are given in English.
(2) Authors' abstracts are used whenever possible. As a rule, popular articles were not abstracted; however their titles are usually given with the notation "Popular article".
(3) As a rule, each paper has been classified into one of 108 numbered subject categories and allocated a serial number within the category. In this way each item is numbered by six figures, the first three of which indicate the number of the category. Three further figures indicate the serial number within the category, which was allocated in the order of the receipt of the abstract. Reference to an abstract in Volume 1 is indicated by "01" before the number of the category; for example, 01.074.028, denotes Volume 1, category 074, abstract 028, Vol. 2 is indicated by "02", etc., Vol. 14 by "14".

A paper may have been classified into more than one category. Then its abstract has been allocated a number in one of the categories involved, and in the other category (or categories) the paper has been indicated by the title and a reference to the abstract number.
Papers whose authors are not named were treated like those with authors' names, with one exception: reports from correspondents of journals whose names were unknown were not numbered.

(4) There are categories which suggest the presentation of the material in subject groups. For instance, a subject group may be formed by all information received on the same solar eclipse, comet, nova, etc. The unsorted presentation of such material in a subject category would be inconvenient for the user, even if the individual comet, etc. were included in the subject index.

The following subject categories are subdivided into subject groups:

008 Observatories, Institutes. The publications of observatories and astronomical institutes are listed in alphabetical order of the towns of the institutions, each town forming a numbered subject group.
010 Societies, Associations, Organizations. The publications of each one form a subject group. The groups are presented in alphabetical order.
079 Solar eclipses. All publications related to one solar eclipse form a subject group.
103 Comets: Listed Objects. All publications related to the same comet form a numbered group.
124 Novae. All publications related to one nova form a subject group.
125 Supernovae. All publications related to one supernova form a subject group.

(5) Border fields of astronomy and astrophysics have been taken into account by presenting titles of papers occasionally without abstracts. The selection of papers for inclusion has been made according to the degree of relevance to astronomical research.

Transliteration of the Russian alphabet

The transliteration of the Russian alphabet in use in Astronomy and Astrophysics Abstracts is presented here.

А	а	a	Р	р	r
Б	б	b	С	с	s
В	в	v	Т	т	t
Г	г	g	У	у	u
Д	д	d	Ф	ф	f
Е	е	e	Х	х	kh
Ё	ё	e	Ц	ц	ts
Ж	ж	zh	Ч	ч	ch
З	з	z	Ш	ш	sh
И	и	i	Щ	щ	shch
Й	й	j	Ъ	ъ	"
К	к	k	Ы	ы	y
Л	л	l	Ь	ь	'
М	м	m	Э	э	eh
Н	н	n	Ю	ю	yu
О	о	o	Я	я	ya
П	п	p			

This transliteration was recommended by the Abstracting Board of the International Council of Scientific Unions in 1969. It is essentially the same as the transliteration proposed by the Academy of Sciences, Moscow, and used by the Referativnyj Zhurnal. It may be noted that the letters can be read and printed by usual data processing machines.
If the names of Russian authors in the literature are transliterated very different from this scheme we present the names in the form in which they are given in the references cited and in addition in round brackets according to our transliteration table.

Sources of information

The majority of sources of information for this volume are given in section **001 Periodicals** and in section **008 Observatories, Institutes.** The term "periodical" has been used in its widest sense for publications in a sequence of undetermined duration, even if the intervals of appearance are not regular. Section 001 records 310 periodicals with their full titles and with abbreviations which are in use in Astronomy and Astrophysics Abstracts. It may be noted that the titles of the periodicals are given in their original languages, and that Russian titles have been transliterated applying the transliteration given above. Section 008 records 138 periodicals; these are publication series of observatories and astronomical institutes which have not been included in section 001. The abbreviations of the titles of the periodicals have been given so that in most cases they permit recognition of the full title without recourse to the key in section 001. The steadily growing number of periodicals makes it necessary to use more extensive abbreviations and to abandon the use of very condensed ones.
Other abstracting journals have been consulted in order to examine the degree of completeness of our service. Occasionally, in particular in Physics Abstracts, Referativnyj Zhurnal, and Bulletin Signalétique abstracts of papers were found which had not come to our attention. In such cases Astronomy and Astrophysics Abstracts cites these paper and gives in general reference to the abstracting service which acted as the source.

Classification into a scheme of subject categories

The subdivision of astronomy and its border fields into subject categories is facilitated by the fact that the astronomical objects appear to be particularly well suited for the formation of categories. Sun, moon, earth, planets, comets, and meteorites, the various kinds of stars, galaxies, radio sources, quasars, and pulsars etc. suggest natural subdivisions. It may be assumed that such subdivisions can be maintained for long periods of time. Experience shows, however, that progress in research may imply changes in the classification scheme, in particular, in fields where the expansion of knowledge is explosive.

A few explanatory remarks may be in order on some of the subject categories. In section 003 books on astronomy and astrophysics and its border fields are listed which came to our notice from August 1975 to March 1976. References to book reviews are given if the review appeared quickly.
For completeness of documentation, personal notes (section 006) and obituaries (section 007) are listed. In section 012 (Proceedings of Colloquia, Congresses, Meetings, and Symposia) the proceedings etc. are listed with titles and editors. The individual papers are classified into their corresponding subject categories, but mostly not included in the subject index. The main subjects of the symposia are cited in the index under section 012.
Errata to papers communicated by the authors are listed at the end of the corresponding subject categories.

Author index and subject index

The subject category and the serial number forming six figures for each abstract have been used as a means of reference in the author index and the subject index. These references are more precise than page references. They offer considerable advantages in indexing by means of data processing machines, and they are more convenient for the user.
The author index of this volume contains 9145 names. A complete reference comprises six figures, three for the subject category and three for the serial number within the category. In the case of more than one reference to abstracts in one category, the number of the category is given only once and not repeated in the immediately following references. The total number of papers (some do not give names of authors) recorded in this volume is about 8130.
We consider the subject index as an approximation to an optimal index covering all fields of astronomy and astrophysics and their border fields. The assigning of one or more key words to a paper is undoubtedly a difficult task. Some journals have started giving key words together with the titles of papers. These key words are chosen by the authors themselves and are in many cases identical with our designations of subject categories with no additional specification. In fact, in some cases it may be more useful to refer to a subject category as a whole than to an item number, in particular, if the total number of abstracts in a category is very small, and if more specific key words do not provide a proper description of the paper.
While each volume is scheduled to contain an author index and a subject index, the magnetic tapes containing the index information will be used to produce separate index volumes (authors and subjects) at intervals of a few years.
The text of the publication was typed on IBM 72 Composers in the editorial office, and it was given to the printer in a form ready for offset reproduction. For the preparation of the indexes a new sorting program has been developed which is based on the IBM SORT-Program. The introduction of small and capital letters in the layout caused some difficulties. Special programs had to code the capital letters into small ones. For the layout a TN chain for a 1403 IBM high speed printer was used. All the programs are written in PL/I. The computations are carried out on an IBM 370/168.

Abbreviations

AAS	American Astronomical Society
AAVSO	American Association of Variable Star Observers
Abh.	Abhandlungen
Abstr.	Abstract
Abt.	Abteilung
Acad.	Academy, etc.
Accad.	Accademia
Adv.	Advances
AG	Astronomische Gesellschaft
AIAA	American Institute of Aeronautics and Astronautics
AJB	Astronomischer Jahresbericht
Akad.	Akademie
An.	Anales, etc.
Ann.	Annals, etc.
Arch.	Archiv, etc.
Ark.	Arkiv
ASA	Astronomical Society of Australia
Asoc.	Asociación
ASP	Astronomical Society of the Pacific
Ass.	Association
ASSA	Astronomical Society of Southern Africa
Astrofis.	Astrofisica, etc.
Astrofiz.	Astrofizika, etc.
Astron.	Astronomy, etc.
Astronaut.	Astronautics, etc.
Astrophys.	Astrophysics, etc.
ASV	Astronomical Society of Victoria
ASWA	Astronomical Society of Western Australia
Atmosph.	Atmosphere, etc.
BA	Bulletin Astronomique
BAA	British Astronomical Association
BAN	Bulletin of the Astronomical Institutes of the Netherlands
Ber.	Berichte
BIH	Bureau International de l'Heure (Paris)
Bol.	Boletin
Boll.	Bolletino
Bull.	Bulletin
Byull.	Byulleten' (Bulletin)
Circ.	Circular
Cl.	Classe
Coll.	Collection
Commun.	Communication
Comun.	Comunicazioni
Contr.	Contributions, etc.
COSPAR	Committee on Space Research
C.S.I.R.O.	Commonwealth Scientific Industrial Research Organization
Dep.	Department
Diss.	Dissertation
Div.	Division
Dokl.	Doklady (Reports)
ESA	European Space Agency
ESO	European Southern Observatory
Fis.	Fisica, etc.
Fiz.	Fizika, etc.
Fys.	Fysica, etc.
Géod.	Géodésie, etc.
Geod.	Geodesy, etc.
Geofis.	Geofisica, etc.
Geofiz.	Geofizika, etc.
Geofys.	Geofysik, etc.
Geol.	Geology, etc.
Geogr.	Geography, etc.
Geophys.	Geophysics, etc.
Ges.	Gesellschaft
Glav.	Glavnyj (Main)
Gos.	Gosudarstvennyj (State)
HRD	Herzsprung-Russell diagram
Hydrogr.	Hydrography, etc.
IAF	International Astronautical Federation
IAU	International Astronomical Union
ICSU	International Council of Scientific Unions
IEEE	Institute of Electrical and Electronics Engineers
Industr.	Industry, etc.
Inform.	Information
Inst.	Institute, etc.
Instn.	Institution
Ionosph.	Ionosphere, etc.
Issled.	Issledovaniya (Research)
Ist.	Istituto
Izv.	Izvestiya (News)
Jb.	Jahrbuch
JO	Journal des Observateurs
Journ.	Journal
Kl.	Klasse
Lab.	Laboratory
Mag.	Magazine
Mat.	Matematica, etc.
Math.	Mathematics, etc.
Mech.	Mechanics, etc.
Med.	Mededelingen
Medd.	Meddelande, Meddelser
Mekhan.	Mekhanika, etc.
Mém.	Mémoires
Mem.	Memoirs, Memorandum, etc.
Meteorol.	Meteorology, etc.
MIT	Massachusetts Institute of Technology
Mitt.	Mitteilungen
MVS Sonneberg	Mitteilungen über Veränderliche Sterne, Sonneberg
Nachr.	Nachrichten
Nat.	Naturwissenschaftlich, etc.
Naut.	Nautics, etc.
NBS	National Bureau of Standards
NRAO	National Radio Astronomy Observatory (Green Bank)
NRL	Naval Research Laboratory (Washington)
NASA	National Aeronautics and Space Administration
Obs.	Observatory, etc.
OSA	Optical Society of America
Oss.	Osservatorio, Osservazioni, etc.
Ped.	Pedagogika, etc. (Pedagogics)
Phil.	Philosophical
Phys.	Physics, etc.
Planet.	Planetary
Priklad.	Prikladnoj (Applied)
Proc.	Proceedings
Progr.	Progress, etc.
Pubbl.	Pubblicazioni
Publ.	Publications
Rap.	Raportoj
RAS	Royal Astronomical Society
RAS Canada	Royal Astronomical Society of Canada
Rech.	Recherches
Rend.	Rendiconti

Rep.	Report
Repr.	Reprint
Res.	Research
Rev.	Review, etc.
Ric.	Ricerche
Roy.	Royal, etc.
SAAO	South African Astronomical Observatory
SAF	Société Astronomique de France
SAI	Società Astronomica Italiana
SAO	Smithsonian Astrophysical Observatory
SAS	Société Astronomique de Suisse
Sci.	Science, etc.
Sect.	Section
Ser.	Series, etc.
S. I. R.	Service International Rapide des Latitudes
Sitzungsber.	Sitzungsberichte
Soc.	Society
Soobshch.	Soobshcheniya (Communications)
Sternw.	Sternwarte
Stud. Cerc.	Studii şi Cercetari
Supl.	Suplemento
Suppl.	Supplement
SuW	Sterne und Weltraum
Techn.	Technics, etc.
Tekhn.	Tekhnika, etc.
Teor.	Teoreticheskij
Terr.	Terrestrial, etc.
TH	Technische Hochschule
Theor.	Theoretical
Tidssk.	Tidsskrift
Trans.	Transactions
Trudy	Trudy (Publications)
Tsentr.	Tsentral'nyj (Central)
Tsirk.	Tsirkulyar (Circular)
TU	Technical University
Uch. Zap.	Uchenye Zapiski (Treatise)
Univ.	University, etc.
URSI	Union Radio Scientifique Internationale
USNO	US Naval Observatory
Verh.	Verhandlungen
Veröff.	Veröffentlichungen
Wet.	Wetenschappen
Wiss.	Wissenschaften, etc.
Zeitschr.	Zeitschrift
ZfA	Zeitschrift für Astrophysik
Zhurn.	Zhurnal (Journal)

Periodicals, Proceedings, Books, Activities

001 Periodicals

AAS Photo-Bull.
AAS (American Astronomical Society) Photo-Bulletin. Published by the Working Group on Photographic Materials. Produced by Eastman Kodak Co., Rochester, N. Y.

Acad. Roy. Belgique, Bull. Cl. Sci.
Académie Royale de Belgique, Bulletin de la Classe des Sciences (Koninklijke Academie van België, Mededelingen van de Klasse der Wetenschappen). 5e Série. Palais des Académies, Bruxelles.

Acta Astron.
Acta Astronomica. An international quarterly journal. Publisher: Polska Akademia Nauk, Komitet Astronomii (Polish Academy of Sciences, Committee of Astronomy), Warszawa – Wrocław.

Acta Astronaut.
Acta Astronautica. Journal of the International Academy of Astronautics. Publisher: Pergamon Press Inc., Elmsford, New York, U.S.A.; Pergamon Press Ltd., Oxford, England.

Acta Astron. Sinica
Acta Astronomica Sinica. Published by Purple Mountain Observatory, Academia Sinica, Nanking, China.

Acta Cosmologica
Acta Cosmologica. Published by Obserwatorium Astronomiczne Universytetu Jagiellońskiego, Kraków, Poland.

Acta Geophys. Sinica
Acta Geophysica Sinica. Chinese Academy of Sciences, Department of Geophysical Research. Published by Science Press, Peking, Peoples Republic of China.

Acta Phys. Austriaca
Acta Physica Austriaca. Publisher: Springer-Verlag, Wien.

Actas Acad. Nacional Cienc. Lima
Actas de la Academia Nacional de Ciencias Exactas, Fisicas y Naturales de Lima. Lima - Peru.

Acta Univ. Carolinae Math. Phys.
Acta Universitatis Carolinae, Mathematica et Physica. Administrace: Matematicko-fyzikálni fakulta University Karlovy, Praha.

Adv. Astron. Astrophys.
Advances in Astronomy and Astrophysics. Publisher: Academic Press, New York – London.

A.F.O.E.V. Bull.
Association Française des Observateurs d'Étoiles Variables. Bulletin. Redaction and publication: M. Émile Schweitzer, 1, rue Beethoven, Strasbourg.

AIAA Journ.
AIAA Journal. A Publication of the American Institute of Aeronautics and Astronautics devoted to Aerospace Research and Development. Published by the American Institute of Aeronautics and Astronautics, New York, N.Y.

American Scient.
American Scientist. Society of Sigma Xi, New Haven, Conn.

Ann. Françaises Chronométrie Micromécanique
Annales Françaises de Chronométrie et de Micromécanique, publication annuelle de l'Observatoire de Besançon, du Centre Technique de l'Industrie Horlogère et de la Société Française de Chronométrie et de Micromécanique. Rédaction et administration: Observatoire de Besançon. Publiées avec le concours du Centre National de la Recherche Scientifique et des organismes corporatifs.

Ann. Géophys.
Annales de Géophysique. Revue Internationale trimestrielle, publiée par le Centre National de la Recherche Scientifique, Paris.

Ann. Obs. Astron. Météorol. Toulouse
Annales de l'Observatoire Astronomique et Météorologique de Toulouse. Publisher: Gauthier-Villars, Paris.

Ann. Physics
Annals of Physics. Publisher: Academic Press Inc., New York, N.Y.

Ann. Physik
Annalen der Physik. 7. Folge. Publisher: Johann Ambrosius Barth, Leipzig.

Ann. Physique
Annales de Physique. Publisher: Masson et Cie., Paris.

Ann. Soc. Sci. Bruxelles
Annales de la Société Scientifique de Bruxelles. Série I: Sciences Mathématiques, Astronomiques et Physiques. Published by Institut de Physique, Heverlé-Louvain.

Annual Rep. Astron. Inst. Greece
Annual Reports of the Astronomical Institutes of Greece. Published by the Greek National Committee for Astronomy. Academy of Athens, Research Center for Astronomy and Applied Mathematics.

Annual Rev. Astron. Astrophys.
Annual Review of Astronomy and Astrophysics. Publisher: Annual Reviews Inc., Palo Alto, California.

Ann. Univ.-Sternw. Wien
Annalen der Universitäts-Sternwarte Wien. In Kommission bei Ferd. Dümmlers Verlag, Bonn.

Anzeiger. Österreich. Akad. Wiss. Math.-Nat. Kl.
Anzeiger. Österreichische Akademie der Wissenschaften. Mathematisch-Naturwissenschaftliche Klasse. Publisher: Springer-Verlag, Wien.

Applied Optics
Applied Optics. A monthly publication of the Optical Society of America. Published for the Optical Society of America by the American Institute of Physics, New York, N. Y.

Arch. Sci. Genève
Archives des Sciences, éditées par la Société de Physique et d'Histoire Naturelle de Genève. Publisher: Imprimerie Kundig, Genève. Subscription address: Librairie Payot, Genève.

Ark. Astron.
Arkiv för Astronomi. Utgivet av Kungliga Svenska Vetenskapsakademien, Stockholm. Printed by Almqvist & Wiksell, Stockholm.

Ark. Geofys.
Arkiv för Geofysik. Kungliga Svenska Vetenskapsakademien, Stockholm. Printed by Almqvist & Wiksell, Stockholm.

Artificial Satellites
Artificial Satellites. Publication of Polish Scientific Institutions. Polish Academy of Sciences, National Committee of Geophysics and Geodesy, National Committee for Space Research, Warsaw. Publishing Office: Palac Kultury i Nauki, Warszawa.

Asoc. Argentina Astron. Bol.
Asociación Argentina de Astronomía. Boletin. Editor: Instituto Argentino de Radioastronomía, Provincia de Buenos Aires, Argentina. Printer: Talleres Gráficos "Renovación", La Plata, República Argentina.

Astrofizika
Astrofizika. Izdatel'stvo Akademii Nauk Armyanskoj SSR, Erevan. [An English translation is published in "Astrophysics".]

Astrofiz. Issled. Izv. Spets. Astrofiz. Obs.
Astrofizicheskie Issledovaniya. Izvestiya Spetsial'noj Astrofizicheskoj Observatorii. Akademiya Nauk SSSR. Publishers: Izdatel'stvo "Nauka", Leningradskoe Otdelenie, Leningrad.

Astron. Astrophys.
Astronomy and Astrophysics. A European Journal. Published by Springer-Verlag, Berlin – Heidelberg– New York.

Astron. Astrophys. Suppl. Ser.
Astronomy and Astrophysics. Supplement Series. A European Journal. Published by the Astronomical Institute Lausanne and Geneva Observatory, Switzerland, on behalf of the Board of Directors.

Astronautik
Astronautik. Organ der Hermann-Oberth-Gesellschaft e.V. Astronautik-Verlag, Druckerei H. Brandt, Delmenhorst (Germany).

Astron. in der Schule
Astronomie in der Schule. Zeitschrift für die Hand des Astronomielehrers. Herausgegeben vom Verlag Volk und Wissen, Berlin. Redaktion: Sternwarte Bautzen.

Astron. Journ.
The Astronomical Journal. Published for the American Astronomical Society by the American Institute of Physics, New York, N. Y. Editorial Office: Department of Astronomy, Columbia University, New York, N. Y.

Astron. Nachr.
Astronomische Nachrichten. Publisher: Akademie-Verlag, Berlin.

Astron. Tidssk.
Astronomisk Tidsskrift. Edited by Astronomisk Selskab, København; Norsk Astronomisk Selskap, Oslo; Svenska Astronomiska Sällskapet, Stockholm. Printed by John Griegs Boktrykkeri, Bergen.

Astron. Tsirk.
Astronomicheskij Tsirkulyar, izdavaemyj Byuro Astronomicheskikh Soobshchenij Akademii Nauk SSSR. Tipografiya Astrosoveta AN SSSR, Moskva.

Astron. Vestn.
Astronomicheskij Vestnik. Publishers: Izdatel'stvo "Nauka", Moskva.

Astron. Zhurn. Akad. Nauk SSSR
Astronomicheskij Zhurnal. Akademiya Nauk SSSR. Publishers: Izdatel'stvo "Nauka", Moskva. [An English translation is published in "Soviet Astronomy AJ"].

Astrophysics
Astrophysics. A cover-to-cover translation of Astrofizika (USSR). Consultants Bureau, New York, N.Y.

Astrophys. Journ.
The Astrophysical Journal. Published for The American Astronomical Society by The University of Chicago Press, Chicago, Illinois.

Astrophys. Journ., (*Letters*)
The Astrophysical Journal. Letters to the Editors. Published for The American Astronomical Society by The University of Chicago Press, Chicago, Illinois.

Astrophys. Journ., Suppl. Ser.
The Astrophysical Journal. Supplement Series. Published for The American Astronomical Society by The University of Chicago Press, Chicago, Illinois.

Astrophys. Letters
Astrophysical Letters. An International *EXPRESS* Journal. Published monthly by Gordon and Breach Science Publishers Ltd., New York – London – Paris.

Astrophys. Space Sci.
Astrophysics and Space Science. An International Journal of Cosmic Physics. Published by D. Reidel Publishing Company, Dordrecht – Holland.

Atti Accad. Nazionale Lincei. Mem.
Atti della Accademia Nazionale dei Lincei. Serie Ottava. Memorie. Classe di Scienze fisiche, matematiche e naturali. Sezione I: Matematica, Meccanica, Astronomia, Geodesia e Geofisica. Published by Accademia Nazionale dei Lincei, Roma.

Atti Accad. Nazionale Lincei. Rend.
Atti della Accademia Nazionale dei Lincei. Serie Ottava. Rendiconti. Classe di Scienze fisiche, matematiche e naturali. Published by Accademia Nazionale dei Lincei, Roma.

Atti Soc. Astron. Italiana
Atti della Società Astronomica Italiana. Publisher: Tipografia Baccini & Chiappi, Firenze (Italy).

Australian Journ. Phys.
Australian Journal of Physics. Published by the Commonwealth Scientific and Industrial Research Organization, East Melbourne, Victoria.

Australian Journ. Phys. Astrophys. Suppl.
Australian Journal of Physics, Astrophysical Supplement. Published by Commonwealth Scientific and Industrial Research Organization, East Melbourne, Victoria.

BAV Rundbrief
BAV Rundbrief. Mitteilungsblatt der Berliner Arbeitsgemeinschaft für Veränderliche Sterne. Editor: BAV Berliner Arbeitsgemeinschaft für Veränderliche Sterne eV., Berlin.

BBSAG Bull.
Bedeckungsveränderlichen Beobachter der Schweizerischen Astronomischen Gesellschaft, [Swiss Astronomical Society's Eclipsing Variable Observers], Bulletin. To be obtained from R. Diethelm, Winterthur, Switzerland.

Bild der Wiss.
Bild der Wissenschaft. Zeitschrift über die Naturwissenschaften und die Technik in unserer Zeit. Publisher: Deutsche Verlagsanstalt, Stuttgart.

Bol. Inst. Mat., Astron., Fis. Univ. Nacional Córdoba
Boletin del Instituto de Matematica, Astronomia y Fisica, Universidad Nacional de Córdoba (R. A.). Dirección General de Publicaciones, Córdoba (Argentina).

Bol. Liga Latinoamericana Astron.
Boletin de la Liga Latinoamericana de Astronomia. Publicado por la Asociacion Argentina Amigos de la Astronomia, Buenos Aires, Argentina.

Boll. Geod. Sci. Affini
Bolletino di Geodesia e Scienze Affini. Pubblicazione dell'Istituto Geografico Militare, Firenze.

British Astron. Ass. Circ.
British Astronomical Association, Circular. Editorial Office: 97 Hawkswood Drive, Hailsham, Sussex.

Bull. American Astron. Soc.
Bulletin of the American Astronomical Society. Published for the American Astronomical Society by the American Institute of Physics Inc., New York, N. Y.

Bull. Astron. Inst. Czechoslovakia
Bulletin of the Astronomical Institutes of Czechoslovakia. Published under the auspices of the Czechoslovak Academy of Sciences by Academia, Praha. Editor: Astronomical Institutes of the Czechoslovak Academy of Sciences, Praha.

Bull. Astron. Soc. India
Bulletin of the Astronomical Society of India. Edited and published by M. S. Vardya, Tata Institute of Fundamental Research, Bombay on behalf of the Astronomical Society of India, Osmania University, Hyderabad.

Bull. Géod.
Bulletin Géodésique, being the Journal of the International Association of Geodesy. Nouvelle Série. Publié par le Bureau Central de l'Association Internationale de Géodésie, Paris.

Bull. Geograph. Survey Inst.
Bulletin of the Geographical Survey Institute. Published by the Geographical Survey Institute, Ministry of Construction, Tokyo, Japan.

Bull. Obs. Astron. Beograd
Bulletin de l'Observatoire Astronomique de Béograd. Editor: Observatoire Astronomique de Béograd. Printed by Naucna delo, Béograd.

Bull. Signal.
Bulletin Signalétique. Section 120: Astronomie, Physique spatiale, Géophysique. Centre de Documentation du Centre Nationale de la Recherche Scientifique, Paris.

Bull. Signal.
Bulletin Signalétique. Bibliographie des Sciences de la Terre. Section 220, Cahier A: Minéralogie, Géochimie, Géologie extraterrestre. Centre de Documentation du C.N.R.S., Paris; Département Documentation du B.R. G.M., Orléans.

Bull. Soc. Roy. Sci. Liège
Bulletin de la Société Royale des Sciences de Liège. L'Université, Liège.

Byull. Abastuman. Astrofiz. Obs.
Abastumanskaya Astrofizicheskaya Observatoriya, Gora Kanobili. Byulleten'. Akademiya Nauk Gruzinskoj SSR. Publishers: Izdatel'stvo "Metsniereba", Tbilisi.

Byull. Stantsij Opitchesk. Nablyud. Iskusstv. Sputnikov Zemli
Byulleten' Stantsij Opticheskogo Nablyudeniya Iskusstvennykh Sputnikov Zemli. Published by Astronomicheskij Sovet Akademii Nauk SSSR, Moskva.
Beginning with number 60 (1971) the title of the publication changed in Nablyudeniya Iskusstvennykh Nebesnykh Tel.

Canadian Journ. Phys.
Canadian Journal of Physics. Published by the National Research Council of Canada, Ottawa. Printed in Canada by the University of Toronto Press, Toronto, Ont.

Celestial Mechanics
Celestial Mechanics. An International Journal of Space Dynamics. Publishers: D. Reidel Publishing Company, Dordrecht–Holland.

Ciel et Terre
Ciel et Terre. Bulletin de la Société Belge d'Astronomie, de Météorologie et de Physique du Globe. Administration: Avenue Circulaire, 3, Bruxelles. Printed by Imprimerie R. Louis, Bruxelles.

Circ. d'Information
Circulaire d'Information. Union Astronomique Internationale. Commission des Etoiles Doubles. Address: Observatoire de Meudon, Meudon, France.

Coelum
Coelum. Periodico bimestrale per la Divulgazione dell' Astronomia. Editor: Osservatorio Astronomico Universitario di Bologna.

Comments Astrophys. Space Phys.
Comments on Astrophysics and Space Physics. A Journal of Critical Discussion of the Current Literature. Comments on Modern Physics: Part C. Publishers: Gordon and Breach Science Publishers, Inc., New York – London

Comptes Rendus Acad. Bulg. Sci.
Comptes Rendus de l'Académie bulgare des Sciences.

(Doklady Bolgarskoj Akademii Nauk). Sofia.

Comptes Rendus Acad. Sci. Paris
Comptes Rendus hebdomadaires des Séances de l'Académie des Sciences, publié avec le concours du Centre National de la Recherche Scientifique. Imprimerie: Gauthier-Villars, Paris.

Contr. Atmosph. Phys.
Contributions to Atmospheric Physics – Beiträge zur Physik der Atmosphäre. Publisher: Friedrich Vieweg & Sohn, Braunschweig.

COSPAR Inform. Bull.
COSPAR. Information Bulletin. Address: COSPAR Secretariat, Paris.

Deutsche Geod. Kommission Bayer. Akad. Wiss.
Deutsche Geodätische Kommission bei der Bayerischen Akademie der Wissenschaften. Reihe A: Höhere Geodäsie; Reihe B: Angewandte Geodäsie; Reihe C: Dissertationen; Reihe D: Tafelwerke; Reihe E: Geschichte und Entwicklung der Geodäsie. Published by Verlag der Bayerischen Akademie der Wissenschaften, München.

Dokl. Akad. Nauk
Doklady Akademii Nauk SSSR. Seriya Matematika, Fizika. Publishers: Izdatel'stvo "Nauka", Moskva.

Dunsink Obs. Publ.
Dunsink Observatory Publications. The Observatory of the School of Cosmic Physics, Dublin Institute for Advanced Studies, Dublin.

Earth Extraterr. Sci.
Earth and Extraterrestrial Sciences. Published by Gordon and Breach Science Publishers, London.

Earth Planet. Sci. Letters
Earth and Planetary Science Letters. A Letter Journal devoted to the Development in Time of the Earth and Planetary System. Publisher: North-Holland Publishing Company, Amsterdam.

Electronics (*USA*)
Electronics. Published by McGraw-Hill Publishing Company, New York, N.Y., U.S.A.

Electronics Letters (*GB*)
Electronics Letters. Published by Institution of Electrical Engineers, London, England.

El Universo
El Universo. Organo de la Sociedad Astronomica de Mexico, Mexico, D. F.

Endeavour
Endeavour. A review of the progress of science, published in four languages by Imperial Chemical Industries Limited, London.

ESO Bull.
European Southern Observatory, Bulletin. Edited by European Southern Observatory. Office of the Director: Hamburg.

ESO Techn. Rep.
European Southern Observatory, (ESO), Technical Report. Published by the European Southern Observatory Telescope Project Division, CERN, Geneva, Switzerland.

Feinwerktechn. & Messtechn.
F & M. Feinwerktechnik und Messtechnik. Fusion of "Feinwerktechnik" and "Messtechnik" (formerly Zeitschrift für Instrumentenkunde) beginning with Jahrgang 82, No. 5 (1974). Publishers: Karl Hanser Verlag, München, Germany.

Gaz. Astron. Mém.
Gazette Astronomique. Mémoires van het Sterrenkundig Genootschap van Antwerpen, (de la Société d'Astronomie d'Anvers), Antwerpen. Printer: «De Voorzorg», A. Van Leuvenhaege, Antwerpen.

General Relativ. Gravitation
General Relativity and Gravitation. Published under the auspices of the International Committee on General Relativity and Gravitation GRG. Publishing Office: Plenum Publishing Company Limited, London.

Geochim. Cosmochim. Acta
Geochimica et Cosmochimica Acta. Journal of the Geochemical Society. Publishing House: Pergamon Press, Ltd., Oxford.

Geodezja Kartografia
Geodezja i Kartografia. Komitet Geodezji Polskiej Akademii Nauk. Publisher: Państwowe Wydawnictwo Naukowe, Warszawa.

Geomagn. Aeronom.
Geomagnetizm i Aehronomiya. Akademiya Nauk SSSR. Izdatel'stvo "Nauka", Moskva [An English translation is published in "Geomagnetism and Aeronomy", American Geophysical Union, Washington, D.C.].

Geophys. Journ. Roy. Astron. Soc.
The Geophysical Journal of the Royal Astronomical Society. Published for the Royal Astronomical Society by Blackwell Scientific Publications, Oxford – Edinburgh.

Geophys. Res. Letters
Geophysical Research Letters. Published monthly by the American Geophysical Union, Washington, D.C., U.S.A.

Gerlands Beiträge Geophys.
Gerlands Beiträge zur Geophysik. Publisher: Akademische Verlagsgesellschaft Geest & Portig K.-G., Leipzig.

Giornale Astron.
Giornale di Astronomia. Pubblicazione trimestrale della Società Astronomica Italiana. Tipolitografia Lodigraf S.p.A. Lodi (Mi).

Giornale dell'A.A.B.
Giornale dell'A.A.B. Notiziario trimestrale delle attività culturali e scientifiche della Associazione Astrofili Bolognesi, Bologna, Italy.

Glasnik Mat.
Glasnik Matematicki. Published by the Society of Mathematicians and Physicists of the S. R. of Croatia. Publisher: Drustvo Matematicara i Fizicara S. R. Hrvatske, Zagreb.

Helvetica Phys. Acta
Helvetica Physica Acta. Schweizerische Physikalische Gesellschaft. Publisher: E. Birkhäuser, Basel.

IAU Circ.
International Astronomical Union, Circular. Central Bureau for Astronomical Telegrams, Smithsonian Astro-

physical Observatory, Cambridge, Mass.

Icarus
Icarus. International Journal of Solar System Studies. Publisher: Academic Press, New York – London.

ICSU Bull.
ICSU Bulletin. International Council of Scientific Unions. Secretariat: 51, Bd de Montmorency, Paris, France.

IEEE Spectrum
IEEE Spectrum. Published monthly by the Institute of Electrical and Electronics Engineers, Inc., New York, N. Y.

IEEE Trans. Aerospace Electron. Systems
IEEE Transactions on Aerospace and Electronic Systems. Published by the Institute of Electrical and Electronics Engineers, New York, N.Y., U.S.A.

IEEE Trans. Antennas Propagation
IEEE Transactions on Antennas and Propagation. Published by the Institute of Electrical and Electronics Engineers, New York, N.Y., U.S.A.

IEEE Trans. Electron Devices
IEEE Transactions on Electron Devices. Published by the Institute of Electrical and Electronics Engineers, New York, N.Y., U.S.A.

IEEE Trans. Microwave Theory Techn.
IEEE Transactions on Microwave Theory and Techniques. Published by the Institute of Electrical and Electronics Engineers, New York, N.Y., U.S.A.

Indian Journ. Pure and Applied Phys.
Indian Journal of Pure and Applied Physics. Council of Scientific and Industrial Research, New Delhi, India.

Inform. Bull. Southern Hemisphere
Information Bulletin for the Southern Hemisphere. Editorial Office: Observatorio Astronómico, La Plata, Argentina.

Inform. Bull. Variable Stars
Commission 27 of the I.A.U. Information Bulletin on Variable Stars. Konkoly Observatory, Budapest.

Infrared Physics
An International Research Journal. Publisher: Pergamon Press Ltd., Oxford – London – New York.

International Journ. Theor. Phys.
International Journal of Theoretical Physics. Publisher: Plenum Publishing Company, Donington House, London.

Irish Astron. Journ.
The Irish Astronomical Journal. A Quarterly Publication under the auspices of the Observatories of Armagh and Dunsink. Subscription address: Managing Editor, Irish Astronomical Journal, Armagh Observatory, Northern Ireland.

Izv. Akad. Nauk Armyan. SSR
Izvestiya Akademii Nauk Armyanskoj SSR. Fizika. Publisher: Izdatel'stvo AN Armyanskoj SSR, Erevan.

Izv. Glav. Astron. Obs. Pulkovo
Izvestiya Glavnoj Astronomicheskoj Observatorii v Pulkove. Akademiya Nauk SSSR. Izdanie Glavnoj astronomicheskoj observatorii v Pulkove, Leningrad.

Izv. Komissii Fiz. Planet
Izvestiya Komissii po Fizike Planet. Akademiya Nauk SSSR. Astronomicheskij Sovet. Moskva.

Izv. Krymskoj Astrofiz. Obs.
Izvestiya Krymskoj Astrofizicheskoj Observatorii. Akademiya Nauk SSR. Publishers: Izdatel'stvo "Nauka", Moskva.

Jenaer Rundschau (Jena Review)
Jenaer Rundschau (Jena Review). Publisher: VEB Verlag Technik, Berlin.

JETP Letters
JETP Letters. A translation of JETP Pis'ma v Redaktsiyu of the Academy of Sciences in the USSR. Published semimonthly by the American Institute of Physics, Lancaster, Pennsylvania.

Journ. American Ass. Variable Star Observers
The Journal of the American Association of Variable Star Observers. Published by The American Association of Variable Star Observers, Cambridge, Mass.

Journ. Astronaut. Sci.
The Journal of the Astronautical Sciences. Published by the American Astronautical Society Inc., Alexandria, VA.

Journ. Astron. Soc. Victoria
The Journal of the Astronomical Society of Victoria. Printed by D. Buscombe Printers, Glen Waverley, Victoria.

Journ. Astron. Soc. Western Australia
The Journal of the Astronomical Society of Western Australia. Edited by the Astronomical Society of Western Australia, Perth, W. A.

Journ. Atmosph. Sci.
Journal of the Atmospheric Sciences. Published by the American Meteorological Society, Boston, Mass.

Journ. Atmosph. Terr. Phys.
Journal of Atmospheric and Terrestrial Physics. Publishers: Pergamon Press, Oxford – London – New York.

Journ. British Astron. Ass.
Journal of the British Astronomical Association. Subscription address: British Astronomical Association, Burlington House, Piccadilly, London.

Journ. British Interplanet. Soc.
Journal of the British Interplanetary Society. Printed by Unwin Brothers Ltd., at the Gresham Press, Old Woking, and published for the British Interplanetary Society Ltd., by Space Educational Aids Ltd., London.

Journ. Fluid Mechanics
Journal of Fluid Mechanics. Published by Cambridge University Press, London – New York.

Journ. Geophys.
Journal of Geophysics / Zeitschrift für Geophysik. Publisher: Springer-Verlag, Berlin–Heidelberg–New York

Journ. Geophys. Res.
Journal of Geophysical Research. An International Scientific Publication. Published three times a month by the American Geophysical Union, Washington, D. C. First section: Space Physics; Second section: Physics and chemistry of the solid earth, planetology, geodesy; Third section: Oceans and atmospheres.

Journ. History Astron.
Journal for the History of Astronomy. Publisher: Science History Publications Ltd., Chalfont St Giles, Buckinghamshire, England. American Representative: Neale Watson Academic Publications, Inc., New York City, U.S.A.

Journ. Navigation
The Journal of Navigation. Published quarterly by The Royal Institute of Navigation at the Royal Geographical Society, London. To be obtained from Scottish Academic Press Ltd., Edinburgh, Scotland.

Journ. Optical Soc. America
Journal of the Optical Society of America. Publisher: American Institute of Physics, New York.

Journ. Phys. A (Math., Nuclear, General)
Journal of Physics A, (Mathematical, Nuclear and General). Europhysics Journal. Published by the Institute of Physics and Physical Society, London, England, in association with the American Institute of Physics, New York.

Journ. Physique
Journal de Physique. Publication de la Société Française de Physique, Paris.

Journ. Plasma Phys.
Journal of Plasma Physics. Publishers: Cambridge University Press, London.

Journ. Proc. Roy. Soc. New South Wales
Journal and Proceedings of the Royal Society of New South Wales. Publisher: Science House, Sydney, N.S.W. (Australia).

Journ. Quant. Spectrosc. Radiat. Transfer
Journal of Quantitative Spectroscopy & Radiative Transfer. Publisher: Pergamon Press, Oxford – New York.

Journ. Roy. Astron. Soc. Canada
The Journal of the Royal Astronomical Society of Canada, devoted to the advancement of astronomy and allied sciences. Printed by the University of Toronto Press, Toronto, Ontario, Canada.

Kometn. Tsirk. *Kiev*
Kometnyj Tsirkulyar. Gruppa po Issledovaniyu Komet Astrosoveta i Mezhduvedomstvennyj Geofizicheskij Komitet. Akademii Nauk SSSR. Kievskij Universitet im. T. G. Shevchenko.

Komety i Meteory
Komety i Meteory. Akademiya Nauk Tadzhikskoj SSR. Astronomicheskij Sovet Akademii Nauk SSSR. Publishers: Izdatel'stvo "Donish", Dushanbe.

Kosmich. Issled.
Kosmicheskie Issledovaniya. Akademiya Nauk SSSR. Publishers: Izdatel'stvo "Nauka", Moskva [An English translation is published as "Cosmic Research", Consultants Bureau, New York, N.Y.].

Kozmos
Kozmos. Popular Astronomical Journal of the Slovak Central Observatory in Hurbanovo. Publisher: Slovenská ústredná hvezdáren v Hurbanove.

L'Astronomie
L'Astronomie et Bulletin de la Société Astronomique de France. Revue mensuelle. Rédaction: Société Astronomique de France, Paris.

L'Universo
L'Universo. Rivista dell'Instituto Geografico Militare. Direzione, Redazione e Amministrazione: Istituto Geografico Militare, Firenze.

Magnitnye Polya Solnech. Pyaten
Magnitnye Polya Solnechnykh Pyaten. (Supplements to Solnechnye Dannye. Byulleten' (*Solar Data*)). Publishers: Izdatel'stvo "Nauka", Leningrad.

Math. Rev.
Mathematical Reviews. Published by the American Mathematical Society, Providence, R. I.

Mem. Fac. Sci. Kyoto Univ.
Memoirs of the Faculty of Science, Kyoto University. Series of Physics, Astrophysics, Geophysics, and Chemistry. Printed by Yamashiro Printing Publishing Co. Ltd., Kamigyo, Kyoto.

Mem. Roy. Astron. Soc.
Memoirs of the Royal Astronomical Society. Published for the Royal Astronomical Society by Blackwell Scientific Publications, Oxford – Edinburgh.

Mem. Soc. Astron. Italiana
Memorie della Società Astronomica Italiana. Nuova Serie. Pubblicate sotto gli auspici del Consiglio Nazionale dell Ricerche. Publisher: Tipografia Baccini & Chiappi, Firenze.

Mercury
Mercury. The Journal of the Astronomical Society of the Pacific. Published by the Astronomical Society of the Pacific, San Francisco, California.

Meteoritics
Meteoritics. The Journal of the Meteoritical Society. Published quarterly by The Meteoritical Society and Arizona State University Bureau of Publications. Editorial address: Center for Meteorite Studies, The Arizona State University, Tempe, Arizona.

Meteoritika
Akademiya Nauk SSSR. Komitet po Meteoritam. Publishers: Izdatel'stvo "Nauka", Moskva.

Microwave Journ. (*USA*)
Microwave Journal. To be obtained from 610 Washington Street, Dedham Plaza, Dedham, Massachusetts, U.S.A.

Microwaves (*USA*)
Microwaves. Published by Hayden Microwaves Corporation, New York, N.Y., U.S.A.

Minor Planet. Bull.
The Minor Planet Bulletin. Bulletin of the Minor Planets Section of the Association of Lunar and Planetary Observers. Editorial Office: R. G. Hodgson, Dordt College, Sioux Center, Iowa, U.S.A.

Mitt. Astron. Ges.
Mitteilungen der Astronomischen Gesellschaft, Hamburg. Printed by G. Braun, GmbH, Karlsruhe.

Monthly Notes Astron. Soc. Southern Africa
Monthly Notes of the Royal Astronomical Society of Southern Africa. Published by the Astronomical Society

of Southern Africa, Royal Observatory, Cape Province, South Africa.

Monthly Notes International Polar Motion Service
Monthly Notes of the International Polar Motion Service. Published by the Central Bureau, International Latitude Observatory of Mizusawa, Mizusawa-shi, Iwate-ken, Japan.

Monthly Notices Roy. Astron. Soc.
Monthly Notices of the Royal Astronomical Society. Published for the Royal Astronomical Society by Blackwell Scientific Publications, Oxford – Edinburgh.

Moon
The Moon. An International Journal of Lunar Studies. Publisher: D. Reidel Publishing Company, Dordrecht – Holland.

MVS Sonneberg
Mitteilungen über Veränderliche Sterne. Edited by Sternwarte Sonneberg (Zentralinstitut für Astrophysik, Bereich Sternphysik) der Deutschen Akademie der Wissenschaften zu Berlin.

Nablyud. Iskusstv. Nebesn. Tel
Nablyudeniya Iskusstvennykh Nebesnykh Tel. Published by Astronomicheskij Sovet Akademii Nauk SSSR, Moskva.

Nachr. Akad. Wiss. Göttingen
Nachrichten der Akademie der Wissenschaften in Göttingen. II. Mathematisch-Physikalische Klasse. Vandenhoeck & Ruprecht, Göttingen.

Nachr. Karten-, Vermessungswesen
Nachrichten aus dem Karten- und Vermessungswesen. Editor: Institut für Angewandte Geodäsie (Abt. II des Deutschen Geodätischen Forschungsinstituts). Published by Verlag des Instituts für Angewandte Geodäsie, Frankfurt a. M.

Nature
Nature. Editorial and Publishing Offices: Macmillan Journals Limited, 4 Little Essex Street, London; 711 National Press Building, Washington, D. C.

Naturwissenschaften
Die Naturwissenschaften. Publisher: Springer-Verlag, Berlin – Heidelberg – New York.

Nauchn. Informatsii
Nauchnye Informatsii. Astronomicheskij Sovet Akademii Nauk SSSR, Moskva.

Nuovo Cimento
Il Nuovo Cimento. Rivista Internazionale e Organo della Società Italiana di Fisica, Series A, B. Publisher: Nicola Zanichelli, Editore, Bologna.

Nuovo Cimento Lettere
Lettere al Nuovo Cimento, a Cura della Società Italiana di Fisica. Editrice Compositori, Bologna.

Nuovo Cimento Suppl.
Supplemento al Nuovo Cimento. Publisher: Nicola Zanichelli, Editore, Bologna.

Observations Artificial Earth Satellites
Observations of Artificial Satellites of the Earth (Nablyudeniya Iskusstvennykh Sputnikov Zemli). Magyar Tudományos Akadémia Csillagvizsgáló Intézete, Budapest.

Observatory
The Observatory. A Review of Astronomy. Publishers: The Editors of "The Observatory", Royal Greenwich Observatory, Herstmonceux Castle, Hailsham, Sussex, England.

Optics News
Optics News. Publication of the Optical Society, Washington, D.C.

Optik
Optik. Zeitschrift für das gesamte Gebiet der Licht- und Elektronenoptik. Publishers: Wissenschaftliche Verlagsgesellschaft mbH., Stuttgart.

Origins of Life
Origins of Life (Formerly Space Life Sciences). An International Journal. Publisher: D. Reidel Publishing Company, Dordrecht–Holland.

Orion
Orion. Zeitschrift der Schweizerischen Astronomischen Gesellschaft (SAG). Bulletin de la Société Astronomique de Suisse (SAS). Printed by A. Schudel & Co. AG, Riehen, Suisse.

Österreich. Zeitschr. Vermessungswesen
Österreichische Zeitschrift für Vermessungswesen und Photogrammetrie. Editor and Publisher: Österreichischer Verein für Vermessungswesen und Photogrammetrie, Wien, Austria.

Peremennye Zvezdy, Byull.
Peremennye Zvezdy, Byulleten', izdavaemyj Astronomicheskim Sovetom Akademii Nauk SSSR. Published by Astronomicheskij Sovet Akademii Nauk SSSR, Moskva.

Peremennye Zvezdy, Prilozhenie
Peremennye Zvezdy, Prilozhenie (The Variable Stars, Supplement). Astronomicheskij Sovet Akademii Nauk SSSR, Moskva.

Phil. Mag.
The Philosophical Magazine. A Journal of Theoretical, Experimental and Applied Physics. Eighth Series. Publisher: Taylor & Francis, Ltd., London.

Phil. Trans. Roy. Soc. London
Philosophical Transactions of the Royal Society of London. Series A, Mathematical and Physical Sciences. Published by the Royal Society, London.

Phys. Abstr.
Physics Abstracts. Science Abstracts, Series A. An INSPEC Publication, published by The Institution of Electrical Engineers, London.

Phys. Ber.
Physikalische Berichte. Herausgegeben von der Deutschen Physikalischen Gesellschaft e.V. und von der Deutschen Akademie der Wissenschaften zu Berlin. Physik-Verlag, Weinheim, West Germany.

Phys. Blätter
Physikalische Blätter. Physik-Verlag, Mosbach/Baden.

Phys. Bull.
Physics Bulletin. Published by the Institute of Physics and the Physical Society, London, England.

Phys. Earth Planet. Interiors
Physics of the Earth and Planetary Interiors. A journal devoted to observational and experimental studies of the Earth and Planetary interiors and their theoretical interpretation by the physical sciences. Publisher: North-Holland Publishing Company, Amsterdam, Netherlands.

Phys. Fluids
The Physics of Fluids. Published by the American Institute of Physics, New York, N.Y.

Phys. Letters
Physics Letters. Volumes A and B. Publisher: North-Holland Publishing Company, Amsterdam.

Phys. Rev. A
Physical Review A, General Physics. Published for the American Physical Society by the American Institute of Physics, Lancaster, Pa., and New York, N.Y.

Phys. Rev. B
Physical Review B, Solid State. Published for the American Physical Society by the American Institute of Physics, Lancaster, Pa., and New York, N. Y.

Phys. Rev. C
Physical Review C, Nuclear Physics. Published for the American Physical Society by the American Institute of Physics, Lancaster, Pa., and New York, N.Y.

Phys. Rev. D
Physical Review D, Particles and Fields. Published for the American Physical Society by the American Institute of Physics, Lancaster, Pa., and New York, N.Y.

Phys. Rev. Letters
Physical Review Letters. Published weekly by The American Physical Society, New York, N. Y.

Phys. Today
Physics Today. Published by the American Institute of Physics, New York, N.Y.

Physica
Physica. Publishers: North-Holland Publishing Company, Amsterdam, The Netherlands, on request of the Foundation "Physica", Utrecht.

Physica Scripta
Physica Scripta. (Formerly Arkiv för Fysik). Published by the Royal Swedish Academy of Sciences, Stockholm.

Pis'ma v Astron. Zhurn.
Pis'ma v Astronomicheskij Zhurnal. Akademiya Nauk SSSR. Publishers: Izdatel'stvo 'Nauka', Moskva.

Planet. Space Sci.
Planetary and Space Science. Pergamon Press, Oxford – London – New York.

Plasma Physics
Plasma Physics. Publisher: Pergamon Press, Oxford, England.

Pokroky
Pokroky matematiky, fyziky a astronomie. Editor: Jednota čs. matematiků a fyziků. Publisher: Academia, Praha.

Postępy Astron.
Postępy Astronomii. Czasopismo Poświecone Upowszechnianiu Wiedzy Astronomicznej. Polskie Towarzystwo Astronomiczne, Warszawa. Printed in Poland by Pánstwowe Wydawnictwo Naukowe, Lódź.

Priroda
Priroda. Publishers: Izdatel'stvo "Nauka", Moskva.

Proc. Astron. Soc. Australia
Proceedings of the Astronomical Society of Australia. Published for the Society by Sydney University Press, Sydney.

Proc. Cambridge Phil. Soc.
Proceedings of the Cambridge Philosophical Society (Mathematical and Physical Sciences). Publishers: Cambridge University Press, London.

Proc. IEEE
Proceedings of the IEEE. Published monthly by the Institute of Electrical and Electronics Engineers, Inc., New York, N. Y.

Proc. IREE Australia
Proceedings of the Institution of Radio and Electronics Engineers, Australia. Science House, Sydney, N.S.W., Australia.

Proc. Koninkl. Nederl. Akad. Wet.
Koninklijke Nederlandse Akademie van Wetenschappen. Proceedings. Series B, Physical Sciences. Publishers: North-Holland Publishing Company, Amsterdam.

Proc. National Acad. Sci. U.S.A.
Proceedings of the National Academy of Sciences of the United States of America. Published monthly by the National Academy of Sciences, Washington, D.C.

Proc. Roy. Soc. London
Proceedings of the Royal Society of London. Series A: Mathematical and Physical Sciences. Published by the Royal Society, London.

Progr. Theor. Phys. Japan
Progress of Theoretical Physics. Published for the Research Institute for Fundamental Physics and the Physical Society of Japan. Publication Office: Progress of Theoretical Physics, Yukawa Hall, Kyoto University, Kyoto, Japan.

Progr. Theor. Phys. Suppl.
Supplement of the Progress of Theoretical Physics. Published for the Research Institute for Fundamental Physics and The Physical Society of Japan. Publication Office: Progress of Theoretical Physics, Yukawa Hall, Kyoto University, Kyoto, Japan.

PTB Mitt.
PTB Mitteilungen. Amts- und Mitteilungsblatt der Physikalisch-Technischen Bundesanstalt, Braunschweig – Berlin.

Publ. Astron. Soc. Japan
Publications of the Astronomical Society of Japan. Published by the Astronomical Society of Japan. Office of the Society: Tokyo Astronomical Observatory, Mitaka,

Tokyo. Agent: Maruzen Co. Ltd. (Export Department), Nihonbashi, Tokyo, Japan.

Publ. Astron. Soc. Pacific
Publications of the Astronomical Society of the Pacific. Published in Provo, Utah, by the Astronomical Society of the Pacific, San Francisco, California. Printed by Brigham Young University Press, Provo, Utah.

Publ. Roy. Obs. Edinburgh
Publications of the Royal Observatory, Edinburgh. Published by The Royal Observatory, Edinburgh, Scotland.

Publ. Tartu Astrofiz. Obs.
W. Struve nimelise Tartu Astrofüüsika Observatooriumi, Publikatsioonid. Eesti NSV Teaduste Akadeemia, Tartu.

Publ. Variable Star Section, Roy. Astron. Soc. New Zealand
Publications of the Variable Star Section, Royal Astronomical Society of New Zealand. Director: F. M. Bateson, Greerton, Tauranga, New Zealand.

Quarterly Journ. Roy. Astron. Soc.
Quarterly Journal of the Royal Astronomical Society. Published for the Royal Astronomical Society by Blackwell Scientific Publications, Oxford.

Radio Sci.
Radio Science. Published by the American Geophysical Union, Richmond, Virginia.

Referativ. Zhurn. 51. Astron.
Referativnyj Zhurnal. 51. Astronomiya. Vsesoyuznyj Institut Nachnoj i Tekhnicheskoj Informatsii. Moskva.

Referativ. Zhurn. 52. Geod. i Aehrosemka
Referativnyj Zhurnal. 52. Geodeziya i Aehrosemka. Vsesoyuznyj Institut Nauchnoj i Tekhnicheskoj Informatsii. Moskva.

Referativ. Zhurn. 62. Issled. kosm. prostranstva
Referativnyj Zhurnal. 62. Issledovanie Kosmicheskogo Prostranstva. Vsesoyuznyj Institut Nauchnoj i Tekhnicheskoj Informatsii. Moskva.

Rep. Progr. Phys.
Reports on Progress in Physics. Published by The Institute of Physics and the Physical Society, London.

Rev. Geophys. Space Phys.
Reviews of Geophysics and Space Physics (formerly Reviews of Geophysics). Published by the American Geophysical Union, Richmond, Virginia.

Revista Astron.
Revista Astronomica. Organo de la Asociación Argentina Amigos de la Astronomia, Buenos Aires.

Rev. Mexicana Astron. Astrofis.
Revista Mexicana de Astronomia y Astrofisica. Dirección: Instituto de Astronomia, Universidad Nacional Autónoma de México, México, D.F.

Rev. Modern Phys.
Reviews of Modern Physics. Published for The American Physical Society by the American Institute of Physics, Lancaster, Pa., and New York, N.Y.

Rev. Sci. Instruments
Reviews of Scientific Instruments. Published by the American Institute of Physics, Lancaster, Pa., and New York, N.Y.

Rezul'taty Nablyud. Iskusstv. Sputnikov Zemli
Rezul'taty Nablyudenij Iskusstvennykh Sputnikov Zemli. Published by Astronomicheskij Sovet Akademii Nauk SSSR, Ryazanskij Gosudarstvennyj Pedagogicheskij Institut, Ryazan'.

Rezul'taty Nablyud. Sovet. Iskusstv. Sputnikov Zemli
Rezul'taty Nablyudenij Sovetskikh Iskusstvennykh Sputnikov Zemli. Published by Astronomicheskij Sovet Akademii Nauk SSSR, Moskva. Replaced after No. 140 by Rezul'taty Nablyudenij Iskusstvennykh Sputnikov Zemli.

Říše hvězd
Říše hvězd. Czechoslovak popular astronomical journal. Publisher: Orbis, Praha.

Roy Astron. Soc. New Zealand Publ.
Publications of Variable Star Section, Royal Astronomical Society of New Zealand. Publication Office: Greerton, Tauranga, New Zealand.

Rumanian Sci. Abstr.
Rumanian Scientific Abstracts. Natural Sciences. Publishers: The Scientific Documentation Centre of the Academy of the Socialist Republic of Romania, Bucureşti.

Sci. American
Scientific American. Published monthly by Scientific American, Inc., New York, N.Y.

Science
Science. American Association for the Advancement of Science, Washington, D.C.

Scient. Sinica
Scientia Sinica. Edited by Editorial Committee of Scientia Sinica, Peking. Published by Science Press, Peking, China.

Sci. Progr. Découverte
Science Progrès Découverte (formerly Science Progrès, La Nature). Revue publiée avec la participation du Palais de la Découverte. Published by Dunod, Editeur, Paris. Imprimerie Bayeusaine, Bayeux.

Sci. Rep. Tôhoku Univ.
The Science Reports of the Tôhuku University. First Series (Physics, Chemistry, Astronomy). Published by the Faculty of Science, Tôhoku University, Sendai, Japan.

Sitzungsber. Akad. Wiss. Berlin
Sitzungsberichte der Akademie der Wissenschaften der DDR. Klasse für Mathematik, Physik und Technik. Publisher: Akademie-Verlag, Berlin.

Sitzungsber. Bayer. Akad. Wiss.
Bayerische Akademie der Wissenschaften. Mathematisch-Naturwissenschaftliche Klasse. Sitzungsberichte. Publisher: Verlag der Bayerischen Akademie der Wissenschaften, München.

Sitzungsber. Heidelberger Akad. Wiss.
Sitzungsberichte der Heidelberger Akademie der Wissenschaften. Mathematisch-Naturwissenschaftliche Klasse. Publisher: Springer-Verlag, Heidelberg.

Sitzungsber. Österreich. Akad. Wiss.
Sitzungsberichte. Österreichische Akademie der Wissenschaften. Mathematisch-Naturwissenschaftliche Klasse. Abteilung II: Mathematik, Astronomie, Meteorologie und Technik. Publisher: Springer-Verlag, Wien.

Sky Telescope
Sky and Telescope. Published by Sky Publishing Corporation, Cambridge, Mass.

Smithsonian Contr. Astrophys.
Smithsonian Contributions to Astrophysics. Smithsonian Institution Astrophysical Observatory, Cambridge, Mass. Printed by Smithsonian Institution Press, City of Washington. For sale by the Superintendent of Documents, U. S. Government Printing Office, Washington, D. C.

Smithsonian Year
Smithsonian Year. Annual Report of the Smithsonian Institution, including the financial report of the Executive Committee of the Boards of Regents. Published by the Smithsonian Institution, Washington, D.C.

Solar Physics
Solar Physics. A Journal for Solar Research and the Study of Solar Terrestrial Physics. Publishers: D. Reidel Publishing Company, Dordrecht–Holland.

Solnechnye Dannye Byull.
Solnechnye Dannye. Byulleten'. *(Solar Data).* Publishers: Izdatel'stvo "Nauka", Leningradskoe Otdelenie, Leningrad.

Soobshch. Byurakan. Obs.
Soobshcheniya Byurakanskoj Observatorii. Akademiya Nauk Armyanskoj SSR, Erevan.

Soobshch. Gos. Astron. Inst. Shternberg
Soobshcheniya Gosudarstvennogo Astronomicheskogo Instituta im P.K. Shternberga. Publishers: Izdatel'stvo Moskovskogo Universiteta, Moskva.

Southern Stars
Southern Stars. The Journal of the Royal Astronomical Society of New Zealand (Inc.). Address of the Society: P.O. Box 3181, Wellington C1, New Zealand.

Soviet Astron.
Soviet Astronomy. A translation of Astronomicheskij Zhurnal (Astronomical Journal). Published by the American Institute of Physics, New York, N.Y.

Spaceflight
Spaceflight. A Publication of the British Interplanetary Society. Printed by Eyre & Spottiswoode Limited at Grosvenor Press, Portsmouth, and published by the British Interplanetary Society, London.

Space Sci. Instrum.
Space Science Instrumentation. An International Journal of Scientific Instruments for Aircraft, Balloons, Sounding Rockets, and Spacecraft. Published by D. Reidel Publishing Company, Dordrecht–Holland.

Space Sci. Rev.
Space Science Reviews. Publishers: D. Reidel Publishing Company, Dordrecht – Holland.

Sterne
Die Sterne. Zeitschrift für alle Gebiete der Himmelskunde. Johann Ambrosius Barth, Leipzig.

Sternenbote
Sternenbote. Monatsschrift für Österreichs Amateurastronomen. Publisher: Astronomisches Büro, Hermann Mucke, Wien.

Stockholms Obs. Ann.
Stockholms Observatoriums Annaler. Printed by Almquist & Wiksell, Stockholm.

Strolling Astronomer
The Strolling Astronomer. The Journal of The Association of Lunar and Planetary Observers, Publication Office: The Strolling Astronomer, Box 3 AZ, University Park, New Mexico.

Stud. Cerc. Astron.
Studii şi Cercetări de Astronomie. Editura Academiei Republicii Socialiste România. Editorial Office: Observatorul Astronomic, Bucureşti.

Stud. Geophys. Geod.
Studia geophysica et geodaetica. Published for the Geophysical Institute of the Czechoslovak Academy of Sciences by Academia, Praha.

Stud. Soc. Sci. Torunensis
Studia Societatis Scientiarum Torunensis, Toruń – Polonia. Sectio F (Astronomia).

Stud. Univ. Babeş-Bolyai
Studia Universitatis Babeş-Bolyai. Series Mathematica-Physica. Publishers: Intreprinderea Poligrafica, Cluj.

SuW
Sterne und Weltraum. Astronomische Monatsschrift. Publisher: Verlag Sterne und Weltraum Dr. Vehrenberg, Düsseldorf, Germany.

Tellus
Tellus, a bi-monthly Journal of Geophysics. Svenska Geofysiska Foreningen. Printed in Sweden by Almqvist & Wiksells Boktryckeri AB, Uppsala.

Trans. Astron. Obs. Yale Univ.
Transactions of the Astronomical Observatory of Yale University. Published by the Observatory, New Haven.

Trans. IAU
Transactions of the International Astronomical Union. Published and distributed for the IAU (UAI) by D. Reidel Publishing Company, Dordrecht – Holland/Boston – U.S.A.

Trans. Roy. Soc. Canada
Transactions of the Royal Society of Canada. Published by the Royal Society of Canada, National Research Building, Ottawa.

Trudy Astrofiz. Inst. Alma-Ata
Trudy Astrofizicheskogo Instituta, Alma-Ata. Akademiya Nauk Kazakhskoj SSR. Publishers: Izdatel'stvo "Nauka" Kazakhskoj SSR, Alma-Ata.

Trudy Glav. Astron. Obs. Pulkovo
Trudy Glavnoj Astronomicheskoj Observatorii v Pulkove. Akademiya Nauk SSSR. Izdanie Glavnoj astronomicheskoj observatorii v Pulkove, Leningrad.

Trudy Inst. Teor. Astron., ***Leningrad***
Trudy Instituta Teoreticheskoj Astronomii. Akademiya Nauk SSSR. Publishers: Izdatel'stvo "Nauka", Leningrad.

Trudy Tashkent. Astron. Obs.
Trudy Tashkentskoj Astronomicheskoj Observatorii. Akademiya Nauk Uzbekskoj SSR. Publishers: Izdatel'-stvo "FAN" Uzbekskoj SSR, Tashkent.

Tsirk. Astron. Inst. Tashkent
Tsirkulyar Astronomicheskogo Instituta. Akademiya Nauk Uzbekskoj SSR. Izdatel'stvo "FAN" Uzbekskoj SSR, Tashkent.

Tsirk. Astron. Obs. L'vov
Tsirkulyar. Astronomicheskaya Observatoriya. L'vovskij Ordena Lenina Gosudarstvennyj Universitet imeni Ivana Franko. Publisher: Izdatel'stvo L'vovskogo Universiteta, L'vov.

Umschau
Umschau in Wissenschaft und Technik, vereinigt mit Weltraumfahrt – Raketentechnik. Umschau Verlag Breidenstein KG, Frankfurt am Main.

Urania Barcelona
Urania. Revista de Astronomia y Ciencias Afines. Organo de la Sociedad Astronómica de España y América, Barcelona; Unión Nacional de Astronomia y Ciencias Afines, Madrid.

Urania Kraków
Urania. Miesiecznik Polskiego Towarzystwa Miłośników Astronomii, Kraków. Publisher: Krakowska Drukarnia Prasowa, Kraków.

Vasiona
Vasiona. Revue d'Astronomie et d'Astronautique. Bulletin de la Société Astronomique "R. Bosković", Beograd.

Veröff. Astron. Rechen-Inst. Heidelberg
Veröffentlichungen des Astronomischen Rechen-Instituts Heidelberg. Verlag G. Braun, Karlsruhe.

Veröff. Sternw. Sonneberg
Deutsche Akademie der Wissenschaften zu Berlin. Institut für Sternphysik. Veröffentlichungen der Sternwarte in Sonneberg. Publisher: Akademie-Verlag, Berlin.

Vesmír
Vesmír. Přírodovědecky časopis Čs. akadmie věd. Publisher: Academia, Praha.

Vestn. Khar'kov. Univ.
Vestnik Khar'kovskogo Universiteta. Seriya Astronomicheskaya. Publishers: Izdatel'stvo Khar'kovskogo Universiteta, Khar'kov.

Vestn. Kiev. Univ.
Vestnik Kievskogo Universiteta. Seriya Astronomii. Publishers: Izdatel'stvo Kievskogo Universiteta, Kiev.

VJS Naturforsch. Ges. Zürich
Vierteljahresschrift der Naturforschenden Gesellschaft in Zürich. Printer and Publisher: Leeman AG, Zürich.

Wiss. Zeitschr. Friedrich-Schiller Univ. Jena
Wissenschaftliche Zeitschrift der Friedrich-Schiller-Universität. Jena. Mathematisch-Naturwissenschaftliche Reihe. Edited by the Rektor der Friedrich-Schiller-Universität Jena.

Wiss. Zeitschr. Humboldt-Univ. Berlin
Wissenschaftliche Zeitschrift der Humboldt-Universität zu Berlin. Mathematisch-Naturwissenschaftliche Reihe. Edited by the Rektor der Humboldt-Universität, Berlin.

Zeitschr. Naturforschung
Zeitschrift für Naturforschung. Europhysics Journal. Teil a: Astrophysik, Physik, Physikalische Chemie. Published by Verlag der Zeitschrift für Naturforschung, Tübingen, Germany.

Zeitschr. Physik A
Atoms and Nuclei. Zeitschrift für Physik A. Springer-Verlag, Berlin–Heidelberg–New York.

Zeitschr. Physik B
Condensed Matter. Zeitschrift für Physik B. Springer-Verlag, Berlin–Heidelberg–New York.

Zemlya i Vselennaya
Zemlya i Vselennaya. Astronomiya, Geofizika, Issledovaniya Kosmicheskogo Prostranstva. Nauchno-Populyarnyj Zhurnal Akademii Nauk SSSR. Publishers: Izdatel'stvo "Nauka", Moskva.

Zenit
Populair wetenschappelijk maandblad over sterrenkunde/weerkunde/ruimtevaart/ruimte-onderzoek/aanverwante wetenschappen en technieken. Bureau: Stichting De Koepel, Utrecht.

Zentralblatt Math. Grenzgebiete – Math. Abstr.
Zentralblatt für Mathematik und ihre Grenzgebiete Mathematics Abstracts. Publisher: Springer-Verlag, Berlin–Heidelberg–New York.

Zvaigžņota Debess
Latvijas PSR Zinātņu Akadēmijas Radioastrofizikas Observatorijas Populārzinatnisks Gadalaiku Izdevums. Izdevnieciba "Zinātne", Riga.

002 Bibliographical Publications

002.001 **The Kepler resources in the archives of the USSR Academy of Sciences.** N. M. Raskin.
Vistas in Astronomy, Vol. 18, (see 012.006), 937 - 943 (1975).

002.002 **Bibliographia Kepleriana, 1967 - 1975.** M. List.
Vistas in Astronomy, Vol. 18, (see 012.006), 957 - 1010 (1975).

002.003 **Annotations on papers on geomagnetism and aeronomy published in "News of Universities. Radiophysics", 1974, Vol. 17, Nos. 2 - 12; 1975, Vol. 18, Nos. 1 - 2.**
Geomagn. Aeronom., Vol. 15, 762 - 765 (1975). In Russian.

002.004 **Hydrographische Bibliographie.** (Ozeanographie, Erdmagnetismus, Nautik).
Selected and arranged by H. Mädler, G. Heise.
Separate prints from Deutsche Hydrographische Zeitschr. [edited by Deutsches Hydrographisches Institut, Hamburg], Jahrgang 27 (1974), 118 pp. (1975). – The hydrographic bibliography is a generalized extract from the "Hydrographische Dokumentation" (hydrographic documentation). It is available on international catalogue cards.

002.005 **Annotations on astrophysical papers published in the journal "Radiofizika", Vol. 17, Nos. 1, 2, 5, 6, 8, 10, 12 for the year 1974.**
Astron. Zhurn. Akad. Nauk, SSSR, Vol. 52, 1340 - 1341 (1975). In Russian. English translation in Soviet Astron., Vol. 19, No. 6.

002.006 **Forschungen und Publikationen zur Geschichte der Astronomie in der Deutschen Demokratischen Republik. Teil II. Eine Bibliographie (1970–1974).**
D. Wattenberg.
Veröff. Archenhold-Sternw., Berlin-Treptow, No. 3, p. 139 - 209 (1975).

002.007 **Index of geophysical publications, series and contributions of the Dominion Observatory and its successor the Earth Physics Branch of the Department of Energy, Mines and Resources, Canada through December, 1974, with addenda.**
Published by the Earth Phys. Branch, Dep. Energy, Mines, Resources, Ottawa, Canada, 2 + 92 pp. (1975).

002.008 **Centre de Données Stellaires. Information Bulletin No. 9.** C. Jaschek (Editor).
Compiled at Observatoire de Strasbourg, Strasbourg, France, 33 pp. (1975). – The individual contributions are included in their corresponding subject categories – see abstracts 041.065, 041.066, 122.023.

002.009 **Astronomy and Astrophysics Abstracts. Vol. 13, Literature 1975, Part I.**
S. Böhme, U. Esser, W. Fricke, U. Güntzel-Lingner, I. Heinrich, F. Henn, D. Krahn, L. Schmadel, H. Scholl, G. Zech (Editors).
Published for Astronomisches Rechen-Institut, Heidelberg by Springer-Verlag, Berlin–Heidelberg–New York. 10 + 632 pp. Price DM 86.00; (US $ 37.00) [Subscription price DM 68.80 (US $ 29.60)] (1975).

002.010 **Rassegna delle riviste e notizie brevi.** P. Maffei.
Coelum, Vol. 43, 168 - 174, 212 - 216, 258 - 263 (1975).

002.011 **Astronomical notebook.** J. S. Griffith.
Journ. British Interplanet. Soc., Vol. 28, 489 - 494 (1975).

002.012 **Bibliography.** Z. Kopal, M. Moutsoulas, J. W. Salisbury, F. B. Waranius (Editors).
The Moon, Vol. 12, 479 - 510; Vol. 13, 495 - 519 (1975).

An index to biographical fragments in unspecialized scientific journals. See Abstr. 003.019.

Observations of Saturn's satellites 1789–1972.
See Abstr. 100.207.

003 Books (Astronomy and Astrophysics)

003.001 **Astrometriya i Astrofizika, Vypusk 26.**
E. P. Fedorov (Editor).
Respublikanskij Mezhvedomstvennyj Sbornik. Akademiya Nauk Ukrainskoj SSR, Glavnaya Astronomicheskaya Observatoriya. Izdatel'stvo "Naukova dumka", Kiev. 124 pp. Price 1 Rbl. 6 Kop. (1975). In Russian. – The papers included are abstracted in their corresponding subject categories – see abstracts 041.020, 062.027, 062.028, 065.063, 066.063, 071.020, 071.025, 073.066, 074.046, 082.044, 094.006, 096.004, 097.024, 114.050, 123.009, 158.112.

003.002 **Astronomy. Volume 10. Equilibrium and stability of gravitating systems.**
I. S. Shcherbina-Samojlova (Editor).
Itogi nauki i tekhniki. Seriya Astronomiya, tom 10, Moskva. 160 pp. Price 99 Kop. (1975). In Russian. – The individual contributions are included in their corresponding subject categories – see abstracts 042.044, 151.042.

003.003 **Problems of cosmic physics. Vypusk 10.**
S. K. Vsekhsvyatskij (Editor).
Mezhvedomstvennyj Nauchnyj Sbornik. Izdatel'skoe Obedinenie 'Vishcha Shkola', Izdatel'stvo pri Kievskom Gosudarstvennom Universitete, Kiev. 184 pp. Price 1 Rbl. 2 Kop. (1975). In Russian. – The papers included are abstracted in their corresponding subject categories - see abstracts 055.013, 062.033, 072.025, 072.026, 074.055, 084.256, 094.138, 102.012, 102.013, 103.115, 103.116, 104.021 - 104.024, 105.031, 107.009, 121.048.

003.004 **Astrometriya i Astrofizika, Vypusk 27.**
I. V. Gavrilov (Editor).
Respublikanskij Mezhvedomstvennyj Sbornik. Akademiya Nauk Ukrainskoj SSR, Glavnaya Astronomicheskaya Observatoriya. Izdatel'stvo "Naukova dumka", Kiev. 128 pp. Price 1 Rbl. 12 Kop. (1975). In Russian. – The papers included are abstracted in their corresponding subject categories – see abstracts 011.032, 041.025, 045.005 - 045.007, 064.037, 071.029, 073.084, 073.085, 082.063, 094.008, 094.140 - 094.142, 102.015, 114.348, 123.021.

003.005 **Annual review of astronomy and astrophysics, Vol. 13.**
G. R. Burbridge, D. Layzer, J. G. Phillips (Editors).
Annual Reviews Inc., Palo Alto, California. 6 + 557 pp. Price $ 12.50 (1975). – The individual contributions are included in their corresponding subject categories – see abstracts 022. 065, 031.409, 032.524, 061.043, 065.117, 065.118, 066.073, 117.051, 131.114, 131.539, 132.044, 141.082, 141.337, 151. 046, 158.139.

003.006 **Annuaire 1976 du Bureau des Longitudes.** Encyclopédie Physique et Spatiale.
Gauthier-Villars Éditeur, Paris. 16 + 946 pp. (1975). – Contents: Part 1. Éphémérides astronomiques; Part 2. Hydrosphère; Part 3. Astrophysique; Part 4. Électromagnétisme; particules, radiations; Part 5. Géographie mondiale; Part 6. Supplément pour l'année 1977; Part 7. Notices; Part 8. Tables analytiques.

003.007 **Results and methods of investigations of cosmic matter and Be-stars.**
G. Sh. Livshits, T. B. Omarov (Editors).
Akademiya Nauk Kazakhskoj SSR; Trudy Astrofizicheskogo Instituta. Tom (Vol.) 25. Izdatel'stvo "Nauka" Kazakhskoj SSR, Alma-Ata. 136 pp. Price 1 Rbl. 27 Kop. (1975). In Russian. – The individual contributions are included in their corresponding subject categories.

003.008 **Investigation of the sun and red stars. 3.**
A. Balklavs (Editor).
Latvijas PSR Zinātņu akadēmija, Radioastrofizikas observatorija. Akademiya nauk Latvijskoj SSR, Radioastrofizicheskaya observatoriya. Izdatel'stvo "Zinatne", Riga. 64 pp. Price 20 Kop. (1975). In Russian. – The individual contributions are included in their corresponding subject categories – see abstracts 021.022, 034.123, 077.062, 113.070, 114.113, 122.213.

003.009 **Introduction to general relativity.**
R. Adler, M. Bazin, M. Schiffer.
International series in pure and applied physics. McGraw-Hill Book Company, New York. 2nd edition. 16 + 550 pp. Price $ 19.95 (1975).

003.010 **Solar-terrestrial physics. II.**
S. I. Akasofu, S. Chapman.
Translated from the English edition. Mir, Moskva. 512 pp. Price 3 Rbl. 56 Kop. (1975). In Russian. – Review in Referativ. Zhurn. 51. Astron., 11.51.450 (1975).

003.011 **Methods in computational physics: advances in research and applications. Vol. 14: Radio astronomy.**
B. Alder, S. Fernbach, M. Rotenberg (Editors).
Academic Press, Inc., New York – London, a subsidiary of Harcourt Brace Jovanovich, Publishers. 10 + 239 pp. Price $ 26.50, £ 12.70 respectively (1975). ISBN 0-12-460814-0.

003.012 **Venus reveals secrets.** V. Alekseev, S. Minchik.
Second revised and enlarged edition. Mashinostroenie, Moskva. 96 pp. Price 51 Kop. (1975). In Russian. – Review in Priroda, No. 12.75, p. 116 (1975).

003.013 **Structure and evolutionary history of the solar system.** H. Alfvén, S. Arrhenius.
D. Reidel Publishing Company, Dordrecht/Holland – Boston/ U.S.A. Geophysics and Astrophysics Monographs, Vol. 5. 16 + 276 pp. Price Dfl. 80.00, US $ 32.00 respectively (1975). ISBN 90-277-0611-5. – Contents: (1) Introduction and survey; (2) The orbits of planets and satellites; (3) The small bodies; (4) Resonance structure; (5) Spin and tides; (6) Post-accretional changes in the solar system; (7) Formation of celestial bodies; (8) Spin and accretion; (9) On the accretion of planets and satellites; (10) Plasma physics and hetegony; (11) Model of the hetegonic plasma; (12) Transfer of angular momentum and condensation of grains; (13) Accretion of the condensation products; (14) Chemical compositions in the solar system; (15) Meteorites and their precursor states; (16) Mass distribution and the critical velocity; (17) The structure of the groups. Summary and conclusions.

003.014 **The planet Mars.** E. M. Antoniadi.
Translated from the French by P. Moore.
Keith Reid, Ltd., Shaldon, Devon. 335 pp. Price £ 15.00 (1975). ISBN 0-904094-14-6.

003.015 **Composition of cosmic radiation.** Topics in astrophysics and space physics, Vol. 11.
K. M. V. Apparao.
Gordon and Breach Science Publishers, Ltd., London – New York – Paris. 10 + 86 pp. Price £ 5.40 (1975). ISBN 0-677-03770-8. – Review in Nature, Vol. 258, 40; 1975 (*H. R. Allan*).

003.016 **Eyes on the universe.** I. Asimov.
Houghton Mifflin Company, Boston. 274 pp. Price

$ 8.95 (1975). – Review in Sky Telescope, Vol. 50, 409 (1975).

003.017 **Gravitation.** H. Aspden.
Sabberton, Southampton, England, 8 + 78 pp. Price cloth $ 7.00, paper $ 3.50 (1975). – (From Science, Vol. 190, 77 (1975)).

003.018 **Die Entstehung der Elemente – Einführung in die Nuklear-Astrophysik.**
J. Audouze, S. Vauclair. Translated from the French edition and revised by R. Engel.
dva-Seminar, Deutsche Verlagsanstalt, Stuttgart. 162 pp. Price öS 98.60 (1974). – Review in Sternenbote, 18. Jahrgang, p. 238 - 239; 1975 (*J. Pfleiderer*).

003.019 **An index to biographical fragments in unspecialized scientific journals.** E. S. Barr.
The University of Alabama Press, University, Alabama. 10 + 294 pp. Price $ 12.50. – Review in Journ. History Astron., Vol. 6, 218; 1975 (*O. Gingerich*).

003.020 **Space navigation.** I. K. Bazhinov, V. I. Aleshin, V. N. Pochukaev, V. S. Polyakov.
Mashinostroenie, Moskva. 352 pp. Price 2 Rbl. 34 Kop. (1975). In Russian. – Review in Referativ. Zhurn. 62. Issled. kosmich. prostranstva, 7.62.328 (1975).

003.021 **Herder Lexikon – Weltraumphysik.**
U. Becker (Editor).
Verlag Herder KG, Freiburg (Breisgau) – Basel – Wien. 239 pp. Price DM 19.80 (1975). – Reviews in Orion, 33. Jahrgang, p. 127; 1975 (*E. Wiedemann*); Sternenbote, 18. Jahrgang, p. 239 - 240 (1975); SuW, Vol. 14, 414; 1975 (*C. Trefzger*).

003.022 **Erforschter Weltraum.** Ein Sachbuch der modernen Astronomie. U. Becker, R. Sauermost.
Verlag Herder KG, Freiburg (Breisgau) – Basel – Wien. 256 pp. Price DM 34.00 (1975). ISBN 3-451-17393-X. – (From Umschau, Vol. 75, 743 (1975)).

003.023 **Motion of a satellite about the center of mass in a gravitational field.** Textbook for students of mech.-mat. and phys.-mat. faculties of universities. V. V. Beletskij.
Mosk. un-t, Moskva. 308 pp. Price 87 Kop. (1975). In Russian. Review in Referativ. Zhurn. 62. Issled. kosmich. prostranstva, 2.62.361 (1976).

003.024 **Optik und optischer Gerätebau: Deutsch-Englisch.**
W. Bindmann.
VEB Verlag Technik, Berlin, DDR. 432 pp. Price M 45.00 (1975). – Review in Sky Telescope, Vol. 50, 254 (1975).

003.025 **Ancient cosmologies.**
C. Blacker, M. Loewe (Editors).
Rowman and Littlefield Publishers, Totowa, N.J. 270 pp. Price $ 17.50 (1975). – Review in Sky Telescope, Vol. 50, 185 (1975).

003.026 **Time and frequency: theory and fundamentals.**
B. E. Blair (Editor).
National Bureau of Standards, Department of Commerce, Washington, D.C., Monograph 140. 470 pp. Price $ 8.65 (1974). Review in Mem. Soc. Astron. Italiana, Vol. 46, 283 - 284 (1975).

003.027 **Space rescue and safety.** P. H. Bolger (Editor).
American Astronautical Society, Publications Office, Tarzana, Calif. 272 pp. Price $ 20.00 (1975). – Review in Sky Telescope, Vol. 50, 328 (1975).

003.028 **Space physics: the study of plasmas in space.**
R. L. F. Boyd.
Clarendon Press, Oxford; Oxford University Press, London – New York. Oxford Physics Series 7. 18 + 100 pp. Price £ 3.50, $ 11.25 respectively (1974/75). – Review in Sky Telescope, Vol. 50, 185 (1975).

003.029 **Astronomy.**
F. M. Branley, M. R. Chartrand III, H. K. Wimmer.
Fitzhenry and Whiteside, Toronto; Donnelly, New York. 10 + 564 pp. Price $ 14.85, $ 13.50 respectively (1975). Reviews in Journ. Roy. Astron. Soc. Canada, Vol. 69, 317 - 318; 1975 (*D. L. DuPuy*); Sky Telescope, Vol. 50, 409 (1975).

003.030 **The earth's density.** K. E. Bullen.
Chapman and Hall Ltd., London; John Wiley and Sons Inc., New York. 13 + 420 pp. Price £ 12.90 (1975).

003.031 **Lunar soil.** I. I. Cherkasov, V. V. Shvarev.
Nauka, Moskva. 144 pp. Price 57 Kop. (1975). In Russian. – Review in Referativ. Zhurn. 62. Issled. kosmich. prostranstva, 1.62.56 (1976).

003.032 **Electrography and astronomical applications.**
G. L. Chincarini, P. J. Griboval, H. J. Smith (Editors).
University of Texas Press, Austin, TX. 364 pp. Price $ 5.50 paperbound (1974). ISBN 0-292-76423-5. – Review in Astrophys. Letters, Vol. 16, 133 - 134 (1975).

003.033 **Strange universe.** W. R. Corliss (Compiler).
The Sourcebook Project, Glen-Arm, Md. 279 pp. Price $ 7.95 (1975). – Review in Sky Telescope, Vol. 50, 409 - 410 (1975).

003.034 **Man and cosmos.** Nine Guggenheim lectures on the solar system.
J. Cornell, E. N. Hayes (Editors).
A Smithsonian Special Publication. W. W. Norton and Co. Ltd., New York. 192 pp. Price $ 8.95 (1975). – Reviews in Sky Telescope, Vol. 50, 254; Vol. 50, 407 - 409; 1975 (*P. Rizzo*).

003.035 **Solar and terrestrial radiation.** Methods and measurements. K. L. Coulson.
Academic Press Inc., New York – London. 12 + 322 pp. Price $ 27.00, £ 12.95 respectively (1975). – (From Science, Vol. 189, 160 (1975)).

003.036 **Statistical properties of scattered light.**
B. Crosignani, P. Di Porto, M. Bertolotti.
Academic Press, Inc., New York. 226 pp. Price $ 19.50 (1975). (From Phys. Today, Vol. 28, No. 12, p. 58 (1975)).

003.037 **Observations of the stellar sky.** M. M. Dagaev.
3rd edition. Nauka, Moskva. 176 pp. Price 33 Kop. (1975). In Russian. – (From Referativ. Zhurn. 51. Astron., 12.51.49 (1975)).

003.038 **Éléments d'astronomie de position.**
M. Danloux-Dumesnils.
Librairie scientifique et technique Albert Blanchard, Paris. 281 pp. Price F 40.00 (1974). – Review in Orion, 33. Jahrgang, p. 126; 1975 (*H. Müller*).

003.039 **La découverte du cosmos par l'astronomie, l'astrophysique et l'astronautique.**
P. de La Cotardière.
Editions Eyrolles, Paris. 96 pp. Price Ffr. 33.00 (1975). Review in Orion, 33. Jahrgang, p. 222; 1975 (*H. Müller*).

003.040 **Choice of sundials.** W. W. Dolan.
The Stephen Greene Press, Brattleboro, Vermont. Price $ 10.00. – Review in Sci. American, Vol. 233, No. 6, p. 136 (1975).

003.041 **Surfaces and interiors of planets and satellites.** A. Dollfus (Editor).
Translated from the English and French editions. Mir, Moskva. 579 pp. Price 3 Rbl. 19 Kop. (1974). In Russian. – (From Referativ. Zhurn. 51. Astron., 7.51.47 (1975)).

003.042 **Experimental and theoretical foundations of astrophysics of cosmic rays.** L. I. Dorman.
Nauka, Moskva. 464 pp. Price 3 Rbl. 17 Kop. (1975). In Russian. – Review in Referativ. Zhurn. 51. Astron., 11.51.54 (1975).

003.043 **Celestial mechanics. Basic problems and methods.** G. N. Duboshin.
3rd enlarged edition. Nauka, Moskva. 800 pp. Price 1 Rbl. 92 Kop. (1975). In Russian. – Review in Referativ. Zhurn. 51. Astron., 1.51.141 (1976).

003.044 **Data reduction: analysing and interpreting statistical data.** A. S. C. Ehrenberg.
John Wiley & Sons Inc., New York. 391 pp. (1975). – (From Phys. Today, Vol. 28, No. 8, p. 76 (1975)).

003.045 **Let's explore outer space.** Elementary research problems in space science.
S. Engelbrektson, P. Greenleaf.
Arco Publishing Co., New York. 128 pp. Price $ 2.25 (1975).

003.046 **Mechanik, Relativität, Gravitation. Ein Lehrbuch.** G. Falk, W. Ruppel.
Springer-Verlag, Berlin–Heidelberg–New York. Second, revised edition. 16 + 445 pp. Price DM 48.00 (1975).

003.047 **Volcanoes of the earth, moon and Mars.** G. Fielder, I. Wilson (Editors).
Elek Science, London. 9 + 126 pp. Price £ 6.00 (1975). ISBN 0-236-31097-6.

003.048 **Lunar mineralogy.** J. W. Frondel.
Wiley-Interscience, New York. 12 + 324 pp. Price $ 18.95 (1975). – (From Science, Vol. 190, 1290 (1975)).

003.049 **Photometric characteristics of Mars according to data of photographic observations during the great opposition of 1971.**
A. R. Gajduk, N. B. Ibragimov, I. K. Koval'.
Akademiya Nauk Ukrainskoj SSR, Glavnaya Astronomicheskaya Observatoriya. Izdatel'stvo Naukova dumka, Kiev. 128 pp. Price 1 Rbl. 1 Kop. (1975). In Russian.

003.050 **Professor Karl Šteins. Bibliography.**
Compiled by G. Gajlit, F. Gejkina, L. Roze, G. Novatskij, S. Skotele.
Latv. un-t. Astron. observ., Riga. 48 pp. Price 20 Kop. (1972). In Russian and Estonian. – (From Referativ. Zhurn. 51. Astron., 7.51.62 (1975)).

003.051 **The pocket encyclopaedia of spaceflight in colour: missiles and rockets.** K. Gatland.
Blandford Press Ltd., London. 256 pp. Price £ 2.40 (1975). Review in Spaceflight, Vol. 17, 341; 1975 (*J. R. Millburn*).

003.052 **Introduction to matrix methods in optics.** A. Gerrard, J. M. Burch.
Wiley-Interscience, New York. Wiley Series in Applied and Pure Optics. 14 + 356 pp. Price $ 29.95 (1975).

003.053 **Dictionary of scientific biography.** Vol. 10: S. G. Navashin–W. Pisco.
C. C. Gillispie (Editor-in-chief).
Charles Scribner's Sons, New York. 12 + 622 pp. Price $ 40.00 (1974). ISBN 0-684-10121-1.

003.054 **Dictionary of scientific biography.** Vol. 11: A.Pitcairn–B.Rush.
C. C. Gillispie (Editor-in-chief).
Charles Scribner's Sons, New York. 13 + 618 pp. Price $ 40.00 (1975). ISBN 0-684-10122-X. – Review in Sky Telescope, Vol. 50, 185 (1975).

003.055 **Dictionary of scientific biography.** Vol. 12: Ibn Rushd–Jean-Servais Stas.
C. C. Gillispie (Editor-in-chief).
Charles Scribner's Sons, New York. 13 + 620 pp. Price $ 40.00 (1975). ISBN 0-684-12924-8.

003.056 **New frontiers in astronomy.**
Readings from Scientific American with introductions by O. Gingerich.
W. H. Freeman and Co., San Francisco. 8 + 369 pp. Price $ 13.00 or $ 7.50 softcover (1975). – (From Sky Telescope, Vol. 50, 179, 254 (1975)).

003.057 **Atmospheres.** R. M. Goody, J. C. G. Walker.
Translated from the English edition by N. N. Tantsova, edited by G. I. Golyshev.
Mir, Moskva. 184 pp. Price 89 Kop. (1975). In Russian. Reviews in Priroda, No. 11.75, p. 128 (1975); Referativ. Zhurn. 51. Astron., 10.51.92 (1975); Zemlya i Vselennaya, 1975, No. 6, p. 90.

003.058 **A source book in medieval science.** E. Grant (Editor).
Harvard University Press, Cambridge, Massachusetts – London. 18 + 864 pp. Price $ 32.50 (1974). – Essay review in Journ. History Astron., Vol. 6, 212 - 217; 1975 (*E. Poulle*).

003.059 **The physics of cosmic X-ray, gamma-ray, and high-energy particles.** K. Greisen.
Translated from the English edition. Mir, Moskva. 191 pp. Price 1 Rbl. 2 Kop. (1975). In Russian. – Review in Referativ. Zhurn. 51. Astron., 11.51.55 (1975).

003.060 **The Jupiter effect: the planets as triggers of devastating earthquakes.**
J. R. Gribbin, S. H. Plagemann.
Vintage, Random House, Inc., New York. 18 + 180 pp. Price $ 1.95 (1975). Reprint of the 1974 edition. – (From Phys. Today, Vol. 28, No. 9, p. 63 (1975)).

003.061 **V hlubinách vesmíru.** J. Grygar.
Edited by Albatros, Praha. 96 pp. Price Kčs 13.00 (1974). In Czech. – Review in Říše hvězd, Vol. 56, 181 (1975); Vesmír, Vol. 54, 261 (1975).

003.062 **The moon.** A. A. Gurshtejn (Editor).
Translated from the English edition. Mir, Moskva. 304 pp. Price 2 Rbl. 12 Kop. (1975). In Russian. – Review in Referativ. Zhurn. 51. Astron., 2.51.386 (1976).

003.063 **Magnetic properties of meteorites.** E. G. Gus'kova.
Nauka, Leningrad. 108 pp. (1972). In Russian. – Review in Meteoritika, vyp. (No.) 34, p. 116; 1975 (*A. A. Yavnel'*).

003.064 **Celestial mechanics.** Y. Hagihara.
Japan Society for the Promotion of Science, Tokyo, Japan. Volume IV, Parts 1 and 2: Periodic and quasi-periodic solutions. 15 + 1243 pp. Price $ 98.00 (1975). – Contents: Volume IV, Part 1: The restricted three-body problem; Poincaré's theory of periodic solutions. Volume IV, Part 2: Asymptotic solutions and second-genus periodic solutions; Infinite

nonlinear analysis; Quasi-periodic solutions.

003.065 **U.S.-Soviet cooperation in space.** D. L. Harvey, L. C. Ciccoritti.
Monographs in International Affairs, Center for Advanced International Studies, University of Miami. 408 pp. (1975). Review in Space Sci. Rev., Vol. 17, 885; 1975 (*G. C. M. Reijnen*).

003.066 **Planetsystemet.** Ø. Hauge.
Universitetsforlaget, Oslo. 157 pp. Price Nkr. 48.00 (1974). – Review in Astron. Tidssk., Årg. 8, 146 - 147; 1975 (*H. Nielsen*).

003.067 **Radio astronomy for the amateur.** D. Heiserman.
Tab books, Blue Ridge Summit, Pa. 252 pp. Price $ 8.95 or $ 5.95 (1975). – (From Science, Vol. 190, 304, (1975)).

003.068 **Albert Einstein.** F. Herneck.
BSB B. G. Teubner Verlagsgesellschaft, Leipzig, 124 pp. Price M 5.00 (1974). – Review in Blick in das Weltall, 23. Jahrgang, p. 59 - 60 (1975).

003.069 **Geschichte der Astronomie von Herschel bis Hertzsprung.** D. B. Herrmann.
VEB Deutscher Verlag der Wissenschaften, Berlin. 282 pp. Price M 12.80 (1975).

003.070 **The radio universe.** J. S. Hey.
Revised second edition. Pergamon Press, Oxford – New York – Toronto – Sydney – Paris – Braunschweig. 8 + 264 pp. (1975). ISBN 0-08-018760-9.

003.071 **Francis Place and the early history of the Greenwich Observatory.** D. Howse.
Science History Publications, New York. 64 pp. Price $ 12.95 (1975). – (From Nature, Vol. 258, No. 5531, p. XII (1975); Science, Vol. 190, 1193 (1975)).

003.072 **Astronomy and cosmology. A modern course.** F. Hoyle.
W. H. Freeman and Co., San Francisco. 14 + 711 pp. Price $ 15.95 (1975). – Contents: (I) A first look at the universe; (II) Basic ideas and instruments; (III) Astrophysics; (IV) The solar system; (V) Radioastronomy; (VI) Cosmology.

003.073 **Astronomy today.** F. Hoyle.
Heineman Educational Books Ltd., London. 179 pp. Price £ 4.50 (1975). ISBN 0-435-54426-8. – (From Nature, Vol. 258, No. 5536, p. VII (1975)).

003.074 **Highlights in astronomy.** F. Hoyle.
W. H. Freeman and Co., San Francisco. 8 + 179 pp. Price $ 10.00 cloth, $ 5.50 paper (1975). – Review in Sky Telescope, Vol. 50, 186 (1975).

003.075 **Infrared detectors.** R. D. Hudson, Jr., J. Wordsworth Hudson (Editors).
John Wiley & Sons, Halsted Division, New York; Dowden, Hutchinson & Ross, Inc., Stroudsburg, Pennsylvania. Benchmark papers in optics, Vol. 2. 18 + 392 pp. Price $ 26.00 (1975). – Reviews in Applied Optics, Vol. 14, 2803; 1975 (*T. Limperis*); Journ. Optical Soc. America, Vol. 65, 1526 - 1527; 1975 (*W. L. Wolfe*); Scientific American, Vol. 233, No. 5, p. 139 - 140 (1975).

003.076 **Sketches on the history of celestial mechanics.** N. I. Idel'son.
Nauka, Moskva. 496 pp. Price 1 Rbl. 91 Kop. (1975). In Russian. – Review in Referativ. Zhurn. 51. Astron., 11.51.51 (1975).

003.077 **White dwarfs. Collection of papers.** V. S. Imshennik (Editor).
Translated from the English edition. Mir. Moskva, 256 pp. Price 1 Rbl. 45 Kop. (1975). In Russian. – (From Referativ. Zhurn. 51. Astron., 12.51.407 (1975)).

003.078 **Atmospheric thermodynamics; geophysics and astrophysics monographs.**
J. V. Iribarne, W. L. Godson.
D. Reidel Publishing Company, Dordrecht, Holland–Boston, USA. Geophysics and Astrophysics Monographs, Vol. 6. 10 + 222 pp. Price Dfl. 65.00 (1973). – Review in Planet. Space Sci., Vol. 23, 1237; 1975 (*R. C. Smith*).

003.079 **Heavy nuclei, superheavy nuclei, and neutron stars.** J. M. Irvine.
Oxford University Press, London; Clarendon Press, Oxford–New York. Oxford Studies in Nuclear Physics. 8 + 164 pp. Price £ 8.00, $ 22.50 respectively (1975). – Review in Nature, Vol. 257, 520; 1975 (*C. J. Batty*).

003.080 **Fall of cosmic matter on the earth's surface.** G. M. Ivanova, Yu. A. L'vov, N. V. Vasil'ev, I. V. Antonov.
Tomsk. universitet, Tomsk. 120 pp. Price 62 Kop. (1975). In Russian. – Review in Referativ. Zhurn. 51. Astron., 10.51.356 (1975).

003.081 **The Webb Society Observers Handbook, Vol. 1: Double stars.** K. G. Jones.
The Webb Society, London. 122 pp. Price £ 2.00 (1975). Review in Sky Telescope, Vol. 50, 410 (1975).

003.082 **Infrared and Raman spectroscopy of lunar and terrestrial minerals.** C. Karr, Jr. (Editor).
Academic Press Inc., New York – London. 14 + 376 pp. Price $ 38.50, £ 18.50 respectively (1975).

003.083 **Comparative meteorology of the planets.** K. Ya. Kondrat'ev.
Gidrometeoizdat, Leningrad. 48 pp. Price 21 Kop. (1975). In Russian. – Review in Referativ. Zhurn. 51. Astron., 7.51.51 (1975).

003.084 **On the long-term (11-year) cyclicity of seismic activity of the earth and of transient events on the visible side of the moon.** O. A. Kulibekov.
Astrofiz. in-t AN KazSSR. Alma-Ata, 1975. 25 pp. In Russian. Review in Referativ. Zhurn. 51. Astron., 11.51.284; 62. Issled. kosmich. prostranstva, 11.62.215 (1975).

003.085 **Dimensional analysis in astrophysics.** R. Kurth.
With a foreword by S. A. Kaplan. Translated from the English edition.
Mir, Moskva. 230 pp. Price 1 Rbl. 2 Kop. (1975). In Russian. Review in Referativ. Zhurn. 51. Astron., 9.51.51 (1975).

003.086 **Hypothesis of the primordially hydride earth (a new global concept).** V. N. Larin.
Nedra, Moskva. 101 pp. Price 74 Kop. (1975). In Russian. Review in Referativ. Zhurn. 51. Astron., 10.51.269 (1975).

003.087 **Radio astronomer's notebook.** J. Lichtman, B. Sickels.
Bob's Electronic Service, Ft. Lauderdale, Fla. 96 pp. Price $ 20.00 (1975). – Review in Sky Telescope, Vol. 50, 409 (1975).

003.088 **Some planetarium programs of 1972–73.**

J. LoGuirato (Editor).
ISPE Special Report No. 5, New York. 25 pp. Price $ 2.00 (1974). – Review in Journ. British Astron. Ass., Vol. 85, 458 - 459 (1975).

003.089 **Plasma-Experimente im Weltraum.** R. Lüst.
Westdeutscher Verlag, Düsseldorf-Opladen, F.R. Germany. 39 pp. (1974). – (From Phys. Today, Vol. 28, No. 12, p. 58 (1975)).

003.090 **A feather for Daedalus.** K. Malville.
Cummings Publishing Co., Menlo Park, Calif. 152 pp. Price $ 3.95 (1975). – Review in Sky Telescope, Vol. 50, 328 (1975).

003.091 **The Mössbauer effect in geochemistry and cosmochemistry.** T. V. Malysheva.
Nauka, Moskva. 168 pp. Price 1 Rbl. 15 Kop. (1975). In Russian. – (From Referativ. Zhurn. 51. Astron., 1.51.88 (1976)).

003.092 **Tables of spectral-line intensities.** Part I – Arranged by elements; Part II – Arranged by wavelengths. 2nd edition.
W. F. Meggers, C. H. Corliss, B. F. Scribner.
National Bureau of Standards, U.S. Department of Commerce. U.S. Government Printing Office, Washington, D.C. NBS Monograph 145. [Available from Superintendent of Documents, Washington, D.C.]. Part I: 16 + 390 pp. Price $ 8.55; Part II: 18 + 216 pp. Price $ 6.80 (1975). – Reviews in Journ. Optical Soc. America, Vol. 65, 1527 - 1528; 1975 (*R. D. Cowan*); Sky Telescope, Vol. 50, 186 (1975).

003.093 **A field guide to the stars and planets.**
D. H. Menzel.
Houghton Mifflin Company, Boston. 16 + 398 pp. Price $ 4.95 (1975). – Review in Science, Vol. 189, 157 (1975).

003.094 **Sternwarten.** P. Müller.
Peter Lang GmbH, Frankfurt/M; Herbert Lang, Bern. 290 pp. Price DM 38.00 or Swiss francs 43.70 (1975). Review in Sky Telescope, Vol. 50, 328 (1975).

003.095 **The rotation of the earth: a geophysical discussion.**
W. H. Munk, G. J. F. MacDonald.
Cambridge University Press, Cambridge – London. 19 + 323 pp. Price £ 6.80, $ 19.95 respectively (1975, first published 1960). ISBN 0-521-20778-9.

003.096 **A history of ancient mathematical astronomy.**
O. Neugebauer.
Studies in History of Mathematics and Physical Sciences 1. Springer-Verlag, Berlin – Heidelberg – New York. In three parts, 1457 pp. Price DM 232.00 (1975). ISBN 3-540-06995-X Set, 0-387-06995-X Set. – Contents: Pt. 1. The Almagest and its direct predecessors. Babylonian astronomy. – Pt. 2. Egypt. Early Greek astronomy. Astronomy during the Roman imperial period and late antiquity. – Pt. 3. Appendices and indices. 1. Astronomy, ancient – history. 2. Astronomy – mathematics – history.

003.097 **Dynamics of gravitating systems of the metagalaxy.**
T. B. Omarov.
Akademiya Nauk Kazakhskoj SSR, Astrofizicheskij Institut. Izdatel'stvo "Nauka" Kazakhskoj SSR, Alma-Ata. 144 pp. Price 1 Rbl. 4 Kop. (1975). In Russian.

003.098 **Pictorial guide to planet earth.**
F. I. Ordway III.
Thomas Y. Crowell Company, New York. 191 pp. Price $ 12.95 (1975). – Review in Sky Telescope, Vol. 50, 185 (1975).

003.099 **A survey of the Almagest.** O. Pedersen.
Odense University Press, Odense, Denmark. 454 pp. Price Dan.kr. 200.00 (about $ 11.00) (1974). – Review in Sky Telescope, Vol. 50, 186 (1975).

003.100 **Physical cosmology.** P. Peebles.
Translated from the English edition. Mir, Moskva. 311 pp. Price 68 Kop. (1975). In Russian. – Review in Referativ. Zhurn. 51. Astron., 9.51.818 (1975).

003.101 **General astrometry.**
V. V. Podobed, V. V. Nesterov.
Izdatel'stvo Nauka, Glavnaya Redaktsiya Fiziko-Matematicheskoj Literatury, Moskva. 552 pp. Price 1 Rbl. 42 Kop. (1975). In Russian. – (From Referativ. Zhurn. 51. Astron., 12.51.44 (1975)).

003.102 **An introduction to the ionosphere and magnetosphere.** J. A. Ratcliffe.
Cambridge University Press, New York. 256 pp. Price $ 6.95 paperback, $ 14.50 hardcover (1975). – (From Phys. Today, Vol. 28, No. 12, p. 59 (1975)).

003.103 **Soyuz – Apollo.** M. F. Rebrov, L. A. Gil'berg.
Edited by V. A. Shatalov.
Mashinostroenie, Moskva. 98 pp. Price 1 Rbl. 52 Kop. (1975). In Russian. – Review in Priroda, No. 11.75, p. 128 (1975).

003.104 **Worlds beyond: a report on the search for life in space.** I. Ridpath.
Wildwood House, Ltd., London. 176 pp. Price £ 2.95 (1975). ISBN 0-7045-0185-6.

003.105 **The nature of the solid earth.**
E. C. Robertson (Editor).
Translated from the English edition. Nauki o Zemle, t. (Vol.) 60. Mir, Moskva. 275 pp. Price 2 Rbl. 85 Kop. (1975). In Russian. – Review in Referativ. Zhurn. 51. Astron., 9.51.52 (1975).

003.106 **Physics and chemistry in space, Vol. 8: Plasma instabilities and nonlinear effects.**
J. G. Roederer, J. T. Wasson (Editors).
Springer-Verlag, Berlin – Heidelberg – New York. Price $ 31.40 (1975). – (From Phys. Today, Vol. 28, No. 10, p. 62 (1975)).

003.107 **Théorie générale de l'univers physique (Réduction à la cinématique). Tome 1: Principes et méthodes.**
L. Romani.
Librairie Scientifique et Technique Albert Blanchard, Paris. 13 + 211 pp. Price Ffr. 48.00 (1975).

003.108 **Growth rhythms and the history of the earth's rotation.** G. D. Rosenberg, S. K. Runcorn (Editors).
Wiley-Interscience, John Wiley and Sons, London – New York. 16 + 559 pp. Price £ 22.50 (1975). ISBN 0-471-73616-3.

003.109 **Origin of the moon.** E. L. Ruskol.
Nauka, Moskva. 188 pp. Price 81 Kop. (1975). In Russian. – Reviews in Referativ. Zhurn. 51. Astron., 11.51. 275; 62. Issled. kosmich. prostranstva, 11.62.196 (1975); Zemlya i Vselennaya, 1975, No. 6, p. 88 - 89.

003.110 **Equilibrium configurations of degenerate gaseous masses.**
G. S. Saakyan, V. A. Ambartsumyan (Editors). Translated from the Russian edition (Moscow, 1972) by C. F. Hall. Halsted (John Wiley & Sons Inc.), New York; Israel Program for Scientific Translations, Jerusalem. 12 + 294 pp. Price

$ 36.00 (1974). – (From Science, Vol. 190, 77 (1975)).

003.111 **Physics of solar cosmic rays.** K. Sakurai. University of Tokyo Press, Tokyo. 428 pp. Price $ 39.50 (1974). – Review in Phys. Today, Vol. 28, No. 9, p. 56 - 57; 1975 (*J. V. Hollveg*).

003.112 **Commonsense celestial navigation.** H. Schlereth. Henry Regnery Company, Chicago, Ill. 231 pp. Price $ 12.95 (1975). – Review in Sky Telescope, Vol. 50, 186 (1975).

003.113 **The Skylab results.** W. C. Schneider, T. E. Hanes (Editors). Advances in the astronautical sciences, Vol. 31. American Astronautical Society, Tarzana, Calif. Part 1: 14 + 602 pp.; Part 2: 14 pp. + p. 603 - 1146. Price $ 60.00 per set (1975). (From Sky Telescope, Vol. 50, 48 (1975)).

003.114 **Gravitation und Kosmologie.** Eine Einführung in die allgemeine Relativitätstheorie. R. U. Sexl, H. K. Urbantke. Bibliographisches Institut AG., Mannheim – Wien – Zürich, B.I.-Wissenschaftsverlag. 335 pp. Price DM 58.00 (1975). ISBN 3-411-01487-3.

003.115 **Planetary geology.** N. M. Short. Prentice-Hall Inc., Englewood Cliffs, N.J. 15 + 361 pp. Price $ 17.95 (1975). – Reviews in Nature, Vol. 256, 685; 1975 (*P. J. Smith*); Sky Telescope, Vol. 50, 185 (1975).

003.116 **15jährige Sonnenhalobeobachtungen von Schwerin/ Meckl. aus klimatologisch-statistischer Sicht.** G. Schubert, W. Hänsch. Akademie-Verlag, Berlin. 53 pp. Price M 26.00 (1974). Review in Gerlands Beiträge Geophys., Vol. 84, 443; 1975 (*D. Spänkuch*).

003.117 **Geodätische Astronomie.** R. Sigl. Sammlung Wichmann, Neue Folge, Band 7. Herbert Wichmann Verlag, Karlsruhe. 11 + 369 pp. Price DM 58.00 (1975). ISBN 3-87907-041-5.

003.118 **Orbital elements of 45 meteorites. Atlas.** A. N. Simonenko. Nauka, Moskva. 68 pp. Price 85 Kop. (1975). In Russian. Review in Referativ. Zhurn. 51. Astron., 2.51.441 (1976).

003.119 **The changing global environment.** S. F. Singer. D. Reidel Publishing Company, Dordrecht, Holland–Boston, U.S.A. 8 + 423 pp. Price Dfl. 95.00 (1975). – Review in Orion, 33. Jahrgang, p. 222 - 223; 1975 (*H. Müller*).

003.120 **Course of theoretical astrophysics.** V. V. Sobolev. Second edition. Nauka, Moskva. 504 pp. Price 1 Rbl. 35 Kop. (1975). In Russian. – (From Referativ. Zhurn. 51. Astron., 11.51.46 (1975)).

003.121 **The Compton effect.** Turning point in physics. R. H. Stuewer. Science History Publications (Neale Watson), New York. 12 + 368 pp. Price $ 25 (1975). – Review in Science, Vol. 189, 711; 1975 (*J. L. Heilbron*).

003.122 **Catalog of solar particle events, 1955 - 1969.** Prepared under the auspices of Working Group 2 of the Inter-Union Commission on Solar-Terrestrial Physics by H. W. Dodson, E. R. Hedeman, R. W. Kreplin, M. J. Martres, V. N. Obridko, M. A. Shea, D. Smart, H. Tanaka. Z. Švestka, P. Simon (Editors), with the assistance of L. Fritzová-Švestková, A. Guitart. Astrophysics and Space Science Library, Vol. 49, D. Reidel Publishing Company, Dordrecht-Holland/Boston-U.S.A. 9 + 428 pp. Price Dfl. 185.00 (1975).

003.123 **Special relativity.** J. G. Taylor. Clarendon Press, Oxford; Oxford University Press, London. 106 pp. Price £ 1.95 (1975). – Review in Nature, Vol. 258, 368; 1975 (*P. C. W. Davies*).

003.124 **Im Banne des Alls.** A. N. Tomilin. Urania-Verlag Leipzig-Jena-Berlin, Leipzig. 296 pp. Price M 8.80 (1974). – Review in Astron. in der Schule, 12. Jahrgang, p. 95; 1975 (*H. Bernhard*).

003.125 **Microwave spectroscopy.** C. H. Townes, A. L. Schawlow. Dover Publications, Inc., New York. 698 pp. Price $ 7.50 (1975). – Review in Astrophys. Letters, Vol. 16, 184; 1975 (*S. P. Maran*).

003.126 **Invito al Firmamento – ciò che si vede / ciò che si conosce.** A. Valetti. Editrice 'La Scuola', Brescia. 64 pp. L. 4500 (1974).

003.127 **Science awakening II: The birth of astronomy.** B. L. van der Waerden. Oxford University Press, New York. 15 + 347 pp. Price $ 35.00 (1975). – Review in Icarus, Vol. 25, 638 - 639; 1975 (*R. Berendzen*).

003.128 **Astronomical photography, from Daguerreotype to electron camera.** G. de Vaucouleurs. Translated from the English edition. Nauka, Moskva, 136 pp. Price 46 Kop. (1975). In Russian. – (From Referativ. Zhurn. 51. Astron., 11.51.52 (1975)).

003.129 **Exobiology of the moon.** G. P. Vdovykin. Akademiya Nauk SSSR, Seriya 'Problemy nauki i tekhnicheskogo progressa'. Izdatel'stvo Nauka, Moskva. 119 pp. Price 40 Kop. (1975). In Russian. – Review in Referativ. Zhurn. 51. Astron., 7.51.359; 62. Issled. kosmich. prostranstva, 7.62.188 (1975).

003.130 **Sterrenkijkers zelf maken.** T. Vermeesch. A. J. G. Strengholt, Naarden (= Strengholt's Handenarbeid-Reeks No. 32). 131 pp. Price f 17.50 (1975). – Review in Zenit, Vol. 2, 308; 1975 (*M. Drummen*).

003.131 **Berliner Mondatlas.** A. Voigt, H. Giebler. Wilhelm-Foerster-Sternwarte Berlin. 2nd edition (1974). – Review in SuW, Vol. 14, 307 - 308; 1975 (*H. Haffner*).

003.132 **History of rocketry and space travel.** W. von Braun, F. I. Ordway III. Thomas Y. Crowell Company, New York. 308 pp. Price $ 25.00 (1975). – Review in Sky Telescope, Vol. 50, 185 (1975).

003.133 **The discovery of our Galaxy.** C. Whitney. Translated from the English edition, edited and with a foreword by T. S. Khromov. Mir, Moskva. 238 pp. Price 1 Rbl. 10 Kop. (1975). In Russian. Review in Referativ. Zhurn. 51. Astron., 1.51.2 (1976).

003.134 **The origin of the planets.** (Monographs on astronomical subjects). I. P. Williams. Adam Hilger, Ltd., Rank Precision Industries, London. 6 + 108 pp. Price £ 8.00 (1975). ISBN 0-85274-258-4.

003.135 **Carl Friedrich Gauß.** H. Wussing.
BSB G. B. Teubner Verlagsgesellschaft, Leipzig. 100 pp. Price M 4.70 (1974). – Review in Blick in das Weltall, 23. Jahrgang, p. 62 (1975).

003.136 **Cosmos satellites.** Yu. I. Zajtsev.
Nauka, Moskva. 136 pp. Price 52 Kop. (1975). In Russian. – Review in Zemlya i Vselennaya, 1975, No. 6, p. 90.

003.137 **Structure and evolution of the universe.**
Ya. B. Zel'dovich, I. D. Novikov.
Nauka, Moskva. 735 pp. Price 3 Rbl. 27 Kop. (1975). In Russian. – Review in Referativ. Zhurn. 51. Astron., 10.51. 729 (1975).

003.138 **Tides and resonances in the solar system.** Collection of articles. V. N. Zharkov (Editor).
Translated from the English edition.
Mir, Moskva. 288 pp. Price 1 Rbl. 92 Kop. (1975). In Russian. Review in Referativ. Zhurn. 51. Astron., 7. 51.44 (1975).

003.139 **Propagation of visible and infrared radiation in the atmosphere.** V. E. Zuev.
Translated from Russian and edited by P. Greenberg.
John Wiley & Sons, New York – Toronto; Israel Program for Scientific Translations, Jerusalem – London. 12 + 405 pp. Price £ 13.25 (1974). – Review in Gerlands Beiträge Geophys., Vol. 84, 444; 1975 (*D. Spänkuch*).

003.140 **Calculation of fastening constructions for precise astronomical mirrors.**
Trudy Mosk. in-ta inzh. geod., aehrofotosemki i kartogr., vyp. (No.) 60. Moskva. 122 pp. Price 45 Kop. (1975). In Russian. – (From Referativ. Zhurn. 51. Astron., 11.51.106 (1975)).

003.141 **Determination of the motion of space probes.**
In-t kosmich. issled. AN SSSR. Nauka, Moskva. 203 pp. Price 87 Kop. (1975). In Russian. – Review in Referativ. Zhurn. 51. Astron., 2. 51. 223; 62. Issled kosmich. prostranstva, 2. 62. 339 (1976).

003.142 **Investigations on cosmic physics and aeronomy.**
AN SSSR. Yakutsk. fil. Sib. otd. In-t kosmofiz. issled. i aehronomii. Yakutsk. 208 pp. Price 1 Rbl. 8 Kop. (1975). In Russian. – Review in Referativ. Zhurn. 62. Issled. kosmich. prostranstva, 10. 62. 181 (1975).

003.143 **Mars as viewed by Mariner 9.**
Mariner 9 Television Team and Planetology Program Principal Investigators.
National Aeronautics and Space Administration, Washington. To be ordered from Superintendent of Documents, U.S. Government Printing Office, Washington, D. C. Stock No. 3300-00577. 225 pp. Price $ 8.15 (1974). – Review in Sky Telescope, Vol. 50, 112 - 115; 1975 (*J. K. Beatty*).

003.144 **Report of the solar astronomy task force.**
Federal Council for Science and Technology, Washington (1975). – Review in Solar Physics, Vol. 43, 513; 1975 (*J. M. Pasachoff*).

003.145 **Some questions of physics of space.**
Vses. astron.-geod. o-vo AN SSSR. Leningr. otd. Astron. sekts., sb. 2. Moskva. 204 pp. Price 1 Rbl. 20 Kop. (1974). In Russian. – (From Referativ. Zhurn. 51. Astron., 10.51.87 (1975)).

003.146 **Sonne, Mond und Sterne – Bunter Kinder-Kosmos.**
Franckh'sche Verlagshandlung/Kosmos-Verlag, Stuttgart. 64 pp. Price DM 16.80 (1975). – Review in SuW, Vol. 14, 307; 1975 (*G. D. Roth*).

003.147 **Transactions of the eighth lectures dedicated to the scientific heritage and development of ideas of K. Eh. Tsiolkovskij**, Kaluga, 14–17 September 1973.
Sekts. Mekh. kosmich. poleta. AN SSSR. Komis. po razrabotke nauch. naslediya K. Eh. Tsiolkovskogo, Gos. muzej istorii kosmonavt. im. K. Eh. Tsiolkovskogo. Moskva. 218 pp. Price 1 Rbl. 7 Kop. (1974). In Russian. – (From Referativ. Zhurn. 62. Issled. kosm. prostranstva, 7.62.93 (1975)).

003.148 **Introduction to astronomical photometry.**
M. Golay.
D. Reidel Publishing Company, Dordrecht, Holland – Boston, U.S.A. Astrophysics and Space Science Library, Vol. 41. 9 + 363 pp. Price Dfl. 120.00, US $ 39.50 respectively (1974). ISBN 90-277-0428-7.

004 History of Astronomy, Chronology

004.001 **Did Chinese cosmology anticipate relativity?**
J. Gribbin.
Nature, Vol. 256, 619 - 620 (1975).
Recent interest in historic novae and supernovae, stimulated by the discovery of pulsars, has shown the value of ancient Chinese astronomical observations. But the Chinese also have a long record of "cosmological" theorising; in some cases their old concepts regarding the universe are remarkably similar to those developed in the West during the twentieth century.

004.002 **Johannes Keplers Beitrag zur Himmelsphysik.**
F. Krafft.
Internationales Kepler-Symposium 1971, (see 012.004), p. 55 - 139 (1973).

004.003 **Methodological aspects of Kepler's theory of refraction.** G. Buchdahl.
Internationales Kepler-Symposium 1971, (see 012.004), p. 141 - 167 (1973).

004.004 **Kepler und Harriot. Ihre Wege zum Brechungsgesetz.** J. A. Lohne.
Internationales Kepler-Symposium 1971, (see 012.004), p. 187 - 214 (1973).

004.005 **Johann Kepler and the development of physical optics.** J. Marek.
Internationales Kepler-Symposium 1971, (see 012.004), p. 215 - 227 (1973).

004.006 **Kepler's treatment of redundant observations or, the computer versus Kepler revisited.**
O. Gingerich.
Internationales Kepler-Symposium 1971, (see 012.004), p. 307 - 314 (1973).

004.007 **The medieval background to Copernicus.**
J. North.
Vistas in Astronomy, Vol. 17, (see 012.005), 3 - 25 (1975).

004.008 **The role of observations in the work of Copernicus.**
J. Dobrzycki.
Vistas in Astronomy, Vol. 17, (see 012.005), 27 - 35 (1975).

004.009 **Copernicus and the quest for certitude: "East" and "West".** B. Nelson.
Vistas in Astronomy, Vol. 17, (see 012.005), 39 - 46 (1975).

004.010 **Some episodes in the history of Copernicanism.**
R. Palter.
Vistas in Astronomy, Vol. 17, (see 012.005), 47 - 61 (1975).

004.011 **Geocentric to heliocentric to galactocentric to acentric: the continuing assault to the egocentric.**
R. Berendzen.
Vistas in Astronomy, Vol. 17, (see 012.005), 65 - 83 (1975).

004.012 **"Crisis" versus aesthetic in the Copernican revolution.** O. Gingerich.
Vistas in Astronomy, Vol. 17, (see 012.005), 85 - 95 (1975).

004.013 **Copernicanism as a scientific revolution.**
D. Shapere.
Vistas in Astronomy, Vol. 17, (see 012.005), 97 - 104 (1975).

004.014 **On the quest of direct proofs of the earth's motions.**
A. A. Mikhailov (*Mikhajlov*).
Vistas in Astronomy, Vol. 17, (see 012.005), 165 - 169 (1975).

004.015 **Copernicanism in Bruno, Kepler, and Galileo.**
S. Drake.
Vistas in Astronomy, Vol. 17, (see 012.005), 177 - 192 (1975).

004.016 **Copernicus and the impact of printing.**
O. Gingerich.
Vistas in Astronomy, Vol. 17, (see 012.005), 201 - 218 (1975).

004.017 **Leon Foucault's heritage to telescope making.**
J. Ashbrook.
Sky Telescope, Vol. 50, 26 - 27, 28 (1975).

004.018 **Kepler's century: prelude to Newton's.**
B. Cohen.
Vistas in Astronomy, Vol. 18, (see 012.006), 3 - 36 (1975).

004.019 **The Copernican system before Kepler.**
A. Hayli.
Vistas in Astronomy, Vol. 18, (see 012.006), 197 - 203 (1975).

004.020 **Copernicus, Rheticus and Kepler.**
W. D. Stahlman.
Vistas in Astronomy, Vol. 18, (see 012.006), 205 - 207 (1975).

004.021 **Kepler and Copernicus.** E. Rybka.
Vistas in Astronomy, Vol. 18, (see 012.006), 209 - 216 (1975).

004.022 **Kepler and the Copernican model.** P. Costabel.
Vistas in Astronomy, Vol. 18, (see 012.006), 217 - 222 (1975).

004.023 **Kepler and Tycho Brahe.** C. D. Hellman.
Vistas in Astronomy, Vol. 18, (see 012.006), 223 - 230 (1975).

004.024 **Kepler, Jost Bürgi and other Swiss friends.**
H. Balmer.
Vistas in Astronomy, Vol. 18, (see 012.006), 231 (1975).
Abstract.

004.025 **Kepler and Jost Bürgi's cross-beat clock.**
H. H. Grötzsch.
Vistas in Astronomy, Vol. 18, (see 012.006), 233 - 235 (1975).

004.026 **Kepler and Galileo.** S. Drake.
Vistas in Astronomy, Vol. 18, (see 012.006), 237 - 247 (1975).

004.027 **Kepler and Galileo.**
I. N. Veselovskii (*Veselovskij*).
Vistas in Astronomy, Vol. 18, (see 012.006), 249 - 253 (1975).

004.028 **Kepler and Galileo.** L. Rosino.
Vistas in Astronomy, Vol. 18, (see 012.006), 255 - 258 (1975).

004.029 **Kepler's place in astronomy.** O. Gingerich.
Vistas in Astronomy, Vol. 18, (see 012.006), 261 - 278 (1975).

004.030 **Kepler's place in the history of science.**
E. Rosen.
Vistas in Astronomy, Vol. 18, (see 012.006), 279 - 285 (1975).

004.031 **Nicolaus Copernicus and Johannes Kepler: new astronomy from old astronomy.** F. Krafft.
Vistas in Astronomy, Vol. 18, (see 012.006), 287 - 306 (1975).

004.032 **Johannes Kepler and the development of science.** P. Urban.
Vistas in Astronomy, Vol. 18, (see 012.006), 307 - 308 (1975).

004.033 **Kepler–a modern view.** W. Voisé.
Vistas in Astronomy, Vol. 18, (see 012.006), 309 - 312 (1975).

004.034 **Kepler and the Copernican revolution.** W. Gerlach.
Vistas in Astronomy, Vol. 18, (see 012.006), 313 (1975). – Abstract.

004.035 **Kepler's Harmonies, between Pansophia and Mathesis Universalis.** R. Haase.
Vistas in Astronomy, Vol. 18, (see 012.006), 519 - 533 (1975).

004.036 **Astronomical eggs–laid and unlaid–in Kepler's planetary theories, 1600 - 1605.**
D. T. Whiteside.
Vistas in Astronomy, Vol. 18, (see 012.006), 553 - 555 (1975).

004.037 **Kepler's contributions to celestial physics.** F. Krafft.
Vistas in Astronomy, Vol. 18, (see 012.006), 567 - 572 (1975).

004.038 **The elliptical orbit and the area law.** E. J. Aiton.
Vistas in Astronomy, Vol. 18, (see 012.006), 573 - 583 (1975).

004.039 **Infinitesimals and the area law.** E. J. Aiton.
Vistas in Astronomy, Vol. 18, (see 012.006), 585 - 586 (1975).

004.040 **Kepler's ellipse and area rule–their derivation from fact and conjecture.** C. A. Wilson.
Vistas in Astronomy, Vol. 18, (see 012.006), 587 - 591 (1975).

004.041 **Cassini and Kepler's second law.** G. Tabarroni.
Vistas in Astronomy, Vol. 18, (see 012.006), 593 (1975). – Abstract.

004.042 **The origins of Kepler's third law.** O. Gingerich.
Vistas in Astronomy, Vol. 18, (see 012.006), 595 - 601 (1975).

004.043 **On the effect of perturbations in the observations of Mars used by Kepler.** F. Schmeidler.
Vistas in Astronomy, Vol. 18, (see 012.006), 603 (1975). Abstract.

004.044 **Kepler's work on the lunar theory.** V. E. Thoren.
Vistas in Astronomy, Vol. 18, (see 012.006), 613 - 616 (1975).

004.045 **Kepler and the theory of gravitation.** H.-J. Treder.
Vistas in Astronomy, Vol. 18, (see 012.006), 617 - 619 (1975).

004.046 **From Kepler to Newton.** J. Lévy.
Vistas in Astronomy, Vol. 18, (see 012.006), 621 - 626 (1975).

004.047 **Kepler–the second great reviver of astronomy.** K. Ferrari d'Occhieppo.
Vistas in Astronomy, Vol. 18, (see 012.006), 627 (1975). Abstract.

004.048 **Methodology of Kepler's work.** R. S. Westman, G. Buchdahl, J. Mittelstrass, G. Alter, J. L. Russell, O. Gingerich, V. Bialas, Y. Maeyama, O. Neugebauer.
Vistas in Astronomy, Vol. 18, (see 012.006), 713 - 785 (1975).

004.049 **Kepler and comets.** C. D. Hellman.
Vistas in Astronomy, Vol. 18, (see 012.006), 789 - 796 (1975).

004.050 **Kepler's supernova of 1604.** F. Zwicky.
Vistas in Astronomy, Vol. 18, (see 012.006), 797 - 798 (1975).

004.051 **Optics and crystallography.** V. P. Linnik, V. L. Chenakal, G. Simon, V. Ronchi, P. Görlich, J. Marek, S. Szpilczynski, J. A. Lohne, H. M. Koelbing, I. I. Shafranovskii (*Shafranovskij*), A. Gawel.
Vistas in Astronomy, Vol. 18, (see 012.006), 809 - 878 (1975).

004.052 **Kepler's inventions, views and visions.** A. Adam, F. Prager, A. Nitschke, F. Seck, D. H. Menzel, C. Sagan, J. Veverka, J. B. Pollack, T. Gehrels.
Vistas in Astronomy, Vol. 18, (see 012.006), 881 - 909 (1975).

004.053 **Kepler's manuscripts.** A. A. Mikhailov (*Mikhajlov*).
Vistas in Astronomy, Vol. 18, (see 012.006), 933 - 935 (1975).

004.054 **On the scientific publications of Johannes Kepler.** E. Rybka.
Vistas in Astronomy, Vol. 18, (see 012.006), 945 - 946 (1975).

004.055 **Old Greenwich Observatory and Flamsteed's well.** J. Ashbrook.
Sky Telescope, Vol. 50, 157 (1975).

004.056 **On the Chinese calendar.** P. A. Startsev.
Istor.-astron. issledovaniya. Vyp. (No.) 12. Moskva, Nauka, 1975, p. 253 - 334. In Russian. – Abstr. in Referativ. Zhurn. 51. Astron., 7.51.2 (1975).

004.057 **Principles of constructing and using the Maya calendar.** V. I. Kozhanchikov.
Istor.-astron. issledovaniya. Vyp. (No.) 12. Moskva, Nauka, 1975, p. 335 - 354. In Russian. – Abstr. in Referativ. Zhurn. 51. Astron., 7.51.3 (1975).

004.058 **Little known Russian medieval chronological sources "Semitysyachniki".** R. A. Simonov.
Istor.-astron. issledovaniya. Vyp. (No.) 12. Moskva, Nauka, 1975, p. 109 - 112. In Russian. – Abstr. in Referativ. Zhurn. 51. Astron., 7.51.4 (1975).

004.059 **Sundial of the Trigorsk park.** O. V. Bobrovnikova, L. E. Majstrov.
Istor.-astron. issledovaniya. Vyp. (No.) 12. Moskva, Nauka, 1975, p. 127 - 137. In Russian. – Abstr. in Referativ. Zhurn. 51. Astron., 7.51.5 (1975).

004.060 **Astronomy in ancient India.** A. I. Volodarskij.
Istor.-astron. issledovaniya. Vyp. (No.) 12. Moskva, Nauka, 1975, p. 237 - 251. In Russian. – Abstr. in Referativ. Zhurn. 51. Astron., 7.51.10 (1975).

004.061 **Abu-Muhmud Chodjeni and Kushar-ibn-Labban–mathematicians and astronomers of the Xth century.**
Kh. F. Abdulla-zade.
Tezisy dokl. Resp. konf. molodykh uchenykh i spetsialistov TadzhSSR. Sekts. fiz. i astrofiz. Dushanbe, Donish, 1974, p.

30 - 31. In Russian. – Abstr. in Referativ. Zhurn. 51. Astron., 7.51.12 (1975).

004.062 **Al-Biruni's doctrine on solar motion.**
Z. G. Dzhalalova.
Istor.-astron. issledovaniya. Vyp. (No.) 12. Moskva, Nauka, 1975, p. 227 - 236. In Russian. – Abstr. in Referativ. Zhurn. 51. Astron., 7.51.13 (1975).

004.063 **Astronomical work of al-Biruni "The book of teaching the elements of the science on stars".**
B. A. Rozenfel'd.
Istor.-astron. issledovaniya. Vyp. (No.) 12. Moskva, Nauka, 1975, p. 205 - 226. In Russian. – Abstr. in Referativ. Zhurn. 51. Astron., 7.51.14 (1975).

004.064 **Astrolabes of the medieval Orient.**
A. K. Tagi-zade, S. A. Vakhabov.
Istor.-astron. issledovaniya. Vyp. (No.) 12. Moskva, Nauka, 1975, p. 169 - 204. In Russian. – Abstr. in Referativ. Zhurn. 51. Astron., 7.51.15 (1975).

004.065 **On the perception of the astronomical treatise "Shestokryl" in the XVth century in Russia.**
V. K. Kuzakov.
Istor.-astron. issledovaniya. Vyp. (No.) 12. Moskva, Nauka, 1975, p. 113 - 120. In Russian. – Abstr. in Referativ. Zhurn. 51. Astron., 7.51.17 (1975).

004.066 **Copernicus, Copernicanism and development of natural sciences.** Yu. A. Belyj.
Istor.-astron. issledovaniya. Vyp. (No.) 12. Moskva, Nauka, 1975, p. 15 - 72. In Russian. – Abstr. in Referativ. Zhurn. 51. Astron., 7.51.18 (1975).

004.067 **Kepler's work in the field of optics.** V. P. Linnik.
Istor.-astron. issledovaniya. Vyp. (No.) 12. Moskva, Nauka, 1975, p. 89 - 100. In Russian. – Abstr. in Referativ. Zhurn. 51. Astron., 7.51.20 (1975).

004.068 **Natural philosophy or physics. Feofan Prokopovich.**
Published by V. M. Nichik, M. D. Rogovich.
Istor.-astron. issledovaniya. Vyp. (No.) 12. Moskva, Nauka, 1975, p. 357 - 368. In Russian. – Abstr. in Referativ. Zhurn. 51. Astron., 7.51.21 (1975).

004.069 **Russian manuscript textbook on navigation of 1703.**
V. Rajen.
Istor.-astron. issledovaniya. Vyp. (No.) 12. Moskva, Nauka, 1975,p. 121 - 126. In Russian. – Abstr. in Referativ. Zhurn. 51. Astron., 7.51.22 (1975).

004.070 **"Sotnya astronomskaya". Lectures on astronomy in the thirthies of the XVIIth century at the Vilnius Academy.** N. D. Bespamyatnykh.
Istor.-astron. issledovaniya. Vyp. (No.) 12. Moskva, Nauka, 1975, p. 101 - 108. In Russian. – Abstr. in Referativ. Zhurn. 51. Astron., 7.51.23 (1975).

004.071 **Teaching of Copernicus at Vilnius University from the XVIIth century to the beginning of the XIXth century.** S. Matulajtite.
Istor.-astron. issledovaniya. Vyp. (No.) 12. Moskva, Nauka, 1975, p. 73 - 82. In Russian. – Abstr. in Referativ. Zhurn. 51. Astron., 7.51.24 (1975).

004.072 **On the history of the question of cosmic dispersion of light.** E. Ya. Vorob'eva.
Istor.-astron. issledovaniya. Vyp. (No.) 12. Moskva, Nauka, 1975, p. 157 - 167. In Russian. – Abstr. in Referativ. Zhurn. 51. Astron., 7.51.26 (1975).

004.073 **A Megalithic observatory on Dartmoor.**
J. E. Wood, A. Penny.
Nature, Vol. 257, 205 - 207 (1975). – Letter.

004.074 **Letters from Einstein to de Sitter on the nature of the universe.** C. Kahn, F. Kahn.
Nature, Vol. 257, 451 - 454 (1975).

The correspondence between Einstein and de Sitter during 1916 and 1918, recently discovered in the archives of the Sterrewacht at Leiden provides a fascinating glimpse of the thinking of the two principal progenitors of modern cosmology.

004.075 **John Flamsteed and the origins of the Greenwich astronomical tradition.** E. G. Forbes.
Journ. Navigation, Vol. 28, 251 - 262 (1975). – Eva G. R. Taylor lecture, Greenwich, 1975 May 21.

004.076 **A centennial tribute to Marcq St. Hilaire.**
C. H. Cotter.
Journ. Navigation, Vol. 28, 449 - 460 (1975).

004.077 **Christopher Wren: astronomy, architecture, and the mathematical sciences.** J. A. Bennett.
Journ. History Astron., Vol. 6, 149 - 184 (1975).

004.078 **The planetary theory of François Viète. 1. The fundamental planetary models.** N. M. Swerdlow.
Journ. History Astron., Vol. 6, 185 - 208 (1975).

004.079 **Der neue Stern von 1572.** M. Griesser.
Orion, 33. Jahrgang, p. 135 - 138 (1975).

004.080 **Adriaan van Maanen's measurements of solar spectra for a general magnetic field.** N. S. Hetherington.
Quarterly Journ. Roy. Astron. Soc., Vol. 16, 235 - 244 (1975).

004.081 **From the history of invention of the astrolabe.**
A. K. Tagi-Zade.
Vopr. istorii estestvozn. i tekhn. Vyp. (No.) 4 (49). Moskva, Nauka, 1975, p. 42 - 54, 101. In Russian. – Abstr. in Referativ. Zhurn. 51. Astron., 9.51.10 (1975).

004.082 **On the prehistory of the Astrophysical Observatory Potsdam (1865 to 1874).** D. B. Herrmann.
Astron. Nachr., Vol. 296, 245 - 259 (1975). In German.

004.083 **Konnten die Völker des Altertums Sonnenfinsternisse voraussagen?** K. Hausberg.
SuW, Vol. 14, 364 - 366 (1975).

004.084 **Al-Farabi. On the occasion of the 1100th anniversary of his birthday.** A. Kubesov.
Priroda, No. 10.75, p. 70 - 75 (1975). In Russian.

004.085 **Giordano Bruno and the present time.**
B. G. Kuznetsov.
Zemlya i Vselennaya, 1975, No. 5, p. 18 - 24. In Russian.

004.086 **300 years Greenwich Observatory.**
A. A. Mikhajlov.
Zemlya i Vselennaya, 1975, No. 5, p. 74 - 80. In Russian.

004.087 **Success and failure in Copernicus' planetary theories.**
K. P. Moesgaard.
Arch. internat. Histoire Sci., Vol. 24, 73 - 111 (1974).

Copernicus' work on the planets is analysed under four headings. (1) The formulation, on the basis of "Almagest" patterns, of long-term generalised models interpretable as heliostatic mechanisms. (2) The agreement, using Ptolemy's observations, of the parameters of these models with those of the "Almagest". (3) The establishment, using sixteenth century observations, of sixteenth century parameters of the models. (4) The comparison of ancient and Renaissance parameters.

004.088 **I presupposti matematici nell' opera di Niccolò Copernico.** F. S. Rossi.
Archimede, Vol. 25, 225 - 234 (1973).
At first a brief survey of Copernicus' heliocentric theory is presented. Then the author tries to give a general and comprehensive account of Copernicus' mathematical activities, especially in the field of plane and spherical trigonometry. At last the well-known lemma III from "De Revolutionibus" (Book I, Ch. XIV), the originality of its proof and its significance are treated.

004.089 **Copernic et les Pays-Bas.** A. De Smet.
Janus, Vol. 60, 13 - 23 (1973).
This paper presents a documented account of the diffusion of knowledge of the Copernican system in the Low Countries.

004.090 **Da Aristotele a Abû Ma'shar, da Richard de Fournival a Guglielmo da Pastrengo. Un'opera astrologica controversa di Alberto Magno.** P. Zambelli.
Physis, Vol. 15, 375 - 400 (1974).

004.091 **History of radio astronomical observation in Czechoslovakia.** J. Olmr.
Kozmos, Vol. 6, 135 - 137 (1975). In Czech.

004.092 **The Accademia del Cimento and Saturn's ring.** A. Van Helden.
Physis, Vol. 15, 237 - 259 (1974).

004.093 **Light, pressure, and rectilinear propagation: Descartes' celestial optics and Newton's hydrostatics.** A. E. Shapiro.
Studies History Philos. Sci., Vol. 5, 239 - 296 (1974).

004.094 **Peder Horrebow's solution of the Kepler-problem.** N. T. Jørgensen.
Centaurus, Vol. 18, 173 - 183 (1974).

004.095 **Argumente für und wider das heliozentrische Weltbild.** K. Møller Pedersen.
Ann. Sci., Vol. 32, 163 - 167 (1975).
It is shown that the general acceptance of the heliocentric system through the work of astronomers resulted not so much from empirical considerations but rather from the fact that this system was suited to a comprehensive, general dynamics.

004.096 **Canada's lost opportunity.** A. V. Douglas.
Journ. Roy. Astron. Soc. Canada, Vol. 69, 248 - 249 (1975). – Abstr. CAS.

004.097 **The development of astronomy in Fredericton, New Brunswick, between 1847 and 1876.**
J. E. Kennedy.
Journ. Roy. Astron. Soc. Canada, Vol. 69, 249 - 250 (1975). Abstr. CAS.

004.098 **Early astronomy at Queen's.** J. N. Clarke.
Journ. Roy. Astron. Soc. Canada, Vol. 69, 250 (1975). – Abstr. CAS.

004.099 **Antique astrolabes.** A. K. Tagi-Zade.
Trudy XVII Nauch. konf. aspirantov i ml. nauch. sotrudn. In-ta istorii estestvozn. i tekhn. AN SSSR. Sekts. istorii mat. i mekh. Moskva, 1975, p. 3 - 17. In Russian. – Abstr. in Referativ. Zhurn. 51. Astron., 12.51.5 (1975).

004.100 **Information on al-Biruni's treatises on astrolabes.** S. A. Vakhabov, A. K. Tagi-Zade.
Trudy XVII Nauch. konf. aspirantov i ml. nauch. sotrudn. In-ta istorii estestvozn. i tekhn. AN SSSR. Sekts. istorii mat. i mekh. Moskva, 1975, p. 18 - 31. In Russian. – Abstr. in Referativ. Zhurn. 51. Astron., 12.51.6 (1975).

004.101 **On Ibn Sine's astronomical instrument.** S. A. Vakhabov.
Trudy XVII Nauch. konf. aspirantov i ml. nauch. sotrudn. In-ta istorii estestvozn. i tekhn. AN SSSR. Sekts. istorii mat. i mekh. Moskva, 1975, p. 32 - 36. In Russian. – Abstr. in Referativ. Zhurn. 51. Astron., 12.51.7 (1975).

004.102 **The astronomy of Jōbir ibn Aflah.** R. P. Lerch.
Centaurus, Vol. 19, 85 - 107 (1975).

004.103 **An analog computer for solving problems of spherical astronomy: The Shakkāzīya quadrant of Jamāl al-Dīn al-Māridīnī.** D. A. King.
Arch. Internationales Histoire Sci., Vol. 24, 219 - 242 (1974/75).

004.104 **Ibn aṣ-Ṣalāḥ. Zur Kritik der Koordinatenüberlieferung im Sternkatalog des Almagest.**
Arabischer Text nebst deutscher Übersetzung, Einleitung und Anhang. P. Kunitzsch (Editor).
Abh. Akad. Wiss. Göttingen, Phil.-hist. Kl., 3. Folge, No. 94. Vandenhoeck & Ruprecht, Göttingen. ISBN 3-525-82370-3. 160 pp. Price DM 88.00 (1975).

004.105 **Ein Exemplar der Chronologie von Newton aus dem Besitz von Pierre Des Maizeaux in der Bibliothèque de la Ville de Colmar.** K. Figala.
Verhandl. Naturforsch. Ges. Basel, Vol. 84, (No. 2), 646 - 697 (1974) = Veröff. Forschungsinst. des Deutschen Museums für die Geschichte der Naturwissesnschaften und der Technik, Reihe C, Quellentexte und Übersetzungen, No. 16 (1975).

004.106 **Zur Namengebung Kairos (al-Qāhir = Mars?).** P. Kunitzsch.
Der Islam [Walter de Gruyter, Berlin – New York], Vol. 52, 209 - 225 (1975).

004.107 **Die arabischen Sternbilder des Südhimmels.** (Fortsetzung und Schluß). P. Kunitzsch.
Der Islam [Walter de Gruyter, Berlin – New York], Vol. 52, 263 - 277 (1975).

004.108 **Giovanni Cassini und die französischen Gradmessungen im 17. und 18. Jahrhundert.** D. Schoeps.
Vermessungstechnik, 23. Jahrgang, p. 471 - 472 (1975).

004.109 **On Kirik Novgorodez's cosmology.** M. F. Mur'yanov.
Vopr. istorii astron. No. 3. Moskva, 1974, p. 12 - 17, 196. In Russian.

004.110 **Copernican theory and the problem of longitudes.** K. F. Ogorodnikov.
Vopr. istorii astron. No. 3. Moskva, 1974, p. 19 - 33, 196. In Russian. – Abstr. in Referativ. Zhurn. 51. Astron., 1.51.12 (1976).

004.111 **Copernican theory and P. Ramus.** G. P. Matvievskaya.
Vopr. istorii astron. No. 3. Moskva, 1974, p. 34 - 42, 196. In Russian. – Abstr. in Referativ. Zhurn. 51. Astron., 1.51.13 (1976).

004.112 **Astronomical-geodetical and cartographic works at**

the Academy of Sciences (Concluding review).
L. S. Khrenov.
Vopr. istorii astron. No. 3. Moskva, 1974, p. 160 - 183, 198. In Russian. – Abstr. in Referativ. Zhurn. 51. Astron., 1.51.15 (1976).

004.113 **On the history of study of telluric lines of the solar spectrum (to the year 1900).**
O. A. Mel'nikov, E. D. Khilov.
Vopr. istorii astron. No. 3, Moskva, 1974, p. 124 - 138, 197. In Russian. – Abstr. in Referativ. Zhurn. 51. Astron., 1.51.27 (1976).

004.114 **Delisle's forgotten article on the oriental astronomy.** N. I. Nevskaya.
Vopr. istorii astron. No. 3. Moskva, 1974, p. 94 - 123, 197. In Russian. – Abstr. in Referativ. Zhurn. 51. Astron., 1.51.41 (1976).

004.115 **A. A. Andronov as historian of science.**
E. S. Bojko.
Vopr. istorii estestvozn. i tekhn. Vyp. (No.) 1 (50). Moskva, Nauka, 1975, p. 51 - 53, 104. In Russian. – Abstr. in Referativ. Zhurn. 51. Astron., 1.51.46 (1976).

004.116 **From the first ideas to the measurements of the earth.** L. Zajdler.
Urania Kraków, Vol. 46, 267 - 271 (1975). In Polish.

004.117 **T. Banachiewicz in Dorpat (1915 - 1918).**
A. W. Szpilewski.
Urania Kraków, Vol. 46, 290 - 295, 322 - 328 (1975). In Polish.

004.118 **Jelgavas Astronomiskā Observatorija.** I. Daube.
Zvaigžņotā debess, 1975. gada pavasaris, p. 36 - 46.

004.119 **Astronomia e calendario dell'antico Messico.**
L. Gratton.
Giornale Astron., Vol. 1, 261 - 282 (1975).

004.120 **Authentisches über die mittelalterliche Almagest-Überlieferung.** K. Ferrari d'Occhieppo.
Sternenbote, 18. Jahrgang, p. 146 - 151 (1975).

004.121 **Man, society and the universe.** N. S. Hetherington.
Mercury, (Journ. Astron. Soc. Pacific), Vol. 4, No. 6, p. 9 - 12 (1975).

004.122 **Astra medicis.** R. R. Reynoso.
El Universo, Vol. 29, No. 112, p. 4 - 6 (1975).

004.123 **The physical doctrine of Aristotle and the Copernican revolution.** V. Hütt.
Tartu Ulikooli toimetised, Uch. zap. Tartus. un-ta, 1975, vyp. (No.) 360, p. 29 - 40. In Russian. – Abstr. in Referativ. Zhurn. 51. Astron. 2.51.3 (1976).

004.124 **Copernicus and the devolopment of the classical concept of time.** Yu. B. Molchanov.
Tartu Ulikooli toimetised, Uch. zap. Tartus. un-ta, 1975, vyp. (No.) 360, p. 41 - 46. In Russian. – Abstr. in Referativ. Zhurn. 51. Astron., 2.51.5 (1976).

004.125 **The way of scientific thinking of Copernicus.**
L. A. Mikeshina.
Tartu Ulikooli toimetised, Uch. zap. Tartus. un-ta, 1975, vyp. (No.) 360, p. 60 - 70. In Russian. – Abstr. in Referativ. Zhurn. 51. Astron., 2.51.6 (1976).

004.126 **Search for the planet X.** T. Sajmon.
Vasiona, Vol. 23, 44 - 51 (1975). In Serbo-Croatian.

004.127 **Astronomers and other scientists on St. Helena.**
W. G. Tatham, K. A. Harwood.
Ann.Sci., (*GB*), Vol. 31, 489 - 509 (1974).
An outline is presented of discoveries of scientific sites on St. Helena, including the site of the observatory of Edmond Halley. The site of a temporary observatory built in order to observe the transit of Venus in 1761, was also identified. Other observatories, for astronomical and meteorological work, which were built on the island, are described.

004.128 **William Stukeley and the early history of the orrery.**
J. R. Millburn.
Ann. Sci., (*GB*), Vol. 31, 511 - 527 (1974).
Mechanical models of the solar system engaged the attention of some well-known scientific instrument makers during the eighteenth century. The history of such models, called orreries, is presented, particularly those accounts of the development of the orrery given in the papers of William Stukeley.

Forschungen und Publikationen zur Geschichte der Astronomie in der Deutschen Demokratischen Republik. Teil II. Eine Bibliographie (1970–1974).
See Abstr. 002.006.

Ancient cosmologies. See Abstr. 003.025.

Geschichte der Astronomie von Herschel bis Hertzsprung. See Abstr. 003.069.

The foundation of the Nikolayev Observatory.
See Abstr. 009.002.

The historical search for stellar parallax.
See Abstr. 111.003.

The historical search for stellar parallax.
See Abstr. 111.008.

005 Biography

005.001 **Johannes Kepler – Homo iste. Selbsterkenntnis – Welterkenntnis – Gotteserkenntnis.** B. Sticker.
Internationales Kepler-Symposium 1971, (see 012.004), p. 455 - 474 (1973).

005.002 **Copernicus as a man and contributor to the advancement of science.** E. Rosen.
Vistas in Astronomy, Vol. 17, (see 012.005), 193 - 199 (1975).

005.003 **Johannes Kepler homo iste.** B. Sticker.
Vistas in Astronomy, Vol. 18, (see 012.006), 55 - 56 (1975).

005.004 **Kepler's life and times.**
W. Gerlach, M. List, H. Tursky, B. Morando, J. Hübner, F. Zagar, C. Agostinelli, E. Bernleithner, A. Beer, P. Beer.
Vistas in Astronomy, Vol. 18, (see 012.006), 73 - 131 (1975).

005.005 **Kepler's wanderings.**
J. Sydow, B. Sutter, Z. Horský, K. Hujer, H. Tursky, H. E. Specker, J. Dobrzycki, S. Färber, A. Menath, K. Rocznik, W. Boll.
Vistas in Astronomy, Vol. 18, (see 012.006), 135 - 194 (1975).

005.006 **Kepler's religious and philosophical beliefs.**
E. Rosen, E. W. Gerdes, J. Hübner, M. List, O. Wenig, N. Schiffers, W. Petri.
Vistas in Astronomy, Vol. 18, (see 012.006), 317 - 396 (1975).

005.007 **Mysticism, astrology and alchemy.**
A. Beer, J. O. Fleckenstein, G. Simon, M. List, G. Colombo, A. Postl, J. A. Hynek, K. Figala.
Vistas in Astronomy, Vol. 18, (see 012.006), 399 - 469 (1975).

005.008 **Kepler as mathematican and physicist.**
P. Costabel, Yu. A. Belyi (*Belyj*), H. S. M. Coxeter, E. J. Aiton, A. E. L. Davis, J. E. Hofmann, S. Bochner, W. Kaunzner, R. J. Seeger, Yu. A. Danilov, Ya. A. Smorodinskii (*Smorodinskij*), H. M. Nobis.
Vistas in Astronomy, Vol. 18, (see 012.006), 631 - 709 (1975).

005.009 **Chronology (*Kepler*).**
Vistas in Astronomy, Vol. 18, (see 012.006), 917 - 919 (1975).

005.010 **The reverend Thomas William Webb: 1807 - 1885.**
E. G. Moore.
Journ. British Astron. Ass., Vol. 85, 426 - 429 (1975).

005.011 **Heinrich Samuel Schwabe, 1789 - 1875.**
M. Bean.
Journ. British Astron. Ass., Vol. 85, 532 - 533 (1975).

005.012 **"...das war mein Weinberg!"** Zur 100. Wiederkehr des Geburtstages von Bruno Hans Bürgel am 14. November 1975. A. Zenker.
Sterne, 51. Jahrgang, p. 148 - 158 (1975).

005.013 **F. W. A. Argelander in Finland (1823–1837).**
C. Gefwert.
Journ. History Astron., Vol. 6, 209 - 211 (1975).

005.014 **Adolphe Quetelet (1796–1874). Ses initiatives créatrices en géophysique.** E. Lahaye.
Ciel et Terre, Vol. 91, 257 - 260 (1975).

005.015 **The founders of American meteoritics.**
E. F. Lange.
Meteoritics, Vol. 10, 241 - 252 (1975).

005.016 **To the 100th anniversary of the birthday of the corresponding member of the USSR Academy of Sciences G. A. Tikhov.**
V. P. Bedenko, Z. S. Parshina, V. S. Sokolova.
Vestn. AN KazSSR, 1975, No. 5, p. 73 - 77. In Russian.
Abstr. in Referativ. Zhurn. 51. Astron., 11.51.9 (1975).

005.017 **G. A. Tikhov (on the occasion of the 100th anniversary of his birthday).** A. I. Eremeeva.
Zemlya i Vselennaya, 1975, No. 6, p. 42 - 45. In Russian.

005.018 **Kopernikus – ein Revolutionär?** W. Kuhn.
Math. naturwiss. Unterricht, Vol. 26, 449 - 457 (1973).

005.019 **Nicolaus Copernicus: Leben und Wirken.**
H. Wussing.
NTM, Schriftenreihe Geschichte Naturw. Techn. Med., Vol. 11, No. 1, p. 98 - 104 (1974).

005.020 **Rayleigh centennial.** J. N. Howard, J. W. Chamberlain, A. N. Wright, C. A. Winkler.
Applied Optics, Vol. 14, 2827 - 2830 (1975).

005.021 **A great Soviet meteoriticist.** E. L. Krinov.
Probl. mineral'n. syr'ya. Moskva, Nauka, 1975, p. 301 - 303. In Russian. – Abstr. in Referativ. Zhurn. 51. Astron., 12.51.14 (1975). – Concerning A. E. Fersman.

005.022 **F. W. A. Argelander (1799 - 1875).**
M. Schmögner.
Kozmos, Vol. 6, 150 (1975). In Slovak.

005.023 **C. Flammarion (1842 - 1925).**
Kozmos, Vol. 6, 151 - 152 (1975). In Slovak.

005.024 **Thaddaeus Hagecius (1525 - 1600).**
Z. Horský.
Říše hvězd, Vol. 56, 228 - 229 (1975). In Czech.

005.025 **Ein Brief des Lichtenberg-Schülers Benzenberg.**
W. Gresky.
Gauss-Ges. Göttingen, Mitt. No. 12, p. 25 - 33 (1975).

005.026 **George Calver – east Anglian telescope maker.**
H. E. Dall.
Journ. British Astron. Ass., Vol. 86, 49 - 52 (1975).

005.027 **Johann Bayer and his star atlas – reconsidered.**
D. J. Warner.
Journ. British Astron. Ass., Vol. 86, 53 - 54 (1975).

005.028 **Friedrich Wilhelm August Argelander.**
I. D. Howarth.
Journ. British Astron. Ass., Vol. 86, 55 - 56 (1975).

005.029 **G. A. Tikhov (to the 100th anniversary of his birthday).** A. F. Bogorodskij.
Vestn. Kiev. Un-ta, Ser. Astron., No. 17, p. 3 - 12 (1975). In Russian.

The article contains a short outline on G. A. Tikhov's life and his scientific activity.

005.030 **Belopol'skij's work in the Commission for Solar Researches.** E. Ya. Vorob'eva.
Vopr. istorii astron. No. 3. Moskva, 1974, p. 139 - 159, 197. In Russian. – Abstr. in Referativ. Zhurn. 51. Astron., 1.51.44 (1976).

005.031 **Leonid Alexandrovich Demin (1897–1973).**
Vopr. istorii astron. No. 3. Moskva, 1974, p. 187 - 189, 198. In Russian. – Abstr. in Referativ. Zhurn. 51. Astron., 1.51.45 (1976).

005.032 **Camille Flammarion** [À l'occasion du 50ᵉ anniversaire de sa mort].
L'Astronomie, 89ᵉ année, p. 399 - 442 (1975).
Included are: Cérémonie à la Sorbonne (*P. de La Cotardière*); Allocution de M. J.-C. Pecker; La voix de Camille Flammarion; La vie de Camille Flammarion (*A. Duplay*), p. 405 - 419; Un siècle de popularisation de la science (*Y. Novozhilov*), p. 420 - 426; Allocution de M. Jean-Pierre Soisson en l'honneur de Camille Flammarion; Une page de Camille Flammarion (*M. Etcheverry*); Témoignages divers; Progrès récents de l'astronomie (see 013.012); Cérémonie Camille Flammarion à Juvisy (*P. de La Cotardière*).

005.033 **Plaque unveiled at Dominion Observatory for W. F. King.** A. E. Covington.
Journ. Roy. Astron. Soc. Canada, Vol. 69, 286 - 288 (1975).

005.034 **Oriental astronomers.** T. Z. Dworak.
Urania Kraków, Vol. 46, 328 - 332 (1975). In Polish.

005.035 **Double jubilee in the Astronomical Observatory of the Jagellonian University.** J. Mietelski, M. Winiarski.
Postępy Astron., Vol. 23, 219 - 221 (1975). In Polish.

A. Bečvář, 1901 - 1965.
Kozmos, Vol. 6, 185 (1975). In Slovak.

Kepler's inventions, views and visions.
See Abstr. 004.052.

006 Personal Notes

E. E. Becklin received AAS Pierce Prize.
Phys. Today, Vol. 28, No. 9, p. 65 (1975).

M. Burbidge, president-elect of AAS.
Physics Today, Vol. 28, No. 10, p. 71 (1975).

G. A. Chebotarev, on the occasion of his 60th birthday. I. D. Zhongolovich.
Byull. Inst. Teoret. Astron., *Leningrad,* Vol. 13, 605 - 608 (1975). In Russian.

G. Duboshin, 70th birthday. L. Dīriķe.
Zvaigžņotā debess, 1975. gada vasara, p. 41 - 42. In Lettish.

R. Gott III received the Trumpler award.
Mercury, (Journ. Astron. Soc. Pacific), Vol. 4, No. 6, p. 2 (1975).

J. L. Greenstein received the Gold Medal of the Royal Astronomical Society. D. E. Blackwell.
Quarterly Journ. Roy. Astron. Soc., Vol. 16, 356 - 357 (1975).

V. Guth, 60th birthday. B. Šternberk.
Pokroky, Vol. 20, 111 - 113 (1975). In Czech.

V. Guth, 60th birthday.
Stud. Geophys. Geod., Vol. 19, 308 - 309 (1975).

S. W. Hawking received the Eddington Medal of the Royal Astronomical Society. D. E. Blackwell.
Quarterly Journ. Roy. Astron. Soc., Vol. 16, 359 - 360 (1975).

A. Hewish wins the Nobel prize for physics.
L. I. Matveenko.
Zemlya i Vselennaya, 1975, No. 5, p. 56 - 61. In Russian.

B. C. Murray, director of Jet Propulsion Laboratory.
Phys. Today, Vol. 28, No. 9, p. 65 (1975).

B. C. Murray, director of the Jet Propulsion Laboratory in Pasadena, California.
Sky Telescope, Vol. 50, 94 (1975).

E. J. Öpik received the Gold Medal of the Royal Astronomical Society.
Irish Astron. Journ., Vol. 12, 70 - 72 (1975).

E. J. Öpik received the Gold Medal of the Royal Astronomical Society. D. E. Blackwell.
Quarterly Journ. Roy. Astron. Soc., Vol. 16, 358 - 359 (1975).

R. Penrose received the Eddington Medal of the Royal Astronomical Society. D. E. Blackwell.
Quarterly Journ. Roy. Astron. Soc., Vol. 16, 359 - 360 (1975).

M. Pupareli, acting director of the Instituto de Astronomía y Física del Espacio, Buenos Aires, Argentina.
Inform. Bull. Southern Hemisphere, No. 25, p. 32 (1974).

H. Rohr, 80th birthday. E. Wiedemann.
Orion, 33. Jahrgang, p. 200 (1975).

M. Ryle wins the Nobel prize for physics.
L. I. Matveenko.
Zemlya i Vselennaya, 1975, No. 5, p. 56 - 61. In Russian.

V. S. Safronov received the O. J. Schmidt prize.
Zemlya i Vselennaya, 1975, No. 4, p. 21. In Russian.

A. Sandage received the Catherine Wolfe Bruce Gold Medal award.
Mercury, (Journ. Astron. Soc. Pacific), Vol. 4, No. 6, p. 2 (1975).

K. A. Strand received la médaille annuelle de l'A.D.I.O.N. P. Couteau, P. Muller.
L'Astronomie, 89e année, p. 330 (1975).

B. Thüring, 70th birthday. A. Fadini.
Wetter–Boden–Mensch, No. 22, p. 1490 (1975).

K. Walter, 70th birthday.
J. Dachs, J. Pfleiderer, M. Reinhardt.
SuW, Vol. 14, 281 (1975).

007 Obituaries

G. Armellini-Conti, 1891 March 19 – 1974 November 8. M. Cimino.
Mem. Soc. Astron. Italiana, Vol. 46, 127 - 128 (1975).

G. S. Badalyan, 1908 - 1972 April 6.
Soobshch. Byurakan. Obs., vyp. (No.) 46, p. 152 - 154 (1975). In Russian.

R. Brandt, died 1975 January 18.
N. Richter.
Sterne, Vol. 51, 247 - 249 (1975).

G. A. Chebotarev, 1913 August 1 – 1975 August 4.
Yu. V. Batrakov.
Byull. Inst. Teoret. Astron., *Leningrad*, Vol. 14, 133 - 135 (1975). In Russian.

G. A. Chebotarev, 1913 August 1 - 1975 August 4.
V. A. Brumberg.
Celestial Mechanics, Vol. 12, 395 - 396 (1975).

G. A. Chebotarev, 1913 - 1975 August 4.
Zemlya i Vselennaya, 1975, No. 6, p. 40 - 41. In Russian.

G. M. Clemence, 1908 August 16 - 1974 November 22. H. S. Hogg.
Journ. Roy. Astron. Soc. Canada, Vol. 69, 136 - 138 (1975).

V. C. A. Ferraro, 1907 April 10 – 1974 January 4.
M. Cutolo.
Mem. Soc. Astron. Italiana, Vol. 46, 123 - 126 (1975).

E.-H. Geneslay, 1892 January 17 - 1975 May 13.
L. Tartois.
L'Astronomie, 89e année, p. 391 - 392 (1975).

A. E. Glancy died 1975 May 19.
Sky Telescope, Vol. 50, 12 (1975).

S. Herrick, 1912 May 29 - 1974 March 24.
D. H. Sadler.
Quarterly Journ. Roy. Astron. Soc., Vol. 16, 321 - 322 (1975).

J. Hopmann died 1975 October 11.
Sternenbote, 18. Jahrgang, p. 189 (1975).

H. Kienle died 1975 February 15.
Orion, 33. Jahrgang, p. 111 (1975).

H. Kienle, 1895 October 22 - 1975 February 15.
D. Labs.
Ruperto Carola, Univ. Heidelberg, 27. Jahrgang, p. 114 - 115 (1975).

K.-O. Kiepenheuer died 1975 May 23.
Observatory, Vol. 95, 227 (1975).

K. O. Kiepenheuer died 1975 May 23.
Sky Telescope, Vol. 50, 230 (1975).

K. O. Kiepenheuer, 1910 November 10 – 1975 May 23. A. Bruzek.
Solar Physics, Vol. 43, 3 - 7 (1975).

K. O. Kiepenheur, 1910 November 10 - 1975 May 23. A. Bruzek.
SuW, Vol. 14, 352 - 353 (1975).

E. M. Lindsay, 1907 - 1974.
Irish Astron. Journ., Vol. 11, 209 (1975).

R. O. Redman, 1905 July 17 - 1975 March 6.
R. H. Stoy.
Monthly Notes Astron. Soc. Southern Africa, Vol. 34, 132 - 138 (1975).

W. M. Smart, died 1975 September 17.
Nature, Vol. 257, 432 (1975).

W. M. Smart died 1975 September 17.
Observatory, Vol. 95, 227 (1975).

W. M. Smart died 1975 September 17.
Sky Telescope, Vol. 50, 380 (1975).

P. S. Sørensen, 1898 - 1975. P. Darnell.
Astron. Tidssk., Årg. 8, 138 - 139 (1975).

W. H. Steavenson died 1975 September 23.
Observatory, Vol. 95, 227 (1975).

W. H. Ward, 1900 - 1973 March 2.
Southern Stars, Vol. 26, 93 - 94 (1975).

K. Wurm, 1899 July – 1975 February 16.
L. Rosino.
Astrophys. Space Sci., Vol. 35, 221 - 222 (1975).

W. Zonn, 1905 - 1975. W. Iwanowska.
Postępy Astron., Vol. 23, 87 - 90 (1975). In Polish.

W. Zonn, 1905 - 1975. S. Piotrowski.
Postępy Astron., Vol. 23, 91 - 95 (1975). In Polish.

008 Observatories, Institutes

Reports, communications and publications of observatories and astronomical institutes are recorded in this section; included are numbered series of reprints. Whenever possible, the numbers of the abstracts referring to the publications are given. Observatories and institutes are listed in alphabetical order of their towns. In some cases observatory publications do not give the name of the town; the following list which gives names and towns of some institutions may serve as an aid in such cases.

Aarne Karjalainen Observatory	**Oulu**, Finland
Algonquin Radio Observatory	**Lake Traverse**, Ontario, Canada
Allegheny Observatory	**Pittsburgh**, Pennsylvania
Archenhold-Sternwarte	**Berlin-Treptow**, Germany
Arthur J. Dyer Observatory	**Nashville**, Tennessee
Astronomical Latitude Station, Polish Academy of Sciences	**Borowiec**, Poland
Bosscha Observatory	**Lembang**, Indonesia
Boyden Observatory	**Bloemfontein**, South Africa
Bureau International de l'Heure	**Paris**, France
Cajigal Observatory	**Caracas**, Venezuela
California Institute of Technology	**Pasadena**, California
Carter Observatory	**Wellington**, New Zealand
Catalina Station	**Tucson**, Arizona
Cavendish Laboratory	**Cambridge**, England
Ceskoslovenská Akademie Ved Astronomický Ustav	**Praha**, Czechoslovakia
Chamberlin Observatory, University of Denver	**Denver**, Colorado
Commonwealth Observatory	**Canberra**, Australia
Corralitos Observatory	**Las Cruces**, New Mexico
David Dunlap Observatory, University of Toronto	**Richmond Hill**, Ontario
Dearborn Observatory	**Evanston**, Illinois
Department of Astronomy and Observatory, Univ. California	**Los Angeles**, California
Department of Astronomy, University of Texas	**Austin**, Texas
Division Radiophysics, C.S.I.R.O. University Grounds	**Sydney**, New South Wales
Dominion Astrophysical Observatory	**Victoria**, British Columbia
Dominion Observatory	**Ottawa**, Ontario
Dominion Radio Astrophysical Observatory	**Penticton**, British Columbia
Dudley Observatory	**Albany**, New York
Dunsink Observatory	**Dublin**, Ireland
Engelhardt Observatory	**Kazan**, U.S.S.R.
European Southern Observatory	**Hamburg**, Federal German Republic
Fernbank Observatory	**Atlanta**, Georgia
Five College Observatories	**Amherst**, Massachusetts
Florida State University Radio Observatory	**Tallahassee**, Florida
Flower and Cook Observatories, University of Pennsylvania	**Philadelphia**, Pennsylvania
Fraunhofer Institut	**Freiburg**, Federal German Republic
Georgetown Observatory	**Washington**, D.C.
Goddard Space Flight Center	**Greenbelt**, Maryland
Goethe Link Observatory, University of Indiana	**Bloomington**, Indiana
Hale Observatories	**Pasadena**, California
Harvard College Observatory	**Cambridge**, Massachusetts
Harvard Radio Astronomy Station	**Cambridge**, Massachusetts
Haystack Observatory	**Westford**, Massachusetts
Heinrich-Hertz-Institut	**Berlin-Adlershof**, Germany
High Altitude Observatory, University of Colorado	**Boulder**, Colorado
Institute for Astronomy, University of Hawaii	**Honolulu**, Hawaii
Institute for Theoretical Astronomy (Institut Teoreticheskoj Astronomii)	**Leningrad**, U.S.S.R.
Institute of Theoretical Astrophysics, Blindern	**Oslo**, Norway
Inter-American Observatory	**Cerro Tololo**, Chile
International Latitude Observatory	**Mizusawa**, Japan
Istituto de Astronomía y Física del Espacio (IAFE)	**Buenos Aires**, Argentina
Joint Institute for Laboratory Astrophysics (JILA)	**Boulder**, Colorado
Kandilli Observatory	**Istanbul**, Turkey
Kansas University Observatory	**Lawrence**, Kansas
Kapteyn Astronomical Laboratory	**Groningen**, Netherlands
Karl-Schwarzschild-Observatorium	**Tautenburg**, German Democratic Republic
Kenneth Mees Observatory	**Rochester**, New York
Kwasan Observatory	**Kyoto**, Japan
Lamont-Hussey Observatory	**Bloemfontein**, South Africa
Leander McCormick Observatory University of Virginia	**Charlottesville**, Virginia
Lee Observatory	**Beirut**, Lebanon
Leopold-Figl-Observatorium	**Wien**, Austria
Leuschner Observatory	**Berkeley**, California
Lick Observatory	**Santa Cruz**, (Mount Hamilton), California
Lindheimer Astronomical Research Center	**Evanston**, Illinois
Lockheed Solar Observatory	**Saugus**, California
Lohrmann-Observatorium für Geodätische Astronomie	**Dresden**, German Democratic Republic
Louisiana State University Observatory	**Baton Rouge**, Louisiana
Lowell Observatory	**Flagstaff**, Arizona
Lunar and Planetary Laboratory	**Tucson**, Arizona
Max-Planck-Institut für Astronomie	**Heidelberg**, Federal German Republic
Max-Planck-Institut für Physik und Astrophysik	**München**, Federal German Republic
Max-Planck-Institut für Radioastronomie	**Bonn**, Federal German Republic
McDonald Observatory	**Fort Davis**, Texas
McMath Hulbert Observatory	**Pontiac**, Michigan
Michigan State University Observatory	**East Lansing**, Michigan
Molonglo Radio Observatory, University of Sydney	**Sydney**, New South Wales
Mount Cuba Observatory	**Wilmington**, Delaware
Mount John Observatory	**Lake Tekapo**, New Zealand
Mount Palomar Observatory	**Pasadena**, California
Mount Wilson Observatory	**Pasadena**, California

Mullard Radio Astronomy Observatory	**Cambridge**, England
Narrabri Observatory, University of Sydney	**Sydney**, New South Wales
National Bureau of Standards	**Washington**, D.C.
National Observatory, USA	**Kitt Peak**, Arizona
National Radio Astronomy Observatory	**Charlottesville**, Virginia **Green Bank**, West Virginia **Tucson**, Arizona
New Mexico State University Observatory	**Las Cruces**, New Mexico
Nizamiah Observatory	**Hyderabad**, India
Nuffield Radio Astronomy Laboratories, Jodrell Bank University of Manchester	**Manchester**, England
Observatoire Royal de Belgique	**Uccle**, Belgium
Observatorio de Cartuja	**Granada**, Spain
Observatorio del Ebro	**Tortosa**, Spain
Observatorio Fabra	**Barcelona**, Spain
Observatory, University of Michigan	**Ann Arbor**, Michigan
Ohio State University Radio Observatory	**Columbus**, Ohio
Ole Roemer-Observatoriet	**Aarhus**, Denmark
Onsala Space Observatory	**Gothenburg**, Sweden
Owens Valley Radio Observatory	**Big Pine**, California
Perkins Observatory, Ohio State and Wesleyan Universities	**Delaware**, Ohio
Purple Mountain Observatory	**Nanking**, China
Radcliffe Observatory	**Pretoria**, South Africa
Remeis-Sternwarte	**Bamberg**, Federal German Republic
Rosemary Hill Observatory	**Gainesville**, Florida
Royal Radar Establishment, Radio Astronomy Division	**Malvern**, England
Sagamore Hill Radio Observatory	**Bedford**, Massachusetts
Saint-Michel, l'Observatoire	**Haute Provence**, France
San Fernando Observatory	**El Segundo**, California
Smithsonian Astrophysical Observatory	**Cambridge**, Massachusetts
Sonnenobservatorium Kanzelhöhe	**Graz**, Austria
South African Astronomical Observatory	**Cape Town**, South Africa
Specola Astronomica Vaticana	**Castel Gandolfo**, Italy
Specola di Padova	**Asiago**, Italy
Sproul Observatory	**Swarthmore**, Pennsylvania
Stellar Data Center	**Strasbourg**, France
Sternberg Observatory	**Moscow**, U.S.S.R.
Steward Observatory, University of Arizona	**Tucson**, Arizona
United States Naval Observatory	**Washington**, D.C.
University of Florida, Radio Observatory	**Gainesville**, Florida
University of Illinois Observatory	**Urbana**, Illinois
University of Michigan Observatories	**Ann Arbor**, Michigan
University of South Florida Observatory	**Tampa**, Florida
Uttar Pradesh State Observatory	**Naini Tal**, India
Van Vleck Observatory	**Middletown**, Connecticut
Wallace Observatory	**Cambridge**, Massachusetts
Warner and Swasey Observatory	**Cleveland**, Ohio
Washburn Observatory	**Madison**, Wisconsin
West Melton Observatory	**Christchurch**, New Zealand
Wilhelm-Förster Sternwarte	**Berlin**, Germany
Yale University Observatory	**New Haven**, Connecticut
Yerkes Observatory	**Williams Bay**, Wisconsin
Zentralinstitut für Astrophysik, Sternwarte Babelsberg, (Fachbereich Kosmische Physik)	**Potsdam-Babelsberg**, German Democratic Republic

008.001 **Abastumani**

Chronicle. [Dissertations at the Abastumani Astrophysical Observatory].
Astron. Tsirk., No. 870, p. 7 - 8 (1975). In Russian.

Chronicle (1974).
Byull. Abastumansk. Astrofiz. Obs., No. 46, p. 257 - 258 (1975). In Russian.

Abastumanskaya Astrofizicheskaya Observatoriya, Gora Kanobili, Byulleten', No. 46 (I. F. Alaniya, 14.114.376; I. F. Alaniya, O. P. Abuladze, 14.114.377; O. P. Abuladze, 14.031.252; V. A. Oshchepkov, 14.121.104; V. Malyuto, 14.114.094; M. D. Metreveli, 14.131.551; E. I. Zajtseva, 14.131.166; G. F. Kevanishvili, 14.155.057; M. G. Saginashvili, 14.151.077; N. G. Kogoshvili, 14.158.204; Ts. S. Khetsuriani, G. N. Salukvadze, Eh. I. Tetruashvili, A. Sh. Khatisov, A. K. Majer, V. K. Lokhov, 14.074.110; Ts. S. Khetsuriani, 14.074.111; Eh. I. Tetruashvili, 14.074.112; Ts. S. Khetsuriani, M. Sh. Gigolashvili, 14.073.129; Ts. S. Khetsuriani, 14.034.108; V. I. Kulidzhanishvili, 14.031.051; V. I. Kulidzhanishvili, 14.079.004; A. Sh. Khatisov, G. N. Salukvadze, 14.097.204; G. G. Mikirtumova, G. V. Rozenberg, 14.082.100; T. I. Toroshelidze, S. P. Chilingarashvili, 14.082.101; E. S. Agapov, Yu. S. Aleksandrin, V. F. Anisimov, S. G. Braunfel'd, R. I. Kiladze, G. D. Matveev, A. Sh. Khatisov, S. M. Chanturiya, 14.034.109).

008.002 **Albany**

Dudley Observatory, *Albany, New York,* **Reports,** No. 10 (I. R. Linscott, J. W. Erkes, N. R. Powell, 14.033.075).

008.003 **Alma-Ata**

Akademiya Nauk Kazakhskoj SSR. **Trudy Astrofizicheskogo Instituta,** *Alma-Ata,* Vol. **25** (G. Sh. Livshits, T. B. Omarov, 14.003.007: D. A. Rozhkovskij, V. S. Matyagin, 14.155.052; A. V. Kurchakov, 14.132.050; L. A. Pavlova, 14.131.161; L. N. Kondrat'eva, 14.133.033; L. M. Sapargalieva, 14.114.373; D. A. Rozhkovskij, 14.036.008; Sh. N. Sabitov, 14.032.037; Z. V. Karyagina, 14.103.101; Eh. S. Eroshevich, Z. V. Karyagina, 14.103.101; Yu. M. Zavarzin, 14.106.036; Yu. M. Zavarzin, 14.106.048; D. A. Rozhkovskij, 14.041.038; Yu. M. Zavarzin, 14.113.050; I. L. Genkin, O. V. Chumak, 14.151.073), **26** (T. B. Omarov, 14.012.033: B. M. Gabdullin, D. I. Gorodetskij, 14.132.051; Yu. M. Zavarzin, 14.106.063; T. S. Kozhanov, T. B. Omarov, 14.151.074; T. S. Kozhanov, T. B. Omarov, 14.151.075; G. S. Minasyants, 14.072.057; L. A. Usol'tseva, 14.032.038; O. V. Chumak, 14.065.148; Z. N. Chumak, O. V. Chumak, 14.151.076; A. N. Aksenov, 14.082.094; A. N. Aksenov, 14.099.072; V. F. Kartashov, 14.099.073; V. F. Kartashov, 14.099.074; A. I. Ivanov, V. A. Molchanov, I. A. Fedulin, 14.082.095; A. I. Ivanov, V. A. Molchanov, I. A. Fedulin, 14.082.096; A. B. Kos'yanenko, 14.082.097; O. D. Tokarev, 14.082.098; A. P. Ten, O. D. Tokarev, T. P. Toropova, 14.082.099; D. I. Gorodetskij, B. M. Gabdullin, 14.034.107; B. Kalzhanov, 14.031.250; B. Kalzhanov, Eh. G. Mychelkin, B. T. Nuraliev, 14.085.015; V. N. Milovanov, 14.032.039; D. A. Rozhkovskij, F. K. Rspaev, 14.031. 251), 27 (L. D. Beshenova, A. V. Kharitonov, 14.114.093).

008.004 **Ankara**

Communications from the Astronomy Department, University of Ankara, Fen Fakültesi, Ankara–Turkey, No. 72 (N. Doğan, 14.072.069).

008.005 **Arcetri**

Osservazioni e Memorie dell'Osservatorio Astrofisico di Arcetri, Fascicolo 104 (G. Righini, 14.012.025), 105 (C. Chiuderi, M. Landini, A. Righini, 14.012.026).

008.006 **Armagh**

Armagh Observatory in 1973 and 1974.
E. J. Öpik.
Irish Astron. Journ., Vol. 11, 212 - 217 (1975).

Armagh Observatory, Leaflet, No. 125, I (E. J. Öpik, 13.104.030),

008.007 **Bamberg**

Dr. Remeis-Sternwarte Bamberg, Astronomisches Institut der Universität Erlangen-Nürnberg. – Report 1974.
W. Strohmeier.
Mitt. Astron. Ges., No. 37, p. 5 - 6 (1975).

Veröffentlichungen der Remeis-Sternwarte Bamberg, Astronomisches Institut der Universität Erlangen–Nürnberg, Vol. 10, Nos. 108 (M. Wälder, 14.121.131), 112 (H. Mauder, 13.117.025), 113 (H. Mauder, F. M. Sosna, 14.123.034), 114 (J. Rahe, 14.102.004), 115 (B. S. Carter, 13.123.024), 116 (W. Strohmeier, R. Knigge, 14.123.059).

008.008 **Beirut**

Lee Observatory, American University of Beirut, Lebanon. Monthly Bulletin, Astronomical Section, 1974 December–1975 May (F. Bruin, H. Hourani, N. G. Bustati, 14.075.018),

008.009 **Belo Horizonte**

Instituto de Ciencias Exatas da Universidade Federal de Minas Gerais (Institute of Exact Sciences of the Federal University of Minas Gerais).
Inform. Bull. Southern Hemisphere, No. 25, p. 7 (1974).
Current research report.

008.010 **Beograd**

Izveštaj o radu Astronomske opservatorije u Beogradu tokom 1973. godine. M. Protić.
Publ. Obs. Astron. Beograd, No. 20, p. 231 - 234 (1975).

Izveštaj o radu Instituta za astronomiju Prirodno–matematičkog fakulteta u Beogradu u 1971., 1972. i 1973. god. B. Ševarlić.
Publ. Obs. Astron. Beograd, No. 20, p. 235 - 253 (1975).

Bulletin de l'Observatoire Astronomique de Belgrade, No. 126 (S. Sadžakov, M. Dačić, D. Šaletić, V. Protić-Benišek, 14.041.067; S. Sadžakov, D. Šaletić, 14.041.068; R. Grujić, 14.045.028; Z. Knežević, 14.032.054; M. Jovanović, D. Vesić, 14.044.043; G. Grujić, M. Djokić, V. Milovanović, L. Djurović, 14.045.029; R. Grujić, M. Djokić, L. Djurović, 14.045.030; M. B. Protitch, V. Protić-Benišek, 14.096.014; V. Protić-Benišek, M. B. Protitch, 14.096.015; V. Protić-Benišek, 14. 095.010; 14.096.016; G. M. Popović, 14.118.027; D. Olević, 14.118.028; G. M. Popović, 14.118.029; D. Olević, 14.118. 030; V. Erceg, 14.118.031; V. Erceg, 14.118.032; J. Arsenijević, A. Kubičela, I. Vince, 14.122.218; J. Arsenijević, A. Kubičela, I. Vince, 14.122.219; I. Vince, 14.034.130; I. Vince, 14.031.278; D. Djurović, 14.045.031).

Publications de l'Observatoire Astronomique de Beograd, Nos. 20 (B. Kilar, G. (*Dj.*) Teleki, V. Vujnović, 14.012.024; M. Protić, 14.008.000; B. Sevarlić, 14.008.000; F. Dominko, 14.009.012; P. Terzić, 14.009.013; V. Petković, 14.009.014; M. Muminović, 14.009.015; M. Muminović, 14.009.016; A Tomić, 14.010.029; G. Kren, T. Vranić-Kren, 14.009.017, G. (*Dj.*) Teleki, V. Milovanović, B. Sevarlić, P. Djurcović, 14.010.037), 21 (S. Sadzakov, D. Saletić, 14.041. 069), 22 (I. Pakvor, 14.032.030), 23 (L. A. Mitić, 14.032.055).

008.011 **Berlin**

Institut für Astrophysik der Technischen Universität. Report 1974. K. Hunger.
Mitt. Astron. Ges., No. 37, p. 6 - 7 (1975).

008.012 **Berlin-Adlershof**

Heinrich-Hertz-Institut, Solare Beobachtungsergebnisse. Akademie der Wissenschaften der DDR, Zentralinstitut für Solar-Terrestrische Physik (Heinrich-Hertz-Institut), Berlin-Adlershof. HHI Solar Data, Vol. 26, 1975 April - September (E. A. Lauter, C.-U. Wagner, A. Böhme, F. Fürstenberg, D. Scholz, S. Böhm, 14.075.017).

Heinrich-Hertz-Institut. Supplement Series of Solar Data. Akademie der Wissenschaften der DDR, Zentralinstitut für Solar-Terrestrische Physik (Heinrich-Hertz-Institut), Berlin-Adlershof. HHI Suppl. Ser. Solar Data, Vol. 4, Nos. 1 (J. Paschke, D. Scholz, W. Voigt, 14.077.064), 2 (J. Paschke, D. Scholz, W. Voigt, 14.077.065).

008.013 **Berlin-Treptow**

Astronomiegeschichte an der Archenhold-Sternwarte im Jahre 1974. D. B. Herrmann.
Sterne, 51. Jahrgang, p. 176 (1975).

Blick in das Weltall, Archenhold-Sternwarte Berlin-Treptow. Astronomische Veranstaltungen und Mitteilungen für Sternfreunde, 23. Jahrgang, Nos. 7 - 12 (1975).

Veröffentlichungen der Archenhold-Sternwarte, Berlin-Treptow, No. 3 (D. Wattenberg, 14.002.006).

Archenhold-Sternwarte, Berlin-Treptow, Vorträge und Schriften No. 47 (D. B. Herrmann, 14.013.015).

008.014 **Bloemfontein**

Boyden Observatory.
Inform. Bull. Southern Hemisphere, No. 25, p. 23 (1974).
Current research report.

008.015 **Bloomington**

Indiana astronomy homecoming.
Sky Telescope, Vol. 50, 149 (1975).

008.016 **Bochum**

Astronomisches Institut. – Report 1974.
T. Schmidt-Kaler.
Mitt. Astron. Ges., No. 37, p. 8 - 13 (1975).

Veröffentlichungen des Astronomischen Instituts der Ruhr-Universität Bochum, No. 3 (M. Reinhardt, 14.142. 149).

Bereich Extraterrestrische Physik in der Abteilung XII der Ruhr-Universität Bochum. – Report 1974.
R. H. Giese.
Mitt. Astron. Ges., No. 37, p. 13 - 14 (1975).

008.017 **Bonn**

Astronomische Institute der Universität Bonn: Sternwarte mit Observatorium Hoher List; Radioastronomisches Institut; Institut für Astrophysik und Extraterrestrische Forschung; Sonderforschungsbereich Radioastronomie.
Reports 1974. W. Priester.
Mitt. Astron. Ges., No. 37, p. 15 - 37 (1975).

Max-Planck-Institut für Radioastronomie. – Report 1974.
Mitt. Astron. Ges., No. 37, p. 37 - 56 (1975).

Max-Planck-Gesellschaft zur Förderung der Wissenschaften, M.P.I. f. Radioastronomie, Bonn, Sonderdruck, Ser. A, Nos. 61 (J. E. Wink, W. J. Altenhoff, W. J. Webster, Jr., 13.131.506), 62 (W. Sieber, L. Oster, 13.141.333), 63 (W. Sieber, R. Reinecke, R. Wielebinski, 13.141.332),64 (A. Greve, 14.031.048), 65 (V. Pankonin, 13.131.514), 66 (E. Churchwell, C. M. Walmsley, 13.131.515), 67 (R. Hills, V. Pankonin, T. L. Landecker, 13.131.065), 68 (F. Chiuderi-Drago, E. Fürst, W. Hirth, P. Lantos, 13.074.044), 69 (A. Wolszczan, K. H. Hesse, W. Sieber, 13.141.306), 70 (E. Fürst, W. Hirth, 13.074.102), 71 (A. Sume, D. Downes, T. L. Wilson, 13.131.091), 72 (A. N. M. Hulsbosch, 13.131.126), 73 (C. G. T. Haslam, N. J. Keen, W. E. Wilson, D. A. Graham, P. Thomasson, 13.125.028), 74 (E. M. Berkhuijsen, 13.157.009), 75 (M. Winnewisser, G. Winnewisser, T. Honda, E. Hirota, 14.131. 021), 76 (A. von Kap-herr, B. B. Jones, R. Wielebinski, 13.158. 128), 77 (G. Winnewisser, E. Churchwell, 14.131.029), 78 (C. M. Walmsley, E. Churchwell, I. Kazès, A. M. Le Squéren, 13.131.539), 80 (T. L. Wilson, P. Thomasson, F. F. Gardner, 14.131.537).

008.018 **Borowiec**

Polish Academy of Sciences, Astronomical Latitude Observatory, Borowiec - Poland, Circular Nos.134 - 135 (14.044.045).

008.019 **Bruxelles**

Astrofysisch Instituut, Vrije Universiteit Brussel, Overdruk Nos. 84 (C. De Loore, H. Hensberge, 13.065.104), 85 (M. Burger, K. A. van der Hucht, 12.114.164), 86 (E. P. J. van den Heuvel, P. Buurman, 14.085.023), 87 (H. Hensberge, C. De Loore, 13.114.305), 88 (H. Gursky, E. P. J. van den Heuvel, 13.142.024), 89 (W. van Rensbergen, E. De Doncker, G. Deridder, 13.071.010), 90 (W. van Rensbergen, 13.004. 043), 91 (E. P. J. van den Heuvel, 14.142.127), 92 (E. P. J. van den Heuvel, 13.126.007), 93 (G. Hensberge, E. P. J. van den Heuvel, 11.117.034), 94 (P. S. Conti, G. Hensberge, E. P. J. van den Heuvel, D. J. Stickland, 12.153.008), 95 (B. P. Flannery, E. P. J. van den Heuvel, 13.141.338), 96 (E. J. Zuiderwijk, E. P. J. van den Heuvel, G. Hensberge, 12.119.014), 97 (H. Hensberge, C. De Loore, 13.114.330), 98 (C. De Loore, J. P. De Grève, J. P. De Cuyper, 14.117.018), 99 (M. A. J. Snijders, H. J. G. L. M. Lamers, 14.114.017), 100 (H. J. G. L. M. Lamers, M. A. J. Snijders, 14.114.018), 101 (C. De Loore, J. P. De Grève, 14.142.107), 102 (J.-P. De Grève, C. De Loore, 14.117.020), 103 (C. De Loore, J. P. De Grève, 14.117.001).

008.020 **Bucarest**

Observations solaires, Académie de la République Socialiste de Roumanie, Observatoire de Bucarest, Secteur Solaire. Rotations 1610–1622 (C. Popovici, E. Ţifrea, V. Dinulescu, A. Dimitriu, S. Dinulescu, G. Mariş, I. D. Niţă, 14.075.010; A. Dimitriu, G. Mariş, I. D. Niţă, 14.073.151).

008.021 **Buenos Aires**

Instituto de Astronomía y Física del Espacio, (IAFE), *Buenos Aires, Argentina,* **Tirada Aparte** Nos. 13 (R. H. Méndez, 13.121.058), 14 (R. H. Méndez, 14.133.001).

008.022 **Byurakan**

Byurakan Astrophysical Observatory, Armenia, USSR, Reprints, Nos. 149 (B. E. Markarian, V. A. Lipovetskij, 12.158.154), 150 (F. W. Baier, M. B. Petrosyan, H. Tiersch, R. K. Shakhbazyan, 12.158.156), 151 (H. M. Tovmassian, G. Haro, J. C. Webber, G. W. Swenson, Jr., K. S. Yang, K. M. Yoss, D. Deming, R. F. Green, 12.122.112), 152 (G. A. Gurzadyan, 12.114.131), 153 (M. B. Petrosyan, 12.158.158), 154 (M. A. Kazaryan, Eh. E. Khachikyan, 12.158.159), 155 (M. A. Arakelyan, 12.158.162), 156 (H. Arp, Eh. E. Khachikyan, N. K. Andreasyan, 12.158.163; Yu. K. Melik-Alaverdyan, 12.158.164; V. G. Malumyan, V. A. Sanamyan, 12.141.108; V. E. Karachentseva, I. P. Kostyuk, 12.158.186).

Soobshcheniya Byurakanskoj Observatorii, vyp. (No.) 46 (Eh. S. Parsamyan, 14.122.104; Eh. S. Parsamyan, O. S. Chavushyan, 14.122.105; M. A. Eritsyan, 14.122. 106; R. Kh. Oganesyan, 14.114.046; R. A. Vardanyan, 14.131. 097; Eh. E. Khachikyan, Dzh. A. Ehjnatyan, 14.132.034; O. V. Oganesyan, 14.133.019; S. G. Iskudaryan, 14.158.109; S. G. Iskudaryan, 14.158.110; V. G. Panadzhyan, 14.106.021; M. A. Mnatsakanyan, 14.063.022; Eh. Kh. Danielyan, M. A. Mnatsakanyan, 14.063.023; M. A. Martirosyan, L. M. Karamyan, 14.034.048; P. A. Kyalyan, 14.034.049; R. A. Sarkisyan, 14.031.224; R. A. Sarkisyan, 14.158.111; 14.007.000).

008.023 **Cambridge, England**

University of Cambridge. Institute of Astronomy. Annual report 1974–75. D. Lynden-Bell.
Separate print Univ. Cambridge, Inst. Astron., 22 pp. (1975).
Report for the year ending 1975 August 31.

008.024 **Cambridge, Massachusetts**

Smithsonian Institution. Astrophysical Observatory. Research in Space Science. SAO Special Reports, Nos. 366 (M. D. Grossi, R. H. Gay, 14.083.051), 367 (R. H. Gay, M. D. Grossi, 14.083.052), 368 (L. G. Jacchia, J. W. Slowey, 14. 082.118), 369 (R. C. Gilman, 14.066.164).

008.025 **Cape Town**

South African Astronomical Observatory (Johannesburg, and **Royal Observatory,** Cape of Good Hope).
Inform. Bull. Southern Hemisphere, No. 25, p. 24 - 26 (1974).
Current research report.

The South African Astronomical Observatory. R. Woolley.
Quarterly Journ. Roy. Astron. Soc., Vol. 16, 478 - 488 (1975).
Report for the year ending 1975 March 31.

Department of Astronomy, University of Cape Town. B. Warner.
Inform. Bull. Southern Hemisphere, No. 25, p. 23 - 24 (1974).
Current research report.

008.026 **Cardiff, Wales**

University College, Cardiff, Astronomical Communications Nos. 15 (S. P. Tarafdar, P. K. Das, 13.064.016), 16 (N. C. Wickramasinghe, 13.131.003), 17 (M. J. Dempsey, N. C. Wickramasinghe, 13.131.082), 18 (V. Vanýsek, N. C. Wickramasinghe, 13.102.021), 19 (S. Ramadurai, N. C. Wickramasinghe, 13.061.024), 20 (S. P. Tarafdar, N. C. Wickramasinghe, 13.131.029).

008.027 **Carloforte**

Pubblicazioni della Stazione Astronomica Internazionale di Latitudine, Carloforte–Cagliari, Nuova Serie, No. 46 (F. Zagar, 13.044.016).

008.028 **Cerro Tololo**

Observatorio Interamericano (Cerro Tololo Interamerican Observatory).
Inform. Bull. Southern Hemisphere, No. 25, p. 14 - 19 (1974).
Current research report.

008.029 **Cincinnati**

Minor Planet Circulars (MPC), Nos. 3829 - 3908 (P. Herget, 14.098.095).

008.030 **Cleveland, Ohio**

Publications of the Warner and Swasey Observatory, Case Western Reserve University, Vol. 2, No. 1 (N. Sanduleak, 14.133.028).

Warner and Swasey Observatory, Case Western Reserve University, Reprints, Nos. 225 (N. J. Irvine, C. E. Irvine, 12.125.036), 230 (C. A. Murray, N. Sanduleak, 07.112. 013), 242 (F. M. Stienon, M. R. Chartrand III, C. Y. Shao, 11.114.061), 247 (L. W. Hartmann, J. F. Dolan, 11.114.002), 249 (C. B. Stephenson, 10.114.184), 250 (J. F. Dolan, 12.031. 018), 251 (P. Pesch, S. W. McCuskey, 11.155.021), 252 (S. W. McCuskey, 11.155.017), 253 (N. Sanduleak, 11.153.010), 254 (N. Sanduleak, J. F. Dolan, 11.141.006), 255 (N. J. Irvine, 11.114.034), 259 (N. Sanduleak, 11.133.012), 260 (W. B. Weaver, 11.113.023), 261 (W. B. Weaver, 11.131.066), 262 (C. B. Stephenson, 12.114.033), 263 (S. W. McCuskey, P. Pesch, G. A. Snyder, 11.155.045), 264 (G. E. Mechler, 11. 118.034), 265 (N. Sanduleak, 12.114.064), 266 (C. B. Stephenson, 12.041.030).

008.031 **Córdoba**

Observatorio Astronómico (Astronomical Observatory, National University of Córdoba).
Inform. Bull. Southern Hemisphere, No. 25, p. 2 - 4 (1974).
Current research report.

Observatorio Astronómico de Córdoba, Córdoba, Argentina, Separata (G. M. Iannini, 14.079.108).

Observatorio Astronómico (*Universidad Nacional de Córdoba, Córdoba, Argentina*), **Tirada Aparte,** Nos. 207 (G. J. Carranza, E. L. Agüero, 11.158.026), 214 (R. F. Sisteró, M. E. Castore de Sisteró, 11.121.046), 215 (C. R. Fourcade, J. R. Laborde, J. C. Arias, 12.154.002).

008.032 **Cracow**

Cracow Observatory, Reprints Nos. 105 (Z. Klimek, J. M. Kreiner, 13.121.017), 106 (J. Machalski, 13.141.037), 107 (Z. Klimek, 12.096.018), 108 (Z. Klimek, 13.162.030), 109 (T. Z. Dworak, 13.121.037).

008.033 **Delaware**

Contributions from the Perkins Observatory, Ohio State – Ohio Wesleyan Universities. Series I, No. 157 (S. J.

Czyzak, L. H. Aller, R. N. Euwema, 12.074.085).

Contributions from the Perkins Observatory, Ohio State – Ohio Wesleyan Universities. Series II, Nos. 50 (S. J. Czyzak, L. H. Aller, R. N. Euwema, 13.074.011), 51 (A. Slettebak, 13.114.024), 52 (G. O. Boeshaar, 13.133.004), 53 (C. A. Dean, 12.121.086), 54 (R. G. Tabak, J. P. Hirth, G. Meyrick, T. P. Roark, 13.131.038), 55 (A. Slettebak, G. W. Collins, II, P. B. Boyce, N. M. White, T. D. Parkinson, 13.116. 012), 56 (S. J. Czyzak, E. G. Buerger, L. H. Aller, 13.133.028).

008.034 **Edinburgh**

Royal Observatory Edinburgh. Annual report 1974–75. V. C. Reddish.
Roy. Obs. Edinburgh, 29 pp. (1975). – Annual Report for the year ending 30 September 1975.

Communications from the Royal Observatory, Edinburgh, Nos. 168 (S. V. M. Clube, 11.155.057), 169 (S. V. M. Clube, W. Nicholson, 11.041.051), 170 (S. V. M. Clube, 11.041.061).

008.035 **Frankfurt**

Astronomisches Institut der Universität. – Report 1974. W. Gleißberg.
Mitt. Astron. Ges., No. 37, p. 56 - 58 (1975).

Veröffentlichungen des Astronomischen Instituts der Universität Frankfurt (Main), Nos. 51 (T. Damboldt, 14.084.323), 52 (W. Gleissberg, 14.072.083), 53 (R. Hartmann, 13.074.114), 54 (W. Gleissberg, 14.072.084), 55 (R. Hartmann, 14.074.127), 56 (W. Gleissberg, 14.072.085), 57 (W. Gleissberg, 14.072.086).

008.036 **Freiburg**

Fraunhofer-Institut mit den Observatorien Schauinsland and Anacapri. – Report 1974. K. O. Kiepenheuer.
Mitt. Astron. Ges., No. 37, p. 58 - 62 (1975).

008.037 **Gent**

Universiteit te Gent, Sterrenkundig Instituut, *Mededelingen*, Nos. 32 (P. Dingens, H. Steyaert, N. Baeck, 14.071.065), 33 (N. Baeck, 14.071.066), 34 (N. Baeck, P. Dingens, H. Steyaert, 14.071.067), 35 (H. Steyaert, N. Baeck, P. Dingens, 14.071.068), 36 (P. Dingens, H. Steyaert, N. Baeck, 14.071.069), 37 (N. Baeck, P. Dingens, H. Steyaert, 14.071.071).

008.038 **Glasgow**

Department of Astronomy, University of Glasgow. P. A. Sweet.
Quarterly Journ. Roy. Astron. Soc., Vol. 16, 432 - 435 (1975). Report for the period 1972 October to 1975 September.

008.039 **Göttingen**

Universitäts-Sternwarte Göttingen und Institut für Sonnenforschung Locarno-Orselina (Tessin). – Report 1974. H. H. Voigt.
Mitt. Astron. Ges., No. 37, p. 63 - 68 (1975).

008.040 **Graz**

Astronomisches Institut (Universitäts-Sternwarte); Sonnenobservatorium Kanzelhöhe. – Reports 1974. H. Haupt.
Mitt. Astron. Ges., No. 37, p. 69 - 71 (1975).

Mitteilungen der Universitätssternwarte Graz Nos. 15 (H. Haupt, 14.101.013), 17 (G. Lustig, R. Dvorak, 14.098.062), 18 (H. J. Schober, G. Lustig, 13.098.073), 20 (G. Hahn, H. F. Haupt, 14.152.004), 22 (H. J. Schober, R. Dvorak, 14.098.028), 23 (H. J. Schober, 14.098.029).

Mitteilungen des Sonnenobservatoriums Kanzelhöhe, Nos. 21 (T. V. Pettauer, 09.034.088), 22 (S. Edelson, F. I. Shimabukuro, E. B. Mayfield, 09.077.062), 23 (T. V. Pettauer, 12.034.027).

008.041 **Green Bank**

National Radio Astronomy Observatory, *Green Bank,* **Reprints,** Series A, Nos. 443 (M. A. Kaftan-Kassim, J. W. Sulentic, G. Sistla, 13.141.007), 444 (N. Kaifu, T. Iguchi, M. Morimoto, 13.131.042), 445 (B. Zuckerman, B. E. Turner, D. R. Johnson, F. O. Clark, F. J. Lovas, N. Fourikis, P. Palmer, M. Morris, A. E. Lilley, J. A. Ball, C. A. Gottlieb, M. M. Litvak, H. Penfield, 13.131.043), 446 (W. W. Roberts, Jr., M. S. Roberts, F. H. Shu, 13.151.008), 447 (R. A. Sramek, H. M. Tovmassian, 13.158.026), 448 (D. S. Heeschen, E. K. Conklin, 13.158.027), 449 (V. Pankonin, A. Parrish, Y. Terzian, 13.131.505), 450 (H. S. Liszt, R. A. Linke, 13.131.041), 451 (P. D. Jackson, F. J. Kerr, 13.157.002), 452 (J. J. Condon, D. C. Backer, 13.141.035), 453 (B. Zuckerman, B. E. Turner, 13.131.047), 454 (L. E. Snyder, D. Buhl, 13.131.053), 455 (D. C. Backer, J. M. Rankin, D. B. Campbell, 13.141.334), 456 (J. W. Erkes, A. G. D. Philip, 13.154.007), 457 (D. R. Johnson, F. O. Clark, 13.131.058), 458 (F. N. Owen, 13.160.011), 459 (J. W. Warner, G. E. Assousa, B. Balick, E. R. Craine, 13.141.047), 460 (M. N. Simon, T. O. M. Simon, 12.103.104), 461 (A. E. Niell, K. I. Kellermann, B. G. Clark, D. B. Shaffer, 13.141.053), 462 (K. I. Kellermann, B. G. Clark, A. E. Niell, D. B. Shaffer, 13.141.054), 463 (D. M. Gibson, 13.141.064), 464 (R. A. Sramek, 13.158.068), 465 (B. E. Turner, R. H. Gammon, 13.131.084), 466 (P. F. Bowers, 14.141.601), 467 (P. L. Baker, W. B. Burton, 13.155.039), 468 (G. S. Shostak, 13.158.104), 469 (H. S. Liszt, R. H. Sanders, W. B. Burton, 13.155.041), 470 (P. T. Giguere, F. O. Clark, 13.103.100), 471 (M. Morris, W. Gilmore, P. Palmer, B. E. Turner, B. Zuckerman, 14.131.008), 472 (N. Kaifu, K. Takagi, T. Kojima, 13.131.120), 473 (B. E. Turner, B. Zuckerman, N. Fourikis, M. Morris, P. Palmer, 13.131.125), 474 (L. E. Snyder, J. M. Hollis, B. L. Ulich, F. J. Lovas, D. R. Johnson, D. Buhl, 13.131.119), 475 (B. G. Clark, K. I. Kellermann, D. Shaffer, 13.121.059), 476 (J. H. Fertel, B. E. Turner, 13.131.123), 477 (A. Parrish, V. Pankonin, 13.132.031), 478 (L. J. Rickard, P. Palmer, M. Morris, B. Zuckerman, B. E. Turner, 14.158. 009), 479 (K. J. Gordon, C. P. Gordon, 13.141.356), 480 (G. W. Swenson, Jr., K. I. Kellermann, 13.033.025), 481 (J. M. Huntley, W. C. Saslaw, 14.151.003), 482 (C. Heiles, M. A. Gordon, 14.131.011), 483 (T. K. Menon, 14.141.016), 484

(R. M. Hjellming, D. M. Gibson, 14.141.311), 485 (N. Z. Scoville, P. M. Solomon, 14.131.016), 486 (E. B. Fomalont, R. A. Sramek, 14.066.037), 487 (D. R. Altschuler, J. F. C. Wardle, 13.158.313), 488 (G. S. Shostak, M. S. Roberts, S. D. Peterson, 14.158.039), 489 (R. M. Hjellming, D. M. Gibson, F. N. Owen, 14.142.017), 490 (J. R. Dickel, H. J. Rood, 14. 158.040), 491 (M. Haynes, R. Sramek, 14.158.113), 492 (R. H. Becker, M. R. Kundu, 14.125.027), 493 (S. von Hoerner, 14.033.002).

National Radio Astronomy Observatory, *Green Bank*, **Reprints**, Series B, Nos. 458 (M. A. Gordon, 12.155. 058), 459 (W. B. Burton, 12.155.064), 460 (C. E. Alissandrakis, M. R. Kundu, 13.077.023).

008.042 **Greenbelt**

Goddard Space Flight Center, Greenbelt, Maryland, GSFC Documents X-660-75-130 (J. H. Trainor, 14.099.064), 192 (F. B. McDonald, B. J. Teegarden, J. H. Trainor, T. T. von Rosenvinge, W. R. Webber, 14.143.044), 661-75-178 (L. J. Kaluzienski, S. S. Holt, E. A. Boldt, P. J. Serlemitsos, G. Eadie; K. A. Pounds, M. J. Ricketts, M. Watson, 14.142.153), 205 (S. S. Holt, E. A. Boldt, P. J. Serlemitsos, L. J. Kaluzienski, 14.142.154), 218 (R. F. Silverberg, 14.143.045), 232 (S. S. Holt, 14.142.201), 233 (E. A. Boldt, S. S. Holt, R. E. Rothschild, P. J. Serlemitsos, 14.142.155), 260 (S. S. Holt, E. A. Boldt, P. J. Serlemitsos, L. J. Kaluzienski, S. H. Pravdo, A. Peacock, M. Elvis, M. G. Watson, K. A. Pounds, 14.142.202), 278 (A. J. Fisher, F. A. Hagan, R. C. Maehl, J. F. Ormes, J. F. Arens, 14.143.066), 303 (S. Holt, E. A. Boldt, P. J. Serlemitsos, L. J. Kaluzienski, 14.142.203), 682-75-167 (C. J. Crannell, R. Ramaty, C. Werntz, 14.073.108), 920-75-220 (P. Musen, 14.052.032), 921-75-110 (J. G. Marsh, S. Vincent, A. T. McClinton, E. S. Chang, 14.081.024), 157 (M. A. Graber, 14.045.013), 179 (C. A. Wagner, 14.081.025), 194 (T. L. Felsentreger, J. G. Marsh, R. W. Agreen, 14.052. 033).

Goddard Space Flight Center, Greenbelt, Maryland, Separate prints (D. E. Smith, R. Kolenkiewicz, R. W. Agreen, P. J. Dunn, 14.046.043; A. T. Serlemitsos, V. K. Balasubrahmanyan, 13.078.012; L. J. Kaluzienski, S. S. Holt, E. A. Boldt, P. J. Serlemitsos, G. Eadie, K. A. Pounds, M. J. Ricketts, M. Watson, 14.142.120; S. H. Pravdo, E. A. Boldt, 14.061. 017; S. S. Holt, E. A. Boldt, L. J. Kaluzienski, P. J. Serlemitsos, 14.142.015).

008.043 **Greenwich**

Det Kongelige Observatorium, Greenwich 1675 - 1975. K. M. Pedersen.
Astron. Tidssk., Årg. 8, p. 97 - 107 (1975).

The Royal Observatory, Greenwich.
Endeavour, No. 123, Vol. 34, 102 (1975).

Royal Observatory, Greenwich.
A. V. Douglas.
Journ. Roy. Astron. Soc. Canada, Vol. 69, 140 - 141 (1975).

300 Jahre Sternwarte Greenwich.
J. G. Growther.
Orion, 33. Jahrgang, p. 114 - 117 (1975).

The Royal Greenwich Observatory. – Report for the year ending 31 December 1974. A. Hunter.
Quarterly Journ. Roy. Astron. Soc., Vol. 16, 444 - 477 (1975).

Royal Greenwich Observatory. Report for the year ending 31 December 1974. A. Hunter.
D. V. Thomas, J. E. Perry (Editors).
Published by the Royal Greenwich Observatory, Herstmonceux Castle, Hailsham, East Sussex, 6 + 43 pp. Price 40 p. (1975).

Royal Greenwich Observatory. Illustrated.
Prepared by D. H. P. Jones, assisted by A. R. Hewerdine, C. M. Lowne, the photography was done by D. A. Calvert.
Published by the Royal Greenwich Observatory, Herstmonceux Castle, Hailsham, East Sussex, 4 + 24 pp. Price 30 p. (1975).

John Flamsteed and the origins of the Greenwich astronomical tradition. See Abstr. 004.075.

300 years Greenwich Observatory.
See Abstr. 004.086.

Royal Observatory Annals, [Royal Greenwich Observatory, Herstmonceux]. No. 10 (P. S. Laurie, G. M. Appleby, C. Y. Hohenkerk, C. M. Ladley, P. J. Rudd, 14.075. 021).

Royal Observatory Bulletins, (Herstmonceux: Royal Greenwich Observatory), No. 179 (K. C. Blackwell, J. V. Carey, R. H. D. Swifte, 14.041.028).

H. M. Nautical Almanac Office, Royal Greenwich Observatory, Library Reprint, Nos. 314 (R. F. Griffin, B. Emerson, 13.119.001), 315 (L. V. Morrison, C. G. Ward, 14.092.007), 316 (14.096.018), 317 (L. V. Morrison, 14.044. 048), 318 (G. E. Taylor, 14.035.012), 319 (A. T. Sinclair, 14.047.042).

008.044 **Groningen**

Nederlandse Vereniging voor Weer- en Sterrenkunde. **Observations of Variable Stars. Report** (Kapteyn Astronomical Laboratory, Groningen–Netherlands), No. 28 (L. Plaut, H. Feijth, 14.123.039).

008.045 **Hamburg**

Hamburger Sternwarte. – Report 1974.
Mitt. Astron. Ges., No. 37, p. 72 - 79 (1975).

Hamburger Sternwarte. Sonderdruck, Serie A, Nos. 31 (J. Materne, G. A. Tammann, 12.158.093), 32 (G. A. Tammann, 12.162.023), 33 (J. Materne, G. A. Tammann, 13.158. 009), 34 (J. W. M. Baars, H. J. Wendker, 12.131.539), 35 (W. K. Huchtmeier, H.-D. Bohnenstengel, 14.158.034).

Deutsches Hydrographisches Institut, Hamburg. **Astronomische Zeit- und Breitenbestimmungen, Empfangszeiten von Zeitsignalen**, 1974 October - December (14.044. 046).

Deutsches Hydrographisches Institut, Hamburg. **Zeit- und Breitendienst,** 1975 January - March (14.044.047).

008.046 **Hannover**

Astronomische Station des Instituts für Theoretische Geodäsie der Technischen Universität. – Report 1974.
G. Seeber.
Mitt. Astron. Ges., No. 37, p. 79 - 81 (1975).

008.047 **Heidelberg**

Astronomisches Rechen-Institut. – Report 1974.
W. Fricke.
Mitt. Astron. Ges., No. 37, p. 81 - 89 (1975).

Astronomy and Astrophysics Abstracts, Vol. 13 (S. Böhme, U. Esser, W. Fricke, U Güntzel-Lingner, I. Heinrich, F. Henn, D. Krahn, L. Schmadel, H. Scholl, G. Zech, 14.002.009).

Astronomisches Rechen-Institut in Heidelberg, Mitteilungen Serie A, Nos. 88 (W. Lohmann, 13.154.001), 89 (W. Fricke, 14.041.023), 90 (H. G. Walter, 13.052.031), 91 (R. Wielen, 14.151.018), 92 (W. Fricke, 14.041.030), 93 (H. Scholl, 13.099.035), 94 (W. Fricke, 14.155.031), 95 (W. Fricke, 14.044.016).

Astronomisches Rechen-Institut in Heidelberg, Mitteilungen Serie B, Nos. 47 (R. Wielen, 12.151.018), 48 (J. Schubart, 13.098.010), 49 (R. Wielen, 13.151.035), 50 (W. Fricke, W. Gliese, 13.155.046), 51 (H. G. Walter, 14.041.049), 52 (W. Gliese, 14.111.010), 53 (W. Fricke, A. Tsioumis, 14.155.022), 54 (H. Scholl, C. Froeschlé, 14.098.019), 55 (R. Wielen, 14.151.056).

Lehrstuhl für Theoretische Astrophysik. – Report 1974. B. Baschek, G. Traving.
Mitt. Astron. Ges., No. 37, p. 89 - 90 (1975).

Landessternwarte und Max-Planck-Institut für Astronomie. – Report 1974. H. Elsässer.
Mitt. Astron. Ges., No. 37, p. 91 - 99 (1975).

Das Max-Planck-Institut für Astronomie – Stand des Aufbaus 1974. H. Elsässer.
Mitt. Astron. Ges., No. 36, p. 53 - 57 (1975).

008.048 **Innsbruck**

Institut für Astronomie und Sternwarte. – Report 1974. J. Pfleiderer.
Mitt. Astron. Ges., No. 37, p. 100 (1975).

008.049 **Jena**

Mitteilungen der Universitäts-Sternwarte zu Jena, Nos. 115 (J. Dorschner, C. Friedemann, J. Gürtler, 10.113.033), 116 (L. Kühn, 10.131.092), 118 (C. Friedemann, 12.113.025), 119 (J. Dorschner, 11.098.016), 120 (J. Dorschner, 11.131.072), 121 (R. Schielicke, 11.034.052).

008.050 **Kazan**

W. J. Struve and the Astronomical Observatory of the Kazan University. V. L. Chenakal.
Trudy Kazan. Gorod. Astron. Obs., No. 39, p. 189 - 198 (1973). In Russian.

Izvestiya Astronomicheskoj Ehngel'gardtovskoj Observatorii, *Kazan',* Nos. **39** (D. I. Stepanov, L. B. Gusakovskaya, 14.104.012; S. S. Tokhtas'ev, 14.113.025; I. A. Dubyago, 14.131.093; R. A. Botsula, 14.117.040; S. G. Valeev, 14.094.135; I. A. Urasina, 14.081.013; O. E. Shornikov, A. S. Mamakov, 14.034.047; A. S. Mamakov, 14.031.223), **40** (A. I. Nefed'eva, 14.082.031; A. A. Krasikova, 14.041.018; D. P. Laas, 14.041.019; S. S. Tokhtas'ev, I. A. Dubyago, 14.113.026; S. S. Tokhtas'ev, 14.113.027; S. S. Tokhtas'ev, 14.131.094; L. A. Urasin, N. V. Lavrova, 14.158.108; L. A. Urasin, 14.131.095; L. A. Urasin, 14.113.028; L. A. Urasin, 14.131.096; L. B. Gusakovskaya, 14.082.032; I. A. Urasina, 14.081.014; S. G. Valeev, 14.044.004).

Trudy Kazanskoj Gorodskoj Astronomicheskoj Observatorii, *Kazan',* Nos. 39 (S. I. Petrusevich, 14.062.022; N. A. Sakhibullin, 14.133.008; V. P. Merezhin, 14.121.029; M. I. Lavrov, 14.121.030; M. P. Volodina, 14.119.009; M. I. Lavrov, N. V. Lavrova, 14.121.031; Yu. V. Evdokimov, 14.104.005; N. A. Belyaev, 14.103.103; N. A. Belyaev, E. A. Reznikov, 14.103.104; L. E. Nikonova, A. Yu. Yatsenko, 14.042.025; Sh. T. Khabibullin, 14.094.129; E. D. Kondrat'eva, 14.103.106; A. A. Nemo, 14.042.026; N. F. Bystrov, N. G. Rizvanov, 14.041.012; D. D. Dzyaman, 14.033.005), 40 (Sh. T. Khabibullin, 14.052.010; Yu. V. Evdokimov, I. Yu. Evdokimov, 14.103.107; L. Ya. Anan'eva, E. D. Kondrat'eva, 14.103.108; L. M. Belous, 14.103.109; L. A. Markelov, 14.103.110; E. E. Belyaeva, 14.064.020; E. E. Belyaeva, 14.064.021; V. P. Merezhin, 14.063.015; D. D. Dzyaman, 14.031.216; D. D. Dzyaman, 14.031.217).

008.051 **Kiel**

Institut für Theoretische Physik und Sternwarte der Universität. – Report 1974. V. Weidemann.
Mitt. Astron. Ges., No. 37, p. 101 (1975).

Sonderdrucke der Sternwarte Kiel, Nos. 202 (A. Unsöld, 13.162.103), 203 (A. Unsöld, T. Gehren, 13.091.006), 204 (R. Mäckle, R. Griffin, R. Griffin, H. Holweger, 13.064.014), 205 (V. Weidemann, 14.065.175), 206 (R. Mäckle, H. Holweger, R. Griffin, R. Griffin, 13.114.316), 207 (T. Gehren, 13.080.004), 208 (D. Reimers, 14.064.034), 209 (A. Unsöld, 14.162.020), 210 (V. Weidemann, 14.011.012).

008.052 **Kiev**

Astrometriya i Astrofizika, *Kiev,* Vyp. (Nos.) 26 (E. P. Fedorov, 14.003.001), 27 (I. V. Gavrilov, 14.003.004).

Kievskij ordena Lenina gosudarstvennyj universitet im. T. G. Shevchenko, Astronomicheskaya Observatoriya (Astron. Obs.), Preprint Nos. 4 (Yu. N. Kudrya, 14.072.059); 5 (P. R. Romanchuk, 14.072.060); 6 (P. R. Romanchuk, T. I. Redyuk, 14.072.061).

008.053 **Kitt Peak**

Das Mayall 4-m-Teleskop auf Kitt Peak.
See Abstr. 032.011.

008.054 **Krim**

Izvestiya Krymskoj Astrofizicheskoj Observatorii, Akademiya Nauk SSSR, Tom (Vol.) 53 (A. M. Zvereva, A. B. Severnyj, Eh. I. Terez, 14.094.152; A. A. Stepanyan, B. M. Vladimirskij, Yu. I. Neshpor, V. P. Fomin, 14.142.150; V. P. Fomin, B. M. Vladimirskij, Yu. I. Neshpor, A. A. Stepanyan, 14.142.151; M. D. Gusejnov, 14.072.035; Eh. A. Baranovskij, 14.073.098; M. B. Ogir', 14.072.036; T. T. Tsap, 14.073.099; S. I. Gopasyuk, 14.062.040; L. G. Kartashova, 14.073.100; N. N. Stepanyan, Z. A. Shcherbakova, 14.073.101; V. A. Efanov, I. G. Moiseev, A. B. Severnyj, 14.077.025; L. I. Yurovskaya, O. Olvares, M. Ferreiro, 14.079.104; L. I. Yurovskaya, 14.077.026; V. A. Efanov, I. I. Zinchenko, A. G. Kislyakov, I. V. Kuznetsov, I. G. Moiseev, A. I. Naumov, 14. 141.097; I. G. Moiseev, G. E. Patston, L. H. Solomon, H. M. Tovmassian, C. S. Higgins, P. F. Chugajnov, 14.122.156; V. I. Krasnobabtsev, R. E. Gershberg, 14.122.157; N. I. Shakhovskaya, 14.122.158; T. M. Rachkovskaya, 14.122.159; G. I. Shanin, A. G. Shcherbakov, 14.122.160; V. A. Efanov, A. G. Kislyakov, I. G. Moiseev, N. S. Nesterov, V. I. Chernyshev, V. B. Shtejnshlejger, 14.132.047; V. I. Pronik, 14.132.048; N. S. Chernykh, L. I. Chernykh, 14.032.031; G. M. Popov, 14.031.028; 14.012.027; A. B. Severnyj, 14.071.035; A. A. Boyarchuk, 14.064.043; Eh. I. Mogilevskij, 14.080.053; V. M. Grigor'ev, I. M. Kats, 14.062.041; S. I. Gopasyuk, V. A. Kotov, A. B. Severnyj, T. T. Tsap, 14.080.054; Z. N. Shukstova, G. V. Lyamova, 14.072.037; M. D. Gusejnov, 14.072. 038; V. L. Khokhlova, 14.064.044; Yu. V. Glagolevskij, 14. 116.009; V. M. Kuvshinov, G. Hildebrandt, W. Schöneich, 14. 116.010; T. A. Ryabchikova, 14.114.356; N. S. Polosukhina, K. K. Chuvaev, 14.114.357; A. Z. Dolginov, 14.116.011).

008.055 **Kyoto**

Contributions from the Department of Astronomy, University of Kyoto, Nos. 53 (S. Kato, 12.121.014), 54 (K.-i. Wakamatsu, S. Oka, K. Sakka, 13.158.038), 55 (Y. Baba, 14.021.023), 56 (T. Okuda, 13.064.027), 57 (T. Ishizawa, 13.062.028), 58 (M. Nishida, T. Ishizawa, 13.151.012), 59 (T. Ishizawa, 12.061.044).

008.056 **La Plata**

Observatorio Astronómico (Astronomical Observatory, National University of La Plata).
Inform. Bull. Southern Hemisphere, No. 25, p. 4 (1974).
Current research report.

008.057 **La Silla**

Observatorio Europeo Austral (European Southern Observatory).
Inform. Bull. Southern Hemisphere, No. 25, p. 19 - 20 (1974).
Current research report.

008.058 **Las Campanas**

Las Campanas Observatory (CARSO) (Carnegie Institution of Washington).
Inform. Bull. Southern Hemisphere, No. 25, p. 20 (1974).
Current research report.

008.059 **Las Cruces**

The spectrum of a city. J. Cuffey.
Sky Telescope, Vol. 50, 155 (1975).

008.060 **Lembang**

Bosscha Observatory.
Inform. Bull. Southern Hemisphere, No. 25, p. 21 - 22 (1974).
Current research report.

Annals of the Bosscha Observatory, Lembang (Java), Indonesia, Vol. IX, Part 4 (P. S. Thé, 14.118.001).

008.061 **Leningrad**

Chronicle.
Byull. Inst. Teoret. Astron., *Leningrad,* Vol. 13, 673 - 674 (1975). In Russian.

Chronicle.
Byull. Inst. Teoret. Astron., *Leningrad,* Vol. 14, 70 (1975).
In Russian.

Chronicle.
Byull. Inst. Teoret. Astron., *Leningrad,* Vol. 14, 132 (1975).
In Russian.

Byulleten' Instituta Teoreticheskoj Astronomii, Akademiya Nauk SSSR, Vol. 13, No. 10 (M. S. Volkov, 14.042.002; N. I. Glebova, 14.097.004; V. L. Mazo, 14.042. 012; V. F. Myachin, 14.042.003; N. M. Bronnikova, L. V. Zhukov, T. P. Kiseleva, A. A. Kiselev, L. S. Koroleva, G. V. Panova, 14.041.004; H. K. Raudsaar, 14.098.002; H. K. Raudsaar, 14.098.003; L. I. Chernykh, B. A. Burnasheva, T. M. Smirnova, N. S. Chernykh, 14.098.004), Vol. 14, No. 1 (N. S. Samojlova-Yakhontova, 14.098.005; A. S. Baranov, 14.151.004; M. A. Dirikis, 14.098.006; I. N. Latyshev, 14. 042.004; B. I. Lokhin, 14.052.004; E. M. Nezhinskij, 14.042. 005; G. A. Merman, 14.042.006; F. B. Khanina, 14.098.007; E. Z. Khotimskaya, 14.042.007; F. F. Kalikhevich, 14.098.008; T. P. Kiseleva, L. S. Koroleva, G. V. Panova, 14.100.202; A. Sh. Khatisov, 14.098.009; L. I. Chernykh, 14.034.004).

Byulleten' Instituta Teoreticheskoj Astronomii, Akademiya Nauk SSSR, Vol. 14, Nos. 2 (N. S. Samojlova-Yakhontova, 14.098.064; M. S. Volkov, 14.042.103; R. A. Zejnalov, 14.052.053; V. I. Orel'skaya, 14.098.065; M. L. Sveshnikov, E. S. Sveshnikova, 14.041.064; A. S. Sochilina, 14.055.026; L. I. Chernykh, T. M. Smirnova, 14.098.066; L. I. Chernykh, N. S. Chernykh, 14.098.067; Yu. V. Batrakov, 14.007.000), 3 (V. M. Amelin, 14.047.032; Yu. V. Batrakov, V. G. Shkodrov, 14.091.035; E. N. Lemekhova, 14.099.239; A. S. Sochilina, L. I. Varenik, 14.054.017; G. A. Chebotarev, N. A. Belyaev, R. P. Eremenko, 14.098.068; E. V. Vojtyuk, Yu. F. Zinchenko, Yu. V. Mikhajlov, 14.098.069; L. V. Zhuravleva, 14.098.070; A. M. Lozinskij, V. P. Osipenko, V. A. Yurevich, 14.103.102).

Ephemerides of minor planets for 1976 (G. A. Chebotarev, 14.098.001).

Trudy Astronomicheskoj Observatorii, (Transactions of the Astronomical Observatory), *Leningrad,* Vol. 31 (D. I. Nagirner, 14.063.021; M. K. Babadzhanyants, S. K. Vinokurov, V. A. Hagen-Thorn, E. V. Semenova, 14. 158.102;

G. V. Khozov, T. N. Khudyakova, L. V. Larionova, 14.113.019; O. S. Shulov, E. N. Kopatskaya, 14.121.043; A. B. Berlin, V. V. Vitkovskij, G. B. Gel'frejkh, R. I. Enikeev, A. V. Ipatov, V. G. Nagnibeda, G. M. Timofeeva, S. P. Yakimov, 14.079.101; T. A. Agekyan, Zh. P. Anosova, 14.117.038; I. V. Petrovskaya, S. P. Yakimov, 14.151.039; E. N. Polyakhova, 14.100.009; O. P. Bykov, 14.052.011; M. P. Mishchenko, A. V. Shiryaev, 14.044.003).

008.062 **Liège**

Institut d'Astrophysique, Université de Liège, Cointe-Ougrée (Belgique), Collection in 4°, Nos. 259 (M. L. Aizenman, J. Perdang, 10.065.075), 260 (T. Danks, C. Arpigny, 10.102.021), 261 (M. Singh, H. Bredohl, F. Remy, I. Dubois, 14.022.111), 262 (M. Gabriel, A. Noels, 11.065.006), 263 (J. P. Swings, 10.133.069), 264 (H. Bredohl, R. Cornet, I. Dubois, F. Remy, 14.022.112), 265 (A. Noels, A. Boury, R. Scuflaire, M. Gabriel, 11.065.020), 266 (M. Singh, H. Bredohl, F. Remy, I. Dubois, 14.022.113), 267 (M. Gabriel, S. Refsdal, H. Ritter, 11.065.065), 268 (M. Gabriel, M. L. Roth, 11.065.074), 269 (J. P. Swings, 12.114.032), 270 (J. M. Vreux, Y. Andrillat, 12.114.031), 271 (R. Scuflaire, 12.065.033), 272 (M. Gabriel, A. Noëls, 12.065.034), 273 (R. Scuflaire, 12.065.111), 274 (M. Maheswaran, 12.116.015), 275 (H. Bredohl, R. Cornet, I. Dubois, 14.022.114), 276 (H. Robe, 13.151.019).

Institut d'Astrophysique, Université de Liège, Cointe-Ougrée (Belgique), Collection in 8°, Nos. 642 (I. Ottelet, 14.081.045), 643(J. P. Swings, J. M.Vreux, 14.103.142), 644 (H. Robe, 12.065.162), 645 (J. Demaret, 12.065.183), 646 (P. Ledoux, 14.013.014), 647 (Y. Andrillat, J. M. Vreux, 14.114.108), 648 (R. Simon, 12.151.005), 649 (F. Dossin, A. Heck, M. Klutz, J. Manfroid, G. Sause, J. P. Swings, J. M. Vreux, 12.103.104), 650 (A. Heck, 14.021.016), 651 (M. Gabriel, R. Scuflaire, A. Noels, A. Boury, 14.065.176), 652 (R. Duysinx, M. Henrist, 12.099.069), 653 (E. Biémont, 12.071.031), 654 (P. Ledoux, 12.065.140), 655 (J. Demaret, 12.065.124), 656 (J. Demaret, 13.065.022).

008.063 **London, Canada**

University of Western Ontario, London, Ontario. A. Wehlau.
Journ. Roy. Astron. Soc. Canada, Vol. 69, 146 - 147 (1975).

008.064 **London, England**

Communications from the University of London Observatory, Nos. 75 (S. L. Wright, J. D. Argyros, 14.021.026), 76 (S. L. Wright, 14.021.027).

Mullard Space Science Laboratory, University College London. R. L. F. Boyd.
Quarterly Journ. Roy. Astron. Soc., Vol. 16, 309 - 317 (1975). Report for the period 1973 October 1 - 1974 September 30.

008.065 **Madrid**

Boletín Astronómico del Observatorio de Madrid, Instituto Geografico y Catastral, Vol. 8, No. 5,(J. F. Lahulla, 14.052.055; M. López Arroyo, J. F. Lahulla, 14.072.078; J. F. Lahulla, 14.031.274; M. de Pascual, 14.098.076; M. de Pascual 14.096.013), Vol. 9, No. 1 (M. Lopez Arroyo, 14.075.009).

Memoria de las actividades del Seminario de Astronomía y Geodesia de la Universidad Complutense de Madrid en 1971. J. M. Torroja.
Univ. Complutense de Madrid, Fac. Ci., Seminario Astron. Geod., Publ. No. 71 (1972).

Memoria de las actividades del Seminario de Astronomía y Geodesia de la Universidad Complutense de Madrid en 1972 y 1973. J. M. Torroja.
Urania Barcelona, Año 59, No. 281–282, p. 431 - 440 (1974).

Universidad Complutense – Facultad de Ciencias, Madrid. Seminario de Astronomía y Geodesia, Publicación, Nos. 71 (J. M. Torroja, 14.008.065), 77 (J. Aranda, 12.065.120), 78 (R. Ortiz, M. J. Sevilla, R. Vieira, 12.041.028), 79 (M. J. Sevilla, 14.046.014), 80 (M. A. Acosta, R. Ortiz, R. Vieira, 14.034.052), 81 (T. J. Vives, C. Morales, J. García-Pelayo, J. Barbero, 14.153.018), 82 (R. Ortiz, R. Vieira, 14.032.016), 83 (J. M. Torroja, 14.008.065).

008.066 **Manchester**

Astronomical Contributions from the University of Manchester, Series III, Nos. 251 (J. Meaburn, 06.133.001), 252 (Z. Kopal, 06.121.008), 253 (F. D. Kahn, 05.134.022), 254 (M. Moutsoulas, 05.094.118), 255 (J. Meaburn, 06.131.091), 256 (F. D. Kahn, 06.131.113), 257 (J. Meaburn, 07.034.001), 258 (Z. Kopal, 14.091.034), 259 (M. Moutsoulas, 08.094.058), 260 (Z. Kopal, 06.117.022), 261 (J. Meaburn, 08.155.038), 262 (M. A. Dopita, A. H. Gibbons, J. Meaburn, 09.132.002), 263 (J. E. Dyson, J. de Vries, 08.132.005), 264 (J. E. Dyson, 09.132.009), 265 (M. Moutsoulas, 08.094.144), 266 (M. A. Dopita, A. H. Gibbons, J. Meaburn, K. Taylor, 09.131.009), 268 (K. H. Elliott, J. Meaburn, 09.131.120), 269 (M. A. Dopita, 09.034.001), 270 (Z. Kopal, 14.094.183), 272 (J. Meaburn, 09.034.030), 273 (K. H. Elliott, 10.132.002), 274 (K. H. Elliott, J. Meaburn, 10.132.005), 275 (G. Tenorio-Tagle, 10.132.024), 276 (M. A. Dopita, 10.132.006), 277 (M. E. Alexander, 10.117.009), 278 (Z. Kopal, 10.065.081), 279 (K. H. Elliott, J. Meaburn, 10.132.014), 280 (E. Budding, 09.121.060), 281 (J. E. Dyson, 10.131.052), 283 (E. Budding, 11.121.016), 284 (G. Tenorio-Tagle, 10.131.022), 285 (F. D. Kahn, 10.131.137), 286 (Z. Kopal, 11.117.005), 287 (M. A. Dopita, 10.132.046), 288 (M. D. Moutsoulas, 10.094.090), 289 (M. A. Dopita, K. H. Elliott, J. Meaburn, 11.132.012), 290 (M. A. Dopita, J. E. Dyson, J. Meaburn, 11.131.531), 291 (E. Budding, 11.117.027), 292 (K. H. Elliott, J. Meaburn, 11.132.020), 293 (C. Goudis, J. Meaburn, 12.131.508), 294 (J. Meaburn, 10.034.085), 296 (J. E. Dyson, 11.131.075), 297 (M. A. Dopita, 11.132.028), 298 (Z. Kopal, M. K. Mahanta, 12.065.093), 299 (M. A. Dopita, 12.131.503), 300 (K. H. Elliott, J. Meaburn, 12.132.012).

008.067 **Mérida**

Instituto de Astronomía (Astronomical Institute).
Inform. Bull. Southern Hemisphere, No. 25, p. 28 (1974).
Current research report.

008.068 **Milano-Merate**

Contributi dell'Osservatorio Astronomico di Milano-Merate, Nuova Serie, Nos. 366 (F. Zagar, 14.044.037), 367 (F. Zagar, 14.112.022).

008.069 **Minneapolis**

University of Minnesota, Minneapolis, Minnesota, Separate prints (W. J. Luyten, 14.112.017; W. J. Luyten, 14. 112.018), (W. J. Luyten, 14.112.019), (W. J. Luyten, 14.112. 020; W. J. Luyten, C. T. Kowal, 14.112.021).

008.070 **Mizusawa**

A history of 75 years at the International Latitude Observatory of Mizusawa. T. Okuda.
Published by the International Latitude Observatory of Mizusawa, Mizusawa-Shi, Iwate-Ken, 10 + 77 pp. (1974).

Annual Report of the International Polar Motion Service 1973, (S. Yumi, 14.045.026).

Bulletins, Time Service of the Mizusawa Observatory, Vol. 18, No. 1 - 12, 1973 (S. Takagi, I. Okamoto, M. Aihara, K. Yokoyama, T. Hara, G. Murakami, T. Okuda, H. Kitago, K. Iwadate, S. Sakai, K. Horiai, T. Sasao, 14.044. 041).

Monthly Notes of the International Polar Motion Service, 1975 Nos. 6 - 11 (14.045.036).

Publications of the International Latitude Observatory of Mizusawa, Vol. 9, No. 2 (K. Sato, Y. Wako, 14.045. 009; S. Yumi, K. Hurukawa, T. Hirayama, 14.041.034; M. Ooe, T. Sasao, 14.045.010; H. Ishii, I. Naito, 14.045.011; S. Takagi, 14.082.076; S. Takagi, 14.032.033; S. Takagi, M. Aihara, H. Kitago, G. Murakami, K. Iwadate, 14.032.034; S. Takagi, H. Ookawa, Y. Goto, 14.045.012).

Annual report of geophysical observations made at the International Latitude Observatory of Mizusawa for the year 1973. T. Okuda.
Published by the International Latitude Observatory of Mizusawa, 42 pp. (1975).

Annual report of the meteorological observations made at the International Latitude Observatory of Mizusawa for the year 1973. T. Okuda.
Published by the International Latitude Observatory of Mizusawa, Japan, 2 + 30 pp. (1974).

Annual report of the meteorological observations made at the International Latitude Observatory of Mizusawa for the year 1974. T. Okuda.
Published by the International Latitude Observatory of Mizusawa, 30 pp. (1975).

008.071 **Mons**

Communications du Département d'Astrophysique de la Faculté des Sciences de Mons. Mons Astrophysical Papers, Nos. 47 (E. Peytremann, 14.064.067), 48 (E. Peytremann, 13.064.045), 49 (Y. Andrillat, L. Houziaux, 14.114. 360), 50 (Y. Andrillat, A. Baranne, L. Houziaux, 13.133.032).

008.072 **Moskva**

Chronicle. [Dissertations at the Sternberg Astronomical Institute]. L. N. Bondarenko.
Astron. Tsirk., No. 858, p. 6 - 8 (1975). In Russian.

Soobshcheniya Gosudarstvennogo Astronomicheskogo Instituta im. P. K. Shternberga. Izdatel'stvo Moskovskogo Universiteta, Nos. 177 (A. S. Sharov, G. I. Polyakova, 14.113. 037; O. D. Dokuchaeva, T. A. Birulya, 14.036.005; N. N. Yakimova, 14.122.138), 181 (V. S. Prokudina, 14.091.024; V. S. Prokudina, 14.072.030; V. I. Kovshov, 14.064.040; L. N. Bondarenko, 14.094.145; L. N. Bondarenko, 14.094. 146), 182 (B. A. Vorontsov-Vel'yaminov, R. I. Noskova, 14.158.144; E. D. Pavlovskaya, D. K. Karimova, 14.112.008; G. A. Starikova, 14.118.014; V. G. Moshkalev, 14.153.026; V. P. Arkhipova, 14.113.038; N. N. Yakimova, 14.122.139; O. D. Dokuchaeva, T. A. Birulya, 14.036.006).

Trudy Gosudarstvennogo Astronomicheskogo Instituta im. P. K. Shternberga. Izdatel'stvo Moskovskogo Universiteta, Vol. 46, vyp. (No.) 2 (I. M. Kalinina, 14.041.042; L. M. Khommik, 14.041.043; L. M. Khommik, 14.041.044; N. M. Artyukhina, P. N. Kholopov, 14.112.013).

008.073 **München**

Institut für Astronomie und Astrophysik der Universität München. Universitäts-Sternwarte. – Report 1974. P. Wellmann.
Mitt. Astron. Ges., No. 37, p. 140 - 143 (1975).

Max-Planck-Institut für Physik und Astrophysik. Institut für Astrophysik und Institut für extraterrestrische Physik. I. Institut für Astrophysik; II. Institut für extraterrestrische Physik. – Reports 1974.
Mitt. Astron. Ges., No. 37, p. 101 - 139 (1975).

Max-Planck-Institut für Physik und Astrophysik, München, MPI-PAE/Astro 73 (J. J. Perry, 14.141.150).

008.074 **Münster**

Astronomisches Institut der Universität. – Report 1974. H. Straßl.
Mitt. Astron. Ges., No. 37, p. 143 - 144 (1975).

008.075 **Oslo**

Institute of Theoretical Astrophysics, Blindern–Oslo. Report, Nos. 43 (P. Iuell, L. Staveland, 14.072.075), 44 (B. M. Rustad, 14.073.147).

008.076 **Ottawa**

National Research Council of Canada, Ottawa, Ontario, Canada, NRC Nos. 13311 (J. A. Galt, H. W. Broten, T. H. Legg, 13.141.331), 14220 (L. W. Avery, B. H. Andrew, 12.103.104), 14270 (M. B. Bell, 12.141.100), 14293 (B. H. Andrew, L. W. Avery, 12.131.071), 14324 (A. E. Covington, 12.077.063), 14328 (B. Balick, R. H. Gammon, L. H. Doherty, 11.132.004), 14376 (L. A. Higgs, 12.133.029), 14658

(B. H. Andrew, M. B. Bell, N. W. Broten, J. M. MacLeod, 13.158.074), 15392 (R. Pike, 12.073.112; A. E. Covington, 12.073.113).

008.077 **Oxford**

The University Observatory, Oxford–founded 1875. V. Reade.
Journ. British Astron. Ass., Vol. 85, 448 - 449 (1975).

Department of Astrophysics,University of Oxford. D. E. Blackwell.
Quarterly Journ. Roy. Astron. Soc., Vol. 16, 436 - 443 (1975).
Report for the year ending 1974 December 31.

008.078 **Paris**

Bureau International de l'Heure, (B.I.H.), Circulars A (14.044.038); **D**105 - **D**109 (14.044.039); E5 (14.044.040).

008.079 **Porto**

Publicações do Observatório Astronómico «Prof. Manuel de Barros» da Faculdade de Ciências do Porto No. 29 (J. P. Osório, 14.052.054).

008.080 **Porto Allegre**

Observatorio Astronomico do Instituto de Física da Universidade Federal do Rio Grande do Sul (Astronomical Observatory, Institute of Physics of the Federal University of Rio Grande do Sul).
Inform. Bull. Southern Hemisphere, No. 25, p. 8 (1974).
Current research report.

008.081 **Potsdam**

Zum 100. Jahrestag der Gründung des Astrophysikalischen Observatoriums Potsdam. J. Wempe.
Sterne, Vol. 51, 193 - 206 (1975).

100 Jahre Astrophysikalisches Observatorium Potsdam – Arbeiten zur Spektroskopie. G. Scholz.
Sterne, Vol. 51, 207 - 218 (1975).

100 Jahre Astrophysikalisches Observatorium Potsdam – Arbeiten zur Theorie der Sternatmosphären.
E. A. Gussmann.
Sterne, Vol. 51, 219 - 227 (1975).

On the prehistory of the Astrophysical Observatory Potsdam (1865 to 1874). See Abstr. 004.082.

Akademie der Wissenschaften der DDR. **Zentralinstitut für Astrophysik, Sternwarte Babelsberg,** Potsdam-Babelsberg, **Mitteilungen.** Neue Folge, Nos. 112 (G. Dautcourt, 12.162.071), 113 (H. Tiersch, 12.141.098), 117 (H.-J. Treder, 13.022.005), 118 (H.-J. Treder, 13.066.004), 119 (H.-J. Treder, 13.162.002), 120 (G. Dautcourt, 13.066.005), 126 (R. W. John, 14.066.145), 127 (H.-H. v. Borzeszkowski, U. Kasper, 14.066.146), 129 (H.-J. Treder, 14.066.147), 130 (H.-J. Treder, 13.081.006), 131 (H.-J. Treder, 14.066.148), 132 (H.-J. Treder, 14.066.149), 133 (G. Dautcourt, 14.066.150), 134 (H.-J. Treder, 14.066.151), 135 (U. Kasper, 14.066.152), 136 (D.-E. Liebscher, 14.066.153), 137 (H.-J. Treder, 14.066.154), 138 (H.-J. Treder, 14.066.155), 139 (H.-J. Treder, 13.066.138), 140 (U. Kasper, 13.066.038), 141 (H.-J. Treder, 13.066.037), 149 (H.-J. Treder, 14.066.156).

008.082 **Praha**

Académie Tchécoslovaque des Sciences, Institut Astronomique, **Station de l'Heure à Prague,** Série 7, Nos. 2 - 6 (L. Webrová, V. Ptáček, 14.044.050).

008.083 **Pretoria**

Radcliffe Observatory, Pretoria.
Inform. Bull. Southern Hemisphere, No. 25, p. 1 (1974).

Radcliffe Observatory.
Inform. Bull. Southern Hemisphere, No. 25, p. 26 (1974).
Current research report.

008.084 **Pulkovo**

Izvestiya Glavnoj Astronomicheskoj Observatorii v Pulkove, No. 193, Astrofizika, Astrometriya i Zvezdnaya Astronomiya (V. A. Krat, E. K. Kokhan, N. I. Pechinskaya, 14.071.064; Yu. I. Vitinskij, 14.072.067; A. N. Dadaev, 14.121.119; A. N. Gerashchenko, 14.131.192; G. G. Lengauer, 14.131.193; V. N. Frolov, 14.153.037; G. A. Plyugin, 14.117.071; A. A. Nemiro, 14.041.056; V. A. Naumov, 14.041.057; G. I. Pinigin, 14.032.046; N. M. Bakhrakh, V. A. Naumov, 14.041.058; G. S. Sheptunov, A. N. Sheptunova, V. P. Kotlyarchuk, I. A. Vladimirova, 14.045.025; L. D. Kostina, 14.041.059; R. Peralta, D. D. Polozhentsev, T. A. Polozhentseva, R. Tajbo, R. Tapia, G. Silva, G. M. Timashkova, 14.032.047; J. Petit, D. D. Polozhentsev, T. A. Polozhentseva, E. Portugal, G. Silva, R. Tajbo, R. Tapia, G. M. Timashkova, 14.032.048; S. Akun'ya, M. P. Varin, V. A. Varina, D. Vivers, G. Leon, R. Tajbo, 14.032.049; N. M. Bronnikova, A. A. Kiselev, 14.041.060; T. P. Kiseleva, 14.041.061; G. K. Gorel', F. F. Kalikhevich, 14.041.062; V. A. Vytnov, 14.044.036; I. I. Brejdo, O. M. Mikhajlova, 14.036.022; T. S. Belorossova, N. V. Merman, M. A. Sosnina, 14.032.050; G. I. Bol'shakova, A. N. Demidova, 14.085.020).

Glavnaya Astronomicheskaya Observatoriya AN SSSR, Katalog Solnechnoj Deyatel'nosti (za 1972 god, R. S. Gnevysheva, 14.075.005; za 1973 god, R. S. Gnevysheva, 14.075.006).

008.085 **Richmond Hill**

David Dunlap Observatory forty years old.
J. F. Heard.
Journ. Roy. Astron. Soc. Canada, Vol. 69, 141 - 144 (1975).

Communications from the David Dunlap Observatory, University of Toronto, Richmond Hill, Ontario, Canada, Nos. 432 (J. P. Vallée, P. P. Kronberg, 12.141.091), 433 (S. van den Bergh, 12.159.006), 434 (R. G. Deupree, 12.122.091), 435 (J. F. C. Wardle, P. P. Kronberg, 12.141.101), 436 (J. D. Fernie, 12.034.073), 437 (E. Honig, K. Lake, R. C.

Roeder, 14.066.157), 438 (J. P. Vallée, P. P. Kronberg, 10. 156.003), 439 (M. J. Clement, 12.065.160), 440 (H. Sawyer Hogg, B. Rutily, A. Terzan, 12.154.014), 441 (S. van den Bergh, G. L. Hagen, 13.153.001), 442 (P. G. Martin, J. R. P. Angel, 13.131.011), 443 (S. van den Bergh, 13.103.103), 444 (J. F. Heard, R. J. Hurkens, 13.119.006), 445 (J. F. Heard, A. Krautter, 13.119.005), 446 (R. G. Conway, P. Haves, P. P. Kronberg, D. Stannard, J. P. Vallée, J. F. C. Wardle, 12.141. 001), 447 (J. R. Percy, C. W. McAlary, 12.122.022), 448 (W. Herbst, 13.132.018), 449 (C. T. Bolton, C. Martin, 12. 036.021), 450 (S. van den Bergh, W. Herbst, 13.132.017).

David Dunlap Observatory, University of Toronto, Richmond Hill, Ontario, Canada. Separate prints (B. F. Madore, S. van den Bergh, 13.122.021), (C. T. Bolton, A. Young, G. Wicks, R. B. Jones, 13.119.015), (S. van den Bergh, 13.031. 260).

008.086 **Rio de Janeiro**

Observatorio Nacional (National Observatory). Inform. Bull. Southern Hemisphere, No. 25, p. 7 - 8 (1974). Current research report.

008.087 **Rochester**

C. E. Kenneth Mees Observatory, University of Rochester, Rochester, N.Y., **Reprint**, No. 53 (J. G. Duthie, 13.008.107)

008.088 **Roma**

Monthly Bulletin. Osservatorio Astronomico di Roma, Nos. 203 - 208 (M. Cimino, M. Torelli, 14.075.020).

Photographic Journal of the Sun. Osservatorio Astronomico di Roma, Nos. 93 - 98 (M. Cimino, 14.075.019).

008.089 **Rosario**

Actividades del Observatorio. Obs. Astron. Municipal Rosario, (*Argentina*), Bol. No. 4, p. 20 - 24 (1974).

Observatorio Astronómico Municipal de Rosario, *Argentina,* Boletín No. 4 (V. Ambartsumyan, 14.065.180; V. Capolongo, 14.092.016; J. R. Garcia, G. A. Galassi, 14.122. 221; J. A. Gutierrez, R. B. Esteve, C. M. Parfait, J. A. Serra, 14.085.022; 14.008.089).

008.090 **Sacramento Peak**

Sacramento Peak Observatory, Contribution Nos. 249 (J. M. Beckers, 14.072.079), 250 (J. M. Beckers, L. Dickson, R. S. Joyce, 14.034.124).

008.091 **San Miguel**

Observatorio Nacional de Física Cósmica (National Observatory of Cosmic Physics). Inform. Bull. Southern Hemisphere, No. 25, p. 4 - 6 (1974). Current research report.

008.092 **Santiago de Chile**

Departamento de Astronomía, Universidad de Chile, Facultad de Ciencias Fisicas y Matematicas, Observatorio Astronómico Nacional, Cerro Calán, Santiago de Chile, **Publicaciones**, Vol. 2, No. 6 (C. Anguita, G. Carrasco, P. Loyola, V. N. Bedin, A. A. Naumova, D. D. Polojentsev (*Polozhentsev*), T. A. Polojentseva (*Polozhentseva*), K. N. Tavastsherna, M. S. Zverev, 14.041.036).

008.093 **São Paulo**

Instituto Astronomico e Geofísico da Universidade de São Paulo (Astronomical and Geophysical Institute of São Paulo University). Inform. Bull. Southern Hemisphere, No. 25, p. 8 - 10 (1974). Current research report.

Centro de Radio Astronomía e Astrofísica da Universidade Mackenzie (CRAAM) (Center of Radio Astronomy and Astrophysics, Mackenzie University). Inform. Bull. Southern Hemisphere, No. 25, p. 10 - 13 (1974). Current research report.

008.094 **Sendai**

Sendai Astronomiaj Raportoj, Nos. 159 (K. Takakubo, 12.155.069), 160 (H. Saio, 14.065.026), 161 (T. Sawa, 14.155.028), 162 (H. Tamura, 14.132.033), 163 (K. Suda, J. Uchida, 14.065.132), 164 (T. Hasegawa, 14.114.359), 165 (O. Kaburaki, 14.062.046).

008.095 **Sonneberg**

Zentralinstitut für Astrophysik. Mitteilungen über Veränderliche Sterne, *Sonneberg,* Vol. 7, Nos. **2** (W. Wenzel, H. Gessner, 14.122.214; V. Novotný, 14.123.066; R. Hudec, 14.142.243; H.-J. Blasberg, 14.123.067; H.-J. Blasberg, 14. 123.068; H.-J. Blasberg, 14.123.069; H. Busch, 14.123.070; M. Heß, 14.121.136; M. Werner, 14.123.071; 14.123.072; P. Ahnert, 14.121.137; P. Ahnert, 14.123.073; E. Splittgerber, 14.122.215; L. Meinunger, 14.158.241; H. Gessner, 14.123. 074; D. Böhme, 14.124.103), **3** (H. Gessner, 14.124.107; H. Gessner, 14.121.138; G. A. Richter, 14.124.104; P. Ahnert, 14.121.139; P. Ahnert, 14.121.140; P. Ahnert, 14.121.141; P. Ahnert, 14.121.142; P. Ahnert, 14.121.143; P. Ahnert, 14.121.144; R. Hudec, 14.114.393; P. Ahnert, 14.121.145; P. Ahnert, 14.121.146).

Akademie der Wissenschaften der DDR, Zentralinstitut für Astrophysik. **Veröffentlichungen der Sternwarte in Sonneberg**, Vol. 8, No. 4 (H. Busch, K. Häußler, C. Thänert, 14.123.065).

008.096 **St. Andrews**

Communications from the University Observatory, St. Andrews, Nos. 7 (D. C. B. Whittet, 12.152.001), 8 (I. C. van Breda, I. S. Glass, D. C. B. Whittet, 12.113.013), 9 (P. W. Hill, D. Kilkenny, I. C. van Breda, 12.113.012), 10 (R. W. Hilditch, 12.119.016), 11 (D. Kilkenny, P. W. Hill, T. Schmidt-Kaler, 13.113.005).

University Observatory, St. Andrews. Reprint Nos. 48 (C. W. Fraser, 07.158.105), 50 (B. D. Kelly, D. Kilkenny, 10.034.004), 51 (D. C. B. Whittet, I. G. van Breda, K. Nandy, 09.131.126), 52 (T. R. Carson, 09.065.111), 53 (D. Kilkenny, 11.112.003), 54 (I. G. van Breda, P. W. Hill, R. J. Campbell, A. E. Lynas-Gray, D. M. Carr, 14.031.412), 55 (R. W. Hilditch, G. Hill, 12.142.029), 56 (P. J. Harvey, R. S. Booth, R. D. Davies, D. C. B. Whittet, W. McLaughlin, 12.131.110), 57 (D. Kilkenny, 13.031.206), 58 (T. R. Carson, D. Ezer, R. Stothers, 12.080.062), 59 (B. D. Kelly, 12.122.008), 60 (R. W. Hilditch, 13.116.009), 61 (D. Kilkenny, C. E. J. Hilditch, R. W. Hilditch, P. W. Hill, A. E. Lynas-Gray, 14.153.002), 62 (R. W. Hilditch, P. W. Hill, 14.119.004).

008.097 **Swarthmore**

Sproul Observatory, Swarthmore, Pennsylvania, **Reprints,** Nos. 230 (S. L. Lippincott, W. D. Heintz, 11.008.123), 231 (W. D. Heintz, 11.118.008), 232 (P. van de Kamp, 11.118.010), 234 (W. D. Heintz, 12.118.001), 235 (S. L. Lippincott, 12.111.005), 236 (W. D. Heintz, 11.111.008), 238 (M. D. Worth, W. D. Heintz, 12.118.010), 239 (S. L. Lippincott, 12.011.016), 240 (W. D. Heintz, 13.118.002), 241 (S. L. Lippincott, W. D. Heintz, 13.008.121), 243 (P. van de Kamp, 14.041.022), 244 (W. D. Heintz, 14.118.008, 14.118.009, 14.118.010), 245 (P. van de Kamp, 14.117.016), 246 (J. L. Hershey, 14.121.021), 247 (P. van de Kamp, 14.117.051), 248 (S. L. Lippincott, 14.118.004), 249 (W. D. Heintz, 14.098.014), 250 (S. L. Lippincott, 14.118.015), 251 (P. J. Bachmann, J. L. Hershey, 14.121.071), 252 (J. L. Hershey, 14.034.141), 253 (P. van de Kamp, 14.011.005), 254 (P. van de Kamp, 14.042.013).

008.098 **Sydney**

Division of Radiophysics, CSIRO, Sydney, Australia, Radiophysics Publication RPP 1593 (J. L. Caswell, W. M. Goss, 11.131.536), 1638 (D. K. Milne, L. H. Aller, 13.133.012), 1722 (J. B. Whiteoak, 12.156.002), 1729 (D. K. Milne, J. R. Dickel, 12.125.053), 1730 (J. L. Caswell, 12.141.623), 1732 (D. K. Milne, L. H. Aller, 12.133.025), 1733 (O. B. Slee, G. A. Dulk, 12.125.054), 1755 (B. J. Robinson, 12.155.062), 1758 (T. Takakura, S. Yousef, 13.077.022), 1772 (J. G. Ables, B. F. C. Cooper, A. J. Hunt, G. G. Moorey, J. W. Brooks, 14.033.021), 1775 (T. W. Cole, 13.031.421), 1788 (J. B. Whiteoak, F. F. Gardner, 13.131.018), 1796 (T. W. Cole, 14.033.022), 1812 (R. T. Stewart, 13.077.021), 1823 (F. F. Gardner, G. Winnewisser, 13.131.059), 1849 (W. M. Goss, I. A. Lockhart, E. B. Fomalont, 13.131.534), 1854 (F. F. Gardner, J. B. Whiteoak, 13.131.528).

Division of Radiophysics, CSIRO, Sydney, Australia. Separate prints (J. A. Roberts, R. S. Roger, J.-C. Ribes, D. J. Cooke, J. D. Murray, B. F. C. Cooper, F. Biraud, 14.141.009; J. L. Caswell, R. F. Haynes, D. H. Clark, 14.141.117; B. J. Wills, 14.141.131; R. F. Haynes, W. K. H. Huchtmeier, B. C. Siegman, A. E. Wright, 14.158.179).

008.099 **Tartu**

B and V photoelectric photometry at Tallinn Astronomical Observatory. See Abstr. 031.238.

Eesti NSV Teaduste Akadeemia (Akademiya Nauk Ehstonskoj SSR), **W. Struwe nimelise Tartu Astrofüüsika Observatooriumi, Publikatsioonid** (Publikatsii Tartuskoj Astrofizicheskoj Observatorii imeni W. Struve), Köide (Tom) 43 (A Kipper, 14.162.043; U. Hänni, L. Luud, 14.114.364; T. Kipper, 14.114.080; T. Kipper, 14.114.081; L. Hänni, M. Kipper, 14.114.365, U. Uus, 14.065.135; M. Jõeveer, 14.155.045; P. Kalv, 14.031.238; P. Kalv, 14.121.084; P. Kalv, I. Pustylnik; 14.121.085; I. Pustylnik, L. Sorgsepp, 14.121.086; I. Pustylnik, L. Sapar, L. Sorgsepp, 14.121.087; V. Malyuto, 14.155.046; Ü. Ibrus, 14.034.097; V. Riives, 14.103.001; J. Jaaniste, E. Saar, 14.155.047; T. Viik, 14.063.036; L. Luud, O. P. Gollandskij, T. Yarygina, 14.114.082; H. Albo, 14.122.164; A. Sapar, T. Feklistova, 14.022.085).

008.100 **Tautenburg**

Zentralinstitut für Astrophysik. Mitteilungen des Karl-Schwarzschild-Observatoriums Tautenburg der Deutschen Akademie der Wissenschaften zu Berlin, Nos. 73 (F. Börngen, N. Richter, P. Notni, H. Oleak, W. Wenzel, 13.032.008), 74 (F. Börngen, 11.032.011), 75 (N. Richter, 11.005.011), 76 (N. Richter, 12.036.010), 77 (S. Marx, 13.131.020).

008.101 **Teramo**

Osservatorio Astronomico di Collurania, Teramo. Contributi Serie II Nos. 1 (P. Tempesti, 06.121.068), 2 (P. Tempesti, 08.124.101), 3 (R. Burchi, L. Milano, 11.098.012).

Osservatorio Astronomico di Collurania, Teramo. Memorie ed Osservazioni, Vol. 2, Fasc. 3 (G. de Mottoni, 14.097.059).

Osservatorio Astronomico di Collurania, Teramo. Note e Comunicationi Nos. 47 (P. Tempesti, 02.121.016), 48 (P. Tempesti, R. Burchi, 02.098.017), 49 (R. Burchi, 07.098.006), 50 (P. Tempesti, R. de Carlo, 07.121.022).

008.102 **Tokyo**

Tokyo Observatory today. F. Moriyama. Sky Telescope, Vol. 50, 276 - 281 (1975).

Annals of the Tokyo Astronomical Observatory, University of Tokyo, Second Series, Vol. 15, No. 3 (M. Kiyokawa, M. Kitamura, 14.121.148).

Tokyo Astronomical Bulletin, Tokyo Astronomical Observatory, Second Series, Nos. 238 (K. Tomita, H. Kosai, 14.103.008), 239 (Y. Niimi, K. Hurukawa, K. Nakajima, 14.098.061), 240 (E. Budding, 14.121.147).

Tokyo Astronomical Observatory, Reprints Nos. 466 (Y. Uchida, T. Sakurai, 13.072.058), 467 (Y. Yamashita, 13.114.078), 468 (S. Okazaki, 13.044.020), 469 (M. Nakagiri, 13.122.114), 471 (M. Miyamoto, 14.155.030), 472 (Y. Yamashita, 14.113.030), 473 (M. A. Dopita, S. Isobe, J. Meaburn, 13.132.024), 474 (T. Iguchi, 14.131.145), 475

(M. Miyamoto, R. Nagai, 14.151.066), 476 (M. Iye, K. Kodaira, S. Kikuchi, H. Ohtani, 14.125.103), 477 (S. Isobe, H. Kosai, H. Hara, R. Fukaya, Y. Adachi, 14.124.104), 478 (K. Nishi, 13.076.060), 479 (S. Kawabata, M. Saito, 14.121.026), 480 (E. Budding, 14.121.027), 481 (Y. Uchida, 14.073.032), 482 (K. Nariai, 14.042.053).

Time and Latitude Bulletins, Tokyo Astronomical Observatory, Vol. 49, Nos. 1 - 4 (K. Osawa, 14.044.049).

Contributions from the Department of Astronomy, University of Tokyo, Nos. 172 (H. Yoshimura, 10.072.078), 173 (F. Kamijo, Y. Nakada, T. Iguchi, M. Fujimoto, M. Takada, S. Takajo, 10.131.175), 174 (M. Hirai, 11.114.153), 175 (M. Takada, M. Hirai, 11.022.070), 176 (K. Kodaira, 11.125.072), 177 (W. Unno, E. Ribes, I. Appenzeller, 11.073.046), 178 (Y. Osaki, 12.122.056), 179 (S. Deguchi, 12.131.085), 180 (T. Nakamura, 12.100.209), 181 (M. Ishikawa, 13.122.017), 182 (O. Kaburaki, 13.073.015), 183 (M. Kurihara, 13.077.018), 184 (W. Unno, 13.080.012), 185 (T. Daishido, 13.142.031), 186 (F. Kamijo, 14.022.123), 187 (Y. Osaki, 13.065.090), 188 (H. Shibahashi, Y. Osaki, W. Unno, 14.080.032), 189 (M. Iye, K. Kodaira, 14.125.026), 190 (T. Mikami, 14.115.017), 191 (T. Takakura, Y. Naito, K.Ohki, 13.077.024), 192 (T. Takakura, 13.076.056), 193 (M. Iye, K. Kodaira, S. Kikuchi, H. Ohtani, 14.125.103), 194 (H. Ando, Y. Osaki, 14.080.058), 195 (W. Unno, 14.065.138).

008.103 **Tonantzintla**

Boletín del Instituto de Tonantzintla, Vol. 1, No. 4 (R. K. Pandey, 14.022.117; B. Iriarte Erro, 14.122.216; L. R. Berriel-Valdos, 14.021.024; D. Malacara H., 14.032.053; J. C. Fouéré, D. Malacara, 14.034.125; D. Malacara, A. Cornejo, M. V. R. K. Murty, 14.034.126; M. L. Celaya, D. Tentori, D. Malacara, 14.031.076; C. Roychoudhuri, A. Cornejo, 14.031.077), No. 5 (E. Sánchez-Sinencio, 14.031.303; R. K. Pandey, G. R. Hernández, 14.022.124; C. Roychoudhuri, 14.031.304; R. I. Mercado, 14.031.080; D. Malacara, A. Cornejo, 14.031.081; L. R. Berriel-Valdos, 14.031.082; A. Cornejo, D. Malacara, 14.031.083; A. S. De Vany, A. De Vany, 14.031.084; A. S. De Vany, 14.031.085; D. Malacara, 14.031.086).

008.104 **Torino**

Contributi dell'Osservatorio Astronomico di Torino (Pino Torinese), Nos. 83 (L. Lorenzi, V. Zappalà, 12.103.104), 84 (R. Pannunzio, V. Zappalà, 12.098.069), 87 (G. De Sanctis, M. A. Vogliotti, V. Zappalà, 13.098.053).

Osservatorio Astronomico di Torino, Pino Torinese. Time Service, Bulletin Nos. 8 - 10 (C. Moranzino, 14.044.042).

Pubblicazioni Varie Fuori Serie dell'Osservatorio Astronomico di Torino (Pino Torinese), No. 58 (M. G. Fracastoro, 14.103.102).

008.105 **Toruń**

Bulletin of the Astronomical Observatory of N. Copernicus University in Toruń, Nos. 52 (A. T. Young, A. Woszczyk, L. G. Young, 11.093.008; A Przybylski, A. Burnicki, 12.114.079), 53 (S. Gorgolewski, 14.035.021; Z. Turło, 14.031.413; B. Wikierski, 14.031.414), 54 (S. Gąska, 14.104.091; W. Wegner, 14.114.394; D. Kuroczkin, 14.114.114).

008.106 **Trieste**

Pubblicazione Osservatorio Astronomico di Trieste, Nos. 484 (R. Faraggiana, K. A. van der Hucht, 13.114.318), 485 (14.075.011), 486 (14.075.011), 487 (R. Stalio, P. L. Selvelli, 14.114.307), 489 (P. L. Bernacca, F. Ciatti, L. Guzzi, G. Sedmak, I. E. Campisi, A. Treves, 13.141.359), 491 (B. Cester, M. Pucillo, 14.121.005), 495 (P. Zlobec, 14.077.030), 496 (P. Zlobec, 14.077.024), 499 (14.075.011), 500 (14.075.011).

008.107 **Tübingen**

Astronomisches Institut der Universität und Lehrstuhl für Theoretische Astrophysik. – Reports 1974.
J. Trümper.
Mitt. Astron. Ges., No. 37, p. 145 - 151 (1975).

008.108 **Uccle**

Bulletin Astronomique. (Astronomisch Bulletin). **Observatoire Royal de Belgique** (Koninklijke Sterrenwacht van België), Vol. 8, Nos. 4 (P. Pâquet, W. De Rop, 14.044.044; H. Debehogne, 14.098.088; H. Debehogne, 14.098.089; H. Debehogne, 14.098.090; H. Debehogne, 14.103.012; C. Gonze, R. Gonze, 14.075.012; J. Dommanget, 14.118.038), 5 (R. Dejaiffe, R. Verbeiren, 14.041.071; H. Debehogne, 14.098.091; H. Debehogne, 14.098.092; C. Gonze, R. Gonze, 14.075.013; O. Nys, 14.118.039).

Observatoire Royal de Belgique (Koninklijke Sterrenwacht van Belgie), **Communications** (Mededelingen), Série B, No. 92 (A. Koeckelenbergh, 13.073.033).

008.109 **Victoria**

The Dominion Astrophysical Observatory 1918–1975. K. O. Wright.
Journ. Roy. Astron. Soc. Canada, Vol. 69, 205 - 211 (1975).
Paper presented at the meeting of the Canadian Astronomical Society in Kingston, Ontario on June 21, 1975.

The Dominion Astrophysical Observatory, Victoria, B.C. J. B. Hutchings.
Journ. Roy. Astron. Soc. Canada, Vol. 69, 257 (1975).

Dominion Astrophysical Observatory Victoria, B.C.
K. O. Wright.
Quarterly Journ. Roy. Astron. Soc., Vol. 16, 298 - 308 (1975).
Report for the year 1974 April 1 to 1975 March 31.

Contributions from the Dominion Astrophysical Observatory, Victoria, B. C.,Nos. 209 (J. B. Hutchings, A. D. Thackeray, B. L. Webster, P. J. Andrews, 10.121.002), 218 (A. D. Thackeray, J. B. Hutchings, 11.121.052), 220 (A. D. Thackeray, G. Hill, 12.119.001), 228 (R. W. Hilditch, G. Hill, 12.142.029), 231 (G. Hill, D. L. Crawford, J. V. Barnes, 12.119.019), 235 (J. M. Marlborough, M. Zamir, 13.116.001), 237 (A. P. Cowley, D. Crampton, J. B. Hutchings, J. M. Marlborough, 13.117.003), 238 (A. P. Cowley, J. B. Hutchings,

12.117.028), 239 (D. Crampton, Y. M. Georgelin, 13.155. 043), 240 (D. Crampton, A. P. Cowley, 13.142.048), 241 (C. R. Cowley, S. J. Adelman, 13.114.004), 242 (J. B. Hutchings, D. Crampton, R. O. Redman, 13.117.010), 243 (C. R. Cowley, G. C. L. Aikman, 13.114.020), 244 (C. D. Scarfe, S. Alers, 13.119.017), 246 (P. S. Conti, A. P. Cowley, G. B. Johnson, 13.119.018), 247 (M. S. Allen, C. R. Cowley, 12.071. 033), 249 (M. Hack, J. B. Hutchings, Y. Kondo, G. E. McCluskey, M. Plavec, R. S. Polidan, 13.121.042), 252 (J. B. Hutchings, 13.121.061), 253 (A. H. Batten, W. A. Fisher, B. W. Baldwin, C. D. Scarfe, 13.117.002), 256 (C. R. Cowley, 13.114.063), 258 (D. Crampton, R. O. Redman, 13.119.012), 259 (D. Crampton, A. P. Cowley, R. M. Humphreys, 13.141. 616).

Publications of the Dominion Astrophysical Observatory, Victoria, B.C., Vol. 14, No. 14 (J. A. Pearce, G. Hill, 14.153.041).

008.110 **Vilnius**

Vilniaus Astronomijos Observatorijos Biuletenis (Bulletin of the Vilnius Astronomical Observatory), Nos. **41** (K. Zdanavičius, 14.113.052; A. Bartkevičius, K. Zdanavičius, 14.113.053; G. Kakaras, 14.113.054), **42** (E. Meištas, K. Zdanavičius, V. Straižys, A. Gurklytė, 14.113.055; V. Straižys, G. Kuriliene, 14.113.056; 14.032.042).

008.111 **Warsaw**

150 years of the Astronomical Observatory of the Warsaw University. L. Zajdler.
Urania Kraków, Vol. 46, 364 - 371 (1975). In Polish.

Politechnika Warszawska, Obserwatorium Astronomiczno-Geodezyjne w Józefosławiu, (Warsaw Technical University, Astronomic-Geodetical Observatory at Józefosław), **Latitude Circular,** Nos. 53, 54 (L. Pieczyński, 14.045. 037).

008.112 **Washington**

Publications of the United States Naval Observatory, *Washington,* Second Series, Vol. 23, Part 1 (R. W. Rhynsburger, F. S. Gauss, 14.041.029); Vol. 24, Part 1 (R. S. Harrington, C. C. Dahn, A. L. Behall, J. B. Priser, J. W. Christy, B. Y. Riepe, H. D. Ables, H. H. Guetter, A. V. Hewitt, R. L. Walker, 14.041.035).

U.S. Naval Observatory, Washington, D.C. Time Service Publications, Series 4, Nos. 439 - 465; Series 7, Nos. 392 - 417; Series 11, No. 223; Series 14, Nos. 17, 19 (14.044. 023 - 14.044.026).

United States Naval Observatory, *Washington, D.C.,* **Circular,** Nos. 151 (G. H. Kaplan, K. F. Pulkkinen, B. Emerson, 14.047), 152 (A. D. Fiala, J. S. Duncombe, 14.079.109).

008.113 **Wellington**

Carter Observatory, Wellington, New Zealand, Astronomical Bulletin, Nos. 83 (B. M. Lewis, 14.047.040),

Report of the Carter Observatory Board and report of the director for the year ended 1975, March 31.
R. P. Gough, B. M. Lewis.
Carter Obs., Wellington, New Zealand, Astron. Bull. No. 84, p. 1 - 8 (1975).

Carter Observatory, *Wellington,* **Reprint Series II,** Nos. 1, (B. M. Lewis, 11.158.027), 2 (B. M. Lewis, 13.162.013).

008.114 **Wien**

Universitäts-Sternwarte mit Leopold Figl-Observatorium für Astrophysik. – Report 1974. J. Meurers.
Mitt. Astron. Ges., No. 37, p. 151 - 157 (1975).

Sternwarten gestern, heute und morgen.
See Abstr. 009.019.

Institut für Theoretische Astronomie. – Report 1974. K. Ferrari d'Occhieppo.
Mitt. Astron. Ges., No. 37, p. 157 - 158 (1975).

008.115 **Wrocław**

Contributions from the Wrocław Astronomical Observatory No. 19 (T. Ciurla, B. Rompolt, 14.073.153).

Wrocław Astronomical Observatory, Reprint No. 96 (M. Jerzykiewicz, 13.122.059).

008.116 **Würzburg**

Institut für Astronomie und Astrophysik. – Report 1974. H. Haffner.
Mitt. Astron. Ges., No. 37, p. 158 - 159 (1975).

008.117 **Zelenchukskaya**

Soobshcheniya Spetsial'noj Astrofizicheskoj Observatorii, Akademiya Nauk SSSR, vyp. (No.) **13** (T. A. Kartasheva, L. I. Snezhko, Kh. F. Khaliullin, N. M. Chunakova, 14.113.064; V. P. Chuprina, 14.126.022; B. P. Artamonov, F. Börngen, A. I. Shapovalova, 14.158.229; I. P. Kostyuk, 14.158.230; I. D. Karachentsev, R. L. Tsarevskaya, A. L. Shcherbanovskij, 14.160.042), **14** (T. B. Pyatunina, 14.131. 556; A. F. Steklov, I. D. Karachentsev, 14.151.085; V. K. Dubrovich, 14.162.061; V. K. Khersonskij, D. A. Varshalovich, 14.131.187; V. K. Khersonskij, D. A. Varshalovich, 14.131. 188; V. K. Khersonskij, D. A. Varshalovich, 14.131.189).

008.118 **Zürich**

Astronomische Mitteilungen der Eidgenössischen Sternwarte Zürich, Nos. 339 (M. Waldmeier, 14.075.014), 342 (H. Pfister, 14.072.080).

Publikationen der Eidgenössischen Sternwarte Zürich, Vol. 14, No. 4 (M. Waldmeier, 14.075.016).

Quarterly Bulletin on Solar activity (Zürich), Nos. 187 - 188 (M. Waldmeier, R. Howard, G. Olivieri, M. Bernot, H. Tanaka, 14.075.015).

009 Notes on Observatories, Planetaria, and Exhibitions

009.001 **The Fiske planetarium in Boulder.** G. L. Verschuur, J. H. Sharp.
Sky Telescope, Vol. 50, 140 - 144 (1975).

009.002 **The foundation of the Nikolayev Observatory.** G. K. Gorel'.
Istor.-astron. issledovaniya. Vyp. (No.) 12. Moskva, Nauka, 1975, p. 139 - 155. In Russian. – Abstr. in Referativ. Zhurn. 51. Astron., 7.51.25 (1975).

009.003 **Le Centre d'Études et de Recherches Géodynamiques et Astronomiques (CERGA).**
J. Kovalevsky.
L'Astronomie, 89e année, p. 271 - 286, 318 (1975).

009.004 **The Mt. Evans Infrared Observatory.** J. R. Lesh.
Bull. American Astron. Soc., Vol. 7, 408 (1975).
Abstr. AAS.

009.005 **"Discovering astronomy": an interactive museum exhibit.** D. L. Schatz, A. J. Friedman.
Bull. American Astron. Soc., Vol. 7, 447 - 448 (1975). – Abstr. AAS.

009.006 **Das Planetarium Freiburg i.Br.** R. Fehrenbach.
Orion, 33. Jahrgang, p. 112 - 114 (1975).

009.007 **Projektionskuppeln für Kleinplanetarien.** P. Ungar.
SuW, Vol. 14, 297 (1975).

009.008 **The Rockford planetarium dream.** C. J. Beaman.
Sky Telescope, Vol. 50, 232 - 233 (1975).

009.009 **La Posta Astrogeophysical Observatory.** – Report from solar institute.
M. P. Bleiweiss, F. L. Wefer.
Solar Physics, Vol. 43, 253 - 259 (1975).

009.010 **Planetarium en astronomisch observatorium 'Im. Mikolaja Kopernika' in Polen.** T. M. Vermeesch.
Zenit, Vol. 2, 330 - 331 (1975).

009.011 **Simon Stevin public observatory in the Netherlands.** B. C. J. van Apeldoorn.
Sky Telescope, Vol. 50, 305 - 307 (1975).

009.012 **Izveštaj o radu Katedre za astronomiju i Astr. sektora Astronomskogeofizičke opservatorije u Ljubljani u godinama 1971–1973.** F. Dominko.
Publ. Obs. Astron. Beograd, No. 20, p. 254 - 257 (1975).

009.013 **Izveštaj o radu Katedre za astronomiju Geodetskog fakulteta u Zagrebu u školskim godinama 1970/71, 1971/72 i 1972/73.** P. Terzić.
Publ. Obs. Astron. Beograd, No. 20, p. 258 - 260 (1975).

009.014 **Opservatorij Hvar Geodetskog fakulteta Sveučilišta u Zagrebu.** V. Petković.
Publ. Obs. Astron. Beograd, No. 20, p. 261 - 271 (1975).

009.015 **Izveštaj o radu Akademskog astronomsko-astronautičkog društva u Sarajevu.** M. Muminović.
Publ. Obs. Astron. Beograd, No. 20, p. 273 - 274 (1975).

009.016 **Projekat snimanja fotografskog zvjezdanog atlasa sa Astronomske opservatorije "Čolina kapa" u Sarajevu.** M. Muminović.
Publ. Obs. Astron. Beograd, No. 20, p. 275 - 276 (1975).

009.017 **Izvještaj o radu Zvjezdarnice u Zagrebu za period 1. I. 1970. do 31. XII. 1973.**
G. Kren, T. Vranić-Kren.
Publ. Obs. Astron. Beograd, No. 20, p. 279 - 281 (1975).

009.018 **C.S.I.R., National Institute for Telecommunications Research.** G. D. Nicolson.
Inform. Bull. Southern Hemisphere, No. 25, p. 27 (1974)
Current research report.

009.019 **Sternwarten gestern, heute und morgen.** J. Meurers.
SuW, Vol. 14, 390 - 394 (1975).

009.020 **The first observatories of the Petersburg Academy of Sciences.** V. L. Chenakal.
Zemlya i Vselennaya, 1975, No. 6, p. 50 - 56. In Russian.

009.021 **Planetaries of the world.** K. A. Portsevskij.
Zemlya i Vselennaya, 1975, No. 6, p. 70 - 77. In Russian.

009.022 **Goth Hill Radio Observatory, 1948–1971.** A. E. Covington.
Journ. Roy. Astron. Soc. Canada, Vol. 69, 251 (1975).
Abstr. CAS.

009.023 **Planetariums in the United States.** L. T. Shapiro.
Bull. American Astron. Soc., Vol. 7, 543 (1975).
Abstr. AAS.

009.024 **University of Hawaii LURE Observatory.** W. E. Carter, J. D. Williams.
Proc. Symposium on the earth's gravitational field and secular variations in position, (see 012.036), p. 433 - 441 (1974).

009.025 **Calar-Alto, Deutsch-Spanisches Astronomisches Zentrum im Aufbau.** M. Lammerer.
Orion, 33. Jahrgang, p. 205 - 211 (1975).

009.026 **Planetarium i obserwatorium astronomiczne im. Mikołaja Kopernika w Chorzowie 1955 - 1975.**
J. Ziętek, H. Chrupała, T. Modrzejowski, J. Palowski, M. Pańków, S. Pańków, I. Sikorska, G. Sitarski, M. Szczepański, W. Szymański, R. Śmiłowski, L. Zajdler.
Urania Kraków, Vol. 46, 225 - 253 (1975).

009.027 **Jūras astronomija planetārijā.** H. Legzdiņš.
Zvaigžņotā debess, 1975. gada pavasaris, p. 47 - 51.

009.028 **Sternwarten – gestern, heute, morgen.** J. Meurers.
Sternenbote, 18. Jahrgang, p. 162 - 179 (1975). – Lecture on the occasion of the 100th anniversary of the Observatory at Vienna.

009.029 **Il nuovo osservatorio di Tignano.**
Giornale Ass. Astrofili Bolognesi, No. 39, p. 13 - 17 (1975).

009.030 **Laboratory for Planetary Studies.** J. L. Elliot, E. K. Elliot.
Lab. Equipment Digest, (*GB*), Vol. 13, No. 4, p. 52 - 53, 55,

57, 59, 63, 65, 67, 69, 71 (1975).

The Laboratory for Planetary Studies comprises four individual laboratories: (1) Mars room; (2) goniometer laboratory; (3) exobiology laboratory; and (4) photometric instrumentation laboratory. Several research projects representative of work being done in the individual departments are described.

Herzberg Institute of Astrophysics formed.
Journ. Roy. Astron. Soc. Canada, Vol. 69, 144 - 145 (1975).

010 Societies, Associations, Organizations

010.001 **American Association of Variable Star Observers (AAVSO)**

Minutes of the general meeting of the AAVSO held at the Fernbank Science Center Atlanta, Georgia, May 24, 1975. C. B. Ford.
Journ. American Ass. Variable Star Observers, Vol. 4, 40 - 48 (1975). – Included are reports of the following committees: Nova Search (*C. V. Borzelli*); Eclipsing Binary (*M. E. Baldwin*); RR Lyrae (*M. E. Baldwin*); Telescope Loans (*C. E. Scovil*); Solar Division (*C. H. Hossfield*); Photoelectric Photometry (*A. J. Stokes*); Variable Star Atlas (*C. B. Ford*); Classical Cepheids (*T. A. Cragg*).

AAVSO meets in Massachusetts. J. Ashbrook.
Sky Telescope, Vol. 50, 384 - 385 (1975).

010.002 **American Astronomical Society (AAS)**

Working Group on Photographic Materials meets at Tucson and Rochester.
AAS Photo-Bull., No. 8, 1975 No. 1 p. 4 - 5.

The 146th meeting of the American Astronomical Society held 18 - 21 August 1975 at San Diego, California. Abstracts of papers presented.
Bull. American Astron. Soc., Vol. 7, 395 - 478 (1975).

Late-paper abstracts presented at the 145th meeting of the American Astronomical Society held 23 - 26 March 1975 at Bloomington, Indiana.
Bull. American Astron. Soc., Vol. 7, 497 - 500 (1975).

The 147th meeting of the American Astronomical Society held 8–10 December 1975 at Chicago, Illinois. Abstracts of papers presented.
Bull. American Astron. Soc., Vol. 7, 503 - 559 (1975).

010.003 **Association of Lunar and Planetary Observers (ALPO)**

Minor Planet Section news.
Minor Planet Bull., Vol. 3, 22, 26 - 28 (1975).

010.004 **Astronomical Society of Australia (ASA)**

Papers presented at the ninth annual general meeting held at ANRAO, Parkes, 14, 15 and 16 May 1975.
Proc. Astron. Soc. Australia, Vol. 2, 315 - 380 (1975). – The individual contributions are included in their corresponding subject categories.

010.005 **Astronomical Society of Czechoslovakia**

No publication received.

010.006 **Astronomical Society of the Pacific (ASP)**

Educational activities of the Astronomical Society of the Pacific. A. G. Fraknoi.
Bull. American Astron. Soc., Vol. 7, 446 (1975). – Abstr. AAS.

Abstracts of papers presented at the Ensenada meeting of the Astronomical Society of the Pacific, 12–15 May 1975.
Publ. Astron. Soc. Pacific, Vol. 87, 499 - 511 (1975).

010.007 **Astronomical Society of Southern Africa (ASSA)**

Notices.
Monthly Notes Astron. Soc. Southern Africa, Vol. 34, 93, 105, 129 (1975).

Proceedings of the annual general meeting, 1975.
Monthly Notes Astron. Soc. Southern Africa, Vol. 34, 106 - 117 (1975). – Included are the reports of the observing sections: Comet and Meteor Section, (*J. C. Bennett*); Occultation Section, (*A. G. F. Morrisby*); Grazing Occultations Section, (*J. Hers*); Variable Star Section, (*M. D. Overbeek*); Nova Search Section, (*J. C. Bennett*).

Centre reports, 1974/5. Cape Centre (*C. R. G. Turk*); Pretoria Centre (*K. J. Sterling*); Natal Centre (*M. A. Gray*); Natal Midlands Centre (*G. R. McNulty*); Transvaal Centre (*M. D. Overbeek*); O. F. S. Centre (*C. J. Muller*).
Monthly Notes Astron. Soc. Southern Africa, Vol. 34, 139 - 146 (1975).

Transvaal Centre occupies new dome.
Monthly Notes Astron. Soc. Southern Africa, Vol. 34, 157 (1975). – News note.

010.008 **Astronomical Society of Victoria (ASV)**

Astrophotographic Section. A. Stern.
Journ. Astron. Soc. Victoria, Vol. 28, 5 (1975).

Auroral Section. T. B. Tregaskis.
Journ. Astron. Soc. Victoria, Vol. 28, 7 (1975).

Current Phenomena Section. J. B. Trainor.
Journ. Astron. Soc. Victoria, Vol. 28, 12 - 13 (1975).

Historical Section. P. Simon.
Journ. Astron. Soc. Victoria, Vol. 28, 13 - 14 (1975).

Instrument Making Section. C. N. Chatfield.
Journ. Astron. Soc. Victoria, Vol. 28, 14 (1975).

Lunar and Planetary Section. B. S. Adcock.
Journ. Astron. Soc. Victoria, Vol. 28, 5 (1975).

Nova Search Section. B. S. Adcock.
Journ. Astron. Soc. Victoria, Vol. 28, 15 (1975).

Variable Stars Section. T. B. Tregaskis.
Journ. Astron. Soc. Victoria, Vol. 28, 11 - 12 (1975).

010.009 **Astronomical Society of Western Australia (ASWA)**

Report on proceedings of ordinary meetings.
Journ. Astron. Soc. Western Australia, Vol. 26 - 29, May, June, August - October (1975).

Report on proceedings of annual general meeting 14th July, 1975.
Journ. Astron. Soc. Western Australia, Vol. 27, July (1975).

010.010 **Astronomische Gesellschaft (AG)**

Wissenschaftliche Tagung der Astronomischen Gesellschaft mit 54. ordentlicher Mitgliederversammlung in Würzburg vom 16. - 20. September 1974.
H. Mauder, with an address by H. H. Voigt.
Mitt. Astron. Ges., No. 36, p. 7 - 13 (1975).

010.011 **Astronomisk Selskab København**

No publication received.

010.012 **British Astronomical Association (BAA)**

Notices.
Journ. British Astron. Ass., Vol. 85, 386 - 387 ; Vol. 86, 1 - 2 (1975).

Meetings of the Association.
Journ. British Astron. Ass., Vol. 85, 387 - 392, 491 - 499 (1975).

New members elected.
Journ. British Astron. Ass., Vol. 85, 459 - 461 (1975).

Historical Section. E. A. Beet.
Journ. British Astron. Ass., Vol. 85, 448 - 449, 532 - 533; Vol. 86, 77 - 79 (1975).

Lunar Section. P. Moore.
Journ. British Astron. Ass., Vol. 85, 511 - 527 (1975).

Mercury and Venus Section. J. H. Robinson.
Journ. British Astron. Ass., Vol. 85, 450 (1975).

Meteor Section. K. B. Hindley, M. A. Houlden.
Journ. British Astron. Ass., Vol. 86, 67 - 73 (1975).

Solar Section. V. Barocas, M. R. Whippey.
Journ. British Astron. Ass., Vol. 85, 433 - 436, 505 - 508; Vol. 86, 61 - 66 (1975).

Variable Star Section. J. E. Isles, M. Taylor.
Journ. British Astron. Ass.,Vol. 85, 438 - 447, 528 - 531; Vol. 86, 74 - 76 (1975).

Report of the council on work during the session 1974 July 1 to 1975 June 30.
Journ. British Astron. Ass., Vol. 85, 466 - 481 (1975). – Presented to members of the Association at the annual general meeting, 1975 October 29. – Included are the reports of the following Sections: Solar Section (*V. Barocas*); Lunar Section (*P. Moore*); Mercury and Venus Section (*J. H. Robinson*) Mars Section (*E. H. Collinson*); Jupiter Section (*W. E. Fox*); Saturn Section (*G. E. Satterthwaite*); Comet Section (*S. W. Milbourn*); Meteor Section (*K. Hindley*); Aurora Section (*B. McInnes*); Variable Star Section (*J. Isles*); Instrument and Observing Methods Section (*E. J. Hysom*); Computing Section (*G. E. Taylor*); Historical Section (*E. A. Beet*); Radio Astronomy Section (*J. R. Smith*); Artificial Satellite Section (*H. Miles*); Education Committee (*V. Barocas*).

Treasurer's accounts.
Journ. British Astron. Ass., Vol. 85, 482 - 489 (1975).

Provincial meeting of the Association, held on 1975 September 25 at Dundee University, Dundee, Scotland.
C. Dinwoodie, G. E. Stone, H. R. Hatfield.
Journ. British Astron. Ass., Vol. 86, 3 - 6 (1975).

010.013 **British Interplanetary Society**

Society news.
Journ. British Interplanet. Soc., Vol. 28, 514 - 516 (1975).

30th annual general meeting.
Spaceflight, Vol. 17, 372 - 375 (1975).

010.014 **Committee on Space Research (COSPAR)**

XVIII meeting plenario del COSPAR – Varna, giugno 1975. L. Cugusi.
Mem. Soc. Astron. Italiana, Vol. 46, 361 - 364 (1975).

XVIIth session of the International Committee on Space Research (COSPAR) from June 17 to July 1, 1974 in Sao Paulo, Brazil. Eh. N. Vaulina, S. S. Yurov.
Radiobiologiya, Vol. 15, 637 (1975). In Russian. – Abstr. in

Referativ. Zhurn. 62. Issled. kosmich. prostranstva, 1.62.9 (1976).

010.015 **European Space Agency (ESA)**

Ten nations reorganize European space efforts.
Phys. Today, Vol. 28, No. 8, p. 62, 64 (1975).

ESA officieel in werking getreden.
G. W. E. Beekman.
Zenit, Vol. 2, 327 (1975).

010.016 **International Astronautical Federation (IAF)**

No publication received.

010.017 **International Astronomical Union (IAU)**

No publication received.

010.018 **Meteoritical Society**

Members of the Meteoritical Society.
Meteoritics, Vol. 10, 159 - 187 (1975).

Abstracts of papers presented at the 38th annual meeting of the Meteoritical Society, July 28 - August 1, 1975, Tours, France.
Meteoritics, Vol. 10, 352 - 509 (1975). – See abstracts 081.046, 081.047, 091.036, 091.037, 094.024, 094.185, 094.187 - 094.196, 094.200, 094.458 - 094.473, 098.074, 104.089, 104.090, 105.094 - 105.199, 107.017, 143.073.

010.019 **Nederlandse Vereniging voor Weer- en Sterrenkunde**

Algemene ledenvergadering en jaarbijeenkomst Jongerenwerkgroep N.V.W.S. L. W. F. Gielingh.
Zenit, Vol. 2, 247 - 248 (1975).

010.020 **Polskie Towarzystwo Astronomiczne (PTA)**

XVII meeting of the Polish Astronomical Association. L. Zajdler.
Urania Kraków, Vol. 46, 377 - 378 (1975). In Polish.

010.021 **Polskie Towarzystwo Miłośników Astronomii (PTMA)**

PTMA chronicle.
Urania Kraków, Vol. 46, 216 - 220, 278 - 284, 343 - 346 (1975). In Polish.

010.022 **Royal Astronomical Society (RAS)**

Meetings of the Society.
Observatory, Vol. 95, 117 - 137, 153 - 161, 161 - 168 (1975).

Summaries of papers presented at RAS specialist discussion on quasars, held on 1975 January 10.
Observatory, Vol. 95, 256 - 263, with introductory comments by M. J. Rees, and concluding remarks by M. S. Longair (1975).

Annual general meeting 1975 March 14.
Quarterly Journ. Roy. Astron. Soc., Vol. 16, 323 - 324 (1975).

Report of the Council to the one hundred and fifty-fifth annual general meeting of the Society.
Quarterly Journ. Roy. Astron. Soc., Vol. 16, 324 - 333 (1975). This report refers to the calendar year 1974.

Report of the honorary auditors for the year 1974.
L. Mestel, M. Rowan-Robinson.
Quarterly Journ. Roy. Astron. Soc., Vol. 16, 333 - 334 (1975).

Treasurer's accounts for the year ended 1974. December 31. A. Hunter.
Quarterly Journ. Roy. Astron. Soc., Vol. 16, 336 - 346 (1975).

The Royal Astronomical Society Education Committee. D. McNally.
Quarterly Journ. Roy. Astron. Soc., Vol. 16, 428 - 431 (1975).

Meetings of the Society.
Quarterly Journ. Roy. Astron. Soc., Vol. 16, 493 - 496 (1975).

010.023 **Royal Astronomical Society of Canada (RAS Canada)**

The Royal Astronomical Society of Canada. Minutes of the annual meeting June 28, 1975.
R. Freeman.
Journ. Roy. Astron. Soc. Canada, Vol. 69, 191 - 193 (1975).

The Canadian Institute, a root of the Royal Astronomical Society of Canada. H. S. Hogg.
Journ. Roy. Astron. Soc. Canada, Vol. 69, 249 (1975). Abstr. CAS.

010.024 **Royal Astronomical Society of New Zealand (RAS New Zealand)**

Report of Section for year ended 30th September, 1974. F. M. Bateson.
Publ. Variable Star Section, Roy. Astron. Soc. New Zealand, No. 3 (C75), p. 50 - 55 (1975).

Report of Section for year ended 30th September 1975. F. M. Bateson.
Publ. Variable Star Section, Roy. Astron. Soc. New Zealand, No. 3 (C75), p. 55 - 60 (1975).

52nd annual general meeting and conference of the Royal Astronomical Society of New Zealand. 1974 conference abstracts.
Southern Stars, Vol. 26, 62 (1975).
Included are abstracts of some papers: The oblateness of the sun, (*K. H. Schatten*); Instrumented occultation observations, (*C. Rowe*); Anti-light pollution measures, (*D. Calder*).

010.025 **Schweizerische Astronomische Gesellschaft (SAG)**

No publication received.

010.026 **Sociedad Astronómica de México**

No publication received.

010.027 **Società Astronomica Italiana (SAI)**

XIX riunione della Società Astronomica Italiana.
S. Ghedini.
Giornale Ass. Astrofili Bolognesi, No. 40, p. 9 (1975).

Notizie della Società.
Giornale Astron., Vol. 1, 399 - 406 (1975).

010.028 **Société Astronomique de France (SAF)**

Les séances de la Société P. de La Cotardière.
L'Astronomie, 89e année, p. 321 - 322, 345 (1975).

Commission du Soleil. M.-J. Martres.
L'Astronomie, 89e année, p. 339 - 342 (1975).

Quelques nouvelles de groupe d'Alsace de la S.A.F. Le deuxième séminaire d'astronomie pratique de Mars 1974.
A. Florsch.
L'Astronomie, 89e année, p. 293 - 302 (1975).

Assemblée générale statutaire du 23 juin 1975.
B. Clouet.
L'Astronomie, 89e année, p. 379 - 385 (1975).

La vie de la Société Astronomique de France.
L. Tartois.
L'Astronomie, 89e année, p. 386 - 390 (1975).

010.029 **Société Astronomique "R. Boškovic"**

O radu Narodne opservatorije Astronomskog društva "Rudjer Bošković" u Beogradu. A. Tomić.
Publ. Obs. Astron. Beograd, No. 20, p. 277 - 278 (1975).

010.030 **Société Chronométrique de France**

No publication received.

010.031 **Société Belge d'Astronomie, de Météorologie et de Physique du Globe**

Réunion mensuelle. M. Ducuroir.
Ciel et Terre, Vol. 91, 313 - 315, 390 - 391, 443 - 446 (1975).

Assemblée Générale Statutaire du 22 mars 1975.
M. Ducuroir.
Ciel et Terre, Vol. 91, 309 - 312 (1975).

010.032 **Svenska Astronomiska Sällskapet**

No publication received.

010.033 **VAGO (Astronomical-Geodetical Society of the USSR)**

Sverdlovsk Department of VAGO is 20 years now.
A. I. Bogolepov.
Geod. raboty na Urale. Sverdlovsk, 1974, p. 5 - 9. In Russian.
Abstr. in Referativ. Zhurn. 51. Astron., 7.51.35 (1975).

Moscow plenum of VAGO Central Council.
V. A. Bronshtehn.
Zemlya i Vselennaya, 1975, No. 4, p. 65 - 67. In Russian.

From the history of the Leningrad VAGO Department. A. I. Stepanov.
Zemlya i Vselennaya, 1975, No. 4, p. 84 - 86. In Russian.

010.034 **Vereniging voor Sterrenkunde, België**

No publication received.

010.035 **Activité de l'A.F.O.E.V. Association Française des Observateurs d'Étoiles Variables.** E. Schweitzer.
L'Astronomie, 89e année, p. 337 - 338 (1975).

010.036 **La vie de l'Association.** E. Schweitzer.
A.F.O.E.V. Bull., Vol. 9, 97 - 98 (1975).

010.037 **Reports on the activity of the Commission for Astronomy of the Union of Societies of mathematicians, physicists and astronomers of Yugoslavia (Yugoslav National Committee of Astronomy) and its working bodies.**
G. *(Dj.)* Teleki, V. Milovanović, B. Ševarlić, P. Djurković.
Publ. Obs. Astron. Beograd, No. 20, p. 283 - 301 (1975).
In Serbo-Croatian.

010.038 **Concise report of CESRA (*Committee of European Solar Radio Astronomers*) meeting on organizational matters.** F. Poumeyrol.
Oss. Mem. Oss. Astrofis. Arcetri, Fasc. 105, (see 012.026), p. 138 - 139 (1975).

010.039 **Sixth meeting of the Canadian Astronomical Society (*CAS*) at Queen's University, Kingston, Ontario June 18–21, 1975.** P. M. Millman.
Journ. Roy. Astron. Soc. Canada, Vol. 69, 242 - 256, with abstracts of papers held during this meeting (1975).

010.040 **The Auckland Astronomical Society.**
W. S. G. Walker.
Southern Stars, Vol. 26, 75 - 82 (1975).

010.041 **Nachrichten der Vereinigung der Sternfreunde e.V.**
SuW, Vol. 14, 254 - 262, 300 - 306, 342 - 345, 375 - 378, 402 - 405 (1975).

011 Reports on Colloquia, Congresses, Meetings, Symposia, and Expeditions

011.001 **Probing the moon: The moon – a new appraisal from space missions and laboratory analyses.**
Nature, Vol. 256, 87 - 88 (1975). – Discussion meeting of the Royal Astron. Soc., held 1975, June 9 - 12.

011.002 **Mercury since Mariner 10.**
Nature, Vol. 256, 455 - 456 (1975). – Report on the first international colloquium on Mercury held at the California Institute of Technology on June 25 - 27, 1975.

011.003 **Far infrared astronomy.** M. Rowan-Robinson.
Nature, Vol. 256, 458 - 459 (1975). – Report on the first international conference devoted to far infrared astronomy held at Cumberland Lodge, Windsor Great Park on July 9 - 11.

011.004 **Hoyle on troubled waters.** J. Faulkner.
Nature, Vol. 256, 535 - 536 (1975). – Report on the conference "The frontiers of astronomy in 1975", held 1975, July 15 - 17.

011.005 **Introduction to the Philadelphia Symposium.**
P. van de Kamp.
Vistas in Astronomy, Vol. 18, (see 012.006), 41 - 48 (1975).

011.006 **Introduction to the Leningrad Symposium.**
A. A. Mikhailov (*Mikhajlov*).
Vistas in Astronomy, Vol. 18, (see 012.006), 49 - 50 (1975).

011.007 **Address at the Leningrad Symposium.** A. Adam.
Vistas in Astronomy, Vol. 18, (see 012.006), 51 - 52 (1975).

011.008 **Discussions at the Weil der Stadt Symposium.**
F. Krafft, K. Meyer, B. Sticker.
Vistas in Astronomy, Vol. 18, (see 012.006), 53 (1975).

011.009 **Continuities in Kepler scholarship: the European Kepler Symposia in historiographical perspective.**
R. S. Westman.
Vistas in Astronomy, Vol. 18, (see 012.006), 57 - 70 (1975).

011.010 **Quantitative data for science and technology. Conference in Tsakhkadzor.** 1974, June 24 - 27.
L. V. Gurvich.
Vestn. AN SSSR, 1975, No. 1, p. 84 - 89. In Russian.
Abstr. in Referativ. Zhurn. 51. Astron., 7.51.31 (1975).

011.011 **Scientific session of the Department of General Physics and Astronomy of the USSR Academy of Sciences, September 25 - 26, 1974.**
Uspekhi fiz. nauk, Vol. 115, 531 - 537 (1975). In Russian.
Abstr. in Referativ. Zhurn. 51. Astron., 7.51.33 (1975).

011.012 **Relativistische Astrophysik.** V. Weidemann.
Naturwissenschaften, 62. Jahrgang, p. 419 - 424 (1975).

Report on the Seventh Texas Symposium on Relativistic Astrophysics, held in Dallas, 16 - 20 December 1974. The topics discussed included pulsars, binary X-ray sources, gamma-ray and neutrino bursts, and theoretical aspects of black hole and neutrino astrophysics, as well as observations of relevance to cosmology and gravitation theory.

011.013 **Astronomie populaire.** P. de La Cotardière.
L'Astronomie, 89e année, p. 327 - 329 (1975).
Rouen, 1975 March 9.

011.014 **The Royal Society lunar conference.**
P. Moore.
Journ. British Astron. Ass., Vol. 85, 500 - 501 (1975).

011.015 **IAU Kolloquium Nr. 31: "Interplanetarer Staub und Zodiakallicht".** H. Link.
SuW, Vol. 14, 280 - 281 (1975).

011.016 **Interstellar dustmen's convention.**
A. Whitworth.
Nature, Vol. 257, 91 (1975).
Workshop on interstellar grains, University College Cardiff, July 11 - 13, 1975.

011.017 **Galaxy formation.** J. E. Jones.
Nature, Vol. 257, 91 - 92 (1975).
Workshop meeting on "The formation and evolution of galaxies" held at the Institute of Astronomy, Cambridge, on August 4 - 8, 1975.

011.018 **European astronomy at Leicester.** A. C. Fabian.
Nature, Vol. 257, 275 - 276 (1975).
Report on the first European conference on astronomy, held at Leicester, UK on August 12 - 14, 1975.

011.019 **The League convenes in Georgia.**
Sky Telescope, Vol. 50, 208 - 211 (1975). – 1975 August 13 - 16, Atlanta, Georgia.

011.020 **Greenwich tercentenary symposium.**
O. Gingerich.
Sky Telescope, Vol. 50, 217 - 218 (1975). – Greenwich, July 14 - 18 (1975).

011.021 **Magnetosphären von Pulsaren.** Seminar der Dr. Wilhelm und Else Heraeus-Stiftung vom 7. - 11. April 1975 in Mönchberg/Spessart. W. Kundt.
Phys. Blätter, 31. Jahrgang, p. 475 - 476 (1975).

011.022 **Scientific sessions of the Department of Physical-Mathematical Sciences of the Academy of Sciences of the Armenian SSR (1974, March 1; 1974, October 7).**
Izv. Akad. Nauk Armyan. SSR, Fizika, Vol. 9, 535 - 541 (1974). In Russian.

011.023 **Fünftes Mondkolloquium.** J. Classen.
Sterne, 51. Jahrgang, p. 140 - 141 (1975).
Houston, Texas, 1974 March 18 - 23.

011.024 **Festkörperastrophysik. – Bericht über ein Symposium.** J. Dorschner.
Sterne, 51. Jahrgang, p. 142 - 144 (1975). – Cardiff, 1974 July 9 - 12.

011.025 **Astrofotografie am Südhimmel.**
E. Alt, E. Brodkorb, R. Mehrmann, K. Rihm.
Orion, 33. Jahrgang, p. 152 - 156, 201 - 204 (1975). – Report on an expedition to South Africa.

011.026 **Resumen de los trabajos presentados en la I Reunión Nacional del Grupo de Astrofísica.**
Urania Barcelona, Año 59, No. 281–282, p. 441 - 453 (1974).

Barcelona, 1973 December 14–15.

011.027 **Symposium No. 67 of the International Astronomical Union: 'Variable stars in stellar systems'.** Moscow, 1974, July 29 - August 3. Yu. N. Efremov.
Astron. Tsirk., No. 842, p. 6 - 8 (1974). In Russian.

011.028 **"Hidden" mass in the universe.** (Results of a conference in Tallin, 1975, January).
B. V. Komberg.
Zemlya i Vselennaya, 1975, No. 4, p. 62 - 64. In Russian.

011.029 **Royal Astronomical Society MIST meeting at Exeter.**
Quarterly Journ. Roy. Astron. Soc., Vol. 16, 282 - 297 (1975).
The tenth of the series of meetings on the magnetosphere, ionosphere and solar-terrestrial environment was held at the University of Exeter on 1975 March 25–26. Report including abstracts of 31 papers dealing with the four meeting topics: (1) Lower ionosphere and lower thermosphere, (2) Solar-terrestrial relationships, (3) Pulsations and VLF emission, (4) Geomagnetism.

011.030 **Icarus meeting report. AAS Division for Planetary Sciences: Fifth (1974) and sixth (1975) annual meetings.** D. Morrison.
Icarus, Vol. 26, 111 - 123 (1975).

011.031 **Frontiers of astronomy: in Venice.** V. Trimble.
Science, Vol. 190, 368 - 369 (1975).
Sir Fred Hoyle celebrated his 60th birthday this year, and a symposium was held to mark the occasion in Venice, 15 to 19 July. Many of Hoyle's colleagues and former students presented current work in some of the fields Hoyle helped to found, including cosmology, the nature of the interstellar medium, and the early history of the solar system.

011.032 **Astrophysics in the system of physical sciences.**
V. S. Kislyuk.
Astrometriya i Astrofizika, *Kiev*, vyp. (No.) 27, (see 003.004), p. 119 - 120 (1975). In Russian. – Session of the Scientific Council of the Main Astronomical Observatory of the Ukrainian Academy of Sciences, 1974, May 23.

011.033 **West coast convention.** D. Milon.
Sky Telescope, Vol. 50, 291 - 295 (1975).

011.034 **What drives the solar cycle?** N. Weiss.
Nature, Vol. 258, 197 - 198 (1975). – IAU symposium No. 71 on 'Basic mechanisms of solar activity', held in Prague, August 25–29, 1975.

011.035 **Conference on "Hidden mass in the universe".** Tallin, 1975, January 28 - 30.
A. G. Doroshkevich, M. M. Jõeveer, J. Einasto.
Astron. Zhurn. Akad. Nauk SSSR, Vol. 52, 1113 - 1115 (1975). In Russian. English translation in Soviet Astron., Vol. 19, No. 5.

011.036 **Conference on mapping the moon and Mars.** Moscow, 1974, December 24 - 27.
G. A. Burba.
Astron. Zhurn. Akad. Nauk SSSR, Vol. 52, 1115 - 1117 (1975). In Russian. English translation in Soviet Astron., Vol. 19, No. 5.

011.037 **Neue Testergebnisse für Gravitationstheorien.**
V. Weidemann.
Phys. Blätter, 31. Jahrgang, p. 502 - 503 (1975).

011.038 **International conference of infrared physics.**
PTB Mitt., 85. Jahrgang, p. 480 - 481 (1975). Zürich 1975 August 8 - 15.

011.039 **IIIrd European astronomical conference.** Tbilisi, 1975, July 1 - 5. – Interview with Prof. E. K. Kharadze.
Priroda, No. 10.75, p. 113 - 115 (1975). In Russian.

011.040 **La Scuola d'estate di Fisica Spaziale di Lannion sulla geodinamica spaziale.** (1974, August 20 - September 13). A. Poma.
Mem. Soc. Astron. Italiana, Vol. 46, 277 - 278 (1975).
Conference abstract.

011.041 **Mehr Energie für die Ionosphäre.**
Umschau, 75. Jahrgang, p. 703 (1975). – San Francisco, 1974 December.
Among the topics for the fall meeting of the American Geophysical Union the determination of the absolute solar extreme ultraviolet fluxes covered a special session for a better understanding of the ionospheric-atmospheric processes. This and most recent measurements forced a correction of the published values towards higher solar energies.

011.042 **IAU symposium on fast transients in X-and gamma rays.** S. V. Damle.
Bull. Astron. Soc. India, Vol. 3, 58 - 59 (1975). – Varna, Bulgaria, 1975 May 29 - 31.

011.043 **The 14th international conference on cosmic rays.**
R. R. Daniel.
Bull. Astron. Soc. India, Vol. 3, 59 - 61 (1975). – Munich, 1975 August 15 - 29.

011.044 **Astrophysical convection.** S. Ramadurai.
Bull. Astron. Soc. India, Vol. 3, 61 - 62 (1975). – Informal meeting of the Fluid Dynamicists and Astrophysicists, held at the Institute of Astronomy, Cambridge, June 16–27, 1975.

011.045 **Symposium 'Variable stars in stellar systems' July 29 - August 3, 1974.** Yu. V. Voroshilov.
Vestn. AN SSSR, 1975, No. 4, p. 113 - 116. In Russian.

011.046 **Imaging in astronomy: OSA topical meeting, Cambridge, 18 - 21 June 1975.** J. C. Wyant.
Applied Optics, Vol. 14, 2322 (1975).

011.047 **Colloquium "Origin and evolution of galaxies", Itkol-Azau, 25 Jan. – 2. Febr. 1974.**
G. M. Vereshkov.
Izv. Sev.-Kavkaz. nauch. tsentra vyssh. shkoly. Ser. estestv. n., 1974, No. 4, p. 131. In Russian.

011.048 **All-Union conference on "Physics of geomagnetic phenomena".** Moscow, 10–12 February 1975.
L. S. Al'perovich.
Geomagn. Aeronom., Vol. 15, 1132 - 1133 (1975). In Russian.

011.049 **Seventh Sikhote-Alin meteoritic expedition.**
E. L. Krinov.
Meteoritika, vyp. (No.) 34, p. 113 (1975). In Russian.

011.050 **Astronomy from the space shuttle.**
J. Gribbin.
Nature, Vol. 258, 663 - 664 (1975).
Report on a discussion meeting of the Royal Astronomical Society and the Institute of Physics – Uses of the space shuttle for astronomy and geophysics – held 1975, December 12 in London.

011.051 **All-Union symposium "Interaction of meteor matter with the earth and estimate of influx of meteor matter on the earth and moon"**. Dushanbe, 1974, September 24 - 28. V. S. Getman.
Izv. AN TadzhSSR. Otd. fiz.-mat. i geol.-khim. n., 1975, No. 1, p. 111 - 112. In Russian.

011.052 **Scientific session of the Department of General Physics and Astronomy of the USSR Academy of Sciences, October 30 - 31, 1974.**
Uspekhi fiz. nauk, Vol. 116, 341 - 342 (1975). In Russian. Abstr. in Referativ. Zhurn. 51. Astron., 11.51.20 (1975).

011.053 **The second European regional astronomical conference, Trieste, Sept. 2–6, 1974.**
O. B. Dluzhnevskaya.
Vestn. AN SSSR, 1975, No. 7, p. 101 - 103. In Russian.

011.054 **Symposium "Theoretical methods of investigation in geophysics, geology and astrophysics", Eisenach, Jan. 13–18, 1975.** L. V. Nikitin.
Vestn. AN SSSR, 1975, No. 7, p. 107 - 108. In Russian.

011.055 **Conference on astrometry, Heidelberg, Oct. 29 - Nov. 1, 1974.** A. A. Nemiro.
Vestn. AN SSSR, 1975, No. 7, p. 103 - 104. In Russian.

011.056 **Scientific session of the Department of General Physics and Astronomy of the USSR Academy of Sciences, Dec. 18, 1974.**
Uspekhi fiz. nauk, Vol. 116, 539 - 540 (1975). In Russian.

011.057 **Scientific session of the Department of General Physics and Astronomy of the USSR Academy of Sciences, Dec. 26, 1974.**
Uspekhi fiz. nauk, Vol. 116, 546 (1975). In Russian.

011.058 **Third European regional meeting in astronomy.** J. Bouška.
Říše hvězd, Vol. 56, 185 - 187 (1975). In Czech.

011.059 **Colloquium about interplanetary dust and zodiacal light.** Heidelberg 10 - 13 June 1975. V. Vanýsek.
Říše hvězd, Vol. 56, 187 - 190 (1975). In Czech.

011.060 **Seminar on geophysical hydrodynamics at the Oceanographical Commission of the USSR Academy of Sciences.** Yu. Z. Miropol'skij.
Okeanologiya, Vol. 15, 759 - 761 (1975). In Russian. – Abstr. in Referativ. Zhurn. 51. Astron., 1.51.57 (1976).

011.061 **Symposium of the Socialist countries on solar-terrestrial physics.** E. V. Ivanov.
Vestn. AN SSSR, 1975, No. 7, p. 104 - 106. In Russian. Abstr. in Referativ. Zhurn. 62. Issled. kosmich. prostranstva, 1.62.10 (1976).

011.062 **Meeting of investigators of variable stars on the problem of R Coronae Borealis type objects.**
Kiev, Goloseevo, 1973, Nov. 19 - 20.
Peremennye Zvezdy, Vol. 19, 499 - 500 (1975). – In Russian.

011.063 **Chronicle.** Report on the 3rd European Astronomical Conference in Tbilisi, 1975, July 1 - 5.
T. M. Borchkhadze.
Astron. Tsirk., No. 890, p. 7 - 8 (1975). In Russian.

011.064 **Royal Astronomical Society specialist discussion on the peculiar and metallic-line stars.**
Observatory, Vol. 95, 229 - 239 (1975). – 1975 February 14.

011.065 **Joint meeting of the Royal Astronomical Society, the Royal Meteorological Society and the Geological Society of London on "Venus and Mercury"**, held on 1975 April 11. G. E. Hunt, J. E. Guest.
Observatory, Vol. 95, 265 - 267 (1975).

011.066 **The nineteenth Herstmonceux conference – Photometry**, 1975 April 9 and 10.
Observatory, Vol. 95, 267 - 282 (1975).

011.067 **Konferences un sanāksmes.**
Zvaigžņotā debess, 1975. gada pavasaris, p. 29 - 35.
Simposzijs par mainzvaigznēm zvaigžnu sistēmās (*A. Alksnis*), p. 29 - 33; Apspriede "Zvaigžņu evolūcijas agrās stadijas" (*J. Francmanis*), p. 34 - 35.

011.068 **Konferences un sanāksmes.**
Zvaigžņotā debess, 1975. gada vasara, p. 43 - 48.
PSRS ZA Astronomiskās padomes plēnumā (*L. Laucenieks*), p. 43 - 46; Saules pētnieku tikšanās Azerbaidžanā (*G. Ozoliņš*), p. 46 - 48.

011.069 **Konferences un sanāksmes.**
Zvaigžņotā debess, 1975. gada rudens, p. 22 - 30.
Zinātniskā darba skate P. Stučkas Latvijas Valsts universitātē (*L. Roze*), p. 22 - 23; Apspriede "Slēptā masa Visumā" (*A. Balklavs*), p. 23 - 27; Starptautiska konference par mezosfēras un jonosferas procesiem (*J. Voss, M. Diriķis*), p. 27 - 30.

011.070 **In memory of P. P. Parenago: 'The hidden mass in the universe' – a conference held in Tallin on January 28 - 30, 1975.** K. Rudnicki.
Postępy Astron., Vol. 23, 215 - 218 (1975). In Polish.

011.071 **Convegno della S.A.It. sul tema: «L'insegnamento dell'astronomia nelle scuole secondarie. Problemi metodologici e didattici»**, Torre del Greco, 25–27 giugno 1975, Soresina, 22–26 settembre 1975. M. A. Santaniello.
Giornale Astron., Vol. 1, 315 - 319 (1975).

011.072 **Die Tagung der VdS in Recklinghausen.**
K. Wälke.
BAV Rundbrief, 24. Jahrgang, p. 50 - 53 (1975).

011.073 **IX congresso dell'U.A.I.** (*Unione degli Astrofili Italiani*). S. Ghedini.
Giornale Ass. Astrofili Bolognesi, No. 40, p. 9 - 10 (1975).

011.074 **Summary of the La Jolla conference on 'Quantitative magnetospheric models'.** W. P. Olson.
EOS Trans. American Geophys. Union, Vol. 56, 607 - 629 (1975).

011.075 **Committee to record history of solar-planetary research.**
EOS Trans. American Geophys. Union, Vol. 56, 699 - 700 (1975).

012 Proceedings of Colloquia, Congresses, Meetings, and Symposia

012.001 **Variable stars and stellar evolution.** International Astronomical Union, Symposium No. 67, held in Moscow, U.S.S.R., July 29–August 4, 1974.
V. E. Sherwood, L. Plaut (Editors).
D. Reidel Publishing Company, Dordrecht–Holland/Boston–U.S.A. 17 + 614 pp. Price Dfl. 210.00 (1975). – The individual contributions are included in their corresponding subject categories – see abstracts 031.201, 064.003, 064.004, 114.007 - 114.012, 114.305, 114.306, 117.003, 121.004, 122.006 - 122.053, 124.002 - 124.005, 124.100, 124.101, 126.002, 141.006, 141.304, 141.305, 141.603, 142.026 - 142.033, 154.003, 158.015 - 158.017.

012.002 **Experimental gravitation.** Proceedings of the International School of Physics 'Enrico Fermi', Course 56, held at Varenna, Italy, 17 - 29 July, 1972.
B. Bertotti (Editor).
Academic Press, London–New York. 15 + 574 pp. Price $ 42.50 (1974). – The individual contributions are included in their corresponding subject categories – see abstracts 022.002, 052.001, 052.002, 066.020 - 066.035, 080.005 - 080.007.

012.003 **Solar gamma-, X-, and EUV radiation.** International Astronomical Union, Symposium No. 68, held in Buenos Aires, Argentina, 11–14 June 1974. Organized by the IAU in cooperation with COSPAR.
S. R. Kane (Editor).
D. Reidel Publishing Company, Dordrecht–Holland/Boston–U.S.A. 12 + 439 pp. Dfl. 145.00 (1975). – The individual contributions are included in their corresponding subject categories – see abstracts 073.005 - 073.021, 074.006 - 074.009, 076.002 - 076.019.

012.004 **Internationales Kepler-Symposium,** Weil der Stadt, 8 - 11 August 1971.
F. Krafft, K. Meyer, B. Sticker (Editors).
Arbor scientiarum: Beiträge zur Wissenschaftsgeschichte, Reihe A: Abhandlungen, Vol. 1. Verlag Dr. H. A. Gerstenberg, Hildesheim, Germany. 12 + 490 pp. Price DM 160.00 (1973). The individual contributions within the subject scope of Astronomy and Astrophysics Abstracts are included in their corresponding categories – see abstracts 004.002 - 004.006, 005.001.

012.005 **Copernicus, yesterday and today.** Proceedings of the Commemorative Conference held in Washington in honour of Nicolaus Copernicus, December 27 - 28, 1972.
A. Beer, K. A. Strand (Editors), with introductory remarks by T. F. Dedman.
Vistas in Astronomy, Vol. 17, Pergamon Press, Oxford–New York–Toronto–Sydney–Braunschweig. 33 + 225 pp. Price $ 38.00 (1975). – The individual contributions within the subject scope of Astronomy and Astrophysics Abstracts are included in their corresponding categories – see abstracts 004.007 - 004.016, 005.002, 042.001.

012.006 **Kepler–four hundred years.** Proceedings of conferences held in honour of Johannes Kepler.
A. Beer, P. Beer (Editors), with a foreword by W. Hartner.
Vistas in Astronomy, Vol. 18. Pergamon Press, Oxford–New York–Toronto–Sydney–Braunschweig. 20 + 1034 pp. Price about $ 162.00 (1975). – The individual papers or in some cases a chapter are included in their corresponding subject categories – see abstracts 002.001, 002.002, 004.018 - 004.054, 005.003 - 005.009, 011.005 - 011.009, 015.001, 015.002, 042.013 - 042.015, 081.002, 091.003, 125.004.

012.007 **Atmospheres of earth and the planets.** Proceedings of the Summer Advanced Study Institute, held at the University of Liège, Belgium, July 29 - August 9, 1974.
B. M. McCormac (Editor).
Astrophys. Space Sci. Library, Vol. 51. D. Reidel Publishing Company, Dordrecht–Holland/Boston–U.S.A. 7 + 454 pp. Price Dfl. 165.00 (1975). – The individual contributions within the subject scope of Astronomy and Astrophysics Abstracts are included in their corresponding categories – see abstracts 082.012 - 082.017, 083.011 - 083.013, 084.005 - 084.010, 091.005, 093.002 - 093.004, 099.011, 099.012, 099.211, 099.212.

012.008 **The magnetospheres of the earth and Jupiter.** Proceedings of the Neil Brice Memorial Symposium, held in Frascati, Italy, May 28 - June 1, 1974.
V. Formisano (Editor).
Astrophys. Space Sci. Library, Vol. 52. D. Reidel Publishing Company, Dordrecht–Holland/Boston – U.S.A. 11 + 485 pp. Price Dfl. 165.00 (1975). – The individual contributions within the subject scope of Astronomy and Astrophysics Abstracts are included in their corresponding categories – see abstracts 074.014, 074.015, 084.011, 084.210 - 084.222, 099.013 - 099.030, 099.213.

012.009 **Dynamics of stellar systems.** International Astronomical Union, Symposium No. 69, held in Besançon, France, September 9 - 13, 1974.
A. Hayli (Editor).
D. Reidel Publishing Company, Dordrecht–Holland/Boston–U.S.A. 10 + 461 pp. Price Dfl 158.00 (1975). – The individual contributions are included in their corresponding subject categories – see abstracts 042.024, 062.021, 117.026, 117.027, 151.009 - 151.030, 153.008, 154.010, 155.013 - 155.015, 158.058 - 158.067, 160.010.

012.010 **Transactions of the VIth international seminar: "Acceleration of particles and nuclear reactions in space", Leningrad, August 19 - 21, 1974.**
Izv. AN SSSR. Ser. fiz.,Vol. 39, 241 - 448 (1975). In Russian.

012.011 **Problems of gravitation.** Collection of papers of the plenary sessions of the 3rd Soviet gravitational conference in Erevan, 1972, October 11 - 14.
Erevan. un-t, Erevan. 228 pp. Price 63 Kop. (1975). In Russian.

012.012 **Origin and evolution of the lunar regolith.** Symposium held at the Lunar Science Institute, Houston, Texas, November 13 - 15, 1974.
H. Alfvén, Z. Kopal, H. C. Urey (Editors).
The Moon, Vol. 13, Nos. 1 - 3, 359 pp. (1975). The individual contributions are included in their corresponding subject categories – see abstracts 094.107 - 094.122, 094.405 - 094.409.

012.013 **Space optics.** Proceedings of the summer school organised by the Commission Internationale d'Optique and the Centre National d'Études Spatiales, Marseille, June 29 - July 4, 1970.
A. Maréchal, G. Courtès (Editors).
Gordon and Breach Science Publishers, London–New York–Paris. 7 + 389 pp. Price $ 26.50 (1974). – Review in Planet. Space Sci., Vol. 23, 1237; 1975 (*B. Bates*). – The individual papers within the subject scope of Astronomy and Astrophysics Abstracts are included in their corresponding categories – see abstracts 031.006 - 031.009, 031.206 - 031.209, 032.501 - 032.509, 034.045, 046.001, 051.005, 051.006.

012.014 **Neutron stars, black holes and X-ray sources.** Articles based on a session of the annual meeting of the American Association for the Advancement of Science, held in San Francisco in February 1974.
H. Gursky, R. Ruffini (Editors).
Astrophys. Space Sci. Library, Vol. 48. D. Reidel Publishing Company, Dordrecht–Holland/Boston–U.S.A. 12 + 441 pp. Price Dfl. 135.00 (1975). – The individual contributions are included in their corresponding subject categories – see abstracts 061.013, 061.014, 066.056, 117.023, 125.015, 141.319, 142.056, 142.057. – Besides these papers a set of "Classic papers" (appendix I) has been reproduced on pp. 259-317 and a set of "Contemporary papers leading to the discovery of gravitationally collapsed stars" (appendix II) on pp. 321 - 437.

012.015 **Soviet-French colloquium concerning the problem 'Physics and evolution of stars'.** Moscow, 1974, January 16 - 21.
Nauchn. Informatsii, vyp. (No.) 33, p. 3 - 35 (1974). In Russian and English. – Abstracts of the individual contributions are presented only. They are included in their corresponding subject categories – see abstracts 064.031, 065.064 - 065.070, 114.051 - 114.056, 117.043, 122.107 - 122.112, 153.015, 153.016.

012.016 **Problèmes d'hydrodynamique stellaire.** Communications présentées au dix-neuvième Colloque International d'Astrophysique, tenu à Liège, les 8, 9 et 10 juillet 1974, with an introduction by P. Ledoux.
Mém. Soc. Roy. Sci. Liège, Coll. 8°, 6ᵉ Sér., Vol. 8, 490 pp. (1975). – The individual contributions are included in their corresponding subject categories – see abstracts 061.035, 064.033 - 064.035, 065.077 - 065.106, 116.003, 117.044, 117.045, 122.113 - 122.118, 124.008 - 124.010, 125.029, 133.020, 142.106, 142.107.

012.017 **Proceedings of the astrometric conference held in connection with the dedication of the 41-inch astrometric reflector of the Observatory of Torino, 1974, May 26 - 28,** with introductory remarks by M. G. Fracastoro, and closing remarks by P. Muller.
Tipografia Baccini & Chiappi, Firenze, Italy. 65 pp. (1975). The individual contributions are included in their corresponding subject categories – see abstracts 031.225, 034.054, 034.055, 041.022, 041.023, 111.004, 117.047, 118.013.

012.018 **Publications of scientific results of the "Intercosmos" cooperation.** Symposium on results of satellite observations, Budapest, October 21 - 24, 1974.
A. Dinescu (Editor).
Observations of artificial earth satellites, No. 14. Editura Academiei Republicii Socialiste România, Bucureşti. 644 pp. (1975). In Russian, English, German and French. – The individual papers within the subject scope of Astronomy and Astrophysics Abstracts are included in their corresponding categories – see abstracts 021.013, 031.408, 046.016 - 046.035, 052.016 - 052.022, 054.005 - 054.007, 055.008 - 055.012, 081.017, 082.055 - 082.061, 091.021, 091.022.

012.019 **XXVIII. Gertsen readings. Theoretical physics and astronomy.**
Nauch. dokl. Leningr. gos. ped. in-t im. A. I. Gertsena. Leningrad. 96 pp. Price 30 Kop. (1975). In Russian.

012.020 **Astrophysics and gravitation.** Proceedings of the sixteenth Solvay Conference on Physics at the University of Brussels, September 1973.
Éditions de l'Université de Bruxelles, Bruxelles, Belgique. 15 + 494 pp. Price FB 1800.00 (1974). – The individual contributions are included in their corresponding subject categories – see abstracts 061.044, 061.045, 065.119, 066.074, 066.075, 141.083 - 141.086, 141.338, 142.124 - 142.130, 158.140.

012.021 **Ondes et radiations gravitationnelles.** Colloques Internationaux du Centre National de la Recherche Scientifique, No. 220. Paris, Institut Henri Poincaré, 18–22 juin 1973, with an introduction by Y. Choquet-Bruhat.
Éditions du Centre National de la Recherche Scientifique, Paris. 392 pp. Price F 120.00 (1974). – The individual papers within the subject scope of Astronomy and Astrophysics Abstracts are included in their corresponding categories – see abstracts 066.076 - 066.094, 162.028.

012.022 **La dynamique des galaxies spirales.** Colloques Internationaux du Centre National de la Recherche Scientifique, No. 241. Institut des Hautes Études Scientifiques, Bures-sur-Yvette, 16–20 septembre 1974.
L. Weliachew (Editor), with concluding remarks by J. H. Oort.
Éditions du Centre National de la Recherche Scientifique, Paris. 534 pp. Price F 120.00 (1975). – The individual contributions are included in their corresponding subject categories – see abstracts 131.541, 151.047 - 151.062, 155.036 - 155.039, 158.147 - 158.168.

012.023 **VIth Leningrad international seminar, August 19 - 21, 1974. Transactions of the international seminar "Particle acceleration and nuclear reactions in space".**
G. E. Kocharov, V. A. Dergachev (Editors).
AN SSSR. Fiz.-tekhn. in-t, NII yader. fiz. Mosk. un-ta. Leningrad. 404 pp. Price 1 Rbl. 50 Kop. (1974). In Russian.

012.024 **Proceedings of the National Conference of Yugoslav Astronomers – 1973, Belgrade, November 6 - 8.**
B. Kilar, G. *(Dj.)* Teleki, V. Vujnović (Editors).
Publ. Obs. Astron. Beograd, No. 20, p. 1 - 227 (1975).
Activity of the conference, p. 1 - 15; The individual contributions are included in their corresponding subject categories - see abstracts 014.011, 014.012, 022.078, 031.029, 032.024 - 032.029, 034.093, 041.031 - 041.033, 042.071, 042.072, 044.008, 044.009, 045.008, 046.039 - 046.041, 061.054, 061.055, 065.130, 072.032, 073.093, 073.094, 080.051, 082.070 - 082.073, 098.034, 098.035, 104.030, 118.017, 118.018, 141.096.

012.025 **Skylab Solar Workshop. Preliminary results from the S-054 X-ray telescope and the correlated ground-based observations (Florence, 21 - 22 March 1974).**
G. Righini (Editor), with introductory remarks by B. Rossi.
Oss. Mem. Oss. Astrofis. Arcetri, Fasc. 104, 10 + 210 pp. (1975). – The individual contributions are included in their corresponding subject categories – see abstracts 022.080, 051.014, 073.102 - 073.107, 074.073 - 074.082, 077.027 - 077.030.

012.026 **Proceedings of the First European Solar Meeting (Florence, February 25 - 27, 1975).**
C. Chiuderi, M. Landini, A. Righini (Editors), with a preface by C. Zwaan.
Oss. Mem. Oss. Astrofis. Arcetri, Fasc. 105, 143 pp. (1975). The individual contributions are included in their corresponding subject categories – see abstracts 010.038, 013.005 - 013.009, 033.012, 033.013, 051.015, 051.016, 071.036, 071.037, 074.083, 077.031, 079.100, 080.055.

012.027 **Conference on the problem "Magnetic fields and chemical composition of the sun and stars".**
Crimean Astrophysical Observatory, 1974, April 9 - 12.
Izv. Krymskoj Astrofiz. Obs., Vol. 53, p. 240 - 257 (1975). In Russian. – The individual contributions are included in their corresponding subject categories – see abstracts 062.041, 064.043, 064.044, 071.035, 072.037, 072.038, 080.

053, 080.054, 114.356, 114.357, 116.009 - 116.011.

012.028 **The rings of Saturn.** Proceedings of the Saturn's rings workshop held at the Jet Propulsion Laboratory Pasadena, California, July 31 and August 1, 1973.
F. D. Palluconi, G. H. Pettengill (Editors), with a foreword by I. Rasool and opening remarks by H. M. Schurmeier.
NASA SP-343. National Aeronautics and Space Administration, Washington, D.C. 10 + 222 pp. Price $ 3.25 (1974).
The individual papers are included in their corresponding subject categories – see abstracts 053.015 - 053.020, 100.017 - 100.029.

012.029 **Time determination, dissemination and synchronization.** Proceedings of the Second Cagliari International Meeting, co-sponsored by the IAU Commission No. 31 and the Faculty of Science of the Cagliari University.
H. Enslin, E. Proverbio (Editors) with opening adresses, by F. Zagar and H. Enslin.
3T Edizioni anastatiche, Cagliari, Italy. 291 pp. (1975).
The individual papers are included in their corresponding subject categories – see abstracts 035.005 - 035.010, 044.010 - 044.022.

012.030 **Evolutionary and physical properties of meteoroids.** The proceedings of the International Astronomical Union's Colloquium No. 13, held at the State University of New York, Albany, N.Y., on June 14–17, 1971.
C. L. Hemenway, P. M. Millman, A. F. Cook (Editors).
NASA SP-319. National Aeronautics and Space Administration, Washington, D.C., 6 + 378 pp. [For sale by the Superintendent of Documents, U.S. Government Printing Office, Washington, D.C.]. Price $ 4.30 (1973). – The individual contributions are included in their corresponding subject categories – see abstracts 051.020, 082.079, 082.080, 094.156, 098.040, 102.017, 104.031 - 104.052, 105.049 - 105.053, 106.037 - 106.040, 107.012, 107.013, 131.137.

012.031 **Conference on interactions of the interplanetary plasma with the modern and ancient moon,** sponsored by the Lunar Science Institute, Houston, Texas and held at Lake Geneva Campus of George Williams College, Wisconsin, 1974 September 30 - October 4, with a summary by D. R. Criswell, J. W. Freeman, Jr.
The Moon, Vol. 14, 3 - 207 (1975). – The individual contributions are included in their corresponding subject categories – see abstracts 084.278 - 084.280, 094.157 - 094.165, 094.417 - 094.419, 106.042.

012.032 **Lunar ranging from a mobile laser station. Part I: Lunar laser ranging.** An informal seminar held at the University of Texas at Austin 18 - 19 December 1975.
J. D. Mulholland (Editor).
Separate print Dep. Astron. and McDonald Obs., Univ. Texas, Austin. 3 + 36 pp. (1975). – The individual contributions are included in their corresponding subject categories – see abstracts 031.253 - 031.255, 094.015, 094.016.

012.033 **Questions of astrophysics and physics of planetary atmospheres.** Conference at the Astrophysical Institute in Alma-Ata, 1974, February 25 - 27.
T. B. Omarov (Editor).
Akademiya Nauk Kazakhskoj SSR; Trudy Astrofizicheskogo Instituta. Tom (Vol.) 26. Izdatel'stvo "Nauka" Kazakhskoj SSR, Alma-Ata. 192 pp. Price 85 Kop. (1974). In Russian.
The individual contributions are included in their corresponding subject categories.

012.034 **On reference coordinate systems for earth dynamics.** International Astronomical Union Colloquium No. 26, held in Toruń, Poland, 26–31 August 1974.
B. Kołaczek, G. Weiffenbach (Editors).
Wykonano w Zakładzie Graficznym Politechniki Warszawskiej, Warsaw, Poland. 478 pp. with a foreword, introduction, summary of proceedings of the colloquium and recommendations, p. 7 - 47 (1975). – The individual contributions are included in their corresponding subject categories – see abstracts 031.266, 032.045, 033.018, 041.030, 041.045, 041.053 - 041.055, 044.034, 044.035, 045.016, 045.022 - 045.024, 046.056 - 046.065, 054.014 - 054.016, 081.036 - 081.040, 094.021, 141.145.

012.035 **Transactions of the XVIIth scientific conference of postgraduate students and junior research workers of the Institute of the History of Natural Sciences and Techniques of the USSR Academy of Sciences.**
Moskva, 1975, 176 pp. In Russian. – Abstr. in Referativ. Zhurn. 51. Astron., 12.51.46 (1975).

012.036 **Proceedings of symposium on the earth's gravitational field and secular variations in position.**
R. S. Mather, P. V. Angus-Leppan (Editors).
Published by the School of Surveying – The University of New South Wales, Sydney, Australia. 726 pp. Price Australian $ 20.00 (1974). – The individual contributions within the subject scope of Astronomy and Astrophysics Abstracts are included in their corresponding categories – see abstracts 009.024, 031.260 - 031.263, 045.017 - 045.020, 046.043, 046.054, 046.055, 055.016, 081.031 - 081.034, 094.020.

012.037 **Space astrometry.** Proceedings of a symposium held at Frascati, Italy, 22 - 23 October 1974.
T. D. Nguyen, B. T. Battrick (Editors).
Published by ESRO Scientific and Technical Information Branch, ESTEC, Noordwijk, The Netherlands. 9 + 137 pp. (1975). – The individual contributions are included in their corresponding subject categories – see abstracts 031.264, 032.044, 032.529 - 032.532, 041.048 - 041.052, 044.033, 045.021, 111.010, 111.011, 117.070, 153.036.

012.038 **Beiträge aus der Bundesrepublik Deutschland zur Vorlage bei der XVI. Generalversammlung der Internationalen Union für Geodäsie und Geophysik, Grenoble 1975.**
Deutsche Geod. Kommission Bayer. Akad. Wiss., München, Reihe B, No. 213, 109 pp. (1975). – The individual contributions within the subject scope of Astronomy and Astrophysics Abstracts are included in their corresponding categories – see abstracts 046.068 - 046.070, 081.041, 081.042, 082.111.

012.039 **Proceedings of the ICO conference on optical methods in scientific and industrial measurements.**
Tokyo, 1974, August 26 - 30. S.-I. Tanaka (Editor).
S.-I. Tanaka (Editor).
Journ. Applied Phys., Vol. 14, Suppl. 14-1, 544 pp. Price 10000 yen (1975). – Review in Optik, Vol. 44, 119 (1975).

012.040 **Convegno della Società Astronomica Italiana sul tema: «Problemi didattici e metodologici dell'insegnamento dell'astronomia»**, Milano, 5–6 aprile 1975.
Giornale Astron., Vol. 1, 321 - 388 (1975). – The individual contributions are included in their corresponding subject categories – see abstracts 014.019 - 014.029.

012.041 **Image processing techniques in astronomy.** Proceedings of a conference held in Utrecht on March 25–27, 1975.
C. de Jager, H. Nieuwenhuijzen (Editors).
D. Reidel Publishing Company, Dordrecht – Holland/Boston – U.S.A. Astrophysics and Space Science Library, Vol. 54 (Proceedings), 11 + 418 pp. Price Dfl. 95.00, US $ 38.00 respectively (1975). – The individual contributions are includ-

ed in their corresponding subject categories – see abstracts 031.280 - 031.296, 031.415 - 031.423, 032.537, 034.131 - 034.152, 041.070, 072.081, 073.154, 113.071.

012.042 **Skylab science experiments.**
G. W. Morgenthaler, G. E. Simonson (Editors).
Science and Technology, Vol. 38. Proceedings of a symposium of the 140th annual meeting of the American Association for the Advancement of Science. American Astronautical Society Publications Office, Tarzana, Calif. 14 + 257 pp. Price $ 20.00 (1975). ISBN 17703-74-X.

012.043 **In the beginning...** Symposium on the origin of planets and life, held as part of the Copernicus 500th birthday celebration at Canberra, 1973 April 27.
Australian Academy of Science, Canberra. 10 + 132 pp. (1974). ISBN 0-85847-0-026-8.

012.044 **Methods in stellar atmosphere and interplanetary plasma research.** Proceedings (Trudy) of the P. N. Lebedev Physics Institute, Vol. 62.
D. V. Skobel'tsyn (Editor).
Consultants Bureau, New York – London. Translated from Russian. 6 + 202 pp. Price $ 42.00 (1974). ISBN 0-306-10905-0.

012.045 **International conference on X rays in space.**
Calgary, Alberta, Canada, August 1974.
D. Venkatesan (Editor).
University of Calgary Department of Physics, Calgary, Canada. Two Volumes. 20 + 1236 pp. + appendices. Price $ 40.00 (1975).

012.046 **Proceedings of the Council "Sun – Earth". 10th issue.**
AN SSSR. Moskva. 28 pp. (1975). In Russian. – (From Referativ. Zhurn., 51. Astron., 12.51.56 (1975)).

012.047 **VIIIth All-Union conference on radioastronomy.**
AN SSSR. Nauch. sovet po kompleks. probl. "Radioastronomiya", Fiz. in-t im. P. N. Lebedeva, Mosk. obl. pravl. NTOREhS im. A. S. Popova. Pushchino. 204 pp. Price 90 Kop. (1975). In Russian. – Rev. in Referativ. Zhurn. 51. Astron., 11.51.50 (1975).

012.048 **Quantum gravity.** Papers from a symposium, Chilton, England, February 1974.
C. J. Isham, R. Penrose, D. W. Sciama (Editors).
Oxford University Press, London; Clarendon Press, Oxford – New York, 12 + 606 pp. Price £ 10.50, $ 29.50 respectively (1975). – Review in Nature, Vol. 258, 25; 1975 (*P. C. W. Davies*).

012.049 **The nature of scientific discovery.** A symposium commemorating the 500th anniversary of the birth of Nicolaus Copernicus. Washington,D.C., April 1973.
O. Gingerich (Editor).
Smithsonian International Symposia Series 5. Smithsonian Institution Press, Washington, D.C. 616 pp. Price $ 15.00 (1975). – Reviews in Sky Telescope, Vol. 50, 186; Vol. 50, 323 - 325; 1975 (*N. S. Hetherington*).

012.050 **Archaeoastronomy in pre-Columbian America.**
Papers from a meeting, Mexico City, June 1973.
A. F. Aveni (Editor).
University of Texas Press, Austin, Tex. 16 + 436 pp. Price $ 16.50 (1975). – Review in Sky Telescope, Vol. 50, 409 (1975).

012.051 **Skylab and Pioneer report.**
P. H. Bolger, P. B. Richards (Editors).
Univelt, Inc., Tarzana, Calif. 135 pp. Price $ 16.00 (1975). Review in Sky Telescope, Vol. 50, 185 - 186 (1975).

013 Reports on Astronomy in Various Countries and Particular Fields, International Cooperation

013.001 **U.S. National Report to the International Union of Geodesy and Geophysics.**
P. M. Bell (Editor).
Rev. Geophys. Space Phys., Vol. 13, No. 3, 9 + 1107 pp. (1975).
This report which is prepared by the American Geophysical Union prior to each quadrennial assembly, reviews the progress in all areas of geophysics and presents an extensive bibliography of the geophysical literature published during the quadrennium. The report also affords an opportunity to focus international attention on the entire U.S. program of research in geophysics. – The individual contributions within the subject scope of Astronomy and Astrophysics Abstracts are included in their corresponding categories – see abstracts 046.003 - 046.007, 074.023 - 074.025, 078.006 - 078.008, 081.010, 082.019, 083.020 - 083.026, 084.014, 084.231 - 084.239, 091.008 - 091.010, 092.004, 094.003, 094.123 - 094.126, 094.410, 094.411, 097.010, 099.042, 105.019 - 105.021, 106.024, 107.005, 143.012.

013.002 **Astronomy in Bulgaria in the years of People's Democracy.** N. Nikolov.
Zemlya i Vselennaya, 1975, No. 4, p. 41 - 43. In Russian.

013.003 **30th anniversary of astronomy in the People's Republic Bulgaria.** N. S. Nikolov.
Priroda (NRB), Vol. 24, No. 1, p. 15 - 18 (1975). In Bulgarian. Abstr. in Referativ. Zhurn. 51. Astron., 9.51.23 (1975).

013.004 **Ergebnisse der astronomischen Forschungen 1974.** K.-H. Schmidt, K.-G. Steinert.
Astron. in der Schule, 12. Jahrgang, p. 100 - 103 (1975).

013.005 **Scientific cooperation of solar physicists in socialistic countries.** V. Bumba.
Oss. Mem. Oss. Astrofis. Arcetri, Fasc. 105, (see 012.026), p. 98 - 101 (1975).

013.006 **Solar physics research at the Appleton Laboratory, Culham.** A. H. Gabriel.
Oss. Mem. Oss. Astrofis. Arcetri, Fasc. 105, (see 012.026), p. 102 - 105 (1975).

013.007 **Summary of report on current solar research at the Lund Observatory.** J. O. Stenflo.
Oss. Mem. Oss. Astrofis. Arcetri, Fasc. 105, (see 012.026), p. 120 (1975).

013.008 **Promotion of solar physics in Europe: introduction to a debate.** J. C. Pecker.
Oss. Mem. Oss. Astrofis. Arcetri, Fasc. 105, (see 012.026), p. 125 - 132 (1975).

013.009 **Report on the session: how to promote solar research?** A. D. Fokker.
Oss. Mem. Oss. Astrofis. Arcetri, Fasc. 105, (see 012.026), p. 133 - 137 (1975).

013.010 **Optics in Switzerland.** W. Lotmar.
Applied Optics, Vol. 14, 2355 - 2370 (1975).
Present work in optics in Switzerland is surveyed, covering universities, research institutes, and industry. Included are short notes on the following astronomical topics: University of Geneva, Observatory, *M. Golay;* Platform for astronomical observations from balloons, *D. Huguenin*; Photometry with an image tube, *F. Pilloud;* PIN diode for high-speed astronomy, *P. Bartholdi;* Spectrophotometer for the determination of stellar radial velocities, *M. Mayor, J. L. Poncet;* International foundation Jungfraujoch and Gornergrat scientific station, *W. Lotmar;* Oxford University solar station at the Gornergrat, *D. E. Blackwell;* Liège University laboratory of solar spectroscopy, *L. Delbouille, G. Roland.*

013.011 **Astronomy at the University of Otago Physics Department.** P. J. Edwards.
Southern Stars, Vol. 26, 90 - 92 (1975). – Presented to Roy. Astron. Soc. New Zealand annual general meeting, 1974 December.

013.012 **Progrès récents de l'astronomie.** J.-C. Pecker.
L'Astronomie, 89ᵉ année, p. 434 - 440 (1975).

013.013 **Current trends in astronomical spectroscopy.** K. O. Wright.
Journ. Roy. Astron. Soc. Canada, Vol. 69, 265 - 285 (1975).
A brief survey is made of improvements in spectroscopic data analysis in recent years. The need for accurate observational material is stressed, especially with the highly efficient receivers now available. A few problems that will be studied for many years are considered: stellar evolution and nucleogenesis, stellar atmospheres and chromospheres, double and variable stars, and galactic, interstellar and extra-galactic astronomy.

013.014 **La nouvelle astronomie.** P. Ledoux.
Acad. Roy. Belgique, Bull. Cl. Sci., 5ᵉ Sér., Vol. 59, 1221 - 1243 = Inst. Astrophys. Univ. Liège, Cointe-Ougrée (*Belgique*), Coll. 8°, No. 646 (1973).

013.015 **Sternforscher und Sternfreunde in der VR Ungarn.** D. B. Herrmann.
Archenhold-Sternw. Berlin-Treptow, Vorträge und Schriften, No. 47, 16 pp. (1975).

013.016 **Astronomisko pētījumu virzieni Latvijas PSR.** K. Šteins.
Zvaigžņotā debess, 1975. gada pavasaris, p. 1 - 3.

013.017 **Zem tālās Bolīvijas debesīm.** K. Lapuska.
Zvaigžņotā debess, 1975. gada vasara, p. 16 - 21.

013.018 **Concept of revolutions in astronomy based on the scheme of Shapley and Rabinovičs.** H. Eelsalu.
Tartu Ulikooli toimetised, Uch. zap. Tartus. un-ta, 1975, vyp. (No.) 360, p. 7 - 13. In Russian. – Abstr. in Referativ. Zhurn. 51. Astron., 2.51.2 (1976).

013.019 **On the characteristics of astronomy as science.** I. S. Alekseev.
Tartu Ulikooli toimetised. Uch. zap. Tartus. un-ta, 1975, vyp. (No.) 360, p. 14 - 20. In Russian. – Abstr. in Referativ. Zhurn. 51. Astron., 2.51.4 (1976).

Solar space research in the Federal Republic of Germany. See Abstr. 051.015.

014 Teaching in Astronomy

014.001 **Astronomy activities for use on the secondary school level.** C. D. Smith, P. H. Knappenberger.
Bull. American Astron. Soc., Vol. 7, 446 - 447 (1975).

014.002 **The open class in astronomy.** D. A. Pierce.
Bull. American Astron. Soc., Vol. 7, 447 (1975).
Abstr. AAS.

014.003 **Use of astronomy in a physics curriculum.** J. T. Dowell.
Bull. American Astron. Soc., Vol. 7, 447 (1975). – Abstr. AAS.

014.004 **Nine semesters of individually paced instruction in astronomy at Oklahoma State University.**
L. W. Schroeder, R. K. Oines, T. M. Jordan.
Bull. American Astron. Soc., Vol. 7, 447 (1975). – Abstr. AAS.

014.005 **Zur Verwendung von Horizont- bzw. Himmelsausschnittskizzen für die kontrollierbare selbständige Schülerbeobachtung.** H. Hirrle.
Astron. in der Schule, 12. Jahrgang, p. 86 - 88 (1975).

014.006 **"Sky service" in the Palace of Young Pioneers.** V. I. Koval'.
Zemlya i Vselennaya, 1975, No. 4, p. 79 - 82. In Russian.

014.007 **Astronomy teaching in the USA.** D. V. Clish.
Quarterly Journ. Roy. Astron. Soc., Vol. 16, 318 - 320 (1975).

014.008 **Certain questions on the methods of exposing the theme 'Internal constitution of the sun' at teachers' training colleges.** A Minbaeva.
Trudy Kirg. un-ta. Ser. fiz. n., 1975, vyp. (No.) 6, ch. (part) 1, p. 51 - 57. In Russian. – Abstr. in Referativ. Zhurn. 51. Astron., 9.51.34 (1975).

014.009 **On teaching of astronomy in the 10-year secondary school (resolution of the board of the Polish Astronomical Society).** J. Stodółkiewicza.
Postępy Astron., Vol. 23, 153 - 154 (1975). In Polish.

014.010 **Die Volksbildungsarbeit Bruno Hans Bürgels.** A. Zenkert.
Astron. in der Schule, 12. Jahrgang, p. 108 - 111 (1975).

014.011 **Nastava astronomije u pojedinim republikama SFRJ s naročitim osvrtom na SR Sloveniju.**
F. Dominko.
Publ. Obs. Astron. Beograd, No. 20, (see 012.024), p. 223 - 225 (1975).

014.012 **Nastava astronomije u srednjim školama S. R. Srbije.** B. Ševarlić.
Publ. Obs. Astron. Beograd, No. 20, (see 012.024), p. 226 - 227 (1975).

014.013 **Astronomy at secondary school.** R. V. Kunitskij, Yu. A. Sklyarov, I. I. Sysoev, F. Yu. Zigel', P. N. Karnitskij, A. N. Leonov.
Zemlya i Vselennaya, 1975, No. 6, p. 64 - 70. In Russian.

014.014 **On the control of laboratory exercises on astronomy at teacher's colleges.** L. S. Bratolyubova.
Probl. programmir. obuch. Vyp. (No.) 3. Vladimir, 1975, p. 85 - 87. In Russian. – Abstr. in Referativ. Zhurn. 51. Astron., 12.51.34 (1975).

014.015 **Teaching astronomy to Navajos.** J. T. Stocke.
Bull. American Astron. Soc., Vol. 7, 526 (1975).

014.016 **Astronomische Beobachtungen und Modellversuche.** A. Werner.
Jahrbuch der Schulphysik, (Aulis-Verlag, Köln), Vol. 2, 1 - 7 (1975).
The article contains suggestions for observation of eclipsing and visual binaries with modest apparatus and instructions for recording of light curves by means of model eclipsing binaries and artificial variable stars.

014.017 **Astronomijas studijas Ļeņingradas universitātē.** J.-I. Straume.
Zvaigžņotā debess, 1975. gada vasara, p. 49 - 52.

014.018 **Zvaigžņu fizika vidusskolā.** J. Francmanis.
Zvaigžņotā debess, 1975. gada rudens, p. 31 - 36.

014.019 **Perché e come la S.A.It. si occupa dell'insegnamento dell'astronomia.** M. Rigutti.
Giornale Astron., Vol. 1, 323 - 338 (1975). – Conference paper.

014.020 **L'astronomia nella didattica delle scienze. Considerazioni generali e rassegna di alcuni corsi stranieri.** V. Zanetti.
Giornale Astron., Vol. 1, 339 - 344 (1975). – Conference paper.

014.021 **Corsi sperimentali di astronomia in Italia. Problemi e possibili sviluppi.** G. Veronese.
Giornale Astron., Vol. 1, 345 - 350 (1975). – Conference paper.

014.022 **Disponibilità per la didattica delle attrezzature degli osservatori astronomici e dei planetari.**
L. Baldinelli.
Giornale Astron., Vol. 1, 351 - 352 (1975). – Conference paper.

014.023 **Insegnamento e divulgazione al planetario.** M. Cavedon.
Giornale Astron., Vol. 1, 353 - 358 (1975). – Conference paper.

014.024 **Il problema della formazione e dell'aggiornamento degli insegnanti di scienze.** P. Tripòdi.
Giornale Astron., Vol. 1, 359 - 362 (1975). – Conference paper.

014.025 **Un'esperienza di astronomia per il II ciclo della scuola elementare.** M. Rigutti, M. A. Santaniello.
Giornale Astron., Vol. 1, 363 - 368 (1975). – Conference paper.

014.026 **Esperienze di insegnamento dell'astronomia nella scuola dell'obbligo alla luce della psicologia piagetiana.** B. Bertotti, A. Terzi Belloni.
Giornale Astron., Vol. 1, 369 - 372 (1975). – Conference paper.

014.027 **Per la formazione di una mentalità e di un atteggiamento scientifico del bambino della scuola primaria.**
C. Pitzalis.
Giornale Astron., Vol. 1, 373 - 376 (1975). – Conference paper.

014.028 **L'insegnamento delle scienze di osservazione: problemi metodologici e didattici.** E. Proverbio.
Giornale Astron., Vol. 1, 377 - 381 (1975). – Conference paper.

014.029 **L'utilizzazione degli audiovisivi.** B. Caccin.
Giornale Astron., Vol. 1, 383 - 388 (1975). – Conference paper.

014.030 **Astrometrie und Astronomieunterricht.**
K.-G. Steinert.
Astron. in der Schule, 12. Jahrgang, p. 128 - 131 (1975).

014.031 **Zur Bedeutung neuerer Erkenntnisse über die Sternentstehung und Sternentwicklung für den Astronomielehrer und den Astronomieunterricht.**
M. Schukowski.
Astron. in der Schule, 12. Jahrgang, p. 131 - 134 (1975).

014.032 **What a teacher in astronomy must know.**
Yu. N. Klevenskij.
Vestn. vyssh. shkoly, 1975, No. 8, p. 78 - 79. In Russian.
Abstr. in Referativ. Zhurn. 51. Astron., 2.51.119 (1976).

014.033 **On the illustration of the method of determination of distances to celestial bodies.**
A. D. Marlenskij, G. S. Yakhno.
Fizika v shkole, 1975, No. 5, p. 77 - 78. In Russian. – Abstr. in Referativ. Zhurn. 51. Astron., 2.51.120 (1976).

014.034 **The conservation principles and Kepler's laws of planetary motion (teaching method).**
L. Motz.
American Journ. Phys., Vol. 43, 575 - 578 (1975).

014.035 **Velocity space and the geometry of planetary orbits (teaching method).**
H. Abelson, A. Di Sessa, L. Rudolph.
American Journ. Phys., Vol. 43, 579 - 589 (1975).

014.036 **Term projects on interstellar comets.** J. E. Mack.
American Journ. Phys., Vol. 43, 590 (1975).

014.037 **Astrophysics: an integrative course.**
G. D. Gutsche.
American Journ. Phys., Vol. 43, 695 - 696 (1975).

014.038 **20th century quasars and 17th century stars.**
D. R. Topper.
Phys. Teacher, (*USA*), Vol. 13, 339 - 342 (1975).
The similarity between the present-day problems concerning quasars and those of the 17th century concerning the stars is discussed. In both cases, 'size' and distance were under debate, 'size' referring to energy in the case of quasars and volume in the case of the stars.

014.039 **Stellar spectrum synthesizer.** G. F. Landegren.
American Journ. Phys., Vol. 43, 821 - 823 (1975).
The author describes an apparatus which may be used in an undergraduate laboratory exercise for observing absorption or emission spectra which may be arbitrarily Doppler shifted, and may therefore be regarded as an artificial stellar spectrum synthesizer. The apparatus is described in some detail and consists essentially of two diffraction gratings and several lenses.

014.040 **Films and the sky. Explorations in space and time.**
M. L. Meeks.
Eight 16-mm. color films. Houghton Mifflin Co., Department M, Boston, Mass. Price $ 95.00 each; three day rental, $ 15.00 each. – Review in Sky Telescope, Vol. 50, 183 - 185; 1975 (*G. S. Mumford*).

015 Miscellanea

015.001 **Kepler's Harmonies–past, present and future.** R. Haase.
Vistas in Astronomy, Vol. 18, (see 012.006), 535 - 536 (1975).

015.002 **Harmonies of the world.** G. Abetti.
Vistas in Astronomy, Vol. 18, (see 012.006), 537 - 539 (1975).

015.003 **Astronomy and music.** C. A. Ronan.
Sky Telescope, Vol. 50, 145 - 148 (1975).

015.004 **Solar systems and life.** I. Ridpath.
Spaceflight, Vol. 17, 323 - 327 (1975).

015.005 **CETI from Copernicus.** A. T. Lawton.
Spaceflight, Vol. 17, 328 - 330 (1975).

015.006 **Interesting astronomical phenomena in 1975.** A. D. Bonov.
Priroda (NRB), Vol. 23, No. 6, p. 3 - 5 (1974). In Bulgarian.

015.007 **The galactic "water hole".** C. L. Seeger.
Bull. American Astron. Soc., Vol. 7, 440 (1975).

015.008 **International legal protection of the "water hole" frequency band for a search for extraterrestrial intelligent life.** M. A. Stull.
Bull. American Astron. Soc., Vol. 7, 440 (1975). – Abstr. AAS.

015.009 **Population explosion and interstellar expansion.** S. von Hoerner.
Journ. British Interplanet. Soc., Vol. 28, 691 - 712 (1975).
This paper, dedicated to C. F. von Weizsäcker on his 60th birthday, appeared in an earlier version, in 1972, in "Einheit und Vielheit", Scheibe and Süssmann (editors), Vandenhoeck and Ruprecht, Göttingen.

015.010 **A long-range perspective and some fundamental aspects of interstellar evolution.** K. A. Ehricke.
Journ. British Interplanet. Soc., Vol. 28, 713 - 734 (1975).

015.011 **Directly interacting extra-terrestrial technological communities.** D. Viewing.
Journ. British Interplanet. Soc., Vol. 28, 735 - 744 (1975).

015.012 **USSR research programme on the problem of Communication with Extraterrestrial Civilisations.**
Journ. British Interplanet. Soc., Vol. 28, 751 - 760 (1975).
This paper is an edited and annotated translation of a research programme on the problem of Communication with Extraterrestrial Civilisations, considered and approved by the Scientific Council on the Problem Area of Radio Astronomy, of the USSR Academy of Sciences. The original text was published in 1974 (Astron. Zhurn. Akad. Nauk SSSR, Vol. 51, 1125 - 1132 – see 12.015.009).

015.013 **Un pianeta per l'uomo.** M. Cantelli.
Coelum, Vol. 43, 160 - 167 (1975).

015.014 **"Stars look downwards" on our aircraft.** R. V. Kunitskij.
Zemlya i Vselennaya, 1975, No. 4, p. 48 - 52. In Russian.

015.015 **500th anniversary of Copernicus in Polish philately.** A. A. Gurshtejn.
Zemlya i Vselennaya, 1975, No. 4, p. 87 - 90. In Russian.

015.016 **The Luneberg lens and the importance of transmission in establishing contact with extraterrestrial civilizations.** H. J. Gerritsen, S. J. McKenna.
Icarus, Vol. 26, 250 - 256 (1975).
A large Luneberg-type antenna with accessories is described as a means of simultaneously monitoring a large number of stars for possible intelligent signals. Various contacting strategies are compared on their probability of success and arguments are advanced favoring a program of transmission at a wavelength of 21 cm to the nearest stars.

015.017 **Project Ozma II.** R. Sheaffer.
Spaceflight, Vol. 17, 421 - 423 (1975).

015.018 **Views of those glorious southern skies.** J. R. Henley, H. Vehrenberg, R. M. Ashby.
Sky Telescope, Vol. 50, 360 - 367 (1975).

015.019 **Sterrenkunde op postzegels.** B. A. Gerritsen.
Zenit, Vol. 2, 404 - 406 (1975).

015.020 **Are we alone?** S. van den Bergh.
Mercury, (Journ. Astron. Soc. Pacific), Vol. 4, No. 4, p. 8 - 10, 26 (1975).

015.021 **Space travel and life beyond earth.** E. J. Öpik.
Irish Astron. Journ., Vol. 11, 220 - 238 (1975).
Popular lecture given in New York on June 10, 1974, with an appendix: 'Our cosmic destiny as it was seen half-a-century ago'.

015.022 **The soviet CETI program.** USSR Academy of Sciences.
Icarus, Vol. 26, 377 - 385 (1975). – Paper originally published in Astron. Zhurn., Vol. 51, 1125 - 1132 (1974). English translation reprinted from Soviet Astron., Vol. 18, 669 - 675 (1975).

015.023 **Space colonies and energy supply to the earth.** G. K. O'Neill.
Science, Vol. 190, 943 - 947 (1975).
Manufacturing facilities in high orbit could be used to build satellite solar power stations from lunar materials.

015.024 **Possible relation between periodic glaciation and the flexure of the Galaxy.** G. E. Williams.
Earth Planet. Sci. Letters, Vol. 26, 361 - 369 (1975).
Employing the following values for parameters of the sun's galactic orbit, present circular valocity $=225 \pm 25$ km sec^{-1}, present galactocentric distance $=9.0 \pm 0.5$ kiloparsec, and eccentricity $= 0.1 \pm 0.01$, it is calculated as a first approximation that the period of the sun's revolution around the galactic centre is 303^{+65}_{-51} m.y. Assuming that the two galactic regions which appear to be tidally flexed by the Magellanic Clouds are stationary with respect to the galactic centre, then the mean period of solar passage between regions of given, diametrically-opposite galactic flexure is 152^{+33}_{-26} m.y. Since the solar system now lies within a little-flexed region of the galactic disc, and the earth is presently subjected to a climatic regime of high-latitude glaciation, it is suggested that passage through such regions may have resulted in the ~155-m.y. rhythm of earth glaciation.

015.025 **High pH, ammonia toxicity, and the search for life on the Jovian planets.**
P. H. Deal, K. A. Souza, H. M. Mack.
Origins of Life, Vol. 6, 561 - 573 (1975).

015.026 **The arrow of time.** D. Layzer.
Sci. American, Vol. 233, No. 6, p. 56 - 69 (1975).

015.027 **Sternwartengründungen, Wissensproduktion und ökonomischer Fortschritt.** D. B. Herrmann.
Sterne, Vol. 51, 228 - 234 (1975).

015.028 **Beitrag zur Diskussion der Möglichkeit von interstellaren Flügen.** G. Müller.
Sterne, Vol. 51, 235 - 246 (1975).

015.029 **The Arecibo message of November, 1974.**
The Staff at the National Astronomy and Ionosphere Center, Arecibo, Puerto Rico.
Icarus, Vol. 26, 462 - 466 (1975).

015.030 **Building Chaucer's astrolabe–I.** S. Eisner.
Journ. British Astron. Ass., Vol. 86, 18 - 29 (1975).

015.031 **The moon illusion.** P. Cope.
Journ. British Astron. Ass., Vol. 86, 44 - 46 (1975).

015.032 **On the relations between astronomy and physics.**
I. Úlehla.
Vesmír, Vol. 54, 101 - 103 (1975). In Czech.

015.033 **Arbeitsschema der methodischen Physik nach Thüring.** A. Fadini.
Wetter–Boden–Mensch, No. 22, p. 1451 - 1455 (1975).

015.034 **Nomlisto de la konstelacioj.** A. Heck.
Sci. Revuo, Vol. 26, 3 - 6 (1975). – List of constellations in Esperanto.

015.035 **Druga fundamentalna kozmička formula.**
Prostorno vremenska konačnost svemira.
M. Hegedušić.
Naučna Misao, Bull. 12, p. 41 - 55 (1975).

015.036 **Pismo društvu naučna misao u vezi s otkrićem druge fundamentalne kozmičke formule.**
M. Hegedušić.
Naučna Misao, Bull. 12, p. 59 - 63 (1975).

015.037 **A new interpretation of the mysterious radioecho.**
A. W. Szpilewski (*Shpilevsky*).
Urania Kraków, Vol. 46, 108 - 116 (1975). In Polish.
Translated from the Russian and with a remark by L. Zajdler.

015.038 **Radioechoes and space probes of Lunan and Shpilevsky.** Z. Paprotny.
Urania Kraków, Vol. 46, 206 - 210 (1975). In Polish.

015.039 **Are we alone in the universe?** S. R. Brzostkiewicz.
Urania Kraków, Vol. 46, 296 - 304 (1975). In Polish.

015.040 **Meklēt, atrast, novērot.** L. Laucenieks.
Zvaigžņotā debess, 1975. gada pavasaris, p. 3 - 9.

015.041 **Uputstvo Medunarodne Astronomske Unije za pisanje tablica i bibliografije.**
Vasiona, Vol. 23, 62 - 64 (1975).

Vikingvluchten naar de planeet Mars. Op zoek naar sporen van het leven. See Abstr. 053.011.

Applied Mathematics, Physics

021 Mathematics, Computing

021.001 **Fourier analysis with unequally-spaced data.**
T. J. Deeming.
Astrophys. Space Sci., Vol. 36, 137 - 158 (1975).
The general problems of Fourier and spectral analysis are discussed. A discrete Fourier transform $F_N(\nu)$ of a function $f(t)$ is presented which (1) is defined for arbitrary data spacing; (2) is equal to the convolution of the true Fourier transform of $f(t)$ with a spectral window. It is shown that the 'pathology' of the data spacing, including aliasing and related effects, is all contained in the spectral window, and the properties of the spectral windows are examined for various kinds of data spacing. The results are applicable to power spectrum analysis of stochastic functions as well as to ordinary Fourier analysis of periodic or quasiperiodic functions.

021.002 **Some remarks on Roche harmonics.**
G. F. Roach.
Astrophys. Space Sci., Vol. 36, 159 - 167 (1975).
In the present paper certain analytical aspects of the separability of the three-dimensional Laplace's equation in Roche harmonics (cf. Kopal, 1969, 1970) are discussed. The intention is to provide a rigorous analytical basis for a further, more detailed investigation of the properties of Roche harmonics which will be developed more fully in subsequent communications. The arguments presented in this paper provide a generalization of some well known techniques in potential theory and as such can be shown to lead to a successively refinable approximation theory.

021.003 **On the subdivision of a spherical surface into elements of equal area.** R. J. Fryer.
Geophys. Journ. Roy. Astron. Soc., Vol. 42, 883 - 891 (1975).
A method is presented which permits the subdivision of a spherical surface into elements of precisely equal area while ensuring a close similarity of shape. The boundaries of individual elements are meridians and parallels. It is demonstrated that certain values of the total number of elements permit close approaches to identity of shape and the parameters necessary for the utilization of these optimum values are tabulated.

021.004 **On the analysis of vector arrays on a spherical surface.** R. J. Fryer.
Geophys. Journ. Roy. Astron. Soc., Vol. 42, 893 - 898 (1975).
The need for, and difficulties commonly experienced during, the analysis of patterns of vectors with superimposed 'noise' upon the surfaces of spheres is discussed. A simple procedure is described whereby many of these difficulties may be circumvented and certain classes of systematic array detected even against a considerable random background. An extension of the same procedure to the analysis of vector patterns in a plane is discussed.

021.005 **Simple graphical solutions of spherical triangles.**
C. H. Cotter.
Journ. Navigation, Vol. 28, 306 - 312 (1975).

021.006 **Service programs for the computer M-222.**
B. M. Shustov, V. I. Ponomareva.
Nauchn. Informatsii, vyp. (No.) 34, p. 3 - 24 (1974). In Russian.

021.007 **A program for calculation of optimum difference grids.** Eh. I. Shustova.
Nauchn. Informatsii, vyp. (No.) 34, p. 25 - 28 (1974). In Russian.
Consideration is given to a program for obtaining nonuniform difference grids for gas-dynamical problems.

021.008 **A system of combined programs for computers of M-20 type.** V. V. Muzylev, V. I. Shipilova.
Nauchn. Informatsii, vyp. (No.) 34, p. 29 - 44 (1974). In Russian.

021.009 **Special procedures of the α-translator for work with supplementary memory of the computer.**
V. V. Muzylev, V. I. Shipilova.
Nauchn. Informatsii, vyp. (No.) 34, p. 45 - 49 (1974). In Russian.

021.010 **A stellar evolution program for the computer M-222.**
B. M. Shustov.
Nauchn. Informatsii, vyp. (No.) 34, p. 50 - 120 (1974). In Russian.
A complicated stellar evolution program is realized for the medium size computer M-222 in ALGOL-60. The algorithm and all necessary technical details are described.

021.011 **Approximation functions to the Emden and associated Emden functions near the first zero.**
A. P. Linnell.
Astrophys. Space Sci., Vol. 37, 57 - 72 (1975).
Analytic expressions are presented which approximate the Emden function θ, and the associated Emden functions, ψ_0 and ψ_2, and their derivatives, near the first zero of θ, ξ_1. The range of accurate representation extends to $\xi = 1.5\,\xi_1$. This range is sufficient to encompass the boundary of a critically rotating polytrope.

021.012 **The structure of the secular system in nonlinear oscillations.** U. Kirchgraber, M. Vitins.
Celestial Mechanics, Vol. 12, 139 - 153 (1975).
If the method of averages is applied to a nearly integrable system of differential equations the so-called secular system is obtained. In the present paper the authors study the structure of the secular system for a large class of problems. It is especially aimed to establish sufficient conditions for the secular system to be solvable by means of elementary functions.

021.013 **The accuracy of programming algorithms in the reduction of satellite photographs.** I. Synek.
Observations of artificial earth satellites, No. 14, (see 012.018), p. 577 - 584 (1975).

021.014 **The method of statistical reduction.** A. Zięba.
Acta Cosmologica, Fasc. 3, p. 75 - 90 (1975).
The method of statistical reduction consists in substituting the population under investigation by the sample of a given

and fixed size. Speaking more exactly we substitute the distribution of interesting property in the population by the expected value of the distribution in the sample, which is treated as a random variable. Such procedure simplifies statistical investigations of very numerous populations as well as allows to compare between each other the distributions in populations of different sizes by means of their reduction to the same size.

021.015 **Imitation of spectral parameters of stars.**
A. B. Belitskij, A. I. Belov, A. V. Dabagyan.
Optich. i ehlektrooptich. obrabotka inform. Moskva, Nauka, 1975, p. 33 - 36. In Russian. – Abstr. in Referativ. Zhurn. 51. Astron., 11.51.235 (1975).

021.016 **A comparative study of maximization methods.**
A. Heck.
Acad. Roy. Belgique, Bull. Cl. Sci. 5e Sér., Vol. 60, 927 - 943 (1974).

A comparative study is made between the following methods: Newton-Raphson method, Davidon-Fletcher-Powell method and modified Gauss-Newton method. These were applied to the maximization of a likelihood function of very general use in astronomy. The analysis shows that the most satisfactory method is the last quoted.

021.017 **Exakte Abschätzung von Maximalfehlern aus vorgegebenen Toleranzen der Beobachtungsgrößen.**
E. Reinhart.
Deutsche Geod. Kommission Bayer. Akad. Wiss., München, Reihe C, No. 211, 54 pp. (1975).

021.018 **On the determination of the period of a variable star with a computer (method of local scanning).**
V. Harvig, P. Kal'v.
Astron. Tsirk., No. 876, p. 3 - 4 (1975). In Russian.

021.019 **Wahrscheinlichkeitsverteilungen für statistische Beurteilungen von Ausgleichungsergebnissen.**
K. R. Koch.
Mitt. Inst. Theor. Geod., Univ. Bonn, No. 38, 3 + 58 pp. (1975).

021.020 **Computers in astronomy.** K. Ziołkowski.
Urania Kraków, Vol. 46, 360 - 364 (1975). In Polish.

021.021 **On the strength of a system of partial differential equations.** B. F. Schutz.
Journ. Math. Phys., *New York*, Vol. 16, 855 - 856 = Separate print Dep. Applied Math. Astron., Univ. College, Cardiff, Wales (1975).

It is shown that Einstein's concept of the "strength" of a system of differential equations is directly related to the number of dynamical degrees of freedom the equations permit. It may become a useful tool for investigating the structure of the system.

021.022 **Application of the theory of events to research of relation between processes.** M. Paupere.
Investigation of the sun and red stars. 3, (see 003.008), p. 21 - 28 (1975). In Russian.

A method for investigation of the relation between stationary normal processes is considered. It consists in the following: by means of experimental data the number of coincidences of events depending on the level is calculated. Then the mean value of coincidences of events of independent processes with chosen probability at the same levels is estimated. If the experimental curve lies above the zone showing the estimate of the number ofcoincidences of events, then it can be assumed that these investigated processes are correlated with the corresponding confidence probability.

021.023 **Statistical test of pseudo-random numbers.**
Y. Baba.
Mem. Fac. Sci., Kyoto Univ., Ser. Phys., Astrophys., Geophys., Chem., Vol. 34, 247 - 259 = Contr. Dep. Astron., Univ. Kyoto, No. 55 (1974).

021.024 **Implementación de la transformada de Fourier rápida para una función segmentada de M variables.**
L. R. Berriel-Valdos.
Bol. Inst. Tonantzintla, Vol. 1, 217 - 220 (1975).

A procedure is presented for determining the fast Fourier transform of a segmented function of M variables. This procedure can be applied to cases in which the number of data is greater than the number of useful words in the computer memory; a common situation in minicomputers.

021.025 **A further ephemeridal A.P.L. computer programme.**
K. Locher.
BBSAG Bull., No. 24, p. 5 - 6 (1975).

021.026 **ATLAS 5** [*a program for the calculation of model stellar atmospheres*] **at the University of London (SAM 1). I: Programming and implementation.**
S. L. Wright, J. D. Argyros.
Commun. Univ. London Obs. No. 75, 63 pp. (1975).

This report is a summary of all the changes that have been made to the model stellar atmospheres computer program, ATLAS 5, at the University of London. The report also describes the implementation of the modified program on the IBM 360/65 computer of the University College London Computer Centre and on the CDC 6600 computer of the University of London Computer Centre.

021.027 **ATLAS 5** [*a program for the calculation of model stellar atmospheres*] **at the University of London (SAM 1). II. User's guide.** S. L. Wright.
Commun. Univ. London Obs. No. 76, 73 pp. (1975).

This report describes all of the facilities available with SAM 1 including a complete description of the output produced by the program. Information is given to help the user select appropriate model parameters, and to suggest ways of recovering from the various error conditions that may arise when running the program. The appendices give a brief description of how the program works, a complete list of all the options available, and details of how to run the program on the computers mentioned above.

On-line computers for telescope control and data handling. See Abstr. 031.409.

On the computation of minimum times of light curves. See Abstr. 120.001.

A nomogram for determination of interstellar absorption according to P. P. Parenago's formula.
See Abstr. 131.193.

Two recent developments concerning the Monte Carlo method. See Abstr. 151.019.

022 Physical Papers Related to Astronomy and Astrophysics

022.001 **The absorption of energetic electrons by molecular hydrogen gas.**
T. E. Cravens, G. A. Victor, A. Dalgarno.
Planet. Space Sci., Vol. 23, 1059 - 1070 (1975).

The processes by which energetic electrons lose energy in a weakly ionized gas of molecular hydrogen are analyzed and calculations are carried out taking into account the discrete nature of the excitation processes. The excitation, ionization and heating efficiencies are computed for electrons with energies up to 100 eV absorbed in a gas with fractional ionizations up to 10^{-2} and the mean energy per neutral hydrogen atom pair is calculated.

022.002 **Experiments to determine the force of gravity on single electrons and positrons.** W. M. Fairbank, F. C. Witteborn, J. M. J. Madey, J. M. Lockhart.
Experimental gravitation, (see 012.002), p. 310 - 330 (1974).

022.003 **Lifetimes and transition probabilities in N V.** Y. Baudinet-Robinet, P. D. Dumont, E. Biémont, N. Grevesse.
Physica Scripta, Vol. 11, 371 - 374 (1975).

022.004 **Study of the electronic transition moment for the CN red band system.** D. C. Jain.
Journ. Quant. Spectrosc. Radiat. Transfer, Vol. 15, 571 - 573 (1975).

022.005 **Transition probabilities for the (0, 0) and (1, 1) bands of the $c\ ^1\Phi - a\ ^1\Delta$ system of the TiO molecule.**
W. Żyrnicki.
Journ. Quant. Spectrosc. Radiat. Transfer, Vol. 15, 575 - 577 (1975).

022.006 **Relative transition probabilities and the electronic transition moment in the $A-X$ system of OH.**
D. R. Crosley, R. K. Lengel.
Journ. Quant. Spectrosc. Radiat. Transfer, Vol. 15, 579 - 591 (1975).

022.007 **Hydrogen broadened half-widths of ammonia.** J. S. Margolis.
Journ. Quant. Spectrosc. Radiat. Transfer, Vol. 15, 637 - 640 (1975).

The hydrogen-broadened ammonia half-widths of 59 lines of the $\nu_1 + \nu_2$ and 21 lines of the $\nu_2 + \nu_3$ bands have been measured. The spectral resolution available for the measurements was <0.06 cm^{-1} full width at half maximum. Lines up to $J = 12$ were measured.

022.008 **Unified Franck–Condon treatment of pressure broadening of spectral lines.**
J. Szudy, W. E. Baylis.
Journ. Quant. Spectrosc. Radiat. Transfer, Vol. 15, 641 - 668 (1975).

022.009 **Measurement of Stark broadened Lyman-lines.** G. Fussmann.
Journ. Quant. Spectrosc. Radiat. Transfer, Vol. 15, 791 - 809 (1975).

022.010 **Measurement of the oscillator strengths of transitions of titanium (I) in the region 3600–4500 Å.**
J. Lotrian, J. Cariou, A. Johannin-Gilles.
Journ. Quant. Spectrosc. Radiat. Transfer, Vol. 15, 815 - 821 (1975).

022.011 **Tentatively identified and predicted H^- Fano resonances.**
C. Ingemann-Hilberg, M. Rudkjøbing.
Journ. Quant. Spectrosc. Radiat. Transfer, Vol. 15, 823 - 830 (1975).

022.012 **Van der Waals broadening for heavy elements.** D. Hofsäss.
Journ. Quant. Spectrosc. Radiat. Transfer, Vol. 15, 901 - 904 (1975).

022.013 **A simple quadrature method for the evaluation of the redistribution functions $R_{\mathrm{III,IV}}(x', \mathbf{n}'; x, \mathbf{n})$.**
A. Reichel, I. M. Vardavas.
Journ. Quant. Spectrosc. Radiat. Transfer, Vol. 15, 929 - 939 (1975).

022.014 **Angular distributions in the elastic scattering and rotational excitation of molecular hydrogen by atomic hydrogen.** S. Chu, A. Dalgarno.
Astrophys. Journ., Vol. 199, 637 - 641 (1975).

The effective close-coupling method of Rabitz is tested and used to calculate the angular distributions of the elastic and inelastic scattering of molecular hydrogen in collision with atomic hydrogen when rotational transitions may occur. The interaction potential adopted is the Porter–Karplus semi-empirical potential surface, which was joined smoothly to the correct long-range potentials. Explicit calculations are reported for the differential elastic and total inelastic cross sections in the energy range from 0.1 to 1.5 eV.

022.015 **Charge transfer reactions in multiply charged ion-atom collisions.** G. Steigman.
Astrophys. Journ., Vol. 199, 642 - 646 (1975).

Charge transfer reactions in collisions between highly charged ions and neutral atoms of hydrogen and/or helium may be rapid at thermal energies. If these reactions are rapid, they will suppress highly charged ions in H I regions and guarantee that the observed absorption features from such ions cannot originate in the interstellar gas. A discussion of such charge transfer reactions is presented and compared with the available experimental data. The possible implications of these reactions for observations of the interstellar medium, H II regions, and planetary nebulae are outlined.

022.016 **Photon emission from suprathermal helium ions in a partially ionized hydrogen gas.**
J. L. Mergenthaler.
Astrophys. Journ., Vol. 199, 770 - 776 (1975).

The relative intensity and profiles of He I λ5876, He II λ4686, and He II λ3203 have been calculated using a suprathermal-ion model for broad emission lines in Seyfert galaxies and quasars. With the results of this calculation it may be possible to explain the observed helium line strengths in 3C 120, NGC 4151, NGC 7469, and I4329A.

022.017 **Observation of subordinate $2p^54d-2p^53p$ transitions in neon-like ions from Ca XI to Ni XIX.**
S. O. Kastner, W. E. Behring, L. Cohen.
Astrophys. Journ., Vol. 199, 777 - 780 (1975).

The authors report the first identification of subordinate $2p^54d-2p^53p$ transitions in the neon-like ions Ca XI through Ni XIX. Such transitions between excited levels have not previously been reported for ionization stages higher than Al IV, though the resonance lines are known through Zn XXI in the neon isoelectronic sequence. Measurement of the solar flux in such transitions may aid in establishing the relative importance of the several cascade processes giving rise to the observed Fe XVII resonance lines.

022.018 **Hyperfine and Λ-doubling splitting in excited rotational levels of CH.**
D. H. Levy, J. Hinze.
Astrophys. Journ., Vol. 200, 236 - 238 (1975).

022.019 **Continuous ultraviolet absorption into double-excited states of H^-.**
V. L. Jacobs, A. K. Bhatia, A. Temkin.
Astrophys. Journ., Vol. 200, 239 - 243 (1975).

022.020 **Stark broadening of N II and N III lines.**
M. V. Popović, M. Platiša, N. Konjević.
Astron. Astrophys., Vol. 41, 463 - 465 (1975).
A pulsed low-pressure arc was the source of a plasma used to measure, on a shot-to-shot basis, profiles of 9 lines of singly ionized and 5 lines of doubly ionized nitrogen. An electron density of 4.8 and 5.5×10^{16} cm^{-3} was determined by laser interferometry while electron temperatures 23150 K and 24300 K were measured using the relative intensities of N II lines. The experimental N II Stark profile half-widths agree well with theory.

022.021 **The continuous absorption coefficient of atomic and molecular negative ions.** T. L. John.
Monthly Notices Roy. Astron. Soc., Vol. 172, 305 - 311 (1975).
The continuous absorption coefficient of the free–free transitions of 17 atomic and molecular negative ions are reported for the temperature and wavelength ranges $100 \leqslant T(K) \leqslant 15000$, $0.5 \leqslant \lambda(\mu) \leqslant 10$. The author gives an analytical expression for some molecular negative ions of astrophysical interest.

022.022 **Dielectronic satellite spectra for highly-charged helium-like ions–II. Improved calculations.**
C. P. Bhalla, A. H. Gabriel, L. P. Presnyakov.
Monthly Notices Roy. Astron. Soc., Vol. 172, 359 - 375 (1975).
Earlier calculations on the intensities of dielectronic satellite spectra for helium-like ion resonance lines (Gabriel) have now been revised. The revision is based upon new and improved calculations for the rates of autoionization of the levels responsible, and of their production rate by direct inner-shell excitation. Recent improved observations of solar active regions and flares and high temperature laboratory plasmas are compared with the calculated spectra.

022.023 **A line-leaking model for silicon monoxide millimetre wavelength emission.** M. Pelling.
Monthly Notices Roy. Astron. Soc., Vol. 172, 421 - 425 (1975).
Radiative transfer theory is applied to the determination of population ratios of the pairs of pure rotational energy levels giving rise to SiO millimetre lines at 43, 86 and 129 GHz. The result is that, within the confines of the model, the required population ratios cannot be found. However, it does emerge that there is inversion in alternate levels J, $J-2$ belonging to a given vibrational state. Most probably each level is inverted relative to the next lower, and this gives fair agreement with observation.

022.024 **On the effect of time-ordering for Lyman-α.**
R. L. Greene, J. Cooper.
Journ. Quant. Spectrosc. Radiat. Transfer, Vol. 15, 991 - 993 (1975).
Numerical calculations of the wings of the hydrogen Lyman-α line due to Stark broadening by electrons have been performed including time ordering effects. It is concluded that two recent calculations, which apparently disagree, are numerically correct within their respective approximations and their differences are explained.

022.025 **Spectroscopic observation of a quasi-steady-state plasma in contact with a neutral gas.**
M. Otsuka, R. Ikee, K. Ishii.
Journ. Quant. Spectrosc. Radiat. Transfer, Vol. 15, 995 - 1005 (1975).
Very intense spectral lines of He^+ were observed when a He plasma was brought into contact with a He gas. The population densities of the excited levels of He^+ ions were found to be much larger than the values expected in the steady state for the given electron temperature and density. The contact mechanism and a method of determining the electron temperature and the electron and ion densities are discussed for the quasi-steady state.

022.026 **Observations on lifetimes and excitation processes for the upper states in atomic helium**
R. T. Thompson, R. G. Fowler.
Journ. Quant. Spectrosc. Radiat. Transfer, Vol. 15, 1017 - 1024 (1975).
Lifetime observations have been made on 49 energy levels of helium. Good overall agreement with theory is obtained, with significant exceptions, the most striking of which are the 3^1D and 3^3D states, where 50% discrepancies with theory and some of the older measurements are found. The new measurements are, however, in harmony with recent results using the Hanle effect. The existence of the upper-state transfer process is clearly delineated by the results for $n > 6$.

022.027 **A unified theory of Stark broadening for hydrogenic ions–I. A general theory (including time ordering).**
R. L. Greene, J. Cooper, E. W. Smith.
Journ. Quant. Spectrosc. Radiat. Transfer, Vol. 15, 1025 - 1036 (1975).
The formalism of the Unified Classical Path Theory of Stark broadening is extended (by use of a technique developed by Lisitsa and Sholin) to include the effects of time-ordering in the calculation of the required time-development operator. In addition, hyperbolic classical paths are considered so that the theory is valid for hydrogenic ions (as well as for the straight line paths of neutral hydrogen).

022.028 **A unified theory of Stark broadening for hydrogenic ions–II. Line wings.** R. L. Greene, J. Cooper.
Journ. Quant. Spectrosc. Radiat. Transfer, Vol. 15, 1037 - 1044 (1975).
The consequences of the attractive Coulomb force between positively charged hydrogenic radiators and electrons on line broadening in the quasistatic wings are examined.

022.029 **A unified theory of Stark broadening for hydrogenic ions–III. Results for He II Lyman-α.**
R. L. Greene, J. Cooper.
Journ. Quant. Spectrosc. Radiat. Transfer, Vol. 15, 1045 - 1053 (1975).
Calculations have been performed for He II Lyman-α using a previously developed theory (see 022.027, 022.028) in which the effect of time-ordering in the required time development operator is retained. Representative results are given for an electron density of 10^{17} cm^{-3} and a temperature of 40,000°K. Compared with non-time ordered results, these profiles are about 17% lower at line center and up to 16% higher in the wings. In addition, since the perturbing electron is speeded up by the Coulomb attraction, in comparison with neutral hydrogen the static results are not obtained until much further into the line wings.

022.030 **On some problems encountered in the theoretical calculation of collisional linewidths.**

A. Levy, C. Boulet.
Journ. Quant. Spectrosc. Radiat. Transfer, Vol. 15, 1067 - 1069 (1975).

This paper relates to the comments recently made by Varanasi and Sarangi about the role of higher multipolar forces in linewidth calculations. While the present authors agree with the broad outline of their analysis, some differences remain concerning the treatment of HCl and N_2O linewidths. The important idea must be retained that the multipole of highest order introduced in the potential expression should be considered as an overall, empirical parameter.

022.031 **Asymptotic formulae for the equation of state of an electron-positron gas.** D. K. Nadezhin.
Nauchn. Informatsii, vyp. (No.) 32, p. 3 - 72 (1974).
In Russian.

The asymptotic behavior of an ideal electron-positron gas is considered in full detail. Special attention is paid to the accuracy of various asymptotic expansions. Except several original results the author has cited well-known data to have all necessary information at hand. The results are represented in a form suitable for using computers.

022.032 **Absolute *f*-values for resonance lines of neutral titanium.** G. D. Bell, L. B. Kalman, E. F. Tubbs.
Astrophys. Journ., Vol. 200, 520 - 523 (1975).

Absolute *f*-values for 13 resonance lines of Ti I have been experimentally measured by the atomic-beam method. The effect of an improved absolute scale for Ti I *gf*-values on the abundance of titanium in the solar photosphere is discussed.

022.033 **The quadratic Zeeman effect in the Ca II H and K lines.** S. B. Kemic.
Astrophys. Space Sci., Vol. 36, 459 - 466 (1975).

The effects of a large magnetic field on the H and K lines of Ca II are investigated incorporating both the linear and quadratic Zeeman effects for fields in the range $10^6 - 10^8$ G. For these fields the linear term is the dominant one, resulting in a Paschen-Back effect with a smaller quadratic Zeeman shift superimposed on all components. Due to the partial cancellation of the quadratic effect between the upper and lower states of the lines, a much smaller quadratic Zeeman shift occurs than in the lines of other atoms previously studied at similar fields.

022.034 **Relative intensities of solar emission lines of ions in the sodium isoelectronic sequence.**
D. R. Flower, H. Nussbaumer.
Astron. Astrophys., Vol. 42, 265 - 268 (1975).

Electron collision cross-sections and radiative transition probabilities are presented for Si^{+3}, S^{+5}, Ca^{+9} and Fe^{+15} of the sodium isoelectronic sequence. An attempt is made to interpret the observed relative intensities of Si IV and Si VI lines in terms of the electron temperatures of the emitting regions.

022.035 **The spectrum of doubly ionized titanium, Ti III.**
B. Edlén, J. W. Swensson.
Physica Scripta, Vol. 12, 21 - 32 (1975).

022.036 **Term analysis of Fe V.** J. O. Ekberg.
Physica Scripta, Vol. 12, 42 - 57 (1975).

022.037 **Lifetimes of excited levels in Ca I–Ca III.**
B. Emmoth, M. Braun, J. Bromander, I. Martinson.
Physica Scripta, Vol. 12, 75 - 79 (1975).

022.038 **Extensions to the spectrum of singly ionized barium (Ba II).** R. A. Roig, G. Tondello.
Journ. Optical Soc. America, Vol. 65, 829 - 830 (1975).

022.039 **Ultraviolet optical constants of water and ammonia ices.** E. V. Browell, R. C. Anderson.
Journ. Optical Soc. America, Vol. 65, 919 - 926 (1975).

022.040 **Doppler and collision-broadening effects in the profile of spectral lines.**
B. W. Fowler, C. C. Sung.
Journ. Optical Soc. America, Vol. 65, 949 - 951 (1975).

The effect of velocity-dependent collision broadening on the shape of spectral lines is investigated. Computations show that the modified line shape has a narrower half-width than the conventional Voigt profile.

022.041 **Unified treatment of perturbed series, continuous spectra and collisions.** U. Fano.
Journ. Optical Soc. America, Vol. 65, 979 - 987 (1975).

022.042 **Theoretical study of the properties of HCO^+ at equilibrium.** P. J. Bruna.
Astrophys. Letters, Vol. 16, 107 - 113 (1975).

022.043 **Cross sections for photoionization of vibrationally excited molecular hydrogen.**
A. L. Ford, K. K. Docken, A. Dalgarno.
Astrophys. Journ., Vol. 200, 788 - 789 (1975).

Cross sections are presented for the photoionization of vibrationally excited molecular hydrogen by radiation with wavelengths longer than 912 Å.

022.044 **Cooling of the 6-centimeter and 2-centimeter doublets of interstellar H_2CO by collision: an accurate quantum-mechanical calculation.**
B. J. Garrison, W. A. Lester, Jr., W. H. Miller, S. Green.
Astrophys. Journ., (*Letters*), Vol. 200, L175 - L177 (1975).

Cooling of the 6-cm and 2-cm levels of interstellar H_2CO by a collisional pump is verified by a series of accurate quantum-mechanical calculations. The $j = 3$ ortho doublet is found to be an integral part of the collisional pumping scheme.

022.045 **Relative Raman cross sections for the *S*(0) through *S*(4) rotational transitions in hydrogen.**
R. C. Harney, J. E. Randolph, F. P. Milanovich.
Astrophys. Journ.,(*Letters*), Vol. 200, L179 - L181 (1975).

The relative Raman scattering cross sections for the *S*(0) through *S*(4) rotational transition in molecular hydrogen have been experimentally determined at two wavelengths, 4880 and 5145 Å. The experimental results are in mild disagreement with theoretical predictions.

022.046 **New identifications and lifetime measurements in Ne VIII.**
L. Barrette, D. J. G. Irwin, R. Drouin.
Physica Scripta, Vol. 12, 113 - 115 (1975).

022.047 **Bremsstrahlung and pair production in the field of free electrons.** E. Haug.
Zeitschr. Naturforschung, Vol. 30a, 1099 - 1113 (1975).

022.048 **Wavelengths of the O VI resonance lines.**
A. N. Ryabtsev.
Pis'ma v Astron. Zhurn., Vol. 1, No. 9, p. 40 - 41 (1975). In Russian.

New measurements of laboratory wavelengths of the O VI resonance lines are reported. They permit to reduce considerably the discrepancy between wavelengths measured recently for stars by the Copernicus satellite and existing laboratory data.

022.049 **Data banks on diatomic molecular absolute intensity constants.** R. A. Hefferlin.
Bull. American Astron. Soc., Vol. 7, 400 - 401 (1975). – Abstr. AAS.

022.050 **The constancy of Planck's constant.**
W. A. Baum, R. Florentin-Nielsen.
Bull. American Astron. Soc., Vol. 7, 412 (1975). – Abstr. AAS.

022.051 **Atomic energy levels data center activities.**
L. Hagan.
Bull. American Astron. Soc., Vol. 7, 470 (1975). – Abstr. AAS.

022.052 **Approximate peak frequency of gyrosynchrotron emissivity.** G. L. Tarnstrom.
Bull. American Astron. Soc., Vol. 7, 475 (1975). – Abstr. AAS.

022.053 **Theoretical cooling coefficients and soft X-ray spectra.**
J. C. Raymond, B. W. Smith, D. P. Cox.
Bull. American Astron. Soc., Vol. 7, 498 (1975). – Abstr. AAS.

022.054 **Cross sections for collisional excitation of interstellar formaldehyde.** N. J. Evans II.
Astrophys. Journ., Vol. 201, 112 - 117 (1975).

The cross sections for excitation of H_2CO in collisions with H_2 have been calculated in a quantum mechanical, close-coupling form. The results lend credence to the collisional model for explaining the observations of anomalous absorption in H_2CO at 6-cm and 2-cm wavelengths.

022.055 **Multiple electron ejection by X-ray photoionization and the abundance of interstellar ions.**
W. D. Watson, A. B. Kunz.
Astrophys. Journ., Vol. 201, 165 - 167 (1975).

Ejection of multiple electrons due to "shake-off" and the double Auger processes is considered in photoionization by X-rays. Calculations using the sudden approximation are presented to support the authors' estimates for multiple ejection in photoionization of nitrogen and oxygen. Multiple ejection is found to be much more probable for initially neutral atoms than for positive ions. Abundances for the various stages of ionization of N and O in the intercloud gas due to the observed X-ray flux are recalculated. The abundances of N V and O VI are increased by an order of magnitude for an electron density 0.02 cm^{-3}.

022.056 **Is there a maser in the silicon monoxide ground state?** D. Buhl, L. E. Snyder, F. J. Lovas, D. R. Johnson.
Astrophys. Journ., (*Letters*), Vol. 201, L29 - L31 (1975).

Observations of the ground state ($v = 0, J = 2–1$) transition of silicon monoxide (SiO) show a rather smooth, low-intensity profile stretching over a large velocity range. The presence of groundstate emission at velocities which are not present in the $v = 1$ or $v = 2$ lines is interpreted as thermal emission in the ground state in regions where there is insufficient excitation to produce the vibrationally excited masers. Only VY CMa shows a profile which could be interpreted as ground-state maser emission.

022.057 **New calculations of atomic data concerning E.U.V. lines of O V.** M. Malinovsky.
Astron. Astrophys., Vol. 43, 101 - 110 (1975).

New values of transition probabilities and cross sections are given for O V. Values of electron density N_e, and electron temperature T_e, are then deduced from appropriate U.V. observed line ratios. The sensitivity of these physical parameters (N_e, T_e) to the remaining uncertainties in atomic data is discussed.

022.058 **Tables related to the equation of state of an electron-positron gas.** D. K. Nadezhin.
Nauchn. Informatsii, vyp. (No.) 33, p. 117 - 142 (1974). In Russian.

Tables related to various asymptotic expansions of the equation of state of an electron-positron gas are given. Five-decimals tables of Fermi-Dirac functions of integer and semi-integer indices are included.

022.059 **Electronic energies including exchange and spin orbit interaction of the interstellar molecules CH^+, CH and OH at large interatomic separations.**
J. F. Bazet, C. Harel, R. McCarroll, A. Riera.
Astron. Astrophys., Vol. 43, 223 - 228 (1975).

A preliminary discussion is given of the problems associated with the use of exchange perturbation expansions for the calculation of molecular potential energy curves at large internuclear distances. Application is made to the interstellar molecules CH^+, CH and OH in view of the treatment of fine structure transitions of the species C^+, C and O arising on collision with atomic hydrogen.

022.060 **Transitions between the fine structure components of C^+ in collision with atomic hydrogen.**
J. F. Bazet, C. Harel, R. McCarroll, A. Riera.
Astron. Astrophys., Vol. 43, 229 - 232 (1975).

The cross section for excitation of the ${}^2P_{1/2}$–${}^2P_{3/2}$ transition in thermal $C^+({}^2P)$–$H({}^2S)$ collisions is calculated in the framework of the semi-classical approximation using the potential energy curves of the ${}^1\Sigma$, ${}^1\Pi$, ${}^3\Pi$, ${}^3\Sigma$ states of CH^+, determined by an exchange perturbation method. The cooling function due to C^+–H collisions in the interstellar clouds is computed as a function of the kinetic temperature of the medium.

022.061 **Proton collisional excitation in the ground configurations of Fe^{+12} and Fe^{+13}. III. Transitions between magnetic sublevels.** D. A. Landman.
Astron. Astrophys., Vol. 43, 285 - 290 (1975).

Values of proton collisional cross-sections at coronal energies have been computed for all transitions between magnetic sublevels of the Fe^{+12} and Fe^{+13} ground configurations. Semi-classical Coulomb excitation theory was used with the resulting truncated Schroedinger equations being directly integrated. The corresponding rate constants for coronal temperatures are presented and comparison is made with previous results.

022.062 **Preparation of fine particles of astrophysical interest by gas evaporation techniques.** F. Kamijo, Y. Nakada, T. Iguchi, M.-K. Fujimoto, M. Takada.
Icarus, Vol. 26, 102 - 110 (1975).

Fine particles of various chemical substances–carbon, iron oxide and silica–which are expected to exist in interstellar space are prepared in argon gas. The size, shape, and crystal structure of the powder particles are studied by electron microscopy and electron diffraction, and may have some bearing on models of comets and of the solar nebula.

022.063 **Rate of the reaction $N_2H^+ + CO \rightarrow HCO^+ + N_2$ and its significance for the interstellar chemistry of N_2H^+.**
E. Herbst, D. K. Bohme, J. D. Payzant, H. I. Schiff.
Astrophys. Journ., Vol. 201, 603 - 606 (1975).

The rate constant for the reaction $N_2H^+ + CO \rightarrow HCO^+ + N_2$ has been determined to be $(8.79 \pm 0.13) \times 10^{-10}$ $cm^3 s^{-1}$ at 297 ± 2 K with an estimated accuracy of ±25 percent. The implications of this rate constant for the interstellar chemistry of N_2H^+ are discussed. Estimates are made for the frequencies of the first rotational transitions in ${}^{15}N^{14}NH^+$, ${}^{14}N^{15}NH^+$, and N_2D^+.

022.064 **Lifetime measurements in the Co I and Fe I spectra using tunable dye laser excitation.**
H. Figger, J. Heldt, K. Siomos, H. Walther.
Astron. Astrophys., Vol. 43, 389 - 394 (1975).

Radiative lifetimes of single fine structure levels of the

multiplets y^4G^0, y^4F^0, y^2G^0 of the Co I configuration $3d^8(a^3F)4p$ were measured. Furthermore, lifetime measurements of Fe I levels belonging to the $3d^7(a^4F)4py^5D^0$ and the $3d^64s(a^4D)4pz^3P^0$ multiplets were performed. The levels were excited by a nitrogen laser pumped dye laser, whose output was frequency doubled; the lifetimes were determined directly from the exponential decay of the fluorescence. The results are used to convert the *gf* values given in the literature to an absolute scale.

022.065 **Thermonuclear reaction rates, II.**
W. A. Fowler, G. R. Caughlan, B. A. Zimmerman.
Annual Rev. Astron. Astrophys., Vol. 13, (see 003.005), 69 - 112 (1975).

022.066 **Measurements of the light flash produced by high velocity particle impact.** G. Eichhorn.
Planet. Space Sci., Vol. 23, 1519 - 1525 (1975).

022.067 **Lifetime measurements for excited levels in Cr II.**
B. Engman, A. Gaupp, L. J. Curtis, I. Martinson.
Phys. Scripta, Vol. 12, 220 - 222 (1975).

Lifetimes for ten odd terms in Cr II which decay by transitions in the region 2000–4000 Å have been measured using the beam-foil technique. The results are compared with previous experimental data and available theoretical calculations.

022.068 **Beam-foil studies of phosphorus in the vacuum ultraviolet.** A. E. Livingston, J. A. Kernahan, D. J. G. Irwin, E. H. Pinnington.
Phys. Scripta, Vol. 12, 223 - 229 (1975).

022.069 **The emission spectrum of SiO: the A–X system in the 2100–2300 Å region.**
J. L. Deutsch, E. W. Deutsch, N. Elander, A. Lagerqvist.
Phys. Scripta, Vol. 12, 248 - 252 (1975).

022.070 **Half-range moment method for solution of the transport equation in a spherically symmetric geometry.** M. N. Özisik, J. Menning, W. Hälg.
Journ. Quant. Spectrosc. Radiat. Transfer, Vol. 15, 1101 - 1106 (1975).

A half-range moment method is presented for solving, in various orders of approximation, a multi-group transport equation subject to generalized boundary conditions in a spherically symmetric geometry. The results for the plane-parallel geometry are obtainable from the present analysis as a special case. The equations and the boundary conditions considered are sufficiently general to characterize a variety of problems in radiative transfer, neutron transport and phonon transport if various coefficients appearing in the equations are properly specified.

022.071 **The thermalization length by the Green function method.** Pham Duc Wuong.
Journ. Quant. Spectrosc. Radiat. Transfer, Vol. 15, 1107 - 1116 (1975).

022.072 **Comment on resonance broadening.**
R. J. Exton.
Journ. Quant. Spectrosc. Radiat. Transfer, Vol. 15, 1141 - 1144 (1975).

The conflicting results describing the relationship between line width and absorber density in resonance broadening are considered. A reexamination of some of these experiments points to a common difficulty – large optical depth. Formulas are given which approximate the widths and dependencies observed under optically thick conditions. It is concluded that a resonance breadth which is linear with density is consistent with all of the observations and that there is no solid basis for a square root dependence at low density.

022.073 **Oscillator strength for the $D\ ^2\Sigma^-$–$X\ ^2\Pi$ transition in OH.** S. Ray, H. P. Kelly.
Astrophys. Journ., (*Letters*), Vol. 202, L57 - L58 (1975).

The oscillator strength $f(0, 0)$ for the $D\ ^2\Sigma^-$–$X\ ^2\Pi$ transition in OH has been calculated by use of the multiconfiguration Hartree-Fock program of Das and Wahl. The authors' length and velocity results are 0.00356 and 0.00369, respectively. These oscillator strengths bring the OH abundance results as determined by the Copernicus satellite into better agreement with the lowest limit measured by Knapp and Kerr for the general interstellar medium.

022.074 **The lower terms in Ni VII and possible identifications in RR Telescopii.** H. F. Henrichs.
Astron. Astrophys., Vol. 44, 41 - 44 (1975).

New observations of Ni VII transitions in the region 205–231 Å have led to a revision of the lower terms in the configurations $3d^4$ and $3d^34p$. Wavelengths of forbidden lines of Ni VII are predicted. Calculation of magnetic dipole transition probabilities in the ground configuration supports possible identifications in the nova-like variable RR Telescopii.

022.075 **Lifetime measurements of some levels of Mn I and Cr I by use of dye laser excitation.** J. Marek.
Astron. Astrophys., Vol. 44, 69 - 71 (1975).

022.076 **Qualitative features of synchrotron radiation in an inhomogeneous magnetic field.**
G. Cavallo, A. Ventura.
Nature, Vol. 258, 309 - 310 (1975). – Letter.

022.077 **Experimental observation of the absorption spectrum of Fe I in the region 1600–900 Å.**
G. Tondello.
Mem. Soc. Astron. Italiana, Vol. 46, 113 - 121 (1975).

022.078 **An analysis of hydrogen spectrum near the Balmer limit.** V. Ruždjak, V. Vujnović.
Publ. Obs. Astron. Beograd, No. 20, (see 012.024), p. 31 - 34 (1975). In Serbo-Croatian.

022.079 **Lifetimes of excited levels in MgIII, CaIII, KIV–V, CaV, and GeI.**
T. Andersen, A. P. Petkov, G. Sørensen.
Physica Scripta, Vol. 12, 283 - 286 (1975).

Radiative lifetimes of excited levels in MgIII, CaIII, KIV, KV, CaV, and GeI have been measured by means of the beam-foil technique in order to test the validity of recent theoretical methods to calculate lifetimes for heavier atoms and ions. Good agreement with theoretical lifetimes is obtained.

022.080 **An experiment of absorption spectroscopy in the XUV of singly ionized species.**
A. M. Cantù, G. Tondello.
Oss. Mem. Oss. Astrofis. Arcetri, Fasc. 104, (see 012.025), p. 202 - 203 (1975).

022.081 **The ionisation balance of C^0 to C^{+4}.**
H. Nussbaumer, P. J. Storey.
Astron. Astrophys., Vol. 44, 321 - 327 (1975).

The ionisation balance for the ions C^0 – C^{+4} has been calculated for $10^8 \leqq N_e\,[\mathrm{cm}^{-3}] \leqq 10^{12}$ and $2 \times 10^4\,\mathrm{K} \leqq T_e \leqq 2.5 \times 10^5\,\mathrm{K}$. The presence of metastable terms is included in the calculation of the collisional dielectronic recombination and ionisation coefficients. The influence of the observed solar radiation field on the ionisation balance is investigated. Changes in that field do strongly influence the results.

022.082 **Transition probabilities for F I lines from wall-stabilized arc measurements.**
G. Baruschka, E. Schulz-Gulde.
Astron. Astrophys., Vol. 44, 335 - 342 (1975).

Relative transition probabilities were obtained for 36 F I lines in the spectral range from 6047 to 7800 Å. Except for a few transitions, the results for the lines comprising the $3s-3p$ transition array are in agreement with the relative transition probabilities obtained from the L, S coupling model. However, the experimental absolute scale for F I line transition probabilities (whose accuracy is ± 25%) is lower by a factor of two than that provided by the Coulomb approximation method.

022.083 **The silicon monoxide radical in M stars – the fundamental rotation-vibration band of SiO.**
P. D. Singh.
Astron. Astrophys., Vol. 44, 411 - 415 (1975).

Line positions of the P and R branches for the 8-μ fundamental rotation-vibration band ($v = 1 \leftarrow 0$ and $2 \leftarrow 1$) in the $X^1\Sigma^+$ electronic ground state of the SiO molecule are derived by a differentiating process invoking the known structures of bands of the $E^1\Sigma^+ - X^1\Sigma^+$ system. A reasonable agreement is found between the computed and observed rotational transitions.

022.084 **Electron capture by C^{2+} from interstellar atomic hydrogen.** R. McCarroll, P. Valiron.
Astron. Astrophys., Vol. 44, 465 - 467 (1975).

The potential energies of the low lying Σ and Π states of the CH^{2+} molecular ion are calculated by a model potential method. The possible charge exchange reactions $C^{2+}+H \rightarrow C^+ + H^+$ are discussed in terms of the calculated energy diagram. It is shown that charge exchange is unlikely to be important in the conversion of C^{2+} to C^+ in the interstellar medium.

022.085 **Investigation of recombination spectra of light ions.**
A. Sapar, T. Feklistova.
Publ. Tartu Astrofiz. Obs., Vol. 43, 262 - 289 (1975). In Russian.

Using the recombination coefficients calculated based on the quantum defect method and the system of calculated transition probabilities in the single configuration the recombination spectra for 13 lower states of NIII and 19 lower states of CIII are calculated for the temperature range 5000–50000 K. Serious discrepancies with observations insist to take into account the interconfigurational transitions and two-electronic recombination.

022.086 **Absorption and emission by hydrogen at low electron temperatures.** M. Salem.
Monthly Notices Roy. Astron. Soc., Vol. 173, 513 - 521 (1975).

The absorption and emission coefficients of ionized hydrogen are computed by exact evaluation of a known classical expression. Several analytical approximation formulae suitable for different physical conditions are presented. A table of a function of one variable, from which the absorption and emission coefficients can be computed, is provided in a form suitable for interpolation.

022.087 **Production of high-energy deuterons in the ^3He + ^3He reaction and the solar neutrino problem.**
R. J. Slobodrian, R. Pigeon, M. Irshad.
Phys. Rev. Letters, Vol. 35, 19 - 22 (1975).

A spectrum of deuterons consistent with the reaction $^3He + {}^3He \rightarrow {}^2H + {}^4He + e^+ + \nu$ has been detected with a system sensitive to cross sections of 300 pb sr^{-1} MeV^{-1}. The measured cross section is too high for a weak-interaction process: 3.4 nb MeV^{-1} sr^{-1}. It might be due to an interaction of intermediate strength and it could explain the low flux of high-energy solar neutrinos.

022.088 **Improved upper limit on the photon rest mass.**
A. Barnes, J. D. Scargle.
Phys. Rev. Letters, Vol. 35, 1117 - 1120 (1975).

By standard arguments, observed hydromagnetic waves in the Crab nebula imply an upper limit to the photon rest mass of $\mu \lesssim 10^{-16}-10^{-15}$ cm^{-1}, $m_{ph} \lesssim 3 \times 10^{-54} - 3 \times 10^{-53}$ g (uncertainty < factor 100), smaller than the previous best limits by a factor $10^4 - 10^5$. However, the standard wave arguments used here and elsewhere are reexamined and found not to be rigorous, although the limits given are probably valid in order of magnitude.

022.089 **Laboratory microwave spectrum of HCO^+.**
R. C. Woods, T. A. Dixon, R. J. Saykally, P. G. Szanto.
Phys. Rev. Letters, Vol. 35, 1269 - 1272 (1975).

The $J = 0 \rightarrow 1$ rotational transition of the HCO^+ ion has been observed with the same apparatus used earlier for detection of CO^+. A discharge in various mixtures of hydrogen and carbon monoxide cooled to near liquid-nitrogen temperature was employed in a slow-flow system. The frequency which has been obtained is 89188.545 ± 0.020 MHz, but because of the ion drift velocity in the dc discharge this exceeds, by an unknown but small amount, the true rest frequency. This observation confirms the long-standing contention that the radio astronomical X-ogen line is in fact due to HCO^+.

022.090 **Spatial distribution of the apparent spectral emissivity of a freezing-point blackbody with Lambertian walls.** W. B. Fussell.
Journ. Optical Soc. America, Vol. 65, 1179 (1975). – Conference abstract.

022.091 **The possible time variations of the fundamental "constants".** M. I. Kalinin, V. N. Mel'nikov.
Probl. teorii gravitatsii i ehlementarn. chastits. Vyp. (No.) 6. Moskva, Atomizdat, 1975, p. 70 - 82. In Russian. – Abstr. in Referativ. Zhurn. 51. Astron., 11.51.817 (1975).

022.092 **Angular properties of blackbody simulators.**
G. Marette.
Applied Optics, Vol. 14, 2665 - 2667 (1975).

Aperture-cavity wall radiative exchange is described for various extended blackbody simulator cavities by means of a simple proper angular parameter in terms of the angle of observation.

022.093 **On the possibility of determination of the value of the magnetic field near pulsars by means of the detection of the "beats" frequency in atomic transitions.**
A. A. Ruzmajkin.
Astron. Zhurn. Akad. Nauk SSSR, Vol. 52, 1173 - 1177 (1975). In Russian. English translation in Soviet Astron., Vol. 19, No. 6.

For the determination of the magnetic field under astrophysical conditions the use of the "beats" effect in atomic transitions is suggested. The frequency of beats under the Zeeman split is proportional to the magnetic field, its Doppler broadening is negligible. Concrete estimates are made for the case of the forbidden lines in the Crab nebula excited by the optical pulse of the pulsar NP 0531.

022.094 **The vibrational spectrum of C_3 in the 5-micron region.** R. R. Treffers, D. P. Gilra.
Astrophys. Journ., Vol. 202, 839 - 843 (1975).

A band near 5 μ has been found in the emission from a carbon tube furnace at 3100 K. It is a broad unresolved feature centered near 1850 cm^{-1} and is about 370 cm^{-1} wide (FWHM). It is shown to be the ν_3 (asymmetric stretch) rotation-vibration band of C_3. A second band at 3200 cm^{-1} is present but weaker; it is possibly consistent with the combination band $\nu_1 + \nu_3$. These bands may be opacity sources in cool carbon stars.

022.095 **Computed $n = 2$ level populations in helium-like ions from C V to Ni XXVII.**
R. Mewe, J. Schrijver.
Astrophys. Space Sci., Vol. 38, 345 - 352 (1975). – Paper presented at the I.A.U. Colloquium No. 27 on 'UV and X-ray spectroscopy of astrophysical and laboratory plasmas', Cambridge, Mass., U.S.A., 9–11 September 1974.

The population densities of all levels with principal quantum number $n = 2$ in a number of helium-like ions with nuclear charge number Z, in the range 6 to 28 have been evaluated as a function of various parameters, i.e., electron temperature, T_e, electron density, N_e, radiation temperature, T_r, dilution factor, W, and of the state of ionization. The spectral line fluxes from all possible radiative transitions from these levels have been calculated for an optically thin plasma. The computations were performed both for stationary and time-varying plasmas. In the latter case strong departures from a stationary ionization equilibrium can significantly alter the line fluxes. A few examples of the results are shown and discussed.

022.096 **Ultraviolet solar identifications based on extended absorption series observed in the laboratory spectrum of Si I.** C. E. Moore, R. Tousey, G. D. Sandlin, C. M. Brown, M. L. Ginter, S. G. Tilford.
Astrophys. Space Sci., Vol. 38, 359 - 364 (1975). – Paper presented at the I.A.U. Colloquium No. 27 on 'UV and X-ray spectroscopy of astrophysical and laboratory plasmas', Cambridge, Mass., U.S.A., 9–11 September 1974.

The absorption spectrum of Si I in the wavelength region 1500–1900 Å has been photographed at high resolution. This laboratory work has provided the basis for extending the identification of silicon lines in the solar spectrum. Nearly all lines found in the laboratory spectrum are also found in rocket spectrograms of the solar chromosphere. More than 300 lines have been attributed to Si I. The excellent correlation between laboratory and solar Si I lines will be illustrated.

022.097 **Density dependence of the line emission of boron-like coronal ions.** G. Elwert, P. K. Raju.
Astrophys. Space Sci., Vol. 38, 369 - 375 (1975). – Paper presented at the I.A.U. Colloquium No. 27 on 'UV and X-ray spectroscopy of astrophysical and laboratory plasmas', Cambridge, Mass., U.S.A., 9–11 September 1974.

Assuming steady state conditions, the authors calculated the occupation of 20 levels of the ion Si X as a function of the electron density and temperature taking into account collisional processes and spontaneous radiative ones. Between the two levels of the ground state also the photoabsorption of the corresponding infrared radiation has been considered. The emission of the most important UV-, X- and IR-lines is determined as a function of the electron density and temperature. Intensity ratios are plotted versus electron density. The line emission as a function of the height of the radiating layers is studied for UV- and X-ray lines. The resulting fluxes in earth distance are compared with observations.

022.098 **Plasma polarization shift: an improved quantum-mechanical treatment.** S. Volonté.
Astrophys. Space Sci., Vol. 38, 377 - 380 (1975). – Paper presented at the I.A.U. Colloquium No. 27 on 'UV and X-ray spectroscopy of astrophysical and laboratory plasmas', Cambridge, Mass., U.S.A., 9–11 September 1974.

Using an improved quantum mechanical treatment of the plasma polarization shift theory, calculations have been made for plasma conditions corresponding to available measurements of the shift of He II resonance lines. Agreement with experimental results at low density is found but large discrepancies still exist at high density. Comparison with other theories leads to agreement with a corrected version of the semi-classical treatment of the effect although discrepancies of one order of magnitude are found with the only quantum treatment presently available.

022.099 **The cascading of cosmic-ray nuclei in various media.** P. S. Freier, C. J. Waddington.
Astrophys. Space Sci., Vol. 38, 419 - 436 (1975).

More than a thousand interactions of primary heavy nuclei of the cosmic radiation with charge $\geqslant 10$ and energy >1 GeV nucleon^{-1} in nuclear emulsion have been studied with emphasis on how the primary nucleus fragments. The fragmentation of $Z>20$ nuclei does depend on the target nucleus to some extent. The fragmentation of $^{16}_{8}O$ at 2.1 GeV nucleon^{-1} measured at the Bevalac shows a similar dependence on target nucleus. By using data from these new interactions combined with published data the authors have simulated on a computer nuclear cascades in both emulsion and air. Results on these cascades are given for both primary silicon and primary iron nuclei. These results are used to discuss the fluctuations expected in extensive air showers produced by heavy primary nuclei.

022.100 **On the extreme ultraviolet solar emission of B-like ions: O IV.** D. R. Flower, H. Nussbaumer.
Astron. Astrophys., Vol. 45, 145 - 150 (1975).

Radiative transition probabilities and electron collision cross-sections have been calculated in intermediate coupling for transitions within the $2s^2\,2p$, $2s2p^2$ and $2p^3$ configurations of O^{+3}, including the intercombination transitions $2s^2\,2p\ ^2P^o_J - 2s2p^2\ ^4P_{J'}$. The relative intensities of the intercombination lines are found to vary with electron density in the range $10^9 \lesssim N_e \lesssim 10^{12}$ cm^{-3}. The results of these calculations, in conjunction with a model of the emitting region, enable the solar EUV observations of O IV to be interpreted.

022.101 **On red shifts and asymmetries of hydrogen spectral lines.** B. Grabowski, J. Halenka.
Astron. Astrophys., Vol. 45, 159 - 166 (1975).

The internal electric field in hydrogen atoms, perturbed by intermolecular interactions in a plasma, is described by the Debye model. Using this model, the authors have determined the energy structure of low-lying levels. The structure obtained is deformed – with respect to the unperturbed structure – in such a way that the spectral lines are shifted towards the red. In so far as the order of magnitude and the direction are concerned, the predicted shifts agree with the recent detailed observations of high-density laboratory plasma. It would appear that the observed asymmetry of hydrogen lines can be also explained with this model.

022.102 **Lifetimes of some excited states in the rare earths: La II, Ce II, Pr II, Nd II, Sm II, Yb I, Yb II, and Lu II.**
T. Andersen, O. Poulsen, P. S. Ramanujam, A. P. Petkov.
Solar Physics, Vol. 44, 257 - 267 (1975).

Radiative lifetimes for excited states in La II, Ce II, Pr II, Nd II, Sm II, Yb I, Yb II, and Lu II have been determined by means of the beam-foil technique or the zero-field level-crossing method. The large discrepancies between the abundance of La, Ce, Pr, Nd, and Sm in the solar photosphere and in meteorites are eliminated or greatly reduced if the abundance determinations of the solar photosphere are based upon the *gf* values of Corliss and Bozman, corrected for by the present lifetimes.

022.103 **Computation of oscillator strengths by a semi-empirical method for some elements of the iron-group and their solar photospheric abundance. III: Results for Mn I.**
E. Biémont.
Solar Physics, Vol. 44, 269 - 273 (1975).

Oscillator strengths are calculated by the STF method for selected transitions of Mn I. The results, applied to the analysis of solar spectra, support the 'high' solar photospheric abundance of manganese recently obtained on the basis of experimental *gf*-values.

022.104 **Laboratory intensity measurements of absorption bands in the long pathlength spectrum of methane.**
B. L. Lutz, T. Owen, R. D. Cess.
Bull. American Astron. Soc., Vol. 7, 508 (1975). – Abstr. AAS.

022.105 **Photoacoustic spectroscopy of the 6190 Å band of CH_4 and the 6450 Å band of NH_3.**
G. Stella, J. Gelfand, W. H. Smith.
Bull. American Astron. Soc., Vol. 7, 522 (1975). – Abstr. AAS.

022.106 **Quantum defects for two-electron ions by Z-expansion perturbation theory.** P. Blanchard.
Bull. American Astron. Soc., Vol. 7, 552 (1975). – Abstr. AAS.

022.107 **Uncertainty in astronomy. Atomic transition probabilities for strong and weak lines, radiative lifetimes; effective temperatures of cooler stars.**
D. E. Blackwell.
Quarterly Journ. Roy. Astron. Soc., Vol. 16, 361 - 377 (1975).
The presidential address delivered at the anniversary meeting of 1975 February 14.

022.108 **Stark broadening of O II and O III lines.**
M. Platiša, M. V. Popović, N. Konjević.
Astron. Astrophys., Vol. 45, 325 - 327 (1975).
The half-widths of twenty one O II and six O III lines have been measured in a pulsed arc operated in an oxygen-nitrogen mixture. The electron density, 5.2×10^{16} cm^{-3}, was determined by laser interferometry at a single wavelength, while the electron temperature, 25900°K, was measured using the relative intensities of O II lines. Some experimental O II Stark profile half-widths are compared with theory.

022.109 **Experimental Stark constants for some Fe I lines.**
J. Moity, P. E. Pieri, J. Richou.
Astron. Astrophys., Vol. 45, 417 - 423 (1975). In French.
Fe I time-resolved spectra were obtained by use of a classical shock-tube, with a copper diaphragm.

022.110 **Profiles of radio recombination lines.** L. A. Higgs.
Journ. Roy. Astron. Soc. Canada, Vol. 69, 289 - 298 (1975).
Graphical relations are presented which allow the determination of Voigt profile parameters of an observed spectral line using its deviations from a best fitting Gaussian profile (in the case where the damping width is $\lesssim$ the Gaussian width of the line). Similar relations are also given for an 'exponential' profile, i.e. one resulting from the convolution of a Gaussian term with an exponential term. An actual observed line is used to demonstrate the application of these relations.

022.111 **The spectrum of SiO^+ and SiN.**
M. Singh, H. Bredohl, F. Remy, I. Dubois.
Journ. Phys. B, (Atomic and Molecular Phys.), Vol. 6, 2656 - 2659 = Inst. Astrophys. Univ. Liège, Cointe-Ougrée (Belgique), Coll. 4°, No. 261 (1973).

022.112 **The a $^3\Pi_r$–X $^1\Sigma^+$ transition of SiO.**
H. Bredohl, R. Cornet, I. Dubois, F. Remy.
Journ. Phys. B, (Atomic and Molecular Phys.), Vol. 7, L66 - L68 = Inst. Astrophys. Univ. Liège, Cointe-Ougrée (Belgique), Coll. 4°, No. 264 (1974). – Letter.

022.113 **The $d^3\Sigma$–$a^3\Pi_r$ transition of SiO.**
M. Singh, H. Bredohl, F. Remy, I. Dubois.
Canadian Journ. Phys., Vol. 52, 569 - 574 = Inst. Astrophys. Univ. Liège, Cointe-Ougrée (Belgique), Col. 4°, No. 266 (1974).

022.114 **A correction to the vibrational numbering of the c $^3\Pi_i$–a $^3\Pi_r$ transition of SiO.**
H. Bredohl, R. Cornet, I. Dubois.
Journ. Phys. B, (Atomic and Molecular Phys.), Vol. 8, L16 - L17 = Inst. Astrophys. Univ. Liège, Cointe-Ougrée (Belgique), Coll. 4°, No. 275 (1975). – Letter.

022.115 **$2p^4$, $4f$, $5f$, and $5g$ configurations of quadruply ionized aluminum (Al V).** W-Ü. L. Brillet, M.-C. Artru.
Journ. Optical Soc. America, Vol. 65, 1399 - 1403 (1975).

022.116 **Absorption spectra of Zn I and Cd I in the 1300–1750 Å region.**
C. M. Brown, S. G. Tilford, M. L. Ginter.
Journ. Optical Soc. America, Vol. 65, 1404 - 1409 (1975).

022.117 **An analytical method for determining the electronic structure of an iron-6.44 at. % silicon alloy.**
R. K. Pandey.
Bol. Inst. Tonantzintla, Vol. 1, 205 - 207 (1975).
The distribution of the 3d and 4s electrons in an iron atom of an iron-silicon alloy having the atomic composition $Fe_{93.56}Si_{6.44}$ is proposed. Primarily the experimental values of the physical parameters like the saturation magnetization, the specific resistivity and the interatomic distance have been used to develop the method. The result explains satisfactorily a few experimental phenomena observed in the alloy.

022.118 **The width of spectral lines.** G. Peach.
Contemporary Phys., (GB), Vol. 16, 17 - 34 (1975).
JDM

022.119 **Potential energy surface for the model unimolecular reaction HNC → HCN.**
P. K. Pearson, H. F. Schaefer, III, U. Wahlgren.
Journ. Chem. Phys., (USA), Vol. 62, 350 - 354 (1975.). – *JDM*

022.120 **Calculation of pressure-broadening effects in oxygen microwave absorption.**
A. H. Barrett, K. S. Lam.
Massachusetts Inst. Technol. Res. Lab. Electronics, RLE Progr. Rep. No. 115, p. 74 - 83 (1975). – *UM*

022.121 **Microwave spectra of molecules of astrophysical interest. 9. Acetaldehyde.** A. Bauder.
U.S. National Bureau of Standards, Washington, D.C., 14 pp. (1975). – *WMG*

022.122 **Microwave spectra of molecules of astrophysical interest. 10. Isocyanic acid.**
G. Winnewisser, W. H. Hocking, M. C. L. Gerry.
U.S. National Bureau of Standards, Washington, D.C., 12 pp. (1975). – *WMG*

022.123 **On the nucleation in a diffusion chamber.**
F. Kamijo.
Japan Journ. Appl. Phys., Vol. 14, 297 - 298 = Contr. Dep. Astron., Univ. Tokyo, No. 186 (1975). – Short note.

022.124 **Techniques for the growth of $Li_2MoO_4 \cdot xGd_2(MoO_4)_3$ crystals.** R. K. Pandey, G. R. Hernández.
Bol. Inst. Tonantzintla, Vol. 1, 255 - 258 (1975).
This communication deals with the crystal growth mechanisms of three members of the $Li_2MoO_4 \cdot x\,Gd_2(MoO_4)_3$ system where x = 1, 2, 3, ... etc.

022.125 **Absolute transition probabilities for Ni I and the solar nickel abundance.** W. N. Lennard.

Thesis, California Inst. Technol., Pasadena. [Available from Univ. Microfilms, Ann Arbor, Mich., USA. Order No. 74-24446], 212 pp. (1974).

Absolute transition probabilities have been measured for 97 lines from 11 levels in Ni I using the lifetime and branching ratio technique. Relative transition probabilities have been determined for 6 lines out of 2 levels whose lifetimes have been measured elsewhere. Equivalent widths have been measured from the Kitt Peak preliminary solar atlas for 48 of the lines for which absolute transition probabilities have been measured. An empirical solar curve of growth has been constructed for Ni I to test the consistency of the transition probability and equivalent width measures. The solar data have been used to calculate a value for the photospheric Ni abundance. A new transition has been observed in the Ni I spectrum at λ 4633.031.

022.126 **One- and two-electron atoms in strong magnetic fields and implications for magnetic white dwarfs and pulsars.** G. L. Surmelian.
Thesis, Louisiana State Univ., Baton Rouge (USA). [Available from Univ. Microfilms, Ann Arbor, Mich., USA. Order No. 75-1954], 131 pp. (1974).

022.127 **On the observability of rotational and vibrational spectra of isotopically asymmetrized homonuclear diatomic ions.** R. B. Bernstein, P. R. Certain.
Comments Atomic Molecular Phys., (*GB*), Vol. 5, 27 - 31 (1975).

022.128 **Experimental consequences of the existence of quarks.** T. R. Mongan, L. Kaufman.
Phys. Rev. D, (Particles and Fields), Vol. 11, 1986 - 1992 (1975).

Assuming the non-observation of free quarks in conventional experiments to be due to very large quark masses, it is argued that free quarks could be produced plentifully by the diffractive disintegration of very high cosmic-ray protons. The production threshold, cross section, quark flux, momentum spectrum and time delay are estimated for quark masses m of 10 and 100 GeV/c^2. The observable quark momentum spectrum at the earth's surface is calculated under assumptions of additivity and small quark inelasticity.

On the extreme ultraviolet solar emission of B-like ions: Na VII, Si X and S XII. See Abstr. 076.058.

Limit on the photon mass deduced from Pioneer-10 observations of Jupiter's magnetic field.
See Abstr. 099.066.

Interstellar HDCO; upper limit and significance as an indicator of interstellar molecule reactions.
See Abstr. 131.087.

Astronomical Instruments and Techniques

031 Astronomical Optics, Methods of Observation and Reduction, Data Processing, Automation

Astronomical Optics

031.001 **Holographische Beugungsgitter, Stand der Entwicklung und neuere Ergebnisse.** G. Schmahl.
Mitt. Astron. Ges., No. 36, p. 123 - 127 (1975).

031.002 **Konstruktion und Eigenschaften holographischer Gitter für weiche Röntgenstrahlung.**
D. Rudolph.
Mitt. Astron. Ges., No. 36, p. 127 - 129 (1975).

031.003 **Untersuchungen über Zonenplattenteleskope zum Einsatz für nicht-solare Quellen im Gebiet der weichen Röntgenstrahlung.** B. Niemann.
Mitt. Astron. Ges., No. 36, p. 129 (1975).

031.004 **Production and testing of the optical elements of the first 2.2 m-telescope for MPIA.** C. Kühne.
Astron. Astrophys., Vol. 41, 345 - 353 (1975).

This is a report on the production and test results of the optics of the first of two 2.2 m-telescopes for the Max-Planck-Institute of Astronomy, Heidelberg. After evaluation of the tests it can be said that the two-mirror systems of the completed telescope give reason to expect a concentration of Hartmann test points of intersection of 96% at 0.3 arcsec dia., and 100% at 0.6 arcsec dia. from the residual production errors alone.

031.005 **The effects on the performance of a spectrograph camera caused by using non-parallel incident light.**
C. F. W. Harmer.
Montly Notices Roy. Astron. Soc., Vol. 172, 325 - 337 (1975).

This theoretical study is related to the practical problems involved in mechanically controlling the focus conditions of image tube-camera systems in astronomical spectrographs, whilst retaining image quality within acceptable limits. The investigation here is for the f/1.6 camera in the AAT intermediate dispersion spectrograph. For the one grating-wavelength arrangement considered here the author derives tolerances on movement which depend on whether (a) the image tube is entirely fixed or (b) the image tube may be tilted to match the focal surface. It is shown that changes in tilt and curvature of the best-fit focal surfaces affect the acceptable limits more than changes in image quality alone, which are relatively slow. Tilt and curvature are shown to be linearly related to collimator defocus.

031.006 **Cours préliminaire d'optique géométrique.**
M. Cagnet.
Space optics, (see 012.013), p. 9 - 42 (1974).

031.007 **Image formation with incoherent light.**
H. H. Hopkins.
Space optics, (see 012.013), p. 43 - 48 (1974).

The basic postulates and definitions associated with the optical transfer function are described. Some of the more important properties of the transfer function curves are then discussed and illustrated graphically.

031.008 **Formation of images in coherent illumination.**
A. Maréchal.
Space optics, (see 012.013), p. 49 - 55 (1974).

031.009 **Two mirror anastigmats.**
G. Monnet, R. Zaharia.
Space optics, (see 012.013), p. 275 - 281 (1974).

This paper considers the full range of two mirror anastigmats giving a real conjugate for an object at infinity. Their optical properties are studied in connection with their usefulness in rocket or satellite experiments.

031.010 **Outlooks on the development of terrestrial optical astronomy.**
Yu. N. Efremov, S. B. Novikov, P. V. Shcheglov.
Uspekhi fiz. nauk, Vol. 115, 301 - 319 (1975).
In Russian. – Abstr. in Referativ. Zhurn. 51. Astron., 7.51.86 (1975).

031.011 **Diffractional image of a double star for the case of central obscuration in the objective.**
N. N. Mikhel'son.
Astron. Zhurn. Akad. Nauk SSSR, Vol. 52, 832 - 838 (1975).
In Russian. English translation in Soviet Astron., Vol. 19, No. 4.

The intensity distribution in the diffractional image of a point at different degrees of central obscuration has been calculated. The following characteristics were determined: the radii of the dark and bright rings, the contrast-transfer function the profiles of diffractional images of double stars at different magnitudes of the components and the dependence of the contrast on the difference between the components.

031.012 **Method of calculation of optical systems with computers.** M. Abele.
Nauchn. Informatsii, vyp. (No.) 28, p. 85 - 102 (1974).
In Russian.

A method of the calculation of optical systems with the aid of computers is described. The basis of the method are the trigonometrical formulas of the passing of beams through the optical system. A method of improvement of the parameters of the optical systems and a method of calculation of the point diagrams for a given system are considered.

031.013 **Holographic interferometer to test optical surfaces.**
F. Broder-Bursztyn, D. Malacara-Hernández.
Applied Optics, Vol. 14, 2280 - 2282 (1975).

A new interferometric method for testing optical surfaces is given. It is related to Murty's Fresnel zone plate interferometer and suggests the fruitful applications of holography to this branch of optics. In this paper a theoretical model that explains both interferometers is developed. Experimental results and pertinent comments are included.

031.014 **Wavefront-error evaluation by mathematical analysis**

of experimental Foucault-test data. R. G. Wilson.
Applied Optics, Vol. 14, 2286 - 2297 (1975).

The diffraction theory of the Foucault test provides an integral formula expressing the complex amplitude and irradiance distribution in the Foucault pattern of a test mirror (lens) as a function of wavefront error. Recent literature presents methods of inverting this formula to express wavefront error in terms of irradiance in the Foucault pattern. The present paper describes a study in which the inversion formulation was applied to photometric Foucault-test measurements on a nearly diffraction-limited mirror to determine wavefront errors for direct comparison with ones determined from scatter-plate interferometer measurements. The results affirm the practicability of the Foucault test for quantitative wavefront analysis of very small errors, and they reveal the fallacy of the prevalent belief that the test is limited to qualitative use only. Implications of the results with regard to optical testing and the potential use of the Foucault test for wavefront analysis in orbital space telescopes are discussed.

031.015 **Compound Schmidt telescope designs with nonzero Petzval curvatures.** R. D. Sigler.
Applied Optics, Vol. 14, 2302 - 2305 (1975).

A variety of aplanatic and anastigmatic Schmidt Cassegrain and Schmidt Gregorian telescope designs with nonzero Petzval curvatures are investigated. Relaxing the Petzval constraint permits the development of high performance photo/visual instruments which are capable of diffraction limited imaging over fields of view of 1–2°.

031.016 **Obtaining information through the atmosphere at the diffraction limit of a large aperture.**
J. B. Breckinridge.
Journ. Optical Soc. America, Vol. 65, 755 - 759 (1975).

An analysis is made which shows that if the pupil plane of a large telescope is relayed into a special achromatic prism interferometer of 180° rotational shear and with point symmetry, the second statistical moment of the pupil irradiance fluctuations reveals information about the source two-dimensional mutual coherence function (MCF) at spatial frequencies that, under ordinary conditions, are severely attenuated by propagation through the atmosphere. To obtain a measure of the MCF at points across the telescope pupil simultaneously, it is necessary to use a two-dimensional photo-electronic array detector, with real-time processing.

031.017 **Far-field diffraction patterns for gaussian beams in the presence of small spherical aberrations.**
D. D. Lowenthal.
Journ. Optical Soc. America, Vol. 65, 853 - 855 (1975).

031.018 **Optimization of telescope images.** A. Martin.
Journ. Optical Soc. America, Vol. 65, 858 (1975).

031.019 **Geometrical optics of holograms.**
H. H. Arsenault.
Journ. Optical Soc. America, Vol. 65, 903 - 908 (1975).

An exact geometrical-optics theory of holography is worked out. A simple derivation of the holographic ray-tracing equations is given; it is used to determine the principal points of a hologram. The well-known paraxial conjugate equations of holography are shown to be exact relations, if the distance and the angles are measured in an appropriate manner. The intersection of all the principal rays determines the position of the aberration-free image.

031.020 **Fangspiegelgröße und Bohrungsdurchmesser bei Cassegrain-Anordnungen.** L. D. Schmadel.
SuW, Vol. 14, 295 (1975).

031.021 **Vorm en plaatsing vangspiegel Newtontelescoop.**
L. Delvoye.
Zenit, Vol. 2, 243 - 245 (1975).

031.022 **Het buigingsbeeldje en wat ermee gebeuren kan.**
H. W. J. Blöte.
Zenit, Vol. 2, 249 - 252 (1975).

031.023 **The application of a laser spherical wave interferometer in astronomical optics.** D. Li, Y. He.
Acta Astron. Sinica, Vol. 16, 51 - 64 (1975). In Chinese.

This paper describes the optical principles, methods of usage, precautions and applications in testing of astronomical optical surfaces by means of an interferometer, the source of which is a He-Ne gas laser tube. Some merits of this method are therein pointed out.

031.024 **Glancing incidence optics in X-ray astronomy: a short review.** J. H. Underwood.
Space Sci. Instrumentation, Vol. 1, 289 - 304 (1975). – Revised version of a talk given at the 'X-ray optics symposium' held at the Mullard Space Science Laboratory of University College, London, April 1973.

This paper reviews the application of the phenomenon of total reflection of X-rays at glancing angles of incidence to the construction of optical devices for use in X-ray astronomy. Examples of instruments which have been used for solar or stellar observations are presented and discussed.

031.025 **Optical materials and processes. Part 2: Grinding to curve.** G. R. Nankivell.
Southern Stars, Vol. 26, 59 - 61 (1975).

031.026 **Spiegel für das Röntgenexperiment Astro 8.**
R. Schlegelmilch.
SuW, Vol. 14, 331 - 332 (1975).

031.027 **Zur Gleichwertigkeit von Kugel- und Parabolspiegel.**
L. D. Schmadel.
SuW, Vol. 14, 338 (1975).

031.028 **The use of properties of pseudo-equivalent systems in calculations of astronomical optics.**
G. M. Popov.
Izv. Krymskoj Astrofiz. Obs., Vol. 53, p. 227 - 239 (1975). In Russian.

The fundamental properties of pseudo-equivalent optical systems are examined and their use for calculations of thin optical systems is discussed. A method for finding elements of a new system from the data of calculations of an initial system is given. Systems of thick lenses are studied too. The method of finding the elements of concentric pseudo-equivalent systems is examined. These results may be used for calculations of lenses for telescopes and collimators of spectrographs, compensating systems.

031.029 **Preliminary testing results of objectives of Belgrade Observatory large meridian circle and large vertical circle.** V. Protić-Benišek, M. Mijatov *(Miyatov).*
Publ. Obs. Astron. Beograd, No. 20, (see 012.024), p. 213 - 217 (1975). In Serbo-Croatian.

031.030 **Zen and the art of big mirror making.** N. Wade.
Science, Vol. 190, 132 - 133 (1975).

031.031 **Aspherisation by elasticity of a 50-cm plate for the Lyon Observatory Schmidt telescope.**
G. Lemaitre.
Astron. Astrophys., Vol. 44, 305 - 313 (1975). In French.

The author describes the development of what he has called the dioptric elasticity method for making aspherics. The method was applied to the construction of the BSC B 16–64

Schmidt corrector plate. The oversize disk is supported on a narrow metal ring which divides the plate into two concentric zones. On each side of the ring the air underneath is partially evacuated and the load-ratio of the inner and outer zones is fixed at a constant value during the grinding and polishing. The elasticity theory used to calculate the flexure of the plate is given, first for the old method invented and used empirically by Schmidt, and then for the flat polishing method used here.

031.032 **Dynamic Hartmann test.** L. J. Golden.
Applied Optics, Vol. 14, 2391 - 2399 (1975).

The dynamic Hartmann test was developed to monitor precisely the alignment and optical figure of a complete telescope system such as the Large Space Telescope by using a real star source. The test was chosen for its large range of operation, sensitivity, and relative simplicity. The dynamic Hartmann test configuration is discussed and analyzed, and the construction of an experimental test instrument is described. A wavefront error simulator is used to determine experimentally the capability of the Hartmann test instrument in detecting low-order aberrations under laboratory conditions. It is found that sensitivities in the $\lambda/50$ to $\lambda/100$ range can be realized.

031.033 **Night myopia and the intermediate dark focus of accommodation.**
H. W. Leibowitz, D. A. Owens.
Journ. Optical Soc. America, Vol. 65, 1121 - 1128 (1975).

031.034 **Multimodule optical systems.**
R. I. Mercado, O. N. Stavroudis.
Journ. Optical Soc. America, Vol. 65, 1133 - 1140 (1975).

Earlier work on modular optical-design techniques is extended to lens systems that consist of an arbitrary number of modules; the work is generalized to include finite-conjugate systems (copy lenses) and afocal systems as well as infinite-conjugate systems (camera lenses). All such modular designs have the property that third-order spherical aberration and astigmatism are identically zero. Equations are derived for obtaining zero values for the remaining third-order monochromatic aberrations as well as the primary chromatic aberrations. Several examples are given.

031.035 **Unobscured-aperture two-mirror systems.**
R. Gelles.
Journ. Optical Soc. America, Vol. 65, 1141 - 1143 (1975).

Two-mirror systems of the type conventionally used as astronomical objectives can be used noncoaxially to avoid the usual central obscuration. Third-order aberration theory is used to derive the conditions under which the primary and secondary mirrors can be tilted and decentered.

031.036 **Reflectance and optical constants for ULE from 250 to 3000 Å.** J. F. Osantowski, A. R. Toft.
Journ. Optical Soc. America, Vol. 65, 1169 (1975). – Conference abstract.

031.037 **High-transparency infrared materials.**
P. A. Miles, R. N. Donadio, R. T. Newberg.
Journ. Optical Soc. America, Vol. 65, 1169 (1975). – Conference abstract.

031.038 **How window emittance degrades the performance of airborne infrared imaging systems.** C. A. Klein.
Journ. Optical Soc. America, Vol. 65, 1169 (1975). – Conference abstract.

031.039 **Measured spatial variations of transmittance and emissivity of infrared windows.**
L. D. Tubbs, R. F. Gasparovic, G. H. Emmons.
Journ. Optical Soc. America, Vol. 65, 1169 (1975). – Conference abstract.

031.040 **Low-temperature thermal-expansion measurements on new optical materials.**
J. S. Browder, S. S. Ballard.
Journ. Optical Soc. America, Vol. 65, 1170 (1975). – Conference abstract.

031.041 **Differentiation of ray-tracing equations.**
B. Tatian.
Journ. Optical Soc. America, Vol. 65, 1176 (1975). – Conference abstract.

031.042 **Optimization of the geometric and diffraction images of concentric-mirror telescopes.**
J. S. Newcomb.
Journ. Optical Soc. America, Vol. 65, 1176 (1975). – Conference abstract.

031.043 **Caustic surface of an optical system.** D. L. Shealy.
Journ. Optical Soc. America, Vol. 65, 1176 (1975). Conference abstract.

031.044 **Simple unequal-path interferometer configurations.**
R. N. Smartt.
Journ. Optical Soc. America, Vol. 65, 1210 (1975). – Conference abstract.

031.045 **Optimum performance of lateral shear interferometer for active wavefront compensation.**
J. C. Wyant.
Journ. Optical Soc. America, Vol. 65, 1211 (1975). – Conference abstract.

031.046 **Degradation of optically finished surfaces caused by stripping of old coatings and re-coating.**
C. V. Muffoletto, V. Williams, B. K. Flint.
Journ. Optical Soc. America, Vol. 65, 1215 (1975). – Conference abstract.

031.047 **Technique for measuring the refractive index and its change with temperature in the infrared.**
B. C. Platt, H. W. Icenogle, J. E. Harvey, R. Korniski, W. L. Wolfe.
Journ. Optical Soc. America, Vol. 65, 1264 - 1266 (1975).

031.048 **Schmidt systems for the near-infrared.**
A. Greve.
Infrared Physics, Vol. 15, 239 - 241 (1975).

The formulae which define the parameters (focal ratio, field) of a seeing-limited Schmidt system are given. Near-infrared (5000 Å $\leqslant \lambda \leqslant$ 15,000 Å) Schmidt systems are discussed for nine different corrector plate materials.

031.049 **On material for the optics of large ground-based and space optical telescopes.**
Yu. N. Kondrat'ev, E. V. Gurkovskij, I. L. Sakin.
Sb. tr. NII stekla, 1974, vyp. (No.) 1, p. 155 - 162. In Russian.
Abstr. in Referativ. Zhurn. 51. Astron., 11.51.107 (1975).

031.050 **Der 3.6-m-Rohling.**
SuW, Vol. 14, 387 - 389 (1975).

031.051 **An investigation of the large coronograph objective.**
V. I. Kulidzhanishvili.
Byull. Abastumansk. Astrofiz. Obs., No. 46, p. 201 - 204 (1975). In Russian.

031.052 **Beryllium mirrors: 10.6-μm characterization.**
M. D. Guadagnoli, T. T. Saito.
Applied Optics, Vol. 14, 2806 - 2808 (1975).

031.053 **Hologram gratings: amplitude and phase compo-**

nents. T. R. Bader.
Applied Optics, Vol. 14, 2818 - 2819 (1975).

031.054 **Relative flatness measurement of uncoated optical flats.** C. F. Bruce, F. P. Sharples.
Applied Optics, Vol. 14, 3082 - 3085 (1975).
A method is described for measuring the relative flatness of uncoated precision optical flats to $\lambda/1000$. The method uses scanning interferometry and can be made automatic fairly readily.

031.055 **Why not design a new telescope optical system?** A. W. Wilkinson.
Journ. British Astron. Ass., Vol. 86, 35 - 39 (1975).

031.056 **Fixing of high-precision astronomical mirrors in their mountings by means of sealing compounds.**
V. S. Plotnikov, D. I. Varfolomeev, L. O. Lebedev.
Trudy Mosk. in-ta inzh. geod., aehrofotosemki i kartogr., 1975, vyp. (No.) 60, p. 7 - 12. In Russian. – Abstr. in Referativ. Zhurn. 51. Astron., 12.51.76 (1975).

031.057 **Calculation of astronomical mirror deformations for vertical position of the optical axis.**
M. K. Troshenkov.
Trudy Mosk. in-ta inzh. geod., aehrofotosemki i kartogr., 1975, vyp. (No.) 60, p. 12 - 23; In Russian. – Abstr. in Referativ. Zhurn. 51. Astron., 12.51.77 (1975).

031.058 **Effect of discreteness of cylindrical layers of the sealing compound on mirror deformations under its own weight.** B. A. Roev, A. V. Skarzhinskaya.
Trudy Mosk. in-ta inzh. geod., aehrofotosemki i kartogr., 1975, vyp. (No.) 60, p. 23 - 30. In Russian. – Abstr. in Referativ. Zhurn. 51. Astron., 12.51.78 (1975).

031.059 **Spherical mirror flexure under its own weight for horizontal position of the optical axis.**
A. V. Skarzhinskaya.
Trudy Mosk. in-ta inzh. geod., aehrofotosemki i kartogr., 1975, vyp. (No.) 60, p. 30 - 40. In Russian. – Abstr. in Referativ. Zhurn. 51. Astron., 12.51.79 (1975).

031.060 **Temperature deformations of spherical mirrors elastically supported along the edge.**
G. V. Arutyunyan, L. O. Lebedev.
Trudy Mosk. in-ta inzh. geod., aehrofotosemki i kartogr., 1975, vyp. (No.) 60, p. 40 - 48. In Russian. – Abstr. in Referativ. Zhurn. 51. Astron., 12.51.80 (1975).

031.061 **Vibrational strength of fixing of astronomical mirrors by harmonic influences.**
B. F. Kaledin, D. I. Varfolomeev.
Trudy Mosk. in-ta inzh. geod., aehrofotosemki i kartogr., 1975, vyp. (No.) 60, p. 49 - 55. In Russian. – Abstr. in Referativ. Zhurn. 51. Astron., 12.51.81 (1975).

031.062 **Non-linear induced oscillations of a mirror with symmetric elastic characteristic.** V. A. Perov.
Trudy Mosk. in-ta inzh. geod., aehrofotosemki i kartogr., 1975, vyp. (No.) 60, p. 55 - 62. In Russian. – Abstr. in Referativ. Zhurn. 51. Astron., 12.51.82 (1975).

031.063 **Investigation of non-linear oscillations of an astronomical mirror by the electron modelling method.**
N. M. Komar, V. A. Perov.
Trudy Mosk. in-ta inzh. geod., aehrofotosemki i kartogr., 1975, vyp. (No.) 60, p. 62 - 72. In Russian. – Abstr. in Referativ. Zhurn. 51. Astron., 12.51.83 (1975).

031.064 **Oscillations of construction elements of composite material.** G. V. Arutyunyan.
Trudy Mosk. in-ta inzh. geod., aehrofotosemki i kartogr., 1975, vyp. (No.) 60, p. 72 - 83. In Russian. – Abstr. in Referativ. Zhurn. 51. Astron., 12.51.84 (1975).

031.065 **Spherical envelope deformations in the presence of accidental technological form distortions.**
V. S. Plotnikov, B. P. Makarov, M. K. Troshenkov.
Trudy Mosk. in-ta inzh. geod., aehrofotosemki i kartogr., 1975, vyp. (No.) 60, p. 84 - 92. In Russian. – Abstr. in Referativ. Zhurn. 51. Astron., 12.51.85 (1975).

031.066 **Non-linear oscillations of an astronomical mirror under the action of accidental pressure forces.**
V. A. Perov, M. K. Troshenkov.
Trudy Mosk. in-ta inzh. geod., aehrofotosemki i kartogr., 1975, vyp. (No.) 60, p. 92 - 100. In Russian. – Abstr. in Referativ. Zhurn. 51. Astron., 12.51.86 (1975).

031.067 **The effect of fluctuations of elastic characteristics of the foundation on the deformation of constructions.** B. A. Roev.
Trudy Mosk. in-ta inzh. geod., aehrofotosemki i kartogr., 1975, vyp. (No.) 60, p. 100 - 108. In Russian. – Abstr. in Referativ. Zhurn. 51. Astron., 12.51.87 (1975).

031.068 **Boundary effects in constructions on the stochastic basis.**
B. P. Makarov, B. A. Roev, A. V. Novozhilov.
Trudy Mosk. in-ta inzh. geod., aehrofotosemki i kartogr., 1975, vyp. (No.) 60, p. 108 - 113. In Russian. – Abstr. in Referativ. Zhurn. 51. Astron., 12.51.88 (1975).

031.069 **Questions arising by measuring small high-frequency deformations.**
V. Yu. Torochkov, A. M. Zhilkin, V. S. Usov, M. D. Krashchin.
Trudy Mosk. in-ta inzh. geod., aehrofotosemki i kartogr., 1975, vyp. (No.) 60, p. 113 - 122. In Russian. – Abstr. in Referativ. Zhurn. 51. Astron., 12.51.89 (1975).

031.070 **Berechnung genauer Prüfwerte für Parabolspiegel.** E. Wiedemann.
Orion, 33. Jahrgang, p. 219 - 220 (1975).

031.071 **Diffraction from a sequence of apertures and disks.** W. A. Kleinhans.
Journ. Optical Soc. America, Vol. 65, 1451 - 1456 (1975).
A simple expression is given for the power diffracted from a sequence of circular apertures and disks. The result is derived by use of the geometrical theory of diffraction, and is thus asymptotically correct in the short-wavelength limit. Several applications to the baffling of optical instruments are presented. An alternate approach based on the Huygens principle is also outlined and compared with the geometrical theory of diffraction.

031.072 **Effects of small rotationally symmetrical aberrations on the irradiance spread function of a system with gaussian apodization over the pupil.** E. Sklar.
Journ. Optical Soc. America, Vol. 65, 1520 - 1521 (1975). Letter.

031.073 **The design and application of interference filters for direct polychromatic MTF measurements.**
M. W. McDowell, D. N. W. Chinnery, E. Theron.
Optik, Vol. 44, 79 - 88 (1975).

031.074 **On estimating the distortions of details caused by the extended wings of the instrumental profile.**
V. N. Karpinskij.
Solnechnye Dannye 1975 Byull., No. 9, p. 93 - 95 (1975). In

Russian.

Far extended wings of the instrumental profile cause certain distortions of the fine details. A method of estimating the degree of these distortions is given.

031.075 **On applying the transfer modulation function method to investigations of optical systems.**
G. S. Minasyants.
Solnechnye Dannye 1975 Byull., No. 9, p. 99 - 102 (1975). In Russian.

A method of measuring the transfer modulation function of a photoheliograph is described. It is shown that the photoheliograph of High Altitude Solar Station at the Alma-Ata Peak may be used for the study of low-contrasting fine structural features on the sun, in particular for bright points in the umbra of sunspots.

031.076 **Some comments about distortion computing in lenses.** M. L. Celaya, D. Tentori, D. Malacara.
Bol. Inst. Tonantzintla, Vol. 1, 241 - 243 (1975).

In this note the authors point out some details regarding the computation of the distortion aberration in optical systems, by means of exact ray tracing and the influence of other aberrations on the result. Of special interest is the fifth order cubic coma.

031.077 **Can a particle scattering analysis explain all interference experiments?**
C. Roychoudhuri, A. Cornejo.
Bol. Inst. Tonantzintla, Vol. 1, 245 - 246 (1975).

Quantum mechanical analysis of particle (photon) scattering by a periodic potential allows only a discrete set of angles of scattering. Such analyses have been used to explain the interference (diffraction) pattern formed by crystals or gratings while, apparently, doing away with the wave-particle dualism. The validity of such a particle-concept approach is questioned by pointing out the difficulties involved in explaining a single slit (or a single pinhole) diffraction pattern.

031.078 **Microwave switching by picosecond photoconductivity.** A. M. Johnson, D. H. Auston.
IEEE Journ. Quantum Electron., Vol. QE-11, 283 - 287 (1975). – *SS*

031.079 **Correction of aberrations due to zonal errors of the main mirror in teleobjectives of Cassegrain systems.**
A. S. Khajkin.
Fiz. in-t AN SSSR. Lab. spektroskop. Preprint No. 79. Moskva, 1975. 10 pp. In Russian. – Abstr. in Referativ. Zhurn. 51. Astron., 2.51.161 (1976).

031.080 **Optical design modules with zero Petzval contribution.** R. I. Mercado.
Bol. Inst. Tonantzintla, Vol. 1, 265 - 276 (1975).

The properties of optical design modules with zero Petzval contribution are re-examined to obtain the exact number of such modules that can be generated from a given selection of three indices of refraction. For refracting systems, two modules can always be obtained from the real-case family and none from the imaginary-case. For reflecting systems, as in earlier work, a real-case and an imaginary-case were obtained. The subdomains of the shape parameter where zero Petzval sum occurs may be isolated depending on the relative values of the three indices of refraction associated with the lens module. A method of selecting glasses to achieve a desired range of the shape parameter is also presented.

031.081 **Shape measurement of optical surfaces with rotational symmetry using Ronchigrams.**
D. Malacara, A. Cornejo.
Bol. Inst. Tonantzintla, Vol. 1, 277 - 283 (1975).

A method of measurement of a Ronchigram in order to compute the shape of an optical surface with rotational symmetry is described. The method is capable of a very high accuracy and can be applied to spherical as well as to aspherical surfaces.

031.082 **Reconstrucción de imagenes con función de densidad de probabilidad a priori discretizada.**
L. R. Berriel-Valdos.
Bol. Inst. Tonantzintla, Vol. 1, 285 - 291 (1975).

Using a discrete prior probability density function for the parameter values associated with an object, positions and irradiances, the estimated values of the object parameters are found. The minimum mean square estimated technique is used.

031.083 **Required accuracy in the radius of curvature of a primary astronomical telescope mirror.**
A. Cornejo, D. Malacara.
Bol. Inst. Tonantzintla, Vol. 1, 293 - 294 (1975). – Note.

031.084 **Patterns of correlation between focograms and Ronchigrams.** A. S. De Vany, A. De Vany.
Bol. Inst. Tonantzintla, Vol. 1, 295 - 297 (1975). – Note.

031.085 **Using Moiré fringes to teach surface testing.**
A. S. De Vany.
Bol. Inst. Tonantzintla, Vol. 1, 299 - 300 (1975). – Note.

031.086 **An observation that demonstrates the chromatic aberration of the eye.** D. Malacara.
Bol. Inst. Tonantzintla, Vol. 1, 301 (1975). – Note.

031.087 **Effect of atmospheric turbulence on the performance of optical systems with generalized annular apertures.** S. D. Gupta.
Japanese Journ. Applied Phys., Vol. 14, 983 - 990 (1975).

The far-field diffraction by generalized annular apertures of light waves rendered partially coherent due to atmospheric turbulence has been examined theoretically. The phase structure function generated by the atmospheric turbulence has been assumed to obey the square law which leads to a gaussian type correlation function. The expressions are worked out for the field irradiance and central irradiance distributions. The field irradiance and the central irradiance distributions, and the two-point resolution have been presented and discussed as functions of transmittance and phase variations in the two parts of the aperture, size of the central part of the aperture and degree of coherence etc.

031.088 **On atmospheric turbulence and the problem of optimization of the telescopic pupil.**
M. De, L. N. Hazra.
Optica Acta, (*GB*), Vol. 22, 853 - 865 (1975).

Presents the report of investigations on a prior apodization of the pupil of a telescope operating in a turbulent medium. The criterion of apodization has been taken to be the maximization of the factor of encircled energy on a pre-specified area on the far-field plane.

Neue Fertigteile für den Fernrohrselbstbau.
Sternenbote, 18. Jahrgang, p. 134 - 138 (1975).

Methods of Observation and Reduction

031.201 **The study of light curves with the aid of smoothing spline-functions.** V. Pop, P. Pop.
IAU Symposium No. 67, (see 012.001), p. 567 - 569 (1975).

By means of smoothing spline functions an attempt is made to approximate cepheid-type light curves. As an example the light curve of XZ Cyg is used.

031.202 **Eine semi-empirische Entfaltungsmethode an scharfen Gradienten einer Strahlungsflußverteilung.**
M. Butz, A. Greve, O. Hachenberg, W. Hirth.
Mitt. Astron. Ges., No. 36, p. 116 - 120 (1975).

031.203 **On the relative importance of phase and frequency information in pulsar timing measurements.**
W. H. Press.
Astrophys. Journ., Vol. 200, 182 - 186 (1975).

Over a given time interval, the phase (time of arrival) of pulsar signals and their frequency can be measured independently. In principle there is also independent information in yet higher derivatives. A common observational practice is to throw away all information except phase; this results in some loss of signal-to-noise over what might optimally have been obtained. This paper calculates quantitatively the signal loss, in the case of a particular model: the detection of a small sinusoidal phase modulation in a perfectly periodic pulsar whose signal is degraded by white noise.

031.204 **Radial velocity and aberration.** R. d'E. Atkinson. Monthly Notices Roy. Astron. Soc., Vol. 172, 465 - 467 = Publ. Goethe Link Obs., Indiana Univ., *Bloomington,* No. 170 (1975).

The observed radial velocity of a star can be corrected to the barycentre of the solar system by using the Besselian Day Numbers C and D, together with two new star-constants c'' and d'' which resemble the familiar c' and d'. Unless corrections for the second-order terms are added, the accuracy of this or any other first-order method is limited in much the same way as with aberration, but it is still sufficient even for the exceptional precision obtainable when atmospheric absorption lines can be used as the comparison spectrum.

031.205 **A short program for the Besselian C and D.** R. d'E. Atkinson.
Monthly Notices Roy. Astron. Soc., Vol. 172, 469 - 471 = Publ. Goethe Link Obs., Indiana Univ., *Bloomington,* No. 171 (1975).

Errors in the standard reductions for aberration can sometimes reach 0."01 (at declinations below 70°) if second-order corrections are not applied. If that is acceptable, the previously-published computer program for C and D can be simplified to $C = -18''804 \cos \odot - 0''00722 \cos M - 0''00787 \cos J$ and $D = -20''496 \sin \odot - 0''00786 \sin M - 0''00858 \sin J$ where $\odot$ contains only four terms, and M and J are the mean longitudes of the moon and Jupiter.

031.206 **Spectrométrie.** P. Bousquet.
Space optics, (see 012.013), p. 85 - 103 (1974).

031.207 **Méthodes de calibration U.V.** D. Labs.
Space optics, (see 012.013), p. 205 - 210 (1974).

The paper concerns the principles and the difficulties of absolute calibration work. As an example, the calibration method is demonstrated using synchrotron-radiation as a standard source.

031.208 **The use of vapours as dispersing systems for spectroscopy.** J.-E. Blamont.
Space optics, (see 012.013), p. 253 - 264 (1974).

031.209 **Fourier spectroscopy in space.** P. Connes.
Space optics, (see 012.013), p. 265 - 273 (1974).

The basic advantages and difficulties of Fourier spectroscopy are briefly reviewed and references to more complete treatments given. The first results gathered from aircraft, balloons or spacecraft are discussed together with proposals for future experiments.

031.210 **Millimetre emission from extraglactic objects–I. Observational techniques and data analysis.**
M. Rowan-Robinson, P. E. Clegg, P. A. R. Ade.
Monthly Notices Roy. Astron. Soc., Vol. 172, 603 - 615 (1975).

A detailed discussion is given of the observing techniques and method of data analysis for observations in the 1–3 mm band. Due to the broadband nature of the observations, the effective frequency depends strongly on the spectral index of the source. The attenuation and noise are well correlated with the precipitable water vapour, as estimated from the ground-level dew point. Fluxes and upper limits are given for 88 extragalactic sources, of which six (3C 84, 120, 147, 273, M82 and NGC 4051) have been detected. The possible detections of 3C 345, 0727–11 and M81 claimed earlier are not confirmed.

031.211 **Algorithm for computation of the ozone concentration from observations of eclipses of artificial earth satellites.** V. N. Lebedinets, Yu. G. Kaufman, V. P. Prorok.
Nauchn. Informatsii, vyp. (No.) 28, p. 3 - 16 (1974).

An algorithm for a more exact calculation of the ozone concentration at heights of 20–50 km from spectrophotometric observations of artificial earth satellite eclipses is suggested. The vertical ozone concentration profile is considered to be a continuous broken line. On calculating the trajectory of the solar rays illuminating the satellite, atmospheric refraction, finite size of the sun and solar limb darkening are taken into account.

031.212 **Sur l'utilisation des observations des satellites artificiels obtenues dans emplacements excentriques.**
A. Dinescu.
Nauchn. Informatsii, vyp. (No.) 28, p. 77 - 84 (1974).
In Russian.

On a établi les formules pour le calcul de réduction des observations des emplacements excentriques au centre de la station en deux cas spéciaux: 1) une station a une camera photographique et un télémètre laser et on doit calculer la distance station-satellite pour la camera et respectivement la direction vers le satellite pour le laser; 2) une station a une ou plusieurs cameras, toutes excentriques et on doit calculer la direction pour le centre de la station. On donne des exemples numériques pour les deux cas et on analyse l'influence de l'erreur de la distance station-satellite sur les coordonnées équatoriales réduites au centre de la station.

031.213 **A possibility to improve the mathematical reduction of satellite tracking data obtained with AFU-75 cameras.** A. V. Fedorova.
Nauchn. Informatsii, vyp. (No.) 30, p. 16 - 18 (1974).
In Russian.

031.214 **The finite Fourier transform definition of an edge on the solar disk.**
H. A. Hill, R. T. Stebbins, J. R. Oleson.
Astrophys. Journ., Vol. 200, 484 - 498 (1975).

A new technique for locating the edge of the sun is proposed. The technique uses a finite Fourier transform of the observed limb darkening function to achieve reduced sensitivity to atmospheric and instrumental effects and heightened sensitivity to the shape of the intrinsic limb darkening function. A theory is developed that predicts these sensitivities.

The location of the edge is shown to be influenced only by solar phenomena down to the milli-arcsecond range.

031.215 **Automatic registration and reduction of the observations made at the Danjon astrolabe of the Milan-Merate Observatory.** L. Buffoni, F. Carta, F. Chlistovsky, A. Manara, F. Mazzoleni.
Boll. Geod. Sci. Affini, Anno 34, p. 249 - 271 (1975). In Italian.

The authors explain the procedure used at the Astronomical Observatory of Milan-Merate in order to automate the recording of the observations made with the Danjon astrolabe. The preliminary reduction program of the observations is also analyzed.

031.216 **The influence of fluctuations of the refractive index of the air on the accuracy of fixing a transit line by the diffraction method.** D. D. Dzyaman.
Trudy Kazan. Gorod. Astron. Obs., No. 40, p. 130 - 135 (1974). In Russian.

031.217 **On the attenuation of optical radiation and the quality of images of geodetic observations in a turbulent atmosphere.** D. D. Dzyaman.
Trudy Kazan. Gorod. Astron. Obs., No. 40, p. 136 - 145 (1974). In Russian.

031.218 **Results of closure phase analysis of VLBI data.** L. K. Hutton, T. A. Clark, I. I. Shapiro, J. J. Wittels, H. F. Hinteregger, C. A. Knight, A. E. E. Rogers, A. R. Whitney, A. E. Niell, B. O. Rönnäng, O. E. H. Rydbeck.
Bull. American Astron. Soc., Vol. 7, 413 (1975). – Abstr. AAS.

031.219 **Digital image centering.** L. H. Auer, W. F. van Altena.
Bull. American Astron. Soc., Vol. 7, 443 (1975). – Abstr. AAS.

031.220 **Special relativity and very long baseline interferometers.** D. S. Robertson.
Nature, Vol. 257, 467 - 468 (1975).

When baseline lengths approach the diameter of the earth, relativistic effects become important, especially in geodetic applications. Failure to take account of these effects results in errors that are equivalent to clock errors of the order of a few parts in 10^{12}. The object of the calculations presented here is to recast an equation involving unobservables into one containing only observables.

031.221 **Optimum entrance apertures for spectroscopy of redshifted galaxies.** D. H. Gudehus.
Publ. Astron. Soc. Pacific, Vol. 87, 541 - 544 (1975).

Optimum entrance apertures, based upon signal-to-noise ratio considerations, are derived for spectroscopy of redshifted galaxies. Signal-to-noise ratio is found to decrease slowly for apertures larger than optimum and to depend strongly on redshift. The optimum apertures decrease with increasing redshift and depend slightly on the deceleration parameter.

031.222 **Velocity selection of cluster members with use of objective-prism plates.**
R. E. White, Jr., D. R. Mosley, I. Furenlid, R. E. White III.
Publ. Astron. Soc. Pacific, Vol. 87, 551 - 556 (1975).

A modification to the method of Fehrenbach for determination of stellar radial velocities from objective-prism plates is described which allows separation of members of discrete stellar systems from the surrounding field stars. The modified procedure, applied to the galactic globular star cluster NGC 3201, uses two objective-prism plates, with the direction of increasing wavelength reversed with respect to the sky orientation from one plate to the other. The use of a digital microdensitometer in the production of scanned spectral data is shown to be necessary in the further application of this method.

031.223 **On the reduction of position angles determined from heliometric measurements.** A. S. Mamakov.
Izv. Astron. Ehngel'gardt. Obs., *Kazan'*, No. 39, p. 127 - 130 (1972). In Russian.

031.224 **The application of methods to photographs obtained with infrared EOT for revealing faint objects.**
R. A. Sarkisyan.
Soobshch. Byurakan. Obs., vyp. (No.) 46, p. 135 - 143 (1975). In Russian.

The application of some methods for photographs is presented which were obtained on the 20″-telescope with infrared electron-optical transformer (EOT) in order to decrease the background sky and remove irregular background of EOT. By superposition of 10 photographs of the central region of the open cluster NGC 1502 an object has been revealed to be $\approx 1^m$ fainter than the limit 16^m of the telescope. So, the discovered limit of a discrete source goes to $\sim 17^m$. An object with $\sim 17^m$ is invisible on the Palomar Sky Survey. Possibly it is an infrared star.

031.225 **A check of the measuring procedures of Torino Observatory, made with the Sproul Observatory plates of Barnard's star.** V. Zappalà.
Proc. Astrometric Conference, Torino 1974, (see 012.017), p. 61 - 62 (1975).

031.226 **Stellar line profiles by tilt-scanned narrow band interference filters.**
D. Clarke, I. S. McLean, T. H. A. Wyllie.
Astron. Astrophys., Vol. 43, 215 - 221 (1975).

The technique of tilt-scanning narrow band interference filters to record stellar line profiles has been explored with success. Indications are that any one filter may have a scanning range of 100 Å or more, allowing profiles to be recorded without severe distortions. The method is both simple and efficient; the whole of the seeing disc may be accepted even with small filters and the total transmittance of the optical components is high. For example, sample data are presented of short term changes which are occurring in the Hβ line of the emission line star, γ Cas.

031.227 **Study of the solar photospheric granulation by Michelson interferometry–first results.**
C. Aime, G. Ricort, G. Grec.
Astron. Astrophys., Vol. 43, 313 - 315 (1975). In French.

A multiple hole aperture experiment has been performed using the Nice coudé solar telescope during January–February 1975. Using very small apertures (< 2 cm), the authors expect that, within each small aperture area, the light coming from a point source is a plane wave, a random phase difference between each aperture being the main atmospheric effect. The authors show the first results in an autocorrelation function curve and compute the contrast of the granulation using a recent shape of the solar granulation power spectrum.

031.228 **Photometric calibration of Schmidt plates by an auxiliary beam splitting prisma.**
W. Bronkalla, P. Notni, J. Wempe, F. Börngen, R. Ziener.
Astron. Nachr., Vol. 296, 261 - 265 (1975). In German.

Plates of the Tautenburg Schmidt mirror can be calibrated with the aid of a beam-splitting prisma only if Schraffierkassette plates are used. The effects of image structure, scintillation and intermittence on in-focus images are different for the main component and the prismatic companion. Therefore, such plates cannot be used to set up a photometric scale.

031.229 **Photoelectric photometry with small telescopes.** W. S. G. Walker.
Southern Stars, Vol. 26, 39 - 42 (1975).

031.230 **Progress on a programme of multicolour photographic sky mapping.** F. P. Andrews.
Southern Stars, Vol. 26, 43 - 53 (1975).

031.231 **Techniques for visual observation of eclipsing binary stars.** M. E. Baldwin.
Journ. American Ass. Variable Star Observers, Vol. 4, 10 - 12 (1975).

031.232 **On the trailing of spectra.** D. H. Gudehus.
Astron. Astrophys., Vol. 44, 113 - 115 (1975).

The ratio of signal-to-noise ratio of trailed to untrailed slit spectra is calculated as a function of the amount of trailing and the magnitude difference between sky and object. An optimum trailed length exists for objects which are bright enough. In general galaxian spectra are not so successfully trailed.

031.233 **On some technical problems of modern astrophotometry.** V. B. Nikonov.
Novaya tekhn. v astron. Vyp. (No.) 5. Leningrad, Nauka, 1975, p. 44 - 46. In Russian. – Abstr. in Referativ. Zhurn. 51. Astron., 8.51.253 (1975).

031.234 **On the photometric calibration of TV photographs of stars.** A. V. Mironov, P. P. Petrov.
Novaya tekhn. v astron. Vyp. (No.) 5. Leningrad, Nauka, 1975, p. 84 - 90. In Russian. – Abstr. in Referativ. Zhurn. 51. Astron., 8.51.266 (1975).

031.235 **On the question of estimating the accuracy of parameters of objects from polarimetric and photometric astrophysical observations in the IR spectral region.**
Yu. K. Bergner, V. G. Lebedev.
Novaya tekhn. v astron. Vyp. (No.) 5. Leningrad, Nauka, 1975, p. 125 - 129. In Russian. – Abstr. in Referativ. Zhurn. 51. Astron., 8.51.275 (1975).

031.236 **Presentazione di una serie di programmi per le identificazioni ottiche.** A. Ficarra, L. Padrielli.
Mem. Soc. Astron. Italiana, Vol. 46, 105 - 111 (1975).

The authors present a series of FORTRAN programmes which permit the optical identification of radiosources and X-sources with a precision up to one second of arc. These programmes can also be utilized for the precise measurement of astronomical coordinates of interesting objects for which a photographic reproduction exists.

031.237 **Some thoughts about randomness.** L. G. Jacchia.
Sky Telescope, Vol. 50, 371 - 374 (1975).

031.238 **B and V photoelectric photometry at Tallinn Astronomical Observatory.** P. Kalv.
Publ. Tartu Astrofiz. Obs., Vol. 43, 96 - 102 (1975).

The eclipsing binary observational program and the photometric technique are described. From numerous observations the mean extinction coefficients and the transformation constants of the photometric system are determined.

031.239 **Total reduction of distorted echelle spectrograms: an automatic procedure.**
R. C. Peterson, A. M. Title.
Applied Optics, Vol. 14, 2527 - 2532 (1975).

The authors describe a semiautomatic procedure for the reduction of high-dispersion echelle spectra recorded with an image tube. The spectra are traced with a computer-controlled microdensitometer that scans along the curved spectral orders. The curvature of each order is calculated approximately by a FORTRAN program from known grating and distortion parameters. A typical spectrum includes 25 orders (covering 1500 Å) and is traced with a slit 0.012 Å wide. To produce an atlas of intensity vs wavelength and to determine the equivalent widths of 300 lines currently require a day. The authors discuss the reduction procedures and time requirements in detail.

031.240 **New perspectives in interferometric spectrometry.** A. Maréchal.
Journ. Optical Soc. America, Vol. 65, 1183 (1975). – Conference abstract.

031.241 **Behavior of Felgett's advantage in multiplex spectroscopy.** T. Hirschfeld.
Journ. Optical Soc. America, Vol. 65, 1186 (1975). – Conference abstract.

031.242 **Astronomical speckle imaging.** D. C. Ehn, P. Nisenson.
Journ. Optical Soc. America, Vol. 65, 1196 (1975). – Conference abstract.

031.243 **Amplitude interferometry for high-angular-resolution astronomy.** D. G. Currie.
Journ. Optical Soc. America, Vol. 65, 1199 (1975). – Conference abstract.

031.244 **Practical stellar amplitude interferometer.** K. M. Liewer, D. G. Currie, S. L. Knapp.
Journ. Optical Soc. America, Vol. 65, 1199 (1975). – Conference abstract.

031.245 **Real-time correction of telescope "seeing".** R. H. Dicke.
Journ. Optical Soc. America, Vol. 65, 1206 (1975). – Conference abstract.

031.246 **Laboratory-simulated speckle interferometry.** A. M. Schneiderman, P. F. Kellen, M. G. Miller.
Journ. Optical Soc. America, Vol. 65, 1287 - 1291 (1975).

The utility of speckle interferometry for overcoming the atmospherically induced resolution limit is investigated with a laboratory simulation. Studies of various geometric object shapes verify reported field measurements for single and binary stars and demonstrate the usefulness of this technique for quantitative measurement of more complex shapes.

031.247 **Apodization for maximum Strehl criterion and specified Sparrow limit of resolution for incoherent illumination.** P. W. Peng, J. E. Wilkins, Jr.
Journ. Optical Soc. America, Vol. 65, 1292 - 1299 (1975).

031.248 **The particular case of a mathematical model of the clock "aging".** A. A. Abgaryan.
Vopr. metrol. i metody optiko-fiz. izmerenij. Moskva, Izd-vo standartov, 1975, p. 36 - 37. In Russian. – Abstr. in Referativ. Zhurn. 51. Astron., 11.51.210 (1975).

031.249 **Large field speckle interferometry.** G. P. Weigelt.
Optik, Vol. 43, 111 - 128 (1975).

A. Labeyrie proposed and implemented a method (speckle interferometry) which is able to measure fine structures of astronomical objects to 0.02 seconds of arc if the 5m-telescope of Mt. Palomar is used. Because of the atmosphere of the earth the resolution with conventional techniques is only one second of arc. Labeyrie obtains with his method only the autocorrelation of the object and not a real image. This article describes 'large field speckle interferometry' (LFSI), a modification of Labeyrie's method. This LFSI method yields real, high resolution images of astronomical objects (double and multiple stars, star disks, fine structures of galaxies etc.).

031.250 **The influence of construction of solar pavilions on the optical pulsation and fluctuation-temperature conditions.** B. Kalzhanov.
Trudy Astrofiz. Inst., *Alma-Ata,* Vol. 26, 147 - 159 (1974). In Russian.

031.251 **On a statistical analysis of star trails.** D. A. Rozhkovskij, F. K. Rspaev.
Trudy Astrofiz. Inst., *Alma-Ata,* Vol. 26, 171 - 178 (1974). In Russian.

031.252 **Differential method of reduction of photoelectric systems (On the compilation of a catalogue of photometric standards in Kapteyn's Selected Areas).**
O. P. Abuladze.
Byull. Abastumansk. Astrofiz. Obs., No. 46, p. 17 - 30 (1975). In Russian.

A differential method of reduction of Abastumani Observatory's instrumental system to the standard UBV system is described. A detailed discussion is given to the technique used for observation and estimation of both the atmospheric extinction and the reduction coefficients needed for the said reduction.

031.253 **Lunar laser observing process and strategy.** E. C. Silverberg.
Lunar laser ranging, (see 012.032), p. 1 - 10 (1975).

031.254 **Data identification, filtering and compression.** P. J. Shelus.
Lunar laser ranging, (see 012.032), p. 11 - 15 (1975).

031.255 **Mathematical modelling of lunar range observations.** J. D. Mulholland.
Lunar laser ranging, (see 012.032), p. 16 - 22 (1975).

031.256 **Astrometry and atmospheric differential color refraction.** W. F. van Altena, H. Moseley.
Bull. American Astron. Soc., Vol. 7, 509 (1975). – Abstr. AAS.

031.257 **A method of radially supporting large mirrors.** K. M. Harrison.
Monthly Notes Astron. Soc. Southern Africa, Vol. 34, 147 - 152 (1975).

031.258 **Treatment of meridian observations by using the M-220 computer.**
V. V. Tel'nyuk-Adamchuk, A. Ya. Gregul'.
Vestn. Kiev. Un-ta, Ser. Astron., No. 17, p. 98 - 105 (1975). In Russian.

031.259 **Untersuchungen zum Nachführverhalten von siderisch nachgeführten Satellitenmeßkammern.**
F. M. Scherer.
Allgemeine Vermessungsnachr., 82. Jahrgang, p. 369 - 370 (1975).

031.260 **Very long baseline interferometry (VLBI) applications to secular geodynamics and earth strain measurement.** P. F. Macdoran.
Proc. Symposium on the earth's gravitational field and secular variations in position, (see 012.036), p. 380 - 395 (1974).

031.261 **Comparison of two intercontinental baselines by VLBI.**
J. S. Gubbay, K. J. W. Lynn, A. J. Legg, D. S. Robertson.
Proc. Symposium on the earth's gravitational field and secular variations in position, (see 012.036), p. 396 - 405 (1974).

031.262 **The national mapping lunar laser program.** J. McK. Luck, M. J. Miller, P. J. Morgan.
Proc. Symposium on the earth's gravitational field and secular variations in position, (see 012.036), p. 413 - 432 (1974).

031.263 **A comparison of estimation methods for the reduction of laser range observations.**
B. D. Tapley, B. E. Schutz.
Proc. Symposium on the earth's gravitational field and secular variations in position, (see 012.036), p. 489 - 508 (1974).

031.264 **Space astrometry project – reduction by multilateration.** P. Bacchus.
Space astrometry, (see 012.037), p. 37 - 42 (1975).

For the spacelab option, the unknowns (positions, parallaxes, proper motions, instrument constants) and the data (extent catalogues, measured arcs) are specified and enumerated. In a simple mathematical model, the minimal number of data needed for the problem to be solvable is specified, and in the case of 'superabundant' data the variance of the unknowns is determined.

031.265 **Photoelectric photometry in the small observatory.** W. S. G. Walker, B. F. Marino.
Publ. Variable Star Section, Roy. Astron. Soc. New Zealand, No. 3 (C75), p. 31 - 39 (1975).

031.266 **The role of VLBI in the establishment of coordinate systems.** I. Zhongolovich.
IAU Colloquium No. 26, (see 012.034), p. 293 - 295 (1975).

031.267 **On calculation of the effective area on the visible hemisphere in optical observations of cosmic objects.**
A. S. Trofimov.
Trudy TsNII geod., aehrosemki i kartogr., 1975, vyp. (No.) 210, p. 20 - 33. In Russian. – Abstr. in Referativ. Zhurn. 62. Issled. kosmich. prostranstva, 1.62.427 (1976).

031.268 **Cross correlation studies of stellar variability using two telescopes.** P. J. Edwards, R. B. Hurst, G. Christie.
Proc. Astron. Soc. Australia, Vol. 2, 349 - 351 (1975).

031.269 **Evaluation of an eclipsing variable minimum with an HP desk calculator and plotter.** Z. Kvíz.
Proc. Astron. Soc. Australia, Vol. 2, 351 - 352 (1975).

031.270 **The limitations of astronomical image reconstruction.** S. P. Worden.
Observatory, Vol. 95, 291 - 293 (1975).

031.271 **On the application of a mirror slit to simultaneous observations of the spectrum and two-dimensional images of the sun.** Yu. V. Platov.
Solnechnye Dannye 1975 Byull., No. 9, p. 53 - 54 (1975). In Russian.

A new method of simultaneous obtaining of the spectrum and two-dimensional images of the sun used in observations with the large extra-eclipse coronograph is described.

031.272 **On the possibility of applying optical heterodyning to astronomical observations.**
S. M. Gorskij, A. L. Poverennov, V. E. Fridman.
Solnechnye Dannye 1975 Byull., No. 9, p. 70 - 76 (1975). In Russian.

The possibility of spectral resolution increase by optical heterodyning when receiving solar radiation is estimated. Lines of laser radiation and those of separate solar spectrum elements are compared. The signal-to-noise relation when using this method is considered.

031.273 **Momentu reģistrācija astronomiskos novērojumos.** L. Roze.
Zvaigžņotā debess, 1975. gada pavasaris, p. 9 - 14.

031.274 **Determincación de hora por observaciones meridianas.** J. F. Lahulla.
Bol. Astron. Obs. Madrid, Vol. 8, No. 5, p. 61 - 74 (1975).
In the present work the clock correction and the constants of Prin's transit instrument are obtained by application of the method of least squares to Mayer's formula.

031.275 **An accurate experimental method for determining the important properties of microstrip transmission lines.** J. G. Richings.
Marconi Rev., Vol. 37, 209 - 216 (1974). – *DNC*

031.276 **The effect of the earth's atmosphere on baseline determinations by very-long-baseline interferometry.**
K. J. W. Lynn, J. S. Gubbay.
Australian Defence Sci. Serv. Weapons Res. Est WRE-Techn. Note-1312, 15 pp. (1975). – *SS*

031.277 **Phase-modulation techniques simplify analysis of phase-locked loops.** R. Rands.
Electronics, Vol. 48, 94 - 98 (1975). – *JBS*

031.278 **The procedure of the treatment of the polarimetric measurements with the photoelectric polarimeter of the Belgrade Astronomical Observatory.** I. Vince.
Bull. Obs. Astron. Belgrade, No. 126, p. 59 - 61 (1975).

031.279 **Esperienze di fotometria fotoelettrica.** L. Baldinelli.
Giornale Astron., Vol. 1, 283 - 297 (1975).

031.280 **Trends in astronomical spectroscopy.** K. O. Wright.
Image processing techniques in astronomy, (see 012.041), p. 31 - 48 (1975).
A brief survey is made of improvements in spectroscopic data analysis in recent years. The need for accurate observational material is stressed, especially with the highly efficient receivers now available. A few problems that will be studied for many years are considered.

031.281 **Information retrieval by microdensitometry.** T. de Groot.
Image processing techniques in astronomy, (see 012.041), p. 51 - 55 (1975).

031.282 **Fourier techniques.** P. Connes.
Image processing techniques in astronomy, (see 012.041), p. 57 - 58 (1975).

031.283 **On line digital correlation of photon counting TV images for stellar interferometry.**
A. Blazit, L. Koechlin, J. L. Oneto.
Image processing techniques in astronomy, (see 012.041), p. 79 - 84 (1975).
Stellar interferometry images using one large or several smaller telescopes require on line statistical analysis of photon events. The authors built a system giving real time two dimensional autocorrelations of photon counting TV images, containing each up to two hundred photon events. The present system is only partly hard wired, a software algorithm being used to derive the autocorrelation. Plans for a fully hard wired digital processor are also given.

031.284 **Mesure des contours de raies par les techniques de télévision.** C. Fehrenbach, A. Vin.
Image processing techniques in astronomy, (see 012.041), p. 109 - 116 (1975).

031.285 **Photometry using electronographic techniques.** J. Ring, S. P. Worswick.
Image processing techniques in astronomy, (see 012.041), p. 117 - 124 (1975).
This paper deals with the photometric performance and limitations of the specific combination of a Spectracon image tube, used with Ilford L4 nuclear track emulsion, with a Joyce Loebl MkIII CS microdensitometer which produces digital output on magnetic tape. The system has been applied to broad and narrow-band surface photometry of extended objects and it is with these types of observations in mind that the evaluation has been made.

031.286 **Problems of oversampling with SEC vidicon television systems.**
E. J. Devinney, D. Fischel, D. Klinglesmith.
Image processing techniques in astronomy, (see 012.041), p. 135 - 139 (1975).

031.287 **Some experiments with an SEC image tube for high precision multi-colour photometry of galactic clusters.** A. Blecha, P. Bartholdi.
Image processing techniques in astronomy, (see 012.041), p. 141 - 155 (1975).
The possibility of high precision multi-colour stellar photometry with an SEC vidicon camera is studied on the galactic cluster NGC 581. In particular, the best exploitation of calibration information is sought. The correlation length is determined and used for data compression. Several calibration techniques are discussed and the best one is selected and tested. A related problem of zero level is examined.

031.288 **Aperture synthesis.** W. N. Brouw.
Image processing techniques in astronomy, (see 012.041), p. 301 - 307 (1975).

031.289 **The design of VAMP software for the measurement and reduction of stellar spectrograms.**
R. J. Rutten, G. P. van Gelder.
Image processing techniques in astronomy, (see 012.041), p. 311 - 314 (1975).

031.290 **Evaluation of raster scans of photographic stellar images.** E. Høg.
Image processing techniques in astronomy, (see 012.041), p. 323 - 328 (1975).
Raster scans of photographic stellar images can be obtained with a microphotometer, or still faster with a photodiode array. A both simple and accurate procedure for evaluating a raster scan is presented. This "sector" method computes the position and size, the non-circular form and the diffuseness of an image. It derives iso-transmission curves for the image, and it can thereby recognize and evaluate partly overlapping images.

031.291 **Handling of 'Galaxy' photometry.** P. A. Wayman, M. J. Stift.
Image processing techniques in astronomy, (see 012.041), p. 335 - 340 (1975).

031.292 **Echelle spectra image processing for the International Ultraviolet Explorer.**
D. A. Klinglesmith, E. Dunford.
Image processing techniques in astronomy, (see 012.041), p. 347 - 352 (1975).

031.293 **Accuracy and results of spectrophotometry with objective prism plates.** A. Cassatella, R. Viotti.
Image processing techniques in astronomy, (see 012.041),

p. 367 - 372 (1975).

031.294 **Removal of vignetting from Apollo low light level photographs,**
G. C. Alvord, D. A. Klinglesmith, L. Dunkelman, R. M. Mercer.
Image processing techniques in astronomy, (see 012.041), p. 391 - 396 (1975).

The comparison of the vignetting function of Apollo 15, 16 and 17 35-mm Nikon cameras is presented. A technique for removing this effect is described, and demonstrated on an image of the L_4 point of the earth–moon system.

031.295 **Lock-in image subtraction: detectability of circumstellar planets with the Large Space Telescope.**
D. Bonneau, M. Josse, A. Labeyrie.
Image processing techniques in astronomy, (see 012.041), p. 403 - 409 (1975).

In this article, the authors discuss on coronographic and lock-in detection techniques applicable in the near future to observations with the Large Space Telescope. Those techniques enhance the visibility of circumstellar features to such a degree that even planets should become marginally detectable in some favorable cases.

031.296 **Digital image centering, I.**
W. F. van Altena, L. H. Auer,
Image processing techniques in astronomy, (see 012.041), p. 411 - 418 (1975).

A series of parallax plates have been measured on a PDS microdensitometer to assess the possibility of using the PDS for precision relative astrometry and to investigate centering algorithms that might be used to analyze digital images obtained with the Large Space Telescope. The basic repeatability of the PDS is found to be $\pm 0.6\ \mu m$ with the potential for reaching $\pm 0.2\ \mu m$. A very efficient centering algorithm has been developed which fits the marginal density distributions of the image with a gaussian profile and a sloping background. The accuracy is comparable with the best results of the Yerkes Observatory photoelectric image bisector.

031.297 **Extrema de l'effet d'une erreur en son point d'application dans la méthode des moindres carrés.**
H. Debehogne.
Acad. Roy. Belgique, Bull. Cl. Sci., 5e Sér., Vol. 61, 31 - 39 (1975).

031.298 **On the prospects of using an economical analysis in astrophysical investigations.**
B. D. Kokarev, V. S. Shevchenko.
Issled. ehkstremal'no molodykh zvezdn. kompleksov. Tashkent, Fan, 1975, p. 148 - 155. In Russian. – Abstr. in Referativ. Zhurn. 51. Astron., 2.51.188 (1976).

031.299 **Tropospheric limits of the accuracy of measuring the coordinates of cosmic sources.**
A. A. Stotskij.
XI Vses. konf. po rasprostr. radiovoln. Chast' (Part) 1. Tezisy dokl. Kazan', Kazan. un-t, 1975, p. 121 - 123. In Russian. Abstr. in Referativ. Zhurn. 51. Astron., 2.51.189 (1976).

031.300 **Spherical harmonics and earth-rotation synthesis in radio astronomy.** R. H. MacPhie, E. H. Okongwu.
IEEE Trans. Antennas Propagation, Vol. AP-23, 386 - 391 (1975). – *BMT*

031.301 **ILTH II [Inverse Laplace Transform–H II region] – analysis of the spectrum of a thermal radioastronomical source.** M. Salem.
Computer Phys. Commun. (*Netherlands*), Vol. 9, 247 - 257 (1975).

It has been shown by Salem and Seaton (1974) that, given the spectrum of a thermal radioastronomical source, information on the structure can be deduced in the form of a function omega (E), which is the total angular area within which the emission measure is greater than or equal to E. The computer program ILTH II (Inverse Laplace Transform–H II region) computes omega (E) and the spherical model from the flux density spectrum, and also compares omega (E) with high-resolution observations.

031.302 **Computation of line and continuum radiation from thermal radioastronomical sources.**
M. Brocklehurst, M. Salem.
Computer Phys. Commun. (*Netherlands*), Vol. 9, 258 - 267 (1975).

The computer program described calculates the intensities and profiles of hydrogen and helium radio recombination lines emitted by a thermal source. The electron temperature is assumed to be constant throughout the source, while the electron density is a known function of position. Spherical or plane parallel symmetry is assumed in the program. The theory of the method used is described in detail by Brocklehurst and Seaton (1972).

031.303 **Generation of first order derivative information for sensitivity computation and optimization of circuits.**
E. Sánchez-Sinencio.
Bol. Inst. Tonantzintla, Vol. 1, 249 - 254 (1975).

This paper deals with the generation of first order derivative information for sensitivity calculations and optimization of circuits in the frequency domain. These techniques are especially applicable to microwave circuits. The method presented here is based on the work of Bordewijk, generalized by Director and Rohrer.

031.304 **Two beam interference experiments and some quantum concepts.** C. Roychoudhuri.
Bol. Inst. Tonantzintla, Vol. 1, 259 - 264 (1975).

Some two-beam interference experiments like the simple double-slit, the holographic double-slit and the Michelson interferometer are described and analysed with a view to demonstrate that there exist conceptual conflicts behind the usual text book assumptions to explain the so-called "wave-particle duality".

031.305 **Atmospheric turbulent layers localisation by optical processing of stellar shadow pattern.**
F. Martin, J. Borgnino, F. Roddier.
Nouvelle Rev. Optique, Vol. 6, 15 - 23 (1975). In French.

An optical information processing technique, commonly used for speckle-pattern analysis, is applied to the study of the motion of shadow patterns associated with stellar scintillation. By fitting the velocity deduced from fringe measurements with the results of meteorological wind soundings, an approximate altitude is obtained for the atmospheric turbulent layers producing stellar scintillation.

031.306 **Cinematography of solar intensity, velocity, and magnetic fields.**
J. M. Beckers, L. Dickson, D. Woodman.
Optical Engineering, (*USA*), Vol. 14, 64 - 68 (1975).

Using a very narrow band (0.004–0.013 nm) universal birefringent filter, rapidly turnable between 410 and 700 nm, and the vacuum telescope at the observatory, the authors have developed photographic and video techniques to observe solar magnetic and velocity fields utilizing the Zeeman and Doppler effects respectively. Both techniques are used to time lapse cinematography of solar activity.

031.307 **Tips für die Astropraxis.**
SuW, Vol. 14, 248 - 250, 293 - 294, 336 - 338, 370 - 372, 406 - 410 (1975).

The method of statistical reduction. See Abstr. 021.014.

Multi-telescope systems for optical astronomy. See Abstr. 032.035.

Transfer functions of space instruments. See Abstr. 032.504.

Spectroradiometric calibration techniques in the far ultraviolet: a stable emission source for the Lyman bands of molecular hydrogen. See Abstr. 032.514.

Observations with a VLB array. I. Introduction and procedures. See Abstr. 033.007.

On aperture correction in scanning interferometry. See Abstr. 034.001.

Extending the visible frontier. New tools of the optical astronomer. See Abstr. 034.115.

Television detector techniques. See Abstr. 034.132.

Precise reduction to the apparent places of stars. See Abstr. 041.034.

Astrometry. See Abstr. 041.070.

On the apparent track of artificial satellites and the tracking frames of satellite cameras. See Abstr. 055.007.

Methoden zur Ausrichtung von siderisch nachgeführten Satellitenmeßkammern und zur Berücksichtigung der dynamischen Refraktionskorrektionen bei der Reduktion von Satellitenaufnahmen. See Abstr. 055.014.

Numerical methods of calculation of stellar opacity. See Abstr. 061.062.

The effect of linearly polarized light from extended stellar atmospheres on interferometer response functions. See Abstr. 064.050.

Astronomical observations and meteorological data (Express information). See Abstr. 082.030.

Der Einfluss der Sternfarbe und einer Inversion des Temperaturprofils der Atmosphäre auf die astronomische Refraktion für Empfänger verschiedener spektraler Empfindlichkeit. See Abstr. 082.111.

The localization of ionospheric irregularities by the holographic method. See Abstr. 083.066.

Comparison of geometrical effects in radio and stellar occultations. See Abstr. 091.025.

Photographic parallaxes of nineteen stars obtained through the central overlap reduction routine. See Abstr. 111.006.

Photometry. See Abstr. 113.071.

Applications of Fourier analysis to broadening of stellar line profiles. I. Three Mn stars. See Abstr. 114.343.

Eine Methode für seeingunabhängige photometrische und astrometrische Messungen der visuellen Doppelsterne. See Abstr. 118.002.

Geodetic and astrometric results of very long-baseline interferometric measurements of natural radio sources. See Abstr. 141.145.

Test observations of galactic nuclei with a television camera. See Abstr. 158.120.

Data Processing, Automation

031.401 **Zur Prozeßrechner-Steuerung der 2,2-m-Teleskope des MPIA.** L. D. Schmadel.
Mitt. Astron. Ges., No. 36, p. 65 - 67 (1975).

031.402 **On an automatic system of satellite data reduction for M-20 type computers.** V. A. Enal'skij.
Nauchn. Informatsii, vyp. (No.) 30, p. 3 - 8 (1974).
In Russian.

This paper is an introduction to the whole complex of problems connected with an automatic system of satellite data reduction programs performed in the Department of Applied Mathematics of the Astronomical Council.

031.403 **A program for processing punched tapes.** Eh. I. Shustova.
Nauchn. Informatsii, vyp. (No.) 30, p. 9 - 15 (1974).
In Russian.

031.404 **Une possibilité de la création d'un complexe astrométrique avec l'utilisation des programmes pour les calculs des coordonnées de satellites artificiels obtenus à l'aide de la camera AFU-75.** L. A. Yurova.
Nauchn. Informatsii, vyp. (No.) 30, p. 19 - 23 (1974).
In Russian.

031.405 **Possibilities of automatization of information transmission from the coordinate-measuring device "Ascorecord" to the electronic computer M-222.**
A. I. Kolpakov, S. A. Nesvedov.
Nauchn. Informatsii, vyp. (No.) 30, p. 40 - 45 (1974).
In Russian.

031.406 **L'automatisation du procédé du calcul des coordon-**

nées de satellites artificiels. L. A. Yurova.
Nauchn. Informatsii, vyp. (No.) 30, p. 46 - 62, [63] - [82] (1974). In Russian.

031.407 **L'instruction pour l'exploitation du complexe K1 des programmes pour calculer des coordonnées de satellites artificiels photographiés à l'aide de la camera AFU-75.** L. A. Yurova.
Nauchn. Informatsii, vyp. (No.) 30, p. 63 - 72 (1974). In Russian.

031.408 **A complex computer program for automatical reduction of satellite observations by the AFU-75 camera.** J. Balodis, K. K. Lapushka.
Observations of artificial earth satellites, No. 14, (see 012.018), p. 623 - 632 (1975). In Russian.

031.409 **On-line computers for telescope control and data handling.** L. B. Robinson.
Annual Rev. Astron. Astrophys., Vol. 13, (see 003.005), 165 - 185 = Lick Obs. Contr., No. 404 (1975).

031.410 **Automatized complex of equipment for reception and treatment of signals during laser location of the moon.** V. N. Zajtsev, V. V. Kurbasov, A. V. Kutsenko, N. M. Lypkan', B. A. Polos'yants.
Pribory i tekhn. ehksperimenta, 1975, No. 2, p. 78 - 80. In Russian. – Abstr. in Referativ. Zhurn. 51. Astron., 8.51.104 (1975).

031.411 **Device for automatic measurements of amplitude characteristics of meteor reflections.**
V. A. Makarov, A. M. Nasyrov.
Meteorn. rasprostr. radiovoln. No. 10 - 11. Kazan', Kazan. un-t, 1975, p. 66 - 70. In Russian. – Abstr. in Referativ. Zhurn. 51. Astron., 12.51.220 (1975).

031.412 **An astronomical application of CAMAC.** I. G. van Breda, P. W. Hill, R. J. Campbell, A. E. Lynas-Gray, D. M. Carr.
Univ. Obs. St. Andrews, Repr. No. 54, 5 pp. (1974).

A single-crate CAMAC system is described. The system is interfaced to a laboratory microdensitometer, for the measurement of photographic plates, and to a photoelectric photometer mounted on a telescope, for on-line measurement of the light fluxes from stars. Data may be recorded on magnetic tape for processing on a large computer.

031.413 **Digital reduction of stellar spectra.** Z. Turło.
Stud. Soc. Sci. Torunensis, Toruń, Poland, Sect. F (Astron.), Vol. 5, 119 - 133 = Bull. Astron. Obs. N. Copernicus Univ. Toruń, No. 53 II (1974).

A digital system for automatic reduction of the stellar spectra recorded on the photographic plates has been constructed and tested. Hardware and software assembled for this purpose are described here with some technical details relevant to the overall performance of the system. Results of the practical tests of this system are also included.

031.414 **Programmed pulse generator for astronomical applications.** B. Wikierski.
Stud Soc. Sci. Torunensis, Toruń, Poland, Sect. F (Astron.), Vol. 5, 135 - 143 = Bull. Astron. Obs. N. Copernicus Univ. Toruń, No. 53 III (1974).

This paper describes an idea of the programmed pulse generator (PPG) intended for synchronization of the astronomical instruments. It is a digital unit based on the programmed binary counter, which is able to give 262 144 different time intervals in 10 microsecond steps. The paper gives a method for pulse programming in the sidereal, mean solar time and widening of stellar spectra obtained with the objective prisms.

031.415 **Four years experience in APMS star plate processing – results and future plans.** J. S. Newcomb.
Image processing techniques in astronomy, (see 012.041), p. 159 - 170 (1975).

031.416 **The fully automatic program-controlled microdensitometer-comparator of the Astronomical Institute at Utrecht.** J. R. W. Heintze, R. L. Porteous, W. Brandie.
Image processing techniques in astronomy, (see 012.041), p. 185 - 198 (1975).

031.417 **Software techniques for image processing.** D. A. Klinglesmith.
Image processing techniques in astronomy, (see 012.041), p. 271 - 283 (1975).

031.418 **Problems of machine independency and interactive matters.** H. Nieuwenhuijzen.
Image processing techniques in astronomy, (see 012.041), p. 285 - 299 (1975).

A review is given of software developed for image processing in astronomy by a number of astronomical institutes. The presented programs are the result of a questionnaire survey and the list is to be considered as preliminary and incomplete.

031.419 **Interactive computer reduction and display of radio supersynthesis maps.** R. D. Ekers, R. J. Allen.
Image processing techniques in astronomy, (see 012.041), p. 309 (1975). – Abstract.

031.420 **Development of digital system for stellar spectra reduction at Toruń Observatory.**
Z. Turło, S. Krawczyk, J. Smoliński.
Image processing techniques in astronomy, (see 012.041), p. 315 - 321 (1975).

031.421 **Image processing systems developed at MIT to handle silicon vidicon images and spectra of astronomical objects.**
T. B. McCord, P. F. Kinnucan, G. Fawcett, Jr.
Image processing techniques in astronomy, (see 012.041), p. 329 - 334 (1975).

031.422 **Semi-automatic determination of equivalent widths.** A. Cassatella, L. Gratton, R. Nesci.
Image processing techniques in astronomy, (see 012.041), p. 353 - 365 (1975).

031.423 **Search of significant features in a direct non parametric pattern recognition method. Application to the classification of a multiwire spark chamber pictures.**
R. Buccheri, P. Coffaro, G. Colomba, V. Di Gesù, S. Salemi.
Image processing techniques in astronomy, (see 012.041), p. 397 - 402 (1975).

Fourier analysis with unequally-spaced data.
See Abstr. 021.001.

On line digital correlation of photon counting TV images for stellar interferometry. See Abstr. 031.283.

A fully automated digitally controlled 30-inch telescope. See Abstr. 032.013.

Hardware and software for the Llanherne low frequency sky surveys. See Abstr. 033.019.

Überlegungen zur Auswahl einer automatischen

Plattenmeßmaschine. See Abstr. 034.007.

Scannermessungen und automatische Äquivalentbreitenbestimmung. See Abstr. 034.012.

The photometry of extended objects: diode-array detectors and interactive data processing. See Abstr. 034.131.

The use of the "GALAXY" machine at the Royal Greenwich Observatory. See Abstr. 034.138.

The COSMOS facility at the Royal Observatory Edinburgh. See Abstr. 034.142.

The ESO S-3000 measuring machine. See Abstr. 034.143.

The computer controlled comparator Astroscan. See Abstr. 034.147.

A simultaneous six-channel microphotometer with computerized data acquisition. See Abstr. 034.148.

Le Centre de Dépouillement de Clichés Astronomiques. See Abstr. 034.152.

A star catalogue on magnetic type and service programs. See Abstr. 041.010.

A possible variant of the construction of a general catalogue of the position of stars by means of an electronic computer. See Abstr. 041.026.

Analysis of the nebulosities near T Tauri using digital computer image processing. See Abstr. 132.049.

032 Astronomical Instruments, Space Instrumentation

Astronomical Instruments

032.001 **The Anglo-Australian telescope.**
S. C. B. Gascoigne.
Endeavour, No. 123, Vol. 34, 131 - 135 (1975).

The Anglo-Australian telescope went into regular operation on 11th December 1974, seven and a half years after the agreement between the Australian and United Kingdom Governments to build it had been announced. With a working aperture of 153.3 inches it is exceeded substantially in size only by the 200-inch telescope at Palomar, California, and it is the largest astronomical project in which either country has so far invested. Technically, it is fulfilling all expectations and enters upon the scene at a time when the outlook for astronomy appears well assured.

032.002 **Das 3,5-m-Teleskop des Max-Planck-Instituts für Astronomie.** K. Bahner.
Mitt. Astron. Ges., No. 36, p. 57 - 61 (1975).

032.003 **Modern optical telescope.**
O. A. Mel'nikov, V. S. Popov.
Nauka i chelovechestvo. 1975. Moskva, Znanie, 1974, p. 307 - 322. In Russian. – Abstr. in Referativ. Zhurn. 51. Astron., 7.51.69 (1975).

032.004 **Investigation of the Zeiss short-focus astrograph of the Kharkov Observatory.** P. P. Pavlenko.
Khar'kov. un-t. Khar'kov, 1975. 24 pp. In Russian. – Abstr. in Referativ. Zhurn. 51. Astron., 7.51.92 (1975).

032.005 **Searching-guiding device of an optical telescope.**
V. A. Malarev, E. M. Neplokhov.
Novaya tekhn. v astron. Vyp. (No.) 5. Leningrad, Nauka, 1975, p. 134 - 139. In Russian. – Abstr. in Referativ. Zhurn. 51. Astron., 7.51.100 (1975).

032.006 **Mounting design for amateur telescopes.**
R. Berry.
Journ. Roy. Astron. Soc. Canada, Vol. 69, L33 - L38 (1975).

032.007 **The temporal and spatial extent of the isoplanatic patch.** A. Title, T. Pope, S. Schoolman.
Bull. American Astron. Soc., Vol. 7, 462 (1975). – Abstr. AAS.

032.008 **The Anglo-Australian telescope project.**
S. C. B. Gascoigne, E. J. Wampler.
Bull. American Astron. Soc., Vol. 7, 497 (1975). – Abstr. AAS.

032.009 **The large German binoculars of World War II.**
J. A. Gould.
Journ. British Astron. Ass., Vol. 85, 393 - 400 (1975).

032.010 **The story of an historical telescope and of some small improvements to it.** R. A. H. Paterson.
Journ. British Astron. Ass., Vol. 85, 412 - 416 (1975).

032.011 **Das Mayall 4-m-Teleskop auf Kitt Peak.**
M. Lammerer, H. Treutner.
Orion, 33. Jahrgang, p. 99 - 102 (1975).

032.012 **Kuppeln für die 2.2-m-Teleskope des Max-Planck-Instituts für Astronomie.**
K. Bahner, H.-L. von Ehr.
SuW, Vol. 14, 229 - 232 (1975).

032.013 **A fully automated digitally controlled 30-inch telescope.** S. A. Colgate, E. P. Moore, R. Carlson.
Publ. Astron. Soc. Pacific, Vol. 87, 565 - 575 (1975).

A fully automated 30-inch telescope has been successfully designed and constructed from a military surplus Nike-Ajax radar mount. Novel features include: closed-loop operation between mountain telescope and campus computer 30 km apart via microwave link, a TV-type sensor which is photon shot-noise limited, a special lightweight primary mirror, and a stepping motor drive capable of slewing and settling one degree in one second or a radian in fifteen seconds.

032.014 **The 153-inch Anglo-Australian telescope.** E. J. Wampler.
Sky Telescope, Vol. 50, 224 - 228 (1975).

032.015 **Kleine Reise-Astrokamera.** D. Bissiri.
Orion, 33. Jahrgang, p. 167 - 169 (1975).

032.016 **Control automático en posición y tiempo de los sistemas de obturación de las cámaras de observación de satélites geodésicos.** R. Ortiz, R. Vieira.
Urania Barcelona, Año 59, No. 281–282, p. 395 - 405 (1974).

032.017 **Study of the sidereal guiding error of the 125-cm reflector of the Sternberg Institute and its image quality.** A. Eh. Gur'yanov.
Astron. Tsirk., No. 855, p. 4 - 7 (1975). In Russian.

032.018 **Results of investigations of thermal properties of a pavilion.** T. R. Kir'yan.
Astron. Tsirk., No. 860, p. 1 - 3 (1975). In Russian.

032.019 **On thermometric investigations of astrometric instruments and pavilions.**
V. V. Kir'yan, T. R. Kir'yan.
Astron. Tsirk., No. 860, p. 3 - 5 (1975). In Russian.

032.020 **On thermal deformations of a tube of an astrometric instrument.** T. R. Kir'yan.
Astron. Tsirk., No. 860, p. 5 - 7 (1975). In Russian.

032.021 **Tableau des grands télescopes en service dans le monde.** P. Kohler.
L'Astronomie, 89e année, p. 370 - 372 (1975).

032.022 **Ein neues Sonnen-Turmteleskop.** C. Kühne.
SuW, Vol. 14, 369 (1975).

032.023 **Some comments on misaligning in double-mirror astronomical telescopes.**
N. V. Merman, M. A. Sosnina.
Astron. Zhurn. Akad. Nauk SSSR, Vol. 52, 1055 - 1058 (1975). In Russian. English translation in Soviet Astron., Vol. 19, No. 5.

Misaligning effects in double-mirror astronomical telescopes by lateral displacement of the secondary mirror and inclination of its optical axis relative to the whole axis of the instrument are considered. The problem of displacements of the image in any double-mirror systems is discussed.

032.024 **Some characteristics of the inclination of the vertical axis of the zenith telescope in Belgrade.**
M. Djokić.
Publ. Obs. Astron. Beograd, No. 20, (see 012.024), p. 84 - 89 (1975). In Serbo-Croatian.

032.025 **Flexure of meridian instruments at day-time conditions.** M. Mijatov *(Miyatov).*
Publ. Obs. Astron. Beograd, No. 20, (see 012.024), p. 195 - 198 (1975). In Serbo-Croatian.

032.026 **Examination of the pivots of the large meridian circle in Belgrade.** M. Mijatov *(Miyatov)*, S. Sadžakov, M. Dačić, D. Šaletić, V. Protić-Benišek.
Publ. Obs. Astron. Beograd, No. 20, (see 012.024), p. 199 - 204 (1975). In Serbo-Croatian.

032.027 **Meridian marks of the large transit instrument of Belgrade Observatory.** L. Pakvor.
Publ. Obs. Astron. Beograd, No. 20, (see 012.024), p. 205 - 207 (1975). In Serbo-Croatian.

032.028 **On the variation of inclination of the south meridian mark pillar of the Belgrade large transit instrument.**
Z. Knežević.
Publ. Obs. Astron. Beograd, No. 20, (see 012.024), p. 208 - 210 (1975). In Serbo-Croatian.

032.029 **The reading accuracy of Belgrade Observatory vacuum meridian marks.** K. N. Tavastsherna.
Publ. Obs. Astron. Beograd, No. 20, (see 012.024), p. 211 - 212 (1975). In Serbo-Croatian.

032.030 **Pivot-examinations of the large transit instrument of Belgrade Observatory.** I. Pakvor.
Publ. Obs. Astron. Beograd, No. 22, 57 pp. (1975). In Serbo-Croatian.

A complete pivot-testing of the large transit instrument of the Belgrade Observatory is shown.

032.031 **The astrometric properties of the 40-cm double astrograph of the Crimean Astrophysical Observatory.** N. S. Chernykh, L. I. Chernykh.
Izv. Krymskoj Astrofiz. Obs., Vol. 53, p. 209 - 226 (1975). In Russian.

A study of the astrometric properties of the 40-cm wide-angle double photographic telescope used for position observations of minor planets and comets was carried out. The limiting magnitude, the coefficient of third-order distortion, the location of the optical center, the spherical aberration and Hartmann's constant, as well as errors of adjustment of the polar axis of the telescope mounting were determined. The curvature of the field, the dependence of the focal length upon temperature, the magnitude equation were investigated. The precision of the determination of minor planets positions at different observing conditions was estimated.

032.032 **A forgotten 41-inch refractor.** B. Warner.
Sky Telescope, Vol. 50, 370 (1975).

032.033 **Practice on the PZT observation.** S. Takagi.
Publ. International Latitude Obs. Mizusawa, Vol. 9, 259 - 282 (1974).

Some discrepancies in the polar motion derived by different observing systems still exist. In order to obtain a much more consistent result, the IAU and IUGG have recommended to establish a new PZT chain on the ILS parallel of 39°08′. This is an exposition on the cooperative PZT observation at the ILS stations, and it gives a detailed explanation of the instrument, a method of observation, and a method of reduction, in that order.

032.034 **Preliminary results of astronomical observations made by the new PZT at the Mizusawa Latitude Observatory.**
S. Takagi, M. Aihara, H. Kitago, G. Murakami, K. Iwadate.
Publ. International Latitude Obs. Mizusawa, Vol. 9, 283 - 289 (1974).

032.035 **Multi-telescope systems for optical astronomy.**
A. Labeyrie.
Journ. Optical Soc. America, Vol. 65, 1191 (1975). – Conference abstract.

032.036 **Rumastrometri.** E. Høg.
Astron. Tidssk., Årg. 8, p. 186 - 188 (1975).

032.037 **Small high-speed Schmidt camera.**
Sh. N. Sabitov.

Trudy Astrofiz. Inst., *Alma-Ata*, Vol. 25, 76 - 90 (1975). In Russian.

Some features and optical properties of a Schmidt camera with focal ratio 1:1 are described. The high effectivity of this instrument for photographing weak nebulae is shown.

032.038 **Estimate of the limiting magnitude of AZT-8 with an opto-electronic TV-system on the bright background of the sky.** L. A. Usol'tseva.
Trudy Astrofiz. Inst., *Alma-Ata*, Vol. 26, 38 - 47 (1974). In Russian.

032.039 **Instrumental polarization and phase characteristics of the ATsU-5 telescope.** V. N. Milovanov.
Trudy Astrofiz. Inst., *Alma-Ata*, Vol. 26, 164 - 171 (1974). In Russian.

032.040 **A study of the instrumental phase polarization of the horizontal solar telescope.**
V. M. Grigor'ev, A. A. Golovko.
Solnechnye Dannye 1975 Byull., No. 8, p. 78 - 85 (1975). In Russian.

The phase polarization of the horizontal solar telescope may be approximately presented as an effect of an equivalent phase plate with a path difference equal to $\lambda/15 - \lambda/30$. The highest phase polarization is observed in the morning and in the evening, and the lowest in the afternoon. The azimuth of the axis of the maximum phase effect varies weakly during the day. The phase polarization of the telescope leads to a considerable distortion of the line profiles observed with a circular polarization analyzer.

032.041 **Home-built mirror cell for a 210 mm Newtonian.** R. W. Stevens.
Journ. British Astron. Ass., Vol. 86, 47 - 48 (1975).

032.042 **The 63-cm reflector of the Molėtai Astronomical Observatory.**
Bull. Vilnius Astron. Obs., No. 42, p. 34 - 36 (1975). In Russian.

032.043 **Simon Stevin-kijker voor Armand Pien.** T. Vermeesch.
Zenit, Vol. 2, 424 - 425 (1975).

032.044 **Present and future meridian astrometry.** E. Høg.
Space astrometry, (see 012.037), p. 59 - 64 (1975).

032.045 **The 65 cm PZT of the US Naval Observatory.** W. J. Klepczyński.
IAU Colloquium No. 26, (see 012.034), p. 309 - 313 (1975).

032.046 **Determination of orientation of the tubes and the mirror of Sukharev's horizontal meridian circle.**
G. I. Pinigin.
Izv. Glav. Astron. Obs. Pulkovo, No. 193, Astrofiz., Astrometr. i Zvezdn. Astron., p. 109 - 116 (1975). In Russian.

Different methods of determining the orientation of the horizontal meridian circle tubes relative to each other and of the mirror relative to the tubes have been worked out. From the reduction of measurements, made during 1968–1970, the daily and seasonal variations of the mirror relative azimuth ($-0\farcs5$ and $+0\farcs2$ at 1°C correspondingly) and also of the reciprocal orientation of the tubes ($+0\farcs08$ and $0\farcs1$ at 1°C) have been detected.

032.047 **Preliminary results of an investigation of the collimation of the Pulkovo large transit instrument in Chile.**
R. Peralta, D. D. Polozhentsev, T. A. Polozhentseva, R. Tajbo, R. Tapia, G. Silva, G. M. Timashkova.
Izv. Glav. Astron. Obs. Pulkovo, No. 193, Astrofiz., Astrometr. i Zvezdn. Astron., p. 141 - 144 (1975). In Russian.

032.048 **On the precision of determination of the parameters of the Pulkovo large transit instrument in Chile.**
J. Petit, D. D. Polozhentsev, T. A. Polozhentseva, E. Portugal, G. Silva, R. Tajbo, R. Tapia, G. M. Timashkova.
Izv. Glav. Astron. Obs. Pulkovo, No. 193, Astrofiz., Astrometr. i Zvezdn. Astron., p. 145 - 148 (1975). In Russian.

032.049 **Investigations of the pivots of the Pulkovo large transit instrument in Chile.**
S. Akun'ya, M. P. Varin, V. A. Varina, D. Vivers, G. Leon, R. Tajbo.
Izv. Glav. Astron. Obs. Pulkovo, No. 193, Astrofiz., Astrometr. i Zvezdn. Astron., p. 149 - 152 (1975). In Russian.

032.050 **The optics of the prime focus of the 6-meter telescope.**
T. S. Belorossova, N. V. Merman, M. A. Sosnina.
Izv. Glav. Astron. Obs. Pulkovo, No. 193, Astrofiz., Astrometr. i Zvezdn. Astron., p. 175 - 177 (1975). In Russian.

An optical scheme of the prime focus of the large azimuthal telescope ($D = 6$ m; $A = 1:4$) is given. The construction elements and aberrations of the system are described. The quality of images is illustrated by dotted diagrams, plotted for three wavelengths C, F and h and also for three values of the field angle W = 0, W = $3\farcm5$ and W = $5\farcm0$. The penetrating power is determined.

032.051 **An observer's-eye view of the Anglo-Australian telescope.** S. C. B. Gascoigne.
Proc. Astron. Soc. Australia, Vol. 2, 345 - 349 (1975).

032.052 **Amatiera telescops Tukumā.** I. Jurgītis.
Zvaigžņotā debess, 1975. gada vasara, p. 53 - 55.

032.053 **Design of a Cassegrain-Maksutov telescope.** D. Malacara H.
Bol. Inst. Tonantzintla, Vol. 1, 221 - 225 (1975).

A design for a small Cassegrain-Maksutov telescope with a focal ratio F/16 is presented. The design demonstrates that this kind of telescope can be of very high quality. Included in the presentation are details for construction of the optical components.

032.054 **Parallel determination of the collimation of the Large Transit Instrument with the vacuum meridian mark and the new OPTON collimator.** Z. Knežević.
Bull. Obs. Astron. Belgrade, No. 126, p. 16 - 18 (1975).

032.055 **Preparation of the large transit instrument of the Belgrade Observatory for regular observation.**
L. A. Mitić.
Publ. Obs. Astron. Beograd, No. 23, 72 pp. (1975). In Serbo-Croatian.

The installation of the large transit instrument and the derivation of right ascensions; Experimental catalog of right ascensions of 40 stars and some preliminary conclusions; Meridian marks of the fundamental instrument.

032.056 **Telescopes for photographic, photoelectric and spectral observations mounted on Mount Majdanak.**
I. S. Isakov.
Issled. ehkstremal'no molodykh zvezdn. kompleksov. Tashkent, Fan, 1975, p. 143 - 148. In Russian. – Abstr. in Referativ. Zhurn. 51. Astron., 2.51.165 (1976).

032.057 **Results of an investigation of the ZTL-180 zenith telescope of the Kitab Observatory.**
N. N. Litvinova.
Vestn. Leningr. un-ta, 1975, No. 13, p. 127 - 130. In Russian.

Abstr. in Referativ. Zhurn. 51. Astron., 2.51.257 (1976).

032.058 **The most accurate telescope in the world.**
C. A. Scholes.
Polytechn. Tijdschr. Elektrotechn. Elektron. (*Netherlands*), Vol. 30, 277 - 281 (1975). In Dutch. – The astronomical telescope in Siding Spring Mountain in N.S.W. (Australia).

032.059 **Portable power supply for telescope clock drives.**
H. M. Babb, D. S. Gale.
American Journ. Phys., Vol. 43, 376 (1975).
A portable power supply for the use of synchronous motors in telescopes is described. The circuit diagram is given and parts used are listed.

Multiple mirror telescope.
Applied Optics, Vol. 14, 3148 (1975).

Calar-Alto, Deutsch-Spanisches Astronomisches Zentrum im Aufbau. See Abstr. 009.025.

Production and testing of the optical elements of the first 2.2 m-telescope for MPIA.
See Abstr. 031.004.

Zen and the art of big mirror making.
See Abstr. 031.030.

Aspherisation by elasticity of a 50-cm plate for the Lyon Observatory Schmidt telescope.
See Abstr. 031.031.

Schmidt systems for the near-infrared.
See Abstr. 031.048.

On atmospheric turbulence and the problem of optimization of the telescopic pupil. See Abstr. 031.088.

Zur Prozeßrechner-Steuerung der 2,2-m-Teleskope des MPIA. See Abstr. 031.401.

On-line computers for telescope control and data handling. See Abstr. 031.409.

Study on the characteristics of a Talcott levels couple. See Abstr. 041.011.

On the use of Tautenburg Schmidt for improvement of astronomical reference system. See Abstr. 041.054.

The U.K. 48-inch (1.2 metre) Schmidt telescope and the SRC IIIaJ Survey. See Abstr. 041.063.

Time and latitude determination using a portable zenith camera. See Abstr. 044.020.

An experiment on dome heat and seeing disk size.
See Abstr. 082.078.

Photometric system of the Schmidt telescope of the Engelhardt Astronomical Observatory and its connection with the UBV-system. See Abstr. 113.026.

Space Instrumentation

032.501 **The role of optics in space research and techniques.**
G. Courtès.
Space optics, (see 012.013), p. 1 - 8 (1974). In English and French.

032.502 **Influence des contraintes au stade initial de l'étude d'une expérience d'optique spatiale.** A. Girard.
Space optics, (see 012.013), p. 147 - 156 (1974).
This paper surveys certain consequences of the constraints applying to an optics space experiment at the very first stage of the feasibility study, up to the more advanced stage of the preliminary design. This study is particularly oriented to infrared optics experiments.

032.503 **Expérience du laboratoire d'astronomie spatiale embarquée à bord du satellite astronomique D 2 B.**
P. Cruvellier.
Space optics, (see 012.013), p. 165 - 188 (1974).
The author presents here a discussion of the optical problems encountered, in the course of the design of the D 2 B satellite, considering the optical (physical, geometric) and space imperatives. The author explains the optical solution used, which, considering the present state of knowledge in optical technology and the constraints of space experiments, can be considered as the best efficiency-resolution compromise.

032.504 **Transfer functions of space instruments.**
D. Malaise.
Space optics, (see 012.013), p. 189 - 203 (1974).
The aim of this paper is to describe a method employed for calibrating an instrument used in space research, the object of which is to measure the UV flux of the stars.

032.505 **Some new optical instruments for space research.**
G. Courtès.
Space optics, (see 012.013), p. 211 - 227 (1974).

032.506 **The Goddard experiment package.** B. W. Jackson.
Space optics, (see 012.013), p. 229 - 234 (1974).
The Goddard experiment package will measure the ultraviolet spectral response of stars and nebulae. It has a spectral resolution of 2 Å in the 1050 Å to 4000 Å band. The telescope has a 36-inch clear aperture and automatically reduces the spectral data to digital form. Guidance accuracy is one arc second.

032.507 **Methods of imaging X-ray astronomy and the X-ray spectroheliograph on ATM.** G. S. Vaiana.
Space optics, (see 012.013), p. 295 - 337 (1974).

032.508 **Les senseurs optiques.** J. Ragot.
Space optics, (see 012.013), p. 349 - 361 (1974).

032.509 **Les satellites d'observation.** C. Fouché.
Space optics, (see 012.013), p. 379 - 389 (1974).

032.510 **Apollo photographic systems and cartography of the moon.** F. J. Doyle.
Nauka i chelovechestvo. 1975. Moskva, Znanie, 1974, p. 229 - 247. In Russian. – Abstr. in Referativ. Zhurn. 51. Astron., 7.51.75 (1975).

032.511 **Magnetic electron multipliers: efficiency changes.** G. Schmidtke, M. Knothe, F. Heidinger.
Applied Optics, Vol. 14, 1645 - 1648 (1975).
During the mission of AEROS-A satellite the recorded calibration data exhibited a strong variation of the magnetic electron multiplier efficiency. Drastic changes do correlate with temperature; however, a clear separation of the temperature effect cannot be derived from the measurements. In space they have been impeded, e.g., by long-term efficiency changes, while in the laboratory ion feedback and other effects interfered with the measurements. The results affect the design of future space experiments especially for solar XUV measurements.

032.512 **Ellipsoid-hyperboloid X-ray imaging instrument for laser-pellet diagnostics.** R. C. Chase, J. K. Silk.
Applied Optics, Vol. 14, 2096 - 2098 (1975).

032.513 **Photomultiplier window materials under electron irradiation: fluorescence and phosphorescence.**
W. Viehmann, A. G. Eubanks, G. F. Pieper, J. H. Bredekamp.
Applied Optics, Vol. 14, 2104 - 2115 (1975).

032.514 **Spectroradiometric calibration techniques in the far ultraviolet: a stable emission source for the Lyman bands of molecular hydrogen.**
W. G. Fastie, D. E. Kerr.
Applied Optics, Vol. 14, 2133 - 2142 (1975).

032.515 **Preliminary results from the Apollo-Soyuz extreme ultraviolet telescope.**
S. Bowyer, M. Lampton, F. Paresce, B. Margon, R. Stern.
Bull. American Astron. Soc., Vol. 7, 405 (1975). – Abstr. AAS.

032.516 **Flight performance of the 102-cm balloon-borne far-infrared telescope.**
G. G. Fazio, E. L. Wright, F. J. Low.
Bull. American Astron. Soc., Vol. 7, 408 (1975). – Abstr. AAS.

032.517 **Filter wedge spectroscopy in the region 3–6 μm.** L. J. Caroff.
Nature, Vol. 257, 650 (1975). – Simulating experiments for Spacelab.

032.518 **The COS-B experiment for gamma-ray astronomy.** G. F. Bignami, G. Boella, J. J. Burger, P. Keirle, H. A. Mayer-Hasselwander, J. A. Paul, E. Pfeffermann, L. Scarsi, B. N. Swanenburg, B. G. Taylor, W. Voges, R. D. Wills.
Space Sci. Instrumentation, Vol. 1, 245 - 268 (1975).
The COS-B satellite will carry a single large experiment, capable of detecting gamma-rays with energies in excess of 20 MeV in order to study in detail the sources of extraterrestrial gamma radiation. The principal objectives of the study will be to investigate the angular structure of gamma-ray emission from the galactic plane, to measure the flux of the isotropic radiation from high galactic latitudes believed to be of extragalactic origin, to examine known or postulated point sources of the radiation and to determine the energy spectra of all identified sources.

032.519 **The capability of COS-B to measure cosmic gamma ray bursts.**
G. Boella, M. Gorisse, J. Paul, B. G. Taylor, R. D. Wills.
Space Sci. Instrumentation, Vol. 1, 269 - 277 (1975).
The COS-B experiment for gamma ray astronomy features a large area (1.1 m^2) plastic scintillator, 1 cm thick, for the detection and rejection of charged particles entering the experiment. Tests with radioactive sources have confirmed that the effective area of this detector is several hundred square centimeters for the detection of gamma rays of energy greater than 100 keV. Burst onset time will be determined to the order of a millisecond which, given the highly eccentric orbit of COS-B, should enable sources to be located to better than one degree using the triangulation technique with other satellite observations. The determination of structure down to 1.6 ms intervals will provide unique data relating to the source mechanisms.

032.520 **The Apollo 15 and 16 X-ray fluorescence experiment.** I. Adler, R. Schmadebeck, J. I. Trombka, P. Gorenstein, P. Bjorkholm.
Space Sci. Instrumentation, Vol. 1, 305 - 316 (1975).
An X-ray fluorescence spectrometer was flown aboard the Apollo 15 and 16 spacecrafts orbiting the moon. The X-ray instrument was used to produce a chemical map of that portion of the moon covered by the projected ground tracks and illuminated by the sun during the period of flight. The instrument includes three thin windowed proportional counters, two of which have selected X-ray filters. The field of view of the surface is determined by a collimator, while a detector on the opposite side of the spacecraft provides a continuous monitoring of the solar X-ray output.

032.521 **Calibration of the UV experiment in ANS.**
J. W. G. Aalders, R. J. van Duinen, W. Luinge, K. J. Wildeman.
Space Sci. Instrumentation, Vol. 1, 343 - 350 (1975).
The paper describes the calibration of the ultraviolet spectrophotometer in the Netherlands Astronomical Satellite, launched 30 August, 1974. The instrument response curves have been determined in a vacuum optical bench facility. The absolute calibration was obtained by comparison with a standard maintained at the Space Astronomy Laboratory in Wisconsin. The final result is believed to be accurate to 30 %.

032.522 **A slotted disk velocity selector for the detection of energetic atoms above the atmosphere.**
J. H. Moore, Jr., C. B. Opal.
Space Sci. Instrumentation, Vol. 1, 377 - 386 (1975).
A slotted disk velocity selector suitable for use in space has been developed to measure the flux of neutral particles with velocities up to 500 km s^{-1}. The instrument is rugged, positively rejects photons and charged particles, and has a very narrow field of view. Possible uses of the instrument are to measure the density and temperature of interstellar helium entering the solar system and to measure the neutral hydrogen flux and H°/H^{+} ratio in a proton aurora.

032.523 **The hard X-ray experiment on the Astronomical Netherlands Satellite.**
H. Gursky, H. Schnopper, D. Parsignault.
Astrophys. Journ., (*Letters*), Vol. 201, L127 - L131 (1975).
The hard X-ray experiment flown on the Astronomical Netherlands Satellite is described. The experiment consists of two parts. One is a large-area detector of about 60 cm^2 total area, sensitive in the energy range 1.5–30 keV. Two counters comprise this detector, each collimated by $10' \times 3°$ and offset in the narrow direction by 4'. The other part is a Bragg crystal assembly consisting of two PET crystals and counters aligned to search for the silicon emission lines near 2 keV. Instrument characteristics and orbital operations are described.

032.524 **Instrumental technique in X-ray astronomy.** L. E. Peterson.
Annual Rev. Astron. Astrophys., Vol. 13, (see 003.005), 423 - 509 (1975).

032.525 **Rocket X-ray studies of supernova remnants.** A. F. Janes.
Journ. British Interplanet. Soc., Vol. 28, 779 - 782 (1975).

032.526 **The use of balloons in astronomical research.** R. R. Hillier.
Journ. British Interplanet. Soc., Vol. 28, 810 - 815 (1975).

032.527 **Computation of the characteristics of a gamma-telescope by the Monte Carlo method.**
N. A. Gun'ko, B. A. Dmitriev.
Kosmich. Issled., Vol. 13, 784 - 786 (1975). In Russian. Brief information.

032.528 **Wavefront error simulator for evaluating optical testing instrumentation.** L. J. Golden.
Applied Optics, Vol. 14, 2756 - 2761 (1975).

A wavefront error simulator has been designed and fabricated to evaluate experimentally test instrumentation for the Large Space Telescope program. The principal operating part of the simulator is an aberration generator that introduces low-order aberrations of several waves magnitude with an incremented adjustment capability of $\lambda/100$. Each aberration type can be introduced independently with any desired spatial orientation.

032.529 **Space astrometry projects.** P. Lacroute.
Space astrometry, (see 012.037), p. 5 - 30 (1975).

This report presents two space-astrometry projects which are specially designed for obtaining a very precise, coherent and complete stellar reference system. The first project utilises a Spacelab-borne 40 cm-diameter telescope which gives positions of 0".001 up to the magnitude 13 and 0".005 up to magnitude 16. The same precision on the parallaxes can be obtained from two years of scanning, and a scan programme of 40000 stars could be envisaged. The second project utilises a simpler guidance device, and could be flown on board a free-flyer similar to the TD-1 satellite. The automatic sky scanning will enable the positions of 300000 stars up to magnitude 11 (approx.) to be determined with random errors in positions and parallaxes of about 0".001 for bright stars and $<$ 0".003 for stars up to magnitude 10.

032.530 **Space astrometry project – optical concept.** P. Bacchus.
Space astrometry, (see 012.037), p. 31 - 35 (1975).

Solutions are proposed for the telescope to be used. In the Spacelab option, an aplanatic system (Ritchey-Chrétien) is sufficient because focals parallel to the grid can be used. In the TD option, an aplanatic system is required. It is possible to use four-reflection optics (plane mirror, large concave mirror, small convex mirror and a fourth reflection from the central part of the large mirror) or three-reflection optics (deformed plane mirror, large mirror, small mirror).

032.531 **Faisabilité d'un satellite astrométrique – resultat d'une étude CNES.** J. C. Husson.
Space astrometry, (see 012.037), p. 43 - 47 (1975).

032.532 **Astrometric experiments for the large space telescope.**
L. W. Fredrick, H. A. McAlister, W. F. van Altena, O. G. Franz.
Space astrometry, (see 012.037), p. 49 - 58 (1975).

This paper reviews the basic astrometry problems to which the LST will offer major contributions, (i) parallaxes, (ii) binary stars, (iii) stellar dimensions, and (iv) fundamental co-ordinate system. Two novel approaches to setting up the fundamental co-ordinate system are discussed: (i) a system of precision gyroscopes, (ii) an orthogonal axis telescope. The relative accuracy of positions over the whole sky should be in the order of ±0".001 of arc after several measurements.

032.533 **On the choice of the spectral sensitivity of optical systems for investigation of the earth from space.**
Yu. M. Chesnokov, V. A. Kottsov.
Uspekhi fiz. nauk, Vol. 116, 746 - 747 (1975). In Russian. Abstr. in Referativ. Zhurn. 62. Issled. kosmich. prostranstva, 1.62.122 (1976).

032.534 **Proportional counters for investigation of high-energy cosmic rays used aboard the Proton scientific stations.** N. L. Grigorov, G. P. Kakhidze, V. I. Lazutin, I. N. Moskalenko, I. D. Rapoport, I. A. Savenko, A. A. Tavrizov, V. A. Tarutin, I. I. Chistov, V. F. Shesterikov, B. M. Yakovlev.
Issled. kosmich. luchej. Moskva, Nauka, 1975, p. 181 - 196. In Russian. – Abstr. in Referativ. Zhurn. 62. Issled. kosmich. prostranstva, 1.62.200 (1976).

032.535 **Mass absorption coefficients of Parylene N at soft X-ray and vacuum-ultraviolet wavelengths.**
R. Stern, F. Paresce.
Journ. Optical Soc. America, Vol. 65, 1515 - 1516 (1975). Letter.

032.536 **Gamma noise in CCDs (charge coupled devices).** G. W. Autio, M. A. Bafico.
Infrared Physics, Vol. 15, 249 - 258 (1975).

032.537 **SEC vidicon system for a Balloon Ultraviolet Stellar Spectrometer.** T. M. Kamperman.
Image processing techniques in astronomy, (see 012.041), p. 127 - 134 (1975).

032.538 **Zone generator for large space telescope technology. Final report.** K. E. Erickson.
NASA Rep. NASA-CR-132471, [Available from NTIS, Springfield, Va., USA], 46 pp. (1974).

A concept is presented for monitoring the optical adjustment and performance of a large space telescope which consists of a 1.2 m diameter turntable with a laser stylus to operate at speeds up to 30 rpm. The focus of the laser stylus is under closed loop control. A technique for scribing zones of suitable depth, width, and uniformity applicable to large telescope mirrors is also reported.

032.539 **Large space telescope oscillations induced by CMG (*control moment gyroscope*) friction.**
S. M. Seltzer.
Journ. Spacecraft Rockets, (*USA*), Vol. 12, No. 2, p. 96 - 105 (1975).

032.540 **An image intensification and video recording system for astronomical observations.**
M. M. Hoffman, W. M. Sanders, D. H. Liebenberg.
Optical Engineering, (*USA*), Vol. 14, 76 - 80 (1975).

An airborne telescope utilizing a scanning interferometer and video recording system has been used by the Los Alamos Scientific Laboratory for coronal-emission-line-profile measurements during the total solar eclipses of 1972 and 1973. A brief description of the experiment, its objectives and methods is given in the first section. The remainder of the paper is devoted to a description and discussion of the image intensification and TV recording system employed for the coronal-emission-line-profile measurements.

032.541 **An Si(Li) detector system for application to X-ray**

astronomy rocket experiments.
R. E. Griffiths, C. Cheron, A. Friant, C. Jehanno, R. Rocchia, R. Rothenflug, O. Testard.
IEEE Trans. Nuclear Sci., Vol. NS-22, 611 - 615 (1975).

The problems associated with the use of Si(Li) detectors in X-ray astronomy rocket experiments are discussed. In particular a detector system is described that can be used at the focus of a grazing-incidence paraboloid telescope for the energy range 0.3–2 keV.

032.542 **Large area soft X-ray imaging system for cosmic X-ray studies from rockets.**
P. Gorenstein, H. Gursky, F. R. Harnden,Jr., A. Decaprio, P. Bjorkholm.
IEEE Trans. Nuclear Sci., Vol. NS-22, 616 - 619 (1975).

032.543 **A balloon-borne infrared telescope cooled by liquid nitrogen.** S. Hayakawa, T. Iijima, K. Ito, T. Matsumoto, T. Ono, K. Uyama, T. Nishimura.
Japanese Journ. Applied Phys., Vol. 14, 1041 - 1048 (1975).

032.544 **Astronomical observations with the University College, London, balloon borne telescope.**
R. E. Jennings,
Rep. Univ. College, London, (*GB*), 9 pp. (1974).

The characteristics of a telescope system which was developed for high-altitude balloon astronomy are discussed. A drawing of the optical system of the telescope is provided. A sample of the signals recorded during one of the flights is included. The correlation between the infrared flux and the radio continuum flux in analyzed. A far infrared map of the radio and infrared peaks of selected stars is developed. The spectrum of the planet Saturn is plotted to show intensity as compared with wavenumber.

032.545 **Design and performance of a 39-cm balloon-borne telescope.** H. S. Tomlinson, W. A. Towlson, T. E. Venis.
Rep. Univ. College, London, (*GB*), 13 pp. (1974).

A system for stabilizing a balloon-borne telescope using a star sensor device is described. Guide stars from –4 to +4 mag can be used.

032.546 **Balloon infrared astronomy platform (BIRAP).**
M. E. Greeb, G. A. True.
Rep. Ball Bros. Res. Corp., Boulder, Colo., USA, 16 pp. (1974).

032.547 **A balloon-borne 1 meter telescope for far-infrared astronomy.** G. G. Fazio, D. E. Kleinmann.
Rep. Univ. Arizona, Tucson, USA. 13 pp. (1974).

The flight of a balloon-borne one-meter telescope for infrared astronomy in the wavelength interval of 40 to 240 microns is discussed. The gyro-stabilized telescope mapped the intensity of the far infrared radiation from NGC 7538, Mars, the Orion nebula, and W3 with a resolution of one minute and from selected regions of these sources with a resolution of 30 seconds. The infrared detection is described and its capabilities are analyzed. The instrumentation, orientation system and modes of observation of the telescope are defined.

032.548 **Airoscope stellar acquisition.**
G. J. Deboo, G. T. Parra, R. C. Hedlund.
Rep. NASA, Moffett Field, Calif., USA. 14 pp. (1975).

The acquisition system which operates in conjunction with a balloon-borne TV system, boresighted to a telescope is described.

032.549 **The Large Space Telescope.** A. B. Wissinger.
Optical Engineering, (*USA*), Vol. 14, 107 - 111 (1975).

The scientific objectives of the Large Space Telescope (LST), currently in the preliminary design phase, are summarized. The conceptual design of the optical telescope assembly and system requirements introduced by the scientific instruments are also described. LST operations are characterized with emphasis on the fine pointing system, star acquisition, and image motion sensing.

032.550 **The Orbiting Astronomical Observatory.**
A. B. Wissinger.
Optical Engineering, (*USA*), Vol. 14, 133 - 140 (1975).

The engineering features of the Perkin-Elmer-built telescope are surveyed. A digest of the scientific results and a bibliography of OAO-C related papers is given, and the nature of the investigations planned at the time of writing (September 1974) is summarized.

Two mirror anastigmats. See Abstr. 031.009.

Space astrometry project – reduction by multilateration. See Abstr. 031.264.

Lock-in image subtraction: detectability of circumstellar planets with the Large Space Telescope. See Abstr. 031.295.

Rocket-photometer for stellar UV-observations. See Abstr. 034.056.

A scanning Fabry-Perot interferometer and its application to space research. III. Piezo-electric scanning Fabry-Perot interferometer. See Abstr. 034.153.

Ergebnisse der Lyman-Alpha Messungen des Satelliten DIAL. See Abstr. 034.168.

Ultraviolet astronomy from Skylark rockets. See Abstr. 051.006.

Radioactivity observed in the sodium iodide gamma-ray spectrometer returned on the Apollo 17 mission. See Abstr. 051.010.

Extreme-ultraviolet astronomy from Apollo-Soyuz. See Abstr. 061.006.

Skylab and the ASE X-ray telescope experiment: a new view of the X-ray corona. See Abstr. 074.073.

Coronal emission line profile observations at total solar eclipses. I: Airborne instrumentation and results. See Abstr. 074.107.

Crystal spectrometer studies of the sun employing a rotation modulation collimator. See Abstr. 076.023.

On the possibility of recording solar γ-emission flares and its realization in extra-atmospheric measurements. See Abstr. 076.040.

Study of the earth-moon system by space techniques. See Abstr. 094.012.

Astronomical far infrared sky survey. See Abstr. 141.645.

033 Radio Telescopes and Equipment

033.001 **Radio telescopes of large resolving power.** M. Ryle.
Rev. Modern Phys., Vol. 47, 557 - 566 (1975). – Lecture delivered on the occasion of the presentation of the 1974 Nobel Prizes in Physics.

033.002 **Radio telescopes for millimeter wavelength.** S. von Hoerner.
Astron. Astrophys., Vol. 41, 301 - 306 (1975).
General rules are discussed for selecting the shortest observational wavelength, and for finding the largest technically possible diameter for the design of steerable radio telescopes. As an example of a highly optimized structure, the NRAO design of a 65-m telescope for 3.5 mm wavelength is briefly described.

033.003 **Some new possibilities of optimum synthetic linear arrays for radioastronomy.**
E. J. Blum, J. C. Ribes, F. Biraud.
Astron. Astrophys., Vol. 41, 409 - 411 (1975).
An array will be called optimum when the set of spacings between its antennas comprises once and only once all multiples of a given increment. Such an array may be synthesized by using a few movable antennas in several successive configurations. The authors show that optimum arrays of 4 antennas used in 5 configurations exist, giving all multiples from 2 through 31 or 3 through 32. There are also quasi optimum solutions: an alternative to the existing Westerbork array is presented.

033.004 **Unique radio telescope at Bonn University.** G. Nestorov.
Priroda (NRB), Vol. 23, No. 6, p. 6 - 7 (1974). In Bulgarian.

033.005 **The influence of reflection of ultrashort waves from an underlying surface on temporary fluctuations of readings of the indicator of a radio range finder due to atmospheric turbulence.** D. D. Dzyaman.
Trudy Kazan. Gorod. Astron. Obs., No. 39, p 176 - 188 (1973). In Russian.

033.006 **The new digitalized solar radio spectrograph in Durnten.** B. von Arx, A. O. Benz, G. L. Tarnstrom.
Bull. American Astron. Soc., Vol. 7, 407 - 408 (1975). – Abstr. AAS.

033.007 **Observations with a VLB array. I. Introduction and procedures.** M. H. Cohen, A. T. Moffet, J. D. Romney, R. T. Schilizzi, D. B. Shaffer, K. I. Kellermann, G. H. Purcell, G. Grove, G. W. Swenson, Jr., J. L. Yen, I. I. K. Pauliny-Toth, E. Preuss, A. Witzel, D. Graham.
Astrophys. Journ., Vol. 201, 249 - 255 (1975).
The authors describe a series of observations intended to investigate the structure and the time development of the structure of compact extragalactic radio sources at $\lambda = 2.8$ cm and resolution 0".001. In this paper the five telescopes, the sensitivity of the system, the calibration procedures, and modeling procedures are discussed.

033.008 **Successful development of a compound radio interferometer ('2 × 16') operating at 450 MHz.**
Radio Astronomy Division, Peking Observatory, Academia Sinica.
Acta Astron. Sinica, Vol. 16, 91 - 92 (1975). In Chinese. – Research note.

033.009 **Radio-interferometer in Nijmegen.** M. Drummen.
Zenit, Vol. 2, 322 (1975).

033.010 **Removal of the influence of the side-lobes of a radio telescope aerial upon the radio map of the sky.**
L. Nowakowski.
Postępy Astron., Vol. 23, 131 - 140 (1975). In Polish.

033.011 **Interferometer for observations of the sun at 127 MHz.**
K. Borkowski, S. Gorgolewski, J. Usowicz.
Postępy Astron., Vol. 23, 141 - 151 (1975). In Polish.
The interferometer for routine observations of the sun at 127 MHz is described. This description contains the aerials, the circuit diagrams and the measured performance parameters of the radiometer.

033.012 **The Nançay radioheliograph.** M. Pick.
Oss. Mem. Oss. Astrofis. Arcetri, Fasc. 105, (see 012.026), p. 106 - 109 (1975).

033.013 **Project of a large collecting array at Nançay.**
M. Aubier, A. Boischot, G. Daigne, Y. Leblanc, A. Lecacheux, J. de la Noë, B. Møller-Pedersen, C. Rosolen.
Oss. Mem. Oss. Astrofis. Arcetri, Fasc. 105, (see 012.026), p. 110 - 114 (1975).

033.014 **Calibration and basic parameters of a solar radiopolarimeter at 9.6 cm wavelength.**
A. M. Karachun,V. A. Kovalev.
Fiz. ionosfery, magnit. pole Zemli i solnechno-zemn. svyazi. Moskva, 1974, p. 74 - 77. In Russian. – Abstr. in Referativ. Zhurn. 51. Astron., 10.51.138 (1975).

033.015 **A high-sensitive 8-mm radiometer with a broadband parametric amplifier.**
Yu. A. Militskij, I. A. Strukov, V. S. Ehtkin.
Izv. vyssh. ucheb. zavedenij. Radiofizika, Vol. 18, 773 - 774 (1975). In Russian. – Abstr. in Referativ. Zhurn. 51. Astron., 11.51.112 (1975).

033.016 **A radiospectrograph for $\lambda = 0.6-1.2$ m.**
Yu. B. Vedeneev, V. O. Kaloshin, N. M. Prytkov.
Solnechnye Dannye 1975 Byull., No. 8, p. 75 - 77 (1975). In Russian.
The main characteristics of a dynamic radio spectrograph for $\lambda = 0.6-1.2$ m are given. The resolving powers are 1 MHz and 1 sec. The spectrograph is used for registrating solar radio bursts with a radio telescope of a 15 m diameter paraboloid.

033.017 **Automatic spectroscopy with the VRO 120-foot radio telescope.** J. C. Webber, K. S. Yang, R. M. Crutcher, G. W. Swenson, Jr., L. E. Snyder.
Bull. American Astron. Soc., Vol. 7, 540 (1975). – Abstr. AAS.

033.018 **The application of radio interferometers in astrometry.** S. Gorgolewski.
IAU Colloquium No. 26, (see 012.034), p. 453 - 456 (1975).

033.019 **Hardware and software for the Llanherne low frequency sky surveys.** P. S. Whitham.
Proc. Astron. Soc. Australia, Vol. 2, 338 - 340 (1975).

033.020 **Radiointerferometru jaunaka paaudze.**
A. Balklavs.
Zvaigžnotā debess, 1975. gada vasara, p. 37 - 41.

033.021 **A 1024-channel digital correlator.**
J. G. Ables, B. F. C. Cooper, A. J. Hunt, G. G. Moorey, J. W. Brooks.
Rev. Sci. Instruments, Vol. 46, 284 - 295 = Div. Radiophys., CSIRO, Sydney, Australia, Radiophys. Publ. RPP 1772 (1975).

A 1024-channel digital correlator designed for spectral analysis of radio astronomy signals is described. The correlator has a maximum sampling rate of 20 MHz, allowing it to analyze noise signals with bandwidths up to 10 MHz.

033.022 **The magnetic delay line processor – a new device.**
T. W. Cole.
Journ. Phys. E, (Sci. Instruments), Vol. 8, 200 - 202 = Div. Radiophys., CSIRO, Sydney, Australia, Radiophys. Publ. RPP 1796 (1975).

A new technique of transversal filtering is introduced for real-time matched filtering, pulse compression, decoding and correlation analysis of electrical signals where the data rates are less than several hundred kilohertz. Experimental results confirm the theory and illustrate the application with band-pass filters.

033.023 **Choice of grid configuration of array antennas.**
B. N. Das, J. S. Rao.
Electronics Letters, Vol. 11, 170 - 171 (1975). – *ACM*

033.024 **Circular aperture distribution with one parameter.**
R. C. Hansen.
Electronics Letters, Vol. 11, 184 (1975). – *ACM*

033.025 **Gunn-effect amplifiers for microwave communication systems in X, KU, and KA bands.**
J. G. de Koning, R. E. Goldwasser, R. J. Hamilton, Jr., F. E. Rosztoczy.
IEEE Trans. Microwave Theory Techn., Vol. MTT-23, 367 - 374 (1975). – *MGM*

033.026 **Analysis and design of a low-noise X-band MIC mixer 1-GHz IF amplifier.**
J. E. Degenford, B. A. Newman.
IEEE Trans. Microwave Theory Techn., Vol. MTT-23, 375 - 376 (1975). – *MGM*

033.027 **Wideband Ga As Gunn reflection amplification for the 18–25.5 GHz waveguide band.**
J. G. de Koning, R. E. Goldwasser, R. J. Hamilton, Jr., F. E. Rosztoczy.
Electronics Letters, Vol. 11, 195 - 196 (1975). – *KJW*

033.028 **The development of an L-band radiometer dual-mode horn.** M. C. Bailey.
IEEE Trans. Antennas Propagation, Vol. AP-23, 439 - 441 (1975). – *BMT*

033.029 **Reflector surface error compensation in Cassegrain antennas.** P. R. Cowles, E. A. Parker.
IEEE Trans. Antennas Propagation, Vol. AP-23, 323 - 328 (1975). – *BMT*

033.030 **8-cm radioheliograph.**
M. Ishiguro, H. Tanaka, S. Énomé, C. Torii, Y. Tsukiji, S. Kobayashi, N. Yoshimi.
Proc. Res. Inst. Atmosph. Nagoya Univ., Vol. 22, 1 - 25 (1975). – *DNC*

033.031 **Noise – confusion in more ways than one. 2-noise temperature and noise generators.** K. L. Smith.
Wireless World, Vol. 81, 169 - 173 (1975). – *SS*

033.032 **A modern loose locked digital frequency synthesizer.** D. V. Stringer.
Marconi Rev., Vol. 37, 181 - 208 (1974). – *SS*

033.033 **The design of coupled microstrip lines.**
S. Akhtarzad, T. R. Rowbotham, P. B. Johns.
IEEE Trans. Microwave Theory Techn., Vol. MTT-23, 486 - 492 (1975). – *JBS*

033.034 **Corrugated rectangular horns for use as microwave feeds.** R. Baldwin, P. A. McInnes.
Proc. Instn. Electr. Engineers (GB), Vol. 122, 465 - 469 (1975).

033.035 **Loss considerations for microstrip resonators.**
E. Belohoubek, E. Denlinger.
IEEE Trans. Microwave Theory Techn., Vol. MTT-23, 522 - 526 (1975). – *JBS*

033.036 **Effects of pump noise in parametric amplifiers.**
J. Clarke.
Proc. Instn. Electr. Engineers (*GB*), Vol. 122, 473 - 478 (1975). – *DJC*

033.037 **Broad-band varactor-tuned impatt-diode oscillator.**
D. H. Claxton, P. T. Greiling.
IEEE Trans. Microwave Theory Techn., Vol. MTT-23, 501 - 504 (1975). – *JBS*

033.038 **Equivalent circuit of the Schottky-barrier field-effect transistor at microwave frequencies.**
R. H. Dawson.
IEEE Trans. Microwave Theory Techn., Vol. MTT-23, 499 - 501 (1975). – *JBS*

033.039 **Conformal antenna synthesis using spherical harmonic wave functions.** J. R. James.
Proc. Instn. Electr. Engineers, (*GB*), Vol. 122, 479 - 486 (1975). – *ACM*

033.040 **Performance of dual-gate Ga As mesfets as gain-controlled low-noise amplifiers and high-speed modulators.** C. A. Liechti.
IEEE Trans. Microwave Theory Techn., Vol. MTT-23, 461 - 469 (1975). – *JBS*

033.041 **Synthesis of a single null response in an otherwise omnidirectional pattern using a circular array.**
J. C. Lim, D. E. N. Davies.
Proc. Instn. Electr. Engineers, (*GB*), Vol. 122, 349 - 352 (1975). – *ACM*

033.042 **Pointing accuracy and electrical performance of Kashima 26 m antenna, obtained from radio astronomical observations.** N. Kawano, M. Inoue.
Journ. Radio Res. Lab., (*Japan*), Vol. 21, 543 - 550 (1974). *BMT*

033.043 **Simple square law circuits of wide dynamic range.**
P. A. McGovern.
Proc. Instn. Radio Electronics Engineers Australia, Vol. 36, 119 - 121 (1975). – *MB*

033.044 **Optimization of an 11 GHz mixer circuit using image recovery.** C. J. Burkley, R. S. O'Brien.
International Journ. Electronics, Vol. 38, 777 - 787 (1975). *TWC*

033.045 **Hybrid parametric amplifier.**
C. S. Aitchison, A. S. Wong.
Electronics Letters, Vol. 11, 338 - 339 (1975). – *GGM*

033.046 **Designing phase-locked oscillators for synchronization.** M. A. Rich.

IEEE Trans. Circuits Systems, Vol. CS-21, 466 - 472 (1974). *GMCC*

033.047 **Alumina microstrip Ga As F.E.T. 11 GHz oscillator.** N. A. Slaymaker, J. A. Turner.
Electronics Letters, Vol. 11, 300 - 301 (1975). – *JWB*

033.048 **Zener-diode noise generators.** P. I. Somlo.
Electronics Letters, Vol. 11, 290 (1975). – *JWB*

033.049 **S-parameters simplify accurate VCO design.** G. Strid.
Microwaves, Vol. 14, No. 5, p. 34, 36, 38, 40 (1975). – *SS*

033.050 **Integrated optics mirror microwave concepts. Part 2.** F. A. Blum.
Microwaves, Vol. 14, No. 5, p. 56 - 57, 59, 60, 62 - 64 (1975).

033.051 **Frequency response of single and multi-stage transistor amplifiers – a simple method of obtaining closed-form solutions.** M. B. Das, H. Farrokh.
International Journ. Electronics, Vol. 38, 657 - 662 (1975).

033.052 **Design method for interferometer arrays: measurement of the elevation angle of waves in the presence of a reflecting terrain.** A. R. Domville, S. R. Moore.
Proc. Instn. Electr. Engineers, (*GB*), Vol. 122, 605 - 608 (1975). – *BMT*

033.053 **Microwave integrated-circuit MIS varactor phase shifter.**
R. L. Camisa, B. F. Hitch, S. Yuan, M. Ettenberg.
RCA Rev., Vol. 36, 296 - 315 (1975). – *MWS*

033.054 **The Schwarzschild reflector antenna with multiple or scanned beams.** B. Claydon.
Marconi Rev., Vol. 38, 14 - 40 (1975). – *DNC*

033.055 **Design of cavity-stabilised microwave oscillators.** R. Knöchel, K. Schünemann.
Electronics Letters, Vol. 11, 405 - 406 (1975). – *DJC*

033.056 **Directional coupler for gigahertz frequencies, based on the coupling properties of two planar comb transmission lines.** D. J. Gunton, E. G. S. Paige.
Electronics Letters, Vol. 11, 406 - 408 (1975). – *DJC*

033.057 **The antenna drive and drive control system of the Helios telecommand station.** R. Schwarz.
Siemens Rev., Vol. 42, 200 - 203 (1975). – *ACM*

033.058 **Open and closed-loop control of drives for ground station antennas.** E. Böhm.
Siemens Rev., Vol. 42, 204 - 209 (1975). – *ACM*

033.059 **Dynamic behaviour and drive control of large antennas simulated by digital computer.**
H. Eschenauer, G. Henning, H. Tröndle.
Siemens Rev., Vol. 42, 209 - 216 (1975). – *ACM*

033.060 **Noise – confusion in more ways than one. 4 – noise figure and the design of front-ends.** K. L. Smith.
Wireless World, Vol. 81, 326 - 330 (1975). – *MB*

033.061 **Tunable X band GaAs F.E.T. amplifier.** R. A. Soares, J. A. Turner.
Electronics Letters, Vol. 11, 474 - 475 (1975). – *JWB*

033.062 **Broadband beam antennas for EHF, circularly-polarized and fan-shaped.**
J. J. Epis, S. N. Watkins.
Microwave Journ., Vol. 18, No. 8, p. 45 - 47, 59 (1975).

033.063 **Low-noise room-temperature and cryogenic mixers for 80 - 120 GHz.** A. R. Kerr.
IEEE Trans. Microwave Theory Techn., Vol. MTT-23, 781 - 787 (1975). – *TWC*

033.064 **A technique for determining the local oscillator waveforms in a microwave mixer.** A. R. Kerr.
IEEE Trans. Microwave Theory Techn., Vol. MTT-23, 828 - 831 (1975). – *TWC*

033.065 **A new algorithm in spectral analysis and band-limited extrapolation.** A. Papoulis.
IEEE Trans. Circuits Systems, Vol. CAS-22, 735 - 742 (1975).

033.066 **Small-signal simulation of impatt diode amplifiers.** J. Christie, J. A. C. Stewart.
International Journ. Electronics, Vol. 39, 61 - 73 (1975).

033.067 **Input admittance of a circular waveguide aperture loaded with an iris.**
M. D. Deshpande, B. N. Das.
Proc. Instn. Electr. Engineers, (*GB*), Vol. 122, 795 - 799 (1975). – *GJM*

033.068 **High-efficiency GaAs mesfet amplifiers.** H. C. Huang, I. Drukier, R. L. Camisa, S. Y. Narayan, S. T. Jolly.
Electronics Letters, Vol. 11, 508 - 509 (1975). – *GJM*

033.069 **Method of testing mirror parallelism of a multi-mirror radio telescope.** A. V. Skuratov.
Trudy TsNII geod., aehrosemki i kartogr., 1975, vyp. (No.) 210, p. 63 - 74. In Russian. – Abstr. in Referativ. Zhurn. 51. Astron., 2.51.175 (1976).

033.070 **Radiometers for superlong-range radio interferometry.** L. R. Kogan, G. D. Kopelyanskij, V. I. Kostenko, L. I. Matveenko, A. N. Timofeev, G. N. Khramov.
In-t kosmich. issled. AN SSSR. Pr-224. Moskva, 1975. 12 pp. In Russian. – Abstr. in Referativ. Zhurn. 51. Astron., 2.51. 176 (1976).

033.071 **Radiometric complex at 927 MHz with a small computer.** V. M. Demarev, S. K. Malyshev, M. I. Metelev, V. A. Mil'chenko, V. I. Projdakov, V. A. Rassadovskij, R. A. Samojlov, V. V. Khrulev, S. P. Chekalev, S. A. Shmulevich.
Izv. vyssh. ucheb. zavedenij. Radiofizika, Vol. 18, 1215 - 1217 (1975). In Russian. – Abstr. in Referativ. Zhurn. 51. Astron., 2.51.177 (1976).

033.072 **Modulation radiometers of the decimeter-wavelength region for study of irregular sporadic radio radiation pulses.** S. K. Malyshev, V. I. Morozov, R. A. Samojlov, B. K. Fedyantsev, V. V. Khrulev, S. P. Chekalev, S. A. Shmulevich.
Izv. vyssh. ucheb. zavedenij. Radiofizika, Vol. 18, 1211 - 1214 (1975). In Russian. – Abstr. in Referativ. Zhurn. 51. Astron., 2.51.178 (1976).

033.073 **Accuracy considerations in the measurement of the power gain of a large microwave antenna.**
M. Kanda.
IEEE Trans. Antennas Propagation, Vol. AP-23, 407 - 411 (1975). – *BMT*

033.074 **The geometrical theory of diffraction for axially symmetric reflectors.**
W. V. T. Rusch, O. Sørensen.

IEEE Trans. Antennas Propagation, Vol. AP-23, 414 - 419 (1975). – *BMT*

033.075 **A fast Fourier transform spectrometer with applications in radio astronomy.**
I. R. Linscott, J. W. Erkes, N. R. Powell.
Dudley Obs. *Albany, New York,* Rep. No. 10, 30 pp. (1975).

A coherent radio-frequency spectrometer, based on the fast Fourier transform (FFT) method, has been designed to study a large class of astrophysical phenomena with higher sensitivity and resolution than is provided by present instrumentation. Thus coupled with an even modest-sized radio telescope, this FFT spectrometer may be used to examine pulsar micro-structure, study pulsar emission mechanisms and the interstellar medium, and search for supernova events and other radio bursts.

033.076 **The mobile meteoric radar of the CNET.**
B. Boulet, H. Gamelon, M. Glass.
Echo Recherche, (*France*), No. 79, p. 20 - 29 (1975). In French.

The CNET has developed a new type of radar to look into the dynamics of the belt between 75 and 105 km in height. This article deals with the electronics of the radar, its automated management by a real time computer, the scientific results already achieved and the prospects for its future utilisation.

033.077 **Receiver for 23 cm.** G. Sinigaglia, G. Tomasetti.
Antenna, (*Italy*), Vol. 47, No. 1, p. 13 - 15 (1975).

The receiver described is intended for radio astronomy work on 1420 MHz, and employs a 1268 MHz oscillator. The circuitry of the receiver is illustrated and described together with some reference to calibrations.

033.078 **Radio telescope project at Frensham Heights school.** J. H. Duncan.
Wireless World, Vol. 81, 289 - 290 (1975).

033.079 **Gravitational deformation and astigmatism of tiltable radio telescopes.**
S. von Hoerner, W.-y. Wong.
IEEE Trans. Antennas Propagation, Vol. AP-23, 689 - 695 (1975).

033.080 **Aperture synthesis.** W. N. Brown.
Methods Comput. Phys., (*USA*), Vol. 14, 131 - 175 (1975).

Théorie de la synthèse d'ouverture; cas de la synthèse d'ouverture rotationnelle. Traitement nécessaire pour convertir les paramètres observés avec un télescope à synthèse d'ouverture en distribution de brillance radio du ciel.

033.081 **Computations in radio-frequency spectroscopy.**
J. A. Ball.
Methods Comput. Phys., (*USA*), Vol. 14, 177 - 219 (1975).

Méthodes d'acquisition et de réduction des données et problèmes concernant l'interprétation des données.

Aperture synthesis. See Abstr. 031.288.

Interactive computer reduction and display of radio supersynthesis maps. See Abstr. 031.419.

Giant radio galaxies. See Abstr. 158.018.

034 Astronomical Accessories (Spectrometers, Photometers, etc.)

034.001 **On aperture correction in scanning interferometry.** P. Nylén.
Physica Scripta, Vol. 11, 356 - 360 (1975).

The instrument functions of scanning Michelson and Fabry-Perot interferometers in the limiting case of coherence are investigated and characteristics such as the scanning aperture correction and resolving power are analyzed. It is shown that there are no essential practical differences in these characteristics from those in the incoherent case.

034.002 **Double-refraction interference filter for coronal line 5303 Å.** B. Valníček.
Bull. Astron. Inst. Czechoslovakia, Vol. 26, 210 - 212 (1975).

For purposes of studying the polarization of the light of the green corona during the eclipse of June 30, 1973 a Lyot type double-refraction filter, equipped with a revolving half-wave plate to change the orientation of the incident light, and a thermostat, was produced.

034.003 **Solar X-ray photometer onboard the Interkosmos-11 satellite and preliminary results of observations.**
F. Fárnik, B. Valníček, B. Komárek.
Bull. Astron. Inst. Czechoslovakia, Vol. 26, 212 - 218 (1975).

The paper describes briefly the principle and design of the new Czechoslovak instrument for measuring solar X rays in the range of 0.4–100 keV, which operated from May to October 1974 onboard the Interkosmos-11 satellite. The detectors used and their effectiveness, the electronics block diagram, the transmission of observed data are described and, to conclude, the preliminary results of the observations, in particular of the flare of July 9, 1974, are given.

034.004 **An investigation of the measuring machine SIP-5.** L. I. Chernykh.
Byull. Inst. Teoret. Astron., *Leningrad*, Vol. 14, 64 - 69 (1975). In Russian.

034.005 **Spektroskopische Ausrüstung der 2,2-m-Teleskope.** J. Solf.
Mitt. Astron. Ges., No. 36, p. 61 - 63 (1975).

034.006 **Die Verwendung eines Lyot-Filterelements zur Intensitätskalibrierung von Spektren.**
C. Trefzger, J. Solf.
Mitt. Astron. Ges., No. 36, p. 64 - 65 (1975).

034.007 **Überlegungen zur Auswahl einer automatischen Plattenmeßmaschine.** G. F. O. Schnur.
Mitt. Astron. Ges., No. 36, p. 121 - 122 (1975).

034.008 **Der "neue" Bonner Komess.** E. Pendl.
Mitt. Astron. Ges., No. 36, p. 122 (1975).
Abstract.

034.009 **Ein neues Chopperprinzip für Infrarotphotometrie.** U. Fahrbach, K. Haussecker, D. Lemke.
Mitt. Astron. Ges., No. 36, p. 130 (1975). – Abstract.

034.010 **Über die Verwendung von elektrostatisch fokussierten Bildverstärkerröhren in der solaren Spektroskopie.** P. N. Brandt, A. Wiesmeier.
Mitt. Astron. Ges., No. 36, p. 130 - 131 (1975).

034.011 **Solare Spektralmessungen im infraroten Bereich.** H. Wöhl.
Mitt. Astron. Ges., No. 36, p. 131 - 132 (1975).

034.012 **Scannermessungen und automatische Äquivalentbreitenbestimmung.** R. Schoembs.
Mitt. Astron. Ges., No. 36, p. 132 - 136 (1975).

034.013 **A dual narrow-band wavelength scanning polarimeter.** D. Clarke, I. S. McLean.
Monthly Notices Roy. Astron. Soc., Vol. 172, 545 - 556 (1975).

A dual-channel polarimeter is described which is capable of making measurements across line profiles of stars to an accuracy limited by photon shot noise statistics. The system is digitally controlled and the photometry performed by pulse counting. Photometric scans can be made by tilting narrow-band interference filters in the orthogonally polarized beams and wavelength positions can be chosen to allow polarimetric measurements to be made at specific points on the line profile. Examples are given of line profile scans and of the statistics associated with polarimetric measurements.

034.014 **Photoguide on a quadrant photomultiplier.** Z. N. Kuteva, R. K. Makarova.
Novaya tekhn. v astron. Vyp. (No.) 5. Leningrad, Nauka, 1975, p. 144 - 146. In Russian. – Abstr. in Referativ. Zhurn. 51. Astron., 7.51.101 (1975).

034.015 **Television astroguide with application of vidicon.** V. A. Malarev.
Novaya tekhn. v astron. Vyp. (No.) 5. Leningrad, Nauka, 1975, p. 139 - 143. In Russian. – Abstr. in Referativ. Zhurn. 51. Astron., 7.51.102 (1975).

034.016 **The solar panorama magnetograph of the Sayan Observatory.**
N. V. Klochek, V. M. Grigor'ev, V. I. Kruglov.
Novaya tekhn. v astron. Vyp. (No.) 5. Leningrad, Nauka, 1975, p. 25 - 30. In Russian. – Abstr. in Referativ. Zhurn. 51. Astron., 7.51.103 (1975).

034.017 **Magnetograph with a photon counter.** N. N. Lebedev.
Novaya tekhn. v astron. Vyp. (No.) 5. Leningrad, Nauka, 1975, p. 30 - 34. In Russian. – Abstr. in Referativ. Zhurn. 51. Astron., 7.51.104 (1975).

034.018 **Electrophotometer for observations of artificial earth satellites.** V. N. Lebedinets, A. A. Gen, L. A. Kapkov, Yu. G. Kaufman, V. P. Prorok.
Nauchn. Informatsii, vyp. (No.) 28, p. 17 - 30 (1974). In Russian.

The requirements to a photometer for artificial earth satellite photometry are given. Methods with application of D.C. amplifier, photon counter and luminous flux modulation are analysed. A description of the photometer model with application of luminous flux modulation and narrow-band amplifier is given.

034.019 **The optical block of the multichannel electrophotometer for satellite observations.**
S. P. Oraevskaya.
Nauchn. Informatsii, vyp. (No.) 28, p. 37 - 50 (1974). In Russian.

034.020 **A comparison of the performances of an insect-eye Fabry-Perot spectrograph and a classical Fabry-Perot monochromator.** J. Meaburn.
Astrophys. Space Sci., Vol. 36, 489 - 493 (1975).

The performance of an image-tube insect-eye Fabry-Perot spectrograph is compared to a classical Fabry-Perot monochromator when used on some topical nebular problems.

034.021 **Photometer calibration problem for extended astronomical sources.** J. A. Muscari.
Applied Optics, Vol. 14, 1491 - 1492 (1975).

034.022 **Temperature and pressure effects on pressure-scanned etalons and gratings.** J. Caplan.
Applied Optics, Vol. 14, 1585 - 1591 (1975).
Various factors affecting the performance of pressure-scanned etalon and grating spectrophotometers are examined. The mechanical properties of the etalon plate, etalon spacer, and grating substrate materials are sometimes important, as is the temperature dependence of the spring constant for the etalon holder springs. The advantage to be gained from invar construction is usually negligible for a properly designed etalon holder.

034.023 **One-dimensional photon-counting detector array for use at EUV and soft X-ray wavelengths.**
J. G. Timothy, R. L. Bybee.
Applied Optics, Vol. 14, 1632 - 1644 (1975).
The performance of a one-dimensional photon-counting detector array based on microchannel array plates has been evaluated in the laboratory over the wavelength range from 461 Å to 1216 Å. The output charge pulses from a set of cascaded microchannel array plates were collected on a proximity-focused anode array consisting of sixty-four linear electrodes 1.3 mm in length and 25 μm in width, set on 50-μm centers. Although the dynamic range of the detector array was severely limited by ion-feedback effects in currently available microchannel array plates, the theoretical spatial resolution of 50 μm was achieved with a detection efficiency at wavelengths below 800 Å equal to that of a conventional channel electron multiplier.

034.024 **Mesh-based semitransparent photocathodes.** G. R. Carruthers.
Applied Optics, Vol. 14, 1667 - 1672 (1975).

034.025 **Focusing crystal polarimeter designed for X-ray polarimetry: construction and performance.**
J. R. Toraskar.
Applied Optics, Vol. 14, 1727 - 1735 (1975).

034.026 **Planar germanium photodiodes.** J. Conradi.
Applied Optics, Vol. 14, 1948 - 1952 (1975).
Planar passivated germanium photodiodes have been developed whose useful response at 22°C extends from 0.6 μm to 1.85 μm at a nominal operating voltage of 20 V. Typical characteristics at 22°C are diode capacitance, 100 pF/cm^2; rise and fall times, 10 nsec; leakage current, 2.5 mA/cm^2; D^* (1.54 μm) $\sim 2.5 \times 10^{10}$ cm/W/$\sqrt{\text{Hz}}$ with a 3-dB frequency of ~45 MHz. Single and multiple element detectors ranging in size from 2 × 0.25-mm^2 elements with 50-μm spacing between elements to 3-cm^2 quadrant detectors have been made successfully for operation at temperatures from 77 K to room temperature.

034.027 **Observing the sun with a fully tunable Lyot-Öhman filter.** J. M. Beckers, L. Dickson, R. S. Joyce.
Applied Optics, Vol. 14, 2061 - 2066 (1975).
The authors describe the properties of the Zeiss universal birefringent filter and its optical and electronic interface with the Sacramento Peak Observatory's vacuum telescope. The instrument permits observations of solar intensities, velocities, and/or magnetic fields in rapid succession in any Fraunhofer line in the 410–700-nm wavelength region with high spectral (0.004–0.013 nm) and high spatial (~1/3 sec of arc) resolution.

034.028 **Astronomical Fourier spectrometer.** P. Connes, G. Michel.
Applied Optics, Vol. 14, 2067 - 2084 (1975).
A high resolution near ir Fourier spectrometer with the same general design as previously described laboratory instruments has been built for astronomical observations at a coudé focus. Present spectral range is 0.8–3.5 μm with PbS and Ge detectors and maximum path difference 1 m. The servo system can accommodate various recording modes: stepping or continuous scan, path difference modulation, sky chopping. A real time computer is incorporated into the system, which has been set up at the Hale 500-cm telescope on Mount Palomar. Samples of the results are given.

034.029 **Infrared Fourier spectrometer for laboratory use and for astronomical studies from aircraft and ground-based telescopes.** H. P. Larson, U. Fink.
Applied Optics, Vol. 14, 2085 - 2095 (1975).
A portable, versatile, ir Fourier spectrometer is described that provides 0.5 cm^{-1} spectral resolution in the 0.87–5.6-μm region. This spectrometer is employed in a varied program of astronomical observations from ground-based telescopes and from the NASA 91.5-cm airborne ir telescope. A number of spectral results are presented to illustrate the performance of this spectrometer in astronomical applications.

034.030 **Temporal- and spatial-intensity-interferometer imaging through a random medium.**
L. E. Estes, R. Boucher.
Journ. Optical Soc. America, Vol. 65, 760 - 764 (1975).

034.031 **Noise-equivalent power and response time of TlSe bolometer at 1.5 K.**
P. S. Nayar, W. O. Hamilton.
Journ. Optical Soc. America, Vol. 65, 831 - 833 (1975).
A far-infrared bolometer detector using *p*-type thallium selenide (TlSe) single crystal as the temperature-sensitive element has been operated at 1.5 K with a noise-equivalent power (NEP) of 8.3×10^{-15} W/Hz$^{1/2}$ and a response time constant of 8×10^{-3} s. The bolometer could be made faster without increasing NEP by reducing the thickness of the element.

034.032 **Image detectors in astronomical observations.** R. C. Smithson.
Bull. American Astron. Soc., Vol. 7, 404 (1975). – Abstr. AAS.

034.033 **High resolution infrared spectroscopy at Kitt Peak.** S. T. Ridgway, D. N. B. Hall.
Bull. American Astron. Soc., Vol. 7, 408 (1975).– Abstr. AAS.

034.034 **Photographic spectrograms in the 3 to 5-micron wavelength region – laboratory results with a broadband infrared upconverter.**
T. R. Gurski, H. W. Epps, S. P. Maran.
Bull. American Astron. Soc., Vol. 7, 408 - 409 (1975). – Abstr. AAS.

034.035 **Multichannel spectrometers of infrared astronomy.** D. M. Rank, J. D. Bregman.
Bull. American Astron. Soc., Vol. 7, 409 (1975). – Abstr. AAS.

034.036 **Composite bolometers for millimeter-wave astronomy.** M. G. Hauser, H. A. Notarys.
Bull. American Astron. Soc., Vol. 7, 409 (1975). – Abstr. AAS.

034.037 **The KPNO auto-phot system.** F. C. Gillett, R. R. Joyce.

Bull. American Astron. Soc., Vol. 7, 409 - 410 (1975). – Abstr. AAS.

034.038 **A new, high resolution ultraviolet solar spectrograph for rocket and satellite applications.**
J.-D. F. Bartoe, G. E. Brueckner.
Bull. American Astron. Soc., Vol. 7, 432 (1975). – Abstr. AAS.

034.039 **Automation of the NASA/SPO solar multichannel spectrometer.** R. W. Hobbs, G. D. Harris, H. P. Jones, G. L. Epstein, D. R. Brown, P. Wassmund.
Bull. American Astron. Soc., Vol. 7, 432 (1975). – Abstr. AAS.

034.040 **Astronomical spectroscopy with a self-scanning diode array.** J. C. Geary, G. R. Gilbert.
Bull. American Astron. Soc., Vol. 7, 432 (1975). – Abstr. AAS.

034.041 **Radial velocity measurement using a PDS microdensitometer.** S. G. Levy, I. Furenlid.
Bull. American Astron. Soc., Vol. 7, 432 (1975). – Abstr. AAS.

034.042 **A centroid-finding analyzer for photon-counting imaging sensors.**
G. Gilbert, S. Grandi, G. Schmidt.
Bull. American Astron. Soc., Vol. 7, 442 - 443 (1975). – Abstr. AAS.

034.043 **A field-grating imaging spectrograph for radial velocity mapping of peculiar galaxies.**
K. Serkowski.
Bull. American Astron. Soc., Vol. 7, 451 - 452 (1975). – Abstr. AAS.

034.044 **Zur lichtelektrischen Photometrie von Veränderlichen, I. II.** A. Schnitzer.
SuW, Vol. 14, 248 - 250, 293 - 294 (1975).

034.045 **Fundamental limits, quality and choice criteria for radiation detectors.** F. Desvignes.
Space optics, (see 012.013), p. 105 - 125 (1974).

034.046 **The Lick astrometric camera.** R. H. Weitbrecht.
Publ. Astron. Soc. Pacific, Vol. 87, 599 - 601 (1975).

A precision camera system was designed and constructed for use with the Lick 92-cm refractor telescope on various astrometric investigations. The camera equipment features photoelectric guiding, and it accepts photographic plates of dimensions up to 20- by 25.4-cm size.

034.047 **Investigation of a dashed measuring semi-automat.** O. E. Shornikov, A. S. Mamakov.
Izv. Astron. Ehngel'gardt. Obs., *Kazan'*, No. 39, p. 122 - 126 (1972). In Russian.

034.048 **A microphotometer with rebuilding structure for registration of spectra in terms of intensities.**
M. A. Martirosyan, L. M. Karamyan.
Soobshch. Byurakan. Obs., vyp. (No.) 46, p. 115 - 120 (1975). In Russian.

An autocompensated system for registration of spectra in terms of intensities, in which as feedback element serves a functional transducer with adjustable output characteristic is discussed. This makes possible to rebuild the instrument's scale, depending on the characteristic curve of the measured object (photographic plate).

034.049 **Some properties of a contactless alternating-current machine (motor). I. Contactless machine of double supply.** P. A. Kyalyan.
Soobshch. Byurakan. Obs., vyp. (No.) 46, p. 121 - 134 (1975). In Russian.

034.050 **Einige Bemerkungen über die Anwendung des Sonnenglases.** W. W. Spangenberg.
Sterne, 51. Jahrgang, p. 167 - 171 (1975).

034.051 **Streulichtbeseitigung in Teleskopen und Okkultationsokular zur Beobachtung von Sternbedeckungen.**
P. Darnell.
Orion, 33. Jahrgang, p. 151 (1975).

034.052 **Diseño y construcción de un fotómetro fotoeléctrico para la observación de ocultaciones de estrellas por la luna.** M. A. Acosta, R. Ortiz, R. Vieira.
Urania Barcelona, Año 59, No. 281–282, p. 365 - 394 (1974).

034.053 **On the time lag of signal formation in astronomical television apparatus with image orthicon.**
A. N. Abramenko, M. N. Naugol'naya.
Astron. Tsirk., No. 840, p. 4 - 6 (1974). In Russian.

034.054 **The multifunction photoelectric photometer of Torino Observatory.** G. Sedmak.
Proc. Astrometric Conference, Torino 1974, (see 012.017), p. 39 - 43 (1975).

034.055 **Un équipment du type «télévision» pour l'observation des étoiles doubles visuelles.**
J. Dommanget.
Proc. Astrometric Conference, Torino 1974, (see 012.017), p. 45 (1975). – Abstract.

034.056 **Rocket-photometer for stellar UV-observations.** G. Boldt, H. Hessberg, J. Niekerke.
Space Sci. Instrumentation, Vol. 1, 317 - 341 (1975).

A rocket-borne spectrophotometer has been developed which allows continuous broad band photometry from λλ3300 Å down to 1100 Å with a bandwidth of 100 Å. The sensitivity is sufficient to measure blue stars with spectral classification $\leqslant B3$ and visible magnitude $m_v \leqslant 2.0$. The instrument can be calibrated to measure in absolute photometric units.

034.057 **Frequenzwandler.** J. Biel.
SuW, Vol. 14, 336 - 338 (1975).

034.058 **A point object guiding system with a circular scanning element.** V. S. Chernyshev, T. P. Gutarko.
Trudy Tomsk. in-ta avtomatizir. sistem upr. i radioehlektron., Vol. 9, 71 - 74 (1974). In Russian. – Abstr. in Referativ. Zhurn. 51. Astron., 8.51.96 (1975).

034.059 **Principal requirement to modern astronomical spectrographs.** A. A. Boyarchuk.
Novaya tekhn. v astron. Vyp. (No.) 5. Leningrad, Nauka, 1975, p. 5 - 7. In Russian. – Abstr. in Referativ. Zhurn. 51. Astron., 8.51.243 (1975).

034.060 **Stellar magnetometer on the basis of a Fabry-Perot interferometer.** Yu. V. Glagolevskij, V. S. Rylov, P. V. Shcheglov, G. A. Chuntonov.
Novaya tekhn. v astron. Vyp. (No.) 5. Leningrad, Nauka, 1975, p. 7 - 9. In Russian. – Abstr. in Referativ. Zhurn. 51. Astron., 8.51.244 (1975).

034.061 **Automatic spectrophotometer.** V. V. Demidov, E. M. Neplokhov, N. A. Lipin, V. G. Andryukov, I. L. Golubchik, G. G. Paritskaya.
Novaya tekhn. v astron. Vyp. (No.) 5. Leningrad, Nauka,

1975, p. 9 - 13. In Russian. – Abstr. in Referativ. Zhurn. 51. Astron., 8.51.245 (1975).

034.062 **Semi-automatic photoelectric spectrophotometer.** N. D. Kalinenkov, A. I. Lobanov.
Novaya tekhn. v astron. Vyp. (No.) 5. Leningrad, Nauka, 1975, p. 13 - 14. In Russian. – Abstr. in Referativ. Zhurn. 51. Astron., 8.51.246 (1975).

034.063 **Scanning electro-spectrophotometer with output to TsPU, PLU-1 and EhPP-09.** S. Z. Omarov, M. S. Gadzhiev, P. N. Shustarev, O. S. Maksumov.
Novaya tekhn. v astron. Vyp. (No.) 5. Leningrad, Nauka, 1975, p. 15 - 17. In Russian. – Abstr. in Referativ. Zhurn. 51. Astron., 8.51.247 (1975).

034.064 **Photoelectric diffraction spectrophotometer.** N. D. Kalinenkov.
Novaya tekhn. v astron. Vyp. (No.) 5. Leningrad, Nauka, 1975, p. 17 - 21. In Russian. – Abstr. in Referativ. Zhurn. 51. Astron., 8.51.248 (1975).

034.065 **Electromagnetic broadening of the spectrum when observing with cascade image converter tubes.** A. I. Smirnov.
Novaya tekhn. v astron. Vyp. (No.) 5. Leningrad, Nauka, 1975, p. 21 - 22. In Russian. – Abstr. in Referativ. Zhurn. 51. Astron., 8.51.249 (1975).

034.066 **Mechanism for spectrum broadening by automatic star guiding on the axis of right ascension.** A. G. Shcherbakov, G. I. Shanin, A. P. Kul'chitskij.
Novaya tekhn. v astron. Vyp. (No.) 5. Leningrad. Nauka, 1975, p. 22 - 24. In Russian. – Abstr. in Referativ. Zhurn. 51. Astron., 8.51.250 (1975).

034.067 **Narrow-band interference-polarization filter with two transmission bands.** S. V. Aleksandrovich, G. N. Domyshev, A. I. Korovkin, V. P. Sadokhin, V. I. Skomorovskij.
Novaya tekhn. v astron. Vyp. (No.) 5. Leningrad, Nauka, 1975, p. 34 - 39. In Russian. – Abstr. in Referativ. Zhurn. 51. Astron., 8.51.251 (1975).

034.068 **Aplanatic two-mirror systems with large aperture ratios as cameras for spectrographs.** G. M. Popov.
Novaya tekhn. v astron. Vyp. (No.) 5. Leningrad, Nauka, 1975, p. 39 - 43. In Russian. – Abstr. in Referativ. Zhurn. 51. Astron., 8.51.252 (1975).

034.069 **Principal requirement to photoelectric astrophotometers.** V. B. Nikonov.
Novaya tekhn. v astron. Vyp. (No.) 5. Leningrad, Nauka, 1975, p. 46 - 49. In Russian. – Abstr. in Referativ. Zhurn. 51. Astron., 8.51.254 (1975).

034.070 **Increasing the maximum capabilities of astrophotometers by a synthesis of their optimum structures.** V. L. Gorokhov.
Novaya tekhn. v astron. Vyp. (No.) 5. Leningrad, Nauka, 1975, p. 49 - 53. In Russian. – Abstr. in Referativ. Zhurn. 51. Astron., 8.51.255 (1975).

034.071 **A scanning electrophotometer.** N. D. Kalinenkov. Novaya tekhn. v astron. Vyp. (No.) 5. Leningrad, Nauka, 1975, p. 53 - 55. In Russian. – Abstr. in Referativ. Zhurn. 51. Astron. 8.51.256 (1975).

034.072 **Efficiency of an extrapolation of threshold photoelectric signals in a photon counting regime.** V. S. Davydov, N. D. Poroshin, V. A. Brasletov.
Novaya tekhn. v astron. Vyp. (No.) 5. Leningrad, Nauka, 1975, p. 56 - 59. In Russian. – Abstr. in Referativ. Zhurn. 51. Astron., 8.51.257 (1975).

034.073 **Photoelectric intensitometer for astrospectrographs.** S. N. Bubnova, A. B. Medvedev, T. N. Grigoritsa.
Novaya tekhn. v astron. Vyp. (No.). 5. Leningrad, Nauka, 1975, p. 59 - 62. In Russian. – Abstr. in Referativ. Zhurn. 51. Astron., 8.51.258 (1975).

034.074 **Cooling of a photoelectric photometer by means of a thermoelectric device.** A. G. Shcherbina, V. V. Demidov, Eh. Kh. Shapiro, N. A. Lipin, I. L. Golubchik, Yu. N. Chistyakov, L. A. Kamionko.
Novaya tekhn. v astron. Vyp. (No.) 5 Leningrad, Nauka, 1975, p. 62 - 64. In Russian. – Abstr. in Referativ. Zhurn. 51. Astron., 8.51.259 (1975).

034.075 **Electrophotometer with automatic subtraction of noise background.** G. V. Spulgis.
Novaya tekhn. v astron. Vyp. (No.) 5. Leningrad, Nauka, 1975, p. 64 - 66. In Russian. – Abstr. in Referativ. Zhurn. 51. Astron., 8.51.260 (1975).

034.076 **Transistor electrometric amplifiers in astronomical electrophotometers.** V. N. Ivanov, E. L. Kovalev.
Novaya tekhn. v astron. Vyp. (No.) 5. Leningrad, Nauka, 1975, p. 67 - 70. In Russian. – Abstr. in Referativ. Zhurn. 51. Astron., 8.51.261 (1975).

034.077 **Calibration and testing bench for radiation receivers.** U. K. Vejsmann.
Novaya tekhn. v astron. Vyp. (No.) 5. Leningrad, Nauka, 1975, p. 70 - 72. In Russian. – Abstr. in Referativ. Zhurn. 51. Astron., 8.51.262 (1975).

034.078 **Dependence of photomultiplier FEhU-79 and photomultiplier FEhU-106 characteristics on operating conditions.** R. Kalitis, Yu. Speraukas.
Novaya tekhn. v astron. Vyp. (No.) 5. Leningrad, Nauka, 1975, p. 72 - 77. In Russian. – Abstr. in Referativ. Zhurn. 51. Astron. 8.51.263 (1975).

034.079 **On the quantum effectiveness of different television systems used in astronomy.** V. V. Prokof'eva.
Novaya tekhn. v astron. Vyp. (No.) 5. Leningrad, Nauka, 1975, p. 78 - 82. In Russian. – Abstr. in Referativ. Zhurn . 51. Astron., 8.51.264 (1975).

034.080 **Experience of using superorthicon and isocon for registration of the spectrum of neon radiation.** N. F. Kuprevich.
Novaya tekhn. v astron. Vyp. (No.) 5. Leningrad, Nauka, 1975, p. 82 - 84. In Russian. – Abstr. in Referativ. Zhurn. 51. Astron., 8.51.265 (1975).

034.081 **Electronographic image converter tubes for astronomical purposes.** M. M. Butslov, S. V. Lipatov.
Novaya tekhn. v astron. Vyp. (No.) 5. Leningrad, Nauka, 1975, p. 90 - 93. In Russian. – Abstr. in Referativ. Zhurn. 51. Astron., 8.51.267 (1975).

034.082 **Laboratory and astronomical testing of an electronographic image converter tube.** I. A. Aslanov, V. F. Esipov, S. V. Lipatov, D. A. Ptitsyn, V. L. Khokhlova, K. K. Chuvaev.
Novaya tekhn. v astron. Vyp. (No.) 5. Leningrad, Nauka, 1975, p. 93 - 94. In Russian. – Abstr. in Referativ. Zhurn. 51. Astron., 8.51.268 (1975).

034.083 **Investigation of the characteristics of electronographic image converter tubes determined to record extreme faint spectra.**
Eh. G. Sil'kis, Yu. G. Vajner, L. P. Malyavkin.
Novaya tekhn. v astron. Vyp. (No.) 5. Leningrad, Nauka, 1975, p. 95 - 98. In Russian. – Abstr. in Referativ. Zhurn. 51. Astron., 8.51.269 (1975).

034.084 **Electron-optical unit with a pulse low-noise amplifier of image brightness for observing astronomical objects.** M. M. Butslov, B. O. Karapetyan, G. E. Smolkin.
Novaya tekhn. v astron. Vyp. (No.) 5. Leningrad, Nauka, 1975, p. 98 - 103. In Russian. – Abstr. in Referativ. Zhurn. 51. Astron., 8.51.270 (1975).

034.085 **Some operating characteristics of the image converter tube UM-92.** M. G. Sosonkin.
Novaya tekhn. v astron. Vyp. (No.) 5. Leningrad, Nauka, 1975, p. 103 - 104. In Russian. – Abstr. in Referativ. Zhurn. 51. Astron., 8.51.271 (1975).

034.086 **Astronomical detectors of infrared radiation.** G. V. Khozov.
Novaya tekhn. v astron. Vyp. (No.) 5. Leningrad, Nauka, 1975, p. 105 - 116. In Russian. – Abstr. in Referativ. Zhurn. 51. Astron., 8.51.272 (1975).

034.087 **Recording of equilibrium thermal radiation by a photon counter.** Ya. Oyaste.
Novaya tekhn. v astron. Vyp. (No.) 5. Leningrad, Nauka, 1975, p. 117 - 121. In Russian. – Abstr. in Referativ. Zhurn. 51. Astron., 8.51.273 (1975).

034.088 **Energetics of photometric IR means for high-altitude astrophysical studies.**
Yu. K. Bergner, V. G. Lebedev.
Novaya tekhn. v astron. Vyp. (No.) 5. Leningrad, Nauka, 1975, p. 121 - 125. In Russian. – Abstr. in Referativ. Zhurn. 51. Astron., 8.51.274 (1975).

034.089 **Observation unit for infrared spectra of stars.** Eh. A. Vitrichenko, G. I. Shanin, V. S. Shevchenko, A. G. Shcherbakov.
Novaya tekhn. v astron. Vyp. (No.) 5. Leningrad, Nauka, 1975, p. 130 - 134. In Russian. – Abstr. in Referativ. Zhurn. 51. Astron., 8.51.276 (1975).

034.090 **The performance of an interferometric system for star guiding.** N. K. Smirnov, F. M. Subbotin.
Optiko-mekh. prom-st', 1975, No. 3, p. 61 - 63. In Russian. Abstr. in Referativ. Zhurn. 51. Astron., 8.51.277 (1975).

034.091 **Amplifiers for neutron supermonitors of and supertelescopes for cosmic rays.** V. F. Lyutenko, A. M. Starodubtsev, A. A. Upol'nikov, N. P. Chirkov.
Pribory i tekhn. ehksperimenta, 1975, No. 2, p. 132 - 134. In Russian. – Abstr. in Referativ. Zhurn. 51. Astron., 8.51. 285 (1975).

034.092 **Radial velocity measurements with the Asiago nebular spectrograph.**
P. Benvenuti, M. Capaccioli, S. D'Odorico.
Mem. Soc. Astron. Italiana, Vol. 46, 69 - 79 (1975).

In connection with current spectroscopic work the accuracy of radial velocity determinations with the nebular image-tube spectrograph of the Asiago Astrophysical Observatory has been investigated. The planetary nebulae NGC 6826 and IC 2149 have been observed as standards. Once the distortions introduced by the optics and the image tube are removed, the measured values of the radial velocities are in good agreement with those previously known. The mean error of a single measurement along the emission line is ± 13 km sec^{-1}.

034.093 **An equatorial solar spectrograph.** A. Kubičela.
Publ. Obs. Astron. Beograd, No. 20, (see 012.024), p. 47 - 51 (1975). In Serbo-Croatian.

034.094 **Mise en service d'un prisme objectif normal de 60 cm de diamètre. Découverte d'une étoile du halo à grande vitesse radiale.** C. Fehrenbach, R. Burnage.
Comptes Rendus Acad. Sci. Paris, Sér. B, Vol. 281, 481 - 483 (1975).

Un prisme objectif de 60 cm de diamètre monté devant un télescope de Schmidt a été mis en service à l'Observatoire de Haute Provence. Le premier cliché a permis la découverte d'une étoile bleue du halo ayant une vitesse radiale de -153 km s^{-1}.

034.095 **Neutron and proton interaction backgrounds in Compton-telescopes used for gamma-ray astronomy.**
R. S. White, V. Schönfelder.
Astrophys. Space Sci., Vol. 38, 19 - 27 (1975).

Gamma-ray background counting rates encountered in astronomy observations are calculated for a double Compton scatter telescope.

034.096 **A photoelectric photometer for astronomical use.** G. Giannì, I. Mazzitelli, G. Natali.
Astron. Astrophys., Vol. 44, 277 - 283 (1975).

A photon counting photometer has been built, which can be used over a wide range of photon fluxes with minimum non linearity. The use of a new photomultiplier (RCA C31034 A) with high quantum efficiency in the near infrared and a test of photometric measurements are described.

034.097 **The effect of cooling on the dark current of a photomultiplier FEhU–83.** Ü. Ibrus.
Publ. Tartu Astrofiz. Obs., Vol. 43, 203 - 207 (1975). In Russian.

The temperature dependence of the dark current of photomultipliers of the type FEhU–83 on cooling down to the temperature of $-50°$ C is investigated. The dark current decreases by about 3 to 5 orders at cooling to $-35°$ to $-40°$ C and remains practically constant ($\sim 10^{-11}$ A) for lower temperatures.

034.098 **Practical multi-spectrum Hadamard transform spectrometer.**
M. H. Tai, D. A. Briotta, Jr., N. S. Kamath, M. Harwit.
Applied Optics, Vol. 14, 2533 - 2536 (1975).

034.099 **Interferometric and conventional star trackers.** J. L. Kreuzer.
Journ. Optical Soc. America, Vol. 65, 1187 (1975). – Conference abstract.

034.100 **Holographically formed gratings for X-ray spectroscopy.** R. J. Speer.
Journ. Optical Soc. America, Vol. 65, 1214 (1975). – Conference abstract.

034.101 **Achromatic focusing device.** F. M. Phelps III, K. B. Newbound.
Journ. Optical Soc. America, Vol. 65, 1283 - 1286 (1975).

An achromatic focusing assembly that uses two plane mirrors and one spherical mirror was designed and built to focus Fabry-Perot interference patterns onto the entrance slit of a stigmatic spectrograph. The various imaging errors are so small, and the device is so simple and inexpensive to make, that it may have wide applicability to other areas of optics, where it could easily replace very expensive achromatic lenses.

The authors discuss the design and calculate numerical values for the blur circles due to astigmatism, coma, and spherical aberration.

034.102 **Automatic scanner for a horizontal solar telescope.** M. Klvaňa.
Bull. Astron. Inst. Czechoslovakia, Vol. 26, 364 - 367 (1975).
The scanner is mainly used for the program-controlled motion of the solar image across the entrance slit of the spectrograph during magnetographic observations. The fundamental technical data, the principle, a description of its function with the use of pointing and without it and measurements, characterizing the accuracy of the whole system are discussed.

034.103 **The CRESS 7m photoelectric scanning grating spectrograph and its potential for solar, atmospheric and laboratory spectroscopy.** W. Fabian, R. W. Nicholls.
Journ. Roy. Astron. Soc. Canada, Vol. 69, 246 (1975).
Abstr. CAS.

034.104 **The DDO reticon system.** B. Campbell.
Journ. Roy. Astron. Soc. Canada, Vol. 69, 246 (1975). – Abstr. CAS.

034.105 **Protuberansinstrumenter.**
Per Darnell, Poul Darnell.
Astron. Tidssk., Årg. 8, p. 189 - 195 (1975).

034.106 **The magneto-optical filter. I: Preliminary observations in Na D lines.**
G. Agnelli, A. Cacciani, M. Fofi.
Solar Physics, Vol. 44, 509 - 518 (1975).
Transmission curves and theoretical calculi of the magneto-optical filter, designed and built by the authors, are shown together with some observed transmission spectra. At present the filter has a total halfwidth of ≈80 mÅ; and the maximum transmission is 25 %. From the analysis of the theoretical curves and from the observed spectra, the authors deduced the possibility of making up a filter with a very narrow passband (less than 30 mÅ) and a very high transmission (up to nearly 100 % apart from losses arising from the glass cell, lenses and polarizers). The article concludes with a selection and discussion of photographs obtained with the filter.

034.107 **Electrophotometer for astronomical purposes on the basis of a cascade image converter.**
D. I. Gorodetskij, B. M. Gabdullin.
Trudy Astrofiz. Inst., *Alma-Ata,* Vol. 26, 133 - 146 (1974).
In Russian.

034.108 **The instrumental contour of the spectrograph ASP-20.** Ts. S. Khetsuriani.
Byull. Abastumansk. Astrofiz. Obs., No. 46, p. 197 - 200 (1975). In Russian.
The instrumental contour of the spectrograph of the horizontal solar telescope has been investigated by measuring the absorption lines of atmospheric oxygen. Voigt parameters of the instrumental contour were determined.

034.109 **Some experiences on experimental testing of the intermediate accumulator in an astronomical TV-system.** E. S. Agapov, Yu. S. Aleksandrin, V. F. Anisimov, S. G. Braunfel'd, R. I. Kiladze, G. D. Matveev, A. Sh. Khatisov, S. M. Chanturiya.
Byull. Abastumansk. Astrofiz. Obs., No. 46, p. 251 - 256 (1975). In Russian.

034.110 **Solar infra-red limb scans using a simple rapid-scanning Michelson interferometer.**
D. A. Burrell, T. A. Clark.
Journ. Roy. Astron. Soc. Canada, Vol. 69, 254 (1975).
Abstr. CAS.

034.111 **Quantum efficiency of a channel electron multiplier in the far ultraviolet.** F. Paresce.
Applied Optics, Vol. 14, 2823 - 2824 (1975).

034.112 **Infrared spectrometer: liquid-helium-cooled rocket-borne circular-variable filter.** C. L. Wyatt.
Applied Optics, Vol. 14, 3086 - 3091 (1975).

034.113 **Goniometric device with numerical phasometer.**
A. M. Nasyrov, K. K. Kostylev.
Meteorn. rasprostr. radiovoln. No. 10 - 11. Kazan', Kazan. un-t, 1975, p. 53 - 60. In Russian.

034.114 **On one of the possible reasons for deformation of spectral lines in the DFS-13 spectrograph.**
V. S. Tyurin.
Trudy Mosk. in-ta inzh. geod., aehrofotosemki i kartogr., 1973, vyp. (No.) 59, p. 33 - 35. In Russian. – Abstr. in Referativ. Zhurn. 51. Astron., 12.51.196 (1975).

034.115 **Extending the visible frontier. New tools of the optical astronomer.** R. Weymann.
Mercury, (Journ. Astron. Soc. Pacific), Vol. 4, No. 5, p. 2 - 9 (1975).

034.116 **Band-netpass filters for the long-wave IR-region.**
N. R. Braverman, L. V. Vorob'ev, V. A. Soglasnova, G. B. Sholomitskij.
In-t kosmich. issled. AN SSSR. Pr-223. Moskva, 1975. 22 pp.
In Russian.

034.117 **The use of a UP-converter for infrared astronomical observations.**
G. N. Agababyan, Yu. K. Melik-Alaverdyan, H. M. Tovmassian.
Astron. Tsirk., No. 876, p. 4 - 6 (1975). In Russian.

034.118 **Investigation of photomultipliers in the near infrared spectral region in the regime of photon counting.**
V. N. Ivanov, A. F. Pereverzentsev, V. V. Galanin.
Astron. Tsirk., No. 877, p. 1 - 2 (1975). In Russian.

034.119 **A motor-micrometer-driven infrared Fourier-transform spectrometer.**
R. I. Thompson, M. A. Reed.
Publ. Astron. Soc. Pacific, Vol. 87, 929 - 932 (1975).
A new design for a Fourier-transform spectrometer (FTS) is presented which uses a motor and micrometer as opposed to magnetic drives. The drive is reliable, simple, and inexpensive. Its use in an astronomical FTS is shown and an example of the data is presented.

034.120 **Use of a Fabry–Perot/monochromator system for enhancement of the spectral line-to-noise power ratio from line and continuum sources.** W. C. Kreye.
Journ. Optical Soc. America, Vol. 65, 1427 - 1435 (1975).

034.121 **Applications of image tubes in astronomy.**
R. G. Bingham.
Observatory, Vol. 95, 283 - 288 (1975).
The best use of image tubes in spectrographic cameras can lead to improvements in light-grasp of about four times compared with existing designs. Possible applications of image tubes to a simple spectrometer and to an autoguider are also described.

034.122 **The double beam chopping photometer-polarimeter.**
V. Piirola.
Ann. Acad. Sci. Fennicae, Ser. A, VI. Phys. 418, 61 pp. (1975).
A photoelectric instrument for both photometric and

polarimetric observations of astronomical objects has been constructed and tested.

034.123 **Preliminary experience in measuring stellar magnitudes with a new experimental version of an iris-diaphragm photometer based on the plate measuring device "Ascorecord".** A. Alksnis, L. Duncāns.
Investigation of the sun and red stars. 3, (see 003.008), p. 56 - 59 (1975). In Russian.

034.124 **A fully tunable Lyot-Öhman filter.**
J. M. Beckers, L. Dickson, R. S. Joyce.
Sacramento Peak Obs., Project ILIR, Air Force Cambridge Res. Lab., *Hanscom, Mass.*, AFCRL-TR-75-0090, Instrument. Papers, No. 227, 27 pp. = Sacramento Peak Obs. Contr. No. 250 (1975).

The authors describe the properties of the Zeiss universal birefringent filter and its optical and electronic interface with the Sacramento Peak Observatory's vacuum telescope. The instrument permits observations of solar intensities, velocities and/or magnetic fields in rapid succession in any Fraunhofer line in the 410 to 700 nm wavelength region with high spectral (0.004 to 0.013 nm) and high spatial (~ 1/3 arc second) resolution.

034.125 **Generalized shearing interferometry.**
J. C. Fouéré, D. Malacara.
Bol. Inst. Tonantzintla, Vol. 1, 227 - 232 (1975).

Radial and lateral shear interferometers are shown to be two special cases of a generalized shear interferometer. The analysis of the concurrent radial and lateral shear is made here and interferograms for third order aberrations are presented.

034.126 **A shearing interferometer for convergent or divergent beams.**
D. Malacara, A. Cornejo, M. V. R. K. Murty.
Bol. Inst. Tonantzintla, Vol. 1, 233 - 239 (1975).

A simple lateral shearing interferometer using a single prismatic lens is described here. This instrument is a modification of the Murty interferometer to permit the testing of convergent or divergent wavefronts.

034.127 **Laser interferometers: opto-electronic yardsticks with microinch resolution.** J. Koch.
Machine Design, Vol. 47, No. 4, p. 92 - 97 (1975). – *KJW*

034.128 **Integrated Ga As F.E.T. mixer performance at X band.** R. A. Pucel, D. Masse, R. Bera.
Electronics Letters, Vol. 11, 199 - 200 (1975). – *KJW*

034.129 **Silicon photodiode arrays.** P. W. Fry.
Journ. Phys. E, (Sci. Instruments), Vol. 8, 337 - 349 (1975). – *GMCC*

034.130 **The determination of the theoretical value of the degree of polarization of the calibrator of the photoelectric polarimeter of the Belgrade Astronomical Observatory.** I. Vince.
Bull. Obs. Astron. Belgrade, No. 126, p. 56 - 59 (1975).

034.131 **The photometry of extended objects: diode-array detectors and interactive data processing.**
W. C. Livingston.
Image processing techniques in astronomy, (see 012.041), p. 21 - 30 (1975).

Monolithic silicon diode-array detector technology is reviewed as applied to astronomical applications. The massive data files that can result from diode-array detectors are most conveniently and efficiently digested by the astronomer through computer-coupled interactive picture processing.

034.132 **Television detector techniques.** A. Boksenberg.
Image processing techniques in astronomy, (see 012.041), p. 59 - 78 (1975).

A review of image detection by television techniques is given in the context of optical and ultraviolet astronomy. The method of image photon counting, now used routinely in optical astronomy, and the detector system of the International Ultraviolet Explorer observatory satellite, are discussed as representative examples in each field.

034.133 **Photon-counting detector arrays based on microchannel array plates.** J. G. Timothy.
Image processing techniques in astronomy, (see 012.041), p. 85 - 90 (1975).

The recent development of the channel electron multiplier (CEM) and its miniaturization into the microchannel array plate (MCP) offers for the first time the possibility of fully combining the advantages of the photographic and photoelectric detection systems. The MCP has an image-intensifying capability and has the potential of being developed to yield signal outputs superior to those of conventional photomultipliers.

034.134 **Performance of the MIT silicon vidicon imaging system at the telescope.**
T. B. McCord, J. P. Bosel, M. J. Frankston.
Image processing techniques in astronomy, (see 012.041), p. 91 - 96 (1975).

A two-dimensional integrating silicon vidicon imagery system has been developed at MIT and used to record images and spectra of astronomical objects. The system is photon noise limited above the ~500 charge carrier preamplifier noise. The response is linear with $\gamma = 1.00$ over a ranging 10^5.

034.135 **A self-scanned silicon diode array for astronomical photometry.** D. Dravins.
Image processing techniques in astronomy, (see 012.041), p. 97 - 102 (1975).

A digital light detector system using a self-scanned silicon photodiode array with 128 elements has been developed for use in a solar spectrograph and a photographic-plate photometer. Noise patterns are reduced in software, making the signal-to-noise level essentially limited by the 10-bit hardware resolution.

034.136 **Image detectors for recording astronomical ultraviolet spectra.** W. M. Burton.
Image processing techniques in astronomy, (see 012.041), p. 103 - 108 (1975).

Various image tube devices have been evaluated as possible detectors for astronomical spectroscopic studies. An electrostatic image intensifier coupled by fibre-optics to photographic film has been used to record echelle format spectra. Ultraviolet sensitivity is obtained by coating the detector faceplate with a fluorescent layer.

034.137 **Integrating T.V. system development at Cambridge.**
C. D. Mackay.
Image processing techniques in astronomy, (see 012.041), p. 125 - 126 (1975).

034.138 **The use of the "GALAXY" machine at the Royal Greenwich Observatory.**
C. A. Murray, W. Nicholson.
Image processing techniques in astronomy, (see 012.041), p. 171 - 184 (1975).

A GALAXY plate measuring machine, similar to that constructed for the Royal Observatory, Edinburgh, has been in use at the Royal Greenwich Observatory, Herstmonceux, since late in 1972. The design and performance of the Edinburgh machine have already been reported; in this paper the current operational procedures at Herstmonceux, including minor modifications to the original design of plateholders, and also the planned enhancement of the system by installation of a NOVA 2/10 Control Computer, are described.

034.139 **The Grant measuring engine.** E. W. Dennison. Image processing techniques in astronomy, (see 012.041), p. 199 - 208 (1975).

034.140 **The U.S. Naval Observatory's automatic measuring machines.** K. A. Strand.
Image processing techniques in astronomy, (see 012.041), p. 209 (1975). – Abstract.

034.141 **The two-coordinate Grant machine in long-focus astrometry.** J. L. Hershey.
Image processing techniques in astronomy, (see 012.041), p. 211 - 216 (1975).

034.142 **The COSMOS facility at the Royal Observatory Edinburgh.**
N. M. Pratt, R. Martin, L. W. G. Alexander, G. S. Walker, P. R. Williams.
Image processing techniques in astronomy, (see 012.041), p. 217 - 222 (1975).

034.143 **The ESO S-3000 measuring machine.** R. M. West. Image processing techniques in astronomy, (see 012.041), p. 223 - 228 (1975).

The ESO S-3000 measuring machine which was installed in 1974 in the Sky Atlas Laboratory is capable of measuring photographic plates up to 14 × 14 inch^2 with a formal accuracy of ±1 μm and ±.02 D in position and density. It is controlled by a 8 k Alpha 16 minicomputer. Software has been written for the ESO-TP HP 2100 computer system, that permits recognition and accurate measurement of stellar and extended images and filtering of plate faults.

034.144 **IRIS – a two axis comparator and microdensitometer using two different scanning modes.**
N. Åslund, N. von Gersdorff, R. Norberg, J. A. Nordin.
Image processing techniques in astronomy, (see 012.041), p. 229 - 236 (1975).

034.145 **Machine à mesurer de l'Observatoire de Haute Provence.** C. Fehrenbach, A. Vin.
Image processing techniques in astronomy, (see 012.041), p. 237 - 243 (1975).

034.146 **The design of the new S.R.C. automated photographic measuring system.**
E. J. Kibblewhite, M. T. Bridgeland, T. Hooley, D. Horne.
Image processing techniques in astronomy, (see 012.041), p. 245 - 246 (1975).

034.147 **The computer controlled comparator Astroscan.** P. W. H. Blansjaar, J. van Kuilenburg.
Image processing techniques in astronomy, (see 012.041), p. 247 - 253 (1975).

A description is given of the computer controlled comparator densitometer "Astroscan". This instrument consists of a monocomparator and a digital computer which controls the comparator stage. The density measurements of a plate or film are performed by using a multi-element diode array, which is also under computer control. The "Astroscan" is fully programmable in a high level programming language.

034.148 **A simultaneous six-channel microphotometer with computerized data acquisition.**
D. H. Andrews, K. O. Wright.
Image processing techniques in astronomy, (see 012.041), p. 255 - 260 (1975).

034.149 **A very simple digital microdensitometer-comparator.** R. J. Rutten, H. J. van Amerongen.
Image processing techniques in astronomy, (see 012.041), p. 261 - 264 (1975).

034.150 **Photoelectric spectroscopic plate measuring machine built at the Meudon Observatory.** F. Launay.
Image processing techniques in astronomy, (see 012.041), p. 265 - 268 (1975).

034.151 **Some preliminary results obtained with a TV camera in the field of the observation of visual double stars.**
J. Dommanget.
Image processing techniques in astronomy, (see 012.041), p. 343 - 345 (1975).

A description of a TV camera used for double star observations is given. Some very preliminary registrations of images of binaries, Saturn and the moon, are shown.

034.152 **Le Centre de Dépouillement de Clichés Astronomiques.**
M. Auvergne, A. Bijaoui, J. L. Heudier, J. Marchal, C. Ounnas, J. Postel, M. Schneider.
Image processing techniques in astronomy, (see 012.041), p. 373 - 378 (1975).

The "Centre de Dépouillement des Clichés Astronomiques" opened during the year 1974 is equiped with a PDS microdensitometer connected to a computer PDP 11/40. The authors develope an interactive software for astronomical purpose such as classical spectrography, astronomical photometry, astrometry.

034.153 **A scanning Fabry-Perot interferometer and its application to space research. III. Piezo-electric scanning Fabry-Perot interferometer.** T. Kohno, N. Yajima.
Journ. Mech. Engineering Lab., (*Japan*), Vol. 29, No. 2, p. 54 - 63 (1975). In Japanese.

A scanning Fabry-Perot interferometer with three piezoelectric discs as spacers was manufactured and tested.

034.154 **Large NaI (TL) crystals as high-energy resolution spectrometers in high-energy physics and in gammy-ray astronomy.** L. H. O'Neill, R. L. Ford, R. Hofstadter, E. B. Hughes, R. Kose, P. Lecoultre, T. W. Martin, R. F. Schilling.
IEEE Trans. Nuclear Sci., Vol. NS-22, 286 - 291 (1975).

The response of 30 in. diameter NaI(TL) crystal detectors to electrons in the GeV range is described, and experience acquired in the operation of such detectors in physics experiments is summarized.

034.155 **A polarimeter to measure the complete state of polarization of scattered solar radiation.**
D. Hannemann.
Contr. Atmosph. Phys., Vol. 48, 76 - 84 (1975).

A new polarimeter for the measurement of the four Stokes parameters of scattered solar radiation is described. Its wavelength range of sensitivity is 600–1000 nm. The four Stokes parameters are obtained by simultaneous measurements of 6 radiances. The maximum number of measurements is 166/sec. The theory of measurements and calibrations has been discussed. One measurement of skylight of a cloudless but turbid atmosphere is described.

034.156 **A large-area lithium-fluoride Bragg spectrometer for stellar X-ray astronomy.**
H. S. Stockman, Jr., J. R. P. Angel, B. E. Woodgate, R. A. Nidey.
Nucl. Instrum. and Methods, (*Netherlands*). Vol. 126, 217 -

220 (1975).

A large-area Bragg spectrometer recently used to search for the Fe XXV X-ray emission lines in Sco X-1 is described.

034.157 **The ANS hard X-ray experiment.** K. Kubierschky, N. Jagoda, G. K. Austin, D. R. Parsignault, R. Frank, J. S. Ballas.
IEEE Trans. Nuclear Sci., Vol. NS-22, 551 - 554 (1975).

The Astronomical Netherlands Satellite hard X-ray experiment is an X-ray spectrometer gathering astronomical data through two pairs of proportional counter. One set of detectors obtains both spectral and spatial resolution by Bragg reflection. The other pair obtains spatial resolution with a compound collimator and spectral resolution by means of pulse-height analysis.

034.158 **An instrument for measuring cosmic X-ray radiation.** A. J. F. den Boggende, H. T. J. A. Lafleur.
IEEE Trans. Nuclear Sci., Vol. NS-22, 555 - 559 (1975).

The instrument is able to measure cosmic X-ray radiation in two channels between 0.15 and 0.46 keV, and in five channels between 0.8 and 8 keV. It consists of detection units together with a data handling unit. The instrument has been prepared for the Astronomical Netherlands Satellite (ANS), which was launched on August 30, 1974.

034.159 **Massachusetts Institute of Technology X-ray Observatory for the Small Astronomy Satellite (SAS-C).** R. S. Taylor.
IEEE Trans. Nuclear Sci., Vol. NS-22, 565 - 569 (1975).

The MIT SAS-C X-ray Observatory is a highly flexible facility for the comprehensive study of the X-ray sky over 0.1 to 50 keV energy range. Event time may be determined to 0.125 ms. An on-board 16 pulse height analyzer allows for fine X-ray spectral studies using the on-board star trackers and rotating modulation collimator.

034.160 **The Pioneer 10/11 and Helios A/B cosmic ray instruments.** D. E. Stilwell, R. M. Joyce, B. J. Teegarden, J. H. Trainor, G. Streeter, J. Bernstein.
IEEE Trans. Nuclear Sci., Vol. NS-22, 570 - 574 (1975).

The design and performance of a set of cosmic-ray particle experiments for the Pioneer 10/11 and Helios A/B space missions are described.

034.161 **An infrared heterodyne radiometer for high-resolution measurements of solar radiation and atmospheric transmission.** B. J. Peyton, A. J. Dinardo, S. C. Cohen, J. H. McElroy, R. J. Coates.
IEEE Journ. Quantum Electronics, Vol. QE-11, 569 - 574 (1975).

A tunable, Dicke-switched, infrared heterodyne radiometer (IHR) has been designed, fabricated, tested, and used to observe solar radiation and determine the atmospheric transmissivity in the 9–11 μm spectral band. Detailed design equations and measured IHR performance are presented. The IHR was used to make solar and atmospheric transmission measurements and a vertical path atmospheric attenuation of 2.3 dB has been established under favorable weather conditions. The attenuation of solar radiation due to cloud cover and haze have also been investigated.

034.162 **A high speed digital autocorrelation spectrometer for millimetre astronomy.** C. Pell, L. T. Little.
Journ. Phys. E, (Sci. Instruments), Vol. 8, 786 - 789 (1975).

034.163 **The effects of the 'ATS-6' satellite television transmission on 2690–2700 MHz radio astronomy band.** J. P. Hagen, P. N. Swanson.
Telecommun. Journ., (*Switzerland*), Vol. 42, 599 - 603 (1975).

Discusses the results of a test which was carried out by the National Aeronautics and Space Administration (NASA) and a group of radio observatories to determine the extent of the interference caused by the ATS-6 television channel in the 2690–2700 MHz band.

034.164 **Solar photometer.** R. Moses, F. van Swearingen.
Phys. Teacher, (*USA*), Vol. 13, 366 - 367 (1975).

The photometer described uses a Si solar cell to measure solar limb darkening and estimate relative temperatures on the sun.

034.165 **A polarimeter for monochromatic studies of solar prominences.** G. Ratier.
Nouvelle Rev. Optique, Vol. 6, 149 - 159 (1975). In French.

The photoelectric polarimeter which is described associates the three major discoveries of Lyot: the coronagraph, the birefringent filter and the polarimeter. However it differs completely from the coronameter.

034.166 **A prototype two channel infrared photometer designed for study and compensation of sky noise.** S. Bensammar, B. De Batz, F. Dampierre.
Nouvelle Rev. Optique, Vol. 6, 161 - 163 (1975). In French.

Atmospheric limitations of ground-based infrared astronomical observations are reviewed, and a prototype photometer used to study them is described.

034.167 **Thin films for X-ray astronomy.** F. Williamson, C. W. Maxson.
Rev. Sci. Instruments, Vol. 46, 51 - 52 (1975).

Description d'une méthode de production de couches plastiques minces, de grande surface, utilisées pour les fenêtres d'entrée de détecteurs proportionnels de rayons X. Obtention de couches ayant des surfaces allant jusqu'à 1800 cm^2.

034.168 **Ergebnisse der Lyman-Alpha Messungen des Satelliten DIAL.** H. D. Schlichter.
Forschungsber. BMFT-FB No. W 75-01, 104 pp. (1975).

The effects on the performance of a spectrograph camera caused by using non-parallel incident light. See Abstr. 031.005.

Stellar line profiles by tilt-scanned narrow band interference filters. See Abstr. 031.226.

Information retrieval by microdensitometry. See Abstr. 031.281.

On line digital correlation of photon counting TV images for stellar interferometry. See Abstr. 031.283.

Mesure des contours de raies par les techniques de télévision. See Abstr. 031.284.

Photometry using electronographic techniques. See Abstr. 031.285.

Problems of oversampling with SEC vidicon television systems. See Abstr. 031.286.

Handling of 'Galaxy' photometry. See Abstr. 031.291.

Two beam interference experiments and some quantum concepts. See Abstr. 031.304.

An astronomical application of CAMAC. See Abstr. 031.412.

The fully automatic program-controlled microdensitometer-comparator of the Astronomical Institute at Utrecht. See Abstr. 031.416.

Calibration of the UV experiment in ANS. See Abstr. 032.521.

SEC vidicon system for a Balloon Ultraviolet Stellar Spectrometer. See Abstr. 032.537.

An Si(Li) detector system for application to X-ray astronomy rocket experiments. See Abstr. 032.541.

On the possibility of the use of circumzenithal in lunar astrometry. See Abstr. 041.002.

The reduction system of signals of a solar magnetograph in true time scale. See Abstr. 072.045.

The work of the diode array: He 10830 observations of spicules and subflares. See Abstr. 073.071.

Flying-spot analysis of solar images. See Abstr. 073.154.

Photoelectric measuring complex for meteor observations. See Abstr. 104.012.

H-beta spectrophotometry of SX Phe. See Abstr. 114.347.

Initial results of spatial interferometry at 5 microns. See Abstr. 141.626.

Stellar and solar X-ray polarimetry. See Abstr. 142.179.

Results from the Ariel-5 All-Sky X-ray Monitor. See Abstr. 142.201.

035 Clocks and Frequency Standards

035.001 **Meteorological influences on Loran-*C* ground wave propagation.** R. H. Doherty, J. R. Johler.
Journ. Atmosph. Terr. Phys., Vol. 37, 1117 - 1124 (1975).

Phase variations of approximately 36° (one microsecond) have been observed and recorded on a 1000 km Loran-*C* overland propagation path. These phase variations are large compared with the receiver resolution of 0.1 μs or better. As a consequence of this analysis, the authors can predict that a phase change will occur before the passage of a warm front and after the passage of a cold front at the surface of the ground.

035.002 **Zur Theorie der "Homogenen Sonnenuhr" (Hybrid Sundial) in moderner Sicht.** W. Hanke.
Sterne, 51. Jahrgang, p. 159 - 166 (1975).

035.003 **Zonnewijzer, gemaakt van oud fietswiel.** G. van Wageningen.
Zenit, Vol. 2, 274 - 276 (1975).

035.004 **A design for an analemmatic standard-time sundial.** P. K. Seidelmann.
Sky Telescope, Vol. 50, 368 - 369 (1975).

035.005 **Loran-C precise time and frequency dissemination.** E. Proverbio.
Time determination, dissemination and synchronization, (see 012.029), p. 93 - 116 (1975).

The Loran-C pulse type system provides very high accuracy in time synchronization. This paper gives the more important features and characteristics of this timing system. The different effects in time synchronization introduced by the propagation medium have been particularly discussed and corrections determined. An accuracy to the order of 0.1–0.2 μs can be ensured when all instrument delay and systematic propagation errors are taken into account.

035.006 **Statistical properties of Loran-C transmissions.** V. Quesada, A. Simoncini.
Time determination, dissemination and synchronization, (see 012.029), p. 117 - 126 (1975).

035.007 **Comparison of clocks by means of TV field synchronizing pulses.** H. Enslin.
Time determination, dissemination and synchronization, (see 012.029), p. 127 - 150 (1975).

Synchronization pulses included in normal television transmissions can be used as common reference markers for the comparison of clocks located at sites far from each other, but within the service area of the same television programme. The method is inexpensive, easy to employ, and of high precision and accuracy. The application of the "TV method" and results obtained from it in different countries are reviewed.

035.008 **Time pulses measurement by digital methods.** J. De Prins, R. Detrie, J.-L. Guisset, F. Guillaume.
Time determination, dissemination and synchronization, (see 012.029), p. 151 - 163 (1975).

035.009 **Clocks synchronised by time signals.** J. Bonanomi.
Time determination, dissemination and synchronization, (see 012.029), p. 165 - 170 (1975).

035.010 **Atomic clocks and the formation of TAI.** H. M. Smith.
Time determination, dissemination and synchronization, (see 012.029), p. 171 - 184 (1975).

035.011 **Schrittregelung für die Mittelung von Zeitsignalen.**

G. Becker.
PTB Mitt., 85. Jahrgang, p. 448 - 457 (1975).

The characteristics of methods using a step control with unity steps (with and without a tolerance) for time signal averaging are given: transient time constant, averaging time, stabilization factor and efficiency. Very high stabilization factors can be achieved by counting the determinations of the sign of the time difference between the received (A) and produced (B) time signals using a counter whose maximum state when reached produces a unity step correction of the B-signals.

035.012 **Equiangular sundials.** G. E. Taylor.
Journ. British Astron. Ass., Vol. 86, 7 - 17 (1975).

035.013 **Hydrogen maser frequency standard.**
P. Petit, J. Viennet, R. Barillet, O. Gheorghiu, M. Desaintfuscien, C. Audoin.
Ann. Françaises Chronométrie Micromécanique, Année 1974, p. A1–1 - 17. In French.

035.014 **X band frequency standards formed by an oscillator with Gunn effect directly stabilized by superconductive cell.** J. J. Jimenez, J. Benard, Nguen Tuong Viet.
Ann. Françaises Chronométrie Micromécanique, Année 1974, p. A2–1 - 8. In French.

035.015 **Standard frequency and time signal broadcasting from amplitude modulation wireless stations.**
A. Gabry.
Ann. Françaises Chronométrie Micromécanique, Année 1974, p. A3–1 - 20. In French.

035.016 **Spectral stability and purity of crystal-controlled oscillators of very high frequency.**
M. Olivier, J. Groslambert.
Ann. Françaises Chronométrie Micromécanique, Année 1974, p. A4–1 - 7. In French.

035.017 **Synchronization in the nanosecond range of distant clocks.** J. Besson, P. Parcelter.
Ann. Françaises Chronométrie Micromécanique, Année 1974, p. A8–1 - 11. In French.

035.018 **Un cadran solaire oublié.** L. Janin.
Orion, 33. Jahrgang, p. 179 - 182 (1975).

035.019 **Über die Bedeutung von Zeitzeichensendern.**
A. Brömme, A. Wörner.
Orion, 33. Jahrgang, p. 214 - 219 (1975).

035.020 **Electronic registration of time (III).** R. Fangor.
Urania Kraków, Vol. 46, 211 - 216 (1975). In Polish.

035.021 **Solar and sidereal digital clock.** S. Gorgolewski.
Stud. Soc. Sci. Torunensis, Toruń, Poland, Sect. F (Astron.), Vol. 5, 111 - 118 = Bull. Astron. Obs. N. Copernicus Univ., Toruń, No. 53 I (1974).

An integrated circuit digital clock was built which derives from a temperature stabilized 100 kHz crystal oscillator, two time scales – solar and sidereal with indication of seconds, minutes and hours.

Choice of sundials. See Abstr. 003.040.

Sundial of the Trigorsk park.
See Abstr. 004.059.

The accuracy of the atomic time scales.
See Abstr. 044.010.

An accuracy algorithm for an atomic time scale.
See Abstr. 044.014.

The IEN time and frequency dissemination.
See Abstr. 044.015.

Chronometry and the variation of the rotation of the earth. See Abstr. 044.028.

036 Photographic Auxiliaries

036.001 **An evaluation of eight films for astrophotography.** G. T. Keene, M. H. Sewell.
Sky Telescope, Vol. 50, 61 - 65 (1975).

036.002 **Empfindlichkeitssteigerungen bei Astro-Emulsionen.** E. Wiedemann.
Orion, 33. Jahrgang, p. 147 (1975).

036.003 **On the hyper-sensitization of photographic plates by baking. II. Preliminary results for Eastman Kodak plates of types 103a-O, 103a-F and IIa-F.**
O. D. Dokuchaeva, T. A. Birulya.
Astron. Tsirk., No. 849, p. 3 - 5 (1974). In Russian.

036.004 **Colour films for astrophotography.** C. Papadopoulos.
Monthly Notes Astron. Soc. Southern Africa, Vol. 34, 118 - 124 (1975). – Presidential address during the annual general meeting of the Astron. Soc. Southern Africa.

036.005 **On the storage of Kodak (London) scientific plates OaO, OaD and OaF at different temperatures.**
O. D. Dokuchaeva, T. A. Birulya.
Soobshch. Gos. Astron. Inst. Shternberga, No. 177, p. 13 - 18 (1973). In Russian.

Kodak (London) plates for astronomical use (emulsion types OaO, OaD and OaF) were kept during 10 months at the temperatures $-21°$ C, $-14°$ C, $\simeq 0°$ C and $+20°$ C. Data were obtained for speed, contrast and fog density. For the plates kept at room temperature it was found that the speed was decreased ($\simeq 30\%$); the fog density was increased for plates except OaF-plates. The emulsion parameters were almost not changed when the plates were kept at temperatures $-21°$ C, $-14°$ C and $0°$ C.

036.006 **On the storage of two kinds of astronomical films.** O. D. Dokuchaeva, T. A. Birulya.
Soobshch. Gos. Astron. Inst. Shternberga, No. 182, p. 37 - 47 (1973). In Russian.

The long-term storage of astronomical films made by Scientific Institute Techphotoproect (Kazan) and by V. V. Kujbishev Chemical Factory was examined. Changes in sensitivity, coefficient of contrast and fog density were determined for the films stored at temperatures -21, -14, $\simeq 0°$ C and at room temperature. The temperature conditions are recommended for which the films should be stored.

036.007 **Photographing of extreme faint stars on different photographic materials.**
I. I. Brejdo, O. M. Mikhajlova.
Zhurn. nauch. i prikl. fotogr. i kinematogr., Vol. 20, No. 2, p. 113 - 121 (1975). In Russian. – Abstr. in Referativ. Zhurn. 51. Astron., 8.51.278 (1975).

036.008 **High-speed astronomical photo emulsions. Treatment in the photo-room, photometrical applications.**
D. A. Rozhkovskij.
Trudy Astrofiz. Inst., *Alma-Ata*, Vol. 25, 38 - 75 (1975). In Russian.

036.009 **Experimental IIIa-J and 127 spectroscopic films.** M. H. Sewell, A. G. Millikan.
AAS Photo-Bull., No. 8, 1975 No. 1, p. 6 - 9.

Hardened versions of the emulsion used for Kodak spectroscopic plate, type IIIa-J, and Kodak special plate type 127-03, were prepared for application on 0.004-inch Estar support. These experimental films have sensitivity characteristics similar to those of the corresponding plates. The susceptibility of the experimental films to abrasion is much less than that of the plates.

036.010 **Photographic experiments in preparation for a test of general relativity during the eclipse of 30 June 1973.** R. A. Matzner, M. P. Ryan, Jr.
AAS Photo-Bull., No. 8, 1975 No. 1, p. 10 - 12.

The optimum combination of photographic emulsion and processing for astrometric observations of stars close to the sun during the total eclipse of 30 June 1973 was selected on the basis of experiments with four types of Kodak plates and four developers. The final choice of Kodak spectroscopic plate, type III-O, developed in Kodak developer D-19, gave an eclipse sky fog density of about 1 and an eclipse limiting magnitude of about 8.5, using an $f/14$ astrographic camera with 165 mm effective aperture in an exposure pattern of 60 seconds on the eclipse field, plus 30 seconds on a nearby (corona-free) comparison field.

036.011 **Preparation of interobservatory densitometer calibration plates.** M. H. Sewell.
AAS Photo-Bull., No.8, 1975 No. 1, p. 13 - 15. – Paper presented at the August 1974 meeting of the AAS Working Group on Photographic Materials.

036.012 **Photographic equidensitometry with Agfacontour film.** M. Gumtau.
AAS Photo-Bull., No. 8, 1975 No. 1, p. 16 - 17.

A relatively new product, called Agfacontour film, is now available for photographic equidensitometry using a single plate and single exposure.

036.013 **Envelopes for the archival storage of processed astronomical photographs.** W. F. van Altena.
AAS Photo-Bull., No. 8, 1975 No. 1, p. 18 - 19.

Several materials have been examined for use in envelopes designed for archival storage of astronomical photographic plates. The most suitable material, based on a variety of considerations, seems to be synthetic Tyvek material, which looks similar to paper. Envelopes made of Tyvek material are now commercially available at a cost comparable to that for envelopes made from high-quality paper.

036.014 **Bending of large photographic plates.** B. H. Rule.
AAS Photo-Bull., No. 8, 1975 No. 1, p. 20 - 21.

036.015 **Rapid reproduction of finding charts on Kodagraph Super-K projection positive paper KPP5.**
M. M. Hanna.
AAS Photo-Bull., No. 8, 1975 No. 1, p. 22 - 23.

Kodagraph Super-K projection positive paper KPP5 is a direct-positive paper of very high contrast and orthochromatic sensitivity designed for high contrast applications such as making copies of line drawings. Normally, this paper has much too high a contrast for making star-finding charts. By preflashing the paper before exposure, and by using low-contrast developers, reasonably good finding charts can be made quickly and inexpensively; and the need for making and filing internegatives is eliminated.

036.016 **Summary of tests with spectroscopic plates baked in nitrogen.** W. C. Miller.
AAS Photo-Bull., No. 9, 1975 No. 2, p. 3 - 7.

036.017 **Influence of some aging and exposure conditions on**

the sensitivity of Kodak film, type SC7.
J. Jamar, A. Cucchiaro.
AAS Photo-Bull., No. 9, 1975 No. 2, p. 8 - 11.

Data for the sensitivity of Kodak film, type SC7 (manufactured by Kodak Pathe) in the 4500 Å wavelength region are presented. The relative influence of the following physical factors has been investigated: age of the latent image, age of the photographic emulsion, age of the developer, temperature of the emulsion at the time of exposure, and humidity or dryness.

036.018 **Long-term space environmental effects on a 3400 type film.** J. T. Gosling, H. H. Avant.
AAS Photo-Bull., No. 9, 1975 No. 2, p. 12 - 13. – Paper presented at the August 1974 meeting of the AAS Working Group on Photographic Materials.

036.019 **Reticulation: effects and cure.**
A. I. Poland, J. T. Gosling.
AAS Photo-Bull., No. 9, 1975 No. 2, p. 14 - 16.

Reticulation can cause severe effects in the image structure of a high-resolution film copy that is made by contact printing with a collimated source. These effects can be reduced by using a diffuse printing light source. Reticulation can be prevented in the original film by reducing the swelling and contraction of the gel during processing.

036.020 **An analytical representation for photographic characteristic curves.**
T. Tsubaki, O. Engvold.
AAS Photo-Bull., No. 9, 1975 No. 2, p. 17 - 19.

An analytical expression is proposed for representing accurately photographic characteristic curves. The function developed can represent various types of characteristic curves almost perfectly.

036.021 **Use of a thin objective prism with Kodak spectroscopic plate, type IIIa-J.**
A. G. D. Philip.
AAS Photo-Bull., No. 9, 1975 No. 2, p. 20 -21. – Paper presented at the August 1974 meeting of the AAS Working Group on Photographic Materials.

036.022 **Influence of the background brightness on registration of extremely faint stars by photographic materials of different types.** I. I. Brejdo, O. M. Mikhajlova.
Izv. Glav. Astron. Obs. Pulkovo, No. 193, Astrofiz., Astrometr. i Zvezdn. Astron., p. 169 - 174 (1975). In Russian.

On models of stars and sky background it is shown that with an increase of the exposure-time stellar magnitudes of the faintest stars reach the limiting values (m_{lim}) at the background density of 0.5-0.8. With the increase of the background brightness the m_{lim} decreases correspondingly. At the given background brightness fine-grain and low-sensitive photographic materials permit to registrate stars essentially fainter than coarse-grain and high-sensitive materials do, though with longer exposures.

036.023 **Results of laboratory tests of Kodak plates.**
I. I. Brejdo, O. M. Mikhajlova.
Astron. Tsirk., No. 872, p. 6 - 8 (1975). In Russian.

036.024 **On a correlation between DQE of photographic material and its ability for reproducing faint stars at the sky background.** I. I. Brejdo.
Astron. Tsirk., No. 876, p. 6 - 8 (1975). In Russian.

036.025 **Effect of nonuniform irradiance, and irradiance fluctuations, upon the response of photographic film.** C. T. Chang, J. L. Bjorkstam.
Journ. Optical Soc. America, Vol. 65, 1495 - 1501 (1975).

036.026 **On the influence of the form of the initial part of the characteristic curve of a photoemulsion on registration of faint stars on the sky background.**
I. I. Brejdo, O. M. Mikhajlova.
Zhurn. nauch. i prikl. fotogr. i kinematogr., Vol. 20, 256 - 261 (1975). In Russian. – Abstr. in Referativ. Zhurn. 51. Astron., 2.51.298 (1976).

Astronomical photography, from Daguerreotype to electron camera. See Abstr. 003.128.

Positional Astronomy. Celestial Mechanics

041 Positional Astronomy, Astrometry, Star Catalogues and Atlases

041.001 **Observations of Saturn obtained with the astrolabe at Paris Observatory during the winter of 1973 – 1974.** S. Débarbat.
Astron. Astrophys., Suppl. Ser., Vol. 21, 1 - 4 (1975). In French.

This paper contains the results of the third campaign at the Paris Observatory. During the 1973–74 winter 20 east transits and 17 west transits were observed.

041.002 **On the possibility of the use of circumzenithal in lunar astrometry.** I. Pešek.
Bull. Astron. Inst. Czechoslovakia, Vol. 26, 234 - 237 (1975).

The principle of a rotational micrometer for a circumzenithal to be used in lunar astrometry is described. The accuracy of the passage times is increased by a considerable acceleration of the motion of the star images in the field of view.

041.003 **An attempt of combining three absolute catalogues of declinations of FKSZ stars.**
Compiled by Ya. S. Yatskiv, A. N. Kur'yanova, edited by A. A. Korsun'.
Glavnaya Astronomicheskaya Observatoriya Akademii Nauk USSR, Kiev. 51 pp. (1975). In Russian.

In connection with the recompilation of the Preliminary General Catalogue of Fundamental Faint Stars (PFKSZ) the methods of comparison of catalogues and of determination of the weights of catalogues are discussed. The results are illustrated by an example of combining the absolute catalogues of Kazan $-50\ \delta$, Goloseevo $-58\ \delta$ and Pulkovo $-58\ \delta$. For the comparison of these catalogues Brosche's method completed with the autocovariance function estimation is used.

041.004 **Precise positions of Venus, Mars and Saturn from photographic observations with the normal astrograph at Pulkovo.** N. M. Bronnikova, L. V. Zhukov, T. P. Kiseleva, A. A. Kiselev, L. S. Koroleva, G. V. Panova.
Byull. Inst. Teoret. Astron., *Leningrad,* Vol. 13, 652 - 655 (1975). In Russian.

041.005 **Astrometric studies of the regions BD+17°4946, BD–5°3071, and Ross 730–1.**
J. Russell, G. Gatewood.
Astron. Journ., Vol. 80, 652 - 654 (1975).

Positions, proper motions, and absolute parallaxes were found for 46 stars in the regions of BD+17°4946, BD–5°3071, and Ross 730–1. BD+17°4946 showed some evidence of irregular variations in position and magnitude. The residuals for Ross 730–1 were used in a preliminary study to test for an appropriate weighting system; a weak weighting system was indicated.

041.006 **On the derivation of absolute coordinates of stars from group observations.** V. S. Gubanov.
Astron. Zhurn. Akad. Nauk SSSR, Vol. 52, 857 - 866 (1975). In Russian. English translation in Soviet Astron., Vol. 19, No. 4.

An extensive application of high-speed electronic computers permits to raise the problem of general adjustment of the results of astrometric observations made according to the group program. A method of adjustment, based on the interrelating of separate nights of observations to the stars common for a transit instrument in the meridian and an astrolabe, has been worked out.

041.007 **On results of an astrometrical reduction of the AFU-75 plates obtained in a compensative regime.**
M. V. Bratijchuk, V. P. Epishev, A. G. Kirichenko, K. A. Kudak, A. V. Martynov, A. S. Oleolenko, V. P. Ryabov.
Nauchn. Informatsii, vyp. (No.) 28, p. 51 - 64 (1974). In Russian.

The accuracy of measurements of star and satellite images for the AFU-75 plates obtained in a compensative regime is determined. The mean error is equal $\pm 1.0\,\mu$ (for reference stars) and $\pm 1.5\,\mu$ (for a satellite). The mean square errors calculated from an adjustment by Turner's method for 340 satellite positions, and topocentric equatorial coordinates for these positions are considered.

041.008 **On the accuracy of a passive or active satellite position obtained by the AFU-75 camera.**
M. V. Bratijchuk, A. G. Kirichenko, M. M. Osipenko.
Nauchn. Informatsii, vyp. (No.) 28, p. 65 - 69 (1974).

400 point images of the Geos B were used for a determination of the positional accuracy and 200 point images were used for a determination of the position and the time errors as well. The plates were obtained in the compensative regime of tracking of the AFU-75. The correlation between the positional error of the satellite flash and the height of a satellite was investigated.

041.009 **Sur la précision des positions photographiques des satellites.** G. Vass.
Nauchn. Informatsii, vyp. (No.) 28, p. 70 - 76 (1974). In Russian.

On présente les méthodes de réduction utilisées à l'Observatoire de Bucarest et on analyse la précision réelle des positions photographiques des satellites en approximant les coordonnées mesurées avec des fonctions polynomiales de temps.

041.010 **A star catalogue on magnetic tape and service programs.** V. A. Enal'skij, S. A. Pavlik.
Nauchn. Informatsii, vyp. (No.) 30, p. 24 - 39 (1974). In Russian.

The paper is devoted to the SAO star catalogue recorded on magnetic tape. Several service programs used for operative access to this information are described.

041.011 **Study on the characteristics of a Talcott levels couple.** L. Cugusi.
Boll. Geod. Sci. Affini, Anno 34, p. 273 - 288 (1975). In Italian.

041.012 **On a new method of photographing the moon with stars on separate plates.**
N. F. Bystrov, N. G. Rizvanov.
Trudy Kazan. Gorod. Astron. Obs., No. 39, p. 156 - 175 (1973). In Russian.

041.013 **Positions of the major planets and the moon.**
G. Soulié, Dupouy, Teulet, Broqua, Dulou.
Astron. Astrophys., Suppl. Ser., Vol. 22, 49 - 61 (1975).
In French.

In order to improve our knowledge of the astronomical constants, great interest must be paid to accurate positions of major planets and their satellites. Likewise, accurate astrometry of the moon on stellar background is essential to determine the ephemeris time. The present work follows up other papers already published in "Journal des Observateurs" and in "Astronomy and Astrophysics Supplement".

041.014 **Individual corrections astrolabe – N30 for 449 southern stars.** F. Noël.
Astron. Astrophys., Suppl. Ser., Vol. 22, 63 - 68 (1975).

Individual corrections in right ascension and declination in the sense astrolabe – catalogue are given for 449 N30 stars between –4° to –62° in declination.

041.015 **On a deduction of an absolute catalogue of right ascensions of stars for the whole sky.**
V. S. Gubanov.
Pis'ma v Astron. Zhurn., Vol. 1, No. 8, p. 43 - 47 (1975). In Russian.

An organization of special observations by an astrolabe at the equator is suggested for determination of the absolute orientation of right ascension catalogues.

041.016 **Observations of Jupiter's Galilean satellites using astrolabes.** H. Davilà, S. Débarbat, A. Journet.
Celestial Mechanics, Vol. 12, 51 - 55 (1975). – Paper presented at IAU Colloquium No. 28, 'Planetary satellites', Ithaca, N.Y., August 1974.

Jupiter's Galilean satellites have been observed at the Quito (1964) and Haute-Provence (1965) Observatories for experimental purposes; the results show that the astrolabe can give an efficient contribution in this field, and that a permanent survey would be useful.

041.017 **Observations of outer planet satellites.**
P. J. Shelus, R. I. Abbot, J. D. Mulholland.
Celestial Mechanics, Vol. 12, 57 (1975). – Abstract of a paper presented at IAU Colloquium No. 28, 'Planetary satellites', Ithaca, N.Y., August 1974.

041.018 **Absolute declinations of stars in the systems of nadir and Polarissima.** A. A. Krasikova.
Izv. Astron. Ehngel'gardt. Obs., *Kazan'*, No. 40, p. 70 - 76 (1973). In Russian.

041.019 **Determination of declinations of stars on the ZTL-180 of the Engelhardt Astronomical Observatory.**
D. P. Laas.
Izv. Astron. Ehngel'gardt. Obs., *Kazan'*, No. 40, p. 77 - 79 (1973). In Russian.

041.020 **Observations of minor planets for improving the positions of stars of a fundamental catalogue.**
V. I. Orel'skaya.
Astrometriya i Astrofizika, *Kiev*, vyp. (No.) 26, (see 003.001), p. 11 - 16 (1975). In Russian.

The state of improving star catalogue positions by minor planets observations is described. A new list of asteroids is suggested for determining the systematic corrections to the star positions of the equatorial zone. The planets trajectories for the period from 1974 to 1990 and for the sky region of ±30° declination are presented.

041.021 **Astrometric observations of Saturn's satellites from McDonald Observatory, 1972.**
R. I. Abbot, J. D. Mulholland, P. J. Shelus.
Astron. Journ., Vol. 80, 723 - 728 (1975).

Observations of Saturn's satellites have been reduced by means of secondary reference stars obtained by reduction of Palomar Sky Survey plates. This involved the use of 29 SAO stars and plate overlap technique to determine the coordinates of 59 fainter stars in the satellite field. Fourteen plate constants were determined for each of the two PSS plates. Comparison of two plate measurement and reduction techniques on the satellite measures appears to demonstrate the existence of a serious background gradient effect and the utility of microdensitometry to eliminate this error source in positional determinations of close satellites.

041.022 **Thoughts on problems of long-focus photographic astrometry.** P. van de Kamp.
Proc. Astrometric Conference, Torino 1974, (see 012.017), p. 13 - 20 (1975).

041.023 **Potentialities of astrometric telescopes for the establishment of reference systems.** W. Fricke.
Proc. Astrometric Conference, Torino 1974, (see 012.017), p. 47 - 51 (1975).

Approaches for extending the fundamental reference coordinate system to faint objects and extragalactic radio sources are discussed. A joint campaign of radio astrometry and photographic astrometry with large telescopes may contribute effectively to the solution of this problem. Data on the accuracy of different methods are given.

041.024 **On magnitude errors and systematic position and proper motion differences between the Bergedorf– and Strasbourg solution of the AGK2/3 catalogue.**
C. de Vegt.
Astron. Astrophys., Vol. 43, 303 - 308 (1975).

From a preliminary proper motion investigation concerning AGK3-members in a small sample of 14 open clusters, a magnitude-dependence of the proper motions has been found in 8 cases. Further on, the comparison of the Bergedorf AGK2/3 catalogue results with an iterative block adjustment solution obtained by Lacroute et al. at the Strasbourg Observatory in these cluster fields has shown, that a similar magnitude equation is present in the Strasbourg solution and systematic differences up to $0''.5$ in position and $0''.016/a$ in proper motion appear. Possible sources for these differences are discussed.

041.025 **Catalogue of declinations of 200 stars of the program of wide-scale pairs in the FK4 system.**
N. P. Krasnenko.
Astrometriya i Astrofizika, *Kiev*, vyp. (No.) 27, (see 003.004), p. 17 - 25 (1975). In Russian.

Declinations of 200 stars of the new Pulkovo program of wide-scale pairs are given. The observations were carried out with the Repsold transit circle at the Odessa Astronomical Observatory in 1973 and were reduced to the FK4 system. The mean error of one determination of declination is $\pm 0''.33$. The catalogue was compared with the AGK3.

041.026 **A possible variant of the construction of a general catalogue of the position of stars by means of an electronic computer.** E. V. Khrutskaya.
Astron. Zhurn. Akad. Nauk SSSR, Vol. 52, 1059 - 1062 (1975). In Russian. English translation in Soviet Astron., Vol. 19, No. 5.

041.027 **On the reduction of astronomical observations to a standard coordinate system.** V. Eh. Brandt.
Astron. Zhurn. Akad. Nauk SSSR, Vol. 52, 1096 - 1107 (1975). In Russian. English translation in Soviet Astron., Vol. 19, No. 5.

Accurate and approximate formulae of the influence of

polar motion on latitude and time are deduced. The zero-points of universal time and of longitudes are considered, and the formulae of reduction of observational differences in longitude to the original pole position are given.

041.028 **Provisional positions of the moon 1958–1971.** K. C. Blackwell, J. V. Carey, R. H. D. Swifte.
Roy. Obs. Bull., [Herstmonceux: Royal Greenwich Obs.], No. 179, p. 301 - 320 (1974).

Provisional positions of the moon deduced from observations of the limbs and the crater Mösting A made with the Cooke Transit Circle at Herstmonceux in the years 1958–1971 are presented. The right ascensions are based on the system of the FK4.

041.029 **Catalog of proper motions for the 5,965 stars of the six-inch transit circle program 1949 – 1956.**
R. W. Rhynsburger, F. S. Gauss.
Publ. United States Naval Obs., *Washington,* Second Ser., Vol. 23, Part 1, 92 pp. (1975).

For slightly more than half of the $W3_{50}$ stars (3,413) the new proper motions are based on positions obtained from a minimum of three catalogs; the bulk of these positions (2,224) are from the AGK2 (or the reference stars AGK2A), the GC, and the $W3_{50}$ catalogs. The remaining 2,552 stars have their proper motions based on but two catalogs, the $W3_{50}$ and one other. Together with the GFH, the AGK2 (including the AGK2A) and the GC supplied 5,898 positions in various combinations. A summary of the catalogs used, excluding the $W3_{50}$, the number of stars for each combination, and the mean errors of the proper motion for those instances where three or more catalogs were involved, are presented.

041.030 **Definition of the celestial reference coordinate system in fundamental catalogues.** W. Fricke.
IAU Colloquium No. 26, (see 012.034), p. 201 - 222 (1975).

041.031 **Fundamental astrometry today.** G. (*Dj.*) Teleki. Publ. Obs. Astron. Beograd, No. 20, (see 012.024), p. 185 - 192 (1975). In Serbo-Croatian.

041.032 **Research in the field of fundamental astrometry at Pulkovo Observatory.** K. N. Tavastsherna.
Publ. Obs. Astron. Beograd, No. 20, (see 012.024), p. 193 - 194 (1975). In Serbo-Croatian.

041.033 **Le programme d'observation sur le circle meridien à Belgrade.** S. Sadžakov, D. Šaletić, M. Dačić.
Publ. Obs. Astron. Beograd, No. 20, (see 012.024), p. 218 - 222 (1975). In Serbo-Croatian.

041.034 **Precise reduction to the apparent places of stars.** S. Yumi, K. Hurukawa, T. Hirayama.
Publ. International Latitude Obs. Mizusawa, Vol. 9, 175 - 222 (1974).

Reduction to the apparent places of stars by the method of rectangular coordinates in a uniform system was investigated as accurately as 0."001 free from all rounding errors over a long time interval such as one or two centuries or more. Coordinates of the sun were carefully examined on the Astronomical Ephemeris, Herget's Solar Coordinates 1800–2000 and Newcomb's expansions. Coordinates of the centre of mass of the solar system were also examined.

041.035 **Third catalog of trigonometric parallaxes of faint stars.** R. S. Harrington, C. C. Dahn, A. L. Behall, J. B. Priser, J. W. Christy, B. Y. Riepe, H. D. Ables, H. H. Guetter, A. V. Hewitt, R. L. Walker.
Publ. United States Naval Obs., *Washington,* Second Ser., Vol. 24, Part 1, 30 pp. (1975).

This publication is a continuation of the series of parallaxes published in the First and Second Catalogs (Publ. U.S. Naval Observatory, Vol. XX, Parts III and VI). The present catalog gives relative parallaxes and UBV photometry for 88 stars in 82 systems.

041.036 **First Santiago-Pulkovo Fundamental Stars Catalogue (SPF 1). A right ascension catalogue of fundamental FK4 stars.** C. Anguita, G. Carrasco, P. Loyola, V. N. Bedin, A. A. Naumova, D. D. Polojentsev (*Polozhentsev*), T. A. Polojentseva (*Polozhentseva*), K. N. Tavastsherna, M. S. Zverev.
Dep. Astron. Univ. Chile, Fac. Cie. Fis. Mat. Obs. Astron. Nacional, Cerro Calan, Santiago de Chile, Publ. Vol. 2, (No. 6), 181 - 214 (1975).

This is the first catalogue, in right ascension observed and prepared jointly by Chilean and Soviet astronomers working under an agreement for astronomical cooperation between Universidad de Chile and the Academy of Sciences of the USSR. The catalogue contains the corrections to the FK4 right ascensions of 1043 stars observed in upper culmination and 82 stars observed in lower culmination. The average number of observations per star is 5.6 for stars observed in upper culmination and 11.4 for the stars in lower culmination. The mean epoch of the catalogue is 1965.024 and 1965.030 respectively.

041.037 **Mikhajlov's star atlases.** P. G. Kulikovskij. Zemlya i Vselennaya, 1975, No. 6, p. 85 - 88. In Russian.

041.038 **New movable chart of the sky with cylindrical projection.** D. A. Rozhkovskij.
Trudy Astrofiz. Inst., *Alma-Ata*, Vol. 25, 119 - 124 (1975). In Russian.

A new movable celestial chart permitting to solve different problems of spherical astronomy is considered.

041.039 **On the possibility of creating a radio-astronomical inertial coordinate system based on the measurement of arcs between QSOs with the aid of VLBI.**
A. F. Dravskikh, A. M. Finkelstein (*Finkel'shtejn*), G. A. Krasinsky (*Krasinskij*).
Astrophys. Space Sci., Vol. 38, 255 - 266 (1975).

The aim of this paper has been to discuss a possibility of the formation of an inertial system of coordinates, precise within 0."001, with the aid of the measures of mutual angular distances of quasars by the VLBI technique used in radioastronomy. Such a system depends on neither the rotation, nor revolution of the earth around the sun – a fact which permits a separation of the astronomical and geophysical aspects of the problem. The proposed system should permit us to resolve some general astrometric problems – such as of the precession and nutation of the earth as well as of the motion of the terrestrial axis of rotation within the earth, or fluctuations in the angular velocity of terrestrial rotation.

041.040 **Shortening of calculation from systematic differences between two catalogues.** A. Valbousquet.
Astron. Astrophys., Vol. 45, 5 - 14 (1975). In French.

For the representation of systematic differences, the products of Fourier terms in right ascension and Legendre polynomials in declination are used. By developing the star density in a series of the same functions, the coefficients of the normal equations can be evaluated with considerably less computational effort.

041.041 **International Information Bureau on Astronomical Ephemerides.**
I.I.B.A.E. (I.A.U.–COSPAR), Information cards Nos. 103 - 111 (1973); Nos. 112 - 125 (1975). – Available from Bureau International d'Information sur les Éphémérides Astrono-

miques, Palais de l'Institut, Paris, France.

041.042 **Catalogue of declinations of the latitude programme stars obtained by the micrometer method.**
I. M. Kalinina.
Trudy Gos. Astron. Inst. Shternberga, Vol. 46, vyp. (No.) 2, p. 3 - 47 (1975). In Russian.

This catalogue is the first part of the declination catalogue of stars of the latitude programmes established at the Moscow Observatory by observing with the ZTL-180. The observations were made at zones 1°30′ wide, symmetrically to the zenith, with micrometric attaching to the stars of AGK3R and FK4. The catalogue contains 2362 stars. The mean epoch of the observations is 1969.0. The mean error of one observation is ± 0″.22. The proper motions obtained by the author are given for the AGK3R reference stars. The comparison of the obtained catalogue with two photographic and three visual catalogues showed that it reproduces the basic system correctly.

041.043 **Individual corrections of the right ascensions of FK4 stars.** L. M. Khommik.
Trudy Gos. Astron. Inst. Shternberga, Vol. 46, vyp. (No.) 2, p. 48 - 52 (1975). In Russian.

The corrections in right ascension of 520 FK4 stars are given. They are the results of re-reduction of PFKSZ stars (1962) in the FK4 system.

041.044 **Individual corrections of right ascensions and declinations of PFKSZ stars.** L. M. Khommik.
Trudy Gos. Astron. Inst. Shternberga, Vol. 46, vyp. (No.) 2, p. 53 - 56 (1975). In Russian.

Corrections in right ascension and declination of 323 PFKSZ stars are given.

041.045 **Some problems related to the definition of reference systems.** J. Kovalevsky.
IAU Colloquium No. 26, (see 012.034), p. 123 - 132 (1975).
Review paper.

041.046 **Differential radioastrometry.**
V. A. Alekseev, B. N. Lipatov, B. V. Shchekotov.
VIII Vses. konf. po radioastron. Tezisy dokl. Pushchino, 1975, p. 185. In Russian. – Abstr. in Referativ. Zhurn. 51. Astron., 12.51.158 (1975).

041.047 **A master list of non-stellar optical objects.**
R. S. Dixon.
Bull. American Astron. Soc., Vol. 7, 538 - 539 (1975).
Abstr. AAS.

041.048 **Contributions possibles de l'astrométrie spatiale à la définition d'un système de référence absolu.**
J. Kovalevsky.
Space astrometry, (see 012.037), p. 67 - 71 (1975).

041.049 **Extension of a stellar reference system by means of spacecraft-observed faint stars.** H. G. Walter.
Space astrometry, (see 012.037), p. 73 - 77 (1975).

Faint stars play a unique role in positional astronomy as intermediaries between the bright stars that define the fundamental system of reference and planetoids, extragalactic nebulae and quasi-stellar radio sources, the latter providing the link for the optical and radio system of reference. Differential observations with 0″.01 accuracy from spacecraft-borne equipment are capable of extending the present reference system down to magnitude 10, or possible 17. Suitably selected faint stars in the vicinity of radio sources facilitate the interconnection of the optical and radio systems.

041.050 **The fundamental reference system and observations from space.** K. A. Strand.
Space astrometry, (see 012.037), p. 79 - 80 (1975).

041.051 **Conséquences de l'amélioration des catalogues sur l'étude du système terre-lune.**
C. Meyer, M. Froeschle.
Space astrometry, (see 012.037), p. 95 - 99 (1975).

041.052 **Astrometric positions of the planet Pluto in the period 1971–1974.**
C. Barbieri, M. Capaccioli, G. Pinto.
Space astrometry, (see 012.037), p. 101 - 102 (1975).

The astrometric positions of the planet Pluto in the years 1971–1974 have been derived from 41 plates taken with the 67/92-cm Schmidt telescope of Asiago Observatory. The precision of the positions, which are referred to an average number of 24 AGK3 stars, is 0″.6 in right ascension and 0″.4 in declination. A comparison is made between the Asiago results and the astrometric positions computed by Kaplan, Seidelmann & Smith (1972). Good agreement is found, with a possible systematic difference of 0″.6 in declination.

041.053 **Celestial reference systems derivable from solar system dynamics.**
R. L. Duncombe, P. K. Seidelmann, T. C. Van Flandern.
IAU Colloquium No. 26, (see 012.034), p. 223 - 233 (1975).

041.054 **On the use of Tautenburg Schmidt for improvement of astronomical reference system.** H. U. Sandig.
IAU Colloquium No. 26, (see 012.034), p. 241 - 246 (1975).

041.055 **The determination of the equator point for observing star declinations.** A. A. Mikhailov.
IAU Colloquium No. 26, (see 012.034), p. 315 - 317 (1975).

041.056 **On the influence of the inclination of the rotation axis of a transit instrument on the results of meridian observations.** A. A. Nemiro.
Izv. Glav. Astron. Obs. Pulkovo, No. 193, Astrofiz., Astrometr. i Zvezdn. Astron., p. 100 - 103 (1975). In Russian.

On the basis of an analysis of the reduction formulae it is shown that the error in determining the inclination of the rotation axis of a transit instrument (Δi) gives rise to an error in determination of the azimuth and clock corrections Δi tg φ and Δi sec φ correspondingly. An influence of these errors on the results of determinations of time, longitude and right ascensions of stars is considered. Corresponding recommendations are given.

041.057 **Deduction of the declination system from observations with vertical circles in both hemispheres.**
V. A. Naumov.
Izv. Glav. Astron. Obs. Pulkovo, No. 193, Astrofiz., Astrometr. i Zvezdn. Astron., p. 104 - 108 (1975). In Russian.

041.058 **Smoothing of declinations and right ascensions of stars from the Pulkovo PZT program made by the free chain method.** N. M. Bakhrakh, V. A. Naumov.
Izv. Glav. Astron. Obs. Pulkovo, No. 193, Astrofiz., Astrometr. i Zvezdn. Astron., p. 117 - 123 (1975). In Russian.

The results of smoothing the declinations and right ascensions of the Pulkovo PZT program stars observed during 1960–1963 are given. The smoothing was made by the free chain method. The catalogue of 178 stars has been compared with the AGK3 catalogues.

041.059 **Comparison of declinations of the Talcott pairs centers obtained from observations with the PZT-135 during 1948.8–1955.0 at Pulkovo with the declinations of the $W2_{50}$ Catalogue.** L. D. Kostina.
Izv. Glav. Astron. Obs. Pulkovo, No. 193, Astrofiz., Astrometr.

i Zvezdn. Astron., p. 139 - 140 (1975). In Russian.

041.060 **Photographic positional observations of Venus at Pulkovo with the 26″ refractor in 1972.**
N. M. Bronnikova, A. A. Kiselev.
Izv. Glav. Astron. Obs. Pulkovo, No. 193, Astrofiz., Astrometr. i Zvezdn. Astron., p. 153 - 157 (1975). In Russian.

041.061 **Positional photographic observations of Venus with the double short-focus astrograph at Pulkovo during 1972.** T. P. Kiseleva.
Izv. Glav. Astron. Obs. Pulkovo, No. 193, Astrofiz., Astrometr. i Zvezdn. Astron., p. 158 - 161 (1975). In Russian.

041.062 **Positions of Mars determined with the zonal astrograph at Nikolaev during 1967–1971.**
G. K. Gorel', F. F. Kalikhevich.
Izv. Glav. Astron. Obs. Pulkovo, No. 193, Astrofiz., Astrometr. i Zvezdn. Astron., p. 162 - 165 (1975). In Russian.

The topocentric coordinates of Mars, obtained during three oppositions in 1967, 1969 and 1971, are tabulated. A graphical comparison of the (O–C) values with the observations made at five different observatories is given.

041.063 **The U.K. 48-inch (1.2 metre) Schmidt telescope and the SRC IIIaJ Survey.** R. D. Cannon.
Proc. Astron. Soc. Australia, Vol. 2, 323 - 324 (1975).

041.064 **Reduction of T. Hornsby's observations of planets between the years 1774 and 1798.**
M. L. Sveshnikov, E. S. Sveshnikova.
Byull. Inst. Teoret. Astron., *Leningrad,* Vol. 14, 102 - 106 (1975). In Russian.

The present paper deals with the planet observations of Hornsby and their reduction to the system of FK4. The values (O–C) have been received by comparing the observations to the theory of Newcomb. The system of (O–C) thus obtained is investigated by statistical methods. A synodical periodic component, being the phase correction, is detected and approximated by a polynomial.

041.065 **AGK2 - AGK3 and the Astrographic Catalog.**
P. Lacroute, A. Valbousquet.
Centre de Données Stellaires, Inform. Bull. No. 9, (see 002. 008), p. 2 (1975).

041.066 **Star catalogs and files available at the Stellar Data Center.**
Centre de Données Stellaires, Inform. Bull. No. 9, (see 002. 008), p. 6 - 31 (1975).

041.067 **On the systematic differences between the Belgrade Catalogue of Latitude Stars and the AGK3R.**
S. Sadžakov, M. Dačić, D. Šaletić, V. Protić-Benišek.
Bull. Obs. Astron. Belgrade, No. 126, p. 3 - 7 (1975).

The comparison of the Belgrade Catalogue of Latitude Stars (KSZ) with the AGK3R was analysed by right ascension, declination, magnitude and spectral type. The change of the constant of refraction in dependence on the colour of the star is pointed out.

041.068 **Investigation of declinations and of proper motions of the Belgrade General Catalogue of Latitude Stars BSKŠZ1 and BSKŠZ2.** S. Sadžakov, D. Šaletić.
Bull. Obs. Astron. Belgrade, No. 126, p. 7 - 11 (1975).

041.069 **Declinations and the proper motions of the stars of the International Latitude Service on the basis of meridian catalogues from 1929 to 1972.**
S. Sadžakov, D. Šaletić.
Publ. Obs. Astron. Beograd, No. 21, 119 pp. (1975).

041.070 **Astrometry.** K. A. Strand.
Image processing techniques in astronomy, (see 012.041), p. 3 - 10 (1975).

An attempt is made to review the present status of some areas in astrometry especially in need of data and to mention the implementation or improvements which might be considered in regard to instrumentation and observations.

041.071 **Observations effectuées au Grand Cercle Méridien Askania pendant la période 1970–1974.**
R. Dejaiffe, R. Verbeiren.
Bull. Astron., Obs. Roy. Belgique, Vol. 8, 250 - 259 (1975).

General astrometry. See Abstr. 003.101.

Space astrometry project – reduction by multilateration. See Abstr. 031.264.

The role of VLBI in the establishment of coordinate systems. See Abstr. 031.266.

Practice on the PZT observation. See Abstr. 032.033.

Rumastrometri. See Abstr. 032.036.

Present and future meridian astrometry. See Abstr. 032.044.

Space astrometry projects. See Abstr. 032.529.

Faisabilité d'un satellite astrométrique – resultat d'une étude CNES. See Abstr. 032.531.

Astrometric experiments for the large space telescope. See Abstr. 032.532.

The application of radio interferometers in astrometry. See Abstr. 033.018.

Überlegungen zur Auswahl einer automatischen Plattenmeßmaschine. See Abstr. 034.007.

Der "neue" Bonner Komess. See Abstr. 034.008.

The two-coordinate Grant machine in long-focus astrometry. See Abstr. 034.141.

Effects of the introduction of a new fundamental reference coordinate system on time scales. See Abstr. 044.016.

La rotation de la terre et le système terre-lune. See Abstr. 044.033.

Research of changes of the Belgrade ***z*****-term for different declination systems in correlation with atmospheric conditions.** See Abstr. 046.040.

Astronomical observations and meteorological data (Express information). See Abstr. 082.030.

Astrometrical observations and atmospherical influences. See Abstr. 082.070.

Recent investigations of astronomical refraction. See Abstr. 082.072.

Chromatic refraction and its influence on astrometrical measurements. See Abstr. 082.073.

On the exclusion of the influence of photographic irradiation on the results of photographic positional observations of Venus. See Abstr. 093.012.

Improvement of the orbital elements of selected asteroids for the correction of star catalogue positions. See Abstr. 098.065.

Statistical properties of the sample "proper motions of the AGK 3". See Abstr. 112.001.

Proper motions and positions of stars in wide surroundings of the Hyades cluster. See Abstr. 112.013.

Optical positions of radio sources. III. Measurements on the Palomar Sky Survey prints. See Abstr. 141.023.

Determination of equatorial coordinates of radio sources with long-baseline radiointerferometers. See Abstr. 141.071.

Le mouvement des étoiles dans la Galaxie. See Abstr. 155.031.

042 Celestial Mechanics, Figure of Celestial Bodies

042.001 **The problem of three bodies: Copernicus and modern dynamical astronomy.** V. Szebehely.
Vistas in Astronomy, Vol. 17, (see 012.005), 171 - 176 (1975).

042.002 **Translational-rotational motion of celestial bodies (quasi-periodic solutions).** M. S. Volkov.
Byull. Inst. Teoret. Astron., *Leningrad*, Vol. 13, 609 - 611 (1975). In Russian.

It has been shown that the differential equations of translational-rotational motion satisfy the conditions of Arnold's existence theorem. The rotation of bodies is close to the inertial one. Two cases of construction of bodies supposed nearly spherically symmetrical are considered. The first case deals with bodies of rotation and the second considers arbitrary bodies. The solutions constructed may be of interest for climate researchers since the amount of the heat on different parts of planetary surfaces is determined by the orientation of a planet with respect to the sun.

042.003 **On the regularization of double encounters in the n-body problem and its application to the numerical integration of equations in celestial mechanics. II.**
V. F. Myachin.
Byull. Inst. Teoret. Astron., *Leningrad*, Vol. 13, 636 - 651 (1975). In Russian.

The theory of regularization developed previously is applied to the simultaneous numerical integration of the system of equations of celestial mechanics along the trajectories corresponding to close double approaches or encounters. The regularization permits to raise considerably the accuracy of computations when passing these trajectories. The continuation of motion after the encounter in the cases of binary collision is secured automatically. The numerical integration is based upon the direct use of Taylor series, the terms being consecutively evaluated following the generalized Steffensen scheme.

042.004 **On the geometry of orbits of distant stellar satellites.** I. N. Latyshev.
Byull. Inst. Teoret. Astron., *Leningrad*, Vol. 14, 16 - 22 (1975). In Russian.

The present paper deals with Hill's planar limit problem for the satellite-star-Galaxy case. The gravitational force of the Galaxy has been assumed to be proportional to the distance from its centre raised to the power k, $-2 \leqslant k < +1$. For the case where the satellite's motion is stable in Hill's sense closed curves exist, the satellite crossing one of these or both. The equations of these curves are given. A simple method has been proposed to determine the stability in Hill's sense, an application to some natural planetary satellites has been made. The particular features of motion changes in the satellite motion direction have been considered.

042.005 **A proof that a spherical shell central configuration is isolated.** E. M. Nezhinskij.
Byull. Inst. Teoret. Astron., *Leningrad*, Vol. 14, 32 - 36 (1975). In Russian.

042.006 **L'instabilité de certains systèmes canoniques à la résonance principale.** G. A. Merman.
Byull. Inst. Teoret. Astron., *Leningrad*, Vol. 14, 37 - 44 (1975). In Russian.

042.007 **Calculation of perturbations in the case of non-intersecting orbits.** E. Z. Khotimskaya.
Byull. Inst. Teoret. Astron., *Leningrad*, Vol. 14, 47 - 53 (1975). In Russian.

A method is suggested for developing the literal expansion of the disturbing function and its derivatives in case of non-intersecting orbits. The expansion has an analytical form with respect to the major axes. Due to the transformation of the hypergeometric functions related to the ratio of the semi-major axes it is possible to take out the poles from the series which also contribute to the convergence of the expansion. The series obtained make it possible to develop the high-order derivatives of the perturbation function which may be used for the application to some problems of celestial mechanics, among them being the group method of calculating minor planet perturbations and high-order planetary perturbations.

042.008 **Die Entstehung der Kirkwood-Lücken im Asteroiden-Gürtel.** H. Scholl.
Mitt. Astron. Ges., No. 36, p. 70 (1975). – Abstract.

042.009 **Bestimmung ganzperiodischer Lösungen eines Bewegungsproblems nach der Methode der unendlich vielen Variablen.** M. Schneider.
Mitt. Astron. Ges., No. 36, p. 136 - 137 (1975).

042.010 **Relativistic spinor linearization of the Kepler motion.** P. E. Kustaanheimo.
Astron. Nachr., Vol. 296, 163 - 170 (1975).

A relativistically conformal mapping between pairs of spinors and relativistic motors with zero trace is given. The corresponding transformation of any equation of motion for a motor with zero trace into equations of motion for two spinors is deduced. It is shown that the spinors execute harmonic oscillations in the case of the unperturbed Kepler motion.

042.011 **Lectures on the dynamics of the system of the earth and the moon.** A. H. Cook.
Boll. Geod. Sci. Affini, Anno 34, p. 289 - 320 (1975).
Lecture held at the International School of Advanced Geodesy, Erice, Sicily, October 1 - 26, 1974.

042.012 **On a precision estimate of the theories of motion of the earth and Mars.** V. L. Mazo.
Byull. Inst. Teoret. Astron., *Leningrad*, Vol. 13, 629 - 635 (1975). In Russian.

The present paper deals with an attempt to estimate the precision of the theory of the earth's motion as given by Newcomb and that of Clemence's theory of Mars.

042.013 **Keplerian motions.** P. van de Kamp.
Vistas in Astronomy, Vol. 18, (see 012.006), 543 - 552 (1975).

042.014 **Johannes Kepler and celestial mechanics.** G. A. Chebotarev.
Vistas in Astronomy, Vol. 18, (see 012.006), 557 - 565 (1975).

042.015 **Kepler's laws and modern celestial mechanics.** J. Kovalevsky.
Vistas in Astronomy, Vol. 18, (see 012.006), 605 - 611 (1975).

042.016 **An investigation of the invariant orbital elements of primary large bodies with parabolic and hyperbolic orbits.** G. F. Sultanov.
Izv. AN AzSSR. Ser. fiz.-tekhn. i mat. n., 1974, No. 5, p. 89 - 91. In Russian. – Abstr. in Referativ. Zhurn. 51. Astron., 7.51.133 (1975).

042.017 **Basis of the selection of the form of Hansen coef-**

ficients. Yu. V. Plakhov.
Izv. vyssh. ucheb. zavedenij. Geod. in aehrofotosemka, 1974, No. 5, p. 43 - 45. In Russian. – Abstr. in Referativ. Zhurn. 51. Astron., 7.51.135 (1975).

042.018 **Iterative method for determination of planetary perturbations.** T. V. Ivanova.
Astron. Zhurn. Akad. Nauk SSSR, Vol. 52, 839 - 843 (1975). In Russian. English translation in Soviet Astron., Vol. 19, No. 4.

A modification of Brouwer's method for determination of general planetary perturbations is proposed. For the right-hand members of the differential equations a new development is suggested which does not depend on the order of perturbations and is convenient for the calculation by computer. The functions of the unperturbed motion to be used in calculation of the perturbations are presented in a symmetric form and have no singularities for eccentricities equal to zero. The integration is performed by an iterative process which allows to take into account at the n-th approximation some terms of higher order than n with respect to the disturbing mass.

042.019 **On the investigation of the motion of a spheroidal planetary satellite.** G. T. Arazov.
Astron. Zhurn. Akad. Nauk SSSR, Vol. 52, 891 - 894 (1975). In Russian. English translation in Soviet Astron., Vol. 19, No. 4.

The solution of the problem of three stationary centres for small inclination orbits is used with special choice of parameters. The investigation of the motion of the fifth Jupiter satellite is given as an example, the gravitational potential of spheroidal Jupiter being approximated by the potential of the problem of three stationary centres. The comparison of the results of calculation with those of observation shows that for investigating the satellite motions the orbits of the problem of three stationary centres may be chosen as intermediate orbits in some cases.

042.020 **Some families of periodic oscillations in the restricted problem with small mass-ratios of three bodies.**
V. V. Markellos.
Astrophys. Space Sci., Vol. 36, 245 - 272 (1975).

This paper reports on the numerical determination of families of periodic oscillations in the case $\mu = 0.00095$ of the restricted problem. The families emanating out of the collinear Lagrangian points are examined as well as some asymmetric periodic oscillations related to them. An effort is made to complete the global picture of simple-periodic symmetric oscillations in the present case of the problem (the S–J case). The phase portrait of asymmetric periodic orbits is also examined.

042.021 **Angular momentum and kinetic energy of rotating deformable bodies.** J. N. Tokis.
Astrophys. Space Sci., Vol. 36, 427 - 438 (1975).

The general equations of angular momentum and kinetic energy of a rotating deformable (or not rigid) body are discussed for a fixed and a rotating coordinate system. A new system of equations is developed for a deformable body of arbitrary form using the Lagrangian (vector) displacement up to the first order terms. The equations are illustrated for a self-gravitating deformable body perturbed by tides.

042.022 **Accurate integration of geostationary orbits with Burdet's focal elements.**
W. Flury, G. Janin.
Astrophys. Space Sci., Vol. 36, 495 - 503 (1975).

The theory of Burdet's focal elements is outlined. The differential equations are presented, and the initial value problem is described together with the transformation to rectangular coordinates and classical elements. For the numerical integration of near-geostationary orbits a comparison of the efficiency is made between focal elements, KS theory and rectangular coordinates. For this class of orbits, a higher accuracy has been obtained by integrating elements than integrating rectangular coordinates.

042.023 **First order perturbations of the four large planets.**
J. L. Simon, P. Bretagnon.
Astron. Astrophys., Vol. 42, 259 - 263 (1975). In French.

The authors present the perturbations, at the first order with respect to masses, of the planets Jupiter, Saturn, Uranus, Neptune under the form of Fourier series with numerical coefficients for the periodic part. The derivatives of the solution with respect to the integration constants are also given. Both the solution and its derivatives are equal to an entirely analytical theory in a field defined around constants chosen for the construction of the solution.

042.024 **Integrals of motion.** G. Contopoulos.
IAU Symposium No. 69, (see 012.009), p. 209 - 225 (1975).

The properties of conservative dynamical systems of two or more degrees of freedom are reviewed. The transition from integrable to ergodic systems is described. Nonintegrability is due to the interaction of two, or more, resonances. Then one sees, on a surface of section, infinite types of islands of various orders, while the asymptotic curves from unstable invariant points intersect each other along homoclinic and heteroclinic points producing an apparent 'dissolution' of the invariant curves. A threshold energy is defined separating near integrable systems from near ergodic ones. The possibility of real ergodicity for large enough energies is discussed. Finally some recent results about the general three-body problem are described. One can extend the families of periodic orbits of the restricted problem to the general three-body problem. Many of these orbits are stable. An empirical study of orbits near the stable periodic orbits indicates the existence of 2 integrals of motion besides the energy.

042.025 **On methods of computation of a cometary ephemeris in parabolic motion.**
L. E. Nikonova, A. Yu. Yatsenko.
Trudy Kazan. Gorod. Astron. Obs., No. 39, p. 116 - 120 (1973). In Russian.

042.026 **On the stability of translatory-rotational motion of an axisymmetric body of spherical structure.**
A. A. Nemo.
Trudy Kazan. Gorod. Astron. Obs., No. 39, p. 143 - 155 (1973). In Russian.

042.027 **Solution of an internal variant of the problem of the three fixed centers.** G. T. Arazov.
Pis'ma v Astron. Zhurn., Vol. 1, No. 7, p. 42 - 46 (1975). In Russian.

Elliptic quadratures describing the general solution for one of the variants of the three fixed centers problem are given.

042.028 **Canonical change of variables in the canonical equations of the quasi Lagrangian dynamical systems.**
J. Meffroy.
Bull. American Astron. Soc., Vol. 7, 466 (1975). – Abstr. AAS.

042.029 **Computer-developed construction of analytic expressions for the coordinates and partial derivatives of Jupiter's Galilean satellites.** J. H. Lieske.
Celestial Mechanics, Vol. 12, 5 - 17 (1975). – Paper presented at IAU Colloquium No. 28, 'Planetary satellites', Ithaca, N. Y., August 1974.

A method which utilizes computer-based algebraic manipulation software has been developed to reconstruct Sampson's theory, to remove existing errors, to introduce neglected effects and to provide analytical expressions for the coordinates as well as for the partial derivatives with respect to orbital parameters, Jupiter and satellite masses, Jupiter's oblateness (J_2, J_4) and Jupiter's pole and period of rotation. The computer-based manipulations enable one to perform, for example, the approximately 10^8 multiplications required in calculating some perturbations (and their partial derivatives) of Satellite II by Satellite III with ease, and provide algebraic expressions which can readily be adjusted to generate theories corresponding to revised constants of integration and physical parameters.

042.030 **A new theory of the motions of the Galilean satellites of Jupiter.** J. L. Sagnier.
Celestial Mechanics, Vol. 12, 19 - 25 (1975). – Paper presented at IAU Colloquium No. 28, 'Planetary satellites', Ithaca, N.Y., August 1974.

The main features of this theory are presented with special emphasis on the most specific of them: choice of the parameters, separation of the problem into subproblems (main problem, generalized main problem, complete problem), special adaptation of the method required by the resonant situation of the Galilean system.

042.031 **Problems on the Galilean satellites of Jupiter.** S. Ferraz Mello.
Celestial Mechanics, Vol. 12, 27 - 37 (1975). – Paper presented at IAU Colloquium No. 28 'Planetary satellites', Ithaca, N.Y., August 1974.

The purpose of this paper is to present a critical review of some problems concerning the dynamics of Jupiter's Galilean system of satellites. Theory, ephemeris and observation are considered. The most recent photographic observations are discussed in order to state guidelines for future work. The problem of the precision of Sampson's tables is discussed on the grounds of the recent observations.

042.032 **The motion of the Martian satellites.** V. A. Shor.
Celestial Mechanics, Vol. 12, 61 - 75 (1975). – Paper presented at IAU Colloquium No. 28, 'Planetary satellites', Ithaca, N. Y., August 1974.

An extensive analysis of the motion of Phobos and Deimos from 1877 to 1973 has been fulfilled. The new values of the parameters of the orbital model first developed by Struve have been determined for both satellites. A secular acceleration in longitude of Phobos is found to be equal to $+(0.107 \pm 0.011) \times 10^{-7}$ deg day^{-2}. The value of the acceleration is little affected when one or another group of oppositions is omitted. The acceleration of Deimos is determined with great uncertainty: $+(0.06 \pm 0.34) \times 10^{-9}$ deg day^{-2}. Values found for the orbital parameters seem to be in good agreement since the mass, oblateness and coordinates of the pole of Mars inferred from the motion of each satellite have similar values in both cases.

042.033 **The orbital resonance amongst the Galilean satellites of Jupiter.** A. T. Sinclair.
Celestial Mechanics, Vol. 12, 89 - 96 (1975). – Paper presented at IAU Colloquium No. 28, 'Planetary satellites', Ithaca, N.Y., August 1974.

The resonance amongst three of the Galilean satellites is described in a way intended to demonstrate the similarities it has to the normal two-satellite resonance. The hypothesis that the resonance was formed by the action of tidal forces is discussed. The problem is too complicated to reach any firm conclusions, but the tidal hypothesis does not seem to be a satisfactory explanation.

042.034 **Density distortion within a rotating body.** P. Lanzano.
Astrophys. Space Sci., Vol. 37, 173 - 181 (1975).

This paper ascertains the distortion of the density distribution within a self-gravitating body in hydrostatic equilibrium under the influence of rotation. The density distortion has been obtained up to third-order terms in a small rotational parameter.

042.035 **Hill regions for the general three-body problem.** C. Marchal, D. G. Saari.
Celestial Mechanics, Vol. 12, 115 - 129 (1975).

For the general three-body problem the authors show the existence and describe regions in physical space and configuration space where motion cannot occur. The description of these regions turns out to be most similar to the Hill regions of the circular restricted problem of three bodies. These regions depend upon the constants of energy and angular momentum.

042.036 **Invariance in von Zeipel method.** R. Cid, M. Calvo, J. M. Correas.
Celestial Mechanics, Vol. 12, 131 - 138 (1975).

In this paper the relation between the von Zeipel generating functions, in two canonical systems, is considered. Explicit recursive formulae between them are given. Also a series of invariants of the generating functions are obtained, so that the equations that appear in the von Zeipel method can be written in terms of Poisson brackets.

042.037 **The continuation of periodic orbits from the restricted to the general three-body problem.**
J. D. Hadjidemetriou.
Celestial Mechanics, Vol. 12, 155 - 174 (1975).

It is proved that a symmetric periodic orbit of the circular planar restricted three-body problem can be continued analytically, when the mass of the third body is small but not negligible, to a periodic motion of the general three-body problem in a rotating frame of reference whose origin coincides with the center of mass of the two bodies with large masses and its x axis always contains these bodies. It is also shown that for a fixed value of the small mass of the third body, a family of symmetric periodic orbits of the restricted problem can be continued analytically to a family of periodic motions of the general three-body problem.

042.038 **Families of periodic orbits in the planar three-body problem.** J. D. Hadjidemetriou, T. Christides.
Celestial Mechanics, Vol. 12, 175 - 187 (1975).

A periodic orbit of the restricted circular three-body problem, selected arbitrarily, is used to generate a family of periodic motions in the general three-body problem in a rotating frame of reference, by varying the mass m_3 of the third body. This family is continued numerically up to a maximum value of the mass of the originally small body, which corresponds to a mass ratio $m_1 : m_2 : m_3 \simeq 5 : 5 : 3$.

042.039 **Collinear equilibria and their characteristic exponents in the restricted three-body problem when the primaries are oblate spheroids.**
R. K. Sharma, P. V. Subba Rao.
Celestial Mechanics, Vol. 12, 189 - 201 (1975).

In this paper location of the collinear libration points is investigated numerically, by taking the oblateness of the primaries into consideration, for 19 systems of astronomical interest. It is found that in some of the systems the shifts are significant. These equilibria are shown to be unstable in general, though the existence of conditional, infinitesimal (linearized) periodic orbits around them can be established, in the usual way. It is shown that the eccentricity and synodic period of these orbits are functions of oblateness too. Numerical study, in this connection, with the above systems, revealed that the orbits around the libration point which is farthest from the primary whose oblateness effect is included, exhibit a different trend from those around the other two points.

042.040 **An extended ideal resonance problem.** B. Garfinkel.
Celestial Mechanics, Vol. 12, 203 - 214, with a correction, p. 524 (1975).

An extended resonance problem is defined by the Hamiltonian, $F = B(y) + 2\mu^2 A(y)[\sin x + \lambda(y)]^2 ; \mu \ll 1$; $\lambda = O(\mu)$. It is noted here that the phase-plane trajectories exhibit a double libration, enclosing two centers, for the initial conditions of motion satisfying the inequality $1 - |\lambda| < |\alpha| < 1 + |\lambda|$, where α is the usual resonance parameter. A first order solution for the case of double libration is constructed.

042.041 **Numerical investigation of the planar restricted three-body problem. III. Closed branches of family *f* and related periodic orbits of the elliptic problem.**
V. V. Markellos.
Celestial Mechanics, Vol. 12, 215 - 224 (1975).

The critical orbits, corresponding to bifurcations of the generating family and its branches, are considered more closely and the part of *f* is investigated that has branches of very high order only. Three families of periodic solutions of the elliptic problem are also determined in an effort to follow the evolution of the stability region around *f* when the eccentricity of the primaries is increased to non-zero values.

042.042 **Numerical accuracy in calculations of class (g) of the restricted problem of three bodies.**
D. K. Leaphart, J. L. Safko.
Astron. Journ., Vol. 80, 747 - 749 (1975).

The Copenhagen problem has been extensively studied numerically with accuracy sufficient to derive the topological properties of classes of the Copenhagen problem. V. Szebehely and P. Nacozy (1967) terminated one of these classes, class (g), with increased accuracy. The authors present a discussion of sample orbits of class (g) with revisions. Various means of determining propagated errors are discussed and applied to class (g).

042.043 **Les variations à longues périodes des éléments de l'orbite terrestre: une question de précision dans la théorie astronomique des paléoclimats.** A. Berger.
Ciel et Terre, Vol. 91, 261 - 277 (1975).

The analytical expression of the "secular" variations of the orbital elements for the earth is derived in a vectorial form for the two basic systems (e, π) and (i, Ω). The accuracy of the solution when expanded in series of trigonometrical terms is discussed.

042.044 **Equilibrium figures.** V. A. Antonov.
Astronomy. Vol. 10. Equilibrium and stability of gravitating systems, (see 003.002), p. 7 - 60 (1975). In Russian.

042.045 **The stability of periodic orbits in the three-body problem.** J. D. Hadjidemetriou.
Celestial Mechanics, Vol. 12, 255 - 276 (1975).

A method is developed to study the stability of periodic motions of the three-body problem in a rotating frame of reference, based on the notion of surface of section. This linear stability analysis is used to study the stability of a family of periodic motions of three bodies with equal masses in a rotating frame of reference. This family represents motion such that two bodies revolve around each other and the third body revolves around this binary system in the same direction to a distance which varies along the members of the family. It was found that a large part of the family, corresponding to the case where the distance of the third body from the binary system is larger than the dimensions of the binary system, represents stable motion.

042.046 **Periodic motions around a collinear equilibrium solution of the general three-body problem.**
G. Bozis, T. Christides.
Celestial Mechanics, Vol. 12, 277 - 295 (1975).

The collinear equilibrium position of the circular restricted problem with the two primaries at unit distance and the massless body at the point L_3 is extended to the planar three-body problem with respect to the mass m_3 of the third body; the mass ratio μ of the two primaries is considered constant and the constant angular velocity of the straight line on which the three masses stay at rest is taken equal to 1. As regards periodic motions 'around' the equilibrium point L_3, four possible extensions from the restricted to the general problem are presented.

042.047 **A comparative study of Newtonian, Kustaanheimo/Stiefel, and Sperling/Burdet optimal trajectories.**
D. J. Jezewski.
Celestial Mechanics, Vol. 12, 297 - 315 (1975). – Paper presented at the AIAA Mechanics and Control of Flight Conference, Anaheim, California, August 1974.

An optimal trajectory problem is formulated in each of three sets of equations, and the resulting solutions are numerically compared. The three formulations are the classical Newtonian (N), the Kustaanheimo/Stiefel (K/S), and the Sperling/Burdet (S/B). The results indicate that, although the K/S and S/B formulations are more difficult to express mathematically than the Newtonian formulation, the transformed solutions are significantly more numerically stable than the Newtonian solution when the perturbing acceleration is less than a minimum value.

042.048 **On redundant variables in Lagrangian mechanics, with applications to perturbation theory and KS regularization.** R. Broucke, H. Lass.
Celestial Mechanics, Vol. 12, 317 - 325 (1975).

It is shown that it is possible to make a change of variables in a Lagrangian in such a way that the number of variables is increased. The Euler-Lagrange equations in the redundant variables are obtained in the standard way (without the use of Lagrange multipliers). These equations are not independent but they are all valid and consistent. In some cases they are simpler than if the minimum number of variables are used. The redundant variables are supposed to be related to each other by several constraints (not necessarily holonomic), but these constraints are not used in the derivation of the equations of motion. The method is illustrated with the well known Kustaanheimo-Stiefel regularization. Some interesting applications to perturbation theory are also described.

042.049 **Sur de nouvelles séries pour le problème de masses critiques de Routh dans le problème restreint plan des trois corps.** J. Roels.
Celestial Mechanics, Vol. 12, 327 - 336 (1975).

The planar restricted 3-body problem, linearized in the neighborhood of Lagrangian equilibria L_4 and L_5, has in general two distinct eigenvalues and their opposites. When they are pure imaginary and not multiples of each other, they generate two families of periodic solutions called long and short periodic families. When the eigenvalues are equal, which is the case for the mass ratio of Routh, the problem was solved by Deprit and Henrard (1968) using formal series in ordinary unnormalized variables. That leads to very complicated series because of the use of variables that are not well adapted to the problem. The convergence of the series was proven by Meyer and Schmidt (1971). In this paper the author solves the problem by using normalized variables.

042.050 **Sur la recherche des mouvements stationnaires dans les systèmes ayant des variables cycliques.**
M. Pascal.
Celestial Mechanics, Vol. 12, 337 - 358 (1975).

A 'constrained system' is associated with a mechanical system having cyclic coordinates. The author proves that the set of steady motions of this mechanical system and the set of steady motions of the constrained system are the same. Investigating the stability of these motions, it is proved that, if the constrained system is stable, then the mechanical system is also stable. As an example, the author considers the problem of the relative equilibrium of gyrostat satellites.

042.051 **The swinging spring – approximate analyses for low and very high energy, II.** D. L. Hitzl.
Celestial Mechanics, Vol. 12, 359 - 382 (1975).

The two families of periodic solutions emanating from the lower equilibrium of a planar spring-pendulum system are analyzed both at and near resonance. Hamiltonian perturbation theory is used to obtain approximate formulas for the characteristic and the period of the motion. As the energy is increased to very high levels, a circulatory and an almost rectilinear periodic motion persist. Expressions for the trace of a stability matrix are determined for both these solutions. Comparisons are made throughout with accurate results from numerical integration.

042.052 **The problem of escape in the *n*-body system.**
R. B. Patnaik.
Celestial Mechanics, Vol. 12, 383 - 390 (1975).

The celestial bodies are considered as mass-points attracting each other according to Newton's law. Khilmi (1961) obtained some sufficient conditions for the occurrence of disintegration and escape in the *n*-body system as time tends to infinity using his theory of continuous induction, inductiveness of inequalities and α-approach. Making an analytic approach, an improved version of Khilmi's result on escape and conditions to avoid disintegration are obtained.

042.053 **The motion of a particle from the L_2 Lagrangian point.** K. Nariai.
Astron. Astrophys., Vol. 43, 309 - 311 (1975).

The author calculates the orbit of a particle in the restricted 3-body problem starting from the L_2 Lagrangian point until a condition is reached in which the particle is far enough from the center for the osculating elements of its orbit to be fairly constant for more than several rotations of the system. The total energy is a function of the mass ratio μ, the initial velocity v_0, and the ejection angle ϕ. When he applies this results to gas flow by averaging over the ejection angle, the author finds that the total energy of the flow is positive for μ less than 0.35, while it is negative for a certain region of v_0 near 0.1 with μ between 0.35 and 0.5.

042.054 **Satellite with high eccentricity. Application to Nereid.** F. Mignard.
Astron. Astrophys., Vol. 43, 359 - 379 (1975). In French.

An analytical solution to the problem of the motion of a satellite with high eccentricity is presented. Perturbation effects which are considered include only those due to the attraction of the sun. Furthermore, the sun, the planet and the satellite are treated as point masses. In a first part, short and intermediate periods terms, respectively terms with a period of the order of the satellite's and the sun's longitudes, are eliminated by means of the Von Zeipel averaging method. In a second part, the problem is integrated in the particular case of Nereid, the inclination of which, referred to the orbit of Neptune, may be considered as a small quantity. The solution is obtained up to the second order of the inclination. In the case of Nereid, the main result is the great stability of the Keplerian elements, although the value of the eccentricity is 0.75. Expressions for secular motions of the "peri-Neptune" and mode are given without development in powers of the eccentricity.

042.055 **A note on the stability of the triangular points of equilibrium in the restricted three-body problem.**
P. V. Subbarao, R. K. Sharma.
Astron. Astrophys., Vol. 43, 381 - 383 (1975).

The planar restricted three-body problem possesses five stationary solutions: three are collinear and the other two form equilateral triangles with the primary ones. The straight line solutions are in general unstable, while the triangular solutions are stable if the mass parameter $\mu < \mu_0 = 0.03852...$ If the bigger primary is an oblate spheroid whose equatorial plane coincides with the plane of motion, the triangular solutions form only nearly equilateral triangles with the primaries, and the range of the mass parameter which leads to stable triangular solutions decreases.

042.056 **On stationary intermediate motions in the stellar problem of three bodies.** N. A. Solovaya.
Vestn. Mosk. un-ta. Fiz., astron., Vol. 16, 206 - 211 (1975). In Russian. – Abstr. in Referativ. Zhurn. 51. Astron., 9.51. 141 (1975).

042.057 **On the evolution of some classes of 'asteroid' orbits in the N-planet system.** M. A. Vashkov'yak.
In-t prikl. mat. AN SSSR. Preprint No. 28. Moskva, 1975. 40 pp. Price 15 Kop. In Russian. – Abstr. in Referativ. Zhurn. 51. Astron., 9.51.144 (1975).

042.058 **On a generalisation of Whittaker's criterion.**
Eh. A. Vagner.
Probl. mekh. upravlyaemogo dvizheniya. Vyp. (No.) 6. Perm', 1974, p. 19 - 21. In Russian. – Abstr. in Referativ. Zhurn. 51. Astron., 9.51.146; 62. Issled. kosmich. prostranstva, 9.62. 314 (1975).

042.059 **On near-equatorial orbits of the satellite of a spheroidal planet.** I. S. Kozlov.
Probl. mekh. upravlyaemogo dvizheniya. Vyp. (No.) 6. Perm', 1974, p. 61 - 65. In Russian. – Abstr. in Referativ. Zhurn. 51. Astron., 9.51.147; 62. Issled. kosmich. prostranstva, 9. 62. 315 (1975).

042.060 **On the Lagrange-Jacobi equation.** G. N. Duboshin.
Izb. probl. prikl. mekh. Moskva, 1974, p. 313 - 321. In Russian. – Abstr. in Referativ. Zhurn. 51. Astron., 9.51. 150 (1975).

042.061 **Stability of planar solutions in the restricted circular twice-averaged three-body problem.**
A. I. Nejshtadt.
Pis'ma v Astron. Zhurn., Vol. 1, No. 10, p. 41 - 45 (1975). In Russian.

By constructing the Lyapunov function it is proved that there is no critical eccentricity in the considered problem, i. e. all planar orbits are stable with respect to the inclination.

042.062 **Results of first order perturbations of the four giant planets.** J. L. Simon, P. Bretagnon.
Astron. Astrophys., Suppl. Ser., Vol. 22, 107 - 160 (1975). In French.

The authors give results dealing with the perturbations of the first order of masses for the four giant planets as well as their derivatives with respect to the integration constants.

042.063 **Analysis of motion in a central field under the action of small acceleration.** E. L. Lukashevich.
Trudy vos'mykh chtenij, posvyashch. razrabotke nauch.

naslediya i razvitiyu idej K. Eh. Tsiolkovskogo, Kaluga, 1973 g. Sekts. Mekh. kosmich. poleta. Moskva, 1974, p. 130 - 140. In Russian. – Abstr. in Referativ. Zhurn. 51. Astron., 8.51. 113; 62. Issled. kosmich. prostranstva, 8.62.281 (1975).

042.064 **On an approach to the problem of spatial motion of a solid body around a fixed point.**
V. S. Aslanov.
Trudy Kujbyshev. aviats. in-t, 1974, vyp. (No.) 69, p. 47 - 52. In Russian. – Abstr. in Referativ. Zhurn. 51. Astron., 8.51. 116; 62. Issled. kosmich. prostranstva, 8.62.295 (1975).

042.065 **The problem of four fixed centres and its applications.** I. S. Kozlov.
Trudy vos'mykh chtenij, posvyashch. razrabotke nauch. naslediya i razvitiyu idej K. Eh. Tsiolkovskogo, Kaluga, 1973 g. Sekts. Mekh. kosmich. poleta. Moskva, 1974, p. 30 - 44. In Russian. – Abstr. in Referativ. Zhurn. 51. Astron., 8.51. 117 (1975).

042.066 **On the stability of orientation of a satellite gyrostat in equilibrium positions in the libration points.**
V. V. Rumyantsev.
Izbr. probl. prikl. mekh., Moskva, 1974, p. 605 - 616. In Russian. – Abstr. in Referativ. Zhurn. 51. Astron., 8.51.119; 62. Issled. kosmich. prostranstva, 8.62.300 (1975).

042.067 **On the evolution of rotational motion of some celestial bodies.** A. A. Khentov.
Astron. Zhurn. Akad. Nauk SSSR, Vol. 52, 1066 - 1075 (1975). In Russian. English translation in Soviet Astron., Vol. 19, No. 5.

The evolution of rotational motion of a celestial body around its centre of mass whose ellipsoid of inertia slightly differs from a sphere is considered. A complete system of non-linear equations taking into account the gravitational torques and tidal friction is constructed. Resonance motions are also studied.

042.068 **Intermediate planar motion of a rigid body in the gravitational field of a sphere.** Yu. V. Barkin.
Astron. Zhurn. Akad. Nauk SSSR, Vol. 52, 1076 - 1083 (1975). In Russian. English translation in Soviet Astron., Vol. 19, No. 5.

Planar translatory-rotational motion of a dual system consisting of a sphere with concentric density distribution and a rigid body with a plane of dynamical symmetry for which the condition of commensurability of orbital and rotational motion periods holds is considered. The intermediate motion described by the Delone scheme averaged equations which assume the whole system of the first integrals is constructed. Translatory and rotational motions are interrelated.

042.069 **On Stäckel's harmonic force function in axisymmetric coordinates.** E. I. Burshtejn.
Astron. Zhurn. Akad. Nauk SSSR, Vol. 52, 1084 - 1088 (1975). In Russian. English translation in Soviet Astron., Vol. 19, No. 5.

The expressions for the force function satisfying Stäckel's theorem are obtained in three-dimensional Euclidean space. These expressions have the form of general solutions of the Laplace equation in axisymmetric coordinates.

042.070 **The symplectic property of matrizants in Störmer's problem.** A. G. Mavraganis.
Astrophys. Space Sci., Vol. 37, 365 - 369 (1975).

The matrizants of periodic solutions in Störmer's problem, because of their symplectic character, can be transformed by means of multiplication with constant matrices into symmetric ones. As a result the six bilinear relations between their elements, existing on account of the symplectic property, are replaced by 14 linear and simple forms. This fact is very useful in numerical integrations where these relations are used as criteria of accuracy.

042.071 **Savremeni pravci razvoja nebeske mehanike.**
B. Popović.
Publ. Obs. Astron. Beograd, No. 20, (see 012.024), p. 129 - 130 (1975).

042.072 **Da li su Lagrange-ova rešenja problema tri tela jedina "egzaktna rešenja" tog problema?**
B. Popović.
Publ. Obs. Astron. Beograd, No. 20, (see 012.024), p. 145 - 158 (1975).

042.073 **Some iterative methods of construction of the three-body problem solution and its realization by a computer.** Yu. A. Ryabov, T. A. Tolmacheva.
Pis'ma v Astron. Zhurn., Vol. 1, No. 11, p 42 - 44 (1975). In Russian.

A construction of a quasiperiodic solution of the three-body problem in numerical - analytical form by a computer is performed. The possibility to construct such a solution by two iterative methods is considered.

042.074 **Kepler's problem in elementary presentation.**
V. Ts. Gurovich, M. K. Karipova.
Trudy Kirg. un-ta. Ser. fiz. n., 1975, vyp. (No.) 6, ch. (part) 1, p. 67 - 78. In Russian. – Abstr. in Referativ. Zhurn. 51. Astron., 10.51.83 (1975).

042.075 **On the satellite capture problem.**
J. Byl, M. W. Ovenden.
Monthly Notices Roy. Astron. Soc., Vol. 173, 579 - 584 (1975).

The possibility that the outer satellites of Jupiter were permanently captured is discussed. The various proposed capture mechanisms are examined briefly, and their shortcomings discussed. It is suggested that all the observed aspects of the outer satellites can be accounted for if the mass of Jupiter, at some time in the past, increased by about 130 earth masses, on a time scale of the order of one sideral period of Jupiter.

042.076 **Concerning the three-body problem.**
A. A. Efimov.
Nekotor. vopr. fiz. kosmosa. Moskva, 1974, p. 115 - 129. In Russian. – Abstr. in Referativ. Zhurn. 62. Issled. kosmich. prostranstva, 10.62.226 (1975).

042.077 **Supplement to the three-body problem.**
A. A. Efimov.
Nekotor. vopr. fiz. kosmosa. Moskva, 1974, p. 197 - 202. In Russian. – Abstr. in Referativ. Zhurn. 62. Issled. kosmich. prostranstva, 10.62.227 (1975).

042.078 **Systematisation of the calculation of orbits.**
J. M. Correas.
Rev. Acad. Ci. exact. fís.-quím. nat. Zaragoza, II. Ser., Vol. 29, 129 - 196 (1974). In Spanish.

This work deals with the problem of orbit determination in Keplerian motion.

042.079 **Series expansion of the eccentricity for near parabolic orbits.** N. X. Vinh.
AIAA Journ., Vol. 12, 1281 - 1282 (1974).

042.080 **Triple collision in the collinear three-body problem.**
R. McGehee.
Inventiones Math., Vol. 27, 191 - 227 (1974).

This paper examines certain motions for the Newtonian gravitational system consisting of three particles moving along the same straight line. The behavior of solutions near triple collision is described. A new proof is given for Sundman's result that orbits ending in triple collision must asymptotically approach a central configuration. A new proof is also given for Siegel's result that the set of all such orbits forms a smooth submanifold of a constant energy surface. Both of these theorems are given a geometric interpretation. The major new result of the paper is that, for certain values of the masses, the system can achieve arbitrarily large kinetic energy after passing close to triple collision.

042.081 **Restrictions on the motion of the three-body problem.** D. G. Saari.
SIAM Journ. Applied Math. Vol. 26, 806 - 815 (1974).
The author considers the general problem of three bodies of Newtonian mechanics and establishes, as a main result, under suitable conditions, the existence of regions of the physical space where the three particles cannot enter. In the case of equal masses and for a special set of initial conditions, these regions are explicitly constructed.

042.082 **On the physical interpretation of some solutions for the Laplace equation.** E. I. Burshtejn.
Astron. Zhurn. Akad. Nauk SSSR, Vol. 52, 1309 - 1313 (1975). In Russian. English translation in Soviet Astron., Vol. 19, No. 6.
The solutions for the Laplace equation in axisymmetric coordinates satisfying Stäckel's theorem are interpreted in real space as potentials of layers of charge and double layers in which the distribution of densities is defined by Green's formula.

042.083 **On equilibrium points of a restricted averaged four-body problem.** S. G. Zhuravlev.
Astron. Zhurn. Akad. Nauk SSSR, Vol. 52, 1314 - 1323 (1975). In Russian. English translation in Soviet Astron., Vol. 19, No. 6.
Equilibrium points of a problem of motion of a mass-point in the gravitational field of an oblate (at least triaxial) planet and two perturbing bodies moving along eccentrical orbits with any inclination are found. The perturbations due to the influence of two perturbing bodies have the same order as those provided by the planet's oblateness.

042.084 **Circular orbits in the problem of two fixed centres. I.** N. P. Pit'ev.
Astron. Zhurn. Akad. Nauk SSSR, Vol. 52, 1324 - 1327 (1975). In Russian. English translation in Soviet Astron., Vol. 19, No. 6.
The existence of circular orbits outside the plane of symmetry is shown and their stability is proved in the problem of fixed centres with equal masses. The stability of the circular orbit in the plane of symmetry at the critical distance $R = c\sqrt{2}$ from the line of centres is proved ($2c$ is the distance between the centres).

042.085 **Algorithms of celestial mechanics. Materials for mathematical providing of computers.**
Institute of Theoretical Astronomy of the USSR Academy of Sciences. Leningrad, 1975. 50 pp. Price 15 Kop. In Russian. Abstr. in Referativ. Zhurn. 51. Astron., 10.51.162 (1975).

042.086 **Periodic orbits in the restricted three-body problem: existence and asymptotic approximation.**
L. M. Perko.
SIAM Journ. Applied Math., Vol. 27, 200 - 237 (1974).
This paper establishes the existence and first-order asymptotic approximation of one parameter families of periodic orbits of the second species of Poincaré for the restricted three-body problem with small mass ratio $\mu > 0$.

042.087 **On averaged equations of the restricted elliptic three-body problem.** V. P. Evteev.
Dokl. AN TadzhSSR, Vol. 18, No. 3, p. 11 - 14 (1975). In Russian. – Abstr. in Referativ. Zhurn. 51. Astron., 12.51.101 (1975).

042.088 **On the stability of some classes of satellite orbits of a nonspherical planet.** M. Kh. Khasanova.
Dokl. AN TadzhSSR, Vol. 18, No. 3, p. 15 - 17 (1975). In Russian. – Abstr. in Referativ. Zhurn. 51. Astron., 12.51.110; 62. Issled. kosmich. prostranstva, 12.62.485 (1975).

042.089 **On the equations of motion of a satellite about its center of mass.** V. A. Sarychev, N. I. Yakovlev.
In-t prikl. mat. AN SSSR. Preprint No. 55. Moskva, 1975. 52 pp. Price 15 Kop. In Russian. – Abstr. in Referativ. Zhurn. 51. Astron., 12.51.118 (1975).

042.090 **Aberration héliocentrique de distance aller-retour à un satellite.** O. Calame.
Bull. Groupe de Recherches de Géod. Spatiale, No. 14/III, 8 pp. (1974).

042.091 **Comments on the application of power series solutions to problems in celestial mechanics.**
C. E. Roberts, Jr.
Celestial Mechanics, Vol. 12, 397 - 407 (1975).
The author relates his experiences in utilizing the power series method to generate trajectories for orbital and sub-orbital vehicles and for the n-body problem.

042.092 **On relative periodic solutions of the planar general three-body problem.** R. Broucke.
Celestial Mechanics, Vol. 12, 439 - 462 (1975).
The author describes two relatively simple reductions to order 6 for the planar general three-body problem. He also shows that this reduction leads to the distinction between two types of periodic solutions: absolute or relative periodic solutions. An algorithm for obtaining relative periodic solutions using heliocentric coordinates is then described. It is concluded from the periodicity conditions that relative periodic solutions must form families with a single parameter. Finally, two such families have been obtained numerically and are described in some detail.

042.093 **On Broucke's velocity-related series expansions in the two-body problem.** A. H. Jupp.
Celestial Mechanics, Vol. 12, 513 - 518 (1975).
This short article supplements a recent paper by Broucke on velocity-related series expansions in the two-body problem. The derivations of the Fourier and Legendre expansions of the functions $F(v)$, $\sqrt{F(v)}$ and $\sqrt{1/F(v)}$ are given, where $F(v) = (1-e^2)/(1+2e\cos v+e^2)$, $e < 1$. In the two-body problem, v is identified with the true anomaly, e the eccentricity and $F(v)$ equals $(an/V)^2$. Some interesting relations involving Legendre polynomials are also noted.

042.094 **A note on the equilibrium points of the restricted problem of three bodies.** G. E. O. Giacaglia.
Celestial Mechanics, Vol. 12, 519 - 520 (1975).
Using Sylvester's theorem on matrices, an elegant expression is obtained for the solutions of the restricted problem of three bodies in the neighborhood of the equilibrium points.

042.095 **A method for determining the proximity of orbits of celestial bodies.** J. L. Simovljević.
Prirodno-matematichki fakultet, Beograd - Glas Srpske akademije nauka i umetnosti 291, 9 - 17 (1974). In Serbo-Croatian. Abstr. in Bull. Sci. Yougoslavie, Sect. A, Vol. 20, 366 (1975).

042.096 **Proximity of orbits of a group of celestial bodies.** J. L. Simovljević.
Prirodno-matematichki fakultet, Beograd - Glas Srpske akademije nauka i umetnosti 291, 19 - 30 (1974). In Serbo-Croatian. – Abstr. in Bull. Sci. Yougoslavie, Sect. A, Vol. 20, 366 (1975).

042.097 **On a type of anomalies in planetary motion.** J. L. Simovljević.
Prirodno-matematichki fakultet, Beograd - Glas Srpske akademije nauka i umetnosti 291, 1 - 8 (1974). In Serbo-Croatian.

042.098 **On periodic solutions of the averaged equations of the restricted elliptic three-body problem.**
V. P. Evteev.
Dokl. An TadzhSSR, Vol. 18, No. 4, p. 12.- 15 (1975). In Russian. – Abstr. in Referativ. Zhurn. 51. Astron., 1.51.144 (1976).

042.099 **On the problem of stability of Lagrangian solutions of the unrestricted problem of three bodies.**
A. L. Kunitsyn.
Probl. analit. mekh., teorij ustojchivosti i upr. Moskva, Nauka, 1975, p. 175 - 179. In Russian. – Abstr. in Referativ. Zhurn. 51. Astron., 1.51.147 (1976).

042.100 **Study of stability by means of polynomials.** G. O. Yaroshevich.
Probl. analit. mekh., teorij ustojchivosti i upr. Moskva, Nauka, 1975, p. 330 - 332. In Russian. – Abstr. in Referativ. Zhurn. 51. Astron., 1.51.148 (1976).

042.101 **On the Moiseev - Stepanov stability of circular solutions in the theory of motion of a small body in the exterior region of a planetary system.** I. P. Tarasashvili.
Trudy Gruz. politekhn. in-t, 1975, No. 3 (176), p. 228 - 234. In Russian. – Abstr. in Referativ. Zhurn. 51. Astron., 1.51. 149 (1976).

042.102 **Effects of the mass ratio on the existence of retrograde satellites in the circular plane restricted problem. II.** D. Benest.
Astron. Astrophys., Vol. 45, 353 - 363 (1975).

The author investigates the extended region of stability for retrograde satellites when the mass of the planet increases. A preliminary study is made of the characteristic and stability properties of the new family of double-periodic orbits, described for μ=0.0540 in a previous paper; it is found to have as much effect as the single periodic family *f* on the region of stability for 0.0540 $\leq\mu\leq$0.150. It is found that no other family of periodic orbits has any significant effect on the region of stability for any μ. Finally the general case of non-periodic orbits is considered; the evolution of the region of stability up to μ=0.5 is presented.

042.103 **Le problème des marées dans le problème des *n* corps. III. Les oscillations élastiques quasi-périodiques.** M. S. Volkov.
Byull. Inst. Teoret. Astron., *Leningrad,* Vol. 14, 81 - 86 (1975). In Russian.

Le problème des marées dans le problème des *n* corps est envisagé. On suppose que les corps non-homogènes possèdent une symmétrie sphérique. Les oscillations élastiques d'un corps se superposent avec les oscillations d'un liquide parfait. Le schéma de la demonstration de l'existence des solutions quasi-périodiques (suivant Arnold) est donné.

042.104 **Détermination du potentiel à partir des trajectoires dans les cas plan.** J. J. Martinez Benjamin.
Thesis, Univ. Paris-VI, 63 pp. (1975).

042.105 **On constancy of the nutation angle in the general three-body problem.** A. A. Efimov.
Vopr. istorii astron. No. 3. Moskva, 1974, p. 192 - 196. In Russian. – Abstr. in Referativ. Zhurn. 51. Astron., 2.51. 192 (1976).

042.106 **A contribution to the general three-body problem.** A. A. Efimov.
Nekotor. vopr. fiz. kosmosa. Moskva, 1974, p. 115 - 129. In Russian. – Abstr. in Referativ. Zhurn. 51. Astron., 2.51. 198 (1976).

042.107 **An addition to the general three-body problem.** A. A. Efimov.
Nekotor. vopr. fiz. kosmosa. Moskva, 1974, p. 197 - 202. In Russian.

042.108 **Circular orbits in the problem of two fixed centres.** N. P. Pit'ev.
Vestn. Leningr. un-ta, 1975, No. 13, p. 131 - 133. In Russian. Abstr. in Referativ. Zhurn. 51. Astron., 2.51.203; 62. Issled. kosmich. prostranstva, 2.62.332 (1976).

042.109 **On quasi-stability of the solutions of differential equations under permanently acting perturbations in normed spaces.** N. I. Lobanova.
Probl. analit. mekh., teorij ustojchivosti i upr. Moskva, Nauka, 1975, p. 179 - 182. In Russian. – Abstr. in Referativ. Zhurn. 51. Astron., 2.51.204 (1976).

042.110 **Estimate of regions of stability "in large" as a problem of nonlinear programming.** M. A. Tagirov.
Probl. analit. mekh., teorij ustojchivosti i upr. Moskva, Nauka, 1975, p. 300 - 303. In Russian. – Abstr. in Referativ. Zhurn. 51. Astron., 2.51.205 (1976).

042.111 **On the property of rigidity of motions and the acceptability of approximate solutions.**
V. N. Skimel'.
Probl. analit. mekh., teorij ustojchivosti i upr. Moskva, Nauka, 1975, p. 284 - 289. In Russian. – Abstr. in Referativ. Zhurn. 51. Astron., 2.51.208 (1976).

042.112 **Oscillations in essentially nonlinear analytical autonomous systems.** V. M. Starzhinskij.
Probl. analit. mekh., teorij ustojchivosti i upr. Moskva, Nauka, 1975, p. 293 - 300. In Russian. – Abstr. in Referativ. Zhurn. 51. Astron., 2.51.209 (1976).

042.113 **On the asymptotical solution of some problems of optimum control.** A. I. Tret'yak.
Probl. analit. mekh., teorij ustojchivosti i upr. Moskva, Nauka, 1975, p. 313 - 317. In Russian. – Abstr. in Referativ. Zhurn. 51. Astron., 2.51.210 (1976).

042.114 **On the motion of a point in an ellipsoidal gravitational field.** N. B. Batueva.
Vestn. Leningr. un-ta, 1975, No. 7, p. 139 - 144. In Russian. Abstr. in Referativ. Zhurn. 51. Astron., 2.51.211 (1976).

042.115 **On the inversion of hyperelliptic quadratures in the problem of motion of a point in an ellipsoidal gravitational field.** N. B. Batueva.
Vestn. Leningr. un-ta, 1975, No. 13, p. 122 - 126. In Russian. Abstr. in Referativ. Zhurn. 51. Astron., 2.51.212; 62. Issled. kosmich. prostranstva, 2.62.334 (1976).

042.116 **Small oscillations of a rigid body about a fixed point and some cases of existence of the "fourth algebraic integral".** F. Kh. Tsel'man.
Probl. analit. mekh., teorij ustojchivosti i upr. Moskva, Nauka,

1975, p. 326 - 329. In Russian. – Abstr. in Referativ. Zhurn. 51. Astron., 2.51.213 (1976).

042.117 **On the stability of regular rotations of a body with a liquid.** A. Ya. Savchenko.
Probl. analit. mekh., teorij ustojchivosti i upr. Moskva, Nauka, 1975, p. 267 - 270. In Russian. – Abstr. in Referativ. Zhurn. 51. Astron., 2.51.214 (1976).

042.118 **Motion of a rigid body with non-holonomic connection around a fixed point in a quasi-potential field of force.** S. M. Matveenko.
Probl. analit. mekh., teorij ustojchivosti i upr. Moskva, Nauka, 1975, p. 195 - 198. In Russian. – Abstr. in Referativ. Zhurn. 51. Astron., 2.51.215 (1976).

042.119 **Some highly perturbed problems of nonlinear mechanics.** E. P. Aksenov, E. A. Grebenikov, V. G. Demin, G. N. Duboshin, A. A. Orlov, Yu. A. Ryabov.
Probl. analit. mekh., teorij ustojchivosti i upr. Moskva, Nauka, 1975, p. 38 - 42. In Russian. – Abstr. in Referativ. Zhurn. 51. Astron., 2.51.218 (1976).

042.120 **Investigations of some instationary problems of celestial mechanics by the method of transformations.** L. M. Berkovich, B. E. Gel'fgat.
Probl. analit. mekh., teorij ustojchivosti i upr. Moskva, Nauka, 1975, p. 54 - 61. In Russian. – Abstr. in Referativ. Zhurn. 51. Astron., 2.51.220; 62. Issled. kosmich. prostranstva, 2.62. 333 (1976).

042.121 **Ephemerides in Chebyshev series.** A. Deprit.
Mech. Res. Commun., (*GB*), Vol. 1, 209 - 211 (1974).

Chebyshev polynomials are used to produce best approximations (in the sense of Chebyshev) of the classical ephemerides of astronomy. Cast in this form the ephemerides are well suited to processing in real time by minicomputers.

042.122 **Resonance capture in certain nearly Hamiltonian systems (planetary spin).** J. A. Murdock.
Journ. Differential Equations, (*USA*), Vol. 17, 361 - 374 (1975).

042.123 **The motion of a particle in orbit around a growing mass.** S. H. Dole.
Journ. Astronaut. Sci., Vol. 21, 252 - 254 (1974).

Use of an iterative solution of Kepler's equation shows how the semimajor axis and the eccentricity of the orbit of a particle moving around a massive body are affected when mass is added to the body incrementally, as by accretion. The net effect of numerous small discrete mass additions to reduce the semimajor axis; the eccentricity, however, remains substantially unchanged.

042.124 **Determination of elements of recurrent and near-recurrent orbits and their application to operational satellites.** K. Takahashi, K. Kawakami.
Rev. Radio Res. Lab., (*Japan*), Vol. 20, 121 - 134 (1974). In Japanese.

Recurrent and near-recurrent orbits are defined, and methods are given to determine the orbital semi-major axis and period when the recurrent number, eccentricity and inclination of these orbits are given, and to determine the orbital inclination, semi-major axis and period when the recurrent number and eccentricity of these orbits with invariable argument of perigee are given. A recurrent perigee orbit and a near-recurrent perigee orbit, whose longitude of perigee or apogee is nearly invariable in spite of change of the argument of perigee, are also defined, and the method of determining orbital elements is shown. Some examples of the recurrent orbit, which are applicable to practical satellites, are considered and compared with stationary satellite and Molniya satellite orbits. The possibility of an extremely low altitude sun-synchronous near-recurrent orbit, is discussed.

042.125 **Problems of classical dynamical systems (lecture notes).** W. Thirring.
Rep. CERN 75-8, CERN, Geneva, Switzerland. 33 pp. (1975).

After a brief survey of Hamiltonian theory and of relevant notions of set theory and manifolds, these lecture notes present some general properties of orbits, paying special attention to integrable systems. This is followed by a discussion of the Kolmogorov-Arnol'd-Moser theorem, dealing with the stability of orbits under small perturbations, and its importance for ergodic theory. Ergodicity and mixing are then treated in detail. In particular, the ergodic theorem of von Neumann is derived, and a specific example is given of a (strongly) mixing system.

Celestial mechanics. Basic problems and methods. See Abstr. 003.043.

Sketches on the history of celestial mechanics. See Abstr. 003.076.

Some remarks on Roche harmonics. See Abstr. 021.002.

The structure of the secular system in nonlinear oscillations. See Abstr. 021.012.

Celestial reference systems derivable from solar system dynamics. See Abstr. 041.053.

Triaxial degenerate dwarfs and pulsing X-ray sources. See Abstr. 065.134.

Celestial-mechanical relativistic experiments in the solar system. See Abstr. 066.144.

On the stability of a spherical gravitating compressible liquid planet without spin. See Abstr. 081.009.

Power series solutions of Laplace's tidal equation. See Abstr. 081.026.

High zonal harmonics of rapidly rotating planets. See Abstr. 091.002.

An analysis of the transits of Mercury: 1677–1973. See Abstr. 092.007.

Lunar daily distance numbers *D* and *D'*. See Abstr. 094.010.

The motions of Phobos and Deimos from Mariner 9 TV data. See Abstr. 097.202.

Asteroidal motion at the 5/2, 7/3 and 2/1 resonances. See Abstr. 098.019.

Sur le probème de proximité des orbites d'astéroïdes. See Abstr. 098.047.

A comparison of some observations of the Galilean satellites with Sampson's tables. See Abstr. 099.219.

Statistical mechanics of Keplerian orbits. See Abstr. 100.014.

Theory of Enceladus and Dione.

See Abstr. 100.205.

On the Laplace relation among the satellites of Uranus. See Abstr. 101.003.

The effects of tidal forces on the stability and dispersal rates of a protoplanet. See Abstr. 107.003.

The three-body problem and eclipsing binaries. Application to Algol and λ Tauri. See Abstr. 117.022.

The fission theory of binary stars for compressible masses. See Abstr. 117.044.

A gravitational three-body scattering experiment. See Abstr. 117.054.

Statistics of three-body experiments: probability of escape and capture. See Abstr. 117.055.

On the stability of congruent Darwin ellipsoids. See Abstr. 117.060.

On coupled second-harmonic oscillations of the congruent Darwin ellipsoids. See Abstr. 117.061.

Numerical experiments on the tensor virial in N-body systems. See Abstr. 151.016.

Resonant stellar orbits in spiral galaxies. III. Applications and numerical tests of the general epicyclic theory. See Abstr. 151.043.

Inner Lindblad resonance in galaxies. Nonlinear theory. I. See Abstr. 151.044.

The theory of resonances in spiral galaxies. See Abstr. 151.047.

Integrals of the motion in the resonance regions. See Abstr. 151.048.

A theorem on the third integral. See Abstr. 151.078.

On the sensitivity of the behaviour of stellar orbits to the parameters of simple galactic mass models. See Abstr. 155.013.

043 Astronomical Constants

043.001 **An analysis of the ancient astronomical observations with the implications for geophysics and cosmology.**
P. M. Muller.
Bull. American Astron. Soc., Vol. 7, 412 - 413 (1975). – Abstr. AAS.

043.002 **Messung der Lichtgeschwindigkeit nach Ole Roemer.** H.-U. Fuchs.
Orion, 33. Jahrgang, p. 120 - 122 (1975) – see Abstr. 13.043.007.

044 Time, Rotation of the Earth

044.001 **La seconde internationale et l'échelle de temps atomique.** B. Morando.
L'Astronomie, 89e année, p. 303 - 304 (1975).

044.002 **Results obtained with the Paris astrolabe. Time and latitude 1971–1974.**
F. Chollet, S. Débarbat.
Astron. Astrophys., Suppl. Ser., Vol. 22, 69 - 106 (1975). In French.

The results are given from 1971 through 1974, for the observations with astrolabes APP and OPL no. 35. The star programme published with the results of 1956 has not been changed since then. No change was made in observing technique. But the computation was modified since 1972 to give directly the results TU0–TUC and TAI–TU0, and to take account of the variation of TU0–TUC during the two hours of observations of each group of stars. The results are in the FK4 system. To express them in a system (A4) corrected for the systematic and accidental errors of the FK4 catalogue, group corrections must be added.

044.003 **Deviations of the clock corrections from the system of standard time observed at the Time Service of the Leningrad University in 1958–1971.**
M. P. Mishchenko, A. V. Shiryaev.
Trudy Astron. Obs., *Leningrad*, Vol. 31 (= Uchenye Zapiski Leningr. Un-ta, No. 381 = Seriya Matem. Nauk, vyp. (No.) 51), p. 228 - 238 (1975). In Russian.

044.004 **Multifactor dispersion analysis of results of positional observations of the moon.** S. G. Valeev.
Izv. Astron. Ehngel'gardt. Obs., *Kazan'*, No. 40, p. 131 - 140 (1973). In Russian.

044.005 **Total effect of any nearly-diurnal wobble of the earth's axis of rotation in latitude and time observations. Application to the Paris astrolabe observations from 1956.6 to 1973.8.** N. Capitaine.
Geophys. Journ. Roy. Astron. Soc., Vol. 43, 573 - 588 (1975).

The existence of a non-spherical ideal liquid core in the earth's interior produces a free nearly-diurnal wobble of the earth's axis of rotation. The effect of this wobble in latitude and time observations is more complex than has generally been considered before. In previous discussions only the variation of latitude and time due to the neglected wobble has been considered and not the errors arising from the complementary neglected nutation in space. The author establishes here the theoretical expression for the total effect and he shows that the first effect is negligible in comparison with the second.

044.006 **The analogue of Zinger's method in radio astrometry.** V. A. Alekseev, N. S. Blinov, B. N. Lipatov, E. N. Fedoseev.
Astron. Zhurn. Akad. Nauk SSSR, Vol. 52, 1089 - 1095 (1975). In Russian. English translation in Soviet Astron., Vol. 19, No. 5.

The problem of determination of universal time from observations with an interferometer with a long base is discussed. A radio interferometer is proposed to be used for this purpose which detects emission of two radio sources with equal zenith distances. An estimate of accuracy of the expected results is given and conditions under which the results could be achieved are discussed.

044.007 **The irregularity of the earth's rotation from data of astronomical observations.** N. S. Sidorenkov.
Astron. Zhurn. Akad. Nauk SSSR, Vol. 52, 1108 - 1112 (1975). In Russian. English translation in Soviet Astron., Vol. 19, No. 5.

Results of calculations of velocity characteristics of the earth's diurnal rotation for 1972–1974 are given. Some peculiarities of variations of the earth's rotation velocity for this period are discussed.

044.008 **Forschungen in der DDR zur Thematik "Rotationsverhalten der Erde".** E. Buschmann.
Publ. Obs. Astron. Beograd, No. 20, (see 012.024), p. 78 - 81 (1975).

044.009 **Analysis of measured values UT0–UTC of Belgrade Observatory time service.** M. Jovanović.
Publ. Obs. Astron. Beograd, No. 20, (see 012.024), p. 90 - 97 (1975). In Serbo-Croatian.

044.010 **The accuracy of the atomic time scale.** B. Guinot.
Time determination, dissemination and synchronization, (see 012.029), p. 15 - 35 (1975).

The author considers the accuracy of atomic time scales (conformity of the scale unit to the second at sea level) and the long term stability ($\tau > 1$ month).

044.011 **The use of Kalman-filter techniques in constructing the UTC (IEN) time scale.** M. T. Angeli.
Time determination, dissemination and synchronization, (see 012.029), p. 37 - 61 (1975).

A preliminar study on the construction of a time scale that uses the Kalman filter approach, is described. Attention is given to different kind of noises affecting the clock readings and an attempt is made to solve the "flicker noise" case.

044.012 **Time scales and relativity.** G. Becker.
Time determination, dissemination and synchronization, (see 012.029), p. 63 - 85 (1975).

First, the different uses of the world "time" are considered. Time scales and time concepts must be distinguished. Next, the differences between time units and time scale units are discussed. Results of measurements of the frequency of the International Atomic Time TAI showing a systematic negative drift are presented. Concerning the relativistic effects on time scales the following topics are treated: clocks at the earth's surface, clocks in a satellite, and clocks orbiting around the sun. It is a question of what kind of co-ordinate time should be the future Ephemeris Time. It is proposed to use TAI as the time variable in astronomy.

044.013 **Time transfer by Timation satellite.** R. L. Easton.
Time determination, dissemination and synchronization, (see 012.029), p. 87 - 91 (1975).

Description of time transfer using Timation satellite; arrangement and results of time comparisons between the United States Naval Observatory and the Royal Greenwich Observatory.

044.014 **An accuracy algorithm for an atomic time scale.** D. W. Allan, H. Hellwig, D. J. Glaze.
Time determination, dissemination and synchronization, (see 012.029), p. 185 - 205 (1975).

The accuracy of the rate (frequency) of an atomic time scale is the degree to which the second (and/or the accumulation of seconds used in its construction) agrees with the definition of the SI second. Typically primary frequency standards are constructed in order to determine as accurately as possible a physical realization of the SI second. These standards are then used to calibrate and/or construct an atomic time scale (reference standard).

044.015 **The IEN time and frequency dissemination.**
E. Angelotti, F. Cordara.
Time determination, dissemination and synchronization, (see 012.029), p. 207 - 233 (1975).
This paper describes how the dissemination of standard time and frequency, and the synchronization of remote clocks are currently achieved at the Istituto Elettrotecnico Nazionale (IEN), which is responsible for the Italian time scale.

044.016 **Effects of the introduction of a new fundamental reference coordinate system on time scales.**
W. Fricke.
Time determination, dissemination and synchronization, (see 012.029), p. 235 - 240 (1975).
As a results of an analysis of determinations of the equinox and equator from observations of the sun and of major and minor planets, the equinox of the FK5 may be expected to differ from the FK4 equinox. It is also very likely that at least a part of the unexplained equinox motion can be eliminated. Attention is drawn to changes in time scales with the introduction of the FK5. Possible contributions of time determinations to future improvements of the fundamental reference system are considered.

044.017 **A review of the development and future use of ephemeris time.** G. A. Wilkins.
Time determination, dissemination and synchronization, (see 012.029), p. 241 - 249 (1975).
Ephemeris time was defined in 1950 in order to provide a time scale that is more uniform than mean solar time, but it is now clear that for current use an atomic time-scale is to be preferred. It is considered, however, that the concept of ephemeris time should be retained for use in long-term ephemerides, but that new definitions of the unit and origin of the scale should be adopted.

044.018 **The measurement of time from radio observations of extragalactic sources.** B. Elsmore.
Time determination, dissemination and synchronization, (see 012.029), p. 251 - 258 (1975).
The determination of time using radio interferometers of baseline length up to a few kilometres, is described. The effects of polar motion and its measurement are considered. Some preliminary results obtained with the Cambridge 5 km radio telescope give a probable error of ±4.6 ms for a single determination of UT1.

044.019 **Scientific problems and time determination.**
G. P. Pilnik.
Time determination, dissemination and synchronization, (see 012.029), p. 259 - 265 (1975).

044.020 **Time and latitude determination using a portable zenith camera.** J. Geßler.
Time determination, dissemination and synchronization, (see 012.029), p. 267 - 272 (1975).
A new portable zenith camera for astronomical time and latitude determination is described in this paper. Remarkable advantages of this instrument are the short observation time, the avoidance of systematic errors and the high accuracy which is similar to first-order field instruments.

044.021 **Results of analysis of A3 (BIH) – UT2.**
S. Takagi.
Time determination, dissemination and synchronization, (see 012.029), p. 273 - 277 (1975).
The data of A3–UT2 in the period 1955–1972 published at the BIH were analysed by the methods of the maximum entropy and the cumulative analysis to see the characteristics of the change in the rate of rotation of the earth. It was concluded: (1) The rate of rotation of the earth has no distinct periodic terms other than the annual and semi-annual ones. (2) Fluctuation of the rate may be considered as a Brownian motion.

044.022 **Concerning an irregularity in the earth's rotation.**
F. Zagar.
Time determination, dissemination and synchronization, (see 012.029), p. 279 - 290 (1975).

044.023 **Daily phase values and time differences.**
U.S. Naval Obs., Washington, D.C., Time Service Publ., Ser. 4, Nos. 439 - 465 (1975). – 1975 July 3 - December 31.

044.024 **Preliminary times and coordinates of the pole.**
U.S. Naval Obs., Washington, D.C., Time Service Publ., Ser. 7, Nos. 392 - 417 (1975). – 1975 July 3 - December 25.

044.025 **Time service report.**
U.S. Naval Obs., Washington, D.C., Time Service Publ., Ser. 11, No. 223, 15 pp. (1975). – Times of emission, UTC(USNO, MC) – UTC(Transmitter); Time scales; Observations of time and latitude; Polar coordinates, corrections for seasonal and polar (longitude) variations. – 1974 January 2 - December 31.

044.026 **Time service announcement.** G. M. R. Winkler.
U.S. Naval Obs., Washington, D.C., Time Service Publ., Ser. 14, Nos. 17 (UTC time step), 19 (OMEGA Reference Epoch) (1975).

044.027 **Change of the rotation rate of the earth in December 1973.** I. Domiński.
Acta Astron., Vol. 25, 439 - 450 (1975).
A rapid increase of the rate of earth rotation which appeared in December 1973 was evaluated. An approximation function of those changes has been developed. Assuming that the perturbing impulse should first disturb the rotation of the outer parts of the earth, the kinetic energy absorbed by the earth as well as the losses of that energy which must have taken place in subsequent stages and the moments of inertia of the outer and inner parts of the earth were calculated.

044.028 **Chronometry and the variation of the rotation of the earth.** A. Stoyko, N. Stoyko.
Ann. Françaises Chronométrie Micromécanique, Année 1974, p. A9–1 - 4. In French.

044.029 **The international atomic time scale.**
M. Granveaud.
Ann. Françaises Chronométrie Micromécanique, Année 1974, p. A10–1 - 10. In French.

044.030 **Le temps est-il connu de ceux qui l'utilisent?**
S. Débarbat.
Ann. Françaises Chronométrie Micromécanique, Année 1974, p. G2–1 - 3.

044.031 **Extrapolation du temps universel à court intervalle.**
A. Stoyko, N. Stoyko.
Ann. Françaises Chronométrie Micromécanique, Année 1974, p. G3–1 - 3.

044.032 **Zeitvergleichsmessungen mittels Fernseh-Bildsynchronisationsimpulsen.** R. Rummel.
Allgemeine Vermessungsnachr., 82. Jahrgang, p. 348 - 357 (1975).

044.033 **La rotation de la terre et le système terre-lune.**
F. Chollet.
Space astrometry, (see 012.037), p. 103 - 106 (1975).

044.034 **Classical methods for the determination of universal time and polar motion.**
N. P. J. O'Hora, B. D. Yallop.
IAU Colloquium No. 26, (see 012.034), p. 303 - 308 (1975).

044.035 **On some problems of BIH and IPMS services.**
P. Melchior.
IAU Colloquium No. 26, (see 012.034), p. 409 - 413 (1975).

044.036 **Transformation of mean time into sidereal time by means of an electronic-mechanical device.**
V. A. Vytnov.
Izv. Glav. Astron. Obs. Pulkovo, No. 193, Astrofiz., Astrometr. i Zvezdn. Astron., p. 166 - 168 (1975). In Russian.

044.037 **Considerazioni su una irregolarità nella rotazione terrestre.** F. Zagar.
Rend. Ist. Lombardo, Accad. Sci. Lettere, Cl. Sci. (A), Vol. 108, 861 - 870 = Contr. Oss. Astron. Milano-Merate, Nuova Ser. No. 366 (1974/75).

044.038 **UT2–UT1 for 1976.**
Bureau International de l'Heure, (B.I.H.), Paris, Circ. A (1975).

044.039 **Universal time and coordinates of the pole; Emission time of time signals; Universal time (coordinated); Independent local atomic time scales AT (i); Information on time signals.**
Bureau International de l'Heure, (B.I.H.), Paris, Circ. D 105 - D109 (1975). – 1975 August - December.

044.040 **UTC time step on the 1st of January 1976.**
B. Guinot.
Bureau International de l'Heure, (B.I.H.), Paris, Circ. E5, 2 pp. (1975). In English and French.

044.041 **Time Service of the Mizusawa Observatory. Bulletins,** Vol. 18, No. 1 - 12, 1973.
S. Takagi, I. Okamoto, M. Aihara, K. Yokoyama, T. Hara, G. Murakami, T. Okuda, H. Kitago, K. Iwadate, S. Sakai, K. Horiai, T. Sasao.
Edited by the International Latitude Observatory of Mizusawa, Mizusawa-Shi, Iwate-Ken, Japan. 2 + 48 pp. (1975).
This Bulletin contains the results of time service and astronomical observations made at the Mizusawa Observatory from 1 January to 31 December, 1973.

044.042 **Time service.** C. Moranzino (Editor).
Oss. Astron. Torino (Pino Torinese), Bull. Nos 8 - 10, 6 + 13 pp. (1974/75). – Results of the time determinations 1974 May - 1975 April.

044.043 **Détermination astronomique de l'heure.**
M. Jovanović, D. Vesić.
Bull. Obs. Astron. Belgrade, No. 126, p. 19 - 22 (1975).

044.044 **Déterminations de l'heure et de la latitude de 1966 à 1972, effectuées à l'astrolabe de Danjon n° 7 (I).**
P. Pâquet, W. De Rop.
Bull. Astron., Obs. Roy. Belgique, Vol. 8, 184 - 197 (1975)

044.045 **Time and latitude service.**
Polish Acad. Sci., Astron. Latitude Obs., Borowiec, Poland, Circ. Nos. 134 - 135, 11 + 11 pp. (1975). – 1975 April–September.

044.046 **Astronomische Zeit- und Breitenbestimmungen. Empfangszeiten von Zeitsignalen.**
Edited by Deutsches Hydrographisches Institut, Hamburg. 5 pp. (1974). – 1974 October - December.

044.047 **Zeit- und Breitendienst.**
Edited by Deutsches Hydrographisches Institut, Hamburg. 9 pp. (1975). – 1975 January - March.

044.048 **Changes in the earth's rotation from astronomical observations.** L. V. Morrison.
H. M. Nautical Almanac Office, *Roy. Greenwich Obs., Herstmonceux,* Library Repr. No. 317 (1975).

044.049 **International Time and Latitude Service at the Tokyo Astronomical Observatory during 1975.**
K. Osawa.
Tokyo Astron. Obs., Time and Latitude Bull., Vol. 49, Nos. 1 - 4, p. i - iv, 1 - 30 (1975). – Results of the time determinations 1975 January - April.

044.050 **Détermination astronomique de l'heure et heures demi-définitives de réception des signaux horaires.**
L. Webrová, V. Ptáček.
Acad. Tchécoslov. Sci., Inst. Astron., Station de l'Heure, Prague, Sér. 7, Nos. 2 - 6, 14 + 14 + 13 + 13 + 13 pp. (1975). 1974 March - December.

The rotation of the earth: a geophysical discussion.
See Abstr. 003.095.

Loran-C precise time and frequency dissemination.
See Abstr. 035.005.

Statistical properties of Loran-C transmissions.
See Abstr. 035.006.

Clocks synchronised by time signals.
See Abstr. 035.009.

Atomic clocks and the formation of TAI.
See Abstr. 035.010.

Chandler wobble, earthquakes, rotation, and geomagnetic changes. See Abstr. 045.001.

The exact solution of the problem of the earth's dynamical equilibrium. See Abstr. 045.015.

Effect of the third product of inertia and of equatorial flattening of the earth on the free nutation of the rotational axis. See Abstr. 081.001.

Effects of tidal dissipation in the oceans on the moon's orbit and the earth's rotation.
See Abstr. 081.003.

Einfluß der Geschwindigkeitsabnahme der Erdrotation auf Figur und Schwere der Erde.
See Abstr. 081.020.

Variations of the angles between the plumb lines of the astronomical instruments. See Abstr. 081.036.

045 Latitude Determination, Polar Motion

045.001 **Chandler wobble, earthquakes, rotation, and geomagnetic changes.** F. Press, P. Briggs.
Nature, Vol. 256, 270 - 273 (1975).
A pattern recognition algorithm can be applied to the seismicity of major earthquake belts, to the amplitudes of Chandler wobble, to changes in the rotational velocity of the earth, and to the drift of the eccentric geomagnetic dipole for the years 1901–64. The patterns which emerged suggest that all of these diverse phenomena are related.

045.002 **On the direction and rate of secular motion of the earth's poles.** Zh. S. Erzhanov, A. A. Kalybaev.
Dokl. Akad. Nauk SSSR. Ser. Mat., Fiz., Vol. 223, 844 - 847 (1975). In Russian.

045.003 **Period, Q_p and amplitude of the pole tide.** R. G. Currie.
Geophys. Journ. Roy. Astron. Soc., Vol. 43, 73 - 86 (1975).
The maximum entropy method of Burg has been applied to monthly mean values of tide gauge records from 126 stations. Principal results are : (1) the Chandlerian induced pole tide signal is detected on a world-wide basis, (2) the spectral character of the signal is a single broad peak, (3) the mean and standard error of the period is 433.16 ± 0.41 mean solar days, (4) the mean and standard deviation of the pole tide Q_p is 22 ± 12, and (5) in agreement with Miller evidence is found for an enhanced pole tide in certain locations outside the shallow seas of North Europe.

045.004 **On the nearly diurnal free wobble – how necessary it is to compare observations with theory.**
E. P. Fedorov.
Astron. Tsirk., No. 861, p. 1 - 3 (1975). In Russian.

045.005 **Does secular polar motion exist?** E. P. Fedorov.
Astrometriya i Astrofizika, *Kiev*, vyp. (No.) 27, (see 003.004), p. 3 - 6 (1975). In Russian.
In connection with the question whether or not variations of mean latitudes are due to secular polar motion the following hypothesis is tested: the rates of the linear trends in observed latitude variations are independent random values. Observations at the International Latitude Stations Mizusawa, Carloforte, Gaithersburg, and Ukiah from 1900 to 1972 are shown not to contradict this hypothesis.

045.006 **On the secular motion of the earth's geographic pole.** L. D. Kostina, V. I. Sakharov.
Astrometriya i Astrofizika, *Kiev*, vyp. (No.) 27, (see 003.004), p. 7 - 13 (1975). In Russian.
The problem of the secular polar motion is considered. The results of the IPMS observations were used as initial data of the analysis. The periodic components of the latitude variations were eliminated by a smoothing process. Using the annual variations of the mean latitudes of IPMS stations an estimation of the velocity $V_p = 0\overset{''}{.}0032$/year in the direction $\lambda_p = 48\overset{\circ}{.}5$ was obtained and the influence of errors of proper motions of stars for different observational programs was evaluated. The results agree with investigations made by other methods.

045.007 **On the excitation of the Chandler polar motion.** Ya. S. Yatskiv.
Astrometriya i Astrofizika, *Kiev*, vyp. (No.) 27, (see 003.004), p. 14 - 16 (1975). In Russian.
Possible mechanisms of excitation of the Chandler polar motion are discussed. The resonance mechanism suggested by Pines and Shaham is preferred.

045.008 **L'examen d'utilité de l'instrument universel Wild-T4 pour la détermination de latitude par diverses méthodes.** M. Dačić.
Publ. Obs. Astron. Beograd, No. 20, (see 012.024), p. 82 - 83 (1975). – Summary. In Serbo-Croatian and French.

045.009 **Latitude star system of the ILS and the Conventional International Origin (CIO).**
K. Sato, Y. Wako.
Publ. International Latitude Obs. Mizusawa, Vol. 9, 167 - 173 (1974).
The essential characteristics of the methods of finding declination corrections for the observed latitudes of the International Latitude Service (ILS) are discussed in this paper.

045.010 **Polar wobble, sway and astronomical latitude–longitude observations.** M. Ooe, T. Sasao.
Publ. International Latitude Obs. Mizusawa, Vol. 9, 223 - 233 (1974).
General formulae relating the polar wobble (geographical motion of the earth's rotation pole), sway (motion of the pole relative to an inertial frame in space) and astronomically observable values with each other are presented.

045.011 **Note on the secular variations of *z*-term.** H. Ishii, I. Naito.
Publ. International Latitude Obs. Mizusawa, Vol. 9, 235 - 240 (1974).
The secular variations of 12-monthly running mean *z*-term during 1900–1973 based on the ILS-system are presented and their several features are discussed. The large-scale changing points appear in 1915 and in 1959.

045.012 **Results of latitude observations with the Floating Zenith Telescope at Mizusawa (1967–1972).**
S. Takagi, H. Ookawa, Y. Goto.
Publ. International Latitude Obs. Mizusawa, Vol. 9, 291 - 331 (1974).
The results of observations made in the period from 1967 to 1972 are given in details as a continuation of the results published by Hattori in 1951, 1953. The scale value of the FZT plate was taken to be variable according to the atmospheric temperature and time in the reduction of the observation. The external and internal errors of observations with the FZT were derived from the results and were compared with those of the Zenith Telescope at the five ILS stations.

045.013 **Polar motion spectra based upon Doppler, I.P.M.S., and B.I.H. data.** M. A. Graber.
Goddard Space Flight Center, Greenbelt, Maryland, GSFC Document X-921-75-157, Preprint, 3 + 24 pp. (1975).
Spectra based upon recent Doppler, International Polar Motion Service, and Bureau International de l'Heure data are compared. An oscillation is found at 1.3 cpy which might be due to an Eulerian motion of the solid inner core. An extended 15-year I.P.M.S. data set is filtered and analyzed to yield a Chandler peak with a period 430.8 solar days and a full width at half-maximum of 0.7 days (Q = 600). The data is reanalyzed in overlapping 3-year segments and indicates that the excitation of the Chandler wobble is a discrete process and that periods as long as three years occur in which the driving mechanism is essentially quiescent.

045.014 **Catalogue of FK4 Horrebow-pairs for latitudes from +30° to +60°.**

F. Reichhart, with a preface by K. Bretterbauer.
Veröff. Inst. Höhere Geod., Techn. Univ. Wien, Geowiss. Mitt., Heft 6, 9 + 103 pp. (1975). In German and English.

Because of the difficulties arising in selecting suitable star pairs for the method of Horrebow-Talcott for latitude determination the author thought it useful to establish a star program generally valid for a larger range in latitude. The catalogue at hand was compiled in accordance with P. Kalaja (1952). In view of the high precision of the HT-method only FK4-stars were incorporated, as for these stars sufficiently accurate coordinates are available. For the selection of star pairs a FORTRAN-program was worked out. The coordinates stated in the catalogue at hand are apparent places 1975.0. They were computed from mean coordinates, making use of a program by Gerstbach (1972).

045.015 **The exact solution of the problem of the earth's dynamical equilibrium.**
Zh. S. Erzhanov, A. A. Kalybaev.
Vestn. AN KazSSR, 1975, No. 3, p. 20 - 29. In Russian – Abstr. in Referativ. Zhurn. 51. Astron., 10.51.193 (1975).

045.016 **Pole coordinates by the IPMS stations.** S. Yumi, K. Yokoyama, H. Ishii.
IAU Colloquium No. 26, (see 012.034), p. 115 - 122 (1975).

045.017 **Astronomical programs for the study of secular variations in position.** W. Markowitz.
Proc. Symposium on the earth's gravitational field and secular variations in position, (see 012.036), p. 11 - 19 (1974).

045.018 **The earth's secular librations.** H. J. M. Abraham.
Proc. Symposium on the earth's gravitational field and secular variations in position, (see 012.036), p. 29 - 35 (1974).

045.019 **Analysis of the first Doppler observations performed at the Royal Observatory of Belgium.**
P. Pâquet, R. Dejaiffe.
Proc. Symposium on the earth's gravitational field and secular variations in position, (see 012.036), p. 347 - 359 (1974).

045.020 **A lod and wobble analysis of generalized Euler-Liouville type based on virial planetary equations.**
E. W. Grafarend.
Proc. Symposium on the earth's gravitational field and secular variations in position, (see 012.036), p. 643 - 659 (1974).

045.021 **Periodic and quasi-periodic fluctuations in the motion of the earth's pole.**
E. Proverbio, S. Uras.
Space astrometry, (see 012.037), p. 83 - 93 (1975).

In the study of the polar motion, periodic or quasi-periodic fluctuations have been emphasised, the polar origin of which has, nevertheless, not been completely clear. Suggestions about the existence of fluctuations in polar motions could be derived from similar variations in latitudes of the international latitude stations. The reality of these fluctuations has been studied in comparing theoretical and observed band-widths for the different fluctuation periods. These same periods have, moreover, been correlated with the typical period of some natural phenomena.

045.022 **Latitude star system of the ILS and the conventional international origin.** K. Sato, Y. Wako.
IAU Colloquium No. 26, (see 012.034), p. 319 (1975).

045.023 **Systematic differences between global or local solutions of polar motion derived from astronomical and spatial measurements.** N. Capitaine, M. Feissel.
IAU Colloquium No. 26, (see 012.034), p. 415 - 432 (1975).

045.024 **Results of latitude observations with the Floating Zenith Telescope at Mizusawa: comparisons between the results obtained from the astronomical and Doppler satellite observations.** S. Takagi.
IAU Colloquium No. 26, (see 012.034), p. 461 - 468 (1975).

045.025 **Variations of the latitude of Blagoveshchensk from observations during 1959–1972.**
G. S. Sheptunov, A. N. Sheptunova, V. P. Kotlyarchuk, I. A. Vladimirova.
Izv. Glav. Astron. Obs. Pulkovo, No. 193, Astrofiz., Astrometr. i Zvezdn. Astron., p. 124 - 138 (1975). In Russian.

045.026 **Annual report of the International Polar Motion Service for the year 1973.** S. Yumi.
Published for the International Council of Scientific Unions by Central Bureau of the International Polar Motion Service, Mizusawa, Japan. 4 + 169 + E7 pp. (1975).

045.027 **Problèmes statistiques liés au mouvement du pôle.** A.-J. Richard.
Thesis, Univ. P. et M. Curie, Paris, 73 pp. (1975).

045.028 **Research in changes of Belgrade *z*-term in different declination systems in correlation with the atmospheric condition.** R. Grujić.
Bull. Obs. Astron. Belgrade, No. 126, p. 11 - 15 (1975).

045.029 **Observations à la lunette zenithale (de 110 mm) du service de latitude de l'Observatoire de Belgrade en 1970, 1971, 1972.**
G. Grujić, M. Djokić, V. Milovanović, L. Djurović.
Bull. Obs. Astron. Belgrade, No. 126, p. 22 - 29 (1975).

045.030 **Observations à la lunette zenithale (de 110 mm) du service de latitude de l'Observatoire de Belgrade 1973.** R. Grujić, M. Djokić, L. Djurović.
Bull. Obs. Astron. Belgrade, No. 126, p. 30 - 31 (1975).

045.031 **Les termes harmoniques dans le mouvement du pôle.** D. Djurović.
Bull. Obs. Astron. Belgrade, No. 126, p. 62 - 71 (1975).

045.032 **On the variations of latitude during night deduced from observations with the Zeiss zenith telescope at Poltava.** R. I. Popova.
Vrashchenie i prilivn. deformatsii Zemli. Vyp. (No.) 7, Kiev, Nauk. dumka, 1975, p. 66 - 70. In Russian. – Abstr. in Referativ. Zhurn. 51. Astron., 2.51.241 (1976).

045.033 **The results of the reduction of latitude observations with the Zeiss zenith telescope at Poltava during 1949.7 - 1956.5.** I. P. Ogorodnik.
Vrashchenie i prilivn. deformatsii Zemli. Vyp. (No.) 7, Kiev, Nauk. dumka, 1975, p. 70 - 74. In Russian. – Abstr. in Referativ. Zhurn. 51. Astron., 2.51.242 (1976).

045.034 **On the variations of dispersion of the accidental errors of latitude observations.**
I. V. Dzhun', I. I. Maksimova.
Vrashchenie i prilivn. deformatsii Zemli. Vyp. (No.) 7, Kiev, Nauk. dumka, 1975, p. 78 - 81. In Russian. – Abstr. in Referativ. Zhurn. 51. Astron., 2.51.243 (1976).

045.035 **A new method for determination of the motion of the earth's poles.** Đ. Teleki.
Vasiona, Vol. 23, 37 - 38 (1975). In Serbo-Croatian.

045.036 **Monthly Notes of the International Polar Motion Service.**
IPMS Monthly Notes, International Latitude Obs. Mizusawa

(Japan). 1975 Nos. 6 - 11, p. 49 - 108 (1975). – Announces the values of latitudes observed at the collaborating stations during 1975 June - November.

045.037 **Results of the determination of latitude in Józefosław by observations of the Horrebow-Talcott pairs.**
L. Pieczyński.
Warsaw Tech. Univ., Astron. - Geod. Obs. Józefosław.
Latitude Circ. Nos. 53, 54 (1975). – 1975 January - June.

Practice on the PZT observation.
See Abstr. 032.033.

Preliminary results of astronomical observations made by the new PZT at the Mizusawa Latitude Observatory.
See Abstr. 032.034.

Comparison of declinations of the Talcott pairs centers obtained from observations with the PZT-135 during 1948.8–1955.0 at Pulkovo with the declinations of the $W2_{50}$ Catalogue. See Abstr. 041.059.

Investigation of declinations and of proper motions of the Belgrade General Catalogue of Latitude Stars BSKŠZ1 and BSKŠZ2. See Abstr. 041.068.

Results obtained with the Paris astrolabe. Time and latitude 1971–1974. See Abstr. 044.002.

Total effect of any nearly-diurnal wobble of the earth's axis of rotation in latitude and time observations. Application to the Paris astrolabe observations from 1956.6 to 1973.8. See Abstr. 044.005.

Time and latitude determination using a portable zenith camera. See Abstr. 044.020.

Time service report. See Abstr. 044.025.

Classical methods for the determination of universal time and polar motion. See Abstr. 044.034.

On some problems of BIH and IPMS services.
See Abstr. 044.035.

Universal time and coordinates of the pole; Emission time of time signals; Universal time (coordinated); Independent local atomic time scales AT (i); Information on time signals. See Abstr. 044.039.

Déterminations de l'heure et de la latitude de 1966 à 1972, effectuées à l'astrolabe de Danjon n°7 (I).
See Abstr. 044.044.

Time and latitude service.
See Abstr. 044.045.

Astronomische Zeit- und Breitenbestimmungen. Empfangszeiten von Zeitsignalen. See Abstr. 044.046.

Zeit- und Breitendienst. See Abstr. 044.047.

International Time and Latitude Service at the Tokyo Astronomical Observatory during 1975.
See Abstr. 044.049.

Accuracy of site coordinates obtainable by a mobile lunar laser station: final report. See Abstr. 046.042.

Dynamic techniques for studies of secular variations in position from ranging to satellites. See Abstr. 046.043.

Variations of the angles between the plumb lines of the astronomical instruments. See Abstr. 081.036.

Refraction anomalies due to wind in the free atmosphere. See Abstr. 082.063.

046 Astronomical Geodesy, Satellite Geodesy, Navigation

046.001 **Les instruments de géodésie spatiale.** J. Kovalevsky.
Space optics, (see 012.013), p. 339 - 347 (1974).

046.002 **On approximate methods for determining astronomical azimuths.** E. M. Voronoj.
Geod. i kartografiya, 1975, No. 2, p. 29 - 32. In Russian. Abstr. in Referativ. Zhurn. 52. Geodeziya i Aehrosemka, 7.52.88 (1975).

046.003 **Introduction to the report on geodesy.** I. I. Mueller.
Rev. Geophys. Space Phys., Vol. 13, No. 3, (see 013.001), p. 242, 276 (1975). – Review paper with bibliographical notes.

046.004 **Geodetic theory.** B. H. Chovitz.
Rev. Geophys. Space Phys., Vol. 13, No. 3, (see 013.001), p. 243 - 245, 276 - 277 (1975). – Review paper with bibliographical notes.

046.005 **Dynamic satellite geodesy.** E. M. Gaposchkin.
Rev. Geophys. Space Phys., Vol. 13, No. 3, (see 013.001), p. 265 - 267, 286 - 287 (1975). – Review paper with bibliographical notes.

046.006 **Positioning by satellites.** I. I. Mueller.
Rev. Geophys. Space Phys., Vol. 13, No. 3, (see 013.001), p. 268 - 270, 288 - 290 (1975). – Review paper with bibliographical notes.

046.007 **Geodesy by very long base line interferometry.** C. C. Counselman III.
Rev. Geophys. Space Phys., Vol. 13, No. 3, (see 013.001), p. 270 - 271, 290 (1975). – Review paper with bibliographical notes.

046.008 **A simple method of the precise determination of any mark azimuth on the base of groups (pairs) of stars observed in symmetrical elongations.** W. Szpunar.
Geod. i Kartografia, Vol. 24, 167 - 175 (1975). In Polish.

This paper contains a very simple proposition of precise azimuth determination on the base of two groups each of n stars observed in symmetrical elongations.

046.009 **A nocturnal.** M. W. G. Webster.
Journ. Navigation, Vol. 28, 372 - 374 (1975).

046.010 **Alignment of geodetic and satellite coordinate systems to the average terrestrial system.**
D. E. Wells, P. Vaniček.
Bull. Géod., Nouvelle Sér., Année 1975, No. 117, p. 241 - 257 (1975).

046.011 **A catalog of station coordinates for GEOS-C orbit determination.**
J. G. Marsh, B. C. Douglas, D. M. Walls.
Bull. Géod., Nouvelle Sér., Année 1975, No. 117, p. 259 - 266 (1975).

046.012 **Solution of the downward continuation problem by collocation.** J. B. Zielinski.
Bull. Géod., Nouvelle Sér., Année 1975, No. 117, p. 267 - 277 (1975).

The modern satellite techniques enable to measure the gravity acceleration in space. The downward continuation of this measurements to the earth's surface is by the closed theory practically impossible. Using the method of collocation we can estimate the values of Δg in the limited region underneath the observation with a little loss of accuracy. Results of some numerical tests for elevation of 300 km are presented.

046.013 **General-altitude transformations between geocentric and geodetic coordinates.** S. A. T. Long.
Celestial Mechanics, Vol. 12, 225 - 230 (1975).

Formulas for the general-altitude (height above the ellipsoid) transformation from geocentric to geodetic coordinates and vice versa are derived. The set of four formulas is expressed in each of two useful forms: series expansions in powers of the earth's flattening and series expansions in powers of the earth's eccentricity. The error incurred in these expansions is of the order of one part in 3×10^7.

046.014 **Método autocorrector para el cálculo de direcciones de satélites geodésicos y análysis de los errores en la restitución de un arco en órbita.** M. J. Sevilla de Lerma.
Urania Barcelona, Año 59, No. 281–282, p. 217 - 364 (1974).

046.015 **Astronomical-geodetical methods for investigation of geodynamical problems.** A. A. Izotov.
Zemlya i Vselennaya, 1975, No. 4, p. 6 - 14. In Russian.

046.016 **Numerical and analytical methods for the solution of surface integrals connected with the simple layer model of the geopotential in satellite geodesy.** H. Fröhlich.
Observations of artificial earth satellites, No. 14, (see 012.018), p. 53 - 60 (1975).

046.017 **Some results of the "Large Chord" international program.** A. G. Masevich, N. P. Erpylev.
Observations of artificial earth satellites, No. 14, (see 012.018), p. 115 - 118 (1975). In Russian.

046.018 **Station coordinates in the standard earth III system derived by using camera data from ISAGEX.**
E. M. Gaposchkin, J. Latimer, G. Mendes.
Observations of artificial earth satellites, No. 14, (see 012.018), p. 119 - 126 (1975).

046.019 **Problems of determination of station-coordinates and of the gravity field with satellite observations and measurements on the earth.** F. Halmos.
Observations of artificial earth satellites, No. 14, (see 012.018), p. 127 - 150 (1975).

046.020 **Global satellite triangulation and trilateration results.** I. I. Mueller.
Observations of artificial earth satellites, No. 14, (see 012.018), p. 151 - 152 (1975).

046.021 **On the application of the method of synchronous planes.** G. Karský, J. Kostelecký.
Observations of artificial earth satellites, No. 14, (see 012.018), p. 153 - 168 (1975).

046.022 **Zur geometrischen Ableitung der Potsdamer Stationskoordinaten aus den Beobachtungen des ISAGEX-Programms.** K.-H. Marek.
Observations of artificial earth satellites, No. 14, (see 012.018), p. 169 - 182 (1975).

046.023 **Analyse der kombinierten kosmisch-geodätischen Beobachtungsmethoden.** F. Halmos.
Observations of artificial earth satellites, No. 14, (see 012.018), p. 183 - 200 (1975).

046.024 **Independent geometrical Doppler geodetic systems.** J. Ádám, G. Tarcsai.
Observations of artificial earth satellites, No. 14, (see 012.018), p. 201 - 214 (1975).

046.025 **Geodetic research in Finland with balloons and satellites.** J. Kakkuri.
Observations of artificial earth satellites, No. 14, (see 012.018), p. 215 - 219 (1975).

046.026 **Analysis of the accuracy of distance determination between two stations on the earth's surface, when also the distance between two geostationary artificial satellites has been measured.** M. Solarić.
Observations of artificial earth satellites, No. 14, (see 012.018), p. 231 - 242 (1975).

046.027 **Quelques utilisations des observations simultanées optiques, laser et Doppler des satellites artificiels dans la géodésie géométrique spatiale.**
F. Barlier, A. Dinescu.
Observations of artificial earth satellites, No. 14, (see 012.018), p. 243 - 262 (1975). In Russian.

046.028 **The adjustment of photographic and laser observations of artificial earth satellites by the group method.** W. Baran, I. Gajderowicz.
Observations of artificial earth satellites, No. 14, (see 012.018), p. 263 - 270 (1975). In Russian.

046.029 **On the use of non-synchronous satellite observations in the geometrical method of cosmic geodesy.**
N. Georgiev.
Observations of artificial earth satellites, No. 14, (see 012.018), p. 271 - 278 (1975). In Russian.

046.030 **On the use of non-synchronous laser and photographic satellite observations by the vector chain arctic–antarctic.** I. Totomanov, N. Georgiev.
Observations of artificial earth satellites, No. 14, (see 012.018), p. 279 - 303 (1975). In Russian.

046.031 **Satellite connections of continental geodetic networks by means of an ellipsoidal method.**
K. A. Czarnecki.
Observations of artificial earth satellites, No. 14, (see 012.018), p. 305 - 312 (1975).

046.032 **Some proposals on the adjustment of cosmic triangulation based on the method of chords.**
M. Daskalova.
Observations of artificial earth satellites, No. 14, (see 012.018), p. 313 - 325 (1975). In Russian.

046.033 **Triangulation by means of high-altitude targets.** I. Kabeláč.
Observations of artificial earth satellites, No. 14, (see 012.018), p. 327 - 347 (1975). In Russian.

046.034 **New consideration concerning the determination of the coordinates of stations of satellite observations in the "Standard Earth" system.**
P. Galsan, O. Baasansuren, L. Luvsan, M. Ehnkhbayar.
Observations of artificial earth satellites, No. 14, (see 012.018), p. 349 - 351 (1975). In Russian.

046.035 **The use of Sumner-lines for position fixing with satellites.** G. Gerstbach.
Observations of artificial earth satellites, No. 14, (see 012.018), p. 571 - 575 (1975).

046.036 **Application de la méthode généralisée des directions de fermeture à la détermination de la position d'une station satellitaire à la base d'observations synchroniques, photographiques et à laser.** W. Baran.
Geod. i Kartografia, Vol. 24, 179 - 190 (1975). In Polish.

046.037 **Satellitengeodäsie.** M. Schneider.
SuW, Vol. 14, 357 - 363 (1975).

046.038 **Investigation of the accuracy of determination of a ship's position by means of the satellite system NNSS.** Yu. A. Peskov.
Sudovozhdenie. Vyp. (No.) 16. Moskva, 1975, p. 19 - 30. In Russian. – Abstr. in Referativ. Zhurn. 51. Astron., 8.51. 160 (1975).

046.039 **Über die Anwendung des sphärischen Bogenschnittes in der geodätischen Astronomie.**
B. Kilar.
Publ. Obs. Astron. Beograd, No. 20, (see 012.024), p. 63 - 75 (1975). In Serbo-Croatian.

046.040 **Research of changes of the Belgrade z-term for different declination systems in correlation with atmospheric conditions.** R. Grujić.
Publ. Obs. Astron. Beograd, No. 20, (see 012.024), p. 76 - 77 (1975). – Summary. In Serbo-Croatian and English.

046.041 **Determination of meridian direction for the Astronomical and Geophysical Observatory in Ljubljana.**
P. Ranzinger.
Publ. Obs. Astron. Beograd, No. 20, (see 012.024), p. 98 - 100 (1975). In Serbo-Croatian.

046.042 **Accuracy of site coordinates obtainable by a mobile lunar laser station: final report.**
G. L. Loumos, J. D. Mulholland, P. J. Shelus, E. C. Silverberg.
Dep. Astron. and McDonald Obs., Univ. of Texas, Austin, Res. Memorandum in Astron., No. 75–007, 19 + 25 pp. (1975).

The accuracy with which a mobile lunar laser station can be located is the subject of a modelling study presented here. The influence of the number and accuracy of fixed lunar ranging stations, the uncertainty in polar motion, and data loss due to weather and similar factors has been considered, and the results are given in cartographic form. In general, all three coordinates (for coordinates to latitude ±60°) are determined to better than the pole uncertainty, given three or more fixed sites and reasonable weather. This result indicates that one or more mobile stations will be extremely suitable for the study of geotectonics.

046.043 **Dynamic techniques for studies of secular variations in position from ranging to satellites.**
D. E. Smith, R. Kolenkiewicz, R. W. Agreen, P. J. Dunn.
Proc. Symposium on the earth's gravitational field and secular variations in position, (see 012.036), p. 291 - 314 (1974).

046.044 **On the gravimetrical connection of basic astronomical points of some towns.** G. A. Aleksashina.
Geod. i kartografiya, 1975, No. 7, p. 15 - 19. In Russian. Abstr. in Referativ. Zhurn. 52. Geodeziya i Aehrosemka, 12. 52.82 (1975).

046.045 **Determination of the approximated azimuth according to Schweizer's method.**
A. V. Vinogradov.
Nauch. trudy Omsk. s.-kh. in-t, 1975, No. 132, p. 34 - 35. In Russian. – Abstr. in Referativ. Zhurn. 52. Geodeziya i Aehrosemka, 12.52.110 (1975).

046.046 **On the determination and applications of gravity gradients in geodetic systems.** E. Groten.
Boll. Geod. Sci. Affini, Anno 34, p. 357 - 395 (1975).– Paper presented at the 3. Martin Hotine-Symposium on mathematical geodesy, Siena 1975.

Results obtained from an ensemble of various geodetic measurements in a local geodetic system are discussed in view of application of the components of the Marussi (Eötvös)-Tensor. The observations consist of recent torsion balance measurements, vertical gravity gradients, astrogeodetic and zenith distance observations and a dense gravity net; first and second derivatives of the well known harmonic functions $r\Delta g$ and T are analyzed. The data lend themselves to a unified approach using collocation procedures but in this context comparison of specific combinations like astrogeodetic and astrogravimetric levellings, boundary value problem solution, integration of gradients etc. is mainly dealt with.

046.047 **Tables K 21 – computed results of altitudes and azimuths for all latitudes and all celestial bodies.**
S. Kotlarić.
Hidrografski Godišnjak 1973, [Hidrografski Institut Jugoslavenske Ratne Mornarice, Split], HI–N–30, p. 49 - 64 (1975). In Serbo-Croatian.

046.048 **Satellitenentfernungsmessungen in Potsdam.**
H. Fischer, R. Neubert, C. Selke, R. Stecher.
Vermessungstechnik, 23. Jahrgang, p. 53 - 56 (1975).

046.049 **Testung und Interpolation von Laserbeobachtungen einer Station mit Hilfe der Bahnmethode.**
H. Montag.
Vermessungstechnik, 23. Jahrgang, p. 60 - 63 (1975).

046.050 **Die topographisch-atmosphärische Reduktion des mittleren Brechungsindex.** M. Schädlich.
Vermessungstechnik, 23. Jahrgang, p. 68 - 71 (1975).

046.051 **Ein Beitrag zur simultanen Ortsbestimmung.**
G. Hemmleb.
Vermessungstechnik, 23. Jahrgang, p. 185 - 187 (1975).

046.052 **Die Methode der „Azimutstandlinien" zur simultanen Bestimmung von λ, φ und α.** H.-U. Sandig.
Vermessungstechnik, 23. Jahrgang, p. 211 - 213 (1975).

046.053 **Nutzung geodätischer Ergebnisse und Methoden in der Geo- und Kosmosforschung.**
H. Kautzleben.
Vermessungstechnik, 23. Jahrgang, p. 332 - 336 (1975).

046.054 **On the scale and orientation of geodetic reference systems from VLBI.** I. D. Lloyd.
Proc. Symposium on the earth's gravitational field and secular variations in position, (see 012.036), p. 406 - 409 (1974).

046.055 **Future applications of laser ranging systems.**
J. W. Siry, F. O. Vonbun.
Proc. Symposium on the earth's gravitational field and secular variations in position, (see 012.036), p. 710 - 715 (1974).

046.056 **Nature of the requirements for reference coordinate systems.** C. A. Lundquist.
IAU Colloquium No. 26, (see 012.034), p. 51 - 62 (1975). Review paper.

046.057 **Magnitudes and spectra of important dynamical phenomena.** E. P. Fedorov.
IAU Colloquium No. 26, (see 012.034), p. 63 - 77 (1975).

046.058 **Reference coordinate system requirements for geodesy and ocean dynamics.** R. S. Mather.
IAU Colloquium No. 26, (see 012.034), p. 93 - 114 (1975).

046.059 **Some problems related to geodetic datum definitions for terrestrial measurements.** E. Groten.
IAU Colloquium No. 26, (see 012.034), p. 247 - 259 (1975).

046.060 **General principles for the realization of reference systems for earth dynamics.** G. Veis.
IAU Colloquium No. 26, (see 012.034), p. 261 - 267 (1975). – Review paper.

046.061 **Review of the classical methods for the determination of geodetic datums.** I. I. Mueller.
IAU Colloquium No. 26, (see 012.034), p. 321 - 339 (1975).

046.062 **On constructing the inertial system of high accuracy by VLBI methods.** C. A. Krasinsky.
IAU Colloquium No. 26, (see 012.034), p. 381 - 393 (1975).

046.063 **Note on selection of coordinate systems for geodynamics.** S. Henriksen.
IAU Colloquium No. 26, (see 012.034), p. 395 - 407 (1975).

046.064 **On the interrelation of classical and space geodetic systems of position reference.** L. Aardoom
IAU Colloquium No. 26, (see 012.034), p. 445 - 452 (1975).

046.065 **Comparison of the Uccle–Brussels coordinates in the BIH, NWL and EUR 50 systems.**
P. Pâquet, R. Dejaiffe, W. De Rop, R. Verbeiren.
IAU Colloquium No. 26, (see 012.034), p. 457 - 459 (1975).

046.066 **On a coordinate system of the astronomical-geodetical net.** M. I. Yurkina.
Geod. i kartografiya, 1975, No. 6, p. 6 - 10. In Russian. Abstr. in Referativ. Zhurn. 52. Geodeziya i Aehrosemka, 1.52.50 (1976).

046.067 **Die Arbeiten des Sonderforschungsbereiches 78 Satellitengeodäsie der Technischen Universität München im Jahre 1974.** Presented by R. Sigl.
Veröff. Bayer. Kommission Internationale Erdmessung, Bayer. Akad. Wiss., München. Astron.-Geod. Arbeiten, No. 33, 193 pp. (1975).

This publication gives the report on the two projects on satellite geodesy at the Technical University of Munich for 1974. Besides reports on activities some papers on research work are presented. The following contributions are within the subject scope of Astronomy and Astrophysics Abstracts: Zur Bestimmung der Verzeichnung ballistischer Kammern aus Sternaufnahmen, *K. Kaniuth*, p. 32 - 41; Genaue Breitenbestimmungen mit dem Universal Theo 002, *R. Sigl, A. Bauch*, p. 42 - 47; Überlegungen zur Darstellung des Einflusses der Polbewegung auf Bahnen künstlicher Erdsatelliten, *G. Dichtl*, p. 105 - 108; Betrachtung möglicher Erdkörpersysteme für geodynamische und geodätische Untersuchungen, *E. Nagel*, p. 117 - 118; Möglichkeiten zur Erstellung von Erdmodellen mit Hilfe der Sampling Funktionen, *H. F. Schmidt*, p. 119 - 121; Auswahl von Quadraturverfahren im Rahmen der Behandlung einer Randwertaufgabe, *K. H. Ilk*, p. 124 - 126; Doppler- und Altimeterdaten in der Bahnbestimmung, *H. Ludwig*, p. 127 - 129; Bahnbestimmung unter Berücksichtigung von Kräften, die nicht modellmässig erfasst werden, *M. Schneider*, p. 130 - 134; Vergleich verschiedener Rechenmodelle zur Ermittlung von Parametern des Erdpotentials, *K. H. Ilk*, p. 137 - 142; Satellitenbahnanalysen zur Ermittlung von Feldparametern der Erde, *C. Reigber*, p. 143 - 145.

046.068 **Latitude and longitude determinations with a transportable zenith camera.**

J. Geßler, G. Seeber.
Deutsche Geod. Kommission Bayer. Akad. Wiss., München, Reihe B, No. 213, p. 23 - 32 (1975).

046.069 **Zum Einfluss der Schichtneigung in der astronomisch-geodätischen Ortsbestimmung.** E. Kuntz.
Deutsche Geod. Kommission Bayer. Akad. Wiss., München, Reihe B, No. 213, p. 45 - 57 (1975).

046.070 **Investigations on the accuracy of transforming the European datum 1950 into a global reference system by means of the minimax-error method.**
R. Sigl, E. Reinhart.
Deutsche Geod. Kommission Bayer. Akad. Wiss., München, Reihe B, No. 213, p. 95 - 109 (1975).

046.071 **Geodätische Astronomie.** K. Ramsayer.
Deutsche Geod. Kommission Bayer. Akad. Wiss., München, Reihe B, No. 212, p. 28 - 36 (1975).

046.072 **Dynamische Satellitengeodäsie.** M. Schneider.
Deutsche Geod. Kommission Bayer. Akad. Wiss., München, Reihe B, No. 212, p. 49 - 58 (1975).

046.073 **On the choice of stars for astronomical correction in a navigational system.** M. A. Sergeev.
Izv. vyssh. ucheb. zavedenij. Priborostroenie, Vol. 18, No. 8, p. 74 - 79 (1975). In Russian. – Abstr. in Referativ. Zhurn. 51. Astron., 2.51.246 (1976).

046.074 **Long term consistency in positions of sites determined from Doppler satellite observations.**
R. J. Anderle.
Naval Surface Weapons Center, Dahlgren Laboratory, Dahlgren, Virginia, NSWC/DL TR-3433, 2 + 9 + A8 + B33 pp. (1975).

Doppler satellite observations have been used to determine positions of mobile equipment for over twelve years. The standard deviation of five day solutions for coordinates of the base stations upon which these computations were based were found to be 70 cm in latitude and 90 cm longitude and height for the period 1973–1975.

Geodätische Astronomie. See Abstr. 003.117.

Astronomical-geodetical and cartographic works at the Academy of Sciences (Concluding review). See Abstr. 004.112.

Exakte Abschätzung von Maximalfehlern aus vorgegebenen Toleranzen der Beobachtungsgrößen. See Abstr. 021.017.

Analysis of the first Doppler observations performed at the Royal Observatory of Belgium. See Abstr. 045.019.

Thirteenth-order orbital resonance on a Diademe 2 fragment. See Abstr. 081.018.

Terrestrial coordinate systems solidly connected with the earth. See Abstr. 081.038.

Gravimetric and astrogeodetic reference systems in space and time. See Abstr. 081.039.

Geophysical implications of lunar laser ranging. See Abstr. 094.020.

047 Ephemerides, Almanacs, Calendars

047.001 **Anuario del Observatorio Astronómico Nacional 1975.**
Published by Observatorio Astronómico Nacional, Univ. Nacional Colombia, Fac. Ciencias, Bogotá, Colombia, S. A. 112 pp. (1975).

047.002 **1976 Nautical Almanac.** Pub. No. 681.
Published by Hydrographic Office of Japan, Tokyo. 4 + 465 + 4 pp. (1975).

047.003 **1976 Abridged Nautical Almanac.** Pub. No. 683.
Published by Hydrographic Office of Japan, Tokyo. 3 + 236 + 7 pp. (1975).

047.004 **Nautisches Jahrbuch oder Ephemeriden und Tafeln für das Jahr 1976,** zur Bestimmung der Zeit, Länge und Breite zur See nach astronomischen Beobachtungen. Edited by "Deutsches Hydrographisches Institut", Hamburg. 125. Jahrgang, 3 + 43 + 366 + 30 pp. (1975).

047.005 **Grundkalendarium 1977.**
Compiled by H.-J. Felber.
Staatsverlag der Deutschen Demokratischen Republik, Berlin. 55 pp. Price M 25.00 (1975).

047.006 **The Handbook of the British Astronomical Association 1976.**
Prepared by the Computing Section of the Association under the supervision of G. E. Taylor.
Office of the Association: Burlington House, Piccadilly, London, 88 pp. Price £ 1.00, $ 3.00 respectively (1975).

047.007 **The Air Almanac 1976, January – April.**
Her Majesty's Stationery Office, London; United States Naval Observatory, Washington. 4 + 244 + A84 + F4 pp. Price £ 3.00 (1975).

047.008 **Annuaire de l'Observatoire Royal de Belgique [Jaarboek van de Koninklijke Sterrenwacht van België] 1976.**
Imprimerie Hayez, Bruxelles. 143e année (jaargang), 225 pp. (1975).

047.009 **Efemérides Astronómicas 1976.**
Published by Instituto y Observatorio de Marina, San Fernando (Cádiz). Printed in Spain by Observatorio de Marina, San Fernando (Cádiz). Vol. 185, 8 + 600 pp. (1975).

047.010 **Efemérides Astronómicas para o ano de 1976.**
Printed by "Secção de impressão, Observatório Astronómico Universidade de Coimbra, 13 + 266 pp. (1975).

047.011 **Der Sternenhimmel 1976.**
Kleines astronomisches Jahrbuch für Sternfreunde für alle Tage des Jahres zum Beobachten von bloßem Auge, mittels Feldstecher und Fernrohr.
R. A. Naef (Editor), under the redaction of P. Wild.
Verlag Sauerländer, Aarau, Switzerland. 36. Jahrgang, 176 pp. Price Fr. 34.00 (1975).

047.012 **Himmelskalender 1976.**
Ein kleines astronomisches Jahrbuch für Österreich.
H. Mucke, K. Mayrhofer (Editors).
Verlag: Astronomisches Büro, H. Mucke, Wien, 86 pp. Price öS 30.00 (1975).

047.013 **Astronomical Ephemeris for 1976.**
Prepared by Shikin-zan Observatory, Chinese Academy of Sciences. Academic Publishing House, Peking. 5 + 367 pp. Price 4.20 Gen (1975). In Chinese.

047.014 **The Observer's Handbook 1976.**
J. R. Percy (Editor).
The Royal Astronomical Society of Canada, Toronto, Ontario, Canada. 68th year of publication. 116 pp. Price $ 3.00 (1975).

047.015 **Explanatory Supplement to the Astronomical Ephemeris and the American Ephemeris and Nautical Almanac.**
Prepared jointly by the Nautical Almanac Offices of the United Kingdom and the United States of America. Issued by H. M. Nautical Almanac Office by order of the Science Research Council. Her Majesty's Stationery Office, London. Third impression (with amendments). 11 + 533 pp. Price £ 3.20 net (1974).

047.016 **Dados Astronómicos para os Almanaques de 1976 (bissexto) para Portugal.**
Compiled by Observatório Astronómico de Lisboa. Printed by Tipografia Duarte, Lda., Lisboa. 50 pp. (1975).

047.017 **The Indian Ephemeris and Nautical Almanac for the Year 1976.**
Office of preparation: Nautical Almanac Unit, Regional Meteorological Centre, Alipore, Calcutta. Printed by the Manager, New Govt. of India Press, Howrah-4 and published by the Controller of Publications, Civil Lines, Delhi. 22 + 475 pp. Price Rs. 25.00, £ 2.91, $ 9.00 respectively (1975).

047.018 **Astronomische Grundlagen für den Kalender 1977.**
Compiled by T. Lederle, edited by Astronomisches Rechen-Institut in Heidelberg.
Verlag G. Braun GmbH, Karlsruhe. 84 pp. Price DM 42.00 (1975).

047.019 **Lunar calender. (The determinator of lunar phases).**
A. V. Mikhajlovskij.
Nekotor. vopr. fiz. kosmosa. Moskva, 1974, p. 130 - 144.
In Russian. – Abstr. in Referativ. Zhurn. 51. Astron., 11.51. 68 (1975).

047.020 **Kalender für Sternfreunde 1976.** Kleines astronomisches Jahrbuch.
P. Ahnert (Editor).
Johann Ambrosius Barth, Leipzig. 192 pp. Price M 4.80 (1975).

047.021 **Nautisches Jahrbuch 1976.**
Edited by Seehydrographischer Dienst der Deutschen Demokratischen Republik, Rostock, 26th year. 29 + 367 pp. (1975).

047.022 **Astronomical calendar of the Observatory in Sofia for the year 1976.** D. Rajkova, Z. Krajcheva, S. Milcheva, R. Radkov, edited by A. Bonov.
Izdatelstvo na Blgarskata Akademiya na Naukite. Sofiya. Godina 23, 119 pp. Price 1 Lv. (1975/76). In Bulgarian.

047.023 **Das Himmelsjahr 1976.** Sonne, Mond und Sterne im Jahreslauf. M. Gerstenberger (Editor).
Franckh'sche Verlagshandlung, Stuttgart. 112 pp. Price DM 8.80 (1975).

047.024 **Heelal: Hemelkalender 1976.** J. Meeus (Editor).
Vereniging voor Sterrenkunde, Bruxelles. 72 pp.

Price FB 100.00 (1975).

047.025 **Hvězdářská ročenka 1976.** J. Bouška, V. Guth, B. Onderlička, J. Ruprecht (Editors).
Ročník 52. Academia, Nakladatelství Československé akademie věd, Praha. 267 pp. Price Kčs 22.00 (1975).

047.026 **The Air Almanac 1976, May - August.**
Her Majesty's Stationery Office, London, United States Naval Observatory, Washington. 248 + 84 + F4 pp. Price £3.00 (1975).

047.027 **Almanac for Geodetic Engineers 1976.**
Prepared under the supervision of R. L. Kintanar. Republic of the Philippines – Department of National Defense, Philippine Atmospheric, Geophysical and Astronomical Services Administration (Weather Bureau), Quezon City. 10 + 30 pp. Price p 3.00 (1975).

047.028 **Tables of Sunrise, Sunset, Twilight, Moonrise, and Moonset 1976.**
Prepared under the supervision of R. L. Kintanar. Republic of the Philippines – Department of National Defense, Philippine Atmospheric, Geophysical and Astronomical Services Administration (Weather Bureau), Quezon City. 11 + 57 pp. (1975).

047.029 **Astronomical Handbook for Southern Africa 1976.**
R. F. Hurly (Editor).
Published by the Astronomical Society of Southern Africa, Cape Town. 2 + 50 pp. (1975).

047.030 **Éphémérides astronomiques pour 1976.**
Publiées par la Société Astronomique de France, with a preface by B. Morando.
L'Astronomie, Vol. 90, Suppl. pour Janvier 1976, 4 + 68 pp. Price F 18.00 (1975).

047.031 **Anuario del Observatorio Astronómico de Madrid para 1976.**
Published by Instituto Geográfico y Catastral, Madrid. 422 pp. Price 100 pesetas (1975).

047.032 **The evolution of contents and form of the Nautical Astronomical Almanac.** V. M. Amelin.
Byull. Inst. Teoret. Astron., *Leningrad,* Vol. 14, 137 - 144 (1975). In Russian.

A brief review of the development of the Soviet Nautical Astronomical Almanac has been given. The main differences are described between the Nautical Astronomical Almanac and the Astronomical Almanac of the USSR. The new form of the Nautical Astronomical Almanac is characterized in detail.

047.033 **1976 Polaris Almanac for azimuth determination.**
Published by Hydrographic Department, Tokyo, Japan, Pub. No. 685, 15 pp. (1975).

047.034 **The earliest sunrise.**
C. R. McLernon, A. C. Gilmore, with an appendix by L. P. Lee.
Southern Stars, Vol. 26, 101 - 108 (1975).

047.035 **Calendar of the Tartu Astronomical Observatory for the year 1976.**
Academy of Sciences of the Estonian S.S.R., Institute of Astrophysics and Atmospheric Physics, W. Struve Tartu Astrophysical Observatory. 52nd year of publication. Publishing House 'Valgus', Tallinn. 103 pp. Price 33 Kop. (1975). In Estonian.

047.036 **Astronomiskais Kalendārs 1976.**
M. Dīriķis (Editor).
Latvijas PSR Zinātnu Akadēmija, Radioastrofizikas Observatorija, Vissavienības Astronomijas un Ģeodēzijas Biedrības Latvijas Nodaļa. Izdevniecība "Zinātne", Riga. 196 pp. Price 41 Kop. (1975).

047.037 **Almanacco Astronomico della Rivista Coelum per l'anno bisestile 1976.** P. Battistini.
Coelum, Suppl. al fasc. 11 - 12, Vol. 43 [Oss. Astron. Univ. Bologna], 28 + 40 pp. Price L. 3.000 (1975).

047.038 **Anuarul Observatorului din Bucureşti – 1976.**
Editura Academiei Republicii Socialiste România, Bucureşti. 267 pp. Price Lei 18.00 (1975).

047.039 **Apparent ephemeris of Mars 1960–1980.**
G. H. Kaplan, K. F. Pulkkinen, B. Emerson.
United States Naval Obs., *Washington,* Circ. No. 151, 128 pp. (1975).

047.040 **Astronomical Handbook for 1975.** B. M. Lewis.
Carter Obs., Wellington, New Zealand, Astron. Bull. No. 83, 40 pp. (1975).

047.041 **Almanaque Nautico 1976.**
Estados Unidos Mexicanos. Secretaria de Marina, Dirección General de Oceanografia y Señalamiento Maritimo, Mexico, D. F. 3 + 280 + 35 pp. Price $ 70.00 (1975).

047.042 **The tabulation of monthly sets of polynomial coefficients for astronomical data in "The Star Almanac for Land Surveyors".** A. T. Sinclair.
Survey Rev., Vol. 23, 151 - 154 = H. M. Nautical Almanac Office, *Roy. Greenwich Obs., Herstmonceux,* Library Repr. No. 319 (1975).

047.043 **The Nautical Almanac for the year 1977.**
Issued by Her Majesty's Nautical Almanac Office, London; and Nautical Almanac Office, United States, Naval Observatory, Washington. Printed and published by Her Majesty's Stationery Office London. A4 + 276 + 36 pp. Price £ 2.20 (1975).

047.044 **Philippine Astronomical Handbook 1976.**
Prepared under the supervision of R. L. Kintanar. Republic of the Philippines-Department of National Defense. Philippine Atmospheric, Geophysical and Astronomical Services Administration (Pagasa (Weather Bureau)), Quezon City. 10 + 59 pp. (1975).

047.045 **The American Ephemeris and Nautical Almanac for the year 1977.**
Issued by the Nautical Almanac Office, United States Naval Observatory, Washington; and Her Majesty's Nautical Almanac Office, Royal Greenwich Observatory, London. U.S. Government Printing Office, Washington. 8 + 572 pp. Price $ 10.20 (1975).

Annuaire 1976 du Bureau des Longitudes. Encyclopédie Physique et Spatiale. See Abstr. 003.006.

Space Research

051 Extraterrestrial Research, Spaceflight Related to Astronomy and Astrophysics

051.001 **Gas chromatographic column for the Viking 1975 molecular analysis experiment.**
M. Novotny, J. M. Hayes, F. Bruner, P. G. Simmonds.
Science, Vol. 189, 215 - 216 (1975).
A gas chromatographic column has been developed for use in the remote analysis of the Martian surface. The column, which utilizes a liquid-modified organic adsorbent (Tenax) as the stationary phase, provides efficient transmission and resolution of nanogram quantities of organic materials in the presence of millionfold excesses of water and carbon dioxide.

051.002 **A voyage to a neutron star.** P. Bielkowicz.
Journ. British Interplanet. Soc., Vol. 28, 523 - 545 (1975).
The possibility of existence of dark bodies in the neighbourhood of the solar system is examined and it is assumed that a neutron star is located at a distance of 1000 AU from the sun. Possible orbits to such a star are investigated, and relativity effects are briefly outlined.

051.003 **Heating and drag at relativistic speeds.**
C. Powell.
Journ. British Interplanet. Soc., Vol. 28, 546 - 552 (1975).
Equations are derived for the drag on a starship and for the forebody surface temperature, as functions of interstellar-medium density, flight speed, forebody half-angle, and forebody surface emissivity. Some numerical results are given, and some implications of interstellar heating for starship design are discussed.

051.004 **Planetary detection from an interstellar probe.**
G. R. Richards.
Journ. British Interplanet. Soc., Vol. 28, 579 - 585 (1975).
Expressions are derived for the distance at which an approaching interstellar probe can detect planets about the target star.

051.005 **Ultraviolet experiments on the Apollo Telescope Mount.** E. M. Reeves.
Space optics, (see 012.013), p. 235 - 251 (1974).

051.006 **Ultraviolet astronomy from Skylark rockets.**
W. M. Burton, N. K. Reay, R. Wilson.
Space optics, (see 012.013), p. 283 - 293 (1974).

051.007 **Investigation of propagation of radio waves in space.** M. A. Kolosov, V. I. Aksenov, N. A. Savich, O. I. Yakovlev.
Issled. v obl. radiotekhn. i ehlektron. 1954 - 1974. Ch. (Part) I. Moskva, 1974, p. 45 - 91. In Russian. – Abstr. in Referativ. Zhurn. 51. Astron., 7.51.64 (1975).

051.008 **Methods of reduction and main results of investigations with ion traps aboard the artificial earth satellite Intercosmos 2.** G. Gdalevich, B. Gorozhankin, Ts. Dachev, I. Kutiev, K. Segafimov.
Izv. Geofiz. in-t. Blg. AN, 1974, Vol. 20, 39 - 53. In Russian. Abstr. in Referativ. Zhurn. 62. Issled. kosmich. prostranstva, 7.62.156 (1975).

051.009 **First expedition of Salyut 4.** A. I. Gorokhov.
Zemlya i Vselennaya, 1975, No. 4, p. 24 - 28. In Russian.

051.010 **Radioactivity observed in the sodium iodide gamma-ray spectrometer returned on the Apollo 17 mission.**
C. S. Dyer, J. I. Trombka, R. L. Schmadebeck, E. Eller, M. J. Bielefeld, G. D. O'Kelley, J. S. Eldridge, K. J. Northcutt, A. E. Metzger, R. C. Reedy, E. Schonfeld, S. M. Seltzer, J. R. Arnold, L. E. Peterson.
Space Sci. Instrumentation, Vol. 1, 279 - 288 (1975).
In order to obtain information on radioactive background induced in the Apollo 15 and 16 gamma-ray spectrometers (7 cm X 7 cm NaI) by particle irradiation during spaceflight, an identical detector was flown and returned to earth on the Apollo 17 mission. The induced radioactivity was monitored both internally and externally from one and a half hours after splashdown. When used in conjunction with a computation scheme for estimating induced activation from calculated trapped proton and cosmic-ray fluences, these results show an important contribution resulting from both thermal and energetic neutrons produced in the heavy spacecraft by cosmic-ray interactions.

051.011 **Space investigations made in the USSR in 1974.**
Nauka, Moskva. 111 pp. (1975). In Russian. – Review in Referativ. Zhurn. 62. Issled. kosmich. prostranstva, 9.62.8 (1975).

051.012 **Our next 25 years in space.** J. K. Beatty.
Sky Telescope, Vol. 50, 288 - 290 (1975). – Interview of the author with NASA's director J. C. Fletcher.

051.013 **Astronautique 1974.** J. Meeus, G. Bodifee.
Ciel et Terre, Vol. 91, 348 - 374 (1975).

051.014 **The USA-Italy collaboration program on the S-054 Skylab experiment.** G. Tagliaferri.
Oss. Mem. Oss. Astrofis. Arcetri, Fasc. 104, (see 012.025), p. 48 - 49 (1975).

051.015 **Solar space research in the Federal Republic of Germany.** K. O. Kiepenheuer.
Oss. Mem. Oss. Astrofis. Arcetri, Fasc. 105, (see 012.026), p. 87 - 89 (1975).

051.016 **Future programs in solar physics from space.**
R. M. Bonnet.
Oss. Mem. Oss. Astrofis. Arcetri, Fasc. 105, (see 012.026), p. 93 - 97 (1975).

051.017 **Second expedition aboard Salyut 4.**
S. A. Nikitin.
Priroda, No. 10.75, p. 19 - 21 (1975). In Russian.

051.018 **The future of astronautics.** V. P. Denisov.

Zemlya i Vselennaya, 1975, No. 5, p. 4 - 10. In Russian.

051.019 **Space report.**
Spaceflight, Vol. 17, 252 - 258, 304 - 311, 368 - 371, 392 - 394, 424 - 426 (1975). – A regular monthly review of space events and technical trends.

051.020 **Spectroscopy of Project Fire I, April 14, 1964.** P. M. Millman.
I.A.U. Colloquium No. 13, (see 012.030), p. 161 - 167 (1973).

051.021 **Experiment "Artificial solar eclipse" during the joint flight of spaceships Soyuz and Apollo.**
G. M. Nikol'skij.
In-t zemn. magnetizma, ionosfery i rasprostr. radiovoln AN SSSR. Pr-232. 1975. 8 pp. In Russian. – Abstr. in Referativ. Zhurn. 62. Issled. kosmich. prostranstva, 1.62.115 (1976).

051.022 **Gravitācijas manevrs starpplanētu lidojumos.** Ė. Mūkins.
Zvaigžņotā debess, 1975. gada rudens, p. 5 - 10.

051.023 **Un programa para vuelos interestelares.** F. Spada S.
El Universo, Vol. 29, No. 112, p. 9 - 12 (1975).

051.024 **The electronics of the induction coil–magnetometers of the Helios solar probe.** G. Dehmel.
Internationale Elektron. Rundschau, (*Germany*), Vol. 29, No. 4, p. 69 - 72 (1975). In German.

051.025 **The international sun-earth Explorer (ISEE) programme.** A. C. Durney.
ESRO/ELDO Bull., (*France*), No. 26, p. 3 - 6 (1974).
Discusses the ISEE programme, a joint NASA/ESRO undertaking involving the launch of three spacecrafts, and the prime aim of the mission to resolve the spatial and temporal ambiguities inherent in magnetospheric measurements from a single platform. This short article outlines the aims of the overall programme, with particular reference to the role of the A and B spacecraft.

051.026 **Space science in retrospect.** H. E. Newell.
EOS Trans. American Geophys. Union, Vol. 56, 213 - 216 (1975).
Reviews the development of our knowledge of the earth's atmosphere, planetary atmospheres and X-ray astronomy due to the development of sounding rockets, artificial satellites and space probes.

051.027 **Space report.**
Journ. British Interplanet. Soc., Vol. 28, 501 - 513, 586 - 587, 681 - 687, 817 - 824 (1975).

Fourier spectroscopy in space.
See Abstr. 031.209.

Soyuz–Apollo space experiment.
See Abstr. 053.014.

Energetic solar proton vs. terrestrially trapped proton fluxes (risk factor). See Abstr. 078.032.

Jupiter's atmosphere: problems and potential of radio occultation. See Abstr. 099.046.

052 Astrodynamics and Navigation of Space Vehicles

052.001 **Lectures on physical and technical problems posed by precision radio tracking.** J. D. Anderson.
Experimental gravitation, (see 012.002), p. 163 - 199 (1974).
Introduction; Orbital equations; Station locations, timing and propagation effects; Estimation theory.

052.002 **Analyse des limitations physiques des accéléromètres pour sonde spatiale et des détecteurs de sonde à traînée compensée « drag-free probe ».** R. Juillerat.
Experimental gravitation, (see 012.002), p. 449 - 489 (1974).

052.003 **Two-impulse minimum-time earth launch-to-rendezvous trajectories.** G. M. Anderson, S. H. Edelman.
Journ. British Interplanet. Soc., Vol. 28, 477 - 484 (1975).
This paper presents an analysis of minimum-time two-impulse ascent trajectories for a single-stage rocket between launch from the earth and rendezvous with a satellite in a circular orbit around the earth.

052.004 **Linear perturbations of the coordinates of satellites in ellipsoidal orbits in the standard gravitational field of the earth.** B. I. Lokhin.
Byull. Inst. Teoret. Astron., *Leningrad*, Vol. 14, 23 - 31 (1975). In Russian.
A method of determination of the linear perturbations of the coordinates of satellites in ellipsoidal orbits in the gravitational field of two fixed centres was obtained in the form of a quadrature. A new form of the linear perturbations of the coordinates of the satellites in non-ellipsoidal orbits is also presented.

052.005 **On the influence of outer forces on the motion of a material point of variable mass.**
B. N. Kiforenko, V. V. Kuz'menko.
Trudy vos'mykh chtenij, posvyashch. razrabotke nauch. naslediya i razvitiyu idej K. Eh. Tsiolkovskogo, Kaluga, 1973 g. Sekts. "Mekh. kosmich. poleta". Moskva, 1974, p. 190 - 198. In Russian. – Abstr. in Referativ. Zhurn. 62. Issled. kosmich. prostranstva, 7.62.294 (1975).

052.006 **Improvement of orbits of artificial earth satellites without using coordinates of terrestrial points.**
A. E. Gusev.
Geod. i kartografiya, 1975, No. 3, p. 11 - 14. In Russian. Abstr. in Referativ. Zhurn. 62. Issled. kosmich. prostranstva, 7.62.297 (1975).

052.007 **Resonance phenomena in rotational motions of artificial and natural celestial bodies.**
V. V. Beletskij.
In-t priklad. mat. AN SSSR Preprint No. 10. Moskva, 1975. 48 pp. Price 15 Kop. In Russian. – Abstr. in Referativ. Zhurn. 62. Issled. kosmich. prostranstva, 7.62.301 (1975).

052.008 **Oscillatory motions of artificial earth satellites in a Keplerian orbit.** G. I. Kudin.
Modelir. i optimiz. sistem. upr. Kiev, Vishcha shkola, 1974, p. 83 - 90. In Russian. – Abstr. in Referativ. Zhurn. 62. Issled. kosmich. prostranstva, 7.62.304 (1975).

052.009 **Long-period perturbations in the motion of an artificial earth satellite caused by the combined influence of atmospheric drag and of the earth's oblateness.**
B. N. Noskov.
Astron. Zhurn. Akad. Nauk SSSR, Vol. 52, 844 - 856 (1975). In Russian. English translation in Soviet Astron., Vol. 19, No. 4.

052.010 **The application of Lindstedt series to problems of oscillations of a satellite around the mass center. I. Plane oscillations of a satellite in synchronous motion in an elliptical orbit.** Sh. T. Khabibullin.
Trudy Kazan. Gorod. Astron. Obs., No. 40, p. 3 - 46 (1974). In Russian.

052.011 **Use of the Laplacian method for determination of instantaneous satellite orbits from a single photograph with many exposures.** O. P. Bykov.
Trudy Astron. Obs., *Leningrad*, Vol. 31 (= Uchenye Zapiski Leningr. Un-ta, No. 381 = Seriya Matem. Nauk, vyp. (No.) 51), p. 217 - 227 (1975). In Russian.

052.012 **Optimization of the generalized linear correction of the trajectory of a space vehicle.**
V. A. Kuz'minykh.
Kosmich. Issled., Vol. 13, 443 - 447 (1975). In Russian.

052.013 **Methods for increasing the effectiveness in applying astronomical measuring equipment to space navigation.** A. G. Nikolaev, N. F. Romanteev, V. I. Sevast'yanov, A. V. Levitskij.
Kosmich. Issled., Vol. 13, 595 - 602 (1975). In Russian.

052.014 **Parameter optimization and optimal attitude response of a passive environmentally controlled space system.** R. Rangarajan, R. C. Flanagan.
Celestial Mechanics, Vol. 12, 231 - 249 (1975).
Numerically generated optimal response curves for a satellite in circular orbit show very rapid damping rates for large disturbances up to ten degrees. Under these conditions, and in the absence of external disturbances, course alignment was reached within several orbits and fine pointing accuracy attainable up to altitudes of 650 kilometers.

052.015 **A calculation method of the perturbation of artificial satellites.** L. Liu.
Acta Astron. Sinica, Vol. 16, 65 - 80 (1975). In Chinese.
This article gives a method for calculation of the perturbation of earth's gravitational field of artificial satellites, in which the singular points at $e = 0$ and critical inclinations ($i = 63° 26'$, $116° 34'$) are eliminated.

052.016 **Erfahrungen aus Bahnberechnungen künstlicher Erdsatelliten am Beispiel von ISAGEX-Beobachtungen.** H. Montag.
Observations of artificial earth satellites, No. 14, (see 012.018), p. 3 - 14 = Mitt. Zentralinst. Phys. d. Erde, Potsdam, No. 442 (1975).

052.017 **Construction of an analytical theory of satellite motion (fundamental problem) on a computer by means of the OPAL programming system.** A. L. Kutuzov.
Observations of artificial earth satellites, No. 14, (see 012.018), p. 25 - 40 (1975). In Russian.

052.018 **Long-periodic perturbations in the elements of an intermediate satellite orbit.** B. N. Noskov.
Observations of artificial earth satellites, No. 14, (see 012.018), p. 41 - 52 (1975). In Russian.

052.019 **Geodetical uses of the resonances of 15th order in the motion of artificial satellites.** J. Klokočník.
Observations of artificial earth satellites, No. 14, (see 012.018), p. 61 - 74 (1975).

052.020 **Improvement of a satellite's orbit from photographic observations of a planet's physical surface.**
V. Ivanova.
Observations of artificial earth satellites, No. 14, (see 012.018), p. 75 - 80 (1975). In Russian.

052.021 **The use of the Intercosmos satellite for astrodynamics.** J. Klokočník, P. Lála, P. Navara, M. Prokeš, L. Sehnal, M. Solarić.
Observations of artificial earth satellites, No. 14, (see 012.018), p. 539 - 560 (1975).

052.022 **Non-classical approach to the estimate of accuracy of orbit determination.** P. E. Ehl'yasberg.
Observations of artificial earth satellites, No. 14, (see 012.018), p. 561 - 570 (1975). In Russian.

052.023 **On the character of extremal motion of a point with variable mass between two configurations.**
I. F. Vereshchagin.
Probl. mekh. upravlyaemogo dvizheniya. Vyp. (No.) 6. Perm', 1974, p. 22 - 24. In Russian. – Abstr. in Referativ. Zhurn. 62. Issled. kosmich. prostranstva, 9.62.310 (1975).

052.024 **Functioning of a system of four satellites.**
G. V. Alferov.
Probl. mekh. upravlyaemogo dvizheniya. Vyp. (No.) 6. Perm', 1974, p. 11 - 15. In Russian. – Abstr. in Referativ. Zhurn. 62. Issled. kosmich. prostranstva, 9.62.312 (1975).

052.025 **Some questions of improvement of the orbits and landmarks in a central gravitational field.**
V. T. Tarushkin.
Probl. mekh. upravlyaemogo dvizheniya. Vyp. (No.) 6. Perm', 1974, p. 103 - 110. In Russian. – Abstr. in Referativ. Zhurn. 62. Issled. kosmich. prostranstva, 9.62.322 (1975).

052.026 **On a class of controlled motions of a solid body around a fixed point in a Newtonian field of forces.**
Yu. M. Loenko.
Probl. mekh. upravlyaemogo dvizheniya. Vyp. (No.) 6. Perm', 1974, p. 66 - 70. In Russian. – Abstr. in Referativ. Zhurn. 62. Issled. kosmich. prostranstva, 9.62.329 (1975).

052.027 **On some motions of a gyrostat of variable mass in the case of Euler-type.** N. N. Makeev.
Probl. mekh. upravlyaemogo dvizheniya. Vyp. (No.) 6. Perm', 1974, p. 71 - 78. In Russian. – Abstr. in Referativ. Zhurn. 62. Issled. kosmich. prostranstva, 9.62.330 (1975).

052.028 **A system for guiding a telescope to the center of a planet in the partial phase.**
G. G. Bil'chenko, S. N. Vasil'ev, V. M. Matrosov.
Upravlenie v prostranstve. Tom (Vol.) 1. Moskva, Nauka, 1975, p. 45 - 57. In Russian. – Abstr. in Referativ. Zhurn. 62. Issled. kosmich. prostranstva, 9.62.335 (1975).

052.029 **Orbital requirements for an Astronomical Roentgen Observatory Satellite.**
G. E. Cook, M. D. Palmer.
Journ. British Interplanet. Soc., Vol. 28, 763 - 778 (1975).

052.030 **Equations of the disturbed rotational motion of a solid body about the center of mass.**
Yu. V. Barkin.
Vestn. Mosk. un-ta. Fiz., astron., Vol. 16, No. 1, p. 46 - 53 (1975). In Russian. – Abstr. in Referativ. Zhurn. 62. Issled. kosmich. prostranstva, 8.62.294 (1975).

052.031 **On the stability of rotational motions of an elastic body.** N. G. Apykhtin.
Probl. mekh. upravlyaemogo dvizheniya. Vyp. (No.) 6. Perm', 1974, p. 16 - 18. In Russian. – Abstr. in Referativ. Zhurn. 62. Issled. kosmich. prostranstva, 8.62.296 (1975).

052.032 **The exterior tidal potential acting on a satellite.**
P. Musen.
Goddard Space Flight Center, Greenbelt, Maryland, GSFC Document X-920-75-220, Preprint, 6 + 26 pp. (1975).

By making use of Dirichlet's theorem the author continues the tidal potential from the surface of the earth into outer space. The final form of the disturbing tidal potential acting on the satellite is obtained in terms of the mean elements of the satellite and in terms of the standard arguments of the lunar theory. The tidal effects in the motion of the satellite are interpreted as perturbations in the orbital elements and as the small periodic variations in the coefficients of the spherical harmonics in the expansion of the geopotential.

052.033 **Analyses of the solid earth and ocean tidal perturbations on the orbits of the GEOS-I and GEOS-II satellites.** T. L. Felsentreger, J. G. Marsh, R. W. Agreen.
Goddard Space Flight Center, Greenbelt, Maryland, GSFC Document X-921-75-194, 7 + 26 pp. (1975).

The luni-solar tidal perturbations in the inclination of the GEOS-I and GEOS-II satellite orbits have been analyzed for the solid earth and ocean tide contributions. Precision reduced camera and TRANET Doppler observations spanning periods of over 600 days for each satellite were used to derive mean orbital elements. Perturbations due to the earth's gravity field, solar radiation pressure, and atmospheric drag were modelled, and the resulting inclination residuals were analyzed for tidal effects.

052.034 **Determination of the coordinates and of the orientation of an apparatus on a planet's surface.**
K. G. Sukhanov.
Kosmich. Issled., Vol. 13, 658 - 668 (1975). In Russian.

052.035 **On the motion of a point in an ellipsoidal gravitational field.** N. B. Batueva.
Vestn. Leningr. un-ta, 1975, No. 7, p. 139 - 144. In Russian. Abstr. in Referativ. Zhurn. 62. Issled. kosmich. prostranstva, 10.62.229 (1975).

052.036 **Theory of passive motion of a space apparatus near the collinear libration point L_2 of the system earth–moon.** M. G. Lidov, M. A. Vashkov'yak, A. P. Markeev.
In-t prikl. mat. AN SSSR. Preprint No. 56. Moskva, 1975. 72 pp. Price 20 Kop. In Russian. – Abstr. in Referativ. Zhurn. 62. Issled. kosmich. prostranstva, 11.62.314 (1975).

052.037 **Determination of the optimum program of trajectory measurements with correlation errors.**
V. B. Britkov.
In-t prikl. mat. AN SSSR. Preprint No. 67. Moskva, 1975. 32 pp. Price 10 Kop. In Russian. – Abstr. in Referativ. Zhurn. 62. Issled. kosmich. prostranstva, 11.62.316 (1975).

052.038 **On the optimum stabilization of circular orbits.**
V. A. Sinitsyn.
(Trudy) Mosk. aviats. in-ta, 1975, vyp. (No.) 321, p. 79 - 83. In Russian. – Abstr. in Referativ. Zhurn. 62. Issled. kosmich. prostranstva, 11.62.320 (1975).

052.039 **On the stability of periodic motions of an asymmetrical satellite in a circular orbit.**
A. P. Markeev, A. G. Sokol'skij.
In-t prikl. mat. AN SSSR. Preprint No. 58. Moskva, 1975, 64 pp. Price 18 Kop. In Russian. – Abstr. in Referativ. Zhurn. 62. Issled. kosmich. prostranstva, 11.62.323 (1975).

052.040 **Some qualitative regularities and estimates of the evolution of orbits of artificial earth satellites of Molniya-1 type.** M. L. Lidov, A. A. Solov'ev.
Kosmich. Issled., Vol. 13, 795 - 803 (1975). In Russian.

052.041 **On the use of simplified calculation schemes for estimating the parameters of motion of space objects.** A. M. Chernitsov.
Kosmich. Issled., Vol. 13, 932 - 935 (1975). In Russian.
Brief information

052.042 **Optimal impulsive transfers between real planetary orbits.** J. P. Gravier, C. Marchal, R. D. Culp.
Journ. Optimization Theory Appl., Vol. 15, 587 - 604 (1975).

052.043 **Dynamical model compensation for near-earth satellite orbit determination.**
K. A. Myers, B. D. Tapley.
AIAA Journ., Vol. 13, 343 - 349 (1975).

Operational requirements in modern space applications often demand orbit determination accuracies which are limited by fundamental mathematical and computational restrictions. It is shown in realistic computer simulation studies how these difficulties can be alleviated for typical near-earth satellites by employing dynamical model compensation and accurate observations in the extended Kalman filter.

052.044 **On asymptotic trajectories of a class of optimization problems.** A. I. Panasyuk, V. I. Panasyuk.
Avtomatika i telemekhanika, 1975, No. 8, p. 5 - 13. In Russian. Abstr. in Referativ. Zhurn. 62. Issled. kosmich. prostranstva, 12.62.486 (1975).

052.045 **Analytical expressions for earth tides perturbations on artificial satellites.** G. Balmino.
Bull. Groupe de Recherches de Géod. Spatiale, No. 14/II, 86 pp. (1974).

An analytical first order formulation of earth tides perturbations on close earth satellites has been derived, generalizing the expansions given by Kaula in 1969. In particular, tesseral harmonics of longitude (and latitude) dependent Love numbers have been considered. The solar and lunar tides perturbations have been treated separately. Some examples of long period perturbations for actual geodetic satellites are given.

052.046 **Poisson equations of rotational motion for a rigid triaxial body with application to a tumbling artificial satellite.** J. J. F. Liu, P. M. Fitzpatrick.
Celestial Mechanics, Vol. 12, 463 - 487 (1975).

Differential equations are derived for studying the effects of either conservative or nonconservative torques on the attitude motion of a tumbling triaxial rigid satellite. These equations, which are analogous to the Lagrange planetary equations for osculating elements, are then used to study the attitude motions of a rapidly spinning, triaxial, rigid satellite about its center of mass, which, in turn, is constrained to move in an elliptic orbit about an attracting point mass.

052.047 **Estimate of the accuracy of prediction of satellite motion.** Yu. Kh. Zhagar.
Vestn. Mosk. un-ta. Fiz., astron., Vol. 16, 266 - 273 (1975). In Russian. – Abstr. in Referativ. Zhurn. 51. Astron., 1.51. 158; 62. Issled. kosmich. prostranstva, 1.62.425 (1976).

052.048 **Methods for taking into account the influence of forces of light pressure in semi-analytical calculation of satellite motion.** M. L. Lidov, E. Ya. Ivanova.
In-t prikl. mat. AN SSSR. Preprint No. 42. Moskva, 1975. 70 pp. Price 22 Kop. In Russian. – Abstr. in Referativ. Zhurn. 62. Issled. kosmich. prostranstva, 1.62.415 (1976).

052.049 **On an approximate method for calculating the trajectory of an orbital transition of space vehicles.**
L. G. Gripp, V. A. Dolodarenko, V. F. Lager'.
Kosmich. issledovaniya na Ukraine, Resp. mezhved. sb., 1975, vyp. (No.) 7, p. 42 - 49. In Russian. – Abstr. in Referativ. Zhurn. 62. Issled. kosmich. prostranstva, 1.62.418 (1976).

052.050 **Analytical method for calculating the regions of dispersion of orbital elements.** E. I. Bushuev.
Kosmich. issledovaniya na Ukraine. Resp. mezhved. sb., 1975, vyp. (No.) 7, p. 60 - 63. In Russian. – Abstr. in Referativ. Zhurn. 62. Issled. kosmich. prostranstva, 1.62.419 (1976).

052.051 **Periodic oscillations of a satellite in the plane of an elliptical orbit.**
V. A. Sarychev, V. A. Zlatoustova.
In-t prikl. mat. AN SSSR. Preprint No. 48. Moskva, 1975. 76 pp. Price 25 Kop. In Russian. – Abstr. in Referativ. Zhurn. 62. Issled. kosmich. prostranstva, 1.62.441 (1976).

052.052 **On equations of motion of a satellite around the centre of mass.**
V. A. Sarychev, N. I. Yakovlev.
In-t prikl. mat. AN SSSR. Preprint No. 55. Moskva, 1975 g. 52 pp. Price 15 Kop. In Russian. – Abstr. in Referativ. Zhurn. 62. Issled. kosmich. prostranstva, 1.62.442 (1976).

052.053 **Method of determination of the circular orbit of an artificial earth satellite from optical observations with unknown moments of time.** R. A. Zejnalov.
Byull. Inst. Teoret. Astron., *Leningrad,* Vol. 14, 87 - 94 (1975). In Russian.

The problem of determination of the circular orbit of an artificial earth satellite from incomplete optical observations with unknown moments of time is considered again, and a new effective method is suggested for the case when there are four observations, i.e. four angular coordinates (α, δ) without moments of time.

052.054 **Perturbações de órbitas de satélites no estudo do campo gravitacional terrestre.** J. P. Osório.
Publ. Obs. Astron. «Prof. Manuel de Barros» Fac. Cí. Porto No. 29, 126 pp. (1973).

052.055 **Estudio de las perturbaciones debidas a los armónicos zonales. Determinación y eliminación de pequeños divisores.** J. F. Lahulla.
Bol. Astron. Obs. Madrid, Vol. 8, No. 5, p. 3 - 25 (1975).

The paper extends the known solution of the main problem of the artificial satellite theory to include the effects of all the higher zonal harmonics of the geopotential. The method of Hori-Lie is used to calculate the secular and the long-periodic variations of orders J_n and J_n/J_2 respectively in Delaunay's elements. The solution includes secular perturbations to the second order in J_2, short period and long-period perturbations to the first order; all coefficients J_p for $p > 2$ are assumed to be of order J_2^2.

052.056 **Correction for the influence of mascons in the semi-analytical method of calculation of lunar artificial satellite motion.** Yu. F. Gordeeva.
Opredelenie dvizheniya kosmich. apparatov. Moskva, Nauka, 1975, p. 43 - 73. In Russian. – Abstr. in Referativ. Zhurn. 51. Astron., 2.51.222; 62. Issled. kosmich. prostranstva, 2.62.336 (1976).

052.057 **A method to construct the fields of isolines for trajectories of gravitational planetary fly-by.**
K. G. Sukhanov.
Opredelenie dvizheniya kosmich. apparatov. Moskva, Nauka, 1975, p. 147 - 154. In Russian. – Abstr. in Referativ. Zhurn.

51. Astron., 2.51.224; 62. Issled. kosmich. prostranstva, 2.62.347 (1976).

052.058 **Trajectories of a solar satellite for radio sounding of the circumsolar space.** G. A. Mersov.
Opredelenie dvizheniya kosmich. apparatov. Moskva, Nauka, 1975, p. 115 - 122. In Russian. – Abstr. in Referativ. Zhurn. 51. Astron., 2.51.225; 62. Issled. kosmich. prostranstva, 2.62.349 (1976).

052.059 **Investigation of the formation of trajectories in the activity sphere of the moon.** L. I. Gusev.
Trudy devyatykh chtenij, posvyashch. razrabotke nauch. naslediya i razvitiyu idej K. Eh. Tsiolkovskogo, Kaluga, 1974 g. Sekts. 'Mekh. kosmich. poleta'. Moskva, 1975, p. 54 - 66. In Russian. – Abstr. in Referativ. Zhurn. 62. Issled. kosmich. prostranstva, 2.62.338 (1976).

052.060 **On deriving the first integrals of equations of motion of space vehicles with the group variation method.** Ya. A. Bolotnikova.
Trudy devyatykh chtenij, posvyashch. razrabotke nauch. naslediya i razvitiyu idej K. Eh. Tsiolkovskogo, Kaluga, 1974 g. Sekts. 'Mekh. kosmich. poleta'. Moskva, 1975, p. 169 - 177. In Russian. – Abstr. in Referativ. Zhurn. 62. Issled. kosmich. prostranstva, 2.62.342 (1976).

052.061 **On spheres of activity in the perturbed motion of a space vehicle.** V. S. Melkumov, A. M. Khabarov, G. G. Fedotov.
Trudy devyatykh chtenij, posvyashch. razrabotke nauch. naslediya i razvitiyu idej K. Eh. Tsiolkovskogo, Kaluga, 1974, g. Sekts. 'Mekh. kosmich. poleta'. Moskva, 1975, p. 47 - 53. In Russian. – Abstr. in Referativ. Zhurn. 62. Issled. kosmich. prostranstva, 2.62.343 (1976).

052.062 **On the calculation of the rated trajectory of a space vehicle.** A. M. Nikulin, R. D. Kulakova, A. T. Danilov.
Trudy devyatykh chtenij, posvyashch. razrabotke nauch. naslediya i razvitiyu idej K. Eh. Tsiolkovskogo, Kaluga, 1974 g. Sekts. 'Mekh. kosmich. poleta'. Moskva, 1975, p. 31 - 46. In Russian. – Abstr. in Referativ. Zhurn. 62. Issled. kosmich prostranstva, 2.62.344 (1976).

052.063 **Choice of the optimum orbit for flights on the course earth – planet – earth.** A. Yu. Kogan.
Opredelenie dvizheniya kosmich. apparatov. Moskva, Nauka, 1975, p. 163 - 168. In Russian. – Abstr. in. Referativ. Zhurn. 62. Issled. kosmich. prostranstva, 2.62.348 (1976).

052.064 **On the problem of motion of a space vehicle about its own center of mass with constant angular velocity.** Yu. M. Loenko.
Trudy devyatykh chtenij, posvyashch. razrabotke nauch. naslediya i razvitiyu idej K. Eh. Tsiolkovskogo, Kaluga, 1974 g. Sekts. 'Mekh. kosmich. poleta'. Moskva, 1975, p. 149 - 156. In Russian. – Abstr. in Referativ. Zhurn. 62. Issled. kosmich. prostranstva, 2.62.363 (1976).

052.065 **On the rotational motion of a space vehicle with generalized precession of the vector of momentum in free space (according to K. Eh. Tsiolkovskij's terminology).** Yu. A. Gartung.
Trudy devyatykh chtenij, posvyashch. razrabotke nauch. naslediya i razvitiyu idej K. Eh. Tsiolkovskogo, Kaluga, 1974 g. Sekts. 'Mekh. kosmich. poleta'. Moskva, 1975, p. 131 - 141. In Russian. – Abstr. in Referativ. Zhurn. 62. Issled. kosmich. prostranstva, 2.62.364 (1976).

052.066 **On spherical motions of space vehicles.** V. V. Lunev.
Trudy devyatykh chtenij, posvyashch. razrabotke nauch. naslediya i razvitiyu idej K. Eh. Tsiolkovskogo, Kaluga, 1974 g. Sekts. 'Mekh. kosmich. poleta'. Moskva, 1975, p. 189 - 196. In Russian. – Abstr. in Referativ. Zhurn. 62. Issled. kosmich. prostranstva, 2.62.365 (1976).

052.067 **On the question of calculation of the program of motion of a space vehicle around the center of mass and its realization on a computer.** A. F. Pavlov, L. I. Klenina.
Trudy devyatykh chtenij, posvyashch. razrabotke nauch. naslediya i razvitiyu idej K. Eh. Tsiolkovskogo, Kaluga, 1974 g. Sekts. 'Mekh. kosmich. poleta'. Moskva, 1975, p. 142 - 148. In Russian.– Abstr. in Referativ. Zhurn. 62. Issled. kosmich. prostranstva, 2.62.366 (1976).

052.068 **Plane motion of a satellite in a toroidal cavity filled with a viscous incompressible liquid.**
L. I. Kuznetsov, V. S. Nerovnaya.
Prikl. mekhanika. Vyp. (No.) 2. Leningrad, Leningr. un-t, 1975, p. 55 - 61. In Russian. – Abstr. in Referativ. Zhurn. 62. Issled. kosmich. prostranstva, 2.62.367 (1976).

052.069 **On oscillations of a magnetized satellite.**
L. N. Romasheva, L. I. Kuznetsov.
Prikl. mekhanika. Vyp. (No.) 2. Leningrad, Leningr. un-t, 1975, p. 43 - 47. In Russian. – Abstr. in Referativ. Zhurn. 62. Issled. kosmich. prostranstva, 2.62.368 (1976).

052.070 **The motion of an artificial satellite in a nonspherical gravitational field and an atmosphere with a quadratic scale height.**
R. E. Willey, V. L. Pisacane.
Journ. Astronaut. Sci., Vol. 21, 230 - 243 (1974).
Presents the development of a solution for satellite orbits in a nonrotating atmosphere and takes into account the second to fourth zonal harmonics of the gravitational potential. It is an extension of the power-law representation of the atmospheric density function in which the linear density scale height is replaced by a quadratic. The secular and mixed secular terms arising from the coupled drag and geopotential effects are complete through the second order in J_2. The periodic terms are complete through the first order in J_2 except for the long period terms in the mean anomaly. For these terms, the result is complete through order B^*/J_2, where B^* is the drag coefficient and of order J_2.

Motion of a satellite about the center of mass in a gravitational field. Textbook for students of mech.-mat. and phys.-mat. faculties of universities. See Abstr. 003.023.

Determination of the motion of space probes. See Abstr. 003.141.

Accurate integration of geostationary orbits with Burdet's focal elements. See Abstr. 042.022.

Sur la recherche des mouvements stationnaires dans les systèmes ayant des variables cycliques. See Abstr. 042.050.

Series expansion of the eccentricity for near parabolic orbits. See Abstr. 042.079.

Détermination du potentiel à partir des trajectoires dans le cas plan. See Abstr. 042.104.

Determination of elements of recurrent and near-recurrent orbits and their application to operational satellites. See Abstr. 042.124.

Artificial lunar satellite. See Abstr. 053.013.

Estimation of unmodeled forces on a lunar satellite. See Abstr. 053.021.

East-west stationkeeping requirements of 24-h satellites due to earth's triaxiality and luni-solar effects. See Abstr. 054.011.

Effect of resonance-oblateness coupling on a satellite orbit. See Abstr. 081.019.

The rotational speed of the upper atmosphere determined from orbital inclinations of Interkosmos satellites. See Abstr. 082.022.

Atmospheric density at heights between 200 and 300 km from drag on Interkosmos satellites. See Abstr. 082.023.

A new atmospheric model and motion of artificial satellites. See Abstr. 082.061.

053 Lunar and Planetary Probes and Satellites

053.001 **Viking to Mars: the orbiter.**
Sky Telescope, Vol. 50, 90 - 92 (1975).

053.002 **Viking to Mars: the lander.**
Sky Telescope, Vol. 50, 150 - 154 (1975).

053.003 **Programm Viking – die Landung auf dem Mars.** H. W. Köhler.
Umschau, 75. Jahrgang, p. 472 - 475 (1975).

053.004 **Nach Apollo/Sojus: Die Zukunft im Weltraum.** J. von Puttkamer.
Umschau, 75. Jahrgang, p. 493 - 500 (1975).

053.005 **Mariner-Venus-Mercury 1973 project history.** D. Baker.
Spaceflight, Vol. 17, 298 - 301 (1975).

053.006 **The galactic orbit of Pioneer 10.** D. W. Keenan.
Bull. American Astron. Soc., Vol. 7, 466 (1975). Abstr. AAS.

053.007 **Das amerikanische Projekt «Viking».** H. Müller.
Orion, 33. Jahrgang, p. 104 - 111 (1975).

053.008 **Planetary spacecraft for the 1980's.** R. C. Parkinson.
Spaceflight, Vol. 17, 346 - 351 (1975).

053.009 **Again to Venus.** S. A. Nikitin.
Priroda, No. 9.75, p. 108 - 109 (1975). In Russian.

053.010 **Spacecraft potential control and plasma diagnostic using electron field emission probes.**
R. J. L. Grard.
Space Sci. Instrumentation, Vol. 1, 363 - 376 (1975).

A new probe is proposed to prevent spacecraft with conductive surfaces which are immersed in energetic plasma environments, from reaching highly negative potentials in case of insufficient photoelectron emission. This probe works on the principle of electron field emission; it is cheap, small in size, light in weight and it uses no power. Information about density and temperature of the surrounding plasma can even be obtained at the cost of very little additional hardware. It is recommended to mount this probe on spacecraft which are venturing in planetary magnetospheres, as a guarantee against failures of subsystems and experimental equipment caused by very negative surface potentials.

053.011 **Vikingvluchten naar de planeet Mars. Op zoek naar sporen van het leven.** C. Titulaer.
Zenit, Vol. 2, 305 - 308 (1975).

053.012 **The Russian Venus probes.** D. Baker.
Spaceflight, Vol. 17, 446 (1975).

053.013 **Artificial lunar satellite.** R. Forga.
Astron. Astrophys., Vol. 44, 25 - 29 (1975). In French.

In the second part of this paper (Forga, 1973), the system (I) supplies the equations of motion. The components $\partial\Omega_1/\partial s$ and $\partial\Omega_1/\partial u$, second member of these equations, hold z in their expression. z may be replaced by a second approximation value, obtained from the z equation with a second member: this value is substituted in the components $\partial\Omega_1/\partial s$ and $\partial\Omega_1/\partial u$.

053.014 **Soyuz–Apollo space experiment.** L. A. Lebedev.
Priroda, No. 10.75, p. 10 - 18 (1975). In Russian.

053.015 **The Mariner Jupiter/Saturn photopolarimeter experiment.** C. F. Lillie.
Rings of Saturn. NASA SP-343, (see 012.028), p. 131 - 139 (1974).

053.016 **The Mariner Jupiter/Saturn visual imaging system.** B. A. Smith.
Rings of Saturn. NASA SP-343, (see 012.028), p. 141 - 143 (1974).

053.017 **The Mariner Jupiter/Saturn ultraviolet experiment.** A. L. Broadfoot.
Rings of Saturn. NASA SP-343, (see 012.028), p. 145 - 146 (1974).

053.018 **The Mariner Jupiter/Saturn IRIS experiment.** T. E. Burke.
Rings of Saturn. NASA SP-343, (see 012.028), p. 147 - 153

(1974).

053.019 **The Mariner Jupiter/Saturn ring occultation experiment.** R. Eshleman.
Rings of Saturn. NASA SP-343, (see 012.028), p. 155 - 166 (1974).

053.020 **The Sisyphus particle detector.** R. K. Soberman.
Rings of Saturn. NASA SP-343, (see 012.028), p. 167 - 173 (1974).

053.021 **Estimation of unmodeled forces on a lunar satellite.** B. D. Tapley, B. E. Schutz.
Celestial Mechanics, Vol. 12, 409 - 424 (1975).

In previous investigations, a procedure for sequentially estimating the state of a lunar orbiting space vehicle acted upon by unmodeled terms in the lunar potential has been developed. Results obtained by processing tracking data from the Apollo 10 and 11 missions indicate that the algorithm provides more precise estimates of the vehicle state than conventional orbit determination procedures and, hence, provides an accurate input for navigation purposes. This investigation considers the question of the accuracy with which the algorithm can estimate the acceleration due to unmodeled lunar surface mascons. It is shown that an accurate estimate of the time history of the unmodeled acceleration can be obtained. The investigation also considers the effects of the magnitude and location of the mascons, as well as the effect of the observation accuracy.

053.022 **Design for a Venus orbital imaging radar mission.** J. R. Rose, L. D. Friedman.
Journ. Spacecraft Rockets, (*USA*), Vol. 12, 106 - 112 (1975).

A planetary exploration mission to map the surface of Venus is discussed. A review of the Venus exploration program provides a basis for determining the probable scientific requirements for resolution, planetary coverage, etc., for such a mission. From these requirements, the range of possible mission, radar, and spacecraft functional requirements is determined and a single 'point' design is investigated in some detail.

Mariner 10: a Boeing's-eye view.
Applied Optics, Vol. 14, A200 - A201 (1975).

Gas chromatographic column for the Viking 1975 molecular analysis experiment. See Abstr. 051.001.

Lectures on physical and technical problems posed by precision radio tracking. See Abstr. 052.001.

Satellite digest. See Abstr. 054.010.

The Mariner 10 Venus encounter – a review.
See Abstr. 093.001.

On the selection of a satellite orbit to express the lunar potential by point masses. See Abstr. 094.184.

Viking to Mars: the mission strategy.
See Abstr. 097.002.

Zweite US-Sonde am Jupiter. Erste Ergebnisse des Vorbeifluges von Pionier 11 an Jupiter.
See Abstr. 099.061.

Saturn's ring particles and space vehicle design.
See Abstr. 100.019.

Impact of Saturn's rings on mission analysis for MJS 77. See Abstr. 100.020.

Saturn's rings investigation from space.
See Abstr. 100.029.

054 Artificial Earth Satellites

054.001 **An assessment of satellite observations from the Russian AFU-75 camera.** H. Hiller.
Journ. British Interplanet. Soc., Vol. 28, 485 - 488 (1975).

The utilisation of observations from the Russian AFU-75 cameras in orbit determination has been assessed by using a total of 137 observations from the cameras at Riga, Zvenigorod and Ulan Bator, with other observations, to determine four orbits of the balloon satellite Explorer 19. The results are enough to show that it will be possible to improve the orbital accuracy of satellites of geophysical interest whenever AFU-75 camera observations of them are available.

054.002 **The Moonwatch era ends.** J. Cornell.
Sky Telescope, Vol. 50, 160 - 163 (1975).

054.003 **Determination of rotational periods of an artificial earth satellite around its axis from electrophotometric observational data.** V. N. Lebedinets, L. A. Kapkov, Yu. G. Kaufman, V. P. Prorok.
Nauchn. Informatsii, vyp. (No.) 28, p. 31 - 36 (1974).
In Russian.

An algorithm for the calculation of the main period of satellite rotation by spectral analysis of the satellite light curve is suggested. The spectral density of Explorer 19 and Pageos light curves calculated on a computer is given. The dependence of spectral density from the phase function, air mass and distance from the observer to the satellite is analysed.

054.004 **Analysis of the orbit of 1970-97B (Cosmos 378 rocket).** H. Hiller.
Planet. Space Sci., Vol. 23, 1369 - 1376 (1975).

The main aim of the analysis was to determine the atmospheric rotation rate from the decrease in orbital inclination, which was determined with a mean standard deviation of 0·0010° and a best standard deviation of 0·0003°. After removal of relevant perturbations, analysis of the variation in inclination between July 1971 and April 1972 yields the surprisingly low average atmospheric rotation rate of 0·75 ± 0·05 rev/day, at a mean height of 250 km.

054.005 **Decay rates of three satellites in 1971 determined from observations of the program "Atmosphere".**
E. Illés-Almár.
Observations of artificial earth satellites, No. 14, (see 012.018), p. 427 - 437 (1975). In Russian.

054.006 **On the changes of the quasi-nodal period of the satellite Samos 2 during May 25 - June 28, 1970.**
T. Oproiu, M. Trifu.
Observations of artificial earth satellites, No. 14, (see 012.018), p. 457 - 472 (1975).

054.007 **Investigation of the period changes of 16 artificial satellites.** S. Nagy.
Observations of artificial earth satellites, No. 14, (see 012.018), p. 483 - 497 (1975). In Russian.

054.008 **Europese COS-B met succes gelanceerd.**
G. W. E. Beekman.
Zenit, Vol. 2, 380 - 382 (1975).

054.009 **The orbital lifetimes of Molniya satellites.**
D. G. King-Hele.
Journ. British Interplanet. Soc., Vol. 28, 783 - 796 (1975).

In this paper an approximate analytical method is developed for 65° Molniyas which shows that the lifetime must be between 1 and 7 years, and successfully defines the domain of permissible lifetimes as a function of the initial right ascension of the node, Ω_0. The lifetimes of all the satellites lie within or very near the boundaries of this domain.

054.010 **Satellite digest.** Compiled by R. D. Christy.
Spaceflight, Vol. 17, 276, 339 - 340, 366 - 367, 406 - 407, 442 - 443 (1975). – A monthly listing of all known artificial satellites and spacecraft.

054.011 **East-west stationkeeping requirements of 24-h satellites due to earth's triaxiality and luni-solar effects.** A. A. Kamel.
Celestial Mechanics, Vol. 12, 425 - 438 (1975).

Ignoring luni-solar perturbations, analytical solution for longitude deviation of the 'stroboscopic' mean center of 24-h satellites ground trace in the neighborhood of a prespecified station is obtained. The initial semimajor axis for two-maneuver east-west stationkeeping is then deduced. Finally, the luni-solar long and short period effects on this initial semimajor axis are discussed.

054.012 **Conversion from geocentric to geodetic coordinates.**
A. Deprit, A. Deprit-Bartholomé.
Celestial Mechanics, Vol. 12, 489 - 493 (1975).

The geodetic latitude and the height of a satellite are obtained as power series in the ellipsoid's eccentricity, the terms being Fourier sums in the geocentric latitude with polynomials in $1/r$ as coefficients, with a view of determining the height to the fullest accuracy required by altimetry and geodesy from satellites.

054.013 **Mean rates of the orbital elements of a satellite perturbed by a lens shaped mass concentration.**
M. Ananda.
Celestial Mechanics, Vol. 12, 495 - 511 (1975).

Long arc gravity analysis of lunar orbiter tracking data in the past has been carried out with the help of averaged equations of motion, in which short period effects have been suppressed. This procedure has required that the harmonic terms in the gravity potential be averaged over an orbital period. In the present paper, the author extends this technique to mass points and mass discs in the gravity field. This required the evaluation of expressions for the mean rates of the orbit elements for a satellite perturbed by a lens shaped mass concentration. Corresponding expressions for the perturbations due to a mass point are obtained in the limit as the lens radius goes to zero.

054.014 **Coordinates used in range or range-rate systems and their extension to a dynamic earth.** R. R. Newton.
IAU Colloquium No. 26, (see 012.034), p. 181 - 200 (1975).

054.015 **Hybrid systems for use in the dynamics of artificial satellites.** Y. Kozai.
IAU Colloquium No. 26, (see 012.034), p. 235 - 240 (1975).

054.016 **Practical realization of a reference system for earth dynamics by satellite methods.**
R. J. Anderle, M. C. Tanenbaum.
IAU Colloquium No. 26, (see 012.034), p. 341 - 380 (1975).

054.017 **On the computation of prediction of visibility of artificial earth satellites.**
A. S. Sochilina, L. I. Varenik.
Byull. Inst. Teoret. Astron., *Leningrad,* Vol. 14, 174 - 177 (1975). In Russian.

A method for computing the visibility predictions for

artificial earth satellites on long time intervals has been developed. The corresponding algorithm of the calculation program and a numerical example are given.

054.018 **Forecast and determination of orbits of artificial earth satellites taking into account the estimate of fluctuations of atmospheric drag.** V. D. Anisimov, L. G. Markova, A. I. Nazarenko, I. G. Pozdnyakov.
Opredelenie dvizheniya kosmich. apparatov. Moskva, Nauka, 1975. p. 3 - 15. In Russian. – Abstr. in Referativ. Zhurn. 62. Issled. kosmich. prostranstva, 2.62.355 (1976).

054.019 **Drag coefficients for astronomical observatory satellites.** R. P. Williams.
Journ. Spacecraft Rockets, (*USA*), Vol. 12, No. 2, p. 74 - 78 (1975).

The effects of satellite attitude history on drag coefficients are evaluated by averaging over the orbital motion. This permits translating the results of a free molecular drag analysis into mission-specific orbit decay rates. The type of mission considered is an astronomical observatory satellite in a circular orbit. Numerical results are given for the high-energy astronomy observatory-B X-ray telescope, and the relation of its drag coefficient to galactic belt mapping is discussed.

All systems "go" for ASTP.
Sky Telescope, Vol. 50, 13 - 14 (1975).

CETI from Copernicus. See Abstr. 015.005.

A possibility to improve the mathematical reduction of satellite tracking data obtained with AFU-75 cameras.
See Abstr. 031.213.

Une possibilité de la création d'un complexe astrométrique avec l'utilisation des programmes pour les calculs des coordonnées de satellites artificiels obtenus à l'aide de la camera AFU-75. See Abstr. 031.404.

L'automatisation du procédé du calcul des coordonnées de satellites artificiels. See Abstr. 031.406.

On results of an astrometrical reduction of the AFU-75 plates obtained in a compensative regime.
See Abstr. 041.007.

Use of the Laplacian method for determination of instantaneous satellite orbits from a single photograph with many exposures. See Abstr. 052.011.

Poisson equations of rotational motion for a rigid triaxial body with application to a tumbling artificial satellite.
See Abstr. 052.046.

Viking to Mars: the lander.
See Abstr. 053.002.

14 years of resonance on Vanguard orbits.
See Abstr. 081.025.

Upper-atmosphere rotation rate from analysis of the orbital inclination of Explorer 1. See Abstr. 082.081.

055 Observations of Earth Satellites, Lunar and Planetary Probes

055.001 **Poljot 1 – 1963-43-1. Oreol 1 – 1971-119-1. Visual observations. Horizontal coordinates (January–November 1973).**
Rezul'taty Nablyud. Iskusstv. Sputnikov Zemli, vyp. (No.) 29 (169), 67 pp. (1974). In Russian.

055.002 **Explorer 39–1968-66-1. Visual observations. Equatorial coordinates (1950.0) August - November 1973.**
Rezul'taty Nablyud. Iskusstv. Sputnikov Zemli, vyp. (No.) 30 (170), 57 pp. (1974). In Russian.

055.003 **Poljot 1–1963-43-1. Visual observations. Equatorial coordinates (1950.0) January - November 1973; January - July 1974.**
Rezul'taty Nablyud. Iskusstv. Sputnikov Zemli, vyp. (No.) 31 (171), 71 pp. (1974). In Russian.

055.004 **Explorer 19–1963-53-1. Visual observations. Horizontal coordinates (January - June 1974).**
Rezul'taty Nablyud. Iskusstv. Sputnikov Zemli, vyp. (No.) 32 (172), 62 pp. (1975). In Russian.

055.005 **Explorer 19–1963-53-1. Visual observations. Horizontal coordinates (June - September 1974).**
Rezul'taty Nalbyud. Iskusstv. Sputnikov Zemli, vyp. (No.) 33 (173), 66 pp. (1975). In Russian.

055.006 **Explorer 19–1963-53-1. Visual observations. Equatorial coordinates (1950.0) January - September 1974.**
Rezul'taty Nablyud. Iskusstv. Sputnikov Zemli, vyp. (No.) 34 (174), 64 pp. (1975). In Russian.

055.007 **On the apparent track of artificial satellites and the tracking frames of satellite cameras.** K. Huang.
Acta Astron. Sinica, Vol. 16, 37 - 50 (1975). In Chinese.

The author discusses the problem of using the arcs of great and small circles of the celestial sphere to approximate the artificial satellite's apparent tracks, and compares the structural features of the triaxial and quadraxial tracking frames of current satellite cameras.

055.008 **Computation of ephemerides of artificial earth satellites for satellite trackers with four-axes mounting.** M. Abele, Ya. Vyater.
Observations of artificial earth satellites, No. 14, (see 012.018), p. 585 - 598 (1975). In Russian.

055.009 **Analysis of the possibility of eliminating observational errors.** H. Hurnik, E. Vnuk.
Observations of artificial earth satellites, No. 14, (see 012.018), p. 599 - 604 (1975). In Russian.

055.010 **Automation of measurements of satellite positions.** Z. Boenigk.
Observations of artificial earth satellites, No. 14, (see 012.018), p. 605 - 608 (1975). In Russian.

055.011 **Use of quasi-synchronous observations of artificial earth satellites.**
G. Vylev, G. Rainov, I. Stanoev.
Observations of artificial earth satellites, No. 14, (see 012.018), p. 609 - 618 (1975). In Russian.

055.012 **Some results of photometric satellite observations from the Mongolian Astronomical Observatory.** S. Sanjav.
Observations of artificial earth satellites, No. 14, (see 012.018), p. 619 - 621 (1975). In Russian.

055.013 **Some results from simultaneous photographic and laser observations.** M. V. Bratijchuk, A. G. Kirichenko, K. A. Kudak, T. J. Laslo, I. I. Motrunich, I. V. Shvalagin.
Problems of cosmic physics. Vyp. (No.) 10, (see 003.003), p. 34 - 39 (1975). In Russian.

The results of simultaneous photographic and laser observations of the satellite 'Geos-A' obtained in Uzhgorod on October 3, 1972, are discussed.

055.014 **Methoden zur Ausrichtung von siderisch nachgeführten Satellitenmeßkammern und zur Berücksichtigung der dynamischen Refraktionskorrektionen bei der Reduktion von Satellitenaufnahmen.** F.-M. Scherer.
Mitt. Geod. Inst., Univ. Bonn, No. 3, 64 pp (1975).

(1.) The hour-axis of a sidereally driven parallactically mounted satellite-camera must be directed parallel to the earth rotation axis with high accuracy. Different methods are discussed which allow to achieve a sufficient parallelity and to improve the direction of the hour-axis. (2.) This topic has led to the problem of the so called dynamic refraction, which until now has partly been ignored, though it may reach the dimension of some seconds of arc. For the reduction of photographic plates, taken with the second generation of parallactically mounted satellite cameras, this effect on the position of the satellite track-points relative to the stars can now be taken in account.

055.015 **Zur Reduktion von Laserentfernungsmessungen nach künstlichen Erdsatelliten.** G. Marek.
Vermessungstechnik, 23. Jahrgang, p. 57 - 60 (1975).

055.016 **Laser technology for high precision satellite tracking.** H. H. Plotkin.
Proc. Symposium on the earth's gravitational field and secular variations in position, (see 012.036), p. 328 - 346 (1974).

055.017 **Poljot 1–1963-43-1; Explorer 19–1963-53-1. Visual observations. Equatorial coordinates (1950.0) August - November 1974.**
Rezul'taty Nablyud. Iskusstv. Sputnikov Zemli, vyp. (No.) 35 (175), 68 pp. (1975). In Russian.

055.018 **Oreol 1–1971-119-1; Intercosmos 10–1973-82-1. Visual observations. Equatorial coordinates (1950.0) January - November 1974.**
Rezul'taty Nablyud. Iskusstv. Sputnikov Zemli, vyp. (No.) 36 (176), 65 pp. (1975). In Russian.

055.019 **Poljot 1–1963-43-1; Intercosmos 10–1973-82-1. Visual observations. Horizontal coordinates (January - November 1974).**
Rezul'taty Nablyud. Iskusstv. Sputnikov Zemli, vyp. (No.) 37 (177), 72 pp. (1975). In Russian.

055.020 **Explorer 39–1968-66-1. Visual observations. Horizontal coordinates (January - June 1974).**
Rezul'taty Nablyud. Iskusstv. Sputnikov Zemli, vyp. (No.) 38 (178), 60 pp. (1975). In Russian.

055.021 **Explorer 39–1968-66-1. Visual observations. Horizontal coordinates (June - September 1974).**
Rezul'taty Nablyud. Iskusstv. Sputnikov Zemli, vyp. (No.) 39

(179), 62 pp. (1975). In Russian.

055.022 **Explorer 19–1963-53-1; Explorer 39–1968-66-1; Oreol –1971-119-1. Visual observations. Horizontal coordinates (October - November 1974; September - November 1974; January - November 1974).**
Rezul'taty Nablyud. Iskusstv. Sputnikov Zemli, vyp. (No.) 40 (180), 66 pp. (1975). In Russian.

055.023 **Explorer 39–1968-66-1. Visual observations. Equatorial coordinates (1950.0) January - April 1974.**
Rezul'taty Nablyud. Iskusstv. Sputnikov Zemli, vyp. (No.) 41 (181), 59 pp. (1975). In Russian.

055.024 **Explorer 39–1968-66-1. Visual observations. Equatorial coordinates (1950.0) May - November 1974.**
Rezul'taty Nablyud. Iskusstv. Sputnikov Zemli, vyp. (No.) 42 (182), 36 pp. (1975). In Russian.

055.025 **Geocentric coordinates of artificial earth satellites calculated from visual observations. Vyp. (No.) 3. Satellite 1963-43-1, September -October 1969. Satellite 1965-53-6, April -October 1969.** V. V. Kondrashin, I. E. Kupriyanova, I. A. Kuznetsov, V. A. Smirnov.
Nablyud. Iskusstv. Nebesn. Tel, No. 66, 110 pp. (1975). In Russian.

055.026 **On ephemeris computation of artificial earth satellites for the AFU tracking cameras.**
A. S. Sochilina.
Byull. Inst. Teoret. Astron., *Leningrad,* Vol. 14, 107 - 112 (1975). In Russian.

The accuracy of the approximation of the apparent orbit of a satellite by a small circle on the celestial sphere has been investigated. The parameters of the small circle are shown to depend essentially on the diurnal rotation of the earth. The possibility of improvement of the AFU camera ephemerides has been investigated. A computation algorithm for special ephemerides for the AFU cameras is proposed.

The accuracy of programming algorithms in the reduction of satellite photographs.
See Abstr. 021.013.

A complex computer program for automatical reduction of satellite observations by the AFU-75 camera.
See Abstr. 031.408.

Electrophotometer for observations of artificial earth satellites. See Abstr. 034.018.

The optical block of the multichannel electrophotometer for satellite observations. See Abstr. 034.019.

On the accuracy of a passive or active satellite position obtained by the AFU-75 camera.
See Abstr. 041.008.

Sur la précision des positions photographiques des satellites. See Abstr. 041.009.

Results of latitude observations with the Floating Zenith Telescope at Mizusawa: comparisons between the results obtained from the astronomical and Doppler satellite observations. See Abstr. 045.024.

Theoretical Astrophysics

061 General Theoretical Problems of Astrophysics, Gravitational Instability, Neutrino Astronomy, Infrared, X-Ray, Gamma-Ray Astronomy, Abundances and Origin of Elements

061.001 **Diffuse X-ray spectrum from 10 to 40 keV measured with rocket-borne scintillation counters.** M. Ikeda.
Astrophys. Space Sci., Vol. 35, 255 - 268 (1975).

The energy spectrum of the diffuse component of cosmic X-rays was measured with rocket-borne scintillation counters. Subtracting the environmental background unambiguously by means of the shutter method, the absolute values of the cosmic X-ray flux are obtained in a few keV band from 10 to 40 keV. The result indicates that the energy spectrum sharply changes its slope around 20–30 keV. The acceptable range of the spectral index in the high energy side is 2.3–2.6, that of the break energy is 20–30 keV and the corresponding transition width is smaller than 50 keV, if the confidence level is to be better than 5%. The authors' result may suggest that current theories on the origin of the diffuse X rays have to be revised.

061.002 **White holes and high energy astrophysics.** J. V. Narlikar, K. M. V. Apparao.
Astrophys. Space Sci., Vol. 35, 321 - 336 (1975).

The role of exploding objects in high-energy astrophysics is discussed. Quantitative results are obtained for the simple case of the canonical white hole which explodes from a singularity and consists of pressure-free matter in the comoving frame of reference. General relativity is used for calculating the dynamical results. Applications to X-ray background, transient X-ray sources, γ-ray bursts and high energy cosmic rays are considered. White holes of more general types are discussed in a qualitative manner.

061.003 **An exact analytic solution for axially-symmetric explosion in a spheroid.** J. B. Bhowmick.
Astrophys. Space Sci., Vol. 35, L35 - L39 (1975).

The Brinkley-Kirkwood theory as modified by Sachdev and Kogure and Osaki, is applied to a point explosion in a spheroid with exponentially decreasing density distribution. Similar results to those of Sakashita are obtained but without the assumption of the mass concentration at the shock surface.

061.004 **Stellar rates for some reactions of interest in astrophysics.**
M. Rios, J. S. Schweitzer, B. D. Anderson.
Astrophys. Journ., Vol. 199, 173 - 178 (1975).

Experimental cross sections for the ^{46}Ti(p, γ)^{47}V, ^{50}Cr(p, γ)^{51}Mn, and ^{58}Ni(α, p)^{61}Cu reactions have been numerically integrated over a Maxwell-Boltzmann velocity distribution to determine thermonuclear reaction rates in the temperature range $2 \leq T_9 \leq 10$ and for the ^{58}Ni(α, γ)^{62}Zn reaction in the range $3 \lesssim T_9 \lesssim 5$. Appropriate analytic expressions for the experimental rates and for stellar rates in which the target nuclei are in a thermal distribution of excited states are presented. Experimental rates are compared with theoretical calculations.

061.005 **Neutrinos in general relativity: four(?) levels of approach.** B. Kuchowicz.
Gen. Rel. Grav., Vol. 5, 201 - 234 (1974).

Studies on the neutrino physics in gravitational fields reveal not only a series of results which are of interest in se, and which may be used as the basis to a unified theory of neutrino and gravitational fields (the Rainich problem for the neutrino). They provide in addition the necessary material for astrophysical and cosmological applications; some of these are outlined in relation to the results presented.

061.006 **Extreme-ultraviolet astronomy from Apollo-Soyuz.** B. Margon, S. Bowyer.
Sky Telescope, Vol. 50, 4 - 9 (1975).

061.007 **^{22}Na, Ne–E, extinct radioactive anomalies and unsupported ^{40}Ar.** D. D. Clayton.
Nature, Vol. 257, 36 - 37 (1975). – Letter.

061.008 **On the information entropy complexity of the chemical composition of some cosmic and terrestrial systems.** T. G. Petrov.
Vestn. Leningr. un-ta, 1974, No. 24, p. 39 - 43. In Russian.
Abstr. in Referativ. Zhurn. 51. Astron., 7.51.246 (1975).

061.009 **On particle acceleration and radiation output in stochastic electromagnetic fields.** I. Lerche.
Astrophys. Space Sci., Vol. 36, 205 - 217 (1975).

The author demonstrates that when charged particles interact with a plane electromagnetic wave which possesses a random amplitude, then the particles are accelerated to high energy because they are pushed along by the wave's Poynting vector. Not only are they so accelerated, as they are carried along by the wave, but also they diffuse at right angles to the direction of the Poynting flux (i.e. in the direction of the wave's electric field). The ultimate energy that such particles can reach is determined when they radiate as much energy per unit time as they receive from the plane wave. For numbers believed typical of the Crab nebula this ultimate energy is of order $10^{10} mc^2$. The author has done these calculations to show that turbulent electromagnetic waves are quite efficient in generating high energy particles from low energy particles.

061.010 **Plasma frequency radiation in astrophysics: what we learn from solar radio observations.**
H. M. Bradford.
Journ. Roy. Astron. Soc. Canada, Vol. 69, 162 - 174 (1975).

Plasma frequency radiation (p.f.r.) accounts for most of the sporadic radio emission from the sun below about one gigahertz, yet radiation from non-solar radio sources is rarely attributed to this mechanism. In this article solar radio observations are reviewed with an emphasis on the questions which a complete p.f.r. theory should answer, and the problems of identifying and interpreting possible p.f.r. radio sources. Among the points discussed are the discrepancy between the observed frequency and the plasma frequency, the relation between the observed spectrum and the intrinsic spectrum, the efficiency of p.f.r., the effects of scattering and absorption, and some remaining gaps between the theory and interpretation of type III bursts.

061.011 **High-energy gamma-ray astronomy.** C. Fichtel, K. Greisen, D. Kniffen.
Phys. Today, Vol. 28, No. 9, p. 42 - 47, 50 (1975).

061.012 **The explosive thermonuclear formation of ^{7}Li and ^{11}B.** M. Arnould, H. Norgaard.
Astron. Astrophys., Vol. 42, 55 - 70 (1975).
In the framework of Arnett's (1969) model, the post-shock explosive nucleosynthesis of hydrogen-rich layers is investigated with particular emphasis on the thermonuclear production of ^{7}Li and ^{11}B. Rather large ranges of peak temperatures and densities are shown to be consistent with the production of ^{7}Li or ^{11}B by this mechanism. It is also demonstrated that there exists a range of temperature-density conditions ($T_0 \cong 4$–6×10^8 °K; $\rho_0 \cong 1$–10^3 g cm^{-3}) for which the enhanced production of both ^{7}Li and ^{11}B – sometimes together with ^{13}C, ^{15}N and ^{17}O – occurs in relative concentrations consistent with solar system abundances (Cameron, 1968, 1973).

061.013 **Pulse astronomy: short time scale phenomena in electromagnetic and gravitational wave astronomy.**
R. B. Partridge.
Astrophys. Space Sci. Library, Vol. 48, (see 012.014), 29 - 45 (1975).

061.014 **Cosmic gamma-ray bursts.** I. B. Strong.
Astrophys. Space Sci. Library, Vol. 48, (see 012.014), 47 - 58 (1975).

061.015 **A photonuclear mechanism for promethium synthesis excluding technetium.**
T. G. Harrison.
Astrophys. Letters, Vol. 16, 87 - 95 (1975).
It is shown where ^{13}C(p, γ)^{14}N photons are able to trigger the photoneutron reaction ^{148}Nd(γ, n)^{147}Nd(β^-)^{147}Pm leading to a synthesis of ^{147}Pm, but are not able to trigger the equivalent reaction leading to the synthesis of ^{99}Tc via ^{100}Mo(γ, n)^{99}Mo(β^-)^{99}Tc. Thus, depending upon the magnitude of a presently unknown photoneutron cross section, this scheme is presented as a means whereby ^{147}Pm is routinely manufactured during CNO burning, but where also, due to differences in the neutron binding energy, the analogous synthesis of all technetium isotopes is discriminated against. Transit time between shell and surface for irradiated material is considered and found to be of little consequence in causing a weakening of atmospheric promethium.

061.016 **Low-energy nucleosynthesis of lithium, beryllium, and boron.** R. Canal, J. Isern, B. Sanahuja.
Astrophys. Journ., Vol. 200, 646 - 652 (1975).
The results from detailed calculations of lithium, beryllium, and boron production by spallation reactions at low energies are reported. Proton and α-induced reactions, destruction, and deceleration effects are included. The abundances and isotopic ratios, for particle spectra analogous to those observed in solar flares (power law in kinetic energy), agree with the solar system values. The authors therefore suggest stellar flares and/or low-energy cosmic rays as the main contributors to the nucleosynthesis of these elements.

061.017 **Kα X-rays from cosmic-ray oxygen.** S. H. Pravdo, E. A. Boldt.
Astrophys. Journ., Vol. 200, 727 - 733 (1975).
Equilibrium charge fractions are calculated for subrelativistic cosmic-ray oxygen ions in the interstellar medium. These are used to determine the expected flux of Kα rays arising from atomic processes for a number of different postulated interstellar oxygen spectra. Relating these results to the diffuse X-ray background measured at the appropriate energy (i.e., ~ 0.6 keV) suggests an observable broadened line feature.

061.018 **Vibration regime for expansion of a gas cloud into a vacuum.** O. I. Bogoyavlenskij.
Pis'ma v Astron. Zhurn., Vol. 1, No. 9, p. 22 - 25 (1975). In Russian.
A vibration regime for expansion of a uniformly rotating gas cloud into vacuum is found, during which the density of gas has any finite number of maxima. A natural approximation of the vibration regime for motion of a cloud with negative total energy, applicable under certain values of parameters, is pointed out.

061.019 **Elementhäufigkeiten im Sonnensystem und Elemententstehung in der Galaxis. Anfängliche Häufigkeiten im Sonnensystem und galaktische Nukleosynthese.** H. Holweger.
SuW, Vol. 14, 233 - 237 (1975).

061.020 **On the theory of a stationary dynamo. I.** Eh. P. Kropachev, S. N. Gorshkov.
Geomagn. Aeronom., Vol. 15, 709 - 714 (1975). In Russian.

061.021 **Observations on the spatial and spectral distribution of diffuse X-ray background from UHURU.**
P. C. Agrawal, F. Cordova, G. Garmire.
Bull. American Astron. Soc., Vol. 7, 415 (1975). – Abstr. AAS.

061.022 **Observation of the galactic X-ray background.** W. A. Wheaton.
Bull. American Astron. Soc., Vol. 7, 415 - 416 (1975). – Abstr. AAS.

061.023 **Search for extreme ultraviolet radiation from selected stars.** G. R. Riegler, G. P. Garmire.
Bull. American Astron. Soc., Vol. 7, 416 (1975). – Abstr. AAS.

061.024 **The low temperature photonuclear metasynthesis of ^{50}V, ^{138}La and ^{180}Ta.** T. G. Harrison.
Bull. American Astron. Soc., Vol. 7, 441 - 442 (1975). – Abstr. AAS.

061.025 **Electron binding effects on scattering contribution to monochromatic and mean opacities.**
R. T. Brown.
Bull. American Astron. Soc., Vol. 7, 445 (1975). – Abstr. AAS.

061.026 **The electromagnetic background: limitations on models of unseen matter.**
D. Eichler, A. Solinger.
Bull. American Astron. Soc., Vol. 7, 454 (1975). – Abstr. AAS.

061.027 **Line blanketed opacities for an expanding gas.** A. H. Karp, G. Lasher, K. L. Chan, E. E. Salpeter.
Bull. American Astron. Soc., Vol. 7, 456 (1975). – Abstr. AAS.

061.028 **A search for weak gamma-ray bursts using a large-area scintillation detector system.**
G. J. Fishman, J. H. Derrickson.
Bull. American Astron. Soc., Vol. 7, 461 - 462 (1975). – Abstr. AAS.

061.029 **Longitudinal photons in the fireball.** G. L. Murphy.
Nature, Vol. 257, 159 - 160 (1975).

061.030 **Production of ^{17}O and ^{18}O by means of the hot CNO tri-cycle.** J. J. Cowan, W. K. Rose.
Astrophys. Journ., (*Letters*), Vol. 201, L45 - L48 (1975).

The authors study high-temperature ($T \geq 10^8$ K) CNO cycle burning. They find an overabundance of both ^{17}O and ^{18}O after times of approximately 10^2 seconds. For temperatures of 150 and 200 million degrees the authors find that the ^{18}O abundance is much larger than the ^{17}O abundance after burning times of about 10^3 seconds. Known uncertainties in the nuclear reaction rates do not alter the conclusions.

061.031 **Isotopes of carbon, nitrogen and oxygen as probes of nucleosynthesis, stellar mass losses and galactic evolution.** J. Audouze, J. Lequeux, L. Vigroux.
Astron. Astrophys., Vol. 43, 71 - 83 (1975).

Recent observations of CO and other interstellar molecules have conclusively shown that differences in the isotopic ratios of C, N and O exist between the interstellar medium and the solar system. In particular, the interstellar medium is very probably enriched in ^{13}C by a factor ~ 2 with respect to ^{12}C, in comparison with the solar system. It is emphasized that ^{13}C, ^{14}N and possibly also ^{17}O are mainly formed in the external regions of stars losing mass in the red giant, Mira, planetary nebula or supernova phase, while ^{15}N is produced in nova outbursts. Simple evolutionary models allow the authors to reconcile the time variation of the $^{12}C/^{13}C$ ratio with the strong spatial nitrogen gradient and with the absence of a spatial gradient for $^{12}C/^{13}C$. Some implications regarding the mass loss by stars and the chemical evolution of the Galaxy are discussed.

061.032 **'Time-integrated' synchrotron radiation in high-energy astrophysics.** D. Falla, A. Evans.
Astrophys. Space Sci., Vol. 37, 11 - 56 (1975).

The authors survey the observational data on infrared, optical and X-radiation sources associated with energetic cosmic events, and note the occurrence of an apparently preferred value of the spectral index, $n = 1$, for the radiation continua. The authors obtain a new form of synchrotron radiation theory, which they describe as time-integrated synchrotron radiation theory, the particular advantage of which is that it is able to give a unique value ($n = 1/2$) of the spectral index for radiation produced by a single high energy electron, independently of the initial electron energy.

061.033 **Observation of the energy spectrum of cosmic diffuse X-rays in the energy range 20 keV–4 MeV.**
F. Makino.
Astrophys. Space Sci., Vol. 37, 115 - 130 (1975).

Balloon observations of the cosmic diffuse component of hard X-rays were conducted with two independent directional counters in two energy bands, from 20 keV to 120 keV and from 90 keV to 4 MeV. If the differential energy spectrum of the photon flux is represented by a power law $E^{-\alpha}$, the value of α is 2.3 up to 100 keV, gradually increases to 2.8 at about 500 keV, and decreases to 2.0 thereabove. The spectrum above 300 keV is in parallel to the Apollo-15 spectrum, whereas the absolute intensity is somewhat smaller. The shape of the spectrum suggests the necessity of a multi-component theory on the origin of cosmic diffuse X-rays.

061.034 **Asymptotic expansions for acoustical modes of the homogeneous compressible model.**
J. Denis, J. Denoyelle, P. Smeyers.
Astrophys. Space Sci., Vol. 37, 221 - 233 (1975).

The behaviour of p-modes of high degree and high order in the homogeneous compressible model is examined. The second-order differential equation of Pekeris is used to construct asymptotic expansions near the centre and near the surface, which are singular points, and near the turning point of that equation. An equation for the frequencies is obtained by requiring the continuity of the asymptotic solutions and of their first derivatives. Numerical applications are considered.

061.035 **On the stability limit of plane-parallel and infinite gravitating media.** D. K. Callebaut.
19th Colloquium International Astrophys. Liège 1974, (see 012.016), p. 467 - 476 (1975).

061.036 **The origin and abundances of the chemical elements.** V. Trimble.
Rev. Modern Phys., Vol. 47, 877 - 976 (1975).

A NATO Advanced Studies Institute on the 'Origins and abundances of the chemical elements' was held at the Institute of Astronomy, Cambridge, England from 23 July to 9 August 1974. The problems discussed there are the basis for this review. Three major areas can be identified. First, the normal distribution of chemical and isotopic abundances and deviations from it must be determined observationally. Second, the relevant nuclear processes and their sites and products must be identified. Third, a model of the evolution of galaxies which includes the relevant processes and produces the observed abundances must be formulated. There has been rapid progress recently in all three of these areas.

061.037 **Neutrino opacities at high temperatures and densities.** D. L. Tubbs, D. N. Schramm.
Astrophys. Journ., Vol. 201, 467 - 488 (1975).

A detailed calculation is made of the major cross sections contributing to neutrino opacities at high temperatures and densities such as those encountered in gravitational collapse. These calculations include the effects of neutral currents, where applicable, and electron degeneracy. The processes considered are electron-neutrino scattering (including both electron and muon neutrinos and antineutrinos), neutrino-nucleon absorption and scattering, and coherent neutrino scattering. Results for these interactions are also given for the average energy transferred by the neutrino as well as the mean scattering angle (thus yielding momentum transfer).

061.038 **Is deuterium of cosmological or of galactic origin?** J. P. Ostriker, B. M. Tinsley.
Astrophys. Journ., (*Letters*), Vol. 201, L51 - L54 (1975).

If deuterium is made primarily in stars, then according to current models of galactic evolution it should have a higher abundance in the metal-rich parts of the Galaxy. If it is produced only in the initial big bang, then astration causes a decrease in D abundance with increasing metal abundance z. Since the sign of the galactic metal abundance gradient is known ($dz/dr < 0$), the authors propose that a determination of the sign of the deuterium abundance gradient would indicate the probable origin of this isotope.

061.039 **Tests of randomness in astronomical objects. I. Along single great circles.** P. S. Wesson.
Australian Journ. Phys., Vol. 28, 453 - 461 (1975).

Results are presented for the first of two extensive computer tests for nonrandomness in the inter-object positions of astronomical objects. Nonrandomness was detected in the inter-object positions of a large class of bright stars (10σ deviation) and of bright galaxies (15σ deviation). However, peculiar objects from the Arp catalogue were found to be uncorrelated in position at the 1σ level (analogous to those just quoted).

061.040 **Tests of randomness in astronomical objects. II. Along two orthogonal great circles.** P. S. Wesson.
Australian Journ. Phys., Vol. 28, 463 - 473 (1975).

A refinement in computing technique and the use of the von Neumann ratio test are shown to improve the detectability of nonrandomness in astronomical object positions. Correlation is confirmed for those objects (bright galaxies and bright stars) shown to be nonrandomly distributed in Part I (Australian Journ. Phys., Vol. 28, 453 - 461 (1975) – see Abstract 061.039). The presence of correlation is also demonstrated, with $\sim 10:1$ likelihood ratio at the 98% confidence level, for the other groups of objects considered in Part I (QSOs, pulsars, planetary nebulae and Arp peculiar objects).

061.041 **Does astronomy need 'new physics'?** V. L. Ginzburg.
Quarterly Journ. Roy. Astron. Soc., Vol. 16, 265 - 281 (1975). George Darwin Lecture delivered on 1975 April 11.

061.042 **Astrophysical discussion of the Compton effect in an outside magnetic field.**
J. Herrmann, B. Kämpfer.
Astron. Nachr., Vol. 296, 267 - 275 (1975). In German.
The authors consider the Compton scattering of a distribution of photons and relativistic electrons in a constant magnetic field. The influence of the magnetic field upon the energy spectrum of the photons scattered by the electrons has been calculated.

061.043 **Equation of state at ultrahigh densities. Part 2.** V. Canuto.
Annual Rev. Astron. Astrophys., Vol. 13, (see 003.005), 335 - 380 (1975). – Review paper.

061.044 **Physics of high density and nuclear matter.** V. R. Pandharipande.
Proc. 16th Solvay Conference Phys., (see 012.020), p. 177 - 218, with a discussion, p. 219 - 220 (1974).

061.045 **Gamma ray bursts of cosmic origin.** R. Hofstadter.
Proc. 16th Solvay Conference Phys., (see 012.020), p. 491 - 493, with a discussion, p. 494 (1974).

061.046 **Extremal states of matter in astrophysics. Part I. Superfluidity and superconductivity: physical aspects.** M. S. Borczuch, B. Kuchowicz.
Postępy Astron., Vol. 23, 111 - 130 (1975). In Polish.
Such states of matter which are produced under very artificial conditions in the physical laboratory may occur often in the astrophysical universe. The main physical ideas associated with the superfluid and superconducting states of matter (which may occur in pulsars and white dwarfs) are outlined.

061.047 **The *r*-process production ratios of long-lived radionuclides.** O. Johns, H. Reeves.
Astrophys. Journ., Vol. 202, 214 - 221 (1975).
The *r*-process production ratios of ^{232}Th, ^{235}U, ^{238}U, and ^{244}Pu, and their uncertainties, are estimated without dependence on specific *r*-process models. The estimates are based on similarities between nuclei of the same nuclear type in the rare-earth and actinide regions. The estimated production ratios and uncertainties are compared with previously published ones.

061.048 **On the production of lithium, beryllium, and boron at low energies.**
D. Bodansky, W. W. Jacobs, D. L. Oberg.
Astrophys. Journ., Vol. 202, 222 - 231 (1975).
Relative yields of LiBeB isotopes are calculated for proton and α-particle induced reactions at energies in the neighborhood of 10 or tens of MeV, and comparisons are made with solar system abundances.

061.049 **The effect of delayed fission in nucleo-cosmochronology.** C.-O. Wene.
Astron. Astrophys., Vol. 44, 233 - 236 (1975).

061.050 **Spectroscopy of highly ionized atoms and some astrophysical questions.** S. L. Mandel'shtam.
Trudy Gos. optich. in-ta, Vol. 42, No. 175, p. 31 - 48 (1974). In Russian. – Abstr. in Referativ. Zhurn. 51. Astron., 8.51. 208 (1975).

061.051 **The role of nuclear reactions in the evolution of matter of the solar system.** A. K. Lavrukhina.
VI Leningr. mezhdunar. seminar, 1974. Uskorenie chastits i yader. reaktsii v kosmose. Leningrad, 1974, p. 251 - 252, 252 - 253, with a discussion, p. 253 - 254. In Russian and English. Abstr. in Referativ. Zhurn. 51. Astron., 8.51.297 (1975).

061.052 **Approximation functions to the Emden and associated Emden functions near the first zero, II.**
A. P. Linnell.
Astrophys. Space Sci., Vol. 37, 401 - 426 (1975).
Improved analytic expressions are presented to approximate the Emden function, θ, and the associated Emden functions ψ_0 and ψ_2 near the first zero of θ at ξ_1. The error in θ and ψ_0 at $\xi = 1.5\,\xi_1$ is not more than 0.000 001. The ψ_2 error at the same point is less than 0.000 05.

061.053 **Cosmic Čerenkov emission.** V. M. Charugin.
Astrophys. Space Sci., Vol. 37, 441 - 447, 449 - 454 (1975). In Russian and English.
In real intensive cosmic sources it is necessary to take into account the Čerenkov reabsorption which restricts the brightness temperature of Čerenkov emission by energies of emitted relativistic electrons. As Čerenkov optical depth is given by $\tau_c \sim N/E$, the brightness temperature of emission on frequencies near gyrofrequency is determined by electron energy E_r in which $\tau_c(E_r) = 1$. It is shown that most of the synchrotron sources (quasars, pulsars) must be powerful sources of Čerenkov emission. In these sources the high density of low frequency Čerenkov emission can give rise to many parametric effects.

061.054 **Physics and astronomy.** V. Vujnović.
Publ. Obs. Astron. Beograd, No. 20, (see 012.024), p. 19 - 26 (1975). In Serbo-Croatian.

061.055 **Theoretical investigations in astrophysics.** M. Vukićević-Karabin.
Publ. Obs. Astron. Beograd, No. 20, (see 012.024), p. 105 - 109 (1975). In Serbo-Croatian.

061.056 **Airborne infrared astronomy.** L. Caroff.
Mercury, (Journ. Astron. Soc. Pacific), Vol. 4, No. 4, p. 2 - 7 (1975).

061.057 **Spontaneous violation of symmetry and its cosmological consequences.** I. Yu. Kobzarev.
Priroda, No. 11.75, p. 79 - 85 (1975). In Russian.

061.058 **The rôle of electrical discharges in astrophysical phenomena.** C. E. R. Bruce.
Observatory, Vol. 95, 204 - 210 (1975).
Electrical discharges may occur on stellar and galactic scales. They almost certainly contribute to the phenomena observed in solar flares; and they offer a ready explanation of some aspects of novae, radio-galaxies, and the 'jet' of M87.

061.059 **The shear modulus of neutron star matter in the free-neutron state.** P. B. Jones.
Astrophys. Space Sci., Vol. 38, 3 - 13 (1975).
The lattice whose properties have been tabulated by Baym et al. (1971) is unstable against nuclear quadrupole deformation for nuclear charge $Z > Z_c$, $120 < Z_c < 201$, if the electron gas is assumed to be of uniform density. Electron screening corrections to the charge in energy density with nuclear deformation are large and it is shown that their inclusion removes the instability.

061.060 **Curvature radiation by bunches of particles.** A. Saggion.
Astron. Astrophys., Vol. 44, 285 - 293 (1975).

A bunch of relativistic particles moving on a curved trajectory is considered. The coherent emission of curvature radiation is described with particular regard to the role played by the "shape" of the bunch as a function of its dimensions. It is found that the length of the bunch along the line of sight strongly effects the spectrum of the radiation emitted, with no effect on its polarization. The combined effect of the three dimensions of the bunch is also described. Some implications of coherent emission are briefly discussed.

061.061 **$\alpha + \alpha$ reaction and the origin of ^{7}Li.**
C. H. King, H. H. Rossner, S. M. Austin, W. S. Chien, G. J. Mathews, V. E. Viola, Jr., R. G. Clark.
Phys. Rev. Letters, Vol. 35, 988 - 991 (1975).
Cross sections are presented for the production of ^{7}Li and ^{7}Be in the $\alpha + \alpha$ reaction between threshold and 140 MeV. Implications of these measurements for the problem of the origin of ^{7}Li in the universe are discussed.

061.062 **Numerical methods of calculation of stellar opacity.**
A. F. Nikiforov, V. B. Uvarov.
II Vses. konf. dinamika izluchayushchego gaza, 1975. Programma i tezisy dokl. Moskva, 1975, p. 34 - 35. In Russian. Abstr. in Referativ. Zhurn. 51. Astron., 11.51.219 (1975).

061.063 **On the possibility of nucleosynthesis of superhigh elements in nature.** Eh. E. Berlovich.
Probl. yader. fiz. i kosmich. luchej. Resp. mezhved. temat. nauch.-tekhn. sb., 1975, vyp. (No.) 2, p. 3 - 17. In Russian. Abstr. in Referativ. Zhurn. 51. Astron. 11.51.758 (1975).

061.064 **N$\bar{\text{N}}$-interaction and matter-antimatter separation in the universe.** L. N. Bogdanova.
Vzaimodejstvie chastits vysok. ehnergii s yadrami i novye yaderno-podob. sistemy. Vyp. (No.) 4. Moskva, Atomizdat, 1974, p. 57 - 60. In Russian. – Abstr. in Referativ. Zhurn. 51. Astron., 11.51.763 (1975).

061.065 **On the theory of a stationary dynamo. II.**
Eh. P. Kropachev, S. N. Gorshkov.
Geomagn. Aeronom., Vol. 15, 1054 - 1061 (1975). In Russian.

061.066 **Materie und Strahlung im interstellaren Raum.**
P. Mezger.
Phys. Blätter, 31. Jahrgang, p. 548 - 557 (1975).

061.067 **Size distribution of cosmic gamma-ray bursts.**
A. Bewick, M. J. Coe, J. S. Mills, J. J. Quenby.
Nature, Vol. 258, 686 - 687 (1975).
The authors report on the combined results of three balloon-borne searches for small cosmic γ-ray bursts, carried out by the Imperial College Space Physics Groups. The authors suggest that their results are incompatible with the extragalactic source postulated for the large events previously seen by satellites.

061.068 **On pair production in intense electromagnetic fields occurring in astrophysical situations.**
J. K. Daugherty, I. Lerche.
Astrophys. Space Sci., Vol. 38, 437 - 445 (1975).
The authors show that the pair production rate in a strong magnetic field is substantially altered when an electric field is also included. They illustrate and emphasize this significant alteration by considering a few special cases. In the vicinity of the polar cap of a rotating magnetized neutron star it is currently thought that both steady electric and magnetic fields must be present. The results presented here then indicate that some considerable degree of caution must be exercised in applying pair production rates calculated under the assumption of zero electric field to the problems of pulsar emission and the generation and maintenance of pulsar magnetospheres.

061.069 **Low-temperature Rosseland opacity tables.**
D. R. Alexander.
Astrophys. Journ., Suppl. Ser., No. 290, Vol. 29, 363 - 374 (1975).
Rosseland opacities are tabulated for five chemical compositions for the temperature range 10,000 K to 700 K. Bound-bound absorption due to molecules has been included whenever possible. For temperatures below 2000 K the absorption and scattering by silicate grains has also been included. Comparison with the Cox-Stewart opacities shows reasonable close agreement down to about 5000 K, where absorption by molecules begins to be important. Water vapor, titanium oxide, and silicate grains are the most significant low-temperature contributors to the Rosseland opacity.

061.070 **Electron-electron bremsstrahlung in a hot plasma.**
E. Haug.
Zeitschr. Naturforschung, Vol. 30a, 1546 - 1552 (1975).
Using the differential cross section for electron-electron (e-e) bremsstrahlung exact to lowest-order perturbation theory the bremsstrahlung spectra and the total rate of energy loss of a hot electron gas are calculated. The results are compared to the nonrelativistic and extreme-relativistic approximations available and to previous work. The e-e contribution to the bremsstrahlung emission of a hydrogenic plasma is found to be not negligible at temperatures above $\approx 10^8$ K, especially in the short-wavelength part of the spectrum.

061.071 **Calculation of energies of autoionization states of helium-like ions with large nuclear charge.**
L. N. Ivanov, U. I. Safronova, I. A. Shavtvalishvili.
AN SSSR. Otd. obshch. fiz. i astron. In-t spektroskopii. Preprint No. 9. Akademgorodok, Mosk. obl., Podol'sk. r-na, 1975 g. 56 pp. In Russian. – Abstr. in Referativ. Zhurn. 51. Astron., 12.51.161 (1975).

061.072 **The mean age of the elements revisited.**
K. L. Hainebach, D. N. Schramm.
Bull. American Astron. Soc., Vol. 7, 520 (1975). – Abstr. AAS.

061.073 **Generalization of the Jeans' instability condition.**
A. E. Glassgold.
Bull. American Astron. Soc., Vol. 7, 541 (1975). – Abstr. AAS.

061.074 **Rotational distortion of polytropes.** A. P. Linnell.
Bull. American Astron. Soc., Vol. 7, 542 (1975). Abstr. AAS.

061.075 **The nonlinear Rayleigh-Taylor instability: some numerical simulations of astrophysical interest.**
I. Easson.
Bull. American Astron. Soc., Vol. 7, 545 (1975). – Abstr. AAS.

061.076 **Neutrino astrophysics at high temperatures and densities.** D. L. Tubbs, D. N. Schramm.
Bull. American Astron. Soc., Vol. 7, 546 (1975). – Abstr. AAS.

061.077 **Dynamic r-process nucleosynthesis.**
E. B. Norman, D. N. Schramm, J. B. Blake.
Bull. American Astron. Soc., Vol. 7, 553 (1975). – Abstr. AAS.

061.078 **Determining the magnetic field under astrophysical conditions from polarization of atomic fluorescence.**
A. A. Ruzmajkin.
In-t prikl. mat. AN SSSR. Preprint No. 62. Moskva, 1975. 26 pp. Price 10 Kop. In Russian. – Abstr. in Referativ. Zhurn. 51. Astron., 1.51.268 (1976).

061.079 **Realistic quark models and astrophysics.**

Ya. B. Zel'dovich, L. B. Okun'.
In-t prikl. mat. AN SSSR. Preprint No. 39. Moskva, 1975. 16 pp. In Russian. – Abstr. in Referativ. Zhurn. 51. Astron., 1.51.891 (1976).

061.080 **The spectrum of relativistic particles accelerated in a Fermi reservoir.** B. J. Burn.
Astron. Astrophys., Vol. 45, 435 - 438 (1975).

The formation of power law spectra of relativistic particles by Fermi acceleration is reconsidered. It is shown that the observed range for the spectral indices of radio sources is readily explained if there is a continuous injection of both relativistic particles and turbulent energy into the accelerating region.

061.081 **Extremal states of matter in astrophysics. Part II. Superfluidity and superconductivity: astrophysical aspects.** M. S. Borczuch, B. Kuchowicz.
Postępy Astron., Vol. 23, 177 - 197 (1975). In Polish.

After a brief discussion of the temperature-density regions in which various kinds of equations of state are relevant, the question of the search for possible extremal states of matter under natural conditions is asked. The possibilities of an appearance of the superfluid and superconducting states of matter in celestial objects (dense plasma in white dwarfs and planets, neutron matter and superconducting protons in pulsar cores) are reviewed.

061.082 **A new theory on the generation of the magnetic field of celestial bodies. II. Spherical model.**
E. Schmutzer.
Experim. Techn. Phys., Vol. 23, 1 - 16 (1975). In German.

The general theory is applied to a spherical model, which is justified for small angular velocities of the rotating body. The complicated system of equations is evaluated by a Taylor expansion with respect to r. Besides a series of detailed results, a formula for the magnetic moment of celestial bodies, a prediction of a toroidal field connected with the meridional field, and a pressure function within the sphere are given. The numerical evaluation of the theoretical formulae for the fluid core of the earth leads to interesting predictions with respect to the conductivity, viscosity, meridional mass currents and differential rotation.

061.083 **Astrophysical applications of atomic physics.**
M. J. Seaton.
Atoms, molecules and lasers. Conference Trieste 1973, [International Atomic Energy Agency, Vienna], p. 261 - 282 (1974).

Discusses the information that can be obtained from the radiation emitted by stellar atmospheres, the solar corona and gaseous nebulae.

061.084 **Channel flow driven by a stationary thermal source.**
D. A. Salstein.
Tellus, Vol. 26, 638 - 651 (1974).

It is shown by analytic means that it is possible to produce mean horizontal flows in a channel from a periodic distribution of stationary heat sources which are placed upon a varying topographic structure. Past analyses and experiments of channel flows are reviewed. The conditions necessary for production of mean horizontal flows are derived using a number of boundary conditions. The heat balance and momentum balance of this fluid system are discussed.

061.085 **Escape of photons from charged bodies.**
K. D. Krori, J. Barua.
Proc. Indian Acad. Sci., Ser. A, Vol. 81, 9 - 14 (1975).

The authors show that the restriction of photons issuing from the surface of an uncharged sphere with Schwarzschild radius to the radial direction is removed for a charged sphere. Also they establish that in the case of a charged infinitely long cylinder the mass per unit length does not impose any restriction on the photons issuing from the surface in non-radial directions unlike in the case of uncharged infinitely long cylinders.

061.086 **High energy X-ray detection opens new window on the universe.** A. R. Engel.
Electron, (*GB*), No. 69, p. 33 - 34, 37 (1975).

Describes the X-ray detector electronics as well as the data storage and transmission facility on the Ariel-5 X-ray satellite.

061.087 **Equation of state at ultra-high densities.**
V. Canuto.
Nordisk Inst. Teoretisk Atomfysik, Copenhagen (Denmark), Lectures Pt. 1, 2, 148 + 100 pp. (1974).

061.088 **Gamma ray astronomy.** D. Kniffen.
IEEE Trans. Nuclear Sci., Vol. NS-22, 45 - 53 (1975).

A survey of the instruments developed for this research is given together with a brief summary of the current status of the observational results. These include the studies of galactic gamma-ray emission, the diffuse, presumably extragalactic gamma radiation and localized gamma-ray sources.

061.089 **Helical structure and galactic gamma-radiation.**
R. Schlickeiser, K. O. Thielheim.
Verhandl. Deutsche Phys. Ges., Reihe 6, Vol. 10, 160 (1975). In German. – Abstract.

061.090 **Cross section calculations of astrophysical interest. Final technical report.** E. Gerjuoy.
NASA Rep. NASA-CR-138600, 7 pp. (1974).

Cross sections are discussed for rotational excitation associated with theories of absorption and emission lines from molecules in space with emphasis on H_2CO, CO, and OH by collisions with neutral particles such H, H_2, and He. The sensitivity of the Thaddeus equation for the H_2CO calculation is examined.

061.091 **The diffuse cosmic gamma radiation.**
V. Schoenfelder.
Fortschr. Phys., Vol. 23, 1 - 59 (1975). In German.

061.092 **Evidence for the production of a new particle in neutrino interactions.**
M. R. Krishnaswamy, M. G. K. Menon, V. S. Narasimham, N. Ito, S. Kawakami, S. Miyake.
Phys. Letters B, (*Netherlands*), Vol. 57B, 105 - 108 (1975).

The authors present evidence, based on 5 events recorded in cosmic ray experiments deep underground, for the production of new, massive (>2 GeV) and long lived particles in neutrino interactions with rock nuclei.

061.093 **Abnormal state of pure neutron matter: effects of correlation.** C.-G. Kallman, S. A. Moszkowski.
Phys. Letters B, (*Netherlands*), Vol. 57B, 183 - 185 (1975).

It is shown that two different quasi-classical approaches to abnormal states of dense nucleons give binding in pure neutron matter. Effects of short range correlations are found to increase the critical density.

061.094 **Instabilities, bifurcations and catastrophes.**
J. M. T. Thompson.
Phys. Letters A, (*Netherlands*), Vol. 51A, 201 - 203 (1975).

Two independently developed bifurcational theories for the instability of a slowly evolving system, such as a stellar mass, are correlated. Applied to the fracture of a mechanically stressed perfect crystal, they predict a sharp cusp on the failure stress locus.

061.095 **Neutron matter in Brueckner theory with the Reid hard-core and Mongan potentials.**
P. Haensel.
Nuclear Phys. A, (*Netherlands*), Vol. A245, 528 - 540 (1975).
The energy per particle in pure neutron matter is calculated in lowest order Brueckner theory with the Reid hard core potential and the Mongan nonlocal separable potential. The integral equation method of Brueckner and Gammel is used for the calculation of the K-matrix for the Reid hard core potential.

061.096 **Magnetic susceptibility of neutron matter.**
P. Haensel.
Phys. Rev. C, (Nuclear Phys.), Vol. 11, 1822 - 1827 (1975).

Die Entstehung der Elemente – Einführung in die Nuklear-Astrophysik. See Abstr. 003.018.

Course of theoretical astrophysics. See Abstr. 003.120.

Relativistische Astrophysik. See Abstr. 011.012.

The quadratic Zeeman effect in the Ca II H and K lines. See Abstr. 022.033.

Glancing incidence optics in X-ray astronomy: a short review. See Abstr. 031.024.

The COS-B experiment for gamma-ray astronomy. See Abstr. 032.518.

The capability of COS-B to measure cosmic gamma ray bursts. See Abstr. 032.519.

The Apollo 15 and 16 X-ray fluorescence experiment. See Abstr. 032.520.

The hard X-ray experiment on the Astronomical Netherlands Satellite. See Abstr. 032.523.

Instrumental technique in X-ray astronomy. See Abstr. 032.524.

Neutron and proton interaction backgrounds in Compton-telescopes used for gamma-ray astronomy. See Abstr. 034.095.

The ANS hard X-ray experiment. See Abstr. 034.157.

Radioactivity observed in the sodium iodide gamma-ray spectrometer returned on the Apollo 17 mission. See Abstr. 051.010.

Wave-wave interactions and gravitational instability. See Abstr. 066.054.

Do comets play a role in galactic chemistry and γ-ray bursts? See Abstr. 102.001.

On the chemical composition of the sun, Jupiter, meteorites and Am stars. See Abstr. 114.073.

062 Hydrodynamics, Magnetohydrodynamics, Plasma

062.001 **Ground state of electron matter in high magnetic fields.** M. L. Glasser.
Astrophys. Journ., Vol. 199, 206 - 207 (1975).

It is shown that for a dense electron plasma, Coulomb interactions, which are negligible in weak magnetic fields, can in strong magnetic fields profoundly modify the properties of the system. They should therefore be considered in speculations regarding the surface properties of pulsars.

062.002 **Dynamics of solar magnetic fields. VI. Force-free magnetic fields and motions of magnetic foot-points.**
B. C. Low, Y. Nakagawa.
Astrophys. Journ., Vol. 199, 237 - 246 (1975).

A mathematical model is developed to consider the evolution of force-free magnetic fields in relation to the displacements of their foot-points. As illustration of the physical process, two specific examples of evolving force-free magnetic fields are examined in detail, one evolving with rising and the other with descending field lines. It is shown that these two contrasting behaviors of the field lines correspond to sheared motions of their foot-points of quite different characters. The physical implications of these two examples of evolving force-free magnetic fields are discussed.

062.003 **Thermodynamics of dense hydrogen-helium fluids.**
H. C. Graboske, Jr., R. J. Olness, A. S. Grossman.
Astrophys. Journ., Vol. 199, 255 - 264 (1975).

The equilibrium thermodynamic properties of hydrogen-helium fluids are studied in the density-temperature regime appropriate to the interiors of evolving very low mass stars and giant planets.

062.004 **A new primary dynamo mechanism.**
M. Reinhardt, A. Rosenblum.
Gen. Rel. Grav., Vol. 6, 71 - 73 (1975). – Presented at the Seventh International Conference on Gravitation and Relativity, June 1974, Tel-Aviv, Israel.

062.005 **Current flow in ion-acoustic and Langmuir turbulence plasma interaction.**
V. N. Tsytovich, L. Stenflo, H. Wilhelmsson.
Physica Scripta, Vol. 11, 251 - 257 (1975).

The purpose of this paper is to consider one of the possible mechanisms of ion-sound and Langmuir turbulence interaction and its consequences.

062.006 **Parameter constraints for plasma turbulent reactors.**
C. A. Norman, D. ter Haar, V. N. Tsytovich.
Physica Scripta, Vol. 11, 323 - 325 (1975).

062.007 **The effects of comptonization and fusion- and decay-processes on the properties of plasma turbulent reactors.** C. A. Norman.
Physica Scripta, Vol. 11, 326 - 329 (1975).

062.008 **The theory of the Stark profile's asymmetry of the hydrogen lines in a dense plasma.**
A. V. Demura, G. V. Sholin.
Journ. Quant. Spectrosc. Radiat. Transfer, Vol. 15, 881 - 899 (1975). In Russian.

062.009 **Thermal equilibrium criteria in a nitrogen plasma.**
W. A. Cilliers, J. D. Hey, J. P. S. Rash.
Journ. Quant. Spectrosc. Radiat. Transfer, Vol. 15, 963 - 978 (1975).

062.010 **Evolution of rotating interstellar clouds. I. Numerical techniques.** D. C. Black, P. Bodenheimer.
Astrophys. Journ., Vol. 199, 619 - 632 (1975).

A method is described for the numerical calculation of the hydrodynamic evolution of a self-gravitating configuration, in two space dimensions with assumed axial symmetry. The calculation is formulated in cylindrical coordinates with respect to a moving Eulerian grid, and is solved using explicit hydrodynamics combined with implicit radiative transfer. The physics included is appropriate for calculation of the collapse of a rotating protostellar cloud. The gravitational field is obtained by means of an alternating-direction iterative technique. Numerical tests to demonstrate the correctness of the method are presented for special cases.

062.011 **Polarization characteristics of synchrotron emission in a plasma.** J. T. Stocke, J. S. Scott.
Monthly Notices Roy. Astron. Soc., Vol. 172, 567 - 576 (1975).

The linear and circular polarization emission characteristics are calculated for a power-law distribution of electron energies in a weak plasma in an attempt to differentiate the Razin–Tsytovich effect from other low-frequency turnover mechanisms in radio sources.

062.012 **Spatial echo oscillations in a magnetoactive plasma.**
V. N. Pavlenko, S. M. Revenchuk.
Kosmich. issled. na Ukraine, Resp. mezhved. sb., 1974, vyp. (No.) 5, p. 28 - 38. In Russian. – Abstr. in Referativ. Zhurn. 51. Astron., 7.51.220 (1975).

062.013 **Numerical solutions of the hydromagnetic dynamo problem.** D. Gubbins.
Geophys. Journ. Roy. Astron. Soc., Vol. 42, 295 - 305 (1975). Paper presented at the Tenth International Symposium on Mathematical Geophysics, Cambridge, England, 1974 June 26 - July 4.

Fluid motions are driven in a sphere by an applied body force. The fluid is electrically conducting and incompressible, and is surrounded by an insulator. The vorticity equation, in which inertial terms have been omitted but which includes magnetic forces, is solved in conjunction with the pre-Maxwell equations which govern the magnetic field. The model allows the magnetic field to grow to a finite limit. The time evolution of the magnetic field is calculated and in most cases the magnetic field is found to approach a steady state in an oscillatory fashion. Possible modifications and improvements to the model, including the addition of rotation, are discussed, and future calculations are outlined.

062.014 **The long time confinement of fast energy particles in a magnetic trap.** A. M. Uralov.
Astron. Zhurn. Akad. Nauk SSSR, Vol. 52, 772 - 777 (1975). In Russian. English translation in Soviet Astron., Vol. 19, No. 4.

The influence of "cone" instability on the time of confinement of fast energy particles captured by a magnetic trap is considered. The cause of the "cone" instability is an anisotropy in the distribution of the thermal ions. In this case the energy of the plasma is transmitted by the whistlers on the transverse mode of the captured particles. This question is closely connected with the problem of confinement of cosmic rays on the sun and in the earth's magnetosphere.

062.015 **The hydrodynamics of accretion discs I: stability.**
J. M. Stewart.
Astron. Astrophys., Vol. 42, 95 - 101 (1975).

This is the first of two papers on the hydrodynamics of

accretion discs. Here the disc is idealised to be a stationary axisymmetric toroidal flow of a compressible fluid. The stability against linearised short wavelength perturbations is discussed. When the Rayleigh and Schwarzschild criteria are satisfied the flow is stable against axisymmetric perturbations. However almost all non-axisymmetric perturbations are not secularly stable, and examples of dynamically unstable modes are given.

062.016 **Stellar winds driven by Alfvén waves.** J. W. Belcher, S. Olbert.
Astrophys. Journ., Vol. 200, 369 - 382 (1975).
The authors consider the properties of stellar winds in which the dynamic expansion of a corona is driven by Alfvén waves propagating outward along radial magnetic field lines. The constants of motion of such winds can be simply expressed in terms of reference-level parameters near the star. The authors discuss in detail the dynamics of the acceleration mechanism which produces these winds.

062.017 **Viscous damping of nonlinear magneto-acoustic waves.** J. A. Adam.
Astrophys. Space Sci., Vol. 36, 479 - 487 (1975).
The nonlinear Burger's equation is derived for wave propagation in a compressible unstratified viscous fluid permeated by a magnetic field (initially constant), using the reductive perturbation method of Taniuti and Wei (1968). An analytic solution of the equation is given (after Sakai, 1972) and the angular behaviour is shown for certain parameters describing the nonlinearity and damping of the system, for both fast- and slow-mode disturbances.

062.018 **Contribution of strong interactions to the wings of hydrogen lines broadened by the charged particles in plasmas.**
M. Caby-Eyraud, G. Coulaud, Nguyen-Hoe.
Journ. Quant. Spectrosc. Radiat. Transfer, Vol. 15, 593 - 614 (1975).

062.019 **The effect of energy generation fluctuations on the onset of thermal instability in a Boussinesq fluid sphere.** A. C. Edwards.
Monthly Notices Roy. Astron. Soc., Vol. 173, 207 - 221 (1975).
The Boussinesq perturbation equations are modified to include the effect of energy generation fluctuations. The equations are then solved for the motion within a sphere of radius R subject to free and rigid boundary conditions. A transcendental equation is derived connecting the growth rate, Rayleigh number, Prandtl number and an energy generation number for each mode. Various cases are considered for the $l = 1$ mode with free boundary conditions and a good approximation is given for the Rayleigh number at high energy generation rates.

062.020 **Electron temperature determination from non-LTE populations in a nitrogen plasma jet.**
A. Catherinot, A. Sy.
Zeitschr. Naturforschung, Vol. 30a, 1143 - 1155 (1975).

062.021 **Crossfertilization between plasma, stellar dynamics and hydrodynamics.** M. R. Feix.
IAU Symposium No. 69, (see 012.009), p. 179 - 194 (1975).
The author presents results on four different mediums characterised by their 'density conservation' in a two dimensional space (phase space for unidimensional plasma and self gravitating systems, configuration space for two dimensional Navier Stokes fluid and guiding center rod plasma). He reviews the different problems (purely collective versus individual) and introduces the distinction between Lagrangian schemes (superparticles) and Eulerian schemes. The steady states structures for these fluids and their open (plasma, hydrodynamic) versus closed (stellar system, accelerator beam, hydrodynamic) topologies are discussed. He then turns to their stability properties and shows the condition for the presence of marginal adiabatic mode. The 'double water bag' system is fully studied and interesting analogies for completely trapped systems (accelerator beam and self gravitating systems) are pointed out.

062.022 **The influence of an inhomogeneous and frozen in magnetic field on the criteria of convective instability of a plasma.** S. I. Petrusevich.
Trudy Kazan. Gorod. Astron. Obs., No. 39, p. 3 - 15 (1973). In Russian.

062.023 **Inverse bremsstrahlung of photons in a strong magnetic field.** G. G. Pavlov, A. D. Kaminker.
Pis'ma v Astron. Zhurn., Vol. 1, No. 9, p. 13 - 18 (1975). In Russian.

062.024 **The pulsation equation for a fully relativistic, differentially rotating system.** M. R. Bernstein.
Bull. American Astron. Soc., Vol. 7, 500 (1975). – Abstr. AAS.

062.025 **The spatial and temperature structure of vacuum spark plasmas.**
U. Feldman, S. Goldsmith, J. L. Schwob, G. A. Doschek.
Astrophys. Journ., Vol. 201, 225 - 237 (1975).
The spatial and temperature structures of the X-ray emitting components of low-inductance vacuum spark plasmas are investigated. The plasmas produced by low-inductance vacuum sparks show striking resemblances to solar flare plasmas. High-resolution photographs show that the spark plasmas are composed of hot ($\sim 30 \times 10^6$ K) compact sources ($< 20\ \mu$) and cooler ($\sim 10 \times 10^6$ K) larger sources ($\sim 100\ \mu$). Both types of sources may be produced by the same plasma at different times. The differences between the solar flare iron-line spectrum near 1.9 Å and the spark iron-line spectrum are discussed and interpreted.

062.026 **On the notion of Thomson scattering in a plasma.** K. Furuya, T. Hamada, S. Kanno.
Publ. Astron. Soc. Japan, Vol. 27, 497 - 500 (1975).
It is argued that the notion of Thomson scattering in a plasma is superfluous. First, a satisfactory theory of radiation transport in a magnetized plasma has been developed based solely on the Thomson scattering in free space. Second, the propagation of the radiation in a plasma is a result of the multiple free-space Thomson scattering. For the sake of generality, the problem is studied under the presence of a magnetic field.

062.027 **On the theory of shock wave dissipation in inhomogeneous stationary moving media.** A. F. Novak.
Astrometriya i Astrofizika, *Kiev*, vyp. (No.) 26, (see 003.001), p. 38 - 46 (1975). In Russian.
The equations are obtained describing the shock wave motion through plane, cylindrical and spherical gravitating atmospheres, with regard to dissipation of energy, steady motions, arbitrary distribution of temperature and density. Particular cases of strong and weak shock wave propagation in isothermal or adiabatic atmospheres are discussed.

062.028 **Cross-modulation of electromagnetic waves in limited plasma volumes.**
A. P. Gorovenko, E. I. Fialko, I. V. Bajrachenko.
Astrometriya i Astrofizika, *Kiev*, vyp. (No.) 26, (see 003.001), p. 108 - 112 (1975).
A possibility of using the non-linear effect of cross-modulation to determine meteor trails is considered. The disturbing effect of a strong impulse on the plasma parameters is estimat-

ed. It is shown that the technique used in the experiments on ionospheric cross-modulation can be a basis for studying heating non-linear effects in meteor trails.

062.029 **The Stark broadening mechanism in an unstable plasma.** D. S. Spicer, J. Davis.
Solar Physics, Vol. 43, 107 - 110 (1975).
The authors consider a beam driven unstable plasma and estimate the turbulent electric fields which may be excited by this beam. The Stark broadening due to such fields is estimated.

062.030 **The two-dimensionality of slow motion in a rotating fluid.** D. R. Inglis.
Rev. Modern Phys., Vol. 47, 841 - 847 (1975).

062.031 **The magnetic thermal wind and differential rotation of ionized, magnetic atmospheres.**
S. D. Gedzelman.
Astrophys. Journ., Vol. 201, 509 - 520 (1975).
Slow motions of an ideal, ionized, rotating fluid in the presence of a nonconstant magnetic field are investigated. A magnetic analog to the thermal wind is found when the fluid is compressible. It is suggested that the magnetic thermal wind can have an important effect on the rotation rate of a stellar atmosphere and may be partially responsible for the apparent slow rotation rate of magnetic A type stars.

062.032 **Criteria for plasma-wave losses for electrons streaming from an acceleration region.**
D. F. Smith.
Astrophys. Journ., Vol. 201, 521 - 525 (1975).
Consistent criteria for determining when losses due to collective excitation of electron plasma waves will be important for electrons streaming from an acceleration region are derived. The analysis is nonrelativistic and one-dimensional, but includes the effect of Coulomb collisions. These criteria are applied to source models for solar hard X-ray bursts, and it is shown that these losses will almost never be important for the observations obtained thus far. The case in which the acceleration region and the X-ray source partially overlap is considered briefly.

062.033 **On the calculation of parameters of nonstationary relativistic shock waves.**
I. A. Klimishin, A. F. Novak.
Problems of cosmic physics. Vyp. (No.) 10, (see 003.003), p.154 - 160 (1975). In Russian.
An approximate method for the calculation of parameters of nonstationary relativistic shock waves moving in a nonhomogeneous medium is derived. It is shown that the maximum shock wave velocity and the path length are essentially dependent on the law of density distribution in a stellar corona.

062.034 **Adiabatic plasma convection in a dipole field: variation of plasma bulk parameters with L.**
S. W. H. Cowley, M. Ashour-Abdalla.
Planet. Space Sci., Vol. 23, 1527 - 1549 (1975).

062.035 **Charged particles spiralling along magnetic field lines in extended gravitational fields and the trapping of plasma by the top of the field lines.**
E. Woyk (Chvojková).
Geomagn. Aeronom., Vol. 15, 799 - 805 (1975).

062.036 **On a type of hydromagnetic waves of finite amplitude in magnetohydrodynamics with anisotropic pressure.** K. G. Ivanov.
Geomagn. Aeronom., Vol. 15, 923 - 925 (1975). In Russian. Brief information.

062.037 **Collective relaxation of ultra-relativistic plasmas.**
P. Stewart, G. S. S. Sweeney.
Astron. Astrophys., Vol. 44, 1 - 7 (1975).
The interaction between ions and strongly radiatively damped electrons is examined for a spatially uniform neutral plasma, and the results are discussed in relation to the Crab pulsar.

062.038 **Wave propagation in rotating plasmas and influence of the Coriolis force.**
E. Engels, F. Verheest.
Astrophys. Space Sci., Vol. 37, 427 - 440 (1975).
When assessing the influence of the Coriolis force on wave propagation in plasmas or other dielectric media, all the equations and relevant physical quantities should be expressed in a rotating reference frame. The resulting wave equation has coefficients depending on the position and this precludes a plane wave solution, even in the slow rotation approximation where the centrifugal force may be neglected in comparison with the Coriolis force. Perturbation analysis then gives a dispersion law as if the system were not rotating. The wave electric field, however, now has a position dependent amplitude, which is not only stretched but also changed in direction compared to the previously known unperturbed or not rotating solution.

062.039 **Equilibrium and stability of the force-free magnetic field. II. Stability.** M. M. Molodensky (*Molodenskij*).
Solar Physics, Vol. 43, 311 - 316 (1975).
The exact limit of stability of force-free fields for the case α=const and spherical region of the disturbances is obtained. The stable fields of active regions must be close to the harmonical ones for length scales which are less than the linear dimensions of spots.

062.040 **Non-uniform spatial conductivity and force-free magnetic fields. II.** S. I. Gopasyuk.
Izv. Krymskoj Astrofiz. Obs., Vol. 53, p. 104 - 107 (1975). In Russian.
The influence of non-uniform plasma conductivity on the structure and instability of a force-free magnetic field is examined. It is shown that for non-uniform plasma conductivity the contribution to the magnetic energy from the toroidal and the poloidal parts of a force-free magnetic field are not equal. Such a force-free magnetic field is not stable against perturbations.

062.041 **Magneto-active lines in a medium with velocity gradient.** V. M. Grigor'ev, I. M. Kats.
Izv. Krymskoj Astrofiz. Obs., Vol. 53, p. 245 - 247 (1975). In Russian. – Abstr. of a paper presented at the conference "Magnetic fields and chemical composition of the sun and stars" at the Crimean Astrophys. Obs., 1974, April 9 - 12, (see 012.027).

062.042 **Continuum radiation of a single ionized helium-plasma in the spectral range 300 nm to 650 nm.**
D. Einfeld, G. Sauerbrey.
Zeitschr. Naturforschung, Vol. 30a, 1413 - 1424 (1975). In German.

062.043 **On an equilibrium magnetohydrodynamical configuration with axial symmetry. I.** V. A. Osherovich.
Solnechnye Dannye 1975 Byull., No. 5, p. 70 - 76 (1975). In Russian.
An axial-symmetrical equilibrium magnetohydrodynamical configuration with a magnetic field intensity periodic along the axis of symmetry has been considered. The physical effects of linear energy finiteness are studied. Simple analytical solutions are obtained.

062.044 **On an equilibrium magnetohydrodynamic configuration with axial symmetry. II. Differential rotation of matter.** V. A. Osherovich.
Solnechnye Dannye 1975 Byull., No. 6, p. 55 - 60 (1975).
In Russian.

An influence of differential rotation of matter on the magnetic configuration is investigated. It is shown that faint differential rotation enlarges the zone of main polarity, whereas a strong one (higher than critical) destroys the configuration entirely.

062.045 **On the cosmic force-free magnetic fields.**
A. A. Korchak, M. M. Molodensky (*Molodenskij*).
Comments Astrophys. Space Phys., Vol. 6, 81 - 85 (1975).

062.046 **On the energy release processes at the neutral sheet.**
O. Kaburaki.
Sci. Rep. Tôhoku Univ., First Ser., Vol. 58, 141 - 149 (1975).

The energy release processes at the neutral sheet are considered, and the qualitative differences between steady and non-steady processes are emphasized, especially in relation to the energy source. The expressions for the energy conversion rate and the total energy to be released are given by taking into account the subsidiary conditions. Comparing the calculated values of these quantities with observational data, the author concludes that the neutral sheet theories (both steady and non-steady) can hardly give any satisfactory account for solar flare phenomena.

062.047 **Ondes de choc, sous des hypothèses générales, en magnétohydrodynamique relativiste.**
A. Lichnerowicz.
Comptes Rendus Acad. Sci. Paris, Sér. A, Vol. 281, 723 - 728 (1975).

On établit, en accord avec une suggestion de Thorne, que les résultats généraux (places des vitesses des fronts d'ondes de choc, inégalités thermodynamiques, propriétés du contour d'Hugoniot) établis sous l'hypothèse $\tau_s' > 0$ (τ volume dynamique, S entropie) sont aussi valables pour un milieu à $\tau_s' < 0$.

062.048 **A statistical theory of electromagnetic waves in turbulent plasmas.**
I. Ben-Israel, T. Piran, A. Eviatar, J. Weinstock.
Astrophys. Space Sci., Vol. 38, 125 - 155 (1975).

A statistical theory, previously developed for electrostatic waves is extended to the case of electromagnetic waves in the presence of a constant magnetic field. The theory is nonlinear and contains the effects of mode-mode coupling. The extension is not straightforward as assumed previously by many authors and involves calculation of perturbed trajectories in both velocity and configuration space. A diffusion equation is derived for the average particle distribution function, the associated diffusion tensor is calculated and a nonlinear wave dispersion relation is found. All these results contain the usual quasilinear theory as the lowest order approximation.

062.049 **The escape of magnetic flux from a turbulent body of gas.** E. N. Parker.
Astrophys. Journ., Vol. 202, 523 - 527 (1975).

Using an idealized two-dimensional model, it is shown through exact calculation of the intermediate field configurations how the magnetic field escapes from surface of a turbulent body of highly conducting fluid.

062.050 **Electric resistivity of electromagnetically turbulent plasma and reconnection rate of magnetic fields.**
S. Ichimaru.
Astrophys. Journ., Vol. 202, 528 - 531 (1975).

The dc electric resistivity of an electromagnetically or magnetohydrodynamically turbulent plasma is expressed in terms of the spectral tensor of the electric-field fluctuations. The result is applied to a calculation of the reconnection velocity u of magnetic fields in the neutral sheet; the calculation, combined with the equipartition of energies between the fluctuating and merging magnetic fields and the plasma, provides a reasonable, microscopic account of Parker's empirical relationship, $u \approx 0.1 \times$(Alfvén speed).

062.051 **On an equilibrium magnetohydrodynamical configuration with axial symmetry. III. Radial and longitudinal motions of matter.** V. A. Osherovich.
Solnechnye Dannye 1975 Byull., No. 7, p. 61 - 64 (1975).
In Russian.

The effect of the longitudinal and radial motions of matter on a magnetic field configuration has been studied. The growth of radial and longitudinal motions is found to lead to destruction of the magnetic field configuration provided fixed slow differential rotation of matter is taking place.

062.052 **The quasi-steady plasma-neutral gas balance in magnetic bottles.** B. Lehnert.
Physica Scripta, Vol. 12, 327 - 336 (1975).

062.053 **On finite amplitude convection in a rotating magnetic system.** P. H. Roberts, K. Stewartson.
Phil. Trans. Roy. Soc. London, Ser. A, Vol. 277, 287 - 315 (1974).

This is a study of instabilities that can arise in a stratified, rapidly rotating, magnetohydrodynamic system. It is motivated by the currently popular view that such instabilities pay a key role in the creation of magnetic fields by cosmic bodies of electrically conducting fluid, such as the earth's core. Such a view was early promulgated by Braginsky, who proposed in 1964 (and elaborated in 1967) a MAC-wave theory that studied motions in which magnetic, Archimedean (bouyancy) and Coriolis forces dominate viscous and inertial forces. A MAC-wave theory encounters difficulties adumbrated by Eltayeb in 1972 and studied in detail in this paper, through a simple model of a plane Bénard layer rotated about the vertical and permeated by a co-rotating uniform horizontal field. The relevance to sunspot formation is discussed.

062.054 **Symmetric periodic orbits in Størmer's problem.**
A. G. Mavraganis.
Astrophys. Space Sci., Vol. 38, 395 - 401 (1975).

The author analyses the motion of a charged particle in a magnetic dipole field, which provides a basis for discussion, first on the symmetry of the orbits and, second, on the reflection of the particle, as it moves on both sides of the equatorial plane of the dipole.

062.055 **Resonant interactions between fast particles and collision-dominated MHD waves.** D. B. Melrose.
Astrophys. Space Sci., Vol. 38, 483 - 493 (1975).

The theory of resonant interactions between particles and MHD waves (in collision-dominated plasmas) is shown to be equivalent to the corresponding theory for waves in collisionless plasmas provided that the wave properties are described in a suitable way. Slow-mode waves (as well as fast mode waves and Alfvén waves) might contribute to the scattering of cosmic rays in the interstellar medium.

062.056 **On the line of force curvature in a force-free magnetic field.** V. A. Osherovich.
Solnechnye Dannye 1975 Byull., No. 8, p. 69 - 71 (1975).
In Russian.

The relations between the line of force curvature in a force-free magnetic field and the magnetic energy distribution are considered. It has been shown that the lines of force straighten near the extremum in the density of the magnetic energy.

062.057 **A method to solve the hydromagnetic dynamo problem.** T. S. Ivanova.
In-t prikl. mat. AN SSSR. Preprint No. 52. Moskva, 1975 g. 25 pp. Price 10 Kop. In Russian. – Abstr. in Referativ. Zhurn. 51. Astron., 12.51.184 (1975).

062.058 **On explosive instability of electromagnetic waves in a non-equilibrium magnetoactive plasma.**
A. N. Karashtin, V. P. Reutov.
Izv. vyssh. ucheb. zavedenij. Radiofizika, Vol. 18, 930 - 934 (1975). In Russian. – Abstr. in Referativ. Zhurn. 51. Astron., 12.51.191 (1975).

062.059 **X-ray spectrum of a hot plasma.**
J. C. Raymond, B. W. Smith, D. P. Cox.
Bull. American Astron. Soc., Vol. 7, 538 (1975). – Abstr. AAS.

062.060 **Magnetic merging in a collisionless plasma.**
T. W. Hill.
Journ. Geophys. Res., Vol. 80, 4689 - 4699 (1975).

062.061 **On the kinetic theory of reflection of an electromagnetic wave from a moving magnetoactive plasma.**
Yu. V. Bogomolov.
Izv. vyssh. ucheb. zavedenij. Radiofizika, Vol. 18, 840 - 846 (1975). In Russian. – Abstr. in Referativ. Zhurn. 51. Astron., 1.51.272 (1976).

062.062 **On plasma perturbation by strong electromagnetic waves.** L. F. Chernogor.
Izv. vyssh. ucheb. zavedenij. Radiofizika, Vol. 18, 847 - 857 (1975). In Russian. – Abstr. in Referativ. Zhurn. 51. Astron., 1.51.273 (1976).

062.063 **On a method of constructing force-free magnetic fields.** V. A. Osherovich.
Astrofizika, Vol. 11, 559 - 561 (1975). – English translation in Astrophysics, Vol. 11, No. 3.

A method of theoretical construction of a force-free magnetic field is discussed. The necessary and sufficient conditions that magnetic fields with α = const. are force-free are formulated. The necessary condition is generalized for α being dependent on the coordinates.

062.064 **Problems on gravitational collapse of interstellar gas clouds. I. Numerical integration of the equations of hydrodynamics: the case of spherical symmetry.**
F. Ferraioli, N. Virgopia.
Mem. Soc. Astron. Italiana, Vol. 46, 313 - 339 (1975).

It is pointed out, through numerical integrations, that the introduction of cooling-heating processes in the dynamics of gravitational collapse of spherical interstellar gas clouds, leads to the formation of shell-shaped regions in the extreme zone of lower density. In the whole, gravitational collapse tends asymptotically to become homologous as the mass of subspheres reaches zero. The cooling effects are in fact such that the interior pressure gradient is strongly reduced against the gravity force and make the collapse to behave almost as in absence of pressure force. A critical time, slightly greater than the free-fall time, exists such that central density attains an infinite value.

062.065 **The propagation of electromagnetic plasma oscillations in the gravitational field.** Yu. G. Ignat'ev.
Acta Phys. Polonica B, Vol. B6, 203 - 221 (1975). In Russian.

062.066 **Particle acceleration by strong waves in ionized media.** A. Ferrari, E. Trussoni.
Phys. Letters A, (*Netherlands*), Vol. 51A, 304 - 306 (1975).

062.067 **Plasma turbulent reactors: an astrophysical paradigm.** C. A. Norman, D. ter Haar.
Phys. Rep., (*Netherlands*), Vol. 17, 307 - 358 (1975).

062.068 **Some remarks on the initiation of inertial Taylor columns.** H. E. Huppert.
Journ. Fluid Mech., Vol. 67, 397 - 412 (1975).

Étude de l'écoulement quasi-géostrophique sur un obstacle placé sur le plan inférieur d'un système de deux plans horizontaux en rotation rapide autour d'un axe vertical dans le cas d'un écoulement général uniforme et à la limite d'une viscosité tendant vers zéro. Les effets obtenus en fluide homogène sont comparés aux effets obtenus en fluide stratifié.

062.069 **The macrodynamics of α-effect dynamos in rotating fluids.** W. V. R. Malkus, M. R. E. Proctor.
Journ. Fluid Mech., Vol. 67, 417 - 443 (1975).

L'étude des conséquences à grande échelle des mouvements forcés à petite échelle dans les fluides conducteurs de l'électricité a conduit aux effets dynamos α. Étude du comportement à amplitude finie des champs magnétiques globaux et des écoulements à grande échelle induits par eux dans les systèmes tournants. Analyse détaillée pour déterminer l'équilibre initial des champs et des écoulements dans une sphère en rotation.

062.070 **Velocity fields which preserve cylindrically symmetric force-free magnetic fields.**
S. R. Sreenivasan, D. L. Thompson.
Physica, Vol. 78, 321 - 339 (1974).

Forme de l'équation des champs de vitesse présentant les champs magnétiques sans force à α constant. Expressions explicites de ces champs de vitesse pour la symétrie cylindrique. Application aux champs en astrophysique.

062.071 **Relativistic effects in magnetohydrodynamic wave propagation.** G. L. Murphy.
Phys. Fluids, Vol. 18, 112 - 114 (1975).

Discussion sur les modifications relativistes d'ondes d'Alfvén, pour les cas où les termes magnétiques, et relativistes sont comparables.

062.072 **Bremsstrahlung in a plasma under the influence of a strong magnetic field.**
J. Virtamo, P. Jauho.
Nuovo Cimento B, Ser. 11, Vol. 26, 537 - 558 (1975).

062.073 **New, almost electrostatic, temperature anisotropy instability.** R. W. Landau.
Journ. Plasma Phys., Vol. 13, 377 - 384 (1975).

Découverte d'une nouvelle instabilité, purement croissante, pour un plasma bi–maxwellien dans un champ magnétique uniforme. L'instabilité ne peut être obtenue qu'à partir de la relation de dispersion complète. Application au vent solaire.

Physics and chemistry in space, Vol. 8: Plasma instabilities and nonlinear effects. See Abstr. 003.106.

Computed n = 2 level populations in helium-like ions from C V to Ni XXVII. See Abstr. 022.095.

Equilibrium figures. See Abstr. 042.044.

Electron-electron bremsstrahlung in a hot plasma. See Abstr. 061.070.

Theory of Thomson scattering in a strong magnetic field. III. See Abstr. 063.040.

Equations for the nonlinear transfer of electromag-

netic radiation and excitation of Langmuir turbulence at decay interaction. See Abstr. 063.048.

The velocity of magnetohydrodynamic waves in general relativity. See Abstr. 066.010.

Equations of hydrodynamics in general relativity in the slow motion approximation. See Abstr. 066.117.

Electromagnetic ion-beam instabilities in the solar wind. See Abstr. 074.090.

Plasma emission due to isotropic fast electrons, and types I, II, and V solar radio bursts. See Abstr. 077.013.

Open and closed magnetospheric tail configurations and their stability. See Abstr. 084.202.

A possible mechanism of stellar flares. See Abstr. 122.013.

Are stellar flares energized by the missing energy in starspots? See Abstr. 122.084.

Magneto-parametric instabilities in the Crab nebula. See Abstr. 134.001.

Superfluidity in neutron stars. I. Steady-state hydrodynamics and frictional heating. See Abstr. 141.317.

Flux limit on cosmic ray acceleration by strong spherical pulsar waves. See Abstr. 141.341.

063 Radiative Transfer, Scattering

063.001 **Parametric conversion and amplification of light in moving media.** I. Lerche.
Astrophys. Space Sci., Vol. 35, 363 - 388 (1975).

The author presents calculations which should be of interest in connection with the problem of radiative transport through the magnetospheres of pulsars and the envelopes of quasars, where not only are particle speeds of the order of the speed of light expected, but also variations of refractive index with frequency and position.The calculations indicate that the production of broad-band, high-intensity radiation from relatively weak monochromatic light is a process which is likely to occur in most astrophysical objects where it is believed, or suspected, that rapid motions and spatial variations occur.

063.002 **Transfer of polarized light in an isotropic medium. Singular eigensolutions of the transfer equation.**
H. Domke.
Journ. Quant. Spectrosc. Radiat. Transfer, Vol. 15, 669 - 679 (1975).

The eigensolutions of the transfer equation for polarized light in a plane parallel isotropic medium with an arbitrary law of single scattering are derived.

063.003 **Transfer of polarized light in an isotropic medium. Biorthogonality and the solution of transfer problems in semi-infinite media.** H. Domke.
Journ. Quant. Spectrosc. Radiat. Transfer, Vol. 15, 681 - 694 (1975).

Sets of adjoint function vectors, which can be interpreted as describing the emergent radiation for generalized Milne's problems, are constructed by means of Case's eigenvectors and the albedo matrix. It is shown that the adjoints allow us to transform the full range orthogonality relation for eigenvectors of the positive half spectrum into a half range biorthogonality relation. A half range closure relation is derived. Half range integral equations are obtained for the eigenvectors, as well as for its adjoints.

063.004 **A modified Henyey method for computing radiative transfer hydrodynamics.** A. H. Karp.
Journ. Quant. Spectrosc. Radiat. Transfer, Vol. 15, 745 - 748 (1975).

The implicit hydrodynamic code of Kutter and Sparks has been modified to include radiative transfer effects in the optically thin zones of the model. This modification is described and the convergence of the method is discussed. Hydrodynamic models of a classical cepheid with a period of 12^d computed using the diffusion approximation and this modified method are compared.

063.005 **Derivation and application of the reciprocity relations for radiative transfer with internal illumination.**
A. C. Cogley.
Journ. Quant. Spectrosc. Radiat. Transfer, Vol. 15, 749 - 760 (1975).

063.006 **A three-parameter analytic phase function for multiple scattering calculations.** G. W. Kattawar.
Journ. Quant. Spectrosc. Radiat. Transfer, Vol. 15, 839 - 849 (1975).

A very simple procedure has been developed to fit the first three moments of an actual phase function with a three parameter analytic phase function. The exact Legendre Polynomial decomposition of this function is known which makes it quite suitable for multiple scattering calculations. The use of this function can be expected to yield excellent flux values at all depths within a medium. Since it is capable of reproducing the glory, it can be used in synthetic spectra computations from planetary atmospheres. Accurate asymptotic radiance values can also be achieved as long as the single scattering albedo $\omega_0 \geq 0.9$.

063.007 **Aspects of frequency redistribution for non-coherent scattering in a turbulent medium.**
C. Magnan.
Journ. Quant. Spectrosc. Radiat. Transfer, Vol. 15, 979 - 989 (1975).

063.008 **Bedingungen für negativen Kontinuums-Absorptionskoeffizienten.** J. Pfleiderer.
Mitt. Astron. Ges., No. 36, p. 72 (1975).

063.009 **Resonance line transfer with partial redistribution. IV. A generalized formulation for lines with common upper states.** R. W. Milkey, R. A. Shine, D. Mihalas.
Astrophys. Journ., Vol. 199, 718 - 723 (1975).

This paper describes a generalized formulation for treating partial redistribution effects in transfer problems in resonance lines with common upper states. The formulation allows explicitly for the possibility that several spectral lines may arise in transitions from a given upper level to several sharp lower levels, including, for example, the ground state and metastable states.

063.010 **Resonance line transfer with partial redistribution. V. The solar Ca II lines.**
R. A. Shine, R. W. Milkey, D. Mihalas.
Astrophys. Journ., Vol. 199, 724 - 733 (1975).

Line profiles for the Ca II H and K lines have been calculated, accounting for the partial frequency coherence of scattered photons. These profiles are compared with calculations made with identical atomic and atmospheric models but assuming complete redistribution. Very significant differences between the profiles obtained using these two different physical descriptions of the scattering process are found.

063.011 **On the diffusion of light in turbulent shearing media.** I. Lerche.
Astrophys. Journ., Vol. 199, 734 - 740 (1975).

The author presents some simple calculations to show that a beam of light is, in general, severely distorted and dispersed from the straight-line path that would obtain in vacuum when passing through a medium possessing a turbulent velocity of order c, the speed of light in vacuum. The motivation underlying these calculations is the problem of radiative transport through the magnetospheres of pulsars where particle energies in excess of about $10^8 mc^2$ are believed to obtain.

063.012 **Comments on the redistribution function of Jefferies and White.** F. Kneer.
Astrophys. Journ., Vol. 200, 367 - 368 (1975).

The simplified form of the angle-averaged redistribution function $R^{II}(\nu', \nu)$ for coherent scattering in the atom's frame, suggested by Jefferies and White (1960), is neither normalized nor symmetric. It is shown how these requirements can be met by a simple change, retaining the idea that photons are completely redistributed in the Doppler core and coherently scattered in the line wings.

063.013 **On induced Compton scattering by relativistic particles.** R. D. Blandford, E. T. Scharlemann.
Astrophys. Space Sci., Vol. 36, 303 - 317 (1975).

Induced Compton scattering by independent relativistic particles in a radiation field is considered from both quantum mechanical and classical viewpoints. Quantum mechanically, the particle recoil causes an imperfect cancellation of the induced scattering rates between pairs of photon states. This results in a net energy transfer from the radiation field to the particle. Classically, the electrodynamical effects are manifestations of non-linear coupling between different wave amplitudes in the equation of motion. The results from these two approaches are related and shown to be equivalent.

063.014 **Non-LTE transfer. $\sqrt{\epsilon}$ revisited.**
U. Frisch, H. Frisch.
Monthly Notices Roy. Astron. Soc., Vol. 173, 167 - 182 (1975).

Assuming a semi-infinite atmosphere, a two-level atom and complete frequency redistribution, the authors give, in the isothermal case, a new derivation of the surface value of the source function based only on the integral equation for the source function. This derivation has a straightforward extension to the case where a uniform velocity gradient is included, to the time-dependent case (thermal source B switched on at time $t = 0$) and to the determination of the mean number of scatterings undergone by an escaping photon (with Ambartsumian's definition).

063.015 **Solution of the equation of radiative transfer in the atmosphere of a double star. Extended stellar atmosphere.** V. P. Merezhin.
Trudy Kazan. Gorod. Astron. Obs., No. 40, p. 119 - 129 (1974). In Russian.

063.016 **Lichtstreuung in der Astronomie und ihre Simulation durch Mikrowellenanalogieversuche.** R. Zerull.
SuW, Vol. 14, 238 - 243 (1975).

063.017 **A tensor formulation of the equation of transfer. I. Application to plane parallel flows.** B. M. Haisch.
Bull. American Astron. Soc., Vol. 7, 449 (1975). – Abstr. AAS.

063.018 **Solution of the co-moving frame equation of transfer in spherically symmetric flows. II. Picket-fence models.** D. Mihalas, P. B. Kunasz, D. G. Hummer.
Bull. American Astron. Soc., Vol. 7, 449 (1975). – Abstr. AAS.

063.019 **Non-LTE radiative transfer in hydrogen with a self-consistent electron energy distribution.**
E. C. Shoub.
Bull. American Astron. Soc., Vol. 7, 450 (1975). – Abstr. AAS.

063.020 **Redistribution perturbations in radiative transfer theory.** C. J. Cannon, P. B. Lopert, C. Magnan.
Astron. Astrophys., Vol. 42, 347 - 353 (1975).

A new technique is developed which enables the fast, efficient and accurate numerical solution of the equation of transfer for spectral line radiation in which the redistribution of both photon angle and frequency at a photon scattering event is incorporated. The authors formulate the technique for both the integro-differential and integral representations of the transfer equation. The essence of the technique is to start with a zero order solution obtained from the equation of radiative transfer in which complete redistribution holds. The first and higher order terms are then obtained from a perturbation series which is shown to converge extremely quickly.

063.021 **The calculation of the profiles of spectral lines forming in a scattering atmosphere.**
D. I. Nagirner.
Trudy Astron. Obs., *Leningrad*, Vol. 31 (= Uchenye Zapiski Leningr. Un-ta, No. 381 = Seriya Matem. Nauk, vyp. (No.) 51), p. 3 - 99 (1975). In Russian.

Multiple scattering of resonance radiation in a semi-infinite atmosphere is considered under the following assumptions: 1) two-level atoms; 2) complete frequency redistribution; 3) the Milne-Eddington model. Tables are given of the functions in terms of which profiles of emission and absorption lines formed in such an atmosphere are expressed.

063.022 **On the solution of the one-dimensional transfer problem.** M. A. Mnatsakanyan.
Soobshch. Byurakan. Obs., vyp. (No.) 46, p. 93 - 100 (1975). In Russian.

Ambartsumyan's shell-adding method can be perfected, if the variation of the method is considered when not a finite but infinite shell is added to the shell which optical characteristics we are interested in. The common case of monochromatic light transfer into a one-dimensional homogeneous medium is discussed as an illustration.

063.023 **On the solution of the light diffusion problem in an optically finite shell.**

Eh. Kh. Danielyan, M. A. Mnatsakanyan.
Soobshch. Byurakan. Obs., vyp. (No.) 46, p. 101 - 114 (1975).
In Russian.

A connection between the problems of monochromatic radiation transfer in optically finite and semi-infinite shells is established. It is realized in terms of a linear integral equation with the kernel expressed by internal characteristics of the semi-inifinite shell.

063.024 **The mean photon path length in extremely opaque media.** T. F. Adams.
Astrophys. Journ., Vol. 201, 350 - 351 (1975).

The mean total path length traversed by a resonance-line photon in escaping from an extremely opaque medium is discussed. Numerical results are presented for damping constants $a = 4.7 \times 10^{-3}$ and $a = 4.7 \times 10^{-4}$. It is shown that the numerical results can be understood in terms of nearly coherent scattering at a frequency x_*. Expressions for x_* for spherical and plane-parallel media are given, and an approximate expression for the thermalization length is derived.

063.025 **Strong scintillations in astrophysics. II. A theory of temporal broadening of pulses.**
L. C. Lee, J. R. Jokipii.
Astrophys. Journ., Vol. 201, 532 - 543 (1975).

The authors develop a theory of temporal broadening of pulse propagating in a turbulent medium, based on the Markov approximation. The theory may be applied to quite general turbulence spectra and to thin or thick turbulent regions. The observed smearing is found to be the combination of three effects which the authors term the dispersion effect, the pure refractive effect, and the diffraction effect. The last of these dominates for typical pulsar parameters. The pulse shapes are explicitly calculated for both Gaussian and Kolmogorov turbulence spectra and it is shown how these scale with the various turbulence parameters.

063.026 **Resonance-line transfer with partial redistribution. VII. Angle-dependent redistribution.**
R. W. Milkey, R. A. Shine, D. Mihalas.
Astrophys. Journ., Vol. 202, 250 - 258 (1975).

A method is presented for treating radiative transfer in resonance lines, allowing for the full angle and frequency dependence of redistribution in the scattering process, as seen in the laboratory frame. The case of an equivalent-two-level-atom source function is considered; the problem to be treated is then linear in the radiation field. The authors apply this method to the Ca II lines in the solar atmosphere. Both the angle-averaged approximation and the full angle-dependent solution were obtained, and were compared to assess, differentially, the effects of angular redistribution upon the computed line profile and its center-to-limb behavior. For the Ca II line in a homogeneous solar chromosphere the angle-dependent effects are found to be negligible.

063.027 **Asymptotic properties of Green's function of the equation of transfer of polarized radiation.**
H. Domke, V. V. Ivanov.
Astron. Zhurn. Akad. Nauk SSSR, Vol. 52, 1034 - 1037 (1975). In Russian. English translation in Soviet Astron., Vol. 19, No. 5.

The asymptotic properties are studied of Green's function of the vector transfer equation for a semi-infinite medium (deep layers, nearly conservative scattering, contributions due to high-order scattering). The scattering matrix is arbitrary.

063.028 **Effects of Thomson scattering on the shape of a spectral line.** M. Missana, A. Piana.
Astrophys. Space Sci., Vol. 37, 263 - 273 (1975).

The effect of Thomson scattering and Compton effect on the shape of a spectral line is studied in the case of a plane-parallel atmosphere. The theoretical values of redshift and intensity at the top of the line, having an equivalent width $5\sqrt{\pi}\ 10^{-10}$ cm, and of the red asymmetry are tabulated for 62 different values of the optical thickness τ'; from $\tau' = 0.2/\sqrt{3}$ to $\tau' = 2.6/\sqrt{3}$. The case of solar atmosphere is considered and it is shown that all main features of the observed redshifts can be interpreted by means of a scattering theory, with an error smaller than one milliangström. The need of further measurements is pointed out.

063.029 **Non-stationary one-dimensional probabilistic radiative transfer in a bounded medium.** R. Grappin.
Astrophys. Space Sci., Vol. 37, 383 - 400 (1975).

Monochromatic time-dependent diffusion of a photon in a homogeneous bounded medium is considered as a one-dimensional probabilistic process in a medium with absorbing boundaries. Simple and intuitive relations are shown to exist between Laplace transforms of time-dependent distributions of photons in infinite medium and inside a bounded medium, as well as for the corresponding emerging intensities. One obtains in particular an explicit and exact expression for the original time-dependent distribution inside a bounded one-dimensional medium.

063.030 **Diffuse reflection of light by an inhomogeneous atmosphere.** V. V. Sobolev, V. M. Loskutov.
Dokl. Akad. Nauk SSSR. Ser. Mat., Fiz., Vol. 224, 1293 - 1296 (1975). In Russian.

063.031 **An approximate estimate of multiple scattering of quanta in the solution of the equations of statistical equilibrium.** E. G. Borodina.
Solnechnye Dannye 1975 Byull., No. 6, p. 80 - 84 (1975).
In Russian.

A simple approximate expression is given for the coefficient which determines the part of quanta escaping from a medium without absorption. The approximation is based on the solution of the integral equation of transfer of line radiation in a plane layer given by Ivanov (1969).

063.032 **A new approach to radiative transfer theory using Jones's vectors. I: Fundamentals.**
A. L. Fymat, R. Vasudevan.
Astrophys. Space Sci., Vol. 38, 95 - 124 (1975).

Radiative transfer of partially polarized radiation in an anisotropically scattering, inhomogeneous atmosphere containing an arbitrary polydispersion of particles is described using Jones's amplitude vectors and matrices. The complete equivalence between the transport equation for Jones's vectors and the classical radiative transfer equation for Stokes's intensity vectors is demonstrated. The present formulation is the most succinct and the most convenient one for both theoretical and experimental studies. It yields a simpler analysis than the classical formulation since it reduces by a factor of two the dimensionality of transfer problems. It preserves information on phases, and thus can be used directly across the entire electromagnetic spectrum without any further conversion into intensities.

063.033 **Time dependent radiative transfer. I. Development of a variational formulation for dispersion relation and stability.** S. J. Wilson, K. K. Sen.
Astron. Astrophys., Vol. 44, 377 - 381 (1975).

A variational formulation of time dependent radiative transfer problem in a plane parallel, homogeneous medium in LTE has been suggested. A dynamical potential has been constructed, the extreme condition of which gives the transfer equation and the cooling rate equation as the Euler-Lagrange equations. This potential has been utilised to determine the dispersion relations and stability criteria for finite, semi-infinite and infinite media under different perturbations.

063.034 **General-relativistic radiative transfer theory in refractive and dispersive media.**
J. Bičák, P. Hadrava.
Astron. Astrophys., Vol. 44, 389 - 399 (1975). – Paper presented at the Czechoslovak Conference on 'Stellar astronomy' in Štiřín on October 7, 1974; and at the International Seminar on 'Relativity' in Georgenthal (E. Germany) on October 22, 1974.

The transfer of radiation in an isotropic, refractive and dispersive medium in a curved spacetime is studied by applying the general-relativistic kinetic theory. The transfer of black-body radiation, the propagation of radiation in a cold plasma in the Friedmann universe, and the derivation of the explicit form of the transfer equation in a spherically symmetric, non-stationary dispersive medium (the generalization of Lindquist's equation) are discussed.

063.035 **Non-LTE line formation in turbulent media.**
H.-P. Gail, E. Sedlmayr, G. Traving.
Astron. Astrophys., Vol. 44, 421 - 429 (1975).

Non-LTE line formation in a turbulent medium is discussed with the assumption that the turbulent field can be described in terms of its stochastic properties. Formulae for the general non-LTE case (multi-level approximation) are given. As an illustration of the method, the non-LTE formation of the Fe I-line λ 5576 in the solar photosphere is calculated.

063.036 **On grey absorption coefficients in radiative transfer.**
T. Viik.
Publ. Tartu Astrofiz. Obs., Vol. 43, 233 - 249 (1975).

The problem of reducing a frequency-dependent radiative transfer equation to an equivalent grey equation is studied using the Pomraning approximation. The respective grey equation in the case of the Elsasser band is solved by the singular eigenfunction method and it is shown that the emerging flux considerably differs from the exact multifrequency result. The best results were obtained by using the Chandrasekhar mean opacity coefficient in the Pomraning approximation.

063.037 **Strong scintillations in astrophysics. III. The fluctuations in intensity.** L. C. Lee, J. R. Jokipii.
Astrophys. Journ., Vol. 202, 439 - 453 (1975).

The authors present a discussion of fluctuations in intensity due to wave propagation in a wide range of turbulent media. An asymptotic analytic solution to the fourth-moment equation is presented, and its range of validity is discussed. Various parameters characterizing the intensity fluctuations, such as scintillation index, correlation scale, etc., are expressed in terms of turbulence parameters for both Gaussian and power-law irregularity spectra and for thick and thin turbulent media.

063.038 **Radiative transfer through a flowing refractive medium.** J. L. Anderson, E. A. Spiegel.
Astrophys. Journ., Vol. 202, 454 - 464 (1975).

The authors discuss the propagation of photons in a non-dispersive refractive medium with arbitrary velocity and gravity fields. Photons move along the geodesics associated with the optical metric given by Gordon in the limit where geometrical optics applies. The known properties of geodesics then simplify the task of finding the world lines of photons in a variety of problems.

063.039 **Solution of the comoving-frame equation of transfer in spherically symmetric flows. I. Computational method for equivalent-two-level-atom source functions.**
D. Mihalas, P. B. Kunasz, D. G. Hummer.
Astrophys. Journ., Vol. 202, 465 - 489 (1975).

The equation of radiative transfer in the comoving frame makes possible an economical solution of the line formation problem in spherical atmospheres expanding with arbitrarily large velocities. A stable differencing scheme and a frequency-by-frequency elimination procedure have been developed to solve the partial differential equations that describe the radiation field in the comoving frame. Numerical results were obtained for a large number of illustrative models involving line formation by two-level atoms, electron scattering, and continuous absorption.

063.040 **Theory of Thomson scattering in a strong magnetic field. III.** S. Kanno, T. Hamada.
Publ. Astron. Soc. Japan, Vol. 27, 545 - 552 (1975).

The Thomson-scattering mean free path is calculated for the two normal modes (ordinary and extraordinary waves) of the electromagnetic wave propagating in a magnetized plasma. For this purpose the Appleton-Hartree dispersion relation is generalized by the inclusion of the radiation damping. Under the presence of a magnetic field, the ordinary wave generally prefers the propagation in the field direction, while the mean free path for the extraordinary wave has a narrower peak in the field direction and, in addition, a broader peak of comparable height in the direction perpendicular to the field.

063.041 **The 'peaking effect' of the emergent intensity in spherical-geometry transfer problems.**
S. J. Wilson, C. T. Tung, K. K. Sen.
Publ. Astron. Soc. Japan, Vol. 27, 611 - 613 (1975).

The paper demonstrates that the modified double interval method of Wilson and Sen (1963) gives the correct outward peaking of emergent intensity in spherical geometry, if it is calculated according to the scheme suggested by the authors.

063.042 **Asymptotic limits of the Mie-scattering characteristics.** P. Chýlek.
Journ. Optical Soc. America, Vol. 65, 1316 - 1318 (1975).

It is proved in a rigorous way that the $x \to \infty$ limits of the Mie normalized scattering cross section $Q(x, m)$ and normalized extinction cross section $Q_{ext}(x, m)$ are equal to $1 + |(m-1)/(m+1)|^2$ and to 2, respectively.

063.043 **Thermal microwave emission from a scattering layer.** A. W. England.
Journ. Geophys. Res., Vol. 80, 4484 - 4496 (1975).

063.044 **Calculation of short-wave radiative transfer in the atmosphere.** V. A. Snegerev, V. V. Kozoderov.
Trudy Gidrometeorol. n.-i. tsentr SSSR, 1975, vyp. (No.) 165, p. 87 - 99. In Russian. – Abstr. in Referativ. Zhurn. 51. Astron., 12.51.171 (1975).

063.045 **Hyperfine structure and line formation in a magnetic field.** E. Landi Degl'Innocenti.
Astron. Astrophys., Vol. 45, 269 - 276 (1975).

The author derives the transfer equations of the Stokes parameters for a line with hyperfine structure in the presence of a magnetic field, both in LTE and NLTE. The effect of unresolved hyperfine structure on magnetographic determinations of solar magnetic fields is considered in the limiting case of small magnetic fields. Finally, an example is given on the determination of the Eu abundance in magnetic stars showing that the combined effects of hyperfine structure and magnetic field may lower the measured abundance by as much as 2.5 dex.

063.046 **Magnetic intensification as a mechanism for broad band linear polarization.**
G. Calamai, E. Landi Degl'Innocenti, M. Landi Degl'Innocenti.
Astron. Astrophys., Vol. 45, 297 - 301 (1975).

The solutions of the transfer equations for polarized light in a Milne-Eddington atmosphere are used to evaluate the importance of magnetic intensification as a possible mechanism for interpreting the net amount of linear polarization observed in broad-band measurements of magnetic regions. Magneto-optical effects are taken into account properly.

063.047 **Eigenfunction solution of the scattering of beam radiation fields by spherical objects.**
W-C. Tsai, R. J. Pogorzelski.
Journ. Optical Soc. America, Vol. 65, 1457 - 1463 (1975).
The theoretical analysis of the scattering of a beam by a spherical object centered on the propagation axis is treated exactly. For both conducting and dielectric bodies, those cases of sphere radii smaller than, equal to, and larger than that of the beam are analyzed numerically and the power radiated in the far zone is obtained. The difference between the scattering of a beam wave and that of a plane wave is discussed.

063.048 **Equations for the nonlinear transfer of electromagnetic radiation and excitation of Langmuir turbulence at decay interaction.** R. D. Lomadze.
Astrofizika, Vol. 11, 409 - 420 (1975). In Russian. – English translation in Astrophysics, Vol. 11, No. 3.
The case when the frequency of radiation is considerably higher than the electronic plasma frequency is considered.

063.049 **Balmer decrement in a medium which is made transparent by very powerful radiation.**
S. A. Kaplan, V. V. Kulinich.
Astrofizika, Vol. 11, 421 - 437 (1975). In Russian. – English translation in Astrophysics, Vol. 11, No. 3.
The nonlinear transfer of radiation in hydrogen lines is considered. It is assumed that the medium is made transparent in the same lines by very strong radiation from a background source. The analytic solution has been found for the case of small optical depth. The emission and absorption decrements of Lyman, Balmer and other series have been found. A numerical algorithm for a more general nonsteady problem has been given.

063.050 **Nonstationary radiative transfer in scattering media.**
N. N. Rogovtsov, A. M. Samson.
Astrofizika, Vol. 11, 439 - 454 (1975). In Russian. – English translation in Astrophysics, Vol. 11, No. 3.
The nonstationary radiation field in stationary homogeneous media is considered. A calculation method for temporal moments of functions defining nonstationary radiation is suggested. Several formulae for moments at arbitrary excitation are obtained. Using these formulae generalized expressions for average time $\bar{u}$ and dispersion σ are found. The quantities $\bar{u}$ and σ for the cases of a flat parallel layer and a sphere at various kinds of excitation are calculated. The influence of time lag due to the finiteness of velocity of light on a nonstationary resonance radiation field is estimated.

063.051 **Non-coherent scattering. V.**
M. S. Gevorkyan, N. B. Engibaryan, A. G. Nikogosyan.
Astrofizika, Vol. 11, 455 - 472 (1975). In Russian. – English translation in Astrophysics, Vol. 11, No. 3.
Some questions are discussed concerning the solution of radiative transfer problems in the general case of non-coherent scattering. Various methods of representation of the redistribution function as a bilinear expansion are considered. Numerical calculations concerning spectral line profiles originating as a result of diffuse reflection from a semi-infinite medium are given. The influence of various approximations of the redistribution law on the final results is shown.

063.052 **X and Y operators for general linear transport processes.** J. Casti.
Proc. National Acad. Sci. U.S.A., Vol. 72, 1210 - 1211 (1975).
The derivation of generalized Ambartsumian-Chandrasekhar X and Y functions (1960) for stationary transfer in a plane-parallel slab is presented. An algebraic formula relating these functions to the usual reflection function is also presented, together with the appropriate generalization of the Chandrasekhar H-equations for the semi-infinite medium. The planetary problem is also briefly discussed.

A line-leaking model for silicon monoxide millimetre wavelength emission. See Abstr. 022.023.

Half-range moment method for solution of the transport equation in a spherically symmetric geometry. See Abstr. 022.070.

The thermalization length by the Green function method. See Abstr. 022.071.

Astrophysical discussion of the Compton effect in an outside magnetic field. See Abstr. 061.042.

On the notion of Thomson scattering in a plasma. See Abstr. 062.026.

Scattering by dust and the photographic appearance of Eta Carinae. See Abstr. 064.046.

An archetype hydrogen atmosphere problem. See Abstr. 064.061.

On the theory of scattering photospheres. See Abstr. 064.066.

Monte Carlo treatment of Lyman-alpha. II. Radiation in a spherical atmosphere. See Abstr. 082.002.

Formation of coupled spectral lines in a planetary atmosphere: II. Numerical evaluation of approximate solutions. See Abstr. 091.001.

Diffuse reflection and transmission of radiation by planetary atmospheres with a diffuse-and-specular reflector. See Abstr. 091.020.

Non-linear radiation transport calculations to obtain mean colors and to verify the Wesselink method. See Abstr. 122.116.

Radiation transport in dense interstellar dust clouds. I. Grain temperature. See Abstr. 131.010.

Uncertainties in the far-infrared absorption of interstellar clouds. See Abstr. 131.014.

Effects of turbulent velocities on maser lines. See Abstr. 131.040.

The transfer of radio recombination line radiation through a cold gas. I. Hydrogen and helium lines in compact H II regions. See Abstr. 131.513.

Radiative transfer in a strong magnetic field and accreting X-ray pulsars. See Abstr. 142.142.

The evolution of the cosmic radiation spectrum under the influence of turbulent heating. I. Theory. See Abstr. 143.008.

The evolution of the cosmic radiation spectrum under the influence of turbulent heating. II. Numerical calculation and application. See Abstr. 143.009.

064 Stellar Atmospheres, Stellar Envelopes, Mass Loss

064.001 **P Cygni stars and mass loss.**
P. Kuan, L. V. Kuhi.
Astrophys. Journ., Vol. 199, 148 - 165 (1975).

The Sobolev approximation is used to compute line profiles formed in rapidly expanding atmospheres. The results are compared to observed hydrogen Balmer series profiles of P Cygni and other stars of similar type. It is found that a decelerating envelope can most adequately account for the observed profiles. It is also found that low momentum flux favors the presence of dust in the envelope, as indicated by the presence of large infrared excess, whereas high momentum fluxes are found for stars having a free-free infrared emission spectrum.

064.002 **Discovery of a corona around an early-type star and the problem of mass loss.**
J. B. Rogerson, Jr., H. J. G. L. M. Lamers.
Nature, Vol. 256, 190 (1975).

The star τ Sco, HD 149438, spectral type B0 V, was observed spectroscopically by the Princeton telescope-spectrometer on the Copernicus satellite during June and July 1973. Apart from the numerous spectral features with a full width at half maximum of about 20 km s^{-1} which originate in the photosphere of the star, and the interstellar absorption lines, the spectrum also shows a few very wide absorption lines. The lines are highly asymmetric with the largest depression close to the laboratory wavelengths and an enhanced short-wavelength wing, which extends to a Doppler velocity of about −1,000 km s^{-1}. This suggests that the lines are formed in the expanding outer layers of the atmosphere.

064.003 **Circumstellar dust shells.**
L. M. Shulman (*Shul'man*).
IAU Symposium No. 67, (see 012.001), p. 165 - 171 (1975).

The optical properties of circumstellar graphite particle shells are considered. In the case of the optically thin shell, the temperature of the grains is calculated. It is shown that the dust shell always has a sharp inner border.

064.004 **Observations of radially expanding stratified atmospheres of Wolf-Rayet stars.** W. Seggewiss.
IAU Symposium No. 67, (see 012.001), p. 285 - 289 (1975).

A correlation between lower excitation potential and radial velocity of violet-displaced absorption components in the spectra of Wolf-Rayet stars is established. The relationship supports the concept of a radially expanding stratified atmosphere in which the expansion velocity increases outwards. The outermost layers of the atmosphere of one star (HD 151932, WN7) display oscillations with a periodicity of about 3.5 days.

064.005 **Dynamik von Sternatmosphären.** – (Karl-Schwarzschild-lecture 1974). C. de Jager.
Mitt. Astron. Ges., No. 36, p. 15 - 30 (1975).

064.006 **Zur Berechnung der Wärmeleitfähigkeit in Sternatmosphären.** T. Nowak, P. Ulmschneider.
Mitt. Astron. Ges., No. 36, p. 72 - 75 (1975).

064.007 **Non-LTE-effects in atmospheres of O-subdwarfs.**
R.-P. Kudritzki.
Mitt. Astron. Ges., No. 36, p. 81 - 83 (1975).

064.008 **Non-LTE line formation of He II in O-stars.**
K. P. Simon.
Mitt. Astron. Ges., No. 36, p. 83 - 84 (1975). – Abstract.

064.009 **Quantitative Feinanalyse von Popper's Stern (HD 124448).** D. Schönberner.
Mitt. Astron. Ges., No. 36, p. 84 (1975). – Abstract.

064.010 **Selbstgravitierender Wind eines rotierenden Sterns.**
R. Weidelt.
Mitt. Astron. Ges., No. 36, p. 87 (1975). – Abstract.

064.011 **Mass loss and circumstellar dust shells at T Tauri stars.** W. Wenzel.
Astron. Nachr., Vol. 296, 183 - 187 (1975).

The mass loss of T Tauri stars leads to the production of dust in circumstellar space. The total amount of lost mass (and therefore of produced dust) is observed to be positive correlated with the intensity of the H and Ca II emission lines of the objects. This fact is used to explain quantitatively another correlation, namely the observation, that the interstellar extinction which is found by star counts in the wider surroundings of a T Tauri star is roughly proportional to the intensity of its Hα emission. By this, the outflow of circumstellar matter into the interstellar space seems to be observable directly.

064.012 **Das Strahlungsfeld zirkumstellarer Staubhüllen.**
J. Schmid-Burgk.
Mitt. Astron. Ges., No. 36, p. 177 (1975). – Abstract.

064.013 **The study of HR 3018, HD 91324 and α Crv and its implication for the age-abundance correlation of the neighbour stars.** L. da Silva.
Astron. Astrophys., Vol. 41, 287 - 300 (1975).

The stars HR 3018, HD 91324 and α Crv have been analyzed using large dispersion spectrograms with a differential model atmosphere analysis. A careful determination of T_{eff}, logg, abundance, micro-turbulence and rotational velocity of the stars has been made. The three stars are metal deficient. The iron abundances of the stars are not in accordance with the ages estimated from the isochrone diagrams. The author discusses these results and analyses their consequences for the galactic structure.

064.014 **On the calculation of diffusion of chemical elements in stellar envelopes.** A. A. Pamyatnykh.
Nauchn. Informatsii, vyp. (No.) 32, p. 104 - 115 (1974).
In Russian.

An approximate formula for computing the diffusion rate of an ion admixture relative to protons in a ternary mixture including electrons is derived by means of the basic equations of diffusion for multicomponent mixtures. The result differs to some extent from that by Aller and Chapman, who utilized the binary mixture approximation. The comparison of results derived with the approximate and the exact formulae is presented for helium diffusion at the bottom of convective zones in envelopes of main sequence A stars.

064.015 **Interstellar bubbles.** J. Castor, R. McCray, R. Weaver.
Astrophys. Journ., (*Letters*), Vol. 200, L107 - L110 (1975).

Early-type stars blow bubbles in the interstellar medium. The radii of the bubbles are typically 30 pc. Typical conditions in their interiors are $T \approx 10^6$ K and $n \approx 0.01$ cm^{-3}. The dense shell of swept-up interstellar gas that surrounds them is likely to trap the ionization front and may also have an outer layer of H_2. The column density of O VI in the interior is in accord with observations by the Copernicus ultraviolet telescope.

064.016 **Time lag effects in the nucleation of particles in stellar atmospheres.** G. H. Walker.
Astrophys. Letters, Vol. 16, 115 - 120 (1975).

When a system goes from a saturated state to a supersaturated one, the classical nucleation rate formula is not instantly valid; rather, the nucleation rate relaxes exponentially (with a characteristic time–the time lag) to the value given by the steady state formula. Under some circumstances, particularly those found in some cool, stellar atmospheres, the time lag can be quite long and a decisive factor in determining the possibility of particle formation. Carbon condensation in the atmospheres of Mira variables is considered and found to be unlikely on the basis of classical nucleation theory because of the long time lags involved unless the parameters describing the physical conditions in these atmospheres are significantly different from current estimates.

064.017 **Shock waves in stellar atmospheres.** P. Kuan.
Astrophys. Journ., Vol. 200, 653 - 659 (1975).

A series solution is found for the propagation of a blast wave in a planar atmosphere with exponentially decreasing density. A perfect gas and adiabatic motion are assumed, and gravity is ignored. The solution is good within about 3 scale heights from the origin of the blast. However, it is within this region that most of the energy of the blast is dissipated in a dense shell, and in particular, approximately 40 percent of it as kinetic energy of the shell. If the T Tauri stars are ejecting mass through blast waves set off in their atmospheres, the maximum mass loss rate is estimated to be about 8×10^{-7} $M_\odot$ yr^{-1}, a value a few times larger than those observed.

064.018 **Stellar model chromospheres. III. Arcturus (K2 III).** T. R. Ayres, J. L. Linsky.
Astrophys. Journ., Vol. 200, 660 - 674 (1975).

The authors construct models for the upper photosphere and chromosphere of Arcturus based on the H, K, and IR triplet lines of Ca II and the h and k lines of Mg II. The chromosphere model is derived from complete redistribution solutions for a five-level Ca II ion and a two-level Mg II ion. A photospheric model is derived from the Ca II wings using first the "traditional" complete redistribution limit and then the more realistic partial redistribution approximation. In particular, the temperature and mass column densities for the temperature minimum region and the chromosphere-transition region boundary are computed and the pressure P_0 in the transition region and corona estimated. The authors find $T_{min}/T_{eff} \approx 0.77$ for Arcturus, Procyon, and the sun and a trend of increasing mass at the temperature minimum with decreasing gravity. They find P_0 to be about 1 percent of the solar value, and on this basis estimate the surface brightness of the Arcturus transition region and coronal spectrum to be much less than for the sun.

064.019 **The $^{12}C/^{13}C$ ratio in stellar atmospheres. V. Twelve K giants and subgiants.**
D. S. P. Dearborn, D. L. Lambert, J. Tomkin.
Astrophys. Journ., Vol. 200, 675 - 681 (1975).

High-resolution photoelectric scans of portions of the CN red system in the spectra of 12 K giants and subgiants have been analyzed for the $^{12}C/^{13}C$ abundance ratio with the following results: > 50 (δ Eri), > 30 (η Cep), 51 ± 5 (ν^2 CMa), 24 ± 3 (γ Cep), 34 ± 4 (53 Eri), 16 ± 2 (ι Cep), 12 ± 2 (θ Cet), 22 ± 2 (λ Sgr), 20 ± 2 (ξ Dra), 15 ± 2 (ρ Boo), 18 ± 2 (δ Sgr), and 25 ± 2 (δ And). Interpertations of the observed range of results, $5 < {}^{12}C/{}^{13}C < 55$, for giants and subgiants are discussed.

064.020 **Investigation of the influence of the chemical composition and acceleration due to gravity on the structure of photospheric models of stars of different spectral classes.** E. E. Belyaeva.
Trudy Kazan. Gorod. Astron. Obs., No. 40, p. 87 - 109 (1974). In Russian.

064.021 **Method for temperature corrections in the theory of stellar atmospheres.** E. E. Belyaeva.
Trudy Kazan. Gorod. Astron. Obs., No. 40, p. 110 - 118 (1974). In Russian.

064.022 **H_2 formation in wind-driven circumstellar shells.** D. J. Hollenbach, R. McCray.
Bull. American Astron. Soc., Vol. 7, 418 (1975). – Abstr. AAS.

064.023 **Spectral line formation in extended stellar atmospheres.** L. R. Doherty.
Bull. American Astron. Soc., Vol. 7, 449 (1975). – Abstr. AAS.

064.024 **Non-LTE calculations for Ca II lines in late B and hot Ap stars.** M. A. J. Snijders.
Bull. American Astron. Soc., Vol. 7, 469 (1975). – Abstr. AAS.

064.025 **A non-linear radiative transfer model of SW Andromedae.** C. G. Davis, Jr., S. S. Bunker.
Bull. American Astron. Soc., Vol. 7, 471 (1975). – Abstr. AAS.

064.026 **A grid of model atmospheres for metal-deficient giant stars. I.** B. Gustafsson, R. A. Bell, K. Eriksson, Å. Nordlund.
Astron. Astrophys., Vol. 42, 407 - 432 (1975).

A grid of flux-constant model atmospheres for stars with $3750\ K \leqq T_{eff} \leqq 6000\ K$, $0.75 \leqq \log g \leqq 3.0$ (cgs units) and $-3.0 \leqq [A/H] \leqq 0.0$ has been constructed. Metal lines and molecular lines, including those from the infrared bands of CO and CN, are taken into account. The variation of the structure of the models with metal abundance and microturbulence parameter is found to be quite regular and not very drastic. The surface cooling produced by CO is important for all the models with $T_{eff} \leqq 5000$ K, while CN mainly causes a backwarming effect but is not very important for solar CNO abundances. The effects of convection, estimated using the mixing-length approximation, are important only for the coldest models. A solar model consistent with the grid models is found to agree satisfactorily with empirical solar model atmospheres. The careful use of scaled solar models for stars with $[A/H] \approx 0$ is justified.

064.027 **Stellar model chromospheres. IV. The formation of the Hϵ feature in the sun (G2 V) and Arcturus (K2 III).** T. R. Ayres, J. L. Linsky.
Astrophys. Journ., Vol. 201, 212 - 221 (1975).

The formation of the Balmer series member Hϵ in the near red wing of the Ca II H line is discussed for two cases: the sun (Hϵ absorption profile) and Arcturus (Hϵ emission profile). It is shown that although the Hϵ source functions in both stars are dominated by the Balmer continuum radiation field through photoionizations, the line formation problems in the two stars are quantitatively different, owing to a substantial difference in the relative importance of the stellar chromosphere temperature inversion compared with the stellar photosphere.

064.028 **Resonance line transfer with partial redistribution. VI. The Ca II K-line in solar-type stars.**
R. A. Shine, R. W. Milkey, D. Mihalas.
Astrophys. Journ., Vol. 201, 222 - 224 (1975).

Using model atmospheres for solar-type stars based on scaled temperature distributions, the authors discuss the effects of partial frequency redistribution on the Ca II K-line profiles. It is shown that the partial redistribution calculations lead to a significantly lower intensity at K_1 than given by calculations based on the assumption of complete redistribution. This implies that fits to observed fluxes with complete redistribution calculations could tend to underestimate systematically the value of the temperature at the chromospheric temperature minimum.

064.029 **A shortcut in lithium abundance determinations.** C. L. Imhoff.
Publ. Astron. Soc. Pacific, Vol. 87, 502 (1975). – Abstr. ASP.

064.030 **Emission envelopes of T Tauri stars.** P. Kuan.
Publ. Astron. Soc. Pacific, Vol. 87, 503 (1975).
Abstr. ASP.

064.031 **Evolution of protostars. On the formation of H II regions around early-type stars.** A. L. Vinogradov.
Nauchn. Informatsii, vyp. (No.) 33, p. 5, 21 - 22 (1974). In Russian and English. – Abstr. of a paper presented at the colloquium 'Physics and evolution of stars', (see 012.015).

064.032 **On computations of hydrogen and helium ionization in main sequence star envelopes.**
Ts. Tsvetkov, A. Pamyatnykh.
Nauchn. Informatsii, vyp. (No.) 33, p. 143 - 145 (1974). In Russian.

It is shown that a non-iterative scheme for treatment of hydrogen and helium partial ionization zones is adequate for computations of model envelopes of main sequence stars with masses about one solar mass and larger.

064.033 **Stellar mass loss.** L. B. Lucy.
19th Colloquium International Astrophys. Liège 1974, (see 012.016), p. 359 - 365 (1975).

Stars losing mass at a rate that may prove of significance for their evolution are discussed, with emphasis on the mechanisms responsible for the outflows and on the possibility of obtaining theoretical mass loss rates. Considerable attention is paid to the mass loss rate for ζ Puppis (O5f) since theoretical estimates differ by a factor of 100.

064.034 **Circumstellar absorption lines and mass loss from red giants.** D. Reimers.
19th Colloquium International Astrophys. Liège 1974, (see 012.016), p. 369 - 382 (1975).

A compilation of all available data shows that mass loss from red giants varies with basic stellar parameters as $\dot{M}$ $[M_\odot/yr] = 4 \times 10^{-13}$ L/gR (solar units) in the range 10^{-5} to 10^{-9} $M_\odot$/yr, with an uncertainty of a factor of at least 2 either way.

064.035 **On mass loss by stellar wind in red giants and supergiants.** F. Fusi-Pecci, A. Renzini.
19th Colloquium International Astrophys. Liège 1974, (see 012.016), p. 383 - 390 (1975).

For a long time a mechanism of mass loss similar to the solar wind has been suspected to be operating in all stars having a convective envelope. The authors have computed for a number of envelope models(with various masses, luminosities, effective temperatures and compositions) the quantity $\dot{\mu}$ = R L_{ac}/GM, where R and M are the stellar radius and mass respectively and L_{ac} is the total energy in acoustic waves generated per second within the convective envelope. The quantity $\dot{\mu}$ represents the mass loss rate.

064.036 **Hydrodynamic models of a cepheid atmosphere. III. Line spectrum and radius determinations.**
A. H. Karp.
Astrophys. Journ., Vol. 201, 641 - 652 (1975).

The hydrodynamic models of Karp have been used to study line profiles in a moving atmosphere. The line profiles are used to generate velocity curves which indicate that the assumptions inherent in the Wesselink method for determining radii of pulsating stars do not introduce large errors. These velocity curves are combined with the light and color curves of Karp to investigate the accuracy of the Wesselink method. It is shown that, while 10 percent accuracy is possible, the errors will probably be 15 percent to 20 percent. This result implies that the empirical period-radius relation cannot be used to detect overtone pulsators as suggested by Fernie.

064.037 **On two-parameter presentations of limb-darkening laws for classic stars.** A. A. Rubashevskij.
Astrometriya i Astrofizika, *Kiev*, vyp. (No.) 27, (see 003.004), p. 94 - 104 (1975). In Russian.

Two-parameter presentations of the limb-darkening law were analysed with the purpose to find the best approximation of the brightness distribution on the star's disk provided the total radiation flux is conserved. An approximation of laws of limb darkening by two straight lines is considered. The calculations show that detection of the effects due to non-linearity of the limb darkening leads to more reliable results than those obtained in previous investigations.

064.038 **Carbon abundances in G and K giants.**
D. Branch, R. A. Bell.
Monthly Notices Roy. Astron. Soc., Vol. 173, 299 - 308 (1975).

Model atmospheres and synthetic spectra are used to calibrate photoelectric measurements of the C_2 Swan bands in spectra of G and K giants. The calculations confirm the anticipated high sensitivity of the measured parameter, the '*c*-index', to the chemical composition. The atmospheric parameters effective temperature, surface gravity and iron-to-hydrogen ratio, derived by others from narrow-band photometry, are used to predict values of the *c*-index for a sample of stars on the 'null hypothesis', [C/Fe] = [O/Fe] = 0.0. The predicted and measured indices are well correlated, and departures from the null hypothesis are evidently small.

064.039 **Magnetic braking by a stellar wind–V. Approximate determination of the poloidal field.** I. Okamoto.
Monthly Notices Roy. Astron. Soc., Vol. 173, 357 - 379 (1975).

The exact equation which determines the steady, axisymmetric structure of the poloidal magnetic field in the stellar wind problem is derived from the transfield component of the equation of motion in the case of an isothermal corona.

064.040 **On the theory of laminar convection in atmospheres.** V. I. Kovshov.
Soobshch. Gos. Astron. Inst. Shternberga, No. 181, p. 33 - 40 (1973). In Russian.

The theory of convective energy transfer given by Lebedinskij (1941) in the case of any motion in an atmosphere is shown to describe in reality only laminar convection. The revision of Schmidt's formula of semi-empirical theory of turbulence in it proves to be ineffective as well as the attempt to find results for small-scale motions. The problem is raised to construct a non-contradictory theory of laminar convection.

064.041 **Objekte mit ausgedehnten Hüllen.** W. H. Kegel.
SuW, Vol. 14, 329 - 331 (1975).

064.042 **Theory of extended stellar atmospheres. II. A grid of static spherical models for O stars and planetary nebula nuclei.** P. B. Kunasz, D. G. Hummer, D. Mihalas.
Astrophys. Journ., Vol. 202, 92 - 113 (1975).

Spherical static non-LTE model atmospheres are presented for stars with $M/M_\odot = 30$ and 60 at various points on their evolutionary tracks, and for some nuclei of planetary nebulae at two points of a modified Harman-Seaton sequence. The temperatures for the non-LTE models are generally lower, at a given depth, than for the corresponding LTE models when $T_{eff} < 45{,}000$ K, while the situation is reversed at higher temperatures. Very severe limb darkening is found for extended models, which have apparent angular sizes significantly smaller than expected from the geometrical size of the star. Extensive tables are given of monochromatic magnitudes, continuum jumps and gradients, Strömgren-system colors, monochromatic extensions, and the profiles and equivalent widths of the hydrogen lines for all models, and of the He II lines for some of the 60 $M_{\odot}$ models.

064.043 **On the main problems in investigating the chemical composition of stellar atmospheres.**
A. A. Boyarchuk.
Izv. Krymskoj Astrofiz. Obs., Vol. 53, p. 242 - 243 (1975).
In Russian. – Abstr. of a paper presented at the conference "Magnetic fields and chemical composition of the sun and stars" at the Crimean Astrophys. Obs., 1974, April 9 - 12, (see 012.027).

064.044 **Method for investigation of the local chemical composition of the atmospheres of magnetic Ap stars.**
V. L. Khokhlova.
Izv. Krymskoj Astrofiz. Obs., Vol. 53, p. 251 - 252 (1975).
In Russian. – Abstr. of a paper presented at the conference "Magnetic fields and chemical composition of the sun and stars" at the Crimean Astrophys. Obs., 1974, April 9 - 12, (see 012.027).

064.045 **Evolutionary variations of helium abundance in the atmospheres of early-type stars.** L. S. Lyubimkov.
Pis'ma v Astron. Zhurn., Vol. 1, No. 11, p. 29 - 31 (1975). In Russian.

Helium abundance is determined in the atmospheres of about 100 stars of types O6.5–B2 which are either members of O-associations and young clusters or field stars. It is shown that the helium abundance tends to increase with the age of the cluster or association.

064.046 **Scattering by dust and the photographic appearance of Eta Carinae.** K. Davidson, M. T. Ruiz.
Astrophys. Journ., Vol. 202, 421 - 424 (1975).

The authors have attempted to construct simple isotropic-scattering models which reproduce the apparent surface brightness of the "halo" of η Carinae, as a function of projected radius.

064.047 **Emission envelopes of T Tauri stars.** P. Kuan.
Astrophys. Journ., Vol. 202, 425 - 432 (1975).

It is shown that the existence of extended moving envelopes around T Tauri stars can most adequately account for the presence of emission lines.

064.048 **Detection of X-ray emission from stellar coronae with ANS.** R. Mewe, J. Heise, E. H. B. M. Gronenschild, A. C. Brinkman, J. Schrijver, A. J. F. den Boggende.
Astrophys. Journ., (*Letters*), Vol. 202, L67 - L71 (1975).

With the soft X-ray detector aboard the Astronomical Netherlands Satellite (ANS) the authors have searched for soft X-ray emission in the range 0.2–0.284 keV from hot stellar coronae. X-ray emission has been detected from Capella (α Aur) and Sirius (α CMa). For 26 other stars (main sequence, subgiants, giants, and supergiants) upper limits to the X-ray luminosity have been obtained. The physical conditions of a corona needed to explain the observations are discussed.

064.049 **Similarity solution for unsteady stellar wind.**
S. Sakashita, M. Yokosawa.
Publ. Astron. Soc. Japan, Vol. 27, 553 - 560 (1975).

A similarity solution for the unsteady stellar wind is obtained and the dynamical effects of the wind on the surrounding interstellar medium are considered. It is shown that the flow patterns can be classified into two types depending on the mass loss rates. If the mass loss rates are small, the stellar wind smoothly adjoins the surrounding interstellar medium through a contact discontinuity, and an outward-facing shock wave proceeds into the interstellar medium. If the mass loss rates exceed a critical value, an inward-facing shock wave appears interior to the contact discontinuity and this front expands together with the contact discontinuity and the outward-facing shock.

064.050 **The effect of linearly polarized light from extended stellar atmospheres on interferometer response functions.** J. P. Cassinelli, N. M. Hoffman.
Monthly Notices Roy. Astron. Soc., Vol. 173, 789 - 800 (1975).

It is shown that the effective radius as determined by an interferometer for a star with an extended atmosphere may change by about 10 per cent or more when measured in orthogonal senses of polarization. Interferometer response functions are determined for four extended electron scattering atmospheres. Two recent expanding model atmospheres for ζ Pup are considered and it is shown that an intensity interferometer could distinguish between them.

064.051 **On the selective absorption in M-type stars. I. Influence of the blanketing effect on stellar magnitudes; the blocking integral coefficient.**
N. S. Komarov, A. V. Dragunova.
Astron. Zhurn. Akad. Nauk SSSR, Vol. 52, 1231 - 1240 (1975). In Russian. English translation in Soviet Astron., Vol. 19, No. 6.

The problem of the influence of the blanketing effect on the atmospherical structure in M-type stars is discussed. The radiation blocking effect by atomic lines and molecular absorption bands, and backwarming effect are considered independently. Calculation data of corrections (conditioned by the blanketing effect) to V and R magnitudes in the Johnson system for 24 M0 – M6-type stars are presented. The integral blocking coefficients and effective temperatures have been obtained as well. The dependence of the coefficients on the spectral type and the luminosity is considered.

064.052 **Radiative equilibrium of strongly non-grey atmospheres. I. General analysis.**
V. V. Ivanov, A. H. Heinlo.
Astron. Zhurn. Akad. Nauk SSSR, Vol. 52, 1252 - 1262 (1975). In Russian. English translation in Soviet Astron., Vol. 19, No. 6.

The radiative equilibrium of atmospheres the non-greyness of which is small-scale in frequency ν, although it may be arbitrarily large, is considered. It is shown that for the description of small-scale dependence of the intensity on ν one may use a statistical approach, retaining at the same time all the information on large-scale variations of the intensity with ν.

064.053 **The subsonic structure of radiatively driven winds of early-type stars.** J. P. Cassinelli, L. Hartmann.
Astrophys. Journ., Vol. 202, 718 - 732 (1975).

Methods are developed for the calculation of extended nongray model atmospheres for early-type stars with radiatively driven mass loss. Models for a central star of a planetary nebula and a Wolf-Rayet star are calculated, and compared with hydrostatic model calculations. It is found that transonic flow cannot occur if only continuum sources of opacity are present. Observations of the Of star ζ Pup and of the Wolf-Rayet star HD 50896 are summarized that indicate the continuum formation regions are spatially extended.

064.054 **An RR Lyrae-type ultraviolet model.**
S. J. Hill, J. L. Hutchinson, C. F. Lillie.
Bull. American Astron. Soc., Vol. 7, 510 - 511 (1975).
Abstr. AAS.

064.055 **Model atmospheres for M stars including the titanium oxide opacity.**
J. G. Collins, H. R. Johnson, B. M. Krupp.
Bull. American Astron. Soc., Vol. 7, 518 (1975). – Abstr. AAS.

064.056 **The effect of carbon enrichment on molecular band strengths in cool stars.** B. M. Krupp, H. R. Johnson.
Bull. American Astron. Soc., Vol. 7, 518 - 519 (1975). Abstr. AAS.

064.057 **On the circumstellar gas shells of the M-supergiants.** A. P. Bernat.
Bull. American Astron. Soc., Vol. 7, 519 (1975). – Abstr. AAS.

064.058 **Laser action in stellar envelopes.** Y. P. Varshni, C. S. Lam.
Bull. American Astron. Soc., Vol. 7, 551 (1975). – Abstr. AAS.

064.059 **Polarization models for early-type stars.** B. M. Haisch, J. P. Cassinelli.
Bull. American Astron. Soc., Vol. 7, 559 (1975). – Abstr. AAS.

064.060 **A method for empirical determination of stellar atmospheric structure.**
L. W. Ramsey, H. R. Johnson.
Solar Physics, Vol. 45, 3 - 14 (1975).

A technique for obtaining information on the temperature structure of a stellar atmosphere from spectral line data where only flux observations are available is discussed. The direct inversion of the flux integral to obtain the line source function can be circumvented by making the physically plausible assumptions of (1) source function equality in a multiplet and (2) the dominance of line absorption over continuum absorption at line center. Using high quality solar spectrum scans obtained from K.P.N.O. the authors demonstrate the accuracy of source function equality for several Fe I multiplets and use one of these multiplets to obtain an empirical outer atmosphere for the sun. The empirical atmosphere agrees well with current solar models.

064.061 **An archetype hydrogen atmosphere problem.** R. G. Athay, D. Mihalas, R. A. Shine.
Solar Physics, Vol. 45, 15 - 23 (1975).

Populations for the first three bound states and the continuum of hydrogen are determined for an isothermal, hydrostatic atmosphere at 20000 K. The atmosphere is treated as being optically thin in the Balmer and Paschen continua and illuminated by continuum radiation at these wavelengths with prescribed radiation temperatures. The atmosphere is optically thick in the 2-1, 3-1, 3-2 and c-1 transitions. The solution of this problem is non-trivial, and presents sufficient difficulty to have caused failure of at least one rather standard technique.

064.062 **On the possibility of constructing model atmospheres of R CrB type stars.** V. E. Panchuk.
Peremennye Zvezdy, Vol. 19, 541 - 548 (1975). In Russian.

A short historical review of the investigation of models of the atmospheres of cool stars is given. The difficulties in constructing models of R CrB atmospheres are discussed on K- and M-type atmospheres taken as an example.

064.063 **Circumstellar dust shells.** L. M. Shul'man
Peremennye Zvezdy, Vol. 19, 549 - 560 (1975). In Russian.

The optical properties of circumstellar dust shells are considered. In the case of an optically thin shell (e.g. the shell of a R CrB star after minimum) the temperature of the graphite grain is calculated. It is shown that the equilibrium temperature of the grain is essentially less than the effective temperature of the star. This result does not remove but reduces the difficulty of the condensation mechanism of the R CrB phenomenon.

064.064 **Model atmospheres of Delta Cephei.** R. Haefner.
Astron. Astrophys., Vol. 45, 287 - 296 = Veröff. Sternw. *München*, Vol. 7, No. 19 (1975).

Equivalent widths and radial velocities of many lines have been measured in several high dispersion spectra of δ Cep. Differential radial velocities at some phases suggest a gradient of the pulsational field in the atmosphere. The author has constructed model atmospheres at two phases –φ = 0.38 (no gradient) and φ = 0.78/0.80 (large gradient)–using the method of fine analysis. The time variations of density and pulsational velocity are similar and in phase. This result indicates that we have to assume a running wave. Using his final model atmospheres the author is able to explain his observations and independent measurements of other authors satisfactorily. The chemical abundances in the atmosphere of δ Cep are quite normal compared to the sun.

064.065 **On the role of He II absorption in the atmospheres of Wolf-Rayet (WR) stars.** I. F. Malov.
Astron. Tsirk., No. 887, p. 3 - 5 (1975). In Russian.

064.066 **On the theory of scattering photospheres.** V. V. Sobolev.
Astrofizika, Vol. 11, 499 - 510 (1975). In Russian. – English translation in Astrophysics, Vol. 11, No. 3.

The problem of the determination of the spectrum of a star in the photosphere of which together with true absorption scattering of radiation takes place is considered. The ratio of the coefficient of true absorption to that of scattering is assumed to depend on depth. The intensity of radiation emerging from the photosphere is expressed in terms of the function $\varphi(\eta, \tau)$ introduced earlier (1956) in solving the problem of diffuse reflection of light from an inhomogeneous atmosphere. Special consideration is given to the case of single scattering albedo exponentially decreasing with optical depth. Tables are given of the auxiliary quantities in terms of which the emergent intensity is expressed in this case. The results may be applied to photospheres of hot and cool stars in which scattering is caused by free electrons and molecules, respectively.

064.067 **Tables of theoretical ultraviolet fluxes and colours, including line opacities.** E. Peytremann.
Bull. Soc. Roy. Sci. Liège, (*Belgium*), Vol. 44, 119 - 156 = Commun. Dép. Astrophys., Fac. Sci. Mons, Mons Astrophys. Papers No. 47 (1975).

The author presents tables of ultraviolet stellar fluxes in the computation of which he has included the opacities of a very large number of atomic lines. In addition, he has used these fluxes to calculate magnitudes and colour indices in the photometric system of the telescope on board the Dutch ANS satellite.

064.068 **Model atmospheres for cool stars.** H. R. Johnson.
NASA Rep. NASA-CR-139752; NCAR-TN/STR-95, 101 pp. (1974).

This report contains an extensive series of model atmospheres for cool stars having a wide range in chemical composition. Model atmospheres are tabulated, along with emergent energy flux distributions, limb darkening, and information on convection for selected models.

064.069 **Temperature, gravity, and metal abundance calibration of the uv by photometric system and its application to blue horizontal-branch stars.** L. T. Matlock.
Thesis, State Univ. New York, Albany (USA). [Available from Univ. Microfilms, Ann Arbor, Mich., *USA*. Order No. 75-1566], 221 pp. (1974).

Theoretical uvby indices are computed from hydrogen-line-blanketed model atmospheres as functions of θ_e and log g for $0.40 \leqq \theta_e \leqq 0.70$, and $2.0 \leqq \log g \leqslant 4.0$. Values of θ_e, log g, and y_{bol} are derived for 19 blue horizontal-branch stars in the globular cluster NGC 6397 and the masses of the stars are computed.

064.070 **Convective envelopes and cooling of degenerate dwarfs.** M. A. Sweeney.
Thesis, Columbia Univ., New York (USA). [Available from Univ. Microfilms, Ann Arbor, Mich., *USA*. Order No. 74-29,669], 176 pp. (1974).

The energy content and thermal structure of degenerate dwarfs is studied for a wide range of masses, luminosities and envelope compositions. The relative importance of different treatment of pressure ionization, crystallization, specific heat, and opacity and of mass, luminosity, and chemical composition in determining the ages is investigated. It is demonstrated that the usual procedure of calculating the adiabatic gradient in high-pressure regions leads to violation of the Maxwell identities if corresponding changes are not made in the entropy and pressure of the gas. As a result of most of the effects considered in this investigation, it is found that convection plays a smaller role in determining the envelope structure and central temperature than previously assumed.

064.071 **Some remarks on supergiant photospheres.** C. de Jager.
Phil. Trans. Roy. Soc. London, Ser. A, Vol. 279, 421 - 427 (1975).

Some characteristic aspects of near ultraviolet spectra of supergiant stars, observed with the Utrecht ultraviolet stellar spectrometer S59 aboard E.S.R.O.'s TD1A satellite are described. A comparison of the observed maximum brightness of supergiants with theoretical computation shows that in the brightest early-type supergiants the outward driving forces due to radiation pressure and to the turbulent pressure gradient must be about equal.

064.072 **Mass loss from the A-type supergiant α Cygni.** H. J. G. L. M. Lamers.
Phil. Trans. Roy. Soc. London, Ser. A, Vol. 279, 445 - 450 (1975).

The Mg II resonance lines in α Cygni (A2Ia) are studied from 22 spectral scans obtained by the experiment S59.

ATLAS 5 [*a program for the calculation of model stellar atmospheres*] **at the University of London (SAM 1). I: Programming and implementation.** See Abstr. 021.026.

ATLAS 5 [*a program for the calculation of model stellar atmospheres*] **at the University of London (SAM 1). II. User's guide.** See Abstr. 021.027.

A photonuclear mechanism for promethium synthesis excluding technetium. See Abstr. 061.015.

Low-temperature Rosseland opacity tables. See Abstr. 061.069.

Stellar winds driven by Alfvén waves. See Abstr. 062.016.

The magnetic thermal wind and differential rotation of ionized, magnetic atmospheres. See Abstr. 062.031.

Aspects of frequency redistribution for non-coherent scattering in a turbulent medium. See Abstr. 063.007.

Resonance line transfer with partial redistribution. V. The solar Ca II lines. See Abstr. 063.010.

Non-LTE transfer. $\sqrt{\epsilon}$ revisited. See Abstr. 063.014.

Solution of the equation of radiative transfer in the atmosphere of a double star. Extended stellar atmosphere. See Abstr. 063.015.

Non-LTE radiative transfer in hydrogen with a self-consistent electron energy distribution. See Abstr. 063.019.

The calculation of the profiles of spectral lines forming in a scattering atmosphere. See Abstr. 063.021.

Non-LTE line formation in turbulent media. See Abstr. 063.035.

Solution of the comoving-frame equation of transfer in spherically symmetric flows. I. Computational method for equivalent-two-level-atom source functions. See Abstr. 063.039.

The 'peaking effect' of the emergent intensity in spherical-geometry transfer problems. See Abstr. 063.041.

Magnetic intensification as a mechanism for broad band linear polarization. See Abstr. 063.046.

On the lower mass limit for the carbon detonation scenario. See Abstr. 065.032.

On the surface gravities of Be stars. See Abstr. 065.049.

On shallow convective envelopes. See Abstr. 065.057.

An explanation for helium-rich and helium-variable stars: diffusion in radial mass-loss flux. See Abstr.065.147.

Macroturbulence broadening of solar lines with implications for stellar studies. See Abstr. 071.006.

Semiempirical upper photosphere models: the sun (G2 V) and Procyon (F5 IV–V). See Abstr. 071.030.

X-ray albedo of stellar atmospheres. See Abstr. 076.033.

The line response function of stellar atmospheres and the effective depth of line formation. See Abstr. 080.052.

Mass,energy and momentum transport in the solar atmosphere. See Abstr. 080.055.

Photometry of O-type stars, interpreted with reference to model atmospheres. See Abstr. 113.009.

The infrared excess of cool giant stars: a chromospheric contribution. See Abstr. 113.010.

Spectroscopic investigations of luminous emission-line objects. See Abstr. 114.010.

Relative abundances of iron-peak, α-, s-, and r-process elements in extremely metal-deficient dwarfs. See Abstr. 114.035.

Physical limits to attainable accuracies in stellar

radial velocities. See Abstr. 114.048.

Absorption line profiles and rotational velocities for 59 stars. See Abstr. 114.049.

The lines He I λ3187 and He II λ3203 in O-type stars. See Abstr. 114.083.

A non–LTE analysis of the CN 3883 Å band head in the upper photosphere of Arcturus. See Abstr. 114.320.

The He II lines in the spectrum of Zeta Puppis. See Abstr. 114.324.

Ultraviolet observations of cool stars. III. Chromospheric and coronal lines in α Tauri, β Geminorum, and α Bootis. See Abstr. 114.350.

On the UV spectrum of Beta Aurigae. See Abstr. 114.353.

A study of the atmospheres of F-type stars. II. A curve of growth analysis of α Canis Minoris. See Abstr. 114.359.

Fine analysis of three R Coronae Borealis stars. See Abstr. 114.362.

Curve-of-growth analysis of M-type giants β And and β Peg. See Abstr. 114.365.

The profile of He I λ5876 in the P-Cygni-type Of star HD 152408. See Abstr. 114.367.

An analysis of the superluminous star HR 8752. See Abstr. 114.371.

Evidence for a temperature rise in the outer layers of α Lyrae, from Copernicus observations of Lyman-α. See Abstr. 114.374.

Evidence for a rotational reddening in early B-type stars. See Abstr. 116.002.

The influence of rotation and stellar winds upon the Be phenomenon. See Abstr. 116.012.

Evolution of massive close binaries. I: The X-ray binary stage. See Abstr. 117.001.

Period changes in binary systems during the Wolf-Rayet stage. See Abstr. 117.020.

Gas flow in cataclysmic variable stars. See Abstr. 117.049.

On the evolution of close binaries with mass and momentum loss from the system. See Abstr. 117.052.

The binary nature of the Of star HD 108. See Abstr. 119.002.

Ultraviolet observations of Alpha Aurigae from Copernicus. See Abstr. 119.003.

Physical interpretation of light curves of R Coronae Borealis-type stars. See Abstr. 122.023.

Hydrodynamic effects in the atmospheres of variable stars. See Abstr. 122.115.

Radial velocities of stars with microwave maser emission. I. Long-period variables. See Abstr. 122.183.

White dwarf external layers. I. Convection and central temperatures. See Abstr. 126.008.

Problems in classifying cool degenerate stars. See Abstr. 126.011.

A numerical calculation of the effect of stellar winds on the interstellar medium. See Abstr. 131.110.

Stellar winds and globules in H II regions. See Abstr. 131.501.

Origin of planetary nebulae. See Abstr. 133.020.

The radio spectra of mass-outflow stars. See Abstr. 141.061.

Mass loss from infrared star CIT 6. See Abstr. 141.625.

Initial results of spatial interferometry at 5 microns. See Abstr. 141.626.

X-ray modulation by stellar winds in binary X-ray sources. See Abstr. 142.070.

A sensitive search for soft X-rays from Arcturus and Procyon. See Abstr. 142.134.

065 Star Formation, Stellar Structure and Evolution, Neutron Stars

065.001 **Rapidly rotating polytropes as a net of singular perturbation problems.** B. L. Smith.
Astrophys. Space Sci., Vol. 35, 223 - 240 (1975).

An analytical method is presented which determines completely the structure of rotating polytropes. It is shown that the basic Chandrasekhar expansion is of local validity only and must, for a comprehensive description, be supplemented by further locally-valid expansions. Although these further expansions have insufficient boundary conditions for unique determination the fact that each is, in principle, derived from the same exact solution leads to an asymptotic matching principle which supplies the missing constraints. The case $n = 1.5$ is considered in detail and third order results are presented which appear to be at most 1/18% in error.

065.002 **Structure of polytropic models of stars containing poloidal magnetic fields.** M. J. Miketinac.
Astrophys. Space Sci., Vol. 35, 349 - 362 (1975).

Polytropic models of axially-symmetric equilibrium stars of infinite conductivity with poloidal magnetic fields are constructed by numerical integration of the exact equations governing internal structure. Detailed results are given only for two polytropes with the index $n = 1.5$ and 3.0, although any other value of n greater than or equal to one could have been selected. Contrary to Monaghan's results, it is found that along the sequence of configurations for $n = 3.0$ the ratio of the magnetic and gravitational energy peaks out before a doughnut-shaped configuration is reached; but this effect does not characterize the $n = 1.5$ sequence. The calculations confirm, however, another result of Monaghan asserting that the magnetic field is a fairly insensitive function of the polytropic index.

065.003 **On the evolution and composition of Sirius and Procyon.**
L. Hartmann, L. M. Garrison, Jr., A. Katz.
Astrophys. Journ., Vol. 199, 127 - 130 (1975).

Stellar models are calculated and compared to the new observational data available for Sirius and Procyon. Sirius can be fitted by normal compositions ($X = 0.7$, $Z = 0.02$) very near the zero-age main sequence. Procyon is more evolved, hence more ambiguous; an anomalous composition (high X or Z) or a smaller mass than the astrometric mass is necessary to fit its position in the H-R diagram.

065.004 **Patterns of convection in the evolution of massive stars.** B. M. Schlesinger.
Astrophys. Journ., Vol. 199, 166 - 172 = Publ. Goethe Link Obs., Indiana Univ., *Bloomington*, No. 167 (1975).

Models for stars of 30 $M_\odot$ have been computed with an algorithm that assumes every point in the star to be radiative or fully convective. The ordinary Schwarzschild criterion is used. Semi-convection is not treated explicitly. In the resulting models, detached convective shells with superadiabatic inner boundaries appear instead of semi-convection. Since these boundaries travel inward by overshoot more rapidly than the core recedes, such detached convective zones would not persist in an actual star but would mix with the core. During shell hydrogen burning, a convective region outside the nuclear burning shell and a semi-conductive region farther outside will arise independent of overshoot. The $^{14}N + \alpha$ reaction may precede triple-α ignition in the core in stars of 30 $M_\odot$.

065.005 **Upper mass limits for stable rotating white dwarfs.**
R. H. Durisen.
Astrophys. Journ., Vol. 199, 179 - 183 (1975).

Models of nonmagnetic, axisymmetric, differentially rotating, zero-temperature white dwarfs are constructed and tested for stability. Upper mass limits, beyond which models become secularly and dynamically unstable to growth of non-axisymmetric perturbations, are found at 2.5 $M_\odot$ and 4.6 $M_\odot$ respectively. Implications of the results for mass transfer binary star systems and degenerate stellar cores are discussed.

065.006 **The surface of a neutron star in superstrong magnetic fields.** M. L. Glasser, J. I. Kaplan.
Astrophys. Journ., Vol. 199, 208 - 219 (1975).

The authors study the effects of electron interactions in the structure of matter in extreme ($B \gtrsim 10^{12}$ gauss) magnetic fields. It is found that interelectronic repulsion tends to produce an anisotropic crystalline structure.

065.007 **Red giants and supergiants with degenerate neutron cores.** K. S. Thorne, A. N. Żytkow.
Astrophys. Journ., (*Letters*), Vol. 199, L19 - L24 (1975).

A new type of stellar model is constructed. It is related to neutron stars as ordinary red giants are related to white dwarfs. Its external appearance is similar to that of an ordinary M supergiant, but its evolutionary lifetime is 10 times longer. The models are constrained to be relativistic but nonrotating, and to contain a degenerate neutron core of mass 1 $M_\odot$ and radius 10 km, surrounded by a nondegenerate, massive, diffuse envelope.

065.008 **The triple-alpha rate, screening factors, and the helium flash.** T. D. Tarbell, R. T. Rood.
Astrophys. Journ., Vol. 199, 443 - 447 (1975).

A simple but accurate stability analysis is presented for a degenerate stellar core containing a temperature-sensitive nuclear fuel. It is used to study the dependence of the helium core mass at the helium flash M_{cf} on the triple-α rate and its screening factor. Updated values of M_{cf} are computed, and a sizable discrepancy between the results of different workers is removed.

065.009 **Hydrodynamic models of a cepheid atmosphere. I. Deep envelope models.** A. H. Karp.
Astrophys. Journ., Vol. 199, 448 - 461 (1975).

The implicit hydrodynamic code of Kutter and Sparks has been modified to include radiative transfer effects. This modified code has been used to compute deep envelope models of a classical cepheid with a period of 12^d. It is shown that in this particular model the hydrogen ionization region plays only a small role in producing the observed phase lag between the light and velocity curves. The cause of the bumps on the model's light curve is examined, and a mechanism is presented to explain those cepheids with two secondary features on their light curves. This mechanism is shown to be consistent with the Hertzsprung sequence only if the evolutionary mass-luminosity law is used.

065.010 **Pion condensed matter at high densities: equation of state and stellar models.**
J. B. Hartle, R. F. Sawyer, D. J. Scalapino.
Astrophys. Journ., Vol. 199, 471 - 481 (1975).

Equations of state for the ground state of matter at nuclear densities and above are derived, including the effects of a condensed charged pion wave. Nonrotating neutron star models are constructed using these equations of state, and their observational significance is discussed.

065.011 **A non-axisymmetric rigid rotator model for magnetic stars.** M. J. Stift.
Monthly Notices Roy. Astron. Soc., Vol. 172, 133 - 139

(1975).

A generalized, non-axisymmetric decentred dipole model is proposed to explain asymmetric magnetic variations and phase shifts between the respective extrema of effective field and surface field of magnetic stars. Detailed models are derived for two stars and it is shown that the model can reproduce the observed magnetic variations of all well-known magnetic stars but one.

065.012 **On fragmentation in a collapsing gas cloud.** R. C. Smith, A. E. Wright.
Monthly Notices Roy. Astron. Soc., Vol. 172, 221 - 225 (1975).

The authors discuss the conditions under which fragmentation in a collapsing gas cloud stops when the cloud ceases to be isothermal, and they present an argument which suggests that these conditions hold in practice.

065.013 **Sterne aus He-reichen Mischungen.** M. Röser.
Mitt. Astron. Ges., No. 36, p. 75 - 79 (1975).

065.014 **Statistics of helium stars.** J. P. Kaufmann.
Mitt. Astron. Ges., No. 36, p. 79 - 80 (1975).

065.015 **Zur Entwicklung der Heliumsterne.** K. Hunger.
Mitt. Astron. Ges., No. 36, p. 80 - 81 (1975).
Abstract.

065.016 **Ein numerisches Verfahren zur Behandlung dynamischer Entwicklungsphasen von Sternen mit Axialsymmetrie.** K. J. Fricke, C. Möllenhoff, W. Tscharnuter.
Mitt. Astron. Ges., No. 36, p. 84 - 85 (1975). – Abstract.

065.017 **Kollaps einer rotierenden protostellaren Wolke.** W. Tscharnuter.
Mitt. Astron. Ges., No. 36, p. 85 (1975). – Abstract.

065.018 **Studies of heavy-element synthesis in the Galaxy. II. A survey of *e*-, *r*-, and *s*-process abundances.**
H. R. Butcher.
Astrophys. Journ., Vol. 199, 710 - 717 (1975).

Results are presented of a survey designed to search for variations in the relative abundances of *e*-, *r*-, and *s*-process elements in G-type dwarfs. Specifically, the abundances of Fe (*e*-process), Eu (*r*-process), and Ba (*s*-process) are examined in some 32 stars with as wide a range of ages, galactic orbits, and compositions as possible. No completely convincing relative abundance variations are found, to a limit of ±25 percent.

065.019 **Neutron star models with realistic high-density equations of state.**
R. C. Malone, M. B. Johnson, H. A. Bethe.
Astrophys. Journ., Vol. 199, 741 - 748 (1975).

The authors calculate neutron star models using four realistic high-density models of the equation of state. It is concluded that the maximum mass of a neutron star is unlikely to exceed 2 $M_\odot$. All of the realistic models are consistent with current estimates of the moment of inertia of the Crab pulsar.

065.020 **On protostellar evolution.**
C. K. Westbrook, C. B. Tarter.
Astrophys. Journ., Vol. 200, 48 - 60 (1975).

An investigation of the evolution of spherically symmetric protostars with initial masses in the range $0.1 \leq M/M_\odot \leq 50$ has been carried out. The authors present detailed numerical results for the evolution of a 1 $M_\odot$ protostar, a summary of the results for other protostellar masses, and a discussion of some observational implications of the work.

065.021 **Vibrational instability of a 1 $M_\odot$ star towards non-radial oscillations.** A. Boury, M. Gabriel, A. Noels, R. Scuflaire, P. Ledoux.
Astron. Astrophys., Vol. 41, 279 - 285 (1975).

The vibrational stability for non-radial oscillations of a 1 $M_\odot$ star during the main-sequence phase, is studied. The star becomes unstable at an age of 2.4×10^8 years. Stability is restored after three billion years.

065.022 **On the structure and stability of a polytrope with an isothermal core.** S. Yabushita.
Monthly Notices Roy. Astron. Soc., Vol. 172, 441 - 453 (1975).

Investigation is made of the structure of a gaseous sphere with a polytropic envelope of index 1.5 and isothermal core. It is shown that if the transition from isothermal state to polytropic equation of state occurs at given density and pressure, the mass-central density diagram of the configuration is very similar to the corresponding one of neutron star models, and that there exists a mass peak in the diagram.

065.023 **On the bubble method of stability analysis.**
R. C. Smith, K. J. Fricke.
Monthly Notices Roy. Astron. Soc., Vol. 172, 577 - 584 (1975).

In a recent paper, Simon has used a 'bubble method' to investigate the secular stability of stellar differential rotation and has obtained a weaker criterion for stability than the Goldreich–Schubert–Fricke (GSF) criterion, which was based on a local (plane-wave) analysis. It is pointed out that Simon's treatment involves a restriction on the class of perturbations considered. A more general bubble method is formulated and the GSF criterion is recovered, thus resolving the contradiction found by Simon.

065.024 **Some peculiarities of formation of population I and II stars.** A. G. Doroshkevich, I. G. Kolesnik.
In-t prikl. mat. AN SSSR. Preprint No. 9. Moskva, 1975. 33 pp. Price 10 Kop. In Russian. – Abstr. in Referativ. Zhurn. 51. Astron., 7.51.498 (1975).

065.025 **Existence and uniqueness of stellar equilibrium models.** J. Perdang.
Astrophys. Space Sci., Vol. 36, 111 - 135 (1975).

The stellar equilibrium equations for given surface pressure $P*$ and temperature $T*$, and in the absence of convection, are translated into a nonlinear integral equation, in which the radius R of the star enters as an eigenvalue. The author shows that under broad mathematical assumptions on the constitutive equations (equation of state, opacity and energy generation) a global existence and uniqueness property can be formulated.

065.026 **On the semiconvective zone in the stars with about one solar mass.** H. Saio.
Sci. Rep. Tôhoku Univ., First Ser., Vol. 58, 9 - 12 (1975).

It is found that during the main sequence evolution of stars with about one solar mass, the situation which requires to introduce the semiconvective zone at the boundary of the convective core arises. This situation is confirmed in terms of opactiy.

065.027 **Evolution of 16 $M_\odot$, 32 $M_\odot$ and 64 $M_\odot$ stars. III. Evolution of 32 $M_\odot$ and 64 $M_\odot$ stars on core carbon– and oxygen-burning stages.**
V. I. Varshavskij, A. V. Tutukov.
Nauchn. Informatsii, vyp. (No.) 32, p. 73 - 103 (1974). In Russian.

The structure and evolution of massive stars on the core carbon– and oxygen-burning stages were investigated by means of computations of evolutionary sequences. The influence of the UFI neutrino energy losses was studied. The computed stellar models allow to estimate changes of helium carbon and oxygen cosmic abundances due to evolution of massive stars. Some possible consequences of the rapid rotation of stellar cores on late stages of evolution and the meridional circulation were considered.

065.028 **Low-velocity *M*-dwarfs explained by very low-mass stars.** R. F. A. Staller.
Astron. Astrophys., Vol. 42, 155 - 157 (1975).

It is suggested that the *M*-dwarfs with a low velocity dispersion (some 50% of the *M*-dwarfs near the galactic north pole) might be contracting stars with masses between 0.01 $M_\odot$ and 0.07 $M_\odot$. Contraction tracks of such stars pass through the *M* spectral region and the total duration of their contraction and cooling (before becoming invisible) is of the order of 5×10^8 to 10^9 yrs. If this hypothesis is correct the number of black dwarfs in the Galaxy must be at least some 10 to 20 times the numer of low velocity dispersion *M*-dwarfs.

065.029 **Transition from a red giant to a blue nucleus after ejction of a planetary nebula.**
R. Härm, M. Schwarzschild.
Astrophys. Journ., Vol. 200, 324 - 329 (1975).

The evolution of two stars, with 0.65 $M_\odot$ and 1.00 $M_\odot$, respectively, has been computed through the transition phase following the rapid dynamical ejection of a planetary nebula by a red giant. This ejection phase was mimicked by a sequence of hydrostatic red-giant models with steadily decreasing mass. The sequence was terminated at the moment of rapid decrease of the previously high pulsational instability. With the starting model thus derived, the constant-mass transition phase was found to last about 6000 years. This time interval is just short enough that the star can start providing the ultraviolet radiation in time to excite the ejected nebula before the nebula completely disperses.

065.030 **Chemical composition and the Hertzsprung gap.**
B. F. Kinahan, R. Härm.
Astrophys. Journ., Vol. 200, 330 - 335 (1975).

The effect of initial chemical composition on the size and position of the Hertzsprung gap is examined. Evolutionary sequences for stars in the mass range 0.8 $M_\odot$ to 2.5 $M_\odot$ with initial chemical composition ranging from $X = 0.7$ to 0.9 and $Z =$ 0.001 to 0.04 have been computed with the Cox-Stewart opacities handled by bicubic interpolation. The models show that chemical composition does not affect the size and position of the Hertzsprung gap to the extent necessary to explain such peculiar color-magnitude diagrams as that of the Large Magellanic Cloud cluster NGC 2209.

065.031 **Stellar cooling in gauge theories of weak interactions.** K. O. Mikaelian.
Astrophys. Journ., Vol. 200, 336 - 338 (1975).

The author considers a large class of processes leading to stellar energy loss, and derives general rules relating the rates in the ordinary V–A theory to those of renormalizable gauge theories of weak and electromagnetic interactions. In these theories energy loss to both electron-type and muon-type neutrinos is allowed. Assuming μ-e universality, the rates attain their minimum for certain values (approximately realized in the Weinberg model) of the neutral-current couplings. More general results, without the assumption of μ-e universality, are also stated.

065.032 **On the lower mass limit for the carbon detonation scenario.** A. S. Endal, W. M. Sparks.
Astrophys. Journ., (*Letters*), Vol. 200, L77 - L78 (1975).

The lower mass limit for carbon detonation ($\sim 4\ M_\odot$) is due to an argument by Paczynski that stars in the 1.4–3.5 $M_\odot$ range will lose their envelopes prior to carbon detonation. This argument is based on two criteria for mass loss: (1) the envelope is unstable to adiabatic oscillations, and (2) the total energy of the envelope is positive. Recent hydrodynamic calculations of red-giant envelopes have shown that the envelopes are highly nonadiabatic, and that the second criterion is not a sufficient condition for mass loss.

065.033 **Discussion of simple stability criteria for stars with central helium burning.**
M. Gabriel, S. Refsdal, H. Ritter.
Astrophys. Space Sci., Vol. 36, 439 - 450 (1975).

The validity of simple stability criteria for stars with central He-burning used in earlier papers is investigated. The determinant of secular stability for $\sigma = 0$ is related to quantities defined as far as possible at the surface of the He-core and at the bottom of the envelope. This relation is used to discuss the validity of two intuitive criteria for secular stability.

065.034 **Line profiles and magnetic intensification in magnetic stars.** O. Havnes, O. K. Moe.
Astron. Astrophys., Vol. 42, 269 - 272 (1975).

Line profiles with Zeeman analyzer and without has been computed for a 8000°K magnetic star using the theory of Unno (1956). Different orientations of the star and its magnetic field and different values of the magnetic field strengths have been considered. For a weak Fe 4260 line the authors find that magnetic intensification can introduce errors of about 0.25 dex in abundance. The effect of the magnetic field on the line profiles is large and the authors find it unlikely that reliable values for micro and macroturbulence can be found by conventional non-magnetic analysis.

065.035 **Sur la stabilité des étoiles supermassives au voisinage de la masse critique.** R. Scuflaire.
Comptes Rendus Acad. Sci. Paris, Sér. B, Vol. 281, 141 - 144 (1975).

Le calcul complet des modes normaux de perturbation linéaire radiale d'étoiles supermassives de masses voisines de la masse critique d'instabilité révèle l'existence d'une transformation continue d'un mode de type dynamique à un mode de type séculaire et réciproquement lorsqu'on fait varier la masse.

065.036 **The effect of composition changes on evolutionary tracks of double-shell models.**
J. M. Scalo, R. K. Ulrich.
Astrophys. Journ., Vol. 200, 682 - 687 (1975).

Evolutionary tracks of double-shell stellar models are computed using a core-envelope fitting technique. Uncertainties in the continuous and molecular contributions to the mean opacity are investigated. A polynomial expression for the Rosseland mean cyanogen opacity is given. The authors find that variations in the carbon and oxygen abundances greatly affect the position of models in the H-R diagram, the effect being much larger for carbon stars than oxygen stars.

065.037 **Finite-amplitude convection in a compressible medium and its application to solar granulation.**
R. Van der Borght.
Monthly Notices Roy. Astron. Soc., Vol. 173, 85 - 95 (1975).

The results of an investigation in the theory of finite-amplitude convection in a compressible medium with polytropic structure are applied to the construction of a model of the solar granulation. A discussion is made of the various parameters involved in the problem and it is shown that a reasonable agreement can be reached between the theoretical predictions and the observed values.

065.038 **On gravity darkening in magnetic stars.**
R. C. Smith.

Monthly Notices Roy. Astron. Soc., Vol. 173, 97 - 102 (1975).

The technique of a previous paper is used to evaluate the gravity darkening in some published models of rotating magnetic stars. For most of the models studied, there is only a small departure from the von Zeipel relation. However, for stars with a large ratio of magnetic energy to rotational energy the variation of emergent flux with gravity could be quite strong. This could be relevant for a few Ap stars.

065.039 **Models of upper main sequence stars containing a toroidal component of magnetic field.**
D. L. Moss.
Monthly Notices Roy. Astron. Soc., Vol. 173, 141 - 160 (1975).

Models appropriate to upper main sequence stars in strict radiative equilibrium with linked toroidal and poloidal magnetic fields are constructed. For most of the calculations the poloidal field is assumed to be strictly dipolar. Models are found with comparable poloidal and toroidal fluxes, and it is plausible that some are stabilized against the instabilities experienced by a purely poloidal field in the neighbourhood of the neutral line. However, it does not seem that models calculated by this method will be dynamically prolate, as may be necessary to explain the large inclination of magnetic and rotational axes in many of the observed magnetic stars.

065.040 **Slowly rotating neutron stars and transient X-ray sources.** A. C. Fabian.
Monthly Notices Roy. Astron. Soc., Vol. 173, 161 - 165 (1975).

A neutron star in a weak stellar wind from a binary companion will spindown because of pulsar emission and the 'propeller' mechanism of Illarionov & Sunyaev. Large density fluctuation in this wind can then turn the neutron star into a transient X-ray source with a relatively long rotation period. The transient X-ray source, Ariel 1118–61, which has a period of 6.75 min is shown to be in general agreement with the above considerations.

065.041 **Time dependent turbulent mixing.**
L. D. Cloutman.
Bull. American Astron. Soc., Vol. 7, 402 - 403 (1975). – Abstr. AAS.

065.042 **Effects of convective overshoot on lithium depletion in main sequence stars.**
J. M. Straus, J. B. Blake, D. N. Schramm.
Bull. American Astron. Soc., Vol. 7, 403 (1975). – Abstr. AAS.

065.043 **Cooling times and parent masses of degenerate dwarfs.** M. A. Sweeney.
Bull. American Astron. Soc., Vol. 7, 403 (1975). – Abstr. AAS.

065.044 **Diffusion of helium in stars.** P. D. Noerdlinger.
Bull. American Astron. Soc., Vol. 7, 403 (1975). Abstr. AAS.

065.045 **The evolution of rotating stars.**
A. S. Endal, S. Sofia.
Bull. American Astron. Soc., Vol. 7, 403 - 404 (1975). – Abstr. AAS.

065.046 **On the relationship between integration and relaxation techniques for the numerical calculation of stellar evolution.** M. J. Newman, R. J. Talbot, Jr.
Bull. American Astron. Soc., Vol. 7, 404 (1975). – Abstr. AAS.

065.047 **A comparison of evolutionary model predictions with statistics of field stars.** B. M. Schlesinger.
Bull. American Astron. Soc., Vol. 7, 404 (1975). – Abstr. AAS.

065.048 **Star formation in NGC 5128 = Centaurus A.**
S. van den Bergh.
Bull. American Astron. Soc., Vol. 7, 412 (1975). – Abstr. AAS.

065.049 **On the surface gravities of Be stars.** G. J. Peters.
Bull. American Astron. Soc., Vol. 7, 433 (1975). Abstr. AAS.

065.050 **The pulsational properties of hot degenerate dwarfs with nuclear reactions near the surface.**
E. M. Sion, S. C. Vila.
Bull. American Astron. Soc., Vol. 7, 442 (1975). – Abstr. AAS.

065.051 **Symmetrical accretion of hydrogen-rich gas onto degenerate carbon dwarfs.**
J. D. Colvin, H. M. Van Horn.
Bull. American Astron. Soc., Vol. 7, 442 (1975). – Abstr. AAS.

065.052 **Hydromagnetostriction and mixing in low-mass stars.**
P. G. Gross.
Bull. American Astron. Soc., Vol. 7, 445 - 446 (1975). – Abstr. AAS.

065.053 **Observational evidence for mass loss in globular clusters: giants.** J. G. Cohen.
Bull. American Astron. Soc., Vol. 7, 446 (1975). – Abstr. AAS.

065.054 **Determining aspect of rotational axis from profiles of He I lines.** T. R. Stoeckley, W. Buscombe.
Bull. American Astron. Soc., Vol. 7, 450 (1975). – Abstr. AAS.

065.055 **Properties of rotating neutron stars.**
M. A. Abramowicz, R. V. Wagoner.
Bull. American Astron. Soc., Vol. 7, 454 - 455 (1975). – Abstr. AAS.

065.056 **A numerical model for the polarized flux from Be stars.** M. W. Johns.
Bull. American Astron. Soc., Vol. 7, 498 (1975). – Abstr. AAS.

065.057 **On shallow convective envelopes.** R. G. Deupree.
Astrophys. Journ., Vol. 201, 183 - 189 (1975).

Three-dimensional (two spatial dimensions and time) partial differential equations of mass, momentum, and energy conservation have been integrated numerically by finite difference techniques over a sufficiently long time interval to allow convective motion to develop from computer round-off error. The effects of several assumptions are discussed, and the method of solution is applied to a particular situation in which most of the assumptions are well satisfied: a shallow convective envelope with extensive convective motion only in the second helium ionization zone. The size of the convective cells, the evidence for the existence of only one growth rate of the convective motion, the pressure perturbations, and the effects of nonadiabaticity are discussed. Finally, the effects of nonlinear terms are discussed, and it is found that these terms tend to drive the convective motion toward a steady state, even though the particle and radiative viscous dissipation is negligible on this scale of motion.

065.058 **Neutronization and thermal disintegration of dense stellar matter.** R. I. Epstein, W. D. Arnett.
Astrophys. Journ., Vol. 201, 202 - 211 (1975).

The properties of matter in nuclear statistical equilibrium are investigated over the range of density, temperature, and electron number $10^9 \lesssim \rho \lesssim 10^{11}$ g cm^{-3}, $2 \times 10^9 \lesssim T \lesssim 4 \times 10^{10}$ and $0.50 \lesssim Y_e \lesssim 0.45$; it is just this range of parameters which is relevant for stellar material subsequent to silicon burning. The ion number, the nuclear binding energy, and the rates of neutronization and energy emission by neutrinos have been calculated. A simple phenomenological picture of the relevant physical processes is presented and is used to develop approximate expressions for the weak interaction rates and the thermodynamic properties of the matter.

065.059 **On the nature and evolution of T Tauri stars.** A. E. Rydgren.
Publ. Astron. Soc. Pacific, Vol. 87, 509 (1975). – Abstr. ASP.

065.060 **Stellar evolution IV: Evolution of a star of 1.5 $M_\odot$ from the main-sequence to the red-giant branch with and without overshooting from convective core.**
A. Maeder.
Astron. Astrophys., Vol. 43, 61 - 69 (1975).

For a star of 1.5 $M_\odot$ with an initial composition given by $X = 0.70$ and $Z = 0.03$, three sets of evolutionary models are computed with different assumptions on the non-local effects characterizing the turbulent motions in the convective core. Some overshooting from the convective core may occur during main-sequence evolution. Some comparisons between models and observational results are attempted. Some comments are made on the estimates of helium and metal abundances derived from features of cluster sequences.

065.061 **Rotational distortion of polytropes by analytic approximation.** A. P. Linnell.
Astrophys. Space Sci., Vol. 37, 73 - 86 (1975).

This paper modifies the first-order perturbation theory of Chandrasekhar, for rotational distortion of polytropes. Comparison with numerical integrations by other authors demonstrates that the present analytic theory is as accurate as other published first-order theories. The present theory is in a form permitting rapid calculation of boundary shapes as a function of the rotation parameter, v, and the polytrope index, n. Results are presented for a critically rotating polytrope, for the case $n = 3$.

065.062 **On the interaction between the URCA process and convection.** O. Regev, G. Shaviv.
Astrophys. Space Sci., Vol. 37, 143 - 150 (1975).

The influence of the URCA process (β-decay, electron capture) on convective stability is investigated and a stability criterion is derived. The criterion contains the Schwarzschild criterion and the thermohaline convection. It is found that URCA may cause convective instability to an otherwise stable medium, but it cannot stabilize an unstable medium, only the growth rate is affected. The evolution of C–O core of an evolved 3.5 $M_\odot$–8 $M_\odot$ star may be affected by formation of convective regions even prior to C^{12} ignition.

065.063 **Gravitational contraction of protostars. IV. Opaque core formation.** I. G. Kolesnik.
Astrometriya i Astrofizika, *Kiev*, vyp. (No.) 26, (see 003.001), p. 24 - 37 (1975). In Russian.

For a protostar with $M = 1\,M_\odot$, $R_0 = 10^6 R_\odot$ and cooling by dust grains the opaque core formation is considered. The changes in the core characteristics with time are calculated. The protostar core mass is about 0.025 M_0. The final mass of the star is supposed not to exceed 0.1 M_0.

065.064 **Hydrodynamics of optically thin collapsing protostars.** I. G. Kolesnik.
Nauchn. Informatsii, vyp. (No.) 33, p. 6, 22 (1974).
In Russian and English. – Abstr. of a paper presented at the colloquium 'Physics and evolution of stars', (see 012.015).

065.065 **Numerical calculation of a collapse.** P. J. Morel.
Nauchn. Informatsii, vyp. (No.) 33, p. 6 - 7, 23 (1974). In Russian and English. – Abstr. of a paper presented at the colloquium 'Physics and evolution of stars' (see 012. 015).

065.066 **Are the cosmic OH and H_2O sources protostars or protoplanets?** V. S. Strel'nitskij.
Nauchn. Informatsii, vyp. (No.) 33, p. 7, 23 - 24 (1974). In Russian and English. – Abstr. of a paper presented at the colloquium 'Physics and evolution of stars', (see 012.015).

065.067 **Late stages of stellar evolution. Pulsed γ-radiation of neutron, collapsing and supernova stars.**
G. S. Bisnovatyj-Kogan, V. S. Imshennik, D. K. Nadezhin, V. M. Chechetkin.
Nauchn. Informatsii, vyp. (No.) 33, p. 15, 31 - 32 (1974). In Russian and English. – Abstr. of a paper presented at the colloquium 'Physics and evolution of stars', (see 012.015).

065.068 **Carbon burning with convective URCA-neutrinos.** B. Paczynski, E. Ergma.
Nauchn. Informatsii, vyp. (No.) 33, p. 16 - 17, 33 (1974). In Russian and English. – Abstr. of a paper presented at the colloquium 'Physics and evolution of stars', (see 012.015).

065.069 **On the problem of blue and red supergiants.** V. I. Varshavskij.
Nauchn. Informatsii, vyp. (No.) 33, p. 17, 33 - 34 (1974). In Russian and English. – Abstr. of a paper presented at the colloquium 'Physics and evolution of stars', (see 012.015).

065.070 **Statistical criteria for stability of arbitrarily rotating stars.** G. S. Bisnovatyj-Kogan, S. I. Blinnikov.
Nauchn. Informatsii, vyp. (No.) 33, p. 18, 34 - 35 (1974). In Russian and English. – Abstr. of a paper presented at the colloquium 'Physics and evolution of stars', (see 012.015).

065.071 **Evolution of intermediate mass stars with rapid differential rotation.** I. Tuominen, V. Muzylev.
Nauchn. Informatsii, vyp. (No.) 33, p. 48 - 71 (1974). In Russian.

The evolution of $4M_\odot$, $6M_\odot$ and $8M_\odot$ stars with rapid differential rotation has been computed under assumptions of spherical symmetry and local conservation of angular momentum. The position of the stellar model on the HR-diagram is determined by its total luminosity and mean effective temperature. Evolutionary tracks and isochrones corresponding to different assumptions on the initial distribution of angular momentum were compared with the HR-diagram of the Pleiades.

065.072 **Evolution of 16 $M_\odot$, 32 $M_\odot$ and 64 $M_\odot$ stars. IV. Evolution of pure hydrogen-helium 16 $M_\odot$ and 32 $M_\odot$ stars.** V. I. Varshavskij.
Nauchn. Informatsii, vyp. (No.) 33, p. 72 - 78 (1974). In Russian.

The structure and evolution of massive pure hydrogen-helium ($Z = 0$) stars were investigated through core hydrogen- and helium-burning stages up to core carbon ignition. The results are compared with those for the evolution of population I stars. The hydrogen-helium stars spend most part of their life time in the O–B supergiant region of the HR-diagram.

065.073 **Investigation of the collapse of protostellar clouds on the transparent stage. I. Intermediate masses.**
I. A. Kharichev.
Nauchn. Informatsii, vyp. (No.) 33, p. 79 - 88 (1974). In Russian.

The influence of the value of relative dust content on main characteristics of the collapse of a protostar were investigated for clouds with mass 5 $M_\odot$. It is shown that for $a_n = N_d/N_H > 10^{-14}$ collapse is nearly isothermal with temperature close to that of the dust. The mass of the central dense core that is formed in the course of the transparent stage of the protostellar evolution depends on the relative dust content.

Is is also shown that for clouds with mass less than ~ $1.5 M_\odot$ the central dense core cannot completely form on the transparent stage of the collapse.

065.074 **Investigation of the collapse of protostellar clouds on the transparent stage. II. Large masses.**
I. A. Kharichev.
Nauchn. Informatsii, vyp. (No.) 33, p. 89 - 100 (1974). In Russian.

The influence of the relative content of heavy elements has been investigated for the collapse of protostellar clouds with mass equal to $50 M_\odot$. It is shown that the influence of heavy elements is relatively weak on the collapse. The relative mass of the central dense core depends on the total mass of the cloud, the initial density and its distribution.

065.075 **La théorie des étoiles jumelles et l'évolution stellaire.** A. Dauvillier.
Ciel et Terre, Vol. 91, 278 - 291 (1975).

065.076 **Die Entwicklung der Sterne.** R. Kippenhahn.
Nova Acta Leopoldina [Deutsche Akademie der Naturforscher Leopoldina, Halle/Saale], New Ser., Vol. 42 (No. 218), 115 - 134 (1975). – Lecture on the occasion of the annual meeting of the Academy, 1973 October 11 - 14.

065.077 **The effect of rotation in stellar structure and evolution.** Introductory report. I. W. Roxburgh.
19th Colloquium International Astrophys. Liège 1974, (see 012.016), p. 15 - 24 (1975).

The development of the present understanding of rotation in stellar structure and evolution is briefly reviewed. More recent work on rotation and convection, solar oblateness, the stability of differential rotation and the effect of rapid rotation on the solar neutrino flux is discussed.

065.078 **The quasidynamic method for rotating stellar models.** A. Kovetz, N. R. Lebovitz, G. Shaviv.
19th Colloquium International Astrophys. Liège 1974, (see 012.016), p. 25 - 30 (1975).

The equations of structure for a rotating star are modified by replacing the dynamic terms with certain quasidynamic terms that vanish when the configuration is in a steady state. The modified equations have the property that the steady-state solution is asymptotically stable if it is dynamically stable in the usual sense, and unstable if it is dynamically unstable in the usual sense. These equations are particularly suited for numerical integration, for the numerical scheme based on them converges if and only if the steady-state solution sought is dynamically stable. This incorporates in the numerical scheme a stability test that would otherwise have to be performed separately.

065.079 **Differential rotation and turbulence in stars.** J.-P. Zahn.
19th Colloquium International Astrophys. Liège 1974, (see 012.016), p. 31 - 34 (1975).

Due to the low viscosity of stellar interiors, shear instabilities are likely to be prominent in the radiative zones of differentially rotating stars. Because of the stabilizing effect of density stratification, the turbulent motions generated by those instabilities are mainly horizontal and they act to smooth out the differential rotation over level surfaces. A much stronger shear is tolerated in the vertical direction, between horizontal layers; the critical conditions for instability are estimated using a modified Richardson criterion. Finally, the interaction of this shear instability with the Eddington-Sweet circulation is briefly discussed.

065.080 **Convection in a rotating sphere.** A. D. Weir.
19th Colloquium International Astrophys. Liège 1974, (see 012.016), p. 37 - 42 (1975).

Numerical calculations of non-linear axisymmetric convection in a uniformly heated self-gravitating sphere of Boussinesq fluid indicate that the law of rotation in the convective core of a rotating star is more like one of uniform angular momentum per unit mass than of uniform rotation.

065.081 **Time scales in stellar spin-down.** A. Clark, Jr.
19th Colloquium International Astrophys. Liège 1974, (see 012.016), p. 43 - 46 (1975).

The distribution of angular momentum in the radiative core of a lower main sequence star will evolve in a manner determined by meridional circulations. Such circulations have two distinct sources: thermal driving throughout the core (Eddington-Sweet currents), and boundary driving associated with the stellar-wind torque on the outer convective envelope (spin-down). Spin-down under stellar conditions is a complex process involving a number of internal time scales. The present work gives a qualitative discussion of these time scales.

065.082 **Bubble stability of differential rotations.** R. Simon.
19th Colloquium International Astrophys. Liège 1974, (see 012.016), p. 57 - 67 (1975).

The stability of stellar differential rotations is investigated by means of the approximate bubble method. Thermal conductivity is taken into account, but viscosity is neglected.

065.083 **Role of semi-convection in the evolution of massive stars.** S. R. Sreenivasan.
19th Colloquium International Astrophys. Liège 1974, (see 012.016), p. 73 (1975). – Summary.

065.084 **Effects of magnetic fields.** Introductory report. L. Mestel.
19th Colloquium International Astrophys. Liège 1974, (see 012.016), p. 79 - 91 (1975). – Review paper.

065.085 **Instabilities of stars containing magnetic fields.** P. Markey, R. J. Tayler.
19th Colloquium International Astrophys. Liège 1974, (see 012.016), p. 93 - 95 (1975).

Many possible magnetic field configurations inside stars can be shown to be unstable and the instabilities have growth times which are very short compared with either the thermal or the nuclear timescale of the star. It seems probable that most purely poloidal and purely toroidal fields are unstable and that the instabilities might lead to a substantial reduction of magnetic flux during a star's lifetime.

065.086 **Meridional circulation in magnetic stars.** D. L. Moss.
19th Colloquium International Astrophys. Liège 1974, (see 012.016), p. 97 - 103 (1975).

Models of rotating magnetic stars with a steady state meridional circulation are found. A first order perturbation technique is used, and the velocity and magnetic fields are expanded in a series of Legendre polynomials.

065.087 **Circulations and fossil magnetic fields in rotating stars.** M. Maheswaran.
19th Colloquium International Astrophys. Liège 1974, (see 012.016), p. 107 - 108 (1975). – Abstract.

065.088 **Some new results on the generation of magnetic fields by dynamo action.** K.-H. Rädler.
19th Colloquium International Astrophys. Liège 1974, (see 012.016), p. 109 - 116 (1975).

The paper deals with spherical dynamo models with both large and small scale motions showing some symmetries as to be expected in stellar objects. A survey is given on the types of magnetic fields which can be excited. Results for a special α-effect dynamo, the α^2-dynamo, are presented. In addition to axisymmetric stationary magnetic fields also non-axisymmetric oscillatory ones were found. The possibility is discussed of superposing different fields which leads to fields as occuring with the oblique rotator.

065.089 **Stellar oscillations, stellar stability and application to variable stars.** J. P. Cox.
19th Colloquium International Astrophys. Liège 1974, (see 012.016), p. 129 - 159 (1975).
A general review of current theories of stellar oscillations is presented, including linear, nonlinear, adiabatic, non-adiabatic, radial, and nonradial oscillations. Applications to various known types of intrinsic variable stars are discussed, and some outstanding problems in our current understanding of these stars are summarized.

065.090 **Vibrational stability of stars in thermal imbalance.** J. Demaret.
19th Colloquium International Astrophys. Liège 1974, (see 012.016), p. 161 - 172 (1975).
The author gives a solution which provides a general integral expression for the vibrational stability coefficient for arbitrary stars in thermal imbalance. The physical interpretation of this stability coefficient is discussed and its generality and its simplicity are stressed. Finally, he describes briefly some applications to simple analytic stellar models and to more realistic configurations.

065.091 **Stability of nonlinear periodic pulsation of stellar envelopes.** K. v. Sengbusch.
19th Colloquium International Astrophys. Liège 1974, (see 012.016), p. 189 - 193 (1975).
The eigenvalue method to calculate models of periodically pulsating stars can be used to study the stability of the resulting oscillations. Results for a series of RR-Lyrae models are discussed.

065.092 **Vibrational stability of relativistic supermassive stars in the pre-main-sequence contraction stage.**
J. Demaret.
19th Colloquium International Astrophys. Liège 1974, (see 012.016), p. 233 - 242 (1975).
The author applies a general relativistic extension of a formula for the vibrational stabiltiy coefficient of classical stars in thermal imbalance to the case of non rotating supermassive stars ($10^5 M_\odot < M < 10^6 M_\odot$) in the pre-main-sequence contraction stage. A vibrational instability appears very near the main-sequence radius for stars of mass smaller than the critical mass for dynamical instability, $M_{CR} \simeq 3.5 \times 10^5 M_\odot$ or very near the critical radius for dynamical instability for stars of mass greater than M_{CR}.

065.093 **The pulsational stability of stars in thermal imbalance.** M. L. Aizenman, J. P. Cox.
19th Colloquium International Astrophys. Liège 1974, (see 012.016), p. 243 - 245 (1975).
In this paper the authors summarize briefly their work on the pulsational stability of stars in thermal imbalance (i.e., stars which in their nonpulsating states are in hydrostatic, but not thermal, equilibrium). This work is all correct to the first order in t_{ff}/t_s, where t_{ff} and t_s are, respectively, the "free fall" time and a "slow", or "secular", time often of the order of the Kelvin time for the star. It is assumed throughout that $t_{ff}/t_s \ll 1$.

065.094 **The vibrational stability of evolving stars.** W. R. Davey.
19th Colloquium International Astrophys. Liège 1974, (see 012.016), p. 247 - 249 (1975).
The vibrational stability of quasi-statically evolving stars (i.e. stars which are in hydrostatic but not thermal equilibrium) has recently been investigated using four separate lines of attack: (1) the small perturbation approach; (2) an energy approach; (3) an asymptotic approach; and (4) direct solutions of the appropriate differential equations. In all cases that the author has examined, each of the four methods produces identical results. Integral expressions are presented in a very simple form for computing stability coefficients.

065.095 **Effect of rigidity on non-radial oscillations.** C. Denis.
19th Colloquium International Astrophys. Liège 1974, (see 012.016), p. 253 - 257 (1975).

065.096 **Non radial oscillations in massive stars during core H-burning phase.** C. Chiosi.
19th Colloquium International Astrophys. Liège 1974, (see 012.016), p. 259 - 272 (1975).
The results of numerical integration of the second order differential equation for the adiabatic non radial oscillations of quadrupole ($l = 2$) for evolutionary models of massive stars during the core H-burning phase are presented. The oscillations have been tested for pulsational stability and several g_+ modes are found to be unstable.

065.097 **Stability of g^+ modes in a 30 $M_\odot$ star.** M. Gabriel, A. Noels, R. Scuflaire, A. Boury.
19th Colloquium International Astrophys. Liège 1974, (see 012.016), p. 273 - 286 (1975).
The evolution of a population I 30 $M_\odot$ star during the main-sequence phases has been computed without semi-convective zone. Three models are tested for stability towards g^+ modes of non-radial oscillation. Whereas the ZAMS model is found stable, evolved models are unstable.

065.098 **Excitation of non-radial oscillations by partial ionization zone mechanism.** W. Dziembowski.
19th Colloquium International Astrophys. Liège 1974, (see 012.016), p. 287 - 297 (1975).
Complete stellar models corresponding to mean states of various pulsating variables were tested for vibrational stability against non-radial modes by means of the linear nonadiabatic analysis. It is shown that only some of pulsating horizontal branch stars are unstable and even then the excitation of non-radial modes is unlikely as the amplitude growth rates are much lower than those for radial modes.

065.099 **Nonadiabatic nonradial oscillations of a solar model.** J. Christensen-Dalsgaard, D. O. Gough.
19th Colloquium International Astrophys. Liège 1974, (see 012.016), p. 309 - 316 (1975).
The stability of a solar model on the main sequence to linearized nonadiabatic nonradial oscillations has been studied. In the inner 99.9% by radius the eigenfunctions were barely distinguishable from those of the adiabatic approximation. The integrated effect of this region is to stabilize the oscillations at zero age, but after a time of order 10^8 yr the destabilizing effect of the $^3He(^3He, 2p)^4He$ reactions exceeds the damping. However there is much more severe damping in a nonadiabatic shell beneath the photosphere about 300 km thick, which dominates the entire effect from the rest of the star interior to this shell.

065.100 **Vibrational stability of g^+ modes towards non-radial oscillations for solar models.**
A. Noels, M. Gabriel, A. Boury, R. Scuflaire, P. Ledoux.
19th Colloquium International Astrophys. Liège 1974, (see 012.016), p. 317 - 323 (1975).
The vibrational stability of solar models towards g^+ modes has been studied from the ZAMS up to the age of 5 billion yr.

The nuclear energy contribution to instability is calculated with the effective sensitivity of the reactions to change of temperature. The ZAMS model and the present sun are stable. However, for intermediate ages, models are unstable with respect to the lowest modes due to the effects of nuclear reactions. If, moreover, the mechanical effects of convection are taken into account, all models are unstable.

065.101 **Numerical models with overshooting of convective cores.** A. Maeder.
19th Colloquium International Astrophys. Liège 1974, (see 012.016), p. 327 - 333 (1975).

Detailed models of stellar interiors have been computed, in which the use of the Schwarzschild criterion for determining the limit of the convective core is replaced by a simultaneous integration of the equations of the stellar structure and of the equation of motion of a mean ascending bubble.

065.102 **Preliminary results on detailed helium shell flash calculations.**
I.-J. Christy-Sackmann, B. Paczynski.
19th Colloquium International Astrophys. Liège 1974, (see 012.016), p. 335 - 353 (1975).

This is a report of work in progress. Up to now, ten successive flashes have been followed for population I stars of 3 and 3.3 $M_\odot$. The flashes are still increasing in strength, consisting of up to 5 subcomponents separated from the main flash on the order of 5, 20, 50 and 800 years, respectively. The core mass M_c of the degenerate carbon-oxygen core seems to determine the interflash period and the intershell mass M_{is}, M_c crowing by an amount M_{is} in 8 flashes for the 3 $M_\odot$ star and 3 flashes for the 6 $M_\odot$ star. The burning of helium during the flash produces primarily ^{12}C and little ^{16}O.

065.103 **Dynamical problems of pre-main sequence evolution.** R. B. Larson.
19th Colloquium International Astrophys. Liège 1974, (see 012.016), p. 451 - 464 (1975).

An overview is given of some of the many dynamical phenomena associated with the earliest stages of stellar evolution, comprising the transition from a dark molecular cloud to a hydrostatic stellar object.

065.104 **The dynamical virial equation as an intuitive aid in problems of proto-stellar gas dynamics.**
A. E. Wright.
19th Colloquium International Astrophys. Liège 1974, (see 012.016), p. 477 - 478 (1975). – Abstract.

065.105 **Gravitational collapse of non-spherical gas clouds.** D. McNally.
19th Colloquium International Astrophys. Liège 1974, (see 012.016), p. 479 - 486 (1975).

Gravitational collapse of non-spherical gas clouds is shown to lead to a flattened system. The effects of rotation are assessed and it is concluded that rotation may not be a severe problem in star formation and may even be advantageous in promoting fragmentation. The consequences for star formation are examined.

065.106 **The angular momentum of the protosun and the equatorial rotational velocities of the early type stars.** J. H. Hunter, R. Fleck.
19th Colloquium International Astrophys. Liège 1974, (see 012.016), p. 487 (1975). – Abstract.

065.107 **Scruttura interna e produzione di energia nelle stelle.** G. Santella.
Coelum, Vol. 43, 141 - 159, 194 - 203 (1975).

065.108 **Pulsation properties of upper-main-sequence stars.** M. L. Aizenman, C. J. Hansen, R. R. Ross.
Astrophys. Journ., Vol. 201, 387 - 391 (1975).

The pulsation properties of zero-age main-sequence stars having an initial chemical composition $X = 0.72$, $Y = 0.25$, $Z = 0.03$ are determined for stars in the mass range $10 \leqslant M/M_\odot \leqslant 110$. Radial and nonradial modes ($l = 1, 2$) are considered in some detail. It is found that for this composition pulsational instability occurs only in the radial fundamental mode for $M \gtrsim 60\, M_\odot$. All nonradial modes examined are found to be stable, as are the radial harmonic oscillations.

065.109 **A variational principle and a stability criterion for the dipole modes of pulsation of stellar models in general relativity.** S. L. Detweiler.
Astrophys. Journ., Vol. 201, 440 - 446 (1975).

A variational principle is found for the frequencies of the dipole modes of pulsation of spherical stellar models in general relativity. This principle reveals that if the relativistic Schwarzschild discriminant, S, is not positive definite, then a stellar model is unstable to dipole perturbations. In addition, if S is positive definite the eigenfrequencies of all of the normal modes are real.

065.110 **The maximum mass of nonrotating neutron stars.** D. J. Hegyi, T.-S. Lee, J. M. Cohen.
Astrophys. Journ., Vol. 201, 462 - 466 (1975).

The authors calculate the maximum mass of a nonrotating neutron star using only general relativity and making no assumptions about the equation of state of matter beyond the limitations of current nuclear physics. They find the upper mass limit of the neutron star to be about 8 $M_\odot$. By including other considerations, it may be possible to obtain even higher masses.

065.111 **Finite amplitude convection in a compressible layer with polytropic structure.** R. Van der Borght.
Australian Journ. Phys., Vol. 28, 437 - 452 (1975).

In this paper are derived the basic equations for finite amplitude convection in a compressible medium with polytropic structure within the framework of the one-mode anelastic approximation. Only the case of non self-interacting planforms is considered, and the effects of viscous dissipation are not taken into account. These equations are solved numerically to illustrate the effects of compressibility and density stratification on the flow pattern and on the thermodynamic variables and their fluctuations.

065.112 **Convective overshooting in main-sequence models.** B. C. Cogan.
Astrophys. Journ., Vol. 201, 637 - 640 (1975).

A simplified treatment of convective overshooting which includes the flux carried by the overshooting elements is presented. It is applied to a 3 $M_\odot$ model and predicts the extent of overshooting to be 0.23 pressure scale heights. This distance is determined by three factors: the convective velocity and efficiency and the gradient of the radiative temperature gradient. Sample calculations indicate that in most main-sequence stars the fractional mass added to the convective core is approximately the same. Finally it is concluded that uncertainties in convection theory make it impossible to assign definitive values to the overshooting distance.

065.113 **Carbon detonations in rapidly rotating stellar cores.** J. H. Mahaffy, C. J. Hansen.
Astrophys. Journ., Vol. 201, 695 - 704 (1975).

Equilibrium models have been constructed for rapidly differentially rotating isothermal stellar cores with central densities ranging from 2×10^9 to 10^{10} g cm^{-3}. The composition, of equal parts by mass of ^{12}C and ^{16}O, is taken to represent that of a pre–carbon-detonation supernova. It is found that the presence of rotation does not lower the central density required for leaving a bound remnant down to the density range at which current evolutionary calculations predict the explosive ignition of carbon. Hence, the results of standard spherically symmetric calculations are not altered in that 4–8 $M_\odot$ stars are totally disrupted after carbon ignition.

065.114 **Linear series and the manifold of stellar equilibrium models.** H. Kähler.
Astron. Astrophys., Vol. 43, 443 - 454 (1975).

Two characteristic properties (multiplicity m and charge c) of stellar equilibrium models are defined. Only models with $m = c = 1$ can be stable. Using these quantities, the general properties of linear series of stellar equilibrium models are investigated. In particular a conservation theorem at arbitrary critical points (turning points, bifurcation points, etc.) is derived. This has implications concerning the manifold of equilibrium models for given mass and composition: under broad assumptions the number N of models is odd; at most $(N+1)/2$ of them can be stable.

065.115 **On a probabilistic model of fragmentation.** Eh. A. Tropp, A. D. Chernin.
Pis'ma v Astron. Zhurn., Vol. 1, No. 10, p. 11 - 13 (1975). In Russian.

It is shown that the Kolmogorov probabilistic scheme of fragmentation contains as a particular case the model proposed by Larson for the star formation process. Comparison of the resulting spectrum of fragments with Salpeter's stellar mass function suggests a possible dependence of the probability of fragmentation on the size or mass of the fragment.

065.116 **On mixing and angular momentum transfer in stars.** Yu. V. Vandakurov.
Pis'ma v Astron. Zhurn., Vol. 1, No. 10, p. 20 - 24 (1975). In Russian.

The possibility of small-scale nonadiabatic instability in a chemically inhomogeneous rotating star is discussed. An estimate of turbulent diffusion due to instability is given. In the case of axisymmetric motions this estimate is also applicable to Goldreich and Schubert's instability of differential rotation in stars.

065.117 **Neutrino processes in stellar interiors.** Z. Barkat.
Annual Rev. Astron. Astrophys., Vol. 13, (see 003.005), 45 - 68 (1975). – Contents: (1) Introduction; (2) Neutrino processes; (3) Late stages of evolution; (4) The evolution of stars in the range $4 \lesssim M/M_\odot \lesssim 10$; (5) Dynamical collapse of stellar cores; (6) Cooling of neutron stars; (7) Concluding remarks.

065.118 **Young stellar objects and dark interstellar clouds.** S. E. Strom, K. M. Strom, G. L. Grasdalen.
Annual Rev. Astron. Astrophys., Vol. 13, (see 003.005), 187 - 216 (1975). – Contents: (1) Introduction; (2) The early phases of stellar evolution; (3) The renewed search for young stellar objects; (4) The Herbig Ae and Be stars associated with nebulosity; (5) The T Tauri stars; (6) The Herbig-Haro objects; (7) A scenario for PMS evolution; (8) The interaction of young stars and dark clouds; (9) Prospects for the future.

065.119 **Neutron stars: general review.** A. G. W. Cameron, V. Canuto.
Proc. 16th Solvay Conference Phys., (see 012.020), p. 221 - 267, with a discussion, p. 268 - 278 (1974).

065.120 **The loss of angular momentum and the evolution of Ap-type stars.** S. C. Wolff.
Astrophys. Journ., Vol. 202, 121 - 126 (1975).

There is a correlation between the radius and period for Ap stars, in the sense that stars of longer period have larger radii. This result is to be expected if the more slowly rotating Ap stars have evolved farther away from the zero-age main sequence than Ap stars with more rapid rotation. It is shown that the period-frequency distribution of the non-Si Ap stars, the correlation of radius with period, and the absence of Si stars with periods longer than 20 days can all be accounted for if the time required for magnetic deceleration to reduce rotational velocities to values less than 10 km s^{-1} is on the order of a few times 10^8 years.

065.121 **Vibrational stability of differentially rotating stars.** M. L. Aizenman, J. P. Cox.
Astrophys. Journ., Vol. 202, 137 - 147 (1975).

The authors have obtained the formal solution, in linear, nonadiabatic theory, of the problem of the vibrational stability of a system possessing arbitrary, but steady, inviscid flows in its unperturbed (nonoscillating) state. The unperturbed system may have any geometry, but no large-scale magnetic fields are assumed. A special case is a (steadily) rotating star, where the rotation may be uniform or differential, slow or fast. This formulation may ultimately provide a new approach to the problem of the interaction (in linear theory) between stellar convection and pulsations. The solution is surprisingly simple, bears considerable resemblance to previously derived formal solutions, and has a straightforward physical interpretation.

065.122 **Atmospheric turbulence measured in stars above the main sequence.** D. F. Gray.
Astrophys. Journ., Vol. 202, 148 - 164 (1975).

Measurements of broadened line profiles are analyzed to identify, measure, and separate macroturbulence from rotation. Evidence is presented for the possibility of more than one macroturbulence velocity distribution.

065.123 **Electrical conductivity and magnetic field decay in neutron stars.**
G. M. Ewart, R. A. Guyer, G. Greenstein.
Astrophys. Journ., Vol. 202, 238 - 247 = Contr. Five-College Astron. Dep., Univ. Mass., *Amherst*, No. 184 (1975).

The authors evaluate the contributions to the electrical resistivity of neutron star crusts due to phonon scattering at high and low temperatures, due to scattering from charge disorders, and due to scattering from dislocations. The results indicate that decay of pulsar magnetic fields can take place only if (1) they are significantly hotter than cooling theory predicts, (2) their crusts are exceedingly impure, or (3) their crusts are made up of an exceedingly fine mosaic of crystallites.

065.124 **Die Endstadien der Sternentwicklung.** H. Zimmermann.
Astron. in der Schule, 12. Jahrgang, p. 103 - 107 (1975).

065.125 **Corequakes of neutron stars and soft γ-ray bursts.** A. I. Tsygan.
Astron. Astrophys., Vol. 44, 21 - 24 (1975).

Within the framework of the model we shall consider, the process by which soft γ-ray bursts are generated can be divided into the following parts: 1. Corequakes in one of the stressed regions within the solid core of the neutron star. 2. Generation of nonradial neutron star vibrations, the amplitude at the surface being much larger than within the star. 3. Generation of a variable electromagnetic field charged particles are detached from the stellar surface and plasma heating results. The presence of a strong magnetic field will ensure that the γ-radiation from this plasma will be collimated: since old neutron stars rotate with a period $P \sim (10\text{–}20)$ s, the observed time structure of γ-bursts will be obtained. The duration of stellar vibrations is determined by gravitational radiation (10–30 s).

065.126 **The peculiar A stars: an adequate method for testing the *r*-processes.** B. Kuchowicz.
VI Leningr. mezhdunar. seminar, 1974. Uskorenie chastits i yader. reaktsii v kosmose. Leningrad, 1974, p. 297 - 308, with a discussion, p. 309 - 310. – Abstr. in Referativ. Zhurn. 51. Astron., 8.51.633 (1975).

065.127 **Stability of a star if there is a phase transition.** G. S. Bisnovatyj-Kogan, S. I. Blinnikov, Eh. Eh. Shnol'.
Astron. Zhurn. Akad. Nauk SSSR, Vol. 52, 920 - 929 (1975). In Russian. English translation in Soviet Astron., Vol. 19, No. 5.

The conditions for stability of a cold star possessing a density jump produced by phase transition are derived. The usual expression for the second variation of the total energy has an additional nonintegral term localized on the density jump. Practical prescriptions for testing the stability are given.

065.128 **Dynamic phases in the evolution of a 1.25 $M_\odot$ C-O star.** A. Kovetz, D. Prialnik, G. Shaviv.
Astrophys. Space Sci., Vol. 37, 463 - 476 (1975).

The evolution of a 1.25 $M_\odot$ carbon and oxygen (equal fractions by mass) homogeneous star is followed by means of a computer code capable of dealing with dynamic evolutionary phases. After carbon ignition at the center, followed by successive shell flashes and the formation of a T-inversion, convection begins at the surface and the model evolves through a very short but strong dynamic phase (viz. a pulsation) after which it settles down to a white dwarf configuration.

065.129 **Population I and population II stellar evolution.** P. Giannone.
Mem. Soc. Astron. Italiana, Vol. 46, 145 - 197 (1975). Lectures for Scuola Superiore di Astronomia, course "Stellar Populations." Erice, May 1974. – Review paper.

065.130 **Contribution to the studies of models of main-sequence stars.** T. Angelov.
Publ. Obs. Astron. Beograd, No. 20, (see 012.024), p. 110 - 122 (1975). In Serbo-Croatian.

065.131 **The life of dying stars.** R. A. Syunyaev, N. I. Shakura.
Priroda, No. 10.75, p. 46 - 53 (1975). In Russian.

065.132 **A method to treat the convective region in the evolving stellar interior. III. Formation of the inhomogeneous zone in the convective core.** K. Suda, J. Uchida.
Sci. Rep. Tôhoku Univ., First Ser., Vol. 58, 113 - 125 (1975).

In connection with the method for generally treating the evolving stellar convective core, the basic problems such as how the evolutionary time step must be taken and how the spatial mesh points must be placed are investigated mainly with regard to the way of computing the second evolutionary model starting directly from the chemically homogeneous initial stage. The background of the investigation is placed to the early evolution of massive main sequence stars and the process through which the inhomogeneous abundance distribution similar to that of semi-convective zone may be formed is described.

065.133 **The influence of the opacity values and mixing length on the evolution of population II main-sequence stars.** R. Stabell.
Astron. Astrophys., Vol. 44, 315 - 319 (1975).

The effects on the main sequence of changes in the opacities and the mixing length is investigated. The slope of the theoretical main sequence is unaffected by changes of the opacity data, but is influenced by the ratio of mixing length to pressure scale height, α. Comparison with observational data seems to favour a small value of this ratio ($\alpha < 0.5$). The effects on the evolutionary tracks and timescales are also discussed.

065.134 **Triaxial degenerate dwarfs and pulsing X-ray sources.** R. H. Durisen.
Astron. Astrophys., Vol. 44, 473 - 476 (1975).

It is argued that triaxial configurations for rotating degenerate dwarfs are not limited to forms analogous to the Jacobi ellipsoids. Some of these alternative forms may exist in mass-transfer binary star systems. Particular consideration is given to the secondaries in Cen X-3 and Her X-1. Plausible triaxial white dwarf candidates for Cen X-3 are found.

065.135 **Location of models of red supergiants with deep mixing in the Hertzsprung-Russell diagram.** U. Uus.
Publ. Tartu Astrofiz. Obs., Vol. 43, 85 - 88 (1975).

A region in the H-R diagram is indicated, where moderately massive red supergiants at the stage of growth of the degenerate carbon-oxygen core may have CN-cycle equilibrium envelope chemical composition as a result of permanent deep mixing in the case of high assumed efficiency of convective heat transport.

065.136 **Stationary elastic rotational deformation of a relativistic neutron star model.** B. Carter, H. Quintana.
Astrophys. Journ., Vol. 202, 511 - 522 (1975).

In order to predict pulsar glitch magnitudes in terms of the Ruderman starquake theory, it is necessary to calculate the deviation from perfect fluid behavior when a stationary rotating partly solid neutron star undergoes an elastic deformation due to a change in angular momentum. Previous calculations have been based on crude Newtonian or semi-Newtonian approximations. The present work describes the way to solve this problem exactly (in the small-deformation limit) using a fully relativistic treatment.

065.137 **Core mass-interflash period relation for double shell source stars.** B. Paczyński.
Astrophys. Journ., Vol. 202, 558 - 560 (1975).

Helium-shell flashes are studied in models of a star of 3 $M_\odot$ with a degenerate core of 0.57, 0.68, 0.97, and 1.374 $M_\odot$. A well-defined core mass-interflash period relation is established. This relation may be combined with the known theoretical core mass-luminosity relation to find a core mass, an intrinsic luminosity, and a distance to an object with known interflash period. An application of this method to FG Sagittae is discussed. A model with a 1.374 $M_\odot$ core developed weak helium-shell flashes every 16 years.

065.138 **Criteria for the thermal instability of the nuclear-burning shell.** W. Unno.
Publ. Astron. Soc. Japan, Vol. 27, 605 - 609 (1975).

Simple criteria are derived for the conditions of the oscillatory and the monotonic thermal instability associated with a nuclear-burning shell. The shell instability begins as the oscillatory type and proceeds to the monotonic type as the star evolves from a thermally stable configuration to a strongly unstable configuration.

065.139 **Some effects of atmospheric opacity on the evolution of a 5 $M_\odot$ star.** H. R. Johnson, R. W. Whitaker.
Monthly Notices Roy. Astron. Soc., Vol. 173, 523 - 526 = Publ. Goethe Link Obs., Indiana Univ., *Bloomington,* No. 172 (1975).

Two 5 $M_\odot$ stellar evolutionary sequences have been computed from the main sequence to core helium exhaustion. One sequence used the standard opacities in the Paczynski code (Cox & Stewart) while in the other sequence the opacities in the atmosphere alone (wherever the temperature was less than

5500 K) were increased to values obtained from model atmosphere calculations. The effect of the increased atmospheric opacity is to extend significantly the blue loop and shift the red giant branch to lower values of effective temperatures.

065.140 **On the internal structure of cold superdense stars with a frozen-in magnetic field.** G. A. Shul'man.
Astron. Zhurn. Akad. Nauk SSSR, Vol. 52, 1166 - 1172 (1975). In Russian. English translation in Soviet Astron., Vol. 19, No. 6.

It is shown that the concentration of protons (electrons) at densities of about the nuclear one depends on the magnetic field, the strength of which has a value $H \gtrsim 10^{18}$ gauss. The limits of application of the obtained relations taking into account an anomalous momentum of the electron are discussed.

065.141 **Non thermal effects in carbon and oxygen burning.** G. Beaudet, G. Shaviv.
Journ. Roy. Astron. Soc. Canada, Vol. 69, 256 (1975). Abstr. CAS.

065.142 **En stjernes død og genfødsel. Hvide dværge, supernovae, pulsarer, sorte huller.** J. O. Petersen.
Astron. Tidssk., Årg. 8, p. 163 - 185 (1975). – Popular review article.

065.143 **Nuclear reactions in carbon-rich stars.** M. Tassoul.
Astrophys. Journ., Vol. 202, 755 - 761 (1975).

The combined effect of neutrino losses and of the detailed network of nuclear reactions on the evolution of low-mass, carbon-rich stars is investigated. It is found that models with $0.93\,M_\odot < M < 0.95\,M_\odot$ ignite in a shell which rapidly extends toward the center. For models with $0.95\,M_\odot < M \leqslant 1.25\,M_\odot$, carbon burning takes place directly in the central region. For both ranges of mass, central combustion is followed by a phase of combustion in an outward-moving shell. For the lower mass range, the end of the evolution is marked by a violent flash; none is observed in the higher mass models which then evolve toward a red giant configuration.

065.144 **The maximum mass of a cold neutron star.** J. Rosen, N. Rosen.
Astrophys. Journ., Vol. 202, 782 - 787 (1975).

The maximum mass of a static cold neutron star is calculated according to the bimetric gravitation theory and according to the general relativity theory for an assumed equation of state containing two parameters. For a certain choice of the parameters the bimetric maximum mass is found to be $8.1\,M_\odot$, while that of general relativity is $1.46\,M_\odot$. It is pointed out that the process of gravitational collapse in the bimetric theory (which does not admit black holes) is quite different from that in general relativity and may lead to considerable energy emission.

065.145 **Thermodynamics of convective Urca cores.** B. Lazareff.
Astron. Astrophys., Vol. 45, 141 - 143 (1975).

The evolution of degenerate carbon cores has recently been computed taking into account convection-driven Urca processes. The treatments used contain the assumption that only the energy balance needs to be taken into account. The aim of the present paper is to show that ignoring the entropy budget leads to an overestimate of the stabilizing properties of convective Urca processes.

065.146 **Estimate of plasticity in the crust of neutron stars.** G. Bambakidis, T. T. Chan.
Astron. Astrophys., Vol. 45, 209 - 211 (1975).

The influence of plasticity in the crust of a neutron star on the frequency of crustquakes has been estimated and found to be small. The resulting rate of creep, assuming it is a classical, thermally activated process, is obtained and compared to the results of others.

065.147 **An explanation for helium-rich and helium-variable stars: diffusion in a radial mass-loss flux.**
S. Vauclair.
Astron. Astrophys., Vol. 45, 233 - 235 (1975).

Diffusion in a radial mass loss flux is shown to be able to account for helium-rich stars, if the mass flux has a value of about 10^{-11} g cm^{-2} s^{-1} (about $10^{-12} M_\odot$/yr). For a given effective temperature, a range of mass loss flux needed for the atmosphere to become helium rich is found. For higher values of the mass loss flux, the star appears normal. For lower values it appears helium weak. Helium variable stars may be understood in the same way within the framework of the oblique rotator model.

065.148 **On the correctness of polytropic and unified polytropic models.** O. V. Chumak.
Trudy Astrofiz. Inst., *Alma-Ata,* Vol. 26, 47 - 55 (1974). In Russian.

065.149 **Linear series of stellar models. V. Hydrogen-helium stars of 0.6, 0.7, 1, 1.5, 2, 3, and 5 $M_\odot$**
M. Kozłowski, B. Paczyński.
Acta Astron., Vol. 25, 321 - 330 (1975).

Linear series of models are constructed for stars of 0.6, 0.7, 1, 1.5, 2, 3, and 5 $M_\odot$. Models consist of hydrogen rich envelopes ($X = 0.7$, $Y = 0.27$, $Z = 0.03$) and helium cores ($X = 0.0$, $Y = 0.97$, $Z = 0.03$), and they have a rectangular hydrogen profile, $X(M_r)$. The linear series are found to be of four types. For stars in the mass range $0.3 < M/M_\odot < 1.4$ two different linear series can be constructed. For a star of about $0.66\,M_\odot$ the two linear series may have a common bifurcation point. In some cases an analogy may be found between the evolutionary tracks and the linear series. Such points can be identified as the Schönberg-Chandrasekhar isothermal core mass limit, the core helium flash, and the minimum mass for helium burning.

065.150 **Thermonuclear explosion of a degenerate carbon stellar core.**
L. N. Ivanova, V. S. Imshennik, V. M. Chechetkin.
In-t prikl. mat. AN SSSR. Preprint No. 31. Moskva, 1975. 54 pp. Price 20 Kop. In Russian. – Abstr. in Referativ. Zhurn. 51. Astron., 12.51.391 (1975).

065.151 **Neutron reactions in stars.** N. A. Vlasov.
Atom. ehnergiya, Vol. 39, No. 2, p. 85, 103 - 108, 114 (1975). In Russian. – Abstr. in Referativ. Zhurn. 51. Astron., 12.51.401 (1975).

065.152 **The origin and evolution of RR Lyrae stars of high metal abundance.**
R. E. Taam, R. P. Kraft, N. Suntzeff.
Bull. American Astron. Soc., Vol. 7, 504 - 505 (1975). Abstr. AAS.

065.153 **Linear and nonlinear models of lower cepheid-strip variable stars.** R. F. Stellingwerf.
Bull. American Astron. Soc., Vol. 7, 511 (1975). – Abstr. AAS.

065.154 **Evolution and secular stability of stellar models crossing the Hertzsprung Gap.**
D. Lauterborn, R. A. Siquig.
Bull. American Astron. Soc., Vol. 7, 512 (1975). – Abstr. AAS.

065.155 **Mixing of CNO-cycle material in red giants.** D. Deming.
Bull. American Astron. Soc., Vol. 7, 518 (1975). – Abstr. AAS.

065.156 **The second convection zone in an A-type star.** J. Latour, E. A. Spiegel, J. Toomre, J.-P. Zahn. Bull. American Astron. Soc., Vol. 7, 526 (1975). – Abstr. AAS.

065.157 **Convection in RR Lyrae stars.** R. G. Deupree. Bull. American Astron. Soc., Vol. 7, 526 - 527 (1975). – Abstr. AAS.

065.158 **Nucleosynthesis and star formation of the Galaxy and the Magellanic Clouds.** G. L. Olson, J. H. Pena. Bull. American Astron. Soc., Vol. 7, 532 (1975). – Abstr. AAS.

065.159 **The Vogt-Russell and the Hayashi theorem.** S. C. Temesvary. Bull. American Astron. Soc., Vol. 7, 541 (1975). – Abstr. AAS.

065.160 **Analytic-numerical calibration of upper main sequence stellar models.** D. L. Hall. Bull. American Astron. Soc., Vol. 7, 542 (1975). – Abstr. AAS.

065.161 **Stellar evolution in the Magellanic Clouds.** P. J. Flower. Bull. American Astron. Soc., Vol. 7, 542 (1975). – Abstr. AAS.

065.162 **Mass loss in early stages of stellar evolution.** D. S. P. Dearborn, M. Koziowskie, D. N. Schramm. Bull. American Astron. Soc., Vol. 7, 544 (1975). – Abstr. AAS.

065.163 **The shape and stability of the magnetospheric boundary of accreting neutron stars.** R. F. Elsner, F. K. Lamb. Bull. American Astron. Soc., Vol. 7, 545 (1975). – Abstr. AAS.

065.164 **Neutron star original spin.** D. Q. Lamb, F. K. Lamb, W. D. Arnett. Bull. American Astron. Soc., Vol. 7, 545 (1975). – Abstr. AAS.

065.165 **Coulomb collisional heating and Compton cooling near the surface of accreting neutron stars.** A. R. Masters. Bull. American Astron. Soc., Vol. 7, 545 - 546 (1975). Abstr. AAS.

065.166 **Obese "neutron" stars.** K. Brecher, G. Caporaso. Bull. American Astron. Soc., Vol. 7, 546 (1975). – Abstr. AAS.

065.167 **Decompression of neutron star matter.** J. M. Lattimer, F. D. Mackie, D. G. Ravenhall, D. N. Schramm. Bull. American Astron. Soc., Vol. 7, 546 - 547 (1975). Abstr. AAS.

065.168 **The boron-11 bypass to carbon-12 production in first generation stars.** J. G. Eoll. Bull. American Astron. Soc., Vol. 7, 558 (1975). – Abstr. AAS.

065.169 **Star formation and early evolution.** M. Cohen. Mercury, (Journ. Astron. Soc. Pacific), Vol. 4, No. 5, p. 10 - 12, 29 (1975). – A report on the May 12 - 15, 1975 scientific meeting of the Astronomical Society of the Pacific, Ensenada, Mexico.

065.170 **Internal structure of helium supergiants.** U. Uus. Peremennye Zvezdy, Vol. 19, 533 - 540 (1975). In Russian.

The main characteristics of computed models of helium supergiants with dense degenerated isothermal carbon-oxygen and oxygen-magnesium cores are discussed. The sensitivity of the model structure to the assumed efficiency of the convective heat transport is investigated.

065.171 **Stochastic stellar evolution. I. The H-burning phase.** G. Bertelli, C. Chiosi, J. Perdang. Astron. Astrophys., Vol. 45, 307 - 314 (1975).

After a short outline of a statistical technique designed to study stochastic processes in stellar evolution, the authors apply this procedure to the analysis of the effect of random abundance perturbations. It is found that this mechanism destroys the tight mass-time relation, and leads to a substantial scatter in the frequency distribution of stars along the isochronous line. In particular the ratio of blue to red supergiants in synthetic young clusters shows an observationally significant dispersion around the usual value.

065.172 **Stellar reaction rates for the $^{36}S(p, \gamma)^{37}Cl$ reaction.** J. Keinonen, A. Anttila, M. Riihonen. Astron. Astrophys., Vol. 45, 315 - 317 (1975).

The $^{36}S(p, \gamma)^{37}Cl$ reaction has been investigated in the proton energy range 0.6–1.9 MeV to obtain data needed for a calculation of the reaction rates in the explosive carbon burning. The γ-ray strengths were determined for 144 resonances. The reaction rates, calculated from the present measurements are given.

065.173 **Evolution of helium rich stars with hydrogen burning.** M. Röser. Astron. Astrophys., Vol. 45, 335 - 341 (1975).

Evolutionary tracks of stars with an initial chemical composition $X = 0.100$, $Y = 0.8790$, $Z = 0.021$ are calculated for masses of $0.35\, M_\odot$, $0.66\, M_\odot$, $1.00\, M_\odot$, $2.00\, M_\odot$, and $5.00\, M_\odot$. The evolution is followed from hydrogen to helium burning, and to carbon burning when the occasion arises. The evolution data are presented and compared with normal Population I-stars. The helium rich stars show higher effective temperatures, much higher luminosities and therefore shorter time scales. They are situated in regions of the HR-diagram where observed helium stars are found.

065.174 **How does rotation change the radius of a low mass stellar model.** J. Aranda, H.-C. Thomas. Astron. Astrophys., Vol. 45, 441 - 442 (1975).

The results of different authors for stellar model computations of chemically homogeneous stars of one solar mass in uniform rotation do not agree near breakup velocity. It is shown that the differences can be explained from the approximations used for the convective envelope. Some qualitative arguments can be given to support the numerical results.

065.175 **Endstadien der Sternentwicklung.** V. Weidemann. Christiana Albertina (18), Neue Folge, Heft 2, Forschungsber. und Halbjahresschrift Univ. Kiel, p. 5 - 10 = Sonderdruck Sternw. Kiel No. 205 (1975).

065.176 **Influence de la convection sur la stabilité des oscillations non radiales des étoiles.** M. Gabriel, R. Scuflaire, A. Noels, A. Boury. Acad. Roy. Belgique, Bull. Cl. Sci., 5ᵉ Sér., Vol. 60, 866 - 887 = Inst. Astrophys. Univ. Liège, Cointe-Ougrée (*Belgique*), Coll. 8°, No. 651 (1974).

The coupling between convection and non-radial oscillations is discussed. The limitation of the theory comes from the lack of a good theory for turbulent convection which implies the use of an approximation equivalent to the Böhm-Vitense theory. In the frame of this approximation an expression which gives the influence of convection on the stability of stellar oscillations is deduced. The result is briefly discussed and compared with previous works.

065.177 **Convective overshooting in main sequence models.** B. C. Cogan.

Proc. Astron. Soc. Australia, Vol. 2, 355 - 356 (1975).

065.178 **Par jaunu zvaigžņu atklāšanu.** Z. Alksne.
Zvaigžņotā debess, 1975. gada vasara, p. 1 - 7.

065.179 **Zvaigžņu evolūcijas agrās stadijas.** J. Francmanis.
Zvaigžņotā debess, 1975. gada vasara, p. 7 - 10.

065.180 **Evolución de las estrellas y galaxias.**
V. Ambartsumyan.
Obs. Astron. Municipal Rosario, (*Argentina*), Bol. No. 4, p. 4 - 6 (1974). – Lecture given 1974 September 16 at Obs. Astron. Municipal Rosario.

065.181 **^{22}Ne (α, n) ^{25}Mg reaction in CONe mantle of massive stars at high temperatures.**
B. N. Kakati, H. L. Duorah.
Pramāņa, (*India*), Vol. 3, 89 - 94 (1974).

065.182 **^{17}O (p, α)^{14}N reaction rate in stars.**
G. W. King, R. E. Brown, D. Dehnhard.
Annual report 1974, Minnesota Univ., Minneapolis (USA). John H. Williams Lab. of Nuclear Physics, p. 38, 42 - 43 (1974). – Summary.

065.183 **Neutrino flow and the collapse of stellar cores.**
J. R. Wilson.
7. Texas symposium on relativistic astrophysics, Dallas, Texas, USA. 16 pp. (1974).

Calculations are repeated with a better model for the neutrino interactions in neutral current theories predicting a scattering cross section for heavy nuclei proportional to the square of the nuclear mass. These calculations show that this high cross section may make it possible for the hot neutron star newly formed in stellar collapse to emit enough neutrinos to blow off the iron envelope which would surround the newly formed neutron star.

065.184 **Cooling of superfluid neutron stars.**
R. C. Malone.
Thesis, Cornell Univ., Ithaca, N.Y. (USA). [Available from Univ. Microfilms, Ann Arbor, Mich. Order No. 74-17,650], 330 pp. (1974).

065.185 **Lower bound on neutrino pair emission.**
K. O. Mikaelian.
Phys. Letters B, (*Netherlands*), Vol. 57B, 181 - 182 (1975).

Weak neutral currents affect the rates with which stars lose energy by emitting neutrino pairs. These rates can be calculated once the parameters of the underlying theory are fixed. It is shown that any new gauge theory will have a minimum rate, equal to half the old rate. The possible existence of new types of neutrinos which may considerably enhance stellar cooling is discussed.

065.186 **Energy production in stars.** H. A. Bethe.
Fiz. Szemle, (*Hungary*), Vol. 25, No. 2, p. 41 - 48 (1975). In Hungarian.

Internal and surface temperatures, brightness and size of stars are summarized on the basis of the Hertzsprung-Russell diagram and equations of hydrostatic equilibrium and radiation propagation. Thermonuclear chain reactions, in particular the C–N and C–N–O cycles are discussed. The development of a normal star into a red giant is traced by Iben's evaluation of temperature and density with respect to time.

065.187 **Neutrino opacity of neutron-star matter.**
R. F. Sawyer.
Phys. Rev. D, (Particles and Fields), Vol. 11, 2740 - 2745 (1975).

Scattering of neutrinos from very hot neutron star matter is calculated in the Weinberg-Salam model of a neutral current interaction, and it is found that the resulting neutrino opacity has an interesting dependence on the equation of state. For an arbitrary equation of state it is possible to estimate the maximum opacity of a thin layer of neutron matter.

065.188 **Solid neutron matter.**
M. T. Takemori, R. A. Guyer.
Phys. Rev. D, (Particles and Fields), Vol. 11, 2696 - 2723 (1975).

The conjectured solid ground state of neutron matter is investigated using a formalism developed for the quantum crystal problem, and neutron interaction potentials of the Reid type. The different computational methods are tested by comparison, and no good evidence is found for the presence of a solid ground state.

065.189 **Hydrogenic atoms in 'superstrong' magnetic fields.**
M. L. Glasser, J. I. Kaplan.
Phys. Letters A, (*Netherlands*), Vol. 53A, 373 - 374 (1975).

An estimate is made of the magnetic field above which many recent non-relativistic treatments of hydrogenic atoms in strong magnetic fields become invalid. The theory is applicable to certain collapsed astronomical objects, such as white dwarfs and neutron stars.

065.190 **Validity of constrained variational methods for dense neutron matter.** V. R. Pandharipande.
Nuclear Phys. A, (*Netherlands*), Vol. A248, 524 - 531 (1975).

The energies of fictitious Boltzmann particles of neutron mass interacting via simple Yukawa potentials are calculated by the method of constrained variation in the lowest order and hypernetted chain approximations. Comparison of calculated energies with exact Monte Carlo results is carried out.

065.191 **Comments on pion condensation in nuclear matter.**
A. Suzuki, Y. Futami, Y. Takahashi.
Progr. Theor. Phys. Japan, Vol. 53, 1204 - 1205 (1975).

If a superdense nuclear matter system is unstable with respect to the presence of pions, the appearance of condensed pions in the ground state is followed by the simultaneous appearance of a macroscopic number of nucleon particle-hole pairs (P–H pairs). It is shown that the condensed state of P–H pairs gives rise to the same ground state energy as the mean field approximation, and a clear image for the ground state is given.

065.192 **Evidence against solidification of a model neutron system.** J. W. Clark, D. G. Sandler.
Phys. Rev. D, (Particles and Fields), Vol. 11, 3365 - 3369 (1975).

The fluid or solid nature of neutron star matter at high density is discussed.

065.193 **Thomas-Fermi nuclei in neutron star matter.**
L. P. Csernai, D. Kisdi, J. Nemeth.
Acta Phys. Acad. Sci. Hungaricae, Vol. 38, 89 - 108 (1975).

The nuclear structure of zero-temperature neutron star matter is investigated in the density region $\rho = 10^{11}$–10^{14} g/cm^3. The energy of a nucleus embedded into a nucleon liquid is calculated in the Thomas-Fermi approximation.

065.194 **Behavior of matter at superhigh density.**
V. Canuto, J. Lodenquai.
Phys. Rev. D, (Particles and Fields), Vol. 11, 233 - 244 (1975).

065.195 **Mass and stability of rigidly rotating relativistic polytropes by energy method.** K. A. van Riper.
Ann. Physics, Vol. 90, 530 - 572 (1975).

065.196 **An equation of state of cold dense matter.**

A. Nishimura, Y. Yamaguchi.
Progr. Theor. Phys. Japan, Vol. 53, 891 - 893 (1975).

065.197 **Superfluid neutron star cores.**
U. de Angelis, L. de Cesare.
Nuovo Cimento Lettere, Ser. 2, Vol. 13, 79 - 82 (1975).

A stellar evolution program for the computer M-222. See Abstr. 021.010.

Approximation functions to the Emden and associated Emden functions near the first zero. See Abstr. 021.011.

Thermonuclear reaction rates, II. See Abstr. 022.065.

Stellar rates for some reactions of interest in astrophysics. See Abstr. 061.004.

On particle acceleration and radiation output in stochastic electromagnetic fields. See Abstr. 061.009.

Production of ^{17}O and ^{18}O by means of the hot CNO tri-cycle. See Abstr. 061.030.

Asymptotic expansions for acoustical modes of the homogeneous compressible model. See Abstr. 061.034.

The origin and abundances of the chemical elements. See Abstr. 061.036.

Equation of state at ultrahigh densities. Part 2. See Abstr. 061.043.

Physics of high density and nuclear matter. See Abstr. 061.044.

Approximation functions to the Emden and associated Emden functions near the first zero, II. See Abstr. 061.052.

The shear modulus of neutron star matter in the free-neutron state. See Abstr. 061.059.

On pair production in intense electromagnetic fields occurring in astrophysical situations. See Abstr. 061.068.

Thermodynamics of dense hydrogen-helium fluids. See Abstr. 062.003.

Evolution of rotating interstellar clouds. I. Numerical techniques. See Abstr. 062.010.

The hydrodynamics of accretion discs I: stability. See Abstr. 062.015.

Mass loss and circumstellar dust shells at T Tauri stars. See Abstr. 064.011.

Evolution of protostars. On the formation of H II regions around early-type stars. See Abstr. 064.031.

Stellar mass loss. See Abstr. 064.033.

Hydrodynamic models of a cepheid atmosphere. III. Line spectrum and radius determinations. See Abstr. 064.036.

Theory of extended stellar atmospheres. II. A grid of static spherical models for O stars and planetary nebula nuclei. See Abstr. 064.042.

On the implications of the nonexistence of unstable outgoing normal modes for the stability of spherical stellar models to nonradial perturbations in general relativity. See Abstr. 066.001.

Gravitational radiation instability in rotating stars. See Abstr. 066.038.

Mass and moment-of-inertia peaks in general relativity. See Abstr. 066.052.

Theories of gravity with structure-dependent γ's. See Abstr. 066.112.

Some magnetic effects in stellar collapse and accretion. See Abstr. 066.165.

Nonradial g-mode oscillations and the stability of the sun. See Abstr. 080.032.

Solar models of low neutrino-counting rate: the central black hole. See Abstr. 080.037.

A solar neutrino loophole: standard solar models. See Abstr. 080.041.

Oscillatory periods in the sun and theoretical models with or without mixing. See Abstr. 080.063.

On Hoyle's theory of the origin of the solar system. See Abstr. 107.010.

Photometry of FG Sagittae. See Abstr. 113.013.

The Wolf-Rayet stars. See Abstr. 114.008.

Photometric variability of magnetic stars. See Abstr. 116.003.

Epsilon Coronae Austrinae: evidence for binary star evolution? See Abstr. 117.012.

The dynamical tide in close binaries. See Abstr. 117.013.

Black holes and neutron stars: evolution of binary systems. See Abstr. 117.023.

On the formation of WR-stars and X-ray sources in close binary systems. See Abstr. 117.043.

The fission theory of binary stars for compressible masses. See Abstr. 117.044.

The age of the W Ursae Majoris stars. See Abstr. 117.056.

Spectroscopic observations of eclipsing binaries. II. Absolute dimensions, evolutionary state, and helium content of RZ Chamaeleontis. See Abstr. 121.080.

An observational approach to stellar evolution (flare stars and related objects). See Abstr. 122.009.

T Tauri and FU Orionis stars. See Abstr. 122.016.

The variable circular polarization of V1057 Cygni. See Abstr. 122.072.

The Hertzsprung progression in cepheid calculations. See Abstr. 122.113.

Comments on the pulsation of double mode cepheids. See Abstr. 122.114.

Period and amplitude changes in δ Scuti stars. See Abstr. 122.117.

On Epstein's weighting function and pulsation constants of cepheid models. See Abstr. 122.118.

Convection in pulsating stars. See Abstr. 122.187.

On presupernova dynamical mass ejection by non-detonated stellar cores. See Abstr. 125.029.

Evolution of crystallizing pure ^{12}C white dwarfs. See Abstr. 126.005.

The demise of the degenerate M stars. See Abstr. 126.007.

Strong molecular line emission associated with small Hα emission regions. See Abstr. 131.056.

Microwave H_2O emission from young stars. See Abstr. 131.092.

H_2O sources in regions of star formation. See Abstr. 131.121.

Bok globules, molecular clouds, infra-red sources and star formation. See Abstr. 131.130.

The structure and stability of shock waves in a multiphase interstellar medium. II. The effect of a magnetic field, thermal conduction, and viscosity. See Abstr. 131.138.

Colliding molecular clouds and star formation in NGC 1333. See Abstr. 131.173.

Stellar winds and globules in H II regions. See Abstr. 131.501.

A correlation of H II regions with H I gas and the rate of star formation in galaxies. See Abstr. 131.545.

Initial mass function for massive stars and integrated spectral properties of H II regions. See Abstr. 131.553.

Herbig-Haro-Objekte und T-Tauri-Sterne, ungelöste Probleme der frühen Sternentwicklung. See Abstr. 132.004.

Relativistic effects in the binary pulsar PSR 1913+16. See Abstr. 141.301.

Pulsar glitches and restlessness as a hard superfluidity phenomenon. See Abstr. 141.303.

Superfluidity in neutron stars. I. Steady-state hydrodynamics and frictional heating. See Abstr. 141.317.

Acceleration of pulsars by asymmetric radiation. See Abstr. 141.331.

Search for continuous X-ray emission from NP 0532. See Abstr. 141.346.

Observations of a bright, variable X-ray source in a globular cluster. See Abstr. 142.010.

Radiative transfer in a strong magnetic field and accreting X-ray pulsars. See Abstr. 142.094.

Hercules X-1: a neutron star with a twisted accretion disk? See Abstr. 142.114.

An interpretation of the general character of the X-ray emission from compact objects in close binary stars. See Abstr. 142.116.

Accretion onto magnetic neutron stars: comments on the position of the Alfvén surface. See Abstr. 142.128.

Observing neutron stars: information on stellar structure from pulsars and compact X-ray sources. See Abstr. 142.129.

Neutron stars, black holes and binary X-ray sources. See Abstr. 142.130.

The Her X-1 spectrum: a measure of the field near the magnetic poles of a neutron star. See Abstr. 142.155.

The tip of the red giant branch (a comparison of theory and observations). See Abstr. 154.006.

Spatial distribution of stars in the globular cluster NGC 5466. See Abstr. 154.038.

Infrared galaxies: evolutionary stages of massive star formation. See Abstr. 158.057.

The distribution of cepheids in the Large Magellanic Cloud and cells of star formation. See Abstr. 159.009.

066 Relativistic Astrophysics (without Cosmology), Background Radiation, Gravitation Theory

066.001 **On the implications of the nonexistence of unstable outgoing normal modes for the stability of spherical stellar models to nonradial perturbations in general relativity.**
J. R. Ipser.
Astrophys. Journ., Vol. 199, 220 - 226 (1975).
This paper studies whether, without tackling the question of completeness of the normal modes, one can relate the question of the nonexistence of unstable normal modes having purely outgoing waves at large radii to the question of the stability of systems to perturbations involving gravitational radiation in general relativity. It is shown that a model is stable to small nonradial perturbations essentially if and only if it does not possess an unstable outgoing normal mode.

066.002 **On ellipsoidal solutions of Liouville's equation in general relativity.** M. Trümper.
Gen. Rel. Grav., Vol. 5, 1 - 12 (1974).
A relativistic, collisionless gas of gravitating particles all having the same proper mass (possibly equal to zero) is studied under the assumption that the one-particle distribution function is locally ellipsoidal in momentum space with respect to some timelike vector field (observer).

066.003 **On relativistic deformable solids and the detection of gravitational waves.** G. A. Maugin.
Gen. Rel. Grav., Vol. 5, 13 - 23 (1974).

066.004 **Shear-free, twisting Einstein-Maxwell metrics in the Newman-Penrose formalism.** R. W. Lind.
Gen. Rel. Grav., Vol. 5, 25 - 47 (1974).

066.005 **The equations of motion of macroscopic bodies in general relativity.** P. A. Hogan, J. D. McCrea.
Gen. Rel. Grav., Vol. 5, 79 - 113 (1974).
A new approach to the problem of motion in general relativity, based upon the systematic approximation procedure of Synge, is presented. The equations of translational motion for a system of spherical bodies moving under their mutual gravitational attractions are derived.

066.006 **A dispersion relation for gravitational radiation in a Schwarzschild background.** J. Madore.
Gen. Rel. Grav., Vol. 5, 169 - 173 (1974).
A dispersion relation is derived for gravitational radiation emitted from an almost spherical object.

066.007 **On electromagnetic and gravitational waves in general relativity. I.** S. R. Roy, V. N. Tripathi.
Gen. Rel. Grav., Vol. 5, 257 - 274 (1974).
In this paper the external field of a bounded source emitting electromagnetic and gravitational radiation has been considered. The field has been assumed to be axially symmetric as well as reflexion symmetric and that it tends to flatness at infinity.

066.008 **Interacting electromagnetic shock waves in general relativity.** P. Bell, P. Szekeres.
Gen. Rel. Grav., Vol. 5, 275 - 286 (1974).
It is shown that the continuity conditions of Lichnerowicz must generally be relaxed in favour of the O'Brien-Synge conditions in the case of shock electromagnetic waves. This is in particular true when two electromagnetic shock waves are in collision. Gravitational impulse waves are produced as a result of the weakened conditions. An exact solution exhibiting this behaviour is derived, and the effect of the impulse waves on a measuring device are compared with experimental results of Weber.

066.009 **A formulation of linearized gravity.**
D. H. Tchrakian.
Gen. Rel. Grav., Vol. 5, 379 - 386 (1974).
The formulation of linearized gravity in terms of the 'electric' and 'magnetic' gravitational fields is extended to take into account the presence of matter. The modes of radiation, the equations of motion and the potential in the static case are given. The relevant components of the superenergy tensor are calculated and a quantity named the superforce is introduced.

066.010 **The velocity of magnetohydrodynamic waves in general relativity.** D. P. Mason.
Gen. Rel. Grav., Vol. 5, 429 - 439 (1974).
The energy-momentum tensor for a perfect fluid with infinite conductivity is modified slightly by attributing a proper energy content to a field of tension, as proposed by Cattaneo, for a perfect nonconducting fluid, and it is shown that with this modification the velocities of the hydrodynamic waves in the fluid do not exceed c. This contrasts with the standard theory in which a lower limit must be placed on the compressibility of the fluid to prevent the velocity of the 'fast' hydrodynamic wave from exceeding c. The velocity of the Alfvén wave is changed only slightly and is always less than c as in the standard theory.

066.011 **Constancy of the local velocity of light in general relativistic gravitational fields.**
G. Horedti (*Horedt*).
Gen. Rel. Grav., Vol. 5, 475 - 477 (1974).
Using the natural, real units (proper time and proper length) for the definition of the spatial velocity, it is shown that the velocity of light during its propagation through gravitational fields is constant and exactly equal to its value in vacuum. Therefore, according to general relativity, it is not correct to speak about a variable velocity of light in gravitational fields.

066.012 **Nongeodesic motion in general relativity.**
B. M. Barker, R. F. O'Connell.
Gen. Rel. Grav., Vol. 5, 539 - 554 (1974).

066.013 **The possibility of superluminal sources and their Doppler effect.** R. Mignani, E. Recami.
Gen. Rel. Grav., Vol. 5, 615 - 620 (1974).
From the special relativity theory, the existence of superluminal reference-frames and faster-than-light objects can be inferred, when its validity is not arbitrarily confined to subluminal velocities. The 'Extended theory of relativity' allows in particular extending the Doppler effect formulae to superluminal sources. This point is here explored, due to its possible astrophysical interest.

066.014 **The second-order gravitational redshift.**
J. Jaffe, R. F. C. Vessot.
Gen. Rel. Grav., Vol. 6, 55 - 61 (1975). – Presented at the Seventh International Conference on Gravitation and Relativity, June 1974, Tel-Aviv, Israel.
It now appears possible that another distinct test of the second-order term of general relativity may be feasible, through the use of very stable atomic clocks. This experiment, which would measure the second-order gravitational redshift,

is a bona fide test of the field equations of gravity, not just a test of the underlying principle of equivalence.

066.015 **A new general covariant approach to the general relativistic two-body problem.** A. Rosenblum.
Gen. Rel. Grav., Vol. 6, 67 - 70 (1975). – Presented at the Seventh International Conference on Gravitation and Relativity, June 1974, Tel-Aviv, Israel.

066.016 **Can quantum gravitational forces stop gravitational collapse?** R. F. O'Connell.
Gen. Rel. Grav., Vol. 6, 99 - 102 (1975). – Presented at the Seventh International Conference on Gravitation and Relativity, June 1974, Tel-Aviv, Israel.

066.017 **Black-and-white holes.** R. Goldoni.
Gen. Rel. Grav., Vol. 6, 103 - 113 (1975). – Presented at the Seventh International Conference on Gravitation and Relativity, June 1974, Tel-Aviv, Israel.

066.018 **Investigation of singularities in general relativity.** K. Bera, B. K. Datta.
Gen. Rel. Grav., Vol. 6, 129 - 139 (1975). – Presented at the Seventh International Conference on Gravitation and Relativity, June 1974, Tel-Aviv, Israel.
The problem of singularities for a non-symmetric and isentropic motion of a perfect fluid under the assumption of adiabatic thermodynamic processes is investigated from the standpoint of a local observer. It is shown that, whatever the distribution of matter might be, there occurred a singularity in the past in the non-rotating parts of the universe and it must occur again in the future if the universe is closed.

066.019 **Massive black holes in globular clusters?** J. N. Bahcall, J. P. Ostriker.
Nature, Vol. 256, 23 - 24 (1975).
Four X-ray sources have been identified with globular star clusters (NGC 1851, NGC 6441, NGC 6624 and M 15). Here the authors suggest that the globular cluster X-ray sources may be a very different phenomenon from the previously studied binary X-ray sources. In particular, they suggest that the observed X-ray emission in globular clusters may be caused by accretion of gas on to a massive black hole at the centre of the cluster.

066.020 **The theoretical tools of experimental gravitation.** C. M. Will.
Experimental gravitation, (see 012.002), p. 1 - 110 (1974).
Theoretical frameworks for testing relativistic gravity; The Dicke framework; Significance of the Eötvös-Dicke-Braginsky experiment; The parametrized post-Newtonian formalism; Metric theories of gravity and their post-Newtonian limit; Equations of motion in the PPN formalism; Observable effects and experimental tests of metric theories of gravity.

066.021 **Lectures on frequency stability and clocks and on the gravitational red-shift experiment.**
R. F. C. Vessot.
Experimental gravitation, (see 012.002), p. 111 - 162 (1974).
Introduction; Frequency and time standards using quartz crystals and cesium, rubidium and hydrogen atoms; The present state of the art; The gravitational red-shift rocket probe experiment.

066.022 **The detection of small accelerations, gravitational antennae, verification of the principle of equivalence.** V. B. Braginsky (*Braginskij*).
Experimental gravitation, (see 012.002), p. 235 - 262 (1974).
The detection of small forces acting on the macroscopic oscillator; Gravitational antennae; Verification of the equivalence of inertial and gravitational mass; Search for gravitational radiation of cosmic origin.

066.023 **Gravitational radiation experiments and instrumentation.** J. Weber.
Experimental gravitation, (see 012.002), p. 263 - 279 (1974).
Gravitational-radiation experiments; Theory of the gravitational radiation detector with ferroelectric crystal coupling; Noise and sensitivity of a gravitational-radiation detector with ferroelectric coupling.

066.024 **The use of low-temperature technology in gravitational experiments.** W. M. Fairbank.
Experimental gravitation, (see 012.002), p. 280 - 293 (1974).
Introduction; Macroscopic quantum phenomena; Other useful low-temperature properties.

066.025 **Search for gravitational radiation using low-temperature techniques.** W. M. Fairbank, S. P. Boughn, H. J. Paik, M. S. McAshan, J. E. Opfer, R. C. Taber, W. O. Hamilton, B. Pipes, T. Bernat, J. M. Reynolds.
Experimental gravitation, (see 012.002), p. 294 - 309 (1974).
Introduction; Theory of gravitational-radiation detector; Experimental program.

066.026 **The gyroscope experiment. - I: general description and analysis of gyroscope performance.**
C. W. F. Everitt.
Experimental gravitation, (see 012.002), p. 331 - 360 (1974).
Introduction; Description of the experiment; Analysis of gyro errors.

066.027 **The gyroscope experiment. - II: development of the London-moment gyroscope and of cryogenic technology for space.** J. A. Lipa, W. M. Fairbank, C. W. F. Everitt.
Experimental gravitation, (see 012.002), p. 361 - 380 (1974).
Background; Design and construction of the gyroscope; The cryogenic system; Laboratory test set-up and flight program.

066.028 **The gyroscope experiment. - III: tests of the equivalence of gravitational and inertial mass based on cryogenic techniques.** P. W. Worden, Jr., C. W. F. Everitt.
Experimental gravitation, (see 012.002), p. 381 - 402 (1974).
Introduction; Design of experiment; Limitations to the accuracy of the experiment.

066.029 **On the detection of gravitational radiation by its interaction with electromagnetic fields.** L. Halpern.
Experimental gravitation, (see 012.002), p. 414 - 424 (1974).

066.030 **Testing gravitation theories by means of a heliocentric probe.** G. Israel, R. Pacault, E. Roth, J. Schniewind, I. F. Croce-Spinelli, S. Larcher.
Experimental gravitation, (see 012.002), p. 425 - 448 (1974).
Introduction; Objectives of the SOREL mission; Description of the mission; Error analysis.

066.031 **Generation of geodesic motion by the twin probe method.** B. Bertotti.
Experimental gravitation, (see 012.002), p. 490 - 495 (1974).

066.032 **Spin, rotation and *C*, *P* and *T* effects in the gravitational interaction and related experiments.**
R. F. O'Connell.
Experimental gravitation, (see 012.002), p. 496 - 514 (1974).

066.033 **Low-temperature gravitational wave detector.** H. Paik.
Experimental gravitation, (see 012.002), p. 515 - 524 (1974).

066.034 **The gravitational deflection of radio waves.** R. Sramek.
Experimental gravitation, (see 012.002), p. 529 - 542 (1974).

066.035 **In search of low-frequency gravitational radiation using cross-correlation of earth-moon vibrations.**
V. S. Tuman.
Experimental gravitation, (see 012.002), p. 543 - 572 (1974).
A brief description of the cryogenic gravity meter; Planetary eigenvibrations and some of the terrestrial results; Proposed experimental plan.

066.036 **Black holes and their astrophysical implications – I, II.** D. L. Block.
Sky Telescope, Vol. 50, 20 - 23, 87 - 90 (1975).

066.037 **A confirmation of Einstein's general theory of relativity by measuring the bending of microwave radiation in the gravitational field of the sun.**
E. B. Fomalont, R. A. Sramek.
Astrophys. Journ., Vol. 199, 749 - 755 (1975).
An experiment to determine the relative bending of microwave radiation from three radio sources (0116+08, 0119+11, and 0111+02) as they passed near the sun was performed with a radio interferometer of baseline 35 km in 1974 March and April. The measured bending in terms of that at the solar limb is 1".775 ± 0".019 (standard error), or 1.015 ± 0.011 times the Einstein prediction of general relativity. The result is not consistent with the scalar-tensor formulation of general relativity with a scalar-coupling constant $\omega < 23$.

066.038 **Gravitational radiation instability in rotating stars.** J. L. Friedman, B. F. Schutz,
Astrophys. Journ., (*Letters*), Vol. 199, L157 - L159 (1975).
The authors have shown elsewhere that gravitational radiation leads to an instability to nonaxisymmetric perturbations in stars that are rotating rapidly enough to be secularly unstable in Newtonian theory, and here some astrophysical consequences are explored. Neutron stars are unstable when $\langle\Omega\rangle$ exceeds $10^3 - 10^4\,s^{-1}$ and lose their excess angular momentum in seconds. Unstable white dwarfs spin down to stability or to collapse in times that vary from weeks to 10^3 years. All stars with ergoregions are unstable; this fact imposes stringent upper limits on the rotation of relativistic configurations.

066.039 **Magnetic viscosity in relativistic accretion disks.** D. M. Eardley, A. P. Lightman.
Astrophys. Journ., Vol. 200, 187 - 203 (1975).
The authors formulate a detailed, self-consistent model of magnetic viscosity in an accretion disk around a rotating, Kerr black hole and give detailed models of stationary disks based on the viscosity of the magnetic field. The observable properties of these models are similar to those of previous models based on less detailed models of viscosity. The authors include the relativistic equations governing the time-evolution of an accretion disk around a Kerr black hole, and some criteria for the existence of a radiation-pressure-dominated region or an optically thin region in a stationary disk.

066.040 **On the stability of relativistic systems.** J. L. Friedman, B. F. Schutz.
Astrophys. Journ., Vol. 200, 204 - 220 (1975).
Using methods applicable to all classical radiative field theories, a necessary and sufficient condition is derived for the onset of instability along a sequence of stationary equilibria in general relativity. A related result is that instability in any mode can set in only when its frequency vanishes. Astrophysical implications are briefly discussed.

066.041 **Experimental test of general relativity using time-delay data from Mariner 6 and Mariner 7.**
J. D. Anderson, P. B. Esposito, W. Martin, C. L. Thornton, D. O. Muhleman.
Astrophys. Journ., Vol. 200, 221 - 233 (1975).
Range and Doppler data from Mariner 6 and Mariner 7 have been analyzed for purposes of measuring the effect of the sun's gravity field on S-band radio transmissions to the spacecraft. The prediction of general relativity, the the round-trip time delays between the station and the spacecraft will be increased by about 200 μs near superior conjunction, has been verified with an uncertainty of 3 percent or less. The value for the parameter $\gamma^* = (1+\gamma)/2$, which multiplies the Einstein prediction of the 'excess' delay in the ranging signal, is $\gamma^* = 1.00 \pm 0.03$, where the uncertainty represents a realistic standard deviation.

066.042 **Some properties of gravitational fields with hydrodynamic fluid and dust.** V. I. Obozov.
Izv. vyssh. ucheb. zavedenij. Fizika, No. 1, p. 27 - 32.
In Russian. – Abstr. in Referativ. Zhurn. 51. Astron., 7.51. 875 (1975).

066.043 **Gravitational radiation from stellar collapse into a disk.** I. D. Novikov.
In-t kosmich. issled. AN SSSR. Pr-215. Moskva, 8 pp. In Russian. – Abstr. in Referativ. Zhurn. 51. Astron., 7.51.879 (1975).

066.044 **Do black holes really exist?** K. Z. Starikov.
Geogr. sb. Kazakhsk. un-t, 1974, vyp. (No.) 1, p. 158 - 160. In Russian. – Abstr. in Referativ. Zhurn. 51. Astron., 7.51.885 (1975).

066.045 **Three types of solutions of Einstein-Maxwell equations.** Yu. A. Khlestkov.
Zhurn. ehksperim. i teor. fiz., Vol. 68, 387 - 391 (1975). In Russian. – Abstr. in Referativ. Zhurn. 51. Astron., 7.51. 888 (1975).

066.046 **Limiting transitions in the Tolman solution of equations of general relativity.**
M. P. Korkina, L. M. Chernyj.
Izv. vyssh. ucheb. zavedenij. Fizika, 1975, No. 1, p. 23 - 26. In Russian. – Abstr. in Referativ. Zhurn. 51. Astron., 7.51. 889 (1975).

066.047 **Ultra-relativistic particles and tetrad theory of some optical effects in a Schwarzschild field.**
A. E. Levashev.
Probl. gravitatsii. Sb. dokl. plenarn. zasedanij 3-j Sov. gravitatsion. konf., Erevan, 1972. Erevan, Erevan. un-t, 1975, p. 113 - 135. In Russian. – Abstr. in Referativ. Zhurn. 51. Astron., 7.51.902 (1975).

066.048 **Deflection of light in the sun's gravitational field.** M. Karipova.
Trudy Kirg. un-ta. Ser. mat. n., 1973 (1974), vyp. (No.) 10, p. 19 - 24. In Russian. – Abstr. in Referativ. Zhurn. 51. Astron., 7.51.905 (1975).

066.049 **On the properties of non-Schwarzschild gravitational lenses.** L. Kh. Ingel'.
Astron. Zhurn. Akad. Nauk SSSR, Vol. 52, 727 - 737 (1975). In Russian. English translation in Soviet Astron., Vol. 19, No. 4.
The lens properties of the simplest non-Schwarzschild gravitational fields are considered. Such gravitational lenses have some interesting features. There are, in particular, two-dimensional caustics, which must be taken into account when calculating the probabilities of gravitational lens effect displays.

066.050 **Relativistic libration of planets.** W. T. Rikow.
Astron. Zhurn. Akad. Nauk SSSR, Vol. 52, 885 - 887 (1975). In Russian. English translation in Soviet Astron., Vol. 19, No. 4.

The method of determination of the angular velocity of proper rotation of test bodies based on the world lines deviation equation is presented. The angular velocity of proper rotation of planets in a central-symmetric field is found. Relativistic physical librations of planets of the solar system are calculated.

066.051 **The generation of gravitational waves. I. Weak-field sources.** K. S. Thorne, S. J. Kovács.
Astrophys. Journ., Vol. 200, 245 - 262 (1975).

This paper derives and summarizes a new "plug-in-and-grind" formalism for calculating the gravitational waves emitted by any system with weak internal gravitational fields. If the internal fields have negligible influence on the system's motions, then the new formalism reduces to standard "linearized theory." Whether or not gravity affects the motions, if the motions are slow and internal stresses are weak, then the new formalism reduces to the standard "quadrupole-moment formalism". In the general case the new formalism expresses the radiation in terms of a retarded Green's function for slightly curved spacetime.

066.052 **Mass and moment-of-inertia peaks in general relativity.**
R. C. Adams, R. D. H. Warburton, J. M. Cohen.
Astrophys. Journ., Vol. 200, 263 - 268 (1975).

The existence of maxima in both the mass and moment of inertia of stellar models is discussed. The moment-of-inertia peak occurs at a lower density than the mass peak. It is shown that general-relativistic effects cause the two peaks to move together as the density increases. Presented is a method (not involving numerical integration) for obtaining the position of the peaks.

066.053 **Rapidly rotating fluid bodies in general relativity.** E. M. Butterworth, J. R. Ipser.
Astrophys. Journ., (*Letters*), Vol. 200, L103 - L106 (1975).

A new self-consistent-field method for numerically constructing rapidly rotating fluid bodies in general relativity has been developed. Results which emerge from using it to build uniformly rotating homogeneous bodies are discussed. The results reveal that, in contrast to the Newtonian sequence of Maclaurin spheroids, the relativistic sequences in most, and probably all, cases terminate at eccentricities < 1 where the centrifugal and gravitational accelerations at the equator become equal. The results also suggest that a highly relativistic rotating body, due to certain effects, might evolve toward a nearly spherical configuration as it contracts.

066.054 **Wave-wave interactions and gravitational instability.** A. Hruška.
Astrophys. Space Sci., Vol. 36, 451 - 454 (1975).

The instability of gravitational waves with $k > k_{\mathrm{Jeans}}$ in a weakly turbulent gas is discussed. Basic features of the processes leading to amplification or damping of gravitational waves are described. Growth rates of the amplitudes are calculated.

066.055 **On the origin of the background radiation.** H. Dehnen, H. Hönl.
Astrophys. Space Sci., Vol. 36, 473 - 478 (1975).

From the assumption of a universal upper limit for the temperature of about 10^{12} K, which is suggested by the experiences of high energy physics, it follows taking additionally into account the strong interaction between hadrons explicitly, that in the earliest stage of the universe the numbers of photons and of nucleons-antinucleons should have been nearly equal, but greater than the present nucleon number by a factor of about 10^7. Accordingly the 3 K background radiation is generated for the most part by annihilation of nearly all primordial heavy particles.

066.056 **The physics of gravitationally collapsed objects.** R. Ruffini.
Astrophys. Space Sci. Library, Vol. 48, (see 012.014), 59 - 118 (1975).

066.057 **Observable aspects of very long wavelength gravitational waves.** J. C. Jackson.
Monthly Notices Roy. Astron. Soc., Vol. 173, 5P - 8P (1975).

The interaction of very long wavelength gravitational waves with clusters of galaxies is considered. These waves induce oscillatory relative motions within such systems, as well as a systematic expansion.

066.058 **A note on the generation of twisting "D" metrics from non-twisting "D" metrics.** T. C. Basey.
Zeitschr. Naturforschung, Vol. 30a, 1200 (1975).

066.059 **Petrov's classification of travelling and standing gravitational waves.**
L. P. Grishchuk, M. V. Sazhin.
Dokl. Akad. Nauk SSSR. Ser. Mat., Fiz., Vol. 223, 72 - 75 (1975). In Russian.

066.060 **Gravitational analogue of the magnetic force.** P. Lorrain, with a reply by D. H. Menzel, W. W. Salisbury.
Nature, Vol. 257, 161 - 162 (1975).

066.061 **Equivalence principle.** G. E. Marsh, C. Nissim-Sabat, with comments by W. G. V. Rosser, and a reply by N. T. Bishop, P. T. Landsberg.
Nature, Vol. 257, 517 - 518 (1975).

066.062 **Thermal instability of disc accretion onto a black hole.** R. A. Syunyaev, N. I. Shakura.
Pis'ma v Astron. Zhurn., Vol. 1, No. 8, p. 6 - 11 (1975). In Russian.

In the inner region of an accretion disc the radiation dominates on the pressure. A stationary regime of accretion is realized only at a fixed value of viscosity. The accretion process is unstable in this region and the radiation of the disc appears to be variable.

066.063 **On a spherically symmetrical solution of the field equations.** A. F. Bogorodskij.
Astrometriya i Astrofizika, *Kiev*, vyp. (No.) 26, (see 003.001), p. 3 - 10 (1975). In Russian.

The non-static solution of the Einstein field equations with spherical symmetry in vacuum is obtained. The matrix determined by the solution is interpreted as a gravitational field inside the expanding or contracting homogeneous sphere. The space represents a three-dimensional spherical continuum, the radius of which is a linear function of time. Some properties of the probe particle motion and of light propagation in this field are considered.

066.064 **On a class of exact spherically symmetric solutions to the Einstein gravitational field equations.**
M. W. Cook.
Australian Journ. Phys., Vol. 28, 413 - 422 (1975).

This paper presents a development of a well-known class of Einstein gravitational fields which describe, in comoving coordinates, spherically symmetric perfect fluids with spatially uniform density but with nonuniform pressure. It is noted that the sign of the spatial curvature for these distributions may vary with time, in contrast with the situation for the uniform model theory used in most cosmological applications. It is shown, however, that this entails physically unacceptable behaviour of the fluid.

066.065 **A relativistic treatment of space–time rays and the Doppler effect in general media.** J. A. Bennett.
Australian Journ. Phys., Vol. 28, 423 - 435 (1975).

For waves in plasmas and similar media, space–time rays with a 4-vector formulation are used to obtain an approximate expression for the rapidly varying part of the wave, on the assumption that the plasma properties vary slowly in space and time. It is shown that the phase function associated with the rapid variations is invariant with respect to Lorentz transformation and that the ray equations may be written in vector form. These results are used to obtain expressions for the Doppler shift that are consistent with special relativity. The possibility that the medium is moving and time-varying is allowed for, and a variety of possible sources are considered.

066.066 **The general cosmological solution of Einstein's equations with time singularity.** V. A. Belinskij, E. M. Lifshits, I. M. Khalatnikov.
Probl. gravitatsii. Sb. dokl. plenarn. zasedanij 3-j Sov. gravitatsion. konf. Erevan, 1972. Erevan, Erevan. un-t, 1975, p. 136 - 164. In Russian. – Abstr. in Referativ. Zhurn. 51. Astron., 9.51.811 (1975).

066.067 **Contemporary state of the theory of superdense celestial bodies.** G. S. Saakyan.
Probl. gravitatsii. Sb. dokl. plenarn. zasedanij 3-j Sov. gravitatsion. konf., Erevan, 1972. Erevan, Erevan. un-t, 1975, p. 208 - 224. In Russian. – Abstr. in Referativ. Zhurn. 51. Astron., 9.51.823 (1975).

066.068 **Contemporary state of the gravitational collapse theory and astrophysical developments.**
I. D. Novikov.
Probl. gravitatsii. Sb. dokl. plenarn. zasedanij 3-j Sov. gravitatsion. konf., Erevan, 1972. Erevan, Erevan. un-t, 1975, p. 37 - 41. In Russian. – Abstr. in Referativ. Zhurn. 51. Astron., 9.51.827 (1975).

066.069 **On the physical interpretation of parameters in the general Kerr-NUT solution. II.** V. P. Semenov.
Izv. vyssh. uchebn. zavedenij. Fizika, 1975, No. 2, p. 11 - 17. In Russian. – Abstr. in Referativ. Zhurn. 51. Astron., 9.51. 832 (1975).

066.070 **Global properties of matter in collapsed state.**
M. A. Markov.
Probl. gravitatsii. Sb. dokl. plenarn. zasedanij 3-j Sov. gravitatsion. konf. Erevan, 1972. Erevan, Erevan. un-t, 1975, p. 165 - 207. In Russian. – Abstr. in Referativ. Zhurn. 51. Astron., 9.51.834 (1975).

066.071 **Arbitrary T-balls in general relativity.**
M. P. Korkina, V. G. Martynenko.
Ukr. fiz. zhurn., Vol. 20, 628 - 632 (1975). In Russian. Abstr. in Referativ. Zhurn. 51. Astron., 9.51.835 (1975).

066.072 **Search for gravitational radiation from cosmic sources.**
V. B. Braginskij, A. B. Manukin, E. I. Popov, V. N. Rudenko.
Probl. gravitatsii. Sb. dokl. plenarn. zasedanij 3-j Sov. gravitatsion. konf., Erevan, 1972. Erevan, Erevan. un-t, 1975, p. 18 - 31. In Russian. – Abstr. in Referativ. Zhurn. 51. Astron., 9.51.840 (1975).

066.073 **Astrophysical processes near black holes.**
D. M. Eardley, W. H. Press.
Annual Rev. Astron. Astrophys., Vol. 13, (see 003.005), 381 - 422 (1975). – Contents: (1) Introduction and overview; (2) Gravitational collapse; (3) Interactions of the hole with its environment; (4) Black holes in binary systems; (5) Exotic possibilities.

066.074 **The black hole.** J. A. Wheeler.
Proc. 16th Solvay Conference Phys., (see 012.020), p. 279 - 311, with a discussion, p. 312 - 316 (1974). – Review paper.

066.075 **Search for observational evidence for black holes.**
I. D. Novikov.
Proc. 16th Solvay Conference Phys., (see 012.020), p. 317 - 340, with a discussion, p. 341 - 347 (1974).

066.076 **New gravitational radiation experiments.**
J. Weber, M. Lee, D. J. Gretz, G. Rydbeck, W. L. Trimble, S. Steppel.
Ondes et radiations gravitationnelles. Colloque International C.N.R.S. No. 220, (see 012.021), p. 23 - 32 (1974).

066.077 **Issues in gravitational wave detection with space missions.** R. W. Davies.
Ondes et radiations gravitationnelles. Colloque International C.N.R.S. No. 220, (see 012.021), p. 33 - 45 (1974).

066.078 **Gravitational wave tails.** W. B. Bonnor.
Ondes et radiations gravitationnelles. Colloque International C.N.R.S. No. 220, (see 012.021), p. 73 - 81 (1974).

066.079 **Meudon gravitational radiation detection experiment.** S. Bonazzola, M. Chevreton, P. Felenbok, G. Herpe, J. Thierry-Mieg.
Ondes et radiations gravitationnelles. Colloque International C.N.R.S. No. 220, (see 012.021), p. 101 - 111 (1974).

066.080 **Experiments and observations with wide band gravity wave detectors.**
R. W. P. Drever, J. Hough, R. Bland, G. W. Lessnoff.
Ondes et radiations gravitationnelles. Colloque International C.N.R.S. No. 220, (see 012.021), p. 113 - 133 (1974).

066.081 **Radiation from highly relativistic geodesics.**
C. W. Misner.
Ondes et radiations gravitationnelles. Colloque International C.N.R.S. No. 220, (see 012.021), p. 145 - 160 (1974).

066.082 **The Einstein-Cartan theory of gravitation.**
A. Trautman.
Ondes et radiations gravitationnelles. Colloque International C.N.R.S. No. 220, (see 012.021), p. 161 - 170 (1974).

066.083 **The gravitational wave detection experiment in Frascati.** K. Maischberger.
Ondes et radiations gravitationnelles. Colloque International C.N.R.S. No. 220, (see 012.021), p. 171 - 180 (1974).

066.084 **On the evaluation of the Munich-Frascati Weber-type experiment.** P. Kafka.
Ondes et radiations gravitationnelles. Colloque International C.N.R.S. No. 220, (see 012.021), p. 181 - 201 (1974).

066.085 **Sources and astrophysical effects of gravitational waves.** M. J. Rees.
Ondes et radiations gravitationnelles. Colloque International C.N.R.S. No. 220, (see 012.021), p. 203 - 212 (1974).

066.086 **How to test gravitation theories by means of gravitational-wave measurements.** K. S. Thorne.

Ondes et radiations gravitationnelles. Colloque International C.N.R.S. No. 220, (see 012.021), p. 213 - 223 (1974).

066.087 **Low temperature gravitational radiation antenna. A progress report.** P. Carelli, M. Cerdonio, U. Giovanardi, G. Lucano, I. Modena.
Ondes et radiations gravitationnelles. Colloque International C.N.R.S. No. 220, (see 012.021), p. 257 - 264 (1974).

066.088 **Gravitational waves and Mach's principle.** D. W. Sciama.
Ondes et radiations gravitationnelles. Colloque International C.N.R.S. No. 220, (see 012.021), p. 267 - 282 (1974).

066.089 **A new covariant theory of gravitation.** N. Rosen. Ondes et radiations gravitationnelles. Colloque International C.N.R.S. No. 220, (see 012.021), p. 285 - 299 (1974).

066.090 **Matching conditions in the design of gravitational detector systems.** D. G. Maeder.
Ondes et radiations gravitationnelles. Colloque International C.N.R.S. No. 220, (see 012.021), p. 359 - 363 (1974).

066.091 **Proposed standardization of gravitational burst detector signals.** D. G. Maeder.
Ondes et radiations gravitationnelles. Colloque International C.N.R.S. No. 220, (see 012.021), p. 365 - 369 (1974).

066.092 **Limits on terrestrial produced gravitational radiation.** L. Halpern.
Ondes et radiations gravitationnelles. Colloque International C.N.R.S. No. 220, (see 012.021), p. 373 - 378 (1974).

066.093 **A radiating Kerr-Schild metric.** P. C. Vaidya. Ondes et radiations gravitationnelles. Colloque International C.N.R.S. No. 220, (see 012.021), p. 379 - 381 (1974).

066.094 **Equations of motion radiation reaction, and gravitational radiation.** P. Havas.
Ondes et radiations gravitationnelles. Colloque International C.N.R.S. No. 220, (see 012.021), p. 383 - 392 (1974).

066.095 **Anisotropic gravity and the binary pulsar PSR 1913+ 16.** K. Nordtvedt, Jr.
Astrophys. Journ., Vol. 202, 248 - 249 (1975).

If anisotropic post-Newtonian metric potentials exist, celestial bodies will possess anisotropic inertial and/or gravitational masses. The binary pulsar system PSR 1913 + 16 puts stringent limits on such anisotropies through its observable perihelion precession rate and hence offers a Hughes-Drever-type experiment for massive celestial bodies.

066.096 **Radiation from spherical accretion onto black holes.** P. Mészáros.
Astron. Astrophys., Vol. 44, 59 - 68 (1975).

The author presents calculations of the radiation properties of black holes accreting in a spherically symmetric manner, both in the case when they are isolated and accreting from the interstellar medium, and when they are in a binary system. Two new effects are included, the heating due to dissipation of turbulent motions and magnetic field line reconnection, and the radiation from cosmic rays captured with the accreted gas. For mass transfer rates comparable to that of binary X-ray source models, the spherically symmetric model gives temperatures above 10^9 K, and therefore can compete with disk models as a source of hard X-ray radiation.

066.097 **Radiation of gravitational waves during captures in black hole clusters.** A. G. Polnarev.
Astron. Zhurn. Akad. Nauk SSSR, Vol. 52, 930 - 936 (1975). In Russian. English translation in Soviet Astron., Vol. 19, No. 5.

A qualitative picture of gravitational radiation during coalescing of two black holes was calculated. This process is characterized by a sequence of repeated bursts of gravitational radiation. The frequency of events was obtained in dependence on the time intervals between the first and second burst, and in dependence on the total number of bursts, supposed that black hole clusters serve as sources of gravitational radiation.

066.098 **On a second region of strong focusing effect created by non-Schwarzschild gravitational lenses.**
L. Kh. Ingel', N. R. Rubakha.
Astron. Zhurn. Akad. Nauk SSSR, Vol. 52, 1049 - 1054 (1975). In Russian. English translation in Soviet Astron., Vol. 19, No. 5.

Non-Schwarzschild gravitational lenses can create a second region of strong focusing effect. However, diffraction being taken into account, the region is reduced considerably. The possibilities of observation of the gravitational focusing effect (especially the focusing of gravitational waves and neutrinos) are discussed.

066.099 **More on gravity waves.** R. L. Garwin, with a reply by J. Weber.
Phys. Today, Vol. 28, No. 11, p. 13, 15, 99 (1975).

066.100 **Some remarks about the concept of cosmic time.** M. Heller.
Acta Cosmologica, Fasc. 3, p. 41 - 44 (1975).

Certain assumptions, which must be postulated to hold on the space-time manifold in order to determine the concept of universal or cosmic time, are discussed.

066.101 **Space-time manifolds and the inertial frames.** M. Heller.
Acta Cosmologica, Fasc. 3, p. 45 - 48 (1975).

It is shown that in any macroscopic physical theory the local inertial frames may be defined, to some extent, absolutely, and "smoothly", on the whole space-time manifold, under certain, very general – and at present indispensable – conditions.

066.102 **Rotating bodies in general relativity.** M. Heller. Acta Cosmologica, Fasc. 3, p. 97 - 107 (1975).

Beginning from the early works of Lense and Thirring, the so-called dragging effects in general relativity are reviewed. Some remarks precede concerning the status of inertial frames in Einstein's theory. "Machian" interpretations of considered effects are discussed.

066.103 **Ondes de choc gravitationnelles et ondes de choc magnétomatérielles extrêmes.** A. Lichnerowicz.
Comptes Rendus Acad. Sci. Paris, Sér. A, Vol. 281, 929 - 934 (1975).

Étude, dans le contexte astrophysique des hautes densités, d'un modèle composé d'un milieu matériel extrême, de conductivité infinie, soumis à un champ magnétique. Les ondes de choc correspondantes engendrent nécessairement, en général, des ondes de choc gravitationnelles.

066.104 **Relativistic spin precession in two-body systems.** G. Börner, J. Ehlers, E. Rudolph.
Astron. Astrophys., Vol. 44, 417 - 420 (1975).

The spin precession of each component of a two-body system is calculated by means of the parametrized post-Newtonian approximation to metric theories of gravity without preferred frame effects. This calculation is carried out in view of an application to the recently discovered binary pulsar

PSR 1913+16. For the case of general relativity the authors find that in the case of arbitrary masses the spin precession rate of the first body is $(M_2 + \mu/3)/(M_1 + M_2)$ times what it would be if the first body were a test particle moving in the field of a fixed body with mass $M_1 + M_2$, μ being the reduced mass.

066.105 **Einstein's gravitational theory in elementary presentation.** M. K. Karipova.
Trudy Kirg. un-ta. Ser. fiz. n., 1975, vyp. (No.) 6, ch. (part) 1, p. 57 - 67. In Russian. – Abstr. in Referativ. Zhurn. 51. Astron., 10.51.82 (1975).

066.106 **Excitation and detection of stationary gravitational waves.** L. P. Grishchuk, M. V. Sazhin.
Zhurn. ehksperim. i teor. fiz., Vol. 68, 1571 - 1582 (1975). In Russian. – Abstr. in Referativ. Zhurn. 51. Astron., 10.51.749 (1975).

066.107 **A new method of testing the effect of light deflection in the gravitational field of the sun.**
M. F. Shirokov, B. V. Bondarev.
Probl. teorii gravitatsii i ehlementarn. chastits. Vyp. (No.) 6. Moskva, Atomizdat, 1975, p. 206 - 212. In Russian. – Abstr. in Referativ. Zhurn. 51. Astron., 10.51.767 (1975).

066.108 **Dual-frequency measurement of the solar gravitational microwave deflection.**
K. W. Weiler, R. D. Ekers, E. Raimond, K. J. Wellington.
Phys. Rev. Letters, Vol. 35, 134 - 137 (1975).
The Westerbork synthesis radio telescope was used to measure the solar gravitational deflection of radio waves over a period of 10 days, in October 1973, centered on the occultation of the quasar 3C 279 by the sun. Simultaneous measurements at two frequencies (5.0 and 1.4 GHz) allowed the removal of the effects of refraction in the solar corona. The gravitational bending measured was 1.038 ± 0.033 times that predicted by general relativity. As a by-product of these observations, values for the parameters at a model of the coronal electron density were also obtained.

066.109 **Search for gravitational radiation at 145 Hz.**
H. Hirakawa, K. Narihara.
Phys. Rev. Letters, Vol. 35, 330 - 334, with a correction, p. 890 (1975).
The authors present the results of gravitational radiation experiments with two 1400-kg antennas at $\nu_0 = 145$ Hz. The square antennas, $1.65 \times 1.65 \times 0.19$ m^3, $Q_M \simeq 2.0 \times 10^5$, with electrostatic transducers of $\beta \simeq 1.8 \times 10^{-3}$, have mean effective noise energy $\bar{D} \simeq 0.034 kT_r$. The observed cross correlation of the outputs of the two detectors gives an upper limit for the relation between the mean energy spectrum density $F(\nu_0)$ and the daily occurrence rate $\dot{N}_G$ of gravitational radiation pulses.

066.110 **Quantum electrodynamical effects in Kerr-Newman geometries.** T. Damour, R. Ruffini.
Phys. Rev. Letters, Vol. 35, 463 - 466 (1975).
Following the classical approach of Sauter, of Heisenberg and Euler and of Schwinger the process of vacuum polarization in the field of a 'bare' Kerr-Newman geometry is studied. The value of the critical strength of the electromagnetic fields is given together with an analysis of the feedback of the discharge on the geometry. The relevance of this analysis for current astrophysical observations is mentioned.

066.111 **Two-detector-coincidence search for bursts of gravitational radiation.** D. H. Douglass, R. Q. Gram, J. A. Tyson, R. W. Lee.
Phys. Rev. Letters, Vol. 35, 480 - 483 (1975).
The authors made a high-sensitivity search with two calibrated detectors in coincidence for bursts of gravitational radiation at 710 Hz of duration < 0.1 sec. Less than one event per day at a flux of 2×10^7 ergs cm^{-2} Hz^{-1} is found.

066.112 **Theories of gravity with structure-dependent γ's.**
M. Camenzind.
Phys. Rev. Letters, Vol. 35, 1188 - 1189 (1975).
A gauge theory for the Lorentz connection of a space-time describes geometries of stars with variable parametrized post-Newtonian parameter γ; its value depends explicitly on the structure of a star: The value of γ for a neutron star is different from that for the solar system ($|\gamma_\odot^{-1}| \leqslant 10^{-2}$). Because of the absence of a "local parity invariance" γ is positive for ordinary stars and γ is negative for sources of gravity consisting of antimatter.

066.113 **The gauge in general relativity.** J. Cohn.
General Relat. Gravitation, Vol. 6, 143 - 150 (1975).

066.114 **'Electric' and 'magnetic' gravitational fields in vacuum.** D. H. Tchrakian.
General Relat. Gravitation, Vol. 6, 151 - 156 (1975).
Gravitational tensor fields E_{ij} and H_{ij} in vacuum are introduced and the case of radiation is formulated, in analogy with electromagnetism.

066.115 **Lagrangian formulation of a geometrical scalar-tensor theory of gravitation.** D. K. Ross.
General Relat. Gravitation, Vol. 6, 157 - 164 (1975).

066.116 **Metric gravitation with a two parameter family of static spherically symmetric space-times.**
M. Camenzind, M. A. Camenzind.
General Relat. Gravitation, Vol. 6, 175 - 195 (1975).

066.117 **Equations of hydrodynamics in general relativity in the slow motion approximation.**
J. L. Anderson, T. C. Decanio.
General Relat. Gravitation, Vol. 6, 197 - 237 (1975).
By means of a formal solution to the Einstein gravitational field equations a slow motion expansion in inverse powers of the speed of light is developed for the metric tensor. The results are compared to those obtained in another gauge by S. Chandrasekhar.

066.118 **A bi-metric theory of gravitation. II.** N. Rosen.
General Relat. Gravitation, Vol. 6, 259 - 267 (1975).
The bi-metric theory of gravitation proposed previously is simplified in that the auxiliary conditions are discarded, the two metric tensors being tied together only by means of the boundary conditions. Some of the properties of the field of a particle are investigated; there is no black hole, and it appears that no gravitational collapse can take place. Although the proposed theory and general relativity are at present observationally indistinguishable, some differences are pointed out which may some day be susceptible of observation. An alternative bi-metric theory is considered which gives for the precession of the perihelion 5/6 of the value given by general relativity; it seems less satisfactory than the present theory from the aesthetic point of view.

066.119 **On Rosen's bi-metric theory of gravitation.**
H. Yilmaz.
General Relat. Gravitation, Vol. 6, 269 - 275 (1975).

066.120 **Bi-metric theories of gravitation.**
D.-E. Liebscher.
General Relat. Gravitation, Vol. 6, 277 - 280 (1975).
The bi-metric theory recently proposed by Rosen is examined in the case of superdense static objects.

066.121 **Gravitational radiation recoil on an N-body system.**

D. D. Dionysiou.
General Relat. Gravitation, Vol. 6, 319 - 323 (1975).

The main purpose of this paper is to present a formula of gravitational radiation recoil of an N-body material system.

066.122 **Effect of gravitational radiation upon electromagnetic waves in a dispersive medium.**
B. Bertotti, R. Catenacci.
General Relat. Gravitation, Vol. 6, 329 - 343 (1975).

066.123 **Relativistic generalization of the Oort distribution.**
L. S. Kuz'menkov.
Vestn. Mosk. un-ta. Fiz., astron., Vol. 16, No. 2, p. 149 - 153 (1975). In Russian. – Abstr. in Referativ. Zhurn. 51. Astron., 11.51.786 (1975).

066.124 **On the theory of super-high density configurations.**
D. D. Ivanenko, N. I. Maksyukov.
Vestn. Mosk. un-ta. Fiz., astron., Vol. 16, No. 2, p. 226 - 229 (1975). In Russian. – Abstr. in Referativ. Zhurn. 51. Astron., 11.51.791 (1975).

066.125 **Problems of gravitational collapse.** V. G. Krechet, V. G. Lapchinskij, V. N. Ponomarev, K. P. Stanyukovich.
Probl. teorii gravitatsii i ehlementarn. chastits. Vyp. (No.) 6. Moskva, Atomizdat, 1975, p. 129 - 139. In Russian. – Abstr. in Referativ. Zhurn. 51. Astron., 11.51.800 (1975).

066.126 **On the radiation-free spherical collapse in the presence of a scalar field.**
N. V. Pavlov, K. A. Bronnikov, M. A. Koval'chuk.
Probl. teorii gravitatsii i ehlementarn. chastits. Vyp. (No.) 6. Moskva, Atomizdat, 1975, p. 146 - 159. In Russian. – Abstr. in Referativ. Zhurn. 51. Astron., 11.51.801 (1975).

066.127 **On some spaces with causality breaking.**
N. A. Voronov.
Pis'ma v ZhurnEhTF, Vol. 21, 605 - 607 (1975). In Russian. Abstr. in Referativ. Zhurn. 51. Astron., 11.51.802 (1975).

066.128 **Scale invariance and gravitational field equations.**
K. P. Stanyukovich.
Probl. teorii gravitatsii i ehlementarn. chastits. Vyp. (No.) 6. Moskva, Atomizdat, 1975, p. 6 - 22. In Russian. – Abstr. in Referativ. Zhurn. 51. Astron., 11.51.806 (1975).

066.129 **On the extension of the solutions of equations of general relativity through singularities.**
M. E. Gertsenshtejn, M. Yu. Konstantinov.
Probl. teorii gravitatsii i ehlementarn. chastits. Vyp. (No.) 6. Moskva, Atomizdat, 1975, p. 23 - 31 . In Russian. – Abstr. in Referativ. Zhurn. 51. Astron., 11.51.807 (1975).

066.130 **On the effect of radiation-reaction force on a particle in a Kerr field.** V. R. Khalilov, G. A. Chizhov.
Izv. vyssh. ucheb. zavedenij. Fizika, 1975, No. 4, p. 87 - 90. In Russian. – Abstr. in Referativ. Zhurn. 51. Astron., 11.51. 815 (1975).

066.131 **Weak cylindrical gravitational waves.**
V. E. Radchenko.
Izv. vyssh. ucheb. zavedenij. Fizika, 1975, No. 5, p. 7 - 11. In Russian. – Abstr. in Referativ. Zhurn. 51. Astron., 11.51. 816 (1975).

066.132 **Black holes – how much do exist?**
P. R. Amnuehl'.
Zemlya i Vselennaya, 1975, No. 6, p. 8 - 12. In Russian.

066.133 **Les principes de la mécanique invariantive.**
O. Onicescu.
Revue Roumaine Sci. techn., Sér. Méc. appl., Vol. 19, 457 - 469 (1974).

066.134 **Weiße Zwerge – schwarze Löcher.** R. Sexl.
Phys. Blätter, 31. Jahrgang, p. 566 - 570 (1975).

066.135 **Accretion of mass and angular momentum onto a non-rotating black hole.**
K. C. Heaton, R. N. Henriksen.
Journ. Roy. Astron. Soc. Canada, Vol. 69, 252 (1975).
Abstr. CAS.

066.136 **Gravitational radiation and gravitational collapse of rotating gaseous ellipsoids.**
W. Y. Chau, T. T. Chia, R. N. Henriksen.
Journ. Roy. Astron. Soc. Canada, Vol. 69, 252 - 253 (1975).
Abstr. CAS.

066.137 **The effects of redshifts and focusing on the spectrum of an accretion disk around a Kerr black hole.**
C. T. Cunningham.
Astrophys. Journ., Vol. 202, 788 - 802 (1975).

The spectrum of X-rays produced by an accretion disk around a black hole is influenced markedly by Doppler shifts, gravitational redshifts, and the gravitational lens effect. These influences can be described by a "transfer function", which one folds into any assumed spectrum emitted on the disk's surface to get the spectrum observed at various locations far from the disk. This paper formulates such a transfer function and tabulates it for Kerr black holes of $a/M = 0$ and 0.9981.

066.138 **Physical properties of a naked singularity.**
F. de Felice.
Astron. Astrophys., Vol. 45, 65 - 68 (1975).

Whenever Kerr metric naked singularities exists, they provide a mechanism of particle acceleration along the axis of symmetry. As a consequence a hypothetical transition from a Kerr metric black-hole to a naked singularity is accompanied by a burst of relativistic particles. Inversely, the time for a naked singularity to become a black hole after accretion along the axis, is evaluated, and varies as $\sim 10^4\ M_\odot/M_\rho$ where ρ is the mean density in the surrounding region.

066.139 **Gravitational collapse in astrophysics and physics of elementary particles ("black holes").** S. Gabrakov.
Fiz.-mat. spisanie, Vol. 17 (50), 262 - 272 (1974). In Bulgarian.

066.140 **Examination of two exact self-similar solutions of general relativity equations.** M. Yu. Ivanov.
Probl. teorii gravitatsii i ehlementarn. chastits. Vyp. (No.) 6. Moskva, Atomizdat, 1975, p. 41 - 48. In Russian. – Abstr. in Referativ. Zhurn. 51. Astron., 12.51.716 (1975).

066.141 **On the speed of test particles near the Schwarzschild sphere.** V. I. Stoyanov.
Dokl. AN TadzhSSR, Vol. 18, No. 5, p. 16 - 18 (1975). In Russian. – Abstr. in Referativ. Zhurn. 51. Astron., 12.51.717 (1975).

066.142 **Gravitational collapse of stars.** W. D. Arnett.
Bull. American Astron. Soc., Vol. 7, 541 (1975).
Abstr. AAS.

066.143 **Zur Geschichte der Ideen über die Wirkung der Schwerkraft auf das Licht.** H.-U. Fuchs.
Orion, 33. Jahrgang, p. 183 - 193 (1975).

066.144 **Celestial-mechanical relativistic experiments in the**

solar system. A. M. Finkel'shtejn.
In-t teor. fiz. AN USSR. Preprint ITF-75-16R. Kiev, 1975. 50 pp. Price 14 Kop. In Russian. – Abstr. in Referativ. Zhurn. 51. Astron., 1.51.151; 62. Issled. kosmich. prostranstva, 1.62. 422 (1976).

066.145 **The geodetic interval in a Riemannian space-time in the second post-Minkowskian approximation.**
R. W. John.
Zeitschr. angew. Math. Mech., Vol. 54, 737 - 740 = Zentralinst. Astrophys. Akad. Wiss. DDR, Potsdam-Babelsberg, Mitt., Neue Folge, No. 126 (1974). In German.

066.146 **Boundaries of space-times and singularities in general relativity.**
H.-H. v. Borzeszkowski, U. Kasper.
Ann. Physik, Ser. 7, Vol. 32, 146 - 158 = Zentralinst. Astrophys. Akad. Wiss. DDR, Potsdam-Babelsberg, Mitt., Neue Folge, No. 127 (1975).

066.147 **Wann kann die Gravitation zu einer starken Wechselwirkung werden?** H.-J. Treder.
Ann. Physik, Ser. 7, Vol. 32, 238 - 240 = Zentralinst. Astrophys. Akad. Wiss. DDR, Potsdam-Babelsberg, Mitt., Neue Folge, No. 129 (1975).

066.148 **Das gravische Wechselwirkungspotential und die Korrekturen der Gaußschen Gravitationskonstanten in der nach-Newtonschen Näherung klassischer und relativistischer Gravitationstheorien.** H.-J. Treder.
Ann. Physik, Ser. 7, Vol. 32, 313 - 320 = Zentralinst. Astrophys. Akad. Wiss. DDR, Potsdam-Babelsberg, Mitt., Neue Folge, No. 131 (1975).

066.149 **Dynamische Äquivalenz, Gaußsche Konstante und Zeit-Skala für die nach-Newtonschen Näherungen klassischer und relativistischer Gravitationstheorien.**
H.-J. Treder.
Ann. Physik, Ser. 7, Vol. 32, 304 - 312 = Zentralinst. Astrophys. Akad. Wiss. DDR, Potsdam-Babelsberg, Mitt., Neue Folge, No. 132 (1975).

066.150 **Schwarzschild-Friedmannsche Lösungen der Einsteinschen Feldgleichungen.** G. Dautcourt.
Math. Nachr., Vol. 66, 211 - 216 = Zentralinst. Astrophys. Akad. Wiss. DDR, Potsdam-Babelsberg, Mitt.,Neue Folge, No. 133 (1975).

066.151 **Unified field theory with Einsteinian photons and heavy bosons as field quants.** H.-J. Treder.
Ann. Physik, Ser. 7, Vol. 32, 375 - 382 = Zentralinst. Astrophys. Akad. Wiss. DDR, Potsdam-Babelsberg, Mitt.,Neue Folge, No. 134 (1975).

066.152 **Corrections and additional remarks to the paper "Kovariante Superpotentiale in der allgemeinen Relativitätstheorie".** U. Kasper.
Ann. Physik, Ser. 7, Vol. 32, 351 - 352 = Zentralinst. Astrophys. Akad. Wiss. DDR, Potsdam-Babelsberg, Mitt.,Neue Folge, No. 135 (1975).

066.153 **Newton's first law and the existence of free tachyons.** D.-E. Liebscher.
Ann. Physik, Ser. 7, Vol. 32, 363 - 365 = Zentralinst. Astrophys. Akad. Wiss. DDR, Potsdam-Babelsberg, Mitt.,Neue Folge, No. 136 (1975).

066.154 **Zur unitarisierten Gravitationstheorie mit lang- und kurzreichweitigen Termen (mit ruhmasselosen und schweren Gravitonen).** H.-J. Treder.
Ann.Physik, Ser. 7, Vol. 32, 383 - 400 = Zentralinst. Astrophys. Akad. Wiss. DDR, Potsdam-Babelsberg, Mitt.,Neue Folge, No. 137 (1975).

066.155 **The post-Newtonian effect of gravitation and the retardation of the gravitational potential in classical and relativistic theories of gravitation.** H.-J. Treder.
Ann. Physik, Ser. 7, Vol. 32, 338 - 350 = Zentralinst. Astrophys. Akad. Wiss. DDR, Potsdam-Babelsberg, Mitt.,Neue Folge, No. 138 (1975).

066.156 **Integral conditions, boundary surfaces and matter in general relativity.** H.-J. Treder.
Ann. Physik, Ser. 7, Vol. 32, 159 - 160 = Zentralinst. Astrophys. Akad. Wiss. DDR, Potsdam-Babelsberg, Mitt.,Neue Folge, No. 149 (1975).

066.157 **Behavior of spacelike geodesics in the extended Schwarzschild manifold.**
E. Honig, K. Lake, R. C. Roeder.
Phys. Rev. D, Particles and Fields, Vol. 10, 3155 - 3161 = Commun. David Dunlap Obs., Richmond Hill, Ontario, Canada, No. 437 (1974).

066.158 **White holes – a coming fashion?** J. Gribbin.
New Scient., Vol. 68, 199 (1975). – *JAR*

066.159 **Energy and momentum transport by neutrino radiation into the mantle of a collapsing star and supernova explosions.** S. S. Gershtejn, V. S. Imshennik, D. K. Nadezhin, V. N. Folomeshkin, M. Yu. Khlopov, V. M. Chechetkin, R. A. Ehramzhyan.
In-t fiz. vysok. ehnerg. IFVEh OTF 75-56. Serpukhov, 1975. 19 pp. Price 9 Kop. In Russian. – Abstr. in Referativ. Zhurn. 51. Astron., 2.51.565 (1976).

066.160 **Solutions of the Einstein field equations for a rotating perfect fluid. II. Properties of the flow-stationary and vortex-homogeneous solutions.** A. Krasinski.
Acta Phys. Polonica B, Vol. B6, 223 - 238 (1975).

066.161 **Solutions of the Einstein field equations for a rotating perfect fluid. III. A survey of models of rotating perfect fluid or dust.** A. Krasinski.
Acta Phys. Polonica B, Vol. B6, 239 - 240 (1975).

066.162 **Black holes.** R. Penrose.
Fiz. Szemle, (*Hungary*), Vol. 25, 15 - 24 (1975).
In Hungarian.

The cycle of stellar development from red giants to black holes is outlined. Topics discussed are the cosmic sensor hypothesis, observation of black holes, space time singularities, gravitational tidal waves and symmetrical gravitational collapse.

066.163 **Two problems in general relativity: 1) Evolving dust shells. 2) Kerr-Newman metrics, spinning charged sources, and their gyromagnetic ratios.** H. P. Siegel.
Thesis, Brandeis Univ., Waltham, Mass. [Available from Univ. Microfilms, Ann Arbor, Mich., USA. Order No. 74-28016], 298 pp. (1974).

The time-evolution of spherically symmetric charged dust shells is explored by study of the detailed 3–geometry of spacelike hypersurfaces embedded in space-time. Under well-defined conditions this 3-geometry 'flips over' from open to closed. The fields generated by charged rotating dust shells are studied, with the intention of eventual comparison to the Kerr-Newman field.

066.164 **Possible implications of the uncertainty principle for the time scale of gravitational collapse.**
R. C. Gilman.
Smithsonian Astrophys. Obs., *Cambridge, Mass.*, Special Rep.

No. 369, 3 + 36 pp. (1975).

Expected quantum fluctuations in the inertial-gravitational field are shown to lead to a loosening of the classical junction conditions of general relativity. This in turn permits an object undergoing gravitational collapse to cross its Schwarzschild horizon in a finite amount of external observer time. Assuming that various problems at the r = 0 singularity can be surmounted, the author shows that the object will re-expand into the original space after a finite amount of external observer time. The characteristic period is about 3.6 ms for a 1 $M_\odot$ object and about 1.4 weeks for a 10^8 $M_\odot$ object. The long term evolution of such objects is briefly discussed.

066.165 **Some magnetic effects in stellar collapse and accretion.** J. R. Wilson.
7. Texas symposium on relativistic astrophysics, Dallas, Texas, USA. UCRL-76317, 21 pp. (1975).

066.166 **Some remarks on the two-body problem in geometrodynamics (black holes).** A. Cadez.
Ann. Physics, Vol. 91, 58 - 74 (1975).

The interest in the two-body problem in geometrodynamics and its present status are discussed. Special consideration is given to time-symmetric initial data. An equation relating the 'bare mass energy', the potential energy and the 'vibrational energy' at the initial moment is proposed. The significance of coordinate conditions in dynamical equations is discussed next. Conditions implying maximal slicings of spacetime and harmonic coordinates in those slices are considered in detail.

066.167 **Black holes with scalar charge.** J. D. Bekenstein.
Ann. Physics, Vol. 91, 75 - 82 (1975).

Previously it has been thought that a stationary black hole with an exterior devoid of matter can be parametrized only by mass, angular momentum, and electric charge. The author shows that scalar charge is also an admissible parameter. The starting point is a new solution of Einstein's equations with stress-energy of electromagnetic and conformal scalar fields. It has a black-hole geometry, and is parametrized by electric and scalar charges.

066.168 **Physical consequences of a solution of the nonsymmetric unified field theory (black holes and magnetic monopoles).** D. H. Boal, J. W. Moffat.
Phys. Rev. D, (Particles and Fields), Vol. 11, 2026 - 2030 (1975).

Physical consequences of a previously published (1975) solution of the nonsymmetric unified field theory are given. The solution contains a universal constant which controls certain fourth-order effects in the propagation of light near charged bodies, and which, if nonzero, forbids the existence of magnetic monopoles.

066.169 **A note on general relativistic dynamo mechanisms.** M. Reinhardt, A. Rosenblum.
Phys. Letters A, (*Netherlands*), Vol. 53A, 269 - 270 (1975).

The authors' previous investigation of a general relativistic primary dynamo mechanism based on the dragging of inertial frames, where the Cohen-Brill metric was used, is generalized to the case of the Kerr metric. It is shown that magnetic fields can be generated without the black hole being charged.

066.170 **The quasi-normal modes of the Schwarzschild black hole.** S. Chandrasekhar, S. Detweiler.
Proc. Roy. Soc. London, Ser. A, Vol. 344, 441 - 452 (1975).

The quasi-normal modes of a black hole represent solutions of the relevant perturbation equations which satisfy the boundary conditions appropriate for purely outgoing (gravitational) waves at infinity and purely ingoing waves at the horizon. For the Schwarzschild black hole the problem reduces to one of finding such solutions for a one-dimensional wave equation (Zerilli's (1970) equation) for a potential which is positive everywhere and is of short-range. The notion of quasi-normal modes of such one-dimensional potential barriers is examined with two illustrative examples; and numerical solutions for Zerilli's potential are obtained by integrating the associated Riccati equation.

066.171 **Vector potential and metric perturbations of a rotating black hole.** P. L. Chrzanowski.
Phys. Rev. D, (Particles and Fields), Vol. 11, 2042 - 2062 (1975).

Homogeneous metric and vector potential perturbations of a Kerr black hole are considered in the framework of the Teukolsky (1972) equations, with the assumption of factorised Green's functions. Analytic solutions are obtained, and lead on to solutions in the presence of sources. Simple formulae for energy and angular momentum at asymptotic distances are found.

066.172 **Testing relativity in the desert: the Texas-Mauritania eclipse expedition.** R. Matzner.
Phys. Teacher, (*USA*), Vol. 13, 215 - 222 (1975).

The difficulties involved in the solar eclipse method of testing the theories of gravitation, which have lead to criticisms in the past are outlined. The experiment undertaken to photograph stars at optical wavelengths during the eclipse of June 30, 1973 is described in detail.

066.173 **On the equations governing the axisymmetric perturbations of the Kerr black hole.**
S. Chandrasekhar, S. Detweiler.
Proc. Roy. Soc. London, Ser. A, Vol. 345, 145 - 167 (1975).

It is shown how Teukolsky's equation, governing the perturbations of the Kerr black hole, can be reduced, in the axisymmetric case, to a one-dimensional wave equation with four possible potentials. In all cases (i.e. whether or not the potentials are real or complex), the problem of the reflexion and the transmission of gravitational waves by the potential barriers can be formulated, consistently, with the known conservation laws. It is, further, shown that all four potentials lead to the same reflexion and transmission coefficients.

066.174 **Spherically symmetric charged dust distributions in general relativity.** A. K. Raychaudhuri.
Ann. Inst. Henri Poincaré A (*France*), Vol. 22, 229 - 240 (1975).

066.175 **Inaccuracy effect of the gravitational parameter.** K. Mison.
Acta Techn. CSAV, (*Czechoslovakia*), Vol. 20, 348 - 365 (1975).

Examines the influence of inaccuracy, while determining the gravitational parameter of the earth, on some quantities concerning the problem of two bodies: semi-major axis, parameter (semi-latus rectum) of the orbit, orbital period and both the circular and escape velocities.

066.176 **Superstars and trapped light. The geometrical optics of gravitationally intense bodies.**
D. J. Baylis.
International Journ. Math. Educ. Sci. Technol., (*GB*), Vol. 6, 223 - 229 (1975).

Starting from the differential equation describing the path of a photon in the Schwarzschild field of a spherically symmetric non rotating star in empty space, the condition for the light emitted from such a star to return to its surface is derived, and some features of its orbit are described. An integral equation for the 'least gravitationally intense star' with the property that its total surface is visible to a fixed distant observer is determined.

066.177 **Axisymmetric quasistationary stability of a charged Kerr black hole.** M. Y. Wang.
Phys. Rev. D, (Particles and Fields), Vol. 11, 3392 - 3394 (1975).

It is proved that a charged Kerr black hole is stable under axisymmetric quasistationary perturbation. The proof utilises the Ernst form of the axisymmetric Einstein-Maxwell equation.

066.178 **Linear-momentum changes in astrophysical processes due to neutrino emission.** K. Tennakone.
Nuovo Cimento Lettere, Ser. 2, Vol. 13, 555 - 556 (1975).

It is pointed out that nonconservation of parity in weak interactions can have an observable macroscopic effect in changing the linear momentum of an astrophysical system emitting neutrinos in the presence of a strong magnetic field.

066.179 **The propagation of gravitational radiation in dense media.** D. K. Ross.
Journ. Phys. A, (Math., Nuclear, General), Vol. 8, 1398 - 1408 (1975).

Einstein's field equations are solved for gravitational radiation propagating in a background metric given by the interior Schwarzschild solution. Expressions for the transmitted fraction as a function of frequency and propagation distance are found.

066.180 **A class of solutions of Einstein-Maxwell equations.** J. F. Plebanski.
Ann. Physics, Vol. 90, 196 - 255 (1975).

066.181 **Relativistic motion of a gyroscope around the moon.** T. Mazeh, J. Katz.
Nuovo Cimento B, Ser. 11, Vol. 25B, 399 - 404 (1975).

066.182 **The rest frame in stationary space-times with axial symmetry.**
R. D. Greene, E. L. Schucking, C. V. Vishveshwara.
Journ. Math. Phys., Vol. 16, 153 - 157 (1975).

066.183 **Instability of intermediate singularities in general relativity.** A. R. King.
Phys. Rev. D, (Particles and Fields), Vol. 11, 763 - 767 (1975).

066.184 **Black-and-white-holes. II. Faster-than-light phenomena and their connection with the complete Schwarzschild manifold.** R. Goldoni.
Acta Phys. Austriaca, Vol. 41, 133 - 139 (1975).

Gravitation und Kosmologie. See Abstr. 003.114.

Special relativity. See Abstr. 003.123.

Relativistische Astrophysik. See Abstr. 011.012.

Neue Testergebnisse für Gravitationstheorien. See Abstr. 011.037.

Experiments to determine the force of gravity on single electrons and positrons. See Abstr. 022.002.

Time scales and relativity. See Abstr. 044.012.

White holes and high energy astrophysics. See Abstr. 061.002.

Neutrinos in general relativity: four(?) levels of approach. See Abstr. 061.005.

Neutrino opacities at high temperatures and densities. See Abstr. 061.037.

Equation of state at ultrahigh densities. Part 2. See Abstr. 061.043.

Physics of high density and nuclear matter. See Abstr. 061.044.

A new primary dynamo mechanism. See Abstr. 062.004.

Ondes de choc, sous des hypothèses générales, en magnétohydrodynamique relativiste. See Abstr. 062.047.

General-relativistic radiative transfer theory in refractive and dispersive media. See Abstr. 063.034.

Vibrational stability of relativistic supermassive stars in the pre-main-sequence contraction stage. See Abstr. 065.092.

Gravitational collapse of non-spherical gas clouds. See Abstr. 065.105.

A variational principle and a stability criterion for the dipole modes of pulsation of stellar models in general relativity. See Abstr. 065.109.

The maximum mass of nonrotating neutron stars. See Abstr. 065.110.

Neutron stars: general review. See Abstr. 065.119.

Corequakes of neutron stars and soft γ-ray bursts. See Abstr. 065.125.

The life of dying stars. See Abstr. 065.131.

Stationary elastic rotational deformation of a relativistic neutron star model. See Abstr. 065.136.

The maximum mass of a cold neutron star. See Abstr. 065.144.

Relativity and the solar oblateness. See Abstr. 080.005.

Solar oblateness and the solar quadrupole moment. See Abstr. 080.007.

The intrinsic visual oblateness of the sun. See Abstr. 080.018.

Effect of the solar activity on the natural oscillations of a torsion pendulum. See Abstr. 085.014.

Black holes and neutron stars: evolution of binary systems. See Abstr. 117.023.

The interrelationship between cosmic dust and the microwave background. See Abstr. 131.050.

Relativistic effects in the binary pulsar PSR 1913+16. See Abstr. 141.301.

On the motion of pulsars and black holes. See Abstr. 141.321.

Magnetospheres and pulsars with net charge. See Abstr. 141.334.

The new binary pulsar and the observation of gravitational spin precession. See Abstr. 141.345.

Can soft γ-ray bursts be emitted by accreting black holes? See Abstr. 142.018.

Cygnus X-1: a two-temperature accretion disk model which explains the observed hard X-ray spectrum. See Abstr. 142.038.

On the nature of the globular cluster X-ray sources. See Abstr. 142.061.

An interpretation of the general character of the X-ray emission from compact objects in close binary stars. See Abstr. 142.116.

The evolutionary origin of massive X-ray binaries and the total number of massive stars with a collapsed close companion in the Galaxy. See Abstr. 142.127.

Neutron stars, black holes and binary X-ray sources. See Abstr. 142.130.

Cygnus X-1, el agujero negro que la devora y los quarks. See Abstr. 142.244.

Thermodynamic stability of relativistic stellar clusters. See Abstr. 151.002.

The distribution of stars in galactic nuclei: loaded polytropes. See Abstr. 151.003.

Dynamics of relativistic stellar systems. See Abstr. 151.028.

The concentric shell method for relativistic star clusters. See Abstr. 151.029.

The star distribution in M15. See Abstr. 154.028.

A new upper limit to the mass loss from the central region of the Galaxy. See Abstr. 155.006.

An experiment to detect the spatial velocity of the earth. See Abstr. 155.012.

Is the nucleus of the Galaxy a black hole? See Abstr. 155.016.

A search for optical pulses from the galactic centre. See Abstr. 155.049.

Galactic nuclei and compact supermassive objects. See Abstr. 158.065.

Antimatter and the particle problem in Einstein's cosmology and field theory of elementary particles (An historical essay on Einstein's work at the Akademie der Wissenschaften zu Berlin). See Abstr. 162.004.

Can collapsed stars close the universe? See Abstr. 162.009.

The primordial black hole mass spectrum. See Abstr. 162.016.

Theoretical determination of the thinning factor in relativistic hierarchical cosmology. See Abstr. 162.017.

Models of the universe in general relativity. See Abstr. 162.029.

Observational evidence for assumptions of cosmology. See Abstr. 162.036.

Déplacements anormaux vers le rouge liés à la traversée des amas de galaxies par la lumière. See Abstr. 162.040.

Variation of the gravitational constant in cosmological models. See Abstr. 162.042.

The universe as a black hole. See Abstr. 162.077.

Errata

066.901 **Erratum: 'A dust model for the cosmic microwave background'** [Astrophys. Space Sci., Vol. 35, L9 - L13 (1975)]. N. C. Wickramasinghe, M. G. Edmunds, S. M. Chitre, J. V. Narlikar, S. Ramadurai.
Astrophys. Space Sci., Vol. 36, 242 (1975).

Sun

071 Solar Photosphere, Spectrum

071.001 **Die Intensitätsfluktuation der Granulation.**
F. L. Deubner, W. Mattig.
Mitt. Astron. Ges., No. 36, p. 148 (1975). – Abstract.

071.002 **Temperature variation with latitude in the upper solar photosphere: relevance to solar oblateness measurements and facular models.**
A. P. Ingersoll, G. A. Chapman.
Solar Physics, Vol. 42, 279 - 288 (1975).

Altrock and Canfield's observations of temperature variation with latitude in the upper solar photosphere refer to higher levels (smaller optical depths) than those to which Dicke and Goldenberg's solar oblateness observations refer. These higher levels account for only 1 % of Dicke and Goldenberg's observed intensity. Thus Altrock and Canfield's observations are not inconsistent with models which have been proposed to account for solar oblateness observations by means of a brightness variation with solar latitude.

071.003 **The formation of Mg I 4571 Å in the solar atmosphere. V. The multi-dimensional structure of the photosphere and low chromosphere.**
R. C. Altrock, C. J. Cannon.
Solar Physics, Vol. 42, 289 - 302 (1975).

The two-dimensional equation of transfer is solved for the case of locally-controlled source function (LTE) and radiationally-controlled ionization. Horizontal fluctuations in electron temperature and macroscopic velocity fields are superposed on the basic one-dimensional model (cf. Altrock and Cannon, 1972). Output intensities are compared with observed rms intensity fluctuations and spatially-averaged intensities in Mg I 4571 Å. The authors find that at least one model can predict the magnitude of the intensity fluctuations in both the continuum and λ4571 Å. The relationship between these results and other multi-dimensional analyses is discussed.

071.004 **Study of the photospheric motion field by the use of weak Fraunhofer lines of metals.**
Eh. A. Gurtovenko.
Astron. Zhurn. Akad. Nauk SSSR, Vol. 52, 804 - 810 (1975). In Russian. English translation in Soviet Astron., Vol. 19, No. 4.

The trial and error method was used for the center-to-limb analysis of 21 weak Fraunhofer lines of Ti I and Fe I. As a result the tangential and radial components of the total photospheric motion field have been deduced within the geometrical depth range -400 km $<h<+50$ km.

071.005 **The solar beryllium abundance.**
Y. Chmielewski, E. A. Müller, J. W. Brault.
Astron. Astrophys., Vol. 42, 37 - 46 (1975).

An interpretation of high resolution center-limb observations of the Be II resonance lines in the solar spectrum is given. It is shown that to account for the observed spectrum it is necessary to allow for (1) a source of additional opacity in the UV continuum which the authors derived empirically for the HSRA and the revised Holweger model and (2) departures from LTE in the Be I–Be II system. The kinetic equilibrium calculations yield a beryllium abundance of $\log \epsilon_{Be} = 1.15 \pm 0.20$ (on the basis of $\log \epsilon_H = 12.00$).

071.006 **Macroturbulence broadening of solar lines with implications for stellar studies.**
J. C. Evans, L. W. Ramsey, L. Testerman.
Astron. Astrophys., Vol. 42, 237 - 241 (1975).

This paper reports the results of numerical studies of macroturbulence broadening on solar LTE intensity profiles with the view toward applying the results toward other solar type stars. We present a comparison of the results for five different velocity models with spacially and temporally averaged center-to-limb observations. The comparisons are made through the curve of line-width correlation and through the curve of line-shape diagram. High accuracy photoelectric data for Procyon used on the curve of line-shape indicates that macroturbulence broadening dominates and the microturbulence component is about 1 km/s.

071.007 **Principe de réalisation d'un hologramme acoustique de la surface du soleil.** F. Roddier.
Comptes Rendus Acad. Sci. Paris, Sér. B, Vol. 281, 93 - 95 (1975).

On décrit le principe d'une méthode permettant d'enregistrer photographiquement en chaque point du soleil l'amplitude complexe des oscillations photosphériques. L'examen, en lumière cohérente, de l'hologramme ainsi obtenu pourrait conduire à une visualisation des sources acoustiques sous-jacentes.

071.008 **The solar spectrum from 1173 to 1324 Å.**
E. Chipman, E. C. Bruner, Jr.
Astrophys. Journ., Vol. 200, 765 - 772 (1975).

The authors present some of the results of a rocket flight which obtained a high-resolution stigmatic spectrum of the sun in the region 1173 – 1324 Å. A list of 179 lines observed in this region is presented, with intensities, widths, and approximate excitation classes for each line. A reproduction and finding chart of the spectrum is included, along with discussion of some individual lines of special interest.

071.009 **Horizontal temperature fluctuations in the low solar photosphere.** R. C. Altrock.
Bull. American Astron. Soc., Vol. 7, 406 (1975). – Abstr. AAS.

071.010 **Photospheric thermal fluctuations in the quiet sun.**
C. A. Lindsey.
Bull. American Astron. Soc., Vol. 7, 406 (1975). – Abstr. AAS.

071.011 **Measurement of the non-flaring solar spectrum from 1.8 - 5.3 Å.** P. B. Landecker, R. S. Wolff.
Bull. American Astron. Soc., Vol. 7, 445 (1975). – Abstr. AAS.

071.012 **Spatial inhomogeneities in the cores of strong lines in the solar spectrum.**
T. J. Schneeberger, H. A. Beebe.
Bull. American Astron. Soc., Vol. 7, 448 - 449 (1975). – Abstr. AAS.

071.013 **The solar profiles of the components of He II 1640 Å from rocket observations.**
J. L. Kohl, W. H. Parkinson.
Bull. American Astron. Soc., Vol. 7, 450 (1975). – Abstr. AAS.

071.014 **NLTE effects and abundance from solar Ba II lines.** K. E. Sedwick, L. H. Aller.
Bull. American Astron. Soc., Vol. 7, 451 (1975). – Abstr. AAS.

071.015 **High resolution disk center and limb observations of the solar ultraviolet beryllium features.**
C. D. Keyes.
Bull. American Astron. Soc., Vol. 7, 451 (1975). – Abstr. AAS.

071.016 **Magnetic field observations with Fe I λ15648 Å.** J. Harvey, D. Hall.
Bull. American Astron. Soc., Vol. 7, 459 (1975). – Abstr. AAS.

071.017 **High resolution observations of photospheric network magnetic fields.**
T. D. Tarbell, A. M. Title.
Bull. American Astron. Soc., Vol. 7, 459 (1975). – Abstr. AAS.

071.018 **Solar infrared spectroscopy at Kitt Peak.** D. N. Hall.
Bull. American Astron. Soc., Vol. 7, 477 (1975). – Abstr. AAS.

071.019 **Photospheric magnetic fields.** J. Harvey. Bull. American Astron. Soc., Vol. 7, 478 (1975).
Abstr. AAS.

071.020 **Centre-to-limb change of the profiles of weak Fraunhofer lines. IV. Interpretation of the width of various lines in the solar disc centre.** Eh. A. Gurtovenko.
Astrometriya i Astrofizika, *Kiev*, vyp. (No.) 26, (see 003.001), p. 78 - 86 (1975). In Russian.

A careful analysis of the line profiles for the elements Ti, Fe, Si, Cr, Ca, O, C, Zn, Mg, N, S, Ni, Sc resulted in deducing the radial component (V^{rad}) for the amplitude of the photospheric velocity field within the optical depth interval of $0.1 \leqslant \tau_s \leqslant 1.9$. The value of V^{rad} increases from 1.7 to 3.5 km/sec to the depth $\tau_s \approx 1.2$, and then decreases smoothly.

071.021 **Forbidden Ca II in the sun unmasked by way of Venus.**
R. A. Schorn, A. T. Young, E. S. Barker.
Solar Physics, Vol. 43, 9 - 14 (1975).

Eleven high-dispersion spectra of Venus, taken with blue Doppler shifts have enabled the authors to unmask the 7323.88 Å forbidden line of Ca II from terrestrial absorption. An equivalent width of 7.4 ± 0.4 mÅ for this line in integrated sunlight is obtained. The value of W_λ is smaller than previous values and much more accurate. The HSRA solar model gives a solar calcium abundance of $A_{\mathrm{Ca}} = 6.21$.

071.022 **The potassium abundance in the solar photosphere.** R. de la Reza, E. A. Müller.
Solar Physics, Vol. 43, 15 - 32 (1975).

High precision center-limb spectrograms of the K I resonance doublet line at λ7699 Å were used to study the line formation and to determine the abundance of potassium in the solar atmosphere. The LTE assumption is not valid for these lines. Synthetic profiles computed in NLTE reproduce very well the observed center-limb line behaviour and yield $\log \epsilon_K = 5.14 \pm 0.10$ for the solar abundance of potassium (on the scale of $\log \epsilon_H = 12$ for hydrogen).

071.023 **A new method for the analysis of the solar photospheric spectrum. The rotational temperature of the C_2 molecule.** P. Persi.
Solar Physics, Vol. 43, 39 - 47 (1975).

A new numerical method for the analysis of the high dispersion photospheric spectrum is described. In particular the method is applied to study the $C_2(0, 0)$ $d\,^3\Pi g$–$a\,^3\Pi u$ molecular band. From measurements of the equivalent widths of C_2 lines, a rotational temperature of 4450 ± 305 K is obtained, and the band intensity $\log W_0/S_0 = -0.051 \pm 0.101$ is found.

071.024 **On temperature inhomogeneities in the solar photosphere.** A. I. Khlystov.
Astron. Tsirk., No. 858, p. 1 - 2 (1975). In Russian.

071.025 **Some results of a study of the photospheric velocity field from moderately strong absorption lines of Fe I.** N. N. Kondrashova.
Astrometriya i Astrofizika, *Kiev*, vyp. (No.) 26, (see 003.001), p. 92 - 96 (1975). In Russian.

The method of fine analysis for some moderately strong absorption lines was used to determine the dependences of turbulent velocity on the depth in the photosphere. The distributions obtained do not coincide with the dependences for the total amplitude of the photospheric velocity field on the depth derived from the study of weak absorption lines. An attempt is made to separate the micro- and macroturbulence, using various effects of the macroturbulence on the lines of different intensity.

071.026 **Photospheric oscillations. I. Large scale observations by optical resonance method.**
E. Fossat, G. Ricort.
Astron. Astrophys., Vol. 43, 243 - 252 (1975).

Intensity fluctuations in the wings of the Fraunhofer line Na D 1 5896 have been recorded for about two hundred hours at the focus of the Nice coudé refractor, using a sodium optical resonance device. The principal results from the analysis of these data are: (1) As shown by White and Cha (1973), the five-minute oscillation has a gaussian random character with a mean lifetime of about 20 min. Its two-dimensional spatial power spectrum is roughly gaussian for every temporal frequency between 2 and 6 MHz. The width of this gaussian spectrum is near 5×10^{-5} km^{-1} (i.e. λ = 20000 km). It is shown that this value corresponds to a cellular pattern having a "coherence scale" near 5000 km. (2) A new oscillatory mode of 10 min period has been discovered, for photospheric apertures larger than 3′.

071.027 **Photospheric oscillations. II. Acoustic filtering in the solar atmosphere and heating of the chromosphere.** E. Fossat, G. Ricort.
Astron. Astrophys., Vol. 43, 253 - 257 (1975).

Observational evidence for acoustic filtering function in the photosphere in the spectral range 2–6 MHz comes from the comparison between the authors' results and some others, which are statistically significant and deduced from observations made with different Fraunhofer lines. The photospheric power spectrum and these filtering functions imply the prediction of a 180 s period oscillation in the chromosphere. The chromosphere acoustic flux is found in good agreement with the value needed for heating the chromosphere, and the value derived of the velocity oscillations amplitude shows that shock waves must begin to appear in the low chromosphere.

071.028 **OSO-8 observations of optically thin lines.** R. A. Shine, D. Rousell-Dupree, E. C. Bruner, Jr., E. G. Chipman, B. W. Lites, G. J. Rottman, R. G. Athay, O. R. White.
Bull. American Astron. Soc., Vol. 7, 552 (1975). – Abstr. AAS.

071.029 **The center-to-limb change of profiles of weak Fraunhofer lines. V. The radial and tangential components of the amplitude of the photospheric velocity field.**
Eh. A. Gurtovenko.
Astrometriya i Astrofizika, *Kiev*, vyp. (No.) 27, (see 003.004), p. 67 - 74 (1975). In Russian.

A center-to-limb analysis of 73 weak Fraunhofer lines was carried out. The lines were supposed to have Voigt profiles. Their damping parameters as well as the thermal velocities of the absorbing particles were calculated for specific optical depths and were taken into account. As a result the tangential and radial components of the photospheric motion field were obtained.

071.030 **Semiempirical upper photosphere models: the sun (G2 V) and Procyon (F5 IV–V).** T. R. Ayres.
Astrophys. Journ., Vol. 201, 799 - 815 (1975).

The author describes a general method for inferring the average upper photosphere temperature structure of a late-type star using calibrated observations of resonance line wings and a partial coherent scattering approximation. The author applies this approach to two problems: (1) He makes a comparison of the excess radiative losses from the upper photospheres of the sun and Procyon by "calibrating" observed profiles of the Ca II K line wings relative to Kurucz's radiative equilibrium models for $\tau_{5000} \gtrsim 0.1$. (2) The author obtains an accurate determination of the Mg II *k* to Ca II K emission core flux ratio for the sun and Procyon by calibrating the Mg II *k* wings relative to the Ca II K wings in each star. He finds that Procyon's chromospheric temperature gradient is probably shallower than the sun's, although the acoustic energy dissipation in Procyon's upper photosphere is very likely larger.

071.031 **A new solar carbon abundance based on non-LTE CN molecular spectra.**
G. H. Mount, J. L. Linsky.
Astrophys. Journ., (*Letters*), Vol. 202, L51 - L54 (1975).

A detailed non-LTE analysis of solar CN spectra strongly suggests a revised carbon abundance for the sun. The authors recommend a value of $\log A_C = 8.35 \pm 0.15$ which is significantly lower than the presently accepted value of $\log A_C = 8.55$.

071.032 **A comparison of modern average characteristics of the solar radiation.**
E. A. Makarova, A. V. Kharitonov.
Astron. Zhurn. Akad. Nauk SSSR, Vol. 52, 965 - 969 (1975).
In Russian. English translation in Soviet Astron., Vol. 19, No. 5.

Various systems of average data on the energy distribution in the solar spectrum and solar constant, derived after 1970 on the basis of averaging results of separate authors, are discussed.

071.033 **Five-minute oscillations of solar equivalent widths.** H. Holweger, L. Testerman.
Solar Physics, Vol. 43, 271 - 284 (1975).

Equivalent widths of weak and moderately strong Fraunhofer lines observed in an area roughly 3500 × 10000 km in extent are found to show small variations of the order of 1 mÅ, on a time scale of minutes. The dependence of amplitude on line strength and excitation closely fits that predicted for a compressional disturbance associated with temperature variations that manifest themselves in the equivalent widths. These equivalent width variations frequently exhibit a 5-min periodicity and are accompanied by prominent oscillations of velocity. Power spectra of both quantities closely resemble each other and demonstrate that most of the equivalent width variations in the range 200 to 500 s are due to the thermal response of the atmosphere to 5-min oscillations. A systematic phase lag between velocity and equivalent width is observed.

071.034 **On the center-to-limb variation of infrared photospheric carbon lines and the infrared continuum intensity around 1.75 μm.** H. Wöhl.
Solar Physics, Vol. 43, 285 - 288 (1975).

The center-to-limb variation of several infrared carbon lines and the infrared continuum intensity around 1.75 μm were measured. The results were compared with theoretical predictions using four different photospheric models. It was found that the model by Holweger and Müller (1974) describes the observations best.

071.035 **Some problems of magnetism and of the chemical composition of the sun and stars.**
A. B. Severnyj.
Izv. Krymskoj Astrofiz. Obs., Vol. 53, p. 240 - 242 (1975).
In Russian. – Abstr. of a paper presented at the conference "Magnetic fields and chemical composition of the sun and stars" at the Crimean Astrophys. Obs., 1974, April 9 - 12, (see 012.027).

071.036 **Photospheric and chromospheric magnetic structures.** F. L. Deubner.
Oss. Mem. Oss. Astrofis. Arcetri, Fasc. 105, (see 012.026), p. 39 - 55 (1975).

The paper deals with the fine structure of the photospheric magnetic field and its chromospheric extension. The author discusses some models of flux concentrations and possible mechanisms to form and maintain such a magnetic structure.

071.037 **The solar brightness temperature in the far infrared.** E. A. Müller, P. Stettler, J. Rast, F. K. Kneubühl, D. Huguenin.
Oss. Mem. Oss. Astrofis. Arcetri, Fasc. 105, (see 012.026), p. 90 - 92 (1975).

071.038 **A study of the equivalent widths of the absorption lines of the solar photosphere.**
P. A. Olijnyk, R. E. Keryk, B. M. Korduba.
Solnechnye Dannye 1975 Byull., No. 4, p. 87 - 91 (1975).
In Russian.

The equivalent widths of the Fe I lines in the 4800–5800 Å region are calculated using the Harvard-Smithsonian model of the solar photosphere. The damping constant for these lines with equivalent width larger than 110 mÅ appears to be 0.01–0.05.

071.039 **On obliteration of the Fraunhofer lines in the spectrum of the inner corona.**
Yu. I. Izotov, N. A. Yakovkin.
Solnechnye Dannye 1975 Byull., No. 5, p. 65 - 70 (1975).
In Russian.

The obliteration of the Fraunhofer lines of the solar photosphere by electron scattering at $T_e = 10^4, 6 \times 10^5, 10^6$ K was calculated. The inner corona spectrum depression was shown to exist at the region λλ3700–4100 Å, which corresponds to concentration of the Fraunhofer lines. The energy distribution in the continuous spectrum of a prominence at temperature $T_e = 10^4$ K has been received.

071.040 **Turbulent velocity determination by Goldberg's method.** G. O. Melikov.
Solnechnye Dannye 1975 Byull., No. 5, p. 83 - 92 (1975).
In Russian.

The profiles of the absorption lines for λ6122.2 and 6162.2 Å Ca I in the centre and the limb of the solar disc have been observed by a photoelectric method. Turbulent velocities have been calculated by Goldberg's method. It is derived that they increase independent of the photospheric model with depth in the centre and the limb. The turbulent velocity on the solar limb is higher than that in the centre and is equal to 270 km/sec.

071.041 **On the profile characteristics of the Fraunhofer lines in the granules and intergranular regions.**
V. M. Sobolev.
Solnechnye Dannye 1975 Byull., No. 6, p. 68 - 71 (1975).
In Russian.

Equivalent widths, half-widths and central depths of the profiles of some Fraunhofer lines averaged over nine granules and intergranular regions were determined as well as their mean ratios.

071.042 **Observations of low wavenumber nonradial eigenmodes of the sun.** F.-L. Deubner.
Astron. Astrophys., Vol. 44, 371 - 375 = Mitt. Fraunhofer-Inst., *Freiburg,* No. 139 (1975).

New photoelectric observations of the photospheric velocity field with high resolution in horizontal wavenumber and frequency have been carried out in the CI 5380 line. Three or four discrete stable modes of the five-minute oscillations are resolved in the k, ω diagram, which agree in many respects with the predicted fundamental modes of subphotospheric standing acoustic waves. Some interesting aspects of the confirmation of a "global" excitation mechanism of the five-minute oscillations are pointed out. Several apparently discordant details of previous observations are rediscussed, and it is shown that they can be reconciled with the present results.

071.043 **On profiles of weak lines of rare earths in the spectrum of the solar photosphere.**
D. M. Kuli-zade, K. I. Gusejnov.
Dokl. AN AzSSR, Vol. 31, No. 1, p. 18 - 22 (1975). In Russian. – Abstr. in Referativ. Zhurn. 51. Astron., 10.51.384 (1975).

071.044 **On the connection of 5-min fluctuations of the solar photosphere with the magnetic field.**
P. A. Golub.
XXVIII Gertsenovsk. chteniya. Teor. fiz. i astron. Leningrad, 1975, p. 78 - 82. In Russian. – Abstr. in Referativ. Zhurn. 51. Astron., 10.51.385 (1975).

071.045 **A study of absorption of the continuous spectrum by Fraunhofer lines. IX. Estimate of the energy reemitted along the spectrum.** P. P. Kozak, I. V. Baranovskij.
Solnechnye Dannye 1975 Byull., No. 7, p. 90 - 96 (1975). In Russian.

On the basis of the solar model by Gingerich et al. the integrals were computed which make it possible to estimate the contribution of all optical layers to the formation of the monochromatic intensity at different depths of the solar atmosphere as well as the effect of the absorption lines on it. The effect of the absorption lines is shown to vary non-monotonically along the spectrum.

071.046 **Estimate of inhomogeneities of objects in calculating the emission line profiles.**
L. N. Kurochka, V. A. Ostapenko.
Solnechnye Dannye 1975 Byull., No. 7, p. 96 - 102 (1975). In Russian.

Estimate of inhomogeneities of emission objects is shown to be a decisive factor in calculating the emission line profiles. Even small macroscopic velocities of emission elements can smooth completely the peculiarities of the line profiles.

071.047 **Differential spectrophotometric measurements of the brightness and temperature of the solar photosphere.** A. A. Kalinyak.
Nekotor. vopr. fiz. kosmosa. Moskva, 1974, p. 33 - 44. In Russian. – Abstr. in Referativ. Zhurn. 51. Astron., 11.51. 399 (1975).

071.048 **The far-infra-red sun–review of recent results and indication of future prospects.** T. A. Clark.
Journ. Roy. Astron. Soc. Canada, Vol. 69, 254 (1975). Abstr. CAS.

071.049 **Line blocking in the solar spectrum.**
A. Ardeberg, B. Virdefors.
Astron. Astrophys., Vol. 45, 19 - 24 (1975).

For the wave-length region λλ 4300–6200, blocking coefficients are derived for the sun. The line-blocking coefficients are given for intervals of 10Å and are reduced for effects of telluric absorption lines. The blocking values of the latter are given separately.

071.050 **New observations of the granular intensity fluctuations.** F. L. Deubner, W. Mattig.
Astron. Astrophys., Vol. 45, 167 - 171 = Mitt. Fraunhofer-Inst., *Freiburg,* No. 140 (1975).

During the partial eclipse of 1973 June 30 white light photographs of the solar granulation were obtained. Two-dimensional power spectra of the intensity distribution were calculated and the instrumental spread function, determined from the lunar limb profile, was used for correcting the data. The mean FWHM of the spread function is 0."52, the corrected intensity fluctuation ($\Delta I/I$) rms = 0.13 for λ 607 nm. The maximum power is found at the wavenumber k=4.7 Mm^{-1}, corresponding to a wavelength Λ=1340 km or 1."85. A remarkable result is the high power at $k \geqq 10$ Mm^{-1}.

071.051 **Th II meanlife and the solar thorium abundance.**
T. Andersen, A. P. Petkov.
Astron. Astrophys., Vol. 45, 237 - 238 (1975).

The meanlife of the 24874 cm^{-1} level in Th II has been measured by the beam-foil technique to be 21 ± 3 ns. The adopted solar thorium abundance is reduced by a factor of four on the basis on this lifetime and is now in agreement with the meteoritic abundance.

071.052 **Evidence for non-radial fields in the sun's photosphere and a possible explanation of the polar magnetic signal.** T. Pope, J. Mosher.
Solar Physics, Vol. 44, 3 - 12 (1975).

The appearance of the Hα fibrils suggests the presence of magnetic fields inclined at noticeably non-radial angles in the sun's chromosphere. The authors present evidence to suggest that these angles continue into the photosphere. They also suggest that the observed polar signal may be nothing more than a geometric effect arising when a balanced but systematically aligned array of bipolar pairs is viewed at an angle.

071.053 **Line intensities in the photosphere-chromosphere transition region. IV: Cr I and metallic ion lines.**
E. L. van Dessel.
Solar Physics, Vol. 44, 13 - 24 (1975).

Eclipse observations of total line intensities are interpreted in terms of departures from LTE. The same method as described in paper III (van Dessel, 1974 – see 12.071.019) is used. The HSRA model is used to compute the LTE populations. A distribution of NLTE-coefficients is obtained for each group of lines. The lines discussed in the present paper are seen to remain in emission down to some 1000 km inside the limb. As a consequence, interpretation is less simple than for the Fe I lines (paper III); population and excitation temperatures are found to be rather divergent from Holweger's temperature, which was used for a first approximation.

071.054 **Global distribution of filaments during solar cycle No. 20.** R. Hansen, S. Hansen.
Solar Physics, Vol. 44, 225 - 230 (1975).

By tracing the positions of filaments on the solar disk for a series of consecutive Carrington rotations, one can make a compact representation of the changes in general topology of photospheric magnetic fields during the course of a solar cycle. Examples are shown for the time period 1964–1974, which may provide some insight into the long-term relationship of the mid-latitude diagonal filaments and the high latitude polar crown.

071.055 **Spectral analyses of solar photospheric fluctuations.**

IV: The low-wavenumber power of granulation brightness fluctuations. F. N. Edmonds, Jr.
Solar Physics, Vol. 44, 293 - 297 (1975).

Synthetic one-dimensional scans of brightness fluctuations are generated from intergranule-granule profiles (IGP), which approximate observed scans of granulation fluctuations except that the local (i.e. IGP) mean intensity is kept constant. Comparing the power spectrum of such scans with the power spectrum of observed scans shows that nearly two thirds of the low-wavenumber ($k < 0.0025\ km^{-1}$) granulation power is due to this variable mean effect. This result favors the interpretation of granulation as turbulent thermal convection but cannot rule out the laminar convection interpretation.

071.056 **Photospheric velocity field associated with moustaches.** R. Kitai, I. Kawaguchi.
Solar Physics, Vol. 44, 403 - 408 (1975).

The photospheric velocity field was observed in an active region which was prolific in moustaches. It is shown that the moustaches occur at the locations where the sign of the line-of-sight velocity changes, and that the extension of the velocity field is large ($\approx 10^4$ km) compared with the dimension of moustaches.

071.057 **Calculation of faint and medium-strong line profiles in the solar photosphere for coherent scattering.** B. T. Babij.
Solnechnye Dannye 1975 Byull., No. 8, p. 72 - 74 (1975).
In Russian.

Methods of determining faint and medium-strong line profiles forming by coherent scattering are proposed. The procedure of calculating the source function for the above lines is shown.

071.058 **Emission conditions for solar He II.**
G. L. Epstein, R. Steinitz.
Bull. American Astron. Soc., Vol. 7, 522 - 523 (1975).
Abstr. AAS.

071.059 **Exit velocities of convective elements on the solar photosphere.** P. R. Romanchuk, M. A. Nuzhdina.
Vestn. Kiev. Un-ta, Ser. Astron., No. 17, p. 57 - 66 (1975).
In Russian.

Convective velocities for 1966 - 1968 from data of area observations of sunspot groups are determined. A model of sunspot groups formation as a result of eruption of large-scale convective elements into the photosphere is presented. The mean velocities are 0.10 ± 0.04 km/sec.

071.060 **The total photospheric motion field.**
E. A. Gurtovenko (*Eh. A. Gurtovenko*).
Solar Physics, Vol. 45, 25 - 33 (1975).

Brief consideration is given to the conception of the total photospheric motion field. A synthesis of the most thorough investigations is made and the radial (V^{rad}) and tangential (V^{tg}) components of the velocity amplitude of the total photospheric motion field are deduced. At depth $\log \tau_5 = -3.0$ V^{rad} and V^{tg} have average values of 1.2 and 1.7 km s^{-1} respectively. They increase smoothly with depth and reach their maximum values of $V^{rad} = 3.0$, $V^{tg} = 3.4$ km s^{-1} at depths $\log \tau_5 = -0.2$ and $\log \tau_5 = +0.4$ respectively. In the deep photospheric layers both components seem to decrease with depth.

071.061 **Visibility of the photospheric granulation in Fe I λ 6569.2.** R. E. Loughhead, R. J. Bray.
Solar Physics, Vol. 45, 35 - 40 (1975).

High-quality photographs of the granulation obtained at seven wavelengths in the Fe I line λ 6569.2 are presented. The granule contrast attains a maximum considerably exceeding the continuum value around Fe+0.05 to Fe+0.10 Å; individual granules remain easily visible to within 5″ of the limb. At line centre only remnants of the granulation pattern are discernible.

071.062 **The ability of current micro-velocity models to represent center-limb line profiles.**
J. C. Evans, L. Testerman.
Solar Physics, Vol. 45, 41 - 58 (1975).

The authors present results on the ability of microturbulence, LTE line formation, and a homogeneous thermal model to realistically represent the center-limb variation of temporally and spacially averaged solar line profiles. The authors have used three somewhat similar semi-empirical thermal models in combination with five current microturbulence models which cover the gamut of homogeneous-isotropic to nonhomogeneous-anisotropic. They find that a microturbulence model, for which the radial and tangential components increase into deeper layers with $\xi_{tan} > \xi_{rad}$, produces a reasonable good center-limb fit for lines less than 100 mÅ. However, for lines stronger than 140 mÅ, microturbulence models with no depth dependence produce the best match between theory and observation.

071.063 **The solar photospheric abundance of iron.**
E. Biémont, N. Grevesse.
Solar Physics, Vol. 45, 59 - 68 (1975).

A new value of the solar photospheric abundance of iron, independent of line-shape parameters, is derived. This analysis is based on a study of 40 weak infrared lines ($0.85 < \lambda < 2.5\ \mu$) for which theoretical oscillator strengths (calculated with configuration interactions taken into account) have recently been computed by Kurucz (1974). The abundance obtained is in agreement with the 'high' solar values recently reported in the literature and with the meteoritic abundance.

071.064 **Spectrophotometry of selected Fraunhofer lines on the solar disk.**
V. A. Krat, E. K. Kokhan, N. I. Pechinskaya.
Izv. Glav. Astron. Obs. Pulkovo, No. 193, Astrofiz. Astrometr. i Zvezdn. Astron., p. 3 - 27 (1975). In Russian.

A special program of observations has been worked out for the study of variations of precise profiles of Fraunhofer lines during a solar cycle. The program includes observations in the center of the solar disk and at the distance of 1.′6 and 0.′6 from the solar limb. The data on corrected profiles of the solar lines are tabulated for the period of observations from 1969 to 1972. An analysis is made for the study of variations of the central intensity and also those of the profiles itself.

071.065 **Analysis of the solar photometric spectrum: a new approach.** P. Dingens, H. Steyaert, N. Baeck.
Acad. Roy. Belgique, Bull. Cl. Sci. 5e Sér., Vol. 61, 582 - 588 = Univ. Gent, Sterrenkundig Inst., Med. No. 32 (1975).

A new very general (computer) method is proposed for decomposing the solar photometric spectrum with the aid of Gaussian components, and without making use of any information about already identified lines. The method has been applied to the wavelength range λ 6105 – λ 6150 of the solar data published by Delbouille, Neven and Roland (1973) and led to the recognition of 239 absorption lines, finding anew the 120 lines mentioned in the S.R.R.T. (Second revision of Rowland's preliminary table of the solar spectrum wavelengths, 1966), and multiplying by a factor 2 the number of detected (weak) lines. An example is given to illustrate the nearly perfect reproduction obtained.

071.066 **Remarks on the presence of some weak lines in the solar photometric spectral range λ 6105 – λ 6150.**
N. Baeck.
Acad. Roy. Belgique, Bull. Cl. Sci., 5e Sér., Vol. 61, 706 – 710 = Univ. Gent, Sterrenkundig Inst., Med. No. 33 (1975).

Using a newly proposed method of analysis of the solar

photometric spectrum in the range λ 6105 – λ 6150, with the spectral data published by Delbouille, Neven and Roland, all lines mentioned in the S.R.R.T. are confirmed with the exception of eight weak lines.

071.067 **The solar photospheric spectrum near wavelength 6138.5 Å.** N. Baeck, P. Dingens, H. Steyaert.
Acad. Roy. Belgique, Bull. Cl. Sci. 5e Sér., Vol. 61, 711 – 714 = Univ. Gent, Sterrenkundig Inst., Med. No. 34 (1975).

The blend mentioned in the S.R.R.T. at 6138.43 Å has been resolved with a new decomposition technique. The presence of Ti I and Y I is confirmed. Besides these two lines the results also reveal a line at 6138.661 Å, corresponding with a line of Ar II.

071.068 **On the presence of Ar and In in the solar photometric spectrum.** H. Steyaert, N. Baeck, P. Dingens.
Acad. Roy. Belgique, Bull. Cl. Sci., 5e Sér., Vol. 61, 761 - 764 = Univ. Gent, Sterrenkundig Inst., Med. No. 35 (1975).

The results obtained with our newly proposed method for analyzing the solar photospheric spectrum are presented for the regions λ 6108.8 – λ 6109.42 and λ 6114.76 – λ 6115.38. Six new weak lines are found, and the analysis suggests the identification of two lines with In II and one with Ar II.

071.069 **Analysis of the solar spectrum: results for the range λ 6105.0 – λ 6116.5.** P. Dingens, H. Steyaert, N. Baeck.
Acad. Roy. Belgique, Bull. Cl. Sci. 5e Sér., Vol. 61, 810 – 814 = Univ. Gent, Sterrenkundig Inst., Med. No. 36 (1975).

Applying a newly proposed method for decomposing the solar spectrum into Gaussian components, the obtained results are given for the interval λ 6105.0 – λ 6116.5. A table lists 61 components, with their central wavelength, equivalent width, (reciprocal) standard deviation and tentative identification.

071.070 **On the nature of dark formations in the solar photosphere.** Eh. V. Kononovich.
Astron. Tsirk., No. 883, p. 6 - 7 (1975). In Russian.

071.071 **Analysis of the solar spectrum: results for the range λ 6116.5 – λ 6132.0.** N. Baeck, P. Dingens, H. Steyaert.
Acad. Roy. Belgique, Bull. Cl. Sci. 5e Sér., Vol. 61, 835 - 840 = Univ. Gent, Sterrenkundig Inst., Med. No. 37 (1975).

The results obtained for the interval λ 6116.5 – λ 6132.0 are given. A table lists 67 components.

071.072 **Models of solar granular structure and the interpretation of photospheric observations.**
S. L. Keil.
Thesis, Boston Univ., Mass. (USA). [Available from Univ. Microfilms, Ann Arbor, Mich., *USA*. Order No. 74-26,459], 235 pp. (1974).

The standard, one-dimensional, calculations are outlined, and the observations which indicate that these calculations can be misleading are reviewed.

Study of the electronic transition moment for the CN red band system. See Abstr. 022.004.

Dielectronic satellite spectra for highly-charged helium-like ions–II. Improved calculations. See Abstr. 022.022.

Absolute *f*-values for resonance lines of neutral titanium. See Abstr. 022.032.

Lifetimes of some excited states in the rare earths: La II, Ce II, Pr II, Nd II, Sm II, Yb I, Yb II, and Lu II. See Abstr. 022.102.

Computation of oscillator strengths by a semi-empirical method for some elements of the iron-group and their solar photospheric abundance. III: Results for Mn I. See Abstr. 022.103.

Absolute transition probabilities for Ni I and the solar nickel abundance. See Abstr. 022.125.

Study of the solar photospheric granulation by Michelson interferometry – first results. See Abstr. 031.227.

Solare Spektralmessungen im infraroten Bereich. See Abstr. 034.011.

Effects of Thomson scattering on the shape of a spectral line. See Abstr. 063.028.

Non-LTE line formation in turbulent media. See Abstr. 063.035.

A model of photospheric faculae deduced from white light high resolution pictures. See Abstr. 072.065.

The work of the diode array: He 10830 observations of spicules and subflares. See Abstr. 073.071.

Optical and magnetic study of the active regions ATM 131 and 137 at the photospheric and chromospheric levels. See Abstr. 073.106.

Asymmetries of the solar Ca II lines. See Abstr. 073.125.

Line profiles resulting from rotating and escaping fine filaments of prominences. See Abstr. 073.153.

The 1640.4 Å Hα line of He II observed from Skylab. See Abstr. 074.004.

Abundance of Fe relative to H at 1.5 solar radii. See Abstr. 074.018.

A study of Fe XVII and Ni XIX coronal lines. See Abstr. 074.101.

Heliographic maps of the photosphere for the year 1974. See Abstr. 075.016.

EUV observations of the active sun from the Harvard experiment on ATM. See Abstr. 076.002.

X-ray solar spectrum in the 1.85 – 1.93 Å region. See Abstr. 076.045.

Spectral distributions of microwave bursts. See Abstr. 077.039.

On the rotation of gas and magnetic fields at the solar photosphere. See Abstr. 080.002.

On the statistical description of inhomogeneities in the quiet solar atmosphere. I. Linear regression analysis and absolute calibration of multichannel observations of the Ca^+ emission network. See Abstr. 080.020.

Heights of formation of non-magnetic solar lines suitable for velocity studies. See Abstr. 080.033.

072 Sunspots, Faculae, Solar Activity Cycles

072.001 **An investigation of material motions in a sunspot.** S. A. Lamb.
Monthly Notices Roy. Astron. Soc., Vol. 172, 205 - 220 (1975).

Sightline velocity components for material within a sunspot are determined for a grid of positions across the spot using the iron lines Fe I λ 6302.499 Å, Fe I λ 6270.231 Å and Fe II λ 6247.562 Å. Assuming that the sunspot velocity field has axial symmetry about the spot centre the author is able to reconstruct all three components of the velocity from the sightline component. A discussion of the results leads to the conclusion that a hydromagnetic wave interpretation may be more promising than an interpretation in terms of streaming motions.

072.002 **Die Feinstruktur von photosphärischen Fackeln.** J. P. Mehltretter.
Mitt. Astron. Ges., No. 36, p. 138 (1975). – Abstract.

072.003 **Eine unerwartete Anomalie in der jährlichen Verteilung der Sonnenfleckenmaxima.**
W. Gleissberg.
Mitt. Astron. Ges., No. 36, p. 139 - 141 (1975).

072.004 **The anomalous splitting of the π-component of a Zeeman triplet in sunspot umbrae and suggestions for its interpretation.** H. Künzel, J. Staude.
Astron. Nachr., Vol. 296, 171 - 176 (1975).

Zeeman spectrograms from 1200 sunspots of the maximum year 1957 are used to investigate the anomalous splitting (i.e., the relative shifting in the $I + V$ and $I - V$ spectra) of the π-component of the line Fe I 6302.5 Å in umbrae. π-component splittings opposite to the side of the main σ-component ((+) effects) are found to appear more than twice as frequently as (–) effects in umbrae with a magnetic northern polarity; the contrary is valid for umbrae with southern polarity. Several disturbing influences such as blends, stray light, and instrumental polarization as well as different possibilities for a theoretical interpretation are discussed.

072.005 **Granulare Geschwindigkeiten in der Nähe von Sonnenflecken.** W. Mattig, A. Nesis.
Mitt. Astron. Ges., No. 36, p. 147 (1975). – Abstract.

072.006 **Zum Problem des Kontinuums-Kontrastes solarer Fackeln.** E. Wiehr, G. Stellmacher.
Mitt. Astron. Ges., No. 36, p. 148 - 149 (1975).

072.007 **Alfvén waves and Meyer's sunspot model.** P. R. Wilson.
Monthly Notices Roy. Astron. Soc., Vol. 172, 535 - 543 (1975).

The recent account of the growth and decay of a sunspot by Meyer, Schmidt, Weiss & Wilson is compared with models exhibiting a large emission of Alfvén flux. It is shown that a composite model which involves both modified convection and Alfvén waves within the sub-surface flux rope can overcome some of the difficulties faced by either type of model independently. It is shown that interactions between convective motions and the flux tubes may give rise to twists in these tubes. The significance of these twists in relation to the decay of the spot magnetic field and the appearance of the moving magnetic features in the moat is discussed.

072.008 **Sunspot and stray light observations during the 1971, February 25 partial solar eclipse.**
J. Köppen.
Solar Physics, Vol. 42, 325 - 332 = Mitt. Fraunhofer Inst., *Freiburg*, No. 134 (1975).

Observations of a sunspot during and after a partial solar eclipse are described. The amount of scattered light confirms the existence of a spread function component with a half width of 10″. The observations also indicate the possibility of severely underestimating this component by aureole measurements. Umbral continuum intensities of 0.10 $I_\odot$ in the red spectral region were directly measured, the correction for scattered light amounts to 0.02 $I_\odot$. Intensities calculated with four umbral models are larger than the observed values, indicating this sunspot to be cooler by some 100 K. The wings of two strong Ca I lines are equally explained by the models of Henoux, Kneer, and Stellmacher/Wiehr. Yun's model can be ruled out because of too high a temperature.

072.009 **Alfvén waves and the sunspot phenomenon.** P. R. Wilson.
Solar Physics, Vol. 42, 333 - 340 (1975).

Parker's explanation of the sunspot phenomenon in terms of the enhanced emission of Alfvén waves (solar vulcanology) is shown to be compatible with observation only if ~90 % of the waves propagate downwards. Further difficulties arise if the region of cooling by Alfvén wave generation is restricted to a depth of 2 Mm. However, it is shown that, if Alfvén wave generation is included in a recent model proposed by Meyer, Schmidt, Weiss and Wilson, these difficulties may be resolved. The problem of the sharp umbra and penumbra boundaries is discussed and it is shown that features of this combined model are relevant to the flare phenomenon.

072.010 **Planetary tides and sunspot cycles.** J. J. Condon, R. R. Schmidt.
Solar Physics, Vol. 42, 529 - 532 (1975).

072.011 **Granular-like pattern in sunspot umbrae.** V. Bumba, L. Hejna, J. Suda.
Bull. Astron. Inst. Czechoslovakia, Vol. 26, 315 - 316 (1975).

A preliminary communication demonstrating the granular-like appearance of the internal parts of all small as well as great umbras of the large complex September 1974 sunspot group. The "bright umbral dots" seem to be the central positions of individual granular-like cells, or even tops of convective elements which may penetrate into the visible layers of the umbra.

072.012 **Analysis of Skylab soft X-ray observations of solar active region 131 (McMath 12379).**
W. Henze, E. J. Reichmann, A. C. Deloach, R. B. Hoover, J. P. McGuire, E. Tandberg-Hanssen, R. M. Wilson, J. B. Smith, Jr., D. M. Speich.
Bull. American Astron. Soc., Vol. 7, 443 - 444 (1975). – Abstr. AAS.

072.013 **Facular models and line profiles.** G. A. Chapman.
Bull. American Astron. Soc., Vol. 7, 449 (1975). Abstr. AAS.

072.014 **The flux of Alfvén waves in sunspots.** J. M. Beckers.
Bull. American Astron. Soc., Vol. 7, 459 (1975). – Abstr. AAS.

072.015 **Comparisons of potential and force-free magnetic-field calculations with Skylab soft X-ray images of a complex solar active region.**
M. J. Hagyard, N. P. Cumings, A. C. Deloach, R. B. Hoover, J. P. McGuire, E. Tandberg-Hanssen, J. R. Watkins, R. M.

Wilson, J. B. Smith, Jr., D. M. Speich, S. T. Wu.
Bull. American Astron. Soc., Vol. 7, 459 (1975). – Abstr. AAS.

072.016 **Physics of an active region loop event.**
R. H. Levine, G. L. Withbroe.
Bull. American Astron. Soc., Vol. 7, 460 (1975). – Abstr. AAS.

072.017 **How to observe faculae.** E. N. Frazier.
Bull. American Astron. Soc., Vol. 7, 472 (1975).
Abstr. AAS.

072.018 **High resolution polarization observations of a solar active region at 6 cm.**
C. E. Alissandrakis, M. R. Kundu.
Bull. American Astron. Soc., Vol. 7, 473 (1975). – Abstr. AAS.

072.019 **Mode of propagation of penumbral waves.**
P. Maltby.
Nature, Vol. 257, 468 - 469 (1975).

The solar atmosphere provides an interesting possibility for the study of low frequency waves both in magnetic and nonmagnetic regions, and the most spectacular result has been the discovery of waves propagating outwards over sunspot penumbrae. The theoretical interpretation of these waves has aroused considerable interest and, in particular, a number of authors have attempted to identify the precise mode of wave propagation involved. The modes proposed include Alfvén waves, acoustic waves, Lamb waves, and magnetogravity waves of the 'plus' type vertically trapped at photospheric levels. The author has explored the conditions of observability as a function of heliocentric angle for purely longitudinal (that is, particle velocity parallel to the propagation vector) and purely transverse (that is, normal to the propagation vector) waves and he presents observations that tend to support a predominantly longitudinal mode of wave propagation.

072.020 **On the periodic structure of plages.**
A. P. Kramynin.
Astron. Tsirk., No. 847, p. 3 - 5 (1974). In Russian.

072.021 **Magnetic fields of small sunspots.**
A. V. Baranov.
Astron. Tsirk., No. 847, p. 5 - 6 (1974). In Russian.

072.022 **Temperature-dimension relation for cores of sunspots.** S. Obashev, G. S. Minasyants.
Astron. Tsirk., No. 850, p. 1 - 2 (1974). In Russian.

072.023 **Magnetic fields of active regions and noise storms.**
M. Ogir', L. Yurovskaya.
Astron. Tsirk., No. 855, p. 1 - 2 (1975). In Russian.

072.024 **The latitudinal zonation and motion of matter on the sun.** N. I. Kozhevnikov.
Astron. Tsirk., No. 865, p. 5 - 6 (1975). In Russian.

072.025 **Resonance phenomena of solar activity.**
P. R. Romanchuk, I. G. D'yachenko.
Problems of cosmic physics. Vyp. (No.) 10, (see 003.003), p. 103 - 108 (1975). In Russian.

Convective element motion is considered under the action of two planets. The solution of the equation of motion admits the existence of resonance strengthening of convective motions. The response of the convective zone is changed from cycle to cycle and reaches a maximum, in the mean, every 80 - 90 years. A model of processes in the convective zone is proposed. The model is confirmed by the observations.

072.026 **Chromospheric flares and rapid changes in sunspots.**
K. G. Chistyakova, V. F. Chistyakov.
Problems of cosmic physics. Vyp. (No.) 10, (see 003.003), p. 108 - 119 (1975). In Russian.

The results of photographic observations of sunspots made in Ussurijsk in 1970 - 1973 are described and discussed.

072.027 **Solar-cycle dynamo wave propagation.**
H. Yoshimura.
Astrophys. Journ., Vol. 201, 740 - 748 (1975).

Dynamo waves, as solutions of the dynamo equations governing the solar cycle, propagate along isorotation surfaces inside the sun. This theorem, universal in most of the dynamo models of the solar cycle, is proven analytically, and the nature of the propagation is analyzed. The effects of meridional circulation and the diffusion process on the propagation nature of the dynamo waves are estimated. The importance of this theorem is stressed for understanding the solar cycle and for inferring the rotational law of the interior of the sun.

072.028 **Numerical modeling of processes of magnetic field skinning in an active solar region.**
S. S. Katsnel'son, V. S. Slavin, V. S. Sokolov.
Aehrofiz. issledovaniya. Vyp. (No.) 3. Novosibirsk, 1974, p. 25 - 27. In Russian. – Abstr. in Referativ. Zhurn. 51. Astron., 9.51.441 (1975).

072.029 **On the latitude periodicity of sunspot origin.**
L. I. Gudzenko, V. E. Chertoprud.
Kratkie soobshch. po fiz., 1975, No. 1, p. 3 - 6. In Russian.
Abstr. in Referativ. Zhurn. 51. Astron., 9.51.464 (1975).

072.030 **Application of the method of planetary configurations to active phenomena on the sun.**
V. S. Prokudina.
Soobshch. Gos. Astron. Inst. Shternberga, No. 181, p. 11 - 32 (1973). In Russian.

The method of planetary configurations is described. Some results of the application of this method to the phenomena of solar activity are given. It is noticed that when the centre of activity passes the direction to the line of the configuration activation is possible: the spot is fractionated, the appearance of a powerful flare is possible.

072.031 **The self-stabilization of the phase of the solar activity cycle.** L. I. Gudzenko, V. E. Chertoprud.
Kratkie soobshch. po fiz., 1975, No. 1, p. 7 - 12. In Russian.
Abstr. in Referativ. Zhurn. 51. Astron., 8.51.412 (1975).

072.032 **Dynamical processes in the development of sunspot groups.** P. Ranzinger.
Publ. Obs. Astron. Beograd, No. 20, (see 012.024), p. 46 (1975). – Summary. In Serbo-Croatian and English.

072.033 **Motion of matter and solar activity cycle.**
N. I. Kozhevnikov.
Astron. Tsirk., No. 864, p. 3 - 5 (1975). In Russian.

072.034 **H and K lines in the sunspot umbra: a source function and Doppler width.** R. B. Teplitskaja (*Teplitskaya*), S. A. Efendieva (*Ehfendieva*).
Solar Physics, Vol. 43, 293 - 309 (1975).

An inversion of line profiles H and K Ca II in the umbra of two sunspots is performed. By means of a numerical experiment, a tentative study of possible application of the method of intercomparison of lines in the multiplet (Goldberg's method) is made in cases when the condition of source function equality in multiplet is not fulfilled. It is found that the source functions of H and K lines in the sunspot umbra vary in a monotonous way. Their ratio in the layers for which measured values of $\Delta\lambda_D$ are available differs from unity. In the region of the lower chromosphere under consideration, the Doppler width is decreasing with height.

072.035 **On the difference of the magnetic field strength in sunspot umbrae determined from neutral vanadium and iron spectral lines.** M. D. Gusejnov.
Izv. Krymskoj Astrofiz. Obs., Vol. 53, p. 66 - 74 (1975). In Russian.

The magnetic field strength *H* was measured in a few sunspots using polarization spectrograms. The values of *H* determined from the spectral lines of Fe I λ6302.5, V I λ6285.2 and λ 6256.9 Å are compared. The magnetic field determined from the vanadium lines was found to be stronger by 200 gauss than that from the iron line.

072.036 **Some characteristics of developing spot groups.** M. B. Ogir'.
Izv. Krymskoj Astrofiz. Obs., Vol. 53, p. 82 - 94 (1975). In Russian.

The development of spot groups is discussed. A photometric study of Hα plages has shown that appearance and development of new spots were accompanied by noticeable changes in plage brightness. During the appearance of a great number of small spots lasting strong pulsations occurred side by side with short ones. Magnetic field maps show that the increasing spot groups could be situated as in isolated magnetic regions, so in magnetic complexes. In spot groups belonging to magnetic complexes plage brightenings and flares were often connected with activisation of the other regions of this complex.

072.037 **On the measurement of magnetic fields of sunspots from different lines.**
Z. N. Shukstova, G. V. Lyamova.
Izv. Krymskoj Astrofiz. Obs., Vol. 53, p. 249 - 250 (1975). In Russian. – Abstr. of a paper presented at the conference "Magnetic fields and chemical composition of the sun and stars" at the Crimean Astrophys. Obs., 1974, April 9 - 12, (see 012.027).

072.038 **Investigation of the magnetic fields in the umbra of sunspots from spectral lines with different temperature sensibility.** M. D. Gusejnov.
Izv. Krymskoj Astrofiz. Obs., Vol. 53, p. 250 - 251 (1975). In Russian. – Abstr. of a paper presented at the conference "Magnetic fields and chemical composition of the sun and stars" at the Crimean Astrophys. Obs., 1974, April 9 - 12, (see 012.027).

072.039 **Magnetic strengthening of the spectral lines of neutral titanium in a sunspot.** E. N. Zemanek.
Solnechnye Dannye 1975 Byull., No. 3, p. 58 - 67 (1975). In Russian.

On the basis of a reduction of the sunspot spectrum obtained on June 22, 1959 the values of magnetic strengthening of the spectral lines of neutral titanium in the region of 4449–6600 Å were calculated. The curve of growth was constructed by means of the equivalent widths of these lines free from the influence of the magnetic field. Physical parameters of the sunspot were determined by the curve-of-growth method.

072.040 **On possibilities of forecasting the solar activity variations.**
G. Ya. Vasil'eva, A. A. Shpital'naya, N. S. Pystina.
Solnechnye Dannye 1975 Byull., No. 3, p. 67 - 75 (1975). In Russian.

Analysis of the solar activity variations caused by the revolution of the earth, Mercury, Venus, and Mars about the sun gives evidence of existence of "active" longitudes which are characterized by increasing the solar activity at the moments of the planets passing these longitudes.

072.041 **On the problem of estimating the validity of forecasts of the mean monthly and quarterly Wolf numbers.** Yu. I. Vitinskij, K. A. Kandaurova.
Solnechnye Dannye 1975 Byull., No. 3, p. 76 - 80 (1975). In Russian.

Meteorological characteristics ρ and Q were used to estimate the validity of forecasts of the average monthly and quarterly Wolf numbers. A comparison of prognoses of the mean monthly Wolf numbers has been made for the methods of Mayot, Kandaurova and Romanchuk.

072.042 **On the spatial scales of the magnetic field of sunspots from observations of change of the polarization in the emission of local sources.** N. G. Peterova.
Solnechnye Dannye 1975 Byull., No. 3, p. 96 - 101 (1975). In Russian.

A method of investigations of the magnetic fields of solar activity regions, based on the known phenomenon of change of the sense of polarization in the emission of local sources of the S-component is described. This method permits to evaluate the total scales of the sunspot magnetic fields or the sizes of the sources of polarized emission.

072.043 **Investigation of solar cycles by the method of differences.** Yu. I. Vitinskij.
Solnechnye Dannye 1975 Byull., No. 4, p. 62 - 70 (1975). In Russian.

Peculiarities of the 11-year, 22-year and 80–90 year solar cycles were analyzed with use of the first and second differences of average annual Wolf numbers for 1749–1973.

072.044 **On the absence of regions with a weak magnetic field in the umbra and penumbra of sunspots.**
Kh. I. Abdusamatov.
Solnechnye Dannye 1975 Byull., No. 4, p. 83 - 86 (1975). In Russian.

The splitting of the Fe I λ 6302.499 Å spectral line was investigated with accounting for the contribution of light and dark features in the umbrae and penumbrae of sunspots. On the basis of the absence, in some cases, of the nonsplitted component in the center of this line it is concluded on the absence of regions with a weak magnetic field in the umbrae and penumbrae of sunspots.

072.045 **The reduction system of signals of a solar magnetograph in true time scale.**
A. M. Morozov, S. G. Bortnik.
Solnechnye Dannye 1975 Byull., No. 4, p. 92 - 97 (1975). In Russian.

A description of the algorithm for calculating the total vector of the magnetic field intensity in sunspots is given. The technical and program equipment of the system providing the operation in a true time scale is also described.

072.046 **Metastable states of sunspots.** A. A. Solov'ev. Solnechnye Dannye 1975 Byull., No. 4, p. 97 - 102 (1975). In Russian.

It is shown that the magnetic flux tube (rope) of a sunspot can have two stable states. The latter correspond to the transparent dense model of a sunspot. The rapid variation of sunspot magnetic fields are connected with the transitions between these states of the magnetic rope.

072.047 **On space-time distribution of large sunspot groups.** Yu. I. Vitinskij.
Solnechnye Dannye 1975 Byull., No. 5, p. 59 - 64 (1975). In Russian.

Peculiarities of longitudinal distribution of sunspot groups with mean area more than 500 millionths of the solar visible hemisphere have been considered.

072.048 **On some peculiarities of the cross-over effect in the**

sunspot spectrum. A. V. Baranov.
Solnechnye Dannye 1975 Byull., No. 5, p. 77 - 82 (1975). In Russian.

A reduction of the polarization spectrograms for the leader penumbra of the large sunspot group taken on August 20, 1971 has been made. It was found that for the cross-over effect the line has a greater equivalent width in the spectrum, when the σ components are better detectable.

072.049 **The magnetic field intensity in sunspots as measured from different absorption lines.**
A. M. Rasulov.
Solnechnye Dannye 1975 Byull., No. 6, p. 61 - 67 (1975). In Russian.

The main characteristics of magnetoactive lines in the λ 5250–6350 Å region are given. For the observed sunspots the magnetic field intensity, measured from the Ti I 6064.626 Å line, is shown to be higher than that from the rest lines by 200 gauss. The distribution of the magnetic field intensity over the sunspot umbra is approximated by Mattig's formula.

072.050 **Solar activity and the solar wind: 1. Eleven-year cycles.** A. J. Hundhausen.
Comments Astrophys. Space Phys., Vol. 6, 63 - 70 (1975).

072.051 **On the nature of solar activity.** G. Ya. Vasil'eva, D. A. Kuznetsov, N. S. Petrova, A. A. Shpital'naya.
Nekotor. vopr. fiz. kosmosa. Moskva, 1974, p. 45 - 62. In Russian. – Abstr. in Referativ. Zhurn. 51. Astron., 10.51. 445 (1975).

072.052 **Theoretical analysis of various statistical indices of sunspot activity.** M. Kopecký.
Bull. Astron. Inst. Czechoslovakia, Vol. 26, 351 - 359 (1975).

Using the physically primary parameters of sunspot groups (number of sunspot groups originated per unit time $f_0(\tau)$, their frequency distribution according to lifetime $F(T, \tau)$, the pattern of the sunspot group area $S(t, T, \tau)$ and the number of sunspots in the group $\sigma(t, T, \tau)$ as a function of the sunspot group age t), of which it is assumed that they depend on time τ, expressions for the time variation of various indices of sunspot activity (Wolf's relative sunspot number, the number of sunspot groups, the total area of the sunspots, the average area of the sunspot groups, the number of new sunspot groups on the solar disc) have been derived.

072.053 **A wave-mechanical forecast of solar activity.**
G. I. Kuz'menko, M. Ya. Postan.
Astron. Zhurn. Akad. Nauk SSSR, Vol. 52, 1334 - 1337 (1975). In Russian. English translation in Soviet Astron., Vol. 19, No. 6.

A superlong-period forecast of increases and decreases of yearly Wolf numbers for the period 1970–1996 is made.

072.054 **The height and duration of sunspot fibrils in the Hα chromosphere.** V. Gaizauskas.
Journ. Roy. Astron. Soc. Canada, Vol. 69, 254 (1975). Abstr. CAS.

072.055 **The deep layers of sunspot umbrae.**
G. Stellmacher, E. Wiehr.
Astron. Astrophys., Vol. 45, 69 - 76 (1975).

An empirical umbral 'working model' is adapted to the 1.67 μ intensities which originate from the deepest observable layers (H^- minimum). The range of possible extrapolations is limited by the two conditions: (1) The continuum near 1.67 μ must not fall below the range of observations. (2) The two C-lines 16888 and 17456 must not come out stronger than the lower observational limit. The calculations show that a $T^4(\tau_R)$ stratification with T_{eff} = 3750° yields the best compromise between (1) and (2).

072.056 **Fine structure and Evershed motions in the sunspot penumbra.** D. J. Galloway.
Solar Physics, Vol. 44, 409 - 415 (1975).

A simple model is put forward in which the interaction of convection rolls with an initially homogeneous magnetohydrostatic sunspot field causes a concentration of flux in the dark filaments. This process drives Evershed motions outwards in these regions; the motions are superficial and shear the lines of force, so that the field appears stronger and more horizontal in the dark filaments. This situation must be time-dependent to avoid rapid destruction of the whole spot.

072.057 **Temperature model of sunspots.**
G. S. Minasyants.
Trudy Astrofiz. Inst., *Alma-Ata,* Vol. 26, 27 - 38 (1974). In Russian.

072.058 **On the character of flare energy release on short-lived sunspot groups.**
V. M. Rossada, L. N. Kurochka.
Solnechnye Dannye 1975 Byull., No. 8, p. 85 - 89 (1975). In Russian.

About 90% of the short-lived sunspot groups are shown to have certain periods (10 - 20 hours) when approximately half of the total energy radiated by flares in the given group is released.

072.059 **The method of superlong-term forecast of Wolf numbers and the 'black box' model.**
Yu. N. Kudrya.
Astron. Obs., *Kiev,* Preprint No. 4, 9 pp. (1974). In Russian.

It is shown that the 'resonance curve' method for a superlong-term forecast of the solar activity is a realization of the mathematical model of the 'black box'.

072.060 **Latitude distribution of sunspots and the common magnetic field of the sun.** P. R. Romanchuk.
Astron. Obs., *Kiev,* Preprint No. 5, 9 pp (1974). In Russian.

By the butterfly diagram it has been established that the equator zone width, almost free from spots, is smaller in even spot cycles than in odd ones. This regularity is explained by the existence of the variable (period 22 years) and the constant poloidal common magnetic field components. According to the author's theory of the generation of the sun's magnetic fields an estimate of the quantity has been obtained and the sign of the constant magnetic field has been determined.

072.061 **The prediction of the 22nd - 26th cycles of solar activity.** P. R. Romanchuk, T. I. Redyuk.
Astron. Obs., *Kiev,* Preprint No. 6, 5 pp. (1974). In Russian.

A prediction of the main characteristics of the 22nd - 26th 11-year cycles of solar activity is given. The epoch of minimum of the current century cycle is supposed on the year 2000.

072.062 **High resolution spectroscopy from Orbiting Solar Observatory VIII: transition zone dynamics over a sunspot.** E. C. Bruner, Jr., E. G. Chipman, B. W. Lites, G. J. Rottman, R. A. Shine, R. G. Athay, O. R. White.
Bull. American Astron. Soc., Vol. 7, 522 (1975). – Abstr. AAS.

072.063 **Analysis of solar active region SFC 158/73.**
M. D. Papagiannis, R. M. Straka, J. A. Kogut.
Bull. American Astron. Soc., Vol. 7, 523 (1975). – Abstr. AAS.

072.064 **On an investigation of the increasing relative rates of areas of sunspot groups.**
P. R. Romanchuk, M. A. Nuzhdina.
Vestn. Kiev. Un-ta, Ser. Astron., No. 17, p. 47 - 57 (1975).

In Russian.

The relative rates of increasing of sunspot group areas during the eighteenth and twentieth cycles of solar activity are considered. The rate of area increasing is correlated with tidal influences by five planets of the solar system.

072.065 **A model of photospheric faculae deduced from white light high resolution pictures.** R. Muller.
Solar Physics, Vol. 45, 105 - 114 (1975).

High resolution pictures (about 0".3) of photospheric faculae near the solar limb have been obtained with the Pic du Midi 50 cm refractor; their granular structure then clearly appears. The microphotometric study of these facular granules shows that the ratio of their intensity to the photospheric intensity, I^f/I^{ph} (cos θ) reaches a maximum near cos $\theta = 0.3$ and then decreases towards the limb. A temperature model of a facular granule is obtained. The temperature excess is 750 K at maximum. This hot region is located over a layer which is cooler than the normal photosphere at the same level.

072.066 **Structure of sunspots. III: A minimum-gradient model atmosphere for umbrae.** C. Zwaan.
Solar Physics, Vol. 45, 115 - 118 (1975). – Research note.

072.067 **On peculiarities of the activity centers during the maximum epoch of the 20th solar cycle.**
Yu. I. Vitinskij.
Izv. Glav. Astron. Obs. Pulkovo, No. 193, Astrofiz., Astrometr. i Zvezdn. Astron., p. 28 - 40 (1975). In Russian.

A catalogue of the activity centers is given, being a continuation of previous works of the author (1965, 1968, 1971). The data are discussed in detail.

072.068 **Discreteness of velocity values of the motion of sunspot groups.** N. I. Kozhevnikov.
Astron. Tsirk., No. 871, p. 5 - 7 (1975). In Russian.

072.069 **Research on the recurrence of sunspot groups during a rotation period of 27.25 days.** N. Doğan.
Commun. Fac. Sci. Univ. Ankara, Sér. A_3: Astron., Vol. 24, 16 + 7 pp. = Comm. Astron; Dep. Univ. Ankara, No. 72 (1975).

The daily values of the sunspot relative numbers in the interval between the beginning of the December 30th, 1903 and December 22nd, 1973 were separately searched at the maximum and the minimum phases of the spot cycles as well as at the rising and declining branches to indicate how sunspot numbers varied in the interval studied. For this purpose 938 rotations of the sun were studied. As a result, it is understood that sunspots prefer to appear at certain longitudes which were previously active or close to previously active ones.

072.070 **On the nature of bright facular dots.**
Eh. V. Kononovich.
Astron. Tsirk., No. 884, p. 4 - 5 (1975). In Russian.

072.071 **On the reality of Spörer's law.**
N. I. Kozhevnikov.
Astron. Tsirk., No. 885, p. 1 - 2 (1975). In Russian.

072.072 **Mechanical model of the structure of the convective zone and sunspot generation.**
N. I. Kozhevnikov.
Astron. Tsirk., No. 885, p. 3 - 5 (1975). In Russian.

072.073 **Phase synchronism of circulation of matter in convective cells of a solar activity region.**
N. I. Kozhevnikov.
Astron. Tsirk., No. 886, p. 3 - 5 (1975). In Russian.

072.074 **What new on the sun?** W. Szymański.
Urania Kraków, Vol. 46, 263 - 267 (1975). In Polish.

072.075 **Correction of an irregular sunspot or facula for stray light.** P. Iuell, L. Staveland.
Inst. Theor. Astrophys., Blindern–Oslo, Rep. No. 43, 2 + 32 pp. (1975).

A formula for correction of the umbra intensity has been developed, the formula is valid even close to the limb. The integrals for the contribution functions for umbra and penumbra of irregular form are solved by a series representation. The blurring spread function is assumed to be Gaussian. The scattering spread function may be a Gaussian, a Lorentzian or a sum of both. Revised formulae for the uncertainty of the relative umbra intensity are given for an elliptical spot.

072.076 **Photoelectric photometry of the facular fields in two regions of the continuum.** I. V. Yudina.
Solnechnye Dannye 1975 Byull., No. 9, p. 55 - 61 (1975). In Russian.

The contrast of the faculae (I_{phot}/I_{fac}) has been measured simultaneously in two wide ($\Delta\lambda = 600$ Å) spectral regions ($\lambda_{eff} =$ 4200 Å, 5300 Å) by the photoelectric method. The results obtained are given in tables. The faculae are shown to have maximum contrast near sin $\theta = 0.95$.

072.077 **Saules aktivitātes cikli. Fakti un prognozes.**
G. Ozoliņš.
Zvaigžņotā debess, 1975. gada vasara, p. 11 - 16.

072.078 **Acerca de la distribución de frecuencias del área de las manchas solares.**
M. López Arroyo, J. F. Lahulla.
Bol. Astron. Obs. Madrid, Vol. 8, No.5, p. 27 - 60 (1975).

In the observed frequency distribution the existence of a discontinuity in the range of area 51 – 90 millionths is shown. Such a discontinuity is evident in every cycle in both hemispheres, North and South. It is suggested that this dicontinuity should be related with the depth of the sunspots which elsewhere has been assumed to be the same for all the sunspots greater than 86 millionths and equal to the total depth of the convective zone.

072.079 **New views of sunspots.** J. M. Beckers.
Sacramento Peak Obs., Project 7649, Air Force Cambridge Res. Lab., *Hanscom, Mass.*, AFCRL-TR-75-0089, Environmental Res. Papers, No. 499, 29 pp. = Sacramento Peak Obs. Contr. No. 249 (1975).

This paper reviews recent theoretical and observational developments in sunspot physics. It concentrates especially on the recent discoveries of rapid changes in sunspots, most of them oscillatory, and their connection with Alfven waves generated in spots.

072.080 **Spezielle Eigenbewegungen in Sonnenfleckengruppen.** H. Pfister.
Astron. Mitt. Eidgenössisch. Sternw. Zürich No. 342, 113 pp. (1975).

In between July 1957 and 1971 120 groups of sunspots were observed which show remarkable proper motions. Similar motions stand out so that it is possible to make a classification. In this work 70 interesting examples are described in detail and divided into 9 classes.

072.081 **Intercomparison of ground based and satellite pictures of the sun.** R. D. Chapman, G. L. Epstein, R. W. Hobbs, W. M. Neupert, R. J. Thomas.
Image processing techniques in astronomy, (see 012.041), p. 379 - 384 (1975).

Using NASA facilities in space (OSO-7) and on the ground

(Goddard Multi-Channel Spectrophotometer at Sacramento Peak, New Mexico) the authors have mapped an active region, and by combining these ultraviolet, X-ray and visible data, have constructed a physical picture of this structured region from the photosphere to the corona, corresponding to temperature regimes over the range 4500 K to 4000000 K.

072.082 **Magnetohydrodynamical dynamo model of the solar cycle.** T. S. Ivanova, A. A. Ruzmajkin.
In-t prikl. mat.AN SSSR. Preprint No. 43. Moskva, 1975. 31 pp. Price 10 Kop. In Russian. – Abstr. in Referativ. Zhurn. 51. Astron., 2.51.526 (1976).

072.083 **Gesetz und Zufall im Ablauf der Sonnentätigkeit.** W. Gleissberg.
Veröff. Astron. Inst. Univ. Frankfurt (Main), No. 52, 4 pp. (1974).

072.084 **An unexpected anomaly in the annual distribution of the maxima of the 11-year sunspot cycle.**
W. Gleissberg.
Journ. Interdiscipl. Cycle Res., Vol. 6, 37 - 40 = Veröff. Astron. Inst. Univ. Frankfurt (Main), No. 54 (1975).

072.085 **Zufall oder Gesetz? Betrachtungen zu einem rätselhaften Verhalten der Sonnenfleckenmaxima.**
W. Gleissberg.
Kleinheubacher Ber., Vol. 18, 161 - 164 = Veröff. Astron. Inst. Univ. Frankfurt (Main), No. 56 (1975).

072.086 **Is there a 179-year recurrence tendency in sunspot frequency?** W. Gleissberg.
Veröff. Astron. Inst. Univ. Frankfurt (Main), No. 57, 13 pp. (1975). In German.

In order to study the character of the 179-year repetition period it will be necessary to consider the relation between this period and the 80-year cycle to which the strengths of the sunspot maxima and all other characteristics of the 11-year sunspot cycles are subject. From a study of these data it follows that the 179-year recurrence tendency in sunspot frequency cannot have been permanent during earlier centuries; thus this repetition period must be considered to be merely a temporary phenomenon.

Magnetic intensification as a mechanism for broad band linear polarization. See Abstr. 063.046.

Temperature variation with latitude in the upper solar photosphere: relevance to solar oblateness measurements and facular models. See Abstr. 071.002.

Exit velocities of convective elements on the solar photosphere. See Abstr. 071.059.

The chromospheric Evershed flow.
See Abstr. 073.068.

On the relation between proton-flares and magnetic configurations of sunspot groups. See Abstr. 073.076.

Emerging flux regions. See Abstr. 073.095.

Birth and development of an active region and the chromospheric network. See Abstr. 073.100.

Investigations according to the programme "Rapid variations of solar magnetic fields", VI. Daily variations of flare activity and variations of magnetic fields of sunspots.
See Abstr. 073.118.

Lack of evidence for solar-cycle variations in high-speed streams in the solar wind. See Abstr. 074.099.

Nord-Süd-Asymmetrie im Sonnenwind und in der Sonnentätigkeit. See Abstr. 074.127.

Actividad solar en 1973. I. Números relativos de Wolf. II. Estadística de manchas y superficie de las mismas.
See Abstr. 075.009.

On the influence of some features of sunspot development on noise storm radio emission.
See Abstr. 077.026.

On possible sources of solar radio emission connected with polar faculae. See Abstr. 077.056.

Constant outflow of low-energy particles from active regions on the sun. See Abstr. 078.026.

On the differential rotation of the sun and cycles of its activity. See Abstr. 080.017.

Differential rotation and the spot zones.
See Abstr. 080.035.

Solar activity and the general absence of hydrostatic equilibrium in an azimuthal magnetic field.
See Abstr. 080.038.

Concerning some actual problems in the observational solar research. See Abstr. 080.051.

Change of low ionosphere characteristics with solar cycle deduced from solar flare effects at VLF.
See Abstr. 083.009.

Geomagnetic disturbances and complex sunspots.
See Abstr. 084.203.

A hundred years of controversy over sunspots and weather. See Abstr. 085.002.

Head echoes and solar activity in the Geminid period. See Abstr. 104.009.

A sunspot as a standard for abundance analyses of cool stars: Ba I spectrum. See Abstr. 114.079.

073 Solar Chromosphere, Flares, Prominences

073.001 **Morphological evolution of X-ray flare structures from the rise through the decay phase.**
S. W. Kahler, A. S. Krieger, G. S. Vaiana.
Astrophys. Journ., (*Letters*), Vol. 199, L57 - L61 (1975).
The morphological evolution of 12 solar X-ray subflares from onset through the decay phase has been studied using photographic X-ray images obtained from Skylab. The spatial configurations are found to vary widely from flare to flare, but they appear to be composed of two basic kinds of structures. The first, which the authors have termed "X-ray kernels", are brightest during the rise phase; and the second, looplike structures, appear during the maximum and decay phases of the event.

073.002 **The helium chromosphere, coronal holes, and stellar X-rays.** H. Zirin.
Astrophys. Journ., (*Letters*), Vol. 199, L63 - L66 (1975).
The solar chromosphere at the limb seen in D3 is an irregular bright band 1000 km thick with a dark band 1000 km thick beneath. The D3 chromosphere disappears in coronal holes. The author demonstrates how the D3 emission, as well as the other He I and He II lines, can be explained quantitatively by photoionization by coronal back-radiation. The chromospheric He emission or absorption is weak in coronal holes because there is no coronal back-radiation. Based on this model, the soft X-ray flux from stars with He λ10830 absorption lines is estimated as proportional to the 10830 equivalent width and the apparent area.

073.003 **The location of the site of energy release in a solar X-ray subflare.** R. D. Petrasso, S. W. Kahler, A. S. Krieger, J. K. Silk, G. S. Vaiana.
Astrophys. Journ., (*Letters*), Vol. 199, L127 - L130 (1975).
A rapid sequence of high-resolution X-ray photographs was obtained by the S-054 X-ray Telescope Experiment on Skylab on 1973 September 1. During the course of this observation, photographs were obtained of a flarelike brightening in a simple, bipolar active region.

073.004 **Comments concerning the photoionization model for excitation of resonance lines of He I and He II in the solar chromosphere.** R. W. Milkey.
Astrophys. Journ., (*Letters*), Vol. 199, L131 - L132 (1975).
Recently there has been much discussion of the production of the solar resonance lines of He I and He II by extreme-ultraviolet photons impinging on the upper chromosphere photoionizing He I and He II with the subsequent recombinations producing the emission lines. In this paper the consequences of this model for the emergent line profiles are examined and are shown to lead to an apparent contradiction with the observed profiles.

073.005 **Ultraviolet emission line profiles of flares and active regions.** G. E. Brueckner.
IAU Symposium No. 68, (see 012.003), p. 135 - 151 (1975).
Review paper.

073.006 **Fe XXIV emission in solar flares observed with the NRL/ATM XUV slitless spectrograph.**
K. G. Widing.
IAU Symposium No. 68, (see 012.003), p. 153 - 163 (1975).

073.007 **X-ray and EUV spectra of solar flares and laboratory plasmas.** G. A. Doschek.
IAU Symposium No. 68, (see 012.003), p. 165 - 181 (1975).
Review paper.

073.008 **Association of X-ray flares with solar coronal active regions.** P. R. Sengupta.
IAU Symposium No. 68, (see 012.003), p. 183 (1975).
Summary.

073.009 **Studies of the dynamic structure and spectra of solar X-ray flares.** S. W. Kahler, A. S. Krieger, J. K. Silk, R. W. Simon, A. F. Timothy, G. Vaiana.
IAU Symposium No. 68, (see 012.003), p. 185 (1975).
Summary.

073.010 **On the thermal structure of the flare-produced plasma.** I. Craig.
IAU Symposium No. 68, (see 012.003), p. 187 - 189 (1975).
Summary.

073.011 **Thermal and nonthermal interpretations of flare X-ray bursts.** S. Kahler.
IAU Symposium No. 68, (see 012.003), p. 211 - 231 (1975).
Review paper.

073.012 **High time resolution analysis of solar flares observed on the ESRO TD-1A satellite.**
P. Hoyng, J. C. Brown, G. Stevens, H. F. van Beek.
IAU Symposium No. 68, (see 012.003), p. 233 - 235 (1975).
Summary.

073.013 **Inference of the hard X-ray source dimensions in the 1972, August 7 white light flare.**
D. M. Rust.
IAU Symposium No. 68, (see 012.003), p. 243 (1975).
Summary of a paper which appeared in Solar Physics, Vol. 40, 141 - 157 (1975). – See 13.073.004.

073.014 **Relation of microwave emission to X-ray emission from solar flares.** T. Takakura.
IAU Symposium No. 68, (see 012.003), p. 299 - 313 (1975).
Review paper.

073.015 **High energy gamma-ray radiation above 300 keV associated with solar activity.**
E. L. Chupp, D. J. Forrest, A. N. Suri.
IAU Symposium No. 68, (see 012.003), p. 341 - 359 (1975).
The present status of our knowledge concerning the production of gamma-ray lines and continuum during the impulsive phase of solar flares is reviewed.

073.016 **Gamma-ray lines from solar flares.**
R. Ramaty, R. E. Lingenfelter.
IAU Symposium No. 68, (see 012.003), p. 363 - 383 (1975).
The theory of gamma-ray line production in solar flares is discussed in detail.

073.017 **Fast electrons in small solar flares.** R. P. Lin.
IAU Symposium No. 68, (see 012.003), p. 385 - 409 (1975).
Because ~5–100 keV electrons are frequently accelerated and emitted by the sun in small flares, it is possible to define a detailed characteristic physical picture of these events. This review summarizes both the direct spacecraft observations of non-relativistic solar electrons, and observations of the X-ray and radio emission generated by these particles at the sun and in the interplanetary medium.

073.018 **Nuclei of heavy elements from solar flares.**
C. Y. Fan, G. Gloeckler, D. Hovestadt.

IAU Symposium No. 68, (see 012.003), p. 411 - 421 (1975). Review paper.

073.019 **Implications of NRL/ATM solar flare observations on flare theories.** C. C. Cheng, D. S. Spicer.
IAU Symposium No. 68, (see 012.003), p. 423 - 424 (1975). Summary.

073.020 **Nonthermal processes in large solar flares.** H. S. Hudson, T. W. Jones, R. P. Lin.
IAU Symposium No. 68, (see 012.003), p. 425 - 426 (1975). Summary.

073.021 **On the acceleration processes in solar flares.** Z. Švestka.
IAU Symposium No. 68, (see 012.003), p. 427 - 439 (1975).
The present day knowledge about the acceleration processes on the sun is summarized.

073.022 **The spatial distribution of X-ray, EUV and GHz emissions in connection with the flare-loop-channel model.** L. Křivský, Š. Pintér.
Bull. Astron. Inst. Czechoslovakia, Vol. 26, 199 - 203 (1975).
The flare-loop-channel model of proton flares is compared with interpretations of measured spatial distribution of some sort of radiation sources, especially with X- and EUV-emissions, carried out by the OSO-7 satellite (Neupert et al., 1974).

073.023 **Trends of flare activity in the July 1974 proton region.** L. Křivský.
Bull. Astron. Inst. Czechoslovakia, Vol. 26, 203 - 205 (1975).
Summation curves of flare activity for the proton region S 15°, CMP 3 July 1974, were constructed in the same manner as for similar cases of occurrence of flares with emissions of sub-cosmic and cosmic radiation. As is known, these curves indicate the main trends of activity which last for several days and which describe the time distribution of the release of flare energy. It was again found that the steep trend of increased activity sets on several tens of hours prior to the occurrence of the first proton flare, which could be used, together with other methods, to forecast proton flares.

073.024 **The appearance of chromospheric structures in Hα.** C. J. Durrant.
Mitt. Astron. Ges., No. 36, p. 139 (1975). – Abstract.

073.025 **Spectral investigation of the chromosphere. V. Observation and analysis of Hβ.**
U. Grossmann-Doerth, M. v. Uexküll.
Solar Physics, Vol. 42, 303 - 309 = Mitt. Fraunhofer Inst., *Freiburg*, No. 137 (1975).
The results of the analysis of 16 Hβ spectra taken with high spatial resolution are presented. The properties of the line profile in the boundary regions of supergranular cells (SGB) are described by the statistical distribution functions and correlation coefficients of four parameters which determine the shape of a suitable fitting curve. Although some of the line profile properties of both Hβ and Hα in the SGB can be interpreted in terms of a model which was recently proposed by Durrant (1975) the agreement between that theory and the author's observations is not entirely satisfactory.

073.026 **Les bulles chromosphériques.** Z. Mouradian, G. Simon.
Solar Physics, Vol. 42, 311 - 323 (1975).
Some characteristics of roundish features which are observed on Hα filtergrams, above the solar chromosphere, at the limb are studied. The 'bulles' – chromospheric bubbles – so called for their apparent shape, move upwards into the low corona and fade out when their material is completely ionized. The physical data for nine observed bubbles are listed. The authors discuss the relation between the ascending velocity and the maximum height reached by the bubbles and compare it with the one obtained by Rush and Roberts for spicules. They also compare the properties of bubbles with some already known chromospheric features and there seems to be a link between bubbles and Becker's grains. Finally the authors propose two hypotheses regarding the origin of bubbles.

073.027 **Flare model chromospheres and photospheres.** M. E. Machado, J. L. Linsky.
Solar Physics, Vol. 42, 395 - 420 (1975).
The authors determine semi-empirical models for different types of flares, and then speculate concerning the nature of the heating mechanism. They describe and reduce observations obtained at Sacramento Peak Observatory of the Ca II lines in a variety of flares. They investigate the photospheric structure of flares, and build a grid of chromospheric flare models which approximately reproduces both the observed line profiles and the estimated parameters such as electron density, temperature and linear thickness. Close agreement between the models and the observations is obtained.

073.028 **White light flares: protons or electrons?** P. Cortellessa, C. Paizis.
Solar Physics, Vol. 42, 421 - 424 (1975).
The estimated values of the energy transported by flare accelerated protons and electrons, based on gamma-ray and hard X-ray measurements in space, are compared with the energy emitted in white light from the 1972, August 7, white light event. The aim is to evaluate the relative contribution of protons and electrons in producing the white light enhancement.

073.029 **Annihilation of magnetic fields and acceleration of particles in solar flares.**
S. B. Pikel'ner, V. N. Tsytovich.
Astron. Zhurn. Akad. Nauk SSSR, Vol. 52, 738 - 751 (1975). In Russian. English translation in Soviet Astron., Vol. 19, No. 4.
The process of annihilation of a field in the neutral layer is considered in the framework of ion-sound turbulence hypotheses. The lower limit of the velocity inflow of the field is in this framework about 10^6 cm s^{-1}. The larger the power of the flare the less the thickness of the layer. The thickness depends on the rate of the gas evacuation from the layer. It is shown that the laminar flow along the layer due to the action of magnetic forces is not enough to explain the power of the flares. The gas can be evacuated by the transversal outbursts due to instability of the gas restrained by the field. The presence of the weak field in the high temperature gas does not stop the outbursts as the field in the turbulent medium reconnects and is detached from the layer. The initial disturbances of the density and of the boundary of the layer are necessary for the formation of the outbursts.

073.030 **Gas dynamical models of solar flares.** N. D. Kostyuk.
Astron. Zhurn. Akad. Nauk SSSR, Vol. 52, 752 - 762 (1975). In Russian. English translation in Soviet Astron., Vol. 19, No. 4.
The flux of electrons with the spectrum $E_0^{-\delta}$ ($E_0 \geqslant 10$ kev) falls from the corona into the chromosphere during 100 sec. The distributions of temperature, density, ionization and velocity of the gas are calculated for the values $\delta = 3, 4, 5, 6$ and for four energy flux values 4×10^9, 10^{10}, 10^{11}, 10^{12} ergs cm^{-2} s^{-1} during the heating time and during the cooling time in the open and closed models of the flare region.

073.031 **Excitation atomique par le rayonnement en présence de champs de vitesse dans la chromosphère solaire.**

Y. Cuny.
Comptes Rendus Acad. Sci. Paris, Sér. B, Vol. 281, 117 - 120 (1975).

Dans la chromosphère solaire le rayonnement photosphérique en présence de la vitesse de microturbulence qui élargit les profils d'absorption des raies, et de la vitesse de macroturbulence qui les décale, produit une excitation atomique élevée. Dans le cas du titane ionisé, la faible dépendance avec le multiplet de l'influence des champs de vitesse et l'ordre de grandeur de l'effet obtenu sont en accord avec les observations d'éclipse solaire.

073.032 **On the possibility of photospheric origin of solar flares.** Y. Uchida.
Astrophys. Letters, Vol. 16, 101 - 106 (1975).

Attention is directed to the possibility that the tiny brightenings observed at the start of a flare in the low chromosphere or the upper photosphere could be the source of an active flare.

073.033 **A theoretical model of the solar chromosphere.** M. Marik.
Bull. Astron. Inst. Czechoslovakia, Vol. 26, 317 - 319 (1975).

A theoretical model of the solar chromosphere is presented. The model has been calculated under the assumption of the equality of the emission coefficient of the solar matter and the absorption coefficient of weak magnetodynamic shock waves generated in the convective zone. The results show an increase of temperature with height.

073.034 **Hard electromagnetic radiation in the flare of August 4, 1972.**
G. Vedrenne, M. I. Kudryavtsev, O. B. Likin, A. S. Melioranskij, N. F. Pisarenko, I. A. Savenko, R. Talon, V. M. Shamolin.
Geomagn. Aeronom., Vol. 15, 729 - 731 (1975). In Russian.
Brief information.

073.035 **A resistive screw instability model of a solar flare.** D. S. Spicer.
Bull. American Astron. Soc., Vol. 7, 397 (1975). – Abstr. AAS.

073.036 **A time dependent study of conductive heat flow in a flaring arch.** P. C. Kepple, D. S. Spicer.
Bull. American Astron. Soc., Vol. 7, 397 - 398 (1975). – Abstr. AAS.

073.037 **Plasma heating and flare X-rays.** J. Davis, P. Kepple, D. Strickland, G. Brueckner.
Bull. American Astron. Soc., Vol. 7, 398 (1975). – Abstr. AAS.

073.038 **Solar flare collisional excitation rate coefficients.** R. Tousey, P. C. Kepple, J. Davis.
Bull. American Astron. Soc., Vol. 7, 398 (1975). – Abstr. AAS.

073.039 **Hydrodynamics of electron beam deposition in solar flares.**
H. Bloomberg, P. C. Kepple, J. Davis, J. Boris, G. Brueckner.
Bull. American Astron. Soc., Vol. 7, 398 (1975). – Abstr. AAS.

073.040 **Acceleration and heating in a solar flare model.** G. Van Hoven.
Bull. American Astron. Soc., Vol. 7, 398 (1975). – Abstr. AAS.

073.041 **Stochastic acceleration of ions by MHD turbulence in solar flares.** C. E. Newman.
Bull. American Astron. Soc., Vol. 7, 398 (1975). – Abstr. AAS.

073.042 **Solar flares.** L. W. Acton.
Bull. American Astron. Soc., Vol. 7, 410 (1975).
Abstr. AAS.

073.043 **The build-up and release processes of major flares.** K. Tanaka.
Bull. American Astron. Soc., Vol. 7, 423 (1975). – Abstr. AAS.

073.044 **Flares in ephemeral active regions.** R. L. Moore, F. Tang.
Bull. American Astron. Soc., Vol. 7, 423 - 424 (1975). – Abstr. AAS.

073.045 **Energy release and thermal structure in solar flares.** C. C. Cheng, K. G. Widing.
Bull. American Astron. Soc., Vol. 7, 424 (1975). – Abstr. AAS.

073.046 **Evolution of XUV plasmas in solar flares.** K. G. Widing, C. C. Cheng.
Bull. American Astron. Soc., Vol. 7, 424 (1975). – Abstr. AAS.

073.047 **A comparison of flares in Hα and D_3 (He I).** H. E. Ramsey, S. F. Martin, K. L. Harvey.
Bull. American Astron. Soc., Vol. 7, 424 (1975). – Abstr. AAS.

073.048 **Observation of the structure of solar flares with a soft X-ray telescope.** E. G. Gibson, P. B. Landecker, D. L. McKenzie, J. H. Underwood, J. A. Vorpahl.
Bull. American Astron. Soc., Vol. 7, 424 - 425 (1975). – Abstr. AAS.

073.049 **S-056 observations of soft X-ray flares–implications of triggering mechanism.** J. A. Vorpahl, E. G. Gibson, P. B. Landecker, D. L. McKenzie, J. H. Underwood.
Bull. American Astron. Soc., Vol. 7, 425 (1975). – Abstr. AAS.

073.050 **Skylab/ATM observations of transient events having the GRF X-ray and microwave character.**
N. R. Sheeley, Jr., J. D. Bohlin, V. E. Scherrer, R. Tousey.
Bull. American Astron. Soc., Vol. 7, 429 - 430 (1975). – Abstr. AAS.

073.051 **X-ray observations of the flare of June 15, 1973.** J. H. Underwood.
Bull. American Astron. Soc., Vol. 7, 438 (1975). – Abstr. AAS.

073.052 **Flares and their relation to the photospheric velocity field.** K. L. Harvey, J. W. Harvey.
Bull. American Astron. Soc., Vol. 7, 438 (1975). – Abstr. AAS.

073.053 **Photometric properties of soft X-ray kernels.** S. Kahler, R. Petrasso, R. Simon.
Bull. American Astron. Soc., Vol. 7, 438 (1975). – Abstr. AAS.

073.054 **An observation of identical hard X-ray bursts from sympathetic flares.** K. Frost, K. Ohki.
Bull. American Astron. Soc., Vol. 7, 438 - 439 (1975). – Abstr. AAS.

073.055 **Radio observations of hard X-ray sympathetic flares and a possible explanation.** K. Ohki, K. Frost.
Bull. American Astron. Soc., Vol. 7, 439 (1975). – Abstr. AAS.

073.056 **On the occurrence of sympathetic flares.** L. Fritzová-Švestková, R. Chase.
Bull. American Astron. Soc., Vol. 7, 439 (1975). – Abstr. AAS.

073.057 **Propagation characteristics of solar flare particles.** L. S. Ma Sung, M. A. Van Hollebeke, F. B. McDonald.
Bull. American Astron. Soc., Vol. 7, 439 (1975). – Abstr. AAS.

073.058 **Transequatorial loops interconnecting McMath 12472 and 12474.** Z. Švestka, A. Krieger.
Bull. American Astron. Soc., Vol. 7, 444 (1975). – Abstr. AAS.

073.059 **Solar He II – a reappraisal.**
G. L. Epstein, R. Steinitz.
Bull. American Astron. Soc., Vol. 7, 450 - 451 (1975). – Abstr. AAS.

073.060 **Extreme ultraviolet solar limb brightening observations.** J. T. Mariska, G. L. Withbroe.
Bull. American Astron. Soc., Vol. 7, 460 (1975). – Abstr. AAS.

073.061 **Helical motion in an eruptive prominence.**
S. F. Martin, R. T. Hansen.
Bull. American Astron. Soc., Vol. 7, 472 (1975). – Abstr. AAS.

073.062 **Thermal instability in loop prominence systems.**
S. K. Antiochos, P. A. Sturrock.
Bull. American Astron. Soc., Vol. 7, 472 - 473 (1975). – Abstr. AAS.

073.063 **Observations at 6 cm of the solar active region.**
M. R. Kundu, C. E. Alissandrakis.
Nature, Vol. 257, 465 - 467 (1975). – Letter.

073.064 **The interpretation of the upper cutoff in the proton energy spectrum of solar flares.** J. Pérez-Peraza.
Publ. Astron. Soc. Pacific, Vol. 87, 508 (1975). – Abstr. ASP.

073.065 **The velocity dependence of the source energy spectrum of solar particles accelerated in the high-density medium of solar flares.**
J. Pérez-Peraza, J. G. Trejo, R. L. Alvarez.
Publ. Astron. Soc. Pacific, Vol. 87, 508 - 509 (1975).
Abstr. ASP.

073.066 **On relative intensities of helium and hydrogen lines in quiescent prominences. I. Homogeneous structure with variable electron density.** N. N. Morozhenko.
Astrometriya i Astrofizika, *Kiev*, vyp. (No.) 26, (see 003.001), p. 69 - 77 (1975). In Russian.

Variation in intensity ratio is considered for helium and hydrogen lines in quiescent prominences of various brightness. The observations are compared with calculations obtained for a prominence model with homogeneous structure and variable electron density. The calculations are in good agreement with the observations of prominences of faint and moderate brightness.

073.067 **The geometry of the chromosphere-corona transition region inferred from the center-to-limb variation of the radio emission.** M. Kanno, R. Tanaka.
Solar Physics, Vol. 43, 63 - 77 (1975).

Based on the observations of the EUV spectroheliograms, the effective chromosphere-corona transition region is assumed to be restricted in a small volume element in the boundaries of the supergranular network. The center-to-limb variation of the quiet sun at cm and dm wavelengths is analyzed to determine where the transition region is located in the network boundaries. Comparison with the observations shows that the spicule-sheath model (Brueckner and Nicolas, 1973) and the hot plagette model (Foukal, 1974) are not compatible with the observations. A new picture is therefore proposed that thin platelet transition regions are placed on top of the chromosphere and scattered between the network boundaries (the platelet transition-region model). This model is in accord with the observed center-to-limb variation of the radio emission.

073.068 **The chromospheric Evershed flow.** P. Maltby.
Solar Physics, Vol. 43, 91 - 105 (1975).

High-resolution filtergrams of a sunspot region observed at seven wavelengths in Hα with a resolution of 1/8 Å have been used to investigate the three-dimensional structure of the chromospheric Evershed flow and its time variation.

073.069 **Observation of possible Fe XVII $2p^5 3p(^1S_0)$–$2p^5 3s(^1P_1, ^3P_1)$ transitions in spectra of a solar active region and flare.** S. O. Kastner, W. M. Neupert, M. Swartz.
Solar Physics, Vol. 43, 111 - 115 (1975).

Expected wavelengths and relative intensities are obtained, from calculations of other workers, for the hitherto unobserved transitions $2p^5 3p(^1S_0)$–$2p^5 3s(^1P_1, ^3P_1)$ in Fe XVII. A candidate pair of lines at 197.05 Å and 242.09 Å was found in laboratory spectra and appears to be present in the spectra of a solar active region observed by the OSO-7 Goddard spectroheliograph, and in the spectrum of a Skylab-observed solar flare.

073.070 **EUV analysis of an active region.**
N. Raghavan, G. L. Withbroe.
Solar Physics, Vol. 43, 117 - 128 (1975).

The present investigation is an attempt to study the coronal structure of an individual active region when it was near the limb. Through use of limb observations, information on the variation of coronal temperature and density with height can be obtained.

073.071 **The work of the diode array: He 10830 observations of spicules and subflares.**
D. M. Rust, C. A. Bridges III.
Solar Physics, Vol. 43, 129 - 145 (1975).

The chromospheric fine structure was studied with an array of 512 silicon photodiodes. The high quantum efficiency of the diodes in the near infrared allowed low noise spectroheliograms to be constructed from observations in the lines of H I, Ca II and He I. Magnetograms of the underlying photosphere were obtained simultaneously. Tachograms in the He 10830 Å line revealed 1″ points and elongated features that are interpreted as spicules seen against the disk. Active regions and filaments at 10830 are compared with Hα and Ca II (8542 Å) features. Twelve subflares were studied and in eight cases, 3–5″ kernels of He I emission appeared over small, growing pores or over 5″ patches of magnetic field emerging through the photosphere. All the subflares showed 10830 emission, contrary to established belief that the 10830 line goes into emission only in the largest flares.

073.072 **Heating and cooling of the thermal X-ray plasma in solar flares.** R. L. Moore, D. W. Datlowe.
Solar Physics, Vol. 43, 189 - 209 (1975).

Temperatures and emission measures derived from UCSD OSO-7 X-ray flare observations are used, and length scales are obtained from Big Bear large-scale Hα filtergrams for 17 small (subflare to Class 1) flares. The empirical values obtained for the characteristic times imply (1) that flares are produced by magnetic field reconnection, (2) that conduction cooling of the thermal X-ray plasma dominates radiative cooling and (3) that reconnection heating and conduction cooling of the thermal X-ray plasma are approximately in balance at thermal X-ray maximum. This model in combination with the data gives estimates for the electron number density (10^{10}–10^{11} cm^{-3}) and the magnetic field strength (10–100 G) in the thermal X-ray plasma and for the total thermal energy generated in a subflare ($\approx 10^{30}$ erg for an Hα area ≈ 1 square degree) which agree with previous observational and theoretical estimates obtained by others.

073.073 **Thermal phase of development of a loop prominence.** P. V. Sasorov.
Astron. Tsirk., No. 849, p. 1 - 3 (1974). In Russian.

073.074 **Variation in prominence height with the solar activity cycle.** V. Dermenjiev.
Astron. Tsirk., No. 868, p. 7 - 8 (1975). In Russian.

073.075 **On the interpretation of bright mottles from ultra-**

violet spectroheliograms. Eh. V. Kononovich.
Astron. Tsirk., No. 870, p. 3 - 6 (1975). In Russian.

073.076 **On the relation between proton-flares and magnetic configurations of sunspot groups.**
Z. Shi, Y. Lin, G. Ai, S. Qian, R. Zhao.
Acta Astron. Sinica, Vol. 16, 12 - 18 (1975). In Chinese.

The magnetic configurations of sunspot groups are separated into two classes, normal and abnormal. The normal class includes typical configurations of ordinary bipolar groups, in which spots are arranged nearly along the equator with preceding spots at lower latitudes, and their polarities conform to the well-known Hale's rule. The abnormal class includes δ-configuration, antipolarity, preceding spots at higher latitude and magnetic axis inclination greater than 30°. Study of the flare statistics in the period 1964 Oct.–1972 Dec. of the 20th cycle shows that 85% of the proton flares occurred in sunspot groups which have abnormal configurations 1–3 days before.

073.077 **An observation of a solar flare in white light on 1974 July 3.** K. J. Medway.
Journ. British Astron. Ass., Vol. 85, 423 (1975).

073.078 **A new method for determining Doppler widths.**
R. Steinitz, V. Bar, S. Wyckoff.
Astrophys. Journ., Vol. 201, 729 - 734 (1975).

An explicit analytical relation between the absorption-line profile and the emergent intensity is developed for a stellar atmosphere in which the absorption profile and the source and sink terms of the source function are depth-independent. Using this relation, a method is described for determining the Doppler width of the solar chromospheric Hα absorption profile. Finally, the method for determining the Doppler width is generalized for application to stellar spectra.

073.079 **Spatial distribution of XUV emission in solar flares.**
C.-C. Cheng, K. G. Widing.
Astrophys. Journ., Vol. 201, 735 - 739 (1975).

The spatial distribution and time evolution of flare plasmas are studied with observations obtained with the NRL XUV spectroheliograph aboard Skylab. Comparison between spectroheliograms and magnetograms shows that the basic magnetic field configuration associated with the flares observed is a low-lying (4000 to 13,000 km) loop or arch structure. The flare apparently results from a plasma instability in the loop which produces localized heating in a small volume, usually near the top of the loop. Geometrical and physical properties of the hot flare plasma are presented, and implications of the observational results are briefly discussed.

073.080 **On the recombination mechanism of 'white' flare emission.** V. A. Kovalev, O. S. Korolev.
Fiz. ionosfery, magnit. pole Zemli i solnechno-zemn. svyazi. Moskva, 1974, p. 54 - 66. In Russian. – Abstr. in Referativ. Zhurn. 51. Astron., 9.51.442 (1975).

073.081 **Anisotropy of solar flare activity in the inertial space and resonance ability of the solar system.**
V. P. Kozelov, G. I. Mingaleva.
Subburi i vosmushcheniya v magnitosfere. Leningrad, Nauka, 1975, p. 264 - 274. In Russian. – Abstr. in Referativ. Zhurn. 51. Astron., 9.51.467 (1975).

073.082 **On the seasonal effect of solar flare activity.**
V. P. Kozelov.
Subburi i vosmushcheniya v magnitosfere. Leningrad, Nauka, 1975, p. 274 - 282. In Russian. – Abstr. in Referativ. Zhurn. 51. Astron., 9.51.468 (1975).

073.083 **Detection of the microwave solar "chromospheric net".** V. M. Bogod, D. V. Korol'kov.
Pis'ma v Astron. Zhurn., Vol. 1, No. 10, p. 25 - 27 (1975). In Russian.

Small-scale brightness fluctuations on the quiet sun are detected on the RATAN-600 radiotelescope at wavelengths 2–4 cm. The typical angular size of fluctuations is about 12″ and the corresponding flux density about 10^{-5} of the total flux from the sun.

073.084 **On the relative intensities of helium and hydrogen lines in quiescent prominences. II. Structure-filamentary model.** N. N. Morozhenko.
Astrometriya i Astrofizika, *Kiev*, vyp. (No.) 27, (see 003.004), p. 75 - 85 (1975). In Russian.

A diffusion penetration of ionized helium radiation in filamentary structured prominences is considered. When consistency was reached of calculations and observations the parameters of thin filamentary structure of prominences were found.

073.085 **Physical conditions in a flare-like prominence.**
E. G. Rudnikova.
Astrometriya i Astrofizika, *Kiev*, vyp. (No.) 27, (see 003.004), p. 86 - 93 (1975). In Russian.

The parameters $T_e = 9700$ K, $n_e = 1.2 \times 10^{11}$ cm^{-3}, $\tau(\mathrm{H}\alpha) = 213$, $\tau(\mathrm{H, Ca^+}) = 31$, $l = 6.6 \times 10^8$ cm for the first two hydrogen levels and 4S, 4P, Ca^+ terms are determined for a flare-like prominence by means of a self-consistent solution of steady-state equations.

073.086 **Die Sonnenaktivität im Juli und August 1975.**
M. Seiler.
SuW, Vol. 14, 373 - 374 (1975).

073.087 **Solar He^3 information from nuclear reactions in flares.** R. Ramaty, B. Kollovsky.
VI Leningr. mezhdunar. seminar, 1974. Uskorenie chastits i yader. reaktsii v kosmose. Leningrad, 1974, p. 58 - 83, with a discussion, p. 84. – Abstr. in Referativ. Zhurn. 51. Astron., 8.51.487 (1975).

073.088 **Gamma-ray lines from solar flares.**
R. Ramaty, R. E. Lingenfelter.
VI Leningr. mezhdunar. seminar, 1974. Uskorenie chastits i yader. reaktsii v kosmose. Leningrad, 1974, p. 25 - 56, with a discussion, p. 56 - 57. – Abstr. in Referativ. Zhurn. 51. Astron., 8.51.490 (1975).

073.089 **Particle acceleration in processes of solar flare type.**
S. I. Syrovatskij.
VI Leningr. mezhdunar. seminar, 1974. Uskorenie chastits i yader. reaktsii v kosmose. Leningrad, 1974, p. 226, 227, with a discussion, p. 227 - 228. In Russian and English. – Abstr. in Referativ. Zhurn. 51. Astron., 8.51.500 (1975).

073.090 **Shock wave with light emission and a burst phase of solar flares.** A. A. Rumyantsev.
VI Leningr. mezhdunar. seminar, 1974. Uskorenie chastits i yader. reaktsii v kosmose. Leningrad, 1974, p. 233, 234, with a discussion, p. 235. In Russian and English. – Abstr. in Referativ. Zhurn. 51. Astron., 8.51.503 (1975).

073.091 **Invariable value of τ_{10830} in plages and helium emission in the EUV radiation field.** M. A. Livshits.
Astron. Zhurn. Akad. Nauk SSSR, Vol. 52, 970 - 974 (1975). In Russian. English translation in Soviet Astron., Vol. 19, No. 5.

The n_2 population of the metastable 2^3S level has been estimated taking into account helium ionisation by radiation with $\lambda < 504$ Å ($T \leqslant 20\,000°$). It has been found that n_2 is limited with increasing electron density n_e, unlike the case of $n_2 \propto n_e$ in the alternative high-temperature model. The data of the absorption line λ 10830, in particular the observed N_{2^3S} value in plages and the fact that it is practically constant for different filaments, gives evidence in favour of a low temperature mechanism of helium emission.

073.092 **Why few Balmer series lines are observed in flares on the disk.** L. N. Kurochka, G. F. Sitnik.
Astron. Zhurn. Akad. Nauk SSSR, Vol. 52, 975 - 977 (1975). In Russian. English translation in Soviet Astron., Vol. 19, No. 5.

It is shown that upper level populations of hydrogen atoms in flares are distributed according to the statistical weights of levels. The optical thicknesses in distant Balmer lines are much less than one. Thanks to these two conditions the intensities of high lines of Balmer series must fall rapidly with decrease of the number, that leads to the small number of observed lines in flares on the disk. The second level population of hydrogen atoms and measures of emission in flares are estimated.

073.093 **Observation of the flare on the Observatory Hvar 7. August 1972.** V. Vujnović, V. Ruždjak.
Publ. Obs. Astron. Beograd, No. 20, (see 012.024), p. 52 - 54 (1975). In Serbo-Croatian.

073.094 **Solar flares in August 1972 followed by atmospheric phenomena of the earth.** M. Vukićević-Karabin.
Publ. Obs. Astron. Beograd, No. 20, (see 012.024), p. 55 - 61 (1975). In Serbo-Croatian.

073.095 **Emerging flux regions.** D. L. Glackin.
Solar Physics, Vol. 43, 317 - 326 (1975).

After a brief review of the properties of emerging flux regions (EFRs), the author reports on new observations of the birth of an EFR and apparent subsequent field reconnection. The author reviews fluxrope theories, predicts the appearance of EFRs, then modifies this picture on the basis of high resolution observations. He arrives at a model of this phenomenon that encompasses relevant aspects such as the axial tilt of spot groups, the observed 'rotation' of EFRs as they emerge, and the fact that EFR fluxtubes are made up of many discrete strands. The relation of the positions of emergence of EFRs to the chromospheric network is investigated.

073.096 **The temperature structure and pressure balance of magnetic loops in active regions.** P. Foukal.
Solar Physics, Vol. 43, 327 - 336 (1975).

EUV observations show many active region loops in lines formed at temperatures between 10^4K and 2×10^6K. The brightest loops are associated with flux tubes leading to the umbrae of sunspots. It is shown that the high visibility of certain loops in transition region lines is due principally to a sharp radial decrease of temperature to chromospheric values toward the loop axis. The plasma density of these cool loops is not significantly greater than in the hot gas immediately surrounding it. When the active region is examined in coronal lines, the hot high pressure plasma shows up as loops that are generally parallel to the cool loops but significantly displaced laterally. In general the loop phenomenon in an active region is the result of temperature variations by two orders of magnitude and density variations of around a factor five between adjacent flux tubes in the corona.

073.097 **The solar-flare infrared continuum.**
K. Ohki, H. S. Hudson.
Solar Physics, Vol. 43, 405 - 414 (1975).

The authors consider potential sources of infrared (1μ to 1 mm) continuum in solar flares. Several mechanisms should produce detectable fluxes: in the 350μ window for ground-based observations, impulsive emission will arise in synchrotron radiation from 1–10 MeV electrons, and possibly thermal (free-free) continuum from the source of the white-light flare; the hot flare plasma responsible for soft X-ray emission will also emit detectable fluxes of free-free continuum in the largest flares. At shorter wavelengths the dominant infrared emission will come from the Hα flare itself. Observations in the infrared wavelengths will help to complete our picture of flare structure in both the impulsive and gradual phases.

073.098 **On the problem of determining the physical conditions in flocculi and the chromospheric net.**
Eh. A. Baranovskij.
Izv. Krymskoj Astrofiz. Obs., Vol. 53, p. 75 - 81 (1975). In Russian.

Temperature and density are determined in a flocculus by comparison of the observed and calculated profiles of Hα and K Ca II lines. It is found that the density in the flocculus (10^{13}–10^{14}) is higher, in the region above the flocculus ($\sim 10^{10}$) lower as compared to the undisturbed chromosphere. The temperature in the flocculus is found to be 6500–7500°. There is an inversion of electron density. It is shown that the appearance of the chromospheric net can be explained in terms of a two-component chromospheric model, the hot component having higher temperature for heights >1000 km.

073.099 **Some features of the chromospheric structure.**
T. T. Tsap.
Izv. Krymskoj Astrofiz. Obs., Vol. 53, p. 95 - 103 (1975). In Russian.

The chromospheric fine-structure in the Hα, Hβ, Hγ, Hδ and K Ca II lines is studied. It is found that there is good accordance between the position of dark and bright details of the chromospheric fine structure in active and quiet regions.

073.100 **Birth and development of an active region and the chromospheric network.** L. G. Kartashova.
Izv. Krymskoj Astrofiz. Obs., Vol. 53, p. 108 - 115 (1975). In Russian.

The birth and development of 15 active regions was studied on the basis of observations in the centre and the wings of the Hα line. The studied regions have simple bipolar configuration with spots without penumbra. Changes in the chromospheric network are apparently connected with the formation of filament structure around active regions.

073.101 **Forecast of plage activity from radio emission at λ 9.1 cm.**
N. N. Stepanyan, Z. A. Shcherbakova.
Izv. Krymskoj Astrofiz. Obs., Vol. 53, p. 116 - 120 (1975). In Russian.

In search of a correspondence between the radio brightness temperature at λ 9.1 cm near the east limb and the power of the plage appearing the next day at that latitude observational data for 1967–1969 were used. The correlation coefficient of these values is found to be 0.84 ± 0.03. An increase of radio emission 1–2 days before the plage birth on the solar disk has been found for 30% of all plages born on the disk. The correlation coefficient between radio emission on the day of plage birth and its power the next day is found to be 0.68 ± 0.07.

073.102 **Hard X-ray coverage of the Skylab period by Utrecht's S-100 instrument on board ESRO TD-1A.**
J. C. Brown.
Oss. Mem. Oss. Astrofis. Arcetri, Fasc. 104, (see 012.025), p. 89 - 92 (1975).

The author indicates the type of data provided by the S-100 hard X-ray spectrometer of the Space Research Laboratory, Utrecht, and the potential importance of these data in theoretical interpretation of flares when combined with observations at other wavelengths.

073.103 **Objective grating studies of X-ray flare spectra.** J. K. Silk, S. Kahler, A. S. Krieger, A. F. Timothy, G. S. Vaiana.
Oss. Mem. Oss. Astrofis. Arcetri, Fasc. 104, (see 012.025), p. 143 - 156 (1975).

The objective grating mode of the A.S.E. X-ray telescope on Skylab provides dispersed and zero order images of coronal features which can be used to determine the temperature of the emitting plasma. Variations of the temperature with time and with position in the flaring region are studied. The M2 flare of 9 August, 1973 is used to illustrate the data obtained with the objective grating. A preliminary analysis of the data suggests a two temperature model for the flare.

073.104 **X-ray and radio emission for the June 15, 1973 solar flare.** R. Pallavicini, S. Kahler, A. S. Krieger, J. K. Silk, G. S. Vaiana.
Oss. Mem. Oss. Astrofis. Arcetri, Fasc. 104, (see 012.025), p. 157 - 168 (1975).

The flare was observed with high spatial and temporal resolution by the grazing-incidence X-ray telescope prepared by American Science and Engineering and mounted on ATM (*SO* 54 experiment; Vaiana et al. 1973). Ground-based radio, optical and magnetogram observations are also available as well as X-ray data from other satellite instruments. Physical parameters for the flare region have been determined.

073.105 **Optical study of June 15, 1973 flare.** G. Godoli, V. Sciuto, R. A. Zappalà.
Oss. Mem. Oss. Astrofis. Arcetri, Fasc. 104, (see 012.025), p. 169 - 174 (1975).

073.106 **Optical and magnetic study of the active regions ATM 131 and 137 at the photospheric and chromospheric levels.** G. Godoli, V. Sciuto, R. A. Zappalà.
Oss. Mem. Oss. Astrofis. Arcetri, Fasc. 104, (see 012.025), p. 192 - 201 (1975).

073.107 **Inclination of solar quiescent prominences and coronal extrapolation of photospheric magnetic fields.** G. Godoli, S. Motta, V. Sciuto, R. A. Zappalà.
Oss. Mem. Oss. Astrofis. Arcetri, Fasc. 104, (see 012.025), p. 204 - 208 (1975).

073.108 **Positron annihilation in solar flares.** C. J. Crannell, R. Ramaty, C. Werntz.
Goddard Space Flight Center, Greenbelt, Maryland, GSFC Document X-682-75-167, 12 pp. (1975).

The various processes that affect the annihilation of positrons are discussed. In particular, since a large fraction of the positrons may annihilate from bound states of positronium, the authors investigate the question of whether positronium annihilation radiation is observable from solar flares.

073.109 **On excitation of the lines of the helium singlet series in prominences and the chromosphere.**
N. N. Morozhenko.
Solnechnye Dannye 1975 Byull., No. 3, p. 81 - 87 (1975). In Russian.

The excitation of the 2^1P, 3^1D, 5^1D, 6^1D and 7^1D levels of helium in quiet prominences and the chromosphere is considered. The lower levels of the singlet system, similar to those of hydrogen, are shown to be excited by resonance (λ 584 Å) and photospheric emissions. The upper levels are populated by the transits from a triplet system so that the relative condition of LTE is established between the systems for $n \geqslant 6$.

073.110 **Birth and motion of spicules.** S. G. Mamedov, Eh. Sh. Orudzhev.
Solnechnye Dannye 1975 Byull., No. 5, p. 92 - 101 (1975). In Russian.

Data on some peculiarities of sudden birth and motion of spicules according to observations in the Hα line are presented. The spicule birth can be explained by the mechanism of condensation of coronal gas. Some arguments in support for the mechanism are given. Sudden variations of radial velocities along the full spicule length are shown to be due to the motion of a spicule as a whole, and not to the plasma motion in the magnetic tube of the spicule. Some peculiarities of the spicule group motion are considered.

073.111 **Determination of n_e, n_1 in the spicules from the Hα emission and spicule optical thickness at the frequencies of this line.** S. G. Mamedov.
Solnechnye Dannye 1975 Byull., No. 6, p. 84 - 91 (1975). In Russian.

A method of estimate of the spicule optical thickness in the Hα line is described. The method is based on the solution of the statistical equilibrium equation from the population of the third hydrogen level. A new method of n_e determination from the intensity of the Hα line only is suggested. An attempt is made to estimate n_1 from the equation of statistical equilibrium for the first hydrogen level.

073.112 **The results of observations of chromospheric spicules in the H_8 and 3889 Å He I lines.**
V. V. Makarova, V. I. Makarov.
Solnechnye Dannye 1975 Byull., No. 6, p. 100 - 106 (1975). In Russian.

From observations of spicules in the 3889 Å line it was found that the intensity of emission of hydrogen and helium varies very strongly and attains the maximum at the height of 1200 km. The spectra of the chromosphere at heights of 700–4000 km were studied. The evolution of the temperature upper limit according to a single spicule gives values of T = 10000°–13000°. A formula of the relation between the intensities of helium and hydrogen has been derived.

073.113 **The Koobi Fora experiment: continuum observations of solar spicules during the 30 June 1973 eclipse.** R. A. Kopp, O. R. White, T. G. Baur.
Astron. Astrophys., Vol. 44, 299 - 304 (1975).

Limb observations of chromospheric spicules in the continuum near 4700 Å were obtained during the solar eclipse of 30 June 1973, in an experiment conducted close to the edge of the path of totality in northern Kenya. The experimental apparatus and observing techniques which were employed are described and some results of the data analysis are presented.

073.114 **Time-dependent 2.2-MeV and 0.5-MeV lines from solar flares.** H. T. Wang, R. Ramaty.
Astrophys. Journ., Vol. 202, 532 - 542 (1975).

The time dependences of the 2.2- and 0.51-MeV γ-ray lines from solar flares are calculated, and the results are compared with observations. The authors' main results are that the calculation is consistent with the observed rise of the 2.2-MeV line and it does not require different time dependences for the accelerated protons and high-energy electrons in the flare region. Positrons of energies greater than about 10 MeV could be detected in interplanetary space following large solar flares.

073.115 **The spectrum of multiple ionized Fe in the 1.7 - 2.0 Å region and its comparison with the solar flare spectrum.** Eh. Ya. Gol'ts, I. A. Zhitnik, Eh. Ya. Kononov, S. L. Mandel'shtam, Yu. V. Sidel'nikov.
IV Vses. konf. po spektroskopii vakuum. ul'trafioleta i vzaimodejstviyu izlucheniya s veshchestvom, 1975. Tezisy dokl. Uzhgorod, 1975, p. 96. In Russian. – Abstr. in Referativ. Zhurn. 51. Astron., 10.51.425 (1975).

073.116 **Plasma instabilities in chromospheric flares and cosmic ray generation.**
Ya. I. Kolesnichenko, A. K. Yukhimchuk.
Geofiz. i astron. Inform. byull., 1975, No. 14, p. 196 - 199. In Russian. – Abstr. in Referativ. Zhurn. 51. Astron., 10. 51. 431 (1975).

073.117 **On applying the pattern identification method to classification of brightness variations in chromospheric objects.** M. I. Fisenko.
Solnechnye Dannye 1975 Byull., No. 7, p. 79 - 86 (1975). In Russian.

The cybernetic pattern identification method has been applied to the classification of brightness variations in chromospheric objects. The method is based on the Tanimoto algorithm. The classification is made in accordance with the degree of similarity to the reference objects.

073.118 **Investigations according to the programme "Rapid variations of solar magnetic fields", VI. Daily variations of flare activity and variations of magnetic fields of sunspots.** A. Krüger, H. Künzel, G. F. Vyal'shin.
Solnechnye Dannye 1975 Byull., No. 7, p. 87 - 90 (1975). In Russian.

Characteristic pulses of solar flare and microwave burst production separated by days of weaker activity have been investigated on the basis of three selected active regions occurring in August 1971, July 1972, and June/July 1973. Typical time intervals between different peaks of activity are of the order of 5–10 days. A satisfactory correspondence of the flare activity of the investigated test regions with that of other regions has been found.

073.119 **Proton flares of July 1974 generating interplanetary shock waves.** L. Křivský, Š. Pinter.
Bull. Astron. Inst. Czechoslovakia, Vol. 26, 360 - 364 (1975).

The paper contains analysis of the proton flare activity in July 1974 especially with respect to interplanetary shock waves. The time of start of shock waves at the sun and of their arrival to the earth are determined; in one case a decelerating wave is traced to the distance of about 5 AU.

073.120 **Resonance phenomena during propagation of magnetosound waves in the chromosphere of the sun.**
M. F. Bakhareva.
Geomagn. Aeronom., Vol. 15, 961 - 969 (1975). In Russian.

073.121 **XUV emission from above the solar west limb near 20:00 UT, January 17, 1974.** R. Tousey.
Astrophys. Space Sci., Vol. 38, 327 - 336 (1975). – Paper presented at the I.A.U. Colloquium No. 27 on 'UV and X-ray spectroscopy of astrophysical and laboratory plasmas', Cambridge, Mass., U.S.A., 9–11 September 1974.

A complex eruptive event took place on January 17, 1974 at approximately 19:20 UT in McMath Hulbert Plage No. 12686, located near N 06° and behind the west limb. The effects were recorded by the XUV spectroheliograph on Skylab from 19:44 UT to 20:46 UT. An eruptive prominence or spray was observed in He II 304 Å to reach $>1\ R_s$. It was also photographed in many emission lines at temperatures reaching 1 MK in Mg IX. There were at least four other parts of the eruption, each changing character differently with temperature. These first appeared with Ne VII at 0.6 MK, and were recorded in lines of ions up to Fe XVI at 2.6 MK.

073.122 **The brightness of the helium D_3 line in the undisturbed chromosphere from eclipse observations.**
R. A. Gulyaev.
Solar Physics, Vol. 44, 25 - 40 (1975).

Slitless spectra of the chromosphere, observed cinematographically at the total solar eclipse of 10 July 1972, were reduced. The surface brightness distribution of the helium D_3 line in the undisturbed chromosphere was obtained in agreement with results by other observers. The available eclipse data on the D_3 absolute brightness was analysed by means of theoretical curves of growth. Intensity data by some observers were found to be certainly too high. A trend was found that the D_3 absolute brightness in the quiet chromosphere decreases with the increasing solar activity (sunspot number). This perhaps indicates a variation of the spicule number over the solar surface during a sunspot cycle.

073.123 **Spectroscopic investigation of the chromosphere. IV: A reassessment of the cloud model.**
C. J. Durrant.
Solar Physics, Vol. 44, 41 - 53 = Mitt. Fraunhofer Inst., *Freiburg,* No. 133 (1975).

The cloud model employed in the analysis of chromospheric contrast profiles is subject to two criticisms. The source function in the cloud may not be varied independently of the Doppler width in the case of Hα and the radiative coupling between the cloud and the underlying atmosphere cannot be ignored. These criticisms are investigated quantitatively with two simple extreme models. It is found that by taking account of both effects the cloud model may be reinstated. Observed chromospheric features may be understood in terms of clouds of varying parameters embedded in the uppermost regions of a generally undisturbed homogeneous atmosphere. The variable cloud parameters are the optical thickness, the Doppler width, the bulk velocity and the angular size viewed from the line forming regions of the underlying atmosphere.

073.124 **Partial analysis of the flare-prominence of 30 April 1974.**
S. T. Wu, M. Dryer, P. S. McIntosh, E. Reichmann.
Solar Physics, Vol. 44, 117 - 133 (1975).

A portion of an east limb flare-prominence observed in Hα is analyzed. Following a rapid (~2 min) achievement of a maximum mass ejection velocity of about 375 km s^{-1}, the ascending prominence reached a height of, at least, 2×10^5 km. The authors use a one-dimensional, time-dependent hydrodynamic theory (Nakagawa et al., 1975) to compute the total mass ($\sim 2 \times 10^{11}$ g) and energy ($\sim 4 \times 10^{26}$ erg) ejected. Theoretical aspects of the coronal response are discussed.

073.125 **Asymmetries of the solar Ca II lines.**
J. N. Heasley.
Solar Physics, Vol. 44, 275 - 292 (1975).

A theoretical study of the influence of propagating acoustic pulses in the solar chromosphere upon the line profiles of the Ca II resonance and infrared triplet lines has been made. The major objective has been to explain the observed asymmetries seen in the cores of the H and K lines and to predict the temporal behavior of the infrared lines caused by passing acoustic or shock pulses. The velocities in the pulses, calculated from weak shock theory, have been included consistently in the non-LTE calculations.

073.126 **Wave systems in the chromosphere.**
R. Giovanelli.
Solar Physics, Vol. 44, 299 - 314 (1975).

Observations are reported of two, possibly three, distinct wave systems in the Hα chromosphere. Alfvén waves: Waves associated with fibrils originating in the network; Waves associated with superpenumbral fibrils; Nature of the wave system. Vertical waves in the chromospheric granulation. Horizontal wave propagation in the chromospheric granulation. Discussion: Manner of generation of the Alfvén waves; The source of coronal heating.

073.127 **A distinctive type of ascending prominence – 'fountain'.**

E. Tandberg-Hanssen, R. T. Hansen, A. C. Riddle.
Solar Physics, Vol. 44, 417 - 427 (1975).

There is a well-defined sub-class of ascending prominences characterized by closed-system transference of chromospheric material along an arch or loop. These prominences are denoted as 'fountains'. Several examples are described. Fountains appear to be well contained by coronal magnetic fields. Their total kinetic energy is in the order of 10^{30} erg but dissipation is typically quite slow (over time periods like 100 min) so that the correlative disturbances (radio bursts, coronal transients, chromospheric brightenings, etc.) are generally unspectacular or non-existent.

073.128 **Prominence eruption accompanied by twist readjustment.** K. Jockers, O. Engvold.
Solar Physics, Vol. 44, 429 - 433 (1975).

Observations of two eruptive prominences are discussed. The authors find evidence for twisting motions in the early stages of the eruption. This twisting may be caused by a mechanism recently discussed by Parker (1974) and by Jockers (1976) wherein a magnetic flux tube, part of which has expanded because of reduction of external pressure, readjusts the magnetic torque along its axis.

073.129 **Observation of a metallic prominence on October 15 and 16, 1969.**
Ts. S. Khetsuriani, M. Sh. Gigolashvili.
Byull. Abastumansk. Astrofiz. Obs., No. 46, p. 189 - 196 (1975). In Russian.

The mean values of electron concentration $N_e = 10^{11}$ cm^{-3}, kinetic and electron temperatures $T_e = T_k = 6000$ K and turbulent velocity $\xi_t = 3$ km/sec computed with different methods are given for a certain height of the prominence.

073.130 **Chromospheric flares in white light.** L. Křivský.
Říše hvězd, Vol. 56, 205 - 206 (1975). In Czech.

073.131 **The isotopic composition of hydrogen and helium in solar particle events.**
J. D. Anglin, W. F. Dietrich, J. A. Simpson.
Bull. American Astron. Soc., Vol. 7, 521 (1975). – Abstr. AAS.

073.132 **Velocities in the solar chromosphere observed in the Si II λ1816 line.** E. G. Chipman, E. C. Bruner, Jr., R. A. Shine, B. W. Lites, G. J. Rottman, O. R. White, R. G. Athay.
Bull. American Astron. Soc., Vol. 7, 522 (1975). – Abstr. AAS.

073.133 **Persistent velocity fields in the middle chromosphere.** B. W. Lites, E. C. Bruner, Jr., E. G. Chipman, R. A. Shine, G. J. Rottman, O. R. White, R. G. Athay.
Bull. American Astron. Soc., Vol. 7, 522 (1975). – Abstr. AAS.

073.134 **The formation of the He I λ584 and He II λ304 resonance lines in the solar transition region.**
K. B. Gebbie, R. Steinitz.
Bull. American Astron. Soc., Vol. 7, 523 (1975). – Abstr. AAS.

073.135 **EUV observations of the prominence-corona interface.** E. J. Schmahl, F. Q. Orrall.
Bull. American Astron. Soc., Vol. 7, 523 (1975). –Abstr. AAS.

073.136 **On the spatial structure of prominence velocity fields.** L. W. Ramsey.
Bull. American Astron. Soc., Vol. 7, 524 (1975). – Abstr. AAS

073.137 **The X-ray spectrum of a solar limb surge.**
R. S. Wolff, G. G. Cohen, H. L. Kestenbaum, M. C. Weisskopf, P. B. Landecker.
Bull. American Astron. Soc., Vol. 7, 536 (1975). – Abstr. AAS.

073.138 **Interpretation of flare-produced proton spectra.**
P. L. Rothwell, L. Katz, G. K. Yates, B. Sellers, F. A. Hanser.
Journ. Geophys. Res., Vol. 80, 4757 - 4761 (1975). – Brief report.

073.139 **Calculation of the radiant energy of chromospheric flares.**
L. N. Kurochka, E. V. Kurochka, V. A. Ostapenko.
Vestn. Kiev. Un-ta, Ser. Astron., No. 17, p. 12 - 25 (1975). In Russian.

For values of optical thicknesses to 10^6, taking into account the broadening action of radiation damping, the self-absorption factor used for determination of chromospheric flares energy in the lines of the Lyman series has been calculated. The self-absorption factor of the Lyman continuum for a number of values of electron temperature has been calculated also. Formulas for estimate of the emission measure in the flares are obtained. For a typical flare of mean power the line intensities of the first four hydrogen series are calculated.

073.140 **The radiant energy and structure of chromospheric flares.**
L. N. Kurochka, E. V. Kurochka, V. A. Ostapenko.
Vestn. Kiev. Un-ta, Ser. Astron., No. 17, p. 25 - 37 (1975). In Russian.

It has been shown that concentration of atomic hydrogen and electron concentration in observed emission regions is 1 - 4 times higher than in the other flare volume. It has been inferred that optical emission is determined mainly by the thin surface layer. It is supposed that the flares arise under action of corpuscular streams.

073.141 **An investigation of the glow of ionized calcium in a large chromospheric flare.** P. N. Polupan.
Vestn. Kiev. Un-ta, Ser. Astron., No. 17, p. 38 - 47 (1975). In Russian.

Some data of a spectrophotometric treatment of ionized calcium lines are presented. It is shown that mechanisms such as excitation by electron impact, own and photospheric radiation, taking account of induced emission, add a small part of energy emitted in calcium lines and emerged from the flare.

073.142 **The solar-flare infrared continuum: observational techniques and upper limits.** H. S. Hudson.
Solar Physics, Vol. 45, 69 - 78 (1975).

Exploratory observations at 20 μ and 350 μ have determined detection thresholds for solar flares in these wavelengths. In the 20 μ range solar atmospheric fluctuations (the 'temperature field') set the basic limits on flare detectability at ~5 K; at 350 μ the extinction in the earth's atmosphere provides the basic limitation of ~30 K. These thresholds are low enough for the successful detection of several infrared-emitting components of large flares. The upper limits obtained thus far, for subflares, indicate that the thickness of the Hα flare region does not exceed ~10 km. This result confirms the conclusion of Suemoto and Hiei (1959) regarding the small effective thickness of the Hα-emitting regions in solar flares.

073.143 **On the possible mechanism of formation of emission rim in hydrogen filaments.**
R. I. Kostik (*Kostyk*), T. V. Orlova.
Solar Physics, Vol. 45, 119 - 124 (1975).

Hα filtergrams of the chromosphere show an emission rim in many hydrogen filaments. The authors suppose that formation of this rim is due to photospheric radiation reflected by the filament in the direction of the chromosphere. The calculations show that: (1) the maximum contrast of the rim relative to the undisturbed chromosphere amounts to 1.4; (2) the larger the optical thickness of the filament and the closer to the solar limb it is situated, the brighter and wider is the rim; (3) the rim was not observed in filaments whose heights exceeds 10000 km above the chromosphere. These results are in close agreement with observations.

073.144 **On the structure of filaments from centimeter and millimeter observations.**
M. Butz, E. Fürst, W. Hirth, M. R. Kundu.
Solar Physics, Vol. 45, 125 - 129 (1975).
From filament observations at centimeter and millimeter wavelengths it is shown that the transition layer between the dense, cool filamentary material and the ambient hot, rarified corona is rather thin, of the order of some hundred kilometers. Hence, the difference in the sizes of filaments in the radio and optical domains is of the order of 1000 km or 2″, thus not detectable by the instruments so far used for observations of filaments.

073.145 **Study of a filament with a circularly polarized beam at 3.8 cm.**
R. M. Straka, M. D. Papagiannis, J. A. Kogut.
Solar Physics, Vol. 45, 131 - 139 (1975).
Extensive observations of left and right circularly polarized emission were carried out with the 120 ft Haystack antenna, which at 3.8 cm has a HPBW of 4.4′. During a very quiet period, September 22–26, 1974, two regions were observed in the southern hemisphere of the sun with brightness temperatures approximately 10% below the surrounding solar disk temperature. Hα photographs show that the main region was associated with a long filament. The observations are in agreement with a filament model consisting of a thin, tall and exceedingly long sheet of enhanced density encaged in a large and equally long tunnellike cavity of lower density.

073.146 **Observations of the structure and evolution of solar flares with a soft X-ray telescope.**
J. A. Vorpahl, E. G. Gibson, P. B. Landecker, D. L. McKenzie, J. H. Underwood.
Solar Physics, Vol. 45, 199 - 216 (1975).
132 soft X-ray flare events have been observed and analyzed and a detailed discussion of the X-ray flare structures is presented. The data indicated that soft X-rays emitted by a flare come primarily from an intense well-defined core surrounded by a region of fainter, more diffuse emission. Size distributions of these core features are presented and a classification scheme describing the brightest flare X-ray features is proposed. Brief comparisons with several theories are presented.

073.147 **The K and H Ca II line emission ratio in prominences.**
B. M. Rustad.
Inst. Theor. Astrophys., Blindern–Oslo, Rep. No. 44, 16 pp. (1975).
The K and H Ca II line emission ratios are studied in a number of prominences observed with the solar tower telescope of Oslo Solar Observatory from August 11, 1972, through June 18, 1974. 87 prominence spectra from 36 prominences are divided into three groups according to the degree of activity. No significant differences are detected in the averaged line ratios for the three groups. The average line emission ratio is 1.28.

073.148 **Large-scale formations of hydrogen filaments.**
V. N. Dermendzhiev.
Solnechnye Dannye 1975 Byull., No. 9, p. 61 - 66 (1975). In Russian.
The surface distribution of hydrogen filaments is studied according to the Meudon catalogue of 1961 in an area including an active longitude of quiet prominences. A statistical method based on the distance between the filaments is used. Indications for the existence of structural formations with dimensions of 10° and 21° are obtained.

073.149 **On motions in quiescent prominence knots.**
V. S. Pyatnitskij.
Solnechnye Dannye 1975 Byull., No. 9, p. 77 - 82 (1975). In Russian.
The results of simultaneous spectral observations of quiescent prominences in the Hα, Hβ, HeD_3, H Ca^+ and K Ca^+ lines with a narrow band interference-polarization filter are given. Rotary and spiral motions are found in the quiescent prominences.

073.150 **Spicule formation caused by the mechanism of coronal gas compression.**
S. G. Mamedov, Eh. Sh. Orudzhev, A. A. Karaev.
Solnechnye Dannye 1975 Byull., No. 9, p. 82 - 87 (1975). In Russian.
The function of plasma emission in the low temperature region for hydrogen ($T \leqslant 2 \times 10^4$ K) has been improved. The process of coronal gas compression has been considered, radiation in the Balmer and Lyman continua being taken into account. It is shown that the mechanism can explain spicule formation.

073.151 **Une éruption chromosphérique remarquable et quelques effets géophysiques associés.**
A. Dimitriu, G. Mariş, I. D. Niţă.
Acad. République Socialiste Roumanie, Obs. Bucarest, Secteur Solaire, p. 43 - 46 (1975).

073.152 **The visible spectrum of quiescent prominences.**
B. Caccin, L. Correra.
Mem. Soc. Astron. Italiana, Vol. 46, 293 - 302 (1975).
The available data on quiescent prominences are critically examined, as well as the attempts to explain the coexistence, in the same spectrum, of lines showing different excitation conditions. The role of XUV radiation from the surrounding regions appears to be fundamental; unfortunately the uncertainties concerning the penetration of such photons within the prominence plasma, whose temperature and density lie within a critical range, make of little use the concept of «average prominence», demanding instead both a detailed analysis of single objects and a careful development of theoretical models.

073.153 **Line profiles resulting from rotating and escaping fine filaments of prominences.**
T. Ciurla, B. Rompolt.
Acta Univ. Wratislav., No. 250, Mat., Fiz. Astron. XVI. 96 pp. = Contr. from the Wrocław Astron. Obs., No. 19 (1975). – Published by Państwowe Wydawnictwo Naukowe, Warszawa–Wrocław. Price zł 35 (1975).
The numerical results of the hydrogen Hα line profiles computed for optically thin rotating and escaping from the solar surface fine filaments, being in the form of a cylinder or a cylindrical shell, are given. The filaments are assumed to scatter the incident photospheric radiation. For an elementary volume element of a filament the Gaussian scattering profile was accepted. In computation the Doppler brightening effect was taken into account.

073.154 **Flying-spot analysis of solar images.**
L. Azzarelli, C. Carlesi, R. Falciani, M. Giordano, R. Panicucci, M. Rigutti, G. Roberti.
Image processing techniques in astronomy, (see 012.041), p. 385 - 390 (1975).

073.155 **Balmer lines in prominence spectra.**
K. I. Selyakov.
Vestn. Leningr. un-ta, 1975, No. 13, p. 134 - 139. In Russian. Abstr. in Referativ. Zhurn. 51. Astron., 2.51.513 (1976).

073.156 **On the connection between solar flares, radio bursts**

of types II and IV and solar cosmic ray events.
M. A. Zel'dovich, B. M. Kuzhevskij, I. A. Savenko, I. A. Yuzefovich.
Issled. kosmich. luchej. Moskva, Nauka, 1975, p. 133 - 153.
In Russian. – Abstr. in Referativ. Zhurn. 51. Astron., 2.51. 520 (1976).

073.157 **Photoelectric solar limb scans for determining mean chromospheric structure.** P. E. Barnhart.
Thesis Ohio State Univ., Columbus, USA. [Available from Univ. Microfilms, Ann Arbor, Mich., USA. Order No. 74-24296], 213 pp. (1974).

An observational technique first described by Mohler is adapted to observe the intensity distribution across any limb of the sun simultaneously in two wavelengths. By selecting one wavelength in the core of a strong chromospheric line and the other in a nearby portion of the continuum it is possible to obtain spatially coherent intensity profiles. The separation of the inflection points, or limbs of the two scans is found to be largely independent of the seeing.

073.158 **An atlas of extreme ultraviolet flashes of solar flares observed via sudden frequency deviations during the ATM-Skylab missions.** R. F. Donnelly, E. L. Berger, J. D. Busman, B. Henson, T. B. Jones.
Rep. UAG-36 NOAA-74110707, World Data Center A for Solar-Terrestrial Physics, Boulder, Colo., USA. [Available from NTIS, Springfield, Va., USA], 103 pp. (1974).

073.159 **Importance of a solar flares electron-electron-bremsstrahlung.** E. Haug.
Verhandl. Deutsche Phys. Ges., Reihe 6, Vol. 10, 195 (1975).
In German. – Abstract.

073.160 **High-energy particles associated with solar flares.** K. Sakurai, A. J. Klimas.
National Aeronautics and Space Administration, Greenbelt, Md. (USA). NASA-TM-X-70655; X-693-74-116, 187 pp. (1974).

High-energy particles are often generated in association with solar flares, and then emitted into interplanetary space. These particles are accelerated in the vicinity of the flare. By studying the temporal and spatial variation of these particles near the earth's orbit, their storage and release mechanisms in the solar corona and their propagation mechanism can be understood. These problems are described in some detail by using observational results on solar cosmic rays and associated phenomena.

073.161 **Acceleration of heavy nuclei in solar flares.** K. Sakurai.
National Aeronautics and Space Administration, Greenbelt, Md. (USA). NASA Rep. NASA-TM-X-7066; X-693-74-142, 20 pp. (1974).

The overabundance of heavy nuclei in solar cosmic rays of energy approximately less than 5 MeV/nucleon is explained by taking into account the pre-flare ionization states of these nuclei in the region where they are accelerated. A model is proposed which considers two-step accelerations associated with the initial development of solar flares.

Relative intensities of solar emission lines of ions in the sodium isoelectronic sequence. See Abstr. 022.034.

The ionisation balance of C^0 to C^{+4}. See Abstr. 022.081.

Ultraviolet solar identifications based on extended absorption series observed in the laboratory spectrum of Si I. See Abstr. 022.096.

The magneto-optical filter. I: Preliminary observations in Na D lines. See Abstr. 034.106.

Low-energy nucleosynthesis of lithium, beryllium, and boron. See Abstr. 061.016.

Asymptotic expansions for acoustical modes of the homogeneous compressible model. See Abstr. 061.034.

The rôle of electrical discharges in astrophysical phenomena. See Abstr. 061.058.

The spatial and temperature structure of vacuum spark plasmas. See Abstr. 062.025.

On the energy release processes at the neutral sheet. See Abstr. 062.046.

Electric resistivity of electromagnetically turbulent plasma and reconnection rate of magnetic fields. See Abstr. 062.050.

Resonance line transfer with partial redistribution. V. The solar Ca II lines. See Abstr. 063.010.

Resonance-line transfer with partial redistribution. VII. Angle-dependent redistribution. See Abstr. 063.026.

Stellar model chromospheres. IV. The formation of the Hϵ feature in the sun (G2 V) and Arcturus (K2 III). See Abstr. 064.027.

The formation of Mg I 4571 Å in the solar atmosphere. V. The multi-dimensional structure of the photosphere and low chromosphere. See Abstr. 071.003.

Photospheric oscillations. II. Acoustic filtering in the solar atmosphere and heating of the chromosphere. See Abstr. 071.027.

Photospheric and chromospheric magnetic structures. See Abstr. 071.036.

Line intensities in the photosphere-chromosphere transition region. IV: Cr I and metallic ion lines. See Abstr. 071.053.

Global distribution of filaments during solar cycle No. 20. See Abstr. 071.054.

Alfvén waves and the sunspot phenomenon. See Abstr. 072.009.

Chromospheric flares and rapid changes in sunspots. See Abstr. 072.026.

Time variations in extreme-ultraviolet emission lines and the problem of coronal heating. See Abstr. 074.005.

Profilberechnung von UV-Linien des Übergangsgebiets Chromosphäre-Korona unter dem Einfluß von Stoßwellen. See Abstr. 074.010.

Evidence for downflow following a coronal transient? See Abstr. 074.017.

Line profile analysis of a coronal formation observed near a quiescent prominence: intensities, temperatures and velocity fields. See Abstr. 074.049.

Coronal X-ray cavities and chromospheric H-alpha prominences correlations. See Abstr. 074.082.

Solar wind parameters and motion of shock waves from solar flares in August 1972. See Abstr. 074.093.

Diffusion effects on the line intensities of He I and He II in the solar transition region. See Abstr. 074.098.

An idealised model of a magnetohydrodynamic spherical blast wave applied to a flare produced shock in the solar wind. See Abstr. 074.102.

Analysis of EUV limb-brightening observations from ATM. I. Model for the transition layer and the corona. See Abstr. 074.103.

Heat transfer in the corona and transition region. See Abstr. 074.106.

EUV observations of the active sun from the Harvard experiment on ATM. See Abstr. 076.002.

Solar activity observed in X-rays and the EUV from OSO 7. See Abstr. 076.004.

Rise time of hard X-ray bursts. See Abstr. 076.014.

The interpretation of spectra, polarization, and directivity of solar hard X-rays. See Abstr. 076.016.

Solar flare X-ray measurements and their relation to microwave bursts. See Abstr. 076.017.

Crystal spectrometer studies of the sun employing a rotation modulation collimator. See Abstr. 076.023.

Slow X-ray bursts and flares with filament disruption. See Abstr. 076.024.

Laboratory reproduction of a solar X-ray flare spectrum. See Abstr. 076.026.

Betatron acceleration in a large solar hard X-ray burst. See Abstr. 076.027.

Association of X-ray arches with chromospheric neutral lines. See Abstr. 076.037.

Measured variation of the XUV line widths and intensities near the solar limb. See Abstr. 076.038.

Gamma-bursts during the flares of 1972, Aug. 4 and 7 observed on satellite Prognoz 2. See Abstr. 076.042.

Observations of solar gamma ray continuum between 360 keV and 7 MeV on August 4, 1972. See Abstr. 076.043.

Solar X-ray flares in September–December 1972. See Abstr. 076.046.

On the directivity of X-ray emission of solar flares. See Abstr. 076.047.

Limb-brightening curves of XUV transition zone lines in the quiet sun and in a polar coronal hole observed from Skylab. See Abstr. 076.051.

Solar gamma rays. See Abstr. 076.055.

Solar flare continua at the metre wavelengths. See Abstr. 077.060.

Enrichment of heavy nuclei in ^{3}He-rich flares. See Abstr. 078.011.

A method for estimation of the quality of the solar proton-event forecast. See Abstr. 078.012.

The spectrum of generation of high-energy particles on the sun during the flare of September 1, 1971. See Abstr. 078.015.

Nuclear composition of energetic solar particles – new source of information on solar flares. See Abstr. 078.019.

Propagation of solar-flare cosmic rays along corotating interplanetary flux-tubes. See Abstr. 078.021.

Propagation of solar cosmic radiation from the flares of October 3 and September 1, 1971 from data of Cosmos 426. See Abstr. 078.025.

Preflare current sheets in the solar atmosphere. See Abstr. 080.036.

Concerning some actual problems in the observational solar research. See Abstr. 080.051.

Effect of solar flares in the lower ionosphere. See Abstr. 083.027.

Solar flare effects in equatorial counter electrojet currents. See Abstr. 084.260.

Short term relationships between solar flares, geomagnetic storms, and tropospheric vorticity patterns. See Abstr. 085.007.

On the possibility of detecting solar flare effects in the zodiacal light. See Abstr. 106.022.

On the propagation of perturbations in the interplanetary plasma caused by solar flares. See Abstr. 106.043.

Micrometeorites and solar flare particles in and out of the ecliptic. See Abstr. 106.055.

Interplanetary scintillation and flare - produced disturbances. See Abstr. 106.070.

Interplanetary shock waves associated with solar flares. See Abstr. 106.077.

Flares of red dwarf stars and solar activity. See Abstr. 122.008.

Comparison of UV Ceti flares with solar flares. See Abstr. 122.015.

Are stellar flares energized by the missing energy in starspots? See Abstr. 122.084.

Upper limits on stellar flare X-ray emission from OSO-3. See Abstr. 122.163.

074 Solar Corona, Solar Wind

074.001 **Columnar content measurements of the solar-wind turbulence near the sun.** P. S. Callahan.
Astrophys. Journ., Vol. 199, 227 - 236 (1975).

The solar-wind turbulence near the sun is investigated with data obtained near the superior conjunction of Mariners 6, 7, and 9. The data are time histories of the change in the electron columnar content between the earth and the spacecraft. It is found that the comoving turbulence spectrum is well represented by a power law [$P(\nu)\alpha\nu^{-\beta}$] of index $\beta = 3.9 \pm 0.2$. It is suggested that there is a region of enhanced turbulence near the sun. The Mariner 9 spectral amplitudes correlate with Zurich sunspot number. The relationship of the present observations to theories of solar-wind heating is discussed.

074.002 **Temperature dependence of emission measure in solar X-ray plasmas. I. Nonflaring active regions.**
K. J. H. Phillips.
Astrophys. Journ., Vol. 199, 247 - 254 (1975).

Previous determinations of the form of the emission measure distribution function $\phi(T)$ for solar active regions are examined. It is found that two analyses in particular are affected by low quality of either observational or atomic data. An alternative, ab initio method of analysis is discussed.

074.003 **Heating of coronal material at X-ray bright points.** W. M. Glencross.
Astrophys. Journ., (*Letters*), Vol. 199, L53 - L56 (1975).

This letter suggests that the magnetic ropes present might be constructed of twisted braids which are wrapped one around the other. These components would be separated by current sheets where the density of material can be far higher than that normally considered appropriate for the corona. If the plasma in these sheets is maintained above 10^6 K, the flux radiated could exceed that from the rest of the magnetic rope. This material would account for the enhanced brightness of the X-ray regions. The conversion of magnetic energy during merging of braids is shown to provide sufficient thermal energy to maintain the dense plasma at coronal temperatures throughout the lifetimes observed, while the development of instabilities in this process accounts for the "flarelike" brightenings.

074.004 **The 1640.4 Å Hα line of He II observed from Skylab.**
U. Feldman, G. A. Doschek, M. E. VanHoosier, R. Tousey.
Astrophys. Journ., (*Letters*), Vol. 199, L67 - L70 (1975).

Profiles obtained from NRL Skylab solar spectra of the 1640.4 Å feature identified as the Hα line of He II are discussed and compared with theoretical profiles derived assuming (a) collisional excitation and (b) radiative recombination as the dominant excitation mechanism. Relative intensities of this line as a function of altitude above and below the solar white-light limb (from $-12''$ to $+8''$) are also given for a representative quiet sun region and for a polar coronal hole.

074.005 **Time variations in extreme-ultraviolet emission lines and the problem of coronal heating.**
J. E. Vernazza, P. V. Foukal, M. C. E. Huber, R. W. Noyes, E. M. Reeves, E. J. Schmahl, J. G. Timothy, G. L. Withbroe.
Astrophys. Journ., (*Letters*), Vol. 199, L123 - L126 (1975).

The authors have analyzed the time structure of the intensity of solar chromospheric and coronal extreme-ultraviolet lines, obtained by the Harvard College Observatory spectrometer aboard Skylab. They find changes in the intensity of up to 50 percent in times as short as 1 minute, but not periodic oscillations. Some evidence is found for the presence of shock waves in the chromosphere and the transition region. It is suggested that the solar chromosphere and corona are heated by nonperiodic waves.

074.006 **Holes in the solar corona.** W. M. Glencross.
IAU Symposium No. 68, (see 012.003), p. 19 - 21 (1975). – Summary.

074.007 **Classification of new spectral lines of Fe XVII observed in solar active regions.** R. J. Hutcheon.
IAU Symposium No. 68, (see 012.003), p. 69 - 70 (1975). Summary.

074.008 **Time variations in coronal active regions.**
A. S. Krieger, R. C. Chase, M. Gerassimenko, S. W. Kahler, A. F. Timothy, G. S. Vaiana.
IAU Symposium No. 68, (see 012.003), p. 103 (1975). Summary.

074.009 **Flare-like ultraviolet spectra of active regions.**
G. E. Brueckner.
IAU Symposium No. 68, (see 012.003), p. 105 - 107 (1975). Summary.

074.010 **Profilberechnung von UV-Linien des Übergangsgebiets Chromosphäre-Korona unter dem Einfluß von Stoßwellen.** L. Elzner.
Mitt. Astron. Ges., No. 36, p. 155 - 157 (1975).

074.011 **Interplanetary gas. XXI. Validity of the Chapman-Enskog description of the solar wind for protons.**
J. C. Price, J. C. Brandt, C. L. Wolff.
Astrophys. Journ., Vol. 199, 756 - 760 (1975).

Measured velocity distribution functions for solar-wind protons are used to estimate the validity of the Chapman-Enskog description, and it is found to be the best description presently in use.

074.012 **On solar coronal temperature.**
U. Narain, S. Chandra.
Astrophys. Journ., Vol. 200, 234 - 235 (1975).

A new calculation based on Shklovskii's method for the determination of the solar coronal temperature from line ratios is reported. The results are in reasonably good agreement with the recent theoretical calculations.

074.013 **On the variation of the solar wind velocity following solar flares.** Y.-N. Huang, Y.-W. Lee.
Journ. Geophys. Res., Vol. 80, 2863 - 2865 (1975).

By use of the superposed epoch method, changes in the solar wind velocity following solar flares have been investigated by using the solar wind velocity data obtained by Pioneer 6 and 7 and Vela 3, 4, and 5 satellites. A significant increase of the solar wind velocity has been found on the second day following importance 3 solar flares and on the third day following importance 2 solar flares. No significant increase of the solar wind velocity has been found for limb flares.

074.014 **The upstream escape of energized solar wind protons from the bow shock.** E. W. Greenstadt.
Astrophys. Space Sci. Library, Vol. 52, (see 012.008), 3 - 10 (1975).

074.015 **Solar wind heating at mid-heliospheric distances due to secondary ions.** H. J. Fahr.
Astrophys. Space Sci. Library, Vol. 52, (see 012.008), 269 - 278 (1975).

074.016 **Precise determination of the orientation of the plane of polarization in the solar corona.**
A. K. Kishonkov, M. M. Molodensky (*Molodenskij*).
Solar Physics, Vol. 42, 341 - 353 (1975).

It has been shown by Molodensky (1973), that precise measurements of the position of the plane of polarization in the corona may allow us to observe overthermal electrons in the solar corona. For such measurements during the eclipse of 10 July 1972, a method has been developed. A technique has also been developed for determining the position of the plane of polarization. The plane of polarization (E-vector position) coincides very exactly with the tangential direction in the region of N-W limb. The corona near the E limb consisted of two 'fans' divided by a thin beam. Calculations have been made of the turn of the plane of polarization caused by an inhomogeneity in the radiation field from the photosphere and due to the presence of spots. The effect qualitatively coincided with that shown by the measurements.

074.017 **Evidence for downflow following a coronal transient?** E. Hildner, W. C. Livingston.
Solar Physics, Vol. 42, 391 - 394 (1975).

On 11 September 1973 a peculiar prominence was observed. The prominence displayed strong (~ 50 km s^{-1}) systematic motions toward and away from the observer. The unusual spectrographic appearance of the prominence might have been due to downflowing material lifted into the corona during an earlier coronal transient.

074.018 **Abundance of Fe relative to H at 1.5 solar radii.**
M. P. Nakada, R. D. Chapman, W. M. Neupert, R. J. Thomas.
Solar Physics, Vol. 42, 487 - 496 (1975).

The purpose of this report is mainly to examine the use of resonance scattering as a method for obtaining abundances relative to H. Abundances of Fe are obtained for these non-simultaneous measurements and some uncertainties in the procedures used are examined.

074.019 **Sheath-limited unipolar induction in the solar wind.**
L. J. Srnka.
Astrophys. Space Sci., Vol. 36, 177 - 204 (1975).

The collection of charged particles by electrodes in plasmas is controlled by the current-voltage characteristics of the plasma sheath which forms at the electrode surface. This principle is applied to the steady-state electromagnetic interaction of the solar wind with moon-like bodies, or 'solid-body' interactions. In some cases the unipolar dynamo response of an electrically conducting body in the solar wind motional electric field can be controlled by sheath effects. This occurs for highly conducting bodies when the body radius R is less than a critical value R_c, with the result that no induced bow shock wave can form.

074.020 **The solar corona on June 30, 1973.**
Sammani A. Jacoub, M. Yu. Nevskij, V. V. Leushin, L. N. Lebedeva.
Astron. Zhurn. Akad. Nauk SSSR, Vol. 52, 778 - 784 (1975). In Russian. English translation in Soviet Astron., Vol. 19, No. 4.

Results of the observation of the total solar eclipse of 1973, June 30, in Sudan are given. Moments of the contacts are determined in the band of totality. The structural features of the corona are discussed. The isophotes of the corona in three colours, the decrease in brightness and flattening of the corona are obtained.

074.021 **The Soviet–French cooperative study of the solar corona. I. The structural peculiarities and photometry of the corona on July 10, 1972.** S. K. Vsekhsvyatskij, N. I. Dzyubenko, A. T. Nesmyanovich, O. S. Popov, S. Koutchmy.
Astron. Zhurn. Akad. Nauk SSSR, Vol. 52, 785 - 791 (1975). In Russian. English translation in Soviet Astron., Vol. 19, No. 4.

The results of an eclipse pictures study obtained by the Kiev University expedition in accordance with the Soviet–French experiment programme "The dynamics of the white corona" are given. Besides the usual details new formations have been detected: "emptinesses" and "sharp boundary".

074.022 **Wave-trains in the solar wind. III: Alfvén waves in the azimuthally-dependent interplanetary medium.**
A. K. Richter.
Astrophys. Space Sci., Vol. 36, 383 - 405 (1975).

Applying an Alfvén-wave-extended-QRH-approximation and the method of characteristics, the author solves the equations of motion for outwardly propagating Alfvén waves analytically for three different cases of an azimuthal dependence of thebackground solar wind, (a) for a pure fast-slow stream configuration, (b) for the situation where the high-speed stream originates from a diverging magnetic field region, and (c) for the case of (b) and an initially decreasing density configuration ('coronal hole'). These three solar wind models are discussed shortly.

074.023 **Plasma processes in the expansion of the solar wind and in the interplanetary medium.** A. Barnes.
Rev. Geophys. Space Phys., Vol. 13, No. 3, (see 013.001), p. 1049 - 1053, 1066 - 1072 (1975). – Review paper with bibliographical notes.

074.024 **Large-scale inhomogeneities in the solar wind of solar origin.** J. T. Gosling.
Rev. Geophys. Space Phys., Vol. 13, No. 3, (see 013.001), p. 1053 - 1058, 1072 - 1076 (1975). – Review paper with bibliographical notes.

074.025 **Composition of the solar wind: present and past.**
J. Hirshberg.
Rev. Geophys. Space Phys., Vol. 13, No. 3, (see 013.001), p. 1059 - 1063, 1077 - 1081 (1975). – Review paper with bibliographical notes.

074.026 **The integral brightness of the corona during the solar eclipse of June 30, 1973.**
M. Rybanský, V. Rušin.
Bull. Astron. Inst. Czechoslovakia, Vol. 26, 206 - 210 (1975).

The total brightness of the corona is determined in the standard region $1.03 \leqq \varrho \leqq 6.00$ for the solar eclipse of June 30, 1973. The absolute photometry of this eclipse formed the basis for computing the integral brightness of the corona. The resultant value is 0.84×10^{-6} of the total brightness of the solar eclipse.

074.027 **Turbulent diffusion of protons in a radial magnetic field of the solar wind.**
M. F. Bakhareva, V. N. Lomonosov, B. A. Tverskoj.
Geomagn. Aeronom., Vol. 15, 593 - 600 (1975). In Russian.

074.028 **On the heliolatitudinal dependence of the velocity of the solar wind.** A. I. Efimov, N. A. Lotova.
Geomagn. Aeronom., Vol. 15, 731 - 732 (1975). In Russian. Brief information.

074.029 **Infrared observations of the solar corona with a silicon vidicon spectrometer.**
J. M. Pasachoff, D. F. Muzyka.
Bull. American Astron. Soc., Vol. 7, 409 (1975). – Abstr. AAS.

074.030 **Skylab soft X-ray observations of coronal condensations.** E. Tandberg-Hanssen, A. C. Deloach,

R. B. Hoover, J. P. McGuire, R. M. Wilson, J. B. Smith, Jr., D. M. Speich, W. Henze, S. T. Wu.
Bull. American Astron. Soc., Vol. 7, 444 (1975). – Abstr. AAS.

074.031 **Skylab soft X-ray observations of magnetic-field reconfiguration.** J. B. Smith, Jr., D. M. Speich, J. P. McGuire, A. C. Deloach, R. B. Hoover, E. Tandberg-Hanssen, R. M. Wilson, S. T. Wu, W. Henze.
Bull. American Astron. Soc., Vol. 7, 444 - 445 (1975). – Abstr. AAS.

074.032 **The sun's polar caps as coronal holes: their sizes, evolution, and phenomenology during the Skylab mission.** J. D. Bohlin, D. M. Rubenstein, N. R. Sheeley, Jr.
Bull. American Astron. Soc., Vol. 7, 457 (1975). – Abstr. AAS.

074.033 **The rigid rotation of coronal holes.** W. J. Wagner.
Bull. American Astron. Soc., Vol. 7, 457 - 458 (1975). – Abstr. AAS.

074.034 **A possible relationship between solar wind density and coronal EUV emission.**
E. J. Rhodes, Jr., E. J. Smith.
Bull. American Astron. Soc., Vol. 7, 458 (1975). – Abstr. AAS.

074.035 **Enhanced ~3 MeV helium and medium fluxes associated with coronal holes.**
R. E. Gold, S. M. Krimigis, E. C. Roelof, A. S. Krieger, J. T. Nolte.
Bull. American Astron. Soc., Vol. 7, 458 - 459 (1975). – Abstr. AAS.

074.036 **The source of material comprising a mass ejection coronal transient.**
E. Hildner, J. T. Gosling, R. T. Hansen, J. D. Bohlin.
Bull. American Astron. Soc., Vol. 7, 473 (1975). – Abstr. AAS.

074.037 **The emission sources and geometry of the coronal transient of 21 August 1973.**
A. I. Poland, R. H. Munro.
Bull. American Astron. Soc., Vol. 7, 473 (1975). – Abstr. AAS.

074.038 **The relative abundance of neon and magnesium in the solar corona.** H. R. Rugge, A. B. C. Walker, Jr.
Bull. American Astron. Soc., Vol. 7, 473 (1975). – Abstr. AAS.

074.039 **Coronal velocity fields derived from line profile analysis of Fe XIV λ5303.** T. Tsubaki.
Bull. American Astron. Soc., Vol. 7, 474 (1975). – Abstr. AAS.

074.040 **Cosmic-ray propagation through the solar corona.**
D. G. Wentzel.
Bull. American Astron. Soc., Vol. 7, 474 (1975). – Abstr. AAS.

074.041 **Extensive spatial distribution of 569.4 nm Ca XV emission in the solar corona during the eclipse of 1973.** M. M. Hoffman, W. M. Sanders, D. H. Liebenberg.
Bull. American Astron. Soc., Vol. 7, 474 (1975). – Abstr. AAS.

074.042 **Coronal magnetic field models.** K. H. Schatten.
Rev. Geophys. Space Phys., Vol. 13, 589 - 592 (1975).

A review of the development of coronal magnetic field is presented. Each model's assumptions are discussed with reference to their validity within the solar corona. It is pointed out how remarkable it is that electrodynamics is applicable on a large scale as well as on the small scale from which it originated.

074.043 **Some peculiarities of investigation of charged particle streams with traps and analyzers. III. Determination of solar wind parameters from energy spectra.**
L. S. Gurin, G. N. Zastenker, V. S. Mokrov, M. Z. Khokhlov.
Kosmich. Issled., Vol. 13, 555 - 562 (1975). In Russian.

074.044 **On the helio-latitudinal dependence of solar wind parameters.** A. I. Efimov, N. A. Lotova.
Kosmich. Issled., Vol. 13, 603 - 605 (1975). In Russian.
Brief information.

074.045 **Solar wind and EUV corona.** T. Watanabe.
Publ. Astron. Soc. Japan, Vol. 27, 385 - 400 (1975).

The relationship between the solar wind velocity and the brightness of the solar EUV corona is examined under the assumption of ideal spiral model for solar wind flow. There exists a general tendency that the wind velocity is high when the EUV coronal brightness at the position of the solar wind source is low and vice versa. Most of the EUV coronal holes are the principal sources of high velocity solar wind faster than 500 km s^{-1}, and the bright EUV coronas formed above the intense active regions with bipolar photospheric magnetic field are the sources of low velocity solar wind slower than 450 km s^{-1}.

074.046 **On cold emission in the solar corona.**
K. V. Alikaeva.
Astrometriya i Astrofizika, *Kiev*, vyp. (No.) 26, (see 003.001), p. 87 - 91 (1975). In Russian.

The faint emission of hydrogen, helium and metals in the corona which appears near an active prominence is studied. The calculations show that the temperature of the emission region is within the limits 11 000 to 30000° and the electron density 10^9–10^{10} cm^{-3}.

074.047 **A loop model of active coronal regions.**
M. Landini, B. C. Monsignori Fossi.
Astron. Astrophys., Vol. 42, 213 - 220 (1975).

The energy balance which is established in a vertical cylinder of constant pressure between the radiated energy, the energy supplied from below and dissipated over one typical scale height $1/k$, and the energy flux conducted along the temperature gradient is investigated: a model of temperature versus height is obtained. A coronal active region is simulated by assembling a large number of such tubes shaped along the lines of force of a dipole field assuming that the flux of the supplied energy is proportional to the magnetic field intensity. Comparison is made with experimental data on several X-ray emission lines.

074.048 **Small-scale inhomogeneities in the solar corona: evidence from meter-λ radio bursts.** D. B. Melrose.
Solar Physics, Vol. 43, 79 - 86 (1975).

A number of inconsistencies between simple theory and observations of solar radio bursts indicate that mode-mode coupling in the solar corona is much stronger than predicted. The inconsistencies include the absence of predicted reversal of the sense of polarization in a type I storm at CMP, and the anomalously weak polarization of type II and type III emission. The strong mode coupling could be explained in terms of small scale inhomogeneities ($L_N \lesssim 100$ km) throughout the relevant regions of the corona. The relevant regions are those with open magnetic field lines overlying active regions. Another observation which may indicate that the corona is inhomogeneous is the amount of Faraday rotation for type III bursts.

074.049 **Line profile analysis of a coronal formation observed near a quiescent prominence: intensities, temperatures and velocity fields.** T. Tsubaki.
Solar Physics, Vol. 43, 147 - 175 (1975).

A technique developed for analyzing line profiles with both speed and high accuracy was used to study the physical conditions of a coronal formation near a quiescent prominence. Detailed analyses of five coronal lines (Fe XIV λ 5303, Fe X λ 6374, Ni XV λ 6702, Fe XV λ 7059, and Fe XI λ 7892) provided total intensities, Doppler width temperatures, ionization temperatures, and velocities.

074.050 **Disturbances in the solar wind from IPS (*interplanetary scintillations*) measurements in August 1972.**
B. J. Rickett.
Solar Physics, Vol. 43, 237 - 247 (1975).

From IPS and spacecraft measurements of the solar wind combined with geomagnetic observations, the author identifies the passage of three main disturbances through the solar wind from solar flares on August 2, 4 and 7. From a detailed study of the IPS data covering the third event, the author concludes that the extent of the disturbance front at 1 AU covered about ±60° in longitude and more than 30° in latitude from the flare normal. If interpreted as a blast wave according to the model of De Young and Hundhausen (1971), the disturbance was ejected from the sun into a cone of half-angle 45° ± 15°.

074.051 **On solar wind helium and heavy ion temperatures.**
J. Feynman (*formerly J. Hirshberg*).
Solar Physics, Vol. 43, 249 - 252 (1975).

Because of theoretical difficulties in predicting the behavior of multi-ionic plasmas, observation must serve as a guide to expectations concerning the results of further measurements of other parameters. Here the author examines the support for the common belief that observations of the helium to hydrogen temperature ratio indicate that the thermal velocities of ionic species in the solar wind tend to be equal.

074.052 **Polarization of the solar corona during the eclipse of June 30, 1973.** A. K. Kishonkov.
Astron. Tsirk., No. 850, p. 3 - 5 (1974). In Russian.

074.053 **Mode decay and evolution of the solar wind Alfvén wave spectrum.** R. H. Cohen.
Journ. Geophys. Res., Vol. 80, 3678 - 3680 (1975).

The treatment of Alfvén wave mode decay by Cohen and Dewar (1974) is extended to a regime of finite temperature and large acoustic wave damping rates for a model (Lorentzian) Alfvén wave power spectrum. When applied to hydromagnetic waves in the solar wind, the theory predicts a degradation of the low-frequency ($\sim$(1 day)$^{-1}$) portion of the spectrum on a time scale of the order of the time required for such waves to travel several AU.

074.054 **Coronale gaten, donkere gebieden in de zonnecorona.** C. de Jager.
Zenit, Vol. 2, 292 - 294 (1975).

074.055 **The structure of the solar corona on March 7, 1970.**
S. K. Vsekhsvyatskij, N. I. Dzyubenko, V. I. Ivanchuk, A. T. Nesmyanovich, G. A. Rubo.
Problems of cosmic physics. Vyp. (No.) 10, (see 003. 003), p. 119 - 144 (1975). In Russian.

The general and detailed structure of the corona up to 3.0 – 3.5 $R_\odot$ have been studied using plates in the photographic region.

074.056 **Analysis of *EUV* observations of regions of the quiet and active corona at the time of the 1970 March 7 eclipse.** A. H. Gabriel, C. Jordan.
Monthly Notices Roy. Astron. Soc., Vol. 173, 397 - 418 (1975).

A series of *EUV* spectra were obtained from a rocket flight during the 1970 March 7 total solar eclipse. The intensities of permitted and forbidden emission lines in both quiet and active regions have been analysed. The densities and temperatures have been derived as a function of position in the active region. In a given line-of-sight the density increases with increasing temperature and overall the central, low parts of the active region contain more material at the higher temperatures than do the more extensive loop systems.

074.057 **Fine structure of solar wind velocity in a model of an exponential spectrum of irregularities.**
N. A. Lotova, I. V. Chashej.
Geomagn. Aeronom., Vol. 15, 769 - 776 (1975). In Russian.

074.058 **On the theory of propagation of discontinuities in the solar wind.** S. V. Sobolev.
Geomagn. Aeronom., Vol. 15, 925 - 927 (1975). In Russian. Brief information.

074.059 **Motion of shocks through interplanetary streams.**
L. F. Burlaga, J. D. Scudder.
Journ. Geophys. Res., Vol. 80, 4004 - 4010 (1975).

A model for the motion of flare-generated shocks through interplanetary streams is presented, illustrating the effects of a stream-shock interaction on the shock strength and geometry. The results show that the Mach number of a shock can decrease appreciably to near unity in the interaction region ahead of streams and that the interaction of a spherically symmetric shock with a spiral-shaped corotating stream can cause significant distortions of the initial shock front geometry.

074.060 **A model of coronal holes.**
W. M. Adams, P. A. Sturrock.
Astrophys. Journ., Vol. 202, 259 - 264 (1975).

It has been noted that coronal holes appear to be associated with regions of diverging magnetic field in the corona. The authors set out to test the hypothesis that coronal holes may be caused by an increased flow of energy into the solar wind resulting directly from this diverging magnetic field pattern. Simple models were devised to approximate the energy flow down into the transition region and up into the solar wind as a function of the temperature, density, and rate of field line divergence in the corona.

074.061 **Observation of the 7892 Å Fe XI line in the corona during the eclipse of June 30, 1973.**
B. Fort, J. P. Picat.
Astron. Astrophys., Vol. 44, 17 - 19 (1975).

Monochromatic images in the 7892 Å line have been obtained by using an interference filter and telescope with a $f/10$ aperture. Local electron densities in fine coronal structures are deduced and compared with previous coronagraphic observations in the red and green coronal lines. As a conclusion it is proposed that the lower non-active corona (i.e. active loops excluded) might consist of more or less abundant aggregates of such fine structures.

074.062 **The effect of binary encounters on the thermal anisotropy of the protons in the solar wind.**
M. Eyni, A. S. Kaufman.
Astron. Astrophys., Vol. 44, 107 - 112 (1975).

The thermal anisotropy is investigated theoretically under conditions where the expansion is both steady-state and adiabatic, and the interplanetary magnetic field is radial. It is assumed further that binary encounters between protons constitute the sole mechanism of limiting the value of the temperature-anisotropy factor K, the ratio of the kinetic temperature in the radial direction to the kinetic temperature in the transverse direction. The predicted value of K at a distance of 1 AU under 'quiet conditions' compares favourably with the measured value.

074.063 **On the acceleration of protons on the boundaries of solar wind with interstellar medium.**
M. F. Bakhareva.

VI Leningr. mezhdunar. seminar, 1974. Uskorenie chastits i yader. reaktsii v kosmose. Leningrad, 1974, p. 175 - 176. In Russian.

074.064 **Influence of the characteristics of analysers of charged particles on the results of measurements of solar wind parameters.** S. A. Romanov.
In-t kosmich. issled. AN SSSR. Pr-211. Moskva, 34 pp. In Russian. – Abstr. in Referativ. Zhurn. 51. Astron., 8.51.467; 62. Issled. kosmich. prostranstva 8.62.196 (1975).

074.065 **OSO-7 results on coronal emission near 304 Å.** M. P. Nakada, R. D. Chapman, W. M. Neupert, R. J. Thomas.
Solar Physics, Vol. 43, 337 - 350 (1975).
The spectral composition and spatial distribution of equatorial coronal emission near 304 Å is examined. Spectral scans indicate that the predominant line is from Si XI. Comparisons of observations with calculations of intensity changes with altitude indicate that collisional excitation is important near the sun but that photoexcitation becomes dominant beyond about $1.3\,R_\odot$ from the solar center. Observed and calculated intensities are in approximate agreement for abundances and electron densities that are within the range of observed values.

074.066 **Investigation of emission lines of the solar corona of 10 July, 1972 using the Fabry-Pérot etalon.**
I. S. Kim, G. M. Nikolsky (*Nikol'skij*).
Solar Physics, Vol. 43, 351 - 358 (1975).
Coronal interferograms in the lines of Fe XIV 5303 Å, He I 5876 Å and Fe X 6374 Å were obtained during the total solar eclipse of 10 July, 1972. He I emission was found in the chromosphere only. A detailed photometry of the 5303 Å line was carried out and the behaviour of the half widths and equivalent widths were studied in different regions of the corona.

074.067 **Energy balance in a magnetically confined coronal structure observed by OSO-7.**
W. M. Neupert, Y. Nakagawa, D. M. Rust.
Solar Physics, Vol. 43, 359 - 376 (1975).
A model of a coronal region of enhanced Fe XV and Fe XVI emission is developed and its energy balance is examined using extreme ultraviolet observations from OSO-7 together with calculations of possible force-free coronal magnetic field configurations. The EUV emissions of Fe XVI and Fe XV are used to estimate electron temperature and density as a function of position in a low coronal arch whose geometry is inferred from the combined observations and force-free field calculations. These data permit estimates of radiative and conductive losses in the coronal enhancement, leading to the conclusion that non-thermal energy deposition is required up to at least 30 000 km if the structure is to maintain the steady state emission pattern as observed.

074.068 **On the reliability of the structure of the low corona as derived from flash spectra.** R. Giovanelli.
Solar Physics, Vol. 43, 377 - 379 (1975).
A study of the photometry of the integrated flash spectrum shows that it is an unreliable method for deriving the structure of the very low corona.

074.069 **Comments on the paper 'On the reliability of the structure of the low corona as derived from flash spectra' by R. Giovanelli.** M. Kanno.
Solar Physics, Vol. 43, 381 - 384 (1975). – See Abstr. 074.068.
In spite of the ambiguity in analyzing the eclipse flash spectrum, pointed out by Giovanelli (1975), the general conclusion of Kanno et al. (1971) is shown to be still valid.

074.070 **The coronal condensation observed at the 1973 eclipse.** H. Kurokawa.
Solar Physics, Vol. 43, 385 - 403 (1975).
In this paper the author has analyzed the monochromatic images of the coronal condensation in Fe XIV 5303, Fe X 6374 and Fe XI 7892 coronal lines and compared its thermal structure with other condensations in the light of the lifetime of the coronal active region.

074.071 **Cyclotron wave instability in the corona and origin of solar radio emission with fine structure. I. Bernstein modes and plasma waves in a hybrid band.**
V. V. Zheleznyakov, E. Ya. Zlotnik.
Solar Physics, Vol. 43, 431 - 451 (1975).
The longitudinal waves (Bernstein modes and plasma waves near the hybrid frequency) in a mixture of equilibrium coronal plasma and a small group of energetic electrons are investigated. The energetic electrons have a nonequilibrium momentum distribution inherent in trapped particles. The frequency dependence of the cyclotron instability increments is studied. Attention is paid to a significant role of the relativistic effects for the cyclotron instability of longitudinal waves. For sufficiently large velocity of nonequilibrium electrons the increments are shown to increase when the hybrid frequency coincides with one of the gyrofrequency harmonics (double plasma resonance).

074.072 **Solar wind heavy ion abundances.**
S. J. Bame, J. R. Asbridge, W. C. Feldman, M. D. Montgomery, P. D. Kearney.
Solar Physics, Vol. 43, 463 - 473 (1975).
During periods in 1969–1971 when interplanetary flow conditions were quiet, 17 determinations of the solar wind Fe and Si abundances and 7 of the O abundance were made. On the average $\langle N(\mathrm{Fe})/N(\mathrm{H})\rangle = 5.3 \times 10^{-5}$, $\langle N(\mathrm{Si})/N(\mathrm{H})\rangle = 7.6 \times 10^{-5}$ and $\langle N(\mathrm{O})/N(\mathrm{H})\rangle = 5.2 \times 10^{-4}$. Variations from the averages over a total range of factors of ~4 for O, ~4 for Si, and ~10 for Fe have been observed. Although Fe and Si abundance variations appear to be correlated, no other element correlation pairs are unambiguously discernible from the present data set.

074.073 **Skylab and the ASE X-ray telescope experiment: a new view of the X-ray corona.**
G. S. Vaiana, R. Chase, J. Davis, M. Gerassimenko, L. Golub, S. Kahler, A. S. Krieger, R. Petrasso, J. K. Silk, R. Simon, A. F. Timothy, M. Zombeck, D. Webb.
Oss. Mem. Oss. Astrofis. Arcetri, Fasc. 104, (see 012.025), p. III, 3 - 47 (1975).
The ATM Solar Observatory on Skylab; The S-054 X-ray telescope; Results.

074.074 **Comparison of Skylab X-ray and ground-based helium observations.**
J. Harvey, A. S. Krieger, A. F. Timothy, G. S. Vaiana.
Oss. Mem. Oss. Astrofis. Arcetri,Fasc. 104, (see 012.025), p. 50 - 58 (1975).
Comparison of soft X-ray images of the sun obtained with the *SO*-54 experiment aboard Skylab and photoelectric helium D_3 spectroheliograms obtained at Kitt Peak yields a good agreement. Distinctively coronal features such as coronal holes and bright points are identified in helium pictures.

074.075 **Models of coronal active regions.** M. Landini, B. C. Monsignori Fossi, A. S. Krieger, G. S. Vaiana.
Oss. Mem. Oss. Astrofis. Arcetri, Fasc. 104, (see 012.025), p. 59 - 66 (1975).
The intensity distribution of a coronal active region observed by means of an X-ray telescope similar to the ATM's one is simulated assuming that temperature and pressure are

radially symmetric and follow a gaussian distribution. A simple way is shown to take into account the instrumental function and to evaluate the temperature and pressure from the comparison of two pictures taken with different filters. Some preliminary results are shown and an evaluation of the magnetic field necessary to confine the structure is given.

074.076 **Coronal active regions evolution.**
M. Landini, B. C. Monsignori Fossi.
Oss. Mem. Oss. Astrofis. Arcetri, Fasc. 104, (see 012.025), p. 67 - 73 (1975).

The evolution of the magnetic structure in an active region is simulated by the sinking and spreading of two magnetic poles below the photosphere.

074.077 **Coronal magnetic flux tubes.** G. Compagno.
Oss. Mem. Oss. Astrofis. Arcetri, Fasc. 104, (see 012. 025), p. 74 - 88 (1975).

A model consisting of magnetic flux tubes, with the direction of the field essentially along the tube's axis and filled with plasma, is assumed to describe quiescent coronal arch structures. A simplified mathematical model is used to study the energy balance of flux tubes.

074.078 **The long term development of the large scale corona and the evolution of coronal holes.**
A. F. Timothy, M. Gerassimenko, L. Golub, A. S. Krieger, R. Petrasso, G. S. Vaiana.
Oss. Mem. Oss. Astrofis. Arcetri, Fasc. 104, (see 012.025), p. 93 - 110 (1975).

In this preliminary discussion on the long term development of the large scale corona and the evolution of coronal holes the authors investigate typical coronal configurations which result in the formation of a hole. They also follow the evolution of one particular feature as a function of time and investigate its rotational characteristics at latitudes ranging from 20° south to 50° north. Finally, they explore the relationship between this feature and a recurrent high velocity stream in the solar wind.

074.079 **Coronal holes as M-regions: correlation between solar features and solar wind disturbances.**
B. Bell, G. Noci.
Oss. Mem. Oss. Astrofis. Arcetri, Fasc. 104, (see 012.025), p. 111 - 119 (1975).

The authors report briefly on a statistical analysis aiming at testing whether coronal holes are the sources of high velocity streams of solar wind as suggested by Noci (1973).

074.080 **A coronal hole observed at X, UV and radio wavelengths.** F. Drago.
Oss. Mem. Oss. Astrofis. Arcetri, Fasc. 104, (see 012.025), p. 120 - 124 (1975).

A large coronal hole, whose central meridian passage took place on May 31, 1973, has been observed by the X-ray ASE telescope mounted on the Skylab, by OSO 7 spacecraft (λ = 284 Å) and by the Nançay interferometer at 408 and 169 MHz.

074.081 **Coronal X-ray structures and coronal magnetic fields.** G. Poletto, A. F. Timothy, A. S. Krieger, G. S. Vaiana.
Oss. Mem. Oss. Astrofis. Arcetri, Fasc. 104, (see 012.025), p. 175 - 191 (1975).

On comparing extrapolated coronal magnetic fields with X-ray structures representative of a variety of situations, a qualitative agreement between the two configurations results, although limited to compact structures. This is particularly verified in absence of flare activity or when large structural changes are not present. In this case the extrapolated potential fields seem to well represent the general topology outlined by the X-ray features, thus revealing a minor contribution of current fields to the total field.

074.082 **Coronal X-ray cavities and chromospheric H-alpha prominences correlations.**
G. Godoli, V. Sciuto, R. A. Zappalà.
Oss. Mem. Oss. Astrofis. Arcetri, Fasc. 104, (see 012.025), p. 209 - 210 (1975).

074.083 **Coronal magnetic structure.** J. L. Steinberg.
Oss. Mem. Oss. Astrofis. Arcetri, Fasc. 105, (see 012.026), p. 56 - 85 (1975). – Review paper.

074.084 **Shock waves of the solar wind and the magnetosphere of the earth.** S. A. Grib.
Subburi i vosmushcheniya v magnitosfere. Leningrad, Nauka, 1975, p. 97 - 103. In Russian. – Abstr. in Referativ. Zhurn. 62. Issled. kosmich. prostranstva, 8.62.212 (1975).

074.085 **On quasi-linear relaxation of electron streams in the solar plasma.** V. V. Zajtsev, V. O. Rapoport.
Pis'ma v Astron. Zhurn., Vol. 1, No. 11, p. 38 - 41 (1975). In Russian.

New data on solar fast electron streams observed by space stations are analysed. These data suggest strong quasi-linear relaxation of non-scattered component of the electron stream caused by excitation of plasma waves by this component in the solar wind plasma.

074.086 **Spectrophotometry of the solar corona taken on July 10, 1972.** K. I. Nikol'skaya.
Solnechnye Dannye 1975 Byull., No. 4, p. 71 - 77 (1975). In Russian.

The main results are given of a study of the solar corona spectra taken during the total solar eclipse on July 10, 1972.

074.087 **On the large-scale brightness distribution in the solar corona on July 10, 1972.**
A. T. Nesmyanovich, O. S. Popov.
Solnechnye Dannye 1975 Byull., No. 4, p. 78 - 83 (1975). In Russian.

074.088 **Asymmetric distant solar wind.**
S. Grzędzielski, R. Ratkiewicz.
Acta Astron., Vol. 25, 177 - 204 (1975).

Distant solar wind deviations from spherical symmetry induced by interaction with neutral interstellar hydrogen (photoionization and charge-exchange) were studied by perturbation technique. For neutral hydrogen density outside the heliosphere $n_H^\infty = 0.2$ cm^{-3} and relative velocity 20 km s^{-1} the temperature decreases along a 150° arc of Neptune's orbit from 3×10^5 K in the apex direction to 1×10^5 K at 30° from the anti-apex and the non-radial component of solar wind velocity increases, respectively, from 0 to 4 km s^{-1}. Possible influence upon magnetospheres of Uranus and Neptune is briefly discussed.

074.089 **Solar wind.** J. Olmr, Š. Pintér.
Pokroky, Vol. 20, 84 - 90 (1975). In Czech.

074.090 **Electromagnetic ion-beam instabilities in the solar wind.** M. D. Montgomery, S. P. Gary, D. W. Forslund, W. C. Feldman.
Phys. Rev. Letters, Vol. 35, 667 - 670, with a correction, p. 890 (1975).

074.091 **Coronal index of solar activity. I. Line 5303 Å, year 1971.** M. Rybanský.
Bull. Astron. Inst. Czechoslovakia, Vol. 26, 367 - 374 (1975).

The paper describes the procedure for computing the energy emitted by the whole corona in emission line Fe XIV 5303 Å. The value of the energy, determined from standard observations of this line at coronal stations, is the proposed index of solar activity in the corona. The results presented cover 1971. In the conclusion the use of the index and of some of the intermediate results of the computation are discussed.

074.092 **Coronal index of solar activity. II. Line 5303 Å, years 1972 and 1973.** M. Rybanský.
Bull. Astron. Inst. Czechoslovakia, Vol. 26, 374 - 377 (1975).
The paper is the second part of a more extensive investigation of the coronal index of solar activity. It contains the results covering the years 1972 and 1973. See also abstr. 074.091.

074.093 **Solar wind parameters and motion of shock waves from solar flares in August 1972.** O. L. Vajsberg, G. N. Zastenker, F. Kambu, V. V. Temnyj, M. Z. Khokhlov.
Kosmich. Issled., Vol. 13, 859 - 865 (1975). In Russian.

074.094 **Labor modelling of the interaction between solar wind and the moon.**
I. M. Podgornyj, Yu. V. Andriyanov, G. B. Kulikov.
Kosmich. Issled., Vol. 13, 914 - 922 (1975). In Russian.

074.095 **The magnetic field of the solar wind and vertical motion of the night ionosphere.**
A. M. Lyatskaya, V. B. Lyatskij.
Geomagn. Aeronom., Vol. 15, 977 - 981 (1975). In Russian.

074.096 **Nonlinear evolution of parallel-propagating hydromagnetic waves.** R. H. Cohen, R. M. Kulsrud.
Phys. Fluids, Vol. 17, 2215 - 2225 (1974).
The authors discuss the non-linear evolution of plane hydromagnetic fluctuations propagating along the unperturbed magnetic field direction. When applied to the solar wind it is shown that the process occurs on a time scale comparable to the time required for Alfvén waves convected by the solar wind to travel one astronomical unit. This can account for the existence of streams of purely Alfvénic fluctuations in the Mariner V data. A class of fluctuations which includes linearly polarized waves is shown to evolve into rotational discontinuities. The theory also explains the conversion of hydromagnetic wave energy into heat in the solar wind.

074.097 **Coronal scattering of solar "plasma frequency" radio bursts.** H. M. Bradford.
Journ. Roy. Astron. Soc. Canada, Vol. 69, 254 (1975).
Abstr. CAS.

074.098 **Diffusion effects on the line intensities of He I and He II in the solar transition region.**
R. Shine, H. Gerola, J. L. Linsky.
Astrophys. Journ., (*Letters*), Vol. 202, L101 - L105 (1975).
The authors develop a heuristic treatment of diffusion in the solar chromosphere-corona transition region. They find that diffusion becomes increasingly important with steeper temperature gradients, in active and quiet regions relative to coronal holes, and with increasing excitation potential. Numerical calculations are made for the resonance lines of He I and He II and show that diffusion can enhance these lines. Thus the helium lines may appear relatively weak in coronal holes due to a weakening of the enhancement mechanism. Most transition region lines will be less affected by diffusion than He I or He II.

074.099 **Lack of evidence for solar-cycle variations in high-speed streams in the solar wind.**
V. M. Vasyliunas.
Astrophys. Journ., (*Letters*), Vol. 202, L159 - L160 (1975).
The observed yearly numbers and durations of high-speed streams in the solar wind previously reported by Intriligator, when normalized to the duration of available observations, do not exhibit a significant systematic variation with the solar cycle, contrary to the inference drawn by Intriligator.

074.100 **Discrepancies between calculated and observed Fe XVII and Ni XIX solar coronal lines.**
M. Loulergue, H. Nussbaumer.
Astrophys. Space Sci., Vol. 38, 365 - 368 (1975). – Paper presented at the I.A.U. Colloquium No. 27 on 'UV and X-ray spectroscopy of astrophysical and laboratory plasmas', Cambridge, Mass., U.S.A., 9–11 September 1974.
The authors extend the 36-levels model atom Fe^{+16} of Loulergue and Nussbaumer (1973) to a 51-levels atom including $n = 4$ configuration. An earlier conclusion that relative intensities among the Fe XVII lines are not sufficiently sensitive to variations in electron densities and electron temperature to deduce these parameters from observed intensity ratios still holds. Discrepancies between calculated and observed relative intensities in Fe XVII and Ni XIX are attributed to blends or misidentifications.

074.101 **A study of Fe XVII and Ni XIX coronal lines.** M. Loulergue, H. Nussbaumer.
Astron. Astrophys., Vol. 45, 125 - 134 (1975).
Collision strengths and radiative transition probabilities for $n = 3$ and $n = 4$ configurations have been calculated in view of the solar Fe XVII and Ni XIX spectra. The overall good agreement between calculated and observed line intensity ratios gives confidence that we now understand the atomic processes involved in forming Fe XVII and Ni XIX. From coronal observations a relative abundance $Fe/Ni \approx 20$ was calculated.

074.102 **An idealised model of a magnetohydrodynamic spherical blast wave applied to a flare produced shock in the solar wind.** D. Summers.
Astron. Astrophys., Vol. 45, 151 - 158 (1975).
A spherically symmetric model incorporating a self-consistent magnetic field is constructed to describe a blast wave in the solar wind caused by the explosive energy release of a solar flare. Three general integrals of the four blast wave equations are found; complete analytic solutions are given for a special choice of parameters and for the case of a strong magnetic field. Numerical solutions are obtained for the magnetic field and gas pressure, velocity and density distributions within the shock, and the energy release required to produce the observed enhancement of the solar wind variables at the orbit of the earth is calculated.

074.103 **Analysis of EUV limb-brightening observations from ATM. I: Model for the transition layer and the corona.** J. T. Mariska, G. L. Withbroe.
Solar Physics, Vol. 44, 55 - 68 (1975).
A comparison of theoretical and observed limb-brightening curves indicates that the lower boundary of the corona where $T_e = 10^6$ K, is at a height of about 8000 km in typical quiet areas. For $1.01\ R_\odot \lesssim r < 1.25\ R_\odot$, the corona can be represented by a homogeneous model in hydrostatic equilibrium. The model for the transition layer is inhomogeneous, with the temperature gradient a factor of 3 shallower in the network than in the intranetwork regions. It appears that spicules should be included in the model in order to account for the penetration into the corona of cool ($T_e < 10^6$ K) EUV-emitting material to heights up to 20000 km above the limb.

074.104 **The coronal structure of active regions.** M. Landini, B. C. Monsignori Fossi, A. Krieger, G. S. Vaiana.
Solar Physics, Vol. 44, 69 - 82 (1975).
A four-parameter model which assumes a Gaussian

dependence of both temperature and pressure on distance from center is used to fit the compact part of coronal active regions as observed in X-ray photographs from a rocket experiment. The four parameters are the maximum temperature, the maximum pressure, the width of the pressure distribution and the width of the temperature distribution.

074.105 **A comparison of coronal X-ray structures of active regions with magnetic fields computed from photospheric observations.** G. Poletto, G. S. Vaiana, M. V. Zombeck, A. S. Krieger, A. F. Timothy.
Solar Physics, Vol. 44, 83 - 99 (1975).

The appearances of several X-ray active regions observed on March 7, 1970 and June 15, 1973 are compared with the corresponding coronal magnetic field topology. Coronal fields have been computed from measurements of the longitudinal component of the underlying magnetic fields, under the current-free hypothesis. An overall correspondence between X-ray structures and calculated field lines is established, and the magnetic counterpart of different X-ray features is also examined. A correspondence between enhanced X-ray emission and the location of compact closed field lines is suggested by this study. Representative magnetic field values calculated under the assumption of current-free fields are given for heights up to 200".

074.106 **Heat transfer in the corona and transition region.** R. G. Giovanelli.
Solar Physics, Vol. 44, 315 - 329 (1975).

Thermal transfer in closed magnetic tubes in the corona and transition region is described on the basis of a static model in which all heat generated is radiated away, though conduction transfers much of the heat to the transition region prior to emission. The rate of conductive transfer depends on the cross-section of the magnetic tube as it passes through the chromosphere and transition region. For the observed rate of heating, the computed densities (or pressures), the ratio of coronal to transition region emissions, and the distribution of radiation in the EUV spectrum agree closely with those observed.

074.107 **Coronal emission line profile observations at total solar eclipses. I: Airborne instrumentation and results.** D. H. Liebenberg.
Solar Physics, Vol. 44, 331 - 344 (1975).

A 25-cm aperture $f/8.0$ telescope, a fixed path Fabry-Perot interferometer spectrograph with 0.008-nm resolution, and a pressure scanned Fabry-Perot interferometer with a photoelectric detector and recorder are described. Photographic interferometer data were obtained out to 1.79 $R_{\odot}$ in 1965 at 530.3 nm. Doppler shifts are determined and indicate generally small velocities of coronal material. Results in Fe XIV and Fe X were obtained in 1966. No Ca XV emission was detected in 1966.

074.108 **Coronal emission line profile observations at total solar eclipses. II: 30 May 1965 results, deconvolution and interpretation.**
D. H. Liebenberg, R. J. Bessey, B. Watson.
Solar Physics, Vol. 44, 345 - 359 (1975).

High resolution spectra of the coronal emission line Fe XIV at 530.3 nm obtained at the 30 May 1965 total solar eclipse are analyzed and interpreted. Deconvolution techniques that preserve the line intensity vs wavelength profile shape are developed to obtain further resolution improvement. Several distinct coronal features are described and discussed.

074.109 **Differential rotation and reconnection as basic causes of some coronal reorientations.**
S. F. Hansen, R. T. Hansen.
Solar Physics, Vol. 44, 503 - 508 (1975).

A process is suggested by which a coronal structure (with underlying filament) may form between a polar crown structure and a low-latitude bipolar region. During the ascending phase of the solar cycle the identifying underlying filament should lie poleward and westward of the active region, but during the descending phase it should appear as an eastward extension of the filament separating leader and follower photospheric fields within the active region.

074.110 **Coronal observations of the July 10, 1972 total solar eclipse.** Ts. S. Khetsuriani, G. N. Salukvadze, Eh. I. Tetruashvili, A. Sh. Khatisov, A. K. Majer, V. K. Lokhov.
Byull. Abastumansk. Astrofiz. Obs., No. 46, p. 163 - 176 (1975).

In order to study the corona of the July 10, 1972 total solar eclipse the following tasks were set: 1.) to study the structure of the solar corona and its photometry; 2.) to measure the polarization of coronal radiation; 3.) a precise measurement of the position of the polarization plane by means of photoelectric methods. The preliminary results of the treatment of the observational material are given.

074.111 **The June 30, 1973 solar corona.**
Ts. S. Khetsuriani.
Byull. Abastumansk. Astrofiz. Obs., No. 46, p. 177 - 184 (1975). In Russian.

The structural peculiarities of the corona are described. The isophotes of the corona are drawn by means of the equidensity method. The shape and structure of the corona are compared with those concerned to previous eclipses.

074.112 **On the distribution of the half-widths of coronal line profiles.** Eh. I. Tetruashvili.
Byull. Abastumansk. Astrofiz. Obs., No. 46, p. 185 - 188 (1975). In Russian.

On the basis of the observational material obtained in 1966–1972 with the help of the Abastumani Astrophysical Observatory coronograph 2530 contours of the green coronal line and 910 for the red coronal line have been treated. In order to study the half-widths distribution of the coronal lines some histograms have been plotted.

074.113 **Electron heat flux instabilities in the solar wind.**
S. P. Gary, W. C. Feldman, D. W. Forslund, M. D. Montgomery.
Geophys. Res. Letters, Vol. 2, 79 - 82 (1975).

There are at least three plasma instabilities associated with the electron heat flux in the solar wind. This letter reports the study of the unstable fast magnetosonic, Alfvén and whistler modes via a computer code which solves the full electromagnetic, linear, Vlasov dispersion relation. Linear theory demonstrates that both the magnetosonic and Alfvén instabilities are candidates for turbulent limitation of the heat flux in the solar wind at 1 A.U.

074.114 **Solar wind electrons.** W. C. Feldman, J. R. Asbridge, S. J. Bame, M. D. Montgomery, S. P. Gary.
Journ. Geophys. Res., Vol. 80, 4181 - 4196 (1975).

Average characteristics of solar wind electron velocity distributions as well as the range and nature of their variations are presented.

074.115 **Heat flux instabilities in the solar wind.**
S. P. Gary, W. C. Feldman, D. W. Forslund, M. D. Montgomery.
Journ. Geophys. Res., Vol. 80, 4197 - 4203 (1975).

The linear dispersion properties of three electron heat flux plasma instabilities are studied in detail by using parameters appropriate to average values in the solar wind at 1 AU. Among the (fast) magnetosonic, Alfvén, and whistler instabilities the first one appears most likely to make a significant contribution to the regulation of the heat flux.

074.116 **An empirical relationship between interplanetary conditions and *Dst*.**
R. K. Burton, R. L. McPherron, C. T. Russell.
Journ. Geophys. Res., Vol. 80, 4204 - 4214 (1975).
An algorithm is presented for predicting the ground-based *Dst* index solely from a knowledge of the velocity and density of the solar wind and the north–south solar magnetospheric component of the interplanetary magnetic field. The algorithm is used to predict the *Dst* signature of seven geomagnetic storm intervals in 1967 and 1968.

074.117 **Unsteady interactions of solar wind disturbances with the magnetosphere.** T. Tamao.
Journ. Geophys. Res., Vol. 80, 4230 - 4236 (1975).
A model for unsteady interactions between solar wind disturbances and the magnetosphere associated with the southward shift of the solar wind magnetic field during the growth phase of substorms is presented. It is emphasized that the presence of the solar wind electric current relating to the southward turning of the magnetic field and its interaction with the magnetosphere are essential in describing the transient process that generates the large-scale magnetospheric convection.

074.118 **The fine structure of solar magnetoplasma.**
Eh. I. Mogilevskij.
Uspekhi fiz. nauk, Vol. 116, 545 - 546 (1975). In Russian.
Abstr. in Referativ. Zhurn. 51. Astron., 12.51.356 (1975).

074.119 **Electric currents in the solar wind.**
P. J. Coleman, Jr.
Journ. Geophys. Res., Vol. 80, 4719 - 4724 (1975).
The electric currents in the solar wind are described for the particular case of Parker's model, in which the sun's magnetic field is a dipole located at the center of the sun and inclined at an arbitrary angle to the sun's axis of rotation. The current consists of two components. The density of one depends only upon the sun's dipole moment, and that of the other depends upon the solar wind velocity and the sun's angular velocity as well as its dipole moment. The latter component flows in heliographic meridional planes. If the angle λ between the sun's spin axis and the dipole axis is different from either 0° or 90°, the electromagnetic force near the equatorial plane has a southward component in the positive sector and a northward component in the negative sector. For a given dipole moment, $\lambda = 90°$ is also the orientation that maximizes the electromagnetic braking. Possible electromagnetic contributions to the sun's differential rotation are also considered.

074.120 **The excitation of the coronal line 5303 Å.**
M. Waldmeier.
Solar Physics, Vol. 45, 147 - 155 = Astron. Mitt. Eidgenössisch. Sternw. Zürich, No. 340 (1975).
The relation between the intensities of the line 5303 Å and that of the nearby continuum has been investigated at the solar eclipse of March 7, 1970. 4 spectrograms have been selected that show the line 5303 Å up to $r = 2$. For distances $r > 1.8$ the ratio of line- to continuum-intensity, Q, becomes constant. This is the region where the excitation by radiation dominates. Towards smaller distances Q increases as the excitation by collisions becomes more and more important. At $r = 1.3$ the contributions from both mechanisms are equal. For $r < 1.3$ Q increases stronger than expected for an isothermal and homogeneous model. In the innermost corona the observed values are twice as high as the theoretical ones. This difference can be accounted for by inhomogeneities of the corona.

074.121 **Interferometric investigation of the line of sight velocities in λ 5303 during the eclipse of 22 September, 1968.** A. B. Delone, E. A. Makarova.
Solar Physics, Vol. 45, 157 - 168 (1975).
Fabry-Perot interferometric observations of the λ5303 Å corona line on 22 September, 1968 were carried out. There are many elements of the interferometric rings with considerable line of sight velocities. The influence on the results of the changes in the brightness of the monochromatic and white-light corona were discussed.

074.122 **Meridional flow in the solar wind in the presence of latitudinally dependent boundary conditions.**
S. F. Nerney, S. T. Suess.
Solar Physics, Vol. 45, 255 - 266 (1975).
The influence of latitudinally dependent boundary conditions on the large radius values of meridional flow in the distant solar wind is examined through a double perturbation expansion of the magnetohydrodynamic equations. A general result is derived for the meridional velocity which allows arbitrary specification of radial velocity, radial magnetic field, and mass flux, as a function of colatitude at some coronal reference surface. Three specific examples are treated.

074.123 **Variations of the solar wind parameters in the solar activity cycle.**
V. A. Kovalenko, V. N. Malyshkin.
Solnechnye Dannye 1975 Byull., No. 9, p. 95 - 99 (1975). In Russian.
The solar wind parameters are compared with solar activity indices during 1962–1973. The average characteristics of the solar wind on the earth's orbit are shown to vary either weakly or in counterphase with the solar activity indices. The fluctuations (characteristic time 24 hours) of the interplanetary magnetic field component perpendicular to the ecliptic increased by about one order from 1964 to 1969.

074.124 **Saules vainaga rentgenstruktūra.** V. Kasinskis.
Zvaigžņotā debess, 1975. gada rudens, p. 1 - 5.

074.125 **Evidence for local ion heating in solar wind high speed streams.** S. J. Bame, J. R. Asbridge, W. C. Feldman, S. P. Gary, M. D. Montgomery.
Geophys. Res. Letters, Vol. 2, 373 - 375 (1975).
The thermal anisotropies near the peaks of proton distributions are observed to vary characteristically across two simple high speed streams. This implies that it may not be necessary to invoke extended ion heating close to the sun to explain enhanced ion temperatures associated with high speed streams and that the T – V relation may be primarily the result of local heating near 1 AU.

074.126 **On coronal ionization cross-section.**
S. Chandra, U. Narain.
Indian Journ. Pure Applied Phys., Vol. 13, 276 (1975).
Coronal ionization cross-sections of Fe ions have been calculated in the energy range 0.5–3.0 keV, using Lotz's (1967) formula. The results support the conclusions drawn by Burgess (1960). The assumption made by Schwartz and Zirin (1959) regarding the cross-sections for these ions has been carefully examined and found incorrect.

074.127 **Nord-Süd-Asymmetrie im Sonnenwind und in der Sonnentätigkeit.** R. Hartmann.
Kleinheubacher Ber., Vol. 18, 165 - 169 = Veröff. Astron. Inst. Univ. Frankfurt (Main), No. 55 (1975).

074.128 **Relative coronal abundances derived from X-ray observations. III. The effect of cascades on the relative intensity of Fe XVII line fluxes, and a revised iron abundance.** A. B. Walker, Jr., H. R. Rugge, K. Weiss.

Rep. TR-0075(9260-02)-1, Aerospace Corp., El Segundo, Calif., USA. [Available from NTIS, Springfield, Va., USA], 43 pp. (1974).

074.129 **HEOS-2 plasma experiment: composition of the solar wind.** H. Gruenwaldt, H. Rosenbauer.
Verhandl. Deutsche Phys. Ges., Reihe 6, Vol. 10, 187 (1975). In German. – Abstract.

074.130 **Effect of Alfvénic fluctuations on the solar wind.** T. H. Chien.
Thesis, Catholic Univ. of America, Washington, D.C. (USA). [Available from Univ. Microfilms, Ann Arbor. Mich., *USA*. Order No. 75-13,624], 131 pp. (1974).

The major source of microscale fluctuations in the interplanetary medium due to outwardly propagating Alfvén waves is considered. The effect of the Alfvén waves on the supersonic expansion of the solar wind is studied under the assumption that the motion of the interplanetary medium can be resolved physically into a comparatively smooth and slowly varying mesoscale flow and field with very irregular disordered incompressible microscale Alfvénic fluctuations superposed on it.

074.131 **The behaviour of minor species in the solar wind.** J. M. Ryan, W. I. Axford.
Journ. Geophys., Vol. 41, 221 - 232 (1975).

The flow of minor species in the solar wind is considered on the basis of a model in which all species are heated in an arbitrary manner as they leave the sun and the minor species interact with the background proton-electron plasma through the radial electric field associated with the latter and by means of Coulomb collisions.

074.132 **Isothermal flow behind a shock wave propagating in the solar wind.** M. P. Ranga Rao, B. V. Ramana.
Zeitschr. Angew. Math. Phys., (*Switzerland*), Vol. 26, 289 - 297 (1975).

A Parker-type blast wave, which is headed by a strong shock, driven out by a propelling contact surface moving into an ambient solar wind having a strictly inverse square law radial decay in density, is studied. Assuming the self-similar flow behind the shock to be isothermal, approximate analytical and exact numerical solutions are obtained.

New calculations of atomic data concerning E.U.V. lines of O V. See Abstr. 022.057.

Proton collisional excitation in the ground configurations of Fe^{+12} and Fe^{+13}. III. Transitions between magnetic sublevels. See Abstr. 022.061.

Density dependence of the line emission of boron-like coronal ions. See Abstr. 022.097.

New, almost electrostatic, temperature anisotropy instability. See Abstr. 062.073.

Dual-frequency measurement of the solar gravitational microwave deflection. See Abstr. 066.108.

On obliteration of the Fraunhofer lines in the spectrum of the inner corona. See Abstr. 071.039.

Solar activity and the solar wind: 1. Eleven-year cycles. See Abstr. 072.050.

The helium chromosphere, coronal holes, and stellar X-rays. See Abstr. 073.002.

Comments concerning the photoionization model for excitation of resonance lines of He I and He II in the solar chromosphere. See Abstr. 073.004.

Association of X-ray flares with solar coronal active regions. See Abstr. 073.008.

Extreme ultraviolet solar limb brightening observations. See Abstr. 073.060.

The geometry of the chromosphere-corona transition region inferred from the center-to-limb variation of the radio emission. See Abstr. 073.067.

EUV analysis of an active region. See Abstr. 073.070.

The temperature structure and pressure balance of magnetic loops in active regions. See Abstr. 073.096.

Inclination of solar quiescent prominences and coronal extrapolation of photospheric magnetic fields. See Abstr. 073.107.

Partial analysis of the flare-prominence of 30 April 1974. See Abstr. 073.124.

The formation of the He I λ584 and He II λ304 resonance lines in the solar transition region. See Abstr. 073.134.

Spicule formation caused by the mechanism of coronal gas compression. See Abstr. 073.150.

Quiet and active solar emission of Fe XV 284 Å radiation. See Abstr. 076.001.

Spatially resolved X-ray spectra of coronal active regions. See Abstr. 076.007.

Moving coronal clouds observed with the Skylab S-054 X-ray telescope. See Abstr. 076.036.

The intensities and profiles of XUV transition zone lines in a quiet sun region compared to a polar coronal hole. See Abstr. 076.050.

Limb-brightening curves of XUV transition zone lines in the quiet sun and in a polar coronal hole observed from Skylab. See Abstr. 076.051.

Analysis of the intensities and profiles of the spectral line Mg XII 8.42 Å in the solar X-ray spectrum. See Abstr. 076.054.

On the extreme ultraviolet solar emission of B-like ions: Na VII, Si X and S XII. See Abstr. 076.058.

Discussion on the coronal structure related to type III bursts. See Abstr. 077.005.

On the decay of type III radio bursts in the solar corona. See Abstr. 077.009.

Spectral characteristics of stria, split pair and triple bursts. See Abstr. 077.017.

Observations of the sun at 408 MHz. See Abstr. 077.027.

On the origin of fast drift absorption bursts. See Abstr. 077.036.

Cyclotron wave instability in the corona and origin of solar radio emission with fine structure. II: Origin of 'tadpoles. See Abstr. 077.044.

Cyclotron wave instability in the corona and origin of solar radio emission with fine structure. III. Origin of zebra-pattern. See Abstr. 077.045.

Fast solar electrons, interplanetary plasma and km-wave type-III radio bursts observed from the IMP-6 spacecraft. See Abstr. 077.047.

Evidence of large scale diverging paths in the solar corona for type III bursts exciters. See Abstr. 077.051.

Unusual emission of iron nuclei from the sun. See Abstr. 078.004.

Cosmic ray intensity variations and solar wind. See Abstr. 078.018.

The Zeeman broadening of high n solar recombination lines. See Abstr. 080.066.

Weak pitch angle scattering of energetic protons in the magnetosphere. See Abstr. 084.208.

Motions of the earth's bow shock. See Abstr. 084.210.

The response of the magnetosphere to the solar wind. See Abstr. 084.213.

Evidence for magnetic field line reconnection in the solar wind. See Abstr. 084.221.

The terrestrial magnetosphere: a half-wave rectifier of the interplanetary electric field. See Abstr. 084.230.

Solar wind–magnetosphere coupling. See Abstr. 084.236.

Magnetic disturbances in the polar caps and parameters of the solar wind. See Abstr. 084.265.

Whistler observations during a magnetospheric sudden impulse. See Abstr. 084.317.

Mercury's helium exosphere. See Abstr. 092.008.

Variations in the solar wind stand-off distance at Mercury. See Abstr. 092.014.

Substorms on Mercury? See Abstr. 092.015.

Formation of the lunar atmosphere. See Abstr. 094.162.

Molecular gas species in the lunar atmosphere. See Abstr. 094.163.

Solar-wind interactions: nature and composition of lunar atmosphere. See Abstr. 094.164.

On the source of lunar limb compressions. See Abstr. 094.180.

The role of sputtering in the accumulation of implanted solar wind species in lunar soils. See Abstr. 094.190.

Solar wind implantation and agglutination in the formation of finely-divided iron in lunar soil. See Abstr. 094.191.

Past record of solar wind, solar flares, and micrometeorites in individual crystals from lunar cores. See Abstr. 094.192.

Experimental detection of the influence of radiation damages upon the diffusion behaviour of solar wind noble gases in single grains of moon dust. See Abstr. 094.478.

On the gas exchange between the Martian upper atmosphere and the solar wind. See Abstr. 097.048.

Pioneer 10 observations of the solar wind and its interaction with Jupiter: plasma electron results. See Abstr. 099.020.

Jupiter's non Io-related decametric radiation. See Abstr. 099.048.

Corrections to the azimuthal component of the interplanetary magnetic field due to meridional flow in the solar wind. See Abstr. 106.013.

Multifrequency techniques for studying interplanetary scintillations. See Abstr. 106.020.

The global solar wind and predictions for Pioneers 10 and 11. See Abstr. 106.053.

Interplanetary scintillation and flare - produced disturbances. See Abstr. 106.070.

Thermoionic emission from hot grains. See Abstr. 131.032.

Observations of Faraday rotation of linearly polarized radio emission from the Crab nebula in the solar supercorona at 4 cm wavelength with the help of the large Pulkovo radio telescope. See Abstr. 134.011.

Cosmic ray intensity gradients in the solar system. See Abstr. 143.012.

Diffusion of cosmic rays in the asymmetrical solar wind. See Abstr. 143.033.

Composition and energy spectra of heavy nuclei of unknown origin detected on Skylab. See Abstr. 143.051.

Cosmic rays and solar wind. See Abstr. 143.054.

Cosmic rays and the solar wind. See Abstr. 143.060.

Errata

074.901 **Erratum: "Low-frequency fluctuations in the solar wind. I. Theory."** [Astrophys. Journ., Vol. 196, 849 - 857 (1975)]. C. S. Wu, J. D. Huba. Astrophys. Journ., Vol. 200, 525 (1975).

075 Solar Patrol

075.001 **Nombres relatifs moyens de l'activité solaire pour l'année 1973.** M. Waldmeier.
L'Astronomie, 89e année, p. 343 (1975).

075.002 **Nombres relatifs moyens de l'activité solaire pour l'année 1974.** M. Waldmeier.
L'Astronomie, 89e année, p. 344 (1975).

075.003 **Solar activity and geomagnetic storms 1974.** A. L. T. Powell, K. Dyson.
Observatory, Vol. 95, 223 - 225 (1975).

075.004 **Visual observation of the sun in Czechoslovakia in the year 1974.** L. Schmied.
Říše hvězd, Vol. 56, 149 - 150 (1975). In Czech.

075.005 **Catalogue of solar activity for the year 1972.** R. S. Gnevysheva.
Glav. Astron. Obs., *Pulkovo,* 128 pp. Price 89 Kop. (1975). In Russian.

075.006 **Catalogue of solar activity for the year 1973.** R. S. Gnevysheva.
Glav. Astron. Obs., *Pulkovo,* 100 pp. Price 86 Kop. (1975). In Russian.

075.007 **Solar and solar system activity. Radio Astronomy Section (BAA).** J. R. Smith, R. J. J. Langton.
Journ. British Astron. Ass., Vol. 85, 436 - 437, 509 - 510 (1975).

075.008 **L'activité solaire.** M.-J. Martres.
L'Astronomie, 89e année, p. 323, 348 - 349, 393 - 394, 442 - 444 (1975). – Rotations 1624 - 1630.

075.009 **Actividad solar en 1973. I. Números relativos de Wolf. II. Estadística de manchas y superficie de las mismas.** M. Lopez Arroyo.
Bol. Astron. Obs. Madrid, Vol. 9, No. 1, 46 pp. (1975).

075.010 **Observations solaires.** C. Popovici, E. Ţifrea, V. Dinulescu, A. Dimitriu, S. Dinulescu, G. Mariş, I. D. Niţă.
Acad. République Socialiste Roumanie, Obs. Bucarest, Secteur Solaire, p. 1 - 42 (1975). – Rotations 1610–1622 (6 janvier–26 décembre 1974).

075.011 **Osservazioni solari.** Nos. 33 - 36, 1974 January - December.
Oss. Astron. Trieste, Pubbl. Nos. 485, 486, 499, 500 (1975).
Flux density and polarization data at 237 MHz; Interferometric and radiometric data at 408 MHz; Distinctive events at 237 and 408 MHz.

075.012 **Observations radioélectriques solaires faites sur 600 MHz en 1973 au laboratoire de radioastronomie de Humain-Rochefort.** C. Gonze, R. Gonze.
Bull. Astron., Obs. Roy. Belgique, Vol. 8, 233 - 243 (1975).
Densité de flux, variabilité; Événements remarquables; Moyennes journalières.

075.013 **Observations radioélectriques solaires faites sur 600 MHz en 1974 au laboratoire de radioastronomie de Humain-Rochefort.** C. Gonze, R. Gonze.
Bull. Astron., Obs. Roy. Belgique, Vol. 8, 278 - 288 (1975).

075.014 **Die Sonnenaktivität im Jahre 1974.** M. Waldmeier.
Vierteljahrsschrift Naturforsch. Ges. Zürich, Jahrgang 120, p. 165 - 180 = Astron. Mitt. Eidgenössisch. Sternw. Zürich No. 339 (1975).
The present paper gives the frequency numbers of sunspots, photospheric faculae and prominences as well as the intensity of the coronal line 5303 Å and of the solar radio emission at the wavelength of 10.7 cm, all characterizing the solar activity in the year 1974.

075.015 **Sunspots** (sunspot relative-numbers and sunspot-areas); **Synoptic charts of solar magnetic fields** (Mount Wilson Observatory); **Eruptions chromosphériques brillantes; Intensité de la couronne solaire; Solar radio emission.**
M. Waldmeier, R. Howard, G. Olivieri, M. Bernot, H. Tanaka.
Quarterly Bull. Solar Activity (published by Eidgenössisch. Sternw. Zürich), Nos. 187 - 188, p. 255 - 340 (1975).
1974 July - December.

075.016 **Heliographic maps of the photosphere for the year 1974.** M. Waldmeier.
Publ. Eidgenössisch. Sternw. Zürich, Vol. 14, (No. 4), 95 - 123 (1975). – Rotations 1610 - 1622.

075.017 **Solare Beobachtungsergebnisse (Solar Data).** E. A. Lauter, C.-U. Wagner, A. Böhme, F. Fürstenberg, D. Scholz, S. Böhm.
Zentralinst. für Solar-Terrestrische Physik (Heinrich-Hertz-Inst.), Akad. Wiss. DDR, HHI Solar Data, Vol. 26, 25 - 73, April - September (1975). – Solar radio emission.

075.018 **Solar photospheric observations.** F. Bruin, H. Hourani, N. G. Bustati.
Lee Obs. American Univ. Beirut, Monthly Bull. Astron. Section, 1974 December–1975 May (1975).
Sunspot relative numbers, Heliographic mean position and classification of the sunspot groups, Number of facular zones.

075.019 **Daily Hα chromosphere pictures, daily K_{232} chromosphere pictures, daily white light photosphere pictures.** M. Cimino (Editor).
Photographic Journ. of the Sun, Oss. Astron. Roma, Nos. 93 - 98 (1974/75). – 1974 November 1 – 1975 April 14. – Rotation 1621 - 1626.

075.020 **Solar phenomena.** M. Cimino, M. Torelli.
Oss. Astron. Roma, Monthly Bull. Nos. 203 - 208 (1975).
1975 March - August: Daily total areas of sunspot-groups; Heliographic position, classification and area of sunspot-groups; Hours of K-line cinematographic patrol; Hours of Hα cinematographic patrol; Sudden cosmic noise absorption S.C.N.A. (18 Mc/s), sudden enhancement of atmospherics S.E.A. (27Kc/s); Explanation.

075.021 **Photoheliographic results 1966.**
Compiled under the supervision of P. S. Laurie by G. M. Appleby, C. Y. Hohenkerk, C. M. Ladley, P. J. Rudd.
Roy. Obs. Ann., [Herstmonceux, Royal Greenwich Obs.], No. 10, 67 pp. Price £ 1.35 net (1974). – The following data are given: Positions and areas of sunspots for each day in the year; General catalogue of groups of sunspots; Total areas of sunspots and faculae; Mean areas of sunspots and faculae; Mean heliographic latitude of sunspots; Summary of solar activity.

075.022 **Daily maps of the sun and geophysical graphs.**
Solnechnye Dannye 1975 Byull., No. 3, p. 1 - 51; No. 4, p. 1 - 46; No. 5, p. 1 - 50; No. 6, p. 1 - 54; No. 7, p. 1 - 60; No. 8, p. 1 - 68; No. 9, p. 1 - 52. In Russian.

075.023 **Magnetic fields of sunspots.**
Prilozhenie k Byulletenyu "Solnechnye Dannye", 1975, Nos. 3 - 9. In Russian.

075.024 **Indices of geomagnetic activity.**
Journ. Atmosph. Terr. Phys., Vol. 37, 1287, 1288 (1975). – 1975 February - March.

075.025 **Geomagnetic and solar data.**
J. V. Lincoln (Editor).
Journ. Geophys. Res., Vol. 80, 2892, 3284, 3732, 4404, 4772 - 4774 (1975). – 1975 March - August.

075.026 **Sunspot numbers.**
Sky Telescope, Vol. 50, 65, 133, 199, 267, 347, 428 (1975).

076 Solar UV, X Rays, Gamma Radiation

076.001 **Quiet and active solar emission of Fe XV 284 Å radiation.** C. W. Allen.
Monthly Notices Roy. Astron. Soc., Vol. 172, 159 - 173 (1975).
Intensity contours of 284 Å radiation from the OSO-7 EUV spectroheliograph observations have been analysed to segregate emissions from the quiet sun (Q) and active regions (AR) during the period 1972 May to 1974 March. Measurements include: central, equatorial limb and polar intensities; also Q, AR and total fluxes. Both Q and AR fluxes vary almost in proportion to sunspot number. The absolute emission of the quiet corona has been compared with expected values based on solar eclipse electron densities. The comparison allows an estimate of the coronal irregularity factor χ which varies from 4 near the surface to 30 at 1.5 solar radii. Coronal densities within well-developed AR's are about five times the Q value.

076.002 **EUV observations of the active sun from the Harvard experiment on ATM.**
R. W. Noyes, P. V. Foukal, M. C. E. Huber, E. M. Reeves, E. J. Schmahl, J. G. Timothy, J. E. Vernazza, G. L. Withbroe.
IAU Symposium No. 68, (see 012.003), p. 3 - 17 (1975).
Review paper.

076.003 **Time variations of solar X-ray bright points.**
L. Golub, A. S. Krieger, J. K. Silk, A. F. Timothy, G. S. Vaiana.
IAU Symposium No. 68, (see 012.003), p. 23 - 24 (1975).
Summary.

076.004 **Solar activity observed in X-rays and the EUV from OSO 7.** R. J. Thomas.
IAU Symposium No. 68, (see 012.003), p. 25 - 42 (1975).

076.005 **X-ray spectra of solar active regions.**
J. H. Parkinson.
IAU Symposium No. 68, (see 012.003), p. 45 - 64 (1975).
The last few years have seen great progress in our understanding of X-ray spectra of solar active regions. This paper demonstrates both the usefulness and the limitations of the techniques, both scientific and instrumental, that have recently become available.

076.006 **X-ray spectroscopy of solar active regions during the third Skylab mission.**
J. P. Pye, R. J. Hutcheon, J. H. Parkinson, K. A. Pounds.
IAU Symposium No. 68, (see 012.003), p. 65 - 66 (1975).
Summary.

076.007 **Spatially resolved X-ray spectra of coronal active regions.** R. C. Catura, L. W. Acton, E. G. Joki, C. G. Rapley, J. L. Culhane.
IAU Symposium No. 68, (see 012.003), p. 67 (1975). – Summary.

076.008 **Statistical methods in the identification and prediction of the solar X-ray spectral lines.** S. Yousef.
IAU Symposium No. 68, (see 012.003), p. 71 - 72 (1975).
Summary.

076.009 **Interpretation of the X-ray spectra of solar active regions.** A. B. C. Walker, Jr.
IAU Symposium No. 68, (see 012.003), p. 73 - 100 (1975).
Review paper.

076.010 **Soft X-radiation from single active regions.**
D. H. Brabban, E. B. Dorling, W. M. Glencross, J. R. H. Herring.
IAU Symposium No. 68, (see 012.003), p. 101 - 102 (1975).
Summary.

076.011 **The structure of solar active regions from EUV and soft X-ray observations.** C. Jordan.
IAU Symposium No. 68, (see 012.003), p. 109 - 131 (1975).
Review paper.

076.012 **The relationship between hard and soft X-ray bursts observed by OSO 7.** D. W. Datlowe.
IAU Symposium No. 68, (see 012.003), p. 191 - 208 (1975).
Review paper.

076.013 **Relationship between hard and soft solar X-ray sources observed by OSO-7.**
D. W. Datlowe, H. S. Hudson.
IAU Symposium No. 68, (see 012.003), p. 209 (1975).
Summary.

076.014 **Rise time of hard X-ray bursts.**
J. Vorpahl, T. Takakura.
IAU Symposium No. 68, (see 012.003), p. 237 - 238 (1975).
Summary.

076.015 **Determination of the height of hard X-ray sources in the solar atmosphere by measurement of photospheric albedo photons.** J. C. Brown, H. F. van Beek.
IAU Symposium No. 68, (see 012.003), p. 239 - 241 (1975).
Summary.

076.016 **The interpretation of spectra, polarization, and directivity of solar hard X-rays.** J. C. Brown.
IAU Symposium No. 68, (see 012.003), p. 245 - 282 (1975).
Review paper.

076.017 **Solar flare X-ray measurements and their relation to microwave bursts.** L. D. de Feiter.
IAU Symposium No. 68, (see 012.003), p. 283 - 297 (1975).
Review paper.

076.018 **X- and γ-ray measurements during the 1972, August 2 and 7 large solar flares.**
R. Talon, G. Vedrenne, A. S. Melioransky (*Melioranskij*), N. F. Pissarenko (*Pisarenko*), V. M. Shamolin, O. B. Likin.
IAU Symposium No. 68, (see 012.003), p. 315 - 339 (1975).
The solar X- and gamma-ray experiment aboard Prognoz 2 recorded solar gamma-radiation bursts during the 1972, August 2, 4 and 7 events. This work analyses the general evolution of the August 2 and 7 events and the possible evidence for γ-ray lines during these two events.

076.019 **Measurements of a gamma-ray burst above 1 MeV.**
R. Koga, G. M. Simnett, R. S. White.
IAU Symposium No. 68, (see 012.003), p. 361 (1975).
Summary.

076.020 **Spectral analysis of the time variations of solar X-rays in the 1–8 Å and 8–20 Å regions.**
E. M. Apostolov, V. Letfus.
Bull. Astron. Inst. Czechoslovakia, Vol. 26, 193 - 198 (1975).
Standard autocorrelation and power spectral analysis has been applied to the solar X-ray measurements made on the Solrad 9 satellite in the spectral regions 1–8 Å and 8–20 Å. Periods of 82, 42, 30.5, 27.3, 22.5, 20 and 11.8 days have been found in the power spectra of the time series of both the

8–20 Å data and of the 1–8 Å daily minima. Strong in power are the periods of 82 days in all power spectra and of 27 days in the 8–20 Å ones. The period of 22.5 days is newly observed and does not occur in the time series of other solar indices. The periods different from the period of the sun's rotation may be due to the structure of the longitudinal distribution of solar activity and its time development.

076.021 **General description of the solar X and γ ray bursts observed aboard Prognoz 2 during August 4 and 7, 1972 events.** A. S. Meliorensky (*Meliorenskij*), N. F. Pissarenko (*Pisarenko*), W. M. Shamolin, O. B. Likin, R. Talon, G. Vedrenne, B. Gourdon.
Astron. Astrophys., Vol. 41, 379 - 384 (1975).
The solar gamma radiation results the authors discuss here were obtained during a Franco-Soviet experiment launched aboard a satellite with a very eccentric orbit, pointed toward the sun. Proton and gamma-ray detector data show a good time correlation between different types of solar emission particularly between gamma and X radiation for the solar events of August 4 and 7, 1972. The authors have therefore been able to determine the different phases of these flares by using these correlations.

076.022 **Determination of the height of hard X-ray sources in the solar atmosphere by measurement of photospheric albedo photons.**
J. C. Brown, H. F. van Beek, A. N. McClymont.
Astron. Astrophys., Vol. 41, 395 - 402 (1975). – Paper presented in summary at IAU/COSPAR Symposium 68 on "Solar γ-, X- and EUV-Radiation", held in Buenos Aires, Argentina, June 11 - 14, 1974.
The importance and difficulties of determining the altitude of hard X-ray sources in the solar atmosphere are discussed.

076.023 **Crystal spectrometer studies of the sun employing a rotation modulation collimator.**
D. H. Brabban, W. M. Glencross, F. D. Rosenberg.
Solar Physics, Vol. 42, 355 - 376 (1975).
The use of rotating modulation collimators in high resolution solar X-ray spectroscopy is discussed with reference to the recent flight of a sounding rocket payload. The observations were made in 1972 October 26 between 0432 UT and 0436 UT. During the flight a scan of the solar X-ray spectrum was made which covered the wavelength range 1.45–1.71 nm. A small flare commenced at about 0432 UT in McMath calcium plage region 12094. X-ray emission from a group of active regions in the SW and the small McMath region 12090 has also been detected. Spectral and spatial observations have been combined to investigate the conditions in both active region and flare plasmas.

076.024 **Slow X-ray bursts and flares with filament disruption.** J.-R. Roy, F. Tang.
Solar Physics, Vol. 42, 425 - 439 (1975).
The data from OGO-5 and OSO-7 X-ray experiments have been compared with optical data from six chromospheric flares with filament disruption associated with slow thermal X-ray bursts. The authors suggest that the soft X-ray emission originates from a growing shell of roughly constant thickness of high-temperature plasma due to the compression of the coronal gas by the expanding prominence.

076.025 **An impulsive solar X-ray burst in the energy range 90–400 keV.**
Y. Fukada, S. Hayakawa, I. Kasahara, F. Makino, I. Suzuki, Y. Tanaka, B. V. Sreekantan.
Solar Physics, Vol. 42, 441 - 444 (1975).

076.026 **Laboratory reproduction of a solar X-ray flare spectrum.** Eh. Ya. Gol'ts, I. A. Zhitnik, Eh. Ya. Kononov, S. D. Mandel'shtam, Yu. V. Sidel'nikov.
In-t spektroskop. AN SSSR. Preprint No. 4. Moskva, 1974. 14 pp. In Russian.

076.027 **Betatron acceleration in a large solar hard X-ray burst.** J. C. Brown, P. Hoyng.
Astrophys. Journ., Vol. 200, 734 - 746 (1975).
The problem of diagnosing flare particle acceleration mechanisms from hard X-ray bursts is discussed, and it is argued that the electron trap model of bursts is more amenable to observational investigation at present than models of thick-target type. Observations show that the burst time profile is not dominated by collisional losses. It is proposed instead that the entire profile is essentially determined by betatron action of the varying trap field on the electrons. This betatron model is then analyzed in detail and shown to predict very well the observed correlation of electron flux and spectral index when it is supposed that the electrons are initially produced by runaway in a direct electric field. Comparison of the model with observations permits inference of the approximate form of magnetic field evolution in the trap. Finally the physics behind this field evolution is briefly considered.

076.028 **Solar hard X-ray bursts phenomena.**
L. E. Peterson, D. W. Datlowe, M. J. Elcan, H. S. Hudson.
Bull. American Astron. Soc., Vol. 7, 399 (1975). – Abstr. AAS.

076.029 **Evidence for long-term variations in the quiet sun emission at EUV wavelengths.** J. G. Timothy.
Bull. American Astron. Soc., Vol. 7, 406 - 407 (1975). – Abstr. AAS.

076.030 **The role of pulse pile-up in the interpretation of solar hard X-ray spectra.** D. W. Datlowe.
Bull. American Astron. Soc., Vol. 7, 422 (1975). – Abstr. AAS.

076.031 **Hard solar X-rays from high temperature plasmas.**
M. J. Elcan.
Bull. American Astron. Soc., Vol. 7, 422 - 423 (1975). – Abstr. AAS.

076.032 **Pulsations in impulsive solar hard X-ray bursts.**
B. J. Lipa, V. Petrosian.
Bull. American Astron. Soc., Vol. 7, 423 (1975). – Abstr. AAS.

076.033 **X-ray albedo of stellar atmospheres.**
S. H. Langer, V. Petrosian.
Bull. American Astron. Soc., Vol. 7, 423 (1975). – Abstr. AAS.

076.034 **X-ray event of August 13 - 15, 1973.**
V. Scherrer, G. Sandlin, N. Sheeley, R. Tousey.
Bull. American Astron. Soc., Vol. 7, 430 (1975). – Abstr. AAS.

076.035 **Coronal X-ray transient events associated with Hα filament disappearances.**
D. Webb, A. Krieger, D. Rust, G. Vaiana.
Bull. American Astron. Soc., Vol. 7, 430 (1975). – Abstr. AAS.

076.036 **Moving coronal clouds observed with the Skylab S-054 X-ray telescope.** D. M. Rust, D. T. Webb.
Bull. American Astron. Soc., Vol. 7, 431 (1975). – Abstr. AAS.

076.037 **Association of X-ray arches with chromospheric neutral lines.**
P. S. McIntosh, A. S. Krieger, J. T. Nolte, G. Vaiana.
Bull. American Astron. Soc., Vol. 7, 444 (1975). – Abstr. AAS.

076.038 **Measured variation of the XUV line widths and intensities near the solar limb.**

O. K. Moe, K. R. Nicolas.
Bull. American Astron. Soc., Vol. 7, 460 (1975). – Abstr. AAS.

076.039 **X-ray bright spots on the sun and the nonequilibrium of a twisted flux rope in a stratified atmosphere.**
E. N. Parker.
Astrophys. Journ., Vol. 201, 494 - 501 (1975).
The calculations are concerned with the hydrostatic equilibrium of a magnetic flux tube whose two ends are anchored at some level in a stratified atmosphere. It is shown that no hydrostatic equilibrium exists if the whole flux tube is twisted more than a modest amount, equivalent to a 45° pitch over only one scale height. The author suggests that the magnetic flux tubes in the sun generally have more than the allowed twisting, so that, when they rise up through the surface, as a consequence of their buoyancy, they are subject to rapid dissipation, causing the phenomenon observed as the X-ray bright spot.

076.040 **On the possibility of recording solar γ-emission flares and its realization in extra-atmospheric measurements.** V. V. Klimenko, R. B. Salimzibarov, Yu. G. Shafer.
Issled. po kosmofiz. i aehronomii. Yakutsk, 1975, p. 5 - 12, 192. In Russian. – Abstr. in Referativ. Zhurn. 51. Astron., 9.51.119 (1975).

076.041 **X-ray spectrum of the sun in the 1.85–1.93 Å region.**
Yu. I. Grineva, U. I. Safronova, A. M. Urnov.
IV Vses. konf. po spektroskopii vakuum. ul'travioleta i vzaimodejstviyu izlucheniya s veshchestvom, 1975. Tezisy dokl. Uzhgorod, 1975, p. 97. In Russian. – Abstr. in Referativ. Zhurn. 62. Issled. kosmich. prostranstva, 9.62.154 (1975).

076.042 **Gamma-bursts during the flares of 1972, Aug. 4 and 7 observed on satellite Prognoz 2.**
Zh. Vedren, O. B. Likin, A. S. Melioranskij, N. F. Pisarenko, I. A. Savenko, R. Talon, V. M. Shamolin.
VI Leningr. mezhdunar. seminar, 1974. Uskorenie chastits i yader. reaktsii v kosmose. Leningrad, 1974, p. 85 - 86, with a discussion, p. 86. In Russian and English. – Abstr. in Referativ. Zhurn. 51. Astron., 8.51.491 (1975).

076.043 **Observations of solar gamma ray continuum between 360 keV and 7 MeV on August 4, 1972.**
A. N. Suri, E. L. Chupp, D. J. Forrest, C. Reppin.
Solar Physics, Vol. 43, 415 - 429 (1975).
The purpose of this paper is to present the gamma ray continuum observations made in the energy interval 0.36–7 MeV during the 3B optical flare of August 4, 1972 and to describe the unfolded continuum high energy photon spectrum, the average electron spectrum that can be deduced from this photon spectrum; and to show the approximate time correlation of the energetic photons with photons < 100 keV and with the radio bursts.

076.044 **Solar X-ray bursts during August, 1972.**
N. N. Evstaf'ev, A. S. Melioranskij, N. I. Nazarova, V. M. Pankov, I. N. Rozantsev, I. A. Savenko.
Solnechnye Dannye 1975 Byull., No. 5, p. 51 - 59 (1975). In Russian.
Prognoz 2 satellite data on the August 1972 solar X-ray bursts in the energy region 2.5–30 keV are presented. The time of the beginning, maximum and end of the bursts is given. The intensity of the bursts is also given. A comparison of the X-ray effects with the associated optical, radio and SID phenomena is made.

076.045 **X-ray solar spectrum in the 1.85 – 1.93 Å region.**
Yu. I. Grineva, U. I. Safronova, A. M. Urnov.
IV Vses. konf. po spektroskopii vakuum. ul'trafioleta i vzaimodejstviyu izlucheniya s veshchestvom, 1975. Tezisy dokl. Uzhgorod, 1975, p. 97. In Russian. – Abstr. in Referativ. Zhurn. 51. Astron., 10.51.426 (1975).

076.046 **Solar X-ray flares in September–December 1972.**
N. N. Evstaf'ev, N. V. Illarionova, A. S. Melioranskij, N. I. Nazarova, V. M. Pankov, I. N. Rozantsev, I. A. Savenko.
Solnechnye Dannye 1975 Byull., No. 7, p. 68 - 79 (1975). In Russian.
Observations of solar X-ray flares for $5 < E < 15$ keV by Prognoz 2 during September–December 1972 are given. These measurements are discussed in connection with different phenomena of solar activity.

076.047 **On the directivity of X-ray emission of solar flares.**
O. M. Kovrizhnykh, I. A. Savenko, L. M. Chupova.
Pis'ma v Astron. Zhurn., Vol. 1, No. 12, p. 32 - 37 (1975). In Russian.

076.048 **On a possibility of recording flares of solar γ-radiation and its realization in extra-atmospherical measurements.**
V. V. Klimenko, R. B. Salimzibarov, Yu. G. Shafer.
Issled. po kosmofiz. i aehronomii, Yakutsk, 1975, p. 5 - 12, 192. In Russian.

076.049 **On streams of short-wave radiation of the quiet sun.**
I. N. Odintsova, L. N. Leshchenko, V. V. Firsov.
Geomagn. Aeronom., Vol. 15, 970 - 972 (1975). In Russian.

076.050 **The intensities and profiles of XUV transition zone lines in a quiet sun region compared to a polar coronal hole.** U. Feldman, G. A. Doschek, R. Tousey.
Astrophys. Journ., (*Letters*), Vol. 202, L147 - L150 (1975).
The intensities of XUV transition zone lines from limb spectra of a quiet sun region and a polar coronal hole are compared. Profiles of selected optically thin transition zone lines from the quiet sun and coronal hole spectra are shown. The lines are broader than expected in ionization equilibrium, and bulk motion velocities are deduced from the widths of the lines. There appears to be little, if any, statistically significant difference in the velocities obtained from the quiet sun and the coronal hole.

076.051 **Limb-brightening curves of XUV transition zone lines in the quiet sun and in a polar coronal hole observed from Skylab.**
G. A. Doschek, U. Feldman, R. Tousey.
Astrophys. Journ., (*Letters*), Vol. 202, L151 - L154 (1975).
Solar limb-brightening curves are discussed for XUV spectral lines formed in the upper chromosphere and transition zone of a quiet sun region and a polar coronal hole. The lines considered are emitted from ions formed within the temperature range from 10^4 K to 2.2×10^5 K. The data are compared with predictions based on both homogeneous and inhomogeneous models of the transition zone. The limb-brightening curve of the O I line at 1355.6 Å indicates that O I is formed in spicules. The limb brightening of the He II line at 1640.4 Å is consistent with a temperature of formation between $\sim 4 \times 10^4$ K and $\sim 9 \times 10^4$ K for He II.

076.052 **Thermal models of flaring region based on observations by the SOLRAD 10 satellite.**
M. Landini, B. C. Monsignori Fossi, R. Pallavicini.
Solar Physics, Vol. 44, 101 - 116 (1975).
A number of solar X-ray bursts, observed with the SOLRAD 10 satellite on 1971 September 14, 17, November 28 and 1972 August 1, have been analysed. From the differential emission measure distribution deduced from the observations the temporal and spatial distribution in a flaring region has

been evaluated and compared with models of instantaneous or continuous energy injection and conductive redistribution. Assuming a power law for the differential emission measure in dependence of electron temperature the ratios of the counting rates observed in different spectral bands have been compared with theoretical predictions so as to test the validity of this particular class of models.

076.053 **Calculated solar X-radiation. II: Spectrum between 61 and 220 Å.** R. Mewe.
Solar Physics, Vol. 44, 383 - 390 (1975).

Earlier calculations on the solar X-ray spectrum between 1 and 61 Å have now been extended by the addition of the intensities of about 90 spectral lines up to 220 Å from ions of the elements O, Ne, Mg, Si, S, Fe, and Ni for electron temperatures between 10^5 and 1.5×10^7 K. The continuum emission has been calculated in the temperature range 10^5–10^8 K for the wavelength region 40–220 Å.

076.054 **Analysis of the intensities and profiles of the spectral line Mg XII 8.42 Å in the solar X-ray spectrum.**
J. Jakimiec, V. V. Korneev, V. V. Krutov, I. A. Zhitnik, S. Płocieniak, B. Sylwester, J. Sylwester.
Solar Physics, Vol. 44, 391 - 401 (1975).

The Mg XII line intensity provides a sensitive indicator of the hot plasma content ($T \gtrsim 3 \times 10^6$ K) in coronal condensations and X-ray flare volumes. For all the investigated coronal condensations the temperature distribution of plasma has been found to be a function monotonically decreasing with temperature. But for some X-ray bursts there occurred a distinct excess of the hot plasma of temperature between 6–10×10^6 K. FWHM values of the Mg XII line profiles have been used to estimate ion temperature in the emitting regions.

076.055 **Solar gamma rays.** R. Ramaty, B. Kozlovsky, R. E. Lingenfelter.
Space Sci. Rev., Vol. 18, 341 - 388 (1975).

The theory of gamma-ray production in solar flares is treated in detail. Both lines and continuum are produced. The strongest line predicted at 2.225 MeV with a width of less than 100 eV and detected at 2.24 ± 0.02 MeV, is due to neutron capture by protons in the photosphere. Its intensity is dependent on the photospheric ^{3}He abundance. The strongest prompt lines are at 4.43 MeV from ^{12}C and at ~ 6.2 from ^{16}O and ^{15}N. The width of the 4.43 MeV line is ~ 100 keV and that of the 6.2 MeV feature is ~ 300 keV. Both these lines have been observed from a solar flare. Other potentially observable lines are predicted at 0.845 and 1.24 MeV from ^{56}Fe, at 1.63 MeV principally from ^{14}N and ^{20}Ne, at 1.78 MeV from ^{28}Si, at ~ 5.3 MeV from ^{15}O and ^{15}N, and at 7.12 MeV from ^{16}O. From the comparison of the observed and calculated intensities of the line at 4.4 MeV to that of the 2.2 MeV line it is possible to obtain information on the spectrum of accelerated nuclei in flares. The authors find that about 10^{33} protons of energies greater than 30 MeV were produced in the 1972, August 4 flare.

076.056 **Relation of the observed far ultraviolet solar irradiance to the solar magnetic sector structure.**
D. F. Heath, J. M. Wilcox, L. Svalgaard, T. L. Duvall.
Solar Physics, Vol. 45, 79 - 82 (1975). – Research note.

Comparison of the observed solar far ultraviolet irradiance and the observed solar sector structure during 1969 through 1972 shows a tendency for EUV maxima to be located near sector boundaries.

076.057 **The influence of shock waves on UV emission lines.** L. R. Elzner.
Solar Physics, Vol. 45, 93 - 103 (1975).

If shock waves are running outwards through the solar atmosphere, the variations of the intensity of optically thin UV-lines should reflect the variations in temperature and pressure of the shocks. In addition, the high fluid velocity produces Doppler shifts in the local line profiles. Assuming vertically propagating shock waves with a period of 200 s, a procedure is specified for calculating the time-dependent line profile for a radial line of sight. The width of the shock pulse which characterizes the shape of the shock, is treated as a free parameter. Because the processes of ionization and recombination are out of equilibrium, the relative ion densities are obtained by solving the continuity equations. Results are presented for the O VI-resonance line 1032 Å.

076.058 **On the extreme ultraviolet solar emission of B-like ions: Na VII, Si X and S XII.**
D. R. Flower, H. Nussbaumer.
Astron. Astrophys., Vol. 45, 349 - 352 (1975).

Radiative and collisional atomic data have been calculated for transitions within the $2s^2\,2p$, $2s2p^2$ and $2p^3$ configurations of Na^{+6}, Si^{+9} and S^{+11}. Configuration interaction, between $2s^2\,2p\,^2P^0$ and $2p^3\,^2P^0$, and spin-orbit interaction within the $2s2p^2$ configuration affect significantly the results for the $2s^2\,2p\,^2P^0 - 2s2p^2\,^2S$, ^2P multiplets. The relative intensities of the allowed transitions between the $2s^2\,2p$ and $2s2p^2$ configurations of Si X and S XII are found to be density dependent for coronal conditions. Calculated intensity ratios are compared with currently available observations.

076.059 **Hard X-rays from the sun.** H. F. van Beek, P. Hoyng, C. de Jager, G. A. Stevens.
Nederl. Tijdschr. Natuurkunde, Vol. 41, 101 - 103 (1975). In Dutch.

076.060 **LF and VLF phase antinomies during solar X-ray flares.** G. E. Perona.
Radio Sci., (*USA*), Vol. 10, 435 - 444 (1975).

Density dependence of the line emission of boron-like coronal ions. See Abstr. 022.097.

On the extreme ultraviolet solar emission of B-like ions: O IV. See Abstr. 022.100.

The Apollo 15 and 16 X-ray fluorescence experiment. See Abstr. 032.520.

Solar X-ray photometer onboard the Interkosmos-11 satellite and preliminary results of observations.
See Abstr. 034.003.

Criteria for plasma-wave losses for electrons streaming from an acceleration region. See Abstr. 062.032.

The solar spectrum from 1173 to 1324 Å.
See Abstr. 071.008.

Analysis of Skylab soft X-ray observations of solar active region 131 (McMath 12379).
See Abstr. Abstr. 072.012.

Morphological evolution of X-ray flare structures from the rise through the decay phase.
See Abstr. 073.001.

The location of the site of energy release in a solar X-ray subflare. See Abstr. 073.003.

Fe XXIV emission in solar flares observed with the NRL/ATM XUV slitless spectrograph. See Abstr. 073.006.

X-ray and EUV spectra of solar flares and laboratory plasmas. See Abstr. 073.007.

Association of X-ray flares with solar coronal active regions. See Abstr. 073.008.

Studies of the dynamic structure and spectra of solar X-ray flares. See Abstr. 073.009.

Thermal and nonthermal interpretations of flare X-ray bursts. See Abstr. 073.011.

High time resolution analysis of solar flares observed on the ESRO TD-1A satellite. See Abstr. 073.012.

Inference of the hard X-ray source dimensions in the 1972, August 7 white light flare. See Abstr. 073.013.

Relation of microwave emission to X-ray emission from solar flares. See Abstr. 073.014.

High energy gamma-ray radiation above 300 keV associated with solar activity. See Abstr. 073.015.

Gamma-ray lines from solar flares. See Abstr. 073.016.

Fast electrons in small solar flares. See Abstr. 073.017.

The spatial distribution of X-ray, EUV and GHz emissions in connection with the flare-loop-channel model. See Abstr. 073.022.

White light flares: protons or electrons? See Abstr. 073.028.

Plasma heating and flare X-rays. See Abstr. 073.037.

S-056 observations of soft X-ray flares–implications of triggering mechanism. See Abstr. 073.049.

X-ray observations of the flare of June 15, 1973. See Abstr. 073.051.

Photometric properties of soft X-ray kernels. See Abstr. 073.053.

An observation of identical hard X-ray bursts from sympathetic flares. See Abstr. 073.054.

Heating and cooling of the thermal X-ray plasma in solar flares. See Abstr. 073.072.

The temperature structure and pressure balance of magnetic loops in active regions. See Abstr. 073.096.

The solar-flare infrared continuum. See Abstr. 073.097.

Hard X-ray coverage of the Skylab period by Utrecht's S-100 instrument on board ESRO TD-1A. See Abstr. 073.102.

Objective grating studies of X-ray flare spectra. See Abstr. 073.103.

X-ray and radio emission for the June 15, 1973 solar flare. See Abstr. 073.104.

Positron annihilation in solar flares. See Abstr. 073.108.

XUV emission from above the solar west limb near 20:00 UT, January 17, 1974. See Abstr. 073.121.

Observations of the structure and evolution of solar flares with a soft X-ray telescope. See Abstr. 073.146.

Temperature dependence of emission measure in solar X-ray plasmas. I. Nonflaring active regions. See Abstr. 074.002.

Heating of coronal material at X-ray bright points. See Abstr. 074.003.

Skylab soft X-ray observations of magnetic-field reconfiguration. See Abstr. 074.031.

A loop model of active coronal regions. See Abstr. 074.050.

Energy balance in a magnetically confined coronal structure observed by OSO-7. See Abstr. 074.067.

Skylab and the ASE X-ray telescope experiment: a new view of the X-ray corona. See Abstr. 074.073.

Comparison of Skylab X-ray and ground-based helium observations. See Abstr. 074.074.

Models of coronal active regions. See Abstr. 074.075.

Coronal active regions evolution. See Abstr. 074.076.

Coronal magnetic flux tubes. See Abstr. 074.077.

The long term development of the large scale corona and the evolution of coronal holes. See Abstr. 074.078.

A coronal hole observed at X, UV and radio wavelengths. See Abstr. 074.080.

Coronal X-ray structures and coronal magnetic fields. See Abstr. 074.081.

Analysis of EUV limb-brightening observations from ATM. I: Model for the transition layer and the corona. See Abstr. 074.103.

The coronal structure of active regions. See Abstr. 074.104.

A comparison of coronal X-ray structures of active regions with magnetic fields computed from photospheric observations. See Abstr. 074.105.

The structure and intensity evolution of a solar burst at 2.8 cm and the relation with the soft X-ray emitting region. See Abstr. 077.037.

Neutron and γ ray emissions from the sun. See Abstr. 078.007.

Solar structure in the extreme ultraviolet. See Abstr. 080.031.

Solar activity and the general absence of hydrostatic equilibrium in an azimuthal magnetic field. See Abstr. 080.038.

On non–LTE H_2^+ as missing solar opacity.

See Abstr. 080.039.

A global model of thermospheric dynamics – II. Wind, density, and temperature fields generated by EUV heating. See Abstr. 082.010.

Improved calculations of electron densities between 110 and 300 km derived from solar EUV fluxes of August 23, 1972. See Abstr. 083.073.

Comment on 'Electron densities between 110 and 300 km derived from solar EUV fluxes of August 23, 1972' by L. Heroux, M. Cohen, and J. E. Higgins. See Abstr. 083.074.

Stellar and solar X-ray polarimetry. See Abstr. 142.179.

077 Solar Radio Radiation

077.001 **Radiation phenomena in type II solar radio bursts and laboratory shock-wave experiments.**
N. A. Krall, D. F. Smith.
Astrophys. Journ., Vol. 199, 500 - 503 (1975).

The known spectrum of turbulence in and behind a collisionless shock wave is used to compute the relaxation of the electron distribution function in the presence of that turbulent spectrum. It is concluded that a nonthermal tail of the sort required to explain type II solar burst radiation does not develop from the turbulent environment of a collisionless shock wave. Alternate mechanisms for the radiation are proposed, based on similar observations in laboratory experiments.

077.002 **Das Problem der Bestimmung der "ruhigen Sonne" im Radiofrequenz-Bereich.**
W. Hirth, E. Fürst.
Mitt. Astron. Ges., No. 36, p. 144 - 147 (1975).

077.003 **On the nature of some active regions in the microwave range.**
M. Felli, G. Tofani, E. Fürst, W. Hirth.
Solar Physics, Vol. 42, 377 - 390 (1975).

The radio emission of a selected number of solar active regions has been investigated with high angular resolution at two frequencies: 10 and 17 GHz. The brightness temperature distribution of an active region is often composed of very bright cores of small dimension (angular extent $\theta \simeq 20''$) imbedded in extended halos of lower brightness. The radio emission of such structures as well as the degree of polarization can be explained with a thermal process.

077.004 **Source regions for type II radio bursts.**
J. C. Dodge.
Solar Physics, Vol. 42, 445 - 459 (1975).

The flares associated with reported type II bursts from 1964–1973 (hereafter abbreviated source-flares) are analyzed with respect to their importance, duration, and heliographic distribution. An analysis of the plage regions which are associated with source-flares shows that relatively few plages ever produce type II bursts, yet certain plages produce type II bursts repeatedly.

077.005 **Discussion on the coronal structure related to type III bursts.**
M. Pick, M. J. Martres, F. Axisa, C. Mercier.
Solar Physics, Vol. 42, 461 - 466 (1975).

The authors think that there is no observational proof for excluding coronal streamers as guides for the escape of type III electrons. At the present time there is no direct conclusive evidence of the association of type III's with known coronal structures.

077.006 **Decameter storm radiation, I.**
T. E. Gergely, W. C. Erickson.
Solar Physics, Vol. 42, 467 - 486 (1975).

Six long duration noise storms have been observed over an eight month period, extending from January to September 1971. The storms are described and their relation to chromospheric active regions and flares, coronal magnetic fields, and meter and decimeter wave activity is discussed. Decametric storms are found to be related to complexes of interacting active regions.

077.007 **Characteristic pairs of type III bursts as the electromagnetic response to a dispersive stream exciter.**
G. Daigne.
Astron. Astrophys., Vol. 42, 71 - 79 (1975).

The concept of characteristic pairs of type III bursts, or "f–h pairs" previously described for the metric range, is extended to other observations in various frequency ranges. It is argued that the two components of a burst pair are radiated in the same mode and from a common source. A review of the properties of type III bursts favours the second harmonic radiation mode at all frequencies. Satellite observations of type III and electron bursts confirm that the exciter is a dispersive electron stream.

077.008 **A search for periodic variations of solar radio emission at 3.7- and 11.1-cm wavelengths.**
M. R. Kundu, C. E. Alissandrakis.
Monthly Notices Roy. Astron. Soc., Vol. 173, 65 - 76 (1975).

The authors have looked for periodic time variations in the emission of quiet solar regions with the NRAO three-element interferometer at 3.7 and 11.1 cm. The estimated size of the time-varying elements is $2\overset{''}{.}5 \pm 0\overset{''}{.}2$ arc at 3.7 cm and $7\overset{''}{.}3 \pm 0\overset{''}{.}4$ arc at 11.1 cm.

077.009 **On the decay of type III radio bursts in the solar corona.** N. N. Gerasimova, I. M. Gordon.
Pis'ma v Astron. Zhurn., Vol. 1, No. 7, p. 34 - 38 (1975). In Russian.

The decay of low frequency type III bursts cannot be explained by Landau damping by the non-thermal tail of a one-dimensional electron velocity distribution in the coronal plasma. The abrupt cutoff of intensity of bursts by Landau damping caused by a density gradient is probably prevented by the induced scattering of plasma waves on ions, which creates a flux of plasmons in the direction of increasing plasma density.

077.010 **Fluctuations of solar radio emission in the noise storm on May 21, 1973.**
A. A. Gnezdilov, T. S. Podstrigach.
Pis'ma v Astron. Zhurn., Vol. 1, No. 8, p. 27 - 31 (1975). In Russian.

Fluctuations of enhanced solar radio emission on May 21, 1973 in the meter range (74 and 100 MHz) are studied. Some quasiperiodic components with a time scale of about 5 minutes are found. There is anticorrelation between total intensity and degree of polarization.

077.011 **Type IVdm bursts vs. the Razin effect.**
A. O. Benz, G. L. Tarnstrom.
Bull. American Astron. Soc., Vol. 7, 439 - 440 (1975). – Abstr. AAS.

077.012 **The radial brightness distribution of the sun at 3.2 mm as determined from the June 30, 1973 total solar eclipse and a reanalysis of the March 7, 1970 total solar eclipse.** P. N. Swanson, J. P. Hagen.
Solar Physics, Vol. 43, 57 - 62 (1975).

The radial brightness distribution of the sun at 3.2 mm is recalculated for the 7 March, 1970 total solar eclipse by an improved method. The results from the first contact of the 30 June, 1973 total eclipse are also analyzed. Limb brightening is apparent for the 1970 eclipse when all four contacts are averaged. The 1973 eclipse indicates strong limb brightening, although the detailed shape differs from the 1970 eclipse.

077.013 **Plasma emission due to isotropic fast electrons, and types I, II, and V solar radio bursts.** D. B. Melrose.
Solar Physics, Vol. 43, 211 - 236 (1975).

The theory of plasma emission is developed under the assumption that the Langmuir waves are generated by an isotropic distribution of fast electrons. Emission from inverse power-law distributions tend to favor emission at the second harmonic with brightness temperatures up to about 10^8K at 100 MHz. The concept of a gap (in velocity space) distribution is developed. Mechanisms by which a gap distribution might be formed are discussed and two effective mechanisms are identified. The theory is applied to the interpretation of radio bursts of types I, II, stationary IV and V. In each case the suggested mechanism appears to be favorable.

077.014 **Observations of dynamical spectra of the sporadic solar radio emission in the 0.6–1.2 m region.**
Yu. B. Vedeneev, V. O. Kaloshin, N. M. Prytkov.
Astron. Tsirk., No. 857, p. 1 - 2 (1975). In Russian.

077.015 **Low-frequency observations of fluctuations of the quasiperiodic component of solar radio radiation.**
V. P. Grigor'eva.
Astron. Tsirk., No. 857, p. 5 - 7 (1975). In Russian.

077.016 **Unusual series of type III-RS solar radio bursts.**
G. L. Tarnstrom, C. Zehntner.
Nature, Vol. 258, 693 - 694 (1975).

Solar radio emission on August 12, 1975, was unusual because of the large number of, and predominance of, type III-RS bursts (' reverse slope', positive frequency-drift type III bursts). These bursts seem to be second branches of U bursts and, as such, are consistent with only one current model. Four burst events have a Y-like profile and may be the first observations of the reflection of the burst exciter at a magnetic mirror.

077.017 **Spectral characteristics of stria, split pair and triple bursts.** J. de la Noë.
Astron. Astrophys., Vol. 43, 201 - 214 (1975).

Spectral parameters of a large number of stria, split pair and triple bursts, grouped in type IIIb bursts, were measured for two consecutive half solar rotations. The results are: the frequency variation of the average duration, the center-to-limb effect of the average frequency and duration, the observing period and longitude independence of the bandwidth and of the frequency splitting in pairs, the random character of the drift rate and that of the circular polarization, the existence of long duration and trailing elements, a broadening effect on the bandwidth of some bursts, the role of an axis of symmetry played by the middle element in a triple burst. All these properties have to be accounted for by a coherent theory of type IIIb bursts.

077.018 **Estimates of magnetic field irregularities over a radio source using emission of solar bursts in the microwave region.** V. A. Kovalev, O. S. Korolev.
Fiz. ionosfery, magnit. pole Zemli i solnechno-zemn. svyazi. Moskva, 1974, p. 67 - 73. In Russian. – Abstr. in Referativ. Zhurn. 51. Astron., 9.51.454 (1975).

077.019 **On induced scattering of plasma waves during generation of solar high-frequency type III radio bursts.** V. V. Zajtsev.
Pis'ma v Astron. Zhurn., Vol. 1, No. 10, p. 28 - 33 (1975). In Russian.

It is shown that induced scattering of plasma waves into electromagnetic ones may be neglected during generation of high-frequency type III radio bursts if their brightness temperature does not exceed 10^{10} K. In this case the fundamental mode radiation results from spontaneous scattering of plasma waves by thermal coronal ions.

077.020 **A three-dimensional trough of an ordinary wave and type I solar radio bursts.**
O. S. Korolev, V. I. Petviashvili.
AN SSSR. In-t zemn. magnetizma, ionosfery i rasprostr. radiovoln. Preprint No. 29 (114). Moskva, 1975. 15 pp. In Russian.
Abstr. in Referativ. Zhurn. 51. Astron., 8.51.512 (1975).

077.021 **Harmonics of electron gyrofrequency in solar radio emission and their interpretation in terms of plasma-beam instability.** V. N. Fedorenko.
Astron. Zhurn. Akad. Nauk SSSR, Vol. 52, 978 - 988 (1975). In Russian. English translation in Soviet Astron., Vol. 19, No. 5.

A possible explanation of the emission lines with frequencies of multiples of the electron gyrofrequency in the dynamical spectrum of solar radiation is given. The electron beam captured in a magnetic trap and interacting with plasma by cone instability is considered. Parameters of the beam are taken the same as for the beams producing type III radio bursts. It is shown that under such conditions the oscillations with frequencies of multiples of the electron gyrofrequency and the shortest wavelength are amplified. Various mechanisms of transformation of these oscillations into electromagnetic waves are considered.

077.022 **Polarization structure of noise storms.**
A. K. Markeev, V. V. Fomichev, G. P. Chernov, I. M. Chertok.
Astron. Zhurn. Akad. Nauk SSSR, Vol. 52, 989 - 992 (1975). In Russian. English translation in Soviet Astron., Vol. 19, No. 5.

The results of simultaneous observations with high time resolution of a spectrum (181–224 MHz) and polarization (204 MHz) of a noise storm performed in 1972 are presented. The polarization structure of isolated bursts and chains of type I bursts are investigated. A detailed comparison of spectral and polarization features of individual bands in the chains of type I bursts with small band-splitting is carried out.

077.023 **On angular dimensions of decameter solar radio burst sources.** Eh. P. Abranin, L. L. Bazelyan, N. Yu. Goncharov, V. V. Zajtsev, V. A. Zinichev, V. O. Rapoport, Ya. G. Tsybko.
Astron. Zhurn. Akad. Nauk SSSR, Vol. 52, 993 - 998 (1975). In Russian. English translation in Soviet Astron., Vol. 19, No. 5.

Data are obtained on the angular structure of type III bursts sources at frequencies of 25 and 12.5 MHz. As a rule, the type III bursts in the decameter range arise in the form of chains of stria-bursts (at the frequency of the fundamental tone) and of diffusive traces (at the second harmonic of the plasma frequency). The angular dimensions of sources of the diffusive traces are 20′ and <40′, respectively, at frequencies of 25 and 12.5 MHz. On the 4th of September 1973, within 24–26 MHz a long-lived ($\tau = 40$ min) narrow band burst ($\Delta f = 100$ kHz) was observed. The mean frequency of this burst shows a systematic variation with solar latitude, whereas the general form of the dynamic spectrum remains invariable in different rays of the antenna.

077.024 **Intermediate polarization of type I bursts.**
P. Zlobec.
Solar Physics, Vol. 43, 453 - 461 (1975).

A constant polarization degree was found in groups of type I bursts having intermediate polarization for all the duration of the group. The same circumstance was verified also for analogous bursts present during a type IV event. The consequences of this behaviour, considering also the results of some recent papers, are discussed. The polarization amount is probably intrinsic to the source itself.

077.025 **The transit of Mercury across the solar disk 9 May 1970 and fluctuations of radio emission at λ = 8 mm.** V. A. Efanov, I. G. Moiseev, A. B. Severnyj.
Izv. Krymskoj Astrofiz. Obs., Vol. 53, p. 121 - 129 (1975). In Russian.

The observations of radio emission at λ = 8 mm of the sun during the transit of Mercury across the disk at 9 May 1970 are described. Discussion of observations and a correlation analysis of the fluctuations of radio emission have permitted to ascribe these fluctuations to the effect of obscuration of the condensations of radio emission by the disk of the planet. Characteristic sizes of condensations of 1."6–3."4 and 17″–40″ and brightness temperatures of 9420–18250°K are obtained. It is possible that the magnetic field plays an important role in formation of condensations of radio emission at λ = 8 mm.

077.026 **On the influence of some features of sunspot development on noise storm radio emission.**
L. I. Yurovskaya.
Izv. Krymskoj Astrofiz. Obs., Vol. 53, p. 139 - 143 (1975). In Russian.

The origin of noise storm sources in the corona most often accompanies fast changes of the largest spot area and the configuration of the magnetic field in the photosphere. It is suggested that the origin of noise storm sources is associated with the nonstationarity of the magnetic flux penetrating into high levels of the solar atmosphere.

077.027 **Observations of the sun at 408 MHz.**
F. Palagi, P. Patriarchi, N. Speroni.
Oss. Mem. Oss. Astrofis. Arcetri, Fasc. 104, (see 012.025), p. 125 - 128 (1975).

In the period from Dec. 14, 1973 to Jan. 24, 1974 absolute measurements of solar radio flux have been made with the *E–W* arm of the "North Cross" radiotelescope, Bologna, Italy, while the third SKYLAB Manned Mission was going on. Calibrations have been made using Cas *A* and Cyg *A* as reference sources, almost daily, to avoid errors due to system-gain variations. There are some indications that a coronal hole has been observed twice at a time-interval of a solar rotation.

077.028 **High resolution solar radio observations at 2.8 cm during the ATM program.**
M. Felli, G. Tofani.
Oss. Mem. Oss. Astrofis. Arcetri, Fasc. 104, (see 012.025), p. 129 - 137 (1975).

The scientific motivations for high resolution observations of active regions in the microwave part of the spectrum, contemporarely to the ATM programs, are briefly described. A description of the type of information which can be derived from the radio data as well as a catalogue of all the radio observations, contemporarely to the second manned mission, is given. Preliminary results and fields of current research are indicated.

077.029 **Results of solar observations during August and September 1973 in Effelsberg.**
E. Fürst, W. Hirth.
Oss. Mem. Oss. Astrofis. Arcetri, Fasc. 104, (see 012.025), p. 138 (1975). – Abstract.

077.030 **Type IV event of January 15, 1974 – polarimetric and interferometric observations respectively at 239 MHz and 408 MHz.** P. Zlobec.
Oss. Mem. Oss. Astrofis. Arcetri, Fasc. 104, (see 012.025), p. 139 - 142 (1975).

077.031 **First digital observation with the Duernten radio spectrograph.**
M. R. Perrenoud, A. O. Benz, G. L. Tarnstrom.
Oss. Mem. Oss. Astrofis. Arcetri, Fasc. 105, (see 012.026), p. 121 - 124 (1975).

077.032 **On the definition of geoeffectivity of solar activity by radioastronomical methods.** S. T. Akin'yan.
Solnechnye Dannye 1975 Byull., No. 3, p. 52 - 58 (1975). In Russian.

The characteristics of the solar radio bursts of type IV are considered in connection with acceleration and escape of fast and slow solar particles.

077.033 **Variations of radio flux and solar activity.**
O. G. Gontarev, Eh. I. Mogilevskij.
Solnechnye Dannye 1975 Byull., No. 6, p. 72 - 79 (1975). In Russian.

A statistical Fourier analysis of short-time variations (STV) of solar radio flux at λ = 3.2 cm has been made for several days with different levels of solar activity. The power spectra of STV is shown to change continuously, i.e. the energy is "pumped over" from the long-time ($\simeq$ 1000 sec) and the short-time ($\simeq$ 100 sec) variations.

077.034 **Analysis of fluctuations in microwave solar radio emission.** Yu. I. Vitinskij.
Solnechnye Dannye 1975 Byull., No. 6, p. 91 - 99 (1975). In Russian.

The values of fluctuation indices of the solar radio emission fluxes and sunspot areas are given. A comparison of these characteristics for the indices of the solar radio emission and sunspots is made. The problem of supposed annual variation of the solar radio emission flux is considered. It is shown that the indices of the solar radio emission cannot be used as equivalents of Wolf numbers.

077.035 **Spectrographic observations of the sun's radio emission at the IZMIRAN in the region of 93 - 186 MHz.** O. S. Korolev.
Astron. Zhurn. Akad. Nauk SSSR, Vol. 52, 1247 - 1251 (1975). In Russian. English translation in Soviet Astron., Vol. 19, No. 6.

The principle of the work of a radio spectrograph with continuous variation of the frequency in the region from 93 MHz to 186 MHz is described in detail. A block scheme of this instrument and its main parameters are given. This spectrograph is used for patrol observations of the radio emission of the sun over the considered frequency interval with registration of the signal on a slowly moving film. Fast-speed observations may be performed simultaneously with slow recording in moments of solar activity increase. There are given dynamical spectra of some phenomena in order to illustrate the work of the spectrograph: bursts and "chains" of type I, a group of bursts of type III, continuous radiation, and their description are presented.

077.036 **On the origin of fast drift absorption bursts.**
V. V. Zaitsev (*Zajtsev*), A. V. Stepanov.
Astron. Astrophys., Vol. 45, 135 - 140 (1975).

The authors present an explanation of the Type-IV continuum with fast-drift absorption radio bursts observed during the event which occurred on the 2nd of March 1970. The continuum emission is assumed to be caused by a high-frequency loss-cone instability in the adiabatic magnetic trap, while the absorption bursts are considered to be a result of hot electrons being injected into the trap, which fill the loss-cone and quench the instability. The emission fringe of absorption bursts at the low-frequency side and the wedge-shaped form of the bursts are also elucidated.

077.037 **The structure and intensity evolution of a solar burst at 2.8 cm and the relation with the soft X-ray emitting region.** M. Felli, R. Pallavicini, G. Tofani.
Solar Physics, Vol. 44, 135 - 147 (1975).

The decay phase of a microwave burst was observed with a one-dimensional angular resolution of 16″ at 2.8 cm. The structure was found to be composed of several bursting regions with different evolution. The event was also observed in soft X-rays by full-disk detectors. The joint analysis of these data suggests that the complete event has thermal origin and that the soft X-ray emission is associated with the rapidly varying component of the radio structure. From these results the average values of the electron temperatures and densities were computed for each component.

077.038 **A radio burst with peculiar polarization behaviour in July 1974.** P. Steffen.
Solar Physics, Vol. 44, 149 - 153 (1975).

Observing the total solar flux at 17 GHz a radio burst with unusual polarization characteristics (e.g. correlation of intensity maxima with maxima of polarization degree) has been recorded. The discussion shows that there is a possibility for a small optical depth of the source for at least one propagation mode during the burst maximum. Furthermore it could be found that the emission has been caused probably by synchrotron radiation with a power law electron energy distribution ($N(\epsilon) \sim \epsilon^{-\gamma}$) with $\gamma \approx 4$.

077.039 **Spectral distributions of microwave bursts.** D. A. Guidice, J. P. Castelli.
Solar Physics, Vol. 44, 155 - 172 (1975).

This paper sets forth various distributions of the peak flux density spectra of large numbers of bursts, based on observations of the Sagamore Hill Radio Observatory at nine discrete frequencies between 245 and 35000 MHz over the years 1968–1971. Emission mechanisms and burst region parameters such as magnetic field strength, electron density and electron energy distribution are determined.

077.040 **Generation of intermediate drift bursts in solar type IV radio continua through coupling of whistlers and Langmuir waves.** J. Kuijpers.
Solar Physics, Vol. 44, 173 - 193 (1975).

The possible generation of intermediate drift bursts ('fiber' bursts) in type IV dm continua through coupling between whistler waves, traveling along the magnetic field, and Langmuir waves, excited by a loss-cone instability in the source region, is elaborated. The generation, propagation and coupling of whistlers are investigated.

077.041 **Solar radio emission at 9.1 cm and sector boundaries.** W. Graf, R. N. Bracewell.
Solar Physics, Vol. 44, 195 - 204 (1975).

The 9.1 cm activity appears to be organized around sector boundaries only in a very limited sense in that the distribution of very strong active regions peaks near the –/+ boundaries. However, this phenomenon is only observed during the most active part of the solar cycle. A peculiar asymmetry is found regarding the length of the positive and negative sectors.

077.042 **Comments on the quiet sun brightness distribution at 1.2 mm wavelength.** M. R. Kundu, S.-Y. Liu.
Solar Physics, Vol. 44, 361 - 364 (1975).

It is concluded that there is no evidence for uniform or limb-brightened distribution at this wavelength.

077.043 **Observations of a complex solar radio burst with fine structure on 3 May 1973.**
G. P. Chernov, O. S. Korolev, A. K. Markeev.
Solar Physics, Vol. 44, 435 - 446 (1975).

The simultaneous high resolution recordings of dynamic spectra in the range 93–220 MHz and polarization at 204 MHz of a complex type II–IV event which started at 08:33 UT on 3 May 1973 have shown a sporadic zebra pattern. In contrast with the unpolarized type II burst, the stripes in the emission and absorption of the zebra pattern were fully polarized and most likely corresponded to the ordinary wave.

077.044 **Cyclotron wave instability in the corona and origin of solar radio emission with fine structure. II: Origin of 'tadpoles'.** V. V. Zheleznyakov, E. Ya. Zlotnik.
Solar Physics, Vol. 44, 447 - 459 (1975).

An interpretation is suggested of the fine structure of tadpole solar radio emission based upon the theory of Bernstein mode cyclotron instability. It is shown that the frequency spectrum of an individual tadpole is similar to the frequency behaviour of the Bernstein mode increment if the velocity of electrons trapped by the magnetic field amounts nearly to 10^9 cm s^{-1}. The appearance of separate tadpoles (interrupted character of emission in time) is attributed to the pulse injection of nonequilibrium electrons into the trap. Estimations are presented for the expected dimensions of emission region and the hot electron density necessary for Bernstein mode excitation.

077.045 **Cyclotron wave instability in the corona and origin of solar radio emission with fine structure. III. Origin of zebra-pattern.** V. V. Zheleznyakov, E. Ya. Zlotnik.
Solar Physics, Vol. 44, 461 - 470 (1975).

The results of investigation of the cyclotron wave instability in the corona for Bernstein modes and plasma waves in a hybrid band have been used to interpret zebra-pattern phenomena. Two models of the generation region of parallel drifting bands are considered: the model of the point source localized at the apex of the magnetic trap and the model of a distributed source extended along the magnetic flux tube. Estimations of the magnetic field and the nonequilibrium component density necessary for the zebra-pattern to generate are presented.

077.046 **Towards a theory for type III solar radio bursts. III: The radiation source including scattering.**
D. F. Smith, A. C. Riddle.
Solar Physics, Vol. 44, 471 - 483 (1975).

The procedure developed in Smith (1974) to model the radiation source for type III bursts is modified to include scattering of radiation in the source itself. The authors use inhomogeneities consistent with the scattering inhomogeneities of Steinberg et al. (1971) and Riddle (1974) and apply the procedure to an archetype 'fundamental-harmonic' pair observed at Culgoora on 28 September, 1973 at 03 19 UT. They find that it is impossible to model this burst with a source which is homogeneous in the sense that every part of the source has the same energy density in plasma waves. Possible solutions are discussed. It is concluded that none of the possibilities are completely satisfactory to explain present observations and suggested that critical observations are missing.

077.047 **Fast solar electrons, interplanetary plasma and km-wave type-III radio bursts observed from the IMP-6 spacecraft.** H. Alvarez, R. P. Lin, S. J. Bame.
Solar Physics, Vol. 44, 485 - 501 (1975).

IMP-6 spacecraft observations of low frequency radio emission, fast electrons, and solar wind plasma are used to examine the dynamics of the fast electron streams which generate solar type-III radio bursts. Of twenty solar electron events observed between April, 1971 and August, 1972, four were found to be amenable to detailed analysis. Observations of the direction of arrival of the radio emission at different frequencies were combined with the solar wind density and velocity measurements at 1 AU to define an Archimedean spiral trajectory for the radio burst exciter. The propagation characteristics of the exciter and of the fast electrons observed at 1 AU were then compared.

077.048 **On some peculiarities of type III solar radio bursts accompanied and unaccompanied by radio bursts of type V.** G. F. Eliseev.
Solnechnye Dannye 1975 Byull., No. 8, p. 90 - 96 (1975).
In Russian.

A statistical analysis of some peculiarities of the intensity fluctuations of type III radio bursts accompanied and unaccompanied by radio bursts of type V at a frequency of 208

MHz is made. The distributions of the time intervals Δt between the neighbouring maxima for type III and III + V radio bursts are shown to be significantly different. It appears that the longer the duration of type III + V radio bursts, the larger is the mean value of the period $\overline{\Delta t}$. The similar peculiarity is absent for type III radio bursts. The difference is connected with different magnetic field configurations in the regions of generation of type III and III + V bursts.

077.049 **Statistical analysis of time properties of 196 broad band solar radio bursts observed in Cracow in the years 1966–1970.** R. Guła, S. Zięba.
Acta Astron., Vol. 25, 427 - 438 (1975).

Statistical analysis of 196 broad band solar radio bursts is carried out. Four time parameters are taken into account, i.e.: duration time, rising time, moment of start and moment of maximum radiation. It seems that the analysed time properties differ in the four frequency intervals into which the whole discussed radio band can be divided. The behaviour of the considered bursts on meter waves suggests that a great part of them cannot be connected with type IV continuous bursts.

077.050 **The quiet sun brightness temperature at 408 MHz.** Y. Avignon, P. Lantos, F. Palagi, P. Patriarchi.
Solar Physics, Vol. 45, 141 - 145 (1975).

The flux of the radio quiet sun and the brightness temperature at 408 MHz (73 cm) are derived from measurements with the E-W Nançay interferometer and the E-W arm of the Medicina North Cross. It is shown that the lowest envelopes, which defined the radio quiet sun, correspond to transits of extended coronal holes across the disk of the sun.

077.051 **Evidence of large scale diverging paths in the solar corona for type III bursts exciters.** C. Mercier.
Solar Physics, Vol. 45, 169 - 179 (1975).

On July 1, 1971, about ten groups of type III radiobursts were observed with the 169 MHz Nançay radioheliograph. Each group consisted of two or several bursts exhibiting short delays from east to west (~0.5 s) and separated by distance of $\sim 1.5 \times 10^5$ km. The author suggests that distinct exciters are simultaneously accelerated at low levels in very close regions and propagate upward along very widely diverging magnetic field lines.

077.052 **The solar elongation distribution of low-frequency radio bursts.** M. L. Kaiser.
Solar Physics, Vol. 45, 181 - 187 (1975).

Over 500 days of low-frequency (<1 MHz) radio observations from the IMP-6 spacecraft have been accumulated to produce a two-dimensional map (frequency vs elongation) of solar type III burst occurrences. This map indicates that most solar bursts in this frequency range are observed at the second harmonic of the plasma frequency rather than the fundamental. The map also shows that the solar wind electron density varies as $R^{-\gamma}$, where γ can be somewhat less than 2 to perhaps 3 or higher.

077.053 **The July 1974 solar events: a possible lower limit for microwave activity.**
P. Kaufmann, P. Iacomo, Jr., E. H. Koppe, P. Marques dos Santos, R. E. Schaal, J. R. Blakey.
Solar Physics, Vol. 45, 189 - 197 (1975).

The active region McMath 10433 was the source of several flares and radio outbursts during the early part of July 1974. This region was tracked continuously, for several periods during the month at 22.2 GHz using a telescope with a 4′ beam. Comparison with the results obtained simultaneously with a normal 7 GHz solar patrol instrument indicate that there is important burst activity occurring at levels below the detection limit of normal solar patrol instruments. A completely new type of event – the fast absorption – has been recognized.

077.054 **A model for the development of a solar outburst based on observations with the Culgoora radio spectrograph and heliograph.** K. Kai.
Solar Physics, Vol. 45, 217 - 226 (1975).

The 80 and 160 MHz radioheliograph observations of a flare-associated outburst in June 1973 reveal three distinctive type IV continuum sources: a flare continuum, an isolated moving continuum and a storm continuum, in that time sequence. The observed characteristics of the three sources and their relations in space and time are described.

077.055 **Radio astronomical observations of wave motions in the solar atmosphere with periods shorter than one minute.**
G. P. Apushkinskij, Yu. V. Baryshev, A. N. Tsyganov.
Izv. vyssh. ucheb. zavedenij. Radiofizika, Vol. 18, 812 - 815 (1975). In Russian. – Abstr. in Referativ. Zhurn. 51. Astron., 1.51.437 (1976).

077.056 **On possible sources of solar radio emission connected with polar faculae.**
Eh. V. Kononovich, N. N. Shakura.
Astron. Tsirk., No. 872, p. 2 - 3 (1975). In Russian.

077.057 **Modulation of the S-component of the solar radio emission by Alfvén waves.** O. G. Gontarev.
Astron. Tsirk., No. 876, p. 1 - 2 (1975). In Russian.

077.058 **Three-frequency radioheliograph observations of solar continuum events.** A. Magun, R. T. Stewart, R. D. Robinson.
Proc. Astron. Soc. Australia, Vol. 2, 367 - 370 (1975).

077.059 **Multi-frequency heliograph observations of type II bursts.** G. J. Nelson, R. D. Robinson.
Proc. Astron. Soc. Australia, Vol. 2, 370 - 373 (1975).

077.060 **Solar flare continua at the metre wavelengths.** R. D. Robinson, S. F. Smerd.
Proc. Astron. Soc. Australia, Vol. 2, 374 - 378 (1975).

077.061 **Solar radioastronomy.** K. Borkowski.
Urania Kraków, Vol. 46, 258 - 263 (1975). In Polish.

077.062 **Observations of quasiperiodic fluctuations of solar radio flux at frequencies of 610 and 780 MHz.**
G. Ozoliņš, M. Paupere, M. Eliass.
Investigation of the sun and red stars. 3, (see 003.008), p. 5 - 20 (1975). In Russian.

Radiometers for investigation of quasiperiodic fluctuations of solar radio flux at frequencies of 610 and 780 MHz and the observation method are described. The observations were carried out in summer of 1973 at the Radioastrophysical Observatory of the Academy of Sciences of the Latvian SSR with a 10-m paraboloidal radiotelescope. As a result of spectral analysis characteristic periods of quasiperiodic fluctuations in the range of 4–30 minutes were obtained.

077.063 **Radio-zračenje Sunca.** M. S. Dimitrijević.
Vasiona, Vol. 23, 38 - 44 (1975).

077.064 **The polarization of solar bursts at the frequencies 775 MHz, 113 MHz, 39.6 MHz, 29.5 MHz and 23.5 MHz based on the measurements of the Zentralinstitut für solar-terrestrische Physik (Heinrich-Hertz-Institut) during the period 1969–1971. Part I: Tables of data.**
J. Paschke, D. Scholz, W. Voigt.
Zentralinst. Solar-Terrestr. Phys. (Heinrich-Hertz-Inst.), Akad. Wiss. DDR, HHI Suppl. Ser. Solar Data, Vol. 4, (No.1), p. 1 - 122 (1975).

By this publication the former publication of polarization data obtained during 1964 – 1968 is continued.

077.065 **The polarization of solar bursts at the frequencies 775 MHz, 113 MHz, 39.6 MHz, 29.5 MHz and 23.5 MHz based on the measurements of the Zentralinstitut für solar-terrestrische Physik (Heinrich-Hertz-Institut) during the period 1969–1971. Part II: Description of instruments and statistical investigations of type III and type V bursts.**
J. Paschke, D. Scholz, W. Voigt.
Zentralinst. Solar-Terrestr. Phys. (Heinrich-Hertz-Inst.), Akad. Wiss. DDR, HHI Suppl. Ser. Solar Data, Vol. 4, (No. 2), p. 123 - 144 (1975).
The publication of polarization measurements of the Observatory for Solar Radioastronomy beginning in 1961 is continued for the period 1969 – 1971. The polarimeters are partly reconstructed and the measurements are extended to the frequencies 775 MHz and 113 MHz. A short description of the instruments is given. No remarkable difference in polarization was observed between individual type III bursts and the following type V burst. Further also it could be stated in 1969–1971 that the percentage of circularly unpolarized bursts is greater for type III–V than for type III without type V.

Plasma frequency radiation in astrophysics: what we learn from solar radio observations.
See Abstr. 061.010.

Fast electrons in small solar flares.
See Abstr. 073.017.

The spatial distribution of X-ray, EUV and GHz emissions in connection with the flare-loop-channel model.
See Abstr. 073.022.

Radio observations of hard X-ray sympathetic flares and a possible explanation. See Abstr. 073.055.

Observations at 6 cm of the solar active region.
See Abstr. 073.063.

Forecast of plage activity from radio emission at λ 9.1 cm. See Abstr. 073.101.

X-ray and radio emission for the June 15, 1973 solar flare. See Abstr. 073.104.

On the structure of filaments from centimeter and millimeter observations. See Abstr. 073.144.

Study of a filament with a circularly polarized beam at 3.8 cm. See Abstr. 073.145.

On the connection between solar flares, radio bursts of types II and IV and solar cosmic ray events.
See Abstr. 073.156.

Small-scale inhomogeneities in the solar corona: evidence from meter-λ radio bursts. See Abstr. 074.048.

Cyclotron wave instability in the corona and origin of solar radio emission with fine structure. I. Bernstein modes and plasma waves in a hybrid band. See Abstr. 074.071.

A coronal hole observed at X, UV and radio wavelengths. See Abstr. 074.080.

Die Mitte-Rand-Variation der solaren Helligkeit im cm-Wellengebiet und die Vorstellungen über die Struktur der unteren Sonnenatmosphäre. See Abstr. 080.012.

Circulation changes in the free atmosphere during proton events associated with type IV radio bursts.
See Abstr. 085.001.

Spectral characteristics of solar radio bursts associated with shortwave fadeouts. See Abstr. 085.003.

078 Solar Cosmic Radiation

078.001 **Pioneer 10 measurements of the charge and energy spectrum of solar cosmic rays during 1972 August.**
W. R. Webber, E. C. Roelof, F. B. McDonald, B. J. Teegarden, J. Trainor.
Astrophys. Journ., Vol. 199, 482 - 493 (1975).
The authors have measured the energy spectrum and charge composition of solar cosmic rays integrated over the giant events of 1972 August on the Pioneer 10 spacecraft located at 2.2 AU. Within the energy range 8–50 MeV nucleon^{-1}, the individual nuclei from He to Si appear to have average differential spectra of the form $dN/dE \sim E^{-2.7}$, except for carbon, which exhibits a slightly steeper spectrum with $\gamma = 3.1$. The data provide further evidence that the He abundance relative to C and N and O nuclei is variable from event to event, and also that this ratio is energy-dependent. The abundance of Na and Al nuclei was determined for the first time in solar cosmic rays. In the light of these data, the question of the relative abundances of solar cosmic rays and solar atmospheric abundances in the $Z = 10$–14 range is re-examined. Results are also presented for the ^{3}He flux.

078.002 **Angular distributions of solar protons and electrons.**
E. Nielsen, M. A. Pomerantz, H. I. West, Jr.
Planet. Space Sci., Vol. 23, 1179 - 1194 (1975).
High angular-resolution measurements of directional fluxes of solar particles in space have been obtained with detectors aboard OGO-5 during the cosmic ray event of 18 November 1968. This is the only case on record for which sharply-defined directional observations of protons and electrons covering a wide rigidity range (0.3 MV to 1.5 GV) are available.

078.003 **Solar proton events: stratospheric sources of nitric oxide.** P. J. Crutzen, I. S. A. Isaksen, G. C. Reid.
Science, Vol. 189, 457 - 459 (1975).
The production of nitric oxide in the stratosphere during each of the solar proton events of November 1960, September 1966, and August 1972 is calculated to have been comparable to or larger than the total average annual production of NO by the action of galactic cosmic rays. It is therefore very important to consider the effect of solar proton events on the temporal and spatial distribution of ozone in the stratosphere.

078.004 **Unusual emission of iron nuclei from the sun.**
G. Gloeckler, D. Hovestadt, O. Vollmer, C. Y. Fan.
Astrophys. Journ., (*Letters*), Vol. 200, L45 - L48 (1975).
Sustained emission of low-energy solar particles with a composition richer in iron than oxygen is observed in the time period 1974 May 7 to 17. Between 0.7 and 4 MeV per nucleon the relative abundances of C:O:Fe are 0.65:1:1.35. The authors suggest that these observations provide indication for effects of heavy-ion enrichment in the lower corona of the sun.

078.005 **Generation of superhigh energy particles on the sun.**
E. V. Kolomeets.
Materialy Itog. nauch. konf. prof.-prepodavat. sostava. Kazakhsk. un-t. Alma-Ata, 1974, p. 262 - 263. In Russian. Abstr. in Referativ. Zhurn. 62. Issled. kosmich. prostranstva, 7.62.245 (1975).

078.006 **Relativistic solar protons.** S. P. Duggal.
Rev. Geophys. Space Phys., Vol. 13, No. 3, (see 013.001), p. 1084 - 1086, 1095 - 1096 (1975). – Review paper with bibliographical notes.

078.007 **Neutron and γ ray emissions from the sun.**
S. P. Duggal.
Rev. Geophys. Space Phys., Vol. 13, No. 3, (see 013.001), p. 1086 - 1087, 1096 - 1098 (1975). – Review paper with bibliographical notes.

078.008 **Low-energy solar cosmic rays: a bibliography.**
E. C. Roelof, S. M. Krimigis.
Rev. Geophys. Space Phys., Vol. 13, No. 3, (see 013.001), p. 1092 - 1094, 1099 - 1104 (1975). – Review paper with bibliographical notes.

078.009 **On the determination of the spectrum of solar protons from the geomagnetic cutoff rigidity along the trajectory of artificial earth satellites.** P. I. Shavrin.
Kosmich. Issled., Vol. 13, 503 - 507 (1975). In Russian.

078.010 **Solar protons on closed magnetospheric field lines after an interplanetary flux decrease.**
M. Scholer.
Planet. Space Sci., Vol. 23, 1445 - 1450 (1975).
Particle measurements from the low altitude polar-orbiting satellite GRS-A/Azur and from Explorer 41 in the magnetosheath during a time period after the sudden commencement at 14:30 UT on 8 March 1970, have been used in order to study the access mode of solar particles into the closed field line region of the magnetosphere. A particle decrease in the magnetosheath and over the central polar cap but not in the stable trapping region indicates that solar particles are temporarily trapped and can complete several drifts around the earth.

078.011 **Enrichment of heavy nuclei in ^{3}He-rich flares.**
G. J. Hurford, R. A. Mewaldt, E. C. Stone, R. E. Vogt.
Astrophys. Journ., (*Letters*), Vol. 201, L95 - L97 (1975).
Five ^{3}He-rich solar-flare particle events were observed in 1973–1974 with ^{3}He/^{4}He ranging from 0.2 to ~8 at 2.9–12.7 MeV per nucleon. In all five events the $(Z \geqslant 6)/^1$H ratio at ~3 MeV per nucleon was enriched by ~10 to ~100 times and the $(Z \geqslant 6)/^4$He ratio was enriched by ~3 to ~30 times, when compared with average solar-particle abundances measured over this 2-year period. It is suggested that the simultaneous enhancement of $Z \geqslant 6$ and ^{3}He nuclei and the absence of ^{2}H and ^{3}H may be partly due to a preferential acceleration process which depends on the Z^2/A of the nuclei.

078.012 **A method for estimation of the quality of the solar proton-event forecast.** Z. Cui.
Acta Astron. Sinica, Vol. 16, 6 - 11 (1975). In Chinese.
The various indicators used to estimate the quality of the solar proton-event forecast are reviewed and their incompleteness is discussed. Introducing the concept of random forecast, the author proposes to adopt a combined indicator, which is more reasonable for estimating the quality of the proton-event forecast.

078.013 **On the origin of the upper cutoff in the solar proton spectrum.** J. Pérez-Peraza.
Journ. Geophys. Res., Vol. 80, 3535 - 3542 (1975).
The problem of the origin of the upper cutoff in the spectrum of multi-GeV solar protons is examined. This analysis concerns the case in which the upper cutoff is determined at the source level by a stochastic mechanism of acceleration. The operation of the random process at the solar source is justified in terms of a small length scale of the fluctuations of the magnetic field. The deviation of the particle spectrum from a perfect power law is explained by a discontinuity in the behavior of the mean free path of particles for acceleration.

078.014 **Solar activity and streams of cosmic ray protons in March - July, 1972.** N. N. Kontor, G. P. Lyubimov, A. G. Nikolaev, N. V. Pereslegina, B. A. Tverskoj, T. G. Khotilovskaya.
Geomagn. Aeronom., Vol. 15, 777 - 786 (1975). In Russian.

078.015 **The spectrum of generation of high-energy particles on the sun during the flare of September 1, 1971.**
I. V. Petrova.
Geomagn. Aeronom., Vol. 15, 930 - 931 (1975). In Russian.

078.016 **Generation of the permanent flux of charged and neutral particles on the sun.** A. B. Bajsakalova, O. A. Bogdanova, V. A. Kobzev, E. V. Kolomeets, A. I. Kupchishin.
Materialy Itog. nauch. konf. prof.-prepodavat. sostava. Kazakhsk. un-t. Alma-Ata, 1974, p. 256. In Russian. – Abstr. in Referativ. Zhurn. 51. Astron., 8.51.455 (1975).

078.017 **Neutrons of solar origin.**
Yu. Dubinskij, Yu. E. Efimov, K. Kudela.
VI Leningr. mezhdunar. seminar, 1974. Uskorenie chastits i yader. reaktsii v kosmose. Leningrad, 1974, p. 89, with a discussion, p. 90. In Russian and English. – Abstr. in Referativ. Zhurn. 51. Astron., 8.51.456 (1975).

078.018 **Cosmic ray intensity variations and solar wind.**
S. N. Vernov, B. A. Tverskoj, G. P. Lyubimov, N. V. Pereslegina, N. N. Kontor, E. A. Chuchkov.
VI Leningr. mezhdunar. seminar, 1974. Uskorenie chastits i yader. reaktsii v kosmose. Leningrad, 1974, p. 159, 160, with a discussion, p. 161 - 162. In Russian and English. – Abstr. in Referativ. Zhurn. 51. Astron., 8.51.458 (1975).

078.019 **Nuclear composition of energetic solar particles – new source of information on solar flares.**
V. N. Lutsenko.
VI. Leningr. mezhdunar. seminar, 1974. Uskorenie chastits i yader. reaktsii v kosmose. Leningrad, 1974, p. 87, with a discussion, p. 88. In Russian and English. – Abstr. in Referativ. Zhurn. 51. Astron., 8.51.492 (1975).

078.020 **Particle acceleration in a neutral line of a magnetic field.** S. V. Bulanov, P. V. Sasorov.
VI Leningr. mezhdunar. seminar, 1974. Uskorenie chastits i yader. reaktsii v kosmose. Leningrad, 1974, p. 231, with a discussion, p. 232. In Russian and English. – Abstr. in Referativ. Zhurn. 51. Astron., 8.51.502 (1975).

078.021 **Propagation of solar-flare cosmic rays along corotating interplanetary flux-tubes.**
C. K. Ng, L. J. Gleeson.
Solar Physics, Vol. 43, 475 - 511 (1975).

The propagation of solar-flare cosmic rays along a corotating spiral mean interplanetary magnetic flux-tube is investigated under a variety of initial, inner and outer boundary conditions, by solving the transport equations numerically. The inner boundary condition is shown to have negligible effects on the redistribution of the cosmic rays in interplanetary space, apart from a proportionality constant. The diffusion coefficients have been classified into type I and type II: for type I the peak-along-spiral stays at a finite heliocentric distance, while for type II it travels to infinity. The general propagation characteristics with a free-escape outer boundary are like those of type I. It is shown that closely exponential decay can be produced without a free-escape boundary. The concept of an 'equilibrium shape' at late time is introduced.

078.022 **Catalogue of solar electron fluxes observed from April 14 to August 16, 1972.**
M. A. Zel'dovich, V. G. Kurt, V. N. Lutsenko, I. A. Savenko.
Solnechnye Dannye 1975 Byull., No. 4, p. 47 - 62 (1975).
In Russian.

The data on solar electron fluxes obtained by the Prognoz satellite during April 14–August 16, 1972 are given. A comparison of these observations with other solar and geomagnetic phenomena is made.

078.023 **Determination of the proton spectrum of the burst of solar cosmic radiation on April 6, 1971 from the geomagnetic cut-off rigidity aboard an artificial earth satellite.**
P. I. Shavrin, E. A. Devicheva, V. S. Tsaplin.
Kosmich. Issled., Vol. 13, 704 - 709 (1975). In Russian.

078.024 **Cosmic ray events on August 4 and 7, 1972.**
N. P. Chirkov, A. T. Filippov, V. I. Kozlov.
Issled. po kosmofiz. i aehronomii. Yakutsk, 1975, p. 38 - 48, 195 - 196. In Russian. - Abstr. in Referativ. Zhurn. 51. Astron., 10.51.415 (1975).

078.025 **Propagation of solar cosmic radiation from the flares of October 3 and September 1, 1971 from data of Cosmos 426.** S. N. Kuznetsov, Yu. I. Logachev, I. V. Petrova, I. A. Savenko.
Kosmich. Issled., Vol. 13, 844 - 850 (1975). In Russian.

078.026 **Constant outflow of low-energy particles from active regions on the sun.** M. A. Zel'dovich, B. M. Kuzhevskij, Yu. I. Logachev, I. A. Savenko, I. P. Shestopalov.
Kosmich. Issled., Vol. 13, 851 - 858 (1975). In Russian.

078.027 **Energetic dependence of the mean free path for scattering of cosmic rays.**
E. V. Gorchakov, G. A. Timofeev, T. I. Morozova.
Geomagn. Aeronom., Vol. 15, 1083 - 1084 (1975). In Russian. Brief information.

078.028 **Solar cosmic rays.** T. N. Charakhch'yan.
Probl. yader. fiz. i kosmich. luchej. Resp. mezhved. temat. nauch.-tekhn. sb., 1975, vyp. (No.) 2, p. 17 - 34. In Russian. – Abstr. in Referativ. Zhurn. 51. Astron., 12.51.345; 62. Issled. kosmich. prostranstva, 12.62.211 (1975).

078.029 **Emission of energetic Fe^{+12} ions from the sun.**
G. Gloeckler, R. Sciambi, C. Y. Fan, D. Hovestadt.
Bull. American Astron. Soc., Vol. 7, 536 (1975). – Abstr. AAS.

078.030 **Simulation of solar flare particle interaction with interplanetary shock waves.**
M. Scholer, G. Morfill.
Solar Physics, Vol. 45, 227 - 240 (1975).

In order to study the propagation of solar cosmic rays in interplanetary space a computer program has been developed using a Monte-Carlo technique, which traces the histories of particles released impulsively at the sun. The particle propagation model considers the adiabatic deceleration during the convective and diffusive transport of the particles, and the model of the interplanetary medium incorporates a radially expanding blast wave which exerts a sweeping action on the particles and accelerates them through the first-order Fermi process.

078.031 **The velocity dependence of the acceleration spectrum of solar protons.**
J. Pérez-Peraza, J. Galindo Trejo.
Rev. Mexicana Astron. Astrofis., Vol. 1, 273 - 281 (1975).

In a preliminary analysis of the source spectrum of solar protons, the authors investigate the case in which energy losses of protons in the source are negligible while they are undergoing Fermi-acceleration. A power law in energy is adopted, under the assumption of an acceleration rate proportional to the energy and dependent on the velocity of the protons. The calculated spectra for several sets of accelerating parameters are compared with the experimental spectrum of the solar proton event of January 28, 1967.

078.032 **Energetic solar proton vs. terrestrially trapped proton fluxes (risk factor).**
J. H. King, E. G. Stassinopoulos.
Journ. Spacecraft Rockets, (*USA*), Vol. 12, 122 - 124 (1975).
Demonstrates the relative importance of solar and trapped proton fluxes in the consideration of shielding requirements for 1977–1983 geocentric space missions.

078.033 **Recordings of solar protons during 26–29 October, 1968.** A. Gjerde, R. Amundsen.
Phys. Norvegica, Vol. 8, 47 - 58 (1975).
The composite events causing increased influx of solar particles during 26–29 October, 1968 are studied, using data from the polar-orbiting ESRO IA satellite. Details on protons of 0.1–10 MeV are presented, and their interplanetary propagation characteristics discussed.

078.034 **Energetic solar particles in interplanetary space. Part 2.** R. Amundsen.
Fra Fys. Verden, Vol. 37, 15 - 22 (1975). In Norwegian.

078.035 **Short time intensity variations of solar cosmic radiation on August 1972.**
G. Kremser, W. Riedler, E. Kirsch.
Verhandl. Deutsche Phys. Ges., Reihe 6, Vol. 10, 174 (1975). In German. – Abstract.

078.036 **Search for solar particle production above approx. 75 GeV.** S. M. Schindler.
Thesis, Colorado State Univ., Fort Collins (USA). [Available from Univ. Microfilms, Ann Arbor, Mich. Order No. 75-4614], 278 pp. (1974).

078.037 **Interplanetary propagation of steady solar injected particles.** S. Cecchini, N.J. Martinic.
Nuovo Cimento B, Ser. 11, Vol. 25B, 191 - 207 (1975).

078.038 **Heavy solar cosmic rays in the January 25, 1971 solar flares.** C. J. Pellerin, Jr.
National Aeronautics and Space Administration, Greenbelt, Md. (USA). NASA Rep. NASA-TM-X-70731; X-662-74-231, 32 pp. (1974).
A detailed study of the charge composition of heavy solar cosmic rays measured in the January 25, 1971 solar flare including differential fluxes for the even charged nuclei from carbon through argon is presented. The measurements are obtained for varying energy intervals for each nuclear species in the energy range from 10 to 35 MeV/nucleon. In addition, abundances relative to oxygen are computed for all the above nuclei in the single energy interval from 15 to 25 MeV/nucleon. An upper limit for the abundance of calcium nuclei is also presented. These measurements, when combined with other experimental results, enable the energy dependence of abundance measurements as a function of nuclear charge to be discussed.

Physics of solar cosmic rays. See Abstr. 003.111.

Catalog of solar particle events, 1955 - 1969. See Abstr. 003.122.

The Pioneer 10/11 and Helios A/B cosmic ray instruments. See Abstr. 034.160.

Plasma instabilities in chromospheric flares and cosmic ray generation. See Abstr. 073.116.

The isotopic composition of hydrogen and helium in solar flare particle events. See Abstr. 073.131.

Interpretation of flare-produced proton spectra. See Abstr. 073.138.

On the connection between solar flares, radio bursts of types II and IV and solar cosmic ray events. See Abstr. 073.156.

Cosmic-ray propagation through the solar corona. See Abstr. 074.040.

Nuclear reactions in the solar atmosphere and solar cosmic rays. See Abstr. 080.076.

Die irdische Magnetopause unter Berücksichtigung einer Winkelverteilung der solaren Teilchen und des interplanetaren Magnetfeldes. See Abstr. 084.324.

A search for solar energetic tritons in lunar samples. See Abstr. 094.471.

Variations of the interplanetary magnetic field and of cosmic rays on August 9, 1972. See Abstr. 106.031.

Cosmic ray propagation in interplanetary space. See Abstr. 143.015.

Long-periodic changes of solar diurnal variation of cosmic rays. See Abstr. 143.026.

The anomalous ^{4}He component in the cosmic-ray spectrum at $\lesssim$50 MeV per nucleon during 1972–1974. See Abstr. 143.034.

An analysis of the connection between the 27-day variations of cosmic rays and various indices of solar activity. See Abstr. 143.036.

The interplanetary acceleration of energetic nucleons. See Abstr. 143.044.

079 Solar Eclipses

079.001 **A note on the periodicity of eclipses.**
C. H. Smiley.
Journ. Roy. Astron. Soc. Canada, Vol. 69, 133 - 135 (1975).

079.002 **Modifications of Minnaert's method of timing solar eclipse contacts.** W. J. Westbrooke.
Bull. American Astron. Soc., Vol. 7, 474 - 475 (1975). – Abstr. AAS.

079.003 **Les éclipses de soleil visibles à Uccle de 1976 à 2075.** J. Meeus.
Ciel et Terre, Vol. 91, 424 - 427 (1975).

079.004 **Relative positions of the solar and lunar disc centres during a total solar eclipse.**
V. I. Kulidzhanishvili.
Byull. Abastumansk. Astrofiz. Obs., No. 46, p. 205 - 216 (1975). In Georgian.

The method for determination of relative positions of the solar and lunar disc centres applied for the July 10, 1972 total solar eclipse is described. The solar disc centre motion is given graphically vs. the "fixed" centre of the lunar disc. The technique of the matching of the solar disc centre with that of the microphotometer rotating table is being described.

079.005 **L'eclisse di sole del 2 Agosto 1133.**
U. Dall'Olmo.
Coelum, Vol. 43, 233 - 243 (1975).

Sky color near the horizon during a total solar eclipse. See Abstr. 082.102.

Sky brightness during eclipses: a review. See Abstr. 082.103.

Cosmic rays and eclipses. See Abstr. 143.089.

079.100 **Solar eclipse 1973 June 30**

Observation of the temperature and pressure changes during the 30 June 1973 solar eclipse.
R. C. Anderson, D. R. Keefer.
Journ. Atmosph. Sci., Vol. 32, 228 - 231 (1975).

Total solar eclipse of 30 June 1973.
J. Sýkora, I. Miko, E. Pittich.
Kozmos, Vol. 6, 118 - 120, 139 - 142, 163 - 166, 182 - 184 (1975). Continued, see Abstr. 13.079.100. In Slovak.

A 16-mm film "Solar spectrum at the extreme limb observed during the 1973 June 30 total eclipse at Atar, Mauritania." J. Houtgast, O. Namba.
Oss. Mem. Oss. Astrofis. Arcetri, Fasc. 105, (see 012.026), p. 115 - 119 (1975).

Testing relativity in the desert: the Texas-Mauritania eclipse expedition. See Abstr. 066.172.

The Koobi Fora experiment: continuum observations of solar spicules during the 30 June 1973 eclipse. See Abstr. 073.113.

The solar corona on June 30, 1973. See Abstr. 074.020.

The integral brightness of the corona during the solar eclipse of June 30, 1973. See Abstr. 074.026.

Observation of the 7892 Å Fe XI line in the corona during the eclipse of June 30, 1973. See Abstr. 074.061.

The coronal condensation observed at the 1973 eclipse. See Abstr. 074.070.

The radial brightness distribution of the sun at 3.2 mm as determined from the June 30, 1973 total solar eclipse and a reanalysis of the March 7, 1970 total solar eclipse. See Abstr. 077.012.

Generation of atmospheric gravity waves by the 30 June 1973 solar eclipse in Africa. See Abstr. 083.018.

079.101 **Solar eclipse 1972 July 10**

Observation of the total solar eclipse of July 10, 1972 at 4.0 cm wavelength. Method and preliminary results.
A. B. Berlin, V. V. Vitkovskij, G. B. Gel'frejkh, R. I. Enikeev, A. V. Ipatov, V. G. Nagnibeda, G. M. Timofeeva, S. P. Yakimov.
Trudy Astron. Obs., *Leningrad*, Vol. 31 (= Uchenye Zapiski Leningr. Un-ta, No. 381 = Seriya Matem. Nauk, vyp. (No.) 51), p. 154 - 173 (1975). In Russian.

The total solar eclipse of July 10, 1972 was observed in the region of Anadyr (Chukotka) at 4.0 cm wavelength. Preliminary results of the observations are presented. The method of observation and reducing of eclipse curves, as well as the sources of errors are discussed in detail. Fine structure of local sources in polarized and unpolarized emission is obtained with angular resolution of 5″. The residual flux at the moment of the central phase was 8 percent of the solar flux out of the eclipse.

The Soviet-French cooperative study of the solar corona. I. The structural peculiarities and photometry of the corona on July 10, 1972. See Abstr. 074.021.

Investigation of emission lines of the solar corona of 10 July, 1972 using the Fabry-Pérot etalon. See Abstr. 074.066.

Spectrophotometry of the solar corona taken on July 10, 1972. See Abstr. 074.086.

On the large-scale brightness distribution in the solar corona on July 10, 1972. See Abstr. 074.087.

079.102 **Solar eclipse 1970 March 7**

Observations of the solar eclipse on March 7, 1970 with the polarimeter of the Havana Radio Astronomical Station at 4.5 cm. J. A. Quiñones, A. N. Korzhavin, N. G. Peterova, J. Santos.
Solnechnye Dannye 1975 Byull., No. 3, p. 87 - 96 (1975). In Russian.

The results of observations of the solar eclipse on March 7, 1970 with the polarimeter of the Havana Radio Astronomical Station at 4.5 cm are given. The data on the seven local sources of radio emission, the regions of the emission enhancement after the burst, have been obtained from the reduction of the eclipse curves; the general magnetic field of the sun close to the southern pole has been evaluated.

The radial brightness distribution of the sun at 3.2 mm as determined from the June 30, 1973 total solar eclipse and a reanalysis of the March 7, 1970 total solar eclipse. See Abstr. 077.012.

079.103 **Solar eclipse 1975 May 11**

Observations of the partial solar eclipse on 11 May 1975 at the Astronomical Observatory of the Kiev University.
A. A. Zhitetskij, G. V. Moskaleva.
Astron. Tsirk., No. 877, p. 7 - 8 (1975). In Russian.

Observation of the partial solar eclipse of 11 May 1975. J. Bouška, J. Stuchlík, L. Hurta, Z. Machovský, V. Roškot, J. Hajfler, L. Kováč.
Říše hvězd, Vol. 56, 151 - 152 (1975). In Czech.

Zonsverduistering van 11 mei 1975.
H. Betlem.
Zenit, Vol. 2, 240 - 242 (1975).

079.104 **Solar eclipse 1973 December 24**

Observations of the solar eclipse in Cuba on the 24th of December 1973 at 1.37 m wavelength.
L. I. Yurovskaya, O. Olvares, M. Ferreiro.
Izv. Krymskoj Astrofiz. Obs., Vol. 53, p. 130 - 138 (1975). In Russian.

Observing the solar eclipse on the 24th of December 1973 in Havana at wavelength 1.37 m the structure of a coronal local source has been obtained. The dimension of the source was $3'.0 \pm 0'.3$. The enhanced background continuum region and that of the bursts were detected simultaneously.

Observations of the annular solar eclipse of 1973, December 24. L. B. Nadeau.
Strolling Astronomer, Vol. 25, 202, 203, 205 - 207 (1975).

079.105 **Solar eclipse 1976 April 29**

L'éclipse de soleil du 29 avril 1976. J. Meeus.
Ciel et Terre, Vol. 91, 387 - 389 (1975).

079.106 **Solar eclipse 1965 May 30**

Coronal emission line profile observations at total solar eclipses. II: 30 May 1965 results, deconvolution and interpretation. See Abstr. 074.108.

079.107 **Solar eclipse 1968 September 22**

Interferometric investigation of the line of sight velocities in λ 5303 during the eclipse of 22 September, 1968.
See Abstr. 074.121.

079.108 **Solar eclipse 1975 November 3**

Eclipse parcial de sol del 3 de noviembre de 1975.
G. M. Iannini.
Obs. Astron., Univ. Nacional Córdoba, Córdoba, Argentina, Separate print, 14 pp. (1975).

079.109 **Solar eclipse 1976 October 23**

Total solar eclipse of 23 October 1976.
A. D. Fiala, J. S. Duncombe.
United States Naval Obs., *Washington,* Circ. No. 152, 23 pp. (1975).

080 Solar Atmosphere, Figure, Internal Constitution, Neutrinos, Magnetic Fields, Rotation, Miscellanea

080.001 **Solar magnetic fields and convection. II: Magnetic diffusivity and flux concentration.** J. H. Piddington.
Astrophys. Space Sci., Vol. 35, 269 - 283 (1975).

It is shown that the observed concentrations of surface fields into strengths of $\gtrsim$ 1000 G cannot be accounted for by observed surface motions. Nor are they accounted for by the numerical models of turbulence of Weiss or Moss whatever values of the magnetic Reynolds number are assumed. A detailed comparison is made between both small-scale and large-scale surface magnetic features and the predictions of the diffuse-field theory. The differences appear irreconcilable and the features only explicable in terms of the twisted flux-rope model.

080.002 **On the rotation of gas and magnetic fields at the solar photosphere.** P. Foukal, J. R. Jokipii.
Astrophys. Journ., (*Letters*), Vol. 199, L71 - L73 (1975).

The authors point out that observations of a 5 percent velocity difference between photospheric gas and magnetic structures at a given latitude may simply result from angular momentum conservation by fluid elements in the convection zone. Estimates of the viscosity and magnetic drag are considered, and it is concluded that they probably are not large enough to enforce strictly rigid rotation.

080.003 **Solar models of low neutrino-counting rate: the depleted Maxwellian tail.**
D. D. Clayton, E. Dwek, M. J. Newman, R. J. Talbot, Jr.
Astrophys. Journ., Vol. 199, 494 - 499 (1975).

The authors present evolutionary sequences for the sun which confirm that the ^{37}Cl counting rate will be much reduced if the high-energy tail of the Maxwellian distribution of relative energies is progressively depleted. If the depletion is characterized by exp $[-\delta(E/kT)^2]$, the counting rate falls below 1 solar neutrino unit (10^{-3} solar neutrino captures per second per target atom) for $\delta \geqslant 10^{-2}$. Observational tests of these ideas are discussed.

080.004 **Chemistry of the solar neutrino problem.**
K. C. Jacobs.
Nature, Vol. 256, 560 - 561 (1975). – Letter.

080.005 **Relativity and the solar oblateness.** R. H. Dicke.
Experimental gravitation, (see 012.002), p. 200 - 234 (1974).

Introduction; The observations; The oblateness and solar physics.

080.006 **A suggested space mission for measuring the angular momentum of the sun.** R. W. Davies.
Experimental gravitation, (see 012.002), p. 405 - 413 (1974).

080.007 **Solar oblateness and the solar quadrupole moment.**
W. Roxburgh.
Experimental gravitation, (see 012.002), p. 525 - 528 (1974).

080.008 **The sun's shape.** J. Ashbrook.
Sky Telescope, Vol. 50, 71, 79 (1975).

080.009 **On a reliable representation of a class of force-free solar magnetic fields.** N. Seehafer.
Astron. Nachr., Vol. 296, 177 - 181 (1975).

It is shown that for a class of force-free magnetic fields the magnetic field cannot be determined uniquely from the observed vertical component of the photospheric magnetic field given in an area of limited extent. Then it is proposed how some functions, the additional knowledge of which permits the magnetic field to be determined uniquely, could be chosen as a first approximation.

080.010 **Differentielle Rotation und meridionale Strömungen des solaren Ca$^+$-networks.**
E. H. Schröter, H. Wöhl.
Mitt. Astron. Ges., No. 36, p. 141 (1975). – Abstract.

080.011 **Entstehung von NaI-Linien in der Sonnenatmosphäre.** T. Gehren.
Mitt. Astron. Ges., No. 36, p. 141 (1975). – Abstract.

080.012 **Die Mitte-Rand-Variation der solaren Helligkeit im cm-Wellengebiet und die Vorstellungen über die Struktur der unteren Sonnenatmosphäre.**
E. Fürst, W. Hirth.
Mitt. Astron. Ges., No. 36, p. 142 - 144 (1975).

080.013 **Acoustic waves and the geometric scale in the solar atmosphere.** F.-L. Deubner.
Mitt. Astron. Ges., No. 36, p. 149 - 150 (1975). – Abstract.

080.014 **Theoretisches Temperaturminimum der Sonne auf Grund der Ausbreitung von strahlungsgedämpften akustischen Wellen.** P. Ulmschneider, W. Kalkofen.
Mitt. Astron. Ges., No. 36, p. 150 - 151 (1975).

080.015 **Entstehung und Ausbreitung von Stoßwellen in der Sonnenatmosphäre.** H. U. Bohn.
Mitt. Astron. Ges., No. 36, p. 151 - 155 (1975).

080.016 **A comparison of the meridional flows in the sun's convection zone, predicted by theories of the solar dynamo and differential rotation.** B. R. Durney.
Astrophys. Journ., Vol. 199, 761 - 764 (1975).

Recently Yoshimura has evaluated the gradient of the sun's angular velocity ($d\Omega/dr$) necessary to give a good fit to the observed solar activity cycle. The author estimates the meridional velocities in the convection zone, implied by this value of $d\Omega/dr$. These meridional velocities are in good agreement with those necessary to explain the sun's surface differential rotation.

080.017 **On the differential rotation of the sun and cycles of its activity.** R. S. Iroshnikov.
Astron. Zhurn. Akad. Nauk SSSR, Vol. 52, 792 - 803 (1975).
In Russian. English translation in Soviet Astron., Vol. 19, No.4.

The size of convective cells is comparable with the thickness of the convective zone. Therefore it is impossible to use Navier-Stokes' or Vasyutinskij's equations for the calculation of the differential rotation of the convective zone; it is necessary to take into account the distortion of the field of accidental velocities by rotation. A basis of acceptability of a previously suggested system of equations for the solution of the problem is given. For a thin convective zone solutions are obtained showing that the equatorial zone rotates faster than the polar zones at all depths.

080.018 **The intrinsic visual oblateness of the sun.**
H. A. Hill, R. T. Stebbins.
Astrophys. Journ., Vol. 200, 471 - 483 (1975).

New solar oblateness observations are reported that yield a value of 18.4 ± 12.5 milli-arcsec for the difference between equatorial and polar diameters. These observations were conducted when the time-varying, excess equatorial brightness was markedly reduced, greatly simplifying the interpretation of the apparent solar oblateness. The experimental scheme for defining the "edge" of the sun and monitoring the shape of the extreme limb darkening function is summarized along with the description of the instrumentation. The result further strengthens the conclusion of Hill et al. that the grave consequence of the Dicke-Goldenberg work for Einstein's general theory of relativity is removed. The new result excludes an oblateness as large as that required by the rapidly rotating interior models used to explain the low neutrino flux.

080.019 **Waves in the solar atmosphere. IV. Magneto-gravity and acoustic-gravity modes.**
R. A. Schwartz, R. F. Stein.
Astrophys. Journ., Vol. 200, 499 - 502 (1975).
The authors consider the properties of waves in a stratified, conducting atmosphere under the influence of an applied magnetic field. Wave normal surfaces are exhibited for acoustic-gravity and magneto-gravity waves, and the propagation of these modes is compared.

080.020 **On the statistical description of inhomogeneities in the quiet solar atmosphere. I. Linear regression analysis and absolute calibration of multichannel observations of the Ca^+ emission network.**
A. Skumanich, C. Smythe, E. N. Frazier.
Astrophys. Journ., Vol. 200, 747 - 764 (1975).
Observations with a 2.″4 square aperture (1968 epoch) of the Ca^+ K core brightness (1.1 Å bandwidth), vertical magnetic field, vertical velocity, and continuum brightness have been analyzed to yield average network and nonnetwork values and dispersions as well as linear regression relations. The average network has a field of 26 gauss, independent of polarity, a Ca^+ brightness 1.27 that of the average nonnetwork or undisturbed chromosphere, a downdraft velocity of 41 m s^{-1}, a monochromatic continuum enhancement of 0.34 percent, and covers 39 percent of the quiet sun. The average outward flux density of network over nonnetwork regions is 2.4×10^5 ergs $cm^{-2} s^{-1}$ for Ca^+K and 3.8×10^8 ergs $cm^{-2} s^{-1}$ for the continuum.

080.021 **Solar five-minute oscillations as a resonance phenomenon.** C. Chiuderi, C. Giovanardi.
Astrophys. Journ., (*Letters*), Vol. 200, L165 - L166 (1975).
The response of a nonisothermal gravitational atmosphere to an oscillatory perturbation is studied in the case of a parabolic temperature profile. The exact analytical solution is then applied to a model solar atmosphere. An interpretation of the 5-minute oscillations as free oscillations of the atmosphere as a whole is proposed.

080.022 **The sun.** E. N. Parker.
Sci. American, Vol. 233, No. 3, p. 42 - 50 (1975).
Popular article.

080.023 **On the nonlinear mechanism of sound pulses generation in the solar atmosphere.**
N. S. Petrukhin, S. M. Fajnshtejn.
Pis'ma v Astron. Zhurn., Vol. 1, No. 8, p. 32 - 36 (1975). In Russian.
The nonlinear interaction of two contrary Alfvén waves effectively generates sound waves with time scale of about 10^3 *s* in the horizontal regions of coronal archs. The suggestion is made to apply this effect to the explanation of quasi-periodic pulsations.

080.024 **Standard solar models: a solar neutrino loophole.**
C. A. Rouse.
Bull. American Astron. Soc., Vol. 7, 406 (1975). – Abstr. AAS.

080.025 **The five minute oscillations as random oscillators.**
D. K. Lynch.
Bull. American Astron. Soc., Vol. 7, 407 (1975). – Abstr. AAS.

080.026 **The character of 300-second oscillators.**
L. J. November, G. W. Simon, S. P. Worden.
Bull. American Astron. Soc., Vol. 7, 407 (1975). – Abstr. AAS.

080.027 **New observational limits on the sun as a pulsating star.** W. Livingston.
Bull. American Astron. Soc., Vol. 7, 407 (1975). – Abstr. AAS.

080.028 **Differential rotation of filaments.**
W. Adams, F. Tang.
Bull. American Astron. Soc., Vol. 7, 458 (1975). – Abstr. AAS.

080.029 **Calculation of solar force-free magnetic field structures.** C. W. Barnes, P. A. Sturrock.
Bull. American Astron. Soc., Vol. 7, 459 (1975). – Abstr. AAS.

080.030 **Recent progress in solar oblateness studies.**
H. A. Hill, R. T. Stebbins, T. M. Brown.
Bull. American Astron. Soc., Vol. 7, 478 (1975). – Abstr. AAS.

080.031 **Solar structure in the extreme ultraviolet.**
G. L. Withbroe.
Bull. American Astron. Soc., Vol. 7, 478 (1975). – Abstr. AAS.

080.032 **Nonradial g-mode oscillations and the stability of the sun.**
H. Shibahashi, Y. Osaki, W. Unno.
Publ. Astron. Soc. Japan, Vol. 27, 401 - 410 (1975).
Numerical study is made of the solar stability against the nonradial g-mode oscillations by using the quasi-adiabatic approximation. The contribution of the outer convection zone to the stability is neglected. All modes become stable in a model closest to the present sun and in more advanced models, with the g_1-mode instability lasting longest.

080.033 **Heights of formation of non-magnetic solar lines suitable for velocity studies.** R. C. Altrock, L. J. November, G. W. Simon, R. W. Milkey, S. P. Worden.
Solar Physics, Vol. 43, 33 - 37 (1975).
Heights of formation of lines that do not exhibit Zeeman splitting are calculated using an LTE, partial non-LTE, and full non-LTE approach. Non-magnetic ($g = 0$) lines are valuable for velocity investigations in quiet-sun magnetic field regions, and a knowledge of their formation heights is useful for obtaining three dimensional velocity profiles in these regions.

080.034 **The five-minute oscillations in the solar atmosphere.**
S. M. Chitre, M. H. Gokhale.
Solar Physics, Vol. 43, 49 - 55 (1975).
The acoustic overstability in a polytropic plane-parallel atmosphere with superadiabatic temperature gradient and radiative dissipation is demonstrated for optically thick disturbances. The periods of oscillation are found to be in the range 250–480 s and the associated wavelength of the order of 4000 km. The five-minute oscillations in the solar surface are attributed to self-excited sound waves in a layer in the sub-photospheric convection zone of about 1000 km thickness.

080.035 **Differential rotation and the spot zones.**
F.-L. Deubner, M. Vazquez.
Solar Physics, Vol. 43, 87 - 90 = Mitt. Fraunhofer Inst., *Freiburg,* No. 136 (1975).
A minimum of angular velocity is not found at the equator as suggested by Gilman, but at a latitude of 6° in both hemispheres.

080.036 **Preflare current sheets in the solar atmosphere.**

E. R. Priest, M. A. Raadu.
Solar Physics, Vol. 43, 177 - 188 (1975).
Neutral current sheets are expected to form in the solar atmosphere when photospheric motions or the emergence of new magnetic flux causes oppositely directed magnetic fields to be pressed together. Magnetic energy may thus be stored slowly in excess of the minimum energy associated with a purely potential field and released suddenly during a solar flare. The authors investigate the neutral sheet which forms between two parallel line dipoles when either the distance between them decreases or their dipole moments increase. It is found that, when the dipoles have approached by an amount equal to a tenth of their original separation distance, the stored energy is comparable with that released in a major flare. A similarity solution for one-dimensional magnetohydrodynamic flow within such a neutral sheet is presented.

080.037 **Solar models of low neutrino-counting rate: the central black hole.**
D. D. Clayton, M. J. Newman, R. J. Talbot, Jr.
Astrophys. Journ., Vol. 201, 489 - 493 (1975).
Partial evolutionary sequences have been calculated for several solar models with central black holes of order $10^{-5} M_{\odot}$. If these are assumed to radiate their Eddington limiting luminosity, the central temperature is depressed to the extent that the predicted count rate for the ^{37}Cl solar neutrino experiment nears the current upper limit of 1 SNU; this occurs when the auxiliary energy source provides about half of the solar luminosity. Count rates below this limit would result from an even larger black-hole luminosity. Consequences for stellar evolution of the occasional presence of black holes inside normal stars are discussed.

080.038 **Solar activity and the general absence of hydrostatic equilibrium in an azimuthal magnetic field.**
E. N. Parker.
Astrophys. Journ., Vol. 201, 502 - 508 (1975).
It is shown that any temperature variation along the axis of symmetry of an azimuthal field causes continual convective activity. There is no static equilibrium configuration, except in very special, and hence improbable, cases. The author suggests that this dynamical effect contributes to the activity associated with the flux tubes extending through the solar photosphere. The X-ray bright spots, evidently caused by the emergence and expansion of small bipolar regions, may be produced in part by the convective effect of an azimuthal field.

080.039 **On non-LTE H_2^+ as missing solar opacity.**
S. P. Tarafdar, M. S. Vardya.
Astrophys. Journ., (*Letters*), Vol. 201, L99 - L102 (1975).
A careful examination has revealed that use of incomplete reaction processes and incorrect rates has lead Krishna Swamy and Stecher to overestimate the H_2^+ opacity by a factor of 10^4 at $\lambda = 2000$ Å. H_2^+ is not a significant source of opacity in the solar atmosphere.

080.040 **The neutrino flux from an updated standard solar model.** A. Bolton.
Monthly Notices Roy. Astron. Soc., Vol. 173, 295 - 298 (1975).
The effect of recent revisions in thermonuclear reaction rates are included in the computation of evolutionary models of the sun. The revised solar neutrino flux is compared to that obtained by other authors and only a slight change is found.

080.041 **A solar neutrino loophole: standard solar models.**
C. A. Rouse.
Astron. Astrophys., Vol. 44, 237 - 240 (1975).
The salient aspects of the existence theorem for a unique solution to a system of linear or nonlinear first-order, ordinary differential equations are given and applied to the equilibrium stellar structure equations. It is shown that values of pressure, temperature, mass and luminosity are needed at one point – and for the sun, the logical point is the solar radius. But without an accurately known chemical composition, an accurate temperature gradient in the convective region, etc., the above conditions are a necessary condition for a unique solution to the solar problem. It is concluded that since standard solar model calculations use split boundary conditions, a solar neutrino loophole still remains: solar model calculations that seek to satisfy the necessary condition for a unique solution to the solar structure equations suggest a solar interior quite different from that deduced in standard models.

080.042 **Some new problems of solar physics based on stratospheric investigations.** V. A. Krat.
VI Leningr. mezhdunar. seminar, 1974. Uskorenie chastits i yader. reaktsii v kosmose. Leningrad, 1974, p. 21 - 22, with a discussion, p. 22 - 24. In Russian and English. – Abstr. in Referativ. Zhurn. 51. Astron., 8.51.411 (1975).

080.043 **Some problems of nuclear astrophysics of the sun.** G. E. Kocharov.
VI Leningr. mezhdunar. seminar, 1974. Uskorenie chastits i yader. reaktsii v kosmose. Leningrad, 1974, p. 17 - 18, 18 - 19, with a discussion, p. 19 - 20. In Russian and English. – Abstr. in Referativ. Zhurn. 51. Astron., 8.51.416 (1975).

080.044 **Report on the Brookhaven solar neutrino experiment.** R. Davis, J. C. Evans, Jr.
VI Leningr. mezhdunar. seminar, 1974. Uskorenie chastits i yader. reaktsii v kosmose. Leningrad, 1974, p. 91 - 107, with a discussion, p. 108 - 110. – Abstr. in Referativ. Zhurn. 51. Astron., 8.51.417 (1975).

080.045 **Feasibility studies on a radiochemical lithium solar neutrino experiment.** J. K. Rowley.
VI Leningr. mezhdunar. seminar, 1974. Uskorenie chastits i yader. reaktsii v kosmose. Leningrad, 1974, p. 111 - 124. Abstr. in Referativ. Zhurn. 51. Astron., 8.51.418 (1975).

080.046 **Bursts of neutrino and antineutrino radiation.** G. E. Kocharov.
VI Leningr. mezhdunar. seminar, 1974. Uskorenie chastits i yader. reaktsii v kosmose. Leningrad, 1974, p. 125 - 130. In Russian. – Abstr. in Referativ. Zhurn. 51. Astron., 8.51. 419 (1975).

080.047 **Some remarks on solar neutrinos.** J. N. Bahcall.
VI Leningr. mezhdunar. seminar, 1974. Uskorenie chastits i yader. reaktsii v kosmose. Leningrad, 1974, p. 131 - 133, with a discussion, p. 133. – Abstr. in Referativ. Zhurn. 51. Astron., 8.51.420 (1975).

080.048 **Fast nonstationary processes in the solar interior and the solar neutrino.** Yu. N. Starbunov.
VI Leningr. mezhdunar. seminar, 1974. Uskorenie chastits i yader. reaktsii v kosmose. Leningrad, 1974, p. 134 - 140. In Russian. – Abstr. in Referativ. Zhurn. 51. Astron., 8.51. 421 (1975).

080.049 **Velocity distribution of plasma ions and the solar neutrino.**
S. S. Vasil'ev, G. E. Kocharov, A. A. Levkovskij.
VI Leningr. mezhdunar. seminar, 1974. Uskorenie chastits i yader. reaktsii v kosmose. Leningrad, 1974, p. 141, with a discussion, p. 142. In Russian and English. – Abstr. in Referativ. Zhurn. 51. Astron., 8.51.422 (1975).

080.050 **Planetary brightness changes: evidence for solar variability.** G. W. Lockwood.
Science, Vol. 190, 560 - 562 (1975).

Brightness changes of Uranus and Neptune at 440 nanometers were highly correlated from 1956 to 1966. Recent observations of Saturn's satellite Titan and of Uranus and Neptune show a steady brightening at 551 and 472 nanometers since 1972. Either the solar constant is slightly variable or solar activity causes correlated changes in the albedos of planetary bodies.

080.051 **Concerning some actual problems in the observational solar research.** V. Bumba.
Publ. Obs. Astron. Beograd, No. 20, (see 012.024), p. 35 - 45 (1975).

080.052 **The line response function of stellar atmospheres and the effective depth of line formation.**
J. M. Beckers, R. W. Milkey.
Solar Physics, Vol. 43, 289 - 292 (1975).
The response function defines the response of line profiles to a depth variation of such atmospheric parameters as velocity, magnetic field and turbulence. The properties of this function are derived and compared with the so-called contribution function.

080.053 **On the problem of investigation of the structure of the solar magnetic field.** Eh. I. Mogilevskij.
Izv. Krymskoj Astrofiz. Obs., Vol. 53, p. 243 - 244 (1975).
In Russian. – Abstr. of a paper presented at the conference "Magnetic fields and chemical composition of the sun and stars" at the Crimean Astrophys. Obs., 1974, April 9 - 12, (see 012.027).

080.054 **Measurements of the magnetic field from different spectral lines.** S. I. Gopasyuk, V. A. Kotov, A. B. Severnyj, T. T. Tsap.
Izv. Krymskoj Astrofiz. Obs., Vol. 53, p. 247 - 249 (1975).
In Russian. – Abstr. of a paper presented at the conference "Magnetic fields and chemical composition of the sun and stars" at the Crimean Astrophys. Obs., 1974, April 9 - 12, (see 012.027).

080.055 **Mass, energy and momentum transport in the solar atmosphere.** L. E. Cram, with an introduction by R. N. Thomas.
Oss. Mem. Oss. Astrofis. Arcetri, Fasc. 105, (see 012.026), p. 11 - 12, 13 - 38 (1975). – Review paper.

080.056 **Solar magnetic fields and convection. III: Recent developments in dynamo and related theories.**
J. H. Piddington.
Astrophys. Space Sci., Vol. 38, 157 - 166 (1975).
Recent developments in solar dynamo and other theories of magnetic fields and convection are discussed and extended. A basic requirement of these theories, that surplus fields are eliminated by turbulent or eddy diffusion, is shown to be invalid. A second basic requirement, that strong surface fields are created by granule or supergranule motions, is shown to be improbable. It is concluded that the dynamo and related diffuse-field theories are unacceptable, that solar magnetic fields are not dominated by convection, and that all the fields emerge as strong, concentrated fields (flux ropes) which were wound and twisted from a permanent, primordial field.

080.057 **The effect of intermediate-scale motions on line formation.** R. A. Shine.
Astrophys. Journ., Vol. 202, 543 - 550 (1975).
The problem of LTE and non-LTE line formation in the presence of velocity fields with geometric scales between the microscopic and macroscopic limits is investigated in the context of periodic sinusoidal and sawtooth waves. The author finds that the presence of intermediate-scale velocities in the solar atmosphere could explain the central intensities of the Na I D lines and other strong absorption lines without the high electron densities suggested in some earlier studies.

080.058 **Nonadiabatic nonradial oscillations: an application to the five-minute oscillation of the sun.**
H. Ando, Y. Osaki.
Publ. Astron. Soc. Japan, Vol. 27, 581 - 603 (1975).
The stability of acoustic modes trapped in the solar convection zone is studied by solving the equations of linear nonadiabatic nonradial oscillations. The radiative transfer of perturbations is treated by the Eddington approximation. The results are shown in the diagnostic (k_x, σ)-diagram, where k_x is the horizontal wave number and σ is the angular frequency of oscillations. It is found that many of the acoustic modes are overstable and that the most unstable modes occupy a long mountain-range-like region in the (k_x,σ)-diagram centered with a period of 300s and with a wide range of horizontal wavelength, which is in good agreement with observations of the five-minute oscillation.

080.059 **Solar neutrinos and variations in the solar luminosity.**
R. K. Ulrich.
Science, Vol. 190, 619 - 624 (1975).
The lack of detectable solar neutrinos has led to the proposal of solar models with variable luminosities.

080.060 **Exploratory journey out of the ecliptic plane.**
D. E. Page.
Science, Vol. 190, 845 - 850 (1975).

080.061 **Structure of the solar atmosphere and 5-minute oscillations.** V. I. Zhukov.
Solnechnye Dannye 1975 Byull., No. 7, p. 64 - 68 (1975).
In Russian.
The solar atmosphere is represented as two plane isothermic layers to be found in the outer horizontal magnetic field. It is shown that under certain conditions a monochromatic surface gravity wave can propagate in the lower layer of the atmosphere, whereas in the upper layer the propagation will have the character of vertical oscillations.

080.062 **Solar neutrinos: A scientific puzzle.**
J. N. Bahcall, R. Davis, Jr.
Phys. Blätter, 31. Jahrgang, p. 609 - 615 (1975).

080.063 **Oscillatory periods in the sun and theoretical models with or without mixing.**
R. Scuflaire, M. Gabriel, A. Noels, A. Boury.
Astron. Astrophys., Vol. 45, 15 - 18 (1975).
Theoretical eigenvalues corresponding to periods less than one hour are presented for a standard solar model and for models of about the solar age which are undergoing a thermal pulse following a fast mixing. The two sets of eigenvalues differ very little, except that the standard solar model presents a quadrupolar oscillation with a 43 min period while the mixed models do not. Attempts to observe that particular period are important since the presence or absence of this 43 min period and its possible identification with a quadrupolar oscillation would be a strong argument in favour of one of the two types of solar models studied here.

080.064 **Scattered light – a comparison between theory and experiments during the 1973 transit of Mercury.**
A. Wittmann, H. Wöhl.
Solar Physics, Vol. 44, 231 - 237 (1975).
The authors check the formalism used to derive stray light corrections from measured aureole intensities and correct an error in the pertinent literature. They solve the alleged problem of appropriately normalizing the spread function by treating blurring and scattering separately. They test the method by comparing stray light corrections derived from both the

aureole and from intensity profiles across Mercury's disc obtained during the transit of November 10, 1973.

080.065 **The representation of magnetic field lines from magnetograph data.** R. H. Levine.
Solar Physics, Vol. 44, 365 - 370 (1975).

Several methods currently used to extrapolate the structure of the solar magnetic field from surface measurements are examined and compared. In particular, the differences between the methods of Schmidt for potential fields and of Nakagawa and Raadu for force-free fields are explained. Suggestions are made regarding the use of these theoretical procedures in making physical conclusions.

080.066 **The Zeeman broadening of high *n* solar recombination lines.** A. Greve.
Solar Physics, Vol. 44, 371 - 382 (1975).

The Zeeman broadening of high *n* lines is derived. While in areas of the quiet sun with field strengths of $\approx$1 G the upper boundary of the observable frequency region of recombination lines is defined by electron impact broadening, in active regions with field strengths $\gtrsim$25 G the Zeeman broadening will shift this boundary to higher frequencies. The frequency region most favourable for observations is derived and the corresponding Doppler, electron impact and Zeeman broadening are discussed.

080.067 **Generation of isochronous sequences of solar models.** M. J. Newman.
Bull. American Astron. Soc., Vol. 7, 526 (1975). – Abstr. AAS.

080.068 **Circulation and redistribution of heat in the solar convective envelope.**
R. S. Bogart, P. J. Gierasch.
Bull. American Astron. Soc., Vol. 7, 526 (1975). – Abstr. AAS.

080.069 **De differentiële rotatie van de zon.** W. van Tend.
Zenit, Vol. 2, 426 - 429 (1975).

080.070 **The sun's magnetic sector structure.**
L. Svalgaard, J. M. Wilcox, P. H. Scherrer, R. Howard.
Solar Physics, Vol. 45, 83 - 91 (1975).

The synoptic appearance of solar magnetic sectors is studied using 454 sector boundaries observed at earth during 1959–1973. The sectors are clearly visible in the photospheric magnetic field. Sector boundaries can be clearly identified as north-south running demarcation lines between regions of persistent magnetic polarity imbalances. These regions extend up to about 35° of latitude on both sides of the equator. They generally do not extend into the polar caps. The polar cap boundary can be identified as an east-west demarcation line marking the poleward limit of the sectors. The typical flux imbalance for a magnetic sector is about 4×10^{21} Mx.

080.071 **Frequency-time spectrum of vibrations of the limb of the solar disk recorded from cinematographic observations.** O. B. Vasil'ev, U. I. Il'yasov.
Izv. AN TurkmSSR. Ser. fiz.-tekhn., khim. i geol. n., 1975, No. 2, p. 56 - 62. In Russian. – Abstr. in Referativ. Zhurn. 51. Astron., 1.51.139 (1976).

080.072 **Active phenomena in the solar atmosphere.**
S. B. Pikel'ner.
Proc. Astron. Soc. Australia, Vol. 2, 316 - 317 (1975). – Invited paper.

080.073 **The solar neutrino problem.**
D. J. Faulkner.
Proc. Astron. Soc. Australia, Vol. 2, 318 - 322 (1975). Invited paper.

080.074 **On calculating the vector of the dipole moment of the magnetic field for a free-current approximation according to the known photospheric surface distribution of the magnetic field intensity.** V. P. Mikhajlutsa.
Solnechnye Dannye 1975 Byull., No. 9, p. 66 - 69 (1975). In Russian.

The vector of the dipole moment is an objective magnetic field characteristic. This characteristic has not been investigated so far as applied to solar active regions. The dipole field, totally determined by the dipole moment vector, could be a satisfactory approximation for the coronal fields. Formulas are derived for calculating the vector components according to the known distribution of the magnetic field intensity over an active region.

080.075 **Is the sun spherical?** A. Krasiński.
Postępy Astron., Vol. 23, 159 - 175 (1975). In Polish.

The article is a review of Dicke's work on the hypothetical quadrupole moment of the sun.

080.076 **Nuclear reactions in the solar atmosphere and solar cosmic rays.** G. M. Blokh, B. M. Kuzhevskij.
Issled. kosmich. luchej. Moskva, Nauka, 1975, p. 114 - 132. In Russian. – Abstr. in Referativ. Zhurn. 51. Astron., 2.51.478 (1976).

080.077 **Inapplicability of Ferraro's theorem to the sun.**
K. H. Schatten.
New Zealand Journ. Sci., Vol. 17, 383 - 385 (1974).

080.078 **Hydromagnetic theory of solar sectors: slow hydromagnetic waves.** S. T. Suess.
AIAA Journ., Vol. 13, 443 - 447 (1975).

Magnetic sectors on the sun are a feature, when the solar dipole field is subtracted, reminiscent of grapefruit sections in terms of the boundaries described by the magnetic field polarity change. One possible suggestion for the origin of these sectors is that they are hydromagnetic waves controlled by the rotation, toroidal magnetic field, and stratification within the convection zone of the sun. The merits of this suggestion are evaluated with respect to the observations and a specific theoretical model.

Production of high-energy deuterons in the ^{3}He + ^{3}He reaction and the solar neutrino problem.
See Abstr. 022.087.

The finite Fourier transform definition of an edge on the solar disk. See Abstr. 031.214.

Cinematography of solar intensity, velocity, and magnetic fields. See Abstr. 031.306.

Dynamics of solar magnetic fields. VI. Force-free magnetic fields and motions of magnetic foot-points.
See Abstr. 062.002.

Equilibrium and stability of the force-free magnetic field. II. Stability. See Abstr. 062.039.

Resonance-line transfer with partial redistribution. VII. Angle-dependent redistribution. See Abstr. 063.026.

Hyperfine structure and line formation in a magnetic field. See Abstr. 063.045.

A method for empirical determination of stellar atmospheric structure. See Abstr. 064.060.

An archetype hydrogen atmosphere problem.

See Abstr. 064.061.

Vibrational instability of a 1 $M_\odot$ star towards non-radial oscillations. See Abstr. 065.021.

Finite-amplitude convection in a compressible medium and its application to solar granulation. See Abstr. 065.037.

The effect of rotation in stellar structure and evolution. See Abstr. 065.077.

Nonadiabatic nonradial oscillations of a solar model. See Abstr. 065.099.

Vibrational stability of g^+ modes towards non-radial oscillations for solar models. See Abstr. 065.100.

Temperature variation with latitude in the upper solar photosphere: relevance to solar oblateness measurements and facular models. See Abstr. 071.002.

The solar beryllium abundance. See Abstr. 071.005.

A comparison of modern average characteristics of the solar radiation. See Abstr. 071.032.

Five-minute oscillations of solar equivalent widths. See Abstr. 071.033.

Evidence for non-radial fields in the sun's photosphere and a possible explanation of the polar magnetic signal. See Abstr. 071.052.

The ability of current micro-velocity models to represent center-limb line profiles. See Abstr. 071.062.

Comparisons of potential and force-free magnetic-field calculations with Skylab soft X-ray images of a complex solar active region. See Abstr. 072.015.

Solar-cycle dynamo wave propagation. See Abstr. 072.027.

Latitude distribution of sunspots and the common magnetic field of the sun. See Abstr. 072.060.

Electric currents in the solar wind. See Abstr. 074.119.

Relation of the observed far ultraviolet solar irradiance to the solar magnetic sector structure. See Abstr. 076.056.

The influence of shock waves on UV emission lines. See Abstr. 076.057.

Radio astronomical observations of wave motions in the solar atmosphere with periods shorter than one minute. See Abstr. 077.055.

Effects of the rotation of the sun on the anisotropy of cosmic rays. See Abstr. 143.014.

The 11-year cycle of galactic cosmic rays in interplanetary space and the general magnetic field of the sun. See Abstr. 143.037.

Earth

081 Figure, Composition, and Gravity of the Earth

081.001 **Effect of the third product of inertia and of equatorial flattening of the earth on the free nutation of the rotational axis.** M. Burša.
Bull. Astron. Inst. Czechoslovakia, Vol. 26, 237 - 241 (1975).
Euler's dynamic equations of the free nutation of the earth's axis of rotation are solved for the case in which the equatorial moments of inertia differ from one another and the third product of inertia differs from zero. The Lagrange method of variation of arbitrary constants was used in combination with the method of successive approximations.

081.002 **From Kepler's heavenly harmony to modern earthly harmonics.** D. G. King-Hele.
Vistas in Astronomy, Vol. 18, (see 012.006), 497 - 517 (1975).

081.003 **Effects of tidal dissipation in the oceans on the moon's orbit and the earth's rotation.**
K. Lambeck.
Journ. Geophys. Res., Vol. 80, 2917 - 2925 (1975).

081.004 **The theory of ideal bodies for gravity interpretation.** R. L. Parker.
Geophys. Journ. Roy. Astron. Soc., Vol. 42, 315 - 334 (1975). Paper presented at the Tenth International Symposium on Mathematical Geophysics, Cambridge, England, 1974 June 26 - July 4.
Ambiguity in gravity interpretation is inevitable because of the fundamental incompleteness of real observations; it is, however, possible to provide rigorous limits on possible solutions even with incomplete data. In this paper a systematic theory is developed for finding such bounds, including an upper bound on depth of burial; the bounds are discovered by constructing the unique body achieving the extreme parameter, e.g. depth; such a body is called the 'ideal' one associated with the given data. General properties of ideal bodies are derived and numerical methods for modest numbers of observations are discussed.

081.005 **Inertial waves and the earth's outer core.** K. D. Aldridge.
Geophys. Journ. Roy. Astron. Soc., Vol. 42, 337 - 345 (1975). Paper presented at the Tenth International Symposium on Mathematical Geophysics, Cambridge, England, 1974 June 26 - July 4.
The existence of wave solutions to the Poincaré equation for a rotating fluid confined between rigid spherical boundaries is investigated. This problem is ill-posed and there appear to be no continuous solutions. However experimental work indicates that solutions do exist under certain conditions and these experiments are at least partially interpreted by an approximate solution found by means of a variational principle.

081.006 **Resolution techniques for density and heterogeneity in the earth.** B. A. Bolt, R. Uhrhammer.
Geophys. Journ. Roy. Astron. Soc., Vol. 42, 419 - 435 (1975). Paper presented at the Tenth International Symposium on Mathematical Geophysics, Cambridge, England, 1974 June 26 - July 4.

081.007 **The static deformation of an earth with a fluid core: a physical approach.** M. A. Chinnery.
Geophys. Journ. Roy. Astron. Soc., Vol. 42, 461 - 475 (1975). Paper presented at the Tenth International Symposium on Mathematical Geophysics, Cambridge, England, 1974 June 26 - July 4.

081.008 **Core undertones with rotation.** D. J. Crossley.
Geophys. Journ. Roy. Astron. Soc., Vol. 42, 477 - 488 (1975). – Paper presented at the Tenth International Symposium on Mathematical Geophysics, Cambridge, England, 1974 June 26 - July 4.
The elastic behaviour of the earth's liquid core at periods longer than those of the free oscillations is shown to be affected by the earth's rotation. In particular, free spheroidal modes of the core, called undertones, exist in a core which is subadiabatic or gravitationally stable. The periods of these undertones are considerably modified by the rotation which serves to couple together modes by the excitation of torsional motion. The results indicate that the undertones will be confined within a period range 0–12 h. Other modes, related to the rotationally-generated inertial modes known in fluid mechanics, may exist at periods greater than 12h.

081.009 **On the stability of a spherical gravitating compressible liquid planet without spin.** I. M. Longman.
Geophys. Journ. Roy. Astron. Soc., Vol. 42, 621 - 635 (1975). Paper presented at the Tenth International Symposium on Mathematical Geophysics, Cambridge, England, 1974 June 26 - July 4.

081.010 **Status of the geopotential.** R. H. Rapp.
Rev. Geophys. Space Phys., Vol. 13, No. 3, (see 013.001), p. 267 - 268, 287 - 288 (1975). – Review paper with bibliographical notes.

081.011 **The earth.** R. Siever.
Sci. American, Vol. 233, No. 3, p. 82 - 90 (1975).
Popular article.

081.012 **Resonant coefficients of the earth's gravity field derived from the inclination of Interkosmos satellites.** J. Klokočnik.
Bull. Astron. Inst. Czechoslovakia, Vol. 26, 293 - 300 (1975).
Certain linear combinations of harmonic coefficients of order 15 and odd degree were determined from resonant perturbations of the inclination of Interkosmos 3, 5 and 9 satellites, which passed through that resonance. Computed lumped coefficients $C(15)$ and $S(15)$ agree well with those of other authors.

081.013 **Investigation of the annual drift of terrestrial inclinations from observations made at the Engelhardt Astronomical Observatory.** I. A. Urasina.
Izv. Astron. Ehngel'gardt. Obs., *Kazan'*, No. 39, p. 111 - 121 (1972). In Russian.

081.014 **The connection of short-period variations of inclinations of the earth's surface with atmospheric pressure from inclination observations at the Engelhardt Astronom-**

ical Observatory. I. A. Urasina.
Izv. Astron. Ehngel'gardt. Obs., *Kazan'*, No. 40, p. 126 - 130 (1973). In Russian.

081.015 **Comparison of the potential coefficient models of the Standard Earth (II and III) and the GEM 5 and GEM 6.** R. H. Rapp.
Bull. Géod., Nouvelle Sér., Année 1975, No. 117, p. 279 - 287 (1975).

The purpose of this paper is to compare, through several procedures, the potential coefficients sets of the Smithsonian Astrophysical Observatory Standard Earth 2 (Gaposchkin and Lambeck, 1970) and 3 (Gaposchkin, 1973) and the sets developed by NASA at the Goddard Space Flight Center and designated GEM 5 and GEM 6 (Richardson and Lerch, 1974).

081.016 **Geopotential harmonics of order 15 and odd degree from analysis of resonant orbits.**
D. G. King-Hele, D. M. C. Walker, R. H. Gooding.
Planet. Space Sci., Vol. 23, 1239 - 1256 (1975).

We have analysed the variations of inclination in 13 satellite orbits as they pass slowly, under the action of air drag, through 15th-order resonance with the geopotential, when successive equatorial crossings are 24° apart and the ground track repeats after 15 rev. Our analysis gives values of the coefficients up to $l = 33$.

081.017 **Evaluation of geopotential coefficients of 15th order and even degree.**
D. G. King-Hele, D. M. C. Walker, R. H. Gooding.
Observations of artificial earth satellites, No. 14, (see 012.018), p. 15 - 24 (1975).

081.018 **Thirteenth-order orbital resonance on a Diademe 2 fragment.** C. A. Wagner.
Journ. Geophys. Res., Vol. 80, 3791 - 3795 (1975).

081.019 **Effect of resonance-oblateness coupling on a satellite orbit.** C. A. Wagner.
Journ. Geophys. Res., Vol. 80, 4089 - 4093 (1975).

Second-order effects of the coupling between geopotential resonance and oblateness on a satellite orbit are calculated. These effects arise from the interaction of resonance with the secular changes of the orbit's node, perigee, and mean anomaly. They have the same period and phase as first-order resonance perturbations. But their amplitudes are proportional to the square of the period and dominate the first-order effects as the orbit becomes commensurate. A striking example of this coupling is seen in the 18-day resonance variation of the node of the orbit of the first earth resources technology satellite.

081.020 **Einfluß der Geschwindigkeitsabnahme der Erdrotation auf Figur und Schwere der Erde.**
V. K. Hristov.
Gerlands Beiträge Geophys., Vol. 84, 355 - 356 (1975).
Short communication.

081.021 **Die Theorie der Erdexpansion.** P. Jordan.
Phys. Blätter, 31. Jahrgang, p. 487 - 495 (1975).

081.022 **Erdexpansion und Kontinentaldrift.** Einige Bemerkungen zur Erdexpansion der Erde und zur Bewegung der Kontinente. G. Angenheister.
Phys. Blätter, 31. Jahrgang, p. 496 - 501 (1975).

081.023 **Marées terrestres.** P. Melchior (Editor).
Bull. d'Informations, (Obs. Roy. Belgique, Bruxelles), No. 72, p. 4085 - 4185 (1975).

081.024 **Detailed gravimetric geoid confirmation of short wavelength features of sea surface topography detected by the Skylab S-193 altimeter in the Atlantic Ocean.**
J. G. Marsh, S. Vincent, A. T. McClinton, E. S. Chang.
Goddard Space Flight Center, Greenbelt, Maryland, GSFC Document X-921-75-110, 5 + 16 pp. (1975).

A new detailed gravimetric geoid has been recently computed for the Northwest Atlantic Ocean and Caribbean Sea area in support of the calibration and evaluation of the GEOS-C altimeter. This paper presents a comparison of the new gravimetric geoid with 10 passes of SKYLAB altimeter data. The agreement of the two data types is quite good with the differences generally less than 2 meters.

081.025 **14 years of resonance on Vanguard orbits.**
C. A. Wagner.
Goddard Space Flight Center, Greenbelt, Maryland, GSFC Document X-921-75-179, 7 + 64 pp. (1975).

081.026 **Power series solutions of Laplace's tidal equation.**
D. Möhlmann.
Gerlands Beiträge Geophys., Vol. 84, 463 - 472 (1975).

Hough's functions are represented in terms of power series of sin for the symmetric and of cos for the antisymmetric modes. The resulting recursion relations are of third, fourth or fifth order. Numerical results are given for a set of interesting tidal modes.

081.027 **Some questions of investigation of the variations of the gravitational field of the earth with time.**
M. E. Khejfets.
Geod. i kartografiya, 1975, No. 7, p. 9 - 15. In Russian.
Abstr. in Referativ. Zhurn. 52. Geodeziya i Aehrosemka, 12. 52.78 (1975).

081.028 **Equations of condition for 13th-order harmonics in the geopotential.** G. Balmino, C. Reigber.
Bull. Groupe de Recherches de Géod. Spatiale, No. 14/I, 10 pp. (1974).

13 different arcs of the satellites GEOS II (680021), BEC (650321), D1D (670141) and ANNA Ib (620601), which are in shallow resonance with the 13th-order harmonics in the geopotential, have been used for an analysis of the long period changes in the mean longitude $\lambda = M + \omega + \Omega$ in the sense of a boundary value problem. By this analysis 47 condition equations for even and odd degree harmonics of order 13 have been derived.

081.029 **The earth, the moon and tidal sediments: an analytical review.** J. G. Ford.
Mercian Geologist, Vol. 5, 205 - 240 (1975).

An attempt is made to relate contemporary tidal sedimentation to that of the past. The causes of the tides are considered at an elementary level, and it is shown that their magnitude depends on a number of factors which may have varied over geological time. The validity of theories in which these variations are implicit are considered, while the scale of the implied fluctuations is investigated semi-quantitatively. One theory is found which dominates all others in its effects on tidal magnitude variation. This is the theory associated with tidal-friction. It is possible that the moon is of recent origin, but no theory fits the facts and this conflict between astronomy and geology remains to the end. The only solution offered is that the radioactive decay constants vary with time, which solves the underlying time-scale difficulty, that prevents the geological and astronomical theories from being compatible.

081.030 **Fundamental geodetic constants.**
H. Moritz.
Bull. Géod., Nouvelle Sér., Année 1975, No. 118, p. 398 - 408 (1975). – Report of Special Study Group N° 5.39 of IAG.

081.031 **An astronomical measurement of the present day drift of the Eurasian and American plates.**
M. Feissel.
Proc. Symposium on the earth's gravitational field and secular variations in position, (see 012.036), p. 20 - 28 (1974).

081.032 **The earth's gravitational field from the combination of satellite and terrestrial data.** R. H. Rapp.
Proc. Symposium on the earth's gravitational field and secular variations in position, (see 012.036), p. 51 - 75 (1974).

081.033 **Adjusted parameters of a mean earth ellipsoid.**
R. H. Rapp.
Proc. Symposium on the earth's gravitational field and secular variations in position, (see 012.036), p. 554 - 558 (1974).

081.034 **Determination of earth and ocean tides from the analysis of satellite orbits.** K. Lambeck.
Proc. Symposium on the earth's gravitational field and secular variations in position, (see 012.036), p. 522 - 528 (1974).

081.035 **Analysis and prediction of tides.** L. P. Lee.
Southern Stars, Vol. 26, 83 - 89 (1975).

081.036 **Variations of the angles between the plumb lines of the astronomical instruments.**
N. T. Mironov.
IAU Colloquium No. 26, (see 012.034), p. 79 - 84 (1975).

081.037 **Reference coordinate system requirements for geophysics.** P. L. Bender.
IAU Colloquium No. 26, (see 012.034), p. 85 - 92 (1975).

081.038 **Terrestrial coordinate systems solidly connected with the earth.** M. Bursa.
IAU Colloquium No. 26, (see 012.034), p. 133 - 159 (1975).

081.039 **Gravimetric and astrogeodetic reference systems in space and time.** H. Moritz.
IAU Colloquium No. 26, (see 012.034), p. 161 - 180 (1975).

081.040 **Prospects for rapid realization of a quasi-earth-fixed coordinate system.** P. L. Bender.
IAU Colloquium No. 26, (see 012.034), p. 297 - 302 (1975).

081.041 **Astrogravimetric levelling with direct regard to topography.** G. Boedecker.
Deutsche Geod. Kommission Bayer. Akad. Wiss., München, Reihe B, No. 213, p. 3 - 13 (1975).

081.042 **A biaxial world geodetic datum WD3.**
B. H. Eitschberger.
Deutsche Geod. Kommission Bayer. Akad. Wiss., München, Reihe B, No. 213, p. 13 - 22 (1975).

081.043 **Physikalische Geodäsie.** K.-R. Koch, H. Wolf.
Deutsche Geod. Kommission Bayer. Akad. Wiss., München, Reihe B, No. 212, p. 72 - 81 (1975).

081.044 **Daily variations in readings of the seismic horizontal pendulum.** N. D. Guzenkova, S. M. Kolesnikov, E. M. Kolesnikova, V. M. Mikheev.
Astron. Tsirk., No. 878, p. 5 - 7 (1975). In Russian.

081.045 **Oscillations sphéroïdales et toroïdales d'un globe élastique homogène en rotation.** I. Ottelet.
Bull. Soc. Roy. Sci. Liège, (*Belgium*), Vol. 42, 417 - 439 = Inst. Astrophys. Univ. Liège, Cointe-Ougrée (*Belgique*), Coll. 8°, No. 642 (1973).

An analytical study is made of the free toroidal and spheroidal oscillations of a slow rotating and gravitating homogeneous elastic globe. It is shown that a strong coupling between spheroidal and toroidal modes may appear.

081.046 **Sulfur in the core of the earth.** R. Brett.
Meteoritics, Vol. 10, 372 - 373 (1975).

081.047 **The earth as a cosmochemical body.**
R. D. Schuiling.
Meteoritics, Vol. 10, 482 (1975).

081.048 **Description of the gravitational field of the earth with series of spherical functions.** Yu. M. Nejman.
Geod. i kartografiya, 1975, No. 9, p. 25 - 29. In Russian.
Abstr. in Referativ. Zhurn. 52. Geodeziya i Aehrosemka, 2.52.88 (1976).

Total effect of any nearly-diurnal wobble of the earth's axis of rotation in latitude and time observations. Application to the Paris astrolabe observations from 1956.6 to 1973.8. See Abstr. 044.005.

Astronomical programs for the study of secular variations in position. See Abstr. 045.017.

Solution of the downward continuation problem by collocation. See Abstr. 046.012.

Numerical and analytical methods for the solution of surface integrals connected with the simple layer model of the geopotential in satellite geodesy.
See Abstr. 046.016.

Problems of determination of station-coordinates and of the gravity field with satellite observations and measurements on the earth. See Abstr. 046.019.

Analyse der kombinierten kosmisch-geodätischen Beobachtungsmethoden. See Abstr. 046.023.

Nature of the requirements for reference coodinate systems. See Abstr. 046.056.

Some problems related to geodetic datum definitions for terrestrial measurements. See Abstr. 046.059.

A new theory on the generation of the magnetic field of celestial bodies. II. Spherical model.
See Abstr. 061.082.

Peculiarities of fields of geomagnetic disturbances and of the geopotential of the stratosphere of the northern hemisphere in 1967–1969. See Abstr. 084.294.

Higher harmonics of the terrestrial and lunar gravitational fields in the theory of the dynamics of the earth's rotation axis. See Abstr. 094.001.

The structure of interplanetary plasma streams and magnetospheric substorms. See Abstr. 106.030.

Die innere Struktur der Erde, des Mondes und der Planeten. See Abstr. 107.016.

082 The Earth's Atmosphere Including Refraction, Scintillation, Extinction, Airglow, Site Testing

082.001 **Earth as an intense planetary radio source: similarities to Jupiter and Saturn.**
M. L. Kaiser, R. G. Stone.
Science, Vol. 189, 285 - 287 (1975).
Observations from spacecraft have revealed naturally occurring radio emission emanating from two regions near earth. The characteristics of these two sources suggest a correlation with areas of known electron precipitation. The possibility of a similar production mechanism for observed nonthermal radio emissions from other planetary magnetospheres permits the polar magnetic field strengths of Jupiter and Saturn to be predicted.

082.002 **Monte Carlo treatment of Lyman-alpha. II. Radiation in a spherical atmosphere.**
S. B. Modali, J. C. Brandt, S. O. Kastner.
Astrophys. Journ., Vol. 199, 530 - 534 (1975).
Intensity and the state of polarization of solar Lα photons as they diffuse through an inhomogeneous, spherically symmetric, and isothermal geocorona are presented. The fine structure of Lα and Doppler redistribution of frequencies are taken into account. The calculations used the Monte Carlo technique involving Stokes vectors. Comparison of the results with OGO-4 and OSO-4 observed intensities at an altitude of 650 km shows good agreement. Calculations of the polarization versus solar zenith angle show a residual polarization at large zenith angles which is mainly due to multiply scattered photons.

082.003 **Ultraviolet laser sounding of the troposphere and lower stratosphere.**
A. J. Gibson, L. Thomas.
Nature, Vol. 256, 561 - 563 (1975). – Letter.

082.004 **Ar isotopes and earth-atmosphere evolution models.**
M. Ozima.
Geochim. Cosmochim. Acta, Vol. 39, 1127 - 1134 (1975).
Earth-atmosphere evolution models are mathematically simulated and the resulting present isotopic ratio ($^{40}Ar/^{36}Ar$) in the mantle is given for each.

082.005 **Absorption coefficients of selected atmospheric water lines.**
J. W. Brault, J. S. Fender, D. N. B. Hall.
Journ. Quant. Spectrosc. Radiat. Transfer, Vol. 15, 549 - 551 (1975).
Eight unblended atmospheric water lines between 4120 cm^{-1} and 14397 cm^{-1} have been selected as having minimum dependence of absorption coefficient on temperature around 300 K. The integrated absorption coefficient of each line has been determined to an accuracy of better than ±10 per cent.

082.006 **High latitude airglow observations of correlated short-term fluctuations in the hydroxyl Meinel 8-3 band intensity and rotational temperature.**
J. W. Meriwether.
Planet. Space Sci., Vol. 23, 1211 - 1221 (1975).

082.007 **Vergleichende Turbulenz- und Seeingmessungen am Pic-du-Midi Observatorium (2855 m).**
K. O. Kiepenheuer.
Mitt. Astron. Ges., No. 36, p. 137 - 138 (1975).

082.008 **Zur Streulichtbehandlung solarer Intensitätsmessungen.** A. Wittmann, H. Wöhl.
Mitt. Astron. Ges., No. 36, p. 138 (1975). – Abstract.

082.009 **Eclipse induced atmospheric gravity waves.**
B. W. Jones, R. S. Bogart.
Journ. Atmosph. Terr. Phys., Vol. 37, 1223 - 1226 (1975).
A null result is reported for eclipse induced atmospheric gravity waves along the centre line of the total solar eclipse of 30 June 1973. This conflicts with the non-null result reported by others for the total solar eclipse of 7 March 1970. A means of reconciliation is suggested.

082.010 **A global model of thermospheric dynamics – II. Wind, density, and temperature fields generated by EUV heating.** J. M. Straus, S. P. Creekmore, R. M. Harris, B. K. Ching, Y. T. Chiu.
Journ. Atmosph. Terr. Phys., Vol. 37, 1245 - 1253 (1975).

082.011 **Diurnal variation of the sodium dayglow.**
M. L. Chanin, J. P. Goutail.
Journ. Geophys. Res., Vol. 80, 2854 - 2858 (1975).

082.012 **Neutral chemistry in the earth's atmosphere.**
H. I. Schiff.
Astrophys. Space Sci. Library, Vol. 51, (see 012.007), 21 - 43 (1975).

082.013 **Atmospheric physics.** J. C. G. Walker.
Astrophys. Space Sci. Library, Vol. 51, (see 012.007), 47 - 58 (1975). – Review paper.

082.014 **Vertical transport in atmospheres.** D. M. Hunten.
Astrophys. Space Sci. Library, Vol. 51, (see 012.007), 59 - 72 (1975).

082.015 **Neutral atmosphere modeling.** G. Kockarts.
Astrophys. Space Sci. Library, Vol. 51, (see 012.007), 235 - 243 (1975).

082.016 **The global pattern of 6300 Å atomic oxygen emission as seen from the ISIS-2 spacecraft.**
G. G. Shepherd.
Astrophys. Space Sci. Library, Vol. 51, (see 012.007), 283 - 288 (1975).

082.017 **OGO-6 observations of 5577 Å.**
T. M. Donahue.
Astrophys. Space Sci. Library, Vol. 51, (see 012.007), 289 - 307 (1975).

082.018 **Nitrogen dioxide in the stratosphere and troposphere measured by ground-based absorption spectroscopy.** J. F. Noxon.
Science, Vol. 189, 547 - 549 (1975).

082.019 **Radiation in the atmosphere.**
K. L. Coulson, R. S. Fraser.
Rev. Geophys. Space Phys., Vol. 13, No. 3, (see 013.001), p. 732 - 737, 826 - 836 (1975). – Review paper with bibliographical notes.

082.020 **Filling in of the Fraunhofer lines by scattering on the ground.** M. L. Chanin.
Journ. Geophys. Res., Vol. 80, 2859 - 2862 (1975).
The author presents experimental evidence of the filling

in of Fraunhofer lines at 5893 and 7699 Å by scattering from terrestrial surfaces. This effect is shown to vary with the scattering angle and with the nature of the surfaces involved. Consequences on dayglow data and on lunar luminescence studies will be considered in future publications.

082.021 **Water vapor absorption spectra of the upper atmosphere (45–185 cm^{-1}).**
G. C. Augason, A. J. Mord, F. C. Witteborn, E. F. Erickson, C. D. Swift, L. J. Caroff, L. W. Kunz.
Applied Optics, Vol. 14, 2146 - 2150 (1975).
The far ir nighttime absorption spectrum of the earth's atmosphere above 14 km is determined from observations of the bright moon. The spectra were obtained using a Michelson interferometer attached to a 30-cm telescope aboard a high-altitude jet aircraft. Comparison with a single-layer model atmosphere implies a vertical column of 3.4 ± 0.4 μm of precipitable water on 30 August 1971 and 2.4 ± 0.3 μm of precipitable water on 6 January 1972.

082.022 **The rotational speed of the upper atmosphere determined from orbital inclinations of Interkosmos satellites.** L. Sehnal.
Bull. Astron. Inst. Czechoslovakia, Vol. 26, 300 - 306 (1975).
The velocity of the rotation of the atmosphere is computed from the secular changes of the inclination of the orbits of the Interkosmos satellites, the elements of which are expressed as polynomials. After numerical integration of the formula for the change of the inclination, the theoretical curves are compared to the observed values. Good results are obtained from the data of the Interkosmos 3, 5 and 9 satellites.

082.023 **Atmospheric density at heights between 200 and 300 km from drag on Interkosmos satellites.**
M. Prokeš.
Bull. Astron. Inst. Czechoslovakia, Vol. 26, 306 - 315 (1975).
Densities derived from the change of orbital periods and density scale heights derived from the decrease of orbital periods are presented for the satellites Interkosmos 3,4,5,7,8 and 9 and compared with existing models. An appraisal of the suitability of the Interkosmos satellites for determining the upper atmosphere characteristics was made.

082.024 **Measurements of the submillimeter radiation of the earth's atmosphere with an artificial earth satellite.**
A. E. Salomonovich, S. V. Solomonov, A. S. Khajkin, V. S. Kovalev, A. A. Kobzev.
Dokl. Akad. Nauk SSSR. Ser. Mat., Fiz., Vol. 223, 852 - 855 (1975). In Russian.

082.025 **Fifty years of atmospheric transmission, water vapor, and ozone: the astronomical implications.**
R. J. Angione, R. G. Roosen.
Bull. American Astron. Soc., Vol. 7, 462 (1975). – Abstr. AAS.

082.026 **The global characteristics of atmospheric emissions in the lower thermosphere and their aeronomic implications.** E. I. Reed, S. Chandra.
Journ. Geophys. Res., Vol. 80, 3053 - 3062 (1975).
The green line (557.7 nm) of atomic oxygen and the Herzberg bands of molecular oxygen (measured between 250 and 280 nm) as observed from the Ogo 4 airglow photometer from August 1967 through January 1968 are discussed in terms of their spatial and temporal distributions and their relation to the atomic oxygen content in the lower thermosphere.

082.027 **The aeronomy of odd nitrogen in the thermosphere.**
E. S. Oran, P. S. Julienne, D. F. Strobel.
Journ. Geophys. Res., Vol. 80, 3068 - 3076 (1975).

082.028 **On fundamental problems in the deduction of atmospheric densities from satellite drag.**
D. R. Imbro, M. M. Moe, K. Moe.
Journ. Geophys. Res., Vol. 80, 3077 - 3086 (1975).
The largest source of error in the deduction of atmospheric densities from orbital data stems from uncertainty in our knowledge of the interaction of atmospheric molecules with satellite surfaces. A method of determining two of the important parameters, the accommodation and drag coefficients, is developed in detail.

082.029 **Laboratory measurements of $H_2O_2^+$ production and loss reactions and implications for mesopheric H_2O_2.** W. Lindinger, D. L. Albritton, C. J. Howard, F. C. Fehsenfeld, E. E. Ferguson.
Journ. Geophys. Res., Vol. 80, 3277 - 3279 (1975).
Rate constants have been measured as a function of kinetic energy for three exothermic reactions: reaction (1), $H_2O_2^+ + NO \rightarrow NO^+ + H_2O_2$; reaction (2), $H_2O_2^+ + H_2O \rightarrow H_3O^+ + HO_2$; and reaction (3), $O_2^+ + H_2O_2 \rightarrow H_2O_2^+ + O_2$.

082.030 **Astronomical observations and meteorological data (Express information).** I. G. Kolchinskij.
Akademiya Nauk Ukrainskoj SSR, Glavnaya astronomicheskaya observatoriya. Izdatel'stvo "Naukova dumka", Kiev. 30 pp. Price 10 Kop. (1974). In Russian.

082.031 **Astronomical refraction. II.** A. I. Nefed'eva.
Izv. Astron. Ehngel'gardt. Obs., *Kazan',* No. 40, p. 3 - 69 (1973). In Russian.

082.032 **Observable brightness of the twilight sky and some errors causing its variations.** L. B. Gusakovskaya.
Izv. Astron. Ehngel'gardt. Obs., *Kazan',* No. 40, p. 112 - 125 (1973). In Russian.

082.033 **Longwave radiation effects of the Harmattan haze.**
P. M. Kuhn, H. K. Weickmann, L. P. Stearns.
Journ. Geophys. Res., Vol. 80, 3419 - 3424 (1975).

082.034 **Applications of the discrete-ordinate method for radiative transfer to inhomogeneous aerosol atmospheres.** K.-N. Liou.
Journ. Geophys. Res., Vol. 80, 3434 - 3440 (1975).

082.035 **A simple variational approach to refraction and related problems in geodesy.** N. Grossman.
Bull. Géod., Nouvelle Sér., Année 1975, No. 117, p. 217 - 240 (1975).

082.036 **Atmospheric density at 169 km from accelerometer measurements and orbital decay of a low altitude satellite.** B. K. Ching, H. R. Rugge.
Planet. Space Sci., Vol. 23, 1301 - 1308 (1975).
Atmospheric densities at 169 km have been obtained for the period 19 August–3 September 1970 from the measurements of an accelerometer on a low altitude satellite and from the orbital decay of the same satellite. Comparisons of the data with the Jacchia-70 and -71 models indicated that these models may give density values which are too small for the conditions and time period corresponding to the data.

082.037 **Variations in the atmospheric neutral density at 145 km.** J. M. Forbes, H. B. Garrett.
Planet. Space Sci., Vol. 23, 1399 - 1404 (1975).
A least-squares multiple linear regression is performed on orbital decay density data obtained from precise orbital analysis of 22 low-perigee (130–160 km) Air Force satellites. Variations related to solar activity, the semi-annual effect, geomagnetic activity, and the zenith angle of the sun are in agreement with the model of Jacchia (1971).

082.038 **Air density at a height of 128 km, from visual observations of 1972-25G.** D. M. Brierley.
Planet. Space Sci., Vol. 23, 1405 - 1411 (1975).

082.039 **Further study of changes in thermospheric molecular oxygen abundance inferred from twilight 6300 Å airglow.** R. C. Schaeffer, J. F. Noxon.
Planet. Space Sci., Vol. 23, 1413 - 1423 (1975).

082.040 **Tropical twilight HeI 10830 emission.** N. R. Teixeira, P. D. Angreji, Y. Sahai, B. A. Tinsley, A. B. Christensen.
Planet. Space Sci., Vol. 23, 1425 - 1430 (1975).

082.041 **Measurements of ozone and minor atmospheric constituents.**
B. Carli, D. H. Martin, E. Puplett, J. E. Harries.
Nature, Vol. 257, 649 (1975). – Simulating experiments for Spacelab.

082.042 **Photography and photometry of the near infrared OH airglow.**
A. W. Peterson, L. M. Kieffaber.
Nature, Vol. 257, 649 - 650 (1975). – Simulating experiments for Spacelab.

082.043 **Airborne television observations of airglow clouds in the near infrared.**
J. Crawford, P. Rothwell, N. Wells.
Nature, Vol. 257, 650 - 651 (1975). – Simulating experiments for Spacelab.

082.044 **Determination of the spectral transmittance of the terrestrial atmosphere on Terskol Peak in October–November 1973.** V. V. Avramchuk, T. P. Semenyuk,
Astrometriya i Astrofizika, *Kiev*, vyp. (No.) 26, (see 003.001), p. 113 - 116 (1975). In Russian.

The results are presented for determinations of the spectral transparency coefficient on Terskol Peak in the interval of λλ3137–7474. The observations show that the terrestrial atmosphere has a high and stable transparency in the ultraviolet region. In the red range the transparency is less stable, but is higher than at other native observatories.

082.045 **Algunos aspectos de la radiación solar global en el Observatorio del Ebro (Tortosa).**
E. Oriol, M. Puigcerver.
Urania Barcelona, Año 59, No. 281–282, p. 406 - 430 (1974).

An 82-month series of continuously recorded horizontal-incidence solar radiation data is analyzed. Hourly means show a multiple-centered maximum of 0.7 kW m^{-2} and a trough near the summar solstice. Decadic means exhibit large fluctuations, particularly in late spring and early summer. Absolute extremes occur in June or July and December or January; the corresponding dates for monthly means are July and December.

082.046 **Analysis of astronomical refraction anomalies.** N. A. Vasilenko.
Astron. Tsirk., No. 840, p. 2 - 4 (1974). In Russian.

082.047 **Observations of Hα-nightglow line profiles.** V. K. Golev, P. V. Shcheglov.
Astron. Tsirk., No. 861, p. 6 - 7 (1975). In Russian.

082.048 **Results of double-beam site testing at the Special Astrophysical Observatory of the USRR Academy of Sciences.** V. V. Leushin, N. F. Nelyubin, V. B. Nebelitskij, S. B. Novikov.
Astron. Tsirk., No. 866, p. 6 - 8 (1975). In Russian.

082.049 **Atmospheric vertical gamma ray flux in the MeV range.** V. Schönfelder, G. Lichti.
Journ. Geophys. Res., Vol. 80, 3681 - 3684 (1975).

On February 27, 1973, gamma ray measurements in the energy range 1.5–10 MeV were performed with a balloon-borne double Compton telescope from Palestine, Texas. The telescope had a mean acceptance cone of 20° half angle. From this measurement the spectra of vertical atmospheric gamma rays could be derived for various depths. These spectra are presented here and compared with other theoretical and experimental work.

082.050 **Présence de HF dans la stratosphère supérieure.** R. Zander.
Comptes Rendus Acad. Sci. Paris, Sér. B, Vol. 281, 213 - 214 (1975).

Des observations solaires effectuées par ballon à 27,5 km d'altitude ont permis d'enregistrer le spectre du soleil au voisinage de 2,5 μ, avec une résolution spectrale égale à 0,04 cm^{-1}. Une raie d'absorption observée à 4038,97 cm^{-1} a été attribuée à HF qui serait présent dans la stratosphère supérieure.

082.051 **Observations crépusculaires du NO_2 stratosphérique.** J. Blamont, J.-P. Pommereau, G. Souchon.
Comptes Rendus Acad. Sci. Paris, Sér. B, Vol. 281, 247 - 252 (1975).

Des mesures absolues du flux solaire visible et ultraviolet au crépuscule ont été conduites à bord de Concorde aux mois de juin et juillet 1973 au-dessus du Golfe de Gascogne. La comparaison du résultat de ces mesures à un modèle théorique montre une absorption entre 350 et 500 nm, que nous interprétons par la présence de NO_2 dans la basse stratosphère dont on observe la variation diurne.

082.052 **Ongewone reflectie.** F. van Loo.
Zenit, Vol. 2, 316 - 317 (1975).

082.053 **Polarisatie van het hemellicht.** G. P. Können.
Zenit, Vol. 2, 318 (1975).

082.054 **Bijzondere halo's. Waarnemingen in 1973 en 1974.** P.-P. H. Verschure.
Zenit, Vol. 2, 364 - 365 (1975).

082.055 **Influence of short-periodic variations of the atmospheric density on the accuracy of calculations of close satellite orbits.**
P. E. Ehl'yasberg, B. V. Kugaenko, M. I. Vojskovskij.
Observations of artificial earth satellites, No. 14, (see 012.018), p. 357 - 388 (1975). In Russian.

082.056 **Determination of the atmospheric density scale height.** M. Ill.
Observations of artificial earth satellites, No. 14, (see 012.018), p. 403 - 412 (1975). In Russian.

082.057 **The rotational speed of the atmosphere from the inclinations of Intercosmos satellites.**
L. Sehnal.
Observations of artificial earth satellites, No. 14, (see 012.018), p. 413 - 415 (1975).

082.058 **Sur les fonctions d'approximation de la densité de l'atmosphère terrestre.** L. Burs, A. Pál.
Observations of artificial earth satellites, No. 14, (see 012.018), p. 439 - 456 (1975). In Russian.

082.059 **Use of the Intercosmos satellites for determination of the upper atmosphere's characteristics from an analysis of their orbits.** M. Prokeš.
Observations of artificial earth satellites, No. 14, (see 012.018),

p. 473 - 482 (1975).

082.060 **On the problem of using satellite microaccelerometers in upper atmospheric research.** F. Veres.
Observations of artificial earth satellites, No. 14, (see 012.018), p. 499 - 508 (1975). In Russian.

082.061 **A new atmospheric model and motion of artificial satellites.** A. M. Fominov.
Observations of artificial earth satellites, No. 14, (see 012.018), p. 509 - 537 (1975). In Russian.

082.062 **On the evaluation of the astronomical refraction.** A. A. Mikhajlov.
Pis'ma v Astron. Zhurn., Vol. 1, No. 10, p. 46 - 48 (1975). In Russian.

Refraction anomalies can reach appreciable amounts at zenith distances greater than 60°, when the condition of the air at high regions many tens and even several hundred kilometers from the point of observation influences the amount of refraction, and may produce an azimuthal anisotropy. In such cases the geography of the surrounding country as well as seasonal factors must be taken into account.

082.063 **Refraction anomalies due to wind in the free atmosphere.** G. S. Tyuterev.
Astrometriya i Astrofizika, *Kiev*, vyp. (No.) 27, (see 003.004), p. 26 - 30 (1975). In Russian.

The author has deduced formulas which enable to derive the inclination of atmospheric layers of equal density from wind velocity components. The atmospheric layer inclinations above the Pulkovo Observatory were calculated. Their effect on the results of latitude observations carried out in 1967 was estimated.

082.064 **Observation of the conjugate sunrise effect on the OI 6300 Å radiation at high latitude.**
M. L. Duboin, G. Weill.
Planet. Space Sci., Vol. 23, 1565 - 1566 (1975).

The paper reports on an experiment to evaluate the high latitude limit of conjugate photoelectron effects on the airglow red line intensity.

082.065 **Doppler temperature and intensity of 6300 Å emission.** A. I. Semenov.
Geomagn. Aeronom., Vol. 15, 876 - 880 (1975). In Russian.

082.066 **Theoretical i.r. emission spectra for the lower stratosphere.** P. Rabache, B. Rebours.
Journ. Quant. Spectrosc. Radiat. Transfer, Vol. 15, 1125 - 1129 (1975).

A direct method of integration of the transfer equation is developed for a concentration distribution model of gaseous constituents. The method is applied to the lower stratosphere and to the constituents H_2O and O_3. The results are presented in the form of synthetic luminance spectra in the far i.r. range from 50 to 100 cm^{-1}.

082.067 **The effective lifetime profiles of the $O(^1S)$ and $N_2^+(B^2\Sigma_u^+, v=0)$ excited species.**
K. Henriksen.
Journ. Atmosph. Terr. Phys., Vol. 37, 1491 - 1495 (1975).

082.068 **Determination of the astronomical refraction and its anomalies from observations of two stars in the common vertical.** V. V. Kirichuk.
Geod., kartogr. i aehrofotosemka. Resp. mezhved. nauch.-tekhn. sb., 1975, vyp. (No.) 21, p. 9 - 13. In Russian. – Abstr. in Referativ. Zhurn. 51. Astron., 8.51.157 (1975).

082.069 **On the secular cycle of variations of the radiocarbon content in the earth's atmosphere.**
G. E. Kocharov, V. A. Dergachev, A. A. Sanadze.
VI Leningr. mezhdunar. seminar, 1974. Uskorenie chastits i yader. reaktsii v kosmose. Leningrad, 1974, p. 163 - 168. In Russian. – Abstr. in Referativ. Zhurn. 51. Astron., 8.51. 550 (1975).

082.070 **Astrometrical observations and atmospherical influences.** G. (*Dj.*) Teleki.
Publ. Obs. Astron. Beograd, No. 20, (see 012.024), p. 101 - 103 (1975). In Serbo-Croatian.

082.071 **On some analogies between phenomena in the atmosphere and in the universe.** M. Čadež.
Publ. Obs. Astron. Beograd, No. 20, (see 012.024), p. 123 - 128 (1975). In Serbo-Croatian.

082.072 **Recent investigations of astronomical refraction.** G. (*Dj.*) Teleki.
Publ. Obs. Astron. Beograd, No. 20, (see 012.024), p. 159 - 162 (1975). In Serbo-Croatian.

082.073 **Chromatic refraction and its influence on astrometrical measurements.** M. Mijatov (*Miyatov*).
Publ. Obs. Astron. Beograd, No. 20, (see 012.024), p. 163 - 165 (1975). In Serbo-Croatian.

082.074 **Latitude and height dependencies of solar atmospheric tides.** G. V. Groves.
Journ. British Interplanet. Soc., Vol. 28, 797 - 809 (1975).

082.075 **Peculiarities of observations of radiation of the earth's upper atmosphere in the twilight zone made from space.**
A. I. Lazarev, A. G. Nikolaev, V. I. Sevast'yanov.
Optiko-mekh. prom-st', 1975, No. 2, p. 3 - 6. In Russian. Abstr. in Referativ. Zhurn. 62. Issled. kosmich. prostranstva, 8.62.252 (1975).

082.076 **Astronomical refraction of the atmosphere.** S. Takagi.
Publ. International Latitude Obs. Mizusawa, Vol. 9, 241 - 258 (1974).

The primitive equation of a light path through refractive mediums are derived by means of the Fermat's principle. These equations are transformed into a form which is easily applied to the numerical calculation of a light path. As an example, a value of the astronomical refraction for a star with declination 30° to be observed at the meridian of Mizusawa was calculated by Runge-Kutta's method of the numerical integration from the data of Radio Sonde observations made at six stations around the Mizusawa Observatory on 1st March 1973.

082.077 **Variations of the height of the λ 5577 Å emission layer.** N. N. Shefov, E. P. Kropotkina.
Kosmich. Issled., Vol. 13, 765 - 770 (1975). In Russian.

082.078 **An experiment on dome heat and seeing disk size.** P. Murdin, R. G. Bingham.
Observatory, Vol. 95, 180 - 187 (1975).

Dome seeing at the Isaac Newton Telescope varies linearly with heat dissipated around the dome (slope 0.05 arc second per kilowatt). This coefficient is apparently inversely proportional to dome radius. The natural seeing at Herstmonceux appears to follow a $(\sec z)^{1/2}$ law.

082.079 **Optical properties of atmospheric dust from twilight observations.**
N. B. Divari, Y. I. Zaginayilo, L. V. Kovalchuk.
I.A.U. Colloquium No. 13, (see 012.030), p. 281 - 285 (1973).

Three methods of approximation are described and are

used to separate the primary twilight brightness from the observed brightness. Photoelectric observations obtained at the Majaky Observatory in 1968–69 (Zaginayilo) are combined with observations from a balloon and from Pic-du-Midi Observatory (Link et al., 1967) to derive the atmospheric scattering phase functions at 0.37 μ and 0.58 μ as a function of height.

082.080 **Unusual chemical compositions of noctilucent-cloud particle nuclei.**
C. L. Hemenway.
I.A.U. Colloquium No. 13, (see 012.030), p. 287 - 290 (1973).

082.081 **Upper-atmosphere rotation rate from analysis of the orbital inclination of Explorer 1.**
D. M. C. Walker.
Planet.Space Sci., Vol. 23, 1573 - 1579 (1975).

Explorer 1, 1958α, the first U.S. artificial satellite, was launched on 1 February 1958 and remained in orbit for 12 years. In this paper theoretical curves have been fitted to the values of inclination, giving three values of the average atmospheric rotation rate at heights of 350–400 km, and latitudes 0–20°. Solar activity was very high in 1958–60, low from 1961 to 1967, and high in 1968–70; so the results strongly suggest that the rotation rate depends on solar activity, being greatest when the sun is most active.

082.082 **Isoplanicity: the translation invariance of the atmospheric Green's function.**
D. Korff, G. Dryden, R. P. Leavitt.
Journ. Optical Soc. America, Vol. 65, 1321 - 1330 (1975).

The isoplanatic condition for imaging through the turbulent atmosphere is investigated, and theoretical expressions are derived for various applications. Explicit calculations are performed for the Pic du Midi 1 m telescope and the Hale 5 m telescope.

082.083 **Spectral analysis of absorption coefficients of H_2O, O_3 and O_2 in the far infrared in the stratosphere.**
P. Rabache, B. Rebours.
Infrared Physics, Vol. 15, 179 - 188 (1975).

082.084 **Electromagnetic beam propagation in turbulent media.** R. L. Fante.
Proc. IEEE, Vol. 63, 1669 - 1692 (1975).

The most recent developments on the propagation of optical beams in a turbulent medium, such as the clear atmosphere, are reviewed. Among the phenomena considered are beam spreading, beam wander, loss of coherence, scintillations, angle-of-arrival variations, and short-pulse effects.

082.085 **On the passage of an Alfvén wave through the separation boundary in an isotropic medium.**
V. A. Pavlov.
Izv. vyssh. ucheb. zavedenij. Radiofizika, Vol. 18, 630 - 635 (1975). In Russian. – Abstr. in Referativ. Zhurn. 62. Issled. kosmich. prostranstva, 10.62.204 (1975).

082.086 **Man investigates the atmosphere.**
V. A. Bronshtehn.
Zemlya i Vselennaya, 1975, No. 6, p. 25 - 32. In Russian.

082.087 **Determination of the height and energetic scale for a photometric analysis of photographs of the atmospheric aureole of the earth from space.** V. A. Shatalov, A. B. Sandomirskij, V. N. Sergeevich, B. M. Stol'berg.
Kosmich. Issled., Vol. 13, 907 - 913 (1975). In Russian.

082.088 **Corpuscular radiation in the upper atmosphere at heights to 100 km above Ostrov Kheysa.**
V. F. Tulinov, L. S. Novikov, V. V. Tulyakov.
Kosmich. Issled., Vol. 13, 937 - 939 (1975). In Russian.
Brief information.

082.089 **Direct investigations of the ion composition in the mesosphere and the upper stratosphere.**
B. F. Shamakhov, Yu. A. Bragin, A. A. Kocheev.
Kosmich. Issled., Vol. 13, 940 - 942 (1975). In Russian.
Brief information.

082.090 **Calculation of meteorological coefficients for a scintillation supertelescope.**
Ya. L. Blokh, L. I. Dorman, I. Ya. Libin, V. G. Yanke.
Geomagn. Aeronom., Vol. 15, 1084 - 1086 (1975). In Russian.
Brief information.

082.091 **Method of determination of the diffusion coefficient and of the initial radius from saturated meteoric traces.** I. A. Delov.
Geomagn. Aeronom., Vol. 15, 1103 - 1105 (1975). In Russian.
Brief information.

082.092 **Variations in air density at heights near 200 km during 1967–1969, from the orbit of 1967-31A.**
B. R. Bowman.
Planet.Space Sci., Vol. 23, 1659 - 1667 (1975).

082.093 **Photometric measurements of the O_2 U.V. nightglow.** J.-C. Gerard.
Planet.Space Sci., Vol. 23, 1681 - 1684 (1975).

Measurements of the $O_2(A^3\Sigma - X^3\Sigma)$ Herzberg system in the night airglow have been made with the ESRO TD-1 satellite in the wavelength range 2400–3100 A. The slant emission rate varies from 3.5 to 15 kR, indicating an irregular structure of the atomic oxygen near the turbopause. A statistical maximum intensity is found near the tropic in the winter hemisphere.

082.094 **Atmospheres of the earth and planets.**
A. N. Aksenov.
Trudy Astrofiz. Inst., *Alma-Ata,* Vol. 26, 67 - 73 (1974).
In Russian.

082.095 **Investigation of the influence of the humidity of the atmosphere on the parameters of the atmospheric aerosol.** A. I. Ivanov, V. A. Molchanov, I. A. Fedulin.
Trudy Astrofiz. Inst., *Alma-Ata,* Vol. 26, 105 - 112 (1974).
In Russian.

082.096 **On some methods of construction of a graduated graph for spectroscopic determination of water vapour in the atmosphere.**
A. I. Ivanov, V. A. Molchanov, I. A. Fedulin.
Trudy Astrofiz. Inst., *Alma-Ata,* Vol. 26, 112 - 117 (1974).
In Russian.

082.097 **On a connection between optical characteristics of the surface layer with meteorological ones.**
A. B. Kos'yanenko.
Trudy Astrofiz. Inst., *Alma-Ata,* Vol. 26, 117 - 121 (1974).
In Russian.

082.098 **Some results of an investigation of the spectral dependence of the scattering indicatrix and the degree of polarization of scattered light for the surface layer of the atmosphere.** O. D. Tokarev.
Trudy Astrofiz. Inst., *Alma-Ata,* Vol. 26, 122 - 126 (1974).
In Russian.

082.099 **On the spectral dependence of transmittance and backscattering of light in the surface layer of the atmosphere.** A. P. Ten, O. D. Tokarev, T. P. Toropova.
Trudy Astrofiz. Inst., *Alma-Ata,* Vol. 26, 127 - 132 (1974).

In Russian.

082.100 **Twilight measurements of height distribution of ozone in the mesosphere.**
G. G. Mikirtumova, G. V. Rozenberg.
Byull. Abastumansk. Astrofiz. Obs., No. 46, p. 225 - 234 (1975). In Russian.

082.101 **An investigation of the sodium layer variations from observations of the D emission in twilight.**
T. I. Toroshelidze, S. P. Chilingarashvili.
Byull. Abastumansk. Astrofiz. Obs., No. 46, p. 235 - 250 (1975). In Russian.

082.102 **Sky color near the horizon during a total solar eclipse.** S. D. Gedzelman.
Applied Optics, Vol. 14, 2831 - 2837 (1975).
A theory for the color of the sky near the horizon for an observer in the umbral region of a total solar eclipse is presented. The model uses a Rayleigh scattering atmosphere.

082.103 **Sky brightness during eclipses: a review.**
S. M. Silverman, E. G. Mullen.
Applied Optics, Vol. 14, 2838 - 2843 (1975).

082.104 **Mass absorption indices of various types of natural aerosol particles in the infrared.** K. Fischer.
Applied Optics, Vol. 14, 2851 - 2856 (1975).

082.105 **Evidence for chemiexcitation as the source of the sodium nightglow.** D. P. Sipler, M. A. Biondi.
Geophys. Res. Letters, Vol. 2, 106 - 108 (1975).
Spectral line profiles of the D_1 and D_2 sodium line emission from the twilight and the nightglow have been obtained by means of a 150 mm aperture Fabry-Perot interferometer.

082.106 **Satellite observations of the nitric oxide nightglow.**
J.-C. Gerard.
Geophys. Res. Letters., Vol. 2, 179 - 182 (1975).
The NO δ-bands have been observed photometrically by the ultraviolet telescope onboard the TD-1 ESRO satellite. The maximum intensity for the system is about 220 R, in good agreement with a previous rocket measurement.

082.107 **Methods of determination and results of measurements of spectral density of kinetic energy in the upper atmosphere in an interval of periods of 5 min–12 h.**
G. M. Teptin, A. N. Fakhrutdinova.
Meteorn. rasprostr. radiovoln., No. 10 - 11. Kazan', Kazan. un-t, 1975, p. 42 - 48. In Russian. – Abstr. in Referativ. Zhurn. 51. Astron., 12.51.98 (1975).

082.108 **Tropospheric odd nitrogen and the atmospheric water vapor cycle.** W. Chameides.
Journ. Geophys. Res., Vol. 80, 4989 - 4996 (1975).

082.109 **Esro 4 gas analyzer results. 3. Spatial and temporal structure of the mid-latitude atmosphere during a geomagnetic storm.** H. Trinks, K. H. Fricke, U. Laux, G. W. Prölss, U. von Zahn.
Journ. Geophys. Res., Vol. 80, 4571 - 4575 (1975).

082.110 **Experiment for determination of the concentration and temperature of atomic oxygen and nitrogen in the upper atmosphere of the earth from a measurement of absorption of ultraviolet radiation during the joint flight of Soyuz–Apollo.** A. A. Vekhov, B. S. Kunashev, V. G. Kurt, V. V. Martynov, V. A. Ol'shevskij.
In-t kosmich. issled. AN SSSR. Pr-230. 1975. In Russian. Abstr. in Referativ. Zhurn. 62. Issled. kosmich. prostranstva, 1.62.375 (1976).

082.111 **Der Einfluss der Sternfarbe und einer Inversion des Temperaturprofils der Atmosphäre auf die astronomische Refraktion für Empfänger verschiedener spektraler Empfindlichkeit.** P. Schott.
Deutsche Geod. Kommission Bayer. Akad. Wiss., München, Reihe B, No. 213, p. 61 - 94 (1975).
The influence of the effective wavelength of starlight on the astronomical refraction as a function of zenith distance is investigated for receivers of various spectral response: human eye for bright and feeble light, photographic plate (Ilford HPS) and photomultiplier tube (spectral response S - 20). For zenith distances $\leqslant 50°$ the corrections due to the effective wavelength of starlight are $\leqslant 0''.07$ for human eye, $\leqslant 0''.44$ and $\leqslant 0''.46$ for the photographic plate respectively the photomultiplier tube. Furthermore the influence of an inversion of the temperature profile in the earth's atmosphere on the astronomical refraction was calculated. Nearly independent of the effective wavelength of starlight and the various receivers these corrections are $\leqslant 0''.05$ for zenith distances $\leqslant 50°$.

082.112 **Geomagnetic activity and air pressure at high latitudes.** R. V. Smirnov, N. Ya. Kuklina.
Astron. Tsirk., No. 875, p. 4 - 6 (1975). In Russian.

082.113 **Penetration of stratospheric disturbances up to the mesopause.** E. P. Kropotkina.
Astron. Tsirk., No. 875, p. 7 - 8 (1975). In Russian.

082.114 **Atmospheric extinction in the Central Kopet-Dag.**
G. A. Nasyrov.
Astron. Tsirk., No. 877, p. 5 - 6 (1975). In Russian.

082.115 **Atmospheric inversions and astroclimate during the Sternberg Astronomical Institute high-altitude expedition.** E. F. Rizov.
Astron. Tsirk., No. 880, p. 6 - 8 (1975). In Russian.

082.116 **The night sky brightness at McDonald Observatory.**
J. K. Kalinowski, R. G. Roosen, J. C. Brandt.
Publ. Astron. Soc. Pacific, Vol. 87, 869 - 876 (1975).
Baseline observations of the night sky brightness in *B* and *V*, obtained in midwinter of 1960, are presented for the McDonald Observatory. In agreement with earlier work by Elvey and Rudnick (1937) and Elvey (1943), significant night-to-night and same-night variations in sky brightness are found. Possible causes for these variations are discussed. The data are used to comment on the accuracy of previously published surface photometry of M 81.

082.117 **Precipitable water above Cerro Tololo between 1971 and 1975.** O. L. Hansen, L. Caimanque.
Publ. Astron. Soc. Pacific, Vol. 87, 935 - 939 (1975).
The precipitable water above Cerro Tololo (CTIO) in Chile has been measured for nearly three years. Daily data and monthly distributions are presented. A pattern of dry winters (April–September) and relatively humid summers appears to repeat itself.

082.118 **A catalog of atmospheric densities from the drag on five balloon satellites.** L. G. Jacchia, J. W. Slowey.
Smithsonian Astrophys. Obs., *Cambridge, Mass.*, Special Report 368, 5 + 416 pp. (1975).
A catalog of atmospheric densities derived from the drag on five balloon satellites is presented. These densities replace and, in two cases, considerably extend previously published results on these same satellites. Much of the catalog is based on precisely reduced Baker-Nunn observations and, for that reason, provides much better time resolution than was possible before. The effect of direct solar radiation pressure has been precisely evaluated, and that of terrestrial radiation pressure has been included in every case. The interval covered for each

satellite varies between 3.1 and 7.6 years, with the data extending from early 1961 to early 1973.

082.119 **Pre-dawn variations of the atmosphere's temperature and the 6300 Å emission intensity.**
A. I. Semenov.
Astron. Tsirk., No. 882, p. 6 - 7 (1975). In Russian.

082.120 **Atmospherically-physical problems of the astroclimate.** E. F. Rizov.
Astron. Tsirk., No. 884, p. 5 - 8 (1975). In Russian.

082.121 **Measurements of the optical depth of the earth's atmosphere in the rotational lines of ozone (λ = 1.23 – 1.4 mm).** I. V. Kuznetsov, Yu. Yu. Kulikov, E. F. Rizov, L. I. Fedoseev, A. A. Shvetsov.
Astron. Tsirk., No. 891, p. 7 - 8 (1975). In Russian.

082.122 **Far infrared atmospheric transmission measurements in North-Norway.**
G. Dall'Oglio, I. Pippi, F. Klokkervoll, S. Sivertsen.
Infrared Physics, Vol. 15, 341 (1975).

082.123 **On the image quality in the surface layer of the atmosphere during daytime.**
Kh. I. Abdusamatov, I. Ferro.
Solnechnye Dannye 1975 Byull., No. 9, p. 87 - 92 (1975). In Russian.

082.124 **Total atmospheric absorption at 22.2 GHz.**
W. G. Fogarty.
IEEE Trans. Antennas Propagation, Vol. AP-23, 441 - 444 (1975). – *BMT*

082.125 **Atmospheric millimeter-wave opacity due to oxygen and water vapor.** P. W. Rosenkranz.
Massachusetts Inst. Technol. Res. Lab. Electronics, RLE Progr. Rep. No. 115, p. 84 - 85 (1975). – *UM*

082.126 **Calculation of the astronomical refraction and its anomalies by means of the M-222 computer.**
F. D. Zabolotskij, V. V. Kirichuk.
Geod., kartogr. i aehrofotosemka. Resp. mezhved. nauch.-tekhn. sb., 1975, vyp. (No.) 22, p. 10 - 14. In Russian. – Abstr. in Referativ. Zhurn. 51. Astron., 2.51.245 (1976).

082.127 **Estimate of the accuracy of constructing a model of the upper atmosphere's density according to deceleration of artificial earth satellites.** P. M. Vingardt.
In-t kosmich. issled. AN SSSR. Pr-218. Moskva, 1975. 20 pp. In Russian. – Abstr. in Referativ. Zhurn. 62. Issled. kosmich. prostranstva, 2.62.283 (1976).

082.128 **Determination of the atmosphere's density from data of deceleration of artificial earth satellites in nearly circular orbits.** E. I. Bushuev, A. A. Krasovskij.
Kosmich. issledovaniya na Ukraine. Resp. mezhved. sb., 1975, vyp. (No.) 7, p. 49 - 54. In Russian. – Abstr. in Referativ. Zhurn. 62. Issled. kosmich. prostranstva, 2.62.284 (1976).

082.129 **An instrument for the measurement of the mean square value of atmospheric star scintillation.**
R. Barletti, L. Paterno.
Rev. Sci. Instruments, Vol. 46, 15 - 17 (1975).
Description d'un appareil photoélectrique donnant directement, sous forme numérique, le rapport de la valeur quadratique moyenne des fluctuations du courant et du carré de la valeur moyenne du courant. Ce rapport est la mesure de la scintillation atmosphérique.

On the history of study of telluric lines of the solar spectrum (to the year 1900). See Abstr. 004.113.

Algorithm for computation of the ozone concentration from observations of eclipses of artificial earth satellites. See Abstr. 031.211.

Cross correlation studies of stellar variability using two telescopes. See Abstr. 031.268.

Atmospheric turbulent layers localisation by optical processing of stellar shadow pattern. See Abstr. 031.305.

Die topographisch-atmosphärische Reduktion des mittleren Brechungsindex. See Abstr. 046.050.

Analysis of the orbit of 1970-97B (Cosmos 378 rocket). See Abstr. 054.004.

Solar proton events: stratospheric sources of nitric oxide. See Abstr. 078.003.

On the U.V. source of the night-time *E* region. See Abstr. 083.006.

Theory of enhanced airglow during ionospheric modifications. See Abstr. 083.067.

Beobachtungen des Zodiakallichtes und des Airglows mit dem Ballonteleskop THISBE. 1. Messungen im nahen Infraroten; 2. Messungen in vier Farbeh des Sichtbaren. See Abstr. 106.008.

On the computation of the first-order scattering of the Milky Way radiation and the zodiacal light in the earth's atmosphere. See Abstr. 106.036.

On the influence of the atmospheric dust on the zodiacal light polarization. See Abstr. 106.039.

Estimate of singly scattered lightfrom bright parts of the Milky Way and zodiacal light. See Abstr. 106.063.

On the dependence of atmospheric extinction coefficients on spectral type, luminosity and interstellar reddening in multicolour photometry. See Abstr. 113.052.

083 Ionosphere

083.001 **The calculated and observed diurnal variation of the ionosphere over Millstone Hill on 23–24 March 1970.** R. G. Roble.
Planet. Space Sci., Vol. 23, 1017 - 1033 (1975).

083.002 **Observed characteristics of polar cleft Hα and OI emissions.** H. Derblom.
Planet. Space Sci., Vol. 23, 1053 - 1058 (1975).

083.003 **A theoretical evaluation of the lunar tidal variations in the ionospheric *F*2-layer.**
M. F. K. Abur-Robb, E. Dunford.
Planet. Space Sci., Vol. 23, 1071 - 1080 (1975).

083.004 **Ionospheric measurements from the ESRO-4 satellite.**
W. J. Raitt, E. B. Dorling, P. H. Sheather, J. Blades.
Planet. Space Sci., Vol. 23, 1085 - 1101 (1975).

083.005 **A comparison of the temperature and density structure in high and low speed thermal proton flows.** W. J. Raitt, R. W. Schunk, P. M. Banks.
Planet. Space Sci., Vol. 23, 1103 - 1117 (1975).

083.006 **On the U.V. source of the night-time *E* region.**
M. P. Gough.
Planet. Space Sci., Vol. 23, 1236 (1975).
The photo-ionization produced in the night-time *E* region by the total stellar u.v. continuum is considered. It is shown to be approximately equal to that produced by the scattered Lyman β above about 115 km.

083.007 **Models of *F*1-region ion composition variations.**
W. L. Oliver.
Journ. Atmosph. Terr. Phys., Vol. 37, 1065 - 1076 (1975).

083.008 **Models for the relationship between electron density and temperature in the upper ionosphere.**
D. Bilitza.
Journ. Atmosph. Terr. Phys., Vol. 37, 1219 - 1222 (1975).
Relationships between electron density profiles and electron temperature profiles in the height range 300–500 km are determined, using incoherent scatter measurements taken at Millstone Hill (U.S.A.), at noon in the years 1965–1968.

083.009 **Change of low ionosphere characteristics with solar cycle deduced from solar flare effects at VLF.**
L. R. Piazza, P. Kaufmann.
Journ. Atmosph. Terr. Phys., Vol. 37, 1281 - 1283 (1975).
Sudden ionospheric disturbances (SIDs) measured at very low frequencies (VLF) during the solar active years are compared with those occurring during solar quiet years. There is a solar cycle dependence of the low altitude ionosphere characteristics related to the behaviour of the ionospheric parameter *L*.

083.010 **Rocket measurement of ion and neutral temperatures in the lower ionosphere.**
S. Sasaki, N. Kawashima.
Journ. Geophys. Res., Vol. 80, 2824 - 2828 (1975).

083.011 **Models of the ionosphere.** J. S. Nisbet.
Astrophys. Space Sci. Library, Vol. 51, (see 012.007), 245 - 258 (1975). – Review paper.

083.012 **The *D* and *E* regions.** W. Swider.
Astrophys. Space Sci. Library, Vol. 51, (see 012.007), 259 - 268 (1975).

083.013 **The *F* region.** P. Stubbe.
Astrophys. Space Sci. Library, Vol. 51, (see 012.007), 269 - 280 (1975).

083.014 **Galactic X-ray sources and the ionospheric D region.**
H. Karszenbaum, D. A. Gagliardini.
Nature, Vol. 257, 34 - 35 (1975).
Several authors have studied the possible influence of galactic point X-ray sources in the ionospheric D region. The observed variabilities in phase and amplitude of very low frequency (v.l.f.) night-time propagation have been attributed to different ionisation sources but the relationship between the appearance of strong X-ray sources in the sky and the recorded v.l.f. phase and intensity variations has been emphasised. Particularly important are the effects of Sco X-1, Cen X-2 and Cen X-4. The contribution of Sco X-1 to the night-time ion production is significant compared with the other major night-time sources, but the effect of the minor X-ray sources included can be neglected. Therefore, these results support claims that there is a Scorpius X-1 effect on the v.l.f. signal propagation records.

083.015 **A simplified *D*-region model and its application to magnetic storm after-effects.**
W. N. Spjeldvik, R. M. Thorne.
Journ. Atmosph. Terr. Phys., Vol. 37, 1313 - 1325 (1975).
Energetic electron precipitation from the earth's radiation belts can provide the dominant *D*-region ionization source at middle latitudes during the recovery phase of geomagnetic storms. To study the ionospheric response, a simplified *D*-region ionic model is developed with emphasis placed on obtaining a reasonably accurate evaluation of the free electron concentration. The model provides an adequate fit with observed quiet-time electron concentration in the *D*-region.

083.016 **Comparison of ionospheric gravity wave periods as measured by different experimental techniques.**
E. A. Essex.
Journ. Atmosph. Terr. Phys., Vol. 37, 1349 - 1356 (1975).

083.017 **Self-heating of the partial reflection experiment.**
W. A. Kissick, A. J. Ferraro.
Journ. Atmosph. Terr. Phys., Vol. 37, 1365 - 1370 (1975).

083.018 **Generation of atmospheric gravity waves by the 30 June 1973 solar eclipse in Africa.**
P. Broche, M. Crochet.
Journ. Atmosph. Terr. Phys., Vol. 37, 1371 - 1374 (1975).
Experimental results of electromagnetic probing of the ionosphere by the HF Doppler technique, obtained at equatorial geomagnetic latitudes in Africa during the 30 June 1973 solar eclipse are presented. Evidence of atmospheric gravity waves generated by the eclipse is discussed, and their properties are compared to the theoretical previsions.

083.019 **On an outer ionosphere hydrostatic model.**
Ja. L. (*Ya. L.*) Al'pert.
Journ. Atmosph. Terr. Phys., Vol. 37, 1375 - 1378 (1975).
Models of electron concentration profiles, $N(z)$, in the outer ionosphere in the altitude range 1000~20,000 km are given. These take into account altitude temperature dependences of the form $T(z) \simeq T_0(z/z_0)^\gamma$, which approximate closely enough those known from the literature on experimental data. It is shown that the theoretical profiles, $N(z)$, are in good agreement with the electron density measurements available to

date.

083.020 **Ionosphere-magnetosphere coupling. 1. Thermal plasma.** C. R. Chappell.
Rev. Geophys. Space Phys., Vol. 13, No. 3, (see 013.001), p. 872 - 873, 904 - 905 (1975). – Review paper with bibliographical notes.

083.021 **Ionosphere-magnetosphere coupling. 2. Electric fields.** P. M. Banks.
Rev. Geophys. Space Phys., Vol. 13, No. 3, (see 013.001), p. 874 - 878, 905 - 908 (1975). – Review paper with bibliographical notes.

083.022 **Ionosphere-magnetosphere coupling. 3. A review of the role of wave-particle interactions.**
R. M. Thorne.
Rev. Geophys. Space Phys., Vol. 13, No. 3, (see 013.001), p. 878 - 881, 908 - 909 (1975). – Review paper with bibliographical notes.

083.023 **Thermal structure of the ionosphere.** L. H. Brace.
Rev. Geophys. Space Phys., Vol. 13, No. 3, (see 013.001), p. 882 - 883, 909 - 910 (1975). – Review paper with bibliographical notes.

083.024 **Ionospheric irregularities.** F. W. Perkins.
Rev. Geophys. Space Phys., Vol. 13, No. 3, (see 013.001), p. 884, 910 - 911 (1975). – Review paper with bibliographical notes.

083.025 **A review of *F* region dynamics.** J. V. Evans.
Rev. Geophys. Space Phys., Vol. 13, No. 3, (see 013.001), p. 887 - 894, 916 - 921 (1975). – Review paper with bibliographical notes.

083.026 **Ionospheric *D* and *E* regions.** C. F. Sechrist, Jr.
Rev. Geophys. Space Phys., Vol. 13, No. 3, (see 013.001), p. 894 - 900, 912 - 916 (1975). – Review paper with bibliographical notes.

083.027 **Effect of solar flares in the lower ionosphere.**
V. Yu. Aksel'rod, B. A. Turkeeva.
Geomagn. Aeronom., Vol. 15, 639 - 642 (1975). In Russian.

083.028 **Positive ions in the D-region. I. Variations of ion composition.** A. D. Danilov, A. G. Simonov.
Geomagn. Aeronom., Vol. 15, 643 - 650 (1975). In Russian.

083.029 **A map of precipitating of proton streams with $E_p > 70$ keV at high latitudes.** O. R. Grigoryan, S. N. Kuznetsov, G. S. Malkiehl', V. G. Stolpovskij.
Geomagn. Aeronom., Vol. 15, 735 - 736 (1975). In Russian. Brief information.

083.030 **A model of seasonal and semi-annual variations of the F2-region of the ionosphere.**
G. S. Ivanov-Kholodnyj, A. V. Mikhajlov.
Geomagn. Aeronom., Vol. 15, 739 - 741 (1975). In Russian. Brief information.

083.031 **On the dependence of the electron concentration in the lower part of the F-region on the solar activity level.** N. M. Boenkova.
Geomagn. Aeronom., Vol. 15, 741 - 743 (1975). In Russian. Brief information.

083.032 **Day-to-day variability of ionospheric electron content at mid-latitudes.** R. P. Kane.
Journ. Geophys. Res., Vol. 80, 3091 - 3099 (1975).

083.033 **Duration of equatorial spread F.** V. V. Somayajulu, K. Sen Gupta, B. V. Krishna Murthy.
Nature, Vol. 257, 112 - 113 (1975). – Letter.

083.034 **Proton streams with $E_p > 70$ keV in the near-equatorial region at small altitudes.** V. D. Butenko, O. R. Grigoryan, S. N. Kuznetsov, G. S. Malkiehl, V. G. Stolpovskij.
Kosmich. Issled., Vol. 13, 508 - 512 (1975). In Russian.

083.035 **On a possibility of determining the collision frequency and temperature of electrons in the outer ionosphere.** G. V. Givishvili, M. D. Fligel'.
Kosmich. Issled., Vol. 13, 528 - 531 (1975). In Russian.

083.036 **On the convective instability of the polar ionosphere.** L. A. Abramov, L. S. Al'perovich.
Kosmich. Issled., Vol. 13, 532 - 538 (1975).

083.037 **On a possible instability of primary photoelectron streams.**
G. G. Butvin, G. V. Popov, G. V. Khazanov.
Kosmich. Issled., Vol. 13, 607 - 609 (1975). In Russian. Brief information.

083.038 **Antarctic substorm events observed by sounding rockets, ionization of the *D*- and *E*-regions by auroral electrons.**
T. Nagata, T. Hirasawa, M. Takizawa, T. Tohmatsu.
Planet. Space Sci., Vol. 23, 1321 - 1327 (1975).

083.039 **Neutral atmospheric composition, plasma density, and electron temperature at *F* region heights.**
G. W. Prölss, U. von Zahn, W. J. Raitt.
Journ. Geophys. Res., Vol. 80, 3715 - 3718 (1975).
This study investigates the connection between neutral composition, plasma density, and electron temperature at *F* region heights and mid-latitudes by using in situ measurements obtained by instruments aboard the Esro 4 satellite.

083.040 **Nalichtend meteoorspoor onthult windsnelheden in de ionosfeer.** P. A. Koning.
Zenit, Vol. 2, 296 - 299 (1975).

083.041 **On the problem of formation of the stationary ionospheric F-region.**
A. G. Khantadze, R. G. Gachechiladze, B. Ya. Chekhoshvili.
Dokl. Akad. Nauk SSSR. Ser. Mat., Fiz., Vol. 224, 332 - 334 (1975). In Russian.

083.042 **Measurements of the electron temperature in the ionosphere by the method of high-frequency sounding aboard the satellites Intercosmos 2 and Intercosmos 8.** Ya. Shmilauehr, K. I. Gringauz, V. V. Afonin.
Geomagn. Aeronom., Vol. 15, 818 - 824 (1975). In Russian.

083.043 **High-latitude ionospheric irregularities from measurements aboard the artificial earth satellite Cosmos 348.** N. V. Dzhordzhio.
Geomagn. Aeronom., Vol. 15, 825 - 831 (1975). In Russian.

083.044 **On intensities of OI 6300 Å, OI 5577 Å and NI 5200 Å emissions by data of measurements of the ionospheric ion composition in the night-time F2-region.**
M. N. Vlasov, Yu. A. Romanovskij.
Geomagn. Aeronom., Vol. 15, 861 - 866 (1975). In Russian.

083.045 **Daily variations of the positive ion concentrations at mean and high latitudes and their connection with solar activity.** B. A. Mirtov, A. G. Starkova.
Geomagn. Aeronom., Vol. 15, 867 - 870 (1975). In Russian.

083.046 **On the distribution of NO and N concentrations at heights of 120–200 km.**
S. I. Belinskaya, V. V. Koshelev.
Geomagn. Aeronom., Vol. 15, 871 - 875 (1975). In Russian.

083.047 **Investigations in the ionosphere with the help of satellite Cosmos 378. N(h)-profiles and the temperature of the F-region by ground and satellite measurements above Khabarovsk.** Eh. G. Mirmovich, B. S. Shapiro.
Geomagn. Aeronom., Vol. 15, 934 - 936 (1975). In Russian. Brief information.

083.048 **Model of Birkeland currents associated with an auroral arc.** R. R. Vondrak.
Journ. Geophys. Res., Vol. 80, 4011 - 4014 (1975).
The ionospheric configuration of the vertical Birkeland currents associated with an auroral arc is calculated from a model based on the assumption that the auroral electrons constitute a net current into the ionosphere. It is further assumed that the return vertical current out of the ionosphere is impeded along the field lines which connect the arc to the magnetosphere. The magnitude, location, and structure of the return Birkeland current calculated from the model agree well with measurements of these parameters by a rocket-borne experiment.

083.049 **Influence of solar cosmic radiation on January 24 and September 1, 1971 on the midlatitude ionosphere.** G. Nestorov, P. Velinov.
Izv. Geofiz. in-t. Blg. AN, Vol. 20, 55 - 71 (1974). In Bulgarian. Abstr. in Referativ. Zhurn. 51. Astron., 8.51.542 (1975).

083.050 **Solar-perturbed variations in the height scale of the lower and mean ionosphere.** K. Serafimov.
Izv. Geofiz. in-t. Blg. AN, Vol. 20, 73 - 84 (1974). In Bulgarian.

083.051 **Doppler measurements of the ionosphere on the occasion of the Apollo-Soyuz Test Project. Part I: Computer simulation of ionospheric-induced Doppler shifts.**
M. D. Grossi, R. H. Gay.
Smithsonian Astrophys. Obs., *Cambridge, Mass.,* Special Report 366, 7 + 54 + A25 pp. (1975).

083.052 **Doppler measurements of the ionosphere on the occasion of the Apollo-Soyuz Test Project. Part II: Inversion of differential and rotating Doppler shifts.**
R. H. Gay, M. D. Grossi.
Smithsonian Astrophys. Obs., *Cambridge, Mass.,* Special Report 367, 5 + 37 pp. (1975).

083.053 **On the origin and composition of excess radiation in the height interval of 200–400 km.**
V. S. Tsaplin.
Trudy I Konf. molodykh uchenykh NIIYaF MGU. Moskva, 1974, p. 53 - 77. In Russian. – Abstr. in Referativ. Zhurn. 62. Issled. kosmich. prostranstva, 8.62.222 (1975).

083.054 **On the anomaly of the ion composition of the equatorial F2 region of the ionosphere.**
Yu. A. Romanovskij, L. I. Pogulyaevskij, I. A. Dubov, E. G. Ul'yanov.
Dokl. Akad. Nauk SSSR. Ser. Mat., Fiz., Vol. 224, 1312 - 1314 (1975). In Russian.

083.055 **High-power radar studies of the ionosphere.**
J. V. Evans.
Proc. IEEE, Vol. 63, 1636 - 1650 (1975).

083.056 **An empirical model of global spread-*F* occurrence.**
D. G. Singleton.
Journ. Atmosph. Terr. Phys., Vol. 37, 1535 - 1544 (1975).
A method of combining models of ionospheric *F*-layer peak electron density and irregularity incremental electron density into a model of the occurrence probability of the frequency-spreading component of spread-*F* is presented.

083.057 **An improved phenomenological model of ionospheric density.** Y. T. Chiu.
Journ. Atmosph. Terr. Phys., Vol. 37, 1563 - 1570 (1975).
A global representation of the large scale variations of ionospheric electron density with the annual, diurnal and solar activity cycles is described.

083.058 **Instationary model of distributions of concentration and temperature of charged particles along the lines of force of the geomagnetic field.**
V. M. Polyakov, M. A. Koen, G. V. Khazanov.
Izv. vyssh. ucheb. zavedenij. Radiofizika, Vol. 18, 510 - 515 (1975). In Russian. – Abstr. in Referativ. Zhurn. 62. Issled. kosmich. prostranstva, 11.62.277 (1975).

083.059 **Positive ions in the D-region. 3. Variations of the effective recombination coefficient.**
A. D. Danilov, A. G. Simonov.
Geomagn. Aeronom., Vol. 15, 1000 - 1004 (1975). In Russian.

083.060 **Ionization of the F-layer of the ionosphere and variations of the magnetic field in the near-polar region.**
Ya. I. Fel'dshtejn, A. M. Lyatskaya, P. V. Sumaruk, N. F. Shevnina.
Geomagn. Aeronom., Vol. 15, 1021 - 1027 (1975). In Russian.

083.061 **On diffusion-recombination parameters of the F-region of the ionosphere.** G. M. Khocholava.
Geomagn. Aeronom., Vol. 15, 1096 - 1098 (1975). In Russian. Brief information.

083.062 **On E_s formation at mean latitudes.**
K. N. Vasil'ev, S. P. Chernysheva, V. M. Sheftel', Eh. G. Shcharenskaya.
Geomagn. Aeronom., Vol. 15, 1098 - 1099 (1975). In Russian. Brief information.

083.063 **Mass spectrometer measurements of the positive ion composition in the *D*- and *E*-regions of the ionosphere.** P. A. Zbinden, M. A. Hidalgo, P. Eberhardt, J. Geiss.
Planet. Space Sci., Vol. 23, 1621 - 1642 (1975).

083.064 **Characteristics of the high-latitude ionospheric irregularity boundary, as monitored by the total ion current probe on ESRO-4.** D. H. Clark, W. J. Raitt.
Planet. Space Sci., Vol. 23, 1643 - 1647 (1975).
The total ion-density probe on the satellite ESRO-4 has been used to monitor the position of the boundary of the high-latitude ionospheric irregularity zone. The diurnal behaviour of this boundary suggests the appearance of a distinct night-time sub-auroral irregularity zone. Possible irregularity production mechanisms are discussed.

083.065 **Narrow band VLF electromagnetic signals generated in the auroral ionosphere by the high-frequency two-stream instability.** E. Ungstrup.
Journ. Geophys. Res., Vol. 80, 4272 - 4278 (1975).

083.066 **The localization of ionospheric irregularities by the holographic method.**
G. Schmidt, A. Tauriainen.
Journ. Geophys. Res., Vol. 80, 4313 - 4324 (1975).
The holographic method has been recently introduced to study satellite scintillations. Only one ground station is needed

to receive the radio hologram. The object wave scattered or diffracted by irregularities is measured in relation to a harmonic reference wave. The resulting amplitude and phase information can be interpreted, by using the reciprocity theorem, as the one-dimensional radio hologram at the height of the satellite orbit. In this paper the mathematical theory is developed to reconstruct the image.

083.067 **Theory of enhanced airglow during ionospheric modifications.** J. Weinstock.
Journ. Geophys. Res., Vol. 80, 4331 - 4345 (1975).

083.068 **Polarization of VHF waves emitted from geostationary satellites.** J. A. Klobuchar.
Journ. Geophys. Res., Vol. 80, 4387 - 4389 (1975).
Brief report.

083.069 **Direct ground-based detection of ionospheric electric fields.** D. Möhlmann.
Gerlands Beiträge Geophys., Vol. 84, 375 - 380 (1975).
The height dependence of the electrostatic potential generated in the ionosphere has been investigated in the region between the earth's surface and the lower boundary of the dynamo region. It is concluded that measurements of the large-scale electrostatic potential in the 10 km level will be possible if measuring points widely spaced in longitude are used.

083.070 **Hall-dominated magnetohydrodynamic waves at and below *E*-region heights.**
H.-R. Lehmann, R. Treumann.
Gerlands Beiträge Geophys., Vol. 84, 449 - 454 (1975).
The dispersion relation of parallel Hall-dominated magnetohydrodynamic waves is discussed. It is shown that short waves with frequencies within the ELF-band can arise in the ionosphere. The waves have a high-frequency cut-off. ELF-hiss in the polar ionosphere is proposed to be explained by such a wave phenomenon. Some remarks on the validity are added.

083.071 **SAR arc event observed during the December 1971 magnetic storm.** E. J. Maier, S. Chandra, L. Brace, J. H. Hoffman, G. G. Shepherd, J. H. Whitteker.
Journ. Geophys. Res., Vol. 80, 4591 - 4597 (1975).

083.072 **Properties of ELF electromagnetic waves in and above the earth's ionosphere deduced from plasma wave experiments on the OV1-17 and Ogo 6 satellites.**
M. C. Kelley, B. T. Tsurutani, F. S. Mozer.
Journ. Geophys. Res., Vol. 80, 4603 - 4611 (1975).
An analysis of ac electric field data obtained on board the OV1-17 satellite and ac magnetic field data obtained on board the Ogo 6 satellite has been made during the northern hemisphere spring and summer of 1969 with the purpose of studying extreme low frequency (ELF) electromagnetic waves above the earth's ionosphere. The results are in basic agreement with a number of previous ground-based and low-altitude satellite experiments.

083.073 **Improved calculations of electron densities between 110 and 300 km derived from solar EUV fluxes of August 23, 1972.** L. Heroux, M. Cohen, J. E. Higgins.
Journ. Geophys. Res., Vol. 80, 4732 - 4734 (1975). – Brief report.

083.074 **Comment on 'Electron densities between 110 and 300 km derived from solar EUV fluxes of August 23, 1972' by L. Heroux, M. Cohen, and J. E. Higgins.**
J. S. Nisbet.
Journ. Geophys. Res., Vol. 80, 4770, with a reply by L. Heroux, M. Cohen, J. E. Higgins, p. 4771 (1975). – Concerning Journ. Geophys. Res., Vol. 79, 5237 - 5244 (1974), (see Abstr. 12.083.107).

083.075 **The effect of solar flares on the ionospheric E-layer.** G. S. Ivanov-Kholodnyj, I. N. Odintsova, L. N. Leshchenko.
XI Vses. konf. po rasprostr. radiovoln. Chast' (Part) 1. Tezisy dokl. Kazan', Kazan. un-t, 1975, p. 66. In Russian. – Abstr. in Referativ. Zhurn. 51. Astron., 1.51.521 (1976).

083.076 **Determination of the total electron content of the ionosphere and plasmasphere.** K. Davies, R. B. Fritz, G. K. Hartmann, R. Leitinger, A. K. Paul, G. Schmidt.
Materialy Mezhdunar. simpoz. 'Issled. struktury ionosf. s pomoshch'yu kogerentn. radiost. mayaka i rezul'taty ehksperimenta na sputnike ATS-6', 1974. Tom 2. Moskva, 1975, p. 146 - 158. – Abstr. in Referativ. Zhurn. 62. Issled. kosmich. prostranstva, 2.62.288 (1976).

083.077 **Ionospheric irregularities as investigated using satellite signals recorded at high and middle latitudes.**
V. A. Misyura, N. P. Svetlitchny, L. V. Bezrodnaya, N. D. Gerasimova, N. D. Zolondkovsky, G. N. Zintchenko, O. F. Tyrnov, A. R. Yagovkin, O. A. Litvinenko, B. V. Zagvozdkin.
Materialy Mezhdunar. simpoz. 'Issled. struktury ionosf. s pomoshch'yu kogerentn. radiost. mayaka i rezul'taty ehksperimenta na sputnike ATS-6', 1974. Tom 1. Moskva, 1975, p. 251 - 260. – Abstr. in Referativ. Zhurn. 62. Issled. kosmich. prostranstva, 2.62.290 (1976).

An introduction to the ionosphere and magnetosphere. See Abstr. 003.102.

Mehr Energie für die Ionosphäre.
See Abstr. 011.041.

The magnetic field of the solar wind and vertical motion of the night ionosphere. See Abstr. 074.095.

The aeronomy of odd nitrogen in the thermosphere.
See Abstr. 082.027.

Auroral heating. See Abstr. 084.005.

Coordinated optical and radar measurements of auroral motions and ionospheric plasma drift.
See Abstr. 084.006.

Microinstabilities and the production of short-wavelength irregularities in the auroral F region.
See Abstr. 084.033.

The role of the magnetosphere in satellite and radio-star scintillation. See Abstr. 084.204.

The electric field in the magnetosphere and ionosphere of the earth. I. See Abstr. 084.224.

The electric field in the magnetosphere and ionosphere of the earth. II. See Abstr. 084.225.

The electric field in the magnetosphere and ionosphere of the earth. III. See Abstr. 084.226.

Field-aligned and ionospheric currents.
See Abstr. 084.248.

Controlled artificial generation of geomagnetic pulsations. See Abstr. 084.262.

A determination of *F* region effective recombination coefficients from the Echo 2 sounding rocket plasma wave and particle measurements. See Abstr. 084.304.

A comparison of electric and magnetic field data from the Ogo 6 spacecraft. See Abstr. 084.313.

Analysis of processes leading to localized electron enhancements in the outer radiation belt. See Abstr. 084.406.

Energy dissipation of the ring current and the state of the ionosphere. See Abstr. 084.408.

Role of meteors in ionospherical processes. II. See Abstr. 104.057.

Influence of the interplanetary magnetic field on the magnetospheric convection and electric currents in the ionosphere. See Abstr. 106.034.

084 Aurorae, Geomagnetic Field, Radiation Belts

Aurorae

084.001 **Comment on the paper: "The reaction of nitric oxide with atomic nitrogen as a possible excitation source of auroral green line"** [Journ. Atmosph. Terr. Phys., Vol. 36, 1437 - 1439 (1974)]. D. W. Rusch, D. G. Torr, W. E. Sharp, T. M. Donahue, K. Henriksen.
Journ. Atmosph. Terr. Phys., Vol. 37, 1173 (1975). – Concerning 12.084.007.

084.002 **The aurora.** B. Hultqvist.
Space Sci. Rev., Vol. 17, 787 - 821 (1975). – Review paper.

084.003 **Direction-finding measurements of auroral kilometric radiation.**
W. S. Kurth, M. M. Baumback, D. A. Gurnett.
Journ. Geophys. Res., Vol. 80, 2764 - 2770 (1975).

084.004 **Auroral electrons and the optical emissions of nitrogen.** P. D. Feldman, J. P. Doering.
Journ. Geophys. Res., Vol. 80, 2808 - 2812 (1975).

084.005 **Auroral heating.** A. Egeland.
Astrophys. Space Sci. Library, Vol. 51, (see 012.007), 99 - 109 (1975).

084.006 **Coordinated optical and radar measurements of auroral motions and ionospheric plasma drift.**
R. D. Sears, J. E. Evans.
Astrophys. Space Sci. Library, Vol. 51, (see 012.007), 125 - 130 (1975).

084.007 **Auroral ultraviolet emissions.** J.-C. Gérard.
Astrophys. Space Sci. Library, Vol. 51, (see 012.007), 309 - 317 (1975).

084.008 **Observed intensity ratios of the N_2 1PG bands in aurora.** R. L. Gattinger, A. V. Jones.
Astrophys. Space Sci. Library, Vol. 51, (see 012.007), 319 - 322 (1975).

084.009 **Rocketborne observations of atmospheric infrared emissions in the auroral region.**
A. T. Stair, Jr., J. C. Ulwick, K. D. Baker, D. J. Baker.
Astrophys. Space Sci. Library, Vol. 51, (see 012.007), 335 - 346 (1975).

084.010 **Summary analysis of 4.3 μm data.** J. B. Kumer.
Astrophys. Space Sci. Library, Vol. 51, (see 012.007), 347 - 358 (1975).

084.011 **The aurora.** B. Hultqvist.
Astrophys. Space Sci. Library, Vol. 52, (see 012.008), 77 - 111 (1975). – Review paper.

084.012 **On the morphology of auroral-zone X-ray events – IV. Substorm-related electron precipitation in the local morning sector.** J. Kangas, L. Lukkari, P. Tanskanen, H. Trefall, J. Stadsnes, G. Kremser, W. Riedler.
Journ. Atmosph. Terr. Phys., Vol. 37, 1289 - 1303 (1975).

084.013 **Simultaneous observations of evening magnetic pulsations and morning electron precipitation events in the auroral zone during the substorm.**
L. Lukkari, J. Kangas, R. R. Heacock.
Journ. Atmosph. Terr. Phys., Vol. 37, 1305 - 1311 (1975).

084.014 **Advances in magnetospheric physics: aurora.**
R. H. Eather.
Rev. Geophys. Space Phys., Vol. 13, No. 3, (see 013.001), p. 925 - 943, 994 - 1002 (1975). – Review paper with bibliographical notes.

084.015 **Investigation of electron precipitation during an auroral substorm by rocket-borne detectors.**
K. K. Vij, D. Venkatesan, C. D. Anger.
Journ. Geophys. Res., Vol. 80, 3205 - 3210 (1975).

084.016 **Spectroscopic determination of the characteristics of precipitating auroral particles.**
S. B. Mende, R. H. Eather.
Journ. Geophys. Res., Vol. 80, 3211 - 3216 (1975).

A large data base obtained from NASA Convair auroral expedition is used to illustrate the variability in $\lambda 6300/\lambda 4278$ and $\lambda 5577/\lambda 4278$ ratios in auroras.

084.017 **The problem of auroral NO^+ concentration and the intensity of 1.27 μ bands of O_2.**
S. S. Prasad, D. R. Furman, A. E. S. Green.
Planet. Space Sci., Vol. 23, 1341 - 1343 (1975). – Research note.

084.018 **Aurora on March 16, 1974 in Moscow.**
V. D. Belyavskaya, A. S. Elokhov, Yu. L. Truttse.
Astron. Tsirk., No. 865, p. 6 - 8 (1975). In Russian.

084.019 **The auroral electrojet and global auroral features.**
Y. Kamide, S.-I. Akasofu.
Journ. Geophys. Res., Vol. 80, 3585 - 3602 (1975).
The distribution of equivalent current vectors of polar magnetic substorms, relative to luminous auroral features of global scale, is examined by comparing a number of DMSP satellite photographs with the simultaneous ground magnetic data.

084.020 **The equatorward boundary of the diffuse aurora and auroral substorms as seen by the Isis 2 auroral scanning photometer.**
A. T. Y. Lui, C. D. Anger, S.-I. Akasofu.
Journ. Geophys. Res., Vol. 80, 3603 - 3614 (1975).

084.021 **Radar observations of auroral electrojet currents.**
R. A. Greenwald, W. L. Ecklund, B. B. Balsley.
Journ. Geophys. Res., Vol. 80, 3635 - 3641 (1975).

084.022 **A new look at radar auroral motions.**
R. A. Greenwald, W. L. Ecklund.
Journ. Geophys. Res., Vol. 80, 3642 - 3648 (1975).

084.023 **Auroral electric field heating: its implications in the altitude profile of ionic composition and emissions.**
S. S. Prasad, D. R. Furman.
Journ. Geophys. Res., Vol. 80, 3701 - 3704 (1975). – Brief report.

084.024 **Simultaneous rocket measurements of structure parameters of the polar thermosphere and of auroral radiation.** V. F. Tulinov, V. M. Fejgin, V. A. Lipovetskij, Yu. M. Zhuchenko, L. S. Novikov, G. F. Tulinov.
Geomagn. Aeronom., Vol. 15, 946 - 947 (1975). In Russian.
Brief information.

084.025 **A measurement of perpendicular current density in an aurora.** E. A. Bering, F. S. Mozer.
Journ. Geophys. Res., Vol. 80, 3961 - 3972 (1975).

084.026 **Similarity between global auroral deformations in DAPP photographs and small scale deformations observed by a TV camera.** T. Oguti.
Journ. Atmosph. Terr. Phys., Vol. 37, 1413 - 1418 (1975).

084.027 **Two-tiered auroral band.** T. Oguti.
Journ. Atmosph. Terr. Phys., Vol. 37, 1501 - 1504 (1975). – Short paper.

084.028 **Auroral activity during 1974.**
D. H. McIntosh, M. Hallissey.
Observatory, Vol. 95, 225 - 227 (1975).

084.029 **Rocket-borne measurements of auroral electrons.**
J. S. Boyd.
Rev. Geophys. Space Phys., Vol. 13, 735 - 740 (1975).
Rocket measurements of the pitch angle and energy distributions of precipitated electrons in the 1- to 20-keV range show a consistent and relatively simple pattern. Features such as 'monoenergetic peaks' and pitch angle distributions showing intense field-aligned components, rather than being anomalous, are characteristic of bright well-defined auroral forms.

084.030 **Magnetospheric substorm energy dissipation in the atmosphere.** M. H. Rees.
Planet. Space Sci., Vol. 23, 1589 - 1596 (1975).
The mechanisms of energy dissipation in the atmosphere for magnetospheric substorm energy sources are examined quantitatively, and the height-integrated energy budget is determined for average auroral conditions.

084.031 **On a magnetostatic model of the auroral electrojet.**
B. L. Shirman, V. N. Khomyak.
Geomagn. Aeronom., Vol. 15, 1112 - 1115 (1975). In Russian.
Brief information.

084.032 **An auroral *F*-region study using in situ measurements by the Atmosphere Explorer-C satellite.**
M. R. Torr, D. G. Torr, R. A. Hoffman, W. B. Hanson, J. H. Hoffman, W. K. Peterson, J. C. G. Walker.
Planet. Space Sci., Vol. 23, 1669 - 1679 (1975).

084.033 **Microinstabilities and the production of short-wavelength irregularities in the auroral F region.**
E. Ott, D. T. Farley.
Journ. Geophys. Res., Vol. 80, 4599 - 4602 (1975).

Model of Birkeland currents associated with an auroral arc. See Abstr. 083.048.

Magnetospheric and auroral plasmas: a short survey of progress. See Abstr. 084.238.

North-south component of the interplanetary magnetic field and large scale auroral dynamics.
See Abstr. 106.002.

The roles of the north–south component of the interplanetary magnetic field on large-scale auroral dynamics observed by the DMSP satellite. See Abstr. 106.023.

Geomagnetic Field

084.201 **Magnetic-field observations in the low-latitude magnetotail during substorms.**
A. Nishida, N. Nagayama.
Planet. Space Sci., Vol. 23, 1119 - 1125 (1975).

084.202 **Open and closed magnetospheric tail configurations and their stability.**
J. Birn, R. Sommer, K. Schindler.
Astrophys. Space Sci., Vol. 35, 389 - 402 (1975).
The asymptotic theory valid for magnetospheric tail configurations that vary only weakly in the antisolar direction is used to derive a number of explicit properties. The conditions under which the magnetopause converges to form a closed magnetosphere or diverges (open magnetosphere) are identified and discussed. It is shown that the presence of the high latitude low pressure tail lobes guarantees the open solution. The large value of the Mach-number of the unperturbed solar wind is the reason for the slow variation of the plasma and field quantities along the tail. Criteria for (two-dimensional) stability are discussed and it is shown that they can be expressed in terms of simple topological properties of the equilibria. Closed magnetospheres turn out to be stable, open magnetospheres with sufficiently stretched field lines are subject to an instability which may be the cause of magnetospheric substorms.

084.203 **Geomagnetic disturbances and complex sunspots.**
J.-T. Horng.
Astrophys. Space Sci., Vol. 35, L33 - L34 (1975).
Statistical result shows that a large geomagnetic disturbance is most likely to occur at the time when a complex sunspot is in 6.7°E–19.9°E from the central meridian of the solar disk during the period 1968–1972.

084.204 **The role of the magnetosphere in satellite and radio-star scintillation.** H. G. Booker.
Journ. Atmosph. Terr. Phys., Vol. 37, 1089 - 1098 (1975).

084.205 **Wave-particle interactions in the outer magnetosphere: a review.** R. W. Fredricks.
Space Sci. Rev., Vol. 17, 741 - 780 (1975).
The paper summarizes the current state of knowledge of the earth's magnetosphere.

084.206 **On the earth's bow shock motion and speed.**
V. Formisano, G. Mastrantonio.
Space Sci. Rev., Vol. 17, 781 - 786 (1975).

084.207 **Heos 2 plasma observations in the distant polar magnetosphere: the plasma mantle.**
H. Rosenbauer, H. Grünwaldt, M. D. Montgomery, G. Paschmann, N. Sckopke.
Journ. Geophys. Res., Vol. 80, 2723 - 2737 (1975).

084.208 **Weak pitch angle scattering of energetic protons in the magnetosphere.**
M. Scholer, G. Morfill, D. Hovestadt.
Journ. Geophys. Res., Vol. 80, 2745 - 2750 (1975).

084.209 **The earth as a radio source: the nonthermal continuum.** D. A. Gurnett.
Journ. Geophys. Res., Vol. 80, 2751 - 2763 (1975).
In addition to the intense and highly variable auroral kilometric radiation the earth also radiates a weak nonthermal continuum from energetic electrons in the outer radiation zone. The intensity of this continuum radiation decreases with increasing frequency and is usually below the cosmic noise level at frequencies above 100 kHz. The author shows that the frequency spectrum of the continuum radiation consists of two components, a trapped component, which is permanently trapped within the magnetosphere at frequencies below the solar wind plasma frequency, and an escaping component, which propagates freely away from the earth at frequencies above the solar wind plasma frequency.

084.210 **Motions of the earth's bow shock.**
R. D. Auer.
Astrophys. Space Sci. Library, Vol. 52, (see 012.008), 11 - 17 (1975).

084.211 **On the earth's bow shock motion and speed.**
V. Formisano, G. Mastrantonio.
Astrophys. Space Sci. Library, Vol. 52, (see 012.008), 19 - 24 (1975).

084.212 **The calm and turbulent state of the earth's magnetosheath from December 1968 through March 1970.**
E. Amata.
Astrophys. Space Sci. Library, Vol. 52, (see 012.008), 25 - 38 (1975).

084.213 **The response of the magnetosphere to the solar wind.** C. T. Russell.
Astrophys. Space Sci. Library, Vol. 52, (see 012.008), 39 - 53 (1975).

084.214 **Substorm effects on the neutral sheet inside 10 earth radii.** B. T. Thomas, P. C. Hedgecock.
Astrophys. Space Sci. Library, Vol. 52, (see 012.008), 55 - 70 (1975).

084.215 **The plasma sheet boundary and K_p.**
J. W. Freeman.
Astrophys. Space Sci. Library, Vol. 52, (see 012.008), 71 - 75 (1975).

084.216 **Wave-particle interactions in the outer magnetosphere: a review.** R. W. Fredricks.
Astrophys. Space Sci. Library, Vol. 52, (see 012.008), 113 - 152 (1975).

084.217 **Hot-cold plasma interactions in the earth's magnetosphere.** D. J. Williams.
Astrophys. Space Sci. Library, Vol. 52, (see 012.008), 161 - 177 (1975).

084.218 **Concepts of magnetospheric convection.**
V. M. Vasyliunas.
Astrophys. Space Sci. Library, Vol. 52, (see 012.008), 179 - 188 (1975).

084.219 **Electric field effects on the access of magnetosheath plasma to the distant magnetotail.** M. K. Bird.
Astrophys. Space Sci. Library, Vol. 52, (see 012.008), 189 - 196 (1975).

084.220 **Neutral sheets.** J. W. Dungey.
Astrophys. Space Sci. Library, Vol. 52, (see 012. 008), 197 - 204 (1975).

084.221 **Evidence for magnetic field line reconnection in the solar wind.** V. Formisano, E. Amata.
Astrophys. Space Sci. Library, Vol. 52, (see 012.008), 205 - 217 (1975).

084.222 **Magnetospheres of earth and Jupiter after Pioneer 10.** A. Prakash, N. Brice.
Astrophys. Space Sci. Library, Vol. 52, (see 012.008), 411 - 423 (1975).

084.223 **Geomagnetic effect associated with X-ray flare from Sco X-1.** H. J. Sastri, B. S. Murthy.
Nature, Vol. 257, 35 - 36 (1975). – Letter.

084.224 **The electric field in the magnetosphere and ionosphere of the earth. I.**
A. V. Gurevich, A. L. Krylov, E. E. Tsedilina.
Issled. po geomagnetizmu, aehron. i fiz. Solntsa, Vyp. (No.) 35. Moskva, Nauka, 1975, p. 85 - 112. In Russian. – Abstr. in Referativ. Zhurn. 62. Issled. kosmich. prostranstva, 7.62.253 (1975).

084.225 **The electric field in the magnetosphere and ionosphere of the earth. II.**
A. V. Gurevich, A. L. Krylov, E. E. Tsedilina.
Issled. po geomagnetizmu, aehron. i fiz. Solntsa. Vyp. (No.) 35. Moskva, Nauka, 1975, p. 113 - 129. In Russian. – Abstr. in Referativ. Zhurn. 62. Issled. kosmich. prostranstva, 7.62.254 (1975).

084.226 **The electric field in the magnetosphere and ionosphere of the earth. III.**
A. V. Gurevich, A. L. Krylov, E. E. Tsedilina.
Issled. po geomagnetizmu, aehron. i fiz. Solntsa. Vyp. (No.) 35. Moskva, Nauka, 1975, p. 130 - 160. In Russian. – Abstr. in Referativ. Zhurn. 62. Issled. kosmich. prostranstva, 7.62.255 (1975).

084.227 **On the helical dynamo of Lortz as a model for the steady main geomagnetic field.** E. R. Benton.
Geophys. Journ. Roy. Astron. Soc., Vol. 42, 385 - 401 (1975). Paper presented at the Tenth International Symposium on Mathematical Geophysics, Cambridge, England, 1974 June 26 - July 4.

An attempt is made to evaluate the infinite cylindrical dynamo of Lortz as a basis for a model of the steady geomagnetic field. Provided it can be deformed from cylindrical to toroidal geometry, when then embedded near the equatorial plane of the liquid outer core, such a dynamo appears to be geophysically relevant.

084.228 **A model of the geodynamo.** F. H. Busse.
Geophys. Journ. Roy. Astron. Soc., Vol. 42, 437 - 459 (1975). – Paper presented at the Tenth International Symposium on Mathematical Geophysics, Cambridge, England, 1974 June 26 - July 4.

The problem of the generation of the geomagnetic field by convection in the earth's core is considered. The complete hydromagnetic problem including the Lorentz force is solved in the case of a cylindrical configuration. Because of the dominant effect of rotation this configuration incorporates the essential features of the dynamo process in the earth's core. The results show general agreement with the known properties of geomagnetism and the earth's core.

084.229 **Effect of drift-resonance broadening on radial diffusion in the magnetosphere.** M. Schulz.
Astrophys. Space Sci., Vol. 36, 455 - 458 (1975).

The purpose of this work is to address a conceptual problem inherent in the theory of magnetospheric radial diffusion.

084.230 **The terrestrial magnetosphere : a half-wave rectifier of the interplanetary electric field.**
R. K. Burton, R. L. McPherron, C. T. Russell.
Science, Vol. 189, 717 - 718 (1975).

A study of geomagnetic disturbances during 1967 and 1968, for which in situ solar wind observations are available, reveals that the magnetosphere acts as a half-wave rectifier of the interplanetary electric field. The rate of injection of energy into the magnetosphere, as inferred from the strength of the disturbance is approximately linearly proportional to the component of the electric field from dawn to dusk but is effectively zero if the electric field has a component from dusk to dawn.

084.231 **The frequency of geomagnetic reversals and the symmetry of the nondipole field.** A. Cox.
Rev. Geophys. Space Phys., Vol. 13, No. 3, (see 013.001), p. 35 - 51 (1975). – Review paper with bibliographical notes.

084.232 **Structure and secular change of the geomagnetic field.** J. C. Cain.
Rev. Geophys. Space Phys., Vol. 13, No. 3, (see 013.001), p. 203 - 206, 239 (1975). – Review paper with bibliographical notes.

084.233 **Core motions and the geodynamo.**
F. H. Busse.
Rev. Geophys. Space Phys., Vol. 13, No. 3, (see 013.001), p. 206 - 208, 240 - 241 (1975). – Review paper with bibliographical notes.

084.234 **Advances in magnetospheric physics: energetic particles.** H. I. West, Jr.
Rev. Geophys. Space Phys., Vol. 13, No. 3, (see 013.001), p. 943 - 951, 1002 - 1012 (1975). – Review paper with bibliographical notes.

084.235 **Magnetospheric physics: magnetic fields.**
C. T. Russell.
Rev. Geophys. Space Phys., Vol. 13, No. 3, (see 013.001), p. 952 - 955, 1012 - 1021 (1975). – Review paper with bibliographical notes.

084.236 **Solar wind–magnetosphere coupling.**
N. U. Crooker.
Rev. Geophys. Space Phys., Vol. 13, No. 3, (see 013.001), p. 955 - 958, 1021 - 1025 (1975). – Review paper with bibliographical notes.

084.237 **Research in magnetospheric wave phenomena.**
J. N. Barfield.
Rev. Geophys. Space Phys., Vol. 13, No. 3, (see 013.001), p. 959 - 967, 1026 - 1032 (1975). – Review paper with bibliographical notes.

084.238 **Magnetospheric and auroral plasmas: a short survey of progress.** L. A. Frank.
Rev. Geophys. Space Phys., Vol. 13, No. 3, (see 013.001), p. 974 - 989, 1036 - 1041 (1975). – Review paper with bibliographical notes.

084.239 **Geomagnetic storms and substorms.**
G. L. Siscoe.
Rev. Geophys. Space Phys., Vol. 13, No. 3, (see 013.001), p. 990 - 993, 1042 - 1048 (1975). – Review paper with bibliographical notes.

084.240 **Reflection of a rarefaction wave from the earth's magnetosphere.** S. A. Grib.
Dokl. Akad. Nauk SSSR. Ser. Mat., Fiz., Vol. 223, 1106 - 1109 (1975). In Russian.

084.241 **Instability of the geomagnetic tail and development of a substorm.** A. P. Kropotkin.
Geomagn. Aeronom., Vol. 15, 687 - 693 (1975). In Russian.

084.242 **An analytical model of the geomagnetic field and its secular variations for the interval of 1950 - 1970 with extrapolation for 1975.**
A. N. Pushkov, T. A. Chernova, M. P. Ivchenko.

Geomagn. Aeronom., Vol. 15, 698 - 704 (1975). In Russian.

084.243 **Space-time structure of the geomagnetic field and of the secular variation for 1500 - 1970.**
A. N. Pushkov, T. A. Chernova.
Geomagn. Aeronom., Vol. 15, 705 - 708 (1975). In Russian.

084.244 **Choice of the normal field of secular variation.**
V. G. Kuznetsova, A. I. Bilinskij, I. M. Rudenskaya.
Geomagn. Aeronom., Vol. 15, 757 - 760 (1975). In Russian. Brief information.

084.245 **The latitudinal morphology of 10-eV to 10-keV electron fluxes during magnetically quiet and disturbed times in the 2100–0300 MLT sector.**
J. D. Winningham, F. Yasuhara, S.-I. Akasofu, W. J. Heikkila.
Journ. Geophys. Res., Vol. 80, 3148 - 3171 (1975).

084.246 **Field-aligned currents and erosion of the dayside magnetosphere.**
Yu. P. Maltsev, W. B. Lyatsky (*V. B. Lyatskij*).
Planet. Space Sci., Vol. 23, 1257 - 1260 (1975).

084.247 **Development of four magnetic storms in February 1972.** L. J. Cahill, Jr., Y. C. Lee.
Planet. Space Sci., Vol. 23, 1279 - 1292 (1975).
Four magnetic storms were observed in February 1972, with instruments on the Explorer 45 satellite in the evening quadrant of the inner magnetosphere. The magnitude of the storms ranged from small, $D_{st} \simeq -40\,\gamma$, to moderate, $D_{st} \simeq -80\,\gamma$.

084.248 **Field-aligned and ionospheric currents.**
F. Yasuhara, Y. Kamide, S.-I. Akasofu.
Planet. Space Sci., Vol. 23, 1355 - 1368 (1975).
Zmuda and Armstrong (1974) showed that the field-aligned currents consist of two pairs; one is located in the morning sector and the other in the evening sector. The authors' analysis of magnetic records from the TRIAD satellite suggests that in each pair the poleward field-aligned current is more intense than the equatorward current, a typical ratio being 2 : 1. This difference has a fundamental importance in understanding the coupling between the magnetosphere and the ionosphere.

084.249 **Magnetic poles of the earth.**
G. A. Fonarev, V. S. Shneer.
Zemlya i Vselennaya, 1975, No. 4, p. 35 - 40. In Russian.

084.250 **Approximations for the study of drift boundaries in the magnetosphere.**
M. G. Kivelson, D. J. Southwood.
Journ. Geophys. Res., Vol. 80, 3528 - 3534 (1975).

084.251 **Magnetotail boundary motion associated with geomagnetic substorms.** K. Maezawa.
Journ. Geophys. Res., Vol. 80, 3543 - 3548 (1975).

084.252 **Protons as the prime contributors to storm time ring current.**
F. W. Berko, L. J. Cahill, Jr., T. A. Fritz.
Journ. Geophys. Res., Vol. 80, 3549 - 3552 (1975).

084.253 **The expansive phase of magnetospheric substorms. 2. The response at synchronous altitude of particles of different energy ranges.**
G. Rostoker, J. L. Kisabeth, R. D. Sharp, E. G. Shelley.
Journ. Geophys. Res., Vol. 80, 3557 - 3570 (1975).

084.254 **Dawn-dusk magnetosheath plasma asymmetries at 60 R_E.** M. A. Fenner, J. W. Freeman, Jr.
Journ. Geophys. Res., Vol. 80, 3693 - 3697 (1975).
A study of data from the Apollo lunar surface suprathermal ion detector experiment package shows that plasma flow and energy parameters in the dusk magnetosheath are much better correlated with geomagnetic activity than those in the dawn magnetosheath. This result is in agreement with a dawn-dusk asymmetry in the magnetosheath magnetic field and in the bow shock configuration. The different orientations between the mean interplanetary magnetic field direction and the shock normal for the magnetosheaths suggest an explanation of the difference in the plasma parameters on the two sides.

084.255 **The dominant mode of standing Alfvén waves at the synchronous orbit.**
W. D. Cummings, C. Countee, D. Lyons, W. Wiley III.
Journ. Geophys. Res., Vol. 80, 3705 - 3708 (1975). – Brief report.

084.256 **On the generation of plasma oscillations and the nature of magnetic fluctuations.**
A. K. Yukhimuk, V. I. Bondarenko.
Problems of cosmic physics. Vyp. (No.) 10, (see 003.003), p. 144 - 154 (1975). In Russian.
Magnetic fluctuations observed by OGO-A near the magnetosphere are analysed. For this purpose instabilities due to the current across the magnetic field as well as instabilities caused by the anisotropic plasma particle velocity distribution are investigated.

084.257 **Time-sequence of polar magnetic substorms.**
S. Kokubun, T. Iijima.
Planet. Space Sci., Vol. 23, 1483 - 1494 (1975).
The time-sequence of polar magnetic substorms is discussed to clarify some controversies on the magnetospheric substorm model including the growth phase. The main purpose of the analyses is to examine magnetic variations in the polar cap and in low latitudes.

084.258 **Three-dimensional polarization characteristics of magnetic variations in the Pc 5 frequency range at conjugate areas near $L = 4$.**
H. Fukunishi, L. J. Lanzerotti, C. G. Maclennan.
Journ. Geophys. Res., Vol. 80, 3973 - 3984 (1975).

084.259 **Destabilization and wave-induced evolution of the magnetospheric plasma clouds.** T. S. T. Young.
Journ. Geophys. Res., Vol. 80, 3995 - 4003 (1975).

084.260 **Solar flare effects in equatorial counter electrojet currents.**
R. G. Rastogi, M. R. Deshpande, N. S. Sastri.
Nature, Vol. 258, 218 - 219 (1975). – Letter.

084.261 **Modulation of geomagnetic lunar daily variation in *H* at Alibag with lunar distance.**
B. R. Arora, D. R. K. Rao.
Geophys. Journ. Roy. Astron. Soc., Vol. 43, 627 - 633 (1975).

084.262 **Controlled artificial generation of geomagnetic pulsations.**
B. J. Lipa, A. C. Fraser-Smith, O. G. Villard, Jr., J. R. Storey.
Nature, Vol. 258, 311 - 313 (1975). – Letter.

084.263 **L'étude internationale de la magnétosphère (IMS – 1976–1978).** E. Aerts.
Ciel et Terre, Vol. 91, 321 - 326 (1975).

084.264 **Substorms and disturbances in the magnetosphere.**
AN SSSR. Kol'sk. fil. Polyarn. geofiz. in-t. Nauka, Leningrad. 328 pp. Price 2 Rbl. 22 Kop. (1975). In Russian.

084.265 **Magnetic disturbances in the polar caps and parameters of the solar wind.** O. A. Troshichev.
Subburi i vosmushcheniya v magnitosfere. Leningrad, Nauka, 1975, p. 66 - 83. In Russian. – Abstr. in Referativ. Zhurn. 62. Issled. kosmich. prostranstva, 8.62.209 (1975).

084.266 **On the theory of the magnetospheric tail.** N. A. Tsyganenko.
Subburi i vosmushcheniya v magnitosfere. Leningrad, Nauka, 1975, p. 83 - 91. In Russian. – Abstr. in Referativ. Zhurn. 62. Issled. kosmich. prostranstva, 8.62.210 (1975).

084.267 **Determination of the effective conductivity of a turbulent magnetospheric plasma.** V. M. Barsukov.
Subburi i vosmushcheniya v magnitosfere. Leningrad, Nauka, 1975, p. 92 - 97. In Russian. – Abstr. in Referativ. Zhurn. 62. Issled. kosmich. prostranstva, 8.62.211 (1975).

084.268 **Regions of influx and shock waves of energy particles in the polar cap.**
A. A. Ostapenko, V. S. Smirnov.
Subburi i vosmushcheniya v magnitosfere. Leningrad, Nauka, 1975, p. 123 - 127. In Russian. – Abstr. in Referativ. Zhurn. 62. Issled. kosmich. prostranstva, 8.62.213 (1975).

084.269 **Dynamics of captured electrons in the energy region of 20 keV – 1.5 MeV during the storm from 21 - 26 November 1971 according to data of satellite Cosmos 426.**
S. N. Kuznetsov, L. L. Lazutin, V. E. Tsirs.
Subburi i vosmushcheniya v magnitosfere. Leningrad, Nauka, 1975, p. 127 - 137. In Russian. – Abstr. in Referativ. Zhurn. 62. Issled. kosmich prostranstva, 8.62.214 (1975).

084.270 **On the dynamics of electrons with energies above 5 keV during the development of a magnetospheric substorm.** M. V. Mal'kov, V. A. Sergeev.
Subburi i vosmushcheniya v magnitosfere. Leningrad, Nauka, 1975, p. 146 - 150. In Russian. – Abstr. in Referativ. Zhurn. 62. Issled. kosmich. prostranstva, 8.62.216 (1975).

084.271 **On the connection between phenomena in the polar ionosphere and particle streams in the tail of the magnetosphere.** L. S. Evlashin, L. M. Evlashina, T. A. Ryzhova, T. A. Kornilova.
Subburi i vosmushcheniya v magnitosfere. Leningrad, Nauka, 1975, p. 151 - 156. In Russian. – Abstr. in Referativ. Zhurn. 62. Issled. kosmich. prostranstva, 8.62.217 (1975).

084.272 **Plasma convection in the evening sector of the magnetosphere and the nature of the plasmapause.**
Yu. I. Gal'perin, V. N. Ponomarev, Yu. N. Ponomarev, A. G. Zosimova.
Kosmich. Issled., Vol. 13, 669 - 686 (1975). In Russian.

084.273 **On effects of external turbulence in an inhomogeneous magnetospheric plasma.**
V. A. Gudkova, V. A. Liperovskij, V. P. Shalimov.
Kosmich. Issled., Vol. 13, 687 - 694 (1975). In Russian.

084.274 **Intensity bursts of low-energy charged particles connected with interplanetary shock waves.**
G. M. Blokh, G. N. Zastenker, B. M. Kuzhevskij, O. B. Likin, N. F. Pisarenko, I. A. Savenko, V. A. Styazhkin.
Kosmich. Issled., Vol. 13, 695 - 703 (1975). In Russian.

084.275 **Observation of electron bursts with energies above 30 keV near the boundary of the magnetosphere aboard the Prognoz satellites.** V. N. Lutsenko, N. S. Nikolaeva, N. F. Pisarenko, I. P. Shestopalov.
Kosmich. Issled., Vol. 13, 710 - 715 (1975). In Russian.

084.276 **What we have learned from the magnetosphere.** C. F. Kennel.
Comments Astrophys. Space Phys., Vol. 6, 71 - 80 (1975). Paper based upon a lecture delivered to the Symposium of the Division of Cosmic Physics, American Physical Society, Berkeley, December 1973.

084.277 **Does the earth's field reverse?** P. J. Smith.
Nature, Vol. 258, 481 - 482 (1975).

084.278 **Bow shock protons in the lunar environment.** J. Bensoñ, J. W. Freeman, H. K. Hills, R. R. Vondrak.
The Moon, Vol. 14, 19 - 25 (1975). – Conference report (see 012.031).

Protons from the earth's bow shock are observed by the Suprathermal Ion Detector Experiment (SIDE) in two regions of the lunar orbit. The dominant contribution to the ion spectra observed by the SIDE in these regions is from particles with energies between ~ 750 eV q^{-1} and 3500 eV q^{-1}. 3500 eV q^{-1} is the upper limit of the energy range of the detector. Analysis of simultaneous data from the Explorer 35 magnetometer and the SIDE indicates that the observability of bow shock protons at the lunar distance is dependent on the configuration of the interplanetary magnetic field.

084.279 **A lunar signature in the geomagnetic Ap-index.** K. Knott.
The Moon, Vol. 14, 49 - 57 (1975). – Conference report (see 012.031).

A study has been performed in an effort to detect a lunar signature in geomagnetic data. Ap-indices available for the period 1932–1972 have been used for this purpose. The data have been divided into some 500 lunar months which were superimposed in synchronism with the phase of the moon. At or shortly after full moon a peak appears which exceeds the average value by about 3 standard deviations. Possible explanations for the generation of geomagnetic disturbances by the moon are given.

084.280 **Magnetospheric protons and electrons encountered by the moon in the plasma sheet.** A. Prakash.
The Moon, Vol. 14, 71 - 78 (1975). – Conference report (see 012.031).

During its passage through the geomagnetic tail, the moon encounters the plasma sheet. Properties of plasma sheet electrons and protons, first detected at lunar distances by Explorer 35, are described. Implications for migration of grains on the lunar surface are also pointed out and it is suggested that strong terrestrial polar winds in the early history of the earth-moon system may have caused some erosion of the earth-facing side of the moon, and that gravitational shielding of interplanetary rock flux by the earth may also be an explanation of the relative smoothness of the front side.

084.281 **Geomagnetic secular variation over the past 5 m.y.** M. W. McElhinny, R. T. Merrill.
Rev. Geophys. Space Phys., Vol. 13, 687 - 708 (1975).

084.282 **A continuous spectrum analyser for analogue analysis of ULF tape recorded data.**
P. W. McNabb, B. J. Fraser.
Australian Journ. Phys., Vol. 28, 647 - 650 (1975). – Short communication.

084.283 **A model of the geomagnetic field at epoch 1975.** D. R. Barraclough, J. M. Harwood, B. R. Leaton, S. R. C. Malin.
Geophys. Journ. Roy. Astron. Soc., Vol. 43, 645 - 659 (1975).

The main magnetic field at epoch 1975 is defined by 168 spherical harmonic coefficients (those representing the internal part to degree and order 12). A further 80 coefficients (degree and order 8) define the secular variation and, in addition, 26 secular acceleration coefficients (those most significantly determined) are included. Details are given of the data used and the methods of reduction.

084.284 **Can precession power the geomagnetic dynamo?** M. G. Rochester, J. A. Jacobs, D. E. Smylie, K. F. Chong.
Geophys. Journ. Roy. Astron. Soc., Vol. 43, 661 - 678 (1975).

084.285 **Electric fields in the high-latitude ionosphere for a model of an open geomagnetosphere.**
M. I. Matveev, V. M. Mishin.
Issled. po geomagnetizmu, aehron. i fiz. Solntsa. Vyp. (No.) 36. Moskva, Nauka, 1975, p. 26 - 33. In Russian. – Abstr. in Referativ. Zhurn. 62. Issled. kosmich.prostranstva, 10.62.183 (1975).

084.286 **Magnetospheric convection and electric currents at high and mean latitudes.** V. M. Mishin, A. D. Bazarzhapov, Eh. I. Nemtsova, A. A. Anistratenko.
Issled. po geomagnetizmu, aehron. i fiz. Solntsa. Vyp. (No.) 36. Moskva, Nauka, 1975, p. 18 - 25. In Russian. – Abstr. in Referativ. Zhurn. 62. Issled. kosmich. prostranstva, 10.62. 184 (1975).

084.287 **Magnetohydrodynamical model of the magnetosphere.** G. F. Krymskij, Yu. A. Romashchenko.
Issled. po geomagnetizmu, aehron. i fiz. Solntsa. Vyp. (No.) 36. Moskva, Nauka, 1975, p. 174 - 199. In Russian. – Abstr. in Referativ. Zhurn. 62. Issled. kosmich. prostranstva, 10.62. 185 (1975).

084.288 **On the origin of magnetospheric electrons of moderate energy.** E. A. Ponomarev.
Issled. po geomagnetizmu, aehron. i fiz. Solntsa. Vyp. (No.) 36. Moskva, Nauka, 1975, p. 96 - 103. In Russian. – Abstr. in Referativ. Zhurn. 62. Issled. kosmich. prostranstva, 10.62. 186 (1975).

084.289 **Artificial and real geomagnetic effects of the sectorial structure of the interplanetary magnetic field.**
V. M. Mishin, V. V. Shelomentsev.
Issled. po geomagnetizmu, aehron. i fiz. Solntsa. Vyp. (No.) 36. Moskva, Nauka, 1975, p. 74 - 83. In Russian. – Abstr. in Referativ. Zhurn. 62. Issled. kosmich. prostranstva, 10.62. 190 (1975).

084.290 **On the influence of the interplanetary magnetic field on the modulation of the planetary geomagnetic activity.** V. V. Shelomentsev.
Issled. po geomagnetizmu, aehron. i fiz. Solntsa. Vyp. (No.) 36. Moskva, Nauka, 1975, p. 70 - 73. In Russian. – Abstr. in Referativ. Zhurn. 62. Issled. kosmich. prostranstva, 10.62. 191 (1975).

084.291 **Secular variations of the geomagnetic field.** A. N. Pushkov.
Zemlya i Vselennaya, 1975, No. 6, p. 18 - 24. In Russian.

084.292 **Geomagnetic disturbances caused by shock waves of large meteoric bodies. I.** Yu. N. Savchenko.
Geomagn. Aeronom., Vol. 15, 1047 - 1053 (1975). In Russian.

084.293 **Pc 2-4 micropulsations of interplanetary origin?** N. D'Angelo.
Geomagn. Aeronom., Vol. 15, 1062 - 1065 (1975). In Russian.

084.294 **Peculiarities of fields of geomagnetic disturbances and of the geopotential of the stratosphere of the northern hemisphere in 1967–1969.** V. V. Mikhnevich, V. G. Kidiyarova, B. N. Trubnikov, I. A. Shcherba.
Geomagn. Aeronom., Vol. 15, 1066 - 1069 (1975). In Russian.

084.295 **The earth's magnetosphere.** S.-I. Akasofu, L. J. Lanzerotti.
Phys. Today, Vol. 28, No. 12, p. 28 - 35 (1975).

084.296 **Radio and X-ray emission as predictor of geomagnetic disturbances.** V. P. Kuleshova, E. V. Lavrova.
Solnechnye Dannye 1975 Byull., No. 8, p. 97 - 103 (1975). In Russian.

The variations of the radio (2800 MHz) and X-ray emission (1–8 Å) are shown to depend on the rate of geomagnetic activity. A distribution of the geoeffective X-ray bursts (1–8 Å) in dependence on their intensity has been derived. The lag of magnetic disturbances as compared to the X-ray bursts is considered.

084.297 **Using the moon to probe the geomagnetic tail lobe plasma.** G. Schubert, C. P. Sonett, B. F. Smith, K. Schwartz, D. S. Colburn.
Geophys. Res. Letters, Vol. 2, 277 - 280 (1975).

The authors have detected the presence of plasma in the lobes of the geomagnetic tail from observations of magnetic induction in the moon forced by time variations of the earth's magnetotail lobe field. The magnitude of the moon's tangential electromagnetic transfer function when the moon is in the lobes of the geomagnetic tail is less than that when the moon is in the solar wind or geomagnetic tail plasma sheet. The waves in the magnetotail lobes which drive the lunar magnetic induction have speeds far less than the speed of light and wavelengths comparable to the size of the moon.

084.298 **Characteristics of the magnetopause energetic electron layer.** C.-I. Meng, K. A. Anderson.
Journ. Geophys. Res., Vol. 80, 4237 - 4243 (1975).

084.299 **Drift-free magnetic geometries in adiabatic motion.** D. P. Stern, P. Palmadesso.
Journ. Geophys. Res., Vol. 80, 4244 - 4248 (1975).

084.300 **VLF line radiation in the earth's magnetosphere and its association with power system radiation.**
R. A. Helliwell, J. P. Katsufrakis, T. F. Bell, R. Raghuram.
Journ. Geophys. Res., Vol. 80, 4249 - 4258 (1975).

084.301 **Electrostatic noise bands associated with the electron gyrofrequency and plasma frequency in the outer magnetosphere.** R. R. Shaw, D. A. Gurnett.
Journ. Geophys. Res., Vol. 80, 4259 - 4271 (1975).

084.302 **A subauroral and mid-latitude view of substorm activity.** D. L. Carpenter, J. C. Foster, T. J. Rosenberg, L. J. Lanzerotti.
Journ. Geophys. Res., Vol. 80, 4279 - 4286 (1975).

084.303 **Explorer 45 (S^3-A) observations of the magnetosphere and magnetopause during the August 4–6, 1972, magnetic storm period.** R. A. Hoffman, L. J. Cahill, Jr., R. R. Anderson, N. C. Maynard, P. H. Smith, T. A. Fritz, D. J. Williams, A. Konradi, D. A. Gurnett.
Journ. Geophys. Res., Vol. 80, 4287 - 4296 (1975).

The Explorer 45 (S^3-A) satellite performed extensive field and particle measurements in the heart of the magnetosphere during the double magnetic storm period of August 4–6, 1972. Both the ground level magnetic records and the magnetic field deformations measured along the orbit by the satellite indicated the existence of only a moderate ring current. Following the period of magnetospheric boundary crossings, Explorer 45 encountered a series of large-amplitude magnetospheric oscillations, seen both in the dc and low-frequency field measurements and in the intensities of the particle fluxes.

084.304 **A determination of *F* region effective recombination coefficients from the Echo 2 sounding rocket plasma wave and particle measurements.**
R. L. Arnoldy, R. A. Hendrickson, J. R. Winckler.
Journ. Geophys. Res., Vol. 80, 4307 - 4312 (1975).

084.305 **Subsonic magnetosheath observations from Explorer 33.** N. U. Crooker, G. L. Siscoe.
Journ. Geophys. Res., Vol. 80, 4368 - 4371 (1975).
Brief report.

084.306 **Independence of geomagnetic polarity intervals.**
J. D. Phillips, R. J. Blakely, A. Cox.
Geophys. Journ. Roy. Astron. Soc., Vol. 43, 747 - 754 (1975).
The statistical independence of sequences of geomagnetic polarity intervals has been investigated using correlation analysis to test the hypothesis that reversals are generated by a renewal process. The present analysis does not support an earlier interpretation that reversals are generated by a Markov process. Instead they are found to be consistent with generation by a renewal process.

084.307 **A model relating undetected geomagnetic polarity intervals to the observed rate of reversals.**
J.-D. Tacier, P. Switzer, A. Cox.
Journ. Geophys. Res., Vol. 80, 4446 - 4448 (1975).
A simple theoretical model is presented to describe the statistical consequences of failing to detect short polarity intervals. A relationship is analytically derived between the true rate of reversals, the observed rate, and a parameter of the model which can be interpreted physically. This is a first step toward unambiguous determination of the true rate of reversals.

084.308 **The polar electrojet (Review).**
V. M. Mishin, A. D. Bazarzhapov, M. I. Matveev, T. I. Sajfudinova, V. V. Shelomentsev.
Issled. po geomagnetizmu, aehron. i fiz. Solntsa. Vyp. (No.) 36. Moskva, Nauka, 1975, p. 46 - 55. In Russian. – Abstr. in Referativ. Zhurn. 62. Issled. kosmich. prostranstva, 12.62. 399 (1975).

084.309 **Some data on the structure of the circumterrestrial shock wave from measurements of α-particles aboard the Prognoz satellite.** A. A. Galeev, O. L. Vajsberg, G. N. Zastenker, A. A. Zertsalov, M. Z. Khokhlov.
Fiz. plazmy, Vol. 1, 441 - 450 (1975). In Russian. – Abstr. in Referativ. Zhurn. 62. Issled. kosmich prostranstva, 12.62.403 (1975).

084.310 **Relationships of the characteristics of magnetohydrodynamic waves to plasma density gradients in the vicinity of the plasmapause.**
L. J. Lanzerotti, H. Fukunishi.
Journ. Geophys. Res., Vol. 80, 4627 - 4634 (1975).

084.311 **Requirements on singly ionized lithium concentrations for magnetospheric seeding experiments.**
S. Cuperman, L. Gomberoff, A. Sternlieb.
Journ. Geophys. Res., Vol. 80, 4643 - 4647 (1975).

084.312 **Complete plasma dropouts at Vela satellites during thinning of the plasma sheet.** A. T. Y. Lui, E. W. Hones, Jr., D. Venkatesan, S.-I. Akasofu, S. J. Bame.
Journ. Geophys. Res., Vol. 80, 4649 - 4659 (1975).
Five satellite years of Vela data are examined for plasma sheet thinnings. Complete proton disappearances (plasma dropouts) are the main subject here. During such times, the Vela satellite is temporarily in the high-latitude tail lobe. The distribution of such plasma dropouts within the magnetotail suggests that the semithickness of the plasma sheet near midnight seldom reaches less than 1 R_E during substorms and that the dawn and dusk portions of the plasma sheet remain thicker than the midnight portion. But it is also shown that the plasma sheet occasionally becomes very thin near the dusk magnetopause. Extension of the magnetopause to greater distances does not seem to be related to magnetically quiet periods nor to the Z component of the interplanetary magnetic field.

084.313 **A comparison of electric and magnetic field data from the Ogo 6 spacecraft.** R. A. Langel.
Journ. Geophys. Res., Vol. 80, 4661 - 4673 (1975).
Previous studies of the Ogo 6 electric field data and of magnetic field magnitude observations have indicated a distinct dependence of disturbance characteristics on interplanetary sector polarity. Examination of simultaneous data below 600 km over the summer polar cap shows that changes in electric field patterns and in the disturbance patterns in magnetic field magnitude ΔB are highly correlated.

084.314 **The paleomagnetosphere.**
G. L. Siscoe, C.-K. Chen.
Journ. Geophys. Res., Vol. 80, 4675 - 4680 (1975).
Paleomagnetic research indicates that the strength of the earth dipole varies over the range 0.1 to 3.0 of the present value. Consequences for magnetospheric physics of changes in the strength of the magnetic dipole are explored in this paper.

084.315 **Spectral analysis of the geomagnetic activity index *aa* over a 103-year interval.**
H. Delouis, P. N. Mayaud.
Journ. Geophys. Res., Vol. 80, 4681 - 4688 (1975).

084.316 **Universal time seasonal variations of auroral zone magnetic activity and VHF scintillations.**
S. Basu.
Journ. Geophys. Res., Vol. 80, 4725 - 4728 (1975).

084.317 **Whistler observations during a magnetospheric sudden impulse.** C. G. Park.
Journ. Geophys. Res., Vol. 80, 4738 - 4740 (1975). – Brief report.

084.318 **Transient occurrence of magnetic loops in the magnetotail.** S. B. Bowling.
Journ. Geophys. Res., Vol. 80, 4741 - 4745 (1975).
A detailed analysis of Explorer 34 magnetometer data indicates that during periods of enhanced geomagnetic activity ($Kp \geqslant 3$) localized regions of the magnetotail are disrupted by magnetic turbulence whose signature is consistent with the presence of closed looplike structures. These turbulent features, which exhibit rapid fluctuations in both the magnitude and the direction of the field, occur at irregular intervals and last no longer than a few hours. They have not been observed for $Kp < 3$ and do not appear to dominate the magnetotail field geometry even during periods of enhanced geomagnetic activity.

084.319 **On the sensitivity of modern inferences of interplanetary magnetic polarity to geomagnetic activity.**
C. T. Russell, B. K. Fleming, P. F. Fougere.
Journ. Geophys. Res., Vol. 80, 4747 - 4750 (1975). – Brief report.

084.320 **Anomalous dissipation of magnetohydrodynamical waves in the magnetosphere at developed ion-sound turbulence.** V. A. Liperovskij, M. Khakimova.
Izv. vyssh. ucheb. zavedenij. Radiofizika, Vol. 18, 833 - 839

(1975). In Russian. – Abstr. in Referativ. Zhurn. 62. Issled. kosmich. prostranstva, 1.62.369 (1976).

084.321 **Propagation of directed waves under cosmic conditions.** I. M. Eremin, M. S. Kovner, E. I. Shkelev. XI Vses. konf. po rasprostr. radiovoln. Chast'(Part) 1. Tezisy dokl. Kazan', Kazan. un-t, 1975, p. 189 - 191. In Russian. – Abstr. in Referativ. Zhurn. 62. Issled. kosmich. prostranstva, 2.62.277 (1976).

084.322 **The magnetospheric contribution to the quiet-time low energy nucleon spectrum in the vicinity of earth.** S. M. Krimigis, J. W. Kohl, T. P. Armstrong. Geophys. Res. Letters, Vol. 2, 457 - 460 (1975).

084.323 **Recurrence tendency in geomagnetic variations and in sunspot frequency.** T. Damboldt. Veröff. Astron. Inst. Univ. Frankfurt (Main), No. 51, 8 + 110 pp. (1974). In German.

An autocorrelation analysis of the Zürich sunspot number and the geomagnetic character figure C9 was made comprising data of 87 years (1884 to 1970). It was found that the 11-year period can be recognized in the recurrence tendency of the geomagnetic figures, whereas this period cannot be recognized in the recurrence tendency of the sunspot numbers.

084.324 **Die irdische Magnetopause unter Berücksichtigung einer Winkelverteilung der solaren Teilchen und des interplanetaren Magnetfeldes.** M. Leubner. Thesis, Univ. Innsbruck, 84 pp. (1975).

An introduction to the ionosphere and magnetosphere. See Abstr. 003.102.

On the energy release processes at the neutral sheet. See Abstr. 062.046.

The upstream escape of energized solar wind protons from the bow shock. See Abstr. 074.014.

Shock waves of the solar wind and the magnetosphere of the earth. See Abstr. 074.084.

An empirical relationship between interplanetary conditions and *Dst*. See Abstr. 074.116.

Unsteady interactions of solar wind disturbances with the magnetosphere. See Abstr. 074.117.

Solar activity and geomagnetic storms 1974. See Abstr. 075.003.

Solar protons on closed magnetospheric field lines after an interplanetary flux decrease. See Abstr. 078.010.

Inertial waves and the earth's outer core. See Abstr. 081.005.

Variations in the atmospheric neutral density at 145 km. See Abstr. 082.037.

Air density at a height of 128 km, from visual observations of 1972-25G. See Abstr. 082.038.

Further study of changes in thermospheric molecular oxygen abundance inferred from twilight 6300 Å airglow. See Abstr. 082.039.

Geomagnetic activity and air pressure at high latitudes. See Abstr. 082.112.

Ionization of the F-layer of the ionosphere and variations of the magnetic field in the near-polar region. See Abstr. 083.060.

Electromagnetic induction in spherical cap current layers under lunar and terrestrial conditions. See Abstr. 094.203.

Magnetospheres of earth and Jupiter after Pioneer 10. See Abstr. 099.007.

Is Jupiter's magnetosphere like a pulsar's or earth's? See Abstr. 099.009.

Relationship between the interplanetary magnetic field and 'isolated substorms'. See Abstr. 106.026.

Influence of the interplanetary magnetic field on the magnetospheric convection and electric currents in the ionosphere. See Abstr. 106.034.

Interplanetary plasma instabilities and geomagnetic pulsations. See Abstr. 106.035.

Energetic lunar nighttime ion events. See Abstr. 106.042.

Interplanetary magnetic fields and geomagnetic variations in the nearpolar region. See Abstr. 106.046.

***Kp* dependence on sectors.** See Abstr. 106.062.

Coordinate transformations for studies of interactions between interplanetary and geomagnetic fields. See Abstr. 106.076.

Splash albedo protons between 4 and 315 MeV at high and low geomagnetic latitudes. See Abstr. 143.021.

Radiation Belts

084.401 **Effects of convection electric field on the distribution of ring current type protons.**
J. M. Grebowsky, A. J. Chen.
Planet. Space Sci., Vol. 23, 1045 - 1052 (1975).

084.402 **Finite temperature effects on the pitch angle diffusion of charged particles in the magnetosphere.**
G. T. Davidson.
Journ. Geophys. Res., Vol. 80, 3172 - 3180 (1975).

084.403 **The intensity of excess radiation from data of Cosmos 137 and Cosmos 219 satellites.**
E. V. Gorchakov, M. V. Ternovskaya, V. I. Severinov.
Kosmich. Issled., Vol. 13, 605 - 606 (1975). In Russian.
Brief information.

084.404 **A coordinated two-satellite study of energetic electron precipitation events.** W. L. Imhof, G. H. Nakano, E. E. Gaines, J. B. Reagan.
Journ. Geophys. Res., Vol. 80, 3622 - 3628 (1975).

084.405 **Satellite observations of X rays associated with energetic electron precipitation near the trapping boundary.** W. L. Imhof, G. H. Nakano, J. B. Reagan.
Journ. Geophys. Res., Vol. 80, 3629 - 3634 (1975).

084.406 **Analysis of processes leading to localized electron enhancements in the outer radiation belt.**
G. Morfill, M. Scholer, D. Hovestadt.
Planet. Space Sci., Vol. 23, 1495 - 1506 (1975).
New information obtained about > 500 keV electron intensity enhancements, which have been observed intermittently close to the outer edge of the electron radiation belt, is used in conjunction with an earlier statistical study by Brown and Stone (1972) to investigate processes which could lead to such structures. The processes investigated quantitatively are distortions of the magnetospheric topology in the equatorial region, wave-particle interactions and the effects of ionospheric currents. It is shown that the various processes which contribute to equatorial field disturbances can explain the observations.

084.407 **The storm and poststorm evolution of energetic (35–560 keV) radiation belt electron distributions.**
L. R. Lyons, D. J. Williams.
Journ. Geophys. Res., Vol. 80, 3985 - 3994 (1975).
Detailed Explorer 45 equatorial observations of the storm and poststorm structure of radiation belt electrons (35–560 keV) for $1.7 \lesssim L \lesssim 5.2$ are presented. Injection during major storms results in electron pitch angle distributions and radial profiles that are greatly distorted from their quiet time equilibrium structure. Following storms the pitch angle distributions return to their prestorm shape over a period of several days. This shape is maintained as electron fluxes decay back to the quiet time levels. The equatorial radial profiles slowly return to their prestorm equilibrium structure over a period of a few weeks and then maintain that structure.

084.408 **Energy dissipation of the ring current and the state of the ionosphere.**
S. A. Zajtseva, N. E. Alekseeva.
Subburi i vosmushcheniya v magnitosfere. Leningrad, Nauka, 1975, p. 156 - 170. In Russian. – Abstr. in Referativ. Zhurn. 62. Issled. kosmich. prostranstva, 8.62.218 (1975).

084.409 **Precipitation of particles at low and middle latitudes.**
G. A. Paulikas.
Rev. Geophys. Space Phys., Vol. 13, 709 - 734 (1975).

084.410 **Dynamics of the outer belt of energetic electrons during moderate magnetic disturbances.**
P. V. Vakulov, L. M. Kovrygina, Yu. V. Mineev, L. V. Tverskaya.
Geomagn. Aeronom., Vol. 15, 1028 - 1032 (1975). In Russian.

084.411 **The ring current and particle precipitation near the plasmapause.** B. Hultqvist.
Ann. Géophys., Vol. 31, 111 - 126 (1975).
Recent observations of ring current particles are reviewed. Strong pitch angle diffusion of ring current protons takes place outside the plasmapause. The magnetosphere is always in a turbulent state in a broad region there.

084.412 **Double structure in the outer electron radiation belt during geomagnetic activity.**
E. Kirsch, J. W. Münch, M. Scholer.
Planet. Space Sci., Vol. 23, 913 - 921 (1975).

Adiabatic plasma convection in a dipole field: variation of plasma bulk parameters with *L*.
See Abstr. 062.034.

A simplified *D*-region model and its application to magnetic storm after-effects. See Abstr. 083.015.

The earth as a radio source: the nonthermal continuum. See Abstr. 084.209.

Dénouement of Jovian radiation belt theory.
See Abstr. 099.008.

High fluxes of low energy (10–25 MeV/amu) C, N, O and heavier ions at 450 km altitude; results from the Skylab experiment. See Abstr. 106.001.

Composition and energy spectra of heavy nuclei of unknown origin detected on Skylab. See Abstr. 143.051.

Errata

084.901 **Correction: 'Self-consistent calculation of the motion of a sheet of ions in the magnetosphere'**
[Journ. Geophys. Res., Vol. 78, 2852 - 2866 (1973)].
R. K. Jaggi, R. A. Wolf.
Journ. Geophys. Res., Vol. 80, 4019 (1975).

085 Solar-Terrestrial Relations

085.001 **Circulation changes in the free atmosphere during proton events associated with type IV radio bursts.**
C. S. Zerefos.
Planet. Space Sci., Vol. 23, 1035 - 1043 (1975).
New evidence of statistically significant circulation changes in the lower stratosphere and in the middle and upper troposphere following type IV proton and ground level events confirms some earlier findings of a relationship between solar activity and weather. A comparison between the effects that followed the flare sample used in this study with earlier results reported for less intense solar particle emission flare samples, implies that if these particles are responsible for triggering the atmospheric circulation in times of enhanced solar activity, their atmospheric effect is enhanced as one moves to the more energetic solar proton events.

085.002 **A hundred years of controversy over sunspots and weather.** A. J. Meadows.
Nature, Vol. 256, 95 - 97 (1975).
After a century of accumulating correlations on the activity of sunspots and atmospheric phenomena around the earth, there exists still no clear idea what exactly the relationships are or what they mean. Much of what is being researched now was already under discussion in the nineteenth century. This article is based on the author's recent Norman Lockyer Lecture at the University of Exeter.

085.003 **Spectral characteristics of solar radio bursts associated with shortwave fadeouts.**
R. R. Babcock, J. T. Parker.
Journ. Atmosph. Terr. Phys., Vol. 37, 1227 - 1230 (1975).
Spectral characteristics of centimeter wavelength solar radio bursts as measured at the Sagamore Hill Radio Observatory were correlated with shortwave fadeouts (SWF). It was found that there is a definite relation between the intensity and shape of the spectrum of a solar radio burst and the occurrence of a SWF. These results imply "threshold" values for the cm-λ component of the radio spectrum if the accompanying X-ray emissions are to be intense enough to cause a SWF.

085.004 **A new look at solar-terrestrial relations.**
J. A. Eddy.
Bull. American Astron. Soc., Vol. 7, 410 (1975). – Abstr. AAS.

085.005 **Solar activity, geomagnetic field and terrestrial weather.** J. W. Knight, P. A. Sturrock.
Bull. American Astron. Soc., Vol. 7, 474 (1975). – Abstr. AAS.

085.006 **Comparison of solar and auroral activity at Glasgow, 1959 to 1973.** R. J. Livesey.
Journ. British Astron. Ass., Vol. 85, 424 - 425 (1975).

085.007 **Short term relationships between solar flares, geomagnetic storms, and tropospheric vorticity patterns.**
R. H. Olson, W. O. Roberts, C. S. Zerefos.
Nature, Vol. 257, 113 - 115 (1975). – Letter.

085.008 **The geomagnetic activity during the decrease of solar cycles.** A. D. Chertkov.
Subburi i vosmushcheniya v magnitosfere. Leningrad, Nauka, 1975, p. 283 - 299. In Russian. – Abstr. in Referativ. Zhurn. 51. Astron., 9.51.465 (1975).

085.009 **Observations of heliogeophysical events in July 1974.** S. I. Avdyushin, R. A. Zevakina, M. N. Nazarova, N. K. Pereyaslova, I. E. Petrenko, N. N. Stepanyan.
VI Leningr. mezhdunar. seminar, 1974. Uskorenie chastits i yader. reaktsii v kosmose. Leningrad, 1974, p. 211 - 220. In Russian. – Abstr. in Referativ. Zhurn. 51. Astron., 8.51. 493 (1975).

085.010 **Reality and nature of a sun–weather correlation.**
C. O. Hines, I. Halevy.
Nature, Vol. 258, 313 - 314 (1975). – Letter.

085.011 **On a relation between secular variations of the mean latitudinal atmospheric pressure and the secular solar cycle.** A. A. Lykov, V. V. Lykova.
Solnechnye Dannye 1975 Byull., No. 3, p. 101 - 105 (1975). In Russian.
Secular variations of Wolf numbers and evolution of anomalies of the mean latitudinal atmospheric pressure are analysed. It has been found that during the epoch of a high level of solar activity the pressure increases in zones with climatic depressions and decreases in zones with high pressures.

085.012 **Results of forecasting of the mean monthly indices of weather during the last 10 years and their dependence on solar activity.** M. Kh. Bajdal.
Trudy Kazakhsk. n.-i. gidrometeorol. in-t, 1975, vyp. (No.) 46, p. 207 - 210. In Russian. – Abstr. in Referativ. Zhurn. 51. Astron., 10.51.464 (1975).

085.013 **Volcanic dust, sunspots, and temperature trends.**
S. H. Schneider, C. Mass.
Science, Vol. 180, 741 - 746 (1975).

085.014 **Effect of the solar activity on the natural oscillations of a torsion pendulum.**
E. M. Kolesnikova, S. M. Kolesnikov.
Probl. teorii gravitatsii i ehlementarn. chastits. Vyp. (No.) 6. Moskva, Atomizdat, 1975, p. 160 - 171. In Russian. – Abstr. in Referativ. Zhurn. 51. Astron., 11.51.820 (1975).

085.015 **Preliminary results of an investigation of the night astroclimate at the Coronal Station of the Astrophysical Institute.**
B. Kalzhanov, Eh. G. Mychelkin, B. T. Nuraliev.
Trudy Astrofiz. Inst., *Alma-Ata,* Vol. 26, 160 - 164 (1974). In Russian.

085.016 **Variability of direct solar radiation with cumulus.**
U.-A. R. Mullamaa, V. K. Pöldmaa.
Oblachnost' i radiatsiya. Tartu, 1975, p. 97 - 109. In Russian. Abstr. in Referativ. Zhurn. 51. Astron., 12.51.380 (1975).

085.017 **The solar situation preceding the generation of the geomagnetic storm of June 17 - 20, 1973.**
B. Bednářová-Nováková.
Stud. Geophys. Geod., Vol. 19, 191 - 197 (1975).

085.018 **Stratospheric thickness variations over the northern hemisphere and their possible relation to solar activity.** C. S. Zerefos, P. J. Crutzen.
Journ. Geophys. Res., Vol. 80, 5041 - 5043 (1975).

085.019 **Electromagnetic processes in the circumterrestrial and interplanetary space during solar flares in August, 1972.** V. I. Bondarenko, A. K. Yukhimuk.
Geofiz. sb. AN USSR, 1975, vyp. (No.) 65, p. 55 - 61. In Russian. – Abstr. in Referativ. Zhurn. 51. Astron., 1.51.516 (1976).

085.020 **Vibration and quality of star images in the Sudak-Alushta region (Crimea).**
G. I. Bol'shakova, A. N. Demidova.
Izv. Glav. Astron. Obs. Pulkovo, No. 193, Astrofiz., Astrometr. i Zvezdn. Astron., p. 178 - 179 (1975). In Russian.

The results of observations of astroclimate made with the AZT-7 telescope in the Sudak-Alushta region in 1962 are given. The data on quality, size and vibration of star images in 4 zones of zenith distance are shown on a graph and in a table.

085.021 **On the influence of the solar (corpuscular) activity upon the variations of the surface atmospheric pressure of the northern hemisphere in summer.**
V. E. Chertoprud.
Astron. Tsirk., No. 883, p. 4 - 6 (1975). In Russian.

085.022 **Estudio sobre una probable relación entre la actividad solar y las variaciones climáticas: el Proyecto Helioklimat.** – Nota preliminar de trabajo.
J. A. Gutierrez, R. B. Esteve, C. M. Parfait, J. A. Serra.
Obs. Astron. Municipal Rosario, (*Argentina*), Bol. No. 4, p. 16 - 17 (1974).

085.023 **Possible causes of glaciations.**
E. P. J. van den Heuvel, P. Buurman.
Marine Geology and Oceanography of the Arctic Seas, p. 359 - 378 = Astrofys. Inst., Vrije Univ. Brussel, Overdruk No. 86 (1974).

085.024 **Ionospheric-magnetospheric relations and the problem of disturbances forecast.** N. P. Ben'kova.
XI Vses. konf. po rasprostr. radiovoln. Chast' (Part) 1. Tezisy dokl. Kazan', Kazan. un-t, 1975, p. 56 - 58. In Russian.
Abstr. in Referativ. Zhurn. 51. Astron., 2.51.543 (1976).

085.025 **Influence of solar activity on geophysical processes.**
P. Velinov.
Blg. geofiz. spisanie, Vol. 1, No. 1, p. 51 - 77 (1975).
In Bulgarian. – Abstr. in Referativ. Zhurn. 51. Astron., 2.51. 544 (1976).

Do solar variations affect earth's weather?
Phys. Today, Vol. 28, No. 9, p. 19 - 20 (1975).

Les variations à longues périodes des éléments de l'orbite terrestre: une question de précision dans la théorie astronomique des paléoclimats. See Abstr. 042.043.

Solar flares in August 1972 followed by atmospheric phenomena of the earth. See Abstr. 073.094.

Une éruption chromosphérique remarquable et quelques effets géophysiques associés. See Abstr. 073.151.

Radio and X-ray emission as predictor of geomagnetic disturbances. See Abstr. 084.296.

Planetary System

091 Physics of the Planetary System (Planetary Atmospheres, Figure, Interior, Magnetic Fields, Rotation, etc.)

091.001 **Formation of coupled spectral lines in a planetary atmosphere: II. Numerical evaluation of approximate solutions.** J. W. Chamberlain, G. R. Smith. Astrophys. Journ., Vol. 199, 282 - 288 (1975).

Approximate solutions to the radiation-transfer problem of two or more resonance lines that share a common upper level have previously been obtained by Chamberlain and Wallace. Essentially exact solutions from a computer program are here compared with those approximate solutions to evaluate the utility of the approximations in solving planetary radiation problems. Isotropic scattering and a semi-infinite atmosphere are adopted throughout. When the atmosphere is homogeneous with depth, the approximation is very good. When it is inhomogeneous, with scattering albedo $\varpi(t)$ gradually decreasing downward, the approximation is good only for single lines.

091.002 **High zonal harmonics of rapidly rotating planets.** W. B. Hubbard, W. L. Slattery, C. L. DeVito. Astrophys. Journ., Vol. 199, 504 - 516 (1975).

A new perturbation expansion is derived for the structure of rotating bodies in hydrostatic equilibrium. The method uses an expansion of the density on Legendre polynomial functions of angle, and can be developed analytically in a manner analogous to the standard level-surface perturbation theory. Sample calculations for polytropic sequences are presented, as well as standard gravity models for Jupiter and Saturn. Mathematical subleties of the theory are discussed in an appendix.

091.003 **Bode's law–truth or consequences?** M. W. Ovenden. Vistas in Astronomy, Vol. 18, (see 012.006), 473 - 496 (1975).

091.004 **Effects of turbulence on radio-occultation scale heights.** W. B. Hubbard, J. R. Jokipii. Astrophys. Journ., (*Letters*), Vol. 199, L193 - L196 (1975).

The authors demonstrate that turbulent scattering can produce a spurious value for the radio-occultation–derived scale height of a planetary atmosphere. Application to the Jovian atmosphere shows that recent measurements obtained during the flyby of Pioneers 10 and 11 can be substantially affected if suitable large amplitude turbulence is present. Other implications of this possibility are briefly discussed.

091.005 **Outer solar-system aeronomy.** D. F. Strobel. Astrophys. Space Sci. Library, Vol. 51, (see 012.007), 401 - 408 (1975).

091.006 **The internal structures of the planets.** A. H. Cook. Geophys. Journ. Roy. Astron. Soc., Vol. 42, 285 - 293 (1975). – Paper presented at the Tenth International Symposium on Mathematical Geophysics, Cambridge, England, 1974 June 26 - July 4.

Some problems of the internal structures of the planets are discussed. The structures must be inferred from the sparse dynamical data (mass, density, J_2) which are all that are available, and must be culled from analogies with the earth for the moon and the terrestrial planets or from theoretical calculations of behaviour of simple materials at high pressures for the outer planets. The importance of the value of the dimensionless moment of inertia, C/Ma^2, is emphasized and the validity of applying Bullen's compressibility-pressure hypothesis to other planets than the earth is considered.

091.007 **A new approach of the adding method for the computations of emergent radiation of an inhomogeneous plane-parallel planetary atmosphere.** T. Takashima. Astrophys. Space Sci., Vol. 36, 319 - 328 (1975).

The procedure of computing the intensity and the polarization parameters of radiation diffusely reflected and transmitted by an inhomogeneous, plane-parallel planetary atmosphere is discussed with the aid of the adding method. If the atmosphere is simulated by a number of homogeneous sublayers, the matrices of radiation diffusely reflected and transmitted by the atmosphere can be expressed in terms of these matrices of sublayers by using only a couple of iterative equations with the polarity effect of radiation.

091.008 **Planetary geodesy.** J. D. Anderson. Rev. Geophys. Space Phys., Vol. 13, No. 3, (see 013.001), p. 274 - 275, 292 - 293 (1975). – Review paper with bibliographical notes.

091.009 **Planetary atmospheres.** P. J. Gierasch. Rev. Geophys. Space Phys., Vol. 13, No. 3, (see 013.001), p. 790 - 793, 862 - 871 (1975). – Review paper with bibliographical notes.

091.010 **Planetary magnetospheres.** T. W. Hill, F. C. Michel. Rev. Geophys. Space Phys., Vol. 13, No. 3, (see 013.001), p. 967 - 974, 1033 - 1036 (1975). – Review paper with bibliographical notes.

091.011 **Resonant fluorescent scattering of solar radiation by the fourth positive band system of CO.** T. Kassal. Applied Optics, Vol. 14, 1513 - 1515 (1975).

The resonant fluorescent scattering equations for a diatomic molecule are developed as a function of temperature, column density, and optical thickness of absorbing rotation lines. The equations are solved numerically for absorption of solar vacuum uv radiation by the carbon monoxide fourth positive band system. Saturation column densities are determined and computer spectra are presented. It is shown that at high column densities, fluorescent scattering of the solar Lyman α line by the (14,0) band constitutes a significant portion of the total absorption.

091.012 **The solar system.** C. Sagan. Sci. American, Vol. 233, No. 3, p. 23 - 31 (1975). Popular article.

091.013 **The reciprocal period rule.** W. E. Greig. Bull. American Astron. Soc., Vol. 7, 448 (1975). Abstr. AAS.

091.014 **Remarks on the distribution of satellites and planets.** W. E. Greig.
Bull. American Astron. Soc., Vol. 7, 499 (1975). – Abstr. AAS.

091.015 **The gravitational field of a feebly-nonuniformly rotating polytrope of index one.** A. V. Kozenko.
Astron. Zhurn. Akad. Nauk SSSR, Vol. 52, 887 - 890 (1975). In Russian. English translation in Soviet Astron., Vol. 19, No. 4.

The expression for the gravitational moments J_2, J_4 and J_6 on the basis of an exact solution of the structure equation for a feebly-nonuniformly rotating polytrope of index $n = 1$ are obtained. The values of gravitational moments for feebly-nonuniformly rotating models of Jupiter are calculated.

091.016 **Laboratory band strengths of methane: planetary abundances revised.**
B. L. Lutz, R. D. Cess, T. Owen.
Publ. Astron. Soc. Pacific, Vol. 87, 504 (1975). – Abstr. ASP.

091.017 **Organic chemistry and biology of planets. Perspectives of experimental investigations.**
L. M. Mukhin.
Kosmich. Issled., Vol. 13, 494 - 502 (1975). In Russian.

091.018 **Motion of large bodies in the atmospheres of planets.** G. I. Petrov, V. P. Stulov.
Kosmich. Issled., Vol. 13, 587 - 594 (1975). In Russian.

091.019 **Thermal equilibrium and hydrostatic equilibrium for excited states in planetary atmospheres.**
G. W. Adams.
Planet. Space Sci., Vol. 23, 1293 - 1300 (1975).

Thermal equilibrium and hydrostatic equilibrium are mutually exclusive for any particular quantum state of an atmospheric constituent in a non-isothermal atmosphere. As a result, there is a flux of rotationally, vibrationally, and electronically excited atoms and molecules down the temperature gradient, balanced by an up-gradient transport of ground-state atoms and molecules, resulting in a net transport of excitation energy, but with no net mass transport.

091.020 **Diffuse reflection and transmission of radiation by planetary atmospheres with a diffuse-and-specular reflector.** S. Mukai.
Publ. Astron. Soc. Japan, Vol. 27, 483 - 495 (1975).

In this paper it is shown how to get the initial value solution of the transfer equation in generalized Chandrasekhar's (1950) planetary atmospheres. Allowing for the effect of the polarity, two systems of integro-differential equations are derived for the diffuse reflection and transmission functions of radiation by an inhomogeneous, anisotropically scattering atmosphere bounded by a reflecting surface at the bottom. As a sample computation, numerical results for the case of a homogeneous isotropic atmosphere bounded by a diffuse-and-specular reflector are obtained, and are compared with those for the case of Lambert's law and specular reflectors, respectively.

091.021 **Model of the gravitational field of a planet to investigate the systematic errors of the potential coefficients.** V. G. Shkodrov.
Observations of artificial earth satellites, No. 14, (see 012.018), p. 81 - 92 (1975). In Russian.

091.022 **Errors of the coefficients of the gravitational potential of a planet caused by non-sphericity of its surface.** V. G. Shkodrov.
Observations of artificial earth satellites, No. 14, (see 012.018), p. 93 - 114 (1975). In Russian.

091.023 **The seven ages of a planet.** W. M. Kaula. Icarus, Vol. 26, 1 - 15 (1975).

091.024 **The motion of the sun about the general center of mass.** V. S. Prokudina.
Soobshch. Gos. Astron. Inst. Shternberga, No. 181, p. 3 - 10 (1973). In Russian.

The question of the motion of center of mass of the solar system is investigated. The curves for the years 1800–1980 are analysed. It is considered that in the 22-year cycle the center of the masses of the solar system describes the curve type of Pascal's limaçon with the parameters $a = R_\odot$, $l = 0.5\ R_\odot$.

091.025 **Comparison of geometrical effects in radio and stellar occultations.** W. B. Hubbard.
Icarus, Vol. 26, 175 - 177 (1975).

An important difference between the radio and stellar occultation techniques for probing planetary atmospheres is that a crucial equation relating refraction angle with ray impact parameter is solved in opposite directions in the two techniques. The solution is tolerant of small geometrical errors for stellar occultations but, unfortunately, not for radio occultations.

091.026 **The development of radar investigations of planets in the Soviet Union.**
V. A. Kotel'nikov, O. N. Rzhiga.
Issled. v obl. radiotekhn. i ehlektron. 1954–1974. Ch. (Part) 1. Moskva, 1974, p. 5 - 43. In Russian. – Abstr. in Referativ. Zhurn. 51. Astron., 8.51.33 (1975).

091.027 **Method for solving the problem of radiative transfer in linear molecular bands in case of LTE disturbances in the upper layers of planetary atmospheres.**
A. A. Kutepov.
Redkollegiya zhurn. 'Vestn. Leningr. un-ta. Ser. fiz., khimiya. Leningrad, 1975. 12 pp. In Russian. – Abstr. in Referativ. Zhurn. 51. Astron., 10.51.234 (1975).

091.028 **Synchronization of satellites.** A. A. Khentov. Dinamika sistem. Vyp. (No.) 4. Gor'kij, 1974, p. 51 - 102. In Russian. – Abstr. in Referativ. Zhurn. 51. Astron., 10.51.271 (1975).

091.029 **Observations of Mercury, Venus, Mars, Jupiter, Saturn and Uranus at 3.92 mm wavelength.**
I. I. Zinchenko, A. G. Kislyakov, V. N. Voronov.
Izv. vyssh. ucheb. zavedenij. Radiofizika, Vol. 18, 501 - 505 (1975). In Russian. – Abstr. in Referativ. Zhurn. 51. Astron., 11.51.247 (1975).

091.030 **Craters on the bodies of the solar system.** P. Příhoda.
Kozmos, Vol. 6, 110 - 116 (1975). Continued, see Abstr. 13.091.044. In Czech.

091.031 **Far infrared photometry of Uranus, Saturn, Jupiter, and Mars.** R. F. Loewenstein, D. A. Harper, H. A. Thronson, C. M. Telesco, S. H. Moseley.
Bull. American Astron. Soc., Vol. 7, 529 (1975). – Abstr. AAS.

091.032 **Giant planets and their satellites, minor planets, meteorites and comets.** R. L. Newburn, S. Gulkis.
Osnovy kosmich. biol. i med. Tom (Vol.) 1. Moskva, Nauka, 1975, p. 198 - 268. In Russian.

091.033 **Terrestrial planets (Mercury, Venus, Mars).** V. D. Davydov, M. Ya. Marov.
Osnovy kosmich. biol. i med. Tom (Vol.) 1. Moskva, Nauka, 1975, p. 139 - 197. In Russian.

091.034 **Inner planets of the solar system: a comparative study.** Z. Kopal.
Ann. New York Acad. Sci., Vol. 187, 108 - 113 = Astron. Contr. Univ. Manchester, Ser. III, No. 258 (1972).

091.035 **Errors of the numerical methods for estimate of integrals determining the coefficients of a planetary gravity field.** Yu. V. Batrakov, V. G. Shkodrov.
Byull. Inst. Teoret. Astron., *Leningrad,* Vol. 14, 145 - 152 (1975). In Russian.

Previous results are explained at least qualitatively by errors in the numerical estimate of integrals determining the coefficients of the planetary gravity field, based on measurements of the gravity force on its surface. Analytic expressions have been derived for the errors in the numerical method, depending on the indices of the harmonics and characterizing the order of magnitude of these errors with respect to a small parameter (the flattening of the planet).

091.036 **The production of atmospheres by impact.** A. Benlow, A. J. Meadows.
Meteoritics, Vol. 10, 360 - 361 (1975).

091.037 **Scaling law for impacts on different planets or satellites.** E. J. H. David.
Meteoritics, Vol. 10, 389 - 390 (1975).

091.038 **Chemical evolution of planets.** B. Kuchowicz. Urania Kraków, Vol. 46, 271 - 275 (1975). In Polish.

091.039 **Gegenseitige Planetenbedeckungen.** J. Meeus. Sternenbote, 18. Jahrgang, p. 222 - 230 (1975).

091.040 **Study of diffraction by radio sounding of planetary atmospheres and radio holography of the surface of Venus.** D. S. Lukin, Yu. G. Spiridonov, V. A. Shkol'nikov, A. S. Mart'yanov, S. I. Fominykh.
XI Vses. konf. po rasprostr. radiovoln. Chast' (Part) 1. Tezisy dokl. Kazan', Kazan. un-t, 1975, p. 151 - 152. In Russian. Abstr. in Referativ. Zhurn. 51. Astron., 2.51.320; 62. Issled. kosmich. prostranstva, 2.62.187 (1976).

091.041 **Fotografisanje površina planeta.** A. Tomić. Vasiona, Vol. 23, 54 - 55 (1975).

091.042 **Natural satellites of the planets.** A. Brahic. Recherche (*France*), Vol. 6, 53 - 56 (1975). In French.

Some recent research results, particularly with reference to Saturn's rings, Titan and Io, are briefly discussed.

091.043 **Radio interferometer studies of solar system objects.** F. H. Briggs.
Thesis, Cornell Univ., Ithaca, N.Y., USA [Available from Univ. Microfilms, Ann Arbor, Mich., USA. Order No. 74-24266], 140 pp. (1974).

The use of a radio interferometer, both in the observation of very weak radio sources and for the derivation of radio brightness distributions, is described. Observations of solar system objects are presented which take advantage of the interferometer's special capabilities. In particular, several observing techniques are discussed which distinguish a moving source from a background of stationary sources. Observations of Mars, Mercury, and Uranus at 21 cm wavelength give disk-average brightness temperatures. High-resolution information about Mars implies a surface dielectric constant and a latitude dependent temperature distribution. Observations of Saturn show that the rings are effective scatterers of microwave radiation but have a very low brightness temperature. Radio detection of thermal emission from the asteroid Ceres is reported. Saturn's moon Titan was detected at 3.7 cm wavelength and must have a surface temperature of less than 200 K.

091.044 **Polar symmetric flow of a viscous compressible atmosphere: an application to Mars.** J. A. Pirraglia.
Journ. Atmosph. Sci., Vol. 32, 60 - 72 (1975).

The atmosphere is assumed to be driven by a polar symmetric temperature field and the equations of motion in pressure ratio coordinates are linearized by considering the zero order in terms of a thermal Rossby number. Using a temperature model for Mars based on Mariner 9 infrared spectral data with a 30 percent increase in the depth-averaged temperature from the winter pole to the subsolar point, results were obtained for the increase in surface pressure from the subsolar point to the winter pole as a function of eddy viscosity and with no-slip conditions imposed at the surface.

091.045 **The hydrostatic equilibrium and its implications on the mechanical properties of planets.**
P. Baldi, M. Caputo.
Ann. Geofis., (*Italy*), Vol. 27, 235 - 244 (1974).

Under the hypothesis of hydrostatic equilibrium the planets of the solar system are represented on the plane (J_2, $w^2 a^3/mG$) (J_2 = second term of the gravitational potential, w = rotation rate of the planet, a = equatorial radius, m = mass of the planet, G = gravitational constant). Since the points representing the planets lie on a straight line it is possible to obtain informations on the gyration moments and the moments of inertia of Uranus and Pluto.

091.046 **Factor group analysis for some molecules of interest in planetary atmospheres.** J. R. Ferraro.
Applied Spectrosc., (*USA*), Vol. 29, 354 - 355 (1975).

091.047 **Sounding of planetary atmospheres: a Fourier analysis of the radiative transfer equation.**
D. Gautier, I. Revah.
Journ. Atmosph. Sci., Vol. 32, 881 - 892 (1975).

In order to determine the information content in the thermal profile of a planetary atmosphere observed by means of remote sensing techniques, Fourier analysis concepts and Shannon's (1949) sampling procedures are applied to the infrared data obtained.

091.048 **Application of the concept of cycles to geological studies in planetology.** J. G. Weihaupt.
Journ. Interdiscipl. Cycle Res., Vol. 6, 25 - 35 (1975).

Surfaces and interiors of planets and satellites.
See Abstr. 003.041.

Hydrogen broadened half-widths of ammonia.
See Abstr. 022.007.

Laboratory intensity measurements of absorption bands in the long pathlength spectrum of methane.
See Abstr. 022.104.

Positions of the major planets and the moon.
See Abstr. 041.013.

Reduction of T. Hornby's observations of planets between the years 1774 and 1798. See Abstr. 041.064.

On the geometry of orbits of distant stellar satellites. See Abstr. 042.004.

Resonance phenomena in rotational motions of artificial and natural celestial bodies. See Abstr. 052.007.

Spacecraft potential control and plasma diagnostic using electron field emission probes. See Abstr. 053.010.

Channel flow driven by a stationary thermal source. See Abstr. 061.084.

Thermodynamics of dense hydrogen-helium fluids. See Abstr. 062.003.

A three-parameter analytic phase function for multiple scattering calculations. See Abstr. 063.006.

X and Y operators for general linear transport processes. See Abstr. 063.052.

Relativistic libration of planets. See Abstr. 066.050.

Application of the method of planetary configurations to active phenomena on the sun. See Abstr. 072.030.

Sheath-limited unipolar induction in the solar wind. See Abstr. 074.019.

Atmospheric physics. See Abstr. 082.013.

Vertical transport in atmospheres. See Abstr. 082.014.

Atmospheres of the earth and planets. See Abstr. 082.094.

The non-homogeneous accumulation model for terrestrial planet formation and the consequences for the atmosphere of Venus. See Abstr. 093.055.

Lunar and planetary seismology. See Abstr. 094.126.

On volatilization in vacuum of the elements from the planet surface soil melts. See Abstr. 094.148.

A review of lunar sample studies and their application to studies of the terrestrial planets. See Abstr. 094.411.

The structure and evolution of Jupiter: the fluid contraction stage. See Abstr. 099.001.

Planetary influences and activation of phenomena on Jupiter. See Abstr. 099.081.

Measuring the magnetic fields of Jupiter and the outer solar system. See Abstr. 099.099.

Characteristics of plasma in the interplanetary medium and in the ionospheres of the planets of the solar system. See Abstr. 106.058.

Rocket and spacecraft studies of ultraviolet emissions from astrophysical targets. See Abstr. 114.398.

Tidal flows in binary systems. See Abstr. 117.017.

092 Mercury

092.001 **Dynamo generation in Mercury.**
D. J. Stevenson.
Nature, Vol. 256, 634 (1975).

Observations during the most recent encounter of the planet Mercury by Mariner 10 seem to have confirmed the existence of a small but intrinsic magnetic dipole of about 3×10^{-3} gauss $R_M{}^3$, where R_M is the radius of Mercury. The author examines whether dynamo generation is possible for the present Mercury.

092.002 **Photographic observations of Mercury's transit across the sun disk on November 10, 1973.**
N. Rizvanov, D. Andrienko, A. Ghezloun, M. Svetchnikov (*Svechnikov*), L. Ourassine (*Urasin*).
Astron. Astrophys. Suppl. Ser., Vol. 21, 171 - 175 (1975). In French.

The results of photographic observations of the Mercury transit across the sun are given.

092.003 **The magnetic field of Mercury, I.** N. F. Ness, K. W. Behannon, R. P. Lepping, Y. C. Whang.
Journ. Geophys. Res., Vol. 80, 2708 - 2716 (1975).

An updated analysis and interpretation are presented of the magnetic field observations obtained during the Mariner 10 encounter with the planet Mercury on March 29, 1974. The combination of data relating to position of the detached bow shock wave and magnetopause and the geometry and magnitude of the magnetic field within the magnetospherelike region surrounding Mercury lead to the conclusion that an internal planetary field exists with dipole moment approximately 5.1×10^{22} G cm^3.

092.004 **Our present view of Mercury and Venus.**
G. E. Danielson, Jr.
Rev. Geophys. Space Phys., Vol. 13, No. 3, (see 013.001), p. 393 - 410, 425 (1975). – Review paper with bibliographical notes.

The discipline of planetology received a quantum jump in new information on the planets Mercury and Venus from the Mariner 10 spacecraft. A detailed look at the ultraviolet markings in the upper atmosphere of Venus and their attendant motion has stimulated a considerable reevaluation of theories concerning the circulation, composition, and nature of this planet's upper atmosphere. The geologic history of

Mercury as revealed by Mariner 10's television science data, from two close flybys, has challenged the prevalent theories of planetary evolution. The Mercury surface morphology and optical properties resemble those of the moon and seem to record a similar sequence of events despite their differences in density and proximity to both the sun and the asteroid belt.

092.005 **Surface properties and effective resolution from ground-based observations of Mercury.**
R. F. Beebe, A. Herzog.
Icarus, Vol. 25, 555 - 560 (1975).

A detailed study to evaluate ground-based photographs of Mercury has been carried out. Models of the surface scattering properties have been assumed and smeared with a Gaussian function for direct comparison with center-to-limb scans along Mercury's intensity equator. Data from a range of phase angles from 31° to 92° have been compared with smeared models assuming a Lambert surface, a surface which obeys the Lommel–Seeliger law and one which is Minnaertian, having a variable coefficient. Within the limits of the observations a lunar Minnaert surface yields the most consistent interpretation. An objective evaluation of the resolution of the photographs is obtained in terms of Gaussian half-widths.

092.006 **Mercury.** B. G. Murray.
Sci. American. Vol. 233, No. 3, p. 58 - 68 (1975).
Popular article.

092.007 **An analysis of the transits of Mercury: 1677–1973.**
L. V. Morrison, C. G. Ward.
Monthly Notices Roy. Astron. Soc., Vol. 173, 183 - 206 (1975).

About 2400 observations of the universal times of internal contact for the transits of Mercury in the period 1677 to 1973 are compared with the ephemeris times calculated from the gravitational theories of the motions of Mercury and the earth. The O–C differences in the times are attributed to two causes: the differences between the ephemeris and universal time scales, ΔT (arising from the non-uniformity of the earth's rate of rotation on its axis) and errors in the values of the orbital elements in the adopted theory of Mercury's motion. The authors find the tidal acceleration of the moon to be $-26'' \pm 2''\,cy^{-2}$, and that the excess of the observed over the Newtonian value of the motion of Mercury's perihelion is $+41''.9 \pm 0''.5\,cy^{-1}$.

092.008 **Mercury's helium exosphere.**
R. E. Hartle, S. A. Curtis, G. E. Thomas.
Journ. Geophys. Res., Vol. 80, 3689 - 3692 (1975).

A model He exosphere of Mercury is developed whose source derives from accretion of the small fraction of solar wind He^{++} that is captured by the magnetosphere and absorbed at the planetary surface. The model is based on a lunar analogy whereby the surface is saturated in He. Surface and radial density distributions are derived, and the day side concentration is determined by requiring the limb intensity at 584 Å to equal the observed intensity. To maintain the He exosphere, the authors find that only the fraction 6.2×10^{-4} of the solar wind He^{++} flux intercepted by the magnetosphere need be captured. This corresponds to a total accretion rate of $6.8 \times 10^{22}\ He^{++}\ s^{-1}$ at the surface, which also represents the upper limit to the total outgassing rate of He from the interior.

092.009 **Mercurius heeft tòch een magnetosfeer.**
C. de Jager.
Zenit, Vol. 2, 345 - 346 (1975).

092.010 **A reinterpretation of the reported energetic particle fluxes in the vicinity of Mercury.**
T. P. Armstrong, S. M. Krimigis, L. J. Lanzerotti.
Journ. Geophys. Res., Vol. 80, 4015 - 4017, with a reply by J. A. Simpson, p. 4018 (1975).

During the Mercury flyby of Mariner 10, observations of large fluxes of energetic electrons ($E > 0.3$ MeV) and protons ($0.53 < E < 1.9$ MeV) have been reported by Simpson et al. (1974). The reported simultaneous enhancements of protons and electrons in the magnetic field of Mercury have raised some perplexing planetology questions. It is shown here that the response of the proton detector in the Mariner 10 experiment is most plausibly attributable to the pileup of low-energy electrons rather than the presence of protons in the vicinity of Mercury. It is concluded that no 'new' acceleration mechanism has been identified at Mercury.

092.011 **The 3.5-micron polarization of Mercury.**
R. Landau.
Icarus, Vol. 26, 243 - 249 (1975).

The polarization of the thermal radiation emitted by Mercury at 3.5 microns was measured through a 4 arcsec aperture on three dates representing two different combinations of phase angle and heliocentric longitude. When fit by a thermal model of the subsurface which takes surface roughness into account, the large measured polarization requires a value of the dielectric constant larger than 2.7 at 3.5 microns with best values around 4.0. These data, together with the smaller values of the dielectric constant derived at longer wavelengths, suggest a surface material, like sand, scree, or talus, which is loosely packed on a scale of centimeters or meters, but compact on a scale of microns.

092.012 **On the distribution of craters on the surface of Mercury.** L. V. Leont'ev.
Kosmich. Issled., Vol. 13, 949 (1975). In Russian. – Brief information.

092.013 **On the theory of Mercury's resonance rotation.**
V. V. Beletskij, Eh. K. Lavrovskij.
Astron. Zhurn. Akad. Nauk SSSR, Vol. 52, 1299 - 1308 (1975). In Russian. English translation in Soviet Astron., Vol. 19, No. 6.

The case of Mercury's 3 : 2 resonance was investigated. The motion under consideration has extended Cassini's general laws for the plane case. Nearly-periodic resonance solutions were found and their stability was considered. The conditions of stability have allowed to find a maximum estimate on the ratio of Mercury's inertia momenta.

092.014 **Variations in the solar wind stand-off distance at Mercury.** G. Siscoe, L. Christopher.
Geophys. Res. Letters, Vol. 2, 158 - 160 (1975).

The intrinsic dipole model of the magnetosphere of Mercury is combined with statistics on the solar wind dynamic pressure from Explorers 33 and 35 to obtain histograms of the solar wind stand-off distance for aphelion, perihelion and median distance of Mercury from the sun. The results show that even at perihelion the stand-off distance lies above the surface the great majority of the time. On rare occasions, direct impact of the solar wind onto the surface will occur.

092.015 **Substorms on Mercury?**
G. L. Siscoe, N. F. Ness, C. M. Yeates.
Journ. Geophys. Res., Vol. 80, 4359 - 4363 (1975).

The Mariner 10 encounter of Mercury provided data showing a strong interaction between the solar wind and the planet similar to a scaled down version of that producing the earth's magnetosphere. Some of the features observed in Mercury's night side magnetosphere suggest time-dependent processes occurring there. Interpreted as temporal events, these features bear striking resemblances to substorm phenomena in the earth's magnetosphere.

092.016 **Investigaciones recientes del planeta Mercurio.**
V. Capolongo.
Obs. Astron. Municipal Rosario, (*Argentina*), Bol. No. 4, p. 7 - 9 (1974).

092.017 **Merkur i Venera jednim letom.**
J. Milogradov-Turin.
Vasiona, Vol. 23, 33 - 36 (1975).

092.018 **What's in a picture. [Some examples of image correction and enhancement from Mariner 10 record of Mercury].** H. J. P. Arnold.
British Journ. Photogr., Vol. 122, 538 - 541 (1975).

Mercury since Mariner 10.
See Abstr. 011.002.

Joint meeting of the Royal Astronomical Society, the Royal Meteorological Society and the Geological Society of London on "Venus and Mercury", held on 1975 April 11.
See Abstr. 011.065.

The transit of Mercury across the solar disk 9 May 1970 and fluctuations of radio emission at $\lambda = 8$ mm.
See Abstr. 077.025.

Mariner unveils Venus and Mercury.
See Abstr. 093.039.

Lunar photoelectron layer dynamics.
See Abstr. 094.160.

The motion of Mercury and variations of the longitude of Jupiter's Red Spot. See Abstr. 099.082.

093 Venus

093.001 **The Mariner 10 Venus encounter – a review.**
J. A. Dunne.
Endeavour, No. 123, Vol. 34, 109 - 114 (1975).

The American spacecraft Mariner 10 encountered Venus on 5 February 1974. The geometry of the encounter was dictated by the role of the planet as a gravitational 'third stage' to deflect the trajectory of the spacecraft towards Mercury. A comet-like plasma tail, extending hundreds of Venus radii downstream, was observed. Ultraviolet spectroscopy revealed the presence of helium in the Venus atmosphere, and ten times the Martian amount of atomic oxygen. Radio propagation and television observations of the upper atmosphere revealed stratification, while near-ultraviolet studies showed in detail a symmetrical but complex system of surface cloud patterns.

093.002 **The atmospheres of Mars and Venus–a comparison.**
J. C. McConnell.
Astrophys. Space Sci. Library, Vol. 51, (see 012.007), 361 - 367 (1975).

093.003 **Venus cloud models.**
S. C. Wofsy, N. D. Sze.
Astrophys. Space Sci. Library, Vol. 51, (see 012.007), 369 - 384 (1975).

093.004 **The ionosphere and upper atmosphere of Venus.**
S. Kumar.
Astrophys. Space Sci. Library, Vol. 51, (see 012.007), 385 - 399 (1975).

093.005 **Analysis of the results of measurements of temperature and pressure in the Venus atmosphere.**
M. Ya. Marov, O. L. Ryabov.
In-t prikl. mat. AN SSSR. Preprint No. 111. Moskva, 1974. 46 pp. In Russian. – Abstr. in Referativ. Zhurn. 51. Astron., 7.51.263; 62. Issled. kosmich. prostranstva, 7.62.190 (1975).

093.006 **Data from direct measurements of temperature and pressure in the Venus atmosphere.**
M. Ya. Marov, O. L. Ryabov.
In-t prikl. mat. AN SSSR. Preprint No. 110. Moskva, 1974. 56 pp. In Russian. – Abstr. in Referativ. Zhurn. 51. Astron., 7.51.264; 62. Issled. kosmich. prostranstva, 7.62.189 (1975).

093.007 **Evidence for the nonuniform distribution of microwave attenuating clouds in the atmosphere of Venus: Mariner 2.** D. E. Jones.
Icarus, Vol. 25, 561 - 568 (1975).

Using the data from Veneras 4–8 and Mariners 5 and 10 related to the composition and structure of the atmosphere of Venus, the three scans obtained with the microwave radiometer on Mariner 2 at a wavelength of 1.9 cm have been reanalyzed. Using more recent data, it has not been possible to match the measured scan ratios with or without any spherically symmetric distribution of microwave cloud absorber. The scan ratios, therefore, require the existence of different average values of microwave cloud opacity for each scan. In addition, the anomalous temperature drop observed in the south polar region of the terminator scan has been found to require a very opaque microwave cloud in the local zenith angle range of 40°–70°. It is suggested that some of the variability seen in the earth-based interferometer data may be a result of changes in the distribution of the microwave clouds over the disc of the planet.

093.008 **Venus.** A. Young, L. Young.
Sci. American, Vol. 233, No. 3, p. 70 - 78 (1975).
Popular article.

093.009 **The Mariner 5 ultraviolet photometer experiment: analysis of Rayleigh-scattered and 1304-Å radiation from Venus.** D. E. Anderson, Jr.
Journ. Geophys. Res., Vol. 80, 3063 - 3067 (1975).

Airglow measurements of the disk of Venus, made by the ultraviolet photometer on Mariner 5 on October 19, 1967, are analyzed to determine the sources of the observed emission. Rayleigh-scattering models for a semi-infinite atmosphere are used to determine the scale height and single-scattering albedo of the scatterer and to determine the 1304-Å emission rate. It is found that (1) the scale height of the Rayleigh-scattered radiation is 4.5 ± 0.5 km, (2) the single-scattering albedo near 2000 Å is 0.8 ± 0.1, and (3) 500 R of 1304-Å radiation is detected at solar zenith angles between 90° and 95°.

093.010 **Internally heated convection in the atmosphere of Venus and in the laboratory.** D. J. Tritton.
Nature, Vol. 257, 110 - 112 (1975). – Letter.

093.011 **Questions of contrast on the surface of Venus.** A. S. Panfilov.
Kosmich. Issled., Vol. 13, 612 - 614 (1975). In Russian. Brief information.

093.012 **On the exclusion of the influence of photographic irradiation on the results of photographic positional observations of Venus.** B. S. Vozdvizhenskij.
Astron. Tsirk., No. 859, p. 1 - 2 (1975). In Russian.

093.013 **Struktuur van de atmosfeer van Venus.** C. de Jager.
Zenit, Vol. 2, 354 - 355 (1975).

093.014 **Photometry of Venus. I. Observation of the brightness distribution over the disk.** A. Dollfus, H. Camichel, C. Boyer, M. Auriere, E. Bowell, J. Nikander.
Icarus, Vol. 26, 53 - 72 (1975).

Isophotes of the luminance distribution over the disk of Venus at several phase angles and in several wavelengths are constructed by means of isodensitometry of high-quality photographic negatives of Venus.

093.015 **Photometry of Venus. II. Theoretical brightness distribution over the disk.** C. Bigourd, C. Devaux, M. Herman, J. Lenoble.
Icarus, Vol. 26, 73 - 84 (1975).

Theoretical brightness distributions over the Venus disk have been computed for homogeneous and multilayered cloud models. If the horizontal inhomogeneities currently seen on Venus are due to a layered structure, with an optical thickness of the upper cloud varying from point to point, large simplifications seem possible, and measured contrasts at various wavelengths should permit a test of such a model.

093.016 **Numerical simulation of the general circulation of the Cytherean lower atmosphere.**
D. V. Chalikov, A. S. Monin, A. S. Safray (*Safraj*), V. G. Turikov, S. S. Zilitinkevich.
Icarus, Vol. 26, 178 - 208 (1975).

The principal features which distinguish the atmosphere on Venus from that of the earth are the slow rotation of the planet, the large mass of the atmosphere, and the opacity of the atmosphere to long-wave radiation. These considerations lead to a much simpler scheme of circulation than that in the earth's atmosphere. The large mass of the atmosphere is the cause of a high thermal and mechanical inertia, which explains why the atmospheric circulation is asymmetrical relative to the solar-antisolar axis. The opacity of the Cytherean atmosphere to long-wave radiation results in vertical heat transfer by turbulence, mesoscale convection, and large-scale currents. This produces adiabatic stratification in the troposphere and a high temperature in the lower layers. These phenomena were studied in a general manner using two- and three-level models.

093.017 **Drie wolkentypen in atmosfeer van Venus.** C. de Jager.
Zenit, Vol. 2, 382 - 383 (1975).

093.018 **Streams of solar radiation in the near infrared spectral region in the lower atmosphere of Venus.**
S. D. Gutshabash, A. S. Safraj.
Kosmich. Issled., Vol. 13, 760 - 764 (1975). In Russian.

093.019 **Numerical modelling of thermal radiative transfer in the lower atmosphere of Venus.** V. P. Shari.
In-t prikl. mat. AN SSSR. Preprint No. 34. Moskva, 1975. 48 pp. Price 16 Kop. In Russian. – Abstr. in Referativ. Zhurn. 51. Astron., 11.51.250 (1975).

093.020 **Calculation of spectral, angular and latitude distribution of long-wave radiation in the Venus atmosphere.** N. I. Moskalenko, A. R. Zakirova.
Izv. AN SSSR. Fiz. atmosf. i okeana, Vol. 11, 599 - 609 (1975) In Russian. – Abstr. in Referativ. Zhurn. 51. Astron., 11.51. 251; 62. Issled. kosmich. prostranstva, 11.62.228 (1975).

093.021 **Infrared limb darkening of the Venus atmosphere.** W. I. Newman.
Icarus, Vol. 26, 451 - 456 (1975).

An analysis of the limb darkening component from the thermal infrared maps of Venus shows that the Cytherean cloud tops were close to radiative equilibrium in 1962. A method for obtaining the optical depth, the extinction coefficient, and the extinction scale height from such data is derived and values are extracted from Marov's standard model of the Venus atmosphere.

093.022 **A model of the Venus ionosphere.**
A. F. Nagy, S. C. Liu, T. M. Donahue, S. K. Atreya, P. M. Banks.
Geophys. Res. Letters, Vol. 2, 83 - 86 (1975).

Results of model calculations of the Venus ionosphere covering the altitude range from 120 km to 300 km are presented. The calculations were carried out for 'low' and 'high' atomic oxygen models, corresponding to an $[O]/[CO_2]$ ratio of 0.4 percent and 4 percent respectively at 140 km. The results of the calculations agree well in both shape and magnitude with the Mariner 5 and 10 occultation results in the chemically controlled region.

093.023 **On the possibility of radio viewing of the Venus surface by bistatic radar.** A. G. Pavel'ev, O. I. Yakovlev, S. S. Matyugov, A. I. Kucheryavenkov, V. I. Kaevitser.
Izv. vyssh. ucheb. zavedenij. Radiofizika, Vol. 18, 816 - 823 (1975). In Russian. – Abstr. in Referativ. Zhurn. 51. Astron., 12.51.234; 62. Issled. kosmich. prostranstva, 12.62.164 (1975).

093.024 **Determination of the relief in the equatorial part of the surface of Venus from radar observations in the USSR.** V. K. Golovkov, B. I. Kuznetsov, G. M. Petrov, A. F. Khasyanov.
XI Vses. konf. po rasprostr. radiovoln. Chast' (Part) 1. Tezisy dokl. Kazan', Kazan. un-t, 1975, p. 157 - 159. In Russian. Abstr. in Referativ. Zhurn. 51. Astron., 2.51.321 (1976).

093.025 **He 584 Å airglow emission from Venus: Mariner 10 observations.** S. Kumar, A. L. Broadfoot.
Geophys. Res. Letters, Vol. 2, 357 - 360 (1975).

Observations were made across the bright limb of Venus by the ultraviolet spectrometer on Mariner 10 as the spacecraft flew by this planet on February 5, 1974. The altitude variation of the brightness of 584 Å emission is interpreted with respect to the helium distribution in the atmosphere. Resonance scattering of solar radiation is considered as the source of this emission. The comparison of model calculations with the data shows that the observed altitude profile is consistent with a He density of $2 \pm 1 \times 10^6$ cm^{-3} at 145 km altitude (mixing ratio $1.0 \pm 5 \times 10^{-5}$) and with an exospheric temperature of

375 ± 105K.

093.026 **The dynamics of the atmosphere of Venus.** P. H. Stone.
Journ. Atmosph. Sci., Vol. 32, 1005 - 1016 (1975).

The wide range of radiative time scales in the Venus atmosphere, together with observations of temperature structure and winds, indicate that the atmosphere contains two distinct regimes. In the deep atmosphere, at altitudes below 40 km, diurnal effects are negligible, motions are weak, and the lapse rate is near adiabatic. In the upper atmosphere, at altitudes above 70 km, diurnal effects are important, strong retrograde zonal motions = 100 m/s occur, and the lapse rate is sub-adiabatic.

093.027 **Further numerical calculations of the circulation of the atmosphere of Venus.** E. Kalnay de Rivas.
Journ. Atmosph. Sci., Vol. 32, 1017 - 1024 (1975).

The results of two-dimensional simulations of the deep circulation of Venus are presented. They prove that the high surface temperature can only be explained by the greenhouse effect, and that Goody and Robinson's dynamical model (1966) is not valid. Very long time integrations, up to a time comparable with the radiative relaxation time, confirm these results.

093.028 **Calculations of the radiative and dynamical state of the Venus atmosphere.**
J. B. Pollack, R. Young.
Journ. Atmosph. Sci., Vol. 32, 1025 - 1037 (1975).

Modelling the atmosphere in accord with recent spacecraft and ground-based observations, the authors have carried out accurate, multiple scattering calculations to determine the solar energy deposition profile in the atmosphere of Venus. Preliminary three-dimensional calculations of the general circulation of the atmosphere are discussed that incorporate the results of the radiative calculations.

093.029 **Meridional circulation and maintenance of the Venus atmospheric rotation.** P. J. Gierasch.
Journ. Atmosph. Sci., Vol. 32, 1038 - 1044 (1975).

A meridional cell, with rising motion near the equator and sinking near the poles, transports angular momentum upward in an atmosphere whenever equatorial regions of the atmosphere have an angular momentum surplus relative to polar regions. This process may contribute to the maintenance of the Venus atmospheric super-rotation. Super-rotation by this process is exhibited in a simple analytical model.

093.030 **Spectroscopic observations of winds on Venus.**
W. A. Traub, N. P. Carleton.
Journ. Atmosph. Sci., Vol. 32, 1045 - 1059 (1975).

The authors have measured the differential Doppler shift between various points on Venus using a high-resolution Pepsios interferometer (three Fabry-Perot etalons in series).

093.031 **Venus: evolution of the upper atmospheric clouds.**
A. Dollfus.
Journ. Atmosph. Sci., Vol. 32, 1060 - 1070 (1975).

The large number of UV photographs taken by different observations throughout the world are grouped for analysis at the IAU planetary photograph center of Meudon Observatory.

093.032 **Comparison of simultaneous CO_2 and H_2O observations of Venus.** E. S. Barker.
Journ. Atmosph. Sci., Vol. 32, 1071 - 1075 (1975).

During the 1972–74 period, 115 pairs of CO_2 and H_2O abundance determinations have been made with the coudé scanner of the 2.7 m reflector at McDonald Observatory. A correlation analysis of the pairs of observations grouping them into eight periods of time which corresponded to telescope observing runs or periods of similar phase angle shows a lack of correlation in all except one period.

093.033 **Dust on Venus.** S. L. Hess.
Journ. Atmosph. Sci., Vol. 32, 1076 - 1078 (1975).

A calculation is performed of the friction velocity needed to lift dust from the surface of Venus. It is found that the most easily lifted grains are 16–17 μm in radius, and a friction velocity of about 1.3 cm/s will suffice.

093.034 **Windblown dust on Venus.** C. Sagan.
Journ. Atmosph. Sci., Vol. 32, 1079 - 1083 (1975).

093.035 **Observations of turbulence in the atmosphere of Venus using Mariner 10 radio occultation measurements.** R. Woo.
Journ. Atmosph. Sci., Vol. 32, 1084 - 1090 (1975).

Develops a technique for estimating the outer scale of turbulence in a planetary atmosphere using dual-frequency radio occultation measurements. This technique is applied to the region of strong turbulence located in the vicinity of 60 km on the dayside of the atmosphere of Venus using the Mariner 10 S/X radio occultation measurements. It is found that, contrary to earlier findings from Mariner 5, the outer scale of turbulence is at least 5 km.

093.036 **Venus: vertical structure of stratospheric hazes from Mariner 10 pictures.** B. O'Leary.
Journ. Atmosph. Sci., Vol. 32, 1091 - 1100 (1975).

High-resolution pictures of the limb of Venus taken by the Mariner 10 television camera indicate the presence of tenuous haze layers high in the stratosphere. At least two distinct layers separated by a few kilometers in altitude appear in pictures taken in both orange and ultraviolet light and extend laterally for several thousand kilometers from the equator to high latitudes. Photometric profiles of these hazes have been analyzed to determine their vertical distribution. An 'optical barometer' technique for determining the altitudes of the hazes is presented.

093.037 **Interpretation of Mariner 10 infrared observations of Venus.** F. W. Taylor.
Journ. Atmosph. Sci., Vol. 32, 1101 - 1106 (1975).

The infrared radiometer experiment on Mariner 10 measured limb darkening curves for Venus in two spectral intervals, one near 11 μm and the other near 45 μm. In this paper, these are analyzed in terms of the vertical opacity profile at each wavelength over a limited altitude range, approximately 60–80 km above the surface of the planet.

093.038 **Cloud structure and heating rates in the atmosphere of Venus.** A. A. Lacis.
Journ. Atmosph. Sci., Vol. 32, 1107 - 1124 (1975).

Ground-based observations and Venera 8 entry probe measurements are used to infer the vertical distribution of cloud particles in the atmosphere of Venus.

093.039 **Mariner unveils Venus and Mercury.**
K. F. Weaver.
National Geogr. Mag., (*USA*), Vol. 147, 858 - 869 (1975).

093.040 **The clouds of Venus.** A. T. Young.
Journ. Atmosph. Sci., Vol. 32, 1125 - 1132 (1975).

The physical and chemical properties of the clouds of Venus are reviewed, with special emphasis on data that are related to cloud dynamics. None of the currently popular interpretations of cloud phenomena on Venus is consistent with all the data.

093.041 **Interpretation of the variation of polarization over**

the disk of Venus. K. Kawabata, J. E. Hansen.
Journ. Atmosph. Sci. Vol. 32, 1133 - 1139 (1975).

The polarization of reflected sunlight is computed for model atmospheres of Venus as a function of location on the apparent planetary disc. The calculations are both homogeneous (uniformly mixed) and layered models, as required to investigate the vertical distribution of particles. The results are compared with available observations. The results for the homogeneous and layered models are also integrated over the planet and compared with whole-disc observations.

093.042 **A determination of the composition of the Venus clouds from aircraft observations in the near infrared.** J. B. Pollack, E. F. Erickson, D. Goorvitch, B. J. Baldwin, D. W. Strecker, F. C. Witteborn, G. C. Augason.
Journ. Atmosph. Sci., Vol. 32, 1140 - 1150 (1975).

The evidence that the first optical depth of the Venus cloud layer is composed of a water solution of sulphuric acid is summarised. Aircraft observations obtained in the 1–4 μm region at a phase angle of 40 degrees and in the 3–6 μm region at a phase angle of 136 degrees are presented.

093.043 **Analysis of spectrophotometric observations of Venus in the 3–4 micron region.**
J. V. Martonchik, R. Beer.
Journ. Atmosph. Sci., Vol. 32, 1151 - 1156 (1975).

Spectrophotometric data of Venus in the 3–4 μm region are analyzed in the context of current ideas concerning the cloud properties. The results show that a homogeneous cloud model, using a unimodal particle size distribution compatible with the results of polarization studies, cannot satisfactorily explain the infrared observations.

093.044 **The infrared polarization of Venus.** R. Landau.
Journ. Atmosph. Sci., Vol. 32, 1157 - 1163 (1975).

The polarization of Venus at 2.2, 3.5 and 4.8 μm, measured through a 4 arc sec aperture at several phase angles is presented and compared with the predictions of multiple scattering calculations using the optical constants of sulfuric acid. To determine the fraction of light which is thermally emitted, drift scans were made across Venus at 3.5, 4.8 and 8.4 μm.

093.045 **Microwave boundary conditions on the atmosphere and clouds of Venus.** W. B. Rossow, C. Sagan.
Journ. Atmosph. Sci., Vol. 32, 1164 - 1176 (1975).

The dielectric properties of H_2O and H_2SO_4 at microwave frequencies have been calculated from the Debye equations. The derived frequency and temperature dependence agrees well with existing data. With the COSPAR model atmosphere and the calculated cloud models the microwave properties of the atmosphere and clouds are determined. Upper limits are also set to the surface density of total cloud constituents and of cloud liquid water. The high surface temperature is discussed in the light of the results.

093.046 **Interpretation of the photometric measurements of Venus by Mariner 10.**
C. Devaux, M. Herman, J. Lenoble.
Journ. Atmosph. Sci., Vol. 32, 1177 - 1189 (1975).

093.047 **On the origin of ultraviolet contrasts on Venus.**
L. D. Travis.
Journ. Atmosph. Sci., Vol. 32, 1190 - 1200 (1975).

Models for the origin of the contrasts in the ultraviolet images of Venus are examined in an attempt to determine the physical differences between light and dark regions fundamental to a clear understanding of the apparent cloud motions.

093.048 **The composition of the ultraviolet dark markings on Venus.** G. T. Sill.
Journ. Atmosph. Sci., Vol. 32, 1201 - 1204 (1975).

The ultraviolet dark clouds are an ephemeral phenomenon in the Venus atmosphere. The UV dark material is moderately abundant, perhaps 10 percent, since the contrasts between light and dark material are some 20 percent.

093.049 **Relative spectrophotometry of Venus from 3067–5960 Å.** E. S. Barker, J. H. Woodman, M. A. Perry, B. A. Hapke, R. Nelson.
Journ. Atmosph. Sci., Vol. 32, 1205 - 1211 (1975).

The relative spectral reflectivity for the integrated disk of Venus is presented.

093.050 **Evidence for an elemental sulfur component of the clouds from Venus spectrophotometry.**
B. Hapke, R. Nelson.
Journ. Atmosph. Sci., Vol. 32, 1212 - 1218 (1975).

The decrease in the reflectivity of Venus in the near UV can be explained if the clouds contain particles of elemental sulfur in addition to sulfuric acid.

093.051 **A numerical model for the dynamics and composition of the Venusian thermosphere.**
R. E. Dickinson, E. C. Ridley.
Journ. Atmosph. Sci., Vol. 32, 1219 - 1231 (1975).

The structure, composition and winds of the mesosphere and thermosphere of Venus are investigated using a nonlinear time-dependent hydrodynamic model.

093.052 **The Mariner 10 radio occultation measurements of the ionosphere of Venus.** G. Fjeldbo, B. Seidel, D. Sweetnam, T. Howard.
Journ. Atmosph. Sci., Vol. 32, 1232 - 1236 (1975).

Data from the Mariner 10 radio occultation experiment have been utilized to determine the vertical electron density distribution in the ionosphere of Venus.

093.053 **Venus: chemical and dynamical processes in the stratosphere and mesosphere.** R. G. Prinn.
Journ. Atmosph. Sci., Vol. 32, 1237 - 1247 (1975).

Photochemical models for the Venus clouds are presented and discussed. Models are illustrated for the sulphuric acid density as a function of altitude. Two reactions are investigated which lead to the production of sulphuric and elemental sulphur respectively.

093.054 **Evolution of the atmosphere of Venus.**
J. C. G. Walker.
Journ. Atmosph. Sci., Vol. 32, 1248 - 1256 (1975).

The runaway greenhouse effect provides the most plausible explanation for the divergent evolution of the atmospheres of Venus and earth.

093.055 **The non-homogeneous accumulation model for terrestrial planet formation and the consequences for the atmosphere of Venus.** K. K. Turekian, S. P. Clarke,Jr.
Journ. Atmosph. Sci., Vol. 32, 1257 - 1261 (1975).

The non-homogeneous accumulation model for the formation of the terrestrial planets is described and its consequences for the formation of the Venusian atmosphere are assayed in the context of our knowledge of the composition of the earth and carbonaceous chondrites.

093.056 **Future exploration of Venus.**
L. D. Friedman, J. L. Lewis.
Astronaut. Aeronaut., (*USA*), Vol. 13, No. 5, p. 46 - 58 (1975).

093.057 **On remote sounding of the upper atmosphere of Venus.** J. T. Houghton, F. W. Taylor.
Journ. Atmosph. Sci., Vol. 32, 620 - 629 (1975).

Some of the possibilities for remote sensing of the upper atmosphere of Venus from an orbiting spacecraft are studied quantitatively.

Veneras 9 and 10 on Venus.
Sky Telescope, Vol. 50, 374 - 375 (1975).

Venus reveals secrets. See Abstr. 003.012.

Joint meeting of the Royal Astronomical Society, the Royal Meteorological Society and the Geological Society of London on "Venus and Mercury", held on 1975 April 11. See Abstr. 011.065.

Photographic positional observations of Venus at Pulkovo with the 26″ refractor in 1972. See Abstr. 041.060.

Positional photographic observations of Venus with the double short-focus astrograph at Pulkovo during 1972. See Abstr. 041.061.

Thermal microwave emission from a scattering layer. See Abstr. 063.043.

Motion of large bodies in the atmospheres of planets. See Abstr. 091.018.

Study of diffraction by radio sounding of planetary atmospheres and radio holography of the surface of Venus. See Abstr. 091.040.

Our present view of Mercury and Venus. See Abstr. 092.004.

Merkur i Venera jednim letom. See Abstr. 092.017.

Occultations of major planets by moon at Belgrade Astronomical Observatory during 1974. See Abstr. 096.015.

CO_2 photodissociation and thermal emission connected with it and some emissions in the thermospheres of Mars and Venus. See Abstr. 097.043.

094 Moon: Dynamics, Global Properties, Local Properties

Moon, Dynamics

094.001 **Higher harmonics of the terrestrial and lunar gravitational fields in the theory of the dynamics of the earth's rotation axis.** M. Burša.
Bull. Astron. Inst. Czechoslovakia, Vol. 26, 218 - 234 (1975).
Euler's dynamic equations, describing the motion of the earth's axis of rotation, are derived with the harmonics of the terrestrial and lunar gravitational fields being retained up to the 4th degree. As regards the terrestrial field Stoke's constants $n = 2, k = 0; n = 2, k = 2; n = 3, k = 0, 1, 2, 3; n = 4, k = 0; n = 4, k = 3$; and as regards the lunar field only the constants $n = 2, k = 0; n = 2, k = 2$ will come to bear.

094.002 **Past orientation of the lunar spin axis.** W. R. Ward.
Science, Vol. 189, 377 - 379 (1975).
The orientation of the lunar spin axis is traced from the early history of the earth-moon system to the present day. Tides raised on the earth by the moon have caused an expansion of the lunar orbit. Tides raised on the moon by the earth have de-spun the moon to synchronous rotation and driven its spin axis to a Cassini state – that is, in a coprecessing configuration, coplanar with the lunar orbit normal and the normal to the Laplacian plane. This combination of events has resulted in a complex history for the lunar spin axis. For much of the period during which its orbital semimajor axis expanded between 30 and 40 earth radii, the obliquity of the moon was of order 25° to 50°. In fact, for a brief period the obliquity periodically attained a value as high as 77°; that is, the spin axis of the moon was only 13° from lying in its orbit plane.

094.003 **Laser measurements of the lunar distance.** P. L. Bender.
Rev. Geophys. Space Phys., Vol. 13, No. 3, (see 013.001), p. 271 - 272, 290 - 291 (1975). – Review paper with bibliographical notes.

094.004 **The topocentric distance of a lunar station as a function of the parameters of the earth-moon system.** M. Burša.
Bull. Astron. Inst. Czechoslovakia, Vol. 26, 275 - 281 (1975).
The topocentric distance of a general point on the moon's surface is derived as a function of the parameters of the terrestrial and lunar co-ordinate system. Terms of the order of 10^{-11} of the distance between the mass centres of the earth and moon are retained.

094.005 **A search for forward scattering of sunlight from lunar libration clouds.** R. H. Munro, J. T. Gosling, E. Hildner, R. M. MacQueen, A. I. Poland, C. L. Ross, A. Hopfield.
Planet. Space Sci., Vol. 23, 1313 - 1319 (1975).
An attempt to determine the radiance of forward scattered sunlight from particles in lunar libration regions was made with the white light coronagraph on Skylab. The libration regions could not be distinguished against the solar $K + F$ coronal background; an upper limit to the libration cloud radiance is determined to be $2.5 \times 10^{-11}\ B_\odot$, where $B_\odot$ is the mean radiance of the solar disk.

094.006 **Derivation of the constants of lunar rotation by the method of adjustment of the amplitude a_3.** A. A. Gorynya.
Astrometriya i Astrofizika, *Kiev*, vyp. (No.) 26, (see 003.001), p. 17 - 21 (1975). In Russian.
The values $f = 0.54$ and $I = 1°32'51''$ are derived from measurement data of 26 pairs of position angles of Mösting A to limb craters (about 4500 measurements).

094.007 **On the accompanying free physical libration of the moon.** K. S. Shakirov.
Astron. Tsirk., No. 853, p. 5 - 7 (1975). In Russian.

094.008 **Determination of the parameters of the moon's rotation from observations of the position angles of crater pairs.** A. A. Gorynya.
Astrometriya i Astrofizika, *Kiev*, vyp. (No.) 27, (see 003.004), p. 43 - 51 (1975). In Russian.

The article deals with the results of treatment of 3540 observations of crater pairs on the moon (Mösting A—regional crater) obtained by means of position angle measuring. The treatment was made by a differential method with initial values $f_0 = 0.85$ and $f_0 = 0.60$. Besides, to clear up on what side from the critical value (0.662) the true value f is situated (when $f = 0.73$ and $f = 0.85$), the amplitude a_3 was determined. Quality of observational material and its treatment have been checked.

094.009 **Accretional capture of the moon.**
S. P. Clark, Jr., D. L. Turcotte, J. C. Nordmann.
Nature, Vol. 258, 219 - 220 (1975).

Hypotheses for the formation of the earth—moon system fall into one of three categories. The moon was either (1) formed by fission from the earth, (2) formed by accretion in earth orbit, or (3) formed in solar orbit and then captured. The authors suggest that hypotheses (2) and (3) may be combined, with the result that many of the constraints on lunar origin are satisfied.

094.010 **Lunar daily distance numbers *D* and *D'*.**
J. C. Gupta.
The Moon, Vol. 14, 247 - 253 (1975).

In analyses of the effect of variation of the earth-moon distance on geophysical phenomena, it is customary to arrange the geophysical data according to the dates of apogee and perigee. However, lunar distances at apogee and especially at perigee vary within wide limits from month to month. A new daily index D' of lunar distance is defined to permit a more precise determination of effects related to lunar distance. It is readily calculated by a computer program.

094.011 **The rotation of the moon.** J. D. Mulholland.
Bull. Cl. Sci. Acad. Roy. Belgique, 5e Sér. Vol. 60, 1088 - 1113 (1974).

The author tries to give a comprehensible demonstration that the Cassini motion is an approximation to a dynamically-consistent system, without being an exact solution to any simplification of the problem. Further, the traditional approaches to the derivation of the physical librations are given, without some of the mystery of the available discussions. Recent studies, motivated by the existence of new high-precision observations and exposing some serious omissions in the available theories, are reviewed. Finally, the present status of the problem is discussed, along with some suggestions for the future.

094.012 **Study of the earth-moon system by space techniques.** J. D. Mulholland.
Separate print Dep. Astron. and McDonald Obs., Univ. Texas, Austin, 59 pp. (1975). – Lectures delivered at the CNES École d'été de physique spatiale, Lannion (France),1974 September 9—14. – Contents: (1) Dynamics and geometry of the lunar orbit; (2) Rotational dynamics of the moon; (3) Lunar gravity field and topography; Résumé of types of space telemetry available for study of the earth-moon system; (4) Laser ranging – description of technique, mathematical modelling for prediction and analysis; (5) Laser ranging – treatment of observations, current results.

094.013 **On equalizing the principal mechanical parameters of the moon.**
G. A. Meshcheryakov, P. M. Zazulyak.
Pis'ma v Astron. Zhurn., Vol. 1, No. 12, p. 43 - 47 (1975). In Russian.

094.014 **On the selection of lunar satellite orbits to determine the spherical harmonics of the gravitational potential of the moon.** N. G. Babaeva.
Pis'ma v Astron. Zhurn., Vol. 1, No. 12, p. 48 - 51 (1975). In Russian.

094.015 **The use of observations in three dimensions for ephemeris improvement.** D. W. Dunham.
Lunar laser ranging, (see 012.032), p. 23 - 27 (1975).

094.016 **Parameter improvement from laser data.**
J. D. Mulholland.
Lunar laser ranging, (see 012.032), p. 28 - 36 (1975).

094.017 **Evolution of the moon: the 1974 model.**
H. H. Schmitt.
Space Sci. Rev., Vol. 18, 259 - 279 = Contr. Div. Geol. Planet. Sci., Cal. Inst. Techn., *Pasadena*, No. 2480 (1975). – This paper is a revision and augmentation of a paper prepared for the Soviet-American Conference on 'Cosmochemistry of the moon and planets', Moscow, U.S.S.R., June 4 - 8, 1974.

The interpretive evolution of the moon can be divided now into seven major stages beginning sometime near the end of the formation of the solar system. These stages and their approximate durations in time are as follows: (1) The beginning – 4.6 billion years ago; (2) The melted shell – 4.6 – 4.4 billion years ago; (3) The cratered highlands – 4.4 – 4.1 billion years ago; (4) The large basins – 4.1– 3.9 billion years ago; (5) The light-colored plains – 3.9 – 3.8 billion years ago; (6) The basaltic maria – 3.8 – 3.0 (?) billion years ago; (7) The quiet crust – 3.0 (?) billion years ago to the present.

094.018 **Lunar laser sounding and celestial mechanics.**
V. K. Abalakin, Yu. L. Kokurin.
Nekotor. vopr. fiz. kosmosa. Moskva, 1974, p. 63 - 79. In Russian. – Abstr. in Referativ. Zhurn. 51. Astron., 12.51.111 (1975).

094.019 **Tables of the lunar diurnal parallax for Kiev.**
A. K. Osipov, E. M. Izhakevich.
Vestn. Kiev. Un-ta, Ser. Astron., No. 17, p. 111 - 112 (1975). In Russian.

094.020 **Geophysical implications of lunar laser ranging.**
W. M. Kaula.
Proc. Symposium on the earth's gravitational field and secular variations in position, (see 012.036), p. 410 - 412 (1974).

094.021 **Coordinate systems in lunar ranging.**
J. D. Mulholland.
IAU Colloquium No. 26, (see 012.034), p. 433 - 443 (1975).

094.022 **The moon and its nature.** H. C. Urey.
Osnovy kosmich. biol. i med. Tom (Vol.) 1. Moskva, Nauka, 1975, p. 119 - 138. In Russian.

094.023 **On the evolution of the moon's rotation.**
K. S. Shakirov.
Astron. Tsirk., No. 874, p. 2 - 4 (1975). In Russian.

094.024 **Geochemical constraints on the origin of the moon.**
H. Wänke, H. Palme.
Meteoritics, Vol. 10, 504 - 505 (1975).

094.025 **The Cracow results of libration studies in view of the lunar laser measurements.** K. Kozieł.

Postępy Astron., Vol. 23, 209 - 211 (1975). In Polish.

The value of the mechanical ellipticity *f* of the moon found recently on the basis of laser measurements confirms earlier results by K. Kozieł derived from the analysis of heliometric observations.

094.026 **Berlage's accretion model of lunar origin and its geochemical consequences. I. II.**
R. W. van Bemmelen.
Proc. Koninkl. Nederl. Akad. Wet., Ser. B, Vol. 78, 168 - 187 (1975).

Origin of the moon. See Abstr. 003.109.

Automatized complex of equipment for reception and treatment of signals during laser location of the moon.
See Abstr. 031.410.

Positions of the major planets and the moon.
See Abstr. 041.013.

Conséquences de l'amélioration des catalogues sur l'étude du système terre-lune. See Abstr. 041.051.

An analysis of the ancient astronomical observations with the implications for geophysics and cosmology.
See Abstr. 043.001.

Multifactor dispersion analysis of results of positional observations of the moon. See Abstr. 044.004.

La rotation de la terre et le système terre-lune.
See Abstr. 044.033.

Effects of tidal dissipation in the oceans on the moon's orbit and the earth's rotation.
See Abstr. 081.003.

The determination of selenodetic coordinates and parameters of the rotational motion of the moon from the lunar surface from measurements of zenith distances.
See Abstr. 094.106.

Lunar evolution: the first 600 million years.
See Abstr. 094.432.

Moon, Global Properties

094.101 **A lunar core and the moon's magnetic field.**
G. C. Kennedy, G. H. Higgins.
The Moon, Vol. 12, 401 - 406 = Inst. Geophys. Planet. Phys., Univ. Calif., Los Angeles, Publ. No. 1394 (1975).

The suggestion that the moon's magnetic field is due to adiabatic magnetohydrodynamic convection of a molten core has been made by a number of recent authors. Considerations based on petrology, mass and rotational inertia limit the size of this hypothetical core to a few 100 km at the most. A proposal has been made that this core is either molten iron or iron sulfide. Fortunately, we know the properties of both molten iron and iron sulfide at lunar core pressures. We can find no way of maintaining circulation in a hypothetical lunar core, as circulation is contingent upon a temperature gradient being greater than the adiabatic gradient, or an internal heat source.

094.102 **Thermal study of the unilluminated surface of the waning moon.** W. L. Raine, W. F. Fountain, J. A. Fountain, M. W. Segewitz, J. van Swearingen, M. K. White.
The Moon, Vol. 12, 407 - 447 (1975).

A program of lunar infrared radiometry which uses large area scanning is described. Procedures for atmospheric attenuation correction and data reduction to temperature by relative radiometry are outlined. The scan data of the waning moon taken during ten evenings in the 10- to 12-μm window are presented as isothermal contour maps of the lunar disc. More than 160 areas of anomalous thermal emission have been found in the lunar darkside data. These are tabulated and are also shown on an accompanying map. An error analysis, derived from accuracy estimates of the independent parameters, is included.

094.103 **Modification of premare impact craters by volcanism and tectonism.** W. J. Brennan.
The Moon, Vol. 12, 449 - 461 (1975).

Many lunar craters greater than 10 km in diam exhibit a variety of morphological characteristics which are not produced by meteorite impact or meteorite erosion. Most such craters are located in or near the margins of the maria. Although some could have resulted from processes such as cauldron resurgence, caldera formation, or ring dike emplacement, most have formed by modification of impact craters by endogenic processes including erosion by flowing lava, fissure volcanism, plutonism and uplift of crater floors along ring fractures of impact origin.

094.104 **Central peaks in lunar craters.** C. C. Allen.
The Moon, Vol. 12, 463 - 474 (1975).

Central peak features in 580 craters on the lunar near side were systematically studied, and detailed studies were made of several features of interest, in an attempt to define the mechanism of central peak formation. The peaks tend to be elongated along the preferred directions of the lunar grid. A weak correlation exists between peak size and crater size.

094.105 **Zeitliche Variation der Einschlagshäufigkeiten großer Körper auf Mond und Erde.**
G. Neukum, B. König, H. Fechtig.
Mitt. Astron. Ges., No. 36, p. 70 (1975). – Abstract.

094.106 **The determination of selenodetic coordinates and parameters of the rotational motion of the moon from the lunar surface from measurements of zenith distances.**
B. Kolaczek, J. Rogowski.
Astron. Zhurn. Akad. Nauk SSSR, Vol. 52, 867 - 874 (1975). In Russian. English translation in Soviet Astron., Vol. 19, No. 4.

The paper presents a theoretical analysis of possible methods for the determination of selenodetic coordinates and the parameters of the moon's rotation on the basis of zenith distance measurements of stars carried out on the lunar surface.

094.107 **Transport and emplacement of crater and basin deposits.**
V. R. Oberbeck, R. H. Morrison, F. Hörz.
The Moon, Vol. 13, (see 012.012), 9 - 26 (1975).

Because lunar mare surfaces contain few large craters, the mare regolith is built up by successive layers of predominantly primary ejecta. In contrast, the lunar highlands are dominated by the effects of large scale craters formed early in lunar history. These effects lead to thick fragmental deposits which are a mixture of primary crater material and local components. These deposits may also properly be named 'regolith' though the term has been traditionally applied only to the relatively thin fine grained surficial deposit on mare and highland terranes generated during the past few billion years. The authors believe that the surficial highland regolith – generated over long periods of time – rests on massive fragmental units that have been produced during the early lunar history.

094.108 **Development of the mare regolith: some model considerations.** W. Quaide, V. Oberbeck.
The Moon, Vol. 13, (see 012.012), 27 - 55 (1975).

094.109 **Shallow lunar structure determined from the passive seismic experiment.** Y. Nakamura, J. Dorman, F. Duennebier, D. Lammlein, G. Latham.
The Moon, Vol. 13, (see 012.012), 57 - 66 (1975).

Data relevant to the shallow structure of the moon obtained at the Apollo seismic stations are compared with previously published results of the active seismic experiments. It is concluded that the lunar surface is covered by a layer of low seismic velocity ($V_p \simeq 100$ m s^{-1}), which appears to be equivalent to the lunar regolith defined previously by geological observations. The regolith thicknesses at the Apollo 11, 12, and 15 sites are estimated from the shear-wave resonance to be 4.4, 3.7, and 4.4 m, respectively.

094.110 **Horizontal transport of the regolith, modification of features, and erosion rates on the lunar surface.**
R. Arvidson, R. J. Drozd, C. M. Hohenberg, C. J. Morgan, G. Poupeau.
The Moon, Vol. 13, (see 012.012), 67 - 79 (1975).

New lunar soils, freshly deposited as impact ejecta, evolve into more mature soils by a complex set of processes involving both near-surface effects and mixing. The deposition rate for material originating more than 1 m away is found to be about 8 mm m.y.$^{-1}$. Material from 10 km away accumulates at a rate of about 0.08 mm m.y.$^{-1}$, providing a steady influx of foreign material.

094.111 **Compositional evidence regarding the influx of interplanetary materials onto the lunar surface.**
J. T. Wasson, W. V. Boynton, C.-L. Chou, P. A. Baedecker.
The Moon, Vol. 13, (see 012.012), 121 - 141 (1975).

Siderophilic element/Ir ratios are higher in mature lunar soils from highlands sites than in those from mare sites. The authors infer that the population of materials responsible for the early intense bombardment of the moon had high ratios, and that the population responsible for the essentially constant flux has low ratios. No group of chondrites has siderophile/Ir ratios identical to those in the mare or highlands soils; CM chondrites are the most similar, and CM-like materials may account for a major fraction of earth-crossing materials during the past 3.7 b.y. The long-lived flux is calculated to be 2.9 g cm^{-2} b.y.$^{-1}$ averaged over the past 3.7 b.y.

094.112 **Stratification in the lunar regolith – a preliminary view.** M. B. Duke, J. S. Nagle.
The Moon, Vol. 13, (see 012.012), 143 - 158 (1975).

094.113 **A Monte Carlo model for the gardening of the lunar regolith.** J. R. Arnold.
The Moon, Vol. 13, (see 012.012), 159 - 172 (1975).
The processes of movement and turnover of the lunar regolith are described by a Monte Carlo model, which includes the effects of collisions by particles from 10^{-7} g to 10^{10} g. The movement of material by the direct cratering process is the dominant mode, but slumping is also included for angles exceeding the static angle of response. Using a group of interrelated computer programs a large number of properties are calculated, including topography, formation of layers, depth of the disturbed layer, nuclear track distributions, cosmogenic nuclides and others. In general the agreement with experiment and observation seems good.

094.114 **Impact ejecta exceeding lunar escape velocity.** E. Schneider.
The Moon, Vol. 13, (see 012.012), 173 - 184 (1975).
Microcrater frequencies caused by fast ($\gtrsim 3$ km s^{-1}) ejecta have been determined using secondary targets in impact experiments. A primary projectile (steel sphere, diam 1.58 mm, mass 1.64×10^{-2} g) was shot in Duran glass with a velocity of 4.1 km s^{-1} by means of a light gas gun. The angular distribution of the secondary crater number densities shows a primary maximum around 25°, and a secondary maximum at about 60° from the primary target surface. The fraction of mass ejected at velocities of $\gtrsim 3$ km s^{-1} is only a factor of 7.5×10^{-5} of the primary projectile mass. A conservative calculation shows that the contribution of secondary microcraters (caused by fast ejecta) to primary microcrater densities on lunar rock surfaces (caused by interplanetary particles) is on the statistical average below 1% for any lunar surface orientation.

094.115 **Track view of the regolithic elephant.** G. Crozaz, R. M. Walker.
The Moon, Vol. 13, (see 012.012), 229 - 234 (1975).
Information gained from track studies in lunar soils is reviewed; soil maturity indices are compared and briefly discussed; track methods to establish the layering chronology of lunar cores are also treated.

094.116 **Catastrophic rupture of lunar rocks: a Monte Carlo simulation.** F. Hörz, E. Schneider, D. E. Gault, J. B. Hartung, D. E. Brownlee.
The Moon, Vol. 13, (see 012.012), 235 - 258 (1975).
A computer model based on Monte Carlo techniques was developed to simulate the destruction of lunar rocks by 'catastrophic rupture' due to meteoroid impact. Energies necessary to accomplish catastrophic rupture were derived from laboratory experiments. A crater-production rate derived from lunar rocks was utilized to calculate absolute time scales. Calculated median survival times for crystalline lunar rocks are 1.9, 4.6, 10.3, and 22 m.y. for rock masses of 10, 10^2, 10^3, and 10^4 g respectively. Corresponding times of 6, 14.5, 32, and 68 $\times$ 10^6 yr are required, before the probability of destruction reaches 0.99.

094.117 **Sintering, crystallization, and breccia formation.** D. R. Uhlmann, L. Klein, R. W. Hopper.
The Moon, Vol. 13, (see 012.012), 277 - 284 (1975).

094.118 **A lunar soil evolution model.** W. W. Mendell, D. S. McKay.
The Moon, Vol. 13, (see 012.012), 285 - 292 (1975).
Comminution, agglutination, and replenishment processes in a lunar soil are modeled by a system of time dependent, linear differential equations. In the model a soil is subdivided into coarse particle, fine particle, and agglutinate fractions. If size data is available for an immature soil and a mature soil of the same system, rates for the various processes can be calculated under certain simplifying assumptions.

094.119 **From regolith to rock by shock.** S. W. Kieffer.
The Moon, Vol. 13, (see 012.012), 301 - 320 (1975).
A model for shock-lithification of terrestrial and lunar regolith is proposed.

094.120 **Distribution, movement, and evolution of the volatile elements in the lunar regolith.**
E. K. Gibson, Jr.
The Moon, Vol. 13, (see 012.012), 321 - 326 (1975).
The abundances and distributions of carbon, nitrogen, and sulfur in lunar soils are reviewed. Carbon and nitrogen have a predominantly extra-lunar origin in lunar soils and breccias, while sulfur is mostly indigeneous to the moon. The lunar processes which effect the movement, distribution, and evolution of carbon, nitrogen, and sulfur, along with the volatile alkali elements sodium, potassium, and rubidium during regolith processes are discussed. Possible mechanisms which may result in the addition to or loss from the moon of these volatile elements are considered.

094.121 **The magnetic characteristics of returned lunar samples and their implications for regolith processes.**
M. Fuller, Y. Wu, P. J. Wasilewski.
The Moon, Vol. 13, (see 012.012), 327 - 338 (1975).

094.122 **Effects of vapor-phase deposition processes on the optical, chemical, and magnetic properties of the lunar regolith.** B. Hapke, W. Cassidy, E. Wells.
The Moon, Vol. 13, (see 012.012), 339 - 353 (1975).
The authors have investigated the physical and chemical properties of materials deposited from vapors generated by hydrogen-ion sputtering and thermal evaporation of lunar and artificial ferrosilicates. Both processes are highly reducing. The deposits are enriched in Fe, have large, nonselective, optical absorptivities, and contain abundant sub-microscopic, superparamagnetic grains of metallic Fe. The hypothesis that the lunar fines contain several percent of materials deposited from the vapor phase accounts in a natural manner for many of the unusual optical, physical and chemical properties of lunar soils.

094.123 **Lunar science: the Apollo legacy.** D. S. Burnett.
Rev. Geophys. Space Phys., Vol. 13, No. 3, (see 013.001), p. 13 - 34 (1975). – Review paper with bibliographical notes.
A general review of lunar science is presented, utilizing two themes: a summary of fundamental problems relating to the composition, structure, and history of the moon and a discussion of some surprising, unanticipated results obtained from Apollo lunar science.

094.124 **Magnetism of the moon and meteorites.** W. A. Gose, R. F. Butler.
Rev. Geophys. Space Phys., Vol. 13, No. 3, (see 013.001), p. 189 - 193, 226 - 228 (1975). – Review paper with bibliographical notes.

094.125 **Selenodesy.** W. M. Kaula.
Rev. Geophys. Space Phys., Vol. 13, No. 3, (see 013.001), p. 272 - 273, 291 - 292 (1975). – Review paper with bibliographical notes.

094.126 **Lunar and planetary seismology.** M. N. Toksöz.
Rev. Geophys. Space Phys., Vol. 13, No. 3, (see 013.001), p. 306 - 311, 322 - 324 (1975). – Review paper with biblio-

graphical notes.

094.127 **The moon after Apollo.** F. El-Baz.
Icarus, Vol. 25, 495 - 537 (1975). – Review paper.

094.128 **The moon.** J. A. Wood.
Sci. American, Vol. 233, No. 3, p. 92 - 102 (1975). Popular article.

094.129 **Correlations between the coefficients of expansion of the potential and relief for some lunar models.**
Sh. T. Khabibullin.
Trudy Kazan. Gorod. Astron. Obs., No. 39, p. 121 - 136 (1973). In Russian.

094.130 **Determination of limits for the seismic velocity structure of the outer layers of the moon by the tau-method.** Eh. N. Bessonova, G. A. Sitnikova, M. N. Toksoez, V. M. Fishman.
Pis'ma v Astron. Zhurn., Vol. 1, No. 8, p. 37 - 42 (1975). In Russian.

The body wave travel time data obtained from the Apollo program are analysed. Using a spherically symmetric model of the moon these seismic data were inverted by the τ-method. The upper and lower bounds for the seismic velocity structure up to depth of 100 km in the Fra Mauro region in Oceanus Procellarum are constructed. One velocity-depth curve within this area fitting travel times is given.

094.131 **The far ultraviolet albedo of the moon.**
R. L. Lucke, R. C. Henry, W. G. Fastie.
Bull. American Astron. Soc., Vol. 7, 467 (1975). – Abstr. AAS.

094.132 **Petrology of lunar soils.** G. Heiken.
Geophys. Space Phys., Vol. 13, 567 - 587 (1975).

The moon is covered with a thin layer of unconsolidated debris called the lunar regolith. A typical sample of this regolith could be characterized as a gray, poorly sorted pebble- or cobble-bearing silty sand having a bulk density of about 1.5 g/cm^3. It consists of lithic and mineral fragments mostly from the local underlying bedrock and glass formed as melt by meteorite impacts. Most investigators conclude that regolith samples are formed primarily by impact-cratering processes, although some may contain a minor pyroclastic component.

094.133 **Transient lunar phenomena – a new approach.**
L. E. Fitton.
Journ. British Astron. Ass., Vol. 85, 511 - 527 (1975).

094.134 **Possible origin of lunar magnetism.** M. Fuller.
Nature, Vol. 257, 295 - 296 (1975). – Letter.

094.135 **Preliminary construction of a secondary selenodetic network.** S. G. Valeev.
Izv. Astron. Ehngel'gardt. Obs., *Kazan'*, No. 39, p. 102 - 110 (1972). In Russian.

094.136 **Mapping of the moon and Mars.** G. A. Burba.
Priroda, No. 9.75, p. 113 (1975). In Russian.
Conference in Moscow, 1974 December.

094.137 **Galactic dust lanes and lunar soil.**
J. F. Lindsay, L. J. Srnka.
Nature, Vol. 257, 776 - 778 (1975).

It has been proposed by McCrea, Shapley and Hoyle and Lyttleton that passages of the solar system through interstellar clouds have appreciable effects on the earth. McCrea argues that the recurrence of ice epochs every ~250 Myr coincides with the passage of the solar system through a galactic spiral arm approximately every 10^8 yr. The authors report here studies on the character and grain-size distribution of texturally-mature lunar soils which support these views.

094.138 **On the connection of colorimetric and spectrophotometric measurements of the lunar surface.**
N. N. Evsyukov.
Problems of cosmic physics. Vyp. (No.) 10, (see 003.003), p. 89 - 95 (1975). In Russian.

094.139 **On the radiogenic atmosphere of the moon.**
Yu. P. Bulashevich, Yu. V. Khachaj.
Geokhimiya, 1975, No. 4, p. 624 - 627. In Russian. – Abstr. in Referativ. Zhurn. 51. Astron., 9.51.360; 62. Issled. kosmich. prostranstva, 9.62.185 (1975).

094.140 **The profile of the far side of the moon from data of a survey by Zond 8.** V. S. Kislyuk.
Astrometriya i Astrofizika, *Kiev*, vyp. (No.) 27, (see 003.004), p. 31 - 36 (1975). In Russian.

Absolutization of the moon's profile obtained by Zond 8 was carried out by means of reference points, the heights of which were determined by selenodetic data of the moon's visible hemisphere, its marginal zone and results of laser altimetry performed by Apollo 15 and Apollo 16. The conclusion is made that the Apollo region on the far side of the moon is a depression with average depth about 6 km relative to a sphere with radius 1738 km. To the north-east of it an elevation with height 4 - 6 km is located.

094.141 **Effect of photographic irradiation on the results of measurements of photographic images of the moon obtained by a ground-based telescope.** A. S. Duma.
Astrometriya i Astrofizika, *Kiev*, vyp. (No.) 27, (see 003.004), p. 37 - 42 (1975). In Russian.

A method is suggested of reducing the measurements of the radius vectors of the moon for the effect of irradiation in the emulsion taking into account photographic density D and contrast of negatives γ. A correlation analysis of deviations ΔR of the mean measured radii from ephemeris values was carried out using measurements of 62 plates. A significant dependence of ΔR on D and γ was found. This dependence was used for taking into account the effect of irradiation while determining the heights in the lunar marginal zone.

094.142 **The relative spectral reflectivity of selected parts of the lunar surface at different phases.**
J. S. Mikhail, L. R. Lisina, V. V. Botvinova.
Astrometriya i Astrofizika, *Kiev*, vyp. (No.) 27, (see 003.004), p. 52 - 58 (1975). In Russian.

17 regions on the moon were investigated for the variation of the ratio of reflectivity with phase and wavelength in the spectral range 0.4–0.8 μ. All parts were divided into five groups depending on the shape of the spectral reflectivity curves. It is shown that the shape of these curves is not changed with phase. This gives an opportunity to compare lunar surface details using the spectral reflectivity curves which were obtained for different phases.

094.143 **Lunar temperature and global heat flux from laboratory electrical conductivity and lunar magnetometer data.** C. P. Sonett, A. Duba.
Nature, Vol. 258, 118 - 121 (1975).

Three-layer monotonic electrical conductivity models for the lunar interior to a depth of 600 km are used in conjunction with laboratory measurements of the electrical conductivity of olivine and pyroxene to estimate a temperature-depth profile. The temperatures calculated for depths of 400–600 km are consistent with attenuation of the seismic shear wave. The temperature calculated at a depth of 100–250 km yields a heat flow that is in good agreement with the directly measured lunar heat flow. The temperature, however, is sufficiently close to melting that mascon anisostasy would not be maintained. Thus a better conductor is required at this depth.

094.144 **Lunar magnetism.** M. L. Goldstein.
Nature, Vol. 258, 175 - 176, with a reply by S. K. Runcorn.

094.145 **Dependence of the degree of polarization of the light reflected by the moon on the phase, on the angles of incidence and reflection and on the azimuth between them.** L. N. Bondarenko.
Soobshch. Gos. Astron. Inst. Shternberga, No. 181, p. 41 - 45 (1973). In Russian.

Phase polarization curves of lunar details at λλ 430, 520 and 600 μm are analysed. Some features were revealed. The polarization of lunar mare depends more on wavelength than polarization of lunar craters. The geometric position of the reflecting surface on the lunar disk slightly influences the degree of polarization.

094.146 **Relative spectrophotometry of lunar details.** L. N. Bondarenko.
Soobshch. Gos. Astron. Inst. Shternberga, No. 181, p. 46 - 55 (1973). In Russian.

The spectral distribution of colour excesses of lunar details is investigated. Colour contrasts are determined. Correlation of colour characteristics with morphological type of the surface, and the time dependence of lunar colour are analysed.

094.147 **Some problems of dynamics of lunar regolith glass particle formation.**
Yu. B. Chernyak, M. D. Nussinov (*Nusinov*).
The Moon, Vol. 13, 363 - 376 (1975).

The objective of this study has been to analyze possibilities of a particular process to occur on the lunar surface, and the relationship of this process with glass particle formation, on the basis of different estimates. A conclusion is reached that condensation mechanism does not appreciably contribute to the formation of small (10^{-1}–10μm) spherical particles of the regolith. Their formation is mainly caused by vibrations and decay of liquid jets in the process of surficial material splashing as a result of lunar surface meteorite bombardment. Large spherical and elongated particle (100–500μm) formation accompanies liquid spraying due to endogeneous events (volcanism, etc.) characterized by significantly lower velocities. Certain notions concerning particle formation mechanisms mentioned in the literature by some authors are shown to be erroneous.

094.148 **On volatilization in vacuum of the elements from the planet surface soil melts.**
M. D. Nussinov (*Nusinov*), Yu. B. Chernyak.
The Moon, Vol. 13, 377 - 394 (1975).

In the present paper a physical model is developed, which is based on experimental data; together with a theory of the process of volatilization of the volatile components of rock melts in vacuum, taking into account an adsorption of the residual atmosphere gases. As a result, some preliminary conclusions have been drawn about such phenomena on the moon and their laboratory simulation.

094.149 **Volatile emission on the moon: possible sources and release mechanisms.** L. J. Friesen.
The Moon, Vol. 13, 425 - 430 (1975).

Evidence for very recent emission of volatiles on the moon is primarily of four types: (1) transient lunar optical events observed by earth-based astronomers; (2) excursions on Apollo SIDE and mass spectrometer instruments; (3) localized Rn^{222}/Po^{210} enhancements on the lunar surface detected by Apollo 15 and 16 orbital alpha spectrometers; (4) presence in lunar fines of retrapped Ar^{40} and other volatiles. Available evidence indicates that the release rate of volatile substances into the lunar atmosphere is not steady, but instead sporadic and episodic.

094.150 **On the petrology and structure of a gravitationally differentiated moon of fission origin.**
A. B. Binder.
The Moon, Vol. 13, 431 - 473 (1975).

This paper presents results obtained from a more detailed analysis of the petrology and structure of a fissioned moon, with emphasis on the fine structure of its upper mantle, the distribution of K, U and Th in the moon and the primary lithology and development of the crust.

094.151 **Temperatures and thermophysical properties of the lunar outermost layer.**
W. P. Jones, J. R. Watkins, T. A. Calvert.
The Moon, Vol. 13, 475 - 494 (1975).

Comparisons of calculated diurnal and eclipse temperatures of the lunar outermost layer are made with earth-based infrared and millimeter data. The thermophysical model upon which the calculations are based incorporates variable physical properties. Laboratory measurements and Apollo sample results are incorporated in the property data. Calculational cases are based largely upon different density profiles. The model is consistent with the data, and the comparisons of theoretical and observational temperatures are very favorable. For such comparisons, further sophistication of the thermophysical model of the outermost layer is probably not justified.

094.152 **Results of an investigation of the lunar sky brightness obtained with the astrophotometer AF-3L aboard Lunokhod 2.**
A. M. Zvereva, A. B. Severnyj, Eh. I. Terez.
Izv. Krymskoj Astrofiz. Obs., Vol. 53, p. 3 - 28 (1975). In Russian.

From the landing of Lunokhod-2 (16 Jan. 1973) 12 telemetric records were obtained: 9 during the lunar day, 1 during the night and 2 in lunar twilight (upper border of the solar disk 1° below the horizon). Excessive brightness as compared with the expected one was recorded during the lunar day and lunar night in UV and especially high in the V region of the spectrum. The excess of brightness shows a characteristic run with the zenith distance of the sun. Being ascribed to some scattering media these measurements show the same ratio of V albedo to UV albedo as for the lunar dust. It is shown that the results can be explained if some very rarefied swarm of dust particles exists around the moon capable to scatter light from sun and earth. The swarm can be formed from secondary ejecta produced by bombardement of the lunar surface by meteorites.

094.153 **Lunar magnetism.** E. G. Eroshenko.
Zemlya i Vselennaya, 1975, No. 5, p. 26 - 32. In Russian.

094.154 **Problems of mapping the moon.** I. A. Kutuzov.
Zemlya i Vselennaya, 1975, No. 5, p. 66 - 67. In Russian.

094.155 **Modern methods and results of mapping the moon and Mars.** Yu. S. Tyuflin.
Zemlya i Vselennaya, 1975, No. 5, p. 68 - 73. In Russian.

094.156 **Lunar rocks as meteoroid detectors.**
J. B. Hartung, F. Hörz, D. E. Gault.
I.A.U. Colloquium No. 13, (see 012.030), p. 227 - 239 (1973).

About 5000 microcraters on seven lunar rocks recovered during the Apollo 12 mission have been systematically studied using a stereomicroscope. Based on comparisons with laboratory cratering experiments, at least 95 percent of all millimeter-sized craters observed were formed by impacts in which the impact velocity exceeded 10 km/s. The dynamics of particle motion near the moon and the distribution of microcraters on the rocks require an extralunar origin for these impacting particles.

094.157 **Lunar remnant magnetic field mapping from orbital observations of mirrored electrons.** J. E. McCoy, K. A. Anderson, R. P. Lin, H. C. Howe, R. E. McGuire.
The Moon, Vol. 14, 35 - 47 (1975). – Conference report (see 012.031).

Areas of lunar surface magnetic field are observed to 'mirror' low energy electrons present in the normal lunar space environment. Preliminary maps of the lunar surface magnetic areas underlying the orbit of the 'Particles and Fields Satellite deployed from Apollo-16' have been generated. The surface field regions observed are generally due to sources smaller than 10–50 km in size, although many individual regions are often so close together as to give much larger regions of effectively continuous mirroring. Absence of consistent mirroring by any global field places an upper limit on the size of any net lunar dipole moment of less than $10^{10}\gamma$ km^3.

094.158 **Lunar topography and the limb compression source regions.** L. J. Srnka, D. R. Criswell, W. R. Wollenhaupt.
The Moon, Vol. 14, 59 - 69 = Contr. Lunar Sci. Inst., *Houston*, Texas, No. 212 (1975). – Conference report (see 012.031).

Data from the Apollo 15, 16, and 17 laser altimeters has been used to study slopes, elevations and roughness in the identifiable regions on the moon which sporadically produce plasma compressions and magnetic field enhancements in the solar wind/lunar void boundary, when those regions are at a flow limb. The results indicate that lunar magnetization is not strongly related to surface features.

094.159 **Lunar electric fields, surface potential and associated plasma sheaths.** J. W. Freeman, M. Ibrahim.
The Moon, Vol. 14, 103 - 114 (1975). – Conference report (see 012.031).

This paper reviews the electric field environment of the moon. Electron and ion densities in the plasma sheath adjacent to each surface potential regime are evaluated and the corresponding Debye length estimated. The electric fields are then approximated by the surface potential over the Debye length.

094.160 **Lunar photoelectron layer dynamics.** E. Walbridge.
The Moon, Vol. 14, 115 - 121 (1975). – Conference report (see 012.031).

Possible waves and oscillations in the lunar photoelectron layer (PEL) are investigated. It is concluded that there are no propagating waves in the PEL other than electromagnetic. However, there is a type of oscillation which appears to be new and which may not be strongly damped. With these oscillations, termed flight-time oscillations, the height of the PEL fluctuates as does the electric field. These oscillations appear to be analogous to the height oscillations of the vertical jet of water in a city park water fountain. For the surfaces of the moon and the planet Mercury, the flight-time oscillation frequency, ω_F, is found to be respectively $\sim 4 \times 10^6$ and $\sim 10^7$ rad s^{-1}. The PEL's of those surfaces may be in a state of continual vertical 'quivering' due to flight-time oscillations, or may be quiescent.

094.161 **The measurements of sky brightness on Lunokhod-2.** A. B. Severny (*Severnyj*), E. I. (*Eh. I.*) Terez, A. M. Zvereva.
The Moon, Vol. 14, 123 - 128 (1975). – Conference report (see 012.031).

An astrophotometer was used for measurements of lunar sky brightness in visible and ultraviolet range during day and night. The data obtained showed unexpectedly high values of brightness during the lunar day in the visible region. From measurements during lunar 'twilight' conditions and from the dependence of excessive flux on cos $Z_\odot$ the authors have concluded that the effect is due to scattering of solar radiation by dust particles above the surface of the moon. Some evidence in favour of dust clouds around the moon is presented.

094.162 **Formation of the lunar atmosphere.** R. R. Hodges, Jr.
The Moon, Vol. 14, 139 - 157 (1975). – Conference report (see 012.031).

Measurements of ^{40}Ar and helium made by the Apollo 17 lunar surface mass-spectrometer are used in the synthesis of atmospheric supply and loss mechanisms. The argon data indicate that about 8% of the ^{40}Ar produced in the moon due to decay of ^{40}K is released to the atmosphere and subsequently lost. Variability of the atmospheric abundance of argon requires that the source be localized, probably in an unfractionated, partially molten core. If so, radiogenic helium released with the argon amounts to 10% of the atmospheric helium supply. The total rate of helium escape from the moon accounts for only 60% of the solar wind α particle influx. This seems to require a nonthermal escape mechanism for trapped solar-wind gases, probably involving weathering of exposed soil grain surfaces by solar wind protons.

094.163 **Molecular gas species in the lunar atmosphere.** J. H. Hoffman, R. R. Hodges, Jr.
The Moon, Vol. 14, 159 - 167 (1975). – Conference report (see 012.031).

There is good evidence for the existence of very small amounts of methane, ammonia and carbon dioxide in the very tenuous lunar atmosphere which consists primarily of the rare gases helium, neon and argon. All of these gases, except ^{40}Ar, originate from solar wind particles which impinge on the lunar surface and are imbedded in the surface material. Methane is the most abundant molecular gas but its concentration is exceedingly low, 1×10^3 mol cm^{-3}, slightly less than ^{36}Ar, whereas the solar wind flux of carbon is approximately 2000 times that of ^{36}Ar. Several reasons are advanced for the very low concentration of methane in the lunar atmosphere.

094.164 **Solar-wind interactions: nature and composition of lunar atmosphere.** N. R. Mukherjee.
The Moon, Vol. 14, 169 - 186 (1975). – Conference report (see 012.031).

The solar wind interacts directly with the lunar surface material resulting in an essentially complete absorption of the corpuscles producing no upstream bowshock but a cavity downstream from the moon. The main source of most neutral species of the atmosphere, except probably ^{40}Ar, is the solar-wind interaction products. On the basis of the solar wind protons, alpha particles and ions of oxygen and carbon, the atmospheric species concentration (cm^{-3}) near the lunar surface at 300 K are as follows: H_2 3.3 to 9.9×10^3; He 2.4 to 4.7×10^3; H 3.7; OH 0.25; H_2O 0.24; and O_2, O, CO, CO_2 and CH_4 in concentrations smaller than H_2.

094.165 **Manifestations and possible sources of lunar transient phenomena (LTP).** W. S. Cameron.
The Moon, Vol. 14, 187 - 199 (1975). – Conference report (see 012.031).

Several different manifestations of lunar transient phenomena (LTP) have been reported. An observing program is being conducted for the Association of Lunar and Planetary Observers which is designed to monitor the LTP sites, the seismic epicenter sites and non-LTP comparison sites. The most extensive new data on albedo versus age are for the crater Dawes. Several LTP effects have been discerned in Dawes.

094.166 **The origin of lunar craters.**
A. Wegener, translated by A. M. C. Şengör.
The Moon, Vol. 14, 211 - 236 (1975). – Translation of 'Die Entstehung der Mondkrater' by A. Wegener, Sammlung Vieweg, Heft 55, Braunschweig, 1921.

094.167 **On the heat flow of a gravitationally-differentiated moon of fission origin.** A. B. Binder.
The Moon, Vol. 14, 237 - 245 (1975).
It is shown that the mean value for the heat flow of a gravitationally-differentiated moon of fission origin is about 13 erg cm^{-2} s^{-1} and that the heat flow varies regionally from about 3 erg cm^{-2} s^{-1} to more than 45 erg cm^{-2} s^{-1}. These regional variations in the heat flow are caused by a non-uniform distribution of K, U and TH in the KREEP zone at the crust-upper mantle boundary and the redistribution of crustal materials and K, U and Th rich KREEP materials by basin-forming impacts. The scale of these regional variations is hundreds of km. The models presented are in accord with the Apollo 15 and 17 heat flow measurements.

094.168 **On the origin of ancient lunar magnetic fields.**
Sh. Sh. Dolginov.
The Moon, Vol. 14, 255 - 261 (1975).
The maximum value of possible lunar dynamo-field in our epoch is estimated in the scheme of precession dynamo model. In the light of our notions on the evolution of the earth-moon system this scheme may account for the size and character of ancient lunar fields that resulted in magnetization of surface lunar rocks, as well as for the absence of lunar dipole field of paleomagnetic origin in our epoch.

094.169 **Interpretation of lunar heat flow data.**
J. E. Conel, J. B. Morton.
The Moon, Vol. 14, 263 - 289 (1975).
Lunar heat flow observations at the Apollo 15 and 17 sites can be interpreted to imply bulk U concentrations for the moon of 5 to 8 times those of normal chondrites and 2 to 4 times terrestrial values inferred from the earth's heat flow and the assumption of thermal steady state between surface heat flow and heat production. A simple model of near-surface structure that takes into account the large difference in (highly insulating) regolith thickness between mare and highland provinces is considered.

094.170 **Large impact craters and the moon's orientation.**
H. J. Melosh.
Earth Planet Sci. Letters, Vol. 26, 353 - 360 = Contr. Div. Geol. Planet. Sci., Cal. Inst. Techn., Pasadena, No. 2551 (1975).
This paper investigates the idea that large impact events have caused the moon to change its orientation in space. It is found that the very largest impact events, such as those which formed Imbrium and Orientale, probably did reorient the moon. This reorientation is primarily due to the change in the moon's moments of inertia consequent upon crater formation. The impulse delivered by the impact can at most unlock the moon's synchronous rotation for a few thousand years, and is thus not of major importance. The moon will attain its new orientation in less than a few times 10^4 years as a result of tidal friction.

094.171 **Deflections of the vertical at lunar mascons.**
M. Burša.
Bull. Astron. Inst. Czechoslovakia, Vol. 26, 346 - 350 (1975).
The lunar mascons Mare Imbrium, Mare Serenitatis, Mare Orientale (Montes Cordillera) are interpreted masswise on the basis of the deflections of the vertical, using the latest data (Michael and Blackshear, 1972). The theory for determining the depth of the centre of gravity and the mass of the mascons has been elaborated for a spherical moon.

094.172 **Results of an investigation of contact-mechanical properties of the lunar ground.** Yu. A. Surkov, V. V. Shandor, Yu. P. Toporov, I. I. Cherkasov, L. T. Zhuravlev.
Kosmich. Issled., Vol. 13, 923 - 931 (1975). In Russian.

094.173 **The moon's geometric figure represented as a spherical- and sampling-function expansion.**
N. A. Chujkova.
Astron. Zhurn. Akad. Nauk SSSR, Vol. 52, 1279 - 1292 (1975). In Russian. English translation in Soviet Astron., Vol. 19, No. 6.
A comparative analysis of the moon's figure representations developed by the author testifies the sampling-function expansion. Comparison of the moon's figure with its gravity field shows a long-wavelength correlation between them.

094.174 **Polarimetric cartography of the moon.**
N. N. Evsyukov, D. I. Shestopalov.
Astron. Zhurn. Akad. Nauk SSSR, Vol. 52, 1293 - 1298 (1975). In Russian. English translation in Soviet Astron., Vol. 19, No. 6.
A polarimetric chart of the east half of the lunar disk in the spectral region 0.40 μ is given; the linearity of connection of polarization p with albedo ρ in coordinates lg ρ, p is shown.

094.175 **Dynamical consequences of meteorite impacts on the moon.** S. J. Peale.
Journ. Geophys. Res., Vol. 80, 4939 - 4946 (1975).
The magnitudes of the excitation of free precession of the lunar spin axis about the position defined by Cassini's laws, free libration in longitude, and free wobble are determined as a function of meteorite angular momentum relative to the lunar center of mass and the position of impact on the lunar surface. Angular momentum conservation suffices for the estimates of precession and libration excitation, but a cratering model for the ejecta distribution is necessary for the estimate of the wobble excitation. For crater sizes scaled as powers of the impact energy, impacts leaving craters as small as a few kilometers in diameter can excite free motions which will ultimately be observable by the lunar laser ranging experiment.

094.176 **Mascons: a two-body solution.**
C. Bowin, B. Simon, W. R. Wollenhaupt.
Journ. Geophys. Res., Vol. 80, 4947 - 4955 (1975).
Almost all of the mass distributions that have been proposed to account for the large positive gravity anomalies associated with lunar mascons have assumed single-body sources of a mass excess. In the case of mare fill with a reasonable density contrast (+0.5 g/cm^3) with crustal material, a fill thickness of about 16 km for Mare Serenitatis is thus required to account for the observed gravity values at 100-km height. Such a great thickness would require a 16-km-deep hole prior to filling, and such a topographic depression is inconsistent with the depths of the topography of the Mare Nectaris and Mare Oriental basins, which have little fill, and with estimates of mare thicknesses based on buried crater dimensions. A two-body mascon solution, however, requires only about a 2-km thickness of fill and a 12-km rise of a lunar Moho beneath Mare Serenitatis to account for observed gravity anomalies.

094.177 **Variation in the number of meteoroid impacts on the moon with lunar phase.**
A. M. Dainty, S. Stein, M. N. Toksöz.
Geophys. Res. Letters, Vol. 2, 273 - 276 (1975).
Data from the ALSEP 12 and 14 seismometers of the

Apollo Passive Seismic Experiment Network are used to determine the statistics of the mass distribution of meteoroids and their direction of approach in near earth space. The authors conclude that, in the mass range studied (> 5 kg), (1) most meteoroid orbits lie near the plane of the ecliptic, (2) the orbits have aphelia between 2 and 5 a.u., and (3) the mass distribution statistics are described by the relation $\log_{10} n = B + G \log_{10} m$, where n = number of meteoroids of mass m grams or greater, $B \sim 0.9$ and $G = -1.5$.

094.178 **Lunar electromagnetic scattering. 3. Propagation at arbitrary angles to the cavity axis.**
B. L. Horning, G. Schubert.
Journ. Geophys. Res., Vol. 80, 4215 - 4229 (1975).
An analytic theory is developed for both the steady state and the time-dependent electric and magnetic fields inside the moon and its downstream cavity for interplanetary electromagnetic field fluctuations incident at arbitrary angles to the cavity axis.

094.179 **Possible origin of the lunar bombardment.**
J. Mottmann.
Bull. American Astron. Soc., Vol. 7, 549 (1975). – Abstr. AAS.

094.180 **On the source of lunar limb compressions.**
C. T. Russell, B. R. Lichtenstein.
Journ. Geophys. Res., Vol. 80, 4700 - 4711 (1975).
Magnetic field measurements made on board the Apollo 15 and 16 subsatellites in low-altitude lunar orbit are used to study the occurrences of compressions of the magnetic field over the terminator regions when the moon is in the solar wind. Where overlap of limb compression occurrence statistics and lunar magnetic field maps exist, the limb compression occurrence rates have a high degree of correlation with the magnetic field strength. Thus the most probable cause of limb compressions is the deflection of the solar wind by the lunar surface magnetic field as the solar wind flows past the lunar terminators.

094.181 **Dependencies of microcrater formation on impact parameters.** K. Nagel, G. Neukum, G. Eichhorn, H. Fechtig, O. Müller, E. Schneider.
Max-Planck-Inst. Kernphysik Heidelberg, MPI H–1975–V 10, p. 171 - 199 (1975). – Contributions to be published in the Proceedings of the Sixth Lunar Science Conference, Houston, Texas, March 17–21, 1975.

094.182 **On the age of Mare Orientale.**
C. S. Beals, R. W. Tanner.
Journ. Roy. Astron. Soc. Canada, Vol. 69, 299 - 306 (1975).
The age of Mare Orientale is reviewed and there is reasonable accord between the results of previous observers and those of the present authors. The catastrophic event producing the basin of Mare Orientale definitely occurred later than that producing the well-known Imbrium basin, making it probably the most recent of the large maria. The age of the central lava flow is comparable with that of the present surface of Mare Imbrium and definitely younger than the rough surfaces of the other parts of the mare basin.

094.183 **The moon in the post Apollo era.** Z. Kopal.
Physics Bull., Vol. 24, 11 - 15 = Astron. Contr. Univ. Manchester, Ser. III, No. 270 (1973).

094.184 **On the selection of a satellite orbit to express the lunar potential by point masses.** N. G. Babaeva.
Astron. Tsirk., No. 883, p. 2 - 4 (1975). In Russian.

094.185 **Comparison of Al/Si intensity ratios and far-ultraviolet albedos of lunar highlands and mare.**
C. G. Andre, M. Hallam, J. R. Weidner, P. Clark.
Meteoritics, Vol. 10, 353 - 354 (1975).

094.186 **Cratering in the earth-moon system – consequences for age determination by crater counting.**
G. Neukum, B. König, H. Fechtig, D. Storzer.
Max-Planck-Inst. Kernphysik Heidelberg, MPI H–1975–V 10, p. 201 - 240 (1975). – Contribution to be published in the Proceedings of the Sixth Lunar Science Conference, Houston, Texas, March 17–21 1975.

094.187 **A Monte-Carlo model for the accumulation of solar flare tracks and spallation products in extraterrestrial regoliths.** J. Borg, J. P. Duraud, Y. Langevin, M. Maurette.
Meteoritics, Vol. 10, 365 - 368 (1975).

094.188 **Lithium in lunar samples, its condensation behaviour and its importance for the estimation of the bulk composition of the moon.**
G. Dreibus, B. Spettel, H. Wänke.
Meteoritics, Vol. 10, 397 - 398 (1975).

094.189 **Are there A 12 mare basalts younger than 3.1 b.y. Unsuccessful search for A 12 mare basalts with crystallization ages below 3.1 b.y.**
P. Horn, T. Kirsten, E. K. Jessberger.
Meteoritics, Vol. 10, 417 - 418 (1975).

094.190 **The role of sputtering in the accumulation of implanted solar wind species in lunar soils.**
A. J. T. Jull, L. R. Gardiner, C. T. Pillinger.
Meteoritics, Vol. 10, 422 - 423 (1975).

094.191 **Solar wind implantation and agglutination in the formation of finely-divided iron in lunar soil.**
C. T. Pillinger, L. R. Gardiner, A. J. T. Jull.
Meteoritics, Vol. 10, 475 (1975).

094.192 **Past record of solar wind, solar flares, and micrometeorites in individual crystals from lunar cores.**
G. Poupeau, E. Zinner, R. M. Walker, D. Morrison.
Meteoritics, Vol. 10, 476 - 477 (1975).

094.193 **Experiments on the stability of FeOOH on the surface of the moon.** L. A. Taylor, J. Burton.
Meteoritics, Vol. 10, 501 - 502 (1975).

094.194 **Age of lunar mantle and crust.**
F. Tera, G. J. Wasserburg.
Meteoritics, Vol. 10, 502 - 503 (1975).

094.195 **The relationship between target and melt rocks in terrestrial craters: implications for the origin of lunar breccia matrices.** S. R. Winzer, R. K. L. Lum.
Meteoritics, Vol. 10, 506 - 507 (1975).

094.196 **Dating of fresh lunar craters by cosmogenic ^{22}Na-^{26}Al studies.** Y. Yokoyama, F. Guichard, J. L. Reyss.
Meteoritics, Vol. 10, 508 (1975).

094.197 **Names of the features on the lunar surface.**
B. Yu. Levin.
Teoriya i praktika toponim. issled. Moskva, 1975, p. 34 - 37. In Russian. – Abstr. in Referativ. Zhurn. 51. Astron., 2.51.7 (1976).

094.198 **Lunar surface study by bistatic radio sounding with a modulated probing signal.** A. L. Zajtsev, V. I. Kaevitser, A. I. Kucheryavenkov, S. S. Matyugov, A. G. Pavel'ev, G. M. Petrov, O. I. Yakovlev.

XI Vses. konf. po rasprostr. radiovoln. Chast' (Part) 1. Tezisy dokl. Kazan', Kazan. un-t, 1975, p. 172 - 176. In Russian. Abstr. in Referativ. Zhurn. 51. Astron., 2.51.398 (1976).

094.199 **Preliminary results of radio sounding of the circumlunar space according to data of Luna 22.**
V. A. Vinogradov, M. B. Vasil'ev, A. S. Vyshlov, M. A. Kolosov, N. A. Savich, V. A. Samovol, L. N. Samoznaev, A. I. Sidorenko, D. Ya. Shtern.
XI Vses. konf. po rasprostr. radiovoln. Chast' (Part) 1. Tezisy dokl. Kazan', Kazan. un-t, 1975, p. 177 - 179. In Russian. – Abstr. in Referativ. Zhurn. 62. Issled. kosmich prostranstva, 2.62.186 (1976).

094.200 **Lunar microcraters and micrometeoroids.**
J. C. Mandeville.
Meteoritics, Vol. 10, 450 - 451 (1975).

094.201 **The gravity and shape of the moon.** W. M. Kaula.
EOS Trans. American Geophys. Union, Vol. 56, 309 - 316 (1975).

Measurements of the geometry and gravity of the moon obtained by satellites and by investigations on the lunar surface are described. Only the second and third degree harmonics of the moon's global gravity field are known reliably. The physical librations and the moments and axes of inertia of the moon are discussed. The tectonic implications of the results are examined on the basis that the moon is chemically differentiated and nearly isostatically compensated. It is emphasised that formation of the mass basins by impact would have left negative gravity anomalies, the present mascons must be attributed to subsequent volcanism.

094.202 **No transform faults on the moon.**
L. Stegena, S. Berczi.
Journ. Geophys., Vol. 40, 775 - 780 (1974).

094.203 **Electromagnetic induction in spherical cap current layers under lunar and terrestrial conditions.**
G. Schubert, K. Schwartz.
Phys. Earth Planet. Interiors, Vol. 10, 55 - 69 (1975).

Presents the theory of electromagnetic induction in spherical cap current sheets of arbitrary angular size, with arbitrary axisymmetric integrated electrical conductivity variations and located at any radial position with respect to the surface of observation. The induced fields are derived for vacuum boundary conditions appropriate to terrestrial induction and plasma confinement boundary conditions relevant to lunar induction. Numerical evaluations show the induced magnetic field as a function of position over the current sheet cap, depth to the current layer, size of the cap, integrated electrical conductivity of the current sheet, and frequency of the fluctuating external field.

094.204 **A determination of the lunar moment of inertia.**
J. P. Gapcynski, W. T. Blackshear, R. H. Tolson, H. R. Compton.
Geophys. Res. Letters, Vol. 2, 353 - 356 (1975).

An estimate of the second zonal coefficient of the spherical harmonic representation of the lunar gravitational field has been obtained from an analysis of particular orbital element variations of the Explorer 35 and Explorer 49 spacecraft. Data from these spacecraft were used because the orbital configurations resulted in variations of the longitude of periapse and node which were, to first order, dependent only on the even zonal harmonics.

The moon. See Abstr. 003.062.

Exobiology of the moon. See Abstr. 003.129.

Berliner Mondatlas. See Abstr. 003.131.

On the analysis of vector arrays on a spherical surface. See Abstr. 021.004.

Apollo photographic systems and cartography of the moon. See Abstr. 032.510.

On a new method of photographing the moon with stars on separate plates. See Abstr. 041.012.

Provisional positions of the moon 1958–1971. See Abstr. 041.028.

Thermal microwave emission from a scattering layer. See Abstr. 063.043.

Labor modelling of the interaction between solar wind and the moon. See Abstr. 074.094.

The earth, the moon and tidal sediments: an analytical review. See Abstr. 081.029.

Bow shock protons in the lunar environment. See Abstr. 084.278.

A lunar signature in the geomagnetic Ap-index. See Abstr. 084.279.

Magnetospheric protons and electrons encountered by the moon in the plasma sheet. See Abstr. 084.280.

Using the moon to probe the geomagnetic tail lobe plasma. See Abstr. 084.297.

Scaling law for impacts on different planets or satellites. See Abstr. 091.037.

Study of the earth-moon system by space techniques. See Abstr. 094.012.

Remarques sur l'utilisation des profils lunaires pour la réduction des observations d'occultations. See Abstr. 096.001.

Abstracts of papers of the All-Union symposium "Interaction of meteor matter with the earth and estimate of the influx of meteor matter on the earth and moon", Dushanbe, 1974, Sept. 24–28. See Abstr. 104.069.

The effect of shock on the magnetic moments and hysteresis properties of natural materials, with special emphasis on rocks from the Lonar meteorite crater, India, and their relationship to lunar paleo-magnetism. See Abstr. 105.115.

Noble gases in grain size fractions of Weston: a comparison with lunar soil data. See Abstr. 105.182.

Energetic lunar nighttime ion events. See Abstr. 106.042.

Die innere Struktur der Erde, des Mondes und der Planeten. See Abstr. 107.016.

Moon, Local Properties

094.401 **Effects of darkening processes on surfaces of airless bodies.** W. Cassidy, B. Hapke.
Icarus, Vol. 25, 371 - 383 (1975). – Presented at I.A.U. Colloquium No. 28, "Planetary Satellites", Ithaca, N.Y., August 18–21, 1974.

There is evidence that lunar soil grains have thin, very light-absorbing coatings that mask absorption bands seen in the reflection spectra of freshly crushed lunar rock, in the same manner as they are masked in the spectra of lunar soils. The authors believe the processes that produce these coatings are (1) deposition of atoms sputtered from lunar soil grains by solar wind particles and (2) deposition of vapor species vaporized from lunar soil grains by micrometeorite impacts.

094.402 **Thicknesses of some lunar mare basalt flows and ejecta blankets based on chemical kinetic data.**
R. Brett.
Geochim. Cosmochim. Acta, Vol. 39, 1135 - 1141 (1975).

Data on the kinetics of some chemical reactions occurring in lunar samples are used to estimate cooling rates over limited temperature ranges. These results, together with data on thermal conductivity and specific heats of lunar samples, have been applied to estimate the thicknesses of the cooling units.

094.403 **Microwave resonance thermomagnetic analysis: a new method for characterizing fine-grained ferromagnetic constituents in lunar materials.**
D. L. Griscom, C. L. Marquardt, E. J. Friebele.
Journ. Geophys. Res., Vol. 80, 2935 - 2946 (1975).

094.404 **Origin of titaniferous lunar basalts.** D. Walker, J. Longhi, E. M. Stolper, T. L. Grove, J. F. Hays.
Geochim. Cosmochim. Acta, Vol. 39, 1219 - 1235 (1975).

Delineation of low pressure phase equilibria in the composition space relevant to titaniferous lunar basalts demonstrates a significant degree of control by those equilibria on the compositions of the basalts. The existence of two distinct chemical groups of basalts (high and low-K) which cannot be related one to the other by fractional crystallization at any pressure, suggests that melting is responsible for the two groups.

094.405 **Light and dark soils at the Apollo 16 landing site.** D. Heymann, J. R. Walton, J. L. Jordan, S. Lakatos, A. Yaniv.
The Moon, Vol. 13, (see 012.012), 81 - 110 (1975).

On the basis of inert gas systematics alone, the soils now near the surface at the Apollo 16 landing site can be divided into three major groups: Group I (North Ray Crater Soils), Group II (Light Soils), and Group III (Dark Soils). Only five soils do not fit this scheme.

094.406 **The source and origin of the exotic component and KREEP-rich materials on the moon.**
V. R. Murthy.
The Moon, Vol. 13, (see 012.012), 111 - 119 (1975).

Trace element concentrations in the lunar soils are inversely correlated to the grain size – i.e., the finer the grain size the higher the trace element content. Based on trace element characteristics, it is suggested that exotic component is finely comminuted KREEP and is dispersed on the lunar surface from the Fra Mauro and other high gamma-ray active regions in the Imbrium-Procellarum regions.

094.407 **Apollo 17 neutron stratigraphy – sedimentation and mixing in the lunar regolith.**
D. B. Curtis, G. J. Wasserburg.
The Moon, Vol. 13, (see 012.012), 185 - 227 (1975).

The authors have measured shifts in the isotopic abundances of Gd and Sm in soils from the Apollo 17 deep drill stem and calculated the neutron fluence from these measurements. The measurements show two well defined regions of nearly constant fluence: (1) a thick deep section with a very large neutron fluence, and (2) a thinner shallow region with a small fluence. This depth dependence is most plausibly described by a model of rapid accumulation in the last 100–200 m.y.

094.408 **Cosmic ray exposure ages of features and events at the Apollo landing sites.** R. Arvidson, G. Crozaz, R. J. Drozd, C. M. Hohenberg, C. J. Morgan.
The Moon, Vol. 13, (see 012.012), 259 - 276 (1975).

094.409 **Effects of maturation on the reflectance of the lunar regolith: Apollo 16 – a case study.**
J. B. Adams, M. P. Charette.
The Moon, Vol. 13, (see 012.012), 293 - 299 (1975).

Darkening of a soil with maturity is due mainly to the increasing proportion of agglutinates. Coating of rock and mineral fragments with thin deposits of glass aids darkening in a minor way, but most of these particles eventually are destroyed by melting as the soils mature.

094.410 **Mineralogy, petrology, and geochemistry of the lunar samples.** J. Warner.
Rev. Geophys. Space Phys., Vol. 13, No. 3, (see 013.001), p. 107 - 113, 163 - 168 (1975). – Review paper with bibliographical notes.

094.411 **A review of lunar sample studies and their application to studies of the terrestrial planets.**
A. L. Albee.
Rev. Geophys. Space Phys., Vol. 13, No. 3, (see 013.001), p. 419 - 422, 426 - 427 (1975). – Review paper with bibliographical notes.

094.412 **Total nitrogen in lunar soils, breccias and rocks.** P. S. Goel, P. N. Shukla, B. K. Kothari, A. N. Garg.
Geochim. Cosmochim. Acta, Vol. 39, 1347 - 1352 (1975).

The total nitrogen contents of a number of lunar samples from Apollo 16 and 17 missions are reported. Solar wind is the main source for the observed excess nitrogen in most fines. Total nitrogen in the soils is found to be proportional to the solar wind rare gases Ar^{36} and Xe^{132}. An inverse correlation between the mean grain diameter and the nitrogen contents is seen, showing that a large fraction of the solar wind nitrogen is surface correlated.

094.413 **Siderophiles and volatiles in Apollo-16 rocks and soils.**
J. T. Wasson, C.-L. Chou, K. L. Robinson, P. A. Baedecker.
Geochim. Cosmochim. Acta, Vol. 39, 1475 - 1485 (1975).

The concentration of the extralunar component in mature Apollo-16 soils is about 3.8 %, substantially higher than values observed at Apollo-14 (2.7 %) or mare landing sites (1.2–1.3 %). It appears that a large amount of local pre-Imbrium regolith was incorporated into Apollo-16 soils: the amount incorporated into Apollo-14 soils was smaller, probably as a result of the combination of two factors: (1) a larger dilution by subregolith materials, and (2) a lower concentration of siderophiles in the ancient Fra Mauro regolith relative to that of the Southern Highlands.

094.414 **Lava flows in Mare Imbrium: an evaluation of anomalously low earth-based radar reflectivity.**
G. G. Schaber, T. W. Thompson, S. H. Zisk.
The Moon, Vol. 13, 395 - 423 (1975).

In the present investigation the authors integrate detailed stratigraphic information for the Imbrium basin with available radar and color data in order to establish the physical and /or chemical relations between distinct mare surfaces, radar reflectivity, and spectral color data.

094.415 **Origin of Mare Frigoris.** J. L. Whitford-Stark, R. J. Fryer.
Icarus, Vol. 26, 231 - 242 (1975).

It is hypothesized that the linear lunar Mare Frigoris originated by the separation, as opposed to the foundering, of blocks of terra-type crust. A computer analysis of possible block motions indicates that an 'initial distribution' of blocks may be postulated which both permits the present distribution to be derived by simple translations, and accounts adequately for certain currently discernible morphological inconsistencies.

094.416 **A blue misty cloud observed in Aristarchus and related matters.** J. R. Smith.
Strolling Astronomer, Vol. 25, 182 - 183 (1975).

094.417 **Magnetochemistry of the Apollo landing sites.** P. J. Wasilewski, M. D. Fuller.
The Moon, Vol. 14, 79 - 101 (1975). – Conference report (see 012.031).

The magnetochemical synthesis, as presented, derives from an integration of the magnetic property systematics developed with corresponding bulk chemistry and petrology, and the results of mixing and reduction responses which feature in the production of the lunar regolith. Magnetochemical properties of the lunar soils are correlated with the Al/Si ratio and telescope spectral reflectivity curves. Magnetic properties of sites not visited can now be reasonably estimated based on telescope reflectivity curves and orbital geochemical experiments.

094.418 **Integrated investigation of the mixed origin of lunar sample 72161,11.**
A. Basu, D. J. DesMarais, J. M. Hayes, W. G. Meinschein.
The Moon, Vol. 14, 129 - 138 (1975). – Conference report (see 012.031).

094.419 **Spectroscopic measurements of surfaces using vacuum ultraviolet radiation.** J. W. Taylor.
The Moon, Vol. 14, 201 - 207 (1975). – Conference report (see 012.031).

Recent research on surface properties of materials using X-ray and vacuum ultraviolet radiation suggest some relationships in common to lunar research. New and intense sources of vacuum ultraviolet continuum radiation may offer the possibility of probing the surface properties of lunar materials and may permit some experimental demonstration of the surface phenomena observed on the moon.

094.420 **Compaction by impact of unconsolidated lunar fines.** T. J. Ahrens.
The Moon, Vol. 14, 291 - 299 = Contr. Div. Geol. Planet. Sci., California Inst. Techn., *Pasadena,* No. 2626 (1975).

New Hugoniot and release adiabate data for 1.8 g cm^{-3} lunar fines (sample, 70051) in the ~2 to ~70 kbar range demonstrate that upon shock compression intrinsic crystal density (~3.1 g cm^{-3}) is achieved under shock stresses of 15 to 20 kbar. Release adiabate determinations indicate that measurable irreversible compaction occurs upon achieving shock pressure above ~4 kbar. If the present data for sample 70051 are taken to be representative of the response to impact of unconsolidated regolith material on the moon, it is inferred that the formation of appreciable quantities of soil breccia can be associated with the impact of meteoroids or ejecta at speeds of as low as ~1 km s^{-1}.

094.421 **The Néel transition and magnetic properties of terrestrial, synthetic, and lunar ilmenites.**
F. E. Senftle, A. N. Thorpe, C. Briggs, C. Alexander, J. Minkin, D. L. Griscom.
Earth Planet. Sci. Letters, Vol. 26, 377 - 386 (1975).

094.422 **Heavy element affinities in Apollo 17 samples.** R. O. Allen, Jr., S. Jovanovic, G. W. Reed Jr.
Earth Planet. Sci. Letters, Vol. 27, 163 - 169 (1975).

094.423 **A note on the magnetic properties of lunar sample 15535,28 and the implications regarding palaeointensity determinations.** A. Stephenson, D. W. Collinson.
Earth Planet. Sci. Letters, Vol. 27, 360 - 361, with a reply by S. K. Banerjee, J. P. Mellema, p. 362 (1975).

094.424 **Specific surface area as a maturity index of lunar fines.** R. B. Gammage, H. F. Holmes.
Earth Planet. Sci. Letters, Vol. 27, 424 - 426 (1975).

Mature surface fines have an equilibrium specific surface area of about 0.6 m^2/g, the equivalent mean particle size being about 3 μm. Apparently in mature soils there is a balance in the forces which cause fining, attrition, pore filling and growth of lunar dust grains. The specific surface area, determined by nitrogen or krypton sorption at 77°K, is a valuable index of soil maturity.

094.425 **Argon-lead isotopic correlation in samples from lunar maria: records from the ancient lunar regolith.**
D. Heymann.
Earth Planet. Sci. Letters, Vol. 27, 445 - 448 (1975).

094.426 **Alteration of an annealed and irradiated lunar fines sample by adsorbed water.**
H. F. Holmes, P. A. Agron, E. Eichler, E. L. Fuller, Jr., G. D. O'Kelley, R. B. Gammage.
Earth Planet. Sci. Letters, Vol. 28, 33 - 36 (1975).

Apollo 12 lunar fines sample 12070,403 was annealed at 1000°C and subsequently irradiated with a beam of 130 MeV Fe ions. Adsorption of nitrogen and water were measured before and after the irradiation. Prior to the irradiation, the fines were non-porous and water had no effect on the physical characteristics of the lunar fines. In contrast, after the irradiation, the interaction with water caused an increase in the specific surface area and created a pore system. These results are definitive evidence that the interaction of water with damage tracks is the prime factor involved in the alteration of lunar fines by adsorbed water.

094.427 **Experimental investigation of reaction coronas on olivine in Apollo 14 high-grade breccias.**
S. E. Kesson.
Earth Planet. Sci. Letters, Vol. 28, 65 - 68 (1975).

094.428 **Search for amino acids in Apollo returned lunar soil.** C. W. Gehrke, R. W. Zumwalt, K. Kuo, C. Ponnamperuma, A. Shimoyama.
Origins of Life, Vol. 6, 541 - 550 (1975).

The lunar samples from Apollo flights 11 through 17 provided the students of chemical evolution with an opportunity of examining extraterrestrial materials for evidence of early prebiological chemistry in the solar system. The search was directed to water-extractable compounds with emphasis on amino acids. The authors conclude that amino acids are not present in the lunar regolith.

094.429 **Metallographic investigation of a fragment of lunar matter returned by the automatic interplanetary station Luna 16 from Mare Foecunditatis.**
R. I. Mints, T. M. Petukhova, V. M. Segal'.
Meteoritika, vyp. (No.) 34, p. 92 -97, 127 (1975). In Russian.

094.430 **Optical detection of Fe^{3+} in lunar plagioclase.** D. J. Telfer, G. Walker.
Nature, Vol. 258, 694 - 695 (1975). – Letter.

094.431 **Lunar shear velocity structure at Apollo sites 12, 14, and 15.** N. Mark, G. H. Sutton.
Journ. Geophys. Res., Vol. 80, 4932 - 4938 (1975).

094.432 **Lunar evolution: the first 600 million years.** L. Husain, O. A. Schaeffer.
Geophys. Res. Letters, Vol. 2, 29 - 32 (1975).

The authors have determined an $^{40}Ar-^{39}Ar$ age of 4.26 ± 0.02 G.y. for the Apollo 17 troctolite 76535. The age of the rock is associated with the Serenitatis Basin forming event. The age of formation of the Serenitatis Basin is then deduced at 4.26 ± 0.02 G.y. ago, about 300 million years before the Imbrium event. It appears that the major basin forming era on the near side of the moon lasted longer than 300 million years. The grouping of many lunar rock ages at 3.9–4.0 G.y. is most likely due to the fact that the Imbrium ejecta predominates all highland landing sites.

094.433 **Ferromagnetic phase-mass equivalence and lunar sample magnetic remanence.** P. J. Wasilewski.
Geophys. Res. Letters, Vol. 2, 93 - 95 (1975).

094.434 **Consortium indomitabile.** J. A. Wood.
The Moon, Vol. 14, 303 - 305 (1975).

When the first polymict, clast-rich breccia sample (12013) was returned from the moon, the Lunar Sample Analysis Planning Team recognized that samples of it could not be parcelled out blindly for study since investigators making different types of measurements might receive samples of totally different types of rock to study. LSAPT's approach was to organize a consortium of scientists to study that particular breccia, whose membership included one specialist for each of the types of study that was deemed essential to a characterization and understanding of the rock. Subsequently, LSAPT set up consortia to study many of the other complex breccias collected in the lunar highlands. Among the last of these created were consortia to study the Apollo 17 boulder samples.

094.435 **Geologic setting of Boulder 1, Station 2, Apollo 17 landing site.** E. W. Wolfe.
The Moon, Vol. 14, 307 - 314 (1975).

Boulder 1 at Station 2 is one of three boulders sampled by Apollo 17 at the base of the South Massif, which rises 2.3 km above the floor of a linear valley interpreted as a graben formed by deformation related to the southern Serenitatis impact. The boulders probably rolled from the upper part of the massif after emplacement of the light mantle. Orbital gravity data and photogeologic reinterpretation suggest that the Apollo 17 area is located approximately on the third ring of the southern Serenitatis basin.

094.436 **The Boulder.** U. B. Marvin.
The Moon, Vol. 14, 315 - 326 (1975).

The external morphologies of Boulder 1, Station 2, and of the four samples taken from it by the Apollo 17 crew, are briefly described. The boulder is a polymict breccia, containing the following principal materials as clasts: gray competent breccias, black competent breccias, anorthositic breccias, pigeonite basalt, coarse norite. All are enclosed in a matrix of light-colored friable breccia.

094.437 **Boulder 1, Station 2, Apollo 17: petrology and petrogenesis.** G. Ryder, D. B. Stoeser, U. B. Marvin, J. F. Bower, J. A. Wood.
The Moon, Vol. 14, 327 - 357 (1975).

Boulder 1, Station 2, Apollo 17 is a stratified boulder containing dark clasts and dark-rimmed light clasts set in a light-gray friable matrix. The gray to black clasts are multigenerational, competent, high-grade metamorphic, and partially melted breccias. The age of the older breccias suggests that the Serenitatis event is younger than 4.01 ± 0.03 b.y.

094.438 **Major and trace element chemistry of Boulder 1 at Station 2, Apollo 17.** D. P. Blanchard, L. A. Haskin, J. W. Jacobs, J. C. Brannon, R. L. Korotev.
The Moon, Vol. 14, 359 - 371 (1975).

Twenty-seven samples of matrix and clast materials from Boulder 1 at Station 2, Apollo 17 have been analyzed for major and trace elements as part of the study of this boulder by Consortium Indomitabile. Both unusual and common types of material have been characterized.

094.439 **Meteoritic material in a boulder from the Apollo 17 site: implications for its origin.**
J. W. Morgan, H. Higuchi, E. Anders.
The Moon, Vol. 14, 373 - 383 (1975).

Sixteen samples of Boulder 1 from Station 2 at the Apollo 17 site were analyzed by radiochemical neutron activation analysis for Ag, Au, Bi, Br, Cd, Cs, Ge, Ir, Ni, Rb, Re, Sb, Te, Tl, U and Zn. Nine grey or black breccia samples contain a unique, Group 3 meteoritic component of Ir/Au ratio 0.65–0.82. Because of the rarity of the Group 3 meteoritic component at the Apollo 17 site, this boulder cannot represent ordinary Serenitatis ejecta. It may represent pre-Serenitatis material excavated from the fringes of the crater during late stages of the Serenitatis impact, but only lightly shocked and hence uncontaminated by the Serenitatis projectile.

094.440 **History of Boulder 1 at Station 2, Apollo 17 based on trace element interrelationships.**
S. Jovanovic, G. W. Reed, Jr.
The Moon, Vol. 14, 385 - 393 (1975).

Correlations among the trace and minor element pairs Cl and Br, Cl and P_2O_5, and Ru and Os, present in parent igneous rocks, generally survived the processes of boulder breccia formation. Fractions of the Cl, Br, and Hg that are mobilized by water leaching and/or volatilization at moderate temperatures ($\leq 450°C$) place constraints on the thermal history of Boulder 1 and its component breccias. Since, and possibly during, consolidation, the boulder has probably not been subjected to temperatures of $\geq 450°C$.

094.441 **Cosmic ray exposure history and compaction age of Boulder 1 from Station 2.**
J. N. Goswami, I. D. Hutcheon.
The Moon, Vol. 14, 395 - 405 (1975).

Fossil track analyses of a ~3 cm section of boulder fragment 72255, collected at the base of the South Massif, yield a surface exposure age for this boulder in its present location of ~40 m.y. This age is in good agreement with the $^{81}Kr-Kr$ exposure age (Leich et al., 1975) suggesting that the boulder was either never exposed to cosmic radiation prior to its emplacement at the foot of the South Massif or that it was heavily shielded during any previous irradiation.

094.442 **Rare gas constraints on the history of Boulder 1, Station 2, Apollo 17.** D. A. Leich, S. B. Kahl, A. R. Kirschbaum, S. Niemeyer, D. Phinney.
The Moon, Vol. 14, 407 - 444 (1975).

Rare gas isotopic analyses have been performed on both pile-irradiated and unirradiated samples from Boulder 1, Station 2. Two samples from rock 72255, the Civet Cat clast and a sample of adjacent breccia, have concordant $^{40}Ar-^{39}Ar$ ages of 3.99 ± 0.03 b.y. and 4.01 ± 0.03 b.y., respectively. The authors regard these ages as upper limits on the time of the Serenitatis basin-forming event.

094.443 **Rb–Sr ages of clasts from within Boulder 1, Station**

2, Apollo 17. W. Compston, J. J. Foster, C. M. Gray.
The Moon, Vol. 14, 445 - 462 (1975).

Rb, Sr and $^{87}Sr/^{86}Sr$ have been determined for fragments of matrix and clasts from three of the hand-specimens of Boulder 1, 72275, 72255, and 72215. Total-rock and certain plagioclase samples from a crushed norite clast (Civet Cat) define an ages of 4.17 ± 0.05 *AE* (2σ) for the pre-Serenitatis igneous differentiation of the norite. Pyroxene and other mineral separates were affected by a later event at about 3.9 ± 0.1 *AE*. An unshocked clast of pigeonite basalt has a well-fitted mineral isochron of 4.01 ± 0.04 *AE*.

094.444 **U–Th–Pb systematics of selected samples from Apollo 17, Boulder 1, Station 2.**
P. D. Nunes, M. Tatsumoto.
The Moon, Vol. 14, 463 - 471 (1975).

Nine U–Th–Pb whole-rock analyses of selected brecciated materials from sample 72215 and one analysis of a pigeonite basalt clast from 72275 are presented. The data suggest that significant differentiation and/or metamorphism occurred ~4.2 b.y. ago. The age of this event, however, is not accurately defined by these data.

094.445 **Natural remanent magnetization studies of a layered breccia boulder from the lunar highland region.**
S. K. Banerjee, G. Swits.
The Moon, Vol. 14, 473 - 481 (1975).

094.446 **Spectral reflectance of highland rock types at Apollo 17: evidence from Boulder 1, Station 2.**
J. B. Adams, M. P. Charette.
The Moon, Vol. 14, 483 - 489 (1975).

Many of the non-mare rock types at Apollo 17 can be identified uniquely by their spectral reflectance properties. Mineralogical and textural information is present in the spectral curves of samples from Boulder 1, Station 2. It should be possible to determine the regional extent of rocks similar to the boulder using reflectance spectra from a spacecraft in lunar orbit.

094.447 **Geological model for Boulder 1 at Station 2, South Massif, Valley of Taurus-Littrow.**
H. H. Schmitt.
The Moon, Vol. 14, 491 - 504 = Contr. Div. Geol. Planet. Sci., California Inst. Techn., *Pasadena*, No. 2631 (1975).

It appears possible to establish a preliminary geological model for the origin and evolution of the breccias of Boulder 1 at Station 2 in the Valley of Taurus-Littrow based on firm and probable geological constraints. During the large basin stage, three major dynamic events profoundly influenced the present character of the Boulder 1 materials. These events probably occurred as follows: (1) formation of gray competent breccia containing ANT-suite clasts in the hot ejecta blanket of an old large basin event, such as Tranquilitatis, that took place about 4.0 b.y. ago; (2) rebrecciation and redeposition of the gray competent breccia, mixed with light friable breccia and pigeonite basalt, in a relatively cool ejecta deposit, possibly produced by the northern Serenitatis event; (3) uplift and exposure of the Boulder 1 materials in the South Massif by the southern Serenitatis event about 3.90 b.y. ago.

094.448 **The nature and origin of Boulder 1, Station 2, Apollo 17.** J. A. Wood.
The Moon, Vol. 14, 505 - 517 (1975).

094.449 **Thermophysical properties of Apollo 14 fines.**
C. J. Cremers.
Journ. Geophys. Res., Vol. 80, 4466 - 4470 (1975).

The vacuum thermal conductivity of lunar fines sample 14163 was measured for the approximate temperature range of 100° - 400° K. Several sample densities from 1100 to 1800 kg/m^3 were used. The temperature dependence of the conductivity was found to be well represented by the relation $k = A + BT^3$, which is predicted by elementary theory. The coefficients A and B were obtained by least squares analysis of the data. The thermal diffusivity was calculated for the various densities by using specific heat data from the literature along with the measured conductivities. The results are compared with those obtained for Apollo 11, Apollo 12, and terrestrial basalt samples.

094.450 **On the names of lunar surface features.**
A. V. Barandeev.
Geod. i kartografiya, 1975, No. 7, p. 59 - 63. In Russian.
Abstr. in Referativ. Zhurn. 51. Astron., 1.51.353 (1976).

094.451 **^{39}Ar–^{40}Ar dating of lunar rocks: effects of grain size and neutron irradiation.**
P. Horn, E. K. Jessberger, T. Kirsten, H. Richter.
Max-Planck-Inst. Kernphysik Heidelberg, MPI H–1975–V 10, p. 1 - 54 (1975). – Contributions to be published in the Proceedings of the Sixth Lunar Science Conference, Houston, Texas, March 17–21, 1975.

094.452 **Rare gases in Apollo 17 soils with emphasis on analysis of size and mineral fractions of soil 74241.**
W. Hübner, T. Kirsten, J. Kiko.
Max-Planck-Inst. Kernphysik Heidelberg, MPI H–1975–V 10, p. 55 - 91 (1975). – Contributions to be published in the Proceedings of the Sixth Lunar Science Conference, Houston, Texas, March 17–21, 1975.

094.453 **Lithophile trace and major elements in Apollo 16 and Apollo 17 lunar samples.** O. Müller.
Max-Planck-Inst. Kernphysik Heidelberg, MPI H–1975–V 10, p. 93 - 110 (1975). – Contributions to be published in the Proceedings of the Sixth Lunar Science Conference, Houston, Texas, March 17–21, 1975.

094.454 **Subsolidus reduction of lunar opaque oxides: textures, assemblages, geochemistry, and evidence for a late stage endogenic gaseous mixture.**
A. El Goresy, P. Ramdohr.
Max-Planck-Inst. Kernphysik Heidelberg, MPI H–1975–V 10, p. 111 - 147 (1975). – Contributions to be published in the Proceedings of the Sixth Lunar Science Conference, Houston, Texas, March 17–21, 1975.

094.455 **The feasibility of ion identification on cosmic ray tracks in lunar feldspars.**
W. Krätschmer, W. Gentner.
Max-Planck-Inst. Kernphysik Heidelberg, MPI H–1975–V 10, p. 149 - 170 (1975). – Contributions to be published in the Proceedings of the Sixth Lunar Science Conference, Houston, Texas, March 17–21, 1975.

094.456 **The grape cluster, metal particle 63344,1.**
J. I. Goldstein, H. J. Axon, S. O. Agrell.
Earth Planet. Sci. Letters, Vol. 28, 217 - 224 (1975).

Metal particle 63344,1 consists of hundreds of metallic globules welded together to form a structure somewhat like a bunch of grapes. It is the largest lunar metal sample found to date, over 5 mm in its longest dimension. The globules are all of the same composition and were produced at the same time and from the same source. Two possible mechanisms are proposed for the formation of the globules of 63344,1. One mechanism involves the primary impact of an iron meteorite which produces a metallic liquid and vapor phase. The second mechanism involves the formation of a liquid pool of metal after impact of an iron meteorite projectile followed by a secondary impact in the liquid metal pool.

094.457 **Lunar glass compositions: Apollo 16 core sections 60002 and 60004.**
H. O. A. Meyer, H.-M. Tsai.
Earth Planet. Sci. Letters, Vol. 28, 234 - 240 (1975).
Approximately 500 glasses between 1 mm and 125 μm in size have been analyzed from fourteen samples from the Apollo 16 core sections 60002 and 60004. The majority of glasses have compositions comparable to those found in previous studies of lunar surface soils; however two new and distinct glass compositions that are probably derived in part from mare material occur in the core samples.

094.458 **Luna 16 and Luna 20 revisited.**
P. H. Cadogan, G. Turner.
Meteoritics, Vol. 10, 375 - 376 (1975).

094.459 **Lunar anorthosite paradox in the light of mineral/ melt equilibria.** M. J. Drake.
Meteoritics, Vol. 10, 396 (1975).

094.460 **A physical and chemical characterization of extraterrestrial dust grains on a submicron scale.**
J. C. Dran, J. Klossa, M. Maurette, Y. Benkheiri, J. P. Duraud.
Meteoritics, Vol. 10, 396 - 397 (1975).

094.461 **The carbon chemistry of lunar glassy agglutinates.**
L. R. Gardiner, A. J. T. Jull, C. T. Pillinger.
Meteoritics, Vol. 10, 405 (1975).

094.462 **The grape cluster, metal particle 63344,1.**
J. I. Goldstein, H. J. Axon, S. O. Agrell.
Meteoritics, Vol. 10, 407 - 408 (1975).

094.463 **Al/Si intensity ratios of crater materials: the depths of Tranquillitatis and Serenitatis basalts.**
M. Hallam, C. G. Andre, J. R. Weidner.
Meteoritics, Vol. 10, 409 (1975).

094.464 **A detailed study of the ^{39}Ar-^{40}Ar ages of two Apollo 11 basalts.** N. Grögler, P. Eberhardt, J. Geiss, S. Guggisberg, P. Maurer, A. Stettler.
Meteoritics, Vol. 10, 411 - 412 (1975).

094.465 **Horizontal transport by impact ejection, an important mixing process in the lunar regolith.**
C. M. Hohenberg.
Meteoritics, Vol. 10, 415 - 416 (1975).

094.466 **Effects of grain size and neutron irradiation in ^{39}Ar-^{40}Ar dating.** T. Kirsten, E. Jessberger, P. Horn.
Meteoritics, Vol. 10, 425 - 427 (1975).

094.467 **Ages of Apollo 16 coarse fines fragments.**
P. Maurer, P. Eberhardt, J. Geiss, N. Grögler, A. Stettler, U. Krähenbühl, A. Peckett, G. M. Brown.
Meteoritics, Vol. 10, 455 - 456 (1975).

094.468 **Ancient meteoritic components in Apollo 17 highland breccias.**
J. W. Morgan, H. Higuchi, E. Anders.
Meteoritics, Vol. 10, 457 - 458 (1975).

094.469 **^{244}Pu fission xenon in structural phases of the Apollo 14 breccia 14305.** M. Mörgeli, P. Eberhardt, O. Eugster, J. Geiss, N. Grögler, F. Meier, W. Kiesl.
Meteoritics, Vol. 10, 458 - 459 (1975).

094.470 **He-depth profiles with high resolution (~100 Å) by the rare gas ion probe technique.**
H. W. Müller, J. Kiko, T. Kirsten.
Meteoritics, Vol. 10, 459 - 460 (1975).

094.471 **A search for solar energetic tritons in lunar samples.**
F. Niederer, M. Wahlen, J. Geiss.
Meteoritics, Vol. 10, 466 - 467 (1975).

094.472 **On the halogen deficiency of lunar apatite.**
C. B. Sclar, J. F. Bauer.
Meteoritics, Vol. 10, 484 - 485 (1975).

094.473 **Pyroxene geothermometry applied to lunar crustal rocks and meteorites.** H. Takeda, T. Ishii.
Meteoritics, Vol. 10, 499 - 500 (1975).

094.474 **Rocks 60618 and 65785: evidence for admixture of KREEP in lunar impact melts.**
K. Keil, R. D. Warner, M. Prinz, E. Dowty.
Geophys. Res. Letters, Vol. 2, 369 - 372 (1975).
The authors present evidence to support the hypothesis that the fine-grained, basaltic-textured portions of rocks 60618 and 65785 were produced by impact melting of the coarse-grained spinel-olivine anorthosite (60618) and spinel troctolite (65785) portions, coupled with admixture of approximately 33% and 63%, respectively, material of alkalic high-alumina basalt (KREEP) composition. The abundance of impact-modified rocks at the Apollo 16 site with compositional similarities to the impact melts reported here, suggests that the relations observed in these rocks are indicative of widespread impact mixing with KREEP in rocks of the lunar highlands.

094.475 **FMR thermomagnetic studies up to 900°C of lunar soils and potential magnetic analogues.**
R. V. Morris, R. V. Gibbons, F. Hörz.
Geophys. Res. Letters, Vol. 2, 461 - 464 (1975).
Using a recently developed furnace, ferromagnetic resonance (FMR) thermomagnetic studies up to 900°C were employed to measure the Curie points of the superparamagnetic (SP) and single domain (SD) particles in lunar soils and potential magnetic analogue materials. Based on measured Curie points of 775±10°C, the SP and SD particles in lunar soils 10084-853, 12070-29, 14161-46, and 67010-4 are essentially pure metallic Fe.

094.476 **Hydrogen and fluorine in the surfaces of lunar samples.** D. A. Leich, R. H. Goldberg, D. S. Burnett, T. A. Tombrello.
California Inst. Technol., Pasadena (USA). NASA-CR-139299, 30 pp. (1974).
The resonant nuclear reaction $F^{19}(p, \alpha, \gamma)O^{16}$ was used to perform depth sensitive analyses for both fluorine and hydrogen in lunar samples. The resonance at 0.83 MeV (center-of-mass) in this reaction was applied to the measurement of the distribution of trapped solar protons in lunar samples to depths of about 1/2 micrometer. These results are interpreted in terms of terrestrial H_2O surface contamination and a redistribution of the implanted solar H. Results are also presented for an experiment to test the penetration of H_2O into laboratory glass samples which have been irradiated with O^{16} to simulate the radiation damaged surfaces of lunar glasses. Fluorine determinations were performed in a 1 pm surface layer on lunar samples using the same $F^{19}(\alpha, \gamma)O^{16}$ resonance. The data are discussed from the standpoint of lunar fluorine and Teflon contamination.

094.477 **Rare gases in Apollo 11 fines.** J. R. Basford.
Thesis, Minnesota Univ., Minneapolis (USA).
[Available from Univ. Microfilms, Ann Arbor, Mich. Order No. 75-2085], 124 pp. (1974).

094.478 **Experimental detection of the influence of radiation damages upon the diffusion behaviour of solar wind**

noble gases in single grains of moon dust. H. Ducati. Thesis, Naturwiss. Gesamtfakultät, Univ. Heidelberg. [Available from the library of the Univ. Heidelberg], 48 pp. (1973). In German.

094.479 **Extra-terrestrial geology.** R. Bowen.
Sci. Progr., (*GB*), Vol. 61, 473 - 494 (1974).
During the Apollo lunar missions rock collections and experiments were made by the astronauts which made possible extra terrestrial geological studies involving in situ field investigations. Recently acquired data are discussed and some comments are made on the two inner and two of the outer planets as well as the Jovian satellite, Io, and several asteroids.

094.480 **Comments on the search for amino acids in Apollo 15, 16 and 17 lunar samples. A study in contamination control.** B. Nagy.
Anal. Chem., (*USA*), Vol. 47, 1718 - 1720 (1975).
The possible presence of amino acids in lunar samples is discussed, along with the question of their origin.

Lunar mineralogy. See Abstr. 003.048.

Probing the moon: The moon – a new appraisal from space missions and laboratory analyses. See Abstr. 011.001.

The moon after Apollo. See Abstr. 094.127.

Manifestations and possible sources of lunar transient phenomena (LTP). See Abstr. 094.165.

Lithium in lunar samples, its condensation behaviour and its importance for the estimation of the bulk composition of the moon. See Abstr. 094.188.

Radiogenic isotope research. See Abstr. 105.019.

Stable isotope geochemistry. See Abstr. 105.020.

Experimental hypervelocity impact into quartz sand: distribution and shock metamorphism of ejecta. See Abstr. 105.040.

Ilmenite-orthopyroxene intergrowths from the moon and the Skaergaard intrusion. See Abstr. 105.055.

The perplexing behavior of niobium in meteorites and lunar samples. See Abstr. 105.161.

Micrometeorites and solar flare particles in and out of the ecliptic. See Abstr. 106.055.

Track length distribution in lunar and meteoritic minerals and the chemical composition of very very heavy cosmic rays. See Abstr. 143.073.

095 Lunar Eclipses

095.001 **The lunar eclipse seen from coast to coast.**
C. A. Federer, Jr.
Sky Telescope, Vol. 50, 72 - 78 (1975). – Concerning the eclipse 1975 May 24 - 25.

095.002 **Gleanings from the mailbag for the lunar eclipse.**
Sky Telescope, Vol. 50, 97 - 98 (1975).

095.003 **Observations de l'éclipse de lune du 4 juin 1974 par des amateurs.** P. de La Cotardière.
L'Astronomie, 89e année, p. 303 - 305 (1975).

095.004 **Some results from May's lunar eclipse.**
J. Ashbrook.
Sky Telescope, Vol. 50, 219 - 223 (1975).

095.005 **Beobachtungen der Mondfinsternis vom 29. November 1974.** P. Rapavý, M. Dujnič.
Sterne, 51. Jahrgang, p. 181 - 182 (1975).

095.006 **L'éclipse totale de lune des 18–19 novembre 1975.**
G. Florsch.
L'Astronomie, 89e année, p. 378 (1975).

095.007 **De maansverduistering van 18/19 november 1975.**
G. Schilling.
Zenit, Vol. 2, 360 - 361 (1975).

095.008 **The total lunar eclipse of November 29, 1974.**
W. H. Haas.
Strolling Astronomer, Vol. 25, 183 (1975).

095.009 **Photographic method of the determination of the enlargement of the shadow during the lunar eclipse.** J. Mach.
Říše hvězd, Vol. 56, 213 - 215 (1975). In Czech.

095.010 **Partial lunar eclipse on June 4/5 1974.**
V. Protić-Benišek.
Bull. Obs. Astron. Belgrade, No. 126, p. 33 (1975).

A note on the periodicity of eclipses. See Abstr. 079.001.

096 Lunar Occultations

096.001 **Remarques sur l'utilisation des profils lunaires pour la réduction des observations d'occultations.**
M. Froeschlé, C. Meyer.
The Moon, Vol. 12, 475 - 478 (1975).
It is shown that a rigorous determination of the argument in the limb charts of the moon leads to a limb correction which, in some cases, may differ perceptibly from the one obtained by the traditional method. In the case of central occultations, the discrepancy may reach a maximum of 0".9 when the moon is near its maximum or minimum declination and introduces a statistical uncertainty of 0".2 in the determination of the position of the moon.

096.002 **Occultations rasantes en France en 1976.** J. Meeus. L'Astronomie, 89e année, p. 331 - 334 (1975).

096.003 **A theory for fading at occultations.** A. J. Jeffries. Journ. British Astron. Ass., Vol. 85, 420 - 422 (1975).

096.004 **Observations of two grazing occultations.**
A. K. Osipov, A. A. Zhitetskij.
Astrometriya i Astrofizika, *Kiev*, vyp. (No.) 26, (see 003.001), p. 22 - 23 (1975). In Russian.
The results of observations of grazing occultations are presented for two stars: 63 κ Aqr (June 22, 1973) and 24 Sgr (September 6, 1973).

096.005 **Photoelectric measurements of lunar occultations VII: further observational results.** J. L. Africano, C. L. Cobb, D. W. Dunham, D. S. Evans, F. C. Fekel, S. S. Vogt. Astron. Journ., Vol. 80, 689 - 697 (1975).
Timings of 397 occultations are reported. Measures needed to correct errors arising from faulty practices and instrumental problems are described. The list includes the first well-authenticated case of distortion of a trace by lunar limb irregularities (run No. 2754), two new determinations of stellar angular diameter (ξ^2 Sagittarii, run No. 3571; RZ Arietis, run No. 3623), and 36 cases, the majority new discoveries, in which there is evidence of duplicity.

096.006 **Grazing occultation of ZC1410, 1975 April 20, Nylstroom, South Africa.** J. Hers.
Monthly Notes Astron. Soc. Southern Africa, Vol. 34, 99 - 101 (1975).

096.007 **Two grazing occultations.** A. G. F. Morrisby. Monthly Notes Astron. Soc. Southern Africa, Vol. 34, 101 - 103 (1975).

096.008 **Okkultasjonsarbeidet i Norge 1974.** H. Brubak. Astron. Tidssk., Årg. 8, p. 133 (1975).

096.009 **Problemen bij het waarnemen van sterbedekkingen.** H. Betlem.
Zenit, Vol. 2, 408 - 409 (1975).

096.010 **Occultations de planètes par la lune, 1976 - 1979.** J. Meeus.
Ciel et Terre, Vol. 91, 419 - 423 (1975).

096.011 **Observations of occultations of stars by the moon at the Astronomical Observatory of the Kiev University in 1973.**
A. K. Osipov, E. M. Izhakevich, A. A. Zhitetskij.
Vestn. Kiev. Un-ta, Ser. Astron., No. 17, p. 106 - 111 (1975). In Russian.

096.012 **First results of photoelectric observations of star occultations by the moon in Poltava.** B. F. Sincheskul.
Astron. Tsirk., No. 871, p. 7 - 8 (1975). In Russian.

096.013 **Ocultaciones de estrellas por la luna observadas en Madrid durante el año 1972.** M. de Pascual.
Bol. Astron. Obs. Madrid, Vol. 8, No. 5, p. 82 (1975).

096.014 **Occultations of stars by the moon, observed at the Belgrade Astronomical Observatory, 1972–1973.**
M. B. Protitch, V. Protić-Benišek.
Bull. Obs. Astron. Belgrade, No. 126, p. 31 - 32 (1975).

096.015 **Occultations of major planets by moon at Belgrade Astronomical Observatory during 1974.**
V. Protić-Benišek, M. B. Protitch.
Bull. Obs. Astron. Belgrade, No. 126, p. 32 - 33 (1975).

096.016 **Lunar occultations 1975 visible at Belgrade.** Bull. Obs. Astron. Belgrade, No. 126, p. 34 - 35 (1975).

096.017 **On determination of refraction in observations of star occultations by the moon.**
V. K. Budz'ko, B. F. Sincheskul.
Vrashchenie i prilivn. deformatsii Zemli. Vyp. (No.) 7. Kiev, Nauk. dumka, 1975, p. 74 - 78. In Russian. – Abstr. in Referativ. Zhurn. 51. Astron., 2.51.244 (1976).

096.018 **1976 occultation supplement.** Predictions for the United States and Canada.
Sky Telescope, Suppl., 16 pp. = H. M. Nautical Almanac Office, *Roy. Greenwich Obs., Herstmonceux,* Library Repr. No. 316 (1975).

Sterbedekkingen januari–maart 1976.
Zenit, Vol. 2, 438 (1975).

Measuring the sizes of Saturn's satellites.
See Abstr. 100.206.

Lunar occultation of Saturn. I. The diameters of Tethys, Dione, Rhea, Titan, and Iapetus. See Abstr. 100.208.

The angular diameter of ψ Virginis.
See Abstr. 115.009.

The angular diameter of TX Piscium.
See Abstr. 115.019.

Lunar occultations of Beta Scorpii in 1975 and 1976. See Abstr. 117.011.

Rocznik Astronomiczny Obserwatorium Krakowskiego 1976. International Supplement No. 47.
See Abstr. 120.005.

Observations of lunar occultations of cosmic radio sources in the decametric wavelength region.
See Abstr. 141.122.

Measurement of the spatial structure of the X-ray source in the Crab nebula. I. Observation of the 1974 November 3 lunar occultation. See Abstr. 142.136.

Measurement of the spatial structure of the X-ray source in the Crab nebula. II. Observation of the 1974 December 28 lunar occultation. See Abstr. 142.137.

097 Mars, Mars Satellites

Mars

097.001 **An analysis of Martian satellite photographic observations of 1967.** D. Pascu.
Icarus, Vol. 25, 479 - 483 (1975). – Presented at I.A.U. Colloquium No. 28, "Planetary Satellites", Ithaca, N.Y. August, 18 - 21, 1974.

Photographic observations of the Martian satellites were made at the opposition of 1967 with the Naval Observatory's 61-inch astrometric reflector. The plates were reduced by the method of plate constants using positions for the faint background stars determined from astrographic field plates. The random mean error of these observations was estimated to be not greater than ±0."10. The main result of the orbital adjustment is a +2° correction to the zero of mean longitude for Phobos. The scale of the orbits of both satellites gave accordant values for the mass of Mars and the combined value of 3099500 ± 2800 (m.e.) is in good agreement with modern determinations.

097.002 **Viking to Mars: the mission strategy.** C. Sagan.
Sky Telescope, Vol. 50, 15 - 19, 23.

097.003 **Determination of the temperature of the Martian atmosphere.** A. Sharma, V. B. Singh.
Journ. Quant. Spectrosc. Radiat. Transfer, Vol. 15, 561 - 570 (1975).

The authors have calculated the temperature distribution of a Martian atmosphere using the high-resolution spectroscopic data of Carleton et al. and the theory of Goody and Gierasch. Models of the Martian atmosphere have been developed and synthetic spectra of the 1.05 μ CO_2 spectrum in the Martian atmosphere have been computed. An effective temperature is obtained from the computed synthetic spectra for a non-isothermal and isothermal Martian atmosphere.

097.004 **Improvement of the orbit of Mars using optical and radar observations.** N. I. Glebova.
Byull. Inst. Teoret. Astron., *Leningrad*, Vol. 13, 612 - 628 (1975). In Russian.

The paper deals with the methods of reduction of optical and radar observations of Mars and gives the preliminary results of the improvement of the Martian orbit based on observations during 1952 - 1972.

097.005 **The isostatic state of Martian topography.**
R. J. Phillips, R. S. Saunders.
Journ. Geophys. Res., Vol. 80, 2893 - 2898 (1975).

As is evidenced by the gravity data, the regional topography of Mars falls into two distinct age groups. The older group is isostatically compensated at relatively shallow depth. All of the authors' analyses indicate that the younger group is only partly compensated and that the depth of partial compensation is also shallow.

097.006 **Mars gravity field based on a short-arc technique.**
W. L. Sjogren, J. Lorell, L. Wong, W. Downs.
Journ. Geophys. Res., Vol. 80, 2899 - 2908 (1975).

The magnitudes of 92 surface mass points at designated locations were estimated from the radio tracking data of the Mariner Mars 1971 (M9) orbiter. This result is the first mass point model of a global field. The derived surface mass distribution correlates positively with the visible topography. The Hellas basin contains a mass deficiency, in contrast to some of the lunar basins which contain mass excesses. The Mars gravity field represented by the four parameters of an optimally located mass point (superimposed on an oblate spheroid) has third- and fourth-degree harmonics comparable to those of the complete model.

097.007 **Martian topography derived from occultation, radar, spectral, and optical measurements.**
E. J. Christensen.
Journ. Geophys. Res., Vol. 80, 2909 - 2913 (1975).

A new global topographic map of the planet Mars, derived primarily from Mariner 9 data, is presented. Spacecraft radio signal occultation times, remote sensing of infrared and ultraviolet spectra, and earth-based radar measurements are included in the data base. Large-scale topographic relief is seen with definite fluid flow patterns established over much of the surface consistent with earlier analyses. An optical flattening of 0.0059 is obtained with a center-of-mass to center-of-figure offset of approximately 2 km to the south and 1 km toward the Tharsis ridge.

097.008 **Mars 3: a narrow-band photoelectric photometry in the 3700–7000 Å region.**
L. V. Ksanfomaliti, V. I. Moroz.
AN SSSR. In-t kosmich. issled. Pr-200. Moskva, 1974, 38 pp. In Russian. – Abstr. in Referativ. Zhurn. 51. Astron., 7.51. 276; 62. Issled. kosmich prostranstva 7.62.208 (1975).

097.009 **Density of matter of the Martian upper layer from data of radiophysical measurements in the 3-cm region.** N. N. Krupenio.
AN SSSR. In-t kosmich. issled. Pr-185. Moskva, 1974. 28 pp. In Russian. – Abstr. in Referativ. Zhurn. 51. Astron., 7.51. 292; 62. Issled. kosmich. prostranstva, 7.62.221 (1975).

097.010 **The geology of Mars: a brief review of some recent results.** T. A. Mutch, J. W. Head.
Rev. Geophys. Space Phys., Vol. 13, No. 3, (see 013.001), p. 411 - 416, 425 (1975). – Review paper with bibliographical notes.

097.011 **Variable features on Mars V: Evidence for crater streaks produced by wind erosion.** J. Veverka.
Icarus, Vol. 25, 595 - 601 (1975).

Evidence is presented to show that the ragged dark streaks which appeared behind many Martian craters several months after the end of the 1971 global duststorm were produced by wind erosion of a thin surface veneer of duststorm fallout.

097.012 **The canals of Mars: an assessment after Mariner 9.**
C. Sagan, P. Fox.
Icarus, Vol. 25, 602 - 612 (1975).

The Lowellian canal network has been compared with the results of Mariner 9 photography of Mars. A small number of canals may correspond to rift valleys, ridge systems, crater chains, and linear surface albedo markings. But the vast bulk of classical canals correspond neither to topographic nor to albedo features, and appear to have no relation to the real Martian surface.

097.013 **Generalized photoclinometry for Mariner 9.**
R. L. Wildey.
Icarus, Vol. 25, 613 - 626 (1975).

The present investigation develops a new theory for the photoclinometric determination of topography when the photometric function of the planetary surface (or that which corresponds to the mean optical depth of emergent scattered solar radiation from an optically thick planetary atmosphere) is not restricted beyond the expectation that it is a function of phase angle, angle of incidence, and angle of emergence.

097.014 **Mars.** J. B. Pollack.
Sci. American, Vol. 233, No. 3, p. 106 - 117 (1975).
Popular article.

097.015 **A new image of Mars.** M. Ya. Marov.
Priroda, No. 8.75, p. 12 - 23 (1975). In Russian.

097.016 **Helium in the Martian atmosphere – a hypothetical estimate.** V. I. Moroz.
Pis'ma v Astron. Zhurn., Vol. 1, No. 7, p. 39 - 41 (1975). In Russian.

Helium is generated, as well as argon, by radioactive decay. The dissipation rate of helium from the atmospheres of Mars and the earth is nearly the same. The presence of argon in the Martian atmosphere means that helium is also present there. At heights of 600–1500 km helium is probably the most abundant component of the Martian atmosphere.

097.017 **On the cause of difference of bright and dark crater streaks on Mars.** R. O. Kuz'min.
Pis'ma v Astron. Zhurn., Vol. 1, No. 9, p. 42 - 44 (1975). In Russian.

Bright and dark crater streaks are interpreted as eolian features with unresolved barchan-like forms of surface relief. The light or dark appearance of streaks is probably related to illumination conditions.

097.018 **Measurement of the intensity and spectral composition of the Martian gamma-radiation aboard the Mars 5 automatic interplanetary station.** A. P. Vinogradov, Yu. A. Surkov, L. P. Moskaleva, F. F. Kirnozov.
Dokl. Akad. Nauk SSSR. Ser. Mat., Fiz., Vol. 223, 1336 - 1339 (1975). In Russian.

097.019 **Elysium and Martian tectonics.**
H. Frey, I. Witkin.
Bull. American Astron. Soc., Vol. 7, 468 (1975). – Abstr. AAS.

097.020 **Topography of the polar layered deposits of Mars.**
D. Dzurisin, K. R. Blasius.
Journ. Geophys. Res., Vol. 80, 3286 - 3306 (1975).

Synthesis of polar topographic data derived from the Mariner 9 radio occultation, ultraviolet spectrometer, and television imaging experiments provides new information on the behavior of polar volatiles and the topographic configuration of the martian polar layered deposits. Total thickness of the deposits is inferred to be 1–2 km in the south and 4–6 km in the north.

097.021 **Naherforschung des Planeten Mars mit Viking.**
H. W. Köhler.
SuW, Vol. 14, 274 - 279 (1975).

097.022 **Pressures and heights from results of a CO_2-altimetry from aboard the automatic interplanetary station Mars 3.** L. V. Ksanfomaliti, V. I. Moroz, B. S. Kunashev, V. S. Zhegulev.
Kosmich. Issled., Vol. 13, 563 - 580 (1975). In Russian.

097.023 **Data of a photometric reduction of photographs obtained by the space apparatus Mars 3 on 28. II. 1972.** V. V. Botvinova, O. I. Bugaenko, I. K. Koval', M. K. Naraeva, A. S. Selivanov.
Kosmich. Issled., Vol. 13, 581 - 586 (1975). In Russian.

097.024 **Results of polarimetric observations of Mars in 1971 and 1973.** A. V. Morozhenko.
Astrometriya i Astrofizika, *Kiev*, vyp. (No.) 26, (see 003.001), p. 97 - 107 (1975). In Russian.

The results of polarimetric observations of Mars within the spectral interval of 0.316–0.798 μ are given. The observations of 1971 show that the polarization of Mars depends on the longitude of the central meridian, the amplitude of this dependence decreases towards the ultraviolet region. The observations of 1973 show the presence of the dust storm on Mars from June 1973 to January 1974.

097.025 **Occultations et appulses planétaires en 1976.**
J. Meeus.
L'Astronomie, 89e année, p. 373 - 376 (1975).

097.026 **The new Martian nomenclature of the International Astronomical Union.** G. de Vaucouleurs, J. Blunck, M. Davies, A. Dollfus, I. K. Koval, G. P. Kuiper, H. Masursky, S. Miyamoto, V. I. Moroz, C. Sagan, B. Smith.
Icarus, Vol. 26, 85 - 98 (1975).

A new nomenclature for Martian regions and topographic features uncovered by Mariner 9, as officially adopted by the International Astronomical Union, is described. About 180 craters, generally of diameters > 100 km, have been named, as well as 13 classes of topographic features designated catena, chasma, dorsum,fossa, labyrinthus, mensa, mons, patera, planitia, planum, tholus, vallis, and vastitas. In addition seven craters and the Kepler Dorsum are named on Phobos, and two craters on Deimos. Coordinates and maps of each named feature are displayed.

097.027 **On the undular character of the hypsographic curve and some hypsometric peculiarities of the Martian surface.** Yu. N. Lipskij, D. A. Kazimirov, Zh. F. Rodionova, V. A. Shuvaeva.
Dokl. Akad. Nauk SSSR. Ser. Mat., Fiz., Vol. 224, 58 - 61 (1975). In Russian.

097.028 **Spectral and spatial structure of the thermal radiation field under conditions of the Martian turbid atmosphere.** K. Ya. Kondrat'ev, N. I. Moskalenko.
Dokl. Akad. Nauk SSSR. Ser. Mat. Fiz., Vol. 224, 316 - 319 (1975). In Russian.

097.029 **Fluid transport on earth and aeolian transport on Mars.** C. Sagan, R. A. Bagnold.
Icarus, Vol. 26, 209 - 218 (1975).

Experimental data on cohesion-free particle transport in fluid beds are applied, via a universal scaling relation, to atmospheric transport of fine grains on Mars. It may be that cohesion due to impact vitrification, vacuum sintering, and adsorbed thin films of water are absent on Mars – in which case the curve of threshold velocity versus grain size may show no turnup to small particle size, and one micron diameter grains may be injected directly by saltation into the Martian atmosphere more readily than 100 micron diameter grains. Suspension of fine grains is significantly easier at low temperatures and high pressures.

097.030 **Geomorphologic analysis of some Martian surface images from the Mars 4 and 5 automatic stations.**
C. P. Florenski, A. T. Basilevski, R. O. Kuzmin, I. M. Chernaya.
Icarus, Vol. 26, 219 - 229, with a comment by D. Pieri, Icarus, Vol. 26, 230 (1975).

The paper deals with the results of geomorphological analysis of some photographs taken by the Mars 4 and 5 probes. The area studied (700 × 900 km, centered at 32° S, 35° W) covers part of Mare Erythraeum. The main landforms of the area are craters which are sorted into four morphological classes. Fluvial, eolian, and tectonic processes as well as downslope displacement of material are identified with different degrees of certainty.

097.031 **Observing Mars. VI – the 1975–76 aphelic apparition.** K. v. Knoepfel.
Strolling Astronomer, Vol. 25, 211 - 215 (1975).

097.032 **On the abundance of NO_2 in the Martian atmosphere.** N. B. Ibragimov, V. A. Bronshtehn.
Pis'ma v Astron. Zhurn., Vol. 1, No. 11, p. 45 - 48 (1975). In Russian.

The analysis of spectrograms of Mars obtained in August 1971 in the coudé focus of the 80-inch reflector of the Shemakha Observatory with a dispersion of 4–6 Å/mm has not shown any evidence of NO_2 absorption. By comparison with lunar spectrograms in four spectral regions the upper limit for the abundance of NO_2 has been evaluated to 0.02 mm atm.

097.033 **Water vapour in the Martian atmosphere according to measurements aboard the automatic interplanetary station Mars 3.** V. I. Moroz, A. Eh. Nadzhip.
Kosmich. Issled., Vol. 13, 738 - 752 (1975). In Russian.

097.034 **On the structure of the Martian atmosphere at heights between 120 and 220 km.**
V. A. Krasnopol'skij.
Kosmich. Issled., Vol. 13, 753 - 759 (1975). In Russian.

097.035 **Radar measurements of Martian topography and surface properties: the 1971 and 1973 oppositions.**
G. S. Downs, P. E. Reichley, R. R. Green.
Icarus, Vol. 26, 273 - 312 (1975).

The Goldstone radar system was used at a wavelength of 12.6 cm to probe the Martian surface during the 1973 opposition. Contours of constant altitude were calculated at 200 m intervals between latitudes $-14°$ and $-22°$. These contours are presented in conjunction with Mars charts derived from Mariner 9 television pictures. Reflected power was measured at angles of incidence between $-5°$ and $+5°$. These measurements were combined with those obtained during the 1971 opposition. Predictions of the reflected power versus the angle of incidence were calculated from the exponential surface model of Hagfors. The smoother regions exhibit a typical reflectivity of 8.2%. A tendency for the reflectivity to decrease with increasing roughness was observed.

097.036 **Geophysical implications of the Martian gravity field.** J. Arkani-Hamed.
Icarus, Vol. 26, 313 - 320 (1975).

The undulations of the Martian gravitational potential indicate lateral density variations in the Mars interior. A gravitating and solid Martian model deforms under the influence of these variations, producing stress differences of about 125 bars at a depth of about 200 km. A strong linear correlation of the gravitational potential and the surface topography indicates that the extensive volcanism at the Tharsis region is a recent phenomenon.

097.037 **Eolian erosion of the Martian surface, part 1: Erosion rate similitude.** J. D. Iversen,
R. Greeley, B. R. White, J. B. Pollack.
Icarus, Vol. 26, 321 - 331 (1975).

A similitude parameter is derived which is based on theoretical considerations of erosion due to sand in saltation. This parameter has been used to correlate wind tunnel experiments of particle flow over model craters. The characteristics of the flow field in the vicinity and downstream of a crater are discussed and it is shown that erosion is initiated in areas lying under a pair of trailing vortices.

097.038 **Mariner 9 ultraviolet spectrometer experiment: scattering properties of Hellas.**
J. M. Ajello, K. D. Pang.
Icarus, Vol. 26, 332 - 340 (1975).

A least-squares analysis of 709 Mariner 9 uv spectra obtained over Hellas showed the extinction optical depth of the atmosphere above Hellas to be 0.5 ± 0.2. This is evidence of lingering dust in the Hellas basin. An atmospheric model, combining dust and Rayleigh scattering, was used to make an apparent pressure map of Hellas.

097.039 **North polar hood observations during Martian dust storms.** L. J. Martin.
Icarus, Vol. 26, 341 - 352 (1975).

The 1956, 1971, and 1973 major dust storms on Mars affected the apparent contrast and extent of the northern polar hood. A survey of photography from seven apparitions indicates that this seasonal feature is consistently prominent and identifiable in the absence of major storms throughout half of the Martian year. Hourly and daily mapping from these photographs indicates that the hood developed a southward protrusion during the first few days of the storm as it began its apparent recession. The hood was seen on at least one side of the planet on every day throughout the duration of the storm, although its normal extent and contrast were greatly reduced.

097.040 **Carbon dioxide–water clathrate as a reservoir of CO_2 on Mars.** A. Dobrovolskis, A. P. Ingersoll.
Icarus, Vol. 26, 353 - 357 (1975).

It has been suggested that the residual polar caps of Mars contain a reservoir of permanently frozen carbon dioxide which is controlling the atmospheric pressure. This paper shows that the permanent polar caps may contain several millibars of CO_2 in the form of clathrate, and discusses the implications of this permanent clathrate reservoir for the present and past atmospheric pressure on Mars.

097.041 **Mass loss from the region of Mars and the asteroid belt.** S. J. Weidenschilling.
Icarus, Vol. 26, 361 - 366 = Contr. Planet. Astron. Lab., Mass. Inst. Techn., *Cambridge,* Mass., No. 118 (1975).

Models of the solar nebula suggest that the mass of solid matter which condensed in the region of Mars and the asteroids was much greater than the amount now present. Bombardment by a primordial population of asteroidal bodies originating near Jupiter's orbit could preferentially remove matter from this region, without significant effects in the earth's zone. The total bombardment flux was at least two orders of magnitude greater in the zone of Mars than in that of the earth. The flux at Venus and Mercury from this source was negligible. The cratering rate for Mars may have differed greatly from those of the other terrestrial planets for a significant fraction of the age of the solar system.

097.042 **Estimate of electric field intensity and charge of Mars.** G. M. Nedyalkova, I. E. Turchinovich.
XXVIII Gertsenovsk. chteniya. Teor. fiz. i astron., Leningrad, 1975, p. 75 - 78. In Russian. – Abstr. in Referativ. Zhurn. 51. Astron., 10.51.280 (1975).

097.043 **CO_2 photodissociation and thermal emission connected with it and some emissions in the thermospheres of Mars and Venus.**
A. V. Dembovskij, M. N. Izakov, O. G. Lisin.
Kosmich. Issled., Vol. 13, 866 - 877 (1975). In Russian.

097.044 **Photometric profiles of the planet in the near infrared spectral region from measurements aboard the automatic interplanetary station Mars 3.**
V. I. Moroz, A. Eh. Nadzhip, V. S. Zhegulev.
Kosmich Issled., Vol. 13, 878 - 892 (1975). In Russian.

097.045 **Photometry of cloud formations at the limb of Mars from photographs obtained on 28.II.1972 with the space apparatus Mars 3.**
V. V. Botvinova, I. K. Koval'.
Kosmich Issled., Vol. 13, 893 - 897 (1975). In Russian.

097.046 **Numerical reduction of photographs of the Martian surface transmitted by the automatic interplanetary stations Mars 4 and Mars 5.** T. P. Belikova, M. A. Kronrod, P. A. Chochia, L. P. Yaroslavskij.
Kosmich. Issled., Vol. 13, 898 - 906 (1975). In Russian.

097.047 **Composition of the Martian dust as derived by infrared spectroscopy from Mariner 9.**
J. R. Aronson, A. G. Emslie.
Journ. Geophys. Res., Vol. 80, 4925 - 4931 (1975).

The Martian dust spectra obtained by the infrared interferometer spectrometer (Iris) experiment on Mariner 9 have been compared with modeling computations for various minerals and rocks. The model used was based upon a recently developed theory of absorption and scattering by particulate materials. The results indicate that the composition of the Martian dust is biased toward silicate structures which involve a high degree of silicate oxygen sharing. A feldspar-rich dust of approximately 1-μm particle size appears to be the best candidate, although sheet structure silicates cannot be ruled out.

097.048 **On the gas exchange between the Martian upper atmosphere and the solar wind.**
A. V. Bogdanov, O. L. Vajsberg, A. P. Kalinin, V. N. Smirnov.
Dokl. Akad. Nauk SSSR. Ser. Mat., Fiz., Vol. 225, 1284 - 1287 (1975). In Russian.

097.049 **Synthesis of Martian colour images according to photomaterials obtained aboard the Mars 5 space vehicle.** D. S. Lebedev, L. P. Yaroslavskij, M. K. Naraeva, A. S. Selivanov, I. S. Fajnberg.
Dokl. Akad. Nauk SSSR. Ser. Mat., Fiz., Vol. 225, 1288 - 1291 (1975). In Russian.

097.050 **Seasonal buffering of atmospheric pressure on Mars.**
D. Dzurisin, A. P. Ingersoll.
Icarus, Vol. 26, 437 - 440 = Contr. Div. Geol. Planet. Sci., California Inst. Technol., *Pasadena*, No. 2620 (1975).

An isothermal reservoir of carbon dioxide in gaseous contact with the Martian atmosphere would reduce the amplitude and advance the phase of global atmospheric pressure fluctuations caused by seasonal growth and decline of polar CO_2 frost caps. Adsorbed carbon dioxide in the upper ~ 10 m of Martian regolith is sufficient to buffer the present atmosphere on a seasonal basis. Available observations and related polar cap models do not confirm or refute the operation of such a mechanism.

097.051 **Dust injection into the Martian atmosphere.**
D. W. Johnson, P. Harteck, R. R. Reeves.
Icarus, Vol. 26, 441 - 443 (1975).

A mechanism for initiating global dust storms on Mars is proposed in which the diurnal variation of surface temperature results in the desorption of adsorbed CO_2, which under certain conditions can inject large amounts of fine (1–10 μm) dust into the atmosphere.

097.052 **The gravity field of Mars.**
R. D. Reasenberg, I. I. Shapiro, R. D. White.
Geophys. Res. Letters, Vol. 2, 89 - 92 (1975).

A complete sixth-degree spherical-harmonic model of the gravitational potential of Mars has been produced from an analysis of Doppler data obtained from the radio tracking of the Mariner-9 spacecraft. The equipotential surface defined by this model is estimated to be accurate to within 100 m below 60° North Latitude and to within 300 m above.

097.053 **An occultation of ϵ Geminorum by Mars on 1976 April 8.** G. E. Taylor.
Journ. British Astron. Ass., Vol. 86, 33 - 34 (1975).

097.054 **Clouds on Mars: some conclusions from the Mars 3 observations.** V. I. Moroz.
In-t kosmich. issled. AN SSSR. Pr-235. Moskva, 1975, 28 pp. In Russian. – Abstr. in Referativ. Zhurn. 51. Astron., 12.51. 241; 62. Issled. kosmich. prostranstva, 12.62.155 (1975).

097.055 **Working model of the Martian atmosphere and surface. Part 1. The atmosphere. Part 2. The surface.**
V. I. Moroz (Editor).
In-t kosmich. issled. AN SSSR. Pr-240, Pr-241, Moskva. 1975. 63 pp., 26 pp. resprectively. In Russian. – Abstr. in Referativ. Zhurn. 51. Astron., 2.51.356 (1976).

097.056 **Peculiarities of spectral and spatial distributions of the thermal radiation field in the Martian atmosphere.** N. I. Moskalenko.
Izv. AN SSSR. Fiz. atmosf. i okeana, Vol. 11, 836 - 844 (1975). In Russian. – Abstr. in Referativ. Zhurn. 51. Astron., 1.51. 324; 62. Issled. kosmich. prostranstva, 1.62.172 (1976).

097.057 **On the structure of Martian soil from optical and infrared observations.** V. I. Moroz.
In-t kosmich. issled. AN SSSR. Pr-234. Moskva, 1975. 28 pp. In Russian. – Abstr. in Referativ. Zhurn. 51. Astron., 1.51. 327; 62. Issled. kosmich. prostranstva, 1.62.171 (1976).

097.058 **Investigation of the Martian ionosphere and circumlunar plasma with the method of a dispersion interferometer.** M. A. Kolosov, N. A. Savich.
XI Vses. konf. po rasprostr. radiovoln. Chast' (Part) 1. Tezisy dokl. Kazan', Kazan. un-t, 1975, p. 147 - 150. In Russian. Abstr. in Referativ. Zhurn. 62. Issled. kosmich. prostranstva, 1.62.169 (1976).

097.059 **Osservazioni del pianeta Marte durante l'opposizione perielica del 1971.** G. de Mottoni.
Oss. Astron. Collurania, Teramo, Mem. ed Oss., Vol. 2, Fasc. 3, 22 pp. (1974).

097.060 **Nominal and effective seasons on Mars.**
G. de Mottoni y Palacios.
Mem. Soc. Astron. Italiana, Vol. 46, 341 - 348 (1975).

Generally the seasonal conditions on Mars are considered to be similar to those of the earth. Several important elements make the Martian seasonal conditions quite different from the terrestrial ones. Neglecting the lower temperatures and the composition of the atmosphere, four other points of differentiation dominate this situation: The lack of oceanic masses; The smallness of the atmospheric density; The large eccentricity of the orbit; The presence of the polar caps.

097.061 **Oposición de Marte (1975).** O. Betancourt.
El Universo, Vol. 29, No. 112, p. 7 - 8 (1975).

097.062 **The electron concentration in the Martian ionosphere according to radio occultation data taking into account the horizontal gradients.** M. A. Kolosov, A. I. Efimov, V. M. Ivanov, D. S. Kukin, Yu. G. Spiridonov, O. I. Yakovlev.
XI Vses. konf. po rasprostr. radiovoln. Chast' (Part) 1. Tezisy dokl. Kazan', Kazan. un-t, 1975, p. 153 - 156. In Russian. Abstr. in Referativ. Zhurn. 51. Astron., 2.51.360; 62. Issled. kosmich. prostranstva, 2.62.194 (1976).

097.063 **The lower ionosphere of Mars.**
R. C. Whitten, I. G. Poppoff.
Geophys. Res. Letters, Vol. 2, 565 - 566 (1975).
It is suggested that a recent observation of an apparent lower ionosphere on Mars may be due to ionization by galactic cosmic rays.

097.064 **Evidence of active and ancient volcanism on Mars. A review. II.** P. Hedervari.
Ann. Geofis.,(*Italy*), Vol. 27, 83 - 90 (1974).

097.065 **Channels on Mars.** R. P. Sharp, M. C. Malin.
Bull. American Soc. Geol., Vol. 86, 593 - 609 (1975)

The planet Mars. See Abstr. 003.014.

Photometric characteristics of Mars according to data of photographic observations during the great opposition of 1971. See Abstr. 003.049.

Mars as viewed by Mariner 9.
See Abstr. 003.143.

Positions of Mars determined with the zonal astrograph at Nikolaev during 1967–1971. See Abstr. 041.062.

On a precision estimate of the theories of motion of the earth and Mars. See Abstr. 042.012.

Apparent ephemeris of Mars 1960–1980.
See Abstr. 047.039.

Planetary magnetospheres. See Abstr. 091.010.

Polar symmetric flow of a viscous compressible atmosphere: an application to Mars. See Abstr. 091.044.

The atmospheres of Mars and Venus–a comparison.
See Abstr. 093.002.

Mapping of the moon and Mars.
See Abstr. 094.136.

Modern methods and results of mapping the moon and Mars. See Abstr. 094.155.

Mars, Satellites

097.201 **Mars: satellite origin and angular momentum.**
W. K. Hartmann, D. R. Davis, C. R. Chapman, S. Soter, R. Greenberg.
Icarus, Vol. 25, 588 - 594 (1975). – Paper presented at IAU Colloquium No. 28, "Planetary satellites", August 18–21, 1974, Ithaca, N.Y.
The origin of Phobos and Deimos is considered with a view to accounting for the existence of very small satellites with circular orbits in the Martian equatorial plane, and simultaneously for the suspected angular momentum deficiency of the Mars system. All models considered failed to satisfy at least one requirement, and the problem is considered more puzzling than is at first apparent. The Martian angular momentum deficiency, if physically significant, may be unrelated to the present satellites' origin, but might relate to a large ancient satellite, long ago destroyed. Accretion onto Mars of large amounts of asteroidal dust brought in by Poynting–Robertson drag may have some bearing on the angular momentum problem.

097.202 **The motions of Phobos and Deimos from Mariner 9 TV data.** G. H. Born, T. C. Duxbury.
Celestial Mechanics, Vol. 12, 77 - 88 (1975). – Paper presented at IAU Colloquium No. 28, 'Planetary satellites', Ithaca, N.Y., August 1974.
Orbit elements for the two Martian satellites Phobos and Deimos have been determined from 80 television photographs of the satellites taken by the imaging system of the Mariner 9 spacecraft. Phobos was found to be within 60 km of its positions predicted by recently published ephemeris theories which include a secular acceleration term in the longitude. Deimos was found to be within 100 km of its position predicted from earth-based observations. Comparison of the satellite's orbits determined from Mariner 9 data are made to these same ephemeric theories which are based on recent processing earth-based observations. In addition, the magnitude of periodic perturbations to the satellite orbits due to Mars' gravity field and solar gravity are discussed and a 110 km long period perturbation in the longitude of Deimos is identified.

097.203 **The orbits of Phobos and Deimos in 1973 as found from observations performed at the Abastumani Astrophysical Observatory.** G. N. Salukvadze, A. Sh. Khatisov, N. K. Sumzina, V. A. Shor.
Astron. Tsirk., No. 857, p. 4 - 5 (1975). In Russian.

097.204 **Results of positional observations of Martian satellites.** A. Sh. Khatisov, G. N. Salukvadze.
Byull. Abastumansk. Astrofiz. Obs., No. 46, p. 217 - 224 (1975). In Russian.
60 relative positions of Phobos and 114 of Deimos in 1973 were measured on 16 plates, the results are presented in tables. The same tables contain the results of comparison of the observed positions with the ephemerides. The mean square errors of one determination for Phobos and Deimos equal to ±0."39 and ±."45, respectively.

097.205 **The first map of Phobos.** S. R. Brzostkiewicz.
Urania Kraków, Vol. 46, 332 - 335 (1975). In Polish.

The motion of the Martian satellites.
See Abstr. 042.032.

An analysis of Martian satellite photographic observations of 1967. See Abstr. 097.001.

The new Martian nomenclature of the International Astronomical Union. See Abstr. 097.026.

The smaller bodies of the solar system.
See Abstr. 098.015.

A coordinated effort for observations and theories of the natural satellites. See Abstr. 099.220.

098 Minor Planets

098.001 **Ephemerides of minor planets for 1976.**
Editor: Institut Teoreticheskoj Astronomii Akademii Nauk SSSR, under the editorship of G. A. Chebotarev. Izdatel'stvo "Nauka", Leningradskoe Otdelenie, Leningrad. 200 pp. Price 2 Rbl. 52 Kop. (1975). In Russian and English. Contents: Introduction, p. 3 - 8; Information on new elements, p. 9 - 11; New elements, p. 12 - 14; Elements, p. 15 - 48; Opposition dates, p. 49 - 60; Ephemerides, p. 61 - 179; Ephemerides of bright planets, p. 180 - 190; Ephemerides of some unusual planets, p. 191 - 198; Critical list, p. 199.

098.002 **Observations of minor planets made at the Tartu Observatory.** H. K. Raudsaar.
Byull. Inst. Teoret. Astron., *Leningrad,* Vol. 13, 656 (1975). In Russian.

098.003 **Photographic positions of minor planets observed at Tartu.** H. K. Raudsaar.
Byull. Inst. Teoret. Astron., *Leningrad,* Vol. 13, 657 (1975). In Russian.

098.004 **Observations of minor planets made at the Crimean Astrophysical Observatory (25th report).**
L. I. Chernykh, B. A. Burnasheva, T. M. Smirnova, N. S. Chernykh.
Byull. Inst. Teoret. Astron., *Leningrad,* Vol. 13, 658 - 672 (1975). In Russian.

098.005 **Minor planets (1972).**
N. S. Samojlova-Yakhontova.
Byull. Inst. Teoret. Astron., *Leningrad,* Vol. 14, 1 - 7 (1975). In Russian.

098.006 **Improved elements of 42 minor planets.**
M. A. Dirikis.
Byull. Inst. Teoret. Astron., *Leningrad,* Vol. 14, 12 - 15 (1975). In Russian.

098.007 **Improvement of orbits of 29 minor planets.**
F. B. Khanina.
Byull. Inst. Teoret. Astron., *Leningrad,* Vol. 14, 45 - 46 (1975). In Russian.

098.008 **Photographic observations of minor planets at the Nikolayev Observatory during 1971.**
F. F. Kalikhevich.
Byull. Inst. Teoret. Astron., *Leningrad,* Vol. 14, 54 - 59 (1975). In Russian.

098.009 **Positional observations of seven minor planets.**
A. Sh. Khatisov.
Byull. Inst. Teoret. Astron., *Leningrad,* Vol. 14, 63 (1975). In Russian.

098.010 **Lichtelektrische Beobachtungen des Kleinen Planeten (675) Ludmilla.** H. J. Schober.
Mitt. Astron. Ges., No. 36, p. 98 - 99 (1975).

098.011 **An observation of Eros with a photographic zenith tube.** D. D. McCarthy.
Astron. Astrophys., Vol. 42, 287 - 288 (1975).
The close approach of minor planet 433 Eros provided the opportunity to obtain the first known position of a minor planet with a photographic zenith tube (PZT). Its position with respect to the PZT catalog was determined to ±0.″2 in declination and ±0.ˢ020 in right ascension.

098.012 **Nouvelles des earth-grazers – 1.**
M.-A. Combes, J. Meeus.
L'Astronomie, 89ᵉ année, p. 305 - 316 (1975).

098.013 **The angular momenta of solar system bodies: implications for asteroid strengths.** J. A. Burns.
Icarus, Vol. 25, 545 - 554 (1975).
The angular momentum H is plotted versus M for the planets and for all asteroids with known rotation rates and shapes. H is approximately proportional to $M^{5/3}$, which shows that the asteroids and most planets spin with nearly the same rate. The very smallest asteroids on the plot deviate from the above relation, usually containing excess angular momentum. This suggests that collisions have transferred substantial angular momentum to the smallest asteroids, perhaps causing their internal stress states to be substantially modified by centrifugal effects. The forces produced by gravitation are then compared to centrifugal effects for a rotating, triaxial ellipsoid of density 3 g cm^{-3}. For all asteroids with known properties the gravitational attraction is shown to be larger than the centrifugal acceleration of a particle on the surface: thus the observed asteroid regoliths are gravitationally bound.

098.014 **Micrometer observations of (433) Eros.**
W. D. Heintz.
Astrophys. Journ., Vol. 200, 787 (1975).
The orientation and elongation of the minor planet Eros was observed in 12 nights of the 1975 apparition.

098.015 **The smaller bodies of the solar system.**
W. K. Hartmann.
Sci. American, Vol. 233, No. 3, p. 142 - 147, 149 - 154, 156, 159 (1975). – Popular article.

098.016 **Coupled librations in the motions of asteroids near the 2 : 1 resonance.** R. Greenberg, F. Franklin.
Monthly Notices Roy. Astron. Soc., Vol. 173, 1 - 8 (1975).
Eight asteroids near the 2 : 1 resonance with Jupiter librate alternately in the secular and the commensurable modes over a time scale of thousands of years. An analytical theory is presented that explains this remarkable behaviour as a coupling between the two modes of libration.

098.017 **Radiometric diameters for an additional 22 asteroids.** D. Morrison, C. R. Chapman.
Bull. American Astron. Soc., Vol. 7, 467 - 468 (1975). – Abstr. AAS.

098.018 **Parallaxenmessung beim Kleinplaneten (433) Eros.**
R. Bendel, F. Frevert, B. Wedel.
SuW, Vol. 14, 301 - 305 (1975).

098.019 **Asteroidal motion at the 5/2, 7/3 and 2/1 resonances.** H. Scholl, C. Froeschlé.
Astron. Astrophys. Vol. 42, 457 - 463 (1975).
The authors test the collision hypothesis for the formation of the Kirkwood gaps at the resonances 5/2, 7/3 and 2/1. The model is based on the plane elliptic restricted three body problem averaged by Schubart's method. Including earlier results obtained for the 3/1 resonance, this model supports the collision hypothesis to a great extent. The model does not explain the complete observed width of the gaps, the absence of almost circular orbits in some cases and it does not favour collisions in the 7/3 gap.

098.020 **Minor planets and comets in libration about the 2 : 1 resonance with Jupiter.**

F. A. Franklin, B. G. Marsden, J. G. Williams, C. M. Bardwell.
Astron. Journ., Vol. 80, 729 - 746 (1975).

The authors examine the orbits of minor planets in a search for objects librating about the 2 : 1 mean-motion resonance with Jupiter. Some 30 candidates are found among the Palomar–Leiden Survey and other unnumbered minor planets. Although almost all the orbits are very uncertain, there does seem to be an indication that librators of low orbital eccentricity do exist, contrary to the hypothesis by Giffen. The authors also tabulate data describing 12 comets that are temporarily librating about the 2 : 1 resonance. Finally, they present a discussion of what are apparently "apocentric" librations.

098.021 **L'anello degli asteroidi.** R. Burchi.
Coelum, Vol. 43, 204 - 211 (1975).

098.022 **Minor planets Zhongshan 1 and Zhongshan 2.**
Planetary Laboratory, Purple Mountain Observatory.
Acta Astron. Sinica, Vol. 16, 88 - 90 (1975). In Chinese.

During the winter of 1974, two asteroids with perihelion distance less than 1.64 a.u. and inclination greater than 20° are found. The observations and orbital elements are listed. Their opposition ephemerides for 1976 and 1977 are given.

098.023 **Iris en Hebe in het voorjaar van 1974.** A. T. Son.
Zenit, Vol. 2, 328 - 329 (1975).

098.024 **Comparison of Krylov-Bogolyubov's and Runge-Kutta's methods in the theory of motion of the Minerva-group asteroids.** V. A Prikhod'ko.
In-t teor. i ehksperim. fiz. ITEhF-80. Moskva, 1974. 28 pp. Price 10 Kop. In Russian. – Abstr. in Referativ. Zhurn. 51. Astron., 9.51.158 (1975).

098.025 **The observational conditions during the great opposition of Eros.** O. K. Ukhova.
XXVIII Gertsenovsk. chteniya. Teor. fiz. i astron. Leningrad, 1975, p. 82 - 87. In Russian. – Abstr. in Referativ. Zhurn. 51. Astron., 9.51.182 (1975).

098.026 **The occultation of κ Geminorum by Eros – a learning experience.**
E. B. Weston, B. G. Marsden, B. O'Leary.
Journ. American Ass. Variable Star Observers, Vol. 4, 34 (1975). – Abstract.

098.027 **Größe und Beschaffenheit der Planetoiden.**
R. Bien.
SuW, Vol. 14, 354 - 356 (1975).

098.028 **Rotation period and photoelectric light curve of the minor planet 675 Ludmilla.**
H. J. Schober, R. Dvorak.
Astron. Astrophys., Vol. 44, 81 - 84 (1975).

The minor planet 675 Ludmilla was observed during the 1973 opposition. The synodic period of rotation comes out to $7^h43^m01^s \pm 16^s$, the maximum amplitude of the light curve is 0.28 mag. The range of the phase angle covered by the observations is 4° to 9°.

098.029 **The minor planet 654 Zelinda: rotation period and light curve.** H. J. Schober.
Astron. Astrophys., Vol. 44, 85 - 89 (1975).

The minor planet 654 Zelinda was observed during the 1973 opposition. The synodic period of rotation comes out to $31^h54^m \pm 3^m$, the maximum amplitude is about 0.3 mag. The range of the phase angle α covered by the observations is 9° to 15°. As the rotation period derived for 654 Zelinda is the longest one measured until now for minor planets, the frequency distribution of rotation periods for 57 asteroids with known periods is given.

098.030 **Photographic positional observations of Pallas in 1973 at the Astronomical Observatory of the Kharkov State University.** P. P. Pavlenko.
Khar'kov. un-t. Khar'kov, 1975. 13 pp. In Russian. – Abstr. in Referativ. Zhurn. 51. Astron., 8.51.167 (1975).

098.031 **The motion of asteroids near the commensurabilities 8/13, 7/12, 4/7.**
G. A. Chebotarev, A. S. Vashkevich.
Astron. Zhurn. Akad. Nauk SSSR, Vol. 52, 1063 - 1065 (1975). In Russian. English translation in Soviet Astron., Vol. 19, No. 5.

The motion of the planets (721) Tabora, (522) Helga and (1144) Oda surrounding the 4/7, 7/12 and 8/13 commensurabilities with Jupiter are examined. It is found that the asteroid (522) moves in a libration mode.

098.032 **Asteroid collisional evolution: evidence for a much larger early population.**
C. R. Chapman, D. R. Davis.
Science, Vol. 190, 553 - 556 (1975).

The present population of asteroids is a remnant of a vastly larger one that contained perhaps a planetary mass, dominantly distributed in planetesimals approximately 500 kilometers or less in diameter. It constituted a large reservoir of objects that plausibly were responsible for cratering the moon, Mars, and Mercury. Much asteroidal dust may have accumulated on Mars and other planets.

098.033 **An ellipsoidal model to interpret the light curve of an asteroid with application to Hector.**
R. De Santis.
Mem. Soc. Astron. Italiana, Vol. 46, 97 - 103 (1975).

A model to determine the ratios of the semi-axes and the coordinates of the pole of an ellipsoidal asteroid is given. This model is applied satisfactorily to 624 Hector.

098.034 **Determination of the perturbed orbit of the minor planet 1952 UV_1.** M. Kuzmanoski.
Publ. Obs. Astron. Beograd, No. 20, (see 012.024), p. 131 - 137 (1975). In Serbo-Croatian.

098.035 **Analyse statistique des lacunes de Kirkwood.**
D. Olević.
Publ. Obs. Astron. Beograd, No. 20, (see 012.024), p. 138 - 144 (1975). In Serbo-Croatian.

098.036 **New findings about Eros.** B. Zellner.
Sky Telescope, Vol. 50, 376 - 379 (1975).

098.037 **Unieke Eros-aktie groot succes.** W. Gielingh.
Zenit, Vol. 2, 384 - 389 (1975).

098.038 **7.5–13.5 micron spectra of Ceres and Vesta.**
F. C. Gillett, K. M. Merrill.
Icarus, Vol. 26, 358 - 360 (1975).

7.5–13.5 μm spectra ($\Delta\lambda/\lambda = 0.015–0.02$) of the asteroids Ceres and Vesta are presented. The observations are consistent with a model in which each element of the surface of the asteroid radiates as a grey body in equilibrium with the insolation.

098.039 **On the alleged collisional origin of the Kirkwood Gaps.** T. A. Heppenheimer.
Icarus, Vol. 26, 367 - 376 = Contr. Div. Geol. Planet. Sci., California Inst. Techn., *Pasadena,* No. 2585 (1975).

This paper examines two proposed mechanisms whereby asteroidal collisions and close approaches may have given rise to the Kirkwood Gaps. The first hypothesis is that asteroids in near-resonant orbits have markedly increased collision probabilities and so are preferentially destroyed, or suffer decay in population density, within the resonance zones. A simple order-of-magnitude analysis shows that this hypothesis is untenable. The second hypothesis is the Brouwer–Jefferys theory that collisions would smooth an asteroidal distribution function, as a function of Jacobi constant, thus forming resonance gaps. This hypothesis is examined by direct numerical integration of 50 asteroid orbits near the 2:1 resonance, with collisions simulated by random variables. No tendency to form a gap was observed.

098.040 **Accumulation of chondrules on asteroids.**
F. L. Whipple.
I.A.U. Colloquium No. 13, (see 012.030), p. 345 (1973). Abstract.

098.041 **Die Kleinen Planeten.** J. Schubart.
Naturwissenschaften, 62. Jahrgang, p. 565 - 571 (1975). – Review paper.

The orbits of 1900 minor planets have been accurately established. The results of a statistical survey of the numerous faint asteroids are available. Special types of orbital motion are revealed in a comparison of the range of variation of heliocentric distance for minor and major planets. New methods are provided for determining asteroid diameters. The resulting low mean density of Ceres is comparable to the densities of some types of carbonaceous chondrites. Measurements of reflectivity as a function of wavelength allow another comparison of asteroids and meteorites. The evolutionary process that led to the present population of minor planets is probably different for objects on regular and unusual orbits.

098.042 **Historique des petites planètes.** M.-A. Combes.
Ciel et Terre, Vol. 91, 393 - 418 (1975).

098.043 **Application of the theory of jet stream to the asteroidal belt.** W.-H. Ip.
Astrophys. Space Sci., Vol. 38, 475 - 481 (1975).

The possibility of incorporating the resonant effect and jet stream formation process into the problems of the Hilda asteroids and Kirkwood gaps is discussed qualitatively. It appears that formation of the precursor jet streams of the resonant asteroids in the main belt would be suppressed due to the collisional perturbation effect of the ambient matter in this region. Together with the biased distribution of near-resonant asteroids, the depletion across the Kirkwood gaps could be understood. Within the context of jet stream theory the existence of Hilda asteroids outside the main belt requires the original limit of the main belt to be not much more extensive than the present value of 3.5 AU. This is suggestive of a cosmogonic origin of the observed outer limit.

098.044 **Infrared spectral observations of asteroid 4 Vesta.**
H. P. Larson, U. Fink.
Icarus, Vol. 26, 420 - 427 (1975).

An ir spectrum of asteroid Vesta, the first of any asteroid, has been recorded at a spectral resolution of 44 cm^{-1} with a Fourier spectrometer. An electronic absorption band is observed that is assigned to an iron-rich pyroxene (pigeonite) spectroscopically similar to that found in certain eucrites. Other important rock-forming minerals such as olivine and plagioclase feldspar are not observed. There is no evidence for compositional variation with rotational phase angle. This spectroscopic picture of Vesta suggests considerable evolution including the melting and differentiation of silicates.

098.045 **Observations of comets and asteroids at the Kleť Observatory in the year 1973.** A. Mrkos.
Acta Univ. Carolinae, Math. Phys., Vol. 16, No. 2, p. 47 - 56 (1975). – Precise positions of the following minor planets are given: 826 Henrika, 897 Lysistrate, 502 Sigune, 167 Urda.

098.046 **Minor planets and related objects. XVIII.** ***UBV*** **photometry and surface composition.**
B. Zellner, W. Z. Wisniewski, L. Andersson, E. Bowell.
Astron. Journ., Vol. 80, 986 - 995 (1975).

New photometric observations on the *UBV* system are presented for 45 asteroids, bringing the total to 91 for which complete *UBV* colors are available. The reduced magnitudes *B* (1,0) scatter about the adopted catalog values by $\pm 0^{m}25$, with negligible systematic difference. The *UBV* colors fall into three distinct groups which correspond almost perfectly to the *S* (silicaceous or stony), *C* (carbonaceous), and *U* (unclassifiable) asteroid types recently identified from polarimetry, infrared radiometry, and narrow-band spectrophotometry. Thus, it becomes possible to classify asteroids from *UBV* observations alone. The *S* asteroids have colors very similar to those of L and LL ordinary chondrites, and some of the *U* objects are consistent in color with nickel–iron or enstatite-rich meteorites, but no meteorites appear to have the *UBV* colors of the *C* asteroids. The asteroid *C* and *S* compositional types have no exact counterparts among the surfaces of natural satellites.

098.047 **Sur le problème de proximité des orbites d'astéroïdes.** V. V. Mišković.
Prirodno-matematichki fakultet, Beograd - Glas Srpske akademije nauka i umetnosti 291, 31 - 54 (1974). In Serbo-Croatian. – Abstr. in Bull. Sci. Yougoslavie, Sect. A, Vol. 20, 366 (1975).

098.048 **Photographic observations of minor planets at Nikolaev in 1972 - 1973.**
F. F. Kalikhevich, V. I. Voronenko, G. K. Gorel'.
Gl. astron. observ. AN SSSR. Leningrad, 1975. 29 pp. In Russian. – Abstr. in Referativ. Zhurn. 51. Astron., 1.51.193 (1976).

098.049 **Photographic observations of Eros.**
V. A. Golubev, A. S. Konyukhov.
Astron. Tsirk., No. 873, p. 7 - 8 (1975). In Russian.

098.050 **Photographic photometry of small asteroids: a light curve survey with Schmidt telescopes.**
C.-I. Lagerkvist.
Astron. Astrophys., Vol. 45, 439 - 440 (1975).

A brief description is given of a photographic survey of light curves of small asteroids with the Kvistaberg and the Uppsala Southern Schmidt (Mount Stromlo) telescopes. Rotation periods and light curve amplitudes of 13 asteroids are presented as a function of absolute magnitude.

098.051 **1973 NA.** J. Sanders, S. J. Bus, E. Helin, T. Kiceniuk.
IAU Circ., No. 2806 (1975).

098.052 **1973 EC.** E. Roemer, C. C. McCarthy.
IAU Circ., No. 2813 (1975).

098.053 **1973 EA.** B. G. Marsden.
IAU Circ., No. 2816 (1975).

098.054 **Ceres-Psyche campaign.** B. Zellner.
IAU Circ., No. 2821 (1975).

098.055 **1973 EC.**
E. Roemer, R. McCallister, C. C. McCarthy.
IAU Circ., No. 2854 (1975).

098.056 **Fast-moving object West.** R. M. West.

IAU Circ., No. 2874 (1975).

098.057 **1975 TB.**
R. M. West, K. Aksnes, B. G. Marsden, Z. Sekanina.
IAU Circ., No. 2886 (1975).

098.058 **1973 EA.** R. E. McCrosky, G. Schwartz, E. Roemer, M. A. Daniel.
IAU Circ., No. 2890 (1975).

098.059 **Fast-moving object Wroblewski.**
C. Torres, H. Wroblewski, B. G. Marsden.
IAU Circ., Nos. 2891, 2893 (1975).

098.060 **Fast-moving object Kowal.**
B. G. Marsden, K. Aksnes.
IAU Circ., No. 2893 (1975).

098.061 **The observation of (433) Eros by Tokyo PZT.**
Y. Niimi, K. Hurukawa, K. Nakajima.
Tokyo Astron. Bull., Second Ser., No. 239, p. 2011 - 2017 (1975).

098.062 **Photometric investigations of the asteroids (43) Ariadne and (71) Niobe.** G. Lustig, R. Dvorak.
Acta Phys. Austriaca, Vol. 43, 89 - 97 = Mitt. Univ. Sternw. Graz No. 17 (1975). In German.

The minor planets (43) Ariadne and (71) Niobe were observed photoelectrically at the Observatoire de Haute Provence (OHP) in August 1972 and February 1974, respectively. The reduction was done by a new computer program. It yielded for the minor planet (43) Ariadne an amplitude of $0^m\!.66$ and led to a rotation period of $5^h\!.760$. The rotation period of (71) Niobe was determined to $11^h\!.213$ with an amplitude of $0^m\!.10$.

098.063 **On the period of light variations of Eros.**
B. N. Firmanyuk.
Astron. Tsirk., No. 881, p. 7 - 8 (1975). In Russian.

098.064 **Minor planets (1973).**
N. S. Samojlova-Yakhontova.
Byull. Inst. Teoret. Astron., *Leningrad*, Vol. 14, 73 - 80 (1975). In Russian.

098.065 **Improvement of the orbital elements of selected asteroids for the correction of star catalogue positions.** V. I. Orel'skaya.
Byull. Inst. Teoret. Astron., *Leningrad*, Vol. 14, 95 - 101 (1975). In Russian.

A brief review of performed and future work on the determination of corrections to catalogue star positions from observations of 10 selected minor planets is given. The orbital elements of these planets are reported. The difficulty of reducing at present the corrected 18000 planet positions to the homogeneous FK4 system is emphasized.

098.066 **Observations of minor planets made at the Crimean Astrophysical Observatory** (26th report).
L. I. Chernykh, T. M. Smirnova.
Byull. Inst. Teoret. Astron., *Leningrad*, Vol. 14, 113 - 126 (1975). In Russian.

098.067 **Observations of minor planets made at the Crimean Astrophysical Observatory** (27th report).
L. I. Chernykh, N. S. Chernykh.
Byull. Inst. Teoret. Astron., *Leningrad*, Vol. 14, 127 - 131 (1975). In Russian.

098.068 **The evolution of orbits of asteroids with small mean motions.**
G. A. Chebotarev, N. A. Belyaev, R. P. Eremenko.
Byull. Inst. Teoret. Astron., *Leningrad*, Vol. 14, 178 - 180 (1975). In Russian.

The evolution of orbits of nine planets with small mean motions has been investigated by the method of special perturbations for a time interval of about 400 years (1660–2059). Four asteroids have been discovered which can passage through Jupiter's sphere of action.

098.069 **Observations of minor planets at Saratov.**
E. V. Vojtyuk, Yu. F. Zinchenko, Yu. V. Mikhajlov.
Byull. Inst. Teoret. Astron., *Leningrad*, Vol. 14, 181 - 182 (1975). In Russian.

098.070 **Observations of minor planets made at the Crimean Astrophysical Observatory** (28th report).
L. V. Zhuravleva.
Byull. Inst. Teoret. Astron., *Leningrad*, Vol. 14, 183 - 194 (1975). In Russian.

098.071 **Photographic observations of Eros.**
G. V. Zhukov, M. I. Kibardina, N. G. Rizvanov.
Kometn. Tsirk., *Kiev*, No. 182 (1975). In Russian.

098.072 **Observations of Eros at the Gissar Astronomical Observatory of the Astrophysical Institute of the Academy of Sciences of the Tadzhik SSR.**
N. Kiselev, N. Narizhnaya, I. Gavrilova, S. Gerasimenko.
Kometn. Tsirk., *Kiev*, No. 184 (1975). In Russian.

098.073 **Photographic observations of Eros during the Alpine expedition of the Sternberg Astronomical Institute.**
V. M. Kovalenko.
Kometn. Tsirk., *Kiev*, No. 186 (1975). In Russian.

098.074 **Analysis of asteroids and small satellites by reflection polarimetry.** A. Dollfus.
Meteoritics, Vol. 10, 394 (1975).

098.075 **Observations: visual and photographic positions of Eros.** R. Fangor, P. Grzędzielski, L. Newelski.
Urania Kraków, Vol. 46, 275 - 278 (1975). In Polish.

098.076 **Posiciones exactas de pequeños planetas obtenidas en el Observatorio Astronómico de Madrid en 1972.**
M. de Pascual.
Bol. Astron. Obs. Madrid, Vol. 8, No. 5, p. 76 - 81 (1975).

098.077 **An ellipsoidal asteroid: 39 Laetitia.**
R. de Santis.
Mem. Soc. Astron. Italiana, Vol. 46, 355 - 360 (1975).

Laetitia is an ellipsoidal asteroid. Its semiaxes ratios are $c/a = 0.40 \pm 0.02$ and $c/b = 0.70 \pm 0.02$, while its ecliptic coordinates of the pole are $\lambda_0 = 121°$ and $\beta_0 = 32°$ with an accuracy of about 3°.

098.078 **Durchmesserbestimmung und Abschätzung der Gesamtmasse der kleinen Planeten.** G. Lustig.
Sternenbote, 18. Jahrgang, p. 114 - 119 (1975).

098.079 **Least distances of Apollo and Amor objects to planetary orbits.** J. Meeus.
Minor Planet Bull., Vol. 3, 1 - 3 (1975).

098.080 **Historical notes: concerning 1 Ceres and 330 Adalberta.** M.-A. Combes.
Minor Planet Bull., Vol. 3, 4 - 5 (1975).

098.081 **Some additional speculations on 330 Adalberta.**
F. Pilcher.
Minor Planet Bull., Vol. 3, 5 - 6 (1975).

098.082 **Guide to "New names of minor planets".** J. LoGuirato.
Minor Planet Bull., Vol. 3, 6 - 7 (1975).

098.083 **Calculation of the magnitude of asteroids.** J. Meeus.
Minor Planet Bull., Vol. 3, 7 - 10 (1975).

098.084 **Photographic photometry of small asteroids.** (Preliminary report). C.-I. Lagerkvist.
Minor Planet Bull., Vol. 3, 11 - 19 (1975).

098.085 **Interpolating minor planet or comet positions between ephemeris dates.** F. Pilcher.
Minor Planet Bull., Vol. 3, 23 - 24 (1975).

098.086 **Minor planets at unusually favorable opposition in 1976.** F. Pilcher.
Minor Planet Bull., Vol. 3, 29 - 32 (1975).

A list is presented of 91 minor planets which are much brighter than usual at their 1976 oppositions. Observations are especially urged for 1580 Betulia and 944 Hidalgo at their coming close approaches to earth, and ephemerides are provided.

098.087 **Planets 1580 Betulia and 1620 Geographos in 1976.** J. Meeus.
Minor Planet Bull., Vol. 3, 33 - 34 (1975).

098.088 **Observations photographiques de petites planètes, effectuées à l'astrographe double de 40 cm au cours de l'année 1972 (2me semestre).** H. Debehogne.
Bull. Astron., Obs. Roy. Belgique, Vol. 8, 198 - 215 (1975).

098.089 **Observations photographiques de petites planètes, effectuées à l'astrographe double de 40 cm au cours de l'année 1973.** H. Debehogne.
Bull. Astron., Obs. Roy. Belgique, Vol. 8, 216 - 218 (1975).

098.090 **Observations photographiques de petites planètes, effectuées en 1973 à l'équatorial Gautier (Carte du Ciel) de l'Observatoire de Santiago du Chili (Cerro Calan).** H. Debehogne.
Bull. Astron., Obs. Roy. Belgique, Vol. 8, 219 - 230 (1975).

098.091 **Observations photographiques de petites planètes, effectuées à l'astrographe double de 40 cm au cours de l'année 1974.** H. Debehogne.
Bull. Astron., Obs. Roy. Belgique, Vol. 8, 260 - 270 (1975).

098.092 **Observations photographiques de petites planètes, effectuées à l'astrographe double de 40 cm au cours de l'année 1975 (1er semestre).** H. Debehogne.
Bull. Astron., Obs. Roy. Belgique, Vol. 8, 271 - 277 (1975).

098.093 **Promatranja prolaza planetoida Eros.** G. Kren.
Vasiona, Vol. 23, 51 - 54 (1975).

098.094 **The construction of intermediate orbits of the asteroids of the Hecuba family with an electronic computer.** G. F. Sultanov, O. A. Belousova, F. I. Isaeva, T. M. Shaposhnikova.
Dokl. AN AzSSR, Vol. 31, No. 7, p. 16 - 20 (1975). In Russian
Abstr. in Referativ. Zhurn. 51. Astron., 2.51.202 (1976).

098.095 **Minor Planet Circulars, (MPC),** Nos. 3829 - 3908 (1975).
Edited by Cincinnati Observatory under the supervision of P. Herget.

A repository of nearly all new data for numbered and unnumbered minor planets: Observations, elements and ephemerides, identifications, newly assigned numbers and names, occultations.

Observations of minor planets for improving the positions of stars of a fundamental catalogue. See Abstr. 041.020.

Die Entstehung der Kirkwood-Lücken im Asteroiden-Gürtel. See Abstr. 042.008.

Giant planets and their satellites, minor planets, meteorites and comets. See Abstr. 091.032.

Extra-terrestrial geology. See Abstr. 094.479.

Occultations et appulses planétaires en 1976. See Abstr. 097.025.

Mass loss from the region of Mars and the asteroid belt. See Abstr. 097.041.

Ultraviolet observations of small bodies in the solar system by OAO-2. See Abstr. 099.202.

099 Jupiter, Jupiter Satellites

Jupiter

099.001 **The structure and evolution of Jupiter: the fluid contraction stage.**
H. C. Graboske, Jr., J. B. Pollack, A. S. Grossman, R. J. Olness.
Astrophys. Journ., Vol. 199, 265 - 281 (1975).

An evolutionary study of a 0.00095 $M_{\odot}$ star, composed of a convective, adiabatic, homogeneous fluid, has been performed using stellar structure methods. An early stellar phase, behaving like a typical low-mass pre-main-sequence object, has high luminosities which have left a record in the structure of the Galilean satellites. This phase also has high internal temperatures, $(T_c)_{max}$ reaching 51,400, which ensures a long subsequent evolution as a fully mixed, convective structure. The second phase is an approach to a degenerate dwarf cooling curve, which gives excellent agreement with the observed radius and luminosity of Jupiter. An analysis of the sensitivity of evolution to chemical composition, solar energy deposition, equation of state, model atmospheres, and superadiabaticity demonstrates that equation of state and superadiabaticity have the strongest influence over planetary time scales.

099.002 **Hydrogen-alpha auroral activity on Jupiter.**
R. S. Clary, J. H. Hunter, Jr.
Astrophys. Journ., Vol. 199, 517 - 529 (1975).

This paper deals with an attempt to observe Hα auroral activity on Jupiter. Photographs in Hα light were taken of the planet in the spring and summer of 1972. This investigation and many others have been undertaken because Jupiter has been shown to have a strong magnetic field and to emit nonthermal radiation at various radio wavelengths which seems to be related to magnetic fields and solar activity.

099.003 **Asymmetries of Jupiter's magnetosphere.**
M. M. Komesaroff, P. M. McCulloch.
Monthly Notices Roy. Astron. Soc., Vol. 172, 91 - 95 (1975).

It is shown that departures from a dipolar configuration cannot, by themselves, completely explain the observed variation of polarization position angle with Jovian longitude and that other effects must be invoked, e.g. planetary shadowing or non-rigid rotation of the magnetosphere. Apart from these effects the position angle measurements are consistent with the symmetry of the dual-dipole model of the planetary field, suggested by Smith et al. on the basis of the Pioneer 10 measurements.

099.004 **Refined close-ups of Jupiter.**
Sky Telescope, Vol. 50, 24 - 25 (1975).

099.005 **A wave driven model of the Jovian equatorial jet.**
T. Maxworthy.
Planet. Space Sci., Vol. 23, 1223 - 1233 (1975).

The author explores the consequences of assuming that the equatorial jet, which is the most prominent feature of the atmospheric motions on Jupiter, is driven by vertically propagating Kelvin and mixed Rossby-gravity waves that are absorbed during their passage through the stratosphere to produce a mean zonal flow. Since even the basic atmospheric parameters on Jupiter are still largely a matter of conjecture, the author is not able to produce a unique model but one that can span a whole range of possible conditions and which can be refined (or rejected) as more information becomes available.

099.006 **Jupiters Magnetfeld und eine obere Grenze für die Ruhemasse des Photons.** I. Khan.
Mitt. Astron. Ges., No. 36, p. 71 (1975). – Abstract.

099.007 **Magnetospheres of earth and Jupiter after Pioneer 10.**
A. Prakash, N. Brice.
Space Sci. Rev., Vol. 17, 823 - 835 (1975).

In this paper the authors point out possible reasons for the remarkable differences observed between the magnetospheres of earth and Jupiter and give a model of Jupiter's magnetosphere which sheds light on the observations of Pioneer 10 mission and gives some idea of the interesting features one should look for during the forthcoming Pioneer 11 flyby of Jupiter.

099.008 **Dénouement of Jovian radiation belt theory.**
F. V. Coroniti.
Space Sci. Rev., Vol. 17, 837 - 856 (1975).

099.009 **Is Jupiter's magnetosphere like a pulsar's or earth's?**
C. F. Kennel, F. V. Coroniti.
Space Sci. Rev., Vol. 17, 857 - 883 (1975).

099.010 **On the 23-micron feature in the spectrum of Jupiter.**
K. Fox.
Astrophys. Journ., (*Letters*), Vol. 199, L189 - L191 (1975).

The spectroscopic activity of gaseous constituents in the Jovian atmosphere is investigated with a view toward absorption at 23 μ. Various possibilities for the source of this feature are excluded, particularly pressure-induced dipole transitions involving collisions between H_2 and either CH_4 or HD.

099.011 **The atmosphere and ionosphere of Jupiter.**
M. B. McElroy.
Astrophys. Space Sci. Library, Vol. 51, (see 012.007), 409 - 423 (1975).

099.012 **Spectroscopy of Jupiter and Saturn.** G. E. Hunt.
Astrophys. Space Sci. Library, Vol. 51, (see 012.007), 425 - 432 (1975).

099.013 **The radioastronomy of Jupiter.**
D. Stannard.
Astrophys. Space Sci. Library, Vol. 52, (see 012.008), 221 - 236 (1975). – Review paper.

099.014 **Long term variations of the decimetric radio emission of Jupiter (and Saturn?).** E. Gerard.
Astrophys. Space Sci. Library, Vol. 52, (see 012.008), 237 - 239 (1975).

099.015 **Upper limits to Jovian X-ray emission from the Uhuru satellite.** K. C. Hurley.
Astrophys. Space Sci. Library, Vol. 52, (see 012.008), 241 - 244 (1975).

099.016 **Upper limits for X-ray emission from Jupiter as measured from the Copernicus satellite.**
J. F. Vesecky, J. L. Culhane, F. J. Hawkins.
Astrophys. Space Sci. Library, Vol. 52, (see 012.008), 245 - 251 (1975).

099.017 **Jupiter and Saturn.** R. Hide.
Astrophys. Space Sci. Library, Vol. 52, (see 012. 008), 253 - 254 (1975). – Abstract of an invited lecture 'The origin of Jupiter's magnetic field' dealing with the dynamics, internal constitution and magnetic fields of the earth, Jupiter and Saturn and the theory of hydromagnetic dynamos in rotating fluids.

099.018 **Photochemistry of hydrocarbons in the Jovian up-**

per atmosphere.
S. S. Prasad, L. A. Capone, L. J. Schneck.
Astrophys. Space Sci. Library, Vol. 52, (see 012.008), 255 - 268 (1975).

099.019 **The Pioneer 10 plasma analyzer results at Jupiter.** J. H. Wolfe.
Astrophys. Space Sci. Library, Vol. 52, (see 012.008), 279 - 296 (1975).

099.020 **Pioneer 10 observations of the solar wind and its interaction with Jupiter: plasma electron results.**
D. S. Intriligator.
Astrophys. Space Sci. Library, Vol. 52, (see 012.008), 297 - 300 (1975).

099.021 **Observations in interplanetary space of relativistic electrons from Jupiter.**
D. L. Chenette, T. F. Conlon, J. A. Simpson.
Astrophys. Space Sci. Library, Vol. 52, (see 012.008), 301 - 306 (1975).

099.022 **On the distortion of the Jovian magnetic field for $R \gtrsim 40 R_J$ as deduced from charged particle studies.**
R. B. McKibben, J. A. Simpson.
Astrophys. Space Sci. Library, Vol. 52, (see 012.008), 307 - 311 (1975).

099.023 **Pioneer 10 observations of the Jovian magnetosphere: plasma electron results.**
D. S. Intriligator.
Astrophys. Space Sci. Library, Vol. 52, (see 012.008), 313 - 316 (1975).

099.024 **Characteristics of Jovian trapped electrons and protons for $R \lesssim 20 R_J$ and their interaction with Io.**
J. A. Simpson, D. C. Hamilton, R. B. McKibben, A. Mogro-Campero, K. R. Pyle, A. J. Tuzzolino.
Astrophys. Space Sci. Library, Vol. 52, (see 012.008), 317 - 324 (1975).

099.025 **Observations of Jovian accelerated particles both inside and outside the Jovian magnetosphere: results from the Goddard-U. of New Hampshire experiment on Pioneer 10.** J. H. Trainor, F. B. McDonald, B. J. Teegarden, W. R. Webber, E. C. Roelof.
Astrophys. Space Sci. Library, Vol. 52, (see 012.008), 325 - 353 (1975).

099.026 **Pioneer 10: observations of energetic electrons in the Jovian magnetosphere.** B. A. Randall.
Astrophys. Space Sci. Library, Vol. 52, (see 012.008), 355 - 373 (1975).

099.027 **Dénouement of Jovian radiation belt theory.** F. V. Coroniti.
Astrophys. Space Sci. Library, Vol. 52, (see 012.008), 391 - 410 (1975).

099.028 **Is Jupiter the cosmic ray source in our solar system?** G. Pizzella.
Astrophys. Space Sci. Library, Vol. 52, (see 012.008), 425 - 432 (1975).

099.029 **The magnetospheres of Jupiter and Saturn.** F. L. Scarf.
Astrophys. Space Sci. Library, Vol. 52, (see 012.008), 433 - 449 (1975).

099.030 **Is Jupiter's magnetosphere like a pulsar's or earth's?** C. F. Kennel, F. V. Coroniti.
Astrophys. Space Sci. Library, Vol. 52, (see 012.008), 451 - 477 (1975).

099.031 **Determination of the abundance of ammonia in the atmosphere of Jupiter.** N. B. Ibragimov.
Astron. Zhurn. Akad. Nauk SSSR, Vol. 52, 895 - 898 (1975). In Russian. English translation in Soviet Astron., Vol. 19, No. 4.

The fine structure of the ammonia absorption band 6450 Å has been studied by spectrograms of Jupiter taken at the 2-m reflector with dispersion of 6 Å/mm. The equivalent widths and half-widths are defined for 20 lines of the band. The comparison of these data with laboratory ones is followed by the estimate of the abundance of ammonia in the atmosphere of Jupiter to be equal to 17 ± 4 m · atm. Comparison of the spectrograms of Jupiter and the moon showed strong change in the ammonia absorption band λ 6450 Å.

099.032 **Jupiter's molecular hydrogen layer and the magnetic field.** R. Smoluchowski.
Astrophys. Journ., (*Letters*), Vol. 200, L119 - L121 (1975).

High pressures and temperatures in the molecular hydrogen layer of Jupiter can lead to electrical conductivity which is high enough to affect the hydromagnetic dynamo mechanism and to help explain the strength of the higher than dipolar components of the magnetic field observed by Pioneer 10 and 11.

099.033 **Jupiter: its infrared spectrum from 16 to 40 micrometers.** J. R. Houck, J. B. Pollack, D. Schaack, R. A. Reed, A. Summers.
Science, Vol. 189, 720 - 722 (1975).

Spectral measurements of the thermal radiation from Jupiter in the band from 16 to 40 micrometers are analyzed under the assumption that pressure-broadened molecular hydrogen transitions are responsible for the bulk of the infrared opacity over most of this spectral interval. Both the vertical pressure-temperature profile and the molecular hydrogen mixing ratio are determined. The derived value of the molecular hydrogen mixing ratio, 0.89 ± 0.11, is consistent with the solar value of 0.86.

099.034 **Cyclic octatomic sulfur: a possible infrared and visible chromophore in the clouds of Jupiter.**
B. N. Khare, C. Sagan.
Science, Vol. 189, 722 - 723 (1975).

A brown polymeric material, produced under simulated Jovian conditions, is composed primarily of cyclic octatomic sulfur, although a range of complex organic compounds is also present. The polymeric sulfur, produced from the ultraviolet photolysis of hydrogen sulfide, exhibits its strongest band at 465 reciprocal centimeters, in fair agreement with the frequency of the unidentified Jovian absorber recently announced by Houck et al. Polymeric octatomic sulfur may be an important constituent of the Jovian clouds.

099.035 **Jupiter vu par Pionnier 11.** B. Morando.
L'Astronomie, 89e année, p. 287 - 292 (1975).

099.036 **Further investigations of Jupiter models.** M. Podolak, A. G. W. Cameron.
Icarus, Vol. 25, 627 - 634 (1975).

In the authors' previous investigation of Jupiter models, subject to the constraint that the hydrogen-to-helium ratio is solar, they found a need to include considerable additional mass in the form of volatilized condensates in the atmosphere and excess mass in a central core. The stricter constraints imposed by the measurement of the gravitational moments of the planet as a result of the Pioneer 10 flyby indicate that the mass of the excess volatilized condensates, assumed to be water, relative to the core mass, assumed to be rock, exceeds the ratio of the relevant elements in the solar composition. In this paper the authors have tested the sensitivity of this conclusion to varying assumptions about the constraints and the equation of state.

099.037 **Motion of Jupiter's decametric sources in Io phase.** J. R. Thieman, A. G. Smith, J. May.
Astrophys. Letters, Vol. 16, 83 - 86 (1975).

Decametric radio observations of Jupiter by the Universities of Florida, Chile, and Texas have been combined to achieve 2° resolution in histograms of occurrence probability vs Io phase for the years 1957 through 1972. Under this high resolution, the centroids and maxima of the Io-related source A, B, and C peaks appear to oscillate sinusoidally with a time scale of the order of a decade, mirroring the drift of source A in central meridian longitude coordinates. The Io-related peaks are persistently asymmetric, being skewed toward earlier positions of Io, and the primary peaks are often preceded by smaller secondary peaks.

099.038 **Relation between Jupiter's decametric radio emission and its activity in the equatorial zone.**
D. Basu, C. J. Banos.
Astrophys. Letters, Vol. 16, 97 - 98 (1975).

Significant correlation has been found between the peculiar optical activity in the equatorial area of the planet Jupiter and the Jovian decametric radiation at 18.0, 22.2 and 27.6 MHz.

099.039 **Detection of carbon monoxide in Jupiter.** R. Beer.
Astrophys. Journ., (*Letters*), Vol. 200, L167 - L169 (1975).

New ground-based, 5-μ spectra of Jupiter have revealed the presence of a trace quantity of carbon monoxide in the lower atmosphere of the planet. A preliminary analysis suggests a number mixing ratio in the neighborhood of 10^{-9} and a column abundance of roughly 5×10^{17} molecules cm^{-2} to a depth of at least 2 atmospheres. It is suggested that the observation might be confirmed by microwave techniques.

099.040 **Possible new Jovian thermospheric models.** S. S. Prasad.
Astrophys. Journ., (*Letters*), Vol. 200, L171 - L174 (1975).

Possibilities of models for the Jovian thermosphere, based upon mechanical heating, have been explored. Under certain reasonable assumptions about the magnitude of the critical Richardson number, the correction to the conventional expression for the eddy heat flux, and the height dependence of the eddy diffusion coefficient, a very hot thermosphere could result, yielding exospheric temperatures more in tune with the bounds being suggested by the recent optical and radio occultation experiments.

099.041 **Jupiter.** J. H. Wolfe.
Sci. American, Vol. 233, No. 3, p. 118 - 126 (1975). Popular article.

099.042 **The outer solar system.** T. Owen.
Rev. Geophys. Space Phys., Vol. 13, No. 3, (see 013.001), p. 416 - 419, 425 - 426 (1975). – Review paper with bibliographical notes.

099.043 **Ethane and acetylene abundances in the Jovian atmosphere.**
A. T. Tokunaga, R. F. Knacke, T. Owen.
Bull. American Astron. Soc., Vol. 7, 468 (1975). – Abstr. AAS.

099.044 **An analysis of Jupiter's decametric emissions as observed with the University of Colorado spectrograph, 1960 - 1974.** A. C. Riddle.
Bull. American Astron. Soc., Vol. 7, 469 (1975). – Abstr. AAS.

099.045 **The Jovian Great Red Spot – its three months oscillation.** C. P. Renschen.
Astron. Nachr., Vol. 296, 241 - 243 (1975).

Fourier-analysis of the motion of Jupiter's Great Red Spot (GRS) yields concealed periodicities in this motion. These periodicities are possible clues for the decision in favour of one of the different models of the GRS.

099.046 **Jupiter's atmosphere: problems and potential of radio occultation.** R. Eshleman.
Science, Vol. 189, 876 - 878 (1975).

The atmospheric temperature-pressure profiles derived from the Pioneer 10–Pioneer 11 radio occultation experiment are mutually consistent but differ markedly from the results of other investigations. Current studies indicate that the occultation interpretation contains errors that were made very large by an inherent magnification effect, and that these errors have both geometrical and equipment sources. The apparent consistency between the Pioneer 10 and the Pioneer 11 results must be considered fortuitous. Despite these difficulties, the occultation technique, when optimally instrumented and carefully interpreted, retains its potential for atmospheric profile measurements of high accuracy and resolution.

099.047 **New studies of Jupiter.**
W. B. Hubbard, J. R. Jokipii.
Sky Telescope, Vol. 50, 212 - 216 (1975).

099.048 **Jupiter's non Io-related decametric radiation.**
G. M. Gruber.
Astrophys. Space Sci., Vol. 37, 169 - 172 (1975).

A correlation study between solar wind activity and Jupiter's decametric radiation seems to indicate that the solar wind has a suppressive effect on the non Io-related emission. A basic model for the non Io-related emission is presented, to explain this and other characteristics of the non Io-related emission.

099.049 **Jupiter: présentation 1974 – opposition 5 septembre 1974.** F. Jetzer.
Orion, 33. Jahrgang, p. 158 - 164 (1975). – Rapport No. 30 du Groupement planétaire SAS.

099.050 **Jupiter's atmospheric circulation.**
G. P. Williams.
Nature, Vol. 257, 778 (1975). – Letter.

099.051 **Gasring om Jupiter in baan van Io.** C. de Jager.
Zenit, Vol. 2, 362 - 363 (1975).

099.052 **The dependence of Jupiter's decametric emissions on the Jovian magnetic latitude of Io.**
H. R. Miller, A. G. Smith.
Icarus, Vol. 26, 16 - 23 (1975).

The present analysis confirms the contention that the majority of Jupiter's decametric radio emissions occur when Io is above the northern magnetic hemisphere of Jupiter. However, a substantial portion of the non-Io-related component of source *A* was found to occur when Io was above southern magnetic latitudes of Jupiter.

099.053 **A planet that is like a star.** F. Zigel'.
Nauka i zhizn', 1975, No. 4, p. 93 - 96. In Russian.

099.054 **On the latitudinal change of the Jovian cloud cover top height.** V. G. Tejfel'.
Pis'ma v Astron. Zhurn., Vol. 1, No. 10, p. 34 - 38 (1975). In Russian.

An analysis of measurements of moderate and strong CH_4 absorption bands has shown that the effective upper boundary of the visible Jovian cloud cover falls down from the equator toward high latitudes to about 15–20 km.

099.055 **Mechanism of the three-month oscillation of the Great Red Spot on Jupiter.** C. P. Renschen.
Astron. Nachr., Vol. 296, 285 - 287 (1975). In German.

The present paper is a test to explain the three-month oscillation of the Great Red Spot on Jupiter as a solution of the Navier-Stokes equation which is coupled with a nonlinear equation of oscillation. Since several parameters resp. functions are unknown it is difficult to find quantitative solutions.

099.056 **Phosphine on Jupiter and implications for the Great Red Spot.** R. G. Prinn, J. S. Lewis.
Science, Vol. 190, 274 - 276 (1975).

A study of the chemistry and photochemistry of the recently discovered phosphine in the atmosphere of Jupiter suggests that the red colorations on this planet result from photochemical production of red phosphorus particles. Chemical-dynamical models of this red phosphorus haze imply that the intensity of the red coloration is a strong function of the strength of vertical turbulent mixing in the atmosphere. If the Jovian Great Red Spot is a region of considerable dynamical activity this model provides a self-consistent explanation for the redness of this region in comparison to the rest of the planet.

099.057 **The thermal structure of Jupiter. I. Implications of Pioneer 10 infrared radiometer data.**
G. S. Orton.
Icarus, Vol. 26, 125 - 141 = Contr. Div. Geol. Planet. Sci., Calif. Inst. Techn., *Pasadena*, No. 2615 (1975).

Temperature profiles for low latitude regions of Jupiter in the 1.0–0.1 bar pressure regime are recovered from Pioneer 10 infrared radiometer data. The temperature near 0.1 bar is 108–117 K. For the South Equatorial Belt, the temperature at 1.0 bar is 170 K. The South Tropical Zone temperature at this level is 155 K. A cloud presence in the STrZ may be correlated with the visible and 5 micron appearance of the planet and with NH_3 saturation just below this position. The molar fraction of H_2 most consistent with the data is 0.91 ± 0.08. The effective temperatures of the SEB and STrZ are 127.6 and 124.2 K, respectively. These temperature profiles are generally consistent with data at other wavelengths and radiative-equilibrium models, but a discrepancy with the preliminary neutral atmosphere inversion of Pioneer 10 radio occultation data remains unexplained.

099.058 **The thermal structure of Jupiter. II. Observations and analysis of 8–14 micron radiation.**
G. S. Orton.
Icarus, Vol. 26, 142 - 158 = Contr. Div. Geol. Planet. Sci., Calif. Inst. Techn., *Pasadena*, No. 2607 (1975).

Observations of Jovian limb structure at 8.11 and 8.45 microns are reported. These are used along with other limb structure and spectral data in the 8–14 micron region to derive a model of the thermal and cloud structure within the 1.0–0.01 bar pressure regime. The temperature is about 165 K at 1.00 bar, 108 K at 0.10 bar, and 143 K at 0.03 bar. In zones, an optically opaque cloud of NH_3 exists near the 143 K (0.60 bar) level. A partly transparent haze of solid NH_3 particles overlies the cloud. Belts are free of the cloud and have a much lower abundance of NH_3 haze than the zones. The data are consistent with an NH_3 gas abundance defined by saturation equilibrium, with a mixing ratio of 1.5×10^{-4} deep in the atmosphere, and with a CH_4 mixing ratio of 2×10^{-3}, about three times the currently accepted value.

099.059 **Spatially resolved absolute spectral reflectivity of Jupiter: 3390–8400 Å.** G. S. Orton.
Icarus, Vol. 26, 159 - 174 = Contr. Div. Geol. Planet. Sci., Calif. Inst. Techn., *Pasadena*, No. 2593 (1975).

Observational determinations of the absolute spectral reflectivities of visually distinct regions of Jupiter are presented. The observations cover the 3390–8400 Å region at 10 Å resolution, and they are compared with observations using 150–200 Å filters in the 3400–6400 Å range. The effective reflectivities for several regions (on the meridian) in the 3400–8400 Å range are: South Tropical Zone, 0.76 ± 0.05; North Tropical Zone, 0.68 ± 0.08; South Equatorial Belt, 0.63 ± 0.08; North Equatorial Belt, 0.62 ± 0.04; and the Great Red Spot, 0.64 ± 0.09.

099.060 **On the tidal influence on Jupiter's gravitational field.**
S. V. Gavrilov, V. N. Zharkov, V. V. Leont'ev.
Astron. Zhurn. Akad. Nauk SSSR, Vol. 52, 1021 - 1028 (1975). In Russian. English translation in Soviet Astron., Vol. 19, No. 5.

The tidal influences on the gravitational fields of the giant planets are computed. The "gravitational noise" due to the tides can influence the determination of J_8 and J_{10} for Jupiter. An idea of tidal sounding of the giant planets has been formulated.

099.061 **Zweite US-Sonde am Jupiter. Erste Ergebnisse des Vorbeifluges von Pionier 11 an Jupiter.**
H. W. Köhler.
Phys. Blätter, 31. Jahrgang, p. 504 - 516 (1975).

099.062 **More on large-scale disturbances on Jupiter.**
W. K. Wacker.
Strolling Astronomer, Vol. 25, 173 (1975).

099.063 **The appearance of Jupiter in 1973 – an interim report.** P. K. Mackal.
Strolling Astronomer, Vol. 25, 173 - 182 (1975).

099.064 **A review of the Jovian magnetosphere based upon Pioneer 10 and 11.** J. H. Trainor.
Goddard Space Flight Center, Greenbelt, Maryland, GSFC Document X-660-75-130, 43 pp. (1975).

This paper reviews the data derived from the plasma, magnetic field and energetic particle experiments on the Pioneer 10 and 11 spacecraft which encountered Jupiter in December of 1973 and 1974, respectively. The major features and phenomena are surveyed with the goal of leading to a consistent picture or model of this giant magnetosphere which is so different from that of our own earth.

099.065 **Jovian sodium cloud geometry.**
A. I. Ershkovich, Y. B. Mekler.
Astrophys. Space Sci., Vol. 38, L1 - L2 (1975). – Letter.

099.066 **Limit on the photon mass deduced from Pioneer-10 observations of Jupiter's magnetic field.**
L. Davis, Jr., A. S. Goldhaber, M. M. Nieto.
Phys. Rev. Letters, Vol. 35, 1402 - 1405 (1975).

The authors report an analysis of the Pioneer-10 data on Jupiter's magnetic field, in which the mass μ of the photon was treated as a free parameter. They set a limit of $\mu \leqslant 2 \times 10^{-11}$ $cm^{-1} \equiv 6 \times 10^{-16}$ eV $\equiv 8 \times 10^{-49}$ g. This is the smallest limit so far obtained from direct measurements.

099.067 **Fluorescence detection of organic molecules in the Jovian atmosphere.** J. S. Levine, R. S. Rogowski.
Origins of Life, Vol. 6, 395 - 399 (1975). – Paper presented at the conference on the 'Chemical evolution of the giant planets', Univ. Maryland, 1974, Oct 23 - 26.

A search for fluorescent emission due to the presence of possible organic molecules in the Jovian atmosphere is presented. The authors consider fluorescent emission excited by a laser system aboard a Jupiter spacecraft. Their calculations indicate that laser-induced fluorescent detection is approximately two orders of magnitude more sensitive than rocket ultraviolet measurements of possible Jovian absorption features at 2600 Å that have been attributed to the presence of adenine or benzene.

099.068 **Vorbeiflug von Pioneer 11 an Jupiter.**
H. W. Köhler.
SuW, Vol. 14, 395 - 397 (1975).

099.069 **On the temperature and pressure in the Jovian atmosphere.** A. D. Kuz'min, T. V. Smirnova.
Astron. Zhurn. Akad. Nauk SSSR, Vol. 52, 1263 - 1267 (1975). In Russian. English translation in Soviet Astron., Vol. 19, No. 6.

The data on the temperature and pressure in the atmosphere of Jupiter which were derived from the Pioneer 10 space probe are compared with the radio astronomical measurements of the planet's brightness temperature. The radio astronomical data are in good agreement with the infrared measurements, but are contrary to the occultation measurements of Pioneer 10. The disagreement between the occultation and infrared measurements is explained by a high-speed vertical motion in the atmosphere of Jupiter. The velocity of vertical motion is estimated to be 1.3 km/sec at the level of $T = 130°K$.

099.070 **Photosynthesis of organic compounds in the atmosphere of Jupiter.** J. P. Ferris, C. T. Chen.
Nature, Vol. 258, 587 - 588 (1975). – Letter.

099.071 **Jupiter etter Pioneer 10 og 11.** Ø. Hauge.
Astron. Tidssk., Årg. 8, p. 153 - 158 (1975).

099.072 **Determination of some parameters of the Jovian atmosphere.** A. N. Aksenov.
Trudy Astrofiz. Inst., *Alma-Ata,* Vol. 26, 74 - 82 (1974). In Russian.

099.073 **Results of a photographic photometry of Jupiter. I. Equatorial zone.** V. F. Kartashov.
Trudy Astrofiz. Inst., *Alma-Ata,* Vol. 26, 83 - 92 (1974). In Russian.

099.074 **Results of a photographic photometry of Jupiter. II. Details of the cloud layer.** V. F. Kartashov.
Trudy Astrofiz. Inst., *Alma-Ata,* Vol. 26, 92 - 104 (1974). In Russian.

099.075 **The influence of ethane and acetylene upon the thermal structure of the Jovian atmosphere.**
R. D. Cess, S. C. Chen.
Icarus, Vol. 26, 444 - 450 (1975).

Ethane and acetylene, both of which possess more efficient emission bands than methane, have been incorporated into a thermal structure model for the atmosphere of Jupiter. Choosing for illustrative purposes the mixing ratios $[C_2H_6]/[H_2] = 10^{-5}$ and $[C_2H_2]/[H_2] = 5 \times 10^{-7}$, it is found that these hydrocarbon gases lower the atmospheric temperature within the thermal inversion region by as much as 20 K, subsequently reducing the emission intensity of the 7.7 μm CH_4 band below the observed result.

099.076 **Modulation of Jovian interplanetary electrons and the longitude variation of decametric emissions.**
V. M. Vasyliunas.
Geophys. Res. Letters, Vol. 2, 87 - 88 (1975).

Interplanetary electrons of Jovian origin show minimum intensity and softest energy spectrum when the subsolar point on Jupiter is near the longitude of most frequent non-Io-related decametric emissions.

099.077 **Effect of the Jovian oblateness on Pioneer 10/11 radio occultations.**
W. B. Hubbard, D. M. Hunten, A. Kliore.
Geophys. Res. Letters, Vol. 2, 265 - 268 (1975).

The oblateness of Jupiter, which was neglected in the standard inversion technique for radio occultation data, has a substantial effect on the refractivity profiles of the neutral Jovian atmosphere derived from the Pioneer 10/11 occultations. The geometry of the Pioneer 10 entry and exit occultations has the effect that oblateness should account for much of the discrepancy of the results with plausible atmospheric models. The effect of the oblateness on Pioneer 11 entry is smaller.

099.078 **An analytical model of the Jovian magnetosphere.**
F. D. Barish, R. A. Smith.
Geophys. Res. Letters, Vol. 2, 269 - 272 (1975).

Using Euler potentials, the authors construct a class of axisymmetric magnetic field models that exhibit the general features of the dayside Jovian field as observed by Pioneers 10 and 11. The particular model used for illustration contains a distended, disk-like field along the equator, a blunt, earth-like magnetosphere boundary, and the possibility of multiple trapping regions along a field line.

099.079 **Pioneer 11 and Jupiter.** J. Bouška.
Vesmír, Vol. 54, 262 - 263 (1975). In Czech.

099.080 **Spatially resolved spectroscopy of the Jovian H_2 quadrupole 4–0 S(1) line.**
W. D. Cochran, J. Gelfand, W. H. Smith.
Bull. American Astron. Soc., Vol. 7, 521 - 522 (1975). Abstr. AAS.

099.081 **Planetary influences and activation of phenomena on Jupiter.** V. S. Prokudina.
Astron. Tsirk., No. 880, p. 1 - 3 (1975). In Russian.

099.082 **The motion of Mercury and variations of the longitude of Jupiter's Red Spot.** V. S. Prokudina.
Astron. Tsirk., No. 880, p. 3 - 5 (1975). In Russian.

099.083 **The occultation of β Scorpii by Jupiter. IV. Divergences with other observers in the derived temperature profiles.** M. Combes, L. Vapillon, J. Lecacheux.
Astron. Astrophys., Vol. 45, 399 - 403 (1975).

Thermal profiles in the Jovian atmosphere derived from the β Sco occultation by the three groups of Texas University, Cornell University and Meudon Observatory are reviewed. There is some disagreement between the published results. It is shown that such divergences are the result of different analysis methods, while no important discrepancies exist in the refractivity profiles themselves. Previous conclusions of Vapillon et al. (1973) are confirmed; in particular the authors' value for the thermal lapse rate ($-1.5°K \times km^{-1}$ below the $n = 5 \times 10^{14}\ cm^{-3}$ level) is reliable if the assumption of spherical symmetry is valid in the Jovian upper atmosphere.

099.084 **Recirculation of energetic particles in Jupiter's magnetosphere.**
D. D. Sentman, J. A. Van Allen, C. K. Goertz.
Geophys. Res. Letters, Vol. 2, 465 - 468 (1975).

A significant new finding from analysis of Pioneer 11 observations in the magnetosphere of Jupiter is that there is net streaming of both electrons $E_e > 40$ keV and $E_e > 560$ keV and protons $0.61 < E_p < 3.41$ MeV away from the planet along high latitude field lines. The recirculation of energetic particles emerges as an important dynamical feature of the Jovian magnetosphere. It is also suggested that emission of energetic particles into interplanetary space occurs at the polar caps rather than at the equator.

099.085 **Observations of Jovian electron events in the vicinity of earth.**
S. M. Krimigis, E. T. Sarris, T. P. Armstrong.
Geophys. Res. Letters, Vol. 2, 561 - 564 (1975).
The purpose of this paper is to report the identification of a number of quiet-time electron increases over the 1972–1974 time interval, with good statistical accuracy and better temporal resolution than was previously possible. It is found that the enhancements occur at times when connection of the interplanetary magnetic field to the Jovian magnetosphere is reasonably favorable and that the spectrum is not unlike that observed by the Pioneer spacecraft in the general vicinity of Jupiter.

099.086 **High-order magnetic multipoles as a source of gross asymmetry in the distant Jovian magnetosphere.**
A. J. Dessler, T. W. Hill.
Geophys. Res. Letters, Vol. 2, 567 - 570 (1975).
The longitudinal asymmetry of the surface magnetic-field strength at Jupiter causes a longitudinal asymmetry in the equatorial plasma mass density within the Jovian magnetosphere. The rotation of these density variations with the planet causes a diurnal variation of the radial distance on the night side at which the centrifugal stress of the magnetospheric plasma exceeds the local magnetic-field tension. This is approximately the distance at which the magnetic field opens to interplanetary space.

099.087 **High-resolution observations of Jovian magnetospheric microwave synchrotron emission. Final technical report.** L. R. D'Addario, S. J. Wernecke.
NASA Rep. NASA-CR-138778, 50 pp. (1974).
High-resolution radio observations of Jupiter were investigated with the use of the Stanford synthesis telescope. The best fit disk was computed from the emission of Jupiter and the residual brightness, due to nonthermal mechanisms, was measured.

099.088 **Jupiter.** E. J. Reese.
IAU Circ., No. 2799 (1975).

099.089 **Jupiter.** Gomez.
IAU Circ., No. 2812 (1975).

099.090 **Jupiter.** S. Ortolani, M. Cortes, Gomez, Reginaldo.
IAU Circ., No. 2821 (1975).

099.091 **Jupiter.** E. J. Reese.
IAU Circ., No. 2844 (1975).

099.092 **Long term variations in Jupiter's 11 cm radio emission.** P. M. McCulloch.
Proc. Astron. Soc. Australia, Vol. 2, 340 - 342 (1975).

099.093 **Jupiter's rotation period from dekametric observations 1951–1975.** R. A. Duncan.
Proc. Astron. Soc. Australia, Vol. 2, 342 - 345 (1975).

099.094 **Phase variation of limb darkening of Jupiter. III. Comparison with numerical calculations.**
V. F. Kartashov.
Astron. Tsirk., No. 887, p. 5 - 7 (1975). In Russian.

099.095 **Jovian activity.** V. A. Zinov'ev.
Astron. Tsirk., No. 890, p. 5 - 6 (1975). In Russian.

099.096 **On the application of Minnaert's formula to the photometry of Jupiter.** V. F. Kartashov.
Astron. Tsirk., No. 891, p. 2 - 4 (1975). In Russian.

099.097 **Observations of Jupiter's sporadic radio emission in the range 7.6 - 80 MHz 10 December 1971 through 21 March 1975.** J. W. Warwick, G. A. Dulk, A. C. Riddle.
World Data Center A for Solar-Terrestrial Phys., Rep. UAG-42, 49 pp. (1975). – *JLC*

099.098 **Jupiter.** W. E. Fox.
British Astron. Ass. Circ., No. 565 (1975).

099.099 **Measuring the magnetic fields of Jupiter and the outer solar system.**
E. J. Smith, B. V. Connor, G. T. Foster, Jr.
IEEE Trans. Magn., Vol. MAG-11, 962 - 980 (1975).
The vector helium magnetometer, one of the Pioneer-Jupiter experiments, has measured the magnetic field of Jupiter and the comprehensive scientific objectives of the investigations are explained and are then translated into the major instrument requirements. The principles of operation of the magnetometer, which involve the optical pumping of metastable helium, are discussed and the Pioneer instrument magnetometer is discussed and principal scientific results obtained thus far by the Pioneer investigation are summarized.

099.100 **Jupiter: interpretation of the Pioneer 10 infrared radiometer and S-band occultation observations.**
R. D. Cess, J. S. Hogan, T. Encrenaz, D. Gautier.
Journ. Atmosph. Sci., Vol. 32, 860 - 862 (1975).
Using recently calculated models of the Jovian atmosphere, a value of 5.0 has been derived for the H_2/He mixing ratio from the Pioneer 10 infrared radiometer data. A far-infrared spectrum corresponding to the thermal profile obtained in the Pioneer S-band occultation experiment has also been computed. The computed spectrum strongly suggests a misinterpretation of the data obtained in that experiment.

The Mariner Jupiter/Saturn photopolarimeter experiment. See Abstr. 053.015.

The Mariner Jupiter/Saturn visual imaging system. See Abstr. 053.016.

The Mariner Jupiter/Saturn ultraviolet experiment. See Abstr. 053.017.

The Mariner Jupiter/Saturn IRIS experiment. See Abstr. 053.018.

Magnetospheres of earth and Jupiter after Pioneer 10. See Abstr. 084.222.

High zonal harmonics of rapidly rotating planets. See Abstr. 091.002.

Effects of turbulence on radio-occultation scale heights. See Abstr. 091.004.

Planetary magnetospheres. See Abstr. 091.010.

The gravitational field of a feebly-nonuniformly rotating polytrope of index one. See Abstr. 091.015.

Io-accelerated electrons and ions. See Abstr. 099.213.

Galilean satellites and Jovian energetic particles. See Abstr. 099.225.

The thermal radiation of Saturn and its rings. See Abstr. 100.011.

Jupiter, Satellites

099.201 **On the secular change in the period of Io, 1668–1926.** S. J. Goldstein, Jr.
Astron. Journ., Vol. 80, 532 - 539 (1975).

A circular orbit in a fixed plane was fitted to 50 eclipse observations of Io made from 1668 to 1678, and separately to 156 observations made from 1908 to 1926. Both models contain the perturbation in longitude produced by Europa. The mean residual of the 17th-century observations is 31.5 sec, and that of the 20th-century observations is 10.7 sec. Comparison of the longitudes from the two models gives an upper limit of 100 sec for the time shift between them, or in other terms the upper limit to the time derivative of Io's period divided by the period is 3×10^{-18} sec^{-1}.

099.202 **Ultraviolet observations of small bodies in the solar system by OAO-2.** J. Caldwell.
Icarus, Vol. 25, 384 - 396 (1975). – Presented at I.A.U. Colloquium No. 28, "Planetary Satellites", Ithaca, N.Y. August 18–21, 1974.

Broadband filter photometry from 2100 to 4300 Å has been obtained by OAO-2 for the following objects: The Galilean satellites; Titan; the rings of Saturn; and three asteroids. Agreement with independent ground-based photometry in the region of overlap is good.

099.203 **New observations of surface markings on Jupiter's satellites.** J. B. Murray.
Icarus, Vol. 25, 397 - 404 (1975). – Presented at I.A.U. Colloquium No. 28, "Planetary Satellites", Ithaca, N.Y. , August 18–21, 1974.

Visual and photographic observations of the Galilean satellites of Jupiter made in September 1973 with the 108 cm reflector at Pic-du-Midi Observatory are presented. Results for Io, Europa, and Ganymede give albedo ranges of 0.28 to 0.67, 0.45 to 0.67, and 0.22 to 0.47, respectively. These are geometric albedos for a phase angle of 5°. The percentages of the disks covered by high albedos are consistent with the conclusions of previous workers regarding the fraction of exposed water ice on the surface.

099.204 **The spectral reflectivity of select areas on Europa.** T. F. Greene, J. R. Vermilion, R. W. Shorthill, R. N. Clark.
Icarus, Vol. 25, 405 - 415 (1975). – Presented at IAU Colloquium No. 28, "Planetary Satellites", Ithaca, N.Y. August 18–21, 1974.

The July 12, 1973, occultation of Europa by Io was observed in 30 wavelength channels in the spectral region λλ3200–11, 000 Å. The data are analyzed to obtain the spectral reflectivity of seven localized areas on the disk of Europa centered on 324°W longitude. The equatorial material is confirmed to be darker than the eastward-skewed bright north polar cap and a hint is evident that the darker material as well may be somewhat redder than the cap material.

099.205 **Optical polarimetry of the Galilean satellites of Jupiter.** A. Dollfus.
Icarus, Vol. 25, 416 - 431 (1975). – Presented at I.A.U. Colloquium No. 28, "Planetary Satellites", Ithaca, N.Y., August 18 - 21, 1974.

New measurements of the amount of polarization of the Galilean satellites are given and, within the context of other data, are interpreted.

099.206 **Io's sodium emission cloud.** W. W. Macy, Jr., L. M. Trafton.
Icarus, Vol. 25, 432 - 438 (1975). – Presented at I.A.U. Colloquium No. 28, "Planetary Satellites", Ithaca, N.Y., August 18 - 21, 1974.

Strong evidence that Io's sodium emission is due to resonant scattering is given by the authors' observations which show a monotonic increase of emission intensity with residual solar intensity. In addition they detected no emission during three eclipse observations of Io. They propose a resonant scattering model with two spacial components comprising an optically thick atmosphere extending 10^3 km above Io's surface surrounded by an optically thin cloud which forms a partial torus around Jupiter.

099.207 **Post-eclipse brightening and non-brightening of Io.** H. Frey.
Icarus, Vol. 25, 439 - 446 (1975).

It may be possible to understand the apparent intermittent nature of the post-eclipse brightenings and nonbrightenings of Io in terms of a nonuniform distribution of blue reflectors grouped in the hemisphere centered at 0° longitude. The dimensions required for such blue mirrors are consistent with very large craters. The high blue albedo of water frost and other ices makes these materials likely candidates for the reflectors.

099.208 **Mid-eclipse brightening of the Galilean satellites.** D. W. Smith.
Icarus, Vol. 25, 447 - 451 (1975).

Brinkmann (1973) has suggested that the Galilean satellites might briefly manifest a brightening at mid-eclipse due to a concentration of light refracted into the geometric umbra of Jupiter by the atmosphere around the terminator. Results obtained using two different models of the Jovian atmosphere indicate that such a brightening is unlikely even for Callisto due to the probable aerosol concentration in the Jovian atmosphere at the relevant altitudes.

099.209 **Further studies of line emission from Io.** R. A. Brown, R. M. Goody, F. J. Murcray, F. H. Chaffee, Jr.
Astrophys. Journ., (*Letters*), Vol. 200, L49 - L53 (1975).

A one-dimensional profile of the sodium emission near Io is presented. The maximum apparent emission rate is about 30 kR for the D_2 line, and the profile falls to half-height at about 4″ from Io. A high-resolution spectrum is analyzed for the thermal line width of the sodium atoms in the near-Io cloud. The FWHM is 68 ± 13 mÅ, which implies a sodium kinetic temperature of 5200 ± 2300 K. An upper limit of 5 kR is set to the apparent emission rate of neutral calcium at 4227 Å.

099.210 **Multicolour photometry of an occultation of Europa by Io.** P. Murdin.
Monthly Notices Roy. Astron. Soc., Vol. 172, 385 - 394 (1975).

Simultaneous *UBV* photometry of an occultation of Jupiter's satellite II (Europa) by satellite I (Io) shows directly the change in colour across Europa's disc. The occultation light curves show structure due to non-uniformity of surface brightness across the disc, particularly in the ultraviolet.

099.211 **Sodium 'airglow' from Io.** T. D. Parkinson.
Astrophys. Space Sci. Library, Vol. 51, (see 012.007), 433 - 440 (1975).

099.212 **Io: recent observations.** Y. L. Yung.
Astrophys. Space Sci. Library, Vol. 51, (see 012.007), 441 - 446 (1975).

099.213 **Io-accelerated electrons and ions.** S. D. Shawhan, C. K. Goertz, R. F. Hubbard, D. A. Gurnett, G. Joyce.
Astrophys. Space Sci. Library, Vol. 52, (see 012.008), 375 -

389 (1975).

099.214 **A model for Io's atmosphere and sodium cloud.** W. Macy, Jr., L. Trafton.
Astrophys. Journ., Vol. 200, 510 - 519 (1975).
Observations of the sodium D-line emissions associated with the Galilean satellite Io suggest that the emitting region has two components. The first is an atmosphere which extends 10^3 km above Io's surface. This is a relatively dense region which is optically thick in the sodium D lines. The second component is a surrounding, optically thin cloud spread out along Io's orbit to form part of a torus. For the atmosphere, measurements of the D-line intensity, D_2/D_1 ratio, and line width indicate a temperature of 500 to 1000 K and a sodium column abundance of 3×10^{13} cm^{-2}. Upper limits to the abundance of Mg, K, Ca, and Li indicate a high sodium abundance ratio in the atmosphere.

099.215 **Le treizième satellite de Jupiter.** B. Morando. L'Astronomie, 89ᵉ année, p. 317 - 318 (1975).

099.216 **Galilean satellite mutual occultation data processing.** T. C. Duxbury, T. V. Johnson, D. L. Matson.
Icarus, Vol. 25, 569 - 584 (1975). – Paper presented at I.A.U. Colloquium No. 28, "Planetary satellites", Ithaca, New York, 18–21 August 1974.
Results of processing seven mutual occultation lightcurves are presented. The lightcurves were obtained using the 60-inch telescope (152 cm) at Mt. Wilson to observe six J1 occulting J2 events and one J3 occulting J2 event. Using a uniformly illuminated disk model, local satellite Jovicentric longitude corrections of 675 ± 150 km, 275 ± 240 km, and 1175 ± 350 km for J1, J2, and J3, respectively, were determined. These corrections enabled the event midpoint times to be computed to ± 5 sec of the observed midpoint times for all seven events.

099.217 **On the secular variations of Io and Titan.** C. Blanco, S. Catalano.
Icarus, Vol. 25, 585 - 587 (1975). – Paper presented at IAU Colloquium No. 28, "Planetary satellites", August 18–21, 1974, Ithaca, N.Y.
A long-term variation of the magnitude at mean opposition is shown to exist for Io and Titan. The variation of Io seems to be correlated with the orbital period of Jupiter, the maximum light occurring at perihelion.

099.218 **Multislit spectra of the sodium emission associated with Io.** G. Münch, J. T. Bergstrahl.
Publ. Astron. Soc. Pacific, Vol. 87, 506 (1975). – Abstr. ASP.

099.219 **A comparison of some observations of the Galilean satellites with Sampson's tables.** J.-E. Arlot.
Celestial Mechanics, Vol. 12, 39 - 50 (1975). – Paper presented at IAU Colloquium No. 28, 'Planetary satellites', Ithaca, N.Y., August 1974.
An analysis is made of two series of photographic observations of the Galilean satellites of Jupiter. In the comparison of theory with observation, the aim of this work is to solve for systematic errors in the observations as well as those in the theory. Both intersatellary and planet-satellite positions were used. The least squares adjustment included as unknowns, corrections to the longitudes, inclinations and scale for both observation types, and an additional periodic term to account for residual phase defect for the planet-satellite coordinates. The validity of the results is discussed in terms of the unknowns introduced, the correlations between them and the reduction of the residuals.

099.220 **A coordinated effort for observations and theories of the natural satellites.** P. K. Seidelmann.
Celestial Mechanics, Vol. 12, 59 (1975). – Abstract of a paper presented at IAU Colloquium No. 28, 'Planetary satellites', Ithaca, N.Y., August 1974.

099.221 **Establishment and evolution of satellite-satellite resonances.** C. F. Yoder.
Celestial Mechanics, Vol. 12, 97 (1975). – Abstract of a paper presented at IAU Colloquium No. 28, 'Planetary satellites', Ithaca, N.Y., August 1974.

099.222 **Foto's van Io uit de ruimte.** C. de Jager. Zenit, Vol. 2, 349 (1975).

099.223 **Infrared albedos and rotation curves of the Galilean satellites.** O. L. Hansen.
Icarus, Vol. 26, 24 - 29 (1975).
Infrared (1.5-5μm) albedos and rotation curves of the Galilean satellites have been obtained. The data suggest that the rotational variation in the infrared is less than ± 10% for all four satellites.

099.224 **Interpretation of surface features of Europa obtained from occultations by Io.**
A. D. Herzog, R. F. Beebe.
Icarus, Vol. 26, 30 - 36 (1975).
Light curves of occultations of Europa by Io have been used to generate a crude map of albedo features on Europa. The best values currently available for the impact parameters and magnitude ratios for each event have been imposed on our model. Residuals between the observed and computed light curves are interpreted as albedo features on Europa. In order to improve the fit between the observations and the model it became necessary to impose a general polar brightening. The effects of additional albedo features and alternate models are discussed.

099.225 **Galilean satellites and Jovian energetic particles.** S. H. Gross.
Science, Vol. 190, 564 - 566 (1975).
The observed infrared temperatures of the four Galilean satellites, Io, Europa, Ganymede, and Callisto, are inconsistent with their equilibrium temperatures. Since these satellites appear to have little or no atmosphere, the discrepancies may be explained as due to the heating of their surfaces by energetic particles from Jupiter's radiation belts. The required energy fluxes are not entirely unreasonable and decrease with the distance of the satellite orbit from Jupiter.

099.226 **de Sitter's theory 'melts' Europa's polar cap.** K. Aksnes, F. A. Franklin.
Nature, Vol. 258, 503 - 505 (1975).
During the favourable passage of the earth through the plane of Jupiter's equator in 1973 many observations of the mutual phenomena of the Galilean satellites were obtained. Since then there have been numerous preliminary analyses arguing for the presence of a bright polar cap on Europa (J2). An inevitable explanation, in view of the finding of ice on J2 and J3, was the suggestion of a bright north polar cap for J2. Alternatively, it may be that at mid-event the projected minimum distance, z, between the centres of the satellites in a direction essentially parallel to Jupiter's polar axis, was smaller than predicted by Sampson's theory. Preliminary analyses of a few occultations of J2 by J1 suggested that to remove the need for a bright polar cap on J2 required Δz corrections too large to be compatible with the observed durations of the occultations. The authors show that the 'Δz explanation' is the more plausible one.

099.227 **High-resolution spectra of Io's sodium emission.** L. Trafton.
Astrophys. Journ., (*Letters*), Vol. 202, L107 - L112 (1975).

The shape of Io's sodium D_2 emission profile varies markedly with Io's orbital phase, and its width suggests that the sodium cloud may be expanding with velocities up to 18 km s^{-1}. This offers an explanation for at least part of the orbital phase asymmetry observed for the D-line emission.

099.228 **Detection of a potassium cloud near Io.** L. Trafton.
Nature, Vol. 258, 690 - 692 (1975).

Jupiter's innermost Galilean satellite Io is surrounded by an extensive cloud of sodium and hydrogen. The hydrogen is observed to extend along a segment of Io's orbit but the sodium cloud envelopes the entire Jovian system, extending tens of arc min from Jupiter. It is brightest, however, near Io and in its orbital plane. Sputtering by charged particles impacting the surface of Io has been the most plausible hypothesis for supplying the sodium. The author reports the first detection of potassium in Io's cloud. Its presence with hydrogen and sodium, along with the absence of more abundant elements such as calcium, might be explained by a rough surface microstructure for Io and a process for multiple impacts of the sputtered products with the surface.

099.229 **Angular momentum transfer to the inner Jovian satellites.** A. Mogro-Campero.
Nature, Vol. 258, 692 - 693 (1975).

The transfer of angular momentum from a central rotating body in the presence of a magnetic field has been discussed in connection with the evolutionary history of the solar system. The author considers angular momentum transfer in the inner Jovian satellite system. Electron flux measurements near Io's flux tube, and theoretical estimates of the electric current flowing through Io's flux tube are used to estimate the angular momentum transfer during the evolutionary history of the Jovian system. He finds that, under certain conditions, those currents are sufficient to produce an angular momentum transfer from Jupiter equal to the present angular momentum of the inner satellites.

099.230 **Jupiters nye måne.** K. Aksnes.
Astron. Tidssk., Årg. 8, p. 159 - 162 (1975).
Paper concerning the recent discovery of the 13th satellite.

099.231 **An oscillating asymmetry to Io's sodium emission cloud.** L. Trafton, W. Macy, Jr.
Astrophys. Journ., (*Letters*), Vol. 202, L155 - L158 (1975).

The spatial distribution of sodium emission north and south of Io was observed to have a time-dependent asymmetry correlated with Io's magnetic latitude. The spatial gradient was greatest when Jupiter's north magnetic pole pointed away from Io; this is a configuration of maximum probability of decametric emission. The authors invoke a hypothesis for explaining the asymmetry and use it to calculate preliminary fluxes of trapped charged particles of low energy in the vicinity.

099.232 **UBV photometry of the Galilean satellites.** R. L. Millis, D. T. Thompson.
Icarus, Vol. 26, 408 - 419 (1975).

UBV observations of the Galilean satellites made at Lowell Observatory and Cerro Tololo Inter-American Observatory during 1973 and 1974 are reported. The dependence of brightness on solar phase angle for various faces of each satellite is determined. Significant differences in this dependence are found between different faces of the same satellite, between satellites, and between the present results and those of previous investigators. Rotational light and color-index curves are presented for the satellites and compared with earlier work. An apparent secular brightening of all four satellites between 1973 and 1974 is discussed.

099.233 **Jupiter XIII.** E. Roemer, C. C. McCarthy.
IAU Circ., No. 2820 (1975).

099.234 **Jupiter XIII.** P. J. Shelus, R. I. Abbot.
IAU Circ., No. 2841 (1975).

099.235 **Probable new satellite of Jupiter.** C. Kowal.
IAU Circ., No. 2845 (1975).

099.236 **Satellites of Jupiter.**
IAU Circ., No. 2846 (1975).

099.237 **Probable new satellite of Jupiter.**
L. M. Vaughn; C. Kowal, E. Roemer, M. A. Daniel, C. C. McCarthy, K. Aksnes, B. G. Marsden.
IAU Circ., Nos. 2855, 2867, 2871 (1975).

099.238 **Jupiter XIII (Leda).**
E. Roemer, C. C. McCarthy.
IAU Circ., No. 2863 (1975).

099.239 **Investigation of the motion of the Galilean satellites of Jupiter.** E. N. Lemekhova.
Byull. Inst. Teoret. Astron., *Leningrad,* Vol. 14, 153 - 173 (1975). In Russian.

The present work is an application of Marsden's theory to the investigation of the motion of the Galilean satellites. The method has been used for the calculations of the ephemerides, for the comparison of the theory with the observations, for obtaining the partial derivatives with respect to the parameters in order to improve them on the basis of the recent observations in the future. The results obtained on the basis observations 1961 - 1971 have been discussed in terms of the correlation between the parameters.

099.240 **Possible discovery of the XIIIth satellite of Jupiter.**
Kometn. Tsirk., *Kiev*, No. 186 (1975). In Russian.

099.241 **Probable new satellite of Jupiter.** C. Kowal.
British Astron. Ass. Circ., No. 569 (1975).

099.242 **Electron impact ionization of Io's sodium emission cloud.**
R. W. Carlson, D. L. Matson, T. V. Johnson.
Geophys. Res. Letters, Vol. 2, 469 - 472 (1975).

The geometry of the sodium cloud associated with Io (Jupiter I) indicates that the lifetime of the neutral sodium atoms is an order of magnitude less than the photoionization lifetime. The authors suggest that ionization by thermal plasma electrons in the Jovian magnetosphere is the dominant Na loss process. Using plasma densities deduced from Pioneer 10 measurements, the lifetime and density distributions are calculated for Na and other species which may be present in the cloud around Io.

High pH, ammonia toxicity, and the search for life on the Jovian planets. See Abstr. 015.025.

Observations of Jupiter's Galilean satellites using astrolabes. See Abstr. 041.016.

Observations of outer planet satellites.
See Abstr. 041.017.

On the investigation of the motion of a spheroidal planetary satellite. See Abstr. 042.019.

Computer-developed construction of analytic expressions for the coordinates and partial derivatives of Jupiter's Galilean satellites. See Abstr. 042.029.

A new theory of the motions of the Galilean

satellites of Jupiter. See Abstr. 042.030.

Problems on the Galilean satellites of Jupiter. See Abstr. 042.031.

The orbital resonance amongst the Galilean satellites of Jupiter. See Abstr. 042.033.

On the satellite capture problem. See Abstr. 042.075.

Extra-terrestrial geology. See Abstr. 094.479.

The smaller bodies of the solar system. See Abstr. 098.015.

Gasring om Jupiter in baan van Io. See Abstr. 099.051.

On the Laplace relation among the satellites of Uranus. See Abstr. 101.003.

100 Saturn, Saturn Satellites

Saturn

100.001 **A numerical study of a gravitating system of colliding particles: applications to the dynamics of Saturn's rings and to the formation of the solar system.** A. Brahic.
Icarus, Vol. 25, 452 - 458 (1975). – Presented at I.A.U. Colloquium No. 28, "Planetary Satellites", Ithaca, N.Y., August 18 - 21, 1974.

The author considers numerically a three-dimensional system of particles moving in the gravitational field of a central mass point and interacting through inelastic collisions. After a very fast flattening, the system reaches a quasiequilibrium state in which there are still collisions and the rings tend to a finite thickness of a few times the mean size of the particles.

100.002 **On a suspected ring external to the visible rings of Saturn.**
B. A. Smith, A. F. Cook II, W. A. Feibelman, R. F. Beebe.
Icarus, Vol. 25, 466 - 469 (1975). – Presented at I.A.U. Colloquium No. 28, "Planetary Satellites", Ithaca, N.Y., August 18 - 21, 1974.

Reexamination of a photograph of Saturn taken on 15 November 1966, when the earth was nearly in the ring plane, indicates that ring material may exist outside the visible rings, extending to more than 6 Saturnian radii. Although the suspected feature on the photograph appears to be real, the possibility of its being a developed pressure mark or a chance alignment of grains cannot be ruled out. The observed brightness in blue light was estimated to be $m_B = 19.5 \pm 0.5$ per linear arcsecond, implying a normal optical thickness, $\tau \simeq 10^{-7}$, for ice-covered particles.

100.003 **A theoretical photometric function of Saturn's rings.** K. A. Hämeen-Anttila, P. Vaaraniemi.
Icarus, Vol. 25, 470 - 478 (1975).

A photometric theory of Saturn's rings is developed on the assumption that partially elastic collisions have brought all the ring particles into the same plane. The resulting photometric function explains the tilt effect of the rings, but the opposition peak must originate in the particles themselves.

100.004 **Saturn, seeing, and Percival Lowell.** W. G. Hoyt.
Sky Telescope, Vol. 50, 28 (1975).

100.005 **On the limb darkening on Saturn's disk.**
V. G. Tejfel'.
Astron. Zhurn. Akad. Nauk SSSR, Vol. 52, 823 - 831 (1975).
In Russian. English translation in Soviet Astron., Vol. 19, No. 4.

On the basis of photometric measurements of the absolute brightness distribution on Saturn's disk and of absolute measurements of the reflectivity of the planet's cloud layer it has been found that the relation between the limb-darkening coefficient and the brightness coefficient in the centre of the disk is in agreement with theoretical calculations for very small particles with nearly Rayleigh scattering as well as for particles with a mean radius $\bar{a}$ of $\sim 1\,\mu$m. The situation for small particles is in contrast to the polarimetric data and to the low brightness temperature of Saturn at λ 8–14 μm. Accordance of all measurements with theory may be achieved for $\bar{a}$ about 1 μm.

100.006 **The outer planets.** D. M. Hunten.
Sci. American, Vol. 233, No. 3, p. 130 - 132, 134, 136, 138, 140 (1975). – Popular article.

100.007 **Azimuthal brightness variations in Saturn's rings.**
H. J. Reitsema, R. F. Beebe, B. A. Smith.
Bull. American Astron. Soc., Vol. 7, 468 (1975). – Abstr. AAS.

100.008 **A tentative identification of C_2H_4 in the spectrum of Saturn.** T. Encrenaz, M. Combes, Y. Zeau, L. Vapillon, J. Berezne.
Astron. Astrophys., Vol. 42, 355 - 356 (1975).

A new spectrum of Saturn has been obtained between 900 and 1000 cm^{-1}, at a resolution of 2.5 cm^{-1}, with a Michelson interferometer. The spectrum shows emission lines which could be attributed to C_2H_4.

100.009 **The influence of the solar radiation pressure on the motion of particles in Saturn's rings.**
E. N. Polyakhova.
Trudy Astron. Obs., *Leningrad*, Vol. 31 (= Uchenye Zapiski Leningr. Un-ta, No. 381 = Seriya Matem. Nauk, vyp. (No.) 51), p. 192 - 217 (1975). In Russian.

The orbital evolution of dust- and microparticles in Saturn's rings under the influence of the direct solar radiation pressure and of the dynamical effect of the radiation is described. It is shown that the secular action of the direct radiation pressure can increase the mean periodical eccentricity value to the hyperbolical one during the time of rings existence.

100.010 **Observations of Saturn in the 0.8 - 1.6 μ region.**
L. V. Zasova.

Astron. Tsirk., No. 845, p. 1 - 2 (1974). In Russian.

100.011 **The thermal radiation of Saturn and its rings.** G. H. Rieke.
Icarus, Vol. 26, 37 - 44 (1975).
Infrared observations of Saturn from 5 to 40 μm are described. There is intense limb brightening at 12.35 μm over the southern polar cap. The C ring is anomalously bright at 10 and 20 μm and has bluer (hotter) colours than the A and B rings. The ring spectra have been extrapolated beyond 40 μm and subtracted from low-resolution far-infrared measurements to show that the far-infrared spectrum of the disk of Saturn is qualitatively similar to that of Jupiter and that Saturn radiates 2.5 ± 0.6 times the energy it absorbs from the sun.

100.012 **Saturn's rings. II. Condensations of light and optical thickness of Cassini's division.** I. R. Ferrín.
Icarus, Vol. 26, 45 - 53 (1975).
'Condensations' of light have been observed when Saturn's rings are seen almost edge on, and the sun and the earth are on opposite sides of the ring plane. These condensations are associated with ring C and Cassini's division. If the relative brightness between the two condensations and the optical thickness of ring C are known, one can calculate the optical thickness of Cassini's division, τ_{CASS}. Using Barnard's and Sekiguchi's measurements, the author has obtained $0.01 \lesssim \tau_{CASS} \lesssim 0.05$. A brightness profile of the condensations which agrees well with visual observations is also presented.

100.013 **Observation of the ν_2 band of PH_3 in the atmosphere of Saturn.**
J. D. Bregman, D. F. Lester, D. M. Rank.
Astrophys. Journ., (*Letters*), Vol. 202, L55 - L56 = Lick Obs. Bull., No. 716 (1975).
Observation of the ν_2 band of PH_3 in the Saturnian atmosphere represents the first positive identification of this compound in the planet. The observed spectrum is consistent with line formation in the planet's troposphere where the dominant sources of opacity are molecular hydrogen and unsaturated phosphine vapor.

100.014 **Statistical mechanics of Keplerian orbits.** K. A. Hämeen-Anttila.
Astrophys. Space Sci., Vol. 37, 309 - 333 (1975).
A statistical theory of Keplerian orbits is constructed for a system of particles, which are subject to partially elastic collisions. If the elasticity decreases with collisional velocity, the system shows an increased tendency to form condensations. Near the central body they are concentric rings, which are separated by gaps void of matter. At larger distances outside the Roche limit, the condensations probably form larger bodies. An application to Saturn's rings suggests that at least rings A and C would consist of separate ringlets.

100.015 **The 1973–74 apparition of Saturn.** J. L. Benton, Jr., C. Sherrod.
Strolling Astronomer, Vol. 25, 183 - 191 (1975).

100.016 **Some systematic observations of Saturn during its 1973–74 apparition.**
E. S. Corsi, P. S. Corsi, A. Fabozzi, G. Fuccillo.
Strolling Astronomer, Vol. 25, 207 - 210 (1975).

100.017 **Structure of Saturn's rings: optical and dynamical considerations.** F. A. Franklin.
Rings of Saturn. NASA SP-343, (see 012.028), p. 3 - 15 (1974).

100.018 **Model of Saturn's rings that satisfies the observed phase curve for optical scattering.**
W. M. Irvine.
Rings of Saturn. NASA SP-343, (see 012.028), p. 17 - 28 (1974).

100.019 **Saturn's ring particles and space vehicle design.** F. D. Palluconi.
Rings of Saturn. NASA SP-343, (see 012.028), p. 29 - 38 (1974).

100.020 **Impact of Saturn's rings on mission analysis for MJS 77.** P. A. Penzo.
Rings of Saturn. NASA SP-343, (see 012.028), p. 39 - 50 (1974).

100.021 **Ring particle sizes and composition derived from eclipse cooling curves and reflection spectra.**
H. H. Kieffer.
Rings of Saturn. NASA SP-343, (see 012.028), p. 51 - 63 (1974).

100.022 **Variations in the infrared brightness temperature of Saturn's rings.** R. E. Murphy.
Rings of Saturn. NASA SP-343, (see 012.028), p. 65 - 72 (1974).

100.023 **Distribution and size of elements of Saturn's rings as inferred from 12-cm radar observations.**
G. A. Morris, Jr.
Rings of Saturn. NASA SP-343, (see 012.028), p. 73 - 82 (1974).

100.024 **Short wavelength radio observations of Saturn's rings.** M. A. Janssen.
Rings of Saturn. NASA SP-343, (see 012.028), p. 83 - 96 (1974).

100.025 **Microwave interferometry of Saturn: application to the brightness temperature of the rings.**
G. L. Berge.
Rings of Saturn. NASA SP-343, (see 012.028), p. 97 - 105 (1974).

100.026 **Estimates of the size of particles in Saturn's rings and their cosmogonic implications.**
J. B. Pollack.
Rings of Saturn. NASA SP-343, (see 012.028), p. 107 - 123 (1974).

100.027 **The "atmosphere" of the rings of Saturn.** J. Blamont.
Rings of Saturn. NASA SP-343, (see 012.028), p. 125 - 129 (1974).

100.028 **Gaps in our knowledge of Saturn's rings.** Rings of Saturn. NASA SP-343, (see 012.028), p. 175 - 196 (1974). – General discussion.

100.029 **Saturn's rings investigation from space.** Rings of Saturn. NASA SP-343, (see 012.028), p. 197 - 219 (1974). – General discussion.

100.030 **Saturne: présentation 1974/75. Opposition: 6 janvier 1975.** F. Jetzer.
Orion, 33. Jahrgang, p. 213 - 214 (1975). – Rapport No. 31 du Groupement planétaire SAS.

Observations of Saturn obtained with the astrolabe at Paris Observatory during the winter of 1973 – 1974. See Abstr.041.001.

The Mariner Jupiter/Saturn photopolarimeter experiment. See Abstr. 053.015.

The Mariner Jupiter/Saturn visual imaging system. See Abstr. 053.016.

The Mariner Jupiter/Saturn ultraviolet experiment. See Abstr. 053.017.

The Mariner Jupiter/Saturn IRIS experiment. See Abstr. 053.018.

The Mariner Jupiter/Saturn ring occultation experiment. See Abstr. 053.019.

The Sisyphus particle detector. See Abstr. 053.020.

High zonal harmonics of rapidly rotating planets. See Abstr. 091.002.

Occultations of major planets by moon at Belgrade Astronomical Observatory during 1974. See Abstr. 096.015.

Spectroscopy of Jupiter and Saturn. See Abstr. 099.012.

Long term variations of the decimetric radio emission of Jupiter (and Saturn?). See Abstr. 099.014.

Jupiter and Saturn. See Abstr. 099.017.

The magnetospheres of Jupiter and Saturn. See Abstr. 099.029.

The outer solar system. See Abstr. 099.042.

Resonant orbits of grains and the formation of satellites. See Abstr. 107.002.

Numerical study of a gravitating system of colliding particles; formation and dynamics of discs. See Abstr. 151.025.

Saturn, Satellites

100.201 **Study of the inner satellites of Saturn by photographic photometry.**
S. Koutchmy, P. L. Lamy.
Icarus, Vol. 25, 459 - 465 (1975). – Presented at I.A.U. Colloquium No. 28, "Planetary Satellites", Ithaca, N.Y., August 18 - 21, 1974.

Photographs of Saturn and its five inner satellites were obtained on January 2, 1974 with the 105 cm telescope at Pic du Midi Observatory with exposure times of 45 sec. The magnitudes, computed by integration, are in good agreement with published values except for Mimas, which is nearly one magnitude fainter then previously believed.

100.202 **Positional photographic observations of the satellites of Saturn with the normal astrograph in 1972 at Pulkovo.** T. P. Kiseleva, L. S. Koroleva, G. V. Panova.
Byull. Inst. Teoret. Astron., *Leningrad,* Vol. 14, 60 - 62 (1975). In Russian.

100.203 **Further observations of the 8–13 micron spectrum of Titan.** F. C. Gillett.
Astrophys. Journ., (*Letters*), Vol. 201, L41 - L43 (1975).

Spectrophotometric observations of Titan, covering the range 7.8–13.3 μ with $\Delta\lambda/\lambda \approx 0.02$, are presented. The resulting spectrum is in excellent agreement with the predicted spectrum based on an inversion model for Titan's atmosphere. In addition to emission bands of methane (CH_4) and ethane (C_2H_6),evidence is found for emission by acetylene (C_2H_2), ethylene (C_2H_4), and monodeuterated methane (CH_3D).

100.204 **Eclipses of natural planetary satellites.**
C. F. Peters.
Celestial Mechanics, Vol. 12, 99 - 110 (1975). – Paper presented at IAU Colloquium No. 28, 'Planetary satellites', N.Y., August 1974.

A general method is given for predicting eclipse events for natural planetary satellites suitable for use on a large scale digital computer. The precision is sufficient to process photometric eclipse observations to improve natural satellite ephemerides. Expected accuracy improvement for Saturn's satellites should be an order of magnitude or better. Predicted eclipse times of satellites by Saturn and the Rings are given for the next decade.

100.205 **Theory of Enceladus and Dione.**
W. H. Jefferys, L. M. Ries.
Astron. Journ., Vol. 80, 876 - 884 (1975).

A new theory of the motion of Enceladus and Dione is under development. The theory is in literal form, with all constants and parameters appearing explicitly, and is being computed with the aid of the algebraic manipulation system TRIGMAN (Jefferys 1970, 1972). In this paper the authors treat the development of the disturbing function; the elimination of the short-period terms; and the resonant Hamiltonian and the constants of integration. The theory has a number of novel features. One important feature is that the programs have been written in such a manner that, with only minor modification, they should be able to develop the theories of Mimas–Tethys and Titan–Hyperion as well.

100.206 **Measuring the sizes of Saturn's satellites.**
J. Veverka, J. Elliot, J. Goguen.
Sky Telescope, Vol. 50, 356 - 359 (1975).

100.207 **Observations of Saturn's satellites 1789–1972.**
D. A. Pierce.
Publ. Astron. Soc. Pacific, Vol. 87, 785 - 787 (1975).

A data-collection and literature search has yielded over 22,000 observations of the known satellites of Saturn, made from 1789 through 1972. The type (photographic, micrometer, etc.) and number of observations are tabulated in ten-year increments, and a complete bibliography is referenced.

100.208 **Lunar occultation of Saturn. I. The diameters of Tethys, Dione, Rhea, Titan, and Iapetus.**
J. L. Elliot, J. Veverka, J. Goguen.
Icarus, Vol. 26, 387 - 407 (1975).

The diameters of Tethys, Dione, Rhea, Titan and Iapetus were determined from observations of their March 30, 1974, lunar occultations. The satellite diameters were determined by least-squares fits of model occultation light curves. In these fits the diameter and degree of limb darkening of the satellite are correlated variables, requiring the limb darkening to be specified before the diameter can be determined, or vice versa. Titan's diameter must be at least 5800 km, much larger than the currently accepted value of 5000 km, making it the largest satellite in the solar system. Except for Titan, the lunar occultation diameters generally agree with previous diskmeter and radiometric determinations.

Observations of outer planet satellites. See Abstr. 041.017.

Astrometric observations of Saturn's satellites from McDonald Observatory, 1972. See Abstr. 041.021.

Planetary brightness changes: evidence for solar variability. See Abstr. 080.050.

The smaller bodies of the solar system. See Abstr. 098.015.

Ultraviolet observations of small bodies in the solar system by OAO-2. See Abstr. 099.202.

On the secular variations of Io and Titan. See Abstr. 099.217.

A coordinated effort for observations and theories of the natural satellites. See Abstr. 099.220.

Establishment and evolution of satellite-satellite resonances. See Abstr. 099.221.

101 Uranus, Neptune, Pluto, Transplutonian Planets

101.001 **When and where were the satellites of Uranus formed?** S. F. Singer.
Icarus, Vol. 25, 484 - 488 (1975). – Presented at I.A.U. Colloquium No. 28, "Planetary Satellites", Ithaca, N.Y., August 18 - 21, 1974.

Uranus exhibits an unusually large obliquity compared to other planets of the solar system. However its five satellites are remarkably regular, with eccentricities and inclinations very nearly zero, but of course with orbit planes that are tilted by $\sim 98^\circ$ to the plane of the ecliptic. This circumstance is used here to relate the formation of satellites to planet formation. Six different cases are discussed.

101.002 **On the thermal structure of Uranus.** L. Wallace.
Icarus, Vol. 25, 538 - 544 (1975).

Uranus has an effective temperature close to the solar equilibrium value and undoubtedly a thermal inversion of at least 140 K at a pressure of ~ 3 dyn cm^{-2}. With the inversion and the thermal opacity provided by a $He-H_2$ mixture in a ratio close to solar abundance, acceptable agreement can be achieved with the available infrared observations. The cause of the inversion is, however, uncertain. The use of the $He-H_2$ opacity for Uranus is justified by the excellent agreement of the frequency variation of that opacity with the thermal spectrum of Jupiter.

101.003 **On the Laplace relation among the satellites of Uranus.** R. Greenberg.
Monthly Notices Roy. Astron. Soc., Vol. 173, 121 - 129 (1975).

Three of Uranus' satellites are involved in an interaction similar to the Laplace relation among the Galilean satellites of Jupiter. However, there are no pair-wise commensurabilities in the Uranus system as there are among the Galilean satellites. Thus analysis by Fourier expansion of disturbing functions becomes too unwieldy for practical calculation. Instead, the unexpanded perturbations are numerically averaged over a time scale short compared to the important variations in the Laplace relation. Perturbations of Miranda's orbit are estimated to be of observable magnitude for reasonable masses of Ariel and Umbriel.

101.004 **The polarization of Uranus as a function of phase angle.** J. J. Michalsky.
Bull. American Astron. Soc., Vol. 7, 468 (1975). – Abstr. AAS.

101.005 **Zur Rotationsperiode des Planeten Neptun.** J. Wempe.
Sterne, 51. Jahrgang, p. 145 - 146 (1975).

101.006 **Riddles of the motion of Uranus and the discovery of Neptune.**
E. A. Grebennikov, Yu. A. Ryabov.
Zemlya i Vselennaya, 1975, No. 4, p. 29 - 34. In Russian.

101.007 **Does the solar system include distant but discoverable infrared dwarfs?** K. Davidson.
Icarus, Vol. 26, 99 - 101 (1975).

It is possible that one or more bodies with masses in the range 0.001 to 0.01 $M_\odot$ may be loosely bound to the solar system, at distances of several thousand astronomical units. Such objects would be extremely difficult to detect at visual wavelengths, but they might be discoverable at infrared wavelengths.

101.008 **Method of constructing an analytical theory of the motion of Pluto.** N. I. Lobkova.
Pis'ma v Astron. Zhurn., Vol. 1, No. 10, p. 39 - 40 (1975). In Russian.

Secular and long-period perturbations of Pluto due to Neptune have been evaluated using a new method of expanding the disturbing function.

101.009 **Spectrophotometry of Uranus in the region $\lambda\lambda$ 5000 - 6800.**
Yu. D. Davudov, N. B. Ibragimov.
Astron. Zhurn. Akad. Nauk SSSR, Vol. 52, 1029 - 1033 (1975). In Russian. English translation in Soviet Astron., Vol. 19, No. 5.

The methane absorption bands and the line $H_2S(0)$ (4–0) 6420 Å have been investigated by spectrograms of Uranus taken at the 2-m reflector with dispersion 6, 12 and 30Å/mm. 45 intensive absorption lines absent in the solar spectrum were discovered in the spectrum of Uranus in the region λλ 6500–6750. Besides, the methane absorption bands 5430, 5840, 6140, 6560 and 6620 Å are registered on one spectrogram of Uranus. The intensities of the methane absorption bands from the centre to the limb are not found to change.

101.010 **Uranian methane abundance, rotational temperature, and effective pressure from the 6800 Å band.**
J. T. Bergstralh.
Astrophys. Journ., Vol. 202, 832 - 838 (1975).

Measurements of the methane absorption features at 6800 Å in two spectrograms of Uranus have been analyzed. Parametric fitting of synthetic reflecting layer (RLM) and homogeneous scattering layer (HSL) spectra to the spectrograms was attempted. The best fit of an HSL spectrum was obtained with a temperature of 100 K and methane abundance in the range $2 \lesssim \eta N \lesssim 8$ km-amagat. Best results for an HSL spectrum ($\varpi_{0c} = 0.88$; $\mu = \mu_0 = 1/\sqrt{3}$) were $T = 100$ K and $0.5 \lesssim Nl \lesssim 2$ km-amagat. These methane abundances, taken with recent estimates of the molecular hydrogen abundance (McKellar 1974), correspond to an average C:H ratio between 1.5 and 25 times larger than the solar ratio. Even greater enrichment of the C:H ratio may occur below the visible cloud boundary.

101.011 **Neptune's atmosphere: the source of the thermal inversion.** W. Macy, Jr., L. Trafton.
Icarus, Vol. 26, 428 - 436 (1975).

The authors present variable-flux models for Neptune's atmosphere which are constrained to fit recent infrared brightness temperature measurements. These models have an inversion much lower ($\leqslant 0.2$ atm) than that inferred from the occultation observation. The calculations indicate that the most probable energy source for the inversion is absorption of sunlight by methane, whose abundance ratio to hydrogen is several times the solar value. The authors conclude that Neptune must have a significant internal heat source because solar-heated models require methane to be supersaturated by an unrealistically large factor.

101.012 **Opet u centru pažnje astronoma i astrofizičara.**
M. Hegedušić.
Naučna Misao, Bull. 12, p. 65 - 70 (1975).

101.013 **Die Suche nach dem zehnten Planeten.**
H. Haupt.
Johannes Kepler, 1571–1971. Gedenkschrift der Universität Graz, p. 29 - 39 = Mitt. Univ. Sternw. Graz No. 15 (1975).

Observations of outer planet satellites.
See Abstr. 041.017.

Astrometric positions of the planet Pluto in the period 1971–1974. See Abstr. 041.052.

Satellite with high eccentricity. Application to Nereid. See Abstr. 042.054.

Asymmetric distant solar wind.
See Abstr. 074.088.

Planetary brightness changes: evidence for solar variability. See Abstr. 080.050.

Occultations et appulses planétaires en 1976.
See Abstr. 097.025.

The outer solar system. See Abstr. 099.042.

A coordinated effort for observations and theories of the natural satellites. See Abstr. 099.220.

The outer planets. See Abstr. 100.006.

The formation of the outer planets.
See Abstr. 107.011.

102 Comets (Origin, Structure, Atmospheres, Dynamics)

102.001 **Do comets play a role in galactic chemistry and γ-ray bursts?** F. L. Whipple.
Astron. Journ., Vol. 80, 525 - 531 (1975).
This paper explores the plausibility of the assumption that enough material from interstellar space is locked up in comets to reduce significantly the apparent growth rate of "heavy" elements (mass greater then He) and, therefore, the present abundance of heavy elements in the interstellar medium and in the disk stars. A related suggestion concerns the influx of comets on neutron stars as a source of γ-ray bursts. Although no interstellar comets have been observed, reasonable upper limits to the observed numbers and masses do not rule out the first suggestion. The γ-ray-burst suggestion appears unlikely. A surprisingly large total mass of comets could be gravitationally bound to the sun and, of course, to other stars, but remain undetected. The nature of the 1908 Tunguska explosion is discussed briefly. It was probably not an encounter with an active comet.

102.002 **Study of comets and meteors.**
O. V. Dobrovol'skij, V. S. Getman.
Nauka i chelovechestvo. 1975. Moskva, Znanie, 1974, p. 257 - 268. In Russian.

102.003 **Ion chemistry in the cometary atmosphere.**
M. Shimizu.
Astrophys. Space Sci., Vol. 36, 353 - 361 (1975).
Dirty ice of a second kind (major components, H_2O, CO, and N_2; minor components less than several percents, NH_3, CH_4, and other organic substances such as HCN, CH_3CN etc.) is assumed for the composition of volatiles in the cometary nucleus. There is a satisfactory agreement for the second kind of dirty-ice model, but the presence of large amounts of CH_4 and NH_3 is found to be rather in contradiction with observational evidence.

102.004 **Natur und Herkunft der Kometen.** J. Rahe.
SuW, Vol. 14, 220 - 228 (1975).– Lecture on the occasion of the annual meeting of the "Astronomische Gesellschaft", Würzburg 1974.

102.005 **Non-gravitational perturbations of long-period comets.** P. Weissman.
Bull. American Astron. Soc., Vol. 7, 448 (1975). – Abstr. AAS.

102.006 **On the existence of interstellar comets and the probability of their encounter with the sun.**
Z. Sekanina.
Bull. American Astron. Soc., Vol. 7, 467 (1975). – Abstr. AAS.

102.007 **The asteroidal planet as the origin of comets.**
T. C. Van Flandern.
Bull. American Astron. Soc., Vol. 7, 467 (1975). – Abstr. AAS.

102.008 **Comet chronology in numbers, AD 200–1882.**
D. J. Schove.
Journ. British Astron. Ass., Vol. 85, 401 - 407 (1975).

102.009 **Comets – witnesses of the history of the solar system.** S. K. Vsekhsvyatskij.
Zemlya i Vselennaya, 1975, No. 4, p. 70 - 73. In Russian.

102.010 **Comètes périodiques.** J. Meeus.
L'Astronomie, 89e année, p. 352 - 354 (1975).

102.011 **Kometen, ijstijden en interstellaire wolken.**
G. W. E. Beekman.
Zenit, Vol. 2, 347 - 348 (1975).

102.012 **Investigation of the motion of some objects with large eccentricities from the Palomar-Leiden Survey.**
G. A. Chebotarev, N. A. Belyaev, R. P. Eremenko.
Problems of cosmic physics. Vyp. (No.) 10, (see 003.003), p. 3 - 9 (1975). In Russian.
The evolution of the orbits of three objects of the Palomar-Leiden Survey (4802, 6889, 7617) with mean daily motions close to those of Jupiter has been studied. The results of this research lead to the conclusion that these objects may be considered as comets.

102.013 **Decay of cometary nuclei. I. Catalogue of disrupted comets.** V. A. Golubev.
Problems of cosmic physics. Vyp. (No.) 10, (see 003.003), p. 23 - 34 (1975). In Russian.

102.014 **Perihelion distribution of near-parabolic comets.**
H. Oja.
Astron. Astrophys., Vol. 43, 317 - 319 (1975).
The distribution of the perihelion directions of 73 comets having great eccentricities in their "original" orbits, has been studied. The perihelion points are found to have a preferred plane inclined at 20° to the plane of the Galaxy. The centre of the distribution is only 7 degrees from the solar apex.

102.015 **Determination of effective accelerations in cometary tails of type II.**
V. P. Konopleva, V. K. Rozenbush, L. M. Shul'man.
Astrometriya i Astrofizika, *Kiev*, vyp. (No.) 27, (see 003.004), p. 59 - 66 (1975). In Russian.
An algorithm of automatic analysis of the acceleration field in type II cometary tails is described. The algorithm was used for calculation of effective accelerations of cometary tail neutral particles and moments of particle emission forming the synchronism of type II tails. As an example the parameter $(1 + \mu)$ determined for comet Bennett 1970 II is taken from a plate obtained in Uzhgorod on March, 30, 1970.

102.016 **Composition of cometary dust: the case against silicates.** D. A. Mendis, N. C. Wickramasinghe.
Astrophys. Space Sci., Vol. 37, L13 - L16 (1975).
It is argued that the infrared emission including 10 and 18 μ features observed in recent comets is unlikely to be due to silicates. The vaporization temperature of the emitting material $\sim$ 500 K is consistent with emission by crystalline polyformaldehyde.

102.017 **New evidence for interplanetary boulders?**
Z. Sekanina.
I.A.U. Colloquium No. 13, (see 012.030), p. 199 - 208 (1973).
The present paper critically discusses the method of detection, the magnitude and the rate of occurrence of sudden disturbances in the motions of some short-period comets. The disturbances have recently been suggested as potential indicators of collisions between the comets and interplanetary boulders – minor objects whose existence was predicted by Harwit in 1967.

102.018 **Dissociation and ionization of CO_2 molecules in cometary atmospheres.** V. I. Cherednichenko.
Geofiz. i astron. Inform. byull., 1975, No. 14, p. 239 - 240. In Russian. – Abstr. in Referativ. Zhurn. 51. Astron., 11.51. 308 (1975).

102.019 **News in physical chemistry of comets.**

V. I. Cherednichenko.
Geofiz. i astron. Inform. byull., 1975, No. 14, p. 241 - 245. In Russian. – Abstr. in Referativ. Zhurn. 51. Astron., 11.51. 309 (1975).

102.020 **The observability of comets.** J. B. Tatum.
Journ. Roy. Astron. Soc. Canada, Vol. 69, 212 - 221 (1975).

102.021 **Orientation of cometary dust grains and polarization observations of comets.**
A. Z. Dolginov, I. G. Mitrofanov.
Astron. Zhurn. Akad. Nauk SSSR, Vol. 52, 1268 - 1278 (1975). In Russian. English translation in Soviet Astron., Vol. 19, No. 6.

As follows from observations, the linear polarization of the light scattered in the cometary coma may be parallel to the plane of scattering. Such a polarization may be explained by the orientation of nonspherical dust grains. The orientation may be caused by the comet's magnetic field, by the solar wind flux or interior gas fluxes and by the solar radiation. These types of orientation are discussed.

102.022 **Dependence of the inclination of cometary orbits on the longitude of the ascending node.**
V. P. Tomanov.
Astron. Zhurn. Akad. Nauk SSSR, Vol. 52, 1332 - 1333 (1975). In Russian. English translation in Soviet Astron., Vol. 19, No. 6.

The paper shows that according to the hypothesis of the origin of comets of Lyttleton there exists a dependence of the inclination i to the ecliptic of the cometary orbits on the longitude of the ascending node Ω: arctg $i = -1.35 \sec \Omega$. Theoretical conclusions are in conformity with the observed data in the system of comets with a period exceeding 500 years.

102.023 **Intensity analysis of the Swan bands of carbon in the spectrum of a comet.** J. B. Tatum.
Journ. Roy. Astron. Soc. Canada, Vol. 69, 255 (1975).
Abstr. CAS.

102.024 **Optical pumping by solar radiation in comets: polarization of the Na resonance lines.** F. H. Mies.
Astrophys. Journ., Vol. 202, 823 - 831 (1975).

At some distance from the nucleus of a comet, the rate of fluorescent scattering of solar radiation by an atom or molecule will be comparable to, or exceed, the rate of collisional relaxation. This is the condition for optical pumping. The pumping causes an alignment of the hyperfine magnetic sublevels along the incident axis, and observations from the earth should exhibit an enhanced linear polarization in the sun-comet-earth plane. Therefore, polarization measurements can be used as a diagnostic for the presence or absence of collisional relaxation.

102.025 **A possible component of cometary nuclei.**
E. A. Kajmakov, S. Aliev.
Dokl. AN TadzhSSR, Vol. 18, No. 4, p. 16 - 18 (1975). In Russian. – Abstr. in Referativ. Zhurn. 51. Astron., 12.51.280 (1975).

102.026 **Cometary outbursts, a brief survey.** D. W. Hughes.
Quarterly Journ. Roy. Astron. Soc., Vol. 16, 410 - 427 (1975).

Cometary outbursts, sudden increases in brightness which last for about a month, are still something of a mystery. The known facts and some of the mechanisms put forward to explain their occurrence are reviewed in this paper.

102.027 **Comets and planets: their interrelated origin.**
E. J. Öpik.
Irish Astron. Journ., Vol. 12, 1 - 48 (1975). – Prepared for the Royal Astronomical Society discussion on 'The structure and origin of comets', University of York, March 26, 1975.

102.028 **Mass accretion and variations of elements of comets' orbits.** S. L. Piotrowski, G. Sitarski.
Polish Acad. Sci., Inst. Astron., Warszawa, Poland, Preprint No. 61, 8 pp. (1975). – This paper will be published in Acta Astron.

102.029 **Comets.** O. Dobrovolsky (*Dobrovol'skij*).
Recherche, (*France*), Vol. 6, 98 - 107 (1975).
In French.

102.030 **Comets in perspective.** K. Hindley.
New Scient., Vol. 66, 120 - 123 (1975).

Structure, composition, processus physiques et chimiques dans les comètes.

On methods of computation of a cometary ephemeris in parabolic motion. See Abstr. 042.025.

Giant planets and their satellites, minor planets, meteorites and comets. See Abstr. 091.032.

Minor planets and comets in libration about the 2 : 1 resonance with Jupiter. See Abstr. 098.020.

103 Comets: Listed Objects

103.001 **Photometry of comets 1969 IX (Tago-Sato-Kosaka) and 1970 XV (Abe).** V. Riives.
Publ. Tartu Astrofiz. Obs., Vol. 43, 208 - 215 (1975). In Russian.

On the basis of photometric data the source intensity of luminous matter in the nucleus varies proportionally r^{-2}. Comet 1969 IX showed outbursts of its luminosity starting on February 6.41, a maximal increase of its brightness of $-1^m.4$, an expanding velocity of the halo of 1.1 km/s and a repulsive acceleration of 0.66 cm/s² or $1 + \mu = 1.5$.

103.002 **Observations of comets and asteroids at the Kleť Observatory in the year 1973.** A. Mrkos.
Acta Univ. Carolinae, Math. Phys., Vol. 16, No. 2, p. 47 - 56 (1975). – Precise positions of the following comets are given: P/Kearns-Kwee (1971c), Sandage (1972h), P/Reinmuth (1972i), Kojima (1972j), P/Schwassmann-Wachmann 1, P/Schwassmann-Wachmann 2, Heck-Sause (1973a), P/Tuttle-Giacobini-Kresák (1973b), Kohoutek (1973e), Kohoutek (1973f), Huchra (1973h), P/Gehrels (1973n).

103.003 **Comets in the year 1974.** J. Bouška.
Vesmír, Vol. 54, 314 (1975). In Czech.

103.004 **Possible comet Antal.** M. Antal.
IAU Circ., No. 2808 (1975).

103.005 **Observations of comets.** R. E. McCrosky, C. Y. Shao, G. Schwartz, J. H. Bulger.
IAU Circ., No. 2863 (1975).

103.006 **Possible comet West.** R. M. West.
IAU Circ., No. 2872 (1975).

103.007 **Observations of southern comets.**
Kometn. Tsirk., *Kiev*, No. 181 (1975). In Russian. Concerning comet du Toit 1 (1944 III); Comet Bennett, 1974h; Comet Lovas, 1974c; Comet Honda-Mrkos-Pajdušáková, 1974f.

103.008 **Photographic observations of comets.** K. Tomita, H. Kosai.
Tokyo Astron. Bull., Second Ser., No. 238, p. 2001 - 2010 (1975).

Precisely reduced positions for 10 comets are presented in this report.

103.009 **Other comets.**
British Astron. Ass. Circ., No. 567 (1975). – Concerning P/Boethin 1975a, P/Wolf 1975f, Longmore 1975g, P/Churyumov-Gerasimenko 1975i.

103.010 **Comet observations.** J. J. Aarnio, M. R. Dykes, P. B. Monk, I. M. Purcell, G. H. Rutter, R. H. H. South, R. L. Waterfield, J. van Maris.
British Astron. Ass. Circ., No. 567 (1975). – Concerning Bradfield 1974b, P/Forbes 1974a, P/Boethin 1975a, P/West-Kohoutek-Ikemura 1975b, P/Smirnova-Chernykh 1975e, Kobayashi-Berger-Milon 1975h.

103.011 **Other comets.**
British Astron. Ass. Circ., No. 569 (1975). – Concerning P/Harrington-Abell 1975*l*, P/Arend 1975m.

103.012 **Observations photographiques de comètes effectuées à l'astrographe double de 40 cm en 1973 et en 1974.** H. Debehogne.
Bull. Astron., Obs. Roy. Belgique, Vol. 8, 231 - 232 (1975). Concerning the comets Heck-Sause (1973a), Kohoutek (1973f), Bradfield (1974b).

103.100 **Comet 1975h Kobayashi-Berger-Milon**

La comète Kobayashi-Berger-Milon.
L'Astronomie, 89ᵉ année, p. 377 - 378 (1975).

New comet Kobayashi-Berger-Milon 1975h.
British Astron. Ass. Circ., No. 564 (1975).

Comet Kobayashi-Berger-Milon 1975h.
H. Broomhall.
British Astron. Ass. Circ., No. 565 (1975).

Comet Kobayashi-Berger-Milon 1975h.
R. J. Buckley
British Astron. Ass. Circ., No. 567 (1975).

Infrared photometry of comet Kobayashi-Berger-Milon. E. P. Ney.
Bull. American Astron. Soc., Vol. 7, 508 (1975). – Abstr. AAS.

Osservazioni fotografiche della cometa Kobayashi-Berger-Milon 1975h. S. Cantù, L. Pansecchi.
Coelum, Vol. 43, 216 - 218 (1975).

Comet Kobayashi-Berger-Milon (1975h).
Matsumoto, Fujii, T. Harada, D. Berger, E. A. Harlan, A. R. Klemola, Yoshida, D. Milon.
IAU Circ., No. 2797 (1975).

Comet Kobayashi-Berger-Milon (1975h).
R. E. McCrosky, C. Y. Shao, A. R. Klemola, G. Schwartz, K. Tomita, B. Milet, Z. Sekanina.
IAU Circ., No. 2798 (1975).

Comet Kobayashi-Berger-Milon (1975h).
T. Seki, N. Kojima, H. Kosai, B. Milet, E. Helin, S. J. Bus, K. Tomita, H. L. Giclas, M. L. Kantz, W. Schlosser, B. G. Marsden.
IAU Circ., No. 2800 (1975).

Comet Kobayashi-Berger-Milon (1975h).
K. Suzuki, T. Urata, M. Hoffmann, R. Lukas, R. E. Millington, P. M. Kilmartin, R. L. Waterfield, T. Seki, E. Roemer, M. A. Daniel, C. McCarthy, K. Tomita.
IAU Circ., No. 2803 (1975).

Comet Kobayashi-Berger-Milon (1975h).
M. Clark, P. M. Kilmartin, A. C. Gilmore, H. Debehogne, B. Milet, T. Seki, K. Tomita, H. L. Giclas, M. L. Kantz, T. Urata, M. Nozaki, R. S. Harrington.
IAU Circ., No. 2806 (1975).

Comet Kobayashi-Berger-Milon (1975h).
E. Devinney, C. McCracken, E. Barker, J. Drake, F. Jenkins, W. Upson.
IAU Circ., No. 2809 (1975).

Comet Kobayashi-Berger-Milon (1975h).
C. Fehrenbach, Y. Andrillat, F. Dossin, A. Woszczyk, E. Devinney, C. McCracken, A. Mrkos, H. Debehogne, G. Roland, R. L. Waterfield, J. Hers, R. S. Harrington, K. Tomita.

IAU Circ., No. 2810 (1975).

Comet Kobayashi-Berger-Milon (1975h).
R. R. D. Austin, P. M. Kilmartin, M. Antal, R. L. Waterfield, J. van Maris, K. Tomita, B. G. Marsden, C. Crane, B. Blakeslee, I. Price, J. Craig, W. Miller, P. Hodge.
IAU Circ., No. 2813 (1975).

Comet Kobayashi-Berger-Milon (1975h).
A. C. Danks, A. Grauer, L. Trafton, R. Tull, J. Woodman, T. Kurosaki, N. S. Chernykh, H. Debehogne, B. Milet, G. Roland.
IAU Circ., No. 2815 (1975).

Comet Kobayashi-Berger-Milon (1975h).
C. Lillie, G. Emerson, P. A. Wehinger, S. Wyckoff, K. Suzuki, T. Urata, T. Taguchi, M. Takeishi, J. J. Aarnio, R. L. Waterfield, G. H. Rutter, J.-A. Bojanowski, P. Collins, H. L. Giclas, M. L. Kantz, S. Kasahara, S. Odagiri.
IAU Circ., No. 2818 (1975).

Comet Kobayashi-Berger-Milon (1975h).
H. Debehogne, R. L. Waterfield, J. J. Aarnio.
IAU Circ., No. 2821 (1975).

Comet Kobayashi-Berger-Milon (1975h).
A. Mrkos, W. Ferreri, A. A. Schoenmaker, H. Debehogne.
IAU Circ., No. 2824 (1975).

Comet Kobayashi-Berger-Milon (1975h).
T. Taguchi, T. Kurosaki, K. Suzuki, S. Kasahara, M. Takeishi, T. Urata, B. Milet, W. Ferreri, E. P. Ney, J. Stoddart, B. Hatfield.
IAU Circ., No. 2831 (1975).

Comet Kobayashi-Berger-Milon (1975h).
N. Kojima, H. Debehogne, E. P. Ney, B. Hatfield, P. Maley, R. Reeves, S. Cantu, L. Mantegazza, L. Pansecchi.
IAU Circ., No. 2836 (1975).

Comet Kobayashi-Berger-Milon (1975h).
L. W. Avery, B. H. Andrew, F. Börngen, P. Lochno, K. Kirsch, J. Bortle.
IAU Circ., No. 2838 (1975).

Comet Kobayashi-Berger-Milon (1975h).
J. M. Codina, N. Torras, T. Kurosaki.
IAU Circ., No. 2841 (1975).

Comet Kobayashi-Berger-Milon (1975h).
H. L. Giclas, M. L. Kantz.
IAU Circ., No. 2844 (1975).

Comet Kobayashi-Berger-Milon (1975h).
J. M. Codina, N. Torras, B. Milet.
IAU Circ., No. 2846 (1975).

Comet Kobayashi-Berger-Milon (1975h).
T. Kurosaki, T. Seki, Mundet, N. Torras, H. L. Giclas, M. L. Kantz, D. Herald, B. Milet, B. G. Marsden.
IAU Circ., No. 2856 (1975).

Comet Kobayashi-Berger-Milon (1975h).
H. K. Raudsaar, J. M. Codina, N. Torras, T. Kurosaki, H. L. Giclas, M. L. Kantz. D. Herald, B. Sumner.
IAU Circ., No. 2872 (1975).

New comet Kobayashi-Berger-Milon (1975h).
Kometn. Tsirk., *Kiev*, No. 182 (1975). In Russian.

Observations of comet 1975h at the Crimean Astrophysical Observatory. N. S. Chernykh, G. R. Kastel'.
Kometn. Tsirk., *Kiev*, No. 182 (1975). In Russian.

Observations of comet 1975h in Kiev.
K. I. Churyumov, F. I. Kravtsov.
Kometn. Tsirk., *Kiev*, No. 182 (1975). In Russian.

Spectrum of comet Kobayashi-Berger-Milon.
V. Doroshenko, V. Tarashchuk.
Kometn. Tsirk., *Kiev*, No. 182 (1975). In Russian.

Comet Kobayashi-Berger-Milon (1975h).
Kometn. Tsirk., *Kiev*, No. 183 (1975). In Russian.

Expansion of the head of comet 1975h on August 9–10, 1975. K. I. Churyumov, F. I. Kravtsov.
Kometn. Tsirk., *Kiev*, No. 183 (1975). In Russian.

Observations of comet 1975h in Vilnius.
K. Chernis.
Kometn. Tsirk., *Kiev*, No. 183 (1975). In Russian.

Observations of comet 1975h at Zailyskiy Alatau.
G. V. Zhukov, M. I. Kibardina.
Kometn. Tsirk., *Kiev*, No. 183 (1975). In Russian.

Observations of comet 1975h at the Kleť Observatory (ČSSR). A. Mrkos.
Kometn. Tsirk., *Kiev*, No. 183 (1975). In Russian.

Spectrum of comet Kobayashi-Berger-Milon (1975h). I. R. Bejtrishvili.
Kometn. Tsirk., *Kiev*, No. 183 (1975). In Russian.

Comet Kobayashi-Berger-Milon, 1975h.
Kometn. Tsirk., *Kiev*, No. 184 (1975). In Russian. – Observations at the Observatory in Mayakakh, Odessa University (*N. A. Mis'kin, N. M. Shiper, G. R. Kastel'*); Observations at the Kleť Observatory (*A. Mrkos*); Observations of comet 1975h at the Saratov University (*L. N. Berdnikov, A. V. Prokhorov*); Observations of comet 1975h at the Ussurijsk Solar Station (*V. A. Golubev, A. S. Konyukhov*); Observations of comet 1975h in Stromville (USA) and Japan (*J. E. Bortle*); Elliptical polarization of radiation of comet 1975h (*R. S. Osherov*); Spectrum of comet 1975h Kobayashi-Berger-Milon (*N. A. Mis'kin, N. M. Shiper*).

Comet Kobayashi-Berger-Milon, 1975h.
Kometn. Tsirk., *Kiev*, No. 185 (1975). In Russian. – Observations at the Gissar Observatory (*O. Dobrovol'skij*); Observations of comet 1975h in Vilnius (*K. Chernis*).

Eine erste Amateuraufnahme des Kometen Kobayashi-Berger-Milon (1975h).
Orion, 33. Jahrgang, p. 124 (1975).

Der Komet Kobayashi-Berger-Milon (1975h).
Orion, 33. Jahrgang, p. 144 - 146 (1975).

Weitere Aufnahmen des Kometen Kobayashi-Berger-Milon (1975h).
Orion, 33. Jahrgang, p. 194 - 195 (1975).

Visual observation of the comet 1975h.
J. Květoň.
Říše hvězd, Vol. 56, 219 (1975). In Czech.

A bright summer comet.
Sky Telescope, Vol. 50, 158 - 159 (1975).

Pictures of comet Kobayashi-Berger-Milon.

Sky Telescope, Vol. 50, 261 - 267 (1975).

The brightness of comet 1975 h. C. S. Morris.
Sky Telescope, Vol. 50, 343 (1975).

The new comet Kobayashi-Berger-Milon 1975h.
D. Milon.
Strolling Astronomer, Vol. 25, 200 - 202, 203 - 205 (1975).

Komet Kobayashi-Berger-Milon 1975h.
R. Lukas.
SuW, Vol. 14, 299 (1975).

Spektrum des Kometen Kobayashi-Berger-Milon (1975 h). G. Schnur.
SuW, Vol. 14, 366 - 367 (1975).

Komeet 1975 h, zichtbaar met prismakijker en blote oog. M. Drummen.
Zenit, Vol. 2, 283 (1975).

Belgische amateur ontdekt komeet.
F. van Loo, J. de Keuster, L. Aerts.
Zenit, Vol. 2, 300 - 301 (1975).

103.101 Comet 1970 II Bennett

Ultraviolet spectrophotometry of comet Bennett (1970 II). I. The 1200 to 1800 A region. C. F. Lillie.
Bull. American Astron. Soc., Vol. 7, 507 (1975). – Abstr. AAS.

Production rates of [OI] and H_2O^+ in comet Bennett (1970 II). A. H. Delsemme, M. R. Combi.
Bull. American Astron. Soc., Vol. 7, 507 - 508 (1975).
Abstr. AAS.

On the motion of matter in the head and tail of comet Bennett. R. S. Osherov.
Dokl. AN TadzhSSR, Vol. 17, No. 10, p. 16 - 19 (1974).
In Russian. – Abstr. in Referativ. Zhurn. 51. Astron., 7.51. 372 (1975).

Spectrophotometric study of the tail of comet Bennett (1969i). Z. V. Karyagina.
Trudy Astrofiz. Inst., *Alma-Ata*, Vol. 25, 91 - 98 (1975).
In Russian.

The results of a spectrophotometric study of the continuous and emission spectrum of the tail of comet Bennett (1969i) are given. The energy distribution in the continuum spectrum of the tail (λ 3800–6000 Å) and the brightness of molecular bands (CN, CO^+, C_2) are obtained in absolute units.

Photographic observations of the tail of comet Bennett (1969i). Eh. S. Eroshevich, Z. V. Karyagina.
Trudy Astrofiz. Inst., *Alma-Ata*, Vol. 25, 99 - 101 (1975).
In Russian.

103.102 Comet 1973 XII Kohoutek

Molecular production rates in comet Kohoutek.
M. F. A'Hearn, J. J. Cowan.
Astron. Journ., Vol. 80, 852 - 860 (1975).

Narrow-band photometry was carried out on comet Kohoutek over a moderately wide range of heliocentric distances (0.66–1.77 AU). Production rates for C_2 and CN seem to vary as $r^{-3.5 \pm 0.5}$. C_3 abundance is comparable to that for CN and C_2. The comet was very dusty at large preperihelion distances and, for comparable heliocentric distances, was less dusty after perihelion than before. There was a major outburst of both dust and molecules on 1 and 2 December. The nuclear radius is of order 1 km.

Spectrophotometry of comet Kohoutek (1973f = 1973XII). M. F. A'Hearn.
Astron. Journ., Vol. 80, 861 - 875 (1975).

Spectrum scans have been obtained of comet Kohoutek with a Fourier spectrometer at heliocentric distances from 0.55 to 1.05 AU. The useful spectral range is roughly 3800–8400 Å and the resolution is about 10–15 Å at 5000 Å. Large variations occurred in the gas-to-dust ratio (pre- versus post-perihelion) and in the relative strength of the NH_2 bands (with heliocentric distance). There is a smaller change in the ratio of CN to C_2 emission. The Swan bands of C_2 show the customary anomaly in strengths.

Komet Kohoutek nok en gang.
B. R. Pettersen.
Astron. Tidssk., Årg. 8, 134 - 136 (1975).

Microwave continuum radiation from comet Kohoutek 1973f: emission from the icy-grain halo?
R. W. Hobbs, S. P. Maran, J. C. Brandt, W. J. Webster, Jr., K. S. Krishna Swamy.
Astrophys. Journ., Vol. 201, 749 - 755 (1975).

Cometary emission was measured at 3.71 cm on 1974 January 10–11. A possible detection at 2.8 cm and an upper limit at 11.1 cm are also reported. Interferometer measurements at 3.71 cm show that the angular diameter is <1.″4 or <850 km at the comet. The corresponding disk brightness temperature is >313 ± 80 K. The radiation presumably originates in the icy-grain halo postulated by Delsemme.

Observations of comet Kohoutek (1973f) with a ground-based Fabry-Perot spectrometer. D. Huppler, R. J. Reynolds, F. L. Roesler, F. Scherb, J. Trauger.
Astrophys. Journ., Vol. 202, 276 - 282 (1975).

Observations of Hα, H_2O^+, and [OI] emission lines from comet Kohoutek were made during a 2-month period around perihelion. Analyses of Hα line profiles and line intensities indicate that the mean outflow velocity of the hydrogen atoms was 7.8 ± 0.2 km s^{-1} and that the hydrogen atom production rate typically varied from about 0.4×10^{29} s^{-1} to about 1.4×10^{29} s^{-1} for comet-sun distances between 1 AU and 0.4 AU, respectively. An upper limit of 1 part in 100 was found for the D/H ratio in the cometary atomic hydrogen cloud.

Continuum observations of comet Kohoutek (1973f). B. H. Andrew, L. W. Avery.
Astrophys. Letters, Vol. 16, 125 - 127 (1975).

Observations of comet Kohoutek on several occasions in December 1973 and January 1974 revealed no detectable continuum radio emission at 2.8 cm wavelength, yet on January 10, 1974 Hobbs et al. (1975) observed 20 mJy at 3.7 cm and 34 mJy at 2.8 cm. The icy grain halo model of the cometary nuclear region seems a likely explanation of the observations, as suggested by Hobbs et al. (1975), but the radio emission they observed may be a relatively rare phenomenon.

Cenni sulle proprietà delle comete e breve relazione sulle osservazioni della cometa Kohoutek (1973f) effettuate all'Osservatorio Astronomico di Torino. M. G. Fracastoro.
Atti Accad. Sci. Torino, Cl. Sci. Fis., Mat., Nat., Vol. 108, 295 - 303 = Pubbl. Varie Fuori Ser. Oss. Astron. Torino (Pino Torinese), No. 58 (1973/74).

Far-ultraviolet observations of carbon and oxygen

in comet 1973 XII. C. B. Opal, T. L. Page, G. R. Carruthers. Bull. American Astron. Soc., Vol. 7, 467 (1975). – Abstr. AAS.

Integral brightness and photometry in the region of emission band C_2 (1–0) and in the adjacent continuum of comet Kohoutek 1973f. J. Svoreň, J. Tremko.
Bull. Astron. Inst. Czechoslovakia, Vol. 26, 342 - 345 (1975).

From photo-electric observations the authors derived the integral brightness of the comet Kohoutek 1973f and the diameters of its coma. The effect of the asymmetry of the coma on determining the integral brightness from photometric sections is discussed. For statistical purposes the photometric parameters in Vsekhsvyatskij's system H_{10}, y, H_y are also derived. The photo-electric observations of the comet in the region of the emission band C_2(1–0) and in the adjacent continuum are given, together with the qualitative indices of radiation changes in these regions, derived from them.

Results of observations of the comet 1973f Kohoutek at Zvenigorod.
A. M. Lozinskij, V. P. Osipenko, V. A. Yurevich.
Byull. Inst. Teoret. Astron., *Leningrad*, Vol. 14, 195 - 196 (1975). In Russian.

Comet Kohoutek (1973 XII). C. Torres.
IAU Circ., No. 2879 (1975).

The cometary magnetic field and its associated electric currents. W-H. Ip, D. A. Mendis.
Icarus, Vol. 26, 457 - 461 (1975).

Two different observations of comet Kohoutek (1973f) seem to suggest the existence of substantial magnetic fields ($\gtrsim 100\ \gamma$) in its coma and tail. The effects of the currents and hydromagnetic waves associated with these magnetic fields are considered and it is shown that while the currents closing through the inner coma may represent an important source of ionization in that region, the dissipation of hydromagnetic waves may also be a significant, if not dominant, source of heating there.

Observations of comet Kohoutek, 1973 XII during the Alpine expedition of the Sternberg Astronomical Institute.
V. M. Kovalenko, L. T. Markova, O. N. Kovalenko.
Kometn. Tsirk., *Kiev*, No. 182 (1975). In Russian.

Physical parameters of comet Kohoutek, 1973 XII.
V. P. Tarashchuk, G. A. Terez, Eh. I. Terez.
Kometn. Tsirk., *Kiev*, No. 182 (1975). In Russian.

Occultation of radio source PKS 2025–15 by comet Kohoutek (1973f). See Abstr. 141.094.

103.103 **Comet 1916 II Neujmin 2**

Elements and ephemeris of comet Neujmin 2, 1916 II.
Kometn. Tsirk., *Kiev*, No. 181 (1975). In Russian.

Unification of phenomena and evolution of the orbit of comet Neujmin 2 (1916 II) for 400 years (1660 - 2060). N. A. Belyaev.
Trudy Kazan. Gorod. Astron. Obs., No. 39, p. 102 - 109 (1973). In Russian.

103.104 **Comet 1969 II Gunn**

Periodic comet Gunn (1969 II). R. E. McCrosky, J. H. Bulger, C. Y. Shao, K. Suzuki, T. Urata, L. Krumenaker.
IAU Circ., No. 2821 (1975).

Periodic comet Gunn.
C. Torres, G. Schwartz, C. Y. Shao.
IAU Circ., No. 2884 (1975).

Periodic comet Gunn.
IAU Circ., No. 2887 (1975).

Investigation of the motion of comet Gunn (1969 II) and Kojima (1970 XII) for 200 years (1800 - 2000).
N. A. Belyaev, E. A. Reznikov.
Trudy Kazan. Gorod. Astron. Obs., No. 39, p. 110 - 115 (1973). In Russian.

103.105 **Comet 1970 XII Kojima**

Investigation of the motion of comet Gunn (1969 II) and Kojima (1970 XII) for 200 years (1800 - 2000).
See Abstr. 103.104.

103.106 **Comet 1866 I Tempel-Tuttle**

On the possibility of division of comet Tempel-Tuttle. E. D. Kondrat'eva.
Trudy Kazan. Gorod. Astron. Obs., No. 39, p. 137 - 142 (1973). In Russian.

103.107 **Comet 1972 VI Giacobini-Zinner**

Evolution of the orbit of comet Giacobini-Zinner for 400 years (1660 - 2060).
Yu. V. Evdokimov, I. Yu. Evdokimov.
Trudy Kazan. Gorod. Astron. Obs., No. 40, p. 47 - 63 (1974). In Russian.

103.108 **Comet 1970 VIII Pons-Winnecke**

Periodic comet Pons-Winnecke. B. G. Marsden.
IAU Circ., No. 2880 (1975).

Investigation of the motion of comet Pons-Winnecke. Communication I. L. Ya. Anan'eva, E. D. Kondrat'eva.
Trudy Kazan. Gorod. Astron. Obs., No. 40, p. 64 - 68 (1974). In Russian.

103.109 **Comet 1913 VI Westphal**

Periodic comet Westphal.
IAU Circ., No. 2876 (1975).

Investigation of the motion of comet Westphal for the period 1852 - 1913. L. M. Belous.
Trudy Kazan. Gorod. Astron. Obs., No. 40, p. 69 - 78 (1974). In Russian.

103.110 **Comet 1964 IX Everhart**

Definitive orbit of comet 1964 IX Everhart.
L. A. Markelov.
Trudy Kazan. Gorod. Astron. Obs., No. 40, p. 79 - 86 (1974). In Russian.

103.111 **Comet 1975e Smirnova-Chernykh**

Periodic comet Smirnova-Chernykh (1975e).
N. Kojima, T. Seki, L. Krumenaker.
IAU Circ., No. 2817 (1975).

Periodic comet Smirnova-Chernykh (1975e).
E. Roemer, C. C. McCarthy, B. G. Marsden.
IAU Circ., No. 2848 (1975).

New comet Smirnova-Chernykh (1975e).
G. R. Kastel'.
Pis'ma v Astron. Zhurn., Vol. 1, No. 7, p. 47 - 48 (1975). In Russian.

A new periodic comet moving in resonance 5/7 with Jupiter has been discovered at the Crimean Astrophysical Observatory. Orbital elements based on 12 observations (March 4 – April 15, 1975) are given. The orbit was transformed into the present one as a result of an approach to Jupiter in 1963. Before 1963 the comet moved in the region between Jupiter and Saturn.

103.112 **Comet 1975f Wolf 1**

Periodic comet Wolf (1975f).
R. E. McCrosky, J. Bulger.
IAU Circ., No. 2851 (1975).

The orbital motion of comets Wolf and Barnard 3.
D. K. Yeomans.
Publ. Astron. Soc. Pacific, Vol. 87, 635 - 637 (1975).

The orbital motion of comet Wolf is investigated over the interval 1884–1976. In order to represent successfully the observations prior to 1925, nongravitational terms are required in the equations of motion. After a close approach to Jupiter in 1922 increased the perihelion distance of comet Wolf from 1.6 to 2.4 astronomical units, the nongravitational effects were found to be negligible. The motion of comet Barnard 3 was investigated and no evidence was found to indicate a common origin with comet Wolf.

103.113 **Comet 1892 V Barnard 3**

The orbital motion of comets Wolf and Barnard 3.
See Abstr. 103.112.

103.114 **Comet 1974b Bradfield**

Positions of comet Bradfield 1974b.
J. M. Kreiner, M. Kurpińska, M. Winiarski.
Acta Astron., Vol. 25, 317 - 319 (1975).

Given are 79 astrometrical positions of comet Bradfield 1974b, obtained in Cracow during the period April 2 – July 1, 1974.

Observations of comet Bradfield, 1974b during the Alpine expedition of the Sternberg Astronomical Institute.
V. M. Kovalenko, O. N. Kovalenko.
Kometn. Tsirk., *Kiev*, No. 185 (1975). In Russian.

Near-infrared observations of comet Bradfield (1974b).
T. Iijima, T. Matsumoto, M. Oishi, H. Okuda, T. Ono.
Publ. Astron. Soc. Japan, Vol. 27, 507 - 510 (1975).

Photometric observations of comet Bradfield (1974b) were performed in the near-infrared region. The energy distribution of the comet was similar to the solar spectrum in 1.03-, 1.25-, and 1.65-μm bands, and a slight infrared excess was observed in 2.2-μm band. The near-infrared spectrum is interpreted as the sum of the scattered sunlight and the thermal radiation from the dust grains whose temperature is ~350 K at 1 AU.

103.115 **Comet 1930 VI Schwassmann-Wachmann 3**

Periodic comet Schwassmann-Wachmann 3 (1930 VI). N. A. Belyaev, S. D. Shaporev.
Problems of cosmic physics. Vyp. (No.) 10, (see 003.003), p. 9 - 16 (1975). In Russian.

The definitive orbit of the lost short-period comet 1930 VI has been calculated from 190 observations. Investigation of the orbit during 200 years (1800–2000) demonstrated that the approach to Jupiter in 1882 resulted in considerable changes of all the orbital elements.

103.116 **Comet 1919 III Brorsen-Metcalf**

An investigation of the motion of the periodic comet Brorsen-Metcalf from 1847 to 1919. L. M. Belous.
Problems of cosmic physics. Vyp. (No.) 10, (see 003.003), p. 16 - 23 (1975). In Russian.

The linking of two apparitions of comet Brorsen-Metcalf is based on 62 observations, the perturbations of all nine planets and nongravitational effects being included.

103.117 **Comet 1972 XI Kearns-Kwee**

Comet Kearns-Kwee 1971c. A correction of the predicted elements. C. Dinwoodie.
Journ. British Astron. Ass., Vol. 86, 40 - 43 (1975).

103.118 **Comet 1954 XI Wirtanen**

The close approach of comet Wirtanen 1947 XIII = 1954 XI to Jupiter in 1972. M. P. Imnadze.
Trudy Vychisl. tsentra. AN GruzSSR, Vol. 13–14, No. 2, p. 28 - 31 (1975). In Russian. – Abstr. in Referativ. Zhurn. 51. Astron., 12.51.122 (1975).

103.119 **Comet 1975d Bradfield**

Comet Bradfield (1975d). J. Hers, A. C. Gilmore, P. M. Kilmartin, S. W. Milbourn.
IAU Circ., No. 2804 (1975).

Comet Bradfield (1975d). B. G. Marsden.
IAU Circ., No. 2842 (1975).

Comet Bradfield (1975d).
C. Torres, S. Barros, H. Moreno.
IAU Circ., No. 2884 (1975).

Comet Bradfield, 1975d.
Kometn. Tsirk., *Kiev*, No. 181 (1975). In Russian.

Ephemeris of comet Bradfield, 1975d.
B. G. Marsden.
Kometn. Tsirk., *Kiev*, No. 186 (1975). In Russian.

Photographic observations of comet Bradfield (1975d). J. Hers.
Monthly Notes Astron. Soc. Southern Africa, Vol. 34, 156 (1975).

103.120 **Comet 1971 II Encke**

Periodic comet Encke.
A. C. Gilmore, P. M. Kilmartin, R. R. D. Austin.
IAU Circ., No. 2795 (1975).

103.121 **Comet 1975m Arend**

Periodic comet Arend.
IAU Circ., No. 2796 (1975).

Periodic comet Arend (1975m).
E. Roemer, M. A. Daniel.
IAU Circ., No. 2848 (1975).

103.122 **Comet 1973*l* Schwassmann-Wachmann 2**

Periodic comet Schwassmann-Wachmann 2 (1973*l*).
R. E. McCrosky, G. Schwartz, J. H. Bulger.
IAU Circ., No. 2797 (1975).

103.123 **Comet 1975a Boethin**

Periodic comet Boethin (1975a).
G. Schwartz, C. Y. Shao, E. Roemer, M. A. Daniel, C. McCarthy.
IAU Circ., No. 2799 (1975).

Periodic comet Boethin (1975a).
R. L. Waterfield, R. H. S. South.
IAU Circ., No. 2825 (1975).

103.124 **Comet 1975*l* Harrington-Abell**

Periodic comet Harrington-Abell.
IAU Circ., No. 2807 (1975).

Periodic comet Harrington-Abell (1975*l*).
E. Roemer, B. G. Marsden.
IAU Circ., No. 2848 (1975).

Elements and ephemeris of the short-periodic comet Harrington-Abell (1954 XIII). B. G. Marsden.
Kometn. Tsirk., *Kiev*, No. 185 (1975). In Russian.

103.125 **Comet 1974c Lovas**

Comet Lovas (1974c). M. Zemelman, B. M. Blanco, A. C. Gilmore, P. M. Kilmartin.
IAU Circ., No. 2809 (1975).

Comet Lovas (1974c). A. C. Gilmore, P. M. Kilmartin.
IAU Circ., No. 2817 (1975).

Comet Lovas (1974c).
A. C. Gilmore, P. M. Kilmartin.
IAU Circ., No. 2842 (1975).

Comet Lovas (1974c).
G. Sanchez, M. R. Cesco, H. Mira.
IAU Circ., No. 2851 (1975).

103.126 **Comet 1975g Longmore**

Comet Longmore (1975g). A. J. Longmore, P. R. Standen, R. D. Cannon, M. E. Sim, B. G. Marsden.
IAU Circ., No. 2812 (1975).

Comet Longmore (1975g).
IAU Circ., No. 2838 (1975).

Periodic comet Longmore (1975g).
T. Hawarden, B. G. Marsden.
IAU Circ., No. 2845 (1975).

Periodic comet Longmore (1975g). R. M. West.
IAU Circ., No. 2846 (1975).

New comet Longmore, 1975g.
Kometn. Tsirk., *Kiev*, No. 181 (1975). In Russian.

Comet Longmore, 1975g.
Kometn. Tsirk., *Kiev*, No. 186 (1975). In Russian.

103.127 **Comet 1975i Churyumov-Gerasimenko**

Periodic comet Churyumov-Gerasimenko (1975i).
E. Roemer.
IAU Circ., No. 2816 (1975).

Periodic comet Churyumov-Gerasimenko (1975i).
E. Roemer, C. C. McCarthy.
IAU Circ., No. 2846 (1975).

Periodic comet Churyumov-Gerasimenko (1975i).
Kometn. Tsirk., *Kiev*, No. 183 (1975). In Russian.

103.128 **Comet 1975b West-Kohoutek-Ikemura**

Periodic comet West-Kohoutek-Ikemura (1975b).
L. Krumenaker, R. E. McCrosky, J. H. Bulger, R. H. S. South, R. L. Waterfield.
IAU Circ., No. 2824 (1975).

Periodic comet West-Kohoutek-Ikemura (1975b).
L. Kohoutek, M. Dieckvoss.
IAU Circ., No. 2843 (1975).

103.129 **Comet 1974f Honda-Mrkos-Pajdušaková**

Periodic comet Honda-Mrkos-Pajdušáková (1974f).
C. Torres.
IAU Circ., No. 2818 (1975).

103.130 **Comet 1975j Mori-Sato-Fujikawa**

New comet Mori-Sato-Fujikawa 1975j.
Z. Sekanina, S. W. Milbourn.
British Astron. Ass. Circ., No. 568 (1975).

Comet Mori-Sato-Fujikawa 1975j. B. G. Marsden.
British Astron. Ass. Circ., No. 569 (1975).

Comet Mori-Sato-Fujikawa 1975j.
B. G. Marsden.
British Astron. Ass. Circ., No. 570 (1975).

Comet Mori-Sato-Fujikawa (1975j).
H. Mori, Y. Sato, S. Fujikawa.
IAU Circ., No. 2847 (1975).

Comet Mori-Sato-Fujikawa (1975j).
R. E. McCrosky, H. L. Giclas, J. E. Bortle.
IAU Circ., No. 2849 (1975).

Comet Mori-Sato-Fujikawa (1975j).
K. Suzuki, T. Urata, E. Moore, E. Tedesco, H. Shibasaki, H. Kosai, N. Kojima, D. Herald, Z. Sekanina.
IAU Circ., No. 2850 (1975).

Comet Mori-Sato-Fujikawa (1975j).
R. E. McCrosky, C. Y. Shao, H. L. Giclas, M. L. Kantz, T. Urata, K. Tomita, Kanda, D. Herald, K. Haddow, R. L. Waterfield, B. Milet.
IAU Circ., No. 2853 (1975).

Comet Mori-Sato-Fujikawa (1975j).
K. Suzuki, T. Urata, A. C. Gilmore, P. M. Kilmartin, B. G. Marsden.
IAU Circ., No. 2854 (1975).

Comet Mori-Sato-Fujikawa (1975j).
M. Koishikawa, T. Kurosaki, A. Kuribayashi, T. Seki, B. Milet, K. Tomita, H. L. Giclas, M. L. Kantz, E. Devinney, C. McCracken.
IAU Circ., Nos. 2862, 2867 (1975).

Comet Mori-Sato-Fujikawa (1975j).
B. Milet, K. Tomita, M. Koishikawa, B. G. Marsden.
IAU Circ., No. 2867 (1975).

Comet Mori-Sato-Fujikawa (1975j).
N. Kojima, T. Kurosaki, T. Seki, A. Mrkos, J. Hers, H. L. Giclas, M. L. Kantz, D. Herald.
IAU Circ., No. 2875 (1975).

Comet Mori-Sato-Fujikawa (1975j).
J. Hers, T. Kurosaki, H. Debehogne, R. R. de Freitas Mourão.
IAU Circ., No. 2886 (1975).

New comet Mori-Sato-Fujikawa, 1975j.
Kometn. Tsirk., *Kiev*, No. 186 (1975). In Russian.

103.131 **Comet 1975k Suzuki-Saigusa-Mori**

New comet Suzuki-Saigusa-Mori 1975k.
R. L. Waterfield, K. Haddow, H. Broomhall.
British Astron. Ass. Circ., No. 568 (1975).

Comet Suzuki-Saigusa-Mori 1975k. B. G. Marsden.
British Astron. Ass. Circ., No. 569 (1975).

Comet Suzuki-Saigusa-Mori (1975k).
S. Suzuki, Y. Saigusa, H. Mori, K. Okazaki, S. Furuyama.
IAU Circ., No. 2847 (1975).

Comet Suzuki-Saigusa-Mori (1975k).
G. Casari, R. E. McCrosky, H. L. Giclas, J. E. Bortle.
IAU Circ., No. 2849 (1975).

Comet Suzuki-Saigusa-Mori (1975k).
K. Suzuki, T. Urata, E. Moore, E. Tedesco, H. Shibasaki, H. Kosai, Z. Sekanina.
IAU Circ., No. 2850 (1975).

Comet Suzuki-Saigusa-Mori (1975k).
R. E. McCrosky, C. Y. Shao, H. L. Giclas, M. L. Kantz, T. Urata, M. F. A'Hearn, B. Milet, B. G. Marsden.
IAU Circ., No. 2852 (1975).

Comet Suzuki-Saigusa-Mori (1975k).
K. Suzuki, T. Kurosaki, M. Koishikawa, E. Moore, E. F. Tedesco, K. A. Haddow, R. L. Waterfield.
IAU Circ., No. 2856 (1975).

Comet Suzuki-Saigusa-Mori (1975k).
V. Guth, I. Hasegawa.
IAU Circ., Nos. 2858, 2877 (1975).

Comet Suzuki-Saigusa-Mori (1975k). T. Seki, B. Milet, M. Koishikawa, M. F. A'Hearn, S. Kasahara, K. Tomita, H. L. Giclas, M. L. Kantz, B. G. Marsden, T. Kurosaki.
IAU Circ., Nos. 2861, 2867 (1975).

Comet Suzuki-Saigusa-Mori (1975k).
A. Mrkos.
IAU Circ., Nos. 2869, 2877 (1975).

Comet Suzuki-Saigusa-Mori (1975k).
T. Kurosaki, J. M. Mundet, J. M. Codina, N. Torras, D. Herald, R. F. Smith, J. Hers, T. Morgan.
IAU Circ., No. 2877 (1975).

Comet Suzuki-Saigusa-Mori (1975k). J. Hers.
IAU Circ., No. 2888 (1975).

New comet Suzuki-Saigusa-Mori, 1975k.
Kometn. Tsirk., *Kiev*, No. 186 (1975). In Russian.

103.132 **Comet 1975c Kohoutek**

Periodic comet Kohoutek (1975c).
L. Kohoutek, M. Dieckvoss.
IAU Circ., No. 2847 (1975).

Periodic comet Kohoutek (1975c).
G. Schwartz, C. Y. Shao, R. E. McCrosky.
IAU Circ., No. 2879 (1975).

103.133 **Comet 1975n West**

New comet West 1975n.
R. W. West, B. G. Marsden.
British Astron. Ass. Circ., No. 570 (1975).

Comet West (1975n).
R. M. West, O. Pizarro, G. Pizarro, B. G. Marsden.
IAU Circ., No. 2860 (1975).

Comet West (1975n). R. M. West, H. E. Schuster, O. Pizarro, G. Pizarro, B. G. Marsden.
IAU Circ., No. 2871 (1975).

Comet West (1975n). D. Herald.
IAU Circ., No. 2878 (1975).

Comet West (1975n). P. M. Kilmartin.
IAU Circ., No. 2882 (1975).

Comet West (1975n).
T. Seki, H. L. Giclas, M. L. Kantz.
IAU Circ., No. 2886 (1975).

103.134 **Comet 1975o Gehrels 3**

New comet Gehrels 1975o.
T. Gehrels, R. J. Buckley.
British Astron. Ass. Circ., No. 570 (1975).

Possible comet Gehrels. T. Gehrels.
IAU Circ., No. 2861 (1975).

Comet Gehrels (1975o).
T. Gehrels, R. Adams, R. Sather, C. Vesely, J. H. Bulger, R. E. McCrosky, C. Y. Shao, B. G. Marsden.
IAU Circ., No. 2865 (1975).

Periodic comet Gehrels 3 (1975o).
J. H. Bulger, R. E. McCrosky, C. Y. Shao.
IAU Circ., No. 2870 (1975).

Periodic comet Gehrels 3 (1975o).
G. Schwartz, C. Y. Shao, E. Roemer, B. G. Marsden.
IAU Circ., No. 2890 (1975).

103.135 **Comet 1974g van den Bergh**

Comet van den Bergh (1974g).
E. Roemer, C. C. McCarthy, L. M. Vaughn, R. A. McCallister.
IAU Circ., No. 2864 (1975).

103.136 **Comet 1975p Bradfield**

New comet Bradfield 1975p.
W. A. Bradfield, R. J. Buckley, S. W. Milbourn.
British Astron. Ass. Circ., No. 570 (1975).

Comet Bradfield (1975p).
W. A. Bradfield, A. B. Muller, H. E. Schuster.
IAU Circ., No. 2866 (1975).

Comet Bradfield (1975p).
E. Moore, E. Tedesco, B. Sumner, D. Herald, B. G. Marsden.
IAU Circ., No. 2868 (1975).

Comet Bradfield (1975p).
C. Jekabsons, M. P. Candy, D. Herald.
IAU Circ., No. 2869 (1975).

Comet Bradfield (1975p).
H. L. Giclas, M. L. Kantz, R. R. D. Austin, P. M. Kilmartin, D. Herald, J. Hers, Z. Sekanina, B. G. Marsden.
IAU Circ., No. 2874 (1975).

Comet Bradfield (1975p).
J. Hers, J. B. Trainor, T. B. Tregaskis.
IAU Circ., No. 2881 (1975).

Comet Bradfield (1975p).
A. C. Gilmore, P. M. Kilmartin.
IAU Circ., No. 2882 (1975).

Comet Bradfield (1975p). J. Hers.
IAU Circ., No. 2889 (1975).

Comet Bradfield (1975p). B. G. Marsden.
IAU Circ., No. 2891 (1975).

103.137 **Comet 1925 II Schwassmann-Wachmann 1**

Periodic comet Schwassmann-Wachmann 1.
A. Mrkos.
IAU Circ., No. 2871 (1975).

103.138 **Comet 1955 VII Perrine-Mrkos**

Periodic comet Perrine-Mrkos.
E. Roemer, C. Kowal.
IAU Circ., No. 2876 (1975).

103.139 **Comet 1973 I Gehrels 1**

Periodic comet Gehrels 1 (1973 I).
E. Roemer, D. J. Granrath, C. C. McCarthy.
IAU Circ., No. 2879 (1975).

103.140 **Comet 1975q Sato**

Comet Sato (1975q). Y. Sato, E. A. Harlan.
IAU Circ., No. 2882 (1975).

Comet Sato (1975q).
T. Seki, I. Hasegawa, Z. Sekanina.
IAU Circ., No. 2883 (1975).

Comet Sato (1975q). E. A. Harlan, A. R. Klemola, H. L. Giclas, M. L. Kantz, M. Koishikawa, R. E. McCrosky, G. Schwartz, C. Y. Shao, B. Jones, B. G. Marsden, M. R. Dykes.
IAU Circ., Nos. 2884, 2890 (1975).

Comet Sato (1975q).
H. L. Giclas, M. L. Kantz.
IAU Circ., No. 2889 (1975).

Comet Sato (1975q).
Y. Hiramatsu, M. Koishikawa, T. Seki, K. Edamatsu.
IAU Circ., No. 2890 (1975).

103.141 **Comet 1970 XI Kopff**

Periodic comet Kopff.
IAU Circ., No. 2885 (1975).

103.142 **Comet 1973b Tuttle-Giacobini-Kresák**

Observations spectrographiques de la comète 1973b immédiatement après son violent sursaut de brillance de juillet 1973. J. P. Swings, J. M. Vreux.
Bull. Soc. Roy. Sci. Liège, (*Belgium*), Vol. 42, 593 - 597 = Inst. Astrophys. Univ. Liège, Cointe-Ougrée (*Belgique*), Coll. 8°, No. 643 (1973).

103.143 **Comet 1965 VII de Vico-Swift**

Short-periodic comet de Vico-Swift, 1965 VII.
S. Shaporev.
Kometn. Tsirk., *Kiev*, No. 185 (1975). In Russian.

104 Meteors, Meteor Streams

104.001 **The dynamics and orbit of an iron meteoroid.**
A. A. Griffin.
Journ. Roy. Astron. Soc. Canada, Vol. 69, 126 - 132 (1975).
Two Super-Schmidt photographs of an unusual meteor are employed to study the orbit, atmospheric trajectory and photometric behaviour of the meteoroid. Previous study of the spectrum had suggested the object was similar to an iron meteorite and the close similarity of its behaviour in the atmosphere to an artificial meteor produced by a steel pellet (Trailblazer 1) strongly supports this interpretation. The orbit is unusually small (a = 0.995 a.u.) but appears to be more closely related to asteroidal rather than cometary orbits. A large deceleration is observed in the late stages of flight and is difficult to explain by the conventional meteor theory.

104.002 **Ablation and shape-density coefficients in meteors.**
Z. Ceplecha.
Bull. Astron. Inst. Czechoslovakia, Vol. 26, 242 - 248 (1975).
Six methods of computing the ablation coefficient and the shape-density coefficient for photographic meteors are given. They are suitable in cases, when a big change of meteor velocity is observed. The methods were applied to PN-fireballs, one of them being used as a numerical example in this paper.

104.003 **Meteor-generated infrasound.**
D. O. ReVelle, with a reply by W. L. Donn, N. K. Balachandran.
Science, Vol. 189, 394 - 396 (1975).

104.004 **Extraterrestrial particles and precipitation.**
J. Rosinski, C. T. Nagamoto, M. Bayard.
Journ. Atmosph. Terr. Phys., Vol. 37, 1231 - 1243 (1975).
Sampling of magnetic particles at different latitudes can be used to detect meteor streams. Chemical composition of surfaces and interiors of magnetic particles is different due to selective fractionation of different chemical compounds between molten and solid phases and due to evaporation into the E- and D-regions during their entry into the atmosphere.

104.005 **The influence of planetary perturbations on the evolution of condensations within meteor streams.**
Yu. V. Evdokimov.
Trudy Kazan. Gorod. Astron. Obs., No. 39, p. 94 - 101 (1973). In Russian.

104.006 **Mass of the grazing fireball of August 1972.**
H. M. Caulk.
Astrophys. Letters, Vol. 16, 99 - 100 (1975).
The mass of the Utah-to-Alberta grazing fireball of August 1972 has been recalculated from momentum considerations to be 2×10^6 g. The reported mass estimates of 10^9 g are not allowed by the published deceleration and altitudes.

104.007 **Enduring meteor trains and chemiluminescence.**
W. J. Baggaley.
Bull. Astron. Inst. Czechoslovakia, Vol. 26, 273 - 275 (1975).
The nitric oxide afterglow and nitrogen dioxide air-glow chemiluminescent processes are examined quantitatively and shown to be quite inadequate as a source of enduring meteor train luminosity.

104.008 **On the association of NO-molecules in meteors.**
J. Rajchl.
Bull. Astron. Inst. Czechoslovakia, Vol. 26, 282 - 288 (1975).
The discussion of the possible processes which may lead to the generation of NO_2-radiation in meteors indicates that the most probable process, in particular with faint meteors, is the process described by the equation $O + (NO)_m \rightarrow (NO)_{m-1} + NO_2 + h\nu$. The chemiluminescence radiation of NO_2 is generated here by means of the $(NO)_m$-cluster, where $m > 1$, most probably $m = 2$. It was found that the atmospheric concentrations of O and NO are sufficient for this process.

104.009 **Head echoes and solar activity in the Geminid period.** P. Pecina.
Bull. Astron. Inst. Czechoslovakia, Vol. 26, 288 - 293 (1975).
The occurrence of head echoes in dependence on solar activity in the Geminid period 1959–1969 is examined. All considerations refer to the interval 22^h–06^h U.T. Solar ionizing radiation is assumed to be of either of X-ray and ultraviolet, or corpuscular radiation nature. A comparison within the 11-year solar cycle does not lead to the conclusion that in the interval 22^h–06^h U.T. the corpuscular radiation acts mainly on the head echo phenomenon.

104.010 **An automated southwestern meteor observatory.**
J. Cuffey, G. A. Harvey.
Bull. American Astron. Soc., Vol. 7, 443 (1975). – Abstr. AAS.

104.011 **Ultraviolet emission of meteors.**
D. D. Meisel, G. A. Harvey, C. B. Opal.
Bull. American Astron. Soc., Vol. 7, 466 - 467 (1975). – Abstr. AAS.

104.012 **Photoelectric measuring complex for meteor observations.** D. I. Stepanov, L. B. Gusakovskaya.
Izv. Astron. Ehngel'gardt. Obs., *Kazan'*, No. 39, p. 3 - 53 (1972). In Russian.

104.013 **Persistent meteor trains.** D. W. Hughes.
Nature, Vol. 257, 533 - 534 (1975).

104.014 **Sodium emission from long enduring meteor trains.**
W. J. Baggaley.
Nature, Vol. 257, 567 - 568 (1975). – Letter.

104.015 **The effect of ionic processes on the characteristics of radio-echoes from meteor trains.**
L. M. G. Poole, T. F. Nicholson.
Planet. Space Sci., Vol. 23, 1261 - 1277 (1975).
The effect on meteoric ionization of known ionic reactions of possible relevance has been assessed. It is shown that oxidation of positive meteoric ions by molecular oxygen and ozone, and the formation of nitride ions, followed in each case by dissociative recombination, can give rise to a pseudo-three-body attachment, which for reasonable ozone concentrations in the meteor region is sufficiently rapid to explain observed deionization.

104.016 **The flux and spatial distribution of micrometeoroids in the near-earth environment.**
D. K. Bedford, N. G. Adams, D. Smith.
Planet. Space Sci., Vol. 23, 1451 - 1456 (1975).
The near-earth micrometeoroid data of Prospero has been re-appraised in the light of recent data from the HEOS satellite and other detectors which sampled the near-earth region. This has enabled a cumulative micrometeoroid flux-mass curve to be defined for this region which shows a pronounced flux enhancement above the similar, more detailed, curve obtained at greater geocentric distances. The latter curve is shown to be consistent with flux data obtained from recent lunar microcrater studies. The trend of increasing flux with decreasing geocentric distance is now positively established.

104.017 **Dust of meteor showers in the mesosphere as**

observed in Abastumani. G. G. Mikirtumova.
Astron. Tsirk., No. 849, p. 5 - 7 (1974). In Russian.

104.018 **On the change of line intensity of a small meteor along the way.**
Kh. Gul'medov, S. Mukhametnazarov, V. A. Smirnov.
Astron. Tsirk., No. 849, p. 7 - 8 (1974). In Russian.

104.019 **On the estimate of masses of meteor bodies from photographic observations in Dushanbe.**
V. I. Musij, I. S. Shestaka.
Astron. Tsirk., No. 850, p. 5 - 8 (1974). In Russian.

104.020 **Jaarverslag 1974 visuele en fotografische meteorensecties.** B. Apeldoorn.
Zenit, Vol. 2, 276 - 282 (1975).

104.021 **Investigation of meteor streams.**
A. K. Terent'eva.
Problems of cosmic physics. Vyp. (No.) 10, (see 003.003), p. 39 - 57 (1975). In Russian.

A short history of meteor radiants study and a review of modern state of minor meteor streams investigations are given. The tasks of future researches in this field are pointed out.

104.022 **Measurements of meteor radiants at different geographical latitudes.**
B. L. Kashcheev, N. V. Novoselova.
Problems of cosmic physics. Vyp. (No.) 10, (see 003.003), p. 58 - 67 (1975). In Russian.

Measurement results of some 4.5 thousand meteor radiants (up to $+8^m$) registered near the equator are presented. The radiant distribution functions on latitude for both the northern and southern hemispheres are obtained.

104.023 **Investigation of the initial radius of meteor trains with the help of a height-measuring system.**
I. A. Delov, V. V. Zhukov.
Problems of cosmic physics. Vyp. (No.) 10, (see 003.003), p. 68 - 76 (1975). In Russian.

Results of an experimental investigation of the initial radii of a meteor train with help of a phase anglemeter are presented. Measurements of initial radii were done by simultaneous radio observations of the meteor trains on two wavelengths: $\lambda_1 = 13.4$ m, $\lambda_2 = 8.13$ m. Height and other relationships of the initial radius are obtained. The dependences of the initial radius on height, meteor velocity, diffusion coefficient appear to be weaker than the theoretical ones.

104.024 **Statistical orthogonal models of interpreting radar observations of the drift of meteor trains.**
V. A. Nechitajlenko.
Problems of cosmic physics. Vyp. (No.) 10, (see 003.003), p. 76 - 84 (1975). In Russian.

The peculiarities of orthogonal models of interpreting the observational data are examined and an algorithm for estimating wind components is presented.

104.025 **On the measurement of meteor parameters by an astronomical TV system.** V. F. Anisimov, M. Begkhanov, Kh. Gul'medov, S. Mukhamednazarov.
AN TurkmSSR. Ser. fiz.-tekhn., khim. i geol. n., 1974, No. 6, p. 108 - 110. In Russian. – Abstr. in Referativ. Zhurn. 51. Astron., 9.51.387 (1975).

104.026 **A quantiative model for the ablation of dustball meteors.** R. L. Hawkes, J. Jones.
Monthly Notices Roy. Astron. Soc., Vol. 173, 339 - 356 (1975).

A quantitative model for the ablation of dustball meteors is presented. The model assumes meteoric bodies are composed of grains (approximately 10^{-6} g mass) held together by a lower boiling point 'glue'. Once the boiling point of the 'glue' is reached grains are detached. Light is assumed to be produced only by detached grains. For meteors fainter than about $+5^M$, the grains are detached above the radiation ceiling, and the heights of such meteors are mass independent. For brighter meteors, grain detachment and ablation occur simultaneously and the model predicts approximately classical light curves.

104.027 **Glimpses of the 1975 Perseid meteor shower.**
Sky Telescope, Vol. 50, 344 - 345 (1975).

104.028 **Role of meteors in ionospheric processes.**
O. Ovezgel'dyev, L. P. Korsunova.
Izv. AN TurkmSSR. Ser. fiz.-tekhn., khim i geol. nauk, Vol. 5, 37 - 41 (1974). In Russian. – Abstr. in Referativ. Zhurn. 51. Astron., 8.51.389 (1975).

104.029 **Determination of the ambipolar diffusion coefficient of unsaturated [meteor] trains from observations in Ashkhabad.** Kh. D. Gul'medov, S. S. Takhautdinova.
Izv. AN TurkmSSR. Ser. fiz.-tekhn., khim. i geol. nauk, Vol. 5, 116 - 117 (1974). In Russian. – Abstr. in Referativ. Zhurn. 51. Astron., 8.51.390 (1975).

104.030 **Experiences and results of meteor expeditions.**
K. Pavlovski.
Publ. Obs. Astron. Beograd, No. 20, (see 012.024), p. 178 - 184 (1975). In Serbo-Croatian.

104.031 **Photographic fireball networks.** I. Halliday.
I.A.U. Colloquium No. 13, (see 012.030), p. 1 - 8 (1973).

Three networks for the photography of bright fireballs are now in operation; in the central United States, central Europe and western Canada. A detailed comparison is made of the parameters which describe the three networks.

104.032 **A determination of meteor mass distribution from meteor echoes.** G. Forti.
I.A.U. Colloquium No. 13, (see 012.030), p. 9 - 12 (1973).

Meteor counts from the observations of the Harvard-Smithsonian Radio Meteor Project at four magnitudes in the range of about +9 to +15 in the visual scale were used to determine the exponent of the mass-distribution law. The echo counts, covering in five successive days 24 hr with overlapping periods, yielded for the exponent a value of 1.95 ± 0.02, close to other determinations. Individual values, however, presented a considerable scatter, probably reflecting changes in radiant and mass distribution over the year.

104.033 **Recombination in radar meteors.**
R. B. Southworth.
I.A.U. Colloquium No. 13, (see 012.030), p. 13 - 21 (1973).

104.034 **Combined observations of meteors by image-orthicon television camera and multi-station radar.**
A. F. Cook, G. Forti, R. E. McCrosky, A. Posen, R. B. Southworth, J. T. Williams.
I.A.U. Colloquium No. 13, (see 012.030), p. 23 - 44 (1973).

104.035 **Mass influx obtained from low-light-level television observations of faint meteors.**
R. J. Naumann, K. S. Clifton.
I.A.U. Colloquium No. 13, (see 012.030), p. 45 - 52 (1973).

Since the advent of low-light-level television systems, it has been recognized that such devices offer the ability to observe meteors as faint as 10th magnitude which allows the extension of optical meteor data to masses as small as 10^{-4} gram. The results of these observations are presented along with an interpretation in terms of mass-flux. This interpretation includes the development of a relationship between peak luminosity of a meteor and mass, velocity, and zenith angle.

104.036 **Determination of meteor parameters using laboratory simulation techniques.**
J. F. Friichtenicht, D. G. Becker.
I.A.U. Colloquium No. 13, (see 012.030), p. 53 - 81 (1973).

104.037 **Laboratory determination of the luminous efficiency of meteor constituents.**
H. F. Savage, C. A. Boitnott.
I.A.U. Colloquium No. 13, (see 012.030), p. 83 - 87 (1973).

104.038 **Evidence from spectra of bright fireballs.**
Z. Ceplecha.
I.A.U. Colloquium No. 13, (see 012.030), p. 89 - 102 (1973).
Spectral data with dispersions from 11 to 94 Å/mm on 4 fireballs of actual brightness of −4 to −12 magnitude and with velocities of about 30 km/s at 70 to 80 km heights are used for studies of meteor radiation problems.

104.039 **Spectral analysis of four meteors.** G. A. Harvey.
I.A.U. Colloquium No. 13, (see 012.030), p. 103 - 129 (1973).
Four meteor spectra from the NASA LRC Faint Meteor Spectra Patrol are analyzed for chemical composition and radiative processes. The chemical compositions of the Taurid, Geminid, and Perseid meteors were found to be similar to that of a typical stony meteorite. The chemical composition of the sporadic meteor was found to be similar to that of a nickel-iron meteorite.

104.040 **NASA-LRC faint meteor spectra.**
G. A. Harvey.
I.A.U. Colloquium No. 13, (see 012.030), p. 131 - 139 (1973).
A brief description is given of the instrumentation, the facilities, and the patrol technique. A classification of 500 meteor spectra obtained in the first 2 1/2 years of the patrol is given. The general characteristics of "typical" spectra are discussed and preliminary conclusions drawn. Examples of unusual spectra are briefly described.

104.041 **The auroral green line in Perseid spectra near sunspot maximum.** J. A. Russell.
I.A.U. Colloquium No. 13, (see 012.030), p. 141 - 146 (1973).
Thirty-one spectra photographed during the Perseid showers of 1969 and 1970 are found to exhibit greater ionization and stronger, more frequent appearance of the forbidden oxygen line at 5577 Å than Perseid spectra obtained with the same instrument in 1961. Data from 13 Perseid showers indicate a relationship between the frequency of occurrence of the oxygen line and solar activity.

104.042 **Image-orthicon spectra of Geminids in 1969.**
P. M. Millman, A. F. Cook, C. L. Hemenway.
I.A.U. Colloquium No. 13, (see 012.030), p. 147 - 151 (1973).
The spectra of 25 meteors, recorded with an image-orthicon technique in December 1969, are studied in relation to similar records made in August of the same year. Of 19 Geminid meteors in the absolute visual magnitude range 0 to +2, only one showed any evidence of the forbidden line of oxygen at 5577 Å, while all Perseid meteors recorded in August exhibited the oxygen line, a result of the large difference in geocentric velocity between the two showers.

104.043 **An unusual meteor spectrum.** A. F. Cook, C. L. Hemenway, P. M. Millman, A. Swider.
I.A.U. Colloquium No. 13, (see 012.030), p. 153 - 159 (1973).

104.044 **Motion of a fragment in disturbed air behind a meteor body.** V. Padevět.
I.A.U. Colloquium No. 13, (see 012.030), p. 169 - 174 (1973).

104.045 **The distribution of 1/*a* in photographic meteor orbits.** B. A. Lindblad.
I.A.U. Colloquium No. 13, (see 012.030), p. 175 - 181 (1973).
The purpose of the present investigation is to study the frequency distribution of the reciprocal of the semi-major axis of precisely reduced meteor orbits and in particular to look for fine structure in the distribution.

104.046 **A working list of meteor streams.** A. F. Cook.
I.A.U. Colloquium No. 13, (see 012.030), p. 183 - 191 (1973).

104.047 **Origin and evolution of recent Leonid meteor showers.** B. A. McIntosh.
I.A.U. Colloquium No. 13, (see 012.030), p. 193 - 198 (1973).
The four most prominent returns of the Leonid shower in the past decade fall into two broad classes. The 1966 and 1969 showers were of short duration, had a high proportion of small particles, and occurred with the longest apparent delay after the perihelion passage of the parent comet Temple-Tuttle. By contrast, the 1961 and 1965 returns were of long duration, and had more large particles. The 1961 return preceded the comet.

104.048 **Cosmic dust in the atmosphere and in the interplanetary space at 1 AU today and in the early solar system.** H. Fechtig.
I.A.U. Colloquium No. 13, (see 012.030), p. 209 - 221 (1973).
The results of cosmic dust research are reviewed. A description of a detector based on ionization during particle impact and of a controlled collection technique is given. Meteoroid fluxes today and in the early solar system are formulated from the distribution of craters in various areas of the moon. Finally, the fluxes in interplanetary space, near the earth, and in the upper atmosphere are discussed.

104.049 **Artificial meteor ablation studies.**
M. B. Blanchard.
I.A.U. Colloquium No. 13, (see 012.030), p. 241 - 254 (1973).

104.050 **Ablation in meteors.** V. N. Lebedinets.
I.A.U. Colloquium No. 13, (see 012.030), p. 259 - 269 (1973).

104.051 **The cometary and asteroidal origins of meteors.**
Ľ. Kresák.
I.A.U. Colloquium No. 13, (see 012.030), p. 331 - 341 (1973).

104.052 **Current evolution of meteoroids.** J. S. Dohnanyi.
I.A.U. Colloquium No. 13, (see 012.030), p. 363 - 374 (1973).
The observed mass distribution of meteoroids at 1 AU from the sun is briefly reviewed in a survey that ranges over the bulk of the mass spectrum from micrometeoroids to meteorite parent objects. The evolution of meteoroids under the influence of collisions, planetary perturbations, the Poynting-Robertson effect and radiation pressure is then discussed.

104.053 **Meteoric spectrosensitometer.** V. A. Smirnov.
Geofiz. i astron. Inform. byull., 1975, No. 14, p. 246 - 250. In Russian. – Abstr. in Referativ. Zhurn. 51. Astron., 10.51.259 (1975).

104.054 **Spectrophotometric properties of the system used for meteor photography.**
V. A. Smirnov, Ya. A. Smirnova.
Geofiz. i astron. Inform. byull., 1975, No. 14, p. 251 - 257. In Russian. – Abstr. in Referativ. Zhurn. 51. Astron., 10.51. 260 (1975).

104.055 **Meteor radioelectronics.** E. I. Fialko.
Geofiz. i astron. Inform. byull., 1975, No. 14, p. 200 - 206. In Russian. – Abstr. in Referativ. Zhurn. 51. Astron., 10.51.340 (1975).

104.056 **Kinetic energy distribution of meteor bodies (Determination of kinetic energy distribution of meteor bodies taking into account the dependence of the diffusion coefficient on height).**
E. I. Fialko, V. F. Romanyuk.
Geofiz. i astron. Inform. byull., 1975, No. 14, p. 207 - 211. In Russian. – Abstr. in Referativ. Zhurn. 51. Astron., 10.51. 341 (1975).

104.057 **Role of meteors in ionospherical processes. II.**
O. Ovezgel'dyev, L. P. Korsunova.
Izv. AN TurkmSSR. Ser. fiz.-tekhn., khim. i geol. n., 1974, No. 6, p. 37 - 42. In Russian. – Abstr. in Referativ. Zhurn. 51. Astron., 10.51.342 (1975).

104.058 **Study of meteors and fireballs.** V. A. Bronshtehn.
II Vses. konf. dinamika izluchayushchego gaza, 1975. Programma i tezisy dokl. Moskva, 1975, p. 16 - 17. In Russian. – Abstr. in Referativ. Zhurn. 51. Astron., 10.51.350 (1975).

104.059 **Meteor trains and chemiluminescent processes.**
W. J. Baggaley.
Monthly Notices Roy. Astron. Soc., Vol. 173, 497 - 512 (1975).

This paper assesses the role of chemiluminescent afterglow in meteor trains and the various mechanisms are reviewed quantitatively. It is shown that only if there were a high degree of atmospheric molecular dissociation in the meteoric column would observable emission be produced. However, the short lifetime of the afterglow precludes its acceptance as the major mechanism operative.

104.060 **On the decay of underdense radio meteor echoes.**
J. Jones.
Monthly Notices Roy. Astron. Soc., Vol. 173, 637 - 647 (1975).

The discrepancy between the theoretical and observed variations of ambipolar diffusion coefficient with height in the meteor region is re-examined and processes which might be relevant are discussed.

104.061 **Southern hemisphere meteor stream determinations.**
G. Gartrell, W. G. Elford.
Australian Journ. Phys., Vol. 28, 591 - 620 (1975).

Meteor orbits have been deduced from radio observations of meteor trails. The orbits of 1667 meteors have been determined down to a limiting radio magnitude of +8. The data have been systematically searched for stream meteors, and the significance of minor associations has been appraised. Altogether 40.4% of the orbits were found to be associated with at least one other orbit, and 29.8% with two or more. Numerous minor streams with high inclination and low eccentricity have been found. In addition to confirming several previously established cometary associations, a comprehensive search has indicated that 34 of the meteor associations found may be related to 17 comets.

104.062 **Mathematical description of distributions of observed apparent and true densities of sporadic meteor radiants over the celestial sphere.**
E. V. Dmitrienko, Yu. A. Pupyshev.
Meteorn. rasprostr. radiovoln. No. 10 - 11, Kazan', Kazan. un-t, 1975, p. 3 - 7. In Russian. – Abstr. in Referativ. Zhurn. 51. Astron., 11.51.317 (1975).

104.063 **Some results of TV observations of meteors.**
M. Begkhanov, Kh. Gul'medov, S. Mukhamednazarov.
Izv. AN TurkmSSR. Ser. fiz.-tekhn., khim. i geol. n., 1975, No. 1, p. 111 - 112. In Russian. – Abstr. in Referativ. Zhurn. 51. Astron., 11.51.318 (1975).

104.064 **Influence of the twilight effects on the observed hourly rates of meteors.** J. Slančiková.
Bull. Astron. Inst. Czechoslovakia, Vol. 26, 321 - 326 (1975).

104.065 **On the structure of the Giacobinid meteor shower.**
L. Kresák, J. Slančiková.
Bull. Astron. Inst. Czechoslovakia, Vol. 26, 327 - 342 (1975).

Using correction procedures for eliminating the background luminance effects and magnitude-dependent incompleteness of meteor counts, visual observations of the 1946 Giacobinid shower, obtained at the Skalnaté Pleso Observatory, are evaluated. The total influx of the Giacobinid meteors into the earth atmosphere, as a function of the zenith magnitude threshold m_0, is estimated at $(1.2 \times 10^6) \times 2.6^{m_0}$, and the total mass infall at 4×10^6 g per magnitude class, or 4×10^7 g for the particle range from 10^{-1} to 10^3 g. The observing geometry of the shower is discussed.

104.066 **High temperature condensates among meteors.**
L. L. Wilkening.
Nature, Vol. 258, 689 - 690 (1975). – Letter.

104.067 **Meteor flux survey.**
A. T. Blackwell, I. Halliday.
Journ. Roy. Astron. Soc. Canada, Vol. 69, 255 (1975). Abstr. CAS.

104.068 **Astronomical factors of the shift of estimates in radar observations of drifts of meteor tracks.**
V. V. Fedynskij, B. L. Kashcheev, V. A. Nechitajlenko.
Dokl. Akad. Nauk SSSR. Ser. Mat., Fiz., Vol. 225, 1292 - 1295 (1975). In Russian.

104.069 **Abstracts of papers of the All-Union symposium "Interaction of meteor matter with the earth and estimate of the influx of meteor matter on the earth and moon", Dushanbe, 1974, Sept. 24–28.**
Astron. sovet AN SSSR, Probl. rabochaya gruppa po issled. meteor. veshchestva, AN TadzhSSR. In-t astrofiz. Dushanbe, 1974. 38 pp. In Russian.

104.070 **Double-frequency direction finder for meteor studies.** N. S. Andrianov.
Meteorn. rasprostr. radiovoln. No. 10 - 11. Kazan', Kazan. un-t, 1975, p. 115 - 119. In Russian.

104.071 **On initial radii of meteor trains from observations of oblique propagation of radio waves.**
N. S. Andrianov, I. V. Lukin.
Meteorn. rasprostr. radiovoln. No. 10 - 11. Kazan', Kazan. un-t, 1975, p. 7 - 12. In Russian. – Abstr. in Referativ. Zhurn. 51. Astron., 12.51.285 (1975).

104.072 **On the measurement of meteor velocities from phase-time registrations before the reflection point.**
N. S. Andrianov, V. V. Sidorov.
Meteorn. rasprostr. radiovoln. No. 10 - 11. Kazan', Kazan. un-t, 1975, p. 29 - 33. In Russian. – Abstr. in Referativ. Zhurn. 51. Astron., 12.51.286 (1975).

104.073 **On the charges of meteor bodies.**
A. A. Dmitrievskij.
Meteorn. rasprostr. radiovoln. No. 10 - 11. Kazan', Kazan. un-t, 1975, p. 12 - 19, 93 - 99. In Russian. – Abstr. in Refera-

tiv. Zhurn. 51. Astron., 12.51.287 (1975).

104.074 **A method to take into account the nonspecular reflection on observations of meteor trains at $\lambda < 2$ m.**
A. N. Aminov, O. I. Bel'kovich.
Meteorn. rasprostr. radiovoln. No. 10 - 11. Kazan', Kazan. un-t, 1975, p. 99 - 103. In Russian. – Abstr. in Referativ. Zhurn. 51. Astron., 12.51.288 (1975).

104.075 **On the accuracy of the phase method of coordinate measurements of reflecting points on meteor trains and on some altitude dependences of meteor phenomena.**
A. N. Fakhrutdinova.
Meteorn. rasprostr. radiovoln. No. 10 - 11. Kazan', Kazan. un-t, 1975, p. 103 - 112. In Russian. – Abstr. in Referativ. Zhurn. 51. Astron., 12.51.289 (1975).

104.076 **On the accuracy of the phase radio method of measurement of meteoroid velocities.**
N. S. Andrianov, V. V. Sidorov, Z. N. Fadeicheva.
Meteorn. rasprostr. radiovoln. No. 10 - 11. Kazan', Kazan. un-t, 1975, p. 112 - 115. In Russian. – Abstr. in Referativ. Zhurn. 51. Astron., 12.51.290 (1975).

104.077 **On the theory of non-linear interaction of electromagnetic waves in a volume of ionized meteor trains.** A. P. Gorovenko, E. I. Fialko, I. V. Bajrachenko.
Meteorn. rasprostr. radiovoln. No. 10 - 11. Kazan', Kazan. un-t, 1975, p. 164 - 169. In Russian. – Abstr. in Referativ. Zhurn. 51. Astron., 12.51.291 (1975).

104.078 **Study of the radial structure of the Quadrantid meteor shower.** Sh. O. Isamutdinov, E. I. Fialko.
Dokl. AN TadzhSSR, Vol. 18, No. 2, p. 17 - 20 (1975). In Russian. – Abstr. in Referativ. Zhurn. 51. Astron., 12.51.293 (1975).

104.079 **The Strathpeffer fireball of 1974 December 14.**
K. B. Hindley, M. A. Houlden.
Journ. British Astron. Ass., Vol. 86, 67 - 73 (1975).

104.080 **Some questions of the physical theory of meteors.**
V. G. Kruchinenko.
Vestn. Kiev. Un-ta, Ser. Astron., No. 17, p. 66 - 75 (1975). In Russian.

Some questions connected with the problem of heating of a meteor body in the earth atmosphere limited with regard to thickness are discussed. On the basis of comparison of the propagation velocity of decrease of body size owing to evaporation, the author concludes on the possibility of "cold" sloughing or thermal destruction of a thin surface layer.

104.081 **Study of meteors from photographs in different spectral regions.**
E. V. Sandakova, E. P. Vashchenko.
Vestn. Kiev. Un-ta, Ser. Astron., No. 17, p. 75 - 79 (1975). In Russian.

An analysis of meteor brightness and mass is given. Change of meteor radiation is different along the way. Velocities of the mass loss calculated in three spectral regions are higher than in integral light.

104.082 **On the question of methods of determination of radiants, heights and velocities of meteors.**
V. V. Kalenichenko.
Vestn. Kiev. Un-ta, Ser. Astron., No. 17, p. 80 - 82 (1975). In Russian.

104.083 **On the question of formation and evolution of meteor streams.** L. M. Sherbaum.
Vestn. Kiev. Un-ta, Ser. Astron., No. 17, p. 83 - 90 (1975). In Russian.

The dependence of the width of a meteor stream on the parameters of the parent comet is considered. The action of planetary perturbations within 200 - 300 years on the evolution of meteor streams is compared with the Poynting-Robertson effect within 10^4 - 10^6 years.

104.084 **Investigation of meteor trails by radar with two coherent frequencies.**
Yu. V. Chumak, R. I. Mojsya.
Vestn. Kiev. Un-ta, Ser. Astron., No. 17, p. 90 - 94 (1975). In Russian.

Some results of meteor radar observations at wavelengths 8.87 m and 13.3 m are presented. The method of obtaining the initial radius and line electron density of the meteor trail is described.

104.085 **On the instrumental errors of a measurement of the ambipolar diffusion coefficients by the radar method.**
G. I. Kolomiets, E. I. Fialko, R. I. Mojsya.
Vestn. Kiev. Un-ta, Ser. Astron., No. 17, p. 95 - 98 (1975). In Russian.

The influence of meteor radar receiver parameters on the accuracy of the ambipolar diffusion coefficient measurement is analysed. The instrumental errors may be of order ± (10 - 40)%.

104.086 **Meteorströme,** Beobachtungen in der Schweiz und in Deutschland Juli–September 1975.
R. Germann.
Orion, 33. Jahrgang, p. 196 - 198 (1975).

104.087 **Resultaten juli/augustusaktie 1975.** H. Betlem.
Zenit, Vol. 2, 446 - 448 (1975).

104.088 **Results of a photometry of meteor trains of the Leonid meteor stream at Ashkhabad during 1965 – 1966.** G. A. Buvidajte, Kh. D. Gul'medov.
Izv. AN TurkmSSR. Ser. fiz.-tekhn., khim. i geol. n., 1975, No. 3, p. 115 - 117. In Russian. – Abstr. in Referativ. Zhurn. 51. Astron., 1.51.383 (1976).

104.089 **Closed-circuit television spectroscopy of meteors.**
P. M. Millman.
Meteoritics, Vol. 10, 456 (1975).

104.090 **Correlations with the forbidden oxygen line at 5577Å in Perseid meteor spectra.** J. A. Russell.
Meteoritics, Vol. 10, 481 - 482 (1975).

104.091 **The relation of a (σ_i) for the Perseid meteor stream.**
S. Gąska.
Stud. Soc. Sci. Torunensis, Toruń, Poland, Sect. F (Astron.), Vol. 5, 147 - 152 = Bull. Astron. Obs. N. Copernicus Univ. Toruń, No. 54 I (1974).

In order to test the method applied by Gąska (1970) to minor planets and other small bodies of the solar system, independent calculations have been performed for the Perseid meteor stream, which originated from the comet 1862 III. The results are more than satisfactory and thus confirm our previous conclusion.

104.092 **Fireball of 1975 September 15.** K. B. Hindley.
British Astron. Ass. Circ., No. 568 (1975).

104.093 **Visual meteoric activity over Waltair during October-February, 1961–1972.**
M. Srirama Rao, A. G. Murthy, S. R. Ratnam.
Current Sci., (*India*), Vol. 43, 647 (1974).

Device for automatic measurements of amplitude

characteristics of meteor reflections. See Abstr. 031.411.

The mobile meteoric radar of the CNET. See Abstr. 033.076.

Spectroscopy of Project Fire I, April 14, 1964. See Abstr. 051.020.

Cross-modulation of electromagnetic waves in limited plasma volumes. See Abstr. 062.028.

Nalichtend meteoorspoor onthult windsnelheden in de ionosfeer. See Abstr. 083.040.

Motion of large bodies in the atmospheres of planets. See Abstr. 091.018.

Lunar rocks as meteoroid detectors. See Abstr. 094.156.

Variation in the number of meteoroid impacts on the moon with lunar phase. See Abstr. 094.177.

Study of comets and meteors. See Abstr. 102.002.

Possible evidence of ablation on cosmic dust particles. See Abstr. 105.049.

Criteria for identification of ablation debris from primitive meteoric bodies. See Abstr. 105.082.

Hyperbolic cosmic dust: its origin and its astrophysical significance. See Abstr. 106.025.

A possible mechanism for the capture of microparticles by the earth and other planets of the solar system. See Abstr. 106.038.

105 Meteorites, Meteorite Craters

105.001 **Giant olivine chondrule as a possible later-stage product in the nebula.**
T. Tanaka, N. Nakamura, A. Masuda, N. Onuma.
Nature, Vol. 256, 27 - 28 (1975).
Abundances of trace elements, in particular, of rare-earth elements (REE) in the Allende chondrite and its inclusions have attracted increasing attention from geochemists. Based on the abundance characteristics of major elements and trace elements (including REE), Martin and Mason distinguished four distinct groups among the aggregates and chondrules of the Allende meteorite. The authors report here on the presence of an olivine chondrule in Allende with REE abundance features quite different from those of any inclusions previously reported.

105.002 **Fracture mechanism of Henbury meteorite by separation along surfaces of shear faulting.**
H. J. Axon, E. M. Steele-Perkins.
Nature, Vol. 256, 635 (1975). – Letter.

105.003 **The Moama eucrite – a pyroxene-plagioclase adcumulate.** J. F. Lovering.
Meteoritics, Vol. 10, 101 - 114 (1975).

105.004 **The Seoni meteorite fall, with a note on its morphological characteristics.**
V. K. Nayak, G. R. Rao.
Meteoritics, Vol. 10, 115 - 120 (1975).

105.005 **Mineralogy, petrology and chemistry of the Isna (C3) meteorite.** R. L. Methot, A. F. Noonan, E. Jarosewich, A. A. deGasparis, D. M. Al-Far.
Meteoritics, Vol. 10, 121 - 131 (1975).
A total chemical analysis of the Isna, Egypt, meteorite is similar to analyses for chondrites of type C3, Ornans sub-type; however, comparison with one group of chemical data indicates that Isna is intermediate between the C3(O)'s and C3(V)'s in terms of total Fe.

105.006 **The Meteoritical Bulletin, No. 53,** sponsored by The Meteoritical Society.
R. S. Clarke, Jr., (Editor), with the assistance of B. H. Mason.
Meteoritics, Vol. 10, 133 - 158 (1975). – Reported are new falls and discoveries of meteorites.

105.007 **Cosmogenic excess of ^{40}K and the flux of fast neutrinos in meteorites.**
W. Stegmann, F. Begemann.
Zeitschr. Naturforschung, Vol. 30a, 968 - 975 (1975). In German.

105.008 **Droplet chondrules.** S. W. Kieffer.
Science, Vol. 189, 333 - 340 (1975).

105.009 **From the past of meteorites.**
B. Yu. Levin, A. N. Simonenko.
Priroda, No. 7.75, p. 39 - 45 (1975). In Russian.

105.010 **Second meteorite from Chukotka.**
A. N. Simonenko, R. L. Khotinok.
Priroda, No. 7.75, p. 94 - 95 (1975). In Russian.

105.011 **The resonant orbit of the Lost City meteorite.**
J. G. Williams.
Journ. Geophys. Res., Vol. 80, 2914 - 2916 (1975).

105.012 **The unmetamorphosed sedimentary fill of the Brent meteorite crater, south-eastern Ontario.**
G. P. Lozei, F. W. Beales.
Canadian Journ. Earth Sci., Vol. 12, 606 - 628 (1975).

105.013 **Water in non-carbonaceous stony meteorites.**
J. R. Ashworth, R. Hutchison.
Nature, Vol. 256, 714 - 715 (1975). – Letter.

105.014 **Preferential orientation in four C3 chondritic meteorites.**
P. M. Martin, A. A. Mills, E. Walker.
Nature, Vol. 257, 37 - 38 (1975). – Letter.

105.015 **On possible short-lived progenitors of fission xenon in carbonaceous chondrites.**
W. M. Howard, M. Arnould, J. W. Truran.
Astrophys. Space Sci., Vol. 36, L1 - L9 (1975).
The authors suggest that the unidentified fission xenon component in carbonaceous chondrites may be attributable to the decay of the extinct radioactivities ^{250}Cm and, to a lesser extent, ^{258}Cm. Two assumptions are fundamental to this picture: (1) these relatively short-lived actinides (1.13×10^4 and 3.7×10^5 yr, respectively) were incorporated into grains in the vicinity of their supernova birthplace (prior to decay); and (2) such grains and their trapped anomalous xenon component were incorporated intact into the meteorites.

105.016 **Some studies of an unusual eucrite: Ibitira.**
L. L. Wilkening, E. Anders.
Geochim. Cosmochim. Acta, Vol. 39, 1205 - 1210 (1975).

105.017 **Critical comments on a proposed cosmothermometer.** M. Blander.
Geochim. Cosmochim. Acta, Vol. 39, 1315 - 1320 (1975).

105.018 **Validity of trace element cosmothermometer.**
E. Anders, J. W. Larimer.
Geochim. Cosmochim. Acta, Vol. 39, 1320 - 1324 (1975).
Reply to the paper of M. Blander – see Abstr. 105.017.

105.019 **Radiogenic isotope research.** S. E. Church.
Rev. Geophys. Space Phys., Vol. 13, No. 3,(see 013.001), p. 98 - 101, 150 - 159 (1975). – Review paper with bibliographical notes.
The review of the literature presented here covers three main areas of research: extraterrestrial geochronology, terrestrial geochronology, and the rapidly growing field of isotopic tracer applications to problems of petrogenesis. Special emphasis is placed on a review of the lunar data.

105.020 **Stable isotope geochemistry.** H. P. Taylor, Jr.
Rev. Geophys. Space Phys., Vol. 13, No. 3, (see 013.001), p. 102 - 107, 159 - 163 (1975). – Review paper with bibliographical notes.

105.021 **Meteorite research.** J. T. Wasson.
Rev. Geophys. Space Phys., Vol. 13, No. 3, (see 013.001), p. 113 - 116, 168 - 173 (1975). – Review paper with bibliographical notes.

105.022 **Tracks of bombardments from space.**
V. Tsvetkov.
Mor. sb., 1975, No. 1, p. 81 - 85. In Russian.

105.023 **Classification and properties of iron meteorites.**
E. R. D. Scott, J. T. Wasson.
Rev. Geophys. Space Phys., Vol. 13, 527 - 546 (1975).

Most (86%) iron meteorites can be assigned to one of 12 genetic groups on the basis of systematic variations in their chemical, mineralogical, and structural properties; the remaining 14% are termed anomalous. The groups are described in detail.

105.024 **Messungen des Mikrometeoritenflusses im Erde–Mond-System mit dem HEOS 2-Satelliten.**
H.-J. Hoffmann, H. Fechtig, E. Grün, J. Kissel.
Mitt. Astron. Ges., No. 36, p. 67 (1975). – Abstract.

105.025 **Pressures of formation of iron meteorites from sphalerite compositions.**
H. P. Schwarcz, S. D. Scott, S. A. Kissin.
Geochim. Cosmochim. Acta, Vol. 39, 1457 - 1466 (1975).

105.026 **Stereoisomers of isovaline in the Murchison meteorite.**
G. E. Pollock, C-N. Cheng, S. E. Cronin, K. A. Kvenvolden.
Geochim. Cosmochim. Acta, Vol. 39, 1571 - 1573 (1975). Note.

105.027 **Contents of eleven trace elements in ureilite achondrites.**
C. M. Binz, M. Ikramuddin, M. E. Lipschutz.
Geochim. Cosmochim. Acta, Vol. 39, 1576 - 1579 (1975). Note.

105.028 **Infrared spectra of carbonaceous chondrites and the composition of interstellar grains.**
A. Zaikowski, R. F. Knacke.
Astrophys. Space Sci., Vol. 37, 3 - 9 (1975).

Infrared spectra (7–40 μ) of the carbonaceous chondrites, Cold Bokkeveld, Murray, and Orgueil are compared with phyllosilicate minerals and interstellar grain spectra. Similarities in the position and shape of the silicate feature near 9.8 μ suggest that similar silicate minerals are present in primitive meteorites and the interstellar dust.

105.029 **Meteorite Egvekinot.** R. L. Khotinok.
Zemlya i Vselennaya, 1975, No. 4, p. 23. In Russian.

105.030 **New hypothesis on the Tunguska meteorite.**
G. I. Petrov, V. P. Stulov.
Zemlya i Vselennaya, 1975, No. 4, p. 74 - 75. In Russian.

105.031 **Materials to estimate the influx of meteoric spherules on the earth's surface.** N. V. Vasil'ev, A. P. Boyarkina, V. D. Gol'din, Yu. A. Grishin, S. N. Gryaznova, T. A. Menyavtseva.
Problems of cosmic physics. Vyp. (No.) 10, (see 003.003), p. 84 - 88 (1975). In Russian.

The content of silicate meteoric spherules in the samples of peatbog from the Podkamennaya Tunguska region has been studied. The relatively monotonous character of the meteoric dust influx in the time 1920 - 1971 is demonstrated. The total amount of this sort of cosmic material is probably 2 - 4 thousand tons per year on the whole surface of the earth.

105.032 **Cosmic flakes in manganese nodules.** J. Jedwab.
Meteoritics, Vol. 10, 191 - 207 (1975).

In addition to the well known cosmic spheres, micron-size metallic flakes have been extracted magnetically from Pacific Ocean manganese nodules. The flakes were individually studied by several techniques and were found to be composed of iron, iron-nickel and nickel-iron. The cosmic origin of these flakes is discussed with regard to chemical composition, crystallinity, agglomeration and comparable terrestrial minerals.

105.033 **On the terrestrial origin of the Tepexitl crater, Mexico.** L. Maupomé, R. Alvarez, S. W. Kieffer, R. S. Dietz.
Meteoritics, Vol. 10, 209 - 214 (1975).

Geophysical studies were initiated in order to test the hypothesis of meteoritical origin (Alvarez et al., 1975) for this crater. The following evidence is against an impact origin for Tepexitl: (1) no meteorites were found; (2) no shatter cones were discovered; (3) the authors could identify no shock metamorphosed rocks; (4) the rim is entirely composed of ash with no uplifted quaquaversally dipping country rock; (5) there are no overturned beds or overturned flap; (6) ejected blocks are of small size, the largest observed being about 1 m^3; (7) the ejected material is entirely volcanic; (8) the volume of the rim exceeds that portion of the crater which is depressed below the regional level, so that new material has been added to the earth's surface; (9) there is no evidence of a central dome or ring syncline; and, (10) there is no suevite (impact-melt rock) as would contain cored bombs, polymict breccia, etc.

105.034 **Sketches in the history of meteoritics 1: The birth of the science.** D. W. Sears.
Meteoritics, Vol. 10, 215 - 225 (1975).

The change of opinion concerning the stories of stones falling from the sky that took place around 1800 is examined.

105.035 **An L6 chondrite found near Parachilna, South Australia.** S. J. B. Reed, D. H. McColl.
Meteoritics, Vol. 10, 227 - 231 (1975).

An 8.82 kg stony meteorite found on Motpena Station, 31°6′ S, 138° 16′ E, near Parachilna, South Australia, is shown to be a somewhat weathered L6 (olivine-hypersthene) chondrite.

105.036 **A method of extraction and analysis of possible cosmic particles from manganese nodules.**
J. Jedwab.
Meteoritics, Vol. 10, 233 - 240 (1975).

105.037 **Factor analysis of elemental abundances in chondritic and achondritic meteorites.**
D. M. Shaw, R. S. Harmon.
Meteoritics, Vol. 10, 253 - 282 (1975).

An explorative study of the use of factor analysis in meteorite geochemistry has been made. Forty-two major and trace elements were sought from analyses of 80 stony meteorites in recent articles.

105.038 **On the depletion of moderately volatile elements in ordinary chondrites.** E. Anders.
Meteoritics, Vol. 10, 283 - 286 (1975).

Within the accuracy of present data, elements of intermediate volatility (condensation temperatures between about 1100 and 600° K) do not show a correlation of abundance with volatility in equilibrated ordinary chondrites.

105.039 **A preliminary study of the ductile-brittle transition under impact conditions in material from an octahedrite.** J. L. Remo, A. A. Johnson.
Journ. Geophys. Res., Vol. 80, 3744 - 3748 (1975).

105.040 **Experimental hypervelocity impact into quartz sand: distribution and shock metamorphism of ejecta.**
D. Stöffler, D. E. Gault, J. Wedekind, G. Polkowski.
Journ. Geophys. Res., Vol. 80, 4062 - 4077 (1975).

The authors report results of vertical impacts of 0.3-g cylindrical Lexan projectiles into noncohesive quartz sand. Material displaced from the upper 15% of the crater depth d is represented within the whole ejecta blanket, material from deeper than 28% of d is deposited inside 2 crater radii, and no material from deeper than 33% of *d* was ejected beyond the crater rim. The shock-metamorphosed particles and the shock-induced change in the grain size distribution of ejected samples have close analogies to the basic characteristics of the lunar regolith. Possible applications to regolith formation and to ejecta formations of large-scale impact craters are discussed.

105.041 **Petrogenesis of eucrite, howardite and diogenite meteorites.** E. Stolper.
Nature, Vol. 258, 220 - 222 (1975). – Letter.

105.042 **Relative chemism of stony meteorites fallen on the territory of Bulgaria.** D. I. Dimov.
Godishn. Sofijsk. un-t Geol.-geogr. fak., Vol. 66, No. 1, p. 209 - 217 (1973/1974). In Bulgarian. – Abstr. in Referativ. Zhurn. 51. Astron., 8.51.406 (1975).

105.043 **Search for traces of heavy and superheavy elements in meteorites.**
G. N. Flerov, O. Otgonsurehn, V. P. Perelygin.
Izv. AN SSSR. Ser. fiz., Vol. 39, 388 - 394 (1975). In Russian.

105.044 **The terrestrial life of explosive meteorite craters.** A. I. Dabizha, M. S. Krass.
Zemlya i Vselennaya, 1975, No. 5, p. 80 - 88. In Russian.

105.045 **Pyroxenes in the Shaw (L-7) chondrite.** R. T. Dodd, J. E. Grover, G. E. Brown.
Geochim. Cosmochim. Acta, Vol. 39, 1585 - 1594 (1975).

105.046 **Calcium-aluminum-rich inclusions in the Allende meteorite: evidence for a liquid origin.**
M. Blander, L. H. Fuchs.
Geochim. Cosmochim. Acta, Vol. 39, 1605 - 1619 (1975).

105.047 **Chromium-bearing olivine in the St. Mesmin chondrite.** R. T. Dodd, D. J. Morrison-Smith, J. V. Heyse.
Geochim. Cosmochim. Acta, Vol. 39, 1621 - 1627 (1975).

105.048 **A final summary of measurements of cosmogenic Al^{26} in stone meteorites.** I. R. Cameron, Z. Top.
Geochim. Cosmochim. Acta, Vol. 39, 1705 - 1707 (1975).
A summary is given of the final results of measurements of cosmogenic Al^{26} activity in a total of 50 stone meteorites, mainly chondrites, carried out using coincidence gamma-spectrometry.

105.049 **Possible evidence of ablation on cosmic dust particles.** C. L. Hemenway.
I.A.U. Colloquium No. 13, (see 012.030), p. 255 - 257 (1973).

105.050 **On the manganese content of cosmic flakes from deep-sea sediments.** K. Utech.
I.A.U. Colloquium No. 13, (see 012.030), p. 297 - 298 (1973).

105.051 **The cosmic dust analyzer: experimental evaluation of an impact ionization model.**
J. F. Friichtenicht, N. L. Roy, D. G. Becker.
I.A.U. Colloquium No. 13, (see 012.030), p. 299 - 310 (1973).

105.052 **On the primordial condensation and accretion environment and the remanent magnetization of meteorites.** A. Brecher.
I.A.U. Colloquium No. 13, (see 012.030), p. 311 - 329 (1973).

105.053 **Cometary vs asteroidal origin of chondritic meteorites.** G. W. Wetherill.
I.A.U. Colloquium No. 13, (see 012.030), p. 343 (1973). Abstract.

105.054 **Indirect evidence of the existence of a new element in meteorites and some of its chemical properties.**
G. N. Goncharov.
Geokhimiya, 1975, No. 6, p. 282 - 294. In Russian. – Abstr. in Referativ. Zhurn. 51. Astron., 10.51.364 (1975).

105.055 **Ilmenite-orthopyroxene intergrowths from the moon and the Skaergaard intrusion.**
J. D. Haselton, W. P. Nash.
Earth Planet. Sci. Letters, Vol. 26, 287 - 291 (1975).

105.056 **Electron petrography of shock-deformed olivine in stony meteorites.** J. R. Ashworth, D. J. Barber.
Earth Planet. Sci. Letters, Vol. 27, 43 - 50 (1975).

105.057 **Search for extinct ^{146}Sm, 1. The isotropic abundance of ^{142}Nd in the Juvinas meteorite.**
G. W. Lugmair, N. B. Scheinin, K. Marti.
Earth Planet. Sci. Letters, Vol. 27, 79 - 84 (1975).

105.058 **Thermomagnetic analysis of meteorites, 2. C2 chondrites.**
D. E. Watson, E. E. Larson, J. M. Herndon, M. W. Rowe.
Earth Planet. Sci. Letters, Vol. 27, 101 - 107 (1975).
Samples of all eighteen of the known C2 chondrites have been analyzed thermomagnetically.

105.059 **$^{40}Ar - ^{39}Ar$ chronology of isolated phases from Bununu and Malvern howardites.**
R. S. Rajan, J. C. Huneke, S. P. Smith, G. J. Wasserburg.
Earth Planet. Sci. Letters, Vol. 27, 181 - 190 = Contr. Div. Geol. Planet. Sci., Cal. Inst. Techn., Pasadena, No. 2604 (1975).

105.060 **Cosmogenic neon produced from sodium in meteoritic minerals.** S. P. Smith, J. C. Huneke.
Earth Planet. Sci. Letters, Vol. 27, 191 - 199 = Contr. Div. Geol. Planet. Sci., Cal. Inst. Techn., Pasadena, No. 2606 (1975).

105.061 **The low-temperature electrical properties of carbonaceous chondrites.**
A. Brecher, P. L. Briggs, G. Simmons.
Earth Planet. Sci. Letters, Vol. 28, 37 - 45 (1975).

105.062 **Astroblemes.** V. L. Masajtis.
Zemlya i Vselennaya, 1975, No. 6, p. 13 - 17. In Russian.

105.063 **Fragmentation of the Sikhote-Alin meteor body.** E. L. Krinov.
Meteoritika, vyp. (No.) 34, p. 3 - 14, 117 - 118 (1975). In Russian.

105.064 **Meteorography of the Sikhote-Alin octahedrite.** L. G. Kvasha.
Meteoritika, vyp. (No.) 34, p. 15 - 20, 119 - 120 (1975). In Russian.

105.065 **On the chemical composition of the Sikhote-Alin meteorite.** A. A. Yavnel'.
Meteoritika, vyp. (No.) 34, p. 21 - 26 (1975). In Russian.

105.066 **Physical properties of samples of the Sikhote-Alin iron meteorite shower.**
Eh. S. Gorshkov, E. G. Gus'kova.
Meteoritika, vyp. (No.) 34, p. 27 - 32 (1975). In Russian.

105.067 **Improvement of the radiant of the Sikhote-Alin meteorite by direct calculation of the squares of closing errors.** I. T. Zotkin, A. N. Chigorin.
Meteoritika, vyp. (No.) 34, p. 33 - 41 (1975).

105.068 **Peculiarities of the chemical composition of type III carbonaceous chondrites.** A. A. Yavnel'.
Meteoritika, vyp. (No.) 34, p. 42 - 48 (1975). In Russian.

105.069 **Cosmogenic isotopes Al^{26} and Na^{22} in stony meteorites.** A. K. Lavrukhina, V. D. Gorin.
Meteoritika, vyp. (No.) 34, p. 49 - 51 (1975). In Russian.

105.070 **On equilibrium conditions in pallasites.** L. N. Kolomejtseva.
Meteoritika, vyp. (No.) 34, p. 52 - 56 (1975). In Russian.

105.071 **Petrographic and chemical investigation of the meteorites Poluyamki and Rakity and comparison with the meteorite Markovka.**
O. A. Kirova, M. I. D'yakonova, V. Ya. Kharitonova.
Meteoritika, vyp. (No.) 34, p. 57 - 64, 121 - 123 (1975). In Russian.

105.072 **Further investigation of the chemical composition of some iron meteorites and pallasite of Thiel Mountains from the collection of the USSR Academy of Sciences.** M. I. D'yakonova, V. Ya. Kharitonova.
Meteoritika, vyp. (No.) 34, p. 65 - 68 (1975). In Russian.

105.073 **Some results of an investigation of the Verkhne-Chirskaya meteorite.** D. P. Shaposhnikov.
Meteoritika, vyp. (No.) 34, p. 69 - 70, 124 (1975). In Russian.

105.074 **Microscopic investigation of the nickel iron of the family of chondrites.** V. P. Semenenko.
Meteoritika, vyp. (No.) 34, p. 71 - 75, 125 (1975). In Russian.

105.075 **Investigation of the porosity of olivines and enstatites of meteoritic and terrestrial origin.**
V. N. Savel'ev, V. D. Kolomenskij, I. A. Yudin.
Meteoritika, vyp. (No.) 34, p. 76 - 82 (1975). In Russian.

105.076 **Asymmetry of destructions in shock and shock-explosion craters.** A. K. Stanyukovich.
Meteoritika, vyp. (No.) 34, p. 83 - 87 (1975). In Russian.

105.077 **Ring structures of the central part of the Russian platform.** A. I. Dabizha, M. S. Anuchin, V. V. Fedynskij, V. R. Melikhov.
Meteoritika, vyp. (No.) 34, p. 88 - 91, 126 (1975). In Russian.

105.078 **Terminology in meteoritics.** E. L. Krinov.
Meteoritika, vyp. (No.) 34, p. 98 - 106 (1975). In Russian.

105.079 **Catalogue of the meteoritic collection of the Museum of Natural Sciences of the Kharkov State University.** T. P. Zholud'.
Meteoritika, vyp. (No.) 34, p. 107 - 110 (1975). In Russian.

105.080 **Meteoritic collection of the Sofiya University in Bulgaria on 31 January 1973.** D. I. Dimov.
Meteoritika, vyp. (No.) 34, p. 111 - 112 (1975). In Russian.

105.081 **Giant plate of the troilite-iron meteorite Mundrabilla-II from Western Australia.** L. G. Kvasha.
Meteoritika, vyp. (No.) 34, p. 113 - 114 (1975). In Russian.

105.082 **Criteria for identification of ablation debris from primitive meteoric bodies.**
D. E. Brownlee, M. B. Blanchard, G. C. Cunningham, R. H. Beauchamp, R. Fruland.
Journ. Geophys. Res., Vol. 80, 4917 - 4924 (1975).

105.083 **Host phase of a strange xenon component in Allende.** R. S. Lewis, B. Srinivasan, E. Anders.
Science, Vol. 190, 1251 - 1262 (1975).

Most of the noble gases reside in three rare minerals comprising less than 0.5 percent of the meteorite.

105.084 **Extinct superheavy element in the Allende meteorite.** E. Anders, H. Higuchi, J. Gros, H. Takahashi, J. W. Morgan.
Science, Vol. 190, 1262 - 1271 (1975).

Element 115 (or 114, 113) may have been present in a rare chromium mineral when the meteorite formed.

105.085 **Der "Stein", der auf Naragh fiel.** I. Eschghi.
Umschau, 75. Jahrgang, p. 733 - 735 (1975).

This report concerns a meteorite (chondrite), which crushed the roof of a school room in Naragh, a village about 200 km to the south of Tehran/Iran. Here is the extremely rare case where an extraterrestrial body could be scientifically analysed immediately after its impact on the earth's surface.

105.086 **Die Forschungsbohrung Nördlingen 1973/74.** Modell eines interdisziplinären Forschungsverbundes.
Umschau, 75. Jahrgang, p. 737 (1975).

Recent results of planetary geology, especially the exploration of the moon, have strengthened the worldwide interest in the well preserved meteorite impact crater "Nördlinger Ries". As a German contribution a drilling down to 1206 m depth has been conducted.

105.087 **On the systematics of uranium and lead isotopes in chondritic meteorites.** S. B. Brandt.
Geokhimiya, 1975, No. 6, p. 939 - 943. In Russian. – Abstr. in Referativ. Zhurn. 51. Astron., 1.51.396 (1976).

105.088 **On the critique of A. P. Kazantsev's hypothesis from the point of view of nuclear energetics.**
B. K. Fedyushin.
Vopr. istorii astron. No. 3. Moskva, 1974, p. 190 - 191. In Russian. – Abstr. in Referativ. Zhurn. 51. Astron., 1.51.425 (1976).

105.089 **The Ijopega chondrite: a new H6 fall.**
A. L. Jaques, P. L. Lowenstein, D. H. Green, E. Kiss, W. B. Nance, S. R. Taylor, N. G. Ware.
Meteoritics, Vol. 10, 289 - 301 (1975).

The Ijopega meteorite is a new H6 group chondrite fall which contains olivine, bronzite, plagioclase, diopside, kamacite, taenite, troilite, chromite and whitlockite. The meteorite is extensively recrystallized and brecciated, and shows evidence of moderate shock deformation.

105.090 **The Nakhlites. Part I: petrography and mineral chemistry.** T. E. Bunch, A. M. Reid.
Meteoritics, Vol. 10, 303 - 315 (1975).

105.091 **The Nakhlites. Part II: where, when, and how.**
A. M. Reid, T. E. Bunch.
Meteoritics, Vol. 10, 317 - 324 (1975).

The Nakhla and Lafayette meteorites are pyroxenite cumulates that show no simple relation to chondrites, basaltic achondrites, or lunar rocks. They present evidence of magmatic activity occurring on a highly differentiated parent body relatively late in the history of the solar system similar in some respects to the earth.

105.092 **Implications of poikilitic textures in LL-group chondrites.** R. V. Fodor, K. Keil.
Meteoritics, Vol. 10, 325 - 339 (1975).

105.093 **Pyroxenes in the Patwar mesosiderite.**
P. W. Weigand.

Meteoritics, Vol. 10, 341 - 351 (1975).

105.094 **^{207}Pb-^{206}Pb isochrons for Bruderheim and Peace River.** M. C. B. Abranches, J. W. Arden, N. H. Gale.
Meteoritics, Vol. 10, 353 (1975).

105.095 **Iron in space condensates.** G. Arrhenius.
Meteoritics, Vol. 10, 354 - 357 (1975).

105.096 **Electron petrography of some chondrites.** J. R. Ashworth, D. J. Barber.
Meteoritics, Vol. 10, 357 - 358 (1975).

105.097 **Evidence for heavily irradiated grains in carbonaceous chondrites.** J. Audouze, J. P. Bibring, J. C. Dran, M. Maurette, R. M. Walker, P. Eberhardt.
Meteoritics, Vol. 10, 358 - 360 (1975).

105.098 **Diffusion losses of spallogenic rare gases from different phases of mesosiderites.** F. Begemann, H. Weber, E. Vilcsek, H. Hintenberger.
Meteoritics, Vol. 10, 360 (1975).

105.099 **The Kenna ureilite: an ultramafic rock with evidence for igneous, metamorphic, tectonic, and shock origin.** J. L. Berkley, H. G. Brown IV, K. Keil, G. Huss.
Meteoritics, Vol. 10, 361 - 362 (1975).

105.100 **Exposure and formation history of the Parsa meteorite based on fossil track and rare gas studies.** N. Bhandari, K. Gopalan, M. N. Rao.
Meteoritics, Vol. 10, 362 - 363 (1975).

105.101 **Primitive features of some unusual metal-rich meteorites.** R. W. Bild.
Meteoritics, Vol. 10, 363 - 364 (1975).

105.102 **^{87}Rb-^{87}Sr chronology of achondrites.** J.-L. Birck, J.-F. Minster, C. J. Allègre.
Meteoritics, Vol. 10, 364 - 365 (1975).

105.103 **The CAI and chondrules – solid or liquid condensates?** M. Blander, L. H. Fuchs.
Meteoritics, Vol. 10, 365 (1975).

105.104 **Guenie and Tillaberi: descriptions of two chondrites fallen in Africa.** M. Bourot, M. Christophe-Michel-Lévy, J. M. Malézieux.
Meteoritics, Vol. 10, 368 - 369 (1975).

105.105 **The magnetic domain structure of iron meteorites.** A. Brecher, M. Cutrera.
Meteoritics, Vol. 10, 372 (1975).

105.106 **On subfossil meteorite craters and the frequency of iron meteorites.** V. F. Buchwald.
Meteoritics, Vol. 10, 373 - 374 (1975).

105.107 **The meteorite of Maragh (Iran).** F. Burragato, G. Cavarretta, R. Funiciello, L. Tolomeo.
Meteoritics, Vol. 10, 374 (1975).

105.108 **The new Brazilian achondrite of Governador Valadares (Minas Gerais).** F. Burragato, G. Cavarretta, R. Funiciello.
Meteoritics, Vol. 10, 374 - 375 (1975).

105.109 **Contribution to the study of the phyllitic silicates of carbonaceous meteorites.** S. Caillère, M. Rautureau.
Meteoritics, Vol. 10, 377 - 379 (1975).

105.110 **Piancaldoli meteorite: chemistry and mineralogy.** M. Carapezza, P. M. Nuccio, M. Valenza.
Meteoritics, Vol. 10, 379 (1975).

105.111 **A new astrobleme: Ile Rouleau, Mistassini Lake, Québec, Canada.** J.-L. Caty, E. H. Chown, D. W. Roy.
Meteoritics, Vol. 10, 379 - 380 (1975).

105.112 **The chondritic shower of Lajeado Ipiranga, Parana', Brazil.** G. Cavarretta, M. Fornaseri, R. Funiciello, L. Tolomeo.
Meteoritics, Vol. 10, 380 (1975).

105.113 **Some primordial lead investigations.** J. H. Chen, G. R. Tilton.
Meteoritics, Vol. 10, 380 - 381 (1975).

105.114 **A few observations on the Tieschitz H3 chondrite.** M. Christophe-Michel-Lévy.
Meteoritics, Vol. 10, 381 - 383 (1975).

105.115 **The effect of shock on the magnetic moments and hysteresis properties of natural materials, with special emphasis on rocks from the Lonar meteorite crater, India, and their relationship to lunar paleo-magnetism.** S. M. Cisowski.
Meteoritics, Vol. 10, 383 (1975).

105.116 **Rare-earth and other elemental abundances in Allende inclusions.** R. L. Conrad, R. A. Schmitt, W. V. Boynton.
Meteoritics, Vol. 10, 384 - 387 (1975).

105.117 **Total sulfur content of ordinary chondrites.** J. D. Cripe, C. B. Moore.
Meteoritics, Vol. 10, 387 - 388 (1975).

105.118 **Composition of individual chondrules in ordinary chondrites.** A. A. deGasparis, K. Fredriksson, P. Brenner.
Meteoritics, Vol. 10, 390 - 392 (1975).

105.119 **Malvern howardite: partial solutions of a puzzle.** C. Desnoyers, D. Y. Jérome.
Meteoritics, Vol. 10, 392 (1975).

105.120 **Oman ring: suspected astrobleme.** R. S. Dietz, J. F. McHone, N. M. Short.
Meteoritics, Vol. 10, 393, 509 (1975).

105.121 **Accretion of the ordinary chondrites.** R. T. Dodd.
Meteoritics, Vol. 10, 394 - 395 (1975).

105.122 **Charged-particle track record of some meteorites and its inferences.** S. A. Durrani, R. K. Bull.
Meteoritics, Vol. 10, 399 - 401 (1975).

105.123 **Ne-E rich phase in Orgueil.** P. Eberhardt.
Meteoritics, Vol. 10, 401 (1975).

105.124 **Parent material for seemingly-exotic, light-colored lithic fragments in LL-group chondrites.** R. V. Fodor, K. Keil.
Meteoritics, Vol. 10, 402 (1975).

105.125 **The matrix in chondrites.**

K. Fredriksson, A. A. deGasparis, E. Rambaldi.
Meteoritics, Vol. 10, 402 - 403 (1975).

105.126 **Volatile movement in Rose City meteorite, and implications concerning the impact and late thermal history of ordinary chondrites.** R. M. Fruland.
Meteoritics, Vol. 10, 403 - 404 (1975).

105.127 **Trace elements in Allende amoeboid aggregates.** R. Ganapathy, L. Grossman, R. L. Methot.
Meteoritics, Vol. 10, 404 (1975).

105.128 **Nature of the carbon and sulfur phases and inorganic gases in the Kenna ureilite.**
E. K. Gibson, Jr.
Meteoritics, Vol. 10, 405 - 407 (1975).

105.129 **Neutron capture profiles in the Mundrabilla iron meteorite.** H. Haag, T. Kirsten.
Meteoritics, Vol. 10, 408 - 409 (1975).

105.130 **Anisotropic taenite in meteorites.** R. Gooley, R. B. Merrill, J. R. Smyth.
Meteoritics, Vol. 10, 410 (1975).

105.131 **Rare gas studies of Allende and St. Severin meteorites and their implications with reference to curium-248.** K. Gopalan, M. N. Rao.
Meteoritics, Vol. 10, 410 - 411 (1975).

105.132 **Kakangari and related meteorites.** A. L. Graham, R. Hutchison.
Meteoritics, Vol. 10, 411 (1975).

105.133 **Factor analysis of chondrite and achondrite elemental abundances.** R. S. Harmon, D. M. Shaw.
Meteoritics, Vol. 10, 412 - 413 (1975).

105.134 **Can enstatite meteorites form from a nebula of solar composition?** J. M. Herndon, H. E. Suess.
Meteoritics, Vol. 10, 413 (1975).

105.135 **Garland: polymict-brecciated unequilibrated hypersthene achondrite.** J. V. Heyse.
Meteoritics, Vol. 10, 413 - 414 (1975).

105.136 **Progenitor of fissiogenic xenon in carbonaceous chondrites: attempted characterization from trace element studies.** H. Higuchi, J. W. Morgan, J. Gros, H. Takahashi, E. Anders.
Meteoritics, Vol. 10, 414 - 415 (1975).

105.137 **Preliminary report on the Bali carbonaceous chondrite.** G. Hoinkes, G. Kurat.
Meteoritics, Vol. 10, 416 - 417 (1975).

105.138 **Trace element variations in artificially heated primitive chondrites.**
M. Ikramuddin, C. M. Binz, M. E. Lipschutz.
Meteoritics, Vol. 10, 418 - 419 (1975).

105.139 **Distribution of minor and trace elements in the elongated troilite inclusions of the Cape York iron Agpalilik.** K. P. Jochum, H. Hintenberger, V. F. Buchwald.
Meteoritics, Vol. 10, 419 - 422 (1975).

105.140 **The chemical history of metal – from condensation to core formation.** W. R. Kelly, J. W. Larimer.
Meteoritics, Vol. 10, 423 - 424 (1975).

105.141 **Mafic silicates in Orgueil.** J. F. Kerridge, J. D. Macdougall.
Meteoritics, Vol. 10, 424 (1975).

105.142 **The Elton, Texas octahedrite.** E A. King.
Meteoritics, Vol. 10, 425 (1975).

105.143 **Oxidation state of iron in the solar nebula.** B. K. Kothari.
Meteoritics, Vol. 10, 427 - 428 (1975).

105.144 **Late compaction of C2 chondrites?** B. K. Kothari, J. D. Macdougall.
Meteoritics, Vol. 10, 428 - 429 (1975).

105.145 **An unusual phosphate-sulfide assemblage in the Cape York iron meteorite.**
A. Kracher, G. Kurat.
Meteoritics, Vol. 10, 429 - 430 (1975).

105.146 **Zoneography of the Rochechouart impact structure and giant crystals in the quartz vein of St.-Paul-la-Roche.** F. Kraut, J. Becker.
Meteoritics, Vol. 10, 430 - 431 (1975).

105.147 **Forms of nonterrestrial dust on the earth.** E. L. Krinov.
Meteoritics, Vol. 10, 431 - 432 (1975).

105.148 **Preliminary report on the Cochabamba carbonaceous chondrite.** G. Kurat, A. Kracher.
Meteoritics, Vol. 10, 432 - 433 (1975).

105.149 **Nickel enrichment of impact melt rocks from Rochechouart. Preliminary results and possibility of meteoritic contamination.** P. Lambert.
Meteoritics, Vol. 10, 433 - 436 (1975).

105.150 **Mass/size frequency distribution of chondrules from four L-chondrites.** B. Lang, Y. I. Stakheav (*Yu. I. Stakheev*), G. V. Baryshnikova, A. K. Lavrukhina.
Meteoritics, Vol. 10, 437 - 439 (1975).

105.151 **Estimation ranges of radiation ages and preatmospheric sizes of meteorites due to cosmic-ray variations.** A. K. Lavrukhina, V. A. Alexeev (*V. A. Alekseev*), G. K. Ustinova.
Meteoritics, Vol. 10, 440 - 442 (1975).

105.152 **Thermal metamorphism: a simulation of effective conditions.** F. A. Levi.
Meteoritics, Vol. 10, 442 (1975).

105.153 **A qualitative comparison of textures in type 4-chondrites chondrules and laser-prepared synthetic chondrule-like spherules.** G. R. Levi-Donati.
Meteoritics, Vol. 10, 442 - 444 (1975).

105.154 **Origin of noble gases in Allende II. Sources of fission xenon.** R. S. Lewis, B. Srinivasan, E. Anders.
Meteoritics, Vol. 10, 444 - 445 (1975).

105.155 **Cosmic-ray tracks in the Lost-City meteorite.** J. C. Lorin, P. Pellas.
Meteoritics, Vol. 10, 445 - 447 (1975).

105.156 **Sm-Nd systematics of the Stannern meteorite.** G. W. Lugmair, N. B. Scheinin.
Meteoritics, Vol. 10, 447 - 448 (1975).

105.157 **Cosmogenic radionuclides in the Allende meteorite.**

H. Mabuchi, Y. Nakamura, H. Takahashi, M. Imamura, Y. Yokoyama, J. L. Reyss.
Meteoritics, Vol. 10, 449 (1975).

105.158 **Pre-compaction irradiation of mafic silicate in C1 and C2 chondrites.** J. D. Macdougall.
Meteoritics, Vol. 10, 449 - 450 (1975).

105.159 **Lead-lead and rubidium-strontium study of the Saint-Severin LL6 chondrite.**
G. Manhes, J.-F. Minster, C. J. Allègre.
Meteoritics, Vol. 10, 451 (1975).

105.160 **Trapped xenon, missing neon: ureilites vs. carbonaceous chondrites.** K. Marti, L. L. Wilkening.
Meteoritics, Vol. 10, 452 (1975).

105.161 **The perplexing behavior of niobium in meteorites and lunar samples.** U. B. Marvin.
Meteoritics, Vol. 10, 452 - 454 (1975).

105.162 **Tepexitl crater, Mexico: not meteoritic.**
L. Maupomé, R. Alvarez, S. W. Kieffer, R. S. Dietz.
Meteoritics, Vol. 10, 454 - 455 (1975).

105.163 **Deposition of ejecta at Lonar crater.**
D. J. Milton, A. Dube, S. Sen Gupta.
Meteoritics, Vol. 10, 456 - 457 (1975).

105.164 **Chemical studies of the Mundrabilla iron meteorite.**
O. Müller.
Meteoritics, Vol. 10, 460 - 462 (1975).

105.165 **Eu fractionation in different kinds of chondrites.**
N. Nakamura, A. Masuda.
Meteoritics, Vol. 10, 462 - 463 (1975).

105.166 **The Renazzo chondrite – a reevaluation.**
J. Nelen, G. Kurat, K. Fredriksson.
Meteoritics, Vol. 10, 464 - 465 (1975).

105.167 **The distribution of Li, Na, and K in the LL-group chondrites.** W. Nichiporuk.
Meteoritics, Vol. 10, 466 (1975).

105.168 **Determination of metallic iron, nickel and cobalt in meteoritic rocks.** P. M. Nuccio, M. Valenza.
Meteoritics, Vol. 10, 467 - 468 (1975).

105.169 **An attempt to recalibrate the pyroxene-pyroxene thermometer (or how to beat a dead horse to death).**
E. J. Olsen, T. Bunch.
Meteoritics, Vol. 10, 468 (1975).

105.170 **Fluid phases related to the formation of crypto-explosion structures as indicated by fluid inclusions.**
M. Pagel.
Meteoritics, Vol. 10, 469 (1975).

105.171 **New data on the chemistry of Allende inclusions.**
H. Palme, B. Spettel, H. Baddenhausen, H. Broo.
Meteoritics, Vol. 10, 469 - 471 (1975).

105.172 **Uranium and plutonium in chondritic phosphates.**
P. Pellas, D. Storzer.
Meteoritics, Vol. 10, 471 - 473 (1975).

105.173 **Tracks of heavy cosmic-ray nuclei in olivines from the Marjalahti and Eagle Station meteorites.**
V. P. Perelygin, O. Otgonsuren (*Otgonsurehn*), S. G. Stetsenko.
Meteoritics, Vol. 10, 473 - 474 (1975).

105.174 **Allende and analogues: genetic implications of lithophile trace element abundances.**
J. A. Philpotts, H. Nagasawa, N. Onuma.
Meteoritics, Vol. 10, 474 - 475 (1975).

105.175 **Search for superheavy elements in meteorites.**
A. G. Popeko.
Meteoritics, Vol. 10, 476 (1975).

105.176 **Trace element distribution between metal and silicates of ordinary chondrites.**
E. Rambaldi, R. Thacker, M. Cendales.
Meteoritics, Vol. 10, 477 (1975).

105.177 **Silicate inclusions in the Mundrabilla meteorite.**
P. Ramdohr, M. Prinz, A. El Goresy.
Meteoritics, Vol. 10, 477 - 479 (1975).

105.178 **Cooling rate determinations for hexahedrites.**
E. Randich, J. I. Goldstein.
Meteoritics, Vol. 10, 479 (1975).

105.179 **Crystallographic and chemical studies of a bronzite and chromite in the Yamato *B* achondrite.**
A. M. Reid, H. Takeda, T. Yamanaka.
Meteoritics, Vol. 10, 479 - 480 (1975).

105.180 **Haughton Dome and Slate Islands: recently discovered Canadian impact structures.**
P. B. Robertson, R. A. F. Grieve.
Meteoritics, Vol. 10, 480 - 481 (1975).

105.181 **Back rotation of shatter-cones from the Charlevoix astrobleme, P. Que., Canada.** D. W. Roy.
Meteoritics, Vol. 10, 481 (1975).

105.182 **Noble gases in grain size fractions of Weston: a comparison with lunar soil data.**
L. Schultz, U. Frick, P. Signer.
Meteoritics, Vol. 10, 483 (1975).

105.183 **Analysis of the Orgueil meteorite by energy-dispersive X-ray fluorescence.**
H. P. Schwarcz, O. Mudroch.
Meteoritics, Vol. 10, 484 (1975).

105.184 **Classification and origin of pallasites.**
E. R. D. Scott, J. T. Wasson.
Meteoritics, Vol. 10, 485 (1975).

105.185 **The metal content of common chondrites.**
D. W. Sears, H. J. Axon.
Meteoritics, Vol. 10, 486 - 487 (1975).

105.186 **The genetic relationships among diogenites, howardites and eucrites.**
N. Shimizu, C. J. Allègre.
Meteoritics, Vol. 10, 488 (1975).

105.187 **Fossil fission track studies in Allende inclusions.**
J. Shirck, R. M. Walker.
Meteoritics, Vol. 10, 488 (1975).

105.188 **Electron microprobe investigation of the howardite Malvern.** A. B. Simpson.
Meteoritics, Vol. 10, 489 - 491 (1975).

105.189 **Origin of noble gases in Allende I. Host phases of fission xenon and other gases.**
B. Srinivasan, R. S. Lewis, E. Anders.
Meteoritics, Vol. 10, 491 - 492 (1975).

105.190 **Melting and reaction behavior of opaques and silicates in very strongly shocked crystalline rocks from the Ries crater.** V. Stähle.
Meteoritics, Vol. 10, 492 - 494 (1975).

105.191 **Cosmogenic excess of ^{40}K and the flux of fast neutrons in meteorites.** W. Stegmann, F. Begemann.
Meteoritics, Vol. 10, 494 - 495 (1975).

105.192 **Ries deep drilling results: implications for the structure of the crater basement and the distribution of excavated masses.** D. Stöffler.
Meteoritics, Vol. 10, 495 - 497 (1975).

105.193 **Experimental petrology of the Juvinas eucrite.** E. Stolper.
Meteoritics, Vol. 10, 498 - 499 (1975).

105.194 **Formation and early brecciation of the Juvinas achondrite inferred from U-Th-Pb systematics.**
M. Tatsumoto, D. M. Unruh.
Meteoritics, Vol. 10, 500 - 501 (1975).

105.195 **Skeleton Lake – a Paleozoic impact crater in Central Canada.** E. D. Wadddington, M. R. Dence.
Meteoritics, Vol. 10, 503 (1975).

105.196 **The age of the Rochechouart impact structure.** G. A. Wagner, D. Storzer.
Meteoritics, Vol. 10, 503 - 504 (1975).

105.197 **Formation locations of meteorites.** J. T. Wasson.
Meteoritics, Vol. 10, 505 - 506 (1975).

105.198 **Primordial rare gases in black inclusions from the ureilite Haverö.** H. W. Weber, F. Begemann, H. Hintenberger.
Meteoritics, Vol. 10, 506 (1975).

105.199 **Silicate inclusions in the El Taco iron meteorite.** F. Wlotzka, E. Jarosewich.
Meteoritics, Vol. 10, 507 (1975).

105.200 **Meteorites in New Zealand.** D. Calder.
Southern Stars, Vol. 26, 95 - 100 (1975). – Presented to the annual conference of the Royal Astronomical Society of New Zealand, Wellington, 1974.

105.201 **Sudbury basin sediments and the meteoritic impact theory of origin for the Sudbury structure.**
F. W. Beales, G. P. Lozej.
Canadian Journ. Earth Sci., Vol. 12, 629 - 635 (1975).

105.202 **Concerning mercury in meteorites.** B. L. K. Somayajulu.
Proc. Indian Acad. Sci., Ser. A, Vol. 81, 137 - 142 (1975).

105.203 **Mesures des taux de refroidissement des chondrites ordinaires à partir des traces de fission du plutonium 244 enregistrées dans les cristaux détecteurs.**
P. Pellas, D. Storzer.
Comptes Rendus Acad. Sci. Paris, Sér. D, Vol. 280, 225 - 228 (1975).

Les traces de fission de ^{244}Pu, enregistrées dans les whitlockites et dans les olivines et pyroxènes au contact des whitlockites, permettent de retracer les histoires thermiques des chondrites ordinaires depuis la rétention du xénon fissiogène (il y a ~4,6 × 10^9 ans).

105.204 **Part of an asteroid?** H. H. Nininger.
Earth Sci., (*USA*), Vol. 28, No, 1, p. 15 - 18 (1975).

Magnetic properties of meteorites.
See Abstr. 003.063.

Fall of cosmic matter on the earth's surface.
See Abstr. 003.080.

Orbital elements of 45 meteorites. Atlas.
See Abstr. 003.118.

The founders of American meteoritics.
See Abstr. 005.015.

Measurements of the light flash produced by high velocity particle impact. See Abstr. 022.066.

Lifetimes of some excited states in the rare earths: La II, Ce II, Pr II, Nd II, Sm II, Yb I, Yb II, and Lu II.
See Abstr. 022.102.

Giant planets and their satellites, minor planets, meteorites and comets. See Abstr. 091.032.

Magnetism of the moon and meteorites.
See Abstr. 094.124.

Lunar rocks as meteoroid detectors.
See Abstr. 094.156.

Variation in the number of meteoroid impacts on the moon with lunar phase. See Abstr. 094.177.

A Monte-Carlo model for the accumulation of solar flare tracks and spallation products in extraterrestrial regoliths.
See Abstr. 094.187.

The relationship between target and melt rocks in terrestrial craters: implications for the origin of lunar breccia matrices. See. Abstr. 094.195.

Lunar microcraters and micrometeoroids.
See Abstr. 094.200.

A physical and chemical characterization of extraterrestrial dust grains on a submicron scale.
See Abstr. 094.460.

Pyroxene geothermometry applied to lunar crustal rocks and meteorites. See Abstr. 094.473.

Minor planets and related objects. XVIII. *UBV* photometry and surface composition. See Abstr. 098.046.

Dust storms in space? See Abstr. 106.037.

The physical nature of interplanetary dust as inferred by particles collected at 35 km. See Abstr. 106.040.

Micrometeorites and solar flare particles in and out of the ecliptic. See Abstr. 106.055.

The application of rare-earth elements to infer gas-grain temperature differentials in the solar nebula.
See Abstr. 107.017.

Experimental investigations of extinction on small meteoritic particles. See Abstr. 131.192.

Track length distribution in lunar and meteoritic minerals and the chemical composition of very heavy cosmic rays. See Abstr. 143.073.

106 Interplanetary Matter, Interplanetary Magnetic Field, Zodiacal Light

106.001 **High fluxes of low energy (10–25 MeV/amu) C, N, O and heavier ions at 450 km altitude; results from the Skylab experiment.** S. Biswas, N. Durgaprasad, J. Nevatia, V. S. Venkatavaradan, J. N. Goswami, U. B. Jayanthi, D. Lal, S. K. Mattoo.
Astrophys. Space Sci., Vol. 35, 337 - 347 (1975).
High fluxes (i.e. the integrated fluxes) of C, N, O group of nuclei and some of the heavier ones, in the energy interval 10–25 MeV/amu, have been identified in a Lexan polycarbonate detector assembly exposed on the exterior of the Skylab for 73 days. The authors have considered two sources for these particles: (1) partly ionized interplanetary ions accelerated within the magnetosphere and (2) heavy nuclei trapped in the earth's radiation belt. The flux and composition of the nuclei observed seem to be significantly different from those in the trapped radiation as known at present; hence it seems likely that the major part of the observed flux may be interpreted, in terms of partly ionized interplanetary ions that are further accelerated in the magnetosphere.

106.002 **North-south component of the interplanetary magnetic field and large scale auroral dynamics.**
S.-I. Akasofu.
Nature, Vol. 256, 191 - 192 (1975). – Letter.

106.003 **On the gradient of zodiacal light with heliocentric distance.** R. Dumont.
Planet. Space Sci., Vol. 23, 1235 (1975).
Space density of interplanetary dust grains is directly related to the gradient of zodiacal light observed, at constant elongation ϵ, by a space photometer moving and aiming in the symmetry plane of the solar system.

106.004 **Die räumliche Anisotropie des interplanetaren Staubflusses.** H. Fechtig, E. Grün.
Mitt. Astron. Ges., No. 36, p. 67 - 68 (1975).

106.005 **Gemessene interplanetare Staubflüsse und optische Zodiakallichtdaten.**
R. H. Giese, E. Grün, C. Leinert.
Mitt. Astron. Ges., No. 36, p. 68 (1975). – Abstract.

106.006 **Zur Dynamik interplanetarer Staubteilchen.**
G. Schwehm, R. Zerull, R. H. Giese.
Mitt. Astron. Ges., No. 36, p. 68 - 69 (1975). – Abstract.

106.007 **Streufunktionen nicht kugelförmiger Partikel, gewonnen aus Mikrowellenanalogieversuchen.**
R. Zerull, W. Lantrewitz, R. H. Giese.
Mitt. Astron. Ges., No. 36, p. 69 (1975). – Abstract.

106.008 **Beobachtungen des Zodiakallichtes und des Airglows mit dem Ballonteleskop THISBE. 1. Messungen im nahen Infraroten; 2. Messungen in vier Farben des Sichtbaren.**
A. Frey, W. Hofmann, D. Lemke, C. Thum.
Mitt. Astron. Ges., No. 36, p. 69 - 70 (1975). – Abstract.

106.009 **Consequences of the inclination of the zodiacal cloud on the ecliptic.** R. Robley.
Astron. Astrophys., Vol. 41, 385 - 389 (1975). In French.
The author shows that the modulation of the radiance $B(90, 0)$ of the zodiacal light observed at the ecliptic pole defines the parameter H as a function of the inclination angle β between the zodiacal cloud and the ecliptic plane; moreover, the experimental value of the ratio $(B(90, 0))/B(90, 90)$ defines the parameter R. It can be deduced that the flatness of the zodiacal cloud, expressed by R/H, is <5 and that the plane of symmetry of the zodiacal cloud is very close to that of the invariant plane of the solar system ($\beta < 2°$).

106.010 **Variations in Gegenschein polarization?**
J. G. Sparrow, J. L. Weinberg.
Astron. Astrophys., Vol. 41, 475 - 476 (1975).
The basis for claims of small scale angular structure and day to day variations in the polarized component of the zodiacal light near the antisolar point has been examined and found to be unsubstantiated.

106.011 **On the use of Godhavn *H* component as an indicator of the interplanetary sector polarity.**
L. Svalgaard.
Journ. Geophys. Res., Vol. 80, 2717 - 2722 (1975).
An objective method of inferring the polarity of the interplanetary magnetic field using the H component at Godhavn is presented. The objectively inferred polarities are compared with a subjective index inferred earlier (Svalgaard, 1972). It is concluded that no significant difference exists between the two methods.

106.012 **Measurements of the interplanetary magnetic field in relation to the modulation of cosmic rays.**
P. C. Hedgecock.
Solar Physics, Vol. 42, 497 - 527 (1975).
Field strength distributions and low frequency power spectra are derived from interplanetary field measurements made by the HEOS-1 and HEOS-2 satellites during the years 1969–1973. The total power level, the power spectral density at zero frequency and the correlation length are found to vary significantly and in a similar way over the solar cycle. The magnitude and phase of these variations are compared with measurements of the cosmic ray neutron monitor rate and the coronal green line intensity and the influence of mid-latitude solar phenomena on the character of the interplanetary field in the ecliptic is demonstrated. The data are interpreted as indicating that the cosmic ray intensity in the Galaxy is some 75 % larger than the intensity recorded by neutron monitors on earth.

106.013 **Corrections to the azimuthal component of the interplanetary magnetic field due to meridional flow in the solar wind.** S. F. Nerney, S. T. Suess.
Astrophys. Journ., Vol. 200, 503 - 509 (1975).
Stellar-wind models which incorporate meridional flows, open magnetic flux tubes in the equatorial plane of the star. The transport of magnetic field to higher latitudes is accompanied by an increase in the azimuthal velocity, increasing the corotation of the plasma at all latitudes. The percentage correction to the azimuthal magnetic field compared to classical models is a 10–20 percent effect at 5 AU and $\theta = 90°$, whereas the azimuthal velocity correction is a 20–40 percent effect.

106.014 **Interplanetary particles and fields.**
J. A. Van Allen.
Sci. American, Vol. 233, No. 3, p. 160 - 161, 164, 166, 169 - 173 (1975). – Popular article.

106.015 **Helios investigates the circumsolar space.**
L. I. Miroshnichenko, E. I. Prutenskaya.
Priroda, No. 12.75, p. 101 (1975). In Russian.

106.016 **Radial dependence of the electron temperature at heliocentric distances from 1 a.u. to 1.5 a.u. according to data obtained by measurements aboard the Mars 3 space vehicle.** K. I. Gringauz, M. I. Verigin.
Dokl. Akad. Nauk SSSR. Ser. Mat., Fiz., Vol. 223, 589 - 592 (1975). In Russian.

106.017 **On the recording of simple waves in the interplanetary plasma.** I. P. Stakhanov.
Dokl. Akad. Nauk SSSR. Ser. Mat., Fiz., Vol. 223, 593 - 595 (1975). In Russian.

106.018 **Some notes on the zodiacal light, zodiacal band and gegenschein.** H. C. Nightingale.
Journ. British Astron. Ass., Vol. 85, 417 - 419 (1975).

106.019 **Pioneer 9 and Ogo 5 observations of an interplanetary multiple shock ensemble on February 2, 1969.**
M. Dryer, Z. K. Smith, T. Unti, J. D. Mihalov, B. F. Smith, J. H. Wolfe, D. S. Colburn, C. P. Sonett.
Journ. Geophys. Res., Vol. 80, 3225 - 3234 (1975).

106.020 **Multifrequency techniques for studying interplanetary scintillations.** R. Woo.
Astrophys. Journ., Vol. 201, 238 - 248 (1975).

Rytov's approximation or the method of smooth perturbations is utilized to derive the temporal frequency spectra of the amplitude and phase fluctuations of multifrequency plane and spherical waves propagating in the interplanetary medium and solar corona. It is shown that multifrequency observations of interplanetary scintillations using either compact radio stars or spacecraft radio signals are desirable because the correlation of the multifrequency waves yields additional independent measurements of the solar wind and turbulence. Measurements of phase fluctuations are also desirable because, unlike amplitude fluctuations, they provide information on the full range of scale sizes for the electron density fluctuations.

106.021 **Frequency correlation of interplanetary scintillations.** V. G. Panadzhyan.
Soobshch. Byurakan. Obs., vyp. (No.) 46, p. 88 - 92 (1975). In Russian.

From simultaneous records of interplanetary scintillations at different frequencies (86, 60, 40 MHz) the frequency correlation functions were computed and from their maximum the correlation factors obtained. The experimental curves of frequency correlation factors are presented.

106.022 **On the possibility of detecting solar flare effects in the zodiacal light.** N. Y. Misconi, M. S. Hanner.
Planet. Space Sci., Vol. 23, 1329 - 1335 (1975).

To evaluate possible effects of solar flares on the brightness of the inner zodiacal light, it is necessary to consider the brightness contribution along the line of sight and as a function of sun-particle distance. For this purpose, models of the brightness contribution along the line of sight are presented for both dielectric and metallic particles with a spatial distribution of the form $r^{-\nu}$, $\nu = 0, 1, 2$. These models are discussed in terms of the geometry of shock front interaction. A reported zodiacal light enhancement following a solar flare (Blackwell and Ingham, 1961) is analyzed on the basis of the shock front geometry.

106.023 **The roles of the north–south component of the interplanetary magnetic field on large-scale auroral dynamics observed by the DMSP satellite.** S.-I. Akasofu.
Planet. Space Sci., Vol. 23, 1349 - 1354 (1975).

106.024 **Interplanetary gas of nonsolar origin.**
G. E. Thomas.
Rev. Geophys. Space Phys., Vol. 13, No. 3, (see 013.001), p. 1063 - 1065, 1081 - 1083 (1975). – Review paper with bibliographical notes.

106.025 **Hyperbolic cosmic dust: its origin and its astrophysical significance.** H. A. Zook.
Planet. Space Sci., Vol. 23, 1391 - 1397 (1975).

The heliocentric radial distribution of the flux of hyperbolic cosmic dust particles, as measured by the Pioneer 8 and 9 spacecraft, is closely related to the radial variation of the spatial density of source or "parent" meteoroids. Within the limits of the experimental and theoretical uncertainties the spatial density of parent meteoroids, as deduced from the hyperbolic cosmic dust data, is found to be increasing with increasing heliocentric distance in the neighborhood of one a.u. Finally it is noted that the solar system is not presently contributing significant amounts of dust to the interstellar medium though it may once have done so.

106.026 **Relationship between the interplanetary magnetic field and 'isolated substorms'.**
J. Hirshberg, T. E. Holzer.
Journ. Geophys. Res., Vol. 80, 3553 - 3556 (1975).

The study of 54 'isolated substorms' by Foster et al. (1971) is reexamined, and five different explanations of the results of this study are found to be consistent with the data. For each of the five explanations the decay time of the *AE* index is found not to be characteristic of internal magnetospheric processes but to be due to data selection. In four of the five explanations the time lag that maximizes the correlation coefficient between B_z and *AE* is found not to represent the time lag between an interplanetary cause and a consequent magnetospheric effect. It is shown that by simple extensions of the Foster et al. study, at least four of the five explanations can be eliminated.

106.027 **Comparison of inferred and observed interplanetary magnetic field polarities, 1970–1972.**
J. M. Wilcox, L. Svalgaard, P. C. Hedgecock.
Journ. Geophys. Res., Vol. 80, 3685 - 3688 (1975).

The inferred polarity (toward or away from the sun) of the interplanetary magnetic field at earth using polar observations of the geomagnetic field has been compared with spacecraft observations. A list published by Svalgaard (1974) of the inferred field polarities during 1970–1972 is found to be correct on 82% of the days. A list of 'well-defined' sector boundaries at earth during 1970–1972 is given.

106.028 **The degree of inhomogeneity of the interplanetary magnetic field in 1958 - 1973.** A. I. Kuz'min, G. V. Skripin, V. A. Filippov, V. M. Migunov.
Issled. po kosmofiz. i aehronomii. Yakutsk, 1975, p. 78 - 83, 199 - 200. In Russian. – Abstr. in Referativ. Zhurn. 51. Astron., 9.51.434; 62. Issled. kosmich. prostranstva, 9.62.226 (1975).

106.029 **Modern aspects of the spectrum of interplanetary plasma irregularities.** N. A. Lotova.
Uspekhi fiz. nauk, Vol. 115, 603 - 620 (1975). In Russian. Abstr. in Referativ. Zhurn. 51. Astron., 9.51.435 (1975).

106.030 **The structure of interplanetary plasma streams and magnetospheric substorms.**
K. G. Ivanov, N. V. Mikerina.
Solnechn. veter i magnitosfera. Moskva, 1974, p. 3 - 30, 94. In Russian. – Abstr. in Referativ. Zhurn. 62. Issled. kosmich. prostranstva, 9.62.213 (1975).

106.031 **Variations of the interplanetary magnetic field and of cosmic rays on August 9, 1972.**
A. V. Belov, L. I. Dorman, N. V. Mikerina.

Geomagn. Aeronom., Vol. 15, 787 - 792 (1975). In Russian.

106.032 **"Magnetic" and "helium" regions of the interplanetary plasma stream and the structure of the main phase of a magnetic storm.** K. G. Ivanov, N. V. Mikerina.
Geomagn. Aeronom., Vol. 15, 881 - 886 (1975). In Russian.

106.033 **On the interpretation of strong discontinuities in the interplanetary plasma.**
V. B. Baranov, M. D. Kartalev.
In-t kosmich. issled. AN SSSR. Pr-210. Moskva, 28 pp. In Russian. – Abstr. in Referativ. Zhurn. 51. Astron., 8.51. 463; 62. Issled. kosmich. prostranstva, 8.62.194 (1975).

106.034 **Influence of the interplanetary magnetic field on the magnetospheric convection and electric currents in the ionosphere.** V. M. Mishin, A. D. Bazarzhapov, Eh. I. Nemtsova, G. V. Popov, V. V. Shelomentsev.
Subburi i vosmushcheniya v magnitosfere. Leningrad, Nauka, 1975, p. 191 - 207. In Russian. – Abstr. in Referativ. Zhurn. 62. Issled. kosmich. prostranstva, 8.62.219 (1975).

106.035 **Interplanetary plasma instabilities and geomagnetic pulsations.** A. V. Gul'el'mi.
Subburi i vosmushcheniya v magnitosfere. Leningrad, Nauka, 1975, p. 214 - 226. In Russian. – Abstr. in Referativ. Zhurn. 62. Issled. kosmich. prostranstva, 8.62.220 (1975).

106.036 **On the computation of the first-order scattering of the Milky Way radiation and the zodiacal light in the earth's atmosphere.** Yu. M. Zavarzin.
Trudy Astrofiz. Inst., *Alma-Ata*, Vol. 25, 102 - 114 (1975). In Russian.

106.037 **Dust storms in space?**
R. G. Roosen, O. E. Berg, N. H. Farlow.
I.A.U. Colloquium No. 13, (see 012.030), p. 223 - 226 (1973).

106.038 **A possible mechanism for the capture of microparticles by the earth and other planets of the solar system.** F. Di Benedetto.
I.A.U. Colloquium No. 13, (see 012.030), p. 271 - 274 (1973).

106.039 **On the influence of the atmospheric dust on the zodiacal light polarization.**
N. B. Divari, S. N. Krylova.
I.A.U. Colloquium No. 13, (see 012.030), p. 275 - 280 (1973).

Three-color observations of the nightglow brightness and polarization were made at the Majaky Observatory in October 1969 to determine the brightness of zodiacal light at the ecliptic pole. On the assumption that the degree of polarization of zodiacal light is 0.20 at the ecliptic pole, the corresponding brightnesses are found to be 68, 70, and 96 G2 stars of the 10th magnitude per square deg. at 0.37, 0.53, and 0.58 μ.

106.040 **The physical nature of interplanetary dust as inferred by particles collected at 35 km.**
D. E. Brownlee, P. W. Hodge, W. Bucher.
I.A.U. Colloquium No. 13, (see 012.030), p. 291 - 295 (1973).

Particles were collected at an altitude of 35 km by two flights of a volume-sampling micrometeorite collector. The collection scheme is very sensitive and is capable of collecting a significant number of particles. Many of the particles collected have chemical compositions similar to solar or to iron meteorites. Morphology of collected particles indicates that both true micrometeorites and ablation products were collected.

106.041 **Influence of non-gravitational effects on the evolution of dust particles in the circumsolar space.**
O. V. Dobrovol'skij, P. E. Egibekov.
Dokl. AN TadzhSSR, Vol. 18, No. 1, p. 16 - 19 (1975). In Russian. – Abstr. in Referativ. Zhurn. 51. Astron., 10.51.351 (1975).

106.042 **Energetic lunar nighttime ion events.**
H. E. Schneider, J. W. Freeman.
The Moon, Vol. 14, 27 - 33 (1975). – Conference report (see 012.031).

The Rice University Suprathermal Ion Detector Experiment regularly observes ion events normally ranging from 250 eV q^{-1} to 1000 eV q^{-1} all through the lunar night. These ion events occur most often 2 to 3 days prior to the sunrise terminator. There is also a secondary activity peak 3 to 4 days after local sunset on the moon. The events are normally of 4 hr or less in duration and the integral flux is 10^6 ions cm^{-2} s^{-1} $ster^{-1}$. This article discusses the character of these events and presents the preliminary findings of a detailed study begun on this subject.

106.043 **On the propagation of perturbations in the interplanetary plasma caused by solar flares.**
L. V. Shidlovskaya.
Dokl. Akad. Nauk SSSR. Ser. Mat., Fiz., Vol. 225, 272 - 275 (1975). In Russian.

106.044 **Possibility of studying the structure of the interplanetary magnetic field using data on the circular polarization of the zodiacal light.**
A. Z. Dolginov, I. G. Mitrofanov.
Pis'ma v Astron. Zhurn., Vol. 1, No. 12, p. 38 - 42 (1975). In Russian.

The analysis of the dependence of circular polarization of the zodiacal light on elongation is shown to confirm the existence of the regular component $B_\perp$ of the interplanetary magnetic field. Direction and dependence of this component on the heliocentric distance are determined.

106.045 **Intensity distribution of interplanetary scintillation at 408 MHz.** R. G. Milne.
Australian Journ. Phys., Vol. 28, 621 - 632 (1975).

It is shown that interplanetary scintillation of small-diameter radio sources at 408 MHz produces intensity fluctuations which are well fitted by a Rice-squared distribution, better so than is usually claimed. The observed distribution can be used to estimate the proportion of flux density in the core of 'core-halo' sources without the need for calibration against known point sources.

106.046 **Interplanetary magnetic fields and geomagnetic variations in the nearpolar region.**
P. V. Sumaruk, Ya. I. Fel'dshtejn.
Uspekhi fiz. nauk, Vol. 116, 342 - 345 (1975). In Russian. Abstr. in Referativ. Zhurn. 62. Issled. kosmich. prostranstva, 10.62.172 (1975).

106.047 **Power spectra of interplanetary scintillations.**
A. V. Pynzar', V. I. Shishov, T. D. Shishova.
Astron. Zhurn. Akad. Nauk SSSR, Vol. 52, 1187 - 1194 (1975). In Russian. English translation in Soviet Astron., Vol. 19, No. 6.

The results of the measurements of the power spectra of the interplanetary scintillations of 3C48 at 86 and 102 MHz are presented. It is shown that there is good agreement of the observational data with the hypothesis on the power law spectra of the interplanetary plasma turbulence: $\Phi_{N_e} \propto q^{-n}$.

106.048 **Observation of the night sky light by means of an SSE high-speed scanning electrophotometer.**
Yu. M. Zavarzin.
Trudy Astrofiz. Inst., *Alma-Ata*, Vol. 25, 115 - 118 (1975). In Russian.

Preliminary results of zodiacal light observations obtained by means of the high-speed scanning electrophotometer are reported.

106.049 **A new test for Alfvén waves in interplanetary space.** A. J. Owens.
Astrophys. Space Sci., Vol. 38, 469 - 473 (1975).

A method of testing for the existence of Alfvén waves using only magnetic-field data is given. It is shown that cross-spectral analysis of the magnetic-field components can be a useful tool in distinguishing between Alfvén waves and magnetosonic waves in interplanetary space.

106.050 **Upper limits for brightness and dust concentration of the hypothetical libration cloud at L_4 in the system earth-moon based on photographic surface photometry.** W. Schlosser, H. M. Maitzen, D. Hablick.
Astron. Astrophys., Vol. 45, 197 - 202 (1975). In German.

Three wide-angle photographs with a field-diameter of 140° were used to derive upper limits for surface-brightness and dust-density of a libration cloud at the Lagrangian point L_4. The plates were selected for small air-mass of L_4 and gegenschein. Kordylewski (1974) gave for the diameter of the hypothetical dust cloud a value around 10°, including some isolated maxima of about 3°. Adopting a diameter of 7°, the quoted scatter gives a statistical uncertainity of 0.9 S_{10}. The authors consider three times this value as the upper limit of the surface-brightness of the hypothetical dust cloud and state that – at least during the observations – it must have been below 3 S_{10}. Since studies of low-level contrast visibility (Weiß, 1972) give 20 S_{10} as limit for detection by human eye, only libration clouds with a density of at least 25 times the interplanetary dust density should be discernible by the unaided observer.

106.051 **The heliographic latitude dependence and sector structure of the interplanetary magnetic field 1969–1974: results from the HEOS satellites.** P. C. Hedgecock.
Solar Physics, Vol. 44, 205 - 224 (1975).

Following conclusions have been made: (1) The interplanetary field was structured into two sectors throughout most of the period but four sectors appeared during at least three intervals each lasting a few solar rotations. (2) These direct measurements show a clear reversal in the sense of the Rosenberg-Coleman dominant polarity effect in the 1970/71 period. (3) The dominant polarity effect is present in 27-day running averages, evaluated along the Archimedian spiral, of both the percentage time coverage and the net flux of outward directed field. (4) A best fit to the rate-of-change of dominant polarity with heliographic latitude indicates that on average the sector boundaries in the interplanetary medium 1 AU from the sun are inclined at approximately 12° to the solar equator. (5) No statistically significant correlation could be found between the heliographic north–south field component and heliographic latitude in the 1969–74 period. (6) Although significant differences between north and south hemispheres may have occurred in 1969, the mean field strength and the amplitude of the directional fluctuation of the interplanetary field showed no significant variation with heliographic latitude.

106.052 **On the radial variation of the interplanetary magnetic field: Pioneer 6.**
U. Villante, F. Mariani.
Geophys. Res. Letters, Vol. 2, 73 - 74 (1975).

Analysis of simultaneous plasma and magnetic field observations by Pioneer 6 suggests, at heliocentric distances of 0.8 to 1.0 AU, significant deviations from Parker's theory: in particular, a radial variation as r^{-2} is estimated for the azimuthal magnetic field component.

106.053 **The global solar wind and predictions for Pioneers 10 and 11.** S. T. Suess, S. F. Nerney.
Geophys. Res. Letters, Vol. 2, 75 - 77 (1975).

The dependence on latitude of the azimuthal component of the interplanetary magnetic field generates a meridional pressure gradient which, in turn, generates a meridional flow. This flow has observable consequences in the equatorial plane, although the meridional velocity itself may be zero in this plane. Analytic estimates of these observables – the mass flux, azimuthal and radial magnetic flux, density, and spiral angle – are made in the range of 1 to 12 AU.

106.054 **Comment on 'Interplanetary magnetic sector structure, 1926–1971' by L. Svalgaard.**
A. Berthelier, C. Guérin.
Journ. Geophys. Res., Vol. 80, 4390 - 4396 (1975).

106.055 **Micrometeorites and solar flare particles in and out of the ecliptic.** I. D. Hutcheon.
Journ. Geophys. Res., Vol. 80, 4471 - 4483 (1975).

Using crystals grown into vugs from oriented lunar rocks as directional detectors of cosmic dust particles and solar flare nuclei, the author has measured the angular distribution of the flux of micrometeoroids of mass $\sim 3 \times 10^{-17}$ to $\sim 8 \times 10^{-14}$ g. Hypervelocity impact craters of diameter 500 Å to 10 μm and tracks from solar flare nuclei of energy $\sim$ 100 keV/amu to $\sim$ 20 MeV/amu were observed in crystals from rock 71055 facing lunar south and in crystals from rock 74255 facing lunar east. The author has found that dust grains both in ecliptic orbits and in orbits inclined to the ecliptic have virtually identical mass-frequency distributions and similar shapes. The flux of particles of mass $\gtrsim 4 \times 10^{-15}$ g confined to the ecliptic is $\sim 1.1 \times 10^{-2}$ /m^2 s 2π sr. The solar flare energy spectra observed in vug crystals from both 71055 and 74255 deviate from the long-term spectrum determined by Hutcheon et al. (1974) at energies below $\sim$ 10 MeV/amu.

106.056 **Zodiacal light - a measure of the interplanetary environment.** C. Leinert.
Space Sci. Rev., Vol. 18, 281 - 339 (1975).

This paper reviews research related to zodiacal light and tries to give the status for the end of 1974. Contents: (1) Introduction; (2) Zodiacal light photometry; (3) Results; (4) Interpretation of the observations; (5) Results of zodiacal light analysis and comparison to other investigations; (6) Conclusion.

106.057 **The influence of the source size and the receiver bandwidth on the coefficient of asymmetry of interplanetary scintillations.** A. V. Pynzar', V. I. Shishov.
Izv. vyssh. ucheb. zavedenij. Radiofizika, Vol. 18, 506 - 509 (1975). In Russian. – Abstr. in Referativ. Zhurn. 51. Astron., 12.51.340 (1975).

106.058 **Characteristics of plasma in the interplanetary medium and in the ionospheres of the planets of the solar system.** K. I. Gringauz, T. K. Breus.
XI Vses. konf. po rasprostr. radiovoln. Chast (Part) I. Tezisy dokl. Kazan', Kazan. un-t, 1975, p. 145 - 146. In Russian. Abstr. in Referativ. Zhurn. 62. Issled. kosmich. prostranstva, 12.62.153 (1975).

106.059 **A measurement of interplanetary turbulence using interferometer visibility scintillations.** R. L. Mutel.
Bull. American Astron. Soc., Vol. 7, 508 (1975). – Abstr. AAS.

106.060 **Statistical properties of the interplanetary microscale fluctuations.** J. W. Belcher.
Journ. Geophys. Res., Vol. 80, 4713 - 4718 (1975).

106.061 **IMF polarity effects on the *Sq* current focus location.** S. Matsushita.

Journ. Geophys. Res., Vol. 80, 4751 - 4754 (1975). – Brief report.

106.062 **Kp dependence on sectors.**
I. B. McDiarmid, E. E. Budzinski.
Journ. Geophys. Res., Vol. 80, 4755 - 4756 (1975). – Brief report.

106.063 **Estimate of singly scattered light from bright parts of the Milky Way and zodiacal light.**
Yu. M. Zavarzin.
Trudy Astrofiz. Inst., *Alma-Ata*, Vol. 26, 10 - 16 (1974). In Russian.

106.064 **Search for long term variations of the interplanetary magnetic field.**
F. Mariani, L. Diodato, G. Moreno.
Solar Physics, Vol. 45, 241 - 254 (1975).

A statistical study is made of the long term variations of the interplanetary magnetic field parameters collected in the years 1964 to 1973 by spacecraft. Although temporal fluctuations are observed on field components and magnitudes no clear solar cycle variation is found. The same conclusion holds for the statistical distributions and variances of these parameters. A search for possible heliographic latitude effects on the field also leads to a negative conclusion.

106.065 **Physical characteristics of the interplanetary space.**
S. N. Vernov, Yu. I. Logachev, N. F. Pisarenko.
Osnovy kosmich. biol. i med. Tom (Vol.) 1. Moskva, Nauka, 1975, p. 47 - 118. In Russian.

106.066 **On the radio interference method for investigation of tangential discontinuities and shock waves of cosmic plasma.**
I. V. Kovalevskij, V. D. Novikov, I. P. Stakhanov.
XI Vses. konf. po rasprostr. radiovoln. Chast' (Part) 1. Tezisy dokl. Kazan', Kazan. un-t, 1975, p. 160 - 162. In Russian. Abstr. in Referativ. Zhurn. 62. Issled. kosmich. prostranstva, 1.62.187 (1976).

106.067 **Excitation and propagation of long-frequency waves in the interplanetary medium.**
N. A. Barkhatov, P. A. Bespalov, A. Sh. Kovner, M. S. Kovner, V. V. Lebedev, V. V. Mishin, T. A. Plyasova-Bakunina, V. A. Troitskaya.
XI Vses. konf. po rasprostr. radiovoln. Chast' (Part) 1. Tezisy dokl. Kazan', Kazan. un-t, 1975, p. 186 - 188. In Russian. Abstr. in Referativ. Zhurn. 62. Issled. kosmich. prostranstva, 1.62.188 (1976).

106.068 **On the estimate of the extension of the circumterrestrial dust cloud.** Sh. N. Irkaeva.
Astron. Tsirk., No. 872, p. 3 - 5 (1975). In Russian.

106.069 **The surface brightness of the northern Milky Way in the red region.** Yu. M. Zavarzin.
Astron. Tsirk., No. 875, p. 1 - 2 (1975). In Russian.

106.070 **Interplanetary scintillation and flare - produced disturbances.** B. D. Ward.
Proc. Astron. Soc. Australia, Vol. 2, 378 - 379 (1975).

106.071 **Electromagnetic processes in the circumterrestrial and interplanetary space in August 1972 during flares on the sun.** V. I. Bondarenko, A. K. Yukhimuk.
Geofiz. sb. AN USSR, 1975, vyp. (No.) 65, p. 55 - 61. In Russian. – Abstr. in Referativ. Zhurn. 62. Issled. kosmich. prostranstva, 2.62.213 (1976).

106.072 **On the anisotropies of interplanetary low-energy proton intensities.** M. E. Pesses, E. T. Sarris.
Geophys. Res. Letters, Vol. 2, 349 - 352 (1975).

Anisotropic fluxes of low-energy protons deviating from the magnetic field direction by as much as 45° during periods of time that the magnetic field vector was along the sun-earth line have been observed under certain conditions of the interplanetary medium.

106.073 **Rotational bursting of interplanetary dust particles.**
S. J. Paddack, J. W. Rhee.
Geophys. Res. Letters, Vol. 2, 365 - 367 (1975).

Solar radiation pressure can cause rotational bursting and eventual elimination from the solar system of asymmetric dust particles by a windmill effect. The life span against this process for metallic particles with radii of 10^{-5} to 10^{-2} cm ranges from 10 to 10,000 years. The effects of magnetic spin damping have been considered in this estimate.

106.074 **Fluctuations and the radial variation of the interplanetary magnetic field.** J. R. Jokipii.
Geophys. Res. Letters, Vol. 2, 473 - 475 (1975).

Correlations between fluctuations in the solar wind velocity and interplanetary magnetic field can influence the observed average interplanetary magnetic field and its variation with heliocentric radius. The average radial magnetic field still varies as $1/r^2$, but, depending on the radial variation of the correlations, the average azimuthal field could depart from the $1/r$ dependence expected for an Archimedian spiral.

106.075 **Interplanetary magnetic fields and the propagation of cosmic rays.** G. Wibberenz.
Journ. Geophys., Vol. 40, 667 - 700 (1974).

Results and limitations of the diffusion-convection model for cosmic-ray propagation in interplanetary space are presented. Magnetic field fluctuations on various scales are discussed and the non-resonant scattering at magnetic field discontinuities. In the second part, how far the diffusion model can carry one to obtain a unified view for the propagation of both galactic and solar cosmic rays is analysed.

106.076 **Coordinate transformations for studies of interactions between interplanetary and geomagnetic fields.** J. L. Burch, R. W. Janetzke.
National Aeronautics and Space Administration, Greenbelt, Md. (USA). NASA-TM-X-70684; X-621-74-179, 19 pp. (1974).

A graphical procedure is provided for performing coordinate transformations between the geocentric-solar-equatorial, geocentric-solar-ecliptic and geocentric-solar-magnetospheric coordinate systems. The procedure is designed to facilitate intercomparison of previous studies of interactions between interplanetary and geomagnetic fields that made use of these coordinate systems. The interaction in the geocentric-solar-magnetosphere system is shown to give the most consistent results.

106.077 **Interplanetary shock waves associated with solar flares.** J. K. Chao, K. Sakurai.
National Aeronautics and Space Administration, Greenbelt, Md. (USA). NASA-TM-X-70683; X-693-74-168, 48 pp. (1974).

The interaction of the earth's magnetic field with the solar wind is discussed with emphasis on the influence of solar flares. The geomagnetic storms are considered to be the result of the arrival of shock waves generated by solar flares in interplanetary space. Basic processes in the solar atmosphere and interplanetary space, and hydromagnetic disturbances associated with the solar flares are discussed along with observational and theoretical problems of interplanetary shock waves. The origin of interplanetary shock waves is also discussed.

106.078 **Out-of-ecliptic dust.** R. H. Giese.

ELDO/ESRO Sci. Techn. Rev., Vol. 7, No. 1, p. 43 - 51 (1975).

Predictions of the number density of interplanetary dust particles out of the ecliptic plane are derived from model computations by adjusting the parameters to the run of brightness of the zodiacal light in the ecliptic plane and at right angles to the sun from the ecliptic plane to the ecliptic pole. Within the limits of the simple models, surfaces of equal number density are presented.

106.079 **Measurement on magnesium isotopes in 'cosmic' dust.** T. Shimamura, O. Arai, K. Kobayashi.
Mass Spectrosc., (*Japan*), Vol. 23, 41 - 47 (1975). In Japanese.

A preliminary study of cosmic dust was carried out with a new ion counting technique for mass spectrometry. The new technique made it possible to measure the isotopic composition of magnesium in less than 0.1 μg-sized particles which had been collected in upper atmosphere or in deep sea sediment.

Measurements of the light flash produced by high velocity particle impact. See Abstr. 022.066.

Proton flares of July 1974 generating interplanetary shock waves. See Abstr. 073.119.

Columnar content measurements of the solar-wind turbulence near the sun. See Abstr. 074.001.

Interplanetary gas. XXI. Validity of the Chapman-Enskog description of the solar wind for protons.
See Abstr. 074.011.

Solar wind heating at mid-heliospheric distances due to secondary ions. See Abstr. 074.015.

Wave-trains in the solar wind. III: Alfvén waves in the azimuthally-dependent interplanetary medium.
See Abstr. 074.022.

Plasma processes in the expansion of the solar wind and in the interplanetary medium. See Abstr. 074.023.

Motion of shocks through interplanetary streams.
See Abstr. 074.059.

The effect of binary encounters on the thermal anisotropy of the protons in the solar wind.
See Abstr. 074.062.

Asymmetric distant solar wind.
See Abstr. 074.088.

An empirical relationship between interplanetary conditions and *Dst*. See Abstr. 074.116.

Solar radio emission at 9.1 cm and sector boundaries.
See Abstr. 077.041.

Fast solar electrons, interplanetary plasma and km-wave type-III radio bursts observed from the IMP-6 spacecraft. See Abstr. 077.047.

Angular distributions of solar protons and electrons.
See Abstr. 078.002.

Unusual emission of iron nuclei from the sun.
See Abstr. 078.004.

Simulation of solar flare particle interaction with interplanetary shock waves. See Abstr. 078.030.

On the earth's bow shock motion and speed.
See Abstr. 084.206.

Heos 2 plasma observations in the distant polar magnetosphere: the plasma mantle.
See Abstr. 084.207.

The calm and turbulent state of the earth's magnetosheath from December 1968 through March 1970.
See Abstr. 084.212.

Solar wind–magnetosphere coupling.
See Abstr. 084.236.

Dawn-dusk magnetosheath plasma asymmetries at 60 R_E. See Abstr. 084.254.

Bow shock protons in the lunar environment.
See Abstr. 084.278.

Artificial and real geomagnetic effects of the sectorial structure of the interplanetary magnetic field.
See Abstr. 084.289.

Complete plasma dropouts at Vela satellites during thinning of the plasma sheet. See Abstr. 084.312.

A comparison of electric and magnetic field data from the Ogo 6 spacecraft. See Abstr. 084.313.

On the sensitivity of modern inferences of interplanetary magnetic polarity to geomagnetic activity.
See Abstr. 084.319.

Die irdische Magnetopause unter Berücksichtigung einer Winkelverteilung der solaren Teilchen und des interplanetaren Magnetfeldes. See Abstr. 084.324.

Electromagnetic processes in the circumterrestrial and interplanetary space during solar flares in August, 1972.
See Abstr. 085.019.

A search for forward scattering of sunlight from lunar libration clouds. See Abstr. 094.005.

Compositional evidence regarding the influx of interplanetary materials onto the lunar surface.
See Abstr. 094.111.

Lunar remnant magnetic field mapping from orbital observations of mirrored electrons. See Abstr. 094.157.

Lunar electromagnetic scattering. 3. Propagation at arbitrary angles to the cavity axis. See Abstr. 094.178.

Preliminary results of radio sounding of the circumlunar space according to data of Luna 22.
See Abstr. 094.199.

Mass loss from the region of Mars and the asteroid belt. See Abstr. 097.041.

Accumulation of chondrules on asteroids.
See Abstr. 098.040.

Observations in interplanetary space of relativistic electrons from Jupiter. See Abstr. 099.021.

Modulation of Jovian interplanetary electrons and the longitude variation of decametric emissions.
See Abstr. 099.076.

New evidence for interplanetary boulders?
See Abstr. 102.017.

Cosmic dust in the atmosphere and in the interplanetary space at 1 AU today and in the early solar system.
See Abstr. 104.048.

The mean colour indices B–V and B–R of integral starlight in the neighbourhood of the North Galactic Pole.
See Abstr. 113.068.

Possible interaction of interstellar particles with the solar and terrestrial environment. See Abstr. 131.137.

Cosmic ray propagation in interplanetary space.
See Abstr. 143.015.

A study of days showing non-field-aligned diffusion vectors of cosmic ray diurnal anisotropy.
See Abstr. 143.019.

The interrelation of cosmic ray intensity increases on August 4 - 5, 1972 with the dynamics of the spectrum of inhomogeneities of the interplanetary magnetic field.
See Abstr. 143.025.

The interplanetary acceleration of energetic nucleons. See Abstr. 143.044.

Errata

106.901 **Correction: 'Response of the plasma sheet at ~ 18 R_E to sudden southward turnings of the interplanetary magnetic field'** [Journ. Geophys. Res., Vol. 80, 929 - 935 (1975)]. A. T. Y. Lui, E. W. Hones, Jr., D. Venkatesan, S.-I. Akasofu, S. J. Bame.
Journ. Geophys. Res., Vol. 80, 3731 (1975).

107 Cosmogony of the Planetary System

107.001 **Extinct radioactivities: trapped residuals of presolar grains.** D. D. Clayton.
Astrophys. Journ., Vol. 199, 765 - 769 (1975).

The xenon anomalies trapped in meteorites and the moon may have first been trapped in circumstellar grains formed in or outside of postexplosive stars. In that case, the initial solar nebula need not have contained most of their radioactive progenitors, and this would necessitate major revision of the history of solar system formation.

107.002 **Resonant orbits of grains and the formation of satellites.** T. Gold.
Icarus, Vol. 25, 489 - 491 (1975). – Paper presented at "Planetary Satellites" Colloquium of I.A.U., Cornell University, August 1974.

Grains, an abundant constituent of the former solar system, will have had a high probability of being driven into orbits resonant with major bodies already formed. Since the mean density in resonant orbits can be built up by such a process without limit, these may become the favored orbits for gravitational contraction to gather material into major bodies. Satellite formation processes may therefore depend upon the buildup of resonant lanes of dust grains around the parent body. Saturn's rings are possibly one example of such lanes, though an unsuitable one for the final step of satellite formation on account of their being too close to Saturn.

107.003 **The effects of tidal forces on the stability and dispersal rates of a protoplanet.**
J. R. Donnison, I. P. Williams.
Monthly Notices Roy. Astron. Soc., Vol. 172, 257 - 269 (1975).

The tidal disruption of a protoplanet by the sun is considered by both analytical and numerical means. It is concluded that the disruption process is very rapid and that this implies that most conventional protoplanetary theories for the formation of planets need to be modified otherwise complete disruption would have occurred before a refractory core could form.

107.004 **Letters of O. J. Schmidt to V. V. Radzievskij.**
Istor.-astron. issledovaniya. Vyp. (No.) 12. Moskva Nauka, 1975, p. 369 - 383. In Russian. – Abstr. in Referativ. Zhurn. 51. Astron., 7.51.28 (1975).

107.005 **Cosmogony of the solar system.** W. R. Ward.
Rev. Geophys. Space Phys., Vol. 13, No. 3, (see 013.001), p. 422 - 424, 427 - 430 (1975). – Review paper with bibliographical notes.

This report is intended to cover research on solar system origin over the last 4 years. Emphasis has been placed on the physics of planetary accumulation mechanisms, although pertinent references on the chemistry of mineral condensation and cometary research are included.

107.006 **The origin and evolution of the solar system.**
A. G. W. Cameron.
Sci. American, Vol. 233, No. 3, p. 32 - 41 (1975). – Popular article.

107.007 **A model of evolution of the planets of the terrestrial group.** V. P. Keondzhyan, A. S. Monin.
Dokl. Akad. Nauk SSSR. Ser. Mat., Fiz., Vol. 223, 599 - 602 (1975). In Russian.

107.008 **Die Entstehung des Planetensystems.**
W. H. McCrea.
Nova Acta Leopoldina [Deutsche Akademie der Naturforscher Leopoldina, Halle/Saale], New Ser., Vol. 42 (No. 218), 135 - 144 (1975). – Lecture on the occasion of the annual meeting of the Academy, 1973 October 11 - 14.

107.009 **Is it possible to clear up the real history of the solar system?** S. K. Vsekhsvyatskij.
Problems of cosmic physics. Vyp. (No.) 10, (see 003.003), p. 96 - 103 (1975). In Russian.

107.010 **On Hoyle's theory of the origin of the solar system.**
N. Dallaporta, L. Secco.
Astrophys. Space Sci., Vol. 37, 335 - 364 (1975).

A quantitative re-formulation of Hoyle's theory on the formation of the solar system is attempted, according to a three-dimensional scheme based on the assumption that the original magnetic field of the star is a dipole field. This allows to obtain analytic expressions for the main parameters describing the different phases of the process. The protostar is assumed to evolve in gravitational contraction along the Hayashi track, along which, for a given value of the radius depending on the total angular momentum, matter begins to be shed at the equator as a consequence of centrifugal instability. The quantitative results are discussed in order to see whether it is possible to deduce the main characteristics of the disk, from which the solar system should have originated, for reasonable values of the main parameters.

107.011 **The formation of the outer planets.**
M. J. Handbury, I. P. Williams.
Astrophys. Space Sci., Vol. 38, 29 - 37 (1975).

One of the outstanding problems in planetary cosmogony is to account for the depletion of hydrogen in the outer planets, Neptune and Uranus. It is suggested that these planets were originally similar to the major planets but that the settling towards the centre of grains, enriched by substances such as methane, ammonia and water because of the low temperatures, released enough energy to cause the evaporation of most of the hydrogen.

107.012 **The early evolution of the solar system.**
A. G. W. Cameron.
I.A.U. Colloquium No. 13, (see 012.030), p. 347 - 353 (1973).

The problems of relating collapse conditions in an interstellar cloud to a model of the primitive solar nebula are discussed. In such a nebula there is a radial force balance between gravity, the pressure gradient, and centrifugal forces due to the rotation.

107.013 **Radial pressure in the solar nebula as affecting the motions of planetesimals.** F. L. Whipple.
I.A.U. Colloquium No. 13, (see 012.030), p. 355 - 361 (1973).

107.014 **Origin of the solar system: problems and perspectives.**
Bibhas R. De.
Bull. Astron. Soc. India, Vol. 3, 44 - 49 (1975).

107.015 **A proto-planetary contribution to interstellar dust.**
D. A. Mendis, N. C. Wickramasinghe.
Astrophys. Space Sci., Vol. 38, L13 - L17 (1975).

It is argued that a significant mass fraction of interstellar dust might be formed in proto-planetary particle streams. Some astrophysical and cosmogonic consequences are discussed.

107.016 **Die innere Struktur der Erde, des Mondes und der Planeten.** H. Stiller.
Sitzungsber. Akad. Wiss. DDR, 1974 No. 13, 32 pp. (1975).

107.017 **The application of rare-earth elements to infer gas-grain temperature differentials in the solar nebula.**
W. V. Boynton.
Meteoritics, Vol. 10, 369 - 371 (1975).

Structure and evolutionary history of the solar system. See Abstr. 003.013.

The role of nuclear reactions in the evolution of matter of the solar system. See Abstr. 061.051.

Are the cosmic OH and H_2O sources protostars or protoplanets? See Abstr. 065.066.

The seven ages of a planet. See Abstr. 091.023.

The production of atmospheres by impact. See Abstr. 091.036.

Possible origin of the lunar bombardment. See Abstr. 094.179.

The angular momenta of solar system bodies: implications for asteroid strengths. See Abstr. 098.013.

Application of the theory of jet stream to the asteroidal belt. See Abstr. 098.043.

A numerical study of a gravitating system of colliding particles: applications to the dynamics of Saturn's rings and to the formation of the solar system. See Abstr. 100.001.

Comets and planets: their interrelated origin. See Abstr. 102.027.

Oxidation state of iron in the solar nebula. See Abstr. 105.143.

Ein magnetohydrodynamisches Modell der Entstehung von Planetensystemen. See Abstr. 117.009.

Peculiar A-stars and planetary systems. See Abstr. 117.010.

Stars

111 Stellar Parallaxes

111.001 **Parallaxes of 18 dwarf members of the Hyades cluster.**
A. R. Klemola, E. A. Harlan, B. McNamara, C. A. Wirtanen.
Astron. Journ., Vol. 80, 642 - 646 = Lick Obs. Bull., No. 705 (1975).

Trigonometric parallaxes measured for 18 dwarf members of the Hyades with magnitudes 9–13 in 16 fields lead to a weighted mean parallax of 0.0230 ± 0.0015 (m.e.) arcsec and distance modulus of 3.19 ± 0.15 (m.e.) mag.

111.002 **Trigonometric parallaxes determined with the Yerkes Observatory 40-in. refractor. IV.**
W. F. van Altena, E. U. Vilkki.
Astron. Journ., Vol. 80, 647 - 651 (1975).

Parallaxes and proper motions based on recent observations with the Yerkes Observatory's 40-in. refractor are presented for 12 stars. Two objects of particular interest are included: G5-28 = EG21b is a supposed red degenerate star based on a photometric parallax of its companion. However, the newly determined parallax of G5-28 casts doubt on its physical association with G5-27. BD+41°3306 is the first subdwarf to have its parallax determined in the new Yerkes subdwarf parallax program. BD+41°3306 is a high ultraviolet excess subdwarf which appears in Sandage's (1970) group D category.

111.003 **The historical search for stellar parallax.**
J. D. Fernie.
Journ. Roy. Astron. Soc. Canada, Vol. 69, 153 - 161 (1975).

111.004 **Trigonometric parallaxes with reflectors.**
K. A. Strand.
Proc. Astrometric Conference, Torino 1974, (see 012.017), p. 21 - 26 (1975).

111.005 **Parallaxes of luminous red stars.** A. R. Upgren.
Astron. Journ., Vol. 80, 828 - 830 (1975).

Three luminous red stars, BN Mon, R Leo, and χ Cyg, have been remeasured for parallax at the Van Vleck Observatory. These are among a total of 23 carbon stars, S stars, and M stars of luminosity class II and brighter for which parallaxes have been found. The mean and distribution of their parallaxes are consistent with distances indicated by their luminosities determined by other methods.

111.006 **Photographic parallaxes of nineteen stars obtained through the central overlap reduction routine.**
H. Eichhorn.
Monthly Notices Roy. Astron. Soc., Vol. 173, 619 - 624 (1975).

Relative trigonometric parallaxes of 19 stars were determined from plates taken at the Van Vleck Observatory. The reductions were carried out based on a statistically rigorous least squares algorithm which makes better use of the available measurements than the conventional dependence method or the standard plate constant method.

111.007 **Parallaxes and proper motions. X.**
A. R. Upgren, E. W. Weis.
Astron. Journ., Vol. 80, 711 - 713 (1975).

Relative parallaxes and proper motions are given for 21 stars, six of which have no previous parallax determination.

111.008 **The historical search for stellar parallax.**
J. D. Fernie.
Journ. Roy. Astron. Soc. Canada, Vol. 69, 222 - 239 (1975).

111.009 **A catalogue of photometric parallaxes of eclipsing binaries.** T. Z. Dworak.
Acta Astron., Vol. 25, 383 - 416 (1975).

On the basis of Gaposchkin's method (1938, 1940) of determination of parallaxes for eclipsing binaries a modified method of photometric parallax determination is derived in the present paper. The modification concerns the mass-luminosity relation. A new, semi-empirical expression for the mass-luminosity relation has been obtained which allows the computation of absolute magnitudes from about $+16^m$ to about -7^m. The best results are obtained for luminosity class V. The modified and approximate methods were then tested on the basis of 44 eclipsing binaries with known parallaxes and the obtained results are satisfactory. The methods were also tested on the basis of 340 eclipsing systems with known normalized brightness of components, L_1 and L_2. As a main result this paper presents a catalogue of absolute magnitudes and parallaxes for 1038 eclipsing binaries.

111.010 **Trigonometric parallaxes – basic data for various problems.** W. Gliese.
Space astrometry, (see 012.037), p. 109 - 113 (1975).

To date trigonometric parallaxes of somewhat more than 7000 stars have been measured, but only for 5% of them are the distances known with probable errors of less than 10%. Parallax work proposed by Lacroute could allow the calibration of luminosity relations, and an accurate determination of the Hyades parallax (an important basic distance for estimating the absolute magnitudes of giants and supergiants). A number of faint-star parallaxes would allow the investigation of both the white-dwarf problems and the red-dwarf population in the vicinity of the sun.

111.011 **Some astrophysical consequences of an improvement in parallaxes.** C. Turon Lacarrieu.
Space astrometry, (see 012.037), p. 115 - 120 (1975).

The projects of space astrometry should yield an important improvement in trigonometric parallaxes for a very large number of stars. Three points are to be emphasised: (i) the limiting apparent magnitude will be much larger, the measurements of trigonometric parallaxes will then extend to the coldest stars and allow calibrations for all types of stars; (ii) the parallax measurements should be much more accurate; this should allow us to reach larger distances from the sun; (iii) many more stars will be measured; this would allow improved sampling of the surrounding stars.

Space astrometry projects. See Abstr. 032.529.

Astrometric studies of the regions BD+17°4946, BD–5°3071, and Ross 730–1. See Abstr. 041.005.

Third catalog of trigonometric parallaxes of faint

stars. See Abstr. 041.035.

Note on an astrometric study of two nearby visual binaries of high space velocity: ADS 9544 and Kuiper 90. See Abstr. 118.004.

Astrometric orbit, eclipsing period changes, and parallax of VW Cephei. See Abstr. 121.021.

Stellar kinematics and the cosmic distance scale. See Abstr. 153.036.

Kinematical properties of young and old disk dwarfs determined from new parallaxes. See Abstr. 155.059.

112 Proper Motions, Radial Velocities, Space Motions

112.001 **Statistical properties of the sample "proper motions of the AGK 3".** W. Dieckvoss.
Mitt. Astron. Ges., No. 36, p. 157 - 159 (1975).

112.002 **Radial velocities of southern early-type stars at intermediate galactic latitudes.**
D. Kilkenny, P. W. Hill.
Monthly Notices Roy. Astron. Soc., Vol. 172, 649 - 665 (1975).

Radial velocities for 68 early-type stars at intermediate galactic latitudes are presented. The velocity-longitude diagram indicates that most of the OB stars out of the plane participate in galactic rotation. An apparent deviation from circular galactic rotation already reported for Ca II, H I and H II regions near the plane is also present in the intermediate latitude observations of interstellar Ca II and is probably due to large scale motions in the $-$I arm. Several high-velocity and run-away stars are briefly discussed.

112.003 **The proper motions of Struve double stars.**
L. Fischer.
Mem. Roy. Astron. Soc., Vol. 80, 93 - 103 (1975).

The proper motions of the components of 103 Struve double stars from Muller's list of neglected stars and of 15 other systems were measured relative to sets of reference stars, for the primary purpose of distinguishing physical from optical pairs. A by-product of this investigation is a list of measured position angles and distances of these objects, at two epochs each.

112.004 **Proper motions of 15 variable stars.**
D. K. Karimova, E. D. Pavlovskaya, V. L. Shtaerman.
Peremennye Zvezdy, Vol. 19, 401 - 402 (1974). In Russian.

112.005 **Reduction of proper motions of stars in the Kapteyn Selected Area 32.** A. V. Bolbochanu.
Astron. Tsirk., No. 864, p. 5 - 7 (1975). In Russian.

112.006 **The proper motion and orbit of HZ Her.**
D. K. Karimova, E. D. Pavlovskaya.
Astron. Tsirk., No. 868, p. 4 - 5 (1975). In Russian.

112.007 **Proper motions of stars in 25 selected regions of the sky.** A. G. Rakhimov.
Fan, Tashkent. 72 pp. Price 80 Kop. (1975). In Russian.
Abstr. in Referativ. Zhurn. 51. Astron., 9.51.676 (1975).

112.008 **Absolute proper motions of 41 Ap and Am stars.**
E. D. Pavlovskaya, D. K. Karimova.
Soobshch. Gos. Astron. Inst. Shternberga, No. 182, p. 14 - 17 (1973). In Russian.

112.009 **Proper motions of faint blue stars near the galactic anticenter. II.** K. M. Cudworth.
Astron. Journ., Vol. 80, 826 - 827 (1975).

Proper motions with standard errors of ± 0.″8/cent. have been measured for additional faint blue stars found by Rubin et al. (1974) in a field near the galactic anticenter. The absence of detectable motion for all the stars on the present list implies that most, if not all, are very distant.

112.010 **Radial velocities of 19 G, K-, and M-type stars.**
M. B. K. Sarma.
Bull. Astron. Soc. India, Vol. 3, 55 - 56 (1975).

112.011 **On the velocity dispersion of nearby M dwarfs.**
K. Krisciunas.
Publ. Astron. Soc. Pacific, Vol. 87, 699 - 703 = Lick Obs. Bull., No. 715 (1975).

It is shown that the velocity dispersion of a large sample of stars can be determined by their distribution on the sky as a function of angular distance from the solar apex. Known or suspected M dwarfs from Gliese's catalog were divided into groups according to distance from the sun, angular distance from the solar apex, whether or not the stars had emission lines, and whether or not they were members of binary or multiple systems. It was found that with this method only a lower limit of the transverse velocity dispersion could be derived. This value was 13 km sec^{-1}. Based on the present technique and the presently available data, the existence of low-velocity M dwarfs cannot be ruled out.

112.012 **The proper motion of CD $-23°$14002.**
G. L. Grasdalen.
Publ. Astron. Soc. Pacific, Vol. 87, 831 (1975).

The large proper motion of the apparent runaway star CD $-23°$14002 is shown to be spurious. The proper motion of this early B star is consistent with its membership and origin in Simeis 188.

112.013 **Proper motions and positions of stars in wide surroundings of the Hyades cluster.**
N. M. Artyukhina, P. N. Kholopov.
Trudy Gos. Astron. Inst. Shternberga, Vol. 46, vyp. (No.) 2,

p. 57 - 121 (1975). In Russian.

Photographic proper motions for 3481 stars brighter than $12^m.5-13^m.0$ pg are derived in three regions of the sky near the Hyades cluster. The proper motions were derived using Astrographic Catalogues (Carte du Ciel) and three plates taken with the wide-angle astrograph of the Sternberg Astronomical Institute in Moscow. The mean square error for one component of proper motion is $\pm 0''.007$. A catalogue of positions and relative proper motions of the stars considered is given. Probable members of the cluster are indicated in a table.

112.014 **HD 134439/40: high space velocity stars.**
J. L. Russell.
Bull. American Astron. Soc., Vol. 7, 510 (1975). – Abstr. AAS.

112.015 **Stars with especially high velocities.**
R. B. Shatsova.
Astron. Tsirk., No. 871, p. 3 - 5 (1975). In Russian.

112.016 **Spectroscopy of five faint M dwarf stars near the north galactic pole.** D. C. Koo, R. G. Kron.
Publ. Astron. Soc. Pacific, Vol. 87, 885 - 889 (1975).

Radial velocities and estimates of absolute magnitudes are presented for five M dwarfs selected from a proper motion study of M stars near the north galactic pole by Murray and Sanduleak. The derived mean distance and velocity dispersion are significantly larger than the corresponding quantities found by Murray and Sanduleak, but the authors have only a small sample limited to the earlier-type stars on their list.

112.017 **Proper Motion Survey with the forty-eight inch Schmidt telescope. XL. Five regions in high galactic latitude.** W. J. Luyten.
Separate print Univ. Minnesota, Minneapolis, Minnesota, 47 pp. (1975).

112.018 **On the proper motion of AT Cancri.**
W. J. Luyten.
Separate print Univ. Minnesota, Minneapolis, Minnesota, June 1975, p. 47.

112.019 **Proper Motion Survey with the 48-inch Schmidt telescope. XLI. Further note on the inaccuracy of the proper motions determined with the Lick automatic measuring machine.** W. J. Luyten.
Separate print Univ. Minnesota, Minneapolis, Minnesota, 4 pp. (1975).

112.020 **Proper Motion Survey with the forty-eight inch Schmidt telescope. XLII. Proper motions for 848 faint stars.** W. J. Luyten.
Separate print Univ. Minnesota, Minneapolis, Minnesota, 8 pp. (1975).

112.021 **Proper Motion Survey with the forty-eight inch Schmidt telescope. XLIII. One hundred and six faint stars with large proper motion.**
W. J. Luyten, C. T. Kowal.
Separate print Univ. Minnesota, Minneapolis, Minnesota, 2 pp. (1975).

112.022 **Le velocità spaziali delle stelle entro 25 parsecs dal sole.** F. Zagar.
Atti Accad. Nazionale Lincei, Ser. Ottava, Rend., Cl. Sci. fis., mat. nat., Vol. 57, 199 - 203 = Contr. Oss. Astron. Milano-Merate, Nuova Ser. No. 367 (1974).

112.023 **Note on radial velocities of southern stars.**
A. D. Thackeray, with a reply by F. Ochsenbein.
Centre de Données Stellaires, Inform. Bull. No. 9, (see 002. 008), p. 3 - 4 (1975).

Radial velocity and aberration.
See Abstr. 031.204.

A short program for the Besselian C and D.
See Abstr. 031.205.

Velocity selection of cluster members with use of objective-prism plates. See Abstr. 031.222.

Four years experience in APMS star plate processing – results and future plans. See Abstr. 031.415.

Mise en service d'un prisme objectif normal de 60 cm de diamètre. Découverte d'une étoile du halo à grande vitesse radiale. See Abstr. 034.094.

Astrometric studies of the regions BD+17°4946, BD–5°3071, and Ross 730–1. See Abstr. 041.005.

Catalog of proper motions for the 5,965 stars of the six-inch transit circle program 1949 – 1956.
See Abstr. 041.029.

Investigation of declinations and of proper motions of the Belgrade General Catalogue of Latitude Stars BSKŠZ1 and BSKŠZ2. See Abstr. 041.068.

Declinations and the proper motions of the stars of the International Latitude Service on the basis of meridian catalogues from 1929 to 1972. See Abstr. 041.069.

Trigonometric parallaxes determined with the Yerkes Observatory 40-in. refractor. IV.
See Abstr. 111.002.

Parallaxes and proper motions. X.
See Abstr. 111.003.

Physical limits to attainable accuracies in stellar radial velocities. See Abstr. 114.048.

On the variations of radial velocities of P Cygni.
See Abstr. 114.082.

Radial velocities of stars with microwave maser emission. I. Long-period variables. See Abstr. 122.183.

Acceleration of pulsars by asymmetric radiation.
See Abstr. 141.331.

Proper motions of two galactic clusters in Cassiopeia and of the classical cepheid CG Cas. See Abstr. 153.021.

Relative proper motions in the open cluster NGC 654. See Abstr. 153.033.

Proper motions and BV photometry of NGC 129 and field stars. See Abstr. 153.037.

A spectroscopic investigation of the Pleiades.
See Abstr. 153.041.

Relative proper motions in M15.
See Abstr. 154.031.

Galactic rotation and local irregularities derived from fundamental proper motions and radial velocities.
See Abstr. 155.022.

Le mouvement des étoiles dans la Galaxie.
See Abstr. 155.031.

113 Stellar Magnitudes, Colors, Photometry

113.001 **Standard stars for photoelectric Hα line photometry.** J. Dachs, T. Schmidt-Kaler.
Astron. Astrophys., Suppl. Ser., Vol. 21, 81 - 97 (1975).

100 bright equatorial and southern stars of spectral types O, B, A, F to G0 and of luminosity classes I through V have been measured with a photoelectric photometer equipped with two interference filters of 13 Å and 157 Å bandwidth centered on Hα. The stars are evenly distributed in right ascension and intended to provide a standard system of Hα line photometry. The observing technique and the method used in establishing the standard system are described.

113.002 **Infrared photometry in the R CrA association.** I. S. Glass, M. V. Penston.
Monthly Notices Roy. Astron. Soc., Vol. 172, 227 - 233 (1975).

JHKL photometry of 44 stars in the area of the R CrA association is reported. No previously undiscovered cases of dust emission were found but several highly reddened stars are identified. These show the dark cloud to have a visual absorption of ~8 mag. The ratio E_{J-H}/E_{H-K} is suspected to be higher in the cloud than predicted by traditional reddening laws. Detailed comments on several stars in the region are given.

113.003 **Zur Transformation ins UBV-Standardsystem.** U. Haug.
Mitt. Astron. Ges., No. 36, p. 88 (1975). – Abstract.

113.004 **Lichtelektrische Beobachtungen quasiperiodischer T-Tau-Sterne.** H. Mauder.
Mitt. Astron. Ges., No. 36, p. 98 (1975). – Abstract.

113.005 **Physikalische Parameter und interstellare Absorption von V 1057 Cyg.**
L. Kohoutek, R. Meinhardt.
Mitt. Astron. Ges., No. 36, p. 101 - 102 (1975).

113.006 **Helligkeiten und Farben von 1030 frühen M-Sternen in der nördlichen Milchstraße.** H. Neckel.
Mitt. Astron. Ges., No. 36, p. 160 - 166 (1975).

113.007 **Photographische UBV-Photometrie von 3 Sternfeldern im Cygnus.** H. J. Hartl.
Mitt. Astron. Ges., No. 36, p. 170 - 171 (1975).

113.008 **Fotografische Fotometrie in Gebieten hoher Sterndichte und mit variablem Hintergrund.**
R. Hering.
Mitt. Astron. Ges., No. 36, p. 171 - 172 (1975).

113.009 **Photometry of O-type stars, interpreted with reference to model atmospheres.** N. D. Morrison.
Astrophys. Journ., Vol. 200, 113 - 121 (1975).

Strömgren-system colors for O-type stars have been compared with colors calculated by Mihalas from non-LTE model atmospheres and by Mihalas and Hummer from models with extended geometry. Within the accuracy of the observations, effective temperatures derived from the photometry for main-sequence stars agree with those derived by Conti from a ratio of intensities of spectral lines; consistency of the calculated colors with the calculated strengths of the lines is demonstrated for the non-LTE plane-parallel models.

113.010 **The infrared excess of cool giant stars: a chromospheric contribution.**
D. L. Lambert, R. L. Snell.
Monthly Notices Roy. Astron. Soc., Vol. 172, 277 - 288 (1975).

The idea that the infrared excesses of evolved M stars may contain a contribution from a chromosphere is explored with α Ori and W Hya as test cases. The α Ori spectrum betweeen 8 and 30 μm can be interpreted satisfactorily in terms of three components: a photosphere, a silicate dust cloud and a cool chromosphere, $T \sim 5000$ K, which is optically thick at 14 μm. A similar modelling for W Hya suggests a hotter chromosphere, $T \sim 8000$ K, with unit optical depth at 30 μm. Some consequences of these chromospheres are briefly discussed.

113.011 **A new photometric metal abundance and luminosity calibration for field G and K giants.**
P. A. Jennens, H. L. Helfer.
Monthly Notices Roy. Astron. Soc., Vol. 172, 667 - 679 (1975).

Photometry of 260 G and K giants, using a fast broad-intermediate band photometric system (*UBViyz* system) is used to calibrate chemical composition, [Fe/H], luminosity, M_V and colour excess, $E(B-V)$.

113.012 **Metal abundances for G and K giants and subgiants at the South Galactic Pole.** P. A. Jennens.
Monthly Notices Roy. Astron. Soc., Vol. 172, 695 - 699 (1975).

[Fe/H], Mv and $E(B-V)$ are determined for 72 G and K stars at the South Galactic Pole for which *UBViyz* photometry has been obtained. Away from the galactic plane the population is dominated by old, low luminosity subgiants. A strong gradient of [Fe/H] with distance from the galactic plane is detected.

113.013 **Photometry of FG Sagittae.** J. D. Fernie.
Astrophys. Journ., Vol. 200, 392 - 394 (1975).

BVRI photometry of FG Sge on seven consecutive nights in 1974 September is reported. The star was definitely variable, but the observations are too concentrated in time to allow suitable comparison with those of Papousek – although, if anything, the present results indicate more rapid variability than found by Papousek. The mean $(B-V)$ of the star has now reached 1.41, which continues the linear trend of $(B-V)$ changes in recent years. The position of the star on the $(B-V)$ versus $(R-I)$ diagram is peculiar: dereddening the star in the usual way does not place it on the intrinsic sequence of supergiants in this diagram.

113.014 **Infrared observations of the association Cyg OB 2.** K. Voelcker.
Astron. Astrophys., Suppl. Ser., Vol. 22, 1 - 7 (1975).

This paper gives observational data of an investigation in the association Cyg OB 2. It is the most highly reddened association in the northern sky with the 1950.0 central coordinates $\alpha = 20^h31^m5$, $\delta = +41°16'$ or $l^{II} = 80°3$, $b^{II} = +0°8$ having an angular diameter of about 0°8 corresponding to ≈30 pc in the distance of 2 kpc. All infrared photometric material for 103 stars is presented.

113.015 **DDO photometry of K giants in Baade's window.** K. A. Janes.
Bull. American Astron. Soc., Vol. 7, 435 (1975). – Abstr. AAS.

113.016 **UBVRI photometry of Ba II stars.** P. A. Ianna, R. B. Culver.
Bull. American Astron. Soc., Vol. 7, 445 (1975). – Abstr. AAS.

113.017 **Separation of main-sequence and supergiant stars of early B type.** I. Furenlid, W. Sanders.
Bull. American Astron. Soc., Vol. 7, 446 (1975). – Abstr. AAS.

113.018 **Pi Aquarii: a pulsating Be star?** J. D. Fernie.
Astrophys. Journ., Vol. 201, 179 - 182 (1975).

Following the discovery by Ringuelet and Machado of large periodic radial-velocity variations in π Aqr, photometric observations of the star have been obtained. It is found that the star has undergone small but significant secular changes in brightness and color since observations made 11 years earlier, and that it also shows instabilities of the order of several hundredths of a magnitude on a time scale of hours or less. Several possible explanations, such as rotation, presence of a binary companion, varying emission lines, and pulsation, are considered. Of these, only pulsation seems plausible.

113.019 **Infrared photometry and polarimetry of cool stars. II.** G. V. Khozov, T. N. Khudyakova, L. V. Larionova.
Trudy Astron. Obs., *Leningrad*, Vol. 31 (= Uchenye Zapiski Leningr. Un-ta, No. 381 = Seriya Matem. Nauk, vyp. (No.) 51), p. 123 - 146 (1975). In Russian.

The results of three-colour infrared photometry of 26 stars in the I, H and K bands are given. Polarimetric observations are reported for 7 of these stars (NML Tau, NML Cyg, IRC+10216, VY CMa, BC Cyg, S Per, χ Cyg). A brief discussion of the results is given.

113.020 **Four-color and Hβ photometry of O-type stars.** D. L. Crawford.
Publ. Astron. Soc. Pacific, Vol. 87, 481 - 494 (1975).

Kitt Peak data on O-type stars have been combined with those of Morrison (1975) to derive interstellar reddening slopes for the *uvby* system. The results are $E(c_1) = 0.20$ $E(b-y)$, $E(m_1) = -0.32\ E(b-y)$, $E(u-b) = 1.5\ E(b-y)$, and $E(b-y) = 0.74\ E(B-V)$.

113.021 **Some photometric characteristics of young stars.** E. E. Mendoza V.
Publ. Astron. Soc. Pacific, Vol. 87, 495 - 497, with a correction, p. 836 (1975). – Paper presented at the Ensenada meeting of the Astronomical Society of the Pacific, 1975 May 12 - 15.

113.022 **Simultaneous visual and infrared monitoring of young stars.** M. Cohen, R. D. Schwartz.
Publ. Astron. Soc. Pacific, Vol. 87, 500 (1975). – Abstr. ASP.

113.023 **Narrow-band photometry of A–F stars.** E. E. Mendoza V.
Publ. Astron. Soc. Pacific, Vol. 87, 505 - 506 (1975).
Abstr. ASP.

113.024 **Be stars with large magnitude changes.** A. Feinstein.
Publ. Astron. Soc. Pacific, Vol. 87, 603 - 605 (1975).

Photoelectric *UBV* observations of 19 Be stars are listed which in a time interval of eleven years exhibit changes in V larger than $\Delta V > 0^m15$. Their light curves display very different shapes; no general rule for the light variation is evident.

113.025 **Catalogue of stellar magnitudes and colour indices of 3937 stars in the surroundings of SA 26.**
S. S. Tokhtas'ev.
Izv. Astron. Ehngel'gardt. Obs., *Kazan'*, No. 39, p. 54 - 88 (1972). In Russian.

113.026 **Photometric system of the Schmidt telescope of the Engelhardt Astronomical Observatory and its connection with the UBV-system.**
S. S. Tokhtas'ev, I. A. Dubyago.
Izv. Astron. Ehngel'gardt. Obs., *Kazan'*, No. 40, p. 80 - 85 (1973). In Russian.

113.027 **On Johnson's Q-method.** S. S. Tokhtas'ev.
Izv. Astron. Ehngel'gardt. Obs., *Kazan'*, No. 40, p. 86 - 90 (1973). In Russian.

113.028 **Three-colour photometric system on a single plate.** L. A. Urasin.
Izv. Astron. Ehngel'gardt. Obs., *Kazan'*, No. 40, p. 103 - 107 (1973). In Russian.

113.029 **Strömgren four-colour observations of northern hemisphere binary systems.**
R. W. Hilditch, G. Hill.
Mem. Roy. Astron. Soc., Vol. 79, 101 - 129 (1975).

A total of 1161 *uvby* observations are presented of 308 close binary systems taken largely from 'A Catalogue of Graded Photometric Studies of Close Binaries' (Koch, Plavec & Wood) and 'A Finding List for Observers of Eclipsing Variables' (Koch, Sobieski & Wood). All the northern stars in the former catalogue are represented.

113.030 **The C-classification and infrared color of carbon stars.** Y. Yamashita.
Publ. Astron. Soc. Japan, Vol. 27, 459 - 467 (1975).

It is shown that a large scatter found in the plots of color temperature of carbon stars deduced from infrared color versus the C-type can be interpreted by the blocking effect due to absorption bands of CN and by interstellar reddening.

113.031 **Pulsation in peculiar A stars.** J. R. Percy.
Astron. Journ., Vol. 80, 698 - 701 (1975).

In order to study the incidence of pulsation in peculiar A (Ap) stars, 15 cool stars of this type were observed for short-period light variations. Only two stars showed significant variations; both were previously suspected to be variable, HD 224801 (HR 9080) showed irregular variations, in blue and yellow light, of up to $0^m.02$. 21 Comae showed regular variations, in blue and yellow light, with an amplitude which varied from $0^m.015$ to $0^m.005$. The primary period was 30 min; a secondary period of 40 min was found in the present observations and also in earlier ones. The present results, and previous photometric observations, indicate that pulsation is rare and/or of very low amplitude in Ap stars which lie in the δ Scuti instability strip.

113.032 **Catalogue of faint O–B5 stars in Centaurus (l = 307°, b = 0°).** C. H. McGruder III.
Astron. Astrophys., Suppl. Ser., Vol. 22, 161 - 170 (1975).

A catalogue of 417 O–B5 stars situated in a field of 5 square degrees in Centaurus (l = 307°, b = 0°) is given. The stars have been selected by a photographic *UBV* photometry. Additionally, individual data for stars belonging to two spatially concentrated groups of B3–B9 stars in the same field are given.

113.033 **Photoelectric multiband photometry of M giants and supergiants in the blue-visual spectral region.**
B. E. Helt, K. Gyldenkerne.
Astron. Astrophys., Suppl. Ser., Vol. 22, 171 - 192 (1975).

Photoelectric photometry of M stars has been carried out in nine narrow bands in the range from 4050 to 5500 Å, and various classification indices based on these observations are discussed. It is shown that six bands define two parameters which clearly separate the spectral type and the luminosity and determine these quantities with an accuracy of about 0.1 subclass and 0.5 mag., respectively. The spectral type parameter is in good agreement with the TiO index measured by Wing, and the luminosity parameter agrees well with $M_V(K)$ determined by the Wilson-Bappu method.

113.034 **$uvby\beta$ photoelectric photometric catalogue (magnetic tape).** B. Hauck, M. Mermilliod.
Astron. Astrophys., Suppl. Ser., Vol. 22, 235 - 237 (1975).
A catalogue on magnetic tape containing the published and homogeneous data in the $uvby\beta$ system is presented. It concerns 9407 stars, being the data published to end 1974.

113.035 **Catalogue of measurements in the Cape *UBV* photometry on magnetic tape.** B. Nicolet.
Astron. Astrophys., Suppl. Ser., Vol. 22, 239 - 241 (1975).
All photoelectric measurements in the Cape *UBV* photometry have been compiled and homogenized on magnetic tape. This tape is available at the Centre de Données Stellaires at Strasbourg.

113.036 **Photoelectric investigation of the four Ap stars HD 9531, HD 10221, HD 27309 and HD 184905.**
G. Hildebrandt.
Astron. Nachr., Vol. 296, 277 - 280 (1975). In German.
The light curves in the UBV-system are given for the four Ap stars HD 9531, HD 10221, HD 27309 and HD 184905 from 1968 to 1971. The measurements are represented by the periods $0^d.6684$, $3^d.16$, $2^d.7098$ and $1^d.845031$ respectively.

113.037 **Photometric systems of the catalogues used for star counts.** A. S. Sharov, G. I. Polyakova.
Soobshch. Gos. Astron. Inst. Shternberga, No. 177, p. 3 - 12 (1973). In Russian.
The magnitudes from the Mount Wilson, Bergedorf and Harvard-Groningen catalogues are compared with the magnitudes in the *B* system. The results are shown in figures.

113.038 **The temperature effect in photoelectric photometry of strong emission-line objects.**
V. P. Arkhipova.
Soobshch. Gos. Astron. Inst. Shternberga, No. 182, p. 24 - 26 (1973). In Russian.
A strong dependence of the *V* magnitudes on the photometer temperature was found in the photoelectric observations of the planetary nebula IC 4997. The temperature effect is due to the variations of *V* filter transparence in the N_1, N_2 emission-line region.

113.039 **A preliminary report on the light variation of the extreme A supergiant CD-33° 12119 (=3U1727 -33?).** A. M. van Genderen, G. Hammerschlag-Hensberge.
Inform. Bull. Variable Stars, (I.A.U. Commission 27), Konkoly Obs., Budapest, No. 1048, 3 pp. (1975).

113.040 **Short-periodic light variations in Wolf-Rayet stars.** H. Lindgren, I. Lundström, B. Stenholm.
Astron. Astrophys., Vol. 44, 219 - 222 (1975).
Three Wolf-Rayet stars were investigated for short-periodic light variations by means of narrow band photometry. Both emission lines and the continuum were observed. No stable variation in the period range 1–10 min was found. The only star observed on several nights, γ^2 Vel, showed definite variations in the emission line strength from night to night.

113.041 **Observations of T Tauri stars and related objects in the Taurus dark cloud.** G. Romano.
Mem. Soc. Astron. Italiana, Vol. 46, 81 - 95 (1975).
This paper gives results of the observation of thirty-two T Tauri and T Tauri-like stars, and six variables of other types in a field around φ Tauri. The mean photovisual amplitude of nine T Tau stars is $1^m.19$ and the mean magnitude is 12.79 while the amplitude of other irregular variables is $1^m.29$ and the mean magnitude 13.62. The characteristics of the light variation are given with the magnitude frequency distribution. Some variables show rapid variations in few hours.

113.042 **List of 506 stars, probable Small Magellanic Cloud members.**
M. Azzopardi, J. Vigneau, M. Macquet.
Astron. Astrophys. Suppl. Ser., Vol. 22, 285 - 321 (1975).
An objective-prism survey in the direction of the Small Magellanic Cloud has enabled the authors to detect 506 stars which show high luminosity spectral characteristics; 193 of them have been confirmed as SMC members by other authors. 1975 coordinates and MK spectral types for all stars, *V* magnitudes, $(B-V)$, $(U-B)$ colour indices and remarks for most of them are presented. A master set and 16 identification astrograph charts are provided.

113.043 **$uvby$ photometry of some southern field stars.** A. Heck, J. Manfroid.
Astron. Astrophys. Suppl. Ser., Vol. 22, 323 - 325 (1975).
Photometric data in the $uvby$ system are given for 38 southern field stars.

113.044 **The ESO/Uppsala survey of the ESO (B) atlas of the southern sky. III.**
E. B. Holmberg, A. Lauberts, H.-E. Schuster, R. M. West.
Astron. Astrophys. Suppl. Ser., Vol. 22, 327 - 402 (1975).
A systematic search for certain objects (NGC + IC galaxies, all galaxies with a diameter larger than about $1'.0$, all disturbed galaxies, all star clusters in the Budapest Catalogue, and all listed planetary nebulae) is being carried out by means of the ESO (B) atlas, covering the southern sky from $-90°$ to $-20°$. The present paper contains the third list of objects in 80 fields. A total of 2280 objects are listed; about 70% of these for the first time. – See also abstracts 12.113.033, 12.113. 034.

113.045 **Infrared observations of southern WC9 stars and He 2-113.** M. Cohen.
Monthly Notices Roy. Astron. Soc., Vol. 173, 489 - 496 (1975).
Observations between 1.57 and 4.7 μ are presented of eight southern WC9 stars. Three of these possess well-defined blackbody components in the near-infrared region suggesting that thermal emission by large graphite grains is observed. None of the eight stars can be identified as a relatively evolved WC9 star like Ve 2-45. New observations are also presented of the object He 2-113 in the range 3.8–19.6 μ, which confirm the suggestion that it represents a very cool WC-type nucleus of a planetary nebula. A very large ratio of infrared to Lα luminosity is found for this star.

113.046 **Photometry of faint blue stars.** D. Kilkenny, P. W. Hill.
Monthly Notices Roy. Astron. Soc., Vol. 173, 625 - 636 (1975).
Photometry on the $uvby$ system is presented for 72 faint blue stars in the southern hemisphere. Classification by means of the Strömgren indices is discussed and the classification criteria are applied to the programme stars. Several new hot subdwarfs and horizontal-branch stars are detected.

113.047 **A finding list of B and A stars near the globular cluster M22.** A. G. D. Philip, B. M. Goodman.
Publ. Astron. Soc. Pacific, Vol. 87, 801 - 803 (1975).
A catalog of 27 B and A stars is presented with spectral types, magnitudes, and positions. The catalog is intended to provide a list of stars to be observed for determination of the color excess in front of M22 (NGC 6656).

113.048 **An analysis of the *Q*-method for determining intrinsic color indices of early-type stars.**
A. Gutiérrez-Moreno.

Publ. Astron. Soc. Pacific, Vol. 87, 805 - 815 (1975).

The *Q*-method for determining intrinsic colors for early-type stars is analyzed; intrinsic colors in the *UBV* system are obtained for early main-sequence stars.

113.049 **The OB stars in northern Monoceros.** D. G. Turner. Journ. Roy. Astron. Soc. Canada, Vol. 69, 245 (1975). – Abstr. CAS.

113.050 **The connection of the photometric system of the high-speed scanning electrophotometer [Alma-Ata] with the *BVR* system.** Yu. M. Zavarzin.
Trudy Astrofiz. Inst., *Alma-Ata,* Vol. 25, 125 - 127 (1975). In Russian.

113.051 **Search for extreme ultraviolet radiation from selected stars.** G. R. Riegler, G. P. Garmire.
Astron. Astrophys., Vol. 45, 213 - 216 (1975).

A rocket-borne extreme ultraviolet photometer was used to search for 140–430 Å radiation from selected objects. Upper limits (2σ) to the flux from α CMa, o' CMa, λGem, γ^2 Vel and NGC 2392 range from 1.1 to 6.3×10^{-8} erg/cm^2 s. Estimates for the flux from these objects near 200 Å are compared with the upper limits obtained in the present experiment.

113.052 **On the dependence of atmospheric extinction coefficients on spectral type, luminosity and interstellar reddening in multicolour photometry.** K. Zdanavičius.
Bull. Vilnius Astron. Obs., No. 41, p. 3 - 29 (1975). In Russian.

The extinction coefficients were computed using the transmission curve of the atmosphere, the stellar energy distribution curves and the response curves of the Vilnius photometric system and the system UBVR. The variations of extinction coefficients are shown in figures.

113.053 **Calculation of reddening-free parameters Q of the Vilnius photometric system by the iteration method.**
A. Bartkevičius, K. Zdanavičius.
Bull. Vilnius Astron. Obs., No. 41, p. 30 - 36 (1975). In Russian.

For the classification purposes in the Vilnius system reddening-free parameters Q are used. To calculate the parameters Q one must know the corresponding colour excess ratios. The correlations between a number of colour excess ratios and parameters Q are obtained on the basis of normal colour indices and colour excess ratios of individual stars. These correlations are further used to compute colour excess ratios and parameters Q with the help of the iteration method.

113.054 **The photometric classification of double stars.** G. Kakaras.
Bull. Vilnius Astron. Obs., No. 41, p. 37 - 43 (1975). In Russian.

A photometric classification of 32 double stars with known MK spectral types is made by the method of comparison of color indices. Total color indices of a double star were compared with those calculated from color indices of single standard stars.

113.055 **Photoelectric photometry of bright stars in the WBVR system.** E. Meištas, K. Zdanavičius, V. Straižys, A. Gurklytė.
Bull. Vilnius Astron. Obs., No. 42, p. 3 - 15 (1975). In Russian.

Photoelectric observations of 101 bright stars of different spectral and luminosity classes are made in the system WBVR, the magnitude W being proposed by Straižys (1973) to replace the ill-defined magnitude U of the system UBV. The indices B–V and V–R are transformed into the standard system but with normalization to zero for unreddened O-type stars. The index W–B is in the new instrumental color system normalized to zero for unreddened O-type stars.

113.056 **Absolute calibration of color indices of three photometric systems.** V. Straižys, G. Kurilienė.
Bull. Vilnius Astron. Obs., No. 42, p. 16 - 33 (1975). In Russian.

The color indices of three photometric systems (the Vilnius system, the uvby and UBV systems) are calibrated in absolute units.

113.057 **Near infrared photometry of S-type stars.** J. Piccirillo.
Bull. American Astron. Soc., Vol. 7, 517 (1975). – Abstr. AAS.

113.058 **Photometric variability of certain red giant stars in the globular cluster NGC 7078 (M15).**
D. R. Mosley, R. E. White.
Bull. American Astron. Soc., Vol. 7, 535 - 536 (1975). Abstr. AAS.

113.059 **An analysis of stars in the Lindemann-Hauck four-color catalogue.** A. G. D. Philip.
Bull. American Astron. Soc., Vol. 7, 542 - 543 (1975). Abstr. AAS.

113.060 **HD 50896: Evidence for an extended continuum formation region.** A. V. Holm, J. P. Cassinelli.
Bull. American Astron. Soc., Vol. 7, 551 (1975). – Abstr. AAS.

113.061 **Apollo 16 far-ultraviolet stellar photometry.** T. L. Page, G. R. Carruthers.
Bull. American Astron. Soc., Vol. 7, 551 (1975). – Abstr. AAS.

113.062 **OAO-2 UV photometry of the Alpha Persei cluster.** M. R. Molnar, T. C. Stephens.
Bull. American Astron. Soc., Vol. 7, 551 - 552 (1975). Abstr. AAS.

113.063 **Empirical calibrations of the *uvby*, β systems. I. The F-type stars.** D. L. Crawford.
Astron. Journ., Vol. 80, 955 - 971 (1975).

Available *uvby* and β photometry for bright stars and cluster members has been used to calibrate observed indices in terms of intrinsic color, absolute magnitude, and [Fe/H]. The mean error, for one star, was calculated to be $\pm$ 0^{m}011 for $(b-y)_0$ and 0^{m}25 for M_V. Checks have been done for possible systematic errors. No significant adverse correlations were found. The calibrations should be valid for stars in the solar neighborhood (Population I) with spectral types F2–G0 and luminosity class III–V.

113.064 **Narrow-band photoelectric observations of the Ap star HD 184905.** T. A. Kartasheva, L. I. Snezhko, Kh. F. Khaliullin, N. M. Chunakova.
Soobshch. Spets. Astrofiz. Obs., *Zelenchukskaya,* vyp. (No.) 13, p. 5 - 8 (1975). In Russian.

Narrow-band photometric results of the Ap star HD 184905 in four intervals of the spectrum with $\lambda_{max} \sim 4240$, 4340, 4440, 4750 Å are presented. The light curves vary both in form and amplitude, showing maximum variability in the band corresponding to the H_γ line. The search for short-time (minute) variability in the H_γ line and in the bands of the continuum has not yielded positive results.

113.065 **Hα and Hβ photoelectric photometry of bright Be stars.** S. L. Baliunas, M. A. Ciccone, E. F. Guinan.
Publ. Astron. Soc. Pacific, Vol. 87, 969 - 974 (1975).

Hα and Hβ photoelectric observations of 33 early-type standard stars and 23 Be stars were obtained. The data permit separation of the stronger-lined Be stars from the normal stars and reveal variations in the Be star γ Cas. When compared with earlier data obtained by others, the data show longer-term variations in five other Be stars as well: ω Ori, ζ Tau, κ Dra, 23 Tau, and ψ Per.

113.066 **Intrinsic properties of high luminosity blue stars from observations of multiple systems.**
M. L. Burnichon.
Astron. Astrophys., Vol. 45, 383 - 398 (1975). In French.

Intrinsic colors and absolute magnitudes of some massive hot stars have been determined from a careful study of wide multiple systems. The brightest member of the system is an O star or a blue supergiant, and the fainter members are less massive main-sequence stars of type B or sometimes *A*. The author assumes that the multiple systems selected are physically connected. He can then derive intrinsic properties of bright components in a way which is probably more certain and more precise than the classical method of cluster or association membership. The intrinsic colors are found to be redder than the "classical" values [the $(B-V)_0$ range from -0.29 to -0.25] Comparison of these results with theory leads to good agreement with both upper main-sequence models and evolutionary time scales (with the adopted T_{eff} and B.C. scales). The author suggests three possible explanations for the derived colors of O stars: (1) the colors of *B* companions are not so well known, (2) there is "residual" circumstellar reddening in the system, (3) these colors are the "true" colors of O stars.

113.067 **Photometry of the Be star Wray 795 and some other stars in the Cen X-3 area.** A. M. van Genderen.
Astron. Astrophys., Vol. 45, 425 - 427 (1975).

Five-colour observations in the Walraven system of five objects in the Cen X-3 area are presented: viz. the Be star Wray 795 which does not show the light-variation as claimed by Vidal et al. (1974), a 14.5 mag variable star known as Liller's star, Cen X-3 (one observation only) and two other non-variable stars A (HD 306373) and No. 14 (finding chart of Brucato et al., 1972).

113.068 **The mean colour indices B–V and B–R of integral starlight in the neighbourhood of the North Galactic Pole.** Yu. M. Zavarzin.
Astron. Tsirk., No. 874, p. 6 - 8 (1975). In Russian.

113.069 **Thirteen-color photometry of 1380 bright stars.**
H. L. Johnson, R. I. Mitchell.
Rev. Mexicana Astron. Astrofis., Vol. 1, 299 - 324 (1975).

In this paper the authors publish the final data from their extended program of the observation on their 13-color medium-narrow-band photometric system. They have observed essentially all of the stars brighter than the fifth visual magnitude north of declination $-20°$, and essentially all of the stars brighter than the fourth visual magnitude south of declination $-20°$. The observations have been carefully tied together and the authors believe that the system is homogeneous to an unusual degree over the entire sky. The effective wavelengths of the filter-bands have been discussed in detail and a new absolute calibration, based upon all of the available data is derived. This photometric system should be useful, therefore, for the standardization of other photometric systems, as well as the narrow-band spectrophotometry.

113.070 **Comparison stars for five variable infrared carbon stars.** A. Alksnis, I. Eglitis.
Investigation of the sun and red stars. 3, (see 003.008), p. 35 - 46 (1975). In Russian.

U, B, V, R photographic magnitudes of comparison stars, as well as identification charts, are given for five carbon stars.

113.071 **Photometry.** J. Borgman.
Image processing techniques in astronomy, (see 012.041), p. 11 - 20 (1975).

The significance of photometry in astronomy is discussed in relation to three basic parameters: wavelength, spectral resolution and spatial resolution.

113.072 **UBVRI photometry of variable red dwarf emission objects.** D. H. Martins.
Thesis, Florida Univ., Gainesville (USA). [Available from Univ. Microfilms, Ann Arbor, Mich., *USA*. Order No. 74-27,880], 153 pp. (1974).

A review of the history and nature of small-scale variability of red-dwarf emission-line stars is presented. Data obtained at the Kitt Peak National Observatory, the Lowell Observatory, and the Rosemary Hill Observatory of the University of Florida are tabulated and analyzed to determine their implications with respect to the known variability of four stars. BY Draconis, GT Pegasi, FF Andromedae, and LP101-15 (an eclipsing binary) are studied. A period of 0.63398 day is supported for LP101-15, and a possible secondary minimum is proposed.

113.073 **Recent results in ultraviolet astronomy obtained by a wide field rocket camera and the French S 183 Skylab experiment.** G. Courtès, M. Laget, J. P. Sivan, M. Viton, A. Vuillemin, H. Atkins.
Phil. Trans. Roy. Soc. London, Ser. A, Vol. 279, 401 - 404 (1975).

113.074 **Preliminary results of absolute U.V. rocket photometry of γ Ori.**
H. Hessberg, J. Niekerke, K.-H. Stephan.
Phil. Trans. Roy. Soc. London, Ser. A. Vol. 279, 457 (1975).

The nineteenth Herstmonceux conference – Photometry, 1975 April 9 and 10. See Abstr. 011.066.

Photometric calibration of Schmidt plates by an auxiliary beam splitting prisma. See Abstr. 031.228.

Differential method of reduction of photoelectric systems (On the compilation of a catalogue of photometric standards in Kapteyn's Selected Areas). See Abstr. 031.252.

Rocket-photometer for stellar UV-observations.
See Abstr. 034.056.

Fifty years of atmospheric transmission, water vapor, and ozone: the astronomical implications.
See Abstr. 082.025.

Spectrophotometric and colorimetric parameters of O–B stars. See Abstr. 114.046.

Possible identification of a circumstellar 33-micron silicate emission band in cool-star spectra.
See Abstr. 114.070.

On the precision of photoelectric absolute spectrophotometry of stars. See Abstr. 114.088.

Spectrophotometry of M stars.
See Abstr. 114.093.

Supergiants in the Milky Way.
See Abstr. 114.105.

Two years of operation of the ultraviolet stellar spectrophotometer S59 in E.S.R.O.'s TD1A satellite.
See Abstr. 114.117.

Absolute rocket photometry of γ-Ori in the VUV-region. See Abstr. 114.335.

On the variability of the peculiar A type star HD 49976. See Abstr. 114.363.

The helium variable HR 7129.
See Abstr. 114.366.

Absolute luminosity calibration of Stroemgren's "intermediate group". See Abstr. 115.016.

Photometric variability of magnetic stars.
See Abstr. 116.003.

Differnetial *uvby* photometry of the spectroscopic binary HD 217061. See Abstr. 119.001.

The light curve of CV Serpentis, the sometimes-eclipsing Wolf-Rayet star. See Abstr. 121.002.

Flare colours and luminosities.
See Abstr. 122.012.

Photographic photometry of V1057 Cygni.
See Abstr. 122.019.

The color excess scale of the longer period cepheids– a discussion. See Abstr. 122.133.

Interstellar reddening and IR-excess of O and B stars.
See Abstr. 131.101.

On the interaction of stars with interstellar matter.
See Abstr. 131.161.

Differential *uvby* photometry of Cygnus X-1.
See Abstr. 142.066.

Hβ photometry in M39. See Abstr. 153.002.

Photometric metal abundances for twenty clusters.
See Abstr. 153.003.

Non-uniform extinction in open star clusters and dispersion of the photometric sequences.
See Abstr. 153.014.

Fotometría fotográfica UBV del cúmulo galáctico King 19. See Abstr. 153.018.

Magnitudes and colour indices of new possible members of the Praesepe cluster. See Abstr. 153.026.

Proper motions and BV photometry of NGC 129 and field stars. See Abstr. 153.037.

On the color indices of giants and subgiants in globular clusters. See Abstr. 154.012.

The interstellar reddening of the globular cluster 47 Tucanae. See Abstr. 154.013.

***UBV*-photometry of three fields in Cygnus.**
See Abstr. 155.008.

Further evidence concerning the local density of red dwarfs. See Abstr. 155.017.

***RGU* photometry of a field to the direction to the galactic centre.** See Abstr. 155.034.

Studies of the LMC stellar content I. A catalogue of 272 new O–B2 stars. See Abstr. 159.001.

UBV photometry for supergiants of the Large Magellanic Cloud. See Abstr. 159.004.

114 Stellar Spectra, Temperatures, Spectroscopy, Spectra of Individual Stars

Stellar Spectra, Temperatures, Spectroscopy

114.001 **Spectral classification of Ap and Am stars.**
M. Floquet.
Astron. Astrophys., Suppl. Ser., Vol. 21, 25 - 32 (1975). In French.

The spectra of 40 Ap stars and 69 Am stars have been analysed. For most of them, the classification is confirmed. This analysis was done on plates taken with a slit spectrograph.

114.002 **Hydrocarbon molecules in carbon stars.**
S. P. Tarafdar, N. C. Wickramasinghe.
Astrophys. Space Sci., Vol. 35, L41 - L45 (1975).

It is shown that the 10–13 μ band observed in the spectra of some carbon stars could arise from simple hydrocarbon molecules, especially of the type C_nH_{2n}. These molecules could be present in circumstellar dust shells which are responsible for the 1000 K black-body excess observed for these stars, and the emission band is produced by collisional excitation.

114.003 **Observations of the Mg II lines near 2800 Å from Copernicus.**
Y. Kondo, J. L. Modisette, G. W. Wolf.
Astrophys. Journ., Vol. 199, 110 - 119 (1975).

Observations of the Mg II resonance doublet lines at 2795 and 2802 Å, and their respective subordinate lines at 2797 and 2790 Å, have been made for some 40 stars in the spectral region O8–B8. Deviations from the theoretical predictions have been found for the earliest spectral types, where the rotational velocity effects, discussed recently by Collins, appear to make the equivalent widths greater. Evidence of emission has been found in a few stars of earliest spectral types studied.

114.004 **Observations of near-infrared C_2 bands in the spectra of carbon stars.** D. P. Gilra.
Astrophys. Journ., Vol. 199, 145 - 147 (1975).

It is pointed out that features due to the (3, 0) Phillips band of C_2 in the 7715–7750 Å region can be used to identify carbon stars at moderate dispersion. Also, the first observations are reported of the (2, 0) Ballik-Ramsay band of C_2 at 1.1726 μ.

114.005 **Skylab ultraviolet stellar spectra: variation of intensity and structure of strong lines with temperature and luminosity.** K. G. Henize, J. D. Wray, S. B. Parsons, G. F. Benedict, F. C. Bruhweiler, P. M. Rybski, F. G. O'Callaghan.
Astrophys. Journ., (*Letters*), Vol. 199, L119 - L121 (1975).

Objective-prism spectra at ultraviolet wavelengths extending to 1300 Å were photographed during the three Skylab missions in a survey covering 9 percent of the sky. Several spectral features, notably resonance lines of C IV and Si IV, show striking variations with stellar temperature and luminosity. Marked P Cygni profiles in C IV and Si IV, indicative of significant outflow of mass, appear in the spectra of all stars with M_{bol} brighter than –8.4.

114.006 **Metal abundances in nearby stars and the chemical history of the solar neighbourhood.**
B. E. J. Pagel, B. E. Patchett.
Monthly Notices Roy. Astron. Soc., Vol. 172, 13 - 40 (1975).

A simple theory of galactic enrichment in heavy elements is described in some detail, for the purpose of indicating clearly how the assumptions in numerical models of galactic evolution are related to the observational constraints imposed by the dependence of stellar chemical composition on time of formation and by the statistical distribution of metal abundances in long-lived stars such as the G dwarfs.

114.007 **Cool carbon stars in open clusters.**
A. Alksnis, Z. Alksne.
IAU Symposium No. 67, (see 012.001), p. 173 (1975). Abstract.

114.008 **The Wolf-Rayet stars.** S. V. Rublev.
IAU Symposium No. 67, (see 012.001), p. 259 - 274 (1975). – Review paper.

114.009 **On the chemical composition of Wolf-Rayet stars.**
T. Nugis.
IAU Symposium No. 67, (see 012.001), p. 291 - 296 (1975).

The chemical composition of 5 WR stars in Cygnus are considered by using the quasi-nebular variant of Beals' model. On the basis of the general considerations which follow from this model it was concluded that the C/He ratio for the WC stars appears to be more than 10 times higher than for normal OB stars and the N/He ratio in the case of the WN stars seems also to be somewhat higher than for normal OB stars.

114.010 **Spectroscopic investigations of luminous emission-line objects.**
A. Cassatella, R. Nesci, R. Viotti.
IAU Symposium No. 67, (see 012.001), p. 297 - 298 (1975). Abstract.

114.011 **Variations of the C III 5696 Å emission line in Wolf-Rayet stars.** J. D. Schumann, W. Seggewiss.
IAU Symposium No. 67, (see 012.001), p. 299 - 302 (1975).

Slow and rapid fluctuations of emission and absorption features of the C III emission line at λ 5696 Å in WC stars are described. They might be attributed to moving clouds in the expanding stellar envelope.

114.012 **Symbiosis in radio stars with infrared excess.**
F. Ciatti, A. Mammano.
IAU Symposium No. 67, (see 012.001), p. 409 - 410 (1975).

114.013 **A spectral survey of the southern Milky Way II. O–B5 stars l = 237° to 281°.** B. Nordström.
Astron. Astrophys. Suppl. Ser., Vol. 21, 193 - 239 (1975).

A catalogue, part II of the Stockholm Observatory Southern Milky Way Survey, is presented, based on objective prism plates with a dispersion of 240 Å mm^{-1} at H_γ as well as blue and visual direct plates. The catalogue contains spectral types, blue magnitudes, positions and finding charts for 562 O–B5 stars and emission-line objects to a limiting magnitude of about $12^m.5$. The region covered is a 3° wide belt along the galactic equator, l = 237° –281°.

114.014 **Spektral- und Leuchtkraftkriterien bei sehr kleinen Dispersionen.** W. Seitter.
Mitt. Astron. Ges., No. 36, p. 159 (1975). – Abstract.

114.015 **Spektralklassifikation und Radialgeschwindigkeiten von ausgewählten Infrarotobjekten.** J. Solf.
Mitt. Astron. Ges., No. 36, p. 160 (1975). – Abstract.

114.016 **Mariner 9 ultraviolet spectrometer experiment: strong stellar lines in OB stars.** M. R. Molnar.

Astrophys. Journ., Vol. 200, 106 - 112 (1975).

Data from the ultraviolet spectrometer on the Mariner 9 Mars orbiter have been used to measure the strong stellar lines of Si III λ1300, Si IV λ1400, and C IV λ1550. The C IV equivalent widths appear to be in good agreement with preliminary OAO-2 results; however, the Mariner 9 Si IV equivalent widths for B1 stars appear to be smaller by a factor of ~ 0.7. The observations are compared with theoretical results. The author suspects that line-blending effects make any interpretation uncertain.

114.017 **Observations and theory of Mg II lines in early-type stars. II. Theory and predicted profiles.**
M. A. J. Snijders, H. J. G. L. M. Lamers.
Astron. Astrophys., Vol. 41, 245 - 257 (1975).

The profiles of the Mg II lines at 2795.5 and 2802.7 Å, 2790.7 and 2798.0 Å, 4481.1 and 4481.3 Å are computed in a grid of model atmospheres with $8000 \leqq T_{eff} \leqq 35000$ K and $2.5 \leqq \log g \leqq 4.0$ by means of LTE and non-LTE theories of line formation. The effects of the various physical assumptions on the strength of the Mg II lines are discussed. The equivalent width, the central depression and the full half-width of the lines are presented in a form useful for a comparison with the observations.

114.018 **Observations and theory of Mg II lines in early-type stars. III. The observations and a comparison with the predictions.** H. J. G. L. M. Lamers, M. A. J. Snijders.
Astron. Astrophys., Vol. 41, 259 - 272 (1975).

The equivalent width of the ultraviolet Mg II lines from the $3s - 3p$ and $3p - 3d$ transitions in the spectra of 106 stars and of the visual lines from the $3d - 4f$ transition in 48 stars of types O 4 to A 3 are presented. The observed equivalent widths in normal stars of class II–V agree with the predictions over the entire range from O 8–A 2 if an abundance of Mg/H = 3×10^{-5} is adopted. In supergiant stars the Mg II lines are at least as strong as predicted if the same microturbulent velocity as derived from the visual spectrum is used.

114.019 **On the rapid spectral variability in Be stars.**
J. D. R. Bahng.
Observatory, Vol. 95, 147 - 148 (1975).

114.020 **Ultraviolet chromospheric lines in the spectra of late-type stars.** G. A. Gurzadyan.
Monthly Notices Roy. Astron. Soc., Vol. 172, 617 - 621 (1975).

The emission lines 2800 Mg II and 3080 Ti II of chromospheric origin have been observed in the ultraviolet Orion-2 spectra of three, G5, K5 and M3 type, stars. The data obtained strongly indicate the existence of stellar chromospheres in the late-type stars with the radiation power much higher than the radiative emissivity of the solar chromosphere. The relative emissivity of the stellar chromospheres itself is independent from the spectral type or effective temperature of the star.

114.021 **Metal abundances for K giants in a field in Scutum.**
H. L. Helfer, P. A. Jennens.
Monthly Notices Roy. Astron. Soc., Vol. 172, 701 - 705 (1975).

UBViyz photometric observations of ~ 100 K giants in a region of Scutum ($l \simeq 25^\circ_{\cdot}, b \simeq -4°$) out to distances of 10 kpc show that the average value of [Fe/H] for the intrinsically brighter ($M_V < 1$) K-giants along the line of sight is just about the same as that found for the average K giant in the solar vicinity. It is also possible that a group of low luminosity metal-rich stars was detected.

114.022 **Study of the F-type stars. I. MK spectral types.**
S. Malaroda.
Astron. Journ., Vol. 80, 637 - 641 (1975).

MK types for 455 F-type stars are provided.

114.023 **The mean energy distribution in stellar spectra. II. Stars of the B, A, F classes.**
O. B. Vasil'ev, L. V. Zyabrikova, E. V. Ruban.
Astron. Zhurn. Akad. Nauk SSSR, Vol. 52, 720 - 726 (1975). In Russian. English translation in Soviet Astron., Vol. 19, No. 4.

The mean spectral energy distributions in B, A, F stars were determined by the method of the authors (1974) at fixed wavelengths. The mean monochromatic fluxes from B, A, F stars, normalized to unity at λ=5556 Å, are given in a table.

114.024 **Interaction of hot stars and of the interstellar medium. VII. The rate and fate of stellar ultraviolet photons.** Y. M. Georgelin, M. C. Lortet-Zuckermann, G. Monnet.
Astron. Astrophys., Vol. 42, 273 - 285 (1975).

The authors use the new set of observational data gathered by Y. M. Georgelin (1975) on exciting stars of galactic nebulae to reanalyze the calibration of spectral type into effective temperature. An empirical calibration of $N_L/\pi F_V$ (the ratio of the ionizing photons flux to the energy emitted in the photometric *V*-band) is obtained as a function of the spectral MK types in the range O 6–B 0, and then the total photon luminosity and excitation parameter. The authors emphasize the advantages of using the ratio $N_L/\pi F_V$ rather than an absolute flux to describe the excitation of the nebula, because this ratio is quite insensitive to gravity, model-atmosphere, absolute magnitude scale and distance.

114.025 **Comments on some of the effects in the quantitative analysis of stellar spectra.** J. Zverko.
Bull. Astron. Inst. Czechoslovakia, Vol. 26, 271 - 273 (1975).

The consequences of inaccuracies in determining the temperature and microturbulence for the resultant values of element abundances are discussed. The effect of the inaccurate determination of the temperature is eliminated if only lines of a more abundant ion of the element are used in the analysis. An inaccurate estimate of the magnitude of the microturbulence is reflected in a large error for heavier elements. The inaccuracies in determining both values may result in apparent overabundances or underabundances of the elements.

114.026 **Quantitative spectral classification of M giants based on low-resolution spectra.** L. D. Frishberg.
Pis'ma v Astron. Zhurn., Vol. 1, No. 8, p. 22 - 26 (1975). In Russian.

A method of spectral classification of M stars based on quantitative estimates of the intensities of TiO absorption bands has been elaborated. The precision of this method is about 0.1–0.2 subclasses. The classification of 126 M giants has been carried out, 80 of which were not classified in the MK-system.

114.027 **A search for ultraviolet objects.**
D. J. Carnochan, M. M. Dworetsky, R. Wilson.
Bull. American Astron. Soc., Vol. 7, 406 (1975). – Abstr. AAS.

114.028 **The Wolf-Rayet stars of M33.** G. J. Corso.
Bull. American Astron. Soc., Vol. 7, 411 (1975). Abstr. AAS.

114.029 **Isotopic carbon abundances in carbon stars.**
J. L. Climenhaga, B. L. Harris, J. T. Holts, J. Smolinski.
Bull. American Astron. Soc., Vol. 7, 433 (1975). – Abstr. AAS.

114.030 **Possible identification of a circumstellar 33-μ silicate emission band in cool star spectra.**
W. Hagen, T. Simon, H. M. Dyck.

Bull. American Astron. Soc., Vol. 7, 434 (1975). – Abstr. AAS.

114.031 **Infrared variations of Wolf-Rayet stars.**
J. A. Hackwell, R. D. Gehrz, J. R. Smith, D. W. Strecker.
Bull. American Astron. Soc., Vol. 7, 441 (1975). – Abstr. AAS.

114.032 **Infrared calcium triplet emission in Be stars.**
R. S. Polidan.
Bull. American Astron. Soc., Vol. 7, 451 (1975). – Abstr. AAS.

114.033 **Calcium II H and K emission in yellow supergiants.**
D. R. Hollars, H. A. Beebe.
Bull. American Astron. Soc., Vol. 7, 469 (1975). – Abstr. AAS.

114.034 **Fe II emission in late-type stars.**
A. M. Boesgaard, H. Boesgaard.
Bull. American Astron. Soc., Vol. 7, 469 - 470 (1975). – Abstr. AAS.

114.035 **Relative abundances of iron-peak, α-, s-, and r-process elements in extremely metal-deficient dwarfs.**
R. C. Peterson.
Bull. American Astron. Soc., Vol. 7, 470 (1975). – Abstr. AAS.

114.036 **Results of 10830 observations in 217 stars.**
H. Zirin.
Bull. American Astron. Soc., Vol. 7, 470 (1975). – Abstr. AAS.

114.037 **Ultraviolet emission lines in F, G, and K giants and supergiants observed from a balloon.**
Y. Kondo, T. H. Morgan, J. L. Modisette.
Bull. American Astron. Soc., Vol. 7, 470 - 471 (1975). – Abstr. AAS.

114.038 **A study of the strength of the 7323 Å line of [Ca II] in late-type stars.** C. J. Webb, W. Sandmann.
Bull. American Astron. Soc., Vol. 7, 499 (1975). – Abstr. AAS.

114.039 **Enhanced abundances of nitrogen and s-process elements in red giant stars in Omega Centauri.**
R. J. Dickens, R. A. Bell.
Bull. American Astron. Soc., Vol. 7, 499 (1975). – Abstr. AAS.

114.040 **A 2- to 4-micron search for molecular features in young stars.** M. Cohen.
Publ. Astron. Soc. Pacific, Vol. 87, 500 (1975). – Abstr. ASP.

114.041 **Ca II K emission in pre-main-sequence stars.**
L. V. Kuhi.
Publ. Astron. Soc. Pacific, Vol. 87, 503 (1975). – Abstr. ASP.

114.042 **A spectroscopic study of stars for the most part behind the Coalsack.**
E. E. Mendoza V, T. Gómez.
Publ. Astron. Soc. Pacific, Vol. 87, 506 (1975). – Abstr. ASP.

114.043 **A study of the λ3984 feature in the mercury-manganese stars.** C. R. Cowley, G. C. L. Aikman.
Publ. Astron. Soc. Pacific, Vol. 87, 513 - 521 (1975).

The isotopic shift of the Hg II λ3984 line is examined in the spectra of a number of strong mercury-manganese stars; the shift cannot be explained by line blending with lines of other elements in these stars, but represents real isotopic variations. An astrophysical log *gf* value of −1.8 is derived for λ3984, based on the Hg I λ4358 line strength in ι CrB and HR 1800. Elemental and (in some cases) isotopic abundances of Hg are derived using a spectrum synthesis technique; overabundances of up to 5.4 dex over the solar system mercury abundance are found; the maximum overabundances may be temperature dependent.

114.044 **Measures of metal line blocking in 38 A-type stars.**
S. Danford.
Publ. Astron. Soc. Pacific, Vol. 87, 577 - 585 (1975).

The fraction of light blocked by metal lines in selected spectral regions is measured from coudé spectra for 38 nearby A-type stars. Results are tabulated over the wavelength range 3300 Å–5681 Å and are useful for photometric systems employing the Oke bandpasses. Comparison with recent metal-line-blanketed stellar atmospheres is made.

114.045 **Spectral types of faint Orion stars.**
M. Breger, P. M. Rybski.
Publ. Astron. Soc. Pacific, Vol. 87, 607 (1975).

Spectral types for eleven Orion stars are presented.

114.046 **Spectrophotometric and colorimetric parameters of O–B stars.** R. Kh. Oganesyan.
Soobshch. Byurakan. Obs., vyp. (No.) 46, p. 28 - 32 (1975). In Russian.

An empirical relation between the spectrophotometric gradient ϕ_1 (for λλ 3900–4700 Å) and B–V colour was obtained using the observed data of ϕ_1 and B–V for a great number of O–B type stars. A relation between the Balmer jump and (U–B) colour exists also. A comparison of the values of ϕ_1 and D obtained both by these formulas and observations are given in tables and figures.

114.047 **MK classification's of some northern hemisphere binary systems.**
G. Hill, R. W. Hilditch, F. Younger, W. A. Fisher.
Mem. Roy. Astron. Soc., Vol. 79, 131 - 144 (1975).

A total of 479 MK spectral classifications of 186 close binary systems taken largely from 'A Catalogue of Graded Photometric Studies of Close Binaries' (Koch, Plavec & Wood) and 'A Finding List for Observers of Eclipsing Variables' (Koch, Sobieski & Wood) are presented.

114.048 **Physical limits to attainable accuracies in stellar radial velocities.** D. Dravins.
Astron. Astrophys., Vol. 43, 45 - 50 (1975).

Although recent instrumental developments allow the determination of radial velocities of stellar spectral lines with a precision of 0.1 km/s, it is not possible to obtain the true stellar radial velocity to better than ≃0.5 km/s unless one has detailed knowledge of the small-scale inhomogeneities in the region of line formation. More accurate velocities might be obtained from strong lines formed sufficiently high in the atmosphere to (hopefully) be above the convective region. The best candidate is the Na I D_1 line.

114.049 **Absorption line profiles and rotational velocities for 59 stars.** W. Buscombe, T. R. Stoeckley.
Astrophys. Space Sci., Vol. 37, 197 - 220 (1975).

Absorption lines of He II, He I, Mg II, and other ions have been measured in coudé spectral tracings for 59 slowly and rapidly rotating southern O, B and A stars on the main sequence. All 668 separate line profiles have been analyzed with a procedure of differential corrections by least-squares to yield the following parameters and their error estimates: half-width, central depth, equivalent width, and shape parameter. A preliminary value of $v \sin i$ is derived for each line by deconvolving it with both the intrinsic and instrumental profiles. The inadequacies of the traditional procedure are discussed, and plans for using calculated non-LTE intrinsic line profiles including line limb darkening and also effects of rotational distortion, will be outlined.

114.050 **Spectral classification of late-type stars of the oxygen sequence.**
N. S. Komarov, N. A. Gladushina, A. V. Dragunova.
Astrometriya i Astrofizika, *Kiev*, vyp. (No.) 26, (see 003.001),

p. 53 - 62 (1975). In Russian.

Peculiarities of infrared radiation were studied for K–M stars from the energy distribution in their spectra. Molecular bands in the spectra are identified, residual intensities in absorption bands of TiO λλ 7150, 7600, 8500 Å are measured and their dependences on the spectral type are considered. The IR index is calculated which enables integral radiation to be compared in the visible and infrared regions, the m_5 index is determined which may serve as a reliable criterion of stellar classification from the temperature within the interval of K2–M4 spectral types.

114.051 **Investigation of F, G and K dwarfs and giant stars.** G. Cayrel.
Nauchn. Informatsii, vyp. (No.) 33, p. 4, 20 (1974). In Russian and English. – Abstr. of a paper presented at the colloquium 'Physics and evolution of stars', (see 012.015).

114.052 **Stars of early spectral types. Evolutionary effect as a possible explanation of the Be phenomenon.**
A. M. Delplace, H. Hubert.
Nauchn. Informatsii, vyp. (No.) 33, p. 7 - 8, 24 - 25 (1974). In Russian and English. – Abstr. of a paper presented at the colloquium 'Physics and evolution of stars', (see 012.015).

114.053 **Some comments on Balmer and Paschen decrements in Be stars.** D. Briot.
Nauchn. Informatsii, vyp. (No.) 33, p. 8 - 9, 25 - 26 (1974). In Russian and English. – Abstr. of a paper presented at the colloquium 'Physics and evolution of stars', (see 012.015).

114.054 **Calculation of emission line intensities in stellar spectra.** L. Luud, M. Ilmas, T. Nugis.
Nauchn. Informatsii, vyp. (No.) 33, p. 9, 26 (1974). In Russian and English. – Abstr. of a paper presented at the colloquium 'Physics and evolution of stars', (see 012.015).

114.055 **The Balmer discontinuity of the Ap-stars.** M. Gerbaldi, B. Hauck, N. Morguleff.
Nauchn. Informatsii, vyp. (No.) 33, p. 11, 27 - 28 (1974). In Russian and English. – Abstr. of a paper presented at the colloquium 'Physics and evolution of stars', (see 012.015).

114.056 **Peculiar stars. A code for investigation of peculiar Ap-stars.** V. L. Khokhlova.
Nauchn. Informatsii, vyp. (No.) 33, p. 11 - 12, 28 - 29 (1974). In Russian and English. – Abstr. of a paper presented at the colloquium 'Physics and evolution of stars', (see 012.015).

114.057 **Chemical abundance of late-type stars.** A. A. Boyarchuk.
Nauchn. Informatsii, vyp. (No.) 33, p. 36 - 47 (1974). In Russian.

A review of observational data concerning the abundances of metals in various groups of giant stars with conspicuous peculiarities in the spectra is presented. The following groups of stars are discussed: CN stars, CH stars, barium stars, S-stars, C-stars and HdC-stars.

114.058 **Search for the TiO band λ 7054 in the spectra of M13 giants.** R. M. Rusev, E. A. Kolotilov.
Astron. Tsirk., No. 841, p. 3 - 4 (1974). In Russian.

114.059 **The effective temperature scale of M giants.** N. S. Komarov.
Astron. Tsirk., No. 842, p. 5 - 6 (1974). In Russian.

114.060 **Comment on the paper of L. N. Kondrat'eva 'New object with broad emission lines'** [Astron. Tsirk., No. 827, p. 1 - 3 (1974)]. L. N. Kondrat'eva.
Astron. Tsirk., No. 844, p. 8 (1974). In Russian.

114.061 **Thirty-five new peculiar stars.** R. Bartaya.
Astron. Tsirk., No. 845, p. 3 - 5 (1974). In Russian.

114.062 **Increasing of accuracy in spectrophotometry of bright stars with small telescopes.**
A. P. Ipatov, P. V. Shcheglov.
Astron. Tsirk., No. 846, p. 5 - 7 (1974). In Russian.

114.063 **Theoretical values of limb-darkening coefficients for classical stars.** A. A. Rubashevskij.
Astron. Tsirk., No. 851, p. 5 - 7 (1974), with a correction in Astron. Tsirk. No. 859, p. 8 (1975). In Russian.

114.064 **On stars – "primary" spectrophotometric standards.** A. V. Kharitonov.
Astron. Tsirk., No. 855, p. 2 - 4 (1975). In Russian.

114.065 **The He I λ 10830 line in the spectrum of the symbiotic star AG Peg.** E. A. Kolotilov.
Astron. Tsirk., No. 865, p. 1 - 3 (1975). In Russian.

114.066 **On the influence of the blanketing effect on theoretical values of darkening coefficients for dwarf stars.** A. A. Rubashevskij.
Astron. Tsirk., No. 866, p. 5 - 6 (1975). In Russian.

114.067 **Ultra-rapid variations of the Hα spectral line of the Be star φ Per. I. Observations.**
A. Kh. Mamatkazina.
Astron. Tsirk., No. 867, p. 1 - 3 (1975). In Russian.

114.068 **Ultra-rapid variations of the Hα line of the Be star φ Per. II. Discussion of observations.**
A. Kh. Mamatkazina.
Astron. Tsirk., No. 867, p. 3 - 4 (1975). In Russian.

114.069 **On the spectral change in the peculiar star Z CMa.** E. A. Kolotilov.
Astron. Tsirk., No. 868, p. 1 - 3 (1975). In Russian.

114.070 **Possible identification of a circumstellar 33-micron silicate emission band in cool-star spectra.**
W. Hagen, T. Simon, H. M. Dyck.
Astrophys. Journ., (*Letters*), Vol. 201, L81 - L84 (1975).

Infrared photometry of S Per, μ Cep, and HD 11979 reveals that these stars have substantially larger flux excesses at 33 μ than do α Ori and o Cet. These prominent excesses may be attributable to a distinct lattice band in circumstellar dust grains composed of silicates such as γ-Ca_2SiO_4, Fe_2SiO_4, or Mn_2SiO_4. The observations show that the abundance of this silicate material varies dramatically from one star to another among the class of M stars.

114.071 **Photometric scans of Hα in early A-stars.** S. Frandsen.
Astron. Astrophys., Vol. 43, 267 - 273 (1975).

Observations of the Hα-line in early A-stars of different luminosity classes are described. About 20 different points in the profiles have been measured. A ratio between the residual intensities in the line core and just outside is shown to be a good indicator of rotational velocity for main sequence stars. The observed ratios are in agreement with theoretically evaluated ratios. For more luminous stars the observed ratio is larger than the theory predicts.

114.072 **Colors and ionization equilibria in K-dwarfs. II. Implications for the lower main sequence.**
M. N. Perrin.
Astron. Astrophys., Vol. 44, 9 - 16 (1975).

The author has performed a detailed analysis of three K-dwarfs. Two of the observed stars, 36 Oph A and B are young population I stars and form with 36 Oph C a triple system. The third star, HD 191408, belongs, on the contrary, to the old disk population. However, no significant differences in chemical composition between the three stars have been found. Masses have been estimated for a small sample of twelve K-dwarfs.

114.073 **On the chemical composition of the sun, Jupiter, meteorites and Am stars.** Eh. M. Drobyshevskij.
VI Leningr. mezhdunar. seminar, 1974. Uskorenie chastits i yader. reaktsii v kosmose. Leningrad, 1974, p. 143 - 152.
In Russian.

114.074 **Spectrophotometric study of stars with composite spectra.** Ya. N. Chkhikvadze, M. A. Hamdi.
Soobshch. AN GruzSSR, Vol. 77, No. 1, p. 61 - 64 (1975).
In Russian. – Abstr. in Referativ. Zhurn. 51. Astron., 8.51. 782 (1975).

114.075 **Possibilities and restrictions of the method of "mapping" the surface of Ap stars from spectral line profiles.** V. L. Khokhlova.
Astron. Zhurn. Akad. Nauk SSSR, Vol. 52, 950 - 955 (1975).
In Russian. English translation in Soviet Astron., Vol. 19, No. 5.

Model calculations of the line profiles in the spectra of a spotty rotating star were made to clear up the most favourable conditions for detailed study of the surface structure of Ap stars.

114.076 **Classification of some bright B8–F0 stars with peculiar spectra.** H. Levato, S. Malaroda.
Inform. Bull. Southern Hemisphere, No. 25, p. 29 - 31 (1974).

114.077 **Spectrophotometric investigations of semiregular and irregular carbon stars.** J. Krempeć.
Acta Astron., Vol. 25, 235 - 264 (1975).

The spectrophotometric observations of 15 semiregular and irrregular carbon stars carried out at the Toruń Observatory between 1966–71 are discussed. The carbon abundance classes are given for all the investigated carbon stars. The comparison of the changes of the C_2^1 Swan system and the CN (7, 1) molecular bands as well as of the sodium lines (NaI λλ5890, 5896 Å) with the light phases for two semiregular carbon stars shows that the absorption lines and molecular bands get stronger with a decrease of temperature for hotter carbon stars. For a low dispersion population criterion, the diagram of central depths of sodium lines against the central depths of the C_2 (0, 1) molecular band is proposed.

114.078 **Late-type giants in Large Magellanic Cloud.** V. M. Blanco, M. F. McCarthy.
Nature, Vol. 258, 407 - 408 (1975).

Stars of class M 6.5 and later seem to form a natural group according to the definitions of Morgan, even when observed with the small dispersions given by the new objective prism. The characteristic feature of this natural group is an apparent drop in the intensity of the continuum at wavelengths longer than 8,400 Å. The present study is limited to stars that belong to this group. The number of late M stars found is 803.

114.079 **A sunspot as a standard for abundance analyses of cool stars: Ba I spectrum.** J. van Paradijs.
Astron. Astrophys., Vol. 44, 295 - 297 (1975).

Lines of neutral barium in the spectra of a sunspot and of α Tau (K5 III), α Boo (K2 III) and ε Peg (K2 Ib) have been analyzed by a curve of growth method, using the *gf*-values of Miles and Wiese (1969). From the sunspot spectrum a solar barium abundance $\log (N_{Ba}/N_H) + 12.00 = 2.07 \pm 0.35$ (s.d.) has been found. The results of previous barium abundance determinations, using strong Ba II lines in the late type stellar spectra, are substantiated.

114.080 **Profiles of strong lines in spectra of late-type stars.** T. Kipper.
Publ. Tartu Astrofiz. Obs., Vol. 43, 39 - 46 (1975).

Observed profiles of strong atomic lines in the spectra of late-type giants do not agree with profiles computed on the basis of model atmospheres. In cases of Na D lines this discrepancy might be caused by absorption due to TiO γ′-system bands. This possibility was tested by approximate calculations.

114.081 **The carbon isotopic ratio in two late-type stars.** T. Kipper.
Publ. Tartu Astrofiz. Obs., Vol. 43, 47 - 50 (1975).

Using high resolution Fourier spectra of R Leo and U Hya obtained by Connes and Michel the lower limit of the carbon isotopic ratio C^{12}/C^{13} is determined. This limit turns out to be 10 and 40 for R Leo and U Hya respectively.

114.082 **On the variations of radial velocities of P Cygni.** L. Luud, O. P. Gollandskij, T. Yarygina.
Publ. Tartu Astrofiz. Obs., Vol. 43, 250 - 255 (1975). In Russian.

Using spectrograms of P Cygni with a dispersion of 14 Å/mm, obtained during 1965–1971, the radial velocities from some selected spectral lines have been determined. Variations of the absorption line radial velocity of hydrogen H_9 and H_{10} lines with a period of 57 days have been found.

114.083 **The lines He I λ3187 and He II λ3203 in O-type stars.** N. D. Morrison.
Astrophys. Journ., Vol. 202, 433 - 438 (1975).

High-dispersion observations of the lines He I λ3187 and He II λ3203 in 19 O and B0.5-type stars are presented. For the purpose of comparing their equivalent widths with the predictions of non-LTE model atmospheres, effective temperatures and gravities of the O-type stars observed are derived from the equivalent widths of other lines of H and He. For supergiants, these temperatures and gravities differ systematically from those derived by the method of spectral classifications, and the reason for the difference is discussed.

114.084 **Some spectroscopic problems of Coronae Borealis stars.** N. Kameswara Rao.
Bull. Astron. Soc. India, Vol. 3, 50 - 54 (1975). – Review paper presented at the special session on 'Stellar spectroscopy' of the Astronomical Society of India held at Kodaikanal on 12 March 1975.

114.085 **A comment on the spectral energy distributions of Ba II stars.** J. D. Fix, J. S. Neff.
Monthly Notices Roy. Astron. Soc., Vol. 173, 83P - 85P (1975).

Contrary to a recent suggestion, the 4000 Å flux depressions of the Ba II stars are shown not to be caused by increased strengths of CN, CH and C_2 bands but rather by a wide, apparently continuous, spectral feature.

114.086 **Anomalies of abundances in Am stars and their relation to anomalies in Ap stars.**
V. L. Khokhlova, D. A. Ptitsyn.
Pis'ma v Astron. Zhurn., Vol. 1, No. 12, p. 26 - 31 (1975). In Russian.

Correlations of abundance anomalies of various elements in Am stars show continuous transition from normal stars through Am to Ap stars. Correlations for light elements are poor, while for iron group and heavier elements they are good.

114.087 **Spectrophotometry of hydrogen lines in the spectrum of Ap stars.** I. A. Aslanov, Yu. S. Rustamov, V. M. Khalilov, A. A. Shakirzade.

Astron. Zhurn. Akad. Nauk SSSR, Vol. 52, 1226 - 1230 (1975). In Russian. English translation in Soviet Astron., Vol. 19, No. 6.

An investigation of the Balmer lines behaviour in the spectra of 13 Ap stars has been carried out. The change of the equivalent line widths, log ($N_{02}H$) and log n_e in all stars is shown. A comparison with the available light curves has been made. It is also shown that in all stars at the moment of light minimum the lines are wider and deeper than at the moment of its maximum. Short-period variations of the radial velocities for two stars (17 Com A and HD 224801) have been found.

114.088 **On the precision of photoelectric absolute spectrophotometry of stars.** A. V. Kharitonov.
Astron. Zhurn. Akad. Nauk SSSR, Vol. 52, 1241 - 1246 (1975). In Russian. English translation in Soviet Astron., Vol. 19, No. 6.

It is shown that the energy distribution in the spectra of eight primary standard stars forms a homogeneous system. The intrinsic accuracy of spectrophotometric observations of stars in dependence of the stellar magnitude, the number of observations and spectral classes are discussed.

114.089 **Turbulence measurements in giants and supergiants.** D. F. Gray.
Journ. Roy. Astron. Soc. Canada, Vol. 69, 245 (1975). Abstr. CAS.

114.090 **Progress in ultraviolet stellar spectrophotometry with S59.**
C. de Jager, H. J. G. L. M. Lamers, K. A. van der Hucht.
Astrophys. Space Sci., Vol. 38, 313 - 325 (1975). – Invited paper presented at the I.A.U. Colloquium No. 27 on 'UV and X-ray spectroscopy of astrophysical and laboratory plasmas', Cambridge, Mass., U.S.A., 9–11 September 1974.

A progress report is given of investigations related to observations of stellar spectra obtained with the ultraviolet stellar spectrophotometer S59 aboard the ESRO TD-1A satellite. Some important groups of lines are dealt with: non-LTE computation of the Mg II lines are presented: apart from the peculiar emission line and shell stars they compare well with the observations; intensity ratios of Fe II and Fe III lines are an important temperature classification criterion. Interstellar UV lines indicate large deficiencies of some metals, as compared to solar values. A new ultraviolet continuous extinction curve is determined.

114.091 **Statistical analysis of velocities and cyanogen indices for G and K giants.** M.-T. C. Peralta.
Astron. Astrophys., Vol. 45, 87 - 95 (1975).

Regressions of the velocity components *U, V, W* and the total velocity *S* with the cyanogen index anomaly ΔCN have been computed for a sample of 631 G and K giants. The resulting curves suggest the establishment of slow radial motions in extensive portions of the galactic gas, after a post-collapse stabilization period. Two kinematical groups with the same metallic abundances have been confirmed. The presence of the abnormal group with high *U* and *V* velocity components and abnormally high abundances is explained here as an effect of the kinematics of the galactic gas.

114.092 **Absolute ultraviolet spectrophotometry from the TD-1 satellite. VI. Statistical study of the 1930 Å Fe III absorption feature in early type stars.** F. Beeckmans.
Astron. Astrophys., Vol. 45, 177 - 183 (1975).

The depth of the 1930 Å feature, attributed to Fe III, has been measured in more than 300 early type star spectra obtained with the S 2/68 U.V. sky survey telescope. The behavior of this feature versus the visual index *Q* is correlated to the ionized equilibrium of Fe but it also shows an important luminosity effect. It increases from dwarfs to supergiants and is usually stronger in rapidly rotating Be stars and shell stars.

114.093 **Spectrophotometry of M stars.** L. D. Beshenova, A. V. Kharitonov.
Trudy Astrofiz. Inst., *Alma-Ata,* Vol. 27, 208 pp. Price 92 Kop. (1975). In Russian.

114.094 **Metal abundance effects in the spectra of F5–G5 dwarfs obtained with an objective prism.**
V. Malyuto.
Byull. Abastumansk. Astrofiz. Obs., No. 46, p. 47 - 58 (1975). In Russian.

Objective prism spectra (D=166 Å/mm at Hγ) for some F5–G5 dwarfs with different chemical composition have been analysed on purpose to study metal abundance effects. There have been measured the central depths of practically all measurable spectral lines and blends within the range of 3700–4500 Å which can be identified with reliability even in the spectra of moderate quality. The influence of metal content on the line depth ratios has been investigated. It has been found that some combinations of line depth ratios can be well used for segregation of stars with various metal content. The luminosity effects on the chosen line depth ratios have been investigated by combining the spectra with those obtained by Shiukashvili. The influence of metal abundance on quantitative spectral class determinations has been estimated.

114.095 **Helium abundances and cepheid pulsation.** E. G. Schmidt.
Bull. American Astron. Soc., Vol. 7, 504 (1975). – Abstr. AAS.

114.096 **On the variations in the spectra of Ap stars.** R. Steinitz, K. B. Gebbie.
Bull. American Astron. Soc., Vol. 7, 512 (1975). – Abstr. AAS.

114.097 **High dispersion observations of Ca II H and K lines in late-type stars.** J. B. Lester, A. K. Dupree.
Bull. American Astron. Soc., Vol. 7, 517 - 518 (1975). Abstr. AAS.

114.098 **Abundances in late-type dwarfs.** L. Hartmann, C. M. Anderson.
Bull. American Astron. Soc., Vol. 7, 518 (1975). – Abstr. AAS.

114.099 **Spectrophotometry of cool carbon stars.** C. E. Gow.
Bull. American Astron. Soc., Vol. 7, 519 (1975). – Abstr. AAS.

114.100 **The spectra of some giant stars in the globular cluster 47 Tuc.**
R. A. Bell, R. J. Dickens, B. Gustafsson.
Bull. American Astron. Soc., Vol. 7, 535 (1975). – Abstr. AAS.

114.101 **The yttrium peak in Hg-Mn stars.** M. S. Allen. Bull. American Astron. Soc., Vol. 7, 544 (1975). Abstr. AAS.

114.102 **Beryllium in main-sequence stars.** A. M. Boesgaard. Bull. American Astron. Soc., Vol. 7, 557 - 558 (1975). – Abstr. AAS.

114.103 **Chlorine abundances in early B stars.** F. C. Bruhweiler, S. B. Parsons.
Bull. American Astron. Soc., Vol. 7, 558 (1975). – Abstr. AAS.

114.104 **Dwarf K and M stars discovered on objective-prism plates.** C. B. Stephenson, N. Sanduleak.
Astron. Journ., Vol. 80, 972 - 975, 1001 - 1007 (1975).

The authors report 100 dwarf K and M stars found on objective-prism plates taken at low galactic latitudes, mostly in the southern sky. Most stars were recognized from the strength of the sodium D lines, and the majority have known proper motions of 0."1 or more per annum. Ninety-nine of these stars lack previous spectral classification; one lacks previous observation of line emission. Most are below the limits of the DM charts, and for these the authors provide identification charts.

114.105 **Supergiants in the Milky Way.** W. Buscombe.
Peremennye Zvezdy, Prilozhenie, Vol. 2, 127 - 157 (1974). In Russian.

To facilitate studies of kinematics in the Galaxy and of the relation of luminous stars to the interstellar gas as observed with radio telescopes, a summary of photometric and spectroscopic data is presented. An attempt is made to include most of the published information about stars of luminosity class I with $M_v < -4$.

114.106 **Investigations of carbon stars in the Radioastrophysical Observatory of the Academy of Sciences of the Latvian SSR.** A. Alksnis.
Peremennye Zvezdy, Vol. 19, 529 - 531 (1975). In Russian.

114.107 **Spectrophotometric gradients of unreddened M-giants.** L. D. Beshenova, A. V. Kharitonov.
Astron. Tsirk., No. 880, p. 5 - 6 (1975). In Russian.

114.108 **Observations à faible dispersion du spectre de 25 étoiles de type Of entre 5850 Å et 8750 Å.**
Y. Andrillat, J. M. Vreux.
Acad. Roy. Belgique, Bull. Cl. Sci., 5ᵉ Sér., Vol. 60, 430 - 435 = Inst. Astrophys. Univ. Liège, Cointe-Ougrée (*Belgique*), Coll. 8°, No. 647 (1974).

Near infra-red spectrograms (5850 Å–8750 Å) of 25 Of type stars have been obtained between 1971 and 1973 at the Haute Provence Observatory. The main features observed in the spectral range are due to H I and He I. Hα is never observed in emission when He II λ 4686 is in absorption, the reverse is not true.

114.109 **On the existence of M-type subgiant stars.**
L. D. Beshenova, A. V. Mironov, A. V. Kharitonov.
Astron. Tsirk., No. 882, p. 3 - 4 (1975). In Russian.

114.110 **Peculiarity of peculiar stars.** B. Kuchowicz.
Urania Kraków, Vol. 46, 354 - 360 (1975). In Polish.

114.111 **Ultraviolet spectrophotometry of a group of hot stars in Vela.**
G. A. Gurzadyan, R. Kh. Oganesyan.
Astrofizika, Vol. 11, 397 - 408 (1975). In Russian. – English translation in Astrophysics, Vol. 11, No. 3.

The results of the measurements of 62 ultraviolet spectrograms, obtained by means of Orion 2, for 12 hot B-type stars in Vela are given. The observed distributions of the energy in the continuous spectrum of these stars are in good accordance with theory in the wavelength region 3700–2200 Å. The equivalent widths of the ultraviolet doublet λ 2800 Mg II for these stars are also presented.

114.112 **The continuous spectra of WR-type stars.**
V. P. Ryl'kov.
Astrofizika, Vol. 11, 473 - 485 (1975). In Russian. – English translation in Astrophysics, Vol. 11, No. 3.

Theoretical WR fluxes received by summing the Planck radiation of the star and recombination radiation of the shell have been compared with observed fluxes for WR stars. Supposing the shell to consist mainly of helium, estimates of electronic temperatures in the shell and ratios of shell to star luminosities have been made. This permitted to evaluate for every WR star the electronic densities at the photospheric boundary and the WR star's mass loss per year.

114.113 **New carbon stars in the zone centered at the galactic longitude $l = 86°$.** Z. Alksne, V. Ozoliņa.
Investigation of the sun and red stars. 3, (see 003.008), p. 29 - 34 (1975). In Russian.

Positions, identification charts and R, N spectral types for 10 newly discovered carbon stars are given. The stars have been found on 4° objective prism plates taken with the Schmidt telescope at the Radioastrophysical Observatory of the Academy of Sciences of the Latvian SSR.

114.114 **Super-metal-rich stars.** D. Kuroczkin.
Stud. Soc. Sci. Torunensis, Toruń, Poland, Sect. F (Astron.), Vol. 5, 175 - 196 = Bull. Astron. Obs. N. Copernicus Univ. Toruń, No. 54 III (1974).

An analysis of spatial distribution and kinematics indicates that SMR stars are strongly concentrated towards the galactic plane with a slight increase of the *U* velocity component. The ages of SMR stars have been subsequently determined in order to discuss the role of these stars in the theory of chemical evolution of the Galaxy.

114.115 **Rapid variations of Balmer line strengths in the spectra of Be stars.** K. B. McBeath.
Thesis, Northwestern Univ., Chicago, Ill. (USA). [Available from Univ. Microfilms, Ann Arbor, Mich., USA. Order No. 75 - 7953], 95 pp. (1974).

Low resolution photoelectric spectrophotometric measurements of the first four members of the Balmer series in the spectra of one Be and five Be (shell) stars were obtained with the 92-cm telescope and image dissecting scanner at Kitt Peak National Observatory in October, 1972. Equivalent widths were computed for each observation and their standard deviations from the mean values examined.

114.116 **Preparation of a U.V. spectrophotometric catalogue of bright stars.** A. Monfils, D. Macau-Hercot.
Phil. Trans. Roy. Soc. London, Ser. A, Vol. 279, 405 - 411 (1975).

114.117 **Two years of operation of the ultraviolet stellar spectrophotometer S59 in E.S.R.O.'s TD1A satellite.** C. de Jager, R. Hoekstra, K. A. van der Hucht, T. M. Kamperman, H. J. G. L. M. Lamers.
Phil. Trans. Roy. Soc. London, Ser. A, Vol. 279, 413 - 420 (1975).

A progress report is given of the current state of the investigation of the ca. 3500 spectra obtained during the life time of the satellite. A summary is presented of the general types of investigation, namely those which involve a study of major parts of the observational material, and the more specific investigations, which refer to selected stars or groups of stars, or to special lines or groups of lines.

114.118 **A search for ultraviolet objects.**
D. J. Carnochan, M. M. Dworetsky, J. J. Todd, A. J. Willis, R. Wilson.
Phil. Trans. Roy. Soc. London, Ser. A, Vol. 279, 479 - 485 (1975).

The observations of the sky-survey telescope in the TD-1A satellite are being searched for ultraviolet objects. The first results, covering about 10 percent of the sky, are presented and rough limits on space density and luminosity are derived. It is concluded that the frequency population of hot objects lying a few magnitudes below the main sequence is probably much greater than had been thought hitherto. Some examples of the observed ultraviolet spectra are presented.

114.119 **The ultraviolet spectrum of B-type supergiants.**
A. B. Underhill.
Phil. Trans. Roy. Soc. London, Ser. A, Vol. 279, 429 - 442 (1975).

The broad, steep-sided absorption lines in B-type supergiant spectra are stronger than the absorption lines in main-sequence stars. A broad, pointed feature is present at 1720 Å which has constant strength in the B-type supergiants regardless of spectral type. The complete identification of this blend is not known. At high resolution the ultraviolet resonance lines of C IV, N V, Si III and Si IV in the spectra of OB supergiants are shortward displaced by velocities up to 1800 km/s indicating the presence of an escaping atmosphere. Evidence is presented of the presence of a stationary shell around the B5Ia supergiant ζ CMa as well as a slowly expanding atmosphere.

114.120 **P-Cygni profiles in hot stars.** D. C. Morton. Phil. Trans. Roy. Soc. London, Ser. A, Vol. 279, 443 (1975).

Royal Astronomical Society specialist discussion on the peculiar and metallic-line stars. See Abstr. 011.064.

The silicon monoxide radical in M stars – the fundamental rotation-vibration band of SiO. See Abstr. 022.083.

Investigation of recombination spectra of light ions. See Abstr. 022.085.

The vibrational spectrum of C_3 in the 5-micron region. See Abstr. 022.094.

Stellar line profiles by tilt-scanned narrow band interference filters. See Abstr. 031.226.

Trends in astronomical spectroscopy. See Abstr. 031.280.

Scannermessungen und automatische Äquivalentbreitenbestimmung. See Abstr. 034.012.

The origin and abundances of the chemical elements. See Abstr. 061.036.

Carbon abundances in G and K giants. See Abstr. 064.038.

Emission envelopes of T Tauri stars. See Abstr. 064.047.

Atmospheric turbulence measured in stars above the main sequence. See Abstr. 065.122.

Some problems of magnetism and of the chemical composition of the sun and stars. See Abstr. 071.035.

A new method for determining Doppler widths. See Abstr. 073.078.

Photoelectric multiband photometry of M giants and supergiants in the blue-visual spectral region. See Abstr. 113.033.

Short-periodic light variations in Wolf-Rayet stars. See Abstr. 113.040.

The demise of the degenerate M stars. See Abstr. 126.007.

Spectrophotometric observations of suspected red subluminous stars in the Hyades and field. See Abstr. 126.012.

UV stars and the interstellar medium: a statistical time-dependent model. See Abstr. 131.089.

Detection of interstellar deuterium Lyman Alpha. See Abstr. 131.183.

Colorimetric measurements of Hα emission stars in the Orion aggregate. See Abstr. 132.038.

The radio spectra of mass-outflow stars. See Abstr. 141.061.

CO in the infrared and radio spectra of carbon stars. See Abstr. 141.623.

NGC 3114: another open cluster possibly rich in peculiar stars. See Abstr. 153.027.

A spectroscopic investigation of the Pleiades. See Abstr. 153.041.

Metal abundances of Draco and eight galactic globular clusters. See Abstr. 154.009.

An atlas for the Armagh Survey of H-alpha emission objects in the Large Magellanic Cloud. See Abstr. 159.006.

Spectra of Individual Stars

114.301 **The spectrum of HD 125823 (a Centauri).**
A. B. Underhill, R. P. Fahey, D. A. Klinglesmith.
Astrophys. Journ., Vol. 199, 120 - 126 (1975).

Equivalent widths and profile observations of representative lines in the spectrum of the helium variable HD 125823 are presented around one cycle of 8.814 days. The equivalent widths of the hydrogen lines and lines of C II, O II, Mg II, Si II, and Ca II do not vary in a systematic way by a significant amount during the He I cycle, while lines of N II, Si III, and Fe III vary in antiphase to the He I lines. Representative profiles are presented of the weak, shallow Fe I lines which appear between phases 0.9 and 0.3 of the He I variation. The origin of these lines is not understood.

114.302 **High-resolution profiles of sodium and potassium lines in Alpha Orionis.**
L. Goldberg, L. Ramsey, L. Testerman, D. Carbon.
Astrophys. Journ., Vol. 199, 427 - 431 (1975).

Profiles of the K I resonance line at 7698.98 Å and of the D_1 and D_2 lines of Na I in the spectrum of α Ori have been recorded photoelectrically. The observed profiles are consistent with a model in which radiation from the stellar photosphere is scattered and reemitted by an expanding envelope or envelopes of radius large compared with that of the photosphere.

114.303 **The $^{12}C/^{13}C$ ratio in stellar atmospheres. IV. Eleven G and K type giants.**
J. Tomkin, D. L. Lambert, R. E. Luck.
Astrophys. Journ., Vol. 199, 436 - 442 (1975).

High-resolution photoelectric scans of portions of the 4–0, 5–1, 2–0, 3–1, and 4–2 bands of the CN red system in the spectra of 11 late-type giants have been obtained.

114.304 **The infrared spectrum and structure of Eta Carinae.**
D. K. Aitken, B. Jones.
Monthly Notices Roy. Astron. Soc., Vol. 172, 141 - 147 (1975).

Maps of η Car at 12 and 20 μ are presented with beam sizes of 2".2 and 3".7 respectively, with 10 μ spectra taken with different beam sizes and a 20 μ spectrum of the complete source. The maps show a remarkable similarity to the visible spatial distribution and also provide evidence for more complex infrared structure. The spectra indicate an optically thin inner region surrounded by cooler thicker dust.

114.305 **Infrared spectra of T Tau stars and related objects.**
G. I. Shanin, V. S. Shevchenko, A. G. Shcherbakov.
IAU Symposium No. 67, (see 012.001), p. 117 - 122 (1975).

Four T Tau stars and related objects (RY Tau, T Tau, AB Aur and V1057 Cyg) have been included in our spectroscopic programme since 1973. The present paper is concerned with the spectroscopic observations made at the Crimea with the single-stage image tube S1. Tentative atomic line identifications are given for programme stars.

114.306 **Observations of the peculiar emission-line objects with infrared excess HD 45677, HD 87643 and GG Carinae.** J.-P. Swings.
IAU Symposium No. 67, (see 012.001), p. 405 - 407 (1975).

114.307 **The ultraviolet spectra of ϵ Ori, B0 Ia and κ Ori, B0.5 Ia.** R. Stalio, P. L. Selvelli.
Astron. Astrophys. Suppl. Ser., Vol. 21, 241 - 277 (1975).

The ultraviolet spectra of the supergiant stars ϵ and κ Ori are described. The observations were made by the Copernicus satellite and cover the spectral ranges 1000–1540 Å and 1620–3200 Å. According to the classification in the visible, the spectral features of the two stars appear very similar.

114.308 **Eine Feinanalyse des Spektrums von Arcturus.**
R. Mäckle, H. Holweger, R. F. Griffin, R. E. M. Griffin.
Mitt. Astron. Ges., No. 36, p. 71 (1975). – Abstract.

114.309 **HD 193793 – Wolf-Rayet-Stern mit Kurzzeitvariationen der C III-Emissionslinie bei 5696 Å.**
J. D. Schumann, W. Seggewiß.
Mitt. Astron. Ges., No. 36, p. 100 (1975). – Abstract.

114.310 **Spectroscopic and photometric investigation of the A 2 Ia-O star HD 160529.**
B. Wolf, L. Campusano, C. Sterken.
Mitt. Astron. Ges., No. 36, p. 100 - 101 (1975). – Abstract.

114.311 **Circumstellar lines in the spectrum of Eta Canis Majoris.** A. B. Underhill.
Astrophys. Journ., Vol. 199, 691 - 693 (1975).

High-resolution scans made with the U1 and V1 scanners of the Copernicus satellite covering the centers of lines noted from low-resolution scans as possibly circumstellar in origin have been examined for the presence of sharp components. Three components are found displaced about -235 km s^{-1} from the line center. They indicate the presence of a cloud moving rapidly out of the star. Other sharp lines due to C II, N I, Mg I, Mg II, Si IV, and Fe II are observed. They originate in a circumstellar shell which is moving with a velocity of about -25 km s^{-1} with respect to the star.

114.312 **Emission lines in the hydrogen-deficient variable MV Sagittarii near maximum light.** G. H. Herbig.
Astrophys. Journ., Vol. 199, 702 - 704 = Lick Obs. Bull., No. 691 (1975).

The H-deficient R CrB type variable MV Sgr near maximum light shows a rich emission spectrum in the 5800–8600 Å region. Most of these lines are due to Fe II, but Si II, N I, O I, and Hα also are present. Such a spectrum has not been observed in any other R CrB star, but there is some resemblance to υ Sgr. The emission lines in MV Sgr probably originate in an extended region of low temperature.

114.313 **Skylab ultraviolet stellar spectra: the Wolf-Rayet stars.**
K. G. Henize, J. D. Wray, S. B. Parsons, G. F. Benedict.
Astrophys. Journ., (*Letters*), Vol. 199, L173 - L175 (1975).

Ultraviolet spectra showing emission lines in the 1300–2000 Å region have been obtained of 12 Wolf-Rayet stars.

114.314 **Short-term variability of γ^2 Velorum.**
J. D. R. Bahng.
Astrophys. Journ., Vol. 200, 128 - 132 (1975).

Photoelectric spectrum scans of γ^2 Velorum in the spectral range 4600–4720 Å were analyzed to study the short-term variability of brightness and of emission line strenghts. The emission strengths of He II λ4686 and C III-IV λ4650 show rms variations of 3 and 2 percent, respectively. These variations have time scales of 1 minute and longer. Night-to-night variations were also found. The continuum radiation also shows small-amplitude variations. No periodic variations attributable to this star were detected on any of the nights on which these observations were made.

114.315 **Spectroscopic observations of the X-ray binary HD 153919 = 3U 1700–37.** P. S. Conti, A. P. Cowley.
Astrophys. Journ., Vol. 200, 133 - 144 = Contr. Dominion Astrophys. Obs., *Victoria,* No. 248 (1975).

Radial velocities, equivalent widths, and line profiles, with emphasis on the lines in the visual region, are discussed for the Of star HD 153919, associated with the binary X-ray source 3U 1700–37.

114.316 **High-speed spectroscopy of the ultrashort-period variable AM Canum Venaticorum (=HZ 29).**
E. L. Robinson, J. Faulkner.
Astrophys. Journ., (*Letters*), Vol. 200, L23 - L25 = Lick Obs. Bull., No. 703 (1975).
The ultrashort period variable AM CVn possesses a most unusual spectrum which may support an interpretation of the object as an accreting, semidetached binary white dwarf system.

114.317 **Spectroscopic and photometric observations of HDE 269006 in the Large Magellanic Cloud.**
B. Wolf.
Astron. Astrophys., Vol. 41, 471 - 473 (1975).
High dispersion spectra of HDE 269006 of the Large Magellanic Cloud are compared with spectra of S Dor of the same dispersion. The visual magnitude of HDE 269006 has monotonously increased during the 23 day interval from Oct. 20 to Nov. 13, 1974 by 0.08 mag. During three full nights HDE 269006 was photometrically monitored. Significant short term variations on a time scale of a few hours were found in V, $B-V$, and $U-B$ on Nov. 12, 1974.

114.318 **Observations of chromospheric and coronal emission lines in F stars.** R. G. Evans, C. Jordan, R. Wilson.
Monthly Notices Roy. Astron. Soc., Vol. 172, 585 - 602 (1975).
The Princeton telescope and ultraviolet spectrograph on the Copernicus satellite has been used to detect chromospheric emission lines of Lyman α, Si III and O VI in the F5 IV star α CMi. The interstellar absorption component of Lyman α leads to a H I column density of $\sim 1.6 \times 10^{17}$ cm^{-2} and a value for the neutral hydrogen density in the solar neighbourhood of $n_H = 0.015$ cm^{-3}. Mg II h and k emission was detected in Procyon and in the F0 Ib star α Car and the observed widths have been used to evaluate the Wilson–Bappu correlation of line width with luminosity. The analysis of the emission line intensities in Procyon leads to two models of its atmosphere in the range 10^4 to 3×10^5 K, for the limiting cases of high and low electron pressure.

114.319 **New absorbing envelope of 59 Cygni.**
V. Doazan, D. Briot, B. Bourdonneau.
Astron. Astrophys., Vol. 42, 161 - 162 (1975). In French.
In June 1973, the spectrum of 59 Cyg shows deep and sharp absorption lines, characteristic of shell stars and of the α Cygni type. It is the first time that such a spectrum is observed for this star. In March 1975, the shell absorption spectrum has considerably faded and in particular, the metallic absorption lines have completely disappeared.

114.320 **A non–LTE analysis of the CN 3883 Å band head in the upper photosphere of Arcturus.**
G. H. Mount, T. R. Ayres, J. L. Linsky.
Astrophys. Journ., Vol. 200, 383 - 391 (1975).
A detailed non–LTE study of the CN(0,0) 3883 Å bandhead spectrum of Arcturus (K2 III) provides an accurate determination of the carbon, nitrogen, and oxygen abundances in Arcturus. Non–LTE effects are significant, and the authors find that the Ayres–Linsky model provides an adequate fit to the observations for $[C, N]_* = 1/3[C, N]_\odot$ and $[O]_* = 0.60[O]_\odot$, or for $[C, N, O]_* = 1/6[C, N, O]_\odot$, but the latter abundances are unlikely. The upper photospheric microturbulence is found to be 2.5 ± 0.2 km s^{-1}.

114.321 **The ultraviolet spectrum of Gamma Cygni.**
H. M. Johnson.
Astrophys. Journ., Vol. 200, 395 - 398 (1975).
These are data of an F8 Ib supergiant in the ranges $\lambda\lambda$2558–2600 and $\lambda\lambda$2765–2805 at a resolution of 0.4 Å. Line identifications are guided by previous identifications in the range $\lambda\lambda$3977–4405. Rare-earth line richness is apparent. Although supernova-companion genesis for rare earths is tenable, there is no demonstrable connection between γ Cyg and the so-called γ-Cygni supernova remnant or nebula. In particular, there is no ultraviolet spectral evidence for mass ejection that might excite the nebula. The Mg II $\lambda\lambda$2795–2802 doublet passes from net absorption to net emission in Ib stars around spectral class G0.

114.322 **The spectra of AE Andromedae and the Hubble-Sandage variables in M31 and M33.**
R. M. Humphreys.
Astrophys. Journ., Vol. 200, 426 - 429 (1975).
Image-tube spectrograms at 83 Å mm^{-1} and 90 Å mm^{-1} have been obtained of AE And and the Hubble-Sandage variables AF And (variable 19 in M31) and variables B, C, and 2 in M33. The spectra of these extragalactic supergiants are described in detail in the paper. The spectra of all five stars have a strong, hot continuum with no Balmer discontinuity. The emission-line spectrum of AE And is very much like η Car, and it is suggested that all five stars may be related to objects like η Car and S Dor.

114.323 **The spectrum of the bright variable A-1 in M31.**
A. Bianchini, L. Rosino.
Astron. Astrophys., Vol. 42, 289 - 291 (1975).
The spectrum of the bright variable A-1 in M31 recently obtained at Asiago with the 182 cm telescope is described. The remarkable similarity of this star to η Car is pointed out.

114.324 **The He II lines in the spectrum of Zeta Puppis.**
M. A. J. Snijders, A. B. Underhill.
Astrophys. Journ., Vol. 200, 634 - 640 (1975).
Equivalent widths of He II lines in the series n = 2, 3, 4, and 5 are compiled and compared with predictions from plane-parallel, static model atmospheres using a non-LTE theory of line formation. The agreement between observation and prediction for a (50,000, 4.0) model atmosphere is good for the upper members of the n = 3 and the n = 5 series, but the two lines of the n = 2 series which are observed and the upper members of the n = 4 series are stronger than predicted. Well-determined profiles of lines from the n = 3 series indicate $v \sin i = 200$ km s^{-1}. Profiles of the higher members of the n = 4 series, however, do not match the predictions, the observed line cores being deeper than predicted. The n = 4 level appears to be more overpopulated at moderate depths in the atmosphere than the non-LTE calculations with plane-parallel layers indicate. This may be due to an overlap of the H and He II lines in the even-even series caused by macroturbulent velocities of the hydrogen atoms and helium atoms.

114.325 **The temperature and spectrum of VX Aquilae.**
A. E. Greene, R. F. Wing.
Astrophys. Journ., Vol. 200, 688 - 693 (1975).
The spectrum of the red variable star VX Aql contains strong CN bands and falls in the intermediate SC category between types S and C. A careful attempt to detect bandheads of ZrO and C_2, involving the comparison of computed CN spectra with tracings of image-tube spectrograms, gave negative results. The authors interpret this to mean that the abundances of oxygen and carbon are more exactly balanced in VX Aql than in most other SC stars, which show either weak ZrO or weak C_2 bands. The color temperature and CN strength of VX Aql have been measured by narrow-band photometry in the near-infrared. The color temperature, after correction for the estimated amount of interstellar reddening, ranges from 2450 to 1920 K, lower than for any other known SC star. The large observed range in temperature and infrared

magnitude, as well as the appearance of hydrogen emission lines, supports the suggestion that the star is a Mira variable.

114.326 **The far UV spectrum of Phi Persei.** M. Plavec.
Bull. American Astron. Soc., Vol. 7, 405 (1975).
Abstr. AAS.

114.327 **On the spectrum of FG Sge, 1972 - 1974.**
J. S. Tenn, E. M. Barsman.
Bull. American Astron. Soc., Vol. 7, 432 (1975). – Abstr. AAS.

114.328 **CO mapping of the V1057 Cygni region.**
K. P. Bechis, K. Y. Lo.
Bull. American Astron. Soc., Vol. 7, 464 (1975). – Abstr. AAS.

114.329 **Interpretation of the Zeta Oph ultraviolet CO spectrum.** K. S. Krishna Swamy, D. C. Morton, A. M. Smith, T. P. Stecher.
Bull. American Astron. Soc., Vol. 7, 464 (1975). – Abstr. AAS.

114.330 **Changes in the K-line of Arcturus observed at high resolution by SEC vidicon spectroscopy.**
H. Y. Chiu, P. Adams, J. L. Linsky, G. S. Basri, S. P. Maran, R. W. Hobbs.
Bull. American Astron. Soc., Vol. 7, 469 (1975). – Abstr. AAS.

114.331 **The ultraviolet spectrum of HR 3089.**
J. B. Lester.
Bull. American Astron. Soc., Vol. 7, 471 (1975). – Abstr. AAS.

114.332 **Observation of the far-ultraviolet (1160 - 1750 Å) spectrum of Capella.**
R. C. Vitz, H. Weiser, A. Weinstein, E. S. Warden, H. W. Moos.
Bull. American Astron. Soc., Vol. 7, 471 (1975). – Abstr. AAS.

114.333 **An analysis of the mildly metal-deficient star HD 28620.** P. Lee, C. L. Perry.
Bull. American Astron. Soc., Vol. 7, 499 (1975). – Abstr. AAS.

114.334 **The infrared spectrum of the carbon stars UU Aur and Y CVn from 4000 to 6800 cm^{-1}.**
M. Querci, F. Querci.
Astron. Astrophys., Vol. 42, 329 - 340 (1975).
The infrared spectra of the carbon stars UU Aur and Y CVn at 0.3 cm^{-1} resolution are presented. The authors have identified the dominant features of the CO, CN and C_2 molecules and some atomic features.

114.335 **Absolute rocket photometry of γ-Ori in the VUV-region.** H. Hessberg, J. Niekerke, K.-H. Stephan.
Astron. Astrophys., Vol. 42, 395 - 406 (1975).
In autumn 1973 the absolute photon flux of γ-Ori (HD No. 35468, B2 III, m_v = 1,64) was measured with two rocket-borne photometers in the wavelength region from $\lambda\lambda$ 1250 to 3000 Å. The bandwidth was 100 resp. 200 Å. Two recent model calculations for line blanketed stellar atmospheres are strongly supported longward of 1500 Å by the presented measurements. Model comparison yields an effective temperature of T_{eff}=24600° K with log g = 4.0. Compared to previous measurements the present results are in general above all other values down to 1500 Å. Below 1500 Å they present the lower limit of all so far measured values and are in disagreement with theories. Detailed critical discussion hints to unknown circumstellar or interstellar variation effects rather than to instrumental failure.

114.336 **BP Tau: new evidence for infall?** L. V. Kuhi.
Publ. Astron. Soc. Pacific, Vol. 87, 502 - 503 (1975). – Abstr. ASP.

114.337 **An interesting new southern peculiar A star – HD 137509.** A. P. Cowley, N. Houk.
Publ. Astron. Soc. Pacific, Vol. 87, 527 - 528 (1975).
A bright southern B star noted as peculiar from objective-prism plates is found to be rich in Si, Fe, Cr, and Ti but deficient in He.

114.338 **Detection of a linear polarization decrease in the Hγ line of ζ Tauri.** D. P. Hayes.
Publ. Astron. Soc. Pacific, Vol. 87, 609 - 611 (1975).
A statistically significant linear polarization decrease is reported in the Hγ line of ζ Tau, an intrinsically polarized shell star. This result reinforces previous observations which indicated that unpolarized line emission could not completely account for the polarization reduction observed in the Hγ line of γ Cas.

114.339 **A new helium spectrum variable and a new helium-rich star.** N. R. Walborn.
Publ. Astron. Soc. Pacific, Vol. 87, 613 - 616 (1975).
HD 184927, known previously as a helium-rich star, has been found to have a variable helium spectrum; the equivalent widths of five He I lines are larger by an average of 46 % on a 1974 spectrogram than on one obtained with the same equipment in 1970. HD 186205 has been found to be a new, pronounced helium-rich star.

114.340 **Detailed analysis of the high-velocity giant HD 6497.**
G. Cayrel de Strobel, L. E. Pasinetti.
Astron. Astrophys., Vol. 43, 127 - 131 (1975).
This new detailed analysis of the high-velocity giant HD 6497 (K 2 III) shows the importance of using really weak lines in abundance determinations. The present analysis has been performed from curves of growth enriched in weak lines. HD 6497 is found normal in iron-content in contrast to some previous determinations. This finding raises questions regarding the star's membership in the Arcturus moving group.

114.341 **Wavelength dependence of polarization XXX. Intrinsic polarization in Phi Persei.**
G. V. Coyne, I. S. McLean.
Astron. Journ., Vol. 80, 702 - 710 (1975).
Narrow, intermediate, and wide-band polarimetric observations obtained with polarimeters at two different observatories are used to analyze the intrinsic polarization mechanisms in the extended atmosphere of Phi Persei. The polarization in the continuum radiation is produced by electron scattering through a scattering optical depth of τ= 0.12 in a homogeneous, isothermal disk of completely ionized hydrogen. The emission flux in Hα and Hβ is partially polarized. If the polarization of the emission flux is also due to electron scattering, then this flux originates at an optical depth estimated to be τ =0.03 corresponding to a geometrical depth of about one-half stellar radius. However, other mechanisms, such as non-LTE resonant scattering, may be responsible for the emission flux polarization.

114.342 **Variation of Balmer lines in the spectrum of Pleione (BU Tau) in 1969 - 1973.**
A. Kh. Mamatkazina, Z. A. Kudryavtseva.
Astron. Tsirk., No. 841, p. 1 - 2 (1974). In Russian.

114.343 **Applications of Fourier analysis to broadening of stellar line profiles. I. Three Mn stars.**
M. A. Smith, S. B. Parsons.
Astrophys. Journ., Suppl. Ser., No. 289, Vol. 29, 341 - 361 (1975).
This paper explores a modified form of Gray's technique of determining classical microturbulence (T) by Fourier analysis of line profiles. The authors' study for three Mn Ap stars shows that the formal errors placed on the solution for T are most sensitive to limited photon statistics, to the choice of

whether rotation (V) or macroturbulence (M) is dominant in broadening the line, and to the total amount of this broadening. On the other hand the solution for T is insensitive to the damping parameter, the model atmosphere, and the assumption of LTE. It is demonstrated that the Fourier technique provides a more convenient and precisely defined test of the value of T than would be obtained from direct comparison of line profiles. For ι CrB $T = 0 \pm 1/4$ if $V = 4$ km s^{-1}, or $T = 3/4 \pm 1/4$ if $M = 3$ km s^{-1} are obtained. For υ Her and ϕ Her, $0 \leqslant T \leqslant 3$ with $V = 12$ is found.

114.344 **A model for VV8.** D. M. Zipoy.
Astrophys. Journ., Vol. 201, 397 - 403 (1975).
The author presents new optical observations for the peculiar emission-line object VV8 and investigates a fairly general model of the object. VV8 probably is not a young planetary; instead, it is a shell star with a rather thick shell that produces the supergiant absorption spectrum characteristic of shell stars and in addition produces considerable reddening of the central star's spectrum.

114.345 **Low-resolution ultraviolet spectroscopy of several hot stars observed from Apollo 17.**
R. C. Henry, A. Weinstein, P. D. Feldman, W. G. Fastie, H. W. Moos.
Astrophys. Journ., Vol. 201, 613 - 623 (1975).
Low-resolution ultraviolet spectra were obtained for six early-type stars. Stars observed, with their spectral classes, were ζ Oph (O9.5), α Vir (B1 V), ζ Tau (B2 IV), η UMa (B3 V), α Eri (B5 IV), and α Gru (B5 V). The absolute intensities are in good general agreement with the observations of others, and with the predictions of stellar model-atmosphere calculations.

114.346 **Observations of the circumstellar gas shells around Betelgeuse and Antares.**
A. P. Bernat, D. L. Lambert.
Astrophys. Journ., (*Letters*), Vol. 201, L153 - L156 (1975).
K I resonance line (λ7699) emission has been detected from the circumstellar gas shell surrounding Betelgeuse (α Ori). A shell diameter of 4″–8″ (80–160 stellar radii) is indicated. New observations of the Ca II infrared triplet line cores are used to obtain a lower limit to the Ca^+ (and, hence, K I) shell radius. The 8542 Å line in α Ori does not show a circumstellar absorption core, and the shell radius limit, $R_{shell} \geqslant 31\ R_{star}$, is estimated. A 108 mÅ circumstellar core is seen in the 8542 Å line in Antares (α Sco), and a shell radius $R_{shell} \approx 4.0\ R_{star}$ is suggested.

114.347 **H-beta spectrophotometry of SX Phe.**
H. J. Wood, R. Albrecht, W. Nees.
Astron. Astrophys., Vol. 43, 455 - 458 (1975).
A new way to use the ESO spectrum scanner is described. The method provides light-curves at the continuum and line center of H-beta and equivalent width curves automatically. Observations of SX Phe show a pre- and post-maximum hump in the equivalent width curve previously undetected in ordinary H-beta photometry.

114.348 **Energy distribution in the spectra of 15 stars of spectral classes O, B, A.**
N. S. Komarov, E. A. Depenchuk, R. I. Chuprina.
Astrometriya i Astrofizika, *Kiev*, vyp. (No.) 27, (see 003.004), p. 113 - 118 (1975). In Russian.
The energy distribution E_λ in 15 stars in the spectral region from 3200 to 7750 Å were calculated from observations. The values of E_λ are given in absolute units.

114.349 **Infrared observations of young stars–VI. A 2- to 4-micron search for molecular features.**
M. Cohen.
Monthly Notices Roy. Astron. Soc., Vol. 173, 279 - 293 (1975).
Narrow-band spectrophotometric observations with a resolution of $\Delta\lambda/\lambda \sim 0.015$ between 2 and 4μ are presented of 21 T Tau and early-type young stars. Virtually all the stars are characterized by smooth featureless continua. The extremely young star HL Tau has an absorption feature centred at 3.1μ which is well matched by extinction from 3×10^{-5} g cm^{-2} of pure ice. Particle radii are $\lesssim 0.3\mu$. FU Ori and V1057 Cyg show absorptions due to water vapour near 2μ.

114.350 **Ultraviolet observations of cool stars. III. Chromospheric and coronal lines in α Tauri, β Geminorum, and α Bootis.** W. McClintock, J. L. Linsky, R. C. Henry, H. W. Moos, H. Gerola.
Astrophys. Journ., Vol. 202, 165 - 182 (1975).
The ultraviolet spectrometer of the Princeton Experiment Package aboard the Copernicus satellite has been used to obtain high-resolution measurements of Lα, the Mg II λ2800 doublet, and upper limits on the Si III λ1206 line in the K giants α Tau and β Gem. The intensities and line shapes are compared with earlier observations of α Boo. The Lα and Mg II profiles for α Tau resemble those for α Boo, in that they are highly asymmetrical, while β Gem shows much more symmetrical profiles. In the case of β Gem only, the O V intercombination line at 1218 Å is observed, suggesting a well-developed corona substantially cooler than that of the sun.

114.351 **The coronal line spectrum of AS 295.**
G. H. Herbig, D. Hoffleit.
Astrophys. Journ., (*Letters*), Vol. 202, L41 - L45 = Lick Obs. Bull., No. 719 (1975).
AS 295 was discovered originally as a twelfth magnitude star having Hα in emission. It has now been found that the twelfth magnitude star at that position is a normal K giant and that the emission spectrum actually originates in a much fainter star 3″ distant (called AS 295 B), which in addition to bright lines of H, He I, He II, and O I shows a number of the stronger coronal lines such as [Fe X] λ6374 and [Fe XIV] λ5303. Provided the Fe ions coexist, the coronal line region of AS 295 B must have a T_e of at least 1.25×10^6 K. AS 295 B was found to have undergone a hitherto undetected nova-like outburst in 1965. However, the persistence of its coronal spectrum 10 years later is unprecedented among the novae that have shown coronal lines, although it may be related to the fact that AS 295 B is the slowest nova of the group.

114.352 **Variable blue object with a peculiar spectrum.**
A. Elvius.
Astron. Astrophys., Vol. 44, 117 - 121 (1975).
A strongly variable object has been found, which exhibits unusual spectral properties. The colour is very blue ($U-B = -0.9$ to -1.0; $B-V = 0.1 \pm 0.2$) and remarkably constant during variations in the V magnitude from 14 to 17. No emission lines and no hydrogen absorption lines were found in these spectra. Comparisons are made with other types of hydrogen-deficient stars like MV Sgr and DB white dwarfs. The nature of the object still remains uncertain.

114.353 **On the UV spectrum of Beta Aurigae.**
D. J. Stickland, K. A. van der Hucht.
Astron. Astrophys., Vol. 44, 139 - 146 (1975).
Theoretically synthesised spectra of Beta Aurigae have been compared with spectrophotometry at 2 Å resolution in three regions between 2000 Å and 3000 Å. Good agreement depends on the completeness of the line list and on the adopted microturbulence rather than the detailed accuracy of the f-values and abundances.

114.354 **An anomalous 2200 Å absorption band in the WR star HD 192163.** A. J. Willis, R. Wilson.

Astron. Astrophys., Vol. 44, 205 - 207 (1975).

New ultraviolet observations of Wolf-Rayet stars show that HD 192163 (WN 6) has a strong 2200 Å absorption band excess. It is proposed that this excess is produced in the ring nebulosity, NGC 6888, surrounding the star. The deduced number density of 2200 Å absorbers in the nebula compared to hydrogen is shown to be 100 times the interstellar space value and indicates a common material.

114.355 **Thorium in the spectrum of HR465.**
C. R. Cowley, M. S. Allen, G. C. L. Aikman.
Nature, Vol. 258, 311 (1975). – Letter.

114.356 **Detailed spectroscopic investigation of the strontium Ap star HD 140160.** T. A. Ryabchikova.
Izv. Krymskoj Astrofiz. Obs., Vol. 53, p. 254 (1975). In Russian. – Abstr. of a paper presented at the conference "Magnetic fields and chemical composition of the sun and stars" at the Crimean Astrophys. Obs., 1974, April 9 - 12, (see 012.027).

114.357 **On rapid changes in the spectrum of the magnetic variable star HD 215441.**
N. S. Polosukhina, K. K. Chuvaev.
Izv. Krymskoj Astrofiz. Obs., Vol. 53, p. 255 (1975). In Russian. – Abstr. of a paper presented at the conference "Magnetic fields and chemical composition of the sun and stars" at the Crimean Astrophys. Obs., 1974, April 9 - 12, (see 012.027).

114.358 **The helium emission-line object G 61-29.** J. Smak.
Acta Astron., Vol. 25, 227 - 233 (1975).

G 61-29 showed no major photometric activity in 1974. The radial velocity variations with a total range of 100 km/sec but with no obvious periodicity were present. The He I emission lines are shown to consist of a double component with a peak-to-peak separation of 1300 km/sec and a single component shifted to the red, with respect to the double component, by 75 km/sec. A tentative interpretation of the observed phenomena is given within the hypothesis of a close binary system.

114.359 **A study of the atmospheres of F-type stars. II. A curve of growth analysis of α Canis Minoris.**
T. Hasegawa.
Sci. Rep. Tôhoku Univ., First Ser., Vol. 58, 126 - 140 (1975).

The absolute curve of growth analysis is carried out for the spectra of α CMi (F5 IV-V) with recently improved *gf*-values. The derived excitation temperature is consistent with the preliminary temperature which is taken from $\tau = 0.22$ in the model atmosphere, but is much lower than effective temperature. The unique value of temperature and turbulent velocity can adequately explain the empirical plots of both neutral and ionized elements for which different values have been adopted by many authors. The relations between *gf*-values, turbulent velocities and abundances are discussed.

114.360 **Spectres infrarouges de quelques étoiles particulières à émission.** Y. Andrillat, L. Houziaux.
Comptes Rendus Acad. Sci. Paris, Sér. B, Vol. 281, 321 - 323 (1975).

Nous décrivons les spectres des objets HD 51585, V 1016 Cyg, HBV 475 et XX Oph, obtenus en 1974 et 1975 entre 8000 et 11000 Å avec une dispersion de 230 Å/mm à l'aide d'un spectrographe à tube image attaché au foyer Cassegrain du télescope de 193 cm de l'Observatoire de Haute Provence.

114.361 **Near-ultraviolet observation of four bright Mn stars.**
R. Faraggiana, K. A. van der Hucht, M. Burger.
Astrophys. Space Sci., Vol. 38, 243 - 249 (1975). – Paper presented at the I.A.U. Colloquium No. 27 on 'UV and X-ray spectroscopy of astrophysical and laboratory plasmas', Cambridge, Mass., U.S.A., 9–11 September 1974.

Near-ultraviolet spectra, obtained with the ESRO TD-1A/S59 experiment, of the four Mn stars β Tau, γ Crv, α And, and μ Lep and six normal stars are compared. The normal stars show good agreement with synthetic spectra. The Mn stars have spectra which become hotter at shorter wavelengths. They also show strong Mn II lines and weak Mg II lines. A time-variation for Ti II λ 2828 was noted in γ Crv and αAnd.

114.362 **Fine analysis of three R Coronae Borealis stars.**
D. Schönberner.
Astron. Astrophys., Vol. 44, 383 - 388 (1975).

The coarse analyses of the R Coronae Borealis stars R CrB, RY Sgr, XX Cham are rechecked with the model atmosphere techniques in order to obtain reliable values of effective temperatures and gravities. The abundances appear to be the same as for the (extreme) helium stars: The hydrogen content n_H is 10^{-4} or less, carbon is the most abundant element after helium, the metals have approximately solar abundances. The conclusion is drawn that the R CrB stars are cool members of the class of the (extreme) helium stars.

114.363 **On the variability of the peculiar A type star HD 49976.** H. M. Maitzen, R. Albrecht.
Astron. Astrophys., Vol. 44, 405 - 410 (1975). In German.

Photoelectric photometry in 11 colors and spectroscopic observations of the Strontium variable HD 49976 were carried out. The authors calculate a rotational period $P = 2.9760$ days. Light variation can be explained by redistribution of flux which is absorbed in the UV and reemitted in the visible part of the spectrum. The Sr variation shows a double wave like the light curves. However, in contrast to the light curves, its secondary maximum is much weaker than its primary. In HD 49976 the light variation is strongly reduced at 5200 Å and noticeable deminished at 4100 Å (v-band). The authors propose a decentered dipole for HD 49976, as found for a number of other Ap stars.

114.364 **The continuous spectrum of P Cygni in 1971–1974.**
U. Hänni, L. Luud.
Publ. Tartu Astrofiz. Obs., Vol. 43, 31 - 38 (1975).

Since 1969 the continuous spectrum of P Cygni has been observed at Tartu Observatory using direct-current scanners. The measurements in 1971 showed a considerable weakening of the ultraviolet continuum with respect to earlier observations. The reality of a flux decrease of about 10 per cent has been confirmed by the latest results, while a further decrease seems to have ceased.

114.365 **Curve-of-growth analysis of M-type giants β And and β Peg.** L. Hänni, M. Kipper.
Publ. Tartu Astrofiz. Obs., Vol. 43, 51 - 84 (1975). In Russian.

A curve-of-growth analysis is carried out with high-dispersion spectrograms in the region λλ 5050 - 6150 for two M-type stars β And and β Peg using the newest available oscillator strengths. The atmospheric parameters and the abundances of elements are determined.

114.366 **The helium variable HR 7129.** L. A. Balona.
Monthly Notices Roy. Astron. Soc., Vol. 173, 605 - 618 (1975).

New spectroscopic and photometric data, including a line identification list, are presented for the spectrum variable HR 7129. The most likely period of the spectrum, light and velocity variations is 7.34 days. The star belongs to the α Scl subgroup of helium-weak stars and closely resembles a Cen in luminosity and effective temperature. The observations are interpreted in terms of the oblique-rotator hypothesis. The position of HR 7129 in the HR diagram relative to other peculiar B-type stars is discussed.

114.367 **The profile of He I λ5876 in the P-Cygni-type Of star HD 152408.** N. R. Walborn.
Publ. Astron. Soc. Pacific, Vol. 87, 735 - 736 (1975).

The blue-shifted absorption component of the P-Cygni profile at He I λ5876 in HD 152408 (O8: Iafpe) has been found to be extremely broad, extending -1000 km sec^{-1} from the emission maximum. This unusual profile is probably due to overpopulation of the lower level of λ5876, which permits it to form throughout a greater extent of the expanding atmosphere than most other lines. The λ5876 profile in HD 152408 is compared to those in the similar but less extreme P-Cygni star HD 151804 (O8 Iaf), and in the Wolf-Rayet star HD 151932 (WN7-A). The similarity between the absorption components in HD 152408 and the WN star is striking.

114.368 **A search for spectrum variability in HD 148688 and ζ^1 Scorpii.** N. R. Walborn.
Publ. Astron. Soc. Pacific, Vol. 87, 759 - 761 (1975).

Equivalent-width measurements are given for lines in the spectra of the luminous, early-B supergiants HD 148688 and ζ^1 Sco. Recently reported large, rapid variations in the spectra of these stars are not observed.

114.369 **Nature of the Wolf-Rayet star in the binary system V 444 Cyg.**
A. M. Cherepashchuk, Kh. F. Khaliullin.
Astron. Zhurn. Akad. Nauk SSSR, Vol. 52, 1214 - 1225 (1975). In Russian. English translation in Soviet Astron., Vol. 19, No. 6.

Using narrow-band light curves of the system V 444 Cyg (WN5 + O6) the photometric structure of the disk WR-star at different wavelengths of the continuum is reconstructed. The empirical distribution of the colour and brightness temperature was obtained. This distribution shows that the temperature of the surface of the WR core exceeds 70000° – 100000°. The colour temperature of the envelope radiation is about 10000°, for the whole disk T_c = 20000°. The WR-star is formed as the result of mass exchange in the binary system and, probably, it is a helium remnant being on the final state of stellar evolution after the main sequence stage. It is possible that it will form a relativistic object – a neutron star or a black hole.

114.370 **Ultraviolet observations of cool stars. IV. Intensities of Lyman-α and Mg II in Epsilon Pegasi and Epsilon Eridani, and line width-luminosity correlations.**
W. McClintock, R. C. Henry, H. W. Moos, J. L. Linsky.
Astrophys. Journ., Vol. 202, 733 - 740 (1975).

The Princeton spectrometer on the Copernicus satellite has been used to confirm the existence of a line width-luminosity relation for the Lα and Mg II λ2800 chromospheric emission lines in K-type stars, by observation of a K2 dwarf (ϵ Eri) and of a K2 supergiant (ϵ Peg).

114.371 **An analysis of the superluminous star HR 8752.** R. E. Luck.
Astrophys. Journ., Vol. 202, 743 - 754 (1975).

A model atmosphere-curve of growth analysis of HR 8752 is presented. An effective temperature of 4000 K is derived from 1973 spectra and is shown to be consistent with photometric data. Gravity-dependent abundances are given, and it is determined that a $\log g \approx -2$ yields the most consistent results. The metal abundances have solar-system values to within the uncertainties; no evidence of *s*-processing is found in the derived abundances. Non-LTE effects are suggested to be important in Na I, O I, and possibly N I.

114.372 **8 Å resolution spectrophotometry of the peculiar A stars HD 110066 and HD 133029.**
R. B. Phillips, J. D. Fix, J. S. Neff.
Astrophys. Journ., (*Letters*), Vol. 202, L145 - L146 (1975).

The authors have made oversampled spectrophotometric observations of the peculiar A stars HD 110066 and HD 133029 using a bandpass of 8 Å and a channel separation of 3.8 Å. Contrary to the suggestion of Adelman, they find no evidence for bound-free discontinuities near 4200 Å in the spectra of these two stars but find that the 4200 Å depressions reported by Adelman are caused primarily by well-known blends of atomic lines of elements whose apparent overabundances are consistent with the classified peculiarities of the two stars.

114.373 **Spectrophotometry of the Be-type star ψ Per.** L. M. Sapargalieva.
Trudy Astrofiz. Inst., *Alma-Ata*, Vol. 25, 26 - 37 (1975). In Russian.

The change of the form and intensity of the H and of the He I profiles are considered. Computed profiles in rest intensities are given. Equivalent widths of the photosphere and of the shell components were measured. All components of the line showed rapid variations during one night and from night to night. The value of the number of the He I atoms on the 2^3P and 2^1P levels over 1 cm^2 of the surfaces is determined. The discontinuity of the intensity near the limit of Balmer series is computed.

114.374 **Evidence for a temperature rise in the outer layers of α Lyrae, from Copernicus observations of Lyman-α.** F. Praderie, E. Simonneau, T. P. Snow, Jr.
Astrophys. Space Sci., Vol. 38, 337 - 343 (1975). – Paper presented at the I.A.U. Colloquium No. 27 on 'UV and X-ray spectroscopy of astrophysical and laboratory plasmas', Cambridge, Mass., U.S.A., 9–11 September 1974.

The wings of Lα in α Lyr (A0 V), observed by the Copernicus satellite, cannot be interpreted if one assumes a classical LTE model photosphere. The authors argue that the electron temperature must increase outwards over the surface value reached in radiative equilibrium.

114.375 **High temperature plasma in β Lyrae, observed from Copernicus.** Y. Kondo, M. Hack, J. B. Hutchings, G. E. McCluskey, Jr., M. Plavec, R. S. Polidan.
Astrophys. Space Sci., Vol. 38, 353 - 358 (1975). – Paper presented at the I.A.U. Colloquium No. 27 on 'UV and X-ray spectroscopy of astrophysical and laboratory plasmas', Cambridge, Mass., U.S.A., 9–11 September 1974.

High-resolution ultraviolet spectrophotometry of the complex close binary system β Lyrae was performed with the Princeton Telescope Spectrometer on Copernicus. Observations were made at phases 0.0, 0.25, 0.5, and 0.75 with the resolution of 0.2 Å (far-ultraviolet) and 0.4 Å (mid-ultraviolet). The far-ultraviolet spectrum is completely dominated by emission lines indicating existence of high temperature plasma in this binary. It is believed that this high temperature plasma results from dynamic mass transfer taking place in this binary. The possibility that the secondary component is a collapsed object is also discussed; the Copernicus observations are consistent with the hypothesis that the spectroscopically invisible secondary component is a black hole.

114.376 **Electrophotometric observations of BT Dra.** I. F. Alaniya.
Byull. Abastumansk. Astrofiz. Obs., No. 46, p. 3 - 8 (1975). In Russian.

About 100 observations of BT Dra in B and V have been carried out. The brightness of the variable in V and the color-indices (B–V) vary in the range of 11.38–12.26 and 0.18–0.40, respectively.

114.377 **Electrophotometry of the magnetic variable 21 Com.** I. F. Alaniya, O. P. Abuladze.
Byull. Abastumansk. Astrofiz. Obs., No. 46, p. 9 - 16 (1975). In Russian.

Electrophotometric observations of the variable star 21 Com in the UBV system were carried out in 1972–1974. The measurements were made relative to 22 Com. The light curves in B and V colors and that of color index Δ(B–V) are given. Rapid light variations with considerable amplitude have been observed during some nights.

114.378 **2.5–1.5 micron spectrum of MWC 349 and Lk Hα101.** R. I. Thompson.
Bull. American Astron. Soc., Vol. 7, 530 (1975). – Abstr. AAS.

114.379 **Mass loss and asymmetries in the lines of ι CrB.** A. H. Karp.
Bull. American Astron. Soc., Vol. 7, 544 (1975). – Abstr. AAS.

114.380 **The far ultraviolet spectrum of Sirius B from Copernicus.** M. P. Savedoff, H. M. van Horn, F. Wesemael, L. H. Auer, T. P. Snow, D. G. York.
Bull. American Astron. Soc., Vol. 7, 552 (1975). – Abstr. AAS.

114.381 **On the accuracy of spectrophotometric standards of α Aql and α Peg.** N. S. Komarov, T. N. Korotkikh.
Astron. Tsirk., No. 877, p. 3 - 5 (1975). In Russian.

114.382 **The composite nature of the F supergiant CD −30°5135.** R. M. Humphreys.
Publ. Astron. Soc. Pacific, Vol. 87, 933 - 934 (1975).
CD −30°5135 has a composite spectrum consisting of an F-type supergiant and a late B-type star (F2ep *Iab* + B8:). The velocity data confirm that it is a double-line spectroscopic binary. P Cygni emission profiles are present at Hγ and Hδ.

114.383 **Spectral variations of the supergiant star 55 Cygni.** P. Granès.
Astron. Astrophys., Vol. 45, 343 - 347 (1975). In French.
Spectral variations of HD 198478 are reported. The radial velocities curves examination shows an oscillation period of 4 to 5 days inside an atmospheric motion of 15 to 16 days. Analysis of the Hα line profiles seem to confirm the radial velocities results.

114.384 **HD 87643.** W. Wamsteker.
IAU Circ., No. 2796 (1975).

114.385 **ο Andromedae.** P. Koubský, Kocourek.
IAU Circ., No. 2802 (1975).

114.386 **ο Andromedae.** B. W. Bopp.
IAU Circ., No. 2810 (1975).

114.387 **ο Andromedae.** G. J. Peters, R. S. Polidan.
IAU Circ., No. 2814 (1975).

114.388 **ο Andromedae.** D. J. Stickland, R. W. Argyle.
IAU Circ., No. 2825 (1975).

114.389 **ο Andromedae.** M. Fracassini, L. E. Pasinetti.
IAU Circ., No. 2881 (1975).

114.390 **ο Andromedae.** P. Koubský.
IAU Circ., No. 2889 (1975).

114.391 **Can the peculiarities of HD 101065 be explained by fission of transuranium elements?** A. Przybylski.
Proc. Astron. Soc. Australia, Vol. 2, 352 - 353 (1975).

114.392 **A nitrogen rich OB supergiant in the Large Magellanic Cloud.** A. R. Hyland, M. S. Bessell.
Proc. Astron. Soc. Australia, Vol. 2, 353 - 355 (1975).

114.393 **Nochmalige Untersuchung von Hα-Sternen in NGC 7000.** R. Hudec.
MVS, *Sonneberg*, Vol. 7, 63 (1975).

114.394 **Interstellar extinction and spectrophotometry of the helium-rich star HD 264111.** W. Wegner.
Stud. Soc. Sci. Torunensis, Toruń, Poland, Sect. F (Astron.), Vol. 5, 153 - 174 = Bull. Astron. Obs. N. Copernicus Univ. Toruń, No. 54 II (1974).
Spectrophotometric measurements of the He-rich star HD 264111 were made on objective prism spectra. The energy distribution and line depths for H, He I, He II, O II, N II, N III, Si II, Si III, C II, C III are obtained. The interstellar extinction from the colour excess of field stars is derived.

114.395 **Fine analysis of the helium-rich star HD 184927.** N. A. Higginbotham.
Thesis, Louisiana State Univ., Baton Rouge (USA). [Available from Univ Microfilms, Ann Arbor, Mich. Order No. 75-1932], 196 pp. (1974).

114.396 **Abundance of sodium in the red giants Alpha Tauri and Beta Geminorum.** J. C. LoPresto.
Thesis, Michigan Univ., Ann Arbor (USA). [Available from Univ. Microfilms, Ann Arbor, Mich. Order No. 74-25,257], 85 pp. (1974).

114.397 **Stellar studies in the balloon ultraviolet.** B. Bates, P. P. D. Carson, P. L. Dufton, C. D. McKeith.
Phil. Trans. Roy. Soc. London, Ser. A, Vol. 279, 379 - 390 (1975).
Observations with a spectral resolution of 0.01 nm have been made of several stars in the balloon ultraviolet. For the two coolest stars studied (β Ori and α Lyr) wavelengths of all observed spectral features in the range 273 - 288 nm are given with an accuracy of approximately ± 0.004 nm. Using these results the velocity field in the atmosphere of β Ori has been investigated and evidence is found for an outward motion in the higher layers and a pulsation-type motion in the deepest layers. The strength of the Mg^+ resonance and subordinate lines near 280 nm for all the stars observed is compared with non-LTE calculations.

114.398 **Rocket and spacecraft studies of ultraviolet emissions from astrophysical targets.** W. G. Fastie, H. W. Moos, R. C. Henry, P. D. Feldman.
Phil. Trans. Roy. Soc. London, Ser. A, Vol. 279, 391 - 400 (1975).

114.399 **Skylark rocket observations of ultraviolet spectra of γ^2 Velorum and η Puppis.** W. M. Burton, R. G. Evans, W. G. Griffin.
Phil. Trans. Roy. Soc. London, Ser. A, Vol. 279, 355 - 369 (1975).
The spectra were obtained in June 1973 by means of three Wadsworth-mounted objective grating spectrographs with a combined wavelength range 90 - 230 nm. Analysis of the spectra provides information about the composition and properties of the interstellar gas in the line of sight. The observations also provide useful data on the stellar atmospheres. P Cygni line profiles are observed in the spectra of both stars indicating that high-velocity material is being ejected.

114.400 **The ultraviolet spectrum of η Tauri.** S. R. Heap.
Phil. Trans. Roy. Soc. London, Ser. A, Vol. 279, 371 - 377 (1975).
The U.V. stellar lines indicate that: the star has an effective temperature as high as 27000 K; the stellar atmosphere appears to have serious abundance deficiencies in carbon and silicon; the velocity field in the atmosphere is

complicated but shows evidence for outward acceleration and differential rotation.

114.401 **The continuum energy distribution of the Wolf-Rayet star in γ^2 Velorum.** K. A. van der Hucht. Phil. Trans. Roy. Soc. London, Ser. A, Vol. 279, 451 - 455 (1975).

Mise en service d'un prisme objectif normal de 60 cm de diamètre. Découverte d'une étoile du halo à grande vitesse radiale. See Abstr. 034.094.

Discovery of a corona around an early-type star and the problem of mass loss. See Abstr. 064.002.

The study of HR 3018, HD 91324 and α Crv and its implication for the age-abundance correlation of the neighbour stars. See Abstr. 064.013.

Model atmospheres of Delta Cephei. See Abstr. 064.064.

HD 50896: evidence for an extended continuum formation region. See Abstr. 113.060.

Metal abundances for K giants in a field in Scutum. See Abstr. 114.021.

Photoelectric investigations of the magnetic fields and brightness of the Am stars 15 Vul and 68 Tau. See Abstr. 116.010.

The binary nature of the Of star HD 108. See Abstr. 119.002.

Ultraviolet observations of Alpha Aurigae from Copernicus. See Abstr. 119.003.

On spectral observations of ζ Gem. See Abstr. 122.032.

Mariner 9 observations of β Cephei stars. See Abstr. 122.096.

The $^{12}C/^{13}C$ ratio in the classical cepheid T Monocerotis. See Abstr. 122.206.

On fluctuations in optical radiation of the carbon star RW LMi (CIT 6). See Abstr. 122.213.

A cool degenerate dwarf with a metal-rich atmosphere. See Abstr. 126.001.

The OB-type subdwarf HD 149382 and the nature of the subdwarf B stars. See Abstr. 126.003.

Detection of interstellar SiS and a study of the IRC+10216 molecular envelope. See Abstr. 131.008.

Differential interstellar polarization in the region of 6284 Å for HD 183143. See Abstr. 131.027.

Velocity component structure of the Zeta Ophiuchi interstellar lines. See Abstr. 131.108.

An upper limit on interstellar C IV in the spectrum of γ^2 Velorum. See Abstr. 131.134.

A study of interstellar lines in the rocket spectrum of δ Scorpii. See Abstr. 131.208.

Spectroscopy of R Monocerotis and NGC 2261. See Abstr. 132.002.

CO mapping and analysis of the V1057 Cygni region. See Abstr. 132.029.

The infrared spectrum of η Carinae: 3–14 microns. See Abstr. 141.640.

A new CH star in the globular cluster Omega Centauri. See Abstr. 154.026.

Errata

114.901 **Erratum: 'Study of the Be stars – the Balmer decrement'** [Astron. Astrophys., Vol. 11, 57 - 64 (1971)]. In French.].
D. Briot.
Astron. Astrophys., Vol. 42, 475 (1975). In French. – See 05.114.012.

114.902 **Errata: 'New variable carbon stars. IV'** [Astron. Tsirk., No. 769, p. 3 - 4 (1972)]. In Russian.
A. Alksnis, Z. Alksne.
Astron. Tsirk., No. 872, p. 8 (1975).

115 Stellar Luminosities, Masses, Diameters, HR-Diagrams and Others

115.001 **Apparent radii and other parameters for 116 A5V–F5V stars of the catalogue of Geneva Observatory.**
M. Fracassini, L. E. Pasinetti, G. Pelagatti.
Astrophys. Space Sci., Vol. 35, 313 - 320 (1975).
Preceding research on 172 B5V–A5V stars (Fracassini et al., 1973) has been extended to cover 116 A5V–F5V stars of the catalogue of Geneva Observatory (Rufener, 1971), to test the validity of the method for later type stars. There is a satisfactory agreement between the authors' results and those obtained by Gaposchkin (1958) using data from 82 eclipsing binary systems. Furthermore, the results also seem to confirm that the sun has an absolute radius larger than that averaged from F5V–G0V stars.

115.002 **The luminosities of the Beta Canis Majoris variables, the zero age main sequence and the distance of the Sco–Cen association.** L. A. Balona, M. W. Feast.
Monthly Notices Roy. Astron. Soc., Vol. 172, 191 - 203 (1975).
The Hγ-luminosity calibration of Balona & Crampton is used to derive the zero age main sequence. The Hγ calibration is applied to the β CMa variables. The results are consistent with a theoretical P–L–C relation. The pulsation constant, Q, is 0.027±0.001 implying that the variables pulsate in the first overtone. The variables fall in the S-bend region of evolution. Contrary to previous work the slope of the instability strip is not inclined to the S-bend locus in the M_{bol}–log T_e diagram. The calibration is consistent with the interferometric absolute magnitude of α Vir and the astrometric distances of β CMa stars in Sco–Cen.

115.003 **Absolute Helligkeiten und Massen der Sterne des Feldriesenastes.** H. Neckel.
Mitt. Astron. Ges., No. 36, p. 166 - 169 (1975).

115.004 **The absolute magnitudes of carbon stars: carbon stars in binary systems.** B. I. Olson, H. B. Richer.
Astrophys. Journ., Vol. 200, 88 - 94 (1975).
Absolute visual magnitudes have been determined for carbon stars belonging to binary systems. These range from about 0 to brighter than −4, with a mean value of −2.3 for N stars. Bolometric corrections are deduced, giving bolometric absolute magnitudes in the range −4 to −8.

115.005 **Bolometric corrections for late-type giants and supergiants.** P. J. Flower.
Astron. Astrophys., Vol. 41, 391 - 394 (1975).
Complete spectral energy distributions of giants, Mira variables, IRC stars, and the supergiants in *h* and χ Persei and in Carina are used to calculate bolometric corrections for M 0 through M 10 stars. The new values agree quite well with those of previous investigators. The results indicate that the absolute values of bolometric corrections for supergiants are smaller than for giants.

115.006 **Physical properties of the red giants in NGC 2420.** W. Osborn.
Monthly Notices Roy. Astron. Soc., Vol. 172, 631 - 637, (1975).
The effective temperatures, surface gravities, and masses of 11 red giant members of the old open cluster NGC 2420 have been derived from existing intermediate-band photometry on the David Dunlap Observatory system. The mean mass of the stars was found to be 0.7 $M_\odot$. The data permit the helium abundance of the cluster to be estimated and result in the value $Y = 0.30$.

115.007 **Les étoiles doubles et la relation masse-luminosité.** P. Baize.
L'Astronomie, 89ᵉ année, p. 306 - 312 (1975).

115.008 **Note on the near-infrared absolute magnitude of irregular carbon variables.** J. H. Baumert.
Astrophys. Journ., (*Letters*), Vol. 200, L141 (1975).
An apparent discrepancy between the results of Peery and Baumert on the mean absolute magnitude at 1.04 μ of irregularly variable carbon stars is resolved. The mean absolute magnitude of this group of stars derived through a statistical analysis of their proper motions and radial velocities is −4.2 ± 0.7 (m.e.).

115.009 **The angular diameter of ψ Virginis.** A. R. Walker.
Monthly Notices Roy. Astron. Soc., Vol. 173, 29P - 32P (1975).
The angular diameter of the M3 III star ψ Virginis was measured on 1975 May 21 by the lunar occultation method. A value of 6.1 ms arc is found, assuming a fully limb-darkened disc, with an error of 5 per cent. The star is a close double, with vector separation 40 ms arc and a red magnitude difference of 4.4.

115.010 **A technique for separating reddening and envelope emission in T Tauri stars.** A. E. Rydgren.
Bull. American Astron. Soc., Vol. 7, 433 (1975). – Abstr. AAS.

115.011 **The distribution of masses, ages and luminosities of population I field giants.** H. Neckel.
Astron. Astrophys., Vol. 42, 379 - 393 (1975).
The initial luminosity function, the time dependence of the stellar birth rate and the lifetimes of the stars on the main sequence as well as in the giant region determine the space density and mass spectrum of the field giants in the solar neighbourhood. The predicted mass spectrum exhibits a very pronounced maximum for stars with masses only slightly exceeding one solar mass, which were born on the main sequence in the first stages of star formation. The resulting space density (6.3×10^{-4} pc^{-3}) is in fair agreement with the values deduced from star counts. The predicted magnitude dispersion as well as the predicted position of the field giant branch in the HR-diagram agree well with the results which follow from a recalibration of the Mt. Wilson spectroscopic absolute magnitudes by means of the Wilson-Bappu effect.

115.012 **A luminosity effect in late-type supergiants.** C. L. Imhoff.
Publ. Astron. Soc. Pacific, Vol. 87, 502 (1975). – Abstr. ASP.

115.013 **The masses and dimensions of early-type binaries.** J. B. Hutchings.
Publ. Astron. Soc. Pacific, Vol. 87, 529 - 535 (1975).
In order to review the basic data on very massive stars, masses and radii are presented for all well-determined binaries of spectral type O and early B. These are derived from spectroscopic and photometric orbital parameters, in many cases using results and methods available in the last few years. The masses are consistent with those recently derived for single stars by Conti and Burnichon, which imply that Of star masses may exceed 60 $M_\odot$. Relationships between masses, temperatures, and radii are given, and compared with those for X-ray binaries.

115.014 **Absolute magnitudes of barium stars based on the width of Hα in absorption.** E. Kemper.
Publ. Astron. Soc. Pacific, Vol. 87, 537 - 540 = Lick Obs. Bull., No. 700 (1975).

Absolute magnitudes of Ba II stars determined from the width of Hα indicate that these stars have, on the average, the luminosity of normal G and K giants. There may, however, be some Ba stars of luminosity class IV. The distribution of stars in absolute magnitude does not seem to depend on Ba line strength.

115.015 **Calibration of the Wilson-Bappu effect using trigonometric parallaxes.** T. E. Lutz, D. H. Kelker.
Publ. Astron. Soc. Pacific, Vol. 87, 617 - 624 (1975).

Corrections are applied to the absolute magnitudes of the stars used by Wilson (1967) to account for the systematic error discussed by Lutz and Kelker (1973). Several methods of fitting straight lines to data are applied to this sample in an effort to determine the relation between M_v and log W_0. The authors conclude that this relation is best represented by the equation $M_v = -17.13 \log W_0 + 30.74$.

115.016 **Absolute luminosity calibration of Stroemgren's "intermediate group".** A. Heck.
Astron. Astrophys., Vol. 43, 111 - 126 (1975).

The author presents here a calibration of the relation $M_v^{(i)} = q_1 a^{(i)} + q_2 r^{(i)} + q_3$ for the intermediate group (A0–A3 stars) as defined by Stroemgren (1966, 1967), by a method based on the principle of maximum likelihood (Heck, 1975). A calibration of the mentioned relation with an index $a_R^{(i)}$ corrected for reddening is also given. In each case, the dispersion in magnitude, the mean velocity components and the corresponding dispersion are also calculated, as well as the precision of each parameter. The author proposes the following relation for main sequence stars $M_v^{(i)} = 1.89(\pm 3.90)a_R^{(i)} - 14.52(\pm 1.93)\, r^{(i)} + 1.62(\pm 0.25)$.

115.017 **Absolute magnitudes of carbon stars in the solar vicinity.** T. Mikami.
Publ. Astron. Soc. Japan, Vol. 27, 445 - 457 (1975).

Absolute magnitudes of carbon stars are studied by the statistical analysis of radial velocities and of proper motions listed in the Preliminary AGK3 Catalogue. The mean visual absolute magnitude of carbon stars varies with the temperature sub-class of C classification from $M_V = -1.3$ at about C2 to $M_V = -2.4$ at about C7. The bolometric absolute magnitudes are $M_{bol} = -2.2$ and -5.5 at C2 and C7, respectively.

115.018 **The angular diameter of Alpha Herculis A.** S. P. Worden.
Astrophys. Journ., (*Letters*), Vol. 201, L69 - L70 (1975).

A digital analysis of speckle interferometry data for the M5 II supergiant α Her A produced for a uniformly illuminated disk diameters of 0".030 ± 0".001 at 5100 Å and 0".028 ± 0".003 at 5200 Å. These results are consistent with two of the three previous diameter determinations.

115.019 **The angular diameter of TX Piscium.** D. W. Dunham, D. S. Evans, E. C. Silverberg, J. R. Wiant.
Monthly Notices Roy. Astron. Soc., Vol. 173, 61P - 62P (1975).

An occultation observation of this star on 1975 January 18 made with the 107-in. telescope gave an angular diameter for a fully darkened disk of 10.2 ± 2.4 arc ms. Taken with the observation by Lasker et al. and that of de Vegt we adopt a mean angular diameter of 9.8 ± 0.5 arc ms with apparently little or no variation.

115.020 **Stellar angular diameters and visual surface brightness: a new calibaration with color index and an application to variable star distances.**
T. G. Barnes III, D. S. Evans, S. B. Parsons.
Bull. American Astron. Soc., Vol. 7, 504 (1975). – Abstr. AAS.

115.021 **Calibration of the Wilson-Bappu effect using statistical parallaxes.** T. E. Lutz, K. M. Yoss.
Bull. American Astron. Soc., Vol. 7, 510 (1975). – Abstr. AAS.

115.022 **Stellar diameter measurements by infrared observation of lunar occultations.**
S. T. Ridgway, D. C. Wells, R. R. Joyce.
Bull. American Astron. Soc., Vol. 7, 510 (1975). – Abstr. AAS.

Photoelectric measurements of lunar occultations VII: further observational results. See Abstr. 096.005.

A new photometric metal abundance and luminosity calibration for field G and K giants. See Abstr. 113.011.

Separation of main-sequence and supergiant stars of early B type. See Abstr. 113.017.

Photoelectric multiband photometry of M giants and supergiants in the blue-visual spectral region.
See Abstr. 113.033.

Intrinsic properties of high luminosity blue stars from observations of multiple systems. See Abstr. 113.066.

Supergiants in the Milky Way.
See Abstr. 114.105.

The helium variable HR 7129.
See Abstr. 114.366.

Distances and luminosities of irregular variables of type N. See Abstr. 122.004.

Flare colours and luminosities.
See Abstr. 122.012.

On the structure of the cepheid instability strip.
See Abstr. 122.030.

Four-color and Hβ photometry of β Cephei stars and their location in the Hertzsprung-Russell diagram.
See Abstr. 122.074.

On the determination of the radii of cepheids.
See Abstr. 122.139.

On fragmentation in a collapsing gas cloud.
See Abstr. 131.154.

HD 219460 and the open cluster Ma 50.
See Abstr. 153.013.

The tip of the red giant branch (a comparison of theory and observations). See Abstr. 154.006.

The colour-magnitude diagram of the globular cluster M4. See Abstr. 154.007.

A preliminary study of the main-sequence luminosity function of the globular cluster M92.
See Abstr. 154.023.

116 Stellar Magnetic Field, Figure, Rotation

116.001 **Distribution of rotational velocities for the Hg-Mn stars.** S. C. Wolff, R. J. Wolff.
Bull. American Astron. Soc., Vol. 7, 445 (1975). – Abstr. AAS.

116.002 **Evidence for a rotational reddening in early B-type stars.** A. Maeder.
Astron. Astrophys., Vol. 42, 471 - 473 (1975).

It is shown that rapid rotation produces an intrinsic reddening of about $0^m.04-0^m.06$ for early B-type stars. The author suggests that this effect can be interpreted in terms of recent models of extended stellar atmospheres.

116.003 **Photometric variability of magnetic stars.** F. A. Catalano.
19th Colloquium International Astrophys. Liège 1974, (see 012.016), p. 117 - 124 (1975).

The observational data on magnetic stars strongly suggest concentrations of the peculiar abundances in definite regions of the stellar surface. Under the hypotheses that: (1) the magnetic axis is inclined by an angle χ with respect to the rotational axis, which, in turn, is inclined of an angle i with respect to the line of sight; (2) the peculiar abundances are concentrated in regions around the magnetic poles; (3) the mechanisms producing the abundances do not depend on the polarity of the magnetic field; (4) the Chandrasekhar's limb darkening law is representative of the radiation field in the V light; the radiative flux emerging from the visible hemisphere of the star has been computed.

116.004 **A search for strong magnetic fields in rapidly rotating Ap stars.**
J. D. Landstreet, E. F. Borra, J. R. P. Angel, R. M. E. Illing.
Astrophys. Journ., Vol. 201, 624 - 629 (1975).

A search for strong magnetic fields ($\gtrsim 10^3$ gauss) in rapidly rotating peculiar A stars has been made using the solar magnetograph technique on Balmer lines. Observations of 16 stars have been made with standard errors typically in the range of 300 to 700 gauss. No definite fields are found, and the results indicate that fields in excess of 10^3 gauss are significantly less common among rapidly rotating Ap stars than among slowly rotating ones.

116.005 **A rotational velocity study of sharp-line B stars.** R. W. Day, B. Warner.
Monthly Notices Roy. Astron. Soc., Vol. 173, 419 - 428 (1975).

Measurements of $v \sin i$ for 74 OB stars, mostly with $v \sin i < 60$ km s^{-1}, have been made. This high-resolution study enables the authors to define the frequency distribution of $v \sin i$ for small values of $v \sin i$ with improved accuracy. Combining their data with that from Slettebak for $v \sin i >$ 60 km s^{-1} they produce new frequency distributions of $v \sin i$ for main sequence and evolved B stars. Bimodal Maxwellian distribution functions are fitted to these curves.

116.006 **Axial rotation: statistics of early B stars and the effect on line intensities.** L. A. Balona.
Monthly Notices Roy. Astron. Soc., Vol. 173, 449 - 464 (1975).

Frequency distributions of $v \sin i$ on a uniform system are presented for O5–B5 stars. Theoretical distributions are given which take into account the effect of observational errors. Some physical implications are discussed. It is found that B stars probably do not rotate as solid bodies. Difficulties are encountered if Be stars are postulated to be rotating near their break-up velocity unless most of them are evolved from the main sequence, or have large differential rotations. A distinct correlation between hydrogen line strength and axial rotation is found.

116.007 **The distribution of periods of the magnetic A-type stars.** S. C. Wolff.
Astrophys. Journ., Vol. 202, 127 - 136 (1975).

Four-color (*uvby*) photoelectric observations are used to obtain periods for several Ap stars with $v \sin i < 10$ km s^{-1}. The present observations, combined with results published previously, make it possible to derive the distribution of periods for the cool (non-Si) Ap-type stars. It is argued that the data support the hypothesis that the period of variation is to be identified in all cases with the period of rotation, and that the existence of Ap stars with extremely long periods is due to the loss of angular momentum through some form of magnetic braking.

116.008 **Circular polarization of radiation of magnetic stars.** G. G. Pavlov, Yu. A. Shibanov, Yu. N. Gnedin.
Astron. Zhurn. Akad. Nauk SSSR, Vol. 52, 956 - 964 (1975). In Russian. English translation in Soviet Astron., Vol. 19, No. 5.

The general expression for the spectral dependence of the degree of circular polarization is obtained using the LTE assumption and taking into account the inhomogeneity of the magnetic field. In the limit cases when the absorption coefficient is smaller or larger than its mean value corresponding to the frequency approximate analytical formulae are derived. The polarization of the radiation by free-free transitions is investigated in detail. The increase of the surface magnetic field inhomogeneity is shown to cause an increase of both the circular polarization in the high frequency region and the frequency of maximum polarization. It is noticed that the maximum of the polarization is shifted relative to the maximum of the flux value toward the high frequency region.

116.009 **Some results of an investigation of magnetic stars at the Special Astrophysical Observatory of the USSR Academy of Sciences.** Yu. V. Glagolevskij.
Izv. Krymskoj Astrofiz. Obs., Vol. 53, p. 252 - 253 (1975). In Russian. – Abstr. of a paper presented at the conference "Magnetic fields and chemical composition of the sun and stars" at the Crimean Astrophys. Obs., 1974, April 9 - 12, (see 012.027).

116.010 **Photoelectric investigations of the magnetic fields and brightness of the Am stars 15 Vul and 68 Tau.**
V. M. Kuvshinov, G. Hildebrandt, W. Schöneich.
Izv. Krymskoj Astrofiz. Obs., Vol. 53, p. 253 - 254 (1975). In Russian. – Abstr. of a paper presented at the conference "Magnetic fields and chemical composition of the sun and stars" at the Crimean Astrophys. Obs., 1974, April 9 - 12, (see 012.027).

116.011 **On some mechanisms of generation of a magnetic field in stars.** A. Z. Dolginov.
Izv. Krymskoj Astrofiz. Obs., Vol. 53, p. 256 - 257 (1975). In Russian. – Abstr. of a paper presented at the conference "Magnetic fields and chemical composition of the sun and stars" at the Crimean Astrophys. Obs., 1974, April 9 - 12, (see 012.027).

116.012 **The influence of rotation and stellar winds upon the Be phenomenon.** D. Massa.
Publ. Astron. Soc. Pacific, Vol. 87, 777 - 784 (1975).

A number of rapidly rotating B stars, not previously known as Be stars, were observed spectroscopically at Hα. These results were then combined with existing data to show that the spectral type of a star and the minimum velocity at which it must rotate in order to become a Be star are related. The trend of this relationship is found to have a natural explanation in terms of stellar winds.

116.013 **Photoelectric observations of the magnetic field of Sirius.** E. F. Borra.
Astrophys. Journ., Vol. 202, 741 - 742 (1975).

Photoelectric observations of the magnetic field of Sirius do not show evidence of longitudinal magnetic fields greater than a few tens of gauss.

116.014 **Discovery of magnetic fields in four southern Ap stars.** E. F. Borra, J. D. Landstreet.
Publ. Astron. Soc. Pacific, Vol. 87, 961 - 963 (1975).

A search for magnetic fields in southern Ap stars has been made using a Balmer-line magnetograph. Observations of nine stars were obtained, and definite or probable evidence for the presence of fields ranging from a few hundred gauss to almost four kilogauss was found in HD 54118, HD 94660, HD 133880, and α Cir.

116.015 **Zeeman observations of new southern magnetic stars.** H. J. Wood, L. B. Campusano.
Astron. Astrophys., Vol. 45, 303 - 305 (1975).

A program of classical Zeeman spectroscopy in the southern hemisphere has resulted in the discovery of several new bright southern magnetic stars. In particular, HR 5463 and HR 5049 of magnitudes 3 and 5 respectively are bright enough for future detailed high-dispersion studies.

The magnetic thermal wind and differential rotation of ionized, magnetic atmospheres. See Abstr. 062.031.

Method for investigation of the local chemical composition of the atmospheres of magnetic Ap stars. See Abstr. 064.044.

A non-axisymmetric rigid rotator model for magnetic stars. See Abstr. 065.011.

Line profiles and magnetic intensification in magnetic stars. See Abstr. 065.034.

On gravity darkening in magnetic stars. See Abstr. 065.038.

Circulations and fossil magnetic fields in rotating stars. See Abstr. 065.087.

The loss of angular momentum and the evolution of Ap-type stars. See Abstr. 065.120.

Absorption line profiles and rotational velocities for 59 stars. See Abstr. 114.049.

Photometric scans of Hα in early A-stars. See Abstr. 114.071.

On rapid changes in the spectrum of the magnetic variable star HD 215441. See Abstr. 114.357.

On the variability of the peculiar A type star HD 49976. See Abstr. 114.363.

Electrophotometry of the magnetic variable 21 Com. See Abstr. 114.377.

117 Binary and Multiple Stars, Planetary Companions, Theory

117.001 **Evolution of massive close binaries. I. The X-ray binary stage.** C. De Loore, J. P. De Grève.
Astrophys. Space Sci., Vol. 35, 241 - 247 (1975).

X-ray binaries such as Cen X-3, Cyg X-1, Vela X-1, 2U1700–37, SMC X-1, Cir X-1, with periods ranging from $2^d.087$ to $12^d.28$, are probably evolution products of massive binary systems. The massive primary starts losing mass after core hydrogen burning and undergoes a supernova explosion. The evolution of binary systems consisting of a 20 $M_\odot$ primary and secondaries of respectively 14, 10, 8 and 6 $M_\odot$ was computed from main sequence stage through the X-ray phase, until the second mass transfer phase. Estimates were performed for the evolution of a secondary of 4 $M_\odot$. It may be seen that some of these systems during their X-ray phase match the observed mass ratios and periods.

117.002 **Linear polarization variations modulo 5.6 days in HDE 226868.** I. G. Nolt, J. C. Kemp, R. J. Rudy, R. G. Southwick, J. V. Radostitz, L. J. Caroff.
Astrophys. Journ., (*Letters*), Vol. 199, L27 - L29 (1975).

The authors have found linear-polarization variations in the ultraviolet in HDE 226868, as folded on the $5^d.6$ orbital period. The amplitudes are $\Delta p = 0.25$ percent and $\Delta\theta = 2^\circ.7$.

117.003 **Nonstationary processes and magnetic field generation in binary stellar systems.** A. Z. Dolginov.
IAU Symposium No. 67, (see 012.001), p. 373 - 374 (1975).

117.004 **Photometric effects in close binary systems including X-ray sources.** Yu. N. Efremov, R. A. Syunyaev, A. M. Cherepashchuk.
Peremennye Zvezdy, Vol. 19, 407 - 419 (1974). In Russian.

117.005 **Statistics of binaries.** C. Jaschek.
Mitt. Astron. Ges., No. 36, p. 45 - 51 (1975).
Review paper.

117.006 **Losses of mass and angular momentum during the evolution towards cataclysmic binaries.**
H. Ritter.
Mitt. Astron. Ges., No. 36, p. 93 - 94 (1975).

117.007 **The importance of tidal interaction during the evolution of close binaries.** H. Ritter.
Mitt. Astron. Ges., No. 36, p. 94 - 95 (1975).

117.008 **Diskussion des Doppelsternsystems AS Eridani.**
S. Refsdal, M. L. Roth, A. Weigert.
Mitt. Astron. Ges., No. 36, p. 95 - 96 (1975). – Abstract.

117.009 **Ein magnetohydrodynamisches Modell der Entstehung von Planetensystemen.** R. Ebert.
Mitt. Astron. Ges., No. 36, p. 115 (1975). – Abstract.

117.010 **Peculiar A-stars and planetary systems.**
E. M. Drobyshevski (*Eh. M. Drobyshevskij*).
Astrophys. Space Sci., Vol. 35, 403 - 408 (1975).

A suggestion is put forward that a number of spectral anomalies observed in some A-stars (Am and Ap-stars) may be due to the existence of planetary systems. Terrestrial type planets condense in matter exchanged between components of close binaries. Exchange of matter takes place at the stage of system formation and in later stages of evolution with both late and early stars. If the outer layers of components in a system do not become renewed by mixing or outflow, then surface anomalies in the chemical composition of stars onto which fall some of the forming planetoids may serve as an evidence of the formation of planets.

117.011 **Lunar occultations of Beta Scorpii in 1975 and 1976.** T. C. Van Flandern, P. Espenschied.
Astrophys. Journ., Vol. 200, 61 - 67 (1975).

The complex multiple star β Scorpii (combined magnitude 2.5) will be occulted by the moon 19 times during 1975-1976. Limiting lines of visibility for these occultations are presented. From modern observations it is apparent that β Sco is at least quadruple, but there is evidence for the existence of three additional components. The 1971 May occultation of the β Sco system by Jupiter and of its C component by Io raised important questions about β Sco and about Jupiter and Io, making the coming occultation series especially significant.

117.012 **Epsilon Coronae Austrinae: evidence for binary star evolution?** S. Tapia, J. Whelan.
Astrophys. Journ., Vol. 200, 98 - 105 (1975).

Spectroscopic radial velocity observations of ϵ Coronae Austrinae, the brightest W UMa type contact binary system, are reported. The parameters, (masses, radii, luminosities) of the components of the system have been determined. The primary, of mass about 1.5 $M_\odot$, is probably evolved and burning hydrogen in its core. The secondary has a very low mass, near 0.15 $M_\odot$.

117.013 **The dynamical tide in close binaries.** J.-P. Zahn.
Astron. Astrophys., Vol. 41, 329 - 344 (1975).

The non-adiabatic oscillations of a star, driven by an outer rotating gravitational field, have been studied by the use of matched asymptotic expansions. The author finds that the resonances of the free gravity modes are damped by radiative dissipation, which operates in a relatively thin region below the surface of the star. Due to that dissipation, some properties of the dynamical tide have observable consequences in close binary systems: (1) A torque is applied to a binary component; this serves to make it corotate with its companion in a time which can be short compared to its nuclear life. (2) Before that synchronization is achieved, the brightness distribution over the surface of the star is in general phase shifted relative to the external driving potential.

117.014 **Stellar stability in close binary systems.**
J. C. B. Papaloizou, G. T. Bath.
Monthly Notices Roy. Astron. Soc., Vol. 172, 339 - 357 (1975).

The linear adiabatic stability of stars in close binary systems is discussed. Both spherical and Roche geometries are considered. Two destabilizing influences are shown to exist. Detailed Roche models are constructed and their stability properties determined. It is concluded that dynamical instabilities may be important in dwarf novae, slow novae and the binary symbiotic stars.

117.015 **Interferometric measurements of binary stars.**
W. C. Wickes.
Astron. Journ., Vol. 80, 655 - 657 (1975).

Ten close binary star systems have been observed with the achromatic interferometer on the Mt. Wilson 60-in. reflecting telescope. No reduction of fringe visibility was observed compared to observations on smaller telescopes. A second resolution of the astrometric binary μ Cas has been obtained. The star α Del A was found to have a previously undetected companion.

117.016 **Astrometric study of Barnard's star from plates**

taken with the Sproul 61-cm refractor.
P. van de Kamp.
Astron. Journ., Vol. 80, 658 - 661 (1975).
To obviate the uncertainty introduced by instrumental equations prior to 1950 an analysis has been made of the Sproul astrometric material of Barnard's star over the interval 1950–1974. All plates were measured on the Grant two-coordinate machine. The "short" perturbation is confirmed, the "long" perturbation is less well determined.

117.017 **Tidal flows in binary systems.**
A. Z. Dolginov, D. G. Yakovlev.
Astrophys. Space Sci., Vol. 36, 31 - 78 (1975).
The expressions of the tidal velocity in not very close binaries (double stars, the sun and a planet, a planet and a satellite) are derived and applied in particular to white dwarfs and the giant planets of the solar system. The magnitude of the velocity on the surface of Jupiter is estimated to be about 0.5 cm s^{-1}. In white dwarfs the velocities of the order of tens m s^{-1} may be encountered, and they can influence their evolution. The symmetry of the tidal flows is noted to be suitable for the magnetic field generation.

117.018 **Evolution of massive close binaries. II: The post X-ray binary stage: origin of run-away and binary pulsars.** C. de Loore, J. P. de Grève, J. P. de Cuyper.
Astrophys. Space Sci., Vol. 36, 219 - 225 (1975).
The further evolution of a massive X-ray binary consisting of a compact object and an OB supergiant is outlined. The supergiant exceeds its critical Roche lobe and a second stage of mass transfer starts. The remnant of the mass losing star – a pure helium star – develops a collapsing iron core and finally undergoes a supernova explosion. If the compact companion is a black hole the system remains bound; if the compact companion is a neutron star the system is disrupted unless an extra kick allowing an asymmetric explosion is given. Computations were performed for the massive binary 22.6 $M_\odot$ + 2 $M_\odot$. The possible final evolutionary products are: (1) a black hole and a compact object, in a binary system, (2) two run-away pulsars, (3) a binary pulsar.

117.019 **A quantitative analysis of the spectrum of Beta Lyrae. IV. Identification of lines of a faint component and star masses of close binaries.** M. Yu. Skul'skij.
Astron. Zhurn. Akad. Nauk SSSR, Vol. 52, 710 - 719 (1975). In Russian. English translation in Soviet Astron., Vol. 19, No. 4.
An investigation of high-dispersion spectrograms reveals wide and nondeep lines in the combination spectrum of Beta Lyrae probably belonging to the faint component. The semi-amplitude of the radial velocity curve from these lines is approximately 48 km/sec. The mass ratio is 0.26, the mass of the bright B9-star is 3.8 $M_\odot$ and the faint component mass 14.6 $M_\odot$.

117.020 **Period changes in binary systems during the Wolf-Rayet stage.** J.-P. De Grève, C. De Loore.
Astron. Astrophys., Vol. 42, 91 - 94 (1975).
Wolf-Rayet stars lose mass at a rate of 10^{-6} to $10^{-5} M_\odot$ yr^{-1}. If the Wolf-Rayet star is a member of a binary system this mass loss can be sufficiently large to cause observable changes in the binary period. Computations were performed in order to estimate these changes (of mass ratios and periods) by assuming a Jeans-like mode of loss of mass and angular momentum. It is argued that Wolf-Rayet binaries originate from a very small region of the HR diagram.

117.021 **The origin of the emission lines shown by RW Tauri and other binary systems.** C.-C. Wu.
Astrophys. Space Sci., Vol. 36, 407 - 426 (1975).
It is proposed that the observed double-component emission lines originate from the triangular Lagrangian points, L4 and L5, of the restricted three-body problem. The light curves of many close binary systems show absorption dips at ±60° of the primary (and sometimes also the secondary) minimum, indicating appreciable accumulation of matter at these points. The orbital velocity of L4 and L5 is derived as a function of period and the masses of the component stars. This equation is an independent relationship for determining the two stellar masses. It also reproduces Struve's empirical finding of $V^3 \propto P^{-1}$. Some related discussion is also given to nova and dwarf nova systems.

117.022 **The three-body problem and eclipsing binaries. Application to Algol and λ Tauri.** S. Söderhjelm.
Astron. Astrophys., Vol. 42, 229 - 236 (1975).
The gravitational perturbations in triple star systems are studied by well-known approximate methods. The major field of application is for a close eclipsing pair with nearly circular orbit, attended by a not too distant companion. (Period in outer orbit less than some 100 times the close one). Formulae are derived for the "long-period" perturbations, with periods of the same order as that in the outer orbit, and for the secular motions of the apse and node. The usually quoted formulae for these latter motions are shown to be valid only for a very massive or distant third body.

117.023 **Black holes and neutron stars: evolution of binary systems.** R. P. Kraft.
Astrophys. Space Sci. Library, Vol. 48, (see 012.014), 235 - 255 (1975).

117.024 **L'origine et l'évolution des étoiles doubles.**
P. Baize.
L'Astronomie, 89e année, p. 319 - 326 (1975).

117.025 **Restricted iterative minimization method for analysing light curves of close binary systems.**
T. B. Horák.
Bull. Astron. Inst. Czechoslovakia, Vol. 26, 257 - 270 (1975).
A modification of the iterative minimization method, which allows the limb darkening coefficients to remain fixed during minimization, is presented. The calculated elements of UV Leo, RX Ari and V 338 Her are the same as the elements obtained by classical methods. New elements of RS Cha were determined: $i = 83.8°$, $R = 0.225$, $r = 0.261$ in the blue and $i = 83.6°$, $R = 0.222$, $r = 0.264$ in the yellow. A three-dimensional presentation of the minimized function Σ demonstrates some basic features of the method.

117.026 **The dynamical evolution of binaries in clusters.**
D. C. Heggie.
IAU Symposium No. 69, (see 012.009), p. 73 - 94 (1975).
Using information on the rates at which binaries suffer encounters in a stellar system (Heggie, 1974), the author studies the effects of such processes on the evolution of the system itself. First considering systems with no binaries initially, he shows that low-energy pairs attain a quasi-equilibrium distribution comparatively quickly. Their effect on the evolution of the cluster is negligible compared with that of two-body relaxation. In small systems energetic pairs may form sufficiently quickly to exercise a substantial effect on its development and on the escape rate, but in large systems their appearance is delayed until the evolution of the core is well advanced. In that case they appear to be responsible for arresting the collapse of the core at some stage.

117.027 **Binary evolution in stellar systems.** R. H. Miller.
IAU Symposium No. 69, (see 012.009), p. 95 - 97 (1975).
Aarseth has shown by means of n-body calculations that, in star systems with a range of particle masses, the most massive stars quickly form a binary which soon takes up a large

fraction of the total binding energy of the cluster. Similar effects appear in other kinds of physical systems as well; mesic atoms behave in much the same way. The phase volumes of two otherwise equivalent stellar systems, each dominated by a tightly bound binary, favor exchange to incorporate the more massive star in the binary by a factor equal to the cube of the ratio of masses.

117.028 **The polarization of optical radiation from close binary systems with X-ray sources.**
Yu. N. Gnedin, N. A. Silant'ev.
Pis'ma v Astron. Zhurn., Vol. 1, No. 7, p. 17 - 22 (1975). In Russian.

The optical radiation of close binary systems becomes linearly polarized due to scattering by electrons of the hot spot and the disc-shaped envelope around the X-ray source. The dependence of polarization on phase for different models is given.

117.029 **Circular polarization of radiation at accretion onto a star with a strong magnetic field.**
Yu. N. Gnedin, G. G. Pavlov.
Pis'ma v Astron. Zhurn., Vol. 1, No. 9, p. 34 - 39 (1975). In Russian.

Spin phase dependence of polarization curves is found for radiation of hot regions that appears as a result of accretion onto a strongly magnetized star in a binary system. Curves obtained are essentially different for various configurations of the radiation region (spot, symmetric or asymmetric column, etc.).

117.030 **On the secular evolution of planetary orbits in a binary system.** S. L. Ziglin.
Pis'ma v Astron. Zhurn., Vol. 1, No. 9, p. 45 - 47 (1975). In Russian.

An integrable limiting case of the restricted elliptical twice-averaged three-body problem is considered. It describes particularly the evolution of planetary orbits in a binary system.

117.031 **On the composite spectra of unresolved late type binaries.** C. L. Bettis, D. R. Branch.
Bull. American Astron. Soc., Vol. 7, 433 (1975). – Abstr. AAS.

117.032 **BD + 40° 4220: a binary on the way?**
B. Bohannan, P. S. Conti.
Bull. American Astron. Soc., Vol. 7, 475 (1975). – Abstr. AAS.

117.033 **Chromospheric emission in binaries with late type components.** A. Young.
Bull. American Astron. Soc., Vol. 7, 475 (1975). – Abstr. AAS.

117.034 **The detection of extrasolar planetary systems.**
G. Gatewood.
Bull. American Astron. Soc., Vol. 7, 475 (1975). – Abstr. AAS.

117.035 **Skylab S-019 ultraviolet spectrograph results: cool stars with hot companions.** S. B. Parsons, J. D. Wray, G. F. Benedict, Y. Kondo, K. G. Henize.
Bull. American Astron. Soc., Vol. 7, 476 (1975). – Abstr. AAS.

117.036 **The stability of triple stars.** R. S. Harrington.
Bull. American Astron. Soc., Vol. 7, 476 (1975). Abstr. AAS.

117.037 **Sternentwicklung in engen Doppelsternsystemen.**
W. Pfau.
Astron. in der Schule, 12. Jahrgang, p. 88 - 92 (1975).

117.038 **The evolution of close binaries in a star field.**
T. A. Agekyan, Zh. P. Anosova.
Trudy Astron. Obs., *Leningrad*, Vol. 31 (= Uchenye Zapiski Leningr. Un-ta, No. 381 = Seriya Matem. Nauk, vyp. (No.) 51), p. 174 - 181 (1975). In Russian.

It is shown by the Monte-Carlo method that the major axis of a close binary system is a nonstationary random function with decreasing mathematical expectation. This is due to the interaction with field stars.

117.039 **The dynamical ages of trapezium systems.**
C. Allen, A. Poveda.
Publ. Astron. Soc. Pacific, Vol. 87, 499 (1975). – Abstr. ASP.

117.040 **On the convergence of the successive approximation process in Kopal's method for determination of photometric orbits.** R. A. Botsula.
Izv. Astron. Ehngel'gardt. Obs., *Kazan'*, No. 39, p. 99 - 101 (1972). In Russian.

117.041 **Models for W Ursae Majoris stars.**
A. Yamasaki.
Publ. Astron. Soc. Japan, Vol. 27, 469 - 481 (1975).

Zero-age models for W Ursae Majoris stars are computed by considering in detail the energy transfer between the components through turbulent motion in the contact region. Four models are constructed with the degree of contact as a parameter, and it is found that the models correspond to the unequal entropy model, and most of the energy is transferred through the deep part of the contact region (the adiabatic part of the surface convection zones). However, the difference in the effective temperature between the components for our models is well outside the range permitted by observations. No W-type systems are obtained with the present method of computation.

117.042 **Eclipses by an elliptical torus.** R. E. Wilson.
Astron. Journ., Vol. 80, 719 - 722 (1975).

The geometry of a torus having elliptical meridian sections is discussed in regard to its eclipsing properties when viewed at arbitrary inclinations. Eclipses involving the hole horizon as well as the outer horizon are considered. Various special cases of such a torus include those of a thin ring or disk, an ellipsoid of revolution, and a section of a right circular cylinder. Thus the relations given here may be used in place of a number of special schemes used previously for these particular cases, as well as for the case of a general torus of finite thickness.

117.043 **On the formation of WR-stars and X-ray sources in close binary systems.**
A. V. Tutukov, L. R. Yungel'son.
Nauchn. Informatsii, vyp. (No.) 33, p. 16, 32 - 33 (1974). In Russian and English. – Abstr. of a paper presented at the colloquium 'Physics and evolution of stars', (see 012.015).

117.044 **The fission theory of binary stars for compressible masses.** N. R. Lebovitz.
19th Colloquium International Astrophys. Liège 1974, (see 012.016), p. 47 - 53 (1975).

A resumé is given of the classical fission theory of binary stars, emphasizing the fundamental assumptions and conjectures, and describing the resolutions of these conjectures that led to the abandonment of the theory. It is pointed out that the formulation of the theory was inconsistent, and that one of the principal conjectures was unnecessary. A reformulated version of the fission theory is described.

117.045 **Theoretical discussion of the binary system AS Eridani.**
S. Refsdal, M. L. Roth, A. Weigert.
19th Colloquium International Astrophys. Liège 1974, (see 012.016), p. 431 - 433 (1975).

Starting from the very accurate observational data of the system, its present evolutionary stage and its past and future evolution are discussed. Special emphasis is given to a possible loss of mass and angular momentum from the system. The present paper summarizes the basic results, while the extensive calculations will be described elsewhere.

117.046 **Polarization of optical radiation of binary systems with X-ray sources.**
Yu. N. Gnedin, N. A. Silant'ev.
Astron. Tsirk., No. 851, p. 1 - 3 (1974). In Russian.

117.047 **Orientation of the orbital planes of the binary systems relative to their respective space velocities.**
F. Zagar.
Proc. Astrometric Conference, Torino 1974, (see 012.017) p. 53 - 60 (1975).

117.048 **Determination of the masses of components of a binary system from the X-ray eclipse duration and the amplitude of regular optical variability.**
N. G. Bochkarev, N. I. Shakura.
Astron. Tsirk., No. 853, p. 1 - 4 (1975). In Russian.

117.049 **Gas flow in cataclysmic variable stars.**
B. P. Flannery.
Astrophys. Journ., Vol. 201, 661 - 694 = Lick Obs. Bull., No. 698 (1975).
A numerical hydrodynamic scheme, applicable to gas flow in binary stars, is presented which incorporates the effects of pressure and gravitational forces, radiative cooling, and shock formation. The method is used to study gas flow originating from Roche lobe overflow through the inner Lagrangian point, L_1, in a binary star modeled to resemble the dwarf nova Z Cam. The calculations confirm the major elements of the "hot spot" model proposed to account for observations of cataclysmic variable stars.

117.050 **Polarization of radiation of double systems with X-ray sources.** Yu. N. Gnedin, N. A. Silant'ev.
AN SSSR. Fiz.-tekhn. in-t. Preprint 483. Leningrad, 1975. 19 pp. In Russian. – Abstr. in Referativ. Zhurn. 51. Astron., 9.51.585 (1975).

117.051 **Unseen astrometric companions of stars.**
P. van de Kamp.
Annual Rev. Astron. Astrophys., Vol. 13, (see 003.005), 295 - 333 (1975).
In this survey the author attempts to gather the available information on unseen companions of stars, revealed either as unresolved astrometric binaries that have been found and studied from perturbations in the proper motion of a star previously considered single, or as an irregularity in the orbital motion of a visual double star, which thus proves to have a third component. The astrometric method is particularly suited for the discovery and subsequent study of perturbations with long periods and large amplitudes. A list is given of well-established as well as of provisional and uncertain results.

117.052 **On the evolution of close binaries with mass and momentum loss from the system.**
A. Massevitch, L. Yungelson.
Mem. Soc. Astron. Italiana, Vol. 46, 217 - 229 (1975).
It is shown that mass and angular momentum loss from a close binary system may take place at different stages of evolution for a broad variety of systems. A simple method to treat this process is presented. It is shown, that mass loss from the system can not seriously influence the evolution of the mass losing component. To obtain agreement between theory and observations for systems with small masses ($M_1 + M_2 \leqslant 6 M_\odot$) it is necessary to assume that about 40–50 per cent of matter lost by the primary is lost by the system.

117.053 **On possible observational tests for some results of the theory of binary stars evolution.**
Z. T. Krajcheva.
Pis'ma v Astron. Zhurn., Vol. 1, No. 11, p. 32 - 37 (1975). In Russian.
Calculations of evolution of massive close binary stars are carried out using Schwarzschild's criterion for convective stability in the zone of variable molecular weight. Comparison with calculations done for Ledoux's criterion permitted to point out observational possibilities to test the validity of these criteria.

117.054 **A gravitational three-body scattering experiment.**
M. J. Valtonen.
Mem. Roy. Astron. Soc., Vol. 80, 61 - 75 (1975).
The gravitational interaction between an equal component circular binary and a third particle of mass equal to the total mass of the binary has been studied in detail for nearly parabolic coplanar orbits. An approximate analytical model for energy transfer has been constructed by harmonic analysis.

117.055 **Statistics of three-body experiments: probability of escape and capture.** M. J. Valtonen.
Mem. Roy. Astron. Soc., Vol. 80, 77 - 91 (1975).
The probability of escape and capture in the gravitational three-body problem has been studied using a large number of numerical orbit calculations. The values of the masses are found to be the most important parameters in both processes. The occurrence of the minimum triple configurations, and their effect on the escape, has also been studied.

117.056 **The age of the W Ursae Majoris stars.**
F. Van't Veer.
Astron. Astrophys., Vol. 44, 437 - 443 (1975).
In a first paper the author tentatively concludes that more than 50% of contact binaries are formed by a fission process of the Jeans-Roxburgh type. In the present paper he discusses in more detail the consequences of this hypothesis. The alignment of the axes of rotation in young galactic clusters and the detectability of W UMa stars are discussed. The author concludes that stars of young clusters generally have a preferential orientation of their rotation axes, which need not coincide with the general orientation of the galactic disk.

117.057 **Fast synchronization of the rotation of binaries by tidally induced turbulence.** G. Horedt.
Astron. Astrophys., Vol. 44, 461 - 463 (1975).
Appreciable synchronization of the spin and orbital angular velocities of binaries takes place during intervals shorter than 10^6 yr, if there occurs energy dissipation by tidally induced turbulence.

117.058 **Magnetic field generation and charged particle acceleration in binary stellar systems.**
A. Z. Dolginov.
VI Leningr. mezhdunar. seminar, 1974. 'Uskorenie chastits i yader. reaktsii v kosmose'. Leningrad, 1974, p. 223 - 224, with a discussion, p. 225. In Russian and English. – Abstr. in Referativ. Zhurn. 51. Astron., 10.51.632 (1975).

117.059 **Evolution of a close binary system: compact object inside moving envelope of normal companion.**
M. E. Alexander, W. Y. Chau, R. N. Henriksen.
Journ. Roy. Astron. Soc. Canada, Vol. 69, 252 (1975). Abstr. CAS.

117.060 **On the stability of congruent Darwin ellipsoids.**
M. Tassoul.
Astrophys. Journ., Vol. 202, 803 - 808 (1975).

The problem of synchronous oscillations of congruent Darwin ellipsoids is reconsidered. Contrary to results obtained by Chandrasekhar, it is shown that along a Darwin sequence the configurations remain stable with respect to ellipsoidal disturbances until they become too closely spaced. The precise limit which separates stable from unstable systems is found. Apart from some minor differences the Darwin sequence is the exact analog of the Roche sequence.

117.061 **On coupled second-harmonic oscillations of the congruent Darwin ellipsoids.** S. Chandrasekhar.
Astrophys. Journ., Vol. 202, 809 - 814 (1975).

A general theory of coupled second-harmonic oscillations of the congruent Darwin ellipsoids is developed. The class of oscillations studied includes the synchronous oscillations as a special degenerate case.

117.062 **Mechanism for inducing synchronous rotation and small eccentricity in close binary systems.**
W. H. Press, P. J. Wiita, L. L. Smarr.
Astrophys. Journ., (*Letters*), Vol. 202, L135 - L137 (1975).

Rotation through a tidal bulge generates a fluid shear which, in stars, implies a very high Reynolds number: fully developed turbulence might result. Turbulent viscosities, synchronization, and circularization time scales are computed. The calculation contains no adjustable dimensional parameters. It explains the observed fact of synchronous rotation for orbital periods $\lesssim 4$ days, nonsynchronous for larger periods. The observed small eccentricity of some short-period binary X-ray sources is also an immediate consequence.

117.063 **The evolution of massive close binaries. III: The possibility of a tidal instability during the post supernova phase.**
J.-P. de Grève, C. de Loore, W. Sutantyo.
Astrophys. Space Sci., Vol. 38, 301 - 312 (1975).

The influence of tidal interaction on the periods of massive X-ray binaries during the post-supernova evolution is investigated. It is assumed that after a certain time the orbit has become circular and synchronous. The tidal effects of subsequent evolutionary changes in the moment of inertia of the massive component are calculated. It is shown that, as is already suggested by Sparks and Stecher (1974), for small mass ratios and short binary periods a tidal instability may occur resulting in an accelerating inward spiral motion. Before the onset of the instability the tidal forces maintain a nearly synchronous orbit. Possibly the orbits of Cen X-3 and 3 U 1700–37 are already unstable at present.

117.064 **Light variations of TT Ari.** J. Smak, K. Stępień.
Acta Astron., Vol. 25, 379 - 381 (1975).

Moments of maxima and minima of the nova-like binary TT Ari observed in 1961–68 are listed. There is an indication for large period variations and, in particular, the periods derived from photometric data differ significantly from the spectroscopic period obtained by Cowley et al. (1975).

117.065 **The evolutionary fate of contact binaries.**
R. F. Webbink.
Bull. American Astron. Soc., Vol. 7, 534 (1975). – Abstr. AAS.

117.066 **Tidal circularization induced in close binaries by turbulent viscosity.**
W. H. Press, P. J. Witta, L. L. Smarr.
Bull. American Astron. Soc., Vol. 7, 534 (1975). – Abstr. AAS

117.067 **Kappa Pegasi – a quadrupole system of low total mass.** W. R. Beardsley, M. W. King.
Bull. American Astron. Soc., Vol. 7, 543 (1975). – Abstr. AAS.

117.068 **A spectroscopic orbit for μ Cassiopeiae.**
T. F. Worek, W. R. Beardsley.
Bull. American Astron. Soc., Vol. 7, 543 (1975). – Abstr. AAS.

117.069 **Dubbelsterren (1). Indeling gebaseerd op waarnemingsmethoden.** E. L. van Dessel.
Zenit, Vol. 2, 419 - 423 (1975).

117.070 **Unseen companions of stars.** P. van de Kamp.
Space astrometry, (see 012.037), p. 121 - 127 (1975).

117.071 **Determination of the parallax and proper motion of the binary Stein 2051.** G. A. Plyugin.
Izv. Glav. Astron. Obs. Pulkovo, No. 193, Astrofiz., Astrometr. i Zvezdn. Astron., p. 92 - 99 (1975). In Russian.

The parallax and proper motion of the binary Stein 2051 have been determined from observations with the 26″-refractor at Pulkovo during 1966–1969. The components of this binary system differ in brightness and colour by about 1.5 magnitudes. This leads to difficulties in observations, measurements and reduction of observational data. The μ and π values were determined with high accuracy.

117.072 **The effects of unresolved duplicity on spectra of solar-type stars.** C. Bettis, D. Branch.
Publ. Astron. Soc. Pacific, Vol. 87, 895 - 899 (1975).

Equivalent widths of selected lines in composite spectra of normal main-sequence stars are calculated. The Na D lines are significantly stronger in many composite spectra than in spectra of single stars of similar color. As a rule, Fe I lines are slightly weaker and Fe II lines are slightly stronger in composite spectra. Effects on spectroscopic determinations of abundance ratios and surface gravities are discussed. It is suggested that spectrophotometric detection of relatively faint companions of solar-type stars is possible.

117.073 **Reflection effect in close binary systems.**
F. G. Vanlandingham.
Thesis, Indiana Univ., Bloomington (USA). [Available from Univ. Microfilms, Ann Arbor, Mich., *USA*. Order No. 75-1767], 176 pp. (1974).

The investigation studies the effects of the irradiation of the hotter component in a close binary system on the atmosphere of the secondary and on the observed flux distribution of the binary system. An existing model atmosphere computer program is modified to include the effects of non-zero incident radiation. Computations reveal that the irradiation can significantly raise the temperature in the upper layers of the atmosphere.

117.074 **Gas flow in cataclysmic variable stars.**
B. P. Flannery.
Thesis, California Univ., Santa Cruz (USA). [Available from Univ. Microfilms, Ann Arbor, Mich., *USA*. Order No. 75-8474], 93 pp. (1974).

A hydrodynamic scheme for numerically simulating gas flow in binary stars is presented. The method includes both gravitational and pressure forces, and allows for shock formation and radiative cooling. Mass transfer originating from thermal evaporation through the inner Lagrangian point is studied in a binary system scaled to resemble the dwarf nova Z Cam. The gas flow is followed for a total time of 1.25 binary orbital periods. The dissipative effect of radiative cooling in shocks results in the rapid formation of a ring in direct rotation around the component receiving mass. Also, an optically thick, relatively confined shock front forms at the intersection of the ring and the stream from the Lagrangian point. The shock's temperature, 23000°K at maximum, and location confirm the major elements of the 'hot spot' model proposed to account for observations of cataclysmic variable stars.

Some remarks on Roche harmonics.
See Abstr. 021.002.

Diffractional image of a double star for the case of central obscuration in the objective. See Abstr. 031.011.

The motion of a particle from the L_2 Lagrangian point. See Abstr. 042.053.

Solution of the equation of radiative transfer in the atmosphere of a double star. Extended stellar atmosphere. See Abstr. 063.015.

La théorie des étoiles jumelles et l'évolution stellaire. See Abstr. 065.075.

Relativistic spin precession in two-body systems. See Abstr. 066.104.

The proper motion and orbit of HZ Her. See Abstr. 112.006.

Strömgren four-colour observations of northern hemisphere binary systems. See Abstr. 113.029.

The photometric classification of double stars. See Abstr. 113.054.

Intrinsic properties of high luminosity blue stars from observations of multiple systems. See Abstr. 113.066.

MK classifications of some northern hemisphere binary systems. See Abstr. 114.047.

The helium emission-line object G 61-29. See Abstr. 114.358.

Nature of the Wolf-Rayet star in the binary system V 444 Cyg. See Abstr. 114.369.

Photometric observations of V470 Cygni (HD 228911). See Abstr. 119.020.

Observational properties of models of semidetached close binaries. II. See Abstr. 121.046.

Infrared observations and a model of β Per. See Abstr. 121.051.

Supernovae in binary systems. See Abstr. 125.006.

The Crab pulsar as a component of a binary system. See Abstr. 141.305.

An interferometric search for the Hulse-Taylor binary pulsar. See Abstr. 141.311.

An upper limit on optical pulsations from PSR 1913+16. See Abstr. 141.312.

The new binary pulsar and the observation of gravitational spin precession. See Abstr. 141.345.

The circularization of close binary orbits: Centaurus X-3. See Abstr. 142.002.

Mass transfer in binary X-ray systems. See Abstr. 142.003.

On light curve analyses of X-ray binaries. See Abstr. 142.013.

X-ray binaries in globular clusters. See Abstr. 142.036.

Tidal capture formation of binary systems and X-ray sources in globular clusters. See Abstr. 142.044.

Ariel 1118–61: a very close binary system? See Abstr. 142.051.

On the possible existence of an extended optical structure around the neutron star in the X-ray binary system HZ Her (Her X-1). See Abstr. 142.054.

On the formation of X-ray binaries. See Abstr. 142.107.

Optical properties of binary X-ray sources. See Abstr. 142.125.

The physics of binary X-ray sources. See Abstr. 142.126.

The evolutionary origin of massive X-ray binaries and the total number of massive stars with a collapsed close companion in the Galaxy. See Abstr. 142.127.

On the long-term behavior of the HZ Herculis/Hercules X-1 system. See Abstr. 142.132.

Some properties of X-ray binaries. See Abstr. 142.143.

Der Röntgendoppelstern HZ Her. See Abstr. 142.149.

The X-ray pulsar 3U 0900–40 with long period – a neutron star with anomalously strong magnetic field. See Abstr. 142.152.

New results from long-term observations of Cyg X-1. See Abstr. 142.154.

Transient X-ray sources. See Abstr. 142.160.

The position and proper motion of HZ Her. See Abstr. 142.191.

N-body simulations. See Abstr. 151.014.

Encounters between binary and single stars and their effect on the dynamical evolution of stellar systems. See Abstr. 151.064.

Binary evolution in stellar dynamics. See Abstr. 151.070.

Statistical identification of binaries in star clusters. See Abstr. 153.029.

118 Visual Double and Multiple Stars

118.001 **Photographic measurements of double stars.** P. S. Thé.
Astron. Astrophys., Suppl. Ser., Vol. 21, 51 - 80 (1975).

In this paper the relative distances and position angles of the components of 170 mostly southern visual double stars are presented. The results are obtained using photographic plates taken with the 60 cm double refractor of the Bosscha Observatory, Lembang, Indonesia.

118.002 **Eine Methode für seeingunabhängige photometrische und astrometrische Messungen der visuellen Doppelsterne.** K. Rakosch.
Mitt. Astron. Ges., No. 36, p. 93 (1975). – Abstract.

118.003 **Orbits and masses of Hyades visual binaries.** W. C. Wickes.
Bull. American Astron. Soc., Vol. 7, 476 (1975). – Abstr. AAS.

118.004 **Note on an astrometric study of two nearby visual binaries of high space velocity: ADS 9544 and Kuiper 90.** S. L. Lippincott.
Publ. Astron. Soc. Pacific, Vol. 87, 557 (1975).

These two binaries were put on the Sproul observing program over 35 years ago with the expectation that orbital elements might be forthcoming within a decade, in view of the obviously short periods. Unfortunately this hope remains unfulfilled. The purpose of this note is to draw attention to the observational needs for these objects; meanwhile precise parallax values have been obtained.

118.005 **Parallaxes, masses, and orbits of the visual binaries η Coronae Borealis and Kuiper 37 (10 Ursae Majoris).** L. A. Breakiron, J. P. Dabrowski, G. Gatewood.
Astron. Journ., Vol. 80, 714 - 718 (1975).

The trigonometric parallaxes, astrometric orbits, and masses of the η CrB and Kuiper 37 binary systems are redetermined. The weighted mean absolute parallaxes of all determinations are $+0\overset{''}{.}060 \pm 0\overset{''}{.}005$ and $+0\overset{''}{.}064 \pm 0\overset{''}{.}004$ (s.e.) for the two stars, respectively. The prallaxes and astrometric orbits, in conjunction with visual elements by Douglass, indicate masses of $0.95 \pm 0.27\ M_\odot$ and $0.94 \pm 0.26\ M_\odot$ for η CrB A and B and $1.13 \pm 0.20\ M_\odot$ and $0.84 \pm 0.15\ M_\odot$ for Kuiper 37 A and B. The masses and absolute magnitudes derived herein agree well with the empirical mass-luminosity relation.

118.006 **CE Cassiopeiae as a double star.** V. N. Frolov.
Astron. Tsirk., No. 848, p. 2 - 3 (1974). In Russian.

118.007 **The results of photographic observations of ADS 7251 at Pulkovo in 1962–1973.** N. A. Shakht.
Astron. Tsirk., No. 848, p. 5 - 7 (1974). In Russian.

118.008 **Micrometer observations of double stars. VIII.** W. D. Heintz.
Astrophys. Journ., Suppl. Ser., No. 288, Vol. 29, 315 - 327 (1975).

Measurements of 935 pairs obtained in 1971.8–1975.0 are tabulated.

118.009 **Photographic double-star measurements.** W. D. Heintz.
Astrophys. Journ., Suppl. Ser., No. 288, Vol. 29, 329 - 330 (1975).

Measurements of eight known and seven new double stars from astrometric plates are listed.

118.010 **Orbits of visual binaries.** W. D. Heintz.
Astrophys. Journ., Suppl. Ser., No. 288, Vol. 29, 331 - 339 (1975).

Elements and dynamical parallaxes have been computed for 17 pairs.

118.011 **Is this pair of binary stars interesting?** R. B. Shatsova, G. B. Shimchuk.
Astron. Tsirk., No. 863, p. 6 - 7 (1975). In Russian. – Concerning Gliese 411, 412 (BD + 36°2147, BD + 44°2051).

118.012 **Visuele waarnemingen van Sirius-B.** L. Aerts.
Zenit, Vol. 2, 295 (1975).

118.013 **Photometric and astrometric observations of close visual binaries.** K. D. Rakos.
Proc. Astrometric Conference, Torino 1974, (see 012.017), p. 27 - 38 (1975).

118.014 **Measurements of double stars with a polarization micrometer. VIII.** G. A. Starikova.
Soobshch. Gos. Astron. Inst. Shternberga, No. 182, p. 18 - 20 (1973). In Russian.

118.015 **Parallax and photographic orbital analysis of the close visual binaries Kuiper 84 and Furuhjelm 54.** S. L. Lippincott.
Astron. Journ., Vol. 80, 831 - 835 (1975).

Analysis of photographs taken with the Sproul 61-cm refractor of Kuiper 84 yield a photocentric orbit with a period of 14.90 yr. The limited number of visual observations at the maximum separation of $0\overset{''}{.}2$ provides a scale of the relative orbit which yields $0.9\,M_\odot$ for the sum of the masses. The Sproul absolute parallax is $+ 0\overset{''}{.}034 \pm 0\overset{''}{.}0002$ (pe). Similar analysis of plates on Furuhjelm 54 yield the first orbit determination for this close pair. A period of $28\overset{y}{.}5$ is found with $0\overset{''}{.}096$ for the semimajor axis of the photocentric orbit. The limited number of visual observations all with separations close to $1\overset{''}{.}0$ provide a scale for the relative orbit. The Sproul absolute parallax is $+ 0\overset{''}{.}058 \pm 0\overset{''}{.}003$ (pe); the mass of the system is $2.0\,M_\odot$, which may indicate the presence of more than two components.

118.016 **The triple system Ψ Sagittarii.** F. C. Fekel.
Astron. Journ., Vol. 80, 844 - 851 (1975).

The visual binary Ψ Sagittarii (ADS 12214) is a spectroscopic triple system with orbital periods of 10.78 days and 20.0 yr. Masses derived from the parameters of the spectroscopic and visual orbits are $2.7\,M_\odot$ for the visual primary, 2.1 $M_\odot$ for the short-period pair primary, and $1.7\,M_\odot$ for the short-period pair secondary. The distance to the system is 100 pc. The orbital inclination of the close pair is nearly the same numerically as that of the visual orbit. Suggested spectral classifications indicate that all three components have evolved off the main sequence. Mass exchange in the close pair has not yet occurred. The system is compared with several other spectroscopic–visual triples.

118.017 **La situation actuelle d'activité de recherche des étoiles doubles et une revue du problème de leur origine et développement.** P. M. Djurković.
Publ. Obs. Astron. Beograd, No. 20, (see 012.024), p. 166 - 169 (1975). In Serbo-Croatian.

118.018 **Untersuchungen in Verbindung mit "The first general catalogue of double-star observations made in Belgrade, 1951–1971".** G. M. Popović.
Publ. Obs. Astron. Beograd, No. 20, (see 012.024), p. 170 -

177 (1975). In Serbo-Croatian.

118.019 **On the evolutionary status of the RS CVn-type binary WW Draconis.** D. S. Hall.
Acta Astron., Vol. 25, 225 - 226 (1975).
It is shown that, because of its F8 V visual companion, WW Dra cannot be in a pre-main-sequence phase of evolution.

118.020 **A synthesis of the solutions of spectroscopic and visual binary orbits.** C. L. Morbey.
Publ. Astron. Soc. Pacific, Vol. 87, 689 - 693 (1975).
A method which simultaneously yields a complete set of orbital elements from spectroscopic and visual binary observations is described. The technique is applied to Burnham 1163.

118.021 **The triple system 20 Leo.**
F. C. Fekel, B. W. Bopp.
Bull. American Astron. Soc., Vol. 7, 512 (1975). – Abstr. AAS.

118.022 **Stellarastronomie, neuere Beobachtungsmethoden bei visuellen Doppelsternen.** A. Werner.
Jahrbuch der Schulphysik, (Aulis-Verlag, Köln), Vol. 2, 227 - 232 (1975).
A survey of the following methods: visual, photographic, by electronic camera (image converter), star occultations by the moon, Michelson interferometer, intensity interferometer, speckle interferometry.

118.023 **Visuelle Doppelsterne, Theorie und Forschungsergebnisse.** A. Werner.
Jahrbuch der Schulphysik, (Aulis-Verlag, Köln), Vol. 2, 250 - 254 (1975).
A review article concerning the differential equation of the two-body problem, the relative orbit, orbital elements and their determination, masses of the stars, Sirius B, proof for invisible companions.

118.024 **Dubbelster Sirius te scheiden met een 10-cm kijker?**
P. B. van der Wal, G. Comello.
Zenit, Vol. 2, 449 - 450 (1975).

118.025 **Visual double stars measured at Las Campanas Observatory, Chile.** F. Holden.
Publ. Astron. Soc. Pacific, Vol. 87, 945 - 948 (1975).

118.026 **The orbits of the visual binaries ϕ316, ADS 9220, ADS 10157, ADS 12973, ADS 14893.**
G. A. Starikova.
Astron. Tsirk., No. 890, p. 6 - 7 (1975). In Russian.

118.027 **The micrometer measures of double stars in Belgrade. (Series 22).** G. M. Popović.
Bull. Obs. Astron. Belgrade, No. 126, p. 35 - 42 (1975).
208 measures of 100 double stars are presented. Among them there are 9 pairs with a parabolic or hyperbolic orbit.

118.028 **Micrometric measurements of 176 pairs of double stars. (Series 23).** D. Olević.
Bull. Obs. Astron. Belgrade, No. 126, p. 42 - 46 (1975).
Information on 192 micrometric measurements of 176 double stars, made with the refractor "Zeiss" 65/1055 cm.

118.029 **The new double stars discovered in Belgrade with the Zeiss refractor 65/1055 cm, Supplement IV.**
G. M. Popović.
Bull. Obs. Astron. Belgrade, No. 126, p. 47 - 49 (1975).
The positions of 23 new double or multiple systems discovered in Belgrade with Zeiss refractor 65/1055 cm are presented together with 58 of their measurements.

118.030 **Orbit of the double star ADS 11311 = STT 353 AB.**
D. Olević.
Bull. Obs. Astron. Belgrade, No. 126, p. 49 - 50 (1975).
Information is given on the elements of the orbit and ephemerides of the pair ADS 11311 = STT 353 as well as on astrophysical quantities resulting from these elements.

118.031 **Les orbites nouvelles de trois étoiles doubles visuelles A162 = ADS 12631, Hu 1268 = ADS 9285, A 1840 AB = ADS 3326.** V. Erceg.
Bull. Obs. Astron. Belgrade, No. 126, p. 50 - 53 (1975).

118.032 **Trajectoire apparente rectiligne de ADS 1322 = A2321.** V. Erceg.
Bull. Obs. Astron. Belgrade, No. 126, p. 53 (1975).

118.033 **Orbites nouvelles.**
W. D. Heintz, Costa-Morales, D. J. Zulević.
Circ. d'Inform. (U.A.I. Commission des Étoiles Doubles), Grasse, France, No. 66 (1975).

118.034 **Étoiles doubles nouvelles, Belgrade** (Lunette de 65 cm), **Nice** (Lunettes de 50 et de 74 cm).
G. M. Popovic, P. Couteau.
Circ. d'Inform. (U.A.I. Commission des Étoiles Doubles), Grasse, France, No. 66 (1975).

118.035 **A revision of the orbit of Capella.** W. S. Finsen.
Circ. d'Inform. (U.A.I. Commission des Étoiles Doubles), Grasse, France, No. 66 (1975).

118.036 **Orbites nouvelles.** V. Erceg, W. D. Heintz, P. Baize, D. J. Zulević, G. M. Popovic.
Circ. d'Inform. (U.A.I. Commission des Étoiles Doubles), Grasse, France, No. 67 (1975).

118.037 **Étoiles doubles nouvelles.**
W. D. Heintz, P. Couteau, P. Muller.
Circ. d'Inform. (U.A.I. Commission des Étoiles Doubles), Grasse, France, No. 67 (1975).

118.038 **Orbite de l'étoile double visuelle ADS 10723 = BDS 8120 = β 1251.** J. Dommanget.
Bull. Astron.,Obs. Roy. Belgique, Vol. 8, 244 - 247 (1975).

118.039 **Trajectoire rectiligne du couple visuel IDS 06163N 1443 = ADS 4973 = Ho 232.** O. Nys.
Bull. Astron., Obs. Roy. Belgique, Vol. 8, 289 - 290 (1975).

118.040 **Un'esercitazione elementare di meccanica: determinazione della massa di una binaria.** L. Stefanini.
Giornale Astron., Vol. 1, 299 - 313 (1975).

The proper motions of Struve double stars.
See Abstr. 112.003.

The absolute magnitudes of carbon stars: carbon stars in binary systems. See Abstr. 115.004.

Les étoiles doubles et la relation masse-luminosité.
See Abstr. 115.007.

Lunar occultations of Beta Scorpii in 1975 and 1976. See Abstr. 117.011.

Unseen astrometric companions of stars.
See Abstr. 117.051.

119 Spectroscopic Binaries

119.001 **Differential *uvby* photometry of the spectroscopic binary HD 217061.**
H. H. Lanning, W. A. James.
Monthly Notices Roy. Astron. Soc., Vol. 172, 1P - 3P (1975).
Strömgren *uvby* differential observations of the spectroscopic binary HD 217061 are presented. No optical variations were detected.

119.002 **The binary nature of the Of star HD 108.**
J. B. Hutchings.
Astrophys. Journ., Vol. 200, 122 - 127 = Contr. Dominion Astrophys. Obs., *Victoria,* No. 260 = NRC No. 14342 (1975).
The peculiar Of star HD 108 is found to vary in velocity with a period of 4.612 days. The star is considered to be a binary with a companion of mass $\lesssim 2.5\ M_\odot$, which may be a collapsed object. The primary has a high-density stellar wind which is partially affected by its companion. Emission lines from the extended envelope show phase-correlated velocity variations, but no marked intensity variations. Comparisons are made with the similar Of binaries HD 153919 and 148937.

119.003 **Ultraviolet observations of Alpha Aurigae from Copernicus.** A. K. Dupree.
Astrophys. Journ., (*Letters*), Vol. 200, L27 - L31 (1975).
Emission lines of Lα (1215.67 Å) and O VI (1031.94 Å) were detected in the spectroscopic binary α Aur with the Princeton experiment on Copernicus. Temperatures of the emitting regions are inferred to be in excess of 3×10^5 K. The temperature and emission measure are consistent with a variable source of soft X-rays. If the emission is attributed to the primary star (G5 III), the atmosphere is expanding with velocities ~20 to 100 km s^{-1}. Such expansion can lead to material within the binary system.

119.004 **Identification of He^3 in the binary system 68u Herculis.** R. W. Hilditch, P. W. Hill.
Monthly Notices Roy. Astron. Soc., Vol. 172, 29P - 34P (1975).
From a radial velocity study of 60 spectrograms of the binary system 68u Herculis, systematic velocity differences have been found to exist between the He I singlet lines and the He I triplet lines of the primary component. These differences are independent of orbital phase, date of observation of spectrograph employed and can be interpreted in terms of the existence of a large fraction of He^3 in the atmosphere of the primary component. A He^3/He^4 ratio of between 4 : 1 and 49 : 1 is suggested.

119.005 **Spectroscopic binary orbits from photoelectric radial velocities. Paper 3: HR 8035.**
G. A. Radford, R. F. Griffin.
Observatory, Vol. 95, 143 - 147 (1975).

119.006 **A spectrographic study of Beta Centauri.**
N. R. Lomb.
Monthly Notices Roy. Astron. Soc., Vol. 172, 639 - 647 (1975).
Using spectra obtained in 1969, it is shown that the radial velocity of β Cen was varying with a period of $0^d.157$ in 1969. The broad spectral line profiles of the star are examined for changes in shape with the above period. This search is considerably complicated by the fact that the star is a single line spectroscopic binary with components of approximately equal brightness. With the help of a series of graphs of predicted blended profiles it is demonstrated that, for the observations under discussion, any variations found are likely to be intrinsic to the star. Variations in the profiles are, in fact, found with the radial velocity period.

119.007 **Properties and nature of shell stars. 5. Re-determination of the velocity curve of 4 Her.** J. F. Heard, R. Hurkens, P. Harmanec, P. Koubský, J. Krpata.
Astron. Astrophys., Vol. 42, 47 - 53 (1975).
A period of 46.194 ± 0.002 days for the radial velocity variation of 4 Her is supported by both visual and oscilloscope measurements of 159 old and new Dunlap, Ondřejov and Victoria spectrograms of the star. It thus appears that the period has remained constant over the past forty years. A period close to one day has been definitely ruled out. The orbital elements presented here are based upon the seemingly justifiable assumption of the binary nature of the star, but distortion of the elements by gas streaming is strongly indicated.

119.008 **A search for optical pulsations from single line spectroscopic binaries.**
I. Nevo, D. Sadeh, A. Frohlich.
Astron. Astrophys., Vol. 42, 247 - 250 (1975).
The authors have obtained optical power spectra from 16 single line binaries that are suspected of harbouring a neutron star or a black hole. The spectra are scanned for both narrow and wide band features. No significant pulsation was detected.

119.009 **Correlation analysis of selected elements of spectrum binaries.** M. P. Volodina.
Trudy Kazan. Gorod. Astron. Obs., No. 39, p. 57 - 82 (1973). In Russian.

119.010 **The double-line spectroscopic binary Ap star HR 2727.** W. K. Bonsack.
Bull. American Astron. Soc., Vol. 7, 476 (1975). – Abstr. AAS.

119.011 **Chemical abundances in the secondary of U Sagittae.**
S. A. Naftilan.
Bull. American Astron. Soc., Vol. 7, 476 - 477 (1975). – Abstr. AAS.

119.012 **Interpretation of a spectroscopic observation of the resolved components of Kappa Pegasi.**
W. R. Beardsley, M. W. King.
Bull. American Astron. Soc., Vol. 7, 477 (1975). – Abstr. AAS.

119.013 **Mass transfer in the symbiotic binary AG Pegasi.**
J. B. Hutchings, A. P. Cowley, R. O. Redman.
Astrophys. Journ., Vol. 201, 404 - 412 = Contr. Dominion Astrophys. Obs., *Victoria,* No. 265=NRC No. 14456 (1975).
Line profiles, equivalent widths, and radial velocities have been derived from high-dispersion spectra of the symbiotic star AG Peg. The model suggested from recent spectra is a hot, approximately 1 $M_\odot$ subluminous star orbiting a 3–4 $M_\odot$ giant. The hot object rotates rapidly and ejects material which streams toward the cool primary. A bright region near the substellar point on the M star is responsible for a 0.3 mag light variation, and is heated either by radiation from the hot star or by collisions from the impinging stream. The exact orbital period remains uncertain because of complications in the interpretation of the velocities from several regions, but a value of 820^d is adopted.

119.014 **Spectroscopic binary orbits from photoelectric radial velocities. Paper 4: HD 204934.**
G. A. Radford, R. F. Griffin.
Observatory, Vol. 95, 187 - 190 (1975).

119.015 **New radial velocities and further photometric observations of λ Sco and κ Sco.**
N. R. Lomb, R. R. Shobbrook.
Monthly Notices Roy. Astron. Soc., Vol. 173, 709 - 728 (1975).

Radial velocity measurements of the southern broad-lined stars λ Sco and κ Sco have confirmed their membership in the β Canis Majoris (or β Cephei) class of variables. For λ Sco the observations show that neither the period nor the amplitude of the short-period variation are constant; the 10-day period is present in each year's observations. The beat period of $7^d.3$ for κ Sco is confirmed, although there are significant variations in the amplitudes which are not accounted for by simple beating periods.

119.016 **A new bright eclipsing star 71 Dra.** W. Fürtig.
Inform. Bull. Variable Stars, (I.A.U. Commission 27), Konkoly Obs., Budapest, No. 1071, 2 pp. (1975).

119.017 **A photometric investigation of the double-lined O-star binary, HD 159176.** J. C. Thomas.
Bull. American Astron. Soc., Vol. 7, 533 (1975). – Abstr. AAS.

119.018 **Hα spectroscopy of the RS CVn star UX Ari (HD 21242).** K. Mulligan, B. W. Bopp.
Inform. Bull. Variable Stars, (I.A.U. Commission 27), Konkoly Obs., Budapest, No. 1075, 2 pp. (1975).

119.019 **The orbit of the BY Draconis variable ADS 16557B.** B. W. Bopp, F. Fekel, Jr.
Publ. Astron. Soc. Pacific, Vol. 87, 891 - 894 (1975).

The BY Draconis-type variable ADS 16557B (HD 218738) is shown to be a double-line spectroscopic binary consisting of two nearly identical K2 dwarfs. The period is 3.03287 days. The system is a prime candidate for photometric spot modeling.

119.020 **Photometric observations of V470 Cygni (HD 228911).** E. G. Ebbighausen, D. Lester, S. Stearns, P. Straton, M. Sykes.
Publ. Astron. Soc. Pacific, Vol. 87, 923 - 927 (1975).

Previous photometric observations of this type B2 spectroscopic binary with a $1^d.87$ period have shown small but not well-defined light variations suggesting that it is an eclipsing binary with a rather small orbital inclination. In the summer of 1974 the authors obtained photoelectrically 2060 individual *V* observations and 2100 *B* observations. The total light range in both colors is nearly $0^m.02$ and the form of the light curve suggests that this object is an ellipsoidal binary. A new epoch of primary minimum and period are derived. All of the data combined are insufficient to confirm Pearce's suggestion of apsidal motion.

119.021 **Spectroscopic binary orbits from photoelectric radial velocities. Paper 5: HR 1970.**
G. A. Radford, R. F. Griffin.
Observatory, Vol. 95, 289 - 291 (1975).

119.022 **Reduction of astrometric observations.** E. Rojas.
Engineering Faculty, Univ. of Zulia, Maracaibo, Venezuela (1975). In Spanish.

The suspected nearby spectroscopic binary Ci 648 = HD 32147 has been studied astrometrically using 165 plates taken on the Sproul refractor between 1912 and 1968. A parallax of $0''.104$ and proper motions of $\mu_x = +0''.5548$ and $\mu_y = -1''.1039$ were derived. The data show no evidence of a second component. The reduction programs are described in some detail.

119.023 **Orbit of the spectroscopic binary B Persei.**
R. Rajamohan, M. Parthasarathy.
Pramāṇa, (*India*), Vol. 4, 153 - 159 (1975).

New spectroscopic orbital elements have been derived for the close binary system B Persei. The period is improved to 1.527360 days. Earlier observations by Cannon (1914), Harper (1930) and Heard (1938) have been reanalysed and the reality of the variations in the orbital elements is established. The change in ω indicates an apsidal motion period of about 60 years.

The composite nature of the F supergiant CD −30°5135. See Abstr. 114.382.

The masses and dimensions of early-type binaries. See Abstr. 115.013.

The triple system Ψ Sagittarii. See Abstr. 118.016.

A synthesis of the solutions of spectroscopic and visual binary orbits. See Abstr. 118.020.

Spectroscopic observations of eclipsing binaries. II. Absolute dimensions, evolutionary state, and helium content of RZ Chamaeleontis. See Abstr. 121.080.

Spectroscopic observations of eclipsing binaries. III. Definitive orbits and effects of line blending in CV Velorum. See Abstr. 121.081.

Spectroscopic observations of eclipsing binaries. I. Description of methods, and results for RS Chamaeleontis and χ^2 Hydrae. See Abstr. 121.082.

Variable radio emission from UX Arietis (HD 21242). See Abstr. 141.028.

The spectroscopic binary Scorpius X-1. See Abstr. 142.115.

Optical studies of UHURU sources. XIII. A photometric analysis of X Persei (= 3U 0352 + 30?). See Abstr. 142.135.

120 Variable Stars: Catalogues, Ephemerides, Miscellanea

120.001 **On the computation of minimum times of light curves.** H. Van Diest.
Astron. Astrophys., Vol. 42, 465 - 466 (1975).
Minimum epochs for periodically variables are derived from cubic Vondrak-spline fits. Obtained results are compared with those from previous methods.

120.002 **Supplement to the Catalogue of elements of RR Lyrae stars.**
B. N. Firmanyuk, G. A. Lange, V. P. Tsesevich.
Astron. Tsirk., No. 853, p. 4 - 5 (1975). In Russian.

120.003 **Complete list of variable or suspected variable Am stars.** M. S. Frolov.
Astron. Tsirk., No. 859, p. 3 - 4 (1975). In Russian.

120.004 **La classification internationale des étoiles variables.** E. Schweitzer.
A.F.O.E.V. Bull., Vol. 9, 59 - 91 (1975).

120.005 **Rocznik Astronomiczny Obserwatorium Krakowskiego 1976. International Supplement No. 47.**
Prepared under the supervision of K. Kozieł.
Komitet Astronomii, Polskiej Akademii Nauk, Kraków, 5 + 137 pp. Price zł 72.00 (1975). – Contents: Eclipsing binaries (K. Kordylewski); Ephemerides of eclipsing binaries among cataclysmic variables 1976 (J. M. Kreiner); RR-Lyrae-type variables (W. Zessewitsch, J. M. Kreiner); Auxiliary tables: Precessional coefficients 1976 (Z. Klimek); Obliquity of the ecliptic; Occultation of stars by the moon 1976 (Z. Klimek); Geocentric ephemeris of the libration points L_4 and L_5 in the earth-moon system for the year 1976 (K. Kordylewski).

120.006 **61st name-list of variable stars.**
B. V. Kukarkin, P. N. Kholopov, N. P. Kukarkina.
Inform. Bull. Variable Stars, (I.A.U. Commission 27), Konkoly Obs., Budapest, No. 1068, 5 pp. (1975).

121 Eclipsing Variables

121.001 **Photometric study of VZ Hydrae.**
T. D. Padalia, R. K. Srivastava.
Astrophys. Space Sci., Vol. 35, 249 - 254 (1975).

The photoelectric elements of the system VZ Hydrae have been redetermined in *U, B* and *V* colours. The authors' results are at variance from those given earlier by Walker. The colour of the system has also been given. The system is a detached one.

121.002 **The light curve of CV Serpentis, the sometimes-eclipsing Wolf-Rayet star.**
R. Schild, W. Liller.
Astrophys. Journ., Vol. 199, 432 - 435 (1975).

New photoelectric observations of the *B*-magnitude of CV Ser made in 1973 and 1974 show no clear evidence of an eclipse, but they establish nigh-to-night variability of several percent, a systematic brightness change of 0.035 mag during a portion of the single orbit observed in 1973, and irregular flaring in 1974. The authors find no evidence of a very deep eclipse such as observed by Hjellming and Hiltner. They present several new light curves and discuss them in the light of the recent results of Cowley et al.

121.003 **Infrared light curves of Algol.**
M. J. Smyth, M. J. Dow, W. McD. Napier.
Monthly Notices Roy. Astron. Soc., Vol. 172, 235 - 244 (1975).

Observations of the light curve of β Per at 2.2 μm are compared with predictions based upon several models of the Algol system; there is no evidence of infrared excess. Published 5-μm observations do not indicate a general infrared excess in the system, but the 5-μm light curve is anomalous near secondary minimum; neither conventional proximity effects nor emission from dust or plasma in the region between Algol A and Algol B can explain the discrepancy.

121.004 **On the eclipsing binaries in the globular clusters NGC 3201 and NGC 5139.** E. H. Geyer.
IAU Symposium No. 67, (see 012.001), p. 559 - 561 (1975).

V 65 is not a member of NGC 3201; there is no certainty about the membership of V 78 to NGC 5139 (ω Cen).

121.005 **Narrow-band photometry of 31 And 32 Cyg.**
B. Cester, M. Pucillo.
Astron. Astrophys. Suppl. Ser., Vol. 21, 189 - 192 (1974).

Photoelectric observations of 31 and 32 Cyg made at Trieste during the international campaigns 1971 and 1972 with three interferential filters are reported. The magnitudes for the components of 31 Cyg are deduced.

121.006 **On the precision of determination of eclipsing binaries elements.**
A. M. Shul'berg, V. P. Murnikova.
Peremennye Zvezdy, Vol. 19, 421 - 430 (1974). In Russian.

A method of computation of errors of eclipsing binaries elements in solving their light curves by the Russell-Merrill method is given. The use of this method is illustrated by two examples: the elements and their errors of the systems S Cnc and YZ Cas are found.

121.007 **Modification of Kopal-Tabachnik's method.**
M. I. Lavrov, N. V. Lavrova.
Peremennye Zvezdy, Vol. 19, 431 - 440 (1974). In Russian.

A new analytic interpretation of the first graphical method of Kopal is described. It is shown that for this interpretation the usual method of least squares may be used. It makes the new version to be less complicated than that of Tabachnik. The method has been tested on five binary systems.

121.008 **Solution of the "stellar" light curves of the Wolf-Rayet type eclipsing binary CQ Cep (Solution of light curves of CQ Cep. II).**
T. A. Kartasheva, M. A. Svechnikov.
Peremennye Zvezdy, Vol. 19, 441 - 459 (1974). In Russian.

The method of solution of the "stellar" light curve of the Wolf-Rayet eclipsing variable CQ Cep is given more explicitly than in the previous work. Nine "stellar" light curves of the system, selected from observations which were obtained in various years by different authors, are solved by this method. The accuracy of the method is estimated.

121.009 **On some peculiarities in brightness variations of AR Lacertae.** M. B. Babaev.
Peremennye Zvezdy, Vol. 19, 461 - 469 (1974). In Russian.

The behaviour of the light variations of AR Lac were studied. The observed peculiarities of the light curve are discussed.

121.010 **Researches of massive close binary systems of early spectral type. 1. Photoelectric observations of the eclipsing binary SX Aurigae.** I. I. Bondarenko.
Peremennye Zvezdy, Prilozhenie, Vol. 2, 51 - 58 (1974). In Russian.

215 observations in a system close to B and 212 observations in a system close to V were carried out by the AVR-2 refractor of the Ural University Observatory in February 1969 until April 1970. The results of the individual observations and mean light curves are given.

121.011 **ES Lac – an eclipsing variable with noticeable eccentricity.** A. F. Pugach.
Peremennye Zvezdy, Prilozhenie, Vol. 2, 59 - 60 (1974). In Russian.

The elements of the primary minimum are: J.D. (min) = $2434240^{d}416 + 4^{d}459349 \times E$. The secondary minimum occurs at phase 0.57.

121.012 **Investigation of an eclipsing variable in the open star cluster NGC 103.** Yu. A. Fadeev.
Peremennye Zvezdy, Prilozhenie, Vol. 2, 71 - 74 (1974). In Russian.

121.013 **Observations of the eclipsing variables AM Aurigae and V411 Cassiopeiae.** S. Yu. Shugarov.
Peremennye Zvezdy, Prilozhenie, Vol. 2, 83 - 89 (1974). In Russian.

121.014 **S10125 = V520 Cas – a W UMa-type variable.**
A. A. Tokovinin.
Peremennye Zvezdy, Prilozhenie, Vol. 2, 101 - 103 (1974). In Russian.

121.015 **Eine Ringscheibe in dem Bedeckungsveränderlichen RX Geminorum.** D. S. Hall, K. Walter.
Mitt. Astron. Ges., No. 36, p. 96 - 97 (1975).

121.016 **Photometric observations of the eclipsing binaries V65 and V78 in the globular clusters NGC 3201 resp. NGC 5139.** E. H. Geyer.
Mitt. Astron. Ges., No. 36, p. 97 (1975). – Abstract.

121.017 **Der Bedeckungsveränderliche VV Orionis.**
H. W. Duerbeck.

Mitt. Astron. Ges., No. 36, p. 98 (1975). – Abstract.

121.018 **Die Gashülle um den bedeckungsveränderlichen Röntgen-Doppelstern HD 153919 = 2U 1700–37.**
J. Dachs.
Mitt. Astron. Ges., No. 36, p. 98 (1975). – Abstract.

121.019 **Search for polarimetric variations in the close binary GK Cephei.** D. F. Lester, N. E. Woods, J. C. Kemp, I. G. Nolt.
Astrophys. Journ., Vol. 200, 95 - 97 (1975).

Polarimetric observations are presented of the close binary system GK Cephei. Based on 27 nights of observations, in the *B* filter, the authors set an upper limit to such effects in GK Cep of $\Delta p = 0.015$ percent, with a phase resolution of 0.10. The interstellar polarization was measured very accurately as $p = 0.20 \pm 0.02$ percent, $\theta = 71° \pm 1°$, in the blue.

121.020 **Observations of rapid blue variables – XVI. VZ Sculptoris.** B. Warner, A. D. Thackeray.
Monthly Notices Roy. Astron. Soc., Vol. 172, 433 - 439 (1975).

High speed photometry of the fifteenth magnitude nova-like variable VZ Scl is reported. The light curve resembles that of DQ Her and U Gem. An improved period of $0\overset{d}{.}14462220$ is derived. Measurements of 200-in. spectra provide a radial velocity curve. Apparent maximum velocity occurs near mid-eclipse, and the velocity curve is markedly non-sinusoidal. These features are interpreted as effects of a hot spot in the system.

121.021 **Astrometric orbit, eclipsing period changes, and parallax of VW Cephei.** J. L. Hershey.
Astron. Journ., Vol. 80, 662 - 667 (1975).

Six hundred ten plates taken from 1942 to 1974 with the 61-cm Sproul refractor have been analyzed for parallax and orbital motion. An orbital period of 30.45 yr and a photocentric semimajor axis of $0\overset{''}{.}130$ have been found, indicating the presence of a third body which has subsequently been observed visually according to prediction. The photographic and visual data yield a direct determination of the mass of the eclipsing pair and of the third star.

121.022 **Fourier analysis of the light curves of eclipsing variables, IV.** Z. Kopal.
Astrophys. Space Sci., Vol. 36, 227 - 237 (1975).

The methods of analysis of the light changes of eclipsing variables in the frequency domain, developed in previous papers (Kopal, 1975) for total or annular eclipses of arbitrarily limb-darkened stars, have now been extended to the case of partial eclipses of occultation as well as transit type. In this paper the even Fourier sine coefficients are formulated – in the guise of the moments A_{2m} of the light curve – in terms of the elements of the eclipse; and their use for a solution for the elements is detailed. A brief appendix containing certain auxiliary tables to facilitate this task concludes the paper.

121.023 **On the light curve of HZ Herculis.**
M. Livio, G. Shaviv.
Astron. Astrophys., Vol. 42, 17 - 23 (1975).

The effect of a low density corona around HZ Her, on the observed light curve is calculated. The agreement with observation, for several assumed density profiles in the corona, is not good. A better agreement is obtained when a cloud of matter around Her X-1 is assumed. The results support the hypothesis of a "third light" in the system HZ Her–Her X-1.

121.024 **A precessional period in U Cephei?** K. Walter.
Astron. Astrophys., Vol. 42, 135 - 138 = Mitt. Astron. Inst. Univ. *Tübingen*, No. 133 (1975).

From photometric observations of the light during total eclipse in primary minimum, obtained by many observers between 1880 and 1970, a period of 12.0 years is derived. It is suggested that this period be identified with the precessional period of the rotational axis of the B component.

121.025 **Eclipse timings of cataclysmic variables.**
G. S. Mumford.
Inform. Bull. Variable Stars, (I.A.U. Commission 27), Konkoly Obs., Budapest, No. 1043, 2 pp. (1975).

121.026 **Spectroscopic observations of Zeta Aurigae during the 1971–72 eclipse. II: Intensity of the chromospheric K-line and radial velocities of chromospheric lines.**
S. Kawabata, M. Saitō.
Astrophys. Space Sci., Vol. 36, 273 - 280 (1975).

The chromospheric phases in the 1971–72 eclipse were observed at wavelength regions containing the K-line of ionized calcium on six nights at ingress and on eight nights at egress at the Okayama Astrophysical Observatory. The spectrograms were measured for line intensities of the K-line and for radial velocities of metallic and hydrogen lines. The results are compared with those obtained in the earlier eclipses.

121.027 **Spectrographic study of the eclipsing binary system SZ Camelopardalis.** E. Budding.
Astrophys. Space Sci., Vol. 36, 329 - 343 (1975).

Thirteen high-dispersion spectrographs of the eclipsing binary star SZ Cam have been studied with a view of determining more accurate information on : (1) the spectral type and luminosity classifications, (2) absolute parameters for the component stars, (3) the stellar environment of SZ Cam. The determined overluminosity of the secondary star and the local concentration of absorbing material are two topics which provide the basis for a discussion section.

121.028 **An unusually strong radio outburst in Algol: flux density, spectral, and polarization observations.**
D. M. Gibson, M. R. Viner, S. D. Peterson.
Astrophys. Journ., (*Letters*), Vol. 200, L143 - L145 (1975).

The authors have monitored the largest radio outburst to date from the binary β Per (Algol). The observations, made at frequencies of 1400, 2695, 6000, and 8085 MHz, mark the first time a non-X-ray radio binary has been observed at flux densities exceeding 1 Jy. The outburst probably consisted of two Algol-type flares. The radio spectrum remained essentially constant throughout the outburst, and no polarization was detected.

121.029 **The monochromatic phase effect in eclipsing variable systems of β Per type.** V. P. Merezhin.
Trudy Kazan. Gorod. Astron. Obs., No. 39, p. 29 - 41 (1973). In Russian.

121.030 **Machine analysis of light curves of S Cancri.**
M. I. Lavrov.
Trudy Kazan. Gorod. Astron. Obs., No. 39, p. 42 - 56 (1973). In Russian.

121.031 **On the period of the eclipsing binary system XX Cephei.** M. I. Lavrov, N. V. Lavrova.
Trudy Kazan. Gorod. Astron. Obs., No. 39, p. 83 - 93 (1973). In Russian.

121.032 **The eclipsing binary system VV Orionis.**
H. W. Duerbeck.
Astron. Astrophys., Suppl Ser., Vol. 22, 19 - 47 (1975).

Simultaneous *UBV*, Hβ and spectroscopic observations yield absolute dimensions and orbital elements of the primary and secondary components of the detached eclipsing binary VV Ori (*B*1*V*+*B*7(?) *V*, *P* = 1.485 days). The rotation of the primary is found to be synchronous with the orbital motion.

Old and new radial velocity measurements lead to improved orbital elements for the third invisible component. Some atmospheric parameters are derived from the equivalent widths of the hydrogen lines. Changes in the equivalent widths and disturbances in the light curves indicate the existence of a gaseous stream with a particle density of about 10^{11} cm^{-3}.

121.033 **Analysis of the light curve of HZ Her.** N. E. Kurochkin.
Pis'ma v Astron. Zhurn., Vol. 1, No. 7, p. 27 - 33 (1975). In Russian.

An analysis of light variation periodicity of HZ Her has been carried out using the author's computer programme and published photoelectric observations. A stationary secondary variation is obtained, its elements being Min I = JD 2440814.824 + $1^{d}_{.}0977 \times E$ and amplitude $\Delta B_1 = 0^{m}_{.}33$. Two further independent variations with $P_2 = 0^{d}_{.}87$ and $P_3 = 1^{d}_{.}62$ are also found. The mean light curve with P_3 is asymmetrical and probably caused by pulsation of the primary star.

121.034 **Far ultraviolet spectrophotometry of β Lyrae.** R. C. Bless, J. Eaton.
Bull. American Astron. Soc., Vol. 7, 405 (1975). – Abstr. AAS.

121.035 **The binary star system CD Tau.** D. B. Wood.
Bull. American Astron. Soc., Vol. 7, 462 (1975). Abstr. AAS.

121.036 **PG 1413+01, a white dwarf–red dwarf eclipsing binary.**
M. Schmidt, R. Green, D. O. Richstone.
Bull. American Astron. Soc., Vol. 7, 462 (1975). – Abstr. AAS.

121.037 **Eclipse observations of H-alpha during the recent outburst of U Cephei.**
R. C. Crawford, M. Plavec.
Bull. American Astron. Soc., Vol. 7, 462 - 463 (1975). – Abstr. AAS.

121.038 **A precision photoelectric analysis of the eclipsing binary WW Aurigae.** P. B. Etzel.
Bull. American Astron. Soc., Vol. 7, 463 (1975). – Abstr. AAS.

121.039 **An extraordinary radio outburst in Algol: flux and structure observations.** D. Gibson, M. Viner, S. Peterson, T. Clark, L. Hutton, C. Ma, W. Webster, H. Hinteregger, A. Rogers, I. Shapiro, A. Whitney, J. Wittels, A. Niell, G. Resch.
Bull. American Astron. Soc., Vol. 7, 498 (1975). – Abstr. AAS.

121.040 **Spectrographic observations of Beta Persei (Algol).** C. T. Bolton, S. N. Shore.
Bull. American Astron. Soc., Vol. 7, 498 (1975). – Abstr. AAS.

121.041 **Minima of eclipsing binaries, 1973–74.**
Journ. British Astron. Ass., Vol. 85, 443 - 447 (1975).

121.042 **Four-colour photometry of eclipsing binaries. I. HS Hya, light curves, photometric elements, and determination of helium content.**
K. Gyldenkerne, H. E. Jørgensen, E. Carstensen.
Astron. Astrophys., Vol. 42, 303 - 309 (1975).

Photometric elements are derived for the eclipsing binary HS Hya. Using published spectroscopic elements the absolute dimensions are derived. A comparison with new stellar model calculations was carried out and the helium content of HS Hya was found to be $Y = 0.22 \pm 0.04$.

121.043 **Effects of intrinsic polarization of eclipsing binaries VW Cep and RW Tau.**
O. S. Shulov, E. N. Kopatskaya.
Trudy Astron. Obs., *Leningrad*, Vol. 31 (= Uchenye Zapiski Leningr. Un-ta, No. 381 = Seriya Matem. Nauk, vyp. (No.) 51), p. 146 - 153 (1975). In Russian.

Variable polarization was observed in the eclipsing binaries VW Cep and RW Tau. In both stars the variations of polarization at primary minima are similar to Sobolev-Chandrasekhar's effect.

121.044 **Photoelectric UBV photometry of the X-ray binary systems HZ Herculis (Her X-1) and V 1357 Cygni (Cyg X-1).** A. M. Cherepashchuk, V. M. Kovalenko, O. N. Kovalenko, A. V. Mironov.
Peremennye Zvezdy, Vol. 19, 305 - 318 (1974). In Russian.

The results of a UBV photometry of HZ Her (Her X-1) and V 1357 Cyg (Cyg X-1) are presented and discussed.

121.045 **On rapid variations in the spectrum of AR Lac.** M. B. Babaev.
Peremennye Zvezdy, Vol. 19, 377 - 379 (1974). In Russian.

The emission bands with variable profiles and intensities have been observed in the region λλ 4300–4870 Å of the spectrum of AR Lac. The observation of the emission lines is assumed to be connected with gaseous streams existing between the components.

121.046 **Observational properties of models of semidetached close binaries. II.** P. Giannone, M. A. Giannuzzi, M. Pucillo.
Astron. Astrophys., Vol. 43, 141 - 148 (1975).

Observational properties and the probability of discovery as eclipsing binaries have been determined for models of semidetached systems. Binaries of cases A and B with intermediate and small masses have been studied. Synthetic light curves are shown to be affected mainly by the assumption concerning the shape of the components. The detection of systems evolving according to case A appears to be favoured in comparison with that of systems of case B. Systems with undersize subgiants result comparatively difficult to observe.

121.047 **On the eclipsing star SVS 1212 Scuti.** V. P. Tsesevich.
Astron. Tsirk., No. 845, p. 7 - 8 (1974). In Russian.

121.048 **Quantitative analysis of the spectrum of Beta Lyrae. III. Investigation of the atmosphere of the bright component.** M. Yu. Skul'skij.
Problems of cosmic physics. Vyp. (No.) 10, (see 003.003), p. 160 - 174 (1975). In Russian.

Quantitative investigation of the atmosphere of Beta Lyrae's bright component for ten different phases of the orbital period has revealed considerable changes of microturbulent velocity with phase. It favours the assumption that the bright component has an unspherical form. The elemental abundance in the atmosphere of the bright component is discussed.

121.049 **Mass flow in the O7f binary UW Canis Majoris.** G. E. McCluskey, Jr., Y. Kondo, D. C. Morton.
Astrophys. Journ., Vol. 201, 607 - 612 (1975).

The ultraviolet spectrum of the eclipsing binary UW CMa (O7f + O–B) has been observed by the Princeton telescope spectrometer on the Copernicus satellite in the wavelength region from about 950–1560 Å at a resolution corresponding to 0.2 Å. A number of stellar and interstellar lines appear in the spectrum. A mass-loss rate of about 3×10^{-6} solar masses per year is estimated. A discussion of the evolution of this system, based on the current observations as well as earlier ground-based spectroscopic and photometric investigations, is given.

121.050 **Binary systems in star clusters. II. HD 162724.**
K.-C. Leung, D. P. Schneider.
Astrophys. Journ., Vol. 201, 792 - 798 (1975).
Over 2000 photoelectric observations of HD 162724 in the yellow, blue, and ultraviolet were analyzed. The period found was 2.785847 days. The depths of primary and secondary minima were about equal. A photometric solution was derived with the Wilson and Devinney model. The binary is found to be a detached system with a mass ratio of 0.88. Both components are at the end of the hydrogen-burning phase of evolution.

121.051 **Infrared observations and a model of β Per.**
A. J. Longmore, R. F. Jameson.
Monthly Notices Roy. Astron. Soc., Vol. 173, 271 - 278 (1975).
Infrared light curves of β Per are presented for wavelengths of 2.2, 3.6, 4.8 and 8.6 μ. From these are deduced a model of the system which has a gas stream flowing from the K0 star to the B8 star. Its temperature and electron density vary from 5 to 20×10^3 K and 30 to 5×10^{11} cm^{-3} respectively. The mass transfer rate is $1.8 \times 10^{-8} M_\odot$/yr.

121.052 **V Sagittae (201520).** E. H. Mayer.
Journ. American Ass. Variable Star Observers, Vol. 4, 7 - 9 (1975).
The light curve is described and visual observations of two recent eclipses are presented. Criteria for future observations are suggested.

121.053 **Evidence for small, short-term period changes in RZ Draconis.** A. D. Mallama.
Journ. American Ass. Variable Star Observers, Vol. 4, 13 - 17 (1975).
An intensive series of timings of minima reveals a quasi-periodic O–C curve. Explanations in terms of a third body or apsidal motion seem unlikely, and mass exchange may be the cause. The importance of timing the secondary minima is stressed.

121.054 **Differential UBV photometry of β Lyrae, V.**
H. J. Landis, L. P. Lovell, D. S. Hall.
Journ. American Ass. Variable Star Observers, Vol. 4, 18 - 21 (1975).
New observations are tabulated.

121.055 **BW Bootis – an Algol type eclipsing variable.**
M. Kurpińska.
Inform. Bull. Variable Stars, (I.A.U. Commission 27), Konkoly Obs., Budapest, No. 1007, 4 pp. (1975).

121.056 **Photographic observations of EE Cephei at the 1969 minimum.**
L. Baldinelli, S. Ghedini, C. Tubertini.
Inform. Bull. Variable Stars, (I.A.U. Commission 27), Konkoly Obs., Budapest, No. 1009 (1975).

121.057 **Photographic observations of EE Cephei at the 1975 minimum.**
L. Baldinelli, S. Ghedini, C. Tubertini.
Inform. Bull. Variable Stars, (I.A.U. Commission 27), Konkoly Obs., Budapest, No. 1009 (1975).

121.058 **Minima of eclipsing variables.**
L. Baldinelli, S. Ghedini, C. Tubertini.
Inform. Bull. Variable Stars, (I.A.U. Commission 27), Konkoly Obs., Budapest, No. 1009 (1975).

121.059 **A. R. Hogg's UBV observations of U Pegasi.**
K. C. Gordon.
Inform. Bull. Variable Stars, (I.A.U. Commission 27), Konkoly Obs., Budapest, No. 1010, 2 pp. (1975).

121.060 **TY Men – determination of new light elements.**
R. R. D. Austin, M. DeGeorge, Jr., R. C. Fleck, Jr.
Inform. Bull. Variable Stars, (I.A.U. Commission 27), Konkoly Obs., Budapest, No. 1013 , 2 pp. = Rosemary Hill Obs., Univ. Florida, Gainesville, Contr. No. 58 (1975).

121.061 **1974 spectrophotometry of six RS Canum Venaticorum-type binaries.** E. J. Weiler.
Inform. Bull. Variable Stars, (I.A.U. Commission 27), Konkoly Obs., Budapest, No. 1014, 2 pp. (1975).

121.062 **Photoelectric times of minima of RZ Cas.**
T. E. Margrave, M. A. Lukes, J. H. Doolittle, R. S. Evenskaas, J. W. MacDonald, N. D. Murray.
Inform. Bull. Variable Stars, (I.A.U. Commission 27), Konkoly Obs., Budapest, No. 1019, 2 pp. (1975).

121.063 **About the variability of the eclipsing variable star NQ Herculis.** S. Bozkurt, C. Ibanoglu, Ö. Gülmen, N. Güdür.
Inform. Bull. Variable Stars, (I.A.U. Commission 27), Konkoly Obs., Budapest, No. 1021, 2 pp. (1975).

121.064 **Minima of the eclipsing variable U Sagittae.**
D. H. McNamara, K. A. Feltz, Jr.
Inform. Bull. Variable Stars, (I.A.U. Commission 27), Konkoly Obs., Budapest, No. 1022 (1975).

121.065 **A critical remark about the determination of minimum times.** H. W. Duerbeck.
Inform. Bull. Variable Stars, (I.A.U. Commission 27), Konkoly Obs., Budapest, No. 1023, 2 pp. (1975).

121.066 **Photoelectric minima of ER Ori and AW Cam.**
F. Zavatti, R. Burchi.
Inform. Bull. Variable Stars, (I.A.U. Commission 27), Konkoly Obs., Budapest, No. 1028 (1975).

121.067 **Photoelectric photometry of WY Cancri.**
C. R. Chambliss.
Inform. Bull. Variable Stars, (I.A.U. Commission 27), Konkoly Obs., Budapest, No. 1029, 2 pp. (1975).

121.068 **The eclipsing binary system X Tri.**
M. I. Kumsiashvili.
Inform. Bull. Variable Stars, (I.A.U. Commission 27), Konkoly Obs., Budapest, No. 1033 (1975).

121.069 **AN Ursae Majoris is probably an exnova?**
B. V. Kukarkin.
Inform. Bull. Variable Stars, (I.A.U. Commission 27), Konkoly Obs., Budapest, No. 1039 (1975).

121.070 **HD 173198, a β Lyrae type star.** F. van Leeuwen.
Inform. Bull. Variable Stars, (I.A.U. Commission 27), Konkoly Obs., Budapest, No. 1041, 2 pp (1975).

121.071 **Orbital analysis of Algol AB, C from combined astrometric, photometric, and radial velocity data.**
P. J. Bachmann, J. L. Hershey.
Astron. Journ., Vol. 80, 836 - 843 (1975).
New astrometric data from the Sproul 61-cm refractor are included in the solutions for the orbit and masses of Algol AB, C. For the first time an orbital determination has been made by combining all astrometric, photometric time-of-minimum, and spectroscopic data in a single solution by simultaneous differential corrections to all elements. Results of separate and combined solutions of the three kinds of data are presented. The errors of all orbital elements have been

computed and carried through the computations for the masses, which have been computed from several sets of observed quantities. Combining elements independent of the parallax, a mass of 5.3 $M_\odot$ is found for Algol AB and 1.8 $M_\odot$ for Algol C.

121.072 **EE Cephei, une algolide à très longue période.**
A.F.O.E.V. Bull., Vol. 9, 95 (1975).

121.073 **Mass transfer in X Trianguli.** A. D. Mallama.
Acta Astron., Vol. 25, 205 - 214 (1975).

The $O-C$ curve for X Tri between 1921 and 1974 is analyzed in terms of the period-change model of Biermann and Hall. Two separate representations of the $O-C$ curve are presented; these place a lower and upper limit on the observable mass transfer inferred from the curve. During each sudden period decrease the observable mass loss is $\gtrsim 6\times10^{-6} M_\odot$. The average observable mass loss rate is $\gtrsim 5\times10^{-7} M_\odot$/yr and corresponds to a thermal time scale. Orbital angular momentum is not conserved.

121.074 **On the period variations and light curve changes in RS Canum Venaticorum.** D. S. Hall.
Acta Astron., Vol. 25, 215 - 224 (1975).

In response to a recent paper by Catalano and Rodono, the problem of spurious $O-C$ shifts generated by light curve asymmetry in RS CVn-type binaries is discussed anew. Three different binaries (RS CVn, SS Cam, CG Cyg) show the same correlation between $O-C$ variation and outside-eclipse wave migration, even after effects of light curve asymmetry have been removed or shown to be negligible. Thus it is premature to dismiss the mass ejection explanation proposed earlier by Hall to explain this correlation. In addition, the tilted ring model of RS CVn, proposed by Catalano and Rodono as an alternative, is criticized.

121.075 **Observations of six southern eclipsing binaries for apsidal motion.** S. Söderhjelm.
Astron. Astrophys. Suppl. Ser., Vol. 22, 263 - 283 (1975).

Photoelectric *UBV* observations are presented for the six eclipsing binaries KX Pup, EO Vel, HH Car, AY Mus, KT Cen and GM Nor. Photometric elements are determined together with rough estimates of masses and radii. KX Pup, EO Vel, KT Cen and GM Nor are found to exhibit apsidal motion in good agreement with theory. HH Car, for which apsidal motion was previously reported, has a circular orbit and shows a distorted light-curve due to circumstellar matter. AY Mus has stopped eclipsing, apparently because of perturbations by a third body in the system.

121.076 **Photoelectric elements of the eclipsing binary XY Ceti.** R. K. Srivastava, T. D. Padalia.
Astrophys. Space Sci., Vol. 38, 79 - 85 (1975).

The absolute elements of the system XY Ceti have been obtained on the basis of the spectroscopic elements given by Popper (1971) and photoelectric elements derived by the authors. The colours of the components have been obtained. Both components are found to lie fairly on the main sequence. The primary component of the system, however, is slightly more evolved as it shows a tendency to drift away from the main sequence. The spectral classes now assigned are A5V (primary) and A7V (secondary). The values of Roche constants indicate that the system is a detached one.

121.077 **The photoelectric elements of AS Camelopardalis.**
T. D. Padalia, R. K. Srivastava.
Astrophys. Space Sci., Vol. 38, 87 - 93 (1975).

Improved geometrical elements of the system AS Cam have been obtained and the colour of the system has been discussed. The absolute dimensions have also been obtained on the basis of the spectroscopic elements given by Hilditch (1972). Both components are found to lie fairly on the main sequence. The system is a detached one.

121.078 **Fourier analysis of the light curves of eclipsing variables. V.** Z. Kopal.
Astrophys. Space Sci., Vol. 38, 191 - 241 (1975).

The methods of analysis of the light changes of eclipsing variables in the frequency-domain, developed in previous papers for an interpretation of mutual eclipses in systems consisting of spherical stars, have now been extended to analyse the light variations – between minima as well as within eclipses – of close binaries whose components are distorted by axial rotation and mutual tidal action.

121.079 **Four-colour photometry of eclipsing binaries. II. RZ Cha, light curves, photometric elements and determination of helium content.**
H. E. Jørgensen, K. Gyldenkerne.
Astron. Astrophys., Vol. 44, 343 - 347 (1975).

Photometric elements are derived for the eclipsing binary RZ Cha. Combined with spectroscopic elements the absolute dimensions are derived. A comparison with new stellar model calculations was carried out and the helium content of RZ Cha was found to be $Y = 0.21 \pm 0.05$.

121.080 **Spectroscopic observations of eclipsing binaries. II. Absolute dimensions, evolutionary state, and helium content of RZ Chamaeleontis.**
J. Andersen, H. Gjerløff, M. Imbert.
Astron. Astrophys., Vol. 44, 349 - 353 (1975).

New spectroscopic observations of the 8-th magnitude F-type eclipsing binary RZ Chamaeleontis are presented and discussed in conjunction with the recent photometric data by Jørgensen and Gyldenkerne (1975). The system is found to consist of two virtually identical components of mass $1.51 \pm 0.03 M_\odot$, radius $2.26 \pm 0.02 R_\odot$, and spectral type F5. Comparison with theoretical stellar models shows that the components are appreciably evolved and presently in the subgiant stage. An age of about 2×10^9 years and a He abundance $Y = 0.21 \pm 0.05$ are derived.

121.081 **Spectroscopic observations of eclipsing binaries. III. Definitive orbits and effects of line blending in CV Velorum.** J. Andersen.
Astron. Astrophys., Vol. 44, 355 - 362 (1975).

New spectroscopic observations of the double-lined eclipsing binary CV Vel (B2.5 V) are discussed. From 32 spectrograms of dispersion 12.3 Å mm^{-1} the following orbital elements have been found: $\gamma = +24.3 \pm 0.5$ km s^{-1}, $K_1 = 127.0 \pm 0.3$ km s^{-1} and $K_2 = 129.2 \pm 0.4$ km s^{-1}, leading to definitive mass values of $m_1 \sin^3 i = 6.07 \pm 0.04 M_\odot$ and $m_2 \sin^3 i = 5.96 \pm 0.04 M_\odot$. Combining with preliminary photometric data, the physical properties of the average component are found to be: mass $= 6.05 M_\odot$, radius $= 4.05 R_\odot$, $Te = 18300$°K, $M_{bol} = -3^m.3$, $M_v = -1^m.5$, distance 600 pc.

121.082 **Spectroscopic observations of eclipsing binaries. I. Description of methods, and results for RS Chamaeleontis and χ^2 Hydrae.** J. Andersen.
Astron. Astrophys., Vol. 44, 445 - 458 (1975).

First results of a spectroscopic observing programme designed to provide an accurate determination of the physical properties of the components of eclipsing binary systems with double-lined spectra are presented and discussed. The system RS Cha is found to consist of two somewhat evolved components of approximately equal spectral type and luminosity, masses of 1.86 ± 0.02 and $1.82 \pm 0.02 M_\odot$, radii of 2.14 ± 0.1 and $2.42 \pm 0.1 R_\odot$, respectively, and an age of some $8-9\times10^8$ years. The system is an example of differential metallicism, the Ca II K line of the less massive component showing moderate weakening. The B-type system χ^2 Hya is observationally difficult because of the large light-ratio of 4:1. The components are found to have masses of 3.71 ± 0.08 and 2.71 ± 0.05 $M_\odot$, radii of 4.73 ± 0.27 and 2.64 ± 0.14 $R_\odot$, respectively, and an age of 1.5–1.7 × 10^8 years

121.083 **The eclipsing binary RS Cha.** H. E. Jørgensen. Astron. Astrophys., Vol. 44, 459 - 460 (1975).

The light curves of RS Cha are rediscussed. The photometric solution shows equally bright components in agreement with new spectroscopic observations.

121.084 **B and V photometry of AO Cassiopeiae,** P. Kalv. Publ. Tartu Astrofiz. Obs., Vol. 43, 103 - 113 (1975).

Photoelectric light curves of the close binary star AO Cas have been obtained from observations during 118 nights from 1967 to 1970 in Tallinn. The light curve shows varying amounts of asymmetry.

121.085 **Photoelectric investigation of the long-period eclipsing binary V367 Cyg based on observations made 1967–1973.** P. Kalv, I. Pustylnik.
Publ. Tartu Astrofiz. Obs., Vol. 43, 114 - 129 (1975). In Russian.

Photoelectric B and V light curves of the long-period eclipsing variable V367 Cyg were obtained at Tallinn Observatory. Orbital elements were determined, the masses of the components were estimated.

121.086 **Analysis of light changes of the contact binary VW Cep in 1957–1960 and 1965–1968.**
I. Pustylnik, L. Sorgsepp.
Publ. Tartu Astrofiz. Obs., Vol. 43, 130 - 175 (1975). In Russian.

Numerous B and V light curves of the contact binary VW Cep have been analyzed based on the observations made by Kwee in 1957–1960, Sorgsepp and Albo in 1965–1968. Rectification constants were determined for individual curves showing considerable time variations. Orbital elements were calculated. The asymmetry of the light curves is explained in terms of a "hot spot" presenting itself in the leading hemisphere of the primary star.

121.087 **On the mass ratio of the components of VW Cephei.** I. Pustylnik, L. Sapar, L. Sorgsepp.
Publ. Tartu Astrofiz. Obs., Vol. 43, 176 - 183 (1975). In Russian.

The method of differential corrections has been applied to analyze the light curves of the contact binary VW Cep. The angle of orbital inclination and the relative luminosities were improved, the mass ratio of the components was estimated.

121.088 **Comments on the tilted ring model for the eclipsing binary RS Canum Venaticorum.** J. P. Oliver.
Publ. Astron. Soc. Pacific, Vol. 88, 695 - 697 = Contr. Rosemary Hill Obs., *Gainesville*, Florida, No. 59 (1975).

The tilted equatorial ring model of Catalano and Rodono is discussed and major weaknesses are pointed out.

121.089 **A spectroscopic orbit for the eclipsing binary IQ Persei.** A. Young.
Publ. Astron. Soc. Pacific, Vol. 87, 717 - 719 (1975).

Spectroscopic orbital elements are derived for the eclipsing binary IQ Per. Faint secondary lines are detected, and a mass ratio and individual masses are inferred. The components are found to be on the main sequence, and the system is detached.

121.090 **Ultraviolet photometry and spectrophotometry of Algol.** J. A. Eaton.
Publ. Astron. Soc. Pacific, Vol. 87, 745 - 751 (1975).

OAO-2 ultraviolet light curves and spectrophotometry of the Algol system are presented and used to establish effective temperatures and luminosities for the three components. The values derived are: (Component, T_e, log $L/L_\odot$): (A, 13,000°K ± 500°K, 2.41 ± 0.2), (B, 4600°K ± 800°K, 0.68 ± 0.3), (C, 8700°K ± 700°K, 1.11 ± 0.2).

121.091 **Photometric study of the eclipsing binary WX Cephei.**
E. G. Ebbighausen, G. Penegor, P. Straton.
Publ. Astron. Soc. Pacific, Vol. 87, 795 - 800 (1975).

The photometric elements of the eclipsing binary WX Cep are determined from 3120 photoelectrically measured observations in each of the *B* and *V* regions. The analysis shows that primary minimum is produced by the occultation of the smaller, fainter type A2 star by the larger brighter star of about type A5, and that during the 10.9-hour duration of the primary eclipse it is total for about 40 minutes. The existing spectroscopic data (of uncertain quality) for this double-line binary are used to estimate the mass and radius of each component. Both components have about one solar mass.

121.092 **Photoelectric minima of eclipsing binaries.** E. Pohl, A. Kizilirmak.
Inform. Bull. Variable Stars, (I.A.U. Commission 27), Konkoly Obs., Budapest, No. 1053, 4 pp. (1975).

121.093 **HR 6611: an interesting new eclipsing binary.** D. W. Kurtz.
Inform. Bull. Variable Stars, (I.A.U. Commission 27), Konkoly Obs., Budapest, No. 1054, 2 pp. (1975).

121.094 **Non-variability of NQ Her.** T. D. Padalia. Inform. Bull. Variable Stars, (I.A.U. Commission 27), Konkoly Obs., Budapest, No. 1059, 2 pp. (1975).

121.095 **About the intrinsic polarization of AW UMa.** V. Piirola.
Inform. Bull. Variable Stars, (I.A.U. Commission 27), Konkoly Obs., Budapest, No. 1060, 3 pp. (1975).

121.096 **Polarimetric observations of U Cephei during the primary eclipse.** V. Piirola.
Inform. Bull. Variable Stars, (I.A.U. Commission 27), Konkoly Obs., Budapest, No. 1061, 6 pp. (1975).

121.097 **Photoelectric minima of eclipsing variables.** L. Patkós.
Inform. Bull. Variable Stars, (I.A.U. Commission 27), Konkoly Obs., Budapest, No. 1065 (1975).

121.098 **Improved elements of the new eclipsing binary BD +60° 2289.** A. A. Wachmann.
Inform. Bull. Variable Stars, (I.A.U. Commission 27), Konkoly Obs., Budapest, No. 1066 (1975).

121.099 **Photometry of the eclipsing binary U Ophiuchi and of nova Cygni 1975.**
G. A. Becker, C. R. Chambliss, A. Kiasat.
Inform. Bull. Variable Stars, (I.A.U. Commission 27), Konkoly Obs., Budapest, No. 1067 (1975).

121.100 **On the period and orbit of the eclipsing binary θ^1 Ori A.** W. A. Feibelman.
Inform. Bull. Variable Stars, (I.A.U. Commission 27), Konkoly Obs., Budapest, No. 1070 (1975).

121.101 **The apparent infrared excess in RT Lacertae.** E. F. Milone.
Journ. Roy. Astron. Soc. Canada, Vol. 69, 244 (1975).
Abstr. CAS.

121.102 **The close O-type eclipsing binary TU Muscae.** J. Andersen, B. Grønbech.
Astron. Astrophys., Vol. 45, 107 - 115 (1975).

New spectroscopic and photometric observations of TU Mus are presented and discussed. From 14 double-lined spectra the authors derive circular orbital elements of $V_0 = +1.8$ km s^{-1}. The spectral type of the primary component is estimated to be O7–O8. The *uvby* light curves (703 observations in each filter) are analysed approximating the stars by triaxial ellipsoids, and are found to be fairly well reproduced with the elements $r_1 = 0.45$, $r_2 = 0.37$, $i = 76°$, $T_1 = 33600$ °K, $T_2 = 29600$ °K. A tentative model of TU Mus is proposed with the following physical properties: masses 23.5 and 15.8 $M_\odot$, radii 8.0 and 6.6 $R_\odot$, $M_{bol} = -7^m.4$ and $-6^m.4$, and $M_v = -4^m.2$ and $-3^m.5$.

121.103 **Spectroscopic observations of eclipsing binaries IV. Absolute dimensions of the giant system SZ Centauri.** J. Andersen.
Astron. Astrophys., Vol. 45, 203 - 208 (1975).

New spectroscopic observations at 12 Å mm^{-1} of the A-type giant eclipsing binary system SZ Centauri are presented and discussed together with new accurate photometric elements by Grønbech et al. (1976). The components of SZ Cen are found to have masses of 2.28 ± 0.02 and 2.32 ± 0.03 $M_\odot$, radii of 3.62 ± 0.02 and 4.53 ± 0.02 $R_\odot$, the slightly more massive component being the larger and cooler of the two. Comparing with a set of theoretical stellar models, an age of 500×10^6 years is derived, and it is estimated that the larger component will fill its Roche lobe in only 9×10^6 years from now.

121.104 **Polarimetric observations of close binary systems of W UMa-type.** V. A. Oshchepkov.
Byull. Abastumansk. Astrofiz. Obs., No. 46, p. 31 - 46 (1975). In Russian. – Concerning i Boo, VW Cep, U Peg, W UMa.

121.105 **Some aspects concerning the determination of minimum times in eclipsing binary systems.**
H. W. Duerbeck.
Acta Astron., Vol. 25, 361 - 369 (1975).

A new method for the determination of minimum times in eclipsing binary systems, the conjunction method, is presented. Several methods which are in general use are applied to visually and photoelectrically observed minima of VV Ori and RZ Cas. In the case of symmetrical minima graphical methods and the Kwee-van Woerden method give better results than polynomial fits. The mean error of a visually determined minimum time is found to be 6 times that of a photoelectrically observed minimum.

121.106 **Light variability of AM CVn (=HZ 29).** J. Smak.
Acta Astron., Vol. 25, 371 - 378 (1975).

Photometric observations on four nights in 1974 show a wide variety of shapes of the 17.5-min. light curve. Significant changes in the shape of the light curve can occur on time-scales as short as 40 - 60 minutes (2 - 3 cycles). On January 23/24, 1974, a transition was observed from a 'standard', double-humped light curve to a 'U Gem-type' light curve with a single, primary minimum. This phenomenon, resembling a similar, occasional behaviour of WZ Sge, provides an additional argument in favour of the binary model for AM CVn.

121.107 **Photoelectric observations of the variable AW UMa.** T. Z. Dworak, M. Kurpińska.
Acta Astron., Vol. 25, 417 - 425 (1975).

This paper contains results of the V and $B-V$ observations of the contact binary AW UMa. The improved photometric elements and the $O-C$ diagram are given. A possible non periodic activity is discussed.

121.108 **Definitive ultraviolet light curves of UW Canis Majoris.** J. A. Eaton.
Bull. American Astron. Soc., Vol. 7, 532 - 533 (1975). Abstr. AAS.

121.109 **Another look at V566 Ophiuchi.** B. B. Bookmyer.
Bull. American Astron. Soc., Vol. 7, 533 (1975). Abstr. AAS.

121.110 **The binary system V566 Oph revisited.** T. A. Nagy.
Bull. American Astron. Soc., Vol. 7, 533 (1975). – Abstr. AAS.

121.111 **A comparison of two computer-generated models of eclipsing binary systems: final results.**
J. S. Shaw.
Bull. American Astron. Soc., Vol. 7, 533 (1975). – Abstr. AAS.

121.112 **Secondary component and mass ratio of β Lyrae.** D. A. Brown, S.-S. Huang.
Bull. American Astron. Soc., Vol. 7, 534 (1975). – Abstr. AAS.

121.113 **Apsidal motion in the Algol system.** J. W. Stein, W. R. Beardsley.
Bull. American Astron. Soc., Vol. 7, 543 (1975). – Abstr. AAS.

121.114 **The 1975 shell spectrum of DV Aquarii.** M. W. King, W. R. Beardsley.
Bull. American Astron. Soc., Vol. 7, 543 - 544 (1975). Abstr. AAS.

121.115 **Photometric study of MY Cygni.** R. M. Williamon.
Astron. Journ., Vol. 80, 976 - 985 (1975).

The eclipsing variable MY Cygni was observed with *UBV* filters during 1972, 1973, and 1974 at the Fernbank Science Center Observatory. The light curves of this metallic line system indicate little interaction between the components and a small eccentricity of orbit. The eclipses are partial, and solutions of the light curves consistent among the three colors and based on the Russell method give $K = 0.96$.

121.116 **V624 Her.** D. Hoffleit.
Inform. Bull. Variable Stars, (I.A.U. Commission 27), Konkoly Obs., Budapest, No. 1074 (1975).

Concerning HR6611 reported by D. W. Kurtz – see Abstr. 121.093.

121.117 **Note on the period variation in the white-dwarf eclipsing binary BD +16°516.** T. J. Herczeg.
Inform. Bull. Variable Stars, (I.A.U. Commission 27), Konkoly Obs., Budapest, No. 1076, 4 pp. (1975).

121.118 **Bedeckungsveränderliche.** A. Werner.
Jahrbuch der Schulphysik, (Aulis-Verlag, Köln), Vol. 2, 7 - 12 (1975).

A review article concerning classification of eclipsing variables, determination of orbital elements (taking into consideration limb-darkening, deformation of the components, reflection effects), new Fourier methods, apsidal motion and internal structure of stars, investigation of stellar atmospheres.

121.119 **The equivalent widths and catalogue of the spectral lines of the β Lyrae primary and secondary. I.**
A. N. Dadaev.
Izv. Glav. Astron. Obs. Pulkovo, No. 193, Astrofiz., Astrometr. i Zvezdn. Astron., p. 41 - 64 (1975). In Russian.

From observations made by the author in 1959 at the Crimean Astrophysical Observatory, using the 120-cm reflector with a glass spectrograph, a quantitative analysis of the light absorption in the spectral lines was developed. The equivalent widths of the absorption lines of a series of elements have been determined from 42 spectrograms covering two eclipse cycles. An increase of absorption in the lines during the primary and secondary eclipses is observed. The velocity curve for the secondary has been plotted and its mass derived ($m_A : m_B = 1 : 1.1$). The luminosity of both components has been determined ($L_A = 0.80$, $L_B = 0.20$).

121.120 **Photometric study of the W UMa type system V523 Cas.** M. I. Lavrov, G. V. Zhukov.
Astron. Tsirk., No. 873, p. 1 - 2 (1975). In Russian.

121.121 **Minimum of EE Cephei.** G. V. Zajtseva, V. M. Lyutyj, N. B. Perova, A. S. Sharov.
Astron. Tsirk., No. 878, p. 1 - 3 (1975). In Russian.

121.122 **Eclipsing binaries found spectroscopically. II. HD 128141.** H. E. Bond.
Publ. Astron. Soc. Pacific, Vol. 87, 877 - 878 = Contr. Louisiana State Univ. Obs., *Baton Rouge*, No. 108 (1975).

The ninth-magnitude G-type star HD 128141 was noted to have diffuse spectral lines on an objective-prism plate. Photometric observations show this object to be an eclipsing binary of the W UMa type, with a period of $0\overset{d}{.}355$.

121.123 **A photometric analysis of the supergiant close binary, V453 Scorpii.**
E. J. Woodward, R. H. Koch.
Publ. Astron. Soc. Pacific, Vol. 87, 901 - 907 (1975).

New yellow and blue photometry of V453 Sco shows the character of the light curve to be completely different from that accepted from photographic estimates. In the new data there are conspicuous photometric complications but it has been possible to solve the light curves by synthesis procedures. Absolute dimensions for the component stars have been derived and an attempt at understanding the systemic evolution is presented.

121.124 **Photoelectric observations of RZ Cassiopeiae.** J. H. Karle, C. Vaucher, B. Gaston.
Publ. Astron. Soc. Pacific, Vol. 87, 909 (1975).

Photoelectric observations for a single night are presented, resulting in a well-observed time of primary minimum light for the eclipsing binary RZ Cas.

121.125 **The eclipsing binary EM Cephei.** G. A. Bakos, J. Tremko.
Journ. Roy. Astron. Soc. Canada, Vol. 69, 307 - 315 (1975).

Photoelectric observations of EM Cep, in the blue spectral region, indicate great irregularities in the shape of the light curve and a change in the period since 1968. Two epochs of minima have been derived.

121.126 **VV Cephei.** K. O. Wright.
IAU Circ., No. 2811 (1975).

121.127 **AZ Cassiopeiae.** P. Tempesti.
IAU Circ., No. 2825 (1975).

121.128 **β Persei.** A. Epstein.
IAU Circ., No. 2859 (1975).

121.129 **θ^1 Orionis A.** W. A. Feibelman.
IAU Circ., No. 2859 (1975).

121.130 **θ^1 Orionis A.**
IAU Circ., No. 2881 (1975).

121.131 **Langfristige Periodenkontrolle für Bedeckungsveränderliche der Bamberger Liste.** M. Wälder.
Veröff. Remeis-Sternw. Bamberg, Astron. Inst. Univ. Erlangen-Nürnberg, Vol. 10, No. 108, 79 pp. (1975).

A period control for 30 eclipsing binaries of the Bamberg list is made. Minima are given for a time interval of ⩾ 70 years. For 18 stars earlier published periods had to be corrected. For some stars period variations have been observed.

121.132 **Revision of the elements of the photometric orbit of TV Cas.** M. I. Lavrov, N. V. Lavrova.
Astron. Tsirk., No. 881, p. 5 - 7 (1975). In Russian.

121.133 **UBV photometry of the primary minimum of U Cep.** L. P. Surkova.
Astron. Tsirk., No. 882, p. 5 - 6 (1975). In Russian.

121.134 **CX Serpentis.** V. P. Tsesevich.
Astron. Tsirk., No. 882, p. 7 - 8 (1975). In Russian.

121.135 **Photoelectric observations of SS Ari.** G. V. Zhukov.
Astron. Tsirk., No. 888, p. 6 - 7 (1975). In Russian.

121.136 **WW Cygni.** M. Heß.
MVS, *Sonneberg,* Vol. 7, 36 (1975).

121.137 **Photographische Reihenbeobachtungen von Bedeckungsveränderlichen.** P. Ahnert.
MVS, *Sonneberg,* Vol. 7, 38 (1975).

121.138 **Minimum des langperiodischen Bedeckungssterns V413 Aquilae.** H. Geßner.
MVS, *Sonneberg,* Vol. 7, 45 (1975).

121.139 **Der Lichtwechsel von XZ Ursae Maioris.** P. Ahnert.
MVS, *Sonneberg,* Vol. 7, 47 - 49 (1975).

121.140 **Konstanz der Periode von V787 Cygni seit 1904.** P. Ahnert.
MVS, *Sonneberg,* Vol. 7, 50 - 51 (1975).

121.141 **Periodenänderungen von BE Vulpeculae.** P. Ahnert.
MVS, *Sonneberg,* Vol. 7, 52 - 54 (1975).

121.142 **Der Lichtwechsel von WW Cancri.** P. Ahnert.
MVS, *Sonneberg,* Vol. 7, 54 - 55 (1975).

121.143 **Der Lichtwechsel von TZ Draconis.** P. Ahnert.
MVS, *Sonneberg,* Vol. 7, 56 - 58 (1975).

121.144 **Verbesserte Lichtwechselelemente von BU Vulpeculae.** P. Ahnert.
MVS, *Sonneberg,* Vol. 7, 59 - 62 (1975).

121.145 **Periodenänderungen von EG Cephei.** P. Ahnert.
MVS, *Sonneberg,* Vol. 7, 63 - 68 (1975).

121.146 **Neue instantane Lichtwechselelemente von VV Ursae Maioris.** P. Ahnert.
MVS, *Sonneberg,* Vol. 7, 68 - 70 (1975).

121.147 **UBV observations of YY Gem.** E. Budding.
Tokyo Astron. Bull., Second Ser., No. 240, p. 2019 - 2053 (1975).

Observations of the dwarf companion to Castor in the UBV system have been carried out with particular attention being paid to the light contamination problem. Results are found to be generally in good agreement with previously quoted values. Absolute parameters and a slightly revised period for the system are given.

121.148 **Investigation of the metallic-line eclipsing binary system WW Aurigae. Part 1. Photoelectric study.**
M. Kiyokawa, M. Kitamura.
Ann. Tokyo Astron. Obs., Second Ser., Vol. 15, (No. 3), 117 - 160 (1975).

More than three thousand photoelectric observations of WW Aur in UBV colours were obtained. The light curves secured in the standard UBV system have been analyzed in detail to obtain accurate informations on the orbital and physical parameters of this metallic-line system.

121.149 **Lists of minima of eclipsing binaries.**
Compiled by T. Bonneville, A. Chetanneau, F. Desprez, R. Diethelm, G. Dumarchi, A. Fries, R. Germann, J.-F. Le Borgne, R. Leydon, K. Locher, A. Marot, N. Mauron, H. Peter, P. Ralincourt, J. Remis, R. Rolland, C. Romoli, A. Royer, W. Sieber, P. Steimer, M. Baumann, R. Boninsegna, J. Bourgeois, P. Carnevali, J.-P. Clovin, M. Kissling, L. Maurin, G. Troispoux, T. Wüthrich, G. Zajàcz.
BBSAG Bull., No. 23, p. 1 - 6; No. 24, p. 1 - 4 (1975). – 56th and 57th list of Swiss Astronomical Society's Eclipsing Variables Observers.

121.150 **A reinterpretation of EP Andromedae.**
K. Locher.
BBSAG Bull., No. 23, p. 6 - 7 (1975).

121.151 **A minimum of V718 Scorpii.** R. Diethelm.
BBSAG Bull., No. 23, p. 7 (1975).

121.152 **The period of V752 Ophiuchi.** K. Locher.
BBSAG Bull., No. 24, p. 4 (1975).

121.153 **La binaria ad eclisse W UMa.**
L. Baldinelli, S. Ghedini, C. Tubertini.
Giornale Ass. Astrofili Bolognesi, No. 39, p. 3 - 12 (1975).

121.154 **Investigation of eclipsing binary stars exhibiting calcium II emission.** J. P. Oliver.
Thesis, California Univ., Los Angeles (USA). [Available from Univ. Microfilms, Ann Arbor, Mich., USA. Order No. 75-9422], 329 pp. (1974).

Three-color photometry of some eclipsing binaries showing calcium II emission is reported. A highly stable and accurate d.c. amplifier, and a new type digital averaging system are described. Past and current light curves of SS Boo, RS CVn, WY Cnc, WW Dra, UV Psc, Z Her, SS Cam, RW UMa, AR Lac, and RT Lac are discussed with particular emphasis on asymmetries in the heights of the maxima and variations in the depths of the minima.

121.155 **Photoelectric observation of Beta Lyrae.**
R. M. Williamon, T. H. Morgan, D. H. Martins, T. F. Collins, H. R. Miller.
Florida Scient., (*USA*), Vol. 37, 100 - 104 (1974).

A further ephemeridal A.P.L. computer programme. See Abstr. 021.025.

Evaluation of an eclipsing variable minimum with an HP desk calculator and plotter. See Abstr. 031.269.

A catalogue of photometric parallaxes of eclipsing binaries. See Abstr. 111.009.

Strömgren four-colour observations of northern hemisphere binary systems. See Abstr. 113.029.

MK classifications of some northern hemisphere binary systems. See Abstr. 114.047.

Spectroscopic observations of the X-ray binary HD 153919 = 3U 1700–37. See Abstr. 114.315.

High-speed spectroscopy of the ultrashort-period variable AM Canum Venaticorum (=HZ 29). See Abstr. 114.316.

On the UV spectrum of Beta Aurigae. See Abstr. 114.353.

High temperature plasma in β Lyrae, observed from Copernicus. See Abstr. 114.375.

The masses and dimensions of early-type binaries. See Abstr. 115.013.

Epsilon Coronae Austrinae: evidence for binary star evolution? See Abstr. 117.012.

The three-body problem and eclipsing binaries. Application to Algol and λ Tauri. See Abstr. 117.022.

Models for W Ursae Majoris stars. See Abstr. 117.041.

Eclipses by an elliptical torus. See Abstr. 117.042.

On the evolutionary status of the RS CVn-type binary WW Draconis. See Abstr. 118.019.

Photometric observations of V470 Cygni (HD 228911). See Abstr. 119.020.

The structure of DQ Herculis. See Abstr. 124.102.

The X-ray sources Sco X-1 and Cyg X-2 as binaries. See Abstr. 142.027.

Optical studies of Uhuru sources. XII. The light curve of Scorpius X-1 = V818 Scorpii, 1889–1974. See Abstr. 142.039.

An extended observation of Cen X-3 with the Ariel-5 sky survey. See Abstr. 142.049.

Parameters of Cen X-3 deduced from observations of its accretion wake. See Abstr. 142.050.

Ariel V sky survey: observations of the accretion wake in Vel X-1. See Abstr. 142.052.

Optical observations of binary X-ray sources. See Abstr. 142.057.

Light-curve analysis of X-ray binaries: a revision. See Abstr. 142.112.

An upper limit to the low-energy X-ray flux from β Persei. See Abstr. 142.159.

122 Intrinsic Variables, Flare Stars, Pulsation Theory

122.001 **Variable stars in clusters of the SMC.**
P. W. Hodge, F. W. Wright.
Astron. Journ., Vol. 80, 510 - 511, 543 (1975).
A search of variable star and star cluster catalogues of the SMC has revealed the presence of 38 known variables that lie within, or very close to, star clusters. About 21 are almost certainly members of the clusters. Of these, the majority are cepheids of short period.

122.002 **Photoelectric *UBV* photometry of cepheids in the Magellanic Clouds and in the southern Milky Way.**
B. F. Madore.
Astrophys. Journ., Suppl. Ser., No. 285, Vol. 29, 219 - 284 (1975).
New photoelectric *UBV* photometry is presented for 46 long-period cepheids in the southern Milky Way. Data are now essentially complete for cepheids with $P > 11$ days and ⟨B⟩ brighter than 14 mag. New *UBV* photometry is also given for 22 cepheids in the Large Magellanic Cloud and for 15 cepheids in the Small Magellanic Cloud.

122.003 **Long-period variables: further work on the correlation of period with OH radial-velocity pattern.**
D. F. Dickinson, E. Kollberg, S. Yngvesson.
Astrophys. Journ., Vol. 199, 131 - 134 (1975).
Recent radio observations at 18 cm have found a number of new OH stars, all of which exhibit the double-peaked spectrum characteristic of these objects. They confirm the correlation between stellar period and the velocity separation of the two emission peaks. A less stringent correlation seems to exist between the $I-K$ color index and the velocity separation.

122.004 **Distances and luminosities of irregular variables of type N.** B. F. Peery, Jr.
Astrophys. Journ., Vol. 199, 135 - 144 = Publ. Goethe Link Obs., Indiana Univ., *Bloomington*, No. 168 (1975).
Near-infrared photometric data of Bäumert are shown to suggest that the variety in the observed (0.78–1.08) μ color index of N irregular variables (L*b* and SR*b*) is primarily a consequence of interstellar reddening. On this hypothesis, color excesses are used to estimate distances of 60 N irregulars. Luminosities follow from photometry relative to Vega and the observed flux distribution of 19 Psc out to 14 μ; the latter star is taken as the prototype N irregular.

122.005 **Light and color variations of the flare star BY Draconis: a critique of starspot properties.**
S. S. Vogt.
Astrophys. Journ., Vol. 199, 418 - 426 (1975).
Periodic light and color variations of the now quiescent flare star BY Dra are presented and shown to be inconsistent with a thermal dark spot model at the present time.

122.006 **Flare stars in star clusters and associations.**
V. A. Ambartsumian (*Ambartsumyan*), L. V. Mirzoyan.
IAU Symposium No. 67, (see 012.001), p. 3 - 14 (1975).
Review paper.

122.007 **Solar neighborhood flare stars – a review.**
W. E. Kunkel.
IAU Symposium No. 67, (see 012.001), p. 15 - 46 (1975).

122.008 **Flares of red dwarf stars and solar activity.**
R. E. Gershberg.
IAU Symposium No. 67, (see 012.001), p. 47 - 64 (1975).
Review paper.

122.009 **An observational approach to stellar evolution (flare stars and related objects).** G. Haro.
IAU Symposium No. 67, (see 012.001), p. 65 - 66 (1975).
Flare stars in the solar neighborhood and in associations belong to the same class of objects but might represent different evolutionary stages. In this report, fundamental contributions, based on observations by various groups, on flare stars ranging from extremely young ones to dMe flare stars are summarized.

122.010 **On variations in the flare activity of V1054 Oph.**
W. E. Kunkel.
IAU Symposium No. 67, (see 012.001), p. 67 - 68 (1975).

122.011 **High-speed photometry of cluster flare stars.**
M. Rodonò.
IAU Symposium No. 67, (see 012.001), p. 69 - 73 (1975).
Recent photoelectric observations made at Catania Observatory show a high occurrence of short-lived faint flares on the Hyades member II Tau (H II 2411). Its activity level turns out to be higher than for UV Cet-type stars of equal absolute luminosity.

122.012 **Flare colours and luminosities.**
S. Cristaldi, M. Rodonò.
IAU Symposium No. 67, (see 012.001), p. 75 - 89 (1975).
Flare colours determined from simultaneous *UBV* observations made at Catania Observatory and from sequential *UBV* observations made at McDonald Observatory are presented. They fit fairly well with the theoretical colours computed according to the Gurzadian's (1970) non-thermal model. From a *B*-band patrol of UV Cet-type stars carried out from 1967 to 1972, some quantitative estimates of flare frequencies and luminosities and of their average contribution to the stellar radiation are given.

122.013 **A possible mechanism of stellar flares.**
Yu. V. Vandakurov.
IAU Symposium No. 67, (see 012.001), p. 91 (1975).

122.014 **The physics of flare stars.** D. S. Evans.
IAU Symposium No. 67, (see 012.001), p. 93 - 98 (1975).
This paper reports the researches of the Texas flare star group. High time resolution and simultaneous photometric, spectroscopic and, if possible, radio observations of flaring stars are essential. Fast and slow flares are distinguished and their distinct optical, spectroscopic, and probably, radio effects are noted. The starspot model satisfactorily accounts for many optical and spectroscopic flare star properties.

122.015 **Comparison of UV Ceti flares with solar flares.**
W. Haupt, W. Schlosser.
IAU Symposium No. 67, (see 012.001), p. 99 - 100 (1975).
Abstract.

122.016 **T Tauri and FU Orionis stars.** G. F. Gahm.
IAU Symposium No. 67, (see 012.001), p. 101 - 107 (1975).
A great deal of information on the continuous radiation of T Tau- and similar type stars has been acquired. The cause of the ultraviolet and infrared excess emission is discussed. The possibility that the blue continuum may arise from numerous line emissions is considered. Various possibilities for the cause of the light variations are described, as well as the nature of the emission lines. The special features of the FU Ori stars are mentioned.

122.017 **Possible groupings of irregular (Orion) variables.** M. V. Dolidze.
IAU Symposium No. 67, (see 012.001), p. 109 - 111 (1975).

122.018 **Spectroscopic and photometric observations of VX Cas.** G. V. Zajtseva, E. A. Kolotilov.
IAU Symposium No. 67, (see 012.001), p. 113 - 115 (1975).

122.019 **Photographic photometry of V1057 Cygni.** O. E. Mandel.
IAU Symposium No. 67, (see 012.001), p. 123 - 125 (1975).
The photographic magnitude of V1057 Cyg is estimated on all plates available at the Odessa plate depositary. The light variation is similar to that of FU Orionis.

122.020 **The R Coronae Borealis type variables.** M. W. Feast.
IAU Symposium No. 67, (see 012.001), p. 129 - 141 (1975).
Review paper.

122.021 **The so-called antiflare stars.** A. F. Pugach.
IAU Symposium No. 67, (see 012.001), p. 143 - 145 (1975). – Abstract.

122.022 **Some statistics of southern R Coronae Borealis variables.** V. E. Sherwood.
IAU Symposium No. 67, (see 012.001), p. 147 (1975). – Abstract.

122.023 **Physical interpretation of light curves of R Coronae Borealis-type stars.** J. Krełowski.
IAU Symposium No. 67, (see 012.001), p. 149 - 157 (1975).
The paper contains preliminary results of theoretical investigations of the light curves of R CrB variables. A circumstellar dust envelope is assumed to be the source of the great visual opacity as well as the infrared and violet excesses. The processes of grain formation and the interaction with the radiation field are discussed briefly. The results of a laboratory experiment concerning the problem of the violet continuum are also described.

122.024 **Spectral and polarimetric investigation of R CrB stars.** M. Ya. Orlov, M. H. Rodríguez.
IAU Symposium No. 67, (see 012.001), p. 159 - 160 (1975).

122.025 **Observations of R CrB and XX Cam.** A. G. Totochava.
IAU Symposium No. 67, (see 012.001), p. 161 - 163 (1975).
The observations of R CrB and XX Cam obtained in 1971–1973 are described. Semiregular light variations with an amplitude of $0^m.15$ and a period of about 30 to 40 days were found during maximum brightness. In addition, fluctuations up to $0^m.2$ for about 2 h were observed.

122.026 **Autocorrelative analysis of the brightness of irregular and semi-regular variable stars.**
F. I. Lukatskaya.
IAU Symposium No. 67, (see 012.001), p. 179 - 182 (1975).
This paper gives an introduction to the use of autocorrelative analyses in the case of irregular and semi-regular brightness variations.

122.027 **A preliminary study of the classical cepheid P-L relation in the *R, I* photometric system.**
J. D. Fernie.
IAU Symposium No. 67, (see 012.001), p. 185 - 190 (1975).
Available *R, I* data for classical cepheids are gathered from the literature, and the P-L, P-C, and P-L-C relations investigated.

122.028 **A photoelectric investigation of the double cepheid CE Cas.** O. G. Franz.
IAU Symposium No. 67, (see 012.001), p. 191 (1975). – Abstract.

122.029 **Some remarks on the absolute magnitudes of cepheids.** A. Opolski.
IAU Symposium No. 67, (see 012.001), p. 193 - 200 (1975).
This paper contains a rediscussion of the period-luminosity relation of type I-cepheids. Special attention is paid to the radius and its variation.

122.030 **On the structure of the cepheid instability strip.** N. N. Yakimova, N. S. Nikolov, G. R. Ivanov.
IAU Symposium No. 67, (see 012.001), p. 201 - 222 (1975).
Review paper.

122.031 **On the presence of variable stars in open clusters.** M. Popova.
IAU Symposium No. 67, (see 012.001), p. 223 (1975).

122.032 **On spectral observations of ζ Gem.** A. G. Shcherbakov.
IAU Symposium No. 67, (see 012.001), p. 225 - 227 (1975).

122.033 **Stars of δ Scuti-type in stellar clusters.** M. Breger.
IAU Symposium No. 67, (see 012.001), p. 231 - 241 (1975).
Review paper.

122.034 **Physical classification of short-period variables.** D. H. P. Jones.
IAU Symposium No. 67, (see 012.001), p. 243 - 246 (1975).

122.035 **Radial pulsations of δ Scuti stars: theory in comparison with observations.**
A. A. Pamjatnykh (*Pamyatnykh*).
IAU Symposium No. 67, (see 012.001), p. 247 - 250 (1975).
In this paper the pulsations of δ Sct star models are compared with observational data. The locations of the blue edges of the instability strips were computed as a function of helium abundance. Linear non-adiabatic pulsations were computed following Dziembowski's programme.

122.036 **The mass and evolutionary status of AI Vel-type variables.** W. Dziembowski.
IAU Symposium No. 67, (see 012.001), p. 251 - 254 (1975).
Arguments are presented that the AI Velorum variables are about 0.2 $M_\odot$ objects evolving with a roughly constant luminosity toward the white dwarf stage.

122.037 **New data on the δ Scuti star 44 Tauri.** N. Morguleff, V. Oskanian (*Oskanyan*), B. Rutily, A. Terzan.
IAU Symposium No. 67, (see 012.001), p. 255 - 256 (1975).
Abstract.

122.038 **S Dor-type variables in other galaxies.** A. S. Sharov.
IAU Symposium No. 67, (see 012.001), p. 275 - 284 (1975).
The variability of these objects, almost the brightest ones found in galaxies, is discussed, followed by a description of colour indices and spectra. In the last sections the classification problems and items for further investigation are mentioned.

122.039 **U Geminorum-type stars.** V. G. Gorbatsky (*Gorbatskij*).
IAU Symposium No. 67, (see 012.001), p. 357 - 367 (1975).
Review paper.

122.040 **VW Hydri, a U Geminorum-type star with periodic**

variations during outbursts. N. Vogt.
IAU Symposium No. 67, (see 012.001), p. 371 (1975). – Abstract.

122.041 **Symbiotic stars.** A. A. Boyarchuk.
IAU Symposium No. 67, (see 012.001), p. 377 - 387 (1975). – Review paper.

122.042 **Physical characteristics of V 1016 Cygni.**
F. Ciatti, A. Mammano, L. Rosino.
IAU Symposium No. 67, (see 012.001), p. 389 - 390 (1975).

122.043 **Photometric properties of V 1329 Cygni.**
V. P. Arhipova (*Arkhipova*), O. E. Mandel.
IAU Symposium No. 67, (see 012.001), p. 391 - 394 (1975).
Light curve and position of V 1329 Cyg in the colour-magnitude diagram are discussed.

122.044 **Some photometric peculiarities of symbiotic stars.**
T. S. Belyakina.
IAU Symposium No. 67, (see 012.001), p. 395 (1975). – Abstract.

122.045 **The outburst of CI Cygni in 1973.**
T. S. Belyakina.
IAU Symposium No. 67, (see 012.001), p. 397 - 399 (1975).
Description of photometric and spectroscopic observations of CI Cyg during the 1973 outburst.

122.046 **Spectroscopic observations of Z Andromedae and CI Cygni from 1968 to 1973.**
A. Mammano, L. Rosino, S. Yildizdogdu.
IAU Symposium No. 67, (see 012.001), p. 401 - 404 (1975).
A description is given of the spectral changes observed in the symbiotic stars Z And and CI Cyg during the nova-like outbursts of 1968 and 1971–73. The observed features can be interpreted by assuming that transient shells are formed around the hot component of a binary system, imbedded in a small nebula.

122.047 **RR Lyrae- and W Virginis-type stars.**
B. V. Kukarkin.
IAU Symposium No. 67, (see 012.001), p. 511 - 529 (1975). Review paper.

122.048 **Red variables of the halo and the globular clusters.**
T. Lloyd Evans.
IAU Symposium No. 67, (see 012.001), p. 531 - 540 (1975).
The present state of knowledge of the red variables in the halo and globular clusters is summarized. Techniques for their discovery are described. The correlations between their properties and those of the globular clusters to which they belong are discussed. The evolutionary state of the variables is considered and stars are assigned tentatively to the first or second ascent of the giant branch or to the excursions from the asymptotic giant branch resulting from instabilities in the He-burning shell.

122.049 **Period changes of RR Lyrae stars in globular clusters.** B. Szeidl.
IAU Symposium No. 67, (see 012.001), p. 545 - 552 (1975).
The dependence of the period changes of RR Lyrae stars, especially in globular clusters, on the evolutionary track through the instability strip has still found no explanation. The present paper is an attempt to give a review of the observed phenomena.

122.050 **Some RR Lyrae variables with Blashko effect.**
G. H. Chis, V. Ureche.
IAU Symposium No. 67, (see 012.001), p. 553 (1975). – Abstract.

122.051 **Spectra of type II cepheids.** T. Lloyd Evans.
IAU Symposium No. 67, (see 012.001), p. 555 (1975). – Abstract.

122.052 **A search for rapid variables and RR Lyrae stars in the LMC on ESO Schmidt plates.** E. H. Geyer.
IAU Symposium No. 67, (see 012.001), p. 557 - 558 (1975).
This is a very preliminary report about a search for rapid variables in extended fields of the Large Magellanic Cloud.

122.053 **The period-luminosity relation for red variables in globular clusters.** R. M. Russev (*Rusev*).
IAU Symposium No. 67, (see 012.001), p. 563 - 565 (1975).
Recent data for red pulsating variables in globular clusters are discussed. There seems to exist a period-luminosity relation in the near-by infrared but not in the visual magnitude.

122.054 **A photometric study of the RR_S-star CY Aqr.**
E. H. Geyer, M. Hoffmann.
Astron. Astrophys. Suppl. Ser., Vol. 21, 177 - 182 (1975).
Fairly high-speed photoelectric photometry in *B* and *V* has been carried out of the ultra-short period RR Lyrae star CY Aqr for five consecutive cycles. The observations cannot be reconciled either by the beat period given by Elst (1972) or by Fitch (1973). The beat phenomenon in this star may therefore be fictitious.

122.055 **A two-colour photometry of the short period RR Lyrae star DY Peg.** E. H. Geyer, M. Hoffmann.
Astron. Astrophys. Suppl. Ser., Vol. 21, 183 - 188 (1975).
By photoelectric photometry in *B* and *V* with a double beam photometer two resp. four consecutive cycles have been observed of the RR_S-star DY Peg. The non-periodic light curve changes suggest overtone pulsation of this c-type variable.

122.056 **Variations of the radial pulsations of XZ Cygni with the Blazhko effect.**
P. Z. Kynchev, Yu. S. Romanov.
Peremennye Zvezdy, Vol. 19, 471 - 475 (1974). In Russian.
The variations of the light gradients for XZ Cyg with the Blazhko effect were studied. On the basis of light gradients the relations of radii at minimum and maximum of light are determined and their variations with phases of the Blazhko effect are constructed.

122.057 **Investigation of variations of the photometric characteristics of XZ Cygni with the Blazhko effect.**
P. Z. Kynchev.
Peremennye Zvezdy, Vol. 19, 477 - 496 (1974). In Russian.
On the basis of photoelectric observations obtained in 1972 variations in photometric characteristics of XZ Cyg with the Blazhko effect have been studied. The light in U, B, V at maximum and minimum, colour index B–V and U–B at maximum are found to change with the Blazhko effect.

122.058 **Photographic observations of DL Comae Berenices.**
N. E. Kurochkin, B. D. Pochinok.
Peremennye Zvezdy, Prilozhenie, Vol. 2, 91 - 95 (1974). In Russian.

122.059 **On the period of CS Monocerotis.**
M. B. Bogdanov.
Peremennye Zvezdy, Prilozhenie, Vol. 2, 97 - 99 (1974). In Russian.

122.060 **V619 Herculis.** A. A. Tokovinin.
Peremennye Zvezdy, Prilozhenie, Vol. 2, 104 - 106 (1974). In Russian.

122.061 **V602 Herculis.** V. S. Malyshev.
Peremennye Zvezdy, Prilozhenie, Vol. 2, 107 - 109 (1974). In Russian.

122.062 **Observations of TV Lyncis and AB Ursae Majoris.** O. E. Mandel'.
Peremennye Zvezdy, Prilozhenie, Vol. 2, 111 - 116 (1974). In Russian.

122.063 **V441 Cassiopeiae.** O. M. Kanalina.
Peremennye Zvezdy, Prilozhenie, Vol. 2, 121 - 122 (1974). In Russian.

122.064 **Investigation of the new variable star SVS 1874 in Ursa Major.** V. P. Goranskij, N. N. Samus'.
Peremennye Zvezdy, Prilozhenie, Vol. 2, 123 - 126 (1974). In Russian.

122.065 **Die Perioden-Leuchtkraft-Farben (PLC) Beziehung in LMC, SMC und der Galaxis.** R. Schröder.
Mitt. Astron. Ges., No. 36, p. 99 (1975). – Abstract.

122.066 **Lichtelektrische Dreifarben-Photometrie an YZ Bootis.** W. Gieren, F. Gieseking, M. Hoffmann.
Mitt. Astron. Ges., No. 36, p. 99 - 100 (1975). – Abstract.

122.067 **Lichtelektrische und spektroskopische Beobachtungen an SU Cassiopeiae.** W. Gieren.
Mitt. Astron. Ges., No. 36, p. 100 (1975). – Abstract.

122.068 **Simultane Beobachtungen an schnell veränderlichen Sternen.** K. Metz.
Mitt. Astron. Ges., No. 36, p. 102 - 104 (1975).

122.069 **Wellenlängenabhängigkeit der Polarisation von Mira-Sternen.** J. Materne.
Mitt. Astron. Ges., No. 36, p. 104 (1975). – Abstract.

122.070 **Beobachtungen von Veränderlichen Sternen im nahen Infrarot.** U. Fahrbach, D. Lemke.
Mitt. Astron. Ges., No. 36, p. 104 - 106 (1975).

122.071 **Nonlinear effects in double-mode cepheids.** R. F. Stellingwerf.
Astrophys. Journ., Vol. 199, 705 - 709 (1975).

The effects of nonlinear mode interaction and period shifts in observed double-mode cepheids are investigated with a detailed analysis of a mixed-mode pulsation model. Derived periods and mean constituent light curves agree very well with the pure-mode nonlinear results and with the linear periods. Complete stabilityanalyses indicate that low-mass cepheid models show aperiodic mixed-mode behavior near the red edge of the instability strip, as well as a mode switch near the center of the strip.

122.072 **The variable circular polarization of V1057 Cygni.** R. D. Wolstencroft, T. Simon.
Astrophys. Journ., (*Letters*), Vol. 199, L169 - L171 (1975).

Circular polarization of the visible light from V1057 Cygni has been detected. The magnitude and variability of this polarization imply that it is intrinsic to the star and not of interstellar origin. The linear polarization shows no evidence for variability and is therefore primarily interstellar.

122.073 **Metal abundances of RR Lyrae stars in galactic globular clusters.** D. Butler.
Astrophys. Journ., Vol. 200, 68 - 81 = Lick Obs. Bull., No.695 (1975).

Image-tube scanner measurements of Preston's metal abundance indicator, ΔS, have been made for 47 RR Lyrae stars in 12 galactic globular clusters, and for 67 field RR Lyrae stars. From curve-of-growth analyses of 13 field RR Lyraes, the relationship $[Fe/H] = -0.16\Delta S - 0.23$ is found, where $[Fe/H] = \log(Fe/H)_* - \log(Fe/H)_\odot$. The authors find metal abundances ranging from $[Fe/H] = -0.39$ for NGC 6712 to $[Fe/H] = -2.18$ for M92.

122.074 **Four-color and Hβ photometry of β Cephei stars and their location in the Hertzsprung-Russell diagram.** J. S. Shaw.
Astron. Astrophys., Vol. 41, 367 - 373 (1975).

New *uvby* – Hβ indices were determined for nine β Cephei stars and 22 suspects. Bolometric magnitudes and effective temperatures were derived for 21 known β Cephei stars and 30 suspects. Almost all β Cephei stars and a few suspects lie in the familiar instability strip in a $M_{bol} - T_e$ diagram. Counts of variables and non-variables in this region of the diagram indicate β Cephei stars are not solely in a secondary contraction or early shell-burning stage of their evolution. The possible use of four-color Hβ photometry and the instability strip to expedite search for new β Cephei stars is proposed.

122.075 **Search for stellar flare activity.** B. N. Andersen, B. R. Pettersen.
Astron. Astrophys., Vol. 41, 459 - 461 (1975).

Criteria for flare star selection are derived, and a list of candidates is presented. The BG 12 filter is found suitable for flare star search with small telescopes. Flare activity is discovered on Gliese 109.

122.076 **The RR Lyrae variable HD 268892.** P. J. Andrews.
Monthly Notices Roy. Astron. Soc., Vol. 172, 271 - 275 (1975).

HD 268892 is shown to be a metal-poor c-type RR Lyrae variable with period 0.328936 days. It has a high radial velocity and lies in front of the Large Magellanic Cloud.

122.077 **The slope of the surface brightness-colour relation for cepheids.** R. J. Thompson.
Monthly Notices Roy. Astron. Soc. Vol. 172, 455 - 463 (1975).

Published *UBV* photometry and radial velocity data for 16 cepheids have been analysed to obtain the slope of the visual surface brightness-intrinsic colour, $[S_V, (B-V)_0]$, relation for each cepheid. The 16 cepheids give a mean slope of 2.11 ± 0.03.

122.078 **Wavelength dependence of polarization. XXIX. Observations of red variable stars.**
S. J. Shawl.
Astron. Journ., Vol. 80, 602 - 624 (1975).

New polarimetric data are presented for 25 stars. Results of comparison between observations and theoretical models of circumstellar clouds are presented. Out of 29 stars that have data good enough to compare with theory, only three are not able to have their observations fitted with either silicate, graphite, or iron grains. The particle sizes are found to be confined to a narrow range of sizes, from Rayleigh particles (radii less than 0.05 μm) up to a maximum of 0.10 μm; the average particle size for silicate cloud models is 0.08 μm (and slightly smaller for iron and graphite); column densities of $1 - 100 \times 10^8$ cm^{-2} are found.

122.079 **A survey of bright variable stars in M33.** S. van den Bergh, E. Herbst, C. T. Kowal.
Astrophys. Journ., Suppl. Ser., No. 287, Vol. 29, 303 - 314 (1975).

Thirty-eight new variable stars have been discovered in M33. Twelve of these objects are red semiregular or irregular variables. These stars typically have light cycles of 1–2 years.

The brightest red variable has $V(\max) \simeq 17.0$, corresponding to $M_V(\max) \simeq -7.6$. Red supergiant variables appear to favor the outer regions of M33. The cepheids in M33 do not seem to be concentrated in the major associations which outline the spiral arms. At present the brightest nonstable star in M33 is the yellow variable No. 83 with $V(\max) \simeq 15.8$, corresponding to $M_V(\max) \simeq -8.8$.

122.080 **Relations between luminosity, period and metallicity in RR Lyrae stars.** A. Heck.
Astron. Astrophys., Vol. 42, 131 - 133 (1975).
The author presents the results of the calibration of relations between the luminosity and the period on the one hand, and between the luminosity and the metallicity index ΔS Preston, 1959) on the other hand for two samples of RR Lyrae stars: one of 134 stars of mixed types and one of 121 stars of Bailey-types *a* and *b* only. All the stars have periods greater than $0\overset{d}{.}2$. The calibration algorithm is based on the principle of maximum likelihood. The best relations, among the considered possibilities, are those which include the metallicity index and the author obtains $M_v^{(i)} = -0.51(\pm 0.21) + 0.21(\pm 0.02)\Delta S^{(i)}$ for the sample of 134 stars and $M_v^{(i)} = -0.26 (\pm 0.19) + 0.17(\pm 0.05)\Delta S^{(i)}$ for the sample of 121 stars of types *a* and *b*.

122.081 **Period-luminosity-color relations and pulsation modes of pulsating variable stars.**
M. Breger, J. N. Bregman.
Astrophys. Journ., Vol. 200, 343 - 353 (1975).
The periods of δ Scuti, RR Lyrae, dwarf cepheid, and W Virginis variables have been investigated for their dependence on luminosity, color, mass, and pulsation modes. A maximum-likelihood method, which includes consideration of the observational errors in each coordinate, has been applied to obtain observational period-luminosity-color (P-L-C) relations. On the radial pulsation hypothesis, the cool variables pulsate in the fundamental mode, while the hotter variables tend to pulsate in the first and second overtones. Analogous to RR Lyrae stars, the division into different pulsation modes seems to be determined by color.

122.082 **Hydrodynamic models of a cepheid atmosphere. II. Continuous spectrum.** A. H. Karp.
Astrophys. Journ., Vol. 200, 354 - 363 (1975).
The hydrodynamic models of a classical cepheid computed by Karp have been used to study the continuous spectrum of cepheids. The structure of each hydrodynamic model was used as a snapshot of the temperature and pressure structure of the atmospheric layers. After using line blocking factors to account for the effect of lines on the spectral energy distributions, broad-band *UBVRI* colors were calculated. The light and color curves of the model reproduce the amplitude and asymmetry observed in cepheids in this period range. It is shown that loops in the $(U-B, B-V)$-diagram can be produced by a dependence of the continuous opacity on the electron pressure. The location of the equilibrium model in the H-R diagram indicates that the zero point of the Sandage and Tammann period-luminosity relation is too high by 0.2 mag.

122.083 **On the absolute magnitudes of RV Tauri variables.**
T. G. Barnes III, D. L. DuPuy.
Astrophys. Journ., Vol. 200, 364 - 366 (1975).
Recently DuPuy suggested that RV Tauri variables show evidence for ~1.5 mag of nearly neutral extinction in the visual region. The suggestion is based upon an apparent discrepancy between photometric absolute magnitudes and spectroscopic luminosity classes. The authors have reexamined the evidence and conclude that no discrepancy exists if the surface gravities of RV Tauri variables are as given by the current calibration of MK luminosity classes. In this case, the photometrically determined period-luminosity relation given by DuPuy correctly represents the luminosities of the variables, and no neutral extinction in the circumstellar envelopes is required.

122.084 **Are stellar flares energized by the missing energy in starspots?** D. J. Mullan.
Astrophys. Journ., Vol. 200, 641 - 645 (1975).
The author suggests that the energy released in stellar flares is provided by a steady accumulation of a fraction of the missing energy in starspots. The mechanism by which the missing energy might be converted to flare-compatible form is not discussed here. However, the principal result of the paper is independent of the (speculative) details of the conversion and storage mechanisms. The formulae are even applicable in an approximate way to sunspots and solar flares, indicating that solar and stellar flares are probably not caused by radically different mechanisms. The lack of flare stars at spectral types earlier than K0 is interpreted to mean that starspots become very small in the late G stars. According to the author's starspot hypothesis, this suggests that the transition from deep to shallow convective envelopes occurs at a much later spectral type than is customarily believed.

122.085 **Observations of rapid blue variables – XVII. V436 Centauri.** B. Warner.
Monthly Notices Roy. Astron. Soc., Vol. 173, 37P - 40P (1975).
High speed photometric observations of V436 Cen during a supermaximum in 1975 June show it to have recurrent humps with a period of 92 min, indicating an orbital period of near this value. Rapid oscillations in brightness, with periods near 20 s, were present over a total of 3 days.

122.086 **Photoelectric photometry of late-type variable stars.** L. Celis S.
Astron. Astrophys., Suppl. Ser., Vol. 22, 9 - 17 (1975).
The author presents 168 photoelectric (*UBV*) observations of 30 late-type variable stars of long period which were made systematically during 8 observing runs from 1972 December to 1973 September. The $B-V$ and $U-B$ colour indices obtained near maximum and minimum in V are plotted in the two-colour diagram.

122.087 **Peculiarities of light and polarization variations of the infrared carbon star CIT 6.**
A. K. Alksnis, G. V. Khozov.
Pis'ma v Astron. Zhurn., Vol. 1, No. 7, p. 23 - 26 (1975). In Russian.
Photometric and polarimetric observations of CIT 6 in the range $0.4 - 2.5\ \mu$ have been carried out. A period of about 640 days appears to satisfy both optical and infrared light variations. Significant deviations from the mean light curve, including fluctuations with a time scale of 1 to 30 days and amplitude of $0\overset{m}{.}4 - 1\overset{m}{.}0$, have been found. Large variations of infrared polarization are present and seem to exhibit no obvious correlation with light variations.

122.088 **Mass loss from stellar flares.**
S. P. Worden, G. D. Coleman.
Bull. American Astron. Soc., Vol. 7, 398 - 399 (1975). – Abstr. AAS.

122.089 **Evidence supporting low cepheid masses.**
N. R. Simon, E. G. Schmidt.
Bull. American Astron. Soc., Vol. 7, 403 (1975). – Abstr. AAS.

122.090 **Pulsation, metallicism, and the nature of 32 Vir.**
S. Evans, M. Breger, D. Kurtz.
Bull. American Astron. Soc., Vol. 7, 404 (1975). – Abstr. AAS.

122.091 **OAO-A2 ultraviolet photometry of RR Lyrae.**

J. L. Hutchinson.
Bull. American Astron. Soc., Vol. 7, 405 (1975). – Abstr. AAS.

122.092 **Properties of cepheids in the Large Magellanic Cloud.**
L. P. Connolly.
Bull. American Astron. Soc., Vol. 7, 411 (1975). – Abstr. AAS.

122.093 **Models of radio flares from UV Ceti stars.**
S. R. Spangler.
Bull. American Astron. Soc., Vol. 7, 477 (1975). – Abstr. AAS.

122.094 **An upper limit to slowly varying radio emission from three UV Ceti-type stars.**
S. R. Spangler, S. D. Shawhan.
Bull. American Astron. Soc., Vol. 7, 477 (1975). – Abstr. AAS.

122.095 **The evolutionary state of the dwarf cepheids.**
D. H. McNamara, W. R. Langford.
Publ. Astron. Soc. Pacific, Vol. 87, 505 (1975). – Abstr. ASP.

122.096 **Mariner 9 observations of β Cephei stars.**
J. R. Lesh, R. C. Bohlin.
Publ. Astron. Soc. Pacific, Vol. 87, 587 - 592 (1975).
Ultraviolet spectra of three β Cep variables and a non-variable star of similar spectral type were observed by the ultraviolet spectrometer on Mariner 9. A significant light variation at 1400 Å may have been detected for β Cep. The variable stars have mean ultraviolet spectra similar to that of the non-variable star and do not show any significant short-period spectral variations.

122.097 **Simultaneous spectroscopic and photometric observations of the β Cephei star HR 6684.**
C. T. Bolton, J. R. Percy, R. E. Shemilt.
Publ. Astron. Soc. Pacific, Vol. 87, 595 - 598 (1975).
The authors have obtained simultaneous radial velocity observations and differential magnitude observations of the β Cep star HR 6684. The full amplitude of the velocity curve is 18 km sec^{-1} and the range in blue light is $0^m\!.033$. The ratio of these values is 550 km sec^{-1} mag^{-1} and maximum light falls at minimum radius. In these respects, HR 6684 is similar to other β Cep stars. There is some evidence for a beat effect. The spectral lines are sharp ($v \sin i$ = 110 km sec^{-1}) and appear to vary in their profiles.

122.098 **Three-colour observations of the variability of Z Camelopardalis.** A. F. Pugach.
Peremennye Zvezdy, Vol. 19, 319 - 326 (1974). In Russian.
Some fragments of four outbursts of Z Cam had been photoelectrically observed in 1970–1972. The variability of Z Cam is considered as a sum of two independent processes: great outbursts ($\Delta m \simeq 3^m - 4^m$) and fast fluctuations of small amplitude (flickering). At the ascending branch of an outburst the star shows some deficiency of U-radiation as compared to B-radiation, performing a loop on the B–V and (U–B)–(B–V) diagrams. There is no correlation between V, B, U magnitudes during flickering. The flickering range increases while the star grows weak.

122.099 **On the value of the factor "p" for the conversion from radial to pulsational velocities of pulsating stars with extended atmospheres.** M. S. Frolov.
Peremennye Zvezdy, Vol. 19, 327 - 329 (1974). In Russian.
On the basis of the theory of stars with extended atmospheres the factor "p" for the conversion from radial to pulsational velocity was found to be equal to 1.28 instead of p = 1.41 (or 24/17) usually assumed in Wesselink's method.

122.100 **On the periods of light variations of variables V 39 and V 42 in the dwarf galaxy IC 1613.**
P. N. Kholopov.
Peremennye Zvezdy, Vol. 19, 331 - 335 (1974). In Russian.
The variable V 39 is a normal cepheid with elements: Max = JD 2426926.946 + $27^d\!.953$ E. The variable V 42 is an interesting cepheid with Blazhko effect and mean elements: Max = JD 2427747.741 + $16^d\!.486$ E. The ranges of light variation for both variables are practically the same.

122.101 **On rapid light variations of a blue star in the vicinity of Cyg X-1.**
A. N. Abramenko, O. P. Gollandskij, V. V. Prokof'eva.
Peremennye Zvezdy, Vol. 19, 357 - 366 (1974). In Russian.
A blue star of 17^m in the vicinity of Cyg X-1 suspected of being pulsating was repeatedly observed with a TV apparatus. 257 TV photographs of a $1'\!.5$ field around the star were taken on August 19, 1972 in order to detect light variations with a period inside the interval between 0.0731 and 0.0754 sec. Histograms showed that there might be rapid light variations of the star with characteristic time $\tau \lesssim 10$ msec. The photographs taken with the resolution period of 0.07395 sec showed quasiperiodic light pulsations.

122.102 **Colorimetric characteristics and the position on the H–R diagram of RV Tauri stars.**
G. E. Erleksova.
Peremennye Zvezdy, Vol. 19, 367 - 375 (1974). In Russian.
The intrinsic colours $\langle B-V \rangle^\circ$ and $(B-V)^\circ_{max}$ of 20 RV Tauri stars were determined. The relation period - intrinsic colour, the position on the H–R diagram and the light gradient diagram of groups A, B, C are different. Detected differences and improved phase shifts between V and B–V curves continue the sequence of the differences which were detected earlier in the light curves and the period changes (Erleksova, 1971, 1972).

122.103 **O–C diagram research of long-period stars.**
A. G. Nudzhenko.
Peremennye Zvezdy, Vol. 19, 381 - 399 (1974). In Russian.
Results of O–C diagram research of 238 long-period variable stars are given. Dependencies $|\Delta P/P|$ and ΔE on the period and $|\Delta P/P|$ on ΔE are considered. The stars which periods change according to Sterne's criterion are distinguished by drastic variations of the period through relatively small intervals of epoch.

122.104 **New flare stars in the Pleiades. II.**
Eh. S. Parsamyan.
Soobshch. Byurakan. Obs., vyp. (No.) 46, p. 3 - 15 (1975). In Russian.
Data about 33 new and 21 repeated flares found during observations of the Pleiades at the end of 1971 and at the beginning of 1972 and re-examined material are given. Among the flare stars there were found stars in the foreground of and behind the Pleiades. It is shown that the flare stars behind are mostly of K type. The distribution of M-type flare stars in the region of the Pleiades confirms the assumption that they are mostly cluster members.

122.105 **Two-colour observations of flare stars in the Pleiades.** Eh. S. Parsamyan, O. S. Chavushyan.
Soobshch. Byurakan. Obs., vyp. (No.) 46, p. 16 - 22 (1975). In Russian.
The results of two-colour observations of rapid and one 'slow' flare-ups in the Pleiades are given. The observations were made in the ultraviolet and photographic region.

122.106 **Polarimetric and photometric observations of EV Lac.** M. A. Eritsyan.
Soobshch. Byurakan. Obs., vyp. (No.) 46, p. 23 - 27 (1975). In Russian.
The results of photometric and polarimetric observation of the star EV Lac are given. During 98 hours of observation

9 flares have been registered. The data concerning these flares are presented in a table and on figures. The total energy of these flares and the luminosity at their maximum are given as well.

122.107 **Optical properties of type-1 OH stars.** R. Foy. Nauchn. Informatsii, vyp. (No.) 33, p. 3 - 4, 19 - 20 (1974). In Russian and English. – Abstr. of a paper presented at the colloquium 'Physics and evolution of stars', (see 012.015).

122.108 **Cepheids in open clusters.** Yu. L. Frantsman. Nauchn. Informatsii, vyp. (No.) 33, p. 5, 21 (1974). In Russian and English. – Abstr. of a paper presented at the colloquium 'Physics and evolution of stars', (see 012.015).

122.109 **On some difficulties in classification of dwarf cepheids and δ Scuti stars.** M. S. Frolov. Nauchn. Informatsii, vyp. (No.) 33, p. 12 - 13, 29 (1974). In Russian and English. – Abstr. of a paper presented at the colloquium 'Physics and evolution of stars', (see 012.015).

122.110 **Relation between δ Scuti- and AI Velorum-type stars.** A. Baglin. Nauchn. Informatsii, vyp. (No.) 33, p. 13, 30 (1974). In Russian and English. – Abstr. of a paper presented at the colloquium 'Physics and evolution of stars', (see 012.015).

122.111 **Nature of the oscillations in δ Scuti stars.** A. Baglin, J. M. Le Contel, J. P. Sareyan, J. C. Valtier. Nauchn. Informatsii, vyp. (No.) 33, p. 13 - 14, 30 (1974). In Russian and English. – Abstr. of a paper presented at the colloquium 'Physics and evolution of stars', (see 012.015).

122.112 **Preliminary results of a study of the pulsational instability of δ Sct stars.** A. A. Pamyatnykh. Nauchn. Informatsii, vyp. (No.) 33, p. 14, 30 - 31 (1974). In Russian and English. – Abstr. of a paper presented at the colloquium 'Physics and evolution of stars', (see 012.015).

122.113 **The Hertzsprung progression in cepheid calculations.** R. F. Christy. 19th Colloquium International Astrophys. Liège 1974, (see 012.016), p. 173 - 181 (1975).

The Hertzsprung bump in cepheid models has been studied for models of a wide range of mass, M, radius, R, and effective temperature, T_e. This study showed that the delay of the bump measured from the time of minimum radius preceding the bump by one period is dependent only on M and R and not on T_e. Over a range of models from $M/M_\odot \approx .5$ and $R/R_\odot = 10$ to $M/M_\odot = 13$ and $R/R_\odot = 160$, the delay of the bump is approximately proportional to the radius. This is consistent with a picture that the bump results from an acoustic signal that traverses the star from the ionization zone in to the center and out again to the surface. The bump and other nonlinear phenomena have been studied over a series of cepheid models from a period of less than one day to a period of 150 days. The results show characteristic changes in amplitude and in the appearance of bumps and are in good correspondence to actual cepheids. The masses in this series are, however, characteristically lower than evelutionary masses.

122.114 **Comments on the pulsation of double mode cepheids.** D. S. King, C. J. Hansen, R. R. Ross, R. F. Stellingwerf, J. P. Cox. 19th Colloquium International Astrophys. Liège 1974, (see 012.016), p. 183 - 187 (1975).

The pulsation properties of stellar envelope models for stars with periods appropriate to the double mode cepheids are investigated. Three different chemical compositions are used with X = 0.70, Z = 0.02; X = 0.80, Z = 0.02 and X = 0.602, Z = 0.044. Certain conclusions can be reached with regard to the masses and radii of these stars on the basis of the calculations. Results for both linear and nonlinear computations are presented.

122.115 **Hydrodynamic effects in the atmospheres of variable stars.** C. G. Davis, Jr. 19th Colloquium International Astrophys. Liège 1974, (see 012.016), p. 197 - 207 (1975).

Numerical models of variable stars are established, using a non-linear radiative transfer coupled hydrodynamics code. The variable Eddington method of radiative transfer is used. Comparisons are for models of W Virginis, β Doradus and η Aquilae. From these models it appears that shocks are formed in the atmospheres of classical cepheids as well as W Virginis stars.

122.116 **Non-linear radiation transport calculations to obtain mean colors and to verify the Wesselink method.** A. N. Cox, C. G. Davis. 19th Colloquium International Astrophys. Liège 1974, (see 012.016), p. 209 - 219 (1975).

Four cepheid models of mass 4, 4, 7, and 15 solar masses, one of the 4 solar mass models representing Beta Doradus, have been numerically integrated. The variable Eddington radiation approximation has been used for the surface layers to produce structures accurate enough for spectrum predictions. Spectra are predicted at about 100 phases over three pulsational cycles. B and V colors are computed at these phases. Mean colors are taken by four different methods that could be applied to colors of an observed cepheid. The cepheid mass discrepancy, with pulsation masses involving the method of color mean, remains, and pulsation masses are still only 70 percent those needed in evolutionary calculations. A discussion of the Wesselink method for pulsating star radius determinations shows that only small positive systematic errors can occur in the derived mean radii.

122.117 **Period and amplitude changes in δ Scuti stars.** J.-C. Valtier, J.-P. Sareyan, J.-M. Le Contel, A Baglin. 19th Colloquium International Astrophys. Liège 1974, (see 012.016), p. 223 - 226 (1975).

Long observational sequences have been performed on four δ Scuti stars at the ESO Observatory in Chile. Narrow band filters on the continuum have been used. The observed light curves are complex: periods and amplitudes vary even on a single night. It seems difficult to interpret these variations in terms of a mixture of different modes as it has been generally proposed up to now. The authors propose that non linear effects taking place in the atmosphere should be responsible for these irregularities. It is stressed that in these objects the time scale of pulsation and convection are comparable and that the coupling between these two velocity fields should be taken into account.

122.118 **On Epstein's weighting function and pulsation constants of cepheid models.** J. O. Petersen. 19th Colloquium International Astrophys. Liège 1974, (see 012.016), p. 299 - 305 (1975).

For a series of cepheid models Epstein's weight functions are discussed, and it is found useful to consider also the normalization functions, which have a similar character. It is shown that the increase in Q_0 with increasing luminosity is due mainly to changes in the structure of the models and effects from the radiation field, while the direct effects of ionizations are small.

122.119 **On the Blazhko effect of AE Ursae Majoris.** V. P. Tsesevich, B. N. Firmanyuk.

Astron. Tsirk., No. 841, p. 7 - 8 (1974). In Russian.

122.120 **Mira variables of class M in the red part of the spectrum.** M. V. Dolidze.
Astron. Tsirk., No. 843, p. 6 - 7 (1974). In Russian.

122.121 **On the Blazhko effect in TV Boo.**
B. N. Firmanyuk.
Astron. Tsirk., No. 843, p. 7 - 8 (1974). In Russian.

122.122 **AS 501.** G. F. Gahm, E. A. Kolotilov, P. P. Petrov, A. G. Shcherbakov, G. I. Shanin.
Astron. Tsirk., No. 852, p. 5 - 6 (1975). In Russian.

122.123 **New elements of light variation of eight variable stars.** S. Yu. Shugarov.
Astron. Tsirk., No. 857, p. 7 - 8 (1975). In Russian.

122.124 **On the relation between UBVRI magnitudes for red variable stars.**
F. I. Lukatskaya, A. Eh. Rozenbush.
Astron. Tsirk., No. 858, p. 3 - 4 (1975). In Russian.

122.125 **Radius and luminosity of the dwarf cepheid EH Lib.**
M. S. Frolov.
Astron. Tsirk., No. 858, p. 6 (1975). In Russian.

122.126 **On spectral peculiarities of SW Andromedae.**
Z. N. Fenina, Yu. S. Romanov.
Astron. Tsirk., No. 862, p. 2 - 4 (1975). In Russian.

122.127 **On the possible variation of spectral peculiarities of XZ Cyg with the Blazhko effect.** Z. N. Fenina.
Astron. Tsirk., No. 862, p. 4 - 5 (1975). In Russian.

122.128 **Recent observations of the Blazhko effect in ST Boo.** G. A. Lange, B. N. Firmanyuk.
Astron. Tsirk., No. 864, p. 7 - 8 (1975). In Russian.

122.129 **Gradient diagram for some types of pulsating variable stars.** T. K. Kiseleva.
Astron. Tsirk., No. 865, p. 3 - 4 (1975). In Russian.

122.130 **Suspected VV Cephei stars in the Large Magellanic Cloud.** N. Sanduleak, P. Pesch.
Astrophys. Journ., Vol. 201, 347 - 349 (1975).
Twelve stars in the Large Magellanic Cloud show both hydrogen emission and TiO absorption features in the red and/or infrared. Five of these are known to be composite, apparently combining an M supergiant and a luminous blue component, and are probably VV Cephei stars. The remaining stars are suspected to be of a similar character.

122.131 **Veränderliche Sterne in Kugelsternhaufen.**
E. H. Geyer.
SuW, Vol. 14, 316 - 322 (1975).

122.132 **A study of Mira variables. Implications for OH stars and galactic evolution.**
R. Foy, A. Heck, M.-O. Mennessier.
Astron. Astrophys., Vol. 43, 175 - 188 (1975).
In order to study properties of type I OH Mira variables and the kinematical evolution of the Galaxy, the authors have analysed a sample of Mira variables, by means of a statistical method based on the principle of maximum likelihood (Heck, 1975). Previous determinations of the relation between absolute visual magnitude M_V and period P relation never did make the distinction between OH and non OH Mira variables. However, since OH stars have been found to be as bright as non OH stars in the visual spectral range, the relation M_V/P is not affected by it, as confirmed by the results. In contrast, in the near infrared (1.04 μm), OH stars have been found to be brighter than non OH stars, by one magnitude. It appears that the only difference between type I and type II OH stars lies in the temperature of the dust grains, respectively $T \simeq 1500$ K and $T \simeq 600$ K.

122.133 **The color excess scale of the longer period cepheids– a discussion.**
R. Canavaggia, P. Mianes, J. Rousseau.
Astron. Astrophys., Vol. 43, 275 - 284 (1975).
In establishing a T_{eff} scale for long period cepheids, Schmidt encountered "color and temperature discrepancies". The authors suggest that these were introduced into his work as a consequence of using Kraft's color excess scale for cepheids with periods longer than about 10 days. Color excesses derived from the six-color photometry (the Lick system) could eliminate the discrepancies.

122.134 **On the variation of the fundamental period of Sigma Scorpii.** C. Sterken.
Astron. Astrophys., Vol. 43, 321 - 322 (1975).
From a set of u, v, b and y lightcurves obtained for σ Sco at the European Southern Observatory in 1972 and 1974 the main period $P_0 = 0^d.2468\,395 \pm 0.0000\,005$ was determined. The results indicate that P_0 increases again with respect to van Hoof's 1962–1965 results. The material and results obtained between 1915 and 1974 show an oscillation of P_0 with period $23^y.2$ and amplitude of 0.7 s.

122.135 **Polarized light from lower main-sequence stars: is it due to synchrotron emission?** D. J. Mullan.
Astrophys. Journ., Vol. 201, 630 - 636 (1975).
Broad-band plane polarization (0.1–10%) has been observed in the light emitted by flare stars outside flares. Here the author examines the possibility that the polarization might be due to synchrotron radiation. On the basis of the electron energy spectrum which would be required, he suggests that synchrotron radiation is probably not responsible for the observed polarization.

122.136 **The nature of dwarf cepheids. I. AD Canis Minoris.**
M. Breger.
Astrophys. Journ., Vol. 201, 653 - 660 (1975).
New $uvby\beta$ photometry of AD CMi covering several cycles is reported. No evidence for shock wave emission is detected in either the Hβ line or in the Wesselink radius solutions. A Wesselink radius of 3 $R_\odot$ is derived. This leads to a population I mass. No evidence for a low-mass population II nature exists for this star. Except for the large amplitude, the star is virtually indistinguishable from a normal δ Scuti variable.

122.137 **The period of CY Aquarii.** J. R. Percy.
Astron. Astrophys., Vol. 43, 469 - 471 (1975).
Published times of maximum light of CY Aquarii have been collected, and used to determine the period of this star by means of the phase-shift method. The period remained constant from 1934 to 1951, changed abruptly at that time, and has remained constant since then. Several light curves of CY Aquarii have been analyzed by the maximum entropy method of spectral analysis. The beat period found by Elst has been found to be probably an artifact of his harmonic analysis method.

122.138 **Structure of the instability strip of cepheids near the center of the Small Magellanic Cloud.**
N. N. Yakimova.
Soobshch. Gos. Astron. Inst. Shternberga, No. 177, p. 19 - 38 (1973). In Russian.
The results of a comparison of the photometric parameters (colours, magnitudes, light and colour amplitudes in the BV-system) for near hundred cepheids in the center of

the Small Magellanic Cloud are given. Van Genderen's conclusions are criticized. The correlations between these parameters along the instability strip (IS) and across it, i.e. along the lines P = const., are qualitatively identical with galactic cepheids although there are some quantitative distinctions. Along the lines of P = const. the largest amplitudes of $(B-V)$ and V are observed on the low-temperature boundary of the IS. The correlations between light and colour amplitudes, between light amplitudes and mean colours (or minimal colours) are not single-valued. This is removed by choosing an additional parameter – the colour in the light maximum. The correlation between light and colour amplitudes (along the lines of P = const.) in the center of SMC is three times weaker than in the Galaxy. In contrary, the correlation between light amplitudes and maximal colours in SMC is more strong than in the Galaxy.

122.139 **On the determination of the radii of cepheids.**
N. N. Yakimova.
Soobshch. Gos. Astron. Inst. Shternberga, No. 182, p. 27 - 36 (1973). In Russian.

The correctness of the Wesselink–Baade method for determination of the radii of cepheids is questioned. A new generalized method is proposed.

122.140 **New infra-red data on Mira variables.**
C. A. Whitney.
Journ. American Ass. Variable Star Observers, Vol. 4, 22 - 24 (1975).

The "IRC" catalog provides photometric data on 5562 objects; most of them are late type stars. Two current projects are described: (1) a search for variables among the IRC objects, (2) construction of infra-red light and color curves.

122.141 **XZ Cygni: recent period study, again.**
M. Taylor.
Journ. American Ass. Variable Star Observers, Vol. 4, 25 - 30 (1975).

This paper analyzes the behavior of the RR Lyrae type variable star XZ Cygni over the past nine years. One hundred visual light curves were analyzed, and forty selected, from which primary and secondary (Blazhko) periods were computed.

122.142 **A photometric analysis of X Persei $\stackrel{?}{=}$ 3U 0352+30.**
E. W. Gottlieb, E. L. Wright, W. Liller.
Journ. American Ass. Variable Star Observers, Vol. 4, 34 (1975). – Abstract.

122.143 **Photoelectric observations of AD Leonis, EV Lacertae and a suspected flare star near HZ Herculis.**
W. A. Christiansen, R. H. Kaitchuck.
Inform. Bull. Variable Stars, (I.A.U. Commission 27), Konkoly Obs., Budapest, No. 1012, 4 pp.(1975).

122.144 **Flares observed from UV Ceti.** A. H. Jarrett, J. B. Gibson.
Inform. Bull. Variable Stars, (I.A.U. Commission 27), Konkoly Obs., Budapest, No. 1017, 4 pp.(1975).

122.145 **Flare photometry of YZ Canis Minoris.**
A. H. Jarrett, J. B. Gibson.
Inform. Bull. Variable Stars, (I.A.U. Commission 27), Konkoly Obs., Budapest, No. 1018, 3 pp. (1975).

122.146 **Photoelectric observations of the star UV Cet in 1972.** M. E. Contadakis, L. N. Mavridis.
Inform. Bull. Variable Stars, (I.A.U. Commission 27), Konkoly Obs., Budapest, No. 1020, 5 pp. (1975).

122.147 **Hβ-photometry of HR 8024 and HR 8102.**
G. C. Kilambi.
Inform. Bull. Variable Stars, (I.A.U. Commission 27), Konkoly Obs., Budapest, No. 1024, 4 pp. (1975).

122.148 **Two possible near-by flare stars in Cancer.**
G. Haro, E. Chavira, G. Gonzalez.
Inform. Bull. Variable Stars, (I.A.U. Commission 27), Konkoly Obs., Budapest, No. 1031 (1975).

122.149 **Photographic observations of UV Cassiopeiae.**
P. Tempesti.
Inform. Bull. Variable Stars, (I.A.U. Commission 27), Konkoly Obs., Budapest, No. 1036, 2 pp. (1975).

122.150 **A revised period for one of the Wesselink-Shuttleworth SMC variables.** A. M. van Genderen.
Inform. Bull. Variable Stars, (I.A.U. Commission 27), Konkoly Obs., Budapest, No. 1037, 2 pp. (1975).

122.151 **Confirmation of the periodicity and the period (about 30 years) of the shell of Omicron Andromedae.** M. Fracassini, L. E. Pasinetti.
Inform. Bull. Variable Stars, (I.A.U. Commission 27), Konkoly Obs., Budapest, No. 1044, 2 pp. (1975).

122.152 **Photoelectric observations of DO Cephei.**
A. J. Nicastro.
Inform. Bull. Variable Stars, (I.A.U. Commission 27), Konkoly Obs., Budapest, No. 1045, 5 pp., with a correction No. 1057 (1975).

122.153 **Photoelectric maxima of RZ Cephei.**
B. Cester, I. Todoran.
Inform. Bull. Variable Stars, (I.A.U. Commission 27), Konkoly Obs., Budapest, No. 1047, 2 pp. (1975).

122.154 **A new determination of the period of the β CMa star γ Pegasi.** J.-P. Sareyan, J.-C. Valtier, J.-M. Le Contel.
Astron. Astrophys., Vol. 44, 215 - 218 (1975).

Spectrographic and photometric observations of γ Peg lead to the determination of a new period P = 0.15 17501 ± 3 × 10^{-7} day. This period is constant to within 0.06 s per century, during the period 1931–1971. A comparison with similar observations of other β CMa variables and with theoretical evolutionary tracks by various authors is made.

122.155 **Possible mechanism of stellar bursts.**
Yu. V. Vandakurov.
VI Leningr. mezhdunar. Seminar, 1974. Uskorenie chastits i yader. reaktsii v kosmose. Leningrad, 1974, p. 236 - 237, 237 - 238, with a discussion, p. 238. In Russian and English. Abstr. in Referativ. Zhurn., 8.51.620 (1975).

122.156 **Radio and optical observations of some flare stars.**
I. G. Moiseev, G. E. Patston, L. H. Solomon, H. M. Tovmassian, C. S. Higgins, P. F. Chugajnov.
Izv. Krymskoj Astrofiz. Obs., Vol. 53, p. 150 - 153 (1975). In Russian.

The results of joint radio-optical observations of flare stars UV Cet, V371 Ori, EQ Peg and GC 22805 made in May–August, 1965, are presented. In total 7 optical flares have been recorded of which 3 were probably accompanied by radio flares. Possible radio flares were observed only at 153 MHz. It is suggested that radio flares have very steep spectra or they occur at lower frequencies only.

122.157 **The distribution of flares of red dwarfs from energies of optical radiation.**
V. I. Krasnobabtsev, R. E. Gershberg.
Izv. Krymskoj Astrofiz. Obs., Vol. 53, p. 154 - 164 (1975).

In Russian.

Assuming the ergodic character of flare activity within an ensemble of stars of similar luminosity the authors have carried out a statistical analysis of photographic observations of flares in the Pleiades and in Orion. A certain community is found for distributions of flares in energy of optical radiation within wide ranges of luminosity and age of flare stars in clusters and in solar vicinity.

122.158 **The dependence between age and flare activity of the red dwarf stars in the solar vicinity.**
N. I. Shakhovskaya.
Izv. Krymskoj Astrofiz. Obs., Vol. 53, p. 165 - 167 (1975).
In Russian.

The investigation of the space motions of 646 red dwarf stars of the solar vicinity shows that the most active UV Cet-type variable stars are members of the young disk population. But less powerful flare activity is observed on members of the old disk population too. Flares are observed more often on young stars, the flare activity decreases when the age of the stars increases, but in some cases flare activity is observed on stars older than 5×10^9 years. This result changes the previous opinion that flare activity is observed only on young stars.

122.159 **Photoelectric observations of EM Cep 1971 – 1972.**
T. M. Rachkovskaya.
Izv. Krymskoj Astrofiz. Obs., Vol. 53, p. 168 - 186 (1975).
In Russian.

Photoelectric observations of EM Cep have been carried out with B filter in the period 1971 – 1972. It has been shown: 1) there are light variations of amplitude 0^m10 – 0^m05; 2) rapid variations of the light are present. The period of the light variations of EM Cep was found to be $0^d.806196$. EM Cep probably belongs to β Cep type variables.

122.160 **The spectrum of the cepheid ζ Geminorum in the $\lambda\lambda$ 10580–11020 Å region.**
G. I. Shanin, A. G. Shcherbakov.
Izv. Krymskoj Astrofiz. Obs., Vol. 53, p. 187 - 195 (1975).
In Russian.

Spectroscopic observations in the region $\lambda\lambda$ 10580–11020 Å of ζ Gem are presented. Numerous atomic line identifications and central line intensities of H, Si I, C I, Fe I, Mg I, Sr II, He I are given for five phases. The λ 10830 Å line of neutral helium appears in absorption and shows variation with phase. The light maximum equivalent width of the helium line is equal or less than 700 mÅ.

122.161 **Beat phenomena in the dwarf cepheid AE UMa.**
P. Broglia, P. Conconi.
Astron. Astrophys. Suppl. Ser., Vol. 22, 243 - 261 (1975).

AE UMa is an ultrashort pulsating variable. 1900 photoelectric B and V observations were carried out at the Merate Observatory during 1974. The main period has been improved: $P_0 = 0^d.086016883$. The residuals computed by means of P_0 and the variation in the brightness of the maxima and of the minima can be accounted for by a beat phenomenon whose period is found to be: $P_b = 0^d.293616$. The secondary period follows: $P_1 = 0^d.066527$ and the ratio $P_1/P_0 = 0.7734$. This ratio and the amplitudes of the oscillations relating to P_0 and to P_1 are compared with those of other double mode AI Vel variables.

122.162 **Current problems on horizontal branch stars. III. An analysis of the evolutionary status of the RR Lyrae stars in the globular cluster ω Cen.**
F. Caputo, V. Castellani.
Astrophys. Space Sci., Vol. 38, 39 - 66 (1975).

Expected characteristics of RR Lyrae stars as a function of the evolutive parameters are reported. Results from both evolutionary and pulsational investigations are collected in a suitable form, to show the general constraints to any interpretative analysis of the observations. It is shown that the spread in luminosity among the RR Lyrae stars results a function of the original chemical composition. It is suggested that variations in the original helium content of the order of $\Delta Y \sim 0.1$ and a correlation $Z=Z(t)$ can account for some well-observed galactic globular clusters, without allowing for mass loss in the redder HB stars belonging to each cluster.

122.163 **Upper limits on stellar flare X-ray emission from OSO-3.** V. Tsikoudi, H. Hudson.
Astron. Astrophys., Vol. 44, 273 - 276 (1975).

The OSO-3 X-ray telescope scanned 82 optically-observed flares on UV Cet, EV Lac, YZ CMi and AD Leo and four radio flares on UV Cet, during 1967–1968. The authors have examined these data individually and by superposed-epoch analysis, and find upper limits to the luminosity ratios $L_X/L_U \lesssim 10^3$, and $L_X/L_{\rm radio} \lesssim 10^7$. Moderate improvement of the X-ray observations should make possible a direct test of the solar-flare analogy in the near future. The integrated X-ray luminosity of stellar flares may contribute significantly to the low-energy diffuse X-ray background in the plane of the Galaxy.

122.164 **Long-period cepheid CP Cephei.** H. Albo.
Publ. Tartu Astrofiz. Obs., Vol. 43, 256 - 261 (1975). In Russian.

Using photographic plates of the Tartu Observatory a mean light curve of CP Cep is obtained. All published maxima in 1899–1960 are collected and a new mean period of CP Cep is derived.

122.165 **The cepheid variables and the stellar populations of the dwarf spheroidal galaxies.**
J. Norris, R. Zinn.
Astrophys. Journ., Vol. 202, 335 - 345 (1975).

A new hypothesis is presented concerning the stellar populations of the dwarf spheroidal galaxies. The authors suggest that these systems have evolved separately from our Galaxy, with initial star formation occurring from an extremely metal-deficient protocloud some few billion years after the formation of the Galaxy. The existence of cepheid variables in the Small Magellanic Cloud similar to those in the dwarf spheroidal galaxies probably implies the existence of young, metal-deficient objects ($Z \lesssim 10^{-4}$) in this system. Finallv. the authors suggest that age is the relevant "second" parameter necessary for the interpretation of the anomalous horizontal-branch morphology of the intergalactic globular clusters and those at the edge of the Galaxy.

122.166 **On the nature of the bright variables in dwarf spheroidal galaxies.**
P. Demarque, A. W. Hirshfeld.
Astrophys. Journ., Vol. 202, 346 - 352 (1975).

Arguments are presented which suggest that the unusual "bright variables" observed in Draco and other dwarf spheroidals are more massive than the RR Lyrae variables in the same systems. Models of very metal-poor stars ($10^{-5} < Z < 10^{-4}$) on the horizontal branch with masses up to $2\,M_\odot$ confirm this interpretation in the sense that both the luminosities and the period-luminosity relation of the models are consistent with the observations of the bright variables. Best fit is obtained for $Y \approx 0.25$ and $Z \approx 7$–8×10^{-5}. Some implications of the presence of stars about 1×10^9 years old in dwarf spheroidal galaxies are briefly discussed.

122.167 **Evidence for X-ray emission from flare stars observed by ANS.**
J. Heise, A. C. Brinkman, J. Schrijver, R. Mewe, E. Gronenschild, A. den Boggende, J. Grindlay.

Astrophys. Journ., (*Letters*), Vol. 202, L73 - L76 (1975).

Observations that detected the first X-ray emission from flare stars are described. An X-ray flare was detected from YZ CMi at 0.28 keV and ~1–7 keV, although no optical or radio coverage was available. During a very large optical flare from UV Ceti, X-ray emission at (only) 0.28 keV was detected. Upper limits for X-ray emission from several small optical flares of UV Ceti are presented. Implications for X-ray flare models, the diffuse X-ray background, and low-energy cosmic-ray flux are mentioned.

122.168 **NT Tel, a halo population S-type Mira variable.** P. J. Andrews.
Monthly Notices Roy. Astron. Soc., Vol. 173, 701 - 707 (1975).

NT Tel, a unique halo-population Se star, is shown to be a Mira-type variable with a period of 252 days and an amplitude A_v= 4.5 mag.

122.169 **Variable stars and the structure of the Galaxy.** L. Plaut.
Priroda, No. 12.75, p. 88 - 95 (1975). In Russian. Translated from English by N. N. Samus'.

122.170 **The behaviour of the fuor (*FU Orionis star*) V 1057 Cygni after its outburst.** A. L. Gyul'budagyan.
Dokl. AN ArmSSR, Vol. 60, No. 2, p. 90 - 93 (1975).
In Russian. – Abstr. in Referativ. Zhurn. 51. Astron., 11.51. 507 (1975).

122.171 **The RR Lyrae stars in the Small Magellanic Cloud.** J. A. Graham.
Publ. Astron. Soc. Pacific, Vol. 87, 641 - 682 (1975).

A field covering an area 1° × 1°.3 in an outlying region of the Small Magellanic Cloud has been searched for variable stars. Ninety-two variables are identified and periods determined for 77 of these. For the RR Lyrae variables, multiple discovery statistics suggest a completeness of 93 % for variables of Bailey class *ab* and 50 % for those of Bailey class *c*. Approximately 30 % of the RR Lyrae variables show variations in amplitude and/or phase of maximum or minimum light. The period distribution is unlike that found anywhere in the Galaxy. The mean, $\langle B \rangle$, for all RR Lyrae variables with known periods and with $\langle B \rangle > 19^m.0$, is $19^m.95$. Adopting a distance modulus of $19^m.0$ for the SMC, a mean absolute visual magnitude of $+0^m.6 \pm 0^m.25$ is found for the RR Lyrae stars in the SMC.

122.172 **A periodogram analysis of 4 Canum Venaticorum.** C. W. McAlary, J. R. Percy.
Publ. Astron. Soc. Pacific, Vol. 87, 789 - 793 (1975).

A periodogram analysis of the δ Scuti star 4 CVn has been carried out using the photometric data of Landolt and the method of Gray and Desikachary. The analysis reveals three frequencies of 3.92, 5.823, and 8.07 cycles day^{-1}; the phases and amplitudes corresponding to these frequencies have been derived.

122.173 **Photoelectric photometry of the δ Scuti star HR 5005.** M. Jerzykiewicz.
Publ. Astron. Soc. Pacific, Vol. 87, 817 - 820 (1975).

About 150 photoelectric observations of HR 5005 are reported. The δ Scuti-type variability of the star is confirmed. On the basis of the light range changes and the distribution of the epochs of maximum light, it is conjectured that HR 5005 has a fairly long beat period.

122.174 **Photoelectric photometry of CW Ser.** W. Gieren, F. Gieseking, M. Hoffmann.
Inform. Bull. Variable Stars, (I.A.U. Commission 27), Konkoly Obs., Budapest, No. 1055, 3 pp. (1975).

122.175 **GS Andromedae.** G. Romano.
Inform. Bull. Variable Stars, (I.A.U. Commission 27), Konkoly Obs., Budapest, No. 1069 (1975).

122.176 **Double mode cepheid V 367 Sct in open cluster NGC 6649.** Yu. N. Efremov, P. N. Kholopov.
Inform. Bull. Variable Stars, (I.A.U. Commission 27), Konkoly Obs., Budapest, No. 1073, 2 pp. (1975).

122.177 **Stellar spots, flares and lithium.** P. F. Chugajnov.
Zemlya i Vselennaya, 1975, No. 6, p. 4 - 8. In Russian.

122.178 **On the period and radius of Zeta Geminorum.** C. D. Scarfe.
Journ. Roy. Astron. Soc. Canada, Vol. 69, 244 - 245 (1975). Abstr. CAS.

122.179 **π Aqr: a pulsating Be star?** J. D. Fernie.
Journ. Roy. Astron. Soc. Canada, Vol. 69, 245 (1975). – Abstr. CAS.

122.180 **Evolution and period changes in pulsating variable stars.** J. R. Percy.
Journ. Roy. Astron. Soc. Canada, Vol. 69, 251 (1975). Abstr. CAS.

122.181 **A search for periodicities in the pulsation of Delta Scuti stars. I. The calculation of a cyclic phenomenon.** T. Dumont, N. Morguleff, B. Rutily, A. Terzan.
Astron. Astrophys., Vol. 45, 57 - 64 (1975). In French.

A new calculation method is proposed to search for periods in vibratory phenomena.

122.182 **The magnetic fields of pulsating variables.** M. J. Stift.
Astron. Astrophys., Vol. 45, 227 - 228 (1975).

Photometric and magnetic observations lend no support to the proposed relation between the periods and the maximum magnetic fields of pulsating variables.

122.183 **Radial velocities of stars with microwave maser emission. I. Long-period variables.**
G. Wallerstein.
Astrophys. Journ., Suppl. Ser., No. 291, Vol. 29, 375 - 396 (1975).

Extensive radial-velocity measurements of high-dispersion optical spectra (λλ 3500–9000) of long-period variables are presented and analyzed so as to establish the photospheric, emission, and circumstellar velocities. The data are compared with microwave emission velocities from the molecules OH, H_2O, and SiO. Correlations among the various velocities are presented and discussed. The question of whether the absorption or emission lines represent the true stellar velocity is carefully discussed.

122.184 **Radial stellar stability and pulsation.** A. N. Cox.
Bull. American Astron. Soc., Vol. 7, 503 (1975). Abstr. AAS.

122.185 **Calculations of periodic, nonlinear pulsation.** R. F. Stellingwerf.
Bull. American Astron. Soc., Vol. 7, 503 (1975). – Abstr. AAS.

122.186 **An analytical theory of periodic variable stars.** P. J. Melvin.
Bull. American Astron. Soc., Vol. 7, 504 (1975). – Abstr. AAS.

122.187 **Convection in pulsating stars.** N. H. Baker, D. O. Gough, R. F. Stellingwerf.
Bull. American Astron. Soc., Vol. 7, 504 (1975). – Abstr. AAS.

122.188 **OAO-2 ultraviolet photometry of RR Lyrae.**
J. L. Hutchinson, C. F. Lillie, S. J. Hill.
Bull. American Astron. Soc., Vol. 7, 505 (1975). – Abstr. AAS.

122.189 **Theoretical colors for cepheids with transfer effects now evident in their light curves.**
C. G. Davis, A. N. Cox, S. S. Bunker.
Bull. American Astron. Soc., Vol. 7, 511 (1975). – Abstr. AAS.

122.190 **On the mass of AC Andromedae.**
D. S. King, A. N. Cox, S. W. Hodson.
Bull. American Astron. Soc., Vol. 7, 511 (1975). – Abstr. AAS.

122.191 **Does the system Theta Virginis represent a new class of pulsation stars?** W. R. Beardsley, F. R. Zizka.
Bull. American Astron. Soc., Vol. 7, 512 (1975). – Abstr. AAS.

122.192 **AM Herculis, an unusual rapid variable.**
R. A. Berg.
Bull. American Astron. Soc., Vol. 7, 512 (1975). – Abstr. AAS.

122.193 **Spectrophotometry of R Coronae Borealis during the minimum of 1974.**
R. S. Patterson, J. D. Fix, J. S. Neff.
Bull. American Astron. Soc., Vol. 7, 512 (1975). – Abstr. AAS.

122.194 **Detection of band polarization in red variable stars.**
J. Landstreet, R. Angel, N. Woolf, E. A. Beaver.
Bull. American Astron. Soc., Vol. 7, 519 (1975). – Abstr. AAS.

122.195 **The pulsation of high luminosity helium stars.**
P. R. Wood.
Bull. American Astron. Soc., Vol. 7, 527 (1975). – Abstr. AAS.

122.196 **Spectrophotometry of the flare star BY Draconis.**
S. R. Spangler, J. D. Fix.
Bull. American Astron. Soc., Vol. 7, 527 (1975). – Abstr. AAS.

122.197 **Flare stars and galactic winds.**
G. D. Coleman, S. P. Worden.
Bull. American Astron. Soc., Vol. 7, 528 (1975). – Abstr. AAS.

122.198 **Observations of some southern variables of the R CrB type.** F. M. Bateson.
Publ. Variable Star Section, Roy. Astron. Soc. New Zealand, No. 3 (C75), p. 1 - 25 (1975).
Light curves are presented for UW Cen; S Aps; GU Sgr and RY Sgr in continuation of those previously published. Light curves are given for lengthy periods for DY Cen; AE Cir; RT Nor; RZ Nor; WX CrA; VZ Sgr; RS Tel and V CrA. All curves are from visual observations. The observations for each star are discussed, and for some stars charts are reproduced.

122.199 **Macro- and microturbulent velocities of the atmosphere of R Sct.** A. A. Aliev.
Dokl. AN AzSSR, Vol. 31, No. 4, p. 11 - 14 (1975). – Abstr. in Referativ. Zhurn. 51. Astron., 1.51.571 (1976).

122.200 **Review of observational results on R Coronae Borealis stars.** M. Ya. Orlov.
Peremennye Zvezdy, Vol. 19, 501 - 513 (1975). In Russian.
Basic observational data on R CrB stars are summarized with emphasis on the spectral behaviour of these stars.

122.201 **On possible mechanisms of brightness variability of R Coronae Borealis type stars.** B. E. Zhilyaev.
Peremennye Zvezdy, Vol. 19, 515 - 522 (1975). In Russian.
Some theoretical aspects of variability of R CrB type stars are discussed. The mechanism of formation of graphite grains in the atmospheres of these stars proposed in the Loreta-O'Keefe hypothesis is considered in detail.

122.202 **A spectroscopic study of the R CrB type star XX Camelopardalis in quiet stage.**
M. Ya. Orlov, M. H. Rodriguez.
Peremennye Zvezdy, Vol. 19, 523 - 524 (1975). In Russian.
A quantitative analysis of the R CrB type star XX Cam has been performed by the curve-of-growth method using coudé-spectrograms obtained at maximum brightness. No conspicuous Balmer lines of hydrogen have been found. A great number of carbon lines were identified, as well as bands of the C_2 and CN molecules. Equivalent widths of 360 lines were measured. The microturbulent velocity was found to be 8.9 km/sec, damping constant a = 0.001, electron pressure lg $p_e = -2.60$. Relative abundances of Ca, Sc, Ti, V, Cr, Mn, and Y turned out to be very close to solar values. The ratio of carbon to iron is 100 times greater in the atmosphere of XX Cam than in the solar atmosphere.

122.203 **Statistical analysis of brightness of R CrB, RY Sgr, SU Tau.** G. U. Koval'chuk, E. I. Lenderman, F. I. Lukatskaya, A. Eh. Rozenbush.
Peremennye Zvezdy, Vol. 19, 525 - 528 (1975). In Russian.

122.204 **Variables desired to be observed.** A. F. Pugach.
Peremennye Zvezdy, Vol. 19, 561 - 564 (1975).
There are some Is(A) variables whose light curves bear a strong resemblance to the R CrB stars, but the brightness variations are displayed in a smaller time-scale. The "normal" light of such a star is interrupted by an Algol-like minimum in a completely irregular fashion. One of the most interesting representatives of these so-called "antiflare" stars is RZ Psc. A table lists four well-determined antiflare stars and some suspected ones.

122.205 **Hα and Hβ lines in the spectrum of the FU Orionis star V1057 Cyg in 1974.** E. A. Kolotilov.
Astron. Tsirk., No. 873, p. 4 - 7 (1975). In Russian.

122.206 **The $^{12}C/^{13}C$ ratio in the classical cepheid T Monocerotis.**
G. L. Loumos, D. L. Lambert, J. Tomkin.
Publ. Astron. Soc. Pacific, Vol. 87, 859 - 861 (1975).
Observations of the CN lines at 8000 Å in the classical cepheid T Mon near minimum light are discussed. Weak ^{13}CN features are identified and an abundance ratio $^{12}C/^{13}C = 9 \pm 3$ is derived. Similar observations for two other cepheids show no detectable ^{13}CN lines; the corresponding limits are $^{12}C/^{13}C > 17$ (ζ Gem) and > 12 (η Aql).

122.207 **G 208-44/45.** S. Cristaldi, M. Rodonò.
IAU Circ., No. 2805 (1975).

122.208 **CoD$-33°$12119.** A. M. van Genderen.
IAU Circ., No. 2814 (1975).

122.209 **AR Serpentis.** V. P. Tsesevich.
Astron. Tsirk., No. 885, p. 7 - 8 (1975). In Russian.

122.210 **AN Ursae Majoris – a probable exnova?**
S. Yu. Shugarov.
Astron. Tsirk., No. 887, p. 7 - 8 (1975). In Russian.

122.211 **Dwarf cepheid LS Herculis.** G. Lange.
Astron. Tsirk., No. 888, p. 4 - 6 (1975). In Russian.

122.212 **New variable star of RR Lyr-type SVS 2146.**
S. Yu. Shugarov.
Astron. Tsirk., No. 888, p. 8 (1975). In Russian.

122.213 **On fluctuations in optical radiation of the carbon star RW LMi (CIT 6).**
A. Alksnis, I. Eglītis.

Investigation of the sun and red stars. 3, (see 003.008), p. 47 - 55 (1975). In Russian.

Photographic observations made with the Schmidt telescope of the Radioastrophysical Observatory of the Academy of Sciences of the Latvian SSR in the R, V and B passbands show that the long-period carbon star RW LMi has light variations with time scale of one day. Changes of 0.4 mag in the R band within seven hours were observed. The light variations in B and V are similar to those in R. The interval between the two last maxima of the long-period variations in RW LMi occurred to be about 100 days longer than the previously assumed period.

122.214 **Neuer veränderlicher Stern mit Nebelanhang.** W. Wenzel, H. Gessner.
MVS, *Sonneberg,* Vol. 7, 23 - 24 (1975).

122.215 **Beobachtungen von Z Andromedae und AX Persei auf Sonneberger Überwachungsplatten.**
E. Splittgerber.
MVS, *Sonneberg,* Vol. 7, 39 - 42 (1975).

122.216 **Spectral types for the brightest flare stars in the Pleiades.** B. Iriarte Erro.
Bol. Inst. Tonantzintla, Vol. 1, 209 - 215 (1975).

Nine of the brightest flare stars in the Pleiades cluster were observed spectroscopically at 121 Å/mm. Among them the earliest spectral type yet known for flare stars in the cluster was found to be K2. This spectral type was assigned to two stars and neither shows Ca II H-K in emission. With different degrees of intensity the remaining flare stars show the H-K lines in emission.

122.217 **On the expected properties of the RR Lyrae pulsators.** F. Caputo, V. Castellani.
Mem. Soc. Astron. Italiana, Vol. 46, 303 - 311 (1975).

The available evolutionary and pulsation theories are combined to give informations on the expected behaviour of the RR Lyrae pulsators, in varying the evolutionary parameters of the cluster stars. In this way the expected dependence of the blue boundaries for instability in the first harmonic and in the fundamental mode as a function of the original chemical composition only is derived. The comparison between typical Oosterhoff II clusters (as M15 and M68) and M3 doesn't give any definitive evidence for a difference in the original helium content.

122.218 **Photoelectric observations of some flare stars during 1972.**
J. Arsenijević, A. Kubičela, I. Vince.
Bull. Obs. Astron. Belgrade, No. 126, p. 54 - 55 (1975).

122.219 **Photoelectric observations of UV Ceti stars during 1973.** J. Arsenijević, A. Kubičela, I. Vince.
Bull. Obs. Astron. Belgrade, No. 126, p. 55 - 56 (1975).

122.220 **Progressi nell'interpretazione e studio delle stelle variabili. XX - Dalle stelle novae alle variabili simbiotiche.** L. Rosino.
Coelum, Vol. 43, 244 - 257 (1975).

122.221 **Sobre el comportamiento de R Doradus.** J. R. Garcia, G. A. Galassi.
Obs. Astron. Municipal Rosario, (*Argentina*), Bol. No. 4, p. 10 - 15 (1974).

122.222 **On the real value of the electron temperature of the atmosphere of R Sct.** A. A. Aliev.
Dokl. AN AzSSR, Vol. 31, No. 6, p. 10 - 12 (1975). In Russian. – Abstr. in Referativ. Zhurn. 51. Astron., 2.51.628 (1976).

122.223 **Metal abundances of RR Lyrae stars in galactic globular clusters.** D. S. Butler.
Thesis, California Univ., Santa Cruz (USA). [Available from Univ. Microfilms, Ann Arbor, Mich., *USA*. Order No. 75-8473], 100 pp. (1974).

Image-tube scanner measuremnts of Preston's (1959) metal abundance indicator have been made for 47 RR Lyraes. Another index that is a measure of metallic-line absorption in the region of the G-band has also been determined. From curve-of-growth analyses of 13 field RR Lyraes, a relationship for Fe/H is found. The problem of the anomalous horizontal-branch distribution functions in M13 and NGC 7006 is discussed in the light of these new observations. It is suggested that differences in the location of the horizontal branch in clusters with the same Fe/H result from different degrees of mixing between the envelope and the hydrogen-burning shell in the horizontal branch progenitor.

122.224 **A study of variable stars in the ultraviolet.** A. V. Holm.
Phil. Trans. Roy. Soc. London, Ser. A, Vol. 279, 459 - 471 (1975).

Some of the OAO-2 ultraviolet observations of intrinsically variable stars are discussed. New data is presented for the Ap variable ι Cassiopeiae. The cepheid variables are illustrated by α Ursae Minoris. The explosive variables nova Serpentis 1970, nova Aquilae 1918, SS Cygni, and the 1972 supernova in NGC 5253 are becoming visually bright only when an extended envelope is formed, is considered and rejected on the basis of observational evidence.

122.225 **A study of the period of HR 6684.** P. Rosenzweig L.
Thesis, Dep. Física, Univ. de Los Andes, Mérida, Venezuela, 49 pp. (1975). In Spanish.

A new period determination for HR 6684 has been carried out using all available observations, both photometric and of radial velocities. The method of Lafler and Kinman was used to search for possible periods in the range 0.11 – 0.21 days. It was found that there are two periods in this range that satisfy all the observations, one being the previously accepted period and the other the period related with this by one in the reciprocal. The available observational material is insufficient for deciding which of the two possibilities is the true period. Nevertheless, the data show that HR 6684 must be pulsating in the first vibrational mode.

On the determination of the period of a variable star with a computer (method of local scanning).
See Abstr. 021.018.

Is there a maser in the silicon monoxide ground state? See Abstr. 022.056.

The lower terms in Ni VII and possible identifications in RR Telescopii. See Abstr. 022.074.

The study of light curves with the aid of smoothing spline-functions. See Abstr. 031.201.

Low-energy nucleosynthesis of lithium, beryllium, and boron. See Abstr. 061.016.

A modified Henyey method for computing radiative transfer hydrodynamics. See Abstr. 063.004.

Circumstellar dust shells. See Abstr. 064.003.

Time lag effects in the nucleation of particles in stellar atmospheres. See Abstr. 064.016.

Hydrodynamic models of a cepheid atmosphere. III. Line spectrum and radius determinations.
See Abstr. 064.036.

An RR Lyrae-type ultraviolet model.
See Abstr. 064.054.

On the possibility of constructing model atmospheres of R CrB type stars. See Abstr. 064.062.

Model atmospheres of Delta Cephei.
See Abstr. 064.064.

Hydrodynamic models of a cepheid atmosphere. I. Deep envelope models. See Abstr. 065.009.

Stellar oscillations, stellar stability and application to variable stars. See Abstr. 065.089.

Stability of nonlinear periodic pulsation of stellar envelopes. See Abstr. 065.091.

Effect of rigidity on non-radial oscillations.
See Abstr. 065.095.

Non radial oscillations in massive stars during core H-burning phase. See Abstr. 065.096.

Excitation of non-radial oscillations by partial ionization zone mechanism. See Abstr. 065.098.

The origin and evolution of RR Lyrae stars of high metal abundance. See Abstr. 065.152.

Linear and nonlinear models of lower cepheid-strip variable stars. See Abstr. 065.153.

Convection in RR Lyrae stars.
See Abstr. 065.157.

Pi Aquarii: a pulsating Be star?
See Abstr. 113.018.

Be stars with large magnitude changes.
See Abstr. 113.024.

Pulsation in peculiar A stars. See Abstr. 113.031.

Observations of T Tauri stars and related objects in the Taurus dark cloud. See Abstr. 113.041.

Narrow-band photoelectric observations of the Ap star HD 184905. See Abstr. 113.064.

Results of 10830 observations in 217 stars.
See Abstr. 114.036.

Some spectroscopic problems of Coronae Borealis stars.
See Abstr. 114.084.

Helium abundances and cepheid pulsation.
See Abstr. 114.095.

Infrared spectra of T Tau stars and related objects.
See Abstr. 114.305.

Emission lines in the hydrogen-deficient variable MV Sagittarii near maximum light. See Abstr. 114.312.

High-speed spectroscopy of the ultrashort-period variable AM Canum Venaticorum (= HZ 29). See Abstr. 114.316.

The temperature and spectrum of VX Aquilae.
See Abstr. 114.325.

Variable blue object with a peculiar spectrum.
See Abstr. 114.352.

Fine analysis of three R Coronae Borealis stars.
See Abstr. 114.362.

The luminosities of the Beta Canis Majoris variables, the zero age main sequence and the distance of the Sco–Cen association. See Abstr. 115.002.

Note on the near-infrared absolute magnitude of irregular carbon variables. See Abstr. 115.008.

A technique for separating reddening and envelope emission in T Tauri stars. See Abstr. 115.010.

Gas flow in cataclysmic variable stars.
See Abstr. 117.049.

Light variations of TT Ari. See Abstr. 117.064.

The triple system 20 Leo. See Abstr. 118.021.

New radial velocities and further photometric observations of λ Sco and κ Sco. See Abstr. 119.015.

Eclipse timings of cataclysmic variables.
See Abstr. 121.025.

Is a dwarf nova really a "dwarf" nova?
See Abstr. 124.008.

The orbital period of U Gem. See Abstr. 124.017.

Eruptive variable in Aquila.
See Abstr. 124.023.

Eruptive variable in Aquila.
See Abstr. 124.025.

Discovery of a nova or eruptive variable with the coordinates $\alpha = 19^h15^m26^s.98$, $\delta = +04°41'44''.3$ (1950).
See Abstr. 124.026.

Time variations in sources of silicon monoxide emission at 7 millimeters. See Abstr. 131.017.

The results of observations of stars with intrinsic polarization. See Abstr. 131.097.

Spectroscopy of R Monocerotis and NGC 2261.
See Abstr. 132.002.

Colorimetric measurements of flare stars in the Orion aggregate. See Abstr. 132.039.

Herbig-Haro objects and postfuors.
See Abstr. 132.060.

Transition radiation and peculiar nebulae.
See Abstr. 132.061.

Radio sources in the vicinity of UV Ceti-type flare stars. See Abstr. 141.136.

On the nature of maser sources in infrared stars.

See Abstr. 141.603.

Combined optical and X-ray observations of variable stars. See Abstr. 142.028.

Polarimetric measurements at Hα and Hβ of X Per. See Abstr. 142.067.

On the possible annihilation origin of gamma-ray bursts and flare stars. See Abstr. 142.146.

Proper motions of two galactic clusters in Cassiopeia and of the classical cepheid CG Cas. See Abstr. 153.021.

General properties of the two groups of globular clusters and their variable stars. See Abstr. 154.003.

Variable stars in the X-ray globular cluster NGC 1851. See Abstr. 154.024.

NGC 6273: a globular cluster with 3 W Virginis type variables. See Abstr. 154.034.

The distribution of cepheids in the Large Magellanic Cloud and cells of star formation. See Abstr. 159.009.

123 Variable Stars: Lists of Observations, Individual Observations

123.001 **Sequences for southern variables.** B. Menzies. Publ. Variable Star Section, Roy. Astron. Soc. New Zealand, No. 3 (C75), p. 39 - 40 (1975).

123.002 **Observations of variable stars in the T-association T1 Cygni.** A. Ya. Filin. Peremennye Zvezdy, Prilozhenie, Vol. 2, 63 - 70 (1974). In Russian.

123.003 **On seven variable stars.** N. B. Perova. Peremennye Zvezdy, Prilozhenie, Vol. 2, 75 - 81 (1974). In Russian.

123.004 **On the variable star V923 Cygni.** E. N. Makarenko. Peremennye Zvezdy, Prilozhenie, Vol. 2, 117 - 119 (1974). In Russian.

123.005 **Z Camelopardalis, 1926–50.** E. H. Collinson. Journ. British Astron. Ass., Vol. 85, 438 - 443 (1975).

123.006 **Two U Geminorum variables: UV Per 1926–69 and SW UMa 1963–69.** P. W. Hornby. Journ. British Astron. Ass., Vol. 85, 528 - 531 (1975).

123.007 **Der langperiodisch veränderliche Stern R Draconis.** R. Lukas. SuW, Vol. 14, 260 (1975).

123.008 **Das Verhalten einiger veränderlicher Sterne über längere Zeiträume.** R. Lukas. Sterne, 51. Jahrgang, p. 177 - 180 (1975). – Concerning UW Dra, VY Dra, μ Cep, ρ Cas, XX Cam and R CrB.

123.009 **Photometry of three variables in the neighbourhood of the galactic cluster NGC 6830.** B. L. Shaganyan, I. I. Vypirajlo. Astrometriya i Astrofizika, *Kiev*, vyp. (No.) 26, (see 003.001), p. 63 - 68 (1975). In Russian.

The results are given of photographic observations of LZ, NS and V 781 Aql in the neighbourhood of the galactic cluster NGC 6830. The maps of neighbourhood of each variable, photographic values of comparison stars and epochs of maximum light are presented. The seasonal curves of light variations in the photographic region are obtained.

123.010 **Information on photoelectric observations of variable stars stored at the Odessa Astronomical Observatory.** V. P. Tsesevich, E. N. Makarenko. Astron. Tsirk., No. 844, p. 7 - 8 (1974). In Russian.

123.011 **Preliminary designations of SVS stars.** Astron. Tsirk., No. 846, p. 7 - 8 (1974). In Russian.

123.012 **New variable of RR Lyrae type SVS 2041 in the central region of the globular cluster M3.** P. N. Kholopov. Astron. Tsirk., No. 848, p. 8 (1974). In Russian.

123.013 **UBV observations of FG Sagittae in 1973–1974.** V. P. Arkhipova. Astron. Tsirk., No. 853, p. 7 - 8 (1975). In Russian.

123.014 **Three new variable stars in the β Aql region.** S. Yu. Shugarov. Astron. Tsirk., No. 858, p. 5 - 6 (1975). In Russian.

123.015 **New variable carbon stars.** I. Daube, L. Duncans. Astron. Tsirk., No. 860, p. 7 - 8 (1975). In Russian.

123.016 **New variable star of U Gem-type SVS 2120.** S. Yu. Shugarov. Astron. Tsirk., No. 862, p. 8 (1975). In Russian.

123.017 **New variable star of RR Lyrae-type SVS 2121.** S. Yu. Shugarov. Astron. Tsirk., No. 863, p. 8 (1975). In Russian.

123.018 **New variables in the Sco–Oph association.** V. Satyvaldiev. Astron. Tsirk., No. 869, p. 6 - 8 (1975). In Russian.

123.019 **New variable stars of spectral type C in the 90° galactic longitude zone.** A. Alksnis, Z. Alksne. Astron. Tsirk., No. 870, p. 6 - 7, with a correction, No. 891, p. 8 (1975). In Russian.

123.020 ***V* photometric observations of AI Vel.** E. W. Elst. Astron. Astrophys., Suppl. Ser., Vol. 22, 221 - 233 (1975).

Photometric observations of AI Vel in the *V* band of the *UBV* system, obtained during November 1972 at the European Southern Observatory, are presented. The fundamental and the beat period are re-estimated.

123.021 **On the variability of some stars in Aql.** B. L. Shaganyan. Astrometriya i Astrofizika, *Kiev*, vyp. (No.) 27, (see 003.004), p. 105 - 112 (1975). In Russian.

The work is a part of the study of all variables in the vicinity of the galactic cluster NGC 6830. Photographic magnitudes of NV, V467 and V700 Aql are given.

123.022 **RY Carinae: a complex star group.** C. Johnson, M. H. Liller. Journ. American Ass. Variable Star Observers, Vol. 4, 31 - 33 (1975).

123.023 **Variable star notes.** J. A. Mattei. Journ. American Ass. Variable Star Observers, Vol. 4, 35 - 39 (1975).

123.024 **New variable stars in a field around 38 o Per.** F. Gieseking. Inform. Bull. Variable Stars, (I.A.U. Commission 27), Konkoly Obs., Budapest, No. 1008, 2 pp. (1975).

123.025 **Observations of the period of Y Leo.** A. Ghezloun, M. A. Svechnikov. Inform. Bull. Variable Stars, (I.A.U. Commission 27), Konkoly Obs., Budapest, No.1015 (1975).

123.026 **Three revised periods and two new variable stars.** D. Hoffleit. Inform. Bull. Variable Stars, (I.A.U. Commission 27), Konkoly Obs., Budapest, No. 1016, 2 pp. (1975).

123.027 **On the variability of θ^1 Ori A.** K. A. Strand. Inform. Bull. Variable Stars, (I.A.U. Commission 27), Konkoly Obs., Budapest, No. 1025, 2 pp. (1975).

123.028 **Photoelectric observations of UV Cas.** F. Zavatti.

Inform. Bull. Variable Stars, (I.A.U. Commission 27), Konkoly Obs., Budapest, No. 1027, 2 pp. (1975).

123.029 **Revised elements of eclipsing stars.** P. Ahnert.
Inform. Bull. Variable Stars, (I.A.U. Commission 27), Konkoly Obs., Budapest, No. 1030 (1975).

123.030 **V3804 Sgr.** M. Brewster.
Inform. Bull. Variable Stars, (I.A.U. Commission 27), Konkoly Obs., Budapest, No. 1032 (1975).

123.031 **Variability of the δ Scuti star 38 Cancri.** G. Guerrero.
Inform. Bull. Variable Stars, (I.A.U. Commission 27), Konkoly Obs., Budapest, No. 1034 (1975).

123.032 **CSV 100778: a variable carbon star probably identical with AS 137.** N. Sanduleak.
Inform. Bull. Variable Stars, (I.A.U. Commission 27), Konkoly Obs., Budapest, No.1035 (1975).

123.033 **Period for BV 1634.** M. Hibbert, D. Hoffleit.
Inform. Bull. Variable Stars, (I.A.U. Commission 27), Konkoly Obs., Budapest, No. 1040, 2 pp. (1975).

123.034 **Light changes of T Tauri stars in the Chamaeleon association.** H. Mauder, F. M. Sosna.
Inform. Bull. Variable Stars, (I.A.U. Commission 27), Konkoly Obs., Budapest, No. 1049, 4 pp. = Veröff. Remeis-Sternw. Bamberg - Astron. Inst. Univ. Erlangen-Nürnberg, Vol. 10, No. 113 (1975).

123.035 **A new red semiregular variable star: BD + 18°4590.** P. Tempesti.
Inform. Bull. Variable Stars, (I.A.U. Commission 27), Konkoly Obs., Budapest, No. 1050, 3 pp. (1975).

123.036 **Photoelectric times of maxima of BW Vul.** T. E. Margrave, J. W. Mefford.
Inform. Bull. Variable Stars, (I.A.U. Commission 27), Konkoly Obs., Budapest, No. 1051 (1975).

123.037 **Amatörobservationer av X Camelopardalis.** H. Bengtsson.
Astron. Tidssk., Årg. 8, 136 - 137 (1975).

123.038 **The variable stars around Omicron Andromedae.** G. Romano, M. Perissinotto.
Mem. Soc. Astron. Italiana, Vol. 46, 259 - 270 (1975).

This paper gives the results of the observations of 22 variable stars in a field around Omicron And. The period of observation is of about 13 years. Three new variable stars have been discovered during the research.

123.039 **Observations of variable stars January - June 1975. Report No. 28.** L. Plaut, H. Feijth.
Nederlandse Vereniging voor Weer- en Sterrenkunde. Kapteyn Astron. Lab. Groningen–Netherlands. 10 pp. (1975).

This report gives 4573 visual observations of 225 Mira stars, 6 semi-regular variables, 3 RV Tauri stars, 15 U Gem and Z Cam stars and novae, and 4 miscellaneous type stars.

123.040 **Tableaux des observations faites par les sociétaires de l'A.F.O.E.V. de mai à août 1975.**
A.F.O.E.V. Bull., Vol. 9, 99 - 128 (1975).

Le nombre des estimations faites par 26 observateurs s'élève à 4961 et porte sur 320 étoiles différentes.

123.041 **Etoiles naines rouges variables.** M. Petit.
Inform. Bull. Variable Stars, (I.A.U. Commission 27), Konkoly Obs., Budapest, No. 1056, 2 pp. (1975).

123.042 **V1057 Cygni.** G. Welin.
Inform. Bull. Variable Stars, (I.A.U. Commission 27), Konkoly Obs., Budapest, No. 1057 (1975).

123.043 **Two peculiar variable stars: OX Cyg and GR 128.** D. Hoffleit.
Inform. Bull. Variable Stars, (I.A.U. Commission 27), Konkoly Obs., Budapest, No. 1063, 2 pp. (1975).

123.044 **Three long-period variables in Sagittarius.** D. Hoffleit.
Inform. Bull. Variable Stars, (I.A.U. Commission 27), Konkoly Obs., Budapest, No. 1064 (1975).

123.045 **SY Muscae.** F. M. Bateson, A. F. Jones, B. Menzies.
Publ. Variable Star Section, Roy. Astron. Soc. New Zealand, No. 3 (C75), p. 40 - 43 (1975).

V and B–V magnitudes are listed for the comparison stars for SY Mus. Visual observations from 2,434,910 to 2, 441,510 are presented in a light curve.

123.046 **U Fornacis.** F. M. Bateson, A. F. Jones, B. Menzies.
Publ. Variable Star Section, Roy. Astron. Soc. New Zealand, No. 3 (C75), p. 44 - 46 (1975).

V and B–V magnitudes for the comparison stars for U For are given, together with visual observations in the form of a light curve.

123.047 **SU Carinae.** F. B. Bateson, B. Menzies.
Publ. Variable Star Section, Roy. Astron. Soc. New Zealand, No. 3 (C75), p. 47 - 49 (1975).

The Mira-type variable, SU Car, is shown to have a period of 575.6 days instead of one of 230.9 days as adopted hitherto. Maximum visual magnitude varies from 8.5 to 10.9, whilst the minima are far below the threshold of instruments used.

123.048 **List of stars.**
Peremennye Zvezdy, Vol. 19, 565 - 570 (1975). In Russian.

This list contains all variables in Peremennye Zvezdy, Vol. 19, ordered with respect to their constellation.

123.049 **Photoelectric observations of Pleione (BU Tau) from August 1973 to March 1975.**
A. S. Sharov, V. M. Lyutyj.
Astron. Tsirk., No. 872, p. 1 - 2 (1975). In Russian.

123.050 **Two new variables in the region of M31.** A. S. Sharov, A. K. Alksnis.
Astron. Tsirk., No. 873, p. 3 - 4 (1975). In Russian.

123.051 **Luminosities of classical cepheids and the value of the factor "p".** G. R. Ivanov.
Astron. Tsirk., No. 875, p. 3 - 4 (1975). In Russian.

123.052 **New variable star of o Ceti type.** E. N. Makarenko.
Astron. Tsirk., No. 877, p. 6 - 7 (1975). In Russian.

123.053 **Preliminary designations of SVS.**
Astron. Tsirk., No. 877, p. 8 (1975). In Russian.

123.054 **CI Cygni.** S. Tapia.
IAU Circ., Nos. 2804, 2825 (1975).

123.055 **R Coronae Borealis.**
IAU Circ., No. 2854 (1975).

123.056 **R Coronae Borealis.**

IAU Circ., No. 2859 (1975).

123.057 **R Coronae Borealis.**
IAU Circ., No. 2873 (1975).

123.058 **CI Cygni.**
IAU Circ., No. 2881 (1975).

123.059 **Mitteilungen über Veränderliche der Bamberger Liste. Neue Veränderliche Sterne am Südhimmel.**
W. Strohmeier, R. Knigge.
Veröff. Remeis-Sternw. Bamberg, Astron. Inst. Univ. Erlangen-Nürnberg, Vol. 10, No. 116, 23 pp. (1975). – Concerning BV 1642 - BV 1719.

123.060 **Information on photoelectric observations of variable stars deposited at the Odessa Astronomical Observatory.** V. P. Tsesevich, E. N. Makarenko.
Astron. Tsirk., No. 883, p. 7 - 8 (1975). In Russian.

123.061 **On three variables in Orion.** V. P. Tsesevich.
Astron. Tsirk., No. 888, p. 2 - 4 (1975). In Russian.

123.062 **New light elements of three variable stars.**
S. Yu. Shugarov.
Astron. Tsirk., No. 888, p. 7 (1975). In Russian.

123.063 **RX Cassiopeiae. An appeal to the observers.**
D. Ya. Martynov.
Astron. Tsirk., No. 891, p. 4 - 5 (1975). In Russian.

123.064 **The moments of light maxima of 5 Mira Ceti-type stars in Hydra.** B. N. Irkaev.
Astron. Tsirk., No. 891, p. 5 - 6 (1975). In Russian.

123.065 **Die veränderlichen Sterne der nördlichen Milchstraße. Teil XVI.**
H. Busch, K. Häußler, C. Thänert.
Veröff. Sternw. Sonneberg, Vol. 8, (No.4), 177 - 242 (1974).
The present paper contains the derivation of the elements of the light variations of variables which have been discovered since 1963 on plates of the Sonneberg 40-cm-astrographs in the fields β Herculis and κ Lyrae. On account of the rather small galactic longitudes and moderate galactic latitudes of both fields the line of sight approaches the inner parts of the galactic halo. Therefore most of the objects belong to population II.

123.066 **Photographische Beobachtungen von R Coronae Borealis 1963 bis 1973.** V. Novotný.
MVS, *Sonneberg,* Vol. 7, 24 - 28 (1975).

123.067 **Visuelle Beobachtungen von R Coronae Borealis 1972 bis 1974.** H.-J. Blasberg.
MVS, *Sonneberg,* Vol. 7, 30 - 31 (1975).

123.068 **Maxima von Mirasternen.** H.-J. Blasberg.
MVS, *Sonneberg,* Vol. 7, 31 (1975).

123.069 **Zwei Normalmaxima von XZ Cygni.**
H.-J. Blasberg.
MVS, *Sonneberg,* Vol. 7, 31(1975).

123.070 **Beobachtungsergebnisse des Arbeitskreises "Veränderliche Sterne" im Kulturbund der DDR.**
Teil II. H. Busch (Compiler).
MVS, *Sonneberg,* Vol. 7, 32 - 36 (1975).
Minima of eclipsing variables; Maxima of RR Lyr- and δ Cep stars; Maxima of Mira stars; Semi- and aperiodic stars.

123.071 **Photographische Schätzungen von TT Aquilae.**
M. Werner.
MVS, *Sonneberg,* Vol. 7, 36 (1975).

123.072 **Photographische Maxima von Mirasternen.**
MVS, *Sonneberg,* Vol. 7, 37 (1975).

123.073 **Visuelle Beobachtungen von Mirasternen und R Sct.**
P. Ahnert.
MVS, *Sonneberg,* Vol. 7, 38 (1975).

123.074 **Bearbeitung von 70 Veränderlichen am Südhimmel.** (Feld η Arae, Teil V). H. Gessner.
MVS, *Sonneberg,* Vol. 7, 43 - 44 (1975). – Concerning S5941-S6010.

123.075 **Variable stars in the globular cluster NGC 5286.**
C. R. Fourcade, J. R. Laborde, A. A. Puch, J. R. Colazo, J. C. Arias.
Inform. Bull. Variable Stars, (I.A.U. Commission 27), Konkoly Obs., Budapest, No. 1077, 3 pp. (1975).

123.076 **Photoelectric observations of VZ Cnc (HD 73857).**
G. Hahn, G. Lustig, H. Jasicek, G. Klement.
Inform. Bull. Variable Stars, (I.A.U. Commission 27), Konkoly Obs., Budapest, No. 1078 (1975).

123.077 **Nova-ähnliches Verhalten von CI Cygni.**
J. Bauer.
BAV Rundbrief, 24. Jahrgang, p. 45 - 49 (1975).

123.078 **R Coronae Borealis.**
M. Hapgood, J. E. Isles.
British Astron. Ass. Circ., No. 569 (1975).

123.079 **Variable star notes.** J. A. Mattei.
Journ. Roy. Astron. Soc. Canada, Vol. 69, 149 - 152, 201 - 204, 261 - 264, 319 - 322 (1975).

The spectrum of the bright variable A-1 in M 31.
See Abstr. 114.323.

Untersuchungen an HDE 245770, Kandidat für die Röntgenquelle A 0535+26. See Abstr. 142.243.

Errata

123.901 **Erratum: "Bright variable stars in the Andromeda M31 and Triangulum M33 nebulae"** [Peremennye Zvezdy, Vol. 19, 3 - 17 (1973)]. In Russian.
A. S. Sharov.
Peremennye Zvezdy, Vol. 19, 403 (1974).

124 Novae

124.001 **Preeruption light curves of novae.**
E. L. Robinson.
Astron. Journ., Vol. 80, 515 - 524 (1975).
A catalogue of all published preeruption photometry of novae is presented, and preeruption light curves are constructed for 11 of the stars. It is found that (1) with the possible exception of BT Mon, the preeruption and posteruption magnitudes are the same for all 18 stars for which both magnitudes are known. (2) Twelve stars have well-defined preeruption light curves. (3) The preeruption and posteruption light curves of one star, V446 Her, are different. The prenova varied irregularly by almost 4 mag. The postnova does not vary by more than 0.4.

124.002 **Novae.**
V. P. Arhipova (*Arkhipova*), E. R. Mustel.
IAU Symposium No. 67, (see 012.001), p. 305 - 333 (1975).
Review paper.

124.003 **Towards a realistic model for nova envelopes.**
M. Friedjung.
IAU Symposium No. 67, (see 012.001), p. 335 - 338 (1975).
Arguments are given, suggesting that the most probable type of a model is one in which ejection continues after maximum light in the visual region, but with rapid time variations.

124.004 **On the origin of nova-like binary systems.**
V. G. Gorbatsky (*Gorbatskij*).
IAU Symposium No. 67, (see 012.001), p. 369 (1975). – Abstract.

124.005 **X-ray nova flares.**
P. R. Amnuel (*Amnuehl'*), O. H. Guseinov (*O. Kh. Gusejnov*).
IAU Symposium No. 67, (see 012.001), p. 501 - 508 (1975).

124.006 **The very slow nova He 3-558.**
K. G. Henize, W. Liller.
Astrophys. Journ., Vol. 200, 694 - 697 (1975).
The emission-line object He 3-558 has been found to be a very slow nova which rose to maximum brightness in 1948 and which is probably still fading. The authors describe its spectrum and light curve and derive a distance of 3.5 kpc. No sign of an earlier outburst was found in the Harvard collection of photographs.

124.007 **Ultraviolet radiation from novae.**
W. M. Sparks, S. G. Starrfield, J. W. Truran.
Bull. American Astron. Soc., Vol. 7, 405 - 406 (1975). – Abstr. AAS.

124.008 **Is a dwarf nova really a "dwarf" nova?**
W. M. Sparks, S. Starrfield.
19th Colloquium International Astrophys. Liège 1974, (see 012.016), p. 407 - 412 (1975).
Since it is observed that the constituents of dwarf and common novae binary systems are similar, it has been commonly assumed that the mechanisms are the same varying only in the intensity of the outburst. The authors' published work shows that a thermonuclear runaway in the hydrogen-rich envelope of a carbon-oxygen white dwarf is responsible for the common nova outburst as long as the CNO nuclei are enhanced. They report here on less energetic thermonuclear runaways which ejects no material and produce a rapid luminosity rise that levels off on a short time scale. However, the luminosity increase is much too larger for a dwarf nova. Several suggestions for reducing the luminosity peak are presented.

124.009 **Accretion and the nova outburst.**
S. G. Starrfield, W. M. Sparks, J. W. Truran.
19th Colloquium International Astrophys. Liège 1974, (see 012.016), p. 413 - 424 (1975).
In previous work the authors studied thermonuclear runaways in the hydrogen rich envelopes of carbon-oxygen white dwarfs and showed that these models reproduced the gross features of the nova outburst. Now new evolutionary sequences have been computed that include the infalling material and allow the initial envelope to be out of equilibrium. These models also reproduce the gross features of the nova outburst.

124.010 **An extremely carbon enhanced "nova" model.**
S. G. Starrfield, W. M. Sparks, J. W. Truran.
19th Colloquium International Astrophys. Liège 1974, (see 012.016), p. 425 - 429 (1975).
Expanding upon the authors' earlier work, Hoyle and Clayton have suggested that a thermonuclear runaway in a white dwarf envelope that consists of equal numbers of protons and ^{12}C nuclei will produce the *s* or *r*-process elements. The present authors have studied such a runaway and find that these initial conditions cause a "super" nova outburst that does not result in any *s* or *r*-process nucleosynthesis. However, the model is very interesting in that it reaches peak temperatures of 1.6×10^9 K and peak burning rates exceeding 2×10^{23} erg/gm/sec. A shock forms and ejects 10^{29} grams moving with speeds up to 60,000 km/sec. The peak bolometric magnitude is $-21\overset{m}{.}2$ and this model will also produce a γ-ray burst.

124.011 **Observations of some novae in M31 discovered by Rosino.** A. S. Sharov, A. K. Alksnis.
Astron. Tsirk., No. 869, p. 3 - 4 (1975). In Russian.

124.012 **Novae in M31 in 1973.**
A. S. Sharov, A. K. Alksnis.
Astron. Tsirk., No. 869, p. 4 - 6 (1975). In Russian.

124.013 **Remarks on the medium-band photometry of novae.** B. Yao, J. Mo.
Acta Astron. Sinica, Vol. 16, 81 - 87 (1975). In Chinese.

124.014 **Novae (1). Kenmerken van hun lichtkrommen en spectra.** G. W. E. Beekman.
Zenit, Vol. 2, 336 - 339 (1975).

124.015 **Helle Novae in unserem Jahrhundert.**
R. Lukas.
SuW, Vol. 14, 339 - 340 (1975).

124.016 **Novae (2).** T. de Groot.
Zenit, Vol. 2, 376 - 377 (1975).

124.017 **The orbital period of U Gem.** J. A. Bailey.
Observatory, Vol. 95, 174 - 176 (1975).
Visual observations of the dwarf nova U Geminorum have been used to derive values for the orbital period and time of eclipse in the nineteenth century. The results are in good agreement with the most recent linear elements and do not support previously-reported large rates of period increase.

124.018 **The effects on emission-line profiles of novae due to variations in thermal broadening and excitation**

conditions. D. R. Bochonko.
Journ. Roy. Astron. Soc. Canada, Vol. 69, 255 (1975).
Abstr. CAS.

124.019 **The rate of decline from dwarf nova outbursts.** J. Bailey.
Journ. British Astron. Ass., Vol. 86, 30 - 32 (1975).

124.020 **Nova infrared curves and the growth of dust.** D. D. Clayton, F. Hoyle, N. C. Wickramasinghe.
Bull. American Astron. Soc., Vol. 7, 553 (1975). – Abstr. AAS.

124.021 **Novae (3). Voortekenen van de eruptie.** G. W. E. Beekman.
Zenit, Vol. 2, 416 - 418 (1975).

124.022 **Nova Aquilae (V1301 Aql); Nova Scuti (V373 Sct); Nova Cygni (V1500 Cyg).** J. A. Mattei.
Journ. Roy. Astron. Soc. Canada, Vol. 69, 319 - 321 (1975).

124.023 **Eruptive variable in Aquila.**
IAU Circ., No. 2795 (1975).

124.024 **V436 Centauri.** B. Warner.
IAU Circ., No. 2795 (1975).

124.025 **Eruptive variable in Aquila.** A. M. van Genderen, G. M. Uiterwaal, G. M. Hurst.
IAU Circ., No. 2802 (1975).

124.026 **Discovery of a nova or eruptive variable with the coordinates $\alpha = 19^h15^m26^s.98$, $\delta = +04°41'44''.3$ (1950).**
Astron. Tsirk., No. 878, p. 7 (1975). In Russian.

The coronal line spectrum of AS 295.
See Abstr. 114.351.

Light variations of TT Ari. See Abstr. 117.064.

Observations of rapid blue variables–XVI. VZ Sculptoris. See Abstr. 121.020.

U Geminorum-type stars. See Abstr. 122.039.

VW Hydri, a U Geminorum-type star with periodic variations during outbursts. See Abstr. 122.040.

Observations of rapid blue variables – XVII. V436 Centauri. See Abstr. 122.085.

Progressi nell'interpretazione e studio delle stelle variabili. XX - Dalle stelle novae alle variabili simbiotiche.
See Abstr. 122.220.

124.100 **Nova Delphini 1967**

Nova Delphini 1967. I. Identification of lines and measurements of radial velocities. E. Bartl, E. Szumiejko.
Acta Astron., Vol. 25, 265 - 303 (1975).

The spectral development of the nova indicates that there were two separate and distinct phases of the outburst. Phase I represents spectra obtained before December 12, 1967, and phase II – remaining spectra. The spectrum in phase I shows lines which are normal for the early stages of a nova, and smoothly changing line profiles and velocities. The spectrum in phase II is characterized by equally rapid and irregularly changing multiple-component spectral line profiles. The observed phases of nova Delphini are difficult to compare with the generally accepted model phases of a typical nova star.

La distance de la nova Delphini 1967.
I. Malakpur.
Astrophys. Space Sci., Vol. 38, 403 - 406 (1975).

The good agreement found between the observed flux of radio emission of the nova Delphini 1967 and that calculated theoretically enables the author to deduce the angular diameter of the nova's envelope. This result, when combined with the known linear diameter of the envelope, gives a distance of the order of 885 pc for nova Delphini.

On the structure of the envelope of nova Delphini 1967 in the nebular stage. R. Tylenda.
IAU Symposium No. 67, (see 012.001), p. 339 - 344 (1975).

The analysis of line profiles in the spectrum of nova Del 1967 in the nebular stage shows that the polar axis of the envelope is longer than the equatorial one. The polar caps are probably cooler and thinner than the equatorial rings.

Nova Delphini 1967: radial velocities and excitation potentials of lines.
E. Szumiejko, T. Jarzębowski, E. Bartl.
IAU Symposium No. 67, (see 012.001), p. 345 - 346 (1975).

Radial velocities were derived from absorption lines of various excitation potentials; there seems to exist a small dependence on this potential.

Observations of nova Delphini 1967.
N. N. Kiselev.
Peremennye Zvezdy, Prilozhenie, Vol. 2, 61 - 62 (1974). In Russian.

124.101 **Nova Cephei 1971**

Spectroscopic observations of nova Cephei 1971.
L. Rosino.
IAU Symposium No. 67, (see 012.001), p. 347 - 354 (1975).

124.102 **Nova Herculis 1934**

The structure of DQ Herculis. J. I. Katz.
Astrophys. Journ., Vol. 200, 298 - 305 (1975).

The old nova (1934) DQ Her is a short-period eclipsing binary which also has a regular 71 s oscillation. A reexamination of the data shows no evidence for change in the orbital period. The 71^s period is extraordinarily stable, with $|1/\dot{P}| \approx 10^{12}$. In analogy with pulsars, this stability indicates that rigid-body rotation is the timing mechanism. A white dwarf with a crystalline core, coupled to the surface by a magnetic field, explains both the timing and the luminosity modulation. The dominant source of optical light is reradiated ultraviolet from the white dwarf. The combination of reradiation by structures rotating with the binary orbit and by matter with a constant configuration gives rise to two terms at differing frequencies whose interference leads to the observed phase shift at eclipse.

Luminosity change in DQ Herculis.
M. R. Nelson, E. C. Olson.
Bull. American Astron. Soc., Vol. 7, 509 (1975). – Abstr. AAS.

124.103 **Nova Scuti 1975**

Le spectre de nova Scuti 1975.
C. Fehrenbach, Y. Andrillat.
Comptes Rendus Acad. Sci. Paris, Sér. B, Vol. 281, 169 - 172 (1975).
Une vingtaine de spectres ont été obtenus jusqu'à 1.13μ: une dizaine d'éléments correspondant à de faibles potentiels d'ionisation ont été identifiés. Dans l'infrarouge, les émissions sont très nombreuses et intenses.

Nova Scuti 1975. A. D. Mallama, D. Wereb, T. Hishikura, K. Tsuchiya, H. Mochizuki, M. Koishikawa, J. Gallagher.
IAU Circ., No. 2795 (1975).

Nova Scuti 1975.
Y. Andrillat, C. Fehrenbach, J. M. Vreux, O. L. Hansen, R. Nishi, T. Harada, K. Osada, Y. Miyamoto, M. Brewster, D. Carmichael, P. Guida, J. Lukas, M. McGrath.
IAU Circ., No. 2798 (1975).

Nova Scuti 1975. Y. Kuwano, M. Honda, J. E. Bortle, H. Huth, M. Jacques, G. M. Hurst, G. M. Uiterwaal, A. M. van Genderen, P. A. Wehinger, S. Wyckoff.
IAU Circ., No. 2801 (1975).

Nova Scuti 1975. G. W. Erickson, H. M. Maitzen, W. Schlosser.
IAU Circ., No. 2803 (1975).

Nova Scuti 1975.
IAU Circ., No. 2811 (1975).

Nova Scuti 1975.
IAU Circ., No. 2821 (1975).

Nova Scuti 1975.
IAU Circ., No. 2835 (1975).

Nova Scuti 1975. P. Chen.
IAU Circ., No. 2840 (1975).

Nova Scuti 1975.
IAU Circ., No. 2858 (1975).

V373 Scuti (nova Scuti 1975).
IAU Circ., No. 2870 (1975).

Spectroscopic observations of nova Scuti 1975.
H. W. Duerbeck, W. C. Seitter.
Inform. Bull. Variable Stars, (I.A.U. Commission 27), Konkoly Obs., Budapest, No. 1026, 3 pp. (1975).

Visuelle Beobachtungen der nova Scuti 1975.
D. Böhme.
MVS, *Sonneberg,* Vol. 7, 44 (1975).

Optical identification of nova Scuti 1975.
M. I. Raff, J. Thorstensen.
Publ. Astron. Soc. Pacific, Vol. 87, 593 - 594, with a correction, p. 836 (1975).
The authors have identified nova Scuti 1975 with a faint star on the Palomar Sky Survey. Estimates of blue and red magnitudes are given for both the nova and suggested prenova.

Die Nova Scuti 1975. R. Lukas.
SuW, Vol. 14, 372 (1975).

Nova Scuti 1975. G. W. E. Beekman.
Zenit, Vol. 2, 272 - 273 (1975).

124.104 **Nova Cygni 1975**

Nova Cygni 1975.
A.F.O.E.V. Bull., Vol. 9, 93 - 94 (1975).

Nova Cygni 1975.
H. Lindgren, L. Lindegren.
Astron. Tidssk., Årg. 8, 137 - 138 (1975).

Nova Cygni 1975 = V1500 Cygni. I.
B. V. Kukarkin, P. N. Kholopov.
Astron. Tsirk., No. 889, p. 1 - 3 (1975). In Russian.

Nova Cygni 1975 = V1500 Cygni. II.
B. V. Kukarkin, P. N. Kholopov.
Astron. Tsirk., No. 889, p. 3 - 6 (1975). In Russian.

Position of nova Cygni 1975.
A. M. Lozinskij, M. O. Sulejman.
Astron. Tsirk., No. 889, p. 6 (1975). In Russian.

Coordinates of N Cyg 1975.
M. I. Kibardina, N. G. Rizvanov, L. A. Urasin.
Astron. Tsirk., No. 889, p. 7 (1975). In Russian.

Observations of nova Cygni 1975.
S. K. Vsekhsvyatskij.
Astron. Tsirk., No. 889, p. 7 (1975). In Russian.

Prediscovery observations of nova Cygni 1975.
Z. Alksne, I. Platais.
Astron. Tsirk., No. 889, p. 8 (1975). In Russian.

Photoelectric observations of nova Cygni 1975 (V1500 Cyg). D. Ya. Martynov.
Astron. Tsirk., No. 890, p. 1 (1975). In Russian.

On the determination of the brightness and coordinates of nova (or supernova) 1975. R. I. Kiladze.
Astron. Tsirk., No. 890, p. 2 - 3 (1975). In Russian.

The spectral development of nova Cygni 1975.
Yu. V. Borisov.
Astron. Tsirk., No. 890, p. 3 - 5 (1975). In Russian.

Coordinates of nova Cyg 1975.
N. M. Evstigneeva, Yu. A. Shokin.
Astron. Tsirk., No. 891, p. 1 (1975). In Russian.

Nova Cygni 1975 (V 1500 Cyg). E. Heiser.
BAV Rundbrief, 24. Jahrgang, p. 41 - 44 (1975).

Erste Beobachtungen der Nova Cygni 1975 an der Schwäbischen Volkssternwarte Stuttgart. H. Marx.
BAV Rundbrief, 24. Jahrgang, p. 44 - 45 (1975).

Nova Cygni 1975.
British Astron. Ass. Circ., Nos. 566, 567 (1975).

ubvy observations of nova Cyg 1975.
G. W. Lockwood, R. L. Millis.
Bull. American Astron. Soc., Vol. 7, 508 (1975). – Abstr. AAS.

Absolute spectrophotometry of nova Cygni 1975.
J. S. Neff, D. A. Ketelsen, V. V. Smith.
Bull. American Astron. Soc., Vol. 7, 508 (1975). – Abstr. AAS.

Infrared photometry and the nature of nova Cygni 1975. J. S. Gallagher, E. P. Ney.
Bull. American Astron. Soc., Vol. 7, 508 - 509 (1975).
Abstr. AAS.

Rapid variations in emission line strengths in the early stages of nova Cygni 1975. W. Weller, S. Jeffers.
Bull. American Astron. Soc., Vol. 7, 509 (1975). – Abstr. AAS.

Nova Cygni: exceptional nova or novel exception?
S. G. Starrfield, J. W. Truran, J. S. Gallagher, W. M. Sparks, P. Strittmatter, H. M. van Horn.
Bull. American Astron. Soc., Vol. 7, 509 (1975). – Abstr. AAS.

SAS-3 observations of nova Cygni 1975.
W. H. G. Lewin, J. Hoffman, and the SAS-3 Group.
Bull. American Astron. Soc., Vol. 7, 524 (1975). – Abstr. AAS.

Una stella nova eccezionale: la nova Cygni 1975.
P. Tempesti.
Coelum, Vol. 43, 185 - 193 (1975).

Nova Cygni 1975. F. Bònoli, L. Rosino.
Coelum, Vol. 43, 221 - 223 (1975).

Le spectre de nova Cygni 1975 au voisinage de son maximum d'éclat. C. Fehrenbach, Y. Andrillat.
Comptes Rendus Acad. Sci. Paris, Sér. B, Vol. 281, 365 - 368 (1975).

Les rapides variations d'éclat et les caractéristiques spectrales montrent que nova Cyg 1975 est une nova très rapide. Nous avons obtenu un spectre dans la phase pré-maximale, très mal connue. Il est caractérisé par l'absence de raies, exceptées quelques larges et diffuses absorptions à peine visibles. Après le maximum, les absorptions ne présentent aucune structure mais les émissions, larges, montrent plusieurs composantes.

Nova Cygni 1975.
IAU Circ., Nos. 2826, 2827, 2828, 2829, 2830, 2832, 2833, 2834, 2837 (1975).

The first announcement on the discovery of the nova was cabled by *M. Honda.* In the above cited Circulars independent discoveries, precise positions, prediscovery, visual, photoelectric, spectroscopic, radio and infrared observations from many observers all over the world are reported.

Nova Cygni 1975. C. de Vegt, R. H. Koch, C. W. Ambruster, A. B. Hull, J. C. Kemp, R. J. Rudy.
IAU Circ., No. 2837 (1975).

Nova Cygni 1975. T. E. Margrave, J. H. Doolittle, H. French, R. H. Koch, C. W. Ambruster, A. B. Hull, G. de Vaucouleurs, J. Tomkin, B. Campbell, S. Jeffers, W. Weller.
IAU Circ., No. 2839 (1975).

Nova Cygni 1975.
W. Gieren, W. Seggewiss, P. Chen, G. R. Gilbert, P. Tempesti.
IAU Circ., No. 2842 (1975).

Nova Cygni 1975.
D. H. Beattie, T. J. Lee, P. M. Williams.
IAU Circ., No. 2846 (1975).

Nova Cygni 1975.
W. Liller, C. Y. Shao, B. Mayer, P. Garnavich.
IAU Circ., No. 2848 (1975).

Nova Cygni 1975.
J. E. Wink, A. von Kap-herr, W. J. Altenhoff.
IAU Circ., No. 2851 (1975).

Nova Cygni 1975.
R. M. Hjellming, T. E. Margrave.
IAU Circ., No. 2853 (1975).

Nova Cygni 1975.
C. Fehrenbach, Y. Andrillat, R. S. Stobie, Purkins, Pickup, O. Citterio, G. Conti, P. Di Benedetto, E. G. Tanzi.
IAU Circ., Nos. 2857, 2858 (1975).

Nova Cygni 1975. P. Kuan.
IAU Circ., No. 2858 (1975).

V1500 Cygni (nova Cygni 1975).
O. Suyarkova, P. A. Whitelock, P. R. Jorden, C. Sanchez Magro.
IAU Circ., No. 2864 (1975).

V1500 Cygni. N. N. Kiselev.
IAU Circ., No. 2873 (1975).

V1500 Cygni.
IAU Circ., No. 2885 (1975).

V1500 Cygni.
P. C. Gregory, W. H. McCutcheon, P. Tempesti.
IAU Circ., No. 2892 (1975).

A short period light variation in nova Cygni 1975.
P. Tempesti.
Inform. Bull. Variable Stars, (I.A.U. Commission 27), Konkoly Obs., Budapetst, No. 1052, 3 pp., with a correction No. 1057 (1975).

Short period variability of nova Cygni 1975.
I. Semeniuk.
Inform. Bull. Variable Stars, (I.A.U. Commission 27), Konkoly Obs., Budapest, No. 1058, 2 pp. (1975).

Observations of nova Cygni 1975 at the Torun Observatory. A. Woszczyk, S. Krawczyk, A. Strobel.
Inform. Bull. Variable Stars, (I.A.U. Commission 27), Konkoly Obs., Budapest, No. 1072, 6 pp. (1975).

UBV observations of nova Cygni 1975.
H. Lindgren, L. Lindegren.
Inform. Bull. Variable Stars, (I.A.U. Commission 27), Konkoly Obs., Budapest, No. 1074, 3 pp. (1975).

A pre-discovery observation of nova Cygni 1975.
A. T. Blackwell, P. A. Feldman, A. A. Griffin, I. Halliday.
Journ. Roy. Astron. Soc. Canada, Vol. 69, 241 (1975).

Nova Cygni 1975. J. C. Bennett.
Monthly Notes Astron. Soc. Southern Africa, Vol. 34, 125 - 126 (1975).

Nova Cygni 1975 im Maximum. G. A. Richter.
MVS, *Sonneberg,* Vol. 7, 46 (1975).

Nova in Cygnus of record brightness.
L. Lindegren, H. Lindgren.
Nature, Vol. 258, 501 - 502 (1975). – Letter.

Die Nova Cygni 1975.
Orion, 33. Jahrgang, p. 142 - 143 (1975).

Spektralaufnahmen der Nova Cygni 1975.
Orion, 33. Jahrgang, p. 212 (1975).

Accurate position of nova Cygni 1975.
S. Isobe, H. Kosai, H. Hara, R. Fukaya, Y. Adachi.
Publ. Astron. Soc. Japan, Vol. 27, 619 - 621 (1975).

An accurate position of nova Cygni 1975 is determined by a meridian circle as R.A. (1950.0) = $21^h 09^m 52^s.822$, Decl. (1950.0) = $+47° 56' 40''.87$. No prenova brighter than 21 mag is found at this position.

The region of nova Cygni 1975 on the Palomar Sky Survey. W. R. Beardsley, M. W. King, J. L. Russell, J. W. Stein. Publ. Astron. Soc. Pacific, Vol. 87, 943 - 944 (1975).

Careful superposition of a blue Palomar Sky Survey print onto a sectored photograph of nova Cygni 1975 obtained with the Thaw 30-inch (76-cm) refractor at the Allegheny Observatory decisively confirms the fact that no star brighter than magnitude 21 appears on the Sky Survey print at that position.

Photometric behaviour of the Nova Cygni 1975. P. Mayer. Říše hvězd, Vol. 56, 225 - 227 (1975). In Czech.

Brightest nova in 33 years. Sky Telescope, Vol. 50, 229, 234 (1975).

More about nova Cygni. Sky Telescope, Vol. 50, 282 - 284 (1975).

Beobachtungen der Nova Cygni. SuW, Vol. 14, 340 - 341 (1975).

Nova Cygni 1975. A. Hänel. SuW, Vol. 14, 411 - 412 (1975).

Nova in Cygnus. D. Ya. Martynov. Zemlya i Vselennaya, 1975, No. 5, p. 96. In Russian.

Nova Cygni 1975. C. de Jager, F. Reijmerink. Zenit, Vol. 2, 324 (1975).

De Zwaan roert zijn staart. J. W. Schippers. Zenit, Vol. 2, 378 - 379 (1975).

Photometry of the eclipsing binary U Ophiuchi and of nova Cygni 1975. See Abstr. 121.099.

124.105 Nova Sagittarii 1974

Nova Sagittarii 1974. A. Sh. Khatisov. Astron. Tsirk., No. 852, p. 7 - 8 (1975). In Russian.

124.106 Nova Persei 1974

Observations of nova Persei 1974 at the Abastumani Observatory. D. M. Depadel'. Astron. Tsirk., No. 855, p. 7 - 8 (1975). In Russian.

Nova Persei 1974. N. V. Narizhnaya, N. N. Kiselev. Astron. Tsirk., No. 888, p. 1 - 2 (1975). In Russian.

Nova Persei 1974. IAU Circ., No. 2837 (1975).

Nova Persei 1974. IAU Circ., No. 2859 (1975).

Observations de nova Per 1974. L. A. Ourassine (*Urasin*), I. A. Ourassina (*Urasina*). Inform. Bull. Variable Stars, (I.A.U. Commission 27), Konkoly Obs., Budapest, No. 1038 (1975).

124.107 Nova Aurigae 1964

Nova Aurigae 1964. N. Sanduleak. Inform. Bull. Variable Stars, (I.A.U. Commission 27), Konkoly Obs., Budapest, No. 1011 (1975).

Nova Aurigae 1964. H. Gessner. Inform. Bull. Variable Stars, (I.A.U. Commission 27), Konkoly Obs., Budapest, No. 1042 (1975).

Lichtkurve der Nova Aurigae 1964. H. Geßner. MVS, *Sonneberg,* Vol. 7, 45 (1975).

124.108 Nova Scuti 1970

Narrow emission lines in the spectrum of nova Scuti 1970. S. Kikuchi, J. Smoliński. Acta Astron., Vol. 25, 305 - 315 (1975).

The spectra obtained about two weeks after the light maximum have been analysed. Narrow emission lines superposed on the broad emission components were observed and supposed to be formed in the condensations within the expanding shell which contributed to the formation of broad emission and absorption features. Some physical conditions for this nova and for condensations were estimated. The distance to the nova is estimated to be 2.0 kpc. The velocity of expansion determined from the absorption lines is 1130 km s^{-1}. The mass of the ejected shell is estimated to more than $10^{-4} M_\odot$

124.109 Nova Sagitarii 1975

Possible nova in Sagittarius. Y. Kuwano. IAU Circ., Nos. 2805, 2808 (1975).

Possible nova in Sagittarius. IAU Circ., No. 2808 (1975).

Nova Sagittarii 1975. IAU Circ., No. 2811 (1975).

Nova Sagittarii 1975. C. Y. Shao, G. Schwartz, P. Chen, L. Loucka, T. Ake, L. Krumenaker. IAU Circ., No. 2813 (1975).

Nova Sagittarii 1975. H. Kosai, N. Kojima. IAU Circ. No. 2819 (1975).

124.110 Nova Aquilae 1975

Nova Aquilae 1975. J. E. Bortle. IAU Circ., No. 2811 (1975).

Nova Aquilae 1975. G. M. Hurst, W. Pennell. IAU Circ., No. 2821 (1975).

Nova Aquilae 1975. P. Pesch. IAU Circ., No. 2835 (1975).

Nova Aquilae 1975. IAU Circ., No. 2837 (1975).

Nova Aquilae 1975. J. Bortle. IAU Circ., No. 2861 (1975).

124.111 **Nova Monocerotis 1975**

A0620–00. Robertson.
IAU Circ., Nos. 2830, 2846 (1975).

Nova Monocerotis 1975 (= A0620–00).
J. B. Oke, J. L. Greenstein, H. French.
IAU Circ., No. 2840 (1975).

Nova Monocerotis 1975 (= A0620–00).
Sheffer, Moskalenko.
IAU Circ., No. 2846 (1975).

Nova Monocerotis 1975.
R. F. Jameson, E. L. Robinson, R. E. Nather.
IAU Circ., No. 2854 (1975).

Nova Monocerotis 1975. J. Bortle.
IAU Circ., No. 2864 (1975).

V616 Monocerotis (nova Monocerotis 1975).
J. Bortle.
IAU Circ., No. 2882 (1975).

A0620–00. See Abstr. 142.216.

A0620–00. See Abstr. 142.221.

A0620–00. See Abstr. 142.222.

125 Supernovae, Supernova Remnants

125.001 **Spectral turnovers of galactic supernova remnants.** G. A. Dulk, O. B. Slee.
Astrophys. Journ., Vol. 199, 61 - 68 (1975).

High-resolution observations of 14 galactic supernova remnants have been made at 80 and 160 MHz. The spectra of 7 to 10 of the sources exhibit a turnover at $\lesssim 100$ MHz which the authors attribute to free-free absorption taking place in the inner regions of the Galaxy. The absorption could occur in large ($\gtrsim 20$ pc), low-density ($n \approx 10\ cm^{-3}$) H II regions created by O5 to O8 stars. Alternatively, absorption could occur in cold dense clouds which are a feature of the steady-state model of the interstellar medium, or, more likely, in cold, partly ionized regions of moderate density which are a feature of the time-dependent model.

125.002 **Numerical calculation of thin shell formation in supernova remnants.** S. A. E. G. Falle.
Monthly Notices Roy. Astron. Soc., Vol. 172, 55 - 84 (1975).

Numerical calculations of the evolution of supernova remnants are carried out from the Sedov phase to the time when they merge with the interstellar medium. It is found that for some energies and densities the radiative cooling is so rapid that secondary shocks are produced. A criterion for this is derived and used as a basis for a dynamical classification of supernova remnants. The interstellar magnetic field is included in the calculations. It is found that the magnetic field may prevent secondary shocks and also that it affects the ultimate fate of the energy of the explosion. The results also suggest that it may make some remnants oblate and some prolate in a uniform field.

125.003 **Supernovae.** B. Baschek.
Mitt. Astron. Ges., No. 36, p. 33 - 44 (1975).
Review paper.

125.004 **The radio remnant of Kepler's supernova of 1604.** C. Hazard.
Vistas in Astronomy, Vol. 18, (see 012.006), 799 - 805 (1975).

125.005 **The X-ray structure of the Vela supernova remnant.** W. E. Moore, G. P. Garmire.
Astrophys. Journ., Vol. 199, 680 - 690 (1975).

The X-ray surface brightness distribution of the Vela supernova remnant is presented in three energy intervals on scales of 1.0 deg^2 and 0.44 deg^2. The 0.15–1.0 keV X-ray emission exhibits two principal regions of enhancement and a general limb brightening which may imply a shell-like source structure.

125.006 **Supernovae in binary systems.** J. C. Wheeler, M. Lecar, C. F. McKee.
Astrophys. Journ., Vol. 200, 145 - 157 (1975).

The general problem of a supernova explosion in a binary star system is studied. The authors show that to a good approximation the effect of the impinging blast wave which strips and ablates mass from the companion can be expressed in terms of a single parameter. The eccentricity imparted to the system and the ratio of final to initial semimajor axes depend on two to six parameters, depending on the level of approximation. Results for calculations of the eccentricity imparted to an initially circular system are presented. It is concluded that neutron stars can be formed in close binary systems but that the resulting orbit must have a finite eccentricity. The relevance of this result to X-ray binary systems is discussed.

125.007 **The soft X-ray spectra of Cassiopeia A and Tycho's supernova remnant.** R. W. Hill, G. A. Burginyon, F. D. Seward.
Astrophys. Journ., Vol. 200, 158 - 162 (1975).

The X-ray spectra of Tycho's supernova remnant and Cas A have been measured with good statistics. Both spectra show evidence for silicon line emission. The thermal model of Tucker and Koren has been used to fit both spectra. To obtain a satisfactory fit to the data with this model, the silicon abundance must be increased about 10 times from the usual cosmic abundance values. Neither spectrum can be fitted with exponential or power-law models without the addition of line emission. Tycho's remnant shows very little soft X-ray absorption .

125.008 **Analysis of X-ray spectra reveals possibility of enhanced silicon in three supernova remnants.**
G. A. Burginyon, R. W. Hill, F. D. Seward.
Astrophys. Journ., Vol. 200, 163 - 170 (1975).

New calculations by Tucker and Koren for X-radiation from a thin hot plasma are applied to observed X-ray spectra of the Cygnus Loop, Vela X, and Puppis A. These calculations cannot satisfactorily describe the spectra of these three old supernova remnants when evaluated at one temperature and for commonly accepted cosmic abundances.

125.009 **The diameter function for supernova remnants and their evolution.** H. Smith, Jr.
Astron. Astrophys., Vol. 41, 273 - 277 (1975).

The data on the diameter function for supernova remnants is re-examined. In contrast to previous studies, the author finds a discrepancy between the adiabatic-expansion prediction and the data. Possible explanations, including error in the surface brightness-diameter ($\Sigma - D$) relation, are examined.

125.010 **Depolarization of five supernova remnants: internal and external Faraday effects.**
T. Velusamy, M. R. Kundu.
Astron. Astrophys., Vol. 41, 307 - 314 (1975).

Depolarization in five supernova remnants Kes 69, W 44, IC 443, 3 C 10 and 3 C 58 has been investigated using the polarization measured at 2.8, 6, 11 and 21 cm with equivalent resolutions. The polarization data for certain regions in Kes 69, W 44 and IC 443 would be best interpreted if it were assumed that the depolarization is caused by an internal Faraday effect in the source. The physical parameters such as the internal and external rotation measures and the degree of inhomogeneity within the source have been derived from source models for internal depolarization.

125.011 **On a model of a type I supernova close to light maximum.** Eh. R. Mustel', N. N. Chugaj.
Astron. Zhurn. Akad. Nauk SSSR, Vol. 52, 673 - 681 (1975).
In Russian. English translation in Soviet Astron., Vol. 19, No.4.

Some photometric and spectrophotometric data on type I supernovae (SN–I) are analyzed. A comparison of colour T_c with spectrophotometric temperatures T for SN–I 1972e permitted to conclude that at light maximum and before it $T_c \approx T$, whereas soon after light maximum $T_c < T$. The behaviour of the "photospheric" radius R_p of SN–I is investigated on the basis of the assumed dependence of T on time. It is found that at light maximum the radius R_p is increasing with time approximately linearly, the average velocity V_p is about 5000 km/sec; the greatest magnitude of R_p is attained in 30–35 days after light maximum. The velocity of expansion of the "photosphere" is approximately two times less than the average velocity of the ejected envelope of SN–I. An extrapolation of the value of $R_p(t)$ to $R_p \approx 0$ for SN–I 1972e and

SN–I 1970j gives for the beginning of the explosion the moments -25^d and -16^d, respectively. The behaviour of the bolometric luminosity of SN–I around light maximum is studied. Computations show that the bolometric maximum of SN–I is attained considerably earlier than the optical maximum.

125.012 **The age of HB 21 and PSR 2021 + 51.**
E. M. Jones.
Astron. Astrophys., Vol. 42, 147 - 150 (1975).
The properties of the observed H I shell around supernova remnant HB 21 are used to derive an age between 0.26 and 0.79 million years. Ages in this range are inconsistent with properties of pulsar PSR 2021 + 51 which has been proposed as the neutron-star product of the HB 21 explosion. HB 21 may show marginal buoyant displacement.

125.013 **An anomaly in the flux of Cassiopeia A at 38 MHz.**
W. C. Erickson, R. A. Perley.
Astrophys. Journ., (*Letters*), Vol. 200, L83 - L87 (1975).
The secular decrease in the flux of the supernova remnant Cassiopeia A has been measured at many radio frequencies. The authors present new observations that indicate that this flux decrease has halted at 38 MHz, and that the flux is increasing at this frequency. This may indicate that a new era of activity has begun in this object.

125.014 **The radio structure of 3 C 358, the remnant of Kepler's supernova.** R. G. Strom, J. Sutton.
Astron. Astrophys., Vol. 42, 299 - 302 (1975).
21 cm synthesis observations of 3 C 358 show the remnant to be a slightly elliptical object, roughly $3'.2 \times 4'.2$ in size with the major axis north–south. The eastern, western and southern edges are sharp while to the north the emission dies away more gradually. There is no clear indication of shell structure. The source is linearly polarized, typically by 3%. 3C358 is compared briefly with the remnant of Tycho's supernova and it is shown that despite their dissimilar structures, in terms of energy content and other physical parameters the two objects are nearly alike.

125.015 **Supernovae.** S. A. Colgate.
Astrophys. Space Sci. Library, Vol. 48, (see 012.014), 13 - 27 (1975).

125.016 **Copernicus: the spatial distribution of the X-ray emission from IC443.**
P. A. Charles, J. L. Culhane, C. G. Rapley.
Astrophys. Letters, Vol. 16, 129 - 132 (1975).
The MSSL X-ray instrumentation on the Copernicus satellite has observed the spatial structure of the supernova remnant IC443. The authors find the soft (0.5–1.4 keV) X-ray emission to be strongly correlated with the bright optical filaments in the north–east. The total intensity is consistent with previous observations of the remnant. An upper limit of $4 \cdot 10^{-11}$ erg cm^{-2} s^{-1} can also be assigned to the flux in the 0.5–1.4 keV band from the nearby pulsar PSR 0611 + 22.

125.017 **The evolution of supernova remnants. IV. The supernova ejecta.** R. A. Chevalier.
Astrophys. Journ., Vol. 200, 698 - 708 (1975).
A model for Cas A based on the properties of the fast moving knots is investigated. The knots form during or soon after the supernova explosion. They become immersed in hot gas, which heats the knots by thermal conduction provided radiative cooling by overabundant heavy elements is not too strong. The heating time scale is of the same order as or less than the age of Cas A. As a supernova remnant evolves, both the heated ejecta and the interstellar medium emit X-rays. The central X-ray source in the Cygnus Loop may be heated ejecta. The relevance of the model for other supernova remnants is considered. The effect of the expansion of a young remnant on oxygen recombination is investigated and found to be important at low densities. Such effects must be taken into account in attempting to derive abundances from X-ray lines, once they can be measured observationally.

125.018 **Detection of the [Fe XIV] 5303 Å coronal line in the supernova remnant Vela X.**
B. E. Woodgate, J. R. P. Angel, R. P. Kirshner.
Astrophys. Journ., Vol. 200, 715 - 718 (1975).
The coronal line [Fe XIV] 5303 Å has been detected in Vela X at a statistical significance above zero of 6.3 σ with a flux $(3.9 \pm 0.6) \times 10^{-12}$ ergs (cm^2 s)$^{-1}$ in a 1° diameter field. This demonstrates that the X-ray flux of Vela X is of thermal origin. The 3 σ upper limits for the [Fe XIV] 5303 Å and [Ar XIV] 4412 Å coronal line fluxes from the supernova remnant Puppis A are respectively 4×10^{-12} and 2.6×10^{-12} ergs (cm^2 s)$^{-1}$.

125.019 **Supernova explosions and the maintenance of a hot component of the interstellar medium.**
C. F. McKee, J. P. Ostriker.
Bull. American Astron. Soc., Vol. 7, 419 (1975). – Abstr. AAS.

125.020 **Detection of X-ray emission from the remnant of the supernova 1006 A.D.**
P. F. Winkler, F. N. Laird.
Bull. American Astron. Soc., Vol. 7, 456 (1975). – Abstr. AAS.

125.021 **Soft X-ray profile and spectrum of the supernova remnant IC 443.**
R. Malina, M. Lampton, S. Bowyer.
Bull. American Astron. Soc., Vol. 7, 456 (1975). – Abstr. AAS.

125.022 **Prospects for deuterium synthesis in supernova shock waves.** T. A. Weaver.
Bull. American Astron. Soc., Vol. 7, 456 - 457 (1975). – Abstr. AAS.

125.023 **Observations of supernovae: 1975a in NGC 2207 and 1975b in the Perseus cluster.**
R. P. Kirshner, H. C. Arp, J. W. Dunlap.
Bull. American Astron. Soc., Vol. 7, 457 (1975). – Abstr. AAS.

125.024 **Isothermal blast wave model of supernova remnants.**
A. Solinger, S. Rappaport, J. Buff.
Bull. American Astron. Soc., Vol. 7, 457 (1975). – Abstr. AAS.

125.025 **A simple model for the early part of type I supernova light curves.** G. Lasher.
Astrophys. Journ., Vol. 201, 194 - 201 (1975).
A simple hydrodynamic and gray radiation diffusion model is used to fit the first 40 days of the type I supernova light curve. The model requires only three input parameters: the density and mass of the envelope of the parent star and the energy which is presumed to be deposited at the center of the envelope by the collapse of the core. A density of 10^{-8} g cm^{-3}, a mass of 2 $M_\odot$, and an energy of 10^{51} ergs gives a model fitting observed values of the maximum magnitude, the duration of the light curve 1 mag below maximum, and a final velocity of the photosphere of 10^9 cm s^{-1}.

125.026 **Distribution of supernovae in spiral galaxies of type Sb–Sd.** M. Iye, K. Kodaira.
Publ. Astron. Soc. Japan, Vol. 27, 411 - 419 (1975).
The derivation of the distribution of extragalactic supernovae by Johnson and MacLeod (1963) is extensively reexamined. The authors present a revised surface number density curve based upon the data of 155 supernovae observed in spiral galaxies of type Sb–Sd during the period 1885–1973.

The resulting curve agrees well with the surface number density curve of the galactic supernova remnants derived by Kodaira (1974), except for the central region, where a slight excess of supernovae is observed relative to supernova remnants.

125.027 **Observations of nine supernova remnants at 10.6 GHz.** R. H. Becker, M. R. Kundu.
Astron. Journ., Vol. 80, 679 - 682 (1975).

Maps of total and linearly polarized intensity over nine supernova remnants as observed with a beam of 3 arcmin at 10.6 GHz are presented in this paper.

125.028 **Les supernovae.** F. R. Stephenson.
Orion, 33. Jahrgang, p. 139 - 141 (1975).

125.029 **On presupernova dynamical mass ejection by non-detonated stellar cores.**
J.-R. Buchler, T. J. Mazurek.
19th Colloquium International Astrophys. Liège 1974, (see 012.016), p. 435 - 445 (1975).

Stars in the range of 4 to 8 solar masses are known to develop degenerate carbon-oxygen cores of about 1.4 $M_\odot$. Arguments are presented why, contrary to past assumption, carbon ignition may not lead to the formation of a detonation wave. Provided that sufficient energy is liberated in the dynamic formation of a neutron star, a supernova event will result. The distended nebula is ideally suited for the production of the visible supernova light curves. Thus, carbon non-detonation may give a viable alternative evolution to supernova that avoids the impasse encountered by the detonation disrupted models.

125.030 **Supernova in an anonymous galaxy.**
L. K. Erastova.
Astron. Tsirk., No. 841, p. 4 - 5 (1974). In Russian. – α_{1950} = $12^h\ 14^m\!.8$, δ_{1950} = +24°12′.

125.031 **On the origin of the anomalous behaviour of the absorption line He I 10830 Å in the spectra of type-I supernovae.** Eh. R. Mustel'.
Astron. Tsirk., No. 846, p. 1 - 3 (1974). In Russian.

125.032 **On the stratification of elements and physical conditions in the shells of type-I supernovae.**
Eh. R. Mustel'.
Astron. Tsirk., No. 870, p. 1 - 3 (1975). In Russian.

125.033 **The distribution of supernova remnants in the Galaxy: evidence for a tunnel system.** E. M. Jones.
Astrophys. Journ., Vol. 201, 377 - 380 (1975).

Counts of supernova remnants are used to derive the supernova rate, average remnant evolution, and probability that a remnant encounters a hot, low-density tunnel in the interstellar medium. The results suggest that tunnels occupy about 80 percent of the galactic volume. The author further finds that the remnants in the inner parts of the Galaxy may be in a later stage of evolution than the average remnant in the outer parts of the Galaxy.

125.034 **Isothermal blast wave model of supernova remnants.** A. Solinger, S. Rappaport, J. Buff.
Astrophys. Journ., Vol. 201, 381 - 386 (1975).

The validity of the "adiabatic" assumption in supernova remnant calculations is examined, and the alternative extreme of an isothermal blast wave is explored. The authors conclude that, because of thermal conductivity, the large temperature gradients predicted by the adiabatic model probably are not maintained in nature. Self-similar solutions to the hydrodynamic equations for an isothermal blast wave have been found and studied. These solutions are then used to determine the relationship between X-ray observations and inferred parameters of supernova remnants. It is concluded that most parameters of supernova remnants inferred from X-ray measurements are relatively insensitive to the specifics of the blast wave model.

125.035 **A high resolution map of the supernova remnant 3C 400.2 at 610 MHz.**
W. M. Goss, S. G. Siddesh, U. J. Schwarz.
Astron. Astrophys., Vol. 43, 459 - 463 (1975).

The large supernova remnant 3C 400.2 has been mapped with the Westerbork synthesis telescope at 610 MHz (49 cm) with a resolution of 1′ X 3′. A comparison with an 11 cm map made by Willis (1973) shows that the spectral index between 49 and 11 cm is constant over most of the source. The spectral index does however steepen in the central depression.

125.036 **Hydrodynamic instabilities and the formation of radio shells in young supernovae remnants.**
L. L. Cowie.
Monthly Notices Roy. Astron. Soc., Vol. 173, 429 - 436 (1975).

Recent calculations by Gull have suggested the possibility that the radio emission in young supernovae remnants (SNRs) may be generated by the Rayleigh–Taylor instability at the interface between material ejected by the supernova (the piston), and the shocked interstellar medium. In this paper the physics of the instability is considered in more detail, and in particular the presence of a reverse shock in the piston and the effects of viscosity are fully included. Finally the predictions of the theory are compared with observation for the cases of Cas A and Tycho. The agreement is reasonable.

125.037 ***p*-process nucleosynthesis in postshock supernova envelope environments.** J. Audouze, J. W. Truran.
Astrophys. Journ., Vol. 202, 204 - 213 (1975).

Nuclear transformations giving rise to the production of the *p*-process (bypassed) elements are examined within the framework established by explosive nucleosynthesis studies. The guiding assumption is that thermonuclear processes occurring in hydrogen-rich matter in supernova envelopes following the passage of a shock wave are responsible for the formation of these nuclei. It is demonstrated that there exists a restrictive range of temperature-density conditions ($T \sim 2 \times 10^9$ K; $\rho \sim 10^4$ g cm^{-3}) for which the enhanced production of most of the *p*-process nuclei occurs in relative concentrations consistent with solar system abundances. The implications of these results for galactic nucleosynthesis are discussed.

125.038 **Can supernovae make white dwarfs?**
K. Hainebach.
Nature, Vol. 258, 268 (1975).

125.039 **Observations of eight supernova remnants at 15 GHz.** A. J. Green, J. R. Baker, T. L. Landecker.
Astron. Astrophys., Vol. 44, 187 - 194 (1975).

Eight supernova remnants have been observed with the 100-m radiotelescope. The spatial resolution is 58″arc. Maps are presented of Cassiopeia A and Taurus A in both linear polarization and total intensity, and of Tycho's SNR, 3 C 58 and W 49 B in total intensity only. Limits to the brightness temperatures of Kepler's SNR, 3 C 391 and 3 C 397 are also given.

125.040 **A magnetohydrodynamical rotational model of a supernova explosion.**
G. S. Bisnovatyj-Kogan, Yu. P. Popov, A. A. Samokhin.
In-t prikl. mat. AN SSSR. Preprint No.16 Moskva, 1975, 64 pp. In Russian. – Abstr. in Referativ. Zhurn. 51. Astron., 8.51. 568 (1975).

125.041 **The cooling wave in supernova shells.**

Eh. K. Grasberg, D. K. Nadezhin.
In-t prikl. mat. AN SSSR. Preprint No. 5. Moskva, 1975. 45 pp. In Russian. – Abstr. in Referativ. Zhurn. 51. Astron., 8.51. 569 (1975).

125.042 **On a possible mechanism of the explosion of a rotating star as a supernova.** A. S. Zentsova.
Astron. Zhurn. Akad. Nauk SSSR, Vol. 52, 905 - 910 (1975). In Russian. English translation in Soviet Astron., Vol. 19, No. 5.

On the basis of results of evolutionary calculations for rotating massive stars the mechanism of a supernova explosion is proposed which is not connected with nuclear detonation. The explosion occurs as a result of kinetic rotating energy transfer from the core of a red supergiant to shock wave energy. By assuming that the core of the red supergiant forms a Jacobi ellipsoid, the flow velocity of the slowly rotating atmosphere caused by the rotation of the core is determined. The flow of the interior parts of the atmosphere is supersonic.

125.043 **The dynamical effects of supernovae on the interstellar medium. I: Spherical shock waves in an adiabatic medium.** J. E. Dyson, P. Gulliford.
Astrophys. Space Sci., Vol. 37, 477 - 498 (1975).

An interstellar cloud suddenly overrun by a supernova blast wave experiences a very rapid increase in boundary pressure. A shock wave propagates into the cloud. As a preliminary investigation, the propagation of spherical shock waves in an adiabatic medium is studied numerically.

125.044 **408 and 5000 MHz observations of 28 new galactic supernova remnants.**
D. H. Clark, J. L. Caswell, A. J. Green.
Australian Journ. Phys.,Astrophys. Suppl. No. 37, p. 1 - 38 (1975).

High-resolution contour maps of surface brightness at 408 and 5000 MHz are presented for 28 new galactic SNRs.

125.045 **408 and 5000 MHz observations of 12 probable supernova remnants.**
J. L. Caswell, D. H. Clark, D. F. Crawford.
Australian Journ. Phys., Astrophys. Suppl. No. 37, p. 39 - 56 (1975).

Twelve radio sources which have been classified as probable SNRs in earlier studies have been observed with high resolution at both 408 and 5000 MHz. Improved values have been determined for their flux densities and spectral indices, and more details can now be seen in their brightness distributions.

125.046 **Improved 408 MHz observations of some galactic supernova remnants.**
D. H. Clark, A. J. Green, J. L. Caswell.
Australian Journ. Phys., Astrophys. Suppl. No. 37, p. 75 - 86 (1975).

High-resolution contour maps of surface brightness at 408 MHz obtained with the Molonglo radio telescope are presented for seven galactic SNRs.

125.047 **The remnants of the supernovae of AD 185 and AD 393.** D. H. Clark, F. R. Stephenson.
Observatory, Vol. 95, 190 - 195 (1975).

125.048 **Studies on supernovae.**
R. Barbon, M. Capaccioli, F. Ciatti.
Astron. Astrophys., Vol. 44, 267 - 271 (1975).

A new calibration of type I supernovae is given through the use of the Hubble diagram. A mean absolute magnitude at maximum $M_{pg} = -18.15 + 5 \log (H/100)$ is found with no clear separation between "fast" and "slow" subtypes. The radial distribution of supernovae within their parent galaxies is considered by using the distance-independent parameter $\vartheta = r/R$. The frequency distribution of the ϑ values, after a proper correction for projection is applied, indicates that about 1/3 of supernovae occur beyond the optical image of galaxies. A surface density of supernovae, consistent with the brightness distribution in the disks of spiral galaxies, is derived.

125.049 **Fe II emission and the presence of hydrogen in the envelopes of type I supernovae.** M. Friedjung.
Astron. Astrophys., Vol. 44, 431 - 436 (1975).

A study of the excitation of features due to Fe II in published tracings of spectra of SN 1972-e supports the identification of a feature by Kirshner et al. (1973), as being due to Hα. An approximate calculation indicates that the total mass of ionized hydrogen above the photosphere is of the order of 0.3 of a solar mass, if most of the emission at Hα is due to recombinations.

125.050 **On the *z*-distribution of supernova remnants.**
K. Henning, H. J. Wendker.
Astron. Astrophys., Vol. 44, 91 - 94 (1975).

Since observational selection effects against the detection of SNR towards high galactic latitudes cannot be ruled out, the authors used the existing data to determine a lower limit of detected SNR at high galactic latitudes over a significant portion of the sky and studied theoretically the dependence of the visibility of SNR on the density of the surrounding medium. These investigations lead to an improved *z*-distribution of SNR and show the dependence of the surface brightness on the gas density.

125.051 **On the presence of helium in the envelopes of type I supernovae.** Eh. R. Mustel'.
Astron. Zhurn. Akad. Nauk SSSR, Vol. 52, 1143 - 1155 (1975). In Russian. English translation in Soviet Astron., Vol. 19, No. 6.

The paper contains a discussion on the problem of the presence of absorption lines of He I in the spectra of type I supernovae (SNI). The changes in the spectrum of SNI 1972e during the period from JD 2441529 to 2441653 are subjected to a special discussion. There are strong arguments that this period was characterized by a noticeable increase of T_{uv} and, possibly, T_{opt}. The increase of T_{uv} was accompanied by strengthening of the absorptions of He I 5876, 7065 Å and simultaneously by the appearance and further strengthening of absorptions of the intense multiplets NN 3, 5 of N II. These facts as well as others confirm the presence of rather strong absorptions of He I in the spectra of SNI 1972e. The present analysis gives also a natural explanation of the absence or weakness of other absorptions of He I.

125.052 **A 10 GHz survey of the γ Cygni region.**
L. A. Higgs.
Journ. Roy. Astron. Soc. Canada, Vol. 69, 248 (1975). Abstr. CAS.

125.053 **On the light curve of 3C273.**
G. G. Fahlman, T. J. Ulrych.
Journ. Roy. Astron. Soc. Canada, Vol. 69, 256 (1975). Abstr. CAS.

125.054 **The next galactic supernova.** S. van den Bergh.
Astrophys. Space Sci., Vol. 38, 447 - 456 (1975).

It is pointed out that there is a better than even chance that a galactic supernova will occur during the lifetime of the present generation of astronomers. A brief discussion is given of a number of types of observations that could be made of such a unique event. It would be particularly important to obtain speckle interferometry of the expanding supernova. High time-resolution spectroscopy and photometry should be obtained during the early phases of the outburst.

125.055 **On the nature of supernovae.** Z. Mikulášek.
Říše hvězd, Vol. 56, 169 - 175, 190 - 194 (1975).
In Czech.

125.056 **A prediction of soft X-ray emission from type I supernovae.** G. Lasher, K. L. Chan.
Bull. American Astron. Soc., Vol. 7, 505 (1975). – Abstr. AAS.

125.057 **Is the Gum nebula an old supernova remnant ionized by ζ Pup and γ^2 Vel?** R. J. Reynolds.
Bull. American Astron. Soc., Vol. 7, 554 - 555 (1975).
Abstr. AAS.

125.058 **A Ni^{56} white dwarf hypothesis for the type I supernova light curve.** M. Leventhal, S. L. McCall.
Bull. American Astron. Soc., Vol. 7, 555 (1975). – Abstr. AAS.

125.059 **Observational support for the white dwarf, nuclear decay hypothesis for type I supernova luminosity.**
B. W. Rust, M. Leventhal, S. L. McCall.
Bull. American Astron. Soc., Vol. 7, 555 (1975). – Abstr. AAS.

125.060 **Supernova in the anonymous galaxy $\alpha = 17^h30^m$, $\delta = +43°25'$.** R. G. Mnatsakanyan.
Astron. Tsirk., No. 871, p. 1 - 3 (1975). In Russian.

125.061 **Discovery of a supernova in the anonymous galaxy with the coordinates $\alpha = 13^h59^m.5$, $\delta = +54°40'$ (1950).**
Astron. Tsirk., No. 878, p. 8 (1975). In Russian.

125.062 **What stars become supernovae?** B. M. Tinsley.
Publ. Astron. Soc. Pacific, Vol. 87, 837 - 848 (1975).

This paper assembles a variety of empirical lines of evidence on the masses and stellar population types of stars that trigger supernova (SN) explosions. Relatively secure conclusions include the following. Statistics of stellar birthrates, SN, pulsars, and SN remnants in the Galaxy show that SN II (or all SN) could arise from stars with masses $>M_s$ where $M_s \sim 4$9 to 12 $M_\odot$. Several methods of estimating the masses of stars that become white dwarfs are consistent with a lower limit, $M_s \gtrsim$ 5 $M_\odot$, so carbon detonation may indeed be avoided, although this conclusion is not secure. Studies of the properties of galaxies in which SN occur, and their distributions within galaxies, support the usual views that SN I have low-mass precursors ($\lesssim 5\ M_\odot$ and typically $\lesssim 1\ M_\odot$) and SN II have massive precursors ($\gtrsim 5\ M_\odot$); the restriction of known SN II to Sc and Sb galaxies, to date, is shown to be consistent, statistically, with massive stars in other galaxies also dying as SN II. Possible implications of the peculiarities of some SN-producing galaxies are discussed.

125.063 **Supernova in anonymous galaxy.**
J. P. Huchra.
IAU Circ., No. 2825 (1975).

125.064 **Supernovae in anonymous galaxies.**
M. Lovas, C. T. Kowal.
IAU Circ., No. 2883 (1975).

125.065 **Supernova in an anonymous galaxy.**
L. K. Erastova.
Astrofizika, Vol. 11, 561 - 562 (1975). In Russian. – English translation in Astrophysics, Vol. 11, No. 3.

Additional data on the light curve of Romano's supernova are presented. The supernova was brighter than the parent galaxy.

125.066 **Calculation of supernova light curves.**
S. W. Falk, Jr.
Thesis, Texas Univ., Austin (USA). [Available from Univ. Microfilms, Ann Arbor, Mich., *USA*. Order No. 75-4361], 170 pp. (1974).

A brief discussion of the observational data pertaining to supernova phenomena is presented, along with a summary of prior attempts to account for various of these features. The physics and numerical techniques appropriate to the treatment of coupled hydrodynamics and radiation transport are discussed. The methods of pseudo-viscosity, equilibrium photon diffusion and frequency dependent photon transport in matter in local thermodynamic equilibrium are used to calculate the passage of strong shock waves through extended red giant-like stellar envelopes, in an attempt to account for the main features of observed supernovae. Models are presented which show good agreement with the general characteristics of observed supernova events.

125.067 **Spectrophotometry of supernovae and their remnants.** R. P. Kirshner.
Thesis, California Inst. of Technol., Pasadena (USA). [Available from Univ. Microfilms, Ann Arbor, Mich., USA. Order No. 75-9530], 149 pp. (1975).

Spectrophotometry of extragalactic supernovae shortly after discovery, and of galactic supernova remnants has been used to derive the masses, temperatures, compositions, and distances of supernovae.

125.068 **Structure of strong shock waves with implications for deuterium synthesis in supernovae.**
T. A. Weaver.
Thesis, California Univ., Livermore (USA). Lawrence Livermore Lab., 180 pp. (1975).

The structure of strong shock waves is calculated over the range of shock energies (1 to 100 MeV/nucleon) and initial number densities (10^{15}–10^{22} cm^{-3}) believed likely to occur in the red-giant-like envelopes of stars undergoing type II supernovae explosions. The general equations governing the structure of such shocks are developed on the basis of a hydrodynamic treatment of a plasma composed of ions, electrons, positrons, and photons, making use of diffusion theory to evaluate the dissipative and transfer terms. Several strong shock structure models are formulated and solved on the basis of these equations and physical processes. A shock model dominated by radiation pressure and transport is considered and criteria for its self-consistency deduced.

Rocket X-ray studies of supernova remnants.
See Abstr. 032.525.

Late stages of stellar evolution. Pulsed γ-radiation of neutron, collapsing and supernova stars.
See Abstr. 065.067.

Carbon detonations in rapidly rotating stellar cores.
See Abstr. 065.113.

En stjernes død og genfødsel. Hvide dværge, supernovae, pulsarer, sorte huller. See Abstr. 065.142.

Energy and momentum transport by neutrino radiation into the mantle of a collapsing star and supernova explosions. See Abstr. 066.159.

An extremely carbon enhanced "nova" model.
See Abstr. 124.010.

Molecular clouds associated with supernova remnants. See Abstr. 131.059.

Behavior of grains in early supernova environments.
See Abstr. 131.190.

Elemental abundances in compact H II regions, planetary nebulae, and supernova remnants in the Magellanic Clouds. See Abstr. 131.524.

New radio map of Cassiopeia A at 5 GHz. See Abstr. 141.056.

Observations of radio sources formerly considered as possible supernova remnants. See Abstr. 141.101.

Neutral hydrogen absorption measurements yielding kinematic distances for 42 continuum sources in the galactic plane. See Abstr. 141.146.

Pulsars: a progress report. See Abstr. 141.338.

Hard emission of young pulsars as the cause of supernovae optical emission. See Abstract 141.342.

A soft X-ray survey of the Scutum-Serpens-Aquila region. See Abstr. 142.059.

Copernicus: X-ray observations of several radio supernova remnants. See Abstr. 142.064.

The soft X-ray enhancement in Eridanus. See Abstr. 142.072.

Transient X-ray sources. See Abstr. 142.160.

X rays from supernova remnants. See Abstr. 142.173.

Where are they now. Cosmic rays from supernovae. See Abstr. 143.081.

Infrared galaxies: evolutionary stages of massive star formation. See Abstr. 158.057.

125.100 **Supernova in NGC 5253**

Five-colour photometry of the supernova in NGC 5253 (1972e). A. M. van Genderen.
Astron. Astrophys., Vol. 45, 429 - 434 (1975).

Five-colour observations (Walraven system) of the bright supernova in NGC 5253 (1972e) are presented. The light-curves show to be very smooth with a slight non-linearity in the U (3620 Å) curve after the end of the steep decline. A steep decline of about 30% in brightness in the L band (3850 Å) which lasts roughly 35 days starts 100 days after maximum. This can probably be attributed to the appearance of a strong absorption feature (Ca II?) visible in the spectral energy distributions measured by Kirshner et al. (1973). The $B-L$ curve shows before this decline a wavy pattern. The photometry presented here is compared with the UBV observations of Ardeberg and de Groot (1973, 1974) and with five-colour observations (also Walraven system) of the SN 1968 d in M83 by Wamsteker (1972).

Supernova 1972e in NGC 5253.
R. P. Kirshner, J. B. Oke.
Astrophys. Journ., Vol. 200, 574 - 581 (1975).

New absolute energy distributions of the type I supernova 1972e in NGC 5253 extending to about 700 days after maximum light have been obtained. A physical model of the expanding envelope, based on the identification of the feature at 6550 Å with Hα, is proposed. More than 200 days after maximum, the spectrum is dominated by four features between 4200 and 5500 Å. Three of these four features match the blended emissions from over 100 lines of [Fe II]. If this identification is correct, the envelope requires about $10^{-2} M_\odot$ of iron, which corresponds to an Fe/H ratio about 20 times higher than the cosmic abundance. Possible identifications of the fourth feature with Mg I λ4571 or permitted lines of Fe II are also discussed.

125.101 **Supernova in NGC 4254**

The spectrum of the type II supernova 1967h in NGC 4254. A. P. Fairall.
Monthly Notes Astron. Soc. Southern Africa, Vol. 34, 94 - 98 (1975). – Research note.

125.102 **Supernova in NGC 4165**

Photographic observations of supernova 1971-g in NGC 4165. S. K. Fomin, L. A. Urasin.
Inform. Bull. Variable Stars, (I.A.U. Commission 27), Konkoly Obs., Budapest, No. 1046 (1975).

125.103 **Supernova in NGC 4414**

Supernova 1974g in NGC 4414.
S. Wyckoff, P. A. Wehinger.
Bull. American Astron. Soc., Vol. 7, 555 (1975). – Abstr. AAS.

Spectra of the supernova in NGC 4414.
M. Iye, K. Kodaira, S. Kikuchi, H. Ohtani.
Publ. Astron. Soc. Japan, Vol. 27, 571 - 579 (1975).

Six spectrograms of the supernova which was detected in NGC 4414 in April 1974 are discussed. These spectra, covering the light-maximum phase, closely resemble the spectra of the type-I fast supernovae 1971i and 1972e, and suggest that the present supernova is also of type-I fast and reached its light maximum around May 1. Special attention is called to the presence of fine spectral features (width $\lesssim$3000 km s^{-1}). The rapid shifts of the wavelengths of some prominent features are found to start already in the pre-maximum phase.

125.104 **Supernova in NGC 7619**

Photographic observations of the supernova 1970-j in NGC 7619. I. Dubjago, S. S. Tokhtas'ev
Inform. Bull. Variable Stars, (I.A.U. Commission 27), Konkoly Obs., Budapest, No. 1062 (1975).

125.105 **Supernova in NGC 5457**

Paschen lines of H and lines of multiplet No. 1 (λ 7773 Å) of OI in the spectrum of SN 1970g.
N. N. Chugaj.
Astron. Tsirk., No. 874, p. 1 - 2 (1975). In Russian.

125.106 **Supernova in NGC 7723**

Supernova in NGC 7723. P. Wild.
IAU Circ., No. 2858 (1975).

Supernova in NGC 7723. V. A. Lipovetskij.
IAU Circ., No. 2859 (1975).

Supernova in NGC 7723.
L. A. Thompson, G. Illingworth, F. Ciatti.
IAU Circ., No. 2866 (1975).

125.107 **Supernova in NGC 2487**

Supernova in NGC 2487.
C. Kowal, A. Boksenberg, L. Searle, R. Kirshner.
IAU Circ., No. 2874 (1975).

125.108 **Supernova in NGC 3583**

Supernova in NGC 3583. C. Kowal.
IAU Circ., No. 2878 (1975).

125.109 **Supernova in NGC 4298**

Possible supernova in NGC 4298.
Natsvlichvili.
IAU Circ., No. 2888 (1975).

125.110 **Supernova in NGC 1325**

Supernova in NGC 1325.
J. Dunlap, Y. Dunlap.
IAU Circ., No. 2893 (1975).

126 Low-luminosity Stars, Subdwarfs, White Dwarfs

126.001 **A cool degenerate dwarf with a metal-rich atmosphere.** P. Hintzen, S. Tapia.
Astrophys. Journ., (*Letters*), Vol. 199, L31 - L33 (1975).

G165-7, a recently discovered cool degenerate dwarf ($B-V = 0.84$), displays a very rich spectrum containing absorption features due to Ca II, Ca I, Mg I, Fe I, and probably CH and Mn I.

126.002 **Discovery of light variability in the white dwarf G 29–38.** O. S. Shulov.
IAU Symposium No. 67, (see 012.001), p. 485 - 486 (1975).

126.003 **The OB-type subdwarf HD 149382 and the nature of the subdwarf B stars.**
B. Baschek, J. Norris.
Astrophys. Journ., Vol. 199, 694 - 701 (1975).

High-dispersion spectra and monochromatic continuum measures of the subdwarf OB star HD 149382 have been obtained and analyzed to yield the atmospheric parameters $T_{\rm eff} = 40{,}000$ K, $\log g = 5.8$. Comparison with other subdwarfs shows that HD 149382 together with Feige 66 represent intermediate cases between the subdwarf B and subdwarf O classes. It is suggested that the Bp stars, the subdwarf B stars, and the DA white dwarfs, all of which are weak-helium-line objects, define a high-temperature boundary in the ($\log g$, $T_{\rm eff}$)-plane hotter than which purely surface peculiarities cannot exist. Consideration of the space motions of 17 subdwarf B stars shows that a large proportion of this class belongs to the old disk population. The authors consider the implications of this result for the evolutionary status of these objects.

126.004 **White dwarf external layers. II. Convection and magnetic fields.** F. D'Antona, I. Mazzitelli.
Astron. Astrophys., Vol. 42, 127 - 129 (1975).

It is shown that, in a white dwarf, a magnetic field of the order of several thousands Gauss can have a strong influence on the behaviour of subatmospheric convection and, for cool stars, even on the cooling time.

126.005 **Evolution of crystallizing pure ^{12}C white dwarfs.**
D. Q. Lamb, H. M. Van Horn.
Astrophys. Journ., Vol. 200, 306 - 323 (1975).

The authors describe the first results of a quantitative exploration of white dwarf evolution for models incorporating an accurate dense plasma equation of state and a full treatment of partial ionization and convection in the envelope. They discuss in detail the results for a 1 $M_\odot$, pure ^{12}C star. The direct effects of crystallization are too small to permit observational detection at present. Debye cooling may offer an indirect test, however, because its effect is dramatic and roughly composition independent. For this, unfortunately, an accurate observational luminosity function down to at least $\log (L/L_\odot) \approx -5$ appears needed.

126.006 **G29-38 and G38-29: two new large-amplitude variable white dwarfs.** J. T. McGraw, E. L. Robinson.
Astrophys. Journ., (*Letters*), Vol. 200, L89 - L93 (1975).

G29-38 and G38-29 are shown to be luminosity-variable white dwarfs whose characteristics are similar to those of HL Tau–76. Power spectrum analysis shows that the light curves of both stars have a complex low-frequency structure that is variable with time. It is suggested that many but not all of the features in the power spectra of these two stars could result from nonradial pulsations in a rotating white dwarf.

126.007 **The demise of the degenerate M stars.**
J. Liebert.
Astrophys. Journ., (*Letters*), Vol. 200, L95 - L98 (1975).

Lick image tube spectrophotometry is presented of suspected degenerate M (DM) stars and of the faintest of the suspected Hyades red subluminous (RSL) star suspects. All appear to be dwarf M's or subdwarfs. It is believed that these results, taken together with recent work by Greenstein and with Naval Observatory parallaxes, leave no remaining degenerate M or strong Hyades RSL candidates.

126.008 **White dwarf external layers. I. Convection and**

central temperatures. F. D'Antona, I. Mazzitelli.
Astron. Astrophys., Vol. 42, 165 - 172 (1975).
Several models for white dwarf atmospheres and sub-atmospheres have been computed for various physical and chemical conditions. It is shown that the equation of state is probably the most critical parameter in determining the structure of the inner layers.

126.009 **Interstellar accretion and the initiation of thermonuclear runaways in the envelopes of low luminosity white dwarfs.**
S. G. Starrfield, J. W. Truran, W. M. Sparks, P. Strittmatter.
Bull. American Astron. Soc., Vol. 7, 442 (1975). – Abstr. AAS.

126.010 **Spectrophotometric studies of metallic-lined white dwarfs.** J. W. Liebert.
Bull. American Astron. Soc., Vol. 7, 471 (1975). – Abstr. AAS.

126.011 **Problems in classifying cool degenerate stars.** P. Hintzen, P. A. Strittmatter.
Astrophys. Journ., (*Letters*), Vol. 201, L37 - L40 (1975).
It is shown from model atmosphere calculations that cool, hydrogen-rich degenerate stars may readily be confused spectroscopically with subdwarfs. Many faint, cool, high-proper-motion stars currently classified as subdwarfs may thus turn out to be cool degenerates. Empirical studies of white-dwarf cooling times which do not take this into account may be in serious error.

126.012 **Spectrophotometric observations of suspected red subluminous stars in the Hyades and field.**
J. W. Leibert.
Publ. Astron. Soc. Pacific, Vol. 87, 504 (1975). – Abstr. ASP.

126.013 **Spectra of white dwarfs.** D. T. Wickramasinghe, P. Hintzen, P. A. Strittmatter, E. M. Burbidge.
Astrophys. Journ., Vol. 202, 191 - 194 (1975).
Results of a spectroscopic survey of white dwarf candidates from the Lowell Observatory lists are reported. *UBV* photometry is shown to be effective in enhancing the detection rate of DB white dwarfs. The spectrum of GD 40 contains lines of Ca II H and K in addition to those of He I; it is the first DB to show lines of any element other than helium.

126.014 **Mass differences and convective mixing in white dwarfs.** K. H. Böhm.
Astrophys. Space Sci., Vol. 37, L7 - L11 (1975).
Very different amounts of convective mixing are predicted for DA and for non-DA white dwarfs if one accepts the recently determined mass differences between these two classes of stars. Using convection zone models calculated by Fontaine (1973) the author shows that this effect contributes strongly to the atmospheric abundance differences provided DA and non-DA stars have originally comparable total hydrogen abundances.

126.015 **Influence of the atmospheric composition on the structure of a white dwarf of 0.45 $M_\odot$.**
F. D'Antona, I. Mazzitelli.
Mem. Soc. Astron. Italiana, Vol. 46, 271 - 275 (1975).
The evolutionary sequences for a helium white dwarf of 0.45 $M_\odot$, with different chemical composition of the external layers, have been computed. The loci of equal central temperature (T_c) are shown in the H–R diagram. A variation $\delta \log L/L_\odot \simeq 0.05$ and $\delta \log T_{eff} \simeq 0.03$ is found for models with the same T_c and pure–He or H–rich external layers constituting 2 ‰ of the total mass. It is concluded that mixing of a surface H–layer with the He–core, due to the sinking of convection at low T_{eff}, cannot be studied by means of static models.

126.016 **White dwarf external layers. III. Accretion and related problems.**
F. D'Antona, I. Mazzitelli.
Astron. Astrophys., Vol. 44, 253 - 257 (1975).
The authors have examined the possibility that the differences in the atmospheric composition of white dwarfs (H- or He-rich) could be ascribed to accretion and/or mixing processes. Models of white dwarf He-envelopes have been constructed, with an H-layer superimposed, making several assumptions about the H-mass, the metal content, the mixing length and the total mass of the star. It is shown that differences between H-rich and He-rich spectra cannot merely be due to accretion and mixing processes, and that very low accretion rates only are not inconsistent with the existence of He-atmospheres.

126.017 **Theoretical polarization and intensity Hα profiles in white dwarfs with strong magnetic fields.**
E. F. Borra.
Journ. Roy. Astron. Soc. Canada, Vol. 69, 253 (1975).
Abstr. CAS.

126.018 **The detection of an Hα Zeeman pattern in the cool magnetic white dwarf G99-47.**
J. Liebert, J. R. P. Angel, J. D. Landstreet.
Astrophys. Journ., (*Letters*), Vol. 202, L139 - L143 (1975).
An apparent Hα absorption feature has been detected in image tube scans of the cool magnetic white dwarf G99-47, previously classified as DC (featureless). The broad ~2.5 Å EW line is centered some 21 Å shortword from the normal Hα position. This line is interpreted as the π component of Hα shifted by the quadratic Zeeman effect in a surface field of $\sim 15 \times 10^6$ gauss. This interpretation is supported by the detection of the characteristic circular polarization of the σ components at ± 350 Å. Both results are fitted well by a simple, centered dipole model with a mean longitudinal field of 5.6×10^6 gauss.

126.019 **Kinematics of newly observed white dwarfs.** E. M. Sion, J. A. Bangert, D. A. Turnshek, J. W. Liebert.
Bull. American Astron. Soc., Vol. 7, 513 (1975). – Abstr. AAS.

126.020 **Accretion of rotating fluid by stars: results for white-dwarf models.** R. H. Durisen.
Bull. American Astron. Soc., Vol. 7, 542 (1975). – Abstr. AAS.

126.021 **On the origin of continuum polarization in white dwarfs.** W. H. Ingham, K. Brecher, I. Wasserman.
Bull. American Astron. Soc., Vol. 7, 558 (1975). – Abstr. AAS.

126.022 **White dwarfs: a brief survey of observational data and theoretical ideas for 1974.** V. P. Chuprina.
Soobshch. Spets. Astrofiz. Obs., *Zelenchukskaya*, vyp. (No.) 13, p. 9 - 31 (1975). In Russian.
This paper is a brief survey of current observational data and theoretical ideas of white dwarfs (WD). Of all the observational aspects special consideration is given to the methods of search for WD's among faint blue objects as well as to the recently discovered phenomena: rapid variability of WD's and the presence of circular polarization in a number of them. As for the theory, possible mechanisms of variability of WD's and the problem of estimate of magnetic fields on their surfaces are discussed in more detail.

126.023 **Time-independent models of white dwarfs accreting mass.** R. Sienkiewicz.
Astron. Astrophys., Vol. 45, 411 - 416 (1975).
Simplified models of 1 $M_\odot$ and 1.39 $M_\odot$ white dwarfs accreting matter have been obtained. These models consist of envelopes, stationary shell source regions and highly degener-

ate isothermal cores. The accreted matter flows from the surface through the hydrogen and helium burning shells into the core. The range of the rates of mass flow for which hydrogen and helium burning shell sources may be stationary, as well as time scales of their formations, have been found.

126.024 **BPM 30551.**
J. E. Hesser, B. M. Lasker, H. E. Neupert.
IAU Circ., No. 2876 (1975).

126.025 **PG 2337+12.** R. F. Green.
IAU Circ., No. 2892 (1975).

126.026 **Rapid rotating white dwarfs.**
G. S. Adzhyan, Yu. L. Vartanyan.
Astrofizika, Vol. 11, 517 - 530 (1975). In Russian. – English translation in Astrophysics, Vol. 11, No. 3.

The dependence of mass and other parameters of differentially rotating white dwarfs on central density have been studied by the energetic method. The effects of general relativity and reverse β-decay are taken into account. It has been shown that rapid rotation depresses these effects and makes possible the existence of stable white dwarfs with masses up to three solar masses and with central densities two orders higher than those of static configurations.

126.027 **On limiting density in white stars.**
J. P. Sharma.
Indian Journ. Phys., Vol. 48, 1143 - 1145 (1974).

The equilibrium mass and the electronic concentration per unit volume is calculated for stars of the white dwarf type.

126.028 **Chemical composition and effective temperatures of metallic line white dwarfs.** G. L. Hammond.
Thesis, Maryland Univ., College Park (USA). [Available from Univ. Microfilms, Ann Arbor, Mich., *USA*. Order No. 74-29,764], 120 pp. (1974).

Model atmosphere techniques have been employed to determine the composition, effective temperatures, radii, masses and surface gravities of white dwarfs Ross 640 and van Maanen 2. The non-gray, LTE, convective, constant flux models employed collisional damping constants for the Ca II H and K lines that were measured in a laboratory device that simulated white dwarf atmospheric conditions. Ross 640 was found to have an extremely helium-rich composition and T_{eff} = 8500 K, while the observed properties of van Maanen 2 were fitted best by a model with 91 percent helium, 9 percent hydrogen and T_{eff} = 6100 K. The laboratory measurements of pressure shifts for the Ca II lines cast some doubt on the interpretation of recent radial velocity determinations for van Maanen 2.

126.029 **Magnetic field and screening effects in condensed and ultradense matter.** K. M. Roussel.
Thesis, Louisiana State Univ., Baton Rouge (USA). [Available from Univ. Microfilms, Ann Arbor, Mich., *USA*. Order No. 75-1951], 209 pp. (1974).

The investigations of three topics are presented: the origin of magnetic fields in white dwarfs and neutron stars, the detection of magnetic fields in white dwarfs, and screening effects due to free charged particles, particularly in semiconductors.

White dwarfs. Collection of papers.
See Abstr. 003.077.

One- and two-electron atoms in strong magnetic fields and implications for magnetic white dwarfs and pulsars.
See Abstr. 022.126.

Non-LTE-effects in atmospheres of O-subdwarfs.
See Abstr. 064.007.

Upper mass limits for stable rotating white dwarfs.
See Abstr. 065.005.

Endstadien der Sternentwicklung.
See Abstr. 065.175.

Gravitational radiation instability in rotating stars.
See Abstr. 066.038.

Trigonometric parallaxes determined with the Yerkes Observatory 40-in. refractor. IV.
See Abstr. 111.002.

A search for ultraviolet objects.
See Abstr. 114.027.

High-speed spectroscopy of the ultrashort-period variable AM Canum Venaticorum (=HZ 29).
See Abstr. 114.316.

Tidal flows in binary systems.
See Abstr. 117.017.

Note on the period variation in the white-dwarf eclipsing binary BD +16°516. See Abstr. 121.117.

Accretion and the nova outburst.
See Abstr. 124.009.

An extremely carbon enhanced "nova" model.
See Abstr. 124.010.

The structure of DQ Herculis.
See Abstr. 124.102.

Can supernovae make white dwarfs?
See Abstr. 125.038.

A Ni^{56} white dwarf hypothesis for the type I supernova light curve. See Abstr. 125.058.

Observational support for the white dwarf, nuclear decay hypothesis for type I supernova luminosity.
See Abstr. 125.059.

Radiative transfer in a strong magnetic field and accreting X-ray pulsars. See Abstr. 142.094.

X-ray sources and the pulsational properties of high luminosity degenerate stars. See Abstr. 142.106.

On the energetics of pulsating X-ray sources.
See Abstr. 142.113.

On the stellar component of galactic coronae.
See Abstr. 155.047.

An observation of the diffuse soft X-ray/extreme ultraviolet background. See Abstr. 155.058.

Suggested observations of supernova remnants in the Large Magellanic Cloud. See Abstr. 159.003.

Interstellar Matter, Gaseous Nebulae, Planetary Nebulae

131 Interstellar Matter, Polarization of Starlight, H I, H II Regions

Interstellar Matter, Polarization of Starlight

131.001 **Carbon monoxide in collapsing interstellar clouds.** T. de Jong, S. Chu, A. Dalgarno.
Astrophys. Journ., Vol. 199, 69 - 78 (1975).
The authors have calculated the energy loss rates, the brightness temperatures and the profiles of lines of carbon monoxide in collapsing interstellar clouds. The density distribution and the velocity field inside the cloud are varied, one model applying to a cloud of uniform density that is collapsing in free fall, and a second to a cloud in isothermal collapse. The line profiles are a sensitive tool for testing the models against the observations. The observed CO lines can be explained by collapsing models in which both density and velocity gradients occur. The authors predict that CO infrared lines from dense interstellar clouds should be observable with instruments of high spectral resolution at wavelengths longward of 200 microns.

131.002 **CO, CS, and HCN in a clustering of reflection nebulae in Monoceros.** M. L. Kutner, K. D. Tucker.
Astrophys. Journ., Vol. 199, 79 - 85 (1975).
Carbon monoxide line emission at $\lambda = 2.6$ mm has been observed over an area of $\sim 3^1/_2{}^\circ \times 3^1/_2{}^\circ$ in L1646, a diffuse dust cloud containing a grouping of reflection nebulae. The H_2 mass is estimated from the CO observations to be $>3.2 \times 10^4 M_\odot$. Five CO emission peaks are observed, each apparently associated with at least one reflection nebula.

131.003 **Observations of the polarization characteristics of galactic H_2O sources.** J. M. Bologna, K. J. Johnston, S. H. Knowles, S. A. Mango, R. M. Sloanaker.
Astrophys. Journ., Vol. 199, 86 - 91 (1975).
Twenty-four H_2O sources associated with H II regions have been examined for circular and linear polarization. No circular polarization was found in any source. Approximately 25 percent have been found to show linear polarization in at least some features. This, together with the fact that 25 percent of the sources show linear polarization, indicates that the cross-relaxation rate of maser-level degenerate substates via infrared line radiation is almost equal to the stimulated emission rate in many H_2O sources.

131.004 **Ultraviolet photometry from the Orbiting Astronomical Observatory XX. The ultraviolet extinction bump.** B. D. Savage.
Astrophys. Journ., Vol. 199, 92 - 109 (1975).
Interstellar extinction curves over the wavelength region 1800–3600 Å are presented for 36 stars. The stars have $E(B-V)$ in the range 0.03 to 0.55, and are mostly confined to the brighter OB associations distributed along the plane of the Galaxy. Every extinction curve exhibits a broad extinction bump peaking near 2175 Å (4.6 μ^{-1}). The position of the peak and the profile of the feature appear remarkably constant among the sample of stars. With only a few exceptions, $E(\lambda_{max} - 3320)$, a measure of the strength of the feature, correlates very well with $E(B-V)$, implying that the bump has an interstellar rather than a circumstellar origin.

131.005 **Interstellar formaldehyde near stars of the Orion population.** N. H. Dieter.
Astrophys. Journ., Vol. 199, 289 - 296 (1975).
Interstellar formaldehyde absorption lines have been detected in the direction of 100 Orion-population stars. The low angular resolution of this survey prevents discerning any direct connection between these young stars and the gas surrounding them, but does allow establishing that the dust clouds in which they have formed are different from quiescent clouds. In these regions of recent star formation, there is a higher percentage of relatively strong H_2CO lines, which can be explained by a difference in their density and/or temperature gradients.

131.006 **On the Orion infrared nebula/molecular cloud.** B. Zuckerman, P. Palmer.
Astrophys. Journ., (*Letters*), Vol. 199, L35 - L38 (1975).
Observational and theoretical studies of the Orion nebula and the associated molecular clouds have greatly increased the understanding of this and other regions in which star formation is taking place. Fundamental questions remain unanswered: (1) the chemical composition of the molecular cloud, (2) its internal motions, and (3) the role of magnetic fields in its evolution. The authors show that the gas phase chemistry and internal motions in one part of the cloud are distinctly different from those in the rest of the cloud, and two recent estimates of the magnetic field strengths are very uncertain.

131.007 **Further evidence for magnetism in the Orion region.** E. J. Chaisson, C. A. Beichman.
Astrophys. Journ., (*Letters*), Vol. 199, L39 - L42 (1975).
Observations of two transitions ($F = 1 \rightarrow 1$, 1665 MHz, and $F = 1 \rightarrow 2$, 1612 MHz) within the ${}^2\Pi_{3/2}$, $J = 3/2$ state of interstellar OH illustrate near classic Zeeman patterns strongly suggestive of a magnetic field of several milligauss, at least within the vicinity of the Orion maser source.

131.008 **Detection of interstellar SiS and a study of the IRC+10216 molecular envelope.** M. Morris, W. Gilmore, P. Palmer, B. E. Turner, B. Zuckerman.
Astrophys. Journ., (*Letters*), Vol. 199, L47 - L51 (1975).
Microwave emission in the $J = 6 \rightarrow 5$ and $J = 5 \rightarrow 4$ transitions of ${}^{28}Si{}^{32}S$ has been detected toward IRC+10216. The $J = 2 \rightarrow 1$ line of ${}^{28}Si{}^{16}O$ and several rotational lines of HC_3N were also detected in the envelope of this unusual carbon star. The SiS/SiO abundance ratio in IRC+10216 is greater than unity, as expected for the atmosphere of a carbon star. The Sgr B2 molecular cloud is probably another source of SiS emission.

131.009 **Carbon-helium charge transfers and the ionization of carbon in the intercloud medium.** G. Steigman.
Astrophys. Journ., Vol. 199, 336 - 339 (1975).
Charge transfer reactions are considered in collisions between carbon and helium ions: $C^{++} + He \rightleftarrows C^+ + He^+$. It is shown that, if these reactions proceed rapidly at high temperatures ($300 \lesssim T \lesssim 10^4$ K), then the degree of carbon

ionization is directly tied to the degree of helium ionization: C III/C II $\gtrsim 10^{-1}$ (He II/He I)$_{T=10^4}$. As a result, even if these charge transfer reactions are fast, it is impossible to account for the C III/C II ratios observed toward several unreddened stars with a hot intercloud medium. More important, it is shown that the presence of the Coulomb barrier in $C^+ + He^+$ collisions prevents the above reactions from proceeding rapidly (in either direction) at thermal energies ($T \lesssim 10^4$ K).

131.010 **Radiation transport in dense interstellar dust clouds. I. Grain temperature.** C. M. Leung.
Astrophys. Journ., Vol. 199, 340 - 360 (1975).

A numerical method is presented for solving the problem of radiation transport in dense interstellar dust clouds in spherical or planar geometry and for general anisotropic scattering. This method is used to determine the grain temperature distribution inside such dark clouds.

131.011 **A comparative study of H_2CO, OH, and H I in the dark Taurus dust cloud.**
C. Heiles, M. A. Gordon.
Astrophys. Journ., Vol. 199, 361 - 378 (1975).

High-sensitivity spectra of H_2CO, OH, and H I, all taken with the same angular resolution, were obtained in the dark dust cloud in Taurus. The authors consider two extreme interpretations of the widely varying relative intensities of the molecular lines. The one, constant excitation temperature of H_2CO within the cloud, leads to large variations of N_{OH}/N_{H_2CO} from point to point within the cloud. The other, a constant value of $N_{OH}/N_{H_2CO} = 20$ deduced from independent observations of other dust clouds, leads to a correlation of OH and H_2CO excitation temperatures–a natural result of collisional excitation of both molecules as previously suggested by others.

131.012 **Line spectra in interstellar clouds. II. CH and CH^+ in Ophiuchus.** F. H. Chaffee, Jr.
Astrophys. Journ., Vol. 199, 379 - 382 (1975).

Observational evidence is presented that suggests that velocity differences of the order of 1 km s^{-1} exist between interstellar CH and CH^+ in four stars in Ophiuchus. Such velocity stratification would be consistent with the formation mechanism proposed by Black and Dalgarno and by Watson in which CH is formed near the center of interstellar clouds where the density of molecular hydrogen, $n(H_2)$, is high, and CH^+ is formed in the outer regions where $n(H_2)$ is negligible.

131.013 **Interstellar H_2CO. II. 2-centimeter emission from the Orion molecular cloud.**
N. J. Evans II, B. Zuckerman, T. Sato, G. Morris.
Astrophys. Journ., Vol. 199, 383 - 397 (1975).

Emission in the $2_{11} \rightarrow 2_{12}$ line of formaldehyde at 2-cm wavelength has been mapped over a $3' \times 6'$ region in the Orion molecular cloud. The region emitting the 2-cm line displays a velocity gradient that is suggestive of rotation about an east-west axis–the direction of flattening. The total mass derived from considerations of excitation and, independently, from dynamical considerations is $\sim 500 M_\odot$.

131.014 **Uncertainties in the far-infrared absorption of interstellar clouds.** M. Harwit.
Astrophys. Journ., Vol. 199, 398 - 400 (1975).

The observed absorption of far-infrared radiation by an interstellar dust cloud is a complicated function of the material properties of grains, of their column density, and of the extent of stimulated emission. At long wavelengths, neglect of stimulated emission would lead to an underestimate of the amount of absorbing material observed along a line of sight. However, if the complex index of refraction for grains is measured at the low temperatures in question, a classical analysis will be correct since the measured refractive index correctly accounts for stimulated emission. Since the grain temperatures may strongly affect the refractive index of a grain, typical interstellar grain temperature fluctuations are analyzed.

131.015 **Complex interstellar calcium line structure in the Carina nebula.** N. R. Walborn, J. E. Hesser.
Astrophys. Journ., Vol. 199, 535 - 543 (1975).

Exceedingly complex structure has been found in the interstellar Ca II lines on 9 Å mm^{-1} coudé spectrograms of stars in the Carina nebula. As many as six components have been observed in a single spectrum; moreover, the profiles show large variations over small angular distances, and a total velocity range of 330 km s^{-1} has been measured. Stars in the outer part of the nebula have single calcium lines, which suggests that the additional components in the inner region are formed there. Measurements of the interstellar Ti II lines, which are double in a few cases, are also presented. A search for similar interstellar calcium-line structure at the center of the 30 Doradus nebula in the Large Magellanic Cloud is briefly discussed.

131.016 **Molecular clouds in the Galaxy.**
N. Z. Scoville, P. M. Solomon.
Astrophys. Journ., (*Letters*), Vol. 199, L105 - L109 (1975).

The $J = 1 \rightarrow 0$ emission of CO has been surveyed in the galactic plane between $l = -10°$ and $+90°$ with a $1'$ beam sampling every degree. Molecular clouds emitting in the CO line are plentiful over the inner region of the Galaxy. Their greatest number occurs in the galactic nucleus and at a radius of 5.5 kpc – a distribution similar to radio H II regions and galactic γ-ray emission but very different from that previously derived for atomic hydrogen. The total mass in molecular clouds is found to be $1-3 \times 10^9 M_\odot$, and each typically has $10^5 M_\odot$, an H_2 density of 700 cm^{-3}, and a temperature near 7 K. These results suggest that most of the interstellar medium in the interior of the Galaxy is molecular H_2.

131.017 **Time variations in sources of silicon monoxide emission at 7 millimeters.**
J. H. Spencer, P. R. Schwartz.
Astrophys. Journ., (*Letters*), Vol. 199, L111 - L113 (1975).

Time variations in SiO $J = 1-0$, $v = 1$ and $v = 2$ intensity with a time scale which probably corresponds to the infrared light curve have been observed in o Cet and R Cas. A short time scale variation of about 1 hour has been detected in the emission of SiO from the Orion infrared source.

131.018 **Line-leaking models for interstellar molecular emission and absorption–II. Masers in the methyl alcohol spectrum.** M. Pelling.
Monthly Notices Roy. Astron. Soc., Vol. 172, 41 - 54 (1975).

A radiative transfer (line-leaking) model is proposed to explain methanol emission at 84 and 36 GHz in Sgr B2, and at 25 GHz in Ori A. The different emission characteristics of the two sources are attributed to differences of hydrogen density and possibly kinetic temperature. The Sgr B2 environment fits the results best if $n_{H_2} \simeq 10^5$ cm^{-3}, $T_K \lesssim 100$ K. In contrast, $n_{H_2} \geqslant 10^6$ cm^{-3}, $T_K \simeq 100$ K are needed in Ori A. It is also shown that a thermal equilibrium explanation for the formation of the lines is unlikely.

131.019 **Giant grains in Orion.** M. Rowan-Robinson.
Monthly Notices Roy. Astron. Soc., Vol. 172, 109 - 116 (1975).

A modification of Larson's model for optically thick dust clouds is used to fit the infrared and submillimetre spectrum of the Orion nebula. For a spherically symmetric, uniform dust cloud in which the grains are composed of silicates and have radius 0.2 μm, the optical depth in the ultraviolet has to exceed 1000, and the infrared cluster would have to lie by chance in front of the centre of the Orion molecular and submillimetre cloud. This conclusion would not appear to be greatly altered if the cloud is flattened or has a density gradient, or if complex ices are used instead of silicates.

131.020 **An explanation of anomalous velocities in pulsar scintillation patterns.** B. J. Uscinski.
Monthly Notices Roy. Astron. Soc., Vol. 172, 117 - 132 (1975).

Velocities of the interstellar medium deduced from the scintillation of pulsar signals frequently exhibit surprising features. They are sometimes very large and can reverse direction in the course of an hour or so. In cases where the pulsar is known to have a high intrinsic velocity the observed pattern velocities are often strangely low. These features are explained in terms of a model of the interstellar medium containing a shear in transverse velocities. Relatively small changes in the medium are required to reverse the high observed velocities and no intrinsic modulation of the pulsar is required.

131.021 **Ground state centrifugal distortion constants of trans-acrolein, $CH_2 = CH-CHO$ from the microwave and millimeter wave rotational spectra.**
M. Winnewisser, G. Winnewisser, T. Honda, E. Hirota.
Zeitschr. Naturforschung, Vol. 30a, 1001 - 1014 (1975).

131.022 **Einfluß der Infrarotlinien auf die Effektivität von OH-Masern.** W. H. Kegel.
Mitt. Astron. Ges., No. 36, p. 72 (1975). – Abstract.

131.023 **The formation of CH^+ in interstellar clouds.**
J. H. Black, A. Dalgarno, M. Oppenheimer.
Astrophys. Journ., Vol. 199, 633 - 636 (1975).

It is suggested that the production of CH^+ in interstellar clouds occurs by a chemical scheme initiated by the radiative association of C^+ and H_2. The scheme leads to CH_2^+ and CH_3^+ which are photodissociated to give CH^+. In contrast to previous formation mechanisms, it is not necessary to suppose that the dissociative recombination of CH^+ is anomalously slow. The same chemical sequence is also able to produce CH in harmony with the measured abundances.

131.024 **Measured extinction of small olivine spheres.**
K. L. Day.
Astrophys. Journ., Vol. 199, 660 - 662 (1975).

Crystalline spheres of the silicate forsterite (Mg_2SiO_4), with a mean diameter of around 600 Å, were prepared in the laboratory. Their infrared extinction in the region 1100–400 cm^{-1} was quantitatively measured and the results were compared with the work of other authors and with Mie theory calculations. The extremely sharp spikes predicted by Mie theory were not realized, though agreement with peak positions and integrated absorption strengths was good.

131.025 **Observations of 1-millimeter continuum radiation from the DR 21 region.** M. W. Werner, J. H. Elias, D. Y. Gezari, M. G. Hauser, W. E. Westbrook.
Astrophys. Journ., (*Letters*), Vol. 199, L185 - L187 (1975).

Continuum emission at 1 mm has been mapped with 1′ resolution over a 3′× 8′ area including the compact H II region DR 21, the maser source W75(S)–OH, and the central portions of the surrounding molecular cloud. Peaks in the 1-mm intensity are observed at the positions of DR 21 and W75(S)–OH, and are roughly coincident with less marked peaks in the intensity of HCN and CO emission. An extended component of 1-mm emission is present as well. The principal source of 1-mm radiation is thought to be thermal emission from dust within the molecular gas. The characteristics of this radiation which make it a promising tool for the study of molecular clouds are briefly discussed.

131.026 **Observations of quasi-thermal and maser phenomena in rotationally excited OH. I.**
L. J. Rickard, B. Zuckerman, P. Palmer.
Astrophys. Journ., Vol. 200, 6 - 21 (1975).

Absorption features arising from Λ-doublet transitions in a very short-lived rotationally excited state of OH have been discovered in the direction of the W3 continuum peak. Analyses of these features and of broad emission features in Sgr B2 indicate the presence of massive regions with very high densities. An extensive search for new sources of maser emission from excited-state OH has resulted in the detection of maser features in NGC 7538, OH 69.5–1.0, M17, W51, and Orion (OH). A rapid time variation has been observed in the maser spectrum of Sgr B2, and evidence is presented for variations in W51 and W3 (continuum).

131.027 **Differential interstellar polarization in the region of 6284 Å for HD 183143.**
G. G. Fahlman, G. A. H. Walker.
Astrophys. Journ., Vol. 200, 22 - 29 (1975).

Differential polarization measurements in the region of the diffuse feature at 6284 Å in the spectrum of HD 183143 are described. The observations were made using a television detector and a Polaroid filter. No systematic change through the diffuse feature was observed, and an upper limit of ± 0.4 percent is placed on the variation of polarization in the spectral region.

131.028 **Absorptive properties of silicate core-mantle grains.** P. A. Aannestad.
Astrophys. Journ., Vol. 200, 30 - 41 (1975).

Absorption and extinction efficiencies for silicate and silicate core-ice mantle grains are presented for olivine-type particles in the wavelength range $0.07\ \mu - 10^3\ \mu$, for fused quartz-type particles in the range $0.056\ \mu - 10^3\ \mu$, and for lunar silicate-type particles in the range $1\ \mu - 10^3\ \mu$. Comparison with the observed near-infrared absorption spectrum in the BN object shows that no single type of these particles can adequately account for the observed 9.8 μ feature, and that a mixture of various types of silicates may be required.

131.029 **Detection of formic acid in Sagittarius B2 by its $2_{11}-2_{12}$ transition.**
G. Winnewisser, E. Churchwell.
Astrophys. Journ., (*Letters*), Vol. 200, L33 - L36 (1975).

Interstellar formic acid, HCOOH, has been detected at 4.9 GHz by its $2_{11}-2_{12}$ transition from the direction of the galactic center radio source Sgr B2. The emission line, which has a peak antenna temperature of 0.04 K and a full width to half-power of 24 km s^{-1}, probably results from an inversion of the $K_a = 1$ doublets. The total column density of HCOOH remains uncertain.

131.030 **On the charge distribution of interstellar dust grains.**
H.-P. Gail, E. Sedlmayr.
Astron. Astrophys., Vol. 41, 359 - 366 (1975).

Spitzer's method to calculate the charge of interstellar dust grains is extended to the case of an ensemble of dust particles. It is shown that the members of the ensemble do not possess a unique charge but show some distribution of the particles over the different possible charge states. The resulting equilibrium charge distribution due to collisions with charged particles and irradiation by photons is derived for different models. Numerical results are presented for dust grains imbedded into a H I-region. For non equilibrium problems the time of relaxation towards equilibrium is derived.

131.031 **A study of the interstellar gas between the sun and the Sagittarius arm.** J. J. Rickard.
Astron. Astrophys., Vol. 41, 403 - 408 (1975).

The velocity field of the interstellar gas is examined in the galactic longitude interval $345° < l < 28°$ by means of Ca II K lines in high dispersion spectra. Two current hypotheses about the structure of the local gas near the sun, one based upon a ring of expanding matter, the other being a linear density-wave perturbation, are compared with the observations. The expanding ring model is in better agreement with the observed K-line velocities.

131.032 **Thermoionic emission from hot grains.**
J. Lefevre.
Astron. Astrophys., Vol. 41, 437 - 440 (1975).

The thermoionic emission from grains heated to a high temperature in the vicinity of stars should be taken into account in the calculation of equilibrium charge. It is always negligible for graphite grains if the star temperature exceeds 14000 °K but it is important in the vicinity of cold stars. The potential of graphite grains is calculated in the solar corona.

131.033 **Absence of the 4400 Å diffuse emission feature.**
A. C. Danks, D. L. Lambert.
Astron. Astrophys., Vol. 41, 455 - 457 (1975).

New spectroscopic observations of 7 stars which exhibit the 4430 Å diffuse interstellar line are presented. A previously reported emission wing at 4400 Å is not confirmed. An alternative mechanism to grains for the production of the 4430 feature is proposed.

131.034 **The structure of the Bok globule B361.**
E. G. Schmidt.
Monthly Notices Roy. Astron. Soc., Vol. 172, 401 - 409 (1975).

Photometry on the *uvbyβ* system has been carried out of stars in and near the Bok globule B361. These data have been supplemented by star counts over the surface of the globule. This object is nearly circularly symmetric and the statistics of the star counts indicate that the distribution of absorption in the globule is quite smooth. The distance of this object is found to be 350 pc and its mass is at least $73 M_\odot$. The variation of extinction with distance from the cloud centre does not agree with models of collapsing clouds but the virial theorem indicates that the object will collapse if the temperature is about 10 K.

131.035 **Whiskers and cosmic millimetre-wave sources.**
M. G. Edmunds, N. C. Wickramasinghe.
Nature, Vol. 256, 713 - 714 (1975). – Letter.

131.036 **High velocity gas in the direction of δ Sco.**
T. R. Hicks, N. K. Reay, J. Ring, R. J. Scaddan.
Monthly Notices Roy. Astron. Soc., Vol. 172, 53P - 55P (1975).

Observations, using a scanning Fabry-Perot monochromator, of the interstellar Na (D2) line in the direction of δ Sco indicate a complex spectral structure. At least two clearly distinguished high velocity components exist, at –47.1 and –63.31 km s^{-1}, in addition to the main interstellar component of –11.8 km s^{-1}.

131.037 **The interstellar medium: UV complex index of refraction of several candidate materials.**
W. G. Egan, T. Hilgeman.
Astron. Journ., Vol. 80, 587 - 594 (1975).

Ultraviolet determinations have been made of the optical complex index of refraction of augite, bytownite, enstatite, diopside, troilite, pyrrhotite, and the Bruderheim meteorite and the results have been used in a Mie scattering program. From this analysis bytownite and meteorite particles of radii between 0.1 and 0.001 μm could account for the interstellar extinction peak at 0.22 μm; however, both of these materials would produce a double peak at 0.22 μm. The splitting for the meteoritic material is stronger than that of the bytownite, and thus harder to reconcile to the interstellar extinction. Enstatite is improbable because the absorption maximum occurs at ~0.26 μm. The other samples also do not have appropriate absorption curves and are ruled out.

131.038 **Wavelength dependence of polarization. XXVIII. Circumstellar scattering: polarization and reddening models.** S. J. Shawl.
Astron. Journ., Vol. 80, 595 - 601 (1975).

Results of calculations of the polarization and reddening produced by circumstellar scattering envelopes are presented. Models are given for scattering by spherical silicate, graphite, and iron grains and by the hydrogen molecule. Various geometries, density distributions, and particle size distributions are included. The following paper contains a comparison of the results of this paper with observation.

131.039 **Neutral hydrogen absorption spectra.**
B. Lazareff.
Astron. Astrophys., Vol. 42, 25 - 35 (1975).

Observations of 21 cm line absorption spectra at intermediate latitudes were made with the Nançay radiotelescope with an effort towards improved sensitivity. The absorption spectra are combined with emission brightness temperature spectra to discuss the spin temperature and other properties of interstellar neutral hydrogen. A correlation is found between the optical depth and the spin temperature of the clouds. The velocity dispersion of the clouds is found to be larger in emission than in absorption. The nature of the diffuse component seen only in emission is discussed. The results are compared with existing models of the interstellar medium, and the observed $\tau - T_s$ correlation seems to favor equilibrium models.

131.040 **Effects of turbulent velocities on maser lines.**
H.-P. Gail, W. H. Kegel, E. Sedlmayr.
Astron. Astrophys., Vol. 42, 81 - 84 (1975).

The theory of line formation in turbulent media (Gail et al., 1974) is applied to the case of an unsaturated maser. It is shown that the emitted maser line depends strongly on the correlation length l of the turbulent velocity field. Both, the total intensity and the line width increase with increasing l. The influence of the finite correlation length has the opposite sign and is much stronger than in the case of a positive absorption coefficient.

131.041 **On the determination of the dust abundance from middle infrared measurements.** N. Panagia.
Astron. Astrophys., Vol. 42, 139 - 143 (1975).

The effects of a temperature range on the effective emissivity of grains have been considered. The author finds that the use of the color temperature as the physical temperature of grains for the interpretation of middle infrared data ($\lambda \approx 10\mu$) leads to serious underestimates of the dust mass inside H II regions (typically by a factor of 20). As a consequence, the corrected abundance of dust relative to the gas turns out to be similar to that of the interstellar medium. It also agrees well with the determinations obtained from far infrared data.

131.042 **The scale height of interstellar ionised hydrogen.**
A. C. S. Readhead, P. J. Duffett-Smith.
Astron. Astrophys., Vol. 42, 151 - 153 (1975)

New observations of interstellar scattering at 81.5 MHz and 151.5 MHz are used to determine the scale height of the free electron layer in the Galaxy. A value of 430 pc to 670pc is derived for the scale height, and the dispersion measure perpendicular to the plane is shown to be between 15 cm^{-3}pc and 19 cm^{-3}pc. The limit imposed by the scattering on observations of extragalactic radio sources is discussed.

131.043 **Copernicus observations of interstellar absorption**

at Lyman Alpha. R. C. Bohlin.
Astrophys. Journ., Vol. 200, 402 - 414 (1975).

Column densities N_H of atomic hydrogen have been derived for 40 OB stars from spectral scans at Lα obtained by the Copernicus (OAO-3) satellite. The stars are all between 60 and 1100 pc away, with a range of mean densities n_H of 0.01 to 2.5 atoms cm^{-3}. In the set of data, the ratio of gas to color excess in clouds varies from 1 to 3 times the mean outside of clouds. The presence of molecular hydrogen correlates with $E(B-V)$, but the best tracer for H_2 is atomic hydrogen. The mean density of the gas for all 40 stars is much smaller than the mean of 0.7 atoms cm^{-3} obtained from 21-cm observations, because the brightest stars with less than average amounts of matter in the line of sight were selected for observation.

131.044 **Sodium depletion within diffuse interstellar clouds.** M. Jura.
Astrophys. Journ., Vol. 200, 415 - 418 (1975).

Having derived estimates of the densities and radiation fields within several interstellar clouds from the observed populations of the H_2 rotational levels, the author estimates the amount of sodium and calcium depletion within different clouds. If the H_2 formation rate R equals 3×10^{-17} cm^3 s^{-1}, and if the hydrogen ionization rate is less than 10^{-16} s^{-1}, he finds that in most clouds with $E(B-V) \lesssim 0.10$, sodium is at most slightly depleted, while calcium is always substantially depleted. This result is important for determining the thermal equilibrium within diffuse interstellar clouds.

131.045 **The formation of carbon monoxide and the thermal balance in interstellar clouds.**
M. Oppenheimer, A. Dalgarno.
Astrophys. Journ., Vol. 200, 419 - 425 (1975).

A chemical scheme is presented for the formation of carbon monoxide in interstellar clouds of low density which suggests that the conversion of C into CO is substantial at depths corresponding to visual extinctions exceeding unity. The production of CO leads to the formation of a cold region in the cloud that may be unstable, and the authors suggest that the contraction of an interstellar cloud depends critically on the formation of CO and of other interstellar molecules.

131.046 **The relative positions of the OH and H_2O masers in W49N and W3(OH).** G. L. Mader, K. J. Johnston, J. M. Moran, S. H. Knowles, S. A. Mango, P. R. Schwartz, W. B. Waltman.
Astrophys. Journ., (*Letters*), Vol. 200, L111 - L114 (1975).

The relative positions of the OH and H_2O masers in W49N and W3(OH) have been found by simultaneously observing both masering transitions with a very-long-baseline interferometer. In both sources the OH and H_2O features are clearly separated and occupy two distinct regions of space, separated by 48,000 AU and 18,000 AU in W49N and W3(OH), respectively. The two masering regions appear unlikely to have the same source of excitation.

131.047 **Population of excited rotational levels of interstellar H_2.** P. Joshi, S. P. Tarafdar.
Astrophys. Space Sci., Vol. 36, 281 - 289 (1975).

The population of the rotational levels of vibrational and electronic ground state of interstellar H_2 has been calculated; different possible excitation and de-excitation mechanisms including the ultraviolet pumping and formation and destruction of molecules have been considered. The values of temperature, density and radiation field towards ζ Oph, ϵ Per and ζ Pup have been obtained by comparing the theoretical distributions with the observations.

131.048 **Population of excited rotational levels of HD and CO in interstellar clouds.**
P. Joshi, S. P. Tarafdar.
Astrophys. Space Sci., Vol. 36, 291 - 301 (1975).

Populations of the rotational levels of vibrational and electronic ground state of interstellar HD and CO have been calculated. Ultraviolet pumping and molecular formation have been found to be very important in the case of HD but not so in CO, the population distribution of which is mainly governed either by 3 K or by collisional excitation. The population inversion has been found to occur in HD but not in CO. The population inversion at large density does not occur in HD, probably due to neglect of multipole collision.

131.049 **Origin of the diffuse interstellar absorption bands. I: The structure of interstellar grains.**
W. W. Duley.
Astrophys. Space Sci., Vol. 36, 345 - 351 (1975).

The influence of crystal structure and surface stresses on the spectrum of small interstellar particles has been investigated. Surface effects are predicted to result in the occurrence of pairs of features in the discrete absorption spectrum of interstellar dust. A simple relationship between the energy separation between lines of these pairs and their widths is derived which is tested against recent observational data on the diffuse interstellar band spectrum. Thirty of the diffuse bands can be accounted for on this basis by assuming that interstellar dust consists of a mixture of components of differing chemical composition.

131.050 **The interrelationship between cosmic dust and the microwave background.** P. S. Wesson.
Astrophys. Space Sci., Vol. 36, 363 - 382 (1975).

Attention is given to the radiation of microwaves by charged dust in space. Presently-used particle distributions do not restrict the presence in space of large numbers of small ($r < 10^{-6}$ cm) silicate grains, but it is shown that such densities ($\varrho \approx 10^{-25} - 10^{-26}$ g cm^{-3}) of small grains would produce a microwave background with an energy density of the same order of magnitude as the energy density of the (presumed) cosmological 3 K background.

131.051 **On the value of $R = A_V/E_{B-V}$.** C. Schalén.
Astron. Astrophys., Vol. 42, 251 - 258 (1975).

The various methods used by H. L. Johnson to determine R and his results are briefly discussed in the first part of the paper. In the second part an attempt is made to derive R from determinations of total extinctions and total colour excesses in dark clouds. A value $R = 3.1$ is suggested. Variations of R seem to be small.

131.052 **Constraints on the molecular properties of interstellar X-ogen derived from radio observations.**
J. M. Hollis, L. E. Snyder, D. Buhl, P. T. Giguere.
Astrophys. Journ., Vol. 200, 584 - 593 (1975).

Twenty-eight X-ogen emission sources at 89.189 GHz have been detected and observed; of these, 20 are new X-ogen sources which also contain other interstellar molecules. No infrared stars were found to be X-ogen sources. A new rest frequency, 89,188.65 MHz, has been derived for X-ogen. The authors show that X-ogen is not SiC, C_2H, or any molecule with spin-doubling and/or hyperfine structure.

131.053 **Interstellar K I and Na I absorption lines near Rho Ophiuchi.** L. M. Hobbs.
Astrophys. Journ., Vol. 200, 621 - 624 (1975).

Interferometric scans of the interstellar K I λ7699 and Na I D1 absorption lines have been obtained toward ρ Oph and o Sco, in the ρ Ophiuchi dark cloud. The scans reveal narrow, multiple lines toward ρ Oph. This multiplicity can explain the difference in radial velocity between the strong atomic absorption lines and the weak molecular lines in this direction, as observed at lower velocity resolution. In both directions the less saturated λ7699 lines yield K I column densities roughly consistent with the reddening.

131.054 **Detection of interstellar nitrogen sulfide.**
C. A. Gottlieb, J. A. Ball, E. W. Gottlieb, C. J. Lada, H. Penfield.
Astrophys. Journ., (*Letters*), Vol. 200, L147 - L149 (1975).
The authors report the detection and positive identification of interstellar NS. They observed emission toward Sgr B2 from the $J = 5/2 \rightarrow 3/2\ ^2\Pi_{1/2}$ *c*-state (115.16 GHz) and *d*-state (115.6 GHz) transitions of $^{14}N^{32}S$. The NS column density is $\sim 10^{14}$ cm^{-2} derived from statistical equilibrium calculations.

131.055 **Detection of 2.6-millimeter radiation probably due to nitrogen sulfide.** T. B. H. Kuiper, B. Zuckerman, R. K. Kakar, E. N. Rodriguez Kuiper.
Astrophys. Journ., (*Letters*), Vol. 200, L151 - L153 (1975).
Millimeter-wavelength radiation was seen in Sgr B2 at two frequencies which correspond to transitions between the $J = 5/2$ and the $J = 3/2$ Λ-doublet sublevels in the $^2\Pi_{1/2}$ state of nitrogen sulfide. If the radiation is due to this molecule, the hyperfine ratios may be anomalous.

131.056 **Strong molecular line emission associated with small Hα emission regions.**
G. N. Blair, W. L. Peters, P. A. Vanden Bout.
Astrophys. Journ., (*Letters*), Vol. 200, L161 - L164 (1975).
A search for new sites of ongoing star formation has been undertaken using Hα emission nebulosity as a guide. The $J = 1 \rightarrow 0$ transition of $^{12}C^{16}O$ is used as a probe of the structure of these regions and an indicator of the position of embedded sources. Of the 60 Hα regions surveyed thus far, 30 distinct areas of CO emission enhanced in one or more localized areas have been detected. Several are very strong molecular emission sources. To a distance of 4 kpc, their galactic distribution is similar to that of known protostellar regions containing powerful radio continuum sources. The data suggest that line broadening is present at the position of peak emission in seven sources and that $^{12}C^{16}O$ may be a poorer probe than $^{13}C^{16}O$ for deeply penetrating into molecular clouds.

131.057 **The relative positions of the OH and H_2O masers in W49N, W3 (OH), and W51.**
G. L. Mader, K. J. Johnston, J. M. Moran.
Bull. American Astron. Soc., Vol. 7, 417 (1975). – Abstr. AAS.

131.058 **Extremely rapid variations of the H_2O maser source near Herbig-Haro object No. 9.**
K. Y. Lo, J. M. Moran, M. Morris, R. Walker, A. H. Haschick.
Bull. American Astron. Soc., Vol. 7, 417 (1975). – Abstr. AAS.

131.059 **Molecular clouds associated with supernova remnants.**
H. A. Wootten, G. N. Blair, P. A. Vanden Bout.
Bull. American Astron. Soc., Vol. 7, 418 (1975). – Abstr. AAS.

131.060 **Limits on the local interstellar atomic hydrogen density determined from Copernicus observations of chromospheric emission profiles of a cool star.**
W. McClintock.
Bull. American Astron. Soc., Vol. 7, 418 (1975). – Abstr. AAS.

131.061 **Neutral hydrogen cloud structure around $l = 140°$, $b = 0°$.** G. L. Verschuur.
Bull. American Astron. Soc., Vol. 7, 418 (1975). – Abstr. AAS.

131.062 **Measurement of the C^{12}/C^{13} ratio in formaldehyde in 4 interstellar clouds.**
M. F. Chui, P. G. Goldsmith, D. M. Matsakis, C. H. Townes.
Bull. American Astron. Soc., Vol. 7, 418 - 419 (1975). – Abstr. AAS.

131.063 **Do X-ray observations provide evidence for abundance anomalies in the interstellar medium?**
S. K. Ride, A. B. C. Walker, Jr.
Bull. American Astron. Soc., Vol. 7, 419 (1975). – Abstr. AAS.

131.064 **Observations of the local interstellar medium with a helium gas absorption cell.**
M. Lampton, J. Freeman, F. Paresce, S. Bowyer.
Bull. American Astron. Soc., Vol. 7, 419 (1975). – Abstr. AAS.

131.065 **Grain heating of a warm intercloud medium.**
J. Lyon.
Bull. American Astron. Soc., Vol. 7, 419 (1975). – Abstr. AAS.

131.066 **Photo-electric heating of interstellar couds.**
M. Jura.
Bull. American Astron. Soc., Vol. 7, 420 (1975). – Abstr. AAS.

131.067 **Recent experimental results on the simulation of the diffuse interstellar lines.** F. M. Johnson.
Bull. American Astron. Soc., Vol. 7, 427 (1975). – Abstr. AAS.

131.068 **Radiation transport and non-LTE analysis of interstellar molecular lines. I. Carbon monoxide.**
C. M. Leung, H. S. Liszt.
Bull. American Astron. Soc., Vol. 7, 427 (1975). – Abstr. AAS.

131.069 **The porosity of the interstellar medium.**
E. M. Jones.
Bull. American Astron. Soc., Vol. 7, 428 (1975). – Abstr. AAS.

131.070 **Interstellar bubbles.**
J. I. Castor, R. McCray, R. Weaver.
Bull. American Astron. Soc., Vol. 7, 428 (1975). – Abstr. AAS.

131.071 **The ionization of cloud and intercloud hydrogen by O and B stars.** B. G. Elmergreen.
Bull. American Astron. Soc., Vol. 7, 428 (1975). – Abstr. AAS.

131.072 **Far infrared emission of interstellar dust.**
C. Ryter, J.-L. Puget.
Bull. American Astron. Soc., Vol. 7, 428 (1975). – Abstr. AAS.

131.073 **Anisotropy of constituents of interstellar grains.**
F. C. Gillett, T. W. Jones, K. M. Merrill, W. A. Stein.
Bull. American Astron. Soc., Vol. 7, 428 (1975). – Abstr. AAS.

131.074 **Grain temperature in giant molecular clouds associated with H II regions.** C. M. Leung.
Bull. American Astron. Soc., Vol. 7, 428 - 429 (1975). – Abstr. AAS.

131.075 **Infrared spectrophotometry of ices and silicates in molecular clouds.**
K. M. Merrill, R. W. Russell, B. T. Soifer.
Bull. American Astron. Soc., Vol. 7, 429 (1975). – Abstr. AAS.

131.076 **10-micron spectrophotometry of compact sources in W3.** S. Willner.
Bull. American Astron. Soc., Vol. 7, 440 (1975). – Abstr. AAS.

131.077 **Near infrared observations of "active" molecular clouds.** G. N. Blair, P. A. Vanden Bout.
Bull. American Astron. Soc., Vol. 7, 441 (1975). – Abstr. AAS.

131.078 **Radio recombination lines at 318 MHz.**
A. Parrish, E. K. Conklin, V. Pankonin.
Bull. American Astron. Soc., Vol. 7, 464 (1975). – Abstr. AAS.

131.079 **Submillimeter observations of NGC 2024, OMC-2, and Mon R-2.** H. S. Hudson, B. T. Soifer.
Bull. American Astron. Soc., Vol. 7, 465 (1975). – Abstr. AAS.

131.080 **On the identification of interstellar X-ogen.** L. E. Snyder, J. M. Hollis, B. L. Ulich, F. J. Lovas, D. Buhl.
Bull. American Astron. Soc., Vol. 7, 497 (1975). – Abstr. AAS.

131.081 **Physics of dense interstellar dust clouds.** C. M. Leung.
Bull. American Astron. Soc., Vol. 7, 498 (1975). – Abstr. AAS.

131.082 **Interstellare Ameisensäure.** G. Winnewisser, E. Churchwell.
SuW, Vol. 14, 288 - 289 (1975).

131.083 **Formaldehyde polymers in interstellar space.** A. H. Fawcett, with a reply by N. C. Wickramasinghe, K. S. V. Santhanan.
Nature, Vol. 257, 159 (1975).

131.084 **The diffuse λ 2200 Å interstellar absorption band and the abundances of neutral interstellar hydrogen and sodium.** C. Friedemann.
Astron. Nachr., Vol. 296, 237 - 239 (1975).

The equivalent width of the diffuse interstellar absorption band W_{2200} and the abundances of neutral interstellar hydrogen and sodium derived from the interstellar absorption lines in the spectra of hot stars are well correlated each other.

131.085 **Radiation density and lifetimes of molecules in interstellar dust clouds.** G. Sandell, K. Mattila.
Astron. Astrophys., Vol. 42, 357 - 364 (1975).

The radiation field in dust clouds as a function of cloud extinction is computed using the Monte Carlo method and recent results for the scattering properties of grains. This is used to recalculate the photodissociation lifetimes for a number of molecules in interstellar dust clouds. The results show that shielding by dust is not as efficient as commonly believed. For molecules in dense dark clouds ($A_V \approx 8^m$) the authors obtain lifetimes which are typically by a factor of $\sim 10^4$ shorter than the lifetimes obtained in the earlier calculations of Stief et al. (1972). Some comments are made concerning the theory by Watson and Salpeter (1972) on the molecule formation on surfaces of interstellar grains.

131.086 **Distribution of NH_3 clouds in the galactic center region.**
N. Kaifu, M. Morris, P. Palmer, B. Zuckerman.
Astrophys. Journ., Vol. 201, 98 - 101 (1975).

The galactic center has been surveyed in three transitions of NH_3 along the galactic plane at $b = -2'$ in the longitude range $358° \lesssim l \lesssim 2.°3$. The velocity-longitude contour map of the (1, 1) inversion transition shows a remarkably asymmetric distribution of dense molecular clouds with respect to the galactic center. Strong NH_3 sources are concentrated in the region of $0° \lesssim l \lesssim 2°$ and $0\ \mathrm{km\ s^{-1}} \lesssim V_r \lesssim 100\ \mathrm{km\ s^{-1}}$. Also weak, extended features have been found in this region and in regions corresponding to the nuclear disk. The total mass of the dense molecular clouds in the northern part of the galactic center region ($0° \lesssim l \lesssim 2°$) is estimated to be $\gtrsim 6 \times 10^7 M_\odot$.

131.087 **Interstellar HDCO; upper limit and significance as an indicator of interstellar molecule reactions.**
W. D. Watson, R. M. Crutcher, J. R. Dickel.
Astrophys. Journ., Vol. 201, 102 - 111 (1975).

Upper limits to the strength of the HDCO ($1_{11}-1_{10}$) transition at 5346 MHz are obtained from observations of Cas A, Cloud 4, NGC 2264, DR 21, Orion B, and Sgr B2. The observations are interpreted to yield an upper limit to the abundance ratio [HDCO]/[H_2CO] that is most likely about 1/170.

131.088 **Interstellar polarization: magnetite dust.** P. R. Shapiro.
Astrophys. Journ., Vol. 201, 151 - 164 (1975).

The polarization and extinction properties of magnetite, or Fe_3O_4, dust grains are calculated based on indirect evidence that magnetite grains might be the interstellar polarizers. The extinction and phase-lag cross sections of magnetite grains are calculated using a computer method for nonspherical grains. From these, the linear and circular polarization which would be produced by a medium of aligned magnetite dust grains is determined. It is found that flat platelets of magnetite can reproduce the wavelength dependence of the polarization observations. It is shown that magnetite grains can produce all the observed polarization at the visual wavelength by consuming roughly 16 percent of the cosmic abundance of iron atoms in the interstellar medium, while contributing roughly 4 percent of the total visual extinction. A general property of interstellar polarization is also discussed based on the Kramers-Kronig dispersion relations. An infrared observational test of this property is proposed.

131.089 **UV stars and the interstellar medium: a statistical time-dependent model.** J. Lyon.
Astrophys. Journ., Vol. 201, 168 - 178 (1975).

A statistical time-dependent model for the effects on the interstellar gas of the ionizing radiation from UV stars is presented. The radiation from the stars is assumed to come out as monochromatic bursts at random times and locations. The thermal and ionization history of the gas is followed in a Monte Carlo simulation. Probability densities as well as mean values for the ionization fraction, temperature, and other observable quantities are derived. The fluctuations in observed quantities due to the stochastic nature of the model are estimated. Models of the interstellar medium perpendicular to the galactic plane are made and compared with observations.

131.090 **Interstellar molecular abundances toward Omicron Persei.** T. P. Snow, Jr.
Astrophys. Journ., (*Letters*), Vol. 201, L21 - L24 (1975).

Extensive scanning of the B1 III star o Per ($V = 3.79$, $E[B-V] = 0.32$ mag) has been done with the Copernicus telescope-spectrometer in an effort to determine column densities or meaningful upper limits for several molecules expected to be relatively abundant in H I regions of moderate density. Molecules detected in ultraviolet absorption are H_2, HD, CO, and possibly OH. The CO rotational temperature is $2.2\ \mathrm{K} \leq T_{rot} \leq 3.8\ \mathrm{K}$. The observed column densities and upper limits are compared with recent models for molecular formation in diffuse clouds.

131.091 **Confirmation of interstellar N_2H^+.** P. Thaddeus, B. E. Turner.
Astrophys. Journ., (*Letters*), Vol. 201, L25 - L26 (1975).

The indentification of the new interstellar lines at 93.174 GHz as the molecular ion H_2H^+ has been confirmed by resolving in the source OMC-2 the predicted hyperfine structure of the inner nitrogen nucleus. The hyperfine constants of N_2H^+ derived from the observational data are eqQ (outer N) = -5.666 ± 0.012 MHz, eqQ(inner N) = -1.426 ± 0.021 MHz, and c(outer N) = 0.0147 ± 0.0023 MHz. The rest center frequency of the $J = 1-0$ rotational transition is determined to be 93173.392 ± 0.031 MHz.

131.092 **Microwave H_2O emission from young stars.** P. R. Schwartz, D. Buhl.
Astrophys. Journ., (*Letters*), Vol. 201, L27 - L28 (1975).

H_2O emission has been detected from the Ae or Be stars HD 250550, LkHα 234, and LkHα 198. Comparison of the H_2O velocity with that of associated interstellar CO indicates that the H_2O originates in an infall region, possibly the contracting parent cloud. H_2O emission has also been detected in OH 205.1–14.1, an OH source near the T Tauri star LkHα 308.

131.093 **Interstellar absorption of light in the direction of SA 28.** I. A. Dubyago.
Izv. Astron. Ehngel'gardt. Obs., *Kazan'*, No. 39, p. 89 - 98 (1972). In Russian.

131.094 **On the ratio of total to selective absorption.** S. S. Tokhtas'ev.
Izv. Astron. Ehngel'gardt. Obs., *Kazan'*, No. 40, p. 91 - 94 (1973). In Russian.

131.095 **Interstellar absorption of light in the direction of NGC 1664.** L. A. Urasin.
Izv. Astron. Ehngel'gardt. Obs., *Kazan'*, No. 40, p. 98 - 102 (1973). In Russian.

131.096 **On the structure of the absorbing complex in Auriga.** L. A. Urasin.
Izv. Astron. Ehngel'gardt. Obs., *Kazan'*, No. 40, p. 108 - 111 (1973). In Russian.

131.097 **The results of observations of stars with intrinsic polarization.** R. A. Vardanyan.
Soobshch. Byurakan. Obs., vyp. (No.) 46, p. 33 - 42 (1975). In Russian.

The results of observation of stars with intrinsic polarization and the conclusions drawn from the statistical investigation of observational data are given.

131.098 **Energy fluctuations of interstellar dust grains.** H.-P. Gail, E. Sedlmayr.
Astron. Astrophys., Vol. 43, 17 - 27 (1975).

The fluctuations of internal temperatures of interstellar dust grains caused by the absorption of photons from the interstellar radiation field and collisions with the hot component of the interstellar gas are treated by means of a probability distribution function. A method is presented to compute this probability for a given heating mechanism and a given model for the dust grains. The resulting integral equation is discussed for the idealized case of particles emitting and absorbing radiation like gray bodies. Numerical results are given for a monochromatic and a black-body radiation field.

131.099 **Radiation from inhomogeneous dusty Strömgren spheres.** B. Baschek, R. Wehrse.
Astron. Astrophys., Vol. 43, 29 - 36 (1975).

The characteristic behavior of the center-to-limb variation and the integrated spectrum of dusty Strömgren spheres is discussed on the basis of simplified models with constant gas-to-dust ratios. It is assumed that the ionization of the gas and the heating of the dust is due to a central star (of type O 6 V). The density distribution is assumed to be gaussian. The spectral region considered ranges from radio frequencies to the near infrared (to about 0.8 μ), including the region in which internal dust absorption becomes important.

131.100 **Observations of the diffuse λ 4430 absorption band in southern O, B, and A stars.** P. Gammelgaard.
Astron. Astrophys., Vol. 43, 85 - 92 (1975).

Photoelectric narrow-band observations of the intensity of the diffuse interstellar λ 4430 absorption band have been carried out for 65 southern O, B and A stars of luminosity classes I and II. The galactic longitudes of the stars are in the interval $184° < l^{II} < 53°$. The dependence of the λ 4430 index on stellar spectral lines is eliminated and the variation of the radio J_c/E_{B-V} of λ 4430 to selective absorption from place to place in the Galaxy is discussed. It is found that J_c/E_{B-V} is significantly lower in the inter-arm region than inside the inner and outer spiral arms. The observed value of J_c/E_{B-V} for HD 153919, the optical component of the binary system containing the X-ray source 2U 1700–37, indicates a distance of at least 1.5 kpc which suggests that it is more luminous and massive than an ordinary O 6 star.

131.101 **Interstellar reddening and IR-excess of O and B stars.** G. V. Schultz, W. Wiemer.
Astron. Astrophys., Vol. 43, 133 - 139 (1975).

350 O and B stars were selected according to specific criteria. Two-colour diagrams were used to derive a mean law of interstellar reddening for the various directions of our Galaxy included in this study. The reddening constant was derived to be $R = A_V/E_{B-V} = 3.14 \pm 0.10$. The authors conclude from their investigation that they are able to distinguish between normally reddened O and B stars and stars which show anomalous infrared excess.

131.102 **Attenuation of whistler mode waves excited in the interstellar medium by pulsars.** R. R. Burman.
Publ. Astron. Soc. Japan, Vol. 27, 511 - 513 (1975).

It is shown that whistler mode radiation that pulsars might excite in the interstellar medium is likely to be absorbed in much less than one light year.

131.103 **Interstellar masers and "antimasers" – thermal machines of space.** V. S. Strel'nitskij.
Zemlya i Vselennaya, 1975, No. 4, p. 15 - 20. In Russian.

131.104 **The molecular cloud Sagittarius B2.** N. Z. Scoville, P. M. Solomon, A. A. Penzias.
Astrophys. Journ., Vol. 201, 352 - 365 (1975).

The structure of the Sgr B2 molecular cloud has been studied by detailed and extensive mappings of the CO, ^{13}CO, CS, and H_2CO (2-cm) transitions. The cloud consists of a dense core at the OH maser position plus a very large envelope 45 pc across. Sagittarius B2 is one of the most massive objects in the Galaxy, containing about $3 \times 10^6 M_{\odot}$ – an estimate arrived at independently from calculated ^{13}CO column densities, from application of the virial theorem, and from far-infrared measurements.

131.105 **Rotational excitation of molecular ions in interstellar clouds.** S. Green.
Astrophys. Journ., Vol. 201, 366 - 372 (1975).

Rate constants for rotational excitation of N_2H^+ by low-energy collisions with He have been obtained from accurate quantum mechanical calculations. Rates for excitation of HCO^+ and rates for excitation by collisions with H_2 are expected to be qualitatively similar. The excitation rate for molecular ions is found to be only slightly faster than that for similar neutral species, and not much faster as had been previously suggested. The implications of this for interpreting microwave observations of interstellar N_2H^+ and HCO^+ are discussed.

131.106 **A semiempirical formula for interstellar birefringence.** P. G. Martin.
Astrophys. Journ., Vol. 201, 373 - 376 (1975).

A formula for the wavelength dependence of interstellar birefringence is found by a semi-empirical technique. This formula should facilitate the reduction and interpretation of observations of interstellar circular polarization, in the same manner in which Serkowski's formula is useful for linear polarization.

131.107 **Interstellarer Staub und seine Bedeutung im kosmogonischen Kreislauf.** E. Sedlmayr.
SuW, Vol. 14, 323 - 327 (1975).

131.108 **Velocity component structure of the Zeta Ophiuchi interstellar lines.** R. M. Crutcher.
Astrophys. Journ., Vol. 200, 625 - 633 (1975).

Multivelocity component structure in the ζ Oph interstellar spectrum has been investigated by fitting theoretical profiles to available high-velocity resolution Na I, Ca II, and K I profiles. Evidence is presented for variable depletion of Ca, Ti, and K with respect to Na, with more severe depletion in the stronger velocity components. Effects of explicitly considering the velocity structure in constructing curves of growth are discussed; many of the interstellar column densities derived from Copernicus ultraviolet data are very sensitive to the exact form of the curve of growth. Either of two models for the distribution of interstellar absorbers toward ζ Oph seems consistent with the data. One model includes significant contributions to the lines of some elements by the H II region toward ζ Oph, while the other calls for severe variable depletion among the neutral-region velocity components.

131.109 **Microwave detection of interstellar cyanamide.** B. E. Turner, A. G. Kislyakov, H. S. Liszt, N. Kaifu.
Astrophys. Journ., (*Letters*), Vol. 201, L149 - L152 (1975).

The authors report the detection of interstellar cyanamide (NH_2CN) based on the observation of line emission in Sgr B2 at 80504.5 and 100629.5 MHz corresponding to the transitions $4_{13}-3_{12}$ and $5_{14}-4_{13}$. Statistical equilibrium calculations indicate a low column density ($\sim 10^{14}$ cm^{-2}) arising from a rather high density region ($\gtrsim 10^5$ cm^{-3}).

131.110 **A numerical calculation of the effect of stellar winds on the interstellar medium.** S. A. E. G. Falle.
Astron. Astrophys., Vol. 43, 323 - 336 (1975).

The effect of a strong stellar wind on the interstellar medium is investigated numerically. It is found that this mechanism can produce the high velocities observed in H II regions provided the density is not greater than 100 protons/cc. The rapid cooling of the gas after it has been heated by a strong shock is found to produce secondary shocks in some cases. These shocks may have interesting observational consequences.

131.111 **Interstellar scattering of pulsar radiation. I. Scintillation.** D. C. Backer.
Astron. Astrophys., Vol. 43, 395 - 404 (1975).

An investigation of the intensity fluctuations of 28 pulsars near 0.4 GHz indicates that spectra of interstellar scintillation are consistent with a gaussian shape, that scintillation indices are near unity, and that the scintillation bandwidth depends linearly on dispersion measure. The variations of scattering parameters with dispersion measure may indicate that the rms deviation of thermal electron density on the scale of 10^{11} cm grows with path length through the Galaxy.

131.112 **Characteristics of the interstellar medium as deduced from low-frequency recombination line observations.** P. A. Shaver.
Astron. Astrophys., Vol. 43, 465 - 468 (1975).

Upper limits on the intensities of recombination lines below 200 MHz imply upper limits on the filling factor for the cold cloud component of the interstellar medium of $\approx 0.005-0.05$, depending on the temperature assumed; for the gas responsible for the observed low-frequency continuum absorption in Sgr A and W49B they imply lower limits on the electron density of 0.2–2 cm^{-3}.

131.113 **Hydrogen recombination lines of cosmological origin.** V. K. Dubrovich.
Pis'ma v Astron. Zhurn., Vol. 1, No. 10, p. 3 - 4 (1975). In Russian.

A possibility of formation of hydrogen recombination lines in transitions between high excitation levels (with numbers $i = 10-15$) at the stage of recombination of cosmological plasma is considered. Their intensities and widths are estimated.

131.114 **Ultraviolet studies of the interstellar gas.** L. Spitzer, Jr., E. B. Jenkins.
Annual Rev. Astron. Astrophys., Vol. 13, (see. 003.005), 133 - 164 (1975). – Contents: (1) Introduction; (2) Atomic hydrogen; (3) Interstellar molecules; (4) Abundances of atoms in H I regions; (5) Properties of ionized gas; (6) Ionization and thermal equilibrium.

131.115 **The correlation between the ultraviolet λ 2200 feature and the diffuse λ 4430 band.**
K. Nandy, G. I. Thompson.
Monthly Notices Roy. Astron. Soc., Vol. 173, 237 - 243 (1975).

Observations of the ultraviolet feature which occurs close to 2200 Å are presented for over 60 stars for which interstellar λ 4430 data are available in the literature. The equivalent widths of the λ 2200 feature have been determined from ultraviolet extinction at 2190 and 2500 Å, and the relation between the equivalent width of the ultraviolet feature and the central depth of the λ 4430 band has been determined. It is found that they are well correlated and the correlation coefficient, including allowance for errors, is greater than 0.9; this indicates that the carriers for the λ 2200 feature and diffuse band λ 4430 coexist in the interstellar medium.

131.116 **Λ-type doubling in SiH molecule.** I. D. L. Wilson, W. G. Richards.
Nature, Vol. 258, 133 - 134 (1975).

In recent years there has been much interest in the detection of the radio-frequency spectra of interstellar molecules. The observed spectra of the molecules OH and CH corresponded to transitions between the Λ-doubling components of the lower rotational levels. While experimental determinations proved very difficult, theoretical calculations on the Λ doubling in the lowest rotational level of CH have been notably successful. If the radioastronomical observation is corrected for the epicentre and the calculated value for centrifugal distortion, then the agreement of theory and experiment for the radioastronomical frequency is within 2 MHz in over 3,000 MHz.

131.117 **The overall distribution of carbon monoxide in the plane of the Galaxy.**
W. B. Burton, M. A. Gordon, T. M. Bania, F. J. Lockman.
Astrophys. Journ., Vol. 202, 30 - 49 (1975).

The authors observed the 2.6-mm line of ^{12}CO at 5° longitude intervals along the galactic equator between $l = 10°$ and 200° with a velocity coverage sufficient to include all velocities expected in terms of differential galactic rotation. CO emission is generally observed at positive velocities at $l < 60°$ up to the sub-central region terminal velocity. The oberservations show that 60 percent of the CO emission observed originates between 4 and 8 kpc from the galactic center. At $R > 8$ kpc and $R < 4$ kpc the accumulated emission decreases rapidly. At directions and velocities corresponding to $R > R_0 =$ 10 kpc, only a few isolated emission peaks are found, generally associated with known H II regions. H II and CO show a rather similar large-scale distribution.

131.118 **A survey of carbon monoxide emission in dark clouds.** R. L. Dickman.
Astrophys. Journ., Vol. 202, 50 - 57 (1975).

Results are reported of a CO and ^{13}CO survey of 68 dark clouds from the Lynds catalog. CO was detected in 63 of the 64 sources in which it was searched for, and the ^{13}CO line was seen in 52 of 55 clouds. There is a rather narrow distribution of CO peak line radiation temperatures about a mean of ~6 K. Typical CO column densities for the clouds surveyed are $N_{CO} \approx 5 \times 10^{17}$ cm^{-2}, assuming a terrestrial carbon isotope ratio. A comparison of line widths shows the ^{13}CO lines to generally be wider than their formaldehyde counterparts. Possible explanations of this fact in terms of internal cloud motions are discussed.

131.119 **Molecular alignment and radiative transport.** M. M. Litvak.
Astrophys. Journ., Vol. 202, 58 - 75 (1975).

Methods and examples are given for the calculation of the unequal populations among magnetic sublevels of interstellar molecules subject to anisotropic and polarized radiation and to elastic and inelastic collisions among the rotational levels. Though as yet unobserved, the most likely source of such radiation that thereby aligns the molecules, such as HCN, is an adjacent layer in a slab or cylinder region that emits as a maser parallel to the layer. The molecular alignment is shown to affect the opacity, source function, and emergent flux, for example, in the hyperfine-split 3-mm HCN components. Polarization properties of masers are briefly discussed.

131.120 **The 9.1-GHz cyanoacetylene line in Sagittarius B2.** R. X. McGee, L. M. Newton, P. W. Butler.
Astrophys. Journ., Vol. 202, 76 - 80 (1975).

Improved observations of the $J = 1 \rightarrow 0$ emission line of cyanoacetylene combined with evidence from other molecular-line observations in Sgr B2 suggest the presence of four separate molecular clouds, close to and in front of the source, with radial velocities near +49.3, +60.5, +68.6, and +80.2 km s^{-1}.

131.121 **H_2O sources in regions of star formation.** K. Y. Lo, B. F. Burke, A. D. Haschick.
Astrophys. Journ., Vol. 202, 81 - 91 (1975).

Regions and objects believed to be in early stages of stellar formation have been searched for H_2O 22-GHz line emission with the Haystack 120-foot (37 m) telescope. The objects include radio condensations, infrared objects in H II regions, and Herbig-Haro objects. Nine new H_2O sources were detected in the vicinity of such objects. The association of H_2O sources with regions of star formation is taken to be real. The spatial relationship of H_2O sources to infrared objects, radio condensations, class I OH sources, and molecular clouds are discussed. The suggestion is made that an H_2O source may be excited by a highly obscured star of extreme youth.

131.122 **"On the relation between CO and gamma-ray observations, cosmic rays, and the thickness of the galactic disk": a comment.** T. C. Mouschovias.
Astrophys. Journ., (*Letters*), Vol. 202, L1 - L3 (1975).

The author points out that on both theoretical and observational grounds molecular clouds are not dynamically decoupled from the rest of the interstellar medium. This removes the complications raised in the paper by Wentzel et al. The author discusses some theoretical difficulties with the stratified model of the interstellar medium, and he cites observational evidence against it. Finally, he comments on the suggestion by Wentzel et al. that the latitude distribution of γ-rays implied by the stratified model is similar to that of H I.

131.123 **Formaldehyde in the Rho Ophiuchi dark cloud.** P. C. Myers, P. T. P. Ho.
Astrophys. Journ., (*Letters*), Vol. 202, L25 - L30 (1975).

The 6-cm transition of H_2CO was mapped in absorption over a region ~1° × ~1° with 6′ resolution toward the ρ Oph dark cloud. The distribution of H_2CO column density agrees well with the extinction due to dust. At least three regions of high H_2CO concentration can be identified. Two regions of reduced H_2CO column density coincide with regions of carbon recombination-line emission, reflection nebulosity, and concentration of infrared sources. A north-south velocity gradient of ~0.5 km s^{-1} pc^{-1} is observed, which is probably not sufficient to stabilize the cloud against large-scale gravitational collapse.

131.124 **CO, dust and H_2 in the molecular cloud near ρ Ophiuchi.** P. J. Encrenaz, E. Falgarone, R. Lucas.
Astron. Astrophys., Vol. 44, 73 - 80 (1975).

A detailed map of the ρ Oph dark cloud in the $J = 1 \rightarrow 0$ line of CO is presented. Both the $^{12}C^{16}O$ and $^{13}C^{16}O$ isotopic molecules were observed. More structure is present on the CO map than on the ^{13}CO map, due to optical depth and temperature effects. The visual extinction A_v has been measured by the method of star counts for a large part of the cloud. N_{CO}/A_v is found to be constant $\simeq 10^{17}$ cm^{-2} mag^{-1} for 2 mag $< A_v <$ 6.3 mag, and much smaller for lower values of A_v in agreement with the Copernicus results. The authors derive a constant n_{CO}/n_{H_2} ratio of 10^{-4}, for the whole observed part of the cloud where $A_v < 6.3$ mag.

131.125 **On the influence of dust on the effectiveness of OH maser sources.** W. H. Kegel.
Astron. Astrophys., Vol. 44, 95 - 100 (1975).

The influence of continuous absorption by dust on the efficiency of OH maser sources is investigated. This influence is an indirect one acting on the infrared lines which connect the maser levels with the higher lying states. It is shown by means of a three level model that under certain conditions population inversion is possible even in regions which are optically thick in the infrared line, which is in contrast to the case of no continuous absorption. The maser condition in this case depends on the intensity in the maser line itself. It is satisfied easier for large intensities than for low ones.

131.126 **Studies of ultraviolet interstellar extinction with the sky-survey telescope of the TD-1 satellite. I. Results for three galactic regions.** K. Nandy, G. I. Thompson, C. Jamar, A. Monfils, R. Wilson.
Astron. Astrophys., Vol. 44, 195 - 203 (1975).

The variation of interstellar extinction in the wavelength range from 2740 Å to 1350 Å has been studied from the observations of about 100 reddened early type stars and a comparable number of comparison stars. It is found that the mean extinction curve derived separately for each of three widely separated galactic regions do not show longitude dependence. The equivalent widths and profile of the strong ultraviolet interstellar absorption band at 2200 Å has also been studied. Its strength is well correlated with the visual colour excess E_{B-V} and the equivalent width per kpc is ~ 130 Å ± 10. It is concluded that separate particles are needed to produce the general extinction (scattering) and the absorption band.

131.127 **The fine structure of the galactic dust cloud distribution near the sun.** P. Turon, M. O. Mennessier.
Astron. Astrophys., Vol. 44, 209 - 213 (1975).

Two very dark clouds in Ophiuchus and Taurus are almost exactly symmetrical around the sun. A study of the color excess of nearby stars confirms the idea of a thin elongated cloud linking the two symmetrical regions. The orientation of this long cloud suggests an aligning mechanism related to spiral density waves.

131.128 **The formation of pancakes from dust matter free of rotation.** I. D. Novikov.
Astron. Zhurn. Akad. Nauk SSSR, Vol. 52, 1038 - 1040 (1975). In Russian. English translation in Soviet Astron., Vol. 19, No. 5.

It is shown that in the framework of the Newtonian theory the formation of the two-dimensional singularity with infinite density, but with finite surface density (pancakes), is inevitable in matter without pressure and without rotation.

131.129 **The abundance of negative hydrogen in small hot interstellar clouds.** M. Rudkjøbing.
Astrophys. Space Sci., Vol. 37, 371 - 381 (1975).

The observed intensities of the diffuse interstellar absorption band at λ 4430 have minimum values, for given stellar distances, that are equivalent to the reddening caused by the small Ambarzumian-type clouds studied by B. Strömgren. This indicates that there is no negative hydrogen present between these clouds, which are then identified with the hot H I phase of interstellar matter. The equilibrium density of H^- inside such clouds is calculated from mean densities of neutral hydrogen atoms and free electrons, derived from radio observations for the local region outside the large Orion-arm clouds. An electron temperature of 3375 K corresponding to the local degree of ionization of the small-cloud hydrogen then leads to a mean density of negative hydrogen equal to 2×10^{13} cm^{-2} kpc^{-1}.

131.130 **Bok globules, molecular clouds, infra-red sources and star formation.** D. McNally.
Irish Astron. Journ., Vol. 11, 239 - 264 (1975). – Lecture given at the meeting of Irish Astronomical Group, Queen's University, Belfast, 1975 Jan. 9.

131.131 **Advances in astrochemistry.** V. S. Strel'nitskij. Zemlya i Vselennaya, 1975, No. 5, p. 32 - 33. In Russian.

131.132 **Laboratory measurements of some ion–molecule reactions related to sulphur chemistry in interstellar clouds.** J. P. Liddy, C. G. Freeman, M. J. McEwan.
Astrophys. Letters, Vol. 16, 155 - 157 (1975).

H_2S and a number of other molecules containing sulphur have been observed in dense interstellar clouds. A scheme for the chemistry of these sulphur-containing species was recently discussed by Oppenheimer and Dalgarno. They included in their scheme a number of ion–molecule reactions of sulphur compounds. The authors have measured in the laboratory rate coefficients of several ion–molecule reactions in the Oppenheimer-Dalgarno scheme (involving H_2S^+, H_2S, SO^+, SO_2 and S^+) and also report measurements of several related reactions.

131.133 **A very high velocity 1,612-MHz OH source near the galactic centre.** B. Baud, H. J. Habing, H. E. Matthews, J. D. O'Sullivan, A. Winnberg.
Nature, Vol. 258, 406 - 407 (1975).

A region of about 12 square degrees around the galactic centre was observed. Here the authors discovered a double-peaked absorption feature close to the galactic centre which appeared in several adjacent grid points. Subsequent analysis showed it to be an emission feature in the reference band and therefore at a high velocity. The most striking characteristic of this source is its high mean radial velocity of -341 km s^{-1}. The source seems to be pointlike with the 7.9′ beam of the 100-m telescope. A search on the Palomar Sky Survey reveals no unambiguous optical counterpart.

131.134 **An upper limit on interstellar C IV in the spectrum of γ^2 Velorum.** D. Lengyel-Frey, T. P. Stecher, D. K. West.
Observatory, Vol. 95, 210 - 211 (1975). – Short note.

131.135 **The ratio of total to selective absorption in dark clouds.** D. C. B. Whittet, I. G. van Breda.
Astrophys. Space Sci., Vol. 38, L3 - L8 (1975).

Evidence relating to the value of $R = A_V/E_{B-V}$ appropriate to discrete dark clouds in the interstellar medium is discussed. A value of $R \sim 4.4$ appears to apply to the most reddened stars in the ρ Oph complex, suggesting the existence of an upper size limit for the grains. If the grains grow by accretion of the heavier elements from the surrounding gas, then the availability of material sets a limitation on mantle growth consistent with the observed increase in R.

131.136 **Radio sources near reflection nebulae in Monoceros.** D. Downes, A. Winnberg, W. M. Goss, L. E. B. Johansson.
Astron. Astrophys., Vol. 44, 243 - 252 (1975).

Compact H II regions, OH and H_2O masers and a molecular cloud have been observed at radio frequencies near the reflection nebula NGC 2170 in Monoceros. The emission at 1612 MHz may be a new type of OH maser emission. The 21-cm line of H I, the 1720 MHz line of OH and the 4830 MHz line of H_2CO appear in absorption against the radio continuum sources, and show that an extensive molecular cloud lies in front of, or surrounds the ionized gas. The compact radio sources are close to a "bar" of optical nebulosity seen only on the red Sky Survey print, and coincide with a cluster of infrared sources. The excitation parameter for the radio sources indicates that the MonR2 group of reflection nebulae contains stars of earlier spectral type than had previously been suspected.

131.137 **Possible interaction of interstellar particles with the solar and terrestrial environment.**
J. M. Greenberg.
I.A.U. Colloquium No. 13, (see 012.030), p. 375 - 377 (1973).

The possibility for detection of interstellar particles in the earth's environment is considered on the basis of the passage of the solar system through the interstellar medium. Among the forces which inhibit interstellar particle penetration, the deflection by the solar magnetic field and the repulsive force due to the radiation from the sun are by far the most important.

131.138 **The structure and stability of shock waves in a multiphase interstellar medium. II. The effect of a magnetic field, thermal conduction, and viscosity.**
S. L. Mufson.
Astrophys. Journ., Vol. 202, 372 - 388 (1975).

The effect of shock waves on a parcel of gas initially in one phase of a multiphase interstellar medium is studied. Here a model interstellar medium is considered which is heated by a constant flux of low-energy cosmic rays and which includes a magnetic field transverse to the direction of the shock and nonzero coefficients of thermal conductivity and viscosity. The stability of the gas in the nonequilibrium conditions behind a shock is investigated in the short wavelength limit. The results show that gas shocked from either the intercloud or cloud phase encounters zones of thermal instability as it moves downstream to equilibrium. The numerical examples demonstrate that shocks in the intercloud gas possess a mechanism to convert intercloud gas to cloud gas and fragment this gas to form clouds and stars.

131.139 **On the Kramers-Kronig relations for interstellar polarization.** P. G. Martin.
Astrophys. Journ., Vol. 202, 389 - 392 (1975).

The difficulties encountered in using the Kramers-Kronig relations to predict the behavior of interstellar polarization are pointed out, while at the same time their value in an interpretive role is acknowledged. Observations of interstellar circular polarization lead to restrictions on the interstellar grain composition, and additional constraints should be possible through measurements of linear polarization in the infrared and the ultraviolet.

131.140 **Some implications of 10-micron interstellar polarization.** P. G. Martin.
Astrophys. Journ., Vol. 202, 393 - 399 (1975).

From observations of the linear polarization at the 10μ

"silicate" feature, such as reported for the Becklin-Neugebauer source, constraints may be placed not only on the shape and degree of alignment of the interstellar grains, but also on the 10 μ band strength of the grain material itself. Preliminary analysis of the available data suggests that the band strength may be smaller than the value usually assumed for interstellar silicates by a factor 2 or 3. Independent knowledge of the band strength is also important for the prediction of the accompanying near-infrared and optical extinction, both of which are needed in the interpretation of the nature of the infrared sources.

131.141 **Carbon monosulfide in dust clouds and globules.**
R. N. Martin, A. H. Barrett.
Astrophys. Journ., (*Letters*), Vol. 202, L83 - L86 (1975).

$C^{32}S$ has been detected in seven out of eight dust clouds searched in the $J = 2 \rightarrow 1$ and $J = 1 \rightarrow 0$ transitions. A search for the $J = 2 \rightarrow 1$ $C^{34}S$ line yielded one possible detection. The authors derive excitation temperatures of 2.9–7.0 K and optical depths greater than 1. Using a simple model, including radiative trapping, molecular hydrogen densities greater than 10^4 cm^{-3} and CS/H_2 abundances of 10^{-8} to 10^{-9} are derived.

131.142 **The depletion of interstellar elements and the interaction between gas and dust in space.**
T. P. Snow, Jr.
Astrophys. Journ., (*Letters*), Vol. 202, L87 - L90 (1975).

Recent data obtained with Copernicus, combined with new results from the literature, indicate that the depletions of interstellar elements may depend on cloud density in a simple way. This is expected if the depletions are due to accretion of gas particles onto grains under presently existing conditions, but is not expected if the depletions take place during the grain formation process, before mixing into the interstellar medium. If the grains do not carry large positive charges, then ion-grain encounters may be important not only in creating the depletions, but also in determining ionization equilibrium, particularly if a large population of very small grains is present.

131.143 **Interstellar magnesium absorption in the direction of four unreddened stars.**
A. Boksenberg, B. Kirkham, M. Pettini, B. Bates, P. P. D. Carson, P. L. Dufton, C. D. McKeith.
Astrophys. Journ., (*Letters*), Vol. 202, L91 - L95 (1975).

Spectral observations of the interstellar lines of Mg^+ at 2795 and 2802 Å and Mg^0 at 2852 Å have been made with a resolution of 0.1 Å, toward several stars, with an objective grating spectrograph mounted on a balloon-borne, star-stabilized platform. The authors discuss the analysis of the spectra of four, nearby, unreddened stars and derive values for the interstellar Mg abundance and electron density for the gas in these directions.

131.144 **Molecular hydrogen in H II region transition zones.**
J. K. Hill, J. Silk.
Astrophys. Journ., (*Letters*), Vol. 202, L97 - L99 (1975).

The authors demonstrate that H II–H I transition zones of moderate density (~10–100 cm^{-3}) may contain column densities of molecular hydrogen of order 10^{13}–10^{15} cm^{-2}.

131.145 **Molecular formation on interstellar grains.** T. Iguchi.
Publ. Astron. Soc. Japan, Vol. 27, 515 - 532 (1975).

The temperature dependence of the relative abundance of complex molecules formed on interstellar grains is examined. H, C, N, and O atoms of cosmic abundance ratio are assumed to stick onto the grain surface from the gas phase and be absorbed semi-chemically. Absorbed atoms migrate on the surface of grain passing over the potential barriers, and grow to more complex radicals or molecules in each encounter with other radicals or atoms. When the dust temperature T_d is below 20K, simple and fully saturated molecules such as H_2, H_2O, NH_3, and CH_4 are formed. Above 50K, complex and unsaturated molecules are formed. The observed variety of molecules is shown to be consistent with the abundance predicted for a grain temperature between 40 and 50K.

131.146 **The formation of large interstellar molecules.**
T. J. Millar, D. A. Williams.
Monthly Notices Roy. Astron. Soc., Vol. 173, 527 - 535 (1975).

A scheme is described for producing interstellar molecules of large size by association reactions of neutral radicals. Molecule densities are calculated for an interstellar cloud of 500$M_\odot$, total hydrogen density 3×10^3 cm^{-3}, in which molecules are destroyed by photodissociation. The significance of these results is discussed.

131.147 **New 1612 MHz OH emission sources.**
J. L. Caswell, R. F. Haynes.
Monthly Notices Roy. Astron. Soc., Vol. 173, 649 - 669 (1975).

A search for OH on the 1612 MHz transition has been made at low galactic latitudes ($|b|<0.°4$) over the longitude range 326° to 340°; 27 emission sources were detected, of which 20 are new discoveries. Fourteen of the sources are apparently masers of the OH/IR star type and many of the remainder represent a new class of source. The survey showed 1612 MHz OH absorption to be prominent in a number of directions and this is discussed briefly. In addition the authors have determined more accurate positions for several known 1612 MHz OH emitters outside the search area.

131.148 **Detection of carbon monoxide in the Large Magellanic Cloud.** P. J. Huggins, A. R. Gillespie, T. G. Phillips, F. Gardner, S. Knowles.
Monthly Notices Roy. Astron. Soc., Vol. 173, 69P - 71P (1975).

CO emission at 2.6 mm has been detected in the region N159 of the Large Magellanic Cloud, using the 3.9-m Anglo-Australian telescope. The molecular cloud observed is comparable in properties with those of large clouds in our Galaxy.

131.149 **Observations of the OH source in W 44 at 18 cm wavelength.** M. I. Pashchenko.
Pis'ma v Astron. Zhurn., Vol. 1, No. 12, p. 23 - 25 (1975). In Russian.

With the large radiotelescope at Nançay the coordinates of the OH emission source W 44 (G34.3 + 0.1) were measured. OH absorption in W 44 was observed at the same radial velocities as OH and H_2O maser emission. The OH and H_2O emission sources seem to be associated with a cold cloud of neutral gas and are situated at a distance of ~3.8 kpc from the sun.

131.150 **Non-linear effects in the emission of interstellar molecules.** G. I. Smirnov.
Zhurn. ehksperim. i teor. fiz., Vol. 69, 3 - 8 (1975). In Russian. Abstr. in Referativ. Zhurn. 51. Astron., 11.51.608 (1975).

131.151 **Polarization of cosmic OH maser emission.**
D. A. Varshalovich, V. V. Burdyuzha.
Astron. Zhurn. Akad. Nauk SSSR, Vol. 52, 1178 - 1186 (1975). In Russian. English translation in Soviet Astron., Vol. 19, No. 6.

A mechanism is proposed which leads to the enhancement of one of the circularly polarized Zeeman components 1665 and 1667 MHz of the OH maser radiation. On the basis of an analysis of observations arguments for the Zeeman interpretation of the circular polarization of OH maser radiation are given.

131.152 **Preliminary interstellar spectral line search in the frequency range 72–78 GHz.**

W. L. H. Shuter, W. H. McCutcheon, R. M. Lees, M. Mahoney, P. Joensen, P. Gregory.
Journ. Roy. Astron. Soc. Canada, Vol. 69, 247 (1975).
Abstr. CAS.

131.153 **A search for a recombination line of ionized helium in W3.** J. N. Clarke, V. A. Hughes, M. R. Viner.
Journ. Roy. Astron. Soc. Canada, Vol. 69, 247 - 248 (1975).
Abstr. CAS.

131.154 **On fragmentation in a collapsing gas cloud.** R. C. Smith, A. E. Wright.
Journ. Roy. Astron. Soc. Canada, Vol. 69, 252 (1975).
Abstr. CAS.

131.155 **Interstellar Na I and Ca II absorption in intercloud regions.** L. M. Hobbs.
Astrophys. Journ., Vol. 202, 628 - 633 (1975).

The weak interstellar absorption lines of Na I, Ca II, and H I formed in the intercloud medium toward eight nearly unreddened stars are analyzed. Compared with the corresponding column densities within a standard cloud, the $N(\text{Na I})/N_H$ ratio is smaller by a factor probably in excess of 17, while the $N(\text{Ca II})/N_H$ ratio is about normal. These results imply that the local intercloud medium extensively contains low-density H I regions in which (1) the temperature $T \gtrsim 2500$ K and (2) the depletion of calcium is less severe than in the clouds by a factor of 20.

131.156 **Observations of interstellar Na I lines at 3302 Å.** R. M. Crutcher.
Astrophys. Journ., Vol. 202, 634 - 637 (1975).

Observations have been made of the equivalent widths of the interstellar Na I λ3302 doublet toward 21 stars which have strongly saturated D-lines. Column densities of Na I have been derived, and the depletion of Na in the interstellar medium is discussed. Toward most stars Na is depleted by about a factor of 6. The distribution of line widths is found to be strongly peaked near 1 km s^{-1}, in agreement with the isothermal sound velocity at 100 K.

131.157 **Interstellar absorption lines toward Gamma Arae.** D. C. Morton, E. M. Hu.
Astrophys. Journ., Vol. 202, 638 - 649 (1975).

The available data on interstellar absorption lines in the spectrum of γ Ara ($E_{B-V} = 0.08$) observed from the ground and with the Copernicus telescope have been analyzed to give column densities of various ion stages of H, C, N, O, Na, Mg, Si, P, S, Ar, Ca, Mn, and Fe. The element abundances with respect to H nuclei in H I regions are similar to the values for ζ Oph, with depletions relative to the sun tending to increase with increasing condensation temperature.

131.158 **Evidence for Zeeman splitting in 1720-MHz OH line emission.** K. Y. Lo, R. C. Walker, B. F. Burke, J. M. Moran, K. J. Johnston, M. S. Ewing.
Astrophys. Journ., Vol. 202, 650 - 654 (1975).

A number of 1720-MHz OH maser sources have relatively simple spectra consisting of doublets with opposite senses of circular polarization. Such doublets could be due to the Zeeman splitting of the $^2\Pi_{3/2}\, F = 2 \rightarrow 1$ transition. A three-station VLBU experiment was conducted to measure the spatial separation of the emitting regions of the components of opposite sense polarization. The doublets in the 1720-MHz OH spectra of W3 (OH), W51, NGC 6334 (N), and NGC 7538 (N) are found to be spatially coincident to within a linear dimension of $\sim 10^{15}$ cm (10^2 AU), suggesting that they originate from the same region in space.

131.159 **Observations of millimeter-wave HCN in four prototype clouds.** C. A. Gottlieb, C. J. Lada, E. W. Gottlieb, A. E. Lilley, M. M. Litvak.
Astrophys. Journ., Vol. 202, 655 - 672 (1975).

High-quality $J = 1 \rightarrow 0$ $H^{12}CN$ and $H^{13}CN$ line profiles at 3.4 mm have been obtained in M17, NGC 2264, Ori A, and NGC 1333. Values for the ratios of the hyperfine-split intensities and the ratio of the integrated intensities of $H^{12}CN$ and $H^{13}CN$ are given. Comparison of calculations with the data indicates that the parts of the collapsing or the turbulent cloud responsible for observed line widths of greater than 1 or 2 km s^{-1} cannot be interacting radiatively via Doppler effects. Conditions of high hydrogen density and moderate HCN column density can be found that are consistent with both CS data and the terrestrial ratio of abundances $H^{12}CN/H^{13}CN$, but not with the observed $H^{13}CN$ hyperfine-split intensity ratios. An isotope ratio smaller than terrestrial by a factor of 4 or more is required.

131.160 **Further investigations of the CO emission from the North America dust cloud and L1630.**
A. S. Milman.
Astrophys. Journ., Vol. 202, 673 - 689 (1975).

This paper reports further observations of the $J = 1-0$ rotational lines of $^{12}C^{16}O$ and $^{13}C^{16}O$ in two dust clouds, the North America nebula dust cloud and L1630. The author has looked for bright CO emission from the directions of individual Hα-emission stars in L1630 and the North America cloud, but the CO emission at these places is not enhanced above that of the surrounding material. He concludes that these stars are too cool to heat the gas appreciably. The shapes of the CO profiles observed near NGC 2023 show that the gas motions there are turbulent, rather than systematic radial motions of the gas. From the observations of the North America nebula dust cloud, the author shows that not more than about 10 percent of the carbon present in the gas is in the form of CO.

131.161 **On the interaction of stars with interstellar matter.** L. A. Pavlova.
Trudy Astrofiz. Inst., *Alma-Ata*, Vol. 25, 18 - 22 (1975).
In Russian.

131.162 **Anisotropy of constituents of interstellar grains.** F. C. Gillett, T. W. Jones, K. M. Merrill, W. A. Stein.
Astron. Astrophys., Vol. 45, 77 - 81 (1975).

New spectrophotometric observations of silicate ($\Delta\lambda/\lambda \approx 0.02$) and ice ($\Delta\lambda/\lambda \approx 0.01$) absorption in the directions of VI Cyg No. 12 and the BN source in Orion are described. The results are discussed in terms of laboratory data on silicate absorption. Calculations indicate that about 5 magnitudes of visual extinction (of the total of 10 magnitudes for VI Cyg No. 12) could be produced by pure silicates that also produce the observed 10 μ absorption band (for a 10 μ silicate opacity of 5000 cm^2 gm^{-1}). This would imply the necessary existence of an additional grain constituent unless the silicate opacity in interstellar space is less than 5000 cm^2 gm^{-1}.

131.163 **H I and OH observations towards W3.** J. Crovisier, R. Fillit, I. Kazès, A. Baudry.
Astron. Astrophys., Vol. 45, 97 - 105 (1975).

The double radio source G 133.7 + 1.2 and G 133.8 + 1.4 in W3 was observed in the H I line and in the ground-state (18 cm) and excited-state (6.3 cm) OH lines. H I and OH (18 cm) absorption features around −20 km s^{-1} are attributed to a cold cloud lying along the line of sight. H I and OH (18 cm) absorption features around −40 km s^{-1} are probably due to a dense ($n \sim 10^3$ cm^{-3}) cold ($T_K \sim 20$ K) cloud associated with the W3 complex (diameter ~ 13 pc). If the observed large H I column density is mainly due to this cloud, then it contains a considerable amount of atomic hydrogen. Several maser-like OH emission features have been observed at 18 cm and 6.3 cm at various velocities. One of them suggests that IRS 5 is an H_2O/OH maser source. Some possible excitation mechanisms are discussed.

131.164 **Observations of methyl formate in the galactic center.** E. Churchwell, G. Winnewisser.
Astron. Astrophys., Vol. 45, 229 - 231 (1975).

The $1_{10} - 1_{11}$ transitions of the A and E internal rotation states of methyl formate ($HCOOCH_3$) have been observed in emission toward Sgr B 2. Both the A- and E-states were clearly detected with essentially the same intensity and the same radial velocity. The mean values are: $T_{AL} = 0.050 \pm 0.009$ K, $\Delta\nu_L = 23.4 \pm 3.0$ km s^{-1}, $V_{LSR} = +59.0 \pm 1.5$ km s^{-1}.

131.165 **Interstellar absorption of light in the dark nebula in Taurus.** M. D. Metreveli.
Byull. Abastumansk. Astrofiz. Obs., No. 46, p. 59 - 80 (1975). In Russian.

The interstellar absorption and distribution of dark matter in the Taurus dark nebula up to an average distance of 1.5–2.0 kpc have been studied on the basis of the author's catalogue (1969) of stellar magnitudes, colour indices and spectral and luminsoity classes. A region of 76 square degrees was investigated, being divided into 10 subregions with homogeneous absorption. The dependence of the total visual absorption upon the distance was plotted for each subregion. The analysis of these curves permitted recognition of individual absorbing clouds. There is a positive correlation observed between the neutral hydrogen distribution and that of absorbing matter.

131.166 **Investigation of interstellar absorption and of spatial distribution of stars in a region near the North Galactic Pole.** E. I. Zajtseva.
Byull. Abastumansk. Astrofiz. Obs., No. 46, p. 81 - 102 (1975) In Russian.

The catalogue of star magnitudes V, colour-indices (B–V), spectral types and luminosity classes of 4268 stars (Bull. Abastumani Astrophys. Obs. 1973, No. 44) provided the basis for an investigation of interstellar absorption and spatial star distribution in one region neighboring the North Galactic Pole.

131.167 **The formation of "pancakes" from dust matter free of rotation.** I. D. Novikov.
In-t kosmich. issled. AN SSSR. Pr-229. Moskva, 10 pp. In Russian. – Abstr. in Referativ. Zhurn. 51. Astron. 12.51.694 (1975).

131.168 **On the scattering properties of graphite grains in circumstellar and interstellar space due to temperature variations.** J. Svatoš, M. Šolc, V. Vanýsek.
Acta Univ. Carolinae Math. Phys., Vol. 16, No. 1, p. 83 - 91 (1975).

The dependence of optical properties of graphite grains on temperature changes was computed. It is shown that this mechanism can play an important role in explaining the time-dependent variations of polarization in some late-type variable stars.

131.169 **Observations of the ^{13}C isotope of H_2CO.**
T. L. Wilson, J. Bieging, D. Downes, F. Gardner.
Bull. American Astron. Soc., Vol. 7, 520 - 521 (1975). Abstr. AAS.

131.170 **Isotopic abundances of interstellar carbon monosulphide.**
P. G. Wannier, R. A. Linke, R. W. Wilson, A. A. Penzias.
Bull. American Astron. Soc., Vol. 7, 521 (1975). – Abstr. AAS.

131.171 **Measurement of the double isotopic abundance ratio: $[^{12}C][^{15}N]/[^{13}C][^{14}N]$ in five interstellar molecular clouds.**
R. A. Linke, A. A. Penzias, R. W. Wilson, P. G. Wannier.
Bull. American Astron. Soc., Vol. 7, 521 (1975). – Abstr. AAS.

131.172 **Digicon spectroscopy of interstellar lines – an attempt to determine the interstellar $^{12}C/^{13}C$ isotopic abundance ratio.**
R. Snell, R. Tull, P. vanden Bout, S. Vogt.
Bull. American Astron. Soc., Vol. 7, 521 (1975). – Abstr. AAS.

131.173 **Colliding molecular clouds and star formation in NGC 1333.** R. B. Loren.
Bull. American Astron. Soc., Vol. 7, 531 - 532 (1975). Abstr. AAS.

131.174 **Radio detection of the formyl radical in molecular interstellar clouds.**
J. M. Hollis, L. E. Snyder, B. L. Ulich.
Bull. American Astron. Soc., Vol. 7, 540 (1975). – Abstr. AAS.

131.175 **Optical properties between 3 and 12 μm of interstellar medium candidates.**
W. G. Egan, T. Hilgeman.
Bull. American Astron. Soc., Vol. 7, 540 (1975). – Abstr. AAS.

131.176 **Calculations of the lower electronic states of CH_3 and the formation of interstellar CH^+.**
R. J. Blint, R. F. Marshall, W. D. Watson.
Bull. American Astron. Soc., Vol. 7, 540 (1975). – Abstr. AAS.

131.177 **Time scales for molecule formation by ion molecule reactions.** W. D. Langer.
Bull. American Astron. Soc., Vol. 7, 540 - 541 (1975). Abstr. AAS.

131.178 **Interstellar Na I and Ca II absorption in intercloud regions.** L. M. Hobbs.
Bull. American Astron. Soc., Vol. 7, 541 (1975). – Abstr. AAS.

131.179 **Condensation of grains in hot cosmic clouds.**
B. Donn.
Bull. American Astron. Soc., Vol. 7, 541 (1975). – Abstr. AAS.

131.180 **Molecular equations of state for dense dust clouds.**
S. Ames.
Bull. American Astron. Soc., Vol. 7, 547 (1975). – Abstr. AAS.

131.181 **Detection of interstellar OH at 3078 Å.**
R. M. Crutcher, W. D. Watson.
Bull. American Astron. Soc., Vol. 7, 547 (1975). – Abstr. AAS.

131.182 **Interstellar material in the direction of χ Oph.**
P. C. Frisch.
Bull. American Astron. Soc., Vol. 7, 547 (1975). – Abstr. AAS.

131.183 **Detection of interstellar deuterium Lyman Alpha.**
W. McClintock, R. C. Henry, H. W. Moos, J. L. Linsky.
Bull. American Astron. Soc., Vol. 7, 547 (1975). – Abstr. AAS.

131.184 **Enhanced effects of starlight on the interstellar medium.** H. Gerola, R. A. Schwartz.
Bull. American Astron. Soc., Vol. 7, 548 (1975). – Abstr. AAS.

131.185 **Theoretical models of interstellar shock waves.**
J. C. Raymond, D. P. Cox.
Bull. American Astron. Soc., Vol. 7, 555 (1975). – Abstr. AAS.

131.186 **Surface brightness profile analysis of a bright dark nebula in Chamaeleon.**
M. P. FitzGerald, T. C. Stephens, A. N. Witt.
Bull. American Astron. Soc., Vol. 7, 556 (1975). – Abstr. AAS.

131.187 **Interstellar molecules. I. Spectral properties of $^{12}C^{16}O$, $^{12}C^{18}O$, $^{12}C^{17}O$, $^{13}C^{16}O$.**
V. K. Khersonskij, D. A. Varshalovich.
Soobshch. Spets. Astrofiz. Obs., *Zelenchukskaya,* vyp. (No.) 14, p. 43 - 58 (1975). In Russian.

Frequencies, line strengths, and probabilities of rotational transitions, $J \rightarrow J-1$, for $J \leqslant 55$ in the vibrational states v = 0, 1, 2, of molecules $^{12}C^{16}O$, $^{12}C^{18}O$, $^{12}C^{17}O$, $^{13}C^{16}O$ are calculated on the basis of the experimental data available. The specific optical depths for more intensive microwave lines at $\lambda = 0.3 - 3.0$ mm are given as a function of temperature.

131.188 **Interstellar molecules. II. Spectral properties of $^{12}C^{32}S$, $^{12}C^{33}S$, $^{12}C^{34}S$, $^{13}C^{32}S$.**
V. K. Khersonskij, D. A. Varshalovich.
Soobshch. Spets. Astrofiz. Obs., *Zelenchukskaya,* vyp. (No.) 14, p. 59 - 77 (1975). In Russian.

Frequencies, line strengths, and probabilities of rotational transitions, $J \rightarrow J-1$, for $J \leqslant 120$ in the vibrational states v = 0, 1, 2, of molecules $^{12}C^{32}S$, $^{12}C^{33}S$, $^{12}C^{34}S$, $^{13}C^{32}S$ are calculated on the basis of the experimental data available. The specific optical depths for more intensive microwave lines at $\lambda = 0.6 - 6.0$ mm are given as a function of temperature.

131.189 **Interstellar molecules. III. Spectral properties of $^{28}Si^{16}O$, $^{29}Si^{16}O$, $^{30}Si^{16}O$.**
V. K. Khersonskij, D. A. Varshalovich.
Soobshch. Spets. Astrofiz. Obs., *Zelenchukskaya,* vyp. (No.) 14, p. 78 - 93 (1975). In Russian.

Frequencies, line strengths, and probabilities of rotational transitions, $J \rightarrow J-1$, for $J \leqslant 140$ in the vibrational states v = 0, 1, 2, of molecules $^{28}Si^{16}O$, $^{29}Si^{16}O$, $^{30}Si^{16}O$ are calculated on the basis of the experimental data available. The specific optical depths for more intensive microwave lines at $\lambda = 0.8 - 7.0$ mm are given as a function of temperature.

131.190 **Behavior of grains in early supernova environments.** S. W. Falk, J. M. Scalo.
Astrophys. Journ., Vol. 202, 690 - 695 (1975).

The efficiency of grain destruction by collisional and radiative heating and sputtering in a supernova environment is examined. It is found that radiative heating rapidly destroys any grains composed of magnesium silicates, graphite, silicon carbide, iron, or ices. Quartz grains without impurities could survive evaporation, but sputtering and fragmentation should eliminate all but the smallest grains. It is concluded that grains have no observable effect on early supernovae, but that small surviving particles might serve as condensation nuclei during the later phases of the expansion.

131.191 **Ultraviolet studies of the interstellar medium.** B. Bates.
Irish Astron. Journ., Vol. 12, 49 - 60 (1975). – Paper based on a talk given at a meeting of the Irish Astronomical Group held in the Queen's University of Belfast, 9th January, 1975.

131.192 **Experimental investigations of extinction on small meteoritic particles.** A. N. Gerashchenko.
Izv. Glav. Astron. Obs. Pulkovo, No. 193, Astrofiz., Astrometr. i Zvezdn. Astron., p. 65 - 75 (1975). In Russian.

The spectral dependence of extinction ($1.3 \leqslant \lambda^{-1} \leqslant 5.0$ mkm^{-1}) was investigated on small particles obtained by fragmentation of stony meteorites. The form and size of particles were investigated simultaneously. The results of the experiment have been compared with theoretical calculations and the results of investigations of extinction from stars in the interstellar space. It is concluded that in the range of the spectra under investigation the curve of the interstellar extinction can be interpreted as the result of extinction on small particles which are similar to stony meteorites in their composition.

131.193 **A nomogram for determination of interstellar absorption according to P. P. Parenago's formula.**
G. G. Lengauer.
Izv. Glav. Astron. Obs. Pulkovo, No. 193, Astrofiz., Astrometr. i Zvezdn. Astron., p. 76 - 79 (1975). In Russian.

131.194 **Faint Hα emission with radial velocity of −180 km/sec in the direction of M31.**
V. K. Golev, P. V. Shcheglov.
Astron. Tsirk., No. 874, p. 4 - 6 (1975). In Russian.

131.195 **On the ratio of extinction to reddening for interstellar matter using galaxies. I. A limit on the neutral extinction from photometry of the 3C 129 group.**
A. Sandage.
Publ. Astron. Soc. Pacific, Vol. 87, 853 - 857 (1975).

The ratio of total extinction to reddening of interstellar matter has been determined by a new method applied to galaxies in the highly obscured 3C 129 and 3C 129.1 group ($l = 160°$, $b = 0°$). The difference between the observed magnitude of the first-ranked group member and the magnitude calculated from the redshift-apparent magnitude (Hubble) diagram (at the known redshift of the group) is the total extinction A_v to within the $\sigma(M_v)$ that applies to the first-ranked cluster member. Comparison of the observed color with the known intrinsic color of giant E galaxies gives $E(B-V)$, and therefore R. Photometry of the 3C 129 group gives $R \equiv A_v/E(B-V) = 3.72 \pm 0.32(\sigma)$. Correction for the finite bandwidth of the BV system gives R (O star base) = $3.35 \pm 0.29(\sigma)$.

131.196 **New OH-emission source W33C.**
M. I. Pashchenko.
Astron. Tsirk., No. 886, p. 1 - 3 (1975). In Russian.

131.197 **Solar system as space-probe.** W. H. McCrea.
Observatory, Vol. 95, 239 - 255 (1975). – Written version of the Halley Lecture, 1975 May 6.

131.198 **A polarimetric search for circumstellar envelopes in the young clusters NGC 6611 and NGC 225.**
L. Carrasco, K. M. Strom, S. E. Strom.
Rev. Mexicana Astron. Astrofis., Vol. 1, 283 - 288 (1975).

Polarimetric observations of stars located in the young clusters NGC 6611 and NGC 225 are used to attempt detection of circumstellar envelopes. While many stars show relatively large polarization values, the observed polarization appears to be primarily interstellar in origin. No evidence is found to support the contention of Reddish (1967) that the most luminous members of young clusters are surrounded by circumstellar envelopes.

131.199 **An atlas of profiles of interstellar calcium and hydrogen lines.** D. A. Beintema.
Thesis, Rijksuniv. Groningen (Netherlands). [Available from Rijksuniv. Groningen, afd. Sterrenkunde], 83 pp. (1975).

131.200 **Electron and ion densities in interstellar clouds.**
A. E. Glassgold, W. D. Langer.
NASA Rep. NASA-CR-138964, 29 pp. (1974).

131.201 **Astronomical water vapor masers.** S. Kwok.
Thesis, Minnesota Univ., Minneapolis (USA). [Available from Univ. Microfilms, Ann Arbor, Mich. Order No. 75-2116], 127 pp. (1974).

131.202 **Fourier analysis of steady-state reaction schemes for interstellar molecules.** P. J. Kuntz.
Hahn-Meitner-Inst. Kernforschung, Berlin. HMI-B-162, 19 pp. (1975).

131.203 **Static equilibria of the interstellar gas in the presence of magnetic and gravitational fields.**
T. C. Mouschovias.
Thesis, California Univ., Berkeley (USA). Lawrence Berkeley Lab., 217 pp. (1975).

No exact self-consistent equilibrium calculations exist for (any model of) the system of the interstellar gas and the frozen-in magnetic field. On a large scale (approximately 1 kpc) this system is affected by the vertical galactic gravitational field, while on a small scale (approximately 1 pc) the self-gravitation of the gas comes into play and is responsible for the collapse of some clouds to form stars. Accessible equilibrium states are determined for the gas–field system on both of these scales.

131.204 **Laboratory determination of the frequencies of possible interstellar lines of NO^+.**
F. Alberti, A. E. Douglas.
Canadian Journ. Phys., Vol. 53, 1179 - 1181 (1975).

131.205 **Relaxation phenomena associated with microwave pumping of rotational transitions.** C. Bottcher.
Chem. Phys. Letters, (*Netherlands*), Vol. 34, 143 - 146 (1975).

It is shown that the Bloch equations describing a two-level system in a radiation field can be extended to take proper account of the degenerate magnetic sublevels.

131.206 **Observations of the interstellar gas with the Copernicus satellite.** D. C. Morton.
Phil. Trans. Roy. Soc. London, Ser. A, Vol. 279, 299 - 302 (1975).

Recent results are summarized on interstellar H I, D I, H_2, HD, and heavy elements whose absorption lines have been measured in the far ultraviolet.

131.207 **Interstellar gas studies in the balloon ultraviolet.**
A. Boksenberg, B. Kirkham, E. Michelson, M. Pettini, B. Bates, P. P. D. Carson, G. R. Courts, P. L. Dufton, C. D. McKeith.
Phil. Trans. Roy. Soc. London, Ser. A, Vol. 279, 303 - 316 (1975).

Spectral observations at a resolution of 0.1 Å covering the region 2870–2740 Å have been made of several stars with an objective grating spectrograph mounted on a balloon-borne star-stabilized platform. Measurements are reported of the interstellar lines of Mg II at 2795.53 Å and 2802.70 Å and Mg I at 2852.13 Å. The interstellar Mg abundance is obtained directly from the strengths of the Mg II lines. Observations of both Mg° and Mg^+ yield information on the ionisation balance and interstellar electron density. The results indicate that the predominant ionising mechanism in cool clouds is photoionisation by starlight.

131.208 **A study of interstellar lines in the rocket spectrum of δ Scorpii.** A. M. Smith.
Phil. Trans. Roy. Soc. London, Ser. A, Vol. 279, 317 - 322 (1975).

Ten interstellar lines in the vacuum ultraviolet spectrum of δ Sco are analyzed. The data were taken from a rocket spectrogram with wavelength coverage extending from 1177 to 1717 Å with a resolution of approximately 0.15 Å. Column densities of C°, C^+, N°, Al^+, Si^+ and Fe^+ are derived, from which abundances relative to atomic hydrogen are determined. Cloud temperatures of 47 and 76 K are obtained from the interstellar H_2 spectrum.

131.209 **Interstellar molecular absorption lines.**
A. Dalgarno.
Phil. Trans. Roy. Soc. London, Ser. A, Vol. 279, 323 - 329 (1975).

The ultraviolet absorption lines produced in interstellar clouds through the absorption of radiation from early type stars by molecular hydrogen H_2 may be analyzed to yield the particle densities and ultraviolet fluxes in the absorbing regions and the absorption by HD may be analyzed to determine the proton densities and the ionizing fluxes. The interpretation carries implications concerning the formation of other interstellar molecules, which may be the source of additional absorption lines.

131.210 **The interstellar ionization balance due to ultraviolet radiation.** P. M. Gondhalekar, R. Wilson.
Phil. Trans. Roy. Soc. London, Ser. A, Vol. 279, 331 - 336 (1975).

A new estimate of the interstellar ultraviolet radiation field is used to analyze some well established observations of interstellar Na I, Ca II, Mg I, II in η, ϵ and δ Ori. Four models of the interstellar medium are used and chemical abundances derived relative to that of hydrogen, obtained from Lyman alpha data. The scatter of the results is discussed together with the apparent anomalous abundance of calcium.

131.211 **Preliminary results of ultraviolet interstellar extinction from TD1 satellite observations.**
K. Nandy, G. I. Thompson, C. Jamar, A. Monfils, R. Wilson.
Phil. Trans. Roy. Soc. London, Ser. A, Vol. 279, 337 - 343 (1975).

The interstellar extinction law in the far ultraviolet has been derived from observations of about 100 stars for three galactic regions in the direction of Cygnus, galactic centre and anticentre. It is concluded that the extinction characteristics will need more than one type of particle for their explanation.

131.212 **The Celescope survey and the galactic distribution of interstellar absorption.** R. J. Davis.
Phil. Trans. Roy. Soc. London, Ser. A, Vol. 279, 345 - 354 (1975).

The Celescope experiment consisted of four 31-cm aperture telescopes equipped with digital television photometers, installed in the Orbiting Astronomical Observatory, launched 7 December 1968. This instrument was used to conduct a survey in four ultraviolet colours: U1 (2100–3200 Å), U2 (1550–3200 Å), U3 (1350–2150 Å) and U4 (1050–2150 Å). The results were used, together with ground-based data, to determine the distribution of interstellar dust and variations of the interstellar extinction law with the position in the Galaxy.

131.213 **A non-linear model for the intensity, linewidth and coherence of astrophysical masers.** R. A. Rosen.
Thesis Columbia Univ., New York, USA. [Available from Univ. Microfilms, Ann Arbor, Mich., USA. Order No. 75-7530] 104 pp. (1974).

The effect of pumping on the steady-state coherent lossless light wave trains which can propagate through a two-level resonant medium is studied. This phenomenon is a form of self-induced transparency.

Angular distributions in the elastic scattering and rotational excitation of molecular hydrogen by atomic hydrogen. See Abstr. 022.014.

Charge transfer reactions in multiply charged ion-atom collisions. See Abstr. 022.015.

Cooling of the 6-centimeter and 2-centimeter doublets of interstellar H_2CO by collision: an accurate quantum-mechanical calculation. See Abstr. 022.044.

Cross sections for collisional excitation of interstellar formaldehyde. See Abstr. 022.054.

Multiple electron ejection by X-ray photoionization and the abundance of interstellar ions. See Abstr. 022.055.

Is there a maser in the silicon monoxide ground state? See Abstr. 022.056.

Electronic energies including exchange and spin orbit interaction of the interstellar molecules CH^+, CH and OH at large interatomic separations. See Abstr. 022.059.

Transitions between the fine structure components of C^+ in collision with atomic hydrogen. See Abstr. 022.060.

Preparation of fine particles of astrophysical interest by gas evaporation techniques. See Abstr. 022.062.

Rate of the reaction $N_2H^+ + CO \rightarrow HCO^+ + N_2$ and its significance for the interstellar chemistry of N_2H^+. See Abstr. 022.063.

Oscillator strength for the $D\ ^2\Sigma^- - X\ ^2\Pi$ transition in OH. See Abstr. 022.073.

Electron capture by C^{2+} from interstellar atomic hydrogen. See Abstr. 022.084.

Laboratory microwave spectrum of HCO^+. See Abstr. 022.089.

$K\alpha$ X-rays from cosmic-ray oxygen. See Abstr. 061.017.

Isotopes of carbon, nitrogen and oxygen as probes of nucleosynthesis, stellar mass losses and galactic evolution. See Abstr. 061.031.

The origin and abundances of the chemical elements See Abstr. 061.036.

Materie und Strahlung im interstellaren Raum. See Abstr. 061.066.

Resonant interactions between fast particles and collision-dominated MHD waves. See Abstr. 062.055.

Problems on gravitational collapse of interstellar gas clouds. I. Numerical integration of the equations of hydrodynamics: the case of spherical symmetry. See Abstr. 062.064.

Strong scintillations in astrophysics. III. The fluctuations in intensity. See Abstr. 063.037.

Interstellar bubbles. See Abstr. 064.015.

Young stellar objects and dark interstellar clouds. See Abstr. 065.118.

On the acceleration of protons on the boundaries of solar wind with interstellar medium. See Abstr. 074.063.

Galactic dust lanes and lunar soil. See Abstr. 094.137.

Infrared spectra of carbonaceous chondrites and the composition of interstellar grains. See Abstr. 105.028.

A proto-planetary contribution to interstellar dust. See Abstr. 107.015.

Infrared photometry in the R CrA association. See Abstr. 113.002.

Interaction of hot stars and of the interstellar medium. VII.The rate and fate of stellar ultraviolet photons. See Abstr. 114.024.

A spectroscopic study of stars for the most part behind the Coalsack. See Abstr. 114.042.

Seach for polarimetric variations in the close binary GK Cephei. See Abstr. 121.019.

Long-period variables: further work on the correlation of period with OH radial-velocity pattern. See Abstr. 122.003.

Wavelength dependence of polarization. XXIX. Observations of red variable stars. See Abstr. 122.078.

Spectral turnovers of galactic supernova remnants. See Abstr. 125.001.

Supernova explosions and the maintenance of a hot component of the interstellar medium. See Abstr. 125.019.

The distribution of supernova remnants in the Galaxy: evidence for a tunnel system. See Abstr. 125.033.

Hydrodynamic instabilities and the formation of radio shells in young supernovae remnants. See Abstr. 125.036.

The dynamical effects of supernovae on the interstellar medium. I: Spherical shock waves in an adiabatic medium. See Abstr. 125.043.

H I observations of dust clouds. See Abstr. 131.563.

40–350 micron emission from NGC 2023. See Abstr. 132.009.

On the iron abundance in interstellar gas. See Abstr. 132.041.

Possible spiral structure in the Orion molecular cloud. See Abstr. 132.042.

A narrow emission line near 1720 MHz in the direction of Sgr A. See Abstr. 141.036.

Infrared sources in obscured regions. See Abstr. 141.602.

Eight new infrared sources associated with OH masers. See Abstr. 141.611.

CO in the infrared and radio spectra of carbon stars. See Abstr. 141.623.

Observations of water line emission near three compact infrared sources. See Abstr. 141.628.

On the ultrasoft X-ray background. See Abstr. 142.073.

Measurement of primary cosmic-ray electrons and positrons from 4 to 50 GeV. See Abstr. 143.002.

The cosmic-ray age deduced from the ^{10}Be abundance. See Abstr. 143.022.

The isotopic composition of galactic cosmic-ray lithium, beryllium, and boron. See Abstr. 143.023.

Cosmic rays and interstellar tunnels. See Abstr. 143.024.

Cosmic ray antiproton and positron production in the interstellar medium. See Abstr. 143.043.

Sterne und interstellares Medium in der Assoziation Cyg OB 2. See Abstr. 152.002.

The stabilizing effects of cloud reacceleration; microturbulence, and rotation on Parker's instability. See Abstr. 155.024.

On the relation between CO and gamma-ray observations, cosmic rays, and the thickness of the galactic disk. See Abstr. 155.026.

Finns det brännvin och saltgurka i Vintergatans centrum? See Abstr. 155.040.

Detection of extragalactic carbon monoxide at millimeter wavelengths. See Abstr. 158.009.

Carbon monoxide in external galaxies. See Abstr. 158.010.

Three-dimensional distribution of interstellar gas in disk-shaped galaxies. See Abstr. 158.128.

Star formation in spiral galaxies. See Abstr. 158.161.

Hot gas in and between galaxies. See Abstr. 161.006.

H I, H II Regions

131.501 **Stellar winds and globules in H II regions.**
J. E. Dyson.
Astrophys. Space Sci., Vol. 35, 299 - 312 (1975).

The interaction of a plane-parallel hypersonic stellar wind with a globule in an H II region is considered in two approximations. In both approximations, the ionization front on the globule remains strong-D type, and a flow pattern containing two oppositely facing shock waves results. In the first approximation, the structure of the shocked region is calculated assuming that globule gas and stellar wind gas mix well and move at the same velocity. In the second approximation, the stellar wind merely applies pressure to balance the momentum flux in the globule gas. The structure of the shocked region is calculated on the assumption that a tangential discontinuity exists between shocked stellar wind and shocked globule gas.

131.502 **A classification of the available astrophysical data of particular H II regions. I: Orion nebula: optical, radio, infrared and molecular mapping.** C. Goudis.
Astrophys. Space Sci., Vol. 35, 409 - 458 (1975).

The available optical, radio, infrared and molecular contour maps of the object are presented. Velocity distribution contour maps are also included. Analytical information for each map is classified in separate tables.

131.503 **G45.5 + 0.1 and G45.1 + 0.1: compact infrared sources.**
M. Zeilik II, D. E. Kleinmann, E. L. Wright.
Astrophys. Journ., Vol. 199, 401 - 405 (1975).

The authors have observed the compact H II regions G45.5 + 0.1 and G45.1 + 0.1 from 2.2–21 μ. The infrared emission from G45.5 + 0.1 appears to come from dust that is well mixed with ionized gas and heated by Lα radiation. For G45.1 + 0.1, the infrared luminosity cannot be accounted for by Lα heating alone. Compared with the conventional interstellar values, the gas-to-dust ratio in G45.5 + 0.1 is substantially depleted, but it is close to normal in G45.1 + 0.1.

131.504 **Die HII-Region W3 im nahen Infrarot.**
M. Beetz, H. Elsässer, R. Weinberger.
Mitt. Astron. Ges., No. 36, p. 172 (1975). – The paper appeared in Astron. Astrophys., Vol. 34, 335 - 337 (1974) – see Abstr. 12.131.505.

131.505 **Beobachtungen an HII-Regionen im nahen Infrarot.**
C. Thum, D. Lemke.
Mitt. Astron. Ges., No. 36, p. 173 (1975).

131.506 **HII-Regionen im fernen Infrarot.**
D. Lemke, C. Thum.
Mitt. Astron. Ges., No. 36, p. 173 (1975). – Abstract.

131.507 **Spectrophotometric observations of ionized hydrogen regions in nearby spiral and irregular galaxies.**
H. E. Smith.
Astrophys. Journ., Vol. 199, 591 - 610 (1975).

A program of direct photography and spectrophotometry has been carried out on H II regions in a selection of nearby spiral and irregular galaxies. Excitation differences among the H II regions of Sbc–Scd–Irr galaxies can best be understood in terms of an abundance sequence which progresses from higher to lower heavy-element enrichment as one progresses from the earlier to later type galaxies.

131.508 **Infrared photometric and spectrophotometric observations of the galactic H II region G29.9–0.0.**
B. T. Soifer, J. L. Pipher.
Astrophys. Journ., Vol. 199, 663 - 668 (1975).

Spatial, photometric, and spectrophotometric observations of the compact H II region G29.9–0.0 are reported. The photometric observations indicate that the infrared object agrees in position and size with the radio source. The photometry indicates that dust competes with the gas for absorbing the ionizing stellar continuum, while there is evidence for a depletion of dust within the ionized region. A comparison of 2.2 μ–10 μ extinction to the source indicates that $A_v/\tau_{9.75\,\mu} > 6$ mag.

131.509 **Observations of 6 GHz absorption lines from the $^2\pi_{3/2}$, $J = 5/2$ state of OH.**
F. F. Gardner, J. B. Whiteoak.
Monthly Notices Roy. Astron. Soc., Vol. 172, 9P - 13P (1975).

Absorption has been detected at the 6035 MHz (F = 3–3) and 6031 MHz (F = 2–2) lines of the $^2\pi_{3/2}$, $J = 5/2$ state of OH in the direction of a number of H II regions, usually of high brightness.

131.510 **The intercloud H I gas in the direction of Cassiopeia A and Cygnus A.**
U. Mebold, D. L. Hills.
Astron. Astrophys., Vol. 42, 187 - 194 (1975).

The authors have detected the absorption of the intercloud-interarm H I-gas in the directions of Cas A and Cyg A with the 100-m telescope in Effelsberg. They derive spin temperatures greater than 3000 K and of about 6000 K, respectively, for these sources. It is shown that there exists no systematic error of the spin temperatures even though the gas is not homogeneously distributed around and in front of discrete radio sources.

131.511 **An 85α recombination line survey of NGC 2024.**
J. M. MacLeod, L. H. Doherty, L. A. Higgs.
Astron. Astrophys., Vol. 42, 195 - 204 (1975).

The 85α recombination lines of hydrogen, helium, and carbon have been observed at a grid of 29 positions, centered on the continuum peak of the H II region NGC 2024 (Orion B, W 12). Contour maps of radial velocity. LTE electron temperature, and the distributions of ionized hydrogen, helium and carbon are presented. A model of NGC 2024 has been constructed, consisting of two exciting stars, (one B0 and the other a B0.5 supergiant) and a layer of differentially absorbing dust lying across the southern half of the nebula. This model accounts satisfactorily for the observed radio parameters.

131.512 **Detection of an additional hydrogen recombination line source at the W49A position.** A. Parrish.
Astrophys. Journ., Vol. 200, 594 - 597 (1975).

New 166α and 248α radio recombination-line observations of the thermal source W49A display not only the expected lines characterized by the $\sim$10 km s^{-1} radial velocity of the nebula, but also an additional hydrogen-carbon line pair that, kinematically, arises in a foreground object. The 248α lines are predominantly stimulated by the background continuum. The presence and intensity of the carbon line suggests that the object contains an H I region. In addition, limited 166α line observations of the W49B position have yielded upper limits for the line intensities there.

131.513 **The transfer of radio recombination line radiation through a cold gas. I. Hydrogen and helium lines in compact H II regions.** R. L. Brown, J. Gómez-González.
Astrophys. Journ., Vol. 200, 598 - 608 (1975).

The authors have considered the transfer of recombination line radiation that is produced in a high emission measure H II region (10^7 cm^{-6} pc) and passed through a cold, partially ionized gas. They conclude first that the cold gas will influence the observed recombination line spectrum at frequencies where the absolute value of the line optical depth in the cold gas is comparable with the ratio of the intensity in the line to the intensity in the continuum as emitted in the H II region. The authors find secondly that the ratio of the peak antenna temperature of the helium α-line T(He) to the peak antenna temperature of the hydrogen α-line T(H) will be a function of principal quantum number n. Thirdly, since the He and H recombination lines are enhanced in different degrees in the cold gas, the ratio of line widths $\Delta\nu$(He)/$\Delta\nu$(H) will also be a function of n when n is large. Finally, the line-to-continuum ratio $(T_L/T_C)\Delta\nu\, \nu^{-2}$ depends on the parameters of the cold gas in a well defined way. The observational constraints and predictions of the model are discussed.

131.514 **The 8–13 μ spectra of compact H II regions.**
F. C. Gillett, W. J. Forrest, K. M. Merrill, R. W. Capps, B. T. Soifer.
Astrophys. Journ., Vol. 200, 609 - 620 (1975).

Observations of the 8–13 μ spectra of four compact H II regions [K3-50, W51–IRS 1, W51–IRS 2, and W49–OH(2)] are reported. All of the spectra show a broad absorption feature centered around 9.7 μ. In addition, a 12.8 μ emission line, attributed to the Ne^+ fine-structure transition, has been found in W51–IRS 1 and probably in W49–OH(2), and is marginally present in K3-50 and W51–IRS 2. The visual extinction A_V (inferred from deficiencies in emission at $\lambda < 5\ \mu$) is compared with the optical depth at 9.7 μ, and it is found that $A_V/\tau_{9.7\,\mu} \gtrsim 10$–15. The derived 9.7 μ optical depths are combined with knowledge of the size of the molecular clouds to derive cloud masses that are consistent with molecular determinations. It is found that the emitting regions have larger gas/dust mass ratios than interstellar values by a factor of 20–200.

131.515 **Recombination line observations of a normal helium abundance in the galactic-center H II region Sagittarius B2.** R. L. Brown, F. J. Lockman.
Astrophys. Journ., (*Letters*), Vol. 200, L155 - L159 (1975).

The authors have detected both the H 76α and He 76α recombination lines in the Sgr B2 H II region; from a comparison of the integrated power in these lines they infer that the ratio of ionized helium to ionized hydrogen in Sgr B2 is $\langle He^+/H^+\rangle = 0.085 \pm 0.015$. The anomalously low He^+/H^+ ratio in Sgr B2 derived from recombination line observations at lower frequencies very likely results from preferential enhancement of the nebular hydrogen lines in the ionized component of the Sgr B2/molecular cloud interface.

131.516 **The 4830 MHz H_2CO absorption in the direction of NGC 6334.** F. F. Gardner, J. B. Whiteoak.
Monthly Notices Roy. Astron. Soc., Vol. 173, 131 - 139 (1975).

The distribution of 4830 MHz H_2CO absorption has been mapped over the northern radio component, G351.4 + 0.7, of NGC 6334. The absorption with an average velocity of –4 km s^{-1} has a general association with a dust lane extending across the H II region. In addition there are three H_2CO concentrations. Two coincide with OH-emission centres, one of which is an infrared source. The third, with the highest absorption and smallest angular extent, is centred on a compact continuum radio component; it is also an infrared source.

131.517 **Observations of the hydrogen radio line H56 α in several H II regions.**
I. I. Berulis, G. T. Smirnov, R. L. Sorochenko.
Pis'ma v Astron. Zhurn., Vol. 1, No. 9, p. 28 - 33 (1975). In Russian.

Measurements of the excited hydrogen radio line H56 α have been carried out for the nebulae Orion A, W3, W12, W51 and Dr-21. Electron temperatures obtained are lower than those deduced from non-LTE analysis of the observations at longer wavelengths for the Orion nebula (T_e = 7000 – 7300 K).

131.518 **Microwave NH_3 emission from selected H II regions.**
P. R. Schwartz, J. M. Bologna, J. H. Spencer,

J. A. Waak, A. C. Cheung, M. F. Chui, D. Matsakis.
Bull. American Astron. Soc., Vol. 7, 401 (1975). – Abstr. AAS.

131.519 **CO observations of compact galactic H II regions.** J. A. Frogel, S. E. Persson, D. F. Dickinson, E. J. Chaisson.
Bull. American Astron. Soc., Vol. 7, 401 (1975). – Abstr. AAS.

131.520 **CO emission around the H II regions S 228 and S 269.** W. J. Wilson, J. D. White.
Bull. American Astron. Soc., Vol. 7, 401 (1975). – Abstr. AAS.

131.521 **Recombination line observations of a normal helium abundance in the galactic center H II region Sgr B2.** R. L. Brown, F. J. Lockman.
Bull. American Astron. Soc., Vol. 7, 401-402 (1975). – Abstr. AAS.

131.522 **Observations of Ne II at 12.8 μ in southern H II regions.** E. R. Wollman, T. R. Geballe, J. H. Lacy, C. H. Townes, D. M. Rank.
Bull. American Astron. Soc., Vol. 7, 402 (1975). – Abstr. AAS.

131.523 **One-millimeter continuum maps of H II-region-molecular cloud complexes.** W. E. Westbrook, J. H. Elias, D. Y. Gezari, M. G. Hauser, G. Neugebauer, M. W. Werner.
Bull. American Astron. Soc., Vol. 7, 402 (1975). – Abstr. AAS.

131.524 **Elemental abundances in compact H II regions, planetary nebulae, and supernova remnants in the Magellanic Clouds.** R. J. Dufour, R. M. Killen.
Bull. American Astron. Soc., Vol. 7, 402 (1975). – Abstr. AAS.

131.525 **The source function for Lyman alpha in a dusty H I region heated by suprathermal ions.**
R. E. Stoner, R. L. Ptak.
Bull. American Astron. Soc., Vol. 7, 413 (1975). – Abstr. AAS.

131.526 **Time variations in 18-cm OH masers in H II regions over the period 1965 - 1972.**
W. T. Sullivan III, J. H. Kerstholt.
Bull. American Astron. Soc., Vol. 7, 417 (1975). – Abstr. AAS.

131.527 **The 10 μ silicate feature in southern H II regions.** S. E. Persson, J. A. Frogel, M. Aaronson.
Bull. American Astron. Soc., Vol. 7, 429 (1975). – Abstr. AAS.

131.528 **21-cm observations of non-planar H I associated with the Perseus spiral arm.** E. J. Grayzeck.
Bull. American Astron. Soc., Vol. 7, 434 (1975). – Abstr. AAS.

131.529 **Radio observations of IC 434.** T. B. H. Kuiper.
Astron. Astrophys., Vol. 42, 323 - 327 (1975).

Continuum radio emission and hydrogen recombination line emission were measured at 2.37 GHz with the 64-m antenna at Goldstone. The radio emission has the same general distribution as the optical nebula. The radial velocity of the nebula agrees with that of the adjacent dust cloud L 1630. The electron temperature is estimated to be approximately 7000 K. The excitation of the nebula is compared with current estimates of the ionizing flux from the exciting star, σ Orionis.

131.530 **An interpretation of the radio- and IR-spectrum of the compact H II regions W 3 (A) and W 3 (OH).**
E. Krügel, P. G. Mezger.
Astron. Astrophys., Vol. 42, 441 - 448 (1975).

Models are fitted to the observed integral characteristics and radio and IR spectra of the compact H II regions W 3 (A) and W 3 (OH). W 3 (OH) appears to be ionized by an O-star, which has not yet reached the main sequence. It is surrounded by a compact, dust-depleted H II region. W 3 (A) is surrounded by a dense, dust-enriched shell of ionized gas which is embedded in a dense cloud of neutral gas and dust. The diameter of the ionized shell is comparable to the diameter of the neutral shell surrounding the compact H II region W 3 (OH). The two H II regions appear to represent two different stages in the dynamical evolution of compact H II regions in which dust plays the dominant role. Some characteristics of the dust are obtained.

131.531 **Observations of faint H II regions in our Galaxy.** M. Peimbert, J. F. Rayo, S. Torres-Peimbert.
Publ. Astron. Soc. Pacific, Vol. 87, 507 - 508 (1975).
Abstr. ASP.

131.532 **Far infrared observations of dark cold clouds and H II regions.** P. Turon, D. Rouan, P. Lena, J. Wijnbergen, J. W. Aalders.
Nature, Vol. 257, 651 (1975). – Simulating experiments for Spacelab.

131.533 **Forbidden lines in a newly formed H II region.** J. Manfroid.
Astron. Astrophys., Vol. 43, 149 - 152 (1975).

The forbidden lines emitted by C^+, N^+, N^{++}, O^+, O^{++}, Ne^+, and Ne^{++} in a newly formed H II region have been calculated during its initial evolution.

131.534 **Characteristics of planetary nebulae and H II regions based on λ = 1.35 cm continuum measurements.**
M. A. Bráz, J. O. Jardim, P. Kaufmann.
Astron. Astrophys., Vol. 43, 153 - 155 (1975).

Physical parameters are derived and discussed for stronger H II regions and planetary nebulae for which continuum radio data at λ = 1.35 cm was obtained. The study includes southern hemisphere planetary nebulae IC 418, NGC 6302, NGC 6369, and H II regions RCW 65, RCW 87, RCW 99, H 2–3 and H 2–6.

131.535 **The rate of star formation in M31 and the Large Magellanic Cloud.** M. Tosa, K. Hamajima.
Publ. Astron. Soc. Japan, Vol. 27, 501 - 505 (1975).

A correlation is obtained between the distribution of H II regions and H I gas in M31 and the LMC: (Surface density of H II regions) $\propto$ (H I surface density)^n with $n = 2.1 \pm 0.2$ in M31 and $n = 1.9 \pm 0.3$ in the LMC. The result for M31 confirms the importance of the beam-smearing effect of the H I 21-cm map on the determination of n.

131.536 **H II zones as indicators of extragalactic distances and the Hubble constant.**
B. A. Vorontsov-Vel'yaminov.
Astron. Tsirk., No. 846, p. 3 - 5 (1974). In Russian.

131.537 **H and C 134α line maps of NGC 2024 and W3.** T. L. Wilson, P. Thomasson, F. F. Gardner.
Astron. Astrophys., Vol. 43, 167 - 173 (1975).

The authors present maps of the 134α (2.7 GHz) lines of hydrogen and carbon for the H II regions NGC 2024 and W3. For the three positions where the authors have detected the narrow ($\Delta V < 10$ km s^{-1}) H 134α line in NGC 2024, they find that the velocity is 0.7 ± 0.8 km s^{-1} lower than the velocity of the C 134α line.

131.538 **Photoelectric photometry of the ionized helium abundance in galactic H II regions.** B. Thomsen.
Astron. Astrophys., Vol. 43, 411 - 418 (1975).

Differential abundances of ionized helium have been derived for twenty-three positions in the Orion nebula and one region in each of the nebulae M 8, M 20, M 16, M 17 and NGC 3327 (the η Carina nebula) through photoelectric measurements. There is no evidence for significant differences in total helium abundance from nebula to nebula when differences in the ionized helium abundances due to variations in the content of neutral helium are accounted for.

131.539 **High-velocity neutral hydrogen.** G. L. Verschuur. Annual Rev. Astron. Astrophys., Vol. 13, (see 003.005), 257 - 293 (1975). – Contents: (1) Introduction; (2) Observations of high-velocity clouds; (3) Models for the general high-velocity hydrogen phenomenon.

131.540 **Detection of the [O III] 88.16 micron line in M17.** D. B. Ward, B. Dennison, G. Gull, M. Harwit.
Astrophys. Journ., (*Letters*), Vol. 202, L31 - L32 (1975).
The [O III] 88.16 μ line has been detected in the galactic H II region M17. The line intensity is 2.2 (+ 1.0, – 0.7) × 10^{-15} W cm^{-2}.

131.541 **The distribution of H II regions in M31.** A. J. Kalnajs.
La dynamique des galaxies spirales. Colloque International C.N.R.S. No. 241, (see 012.022), p. 103 - 107, with a discussion, p. 108 - 109 (1975).

131.542 **The radio spectra of four H II regions (Orion B, M8, M16 and M17).** C. Goudis.
Astrophys. Space Sci., Vol. 37, 455 - 462 (1975).
The radio spectra of Orion B, M8, M16 and M17 are established by using more data over a wider range of frequencies than previous authors. The nature of these objects is discussed. One of these H II regions (M8) may be mixed with non-thermal components.

131.543 **The gas dynamics of contracting ionization fronts: a numerical solution.** J. S. Berry.
Astron. Astrophys., Vol. 44, 401 - 404 (1975).
The flow of the ionized gas behind a contracting ionization front is investigated for spherical symmetry. A numerical solution is given when the globule of neutral hydrogen is in isothermal, gravitational equilibrium and subject to ionizing radiation from outside. It is seen that large velocities do not occur within the H II region as suggested by some authors.

131.544 **H II-Regionen und assoziierte Punktquellen: Infrarot-Beobachtungen und Modelle.** C. Thum.
Diss. Nat. Gesamtfakultät, Ruprecht-Karl Univ., Heidelberg. 6 + 119 pp. (1975).

131.545 **A correlation of H II regions with H I gas and the rate of star formation in galaxies.**
K. Hamajima, M. Tosa.
Publ. Astron. Soc. Japan, Vol. 27, 561 - 569 (1975).
A correlation analysis is made between the distributions of H II regions and neutral hydrogen is seven galaxies, LMC, SMC, M31, NGC 2403, M101, M51, and NGC 6946. A detailed analysis is made of the distribution of H II regions in the LMC and it is found that the H II regions have a tendency to form a clustering which cannot be explained only by the irregular distribution of H I gas. This suggests the presence of another factor for the rate of star formation besides the gas density.

131.546 **Recombination line observations of the compact H II region G32.8+0.2.**
M. R. Viner, V. A. Hughes.
Journ. Roy. Astron. Soc. Canada, Vol. 69, 246 (1975).
Abstr. CAS.

131.547 **Recombination line observations of W3(OH): a model for OH and H_2O emission.**
V. A. Hughes, M. R. Viner.
Journ. Roy. Astron. Soc. Canada, Vol. 69, 246 (1975).
Abstr. CAS.

131.548 **Compact H II regions in the Ophiuchus and R Coronae Austrinae dark clouds.**
R. L. Brown, B. Zuckerman.
Astrophys. Journ., (*Letters*), Vol. 202, L125 - L128 (1975).
The authors report the identification of six compact H II regions in the Ophiuchus dark cloud and two H II regions in the R CrA dark cloud. Three of the H II regions in Ophiuchus appear to be excited by highly obscured stars that have been detected in recent 2 μ surveys of this cloud; using the radio and infrared results, the authors infer that the ratio of total to selective extinction $R = 4.1$ in the dense regions of the cloud. Further observations suggest that the density contrast between inner and outer parts of the cloud is at least a factor of 100 and that the total visual extinction through the most dense parts of the cloud $A_v > 100$ mag.

131.549 **Forbidden sulfur II structures in the Carina nebula.** N. R. Walborn.
Astrophys. Journ., (*Letters*), Vol. 202, L129 - L130 (1975).
[S II] and [O III] interference-filter photographs of the northern part of the Carina nebula, obtained at the prime focus of the CTIO 4-meter reflector, are presented. Many of the principal features of each are absent in the other. The most prominent feature in the light of [S II] is the curious ring associated with the radio source Carina II, which shows intricate filamentary structure. Also conspicuous are three small clouds associated with dust, one of them spiral in shape, which glow brightly in [S II] but are invisible in [O III].

131.550 **A classification of the available astrophysical data of particular H II regions. III. NGC 2024: mapping and physical parameters of the object.** C. Goudis.
Astrophys. Space Sci., Vol. 38, 283 - 300 (1975).
Optical and radio contour maps of the object are presented with the analytical information concerned classified in separate tables. Estimations of the various physical parameters of the object are properly classified and presented either in separate tables or in contour map form. The latest work in the infrared range is also classified and presented.

131.551 **Étude à haute résolution spatiale et spectrale du champ de vitesse de la région H II compacte Sh2-106.** A. J. Maucherat.
Astron. Astrophys., Vol. 45, 193 - 196 (1975).
This paper presents the interferometric Hα radial velocities of the compact H II region Sh 2-106. The line is clearly splitting in the south intense region, which can be interpreted as an expanding shell with an expansion velocity of 36 km s^{-1}. This behaviour is unusual in an optical compact H II region.

131.552 **Infrared photometric and spectrophotometric observations of the compact H II region G75.84 + 0.40.** J. L. Pipher, B. T. Soifer.
Bull. American Astron. Soc., Vol. 7, 530 (1975). – Abstr. AAS.

131.553 **Initial mass function for massive stars and integrated spectral properties of H II regions.** M. Kaufman.
Bull. American Astron. Soc., Vol. 7, 532 (1975). – Abstr. AAS.

131.554 **Self-absorbed CO profiles in molecular complexes.** H. R. Dickel, J. R. Dickel, W. J. Wilson, R. B. Pomphrey.
Bull. American Astron. Soc., Vol. 7, 556 - 557 (1975).
Abstr. AAS.

131.555 **On the structure of the H II regions W49A and W51.** S. L. Mufson, H. S. Liszt.
Bull. American Astron. Soc., Vol. 7, 557 (1975). – Abstr. AAS.

131.556 **Observations of some compact H II regions at wavelengths of 4 and 2 cm at Pulkovo.**
T. B. Pyatunina.
Soobshch. Spets. Astrofiz. Obs., *Zelenchukskaya,* vyp. (No.) 14, p. 5 - 19 (1975). In Russian.

Observations of 8 compact H II regions, W 12, W 29, W 31, W 33A, W 43, W 49, W 51, W 75, at 4 cm wavelength and one H II region, W 51, at 2 cm with the large Pulkovo radio telescope are presented. The connection between small-scale and large-scale inhomogeneities in density distributions is discussed.

131.557 **Spectrophotometric studies of galactic nebulae. XII. New compact H II regions.**
Yu. I. Glushkov, Eh. K. Denisyuk, Z. V. Karyagina.
Astron. Tsirk., No. 879, p. 1 - 2 (1975). In Russian.

131.558 **Spectrophotometric studies of galactic nebulae. XIII. Compact H II regions in SH 146, Sh 149, Sh 152.** Yu. I. Glushkov, Eh. K. Denisyuk, Z. V. Karyagina.
Astron. Tsirk., No. 879, p. 2 - 5 (1975). In Russian.

131.559 **Spectrophotometric studies of galactic nebulae. XIV. Compact H II regions in Sh 307 and NGC 1491**
Yu. I. Glushkov, Eh. K. Denisyuk, Z. V. Karyagina.
Astron. Tsirk., No. 879, p. 5 - 6 (1975). In Russian.

131.560 **A comparison of photometric and kinematic distances for southern H II regions.**
N. Vogt, A. F. J. Moffat.
Astron. Astrophys., Vol. 45, 405 - 409 (1975).

Candidates for exciting stars have been examined in 49 southern H II regions by means of photoelectric *UBV* and Hβ photometry. The photometric distances of the related stars are compared with the kinematic distances of the corresponding H II regions as given by Georgelin et al. (1973). In the range $l = 250°...360°...30°$ there are no systematic differences; only a region in Puppis at $l \approx 230°$ yields photometric distances which systematically exceed the kinematic ones. It is shown that an abnormal R-value cannot account for the observed discrepancy in this region.

131.561 **Observations of faint H II regions in our Galaxy.**
M. Peimbert, J. F. Rayo, S. Torres-Peimbert.
Rev. Mexicana Astron. Astrofis., Vol. 1, 289 - 297 (1975).

The authors present photoelectric observations of faint H II regions of emission measure lower than 500 cm^{-6} pc at Hα. From these observations it is found that the limiting surface brightness in Hα for detection of an H II region in the Red Palomar Sky Survey (PSS) is smaller than 2.75×10^{-6} erg $cm^{-2}\,s^{-1}\,sr^{-1}$. By comparing the observations by Reynolds et al. (1974) with the Red PSS it is found that the limiting value lies in the 2.5×10^{-6} to 5.5×10^{-6} erg $cm^{-2}\,s^{-1}\,sr^{-1}$ range which for an unreddened nebula at $T_e = 6000$ °K corresponds to an emission measure between 17 and 40 cm^{-6} pc. The mechanism of ionization and the chemical abundances of two of the observed regions, belonging to the Barnard Loop nebula and located about 4.5° away from the Orion nebula, are studied.

131.562 **Carbon monoxide studies of H II regions.**
H. S. Liszt.
Thesis, Princeton Univ., N.J., USA., 84 pp. (1974).

The author has made a survey of the $^{12}C\,^{16}O$ and $^{13}C\,^{16}O$ J = 1–0 line emission at 2.6 mm from some 25 H II regions and OH sources outside the galactic center. A new interpretation of carbon monoxide spectra is presented which indicates that collapse, rather than turbulence, is the primary source of the line profiles. The kinematics of molecule clouds are discussed and kinetic temperatures, carbon monoxide column densities and masses are derived for the regions surveyed.

131.563 **H I observations of dust clouds.** K. W. Chu.
Thesis, California Inst. of Technol. Pasadena (USA). [Available from Univ. Microfilms, Ann Arbor, Mich., *USA*. Order No. 75-9520], 127 pp. (1975).

Aperture synthesis observations of three dust clouds, L134, L1495 and L1517, were made at 21 cm. A 7 arc min diameter beam was synthesized, and the velocity resolution was 0.2 km/sec. In addition to interferometer observations, single dish measurements were carried out to provide large scale information.

Grain temperature in giant molecular clouds associated with H II regions. See Abstr. 131.074.

Radiation from inhomogeneous dusty Strömgren spheres. See Abstr. 131.099.

Ultraviolet studies of the interstellar gas. See Abstr. 131.114.

Radio sources near reflection nebulae in Monoceros. See Abstr. 131.136.

Molecular hydrogen in H II region transition zones. See Abstr. 131.144.

H I and OH observations towards W3. See Abstr. 131.163.

A microwave investigation of the Trifid nebula and its surrounding environment. See Abstr. 132.005.

Large scale line splitting associated with the dense neutral intrusions and ionization fronts in the core of M8. See Abstr. 132.010.

A classification of the available astrophysical data of particular H II regions. II: Orion nebula: physical parameters. See Abstr. 132.012.

Kinematics of the Gum nebula. See Abstr. 132.043.

Chemical composition of extragalactic gaseous nebulae. See Abstr. 132.044.

8–13 micron maps of the Trapezium region of the Orion nebula. See Abstr. 132.046.

Interstellar ionized nebulae. See Abstr. 132.063.

Observable effects of dust on the ionization of nebulae. See Abstr. 133.022.

The physical parameters and exciting stars of six H II regions. See Abstr. 141.104.

Neutral hydrogen absorption measurements yielding kinematic distances for 42 continuum sources in the galactic plane. See Abstr. 141.146.

A high-resolution map of the W3 region at far-infrared wavelengths. See Abstr. 141.605.

Infrared measurements on several sources in the DR-21 region. See Abstr. 141.607.

40–350 micron observations of galactic sources. See Abstr. 141.629.

Near infrared sources in Sgr B2. See Abstr. 141.634.

The infrared emission from Sgr B2. See Abstr. 141.636.

Infrared star in RCW 113. See Abstr. 141.641.

NGC 2175: the cluster age and the nature of the nebulosity surrounding S 252a. See Abstr. 153.024.

H 134α recombination lines from the group of sources near galactic longitude 30°. See Abstr. 155.055.

Ionized gas in the direction of the galactic center. II. The line of sight problem. See Abstr. 157.001.

Beobachtungen von neutralem Wasserstoff in Galaxien des Virgo-Haufens. See Abstr. 158.024.

Limits to the neutral hydrogen content of elliptical galaxies. See Abstr. 158.039.

The interdependence of H I, galaxy counts, reddening, and galactic latitude. See Abstr. 158.083.

H II regions in peculiar galaxies. See Abstr. 158.132.

H I in early-type galaxies. I. Observations. See Abstr. 158.146.

Radio observations of spiral structures in galaxies. See Abstr. 158.147.

Distribution and kinematics of neutral hydrogen in M81. See Abstr. 158.149.

High resolution observations of neutral hydrogen in Messier 51. See Abstr. 158.151.

High resolution observations of the H I in M31. See Abstr. 158.153.

High resolution observations of neutral hydrogen in Messier 31. See Abstr. 158.154.

Neutral hydrogen, spiral structure and density waves in M31. See Abstr. 158.155.

An optical study of M33. See Abstr. 158.157.

Comments on the association of the H I and H II distributions in M101. See Abstr. 158.158.

Neutral hydrogen line and radio continuum emission from the edge-on galaxy NGC 891. See Abstr. 158.160.

High resolution observations of neutral hydrogen in NGC 1300. See Abstr. 158.165.

The groundstate OH spectrum of the galaxy NGC 253. See Abstr. 158.192.

H II regions in double and interacting galaxies. See Abstr. 158.198.

Rotation-curves of galaxies from 21 cm-line observations. See Abstr. 158.235.

Nuclei and H II regions. See Abstr. 158.240.

Errata

131.901 **Erratum: "Effects of a free-free radio continuum on the populations of high atomic levels at low temperatures and densities"** [Astrophys. Journ., Vol. 194, 503 - 512 (1974)]. I. A. Ahmad.
Astrophys. Journ., Vol. 199, 781 (1975).

131.902 **Erratum: "Temperature and emission-line structure at the edges of H II regions"** [Astrophys. Journ., Vol. 197, 355 - 363 (1975)]. D. C. V. Mallik.
Astrophys. Journ., Vol. 200, 803 (1975).

131.903 **Erratum: 'Interstellar clouds containing optically thin H_2"** [Astrophys. Journ., Vol. 197, 575 - 580 (1975)]. M. Jura.
Astrophys. Journ., Vol. 202, 561 (1975).

132 Emission Nebulae, Reflection Nebulae

132.001 **Observations of starlight-excited lines in the Orion nebula.** S. A. Grandi.
Astrophys. Journ., (*Letters*), Vol. 199, L43 - L46 (1975).
Image tube spectra have been obtained of the Orion nebula for wavelengths between 5000 and 9000 Å. Permitted lines of N I and O I are identified and are explained by direct starlight excitation.

132.002 **Spectroscopy of R Monocerotis and NGC 2261.** A. Stockton, D. Chesley, S. Chesley.
Astrophys. Journ., Vol. 199, 406 - 410 (1975).
Spectrograms obtained in early 1974 of NGC 2261, the cometary nebula associated with R Mon, show a variation in the velocity of the reflected absorption spectrum as a function of distance from R Mon. All of the differences observed between the spectra of R Mon and NGC 2261 can be interpreted in terms of (1) variations in the spectrum of R Mon and time delay in the light reaching us via NGC 2261, or (2) contamination of the spectrum of NGC 2261 by lines from the H II region around S Mon. Thus, the interpretation of NGC 2261 as a pure reflection nebula is reaffirmed.

132.003 **Spectroscopy of the Orion nebula from 80 to 135 microns.** D. Brandshaft, R. A. McLaren, M. W. Werner,
Astrophys. Journ., (*Letters*), Vol. 199, L115 - L117 (1975).
Airborne observations have been made of the spectrum of the Orion nebula from 80 to 135 μ with 6-μ resolution and with a 7:4 diameter beam. The spectrum is consistent with that of a smooth continuum source having a flux density which decreases linearly by approximately a factor of 1.7 from 80 to 135 μ.

132.004 **Herbig-Haro-Objekte und T-Tauri-Sterne, ungelöste Probleme der frühen Sternentwicklung.**
K.-H. Böhm.
Mitt. Astron. Ges., No. 36, p. 31 - 32 (1975). – Review paper.

132.005 **A microwave investigation of the Trifid nebula and its surrounding environment.**
E. J. Chaisson, R. F. Willson.
Astrophys. Journ., Vol. 199, 647 - 659 (1975).
The M20 gaseous nebula and the nearby environment in which it resides have been studied extensively at microwave wavelengths. New maps of thermal bremsstrahlung radiation are used to delineate the full extent of the ionized gas relative to the partially obscuring matter of an adjacent molecular cloud. The surrounding environment is briefly surveyed by means of 18-cm OH absorption, which is shown to peak to the west of the M20 nebula where optical obscuration is strongest. An empirical model for the evolution of an interstellar cloud, from a quiescent to a collapsing to a nebular phase, is proposed and discussed.

132.006 **The spectrum and structure of 'Minkowski's footprint': M 1–92.** G. H. Herbig.
Astrophys. Journ., Vol. 200, 1 - 5 = Lick Obs. Bull., No. 692 (1975).
M 1–92 is a small elliptical nebulosity of quite high surface brightness, discovered originally by Minkowski. It is divided unequally by a dark gap so as to have the appearance of a human footprint. On infrared photographs a faint red star appears in the gap between the two lobes. The spectroscopic and photometric data can be explained if the emission spectrum observed in the nebulae actually originates near the central star, and is only scattered off dust being driven radially outward through the two lobes at a velocity of approximately 30 km s^{-1}. The central star is believed to be surrounded by a dense, flattened, circumstellar dust disk and by a much larger and more tenuous cloud.

132.007 **Far-infrared spectoscopy of the Orion nebula.** D. B. Ward.
Astrophys. Journ., (*Letters*), Vol. 200, L41 - L43 (1975).
A coarse-resolution spectrum of the Orion nebula between 75 and 120 μ closely fits the color dependence of a 75 K blackbody. A high-resolution search for O III at 88.16 μ yields an upper limit below theoretical predictions.

132.008 **Concerning the nature of the nebula V–V 1–7 surrounding BD –18°1967.**
L. Kohoutek, R. Wehmeyer.
Astron. Astrophys., Vol. 41, 451 - 453 (1975).
BD –18°1967 has been identified as a normal A 0–1 V star of M_{bol} = +1.0, at a distance of 213 pc, and with a rather high velocity of v_r = +73.4 km/s. Although the nebula V–V 1–7 is very probably associated with the star, it should not be regarded as a planetary nebula; a reflection type of the nebula is most probable.

132.009 **40–350 micron emission from NGC 2023.** J. P. Emerson, I. Furniss, R. E. Jennings.
Monthly Notices Roy. Astron. Soc., Vol. 172, 411 - 420 (1975).
40–350 μm photometric measurement of the reflection nebula and molecular cloud NGC 2023 is reported. Absorption by dust of ultraviolet photons from the B star which illuminates the reflection nebula is capable of powering the far-infrared emission. Comparison of this source with OMC2 and the KL nebula suggests that the dust temperature is roughly proportional to the gas temperature (= CO brightness temperature) for ice, silicate or graphite grains. If the dust and gas temperatures are equal and the grains are made of ice the gas to dust ratios in all three objects are close to the value in the interstellar medium.

132.010 **Large scale line splitting associated with the dense neutral intrusions and ionization fronts in the core of M8.** K. H. Elliott, J. Meaburn.
Monthly Notices Roy. Astron. Soc., Vol. 172, 427 - 431 (1975).
A large area where the [O III] line is distinctly split was discovered and mapped in the very core of M8. This was carried out with a Fabry-Perot monochromator. These motions appear to be related to the ionization fronts which are moving into the obvious neutral volumes in this nebula. A model is presented which seeks to explain many of the observed phenomena of this complex of H II, H I, dust, molecules and stars.

132.011 **Nickel and copper in the Orion nebula.** A. D. Thackeray.
Monthly Notices Roy. Astron. Soc., Vol. 172, 49P - 51P, with an addendum, Vol. 173, 67P (1975).
The probable presence of three lines of Ni II and one of Cu II in published spectra of the Orion nebula is pointed out.

132.012 **A classification of the available astrophysical data of particular H II regions. II: Orion nebula: physical parameters.** C. Goudis.
Astrophys. Space Sci., Vol. 36, 79 - 104 (1975).
The physical parameters of the object compiled from various authors are classified in a separate table. Line ratios are presented for their significance in determining the density and

temperature of the nebula. Data for the electron temperature of the object are also presented. A correlation of several interesting features of the core of the nebular complex is also shown.

132.013 **The radio spectrum of the Orion nebula.** C. Goudis.
Astrophys. Space Sci., Vol. 36, 105 - 109 (1975).

The radio spectrum of the Orion A (combined spectrum of both the M42 and M43 sources) is established by using more data than previous authors. The frequency range is also substantially expanded by using recent results. A peculiar feature appearing in the produced radio spectrum is discussed.

132.014 **Spectral and interferometric observations of four emission nebulae.** T. A. Lozinskaya, A. Yu. Klement'eva, G. V. Zhukov, V. I. Shenavrin.
Astron. Zhurn. Akad. Nauk SSSR, Vol. 52, 682 - 691 (1975). In Russian. English translation in Soviet Astron., Vol. 19, No. 4.

The results of spectrophotometric and interferometric observations of four emission nebulae are presented. The evaluations of T_e, N_e, the mean radial velocity and the velocity of internal motions in the nebulae are carried out. The following nebulae are investigated: 1) a bright compact nebula of unknown nature identified with the nonthermal radio source G 6.4–0.5 in the region W28, 2) the nebula RCW 171 identified with the radio source G23.1 + 0.6, 3) the nebula Simeiz 34 (Sharpless 261), 4) the extended complex of the bright emission filaments in Cygnus, partly projected on the possible supernova remnant W 63.

132.015 **Radio observations of small galactic nebulae.** I. Kazès, A. M. Le Squéren, F. Gadéa.
Astron. Astrophys., Vol. 42, 9 - 16 (1975).

Out of 97 Sharpless small nebulae 45 have been detected at 1416 and 4850 MHz; they mostly show flat spectral indices. Optical positions and diameters have been measured and data for the radio counterparts are presented. In some cases radio and optical excitation parameters are compared and their apparent differences are discussed.

132.016 **Asymmetric HCN line profiles from the Orion nebula.** F. O. Clark.
Astrophys. Journ., (*Letters*), Vol. 200, L115 - L117 (1975).

The hyperfine components of the HCN $J = 1-0$ transition in Orion are observed to be asymmetric. This asymmetry apparently indicates some inhomogeneity of the molecular cloud and is interpreted as due to radial motions in the source. The magnitude of these implied motions is discussed and the sense examined. Alternate interpretations are also discussed.

132.017 **High resolution radio and optical observations of NGC 6888.** H. J. Wendker, L. F. Smith, F. P. Israel, H. J. Habing, H. R. Dickel.
Astron. Astrophys., Vol. 42, 173 - 185 (1975).

NGC 6888 is a filamentary nebulosity associated with the Wolf-Rayet star HD 192163. Maps are presented with a resolution of 5 arcmin at 11 cm and better than 1.5 arcmin at Hα and at 6 cm, 21 cm and 49 cm (continuum). There is a smooth component to the Hα radiation which presumably originates from gas between the filaments. From comparison of the line and continuum intensities, it is found that the smooth component has essentially the same temperature as the filaments. The mechanism by which the filaments are created and maintained is not clear.

132.018 **Investigation of OH absorption in the Trifid nebula (M20).** V. I. Slysh.
Pis'ma v Astron. Zhurn., Vol. 1, No. 8, p. 12 - 15 (1975). In Russian.

OH absorption measurements of the Trifid nebula were made with angular resolution 3'.4 × 18'. Two spectral features were detected, the region where the strongest feature at the LSR velocity 18.7 km/s was observed spatially coincides with the M20 SW formaldehyde cloud discovered by Chaisson and Willson. This cloud is a part of the molecular cloud adjacent to the supernova remnant W28.

132.019 **On the electron density in the cometary nebula NGC 2261.**
J. L. Greenstein, M. A. Kazaryan, Eh. E. Khachikyan.
Pis'ma v Astron. Zhurn., Vol. 1, No. 9, p. 26 - 27 (1975). In Russian.

The electron density in the narrow strip of NGC 2261 located at 10″ to the north from R Mon was determined from the components intensities of the [O II] doublet λλ 3729, 3726 Å. The electron density is shown to change more than 10 times from one point to another.

132.020 **The Gum nebula – arguments for the fossil Strömgren sphere theory.**
M. Kafatos, J. C. Brandt, S. P. Maran, T. P. Stecher.
Bull. American Astron. Soc., Vol. 7, 400 (1975). – Abstr. AAS.

132.021 **A study of small diffuse nebulae.** C. J. Matthews.
Bull. American Astron. Soc., Vol. 7, 463 (1975). Abstr. AAS.

132.022 **The nature of Herbig-Haro objects Nos. 1 and 2: compact emission nebulae.**
G. D. Schmidt, F. J. Vrba.
Bull. American Astron. Soc., Vol. 7, 463 (1975). – Abstr. AAS.

132.023 **Excitation of heavy element permitted lines in the spectra of gaseous nebulae.** S. A. Grandi.
Bull. American Astron. Soc., Vol. 7, 463 (1975). – Abstr. AAS.

132.024 **Radio recombination line mapping of M8.**
L. F. Rodriguez, E. J. Chaisson.
Bull. American Astron. Soc., Vol. 7, 464 - 465 (1975). – Abstr. AAS.

132.025 **27-110μ observations of the Kleinmann-Low nebula.** J. P. Simpson, G. C. Augason, L. J. Caroff, E. F. Erickson, D. Goorvitch, J. D. Scargle, D. W. Strecker, F. C. Witteborn.
Bull. American Astron. Soc., Vol. 7, 465 (1975). – Abstr. AAS.

132.026 **A dual radio-optical spectroscopic study of the Orion nebula.** E. J. Chaisson, M. A. Dopita.
Bull. American Astron. Soc., Vol. 7, 465 - 466 (1975). – Abstr. AAS.

132.027 **Infrared maps of the Orion nebula.**
J. R. Smith, R. D. Gehrz, J. A. Hackwell.
Bull. American Astron. Soc., Vol. 7, 466 (1975). – Abstr. AAS.

132.028 **Polarization of the Egg nebula.**
S. J. Shawl, M. Tarenghi.
Bull. American Astron. Soc., Vol. 7, 498 (1975). – Abstr. AAS.

132.029 **CO mapping and analysis of the V1057 Cygni region.** K. P. Bechis, K. Y. Lo.
Astrophys. Journ., Vol. 201, 118 - 133 (1975).

The authors have mapped emission lines from the $J = 1-0$ rotational transitions of $^{12}C^{16}O$ and $^{13}C^{16}O$ in the vicinity of the unusual star V1057 Cygni and its reflection nebula. They interpret the reflection nebula as a cometary- or fan-shaped nebula aligned with its apex (close to which the star is located) facing the earth, and its major axis nearly along the line of sight. The cometary shape of the nebula may be caused in part by interaction with a magnetic field of up to 1.5×10^{-3} gauss in this region.

132.030 **The radio recombination line spectrum of Orion A: observations and analysis.**
F. J. Lockman, R. L. Brown.
Astrophys. Journ., Vol. 201, 134 - 150 (1975).
The authors have considered the entire body of radio observations of Orion A, and examine how the temperature and density can be derived from the ratio T_L/T_C and the line width $\Delta\upsilon$ of the hydrogen radio recombination lines. New observations support the following conclusions: (1) T_L/T_C and $\Delta\upsilon$ can be accurately determined in the low-frequency observations even when erroneous baselines are removed; (2) in accord with other studies, a density gradient must exist in the nebula; (3) isothermal models cannot reproduce the observed line strengths – the most dense parts of the nebula must be somewhat cooler than the surrounding gas.

132.031 **The nature of Herbig-Haro objects 1 and 2: compact emission nebulae.** G. D. Schmidt, F. J. Vrba.
Astrophys. Journ., (*Letters*), Vol. 201, L33 - L35 (1975).
Optical polarization and infrared photometric observations have been obtained for Herbig-Haro objects 1 and 2. The results indicate that these objects are emission rather than the reflection nebulae predicted by the Strom interpretation of H–H objects.

132.032 **Optical identification of the supersonic emission features in the Orion nebula.**
K. Taylor, G. Münch.
Publ. Astron. Soc. Pacific, Vol. 87, 509 - 510 (1975).
Abstr. ASP.

132.033 **A model of the Orion nebula.** H. Tamura.
Sci. Rep. Tôhoku Univ., First Ser., Vol. 58, 23 - 39 (1975).
A model of the Orion nebula is constructed basing on an assumption that the nebula is composed of gaseous clumps surrounded by vacuum space. A set of values of electron density, electron temperature, extinction by dust particles, ionization degrees of helium and oxygen, oxygen abundance, volume occupied by clumps, and probability of seeing clumps are estimated using the available data of H and He radio recombination lines, radio continuum, Hα, [O III] 5007/4363, [O II] 3729/3726, [O II] 3727/Hβ, [O II] 3727/[O III] $(N_1 + N_2)$, and [S II] 6717/6731.

132.034 **Colorimetry of the nebulae NGC 6914b and Parsamian 22.**
Eh. E. Khachikyan, Dzh. A. Ehjnatyan.
Soobshch. Byurakan. Obs., vyp. (No.) 46, p. 43 - 54 (1975).
In Russian.
The results of colour observations of the two diffuse nebulae NGC 6914b and Parsamian 22 are presented. The surface brightnesses and (U–B), (B–V) colours of separate parts of these nebulae are measured. The colours of these nebulae differ strongly from each other, though both nebulae are located in the same region of the sky: NGC 6914b is very blue, but Parsamian 22 is very red.

132.035 **Forbidden lines from inhomogeneous nebulae: the case of a density distribution $N \propto r^{-2}$.**
N. Panagia, A. Preite-Martinez.
Astron. Astrophys., Vol. 43, 93 - 100 (1975).
The formation of forbidden lines in a nebula with density distribution $N \propto r^{-2}$ is considered. The authors show that the forbidden-line-intensity-ratios (FLIR) may mimic those of constant density nebulae, whose physical conditions differ considerably from the actual ones. The limits of a FLIR analysis in terms of a constant density model are discussed. Some considerations about the abundance determination from emission line measurements and the case of a more general density distribution are also presented.

132.036 **Spectrophotometric studies of galactic nebulae. XI. The spectrum of the high-excitation nebula Sh2-71.** Yu. I. Glushkov, Eh. K. Denisyuk, Z. V. Karyagina.
Astron. Tsirk., No. 852, p. 3 - 4 (1975). In Russian.

132.037 **Polarimetry of the faint nebula near the Pleiades star cluster.** Sh. N. Sabitov.
Astron. Tsirk., No. 864, p. 1 - 2 (1975). In Russian.

132.038 **Colorimetric measurements of Hα emission stars in the Orion aggregate.** K. G. Gasparyan.
Astron. Tsirk., No. 867, p. 6 - 8 (1975). In Russian.

132.039 **Colorimetric measurements of flare stars in the Orion aggregate.** K. G. Gasparyan.
Astron. Tsirk., No. 868, p. 5 - 7 (1975). In Russian.

132.040 **New polarization measurements of the Orion nebula at the wavelengths 5765, 6560, 6930 Å.**
D. I. Gorodetskij, K. G. Dzhakusheva.
Astron. Tsirk., No. 869, p. 1 - 3 (1975). In Russian.

132.041 **On the iron abundance in interstellar gas.**
H. Olthof, S. R. Pottasch.
Astron. Astrophys., Vol. 43, 291 - 295 (1975).
The iron abundance is determined in the Orion nebula from the observed emission lines of Fe^+ and Fe^{++}. A comparison of the value obtained, with the iron abundance in the sun and the interstellar medium is made.

132.042 **Possible spiral structure in the Orion molecular cloud.** T. G. Phillips, K. B. Jefferts, P. J. Huggins, P. A. R. Ade, P. E. Clegg.
Astrophys. Journ., (*Letters*), Vol. 201, L157 - L159 (1975).
Results are presented of extensive mapping of the Orion molecular cloud in the CO ($J = 2-1$) line indicating the existence within the cloud of large-scale ordered structure of ring or spiral type.

132.043 **Kinematics of the Gum nebula.**
H. H. Hippelein, R. Weinberger.
Astron. Astrophys., Vol. 43, 405 - 409 (1975).
Radial velocities for the Gum nebula have been measured with a Fabry-Pérot interferometer. There is no significant evidence for an expansion of the H II associated with the Gum nebula. The electron temperature agrees with the temperatures in normal H II regions.

132.044 **Chemical composition of extragalactic gaseous nebulae.** M. Peimbert.
Annual Rev. Astron. Astrophys., Vol. 13, (see 003.005), 113 - 131 (1975).

132.045 **The visual Orion nebula.** J. Ashbrook.
Sky Telescope, Vol. 50, 299 - 301 (1975).

132.046 **8–13 micron maps of the Trapezium region of the Orion nebula.**
R. D. Gehrz, J. A. Hackwell, J. R. Smith.
Astrophys. Journ., (*Letters*), Vol. 202, L33 - L36 (1975).
The Trapezium region of the Orion nebula has been mapped at 8.6, 10.1, 11.2, 12.2, and 13.1 μ with an infrared array spectrometer. The 8–13 μ spectral structure of the entire region is characterized by emission and absorption by silicate grains. The grain emission is most prominent in the vicinity of the Trapezium. Absorption by silicate grains is evident only in the Becklin-Neugebauer–Kleinmann-Low region. It is

concluded that the most likely grain heating mechanism in the Ney-Allen sources is the absorption by the grains of Lα radiation resonantly trapped within the H II region.

132.047 **On the radio emission of thermal sources in the Omega and Orion nebulae at 3.9 mm and 13.5 mm wavelengths.** V. A. Efanov, A. G. Kislyakov, I. G. Moiseev, N. S. Nesterov, V. I. Chernyshev, V. B. Shtejnshlejger.
Izv. Krymskoj Astrofiz. Obs., Vol. 53, p. 196 - 201 (1975). In Russian.

The results of radio brightness distribution investigations of the sources M 17, M 42 and M 43 at wavelengths 3.9 mm and 13.5 mm are presented. Radio maps of the sources with angular resolution 50″ and 2!5, respectively, are obtained. Radio-emission flux densities of M 17, M 42, M 43 and of the source in Orion DNCL 1640 have been measured. The spectra of the sources at millimeter wavelengths are discussed.

132.048 **Determination of relative helium and oxygen abundances in gaseous nebulae.** V. I. Pronik.
Izv. Krymskoj Astrofiz. Obs., Vol. 53, p. 202 - 208 (1975). In Russian.

On the basis of observed numbers of O II and O III ions a method of determination of oxygen abundance is proposed. Graphs for He/H and O/H abundances are presented.

132.049 **Analysis of the nebulosities near T Tauri using digital computer image processing.** J. J. Lorre.
Astrophys. Journ., Vol. 202, 696 - 717 (1975).

Direct plates of T Tauri taken with the 120-inch (3 m) and 36-inch (91 cm) Lick reflectors were digitized and analyzed using digital computer image processing techniques. Luminous emission protrusions extending to as far as 13″ from T Tauri in position angles 170°, 210°, and 330° are shown. These features are variable and may contain layered structure. The complex reflection nebula west of T Tauri (NGC 1555) appears to be an illuminated portion of a much larger dark nebula whose variability is due to obscuring material near the star.

132.050 **Photometric study of the Ced 90 nebula.**
A. V. Kurchakov.
Trudy Astrofiz. Inst., *Alma-Ata*, Vol. 25, 12 - 17 (1975). In Russian.

132.051 **Photoelectric observations of the polarization of radiation in the Orion nebula.**
B. M. Gabdullin, D. I. Gorodetskij.
Trudy Astrofiz. Inst., *Alma-Ata,* Vol. 26, 3 - 9 (1974). In Russian.

132.052 **On the reddening law in the Carina nebula.**
W. Herbst.
Bull. American Astron. Soc., Vol. 7, 530 (1975). – Abstr. AAS.

132.053 **The rocket ultraviolet spectrum of NGC 1976, the Orion nebula.** R. C. Bohlin, T. P. Stecher.
Bull. American Astron. Soc., Vol. 7, 547 - 548 (1975). Abstr. AAS.

132.054 **Structural variations of the emission line intensities in Herbig-Haro object No. 1.** R. D. Schwartz.
Bull. American Astron. Soc., Vol. 7, 548 (1975). – Abstr. AAS.

132.055 **Apollo-16 far-ultraviolet nebular photometry.**
G. R. Carruthers, T. L. Page.
Bull. American Astron. Soc., Vol. 7, 548 (1975). – Abstr. AAS.

132.056 **Study of a triple nebula in Orion.**
P. Pişmiş, I. Hasse.
Bull. American Astron. Soc., Vol. 7, 556 (1975). – Abstr. AAS.

132.057 **Asymmetric HCN line profiles from the Orion nebula.** F. O. Clark.
Bull. American Astron. Soc., Vol. 7, 556 (1975). – Abstr. AAS.

132.058 **Central star temperatures of emission nebulae.**
J. B. Kaler.
Bull. American Astron. Soc., Vol. 7, 557 (1975). – Abstr. AAS.

132.059 **Spectrophotometric studies of galactic nebulae. XV. Optical data for NGC 7635 (C), IC 1470, Sh 159.** [Yu. I. Glushkov, Eh. K. Denisyuk, Z. V. Karyagina].
Astron. Tsirk., No. 879, p. 7 - 8 (1975). In Russian.

132.060 **Herbig-Haro objects and postfuors.**
A. L. Gyul'budagyan.
Astrofizika, Vol. 11, 511 - 516 (1975). In Russian. – English translation in Astrophysics, Vol. 11, No. 3.

It is suggested to divide the Herbig-Haro objects into two groups. The first group will involve objects the luminosity of which can be explained by the reflection of the light of a nearby source and the second group will involve objects which are not illuminated from outside. A hypothesis is proposed for the second group, which asserts that there is an analogy between the nature of these objects and postfuors. Several observational facts have been suggested in favor of this hypothesis.

132.061 **Transition radiation and peculiar nebulae.**
G. A. Gurzadyan.
Astrofizika, Vol. 11, 531 - 550 (1975). In Russian. – English translation in Astrophysics, Vol. 11, No. 3.

Transition radiation arising as a result of electrodynamic interaction of high energy electrons with dust particles may explain some observed properties of galactic peculiar objects, containing dust particles in large amounts – Herbig-Haro objects, cometary and Barnard 10 type nebulae, T Tauri and FU Ori type stars etc. The general relationships for the calculation of the spectral distribution of the transition radiation are derived for arbitrary dielectric properties of dust particles. The probable value of the energy of fast electrons, 1.5 MeV, is found by comparing the theory with observations. The problem of the excitation of emission lines, colour indexes and the contribution of the various types of the radiation have also been examined. A qualitative application of the theory to the FU Ori-type stars (fuors) is made.

132.062 **Optical colours and polarization of a model reflection nebula. I. Star behind the nebula.**
G. A. Shah.
Pramāṇa, (*India*), Vol. 3, 338 - 353 (1974).

Simple models of a reflection nebula in the form of a plane-parallel slab containing smooth spherical solid particles in submicron-size range have been considered. Single scattering has been assumed. The effect of varying the composition and size distribution function of the grains have been brought out in the calculations using Mie theory of scattering. The analytical part of the geometry of the problem has been treated quite rigorously and the resulting expression for nebular intensity has been presented in a somewhat new form. The case of the star behind the nebula has been examined.

132.063 **Interstellar ionized nebulae.** Y. Terzian.
Fundam. Cosmic Phys., Vol. 1, 301 - 386 (1974).

The discussion of interstellar ionized nebula covers the physical structure, emission processes, prominent nebulae, the distribution of ionized nebulae in the Galaxy, chemical abundances, and the evolution of H II regions.

Young stellar objects and dark interstellar clouds.
See Abstr. 065.118.

Interaction of hot stars and of the interstellar medium. VII. The rate and fate of stellar ultraviolet photons. See Abstr. 114.024.

Spectral types of faint Orion stars. See Abstr. 114.045.

CO mapping in the V1057 Cygni region. See Abstr. 114.328.

An anomalous 2200 Å absorption band in the WR star HD 192163. See Abstr. 114.354.

Is the Gum nebula an old supernova remnant ionized by ζ Pup and γ^2 Vel? See Abstr. 125.057.

CO, CS, and HCN in a clustering of reflection nebulae in Monoceros. See Abstr. 131.002.

Interstellar formaldehyde near stars of the Orion population. See Abstr. 131.005.

On the Orion infrared nebula/molecular cloud. See Abstr. 131.006.

Further evidence for magnetism in the Orion region. See Abstr. 131.007.

Interstellar H_2CO. II. 2-centimeter emission from the Orion molecular cloud. See Abstr. 131.013.

Complex interstellar calcium line structure in the Carina nebula. See Abstr. 131.015.

Time variations in sources of silicon monoxide emission at 7 millimeters. See Abstr. 131.017.

Giant grains in Orion. See Abstr. 131.019.

Submillimeter observations of NGC 2024, OMC-2, and Mon R-2. See Abstr. 131.079.

Radio sources near reflection nebulae in Monoceros. See Abstr. 131.136.

Further investigations of the CO emission from the North America dust cloud and L1630. See Abstr. 131.160.

A classification of the available astrophysical data of particular H II regions. I: Orion nebula: optical, radio, infrared and molecular mapping. See Abstr. 131.502.

Photoelectric photometry of the ionized helium abundance in galactic H II regions. See Abstr. 131.538.

Detection of the [O III] 88.16 micron line in M17. See Abstr. 131.540.

The radio spectra of four H II regions (Orion B, M8, M16 and M17). See Abstr. 131.542.

Forbidden sulfur II structures in the Carina nebula. See Abstr. 131.549.

Spectrophotometric studies of galactic nebulae. XII. New compact H II regions. See Abstr. 131.557.

Spectrophotometric studies of galactic nebulae. XIII. Compact H II regions in Sh 146, Sh 149, Sh 152. See Abstr. 131.558.

Spectrophotometric studies of galactic nebulae. XIV. Compact H II regions in Sh 307 and NGC 1491. See Abstr. 131.559.

Observations of faint H II regions in our Galaxy. See Abstr. 131.561.

Spectrophotometry of CRL 2688 from 2 to 24 microns. See Abstr. 141.606.

Haro 13a: a luminous, heavily obscured star in Orion? See Abstr. 141.608.

An unusual nebula near UOA 27. See Abstr. 141.624.

A search for optical pulsations from X-ray sources near λλ4640–4650 and λ4686. See Abstr. 142.098.

Soft X-ray search of centre of Cygnus Loop. See Abstr. 142.139.

R-associations IV. Vela R2, a young stellar group. See Abstr. 152.005.

NGC 2175: the cluster age and the nature of the nebulosity surrounding S 252a. See Abstr. 153.024.

Possible variability of IV Zw67 – the "Egg nebula" in Cygnus. See Abstr. 158.124.

133 Planetary Nebulae

133.001 **The relationship between HD 87892 and NGC 3132.** R. H. Méndez.
Astrophys. Journ., Vol. 199, 411 - 417 (1975).

The observations suggest that the central star of NGC 3132 is a double system, the visually brighter component being HD 87892. The nebula is ionized by a very hot object whose characteristics are briefly discussed. The assumption that NGC 3132 lies at the same distance as HD 87892 leads to a very small ionized nebular mass.

133.002 **The ultraviolet spectrum of planetary nebula IC 2149.** G. A. Gurzadyan.
Monthly Notices Roy. Astron. Soc., Vol. 172, 249 - 256 (1975).

An ultraviolet spectrogram of the planetary nebula NGC IC 2149 and its nucleus has been obtained down to a short wavelength limit of 2400 Å by means of the space orbital observatory 'Orion-2'. The ultraviolet emission lines of the nebula are measured for intensity and allocated with their identifications. The chemical elements aluminium and titanium are observed in planetary nebula for the first time. The distribution of the energy in the continuous spectrum of the central star of the nebula is derived and analysed; as a result an excess of radiation near 2500 Å is obtained.

133.003 **On the occurrence of planetary nebulae in open star clusters.** J. Žižňovský.
Bull. Astron. Inst. Czechoslovakia, Vol. 26, 248 - 251 (1975).

The paper investigates the association of planetary nebulae with open star clusters on the basis of their apparent and space distribution between +15 and −15° galactic latitude. No association was observed.

133.004 **Oxygen abundances in planetary nebulae.** C. T. Daub.
Astrophys. Journ., Vol. 200, 82 - 87 (1975).

The abundance of oxygen in planetaries is examined for three different physical descriptions of the density and temperature structure. A comparison of these abundances with expectations derived from an energy-balance perspective leads to the conclusion that, for the models considered here, the most satisfactory oxygen abundance is derived by assuming a uniform density and temperature for a planetary.

133.005 **The 2–4 micron spectrum of NGC 7027.** K. M. Merrill, B. T. Soifer, R. W. Russell.
Astrophys. Journ., (*Letters*), Vol. 200, L37 - L39 (1975).

Spectrophotometric observations of NGC 7027 from 2.1 to 4.1 μ with a resolution $\Delta\lambda/\lambda \sim 0.015$ are reported. The spectrum is rich in features, including three hydrogen recombination lines and several unidentified features. The 2-μ continuum level and the hydrogen recombination-line fluxes agree well with predictions based on the observed radio and Hβ fluxes.

133.006 **The far-ultraviolet spectrum of NGC 6543.** A. Boksenberg, D. Carnochan, J. Cahn, S. P. Wyatt.
Monthly Notices Roy. Astron. Soc., Vol. 172, 395 - 400 (1975).

The far-ultraviolet spectrum (1350–2550 Å) of the central star of the planetary nebula, NGC 6543, has been obtained by the S2/68 sky-survey telescope in the TD-1A European satellite. It is possible to fit a blackbody function at 35000 K to this spectrum. Attempts to fit the observed spectrum to Cassinelli's spherical models of planetary central stars suggest temperatures in the range $40000\ K \leqslant T(\tau = 2/3) \leqslant 50000\ K$. The temperature and range of likely luminosities of the central star place it near the beginning of the Harman-Seaton evolutionary sequence, consistent with an enveloping planetary nebula that is young and optically thick.

133.007 **On the kinematics and orientations of planetary nebulae.** K. Cudworth.
Monthly Notices Roy. Astron. Soc., Vol. 172, 57P - 58P (1975).

The kinematics of planetary nebulae are shown to be inconsistent with the hypothesis that their orientations (long axis parallel to the galactic plane) are due to compression along the direction of motion through the ambient interstellar medium.

133.008 **The theoretical spectrum of nuclei of planetary nebulae in the region of vacuum ultraviolet.**
N. A. Sakhibullin.
Trudy Kazan. Gorod. Astron. Obs., No. 39, p. 16 - 28 (1973).
In Russian.

133.009 **Infrared radiation and surface brightness of planetary nebulae; evolutionary effects.**
G. S. Khromov.
Pis'ma v Astron. Zhurn., Vol. 1, No. 8, p. 16 - 21 (1975).
In Russian.

It is shown that the intensity of infrared ($\lambda \sim 11\mu$) continuous emission of planetary nebulae is connected with their surface brightness in hydrogen emission. This connection is explained as a result of the evolutionary decrease of dust grains radiative power.

133.010 **On the number of planetary nebulae in our Galaxy.** D. Alloin, C. Cruz-Gonzalez, M. Peimbert.
Bull. American Astron. Soc., Vol. 7, 399 (1975). – Abstr. AAS.

133.011 **Planetary nebula composition and central star spectral class.** L. H. Aller, S. J. Czyzak.
Bull. American Astron. Soc., Vol. 7, 399 (1975). – Abstr. AAS.

133.012 **The rocket ultraviolet spectrum of the planetary nebula NGC 7027 II.**
R. C. Bohlin, P. A. Marionni, T. P. Stecher.
Bull. American Astron. Soc., Vol. 7, 399 (1975). – Abstr. AAS.

133.013 **The 3.3 μ emission feature in nebulae and galaxies.** K. M. Merrill, R. W. Russell, B. T. Soifer.
Bull. American Astron. Soc., Vol. 7, 399 - 400 (1975). – Abstr. AAS.

133.014 **S IV and Ne II in NGC 7027.** J. D. Bregman, D. M. Rank.
Bull. American Astron. Soc., Vol. 7, 400 (1975). – Abstr. AAS.

133.015 **The detection of carbon monoxide emission in planetary nebulae.**
S. L. Mufson, J. G. Lyon, P. Marionni.
Bull. American Astron. Soc., Vol. 7, 400 (1975). – Abstr. AAS.

133.016 **The diffuse gas in planetary nebulae: evidence for stellar winds?** L. Goad.
Bull. American Astron. Soc., Vol. 7, 400 (1975). – Abstr. AAS.

133.017 **Identifications, positions, number, and production rate of planetary nebulae in NGC 221.** H. C. Ford.
Bull. American Astron. Soc., Vol. 7, 411-412 (1975). – Abstr. AAS.

133.018 **The peculiar object Mz-3.** M. Cohen.
Publ. Astron. Soc. Pacific, Vol. 87, 500 (1975).
Abstr. ASP.

133.019 **Photographic photometry of some bipolar planetary nebulae.** O. V. Oganesyan.
Soobshch. Byurakan. Obs., vyp. (No.) 46, p. 55 - 61 (1975).
In Russian.

A photographic photometry of four bipolar planetary nebulae –NGC 650-1, 6905, 7008 and anon. 07^h07^m – has been carried out. The results of the observations and measurements are given in tables. The isophotes in photographic light for the nebulae NGC 650-1, 6905 and 7008 are shown in figures in arbitrary units of the intensity. The distance of the nebula NGC 650-1 is 400 pc, the mass 0.065 $M_\odot$. The isophotes in Hα for anon. 07^h07^m are derived also.

133.020 **Origin of planetary nebulae.** S. C. Vila.
19th Colloquium International Astrophys. Liège 1974, (see 012.016), p. 391 - 395 (1975).

Static models have been constructed for giant star envelopes of luminosities greater than 10^4 $L_\odot$ using the mixing length, pressure scale height theory of convection. A density inversion appears in the convective zone. This inversion increases with increasing luminosity of the model and eventually produces a detached shell. The mass of the shell is similar to the observed mass of planetary nebulae shells. The remnant star has a luminosity and effective temperature similar to the one observed for the central stars of planetary nebulae. It is suggested that planetary nebulae might be produced by this process.

133.021 **The detection of carbon monoxide emission in planetary nebulae.**
S. L. Mufson, J. Lyon, P. A. Marionni.
Astrophys. Journ., (*Letters*), Vol. 201, L85 - L89 (1975).

The authors report the first detection of a molecule from planetary nebulae. They have found carbon monoxide emission in the planetary nebulae NGC 7027, IC 418, and NGC 6543. Compared with the optical nebula, the carbon monoxide emission about NGC 7027 is extended. The authors interpret their observations of NGC 7027 as a shell of gas ejected from the preplanetary nucleus as it evolved into a planetary.

133.022 **Observable effects of dust on the ionization of nebulae.** B. Balick.
Astrophys. Journ., Vol. 201, 705 - 718 = Lick Obs. Bull., No. 702 (1975).

The effects of dust on the ionization and thermal structure of a photoionized nebula are explored by numerical model calculations. Particular emphasis is placed on the calculation of observable quantities such as the radio and infrared continua and the optical emission-line spectrum and the effects that dust might have on these quantities. The role of the dust in the nebular structure is summarized. Model calculations of planetary nebulae are shown to agree very well with much of the radio, infrared, and optical data. The importance of dust in the case of H II regions is called into question, especially as an explanation for the low abundance of ionized helium observed in several nebulae.

133.023 **Spectrophotometric study of compact planetary nebulae.** L. N. Kondrat'eva.
Pis'ma v Astron. Zhurn., Vol. 1, No. 10, p. 14 - 19 (1975). In Russian.

Some relative line intensities of 14 planetary nebulae are presented in the 3700–7300 Å region. Electron temperatures and densities are determined.

133.024 **A survey of planetary nebulae towards the galactic bulge.** B. L. Webster.
Monthly Notices Roy. Astron. Soc., Vol. 173, 437 - 447 (1975).

A spectroscopic survey has been made of 50 possible planetary nebulae towards the galactic bulge and of some stellar planetaries in the southern Milky Way. Possible emission-line stars have been described. The nebulae have been assigned excitation classes and the distribution of these among the galactic bulge objects is compared with the distributions in the Large and Small Magellanic Clouds. The galactic bulge planetaries do not apparently relate to either Cloud, though selection effects may distort the distribution.

133.025 **A study of the compact nebulae VV8 and M3-27.** T. F. Adams.
Astrophys. Journ., Vol. 202, 114 - 120 (1975).

New photometric observations of the lines and continuum in the compact nebulae VV8 and M3-27 are presented. Both nebulae have high electron densities and are self-absorbed in Hα. Parameters describing the physical conditions are estimated using the observed Balmer and [O III] line strengths. By comparing the observations with suitable models for young planetary nebulae, the abundances of helium, oxygen, and neon are shown to be normal. The [N II] lines are stronger than predicted. The similarity between the line spectra and inferred properties of the nebulae suggests that the optical continuum in VV8 is unrelated to the nebula, and may come from a late-type companion in a binary. Serious difficulties remain, however, concerning the absolute magnitude and color of the companion in the binary model.

133.026 **Models of planetary nebulae with dust.** R. L. Ferch, E. E. Salpeter.
Astrophys. Journ., Vol. 202, 195 - 203 (1975).

Computer models have been constructed of planetary nebulae of 0.15 and 0.5 $M_\odot$ total mass, containing varying amounts of absorbing dust grains. In all cases the dust remained well mixed with the gas, and was responsible for the maintenance of a well-defined inner edge for the nebular shell. Once the nebula was ionized, the expansion in the outer regions accelerated due to pressure gradients, so that the observed mass decreased with time. The more massive nebulae were found to expand more slowly. Even when all of the Lα radiation was absorbed by grains, it was not as important for grain heating and acceleration as the direct stellar radiation. Agreement with several aspects of recent infrared observations was obtained.

133.027 **Spectroscopic observations of the planetary nebula 283 + 25° 1.** M. G. Smith, T. R. Gull.
Astron. Astrophys., Vol. 44, 223 - 225 (1975).

Spectroscopic data on the planetary nebula K1-22 is presented. A centrally-located star with a very blue spectrum is identified as the true nucleus. The nebula has an expansion velocity of 28 km s^{-1} in the [O III] λ 5007 line and a most probable heliocentric velocity of -14 km s^{-1}.

133.028 **A catalog of confirmed planetary nebulae in the southern Milky Way noted on low-dispersion objective-prism plates.** N. Sanduleak.
Publ. Warner and Swasey Obs., Cleveland, Ohio, Vol. 2, No. 1, 17 pp. (1975).

The author has made a survey for planetary nebulae in the southern Milky Way on objective-prism plates which can detect the presence of the [O III] doublet at λ 5007 and λ 4959 (the N_1 and N_2 nebular lines) which are a defining characteristic in the spectra of gaseous nebulae. The catalog, that has been compiled as a result of this survey, contains only objects found to show very strong nebular line emission (i.e. at least the N_1 line was seen in all cases) and which, insofar as can be determined, have the spectral characteristics of conventional planetary nebulae as described for instance by Aller

(1956). Planetary-like emission-line objects showing pathological spectral features are excluded and have been or will be discussed elsewhere (see Sanduleak and Stephenson, 1972, 1973).

133.029 **Infrared observations of three unusual nebulae.** M. Cohen, M. J. Barlow.
Astrophys. Letters, Vol. 16, 165 - 171 (1975).

Optical, infrared and radio data are assembled on the three objects M1–67, Sh2-266 and NGC 2346 which previously have been classified as planetary nebulae. It is argued that none of these nebulae is in fact a planetary. M1–67 is almost certainly a ring nebula containing a population I WN star; Sh2-266 is probably a nebulosity containing an emission-line B star; while the nature of NGC 2346 is presently indeterminate.

133.030 **A new southern planetary nebula.** A. Blaauw, I. J. Danziger, H. E. Schuster.
Astron. Astrophys., Vol. 44, 469 - 471 (1975).

A new southern planetary nebula of low surface brightness has been discovered on ESO Schmidt plates. Spectral characteristics, velocity, the central star, and estimates of the distance are discussed.

133.031 **The rocket-ultraviolet spectrum of the planetary nebula NGC 7027.**
R. C. Bohlin, P. A. Marionni, T. P. Stecher.
Astrophys. Journ., Vol. 202, 415 - 420 (1975).

An ultraviolet spectrum of NGC 7027 was obtained with a rocket-borne telescope. When corrected for extinction, the C IV resonance line at 1549 Å is in good agreement with the intensity calculated from models, but the C III] intercombination line at 1909 Å is a factor of 10 too bright. The abundance of carbon is assumed to be 2×10^{-4} that of hydrogen. Using carbon abundances for the sun, which are up to 3 times higher, this discrepancy disappears, and there must be attenuation in the C IV line. Since the optical depth is $\sim 10^4$ at the line center, no appreciable number of absorbing grains can exist in the C IV-producing region of the nebula.

133.032 **A new planetary nebula in Puppis.** N. Sanduleak.
Publ. Astron. Soc. Pacific, Vol. 87, 705 (1975).

The discovery of a new planetary nebula is reported.

133.033 **Two emission objects with broad hydrogen lines.** L. N. Kondrat'eva.
Trudy Astrofiz. Inst., *Alma-Ata*, Vol. 25, 23 - 25 (1975).
In Russian.

Forbidden emission lines and bright broad hydrogen lines have been detected in the spectra of the objects He 2-446 and Sh 2-266. The half-width of Hα-lines profiles were equal to 6.62 ± 0.05 Å and 5.6 ± 0.1 Å, respectively.

133.034 **The birth rate of planetary nebulae.** J. H. Cahn, S. P. Wyatt.
Bull. American Astron. Soc., Vol. 7, 532 (1975). – Abstr. AAS.

133.035 **Observations of the central stars of planetary nebulae.** J. H. Lutz.
Bull. American Astron. Soc., Vol. 7, 552 (1975). – Abstr. AAS.

133.036 **Formation of planetary nebulae.** P. E. Stry.
Thesis, Colorado Univ., Boulder (USA). [Available from Univ. Microfilms, Ann Arbor, Mich., USA. Order No. 75-3838], 84 pp. (1974).

133.037 **Chemical abundances and physical conditions in planetary nebulae.** P. T. Barker.
Thesis, California Univ. Santa Cruz (USA). [Available from Univ. Microfilms, Ann Arbor, Mich., *USA*. Order No. 75-8472], 117 pp. (1974).

Photoelectric scanner observations have been made of the intensities of a number of emission lines between λ 3400 Å and λ 7400 Å in thirty planetary nebulae. The electron temperatures and densities implied by different forbidden line intensity ratios agree well for most nebulae. The relative abundances of H, He, Ne, O, and, less reliably, N and S, were determined for thirty planetaries and for several planetaries observed by others.

Charge transfer reactions in multiply charged ion-atom collisions. See Abstr. 022.015.

Radial velocity measurements with the Asiago nebular spectrograph. See Abstr. 034.092.

Theory of extended stellar atmospheres. II. A grid of static spherical models for O stars and planetary nebula nuclei. See Abstr. 064.042.

The subsonic structure of radiatively driven winds of early-type stars. See Abstr. 064.053.

Transition from a red giant to a blue nucleus after ejection of a planetary nebula. See Abstr. 065.029.

Dynamic phases in the evolution of a 1.25 $M_\odot$ C-O star. See Abstr. 065.128.

The temperature effect in photoelectric photometry of strong emission-line objects. See Abstr. 113.038.

The ESO/Uppsala survey of the ESO (B) atlas of the southern sky. III. See Abstr. 113.044.

UV stars and the interstellar medium: a statistical time-dependent model. See Abstr. 131.089.

Characteristics of planetary nebulae and H II regions based on λ = 1.35 cm continuum measurements. See Abstr. 131.534.

Concerning the nature of the nebula V–V 1–7 surrounding BD –18°1967. See Abstr. 132.008.

New southern radio sources associated with planetary nebulae. See Abstr. 141.103.

Observations of an isolated compact infrared source in Perseus. See Abstr. 141.630.

X-ray sources and planetary nebulae. See Abstr. 142.148.

Pregalactic helium abundance and abundance gradients across our Galaxy derived from planetary nebulae. See Abstr. 155.066.

Planetary nebulae in local group galaxies. II. Identifications, positions, number, and production rate of nebulae in NGC 221. See Abstr. 158.189.

134 Crab Nebula

134.001 **Magneto-parametric instabilities in the Crab nebula.** G. S. S. Sweeney, P. Stewart.
Astron. Astrophys., Vol. 41, 431 - 436 (1975).
An analysis is given of the plasma instabilities associated with a strong low frequency electromagnetic wave propagating through a magnetized plasma. The results are discussed in relation to the Crab nebula.

134.002 **The outer structure of the Crab nebula.** R. A. Chevalier, T. R. Gull.
Astrophys. Journ., Vol. 200, 399 - 401 (1975).
Observations of the line emission from the Crab nebula indicate a structure consisting of dense filaments from which thin clumpy sheets of gas fan out radially. The jet feature noted by van den Bergh is one of these sheets. The sheets may have been created by the differential acceleration of filaments, or they may be the result of the formation of the dense filaments by the Rayleigh–Taylor instability.

134.003 **On some peculiaritis of the radio emission of the Crab nebula.** L. I. Matveenko.
Pis'ma v Astron. Zhurn., Vol. 1, No. 7, p. 13 - 16 (1975). In Russian.
An 8-cm radio burst of the Crab nebula emission has been observed on June 9, 1964. Possible causes of apparition of bright flares are discussed.

134.004 **A possible extended soft X-ray component of the Crab nebula.**
A. Toor, T. M. Palmieri, F. D. Seward.
Bull. American Astron. Soc., Vol. 7, 456 (1975). – Abstr. AAS.

134.005 **No indication of excess MeV gamma radiation from the Crab nebula.**
V. Schönfelder, G. Lichti, C. Moyano.
Nature, Vol. 257, 375 - 376 (1975). – Letter.

134.006 **The Crab nebula: high energy X-ray observation of a lunar occultation.** R. Staubert, E. Kendziorra, J. Trümper, C. Reppin, J. A. Hoffman, K. A. Pounds, A. B. Giles, L. V. Morrison.
Astrophys. Journ., (*Letters*), Vol. 201, L15 - L19 (1975).
A lunar occultation of the Crab nebula was observed at X-ray energies above 15 keV by a scintillation counter. The angular extent of the region of emission of 75 percent of the total flux between 15 and 150 keV (average photon energy 39 keV) along position angle 263° is found to be 110″± 25″. The offset of the centroid of the extended source is such that its projection to the west of NP 0532 is 8″± 6″. The variation of size with energy, as found from this and other measurements, sets constraints on models for energy transfer along magnetic field lines in the Crab.

134.007 **On the production of power-law spectra and the evolution of cosmic synchrotron sources: a model for the Crab nebula.** W. J. Cocke.
Astrophys. Journ., Vol. 202, 773 - 781 (1975).
The author investigates new solutions of the energy diffusion equation for ultrarelativistic electrons undergoing Fermi acceleration. Two special cases are considered, which produce power-law electron spectra in a natural way, and which are applicable to models of expanding synchrotron sources. These are applied to the Crab nebula. The model gives a satisfactory fit to the overall nebular spectrum. The author predicts magnetic fields of 2×10^{-4} gauss in the outer reaches of the nebula and 3×10^{-5} gauss near the pulsar, where Fermi acceleration is believed to occur, and where hydromagnetic activity is observed. He further predicts a cutoff in the nebular γ-ray spectrum between 1 and 10 MeV.

134.008 **Extended soft X-ray emission from the Crab nebula.** P. A. Charles, J. L. Culhane.
Bull. American Astron. Soc., Vol. 7, 505 (1975). – Abstr. AAS.

134.009 **On the mass of the Crab nebula.** D. Kazanas, D. N. Schramm.
Bull. American Astron. Soc., Vol. 7, 554 (1975). – Abstr. AAS.

134.010 **Energy dependence of the size of the Crab nebula emission volume.** R. S. Wolff, R. Novick.
Bull. American Astron. Soc., Vol. 7, 554 (1975). – Abstr. AAS.

134.011 **Observations of Faraday rotation of linearly polarized radio emission from the Crab nebula in the solar supercorona at 4 cm wavelength with the help of the large Pulkovo radio telescope.** N. A. Soboleva.
Astron. Tsirk., No. 881, p. 1 - 3 (1975). In Russian.

134.012 **Rocket observation of a Crab nebula lunar occultation.** E. Kendziorra, R. Staubert, J. Truemper, C. Reppin, J. A. Hoffmann, K. A. Pounds.
Verhandl. Deutsche Phys. Ges., Reihe 6, Vol. 10, 158 (1975). In German. – Abstract.

134.013 **Observation of pulsed and non-pulsed gamma radiation from the Crab nebula near 1 MeV.**
D. E. Gruber.
Thesis, California Univ., San Diego (USA). [Available from Univ. Microfilms, Ann Arbor, Mich., *USA*. Order No. 75-8259], 99 pp. (1974).
Gamma-rays in the energy range 0.2 to 10 MeV have been observed from the Crab nebula and its central pulsar, NP 0532, on three occasions in 1971. Besides confirming the continuous and pulsed emissions at low energy, the new observations indicate an additional flux above 1 MeV.

Improved upper limit on the photon rest mass.
See Abstr. 022.088.

Collective relaxation of ultra-relativistic plasmas.
See Abstr. 062.037.

Meter-wavelength VLBI. II. The observations.
See Abstr. 141.143.

Anomalous scattering in the Crab nebula.
See Abstr. 141.302.

The pulse shape of the Crab nebula pulsar, PSR 0531+21, in the *V* and *I* bands. See Abstr. 141.316.

A high time resolution X-ray light curve for the Crab pulsar, NP 0532. See Abstr. 141.326.

Crab pulsar optical and X-ray radiation through pinching instabilities. See Abstr. 141.328.

On the pulsar emission mechanisms.
See Abstr. 141.337.

Pulsars: a progress report.
See Abstr. 141.338.

Search for continuous X-ray emission from NP 0532.

See Abstr. 141.346.

Third speed-up of the Crab pulsar. See Abstr. 141.348.

The temporal behaviour of Taurus X-1 (the Crab nebula). See Abstr. 142.063.

Measurement of the spatial structure of the X-ray source in the Crab nebula. I. Observation of the 1974 November 3 lunar occultation. See Abstr. 142.136.

Measurement of the spatial structure of the X-ray source in the Crab nebula. II. Observation of the 1974 December 28 lunar occultation. See Abstr. 142.137.

Spatial distribution of X-rays in the Crab nebula. See Abstr. 142.164.

Measurements of low-energy gamma radiation from the Crab nebula and from pulsar NP 0532. See Abstr. 142.165.

The X-ray pulsed fraction of NP0532 and interstellar grains. See Abstr. 142.241.

Diffusion of cosmic rays in non-thermal radio sources. See Abstr. 143.018.

Radio Sources, Quasars, Pulsars, Infrared, X-Ray, Gamma-Ray Sources, Cosmic Radiation

141 Radio Sources, Quasars, Pulsars, Infrared Sources

Radio Sources, Quasars

141.001 **Quasars flare sharply: explaining the energy gets harder.** W. D. Metz.
Science, Vol. 189, 129 (1975).

141.002 **Compton-synchrotron models for the nucleus of Centaurus A (NGC 5128).** J. E. Grindlay.
Astrophys. Journ., Vol. 199, 49 - 53 (1975).

Compton-synchrotron models are constructed to account for the observed microwave through X-ray spectrum and the recently reported flux of γ-rays with $E_\gamma \gtrsim 3 \times 10^{11}$ eV from NGC 5128 (Cen A). The spectrum suggests a double source structure in the nucleus of NGC 5128. The γ-rays arise from inverse Compton scattering of a synchrotron photon spectrum that must extend through X-ray energies. The observed γ-ray flux and the self-absorbed synchrotron spectra determine the source angular sizes, energy densities, and magnetic fields. Such parameters for the sources in the nuclei of active galaxies may enable the eventual description of the source mechanisms. Further implications of the model and possible observations are discussed.

141.003 **3C 123: a distant first-ranked cluster galaxy at $z = 0.637$.** H. Spinrad.
Astrophys. Journ., (*Letters*), Vol. 199, L3 - L4 (1975).

Spectra of the very faint cluster galaxy and radio source 3C 123 yield a redshift $z = 0.637$. The optical spectrum, morphological type, approximate absolute magnitude, and cluster membership strongly suggest 3C 123 to be an optical object more distant, but very similar to, 3C 295. The radio source is a little larger and intrinsically more powerful than Cygnus A.

141.004 **The optical jet in the quasar 3C 273.** B. N. G. Guthrie, W. McD. Napier.
Monthly Notices Roy. Astron. Soc., Vol. 172, 85 - 89 (1975).

Various models of the optical jet in 3C 273 are discussed. The optical emission is probably due to inverse Compton scattering of infrared radiation from the quasar by a mass $\gtrsim 6 \times 10^7 M_\odot$. It seems that the jet was relativistically ejected along the polar axis of the quasar $\sim 10^6$ yr ago. Difficulties are encountered in applying current models of extended radio sources to the jet.

141.005 **3C 219: a double radio source with a jet.** B. D. Turland.
Monthly Notices Roy. Astron. Soc., Vol. 172, 181 - 189 (1975).

3C 219 has been mapped at 5 GHz with the 5-km telescope. Besides the two extended outer components and a component coincident with the nucleus, a jet was found which lies along the source axis between the nucleus and a compact 'head' in the Sp component. This jet is compared with jets found in other radio sources and the possibility that such jets are associated with the flow of energy from the nucleus to the outer components is discussed.

141.006 **Variable quasi-stellar sources with particular emphasis on objects of the BL Lac type.**
T. D. Kinman.
IAU Symposium No. 67, (see 012.001), p. 573 - 590 (1975).
Review paper.

141.007 **Observations of extended sources at 408 MHz – I. The radio structures.**
R. T. Schilizzi, W. B. McAdam.
Mem. Roy. Astron. Soc., Vol. 79, 1 - 73 (1975).

The 1.6 km Molonglo cross radio telescope has been used to study the structure at 408 Mhz of 116 extragalactic sources between declinations +18° and −80°. The sources are selected to have flux density greater than 1.5 fu and angular size between 2 and 15 arcmin. Contour maps are presented together with centroid and peak positions, size, orientation and flux density for each resolved component.

141.008 **Observations of extended sources at 408 MHz – II. The optical fields.** R. T. Schilizzi.
Mem. Roy. Astron. Soc., Vol. 79, 75 - 100 (1975).

The optical fields of 114 of the radio sources described in the companion paper I (see 141.007) have been examined. Selection criteria for identification of extended or multiple radio sources are developed, and 34 new identifications are proposed based on these criteria. All of the identifications with magnitudes brighter than or equal to 18^m are believed to be real associations. Optical positions have been measured for 234 objects in 86 radio fields north of −45°. The internal errors in these measurements are 0.5 arcsec in each coordinate; comparisons with measurements by other observers confirm the present accuracy.

141.009 **Measurements of the circular polarization of radio sources at frequencies of 0.63, 1.4, 5.0 and 8.9 GHz.**
J. A. Roberts, R. S. Roger, J.-C. Ribes, D. J. Cooke, J. D. Murray, B. F. C. Cooper, F. Biraud.
Australian Journ. Phys., Vol. 28, 325 - 351 (1975).

Circular polarization measurements were made with the Parkes 64 m telescope on 66 mainly extragalactic sources, 33 of which were studied at two or more frequencies. Circular polarization was definitely found in eight quasi-stellar sources. For several of the sources the degree of circular polarization changed by a factor $\gtrsim 2$ between observing sessions, while any accompanying change in total flux density was $\lesssim 20\%$. Instrumental effects are discussed in some detail.

141.010 **Hochauflösende Interferometrie kompakter extragalaktischer Radioquellen.**
I. I. K. Pauliny-Toth, E. Preuss, A. Witzel, D. Graham.
Mitt. Astron. Ges., No. 36, p. 107 - 110 (1975).

141.011 **Strahlungsbeschleunigung in der Nachbarschaft eines Quasars.**
R. Kippenhahn, J. J. Perry, H.-J. Röser.

Mitt. Astron. Ges., No. 36, p. 112 (1975). – The paper appeared in Astron. Astrophys., Vol. 34, 211 - 224 (1974) – see Abstr. 12.141.016.

141.012 **Models for radiation pressure driven absorption layers around quasars.**
M. Grewing, P.-R. Preussner, G. Terstiege.
Mitt. Astron. Ges., No. 36, p. 113 - 115 (1975).

141.013 **The absorption spectrum of 4C 25.05.**
A. Oemler, Jr., C. R. Lynds.
Astrophys. Journ., Vol. 199, 558 - 564 (1975).

The QSO 4C 25.05 (z_{em} = 2.360) has a very rich absorption spectrum, especially shortward of the Lα emission feature at 4086 Å. Only one redshift system can be found which identifies enough (11 of 83) absorption lines to be considered unquestionably real. The hypothesis is considered that many of the remaining strong lines are due to Lα in absorption systems in which most of the lines of other elements are too weak to be seen. Thirteen redshift systems were found which explain most of the observed absorption lines.

141.014 **Positions and flux densities of 5C 3 sources near M31.** J. H. Spencer, B. F. Burke.
Astrophys. Journ., Vol. 199, 611 - 618 (1975).

Observations of selected regions in the direction of M31 provide information on the flux densities and spectral indices between 408 and 8085 MHz for six sources from the 5C 3 catalog. A total of 12 sources were detected at 2695 MHz, and upper limits on flux density and spectral index are provided for 33 undetected sources. Improved positions are given for the detected sources. Spectral indices of the detected 5C 3 sources have steep spectra, which provide an argument that the 5C 3 sources are not associated with M31.

141.015 **Photometric histories of QSOs: two QSOs with large light amplitude.**
M. H. Liller, W. Liller.
Astrophys. Journ., (*Letters*), Vol. 199, L133 - L135 (1975).

PKS 1510–089, a QSO with a redshift z = 0.361, has varied in B from 11.8 to 17.8 mag since 1899.6. At its brightest in 1948, PKS 1510–089 was optically one of the most luminous known objects in the universe if its redshift is cosmological. Its observed range in brightness, 5.4 mag, is larger than that known for any other QSO. MA 0829 + 047 is a BL Lac object which has varied in B from 14.2 to 17.8 mag since 1934.1. In one 23.8 hour period it increased in brightness by more than a factor of 2, implying a line-of-sight dimension of $\lesssim 8 \times 10^{-4}$ pc. The authors compare MA 0829 + 047 and PKS 1510–089 with other similar objects and discuss the significance of their findings.

141.016 **High-frequency structure of 3C 154.** T. K. Menon.
Astrophys. Journ., (*Letters*), Vol. 199, L161 - L163 (1975).

The structure of the source 3C 154 at 2695 MHz and 8085 MHz is compared with its structure at 327 MHz obtained with the Ooty occultation telescope . It is found that the fraction of the total flux in a central compact component of the basic triple source increases with frequency and reaches nearly 40 percent at 8085 MHz. There is some indication that the fraction in central components may be increasing in fainter triple sources.

141.017 **Splittings in quasar absorption lines.**
J. N. Bahcall.
Astrophys. Journ., (*Letters*), Vol. 200, L1 - L3 (1975).

It is shown that the apparent constancy of the splitting between quasar absorption lines observed by Boksenberg and Sargent could be understood if the typical velocity in the absorbing material were somewhat less than the observing resolution (estimated to be 100 km s^{-1}). If this interpretation were correct, then there sould be many more redshifts that are split with velocity separations that are less than 100 km s^{-1} than have been so far observed at velocity separations of the order of 150 km s^{-1}. Some results are presented for a specific model in which the splittings are supposed due to absorption in halos or clouds surrounding galaxies or small groups of galaxies and some tests of this model are suggested. The model parameters are consistent with recent radio observations of H I clouds or halos that extend far from their parent galaxies.

141.018 **New spectrometric results on the quasar 3C 273.**
A. Boksenberg, K. Shortridge, R. A. E. Fosbury, M. V. Penston, A. Savage.
Monthly Notices Roy. Astron. Soc., Vol. 172, 289 - 303 (1975).

A systematic spectrometric study of 3C 273 has been performed using a digital image tube. The forbidden lines are well fitted by a model with an ion number density of 2×10^8 cm^{-3} and our upper limit on the intensity ratio of [Ne V] λ3426 to [Ne III] λ3868 gives indirect evidence that the helium to hydrogen ratio is greater than 0.01 by number. The Balmer decrement is considerably steeper than predicted. The emission feature previously identified as the sodium D-lines by Wampler & Oke should now probably be attributed to He I.

141.019 **Optical activity and absorption lines in quasars.**
E. F. Borra.
Observatory, Vol. 95, 141 - 143 (1975).

141.020 **Angular size–flux density relation for extraglactic radio sources.** G. Swarup.
Monthly Notices Roy. Astron. Soc., Vol. 172, 501 - 512 (1975).

A comparison has been made of the maximum angular extents of 163 weak extragalactic radio sources observed by the method of lunar occultation at Ootacamund with those for 199 strong sources from the 3CR complete sample. The angular sizes of the 62 sources with flux densities $S_{408} \geqslant 16.5$ Jy (1 Jy = 10^{-20} W m^{-2} Hz^{-1}) in the All-sky catalogue of Robertson were also considered.

141.021 **Gravitational redshifts and angular diameters of collapsed objects.** P. K. Das.
Monthly Notices Roy. Astron. Soc., Vol. 172, 623 - 630 (1975).

Light propagation from the surfaces and the interiors of static, spherically symmetric collapsed objects with isothermal cores and adiabatically stable envelopes capable of giving high internal gravitational redshifts is studied within the framework of general relativity. Optical appearance and variation of angular diameter are discussed in relation to the equation of state. A method to estimate the mass and the distance of the object from the angular diameter or the period of flux variation has been outlined and illustrated with the example of the QSO 3C9.

141.022 **Study of the brightness and polarization structure of extragalactic radio sources.**
M. A. Stull, K. M. Price, L. R. D'Addario, S. J. Wernecke, W. Graf, C. J. Grebenkemper.
Astron. Journ., Vol. 80, 559 - 580 (1975).

The authors have determined the brightness and linear polarization structure of 19 extended extragalactic radio sources at a frequency of 10.7 GHz and with a synthesized beam 17″ × 17″ csc δ. Compact components with self-absorption spectra coincident with the nucleus of an associated galaxy are fairly common, while many sources display regions of high-percentage polarization, sometimes greater than 40%. Comparison of the results with other studies, all at lower frequencies (generally ⩽ 5 GHz), has yielded information on spectra, depolarization rates, and Faraday rotation of the polarization plane in various source regions.

141.023 **Optical positions of radio sources. III. Measurements on the Palomar Sky Survey prints.**
M. P. Véron, P. Véron.
Astron. Astrophys., Vol. 42, 1 - 8 (1975).
Using the Schlesinger's method (Schlesinger, 1926) with the AGK3 stars as reference, the authors have measured the optical position of 132 quasars and radio galaxies with an r.m.s. error of the order of 0.5–0.6 arcsec. They show that there is no systematic difference between the AGK3 and SAO catalogues but that the accuracy of the AGK3 positions is much better so that the AGK3, when available, should always be preferred to the SAO.

141.024 **Detection of a 21 cm absorption line against the suggested galaxy/QSO pair NGC 3067/3C 232.**
M. Grewing, U. Mebold.
Astron. Astrophys., Vol. 42, 119 - 125 (1975).
The 21 cm line emission of the four suggested galaxy/QSO pairs NGC 3067/3C 232, NGC 4138/3C 268.4, NGC 4651/3C 275.1 and NGC 5832/3C 309.1 has been studied with the Effelsberg 100-m telescope. In the case of the first pair, a narrow absorption feature at v_{LSR} = 1411 km s^{-1} was found. The physical implications of the results as to the reality of the galaxy/QSO associations are discussed.

141.025 **A radio source and cluster of galaxies in 3 U 1252-28.** N. V. Vidal.
Astron. Astrophys., Vol. 42, 145 - 146 (1975).
The optical identification of 3 U 1252–28 with a cluster of galaxies is discussed.

141.026 **A correlation test of the intrinsic interpretation of QSO absorption redshifts.** R. Opher.
Astrophys. Journ., Vol. 200, 450 - 453 (1975).
The general intrinsic interpretation of QSO absorption redshifts (clouds expelled from the QSO core) predicts that the expulsion velocities of the clouds have a high probability of being clustered together. A correlation function, $\bar{C}$, is defined which depends on the probability of three or more expulsion velocities of a QSO being clustered. The data of pairs of QSOs of similar z_{em} are used. Of prime importance, $\bar{C}$ is insensitive to cosmological models. If the absorption systems are intrinsic and correlated, $\bar{C}$ should be positive with a maximum value $\bar{C}$ = +100 percent; if there is no physical connection between the absorption systems, then $\bar{C} \approx 0$; a strong anticorrelation, or a tendency for QSOs to emit one or two clouds at well separated characteristic velocities, results in $\bar{C} \approx -100$ percent. Data from the QSOs Ton 1530, 1331 + 170, PKS 0237, PHL 957, and 4C 05.34 are used to evaluate $\bar{C}$. A large positive value for $\bar{C}$ is found.

141.027 **Photometric histories of QSOs: 3C 279, the most variable and possibly most luminous QSO yet studied.** L. J. Eachus, W. Liller.
Astrophys. Journ., (*Letters*), Vol. 200, L61 - L62 (1975).
3C 279, a QSO with a red shift z = 0.536, reached a B magnitude of 11.27 ± 0.07 in 1937 April, and several times has become brighter than B = 13.0. The authors calculate that at its brightest, the absolute magnitude of 3C 279 was –31.4, making it optically one of the most luminous known objects to occur in the universe, assuming its redshift is cosmological. With a minimum brightness recorded by Sandage et al. at B = 18.0, 3C 279 exhibits a range of >6.7 mag, and thus becomes the most variable QSO or BL Lacertae object yet studied. Its light curve suggests a tendency to repeat outbursts at ~7 year intervals.

141.028 **Variable radio emission from UX Arietis (HD 21242).** D. M. Gibson, R. M. Hjellming, F. N. Owen.
Astrophys. Journ., (*Letters*), Vol. 200, L99 - L102 (1975).
Variable radio emission from the spectroscopic binary UX Ari has been detected at frequencies of 2695, 5010, and 8085 MHz. The UX Ari radio source was present on all 13 days the source was observed with the interferometer, and was probably present on six of the 10 days the source was observed at 5010 MHz. The radio behavior included at least two Algol-type flares, and during the decay of the flare of 1974 August 11 - 14 the spectrum became inverted, indicating the nonthermal nature of the radio emission.

141.029 **Identification and photometry of radiosources of the 3 CR catalogue. III. The case of the source 3C303.** G. Lelièvre, G. Wlérick.
Astron. Astrophys., Vol. 42, 293 - 297 (1975). In French.
Radio observations of Pooley and Henbest (1974) at 5 GHz, show that the source 3C 303 is double with 19″ separation. Using new optical positions, the authors find, on deep electronographic plates that there exists an optical object in coincidence with each radiocomponent. At the position of the preceding one, is a faint object showing a strong U.V. excess. The following component is unresoved at 5 GHz and coincides with the nucleus of an N-galaxy. These results open at least three possibilities for the optical identification: (1) two independent sources: one quasar and an N-galaxy; (2) a multiple source associated with the N-galaxy; (3) two radiocomponents of the same source each having an identification with an U.V. excess object.

141.030 **Physical conditions and the locations of the absorption-line systems in the QSO 3C 286 and the BL Lacertae object PKS 0735 + 178.**
A. M. Wolfe, G. R. Burbidge.
Astrophys. Journ., Vol. 200, 548 - 557 (1975).
The redshifted 21-cm and Mg II absorption lines observed in the QSO 3C 286 and the BL Lacertae object 0735 + 178, respectively, are analyzed to determine the location of the absorbing gas. The authors first describe evidence which supports the presence of neutral hydrogen intrinsic to QSOs, and then examine excitation conditions influencing the formation of 21-cm absorption lines. An analysis of the Mg II absorption lines seen in 0735 + 178 leads to two possibilities: (1) a large column density of Mg^+ ions and low velocity dispersion; or (2) a low column density and large velocity dispersion. Whereas (2) is kinematically inconsistent with an intervening galaxy, (1) is not. The authors propose that 21-cm absorption features be searched for in the 30-cm continuum of 0735 + 178. If no absorption is seen, and (1) applies, then the spin temperature must be much higher than that observed in absorbing clouds in our Galaxy, thus making an intervening galaxy implausible.

141.031 **Variable radio source structure on a scale of several minutes of arc.** H. N. Ross.
Astrophys. Journ., Vol. 200, 790 - 802 (1975).
During the course of high-frequency flux-density measurements on sources exhibiting low-frequency cutoffs, at least three such sources, 4C 79.4, 4C 79.15, and PKS 2344 + 09, revealed structure on a scale ranging from a fraction of a minute of arc to as much as 15 minutes of arc. This structure appears to change significantly on a time scale of several months. It is suggested that this variable structure arises from the existence of several centers that are highly variable in flux density, but fixed in position.

141.032 **Neutral hydrogen absorption in the spectrum of the quasar-galaxy pair 4C32.33/NGC 3067.**
A. D. Haschick, B. F. Burke.
Astrophys. Journ., (*Letters*), Vol. 200, L137 - L140 (1975).

Radio spectra near 21 cm wavelength were taken of five close quasar-galaxy pairs. Neutral hydrogen absorption of radiation from the quasar 4C32.33 (3C 232) has been detected at a velocity ($C\Delta\lambda/\lambda_0$) of 1418 km s^{-1}. This feature lies 76 km s^{-1} away from the velocity of the galaxy NGC 3067 which is separated from the position of 4C 32.33 by less than 2 Holmberg radii. The similarity of the absorption feature in 4C 32.33 to that discovered by Brown and Roberts in 3C 286 lends some support to the suggestion by Brown and Roberts that the feature in 3C 286 is caused by a cloud of neutral hydrogen associated with a galaxy somewhere along the line of sight to the quasar.

141.033 **Measurements of the linear and circular polarization of some compact radio sources at 5 GHz.**
M. Ryle, D. M. Odell, P. C. Waggett.
Monthly Notices Roy. Astron. Soc., Vol. 173, 9 - 20 (1975).

Observations have been made at λ 6 cm with the 5-km telescope to determine the linear polarization of 60 compact sources and the circular polarization of five of them. An attempt to find physically useful relationships between the magnitude of the linear polarization and other properties of the sources has proved no more successful than previous analyses using a different source sample. No circularly-polarized emission was detected in any of the five sources studied, and the new upper limits avoid the theoretical difficulties associated with the earlier reported values.

141.034 **Observations of 3C33, 61.1, 379.1 and 390.3 at 5 GHz with the 5-km radio telescope.**
P. J. Hargrave, M. McEllin.
Monthly Notices Roy. Astron. Soc., Vol. 173, 37 - 63 (1975).

The Cambridge 5-km radio telescope has been used to map the sources 3C33, 61.1, 379.1 and 390.3 in both total intensity and linear polarization. The maps, which have a resolution of 2''0 arc in right ascension and 2''0 cosec δ in declination, reveal that the components of these sources contain features of high surface brightness, some of which show striking alignment with the associated optical objects. These features are in many cases highly polarized (up to 40 per cent). The results of the observations are discussed in the light of the various theoretical models proposed to explain the existence of double extragalactic radio sources.

141.035 **Radio source models with very small pitch angle particles.** P. S. Callahan.
Monthly Notices Roy. Astron. Soc., Vol. 173, 111 - 120 (1975).

Formulae for the emission by very small pitch angle electrons are used to derive the parameters of self-consistent radio source models from observational data. It is found that while no observed radio source requires emission from very small pitch angle particles, emission from small pitch angle particles may contribute in some source.

141.036 **A narrow emission line near 1720 MHz in the direction of Sgr A.** J. B. Whiteoak, F. F. Gardner.
Monthly Notices Roy. Astron. Soc., Vol. 173, 25P - 28P = Div. Radiophys., C.S.I.R.O., Radiophys. Publ. RPP 1851 (1975).

A narrow emission feature (of half-width 4.5 km s^{-1} in velocity) has been detected in the 1720 MHz OH-line spectrum of Sgr A at a frequency that would correspond to an OH velocity of +132 km s^{-1} (lsr). However, the identification as an OH transition is questioned because there are no features of similar velocity in the line profiles of either the other OH ground-state transitions or transitions for other molecules.

141.037 **Apparent distribution of radio sources.**
A. G. Gorshkov, M. G. Larionov, M. V. Popov.
Pis'ma v Astron. Zhurn., Vol. 1, No. 9, p. 19 - 21 (1975). In Russian.

Analysis of the sky survey obtained at frequency 8550 MHz has revealed the trend of radio sources to form groups at a scale ~1° and clusters with typical dimensions ~10°.

141.038 **Transfer of polarized radiation in opaque nonthermal radio sources.** T. W. Jones, S. L. O'Dell.
Bull. American Astron. Soc., Vol. 7, 413 (1975). – Abstr. AAS.

141.039 **A test of some extended extragalactic radio source models.** D. S. De Young.
Bull. American Astron. Soc., Vol. 7, 413 - 414 (1975). – Abstr. AAS.

141.040 **Strong millimeter wave components in extended radio sources.** B. J. Wills.
Bull. American Astron. Soc., Vol. 7, 414 (1975). – Abstr. AAS.

141.041 **Radio frequency spectra of components of extragalactic sources.** T. K. Menon.
Bull. American Astron. Soc., Vol. 7, 414 (1975). – Abstr. AAS.

141.042 **Arc-second source structure from IPS observations at a decametric wavelength.**
W. M. Cronyn, F. T. Erskine, S. D. Shawhan.
Bull. American Astron. Soc., Vol. 7, 414 (1975). – Abstr. AAS.

141.043 **Physical conditions and the locations of the absorption-line systems in the QSO, 3C 286, and the BL Lac object PKS 0735+178.** A. M. Wolfe, G. R. Burbidge.
Bull. American Astron. Soc., Vol. 7, 422 (1975). – Abstr. AAS.

141.044 **The abundance of nitrogen in QSOs.** G. A. Shields.
Bull. American Astron. Soc., Vol. 7, 426 (1975). Abstr. AAS.

141.045 **Observations of QSO pairs.**
B. J. Wills, D. Wills.
Bull. American Astron. Soc., Vol. 7, 426 (1975). – Abstr. AAS.

141.046 **On the reproducibility of periodicities in the distribution of redshifts of QSOs.**
R. F. Green, D. O. Richstone.
Bull. American Astron. Soc., Vol. 7, 426 - 427 (1975). – Abstr. AAS.

141.047 **One-millimeter continuum observations of extragalactic objects.** J. H. Elias, D. Y. Gezari, M. G. Hauser, G. Neugebauer, M. W. Werner, W. E. Westbrook.
Bull. American Astron. Soc., Vol. 7, 436 - 437 (1975). – Abstr. AAS.

141.048 **New variable radio sources at 430 MHz.**
E. K. Conklin.
Bull. American Astron. Soc., Vol. 7, 437 (1975). – Abstr. AAS.

141.049 **Comparison between VLBI and IPS observations of compact sources.**
G. M. Resch, N. R. Vandenberg, W. C. Erickson, T. A. Clark.
Bull. American Astron. Soc., Vol. 7, 437 (1975). – Abstr. AAS.

141.050 **The intrinsic sizes of compact radio sources at meter wavelengths.**
G. M. Resch, W. C. Erickson, N. R. Vandenberg, T. A. Clark.
Bull. American Astron. Soc., Vol. 7, 437 - 438 (1975). – Abstr. AAS.

141.051 **Absorption lines in QSOs and the line-locking hypothesis.** G. Burbidge, M. Burbidge.
Bull. American Astron. Soc., Vol. 7, 453 (1975). – Abstr. AAS.

141.052 **A discovery program for bright quasars.** R. F. Green.
Bull. American Astron. Soc., Vol. 7, 454 (1975). – Abstr. AAS.

141.053 **Local quasars and radio galaxies.** J. Terrell. Bull. American Astron. Soc., Vol. 7, 454 (1975). Abstr. AAS.

141.054 **A sudden increase in the X-ray flux from Centaurus A.** P. F. Winkler, Jr., A. E. White.
Bull. American Astron. Soc., Vol. 7, 498 - 499 (1975). – Abstr. AAS.

141.055 **Structure of the compact radio component in 3C236.** E. B. Fomalount, G. K. Miley.
Nature, Vol. 257, 99 - 103 (1975).

The compact component in the giant radio galaxy 3C236 has been studied at 0.4, 2.7 and 8.1 GHz, with resolution of up to 0.13″. The overall structure is double and asymmetric. No difference between the spectra of the subcomponents was detected. On a scale of 0.2″ there is fine structure. The radio subcomponents are almost certainly not confined and are thus the results of a relatively recent event in the nucleus of 3C236.

141.056 **New radio map of Cassiopeia A at 5 GHz.** A. R. Bell, S. F. Gull, S. Kenderdine.
Nature, Vol. 257, 463 - 465 (1975).

The radio source Cassiopeia A is believed to be the remnant of a type II supernova that occurred in our galaxy about 280 yr ago. The radio remnant forms an irregular shell of radius ~2′, corresponding to ~1.6 pc at its distance of 2.8 kpc, which is broadly coincident with an incomplete shell of fast-moving optical knots. The authors present here a new radio map of the source with a resolution three times greater than has previously been achieved.

141.057 **Search for optical identifications in the 5C1 radio survey.** P. Notni, H.-E. Fröhlich.
Astron. Nachr., Vol. 296, 197 - 219 (1975).

Identifications of 5C1 radio sources with optical objects are given. The identification rates amount to 44 and 30 per cent up to the limiting magnitude of about 21^m in R and B, respectively. The objects are to about equal parts galaxies, blue starlike objects, and "unclassified objects" (presumably mostly faint galaxies and less pronounced blue objects). For every proposed identification a reliability and for the identification samples the reliability and the completeness are given.

141.058 **Source count analysis for a homogeneous sample of faint radio quasars.** C. Fanti, R. Fanti, A. Ficarra, L. Formiggini, G. Giovannini, C. Lari, L. Padrielli.
Astron. Astrophys., Vol. 42, 365 - 377 (1975).

The statistical properties of a sample of faint radio quasars are examined, and a comparison is made with the stronger quasar samples from the 3 CR and 4 C catalogues. The authors find a significant correlation between the radio spectral index distribution and the apparent magnitude. A second, but less significant, correlation is also found between the spectral index distribution and the ratio R between radio and optical flux. It is found that: a) the quasar counts are explained for a wide range of values of R by a unique evolution law similar to the one already found for 3 CR and 4 C samples; b) quasars at low R, a large fraction of which may have prominent compact flat spectrum components, evolve significantly lower in number density with increasing z; c) locally ($z \ll 1.0$) a large fraction of all quasars (30% to 100%) are radio emitters with $\log R > 2.0$; at higher redshifts the bulk of quasars tend to have lower radio luminosity ($\log R < 2.5$) or to become radio quiet; and d) locally the radio luminosity of quasars matches that of radio nuclei of elliptical galaxies, suggesting some continuity between the nuclear phenomena in ellipticals and the quasar phenomena.

141.059 **Spectral intensity dependence and isotropy of sources stronger than 0.1 Jy at 2700 MHz.**
T. J. Balonek, J. J. Broderick, J. J. Condon, D. F. Crawford, D. L. Jauncey.
Astrophys. Journ., Vol. 201, 20 - 25 (1975).

The 1000-foot telescope of the National Astronomy and Ionosphere Center was used to measure 430 MHz flux densities of sources stronger than 0.1 Jy at 2700 MHz. Distributions of the resulting two-point spectral indices $\alpha(430, 2700)$ of sources in the intensity range $0.1 \leq S < 0.35$ Jy were compared with $\alpha(318, 2700)$ distributions of sources stronger than 0.35 Jy at 2700 MHz. The median normal-component spectral index and fraction of flat-spectrum sources in the faintest sample do not continue the previously discovered trend toward increased spectral steepening of faint sources. This result differs from the prediction of simple evolutionary cosmological models and therefore favors the alternative explanation that local source-density inhomogeneities are responsible for the observed intensity dependence of spectral indices.

141.060 **Spectrophotometry of low-redshift quasi-stellar objects.** J. A. Baldwin.
Astrophys. Journ., Vol. 201, 26 - 44 = Lick Obs. Bull., No. 699 (1975).

Emission line strengths, low-resolution line profiles, and continuum energy distributions were measured for 14 low-redshift QSOs. The profiles of many of the broad emission lines are quite symmetrical and are in excellent agreement either with electron-scattering profiles, or, as has been shown by Blumenthal and Mathews, with the profiles expected for mass loss driven by radiation pressure, but some of these QSOs have asymmetrical broad lines which necessarily involve mass motions at high velocities. The Balmer decrements are considerably steeper than is predicted by recombination theory, implying that there is either reddening or at least some collisional excitation.

141.061 **The radio spectra of mass-outflow stars.** K. A. Marsh.
Astrophys. Journ., Vol. 201, 190 - 193 (1975).

The occurrence of radio spectra of the approximate form $S_\nu \propto \nu^{+1}$ in emission-line stars with infrared excesses can be explained by a model in which mass outflow is the dominant characteristic of the extended stellar atmosphere. An unbounded spherically-symmetric inverse-square density distribution of ionized gas would produce a spectrum of the form $S_\nu \propto \nu^{0.6}$. The fact that the measured radio spectral indices are significantly greater than 0.6 can be explained in terms of an inverse-square distribution which has been truncated at some finite outer radius. This simple model provides a good fit to the observed spectrum of the three radio sources considered here, namely Vy 2–2, Hb 12, and V1016 Cygni.

141.062 **Photographs of 3C 120 and NGC 7603 compared to spectracon images.** H. Arp.
Publ. Astron. Soc. Pacific, Vol. 87, 545 - 549 (1975).

Photographs on IIIa-J plates show more detail and fainter features than do recent observations of the same objects with the spectracon. It is suggested that the radio source 3C 120 is some form of a spiral galaxy. This conclusion is in contradiction to that of Walker, Pike, and McGee (1974).

141.063 **Accurate positions and identifications for a complete sample of 341 radio sources from the Parkes ± 4° survey.**
N. J. McEwan, I. W. A. Browne, J. H. Crowther.
Mem. Roy. Astron. Soc., Vol. 80, 1 - 59 (1975).

Accurate radio positions and optical identifications are

presented for a complete sample of 341 radio sources taken from the 2700 MHz Parkes survey 'PKS ± 4°'. For all sufficiently compact sources in this sample, radio positions have been measured with the RRE Malvern interferometer to an accuracy of about 1.5 arcsec rms, and frequently to about 0.6 arcsec rms. Optical positions have been measured for candidate identifications to an rms accuracy of 1.7 arcsec. The combination of the high accuracy radio and optical positions enables reliable identifications to be made down to the limit of the Palomar Sky Survey prints. Seventy-one new identifications are proposed; 34 quasar-like objects and 37, mostly faint, galaxies.

141.064 **Interpretation of the magnitude-redshift diagram for quasars.** A. Evans, D. Hart.
Astron. Astrophys., Vol. 43, 13 - 15 (1975).

It is shown that the slope of the magnitude-redshift relationship for cosmologically distant quasars need not have the usual value of 5 mag per decade, provided that the luminosity of an individual quasar decreases over a period $\sim 10^9$ years and the effects of light travel time are taken into account The possibility of detecting quasars of very high redshift is discussed.

141.065 **Alternative explanation for the spectral lines observed in quasars.** Y. P. Varshni.
Astrophys. Space Sci., Vol. 37, L1 - L6 (1975).

It is shown that the emission lines observed in quasars can be satisfactorily explained as being due to laser action in certain atomic species in the expanding envelope of a star. There is no need to assume a redshift.

141.066 **High-time-resolution photometry of quasistellar and related objects.** A. L. Kiplinger
Astron. Journ., Vol. 80, 669 - 672 (1975).

A total of 30 h of photometry on eight quasistellar and related objects indicate stable luminosity over time scales of hours to seconds.

141.067 **Observations of some discrete radio sources at 1.35 cm wavelength.** V. A. Efanov, V. I. Zagatin, I. G. Moiseev, N. S. Nesterov, V. B. Shtejnshlejger.
Astron. Tsirk., No. 842, p. 1 - 3 (1974). In Russian.

141.068 **Observations of quasar 3C 345.**
L. T. Markova, G. V. Zhukov.
Astron. Tsirk., No. 843, p. 1 - 2 (1974). In Russian.

141.069 **New flare in the quasar 3C 446.**
N. E. Kurochkin.
Astron. Tsirk., No. 844, p. 1 - 2 (1974). In Russian.

141.070 **Hypothesis on a 'geyser' model for big flares of quasars.** N. E. Kurochkin.
Astron. Tsirk., No. 844, p. 2 - 4 (1974). In Russian.

141.071 **Determination of equatorial coordinates of radio sources with long-baseline radiointerferometers.**
N. S. Blinov.
Astron. Tsirk., No. 857, p. 2 - 4 (1975). In Russian.

141.072 **Observations of PKS 2135-147.**
L. T. Markova, G. V. Zhukov.
Astron. Tsirk., No. 862, p. 1 - 2 (1975). In Russian.

141.073 **Observations with a VLB array. II. The sources 4C 39.25, NRAO 150, VRO 42.22.01, 3C 345, and 3C 454.3.** D. B. Shaffer, M. H. Cohen, J. D. Romney, R. T. Schilizzi, K. I. Kellermann, G. W. Swenson, Jr., J. L. Yen, R. Rinehart.
Astrophys. Journ., Vol. 201, 256 - 262 (1975).

Three-antenna long-baseline interferometry at 10.7 GHz is reported for NRAO 150, 4C 39.25, 3C 345, VRO 42.22.01, and 3C 454.3 for several epochs between 1972 April and 1974 February. Simple models for the source brightness distributions fit the observations and indicate that most of the radio emission comes from one or two components. Component separations range up to ~50 lt-yr, and component sizes are on the order of 10 lt-yr. Small variations in source structure have been observed.

141.074 **Observations with a VLB array. III. The sources 3C 120, 3C 273B, 2134 + 004, and 3C 84.**
R. T. Schilizzi, M. H. Cohen, J. D. Romney, D. B. Shaffer, K. I. Kellermann, G. W. Swenson, Jr., J. L. Yen, R. Rinehart.
Astrophys. Journ., Vol. 201, 263 - 274 (1975).

The sources 3C 120, 3C 273B, 2134 + 004, and 3C 84 have been observed at several epochs at 2.8 cm using a multielement very-long-baseline interferometer. The resulting visibility data show that the brightness distributions of each source are variable. There are large changes in the visibilities of 3C 120 and 3C 273B on a time scale of months. For 3C 120, and perhaps 3C 273B, the changes can be explained by source distributions in which the components are separating at high (probably relativistic) speeds.

141.075 **AO 0235 + 164 – a red BL Lacertae object.**
H. Spinrad, H. E. Smith.
Astrophys. Journ., Vol. 201, 275 - 276 (1975).

AO 0235 + 164, an inverted-spectrum radio source identified with a faint red stellar object, is shown to be an optical/radio variable. This plus its featureless optical spectrum lead us to classify AO 0235 + 164 as an object similar to BL Lacertae.

141.076 **On the light curve of 3C 273.**
G. G. Fahlman, T. J. Ulrych.
Astrophys. Journ., Vol. 201, 277 - 286 (1975).

The historical light curve of 3C 273 is analyzed using the technique of autoregressive decomposition. The observed optical fluctuations are regarded as the output of a linear system whose input is shot noise. The application to the observed light curve of 3C 273 indicates that (1) no low-frequency cyclic component is present and (2) the characteristic optical pulse responsible for the observed variability qualitatively resembles the light curve of a supernova. However, the time scale is one order of magnitude longer and, at the cosmological distance of 3C 273, the total energy release in the pulse is some three orders of magnitude greater than for conventional supernovae.

141.077 **Comments on the use of the limiting volume test.**
J. M. Barnothy.
Astrophys. Journ., Vol. 201, 287 - 288 (1975).

It is shown that inferences reached from the application of the limiting volume test to quasars, presupposes the knowledge of the cosmological model of the universe.

141.078 **The origin of the nonthermal radiation of quasistellar objects and galactic nuclei.** R. Opher.
Astrophys. Journ., Vol. 201, 526 - 531 (1975).

The supermassive star model for the origin of the nonthermal luminosity of QSOs and galactic nuclei is discussed. The basic model analyzed is the combined model of moving magnetic fields for the accelerating mechanism of ultrarelativistic particles and gravitational collapse for the energy source. The accelerating efficiency, η, has in general been assumed to be close to unity. Various previously proposed magnetic accelerating field models are analyzed in detail. It is shown for these models that observational and theoretical evidence indicate that η is not on the order of unity but is most likely less than 1 percent. The implication of this result to the rate of mass loss of QSOs and galactic nuclei is discussed.

141.079 **A full synthesis map of Sgr A at 5 GHz.**
R. D. Ekers, W. M. Goss, U. J. Schwarz, D. Downes, D. H. Rogstad.
Astron. Astrophys., Vol. 43, 159 - 166 (1975).
A full-synthesis map of Sgr A has been made at 5 GHz by combining data taken with the Westerbork and Owens Valley radio telescopes. The sidelobe level is less than 10% and the angular resolution is 6.3″ × 34″, which corresponds to a linear resolution of 0.3 × 1.7 pc at a distance of 10 kpc. The map shows the two major components, Sgr A West and Sgr A East, as well as structural details within these components.

141.080 **The rotation measures of radio sources and their interpretation.** J. P. Vallée, P. P. Kronberg.
Astron. Astrophys., Vol. 43, 233 - 242 (1975).
Rotation measures of 251 discrete radio sources have been determined after incorporating new polarization data at short wavelengths. These have been applied to a "slab" model-fitting technique to determine the most likely spiral arm magnetic field structure. The best agreement is obtained for a longitudinal spiral arm magnetic field, directed toward ($l^{II} \approx 90°$, $b^{II} \approx 0°$), but perturbed by an anomaly towards the north galactic spur.

141.081 **Analysis of ghost spectra which simulate the absorption-line spectrum of 4C 05.34.** Y. P. Varshni.
Astrophys. Journ., Vol. 201, 547 - 550 (1975).
Ten ghost spectra which simulate the absorption-line spectrum of the QSO 4C 05.34 are generated on a computer and analyzed according to the rules of Bahcall and Goldsmith. The results conclusively show that the number and properties of the absorption redshift systems proposed by Bahcall and Goldsmith are insignificantly different from those that would be expected from chance coincidences.

141.082 **Radio surveys and source counts.** D. L. Jauncey.
Annual Rev. Astron. Astrophys., Vol. 13, (see 003.005), 23 - 44 (1975).

141.083 **Theories of quasars.** L. Woltjer.
Proc. 16th Solvay Conference Phys., (see 012.020), p. 429 - 439, with a discussion, p. 440 (1974).

141.084 **Distances of the quasars and evidence for nonvelocity redshifts.** H. Arp.
Proc. 16th Solvay Conference Phys., (see 012.020), p. 443 - 450, with a discussion, p. 461 - 462 (1974).

141.085 **Recent observations of quasi-stellar objects.**
E. M. Burbidge.
Proc. 16th Solvay Conference Phys., (see 012.020), p. 451 - 460, with a discussion, p. 461 - 462 (1974).

141.086 **Distribution of quasars in the universe.**
M. Schmidt.
Proc. 16th Solvay Conference Phys., (see 012.020), p. 463 - 470, with a discussion, p. 471 (1974).

141.087 **High resolution radio observations fo 10 unidentified radio sources.** M. S. Longair.
Monthly Notices Roy. Astron. Soc., Vol. 173, 309 - 326 (1975).
Radio observations with angular resolution 2″ × 2″ cosec δ have been made with the Cambridge 5-km telescope of 10 unidentified radio sources. Deep plates of the optical fields of these sources were available and the survey resulted in one new probable identification, 3C 41. None of the unidentified radio sources possesses a compact central radio component, in contrast to many of the most powerful identified radio sources. It is argued that for compact central objects a correlation exists between non-thermal optical and radio luminosity.

141.088 **The orientation of the magnetic field in radio sources.** P. Haves, R. G. Conway.
Monthly Notices Roy. Astron. Soc., Vol. 173, 53P - 56P (1975).
Recent data on the polarization of extragalactic radio sources are used to investigate the distribution of Δ, the angle between the major axis of a source and the intrinsic position angle of the E vector of linear polarization. The authors find that sources of high radio luminosity usually have Δ near 90°, implying that the magnetic fields in such sources are oriented along the major axis. For radio galaxies with low luminosity, on the other hand, Δ tends to lie nearer 0°.

141.089 **Radio galaxies and local quasars.**
J. Terrell.
Nature, Vol. 258, 132 - 133 (1975). – Letter.

141.090 **18-cm visibility functions of high-frequency compact sources.** D. B. Shaffer, R. T. Schilizzi.
Astron. Journ., Vol. 80, 753 - 758 (1975).
Observations at 18 cm, with an angular resolution of ~0.″01, of the strongest compact radio sources with milliarcsecond structure at shorter wavelengths (6–2 cm) show that most of the sources are only slightly resolved, and most of the 18-cm radiation comes from quite small components, probably the same components seen at the short wavelengths. Some sources have brightness temperatures $>10^{12}$ K.

141.091 **Ohio Survey supplement 2.**
C. P. Rinsland, R. S. Dixon, J. D. Kraus.
Astron. Journ., Vol. 80, 759 - 770 (1975).
Results are presented for a 1415-MHz continuum survey made with the Ohio State University 110 × 21-m radio telescope. The survey covers 312 deg^2 between the declinations of 1° N and 11° S and includes four maps and a list of 239 sources at or above 0.25 Jy. Of these sources, 170 are previously uncataloged. The seven installments and two supplements of the Ohio Survey include a total of over 19000 sources in 8.7 sr of sky.

141.092 **Expanding quasar envelopes. I. Steady radiation-driven winds.** R. Kippenhahn, L. Mestel, J. J. Perry.
Astron. Astrophys., Vol. 44, 123 - 138 (1975).
The authors have investigated spherically symmetric expanding envelopes which are in ionization and thermal equilibrium in non-thermal radiation field of a QSO, and which are accelerated by momentum absorption from this radiation field as well as by the thermal pressure gradient and gravity. They find that the winds are parameterized by the luminosity to mass ratio (L_{vis}/M) and by the fractional mass loss rate ($-\dot{M}/M$).

141.093 **Modern models of the source of activity in quasars and galactic nuclei.** L. M. Ozernoj.
Probl. gravitatsii. Sb. dokl. plenarn. zasedanij 3-j Sov. gravitatsion. konf. Erevan, 1972. Erevan, Erevan. un-t, 1975, p. 42 - 70. In Russian. – Abstr. in Referativ. Zhurn. 51. Astron., 8.51.882 (1975).

141.094 **Occultation of radio source PKS 2025–15 by comet Kohoutek (1973f).**
S. Ananthakrishnan, S. M. Bhandari, A. Pramesh Rao.
Astrophys. Space Sci., Vol. 37, 275 - 282 (1975).
This paper describes the strong fluctuations in the intensity of radio source PKS 2025–15, observed at 327 MHz, during its occultation by comet Kohoutek (1973f) in 1974, January 5. Possible mechanisms which could produce the

observed fluctuations are examined. It is difficult to explain the fluctuations in terms of scintillation produced due to the passage of radio waves through the irregular cometary plasma. No detectable radio emission was observed from the comet at 327 MHz.

141.095 **Two luminosity classes of quasars.**
T. Jaakkola, K. J. Donner, P. Teerikorpi.
Astrophys. Space Sci., Vol. 37, 301 - 308 (1975).

Observations of quasars suggest that compact quasars with flat radio spectra belong to the class of dwarf galaxies and extended quasars with steep spectra to the one of bright galaxies. This can be seen also in the associations of quasars with clusters of galaxies, compact quasars occurring in nearby clusters and extended quasars in distant clusters.

141.096 **The survey of the region $6^h > \alpha > 19^h$, $-25° \leqslant \delta \leqslant 70°$ at 38 MHz.** J. Milogradov-Turin.
Publ. Obs. Astron. Beograd, No. 20, (see 012.024), p. 27 - 30 (1975). – Summary. In Serbo-Croatian and English.

141.097 **Observations of cosmic radio sources at 3.9 mm.**
V. A. Efanov, I. I. Zinchenko, A. G. Kislyakov, I. V. Kuznetsov, I. G. Moiseev, A. I. Naumov.
Izv. Krymskoj Astrofiz. Obs., Vol. 53, p. 144 - 149 (1975). In Russian.

The results of measurements of the flux density of 21 radio sources at 3.9 mm are given. The observations were made with the 22-m radiotelescope of the Crimean Observatory in August–September 1973. The sensitivity of the radiometer is 0.5°K at τ = 1 sec.

141.098 **Radio emission from galaxies and quasars.**
B. V. Komberg.
Priroda, No. 11.75, p. 23 - 31 (1975). In Russian.

141.099 **The large scale distribution of quasi-stellar radio sources.** M. Kurpińska, S. Zięba.
Acta Cosmologica, Fasc. 3, p. 55 - 60 (1975).

Based on the homogeneous sample of QSS taken from the Agnew's and Arp's catalogue and using the method of statistical reduction the distribution of QSS is investigated. It seems that the QSS are randomly distributed although there are some systematical tendencies towards clustering.

141.100 **On the problems associated with the rapidly varying compact nonthermal sources.**
G. R. Burbidge, W. A. Stein.
Comments Astrophys. Space Phys., Vol. 6, 87 - 94 (1975).

141.101 **Observations of radio sources formerly considered as possible supernova remnants.**
J. L. Caswell, D. H. Clark.
Australian Journ. Phys., Astrophys. Suppl. No. 37, p. 57 - 73 (1975).

A number of radio sources which have at some time been regarded as possible SNRs, or parts of SNRs, have been investigated. The measurements of the spectrum and, in some instances, the detection of hydrogen recombination line emission, show that most of these sources are probably not SNRs. In each of two cases a proposed large-diameter SNR appears to be predominantly thermal but contains a source which may be an SNR of much smaller size than originally suggested.

141.102 **The quasars from the Parkes 2700-MHz survey.**
J. V. Wall.
Observatory, Vol. 95, 196 - 204 (1975). – Review paper.

141.103 **New southern radio sources associated with planetary nebulae.** M. A. Bráz.
Observatory, Vol. 95, 211 - 212 (1975). – Short note.

141.104 **The physical parameters and exciting stars of six H II regions.** C. Goudis.
Astrophys. Space Sci., Vol. 38, 13 - 18 (1975).

The radio spectra of particular radio sources (mainly H II regions) are plotted on the same graph for comparison. Various physical parameters of these objects are obtained from the radio data through a model. Predictions about the number and the nature of the exciting stars of six H II regions are made. A comparison with the so far identified exciting stars is also presented.

141.105 **Radio observations of early-type stars.**
G. Sistla, S. S. Hong.
Astron. Astrophys., Vol. 44, 477 - 478 (1975).

Radio continuum observations of 18 early-type stars are reported. None of the observed stars show emission higher than the 3σ level of detection.

141.106 **Empirical evidence concerning absorption lines and radiation pressure in quasi-stellar objects.**
E. M. Burbidge, G. R. Burbidge.
Astrophys. Journ., Vol. 202, 287 - 295 (1975).

A critical compilation is given of all absorption-line systems which have so far been identified in quasi-stellar objects. For each QSO all values of R greater than unity were computed, where $R = (1 + z_i)/(1 + z_j)$ and z_i (but not z_j) includes z_{em}. If the absorption-line systems are being driven or held by line-locking or edge-locking mechanisms, the ratios $(1 + z_i)/(1 + z_j)$ should correspond approximately to the ratio λ_m/λ_n where λ_m and λ_n are the wavelengths of the lines or edges which are locking together. A discussion is given of the observable consequences of the line- and edge-locking hypothesis.

141.107 **Splitting of absorption lines in 3C 191.**
R. E. Williams, P. A. Strittmatter, R. F. Carswell, E. R. Craine.
Astrophys. Journ., Vol. 202, 296 - 302 (1975).

Spectroscopic observations of the quasar 3C 191 have been made at a higher dispersion than previous studies in an attempt to detect absorption from the excited fine-structure level of the ground state of C II. It was found that the previously studied absorption system is split into two systems (z_A = 1.945, z_B = 1.949), which show apparent line-locking on the C IV λ1550 doublet. Both systems show a similar and wide range of ionization. The results are interpreted in terms of a multiple cloud model which is photoionized by a central source, and the possible implications for the ionization structure in QSOs generally are discussed.

141.108 **A quasi-statistical analysis of absorption-line redshifts in 4C 05.34.** J. D. Colvin.
Astrophys. Journ., Vol. 202, 303 - 305 (1975).

The quasi-stellar object 4C 05.34 has a particularly rich absorption-line spectrum. An analysis of this spectrum by Bahcall and Goldsmith identified eight absorption-line redshift systems, ranging in value from z = 1.7758 to z = 2.8751 (the emission-line redshift is 2.877), and they concluded from an analysis of uniform random number spectra that the identifications are statistically significant. This finding has been questioned. The author shows that the absorption-line redshifts in the real spectrum are statistically significant, confirming the finding of Bahcall and Goldsmith.

141.109 **Radiation pressure in quasar clouds.**
C. F. McKee, C. B. Tarter.
Astrophys. Journ., Vol. 202, 306 - 318 (1975).

The effects of ionizing radiation pressure on optically thick clouds are analyzed and the results applied to quasars. The authors' analysis shows that the radiative acceleration of a cloud is $\gtrsim$10 percent of the fully optically thin value unless the continuum optical depth is very large. It is concluded that if quasar clouds are radiatively accelerated, they are optically thin; and the authors show that this result is consistent with the observations. Confinement of the clouds is discussed, and it is shown that photoionized clouds cannot be confined by radiation pressure alone. A model for absorption lines produced by numerous small clouds is developed, and they argue that the absorption lines most likely arise outside the emission-line region.

141.110 **VLBI observations of the compact radio source in the center of the Galaxy.**
K. Y. Lo, R. T. Schilizzi, M. H. Cohen, H. N. Ross.
Astrophys. Journ., (*Letters*), Vol. 202, L63 - L65 (1975).

A compact radio source at the galactic center has been detected at 3.7 cm on a very long baseline interferometer. The source has an angular size, θ_s, on the order of or less than 0".02, corresponding to a linear size $\lesssim$200 AU or 10^{-3} pc, at the distance of the galactic center (10 kpc). The measured flux density is 0.6 ± 0.1 Jy. The brightness temperature of the source is 3×10^7 K if $\theta_s \approx 0''.02$. Comparison with previous high-resolution observations of the galactic center suggests time variation in the flux density. The nature of the compact radio source is discussed.

141.111 **Polarization parameters of 183 extragalactic radio sources.** P. Haves.
Monthly Notices Roy. Astron. Soc., Vol. 173, 553 - 568 (1975).

Estimates are presented of the rotation measures, intrinsic position angles, maximum degrees of polarization and depolarization wavelengths of 183 extragalactic radio sources. A procedure is developed for classifying rotation measures so as to eliminate as far as possible incorrect or ambiguous values, and to estimate the realiability of the values retained. For each source in the list presented the quoted rotation measure is the only good fit in the range −300 to +300 rad m^{-2}. It is estimated that the correct choice of position angle has been made for all but ~1 per cent of the list, and that for most sources excluded there are insufficient data to define their rotation measures unambiguously.

141.112 **Symmetry parameters for double radio sources.**
W. Ingham, P. Morrison.
Monthly Notices Roy. Astron. Soc., Vol. 173, 569 - 577 (1975).

Two simple symmetry parameters are introduced to describe those double radio sources with an observed central object. An appropriate subsample of 21 sources is selected from a pool of 223 3C sources, and the distribution of the symmetry parameters is examined. It is concluded that the data are consistent with the existence of a single class of double sources with rms intrinsic bending of 4–18° and with arms which typically differ in actual length by 15–30 per cent or less. An upper limit of 0.15c on typical component speeds is obtained.

141.113 **The companion object to W3 (OH).**
V. A. Hughes, M. R. Viner.
Monthly Notices Roy. Astron. Soc., Vol. 173, 73P - 76P (1975).

A mapping programme of the W3 complex at 2.8 cm has shown the presence of a new radio object, close to W3 (OH) and coincident in position with a 69 μm infrared object discovered by Fazio et al. The object is discussed in terms of a recent model for W3 (OH) by Hughes & Viner.

141.114 **Redshifts for southern quasars.**
I. W. A. Browne, A. Savage, J. G. Bolton.
Monthly Notices Roy. Astron. Soc., Vol. 173, 87P - 91P (1975).

Spectroscopic observations of 21 quasi-stellar objects have been made with the Anglo-Australian 3.9-m telescope. Redshifts have been determined for 15 of the QSOs. Two other objects may have a single emission line while four show no features at all.

141.115 **High resolution observations of extended radio sources at 1666 MHz.**
M. Bentley, P. Haves, R. E. Spencer, D. Stannard.
Monthly Notices Roy. Astron. Soc., Vol. 173, 93P - 97P (1975).

Compact emission regions in extended radio sources have been observed with an interferometer of baseline 127 km. Of the 49 sources studied, ten were found to have small diameter components coincident with the central optical nucleus. Compact regions of emission were also detected in the outer components of Cygnus A.

141.116 **Centaurus-A: observations of the central region at cm-wavelengths by the radio telescope RATAN-600.**
A. B. Berlin, V. Ya. Gol'nev, N. A. Esepkina, Yu. K. Zverev, A. V. Ipatov, N. L. Kajdanovskij, D. V. Korol'kov, A. P. Lavrov, Yu. N. Parijskij, N. S. Soboleva, A. A. Stotskij, G. M. Timofeeva, O. N. Shivris.
Pis'ma v Astron. Zhurn., Vol. 1, No. 12, p. 3 - 8 (1975). In Russian.

Results of observations of the central region around the nucleus of NGC 5128 at 2.08, 3.9, 6.52 cm with resolutions of 15", 32", 50" are given. The spectrum of the compact non-thermal source in the core is shown to differ from that at epoch 1970.

141.117 **5000 MHz flux densities and spectra for 325 small-diameter radio sources at low galactic latitudes.**
J. L. Caswell, R. F. Haynes, D. H. Clark.
Australian Journ. Phys., Vol. 28, 633 - 646 (1975).

Flux density measurements at 5000 MHz are presented for 325 of the small-diameter sources detected in the Molonglo 408 MHz galactic survey. By investigating the spectra the authors show that probably between 10% and 15% of the sources are galactic and the remainder extragalactic.

141.118 **Observations of 31 extragalactic radio sources with the Cambridge 5-km telescope at 5 GHz.**
J. M. Riley, G. G. Pooley.
Mem. Roy. Astron. Soc., Vol. 80, 105 - 137 (1975).

Thirty-one radio sources from the 3C and 4C catalogues have been mapped with a resolution of 2" in RA and 2" cosec δ in Dec. The results are presented as contour maps and in tabular form, with accurate measurements of the positions of optical objects in the fields.

141.119 ***UBV* observations of 3C 273. VI.**
M. S. Burkhead, R. K. Hill.
Publ. Astron. Soc. Pacific, Vol. 87, 821 (1975).

141.120 **On the dependence of luminosity and spatial density of quasars on the redshift.**
M. F. Khodyachikh.
Astron. Zhurn. Akad. Nauk SSSR, Vol. 52, 1132 - 1138 (1975). In Russian. English translation in Soviet Astron., Vol. 19, No. 6.

The Hubble diagram for the brightest quasars with slight dispersion is constructed. The slope of the redshift-magnitude relation is 2.24 ± 0.09. For coordination of this slope with the theoretical dependence $m(\lg z)$ in the standard cosmological models, a decrease of quasars luminosities is required of 1 mag in 2.5×10^9 years. The condition of constancy of quasars spatial density in the comoving space is fulfilled by the deceleration parameter $q_0 = 4.1 \pm 1.3$.

141.121 **Flux densities for radio sources selected from 3CR**

and PKS catalogues.
A. G. Gorshkov, V. K. Konnikova.
Astron. Zhurn. Akad. Nauk SSSR, Vol. 52, 1195 - 1205 (1975). In Russian. English translation in Soviet Astron., Vol. 19, No. 6.

Flux densities at 8550 MHz are given for 246 radio sources mainly selected from 3CR and PKS catalogues. Observations have been carried out with the 22-m radio telescope at the Crimean Astrophysical Observatory in the period from 1968 to 1971. Spectral indices $\alpha_{1400-8550}$ for quasars and radio galaxies were computed.

141.122 **Observations of lunar occultations of cosmic radio sources in the decametric wavelength region.**
V. P. Bovkun.
Astron. Zhurn. Akad. Nauk SSSR, Vol. 52, 1206 - 1213 (1975). In Russian. English translation in Soviet Astron., Vol. 19, No. 6.

Some peculiarities of the observations and processing of the lunar occultations of cosmic radio sources in decametric wavelength region are described. Results of measurements of the occultation of 3C 212 are presented.

141.123 **B0924 + 30 an unusual radio galaxy.**
R. D. Ekers, R. Fanti, C. Lari, M.-H. Ulrich.
Nature, Vol. 258, 584 - 586 (1975).

The radio source B0924 + 30 has been studied as part of a systematic survey of radio galaxies of relatively low ($<10^{41}$ erg s^{-1}) radio luminosity. The radio source is identified with a 14 mag elliptical galaxy with a redshift $z = 0.0266$. A 12-h observation of this radio source was made with the Westerbork Synthesis Radio Telescope at 1.4 GHz in September 1974, and a contour plot of the brightness distribution is obtained.

141.124 **Selection of quasars by variability.** N. Sanitt.
Nature, Vol. 258, 586 - 587 (1975).

For the majority of quasars, which are radio quiet existing samples based on ultraviolet excess are not only incomplete, but are dominated by colour selection effects. The relative importance of these effects depends on redshift, and, in particular the detection of high redshift quasars is extremely difficult; consequently other methods of selection of radio-quiet quasars and BL Lac-type objects have been suggested. Here, one such method based on optical variability is considered, with particular regard to a sample of 13 objects observed by van den Bergh et al.

141.125 **Millimetre wavelength observations of the circumstellar shells of some mass-loss stars.**
C. R. Purton, K. A. Marsh, P. A. Feldman.
Journ. Roy. Astron. Soc. Canada, Vol. 69, 247 (1975).
Abstr. CAS.

141.126 **Variable radio source structure on a scale of several minutes of arc.** H. N. Ross.
Journ. Roy. Astron. Soc. Canada, Vol. 69, 248 (1975).
Abstr. CAS.

141.127 **On the largest angular size-redshift diagram for quasars.** Y. P. Varshni.
Journ. Roy. Astron. Soc. Canada, Vol. 69, 256 (1975).
Abstr. CAS.

141.128 **Compact components in a complete sample of extragalactic radio sources.**
J. J. Broderick, J. J. Condon.
Astrophys. Journ., Vol. 202, 596 - 602 (1975).

A statistically complete sample of 100 extragalactic sources stronger than 2 Jy at 1400 MHz was observed on a 3.5×10^6 wavelength baseline at 430 MHz. In general, sources with flat spectra ($\alpha < +0.5$) show fringes, while steep-spectrum sources do not, confirming that flat spectra are produced by synchrotron self-absorption rather than unusual electron-energy distributions. The spectral index-luminosity relation for radio galaxies is shown to reflect differing electron-energy distributions rather than self-absorption. Nearly all quasars, even those with steep spectra, contain components brighter than 10^{10}K.

141.129 **Spectral index dispersion in extragalactic radio sources.** F. W. Peterson.
Astrophys. Journ., Vol. 202, 603 - 609 (1975).

By assuming synchrotron radiation losses to be an important process, even in the straight spectra of extragalactic radio sources with spectral index between -0.5 and -1.1, and by introducing a different distribution in electron energies, the observed dispersion in spectral indices can be related to a dispersion in magnetic field strengths of the sources.

141.130 **Reexamination of the correlation of galaxies and QSO's.** P. D. Noerdlinger.
Astrophys. Space Sci., Vol. 38, 457 - 468 (1975).

It is shown that the correlation between the separation on the sky of certain QSO's and the nearest bright galaxy to them with the galaxy's redshift could be an artifact of the sampling procedure or small number statistics. The overall association between 3C QSO's and bright galaxies remains as suggestive, but not compelling, evidence for a physical relationship.

141.131 **Accurate spectra of 300 radio sources from the Parkes catalogue: observational results.**
B. J. Wills.
Australian Journ. Phys., Astrophys. Suppl. No. 38, 65 pp. = Separate print Div. Radiophys. C.S.I.R.O. Sydney (1975).

Accurate measurements of relative flux density have been made between frequencies of 468 and 5009 MHz, for 300 radio sources selected from the Parkes catalogue. The observations, their analysis and accuracy are described, and the flux densities presented together with their errors. Excluding the uncertainties in scaling these flux densities to an absolute system, r.m.s. errors vary from 0.24 Jy and 2.4% at 635 MHz to 1.8% at 2700 MHz; uncertainties in 'absolute' scaling are thought to be about 5%. The present flux densities are compared with accurate values published elsewhere. The data have been combined with other published flux densities between 10 and 10000 MHz to derive spectra, which are presented graphically. Some sources whose spectra may be useful for flux density calibration are noted.

141.132 **The influence of the interstellar medium on observed dimensions of radio sources.**
N. A. Lotova, A. V. Pynzar'.
Izv. vyssh. ucheb. zavedenij. Radiofizika, Vol. 18, 919 - 925 (1975). In Russian. – Abstr. in Referativ. Zhurn. 51. Astron., 12.51.667 (1975).

141.133 **QSO redshifts and the steady state cosmology.**
J. V. Narlikar, P. A. Das.
Nature, Vol. 258, 660 (1975). – Letter.

141.134 **Are there two quasar populations?**
M. F. Barnothy, J. M. Barnothy.
Bull. American Astron. Soc., Vol. 7, 515 (1975). – Abstr. AAS.

141.135 **JPL catalog of VLBI radio sources.** R. A. Preston, A. W. Harris, M. A. Slade, J. G. Williams, J. L. Fanselow, J. B. Thomas, D. D. Morabito, D. J. Spitzmesser, L. J. Skjerve, B. Johnson, D. Jauncey.
Bull. American Astron. Soc., Vol. 7, 517 (1975). – Abstr. AAS.

141.136 **Radio sources in the vicinity of UV Ceti-type flare**

stars. S. R. Spangler, R. L. Mutel.
Bull. American Astron. Soc., Vol. 7, 527 (1975). – Abstr. AAS.

141.137 **Linear polarization and flux density variations in 3C120 at λ3.75 cm.**
H. D. Aller, E. T. Olsen, M. F. Aller.
Bull. American Astron. Soc., Vol. 7, 528 (1975). – Abstr. AAS.

141.138 **Apparent rotational axes of 3C405 (Cyg A) and 3C33.** S. M. Simkin.
Bull. American Astron. Soc., Vol. 7, 528 (1975). – Abstr. AAS.

141.139 **Photographic monitoring of 3C273 and 3C351.** G. J. Corso.
Bull. American Astron. Soc., Vol. 7, 528 (1975). – Abstr. AAS.

141.140 **Low frequency variability in extragalactic radio sources.** W. D. Cotton.
Bull. American Astron. Soc., Vol. 7, 528 - 529 (1975). Abstr. AAS.

141.141 **Optical identifications of sources in the NRAO 5-GHz deep survey.**
J. J. Condon, T. J. Balonek, D. L. Jauncey.
Astron. Journ., Vol. 80, 887 - 894, 997 - 999 (1975).

Accurate radio positions of extragalactic radio sources forming a sample complete to 0.1 Jy at 5 GHz were measured with the NRAO dual-frequency interferometer. Comparably accurate positions of all optical objects within 30 arcsec of each source were determined. Identifications were made on the basis of position coincidence without regard to color or morphology.

141.142 **Optical behavior of 20 violently variable extragalactic radio sources.**
B. Q. McGimsey, A. G. Smith, R. L. Scott, R. J. Leacock, P. L. Edwards, R. L. Hackney, K. R. Hackney.
Astron. Journ., Vol. 80, 895 - 922 = Contr. Rosemary Hill Obs., *Gainesville*, Florida, No. 60 (1975).

A 7-yr photometric study of nearly 200 extragalactic radio sources has confirmed 20 of these objects as violent optical variables. Light curves and photometric data are given for these 20 sources, which are placed in one of four categories according to the type of variability displayed.

141.143 **Meter-wavelength VLBI. II. The observations.**
T. A. Clark, W. C. Erickson, L. K. Hutton, G. M. Resch, N. R. Vandenberg, J. J. Broderick, S. H. Knowles, A. B. Youmans.
Astron. Journ., Vol. 80, 923 - 930 (1975).

Observations made during a series of meter-wavelength very-long-baseline (VLBI) experiments conducted during 1971–1973 are reported. A wide variety of objects was observed, including many extragalactic sources known to have compact components, several strong pulsars (especially the Crab nebula pulsar), and a selection of supernova remnants. The experiments are discussed in detail. The results show that all sources except the pulsars, which are known to have intrinsically small sizes, are resolved on the longest baseline. The authors conclude that no supernova remnants except the Crab nebula and Cassiopeia A gave evidence of compact components.

141.144 **26.3-MHz radio source survey. II. Radio source positions and fluxes.** M. R. Viner, W. C. Erickson.
Astron. Journ., Vol. 80, 931 - 954 (1975).

The results of the Clark Lake 26.3-MHz Sky Survey are presented. Data concerning the apparent fluxes and positions of 471 radio sources in the declination range $3° < \delta < 63°$ are given. Almost all of these sources are identified with 3CR or 4C sources.

141.145 **Geodetic and astrometric results of very long-baseline interferometric measurements of natural radio sources.** J. M. Moran.
IAU Colloquium No. 26, (see 012.034), p. 269 - 292 (1975).

141.146 **Neutral hydrogen absorption measurements yielding kinematic distances for 42 continuum sources in the galactic plane.** J. L. Caswell, J. D. Murray, R. S. Roger, D. J. Cole, D. J. Cooke.
Astron. Astrophys., Vol. 45, 239 - 258 = Div. Radiophys., C.S.I.R.O., Publ. RPP 1790 (1975).

The Parkes hydrogen line interferometer has been used to measure 21 cm absorption profiles for 42 radio sources: approximately half are supernova remnants and most of the remainder are H II regions. The authors have used the results to determine kinematic distances to the sources and have thus greatly increased the number of galactic radio sources with well-determined distances. For supernova remnants this is of special importance, since very few reliable distance calibrators have previously been available.

141.147 **A 408 MHz fan-beam survey at low galactic latitudes. IV. The nature of the small diameter sources.**
C. Fanti, M. Felli, N. Panagia, G. Tofani, P. Tomasi.
Astron. Astrophys., Vol. 45, 277 - 286 (1975).

The statistical properties of the 586 small diameter radio sources in the BG survey of the galactic plane (Fanti et al., 1974) are discussed. The number flux density relation, the spatial distribution and an analysis of spectral indices indicate that the sample of unresolved sources has statistical properties similar to those of an extragalactic population, while the more extended sources are mostly of galactic nature.

141.148 **QSO absorption spectra caused by protogalaxies.** H.-J. Röser.
Astron. Astrophys., Vol. 45, 329 - 333 (1975).

The absorption spectra of QSOs are interpreted in terms of cosmological foreground objects. The increase of the number of absorption systems with z_{em} of the QSO can be explained, if protogalaxies of $10^{12} M_{\odot}$ and radius of about 70 kpc ($0.01 < q_0 < 1$ for $\Lambda = 0$) are believed to be responsible for the absorption. This also gives the observed order of magnitude for column densities of C IV. The high degree of ionization may be maintained by the intergalactic flux of QSO radiation.

141.149 **OX-192.** R. F. Willson.
IAU Circ., No. 2861 (1975).

141.150 **On the correlation between the absolute magnitudes of quasi-stellar objects and the velocities of ejected gas.** J. J. Perry.
Max-Planck-Inst. Phys. Astrophys., MPI-PAE/Astro 73, 2 + 12 pp. (1975).

For all QSOs which show absorption-line systems $\Delta z = (z_{em} - z_{abs})/(1 + z_{em})$ is plotted against the absolute visual magnitude (evaluated assuming that the emission line redshifts are of cosmological origin, and for $q_0 = 0.1$). It is shown that there is a weak correlation between these two parameters. This lends support to the hypothesis that the QSO's are at their redshift distance and that the absorption lines arise in gas which is intrinsic to the QSOs and is being driven out from the central object, at velocities $c\Delta z$, by radiation pressure.

141.151 **Early-type emission line radio stars.** A. E. Wright.
Proc. Astron. Soc. Australia, Vol. 2, 329 - 330 (1975).

141.152 **The Parkes 408 MHz southern survey.** C. G. T. Haslam, W. E. Wilson, D. J. Cooke, M. N. Cleary, D. A. Graham, R. Wielebinski, G. A. Day.
Proc. Astron. Soc. Australia, Vol. 2, 331 - 334 (1975).

141.153 **High-resolution observations of Centaurus-A and Pictor-A at 160 MHz.** O. B. Slee, K. V. Sheridan.
Proc. Astron. Soc. Australia, Vol. 2, 336 - 338 (1975).

141.154 **The spectra of Sgr A and Sgr B2 between 1636 and 1640 MHz.** J. B. Whiteoak, F. F. Gardner.
Proc. Astron. Soc. Australia, Vol. 2, 360 - 362 (1975).

141.155 **A search for HI absorption in the radio spectra of the quasi-stellar objects 4C 05.34 and PKS 0237-23.**
B. A. Peterson, J. D. Murray, A. E. Wright.
Proc. Astron. Soc. Australia, Vol. 2, 365 - 366 (1975).

141.156 **Centaurus A : spectra of the optical filament.** B. A. Peterson, R. J. Dickens, R. D. Cannon.
Proc. Astron. Soc. Australia, Vol. 2, 366 - 367 (1975).

141.157 **Group characteristics of radio sources of the Simeis survey.** M. G. Larionov.
Astron. Tsirk., No. 882, p. 1 - 3 (1975). In Russian.

141.158 **The relation between discrete radio sources and the fine structure of the galactic background.**
M. G. Larionov.
Astron. Tsirk., No. 884, p. 1 - 4 (1975). In Russian.

141.159 **Recent observations of quasar spectra.** A. Boksenberg.
Observatory, Vol. 95, 257 (1975). – Summary of a paper presented at RAS specialist discussion on quasars.

141.160 **Absorption lines in QSOs – evidence for radiation pressure.** R. Carswell.
Observatory, Vol. 95, 257 - 258 (1975). – Summary of a paper presented at RAS specialist discussion on quasars.

141.161 **The emission spectrum of OH 471.** A. R. G. Jackson.
Observatory, Vol. 95, 258 - 259 (1975). – Summary of a paper presented at RAS specialist discussion on quasars.

141.162 **The Balmer decrement in quasars.** H. Netzer. Observatory, Vol. 95, 259 - 260 (1975). – Summary of a paper presented at RAS specialist discussion on quasars.

141.163 **Radiation pressure and winds.** J. J. Perry. Observatory, Vol. 95, 260 - 261 (1975). – Summary of a paper presented at RAS specialist discussion on quasars.

141.164 **Recent work on quasars at Jodrell Bank.** D. Walsh.
Observatory, Vol. 95, 261 - 263 (1975). – Summary of a paper presented at RAS specialist discussion on quasars.

141.165 **The minimum size and mass of some quasi-stellar objects.** M. C. Durgapal.
Journ. Phys. A (Math., Nuclear, General), Vol. 8, 1014 - 1018 (1975). *–JAR*

141.166 **Calibration radio sources.** K. M. Borkowski. Postepy Astron., Vol. 23, 199 - 208 (1975). In Polish.

In this paper a short review of methods of measuring the absolute flux of cosmic radio sources is given. Also, results of such measurements are discussed. The widely used flux scales and recently published results are considered. The list of radio sources suitable for calibration and measurements of large antennas together with their parameters is presented and an absolute flux scale at 127 MHz is included.

141.167 **The effect of large fields on the lifetime of the forbidden lines seen in quasars.**
H. Ezawa, M. Leventhal.
Journ. Phys. B, (Atomic and Molecular), Vol. 8, 1824 - 1830 (1975).

141.168 **Nature of extended radio sources in clusters of galaxies.** K. Aizu.
Progr. Theor. Phys. Japan, Vol. 53, 893 - 895 (1975).

141.169 **Spectrophotometry of low-redshift quasistellar objects.** J. A. Baldwin.
Thesis, California Univ., Santa Cruz (USA). [Available from Univ. Microfilms, Ann Arbor, Mich., *USA*. Order No. 75-8471], 101 pp. (1974).

Emission line strengths, line profiles, and continuum energy distributions have been measured for fourteen quasi-stellar objects with redshifts z less than 0.3. The results are discussed in detail.

The radio universe. See Abstr. 003.070.

Millimetre emission from extragalactic objects–I. Observational techniques and data analysis.
See Abstr. 031.210.

Presentazione di una serie di programmi per le identificazioni ottiche. See Abstr. 031.236.

Observations with a VLB array. I. Introduction and procedures. See Abstr. 033.007.

The measurement of time from radio observations of extragalactic sources. See Abstr. 044.018.

Cosmic Čerenkov emission.
See Abstr. 061.053.

The spectrum of relativistic particles accelerated in a Fermi reservoir. See Abstr. 061.080.

Polarization characteristics of synchrotron emission in a plasma. See Abstr. 062.011.

Intensity distribution of interplanetary scintillation at 408 MHz. See Abstr. 106.045.

Power spectra of interplanetary scintillations.
See Abstr. 106.047.

On the proper motion of AT Cancri.
See Abstr. 112.018.

Symbiosis in radio stars with infrared excess.
See Abstr. 114.012.

An unusually strong radio outburst in Algol: flux density, spectral, and polarization observations.
See Abstr. 121.028.

An extraordinary radio outburst in Algol: flux and structure observations. See Abstr. 121.039.

Models of radio flares from UV Ceti stars.
See Abstr. 122.093.

An upper limit to slowly varying radio emission from three UV Ceti-type stars. See Abstr. 122.094.

Radio and optical observations of some flare stars.
See Abstr. 122.156.

The radio remnant of Kepler's supernova of 1604. See Abstr. 125.004.

An anomaly in the flux of Cassiopeia A at 38 MHz. See Abstr. 125.013.

A high resolution map of the supernova remnant 3C 400.2 at 610 MHz. See Abstr. 125.035.

Observations of eight supernova remnants at 15 GHz. See Abstr. 125.039.

408 and 5000 MHz observations of 28 new galactic supernova remnants. See Abstr. 125.044.

408 and 5000 MHz observations of 12 probable supernova remnants. See Abstr. 125.045.

On the light curve of 3C 273. See Abstr. 125.053.

Characteristics of the interstellar medium as deduced from low-frequency recombination line observations. See Abstr. 131.112.

The radio spectra of four H II regions (Orion B, M8, M16 and M17). See Abstr. 131.542.

High resolution radio and optical observations of NGC 6888. See Abstr. 132.017.

The radio recombination line spectrum of Orion A: observations and analysis. See Abstr. 132.030.

On some peculiarities of the radio emission of the Crab nebula. See Abstr. 134.003.

On the production of power-law spectra and the evolution of cosmic synchrotron sources: a model for the Crab nebula. See Abstr. 134.007.

Optical studies of 10 high galactic latitude X-ray sources. See Abstr. 142.005.

Radio observations of Cygnus X-3 from 1973 December to 1974 June. See Abstr. 142.007.

Another major change in the radio source associated with Cyg X-1. See Abstr. 142.017.

Radio emission from Cyg X-3. See Abstr. 142.020.

An interpretation of the radio outbursts of Cygnus X-3. See Abstr. 142.060.

Copernicus: X-ray observations of several radio supernova remnants. See Abstr. 142.064.

Possible identification of a high-latitude X-ray source with a QSO by the Astronomical Netherlands Satellite. See Abstr. 142.122.

X-ray stars and quasars: variables that need watching. See Abstr. 142.138.

A0620–00. See Abstr. 142.215.

Diffusion of cosmic rays in non-thermal radio sources. See Abstr. 143.018.

The redshifts of 3C 129 and 3C 129.1, the radio-tail galaxies. See Abstr. 158.007.

Optical variability of the *N* galaxy 3C 371. See Abstr. 158.017.

Observations of galaxies at high frequencies. See Abstr. 158.022.

Isophotal tracings of galaxies near quasars. See Abstr. 158.029.

A jet in the galaxy identified with the radio source DA 240. See Abstr. 158.031.

Symmetric ejection of massive objects from galactic nuclei. See Abstr. 158.066.

On the energy source of active galactic nuclei and quasars. See Abstr. 158.068.

Spectrophotometric observations of N systems at large redshift – PKS 0353+027, 3C99, 3C467. See Abstr. 158.085.

Optical emission-line spectra of 3C 327 and other narrow-line radio galaxies. See Abstr. 158.091.

A new approach to the variability of radio galaxies, Seyfert nuclei, and QSO's. See Abstr. 158.099.

Photographic photometry of compact extragalactic objects. III. See Abstr. 158.102.

Gradient diagram for optically variable nuclei of galaxies and quasars. See Abstr. 158.112.

Interferometer survey of normal galaxies. See Abstr. 158.113.

A dynamical model of the tailed radio galaxies. See Abstr. 158.134.

New observations of the optical spectrum of the Seyfert galaxy NGC 4151. See Abstr. 158.141.

Activity of the nucleus of the southern Seyfert galaxy NGC 1566. See Abstr. 158.143.

5-GHz survey of bright galaxies. See Abstr. 158.169.

Brightness distributions across six strong radio galaxies at 15 GHz. See Abstr. 158.173.

A compendium of radio measurements of bright galaxies. See Abstr. 158.179.

The all scales distribution of different extragalactic objects. See Abstr. 158.180.

A comparison of the radio and optical orientations of elliptical galaxies identified with double radio sources. See Abstr. 158.181.

A strong millimeter wave component in the radio galaxy 3C 111. See Abstr. 158.190.

The groundstate OH spectrum of the galacy NGC 253. See Abstr. 158.192.

Emission-line galaxies and quasars in the southern

hemisphere. I. Description and applications of an objective-prism survey. See Abstr. 158.197.

The large scale radio structure of NGC 1275. See Abstr. 158.203.

Hydrogen line luminosities for Seyfert galaxies and QSO's. See Abstr. 158.215.

The optical emission-line spectrum of 3C 120. See Abstr. 158.233.

Polarimetric observations of compact extragalactic objects. See Abstr. 158.239.

The composite nature of the N galaxy 3C 371. See Abstr. 158.302.

Photometric histories of BL Lac objects and QSOs. See Abstr. 158.307.

B2 1101+38: a BL Lacertae object. See Abstr. 158.315.

Radio galaxies in the Coma cluster. See Abstr. 160.002.

A Westerbork survey of clusters of galaxies. I. 21 cm observations of A1656, A2147, A2151, A2197 and A2199: radio data and identifications. See Abstr. 160.005.

On the redshifts of clusters of galaxies and quasars. See Abstr. 160.018.

Clustering of faint blue objects. See Abstr. 160.020.

Cosmology from angular size counts of extragalactic radio sources. See Abstr. 162.006.

The composite Hubble diagram. See Abstr. 162.048.

Pulsars

141.301 **Relativistic effects in the binary pulsar PSR 1913+16.**
B. M. Barker, R. F. O'Connell.
Astrophys. Journ., (*Letters*), Vol. 199, L25 - L26 (1975).

The relativistic precession of both the spin and the periastron of the pulsar 1913+16, in the gravitational field of its companion, are analyzed.

141.302 **Anomalous scattering in the Crab nebula.**
A. G. Lyne, D. J. Thorne.
Monthly Notices Roy. Astron. Soc., Vol. 172, 97 - 108 (1975).

During 1974, the width of the pulses from the Crab nebula pulsar, PSR 0531 + 21, has increased by a factor of 10 at 408 MHz. It is shown that this is probably due to multipath scattering causing a 100-fold increase in the pulse broadening. The form of the broadening function is irregular and changes on a time scale of a few days. There is evidence that the scattering is due to an increase in turbulence within the Crab nebula. Over this same period the radio intensity of the pulses has been remarkably low and it is suggested that the enhanced scattering is connected with this.

141.303 **Pulsar glitches and restlessness as a hard superfluidity phenomenon.** P. W. Anderson, N. Itoh.
Nature, Vol. 256, 25 - 27 (1975).

Several pulsars have "restless" behaviour of the period similar to that of the Crab and Vela pulsars which has been explained as due to "microquakes" in the crust. A glitch has also been observed with, possibly, a new type of signature, in PSR 1508+55. The authors suggest that these phenomena can be explained at least equally well by noisiness in the rate of creep of vorticity through the neutron superfluid in the crust, in almost precise analogy to the well known noisy behaviour of flux creep in hard superconducting magnets. In addition, even the macroglitches, especially those in the Vela pulsar, may be caused by the catastrophic release of pinned vorticity.

141.304 **A model for the double system containing the X-ray pulsar Her X-1.**
G. S. Bisnovatyi-Kogan (*Bisnovatyj-Kogan*), B. V. Komberg.
IAU Symposium No. 67, (see 012.001), p. 497 (1975). – Abstract.

141.305 **The Crab pulsar as a component of a binary system.**
A. I. Tsygan.
IAU Symposium No. 67, (see 012.001), p. 499 - 500 (1975).

141.306 **Relativistic solitons and pulsars.**
V. I. Karpman, C. A. Norman, D. ter Haar, V. N. Tsytovich.
Physica Scripta, Vol. 11, 271 - 274 (1975).

The authors consider solitons in which the electrons and/or the positive ions have relativistic random velocities and discuss possible applications of such solitons to pulsar phenomena.

141.307 **Pulsars and high density physics.** A. Hewish.
Rev. Modern Phys., Vol. 47, 567 - 572 (1975).
Lecture delivered on the occasion of receiving the 1974 Nobel Prize in Physics.

141.308 **Discovery of pulsars: a graduate student's story.**
N. Wade.
Science, Vol. 189, 358 - 364 (1975).

141.309 **Pulsar emission patterns: a geometrical model.**

L. Oster.
Mitt. Astron. Ges., No. 36, p. 87 (1975). – Abstract.

141.310 **PSR 0943 + 10 und 2303 + 30, zwei Pulsare mit driftenden Unterpulssystemen.**
W. Sieber, L. Oster.
Mitt. Astron. Ges., No. 36, p. 87 - 88 (1975). – Abstract.

141.311 **An interferometric search for the Hulse-Taylor binary pulsar.**
R. M. Hjellming, D. M. Gibson.
Astrophys. Journ., (*Letters*), Vol. 199, L165 - L166 (1975).

An interferometric search for continuum radio emission from the Hulse-Taylor binary pulsar made with the NRAO interferometer at 2695 and 8085 MHz has resulted in the detection of a weak source well within the positional error box of the binary pulsar and a relatively strong source several arc minutes outside the error box.

141.312 **An upper limit on optical pulsations from PSR 1913+16.**
G. Chanan, J. Middleditch, J. Nelson.
Astrophys. Journ., (*Letters*), Vol. 199, L167 - L168 (1975).

A region of the sky covering the 2′ by 2′ error box of PSR 1913+16, a radio pulsar in a binary system, was searched for optical pulsations at the known pulsar frequency, and none were observed. The authors place an upper limit of 18.2 mag on any such pulsations. Within the smaller error box which has recently been suggested for this object, the limit is 19.2 mag.

141.313 **Optical and X-ray observations of the PSR 1913+16 field.** A. Davidsen, B. Margon, J. Liebert, H. Spinrad, J. Middleditch, G. Chanan, K. O. Mason, P. W. Sanford.
Astrophys. Journ., (*Letters*), Vol. 200, L19 - L21 (1975).

The authors have obtained medium-resolution spectra in the yellow of seven stars in the error box of the binary radio pulsar PSR 1913+16. None of the objects is spectroscopically abnormal and thus a strong candidate for membership in the system. The data are used to establish the extinction in the field, which is modest for this low galactic latitude. For three of the stars, limits of order $V > 23$ mag on optical pulsations with the frequency of the radio pulsar are derived. X-ray observations of the field have been obtained in the 2.5–7.5 keV band, using the UCL experiment aboard OAO Copernicus. No source is detected.

141.314 **The pulsars.** F. G. Smith.
Observatory, Vol. 95, 137 - 141 (1975). – Review paper.

141.315 **Probability distributions related to power spectra.**
E. J. Groth.
Astrophys. Journ., Suppl. Ser., No. 286, Vol. 29, 285 - 302 (1975).

Power-spectrum techniques are becoming an increasingly important tool of the astronomer. Probability distributions for the power obtained from a power-spectrum analysis are derived for the case where the power consists of a signal component as well as a noise component. Knowledge of these distributions is necessary when obtaining limits on threshold signals. Figures giving percentage points of the distributions are presented for selected values of the parameters.

141.316 **The pulse shape of the Crab nebula pulsar, PSR 0531+21, in the *V* and *I* bands.** E. J. Groth.
Astrophys. Journ., Vol. 200, 278 - 280 (1975).

Data obtained through *V* and *I* filters indicate that the leading edge of the interpulse of the Crab pulsar is brighter in *V* than in *I* relative to the rest of the pulse shape. In terms of amplitude, the excess is ~2 percent of the main pulse peak, and in terms of energy the excess is 4.4 ± 0.9 percent of the energy contained in the interpulse. The *V* and *I* pulses arrive simultaneously within 12 μs at 95 percent confidence.

141.317 **Superfluidity in neutron stars. I. Steady-state hydrodynamics and frictional heating.** G. Greenstein.
Astrophys. Journ., Vol. 200, 281 - 297 = Contr. Five College Astron. Dep., *Amherst, Mass.*, No. 200 (1975).

This paper describes the steady-state hydrodynamic response of the superfluid interior of a neutron star to the decelerating torque responsible for the slowing-down of pulsars. This torque throws the superfluid into a differential rotation, constant on cylinders but varying with distance from the axis of rotation, plus an exceedingly slow circulation in the $\hat{r}$ and $\hat{z}$ directions. This circulation is qualitatively similar to the meridional circulation found in "ordinary" stars. The motion of the superfluid does not lead to a tangling of the vortex lines threading it. The superfluid slows at the same rate as the pulsar.

141.318 **SAS-2 high-energy gamma-ray observations of the Vela pulsar.** D. J. Thompson, C. E. Fichtel, D. A. Kniffen, H. B. Ögelman.
Astrophys. Journ., (*Letters*), Vol. 200, L79 - L82 (1975).

The Second Small Astronomy Satellite (SAS-2) high-energy (> 35 MeV) γ-ray telescope has detected pulsed γ-ray emission at the radio period from PSR 0833–45, the Vela pulsar, as well as an unpulsed flux from the Vela region. The pulsed emission consists of two peaks, following the single radio peak by about 13 ms and 48 ms. The relatively high intensity of the pulsed γ-ray emission, and the double peak structure, compared with the single pulse in the radio emission, suggest that the high-energy γ-ray pulsar emission may be produced under different conditions from those at lower energies.

141.319 **Observational properties of pulsars.**
E. J. Groth.
Astrophys. Space Sci. Library, Vol. 48, (see 012.014), 119 - 173 (1975).

141.320 **Electrical conductivity and emissivity of the pulsar surface.** N. Itoh.
Monthly Notices Roy. Astron. Soc., Vol. 173, 1P - 3P (1975).

The electrical conductivity of the pulsar surface in the presence of a strong magnetic field is estimated. It is found that the conductivity perpendicular to the magnetic field is $\sim 10^4$ times smaller than that parallel to the magnetic field. The emissivity of the surface is found to be strongly affected by the magnetic field, and a strong polarization in the thermal radiation is predicted.

141.321 **On the motion of pulsars and black holes.**
S. L. Galkin.
Dokl. Akad. Nauk SSSR. Ser. Mat., Fiz., Vol. 223, 827 - 829 (1975). In Russian.

141.322 **Pulsar electrosphere and 'checkerboard plasma ejection'.** R. N. Henriksen, J. A. Norton.
Bull. American Astron. Soc., Vol. 7, 455 (1975). – Abstr. AAS.

141.323 **Detection of pulsar proper motion by radio interferometry.** D. C. Backer, R. A. Sramek.
Bull. American Astron. Soc., Vol. 7, 455 (1975). – Abstr. AAS.

141.324 **Pulsar microstructure: a search for longitude dependence.** B. J. Rickett.
Bull. American Astron. Soc., Vol. 7, 455 (1975). – Abstr. AAS.

141.325 **Clusters in the distribution of pulsars in period, pulse-width and age.**
K. B. Baker, P. A. Sturrock.
Bull. American Astron. Soc., Vol. 7, 455 (1975). – Abstr. AAS.

141.326 **A high time resolution X-ray light curve for the Crab pulsar, NP 0532.** F. Primini, D. Hearn, W. Lewin, F. Li, W. Mayer, J. McClintock, S. Rappaport, J. A. Richardson. Bull. American Astron. Soc., Vol. 7, 455 - 456 (1975). – Abstr. AAS.

141.327 **Drifting subpulses and pulsar slowdown rates.** R. T. Ritchings, A. G. Lyne.
Nature, Vol. 257, 293 - 294 (1975).
The authors present here evidence for a close relationship between the direction of drifting of subpulses from a pulsar and its position in a P–$\dot{P}$ diagram – the plot of the derivative of a pulsar's period, $\dot{P}$, against period, P.

141.328 **Crab pulsar optical and X-ray radiation through pinching instabilities.** G. Benford.
Astrophys. Journ., Vol. 201, 419 - 424 (1975).
Interpenetrating charged streams carrying no net current are unstable to formation of filaments of alternating positive and negative current. The electron-positron streams formed above the Crab pulsar gap become liable to this mode near the light cylinder distance from the pulsar surface, if they have differing average energies. The author calculates the growth rate and nonlinearly saturated amplitude of the magnetic turbulence generated. Particles emit synchrotron radiation incoherently in this turbulence. Power spectrum, polarization, and luminosity agree with observations, and indicate that the optical X-ray and γ-ray spectra may stem from this same mechanism.

141.329 **The radio spectrum of micropulses from pulsar PSR 0950 +08.**
B. J. Rickett, T. H. Hankins, J. M. Cordes.
Astrophys. Journ., Vol. 201, 425 - 430 (1975).
The microstructure in the radio pulses from PSR 0950 + 08 has been studied over both wide and narrow frequency ranges. The micropulses characterized by 0.17 ms duration are broad-band, exhibiting 50 percent correlation between 111 and 318 MHz. Over 125 kHz centered on 111 MHz the dynamic spectra of individual micropulses show deep modulation over intervals equal to the reciprocal of the pulse durations. These results are discussed in the light of the coherent radiation which is necessary to explain the large apparent brightness temperature seen in pulsars.

141.330 **Does a steady axisymmetric pulsar magnetosphere accelerate charges?** R. N. Henriksen, J. A. Norton.
Astrophys. Journ., Vol. 201, 431 - 439 (1975).
The force-free axisymmetric pulsar magnetosphere is examined, including the inertial effects of a two-fluid plasma. Strictly speaking, only electrostatic acceleration is possible. This is not the case, however, for charged particles which are constrained to flow along rigidly corotating, force-free magnetic field lines. The integrals of the motion and the electromagnetic constraints together yield an algebraic set that describes completely this constrained particle acceleration. The general theory is applied numerically and analytically to a two-fluid rotating monopole. Electrostatic acceleration is found not to be significantly responsible for particle acceleration in this case.

141.331 **Acceleration of pulsars by asymmetric radiation.** E. R. Harrison, E. Tademaru.
Astrophys. Journ., Vol. 201, 447 - 461 (1975).
Examination of the observational data for 145 pulsars confirms previous conclusions that pulsars have velocities generally in excess of 100 km s^{-1}. Earlier theories on the origin of these high velocities are discussed and found to be not entirely satisfactory. A new theory is proposed in which it is suggested that pulsars, and perhaps all rotating stellar objects, have external magnetic fields resembling to a first approximation that of an oblique dipole moment displaced from the center of the star.

141.332 **A deep sample of new pulsars and their spatial extent in the Galaxy.**
R. A. Hulse, J. H. Taylor.
Astrophys. Journ., (*Letters*), Vol. 201, L55 - L59 (1975).
The authors present parameters for 40 pulsars discovered in a survey conducted at the Arecibo Observatory, and list the 10 previously known pulsars which were also detected. Of this total of 50 pulsars, 47 lie in the main search region $42° \lesssim l \lesssim 60°$, $|b| \lesssim 4°$. The survey reached a limiting average flux density of ~1 mJy, and appears to have detected a limit to the spatial distribution of pulsars in this direction, at a distance corresponding to DM = 260 cm^{-3} pc. This limit is consistent with a markedly reduced density of pulsars at distances beyond 8–10 kpc from the galactic center.

141.333 **Oblique rotating pulsar magnetospheres with wave zones.** R. N. Henriksen, J. A. Norton.
Astrophys. Journ., Vol. 201, 719 - 728 (1975).
The equations for the oblique force-free pulsar magnetosphere are formulated and solved in the linear case. The total solution is a superposition of fundamental modes, subject to boundary conditions at the star. The eigensolutions are wavelike in the far zone. For a perpendicular oblique rotator, the resulting limiting polar field structure near the light cylinder is cusplike, providing a possible efficient emission zone.

141.334 **Magnetospheres and pulsars with net charge.** J. M. Cohen, L. S. Kegeles, A. Rosenblum.
Astrophys. Journ., Vol. 201, 783 - 791 (1975).
It is shown that a pulsar surrounded by a corotating magnetosphere region can possess a net charge. If there is no streaming, the star must be charged. The result is shown for the special-relativistic case as well as for the fully general-relativistic model, including the effects of strong gravitation and the dragging of inertial frames. When general relativity is included, the magnetosphere possesses net charge which can be comparable to that of the star itself. This is a new general-relativistic effect. Because of their net charge, pulsars possessing this type of corotating magnetosphere region are sources of charged particles.

141.335 **Interpretation of the pulsed gamma-ray emission from Vela.** D. J. Thompson.
Astrophys. Journ., (*Letters*), Vol. 201, L117 - L120 (1975).
A model is proposed for the Vela pulsar in which the radio emission originates near the surface of the neutron star while the pulsed γ-ray emission is produced by synchrotron radiation near the speed-of-light cylinder. This model can explain the energy flux, double pulse structure, and phase shift (with respect to the radio) of the γ-ray emission, and offers approximate quantitative predictions for other X- and γ-ray fluxes.

141.336 **On the possibility of determining the value of the magnetic field near pulsars by recording the frequency of beats in atomic transitions.** A. A. Ruzmajkin. Moskva, 1975. 14 pp. In Russian. – Abstr. in Referativ. Zhurn. 51. Astron., 9.51.245 (1975).

141.337 **On the pulsar emission mechanisms.** V. L. Ginzburg, V. V. Zheleznyakov.
Annual Rev. Astron. Astrophys., Vol. 13, (see 003.005), 511 - 535 (1975). – Contents: (1) Introduction; (2) The pulsar magnetosphere; (3) Mechanisms of electromagnetic radiation; (4) Formation of the radiation polar diagram (beaming mechanism); (5) Mechanism of optical, X-ray and gamma-ray emission of the pulsar in the Crab; (6) Mechanisms

of pulsars' radio emission; (7) Conclusion.

141.338 **Pulsars: a progress report.** F. Pacini.
Proc. 16th Solvay Conference Phys., (see 012.020), p. 1 - 23, with a discussion, p. 24 - 26 (1974).

141.339 **Proper motions of six pulsars.**
B. Anderson, A. G. Lyne, R. J. Peckham.
Nature, Vol. 258, 215 - 217 (1975).

The authors report measurements on six pulsars which seem to have motion velocities at least five times greater than the formal standard errors. The data show typical transverse velocity components in the range 100–200 km s^{-1}. Thus the authors deduce that the resultant velocities are high, typically 150–300 km s^{-1}, although none is as high as the 1,000 km s^{-1} suggested by several authors.

141.340 **Observed properties of 147 pulsars.**
J. H. Taylor, R. N. Manchester.
Astron. Journ., Vol. 80, 794 - 806 (1975).

The authors have compiled the principal parameters of all known pulsars and list them in tabular form, together with references to the sources of information. A number of quantities derived from the observations are also tabulated, including distance, height above the galactic plane, spectral index, radio luminosity, rate of kinetic energy loss, characteristic age, and surface magnetic field.

141.341 **Flux limit on cosmic ray acceleration by strong spherical pulsar waves.** E. Asseo, F. C. Kennel, R. Pellat.
Astron. Astrophys., Vol. 44, 31 - 40 (1975).

In this paper the authors investigate acceleration of cosmic rays by super-relativistic plasma waves. The pulsar wave field imposed upon the particles is that of a magnetic dipole wave including azimuthal dependence and near field terms. The authors first suppose the refractive index n of the plasma to be constant everywhere. Two distinct zones appear: in the 'near' zone spherical effects dominate, particles may be decelerated; in the 'far' zone the results for the motion differ noticeably from the vacuum case. From a WBK-like treatment for a spherical 'S' wave, it is concluded that for a wave solution to be valid even near the light cylinder, the wave can carry a sufficiently small particle flux.

141.342 **Hard emission of young pulsars as the cause of supernovae optical emission.** I. S. Shklovskij.
Astron. Zhurn. Akad. Nauk SSSR, Vol. 52, 911 - 919 (1975). In Russian. English translation in Soviet Astron., Vol. 19, No. 5.

The theory of energy pumping into the envelope by X-ray emission of a young pulsar is developed. A model for a source of such an emission is constructed that defines the frequency drift of its spectral distribution maximum. The light curves for supernovae of type I and type II after maximum should obey the power law $L \propto t^{-2.5}$.

141.343 **Production of fast particles by pulsars.**
M. Salvati.
Mem. Soc. Astron. Italiana, Vol. 46, 241 - 257 (1975).

141.344 **Relativistic effect of variability of the pulsar rotation on an elliptical orbit.**
Ya. B. Zel'dovich, N. I. Shakura.
Pis'ma v Astron. Zhurn., Vol. 1, No. 11, p. 19 - 22 (1975). In Russian.

A small periodic change of the observable period of pulsar rotation due to variable gravimagnetic interaction with the companion star on an elongated elliptical orbit is estimated.

141.345 **The new binary pulsar and the observation of gravitational spin precession.**
N. D. Hari Dass, V. Radhakrishnan.
Astrophys. Letters, Vol. 16, 135 - 139 (1975).

The recent discovery by Hulse and Taylor (1975) of a pulsar in a binary system has led to considerable excitement among astrophysicists; for the first time, as pointed out by these authors, a study of the gravitational interactions of the pulsar with the parent object, as well as a study of the astrophysics of close binary systems, should be possible. The scale of the binary system is such as to make relativistic gravitational effects on the orbital motion rather large. The precession of the periastron, for example, is of the order of magnitude of roughly 2π per century which should be compared with a similar effect of only 46″/century in our solar system.

141.346 **Search for continuous X-ray emission from NP 0532.**
R. S. Wolff, H. L. Kestenbaum, W. Ku, R. Novick.
Astrophys. Journ., (*Letters*), Vol. 202, L77 - L81 (1975).

A search for continuous X-ray emission from the pulsar NP 0532 has been performed using lunar occultations and high temporal resolution. No increase in the nonpulsed phase of the flux from the Crab nebula was observed when NP 0532 was exposed. The result is used to establish an upper limit of 4.7×10^6 K (99 % confidence) on the surface temperature of the neutron star.

141.347 **Spectral peculiarities of pulsars 0329, 0834 and 1919 radiation.**
V. N. Zlobin, V. A. Udal'tsov.
Astron. Zhurn. Akad. Nauk SSSR, Vol. 52, 1139 - 1142 (1975). In Russian. English translation in Soviet Astron., Vol. 19, No. 6.

Investigations of changes in the spectra of adjacent pulses of pulsars 0329, 0834 and 1919 are made. A result of the investigations is the detection of sporadic intense narrow-band radio emission of short life-time similar to that observed from PSR 0950. It is shown that the radio emission is not a consequence of influence of the medium between observer and pulsar. The mechanism suggested by Smith explains the observed properties of the received narrow-band radiation. Coherent radio emission of this mechanism results from the bunch of relativistic electrons in the magnetic field and it happens at a gyrofrequency in a narrow frequency band.

141.348 **Third speed-up of the Crab pulsar.** E. Lohsen.
Nature, Vol. 258, 688 - 689 (1975).

Optical timing of the pulsar in the Crab nebula revealed a further speed-up by $\Delta\nu = 0.097$ periods $d^{-1} = 3.7 \times 10^{-8} \nu$, on 1975 February 4. The initial jump in pulse frequency was at least three times larger than that reported for the first glitch, and allows a better estimate of the glitch parameters.

141.349 **On the original binary system of PSR 1913 + 16.**
R. Mitalas.
Journ. Roy. Astron. Soc. Canada, Vol. 69, 251 - 252 (1975). Abstr. CAS.

141.350 **Pulsar electrospheres and "checkerboard" plasma ejection.** R. N. Henriksen, J. A. Norton.
Journ. Roy. Astron. Soc. Canada, Vol. 69, 253 (1975). Abstr. CAS.

141.351 **Numerical studies of the axisymmetric pulsar magnetosphere.**
L. G. Kuo-Petravic, M. Petravic, K. V. Roberts.
Astrophys. Journ., Vol. 202, 762 - 772 (1975).

A numerical model based on the assumption that pulsars are spinning magnetized neutron stars is studied. A two-fluid model which includes inertial terms coupled self-consistently with Maxwell's equations is used. In this paper the case with aligned rotation and magnetic axes is examined with the assumption that the only source of charged particles is the star itself. The authors find that under these conditions the magnetic field lines are closed everywhere and the particles in the equatorial region close to the light cylinder reach sufficiently high energies to escape from the magnetosphere.

141.352 **Pulsar scintillation revisited – the physically thin screen model.** S. Bonazzola, L. M. Celnikier.
Astron. Astrophys., Vol. 45, 185 - 192 (1975).

In this paper, the authors have reconsidered the physically thin screen model of pulsar scintillation. They show that 'canonical' theories in which the decorrelation bandwidth is proportional to λ^{-4} or λ^{-2} are limiting cases of a more general formulation. Pulsars have been observed for which the decorrelation bandwidth varies as $\lambda^{-\alpha}$, with $2 \leqq \alpha \leqq 4$, and their result leads to a natural and unified interpretation of this phenomenon without ad hoc assumptions. The authors show in particular that the observed amplitude fluctuations at high frequencies of certain pulsars are not inconsistent with values of α approaching the lower limit 2.

141.353 **High-energy X-ray balloon observations of NP0532, Cyg X-1, Cyg X-3 and A0535 + 26.**
G. Ricker, A. Scheepmaker, J. Ballintine, J. Doty, G. Kriss, S. Ryckman, W. Lewin.
Bull. American Astron. Soc., Vol. 7, 524 (1975). – Abstr. AAS.

141.354 **Observational constraints on pulsar binary motion.** F. K. Lamb, D. Q. Lamb.
Bull. American Astron. Soc., Vol. 7, 545 (1975). – Abstr. AAS.

141.355 **Plasma collective effects and the origin of pulsar radiation.** P. E. Hardee, W. K. Rose.
Bull. American Astron. Soc., Vol. 7, 553 (1975). – Abstr. AAS.

141.356 **Orthogonal polarization in pulsar radio emission.** R. N. Manchester.
Proc. Astron. Soc. Australia, Vol. 2, 334 - 336 (1975).

141.357 **Pulsars.** H. K. Olsen.
Fys. Tidssk., Vol. 73, 3 - 9 (1975). In Danish.

141.358 **Meter-wavelength observations of pulsars using very long baseline interferometry.** N. R. Vandenberg.
Thesis, Maryland Univ., College Park (USA), 200 pp. (1974).

141.359 **Some time-dependent phenomena of pulsars.** K. H. Hesse.
Thesis, Math.-Naturwiss. Fakultät, Univ. Bonn. [Available from the library of the Bonn Univ.], 124 pp. (1974).

141.360 **Pulsars. II. Theories.** H. K. Olsen.
Fys. Tidssk., Vol. 73, No. 2, p. 49 - 55 (1975). In Danish.

141.361 **Pulsars.** D. ter Haar.
Contemporary Phys., (*GB*), Vol. 16, 243 - 256 (1975).

Review paper of observational data on pulsars and their theoretical interpretation.

141.362 **Acceleration of charged particles and radiation reaction. III. Motion in pulsar vacuum fields.**
M. Grewing, E. Schrüfer, H. Heintzmann.
Zeitschr. Physik, Ser. B, Vol. 20, 123 - 129 (1975).

Photon emission from suprathermal helium ions in a partially ionized hydrogen gas. See Abstr. 022.016.

On the possibility of determination of the value of the magnetic field near pulsars by means of the detection of the "beats" frequency in atomic transitions.
See Abstr. 022.093.

One- and two-electron atoms in strong magnetic fields and implications for magnetic white dwarfs and pulsars.
See Abstr. 022.126.

On the relative importance of phase and frequency information in pulsar timing measurements.
See Abstr. 031.203.

On particle acceleration and radiation output in stochastic electromagnetic fields. See Abstr. 061.009.

Cosmic Čerenkov emission.
See Abstr. 061.053.

Curvature radiation by bunches of particles.
See Abstr. 061.060.

On pair production in intense electromagnetic fields occurring in astrophysical situations. See Abstr. 061.068.

Ground state of electron matter in high magnetic fields. See Abstr. 062.001.

Collective relaxation of ultra-relativistic plasmas.
See Abstr. 062.037.

Particle acceleration by strong waves in ionized media. See Abstr. 062.074.

Parametric conversion and amplification of light in moving media. See Abstr. 063.001.

On the diffusion of light in turbulent shearing media. See Abstr. 063.011.

Strong scintillations in astrophysics. II. A theory of temporal broadening of pulses. See Abstr. 063.025.

Neutron star models with realistic high-density equations of state. See Abstr. 065.019.

Electrical conductivity and magnetic field decay in neutron stars. See Abstr. 065.123.

Stationary elastic rotational deformation of a relativistic neutron star model. See Abstr. 065.136.

En stjernes død og genfødsel. Hvide dværge, supernovae, pulsarer, sorte huller. See Abstr. 065.142.

Estimate of plasticity in the crust of neutron stars.
See Abstr. 065.146.

Anisotropic gravity and the binary pulsar PSR 1913 + 16. See Abstr. 066.095.

Relativistic spin precession in two-body systems.
See Abstr. 066.104.

Is Jupiter's magnetosphere like a pulsar's or earth's?
See Abstr. 099.009.

Is Jupiter's magnetosphere like a pulsar's or earth's?
See Abstr. 099.030.

Evolution of massive close binaries. II: The post X-ray binary stage: origin of run-away and binary pulsars.
See Abstr. 117.018.

The age of HB 21 and PSR 2021 + 51. See Abstr. 125.012.

Copernicus: the spatial distribution of the X-ray emission from IC443. See Abstr. 125.016.

An explanation of anomalous velocities in pulsar scintillation patterns. See Abstr. 131.020.

Attenuation of whistler mode waves excited in the interstellar medium by pulsars. See Abstr. 131.102.

Interstellar scattering of pulsar radiation. I. Scintillation. See Abstr. 131.111.

Meter-wavelength VLBI. II. The observations. See Abstr. 141.143.

The temporal behaviour of Taurus X-1 (the Crab nebula). See Abstr. 142.063.

Results of a southern-hemisphere search for gamma-ray sources at $E_\gamma \geq 3 \times 10^{11}$ eV. See Abstr. 142.095.

Observing neutron stars: information on stellar structure from pulsars and compact X-ray sources. See Abstr. 142.129.

Measurements of low-energy gamma radiation from the Crab nebula and from pulsar NP 0532. See Abstr. 142.165.

The X-ray pulsed fraction of NP0532 and interstellar grains. See Abstr. 142.241.

Infrared Sources

141.601 **1612-MHz survey of M supergiants.**
P. F. Bowers.
Astron. Journ., Vol. 80, 512 - 514 (1975).

A total of 66 M supergiants have been surveyed for OH emission at 1612 MHz. One new source, S Per, has been detected and is the fifth known supergiant OH/IR star. It is suggested that supergiant OH/IR stars are a small percentage ($\lesssim$ 10%) of the M-supergiant population.

141.602 **Infrared sources in obscured regions.**
D. A. Allen, M. V. Penston.
Monthly Notices Roy. Astron. Soc., Vol. 172, 245 - 248 (1975).

Two new infrared sources in the direction of dark nebulae in Taurus and Cygnus are reported and one known source re-examined. The new source in Taurus may be suitable for determining the infrared reddening law.

141.603 **On the nature of maser sources in infrared stars.**
V. V. Burdjuzha (*Burdyuzha*), T. V. Ruzmaikina (*Ruzmajkina*).
IAU Symposium No. 67, (see 012.001), p. 175 - 177 (1975).

Supergiant infrared stars with OH maser emissions, like NML Cyg, VY CMa, and VX Sgr, are discussed assuming a model consisting of a star with a photospheric temperature of about 2000 K and a radius of $R \sim 2 \times 10^{14}$ cm surrounded by a gas and dust shell. A possible similarity with OH Miras is mentioned.

141.604 **16 neue IR-Objekte, deren Positionen mit OH-Quellen zusammenfallen.**
G. V. Schultz, H. W. Humberg, E. Kreysa.
Mitt. Astron. Ges., No. 36, p. 174 - 177 (1975).

141.605 **A high-resolution map of the W3 region at far-infrared wavelengths.** G. G. Fazio, D. E. Kleinmann, R. W. Noyes, E. L. Wright, M. Zeilik II, F. J. Low.
Astrophys. Journ., (*Letters*), Vol. 199, L177 - L179 (1975).

The authors have mapped the area in the vicinity of the H II region W3 at an effective wavelength of 69 μ, with high angular resolution (1′), using the 102-cm balloon-borne telescope. The main continuum source, W3, shows a partially resolved, complex structure with a FWHM size of 1′.8. The radio sources W3 (OH) and G133.8 + 1.4 are also observed, as well as three faint sources not associated with known radio emission.

141.606 **Spectrophotometry of CRL 2688 from 2 to 24 microns.**
W. J. Forrest, K. M. Merrill, R. W. Russell, B. T. Soifer.
Astrophys. Journ., (*Letters*), Vol. 199, L181 - L183 (1975).

Medium-resolution spectrophotometry ($\Delta\lambda/\lambda \approx 0.01$) of CRL 2688 from 2 μ to 24 μ are reported. No significant features arefound from 3 μ to 24 μ, although the spectral flux distribution is broader than that of a single-temperature blackbody.

141.607 **Infrared measurements on several sources in the DR-21 region.** C. Thum, D. Lemke.
Astron. Astrophys., Vol. 41, 467 - 469 (1975).

2.2 and 3.5 μ measurements on DR-21 and three nearby pointlike sources are reported. A beam size effect was found on DR-21. W 75 (S)-OH was detected for the first time in the near infrared. A physical model for W 75/IRS 1 is proposed. According to it the source might be a heavily obscured B-star close to the main sequence.

141.608 **Haro 13a: a luminous, heavily obscured star in Orion?** D. A. Allen, K. M. Strom, G. L. Grasdalen, S. E. Strom, K. M. Merrill.
Monthly Notices Roy. Astron. Soc., Vol. 173, 47P - 50P (1975).

Haro 13a may be a luminous pre-main-sequence star of the Orion population obscured by over 20 mag of visual extinction.

141.609 **Initial results of spatial interferometry at 5 μ.**
D. W. McCarthy, F. J. Low.
Bull. American Astron. Soc., Vol. 7, 410 (1975). – Abstr. AAS.

141.610 **New AFCRL IR sources.**
M. J. Lebofsky, S. G. Kleinmann, G. H. Rieke.
Bull. American Astron. Soc., Vol. 7, 410 (1975). – Abstr. AAS.

141.611 **Eight new infrared sources associated with OH masers.** S. Beckwith, N. J. Evans II.
Bull. American Astron. Soc., Vol. 7, 417 - 418 (1975). – Abstr. AAS.

141.612 **Accurate positions for some new sources in the AFCRL catalogue.**
S. G. Kleinmann, M. J. Lebofsky.
Bull. American Astron. Soc., Vol. 7, 433 (1975). – Abstr. AAS.

141.613 **Infrared spectrophotometry of compact sources from the AFCRL survey.** K. M. Merrill.
Bull. American Astron. Soc., Vol. 7, 433 - 434 (1975). – Abstr. AAS.

141.614 **Objective-prism spectra of some suggested identifications of AFCRL infrared sources.**
T. Ake, P. Pesch.
Bull. American Astron. Soc., Vol. 7, 434 (1975). – Abstr. AAS.

141.615 **A far-infrared source in the Rho Ophiuchus dark cloud.**
G. G. Fazio, E. L. Wright, M. Zeilik II, F. J. Low.
Bull. American Astron. Soc., Vol. 7, 440 - 441 (1975). – Abstr. AAS.

141.616 **OH maser emission from an F-supergiant.**
P. T. Giguere, N. J. Woolf.
Bull. American Astron. Soc., Vol. 7, 445 (1975). – Abstr. AAS.

141.617 **Cool carbon monoxide around VY Canis Majoris and NML Cygni.**
T. R. Geballe, E. R. Wollman, J. H. Lacy, D. M. Rank.
Bull. American Astron. Soc., Vol. 7, 464 (1975). – Abstr. AAS.

141.618 **CO observations of high-velocity gas towards the Orion infrared source.**
T. B. H. Kuiper, B. Zuckerman, E. N. R. Kuiper.
Bull. American Astron. Soc., Vol. 7, 466 (1975). – Abstr. AAS.

141.619 **Infrared stars and the morphology of dusty regions.**
E. P. Ney.
Publ. Astron. Soc. Pacific, Vol. 87, 507 (1975). – Abstr. ASP.

141.620 **10 micron spectrophotometry of the compact sources in NGC 7538.** S. P. Willner.
Publ. Astron. Soc. Pacific, Vol. 87, 510 (1975). – Abstr. ASP.

141.621 **Die hellsten Sterne am Infrarothimmel bei 2.2 μm.**
S. Rössiger.
Sterne, 51. Jahrgang, p. 136 - 139 (1975).

141.622 **The nature of Kleinmann-Low (KL) and Becklin-Neugebauer (BN) infrared sources.**
Yu. N. Gnedin, I. G. Mitrofanov.
Astron. Tsirk., No. 852, p. 1 - 3 (1975). In Russian.

141.623 **CO in the infrared and radio spectra of carbon stars.**
J. A. Frogel, D. F. Dickinson, A. R. Hyland.
Astrophys. Journ., Vol. 201, 392 - 396 (1975).

Spectral scans from 2.0 to 2.5 μ of many of the reddest sources in the Two-Micron Sky Survey have resulted in the identification of eight new carbon stars, five of which have subsequently been identified by other observers. These and other known carbon stars have been searched for 2.6-mm CO emission. A broad emission line, similar to that in IRC + 10216, has been found in IRC + 40540. IRC–20131 and IRC + 50338 (V Cyg) show narrow CO lines whose origin is probably interstellar.

141.624 **An unusual nebula near UOA 27.**
S. G. Kleinmann, M. J. Lebofsky.
Astrophys. Journ., (*Letters*), Vol. 201, L91 - L93 (1975).

Near-infrared plates of a region in Cygnus show an extremely red nebula, ~25″ in diameter, which is displaced by 18″ ± 6″ from UOA 27, an object with infrared and microwave properties similar to those of the Becklin-Neugebauer point source in Orion. The authors suggest that this nebula may be reflecting light from the same obscured star which is the energy source for UOA 27, and could therefore provide a unique, though indirect, means of investigating the nature of stars which heat the BN point source and related objects.

141.625 **Mass loss from the infrared star CIT 6.**
S. L. Mufson, H. S. Liszt.
Astrophys. Journ., Vol. 202, 183 - 190 (1975).

The authors have observed 2.6 mm carbon monoxide emission from a small (~20″) region surrounding CIT 6. The line profiles are doubly peaked, and are interpreted as arising from a disk which is rotationally ejected from the outer layers of the stellar atmosphere. The mathematics of rotationally forced ejection are presented.

141.626 **Initial results of spatial interferometry at 5 microns.**
D. W. McCarthy, F. J. Low.
Astrophys. Journ., (*Letters*), Vol. 202, L37 - L40 (1975).

A new type of infrared spatial interferometer has been developed to measure the angular sizes and shapes of circumstellar dust shells. At a wavelength of 5 μ the diameter of VY CMa is approximately 0″.45. The size and structure of IRC +10216 are consistent with previous lunar occultation results. Upper limits are placed on the angular diameters of nine other stars.

141.627 **Narrow band photometry of infrared objects at 10.7 microns.** J. P. Baluteau, E. Bussoletti, C. Delage, N. Epchtein, D. Stefanovitch, B. Talurau.
Mem. Soc. Astron. Italiana, Vol. 46, 231 - 239 (1975).
Presented at the SIF Conference, Bologna, 1974 October 29, November 2.

The authors present preliminary results of narrow band photometry performed at 10.7 microns at the 193 cm telescope of the Observatory of Haute Provence. A schematic description of the instrument is given. Besides planets and stellar standards, three infrared objects have been observed: IRC + 10216, VY CMa, NML Cyg. NML Cyg presents an asymmetric shape never observed before which is related to the OH maser emitting region coincident with the IR source. The nature of this asymmetry is unexplained at present.

141.628 **Observations of water line emission near three compact infrared sources.**
G. J. White, L. T. Little.
Astrophys. Letters, Vol. 16, 151 - 154 (1975).

Observations of water line emission from regions near three compact infrared sources are reported. A new water source has been discovered in S140. New components have been observed in NGC 1333, one of which decreased in intensity by a factor of 2.5 over three days. The position of the W51(N) water source differs by 9″±10″ from that of the compact H II/infrared source IRS-2 in W51.

141.629 **40–350 micron observations of galactic sources.**
I. Furniss, R. E. Jennings, A. F. M. Moorwood.
Astrophys. Journ., Vol. 202, 400 - 406 (1975).

Broad-band 40–350 μ measurements are reported for 31 sources observed with a balloon-borne telescope. All but one of the sources have been identified with galactic H II regions, of which about one-third are optically visible. The objects include W3 (three components), W49, RCW 36, RCW 38, and 24 sources lying along the galactic plane between $l^{II} = 327°$ and 349°. Most of these sources have not previously been observed in the far-infrared. A contour map of the W3 region shows an emission arm extending from the main continuum component.

141.630 **Observations of an isolated compact infrared source in Perseus.** W. E. Westbrook, E. E. Becklin, K. M. Merrill, G. Neugebauer, M. Schmidt, S. P. Willner, C. G. Wynn-Williams.
Astrophys. Journ., Vol. 202, 407 - 414 (1975).

Infrared, optical, and radio observations are described of an unusual compact infrared source discovered during the

AFCRL survey. It is suggested that the object may be a planetary nebula seen at an unprecedentedly early phase in its evolution.

141.631 **The nature of the Kleinmann-Low (KL) and Becklin-Neugebauer (BN) infrared sources.**
Yu. N. Gnedin, I. G. Mitrofanov.
Astron. Zhurn. Akad. Nauk SSSR, Vol. 52, 1121 - 1131 (1975). In Russian. English translation in Soviet Astron., Vol. 19, No. 6.

The polarization observations of the Becklin-Neugebauer infrared point source in the interior of the Kleinmann-Low extended source are discussed. The light propagated through the ensemble of absorbing nonspherical oriented dust grains in the periphery of the KL object becomes polarized. The KL object as the optically thick dust cloud with hot O and B stars in the interior is also considered. This cloud seems to reradiate their radiation in the infrared. The dust grains accelerated by the light pressure are oriented in the region with concentration $n_H \lesssim 300$ cm^{-3} due to collisions with gas. The existence of a transversal magnetic field is the necessary condition for the light polarization. The KL mass may be estimated to $M_{KL} \sim 10^4 M_\odot$ from the analysis of the polarization data.

141.632 **Maser emission from infrared stars.** S. Kwok.
Journ. Roy. Astron. Soc. Canada, Vol. 69, 244 (1975). – Abstr. CAS.

141.633 **Radio spectral line observations of mass-loss: the expanding envelope of IRC+10216.**
T. B. H. Kuiper, G. R. Knapp, S. L. Knapp, R. L. Brown.
Journ. Roy. Astron. Soc. Canada, Vol. 69, 247 (1975). Abstr. CAS.

141.634 **Near infrared sources in Sgr B2.**
C. Thum, D. Lemke.
Astron. Astrophys., Vol. 45, 83 - 85 (1975).

Two 2.2 μ sources were discovered in the Sgr B2 region. They appear to be coincident with two compact H II regions detected earlier at radio wavelengths.

141.635 **Infrared observations of stars in NGC 2264.**
J. W. Warner.
Bull. American Astron. Soc., Vol. 7, 519 (1975). – Abstr. AAS.

141.636 **The infrared emission from Sgr B2.**
D. A. Harper, H. A. Thronson, S. Whitcomb, R. Hildebrand, R. Stiening, R. Winston.
Bull. American Astron. Soc., Vol. 7, 529 (1975). – Abstr. AAS.

141.637 **Far infrared maps of Sagittarius B2.**
M. W. Werner, E. E. Becklin, I. Gatley.
Bull. American Astron. Soc., Vol. 7, 529 - 530 (1975). Abstr. AAS.

141.638 **Far infrared maps of the Sagittarius A region.**
I. Gatley, E. E. Becklin, M. W. Werner, C. G. Wynn-Williams.
Bull. American Astron. Soc., Vol. 7, 530 (1975). – Abstr. AAS.

141.639 **Far infrared observations of NGC 7538.**
H. A. Thronson, I. Gatley, D. A. Harper, E. Becklin, R. Loewenstein, S. H. Moseley, G. Neugebauer, G. Wynn-Williams.
Bull. American Astron. Soc., Vol. 7, 530 (1975). – Abstr. AAS.

141.640 **The infrared spectrum of η Carinae: 3–14 microns.**
R. R. Joyce.
Publ. Astron. Soc. Pacific, Vol. 87, 917 - 921 (1975).

Intermediate resolution spectral scans of η Car in the range 3.2–14 μm provide strong support for the model proposed by Robinson, Hyland, and Thomas (1973) in which the infrared emission arises from two concentric dust shells heated by the central object. No fine-structure or recombination emission lines were observed over this spectral range.

141.641 **Infrared star in RCW 113.** A. C. Danks.
Publ. Astron. Soc. Pacific, Vol. 87, 941 - 942 (1975).

Photometric observations of an infrared star in the *H*, *K*, and *L* bands are presented. The star was detected by drift scans of the H II region RCW 113 at 2.37 μ.

141.642 **The galactic distribution and kinematics of OH stars (a preliminary discussion).** P. F. Bowers.
Proc. Astron. Soc. Australia, Vol. 2, 362 - 363 (1975).

141.643 **A model for the 1612 MHz masers in OH-IR stars.**
M. Elitzur, P. Goldreich, N. Scoville.
Rep. 1974-13, Owens Valley Radio Obs., Pasadena, Calif., USA. [Available from NTIS, Springfield, Va., USA.], 51 pp. (1974).

Attempts to bring into sharper focus the essential features responsible for the inversion of the 1612 MHz transition in OH-IR stars.

141.644 **VLBI observations of OH/IR stars.**
M. J. Reid, D. O. Muhleman.
Rep. 1974-14, Owens Valley Radio Obs., Pasadena, Calif., USA. [Available from NTIS, Springfield, Va., USA.], 13 pp. (1974).

Very-long-baseline interferometric observations of 5 type II OH/IR objects indicate that the apparent size of the masering regions are about or greater than 0.1 seconds of arc. The observations, combined with current theories, indicate that the 1612 MHz OH maser associated with Mira variables originates in the outer regions of the circumstellar dust.

141.645 **Astronomical far infrared sky survey.**
J. H. Goebel.
Thesis Washington Univ., St. Louis, Mo., USA. [Available from Univ. Microfilms, Ann Arbor, Mich., USA. Order No. 75-6596], 223 pp. (1974).

Reports the instrumentation used and the results obtained by a high-altitude balloon-borne far-infrared telescope. Approximately one-half of the sky was examined in the waveband 50–500 microns. The infrared detectors are discussed, and new findings on the factors determining sensitivity are presented. The telescope design and its orientation control are described. Possible improvements are indicated. Twelve objects were detected. Five of them are probably associated with stars within 20 pc. The sources also appear to be significantly concentrated to the ecliptic. Possible radiation mechanisms are discussed.

Young stellar objects and dark interstellar clouds.
See Abstr. 065.118.

Precipitable water above Cerro Tololo between 1971 and 1975. See Abstr. 082.117.

Infrared photometry and polarimetry of cool stars. II.
See Abstr. 113.019.

The infrared spectrum and structure of Eta Carinae.
See Abstr. 114.304.

Infrared observations of young stars–VI. A 2- to 4-micron search for molecular features.
See Abstr. 114.349.

The variable circular polarization of V1057 Cygni.
See Abstr. 122.072.

Peculiarities of light and polarization variations of the infrared carbon star CIT 6. See Abstr. 122.087.

Detection of interstellar SiS and a study of the IRC+10216 molecular envelope. See Abstr. 131.008

Giant grains in Orion. See Abstr. 131.019.

Observations of 1-millimeter continuum radiation from the DR 21 region. See Abstr. 131.025.

Observations of quasi-thermal and maser phenomena in rotationally excited OH. I. See Abstr. 131.026.

Absorptive properties of silicate core-mantle grains. See Abstr. 131.028.

Whiskers and cosmic millimetre-wave sources. See Abstr. 131.035.

10-micron spectrophotometry of compact sources in W3. See Abstr. 131.076.

Near infrared observations of "active" molecular clouds. See Abstr. 131.077.

Radiation from inhomogeneous dusty Strömgren spheres. See Abstr. 131.099.

The molecular cloud Sagittarius B2. See Abstr. 131.104.

H_2O sources in regions of star formation. See Abstr. 131.121.

On the influence of dust on the effectiveness of OH maser sources. See Abstr. 131.125.

Bok globules, molecular clouds, infra-red sources and star formation. See Abstr. 131.130.

Some implications of 10-micron interstellar polarization. See Abstr. 131.140.

New 1612 MHz OH emission sources. See Abstr. 131.147.

Anisotropy of constituents of interstellar grains. See Abstr. 131.162.

H I and OH observations towards W3. See Abstr. 131.163.

G45.5 + 0.1 and G45.1 + 0.1: compact infrared sources. See Abstr. 131.503.

Infrared photometric and spectrophotometric observations of the galactic H II region G29.9–0.0. See Abstr. 131.508.

Observations of 6 GHz absorption lines from the $^2\pi_{3/2}$, $J = 5/2$ state of OH. See Abstr. 131.509.

Detection of an additional hydrogen recombination line source at the W49A position. See Abstr. 131.512.

The 8–13 μ spectra of compact H II regions. See Abstr. 131.514.

The 4830 MHz H_2CO absorption in the direction of NGC 6334. See Abstr. 131.516.

Far infrared observations of dark cold clouds and HII regions. See Abstr. 131.532.

Detection of the [O III] 88.16 micron line in M17. See Abstr. 131.540.

H II-Regionen und assoziierte Punktquellen: Infrarot-Beobachtungen und Modelle. See Abstr. 131.544.

Compact H II regions in the Ophiuchus and R Coronae Austrinae dark clouds. See Abstr. 131.548.

Spectroscopy of the Orion nebula from 80 to 135 microns. See Abstr. 132.003.

The spectrum and structure of 'Minkowski's footprint': M 1–92. See Abstr. 132.006.

40–350 micron emission from NGC 2023. See Abstr. 132.009.

The nature of Herbig-Haro objects 1 and 2: compact emission nebulae. See Abstr. 132.031.

The peculiar object Mz-3. See Abstr. 133.018.

The companion object to W3 (OH). See Abstr. 141.113.

Infrared observations of the field cluster NGC 7419. See Abstr. 153.040.

Measurements of galactic nuclei at 34 microns. See Abstr. 158.048.

Errata

141.901 **Errata: 'PSR 1911+03: a new pulsar with a large duty-cycle'** [Monthly Notices Roy. Astron. Soc., Vol. 171, 17P - 18P (1975)]. D. K. Mohanty, V. Balasubramanian.
Monthly Notices Roy. Astron. Soc., Vol. 172, 707 (1975).

141.902 **Erratum: 'A comparison of some radio and optical properties of quasi-stellar sources and BL Lacertae objects'** [Astrophys. Journ., (*Letters*), Vol. 198, L57 - L60 (1975)]. P. D. Usher.
Astrophys. Journ., (*Letters*), Vol. 201, L49 (1975).

142 X-Ray, Gamma-Ray Sources

142.001 **Further observations of Cygnus X-3 with the Uhuru satellite.** R. W. Leach, S. S. Murray, E. J. Schreier, H. D. Tananbaum, M. P. Ulmer, D. R. Parsignault. Astrophys. Journ., Vol. 199, 184 - 188 (1975).

The authors present observations of the periodically varying X-ray source Cygnus X-3. The results are: 1. A value for the period. $P = 0^d\!.1996811 \pm 0^d\!.0000016$. 2. No systematic change is observed in the effective low energy absorption across the $0^d\!.2$ cycle. 3. Cyg X-3 exhibits different average intensity levels which have been seen to persist for as long as 9 days at a time. The hardest spectra (in the 2–10 keV range) are derived from the times of lowest 2–6 keV intensity. 4. Cygnus X-3 was observed during the large radio outburst first noted on 1972 September 3. The spectral parameter kT for thermal bremsstrahlung emission increased systematically from ~2 keV to ~3 keV from ~August 30 to ~September 3, and then decreased back to ~2 keV by September 7.

142.002 **The circularization of close binary orbits: Centaurus X-3.** R. A. Chevalier. Astrophys. Journ., Vol. 199, 189 - 195 (1975).

If the neutron star in Cen X-3 resulted from a normal supernova explosion, the resulting binary system must have had considerable eccentricity. As the present eccentricity is small, mechanisms for circularizing the orbit are examined with the aim of setting constraints on the evolution of the binary system. Tidal interaction does not yield an explanation for the observed period changes in Cen X-3. Thus it may be necessary to go to a combination of mass loss and mass transfer as the circularization mechanism, although a delicate balance between the two is required.

142.003 **Mass transfer in binary X-ray systems.** R. McCray, S. Hatchett. Astrophys. Journ., Vol. 199, 196 - 205 (1975).

The influence of X-ray heating on gas flows in binary X-ray systems is examined. A simple estimate is obtained for the evaporative wind flux from a stellar atmosphere due to X-ray heating. The wind flux is sensitive to the soft X-ray spectrum. The self-excited wind mechanism does not work. Mass transfer in the Hercules system probably occurs by flow of the atmosphere of HZ Her through the gravitational saddle point of the system. The accretion gas stream is probably opaque with atomic density $n \gtrsim 10^{15}$ cm^{-3} and is confined to a small fraction of 4π steradians. Other binary X-ray systems are briefly discussed.

142.004 **Hard X-ray observations of NGC 4151.** W. A. Baity, T. W. Jones, W. A. Wheaton, L. E. Peterson. Astrophys. Journ., (*Letters*), Vol. 199, L5 - L8 (1975).

The X-ray source 3U 1207+39, associated with the Seyfert galaxy NGC 4151, has been scanned repeatedly by the UCSD X-ray telescope on board the OSO-7 satellite. A weak source with a hard spectrum has been unambiguously detected in this region. The hard spectrum observed and very short electron lifetimes implied severely constrain nonthermal models for the emission.

142.005 **Optical studies of 10 high galactic latitude X-ray sources.** J. N. Bahcall, N. A. Bahcall, S. S. Murray, M. Schmidt. Astrophys. Journ., (*Letters*), Vol. 199, L9 - L11 (1975).

Optical studies are reported for a representative sample of 10 unidentified high galactic latitude ($|b| \geqslant 20°$) X-ray sources that are north of declination −25° and that have 3U error boxes less than or of the order of 0.5 square degrees. The source 3U 1410–03 is identified with a peculiar emission-line spiral galaxy, NGC 5506, and a distance group 4 rich cluster of galaxies, Abell 2142, is suggested as a possible identification for 3U 1555+27. It is shown that very few of the unidentified high galactic latitude sources are quasi-stellar sources brighter than 17.5 mag, although one possible X-ray quasar, NAB 0137–01, is reported.

142.006 **Period changes in Centaurus X-3.** W. M. Sparks. Astrophys. Journ., Vol. 199, 462 - 466 (1975).

A number of mechanisms which can change the orbital period are applied to various models of Cen X-3. Only four models which give rise to feasible mechanisms are found. Possible observations which could distinguish between these models are suggested.

142.007 **Radio observations of Cygnus X-3 from 1973 December to 1974 June.** P. C. Gregory, E. R. Seaquist, V. A. Hughes, A. Woodsworth. Astrophys. Journ., Vol. 199, 467 - 470 (1975).

Observations of Cygnus X-3 on 79 days during the period 1973 December to 1974 June at 10.5 or 22 GHz are presented. During this period several large outbursts (>5 Jy) were recorded, together with many smaller ones. From an analysis of a total of 253 days of observations, the authors find that the flux density of Cygnus X-3 at centimeter wavelengths exceeded 10 Jy on ~2 percent of the days and exceeded 1 Jy on about one-third of the days observed.

142.008 **The unidentified high galactic latitude X-ray sources: bright galaxies or rich clusters?** J. N. Bahcall, N. A. Bahcall. Astrophys. Journ., (*Letters*), Vol. 199, L89 - L92 (1975).

The authors analyze the positional correlations between the unidentified high galactic latitude 3U X-ray sources, as well as (1) all 2712 rich clusters of galaxies in the Abell catalog and (2) all bright galaxies ($m_p \leqslant 15.7$ mag) in the Zwicky catalog. Computer studies using simulated Uhuru catalogs are used to evaluate the chance coincidence rate and its dispersion. They confirm the validity of practically all of the identifications previously suggested by Kellogg et al. of 3U sources with nearby rich Abell clusters.

142.009 **Observations of variable X-ray sources in globular clusters.** G. W. Clark, T. H. Markert, F. K. Li. Astrophys. Journ., (*Letters*), Vol. 199, L93 - L96 (1975).

Observations with the MIT detectors on the OSO-7 satellite have revealed a previously unreported variable hard X-ray source at midgalactic latitude in the globular cluster NGC 1851. The source, designated MX 0513–40, varied in intensity by a factor $\gtrsim 5$ in 16 days and had a peak 12-day average energy flux of 4.7×10^{-10} ergs cm^{-2} s^{-1} in the energy range 1–10 keV, with $dN/dE \sim E^{-1.4}$. Observations of the X-ray sources 3U 2131+11 and 3U 1746–37 are also reported which confirm their locations in globular clusters NGC 7078 (M15) and NGC 6441, respectively, and demonstrate that they too are variable.

142.010 **Observations of a bright, variable X-ray source in a globular cluster.** C. R. Canizares, J. E. Neighbours. Astrophys. Journ., (*Letters*), Vol. 199, L97 - L99 (1975).

The source 3U 1820–30 in NGC 6624 has been observed with the OSO-7 satellite. It shows variability of up to a factor of 5 in intensity over days and months and of a factor of 2 in ~5–10 minutes, but there are no evident periodicities. The luminosity of the source is 1.0×10^{38} ergs s^{-1} (1–10 keV). The authors review arguments which suggest that mass accretion onto a neutron star or black hole provides the most likely mechanism of X-ray production.

142.011 **Soft X-ray emission from Hercules X-1.**
S. Shulman, H. Friedman, G. Fritz, R. C. Henry, D. J. Yentis.
Astrophys. Journ., (*Letters*), Vol. 199, L101 - L104 (1975).

A rocket observation of Her X-1 has detected an intense flux of X-rays at energies below 1 keV. The 0.18–0.28 keV flux was 2×10^{-9} ergs cm^{-2} s^{-1} compared with a 1–10 keV flux of 3.3×10^{-9} ergs cm^{-2} s^{-1}. The low-energy flux was pulsating with the $1^s.24$ period, although the pulse profile differed markedly from that at higher energies.

142.012 **X-ray source counts and the X-ray background from Uhuru.** A. C. Fabian.
Monthly Notices Roy. Astron. Soc., Vol. 172, 149 - 157 (1975).

Fluctuations in the X-ray background observed from the Uhuru satellite have been used to estimate the number counts of X-ray sources at high galactic latitude ($|b| > 20°$). Differential counts represented by $N(S)\,dS = 25(\pm 10)\,S^{-2.5 \pm 0.2}\,dS$ fit these data adequately. It is suggested that the previous counts by Matilsky et al. and Holt et al. have been overestimated.

142.013 **On light curve analyses of X-ray binaries.**
D. T. Wickramasinghe, J. Whelan.
Monthly Notices Roy. Astron. Soc., Vol. 172, 175 - 179 (1975).

A comparison is made between the results of a number of programs currently in use for computing theoretical X-ray binary optical light curves. For four programs excellent agreement is found, and the small differences between them are the result of different treatments of limb darkening. The use of approximate limb-darkening laws is discussed and it is shown that the commonly used linear approximation generally tends to overestimate the 'larger' light amplitude. A discrepancy between the results obtained by Hutchings and other investigators is pointed out.

142.014 **X-ray observations of Cyg X-1 with ANS.**
J. Heise, A. C. Brinkman, J. Schrijver, R. Mewe, A. Den Boggende, E. Gronenschild, D. Parsignault, J. Grindlay, H. Schnopper, E. Schreier, H. Gursky.
Nature, Vol. 256, 107 - 108 (1975).

From May 1 to May 8, 1975, Cyg X-1 was observed with the instruments on board the Astronomical Netherlands Satellite (ANS). The authors report here the results of the measurements performed with soft and hard X-ray detectors. Cyg X-1 underwent an upward transition in its intensity, that seems to be the inverse of the downward transition seen by Uhuru in April 1971. The bulk of the increase occurs at low energy: between 1 and 2 keV, the intensity increased by a factor of 10 over the November 1974 intensity observed by ANS, while above 8 keV there is no significant change.

142.015 **Observations of a new transition in the emission from Cyg X-1.**
S. S. Holt, E. A. Boldt, L. J. Kaluzienski, P. J. Serlemitsos.
Nature, Vol. 256, 108 - 109 (1975). – Letter.

142.016 **Ariel V and Copernicus measurements of the X-ray variability of Cyg X-1.** P. W. Sanford, J. C. Ives, S. J. Bell Burnell, K. O. Mason, P. Murdin.
Nature, Vol. 256, 109 - 111 (1975).

The collimated proportional counter (experiment C of MSSL) on Ariel V viewed Cyg X-1 in 1975 from May 3 1400 UT until May 15 1000 UT. The authors report here observations of the intensity and spectrum obtained with an integration time of ~30 min. They show that, at the beginning of this period, Cyg X-1 was more intense, had a steeper spectrum and was generally more active than has been observed during the past $2^1/_2$ yr by Copernicus. Towards the end of the period the source seemed to be returning to its lower state.

142.017 **Another major change in the radio source associated with Cyg X-1.**
R. M. Hjellming, D. M. Gibson, F. N. Owen.
Nature, Vol. 256, 111 - 112 (1975).

The sudden appearance of a faint radio source inside the probable position region for the X-ray source Cyg X-1 was associated with major changes in the X-ray emission observed during the period February–May 1971. This coincidence established a probable association with the X-ray source. The accurate position of the radio source led to an identification with the bright, blue star HDE 226868. Subsequent optical investigations have established this object to be a binary system with a massive, invisible component that may be a black hole. The authors now report further observations of the Cyg X-1 radio source, the most recent of which were made after learning that Cyg X-1 had undergone another major change in X-ray state in, or before, early May 1975.

142.018 **Can soft γ-ray bursts be emitted by accreting black holes?** T. Piran, J. Shaham.
Nature, Vol. 256, 112 - 113 (1975).

There are many models for the observed γ-ray bursts, but none has proved entirely satisfactory. Here the authors suggest another possible model: that the γ rays are produced by inverse Compton processes occurring deep in the ergosphere of a fast rotating, accreting, black hole. In this model the black hole itself is probably a member of a binary system, and, depending on its steady-state accretion rate, may or may not be observable as an X-ray source.

142.019 **Hard X-ray observations near Ariel 1118–61.**
G. F. Carpenter, M. J. Coe, A. R. Engel, J. J. Quenby.
Nature, Vol. 256, 292 - 293 (1975).

Observations made by Ariel V in the region of Cen X-3 are discussed. The authors conclude that the addition of scintillator X-ray data to the A1118–61 observations suggests that if a dominant, single source mechanism is sought, black body radiation from a low skin depth gas is the most likely.

142.020 **Radio emission from Cyg X-3.**
H.-D. Bohnenstengel, K. Henning, H. J. Wendker.
Nature, Vol. 256, 294 - 295 (1975).

The most recent radio outburst of Cyg X-3 reported occurred in May, 1974. Following the announcement by Ryle of high activity the authors observed part of this outburst at 4.8 GHz with the 100-m telescope of the MPIfR at Effelsberg from May 22 to May 31, during a general survey of the Cygnus X region. The observed flux densities are given. The authors conclude that ten years ago the source was not so active as today.

142.021 **Number counts of γ-ray bursts.**
A. C. Fabian, with a reply by I. B. Strong and R. W. Klebesadel.
Nature, Vol. 256, 347 - 348 (1975).

142.022 **Observations of a transient X-ray source with a period of 104 s.**
F. D. Rosenberg, C. J. Eyles, G. K. Skinner, A. P. Willmore.
Nature, Vol. 256, 628 - 630 (1975).

During routine observations of the Crab nebula by Ariel V, experiment A, the rotation modulation collimator discovered a new X-ray source A0535+26. The source was just detectable during the first observation on April 13, and during the 16 d of observation was seen to brighten to nearly twice the Crab's intensity (in the 3–7 keV range). The source varies periodically, with a period of 104 s, and has a very flat spectrum in the 3–7 keV range.

142.023 **Hard X-ray measurements of nova A0535+26 in Taurus.**
M. J. Coe, G. F. Carpenter, A. R. Engel, J. J. Quenby.
Nature, Vol. 256, 630 - 631 (1975).
Measurements of high energy X rays in the region of the Crab nebula were made with the Imperial College detector on Ariel V between April 13 and April 29, 1975. During that time the X-ray nova A0535+26 was first detected and observed to rise to its maximum intensity. At 30 keV this nova reached an intensity considerably brighter than the Crab nebula, making it the brightest known source in the sky.

142.024 **Observations of A0535+26 with the Leicester sky survey experiment.**
M. J. Ricketts, M. J. L. Turner, C. G. Page, K. A. Pounds.
Nature, Vol. 256, 631 - 633 (1975). – Letter.

142.025 **Decay of X-ray source A0535+26.**
L. J. Kaluzienski, S. S. Holt, E. A. Boldt, P. J. Serlemitsos.
Nature, Vol. 256, 633 - 634 (1975). – Letter.

142.026 **The binary X-ray stars – the observational picture.**
H. Gursky, E. Schreier.
IAU Symposium No. 67, (see 012.001), p. 413 - 464 (1975). Review paper.

142.027 **The X-ray sources Sco X-1 and Cyg X-2 as binaries.**
V. M. Lyuty (*Lyutyj*).
IAU Symposium No. 67, (see 012.001), p. 465 - 466 (1975). Abstract.

142.028 **Combined optical and X-ray observations of variable stars.** C. S. Bowyer.
IAU Symposium No. 67, (see 012.001), p. 467 - 473 (1975).
Two problems of X-ray astronomy are discussed: (1) Optical identifications of X-ray sources, and (2) optical and X-ray observations of U Geminorum.

142.029 **The identification and UBV photometry of the visible component of the Centaurus X-3 binary system.** W. Krzemiński.
IAU Symposium No. 67, (see 012.001), p. 475 (1975). – Abstract.

142.030 **A search for super-rapid variability of X-ray sources and other peculiar objects.**
O. A. Evseyev (*Evseev*), V. N. Mansurov, V. S. Pavlova, L. A. Pustil'nik, V. F. Shvartsman, G. S. Tsarevsky (*Tsarevskij*), A. K. Yakushev.
IAU Symposium No. 67, (see 012.001), p. 477 - 480 (1975).

142.031 **Circular polarization of Cyg X-1, Sco X-1 and 3C 273.**
A. B. Severny (*Severnyj*), V. M. Kuvshinov.
IAU Symposium No. 67, (see 012.001), p. 481 - 482 (1975).

142.032 **Observations of circular polarization in V 1357 Cyg = Cyg X-1.** O. S. Shulov.
IAU Symposium No. 67, (see 012.001), p. 483 - 484 (1975).

142.033 **Linear polarization observations of some X-ray sources.**
N. M. Shakhovskoy (*Shakhovskoj*), Yu. S. Efimov.
IAU Symposium No. 67, (see 012.001), p. 487 - 495 (1975).
Multicolour linear polarization of optical radiation of the X-ray sources Sco X-1, Cyg X-2, Cyg X-1 and Her X-1 was measured at the Crimean Astrophysical Observatory in 1970–1973. These observations indicate that polarization of Sco X-1 in the ultraviolet, blue and red spectral regions appears to be variable. No statistically significant variations of polarization were found for the other three sources observed.

142.034 **Die optische Lichtkurve von Cen X-3.**
H. Mauder.
Mitt. Astron. Ges., No. 36, p. 98 (1975). – This paper appeared in Astrophys. Journ., (*Letters*), Vol. 195, L27 - L30 (1975) – see Abstr. 13.142.005.

142.035 **A sudden increase in the X-ray flux from Centaurus A.** P. F. Winkler, Jr., A. E. White.
Astrophys. Journ., (*Letters*), Vol. 199, L139 - L142 (1975).
Observations from OSO-7 show that the X-ray flux from Cen A increased by a factor of at least 1.6 over a six-day period in 1973 April. Long-term observations indicate greater increases and a hardening of the spectrum. The maximum flux exceeded that measured by Tucker et al. and Lampton et al. in 1970–1971 by factors of 6.7 at 2–10 keV and 14 at 10–50 keV. Both rapid variability and a harder spectrum are consistent with a model proposed by Grindlay. At maximum brightness the best-fit spectrum leads to a 2–50 keV luminosity of 1.1×10^{43} ergs s^{-1}.

142.036 **X-ray binaries in globular clusters.**
G. W. Clark.
Astrophys. Journ., (*Letters*), Vol. 199, L143 - L145 (1975).
It appears to be very unlikely that primordial binaries in globular clusters have evolved to produce high-luminosity X-ray sources like the four variable sources detected in observations by the Uhuru and OSO-7 satellites. Therefore, it is suggested that the cluster X-ray sources are binaries formed by capture from the remnants of massive single stars that exploded with sufficient isotropy to remain bound in the cluster. A case is made for the conclusion that there is approximately one potential binary X-ray source near the center of each of the globular clusters.

142.037 **Short-term time variability of Cygnus X-1.**
M. C. Weisskopf, S. M. Kahn, P. G. Sutherland.
Astrophys. Journ., (*Letters*), Vol. 199, L147 - L151 (1975).
Autocorrelation functions, power density spectra, and cross-correlation functions were calculated for 71 observations of Cyg X-1 obtained by the Uhuru satellite. The average autocorrelation function and power density spectrum were found to be independent of energy and consistent with a shot-noise source with a characteristic time constant of 0.45 s. The data in different energy bands were found to be 100 percent correlated.

142.038 **Cygnus X-1: a two-temperature accretion disk model which explains the observed hard X-ray spectrum.** D. M. Eardley, A. P. Lightman, S. L. Shapiro.
Astrophys. Journ., (*Letters*), Vol. 199, L153 - L155 (1975).
The authors present a model for Cygnus X-1, involving an accretion disk around a black hole, which fits the observed X-ray spectrum from 8 to 500 keV. The hard component of the spectrum ($E > 8$ keV) is produced by an optically thin inner portion of the disk in which the electrons are 10^9 K and the ions are 3–300 times hotter.

142.039 **Optical studies of Uhuru sources. XII. The light curve of Scorpius X-1 = V818 Scorpii, 1889–1974.**
E. L. Wright, E. W. Gottlieb, W. Liller.
Astrophys. Journ., Vol. 200, 171 - 176 (1975).
An analysis of 1068 photographic observations of V818 Sco = Sco X-1 from the Harvard plate collection, as well as 698 post discovery magnitudes, shows that its B magnitude ranges between 11.8 and 13.7. The radial velocity curve, based on published data, and the variations in color make it possible to draw new conclusions on plausible models for Sco X-1. These are described, and the authors conclude by making recommendations for further observations.

142.040 **The X-ray source MX 0513–40 and the globular cluster NGC 1851.** N. V. Vidal, K. C. Freeman.
Astrophys. Journ., (*Letters*), Vol. 200, L9 - L11 (1975).

A hot low-gravity "UV-bright" star is found near the edge of the 1 σ error box of MX 0513–40 and about 0'.5 from the nucleus of the globular cluster NGC 1851. Its radial velocity is consistent with cluster membership and may be slightly variable. On some spectra a blueward emission component is found at Hβ.

142.041 **Circular polarization of the X-ray sources Scorpius X-1, Cygnus X-1, and 3C 273.**
A. B. Severny, V. M. Kuvshinov.
Astrophys. Journ., (*Letters*), Vol. 200, L13 - L17 (1975).

Systematic *B*- and *V*-color observations during 1972–1974 of the circular polarization of Sco X-1 (12 nights), Cyg X-1 (28 nights), and 3C 273 (31 nights) showed the occasional appearance of appreciable and variable polarization.

142.042 **Optical observations of stars near southern galactic X-ray sources.**
M. V. Penston, M. J. Penston, P. Murdin, W. L. Martin.
Monthly Notices Roy. Astron. Soc., Vol. 172, 313 - 324 (1975).

The authors report the results of spectroscopic and photometric observations of stars in the fields of 3U 0115–73, 0900–40, 1223–62, 1254–69, 1516–56, 1543–47, 1543–62, 1556–60, 1626–67, 1630–47, 1636–53, 1642–45, 1658–48, 1702–36, 1705–44, 1728–24, 1758–25, 1813–14, 1820–30 and 1837 + 04. Assuming that flickering in the optical counterpart is only seen if the ratio of optical to X-ray flux is sufficiently small the authors are able to eliminate some stars as identifications.

142.043 **Optical photometry of stars near Cen X-3.**
M. J. Penston, P. Murdin.
Monthly Notices Roy. Astron. Soc., Vol. 172, 377 - 384 (1975).

The authors report upper limits to regular optical pulsations in Cen X-3 of 0.006 mag, and to flickering of 0.01 mag. The large reddening of the source is consistent with the field reddening at a distance of 5–10 kpc, but in view of observational selection in such data no firm conclusions about the absolute magnitude of the primary are made.

142.044 **Tidal capture formation of binary systems and X-ray sources in globular clusters.**
A. C. Fabian, J. E. Pringle, M. J. Rees.
Monthly Notices Roy. Astron. Soc., Vol. 172, 15P - 18P (1975).

The authors propose that the X-ray sources associated with globular clusters are very close binary systems. These result from the tidal capture of compact objects by ordinary stars in the core of the cluster.

142.045 **Soft X rays from Sirius.** R. Mewe, J. Heise, E. H. B. M. Gronenschild, A. C. Brinkman, J. Schrijver, A. J. F. Den Boggende.
Nature, Vol. 256, 711 - 712 (1975).

On April 3, 4 and 5, 1975 the star Sirius (α CMa) was observed with the soft X-ray detector aboard the Astronomical Netherlands Satellite (ANS). The observed counting rate and the instrumental sensitivity give for Sirius an X-ray luminosity in the energy band 0.284–0.2 keV: $L_x = (9.1 \pm 1.6) \times 10^{27}$ erg s^{-1}. The observations could be explained by invoking X-ray emission from a hot corona around Sirius A.

142.046 **Anomalous temporal behaviour of Her X-1.**
B. A. Cooke, C. G. Page.
Nature, Vol. 256, 712 - 713 (1975). – Letter.

142.047 **Observations of X rays from near NGC6440.**
T. H. Markert, D. E. Backman, C. R. Canizares, G. W. Clark, A. M. Levine.
Nature, Vol. 257, 32 - 33 (1975). – Letter.

142.048 **Abell 478 and X-ray source 3U0405 + 10.**
M. Elvis, B. A. Cooke, K. A. Pounds, M. J. L. Turner.
Nature, Vol. 257, 33 - 34 (1975).

Three sightings of the faint X-ray source 3U0405+10 with the Ariel V sky survey instrument have produced an improved position, with the error box now containing Abell 478. This distance class 6 Abell cluster is the most distant X-ray source yet detected. It has a (2–20 keV) X-ray luminosity of $(5.4 \pm 1.1) \times 10^{45}$ erg s^{-1}.

142.049 **An extended observation of Cen X-3 with the Ariel-5 sky survey.** K. A. Pounds, B. A. Cooke, M. J. Ricketts, M. J. Turner, M. Elvis.
Monthly Notices Roy. Astron. Soc., Vol. 172, 473 - 481 (1975).

A continuous observation of Cen X-3 over 8 cycles of the binary period was carried out in 1974 November with the Leicester Sky Survey experiment on board the Ariel-5 satellite. During these observations the source was seen to change from a low to a high intensity state and evidence is found for X-ray attenuation by both the primary stellar wind and the accretion wake formed beyond the neutron star.

142.050 **Parameters of Cen X-3 deduced from observations of its accretion wake.** J. C. Jackson.
Monthly Notices Roy. Astron. Soc., Vol. 172, 483 - 492 (1975).

Recent observations of Cen X-3 show a dip in its light curve around phase 0.6. This is interpreted as being due to electron scattering in a dense wake, via which the secondary accretes material from a primary stellar wind. The phase and width of the dip yield estimates of the wind velocity and temperature. The optical depth in the wake is a function of the accretion rate to secondary mass ratio, enabling this quantity to be determined.

142.051 **Ariel 1118–61: a very close binary system?**
J. E. Pringle, R. F. Webbink.
Monthly Notices Roy. Astron. Soc., Vol. 172, 493 - 500 (1975).

The hypothesis is considered that the 6.75-min periodicity observed in the transient X-ray source Ariel 1118–61 is due to orbital motion. Such a binary system would consist of a white dwarf transferring matter on to a compact object. The average mass transfer rate is $\sim 10^{-7} M_\odot$ yr^{-1} and is driven almost entirely by gravitational radiation. This rate of mass transfer obscures the X-ray source for most of the time and a transient X-ray source is seen when the mass transfer temporarily drops below this level.

142.052 **Ariel V sky survey: observations of the accretion wake in Vel X-1.** G. Eadie, A. Peacock, K. A. Pounds, M. Watson, J. C. Jackson, R. Hunt.
Monthly Notices Roy. Astron. Soc., Vol. 172, 35P - 39P (1975)

Extended observations of the binary X-ray source Vel X-1 (3U0900–40) have shown a pattern of absorption dips in the X-ray light curve. These are interpreted in terms of attenuation near to shock fronts in the accretion wake formed by gravitational focusing of the primary stellar wind by the compact X-ray secondary. Evidence is seen for both the primary shock, as previously reported by Ariel V observations of Cen X-3, and for a secondary shock lying on the up-wind side of the compact star.

142.053 **Photometry of LMC X-ray candidates.**

P. R. Warren, J. E. Penfold.
Monthly Notices Roy. Astron. Soc., Vol. 172, 41P - 47P (1975).

Stars in the fields of the LMC X-ray sources have been examined in *UBV*. No statistically significant variations were found for LMC X-1 and LMC X-2 but variability of the LMC X-3 candidate is suggested by the data.

142.054 **On the possible existence of an extended optical structure around the neutron star in the X-ray binary system HZ Her (Her X-1).** A. M. Cherepashchuk.
Astron. Zhurn. Akad. Nauk SSSR, Vol. 52, 883 - 885 (1975). In Russian. English translation in Soviet Astron., Vol. 19, No. 4.

It is possible that in the HZ Her system during all phases of the 35-day cycle exists an extended optical accreting structure. Its radius is about 3–4 $R_\odot$ and its optical luminosity is $\leqslant 10^{35}$ erg/sec.

142.055 **Orbital elements and an analysis of models for HDE 226868 = Cygnus X-1.** C. T. Bolton.
Astrophys. Journ., Vol. 200, 269 - 277 (1975).

Radial velocities from 21 new high-dispersion spectrograms of HDE 226868 are presented. These are combined with previously published data to calculate a "definitive" set of orbital elements for the binary system. In particular, archival data are used to obtain a precise period. The ellipsoidal light curve is analyzed using both a Roche model and an ellipsoidal model. A qualitative model for the mass transfer is proposed that explains the intensity and velocity variations of the optical emission lines and the variations in the X-ray intensity – including the low-energy X-ray absorption events sometimes seen near superior conjunction of the secondary. Tests of this model are also proposed. Models in which the secondary is a black hole are consistent with all available observations.

142.056 **The galactic X-ray sources.**
H. Gursky, E. Schreier.
Astrophys. Space Sci. Library, Vol. 48, (see 012.014), 175 - 220 (1975).

142.057 **Optical observations of binary X-ray sources.**
P. E. Boynton.
Astrophys. Space Sci. Library, Vol. 48, (see 012.014), 221 - 234 (1975).

142.058 **Absorption anomalies in compact X-ray sources.**
G. P. Garmire, C. E. Ryter.
Astrophys. Letters, Vol. 16, 121 - 123 (1975).

It is suggested that apparently deficient interstellar X-ray absorption on the line of sight to compact X-ray sources must be due to a soft X-ray emission excess. Suitable multiple emission regions are predicted to exist in accretion discs (Thorne and Price, 1975). Examples of this feature are found in Cyg X-1 and Sco X-1.

142.059 **A soft X-ray survey of the Scutum-Serpens-Aquila region.**
R. W. Hill, G. A. Burginyon, F. D. Seward.
Astrophys. Journ., Vol. 200, 709 - 714 (1975).

Data from a 1973 June 23 rocket flight are presented and discussed. Soft X-ray spectra of the sources GX 17 + 2 and Ser X-1 together with the best-fit model spectra are given. The authors derive a location of Ser X-1 which confirms that listed in the 3U catalog. Upper limits are set to the X-ray flux from eight supernova remnants, and upper limits to their X-ray luminosity are calculated. Spatial structure in the diffuse soft X-ray background was observed. A map of the diffuse background in this region is derived.

142.060 **An interpretation of the radio outbursts of Cygnus X-3.** A. P. Marscher, R. L. Brown.
Astrophys. Journ., Vol. 200, 719 - 726 (1975).

The authors describe the radio bursts from Cyg X-3 by an expanding cloud of relativistic particles and magnetic fields that is uniformly mixed with an ionized thermal gas. The agreement obtained with this model between the observed and calculated radio bursts is very satisfactory.

142.061 **On the nature of the globular cluster X-ray sources.**
J. Silk, J. Arons.
Astrophys. Journ., (*Letters*), Vol. 200, L131 - L135 (1975).

It is suggested that the globular cluster X-ray sources can be interpreted in terms of mass shed by evolving post-main-sequence stars in the central regions of the globular cluster and accreting onto a massive ($\sim 100-1000\ M_\odot$) central black hole. Tentative indications that support this hypothesis include the high central escape velocities and the distribution of metallicities among the four globular clusters containing luminous X-ray sources. The authors argue that these X-ray sources cannot be explained as a consequence of the close capture of a nuclear burning star by a compact stellar remnant.

142.062 **The effects of a departure from co-rotation in X-ray binaries.** D. T. Wickramasinghe.
Monthly Notices Roy. Astron. Soc., Vol. 173, 21 - 31 (1975).

The effects of a departure from co-rotation on the light curves of X-ray binaries are computed. The results are used to discuss the observations of 3U 0900–40.

142.063 **The temporal behaviour of Taurus X-1 (the Crab nebula).** P. J. N. Davison.
Monthly Notices Roy. Astron. Soc., Vol. 173, 77 - 83 (1975).

Copernicus data on Taurus X-1 and the Crab pulsar extending over a 2 ½-yr period indicate that under normal conditions the source has a flux that is constant to within 2.5 per cent at the 90 per cent confidence level. The pulsed/total flux ratio also shows no significant changes during the same time.

142.064 **Copernicus: X-ray observations of several radio supernova remnants.** J. C. Zarnecki, J. P. Stark, P. A. Charles, J. L. Culhane.
Monthly Notices Roy. Astron. Soc., Vol. 173, 103 - 109 (1975).

The MSSL X-ray detectors on Copernicus have observed five radio sources, 3C392, DR4, MSH15-52A and Downes Nos 83 and 84 in the energy range 0.5–7.5 keV. The first three objects are thought to be supernova remnants. The authors present here upper limits from their observations and in the case of 3C392 a possible positive detection of X-ray emission. Some comments on the variability of the X-ray source 3U 1901 + 03 are also given.

142.065 **Observation of a line feature in the X-ray spectrum of Cygnus X-3.**
P. Sanford, K. O. Mason, J. Ives.
Monthly Notices Roy. Astron. Soc., Vol. 173, 9P - 14P (1975).

A narrow emission feature at ≈6.5 keV has been detected in the spectrum of Cygnus X-3, using instrumentation on the Ariel-5 satellite. The average strength of the feature is 0.027 ± 0.004 photon $cm^{-2}\ s^{-1}$. The strength varies with the phase of the 4.8-hr binary cycle, but its the same as that seen by Serlemitsos et al. when the overall source intensity was much lower.

142.066 **Differential *uvby* photometry of Cygnus X-1.**
H. H. Lanning.
Monthly Notices Roy. Astron. Soc., Vol. 173, 15P - 19P (1975).

Strömgren *uvby* differential photometry of the X-ray source Cygnus X-1 is presented. Variations in the light curve as large as $0^m.05$ at primary and $0^m.03$ at secondary are noted.

142.067 **Polarimetric measurements at Hα and Hβ of X Per.**
D. Clarke, I. S. McLean.
Monthly Notices Roy. Astron. Soc., Vol. 173, 21P - 24P (1975).

Polarimetric measurements have been made of the star X Per at Hα and Hβ. No reduction of polarization was observed at these emission lines on a scale which has been seen in other early-type emission line stars. There is conflicting evidence on the origin of the polarization associated with this star, whether it is intrinsic to the star or has an interstellar origin.

142.068 **Copernicus observations of Circinus X-1.**
P. J. N. Davison, I. R. Tuohy.
Monthly Notices Roy. Astron. Soc., Vol. 173, 33P - 35P (1975).

The authors present Copernicus X-ray observations of Circinus X-1 made during August 1973 and April 1974. Their results confirm that the reported 12.3-day periodicity is spurious, and in addition, the data show a clear correlation between the low energy X-ray intensity and spectral slope.

142.069 **Galactic super halos, globular clusters, and the nature of the high latitude X-ray sources.**
T. Matilsky.
Bull. American Astron. Soc., Vol. 7, 396 (1975). – Abstr. AAS.

142.070 **X-ray modulation by stellar winds in binary X-ray sources.** S. P. Hatchett, R. McCray.
Bull. American Astron. Soc., Vol. 7, 415 (1975). – Abstr. AAS.

142.071 **X-ray sky fluctuations measured by UHURU.**
D. Schwartz, S. Murray, H. Gursky.
Bull. American Astron. Soc., Vol. 7, 415 (1975). – Abstr. AAS.

142.072 **The soft X-ray enhancement in Eridanus.**
K. S. Long, P. C. Agrawal, G. P. Garmire.
Bull. American Astron. Soc., Vol. 7, 416 (1975). – Abstr. AAS.

142.073 **On the ultrasoft X-ray background.** A. Levine, S. Rappaport, R. Doxsey, G. Jernigan.
Bull. American Astron. Soc., Vol. 7, 416 (1975). – Abstr. AAS.

142.074 **Spatial extent and spectrum of Gemini-Monoceros X-ray source.**
J. R. Patterson, K. S. Long, W. E. Moore, G. P. Garmire.
Bull. American Astron. Soc., Vol. 7, 416 (1975). – Abstr. AAS.

142.075 **Precise positions of galactic X-ray sources.**
R. Doxsey, H. Bradt, J. Buff, J. Delvaille, G. Jernigan, A. Levine, W. Mayer, S. Rappaport, H. Schnopper.
Bull. American Astron. Soc., Vol. 7, 416 (1975). – Abstr. AAS.

142.076 **Precise positions of several high galactic latitude X-ray sources.**
A. Epstein, G. Clark, J. P. Delvaille, R. Doxsey, G. Jernigan, W. Mayer, F. Primini, H. W. Schnopper.
Bull. American Astron. Soc., Vol. 7, 416 (1975). – Abstr. AAS.

142.077 **Intensity and spectral variations of Cyg X-1 observed from balloons.**
J. L. Matteson, R. F. Mushotzky, W. S. Paciesas.
Bull. American Astron. Soc., Vol. 7, 429 (1975). – Abstr. AAS.

142.078 **A twenty-fold increase in the intensity of Aql X-1.**
J. Buff, H. Bradt, G. Clark, G. Jernigan, W. Lewin, T. Matilsky, W. Mayer, J. McClintock, F. Primini.
Bull. American Astron. Soc., Vol. 7, 429 (1975). – Abstr. AAS.

142.079 **Intense soft X-ray flux from Her X-1.**
R. C. Catura, L. W. Acton.
Bull. American Astron. Soc., Vol. 7, 430 (1975). – Abstr. AAS.

142.080 **High-energy X-ray observations of the transient source A0535+26 from a balloon borne telescope.**
G. Ricker, A. Scheepmaker, J. Ballintine, J. Doty, G. Kriss, S. Ryckman, W. H. G. Lewin.
Bull. American Astron. Soc., Vol. 7, 431 (1975). – Abstr. AAS.

142.081 **The transient periodic X-ray source in Taurus, A0535+26.** J. McClintock, H. Bradt, J. Buff, G. Clark, R. Doxsey, D. Hearn, G. Jernigan, W. Lewin, F. Li, T. Matilsky, W. Mayer, F. Primini, S. Rappaport, J. Richardson, H. Schnopper.
Bull. American Astron. Soc., Vol. 7, 431 (1975). – Abstr. AAS.

142.082 **Observations of variable X-ray sources in Centaurus.**
W. A. Baity, W. A. Wheaton, L. E. Peterson.
Bull. American Astron. Soc., Vol. 7, 431 (1975). – Abstr. AAS.

142.083 **X-ray observations of Cen A.**
J. P. Delvaille, H. Bradt, J. Buff, A. Epstein, W. Mayer, J. McClintock, S. Rappaport, H. W. Schnopper.
Bull. American Astron. Soc., Vol. 7, 443 (1975). – Abstr. AAS.

142.084 **Precise location of the X-ray source 3U1820-30 associated with the globular cluster NGC 6624.**
G. Jernigan, C. Canizares, G. Clark, R. Doxsey, A. Epstein, T. Matilsky, W. Mayer, J. McClintock.
Bull. American Astron. Soc., Vol. 7, 443 (1975). – Abstr. AAS.

142.085 **Search for Si XIV and Si XIII line emission from cosmic X-ray sources.**
H. W. Schnopper, J. P. Delvaille, A. Epstein, H. Gursky, K. Kalata, A. R. Sohval, D. R. Parsignault.
Bull. American Astron. Soc., Vol. 7, 460 (1975). – Abstr. AAS.

142.086 **Analysis of a Scorpius X-1 X-ray spectrum obtained with cooled silicon (Li) detectors.**
J. G. Laros, S. Singer.
Bull. American Astron. Soc., Vol. 7, 460 - 461 (1975). – Abstr. AAS.

142.087 **UCSD OSO-7 observations of four extragalactic X-ray sources.**
R. F. Mushotzky, W. A. Baity, W. A. Wheaton, L. E. Peterson.
Bull. American Astron. Soc., Vol. 7, 461 (1975). – Abstr. AAS.

142.088 **Observations of extragalactic X-ray sources with ANS.**
J. E. Grindlay, E. J. Schreier, H. W. Schnopper, H. Gursky, D. Parsignault.
Bull. American Astron. Soc., Vol. 7, 461 (1975). – Abstr. AAS.

142.089 **Discovery of a soft (0.1 - 0.28 keV) X-ray source in Coma.**
J. A. Richardson, C. Canizares, G. Clark, D. Hearn, W. Lewin, W. Mayer, J. McClintock, F. Primini, S. Rappaport.
Bull. American Astron. Soc., Vol. 7, 461 (1975). – Abstr. AAS.

142.090 **A triaxial white dwarf model for Cen X-3.**
R. H. Durisen.
Bull. American Astron. Soc., Vol. 7, 477 (1975). – Abstr. AAS.

142.091 **Candidate for LMC X-5.** P. B. Byrne.
Nature, Vol. 257, 110 (1975).

A new X-ray source in the Large Magellanic Cloud, LMC X-5, has been reported. The author draws attention to the close proximity of the radio sources MC32 and MC33 to the best X-ray position.

142.092 **Precession in transient X-ray sources.**

K. Brecher.
Nature, Vol. 257, 203 (1975).

The origin of both the relatively long period and the transient nature of the recently discovered X-ray sources Ariel 1118–61 and Ariel 0535+26, with pulse periods of about 405 and 104 s, respectively, and decay times of days, is unclear. The author reiterates here an earlier suggestion that free precession may be important in such galactic X-ray sources. Such precession, which provides a plausible explanation for the 35-d quasi-sinusoidal period and intensity variation of Her X-1 (which has a 1.25-s pulse period) could directly account for the shape of the light curves of the transient X-ray sources.

142.093 **X-ray outburst from the direction of the galactic centre.** C. J. Eyles, G. K. Skinner, A. P. Willmore, F. D. Rosenberg.
Nature, Vol. 257, 291 - 293 (1975).

When the Uhuru satellite examined the X-ray emission from the region of the galactic centre in 1971 an extended region of emission centred on the radio source Sgr A and about 2° in the extent was observed. The authors report details of a transient point X-ray source, A1742–28, which has been observed at the centre of this region by the Ariel V rotation modulation collimator (RMC) experiment.

142.094 **Radiative transfer in a strong magnetic field and accreting X-ray pulsars.**
M. M. Basko, R. A. Sunyaev (*Syunyaev*).
Astron. Astrophys., Vol. 42, 311 - 321 (1975).

When matter accretes onto a neutron star endowed with a strong magnetic field $H \gtrsim 10^{12}$-10^{13} Gauss, its kinetic energy is liberated in the vicinity of the magnetic poles. The accreting protons with energy ~ 140 MeV are stopped by nuclear encounters in a layer as deep as 50 g cm^{-2}. As a result of high anisotropy in the scattering cross-section of photons by electrons in a magnetized plasma, the radiation emerging from the surface at frequencies $\nu \ll \nu_H$ has a 'pencil' beam pattern and is highly polarized. The problem of radiative transfer in an atmosphere with a strong magnetic field is solved. The mechanism being discussed can account also for the beamed and polarized radiation of white dwarfs having a strong magnetic field. In the framework of the theoretical model adopted, an upper limit is obtained for the luminosity of X-ray pulsars, which can greatly exceed the Eddington critical value per unit surface area. The luminosity of X-ray pulsars is limited by distortion of the beam pattern due to the scattering of emitted X-rays by the electrons of the infalling gas.

142.095 **Results of a southern-hemisphere search for gamma-ray sources at $E_\gamma \geq 3 \times 10^{11}$ eV.**
J. E. Grindlay, H. F. Helmken, R. Hanbury Brown, J. Davis, L. R. Allen.
Astrophys. Journ., Vol. 201, 82 - 89 (1975).

The optical intensity interferometer operated by Sydney University at Narrabri, NSW, Australia, was converted in 1972 April to a sensitive detector of atmospheric Cerenkov light from extensive air showers (EAS) with $E_0 > 2 \times 10^{11}$ eV. Possible ($\gtrsim 3\ \sigma$) sources were found in 1972 in the directions of Cen A (NGC 5128), the Vela pulsar, and MP 1451, out of 11 candidate objects studied. The objects included pulsars, unusual X-ray sources, active galaxies, a supernova, and lower energy (~100 MeV) γ-ray sources such as the galactic center. Confirmation of the 1972 results was attempted in an improved series of observations in 1973 April–June and 1974 March–April. The Cen A result was confirmed. Upper limits are given here for the other objects, together with evidence for a variable pulsed flux from the Vela pulsar. Implications of the results are discussed.

142.096 **Molecular hydrogen in the Galaxy and galactic gamma rays.**
F. W. Stecker, P. M. Solomon, N. Z. Scoville, C. E. Ryter.
Astrophys. Journ., Vol. 201, 90 - 97 (1975).

Recent surveys of 2.6-cm CO emission and 100-MeV γ-radiation in the galactic plane reveal a striking correlation suggesting that both emissions may be primarily proportional to the line-of-sight column density of H_2 in the inner Galaxy. Both the γ-ray and CO data suggest a prominent ring or arm consisting of cool clouds of H_2 at a galactocentric distance of ~5 kpc with a mean total hydrogen density equivalent to ~5 atoms cm^{-3}. The deduced cosmic-ray distribution inferred from the calculations is similar to that of galactic supernova remnants, suggesting a galactic origin for most cosmic rays. A detailed calculation of the γ-ray flux distribution in the 0°–180° longitude range using the CO data to obtain the average distribution of molecular clouds in the Galaxy shows that most of the enhancement in the inner Galaxy is due to π^0-decay radiation and the 5-kpc ring plays a major role. Detailed agreement with the γ-ray data is obtained with the additional inclusion of contributions from bremsstrahlung and Compton radiation of secondary electrons and Compton radiation from the intense radiation field near the galactic center.

142.097 **Spectral variability of Cygnus X-3.**
P. J. Serlemitsos, E. A. Boldt, S. S. Holt, R. E. Rothschild, J. L. R. Saba.
Astrophys. Journ., (*Letters*), Vol. 201, L9 - L13 (1975).

Spectra of Cyg X-3 from 1.7 to 40 keV obtained about a year apart using the same rocket payload show large spectral differences. The two observations suggest that while the luminosity of this source remains roughly the same, its spectrum can vary from a featureless blackbody distribution to a flat spectrum which includes strong iron line emission at ~6.7 keV. The flux in the line is 0.018 ±0.004 photons cm^{-2} s^{-1}, corresponding to an equivalent continuum width of 1.2 keV.

142.098 **A search for optical pulsations from X-ray sources near λλ4640–4650 and λ4686.**
C. R. Canizares, J. E. McClintock.
Astrophys. Journ., Vol. 200, 177 - 181 (1975).

The authors have searched for periodic optical pulsations from Cyg X-1, Cyg X-2, 3U 1700–37, Her X-1, Sco X-1, Vela X-1, SMC X-1, WRA 977, and from the central region of the Cygnus Loop. Most observations were confined to the spectral features around λλ4640–4650 and λ4686, which are seen in most X-ray source counterparts, are known to be variable, and are likely to be associated with the X-ray emission. No optical pulsations were detected, but upper limits are established and related to the X-ray behavior.

142.099 **Clusters of galaxies within X-ray error boxes in the southern sky.** N. V. Vidal.
Publ. Astron. Soc. Pacific, Vol. 87, 625 - 633 (1975).

A photographic survey of high galactic latitudes and southern X-ray error boxes revealed five possible identifications of X-ray sources with clusters of galaxies. Spectra of galaxies taken in these clusters yielded distances and X-ray luminosities.

142.100 **Astronomy from space: new class of X-ray sources found.** W. D. Metz.
Science, Vol. 189, 1073 - 1074 (1975). – Research news.

142.101 **Discovery of powerful transient X-ray source A0620–00 with Ariel V Sky Survey Experiment.**
M. Elvis, C. G. Page, K. A. Pounds, M. J. Ricketts, M. J. L. Turner.
Nature, Vol. 257, 656 - 657 (1975).

On August 3 a faint new X-ray source was discovered near the boundary of the Monoceros and Orion constellations and 6° from the galactic plane, during routine monitoring of the Milky Way with the Ariel V Sky Survey Experiment. During the following week, the source-increased rapidly in intensity to become the brightest in the sky (at 2–18 keV).

142.102 **The X-ray spectrum of A0620–00.**
M. J. Ricketts, K. A. Pounds, M. J. L. Turner.
Nature, Vol. 257, 657 - 658 (1975). – Letter.

142.103 **Observations of A0620–00 at 962 and 151 MHz.**
R. J. Davis, M. R. Edwards, I. Morison, R. E. Spencer.
Nature, Vol. 257, 659 (1975). – Letter.

142.104 **The optical counterpart of A0620–00 before its eruption.** M. J. Ward, M. V. Penston, C. A. Murray, E. D. Clements.
Nature, Vol. 257, 659 - 660 (1975). – Letter.

142.105 **Galactic superhaloes, globular clusters and high latitude X-ray sources.** T. A. Matilsky.
Nature, Vol. 257, 660 - 661 (1975).
The author shows here that the observational features of unidentified, high latitude, X-ray sources are consistent with the suggestion that they are a population of remote, optically undetectable, globular clusters.

142.106 **X-ray sources and the pulsational properties of high luminosity degenerate stars.** E. M. Sion, S. C. Vila.
19th Colloquium International Astrophys. Liège 1974, (see 012.016), p. 227 - 231 (1975).
Models of high luminosity degenerate stars have been calculated in which nuclear burning at the base of the envelope provides the total luminosity of the model. The pulsation periods, radial eigenfunctions, pulsational properties and stability have been calculated for the fundamental and first harmonic modes. The applicability of the results to Cen X-3 and Her X-1 as accreting, pulsating white dwarf binary components is discussed.

142.107 **On the formation of X-ray binaries.**
C. De Loore, J. P. De Greve.
19th Colloquium International Astrophys. Liège 1974, (see 012.016), p. 399 - 405 (1975).
It is assumed that massive X-ray binaries such as Cen X-3, 2U 1700-37, Cyg X-1, SMC X-1, Vela X-1, with periods between 2 and 9 days, are evolution products of initial binary systems where mass loss from a massive primary occurred after hydrogen exhaustion in the core, followed by a supernova explosion. Computations are performed for a binary system with a 20 $M_\odot$ primary, accompanied by secondaries of 14, 10, 8, 6 and 4 $M_\odot$. It is argued that some of these initial systems might evolve towards final models that satisfy the observed mass ratios and periods for the mentioned X-ray binaries.

142.108 **Analysis of distribution of γ-ray bursts.**
O. F. Prilutskij.
Astron. Tsirk., No. 840, p. 6 - 8 (1974). In Russian.

142.109 **Polarization observations of Sco X-1 = V818 Sco and estimate of the mass of the optical component.**
Yu. N. Gnedin, N. A. Silant'ev.
Astron. Tsirk., No. 851, p. 3 - 5 (1974). In Russian.

142.110 **Preliminary results of simultaneous observations of Sco X-1 in infrared and UBV.**
V. M. Lyutyj, V. I. Moroz, V. I. Pronik, A. A. Liberman. V. I. Shenavrin.
Astron. Tsirk., No. 861, p. 3 - 5 (1975). In Russian.

142.111 **On extended X-ray sources of the metagalaxy.**
V. N. Kuril'chik.
Astron. Tsirk., No. 863, p. 1 - 3 (1975). In Russian.

142.112 **Light-curve analysis of X-ray binaries: a revision.**
J. B. Hutchings.
Astrophys. Journ., Vol. 201, 413 - 414 = Contr. Dominion Astrophys. Obs., *Victoria*, No. 264 (1975).
A correction to the light-curve analyses previously published shows that most systems appear to fill their Roche lobes. The mass estimates are not significantly changed.

142.113 **On the energetics of pulsating X-ray sources.**
P. D. Guthrie, E. Tademaru.
Astrophys. Journ., Vol. 201, 415 - 418 = Contr. Five College Obs., *Amherst*, Mass., No. 200 (1975).
The authors consider nuclear and kinetic energy of accreting matter as possible energy sources for white dwarf models of the pulsating X-ray sources. It is argued that the rate of kinetic energy accretion determines the source luminosity, and the equilibrium velocity amplitude of the pulsation is determined. Inverse Compton cooling of accretion shocks is also considered and it is found that it is generally a small effect.

142.114 **Hercules X-1: a neutron star with a twisted accretion disk?** J. A. Petterson.
Astrophys. Journ., (*Letters*), Vol. 201, L61 - L64 (1975).
A model is proposed for the binary X-ray source Her X-1, in order to explain its 35-day periodicity. It consists of a normal star and a neutron star with a twisted accretion disk. The 35-day clock in the model is provided by a precession of the normal star. The disk is responsible for a periodic obscuration of the pulsed X-rays, causing the on-off behavior. The twisted shape of the disk, which arises naturally from the precession of the normal star, is shown to explain a considerable number of X-ray and optical observations.

142.115 **The spectroscopic binary Scorpius X-1.**
A. P. Cowley, D. Crampton.
Astrophys. Journ., (*Letters*), Vol. 201, L65 - L68 (1975).
Recent spectroscopic observations of Sco X-1 show conclusively that orbital velocity variations of $P = 0^d.787 \pm 0^d.006$ and a range of ~ 120 km s^{-1} are present in both the He II and H emission lines. The most probable masses are less than 2 $M_\odot$ for each star. The authors suggest that the non-degenerate companion is somewhat evolved from the main sequence.

142.116 **An interpretation of the general character of the X-ray emission from compact objects in close binary stars.** L. Fang.
Acta Astron. Sinica, Vol. 16, 19 - 26 (1975). In Chinese.
The emission mechanism from X-ray sources in close binary stars is studied. It is suggested that pair production may play an important role in the X-ray emission region which is formed by the accretion of a compact object (a neutron star or a black hole). The present model accounts qualitatively for the following observed general properties in X-ray binaries: 1.) The X-ray luminosity lies within the range 10^{36} –10^{38} erg/sec. 2.) The X-ray energy spectrum can in the majority of cases be represented either as power law or in the thermal bremsstrahlung form. 3.) The temperature of the thermal bremsstrahlung spectrum is of the order of 10^8 K. 4.) The form of the energy spectrum does not change appreciably when the luminosity fluctuates.

142.117 **A comparison of X-ray sources in globular clusters and in the galactic bulge.** C. R. Canizares.
Astrophys. Journ., Vol. 201, 589 - 592 (1975).
The author suggests the hypothesis that the X-ray sources

of the galactic bulge and those in globular clusters are of the same type. This is supported by the similarity of the respective stellar populations and the observed incidence of luminous sources. Evolutionary considerations suggest that it is the high stellar density that favors X-ray source formation in those regions. If the correspondence is correct, there must be ~100 additional weak sources in the bulge which have so far eluded detection because of source confusion.

142.118 **Detection of nuclear gamma rays from the galactic center region.** R. C. Haymes, G. D. Walraven, C. A. Meegan, R. D. Hall, F. T. Djuth, D. H. Shelton. Astrophys. Journ., Vol. 201, 593 - 602 (1975).

On 1974 April 2 a balloon-borne measurement was undertaken of γ-rays in the 0.02–12.27 MeV energy interval from the galactic center region. A power-law continuum was observed and several spectral lines were detected for the first time, at energies close to the theoretically expected nuclear γ-rays. The nuclear γ-ray data are consistent with several models of the longitudinal distribution of low-energy cosmic rays in the Galaxy. In these models, the nuclear γ-radiation arises from the interactions of low-energy cosmic rays with the interstellar medium.

142.119 **Why are no radio pulses associated with the bursts of celestial gamma-rays?**
G. Cavallo, J. V. Jelley.
Astrophys. Journ., (*Letters*), Vol. 201, L113 - L115 (1975).

The authors discuss various interpretations of the apparent absence of radio pulses at 151 MHz in time coincidence with the Vela satellite γ-ray bursts. While the negative results are consistent with either a galactic or an extragalactic origin for the γ-ray bursts, the constraints are in most cases less stringent for models based on a local or galactic origin. The absence of VHF radio pulses suggests, however, that the γ-ray bursts do not arise from stellar superflares, on the restrictive assumption that they are directly scaled-up events of solar type.

142.120 **The light curve of a transient X-ray source.** L. J. Kaluzienski, S. S. Holt, E. A. Boldt, P. J. Serlemitsos, G. Eadie, K. A. Pounds, M. J. Ricketts, M. Watson.
Astrophys. Journ., (*Letters*), Vol. 201, L121 - L124 (1975).

The Ariel-5 satellite has monitored the X-ray light curve of A1524–62 almost continuously from 40 days prior to maximum light until its disappearance below the effective experimental sensitivity. Although similar to previously reported transient sources with a decay time constant of ~2 months, the source exhibited an extended, variable preflare "on"-state of ~1 month at a level of $\gtrsim$ 0.1 maximum light.

142.121 **The location and intensity of the X-ray source Centaurus A observed by the Astronomical Netherlands Satellite.** J. E. Grindlay, H. Schnopper, E. J. Schreier, H. Gursky, D. R. Parsignault.
Astrophys. Journ., (*Letters*), Vol. 201, L133 - L136 (1975).

The HXX X-ray detectors on ANS have observed the source in NGC 5128, Cen A, with 10′ collimators and high time resolution during 1975 January 26–29. A source position is obtained that is consistent with the nucleus of NGC 5128. No rapid source variability was detected and limits on pulsations are given. The source variability (1969–1975) is related to models for the nucleus of NGC 5128.

142.122 **Possible identification of a high-latitude X-ray source with a QSO by the Astronomical Netherlands Satellite.**
E. Schreier, H. Schnopper, H. Gursky, D. Parsignault.
Astrophys. Journ., (*Letters*), Vol. 201, L137 - L139 (1975).

A recently discovered QSO, NAB 0137–01, located in proximity to the high galactic latitude X-ray source 3U 0138–01 was observed using the hard X-ray experiment on the Astronomical Netherlands Satellite. A flux of 0.33 ± 0.14 counts s^{-1} (1.3–7.0 keV) was detected, consistent with the intensity of 3U 0138–01. The error box for the X-ray source is thus reduced to 0.02 square degrees, greatly increasing the likelihood that the association of the X-ray source with the QSO is not accidental. The intrinsic luminosity of the QSO is about 2×10^{46} ergs s^{-1}, with an X-ray to optical luminosity ratio of 50.

142.123 **On the shape and variability of the X-ray pulses from Her X-1.** G. S. Bisnovatyi-Kogan (*Bisnovatyj-Kogan*), B. V. Komberg, M. Reinhardt.
Astron. Astrophys., Vol. 43, 385 - 388 (1975).

In the first section the authors compile the observed properties of the X-ray pulse profile. Then they discuss the expected radiation pattern and argue that it should be "cup"-shaped. It is shown qualitatively that such a radiation characteristic can explain all the salient features of the pulse profile, if one allows for variations of the angular width of the radiation pattern due to changes in the accretion rate. Some of the relevant parameters are estimated. Several observational predictions are made.

142.124 **Observational results on compact galactic X-ray sources.** R. Giacconi.
Proc. 16th Solvay Conference Phys., (see 012.020), p. 27 - 71, with a discussion, p. 72 (1974).

142.125 **Optical properties of binary X-ray sources.** J. N. Bahcall, N. A. Bahcall.
Proc. 16th Solvay Conference Phys., (see 012.020), p. 73 - 94, with a discussion, p. 95 - 96 (1974).

142.126 **The physics of binary X-ray sources.** M. J. Rees. Proc. 16th Solvay Conference Phys., (see 012.020), p. 97 - 114, with a discussion, p. 115 - 117 (1974).

142.127 **The evolutionary origin of massive X-ray binaries and the total number of massive stars with a collapsed close companion in the Galaxy.**
E. P. J. van den Heuvel.
Proc. 16th Solvay Conference Phys., (see 012.020), p. 119 - 130, with a discussion, p. 131 - 133 (1974).

142.128 **Accretion onto magnetic neutron stars: comments on the position of the Alfvén surface.**
F. K. Lamb, C. J. Pethick.
Proc. 16th Solvay Conference Phys., (see 012.020), p. 135 - 141, with a communication by A. G. W. Cameron, p. 143 - 144, and a discussion of that paper, p. 145 - 146 (1974).

142.129 **Observing neutron stars: information on stellar structure from pulsars and compact X-ray sources.**
D. Pines.
Proc. 16th Solvay Conference Phys., (see 012.020), p. 147 - 173, with a discussion, p. 174 - 175 (1974).

142.130 **Neutron stars, black holes and binary X-ray sources.** R. Ruffini.
Proc. 16th Solvay Conference Phys., (see 012.020), p. 349 - 424, with a short communication by M. J. Rees, p. 425 - 426, and a discussion of that paper, p. 427 (1974). – Review paper.

142.131 **Observations of the 1975 May X-ray outburst of Cygnus X-1.**
C. J. Eyles, G. K. Skinner, A. P. Willmore, F. D. Rosenberg. Monthly Notices Roy. Astron. Soc., Vol. 173, 63P - 66P (1975).

Measurements of the intensity and spectrum of Cygnus

X-1 were made with the Ariel V rotation-modulation-collimator experiment over a period of one month during the recent outburst. The data have been examined for variations with the 5.6-day binary period.

142.132 **On the long-term behavior of the HZ Herculis/ Hercules X-1 system.** L. Anderson.
Astrophys. Journ., Vol. 202, 232 - 237 (1975).
In accretion models for binary X-ray sources based on Roche-lobe overflow, mass transfer from an otherwise normal, expanding star to a collapsed companion is responsible for the production of X-rays. For a range of time-averaged rates of mass transfer, the instantaneous rate of transfer depends on the X-ray luminosity. Thus the rate may be subject to a feedback instability. The author proposes that such an instability is responsible for the long-period on-off behavior of HZ Herculis.

142.133 **Intense soft X-ray flux from Hercules X-1.** R. C. Catura, L. W. Acton.
Astrophys. Journ., (*Letters*), Vol. 202, L5 - L7 (1975).
An intense flux of soft X-rays extending up to at least 1 keV has been observed from Her X-1. If the soft X-ray intensity is corrected for interstellar absorption, the luminosity in the 0.16–0.28 keV band is comparable to that from 2 to 10 keV. This confirms the conclusion of Shulman et al. but extends the detection of this flux to higher energy, a different binary phase, and to a time 19 periods later in the 35-day cycle. These observations suggest that this soft emission may be a stable feature in the spectrum of Her X-1 during its X-ray ON state. It appears difficult to account for the high soft X-ray luminosity by current simple models of the X-ray source.

142.134 **A sensitive search for soft X-rays from Arcturus and Procyon.** R. Cruddace, S. Bowyer, R. Malina, B. Margon, M. Lampton.
Astrophys. Journ., (*Letters*), Vol. 202, L9 - L12 (1975).
Sensitive soft X-ray observations have been obtained of α Boo and α CMi using rocket-borne large area proportional counters. No flux was detected from either star. If these stars are typical, then giants and subgiants cannot be responsible for a significant fraction of the soft component of the diffuse X-ray background.

142.135 **Optical studies of UHURU sources. XIII. A photometric analysis of X Persei (= 3U 0352 + 30?).**
E. W. Gottlieb, E. L. Wright, W. Liller.
Astrophys. Journ., (*Letters*), Vol. 202, L13 - L14 (1975).
2677 magnitudes of the hot massive variable star X Persei, the prime optical candidate for the X-ray source 3U 0352 + 30, have been obtained from blue plates in the Harvard collection taken between 1894.9 and 1975.1. The authors have analyzed these data for periods from 0.5 to 180 days and find no periodic component of brightness with full sinusoidal amplitude ≥ 0.05 mag.

142.136 **Measurement of the spatial structure of the X-ray source in the Crab nebula. I. Observation of the 1974 November 3 lunar occultation.**
R. S. Wolff, H. L. Kestenbaum, W. Ku, R. Novick.
Astrophys. Journ., (*Letters*), Vol. 202, L15 - L19 (1975).
The 1974 November 3 (UT) lunar occultation of the Crab nebula has been observed in the 1.5–17.5 keV X-ray range, at a position angle of 255°. The western portion of the nebula and the pulsar NP 0532 were seen in reappearance. Analysis of the data reveals that the western extent of the X-ray volume is approximately 48″ west of the pulsar. The centroid of the nonpulsed X-ray flux for the portion of the nebula observed is located about 20″ west of the pulsar, and the FWHM of the best-fitting Gaussian distribution is 42″ ± 2″ in the observed energy range.

142.137 **Measurement of the spatial structure of the X-ray source in the Crab nebula. II. Observation of the 1974 December 28 lunar occultation.**
H. L. Kestenbaum, W. Ku, R. Novick, R. S. Wolff.
Astrophys. Journ., (*Letters*), Vol. 202, L21 - L24 (1975).
The 1974 December 28 (UT) lunar occultation of the Crab nebula has been observed in the 0.6–23 keV X-ray range at a position angle of 300°. Reappearance of the region from 62″ west of the pulsar NP 0532 to 62″ east was observed. Analysis of the nonpulsed portion of the data reveals that the X-ray centroid is 12.″8 ± 0.″3 west of the pulsar, averaged over the 0.6–23 keV range. The total width of the X-ray distribution has a mean value of 60.″5 ± 1.″5 and decreases with increasing photon energy.

142.138 **X-ray stars and quasars: variables that need watching.** W. Liller.
Journ. American Ass. Variable Star Observers, Vol. 4, 2 - 6 (1975).

142.139 **Soft X-ray search of centre of Cygnus Loop.** W. A. Snyder, R. C. Henry, P. A. Charles, J. L. Culhane, P. W. Sanford, R. Bleach, J. Drake.
Nature, Vol. 258, 214 - 215 (1975).
The Mullard Space Science Laboratory equipment on the Copernicus satellite has been used to search, with negative results, for evidence of a compact object in the centre of the Cygnus Loop supernova remnant.

142.140 **The formation of globular cluster X-ray sources through neutron star – giant collisions.**
W. Sutantyo.
Astron. Astrophys., Vol. 44, 227 - 230 (1975).
The possibility is examined that the globular cluster X-ray sources are binaries with a compact (neutron star or black hole) component which were formed by direct collisions between compact stars and giant branch stars in the clusters. The anticipated number of $\sim 2 \times 10^4$ compact stars in the globular cluster M3 produces a few collisions with giant branch stars in 10^{12} yr. In the densest globular clusters this frequency can be a few collisions per 10^{10} yr. Such a frequency is sufficient to explain the observed frequency of occurrence of globular cluster X-ray sources.

142.141 **Discrete X- and γ-ray sources.** Yu. N. Gnedin.
VI Leningr. mezhdunar. seminar, 1974. Uskorenie chastits i yader. reaktsii v kosmose. Leningrad, 1974, p. 341, with a discussion, p. 342. In Russian. – Abstr. in Referativ. Zhurn. 51. Astron., 8.51.671 (1975).

142.142 **Radiative transfer in a strong magnetic field and accreting X-ray pulsars.**
M. M. Basko, R. A. Syunyaev.
AN SSSR. In-t kosmich. issled. Pr-208. Moskva, 1974, 40 pp. In Russian. – Abstr. in Referativ. Zhurn. 51. Astron., 8.51. 672 (1975).

142.143 **Some properties of X-ray binaries.** J. N. Bahcall.
VI Leningr. mezhdunar. seminar, 1974. Uskorenie chastits i yader. reaktsii v kosmose. Leningrad, 1974, p. 343 - 354. – Abstr. in Referativ. Zhurn. 51. Astron., 8.51.676 (1975).

142.144 **Variability of the form of X-ray pulses of the X-ray pulsar Her X-1.**
M. I. Rejnkhardt, G. S. Bisnovatyj-Kogan, B. V. Komberg.
AN SSSR. In-t kosmich. issled. Pr-216. Moskva, 12 pp. In Russian. – Abstr. in Referativ. Zhurn. 51. Astron., 8.51. 682 (1975).

142.145 **On the possibility of determination of the direction**

of a cosmic γ-ray burst from measurement of its intensity. E. M. Kruglov, G. M. Gorodinskij, V. L. Bokov. VI Leningr. mezhdunar. seminar, 1974. Uskorenie chastits i yader. reaktsii v kosmose. Leningrad, 1974, p. 395 - 403. In Russian. – Abstr. in Referativ. Zhurn. 51. Astron., 8.51. 697 (1975).

142.146 **On the possible annihilation origin of gamma-ray bursts and flare stars.** P. I. Fomin.
VI Leningr. mezhdunar. seminar, 1974. Uskorenie chastits i yader. reaktsii v kosmose. Leningrad, 1974, p. 385 - 392. In Russian. – Abstr. in Referativ. Zhurn. 51. Astron., 8.51. 699 (1975).

142.147 **Improved declination for transient X-ray source A0620–00.** J. Bieging, D. Downes.
Nature, Vol. 258, 307 - 308 (1975). – Letter.

142.148 **X-ray sources and planetary nebulae.** M. Livio, G. Shaviv.
Nature, Vol. 258, 308 - 309 (1975).
Current models of X-ray sources in binary systems generally assume accretion on to a neutron star, a black hole or a white dwarf as the mechanism of X-ray production. The accretion takes place from either a stellar wind or the inner Lagrangian point. Here the authors explore the possible existence of a compact X-ray source in a binary system, where the companion star ejects its envelope to become a planetary nebula.

142.149 **Der Röntgendoppelstern HZ Her.** M. Reinhardt.
Veröff. Astron. Inst. Ruhr Univ. Bochum No. 3, 83 pp. (1975).
An introduction is given to X-ray binaries in general, the history of their discovery and the astrophysical problems involved such as the nature of the compact companions, the radiation mechanisms in the X-ray region, and the origin and evolution of these systems. The observational data on the system Her X-1 are summarized and discussed. A model for Her X-1 is developed. The author shows that his model of a supercorotating activity centre can explain several salient features of the optical behaviour like the wandering humps and the phase-lock of the optical pulsations, as well as the disturbances of the 35 d cycle and the quiet years. Observational tests of the model proposed are discussed.

142.150 **Results of the search for high-energy gamma-ray sources carried out at the Crimean Astrophysical Observatory in the years 1969–1973.**
A. A. Stepanyan, B. M. Vladimirskij, Yu. I. Neshpor, V. P. Fomin.
Izv. Krymskoj Astrofiz. Obs., Vol. 53, p. 29 - 58 (1975). In Russian.
The results of observations of 43 objects of different type are presented. A comparison with published data has been made. It is shown that, besides NP 0532, Cyg X-3, Cas γ-1, some objects may be considered as gamma-ray sources. Some of them are pulsars, others are gamma-ray sources with energy of 10^8 eV. Some of the radio galaxies and quasars may also be considered as gamma-ray sources, but its statistical significance is not reliable.

142.151 **On the variable character of the gamma-ray emission of Cyg X-3 and Cas γ-1 in the energy region 10^{12} eV.**
V. P. Fomin, B. M. Vladimirskij, Yu. I. Neshpor, A. A. Stepanyan.
Izv. Krymskoj Astrofiz. Obs., Vol. 53, p. 59 - 65 (1975). In Russian.
During 1972–1973 observation of Cas γ-1 and Cyg X-3 in the energy region 10^{12} eV was continued by means of ground-based atmospheric Čerenkov technique. The variability of these gamma-ray sources was confirmed. Flux values are presented.

142.152 **The X-ray pulsar 3U 0900–40 with long period – a neutron star with anomalously strong magnetic field.**
N. I. Shakura.
Pis'ma v Astron. Zhurn., Vol. 1, No. 11, p. 23 - 28 (1975). In Russian.
It is shown that a magnetic field $H \approx 10^{14}-10^{15}$ gauss is necessary for the neutron star in the binary X-ray source 3U 0900–40 to explain the observed period of pulses.

142.153 **The light curve of a transient X-ray source.** L. J. Kaluzienski, S. S. Holt, E. A. Boldt, P. J. Serlemitsos, G. Eadie, K. A. Pounds, M. J. Ricketts, M. Watson.
Goddard Space Flight Center, Greenbelt, Maryland, GSFC Document X-661-75-178, 13 pp. (1975).
The Ariel-V satellite has monitored the X-ray light curve of A1524-62 almost continuously from 40 days prior to maximum light until its disappearance below the effective experimental sensitivity. The source exhibited maximum light on ~4 December 1974, at a level of 0.9 the apparent magnitude of the Crab nebula in the energy band 3–6 keV. The source exhibited an extended, variable pre-flare on-state of ~1 month at a level of $\gtrsim$0.1 maximum light.

142.154 **New results from long-term observations of Cyg X-1.** S. S. Holt, E. A. Boldt, P. J. Serlemitsos, L. J. Kaluzienski.
Goddard Space Flight Center, Greenbelt, Maryland, GSFC Document X-661-75-205, 13 pp. (1975).
Observations of Cyg X-1 between October 1974 and July 1975 reveal a persistent 5.6 day modulation of the 3–6 keV X-ray intensity, having a minimum in phase with superior conjunction of the HDE 226868 binary system. The data imply that the X-ray emission from Cyg X-1 arises from the compact member of HDE 226868.

142.155 **The Her X-1 spectrum: a measure of the field near the magnetic poles of a neutron star.**
E. A. Boldt, S. S. Holt, R. E. Rothschild, P. J. Serlemitsos.
Goddard Space Flight Center, Greenbelt, Maryland, GSFC Document X-661-75-233, 9 pp. (1975).
The steep high energy cutoff observed in the spectrum for Her X-1 is analyzed in terms of the severely modified Thomson scattering that dominates the radiative transfer in a highly magnetized plasma near the surface of a neutron star. The data are shown to indicate a field of about 10^{13} G near the magnetic poles and the stopping of accreting matter by nuclear collisions in the neighboring plasma.

142.156 **How the sky looks like in X-rays.** G. Gurskij (*H. Gursky*).
Priroda, No. 10.75, p. 34 - 45 (1975). In Russian.

142.157 **Optical variability of X-ray sources.** V. M. Lyutyj.
Zemlya i Vselennaya, 1975, No. 5, p. 34 - 38. In Russian.

142.158 **Copernicus: The position and nature of Scutum X-1.** P. A. Charles, K. O. Mason, P. J. N. Davison.
Astrophys. Letters, Vol. 16, 145 - 149 (1975).
X-ray observations of Scutum X-1 by the Copernicus satellite have been combined with the published results of a rocket scan along the galactic plane (Hill et al., 1974) to derive the position of the source. The most likely position is R.A. $18^h 35^m 11^s$, Dec $-07° 48' 00''$ (1950), and the authors suggest possible optical counterparts. The cut-off in the spectrum is equivalent to a column density of between 10^{23} and 10^{24} atoms cm^{-2}. Upper limits to the time variability of the source are computed.

142.159 **An upper limit to the low-energy X-ray flux from β Persei.**
T. Kifune, R. S. Wolff, M. C. Weisskopf.
Astrophys. Letters, Vol. 16, 179 - 181 (1975).

A region of the sky including the binary system Algol (β Persei) was observed with a one-dimensional, grazing-incidence X-ray telescope sensitive in the energy range from 0.5 to 4.0 keV. No statistically significant flux was detected above that attributable to the diffuse and cosmic-ray induced backgrounds. This result allows to set an upper limit of 2×10^{-11} ergs (cm^2 sec keV)$^{-1}$ at 1.6 keV to the X-ray emission from this source at the epoch of observation. The authors conclude that the absense of detectable low-energy X-rays is consistent with a thermal radio source, on condition that no flares occurred at the time of observation.

142.160 **Transient X-ray sources.**
D. H. Clark, J. H. Parkinson.
Nature, Vol. 258, 408 - 409 (1975).

In the eccentric-orbit binary model, a compact object is in an eccentric orbit around a more massive, slightly evolved star. If the compact object accretes mass from the stellar wind of the normal star, the accretion rate is time dependent, so that the resulting X-ray emission would show a maximum at periastron (in the absence of absorption extinction) and a minimum at apastron. It is also possible that accretion might arise from direct mass transfer resulting from contraction of the Roche lobe with decreasing primary to secondary distance. The authors show here that the stellar wind model, is adequate to describe the majority of transient X-ray sources.

142.161 **The periodic transient X-ray sources.**
D. T. Wickramasinghe, J. A. J. Whelan.
Nature, Vol. 258, 502 - 503 (1975).

The transient X-ray sources Ariel 1118–61 and Ariel 0535+26 have periods of 405 s and 114 s respectively. Rappaport and McClintock reported the discovery of a 283-s period in the X-ray binary 3 U0900–40 (a 'continuous' source) which is known to have an orbital period $P_{orb} = 8.95$ d. This raises the possibility that the periodic transients may be a particular stage of evolution of the more familiar class of X-ray binary of the 3 U0900–40 variety. The authors propose that the transient phenomenon occurs during the 'sleeping phase' of the binary system when the early type star is still essentially on the main sequence, significantly underfilling its Roche lobe.

142.162 **On the distance to Centaurus X-3.**
R. M. Humphreys, J. Whelan.
Observatory, Vol. 95, 171 - 173 (1975).

From considerations of spectral type, extinction and galactic kinematics, the distance to Cen X-3 is estimated to be between 6 and 9 kpc.

142.163 **The distribution of X-ray sources in the Galaxy. II.**
H. M. Johnson.
Astrophys. Journ., Vol. 202, 490 - 493 (1975).

The distribution of the strongest X-ray sources near the galactic nucleus is consistent with evolved products of massive stars condensed in the past from such rings of dense interstellar matter as are presently observed to be expanding in the vicinity of the galactic plane from the radii of a few hundred parsecs. Some of the weaker and more dispersed field sources may be aging ex-strong sources. X-ray sources in globular clusters may be derived from stars recently condensed out of the ejecta of evolving red giants in the clusters.

142.164 **Spatial distribution of X-rays in the Crab nebula.**
T. M. Palmieri, F. D. Seward, A. Toor, T. C. Van Flandern.
Astrophys. Journ., Vol. 202, 494 - 501 (1975).

X-ray measurements made during a lunar occulation of the Crab nebula show that the spatial extent of the source is about 100″ in the east-west direction. The center of the X-ray emitting region is about 8″ west of the pulsar. Optical and X-ray intensity distributions are compared, and no optical features other than the pulsar are obvious in the X-ray data.

142.165 **Measurements of low-energy gamma radiation from the Crab nebula and from pulsar NP 0532.**
G. D. Walraven, R. D. Hall, C. A. Meegan, P. L. Coleman, D. H. Shelton, R. C. Haymes.
Astrophys. Journ., Vol. 202, 502 - 510 (1975).

A balloon-altitude measurement of the energy spectra from the Crab nebula and pulsar NP 0532 was conducted in the 0.056–12.110 MeV energy interval. A power law with a photon-number spectral index of 2.13 ± 0.04 fits the nebular spectrum and a prolongation of the X-ray spectrum observed by others. The pulsar spectrum is also well approximated by a power law, with a number spectral index of 2.2 ±0.2. No statistically significant evidence for departures from these power laws was detected.

142.166 **Soft X-ray emission from the direction of the Coma cluster.** S. Hayakawa, Y. Tanaka, K. Yamashita, J. A. M. Bleeker, A. J. M. Deerenberg, P. A. J. de Korte.
Publ. Astron. Soc. Japan, Vol. 27, 615 - 617 (1975).

A soft X-ray source was observed in the direction of the Coma cluster. The flux in the energy range 0.2–0.4 keV was found to change within a time scale shorter than 80 s. The fast transient and the energy spectrum prohibit identification of this source with the Coma cluster. It is suggested that this source belongs to a class of nearby transient soft X-ray sources.

142.167 **On the optical stability of Cygnus X-1.**
A. Frohlich.
Monthly Notices Roy. Astron. Soc., Vol. 173, 821 - 825 (1975).

High resolution optical power spectra of Cygnus X-1 were measured in 1973 October. No evidence for any coherent oscillations was found. The attainable sensitivity closely approached the theoretical predictions for the optical conversion of the known X-ray shot noise.

142.168 **Observations of hard X-ray emission of 3U 1700-37 by the satellite Cosmos 428.**
L. S. Bratolyubova-Tsulukidze, M. I. Kudryavtsev, A. S. Melioranskij, I. A. Savenko, B. Yu. Yushkov.
Pis'ma v Astron. Zhurn., Vol. 1, No. 12, p. 9 - 12 (1975). In Russian.

The satellite Cosmos 428 had registered the source of hard ($\geqslant$30 kev) X-ray emission (named Cosmos HX No. 9) that could be identified with the source 3U 1700-37. Data on the position, time-dependence of intensity and mean spectrum of the X-ray source in the 30–300 keV region are reported.

142.169 **The parameters of the binary X-ray system 3U 0900-40.**
N. G. Bochkarev, E. A. Karitskaya, N. I. Shakura.
Pis'ma v Astron. Zhurn., Vol. 1, No. 12, p. 13 - 17 (1975). In Russian.

By means of comparison of observed and theoretically predicted light curves and durations of X-ray eclipses the permitted parameters for 3U 0900-40 are found.

142.170 **The effect of induced Compton scattering on radio-bursts of the X-ray source Cyg X-3.** M. V. Popov.
Pis'ma v Astron. Zhurn., Vol. 1, No. 12, p. 18 - 22 (1975). In Russian.

Peculiarities of the radiospectrum of Cyg X-3 during the burst of September 2–10, 1972 are explained as possibly due to induced Compton scattering on thermal electrons inside the emitting region. The electron concentration (2×10^6 cm^{-3})

and temperature (10^6 K) are estimated for the adiabatic model of an expanding electron cloud with synchrotron radiation.

142.171 **Search for cosmic γ-ray bursts at $E_\gamma \geq 100$ MeV.**
A. I. Belyaevskij, V. L. Bokov, V. K. Bocharkin, I. F. Bugakov, G. M. Gorodinskij, Yu. G. Derevitskij, E. M. Kruglov, G. A. Pyatigorskij, E. I. Chujkin.
Pis'ma v ZhurnEhTF, Vol. 21, 729 - 733 (1975). In Russian. Abstr. in Referativ. Zhurn. 51. Astron., 11.51.573 (1975).

142.172 ***UBV* photometry of stars in the direction of three LMC X-ray sources.** R. J. Dufour, J. E. Duval.
Publ. Astron. Soc. Pacific, Vol. 87, 769 - 776 (1975).
The results of *UBV* photometry of stars within or near the error boxes of LMC X-1, LMC X-2, and LMC X-3 are presented. The data on stars observed on several nights were analyzed for variability in their *U, B,* and *V* magnitudes. Suggestions are made concerning which stars around LMC X-1 and LMC X-3 are possibly variable and are promising candidates for the optical counterparts to these sources.

142.173 **X rays from supernova remnants.**
P. A. Charles, J. L. Culhane.
Sci. American, Vol. 233, No. 6, p. 38 - 46 (1975).
An exploding supernova hurls a shell of gas into the interstellar medium. In the region where the gas of the shell meets the gas of the medium, X rays are generated that yield information about both.

142.174 **A spherical accretion model of Cyg X-1.**
P. Mészáros.
Nature, Vol. 258, 583 - 584 (1975).
A spherical model may be necessary in Cyg X-1 if the mass transfer in the binary system is not fed by Roche lobe overflow, but rather by a wind. Even if there was enough angular momentum to form a disk, or even a marginal, unstable disk, the thin disk approximation used would not be valid in the hard X-ray emitting regions, since height ~ radius. In such circumstances, spherical geometry seems a closer representation of reality than a disk, and furthermore does not require spin flipping of a specialised nature, extraneous photon sources or uncoupled electron and proton temperatures (although these last two could be accommodated). A spherical model therefore seems to afford a certain economy of assumptions.

142.175 **A dynamical model for Cygnus X-3.**
A. Woodsworth, V. A. Hughes.
Journ. Roy. Astron. Soc. Canada, Vol. 69, 255 (1975). Abstr. CAS.

142.176 **Masses for Vela X-1 and other X-ray binaries.**
Y. Avni, J. N. Bahcall.
Astrophys. Journ., (*Letters*), Vol. 202, L131 - L134 (1975).
The authors report a detailed analysis of the optical light curve and other previously known observations of the X-ray binary Vela X-1 (3U 0900–40). All constraints are consistent only with a projected orbital velocity of the X-ray source of between 200 and 300 km s^{-1}. The allowed ranges of the masses and other parameters are estimated. If the mass of the X-ray source is less than the Chandrasekhar limit for white dwarfs, then the optical star must have an unusually small mass-to-radius ratio for its spectral type. Some new results are also presented from analyses of the observations of 3U 1700–37 and other X-ray binaries.

142.177 **A search for discrete gamma-ray sources of energy greater than 2×10^{12} eV.**
A. A. Stepanian (*Stepanyan*), B. M. Vladimirsky (*Vladimirskij*), Yu. I. Neshpor, V. P. Fomin.
Astrophys. Space Sci., Vol. 38, 267 - 282 (1975).
The results of the observations to search gamma-ray sources with the energy greater than 2×10^{12} eV, which were made in Crimean Astrophysical Observatory during the years 1969–73 are presented. A technique of the detection of the EAS Cerenkov flashes was used. It was shown that two objects, namely, Cyg X-3 and Cas γ-1, may be the sources of high-energy gamma quanta. It is probable that the object with the coordinates $\alpha = 05^h 15^m$, $\delta = +1°$ is the source of gamma-rays as well. An unidentified object Cas γ-1 is variable: gamma-ray flux was observed twice – in September–October 1971 and in December 1972. It is possible that the flux from Cyg X-3 has a period of 4.8 hr.

142.178 **Search for soft X-ray emission from α CMa and γ^2 Vel.** J. R. Patterson, W. E. Moore, G. P. Garmire.
Astron. Astrophys., Vol. 45, 217 - 218 (1975).
A rocket borne soft X-ray experiment with a field of view of $5° \times 10°$ full width at half maximum (FWHM) and an effective area of 180 cm^2 viewed α CMa for 19 s on March 23, 1973. A second detector with area 190 cm^2 and a $0\overset{\circ}{.}4 \times 10°$ (FWHM) aperture viewed γ^2 Vel for 5 s on the same flight. The 3σ upper limits of $F_x \sim 1.0 \times 10^{-10}$ and 1.6×10^{-10} erg $s^{-1} cm^{-2}$ respectively were established for the fluxes at the top of the atmosphere assuming $kT \lesssim 0.3$ keV. Corresponding luminosity upper limits are $L_x \sim 1.0 \times 10^{29}$ erg s^{-1} and 2.1×10^{34} erg s^{-1}.

142.179 **Stellar and solar X-ray polarimetry.** R. Novick.
Space Sci. Rev., Vol. 18, 389 - 408 (1975). – Paper presented at the International Conference on 'X-Rays in Space', held at the University of Calgary, Calgary, Canada, 14 - 21 August 1974.
The scientific motivation for X-ray polarimetry is discussed with particular emphasis on the information that might be obtained on the binary X-ray pulsars in addition to a number of other classes of objects including solar flares. Detailed discussions are given for Thomson-scattering and Bragg-crystal polarimeters with numerical estimates for the sensitivity of various existing and proposed instruments.

142.180 **Hard X-ray bursts of 25 and 26 June, 1971.**
O. P. Babushkina, L. S. Bratolyubova-Tsulukidze, M. I. Kudryavtsev, A. S. Melioranskij, I. A. Savenko, B. Yu. Yushkov.
In-t kosmich. issled. AN SSSR. Pr-220. Moskva, 1975. 10 pp. In Russian.

142.181 **Intense outbursts of cosmic gamma radiation.**
O. F. Prilutskij, I. L. Rozental'.
Uspekhi fiz. nauk, Vol. 116, 517 - 538 (1975). In Russian. Abstr. in Referativ. Zhurn. 51. Astron., 12.51.508; 62. Issled. kosmich. prostranstva, 12.62.120 (1975).

142.182 **A soft X-ray observation of Cyg X-2.**
W. Snyder, A. Davidsen, R. Henry, S. Shulman, G. Fritz, H. Friedman.
Bull. American Astron. Soc., Vol. 7, 505 - 506 (1975). Abstr. AAS.

142.183 **Soft X-ray observations of several sources in Cygnus.**
A. Davidsen, R. Henry, W. Snyder, S. Shulman, G. Fritz, H. Friedman.
Bull. American Astron. Soc., Vol. 7, 506 (1975). – Abstr. AAS.

142.184 **Calculation of the cross section for C IV–H charge-exchange: significance for interstellar X-rays/cosmic ray particles.** R. J. Blint, W. D. Watson, R. B. Christensen.
Bull. American Astron. Soc., Vol. 7, 519 - 520 (1975). Abstr. AAS.

142.185 **X-ray observations from SAS-3.** G. W. Clark.
Bull. American Astron. Soc., Vol. 7, 524 (1975).

Abstr. AAS.

142.186 **SAS-3 observations of GX 301−2 (3U1223−62).** W. H. G. Lewin, J. Hoffman, and the SAS-3 Group.
Bull. American Astron. Soc., Vol. 7, 524 (1975). – Abstr. AAS.

142.187 **SAS-3 observations of an X-ray outburst from Cygnus X-1.**
C. Canizares, H. Bradt, J. Buff, B. Laufer.
Bull. American Astron. Soc., Vol. 7, 524 (1975). – Abstr. AAS.

142.188 **The X-ray light curve and spectrum of A0620−00 using SAS-3.** T. Matilsky, H. Bradt, J. Buff, D. Hearn, J. Richardson.
Bull. American Astron. Soc., Vol. 7, 524 - 525 (1975). Abstr. AAS.

142.189 **The 3U0900−40 binary system: orbital elements and masses and their theoretical implications.**
P. C. Joss, J. E. McClintock, S. A. Rappaport.
Bull. American Astron. Soc., Vol. 7, 525 (1975). – Abstr. AAS.

142.190 **Observations of Her X-1 by OSO-8.** R. H. Becker, E. A. Boldt, S. S. Holt, S. H. Pravdo, R. E. Rothchild, P. J. Serlemitsos, J. H. Swank.
Bull. American Astron. Soc., Vol. 7, 525 (1975). – Abstr. AAS.

142.191 **The position and proper motion of HZ Her.** G. D. Gatewood, S. Sofia.
Bull. American Astron. Soc., Vol. 7, 525 (1975). – Abstr. AAS.

142.192 **Optical search for GX17 + 2.** A. Hoag, J. Weisberg.
Bull. American Astron. Soc., Vol. 7, 525 (1975). – Abstr. AAS.

142.193 **Photometric studies of the X-ray globular cluster NGC 6624 = 3U 1820−30.**
M. H. Liller, W. Liller.
Bull. American Astron. Soc., Vol. 7, 536 (1975). – Abstr. AAS.

142.194 **Iron line emission in the X-ray spectrum of Cygnus X-3.** D. R. Parsignault, J. Grindlay, E. Schreier, H. Schnopper, H. Gursky.
Bull. American Astron. Soc., Vol. 7, 537 (1975). – Abstr. AAS.

142.195 **Stability of the magnetic field in Cygnus X-3: observations at 8 and 14.5 GHz.**
J. E. Ledden, P. E. Hodge, H. D. Aller.
Bull. American Astron. Soc., Vol. 7, 537 (1975). – Abstr. AAS.

142.196 **Observations of Cen X-3 by OSO-8.** J. H. Swank, R. Becker, E. Boldt, S. Holt, S. Pravdo, R. Rothschild, P. J. Serlemitsos.
Bull. American Astron. Soc., Vol. 7, 537 (1975). – Abstr. AAS.

142.197 **Period variations in Her X-1 and Cen X-3: shot noise or Tkachenko oscillation?**
F. K. Lamb, D. Pines, J. Shaham.
Bull. American Astron. Soc., Vol. 7, 537 (1975). – Abstr. AAS.

142.198 **Emission line blending effects and the black hole candidacy of Cygnus X-1.** M. Moulding.
Bull. American Astron. Soc., Vol. 7, 537 - 538 (1975). Abstr. AAS.

142.199 **The nature of transient X-ray sources.** K. Brecher, P. Morrison.
Bull. American Astron. Soc., Vol. 7, 538 (1975). – Abstr. AAS.

142.200 **The origin of X-ray sources in clusters of galaxies.** J. I. Katz.
Bull. American Astron. Soc., Vol. 7, 539 (1975). – Abstr. AAS.

142.201 **Results from the Ariel-5 All-Sky X-ray Monitor.** S. S. Holt.
Goddard Space Flight Center, Greenbelt, Maryland, GSFC Document X-661-75-232, Preprint, 30 pp. (1975).

A summary of results obtained from the first year of Ariel-5 All-Sky Monitor operation is presented. Transient source observations, as well as the results of long-term studies of Sco X-1, Cyg X-3 and Cyg X-1 are described. By example, the included results are indicative of the temporal effects to which the All-Sky Monitor remains sensitive as it begins its second year of observation.

142.202 **Evidence for a 17-day periodicity from Cyg X-3.** S. S. Holt, E. A. Boldt, P. J. Serlemitsos, L. J. Kaluzienski, S. H. Pravdo, A. Peacock, M. Elvis, M. G. Watson, K. A. Pounds.
Goddard Space Flight Center, Greenbelt, Maryland, GSFC Document X-661-75-260, Preprint, 10 pp. (1975).

142.203 **Long-term X-ray studies of Sco X-1.** S. Holt, E. A. Boldt, P. J. Serlemitsos, L. J. Kaluzienski.
Goddard Space Flight Center, Greenbelt, Maryland, GSFC Document X-661-75-303, Preprint, 17 pp. (1975).

No modulation of the 3 - 6 keV X-ray intensity of Sco X-1 at a level of excess of 1% is observed at the optical period of 0.787313d. Evidence is found for shot-noise character in a large fraction or the X-ray emission. Almost all of the Sco X-1 emission can be synthesized in terms of ~200 shots per day, each with a duration of ~1/3 day.

142.204 **HDE 226868.** B. Campbell.
IAU Circ., No. 2795 (1975).

142.205 **New soft X-ray source.** G. W. Clark.
IAU Circ., No. 2799 (1975).

142.206 **HDE 226868.** G. Auriemma, D. Cardini, E. Costa, F. Giovannelli, M. Orciuolo, M. Ranieri.
IAU Circ., No. 2802 (1975).

142.207 **HZ 43.** C. C. Wu, P. R. Wesselius, R. J. van Duinen, J. W. G. Aalders, K. S. de Boer.
IAU Circ., No. 2805 (1975).

142.208 **MX1401−45.** D. Hearn.
IAU Circ., No. 2808 (1975).

142.209 **A1524−61.** M. Watson.
IAU Circ., No. 2808 (1975).

142.210 **A0601−00.** M. Elvis, C. G. Griffiths, M. J. L. Turner, C. G. Page.
IAU Circ., No. 2814 (1975).

142.211 **Transient X-ray sources.** G. Jernigan, M. J. L. Turner.
IAU Circ., No. 2817 (1975).

142.212 **A0621−00 and probable optical identification.** T. Matilsky, E. Boley, R. Wolfson.
IAU Circ., No. 2819 (1975).

142.213 **Cygnus X-3.** N. Kawajiri.
IAU Circ., No. 2819 (1975).

142.214 **Positions of X-ray sources.** R. Doxsey.
IAU Circ., No. 2820 (1975).

142.215 **A0620–00.** C. J. Eyles, G. K. Skinner, A. P. Willmore, H. D. Craft, Jr., L. Bottinelli, P. Encrenaz, L. Gouguenheim, R. Lucas, Davis, Edwards, Morison, Spencer, G. Kojoian, C. R. Purton.
IAU Circ., No. 2822 (1975).

142.216 **A0620–00.** L. J. Eachus, F. Owen, T. Balonek, J. Dickey, Y. Terzian, S. Gottesman, P. F. Scott.
IAU Circ., No. 2823 (1975).

142.217 **Hercules X-1.** P. J. Serlemitsos.
IAU Circ., No. 2824 (1975).

142.218 **Serpens X-1.** A. Davidsen.
IAU Circ., No. 2824 (1975).

142.219 **Cygnus X-1.** F. Primini.
IAU Circ., No. 2833 (1975).

142.220 **3U 0900–40.** S. Rappaport, J. McClintock.
IAU Circ., No. 2833 (1975).

142.221 **A0620–00.**
H. French, J. F. Dolan, S. Wyckoff, P. A. Wehinger.
IAU Circ., No. 2835 (1975).

142.222 **A0620–00.** B. Peterson, D. Jauncey, A. E. Wright.
IAU Circ., No. 2837 (1975).

142.223 **HD 77581.** J. A. van Paradijs, G. Hammerschlag-Hensberge, E. P. J. van den Heuvel, E. J. Zuiderwijk, R. J. Takens, C. De Loore.
IAU Circ., No. 2841 (1975).

142.224 **MX0656–07.**
G. W. Clark, G. D. Schmidt, J. R. P. Angel.
IAU Circ., No. 2843 (1975).

142.225 **MX0656–07.** G. F. Carpenter, C. J. Eyles, G. K. Skinner, A. P. Willmore, A. M. Wilson.
IAU Circ., No. 2852 (1975).

142.226 **3U 0352+30.**
N. E. White, K. O. Mason, P. W. Sanford.
IAU Circ., No. 2854 (1975).

142.227 **X-ray flare in Norma.** L. J. Kaluzienski, S. S. Holt, E. A. Boldt, P. J. Serlemitsos.
IAU Circ., No. 2859 (1975).

142.228 **Cygnus X-2.** C. Chevalier, S. A. Ilovaisky, G. Branduardi, P. W. Sanford.
IAU Circ., No. 2859 (1975).

142.229 **A0535+26.** P. C. Joss.
IAU Circ., No. 2863 (1975).

142.230 **Cygnus X-1.** L. J. Kaluzienski, S. S. Holt, E. A. Boldt, P. J. Serlemitsos, J. Grindlay, E. Schreier, A. den Boggende, A. Brinkman.
IAU Circ., No. 2863 (1975).

142.231 **Transient high-latitude X-ray source.**
S. Rappaport.
IAU Circ., No. 2869 (1975).

142.232 **X Persei.** E. L. Robinson, J. Africano.
IAU Circ., No. 2869 (1975).

142.233 **X-ray sources.** N. E. White, H. E. Huckle, K. O. Mason, P. A. Charles, G. Pollard, J. L. Culhane, P. W. Sanford.
IAU Circ., No. 2870 (1975).

142.234 **B2 1101+38.** K. Pounds.
IAU Circ., No. 2873 (1975).

142.235 **HDE 245770.**
G. D. Schmidt, W. Romanishin.
IAU Circ., No. 2873 (1975).

142.236 **Transient high-latitude X-ray source.**
A. P. Willmore.
IAU Circ., No. 2876 (1975).

142.237 **SMC X-1.** R. Lucke, D. Yentis, H. Friedman, G. Fritz, S. Shulman.
IAU Circ., No. 2878 (1975).

142.238 **3U 0352+30.** N. V. Vidal, I. Kupo.
IAU Circ., No. 2878 (1975).

142.239 **Cygnus X-1.** S. S. Holt, L. J. Kaluzienski, E. A. Boldt, P. J. Serlemitsos.
IAU Circ., No. 2879 (1975).

142.240 **X Persei.** W. Liller.
IAU Circ., No. 2888 (1975).

142.241 **The X-ray pulsed fraction of NP0532 and interstellar grains.** R. M. Thomas.
Proc. Astron. Soc. Australia, Vol. 2, 325 - 327 (1975).

In this paper the author reports that the available data on the measured pulsed fraction of X-rays from the crab nebula between the energies 0.7 and 300 keV indicate that it is no longer necessary to postulate an intrinsic change in the pulsar spectral index near 30 keV. Furthermore the spectral flattening which is evident at low energies can be entirely attributed, within the precision of the available statistics to the destruction of time coherence of the pulse structure by the scattering action of interstellar dust grains.

142.242 **The motion of Cyg X-1.**
D. K. Karimova, K. V. Kuimov, E. D. Pavlovskaya.
Astron. Tsirk., No. 883, p. 1 (1975). In Russian.

142.243 **Untersuchungen an HDE 245770, Kandidat für die Röntgenquelle A 0535+26.** R. Hudec.
MVS, *Sonneberg,* Vol. 7, 29 - 30 (1975).

142.244 **Cygnus X-1, el agujero negro que la devora y los quarks.** R. R. Gil.
El Universo, Vol. 29, No. 112, p. 17 - 19 (1975).

142.245 **A high energy gamma-ray survey of Cygnus and Cassiopeia.** M. F. Campbell.
Thesis, Cornell Univ., Ithaca, N.Y. [Available from Univ. Microfilms, Ann Arbor, Mich., USA. Order No. 74-24270], 258 pp. (1974).

A unique gamma-ray telescope has been used to search for celestial sources at energies above several hundred MeV. Upper limits for the gamma-ray emission of six strong X-ray sources are given. The atmospheric generation rate of secondary gamma-rays is also determined. The balloon flight analysed provided exposure to the strong X-ray sources Cyg X-1, Cyg X-2, Cyg X-3, Cas A, Cas B and Per X-1.

142.246 **High-energy astronomy.** L. Koch-Miramond.
Recherche, (*France*), Vol. 6, 25 - 35 (1975). In French.

A review for the nonspecialist of recent developments in astrophysical research, including discussions of: X-ray sources and their lunar occultations; absorption of photons in matter; the Uhuru program.

142.247 **A model for gamma emission from thermal X sources.** S. Serio.
Nuovo Cimento B, Ser. 11, Vol. 28B, 143 - 151 (1975).

The cocoon model for X sources is examined. It is shown that inverse Compton scattering of relativistic electrons which are emitted from the central pulsar causes an appreciable flux of gamma radiation emitted in the MeV range. The same electrons give, by collisions with the plasma, the energy necessary to maintain the X flux. The ratio of X to gamma luminosity, dependent only on the shape of the electron spectrum, is computed for both power law and monochromatic spectra. The results are consistent with the gamma observations from the Cygnus region.

IAU symposium on fast transients in X-and gamma rays. See Abstr. 011.042.

Presentazione di una serie di programmi per le identificazioni ottiche. See Abstr. 031.236.

Instrumental technique in X-ray astronomy. See Abstr. 032.524.

Diffuse X-ray spectrum from 10 to 40 keV measured with rocket-borne scintillation counters. See Abstr. 061.001.

Cosmic gamma-ray bursts. See Abstr. 061.014.

Observation of the galactic X-ray background. See Abstr. 061.022.

Size distribution of cosmic gamma-ray bursts. See Abstr. 061.067.

The hydrodynamics of accretion discs I: stability. See Abstr. 062.015.

Detection of X-ray emission from stellar coronae with ANS. See Abstr. 064.048.

Slowly rotating neutron stars and transient X-ray sources. See Abstr. 065.040.

Late stages of stellar evolution. Pulsed γ-radiation of neutron, collapsing and supernova stars. See Abstr. 065.067.

Triaxial degenerate dwarfs and pulsing X-ray sources. See Abstr. 065.134.

Massive black holes in globular clusters? See Abstr. 066.019.

Radiation from spherical accretion onto black holes. See Abstr. 066.096.

The helium chromosphere, coronal holes, and stellar X-rays. See Abstr. 073.002.

Galactic X-ray sources and the ionospheric D region. See Abstr. 083.014.

Geomagnetic effect associated with X-ray flare from Sco X-1. See Abstr. 084.223.

A preliminary report on the light variation of the extreme A supergiant CD−33°12119 (=3U1727−33?). See Abstr. 113.039.

Photometry of the Be star Wray 795 and some other stars in the Cen X-3 area. See Abstr. 113.067.

Spectroscopic observations of the X-ray binary HD 153919 = 3U 1700−37. See Abstr. 114.315.

Evolution of massive close binaries. I: The X-ray binary stage. See Abstr. 117.001.

Linear polarization variations modulo 5.6 days in HDE 226868. See Abstr. 117.002.

Photometric effects in close binary systems including X-ray sources. See Abstr. 117.004.

Evolution of massive close binaries. II: The post X-ray binary stage: origin of run-away and binary pulsars. See Abstr. 117.018.

The polarization of optical radiation from close binary systems with X-ray sources. See Abstr. 117.028.

On the formation of WR-stars and X-ray sources in close binary systems. See Abstr. 117.043.

Polarization of optical radiation of binary systems with X-ray sources. See Abstr. 117.046.

Determination of the masses of components of a binary X-ray system from the X-ray eclipse duration and the amplitude of regular optical variability. See Abstr. 117.048.

Polarization of radiation of double systems with X-ray sources. See Abstr. 117.050.

Mechanism for inducing synchronous rotation and small eccentricity in close binary systems. See Abstr. 117.062.

The evolution of massive close binaries. III: The possibility of a tidal instability during the post supernova phase. See Abstr. 117.063.

Die Gashülle um den bedeckungsveränderlichen Röntgen-Doppelstern HD 153919 = 2U 1700−37. See Abstr. 121.018.

On the light curve of HZ Herculis. See Abstr. 121.023.

Photoelectric UBV photometry of the X-ray binary systems HZ Herculis (Her X-1) and V 1357 Cygni (Cyg X-1). See Abstr. 121.044.

A photometric analysis of X Persei $\stackrel{?}{=}$ 3U 0352+30. See Abstr. 122.142.

Upper limits on stellar flare X-ray emission from OSO-3. See Abstr. 122.163.

Evidence for X-ray emission from flare stars observed by ANS. See Abstr. 122.167.

CoD−33°12.119. See Abstr. 122.208.

X-ray nova flares. See Abstr. 124.005.

The X-ray structure of the Vela supernova remnant. See Abstr. 125.005.

Supernovae in binary systems. See Abstr. 125.006.

The soft X-ray spectra of Cassiopeia A and Tycho's supernova remnant. See Abstr. 125.007.

Analysis of X-ray spectra reveals possibility of enhanced silicon in three supernova remnants. See Abstr. 125.008.

Copernicus: the spatial distribution of the X-ray emission from IC443. See Abstr. 125.016.

The evolution of supernova remnants. IV. The supernova ejecta. See Abstr. 125.017.

Detection of the [Fe XIV] 5303 Å coronal line in the supernova remnant Vela X. See Abstr. 125.018.

Detection of X-ray emission from the remnant of the supernova 1006 A.D. See Abstr. 125.020.

Soft X-ray profile and spectrum of the supernova remnant IC 443. See Abstr. 125.021.

Isothermal blast wave model of supernova remnants. See Abstr. 125.034.

Observations of the diffuse λ 4430 absorption band in southern O, B, and A stars. See Abstr. 131.100.

"On the relation between CO and gamma-ray observations, cosmic rays, and the thickness of the galactic disk": a comment. See Abstr. 131.122.

A possible extended soft X-ray component of the Crab nebula. See Abstr. 134.004.

No indication of excess MeV gamma radiation from the Crab nebula. See Abstr. 134.005.

The Crab nebula: high energy X-ray observation of a lunar occultation. See Abstr. 134.006.

On the production of power-law spectra and the evolution of cosmic synchrotron sources: a model for the Crab nebula. See Abstr. 134.007.

Compton-synchrotron models for the nucleus of Centaurus A (NGC 5128). See Abstr. 141.002.

A radio source and cluster of galaxies in 3 U 1252-28. See Abstr. 141.025.

A model for the double system containing the X-ray pulsar Her X-1. See Abstr. 141.304.

Optical and X-ray observations of the PSR 1913+16 field. See Abstr. 141.313.

SAS-2 high-energy gamma-ray observations of the Vela pulsar. See Abstr. 141.318.

A high time resolution X-ray light curve for the Crab pulsar, NP 0532. See Abstr. 141.326.

Crab pulsar optical and X-ray radiation through pinching instabilities. See Abstr. 141.328.

Interpretation of the pulsed gamma-ray emission from Vela. See Abstr. 141.335.

Search for continuous X-ray emission from NP 0532. See Abstr. 141.346.

High-energy X-ray balloon observations of NP0532, Cyg X-1, Cyg X-3 and A0535 + 26. See Abstr. 141.353.

Origin of cosmic rays. See Abstr. 143.049.

Variable stars in the X-ray globular cluster NGC 1851. See Abstr. 154.024.

High-energy galactic gamma radiation from cosmic rays concentrated in spiral arms. See Abstr. 155.003.

Activity of the nucleus of the southern Seyfert galaxy NGC 1566. See Abstr. 158.143.

A strong millimeter wave component in the radio galaxy 3C 111. See Abstr. 158.190.

An X-ray survey of BL Lac objects. See Abstr. 158.309.

Studies of cluster X-ray sources. Energy spectra for the Perseus, Virgo, and Coma clusters. See Abstr. 160.004.

X-ray emission from the Centaurus cluster. See Abstr. 160.006.

Synchrotron and inverse Compton processes in X-ray clusters of galaxies. See Abstr. 160.045.

Accretion to the cluster of galaxies and X-ray emission. See Abstr. 160.048.

Accretion of intergalactic gas by a realistic model of the Galaxy and its implications. See Abstr. 161.004.

Explosive events in the early universe. See Abstr. 162.030.

Antimatter bodies and closure of the universe. See Abstr. 162.049.

143 Cosmic Radiation

143.001 **The abundance of nuclei from sodium to silicon in cosmic rays.** R. Andersson.
Physica Scripta, Vol. 11, 394 - 398 (1975).
The expected charge spectrum at the top of the atmosphere has been calculated for the charge interval $11 \leqslant Z \leqslant 14$. The calculations have been based on the assumption that cosmic ray particles are selected by an ionization process from matter having 'normal' composition.

143.002 **Measurement of primary cosmic-ray electrons and positrons from 4 to 50 GeV.**
A. Buffington, C. D. Orth, G. F. Smoot.
Astrophys. Journ., Vol. 199, 669 - 679 (1975).
The authors have used a bremsstrahlung-identification technique to measure separate e^- and e^+ spectra in the primary cosmic rays from 4 to 50 GeV. Above the average geomagnetic cutoff of 4.0 GV/c, the authors find fluxes of 4.4 ± 0.3 e^- and 0.37 ± 0.09 e^+ $m^{-2} sr^{-1} s^{-1}$. The e^- events follow a differential power law whose spectral index is 2.8 ± 0.1. The ratio $e^+/(e^+ + e^-) = 0.08 \pm 0.02$ implies traversal of an equivalent slab thickness of 4.3 ± 1.5 g cm^{-2} of interstellar and source material.

143.003 **Fokker-Planck theory for cosmic ray diffusion in the presence of Alfvén waves. 1. Theory.**
J. V. Hollweg, G. Skadron.
Journ. Geophys. Res., Vol. 80, 2701 - 2707 (1975).

143.004 **Cosmic ray streaming–I. Effect of Alfvén waves on particles.** J. Skilling.
Monthly Notices Roy. Astron. Soc., Vol. 172, 557 - 566 (1975).
The motion of cosmic rays through a moving magnetized medium supporting a spectrum of hydromagnetic waves is reexamined. In different approximations, Fermi acceleration, convective motion through space, diffusive motion, and free streaming are recovered.

143.005 **Metagalactic protons of ultra-high energies.**
V. S. Berezinsky (*Berezinskij*), S. I. Grigor'eva, G. T. Zatsepin.
Astrophys. Space Sci., Vol. 36, 3 - 15, 17 - 30 (1975).
In Russian and English.
The flux density of ultra-high-energy metagalactic protons generated by normal galaxies is calculated. Energetic losses of protons at interaction with the 2.7° relict and infrared radiations are revised. The time change of the energy of protons with different energies is estimated. The influence of fluctuations of energetic losses of protons on their spectrum is evaluated. The measured spectrum of cosmic rays can have a steepening ($\Delta\gamma \approx 1.6$) at $E \sim 3 \times 10^{19}$ eV instead of a sharp break expected previously.

143.006 **The energy spectrum of cosmic ray electrons between 10 and 1000 GeV.**
K. C. Anand, R. R. Daniel, S. A. Stephens.
Astrophys. Space Sci., Vol. 36, 169 - 175 (1975).
Measurements made by the Bombay Group on the fluxes of cosmic ray electrons in the energy range 10–1000 GeV have been compared with those of other workers in the same energy domain with a view to understand the present confused situation on the existing observations at these high energies.

143.007 **Energetic spectrum of particles accelerated in the neighbourhood of the zero line of the magnetic field.** S. V. Bulanov, P. V. Sasorov.
Astron. Zhurn. Akad. Nauk SSSR, Vol. 52, 763 - 771 (1975).
In Russian. English translation in Soviet Astron., Vol. 19, No. 4.
The acceleration of charged particles in the electric and magnetic fields arising in the decay of neutral sheets is described. The process of acceleration consists of the direct acceleration by the electric field in the nonadiabatic region near the zero (neutral) line and betatronic acceleration in the drift region. Characteristic energies of the accelerated particles in nonrelativistic and relativistic limits are defined.

143.008 **The evolution of the cosmic radiation spectrum under the influence of turbulent heating. I. Theory.**
K. L. Chan, B. J. T. Jones.
Astrophys. Journ., Vol. 200, 454 - 460 (1975).
Assuming that a radiative fluid is incompressible and has rather homogeneous physical conditions, the authors show that the evolution of the spectrum of the radiation field under viscous heating is described by an equation similar to that derived by Kompaneets. Since they include the effects of stimulated emission and keep the accuracy of the calculation to second order, they are able to consider the effect of the thermal motion as well as the effect of fluid motion on the radiation spectrum simultaneously.

143.009 **The evolution of the cosmic radiation spectrum under the influence of turbulent heating. II. Numerical calculation and application.**
K. L. Chan, B. J. T. Jones.
Astrophys. Journ., Vol. 200, 461 - 470 (1975).
The authors study the distortions of the cosmic microwave background radiation spectrum resulting from fluid energy input to the radiation field by analytical approximations and numerical integrations. The results are then used to reevaluate the constraints on the theory of cosmic turbulence imposed by the observations in the Rayleigh–Jeans part of the radiation spectrum.

143.010 **Time delay between explosive nucleosynthesis and cosmic-ray acceleration.** M. Cassé, A. Soutoul.
Astrophys. Journ., (*Letters*), Vol. 200, L75 - L76 (1975).
Pure electron capture nuclei synthesized during explosive events could, under special conditions, be present in the relativistic cosmic radiation, providing an opportunity to set limits on the time elapsed between nucleosynthesis and acceleration of cosmic-ray nuclei. In this respect, measurement of the cobalt abundance in relativistic cosmic rays seems the more promising test for the near future.

143.011 **On the acceleration of cosmic rays.** E. T. Sarris.
Astrophys. Space Sci., Vol. 36, 467 - 472 (1975).
The intensive acceleration of energetic charged particles in perpendicular shock waves which has been known to take place in the interplanetary medium has been utilized in this work in order to account for the energization of cosmic rays. It is proposed that cosmic rays can be accelerated up to 10^{14}–10^{15} eV in successive perpendicular shock waves which appear inside supernova shells in our Galaxy.

143.012 **Cosmic ray intensity gradients in the solar system.**
R. B. McKibben.
Rev. Geophys. Space Phys., Vol. 13, No. 3, (see 013.001), p. 1088 - 1092, 1098 - 1099 (1975). – Review paper with bibliographical notes.

143.013 **Origin and composition of heavy nuclei between 10 and 60 MeV per nucleon during interplanetary quiet times in 1968–1972.**

A. Mogro-Campero, J. A. Simpson.
Astrophys. Journ., Vol. 200, 773 - 786 (1975).

The relative abundances of the nuclei B–Si, the group P–V, and the group Cr–Ni (the iron group) have been measured with an instrument on the OGO-5 satellite in the energy range ~ 10 to 60 MeV nucleon^{-1} in the period of changing solar modulation in 1968–1971. Stringent criteria were applied for the selection of interplanetary quiet-time data in order to eliminate the highly variable solar component. The purpose of the investigation was to obtain evidence for the solar or galactic origin of the nuclei in this energy range where it is known that the H and He spectra display striking differences in shape from the galactic cosmic-ray spectra at higher energies. It is concluded that the experimental evidence favors a galactic origin for the nuclei in this energy range.

143.014 **Effects of the rotation of the sun on the anisotropy of cosmic rays.**
P. A. Krivoshapkin, G. F. Krymskij, A. I. Kuz'min, V. P. Mamrukova, Eh. A. Metlyaeva, G. V. Skripin.
Geomagn. Aeronom., Vol. 15, 601 - 604 (1975). In Russian.

143.015 **Cosmic ray propagation in interplanetary space.**
H. J. Völk.
Rev. Geophys. Space Phys., Vol. 13, 547 - 566 (1975).

The validity of the test particle picture and the approximation of static fields, as well as the spatial diffusion approximation, are discussed in a general way before specific technical assumptions are introduced. It is argued that the spatial diffusion equation for the intensity per unit energy has a much wider range of applicability than the kinetic (Fokker-Planck) equation it is derived from. This gives a strong weight to the phenomenological propagation theory. Its general success (and possible failure at small energies) for the modulation of galactic cosmic rays and solar events is described.

143.016 **The force field approach to the 11-year cosmic ray modulation cycle as observed by stationary neutron monitors.** H. Moraal, P. H. Stoker.
Journ. Geophys. Res., Vol. 80, 3253 - 3264 (1975).

Observations from five stationary neutron monitors are used to demonstrate certain interesting effects in the 11-year modulation cycle of galactic cosmic rays. These effects are large-scale rigidity-dependent hysteresis structures, steplike modulation changes, and double minima in cosmic ray intensities over the modulation cycle. A technique is developed to interpret these effects in terms of the force field solution of the cosmic ray equation of streaming.

143.017 **A measurement of abundances of VVH nuclei above 0.6 GeV per nucleon.** E. Juliusson, P. Meyer.
Astrophys. Journ., Vol. 201, 76 - 81 (1975).

The abundance distribution of very very heavy (VVH) nuclei just beyond the iron group has been measured with a balloon-borne counter telescope flown in 1971 and 1972. While the statistical accuracy of these observations is limited, they are nevertheless of considerable interest, since very little work has so far been done to determine the abundance of nuclei with charge between 27 and 36. It is found that the element distribution in the cosmic rays arriving at the top of the atmosphere is quite similar to the solar system abundance distribution in this interval.

143.018 **Diffusion of cosmic rays in non-thermal radio sources.** A. S. Wilson.
Astron. Astrophys., Vol. 43, 1 - 12 (1975).

The author investigates spherically-symmetric models of extended non-thermal radio sources in which relativistic particles are generated by a compact source (e.g. pulsar) in the centre ($r = 0$) and then diffuse away, losing energy by various processes. The main aim is to calculate the density of relativistic electrons $N(E, r, t)$ taking account of diffusion coefficients D and energy loss rates $d\epsilon/d\tau$ which may be functions of energy, position and time. Detailed computations of the synchrotron spectra as a function of radius are made for representative dependences of D and $d\epsilon/d\tau$ on energy and position, and the form of the spectrum of the integrated emission is calculated. The integrated spectrum of the Crab nebula from radio to X rays may be explained in terms of continuous injection of relativistic electrons having a power-law energy spectrum over the whole range of energies (10^8 eV to 10^{14} eV) into a magnetic field which decreases as r^{-1}.

143.019 **A study of days showing non-field-aligned diffusion vectors of cosmic ray diurnal anisotropy.**
R. P. Kane.
Journ. Geophys. Res., Vol. 80, 3509 - 3518 (1975).

Days when significant non-field-aligned diffusion of cosmic ray diurnal anisotropy occurred during 1965–1968 are studied in detail. It is shown that whereas a large hour-to-hour and/or day-to-day variability of the interplanetary ecliptic magnetic field direction may favor non-field-aligned diffusion, it is not a necessary condition. Some characteristics of the diurnal anisotropy seem to indicate the possibility of more than one mechanism (sources and/or sinks) operating simultaneously, distorting the usually assumed simple picture of a radially outward convection and a field-aligned inward diffusion giving the observed vector along about 1800 LT in space.

143.020 **The energy distribution of very heavy cosmic ray nuclei within the penumbra at Palestine, Texas.**
O. Mathiesen, L. Larsson, R. Andersson.
Journ. Geophys. Res., Vol. 80, 3519 - 3524 (1975).

The low-energy end of the rigidity spectrum of cosmic ray nuclei with atomic numbers $Z \geqslant 18$, recorded in a balloon flight conducted from Palestine, Texas, has been compared with a corresponding theoretical spectrum based on the computations by Shea and Smart. The two distributions are found to be in qualitative agreement with each other.

143.021 **Splash albedo protons between 4 and 315 MeV at high and low geomagnetic latitudes.**
K.-P. Wenzel, E. C. Stone, R. E. Vogt.
Journ. Geophys. Res., Vol. 80, 3580 - 3584 (1975).

143.022 **The cosmic-ray age deduced from the ^{10}Be abundance.**
M. Garcia-Munoz, G. M. Mason, J. A. Simpson.
Astrophys. Journ., (*Letters*), Vol. 201, L141 - L144 (1975).

The authors report the first satellite measurements of the radioactive cosmic-ray species ^{10}Be. The abundance was measured near 100 MeV per nucleon with high-resolution solid-state telescopes on the IMP-7 and IMP-8 satellites during 1973–1974. These two independent measurements yield ^{10}Be/Be ratios of less than 10 percent. Taken together with the galactic propagation model these results show that the cosmic-ray "clock" lifetime (most probable value of $\sim 2 \times 10^7$ years) is significantly greater than that deduced for a nominal interstellar gas density of ~1–3 atoms cm^{-3}.

143.023 **The isotopic composition of galactic cosmic-ray lithium, beryllium, and boron.**
M. Garcia-Munoz, G. M. Mason, J. A. Simpson.
Astrophys. Journ., (*Letters*), Vol. 201, L145 - L148 (1975).

The isotopes of cosmic-ray Li, Be, and B near 100 MeV per nucleon have been measured with cosmic-ray telescopes on board the IMP-7 and IMP-8 satellites during 1973–1974. The measured isotopic abundances provide a stringent test for models of interstellar propagation and solar modulation. It is found that the isotopic abundances can be explained using a steady-state interstellar propagation model with a 5 g cm^{-2}

leakage mean free path.

143.024 **Cosmic rays and interstellar tunnels.** J. S. Scott.
Nature, Vol. 258, 58 (1975). – Letter.

143.025 **The interrelation of cosmic ray intensity increases on August 4 - 5, 1972 with the dynamics of the spectrum of inhomogeneities of the interplanetary magnetic field.** V. I. Kozlov.
Issled. po kosmofiz. i aehronomii. Yakutsk, 1975, p. 49 - 61, 196 - 197. In Russian. – Abstr. in Referativ. Zhurn. 51. Astron., 9.51.433; 62. Issled. kosmich. prostranstva, 9.62.232 (1975).

143.026 **Long-periodic changes of solar diurnal variation of cosmic rays.** G. V. Skripin, A. I. Gavril'ev, V. A. Filippov, A. N. Prikhod'ko, I. P. Karmadonov.
Issled. po kosmofiz. i aehronomii. Yakutsk, 1975, p. 21 - 29, 194. In Russian. – Abstr. in Referativ. Zhurn. 51. Astron., 9.51.473 (1975).

143.027 **Transverse density gradient and cosmic ray currents.** I. S. Samsonov, A. M. Altukhov, G. F. Krymskij, Z. N. Samsonova, N. P. Chirkov.
Issled. po kosmofiz. i aehronomii. Yakutsk, 1975, p. 30 - 37, 195. In Russian. – Abstr. in Referativ. Zhurn. 51. Astron., 9.51.474 (1975).

143.028 **Observation of the high-energy primary nuclei with balloon-borne emulsion chamber.**
T. Saito, Y. Saito, H. Sugimoto, T. Matsubayashi, M. Noma.
Journ. Phys. Soc., (*Japan*), Vol. 37, 1462 (1974).

143.029 **Cosmic rays and eclipses.** S. A. Korff.
Phys. Teacher, (*USA*), Vol. 13, 333 - 338 (1975).

After an introduction to cosmic rays, the phenomena of solar corona, solar eclipses and solar wind are discussed. Interactions with the earth are considered. Finally, evidence about long-term solar variations which comes from comparisons of dendrochronological dates with radiocarbon dates is described.

143.030 **Cosmic ray streaming–II. Effect of particles on Alfvén waves.** J. Skilling.
Monthly Notices Roy. Astron. Soc., Vol. 173, 245 - 254 (1975).

Alfvén waves, generated by streaming cosmic rays, may decay via a cascade of forward and backward waves, caused by an interaction with sound waves. The analysis of this process by Chin & Wentzel is extended, and limits are placed on the validity of various approximate solutions.

143.031 **Cosmic ray streaming–III. Self-consistent solutions.** J. Skilling.
Monthly Notices Roy. Astron. Soc., Vol. 173, 255 - 269 (1975).

Self-consistent solutions are presented, in which cosmic rays are confined by Alfvén waves which are dissipated through a cascade involving sound waves. There are solutions in which Fermi acceleration of particles by the Alfvén waves can reduce or even invert the adiabatic energy losses associated with motion down density gradients. For galactic parameters, such acceleration does not occur, but a very high degree of confinement is predicted, with particles of energy less than some tens of GeV being restricted to stream less quickly than ~50 km s^{-1} (twice the Alfvén speed).

143.032 **Adiabatic approximation and Fokker-Planck formalism in the theory of cosmic ray transport.**
V. F. Zakharchenko.
Geomagn. Aeronom., Vol. 15, 793 - 798 (1975). In Russian.

143.033 **Diffusion of cosmic rays in the asymmetrical solar wind.** E. Ya. Gidalevich.
Geomagn. Aeronom., Vol. 15, 931 - 934 (1975). In Russian. Brief information.

143.034 **The anomalous ^{4}He component in the cosmic-ray spectrum at $\lesssim$50 MeV per nucleon during 1972–1974.** M. Garcia-Munoz, G. M. Mason, J. A. Simpson.
Astrophys. Journ., Vol. 202, 265 - 275 (1975).

When the current solar cycle approached near minimum modulation in 1972, the authors found that during solar quiet times the low-energy cosmic-ray proton spectrum had a typically steep form, while the helium spectrum had evolved into a "flat" form not observed before, such that at low energies the ratio ^{4}He/^{1}H became $\gtrsim$1. This paper reports a further study of these spectra by comparing them with the spectra of several other cosmic-ray species, including ^{3}He and B which are of known galactic origin. The paper includes a comparison of the ^{1}H and ^{4}He behavior in the phase-lag effect of the low-energy galactic cosmic-ray fluxes versus the Climax neutron monitor intensities.

143.035 **Energy spectrum of ultra high energy cosmic rays.** J. Wdowczyk, A. W. Wolfendale.
Nature, Vol. 258, 217 - 218 (1975). – Letter.

143.036 **An analysis of the connection between the 27-day variations of cosmic rays and various indices of solar activity.** V. P. Okhlopkov.
Trudy I Konf. molodykh uchenykh NIIYaF MGU, Moskva, 1974, p. 7 - 52. In Russian. – Abstr. in Referativ. Zhurn. 51. Astron., 8.51.541 (1975).

143.037 **The 11-year cycle of galactic cosmic rays in interplanetary space and the general magnetic field of the sun.** S. N. Vernov, A. N. Charakhch'yan, Yu. I. Stozhkov, T. N. Charakhch'yan.
VI Leningr. mezhdunar. seminar, 1974. Uskorenie chastits i yader. reaktsii v kosmose. Leningrad, 1974, p. 153, 154, with a discussion, p. 155 - 158. In Russian and English. – Abstr. in Referativ. Zhurn. 51. Astron., 8.51.756 (1975).

143.038 **Evidence for the *r*-processes of nucleosynthesis from ultra-heavy cosmic rays.**
B. Kuchowicz, D. N. Schramm.
VI Leningr. mezhdunar. seminar, 1974. Uskorenie chastits i yader. reaktsii v kosmose. Leningrad, 1974, p. 287 - 296. Abstr. in Referativ. Zhurn. 51. Astron., 8.51.757 (1975).

143.039 **On the chemical composition of cosmic rays in the Galaxy.** V. S. Ptuskin.
VI Leningr. mezhdunar. seminar, 1974. Uskorenie chastits i yader. reaktsii v kosmose. Leningrad, 1974, p. 339, with a discussion, p. 340. In Russian and English. – Abstr. in Referativ. Zhurn. 51. Astron., 8.51.758 (1975).

143.040 **Anisotropy of charged particles with energies between 10^{11} and 10^{14} eV in the Galaxy.**
A. J. Somogyi.
VI Leningr. mezhdunar. seminar, 1974. Uskorenie chastits i yader. reaktsii v kosmose. Leningrad, 1974, p. 311 - 323, with a discussion, p. 324. – Abstr. in Referativ. Zhurn. 51. Astron., 8.51.763 (1975).

143.041 **The problem of subcosmic rays. Modulation, spectrum, composition, nuclear interactions, acceleration, role in space.** L. I. Dorman.
VI Leningr. mezhdunar. seminar, 1974. Uskorenie chastits i yader. reaktsii v kosmose. Leningrad, 1974, p. 169 - 170, 171, with a discussion, p. 172 - 174. In Russian and English. – Abstr. in Referativ. Zhurn. 51. Astron., 8.51.765 (1975).

143.042 **On the second spherical harmonics of the cosmic ray angular distribution.** J. Kota.
Journ. Phys. A, (Math., Nuclear, General), Vol. 8, 1349 - 1360 (1975).

Explicit expressions describing the semi-diurnal variation of cosmic ray intensity are obtained in the frame of the convection-diffusion theory. Convection-diffusion equations are extended by introducing a symmetric traceless tensor accounting for the second harmonics of the cosmic ray angular distribution. Starting from the statistical Boltzmann equation a new transport equation is deduced which relates the second harmonics of the angular distribution to the gradients of cosmic ray streaming.

143.043 **Cosmic ray antiproton and positron production in the interstellar medium.**
G. D. Badhwar, R. L. Golden, M. L. Brown, J. L. Lacy.
Astrophys. Space Sci., Vol. 37, 283 - 300 (1975).

The authors have calculated the energy spectra of cosmic ray secondary antiprotons and positrons using the latest available data on inclusive reactions. Using the measured positron spectrum, they have found that the amount of matter traversed by the cosmic rays in the few GeV region to be $m \simeq 4.7 \pm 1.5$ g cm^{-2} of interstellar hydrogen. The computed antiproton to proton ratio is about 4×10^{-4} for energies 5–10 GeV.

143.044 **The interplanetary acceleration of energetic nucleons.** F. B. McDonald, B. J. Teegarden, J. H. Trainor, T. T. von Rosenvinge, W. R. Webber.
Goddard Space Flight Center, Greenbelt, Maryland, GSFC Document X-660-75-192, 17 pp. (1975).

Co-rotating proton and electron streams are the dominant type of low-energy (i.e. 0.1–10 MeV/nucleon) particle event observed at 1 A.U. The radial dependence of these events has been studied between 1 and 4.6 A.U. using essentially identical low-energy detector systems on IMP 7, Pioneer 10 and Pioneer 11. It is observed that from event to event the intensity either remains roughly constant or increases significantly (more than an order of magnitude) between 1 and 3 A.U. It appears that interplanetary acceleration processes are the most plausible explanation. Several possible acceleration models are explored.

143.045 **Measurement of the primary cosmic electron spectrum from 10 to about 250-GeV.**
R. F. Silverberg.
Goddard Space Flight Center, Greenbelt, Maryland, GSFC Document X-661-75-218, 2 + 37 pp. (1975).

The intensity and energy spectrum of primary cosmic electrons from 10 to ~250 GeV has been studied using balloon-borne detectors. The results indicate that the cosmic electron spectrum is steeper than the cosmic ray proton spectrum. It is shown that these data are consistent with the leakage lifetime model for the propagation of cosmic electrons in the Galaxy, although other more complex models cannot be excluded on the basis of these data.

143.046 **The electron component of cosmic rays.**
S. V. Bulanov, V. A. Dogel', S. I. Syrovatskij.
Kosmich. Issled., Vol. 13, 787 - 791 (1975). In Russian. Letter.

143.047 **Enhanced abundances of Ca, Ti and Cr in low energy cosmic rays in Skylab experiment.**
S. Biswas, J. Nevatia, N. Durgaprasad, V. S. Venkatavaradan.
Nature, Vol. 258, 409 - 412 (1975).

The authors report the results of the measurements of the abundances of Ca, Ti and Cr nuclei relative to Fe, in low energy cosmic rays in 25–100 MeV/a.m.u. interval, which were measured in a dielectric detector assembly exposed on the outside of Skylab III for 73 d. The authors conclude that the observed heavy nuclei in Skylab are mainly interplanetary cosmic-ray particles. The enhanced abundances of these heavy nuclei at low energies as compared to those at high energy (> 1 GeV/a.m.u.) seem to be an important property of cosmic ray nuclei.

143.048 **The nuclear composition of cosmic rays at their source in a closed galaxy.**
I. L. Rasmussen, B. Peters.
Nature, Vol. 258, 412 - 413 (1975).

The authors construct an alternative model to explain the observed nuclear composition, which starts with the assumption that all cosmic ray particles are of galactic origin and, once created, remain confined to the galaxy by a suitable large scale magnetic field, even at the highest energies. The main advantage of the model seems to be that it makes a galactic origin of cosmic rays possible at all energies, and thus abolishes the need for the extragalactic cosmic ray component which was supposed to dominate in the very high energy region.

143.049 **Origin of cosmic rays.** F. W. Stecker.
Phys. Rev. Letters, Vol. 35, 188 - 191 (1975).

With the use of recent observations of the galactic gas and γ-ray distributions, the galactic cosmic-ray distribution is deduced. This distribution is identical to that of supernova remnants (within experimental error), strongly supporting the hypothesis that most observed cosmic rays are produced by supernovae in our own Galaxy. The average age of the cosmic-ray sources is suggested from the character of their distribution to be about 30 million years.

143.050 **Search for cosmic-ray antimatter.**
G. F. Smoot, A. Buffington, C. D. Orth.
Phys. Rev. Letters, Vol. 35, 258 - 261 (1975).

In a sample of 1.5×10^4 helium and 4.0×10^4 higher-charged nuclei, obtained with balloon-borne superconducting magnetic spectrometers, the authors find the ratio of antinuclei to nuclei in the cosmic rays to be less than 8×10^{-5} for rigidities (momentum/charge) between 4 and 33 GV/c and less than 10^{-2} between 33 and 100 GV/c, at the 95 % confidence level.

143.051 **Composition and energy spectra of heavy nuclei of unknown origin detected on Skylab.**
J. H. Chan, P. B. Price.
Phys. Rev. Letters, Vol. 35, 539 - 542 (1975).

At the orbit of Skylab the authors observed steeply falling energy spectra of nuclei with $Z \geqslant 8$ and $10 \lesssim E \lesssim 40$ MeV/amu at intensities much higher than seen outside the magnetosphere. The composition is O: Ne: [Na-Si] : Fe = 100:33:13: (~ 2), consistent with that of the solar corona. It is suggested that heavy solar-wind ions enter the magnetosphere, are accelerated, and populate the inner radiation belt.

143.052 **Isotopic composition of cosmic-ray nitrogen at 1.5 GeV/amu.** R. Dwyer, P. Meyer.
Phys. Rev. Letters, Vol. 35, 601 - 604 (1975).

143.053 **Energy spectrum of primary cosmic rays (in the region of $10^{11} - 10^{20}$ eV).**
G. B. Khristiansen, G. V. Kulikov.
Izv. AN SSSR. Ser. fiz., Vol. 39, 1130 - 1138 (1975). In Russian. – Abstr. in Referativ. Zhurn. 51. Astron., 11.51.625 (1975).

143.054 **Cosmic rays and solar wind.** L. I. Dorman.
Uspekhi fiz. nauk, Vol. 116, 548 - 550 (1975). In Russian. – Abstr. in Referativ. Zhurn. 62. Issled. kosmich. prostranstva, 11.62.242 (1975).

143.055 **Isotropic diffusion of cosmic rays in arbitrary de-**

pendence of the diffusion coefficient on energy.
E. Ya. Gidalevich.
Geomagn. Aeronom., Vol. 15, 973 - 976 (1975). In Russian.

143.056 **Discovery of the source Z-1 in Taurus.**
L. Liebovitch.
Zemlya i Vselennaya, 1975, No. 6, p. 91 - 92. In Russian.

143.057 **Cosmic ray streaming.** J. Skilling.
Nature, Vol. 258, 687 - 688 (1975).

Taking the confinement mechanism for cosmic rays in the Galaxy to be resonant scattering by Alfvén or hydromagnetic waves generated by the streaming of the cosmic rays themselves, the author shows that the abundant low energy cosmic rays can generate, indirectly, enough Alfvén waves of long wavelength to scatter cosmic rays of considerably higher energy. Cosmic rays of energy up to around 1,000 GeV may thus be confined within the Galaxy more efficiently than has previously been supposed.

143.058 **Measurement of the fluxes of galactic cosmic-ray ^{2}H and ^{3}He in 1972–1973.**
B. J. Teegarden, T. T. von Rosenvinge, F. B. McDonald, J. H. Trainor, W. R. Webber.
Astrophys. Journ., Vol. 202, 815 - 822 (1975).

This paper reports measurements of the fluxes of cosmic-ray ^{2}H and ^{3}He during 1972–1973 carried out by instruments on board the IMP-7 and Pioneer 10 spacecraft. Energy spectra are generally consistent with a positive slope of unity as would be expected for modulated secondary components. Ratios of the form ^{2}H/^{4}He and ^{3}He/^{4}He are examined and compared with calculated values based on an assumed exponential path-length distribution for galactic propagation. It is concluded that unless extreme assumptions are made about the modulation (e.g., no energy loss), it is impossible to reconcile the simple theory and observation.

143.059 **Neutron-rich isotopes of cosmic rays with $9 \leq Z \leq 16$.**
R. C. Maehl, A. J. Fisher, F. A. Hagen, J. F. Ormes.
Astrophys. Journ., (*Letters*), Vol. 202, L119 - L123 (1975).

The authors report here new data on the isotopic composition of cosmic rays with $9 \leq Z \leq 16$ at $T \approx 450$ MeV per amu. The data show the presence of a considerable neutron-rich component of Mg and Ne, the measured mean masses being 24.3 ± 0.1 and 20.8 ± 0.1, respectively. Coupled with a calculation of the evolution of cosmic-ray isotopic abundances during propagation, these masses indicate that there is probably some neutron-rich material in this charge range at the cosmic-ray source.

143.060 **Cosmic rays and the solar wind.** L. I. Dorman.
Uspekhi fiz. nauk, Vol. 116, 548 - 550 (1975). In Russian. – Abstr. in Referativ. Zhurn. 51. Astron., 12.51.344 (1975).

143.061 **Cosmic-ray chronology.**
M. M. Shapiro, R. Silberberg.
Bull. American Astron. Soc., Vol. 7, 520 (1975). – Abstr. AAS.

143.062 **Measurements of the cosmic ray lifetime determined from the isotopic abundances of Li, Be and B in the cosmic radiation.**
M. Garcia-Munoz, G. M. Mason, J. A. Simpson.
Bull. American Astron. Soc., Vol. 7, 520 (1975). – Abstr. AAS.

143.063 **Astrophysical implications of a cosmic ray lifetime $\geqslant 2 \times 10^7$ years obtained from the Be^{10} abundance measurements.**
M. Garcia-Munoz, G. M. Mason, J. A. Simpson.
Bull. American Astron. Soc., Vol. 7, 520 (1975). – Abstr. AAS.

143.064 **On the question of the supernova origin of cosmic rays.** E. G. Zweibel, R. M. Kulsrud.
Bull. American Astron. Soc., Vol. 7, 553 - 554 (1975). – Abstr. AAS.

143.065 **Cosmic ray acceleration and the radio evolution of Cassiopeia A.**
R. A. Chevalier, J. W. Robertson, J. S. Scott.
Bull. American Astron. Soc., Vol. 7, 554 (1975). – Abstr. AAS.

143.066 **The isotopic composition of cosmic rays with $5 \leq Z \leq 26$.** A. J. Fisher, F. A. Hagan, R. C. Maehl, J. F. Ormes, J. F. Arens.
Goddard Space Flight Center, Greenbelt, Maryland, GSFC Document X-661-75-278, Preprint, 30 pp. (1975).

The authors report here results obtained from a high altitude balloon flight from Thompson, Canada in August, 1973. In the energy range 350 MeV/amu $\lesssim T \lesssim$ 600 MeV/amu they find ^{16}O/O > 0.9, ^{12}C/C > 0.9, ^{14}N/N = 0.6 ± 0.2. Additionally they find even-Z elements in the $12 \leq Z \leq 16$ group to be predominantly α-particle nuclei. In the $17 \leq Z \leq 25$ region, the data appears to be consistent with current cosmic-ray propagation models.

143.067 **Polarization of cosmic maser radiation.**
D. A. Varshalovich, V. V. Burdyuzha.
In-t kosmich. issled. AN SSSR. Pr-213. Moskva, 1975. 27 pp. In Russian. – Abstr. in Referativ. Zhurn. 51. Astron., 1.51. 760 (1976).

143.068 **Nuclear-physical investigations of cosmic rays.**
S. A. Slavatinskij.
Vestn. AN SSSR, 1975, No. 7, p. 82 - 85. In Russian. – Abstr. in Referativ. Zhurn. 51. Astron., 1.51.762 (1976).

143.069 **Energy spectrum of the α-particles of primary cosmic rays in the high-energy region from measurements aboard the Proton 3 artificial earth satellite.**
N. L. Grigorov, V. E. Nesterov, V. L. Prokhin, I. D. Rapoport, I. A. Savenko.
Issled. kosmich. luchej. Moskva, Nauka, 1975, p. 20 - 25. In Russian. – Abstr. in Referativ. Zhurn. 62. Issled. kosmich. prostranstva, 1.62.197 (1976).

143.070 **The effectiveness of recording inelastic interactions of protons with energies of 5 GeV with carbon nuclei with the instrument SEhZ-14 used aboard the Proton 1, 2, 3 space stations.** V. V. Akimov, V. S. Barashenkov, G. V. Veselova, A. S. Il'inov, V. E. Nesterov.
Issled. kosmich. luchej. Moskva, Nauka, 1975, p. 178 - 180. In Russian. – Abstr. in Referativ. Zhurn. 62. Issled. kosmich. prostranstva, 1.62.198 (1976).

143.071 **Search for quarks beyond the atmosphere aboard the Proton 3 artificial earth satellite.**
P. L. Grigorov, G. P. Kakhidze, L. V. Kuleshova, I. D. Rapoport, Yu. V. Trigubov, I. A. Savenko, V. F. Shesterikov, B. A. Yakovlev.
Issled. kosmich. luchej. Moskva, Nauka, 1975, p. 26 - 33. In Russian. – Abstr. in Referativ. Zhurn. 62. Issled. kosmich. prostranstva, 1.62.199 (1976).

143.072 **Anisotropy of cosmic rays with energies higher than 10^{19} eV.** D. D. Krasil'nikov, T. A. Egorov, N. N. Efimov, A. A. Ivanov, V. A. Kolosov, A. I. Kuz'min, K. N. Makarov, S. V. Maksimov, V. A. Orlov, I. E. Sleptsov, S. I. Nikol'skij, S. N. Vernov, G. B. Khristiansen.
Izv. AN SSSR. Ser. fiz., Vol. 39, 1245 - 1248 (1975). In Russian. – Abstr. in Referativ. Zhurn. 62. Issled. kosmich. prostranstva, 1.62.315 (1976).

143.073 **Track length distribution in lunar and meteoritic minerals and the chemical composition of very very heavy cosmic rays.** J. P. Duraud, J. C. Dran, Y. Langevin, M. Maurette, E. Dartyge.
Meteoritics, Vol. 10, 398 - 399 (1975).

143.074 **High-energy gamma quanta in primary cosmic radiation.** L. S. Bratolyubova-Tsulukidze, N. L. Grigorov, L. F. Kalinkin, A. S. Melioranskij, E. A. Pryakhin, I. A. Savenko, V. Ya. Yufarkin.
Issled. kosmich. luchej. Moskva, Nauka, 1975, p. 52 - 66.
In Russian. – Abstr. in Referativ. Zhurn. 51. Astron., 2.51. 789; 62. Issled. kosmich. prostranstva, 2.62.220 (1976).

143.075 **Interplanetary magnetic field fluctuations and the diurnal variation of cosmic ray intensity.**
J. W. Sari, L. J. Lanzerotti, C. G. Maclennan, D. Venkatesan.
Geophys. Res. Letters, Vol. 2, 571 - 574 (1975).

Predictions of the amplitudes of the diurnal variation of cosmic ray intensity, based on calculations of the power spectra of interplanetary magnetic field fluctuations, are compared with measurements of the magnitude of the diurnal variations from Sulphur Mountain neutron monitor data for five solar rotations during mid-December 1965 - April 1966, a period during which the interplanetary magnetic field data were also available. The nature of the general agreement between the derived theoretical predictions and the average observed variations over this limited period is pointed out.

143.076 **Study of cosmic ray diurnal variation on a day-to-day basis.** A. G. Ananth, S. P. Agrawal, U. R. Rao.
Pramāṇa (*India*), Vol. 3, No. 2, p. 74 - 88 (1974).

From a careful examination of the diurnal variation of cosmic ray intensity at high energies and the interplanetary field characteristics, the average characteristics of diurnal variation were recently explained by the authors in terms of a balance between outward convection and field aligned diffusion, the latter arising out of a positive radial density gradient. The authors extend this new concept to explain the large variability observed in the diurnal variation on a day-to-day basis and further demonstrate that the measurement of diurnal anisotropy characteristics of cosmic ray particles on a day-to-day basis can be used directly to infer the nature of scale sizes and interplanetary field parameters. By comparison with the magnetic field vector it is shown that this simple concept holds good on more than 80 percent of days.

143.077 **Measurement of the charge composition of primary cosmic radiation.** K. Atallah.
Thesis, Fakultät für Allgemeine Wissenschaften, Techn. Univ. München. [Available from the library of the TU München], 104 pp. (1973). In German.

143.078 **The importance of nuclear collisions in a counter detector telescope in the determination of the charge spectrum of the iron group of the cosmic radiation.**
S. Lindstam.
Lund Univ., Fys. Instn., LUIP-CR-74-16, 29 pp. (1974).

143.079 **Two views of cosmic-ray propagation in the solar system.** E. Barouch.
National Aeronautics and Space Administration, Greenbelt, Md. (USA). NASA-TM-X-70664; X-693-74-155, 15 pp. (1974).

Diffusion and scatter-free cosmic-ray propagation theories are discussed in terms of the quality of their alternate viewpoints. A hypothetical model of conditions in interplanetary space is described which aids in the comparison. A plot of the intensity of the interplanetary magnetic field over a long period is presented, and the association of these regions with high-velocity streams is shown to support the scatter-free viewpoint.

143.080 **Power spectrum analysis of cosmic-ray fluctuations. Pt. 3. The power spectral density in the region (10^{-7}–10^{-4}) Hz.**
M. R. Attolini, S. Cecchini, I. Guidi, M. Galli.
Nuovo Cimento Lettere, Ser. 2, Vol. 11, 391 - 399 (1974).

143.081 **Where are they now. Cosmic rays from supernovae.** J. C. Higdon.
Thesis, California Univ., Los Angeles (USA). [Available from Univ. Microfilms, Ann Arbor, Mich., USA. Order No. 75-9395], 144 pp. (1974).

This thesis investigates whether it is possible to observe the early stages of cosmic ray propagation near young supernova remnants. Various radiation processes are investigated: nuclear interaction, bremsstrahlung, and Compton scattering. It is found that the most likely process by which such cosmic ray cloud could be detected is γ-ray emission through pion production. Diffusive models for propagation are used to calculate the γ-ray radiation from various possible diffuse cosmic ray clouds in the vicinity of known supernova remnants.

143.082 **Some recent developments in cosmic rays.** A. M. Hillas.
Phys. Rep., (*Netherlands*), Vol. 20, 59 - 136 (1975).

143.083 **Power spectrum analysis of cosmic-ray fluctuations. Pt. 1. The bias due to data averaging.**
M. R. Attolini, S. Cecchini, I. Guidi, M. Galli.
Nuovo Cimento Lettere, Ser. 2, Vol. 11, 377 - 385 (1974).

143.084 **Power spectrum analysis of cosmic-ray fluctuations. Pt. 2. The measurement error effect.**
M. R. Attolini, S. Cecchini, I. Guidi, M. Galli.
Nuovo Cimento Lettere, Ser. 2, Vol. 11, 386 - 390 (1974).

143.085 **On the propagation of cosmic rays in the Galaxy.** G. J. Dickinson.
Thesis, Univ. Durham, England (1975).

The experimental observations of the cosmic rays, i.e. the energy spectrum, anisotropy, composition, gamma-ray and long-term variations, are reviewed. Parameters of the interstellar medium relevant to the propagation of cosmic rays are discussed. Theories about galactic cosmic-ray sources and cosmic-ray acceleration are also discussed.

143.086 **The anisotropy of ultra-high-energy cosmic rays. I. Large-scale features.** P. Kiraly, M. White.
Journ. Phys. A, (Math., Nuclear, General), Vol. 8, 1336 - 1348 (1975).

A detailed discussion is given of the statistical selection effects involved and it is concluded that the anisotropy is not proven. Some astrophysical implications of the anisotropy, if genuine, are also considered.

143.087 **Extragalactic heavy nuclei in cosmic rays.** W. Tkaczyk, J. Wdowczyk, A. W. Wolfendale.
Journ. Phys. A, (Math., Nuclear, General), Vol. 8, 1518 - 1529 (1975).

It might be thought that if the energetic particles in the cosmic radiation are of extragalactic origin then they must be protons because heavy nuclei would be completely fragmented by interactions with the radiation fields in space. It is argued that this is not the case and that it is, in principle, possible to choose an origin model in which the very energetic particles commence their life as heavy nuclei; these nuclei then interact with extragalactic photons and give rise to the observed primary spectrum. Such a model would relate mainly to energies above about 10^{18} eV where there is the well known need to postulate extragalactic origin on the grounds of the observed near-isotropy of arrival directions.

Cosmic background radiation reveals its blackbody shape.
Physics Today, Vol. 28, No. 7, p. 17, 20 (1975).

Experimental and theoretical foundations of astrophysics of cosmic rays. See Abstr. 003.042.

The cascading of cosmic-ray nuclei in various media. See Abstr. 022.099.

Proportional counters for investigation of high-energy cosmic rays used aboard the Proton scientific stations. See Abstr. 032.534.

The Pioneer 10/11 and Helios A/B cosmic ray instruments. See Abstr. 034.160.

Kα X-rays from cosmic-ray oxygen. See Abstr. 061.017.

The spectrum of relativistic particles accelerated in a Fermi reservoir. See Abstr. 061.080.

The long time confinement of fast energy particles in a magnetic trap. See Abstr. 062.014.

Resonant interactions between fast particles and collision-dominated MHD waves. See Abstr. 062.055.

Is Jupiter the cosmic ray source in our solar system? See Abstr. 099.028.

Estimation ranges of radiation ages and preatmospheric sizes of meteorites due to cosmic-ray variations. See Abstr. 105.151.

Tracks of heavy cosmic-ray nuclei in olivines from the Marjalahti and Eagle Station meteorites. See Abstr. 105.173.

Measurements of the interplanetary magnetic field in relation to the modulation of cosmic rays. See Abstr. 106.012.

Interplanetary magnetic fields and the propagation of cosmic rays. See Abstr. 106.075.

"On the relation between CO and gamma-ray observations, cosmic rays, and the thickness of the galactic disk": a comment. See Abstr. 131.122.

Flux limit on cosmic ray acceleration by strong spherical pulsar waves. See Abstr. 141.341.

Molecular hydrogen in the Galaxy and galactic gamma rays. See Abstr. 142.096.

Calculation of the cross section for C IV–H charge-exchange: significance for interstellar X-rays/cosmic ray particles. See Abstr. 142.184.

High-energy galactic gamma radiation from cosmic rays concentrated in spiral arms. See Abstr. 155.003.

On the relation between CO and gamma-ray observations, cosmic rays, and the thickness of the galactic disk. See Abstr. 155.026.

Galactic magnetic fields and their interactions with the gas and cosmic rays. See Abstr. 156.002.

Cosmic ray acceleration of gas in active galactic nuclei. See Abstr. 158.097.

Stellar Systems

151 Kinematics and Dynamics of Stellar Systems

151.001 **On the stability of galactic shocks.**
Yu. N. Mishurov, A. A. Suchkov.
Astrophys. Space Sci., Vol. 35, 285 - 297 (1975).

The stability of galactic spiral shocks is considered. A steady-state shock should be checked to see (1) if it is evolutionary; (2) if its front is stable against bending and torsion; and (3) if the gas flow far from the front is stable. In the present paper the evolutionary criterion is obtained, which implies that conditions in galaxies may lead to the evolutionary spiral shocks as well as to the nonevolutionary ones. In the latter case a galactic shock cannot persist – it instantly decays, emitting spontaneously spiral waves. This leads to a plausible stratification of the spiral arms, to the formation of the secondary arms, 'spurs' and other secondary features.

151.002 **Thermodynamic stability of relativistic stellar clusters.**
J. Katz, G. Horwitz, M. Klapisch.
Astrophys. Journ., Vol. 199, 307 - 321 (1975).

Equilibrium configurations of isothermal relativistic stellar clusters with truncated isotropic Maxwell-Boltzmann distribution of energies have been studied. An argument is given, showing that the energy cutoff should be equal to the gravitochemical potential as seen from infinity. Solutions of Einstein's equations are obtained for a wide range of values of temperature and chemical potential as seen from infinity. Conditions of stability are determined by thermodynamic criteria. It is shown that the (temperature, chemical potential)-plane contains two possible regions of stability in the limits of our calculations. One region is related to the stable configurations found by dynamical means. The second zone is new and is associated with values of the central redshift ranging from 1 to 4.

151.003 **The distribution of stars in galactic nuclei: loaded polytropes.** J. M. Huntley, W. C. Saslaw.
Astrophys. Journ., Vol. 199, 328 - 335 (1975).

The authors consider the structure of a galactic nucleus in which a compact supermassive object (e.g., a black hole or spinar) is embedded. The surrounding stars are assumed to satisfy a polytropic or isothermal equation of state. The density distribution of these loaded polytropes has a sharp central spike in which the time scale for stellar collisions can be short. Other properties of loaded polytropes, such as their masses, radii, projected density distribution, relaxation times, and stability, can also differ significantly from those of normal polytropes.

151.004 **Capture of a single star by a homogeneous spherical cluster.** A. S. Baranov.
Byull. Inst. Teoret. Astron., *Leningrad,* Vol. 14, 8 - 11 (1975). In Russian.

Changes in the orbit of a single star passing through a homogeneous spherical cluster have been estimated. Before entering the cluster the star is supposed to move in a Keplerian parabolic orbit. When passing through the cluster the star is subjected both to attraction of the cluster as a whole and to dynamical friction. Due to dynamical friction the energy of the star becomes negative which results in elliptic-type motion of the star after leaving the cluster and in capturing the star by the cluster. The formulae for changes of the star orbit per one passage through the cluster are given. Numerical estimates show that open clusters are more powerful transformers of the star orbits than globular ones.

151.005 **On the physical sense of some kinds of mixing in stellar systems.** L. P. Osipkov.
Astron. Zhurn. Akad. Nauk SSSR, Vol. 52, 875 - 877 (1975). In Russian. English translation in Soviet Astron., Vol. 19, No. 4.

The role of divergent and circulating mixings for stellar systems with continuate star formation is discussed.

151.006 **A qualitative analysis of forms of motion of a star in a stationary stellar system with axisymmetric nucleus. I.** V. K. Kajsin.
Astron. Zhurn. Akad. Nauk SSSR, Vol. 52, 878 - 882 (1975). In Russian. English translation in Soviet Astron., Vol. 19, No. 4.

A solution of the motion of a star in a stellar system corona with axisymmetric nucleus is given. A qualitative analysis of various forms of motion in the system with a flattened nucleus is given.

151.007 **Zur Interpretation des K-Effektes.**
J. Feitzinger.
Mitt. Astron. Ges., No. 36, p. 169 - 170 (1975). – Abstract.

151.008 **Random gravitational encounters and the evolution of spherical systems. VI. Plummer's model.**
L. Spitzer, Jr., J. M. Shull.
Astrophys. Journ., Vol. 200, 339 - 342 (1975).

Monte Carlo evolutionary computations for spherical clusters have been extended to a system whose initial state is an $n = 5$ polytrope (Plummer's model). The initial halo of this model is intermediate between that of the previous models D and F, and as a result both the time to collapse and the rate of growth of the escape rate are intermediate between the corresponding values found for these earlier models. The rate of collapse in all these one-component models is consistent with a relative energy loss of 0.2 percent of the core kinetic energy per relaxation time at the center.

151.009 **Dynamical theory of spherical stellar systems with large** N. L. Spitzer, Jr.
IAU Symposium No. 69, (see 012.009), p. 3 - 26 (1975).

151.010 **Homology in the evolution of cluster cores.**
D. Lynden-Bell.
IAU Symposium No. 69, (see 012.009), p. 27 - 32 (1975).

It is pointed out that in advanced phases of evolution the cores of clusters ought to evolve homologously and reasons are sought as to why this evolution should be at constant core energy. It is pointed out that continued evolution after infinite core densities have been achieved is an important area for future research.

151.011 **Correlation functions for the gravitational force.**
L. Cohen.
IAU Symposium No. 69, (see 012.009), p. 33 - 45 (1975).

Spatial and time correlations of the force acting on a star are derived for finite gravitational systems. It is shown that their behavior is qualitatively different than that for infinite mediums.

151.012 **Vibrations of non-uniform spherical systems.** G. Severne, A. Kuszell.
IAU Symposium No. 69, (see 012.009), p. 47 - 52 (1975).
Dispersion relations have been obtained and analyzed for a non-uniform, non-rotating spherical gravitational system. A restriction to short wavelengths makes it possible to consider a linearized form of the collisionless Boltzmann equation differing from that for homogeneous systems by the appearance of a term expressing the effect of the mean self-gravitational field upon the motion. The mean field affects the radially directed wave perturbations, with a breaking of symmetry.

151.013 **Tidal effects on spherical stellar systems.** D. W. Keenan, K. A. Innanen.
IAU Symposium No. 69, (see 012.009), p. 53 - 55 (1975).

151.014 ***N*-body simulations.** S. J. Aarseth.
IAU Symposium No. 69, (see 012.009), p. 57 - 60 (1975).
Numerical results on escape and binary evolution in N-body systems are summarized and some results of a new calculation with initial binaries are described.

151.015 **The simulation of non-isolated expanding gravitational systems.** G. Janin, M. J. Haggerty.
IAU Symposium No. 69, (see 012.009), p. 61 - 63 (1975).
Numerical experiments on the gravitational N-body problem are reviewed for the case of non-isolated systems. The effect of an external field on expanding cubical systems is discussed.

151.016 **Numerical experiments on the tensor virial in *N*-body systems.** R. H. Miller.
IAU Symposium No. 69, (see 012.009), p. 65 - 72 (1975).
Tensor generalizations of the virial theorem were checked in a 100-body integration. The virial theorem was remarkably well satisfied, and the claculation confirmed the generalized Lagrange-Jacobi identities. The potential energy tensor, the kinetic energy tensor, and the virial tensor showed surprisingly long correlation times of about $^1/_5$ of a crossing time.

151.017 **The structure of round stellar systems: observation and theory.** I. R. King.
IAU Symposium No. 69, (see 012.009), p. 99 - 117 (1975).
In addition to its traditional roles of checking theories and applying them, observation also plays an important role in guiding theory and indicating the form that is should take. This paper reviews types of observation, types of stellar system (omitting the highly flattened ones), and the relations between them. Observational limitations are indicated, as well as places where observations of particular sorts are needed, either for guidance of theories or for their application.

151.018 **Dynamics of star clusters: comparison of theory with observations and simulations.** R. Wielen.
IAU Symposium No. 69, (see 012.009), p. 119 - 131 (1975).
Stellar dynamical theories for collisional systems are tested by numerical N-body simulations of isolated star clusters with N up to 500. For the dynamical evolution of the density distribution in star clusters, good agreement is found between Monte Carlo results and N-body models. There remains a discrepancy between theory and simulations in the rate of stellar escape from clusters and/or in the mechanism by which escapers are produced. Simulations of non-isolated star clusters are compared with observations of open clusters. The observed age distribution of open clusters can be explained by the dynamical dissolution of clusters. Stars of low mass do not escape relatively more frequently than stars of average mass.

151.019 **Two recent developments concerning the Monte Carlo method.** M. Hénon.
IAU Symposium No. 69, (see 012.009), p. 133 - 149 (1975).
This paper consists of two independent parts. (1) The Monte Carlo method for computing the evolution of spherical stellar systems has been modified so that the computation can be continued after the time of formation of the central singularity. Results are presented for systems with equal and unequal star masses. The initial core-halo formation is followed by a general expansion of the cluster, while the central singularity absorbs a growing fraction of the total negative energy. (2) Theoretical expressions of the 'diffusion coefficients', which describe the effect of encounters in a stellar system, contain a factor $\ln(\gamma N)$ where N is the number of stars and γ is a constant usually taken to be of the order of 0.4. A reconsideration of the 'non-dominant terms' leads to a substantially lower value, of the order of 0.15 for equal masses and 0.075 for unequal masses with a typical distribution. This correction improves the agreement between N-body and Monte Carlo simulations of spherical systems.

151.020 **Formation of double galaxies by tidal capture.** S. M. Alladin, A. Potdar, K. S. Sastry.
IAU Symposium No. 69, (see 012.009), p. 167 - 175 (1975).
The conditions under which double galaxies may be formed by tidal capture are considered. Estimates for the increase in the internal energy of colliding galaxies due to tidal effects are used to determine the magnitudes V_{cap} and V_{dis} of the maximum relative velocities at infinite separation required for tidal capture and tidal disruption respectively. A double galaxy will be formed by tidal capture without tidal disruption of a component if $V_{cap} > V_i$ and $V_{cap} > V_{dis}$ where V_i is the initial relative speed of the two galaxies at infinite separation. If one assumes, as statistics suggest, that the mass of a galaxy is proportional to the square of its radius, it follows that the probability of the formation of double galaxies by tidal capture increases with the increase in mass of the galaxies and tidal disruption does not occur in a single collision for any distance of closest approach of the two galaxies.

151.021 **Ellipsoidal stellar systems.** C. Hunter.
IAU Symposium No. 69, (see 012.009), p. 195 - 203 (1975).
This paper gives a unified account of exact solutions for ellipsoids of uniform density. An approximate solution for a slightly spheroidal system of non-uniform density is given in the final section.

151.022 **A method for solving Poisson's equation for systems with axial symmetry.** T. S. van Albada.
IAU Symposium No. 69, (see 012.009), p. 205 - 206 (1975).
In computer experiments on the dynamics of stellar systems special methods are often required for the computation of the forces to keep the problem manageable. For collisionless systems a method based on solving Poisson's equation with Fourier-series expansions has been used with success (Miller et al., 1970; Hohl, 1973). For three-dimensional systems with axial symmetry a similar method can be used, based on the expansion of density and potential in Legendre polynomials (cf. Prendergast and Tomer, 1970). The author has used this later scheme to study the evolution of a system of 4000 particles; its numerical behavior is quite satisfactory.

151.023 **Resonant stellar orbits in spiral galaxies.** P. O. Vandervoort.

IAU Symposium No. 69, (see 012.009), p. 237 - 244 (1975).
This paper reviews a series of investigations of the orbits of stars in the regions of the Lindblad resonances of a spiral galaxy. The analysis is formulated in an epicyclic approximation. Analytic solutions of the epicyclic equations of motion are obtained by the method of harmonic balance of Bogoliubov and Mitropolsky. These solutions represent the resonance phenomena exhibited by the orbits in generally excellent agreement with numerical solutions.

151.024 **Is there a third integral of motion?** P. Pişmiş.
IAU Symposium No. 69, (see 012.009), p. 245 (1975). – Abstract.

151.025 **Numerical study of a gravitating system of colliding particles; formation and dynamics of discs.**
A. Brahic.
IAU Symposium No. 69, (see 012.009), p. 287 - 295 (1975).
The study of gravitating systems of colliding particles has many potential astrophysical applications, for instance the dynamics of Saturn's ring, the formation of the solar system, the flattening of protogalaxies and the evolution of galactic nuclei. We consider numerically a three-dimensional system of particles moving in the gravitational field of a central mass point and interacting through inelastic collisions. After a very fast flattening, the system forms a disc of finite thickness: this disc spreads slowly, and collisions still occur. A central condensation is formed and there is an outward flux of angular momentum. The energy which is continually lost in the inelastic collisions is obtained at the expense of the bodies which fall inwards.

151.026 **Global instabilities of disks.** J. M. Bardeen.
IAU Symposium No. 69, (see 012.009), p. 297 - 320 (1975).
Current understanding of the stability of gas and stellar disks suggests very strongly that local stability to axisymmetric modes is not sufficient for global stability. A global instability to a bar mode will develop unless the rotational kinetic energy is sufficiently small compared with the random kinetic energy for the system as a whole. A disk as cool as the galactic disk near the sun can survive only if most of the mass of the Galaxy is in a 'hot' component, such as a central bulge and/or an extended halo. The author reviews the theoretical evidence for this conclusion coming from analytic results for simple gas and stellar disks, from numerical simulations of stellar disks, and from numerical calculations of the stability of gas disks. Some new results on the precise form of dynamic bar instabilities of gas disks with and without halos are reported.

151.027 ***N*-body simulations of disks.** F. Hohl.
IAU Symposium No. 69, (see 012.009), p. 349 - 366 (1975).
The methods used in the large-scale n-body simulations are discussed. However, the present review concentrates on the results already obtained in the n-body simulations using systems containing up to 200000 simulation stars. Results are presented which show that the stability criterion developed for flattened systems applies only to truly axisymmetric instabilities. Purely stellar disks acquire rather large velocity dispersions, generally two or more times the velocity dispersion required by Toomre for axisymmetric stability. Experiments performed to determine the collisional relaxation time for the large-scale gravitational n-body calculations show that these models are indeed 'collisionless'.

151.028 **Dynamics of relativistic stellar systems.**
J. R. Ipser.
IAU Symposium No. 69, (see 012.009), p. 423 - 431 (1975).
Relativistic stellar-dynamical systems and their possible occurrence in nature are discussed. Features of the equilibrium models that have been constructed for spherical star clusters in general relativity are delineated. The results of studies of the stability of relativistic spherical clusters are reviewed. It is noted that the results, while not conclusive, indicate that realistic spherical clusters are stable against gravitational collapse if their central redshifts $z_c \lesssim 0.55$ and unstable if $z_c \gtrsim 0.55$.

151.029 **The concentric shell method for relativistic star clusters.** E. D. Fackerell.
IAU Symposium No. 69, (see 012.009), p. 433 - 439 (1975).
The analytic aspects of the Campbell-Hénon method of concentric spherical shells are generalized for application to relativistic spherically symmetric star clusters.

151.030 **Do stellar systems agree with the singular evolution of *N*-body systems?** D. Lynden-Bell.
IAU Symposium No. 69, (see 012.009), p. 443 - 446, 447 - 451 (1975). – Concluding remarks with a final discussion.

151.031 **Numerical study of a random dynamical system with two degrees of freedom.** C. Froeschlé.
Astrophys. Space Sci., Vol. 37, 87 - 100 (1975).
The classical situation of dynamical systems with two degrees of freedom is completely modified by the introduction of a stochastic element which provides a slow dissolution of the zone of stability. Using a Fokker-Planck method it has been shown that the Gambler's ruin problem is a rather good approximation of the diffusion process. Furthermore the orbits exhibit a *C*-system character even during the diffusion process.

151.032 **Models of disk systems of gravitating point masses with power distribution of surface density.**
G. S. Bisnovatyj-Kogan.
Pis'ma v Astron. Zhurn., Vol. 1, No. 9, p. 3 - 7 (1975). In Russian.
Exact self-consistent solutions are obtained for disk systems of gravitating point masses with power dependence of surface density and gravitational potential on the cylindrical radius.

151.033 **Recent developments in the formation and evolution of galaxies.** S. M. Faber.
Bull. American Astron. Soc., Vol. 7, 395 (1975). – Abstr. AAS.

151.034 **Reflection, amplification and maintenance of spiral density waves in galaxies with a bulge-halo subsystem.** J. W.-K. Mark.
Bull. American Astron. Soc., Vol. 7, 395 (1975). – Abstr. AAS.

151.035 **NGC 5364 and the density-wave model.**
J. W. Goad, S. E. Strom, L. E. Goad.
Bull. American Astron. Soc., Vol. 7, 395 - 396 (1975). – Abstr. AAS.

151.036 **N-body studies of the clustering of galaxies.**
E. J. Groth, P. J. E. Peebles.
Bull. American Astron. Soc., Vol. 7, 425 - 426 (1975). – Abstr. AAS.

151.037 **Conditions of persistence for old stellar groups.**
A. E. Gómez, F. Mayer.
Astron. Astrophys., Vol. 42, 341 - 346 (1975).
The proposition given by Innanen and House (1970) to explain the persistence of old stellar groups in terms of successive close mutual approaches of stars having a common origin is tested systematically by numerical integration of orbits in Schmidt's (1965) potential. The authors used the method of the surface of section (ϖ, $\dot{\varpi}$) to distinguish between the various types of orbits: tube, box, semi-ergodic. The condi-

tions for the persistence of approaches are very restrictive: the dispersion of the energy and of angular momentum among the stars of a group must be very small, and close approaches are present only for particular initial conditions, related with the 'third' integral of motion. Observed old groups do not fill these conditions.

151.038 **Resonant stellar orbits in spiral galaxies. II. A general epicyclic theory.** P. O. Vandervoort.
Astrophys. Journ., Vol. 201, 50 - 62 (1975).

An earlier theory of the epicyclic motions of a star near the Lindblad resonances of a spiral galaxy is generalized for situations in which the pitch angle of the spiral pattern is arbitrary and the wavenumber of the pattern is complex. The epicyclic equations of motion are solved by the method of harmonic balance, and two integrals of the basic equations of the method are derived. In terms of these integrals, Jacobi's integral is recovered, and a second isolating integral of the motion can be constructed approximately. Equations are derived which isolate the periodic orbits and determine their stability.

151.039 **Numerical experiments for investigating the effect of multiplicity of a stellar encounter.**
I. V. Petrovskaya, S. P. Yakimov.
Trudy Astron. Obs., *Leningrad*, Vol. 31 (= Uchenye Zapiski Leningr. Un-ta, No. 381 = Seriya Matem. Nauk, vyp. (No.) 51), p. 182 - 191 (1975). In Russian.

A method is proposed for investigation of the effect of multiplicity of stellar encounters by numerical experiments. The method is used to find the coefficient of multiplicity of stellar encounters in open clusters.

151.040 **Periodic orbits in barred galaxies.**
M. Michalodimitrakis.
Astrophys. Space Sci., Vol. 37, 131 - 142 (1975).

A quantitative study of periodic orbits and particle trapping near them in the gravitational field of a uniformly rotating homogeneous solid parallelepiped is made. It is found that particle trapping leads to the possibility of formation of ring-like structures around the parallelepiped and blob-like concentrations around the equilibrium points of the parallelepiped.

151.041 **Amplification of galactic density waves by non-adiabatic processes in gas.** S. Kato.
Publ. Astron. Soc. Japan, Vol. 27, 421 - 429 (1975).

In this paper it is suggested that non-adiabatic increase of clouds' random motion in the arm region acts so as to maintain a density wave against the damping due to galactic shocks. To demonstrate this, the order of magnitude of the work done on the wave by the non-adiabatic pressure increase is estimated for a tightly wrapped wave case by using simple model of pressure variation. The result is compared with the shock damping.

151.042 **Equilibrium and stability of collisionless gravitating systems.** A. M. Fridman.
Astronomy. Vol. 10. Equilibrium and stability of gravitating systems, (see 003.002), p. 61 - 160 (1975). In Russian.

151.043 **Resonant stellar orbits in spiral galaxies. III. Applications and numerical tests of the general epicyclic theory.** P. O. Vandervoort, D. G. Monet.
Astrophys. Journ., Vol. 201, 311 - 326 (1975).

This paper describes applications and numerical tests of the epicyclic theory for the orbits of stars near the Lindblad resonances of a spiral galaxy. The applications consist of a survey of the resonance phenomena exhibited by orbits near the inner Lindblad resonance. The resonance phenomena are described in terms of solutions for the periodic orbits and in terms of surfaces of section derived from the two isolating integrals of the motion in the epicyclic theory. The numerical tests of the epicyclic theory consist of comparisons of surfaces of section derived from analytic solutions and from numerical solutions of the epicyclic equations of motion. Analytic and numerical versions of the surfaces of section agree very well.

151.044 **Inner Lindblad resonance in galaxies. Nonlinear theory. I.** G. Contopoulos.
Astrophys. Journ., Vol. 201, 566 - 584 (1975).

In order to attack the self-consistent nonlinear problem near the inner Lindblad resonance, a theory of resonant orbits is as simple as possible is required, but which at the same time gives approximately correct numerical results. In the present paper the author uses the action-angle variables $(I_1, I_2, \theta_1, \theta_2)$ introduced in the galactic problem by Kalnajs and Lynden-Bell. He develops the spiral Hamiltonian in these variables and eliminates all trigonometric terms except the resonant ones, that contain the combination $\theta_1 - 2\theta_2$. Then it is known that the combination of the new actions $I_2^* + 2I_1^*$ is an integral of motion. Using this integral, together with the Jacobi integral, the author can explain the properties of the orbits and their invariant curves near resonance, as well as far from it. The comparison of theoretical and numerically calculated orbits gives in most cases a good quantitative agreement.

151.045 **Random gravitational encounters and the evolution of spherical systems. VII. Systems with several mass groups.** L. Spitzer, Jr., J. M. Shull.
Astrophys. Journ., Vol. 201, 773 - 782 (1975).

Monte Carlo computations for model clusters have given quantitative information on the dynamical evolution of systems in which stars of three different mass groups are present, with masses in the ratios 1 : 2.5 : 6.25. The seven systems analyzed all show pronounced mass stratification developing in a time of 1 to 2 times t_{rh}, the reference relaxation time. Both analytical theory and special Monte Carlo runs verify that the tendency toward equipartition does not play the dominant part in the late collapse of these three-component systems. During the life of the clusters before collapse, at most three-fourths of the lighter stars escape, if the maximum shock heating experienced by present clusters is assumed. Comparison of actual globular clusters with the present models suggests that the observed heavier stars, each of mass about 0.7 $M_\odot$, may be embedded in a more massive, more extended system of very faint lighter stars.

151.046 **Computer simulations of stellar systems.**
S. J. Aarseth, M. Lecar.
Annual Rev. Astron. Astrophys., Vol. 13, (see 003.005), 1 - 21 (1975). – Contents: (1) Introduction; (2) The N-body problem; (3) Galactic clusters; (4) Globular clusters; (5) Galaxies and clusters of galaxies; (6) Collisionless relaxation.

151.047 **The theory of resonances in spiral galaxies.**
G. Contopoulos.
La dynamique des galaxies spirales. Colloque International C.N.R.S. No. 241, (see 012.022), p. 19 - 37, with a discussion, p. 41 (1975).

The role of resonances in a galaxy is best illustrated if action-angle variables in the Hamiltonian are used. The Hamiltonian is separated into various components; besides the axisymmetric component the most important components are the bar, the inner resonant and the outer resonant components. At each resonance a special form of a new integral is derived, which permits to explain the forms of the trapped and untrapped orbits. The new integrals are used in the distribution function to calculate the amount of trapped matter near each resonance. A self-consistent trailing spiral can exist only outside the inner Lindblad resonance, while a barred spiral can exist only inside the inner Lindblad resonance.

151.048 **Integrals of the motion in the resonance regions.** P. O. Vandervoort.
La dynamique des galaxies spirales. Colloque International C.N.R.S. No. 241, (see 012.022), p. 39 - 40, with a discussion, p. 41 (1975).

151.049 **Numerical experiments on spiral structure.** R. H. Miller.
La dynamique des galaxies spirales. Colloque International C.N.R.S. No. 241, (see 012.022), p. 43 - 53, with a discussion, p. 64 - 66 (1975). – Review paper.

151.050 **Effect of halo component on bar-formation in disk galaxies.** F. Hohl.
La dynamique des galaxies spirales. Colloque International C.N.R.S. No. 241, (see 012.022), p. 55 - 63, with a discussion, p. 64 - 66 (1975).

Numerical experiments are performed to determine the effect of a fixed halo component of mass on the stability of purely stellar disks. The rotation curve of the fixed halo component corresponds to the Schmidt model of the galaxy. It is found that when the stellar disk contains less than 50 % of the total mass, the large-scale bar-making instability is effectively suppressed. For disks containing 50 % or more of the total mass, a bar structure quickly forms.

151.051 **Wave-wave and wave-particle interactions in mechanisms for the maintenance of spiral structure.**
J. W-K. Mark.
La dynamique des galaxies spirales. Colloque International C.N.R.S. No. 241, (see 012.022), p. 67 - 79 (1975).

Due to the presence of dissipation and the frequent possibility of weak driving influences, the amplification of density waves is essential to the maintenance of spiral structure in galaxies. At the corotation region of density waves in the disk, wave-wave and wave-particle exchanges of angular momentum amplifies these waves. The slowly rotating bulge-halo subsystem is effective in removing angular momentum from the disk. This effect actually amplifies those density waves inside corotation. Order of magnitude amplification of wave angular momentum is possible. If proper wave reflection occurs at the inner regions of galaxies, then standing wave systems with temporal growth can also be obtained.

151.052 **Note on the persistence of spiral structure.** E. Dekker.
La dynamique des galaxies spirales. Colloque International C.N.R.S. No. 241, (see 012.022), p. 81 - 89, with a discussion, p. 90 (1975).

151.053 **What is essential for spiral structure?** D. Lynden-Bell.
La dynamique des galaxies spirales. Colloque International C.N.R.S. No. 241, (see 012.022), p. 91 - 101, with a discussion, p. 102 (1975).

151.054 **Spiral structure of the gas in spiral galaxies. – Theory.** W. W. Roberts, Jr.
La dynamique des galaxies spirales. Colloque International C.N.R.S. No. 241, (see 012.022), p. 113 - 149, with a discussion, p. 150 - 155 (1975). – Review paper.

151.055 **Spiral structure produced by tidal interaction of galaxies.** F. Hohl.
La dynamique des galaxies spirales. Colloque International C.N.R.S. No. 241, (see 012.022), p. 237 - 241 (1975).

A simulation of a close passage of two galaxies where the self-gravity of the disk stars is taken into account was performed using the large-scale n-body model (Hohl and Hockney, 1969; Hohl, 1972).

151.056 **On the age dependence of the spatial distribution and velocity dispersion of stars in a spiral galaxy.**
R. Wielen.
La dynamique des galaxies spirales. Colloque International C.N.R.S. No. 241, (see 012.022), p. 357 - 369, with a discussion, p. 370 (1975).

The author discusses whether star migration in the gravitational field of a density wave may explain the observed dissolution of spiral arms and the increase in velocity dispersion with increasing age of stars. Some preliminary results of orbit calculations, mainly concerning periodic orbits, are presented.

151.057 **Elliptical solutions of the Vlasov equation.** J. P. Petit, G. Monnet.
La dynamique des galaxies spirales. Colloque International C.N.R.S. No. 241, (see 012.022), p. 391 - 399 (1975).

151.058 **On the nature of barred spirals.** I. R. King.
La dynamique des galaxies spirales. Colloque International C.N.R.S. No. 241, (see 012.022), p. 417 - 418 (1975).

151.059 **Numerical experiments on the structure of barred spiral galaxies.** A. Duus, K. C. Freeman.
La dynamique des galaxies spirales. Colloque International C.N.R.S. No. 241, (see 012.022), p. 419 - 422, with a discussion, p. 423 (1975).

The authors report preliminary results of some numerical studies designed to see what sort of structures and motions are induced by the presence of a bar in the center of a disk galaxy. Two kinds of characteristic structures appear in these results: (1) the luminous rings which encircle the bars of SB(r) galaxies, and (2) the complicated split nuclei which are often found in SB systems.

151.060 **A kinematic study of several barred spirals.** M. F. Cheriguène.
La dynamique des galaxies spirales. Colloque International C.N.R.S. No. 241, (see 012.022), p. 439 - 480, with a discussion, p. 481 (1975).

151.061 **On the origin and long term maintenance of spiral structure.** C. C. Lin.
La dynamique des galaxies spirales. Colloque International C.N.R.S. No. 241, (see 012.022), p. 491 - 503, with a discussion, p. 504 - 505 (1975). – Review paper.

151.062 **The dynamics of interacting galaxies.** A. E. Wright.
La dynamique des galaxies spirales. Colloque International C.N.R.S. No. 241, (see 012.022), p. 507 - 508, with a discussion, p. 509 - 510 (1975). – Summary.

151.063 **Determination of the distribution function of an elliptical galaxy.** C. Hunter.
Astron. Journ., Vol. 80, 783 - 793 (1975).

A new method is presented for constructing a collisionless distribution function for an elliptical galaxy whose mass distribution is known. The method, which involves the use of Stieltjes transforms, is used in the study of a new set of mass distributions that have two dimensionless parameters.

151.064 **Encounters between binary and single stars and their effect on the dynamical evolution of stellar systems.** J. G. Hills.
Astron. Journ., Vol. 80, 809 - 825 (1975).

More than 1.4×10^4 computer-simulated encounters between binaries and field stars have been run for various pre-encounter relative velocities V_f and impact parameters p. At low values of V_f, the encounter causes the binary orbit to shrink and become more tightly bound with the two more massive of the three stars involved in the encounter remaining in the binary and the third being ejected from the binary at a relatively high velocity. As V_f increases, the amount of orbit shrinking decreases until, for V_f greater than a well-defined velocity V_E, the encounter tends to increase the semimajor axis of the binary and eventually to break it apart. The break-up of loosely bound binaries has relatively little effect on the dynamical evolution of the stellar system. However, the shrinking of the more tightly bound binary orbits, which feeds kinetic energy into the cluster, will destroy an open cluster in a few times 10^8 yr unless it is either unusually massive or unusually deficient in binaries. While the lifetime of a globular cluster due to this mechanism is about 10^{11} yr, it is likely that binary stars limit the maximum density attainable in the nucleus of a globular cluster. Binaries are not likely to have any significant effect on the dynamical evolution of a dense galactic nucleus.

151.065 **Motion in the field of a rotational-symmetrical potential. II. Orbits adjoining the plane of symmetry.** T. A. Agekyan, S. P. Yakimov.
Vestn. Leningr. un-ta, 1975, No. 1, p. 184 - 187. In Russian. Abstr. in Referativ. Zhurn. 51. Astron., 8.51.818 (1975).

151.066 **Three-dimensional models for the distribution of mass in galaxies.** M. Miyamoto, R. Nagai.
Publ. Astron. Soc. Japan, Vol. 27, 533 - 543 (1975).

From Toomre's (1963) disk models of galaxies, the authors have generalized other convenient pairs of functions to represent the axisymmetric potential and density of models in which the mass is no longer confined to a single plane. The models are free from singularities and involve functions that are quite explicit and elementary. Even without superposing any of them, the present density laws seem to mimic remarkably well the stratifications of mass both in the central bulges and in the disk parts of real galaxies.

151.067 **On constructing a limited self-gravitating model of a galaxy.** S. A. Kutuzov.
Vestn. Leningr. un-ta, 1975, No. 7, p. 145 - 150. In Russian. – Abstr. in Referativ. Zhurn. 51. Astron., 10.51.663 (1975).

151.068 **On the applicability of the third quadratic integral for some models of mass distribution in stellar systems. I.** L. P. Osipkov.
Vestn. Leningr. un-ta, 1975, No. 7, p. 151 - 158. In Russian. – Abstr. in Referativ. Zhurn. 51. Astron., 10.51.664 (1975).

151.069 **Models for the formation of elliptical galaxies.** R. B. Larson.
Monthly Notices Roy. Astron. Soc., Vol. 173, 671 - 699 (1975).

Previous models for the formation and evolution of spherical galaxies are generalized to the case of axisymmetric rotating systems. It is found that a completely inviscid collapse does not yield a realistic model, but if a plausible turbulent viscosity is assumed it produces significant outward transfer of angular momentum during the collapse and leads to models which closely resemble elliptical galaxies.

151.070 **Binary evolution in stellar dynamics.** D. C. Heggie.
Monthly Notices Roy. Astron. Soc., Vol. 173, 729 - 787 (1975).

Numerical experiments have demonstrated something of the importance of binaries in *N*-body systems, and this paper aims to provide a comprehensive theoretical picture of their behaviour. Using an impulsive approximation, it is shown that pairs with low binding energies, i.e. much less than the average kinetic energy of single cluster members, tend to be disrupted by encounters. Especially difficult to treat are close encounters resulting in the formation of temporarily bound triple systems; a new hypothesis on the three-body problem deals successfully with this important case. The result of all types of encounter is that energetic pairs tend to become more energetic, at an average rate which is approximately independent of their binding energy.

151.071 **Kinetics of stars in flat systems. I. Dynamical friction.** I. L. Genkin, Z. N. Chumak.
Astron. Zhurn. Akad. Nauk SSSR, Vol. 52, 1329 - 1331 (1975). In Russian. English translation in Soviet Astron., Vol. 19, No. 6.

The dynamical friction in flat stellar systems is studied. It is shown that the force of friction which acts on a rapidly moving star is equal to the regular force acting on external objects by the flatness. All formulae, in contrast to the appropriate three-dimentional ones, do not have a parameter analogous to the radius of the Debye sphere.

151.072 **Numerical experiments on dynamics of axially symmetric stellar systems.** A. Ahmad.
Journ. Roy. Astron. Soc. Canada, Vol. 69, 253 (1975). Abstr. CAS.

151.073 **Azimuthal perturbations in a homogeneously gravitating medium.** I. L. Genkin, O. V. Chumak.
Trudy Astrofiz. Inst., *Alma-Ata*, Vol. 25, 128 - 131 (1975). In Russian.

Properties of sectorial waves in a homogeneous gravitating medium are investigated. It is shown that propagation of perturbations is possible only with sectorial number m = 2.

151.074 **On the behaviour of peculiar motions in the metagalaxy.** T. S. Kozhanov, T. B. Omarov.
Trudy Astrofiz. Inst., *Alma-Ata,* Vol. 26, 16 - 20 (1974). In Russian.

151.075 **On the dynamics of a large-scale gravitating system.** T. S. Kozhanov, T. B. Omarov.
Trudy Astrofiz. Inst., *Alma-Ata,* Vol. 26, 21 - 27 (1974). In Russian.

151.076 **On the stability of a differentially rotating cylinder.** Z. N. Chumak, O. V. Chumak.
Trudy Astrofiz. Inst., *Alma-Ata,* Vol. 26, 55 - 66 (1974). In Russian.

151.077 **The third intragral of motion and the velocity field for a quasi-Hookian potential.** M. G. Saginashvili.
Byull. Abastumansk. Astrofiz. Obs., No. 46, p. 125 - 132 (1975). In Russian.

The geometry of orbits is investigated for the Hooke potential in the meridional plane. The velocity field is being constructed in the same plane, its two-valuedness shown. The form found for the third integral of motion is also true for a quasi-Hookian field. The semiaxes expressions of the orbit ellipse and inclination are obtained.

151.078 **A theorem on the third integral.** J. S. Stodółkiewicz.
Acta Astron., Vol. 25, 331 - 360 (1975).

The theorem on the relationship between alterations of the kinetic energy function and the response of the third integral caused by them in axially symmetrical stationary stellar systems is given. It is shown that the only third integrals satisfying quite general assumptions of the theorem are integrals quadratic in velocities. The possibility of application of these integrals to the problem of stellar motions in the Galaxy is discussed.

151.079 **Magnetogasdynamic mechanism of rotation of galaxies with spiral structure.**

V. A. Krol', S. A. Silich, P. I. Fomin.
In-t teor. fiz. AN USSR. Preprint ITF-75–17R. Kiev, 1975. 20 pp. Price 6 Kop. In Russian. – Abstr. in Referativ. Zhurn. 51. Astron., 12.51.617 (1975).

151.080 **On mathematical modelling of interacting galaxies.** N. N. Kozlov.
In-t prikl. mat. AN SSSR. Preprint No. 66. Moskva, 1975. 36 pp. Price 12 Kop. In Russian. – Abstr. in Referativ. Zhurn. 51. Astron., 12.51.660 (1975).

151.081 **Gas response to barlike perturbations.** R. H. Sanders, J. M. Huntley.
Bull. American Astron. Soc., Vol. 7, 513 (1975). – Abstr. AAS.

151.082 **Stellar orbits near the Lindblad resonances of a spiral galaxy.** D. G. Monet.
Bull. American Astron. Soc., Vol. 7, 513 - 514 (1975). Abstr. AAS.

151.083 **On the structure of a spiral galaxy in the regions of the Lindblad resonances.** P. O. Vandervoort.
Bull. American Astron. Soc., Vol. 7, 514 (1975). – Abstr. AAS.

151.084 **Evolution of spiral galaxies.** R. J. Talbot, Jr.
Bull. American Astron. Soc., Vol. 7, 550 (1975). Abstr. AAS.

151.085 **Numerical calculation of the dynamical evolution of chains of galaxies.**
A. F. Steklov, I. D. Karachentsev.
Soobshch. Spets. Astrofiz. Obs., *Zelenchukskaya*, vyp. (No.) 14, p. 20 - 32 (1975). In Russian.

Results of numerical calculation of evolution of rotating and oscillating chains are presented. It is shown that the action of irregular forces leads to disruption of the initial linear form of a chain for a time of the order of one to two rotation (or oscillation) periods. It is pointed out that there is a possibility to determine the type of motions in chains using the distribution of the radial velocity dispersion across the main axis of the chain.

151.086 **Non-linear oscillations of some homogeneous models of stellar systems. I. Case of radial oscillations.**
V. A. Antonov, S. N. Nuritdinov.
Vestn. Leningr. un-ta, 1975, No. 7, p. 133 - 138. In Russian. Abstr. in Referativ. Zhurn. 51. Astron., 1.51.824 (1976).

151.087 **The absence of autonomy of stellar systems leads to underestimating of their masses by the dynamical method.** F. A. Tsitsin.
Astron. Tsirk., No. 878, p. 3 - 5 (1975). In Russian.

151.088 **Ellipsoidal figures of equilibrium of interstellar medium inside spheroidal galaxies.**
M. G. Abramyan.
Astrofizika, Vol. 11, 487 - 498 (1975). In Russian. – English translation in Astrophysics, Vol. 11, No. 3.

The question of possible ellipsoidal equilibrium figures of a stratum of the interstellar medium (stellar subsets) inside spheroidal galaxies is considered. It is shown that beside spheroidal figures triaxial ellipsoids are also possible figures not being ellipsoids of Jacobi. The question of stability of spheroidal figures and their transition to triaxial ellipsoidal forms of equilibrium is also taken into consideration.

151.089 **Dynamical evolution and instability of the Trapezium type systems.**
L. V. Mirzoyan, M. A. Mnatsakanyan.
Astrofizika, Vol. 11, 551 - 553 (1975). In Russian. – English translation in Astrophysics, Vol. 11, No. 3.

It has been shown that the calculations of the dynamical evolution of the Trapezium type systems made by Allen and Poveda (1974) are in complete agreement with Ambartsumyan's conclusion (1960) on the dynamical instability and the disintegration of these systems.

151.090 **Motion in the field of a rotationally symmetric potential. III. Symmetry of the surface density and orthogonality of box-type orbits.**
T. A. Agekyan, S. P. Yakimov.
Vestn. Leningr. un-ta, 1975, No. 13, p. 120 - 121. In Russian. Abstr. in Referativ. Zhurn. 51. Astron., 2.51.849 (1976).

151.091 **Galaxies with ellipsoidal distribution cannot be stationary.** N.-g. Han.
Acta Phys. Sinica, Vol. 24, 72 - 78 (1975). In Chinese.

A revised expression is given for the stellar dynamic basic equation. The nonexistence of a self-rotating stationary galaxy satisfying the ellipsoidal distribution law is verified. The author also proves that certain quasi-stationary galaxies must possess a spiral structure.

151.092 **Spiral structure and the dynamics of flat stellar systems.** E. Dekker.
Thesis, Rijksuniv. Leiden (Netherlands), 148 pp. (1975).

151.093 **Chemical evolution of a spiral galaxy.**
T. Ohnishi.
Progr. Theor. Phys. Japan, Vol. 53, 1042 - 1057 (1975).

The time variation and the spatial variation of the gas density and the metal abundance Z on the disk of a spiral galaxy are investigated by introducing the density wave theory. It becomes clear that stars are formed in the region $R<12.5$ kpc mainly through the gas compression by the galactic shocks which take place with the passage of a density wave, and only through the inelastic collision of clouds in $R>12.5$ kpc, R being the distance from the galactic center.

151.094 **Collective instabilities of self-gravitating systems. II. Instabilities due to temperature anisotropy.**
T. Nakamura, F. Takahara, S. Ikeuchi.
Progr. Theor. Phys. Japan, Vol. 53, 1348 - 1359 (1975).

The instability modes of rotating self-gravitating systems are investigated on the assumption of infinitely long cylinders. The systems under consideration are collisionless stellar systems with anisotropic velocity dispersion and a gaseous system with anisotropic pressure. In the collisionless stellar system, the Jeans instability mode and the Harris instability mode exist. The dispersion relation is solved numerically.

A comparative study of maximization methods. See Abstr. 021.016.

Integrals of motion. See Abstr. 042.024.

The problem of escape in the *n*-body system. See Abstr. 042.052.

Détermination du potentiel à partir des trajectoires dans le cas plan. See Abstr. 042.104.

Crossfertilization between plasma, stellar dynamics and hydrodynamics. See Abstr. 062.021.

Problems on gravitational collapse of interstellar gas clouds. I. Numerical integration of the equations of hydrodynamics: the case of spherical symmetry.
See Abstr. 062.064.

The dynamical evolution of binaries in clusters.
See Abstr. 117.026.

Binary evolution in stellar systems. See Abstr. 117.027.

A gravitational three-body scattering experiment. See Abstr. 117.054.

Statistics of three-body experiments: probability of escape and capture. See Abstr. 117.055.

New members of the Hyades cluster and a discussion of its structure. See Abstr. 153.025.

Stellar collisions in globular clusters. See Abstr. 154.032.

Theoretical limits on the properties of low-velocity M-dwarfs. See Abstr. 155.014.

The density distribution of the interstellar gas near the galactic shock region. See Abstr. 155.028.

Self-gravitating, disk-like, galactic models with stars having velocity dispersions. See Abstr. 155.030.

Galactic evolution. I. Single-zone models. See Abstr. 155.032.

Some remarks on local kinematical disturbances in our Galaxy and their interpretations. See Abstr. 155.038.

Galactic orbits and integrals of motion for stars of old galactic populations. III. Conclusions and applications. See Abstr. 155.042.

On the stellar component of galactic coronae. See Abstr. 155.047.

Galactic evolution. II. Disk galaxies with massive halos. See Abstr. 155.048.

On the origin of S0 galaxies. See Abstr. 158.033.

The bar-like objects as a possible generator of spiral density waves. II. The wave pattern in a differentially rotating disk. See Abstr. 158.043.

The collapse and formation of galaxies. See Abstr. 158.059.

The development of the spiral structure in M51. See Abstr. 158.152.

Origine explosive. See Abstr. 158.168.

Collisions of galaxies in dense clusters. I. Dynamics of collisions of two galaxies. See Abstr. 160.008.

152 Stellar Associations

152.001 **R-associations II. The ratio of total to selective extinction.** W. Herbst.
Astron. Journ., Vol. 80, 498 - 502 (1975).

The variable extinction method for determining the ratio of total to selective extinction, R, has been applied to 19 R-associations. A value of $\langle R \rangle = 3.31 \pm 0.13$ is derived for "normal" regions of the Galaxy. Combination of this result with recent determinations by other authors leads to a mean value of $\langle R \rangle = 3.3 \pm 0.1$. Two R-associations (Car R1 and Cen R2) are found in addition to two previously reported [Ara R1 by Herbst (1974); vdB 130 by Racine (1974)] which have anomalously large values of R and are located in regions of extensive emission nebulosity.

152.002 **Sterne und interstellares Medium in der Assoziation Cyg OB 2.** K. Voelcker, H. Elsässer.
Mitt. Astron. Ges., No. 36, p. 173 - 174 (1975).

152.003 **Rings and chains in simulated stellar fields.** E. Lindemann, G. Burki.
Astron. Astrophys., Vol. 41, 355 - 358 (1975).

Stellar fields are simulated by means of coordinates generated by random numbers. The diameters of stars in those simulated fields are defined according to a magnitude-like distribution. An equivalent surface of about 8.3 sq.deg. of sky at low galactic latitude is plotted. The authors obtain frequencies of stellar rings and chains of the same order of magnitude as in real fields. This suggests that at least a large part of the stellar rings and chains could be due to random arrangements of stars.

152.004 **Physical investigations of star chains.** G. Hahn, H. F. Haupt.
Astron. Astrophys., Vol. 41, 447 - 450 (1975).

The phenomenon of the star chains as described by Oberguggenberger has been investigated in order to decide on the chains' physical reality. For 8 chains the results show clearly no physical correlation between the single chain stars, for one chain the data are insufficient to decide whether such a correlation exists. For the star chains as a whole the conclusion is drawn that they are a product of chance produced by projection to the sphere.

152.005 **R-associations IV. Vela R2, a young stellar group.** W. Herbst.
Astron. Journ., Vol. 80, 683 - 688, 751 - 752 (1975).

The R-association Vela R2 is investigated in detail. Evidence is presented that it is a very young group and includes several probable pre-main-sequence stars of B spectral type. The exciting star of the peculiar emission nebula RCW 34 is an apparent member of this group and may be an O-type analog of the Herbig Be- and Ae-type stars. A list of other R-associations which appear to be generically associated groups of young stellar objects is given.

152.006 **On the nature of stellar associations.** P. N. Kholopov.
Astron. Tsirk., No. 847, p. 7 - 8 (1974). In Russian.

Population I and population II stellar evolution. See Abstr. 065.129.

Infrared photometry in the R CrA association. See Abstr. 113.002.

Infrared observations of the association Cyg OB 2. See Abstr. 113.014.

The luminosities of the Beta Canis Majoris variables, the zero age main sequence and the distance of the Sco–Cen association. See Abstr. 115.002.

Flare stars in star clusters and associations. See Abstr. 122.006.

Observations of variable stars in the T-association T1 Cygni. See Abstr. 123.002.

On the evidence for circumstellar obscuration of OB stars. See Abstr. 153.004.

R-associations III. Local optical spiral structure. See Abstr. 155.002.

Evidence of density wave motions from the ages of stars and associations in M33: Lallemand electronic camera stellar photometry. See Abstr. 158.163.

153 Galactic Clusters

153.001 **Three-colour photometry of the two galactic star clusters NGC 559 and NGC 637.** C. Grubissich.
Astron. Astrophys., Suppl. Ser., Vol. 21, 99 - 107 (1975). In German.

Three-colour photometry of the clusters NGC 559 and NGC 637 in the *RGU* system has been carried out. The first cluster is a heavily reddened old cluster with many late-type giants at a distance of 1.2 kpc or slightly more. The second cluster is a young one at a distance of 2.45 kpc located in the spiral arm + I. The field stars in the area of the two clusters show colour magnitude diagrams not very different from cluster diagrams. They indicate a space distribution in the form of star clouds. The old cluster NGC 559 seems to be embedded in the nearer cloud (1.2 kpc) which consists of young and old stars.

153.002 **Hβ photometry in M39.** D. Kilkenny, C. E. J. Hilditch, R. W. Hilditch, P. W. Hill, A. E. Lynas-Gray.
Monthly Notices Roy. Astron. Soc., Vol. 172, 5P - 8P (1975).

Hβ photometry of 24 stars in the region of the galactic cluster M39 is reported. The observations are combined with published data to derive a distance of 235 ± 30 pc for the cluster. An estimate of the cluster age is made.

153.003 **Photometric metal abundances for twenty clusters.** P. A. Jennens, H. L. Helfer.
Monthly Notices Roy. Astron. Soc., Vol. 172, 681 - 694 (1975).

Metal abundances, colour excesses and distance moduli have been determined for individual giant stars, using *UBViyz* photometry in 19 distant galactic clusters and one globular cluster. The purpose of this study is to establish metal abundances for clusters for which ages can be assigned in order to see if there has been significant metal enhancement in the galactic disc in the last 5×10^9 yr.

153.004 **On the evidence for circumstellar obscuration of OB stars.** B. Bohannan.
Astron. Journ., Vol. 80, 625 - 636 (1975).

Reddish found a strong increase of color excess with increasing stellar luminosity – here referred to as the Reddish effect – in ten young galactic clusters and associations. New photometric and spectroscopic observations drawn from the literature are combined with data used by Reddish to reconsider the color excess versus intrinsic magnitude diagrams of the same ten stellar groups. The reality of the effect is questioned in all but one of the systems; this was accomplished by identifying possible foreground stars and by recognizing some of the brightest stars as supergiants, then applying correct intrinsic color indices. After careful reanalysis, the one stellar group to retain an indication of the Reddish effect is Cygnus OB2. No correlation of reddening with luminosity was found for five additional very young stellar groups in the southern hemisphere; these groups should exhibit the effect if it is a natural consequence of stellar evolution.

153.005 **Individual characteristics of stars in open clusters. Mass distribution of cluster members.**
V. V. Muzylev, A. Eh. Piskunov.
Nauchn. Informatsii, vyp. (No.) 32, p. 116 - 122 (1974). In Russian.

The statistical distribution of stellar masses in 19 open clusters has been studied. A special computing procedure is used for determination of masses and ages of each cluster member depending on its location on the H–R diagram. It has been shown that the initial mass function for all clusters is age-independent. Nevertheless, observed mass functions for old clusters considerably differ from the initial one, thus evidencing that the process of stellar formation in open clusters was relatively long.

153.006 **Mass loss from red giants in old open clusters.** T. G. Hawarden.
Monthly Notices Roy. Astron. Soc., Vol. 173, 223 - 229 (1975).

The stars of the giant branch clump or red horizontal branch in all members of a sample of six open clusters older than the Hyades are found to be less concentrated towards the cluster centre than the stars of the upper main sequence. In the cluster NGC 2158 the effect is very prominent and is shown to be a real physical property of the cluster. A similar apparently strong effect in NGC 2477 may, however, be a consequence of a systematic variation of reddening with radial position in the cluster. The fact that at least five of the six old clusters studied show the deconcentration phenomenon in the same sense is taken as evidence that the phenomenon is a general property of the horizontal branches of population I systems. It is argued that the phenomenon is a consequence of mass loss during the most luminous phases of red giant evolution, probably by a stellar wind mechanism.

153.007 **A search for old southern open clusters.** T. G. Hawarden.
Monthly Notices Roy. Astron. Soc., Vol. 173, 231 - 235 (1975).

The methods used to identify old open clusters are described and results are given for the clusters studied. A catalogue of clusters older than $\sim 4 \times 10^8$ yr is presented. There appears to be a deficiency of old clusters in the direction of the galactic centre and an excess in the anticentre direction.

153.008 **Structure of the Hyades cluster.** G. Pels, J. H. Oort, H. A. Pels-Kluyver.
IAU Symposium No. 69, (see 012.009), p. 159 (1975).

153.009 **Luminosity and mass functions of stars in the open cluster NGC 129.** V. N. Frolov.
Pis'ma v Astron. Zhurn., Vol. 1, No. 8, p. 3 - 5 (1975). In Russian.

Luminosity and mass functions for the open cluster NGC 129 were obtained from new photometric and proper motion data on members of this cluster. In the range of $2 M_\odot < M < 8 M_\odot$ the mass function may be represented by a straight line with the slope −2.54. This value is in a satisfactory agreement with the slope of the initial mass function by Salpeter (−2.35).

153.010 **Distances, motions, and space distribution of the brighter Hyades cluster stars.**
M. S. Carpenter, T. E. Corbin, R. E. Schmidt.
Bull. American Astron. Soc., Vol. 7, 435 (1975). – Abstr. AAS.

153.011 **An astrometric study of the galactic cluster NGC 1039.** P. A. Ianna, H. A. McAlister.
Bull. American Astron. Soc., Vol. 7, 435 (1975). – Abstr. AAS.

153.012 **Galactic star clusters.** S. J. Hynes.
Journ. British Astron. Ass., Vol. 85, 502 - 504 (1975).

153.013 **HD 219460 and the open cluster Ma 50.** D. Crampton.
Publ. Astron. Soc. Pacific, Vol. 87, 523 - 525 (1975).

The distance modulus of the young compact cluster Ma 50 is estimated to be $12^m_.0 \pm 0.1$ from spectroscopic observations of the brighter members. The radial velocity of the cluster is -81 ± 1 km s^{-1}. The absolute magnitude of the W-N 4.5 component of HD 219460 is $M_v = -3^m_.9 \pm 0.2$; it is slightly less luminous than three stars of spectral type ~B0.5 which have evolved to luminosity class III stars.

153.014 **Non-uniform extinction in open star clusters and dispersion of the photometric sequences.**
G. Burki.
Astron. Astrophys., Vol. 43, 37 - 44 (1975).

The importance of non-uniform extinction in open star clusters of intermediate age is studied on the basis of photoelectric *UBV* measurements in open clusters catalogued by Mermilliod (1975) and of Wallenquist's study (1975) of the presence of absorbing matter by star counts in the fields of open clusters. It is found that clusters having less uniform projected stellar density on the sky are also those which present the largest dispersion in the two colour diagram.

153.015 **Investigation of open clusters by means of a theoretical HR-diagram.**
O. B. Dluzhnevskaya, A. Eh. Piskunov.
Nauchn. Informatsii, vyp. (No.) 33, p. 10, 27 (1974). In Russian and English. – Abstr. of a paper presented at the colloquium 'Physcis and evolution of stars', (see 012.015).

153.016 **Integral parameters and the total mass of open clusters.** A. Eh. Piskunov.
Nauchn. Informatsii, vyp. (No.) 33, p. 10, 27 (1974).
In Russian and English. – Abstr. of a paper presented at the colloquium 'Physics and evolution of stars', (see 012.015).

153.017 **Integral parameters and masses of open clusters.** A. Eh. Piskunov.
Nauchn. Informatsii, vyp. (No.) 33, p. 101 - 116 (1974). In Russian.

Integral parameters of 22 open clusters with well selected members are derived. The data obtained allow to determine the masses of open clusters and to compare them with the theoretical relations obtained previously (1972).

153.018 **Fotometría fotográfica UBV del cúmulo galáctico King 19.**
T. J. Vives, C. Morales, J. García-Pelayo, J. Barbero.
Urania Barcelona, Año 59, No. 281–282, p. 205 - 216 (1974).

153.019 **The dependence of the orbital elements of open clusters on their age.**
K. A. Barkhatova, E. D. Pavlovskaya.
Astron. Tsirk., No. 841, p. 5 - 7 (1974). In Russian.

153.020 **Proper motions and photometry of the galactic clusters NGC 7788 and NGC 7790.** V. N. Frolov.
Astron. Tsirk., No. 845, p. 5 - 7 (1974). In Russian.

153.021 **Proper motions of two galactic clusters in Cassiopeia and of the classical cepheid CG Cas.**
V. N. Frolov.
Astron. Tsirk., No. 848, p. 1 - 2 (1974). In Russian.

153.022 **Determination of the proper motion of the cluster M44 (Praesepe) from 26″ refractor plates.**
I. A. Zykov.
Astron. Tsirk., No. 863, p. 3 - 6 (1975). In Russian.

153.023 **Changes of magnitudes of bright stars in the cluster NGC 6913 from 1967 to 1974.** R. M. Raznik.
Astron. Tsirk., No. 866, p. 3 - 5 (1975). In Russian.

153.024 **NGC 2175: the cluster age and the nature of the nebulosity surrounding S 252a.**
G. L. Grasdalen, L. Carrasco.
Astron. Astrophys., Vol. 43, 259 - 265 (1975).

The galactic cluster NGC 2175 has been proposed as an extraordinarily young stellar association. The authors have derived an age for the cluster of 2×10^6 years. The physical nature of an optical nebulosity associated with the B star S 252a was investigated. It is concluded that the object is primarily a reflection nebula due to a dense ($n \gtrsim 10^3$ cm^{-3}) neutral cloud. This cloud is surrounded by a thin, high density, low excitation ionized zone. An evolutionary model for the cloud is proposed.

153.025 **New members of the Hyades cluster and a discussion of its structure.**
G. Pels, J. H. Oort, H. A. Pels-Kluyver.
Astron. Astrophys., Vol. 43, 423 - 441 (1975).

By means of new proper motion data a search for fainter members of the Hyades cluster has been made over a region which should comprise the entire cluster. 119 new members have been found. Surface and space density distributions were derived for bright ($V < 9$) and faint ($V > 9$) stars separately. Within about 3 pc the brighter stars are strikingly more concentrated than the fainter ones. Total masses and gravitational potential have been computed. The density distribution was compared with the expected distribution in a cluster in dynamical equilibrium, with a truncated Maxwellian velocity distribution. The agreement is satisfactory. At $r = 4$ pc the average internal space velocities are 0.36 and 0.40 km/s for the two groups. If the boundary of the cluster is taken at 14 pc the total mass is about 300 $M_\odot$, half of which lies within 3.6 pc. A tentative determination of the luminosity function shows it to be roughly constant between $M_{pg} = +3$ and +9, and perhaps also beyond +9.

153.026 **Magnitudes and colour indices of new possible members of the Praesepe cluster.**
V. G. Moshkalev.
Soobshch. Gos. Astron. Inst. Shternberga, No. 182, p. 21 - 23 (1973). In Russian.

The magnitudes and colour indices of 13 stars detected as new possible members of Praesepe cluster are given. It is shown that 60% of them may belong to the cluster.

153.027 **NGC 3114: another open cluster possibly rich in peculiar stars.** H. Levato, S. Malaroda.
Astron. Journ., Vol. 80, 807 - 808 (1975).

MK types for 22 stars in the field of NGC 3114 are provided. Two new silicon stars were found in addition to the two already known. Another three anomalous spectra were discovered; one Hg–Mn star, one with O II lines enhanced, and the other with N II and λ4200 enhanced. A comparison of descriptions of the peculiar stars in NGC 2516 and IC 2602 is made.

153.028 **The old, metal-poor open cluster NGC 2243.** T. G. Hawarden.
Monthly Notices Roy. Astron. Soc., Vol. 173, 801 - 820 (1975).

Photoelectric *UBV* and photographic *BV* photometry of the southern open cluster NGC 2243 is presented. The reddening and ultraviolet excess are determined by a general method which has been tested on several better-observed clusters. The adopted values for NGC 2243 are $E(B-V) = 0^m_.06$, $\delta(U-B)_{0.6} = 0.15 \pm 0^m_.035$. The metal abundance is estimated to be [Fe/H] = –0.5 relative to the sun. The unusually large age $(5.0 \pm 0.8) \times 10^9$ yr is deduced from the colour of the main sequence turn-off for an assumed helium abundance $Y = 0.30$. A distance modulus $(m-M)_0 = 12^m_.8$ is derived.

153.029 **Statistical identification of binaries in star clusters.** C. Bettis.
Publ. Astron. Soc. Pacific, Vol. 87, 707 - 715 (1975).
A technique is presented for identifying binary systems using computer curve-fitting methods applied to the color-magnitude diagrams of individual star clusters (Hyades, Pleiades, and Praesepe). The frequency of binaries in the analyzed clusters is discussed, and it appears that the Hyades has a somewhat higher percentage of binaries than the other two clusters. The effect of unresolved binaries on the distance determination of galactic clusters by main sequence fitting is analyzed.

153.030 **Four-color and Hβ photometry for open clusters I. NGC 2516.** M. S. Snowden.
Publ. Astron. Soc. Pacific, Vol. 87, 721 - 734 (1975).
Extensive *uvby* and Hβ photometry was obtained for stars in the region of the open cluster NGC 2516. A photometric analysis revealed variable reddening and a mean reddening of $E(b-y) = 0^m.088$. In addition to determining a new age of 137×10^6 years and a new adopted distance modulus of $8^m.01$, several possible new variable stars were discovered.

153.031 ***UBV* photometry of Collinder 463.** R. E. Townsend.
Publ. Astron. Soc. Pacific, Vol. 87, 753 - 757 (1975).
Collinder 463 is a relatively poor galactic cluster located near the edge of the Orion arm at a distance of 650 parsecs. The cluster is found to have an age of 1.5×10^8 years and probably contains at least three giants. Differential reddening of $0^m.8$ across the cluster yields an estimate for R which is consistent with the currently adopted value.

153.032 **Spectral types in the open cluster NGC 2451.** H. Levato, S. Malaroda.
Publ. Astron. Soc. Pacific, Vol. 87, 823 - 825 (1975).
MK spectral types are provided for 23 stars in the field of the open cluster NGC 2451. Two peculiar stars were found.

153.033 **Relative proper motions in the open cluster NGC 654.** R. C. Stone.
Bull. American Astron. Soc., Vol. 7, 509 - 510 (1975).
Abstr. AAS.

153.034 **The open cluster Trumpler 5.** J. K. Kalinowski.
Bull. American Astron. Soc., Vol. 7, 542 (1975).
Abstr. AAS.

153.035 **A brief history of open-cluster research.** M. S. Snowden.
Southern Stars, Vol. 26, 63 - 74 (1975).

153.036 **Stellar kinematics and the cosmic distance scale.** C. A. Murray.
Space astrometry, (see 012.037), p. 129 - 134 (1975).
The present distance scale relies heavily on kinematic distances measured for members of the Hyades cluster and Sco-Cen association. Part of the uncertainty in the scale arises from suspected errors in the proper motion system over relatively large areas of sky; an independent determination of absolute proper motions of member stars from space is very desirable. Higher systematic accuracy over the whole sky should also improve the determination of statistical and secular parallaxes of variable stars, particularly RR Lyrae variables, which are of extreme importance in the extension of the scale to extra-galactic objects.

153.037 **Proper motions and BV photometry of NGC 129 and field stars.** V. N. Frolov.
Izv. Glav. Astron. Obs. Pulkovo, No. 193, Astrofiz., Astrometr. i Zvezdn. Astron., p. 80 - 91 (1975). In Russian.
Proper motions of 326 NGC 129 and field stars relative to reference stars of the 14th magnitude were obtained from plates taken with the Pulkovo normal astrograph. Three plate pairs with a mean epoch difference of 65.5 years were used and all stars brighter than $B = 15.65$ measured. The coordinates of the center of the measured area ($25' \times 25'$) are: $\alpha = 0^h27^m.1$, $\delta = +59°57'$ (1950.0). Photographic B and V magnitudes were determined from plates taken with the normal astrograph and the 26″ refractor of the Pulkovo Observatory. All 141 cluster members were found using proper motion and B, V photometric criteria. The classical cepheid DL Cas and one supergiant (class Ib) are members of the cluster according to proper motion data. A catalogue of the positions, magnitudes and proper motions of the stars measured is given.

153.038 **Spectral types in the open cluster IC 4665.** H. A. Abt, H. Levato.
Publ. Astron. Soc. Pacific, Vol. 87, 849 - 851 (1975).
Spectral types for 18 bright cluster members have been derived from spectra of several dispersions. Among the peculiar spectra are (1) two B6 Vp stars with enhanced C II, (2) one Hg-Mn Ap star, (3) a shell star, and (4) a classical Am star.

153.039 **Intermediate- and wide-band photometry of NGC 6025.** G. C. Kilambi.
Publ. Astron. Soc. Pacific, Vol. 87, 975 - 981 (1975).
The *uvby* photoelectric photometry and *UBV* photographic photometry of NGC 6025 have yielded a distance modulus of 9.7 ± 0.1 and an age of 90 million years. No major deviation from a normal interstellar reddening law has been noted. A gap in the $(B-V)$ distribution of stars has been detected and the luminosity function has been derived.

153.040 **Infrared observations of the field cluster NGC 7419.** R. A. Vardanyan, L. G. Akhverdyan.
Astrofizika, Vol. 11, 553 - 554 (1975). In Russian. – English translation in Astrophysics, Vol. 11, No. 3.
The field of the star cluster NGC 7419 is observed by means of an image tube in the infrared (0.9 μ). Five very red stars are detected.

153.041 **A spectroscopic investigation of the Pleiades.** J. A. Pearce, G. Hill.
Publ. Dominion Astrophys. Obs., Victoria, B.C., Vol. 14, (No. 14), 319 - 343 = National Res. Council, Canada, NRC No. 14569 (1975).
A total of 469 radial velocities of 115 stars located in, and around, the Pleiades is presented. The authors have determined the velocity differences between the Victoria material and other extensive sources of data. An investigation of the binary frequency preceded a membership study based on proper motion, radial velocity and photometric criteria. New and revised orbits have been obtained for the stars HD 22637, HD 22805, HD 23964 and HD 24769. From members thought to be constant velocity objects a mean cluster velocity of $+7.6 \pm 2.3$ (s.d. 45 stars) km/sec has been found.

153.042 **Photoelectric photometry of the open cluster NGC 1778.** U. C. Joshi, R. Sagar, P. Pandey.
Pramāṇa, (*India*), Vol. 4, 160 - 163 (1975).
Photoelectric U, B, V magnitudes and colours for 48 stars in the open cluster NGC 1778 have been determined. The distance to the cluster is 1062 pc, and the apparent diameter of the cluster 4.5 minutes of arc. The age is estimated at 1.6×10^8 years.

Some experiments with an SEC image tube for high precision multi-colour photometry of galactic clusters.
See Abstr. 031.287.

Stellar evolution IV: Evolution of a star of 1.5 $M_\odot$

from the main-sequence to the red-giant branch with and without overshooting from convective core.
See Abstr. 065.060.

La théorie des étoiles jumelles et l'évolution stellaire.
See Abstr. 065.075.

Population I and population II stellar evolution.
See Abstr. 065.129.

Parallaxes of 18 dwarf members of the Hyades cluster. See Abstr. 111.001.

Proper motions and positions of stars in wide surroundings of the Hyades cluster. See Abstr. 112.013.

The ESO/Uppsala survey of the ESO (B) atlas of the southern sky. III. See Abstr. 113.044.

Cool carbon stars in open clusters.
See Abstr. 114.007.

Physical properties of the red giants in NGC 2420.
See Abstr. 115.006.

The age of the W Ursae Majoris stars.
See Abstr. 117.056.

Orbits and masses of Hyades visual binaries.
See Abstr. 118.003.

Investigation of an eclipsing variable in the open star cluster NGC 103. See Abstr. 121.012.

Binary systems in star clusters. II. HD 162724.
See Abstr. 121.050.

Flare stars in star clusters and associations.
See Abstr. 122.006.

On the presence of variable stars in open clusters.
See Abstr. 122.031.

Stars of δ Scuti-type in stellar clusters.
See Abstr. 122.033.

New flare stars in the Pleiades. II.
See Abstr. 122.104.

Two-colour observations of flare stars in the Pleiades. See Abstr. 122.105.

Cepheids in open clusters.
See Abstr. 122.108.

The distribution of flares of red dwarfs from energies of optical radiation. See Abstr. 122.157.

Double mode cepheid V367 Sct in open cluster NGC 6649. See Abstr. 122.176.

Spectral types for the brightest flare stars in the Pleiades. See Abstr. 122.216.

Spectrophotometric observations of suspected red subluminous stars in the Hyades and field.
See Abstr. 126.012.

A polarimetric search for circumstellar envelopes in the young clusters NGC 6611 and NGC 225.
See Abstr. 131.198.

Polarimetry of the faint nebula near the Pleiades star cluster. See Abstr. 132.037.

On the occurrence of planetary nebulae in open star clusters. See Abstr. 133.003.

Infrared observations of stars in NGC 2264.
See Abstr. 141.635.

Dynamics of star clusters: comparison of theory with observations and simulations. See Abstr. 151.018.

Numerical experiments for investigating the effect of multiplicity of a stellar encounter. See Abstr. 151.039.

Computer simulations of stellar systems.
See Abstr. 151.046.

Atlas of star clusters with colour magnitude diagrams in the Magellanic Clouds. See Abstr. 154.004.

Is NGC 6528 an intermediate age cluster?
See Abstr. 154.017.

The cluster system in the Small Magellanic Cloud.
See Abstr. 159.005.

Errata

153.901 **Erratum: "Spectroscopic study of the open cluster NGC 2422"** [Astron. Journ., Vol. 80, 131 - 133 (1975)]. M. M. Dworetsky.
Astron. Journ., Vol. 80, 668 (1975).

154 Globular Clusters

154.001 **M4 the closest globular cluster.** G. Alcaino.
Astron. Astrophys., Suppl. Ser., Vol. 21, 5 - 14 (1975).

A *BV* photometric investigation of the bright southern globular cluster M4 was carried out. The apparent distance modulus from the mean visual magnitude of 30 horizontal branch RR Lyrae stars within a radius of 9'.3 from the cluster centre and adopting $M_v = +0.6$ for their absolute magnitude is $(m-M)$app.= 12.86. The adopted distance is 1.75 kpc making it the closest globular cluster to the sun, highly reddened by $E(B-V)=0^m55$ because of the dark nebulosity in Scorpio-Ophiuchus. The giant branch rises 2^m45 above the horizontal branch at $(B-V)_0=1^m4$, characteristic of an object of medium metallicity.

154.002 **The globular cluster NGC 288.** G. Alcaino.
Astron. Astrophys., Suppl. Ser., Vol. 21, 15 - 24 (1975).

A *BV* photometric investigation of the globular cluster NGC 288 situated in the direction of the south galactic pole, was carried out. The derived apparent distance modulus is $(m-M)$app. = 14^m80. The reddening $E(B-V)=0^m05$. The true distance modulus is $(m-M)_0 = 14.65$. The distance is 8.5 kpc from the sun and the galactic plane and 12 kpc from the galactic centre. The main feature of the colour magnitude diagram is the presence of a large sample of blue horizontal branch stars in the range $-0.23 < (B-V) < 0.03$. Due to the undefined position of the horizontal branch it is difficult to determine its metallicity from the ΔV method; but the average steepness of the giant branch suggests that this cluster ranks among those of medium metallicity richness.

154.003 **General properties of the two groups of globular clusters and their variable stars.**
A. V. Mironov, N. N. Samus'.
IAU Symposium No. 67, (see 012.001), p. 541 - 543 (1975).

The dependences of the numbers of variable stars in globular clusters on the chemical composition are studied. For given metallicity the numbers of RR Lyrae stars reduced to some definite total number of stars in the cluster are different for the two groups of globular clusters introduced by Mironov.

154.004 **Atlas of star clusters with colour magnitude diagrams in the Magellanic Clouds.** G. Alcaino.
Astron. Astrophys., Suppl. Ser., Vol. 21, 279 - 406 (1975).

The author presents on a uniform scale the colour-magnitude diagrams of 30 of the most spectacular star clusters in the LMC and 6 in the SMC together with the magnitudes and colours of the stars with their identification charts. He includes the photometric data and finding maps of 13 selected photoelectric sequences in the LMC and 9 in the SMC. Tables with basic data are listed for 162 of the most conspicuous star clusters in the LMC and 116 in the SMC, and for 42 photoelectric sequences in the LMC and 22 in the SMC.

154.005 **UBV photometry of the LMC cluster NGC 1466.**
A. J. Penny.
Monthly Notices Roy. Astron. Soc., Vol. 172, 65P - 67P (1975).

UBV electronographic photometry of stars in the LMC globular cluster NGC 1466 down to $V = 21^m$ indicates that the giant branch stars have an ultraviolet excess of $0^m32 \pm 0^m09$, showing that the cluster is very metal poor. The horizontal branch has a magnitude of $V = 18^m8$, implying a distance modulus of 18^m2.

154.006 **The tip of the red giant branch (a comparison of theory and observations).** Z. I. Kadla.
Astron. Zhurn. Akad. Nauk SSSR, Vol. 52, 811 - 814 (1975). In Russian. English translation in Soviet Astron., Vol. 19, No. 4.

Observational data on the five brightest giants in the globular clusters M3, M5, M13, M15 and M92 are compared with theoretically computed data on the helium flash. The considered theoretical models are in agreement with observations.

154.007 **The colour-magnitude diagram of the globular cluster M4.** V. G. Moshkalev.
Astron. Zhurn. Akad. Nauk SSSR, Vol. 52, 815 - 822 (1975). In Russian. English translation in Soviet Astron., Vol. 19, No. 4.

The $V-(B-V)$ diagram of the globular star cluster M4 = NGC 6121 was obtained photographically on the basis of the author's photoelectric sequence. The basic parameters of the cluster are discussed. The questions of the chemical composition of the cluster stars and of the shape of its luminosity function are briefly considered.

154.008 **The galactic halo: globular clusters.** L. Woltjer.
Astron. Astrophys., Vol. 42, 109 - 118 (1975).

The distribution of globular clusters is studied. Individual distances are redetermined on the basis of available data and an adopted $M_V = +0^m6$ for RR Lyrae stars. Both metal poor (H) and comparatively metal rich (D) clusters appear to have an essentially spherical distribution with the D clusters concentrated close to the galactic center, except for a few cases. Kinematical data are consistent with the spatial density distributions if a value of Θ_c for the LSR of 200–225 km/s (for $R_0 = 9$ kpc) is adopted. The mass density in the globular cluster system declines sharply beyond 20 kpc from the center. Estimates of the mass of the halo are made.

154.009 **Metal abundances of Draco and eight galactic globular clusters.** R. Canterna.
Astrophys. Journ., (*Letters*), Vol. 200, L63 - L66 (1975).

Photometric metal abundances of individual giants in the Draco galaxy and eight globular clusters have been obtained using the C, M, T_1, T_2 photometric system. This new broad-band system was designed specifically for photometric abundance determinations of G and K giants. The results show that [Fe/H] = –3.0 for the Draco giants, much lower than M92. M22 is metal-deficient, [Fe/H] = –2.0, in agreement with its horizontal-branch morphology, but inconsistent with the slope of the red-giant branch. [Fe/H] of NGC 4147 and NGC 7006 is similar to M22.

154.010 **Dynamical masses and mass-to-light ratios for globular clusters.** G. Illingworth.
IAU Symposium No. 69, (see 012.009), p. 151 - 157 (1975).

Dynamical masses have been determined for 10 globular clusters. Comparison of the dynamically determined M/L values with those calculated from extrapolations of the observed upper main sequence/giant branch luminosity functions for M3 and M5 indicate that low mass stars ($\sim 0.2 M_\odot$) comprise a large fraction of the total number.

154.011 **Early-type stars near globular clusters. NGC 104, M 4, M 22, and M 55.** A. G. D. Philip.
Bull. American Astron. Soc., Vol. 7, 446 (1975). – Abstr. AAS.

154.012 **On the color indices of giants and subgiants in globular clusters.** D. Butler.
Publ. Astron. Soc. Pacific, Vol. 87, 559 - 560 (1975).

The globular-cluster metal-abundance parameter

$(B-V)_{o,g}$, which is the intrinsic color index at the intersection of the horizontal and subgiant branch in the color-magnitude diagram, is shown to be tightly correlated with [Fe/H].

154.013 **The interstellar reddening of the globular cluster 47 Tucanae.** D. L. Crawford, M. S. Snowden.
Publ. Astron. Soc. Pacific, Vol. 87, 561 - 563 (1975).

The authors have estimated the interstellar reddening in front of 47 Tuc by observations of 26 foreground A- and F-type stars. The result is $E(b-y) = 0^m.021$. The zero point may be uncertain by as much as $0^m.013$.

154.014 **On the chemical composition and the origin of the globular clusters of our Galaxy.**
A. V. Mironov, N. N. Samus'.
Peremennye Zvezdy, Vol. 19, 337 - 356 (1974). In Russian.

16 morphological parameters characterizing the shape of the horizontal branches of globular clusters were introduced. The values of these parameters have been determined for 47 globular clusters of our Galaxy. A number of correlations between morphological parameters is found. The helium abundance for stars of 39 globular clusters has been estimated.

154.015 **The globular cluster IC 4499.**
C. R. Fourcade, J. R. Laborde, J. C. Arias.
Peremennye Zvezdy, Prilozhenie, Vol. 2, No. 7, p. 3 - 17 (1974). In Russian.

The colour-magnitude diagram of the globular cluster IC 4499 was constructed on the basis of 6 B– and 6 V–plates. 171 stars suspected of variability were measured. Adopting for the absolute magnitude of the RR Lyrae stars the value Mv = $0^m.5$ the apparent modulus is $17^m.26$. Since the adopted absorption is Av = $1^m.05$ the real modulus is $16^m.21$, which corresponds to a distance of 17 kpc and a height above the galactic plane of 6.1 kpc. It is characteristic for a globular cluster of low metallicity.

154.016 **Study of the globular cluster NGC 6362.**
C. R. Fourcade.
Peremennye Zvezdy, Prilozhenie, Vol. 2, No. 7, p. 18 - 30 (1974). In Russian.

Three-colour photometry of the globular cluster NGC 6362 was made on the basis of 5 U-, 10 B- and 17 V-plates obtained with the 154-cm reflector at the Astronomical station Bosque Alegre, Cordoba, Argentina. The colour-magnitude and colour-colour diagrams are shown on figures. Adopting the absolute magnitude of RR Lyrae stars Mv = $+0^m.5$ and the apparent magnitude mv = $15^m.30$ the apparent modulus is $14^m.80$. Since the adopted absorption Av = $0^m.24$, the true modulus is $14^m.56$, which corresponds to a distance of 8.17 kpc and a height above the galactic plane of 2.46 kpc. The cluster NGC 6362 is of medium-high metallicity.

154.017 **Is NGC 6528 an intermediate age cluster?**
C. R. Fourcade.
Peremennye Zvezdy, Prilozhenie, Vol. 2, No. 7, p. 31 - 41 (1974). In Russian.

On the basis of 4 B–plates and 3 V–plates obtained with the 154-cm reflector of the observatory Bosque Alegre (Argentina) the colour-magnitude diagram of the globular cluster NGC 6528 is constructed. The distance of NGC 6528 is 6.4 kpc from the sun and 467 pc from the galactic plane. Observations show that NGC 6528 would be an intermediate cluster between a normal globular cluster and an old galactic cluster.

154.018 **Some comments on the three globular clusters IC 4499, NGC 6362 and NGC 6528.**
B. V. Kukarkin.
Peremennye Zvezdy, Prilozhenie, Vol. 2, No. 7, p. 42 - 47 (1974). In Russian.

154.019 **Observations of eight globular clusters at 2.3 and 4.7 μm.** O. L. Hansen, J. E. Hesser.
Nature, Vol. 257, 568 - 569 (1975). – Letter.

154.020 **The axial ratio and position angle of the semi-major axis of some globular clusters.** Z. I. Kadla, A. A. Strugatskaya, N. Richter, W. Högner.
Astron. Tsirk., No. 843, p. 2 - 5 (1974). In Russian.

154.021 **The chemical inhomogeneity of Omega Centauri.**
K. C. Freeman, A. W. Rodgers.
Astrophys. Journ., (*Letters*), Vol. 201, L71 - L74 (1975).

Observations of 25 RR Lyrae stars in ω Cen have revealed a diversity in the composition of these stars. The composition spread explains the ill-defined giant branch and the existence of M stars in the cluster. The diversity may be due either to variable amounts of giant-branch mixing or to metal enrichment of protostellar gas during the cluster collapse. Among other considerations, the uniqueness of ω Cen leads the authors to prefer the latter hypothesis.

154.022 **Abundance variations on the lower giant branch of Omega Centauri.** J. Norris, M. S. Bessell.
Astrophys. Journ., (*Letters*), Vol. 201, L75 - L79 (1975).

Photoelectric photometry on the *UBV, RI,* and DDO systems is presented for stars on the lower giant branch of ω Centauri. Analysis of these data leads to the following conclusions: (1) Differential reddening is small across the cluster ($\Delta E(B-V) \lesssim 0.04$). (2) There exists a range of surface composition in the giants at the level of the horizontal branch which may be due either to abundance inhomogeneities within the cluster (which presumably existed at the end of the epoch of star formation) or to the mixing of processed materials from the interiors to the surfaces of these objects. Low-dispersion spectra (50 Å mm^{-1}) of some of these objects show relatively strong CN features indicating that the mixing hypothesis is probably the correct explanation.

154.023 **A preliminary study of the main-sequence luminosity function of the globular cluster M92.**
S. van den Bergh.
Astrophys. Journ., Vol. 201, 585 - 588 (1975).

The Hale 5-m telescope has been used to study the luminosity function of M92 down to $V \approx 23.25$. The main-sequence luminosity function is found to rise very steeply from $M_V = +4$ to $M_V = +6$. Between $M_V = +6$ and $M_V = +8$ the luminosity function appears to remain essentially constant.

154.024 **Variable stars in the X-ray globular cluster NGC 1851.** M. H. Liller.
Astrophys. Journ., (*Letters*), Vol. 201, L125 - L126 (1975).

Ten known and four new variable stars in the X-ray globular cluster NGC 1851 have been studied. None of these variables show obvious unusual characteristics that would make them candidates for the X-ray emission. The characteristics of the variable stars are presented.

154.025 **The globular cluster NGC 6809.** G. Alcaino.
Astron. Astrophys., Suppl. Ser., Vol. 22, 193 - 205 (1975).

A *BV* photometric investigation of the southern globular cluster NGC 6809 was carried out. The derived apparent distance modulus is $(m-M)_{app} = 13^m.65$. The reddening is $E(B-V) = 0^m.08$. The true distance modulus is $(m-M)_0 = 13^m.41$. The distance is 4.81 kpc from the sun and 1.90 kpc from the galactic plane. The main features of the colour-magnitude diagram are: the presence of a large sample of blue horizontal branch stars, a well defined gap in the horizontal branch, a steep and richly populated subgiant branch, and a deficiency of bright red giants. The giant branch rises $2^m.8$ above the horizontal branch at $(B-V)_0 = 1^m.4$, characteristic of a rather metal poor object.

154.026 **A new CH star in the globular cluster Omega Centauri.** H. E. Bond.
Astrophys. Journ., (*Letters*), Vol. 202, L47 - L49 = Contr. Louisiana State Univ. Obs., *Baton Rouge*, No. 109 (1975).

A photograph of ω Centauri made with a transmission-grating slitless spectrograph has revealed a new CH star near the center of this globular cluster. The star's radial velocity is found to be consistent with cluster membership, and a visual absolute magnitude $M_v = -1.8$ is derived. The total number of carbon stars now known in globular clusters is three, all of which are in ω Cen.

154.027 **On the distribution function of a number of stars according to velocities and coordinates in the model of a globular cluster with discrete stellar mass distribution.**
V. M. Bagin.
Astron. Zhurn. Akad. Nauk SSSR, Vol. 52, 1011 - 1016 (1975). In Russian. English translation in Soviet Astron., Vol. 19, No. 5.

The distribution function of a number of stars according to velocities and coordinates for each group of stars is obtained according to a given distribution of an average space density of a number of stars in a globular cluster model with discrete stellar mass distribution in the approximation of a weak interaction of stars. Relative densities of the stellar number distribution according to the velocity modulus were calculated and constructed as an example for the case of two stellar groups in the cluster with stellar mass ratio 2 and 3.

154.028 **The star distribution in M15.**
J. N. Bahcall, N. A. Bahcall, D. Weistrop.
Astrophys. Letters, Vol. 16, 159 - 163 (1975).

Star counts are reported that extend to within about 0.5 pc of the center of the globular star cluster M 15. Combining the results with previous surface brightness measurements, the authors derive the best-fitting parameters for the star distribution (core radius = 0.4 pc, central density = $4.5 \times 10^4 M_\odot/pc^3$) and the limits that can be placed on the mass of a central massive black hole ($M_{BH} \lesssim 8 \times 10^3 M_\odot$).

154.029 **Inside a globular star cluster.** A. D. Boyko.
Zemlya i Vselennaya, 1975, No. 6, p. 34 - 39. In Russian.

154.030 **The reddening of globular clusters.** R. Racine.
Journ. Roy. Astron. Soc. Canada, Vol. 69, 244 (1975). – Abstr. CAS.

154.031 **Relative proper motions in M15.** K. M. Cudworth.
Bull. American Astron. Soc., Vol. 7, 510 (1975). Abstr. AAS.

154.032 **Stellar collisions in globular clusters.**
J. G. Hills, C. A. Day.
Bull. American Astron. Soc., Vol. 7, 535 (1975). – Abstr. AAS.

154.033 **Surface photometry of globular clusters.**
T. W. Rettig.
Bull. American Astron. Soc., Vol. 7, 535 (1975). – Abstr. AAS.

154.034 **NGC 6273: a globular cluster with 3 W Virginis type variables.** C. M. Coutts, H. S. Hogg, D. C. Thompson.
Bull. American Astron. Soc., Vol. 7, 536 (1975). – Abstr. AAS.

154.035 **An optical search for ionized hydrogen in globular clusters.**
M. G. Smith, J. E. Hesser, S. J. Shawl.
Bull. American Astron. Soc., Vol. 7, 536 (1975). – Abstr. AAS.

154.036 **Intense X-ray bursts from a globular cluster.**
J. Grindlay, J. Heise.
IAU Circ., No. 2879 (1975).

154.037 **Globular cluster winds.** D. J. Faulkner, K. C. Freeman.
Proc. Astron. Soc. Australia, Vol. 2, 356 - 358 (1975).

154.038 **Spatial distribution of stars in the globular cluster NGC 5466.**
R. Buonanno, F. D'Antona, R. De Amicis.
Mem. Soc. Astron. Italiana, Vol. 46, 349 - 354 (1975).

The apparent spatial distribution of the stars in NGC 5466, taken from the Palomar Sky Survey prints, is examined and compared to the previous results by Kholopov (1962). The physical behaviour of differently evolving stars is discussed.

Velocity selection of cluster members with use of objective-prism plates. See Abstr. 031.222.

Observational evidence for mass loss in globular clusters: giants. See Abstr. 065.053.

Population I and population II stellar evolution. See Abstr. 065.129.

Massive black holes in globular clusters? See Abstr. 066.019.

The ESO/Uppsala survey of the ESO (B) atlas of the southern sky. III. See Abstr. 113.044.

A finding list of B and A stars near the globular cluster M22. See Abstr. 113.047.

Photometric variability of certain red giant stars in the globular cluster NGC 7078 (M15). See Abstr. 113.058.

Enhanced abundances of nitrogen and s-process elements in red giant stars in Omega Centauri. See Abstr. 114.039.

Search for the TiO band λ 7054 in the spectra of M13 giants. See Abstr. 114.058.

The spectra of some giant stars in the globular cluster 47 Tuc. See Abstr. 114.100.

On the eclipsing binaries in the globular clusters NGC 3201 and NGC 5139. See Abstr. 121.004.

Photometric observations of the eclipsing binaries V65 and V78 in the globular clusters NGC 3201 resp. NGC 5139. See Abstr. 121.016.

Red variables of the halo and the globular clusters. See Abstr. 122.048.

Period changes of RR Lyrae stars in globular clusters. See Abstr. 122.049.

The period-luminosity relation for red variables in globular clusters. See Abstr. 122.053.

Metal abundances of RR Lyrae stars in galactic globular clusters. See Abstr. 122.073.

Veränderliche Sterne in Kugelsternhaufen. See Abstr. 122.131.

Current problems on horizontal branch stars.

III. An analysis of the evolutionary status of the RR Lyrae stars in the globular cluster ω Cen. See Abstr. 122.162.

On the expected properties of the RR Lyrae pulsators. See Abstr. 122.217.

Metal abundances of RR Lyrae stars in galactic globular clusters. See Abstr. 122.223.

Variable stars in the globular cluster NGC 5286. See Abstr. 123.075.

Observations of variable X-ray sources in globular clusters. See Abstr. 142.009.

Observations of a bright, variable X-ray source in a globular cluster. See Abstr. 142.010.

X-ray binaries in globular clusters. See Abstr. 142.036.

The X-ray source MX 0513–40 and the globular cluster NGC 1851. See Abstr. 142.040.

Tidal capture formation of binary systems and X-ray sources in globular clusters. See Abstr. 142.044.

Observations of X rays from near NGC6440. See Abstr. 142.047.

On the nature of the globular cluster X-ray sources. See Abstr. 142.061.

Galactic super halos, globular clusters, and the nature of the high latitude X-ray sources. See Abstr. 142.069.

Precise location of the X-ray source 3U1820-30 associated with the globular cluster NGC 6624. See Abstr. 142.084.

Astronomy from space: new class of X-ray sources found. See Abstr. 142.100.

Galactic superhaloes, globular clusters and high latitude X-ray sources. See Abstr. 142.105.

A comparison of X-ray sources in globular clusters and in the galactic bulge. See Abstr. 142.117.

The formation of globular cluster X-ray sources through neutron star – giant collisions. See Abstr. 142.140.

The distribution of X-ray sources in the Galaxy. II. See Abstr. 142.163.

Photometric studies of the X-ray globular cluster NGC 6624 = 3U 1820–30. See Abstr. 142.193.

Random gravitational encounters and the evolution of spherical systems. VI. Plummer's model. See Abstr. 151.008.

Homology in the evolution of cluster cores. See Abstr. 151.010.

The structure of round stellar systems: observation and theory. See Abstr. 151.017.

Random gravitational encounters and the evolution of spherical systems. VII. Systems with several mass groups. See Abstr. 151.045.

Computer simulations of stellar systems. See Abstr. 151.046.

Encounters between binary and single stars and their effect on the dynamical evolution of stellar systems. See Abstr. 151.064.

Photometric metal abundances for twenty clusters. See Abstr. 153.003.

Stellar kinematics and the cosmic distance scale. See Abstr. 153.036.

The origin of the Galaxy. See Abstr. 155.043.

Stellar populations in galaxies. See Abstr. 158.139.

4000 globular clusters in M87. See Abstr. 158.221.

The cluster system in the Small Magellanic Cloud. See Abstr. 159.005.

Errata

154.901 **Errata: 'Distribuzione spaziale delle stelle nell'ammasso globulare M13'** [Mem. Soc. Astron. Italiana, Vol. 42, 23 - 32 (1971)]. A. Blaghikh, V. Castellani. Mem. Soc. Astron. Italiana, Vol. 46, 143 (1975).

155 Structure and Evolution of the Galaxy

155.001 **The space distribution of M giants in the Warner and Swasey Luminosity Function Field LF 13.**
G. Vleeming, P. S. Thé.
Astron. Astrophys., Suppl. Ser., Vol. 21, 33 - 50 (1975).
The present paper is the third in a series, concerning the space distribution of M giants in southern Milky Way regions. A survey of M-type stars has been made in a square field of 3°× 3° in LF 13, centered at RA= 10^h18^m, Dec.= −52° 18′ (1950).

155.002 **R-associations III. Local optical and spiral structure.** W. Herbst.
Astron. Journ., Vol. 80, 503 - 509 (1975).
The galactic distribution of R-associations is investigated. They show that the local (Cygnus) spiral feature extends over 2 kpc in the direction of Vela ($l \sim 265°$), and is clearly separated from the next inner (Carina-Sagittarius) spiral feature. There is some evidence for a concentration of the dust towards the inner edge of the local optical spiral feature, a phenomenon which is observed in external galaxies (Lynds 1970, 1972). The R-associations are combined with other good spiral tracers to produce a new map of optical spiral structure. A possible interpretation of this map is suggested in which the angle of inclination of the spiral arms is 13°±4°.

155.003 **High-energy galactic gamma radiation from cosmic rays concentrated in spiral arms.**
G. F. Bignami, C. E. Fichtel, D. A. Kniffen, D. J. Thompson.
Astrophys. Journ., Vol. 199, 54 - 60 (1975).
A model for the emission of high-energy (> 100 MeV) γ-rays from the galactic disk has been developed and compared with recent SAS-2 observations. In the calculation, it is assumed that (1) the high-energy galactic γ-rays result primarily from the interaction of cosmic rays with galactic matter, (2) on the basis of theoretical and experimental arguments the cosmic-ray density is proportional to the matter density on the scale of galactic arms, and (3) the matter in the Galaxy, atomic and molecular, is distributed in a spiral pattern consistent with density wave theory.

155.004 **Die Feinstruktur des Lokalen Spiralarms und ihre Beziehung zum Gouldschen Gürtel.**
W. Schlosser, T. Schmidt-Kaler.
Mitt. Astron. Ges., No. 36, p. 93 (1975). – Abstract.

155.005 **Die galaktische Struktur im Bereich l = 310° bis 325°.** W. Klinkmann.
Mitt. Astron. Ges., No. 36, p. 170 (1975). – Abstract.

155.006 **A new upper limit to the mass loss from the central region of the Galaxy.**
A. Poveda, C. Allen.
Astrophys. Journ., Vol. 200, 42 - 47 (1975).
A mass loss of 200 $M_\odot$ yr^{-1}, as has been conservatively suggested if Weber is detecting gravitational waves from an isotropic source at the galactic center, is shown to be incompatible with (*a*) the existence of old wide binaries, (*b*) the existence of globular clusters if this loss rate has been constant over the lifetime of the Galaxy, and (*c*) the observed *K*-term, even if this loss has occurred only during the last 10^8 years. From the orbit of ω Centauri in the galactic field, its observed mass distribution, and its tidal radius, it is found that an upper limit to the mass loss from a region of ~1.3 kpc radius around the galactic center is 1 $M_\odot$ yr^{-1}, averaged over the lifetime of the Galaxy. The alternatives left open by this limit are briefly discussed.

155.007 **An empirical determination of the star density in the galactic halo with three-colour-photometric methods in the direction to M 13.**
R. P. Fenkart, R. Wagner.
Astron. Astrophys., Vol. 41, 315 - 320 (1975).
With the help of an RGU photometry the authors separated the disk- and halo-populations in a starfield in the direction to M 13 (l = 58°, b = +40°) and determined their density functions for different intervals in absolute magnitude. The assumption of rotational symmetry of the halo density distribution with respect to the galactic polar axis can be tested by a comparison of the results obtained in the present field with those obtained in the direction to SA 82 (l = 6°, b = +66°). The substantial discrepancies do not support the assumed symmetry, at least not for the neighbourhood of the test-plane covered by this investigation.

155.008 ***UBV*-photometry of three fields in Cygnus.**
H. Hartl.
Astron. Astrophys., Vol. 41, 321 - 327 (1975).
1682 stars in three fields in Cygnus (l = 90°, b = −1°) with a total area of 1 square degree were measured from Asiago Schmidt *UBV*-plates. One of the fields showed large and varying absorption and did not give results that could be discussed in detail. The other two give an interstellar absorption that consists of two absorption shields, one at about 450 pc distance, the other one beginning between 4 and 5 kpc distance. They can probably be identified with the Orion and the Perseus spiral arm. The luminosity functions for the distance intervals 0.2–1.4 kpc and 1.4–2.6 kpc have the same slope as the one for the solar neighbourhood but the more distant one has lower values.

155.009 **The bright star cloud in Sagittarius at l = 12°, b = −1°.** A. F. J. Moffat, N. Vogt.
Astron. Astrophys., Vol. 41, 413 - 422 (1975).
Three-colour photometry shows that this bright 1°× 2° stellar cloud is composed of a spatially isolated group of stars formed about 10^7 years ago at a mean distance of 1.5 kpc. Its luminosity function appears slightly evolved compared to that of *h* and χ Persei. The cloud forms an integral part of one of the "shingles" proposed by Schmidt-Kaler and Schlosser to account for the three-dimensional structure of spiral arms.

155.010 **A model for the galactic centre.**
V. de Sabbata, P. Fortini, C. Gualdi.
Probl. gravitatsii. Sb. dokl. plenarn. zasedanij 3–j Sov. gravitatsion. konf., Erevan, 1972. Erevan, Erevan.un-t, 1975, p. 32 - 36. In Russian. – Abstr. in Referativ. Zhurn. 51. Astron., 7.51.747 (1975).

155.011 **High-resolution maps of the galactic center at 2.2 and 10 microns.** E. E. Becklin, G. Neugebauer.
Astrophys. Journ., (*Letters*), Vol. 200, L71 - L74 (1975).
Photometric maps of the central 1′ of the galactic center are presented at 2.2 and 10 μ with a spatial resolution of ~ 2.″5. Most of the 2.2-μ radiation within the central 2 pc comes from discrete sources with absolute 2.2-μ magnitudes brighter than −8. At 10 μ, nine discrete sources plus an extended ridge of emission are seen. Five of the discrete sources are located within the extended ridge and have linear dimensions and 10-μ luminosities similar to the planetary nebula NGC 7027.

155.012 **An experiment to detect the spatial velocity of the earth.** D. J. Mora.

Monthly Notices Roy. Astron. Soc., Vol. 173, 33 - 36 (1975).

The purpose of this paper is to suggest that the knowledge which is now available concerning the solar galactic velocity makes it possible to perform an astronomical alternative to physical experiments for testing special relativity. This astronomical method could also be used to check some observations made in a previous Michelson-Morley type experiment which were at variance with the null results generally obtained.

155.013 **On the sensitivity of the behaviour of stellar orbits to the parameters of simple galactic mass models** L. Martinet.
IAU Symposium No. 69, (see 012.009), p. 227 - 235 (1975).

Some conjectures are given about the sensitivity of the behaviour of the stellar orbits (appearance of resonant orbits, possible interaction of resonances etc.) to the set of parameters characterizing a simple realistic 'spheroid + nucleus' mass model of our galaxy. Connections with a systematic exploration of orbits in Schmidt mass models are mentioned. Finally the author presents arguments in favour of the existence of a third integral for the whole solar neighbourhood old stellar populations, independently of the chosen model.

155.014 **Theoretical limits on the properties of low-velocity M-dwarfs.** P. Biermann.
IAU Symposium No. 69, (see 012.009), p. 321 - 324 (1975).

Using a version of the Toomre-criterion (Toomre, 1964) for a mixture of stellar populations, limits are derived for the kinematical properties of the newly discovered population of low-velocity M-dwarfs (Weistrop, 1972; Murray and Sanduleak, 1972); these stars are claimed to be a large mass-fraction ($\gtrsim 0.5$) of all stars in the solar neighbourhood and to have only 10 km s^{-1} velocity dispersion. They could make up for the locally missing mass (Oort, 1965; Schmidt, 1974). It is found that a configuration of stellar populations with the M-dwarfs included (having such properties) is not stable in the sense of the Toomre-criterion. The limiting stability properties are given.

155.015 **Problems associated with the M-dwarf population.** M. Schmidt.
IAU Symposium No. 69, (see 012.009), p. 325 - 329 (1975).

The existence of a rich population of late M-type dwarfs with small peculiar velocities is discussed on the basis of the work of Weistrop, Murray and Sanduleak, Pesch and others. These stars probably supply the 'missing' mass in the solar neighborhood. Their small peculiar velocities create problems related to the stability of the galactic disk.

155.016 **Is the nucleus of the Galaxy a black hole?** I. S. Shklovskij.
Pis'ma v Astron. Zhurn., Vol. 1, No. 7, p. 3 - 7 (1975). In Russian.

The source located in the Galaxy nucleus has synchrotron nature. The power of relativistic particles generated inside of this source is about 10^{40} erg/s. Its source of energy is probably the accretion of interstellar gas onto a massive ($M = 3 \times 10^4\ M_\odot$) black hole – the nucleus of the Galaxy.

155.017 **Further evidence concerning the local density of red dwarfs.** D. Weistrop.
Bull. American Astron. Soc., Vol. 7, 411 (1975). – Abstr. AAS.

155.018 **Warping of the galactic disc by the LMC.** L. D. Spight, E. J. Grayzeck.
Bull. American Astron. Soc., Vol. 7, 435 (1975). – Abstr. AAS.

155.019 **Distance to the galactic center measured from globular clusters.** W. E. Harris.
Bull. American Astron. Soc., Vol. 7, 436 (1975). – Abstr. AAS.

155.020 **Infrared polarization of the galactic nucleus.** R. F. Knacke, R. W. Capps.
Bull. American Astron. Soc., Vol. 7, 441 (1975). – Abstr. AAS.

155.021 **Sternpopulationen in Galaxien.** C. Trefzger.
SuW, Vol. 14, 282 - 287 (1975).

155.022 **Galactic rotation and local irregularities derived from fundamental proper motions and radial velocites.** W. Fricke, A. Tsioumis.
Astron. Astrophys., Vol. 42, 449 - 455 (1975).

Proper motions and radial velocities of FK 4/FK 4 Sup stars of the disk population have been analysed in the distance intervals 70 to 300 pc and 300 to 1300 pc. The stars at large distances are concentrated to the galactic plane; their motions confirm the values of Oort's constants A and B ($+14.2 \pm 2.8$, -11.8 ± 2.8 km s^{-1} kpc^{-1}) derived by Fricke (1967). The stars nearer than 300 pc are concentrated towards the Gould Belt; they all together do not yield significantly different rotation parameters than the stars at large distances. The motions of the nearby stars younger than 4×10^7 years deviate strongly from galactic rotation. The explanation of a significant K term remains uncertain; the occurrence of nearby groups with diverging motions appears to offer a more realistic explanation than an expansion from a point source.

155.023 **Rotational interpretation of recombination line widths in the galactic nucleus.** S. A. Kellman, R. H. Durisen.
Astron. Astrophys., Vol. 42, 467 - 469 (1975).

The widths of the radio recombination lines observed in the direction of the galactic center are interpreted as due to pure rotation. This interpretation leads to a mass estimate for the inner 1.6 pc region of our galaxy. It is noted that recent observations of M 31 suggest the existence of a distinct dynamic entity at its center–the nucleus. The rotational interpretation of the recombination line widths is shown to be consistent with the existence of a nucleus at the center of our own galaxy which is similar in nature to the nucleus of M 31.

155.024 **The stabilizing effects of cloud reacceleration; microturbulence, and rotation on Parker's instability.** E. G. Zweibel, R. M. Kulsrud.
Astrophys. Journ., Vol. 201, 63 - 73 (1975).

The authors reconsider Parker's instability including a model equation of state for the gas of interstellar clouds, a tangled interstellar field, and galactic rotation. They find that for most choices of interstellar parameters consistent with observation the instability is stabilized. For the remaining choices which can lead to instability the growth time is longer than 10^8 years. Thus, the authors conclude the instability can have little to do with cosmic-ray escape and probably cannot destroy a primordial galactic field.

155.025 **Comment on "The stabilizing effects of cloud reacceleration, microturbulence, and rotation on Parker's instability".** E. N. Parker.
Astrophys. Journ., Vol. 201, 74 - 75 (1975).

It is pointed out that the calculations in the paper by Zweibel and Kulsrud do not speak to the question of the instability and evolution of the gaseous disk of the Galaxy. Rather, they seem to treat the question of the statistical stability of the present turbulent dynamical state. (See 155.024).

155.026 **On the relation between CO and gamma-ray observations, cosmic rays, and the thickness of the galactic disk.** D. G. Wentzel, P. D. Jackson, W. K. Rose, R. P. Sinha.

Astrophys. Journ., (*Letters*), Vol. 201, L5 - L8 (1975).

Recent CO observations suggest that dense clouds provide a high mean gas density (mostly H_2) at approximately 5 kpc from the galactic center. Gamma-ray observations have been explained as resulting primarily from the interaction of cosmic rays with interstellar matter. The authors discuss the relationship between cosmic-ray density and mean gas density, and argue that the CO clouds must be dynamically independent of the gas that is observed in 21-cm line radiation. They derive a plausible upper limit to the cosmic-ray density at $R = 5$ kpc that is based on measurements of galactic disk thickness, hydrogen turbulent velocities from 21-cm line data, and an estimate of the "distributed" gas density which confines the cosmic rays to the disk. This upper limit is approximately 5 times the cosmic-ray density near the sun. Observations of the thickness of the galactic disk suggest that the cosmic-ray density increases with distributed gas density.

155.027 **The spiral pattern of our Galaxy from the terminal velocities.** C. Cruz-González.
Publ. Astron. Soc. Pacific, Vol. 87, 500 - 501 (1975).
Abstr. ASP.

155.028 **The density distribution of the interstellar gas near the galactic shock region.** T. Sawa.
Sci. Rep. Tôhoku Univ., First Ser., Vol. 58, 13 - 22 (1975).

The cloud motions and the distribution of the interstellar gas density in the galactic shock region are calculated, considering the drag force due to the turbulent viscosity of the intercloud medium. The galactic shock waves are assumed to be produced by the intercloud medium and not to be produced by the clouds. It is shown that the interstellar gas density has a peak at the downstream of the shock front and its value is only 2.5 - 4 times larger than the value at the interarm region for the case the mean turbulent velocity is 7 km/sec and the effective size of the turbulent eddies is 1 - 3 pc, and that the distribution of it is considerably smooth at the arm region.

155.029 **Photographic *UBV* observations and the galactic structure in Centaurus ($l = 307°$, $b = 0°$).**
C. H. McGruder III.
Astron. Astrophys., Vol. 43, 51 - 60 (1975).

An area of 5□° centered at $l = 307°$, $b = 0°$ has been studied by means of photographic UBV-photometry. 38 094 stars were measured by the GALAXY machine. Color excesses, frequencies and space densities of O, B, A, F main sequence stars and G, K giants are derived. A maximum in the space densities is found at $r \approx 0.7$ kpc; other maxima are detected at $r \approx 1.8$ kpc and $r \approx 3.5$ kpc, coinciding with the distances of spiral arm –I and a possible spiral arm –II. The presence of spiral arm –I at $l = 307°$ means that this arm probably extends from Carina through Sagittarius.

155.030 **Self-gravitating, disk-like, galactic models with stars having velocity dispersions.** M. Miyamoto.
Publ. Astron. Soc. Japan, Vol. 27, 431 - 443 (1975).

This paper gives numerical results for disk-like models of the self-gravitating stellar system, according to the construction scheme which has been proposed previously (Miyamoto 1974). Particular attention is devoted to a density distribution (a superposition of two Toomre's [1963] mass models) for our Galaxy. For this density distribution, four characteristic models, the central parts of which are essentially pressure-supported, are investigated. The radial velocity dispersion at the center of the four models is almost independent of the models considered. The radial velocity dispersion of stars at the galactic center is estimated to be 225 km s^{-1}, which is much larger than the one deduced from Toomre's (1964) criterion for the local stability of the stellar disk.

155.031 **Le mouvement des étoiles dans la Galaxie.**
W. Fricke.
L'Astronomie, 89ᵉ année, p. 355 - 370 (1975).

155.032 **Galactic evolution. I. Single-zone models.**
T. X. Thuan, M. H. Hart, J. P. Ostriker.
Astrophys. Journ., Vol. 201, 756 - 772 (1975).

The authors calculate possible evolutionary histories for a cylindrical element (1 pc^2 cross section) perpendicular to the plane of the Galaxy including the sun, under the following restrictive assumptions: (1) the total mass in the cylinder, which is constant in time, is initially gaseous and metal free; (2) the gaseous component is always well mixed; (3) the birthrates of various stellar types are declining (or increasing) exponential functions of time. Although this procedure allows a great variety of evolutionary paths, no satisfactory model could be constructed when it was required that the calculated models, after $\sim 12 \times 10^9$ years, match present observations in the solar neighborhood. After demonstrating the failure of one-zone galactic models, the authors suggest that an alternative scheme, in which material shed from dying halo stars makes a substantial contribution to the disk mass, will solve or alleviate the problems described in the present investigation.

155.033 **A new Milky Way satellite found in 21-centimeter line observations.** S. C. Simonson III.
Astrophys. Journ., (*Letters*), Vol. 201, L103 - L108 (1975).

An anomalous feature in the 21-cm line observations has the characteristics of an external galaxy passing near the Milky Way. The galaxy's center is at $l = 197.°3 \pm 0.°5$, $b = +2.°1 \pm 0.°5$, and its angular size is $7° \times 2°$. Tidal debris extends to $l = 155°$, $b = +8°$, forming a stream of H I high-velocity clouds that crosses $l = 180°$ at $V = -90$ km s^{-1}. The observations together with dynamical models of the tidal interaction using Schmidt's rotation parameters indicate a distance from the sun of 17 ± 4 kpc, a perigalactic distance of 22 ± 2 kpc, and a mass of $1.0 \pm 0.2 \times 10^8 M_\odot$.

155.034 ***RGU* photometry of a field to the direction to the galactic centre.** L. Topaktas.
Astron. Astrophys., Suppl. Ser., Vol. 22, 207 - 219 (1975).

A galactic centre field of 0.065 square degrees with 1608 stars in the Small Sagittarius Cloud has been studied photometrically in the *RGU* system down to a limiting magnitude of 17.2 in *G*. The interstellar reddening function, the density functions for different groups of stars and the luminosity functions have been determined. The reddening is caused by three interstellar clouds with distances of 725, 3160 and 5890 pc. The density functions of the late-type giants and of the main-sequence stars with $0 \leq M(G) < 2$ show maxima between 1 and 2 kpc which agree well with the distance of the next inner spiral arm (–I). The luminosity function for the distance intervals of 0.4 to 0.6 and of 1.0 to 2.0 kpc corresponding to the interarm region and to the distance of the next inner spiral arm, reflect the behaviour of the density functions.

155.035 **The mass of the galactic halo derived from the luminosity function of high-velocity stars.**
M. Schmidt.
Astrophys. Journ., Vol. 202, 22 - 29 (1975).

The local luminosity function of high-velocity stars is derived on the basis of a complete sample of stars of large proper motion with trigonometric parallaxes. The fraction of these stars belonging to a roughly spherical galactic halo is deduced by using a kinematical criterion based on space velocities of metal-poor RR Lyrae variables. The local mass density of halo stars is about 1.7×10^{-4} solar masses per cubic parsec, corresponding to a fractional mass within the sun's distance from the center of 6 percent.

155.036 **A density-wave map of the galactic spiral structure.** S. C. Simonson III.
La dynamique des galaxies spirales. Colloque International C.N.R.S. No. 241, (see 012.022), p. 303 - 305 (1975).

The large-scale spiral structure is mapped by constructing a model based on density-wave kinematics that reproduces the main features of the 21-cm H I observations. The map shows a basically two-armed spiral pattern with a pitch angle of $6°-8°$ between the 4-kpc dispersion ring and the 4-kpc dispersion ring and the solar circle. Near the solar circle two additional major arms originate, and the pattern outside is multiple-armed, with pitch angles of $\sim 16°$. Comparison of the gas flow in both a linear and a shock solution shows that the shock solution gives a better representation of the tangential directions of spiral arms.

155.037 **Mise en évidence de la structure spirale à partir de la cinématique des étoiles OB.**
M. O. Mennessier, M. Crèzé.
La dynamique des galaxies spirales. Colloque International C.N.R.S. No. 241, (see 012.022), p. 325 - 331, with a discussion, p. 332 - 333 (1975).

In order to point out the effects of the large scale spiral structure of our galaxy on the kinematics of the youngest stars, the authors derive the components of the velocity field of O-B3 stars within 3 kpc of the sun. The local region is found to be crossed by two arms of a general spiral pattern with a pitch angle of about $-7°$, one of them crosses the line galactic center-sun at about 9 kpc from the center and the other at 13 kpc.

155.038 **Some remarks on local kinematical disturbances in our Galaxy and their interpretations.**
M. Mayor.
La dynamique des galaxies spirales. Colloque International C.N.R.S. No. 241, (see 012.022), p. 383 - 387, with a discussion, p. 388 - 389 (1975).

155.039 **Dynamics of the Magellanic stream.**
A. E. Wright.
La dynamique des galaxies spirales. Colloque International C.N.R.S. No. 241, (see 012.022), p. 511 - 512, with a discussion, p. 513 - 515 (1975). – Summary.

155.040 **Finns det brännvin och saltgurka i Vintergatans centrum?** A. Sandqvist.
Astron. Tidssk., Årg. 8, 108 - 126 (1975).

155.041 **The exploration of our Galaxy.** V. Icke.
Spaceflight, Vol. 17, 414 - 420 (1975).

155.042 **Galactic orbits and integrals of motion for stars of old galactic populations. III. Conclusions and applications.** L. Martinet, F. Mayer.
Astron. Astrophys., Vol. 44, 45 - 57 (1975).

The conclusions of a comparative study of the orbital behaviour in different mass models of our Galaxy are presented. Extensive statistics of resonant periodic orbits for the Schmidt mass models are obtained. Applications of the authors' systematic exploration of orbits to the dynamics of the solar neighbourhood and of more central regions of the Galaxy are proposed. In particular, for the stellar populations near the sun, the authors delimit the regions of the residual velocity plane (Π, $\Theta - \Theta_c$) for which the existence of a third integral is certain, independently of the model. Finally, some conjectures related to the question of a general connection between the parameters characterizing a simple galactic mass model and the orbital behaviour, especially the appearance of peculiar resonances are presented.

155.043 **The origin of the Galaxy.** S. van den Bergh.
Astron. Astrophys., Vol. 44, 231 - 232 (1975).

Observations of galactic globular clusters strongly favour the view that the Galaxy formed from a single protogalaxy rather than by coalescence of smaller ancestral galaxies.

155.044 **The kinematic inhomogeneity of G-type dwarfs.**
T. G. Gotska.
Astron. Zhurn. Akad. Nauk SSSR, Vol. 52, 1017 - 1020 (1975). In Russian. English translation in Soviet Astron., Vol. 19, No. 5.

From the statistics of the velocities and of the elements of the galactocentric orbits of the near stars the G-type dwarfs are related to the disk [$\delta\,(U-B) \lesssim 0\overset{m}{.}015$] and intermediate [$\delta\,(U-B) \gtrsim 0\overset{m}{.}115$] subsystems of the Galaxy. The stars with ultraviolet excess $0\overset{m}{.}015 < \delta\,(U-B) < 0\overset{m}{.}115$ are constituting an inhomogeneous group.

155.045 **Determination of the local mass density on the basis of photographic B, V photometry.** M. Jõeveer.
Publ. Tartu Astrofiz. Obs., Vol. 43, 89 - 95 (1975). In Russian.

It has been shown that the luminosity function of intrinsically faint stars derived by Weistrop is fraught with systematic errors in the colour system of her photoelectric standards, so that the number of red dwarfs has to be reduced significantly. For the local mass density corresponding to the stars of the main sequence an estimate of 0.05 solar masses per pc^{-3} is made.

155.046 **A spectral study of F and G stars near the North Galactic Pole in connection with aging effects. I. Statement of the problem.** V. Malyuto.
Publ. Tartu Astrofiz. Obs., Vol. 43, 184 - 202 (1975). In Russian.

The paper gives a brief critical review of works in which the chemical composition and the kinematics of F and G stars have been studied in connection with their ages. To obtain additional data for the study, the spectral investigation of the F5–G5 stars near NGP is being carried out at Abastumani and Tartu with the aid of an 8° objective prism. The main features of this investigation are described.

155.047 **On the stellar component of galactic coronae.**
J. Jaaniste, E. Saar.
Publ. Tartu Astrofiz. Obs., Vol. 43, 216 - 232 (1975).

The velocity distributions of stars in galactic coronae are compared with available data on stellar motions in the solar neighbourhood. As a result, it is shown that the corona of our Galaxy can be composed of non-luminous objects (planets, black dwarfs) or large groups, containing about 10^5 dwarf stars.

155.048 **Galactic evolution. II. Disk galaxies with massive halos.** J. P. Ostriker, T. X. Thuan.
Astrophys. Journ., Vol. 202, 353 - 364 (1975).

Models of galactic evolution are computed in which matter shed by dying halo stars accumulates in a smaller, more rapidly rotating disk. The models are both simpler and more successful than one-zone (pure disk) models in that (1) the observed absence of low-metal-abundance, low-mass dwarfs is expected, not anomalous, and (2) the relative birthrate function need not be a strongly variable function of time. Models for the spherical (halo) component of the Galaxy produced as a by-product of this investigation agree fairly well with those of Tinsley for elliptical galaxies but tend to produce somewhat smaller evolutionary corrections to the luminosity of old or cosmologically distant galaxies.

155.049 **A search for optical pulses from the galactic centre.**
P. B. Byrne, P. A. Wayman.

Monthly Notices Roy. Astron. Soc., Vol. 173, 537 - 552 (1975).

A search has been carried out for optical flashes in the direction of the centre of the Galaxy. Most of the flashes recorded are adequately accounted for by faint meteors crossing the area of the sky under examination. There remain a few multiple optical pulses, closely grouped in time, which are not readily explained. Possible origins are discussed.

155.050 **Gaussian components of velocity distributions of giant stars.** R. L. Dessureau, A. R. Upgren.
Publ. Astron. Soc. Pacific, Vol. 87, 737 - 743 (1975).

An investigation of the velocity distribution of giant stars perpendicular to the galactic plane has been made using a new determination of their space densities in the region of the north galactic pole. These densities, together with the $K(z)$ force as determined by Oort, give rise to a total velocity distribution which is well represented by three Gaussian distributions. A stable and realistic model is achieved with no constraints imposed upon the parameters of each velocity distribution. It is concluded that Oort's values of $K(z)$ cannot be reconciled with his distributions.

155.051 **On the production of deuterium and helium-3 in the active galactic nucleus.**
L. M. Ozernoj, V. V. Chernomordik.
Astron. Zhurn. Akad. Nauk SSSR, Vol. 52, 1156 - 1165 (1975). In Russian. English translation in Soviet Astron., Vol. 19, No. 6.

The possibility to explain the production of observed abundances of deuterium and helium-3 by cosmic rays and γ-rays produced by the nucleus of the Galaxy at the stage of activity and interacting with the envelope opaque for them around the nucleus is discussed.

155.052 **The diffuse glow of galactic dust under the action of extragalactic radiation sources.**
D. A. Rozhkovskij, V. S. Matyagin.
Trudy Astrofiz. Inst., *Alma-Ata,* Vol. 25, 3 - 11 (1975). In Russian.

155.053 **Observations of the soft X-ray diffuse background.**
W. T. Sanders.
Bull. American Astron. Soc., Vol. 7, 505 (1975). – Abstr. AAS

155.054 **An anomalous velocity neutral hydrogen structure near the galactic center.**
I. F. Mirabel, K. C. Turner.
Astrophys. Space Sci., Vol. 38, 381 - 394 (1975).

An extensive concentration of neutral hydrogen has been observed in the fourth galactic quadrant, with a mean radial velocity of +44 km s^{-1} referred to the local standard of rest. At a distance of R kpc from the sun this structure would contain $2.5 \times 10^4 R^2$ solar masses of neutral hydrogen. Five possible interpretations of this extensive concentration are considered: (1) part of the shell of a nearby explosive event; (2) a distant spiral arm of the Galaxy; (3) an extragalactic object; (4) material falling into our Galaxy; (5) gas expelled from the galactic center. Arguments are offered against the first three possibilities.

155.055 **H 134α recombination lines from the group of sources near galactic longitude 30°.**
F. F. Gardner, P. Thomasson.
Astron. Astrophys., Vol. 45, 1 - 4 (1975).

Observations of the H 134α line (2702.799 MHz) have been made with a 4.8′ beam at 32 source positions near galactic longitude l = 30°. The velocities are constant to within 10 km s^{-1} over a region almost 1.5 degrees in extent. The derived values of electron temperature lie in the range 5000 – 6000 K. Comparison with H_2CO absorption suggests that all the sources are within a single spiral arm near, but in front of the tangential point.

155.056 **Non-circular motions in the Perseus and Carina spiral arms.** R. M. Humphreys.
Bull. American Astron. Soc., Vol. 7, 513 (1975). – Abstr. AAS.

155.057 **Investigation of A-type star condensations in Perseus and Cassiopeia on the basis of a proper motion analysis.** G. F. Kevanishvili.
Byull. Abastumansk. Astrofiz. Obs., No. 46, p. 101 - 124 (1975). In Russian.

The existence of A-type stars grouping in Perseus and Cassiopeia has been revealed on the basis of their space distribution. The reality of the groupings and their expansion in the direction of galactic longitude have been ascertained by applying methods of mathematical statistics and proper motion analysis. The expansions of both Perseus and Cassiopeia condensations proceed at a speed of 0.″004 ± 0.″0018 and 0.″0031 ± 0.″0024 per year, respectively. No effect of the groupings expansion is observed in the direction of galactic latitude.

155.058 **An observation of the diffuse soft X-ray/extreme ultraviolet background.**
W. Cash, R. Malina, R. Stern, S. Bowyer.
Bull. American Astron. Soc., Vol. 7, 505 (1975). – Abstr. AAS.

155.059 **Kinematical properties of young and old disk dwarfs determined from new parallaxes.** A. R. Upgren.
Bull. American Astron. Soc., Vol. 7, 513 (1975). – Abstr. AAS.

155.060 **Observations of carbon monoxide in the inner Galaxy.** T. M. Bania.
Bull. American Astron. Soc., Vol. 7, 514 (1975). – Abstr. AAS.

155.061 **Observations of galactic CO emission in longitudes 10° to 36°.** M. A. Gordon, W. B. Burton.
Bull. American Astron. Soc., Vol. 7, 514 (1975). – Abstr. AAS.

155.062 **The distribution of ionized hydrogen in the central region of the northern Milky Way.** F. J. Lockman.
Bull. American Astron. Soc., Vol. 7, 514 - 515 (1975). Abstr. AAS.

155.063 **Relative galactic morphology of hydrogen and other tracers.** W. B. Burton.
Bull. American Astron. Soc., Vol. 7, 515 (1975). – Abstr. AAS.

155.064 **A 60μ map of the galactic center with 30″ resolution.** C. M. Telesco, D. A. Harper.
Bull. American Astron. Soc., Vol. 7, 530 (1975). – Abstr. AAS.

155.065 **Dynamics of CO molecular clouds in the Galaxy.**
F. N. Bash, W. L. Peters.
Bull. American Astron. Soc., Vol. 7, 531 (1975). – Abstr. AAS.

155.066 **Pregalactic helium abundance and abundance gradients across our Galaxy derived from planetary nebulae.** S. d'Odorico, M. Peimbert, F. Sabbadin.
Bull. American Astron. Soc., Vol. 7, 531 (1975). – Abstr. AAS.

155.067 **Steps toward understanding the large-scale structure of the Milky Way – Part I.** H. Weaver.
Mercury, (Journ. Astron. Soc. Pacific), Vol. 4, No. 5, p. 18 - 24 (1975).

155.068 **Model calculations on the dynamics of the Milky Way under the influence of spiral arms and the Magellanic Clouds.** J. V. Feitzinger.
Astron. Astrophys., Vol. 45, 377 - 382 (1975).

The aim of this paper is to calculate velocity fields for

our stellar system in order to understand the north–south asymmetry of the rotation curves. To do this, model calculations were performed assuming that the velocity field is disturbed by the spiral arms and the Magellanic Clouds. The Magellanic Clouds are revolving about the Galaxy in circular orbits. In the author's model, these calculations show that the difference between the galactic rotation curves for the northern and southern side of our stellar system is the result of tidal action of the Magellanic Clouds.

155.069 **Progress on 21 cm studies of the Southern Milky Way.** F. J. Kerr.
Proc. Astron. Soc. Australia, Vol. 2, 364 - 365 (1975).

155.070 **Galactic rotation of supergiants.** A. V. Charikov.
Astron. Tsirk., No. 885, p. 5 - 7 (1975). In Russian.

155.071 **Steps toward understanding the large scale structure of the Milky Way. – Part II.** H. Weaver.
Mercury, (Journ. Astron. Soc. Pacific), Vol. 4, No. 6, p. 18 - 29 (1975).

155.072 **Derivation of γ-ray emissivity in the Galaxy from satellite data.** A. W. Strong.
Journ. Phys. A (Math., Nuclear, General), Vol. 8, 617 - 623 (1975).

On propose une méthode pour déduire l'émissivité de la distribution longitudinale de l'intensité R_γ et on l'applique aux données des satellites SAS-II et OSO-III. On trouve une augmentation importante de l'émission de 3 à 8 kpc du centre galactique, et une chute au centre ($r < 3$ kpc).

155.073 **An assessment of models for the γ-ray flux from the galactic plane.** D. Dodds, A. W. Strong, A. W. Wolfendale, J. Wdowczyk.
Journ. Phys. A (Math., Nuclear, General), Vol. 8, 624 - 635 (1975).

La discussion concerne 5 modèles pour la production des rayons γ galactiques qui impliquent tous l'interaction des protons du rayonnement cosmique et l'hydrogène neutre. Aucun de ces modèles n'est entièrement satisfaisant.

The discovery of our Galaxy.
See Abstr. 003.133.

Possible relation between periodic glaciation and the flexure of the Galaxy. See Abstr. 015.024.

Keplerian motions. See Abstr. 042.013.

Elementhäufigkeiten im Sonnensystem und Elemententstehung in der Galaxis. Anfängliche Häufigkeiten im Sonnensystem und galaktische Nukleosynthese.
See Abstr. 061.019.

Observation of the galactic X-ray background.
See Abstr. 061.022.

Isotopes of carbon, nitrogen and oxygen as probes of nucleosynthesis, stellar mass losses and galactic evolution.
See Abstr. 061.031.

Is deuterium of cosmological or of galactic origin?
See Abstr. 061.038.

The study of HR 3018, HD 91324 and α Crv and its implication for the age-abundance correlation of the neighbour stars. See Abstr. 064.013.

The surface brightness of the northern Milky Way in the red region. See Abstr. 106.069.

Proper motions of faint blue stars near the galactic anticenter. II. See Abstr. 112.009.

Spectroscopy of five faint M dwarf stars near the north galactic pole. See Abstr. 112.016.

Catalogue of faint O–B5 stars in Centaurus (l = 307°, b = 0°). See Abstr. 113.032.

The mean colour indices B–V and B–R of integral starlight in the neighbourhood of the North Galactic Pole.
See Abstr. 113.068.

Metal abundances in nearby stars and the chemical history of the solar neighbourhood. See Abstr. 114.006.

Statistical analysis of velocities and cyanogen indices for G and K giants. See Abstr. 114.091.

Super-metal-rich stars. See Abstr. 114.114.

A study of Mira variables. Implications for OH stars and galactic evolution. See Abstr. 122.132.

The dependence between age and flare activity of the red dwarf stars in the solar vicinity.
See Abstr. 122.158.

Variable stars and the structure of the Galaxy.
See Abstr. 122.169.

Spectral turnovers of galactic supernova remnants.
See Abstr. 125.001.

The distribution of supernova remnants in the Galaxy: evidence for a tunnel system. See Abstr. 125.033.

On the z-distribution of supernova remnants.
See Abstr. 125.038.

Molecular clouds in the Galaxy.
See Abstr. 131.016.

A study of the interstellar gas between the sun and the Sagittarius arm. See Abstr. 131.031.

The scale height of interstellar ionised hydrogen.
See Abstr. 131.042.

Distribution of NH_3 clouds in the galactic center region. See Abstr. 131.086.

Observations of the diffuse λ 4430 absorption band in southern O, B, and A stars. See Abstr. 131.100.

The overall distribution of carbon monoxide in the plane of the Galaxy. See Abstr. 131.117.

Studies of ultraviolet interstellar extinction with the sky-survey telescope of the TD-1 satellite. I. Results for three galactic regions. See Abstr. 131.126.

The fine structure of the galactic dust cloud distribution near the sun. See Abstr. 131.127.

Investigation of interstellar absorption and of spatial distribution of stars in a region near the North Galactic Pole.
See Abstr. 131.166.

Solar system as space-probe. See Abstr. 131.197.

The Celescope survey and the galactic distribution of interstellar absorption. See Abstr. 131.212.

G45.5 + 0.1 and G45.1 + 0.1: compact infrared sources. See Abstr. 131.503.

21-cm observations of non-planar H I associated with the Perseus spiral arm. See Abstr. 131.528.

High-velocity neutral hydrogen. See Abstr. 131.539.

A full synthesis map of Sgr A at 5 GHz. See Abstr. 141.079.

The relation between discrete radio sources and the fine structure of the galactic background. See Abstr. 141.158.

A deep sample of new pulsars and their spatial extent in the Galaxy. See Abstr. 141.332.

The galactic distribution and kinematics of OH stars (a preliminary discussion). See Abstr. 141.642.

Molecular hydrogen in the Galaxy and galactic gamma rays. See Abstr. 142.096.

Detection of nuclear gamma rays from the galactic center region. See Abstr. 142.118.

The distribution of X-ray sources in the Galaxy. II. See Abstr. 142.163.

Global instabilities of disks. See Abstr. 151.026.

Amplification of galactic density waves by non-adiabatic processes in gas. See Abstr. 151.041.

Note on the persistence of spiral structure. See Abstr. 151.052.

Spiral structure of the gas in spiral galaxies. – Theory. See Abstr. 151.054.

On the age dependence of the spatial distribution and velocity dispersion of stars in a spiral galaxy. See Abstr. 151.056.

A theorem on the third integral. See Abstr. 151.078.

Rings and chains in simulated stellar fields. See Abstr. 152.003.

The galactic halo: globular clusters. See Abstr. 154.008.

On the chemical composition and the origin of the globular clusters of our Galaxy. See Abstr. 154.014.

Three-dimensional distribution of interstellar gas in disk-shaped galaxies. See Abstr. 158.128.

Stellar populations in galaxies. See Abstr. 158.139.

Star formation in spiral galaxies. See Abstr. 158.161.

Synthesis of stellar populations. See Abstr. 158.175.

Accretion of intergalactic gas by a realistic model of the Galaxy and its implications. See Abstr. 161.004.

Acoustic modes and gravitational instabilities in the early Universe. See Abstr. 162.023.

156 Galactic Magnetic Field

156.001 **The galactic dynamo.** M. Stix.
Astron. Astrophys., Vol. 42, 85 - 89 (1975).

The generation of mean magnetic fields by an $\alpha\omega$-dynamo is studied in the geometry of an oblate spheriod. Applied to the galaxy, the model has the steady, quadrupolar (even) mode as the most easily excited mean field, in agreement with results obtained by Parker from a plan-parallel model. However, oscillatory modes which were reported by Parker to occur under similar excitation conditions, exist in the present model only at substantially larger dynamo numbers.

156.002 **Galactic magnetic fields and their interactions with the gas and cosmic rays.** J. H. Piddington.
Astrophys. Space Sci., Vol. 37, 183 - 196 (1975).

(1) The controversy of dynamo or primordial origin of galactic magnetic fields is summarized and extended to show that the dynamo theory appears to fail. (2) A passive magnetic field frozen into a turbulent plasma is not dissipated like a cloud of smoke ('turbulent or eddy diffusion') as believed previously. On the contrary it is amplified exponentially until, within a few eddy periods, either the growing magnetic stresses halt the turbulence or the field becomes chaotic. (3) It is concluded that the gas motions observed in our Galaxy are not convective but are mainly hydromagnetic waves or oscillations, with the magnetic field in control. The significance of this result is discussed in connection with the overall gas velocity field, the creation of stars and stellar systems, and with the origin and distribution of cosmic rays.

156.003 **The effective magnetic field of the corona of our Galaxy.** N. M. Lipovka.
Astron. Tsirk., No. 887, p. 1 - 3 (1975). In Russian.

156.004 **Structure of irregular galactic magnetic fields determined on the basis of cosmic ray measurements.**
A. Somogyi.
Hungarian Acad. Sci., Budapest. KFKL-75-6, 6 pp. (1975).

Based on galactic cosmic ray anisotropy measurements and theoretical results concerning the relation of particle diffusion to the structure of the irregular magnetic field, the power spectrum of the galactic magnetic field irregularities is estimated.

The rotation measures of radio sources and their interpretation. See Abstr. 141.080.

157 Galactic Radio Radiation

157.001 **Ionized gas in the direction of the galactic center. II. The line of sight problem.**
T. Pauls, P. G. Mezger.
Astron. Astrophys., Vol. 44, 259 - 266 (1975).

It is shown that 6 cm radio recombination line observations can be used to decide whether the emission originates in the galactic center or along the line of sight. The lines originating in the galactic center are broad and have large radial velocities, while the lines originating along the line of sight are narrow and have radial velocities near zero. The lines originating along the line of sight show a velocity asymmetry when plotted in a velocity-latitude diagram. This asymmetry is also present in recombination-line data taken at 18 cm and in observations of neutral hydrogen.

157.002 **Low frequency maps of the Galaxy.** H. V. Cane.
Proc. Astron. Soc. Australia, Vol. 2, 330 - 331 (1975).

This paper is a preliminary report on the sky mapping being undertaken at the University of Tasmania using the Llanherne low frequency array (Ellis 1972).

157.003 **Radio emission of the spherical component of our Galaxy from observations at 4 centimeter and decametric wavelengths.** A. V. Ipatov, N. M. Lipovka.
Astron. Tsirk., No. 886, p. 6 - 8 (1975). In Russian.

157.004 **Linear polarization of the galactic radio emission at 102 MHz near $l = 140°$, $b = +8°$.**
A. M. Paseka, L. V. Popova, V. A. Razin, V. G. Arkhangel'skij, Yu. E. Samokhvalov.
Izv. vyssh. ucheb. zavedenij. Radiofizika, Vol. 18, 926 - 929 (1975). In Russian. – Abstr. in Referativ. Zhurn. 51. Astron., 2.51.801 (1976).

157.005 **Interpretation of the intensity distribution of galactic synchrotron radiation.**
C. Jaekel, K. O. Thielheim.
Verhandl. Deutsche Phys. Ges., Reihe 6, Vol. 10, 161 (1975). In German. – Abstract.

A full synthesis map of Sgr A at 5 GHz.
See Abstr. 141.079.

VLBI observations of the compact radio source in the center of the Galaxy. See Abstr. 141.110.

A 408 MHz fan-beam survey at low galactic latitudes. IV. The nature of the small diameter sources.
See Abstr. 141.147.

The Parkes 408 MHz southern survey.
See Abstr. 141.152.

The relation between discrete radio sources and the fine structure of the galactic background.
See Abstr. 141.158.

Finns det brännvin och saltgurka i Vintergatans centrum? See Abstr. 155.040.

Progress on 21 cm studies of the Southern Milky Way. See Abstr. 155.069.

158 Single and Multiple Galaxies, Peculiar Objects

Single and Multiple Galaxies

158.001 **Distribution of observed orientations of galaxies.** D. L. Hawley, P. J. E. Peebles.
Astron. Journ., Vol. 80, 477 - 491 (1975).
The authors report on the results of a search for anisotropy in the orientations of observed projected images of galaxies. They find small but significant departure from isotropy, but there is no reason to believe the effect is real because it does not satisfy tests of reproducibility. The authors find no evidence of anisotropy in the relative orientations of pairs of galaxies separated by less than about 1 Mpc (H = 100 km sec^{-1} Mpc^{-1}). For the galaxies in the Coma cluster proper they find no evidence for anisotropy relative to a fixed direction in space. There is a possible indication that the galaxies are preferentially aligned along the radius vector to the center of the cluster.

158.002 **Photoelectric photometry and calibrated isophotometry of NGC 520.** O. K. Krienke, Jr.
Astron. Journ., Vol. 80, 492 - 497, 541 - 542 (1975).
BVr photoelectric observations were made of an 18 spot rectangular grid fitted to the Irr II galaxy, NGC 520, and photographic isophotes were constructed. An isophote calibration procedure is demonstrated for the utilization of photoelectric spots of diameter several times of scale of resolved structure. Luminosity profiles for portions of NGC 520 show the presence of an exponential disc and a spheroidal component. A comparison of models for NGC 520 favors the conclusion that NGC 520 is best understood as a close pair of tidally interacting galaxies.

158.003 **Southern galaxies. VII. The remarkable lenticular barred galaxy NGC 1291.** G. de Vaucouleurs.
Astrophys. Journ., Suppl. Ser., No. 284, Vol. 29, 193 - 218 (1975).
The lenticular barred galaxy NGC 1291 is the largest example of revised type (R) SB(s)0/a at the transition stage between late lenticulars and early spirals. The central lens (12 × 10 kpc) is surrounded by a large, semi-detached outer ring (21 × 19 kpc) from which emerge two faint spiral arms showing incipient resolution at $m_B \approx 20.5$. In the nuclear region at the center of the primary bar (10 × 2.5 kpc), a secondary bar (1.8 × 0.2 kpc) and secondary nucleus (0.5 × 0.4 kpc) have been detected, repeating the primary lens-bar-nucleus pattern on a smaller scale approximating the ratio of the Jeans lengths. Other examples suggest that this repetitive pattern may be typical. Detailed surface photometry is presented (isophotes, luminosity profiles, and photometric parameters).

158.004 **Emission-line luminosities of Seyfert galaxies.** T. F. Adams, D. W. Weedman.
Astrophys. Journ., Vol. 199, 19 - 30 (1975).
New observations of emission-line intensities are presented for 32 Seyfert galaxies. On the basis of correlations among these intensities and continuum properties inferred from *UBV* photometry, the authors conclude that class 1 Seyfert galaxies (those with broad Balmer line wings) are ionized by a strong radiative source and are not significantly reddened by dust. For class 2 Seyfert galaxies (those resembling NGC 1068), the intrinsic $L(H\alpha)$ after correction for dust reddening are comparable to those of class 1, but the forbidden line luminosities are typically 30 times stronger in class 2. It is concluded that the properties of class 2 nuclei could be explained by the presence of large numbers of very hot, luminous stars substantially reddened by dust.

158.005 **Observations of NGC 6764, a barred Seyfert galaxy.** V. C. Rubin, N. Thonnard, W. K. Ford, Jr.
Astrophys. Journ., Vol. 199, 31 - 38 (1975).
Spectra of NGC 6764, a previously unstudied barred spiral galaxy, reveal strong asymmetrical emission lines from the semistellar nucleus. Permitted and forbidden lines are about 750 km s^{-1} wide, which identify the galaxy as a Seyfert galaxy. Spectra in several position angles suggest that discrete gas clouds just outside the nucleus have both tangential and radial components of motion.

158.006 **Evidence for contraction in the nuclear ring of the barred spiral galaxy NGC 3351.**
V. C. Rubin, W. K. Ford, Jr., C. J. Peterson.
Astrophys. Journ., Vol. 199, 39 - 48 (1975).
The nucleus of the barred spiral galaxy NGC 3351 consists of a ring of low-excitation H II regions. Line-of-sight velocities around the ring indicate a well-defined sinusoid, but with maximum and minimum velocities displaced about 20° from that predicted from the geometry of the outer galaxy. One simple interpretation is that the nuclear ring is a circle in the plane of the galaxy which is rotating and contracting with V_{rot} = 126 ± 16 km s^{-1} and V_{con} = 34 ± 11 km s^{-1} at R = 340 pc. Other possible interpretations are also discussed.

158.007 **The redshifts of 3C 129 and 3C 129.1, the radio-tail galaxies.** H. Spinrad.
Astrophys. Journ., (*Letters*), Vol. 199, L1 (1975).
The average redshift of the "head-tail" radio galaxies 3C 129 and 3C 129.1 is found to be +6400 km s^{-1}; the projected length of the radio tail for 3C 129 is then derived as at least 625 kpc.

158.008 **The infrared spectrum of NGC 1068.** G. H. Rieke, F. J. Low.
Astrophys. Journ., (*Letters*), Vol. 199, L13 - L17 (1975).
New infrared photometry of NGC 1068 shows that the spectrum continues to rise smoothly through 34 μ and that there is a shallow absorption feature near 10 μ. The total infrared flux from the nuclear source is about 2.3 × 10^{-11} W m^{-2}, which corresponds to a luminosity of 10^{11} $L_\odot$ at a distance of 11 Mpc.

158.009 **Detection of extragalactic carbon monoxide at millimeter wavelengths.** L. J. Rickard, P. Palmer, M. Morris, B. Zuckerman, B. E. Turner.
Astrophys. Journ.,(*Letters*), Vol. 199, L75 - L78 (1975).
The authors report the discovery of emission from the $J = 1 \rightarrow 0$ transition of $^{12}C^{16}O$ toward the central regions of external galaxies. In this letter, they present preliminary results for two galaxies with active nuclei, M82 and NGC 253, in which the observed emission features are surprisingly strong. In the directions of the nuclear continuum peaks, the velocity range of CO emission is roughly the same as that of 21-cm absorption. For each galaxy, the CO features are spatially extended, and the velocity shift of the emission centroid along the major axis is in the sense of the galactic rotation.

158.010 **Carbon monoxide in external galaxies.** P. M. Solomon, R. de Zafra.
Astrophys. Journ., (*Letters*), Vol. 199, L79 - L83 (1975).
The authors have observed emission lines from the $J = 1 \rightarrow 0$ carbon monoxide rotational transition at 2.6 mm in the central regions of five external galaxies, including two at a distance of ~9 Mpc. They have also detected ^{13}CO emission in one of these (M82) with an observed $^{13}CO/^{12}CO$ intensity ratio of ~0.23 ± 0.1. The total mass in the form of molecular

clouds is typically of order $2 \times 10^8 M_\odot$ in the central 1′ and probably a few times $10^9 M_\odot$ for an entire galaxy.

158.011 **Permitted iron lines in NGC 7469.**
S. A. Shectman, G. M. MacAlpine.
Astrophys. Journ., *(Letters)*, Vol. 199, L85 - L87 (1975).
Relatively strong Fe II emission has been detected in the Seyfert galaxy NGC 7469. This object is different from the handful of others with permitted iron lines in that it also exhibits intense forbidden lines of oxygen. The observations do not show the strong He II λ4686 reported by previous investigators.

158.012 **A study of the neutral hydrogen near the interacting galaxies NGC 4631 and NGC 4656.**
A. J. B. Winter.
Monthly Notices Roy. Astron. Soc., Vol. 172, 1 - 11 (1975).
High-resolution 21-cm maps of the pair of interacting galaxies NGC 4631/NGC 4656 have been made with the Half-Mile telescope at Cambridge. Large amounts of neutral hydrogen are found at distances up to 12 kpc above the plane of NGC 4631, together with complex structure in the bridge joining the two galaxies. The bearing of these results on the nature of the interaction is discussed.

158.013 **The classification of active galaxies.**
S. van den Bergh.
Journ. Roy. Astron. Soc. Canada, Vol. 69, 105 - 125 (1975).
This paper discusses the classification of and the interrelations between Haro galaxies, Markarian galaxies, compact galaxies, N galaxies, Seyfert galaxies, and quasars.

158.014 **Anisotropic redshift distribution for compact galaxies with absorption spectra.**
T. Jaakkola, H. Karoji, M. Moles, J. P. Vigier.
Nature, Vol. 256, 24 - 25 (1975).
Rubin et al. have found that in one hemisphere of the sky ScI galaxies with $14.0 \leqslant m \leqslant 15.0$ have a mean velocity $1{,}365 \pm 200$ km s^{-1} larger than in the other hemisphere. Anisotropies, corresponding to this, have also been found for brightest cluster galaxies, supernovae, and Markarian galaxies. Here the authors present a similar result for absorption line compact galaxies.

158.015 **Optical variability of the nuclei of Seyfert galaxies.**
V. M. Lyuty (*Lyutyj*), V. I. Pronik.
IAU Symposium No. 67, (see 012.001), p. 591 - 604 (1975).
The continuum variability (*UBV* observations), the variations in the emission lines, and the properties of the central source in the nuclei of Seyfert galaxies are discussed.

158.016 **Variability of the emission-line spectrum of the nucleus of the Seyfert galaxy NGC 7469.**
I. I. Pronik.
IAU Symposium No. 67, (see 012.001), p. 605 - 609 (1975).
It is shown that the emission lines of the NGC 7469 nucleus spectrum are variable. Time variability of hydrogen lines is less than 20 days. The Hα emission line is at the profile base about a factor of two narrower than the Hγ line. Estimation of the density of hydrogen envelope, where Balmer lines wings were emitted, gives the value of about 10^8–10^9 cm^{-3}.

158.017 **Optical variability of the *N* galaxy 3C 371.**
M. K. Babadzanjanz (*Babadzhanyants*), E. T. Belokon'.
IAU Symposium No. 67, (see 012.001), p. 611 - 614 (1975).
On the basis of a uniform series of photographic observations a 163 day period for the light variations of the *N* galaxy 3C 371 is found. The similarity of the form of the light curve for the different components of the light variations with the different time scales is noted. The form of the light variations was observed to be identical to that of the quasars 3C 345 and 3C 454.3.

158.018 **Giant radio galaxies.**
R. G. Strom, G. K. Miley, J. Oort.
Sci. American, Vol. 233, No. 2, p. 26 - 35 (1975).

158.019 **Aktive Galaxien.** K. J. Fricke, M. Reinhardt.
Naturwissenschaften, 62. Jahrgang, p. 309 - 320 (1975).
The different forms of active galaxies are reviewed: M 82-type galaxies, Seyfert galaxies, the NGC 4258 phenomenon, N-galaxies, active elliptical galaxies, and tadpole galaxies. The objects M 82, NGC 1275, NGC 4258, and M 87 are discussed in detail. One section deals with the models proposed so far to explain these phenomena.

158.020 **On the origin of the mass expulsion in the galaxy NGC 4258.** K. J. Fricke.
Mitt. Astron. Ges., No. 36, p. 86 (1975).

158.021 **Noncircular motions in NGC 4151.**
K. J. Fricke, M. Reinhardt.
Mitt. Astron. Ges., No. 36, p. 91 - 93 (1975).

158.022 **Observations of galaxies at high frequencies.**
R. Wielebinski.
Mitt. Astron. Ges., No. 36, p. 106 - 107 (1975).

158.023 **Rotationskurven und Gesamtmassen von Galaxien.**
W. K. Huchtmeier.
Mitt. Astron. Ges., No. 36, p. 110 - 112 (1975).

158.024 **Beobachtungen von neutralem Wasserstoff in Galaxien des Virgo-Haufens.**
W. K. Huchtmeier, G. A. Tammann, H. J. Wendker.
Mitt. Astron. Ges., No. 36, p. 112 (1975). – Abstract.

158.025 **Radiobeobachtungen optischer Galaxien des Virgo-Haufens.** I. Khan, A. von Kap-herr, J. Pfleiderer.
Mitt. Astron. Ges., No. 36, p. 112 (1975). – Abstract.

158.026 **Missing mass in galactic coronae?** J. Dorschner, C. Friedemann, J. Gürtler, K.-H. Schmidt.
Astron. Nachr., Vol. 296, 189 - 191 (1975).
The concept of massive galactic coronae proposed by Einasto et al. is discussed. Only hot gas and under certain circumstances black holes have a good chance to be main constituents of the galactic coronae proposed by Einasto et al.

158.027 **cD galaxies in poor clusters.**
W. W. Morgan, S. Kayser, R. A. White.
Astrophys. Journ., Vol. 199, 545 - 548 (1975).
An investigation has been made to determine whether the supergiant cD galaxies, which are at present only known in rich galaxy clusters, are also present in galaxy clusterings poorer than those of the Abell catalog. A search was carried out over 73 fields of the Palomar Observatory Sky Survey; the fields were located in galactic latitudes greater than +30°. Eight clusters were found which probably contain galaxies of the cD type. While it will be necessary to obtain redshifts before cluster membership is definitely established, it seems likely that cD galaxies do occur in some poor clusters.

158.028 **Empirical properties of the mass discrepancy in groups and clusters of galaxies. VI. Projection effects.** H. J. Rood.
Astrophys. Journ., Vol. 199, 549 - 554 (1975).
Study of a hypothetical sample of identical double galaxies with circular orbits and randomly distributed relative velocity and separation vectors indicates that the empirical

(M_{vt}/M, V)-, (M_{vt}/M, R)-, and (V, R)-relations for double galaxies are almost certainly caused by projection effects. Dynamical properties of loose groups of galaxies and spectroscopic double stars are discussed.

158.029 **Isophotal tracings of galaxies near quasars.**
H. Arp, N. M. Pratt, J. W. Sulentic.
Astrophys. Journ., Vol. 199, 565 - 585 (1975).
Three cases where quasars fall close to large galaxies on the plane of the sky are investigated. In all three cases, disturbances in the inner isophotes of the galaxies are found. In all cases the isophotal disturbances are extended fairly closely in the direction of the quasar and could be interpreted as due to an event that occurred in the galaxy at a time of the order of 10^7 years previously.

158.030 **The rotation curve of NGC 3115.** T. B. Williams.
Astrophys. Journ., Vol. 199, 586 - 590 (1975).
A rotation curve is derived for the S0 galaxy NGC 3115 from its absorption-line spectrum. The central velocity is 728 ± 9 km s^{-1}. The rotation curve rises to 260 km s^{-1} at 25″ from the nucleus, and remains constant at that value to the limit of the measurements. A rough computation of the mass gives $2.4 \times 10^{10} M_\odot$ out to 3.4 kpc, and a mass-to-light ratio of 20 to the same radius.

158.031 **A jet in the galaxy identified with the radio source DA 240.**
E. M. Burbidge, H. E. Smith, G. R. Burbidge.
Astrophys. Journ., (*Letters*), Vol. 199, L137 - L138 (1975).
The galaxy MCG 9-13-66 associated with the large radio source DA 240 has been shown to contain a large jet with a featureless spectrum somewhat bluer in color than the galaxy, oriented at 62° to the axis of radio emission. Observations made at Kitt Peak National Observatory and Lick Observatory are described.

158.032 **Radial velocities of compact galaxies in pairs.**
I. D. Karachentsev, V. I. Pronik, K. K. Chuvaev.
Astron. Astrophys., Vol. 41, 375 - 377 (1975).
Spectral observations of 80 compact galaxies in pairs have been carried out. In 48 of them, emission lines were found and radial velocities were measured. For 18 pairs, orbital masses were determined assuming that the galaxies have circular orbits. The mean mass-to-luminosity ratio for them is $\langle f \rangle = (20.6 \mp 5.4) f_\odot$.

158.033 **On the origin of S0 galaxies.**
P. Biermann, B. M. Tinsley.
Astron. Astrophys., Vol. 41, 441 - 446 (1975).
The authors discuss the hypothesis that S0 galaxies are former spirals, in which star formation has been cut off by loss of gas. Observed *UBV* colors, compiled from the literature, are compared with those of model galaxies; they are consistent with the cutoff hypothesis, with the reddest S0's having ceased star formation at a very early epoch, and the bluest having made stars up to 1 or 2×10^9 years ago. Several mechanisms for cutting off star formation are mentioned, and it is suggested that galaxies classified as S0 may have had a variety of origins.

158.034 **Radial velocities and neutral hydrogen content of the tails of the interacting galaxies NGC 4038/39.**
W. K. Huchtmeier, H.-D. Bohnenstengel.
Astron. Astrophys., Vol. 41, 477 - 479 (1975).
Neutral hydrogen has been observed in the double galaxy NGC 4038/39 (Arp 244). A considerable amount of neutral hydrogen was found along the two arms (antennas). The observed radial velocities strongly support the view that the antennas are part of the double system. More than a third of all H I is connected with the tails.

158.035 **Infrared photometry of NGC 1068 at 25 and 33 μ**
T. Simon, H. M. Dyck.
Monthly Notices Roy. Astron. Soc., Vol. 172, 19P - 21P (1975).
Ground-based broadband photometry centred at wavelengths of 25 μ and 33 μ confirms a previously-reported turnover in the infrared spectrum of the Seyfert galaxy NGC 1068.

158.036 **The ultraviolet fluxes from some nearby galaxies.**
D. J. Carnochan, C. Navach, R. Wilson.
Monthly Notices Roy. Astron. Soc., Vol. 172, 27P - 28P (1975).
Ultraviolet fluxes in four wavebands are presented for some bright galaxies as observed with the sky-survey telescope (S2-68) in the ESRO TD-1 satellite. The data do not show the pronounced increase in the far ultraviolet that is a characteristic of the observations of similar galaxies made with OAO-2.

158.037 **On the age of the Draco dwarf galaxy.**
V. Castellani.
Monthly Notices Roy. Astron. Soc., Vol. 172, 59P - 63P (1975).
By adopting low Z values, as recently confirmed by Hartwick & McClure, it is possible to show that the age of the dwarf elliptical galaxy Draco is expected to be definitely lower in comparison with the ages of galactic globular clusters like M 15. The agreement with previous suggestions is reported and some evolutionary consequences are briefly discussed.

158.038 **Eighth list of compact galaxies.**
F. Zwicky, W. L. W. Sargent, C. T. Kowal.
Astron. Journ., Vol. 80, 545 - 558 (1975).
Positions and descriptions are given for 496 compact galaxies discovered by Zwicky after the completion of his catalog of such objects.

158.039 **Limits to the neutral hydrogen content of elliptical galaxies.**
G. S. Shostak, M. S. Roberts, S. D. Peterson.
Astron. Journ., Vol. 80, 581 - 583 (1975).
Upper limits to the H I content of 15 elliptical galaxies are presented. These limits are typically $\sim 10^8 M_\odot$, assuming a rectangular 21-cm profile with 330-km sec^{-1} width. The effects of alternative line shapes and widths on the derived hydrogen mass are considered, and explanations for the lack of detected H I are reviewed.

158.040 **Comparison of new 21-cm radial velocities with VVO optical radial velocities of galaxies.**
J. R. Dickel, H. J. Rood.
Astron. Journ., Vol. 80, 584 - 586 (1975).
21-cm radial velocities (V_{21}) are given for nine galaxies with optical radial velocities (V_{CR}) published by Chincarini and Rood (1972). The statistic, $V_{CR} - V_{21}$, has an average value of 4 ± 19 km sec^{-1} and a standard derivation of 52 ± 13 km sec^{-1}. The 21-cm radial velocity for NGC 4670 is 106 km sec^{-1} smaller than the optical radial velocity derived by Kintner (1971).

158.041 **On the physical nature of coronae around spiral galaxies.** B. V. Komberg, I. D. Novikov.
In-t kosmich. issled. AN SSSR. Pr-214. Moskva, 10 pp.
In Russian. – Abstr. in Referativ. Zhurn. 51. Astron., 7.51. 762 (1975).

158.042 **160 systems of galaxies appearing as M51.**
B. A. Vorontsov-Vel'yaminov.
Astron. Zhurn. Akad. Nauk SSSR, Vol. 52, 692 - 699 (1975).
In Russian. English translation in Soviet Astron., Vol. 19, No. 4.
When the Morphological Catalogue of Galaxies (MCG) was

completed, among the interacting galaxies some 160 pairs of M51 appearance were described. Here are given the data for 18 systems of known redshift and for 21 systems which have a satellite on each of its spiral arms. The latter fact completely rules out the origin of the spiral arms by tidal disturbances. The ratio of the diameters of both components and the magnitude difference ΔM are presented for 190 companions. Pairs with nearly equal components were not included in the list of M51-like systems.

158.043 **The bar-like objects as a possible generator of spiral density waves. II. The wave pattern in a differentially rotating disk.** V. I. Korchagin, L. S. Marochnik.
Astron. Zhurn. Akad. Nauk SSSR, Vol. 52, 700 - 709 (1975). In Russian. English translation in Soviet Astron., Vol. 19, No. 4.

It is shown that a "narrow" bar can generate the two-armed spiral pattern in a differentially rotating disk modelling the flat subsystem of a galaxy. The actual wave pattern is determined by the properties of generator and system. The trailing spiral structure is generated by the bar rotating more rapidly than the central parts of the flat subsystem of the galaxy. A bar rotating more slowly than the edge of the galaxy generates a leading spiral structure. A bar with a middle value of angular speed generates the "multistoreyed" spiral structure.

158.044 **Photographic photometry of bright galaxies. III. NGC 1023.** R. Barbon, M. Capaccioli.
Astron. Astrophys., Vol. 42, 103 - 108 (1975).

The galaxy NGC 1023 has been investigated by means of photographic photometry up to a threshold surface brightness $\mu_B = 26.6$ mag arcsec^{-2}. All the standard photometric parameters are derived. The morphology of the galaxy is also discussed within the context of the different classifications made till now. The faint condensation on the eastern end of the major axis of the galaxy has also been studied. The authors conclude that it is an individual object, very probably a peculiar dwarf member of the NGC 1023 group.

158.045 **High-resolution, multifrequency radio observations of M82.** P. P. Kronberg, P. N. Wilkinson.
Astrophys. Journ., Vol. 200, 430 - 438 (1975).

A supersynthesis radio map of the 8085 MHz emission in M82 is presented. The radio structure is complex, and is dominated by a bright unresolved source which is unpolarized (<0.7 %), becomes optically thick at $\nu \approx 650$ MHz, and emits 65 percent of its 1666 MHz flux from within a region <1.2 pc in diameter. Nine other discrete radio components near the M82 nucleus have been isolated, and no other radio sources with surface brightness >5 K exist in the galaxy but outside the main radio source. The compact radio components in M82 may be supernova remnants, the brightest of which (source 41.9 + 58) is very recent ($\lesssim 150$ years) and has brightness and size limits which are consistent with the Σ-D relationship for known supernova remnants.

158.046 **Dynamics of early type galaxies. I. The rotation curve of the elliptical galaxy NGC 4697.**
F. Bertola, M. Capaccioli.
Astrophys. Journ., Vol. 200, 439 - 445 (1975).

The rotation curve, up to 0.′6 from the center, has been derived from Ca II H and K lines for the elliptical galaxy NGC 4697. A maximum seems to be reached at 15″ with 65 km s^{-1}. The same galaxy has been investigated by means of photographic photometry in the B system. The luminosity profile follows the $r^{1/4}$ law,and the mean ellipticity is 0.4. The total apparent magnitude is $B = 10.30$ mag, and the corresponding absolute magnitude reaches $M_B = -20.5$ mag, assuming a distance of 14.8 Mpc. The spatial velocity curve and the light density profile are derived.

158.047 **Hα emission-line galaxies in a south galactic polar region.** N. Sanduleak, P. Pesch.
Astrophys. Journ., Vol. 200, 446 - 449 (1975).

Nine new Hα emission-line galaxies have been found in a region near the south galactic pole.

158.048 **Measurements of galactic nuclei at 34 microns.** G. H. Rieke, F. J. Low.
Astrophys. Journ., (*Letters*), Vol. 200, L67 - L69 (1975).

NGC 1275, NGC 4151, Markarian 231, NGC 5195, and NGC 5253 have been observed at 34 μ. The measurements provide improved estimates for the total luminosities of these sources, and show that Markarian 231 emits nearly 10^{13} $L_\odot$.

158.049 **The neutral hydrogen content of elliptical galaxies.** W. K. Huchtmeier, G. A. Tammann, H. J. Wendker.
Astron. Astrophys., Vol. 42, 205 - 211 (1975).

Nine elliptical galaxies have been searched for 21 cm line emission of neutral hydrogen with the 100 m radio telescope in Effelsberg. Using measurements of 6 elliptical galaxies we derive for the mean hydrogen mass of an elliptical $M_{\rm HI} < 3 \times 10^7 M_\odot$. The upper limits imply $M_{\rm HI}/M_T < 5 \times 10^{-4}$ and $M_{\rm HI}/L < 0.01$. For two elliptical galaxies. NGC 4278 and 4636, the authors suspect a "marginal line" which would correspond to a value of $10^8 M_\odot$ of neutral hydrogen in these galaxies. Implications for the evolution of elliptical galaxies are discussed.

158.050 **Evolution der Galaxien.** J. H. Oort.
Nova Acta Leopoldina [Deutsche Akademie der Naturforscher Leopoldina, Halle/Saale], New Ser., Vol. 42 (No. 218), 97 - 114 (1975). – Lecture on the occasion of the annual meeting of the Academy, 1973 October 11 - 14.

158.051 **Photographic photometry of bright galaxies. IV. NGC 2683.** R. Barbon, M. Capaccioli.
Astron. Astrophys., Vol. 42, 221 - 228 (1975).

Surface photographic photometry in the B system of the spiral galaxy NGC 2683 is presented. Standard photometric parameters are derived. It is shown that the equivalent luminosity profile does follow an exponential law, thus confirming that the luminosity gradients in the outer regions of spirals are different from those in ellipticals. The direction of rotation of the spiral arms is discussed, and they are found to be trailing. From rough estimates of the mass and the mass-to-light ratio, the distance of the galaxy is discussed.

158.052 **Remarques sur une faible valeur de la constante de Hubble.** T. Jaakkola.
Comptes Rendus Acad. Sci. Paris, Sér. B, Vol. 281, 49 - 51 (1975).

Plusieurs remarques faites sur les données fournies par Sandage et Tamman suggèrent que ces auteurs ont surestimé les distances des galaxies les plus lumineuses et ont ainsi obtenu une valeur de H, constante de Hubble, trop faible.

158.053 **The effect of suprathermal protons on the physical conditions in Seyfert galaxy nuclei.**
R. Ptak, R. Stoner.
Astrophys. Journ., Vol. 200, 558 - 566 (1975).

The physical conditions in a high-density hydrogen gas heated and ionized by suprathermal protons are investigated, with application to the gas in the nuclear region of Seyfert galaxies. The gas is assumed to be optically thick to Lyman and Balmer line radiation. Mechanisms by which the radiation from the gas can balance the heating by the fast protons are investigated, and minimum values for the mass of gas are estimated. Under certain conditions, the suprathermal atoms themselves can cool the ambient gas by rescattering the line radiation into the optically thin region in the wings of the line. This mechanism, which the authors call optical reverberation, can enhance the broad component of the hydrogen lines produced by inelastic atomic collisions and yield line widths consistent with those observed in Seyfert and quasar spectra.

158.054 **The disk of the Seyfert galaxy NGC 4151.**
S. M. Simkin.
Astrophys. Journ., Vol. 200, 567 - 573 (1975).

Spectroscopic observations of the "bar" in NGC 4151 show its continuous spectrum to have a large ultraviolet excess and no detectable stellar absorption features. The λ3727 [O II] line in emission leads to line-of-sight velocities for the ends of the bar of 35–55 km s^{-1} relative to the nucleus. The most probable interpretation of the observations suggest that both large circular velocities (300–500 km s^{-1}) and radial streaming out into the disk (100–300 km s^{-1}) occur near the nucleus, with outward radial streaming of 60–100 km s^{-1} occurring out to a distance of 4 kpc in the bar.

158.055 **A comparison of colors and line strengths at and near the nucleus of M 31.**
H. Spinrad, J. W. Liebert.
Astrophys. Journ., Vol. 200, 582 - 583 (1975).

High–angular-resolution spectrophotometry shows that the energy distribution of the nucleus of M 31 ($r \leq 2''$) is only slightly redder and marginally more strong-lined than the surrounding nuclear bulge (measured 4″ E). No discontinuity in metallicity seems to be present.

158.056 **The $V - [2.2\mu]$ colors of elliptical galaxies.**
J. A. Frogel, S. E. Persson, M. Aaronson, E. E. Becklin, K. Matthews, G. Neugebauer.
Astrophys. Journ., (*Letters*), Vol. 200, L123 - L126 (1975).

Infrared photometric measurements of 19 normal and giant elliptical galaxies are presented. These measurements have been made with apertures of diameters ranging from 0.1 to 1.3 times the visual diameter. The derived $V - [2.2\mu]$ colors range from 3.15 to 3.60 magnitudes, and show no significant radial dependence. These two results differ markedly from those of Grasdalen.

158.057 **Infrared galaxies: evolutionary stages of massive star formation.** M. Harwit, F. Pacini.
Astrophys. Journ., (*Letters*), Vol. 200, L127 - L129 (1975).

The authors cite evidence which indicates that infrared galaxies may represent evolutionary stages during which a large number of massive stars are being formed. The lifetime of these stars would be rather short (10^6–10^7 years), and the resulting supernova explosions could account for the level of nonthermal activity which often accompanies the thermal infrared emission.

158.058 **Dynamical models of elliptical galaxies.**
C. P. Wilson.
IAU Symposium No. 69, (see 012.009), p. 207 - 208 (1975).

Self-consistent dynamical models of elliptical galaxies have been constructed using a modification of the Prendergast-Tomer algorithm. They reproduce many of the observed properties of elliptical galaxies.

158.059 **The collapse and formation of galaxies.**
R. B. Larson.
IAU Symposium No. 69, (see 012.009), p. 247 - 269 (1975).
Review paper.

158.060 **On the formation of elliptical galaxies.**
J. R. Gott III.
IAU Symposium No. 69, (see 012.009), p. 271 - 285 (1975).

Simple collapse and violent relaxation models produce sharply cut off envelopes with $\rho(r) \propto r^{-4}$, whereas actual elliptical galaxies have more extended envelopes approximately given by $\rho(r) \propto r^{-3}$ as implied by Hubble's law. Numerical models are presented, showing that when cosmological infall effects are included, galaxies are produced with more extended envelopes having a $\rho(r) \propto^{-2.8}$ dependence in excellent agreement with Hubble's law. The author has also computed a realistic rotating model of an E5 galaxy which includes infall effects and where the angular momentum of the galaxy is gained through tidal interactions with neighboring protogalaxies. The envelope displays a Hubble dependence but at great distances shows a cutoff due to the tidal effects. The model is compared to the E5 galaxy NGC 4697 and provides an outstanding good fit to the observations. Except for simple scaling, this excellent fit is produced without the use of any free fitting parameters.

158.061 **The rotation curves of galaxies.** M. S. Roberts.
IAU Symposium No. 69, (see 012.009), p. 331 - 340 (1975).

Currently available data on rotation curves are reviewed. The assumption that a rotation curve approaches a Keplerian curve after passing its peak rotational velocity implies a strongly concentrated and limited extent of the mass distribution within a galaxy. This assumption is not supported by 21-cm observations of the velocity field within a galaxy. Because of the greater extent of H I compared to measurable optical (blue) surface brightness, rotation curves may be defined to much larger radii from 21-cm observations.

158.062 ***M/L* ratios in galactic disks.** J. E. Baldwin.
IAU Symposium No. 69, (see 012.009), p. 341 - 348 (1975).

Models of galactic disks based on optical photometry and uniform *M/L* ratios have rotation curves which are indistinguishable from those derived from 21 cm hydrogen line observations.

158.063 **Comparison with observations of disk galaxies.**
K. C. Freeman.
IAU Symposium No. 69, (see 012.009), p. 367 - 372 (1975).

The author discusses some observational aspects of the structure and dynamics of disk galaxies. Topics include the disk, the bulge, the lens, and the *z*-structure of the disk.

158.064 **Dynamics of early type galaxies.**
F. Bertola, M. Capaccioli.
IAU Symposium No. 69, (see 012.009), p. 373 - 375 (1975).

An attempt has been made to compute the angular momenta of the early type galaxies NGC 4697 and NGC 4762. Comparison with spiral galaxies indicates that there is a correlation between angular momentum per unit mass and morphological class.

158.065 **Galactic nuclei and compact supermassive objects.**
W. C. Saslaw.
IAU Symposium No. 69, (see 012.009), p. 379 - 393 (1975).
Review paper.

158.066 **Symmetric ejection of massive objects from galactic nuclei.** M. J. Valtonen.
IAU Symposium No. 69, (see 012.009), p. 395 - 403 (1975).

It is shown that massive objects can be ejected symmetrically in two opposite directions from the nucleus of a giant elliptical galaxy only under special initial conditions. A way of obtaining these conditions is outlined, and the resulting picture of evolution galactic nuclei, including a quasar or a Seyfert galaxy phase, is shown to be compatible with observations.

158.067 **Expériences numeriques sur l'éjection de masses depuis le centre d'un disque d'étoiles.**
S. Clairemidi, A. Hayli.

IAU Symposium No. 69, (see 012.009), p. 405 - 419 (1975).

Numerical experiments on simple or symmetric ejections of massive bodies from the nucleus of a disk of stars are presented. The ejected masses range from 0.04 to 0.1 of the total mass. Either spiral features or a barred structure appear but do not last for a long time. Material arms start leading and become trailing.

158.068 **On the energy source of active galactic nuclei and quasars.** Eh. Ya. Vil'koviskij.
Pis'ma v Astron. Zhurn., Vol. 1, No. 7, p. 8 - 12 (1975). In Russian.

The interaction of stars with the central gas body in galaxies permits to explain the activity of galactic nuclei and quasars.

158.069 **Beam-switched surface photometry of galaxies.**
J. S. Gallagher, H. S. Hudson.
Bull. American Astron. Soc., Vol. 7, 396 (1975). – Abstr. AAS.

158.070 **The cool stellar component of elliptical galaxies.**
J. A. Frogel, S. E. Persson, M. Aaronson, E. E. Becklin, K. Matthews, G. Neugebauer.
Bull. American Astron. Soc., Vol. 7, 396 (1975). – Abstr. AAS.

158.071 **Composition and composition gradients in disk galaxies. I.**
E. B. Jensen, S. E. Strom, K. M. Strom, S. P. Worden.
Bull. American Astron. Soc., Vol. 7, 396 (1975). – Abstr. AAS.

158.072 **Composition and composition gradients in disk galaxies. II.**
S. E. Strom, K. M. Strom, E. B. Jensen.
Bull. American Astron. Soc., Vol. 7, 397 (1975). – Abstr. AAS.

158.073 **Galaxy population synthesis from image-tube spectra.** T. B. Williams.
Bull. American Astron. Soc., Vol. 7, 397 (1975). – Abstr. AAS.

158.074 **Near infrared energy distributions of elliptical galaxies.** R. W. O'Connell.
Bull. American Astron. Soc., Vol. 7, 411 (1975). – Abstr. AAS.

158.075 **Spectroscopy of the center of M31.**
D. C. Morton, B. G. Elmergreen, C. D. Andereck.
Bull. American Astron. Soc., Vol. 7, 414 (1975). – Abstr. AAS.

158.076 **Observations of the optical jet in NGC 5128.**
J. A. Graham.
Bull. American Astron. Soc., Vol. 7, 414 - 415 (1975). – Abstr. AAS.

158.077 **Galaxy correlations as a function of morphological type.** M. J. Geller, M. Davis.
Bull. American Astron. Soc., Vol. 7, 415 (1975). – Abstr. AAS.

158.078 **The stellar content and BL Lac characteristics of the variable N galaxy 3C 371.** J. S. Miller.
Bull. American Astron. Soc., Vol. 7, 422 (1975). – Abstr. AAS.

158.079 **A critique of the luminosity function for galaxies.**
J. E. Felten.
Bull. American Astron. Soc., Vol. 7, 425 (1975). – Abstr. AAS.

158.080 **Angular momentum properties of galaxies in rich clusters.** L. A. Thompson.
Bull. American Astron. Soc., Vol. 7, 425 (1975). – Abstr. AAS.

158.081 **A new method of determining distances to galaxies.**
R. B. Tully, J. R. Fisher.
Bull. American Astron. Soc., Vol. 7, 426 (1975). – Abstr. AAS.

158.082 **A faint galaxy redshift program.**
R. J. Harms, P. A. Strittmatter.
Bull. American Astron. Soc., Vol. 7, 427 (1975). – Abstr. AAS.

158.083 **The interdependence of H I, galaxy counts, reddening, and galactic latitude.** C. Heiles.
Bull. American Astron. Soc., Vol. 7, 427 - 428 (1975). – Abstr. AAS.

158.084 **Gas motions around bars.**
R. H. Sanders, J. Huntley.
Bull. American Astron. Soc., Vol. 7, 434 - 435 (1975). – Abstr. AAS.

158.085 **Spectrophotometric observations of N systems at large redshift – PKS 0353+027, 3C99, 3C467.**
H. E. Smith, H. Spinrad, R. Hunstead.
Bull. American Astron. Soc., Vol. 7, 436 (1975). – Abstr. AAS.

158.086 **The 37 to 350μ flux from NGC 1068.**
C. M. Telesco, D. A. Harper.
Bull. American Astron. Soc., Vol. 7, 436 (1975). – Abstr. AAS.

158.087 **The 8 - 13μ spectrum of NGC 1068.**
D. E. Kleinmann, F. C. Gillett, E. L. Wright.
Bull. American Astron. Soc., Vol. 7, 436 (1975). – Abstr. AAS.

158.088 **The variable radio nucleus of M81.**
P. C. Crane, T. S. Giuffrida, J. B. Carlson.
Bull. American Astron. Soc., Vol. 7, 437 (1975). – Abstr. AAS.

158.089 **Near infrared and visual observations of NGC 4565.**
R. Schild, M. Frankston, T. McCord.
Bull. American Astron. Soc., Vol. 7, 441 (1975). – Abstr. AAS.

158.090 **Winds in elliptical galaxies.** L. L. Cowie.
Bull. American Astron. Soc., Vol. 7, 452 (1975). Abstr. AAS.

158.091 **Optical emission-line spectra of 3C 327 and other narrow-line radio galaxies.**
R. Costero, D. E. Osterbrock.
Bull. American Astron. Soc., Vol. 7, 452 (1975). – Abstr. AAS.

158.092 **The optical emission-line spectrum of 3C 120.**
M. M. Phillips, D. E. Osterbrock.
Bull. American Astron. Soc., Vol. 7, 452 (1975). – Abstr. AAS.

158.093 **Optical spectra of 3C 227 and other broad-line radio galaxies.**
D. E. Osterbrock, A. T. Koski, M. M. Phillips.
Bull. American Astron. Soc., Vol. 7, 452 - 453 (1975). – Abstr. AAS.

158.094 **Electron temperature of NGC 1052.**
A. T. Koski, D. E. Osterbrock.
Bull. American Astron. Soc., Vol. 7, 453 (1975). – Abstr. AAS.

158.095 **Detection of wavelength dependence of optical polarization in Markarian 421 and Markarian 501.**
K. Serkowski, S. Tapia, M. Tarenghi.
Bull. American Astron. Soc., Vol. 7, 453 (1975). – Abstr. AAS.

158.096 **Models of star formation in Markarian galaxies.**
J. P. Huchra.
Bull. American Astron. Soc., Vol. 7, 453 (1975). – Abstr. AAS.

158.097 **Cosmic ray acceleration of gas in active galactic nuclei.** J. A. Eilek.
Bull. American Astron. Soc., Vol. 7, 500 (1975). – Abstr. AAS.

158.098 **Detection of M104 to a diameter of half a degree – preliminary results.**
M. S. Burkhead, Z. G. Levay.
Bull. American Astron. Soc., Vol. 7, 500 (1975). – Abstr. AAS.

158.099 **A new approach to the variability of radio galaxies, Seyfert nuclei, and QSO's.** J. D. Scargle.
Bull. American Astron. Soc., Vol. 7, 500 (1975). – Abstr. AAS.

158.100 **Spectral observations of some double galaxies.**
V. Afanasjev (*Afanas'ev*), I. Karachentsev, P. Notni.
Astron. Nachr., Vol. 296, 233 - 235 (1975).

Estimates of radial velocities for six components of isolated pairs of galaxies are presented. Orbital masses are determined for two pairs with Markarian galaxies.

158.101 **High resolution neutral hydrogen observations of NGC 4258.** G. D. van Albada, W. W. Shane.
Astron. Astrophys., Vol. 42, 433 - 439 (1975).

Preliminary results of 21 cm line observations of NGC 4258 obtained with the Westerbork Synthesis Radio Telescope are presented. Disturbances in both the H I density distribution and the velocity field in the neighbourhood of the anomalous arms tend to support the expulsion model of Van der Kruit et al. (1972). Outside the region of the anomalous arms the galaxy behaves like a normal spiral galaxy.

158.102 **Photographic photometry of compact extragalactic objects. III.** M. K. Babadzhanyants, S. K. Vinokurov, V. A. Hagen-Thorn, E. V. Semenova.
Trudy Astron. Obs., *Leningrad*, Vol. 31 (= Uchenye Zapiski Leningr. Un-ta, No. 381 = Seriya Matem. Nauk, vyp. (No.) 51), p. 100 - 122 (1975). In Russian.

The results of photographic observations of compact extragalactic objects made in 1972 are presented. The nature of the variability of 3C 345, 3C 371 and 3C 390.3 is discussed.

158.103 **Hα emission from disks of spiral galaxies and implications for galactic luminosity functions.**
J. G. Cohen.
Publ. Astron. Soc. Pacific, Vol. 87, 499 - 500 (1975).
Abstr. ASP.

158.104 **On the formation of spiral and elliptical galaxies.**
J. R. Gott III, T. X. Thuan.
Publ. Astron. Soc. Pacific, Vol. 87, 501 (1975). – Abstr. ASP.

158.105 **Photometric properties of Markarian galaxies.**
J. P. Huchra.
Publ. Astron. Soc. Pacific, Vol. 87, 501 - 502 (1975).
Abstr. ASP.

158.106 **The variable galaxy X Comae.**
J. P. Huchra, R. F. Green.
Publ. Astron. Soc. Pacific, Vol. 87, 502 (1975). – Abstr. ASP.

158.107 **Markarian 376 – a Seyfert galaxy with strong Fe II emission.** D. E. Osterbrock.
Publ. Astron. Soc. Pacific, Vol. 87, 507 (1975). – Abstr. ASP.

158.108 **Photographic observations of 3C 371 and 3C 390.3.**
L. A. Urasin, N. V. Lavrova.
Izv. Astron. Ehngel'gardt. Obs., *Kazan'*, No. 40, p. 95 - 97 (1973). In Russian.

158.109 **The colours of nuclei of Hubble Sc galaxies.**
S. G. Iskudaryan.
Soobshch. Byurakan. Obs., vyp. (No.) 46, p. 62 - 72 (1975). In Russian.

The results of colorimetry of compact and some non-compact nuclei of Sc galaxies are given. A comparison is made with the other Hubble type spirals. Correlations between the following parameters: colour of the nucleus, integral colour of the galaxy, Byurakan class, spectral type are investigated.

158.110 **On the connection of the mechanism of formation of galaxies and the way of origin of their structural features.** S. G. Iskudaryan.
Soobshch. Byurakan. Obs., vyp. (No.) 46, p. 73 - 87 (1975). In Russian.

There is some evidence in favour of the idea that super-associations and the brightest associations can be considered as secondary local centres of active processes in galaxies. The origin of some structural features, which are observed in galaxies, apparently, are due to such local active processes.

158.111 **Application of the superposition method of negatives to the galaxy NGC 5195.** R. A. Sarkisyan.
Soobshch. Byurakan. Obs., vyp. (No.) 46, p. 144 - 151 (1975). In Russian.

The results of superposition of ten negatives obtained with the 40″-Schmidt camera for the galaxy NGC 5195 are presented. The results are represented by isophotes. Arguments are presented in favour of the fact that the arm of the galaxy NGC 5194 is connecting both components of M 51.

158.112 **Gradient diagram for optically variable nuclei of galaxies and quasars.** F. I. Lukatskaya.
Astrometriya i Astrofizika, *Kiev*, vyp. (No.) 26, (see 003.001), p. 47 - 52 (1975). In Russian.

The *V, B* and *U, B* dependences are linear regressions for simultaneous photoelectric observations of the optically variable nuclei of galaxies. The angular coefficients of those dependences or brightness gradients are determined for four N-galaxies, six Seyfert galaxies and the objects OJ 287 and PKS 0537-441.

158.113 **Interferometer survey of normal galaxies.**
M. Haynes, R. Sramek.
Astron. Journ., Vol. 80, 673 - 678 (1975).

Flux densities at 2.7 and 8.1 GHz and the basic radio structures are presented for 71 normal galaxies; accurate radio positions are given for 29 of these galaxies. About 20% of the galaxies have compact components.

158.114 **The mass of the local system of galaxies.**
J. Einasto.
Astron. Tsirk., No. 840, p. 1 - 2 (1974). In Russian.

158.115 **On the evolution of protogalaxies and protoclusters of galaxies.** A. A. Suchkov, Yu. A. Shchekinov.
Astron. Tsirk., No. 842, p. 3 - 5 (1974). In Russian.

158.116 **On the reality of a non-linear relation between virial and luminous masses in groups of galaxies.**
L. M. Ozernoj.
Astron. Tsirk., No. 844, p. 4 - 7 (1974). In Russian.

158.117 **Photographic observations of 3C 371.**
I. A. Dubyago, S. S. Tokhtas'ev, S. K. Fomin.
Astron. Tsirk., No. 845, p. 2 - 3 (1974). In Russian.

158.118 **Test for choosing between hypotheses on the localization of a 'hidden' mass.** L. M. Ozernoj.
Astron. Tsirk., No. 847, p. 1 - 3 (1974). In Russian.

158.119 **Chains of galaxies.** B. A. Vorontsov-Vel'yaminov.
Astron. Tsirk., No. 848, p. 3 - 5 (1974). In Russian.

158.120 **Test observations of galactic nuclei with a television camera.** B. A. Vorontsov-Vel'yaminov, V. V. Prokof'eva, M. V. Savel'eva.

Astron. Tsirk., No. 851, p. 7 - 8 (1974). In Russian.

158.121 **Optical variability of seven Markarian galaxies.** K. A. Saakyan.
Astron. Tsirk., No. 861, p. 7 - 8 (1975). In Russian.

158.122 **Interaction of galaxies with gaseous objects in their centres.** Eh. Ya. Vil'koviskij.
Astron. Tsirk., No. 862, p. 6 - 7 (1975). In Russian.

158.123 **Dynamical indications concerning the nature of the "hidden" mass.** L. M. Ozernoj.
Astron. Tsirk., No. 866, p. 1 - 3 (1975). In Russian.

158.124 **Possible variability of IV Zw67 – the "Egg nebula" in Cygnus.** Yu. N. Efremov, N. E. Kurochkin.
Astron. Tsirk., No. 867, p. 4 - 5 (1975). In Russian.

158.125 **Velocity dispersions in galaxies. IV. The nucleus of NGC 1068.** D. O. Richstone, D. C. Morton.
Astrophys. Journ., Vol. 201, 289 - 295 (1975).

The velocity dispersion of the stars in the nucleus of the Seyfert galaxy NGC 1068 was found to be 150 ± 50 km s^{-1} from an Fe I absorption line in a spectrum obtained with an integrating television sensor. If the structural length R_0 <140 pc in the nucleus and H = 50 km s^{-1} Mpc^{-1}, the mass within this radius is 14(+10, –8) × $10^8 M_\odot$, and less if $R_0 \gtrsim 140$ pc.

158.126 **On the formation of elliptical galaxies.** J. R. Gott III.
Astrophys. Journ., Vol. 201, 296 - 310 (1975).

Simple collapse and violent relaxation models produce sharply cut off envelopes with $\rho(r) \propto r^{-4}$, whereas actual elliptical galaxies have more extended envelopes approximately given by $\rho(r) \propto r^{-3}$ as implied by Hubble's law. Numerical models are presented, showing that when cosmological infall effects are included, galaxies are produced with envelopes having a $\rho(r) \propto r^{-2.8}$ dependence in excellent agreement with Hubble's law. The author has also computed a realistic rotating model of an E5 galaxy.

158.127 **The rotation curve and geometry of M31 at large galactocentric distances.**
M. S. Roberts, R. N. Whitehurst.
Astrophys. Journ., Vol. 201, 327 - 346 (1975).

New 21-cm observations of the southern end of M31 indicate (1) that the plane of H I is bent away from the conventional plane by up to ~5 kpc and (2) that the rotational velocity is essentially constant over the outer 10 kpc, i.e., from 20 to 30 kpc radius. The latter implies a mass that increases linearly with R over this range and a mass-to-luminosity ratio of ≳200 for this outer region. At 30 kpc the surface density is ~50 $M_\odot$ pc^{-2}. Dwarf M stars, the most common type of star in the solar vicinity, of number density of a few tenths pc^{-3} are adequate to explain the required mass and the mass-to-luminosity ratio. At these distances H I is ~1 percent of the total mass.

158.128 **Three-dimensional distribution of interstellar gas in disk-shaped galaxies.** W. Hu.
Acta Astron. Sinica, Vol. 16, 27 - 36 (1975). In Chinese.

The three-dimensional structure of the large-scale interstellar gas under the action of the resultant gravitational field of stars and gas is discussed on the basis of the fundamental equations of gas dynamics. The galactic mass distribution and the kinetic characteristics of gas (such as the rotation curve) being given, an exact solution of the spatial distribution of interstellar gas is obtained, and the peak distribution of gas in disk-shaped galaxies is discussed. A general discussion about the origin of the high-velocity characteristic gaseous clouds in the galactic wing is also given.

158.129 **Why there are no elliptical galaxies flatter than E7.** T. X. Thuan, J. R. Gott III.
Nature, Vol. 257, 774 - 776 (1975).

It has long been known that the flattest elliptical galaxies observed are of type E7, corresponding to an axial ratio of 3.33:1 whereas S0 and spiral galaxies are seen with axial ratios up to 20:1. The authors show, using Maclaurin spheroid models, that an initially spherical protogalaxy, composed entirely of stars, will always relax after a dissipationless collapse to an equilibrium configuration with $\epsilon \equiv 1-(c/a) \leqslant 0.70925$, where a, c are the semi-major, semi-minor axes. This happens regardless of original angular momentum of the galaxy, and is in remarkable agreement with the maximum observed flattening of elliptical galaxies.

158.130 **M82: nog steeds een raadsel.** M. Drummen.
Zenit, Vol. 2, 263 - 267 (1975).

158.131 **Is the chain of galaxies near NGC 247 anomalous?** C. Balkowski, P. Chamaraux.
Astron. Astrophys., Vol. 43, 297 - 301 (1975).

Neutral hydrogen 21-cm observations of the chain of galaxies near NGC 247 are reported. From H I and optical data, the chain's distance is found to be 62^{+41}_{-25} Mpc, in agreement with the redshift. This result rules out Arp's assumption of a physical association between NGC 247 and the chain at a confidence level of better than 3.9 times the r.m.s. dispersion. The H I content of the chain is quite normal, and it is likely to be dynamically stable: these results would argue against the proposal made by Burbidge et al. that the chain was formed recently.

158.132 **H II regions in peculiar galaxies.** P. W. Hodge.
Astrophys. Journ., Vol. 201, 556 - 562 (1975).

A survey of 32 galaxies, either classified as peculiar or having peculiar distributions of H II regions, is presented in order to ascertain the range of the peculiarities and the degree to which they are similar. Ten different types are recognized, several of which are interrelated. They include distortions of spiral patterns, bright diffuse emission, filaments, central core features, and spiral-arm anomalies of various sorts, and are probably caused by either gravitational or explosive disturbances of the galaxies.

158.133 **An attempt to measure the metallicity of galaxies far into their halos.** H. Spinrad, R. P. S. Stone.
Astrophys. Journ., Vol. 201, 563 - 565 (1975).

Attempts were made to extend photoelectric CN measures to large radial distances in three galaxies. Low surface brightnesses preclude very definite conclusions, but the E0 galaxy NGC 1399 appears to be metal-rich out to 20 kpc from its center, while the S0 galaxy NGC 1553 and the Sa galaxy NGC 7814 show a definite CN decrease to a likely extreme population II level. Measurements of the CN strength were also made for several southern halo globular clusters of our Galaxy.

158.134 **A dynamical model of the tailed radio galaxies.** L. L. Cowie, C. F. McKee.
Astron. Astrophys., Vol. 43, 337 - 343 (1975).

A dynamical model of the tailed radio galaxies (TRGs) is developed in which mildly supersonic plasmoids (consisting of relativistic particles, magnetic fields and cold plasma) are ejected by the galaxy, stopped by the ram pressure of the intracluster medium, and subsequently leak electrons into a surrounding magnetic field. A simple theory for the paths of the plasmoids is given, and the effects of hydrodynamic instabilities on their evolution is discussed. The physical theory is applied to the radio source 3C 129, and shown to be consistent with observation. The authors briefly discuss the application of the theory to other TRGs.

158.135 **Spectra of nuclei of the Seyfert galaxies between 8000 and 11000 Å.**
Y. Andrillat, S. Collin-Souffrin.
Astron. Astrophys., Vol. 43, 419 - 422 (1975). In French.

The authors have obtained spectra in the range 7000–11000 Å of the nuclei of the Seyfert galaxies NGC 1068, NGC 3516, NGC 4051, NGC 4151 and of the nucleus of NGC 3031 (M81). The spectra of NGC 1068 and NGC 4151 are characterized by the high intensity of the lines [S III] 9069 and 9532 and He I 10830, while these lines are absent or weak in the other spectra. An interpretation is given in terms of a higher continuum contribution in these latter spectra, probably due to stellar emission.

158.136 **On the nuclei of galaxies and their activity.**
V. A. Ambartsumyan.
Probl. gravitatsii. Sb. dokl. plenarn. zasedanij 3-j Sov. gravitatsion. konf. Erevan, 1972. Erevan, Erevan. un-t, 1975, p. 5 - 17. In Russian. – Abstr. in Referativ. Zhurn. 51. Astron., 9.51. 761 (1975).

158.137 **Accuracy of determination of space densities of galaxies calculated from rotation curves.**
V. G. Metlov.
Pis'ma v Astron. Zhurn., Vol. 1, No. 10, p. 5 - 10 (1975). In Russian.

Accuracy of space densities calculated from rotation curves of galaxies is estimated. It is shown that local density maxima calculated from "waves" on the rotation curve are not reliable, even in the case of accurate velocity measurements. The discussed method of calculation of masses, densities and their uncertainties is applied to several galaxies with known characteristics.

158.138 **On the evolution of galaxies considering collisions of gas clouds.** F. W. Baier.
Astron. Nachr., Vol. 296, 281 - 284 (1975). In German.

Brosche's investigations concerning the evolution of galaxies are repeated using values for the starting-point which deviate from that of Brosche. Especially, a protogalaxy is assumed to consist of gas clouds each of them having a mass of $10^5\ M_\odot$. A comparison of the results with the observations is satisfactory. Brosche's interpretation of the Hubble sequence as a sequence depending on the angular momentum seems to be questionable.

158.139 **Stellar populations in galaxies.** S. van den Bergh.
Annual Rev. Astron. Astrophys., Vol. 13, (see 003.005), 217 - 255 (1975). – Contents: (1) Introduction; (2) Stellar populations in the Galaxy; (3) Stellar populations in the Magellanic Clouds; (4) Globular clusters; (5) Stellar populations in M31 and its companions; (6) Dwarf spheroidal galaxies; (7) The missing mass in galaxy halos and in clusters of galaxies; (8) The evolution of elliptical galaxies.

158.140 **The masses of the galaxies and the mass-energy in the universe.** G. R. Burbidge.
Proc. 16th Solvay Conference Phys., (see 012.020), p. 473 - 486, with a communication by M. J. Rees, p. 489 (1974).

158.141 **New observations of the optical spectrum of the Seyfert galaxy NGC 4151.** A. Boksenberg, K. Shortridge, D. A. Allen, R. A. E. Fosbury, M. V. Penston, A. Savage.
Monthly Notices Roy. Astron. Soc., Vol. 173, 381 - 396 (1975).

Spectra and scans of NGC 4151 are found to show broad lines of He I and Fe II. The He I lines are twice the width of the Balmer line wings and the broad line spectrum closely resembles that of 3C 273. Limits to the density and temperature for the region emitting the broad lines are deduced. The narrow forbidden lines are emitted in a region with log T = 4.07 ± 0.03 and density (possibly stratified) between log $N_e \sim 4.3$ and 3.7. How the hydrogen and helium line widths can differ is briefly discussed. Possible reasons for the difference between quasars and Seyferts with Fe II emission and those with strong forbidden line spectra are advanced.

158.142 **The peculiar galaxy NGC 1097.**
R. D. Wolstencroft, W. J. Zealey.
Monthly Notices Roy. Astron. Soc., Vol. 173, 51P - 52P (1975).

Surface photometry reveals low intensity radial features extending up to 55 kpc from the SB galaxy NGC 1097. It is suggested that these are the result of ejection.

158.143 **Activity of the nucleus of the southern Seyfert galaxy NGC 1566.**
H. Quintana, P. Kaufmann, J. L. Sérsic.
Monthly Notices Roy. Astron. Soc., Vol. 173, 57P - 60P (1975).

Millimetre-wave and optical observations of the nucleus of NGC 1566 are reported. Measurements of the blue luminosity over the last 20 yr confirm the optical variability of the nucleus, suggested previously. At millimetric wavelengths NGC 1566 is detected at 20 GHz and 22.2 GHz with a flux density of 0.52 ± 0.19 Jy and a spectral index $\alpha \simeq -0.5$. This implies a total radio luminosity of 1.6×10^{41} erg s^{-1} or higher. It is tentatively suggested that this Seyfert nucleus could be associated with the X-ray source 3U 0400–59.

158.144 **Catalogue of the photometric parameters of flat galaxies.**
B. A. Vorontsov-Vel'yaminov, R. I. Noskova.
Soobshch. Gos. Astron. Inst. Shternberga, No. 182, p. 3 - 13 (1973). In Russian.

The catalogue contains the fundamental photometric parameters of 130 flat galaxies of the S0, S and Irr types. The limit of galaxies is set at the surface brightness $26^m\!.5$ per 1″ square taking account of galactic absorption and inclination effect of galaxies.

158.145 **The mass of M31.** B. M. Lewis.
Southern Stars, Vol. 26, 54 - 58 (1975).

The limiting mass of M31 is estimated to be (3.5 ± 0.5) $\times 10^{11} M_{solar}$ for a distance of 690 kpc, from the velocities of its globular clusters, dwarf satellite galaxies and from a treatment of M31-M33 as a binary system. The existing observations are consistent with most of the mass being associated with the disk.

158.146 **H I in early-type galaxies. I. Observations.**
J. S. Gallagher, S. M. Faber, B. Balick.
Astrophys. Journ., Vol. 202, 7 - 21 (1975).

New 21-cm observations of 25 E and S0 galaxies provide a basis for an observational study of relationships between H I content and other properties of early-type systems. All available H I observations of early-type galaxies are used to generate a list (containing four normal S0's and 12 peculiar galaxies) of high-probability detections. From these data and new measurements, the authors conclude: (1) Elliptical galaxies contain at most 0.1 percent H I by mass. (2) Some structurally normal S0 galaxies exist with $M_{H\,I}/L_{pg} \gtrsim 0.05$, while other, apparently similar, systems have upper limits $M_{H\,I}/L_{pg} <$ 0.03. (3) In terms of H I content, the S0 class appears to be a true transition between the H I properties of spirals and ellipticals. (4) In some galaxies an extensive interstellar medium is obviously present, but has different properties from those usually found in later type spirals. (5) Among the morphologically peculiar early-type galaxies, those which exhibit violent activity are rich in H I.

158.147 **Radio observations of spiral structures in galaxies.** R. J. Allen.
La dynamique des galaxies spirales. Colloque International C.N.R.S. No. 241, (see 012.022), p. 157 - 195, with a discussion, p. 196 - 197 (1975). – Review paper.

158.148 **A neutral hydrogen study of the galaxy M81.** S. T. Gottesman, L. Weliachew.
La dynamique des galaxies spirales. Colloque International C.N.R.S. No. 241, (see 012.022), p. 199, with a discussion, p. 200 (1975), – Abstract of a paper which has been published in Astrophys. Journ., Vol. 195, 23 - 45 (1975) – see abstract 13.158.003.

158.149 **Distribution and kinematics of neutral hydrogen in M81.** A. H. Rots.
La dynamique des galaxies spirales. Colloque International C.N.R.S. No. 241, (see 012.022), p. 201 - 208, with a discussion, p. 209 (1975).

158.150 **Non-linear gas flow in M81.** H. C. D. Visser.
La dynamique des galaxies spirales. Colloque International C.N.R.S. No. 241, (see 012.022), p. 211 - 215, with a discussion, p. 216 (1975).

158.151 **High resolution observations of neutral hydrogen in Messier 51.** W. W. Shane.
La dynamique des galaxies spirales. Colloque International C.N.R.S. No. 241, (see 012.022), p. 217 - 225, with a discussion, p. 226 - 227 (1975).

158.152 **The development of the spiral structure in M51.** R. B. Tully.
La dynamique des galaxies spirales. Colloque International C.N.R.S. No. 241, (see 012.022), p. 229 - 234, with a discussion, p. 235 (1975).

158.153 **High resolution observations of the H I in M31.** D. T. Emerson.
La dynamique des galaxies spirales. Colloque International C.N.R.S. No. 241, (see 012.022), p. 243 - 254, with a discussion, p. 255 - 256 (1975).

158.154 **High resolution observations of neutral hydrogen in Messier 31.** W. W. Shane.
La dynamique des galaxies spirales. Colloque International C.N.R.S. No. 241, (see 012.022), p. 257 - 261, with a discussion, p. 262 (1975).

158.155 **Neutral hydrogen, spiral structure and density waves in M31.** J. Guibert.
La dynamique des galaxies spirales. Colloque International C.N.R.S. No. 241, (see 012.022), p. 263 - 271, with a discussion, p. 271 - 272 (1975).

158.156 **Two anonymous galaxies with interesting spiral structure.** R. B. Tully.
La dynamique des galaxies spirales. Colloque International C.N.R.S. No. 241, (see 012.022), p. 273 - 277, with a discussion, p. 278 (1975).

158.157 **An optical study of M33.** J. Boulesteix, G. Monnet.
La dynamique des galaxies spirales. Colloque International C.N.R.S. No. 241, (see 012.022), p. 279 - 283 (1975).

158.158 **Comments on the association of the H I and H II distributions in M101.** R. J. Allen.
La dynamique des galaxies spirales. Colloque International C.N.R.S. No. 241, (see 012.022), p. 285 - 288, with a discussion, p. 289 (1975).

158.159 **Observations of a warp in the edge-on galaxy NGC 5907.**
M. Guélin, R. Sancisi, L. Weliachew, H. van Woerden.
La dynamique des galaxies spirales. Colloque International C.N.R.S. No. 241, (see 012.022), p. 291 - 294 (1975).

Observations of neutral hydrogen in the edge-on galaxy NGC 5907 show a large-scale distortion in the northern part of the disk. The deviation of the hydrogen layer from the apparent plane of the galaxy amounts to 1.5 arcminutes, i.e. about 7 kpc at a distance of 16 Mpc. It is not yet known whether a similar distortion is also present on the southern side of the system. NGC 5907 is not in close interaction with any bright galaxy.

158.160 **Neutral hydrogen line and radio continuum emission from the edge-on galaxy NGC 891.**
R. Sancisi, R. J. Allen, T. S. van Albada.
La dynamique des galaxies spirales. Colloque International C.N.R.S. No. 241, (see 012.022), p. 295 - 301, with a discussion, p. 302 (1975).

The distribution and kinematics of neutral hydrogen in the edge-on galaxy NGC 891 are briefly described. A comparison with the radio-continuum shows a significant difference between the thickness and radial extent of the H I and continuum disks.

158.161 **Star formation in spiral galaxies.** F. H. Shu.
La dynamique des galaxies spirales. Colloque International C.N.R.S. No. 241, (see 012.022), p. 309 - 322, with a discussion, p. 323 - 324 (1975). – Review paper.

158.162 **The spiral distribution of stars in galaxies.** F. Schweizer.
La dynamique des galaxies spirales. Colloque International C.N.R.S. No. 241, (see 012.022), p. 337 - 355, with a discussion, p. 355 - 356 (1975). – Review paper.

158.163 **Evidence of density wave motions from the ages of stars and associations in M33: Lallemand electronic camera stellar photometry.** R. Dubout.
La dynamique des galaxies spirales. Colloque International C.N.R.S. No. 241, (see 012.022), p. 371 - 380, with a discussion, p. 381 (1975).

158.164 **Barred spirals.** R. Sancisi.
La dynamique des galaxies spirales. Colloque International C.N.R.S. No. 241, (see 012.022), p. 403 - 411, with a discussion, p. 412 (1975). – Review paper.

158.165 **High resolution observations of neutral hydrogen in NGC 1300.** W. W. Shane.
La dynamique des galaxies spirales. Colloque International C.N.R.S. No. 241, (see 012.022), p. 413 - 416, with a discussion, p. 416 (1975).

158.166 **Neutral hydrogen observations of the barred spiral galaxy NGC 3359.**
P. T. Siefert, S. T. Gottesman, M. C. H. Wright.
La dynamique des galaxies spirales. Colloque International C.N.R.S. No. 241, (see 012.022), p. 425 - 436, with a discussion, p. 437 (1975).

158.167 **Distribution and motions of neutral hydrogen in the giant irregular galaxy NGC 4449.**
H. van Woerden, A. Bosma, U. Mebold.
La dynamique des galaxies spirales. Colloque International C.N.R.S. No. 241, (see 012.022), p. 483 - 487, with a discussion, p. 488 (1975).

Westerbork observations of NGC 4449, a barred magellanic irregular galaxy, show a strong ridge of neutral hydrogen, coinciding with a chain of H II regions at the NE end of the bar, and some rudimentary spiral features. At Effelsberg, the extent of hydrogen has been found to exceed ten times the optical diameter. The velocity gradient along the bar is opposite to the one outside.

158.168 **Origine explosive.** J. H. Oort.
La dynamique des galaxies spirales. Colloque International C.N.R.S. No. 241, (see 012.022), p. 517 - 527, with a discussion, p. 528 - 529 (1975).

The possibility of formation of spiral arms by expulsion of gas from a galactic nucleus is investigated on the basis of phenomena observed in NGC 4258. A brief discussion is given on the frequency of large-scale eruptive activity in spiral galaxies.

158.169 **5-GHz survey of bright galaxies.**
R. Sramek.
Astron. Journ., Vol. 80, 771 - 777 (1975).

The 1135 galaxies in the Reference Catalogue of Bright Galaxies with declinations between +10° and +70° were surveyed at 6-cm wavelength with a 3-σ detection limit of about 30 mf.u.; of these 149 were detected. For 90 of these galaxies, spectral indices were obtained. There is a suggestion that the elliptical galaxies have flat spectra more often than the spiral galaxies.

158.170 **Radio flux measurements of normal galaxies at 80, 160 and 8870 MHz.** W. K. Huchtmeier.
Astron. Astrophys., Vol. 44, 101 - 106 (1975).

Radio flux measurements of 76 galaxies are given. Each galaxy was observed at least at one of the frequencies of 80, 160 and 8870 MHz. A low-frequency cut-off in the spectra of optically bright radio weak galaxies is a common feature. Free-free absorption seems to be a likely mechanism for the observed low frequency cut-off. The change in spectral index over the cut-off point is higher at high spectral indices. It is greatest for Sc and Sbc galaxies and becomes smaller for earlier type galaxies.

158.171 **The inclination of NGC 1058.** B. M. Lewis.
Astron. Astrophys., Vol. 44, 147 - 149 (1975).

Model profiles fitted to the integrated H I profile of NGC 1058 show that the upper limit to the inclination of this galaxy is 11°, and that the velocity dispersion perpendicular to the plane of the galaxy is between 7 and 11 km s^{-1}. Models which calculate the local value of σ_v from the rotation curve suggest that $i \simeq 4.5° \pm 0.5°$.

158.172 **Neutral hydrogen observations of DDO dwarf galaxies.** J. R. Fisher, R. B. Tully.
Astron. Astrophys., Vol. 44, 151 - 171 (1975).

All but 2 of the list of 243 David Dunlap Observatory (DDO) dwarf galaxy candidates have been observed in the H I 21-cm line and 179 have been detected. Among these, there are objects which are particularly interesting because of their low surface brightness. There are others of interest because they are intrinsically the smallest known extragalactic hydrogen-bearing systems. In this latter category are several new candidates for membership in the Local Group. For all the detected systems, the radio observations provide redshifts, neutral hydrogen masses, and a 'total mass index'.

158.173 **Brightness distributions across six strong radio galaxies at 15 GHz.** J. R. Baker, A. J. Green, T. L. Landecker.
Astron. Astrophys., Vol. 44, 173 - 185 (1975).

Observations of six strong radio galaxies and a number of small diameter calibration sources have been made with the 100-m radiotelescope at a frequency of 15.0 GHz. Maps are presented of Virgo A and Cygnus A in both linear polarization and total intensity, and of 3C 111, 3C 270, Hercules A and 3C 353 in total intensity only.

158.174 **On dimensions and stability of galaxy coronas.**
E. M. Nezhinskij.
Astron. Zhurn. Akad. Nauk SSSR, Vol. 52, 1007 - 1010 (1975). In Russian. English translation in Soviet Astron., Vol. 19, No. 5.

If one supposes that each galaxy is surrounded by a wide and massive corona, then from dynamical aspects it follows that coronas are stable and their mean dimensions exceed those of galaxies by more than one order.

158.175 **Synthesis of stellar populations.** B. M. Tinsley.
Mem. Soc. Astron. Italiana, Vol. 46, 3 - 68 (1975).
Lectures given at the Scuola Superiore di Astronomia, "Stellar populations," Erice, Sicily, May 20–31, 1974. – Review paper concerning the following topics: (1) Concepts and techniques of population synthesis; (2) Evolutionary synthesis of single-generation populations; (3) Evolutionary models for the solar neighborhood; (4) Evolutionary synthesis with continued star formation; (5) Some other applications of evolutionary population synthesis.

158.176 **Infrared observations of galaxies.**
A. G. Pacholczyk, M. Tarenghi.
Mem. Soc. Astron. Italiana, Vol. 46, 199 - 216 (1975).

New K and UBV data for a number of galaxies are presented, together with a discussion of near infrared colors for all galaxies for which useful data were reported in the literature.

158.177 **Interacting galaxies and tidal theory of intergalactic bridges.** B. A. Vorontsov-Vel'yaminov.
Pis'ma v Astron. Zhurn., Vol. 1, No. 11, p. 3 - 7 (1975). In Russian.

Photographs of galaxies with blue satellites at the ends of short straight bridges are presented. A similar phenomenon is present in the nuclei of M51 and 30 Dor in the LMC. Examples of pairs of galaxies are also given where, besides the curved branches, components are connected direclty by a straight bridge not predicted by tidal theory. Very similar pairs which have either only a tail or only a bridge are also demonstrated.

158.178 **Rotation, mass and density of the galaxy NGC 4096.**
V. G. Metlov.
Pis'ma v Astron. Zhurn., Vol. 1, No. 11, p. 15 - 18 (1975). In Russian.

The redshift and rotation velocities of the spiral galaxy NGC 4096 are determined. The rotation curve is obtained for distances up to $\sim ^3/_4$ of the observable radius of the galaxy. The mass contained within this region is about $1.3 \times 10^{10} M_\odot$, and the density is varying from 10^{-20} g/cm^3 in the centre to 10^{-24} g/cm^3 in the outer regions.

158.179 **A compendium of radio measurements of bright galaxies.** R. F. Haynes, W. K. H. Huchtmeier, B. C. Siegman, A. E. Wright.
Separate print C.S.I.R.O. Division Radiophys., Sydney. [ISBN 0 643 00144 1], 6 + 200 pp. (1975).

This compendium contains all radio measurements of optically bright 'normal' galaxies available to date. It considerably simplifies the statistical analysis of the radio properties of normal galaxies and indicates where improvements in the available radio data are needed. Statistical analysis is to be published elsewhere.

158.180 **The all scales distribution of different extragalactic objects.** A. Zięba, S. Zięba.
Acta Cosmologica, Fasc. 3, p. 91 - 94 (1975).

Using the method of statistical reduction the distributions of various samples of extragalactic objects are investigated. It is evidently seen that the distributions of radio sources and galaxies are quite different. The surface distribution of galaxies is found to be highly nonrandom and it is dependent on physical parameters of the discussed sample.

158.181 **A comparison of the radio and optical orientations of elliptical galaxies identified with double radio sources.** W. T. Sullivan III, L. A. Sinn.
Astrophys. Letters, Vol. 16, 173 - 177 (1975).

The position angles of the major axes of elliptical galaxies (θ_g) are compared with the position angles of the associated double radio sources (θ_r) for a sample of 15 galaxies; in each case the galaxy is centrally located with respect to the radio components. The distribution of $|\theta_g - \theta_r|$ is most simply interpreted as arising from ejection of the radio components in random directions with respect to the distribution of light. The distribution is statistically consistent, if galaxies are oblate rotators, with only the gravitational slingshot model. If galaxies are prolate rotators, the distribution is inconsistent with magnetohydrodynamic models which predict ejection along the axis of rotation.

158.182 **The redshifts of compact nuclei.** B. M. Lewis.
Observatory, Vol. 95, 168 - 171 (1975).

The Byurakan classification of galaxies is shown to be correlated with the difference between optical and 21-cm estimates of the systemic velocity. Analysis shows (1) that galaxies with compact nuclei and $i < 53°$ have large outflows in their nuclei and (2) that the optical velocities of all other galaxies are overestimated by about 13 km/s. No evidence is found in support of Pecker's anomalous redshifting mechanism.

158.183 **The systemic velocity of NGC 1313.**
D. S. Mathewson, V. L. Ford, J. D. Murray.
Observatory, Vol. 95, 176 - 177 (1975). – Short contribution.

158.184 **The radial velocity of NGC 1313: a correction.**
G. de Vaucouleurs, A. de Vaucouleurs.
Observatory, Vol. 95, 178 - 179 (1975).

158.185 **Kinematics of NGC 1313: a correction and new data.** E. Agüero, G. Carranza.
Observatory, Vol. 95, 179 - 180 (1975).

158.186 **A catalogue of southern peculiar galaxies from the U.K. Schmidt survey.** H. Arp, B. F. Madore.
Observatory, Vol. 95, 212 - 214 (1975).

158.187 **On the relative importance of the bulge to the disk of spiral and S0 galaxies.**
M. Yoshizawa, K. Wakamatsu.
Astron. Astrophys., Vol. 44, 363 - 370 (1975).

Photometric parameters, which characterize the spherical central bulge and the flat disk of spiral and S0 galaxies, are defined and the relations among the parameters are examined statistically for twenty-four objects, together with the trends of the parameters along the morphologic sequence. The ratio of the total luminosity of the bulge (L_b) to that of the disk (L_d), i.e. $\gamma \equiv L_b/L_d$, which ranges from 0.007 to 0.8, is found to signify the type of galaxies along the morphologic sequence. The influence of the bulge on the kinematics of the disk is discussed in relation to the exponential nature of the surface mass distribution in the disk.

158.188 **Neutral hydrogen in some peculiar galaxies.**
W. K. Huchtmeier, H.-D. Bohnenstengel.
Astron. Astrophys., Vol. 44, 479 - 484 (1975).

Neutral hydrogen has been observed in five peculiar galaxies: NGC 3227, 3448, 3718, 4438 and 4651. Four of these – galaxies with companions – show indications of gravitational interaction in their H I-distribution. The relative H I-content of NGC 3448 and 3718 is relatively high.

158.189 **Planetary nebulae in local group galaxies. II. Identifications, positions, number, and production rate of nebulae in NGC 221.** H. C. Ford, D. C. Jenner.
Astrophys. Journ., Vol. 202, 365 - 371 (1975).

The authors extend the identifications of planetary nebulae in NGC 221 to 21 nebulae. The observed number of nebulae is corrected for the losses in the saturated image of the galaxy and for the loss of limiting magnitude in the envelope of the galaxy. With these corrections the authors estimate that there are 34 planetary nebulae which are brighter than the Hα limiting magnitude. It is estimated that the production rate of planetaries in NGC 221 is $\chi_{pn} \geq 2.8 \times 10^{-3}$ nebulae yr^{-1} and that the rate of hydrogen mass loss from planetaries is $\sim 6 \times 10^{-4} M_\odot$ yr^{-1}. Assuming a constant production rate of planetaries for the past 10^{10} years, at least $6 \times 10^6 M_\odot$ of hydrogen has been returned to the interstellar medium in NGC 221.

158.190 **A strong millimeter wave component in the radio galaxy 3C 111.** B. J. Wills.
Astrophys. Journ., (*Letters*), Vol. 202, L59 - L62 (1975).

Observations of the radio galaxy 3C 111 show this source to have a very strong and very compact component at millimeter wavelengths, 3000 times more luminous than that recently found in Cen A. This component is coincident with the optical galaxy and may also be related to the X-ray source 3U 0430+37. The spectrum of the north-following, extended outer component remains power-law even at 3.3 mm wavelength. At least five spectral components of 3C 111 may be distinguished.

158.191 **Non-Dopplerain explanations of the redshift in spectra of distant galaxies.**
O. A. Mel'nikov, V. S. Popov.
Nekotor. vopr. fiz. kosmosa. Moskva, 1974, p. 9 - 32. In Russian. – Abstr. in Referativ. Zhurn. 51. Astron., 10.51.709 (1975).

158.192 **The groundstate OH spectrum of the galaxy NGC 253.** F. F. Gardner, J. B. Whiteoak.
Monthly Notices Roy. Astron. Soc., Vol. 173, 77P - 81P (1975).

The 1665 and 1667 MHz OH profiles in NGC 253 show several narrow emission features lying in the middle of a broad absorption line. The emission is stronger at 1667 MHz and is unpolarized. It is more likely to be due to maser amplification of the radiation from the continuum source in the nucleus than to OH sources of the type associated with H II regions in our Galaxy.

158.193 **Further infrared and millimeter observations of Markarian 231.**
R. R. Joyce, R. F. Knacke, M. Simon, E. Young.
Publ. Astron. Soc. Pacific, Vol. 87, 683 - 687 (1975).

New infrared and 3.3 millimeter observations of the Seyfert galaxy, Markarian 231, are reported. Thermal and nonthermal models of the radiation source are discussed although it is not possible to specify a unique model based on these data.

158.194 **On the mass of the Local Group.** W. Herbst.
Publ. Astron. Soc. Pacific, Vol. 87, 827 - 829 (1975).

The effect of varying the rotation velocity of the local standard of rest on the stability of the Local Group is investigated. It is concluded that, unless $\theta_0 > \sim 290$ km sec^{-1}, unseen mass is probably required for stability.

158.195 **The most distant galaxy.** B. V. Komberg.
Zemlya i Vselennaya, 1975, No. 6, p. 32 - 33. In Russian.

158.196 **On the colour differences between ordinary and**

barred galaxies. V. Ya. Solov'ev.
Astron. Zhurn. Akad. Nauk SSSR, Vol. 52, 1328 - 1329 (1975). In Russian. English translation in Soviet Astron., Vol. 19, No. 6.

The colour indices of A, AB, B families of galaxies of all morphological types were considered statistically on the basis of de Vaucouleurs' catalogue. The integral $(U-B)_{oc}$ of SBc–SBm was less by $0^m_.08 \pm 0^m_.03$ than that of SAc–SAm. The outer regions of SBc–SBd galaxies give the main contribution to integral ultraviolet excesses of these galaxies as follows from the radial $\overline{(U-B)}_0$ distributions.

158.197 **Emission-line galaxies and quasars in the southern hemisphere. I. Description and applications of an objective-prism survey.** M. G. Smith.
Astrophys. Journ., Vol. 202, 591 - 595 (1975).

A selection of objects from the first plates of a low-dispersion, objective-prism survey for emission-line galaxies and quasars is used to illustrate the application of the survey to the following lines of research in extragalactic astronomy: (a) quasi-stellar objects, (b) Seyfert galaxies, (c) instabilities in galaxies produced by tidal interaction or explosive events, and (d) rates of star formation and the general chemical evolution of galaxies. Included in the discussion is a description of how the survey provides a new, purely optical, color-independent method for the direct isolation of bright, high-redshift QSOs with strong emission lines (Lα is often directly visible on the Schmidt-survey plates).

158.198 **H II regions in double and interacting galaxies.** P. W. Hodge.
Astrophys. Journ., Vol. 202, 619 - 627 (1975).

The distributions of H II regions in 14 systems of double and interacting galaxies are examined to detect any possible distortion of the gas that may have been caused by tidal interaction between the galaxies. Most systems show no significant effect; but some, especially NGC 4038–4039, show a conspicuously anomalous H II region distribution that is probably caused by tidal effects. The author looks statistically at Arp's hypothesis that the H II regions are more numerous and more conspicuous on the side of a galaxy facing its companion. No such systematic tendency is found.

158.199 **Gravitational potential energy of a disk-sphere pair of galaxies.** G. M. Ballabh.
Astrophys. Space Sci., Vol. 38, 407 - 418 (1975).

Algebraic expressions are obtained for the interaction potential energy of a pair of galaxies in which one is disk shaped and the other spherical. The density distribution in the disk galaxy is represented by a polynomial in ascending powers of the distance from the centre of the disk while the density distribution in the spherical galaxy is represented by the superposition of spherical polytropes of integral indices. The basic functions required for obtaining the interaction potential energy of a coplanar disk-sphere pair of galaxies are tabulated.

158.200 **Distribution and kinematics of neutral hydrogen in the spiral galaxy M 81. I. Observations.**
A. H. Rots, W. W. Shane.
Astron. Astrophys., Vol. 45, 25 - 42 (1975).

The authors present observations of the early type spiral galaxy M 81 made with the Westerbork Synthesis Radio Telescope in the 21-cm line of neutral hydrogen. A velocity range from −300 km/s to +320 km/s is covered at a resolution of 27 km/s. The central regions exhibit a marked lack of H I. Between 3 and 10 kpc radius the distribution of H I shows a distinct regular two-armed spiral pattern. Outside 10 kpc two outer arm features develop. At larger radii, on the eastern side of the galaxy, two large H I complexes are found. These may be part of a bridge connecting M 81 and M 82. Inside a radius of 10 kpc the velocity field is on the whole fairly regular. Outside 10 kpc the velocity field is dominated by the non-circular motions of the two large H I complexes and the northern outer arm.

158.201 **Distribution and kinematics of neutral hydrogen in the spiral galaxy M 81. II. Analysis.** A. H. Rots.
Astron. Astrophys., Vol. 45, 43 - 55 (1975).

In this paper the author analyzes the data presented in Paper I (see Abstr. 158.200). From the velocity field a rotation curve and several geometrical parameters are derived. A mass model based on the rotation curve, and the light distribution obtained from the data by Brandt et. al. (1972) yield M/L ratios between 1.5 and 8 (corrected for internal absorption). The distribution of H I shows a distinct regular two-armed spiral pattern which can be well approximated by a logarithmic spiral with a pitch angle of 15°. There is a close correspondence with the optical spiral structure. From the velocity field and the rotation curve a residual velocity field is constructed. This velocity field is compared with similar fields predicted by linear density-wave-theory models. Given the limitations of linear density wave theory, a reasonable agreement can be achieved by assuming a phase shift of 20° between the H I spiral arms and the potential minimum.

158.202 **Diameter-luminosity relation. II. The D_{25} diameters system.** G. Paturel.
Astron. Astrophys., Vol. 45, 173 - 175 (1975).

In order to use the diameter-luminosity relation, it is necessary to obtain a diameter system free from systematic effects. The author has therefore demonstrated the existence of 'diameter' and 'morphological type' effects on the limiting surface brightness which defines the optical diameters in the catalogues of Nilson and Vorontsov–Vel'yaminov. These results allow us to understand the mathematical form of the relation-ship between two systems of diameters. Using known systems of diameters the parameter on this relation is calculated: this is then applied to obtain a system at the 25 magnitudes B arc s^{-2} brightness level. This D_{25} diameter system agrees with photometric diameters at the same 25 magnitude level.

158.203 **The large scale radio structure of NGC 1275.**
G. K. Miley, G. C. Perola.
Astron. Astrophys., Vol. 45, 223 - 226 (1975).

New data at 1.4 GHz prove the existence of a component in NGC 1275 on a scale of $\sim 30''$ and give structural information about the previously known $5'$ halo.

158.204 **Catalogue of galaxies brighter than 15 mag on magnetic tape of the computer M-220.**
N. G. Kogoshvili.
Byull. Abastumansk. Astrofiz. Obs., No. 46, p. 133 - 162 (1975). In Russian.

158.205 **M31 as a warped galaxy.** G. G. Byrd.
Bull. American Astron. Soc., Vol. 7, 506 (1975). Abstr. AAS.

158.206 **Nuclear bulge of M31 in UBVI.**
R. D. Burgess, M. S. Burkhead.
Bull. American Astron. Soc., Vol. 7, 506 (1975). – Abstr. AAS.

158.207 **The mass of M31.** P. W. Hodge.
Bull. American Astron. Soc., Vol. 7, 506 (1975). Abstr. AAS.

158.208 **Neutral hydrogen in M81.** A. H. Rots.
Bull. American Astron. Soc., Vol. 7, 506 - 507 (1975). – Abstr. AAS.

158.209 **A reinvestigation of the scattering halo of M82 based on polarimetric and isophotal maps.**
G. D. Schmidt, J. R. P. Angel, R. H. Cromwell.
Bull. American Astron. Soc., Vol. 7, 507 (1975). – Abstr. AAS.

158.210 **Absorption-line spectra of model galactic coronae.**
J. C. Weisheit, L. A. Collins.
Bull. American Astron. Soc., Vol. 7, 507 (1975). – Abstr. AAS.

158.211 **A preliminary discussion of the morphology of Seyfert galaxies.** T. F. Adams.
Bull. American Astron. Soc., Vol. 7, 516 (1975). – Abstr. AAS.

158.212 **Electronographic observations of Seyfert galaxies.**
P. A. Wehinger, S. Wyckoff.
Bull. American Astron. Soc., Vol. 7, 516 (1975). – Abstr. AAS.

158.213 **Spectrophotometry of Seyfert 1 galaxies.**
D. E. Osterbrock.
Bull. American Astron. Soc., Vol. 7, 516 (1975). – Abstr. AAS.

158.214 **Spectroscopic observations of Seyfert galaxies using a self-scanned silicon diode array.**
J. W. Glaspey, G. A. H. Walker, A. Stockton.
Bull. American Astron. Soc., Vol. 7, 516 (1975). – Abstr. AAS.

158.215 **Hydrogen line luminosities for Seyfert galaxies and QSO's.** D. W. Weedman.
Bull. American Astron. Soc., Vol. 7, 517 (1975). – Abstr. AAS.

158.216 **Collisional enhancement of HeI lines in Seyfert galaxy nuclei.** G. M. MacAlpine.
Bull. American Astron. Soc., Vol. 7, 517 (1975). – Abstr. AAS.

158.217 **Ultraviolet spectrum of early-type galaxies..**
A. D. Code, G. A. Welch.
Bull. American Astron. Soc., Vol. 7, 527 - 528 (1975).
Abstr. AAS.

158.218 **Galaxy population analysis.** M. MacFarlane.
Bull. American Astron. Soc., Vol. 7, 529 (1975).
Abstr. AAS.

158.219 **Observations of extragalactic carbon monoxide emission.** L. J. Rickard, P. Palmer,
M. Morris, B. Zuckerman, B. E. Turner.
Bull. American Astron. Soc., Vol. 7, 529 (1975). – Abstr. AAS.

158.220 **A galaxy of coefficients: A_b, A_i, A_d, A_{LC}, D_b, and D_z.** V. C. Rubin, N. Thonnard, W. K. Ford, Jr.,
J. A. Graham, M. S. Roberts.
Bull. American Astron. Soc., Vol. 7, 531 (1975). – Abstr. AAS.

158.221 **4000 globular clusters in M87.**
W. E. Harris, M. G. Smith.
Bull. American Astron. Soc., Vol. 7, 534 - 535 (1975).
Abstr. AAS.

158.222 **Motions of the stars and excited gas in the barred spiral galaxy NGC 3351.**
C. J. Peterson, V. C. Rubin, W. K. Ford, Jr., N. Thonnard.
Bull. American Astron. Soc., Vol. 7, 538 (1975). – Abstr. AAS.

158.223 **Ring galaxies in rich clusters and the intergalactic medium.** L. A. Thompson.
Bull. American Astron. Soc., Vol. 7, 549 (1975). – Abstr. AAS.

158.224 **Head-tail sources: new and old.**
L. Rudnick, F. N. Owen.
Bull. American Astron. Soc., Vol. 7, 549 (1975). – Abstr. AAS.

158.225 **Radio galaxies in rich clusters with wide radio tails.**
F. N. Owen, L. Rudnick.
Bull. American Astron. Soc., Vol. 7, 549 (1975). – Abstr. AAS.

158.226 **Particle acceleration in radio tail galaxies.**
A. G. Pacholczyk, J. S. Scott.
Bull. American Astron. Soc., Vol. 7, 549 (1975). – Abstr. AAS.

158.227 **The 21-cm structure of WLM and its surroundings.**
N. Thonnard.
Bull. American Astron. Soc., Vol. 7, 550 (1975). – Abstr. AAS.

158.228 **Distributions of metals in spheroidal galaxies.**
J. H. Mahaffy, B. A. Fryxell, W. D. Arnett.
Bull. American Astron. Soc., Vol. 7, 550 - 551 (1975).
Abstr. AAS.

158.229 **U, B, V photometry of the galaxy NGC 520.**
B. P. Artamonov, F. Börngen, A. I. Shapovalova.
Soobshch. Spets. Astrofiz. Obs., *Zelenchukskaya,* vyp. (No.) 13, p. 32 - 44 (1975). In Russian.

Results are presented of photographic U, B, V photometry of the galaxy NGC 520 from plates obtained with the 2-meter Schmidt telescope of Tautenburg Observatory. Distributions of brightness in B as well as of B – V, U – B color indices along five directions in the galaxy are obtained in relative units. Zero points of B – V, U – B are fitted to the photoelectric determinations by Hodge with an error of $\sim \pm 0^m\!.2$. It is shown that by color and spectral characteristics NGC 520 is probably a single physical system. A similarity between NGC 520 and M 82 in structure, possible stellar composition, spectral and kinematic characteristics is noted.

158.230 **A list of galaxies with outer ring-shaped structure.**
I. P. Kostyuk.
Soobshch. Spets. Astrofiz. Obs., *Zelenchukskaya,* vyp. (No.) 13, p. 45 - 62 (1975). In Russian.

The list contains data on 143 ring-shaped galaxies with apparent magnitudes $m \leqslant 15.7$ and declinations $\delta \geqslant -3°$.

158.231 **On the redshift in the spectra of distant galaxies.**
O. A. Mel'nikov, V. S. Popov.
Gl. astron. observ. AN SSSR. Leningrad, 1975. 16 pp. In Russian. – Abstr. in Referativ. Zhurn. 51. Astron., 1.51.843 (1976).

158.232 **The spectral energy distribution of NGC 1275.**
G. A. Shields, J. B. Oke.
Publ. Astron. Soc. Pacific, Vol. 87, 879 - 883 (1975).

An analysis of absolute spectral energy distributions for NGC 1275 (Per A) covering the wavelength interval λλ3300–10,800 is presented. The data are consistent with the heavy reddening discovered by Wampler. The line-emitting region has a characteristic density $\sim 10^{4.5}$ cm^{-3}, a mass $\sim 10^{5.5} M_{\odot}$, and a volume filling factor $\sim 10^{-6}$. The gas may be ionized by shock waves or by nonthermal or stellar radiation. It is again suggested, in the vein of Minkowski's original proposal, that the high-velocity, emission-line knots described by Minkowski are H II regions in a Perseus cluster galaxy or intergalactic gas cloud seen in projection against NGC 1275.

158.233 **The optical emission-line spectrum of 3C 120.**
M. M. Phillips, D. E. Osterbrock.
Publ. Astron. Soc. Pacific, Vol. 87, 949 - 959 = Lick Obs. Bull., No. 720 (1975).

Spectrophotometric measurements are reported of the nuclear region of 3C 120. The relative intensities of the individual emission lines and the continuous spectrum were measured in the wavelength region λλ3346–7065 (in the rest system of 3C 120). The Balmer decrement is quite steep and may be caused by interstellar extinction of an amount $E_{B-V} = 0.50$.

The relative strengths of the He I lines are inconsistent with pure recombination theory and suggest that collisional or radiative transfer processes are important. Broad emission features of Fe II are present in the spectrum. The profiles of the H I, He I, and He II lines are fairly symmetrical, with the full width at half maximum of the H I and He I lines approximately 2000 km s^{-1}, while He II λ4686 appears to have somewhat wider wings. The narrower forbidden lines have widths of roughly 300 km s^{-1}. A comparison of the relative forbidden line strengths to those of Cygnus A suggests the presence of a rather complicated distribution of density with degree of ionization in 3C 120.

158.234 **Velocity gradient of the gas along the apparent minor axis of the galaxy NGC 2685.**
M.-H. Ulrich.
Publ. Astron. Soc. Pacific, Vol. 87, 965 - 967 (1975).

Spectrograms were taken in the direction of the apparent minor axis of the peculiar galaxy NGC 2685. Measurements along the [O II] λ3727 emission feature within 10″ from the center show that the line-of-sight velocity of the ionized gas increases by ≃ 250 km s^{-1} from the NW side to SE side of the inner region of the galaxy.

158.235 **Rotation-curves of galaxies from 21 cm-line observations.** W. K. Huchtmeier.
Astron. Astrophys., Vol. 45, 259 - 268 (1975).

High-sensitive single dish observations of the 21 cm-line emission of the neutral hydrogen in the fainter outer regions of the H I distribution yielded extended rotation curves for 8 nearby galaxies. The ratio of the radius (R_{max}) of the maximum rotational velocity to the Holmberg radius (R_{Ho}) shows a correlation with the galaxy's morphological type in the sense that this parameter is an increasing function with morphological type: $R_{max}/R_{Ho} \sim 0.3$ for Sab to $R_{max}/R_{Ho} \sim 2$ for Ir type galaxies.

158.236 **Hypergalaxies.** K. Rudnicki.
Urania Kraków, Vol. 46, 194 - 197 (1975). In Polish.

158.237 **Morphological investigation of 40 Markarian galaxies.**
F. Börngen, A. T. Kalloglyan.
Astrofizika, Vol. 11, 369 - 375 (1975). In Russian. – English translation in Astrophysics, Vol. 11, No. 3.

Morphological types are determined and structural details are described for 40 Markarian galaxies on plates obtained with the 2-m Tautenburg telescope. More than 50 % of the studied galaxies have spiral structure. Most of the galaxies have structural peculiarities.

158.238 **Spectral observations of galaxies of high surface brightness. II.**
M. A. Arakelyan, Eh. A. Dibaj, V. F. Esipov.
Astrofizika, Vol. 11, 377 - 384 (1975). In Russian. – English translation in Astrophysics, Vol. 11, No. 3.

The results of spectral observations of 98 objects from the list of galaxies of high surface brightness (Arakelyan, 1975) are presented. Emission lines are detected and redshifts are measured in the spectra of 50 galaxies. The objects No. 42, 79, 81 and 120 reveal the spectral properties of nuclei of Seyfert galaxies. The object No. 80 has possibly less prominent Seyfert-type features. If so, the objects No. 80 and 81 form an isolated pair of Seyfert-type objects with mutual distance of about 30 kpc. The abundance of objects of high luminosity between galaxies of high surface brightness having emission lines is emphasized.

158.239 **Polarimetric observations of compact extragalactic objects.**
M. K. Babadzhanyants, V. A. Hagen-Thorn.
Astrofizika, Vol. 11, 385 - 395 (1975). In Russian. – English translation in Astrophysics, Vol. 11, No. 3.

Results are given of the polarimetric observations made in 1971–1972 of nuclei of Seyfert galaxies, N-galaxies 3C 371 and 3C 390.3 and QSS 3C 345. The variations of the parameters of polarization for the last three objects are compared with brightness variations. For 3C 371 the existence of a correlation between the variations of the degree of polarization and brightness in August 1972 is established. For 3C 390.3 all available data on polarization and brightness show that there is a correlation on a time-scale of years. According to its optical properties 3C 390.3 seems to be closer to Seyfert galaxies than to 3C 371.

158.240 **Nuclei and H II regions.** S. G. Iskudaryan.
Astrofizika, Vol. 11, 555 - 558 (1975). In Russian. English translation in Astrophysics, Vol. 11, No. 3.

Besides class 3, galaxies of all the other Byurakan classes having low luminous nuclei are rich in H II regions. Among them normal spiral galaxies are richer in H II regions than barred ones having nuclei of the same luminosity.

158.241 **Galaxie AU Leonis.** L. Meinunger.
MVS, *Sonneberg,* Vol. 7, 42 (1975).

AU Leo is a nonvariable irregular galaxy.

158.242 **Formación y evolución de las galaxias.**
J. Mandujano.
El Universo, Vol. 29, No. 112, p. 20 - 22 (1975).

158.243 **Apparent luminosity function of galaxies to the twenty-first magnitude.** G. S. Brown.
Thesis, Texas Univ., Austin (USA). [Available from Univ. Microfilms, Ann Arbor, Mich. Order No. 75-4335], 161 pp. (1974).

158.244 **Spiral galaxy dynamics.** L. Weliachew.
Recherche, (*France*), Vol. 6, 319 - 327 (1975). In French.

Discoveries in NGC 5128.
Sky Telescope, Vol. 50, 84 - 86 (1975).

Photon emission from suprathermal helium ions in a partially ionized hydrogen gas. See Abstr. 022.016.

Millimetre emission from extragalactic objects–I. Observational techniques and data analysis.
See Abstr. 031.210.

Optimum entrance apertures for spectroscopy of redshifted galaxies. See Abstr. 031.221.

A field-grating imaging spectrograph for radial velocity mapping of peculiar galaxies.
See Abstr. 034.043.

The origin and abundances of the chemical elements.
See Abstr. 061.036.

Tests of randomness in astronomical objects. I. Along single great circles. See Abstr. 061.039.

Tests of randomness in astronomical objects. II. Along two orthogonal great circles. See Abstr. 061.040.

The rôle of electrical discharges in astrophysical phenomena. See Abstr. 061.058.

Evolución de las estrellas y galaxias.
See Abstr. 065.180.

The ESO/Uppsala survey of the ESO (B) atlas of the southern sky. III. See Abstr. 113.044.

The Wolf-Rayet stars of M33. See Abstr. 114.028.

The spectrum of the bright variable A-1 in M31. See Abstr. 114.323.

S Dor-type variables in other galaxies. See Abstr. 122.038.

A survey of bright variable stars in M33. See Abstr. 122.079.

On the periods of light variations of variables V 39 and V 42 in the dwarf galaxy IC 1613. See Abstr. 122.100.

The cepheid variables and the stellar populations of the dwarf spheroidal galaxies. See Abstr. 122.165.

On the nature of the bright variables in dwarf spheroidal galaxies. See Abstr. 122.166.

Distribution of supernovae in spiral galaxies of type Sb–Sd. See Abstr. 125.026.

Studies on supernovae. See Abstr. 125.048.

Whiskers and cosmic millimetre-wave sources. See Abstr. 131.035.

On the ratio of extinction to reddening for interstellar matter using galaxies. I. A limit on the neutral extinction from photometry of the 3C 129 group. See Abstr. 131.195.

Spectrophotometric observations of ionized hydrogen regions in nearby spiral and irregular galaxies. See Abstr. 131.507.

The rate of star formation in M31 and the Large Magellanic Cloud. See Abstr. 131.535.

H II zones as indicators of extragalactic distances and the Hubble constant. See Abstr. 131.536.

The distribution of H II regions in M31. See Abstr. 131.541.

A correlation of H II regions with H I gas and the rate of star formation in galaxies. See Abstr. 131.545.

The 3.3 μ emission feature in nebulae and galaxies. See Abstr. 133.013.

Identifications, positions, number, and production rate of planetary nebulae in NGC 221. See Abstr. 133.017.

Compton-synchrotron models for the nucleus of Centaurus A (NGC 5128). See Abstr. 141.002.

3C 123: a distant first-ranked cluster galaxy at $z = 0.637$. See Abstr. 141.003.

Positions and flux densities of 5C 3 sources near M31. See Abstr. 141.014.

Detection of a 21 cm absorption line against the suggested galaxy/QSO pair NGC 3067/3C 232. See Abstr. 141.024.

Neutral hydrogen absorption in the spectrum of the quasar-galaxy pair 4C32.33/NGC 3067. See Abstr. 141.032.

Local quasars and radio galaxies. See Abstr. 141.053.

Structure of the compact radio component in 3C236. See Abstr. 141.055.

Photographs of 3C 120 and NGC 7603 compared to spectracon images. See Abstr. 141.062.

Observations with a VLB array. III. The sources 3C 120, 3C 273B, 2134 + 004, and 3C 84. See Abstr. 141.074.

The origin of the nonthermal radiation of quasi-stellar objects and galactic nuclei. See Abstr. 141.078.

The rotation measures of radio sources and their interpretation. See Abstr. 141.080.

Radio galaxies and local quasars. See Abstr. 141.089.

Modern models of the source of activity in quasars and galactic nuclei. See Abstr. 141.093.

Radio emission from galaxies and quasars. See Abstr. 141.098.

On the problems associated with the rapidly varying compact nonthermal sources. See Abstr. 141.100.

Centaurus-A: observations of the central region at cm-wavelengths by the radio telescope RATAN-600. See Abstr. 141.116.

B0924 + 30 an unusual radio galaxy. See Abstr. 141.123.

Reexamination of the correlation of galaxies and QSO's. See Abstr. 141.130.

Hard X-ray observations of NGC 4151. See Abstr. 142.004.

Optical studies of 10 high galactic latitude X-ray sources. See Abstr. 142.005.

The location and intensity of the X-ray source Centaurus A observed by the Astronomical Netherlands Satellite. See Abstr. 142.121.

Metagalactic protons of ultra-high energies. See Abstr. 143.005.

Formation of double galaxies by tidal capture. See Abstr. 151.020.

Resonant stellar orbits in spiral galaxies. See Abstr. 151.023.

Numerical study of a gravitating system of colliding particles; formation and dynamics of discs. See Abstr. 151.025.

N-body simulations of disks. See Abstr. 151.027.

Resonant stellar orbits in spiral galaxies. II. A general epicyclic theory. See Abstr. 151.038.

Periodic orbits in barred galaxies. See Abstr. 151.040.

Resonant stellar orbits in spiral galaxies. III. Applications and numerical tests of the general epicyclic theory. See Abstr. 151.043.

The theory of resonances in spiral galaxies. See Abstr. 151.047.

Numerical experiments on spiral structure. See Abstr. 151.049.

Effect of halo component on bar-formation in disk galaxies. See Abstr. 151.050.

Wave-wave and wave-particle interactions in mechanisms for the maintenance of spiral structure. See Abstr. 151.051.

Note on the persistence of spiral structure. See Abstr. 151.052.

What is essential for spiral structure? See Abstr. 151.053.

Spiral structure of the gas in spiral galaxies. – Theory. See Abstr. 151.054.

Spiral structure produced by tidal interaction of galaxies. See Abstr. 151.055.

On the age dependence of the spatial distribution and velocity dispersion of stars in a spiral galaxy. See Abstr. 151.056.

On the nature of barred spirals. See Abstr. 151.058.

Numerical experiments on the structure of barred spiral galaxies. See Abstr. 151.059.

A kinematic study of several barred spirals. See Abstr. 151.060.

The dynamics of interacting galaxies. See Abstr. 151.062.

Determination of the distribution function of an elliptical galaxy. See Abstr. 151.063.

Three-dimensional models for the distribution of mass in galaxies. See Abstr. 151.066.

Models for the formation of elliptical galaxies. See Abstr. 151.069.

Numerical calculation of the dynamical evolution of chains of galaxies. See Abstr. 151.085.

Metal abundances of Draco and eight galactic globular clusters. See Abstr. 154.009.

Sternpopulationen in Galaxien. See Abstr. 155.021.

Rotational interpretation of recombination line widths in the galactic nucleus. See Abstr. 155.023.

A new Milky Way satellite found in 21-centimeter line observations. See Abstr. 155.033.

Galactic evolution. II. Disk galaxies with massive halos. See Abstr. 155.048.

Galactic magnetic fields and their interactions with the gas and cosmic rays. See Abstr. 156.002.

Spectroscopy of selected Coma cluster galaxies. See Abstr. 160.032.

Another evolutionary correction to the luminosity of giant galaxies. See Abstr. 160.035.

A theory of galaxy formation and clustering. See Abstr. 160.044.

On galactic rotation and the intergalactic medium. See Abstr. 161.001.

Gas loss in groups of galaxies. See Abstr. 161.010.

The mean luminosity and mass densities in the universe. See Abstr. 162.011.

On the origin of rotation of galaxies in Ambartsumyan's cosmogony. See Abstr. 162.026.

Sur les dangers d'une approximation classique dans l'analyse des décalages spectraux vers le rouge. See Abstr. 162.039.

Déplacements anormaux vers le rouge liés à la traversée des amas de galaxies par la lumière. See Abstr. 162.040.

The redshift-distance relation. VIII. Magnitudes and redshifts of southern galaxies in groups: a further mapping of the local velocity field and an estimate of q_0. See Abstr. 162.047.

The composite Hubble diagram. See Abstr. 162.048.

On the effect of the luminosity selection on the redshift-distance relationship. See Abstr. 162.050.

Decrease of the Hubble constant with distance in the Local Supercluster anticenter direction. See Abstr. 162.051.

Peculiar Objects

158.301 **New photographic observations of BL Lacertae.** M. P. Véron.
Astron. Astrophys., Vol. 41, 423 - 430 (1975).

The author presents new photographic observations of BL Lacertae, made between May 1973 and September 1974. On September 25, 1973, BL Lac was as faint as $B = 16.76$ which is the smallest published flux. Variations of one magnitude per day and 0.25 magnitude in 3 hours are observed. The author discusses all the published variations of BL Lac, having a time scale of the order of one day, or less. He concludes that variations of 2 magnitudes per day are well established, corresponding to a maximum linear diameter $D = 7$ light-days, in absence of highly relativistic effects. Faster variations need confirmations.

158.302 **The composite nature of the N galaxy 3C 371.** J. S. Miller.
Astrophys. Journ., (*Letters*), Vol. 200, L55 - L59 = Lick Obs. Bull., No. 704 (1975).

Observations of the N galaxy 3C 371 with the Lick Observatory image-tube scanner show that the spectrum has three components: the stellar absorption lines of an E galaxy, a power-law continuum of index − 1.35, and a weak emission-line spectrum qualitatively similar to those observed in some giant E and radio galaxies. It is concluded that 3C 371 is an intermediate object, with features observed in giant E galaxies, highly variable QSOs, and BL Lacertae objects.

158.303 **BL Lac and related objects.** T. D. Kinman.
Bull. American Astron. Soc., Vol. 7, 420 (1975). Abstr. AAS.

158.304 **Long baseline interferometry of VRO 42.22.01 (BL Lac) and OJ 287.** D. B. Shaffer.
Bull. American Astron. Soc., Vol. 7, 420 (1975). – Abstr. AAS.

158.305 **Analysis of the variability of BL Lacertae and OJ 287 at 2.8 cm and 4.5 cm.** J. D. Scargle.
Bull. American Astron. Soc., Vol. 7, 420 (1975). – Abstr. AAS.

158.306 **Further observations of BL Lac at 2.8 cm.** B. H. Andrew, G. A. Harvey, J. M. MacLeod.
Bull. American Astron. Soc., Vol. 7, 420 (1975). – Abstr. AAS.

158.307 **Photometric histories of BL Lac objects and QSOs.** W. Liller.
Bull. American Astron. Soc., Vol. 7, 421 (1975). – Abstr. AAS.

158.308 **A suspected BL Lacertae object near a compact galaxy.** E. R. Craine, J. W. Warner.
Bull. American Astron. Soc., Vol. 7, 421 (1975). – Abstr. AAS.

158.309 **An X-ray survey of BL Lac objects.** B. Margon, A. Davidsen, S. Bowyer, T. Jones, P. Sanford, K. Mason.
Bull. American Astron. Soc., Vol. 7, 421 (1975). – Abstr. AAS.

158.310 **Linear polarization and flux of OJ 287 and BL Lac at 8 and 14.5 GHz.** H. D. Aller, J. E. Ledden, T. V. Seling.
Bull. American Astron. Soc., Vol. 7, 421 (1975). – Abstr. AAS.

158.311 **B2 1101 + 38: a new BL Lacertae-type object.** H. R. Miller.
Bull. American Astron. Soc., Vol. 7, 421 - 422 (1975). – Abstr. AAS.

158.312 **Further observations of BL Lacertae.** T. X. Thuan, J. B. Oke, J. E. Gunn.
Bull. American Astron. Soc., Vol. 7, 499 (1975). – Abstr. AAS.

158.313 **New optical polarization data of BL Lacertae objects.** K. Serkowski, S. Tapia.
Bull. American Astron. Soc., Vol. 7, 499 - 500 (1975). – Abstr. AAS.

158.314 **Further observations of BL Lacertae.** T. X. Thuan, J. B. Oke, J. E. Gunn.
Astrophys. Journ., Vol. 201, 45 - 49 (1975).

New spectrophotometric observations of BL Lac, using either a 7″ diameter aperture or a 3″.7–10″ diameter annulus, have been obtained under conditions of fair to excellent seeing and with BL Lac near the zenith. The nucleus was faint for all observations. Four absorption features identified as the Mg I and MgH band, the λ4410 band, the G-band, and H and K of Ca II, are observed. The redshift is $z = 0.070 \pm 0.005$. The spectral energy distribution is found to be very similar to that of a standard giant elliptical galaxy redshifted by $z = 0.070$ and reddened corresponding to $E_{B-V} = 0.28$ to 0.40. The absolute visual magnitude of the underlying galaxy is −22.6 if $H_0 = 60$ km s^{-1} Mpc^{-1} and $E_{B-V} = 0.28$.

158.315 **B2 1101+38: a BL Lacertae object.** H. R. Miller.
Astrophys. Journ., (*Letters*), Vol. 201, L109 - L111 (1975).

The archive plate collection of the Harvard College Observatory has been searched to determine the long-term optical behavior of the radio source B2 1101+38. B2 1101+38 was found to be a large-amplitude, rapid variable with a total range of ⩾4.7 mag. This is the largest range observed for any BL Lacertae object.

158.316 **New non-stellar BL Lacertae objects OY091 and PKS2335 + 03.** E. R. Craine, S. Tapia, M. Tarenghi.
Nature, Vol. 258, 56 - 57 (1975). – Letter.

158.317 **BL Lac at 8 and 14.5 GHz: evidence of multiple epoch particle injection from a single location.** H. D. Aller, J. E. Ledden.
Bull. American Astron. Soc., Vol. 7, 515 (1975). – Abstr. AAS.

158.318 **OY 091 and PKS 2335+03.** E. R. Craine, S. Tapia, M. Tarenghi.
IAU Circ., No. 2813 (1975).

158.319 **Wild's variable object.** D. Wills, H. J. Smith.
IAU Circ., No. 2820 (1975).

158.320 **A0 0235+164.** T. D. Kinman, G. H. Rieke.
IAU Circ., No. 2867 (1975).

158.321 **A0 0235+164.** L. J. Eachus.
IAU Circ., No. 2892 (1975).

Variable quasi-stellar sources with particular emphasis on objects of the BL Lac type. See Abstr. 141.006.

Photometric histories of QSOs: two QSOs with large light amplitude. See Abstr. 141.015.

Photometric histories of QSOs: 3C 279, the most variable and possibly most luminous QSO yet studied. See Abstr. 141.027.

Physical conditions and the locations of the absorption-line systems in the QSO 3C 286 and the BL Lacertae object PKS 0735 + 178. See Abstr. 141.030.

Physical conditions and the locations of the absorption-line systems in the QSO, 3C 286, and the BL Lac object PKS 0735+178. See Abstr. 141.043.

One-millimeter continuum observations of extragalactic objects. See Abstr. 141.047.

High-time-resolution photometry of quasistellar and related objects. See Abstr. 141.066.

Observations with a VLB array. II. The sources 4C 39.25, NRAO 150, VRO 42.22.01, 3C 345, and 3C 454.3. See Abstr. 141.073.

AO 0235 + 164 – a red BL Lacertae object. See Abstr. 141.075.

On the problems associated with the rapidly varying compact nonthermal sources. See Abstr. 141.100.

A search for super-rapid variability of X-ray sources and other peculiar objects. See Abstr. 142.030.

159 Magellanic Clouds

159.001 **Studies of the LMC stellar content I. A catalogue of 272 new O–B2 stars.** J. P. Brunet, M. Imbert, N. Martin, P. Mianes, L. Prévot, E. Rebeirot, J. Rousseau.
Astron. Astrophys., Suppl. Ser., Vol. 21, 109 - 136 (1975). In French.

From plates obtained with the objective-prism astrograph (dispersion 95 Å mm^{-1}) equipped with an intermediate-band interference filter, at La Silla, the authors have detected 272 O–B2 stars belonging to the Large Magellanic Cloud. They give finding charts, 1975 equatorial coordinates, magnitudes and colours in *UBV* photometry, and spectral classifications for these objects. Among them 105 are classified as O–B0 stars, 135 are classified as B1–2 stars, and for 32 stars, the authors could not assign any accurate spectral type.

159.002 **Die Magnetfeldstruktur im Bereich der Magellanschen Wolken.** T. Schmidt.
Mitt. Astron. Ges., No. 36, p. 88 - 90 (1975).

159.003 **Suggested observations of supernova remnants in the Large Magellanic Cloud.** E. M. Jones.
Astron. Astrophys., Vol. 42, 159 - 160 (1975).

Observers are urged to attempt radial velocity measurements of the SNRs in the LMC so that estimates can be made of the gas density around each remnant. One purpose is to check for a possible gas density effect on the radio-continuum surface-brightness vs linear diameter relation used to find the distance to galactic supernova remnants.

159.004 ***UBV* photometry for supergiants of the Large Magellanic Cloud.** J. P. Brunet.
Astron. Astrophys., Vol. 43, 345 - 358 (1975).

The author presents *UBV* photometric results from the study of 603 stars in the direction of the Large Magellanic Cloud (LMC). A detailed study of the local variations of the reddening enables the author to determine the mean $(B-V)$ excess inside the LMC, equal to 0.05 ± 0.02. Finally, the LMC evolutionary status is compared with the one of the $h + \chi$ Per galactic cluster.

159.005 **The cluster system in the Small Magellanic Cloud.** M. T. Brück.
Monthly Notices Roy. Astron. Soc., Vol. 173, 327 - 337 (1975).

Visual examination of 330 star clusters in the SMC on *UBV* plates taken with the UK 48-in. Schmidt telescope allows clusters to be classified by colour and structure into six types. A distinct pattern in the spatial distribution of clusters reported by Hodge is confirmed, though the effective size of the cluster system is the same as that defined by integrated star light. Cluster sizes, excluding red globulars, are on the whole larger than their galactic counterparts.

159.006 **An atlas for the Armagh Survey of H-alpha emission objects in the Large Magellanic Cloud.**
J. McFarland, E. M. Lindsay, A. D. Andrews.
Armagh Obs., Separate print, 56 pp. (1975). – Containing finding charts for 800 emission objects of the 2nd and 3rd Armagh-lists.

The 2nd and 3rd lists contain a total of 800 emission objects of stellar appearance. A preliminary comparison with the recent objective-prism survey of the LMC by Bohannan and Epps (1974) in which 625 emission objects were found within 25 square degrees of the Cloud centre shows that there are only about 296 objects in common. The Armagh survey yielded about 723 objects within this common area. Finding charts have been prepared of the emission objects of the Armagh survey from blue ADH plates (103a–0 with or without filter) and the limiting photographic magnitude of the charts is about 18.

159.007 **The Magellanic stream: further investigations.** W. E. Kunkel, S. Demers.
Bull. American Astron. Soc., Vol. 7, 550 (1975). – Abstr. AAS.

159.008 **The [SII]/(Hα+ [NII]) line intensity ratio for selected emission regions in the Large Magellanic Cloud.** P. S. Hoover.
Bull. American Astron. Soc., Vol. 7, 557 (1975). – Abstr. AAS.

159.009 **The distribution of cepheids in the Large Magellanic Cloud and cells of star formation.**
Yu. N. Efremov.
Astron. Tsirk., No. 881, p. 3 - 5 (1975). In Russian.

159.010 **Preliminary observations of the Magellanic Clouds with the ultraviolet sky survey telescope.**
R. G. Evans, K. Nandy, R. Wilson.
Phil. Trans. Roy. Soc. London, Ser, A, Vol. 279, 473 - 477 (1975).

Observations of the Magellanic Clouds have been made

with the ultraviolet sky survey telescope (S2/68) on board the E.S.R.O. satellite TD-1A. From the data so far reduced, the ultraviolet surface brightness has been derived for four wavelength bands centred near 1550, 1950, 2350 and 2740 Å. Ultraviolet maps of the Large Magellanic Cloud are presented and the flux distributions in each Cloud are compared with that expected from a galactic luminosity function.

Nucleosynthesis and star formation of the Galaxy and the Magellanic Clouds. See Abstr. 065.158.

Stellar evolution in the Magellanic Clouds. See Abstr. 065.161.

List of 506 stars, probable Small Magellanic Cloud members. See Abstr. 113.042.

Late-type giants in Large Magellanic Cloud. See Abstr. 114.078.

Spectroscopic and photometric observations of HDE 269006 in the Large Magellanic Cloud. See Abstr. 114.317.

A nitrogen rich OB supergiant in the Large Magellanic Cloud. See Abstr. 114.392.

Variable stars in clusters of the SMC. See Abstr. 122.001.

Photoelectric *UBV* photometry of cepheids in the Magellanic Clouds and in the southern Milky Way. See Abstr. 122.002.

A search for rapid variables and RR Lyrae stars in the LMC on ESO Schmidt plates. See Abstr. 122.052.

Die Perioden-Leuchtkraft-Farben (PLC) Beziehung in LMC, SMC und der Galaxis. See Abstr. 122.065.

Properties of cepheids in the Large Magellanic Cloud. See Abstr. 122.092.

Suspected VV Cephei stars in the Large Magellanic Cloud. See Abstr. 122.130.

Structure of the instability strip of cepheids near the center of the Small Magellanic Cloud. See Abstr. 122.138.

A revised period for one of the Wesselink-Shuttleworth SMC variables. See Abstr. 122.150.

The cepheid variables and the stellar populations of the dwarf spheroidal galaxies. See Abstr. 122.165.

The RR Lyrae stars in the Small Magellanic Cloud. See Abstr. 122.171.

Detection of carbon monoxide in the Large Magellanic Cloud. See Abstr. 131.148.

Elemental abundances in compact H II regions, planetary nebulae, and supernova remnants in the Magellanic Clouds. See Abstr. 131.524.

The rate of star formation in M31 and the Large Magellanic Cloud. See Abstr. 131.535.

Candidate for LMC X-5. See Abstr. 142.091.

Photometry of LMC X-ray candidates. See Abstr. 142.053.

***UBV* photometry of stars in the direction of three LMC X-ray sources.** See Abstr. 142.172.

Atlas of star clusters with colour magnitude diagrams in the Magellanic Clouds. See Abstr. 154.004.

UBV photometry of the LMC cluster NGC 1466. See Abstr. 154.005.

Dynamics of the Magellanic stream. See Abstr. 155.039.

Model calculations on the dynamics of the Milky Way under the influence of spiral arms and the Magellanic Clouds. See Abstr. 155.068.

Stellar populations in galaxies. See Abstr. 158.139.

Supergalactic studies. II. Supergalactic distribution of the nearest intergalactic gas clouds. See Abstr. 161.007.

Errata

159.901 **Erratum: "Ages of star clusters in a section of the Large Magellanic Cloud"** [Astron. Journ., Vol. 79, 1365 - 1368, 1489 - 1491 (1974)].
S. Baird, P. Flower, P. Hodge, P. Szkody.
Astron. Journ., Vol. 80, 540 (1975).

160 Clusters of Galaxies

160.001 **Redshifts and morphology of galaxies in the Coma cluster.** S. A. Gregory.
Astrophys. Journ., Vol. 199, 1 - 9 (1975).

By new observations of redshift and morphology, a sample of galaxies brighter than $m_p = 15.8$ is nearly completed out to a radial distance of 2°79 from the center of the Coma cluster. The redshift results show that the line-of-sight velocity dispersion first decreases and then increases again with distance from the cluster center with a mean value of $\sigma_v = 896$ km s^{-1}. A marginally positive test for solid body rotation of 85 ± 66 km s^{-1} degree^{-1} is also found. These results are consistent with numerical models of partially relaxed systems.

160.002 **Radio galaxies in the Coma cluster.** W. G. Tifft, M. Tarenghi.
Astrophys. Journ., Vol. 199, 10 - 15 (1975).

Redshifts for 23 possible radio galaxies with $m_p < 17$ in the field of the Coma cluster are reported and discussed. Except for two of the weakest sources, the galaxies have redshifts which average 600 km s^{-1} greater than the cluster mean. The non-emission-line galaxies are brighter both optically and at radio wavelengths than the emission-line objects. A direct correlation of radio magnitude and redshift appears to exist for the non-emission-line radio sources.

160.003 **The NGC 507 cluster of galaxies.** W. G. Tifft, K. A. Hilsman, L. C. Corrado.
Astrophys. Journ., Vol. 199, 16 - 18 (1975).

Redshifts and accurate positions for galaxies in the core of the NGC 507 group are presented. The group is shown to be closely similar to other groups of galaxies located nearby, including the Perseus cluster. A conventional virial analysis shows the usual excess kinetic energy. The redshift-magnitude diagram is consistent with the presence of bands.

160.004 **Studies of cluster X-ray sources. Energy spectra for the Perseus, Virgo, and Coma clusters.**
E. Kellogg, J. R. Baldwin, D. Koch.
Astrophys. Journ., Vol. 199, 299 - 306 (1975).

The authors present the final Uhuru X-ray differential-energy spectra for the Perseus, Virgo, and Coma clusters of galaxies. The power-law and isothermal bremsstrahlung model forms, both with a low-energy cutoff, are given. The spectra, best fits to the Uhuru 2–10 keV data, are also compared with other observations of these sources in the energy range 0.1–100 keV. For Perseus, the data above 20 keV favor the bremsstrahlung fit marginally. For Virgo, the data of Catura et al. between 0.25 and 1.0 keV clearly favor the bremsstrahlung curve. For Coma, the weakest of the three sources, the data are less precise, but there is some evidence for a low-energy turnover or cutoff. The implications of such a cutoff are discussed briefly.

160.005 **A Westerbork survey of clusters of galaxies. I. 21 cm observations of A1656, A2147, A2151, A2197 and A2199: radio data and identifications.**
W. J. Jaffe, G. C. Perola.
Astron. Astrophys. Suppl. Ser., Vol. 21, 137 - 169 (1975).

As a part of a programme of radio observations of rich clusters of galaxies, the authors have observed 5 nearby Abell clusters at 21 cm wavelength using the Westerbork Synthesis Radio Telescope. Here they describe the observing and reduction techniques, and present a list of all detected radio sources, along with contour maps of the more extended sources. An identification search was carried out on the Palomar Sky Survey prints and on deep 48″ IIIaJ Schmidt plates. For each identification optical data including a finding chart are presented.

160.006 **X-ray emission from the Centaurus cluster.** R. J. Mitchell, P. A. Charles, J. L. Culhane, P. J. N. Davison, A. C. Fabian.
Astrophys. Journ., (*Letters*), Vol. 200, L5 - L8 (1975).

X-ray emission from the Centaurus cluster of galaxies has been studied in the energy range 0.6–7.5 keV. Exponential and power-law spectra fit the data equally well. A value of the interstellar column density is found which is in good agreement with radio observations.

160.007 **The density distribution in clusters of galaxies.** L. G. Taff.
Astron. Astrophys., Vol. 42, 243 - 246 (1975).

The number counts for thirteen clusters of galaxies (thirty-five separate counts) have been examined. For these rich, symmetric clusters there always appears to be spherically symmetric departures from monotonicity. There may be evidence to document the existence of secondary maxima in the projected surface densities. If real such maxima would be a cosmological distance indicator. The publication of morphologically separate number counts might help to further clarify the situation.

160.008 **Collisions of galaxies in dense clusters. I. Dynamics of collisions of two galaxies.** D. O. Richstone.
Astrophys. Journ., Vol. 200, 535 - 547 (1975).

Collisions of spherical galaxies composed only of stars were simulated numerically. Test stars representative of the model galaxy were selected by a Monte Carlo technique and followed through the collision by integrating their equations of motion, within the approximations that the stars react only to the smoothed-out mass distributions of the galaxies, and that those distributions are unchanged during the collision. Approximate analytic arguments based on the impulsive approximation have been used to extend the numerical results. These computer and analytic studies have been used to obtain cross sections for certain processes which may be important in dense clusters of galaxies. The author notes that if the galaxies in the center of the Coma cluster were initially 50 kpc or more in radius, then they have suffered substantial changes in mass and radius over the age of the universe.

160.009 **Are nearby small groups of galaxies expanding with the universe?** J. C. Jackson.
Monthly Notices Roy. Astron. Soc., Vol. 173, 41P - 46P (1975).

Turner and Sargent have argued that many of the small groups of galaxies catalogued by de Vaucouleurs have crossing-times of the order of the Hubble time, and hence are probably unbound systems expanding as the universe expands. The author shows that they used an inappropriate definition of crossing-time, and that the properly defined crossing-times are typically about 20 per cent of the Hubble time.

160.010 **Dynamical friction in the Coma cluster.** M. Lecar.
IAU Symposium No. 69, (see 012.009), p. 161 - 166 (1975).

If the galaxies composing the Coma cluster were initially 'Ostriker-Peebles' galaxies (i.e., isothermal spheres extending to 250–500 kpc), the hidden mass is likely to be tidally torn-off halos. Owing to dynamical friction, the remnant cores tend to spiral in to the center of the cluster. This happens as long as the hidden mass, whatever its origin, consists of objects much less massive than galaxies.

160.011 **Optical observations of diffuse intergalactic light in the Coma cluster of galaxies.**
J. E. Gunn, J. Melnick.
Bull. American Astron. Soc., Vol. 7, 412 (1975). – Abstr. AAS.

160.012 **Nature and significance of halos around clusters of galaxies.**
P. J. E. Peebles, M. Seldner, E. J. Groth.
Bull. American Astron. Soc., Vol. 7, 426 (1975). – Abstr. AAS.

160.013 **The Virgo clusters of galaxies.**
G. O. Abell, T. S. Eastmond.
Bull. American Astron. Soc., Vol. 7, 426 (1975). – Abstr. AAS.

160.014 **Digital enhancement of selected clusters of galaxies by image processing.** J. J. Lorre.
Bull. American Astron. Soc., Vol. 7, 427 (1975). – Abstr. AAS.

160.015 **Size of the Coma cluster.**
G. Chincarini, H. J. Rood.
Nature, Vol. 257, 294 - 295 (1975). – Letter.

160.016 **Structural investigations in the central part of the Coma cluster on the base of equidensity diagrams. I.**
N. Richter, W. Högner.
Astron. Nachr., Vol. 296, 221 - 231 (1975).

On the base of a photographic plate in system V, maximal diameters of galaxies in the centre of the Coma cluster are determined by photographic equidensitometry. Another system of diameters is determined for medium densities of the objects. The results are compared with Rood and Baum's who derived the diameters by visual measurements. It is pointed out that the Tautenburg diameters are in the mean 3 times larger than those derived by Rood and Baum. Diameter-magnitude diagrams and ellipticities of the objects are derived, further photometric corrections for tilted objects. For describing the general structure of the cluster diameter functions are derived instead of magnitude functions. The accuracy of a brightness determination by the equidensity method from the diameter-magnitude diagrams is nearly twice better than according to Rood and Baum.

160.017 **Kompakte Gruppen von kompakten Galaxien.**
F. W. Baier.
Sterne, 51. Jahrgang, p. 129 - 135 (1975).

160.018 **On the redshifts of clusters of galaxies and quasars.**
B. I. Gorbachev.
Astron. Tsirk., No. 859, p. 5 - 6 (1975). In Russian.

160.019 **On the luminosity function of galaxies in clusters.**
B. I. Gorbachev.
Astron. Tsirk., No. 859, p. 6 - 8 (1975). In Russian.

160.020 **Clustering of faint blue objects.**
M. R. S. Hawkins, V. C. Reddish.
Nature, Vol. 257, 772 - 773 (1975).

Photographs taken with the UK 48-inch Schmidt telescope at Siding Spring, Australia using hypersensitised Kodak IIIaJ emulsions, record objects as faint as magnitude 23 in exposures of an hour. Most galaxies on the photographs are very faint and at, say 22 mag probably have redshifts $Z \sim 0.5$. At such large redshifts the K term is responsible for considerable reddening $(B-V) \simeq +2$ mag. The number of objects does seem to continue to increase significantly with limiting blue magnitude and the question therefore arises as to whether or not a substantial number of faint images are blue and if so, what objects do they represent and how are they distributed. The blueness suggests that they are not ordinary galaxies but lie in the range from very blue compact galaxies to quasars. Whatever they are the high degree of clustering, and especially the preponderance of pairs, appears to be significant. Taken together with a recent indication of an increase in the degree of clustering of galaxies with distance this may be further direct evidence of an evolving universe.

160.021 **The structure of the outskirts of the Coma cluster.**
H. J. Rood.
Astrophys. Journ., Vol. 201, 551 - 555 (1975).

If the line-of-sight velocities of a large complete sample of galaxies brighter than a given apparent magnitude limit in a region of a cluster are known, then the number of field galaxies within $\pm 3\, \sigma_v$ of the mean velocity of the cluster can be estimated statistically. The method is applied to samples of galaxies in an outer region of the Coma cluster.

160.022 **Photometry of the cluster of compacts ZwCl 0152 + 3337.** P. L. Massey.
Astron. Journ., Vol. 80, 778 - 782, 885 - 886 (1975).

ZwCl 0152 + 3337 contains over 20 compact galaxies arranged in a linear "dorsal spine". Brightness profiles of several of the compacts indicate that they are S0s with one or more ellipticals. The cluster has yet to reach dynamical equilibrium and we probably observe both relaxation due to binary encounters and the first collapse. The cluster is otherwise usual, as evidenced from its luminosity function.

160.023 **On the existence of multiple clusters of galaxies.**
I. D. Karachentsev, R. L. Tsarevskaya, A. L. Shcherbanovskij.
Astron. Zhurn. Akad. Nauk SSSR, Vol. 52, 999 - 1006 (1975). In Russian. English translation in Soviet Astron., Vol. 19, No. 5.

The distribution of Abell's clusters of galaxies with different classes of distance and richness was analysed by the method of counts of number of clusters within distance limits $r \leqslant 6°$ round clusters and points chosen randomly. Some evidence confirming the reality of space clustering of clusters is obtained. The distribution of clusters may be represented by a model of simple clustering.

160.024 **Expected distribution of 'hidden' mass inside rich clusters of galaxies.** N. A. Bobrova, L. M. Ozernoj.
Pis'ma v Astron. Zhurn., Vol. 1, No. 11, p. 8 - 14 (1975). In Russian.

The expected distribution of 'hidden' mass necessary for stationarity of a spherically symmetric galaxy cluster is calculated using the known distribution of visible galaxies in the Coma cluster. Calculations are made both for the case of localization of hidden mass inside massive galactic coronae and for the case of collectivized distribution over the intergalactic space. In both cases the distribution of hidden mass can differ appreciably from that of visible galaxies.

160.025 **Grouping of galaxies. I.** H. Bereś.
Acta Cosmologica, Fasc. 3, p. 7 - 31 (1975).

The application of a mathematical method to the investigation of the distribution of objects in a finite region and of the structure of this distribution is presented. The clusters of galaxies: Hydra I and Zwicky 97–8 = Zwicky 127–2 constituted the basic material in this paper. They served as natural models in the presentation of the method. Apart from them artificial models were constructed and investigated.

160.026 **The estimation of errors in "A catalogue of 15650 Galaxies in the Jagellonian Field".** P. Flin.
Acta Cosmologica, Fasc. 3, p. 33 - 37 (1975).

The comparison of independently elaborated parts of the Jagellonian Field leads to conclusion that the average error of the class of visibility given in the Catalogue is equal almost

1 class per galaxy. It was also checked that the "loosing of galaxies" does not depend on the place where galaxy is located on the Field; the number of "loosing galaxies" is proportional to the number of galaxies in the vicinity of the "loosing galaxy".

160.027 **On the equilibrium distribution of gas in clusters of galaxies.** S. M. Lea.
Astrophys. Letters, Vol. 16, 141 - 144 (1975).

The hydrostatic equilibrium solution for a polytropic gas in a cluster of galaxies has been calculated. The gas has a wider distribution than the galaxies. If the central gas density is 8×10^{-4} particles cm^{-3} for $\gamma = 5/3$ or 2×10^{-3} cm^{-3} for $\gamma = 1.35$ the total X-ray emission from a model Coma cluster is equal to the observed emission. The spatial extent and spectrum for the source are also well-represented by the model, if $\gamma = 1.35$.

160.028 **Supergalactic studies. I. Supergalactic distribution of the nearest galaxies.** G. de Vaucouleurs.
Astrophys. Journ., Vol. 202, 319 - 326 (1975).

The supergalactic distribution of the nearest galaxies is investigated to test the nature of the Local Supercluster and to determine whether the Local Group is inside or outside its boundaries. It is concluded that (1) the Local Supercluster is a disklike physical and dynamical system, and (2) the Local Group is well within the borders of the system. The alternative hypothesis that it is an appearance resulting from a random clumping accident has negligibly small probability.

160.029 **The mass-luminosity ratio in the Coma cluster of galaxies from two-body interactions.**
T. W. Noonan.
Astrophys. Journ., Vol. 202, 551 - 557 (1975).

An analysis of galaxies in the central region of the Coma cluster with available redshifts indicates that galaxies with nearer neighbors tend to have a larger velocity dispersion. Interpretation of the effect in terms of two-body encounters indicates a mass-luminosity ratio of several hundred.

160.030 **Hot gas in clusters of galaxies.**
S. F. Gull, K. J. E. Northover.
Monthly Notices Roy. Astron. Soc., Vol. 173, 585 - 603 (1975).

The authors present a series of models which describe adiabatic, hydrostatic atmospheres of hot gas in the gravitational potential wells of clusters of galaxies. The models can account for the observed X-ray spectra and surface brightness distributions of clusters. The authors show how adiabatic atmospheres arise naturally from the infall of primordial matter into the cluster and consider their temporal evolution as cooling begins.

160.031 **On the surface brightnesses of first-ranked cluster galaxies.** D. H. Gudehus.
Publ. Astron. Soc. Pacific, Vol. 87, 763 - 768 (1975).

It is shown that Sandage's arguments for a constant mean surface brightness of first-ranked cluster ellipticals are incorrect. In addition, a cosmic dispersion of 0.29 mag is found and first-ranked galaxies in irregular clusters appear to have brighter than average surface brightness. The potential usefulness of a surface brightness-redshift plot as an important cosmological test is discussed.

160.032 **Spectroscopy of selected Coma cluster galaxies.**
S. A. Gregory.
Publ. Astron. Soc. Pacific, Vol. 87, 833 - 834 (1975).

New determinations of redshift are given for 14 galaxies in the direction of the Coma cluster. Nine of these galaxies have not been studied spectroscopically before. The comparison of the new redshifts of the remaining five galaxies with those published previously is good with $\langle V_{new} - V_{old} \rangle = -29$ km sec^{-1}.

160.033 **Supergalactic studies. III. The supergalactic distribution of nearby groups of galaxies.**
G. de Vaucouleurs.
Astrophys. Journ., Vol. 202, 610 - 615 (1975).

The probability that the observed concentration of nearby groups of galaxies within 10°, 20°, or 30° from the same great circle (supergalactic equator SGE) in both galactic hemispheres could arise by chance in a random (Poisson) distribution of cluster centers is less than 2×10^{-2} for the 17 nearer groups ($\Delta < 10$ Mpc) and less than 3×10^{-2} for the 46 groups in the next distance range ($10 < \Delta < 16$ Mpc on 1965 distance scale). The probability that both sets could fall by chance within 30° from the same circle with the observed frequencies is less than 5×10^{-4}. The variations of the supergalactic concentration index C as a function of magnitude and of supergalactic latitude and longitude define unequivocally the direction of the center of the Local Supercluster even when a large sector, including the Virgo cluster, is excluded from the statistics. It is clear that the "supercluster effect" cannot be explained by the combined effects of galactic extinction, small-scale clustering and a hypothetical "large" Virgo cluster.

160.034 **Supergalactic studies. IV. Systematic orientation of galaxy clouds relative to the supergalactic plane.**
G. de Vaucouleurs.
Astrophys. Journ., Vol. 202, 616 - 618 (1975).

The major nearby clouds of galaxies tend to be elongated in directions roughly parallel to the supergalactic plane. Out of 11 clouds (or 12 including the Local Cloud) with major axes $a > 10°$ and sufficiently elongated ($b/a < 0.8$) to give an unambiguous direction of elongation, none is inclined to the supergalactic equator by more than 35°. If the space orientations of the cloud axes were random and uncorrelated, the probability of this happening by chance is 3×10^{-5} (or 1×10^{-5} if the Local Cloud is included). Several possible mechanisms leading to this systematic orientation in a flat supercluster are noted. This effect adds greatly to the difficulty of explaining the "supercluster effect" as a random clumping accident.

160.035 **Another evolutionary correction to the luminosity of giant galaxies.** J. P. Ostriker, S. D. Tremaine.
Astrophys. Journ., (*Letters*), Vol. 202, L113 - L117 (1975).

Isolated giant galaxies may accrete their dwarf companions, and those in clusters their neighbors, by the process of dynamical friction. The computed effect is of the opposite sign and similar magnitude to the luminosity change caused by normal stellar evolution. Cosmological investigations could compensate for the postulated effect, if fainter galaxies were used as well as first brightest cluster galaxies in the magnitude-redshift test.

160.036 **Systematic redshifts in the outer parts of the Coma cluster.** S. A. Gregory, W. G. Tifft.
Bull. American Astron. Soc., Vol. 7, 539 (1975). – Abstr. AAS.

160.037 **Statistical uncertainties in determinations of effective radii of clusters of galaxies.** P. Schechter.
Bull. American Astron. Soc., Vol. 7, 539 (1975). – Abstr. AAS.

160.038 **Surface photometry of Virgo cluster barred galaxies.**
G. F. Benedict.
Bull. American Astron. Soc., Vol. 7, 539 (1975). – Abstr. AAS.

160.039 **A model of X-ray and radio emission from the Perseus cluster of galaxies.**
G. D. Holman, W. A. Christiansen.
Bull. American Astron. Soc., Vol. 7, 539 (1975). – Abstr. AAS.

160.040 **Velocity dispersions in the X-ray clusters of galaxies, Abell 262 and Abell 1367.**
R. J. Dickens, C. Moss.
Bull. American Astron. Soc., Vol. 7, 548 - 549 (1975).
Abstr. AAS.

160.041 **Position angle frequencies in the Virgo cluster of galaxies.** J. F. Dean.
Monthly Notes Astron. Soc. Southern Africa, Vol. 34, 153 - 155 (1975).

160.042 **Comparison of Abell's and Zwicky's catalogues of galaxies.** I. D. Karachentsev, R. L. Tsarevskaya, A. L. Shcherbanovskij.
Soobshch. Spets. Astrofiz. Obs., *Zelenchukskaya,* vyp. (No.) 13, p. 63 - 83 (1975). In Russian.

A list of 1524 clusters of galaxies from Abell's catalogue which are identified with those from Zwicky's catalogue is presented. The extent of overlapping of the two catalogues, errors of determining distances, richnesses, and coordinates of the clusters in each catalogue are studied.

160.043 **Calán 129-1: a southern double cluster of galaxies?** H. Quintana, J. Melnick.
Publ. Astron. Soc. Pacific, Vol. 87, 863 - 867 (1975).

Observations of a peculiar southern cluster with Abell richness class O and a redshift of 0.015 are presented. A population and morphology analysis leads to the interpretation that this is a system of two clusters. Some evidence suggesting an interaction due to intracluster gas is also briefly discussed.

160.044 **A theory of galaxy formation and clustering.** J. R. Gott III, M. J. Rees.
Astron. Astrophys., Vol. 45, 365 - 376 (1975).

The authors discuss how galaxies and clusters might evolve from primordial density fluctuations which, at the recombination epoch t_{rec}, have small amplitudes $\delta\rho/\rho \propto M^{-1/2-n/6}$. This approach is stimulated by Peebles' claim that the spatial distribution of galaxies can be described by a power law covariance function $\xi(r)$ of the approximate form $r^{-\gamma}$ with $\gamma \simeq 1.77$, which shows no well-defined preferred scale of clustering. The data indicate $n = -1$ i.e. $\delta\rho/\rho \propto M^{-1/3}$. The authors show that the desired $M^{-1/3}$ spectrum can arise naturally from scale-independent metric fluctuations involving inhomogeneities in the baryon distribution. These (isothermal) fluctuations, unlike adiabatic fluctuations, survive undamped through recombination with a simple $M^{-1/3}$ spectrum on relevant scales.

160.045 **Synchrotron and inverse Compton processes in X-ray clusters of galaxies.** N. V. Vidal.
Proc. Astron. Soc. Australia, Vol. 2, 327 - 329 (1975).

160.046 **Distribution of elliptical galaxies in the Virgo cluster.** B. M. Lewis, R. G. Gibb.
Proc. Astron. Soc. Australia, Vol. 2, 358 - 359 (1975).

160.047 **Evolutionary effects on radially varying properties of Coma cluster galaxies.** S. A. Gregory.
Thesis, Arizona Univ., Tucson (USA). [Available from Univ. Microfilms, Ann Arbor, Mich. Order No. 75-4111], 80 pp. (1974).

160.048 **Accretion to the cluster of galaxies and X-ray emission.** F. Takahara, S. Ikeuchi.
Progr. Theor. Phys. Japan, Vol. 53, 646 - 655 (1975).

Observable aspects of very long wavelength gravitational waves. See Abstr. 066.057.

3C 123: a distant first-ranked cluster galaxy at $z = 0.637$. See Abstr. 141.003.

A radio source and cluster of galaxies in 3 U 1252-28. See Abstr. 141.025.

Two luminosity classes of quasars. See Abstr. 141.095.

Nature of extended radio sources in clusters of galaxies. See Abstr. 141.168.

The unidentified high galactic latitude X-ray sources: bright galaxies or rich clusters? See Abstr. 142.008.

Clusters of galaxies within X-ray error boxes in the southern sky. See Abstr. 142.099.

Soft X-ray emission from the direction of the Coma cluster. See Abstr. 142.166.

The origin of X-ray sources in clusters of galaxies. See Abstr. 142.200.

Computer simulations of stellar systems. See Abstr. 151.046.

Distribution of observed orientations of galaxies. See Abstr. 158.001.

cD galaxies in poor clusters. See Abstr. 158.027.

Empirical properties of the mass discrepancy in groups and clusters of galaxies. VI. Projection effects. See Abstr. 158.028.

Angular momentum properties of galaxies in rich clusters. See Abstr. 158.080.

On the evolution of protogalaxies and protoclusters of galaxies. See Abstr. 158.115.

The masses of the galaxies and the mass-energy in the universe. See Abstr. 158.140.

On the mass of the Local Group. See Abstr. 158.194.

Ring galaxies in rich clusters and the intergalactic medium. See Abstr. 158.223.

Radio galaxies in rich clusters with wide radio tails. See Abstr. 158.225.

Ionized intracluster gas. See Abstr. 161.003.

Hot gas in and between galaxies. See Abstr. 161.006.

Supergalactic studies. II. Supergalactic distribution of the nearest intergalactic gas clouds. See Abstr. 161.007.

Gas loss in groups of galaxies. See Abstr. 161.010.

Selective extinction in the local supercluster. See Abstr. 161.011.

The scale of galaxy clustering and the mean matter density of the universe. See Abstr. 162.005.

Sur les dangers d'une approximation classique dans l'analyse des décalages spectraux vers le rouge.
See Abstr. 162.039.

Déplacements anormaux vers le rouge liés à la traversée des amas de galaxies par la lumière.
See Abstr. 162.040.

On the Tifft effect and possibilities to explain it.
See Abstr. 162.043.

The redshift-distance relation. VIII. Magnitudes and redshifts of southern galaxies in groups: a further mapping of the local velocity field and an estimate of q_0.
See Abstr. 162.047.

Errata

160.901 **Erratum: "Study of three southern groups of galaxies"** [Astron. Journ., Vol. 80, 77 - 82, 163 - 164 (1975)].
G. A. Welch, G. Chincarini, H. J. Rood.
Astron. Journ., Vol. 80, 668 (1975).

161 Intergalactic Matter

161.001 **On galactic rotation and the intergalactic medium.**
R. Cowsik, I. Lerche.
Astrophys. Journ., Vol. 199, 555 - 557 (1975).

The observed spectrum of the isotropic diffuse X-ray background has often been invoked to infer the possible existence of a hot intergalactic medium at a density of the order of that required for closure of the universe. The authors consider here the effect of such a hot isotropic gas on the rotation of galaxies, with the coupling between the intergalactic medium and the galaxies being provided by galactic magnetic fields. The calculation provides the basis for an indirect observational test which bears on the existence of the hot intergalactic medium. It is pointed out that the basic physics involved should also pertain to variable X-ray stars emitting by mass accretion and, perhaps, to the Coma cluster of galaxies.

161.002 **Intergalactic matter in Abell 194 and the radio source 3C40.** S. M. Simkin.
Bull. American Astron. Soc., Vol. 7, 500 (1975). – Abstr. AAS.

161.003 **Ionized intracluster gas.**
P. Crane, J. A. Tyson.
Astrophys. Journ., (*Letters*), Vol. 201, L1 - L4 (1975).

Stringent upper limits for Hβ emission from ionized intergalactic gas are reported for four rich clusters of galaxies with a wide range of X-ray luminosities. For the clusters Abell 779, Abell 1367, Abell 1656 (Coma), and Abell 2199, the 2σ limit gives $F(\mathrm{H}\beta) \lesssim 2 \times 10^{-18}$ ergs s^{-1} cm^{-2} arcsec^{-2} and implies that it is unlikely that these clusters are bound by this type of gas at temperatures less than 2×10^6 K. Combined with the X-ray results, these limits imply that it is unlikely that these clusters are bound by ionized gas at any temperature for H_0 = 55 km s^{-1} Mpc^{-1}.

161.004 **Accretion of intergalactic gas by a realistic model of the Galaxy and its implications.** R. Hunt.
Monthly Notices Roy. Astron. Soc., Vol. 173, 465 - 487 (1975).

A numerical solution is found for the accretion of an intergalactic gas by a stationary, realistic model of the Galaxy in which the flow is unimpeded by the interstellar pressure of the disc, in which case the accretion rate is a maximum. The calculation was performed for a range of initial temperatures for the intergalactic gas and for three values of its specific heat ratio including both adiabatic and isothermal cases. The accretion rates derived are quite large being, in every case, greater than 5.6 $M_\odot$ yr^{-1} if the initial density of the gas is 3×10^{-4} cm^{-3}. Soft X-radiation from the inflowing gas having initial density $\sim 3 \times 10^{-4}$ cm^{-3} and initial temperature greater than $\sim 2.5 \times 10^5$ K would constitute an appreciable fraction of the observed soft X-ray flux.

161.005 **Remarks to B. Takase's paper "Intergalactic reddening of galaxies".**
R. Guła, K. Rudnicki, I. Tarraro.
Acta Cosmologica, Fasc. 3, p. 39 - 40 (1975).

It is shown that a part of the reddening effect of galaxies explained by intergalactic extinction (Takase 1972) can also be accounted for by an inaccuracy of the galactic extinction coefficient used in preparing de Vaucouleurs' 1964 catalogue.

161.006 **Hot gas in and between galaxies.** G. B. Field.
Astrophys. Space Sci., Vol. 38, 167 - 190 (1975).
Invited paper presented at I.A.U. Colloquium No. 27 on 'UV and X-ray spectroscopy of astrophysical and laboratory plasmas', Cambridge, Mass., U.S.A., September 9–11, 1974.

The fate of interstellar gas heated by energetic radiation and far from the galactic plane is considered. It is shown that for plausible heating rates a very hot corona forms, separated from the cooler gas in the disk by a conductive boundary layer. For heating rates exceeding a critical value a galactic wind is set up. Such winds may fill the intracluster space in clusters of galaxies; further heating may drive a cluster wind. It is believed that the flow problem of intracluster gas in the presence of intercluster gas is not yet adequately solved. A variety of observations at X-ray and ultraviolet wavelengths are needed to resolve these questions.

161.007 **Supergalactic studies. II. Supergalactic distribution of the nearest intergalactic gas clouds.**
G. de Vaucouleurs, H. G. Corwin, Jr.
Astrophys. Journ., Vol. 202, 327 - 334 (1975).

The authors confirm the report by Mathewson, Cleary, and Murray that the nearby "high velocity" H I clouds, and in particular the Magellanic Stream, are strongly concentrated toward the supergalactic plane. It is found that actually the high-velocity clouds are not so much concentrated toward the supergalactic equator as toward the equator of the "Local Cloud" of galaxies inclined 14° to the main supergalactic plane. Both galaxies and H I clouds define the same small circle of maximum concentration and exhibit the same standard deviation (15°) from it, demonstrating closely related space distributions. The authors conclude that, with the possible exception of a few of the largest and probably nearest cloud complexes, most of the high-velocity clouds are truly intergalactic and associated with the Local Group and nearer groups of galaxies.

161.008 **Heating of the universe by quasars.** G. B. Field.
Bull. American Astron. Soc., Vol. 7, 516 (1975).
Abstr. AAS.

161.009 **Intergalactic dust and possible cosmological consequences.** S. Margolis, D. N. Schramm.
Bull. American Astron. Soc., Vol. 7, 550 (1975). – Abstr. AAS.

161.010 **Gas loss in groups of galaxies.**
R. B. Larson, H. L. Dinerstein.
Publ. Astron. Soc. Pacific, Vol. 87, 911 - 915 (1975).

Estimates are made of the total amount of mass and the total amount of heavy elements expected to be lost during the early evolution of the galaxies in a typical group or cluster. The estimated amount of gas lost is roughly one quarter of the mass present in galaxies, and the metal abundance of the lost gas is essentially solar. Several possibilities concerning the fate of the gas lost from galaxies are discussed.

161.011 **Selective extinction in the local supercluster.**
H. I. Abadi, M. G. Edmunds.
Astron. Astrophys., Vol. 45, 319 - 323 (1975).

A re-examination of the colours of galaxies in the local supercluster does not support previous suggestions of a concentration of dust in the supergalactic plane. The authors show that the observed reddening is due to a systematic difference between the colours of galaxies in large clusters and those in the field.

Missing mass in galactic coronae?
See Abstr. 158.026.

Ring galaxies in rich clusters and the intergalactic medium. See Abstr. 158.223.

Hot gas in clusters of galaxies.
See Abstr. 160.030.

Explosive events in the early universe.
See Abstr. 162.030.

162 Structure and Evolution of the Universe, Cosmology

162.001 **Uniform model universes containing interacting blackbody radiation and matter with internal energy.** T. L. May.
Astrophys. Journ., Vol. 199, 322 - 327 (1975).

Techniques previously developed for the production of general-relativistic models of the universe which contain interacting matter and blackbody radiation are extended to include models with internal energy and with nonzero cosmological constant. The material content of the model obeys a polytropic equation of state. The nature of the interaction between the matter and radiation is elucidated, and an alternate interpretation of the rate of interaction is presented.

162.002 **Correction of a misunderstanding about Olbers' paradox.** F. J. Belinfante.
Gen. Rel. Grav., Vol. 6, 9 - 12 (1975). – Presented at the Seventh International Conference on Gravitation and Relativity, June 1974, Tel-Aviv, Israel.

162.003 **Cosmological model imposed by a unified theory.** F. Goded.
Gen. Rel. Grav., Vol. 6, 119 - 121 (1975). – Presented at the Seventh International Conference on Gravitation and Relativity, June 1974, Tel-Aviv, Israel.

162.004 **Antimatter and the particle problem in Einstein's cosmology and field theory of elementary particles (An historical essay on Einstein's work at the Akademie der Wissenschaften zu Berlin).** H.-J. Treder.
Astron. Nachr., Vol. 296, 149 - 161 (1975).

The history of Einstein's Berlin work about the 'particle problem' and the 'quantum riddle' in general relativity and their connections with cosmology and unified field theory is outlined. The author's historico-critical discussion centers on Einstein's foremost proof of the existence of 'antimatter' in each general-relativistic universe.

162.005 **The scale of galaxy clustering and the mean matter density of the universe.** S. M. Fall.
Monthly Notices Roy. Astron. Soc., Vol. 172, 23P - 26P (1975).

The correlation energy computed from Peebles' power-law galaxy covariance function suggests that, if the gravitational instability picture is correct, then there must be a feature in the covariance function where it steepens. The scale of this cut-off cannot be accurately determined by this argument but a comparison of the random kinetic energies of galaxies with their correlation energy gives an estimate of the mean matter density of the universe. The suggested value is less than 5 per cent of the closure density if most matter is distributed as galaxies are.

162.006 **Cosmology from angular size counts of extragalactic radio sources.** V. K. Kapahi.
Monthly Notices Roy. Astron. Soc., Vol. 172, 513 - 533 (1975).

In this paper the cosmological implications of the observed angular sizes of extragalactic radio sources are investigated using (1) the $\log N$–$\log \theta$ relation, where N is the number of sources with an angular size greater than a value θ, for the complete sample of 3CR sources, and (2) the θ_{median} *vs* flux density (S) relation derived from the 3CR, the All-sky, and the Ooty occultation surveys, spanning a flux density range of about 300 : 1. The method of estimating the expected $N(\theta)$ and $\theta_m(S)$ relations for a uniform distribution of sources in space is outlined. Since values of $\theta \gtrsim 100''$ arc in the 3C sample arise from sources of small z, the slope of the $N(\theta)$ relation in this range is practically independent of the world model and the distribution of source sizes, but depends strongly on the radio luminosity function.

162.007 **On the origin of the metagalaxy.** L. Eh. Gurevich.
In-t teor. fiz. AN USSSR. Preprint ITF-74-104R. Kiev, 1974. 42 pp. Price 11 Kop. In Russian. – Abstr. in Referativ. Zhurn. 51. Astron., 7.51.849 (1975).

162.008 **On the effect of viscosity on the character of the cosmological singularity.**
V. A. Belinskij, I. M. Khalatnikov.
Pis'ma v ZhurnEHTF, Vol. 21, 223 - 225 (1975). In Russian. Abstr. in Referativ. Zhurn. 51. Astron., 7.51.863 (1975).

162.009 **Can collapsed stars close the universe?** J. R. Thorstensen, R. B. Partridge.
Astrophys. Journ., Vol. 200, 527 - 534 (1975).

Can black holes and other collapsed objects contribute enough mass to close the universe (i.e., to bring the density of the universe up to the critical density, ρ_c)? If the black holes are the collapsed remnants of an early generation of massive stars, current observations of the nightsky brightness may be used to place an upper limit on the mass contributed by such objects. The authors show that the density contributed by collapsed objects must be less than 0.01 ρ_c unless the stars formed before an epoch of 10^9 yr ($z > 5$). Even more stringent limits may be set for many of the models considered below.

162.010 **Fermion pair creation near the Friedmann singularity.**
S. G. Mamaev, V. M. Mostepanenko, V. M. Frolov.
Pis'ma v Astron. Zhurn., Vol. 1, No. 9, p. 8 - 12 (1975). In Russian.

Estimates are obtained for the density of fermions created near the singularity in homogeneous isotropic cosmological models. The effect is shown to depend essentially upon the sign of the curvature of the model, in contrast to the boson case.

162.011 **The mean luminosity and mass densities in the universe.** J. R. Gott III, E. L. Turner.
Bull. American Astron. Soc., Vol. 7, 412 (1975). – Abstr. AAS.

162.012 **Quelle est l'origine de l'Univers?** E. I. Robson.
Orion, 33. Jahrgang, p. 103 - 104 (1975).

162.013 **Crystal structures, the earth and Dirac's large numbers hypothesis.** K. M. Towe.
Nature, Vol. 257, 115 - 116 (1975). – Letter.

162.014 **Possible cosmological origin of spontaneous symmetry breaking.**
G. Domokos, M. M. Janson, S. Kovesi-Domokos.
Nature, Vol. 257, 203 - 205 (1975). – Letter.

162.015 **An accelerating universe.** J. E. Gunn, B. M. Tinsley.
Nature, Vol. 257, 454 - 457 (1975).

New data on the Hubble diagram, combined with constraints on the density of the universe and the ages of galaxies, suggest that the most plausible cosmological models have a positive cosmological constant, are closed, too dense to make deuterium in the big bang, and will expand for ever. Possible errors in the supporting arguments are discussed.

162.016 **The primordial black hole mass spectrum.** B. J. Carr.
Astrophys. Journ., Vol. 201, 1 - 19 (1975).

The author examines what mass spectrum of primordial black holes should result if the early universe consisted of small density fluctuations superposed on a Friedmann background. It is shown that only a certain type of fluctuation favors the formation of primordial black holes and that, consequently, their spectrum should always have a particular form. Since both the fluctuations which arise naturally and the fluctuations which are often invoked to explain galaxy formation are of the required type, primordial black holes could have had an important effect on the evolution of the universe. Any model in which the early universe has a soft equation of state for a prolonged period is shown to be suspect, since primordial black holes probably form too prolifically in such a situation to be consistent with observation.

162.017 **Theoretical determination of the thinning factor in relativistic hierarchical cosmology.** P. S. Wesson.
Astrophys. Space Sci., Vol. 37, 101 - 114 (1975).

Recent computer simulations by Haggerty (1974) and Haggerty and Janin (1974) on the gravitational clumping of bodies into clusters and superclusters in Newtonian cosmology have given an approximate value of $\theta = 1.9$ for the thinning factor (defined for two systems of sizes R_1, R_2 and densities ρ_1, ρ_2 by $\rho_1/\rho_2 = (R_2/R_1)^\theta$), close to the observed value of $\theta = 1.7$. To get an almost exact value of θ in a general relativistic hierarchy, algebraic conditions on the metric tensor are employed: the result is $\theta = 2$.

162.018 **On the possibility of nucleosynthesis in any matter-antimatter symmetric cosmology.**
F. Combes, O. Fassi-Fehri, B. Leroy.
Astrophys. Space Sci., Vol. 37, 151 - 167 (1975).

In any matter-antimatter symmetric cosmology the primeval nucleosynthesis imposes a typical size of a region of matter (antimatter) much larger than the diffusion length of neutrons. Observational results on the abundances of cosmic deuterium and helium are used to evaluate how large it has to be. As a result the authors find that a minimal size $L_1 \sim 1.5 \times 10^8$ cm at the temperature $T = 1$ MeV is compatible with the standard abundances of primeval helium and deuterium.

162.019 **Discrete source counts and the rotation of the local universe in hierarchical cosmology.** P. S. Wesson.
Astrophys. Space Sci., Vol. 37, 235 - 256 (1975).

Attention is given to reasons for believing that the upper limit on the rotation of the universe ω set by isotropy of the 3 K background may not be appropriate to the local system because of its hierarchical structure. In particular, recent work of Rubin et al. (1973) on the anisotropy of Hubble's parameter (H) as determined by certain galaxies is examined. It seems likely that $\omega_1 r \simeq (\text{const})\, r^{-n}$ with $0 \lesssim n \lesssim 1$ over the interval 500–1000 Mpc ($H = 60$ km s^{-1} Mpc^{-1}) with a total rotation of $\omega < 2\,\omega_1$, and $\omega_1 = 1.2\,(\pm 0.25) \times 10^{-18}$ s^{-1} evaluated on data out to 10^3 Mpc. Strictly, the quoted value of the rotation only applies to a region of space that in some sense has an isotropic limit: if the actual hierarchy has a large density-dependence away from a local origin, then the numerical value of the rotation is smaller than the quoted value but still finite and significant.

162.020 **Evolution von Kosmos und Erde. Einführung und Koordination.** A. Unsöld.
Nova Acta Leopoldina [Deutsche Akademie der Naturforscher Leopoldina, Halle/Saale], New Ser., Vol. 42 (No. 218), 91 - 95 (1975). – Lecture on the occasion of the annual meeting of the Academy, 1973 October 11 - 14.

162.021 **Possible local variation of the Hubble constant in van den Bergh's calibration of Sc-type galaxies.**
G. Le Denmat, M. Moles, J. P. Vigier, J. L. Nieto.
Nature, Vol. 257, 773 - 774 (1975). – Letter.

162.022 **Cosmology and verifiability.** G. F. R. Ellis.
Quarterly Journ. Roy. Astron. Soc., Vol. 16, 245 - 264 (1975). – Paper delivered at the Boston Colloquium for the Philosophy of Science, 1973 November 6.

The way in which one obtains a cosmological model from basic principles is discussed. It is shown that some of these principles are of necessity unverifiable. The observed isotropy of the universe leads to a uniformity assumption which can be codified in various ways, leading to slightly different world models. The overall structure of these world models is discussed in detail, and the different ways in which different parts of the universe are accessible to observation can then be made explicit.

162.023 **Acoustic modes and gravitational instabilities in the early Universe.** S. A. Bonometto, F. Lucchin.
Astron. Astrophys., Vol. 43, 189 - 199 (1975).

As several authors (e.g. Peebles, Silk, Field) have outlined, two acoustic modes are possible in the Universe before cosmological hydrogen decoupling occurs: they involve fluctuations which can be roughly described as either adiabatic or isothermal. In this article the two modes are studied in detail, together with the different kinds of gravitational instabilities connected with them. A number of peculiar features of the two acoustic regimes are also emphasized: in particular, we show that there exist "zero speed" bands for the "adiabatic" mode and exact formulae are given, which enable one to calculate to any approximation the rate of occurrence of several effects some of which were previously known.

162.024 **On defining "rest" in Robertson-Walker cosmologies.**
J. G. Fox, K. C. Jacobs, B. Thomsen, S. von Hoerner.
Astrophys. Journ., Vol. 201, 545 - 546 (1975).

An earlier conclusion, that peculiar motions of test particles in empty Robertson-Walker cosmologies are potentially observable, is here shown to be incorrect.

162.025 **Nuclear processes in the early universe.**
T. de Graaf.
II Leningr. mezhdunar. seminar, 1974. Uskorenie chastits i yader. reaktsii v kosmose. Leningrad, 1974, p. 239 - 250. Abstr. in Referativ. Zhurn. 51. Astron., 9.51.782 (1975).

162.026 **On the origin of rotation of galaxies in Ambartsumyan's cosmogony.** R. M. Muradyan.
Obedin. in-t yader. issled. Preprint R2-8585. Dubna, 1975. 18 pp. In Russian. – Abstr. in Referativ. Zhurn. 51. Astron., 9.51.792 (1975).

162.027 **Anisotropic cosmological model of Bianchi type V in the general (axisymmetric) case with moving matter.** I. S. Shikin.
Zhurn. ehksperim. i teor. fiz., Vol. 68, 1583 - 1596 (1975). In Russian. – Abstr. in Referativ. Zhurn. 51. Astron., 9.51.810 (1975).

162.028 **Développements récents en cosmologie de l'antimatière.** R. Omnes, S. Caser.
Ondes et radiations gravitationnelles. Colloque International C.N.R.S. No. 220, (see 012.021), p. 225 - 233 (1974).

162.029 **Models of the universe in general relativity.**
A. Krasiński.
Postępy Astron., Vol. 23, 97 - 110 (1975). In Polish.

Some difficulties of the Newtonian theory of gravitation which led to the formulation of the general relativity theory are presented. Then the basic ideas of general relativity are introduced at an elementary level. The properties of a few most important cosmological models are briefly discussed.

162.030 **Explosive events in the early universe.**
J. Schwarz, J. P. Ostriker, A. Yahil.
Astrophys. Journ., Vol. 202, 1 - 6 (1975).

Large-scale explosions, possibly the cumulative effects of supernovae, may have occurred in the early history of the universe ($10 \geq z \geq 2$). The energy output of any galaxy can be estimated at 5×10^{51} ergs per present visual solar luminosity. Blast waves from such a large input of energy could heat and ionize an intergalactic medium of closure density to a degree that would be consistent with the limits on both the 21-cm absorption in nearby radio sources and the Lα absorption in quasars. The bremsstrahlung radiation emitted by the ionized gas might explain the possible excess observed in the diffuse X-ray background below 1 keV.

162.031 **Er universet uendelig?** R. Stabell.
Astron. Tidssk., Årg. 8, 128 - 130 (1975).

162.032 **On the evolution of cosmological turbulence. IV. Gravitational growth of inhomogeneities, isolation of galaxies and their systems.**
A. A. Kurskov, L. M. Ozernoj.
Astron. Zhurn. Akad. Nauk SSSR, Vol. 52, 937 - 949 (1975). In Russian. English translation in Soviet Astron., Vol. 19, No. 5.

The theory of gravitational growth and isolation of inhomogeneities in Friedmann universes with density less than critical is presented. The authors suggest three regimes of the growth and isolation of inhomogeneities, in dependence on the interrelation between their initial amplitude and the critical value. The results are applied to the density inhomogeneities produced by cosmological turbulence in the course of its evolution both before and after recombination.

162.033 **Globally inhomogeneous "glued together" universes.**
D. D. Sokolov, A. A. Starobinskij.
Astron. Zhurn. Akad. Nauk SSSR, Vol. 52, 1041 - 1048 (1975). In Russian. English translation in Soviet Astron., Vol. 19, No. 5.

Topologically "glued together" Friedmann universes with density below the critical one are considered. The notion of G (ghost)-domain is introduced in which the minimal "gluing parameter" is less than the characteristic distance between clusters of galaxies. Small adiabatic perturbations of such spaces are investigated and it is shown that the formation of clusters of galaxies (as well as galaxies themselves and QSOs) in the G-domains of definite type is forbidden. The hypothesis is assumed that all perturbation modes are uncorrelated and all modes with the same scale are equally probable.

162.034 **The meaning of curved space.** D. Albers.
Mercury, (Journ. Astron. Soc. Pacific), Vol. 4, No. 4, p. 16 - 19 (1975).

162.035 **Dissipative processes in the Friedmannian cosmology.** Z. Klimek.
Acta Cosmologica, Fasc. 3, p. 49 - 53 (1975).

The paper presents certain qualitative results concerning dissipative processes in the homogeneous isotropic hot universes. Two results seem to be interesting: 1. Maximal heating of the substratum is much smaller than the rate of adiabatic cooling in the expanding radiative model. 2. For the flat Friedmann model, maximal amount of radiation entropy per particle Σ which could be generated in periods when bulk viscosity was most efficient (mixture of particles and radiation), cannot explain the high value of the observed at the present epoch $\Sigma = 10^8$. It suggests that in the Friedmannian models high entropy per particle is connected with the initial conditions and may constitute an essential property of the initial singularity.

162.036 **Observational evidence for assumptions of cosmology.** E. Skarżyński.
Acta Cosmologica, Fasc. 3, p. 61 - 67 (1975).

After comments on recent proposals (Heller, Klimek, Rudnicki 1973) concerning empirical support for assumptions of the Friedmann-Lemaître cosmology, Hawking and Collins' arguments upholding the thesis that the universe is isotropic, are discussed. Some remarks about four criteria of spatial homogeneity are made. The cosmological principle in its weak formulation (= space homogeneity in Bianchi sense) implies homogeneity of the substratum in general relativity and in Dicke–Brans theory.

162.037 **Steady-state cosmology with the creation of photons.**
E. Skarżyński.
Acta Cosmologica, Fasc. 3, p. 69 - 73 (1975).

The hypothesis of the continuous creation of radiation follows from the so-called perfect cosmological principle. The steady-state cosmology with the continuous creation of radiation seems to agree with all at present known observational data.

162.038 **Cosmology with spin and torsion. Part I. Physical and mathematical foundations.** B. Kuchowicz.
Acta Cosmologica, Fasc. 3, p. 109 - 129 (1975). – Review paper.

162.039 **Sur les dangers d'une approximation classique dans l'analyse des décalages spectraux vers le rouge.**
J.-C. Pecker, J.-P. Vigier.
Comptes Rendus Acad. Sci. Paris, Sér. B, Vol. 281, 369 - 372 (1975).

La vitesse du soleil par rapport à l'ensemble des galaxies assez proches (et liée aux mouvements internes aux galaxies du groupe local) doit être prise en considération pour interpréter la variation éventuelle avec la distance de la «constante de Hubble». Toutefois, les résultats récents obtenus à partir des galaxies proches et des radiogalaxies proches, et qui montrent que dH/dr est négatif, ne sont affectés que quantitativement, non qualitativement, par cette remarque.

162.040 **Déplacements anormaux vers le rouge liés à la traversée des amas de galaxies par la lumière.**
H. Karoji, L. Nottale, J.-P. Vigier.
Comptes Rendus Acad. Sci. Paris, Sér. B, Vol. 281, 409 - 412 (1975).

Partant d'une analyse de la distribution dans l'espace des amas de galaxies proches et de l'anisotropie angulaire de la constante de Hubble découverte par Rubin, Ford et Rubin, on peut vérifier avec une très forte signification statistique l'hypothèse de Roberts-Pecker et Vigier suivant laquelle la lumière provenant de sources lointaines subit une interaction d'un type nouveau en traversant le champ de radiation des amas de galaxies.

162.041 **On the origin of the metagalaxy.**
L. E. (*L. Eh.*) Gurevich.
Astrophys. Space Sci., Vol. 38, 67 - 78 (1975).

An attempt is made to explain two properties of the metagalaxy: its expansion and the absence of causal connection of its distant regions during a large part of its initial history. It is postulated that the gravitating matter was born from the 'cosmological field', i.e., a medium with equation of state $p = -\epsilon$. It is shown that such a postulate explains the expansion of metagalaxy and leads to a correct estimate of the entropy per baryon. The problem of causal connection can also be solved on this basis.

162.042 **Variation of the gravitational constant in cosmological models.** F. Occhionero, F. Vagnetti.
Astron. Astrophys., Vol. 44, 329 - 333 (1975).

The field equations of the scalar-tensor Brans-Dicke theory of gravitation are generalized to the case of a non-vanishing cosmological constant and applied to the dust models of the canonical Friedmann type. In particular, results are given (1) for the present fractional rate of change of the gravitational constant, (2) for the age of the universe and (3) for the evolution with redshift of the gravitational constant.

162.043 **On the Tifft effect and possibilities to explain it.** A. Kipper.
Publ. Tartu Astrofiz. Obs., Vol. 43, 3 - 30 (1975).

In the present paper an attempt is made to give a theoretical explanation of the Tifft effect – the dependence of the redshift of a galaxy on the luminosity of its nucleus. A theory of the author is used, according to which the cosmological redshift is a consequence of a continuous loss of energy of a light quantum in cosmological space. This hypothesis enables to predict the possible dependence of the value of cosmological redshift z on the luminosity of the object.

162.044 **Boundary conditions in closed universes.** J. K. Lawrence.
General Relat. Gravitation, Vol. 6, 253 - 258 (1975).

The author considers the problem of a particle in a square potential well in a one-dimensional closed universe. Imposition of periodic boundary conditions on the wave function changes the usual discrete energy levels to narrow but finite energy bands. Effects on the properties of atoms in a closed universe are estimated to be undetectable.

162.045 **Non-singular cosmological models in scalar-tensor theory of gravitation.**
V. I. Antonov, Yu. S. Vladimirov.
Probl. gravitatsii i ehlementarn. chastits. Vyp. (No.) 6. Moskva, Atomizdat, 1975, p. 121 - 129. In Russian. – Abstr. in Referativ. Zhurn. 51. Astron., 11.51.780 (1975).

162.046 **Quantization of homogeneous space cosmologies.** V. S. Gusakov.
Probl. teorii gravitatsii i ehlementarn. chastits. Vyp. (No.) 6. Moskva, Atomizdat, 1975, p. 66 - 70. In Russian. – Abstr. in Referativ. Zhurn. 51. Astron., 11.51.785 (1975).

162.047 **The redshift-distance relation. VIII. Magnitudes and redshifts of southern galaxies in groups: a further mapping of the local velocity field and an estimate of q_0.**
A. Sandage.
Astrophys. Journ., Vol. 202, 563 - 582 (1975).

Redshifts of 138 galaxies in 22 southern multiplets, groups, and clusters, and photoelectric *UBVR* magnitudes for 74 of the first few brightest galaxies in each aggregate give the following results. (1) The color excess as a function of galactic latitude for southern galaxies increases less steeply toward the galactic plane than expected from galaxy counts. (2) The absolute magnitude for first-ranked E galaxies is a very shallow function of richness for $N_c^{48} \gtrsim 20$, but steepens for $N_c^{48} < 20$. (3) Deviations from the redshift-magnitude (Hubble) diagram are uncorrelated with either supergalactic longitude, with Rubin-Ford-Rubin (RFR) regions, or with the apex of the Rubin-Ford proposed motion of the Local Group. (4) A crude estimate of q_0, found from the upper limit of $\delta H/H_0 < 0.08\,(\sigma)$ for any possible velocity perturbation for thc E groups in the presence of a density contrast of $\delta\rho/\rho = 3.4 \pm 0.2\,(\sigma)$, is $q_0 \leqslant 0.02 \pm 0.02$, from the idealized model of Silk. But the E groups here are generally beyond the Virgo cluster, giving this determination low weight; reanalysis of the spiral data puts an upper limit of $\delta H/H_0 = 0.40 \pm 0.09\,(\sigma)$, if the uncertain data for field spirals are adopted uncritically, giving $q_0 \leqslant 0.28 \pm 0.08\,(\sigma)$. These low values of q_0 agree with similar low values from five other methods summarized in the paper.

162.048 **The composite Hubble diagram.** K. R. Lang, S. D. Lord, J. M. Johanson, P. D. Savage.
Astrophys. Journ., Vol. 202, 583 - 590 (1975).

The composite Hubble diagram for 663 normal galaxies, 235 radio galaxies, and 265 quasi-stellar objects is given. The least squares solution to the Hubble diagram slope for each of the three classes of objects is compatible, within the errors, with that expected for a homogeneous, isotropic expanding universe. It is shown that the uncertainty in the slope of the Hubble diagram for QSOs is comparable to that obtained for normal galaxies over a similar range in magnitude. This means that the scatter in the QSO Hubble diagram should not be used in arguments which question their cosmological nature. It is shown that the absolute optical luminosity of extragalactic objects increases with increasing redshift, and that any given object class occurs infrequently in the magnitude and redshift range of the others.

162.049 **Antimatter bodies and closure of the universe.** R. E. Wilson.
Astrophys. Space Sci., Vol. 38, L9 - L11 (1975).

One of the hypotheses to account for the cosmic gamma-ray bursts invokes collisions between antimatter bodies of asteroidal mass and stars. This paper points out that, should this idea be correct, the numbers suggest that the antimatter is cosmological, and that a locally measurable ratio might be used to determine the effective gain in mean mass density relevant to the possible closure of the universe. The effective mass gain factor would have a minimum value of 2 and could be much larger.

162.050 **On the effect of the luminosity selection on the redshift-distance relationship.** P. Teerikorpi.
Astron. Astrophys., Vol. 45, 117 - 124 (1975).

Non-linearities in the redshift-distance relationship claimed for nearby ($V \lesssim 4000$ km s^{-1}) galaxies are interpreted in terms of luminosity selection in the magnitude limited data used. The effect, as seen in the $V/R-V$ diagram (V = radial velocity, R = calculated distance) for a sample of luminosity classified field galaxies is interpreted by taking into account the selection function (completeness of the data) and the luminosity function of the galaxies. In addition, theeffect of the selection on the de Vaucouleurs groups is pointed out. A selection free value for H is derived using nearby field galaxies. The present effect probably contributes to the dispersion found in published values of the Hubble constant; it gives rise to excessively large values.

162.051 **Decrease of the Hubble constant with distance in the Local Supercluster anticenter direction.**
G.Le Denmat, M. Moles, J. L. Nieto.
Astron. Astrophys., Vol. 45, 219 - 221 (1975).

If one recalibrates the sample of unbiased galaxies given by Sandage and Tammann using the absolute magnitudes given by Van den Bergh, one obtains a significant decrease of the Hubble constant in the Local Supercluster anticenter direction.

162.052 **Symmetry group of a closed universe.** F. Vollendorf, Zeitschr. Naturforschung, Vol. 30a, 1510 - 1515 (1975). In German.

The fourteen degrees of freedom of a free falling and rotating body are described by means of a non-compact simple 14-parameter Lie group G^{14}. It is shown that this group can be interpreted as the symmetry group of a spherically closed space. Finally an outlook is made to the combination of an internal and an external symmetry group in a nontrivial way.

162.053 **Spontaneously violated symmetries and cosmology.**

I. Yu. Kobzarev.
Ehlementar. chastitsy. Vyp. (No.) 3. Moskva, Atomizdat, 1975, p. 44 - 54. In Russian. – Abstr. in Referativ. Zhurn. 51. Astron., 12.51.700 (1975).

162.054 **Present-day trends in cosmology.**
Ya. B. Zel'dovich, I. D. Novikov.
Vopr. filosofii, 1975, No. 6, p. 51 - 62, 184. In Russian. Abstr. in Referativ. Zhurn. 51. Astron., 12.51.703 (1975).

162.055 **Cosmological models of the universe.** J. Štohl.
Kozmos, Vol. 6, 100 - 103 (1975). In Slovak.

162.056 **On the usefulness and reliability of some cosmological tests.** J. M. Barnothy.
Bull. American Astron. Soc., Vol. 7, 515 (1975). – Abstr. AAS.

162.057 **A "foil" for gravitational clustering investigations.**
W. H. Press.
Bull. American Astron. Soc., Vol. 7, 539 (1975). – Abstr. AAS.

162.058 **A zero-energy cosmology.** J. G. Dodd.
Bull. American Astron. Soc., Vol. 7, 550 (1975). Abstr. AAS.

162.059 **The number of Class A phenomena characterizing the universe.** M. Harwit.
Quarterly Journ. Roy. Astron. Soc., Vol. 16, 378 - 409 (1975).
The author constructs a phase space of observations that contains all conceivable astronomical observations and estimates its volume for the universe. He shows which observations are currently accessible and those that remain to be made, and he limits the discussion to observations, as distinct from experimentation. One way of describing the universe is in terms of sharply distinct observable patterns of events that constitute different phenomena. To fix ideas the author arbitrarily defines a class of phenomena that he calls Class *A*. Any given Class *A* phenomenon differs from all other phenomena in the class by a factor of $\gtrsim 10^3$ in at least one observational trait. The phenomenon can be identified by that trait. The author estimates that the total number of Class *A* phenomena characterizing our universe is of order one hundred. Phenomena that have not yet been discovered are most likely to be found by searching with observational techniques in the currently inaccessible domains of the phase space mentioned above.

162.060 **The origin of the universe.** Summary of the first "J. C. Bentley memorial lecture". E. E. Baart.
Monthly Notes Astron. Soc. Southern Africa, Vol. 34, 130 - 131 (1975).

162.061 **Some features of the relict radiation spectrum.**
V. K. Dubrovich.
Soobshch. Spets. Astrofiz. Obs., *Zelenchukskaya,* vyp. (No.) 14, p. 33 - 41 (1975). In Russian.
A possibility of hydrogen recombination line formation in transition between highly excited levels (with numbers $i \approx 10 - 15$) is considered. Their intensities and widths are estimated. It is shown that from the results of observations of these lines it will be possible to determine the density of hydrogen in the universe as well as the density of hidden matter.

162.062 **The transactinium elements in the developing universe.** V. P. Chechev, Ya. M. Kramarovskij.
Uspekhi fiz. nauk, Vol. 116, 687 - 708 (1975). In Russian. Abstr. in Referativ. Zhurn. 51. Astron., 1.51.886 (1976).

162.063 **Creation of particles in vortical cosmology.**
V. N. Lukash, I. D. Novikov, A. A. Starobinskij.
In-t kosmich. issled. AN SSSR. Pr-233. Moskva, 1975. 40 pp. In Russian. – Abstr. in Referativ. Zhurn. 51. Astron., 1.51.890 (1976).

162.064 **The theory of origin and nature of the universe.**
A. G. W. Cameron.
Osnovy kosmich. biol. i med. Tom (Vol.) 1. Moskva, Nauka, 1975, p. 17 - 46. In Russian. – Abstr. in Referativ. Zhurn. 51. Astron., 1.51.894 (1976).

162.065 **On the effect of viscosity on the nature of cosmological evolution.**
V. A. Belinskij, I. M. Khalatnikov.
Zhurn. ehksperim. i teor. fiz., Vol. 69, 401 - 413 (1975). In Russian. – Abstr. in Referativ. Zhurn. 51. Astron., 1.51.903 (1976).

162.066 **Will the universe expand forever?** J. E. Gunn.
Mercury, (Journ. Astron. Soc. Pacific), Vol. 4, No. 6, p. 4 - 8 (1975).

162.067 **The structure of the universe. A survey of contemporary cosmology.** Ya. B. Zel'dovich.
Fiz. Szemle, (*Hungary*), Vol. 24, 304 - 308 (1974).
In Hungarian.
Experimental results of the past ten years on the composition of the universe have shown the largest component to be primary photons in thermal equilibrium at 2.7 K. Earlier structures of the universe are arrived at by statistical mechanical considerations and extrapolating the present photon population. The hot universe theory explains only the formation of helium and neutrons, and protons are thought to have been formed before the last hot state of the universe.

162.068 **Radiation in cosmological backgrounds.**
S. Chang.
Thesis, Univ. Pittsburgh, Pa. [Available from Univ. Microfilms, Ann Arbor, Mich., USA. Order No. 74-24175], 129 pp. (1974).
The author studies scalar and gravitational radiation in Friedmann backgrounds with positive, negative and zero spatial curvature, and attempts to answer the question under what circumstances a wave propagates with or without a tail. Based on Newman-Penrose formalism, a basic wave equation is derived for pure gravitational radiation in Friedmann backgrounds. This wave equation coupled with a scalar wave equation serves as the basis of the study.

162.069 **Some magnetohydrodynamic Gödel type universes.**
M. Bray.
Ann. Inst. Henri Poincaré A (*France*), Vol. 22, 29 - 42 (1975).
In French.
Given a Gödel type metric tensor $G_{\alpha\beta}$ some solutions of the Einstein gravitational field equations are obtained for a total energy tensor corresponding to the magnetohydrodynamic case.

162.070 **Bianchi type I cosmological models with pure magnetic field.** U. K. De.
Acta Phys. Polonica B, Vol. B6, 341 - 346 (1975).
Presents the possible solutions for electromagnetic fields in a source-free region when the space-time admits an Abelian group of motions with three space-like generators (Bianchi type I group). The entire set of solutions can be recovered by a simple transformation of an already known group of solutions.

162.071 **An indirect confirmation of hierarchical cosmological models.** M. Kalinkov, K. Stavrev.
Comptes Rendus Acad. Bulg. Sci., Vol. 28, 13 - 16 (1975).
Discusses galaxy counts from the Palomar Sky Survey plates. The results presented may be viewed as an indirect con-

firmation of hierarchical cosmological models. There are areas on the celestial sphere for which the galaxy counts are in agreement with a model of a hierarchical universe.

162.072 **The amplification of gravitational waves and creation of gravitons in isotropic universes.**
L. P. Grishchuk.
Nuovo Cimento Lettere, Ser. 2, Vol. 12, 60 - 64 (1975).

162.073 **Antimatter and cosmology.** E. M. Svaton.
Inst. Teoret. Fys., Stockholm Univ. (Sweden), USIP–75-01, 37 pp. (1975).

162.074 **Cosmological singularity in Yilmaz-Tupper theory of gravitation.** Z. Klimek.
Acta Phys. Polonica B, Vol. 4 B, 153 - 156 (1973). – Letter.

162.075 **A class of inhomogeneous cosmological models.**
P. Szekeres.
Commun. Math. Phys., Vol. 41, 55 - 64 (1975).
All solutions of Einstein's field equations for vortex-free dust and a defined metric are presented. The new metric degeneralizes the cosmological models of Bondi-Tolman, Eardley-Liang-Sachs and Kantowski-Sachs.

162.076 **Hadronic matter.** K. Koebke.
Thesis, Naturwiss. Fakultät, Univ. Würzburg. [Available from the library of the Univ. Würzburg], 56 pp. (1973). In German.

162.077 **The universe as a black hole.** I. M. Freeman.
American Journ. Phys., Vol. 43, 644 (1975).

162.078 **Irreversible processes in homogeneous isotropic cosmological models.** W. Meier.
Thesis A, Friedrich-Schiller-Univ., Jena. [Available from Univ. Library, Berlin (GDR)], 89 pp. (1974). In German.

162.079 **Static fluid spheres in Einstein-Cartan theory.**
A. R. Prasanna.
Phys. Rev. D, (Particles and Fields), Vol. 11, 2076 - 2082 (1975).
Static fluid spheres in Einstein-Cartan theory are considered, following Trautman. In the presence of radially-aligned spins preventing collapse, the interior Schwarzschild solution is inappropriate and the modified equation of state is found. The hydrostatic pressure is discontinuous at the surface.

162.080 **Static cylinder of perfect fluid with nonzero spin density.** A. R. Prasanna.
Phys. Rev. D, (Particles and Fields), Vol. 11, 2083 - 2087 (1975).
The interior field of an infinite static cylinder of perfect fluid is studied, when the individual particles have spins aligned along the axis. A solution is found with an assumed equation of state, where the pressure is discontinuous at the surface.

162.081 **Spherical mass immersed in a cosmological universe: a class of solutions.** J. Eisenstaedt.
Phys. Rev. D, (Particles and Fields), Vol. 11, 2021 - 2025 (1975).
A model of a spherical mass immersed in a cosmological universe is constructed, illustrating a new exact solution of the Einstein field equations in the spherical case.

162.082 **Cosmological models of the Kantowski-Sachs type.**
I. M. Goethals.
Ann. Soc. Sci. Bruxelles I (*Belgium*), Vol. 89, 50 - 62 (1975). In French.
The author considers axially symmetric cosmological models with a metric of Kantowski-Sachs type. Einstein's equations are reduced to a system of two ordinary differential equations of first order, and the behaviour of the solutions in the neighborhood of the singular points of the system are classified. The implications for the cosmological models are discussed.

162.083 **Mach's principle in cosmology.** E. Ihrig.
International Journ. Theor. Phys., (*GB*), Vol. 12, 401 - 406 (1975).
In a certain class of cosmological models it is shown that every Ricci flat complete spacetime must be locally flat.

162.084 **Quantum cosmology: exact solutions for the Gowdy T^3 model.** B. K. Berger.
Phys. Rev. D, (Particles and Fields), Vol. 11, 2770–2780 (1975).
For each mode of the quantised Gowdy T^3 universe a complete set of orthonormal wave functions is obtained. They are valid at all times, including that of the classical singularity, where they reduce to a given initial state. The consistency of the quantum theory is checked by comparing quantum and (semi) classical expectation values.

162.085 **Role of F-gravity in cosmological models.**
C. Sivaram, K. P. Sinha, E. A. Lord.
Current Sci., (*India*), Vol. 44, 143 (1975).
The role of a massive scalar-tensor of F-gravity in cosmology is clarified to remove earlier ambiguities.

162.086 **Non-singular homogeneous universes with four-parameter Lie groups.** B. Kuchowicz.
Phys. Letters A, (*Netherlands*), Vol. 54A, 13 - 14 (1975).
It is proved, within the framework of the Einstein-Cartan theory, that singularities may be prevented in homogeneous cosmological models which do not permit any simply transitive group of motions. An oscillating universe with a minimum non-zero radius is possible.

162.087 **Vacuum fluctuations of a quantised scalar field in a Robertson-Walker universe.** E. Streeruwitz.
Phys. Rev. D, (Particles and Fields), Vol. 11, 3378 - 3383 (1975).
In a suitably regularised calculation of the vacuum expectation value of the energy-momentum tensor of a quantised scalar field in a Robertson-Walker universe, three contributions are found up to higher orders in time derivatives and the universe's inverse radius. An estimation of these contributions yields interesting results on the behaviour of a Robertson-Walker universe under collapse. A discussion is given on the role of quantum corrections in general relativity and their possible importance for highly collapsed states of the early universe.

162.088 **Cosmological models in Rosen's theory of gravitation.** D. Babala.
Journ. Phys. A, (Math.,Nuclear, General), Vol. 8, 1409 - 1412 (1975).
Solutions of Rosen's equations of gravitation for a homogeneous isotropic universe are obtained. The solutions resemble the Lemaître universe. A universal repulsive force, introduced in general relativity by a positive cosmological constant, appears directly as a consequence of the (unmodified) field equations.

162.089 **Nature of the singularity in some Brans-Dicke universes.** A. K. Raychaudhuri.
Progr. Theor. Phys. Japan, Vol. 53, 1360 - 1364 (1975).
Against a background of somewhat conflicting results about the possibility of nonsingular cosmological solutions of the Brans-Dicke equations, the author investigates some explicit solutions of these equations.

162.090 **Mach's principle and matter production process in non-steady state cosmological model.**
F. M. Gomide, M. Uehera.
Progr. Theor. Phys. Japan, Vol. 53, 1365 - 1371 (1975).
A Hoyle-Pryce tensor can be introduced into the Einstein field equations of general relativity as a possible inference from Mach's principle as formulated in Whitrow's relation. The set of equations obtained can be completely determined for the matter injection field, pressure, density and scaling functions in a closed model universe with asymptotically limited expansion. This model presents a cosmic density higher than the usually accepted value for the smearing out of galactic matter.

162.091 **Conformal invariance and scalar-tensor theories.**
E. A. Lord.
Pramana, (*India*), Vol. 4, 164 - 170 (1975).
A new generalisation of Einstein's theory is proposed which is invariant under conformal mappings. Two scalar fields are introduced in addition to the metric tensor field, so that two special choices of gauge are available for physical interpretation, the 'Einstein gauge' and the 'atomic gauge'.

162.092 **Cylindrically symmetric cosmological solutions of the Lyttleton-Bondi universe.**
J. R. Rao, H. S. Panda.
Journ. Phys. A, (Math., Nuclear, General), Vol. 8, 1413 - 1422 (1975).
The exterior field of the Einstein-Rosen cylindrically symmetric metric in the Lyttleton-Bondi universe is considered and exact solutions are obtained in some specific situations.

162.093 **Charge conjugation and Segal's cosmology.**
S. Sternberg, J. A. Wolf.
Nuovo Cimento A, Ser. 11, Vol. 28A, 253 - 271 (1975).
It is shown that the operation of charge conjugation can be realized as an actual geometric transformation within the framework of Segal's chronogeometric theory. Some geometric questions connected to the foundations of the chronogeometric theory are also discussed.

162.094 **A geometric aspect of Hubble's phenomenon.**
H. Konitz.
Foundation Phys., (*GB*), Vol. 5, 185 - 191 (1975).
The conventional interpretation of the Hubble effect as a Doppler effect, based upon the concept of an expanding universe or based upon the idea of a continuously increasing radius of curvature of space, leads to some difficulties. It seems possible to avoid these difficulties by ascribing the red-shift of light coming from remote galaxies to the fact that the non-Euclidean structure of the universe gets more and more important as observation extends to regions extremely distant from the point of observation. Hubble's phenomenon thus seems based upon the structure of space and time as judged from the point of observation; this means that it appears as a consequence of the 'geometria intrinsica' of the universe.

Ancient cosmologies. See Abstr. 003.025.

Gravitation und Kosmologie. See Abstr. 003.114.

Structure and evolution of the universe. See Abstr. 003.137.

The arrow of time. See Abstr. 015.026.

An analysis of the ancient astronomical observations with the implications for geophysics and cosmology. See Abstr. 043.001.

The electromagnetic background: limitations on models of unseen matter. See Abstr. 061.026.

Is deuterium of cosmological or of galactic origin? See Abstr. 061.038.

Tests of randomness in astronomical objects. I. Along single great circles. See Abstr. 061.039.

Tests of randomness in astronomical objects. II. Along two orthogonal great circles. See Abstr. 061.040.

Spontaneous violation of symmetry and its cosmological consequences. See Abstr. 061.057.

$N\bar{N}$-interaction and matter-antimatter separation in the universe. See Abstr. 061.064.

On ellipsoidal solutions of Liouville's equation in general relativity. See Abstr. 066.002.

On a class of exact spherically symmetric solutions to the Einstein gravitational field equations. See Abstr. 066.064.

Quantum electrodynamical effects in Kerr-Newman geometries. See Abstr. 066.110.

A bi-metric theory of gravitation. II. See Abstr. 066.118.

On Rosen's bi-metric theory of gravitation. See Abstr. 066.119.

Bi-metric theories of gravitation. See Abstr. 066.120.

Schwarzschild-Friedmannsche Lösungen der Einsteinschen Feldgleichungen. See Abstr. 066.150.

Interstellarer Staub und seine Bedeutung im kosmogonischen Kreislauf. See Abstr. 131.107.

A correlation test of the intrinsic interpretation of QSO absorption redshifts. See Abstr. 141.026.

Spectral intensity dependence and isotropy of sources stronger than 0.1 Jy at 2700 MHz. See Abstr. 141.059.

Interpretation of the magnitude-redshift diagram for quasars. See Abstr. 141.064.

Comments on the use of the limiting volume test. See Abstr. 141.077.

Distribution of quasars in the universe. See Abstr. 141.086.

QSO redshifts and the steady state cosmology. See Abstr. 141.133.

QSO absorption spectra caused by protogalaxies. See Abstr. 141.148.

On the formation of elliptical galaxies. See Abstr. 158.126.

Clustering of faint blue objects. See Abstr. 160.020.

On the surface brightnesses of first-ranked cluster galaxies. See Abstr. 160.031.

Another evolutionary correction to the luminosity of giant galaxies. See Abstr. 160.035.

A theory of galaxy formation and clustering. See Abstr. 160.044.

Author Index

The authors are listed in alphabetical order according to the initial letter following the first names.

Aalders, J. W.
131.532
Aalders, J. W. G.
032.521
142.207
Aannestad, P. A.
131.028
Aardoom, L.
046.064
Aarnio, J. J.
103.010 .100
Aaronson, M.
131.527
158.056 .070
Aarseth, S. J.
151.014 .046
Abadi, H. I.
161.011
Abalakin, V. K.
094.018
Abbot, R. I.
041.017 .021
099.234
Abdulla-zade, Kh. F.
004.061
Abdusamatov, Kh. I.
072.044
082.123
Abele, M.
031.012
055.008
Abell, G. O.
160.013
Abelson, H.
014.035
Abetti, G.
015.002
Abgaryan, A. A.
031.248
Ables, H. D.
041.035
Ables, J. G.
033.021
Abraham, H. J. M.
045.018
Abramenko, A. N.
034.053
122.101
Abramov, L. A.
083.036
Abramowicz, M. A.
065.055
Abramyan, M. G.
151.088
Abranches, M. C. B.
105.094
Abranin, Eh. P.
077.023
Abt, H. A.
153.038
Abuladze, O. P.
031.252
114.377
Abur-Robb, M. F. K.
083.003
Acosta, M. A.
034.052
Acton, L. W.
073.042
076.007
142.079 .133
Adachi, Y.
124.104
Adam, A.
004.052
011.007
Adam, J.
046.024
Adam, J. A.
062.017
Adams, G. W.
091.019
Adams, J. B.
094.409 .446
Adams, N. G.
104.016
Adams, P.
114.330
Adams, R.
103.134
Adams, R. C.
066.052
Adams, T. F.
063.024
133.025
158.004 .211
Adams, W.
080.028
Adams, W. M.
074.060
Adcock, B. S.
010.008
Ade, P. A. R.
031.210
132.042
Adler, I.
032.520
Adler, R.
003.009
Adzhyan, G. S.
126.026
Aerts, E.
084.263
Aerts, L.
103.100
118.012
Afanas'ev, V.
158.100
Afanasjev, V.
See Afanas'ev, V.
Afonin, V. V.
083.042
Africano, J.
142.232
Africano, J. L.
096.005
Agababyan, G. N.
034.117
Agapov, E. S.
034.109
Agekyan, T. A.
117.038
151.065 .090
Agnelli, G.
034.106

Agostinelli, C.
005.004
Agrawal, P. C.
061.021
142.072
Agrawal, S. P.
143.076
Agreen, R. W.
046.043
052.033
Agrell, S. O.
094.456 .462
Agron, P. A.
094.426
Agueero, E.
158.185
A'Hearn, M. F.
103.102 .131
Ahmad, A.
151.072
Ahmad, I. A.
131.901
Ahnert, P.
047.020
121.137 .139 .140 .141 .142 .143 .144 .145 .146
123.029 .073
Ahrens, T. J.
094.420
Ai, G.
073.076
Aihara, M.
032.034
044.041
Aikman, G. C. L.
114.043 .355
Aime, C.
031.227
Aitchison, C. S.
033.045
Aitken, D. K.
114.304
Aiton, E. J.
004.038 .039
005.008
Aizenman, M. L.
065.093 .108 .121
Aizu, K.
141.168
Ajello, J. M.
097.038
Akasofu, S.-I.
003.010
084.019 .020 .245 .248 .295 .312
106.002 .023 .901
Ake, T.
124.109
141.614
Akhtarzad, S.
033.033
Akhverdyan, L. G.
153.040
Akimov, V. V.
143.070
Akin'yan, S. T.
077.032
Aksel'rod, V. Yu.
083.027
Aksenov, A. N.
082.094
099.072
Aksenov, E. P.
042.119
Aksenov, V. I.
051.007
Aksnes, K.
098.057 .060
099.226 .230 .237
Akun'ya, S.
032.049
Al-Far, D. M.
105.005
Alaniya, I. F.
114.376 .377
Albee, A. L.
094.411
Albers, D.
162.034
Alberti, F.
131.204
Albo, H.
122.164
Albrecht, R.
114.347 .363
Albritton, D. L.
082.029
Alcaino, G.
154.001 .002 .004 .025
Alder, B.
003.011
Aldridge, K. D.
081.005
Aleksandrin, Yu. S.
034.109
Aleksandrovich, S. V.
034.067
Aleksashina, G. A.
046.044
Alekseev, I. S.
013.019
Alekseev, V.
003.012
Alekseev, V. A.
041.046
044.006
105.151
Alekseeva, N. E.
084.408
Aleshin, V. I.
003.020
Alexander, C.
094.421
Alexander, D. R.
061.069
Alexander, L. W. G.
034.142
Alexander, M. E.
117.059
Alexeev, V. A.
See Alekseev, V. A.
Alferov, G. V.
052.024
Alfven, H.
003.013
012.012
Aliev, A. A.
122.199 .222
Aliev, S.
102.025
Alikaeva, K. V.
074.046
Alissandrakis, C. E.
072.018
073.063
077.008
Alksne, Z.
065.178
114.007 .113 .902
123.019
124.104
Alksnis, A.
011.067
034.123
113.070
114.007 .106 .902
122.213
123.019
Alksnis, A. K.
122.087
123.050
124.011 .012
Alladin, S. M.
151.020
Allan, D. W.
044.014
Allegre, C. J.
105.102 .159 .186
Allen, C.
117.039
155.006
Allen, C. C.
094.104
Allen, C. W.
076.001
Allen, D. A.
141.602 .608
158.141
Allen, L. R.
142.095
Allen, M. S.
114.101 .355
Allen, R. J.
031.419
158.147 .158 .160
Allen Jr., R. O.
094.422
Aller, H. D.
141.137
142.195
158.310 .317
Aller, L. H.
071.014
133.011
Aller, M. F.
141.137
Alloin, D.
133.010
Al'perovich, L. S.
011.048
083.036
Al'pert, Ya. L.
083.019
Alt, E.
011.025
Altenhoff, W. J.
124.104
Alter, G.
004.048
Altrock, R. C.
071.003 .009

Altrock, R. C.
080.033
Altukhov, A. M.
143.027
Alvarez, H.
077.047
Alvarez, R.
105.033 .162
Alvarez, R. L.
073.065
Alvord, G. C.
031.294
Amata, E.
084.212 .221
Ambartsumian, V. A.
See Ambartsumyan, V. A.
Ambartsumyan, V.
065.180
Ambartsumyan, V. A.
003.110
122.006
158.136
Ambruster, C. W.
124.104
Amelin, V. M.
047.032
Ames, S.
131.180
Aminov, A. N.
104.074
Amnuehl', P. R.
066.132
124.005
Amnuel, P. R.
See Amnuehl', P. R.
Amundsen, R.
078.033 .034
Anand, K. C.
143.006
Ananda, M.
054.013
Anan'eva, L. Ya.
103.108
Ananth, A. G.
143.076
Ananthakrishnan, A.
141.094
Andereck, C. D.
158.075
Anderle, R. J.
046.074
054.016
Anders, E.
094.439 .468
105.016 .018 .038 .083
.084 .136 .154 .189
Andersen, B. N.
122.075
Andersen, J.
121.080 .081 .082 .102
.103
Andersen, T.
022.079 .102
071.051
Anderson, B.
141.339
Anderson, B. D.
061.004
Anderson, C. M.
114.098
Anderson, G. M.
052.003
Anderson, J. D.
052.001
066.041
091.008
Anderson, J. L.
063.038
066.117
Anderson, K. A.
084.298
094.157
Anderson, L.
142.132
Anderson, P. W.
141.303
Anderson, R. C.
022.039
079.100
Anderson, R. R.
084.303
Anderson Jr., D. E.
093.009
Andersson, L.
098.046
Andersson, R.
143.001 .020
Ando, H.
080.058
Andre, C. G.
094.185 .463
Andrew, B. H.
103.100 .102
158.306
Andrews, A. D.
159.006
Andrews, D. H.
034.148
Andrews, F. P.
031.230
Andrews, P. J.
122.076 .168
Andrianov, N. S.
104.070 .071 .072 .076
Andrienko, D.
092.002
Andrillat, Y.
103.100
114.108 .360
124.103 .104
158.135
Andriyanov, Yu. V.
074.094
Andryukov, V. G.
034.061
Angel, J. R. P.
034.156
116.004
125.018
126.018
142.224
158.209
Angel, R.
122.194
Angeli, M. T.
044.011
Angelotti, E.
044.015
Angelov, T.
065.130
Angenheister, G.
081.022
Anger, C. D.
084.015 .020
Angione, R. J.
082.025
Anglin, J. D.
073.131
Angreji, P. D.
082.040
Anguita, C.
041.036
Angus-Leppan, P. V.
012.036
Anisimov, V. D.
054.018
Anisimov, V. F.
034.109
104.025
Anistratenko, A. A.
084.286
Anosova, Zh. P.
117.038
Antal, M.
103.004 .100
Antiochos, S. K.
073.062
Antoniadi, E. M.
003.014
Antonov, I. V.
003.080
Antonov, V. A.
042.044
151.086
Antonov, V. I.
162.045
Anttila, A.
065.172
Anuchin, M. S.
105.077
Apeldoorn, B.
104.020
Apostolov, E. M.
076.020
Apparao, K. M. V.
003.015
061.002
Appleby, G. M.
075.021
Apushkinskij, G. P.
077.055
Apykhtin, N. G.
052.031
Arai, O.
106.079
Arakelyan, M. A.
158.238
Aranda, J.
065.174
Arazov, G. T.
042.019 .027
Ardeberg, A.
071.049
Arden, J. W.
105.094
Arens, J. F.
143.066
Argyle, R. W.
114.388
Argyros, J. D.
021.026

Arhipova, V. P.
See Arkhipova, V. P.
Arias, J. C.
123.075
154.015
Arkani-Hamed, J.
097.036
Arkhangel'skij, V. G.
157.004
Arkhipova, V. P.
113.038
122.043
123.013
124.002
Arlot, J.-E.
099.219
Armstrong, T. P.
084.322
092.010
099.085
Arnett, W. D.
065.058 .164
066.142
158.228
Arnold, H. J. P.
092.018
Arnold, J. R.
051.010
094.113
Arnoldy, R. L.
084.304
Arnould, M.
061.012
105.015
Arons, J.
142.061
Aronson, J. R.
097.047
Arora, B. R.
084.261
Arp, H.
141.062 .084
158.029 .186
Arp, H. C.
125.023
Arrhenius, G.
105.095
Arrhenius, S.
003.013
Arsenault, H. H.
031.019
Arsenijevic, J.
122.218 .219
Artamonov, B. P.
158.229
Artru, M.-C.
022.115
Artyukhina, N. M.
112.013
Arutyunyan, G. V.
031.060 .064
Arvidson, R.
094.110 .408
Arx, B. Von
See Von Arx, B.
Asbridge, J. R.
074.072 .114 .125
Ashbrook, J.
004.017 .055
010.001
080.008
Ashbrook, J.
095.004
132.045
Ashby, R. M.
015.018
Ashour-Abdalla, M.
062.034
Ashworth, J. R.
105.013 .056 .096
Asimov, I.
003.016
Aslanov, I. A.
034.082
114.087
Aslanov, V. S.
042.064
Aslund, N.
034.144
Aspden, H.
003.017
Asseo, E.
141.341
Atallah, K.
143.077
Athay, R. G.
064.061
071.028
072.062
073.132 .133
Atkins, H.
113.073
Atkinson, R. D'E.
031.204 .205
Atreya, S. K.
093.022
Attolini, M. R.
143.080 .083 .084
Aubier, M.
033.013
Audoin, C.
035.013
Audouze, J.
003.018
061.031
105.097
125.037
Auer, L. H.
031.219 .296
114.380
Auer, R. D.
084.210
Augason, G. C.
082.021
093.042
132.025
Auriemma, G.
142.206
Auriere, M.
093.014
Austin, G. K.
034.157
Austin, R. R. D.
103.100 .120 .136
121.060
Austin, S. M.
061.061
Auston, D. H.
031.078
Autio, G. W.
032.536
Auvergne, M.
034.152
Avant, H. H.
036.018
Avdyushin, S. I.
085.009
Aveni, A. F.
012.050
Avery, L. W.
103.100 .102
Avignon, Y.
077.050
Avni, Y.
142.176
Avramchuk, V. V.
082.044
Axford, W. I.
074.131
Axisa, F.
077.005
Axon, H. J.
094.456 .462
105.002 .185
Ayres, T. R.
064.018 .027
071.030
114.320
Azzarelli, L.
073.154
Azzopardi, M.
113.042

Baart, E. E.
162.060
Baasansuren, O.
046.034
Baba, Y.
021.023
Babadzanjanz, M. K.
See Babadzhanyants, M.
Babadzhanyants, M. K.
158.017 .102 .239
Babaev, M. B.
121.009 .045
Babaeva, N. G.
094.014 .184
Babala, D.
162.088
Babb, H. M.
032.059
Babcock, R. R.
085.003
Babij, B. T.
071.057
Babushkina, O. P.
142.180
Bacchus, P.
031.264
032.530
Bachmann, P. J.
121.071
Backer, D. C.
131.111
141.323
Backman, D. E.
142.047
Baddenhausen, H.
105.171
Bader, T. R.
031.053

Badhwar, G. D.
143.043
Baeck, N.
071.065 .066 .067 .068 .069 .071
Baedecker, P. A.
094.111 .413
Bafico, M. A.
032.536
Baggaley, W. J.
104.007 .014 .059
Bagin, V. M.
154.027
Baglin, A.
122.110 .111 .117
Bagnold, R. A.
097.029
Bahcall, J. N.
066.019
080.047 .062
141.017
142.005 .008 .125 .143 .176
154.028
Bahcall, N. A.
142.005 .008 .125
154.028
Bahner, K.
032.002 .012
Bahng, J. D. R.
114.019 .314
Baier, F. W.
158.138
160.017
Bailey, J.
124.019
Bailey, J. A.
124.017
Bailey, M. C.
033.028
Baird, S.
159.901
Baity, W. A.
142.004 .082 .087
Baize, P.
115.007
117.024
118.036
Bajdal, M. Kh.
085.012
Bajrachenko, I. V.
062.028
104.077
Bajsakalova, A. B.
078.016
Baker, D.
053.005 .012
Baker, D. J.
084.009
Baker, J. R.
125.039
158.173
Baker, K. B.
141.325
Baker, K. D.
084.009
Baker, N. H.
122.187
Bakhareva, M. F.
073.120
074.027 .063
Bakhrakh, N. M.
041.058
Bakos, G. A.
121.125
Balachandran, N. K.
104.003
Balasubramanian, V.
141.901
Baldi, P.
091.045
Baldinelli, L.
014.022
031.279
121.056 .057 .058 .153
Baldwin, B. J.
093.042
Baldwin, J. A.
141.060 .169
Baldwin, J. E.
158.062
Baldwin, J. R.
160.004
Baldwin, M. E.
010.001
031.231
Baldwin, R.
033.034
Balick, B.
133.022
158.146
Baliunas, S. L.
113.065
Balklavs, A.
003.008
011.069
033.020
Balkowski, C.
158.131
Ball, J. A.
033.081
131.054
Ballabh, G. M.
158.199
Ballard, S. S.
031.040
Ballas, J. S.
034.157
Ballintine, J.
141.353
142.080
Balmer, H.
004.024
Balmino, G.
052.045
081.028
Balodis, J.
031.408
Balona, L. A.
114.366
115.002
116.006
Balonek, T.
142.216
Balonek, T. J.
141.059 .141
Balsley, B. B.
084.021
Baluteau, J. P.
141.627
Bambakidis, G.
065.146
Bame, S. J.
074.072 .114 .125
077.047
084.312
106.901
Banerjee, S. K.
094.423 .445
Bangert, J. A.
126.019
Bania, T. M.
131.117
155.060
Banks, P. M.
083.005 .021
093.022
Banos, C. J.
099.038
Bar, V.
073.078
Baran, W.
046.028 .036
Barandeev, A. V.
094.450
Baranov, A. S.
151.004
Baranov, A. V.
072.021 .048
Baranov, V. B.
106.033
Baranovskij, Eh. A.
073.098
Baranovskij, I. V.
071.045
Barashenkov, V. S.
143.070
Barber, D. J.
105.056 .096
Barbero, J.
153.018
Barbieri, C.
041.052
Barbon, R.
125.048
158.044 .051
Bardeen, J. M.
151.026
Bardwell, C. M.
098.020
Barfield, J. N.
084.237
Barillet, R.
035.013
Barish, F. D.
099.078
Barkat, Z.
065.117
Barker, B. M.
066.012
141.301
Barker, E.
103.100
Barker, E. S.
071.021
093.032 .049
Barker, P. T.
133.037
Barkhatov, N. A.
106.067
Barkhatova, K. A.
153.019

Barkin, Yu. V.
042.068
052.030
Barletti, R.
082.129
Barlier, F.
046.027
Barlow, M. J.
133.029
Barnes, A.
022.088
074.023
Barnes, C. W.
080.029
Barnes III, T. G.
115.020
122.083
Barnhart, P. E.
073.157
Barnothy, J. M.
141.077 .134
162.056
Barnothy, M. F.
141.134
Barocas, V.
010.012
Barouch, E.
143.079
Barr, E. S.
003.019
Barraclough, D. R.
084.283
Barrett, A. H.
022.120
131.141
Barrette, L.
022.046
Barros, S.
103.119
Barsman, E. M.
114.327
Barsukov, V. M.
084.267
Bartaya, R.
114.061
Bartholdi, P.
013.010
031.287
Bartkevicius, A.
113.053
Bartl, E.
124.100
Bartoe, J.-D. F.
034.038
Barua, J.
061.085
Baruschka, G.
022.082
Baryshev, Yu. V.
077.055
Baryshnikova, G. V.
105.150
Baschek, B.
008.047
125.003
126.003
131.099
Basey, T. C.
066.058
Basford, J. R.
094.477
Bash, F. N.
155.065
Basilevski, A. T.
097.030
Basko, M. M.
142.094 .142
Basri, G. S.
114.330
Basu, A.
094.418
Basu, D.
099.038
Basu, S.
084.316
Bates, B.
114.397
131.143 .191 .207
Bateson, F. B.
123.047
Bateson, F. M.
010.024
122.198
123.045 .046
Bath, G. T.
117.014
Batrakov, Yu. V.
007.000
091.035
Battistini, P.
047.037
Battrick, B. T.
012.037
Batueva, N. B.
042.114 .115
052.035
Bauch, A.
046.067
Baud, B.
131.133
Bauder, A.
022.121
Baudinet-Robinet, Y.
022.003
Baudry, A.
131.163
Bauer, J.
123.077
Bauer, J. F.
094.472
Baum, W. A.
022.050
Baumann, M.
121.149
Baumback, M. M.
084.003
Baumert, J. H.
115.008
Baur, T. G.
073.113
Bayard, M.
104.004
Baylis, D. J.
066.176
Baylis, W. E.
022.008
Bazarzhapov, A. D.
084.286 .308
106.034
Bazelyan, L. L.
077.023
Bazet, J. F.
022.059 .060
Bazhinov, I. K.
003.020
Bazin, M.
003.009
Beales, F. W.
105.012 .201
Beals, C. S.
094.182
Beaman, C. J.
009.008
Bean, M.
005.011
Beardsley, W. R.
117.067 .068
119.012
121.113 .114
122.191
124.104
Beattie, D. H.
124.104
Beatty, J. K.
051.012
Beauchamp, R. H.
105.082
Beaudet, G.
065.141
Beaver, E. A.
122.194
Bechis, K. P.
114.328
132.029
Becker, D. G.
104.036
105.051
Becker, G.
035.011
044.012
Becker, G. A.
121.099
Becker, J.
105.146
Becker, R.
142.196
Becker, R. H.
125.027
142.190
Becker, U.
003.021 .022
Beckers, J. M.
031.306
034.027 .124
072.014 .079
080.052
Becklin, E. E.
141.630 .637 .638 .639
155.011
158.056 .070
Beckwith, S.
141.611
Bedenko, V. P.
005.016
Bedford, D. K.
104.016
Bedin, V. N.
041.036
Bednarova-Novakova, B.
085.017
Beebe, H. A.
071.012

Beebe, H. A.
114.033
Beebe, R. F.
092.005
099.224
100.002 .007
Beeckmans, F.
114.092
Beekman, G. W. E.
010.015
054.008
102.011
124.014 .021 .103
Beer, A.
005.004 .007
012.005 .006
Beer, P.
005.004
012.006
Beer, R.
093.043
099.039
Beet, E. A.
010.012
Beetz, M.
131.504
Begemann, F.
105.007 .098 .191 .198
Begkhanov, M.
104.025 .063
Behall, A. L.
041.035
Behannon, K. W.
092.003
Behring, W. E.
022.017
Beichman, C. A.
131.007
Beintema, D. A.
131.199
Bejtrishvili, I. R.
103.100
Bekenstein, J. D.
066.167
Belcher, J. W.
062.016
106.060
Beletskij, V. V.
003.023
052.007
092.013
Belikova, T. P.
097.046
Belinfante, F. J.
162.002
Belinskaya, S. I.
083.046
Belinskij, V. A.
066.066
162.008 .065
Belitskij, A. B.
021.015
Bel'kovich, O. I.
104.074
Bell, A. R.
141.056
Bell, B.
074.079
Bell, G. D.
022.032
Bell, P.
066.008
Bell, P. M.
013.001
Bell, R. A.
064.026 .038
114.039 .100
Bell, T. F.
084.300
Bell Burnell, S. J.
142.016
Belohoubek, E.
033.035
Belokon', E. T.
158.017
Belorossova, T. S.
032.050
Belous, L. M.
103.109 .116
Belousova, O. A.
098.094
Belov, A. I.
021.015
Belov, A. V.
106.031
Belyaev, N. A.
098.068
102.012
103.103 .104 .115
Belyaeva, E. E.
064.020 .021
Belyaevskij, A. I.
142.171
Belyakina, T. S.
122.044 .045
Belyavskaya, V. D.
084.018
Belyj, Yu. A.
004.066
005.008
Ben-Israel, I.
062.048
Benard, J.
035.014
Bendel, R.
098.018
Bender, P. L.
081.037 .040
094.003
Benedict, G. F.
114.005 .313
117.035
160.038
Benest, D.
042.102
Benford, G.
141.328
Bengtsson, H.
123.037
Benkheiri, Y.
094.460
Ben'kova, N. P.
085.024
Benlow, A.
091.036
Bennett, J. A.
004.077
066.065
Bennett, J. C.
010.007
124.104
Bensammar, S.
034.166
Benson, J.
084.278
Bentley, M.
141.115
Benton, E. R.
084.227
Benton Jr., J. L.
100.015
Benvenuti, P.
034.092
Benz, A. O.
033.006
077.011 .031
Bera, K.
066.018
Bera, R.
034.128
Berczi, S.
094.202
Berdnikov, L. N.
103.100
Berendzen, R.
004.011
Beres, H.
160.025
Berezinskij, V. S.
143.005
Berezne, J.
100.008
Berg, O. E.
106.037
Berg, R. A.
122.192
Berge, G. L.
100.025
Berger, A.
042.043
Berger, B. K.
162.084
Berger, D.
103.100
Berger, E. L.
073.158
Bergner, Yu. K.
031.235
034.088
Bergstrahl, J. T.
099.218
Bergstralh, J. T.
101.010
Bering, E. A.
084.025
Berkley, J. L.
105.099
Berko, F. W.
084.252
Berkovich, L. M.
042.120
Berlin, A. B.
079.101
141.116
Berlovich, Eh. E.
061.063
Bernat, A. P.
064.057
114.346
Bernat, T.
066.025

Bernleithner, E.
005.004
Bernot, M.
075.015
Bernstein, J.
034.160
Bernstein, R. B.
022.127
Berriel-Valdos, L. R.
021.024
031.082
Berry, J. S.
131.543
Berry, R.
032.006
Berstein, M. R.
062.024
Bertelli, G.
065.171
Berthelier, A.
106.054
Bertola, F.
158.046 .064
Bertolotti, M.
003.036
Bertotti, B.
012.002
014.026
066.031 .122
Berulis, I. I.
131.517
Beshenova, L. D.
114.093 .107 .109
Bespalov, P. A.
106.067
Bespamyatnykh, N. D.
004.070
Bessell, M. S.
114.392
154.022
Bessey, R. J.
074.108
Besson, J.
035.017
Bessonova, Eh. N.
094.130
Betancourt, O.
097.061
Bethe, H. A.
065.019 .186
Betlem, H.
079.103
096.009
104.087
Bettis, C.
117.072
153.029
Bettis, C. L.
117.031
Bewick, A.
061.067
Bezrodnaya, L. V.
083.077
Bhalla, C. P.
022.022
Bhandari, N.
105.100
Bhandari, S. M.
141.094
Bhatia, A. K.
022.019
Bhowmick, J. B.
061.003
Bialas, V.
004.048
Bianchini, A.
114.323
Bibring, J. P.
105.097
Bicak, J.
063.034
Bieging, J.
131.169
142.147
Biel, J.
034.057
Bielefeld, M. J.
051.010
Bielkowicz, P.
051.002
Biemont, E.
022.003 .103
071.063
Bien, R.
098.027
Biermann, P.
155.014
158.033
Bignami, G. F.
032.518
155.003
Bigourd, C.
093.015
Bijaoui, A.
034.152
Bil'chenko, G. G.
052.028
Bild, R. W.
105.101
Bilinskij, A. I.
084.244
Bilitza, D.
083.008
Binder, A. B.
094.150 .167
Bindmann, W.
003.024
Bingham, R. G.
034.121
082.078
Binz, C. M.
105.027 .138
Biondi, M. A.
082.105
Biraud, F.
033.003
141.009
Birck, J.-L.
105.102
Bird, M. K.
084.219
Birn, J.
084.202
Birulya, T. A.
036.003 .005 .006
Bishop, N. T.
066.061
Bisnovatyj-Kogan, G. S.
065.067 .070 .127
125.040
141.304
142.123 .144
Bisnovatyj-Kogan, G. S.
151.032
Bissiri, D.
032.015
Biswas, S.
106.001
143.047
Bjorkholm, P.
032.520 .542
Bjorkstam, J. L.
036.025
Blaauw, A.
133.030
Black, D. C.
062.010
Black, J. H.
131.023
Blacker, C.
003.025
Blackshear, W. T.
094.204
Blackwell, A. T.
104.067
124.104
Blackwell, D. E.
006.000
008.077
013.010
022.107
Blackwell, K. C.
041.028
Blades, J.
083.004
Blaghikh, A.
154.901
Blair, B. E.
003.026
Blair, G. N.
131.056 .059 .077
Blake, J. B.
061.077
065.042
Blakely, R. J.
084.306
Blakeslee, B.
103.100
Blakey, J. R.
077.053
Blamont, J.
082.051
100.027
Blamont, J.-E.
031.208
Blanchard, D. P.
094.438
Blanchard, M. B.
104.049
105.082
Blanchard, P.
022.106
Blanco, B. M.
103.125
Blanco, C.
099.217
Blanco, V. M.
114.078
Bland, R.
066.080
Blander, M.
105.017 .046 .103

Blandford, R. D.
063.013
Blansjaar, P. W. H.
034.147
Blasberg, H.-J.
123.067 .068 .069
Blasius, K. R.
097.020
Blazit, A.
031.283
Bleach, R.
142.139
Blecha, A.
031.287
Bleeker, J. A. M.
142.166
Bleiweiss, M. P.
009.009
Bless, R. C.
121.034
Blinnikov, S. I.
065.070 .127
Blinov, N. S.
044.006
141.071
Blint, R. J.
131.176
142.184
Block, D. L.
066.036
Bloete, H. W. J.
031.022
Blokh, G. M.
080.076
084.274
Blokh, Ya. L.
082.090
Bloomberg, H.
073.039
Blum, E. J.
033.003
Blum, F. A.
033.050
Blunck, J.
097.026
Boal, D. H.
066.168
Bobrova, N. A.
160.024
Bobrovnikova, O. V.
004.059
Bocharkin, V. K.
142.171
Bochkarev, N. G.
117.048
142.169
Bochner, S.
005.008
Bochonko, D. R.
124.018
Bodansky, D.
061.048
Bodenheimer, P.
062.010
Bodifee, G.
051.013
Boedecker, G.
081.041
Boehm, E.
033.058
Boehm, K. H.
126.014
132.004
Boehm, S.
075.017
Boehme, A.
075.017
Boehme, D.
124.103
Boehme, S.
002.009
Boella, G.
032.518 .519
Boenigk, Z.
055.010
Boenkova, N. M.
083.031
Boerner, G.
066.104
Boerngen, F.
031.228
103.100
158.229 .237
Boesgaard, A. M.
114.034 .102
Boesgaard, H.
114.034
Bogart, R. S.
080.068
082.009
Bogdanov, A. V.
097.048
Bogdanov, M. B.
122.059
Bogdanova, L. N.
061.064
Bogdanova, O. A.
078.016
Bogod, V. M.
073.083
Bogolepov, A. I.
010.033
Bogomolov, Yu. V.
062.061
Bogorodskij, A. F.
005.029
066.063
Bogoyavlenskij, O. I.
061.018
Bohannan, B.
117.032
153.004
Bohlin, J. D.
073.050
074.032 .036
Bohlin, R. C.
122.096
131.043
132.053
133.012 .031
Bohme, D. K.
022.063
Bohn, H. U.
080.015
Bohnenstengel, H.-D.
142.020
158.034 .188
Boischot, A.
033.013
Boitnott, C. A.
104.037
Bojanowski, J.-A.
103.100
Bojko, E. S.
004.115
Bokov, V. L.
142.145 .171
Boksenberg, A.
034.132
125.107
131.143 .207
133.006
141.018 .159
158.141
Bolbochanu, A. V.
112.005
Boldt, E.
142.196
Boldt, E. A.
061.017
142.015 .025 .097 .120 .153 .154 .155 .190 .202 .203 .227 .230 .239
Boldt, G.
034.056
Boley, E.
142.212
Bolger, P. H.
003.027
012.051
Boll, W.
005.005
Bologna, J. M.
131.003 .518
Bolotnikova, Ya. A.
052.060
Bol'shakova, G. I.
085.020
Bolt, B. A.
081.006
Bolton, A.
080.040
Bolton, C. T.
121.040
122.097
142.055
Bolton, J. G.
141.114
Bonanomi, J.
035.009
Bonazzola, S.
066.079
141.352
Bond, H. E.
121.122
154.026
Bondarenko, I. I.
121.010
Bondarenko, L. N.
008.072
094.145 .146
Bondarenko, V. I.
084.256
085.019
106.071
Bondarev, B. V.
066.107
Boninsegna, R.
121.149
Bonneau, D.
031.295

Bonnet, R. M.
051.016
Bonneville, T.
121.149
Bonnor, W. B.
066.078
Bonoli, F.
124.104
Bonometto, S. A.
162.023
Bonov, A.
047.022
Bonov, A. D.
015.006
Bonsack, W. K.
119.010
Booker, H. G.
084.204
Bookmyer, B. B.
121.109
Bopp, B. W.
114.386
118.021
119.018 .019
Borchkhadze, T. M.
011.063
Borczuch, M. S.
061.046 .081
Borg, J.
094.187
Borgman, J.
113.071
Borgnino, J.
031.305
Boris, J.
073.039
Borisov, Yu. V.
124.104
Borkowski, K.
033.011
077.061
Borkowski, K. M.
141.166
Born, G. H.
097.202
Borodina, E. G.
063.031
Borra, E. F.
116.004 .013 .014
126.017
141.019
Bortle, J. E.
103.100 .130 .131
124.103 .110 .111
Bortnik, S. G.
072.045
Borzelli, C. V.
010.001
Borzeszkowski, H.-H. V.
See Von Borzeszkowski, H.-H.
Bosel, J. P.
034.134
Bosma, A.
158.167
Botsula, R. A.
117.040
Bottcher, C.
131.205
Bottinelli, L.
142.215
Botvinova, V. V.
094.142
097.023 .045
Boucher, R.
034.030
Boughn, S. P.
066.025
Boulesteix, J.
158.157
Boulet, B.
033.076
Boulet, C.
022.030
Bourdonneau, B.
114.319
Bourgeois, J.
121.149
Bourot, M.
105.104
Boury, A.
065.021 .097 .100 .176
080.063
Bouska, J.
011.058
047.025
079.103
099.079
103.003
Bousquet, P.
031.206
Bovkun, V. P.
141.122
Bowell, E.
093.014
098.046
Bowen, R.
094.479
Bower, J. F.
094.437
Bowers, P. F.
141.601 .642
Bowin, C.
094.176
Bowling, S. B.
084.318
Bowman, B. R.
082.092
Bowyer, C. S.
142.028
Bowyer, S.
032.515
061.006
125.021
131.064
142.134
155.058
158.309
Boyarchuk, A. A.
034.059
064.043
114.057
122.041
Boyarkina, A. P.
105.031
Boyd, J. S.
084.029
Boyd, R. L. F.
003.028
008.064
Boyer, C.
093.014
Boyko, A. D.
154.029
Boynton, P. E.
142.057
Boynton, W. V.
094.111
105.116
107.017
Bozis, G.
042.046
Bozkurt, S.
121.063
Brabban, D. H.
076.010 .023
Brace, L.
083.071
Brace, L. H.
083.023
Bracewell, R. N.
077.041
Bradfield, W. A.
103.136
Bradford, H. M.
061.010
074.097
Bradt, H.
142.075 .078 .081 .083
.187 .188
Bragin, Yu. A.
082.089
Braginskij, V. B.
066.022 .072
Brahic, A.
091.042
100.001
151.025
Branch, D.
064.038
117.072
Branch, D. R.
117.031
Brandie, W.
031.416
Brandshaft, D.
132.003
Brandt, J. C.
074.011
082.002 .116
103.102
132.020
Brandt, P. N.
034.010
Brandt, S. B.
105.087
Brandt, V. Eh.
041.027
Branduardi, G.
142.228
Branley, F. M.
003.029
Brannon, J. C.
094.438
Brasletov, V. A.
034.072
Bratijchuk, M. V.
041.007 .008
055.013
Bratolyubova, L. S.
014.014

Bratolyubova-Tsulukidze, L. S.
142.168 .180
143.074
Brault, J. W.
071.005
082.005
Braun, M.
022.037
Braun, W. Von
See Von Braun, W.
Braunfel'd, S. G.
034.109
Braverman, N. R.
034.116
Bray, M.
162.069
Bray, R. J.
071.061
Braz, M. A.
131.534
141.103
Breakiron, L. A.
118.005
Brecher, A.
105.052 .061 .105
Brecher, K.
065.166
126.021
142.092 .199
Breckinridge, J. B.
031.016
Bredekamp, J. H.
032.513
Bredohl, H.
022.111 .112 .113 .114
Breger, M.
114.045
122.033 .081 .090 .136
Bregman, J. D.
034.035
100.013
133.014
Bregman, J. N.
122.081
Brejdo, I. I.
036.007 .022 .023 .024 .026
Brennan, W. J.
094.103
Brenner, P.
105.118
Bretagnon, P.
042.023 .062
Brett, R.
081.046
094.402
Bretterbauer, K.
045.014
Breus, T. K.
106.058
Brewster, M.
123.030
124.103
Brice, N.
084.222
099.007
Bridgeland, M. T.
034.146
Bridges III, C. A.
073.071
Brierley, D. M.
082.038
Briggs, C.
094.421
Briggs, F. H.
091.043
Briggs, P.
045.001
Briggs, P. L.
105.061
Brillet, W-Ue. L.
022.115
Brinkman, A.
142.230
Brinkman, A. C.
064.048
122.167
142.014 .045
Briot, D.
114.053 .319 .901
Briotta Jr., D. A.
034.098
Britkov, V. B.
052.037
Broadfoot, A. L.
053.017
093.025
Broche, P.
083.018
Brocklehurst, M.
031.302
Broder-Bursztyn, F.
031.013
Broderick, J. J.
141.059 .128 .143
Brodkorb, E.
011.025
Broemme, A.
035.019
Broglia, P.
122.161
Bromander, J.
022.037
Bronkalla, W.
031.228
Bronnikov, K. A.
066.126
Bronnikova, N. M.
041.004 .060
Bronshtehn, V. A.
010.033
082.086
097.032
104.058
Broo, H.
105.171
Brooks, J. W.
033.021
Broomhall, H.
103.100 .131
Broqua
041.013
Broucke, R.
042.048 .092
Brouw, W. N.
031.288
Browder, J. S.
031.040
Browell, E. V.
022.039
Brown, C. M.
022.096 .116
Brown, D. A.
121.112
Brown, D. R.
034.039
Brown, G. E.
105.045
Brown, G. M.
094.467
Brown, G. S.
158.243
Brown, J. C.
073.012 .102
076.015 .016 .022 .027
Brown, M. L.
143.043
Brown, R. A.
099.209
Brown, R. E.
065.182
Brown, R. Hanbury
See Hanbury Brown, R.
Brown, R. L.
131.513 .515 .521 .548
132.030
141.633
142.060
Brown, R. T.
061.025
Brown, T. M.
080.030
Brown, W. N.
033.080
Brown IV, H. G.
105.099
Browne, I. W. A.
141.063 .114
Brownlee, D. E.
094.116
105.082
106.040
Brubak, H.
096.008
Bruce, C. E. R.
061.058
Bruce, C. F.
031.054
Brueck, M. T.
159.005
Brueckner, G.
073.037 .039
Brueckner, G. E.
034.038
073.005
074.009
Bruhweiler, F. C.
114.005 .103
Bruin, F.
075.018
Brumberg, V. A.
007.000
Bruna, P. J.
022.042
Bruner, F.
051.001
Bruner Jr., E. C.
071.008 .028
072.062
073.132 .133

Brunet, J. P.
159.001 .004
Bruzek, A.
007.000
Brzostkiewicz, S. R.
015.039
097.205
Bubnova, S. N.
034.073
Buccheri, R.
031.423
Buchdahl, G.
004.003 .048
Bucher, W.
106.040
Buchler, J.-R.
125.029
Buchwald, V. F.
105.106 .139
Buckley, R. J.
103.100 .134 .136
Budding, E.
121.027 .147
Budzinski, E. E.
106.062
Budz'ko, V. K.
096.017
Buff, J.
125.024 .034
142.075 .078 .081 .083
.187 .188
Buffington, A.
143.002 .050
Buffoni, L.
031.215
Bugaenko, O. I.
097.023
Bugakov, I. F.
142.171
Buhl, D.
022.056
131.052 .080 .092
Bulanov, S. V.
078.020
143.007 .046
Bulashevich, Yu. P.
094.139
Bulger, J.
103.112
Bulger, J. H.
103.005 .104 .122 .128
.134
Bull, R. K.
105.122
Bullen, K. E.
003.030
Bumba, V.
013.005
072.011
080.051
Bunch, T.
105.169
Bunch, T. E.
105.090 .091
Bunker, S. S.
064.025
122.189
Buonanno, R.
154.038
Burba, G. A.
011.036
Burba, G. A.
094.136
Burbidge, E. M.
126.013
141.085 .106
158.031
Burbidge, G.
141.051
Burbidge, G. R.
003.005
141.030 .043 .100 .106
158.031 .140
Burbidge, M.
141.051
Burch, J. L.
106.076
Burch, J. M.
003.052
Burchi, R.
098.021
121.066
Burdjuzha, V. V.
See Burdyuzha, V. V.
Burdyuzha, V. V.
131.151
141.603
143.067
Burger, J. J.
032.518
Burger, M.
114.361
Burgess, R. D.
158.206
Burginyon, G. A.
125.007 .008
142.059
Burke, B. F.
131.121 .158
141.014 .032
Burke, T. E.
053.018
Burkhead, M. S.
141.119
158.098 .206
Burki, G.
152.003
153.014
Burkley, C. J.
033.044
Burlaga, L. F.
074.059
Burman, R. R.
131.102
Burn, B. J.
061.080
Burnage, R.
034.094
Burnasheva, B. A.
098.004
Burnett, D. S.
094.123 .476
Burnichon, M. L.
113.066
Burns, J. A.
098.013
Burragato, F.
105.107 .108
Burrell, D. A.
034.110
Burs, L.
082.058
Bursa, M.
081.001 .038
094.001 .004 .171
Burshtejn, E. I.
042.069 .082
Burton, J.
094.193
Burton, R. K.
074.116
084.230
Burton, W. B.
131.117
155.061 .063
Burton, W. M.
034.136
051.006
114.399
Bus, S. J.
098.051
103.100
Busch, H.
123.065 .070
Buschmann, E.
044.008
Buscombe, W.
065.054
114.049 .105
Bushuev, E. I.
052.050
082.128
Busman, J. D.
073.158
Busse, F. H.
084.228 .233
Bussoletti, E.
141.627
Bustati, N. G.
075.018
Butcher, H. R.
065.018
Butenko, V. D.
083.034
Butler, D.
122.073
154.012
Butler, D. S.
122.223
Butler, P. W.
131.120
Butler, R. F.
094.124
Butslov, M. M.
034.081 .084
Butterworth, E. M.
066.053
Butvin, G. G.
083.037
Butz, M.
031.202
073.144
Buurman, P.
085.023
Buvidajte, G. A.
104.088
Bybee, R. L.
034.023
Bykov, O. P.
052.011
Byl, J.
042.075

Byrd, G. G.
158.205
Byrne, P. B.
142.091
155.049
Bystrov, N. F.
041.012

Caby-Eyraud, M.
062.018
Cacciani, A.
034.106
Caccin, B.
014.029
073.152
Cadez, A.
066.166
Cadez, M.
082.071
Cadogan, P. H.
094.458
Cagnet, M.
031.006
Cahill Jr., L. J.
084.247 .252 .303
Cahn, J.
133.006
Cahn, J. H.
133.034
Caillere, S.
105.109
Caimanque, L.
082.117
Cain, J. C.
084.232
Calamai, G.
063.046
Calame, O.
042.090
Calder, D.
010.024
105.200
Caldwell, J.
099.202
Callahan, P. S.
074.001
141.035
Callebaut, D. K.
061.035
Calvert, D. A.
008.043
Calvert, T. A.
094.151
Calvo, M.
042.036
Camenzind, M.
066.112 .116
Camenzind, M. A.
066.116
Cameron, A. G. W.
065.119
099.036
107.006 .012
142.128
162.064
Cameron, I. R.
105.048
Cameron, W. S.
094.165
Camichel, H.
093.014
Camisa, R. L.
033.053 .068
Campbell, B.
034.104
124.104
142.204
Campbell, M. F.
142.245
Campbell, R. J.
031.412
Campusano, L.
114.310
Campusano, L. B.
116.015
Canal, R.
061.016
Canavaggia, R.
122.133
Candy, M. P.
103.136
Cane, H. V.
157.002
Canizares, C.
142.084 .089 .187
Canizares, C. R.
142.010 .047 .098 .117
Cannon, C. J.
063.020
071.003
Cannon, R. D.
041.063
103.126
141.156
Cantelli, M.
015.013
Canterna, R.
154.009
Cantu, A. M.
022.080
Cantu, S.
103.100
Canuto, V.
061.043 .087
065.119 .194
Capaccioli, M.
034.092
041.052
125.048
158.044 .046 .051 .064
Capitaine, N.
044.005
045.023
Caplan, J.
034.022
Capolongo, V.
092.016
Capone, L. A.
099.018
Caporaso, G.
065.166
Capps, R. W.
131.514
155.020
Caputo, F.
122.162 .217
Caputo, M.
091.045
Carapezza, M.
105.110
Carbon, D.
114.302
Cardini, D.
142.206
Carelli, P.
066.087
Carey, J. V.
041.028
Cariou, J.
022.010
Carlesi, C.
073.154
Carleton, N. P.
093.030
Carli, B.
082.041
Carlson, J. B.
158.088
Carlson, R.
032.013
Carlson, R. W.
099.242
Carmichael, D.
124.103
Carnevali, P.
121.149
Carnochan, D.
133.006
Carnochan, D. J.
114.027 .118
158.036
Caroff, L.
061.056
Caroff, L. J.
032.517
082.021
117.002
132.025
Carpenter, D. L.
084.302
Carpenter, G. F.
142.019 .023 .225
Carpenter, M. S.
153.010
Carr, B. J.
162.016
Carr, D. M.
031.412
Carranza, G.
158.185
Carrasco, G.
041.036
Carrasco, L.
131.198
153.024
Carruthers, G. R.
034.024
103.102
113.061
132.055
Carson, P. P. D.
114.397
131.143 .207
Carstensen, E.
121.042
Carswell, R.
141.160
Carswell, R. F.
141.107
Carta, F.
031.215

Carter, B.
065.136
Carter, W. E.
009.024
Casari, G.
103.131
Caser, S.
162.028
Cash, W.
155.058
Cassatella, A.
031.293 .422
114.010
Casse, M.
143.010
Cassidy, W.
094.122 .401
Cassinelli, J. P.
064.050 .053 .059
113.060
Castellani, V.
122.162 .217
154.901
158.037
Castelli, J. P.
077.039
Casti, J.
063.052
Castor, J.
064.015
Castor, J. I.
131.070
Caswell, J. L.
125.044 .045 .046
131.147
141.101 .117 .146
Catalano, F. A.
116.003
Catalano, S.
099.217
Catenacci, R.
066.122
Catherinot, A.
062.020
Catura, R. C.
076.007
142.079 .133
Caty, J.-L.
105.111
Caughlan, G. R.
022.065
Caulk, H. M.
104.006
Cavallo, G.
022.076
142.119
Cavarretta, G.
105.107 .108 .112
Cavedon, M.
014.023
Cayrel, G.
114.051
Cayrel De Strobel, G.
114.340
Cecchini, S.
078.037
143.080 .083 .084
Celaya, M. L.
031.076
Celis S., L.
122.086
Celnikier, L. M.
141.352
Cendales, M.
105.176
Ceplecha, Z.
104.002 .038
Cerdonio, M.
066.087
Certain, P. R.
022.127
Cesco, M. R.
103.125
Cess, R. D.
022.104
091.016
099.075 .100
Cester, B.
121.005
122.153
Chaffee Jr., F. H.
099.209
131.012
Chaisson, E. J.
131.007 .519
132.005 .024 .026
Chalikov, D. V.
093.016
Chamaraux, P.
158.131
Chamberlain, J. W.
005.020
091.001
Chambliss, C. R.
121.067 .099
Chameides, W.
082.108
Chan, J. H.
143.051
Chan, K. L.
061.027
125.056
143.008 .009
Chan, T. T.
065.146
Chanan, G.
141.312 .313
Chandra, S.
074.012 .126
082.026
083.071
Chandrasekhar, S.
066.170 .173
117.061
Chang, C. T.
036.025
Chang, E. S.
081.024
Chang, S.
162.068
Chanin, M. L.
082.011 .020
Chanturiya, S. M.
034.109
Chao, J. K.
106.077
Chapman, C. R.
097.201
098.017 .032
Chapman, G. A.
071.002
072.013
Chapman, R. D.
072.081
074.018 .065
Chapman, S.
003.010
Chappell, C. R.
083.020
Charakhch'yan, A. N.
143.037
Charakhch'yan, T. N.
078.028
143.037
Charette, M. P.
094.409 .446
Charikov, A. V.
155.070
Charles, P. A.
125.016
134.008
142.064 .139 .158 .173
.233
160.006
Chartrand III, M. R.
003.029
Charugin, V. M.
061.053
Chase, R.
073.056
074.073
Chase, R. C.
032.512
074.008
Chashej, I. V.
074.057
Chatfield, C. N.
010.008
Chau, W. Y.
066.136
117.059
Chavira, E.
122.148
Chavushyan, O. S.
122.105
Chebotarev, G. A.
042.014
098.001 .031 .068
102.012
Chechetkin, V. M.
065.067 .150
066.159
Chechev, V. P.
162.062
Chekalev, S. P.
033.071 .072
Chekhoshvili, B. Ya.
083.041
Chen, A. J.
084.401
Chen, C. T.
099.070
Chen, C.-K.
084.314
Chen, J. H.
105.113
Chen, P.
124.103 .104 .109
Chen, S. C.
099.075
Chenakal, V. L.
004.051
008.050

Chenakal, V. L.
009.020
Chenette, D. L.
099.021
Cheng, C. C.
073.019 .045 .046 .079
Cheng, C-N.
105.026
Cherednichenko, V. I.
102.018 .019
Cherepashchuk, A. M.
114.369
117.004
121.044
142.054
Cheriguene, M. F.
151.060
Cherkasov, I. I.
003.031
094.172
Chernaya, I. M.
097.030
Chernin, A. D.
065.115
Chernis, K.
103.100
Chernitsov, A. M.
052.041
Chernogor, L. F.
062.062
Chernomordik, V. V.
155.051
Chernov, G. P.
077.022 .043
Chernova, T. A.
084.242 .243
Chernyak, Yu. B.
094.147 .148
Chernyj, M. L.
066.046
Chernykh, L. I.
032.031
034.004
098.004 .066 .067
Chernykh, N. S.
032.031
098.004 .067
103.100
Chernyshev, V. I.
132.047
Chernyshev, V. S.
034.058
Chernysheva, S. P.
083.062
Cheron, C.
032.541
Chertkov, A. D.
085.008
Chertok, I. M.
077.022
Chertoprud, V. E.
072.029 .031
085.021
Chesley, D.
132.002
Chesley, S.
132.002
Chesnokov, Yu. M.
032.533
Chetanneau, A.
121.149
Cheung, A. C.
131.518
Chevalier, C.
142.228
Chevalier, R. A.
125.017
134.002
142.002
143.065
Chevreton, M.
066.079
Chia, T. T.
066.136
Chien, T. H.
074.130
Chien, W. S.
061.061
Chigorin, A. N.
105.067
Chilingarashvili, S. P.
082.101
Chincarini, G.
160.015 .901
Chincarini, G. L.
003.032
Ching, B. K.
082.010 .036
Chinnery, D. N. W.
031.073
Chinnery, M. A.
081.007
Chiosi, C.
065.096 .171
Chipman, E.
071.008
Chipman, E. G.
071.028
072.062
073.132 .133
Chirkov, N. P.
034.091
078.024
143.027
Chis, G. H.
122.050
Chistov, I. I.
032.534
Chistyakov, V. F.
072.026
Chistyakov, Yu. N.
034.074
Chistyakova, K. G.
072.026
Chitre, S. M.
066.901
080.034
Chiu, H. Y.
114.330
Chiu, Y. T.
082.010
083.057
Chiuderi, C.
012.026
080.021
Chizhov, G. A.
066.130
Chkhikvadze, Ya. N.
114.074
Chlistovsky, F.
031.215
Chmielewski, Y.
071.005
Chochia, P. A.
097.046
Chollet, F.
044.002 .033
Chong, K. F.
084.284
Choquet-Bruhat, Y.
012.021
Chou, C.-L.
094.111 .413
Chovitz, B. H.
046.004
Chown, E. H.
105.111
Christensen, A. B.
082.040
Christensen, E. J.
097.007
Christensen, R. B.
142.184
Christensen-Dalsgaard, J.
065.099
Christiansen, W. A.
122.143
160.039
Christides, T.
042.038 .046
Christie, G.
031.268
Christie, J.
033.066
Christophe-Michel-Levy, M.
105.104 .114
Christopher, L.
092.014
Christy, J. W.
041.035
Christy, R. D.
054.010
Christy, R. F.
122.113
Christy-Sackmann, I.-J.
065.102
Chrupala, H.
009.026
Chrzanowski, P. L.
066.171
Chu, K. W.
131.563
Chu, S.
022.014
131.001
Chuchkov, E. A.
078.018
Chugaj, N. N.
125.011 .105
Chugajnov, P. F.
122.156 .177
Chui, M. F.
131.062 .518
Chujkin, E. I.
142.171
Chujkova, N. A.
094.173
Chumak, O. V.
065.148
151.073 .076

Chumak, Yu. V.
104.084
Chumak, Z. N.
151.071 .076
Chunakova, N. M.
113.064
Chuntonov, G. A.
034.060
Chupova, L. M.
076.047
Chupp, E. L.
073.015
076.043
Chuprina, R. I.
114.348
Chuprina, V. P.
126.022
Church, S. E.
105.019
Churchwell, E.
131.029 .082 .164
Churyumov, K. I.
103.100
Chuvaev, K. K.
034.082
114.357
158.032
Chylek, P.
063.042
Ciatti, F.
114.012
122.042
125.048 .106
Ciccone, M. A.
113.065
Ciccoritti, L. C.
003.065
Cid, R.
042.036
Cilliers, W. A.
062.009
Cimino, M.
007.000
075.019 .020
Cisowski, S. M.
105.115
Citterio, O.
124.104
Ciurla, T.
073.153
Clairemidi, S.
158.067
Clark, D. G.
141.117
Clark, D. H.
083.064
125.044 .045 .046 .047
141.101
142.160
Clark, F. O.
132.016 .057
Clark, G.
142.076 .078 .081 .084 .089
Clark, G. W.
142.009 .036 .047 .185 .205 .224
Clark, J. W.
065.192
Clark, M.
103.100
Clark, P.
094.185
Clark, R. G.
061.061
Clark, R. N.
099.204
Clark, T.
121.039
Clark, T. A.
031.218
034.110
071.048
141.049 .050 .143
Clark Jr., A.
065.081
Clark Jr., S. P.
094.009
Clarke, D.
031.226
034.013
142.067
Clarke, J.
033.036
Clarke, J. N.
004.098
131.153
Clarke Jr., R. S.
105.006
Clarke Jr., S. P.
093.055
Clary, R. S.
099.002
Classen, J.
011.023
Claxton, D. H.
033.037
Claydon, B.
033.054
Clayton, D. D.
061.007
080.003 .037
107.001
124.020
Cleary, M. N.
141.152
Clegg, P. E.
031.210
132.042
Clements, E. D.
142.104
Clifton, K. S.
104.035
Climenhaga, J. L.
114.029
Clish, D. V.
014.007
Clouet, B.
010.028
Cloutman, L. D.
065.041
Clovin, J.-P.
121.149
Coates, R. J.
034.161
Cobb, C. L.
096.005
Cochran, W. D.
099.080
Cocke, W. J.
134.007
Code, A. D.
158.217
Codina, J. M.
103.100 .131
Coe, M. J.
061.067
142.019 .023
Coffaro, P.
031.423
Cogan, B. C.
065.112 .177
Cogley, A. C.
063.005
Cohen, B.
004.018
Cohen, G. G.
073.137
Cohen, J. G.
065.053
158.103
Cohen, J. M.
065.110
066.052
141.334
Cohen, L.
022.017
151.001
Cohen, M.
065.169
083.073 .074
113.022 .045
114.040 .349
133.018 .029
Cohen, M. H.
033.007
141.073 .074 .110
Cohen, R. H.
074.053 .096
Cohen, S. C.
034.161
Cohn, J.
066.113
Colazo, J. R.
123.075
Colburn, D. S.
084.297
106.019
Cole, D. J.
141.146
Cole, T. W.
033.022
Coleman, G. D.
122.088 .197
Coleman, P. L.
142.165
Coleman Jr., P. J.
074.119
Colgate, S. A.
032.013
125.015
Collin-Souffrin, S.
158.135
Collins, J. G.
064.055
Collins, L. A.
158.210
Collins, P.
103.100
Collins, T. F.
121.155

Collinson, D. W.
094.423
Collinson, E. H.
010.012
123.005
Colomba, G.
031.423
Colombo, G.
005.007
Colvin, J. D.
065.051
141.108
Combes, F.
162.018
Combes, M.
099.083
100.008
Combes, M.-A.
098.012 .042 .080
Combi, M. R.
103.101
Comello, G.
118.024
Compagno, G.
074.077
Compston, W.
094.443
Compton, H. R.
094.204
Conconi, P.
122.161
Condon, J. J.
072.010
141.059 .128 .141
Conel, J. E.
094.169
Conklin, E. K.
131.078
141.048
Conlon, T. F.
099.021
Connes, P.
031.209 .282
034.028
Connolly, L. P.
122.092
Connor, B. V.
099.099
Conrad, R. L.
105.116
Conradi, J.
034.026
Contadakis, M. E.
122.146
Conti, G.
124.104
Conti, P. S.
114.315
117.032
Contopoulos, G.
042.024
151.044 .047
Conway, R. G.
141.088
Cook, A. F.
012.030
104.034 .042 .043 .046
Cook, A. H.
042.011
091.006
Cook, G. E.
052.029
Cook, M. W.
066.064
Cook II, A. F.
100.002
Cooke, B. A.
142.046 .048 .049
Cooke, D. J.
141.009 .146 .152
Cooper, B. F. C.
033.021
141.009
Cooper, J.
022.024 .027 .028 .029
Cope, P.
015.031
Corbin, T. E.
153.010
Cordara, F.
044.015
Cordes, J. M.
141.329
Cordova, F.
061.021
Corliss, C. H.
003.092
Corliss, W. R.
003.033
Cornejo, A.
031.077 .081 .083
034.126
Cornell, J.
003.034
054.002
Cornet, R.
022.112 .114
Coroniti, F. V.
099.008 .009 .027 .030
Corrado, L. C.
160.003
Correas, J. M.
042.036 .078
Correra, L.
073.152
Corsi, E. S.
100.016
Corsi, P. S.
100.016
Corso, G. J.
114.028
141.139
Cortellessa, P.
073.028
Cortes, M.
099.090
Corwin Jr., H. G.
161.007
Costa, E.
142.206
Costa-Morales
118.033
Costabel, P.
004.022
005.008
Costero, R.
158.091
Cotter, C. H.
004.076
021.005
Cotton, W. D.
141.140
Coulaud, G.
062.018
Coulson, K. L.
003.035
082.019
Counselman III, C. C.
046.007
Countee, C.
084.255
Courtes, G.
012.013
032.501 .505
113.073
Courts, G. R.
131.207
Couteau, P.
006.000
118.034 .037
Coutts, C. M.
154.034
Covington, A. E.
005.033
009.022
Cowan, J. J.
061.030
103.102
Cowie, L. L.
125.036
158.090 .134
Cowles, P. R.
033.029
Cowley, A. P.
114.315 .337
119.013
142.115
Cowley, C. R.
114.043 .355
Cowley, S. W. H.
062.034
Cowsik, R.
161.001
Cox, A.
084.231 .306 .307
Cox, A. N.
122.116 .184 .189 .190
Cox, D. P.
022.053
062.059
131.185
Cox, J. P.
065.089 .093 .121
122.114
Coxeter, H. S. M.
005.008
Coyne, G. V.
114.341
Craft Jr., H. D.
142.215
Cragg, T. A.
010.001
Craig, I.
073.010
Craig, J.
103.100
Craine, E. R.
141.107
158.308 .316 .318
Cram, L. E.
080.055

Crampton, D.
142.115
153.013
Crane, C.
103.100
Crane, P.
161.003
Crane, P. C.
158.088
Crannell, C. J.
073.108
Cravens, T. E.
022.001
Crawford, D. F.
125.045
141.059
Crawford, D. L.
113.020 .063
154.013
Crawford, J.
082.043
Crawford, R. C.
121.037
Creekmore, S. P.
082.010
Cremers, C. J.
094.449
Creze, M.
155.037
Cripe, J. D.
105.117
Cristaldi, S.
122.012 .207
Criswell, D. R.
012.031
094.158
Croce-Spinelli, I. F.
066.030
Crochet, M.
083.018
Cromwell, R. H.
158.209
Cronin, S. E.
105.026
Cronyn, W. M.
141.042
Crooker, N. U.
084.236 .305
Crosignani, B.
003.036
Crosley, D. R.
022.006
Crossley, D. J.
081.008
Crovisier, J.
131.163
Crowther, J. H.
141.063
Crozaz, G.
094.115 .408
Cruddace, R.
142.134
Crutcher, R. M.
033.017
131.087 .108 .156 .181
Crutzen, P. J.
078.003
085.018
Cruvellier, P.
032.503
Cruz-Gonzalez, C.
133.010
155.027
Csernai, L. P.
065.193
Cucchiaro, A.
036.017
Cudworth, K.
133.007
Cudworth, K. M.
112.009
154.031
Cuffey, J.
008.059
104.010
Cugusi, L.
010.014
041.011
Cui, Z.
078.012
Culhane, J. L.
076.007
099.016
125.016
134.008
142.064 .139 .173 .233
160.006
Culp, R. D.
052.042
Culver, R. B.
113.016
Cumings, N. P.
072.015
Cummings, W. D.
084.255
Cunningham, C. T.
066.137
Cunningham, G. C.
105.082
Cuny, Y.
073.031
Cuperman, S.
084.311
Currie, D. G.
031.243 .244
Currie, R. G.
045.003
Curtis, D. B.
094.407
Curtis, L. J.
022.067
Curtis, S. A.
092.008
Cutolo, M.
007.000
Cutrera, M.
105.105
Czarnecki, K. A.
046.031
Czyzak, S. J.
133.011
Da Silva, L.
064.013
Dabagyan, A. V.
021.015
Dabizha, A. I.
105.044 .077
Dabrowski, J. P.
118.005
Dachev, Ts.
051.008
Dachs, J.
006.000
113.001
121.018
Dacic, M.
032.026
041.033 .067
045.008
Dadaev, A. N.
121.119
D'Addario, L. R.
099.087
141.022
Dagaev, M. M.
003.037
Dahn, C. C.
041.035
Daigne, G.
033.013
077.007
Dainty, A. M.
094.177
Dalgarno, A.
022.001 .014 .043
131.001 .023 .045 .209
Dall, H. E.
005.026
Dallaporta, N.
107.010
Dall'Oglio, G.
082.122
Dall'Olmo, U.
079.005
Damboldt, T.
084.323
Damle, S. V.
011.042
Damour, T.
066.110
Dampierre, F.
034.166
Danford, S.
114.044
D'Angelo, N.
084.293
Daniel, M. A.
098.058
099.237
103.100 .121 .123
Daniel, R. R.
011.043
143.006
Danielson Jr., G. E.
092.004
Danielyan, Eh. Kh.
063.023
Danilov, A. D.
083.028 .059
Danilov, A. T.
052.062
Danilov, Yu. A.
005.008
Danks, A. C.
103.100
131.033
141.641
Danloux-Dumesnils, M.
003.038

D'Antona, F.
126.004 .008 .015 .016
154.038
Danziger, I. J.
133.030
Darnell, P.
007.000
034.051
Darnell, Per
034.105
Darnell, Poul
034.105
Dartyge, E.
143.073
Das, B. N.
033.023 .067
Das, M. B.
033.051
Das, P. A.
141.133
Das, P. K.
141.021
Daskalova, M.
046.032
Datlowe, D. W.
073.072
076.012 .013 .028 .030
Datta, B. K.
066.018
Daub, C. T.
133.004
Daube, I.
004.118
123.015
Daugherty, J. K.
061.068
Dautcourt, G.
066.150
Dauvillier, A.
065.075
Davey, W. R.
065.094
David, E. J. H.
091.037
Davidsen, A.
141.313
142.182 .183 .218
158.309
Davidson, G. T.
084.402
Davidson, K.
064.046
101.007
Davies, D. E. N.
033.041
Davies, K.
083.076
Davies, M.
097.026
Davies, R. W.
066.077
080.006
Davila, H.
041.016
Davis
142.215
Davis, A. E. L.
005.008
Davis, C. G.
122.116 .189
Davis, D. R.
097.201
098.032
Davis, J.
062.029
073.037 .038 .039
074.073
142.095
Davis, M.
158.077
Davis, R.
080.044
Davis, R. J.
131.212
142.103
Davis Jr., C. G.
064.025
122.115
Davis Jr., L.
099.066
Davis Jr., R.
080.062
Davison, P. J. N.
142.063 .068 .158
160.006
Davudov, Yu. D.
101.009
Davydov, V. D.
091.033
Davydov, V. S.
034.072
Dawson, R. H.
033.038
Day, C. A.
154.032
Day, G. A.
141.152
Day, K. L.
131.024
Day, R. W.
116.005
De, Bibhas R.
107.014
De, M.
031.088
De, U. K.
162.070
De Amicis, R.
154.038
De Angelis, U.
065.197
De Batz, B.
034.166
De Boer, K. S.
142.207
De Cesare, L.
065.197
De Cuyper, J. P.
117.018
De Feiter, L. D.
076.017
De Felice, F.
066.138
De Freitas Mourao, R. R.
103.130
De Graaf, T.
162.025
De Greve, J. P.
117.001 .018 .020 .063
142.107
De Groot, T.
031.281
124.016
De Jager, C.
012.041
064.005 .071
074.054
076.059
092.009
093.013 .017
099.051 .222
114.090 .117
124.104
De Jong, T.
131.001
De Keuster, J.
103.100
De Koning, J. G.
033.025 .027
De Korte, P. A. J.
142.166
De La Cotardiere, P.
003.039
005.032
010.028
011.013
095.003
De La Noe, J.
033.013
077.017
De La Reza, R.
071.022
De Loore, C.
117.001 .018 .020 .063
142.107 .223
De Mottoni, G.
097.059
De Mottoni y Palacios, G.
097.060
De Pascual, M.
096.013
098.076
De Prins, J.
035.008
De Rop, W.
044.044
046.065
De Sabbata, V.
155.010
De Santis, R.
098.033 .077
De Smet, A.
004.089
De Vany, A.
031.084
De Vany, A. S.
031.084 .085
De Vaucouleurs, A.
158.184
De Vaucouleurs, G.
003.128
097.026
124.104
158.003 .184
160.028 .033 .034
161.007
De Vegt, C.
041.024
124.104

De Young, D. S.
141.039
De Zafra, R.
158.010
Deal, P. H.
015.025
Dean, J. F.
160.041
Dearborn, D. S. P.
064.019
065.162
Debarbat, S.
041.001 .016
044.002 .030
Debehogne, H.
031.297
098.088 .089 .090 .091 .092
103.012 .100 .130
Deboo, G. J.
032.548
Decanio, T. C.
066.117
Decaprio, A.
032.542
Dedman, T. F.
012.005
Deeming, T. J.
021.001
Deerenberg, A. J. M.
142.166
DeGasparis, A. A.
105.005 .118 .125
Degenford, J. E.
033.026
DeGeorge Jr., M.
121.060
Dehmel, G.
051.024
Dehnen, H.
066.055
Dehnhard, D.
065.182
Dejaiffe, R.
041.071
045.019
046.065
Dekker, E.
151.052 .092
Delage, C.
141.627
Delbouille, L.
013.010
Deloach, A. C.
072.012 .015
074.030 .031
Delone, A. B.
074.121
Delouis, H.
084.315
Delov, I. A.
082.091
104.023
Delplace, A. M.
114.052
Delsemme, A. H.
103.101
Delvaille, J.
142.075
Delvaille, J. P.
142.076 .083 .085
Delvoye, L.
031.021
Demaret, J.
065.090 .092
Demarev, V. M.
033.071
Demarque, P.
122.166
Dembovskij, A. V.
097.043
Demers, S.
159.007
Demidov, V. V.
034.061 .074
Demidova, A. N.
085.020
Demin, V. G.
042.119
Deming, D.
065.155
Demura, A. V.
062.008
Den Boggende, A.
122.167
142.014 .230
Den Boggende, A. J. F.
034.158
064.048
142.045
Dence, M. R.
105.195
Denis, C.
065.095
Denis, J.
061.034
Denisov, V. P.
051.018
Denisyuk, Eh. K.
131.557 .558 .559
132.036 .059
Denlinger, E.
033.035
Dennison, B.
131.540
Dennison, E. W.
034.139
Denoyelle, J.
061.034
Depadel', D. M.
124.106
Depenchuk, E. A.
114.348
Deprit, A.
042.121
054.012
Deprit-Bartholome, A.
054.012
Derblom, H.
083.002
Derevitskij, Yu. G.
142.171
Dergachev, V. A.
012.023
082.069
Dermendzhiev, V. N.
073.148
Dermenjiev, V.
073.074
Derrickson, J. H.
061.028
Desaintfuscien, M.
035.013
Deshpande, M. D.
033.067
Deshpande, M. R.
084.260
DesMarais, D. J.
094.418
Desnoyers, C.
105.119
Desprez, F.
121.149
Dessler, A. J.
099.086
Dessureau, R. L.
155.050
Desvignes, F.
034.045
Detrie, R.
035.008
Detweiler, S.
066.170 .173
Detweiler, S. L.
065.109
Deubner, F.-L.
071.001 .036 .042 .050
080.013 .035
Deupree, R. G.
065.057 .157
Deutsch, E. W.
022.069
Deutsch, J. L.
022.069
Devaux, C.
093.015 .046
Devicheva, E. A.
078.023
Devinney, E.
103.100 .130
Devinney, E. J.
031.286
DeVito, C. L.
091.002
Di Benedetto, F.
106.038
Di Benedetto, P.
124.104
Di Gesu, V.
031.423
Di Porto, P.
003.036
Di Sessa, A.
014.035
Dibaj, Eh. A.
158.238
Dichtl, G.
046.067
Dicke, R. H.
031.245
080.005
Dickel, H. R.
131.554
132.017
Dickel, J. R.
131.087 .554
158.040
Dickens, R. J.
114.039 .100
141.156
160.040

Dickey, J.
142.216
Dickinson, D. F.
122.003
131.519
141.623
Dickinson, G. J.
143.085
Dickinson, R. E.
093.051
Dickman, R. L.
131.118
Dickson, L.
031.306
034.027 .124
Dieckvoss, M.
103.128 .132
Dieckvoss, W.
112.001
Dieter, N. H.
131.005
Diethelm, R.
121.149 .151
Dietrich, W. F.
073.131
Dietz, R. S.
105.033 .120 .162
Dimitrijevic, M. S.
077.063
Dimitriu, A.
073.151
075.010
Dimov, D. I.
105.042 .080
Dinardo, A. J.
034.161
Dinerstein, H. L.
161.010
Dinescu, A.
012.018
031.212
046.027
Dingens, P.
071.065 .067 .068 .069 .071
Dinulescu, S.
075.010
Dinulescu, V.
075.010
Dinwoodie, C.
010.012
103.117
Diodato, L.
106.064
Dionysiou, D. D.
066.121
Dirike, L.
006.000
Dirikis, M.
011.069
047.036
Dirikis, M. A.
098.006
Divari, N. B.
082.079
106.039
Dixon, R. S.
041.047
141.091
Dixon, T. A.
022.089
Djokic, M.
032.024
045.029 .030
Djurkovic, P.
010.037
Djurkovic, P. M.
118.017
Djurovic, D.
045.031
Djurovic, L.
045.029 .030
Djuth, F. T.
142.118
Dluzhnevskaya, O. B.
011.053
153.015
Dmitrienko, E. V.
104.062
Dmitriev, B. A.
032.527
Dmitrievskij, A. A.
104.073
Doazan, V.
114.319
Dobrovol'skij, O.
102.029
103.100
Dobrovol'skij, O. V.
102.002
106.041
Dobrovolskis, A.
097.040
Dobrzycki, J.
004.008
005.005
Docken, K. K.
022.043
Dodd, J. G.
162.058
Dodd, R. T.
105.045 .047 .121
Dodge, J. C.
077.004
D'Odorico, S.
034.092
155.066
Dodson, H. W.
003.122
Doering, J. P.
084.004
Dogan, N.
072.069
Dogel', V. A.
143.046
Doherty, L. H.
131.511
Doherty, L. R.
064.023
Doherty, R. H.
035.001
Dohnanyi, J. S.
104.052
Dokuchaeva, O. D.
036.003 .005 .006
Dolan, J. F.
142.221
Dolan, W. W.
003.040
Dole, S. H.
042.123
Dolginov, A. Z.
102.021
106.044
116.011
117.003 .017 .058
Dolginov, Sh. Sh.
094.168
Dolidze, M. V.
122.017 .120
Dollfus, A.
003.041
093.014 .031
097.026
098.074
099.205
Dolodarenko, V. A.
052.049
Dominko, F.
009.012
014.011
Dominski, I.
044.027
Domke, H.
063.002 .003 .027
Dommanget, J.
034.055 .151
118.038
Domokos, G.
162.014
Domville, A. R.
033.052
Domyshev, G. N.
034.067
Donadio, R. N.
031.037
Donahue, T. M.
082.017
084.001
093.022
Donn, B.
131.179
Donn, W. L.
104.003
Donnelly, R. F.
073.158
Donner, K. J.
141.095
Donnison, J. R.
107.003
Doods, D.
155.073
Doolittle, J. H.
121.062
124.104
Dopita, M. A.
132.026
Dorling, E. B.
076.010
083.004
Dorman, J.
094.109
Dorman, L. I.
003.042
082.090
106.031
143.041 .054 .060
Doroshenko, V.
103.100
Doroshkevich, A. G.
011.035
065.024

Dorschner, J.
011.024
158.026
Doschek, G. A.
062.025
073.007
074.004
076.050 .051
Dossin, F.
103.100
Doty, J.
141.353
142.080
Douglas, A. E.
131.204
Douglas, A. V.
004.096
008.043
Douglas, B. C.
046.011
Douglass, D. H.
066.111
Dow, M. J.
121.003
Dowell, J. T.
014.003
Downes, D.
131.136 .169
141.079
142.147
Downs, G. S.
097.035
Downs, W.
097.006
Dowty, E.
094.474
Doxsey, R.
142.073 .075 .076 .081
.084 .214
Doyle, F. J.
032.510
Drago, F.
074.080
Dragunova, A. V.
064.051
114.050
Drake, J.
103.100
142.139
Drake, M. J.
094.459
Drake, S.
004.015 .026
Dran, J. C.
094.460
105.097
143.073
Dravins, D.
034.135
114.048
Dravskikh, A. F.
041.039
Dreibus, G.
094.188
Drever, R. W. P.
066.080
Drobyshevskij, Eh. M.
114.073
117.010
Drouin, R.
022.046
Drozd, R. J.
094.110 .408
Drukier, I.
033.068
Drummen, M.
033.009
103.100
158.130
Dryden, G.
082.082
Dryer, M.
073.124
106.019
Duba, A.
094.143
Dube, A.
105.163
Dubinskij, Yu.
078.017
Dubjago, I.
125.104
Duboin, M. L.
082.064
Dubois, I.
022.111 .112 .113 .114
Duboshin, G. N.
003.043
042.060 .119
Dubout, R.
158.163
Dubov, I. A.
083.054
Dubrovich, V. K.
131.113
162.061
Dubyago, I. A.
113.026
131.093
158.117
Ducati, H.
094.478
Ducuroir, M.
010.031
Duennebier, F.
094.109
Duerbeck, H. W.
121.017 .032 .065 .105
124.103
Duffett-Smith, P. J.
131.042
Dufour, R. J.
131.524
142.172
Dufton, P. L.
114.397
131.143 .207
Duggal, S. P.
078.006 .007
Dujnic, M.
095.005
Duke, M. B.
094.112
Duley, W. W.
131.049
Dulk, G. A.
099.097
125.001
Dulou
041.013
Duma, A. S.
094.141
Dumarchi, G.
121.149
Dumont, P. D.
022.003
Dumont, R.
106.003
Dumont, T.
122.181
Duncan, J. H.
033.078
Duncan, R. A.
099.093
Duncans, L.
034.123
123.015
Duncombe, J. S.
079.109
Duncombe, R. L.
041.053
Dunford, E.
031.292
083.003
Dungey, J. W.
084.220
Dunham, D. W.
094.015
096.005
115.019
Dunkelman, L.
031.294
Dunlap, J.
125.110
Dunlap, J. W.
125.023
Dunlap, Y.
125.110
Dunn, P. J.
046.043
Dunne, J. A.
093.001
Duorah, H. L.
065.181
Duplay, A.
005.032
Dupouy
041.013
Dupre, A. K.
114.097
Dupree, A. K.
119.003
DuPuy, D. L.
122.083
Duraud, J. P.
094.187 .460
143.073
Durgapal, M. C.
141.165
Durgaprasad, N.
106.001
143.047
Durisen, R. H.
065.005 .134
126.020
142.090
155.023
Durney, A. C.
051.025
Durney, B. R.
080.016
Durrani, S. A.
105.122

Durrant, C. J.
073.024 .123
Duus, A.
151.059
Duval, J. E.
142.172
Duvall, T. L.
076.056
Duxbury, T. C.
097.202
099.216
Dvorak, R.
098.028 .062
Dwek, E.
080.003
Dworak, T. Z.
005.034
111.009
121.107
Dworetsky, M. M.
114.027 .118
153.901
Dwyer, R.
143.052
D'yachenko, I. G.
072.025
D'yakonova, M. I.
105.071 .072
Dyck, H. M.
114.030 .070
158.035
Dyer, C. S.
051.010
Dykes, M. R.
103.010 .140
Dyson, J. E.
125.043
131.501
Dyson, K.
075.003
Dzhakusheva, K. G.
132.040
Dzhalalova, Z. G.
004.062
Dzhordzhio, N. V.
083.043
Dzhun', I. V.
045.034
Dziembowski, W.
065.098
122.036
Dzurisin, D.
097.020 .050
Dzyaman, D. D.
031.216 .217
033.005
Dzyubenko, N. I.
074.021 .055

Eachus, L. J.
141.027
142.216
158.321
Eadie, G.
142.052 .120 .153
Eardley, D. M.
066.039 .073
142.038
Easson, I.
061.075
Eastmond, T. S.
160.013
Easton, R. L.
044.013
Eather, R. H.
084.014 .016
Eaton, J.
121.034
Eaton, J. A.
121.090 .108
Ebbighausen, E. G.
119.020
121.091
Eberhardt, P.
083.063
094.464 .467 .469
105.097 .123
Ebert, R.
117.009
Ecklund, W. L.
084.021 .022
Edamatsu, K.
103.140
Eddy, J. A.
085.004
Edelman, S. H.
052.003
Edlen, B.
022.035
Edmonds Jr., F. N.
071.055
Edmunds, M. G.
066.901
131.035
161.011
Edwards
142.215
Edwards, A. C.
062.019
Edwards, M. R.
142.103
Edwards, P. J.
013.011
031.268
Edwards, P. L.
141.142
Eelsalu, H.
013.018
Efanov, V. A.
077.025
132.047
141.067 .097
Efendieva
See Ehfendieva
Efimov, A. A.
042.076 .077 .105 .106 .107
Efimov, A. I.
074.028 .044
097.062
Efimov, N. N.
143.072
Efimov, Yu. E.
078.017
Efimov, Yu. S.
142.033
Efremov, Yu. N.
011.027
031.010
117.004
122.176
Efremov, Yu. N.
158.124
159.009
Egan, W. G.
131.037 .175
Egeland, A.
084.005
Egibekov, P. E.
106.041
Eglitis, I.
113.070
122.213
Egorov, T. A.
143.072
Ehfendieva, S. A.
072.034
Ehjnatyan, Dzh. A.
132.034
Ehlers, J.
066.104
Ehl'yasberg, P. E.
052.022
082.055
Ehn, D. C.
031.242
Ehnkhbayar, M.
046.034
Ehr, H.-L. Von
See Von Ehr, H. L.
Ehramzhyan, R. A.
066.159
Ehrenberg, A. S. C.
003.044
Ehricke, K. A.
015.010
Ehtkin, V. S.
033.015
Eichhorn, G.
022.066
094.181
Eichhorn, H.
111.006
Eichler, D.
061.026
Eichler, E.
094.426
Eilek, J. A.
158.097
Einasto, J.
011.035
158.114
Einfeld, D.
062.042
Eisenstaedt, J.
162.081
Eisner, S.
015.030
Eitschberger, B. H.
081.042
Ekberg, J. O.
022.036
Ekers, R. D.
031.419
066.108
141.079 .123
El Goresy, A.
094.454
105.177
El-Baz, F.
094.127

Elander, N.
022.069
Elcan, M. J.
076.028 .031
Eldridge, J. S.
051.010
Elford, W. G.
104.061
Elias, J. H.
131.025 .523
141.047
Eliass, M.
077.062
Eliseev, G. F.
077.048
Elitzur, M.
141.643
Eller, E.
051.010
Elliot, E. K.
009.030
Elliot, J.
100.206
Elliot, J. L.
009.030
100.208
Elliott, K. H.
132.010
Ellis, G. F. R.
162.022
Elmergreen, B. G.
131.071
158.075
Elokhov, A. S.
084.018
Elsaesser, H.
008.047
131.504
152.002
Elsmore, B.
044.018
Elsner, R. F.
065.163
Elst, E. W.
123.020
Elvis, M.
142.048 .049 .101 .202
.210
Elvius, A.
114.352
Elwert, G.
022.097
Elzner, L.
074.010
Elzner, L. R.
076.057
Emerson, B.
047.039
Emerson, D. T.
158.153
Emerson, G.
103.100
Emerson, J. P.
132.009
Emmons, G. H.
031.039
Emmoth, B.
022.037
Emslie, A. G.
097.047
Enal'skij, V. A.
031.402
041.010
Encrenaz, P.
142.215
Encrenaz, P. J.
131.124
Encrenaz, T.
099.100
100.008
Endal, A. S.
065.032 .045
Engel, A. R.
061.086
142.019 .023
Engel, R.
003.018
Engelbrektson, S.
003.045
Engels, E.
062.038
Engibaryan, N. B.
063.051
England, A. W.
063.043
Engman, B.
022.067
Engvold, O.
036.020
073.128
Enikeev, R. I.
079.101
Enome, S.
033.030
Enslin, H.
012.029
035.007
Eoll, J. G.
065.168
Epchtein, N.
141.627
Epis, J. J.
033.062
Epishev, V. P.
041.007
Epps, H. W.
034.034
Epstein, A.
121.128
142.076 .083 .084 .085
Epstein, G. L.
034.039
071.058
072.081
073.059
Epstein, R. I.
065.058
Erastova, L. K.
125.030 .065
Erceg, V.
118.031 .032 .036
Eremeeva, A. I.
005.017
Eremenko, R. P.
098.068
102.012
Eremin, I. M.
084.321
Ergma, E.
065.068
Erickson, E. F.
082.021
093.042
132.025
Erickson, G. W.
124.103
Erickson, K. E.
032.538
Erickson, W. C.
077.006
125.013
141.049 .050 .143 .144
Eriksson, K.
064.026
Eritsyan, M. A.
122.106
Erkes, J. W.
033.075
Erleksova, G. E.
122.102
Eroshenko, E. G.
094.153
Eroshevich, Eh. S.
103.101
Erpylev, N. P.
046.017
Ershkovich, A. I.
099.065
Erskine, F. T.
141.042
Erzhanov, Zh. S.
045.002 .015
Eschenauer, H.
033.059
Eschghi, I.
105.085
Esepkina, N. A.
141.116
Eshleman, R.
053.019
099.046
Esipov, V. F.
034.082
158.238
Espenschied, P.
117.011
Esposito, P. B.
066.041
Esser, U.
002.009
Essex, E. A.
083.016
Estes, L. E.
034.030
Esteve, R. B.
085.022
Etcheverry, M.
005.032
Ettenberg, M.
033.053
Etzel, P. B.
121.038
Eubanks, A. G.
032.513
Eugster, O.
094.469
Evans, A.
061.032
141.064
Evans, D. S.
096.055

Evans, D. S.
115.019 .020
122.014
Evans, J. C.
071.006 .062
Evans, J. E.
084.006
Evans, J. V.
083.025 .055
Evans, R. G.
114.318 .399
159.010
Evans, S.
122.090
Evans, T. L.
122.048 .051
Evans II, N. J.
022.054
131.013
141.611
Evans Jr., J. C.
080.044
Evdokimov, I. Yu.
103.107
Evdokimov, Yu. V.
103.107
104.005
Evenskaas, R. S.
121.062
Everitt, C. W. F.
066.026 .027 .028
Eviatar, A.
062.048
Evlashin, L. S.
084.271
Evlashina, L. M.
084.271
Evseev, O. A.
142.030
Evseyev, O. A.
See Evseev, O. A.
Evstaf'ev, N. N.
076.044 .046
Evstigneeva, N. M.
124.104
Evsyukov, N. N.
094.138 .174
Evteev, V. P.
042.087 .098
Ewart, G. M.
065.123
Ewing, M. S.
131.158
Exton, R. J.
022.072
Eyles, C. J.
142.022 .093 .131 .215 .225
Eyni, M.
074.062
Ezawa, H.
141.167

Faber, S. M.
151.033
158.146
Fabian, A. C.
011.018
065.040
142.012 .021 .044
Fabian, A. C.
160.006
Fabian, W.
034.103
Fabozzi, A.
100.016
Fackerell, E. D.
151.029
Fadeev, Yu. A.
121.012
Fadeicheva, Z. N.
104.076
Fadini, A.
006.000
015.033
Faerber, S.
005.005
Fahey, R. P.
114.301
Fahlman, G. G.
125.053
131.027
141.076
Fahr, H. J.
074.015
Fahrbach, U.
034.009
122.070
Fairall, A. P.
125.101
Fairbank, W. M.
022.002
066.024 .025 .027
Fajnberg, I. S.
097.049
Fajnshtejn, S. M.
080.023
Fakhrutdinova, A. N.
082.107
104.075
Falgarone, E.
131.124
Falk, G.
003.046
Falk, S. W.
131.190
Falk Jr., S. W.
125.066
Fall, S. M.
162.005
Falla, D.
061.032
Falle, S. A. E. G.
125.002
131.110
Fan, C. Y.
073.018
078.004 .029
Fang, L.
142.116
Fangor, R.
035.020
098.075
Fano, U.
022.041
Fanselow, J. L.
141.135
Fante, R. L.
082.084
Fanti, C.
141.058 .147
Fanti, R.
141.058 .123
Faraggiana, R.
114.361
Farley, D. T.
084.033
Farlow, N. H.
106.037
Farnik, F.
034.003
Farrokh, H.
033.051
Fassi-Fehri, O.
162.018
Fastie, W. G.
032.514
094.131
114.345 .398
Faulkner, D. J.
080.073
154.037
Faulkner, J.
011.004
114.316
Fawcett, A. H.
131.083
Fawcett Jr., G.
031.421
Fazio, G. G.
032.516 .547
141.605 .615
Feast, M. W.
115.002
122.020
Fechtig, H.
094.105 .181 .186
104.048
105.024
106.004
Federer Jr., C. A.
095.001
Fedorenko, V. N.
077.021
Fedorov, E. P.
003.001
045.004 .005
046.057
Fedorova, A. V.
031.213
Fedoseev, E. N.
044.006
Fedoseev, L. I.
082.121
Fedotov, G. G.
052.061
Fedulin, I. A.
082.095 .096
Fedyantsev, B. K.
033.072
Fedynskij, V. V.
104.068
105.077
Fedyushin, B. K.
105.088
Fehrenbach, C.
031.284
034.094 .145
103.100
124.103 .104
Fehrenbach, R.
009.006

Fehsenfeld, F. C.
082.029
Feibelman, W. A.
100.002
121.100 .129
Feijth, H.
123.039
Feinstein, A.
113.024
Feissel, M.
045.023
081.031
Feitzinger, J.
151.007
Feitzinger, J. V.
155.068
Feix, M. R.
062.021
Fejgin, V. M.
084.024
Fekel, F. C.
096.005
118.016 .021
Fekel Jr., F.
119.019
Feklistova, T.
022.085
Felber, H.-J.
047.005
Feldman, P. A.
124.104
141.125
Feldman, P. D.
084.004
114.345 .398
Feldman, U.
062.025
074.004
076.050 .051
Feldman, W. C.
074.072 .090 .113 .114
.115 .125
Fel'dshtejn, Ya. I.
083.060
106.046
Felenbok, P.
066.079
Felli, M.
077.003 .028 .037
141.147
Felsentreger, T. L.
052.033
Felten, J. E.
158.079
Feltz Jr., K. A.
121.064
Fender, J. S.
082.005
Fenina, Z. N.
122.126 .127
Fenkart, R. P.
155.007
Fenner, M. A.
084.254
Ferch, R. L.
133.026
Ferguson, E. E.
082.029
Fernbach, S.
003.011
Fernie, J. D.
111.003 .008
113.013 .018
122.027 .179
Ferraioli, F.
062.064
Ferrari, A.
062.066
Ferrari D'Occhieppo, K.
004.047 .120
008.114
Ferraro, A. J.
083.017
Ferraro, J. R.
091.046
Ferraz Mello, S.
042.031
Ferreiro, M.
079.104
Ferreri, W.
103.100
Ferrin, I. R.
100.012
Ferris, J. P.
099.070
Ferro, I.
082.123
Feynman, J.
074.051
Fiala, A. D.
079.109
Fialko, E. I.
062.028
104.055 .056 .077 .078
.085
Ficarra, A.
031.236
141.058
Fichtel, C.
061.011
Fichtel, C. E.
141.318
155.003
Field, G. B.
161.006 .008
Fielder, G.
003.047
Figala, K.
004.105
005.007
Figger, H.
022.064
Filin, A. Ya.
123.002
Filippov, A. T.
078.024
Filippov, V. A.
106.028
143.026
Fillit, R.
131.163
Fink, U.
034.029
098.044
Finkel'shtejn, A. M.
041.039
066.144
Finkelstein, A. M.
See Finkel'shtejn, A. M.
Finsen, W. S.
118.035
Firmanyuk, B. N.
098.063
120.002
122.119 .121 .128
Firsov, V. V.
076.049
Fischel, D.
031.286
Fischer, H.
046.048
Fischer, K.
082.104
Fischer, L.
112.003
Fisenko, M. I.
073.117
Fisher, A. J.
143.059 .066
Fisher, J. R.
158.081 .172
Fisher, W. A.
114.047
Fishman, G. J.
061.028
Fishman, V. M.
094.130
Fitton, L. E.
094.133
FitzGerald, M. P.
131.186
Fitzpatrick, P. M.
052.046
Fix, J. D.
114.085 .372
122.193 .196
Fjeldbo, G.
093.052
Flaciani, R.
073.154
Flanagan, R. C.
052.014
Flannery, B. P.
117.049 .074
Fleck, R.
065.106
Fleck Jr., R. C.
121.060
Fleckenstein, J. O.
005.007
Fleming, B. K.
084.319
Flerov, G. N.
105.043
Fligel', M. D.
083.035
Flin, P.
160.026
Flint, B. K.
031.046
Floquet, M.
114.001
Florenski, C. P.
097.030
Florentin-Nielsen, R.
022.050
Florsch, A.
010.028
Florsch, G.
095.006
Flower, D. R.
022.034 .100

Flower, D. R.
076.058
Flower, P.
159.901
Flower, P. J.
065.161
115.005
Flury, W.
042.022
Fodor, R. V.
105.092 .124
Fofi, M.
034.106
Fogarty, W. G.
082.124
Fokker, A. D.
013.009
Folomeshkin, V. N.
066.159
Fomalont, E. B.
066.037
Fomalount, E. B.
141.055
Fomichev, V. V.
077.022
Fomin, P. I.
142.146
151.079
Fomin, S. K.
125.102
158.117
Fomin, V. P.
142.150 .151 .177
Fominov, A. M.
082.061
Fominykh, S. I.
091.040
Fonarev, G. A.
084.249
Forbes, E. G.
004.075
Forbes, J. M.
082.037
Ford, A. L.
022.043
Ford, C. B.
010.001
Ford, H. C.
133.017
158.189
Ford, J. G.
081.029
Ford, R. L.
034.154
Ford, V. L.
158.183
Ford Jr., W. K.
158.005 .006 .220 .222
Forga, R.
053.013
Formiggini, L.
141.058
Formisano, V.
012.008
084.206 .211 .221
Fornaseri, M.
105.112
Forrest, D. J.
073.015
076.043
Forrest, W. J.
131.514
141.606
Forslund, D. W.
074.090 .113 .115
Fort, B.
074.061
Forti, G.
104.032 .034
Fortini, P.
155.010
Fosbury, R. A. E.
141.018
158.141
Fossat, E.
071.026 .027
Foster, J. C.
084.302
Foster, J. J.
094.443
Foster Jr., G. T.
099.099
Fouche, C.
032.509
Fouere, J. C.
034.125
Fougere, P. F.
084.319
Foukal, P.
073.096
080.002
Foukal, P. V.
074.005
076.002
Fountain, J. A.
094.102
Fountain, W. F.
094.102
Fourcade, C. R.
123.075
154.015 .016 .017
Fowler, B. W.
022.040
Fowler, R. G.
022.026
Fowler, W. A.
022.065
Fox, J. G.
162.024
Fox, K.
099.010
Fox, P.
097.012
Fox, W. E.
010.012
099.098
Foy, R.
122.107 .132
Fracassini, M.
114.389
115.001
122.151
Fracastoro, M. G.
012.017
103.102
Fraknoi, A. G.
010.006
Francmanis, J.
011.067
014.018
065.179
Frandsen, S.
114.071
Frank, L. A.
084.238
Frank, R.
034.157
Franklin, F.
098.016
Franklin, F. A.
098.020
099.226
100.017
Frankston, M.
158.089
Frankston, M. J.
034.134
Frantsman, Yu. L.
122.108
Franz, O. G.
032.532
122.028
Fraser, B. J.
084.282
Fraser, R. S.
082.019
Fraser-Smith, A. C.
084.262
Frazier, E. N.
072.017
080.020
Fredrick, L. W.
032.532
Fredricks, R. W.
084.205 .216
Fredriksson, K.
105.118 .125 .166
Freeman, C. G.
131.132
Freeman, I. M.
162.077
Freeman, J.
131.064
Freeman, J. W.
084.215 .278
094.159
106.042
Freeman, K. C.
142.040
151.059
154.021 .037
158.063
Freeman, R.
010.023
Freeman Jr., J. W.
012.031
084.254
Freier, P. S.
022.099
French, H.
124.104 .111
142.221
Frevert, F.
098.018
Frey, A.
106.008
Frey, H.
097.019
099.207
Friant, A.
032.541

Frick, U.
105.182
Fricke, K. H.
082.109
Fricke, K. J.
065.016 .023
158.019 .020 .021
Fricke, W.
002.009
008.047
041.023 .030
044.016
155.022 .031
Fridman, A. M.
151.042
Fridman, V. E.
031.272
Friebele, E. J.
094.403
Friedemann, C.
131.084
158.026
Friedjung, M.
124.003
125.049
Friedman, A. J.
009.005
Friedman, H.
142.011 .182 .183 .237
Friedman, J. L.
066.038 .040
Friedman, L. D.
053.022
093.056
Fries, A.
121.149
Friesen, L. J.
094.149
Friichtenicht, J. F.
104.036
105.051
Frisch, H.
063.014
Frisch, P. C.
131.182
Frisch, U.
063.014
Frishberg, L. D.
114.026
Fritz, G.
142.011 .182 .183 .237
Fritz, R. B.
083.076
Fritz, T. A.
084.252 .303
Fritzova-Svestkova, L.
003.122
073.056
Froehlich, H.
046.016
Froehlich, H.-E.
141.057
Froeschle, C.
098.019
151.031
Froeschle, M.
041.051
096.001
Frogel, J. A.
131.519 .527
141.623
Frogel, J. A.
158.056 .070
Frohlich, A.
119.008
142.167
Frolov, M. S.
120.003
122.099 .109 .125
Frolov, V. M.
162.010
Frolov, V. N.
118.006
153.009 .020 .021 .037
Frondel, J. W.
003.048
Frost, K.
073.054 .055
Fruland, R.
105.082
Fruland, R. M.
105.126
Fry, P. W.
034.129
Fryer, R. J.
021.003 .004
094.415
Fryxell, B. A.
158.228
Fuccillo, G.
100.016
Fuchs, H.-U.
043.002
066.143
Fuchs, L. H.
105.046 .103
Fuerst, E.
073.144
077.002 .003 .029
080.012
Fuerstenberg, F.
075.017
Fuertig, W.
119.016
Fujii
103.100
Fujikawa, S.
103.130
Fujimoto, M.-K.
022.062
Fukada, Y.
076.025
Fukaya, R.
124.104
Fukunishi, H.
084.258 .310
Fuller, M.
094.121 .134
Fuller, M. D.
094.417
Fuller Jr., E. L.
094.426
Funiciello, R.
105.107 .108 .112
Furenlid, I.
031.222
034.041
113.017
Furman, D. R.
084.017 .023
Furniss, I.
132.009
Furniss, I.
141.629
Furuya, K.
062.026
Furuyama, S.
103.131
Fusi-Pecci, F.
064.035
Fussell, W. B.
022.090
Fussmann, G.
022.009
Futami, Y.
065.191
Fymat, A. L.
063.032

Gabdullin, B. M.
034.107
132.051
Gabrakov, S.
066.139
Gabriel, A. H.
013.006
022.022
074.056
Gabriel, M.
065.021 .033 .097 .100
.176
080.063
Gabry, A.
035.015
Gachechiladze, R. G.
083.041
Gadea, F.
132.015
Gadzhiev, M. S.
034.063
Gagliardini, D. A.
083.014
Gahm, G. F.
122.016 .122
Gail, H.-P.
063.035
131.030 .040 .098
Gaines, E. E.
084.404
Gaizauskas, V.
072.054
Gajderowicz, I.
046.028
Gajduk, A. R.
003.049
Gajlit, G.
003.050
Galanin, V. V.
034.118
Galassi, G. A.
122.221
Gale, D. S.
032.059
Gale, N. H.
105.094
Galeev, A. A.
084.309
Galindo Trejo, J.
078.031
Galkin, S. L.
141.321

Gallagher, J.
124.103
Gallagher, J. S.
124.104
158.069 .146
Galli, M.
143.080 .083 .084
Galloway, D. J.
072.056
Gal'perin, Yu. I.
084.272
Galsan, P.
046.034
Gamelon, H.
033.076
Gammage, R. B.
094.424 .426
Gammelgaard, P.
131.100
Ganapathy, R.
105.127
Gapcynski, J. P.
094.204
Gaposchkin, E. M.
046.005 .018
Garcia, J. R.
122.221
Garcia-Munoz, M.
143.022 .023 .034 .062 .063
Garcia-Pelayo, J.
153.018
Gardiner, L. R.
094.190 .191 .461
Gardner, F.
131.148 .169
Gardner, F. F.
131.509 .516 .537
141.036 .154
155.055
158.192
Garfinkel, B.
042.040
Garg, A. N.
094.412
Garmire, G.
061.021
Garmire, G. P.
061.023
113.051
125.005
142.058 .072 .074 .178
Garnavich, P.
124.104
Garrett, H. B.
082.037
Garrison, B. J.
022.044
Garrison Jr., L. M.
065.003
Gartrell, G.
104.061
Gartung, Yu. A.
052.065
Garwin, R. L.
066.099
Gary, S. P.
074.090 .113 .114 .115 .125
Gascoigne, S. C. B.
032.001 .008 .051
Gaska, S.
104.091
Gasparovic, R. F.
031.039
Gasparyan, K. G.
132.038 .039
Gaston, B.
121.124
Gatewood, G.
041.005
117.034
118.005
Gatewood, G. D.
142.191
Gatland, K.
003.051
Gatley, I.
141.637 .638 .639
Gattinger, R. L.
084.008
Gault, D. E.
094.116 .156
105.040
Gaupp, A.
022.067
Gauss, F. S.
041.029
Gautier, D.
091.047
099.100
Gavril'ev, A. I.
143.026
Gavrilov, I. V.
003.004
Gavrilov, S. V.
099.060
Gavrilova, I.
098.072
Gawel, A.
004.051
Gay, R. H.
083.051 .052
Gdalevich, G.
051.008
Geary, J. C.
034.040
Geballe, T. R.
131.522
141.617
Gebbie, K. B.
073.134
114.096
Gedzelman, S. D.
062.031
082.102
Gefwert, C.
005.013
Gehrels, T.
004.052
103.134
Gehren, T.
080.011
Gehrke, C. W.
094.428
Gehrz, R. D.
114.031
132.027 .046
Geiss, J.
083.063
094.464 .467 .469 .471
Gejkina, F.
003.050
Gelfand, J.
022.105
099.080
Gel'fgat, B. E.
042.120
Gel'frejkh, G. B.
079.101
Geller, M. J.
158.077
Gelles, R.
031.035
Gen, A. A.
034.018
Genkin, I. L.
151.071 .073
Gentner, W.
094.455
Georgelin, Y. M.
114.024
Georgiev, N.
046.029 .030
Gerard, E.
099.014
Gerard, J.-C.
082.093 .106
084.007
Gerashchenko, A. N.
131.192
Gerasimenko, S.
098.072
Gerasimova, N. D.
083.077
Gerasimova, N. N.
077.009
Gerassimenko, M.
074.008 .073 .078
Gerbaldi, M.
114.055
Gerdes, E. W.
005.006
Gergely, T. E.
077.006
Gerjuoy, E.
061.090
Gerlach, W.
004.034
005.004
Germann, R.
104.086
121.149
Gerola, H.
074.098
114.350
131.184
Gerrard, A.
003.052
Gerritsen, B. A.
015.019
Gerritsen, H. J.
015.016
Gerry, M. C. L.
022.122
Gersdorff, N. Von
See Von Gersdorff, N.
Gershberg, R. E.
122.008 .157
Gershtejn, S. S.
066.159

Gerstbach, G.
046.035
Gerstenberger, M.
047.023
Gertsenshtejn, M. E.
066.129
Gessler, J.
044.020
046.068
Gessner, H.
121.138
122.214
123.074
124.107
Getman, V. S.
011.051
102.002
Gevorkyan, M. S.
063.051
Geyer, E. H.
121.004 .016
122.052 .054 .055 .131
Gezari, D. Y.
131.025 .523
141.047
Ghedini, S.
010.027
011.073
121.056 .057 .058 .153
Gheorghiu, O.
035.013
Ghezloun, A.
092.002
123.025
Giacaglia, G. E. O.
042.094
Giacconi, R.
142.124
Gianni, G.
034.096
Giannone, P.
065.129
121.046
Giannuzzi, M. A.
121.046
Gibb, R. G.
160.046
Gibbons, R. V.
094.475
Gibson, A. J.
082.003
Gibson, D.
121.039
Gibson, D. M.
121.028
141.028 .311
142.017
Gibson, E. G.
073.048 .049 .146
Gibson, J. B.
122.144 .145
Gibson Jr., E. K.
094.120
105.128
Giclas, H. L.
103.100 .130 .131 .133
.136 .140
Gidalevich, E. Ya.
143.033 .055
Giebler, H.
003.131
Gielingh, L. W. F.
010.019
Gielingh, W.
098.037
Gierasch, P. J.
080.068
091.009
093.029
Gieren, W.
122.066 .067 .174
124.104
Giese, R. H.
008.016
106.005 .006 .007 .078
Gieseking, F.
122.066 .174
123.024
Gigolashvili, M. Sh.
073.129
Giguere, P. T.
131.052
141.616
Gil, R. R.
142.244
Gil'berg, L. A.
003.103
Gilbert, G.
034.042
Gilbert, G. R.
034.040
124.104
Giles, A. B.
134.006
Gillespie, A. R.
131.148
Gillett, F. C.
034.037
098.038
100.203
131.073 .162 .514
158.087
Gillispie, C. C.
003.053 .054 .055
Gilman, R. C.
066.164
Gilmore, A. C.
047.034
103.100 .119 .120 .125
.130 .136
Gilmore, W.
131.008
Gilra, D. P.
022.094
114.004
Gingerich, O.
003.056
004.006 .012 .016 .029
.042 .048
011.020
012.049
Ginter, M. L.
022.096 .116
Ginzburg, V. L.
061.041
141.337
Giordano, M.
073.154
Giovanardi, C.
080.021
Giovanardi, U.
066.087
Giovanelli, R.
073.126
074.068
Giovanelli, R. G.
074.106
Giovannelli, F.
142.206
Giovannini, G.
141.058
Girard, A.
032.502
Giuffrida, T. S.
158.088
Givishvili, G. V.
083.035
Gjerde, A.
078.033
Gjerloeff, H.
121.080
Glackin, D. L.
073.095
Gladushina, N. A.
114.050
Glagolevskij, Yu. V.
034.060
116.009
Glaspey, J. W.
158.214
Glass, I. S.
113.002
Glass, M.
033.076
Glasser, M. L.
062.001
065.006 .189
Glassgold, A. E.
061.073
131.200
Glaze, D. J.
044.014
Glebova, N. I.
097.004
Gleeson, L. J.
078.021
Gleissberg, W.
008.035
072.003 .083 .084 .085
.086
Glencross, W. M.
074.003 .006
076.010 .023
Gliese, W.
111.010
Gloeckler, G.
073.018
078.004 .029
Glushkov, Yu. I.
131.557 .558 .559
132.036 .059
Gnedin, Yu. N.
116.008
117.028 .029 .046 .050
141.622 .631
142.109 .141
Gnevysheva, R. S.
075.005 .006
Gnezdilov, A. A.
077.010
Goad, J. W.
151.035

Goad, L.
133.016
Goad, L. E.
151.035
Goded, F.
162.003
Godoli, G.
073.105 .106 .107
074.082
Godson, W. L.
003.078
Goebel, J. H.
141.645
Goel, P. S.
094.412
Goerlich, P.
004.051
Goertz, C. K.
099.084 .213
Goethals, M.
162.082
Goguen, J.
100.206 .208
Gokhale, M. H.
080.034
Golay, M.
003.148
013.010
Gold, R. E.
074.035
Gold, T.
107.002
Goldberg, L.
114.302
Goldberg, R. H.
094.476
Golden, L. J.
031.032
032.528
Golden, R. L.
143.043
Goldhaber, A. S.
099.066
Gol'din, V. D.
105.031
Goldoni, R.
066.017 .184
Goldreich, P.
141.643
Goldsmith, P. G.
131.062
Goldsmith, S.
062.025
Goldstein, J. I.
094.456 .462
105.178
Goldstein, M. L.
094.144
Goldstein Jr., S. J.
099.201
Goldwasser, R. E.
033.025 .027
Golev, V. K.
082.047
131.194
Gollandskij, O. P.
114.082
122.101
Gol'nev, V. Ya.
141.116
Golovko, A. A.
032.040
Golovkov, V. N.
093.024
Gol'ts, Eh. Ya.
073.115
076.026
Golub, L.
074.073 .078
076.003
Golub, P. A.
071.044
Golubchik, I. L.
034.061 .074
Golubev, V. A.
098.049
102.013
103.100
Golyshev, G. I.
003.057
Gomberoff, L.
084.311
Gomez
099.089 .090
Gomez, A. E.
151.037
Gomez, T.
114.042
Gomez-Gonzalez, J.
131.513
Gomide, F. M.
162.090
Goncharov, G. N.
105.054
Goncharov, N. Yu.
077.023
Gondhalekar, P. M.
131.210
Gontarev, O. G.
077.033 .057
Gonzalez, G.
122.148
Gonze, C.
075.012 .013
Gonze, R.
075.012 .013
Gooding, R. H.
081.016 .017
Goodman, B. M.
113.047
Goody, R. M.
003.057
099.209
Gooley, R.
105.130
Goorvitch, D.
093.042
132.025
Gopalan, K.
105.100 .131
Gopasyuk, S. I.
062.040
080.054
Goranskij, V. P.
122.064
Gorbachev, B. I.
160.018 .019
Gorbatskij, V. G.
122.039
124.004
Gorchakov, E. V.
078.027
084.403
Gordeeva, Yu. F.
052.056
Gordon, I. M.
077.009
Gordon, K. C.
121.059
Gordon, M. A.
131.011 .117
155.061
Gorel', G. K.
009.002
041.062
098.048
Gorenstein, P.
032.520 .542
Gorgolewski, S.
033.011 .018
035.021
Gorin, V. D.
105.069
Gorisse, M.
032.519
Gorodetskij, D. I.
034.107
132.040 .051
Gorodinskij, G. M.
142.145 .171
Gorokhov, A. I.
051.009
Gorokhov, V. L.
034.070
Gorovenko, A. P.
062.028
104.077
Gorozhankin, B.
051.008
Gorshkov, A. G.
141.121
Gorshkov, Eh. S.
105.066
Gorshkov, S. N.
061.065
Gorskij, S. M.
031.272
Gorynya, A. A.
094.006 .008
Gose, W. A.
094.124
Gosling, J. T.
036.018 .019
074.024 .036
094.005
Goss, W. M.
125.035
131.136
141.079
Goswami, J. N.
094.441
106.001
Goto, Y.
045.012
Gotska, T. G.
155.044
Gott III, J. R.
158.060 .104 .126 .129
160.044
162.011

Gottesman, S.
142.216
Gottesman, S. T.
158.148 .166
Gottlieb, C. A.
131.054 .159
Gottlieb, E. W.
122.142
131.054 .159
142.039 .135
Goudis, C.
131.502 .542 .550
132.012 .013
141.104
Gough, D. O.
065.099
122.187
Gough, M. P.
083.006
Gough, R. P.
008.113
Gouguenheim, L.
142.215
Gould, J. A.
032.009
Gourdon, B.
076.021
Goutail, J. P.
082.011
Gow, C. E.
114.099
Graber, M. A.
045.013
Graboske Jr., H. C.
062.003
099.001
Grabowski, B.
022.101
Graf, W.
077.041
141.022
Grafarend, E. W.
045.020
Graham, A. L.
105.132
Graham, D.
033.007
141.010
Graham, D. A.
141.152
Graham, J. A.
122.171
158.076 .220
Gram, R. Q.
066.111
Grandi, S.
034.042
Grandi, S. A.
132.001 .023
Granes, P.
114.383
Granrath, D. J.
103.139
Grant, E.
003.058
Granveaud, M.
044.029
Grappin, R.
063.029
Grard, R. J. L.
053.010
Grasberg, Eh. K.
125.041
Grasdalen, G. L.
065.118
112.012
141.608
153.024
Gratton, L.
004.119
031.422
Grauer, A.
103.100
Gravier, J. P.
052.042
Gray, C. M.
094.443
Gray, D. F.
065.122
114.089
Gray, M. A.
010.007
Grayzeck, E. J.
131.528
155.018
Grebenikov, E. A.
042.119
Grebenkemper, C. J.
141.022
Grebennikov, E. A.
101.006
Grebowsky, J. M.
084.401
Grec, G.
031.227
Greeb, M. E.
032.546
Greeley, R.
097.037
Green, A. E. S.
084.017
Green, A. J.
125.039 .044 .046
158.173
Green, D. H.
105.089
Green, R.
121.036
Green, R. F.
126.025
141.046 .052
158.106
Green, R. R.
097.035
Green, S.
022.044
131.105
Greenberg, J. M.
131.137
Greenberg, P.
003.139
Greenberg, R.
097.201
098.016
101.003
Greene, A. E.
114.325
Greene, R. D.
066.182
Greene, R. L.
022.024 .027 .028 .029
Greene, T. F.
099.204
Greenleaf, P.
003.045
Greenstadt, E. W.
074.014
Greenstein, G.
065.123
141.317
Greenstein, J. L.
124.111
132.019
Greenwald, R. A.
084.021 .022
Gregory, P.
131.152
Gregory, P. C.
124.104
142.007
Gregory, S. A.
160.001 .032 .036 .047
Gregul', A. Ya.
031.258
Greig, W. E.
091.013 .014
Greiling, P. T.
033.037
Greisen, K.
003.059
061.011
Gresky, W.
005.025
Gretz, D. J.
066.076
Greve, A.
031.048 .202
080.066
Grevesse, N.
022.003
071.063
Grewing, M.
141.012 .024 .362
Grib, S. A.
074.084
084.240
Gribbin, J.
004.001
011.050
066.158
Gribbin, J. R.
003.060
Griboval, P. J.
003.032
Griesser, M.
004.079
Grieve, R. A. F.
105.180
Griffin, A. A.
104.001
124.104
Griffin, R. E. M.
114.308
Griffin, R. F.
114.308
119.005 .014 .021
Griffin, W. G.
114.399
Griffith, J. S.
002.011
Griffiths, C. G.
142.210

Griffiths, R. E.
032.541
Grigor'ev, V. M.
032.040
034.016
062.041
Grigor'eva, S. I.
143.005
Grigor'eva, V. P.
077.015
Grigoritsa, T. N.
034.073
Grigorov, N. L.
032.534
143.069 .074
Grigorov, P. L.
143.071
Grigoryan, O. R.
083.029 .034
Grindlay, J.
122.167
142.014 .194 .230
154.036
Grindlay, J. E.
141.002
142.088 .095 .121
Grineva, Yu. I.
076.041 .045
Gringauz, K. I.
083.042
106.016 .058
Gripp, L. G.
052.049
Griscom, D. L.
094.403 .421
Grishchuk, L. P.
066.059 .106
162.072
Grishin, Yu. A.
105.031
Groegler, N.
094.464 .467 .469
Groenbech, B.
121.102
Groetzsch, H. H.
004.025
Gronenschild, E.
122.167
142.014
Gronenschild,
E. H. B. M.
064.048
142.045
Gros, J.
105.084 .136
Groshkov, A. G.
141.037
Groshkov, S. N.
061.020
Groslambert, J.
035.016
Gross, P. G.
065.052
Gross, S. H.
099.225
Grossi, M. D.
083.051 .052
Grossman, A. S.
062.003
099.001
Grossman, L.
105.127
Grossman, N.
082.035
Grossmann-Doerth, U.
073.025
Groten, E.
046.046 .059
Groth, E. J.
141.315 .316 .319
151.036
160.012
Grove, G.
033.007
Grove, T. L.
094.404
Grover, J. E.
105.045
Groves, G. V.
082.074
Growther, J. G.
008.043
Gruber, D. E.
134.013
Gruber, G. M.
099.048
Grubissich, C.
153.001
Gruen, E.
105.024
106.004 .005
Gruenwaldt, H.
074.129
084.207
Grujic, G.
045.029
Grujic, R.
045.028 .030
046.040
Gryaznova, S. N.
105.031
Grygar, J.
003.061
Grzedzielski, P.
098.075
Grzedzielski, S.
074.088
Guadagnoli, M. D.
031.052
Gualdi, C.
155.010
Gubanov, V. S.
041.006 .015
Gubbay, J. S.
031.261 .276
Gubbins, D.
062.013
Gudehus, D. H.
031.221 .232
160.031
Gudkova, V. A.
084.273
Gudzenko, L. I.
072.029 .031
Gueduer, N.
121.063
Guelin, M.
158.159
Guelmen, Oe.
121.063
Guentzel-Lingner, U.
002.009
Guerin, C.
106.054
Guerrero, G.
123.031
Guertler, J.
158.026
Guest, J. E.
011.065
Guetter, H. H.
041.035
Guggisberg, S.
094.464
Guibert, J.
158.155
Guichard, F.
094.196
Guida, P.
124.103
Guidi, I.
143.080 .083 .084
Guidice, D. A.
077.039
Guillaume, F.
035.008
Guinan, E. F.
113.065
Guinot, B.
044.010 .040
Guisset, J.-L.
035.008
Guitart, A.
003.122
Gula, R.
077.049
161.005
Gul'el'mi, A. V.
106.035
Gulkis, S.
091.032
Gull, G.
131.540
Gull, S. F.
141.056
160.030
Gull, T. R.
133.027
134.002
Gulliford, P.
125.043
Gul'medov, Kh.
104.018 .025 .063
Gul'medov, Kh. D.
104.029 .088
Gulyaev, R. A.
073.122
Gumtau, M.
036.012
Gun'ko, N. A.
032.527
Gunn, J. E.
158.312 .314
160.011
162.015 .066
Gunton, D. J.
033.056
Gupta, J. C.
094.010
Gupta, K. Sen
See Sen Gupta, K.

Gupta, S. D.
031.087
Gupta, S. Sen
See Sen Gupta, S.
Gurevich, A. V.
084.224 .225 .226
Gurevich, L. Eh.
162.007 .041
Gurin, L. S.
074.043
Gurklyte, A.
113.055
Gurkovskij, E. V.
031.049
Gurnett, D. A.
084.003 .209 .301 .303
099.213
Gurovich, V. Ts.
042.074
Gurshtejn, A. A.
003.062
015.015
Gurski, T. R.
034.034
Gursky, H.
012.014
032.523 .542
142.014 .026 .056 .071
.085 .088 .121 .122
.156 .194
Gurtovenko, Eh. A.
071.004 .020 .029 .060
Gurvich, L. V.
011.010
Gur'yanov, A. Eh.
032.017
Gurzadyan, G. A.
114.020 .111
132.061
133.002
Gusakov, V. S.
162.046
Gusakovskaya, L. B.
082.032
104.012
Guseinov, O. H.
See Gusejnov, O. Kh.
Gusejnov, K. I.
071.043
Gusejnov, M. D.
072.035 .038
Gusejnov, O. Kh.
124.005
Gusev, A. E.
052.006
Gusev, L. I.
052.059
Gus'kova, E. G.
003.063
105.066
Gussmann, E. A.
008.081
Gustafsson, B.
064.026
114.100
Gutarko, T. P.
034.058
Guth, V.
047.025
103.131
Guthrie, B. N. G.
141.004
Guthrie, P. D.
142.113
Gutierrez, J. A.
085.022
Gutierrez-Moreno, A.
113.048
Gutsche, G. D.
014.037
Gutshabash, S. D.
093.018
Guyer, R. A.
065.123 .188
Guzenkova, N. D.
081.044
Gyldenkerne, K.
113.033
121.042 .079
Gyul'budagyan, A. L.
122.170
132.060

Haag, H.
105.129
Haar, D. Ter
See Ter Haar, D.
Haas, W. H.
095.008
Haase, R.
004.035
015.001
Habing, H. J.
131.133
132.017
Hablick, D.
106.050
Hachenberg, O.
031.202
Hack, M.
114.375
Hackney, K. R.
141.142
Hackney, R. L.
141.142
Hackwell, J. A.
114.031
132.027 .046
Haddow, K. A.
103.130 .131
Hadjidemetriou, J. D.
042.037 .038 .045
Hadrava, P.
063.034
Haefner, R.
064.064
Haelg, W.
022.070
Haemeen-Anttila, K. A.
100.003 .014
Haenel, A.
124.104
Haenni, L.
114.365
Haenni, U.
114.364
Haensch, W.
003.116
Haensel, P.
061.095 .096
Haerm, R.
065.029 .030
Haeussler, K.
123.065
Haffner, H.
008.116
Hagan, F. A.
143.066
Hagan, L.
022.051
Hagen, F. A.
143.059
Hagen, J. P.
034.163
077.012
Hagen, W.
114.030 .070
Hagen-Thorn, V. A.
158.102 .239
Haggerty, M. J.
151.015
Hagihara, Y.
003.064
Hagyard, M. J.
072.015
Hahn, G.
123.076
152.004
Hainebach, K.
125.038
Hainebach, K. L.
061.072
Haisch, B. M.
063.017
064.059
Hajfler, J.
079.103
Halenka, J.
022.101
Halevy, I.
085.010
Hall, C. F.
003.110
Hall, D.
071.016
Hall, D. L.
065.160
Hall, D. N.
071.018
Hall, D. N. B.
034.033
082.005
Hall, D. S.
118.019
121.015 .054 .074
Hall, R. D.
142.118 .165
Hallam, M.
094.185 .463
Halliday, I.
104.031 .067
124.104
Hallissey, M.
084.028
Halmos, F.
046.019 .023
Halpern, L.
066.029 .092
Hamada, T.
062.026
063.040

Hamajima, K.
131.535 .545
Hamdi, M. A.
114.074
Hamilton, D. C.
099.024
Hamilton, W. O.
034.031
066.025
Hamilton Jr., R. J.
033.025 .027
Hammerschlag-Hensberge, G.
113.039
142.223
Hammond, G. L.
126.028
Han, N.
151.091
Hanbury Brown, R.
142.095
Handbury, M. J.
107.011
Hanes, T. E.
003.113
Hanke, W.
035.002
Hankins, T. H.
141.329
Hanna, M. M.
036.015
Hannemann, D.
034.155
Hanner, M. S.
106.022
Hansen, C. J.
065.108 .113
122.114
Hansen, J. E.
093.041
Hansen, O. L.
082.117
099.223
124.103
154.019
Hansen, R.
071.054
Hansen, R. C.
033.024
Hansen, R. T.
073.061 .127
074.036 .109
Hansen, S.
071.054
Hansen, S. F.
074.109
Hanser, F. A.
073.138
Hanson, W. B.
084.032
Hapgood, M.
123.078
Hapke, B.
093.050
094.122 .401
Hapke, B. A.
093.049
Hara, H.
124.104
Hara, T.
044.041
Harada, T.
103.100
124.103
Hardee, P. E.
141.355
Harel, C.
022.059 .060
Hargrave, P. J.
141.034
Hari Dass, N. D.
141.345
Harlan, E. A.
103.100 .140
111.001
Harmanec, P.
119.007
Harmer, C. F. W.
031.005
Harmon, R. S.
105.037 .133
Harms, R. J.
158.082
Harnden Jr., F. R.
032.542
Harney, R. C.
022.045
Haro, G.
122.009 .148
Harper, D. A.
091.031
141.636 .639
155.064
158.086
Harries, J. E.
082.041
Harrington, R. S.
041.035
103.100
117.036
Harris, A. W.
141.135
Harris, B. L.
114.029
Harris, G. D.
034.039
Harris, R. M.
082.010
Harris, W. E.
155.019
158.221
Harrison, E. R.
141.331
Harrison, K. M.
031.257
Harrison, T. G.
061.015 .024
Hart, D.
141.064
Hart, H. M.
155.032
Harteck, P.
097.051
Hartl, H.
155.008
Hartl, H. J.
113.007
Hartle, J. B.
065.010
Hartle, R. E.
092.008
Hartmann, G. K.
083.076
Hartmann, L.
064.053
065.003
114.098
Hartmann, R.
074.127
Hartmann, W. K.
097.201
098.015
Hartner, W.
012.006
Hartung, J. B.
094.116 .156
Harvey, D. L.
003.065
Harvey, G. A.
104.010 .011 .039 .040
158.306
Harvey, J.
071.016 .019
074.074
Harvey, J. E.
031.047
Harvey, J. W.
073.052
Harvey, K. L.
073.047 .052
Harvig, V.
021.018
Harwit, M.
034.098
131.014 .540
158.057
162.059
Harwood, J. M.
084.283
Harwood, K. A.
004.127
Haschick, A. D.
131.121
141.032
Haschick, A. H.
131.058
Hasegawa, I.
103.131 .140
Hasegawa, T.
114.359
Haselton, J. D.
105.055
Haskin, L. A.
094.438
Haslam, C. G. T.
141.152
Hasse, I.
132.056
Hatchett, S.
142.003
Hatchett, S. P.
142.070
Hatfield, B.
103.100
Hatfield, H. R.
010.012
Hauck, B.
113.034
114.055
Haug, E.
022.047
061.070

Haug, E.
073.159
Haug, U.
113.003
Hauge, Oe.
003.066
099.071
Haupt, H.
008.040
101.013
Haupt, H. F.
152.004
Haupt, W.
122.015
Hausberg, K.
004.083
Hauser, M. G.
034.036
131.025 .523
141.047
Haussecker, K.
034.009
Havas, P.
066.094
Haves, P.
141.088 .111 .115
Havnes, O.
065.034
Hawarden, T.
103.126
Hawarden, T. G.
153.006 .007 .028
Hawkes, R. L.
104.026
Hawkins, F. J.
099.016
Hawkins, M. R. S.
160.020
Hawlęy, D. L.
158.001
Hayakawa, S.
032.543
076.025
142.166
Hayes, D. P.
114.338
Hayes, E. N.
003.034
Hayes, J. M.
051.001
094.418
Hayli, A.
004.019
012.009
158.067
Haymes, R. C.
142.118 .165
Haynes, M.
158.113
Haynes, R. F.
131.147
141.117
158.179
Hays, J. F.
094.404
Hazard, C.
125.004
Hazra, L. N.
031.088
He, Y.
031.023
Heacock, R. R.
084.013
Head, J. W.
097.010
Heap, S. R.
114.400
Heard, J. F.
008.085
119.007
Hearn, D.
141.326
142.081 .089 .188 .208
Heasley, J. N.
073.125
Heath, D. F.
076.056
Heaton, K. C.
066.135
Heck, A.
015.034
021.016
113.043
115.016
122.080 .132
Hedeman, E. R.
003.122
Hedervari, P.
097.064
Hedgecock, P. C.
084.214
106.012 .027 .051
Hedlund, R. C.
032.548
Hefferlin, R. A.
022.049
Hegedusic, M.
015.035 .036
101.012
Heggie, D. C.
117.026
151.070
Hegyi, D. J.
065.110
Heidinger, F.
032.511
Heiken, G.
094.132
Heikkila, W. J.
084.245
Heiles, C.
131.011
158.083
Heinlo, A. H.
064.052
Heinrich, I.
002.009
Heintz, W. D.
098.014
118.008 .009 .010 .033 .036 .037
Heintze, J. R. W.
031.416
Heintzmann, H.
141.362
Heise, G.
002.004
Heise, J.
064.048
122.167
142.014 .045
154.036
Heiser, E.
124.104
Heiserman, D.
003.067
Hejna, L.
072.011
Heldt, J.
022.064
Helfer, H. L.
113.011
114.021
153.003
Helin, E.
098.051
103.100
Heller, M.
066.100 .101 .102
Helliwell, R. A.
084.300
Hellman, C. D.
004.023 .049
Hellwig, H.
044.014
Helmken, H. F.
142.095
Helt, B. E.
113.033
Hemenway, C. L.
012.030
082.080
104.042 .043
105.049
Hemmleb, G.
046.051
Hendrickson, R. A.
084.304
Henize, K. G.
114.005 .313
117.035
124.006
Henley, J. R.
015.018
Henn, F.
002.009
Henning, G.
033.059
Henning, K.
125.050
142.020
Henon, M.
151.019
Henrichs, H. F.
022.074
Henriksen, K.
082.067
084.001
Henriksen, R. N.
066.135 .136
117.059
141.322 .330 .333 .350
Henriksen, S.
046.063
Henry, R.
142.182 .183
Henry, R. C.
094.131
114.345 .350 .370 .398
131.183
142.011 .139
Henson, B.
073.158

Henze, W.
072.012
074.030 .031
Heppenheimer, T. A.
098.039
Herald, D.
103.100 .130 .131 .133 .136
Herbig, G. H.
114.312 .351
132.006
Herbst, E.
022.063
122.079
Herbst, W
152.001
Herbst, W.
132.052
152.005
155.002
158.194
Herczeg, T. J.
121.117
Herget, P.
098.095
Hering, R.
113.008
Herman, M.
093.015 .046
Hernandez, G. R.
022.124
Herndon, J. M.
105.058 .134
Herneck, F.
003.068
Heroux, L.
083.073 .074
Herpe, G.
066.079
Herring, J. R. H.
076.010
Herrmann, D. B.
003.069
004.082
008.013
013.015
015.027
Herrmann, J.
061.042
Hers, J.
010.007
096.006
103.100 .119 .130 .131 .136
Hershey, J. L.
034.141
121.021 .071
Herzog, A.
092.005
Herzog, A. D.
099.224
Hess, M.
121.136
Hess, S. L.
093.033
Hessberg, H.
034.056
113.074
114.335
Hesse, K. H.
141.359
Hesser, J. E.
126.024
131.015
154.019 .035
Hetherington, N. S.
004.080 .121
Heudier, J. L.
034.152
Hewerdine, A. R.
008.043
Hewish, A.
141.307
Hewitt, A. V.
041.035
Hey, J. D.
062.009
Hey, J. S.
003.070
Heymann, D.
094.405 .425
Heyse, J. V.
105.047 .135
Hibbert, M.
123.033
Hicks, T. R.
131.036
Hidalgo, M. A.
083.063
Hide, R.
099.017
Higdon, J. C.
143.081
Higginbotham, N. A.
114.395
Higgins, C. S.
122.156
Higgins, G. H.
094.101
Higgins, J. E.
083.073 .074
Higgs, L. A.
022.110
125.052
131.511
Higuchi, H.
094.439 .468
105.084 .136
Hildebrand, R.
141.636
Hildebrandt, G.
113.036
116.010
Hilditch, C. E. J.
153.002
Hilditch, R. W.
113.029
114.047
119.004
153.002
Hildner, E.
074.017 .036
094.005
Hilgeman, T.
131.037 .175
Hill, G.
113.029
114.047
153.041
Hill, H. A.
031.214
080.018 .030
Hill, J. K.
131.144
Hill, P. W.
031.412
112.002
113.046
119.004
153.002
Hill, R. K.
141.119
Hill, R. W.
125.007 .008
142.059
Hill, S. J.
064.054
122.188
Hill, T. W.
062.060
091.010
099.086
Hillas, A. M.
143.082
Hiller, H.
054.001 .004
Hillier, R. R.
032.526
Hills, D. L.
131.510
Hills, H. K.
084.278
Hills, J. G.
151.064
154.032
Hilsman, K. A.
160.003
Hindley, K.
010.012
102.030
Hindley, K. B.
010.012
104.079 .092
Hines, C. O.
085.010
Hintenberger, H.
105.098 .139 .198
Hinteregger, H.
121.039
Hinteregger, H. F.
031.218
Hintzen, P.
126.001 .011 .013
Hinze, J.
022.018
Hippelein, H. H.
132.043
Hirakawa, H.
066.109
Hiramatsu, Y.
103.140
Hirasawa, T.
083.038
Hirayama, T.
041.034
Hirota, E.
131.021
Hirrle, H.
014.005
Hirschfeld, T.
031.241
Hirshberg, J.
074.025

Hirshberg, J.
106.026
Hirshfeld, A. W.
122.166
Hirth, W.
031.202
073.144
077.002 .003 .029
080.012
Hishikura, T.
124.103
Hitch, B. F.
033.053
Hitzl, D. L.
042.051
Hjellming, R. M.
124.104
141.028 .311
142.017
Ho, P. T. P.
131.123
Hoag, A.
142.192
Hobbs, L. M.
131.053 .155 .178
Hobbs, R. W.
034.039
072.081
103.102
114.330
Hocking, W. H.
022.122
Hodge, P.
103.100
159.901
Hodge, P. E.
142.195
Hodge, P. W.
106.040
122.001
158.132 .198 .207
Hodges Jr., R. R.
094.162 .163
Hodson, S. W.
122.190
Hoeg, E.
031.290
032.036 .044
Hoegner, W.
154.020
160.016
Hoekstra, R.
114.117
Hoenl, H.
066.055
Hoerner, S. Von
See Von Hoerner, S.
Hoerz, F.
094.107 .116 .156 .475
Hoffleit, D.
114.351
121.116
123.026 .033 .043 .044
Hoffman, J.
124.104
142.186
Hoffman, J. A.
134.006
Hoffman, J. H.
083.071
084.032
Hoffman, J. H.
094.163
Hoffman, K. R. A.
084.303
Hoffman, M. M.
032.540
074.041
Hoffman, N. M.
064.050
Hoffman, R. A.
084.032
Hoffmann, H.-J.
105.024
Hoffmann, J. A.
134.012
Hoffmann, M.
103.100
122.054 .055 .066 .174
Hofmann, J. E.
005.008
Hofmann, W.
106.008
Hofsaess, D.
022.012
Hofstadter, R.
034.154
061.045
Hogan, J. S.
099.100
Hogan, P. A.
066.005
Hogg, H. S.
007.000
010.032
154.034
Hohenberg, C. M.
094.110 .408 .465
Hohenkerk, C. Y.
075.021
Hohl, F.
151.027 .050 .055
Hoinkes, G.
105.137
Holden, F.
118.025
Hollars, D. R.
114.033
Hollenbach, D. J.
064.022
Hollis, J. M.
131.052 .080 .174
Hollweg, J. V.
143.003
Holm, A. V.
113.060
122.224
Holman, G. D.
160.039
Holmberg, E. B.
113.044
Holmes, H. F.
094.424 .426
Holt, S.
142.196
Holt, S. S.
142.015 .025 .097 .120
.153 .154 .155 .190
.201 .202 .203 .227
.230 .239
Holts, J. T.
114.029
Holweger, H.
061.019
071.033
114.308
Holzer, T. E.
106.026
Honda, M.
124.103 .104
Honda, T.
131.021
Hones Jr., E. W.
084.312
106.901
Hong, S. S.
141.105
Honig, E.
066.157
Hooley, T.
034.146
Hoover, P. S.
159.008
Hoover, R. B.
072.012 .015
074.030 .031
Hopfield, A.
094.005
Hopkins, H. H.
031.007
Hopper, R. W.
094.117
Horak, T. B.
117.025
Horedt, G.
066.011
117.057
Horiai, K.
044.041
Horn, P.
094.189 .451 .466
Hornby, P. W.
123.006
Horne, D.
034.146
Horng, J.-T.
084.203
Horning, B. L.
094.178
Horsky, Z.
005.005 .024
Horwitz, G.
151.002
Hossfield, C. H.
010.001
Houck, J. R.
099.033
Hough, J.
066.080
Houghton, J. T.
093.057
Houk, N.
114.337
Houlden, M. A.
010.012
104.079
Hourani, H.
075.018
Houtgast, J.
079.100
Houziaux, L.
114.360

Hovestadt, D.
073.018
078.004 .029
084.208 .406
Howard, C. J.
082.029
Howard, J. N.
005.020
Howard, R.
075.015
080.070
Howard, T.
093.052
Howard, W. M.
105.015
Howarth, I. D.
005.028
Howe, H. C.
094.157
Howse, D.
003.071
Hoyle, F.
003.072 .073 .074
124.020
Hoyng, P.
073.012
076.027 .059
Hoyt, W. G.
100.004
Hristov, V. K.
081.020
Hruska, A.
066.054
Hu, E. M.
131.157
Hu, W.
158.128
Huang, H. C.
033.068
Huang, K.
055.007
Huang, S.-S.
121.112
Huang, Y.-N.
074.013
Huba, J. D.
074.901
Hubbard, R. F.
099.213
Hubbard, W. B.
091.002 .004 .025
099.047 .077
Huber, M. C. E.
074.005
076.002
Hubert, H.
114.052
Huchra, J. P.
125.063
158.096 .105 .106
Huchtmeier, W. K.
158.023 .024 .034 .049
.170 .188 .235
Huchtmeier, W. K. H.
158.179
Huckle, H. E.
142.233
Hudec, R.
114.393
142.243
Hudson, H.
122.163
Hudson, H. S.
073.020 .097 .142
076.013 .028
131.079
158.069
Hudson Jr., R. D.
003.075
Huebner, J.
005.004 .006
Huebner, W.
094.452
Huett, V.
004.123
Huggins, P. J.
131.148
132.042
Hughes, D. W.
102.026
104.013
Hughes, E. B.
034.154
Hughes, V. A.
131.153 .546 .547
141.113
142.007 .175
Huguenin, D.
013.010
071.037
Hujer, K.
005.005
Hull, A. B.
124.104
Hulse, R. A.
141.332
Hultqvist, B.
084.002 .011 .411
Humberg, H. W.
141.604
Hummer, D. G.
063.018 .039
064.042
Humphreys, R. M.
114.322 .382
142.162
155.056
Hundhausen, A. J.
072.050
Huneke, J. C.
105.059 .060
Hunger, K.
008.011
065.015
Hunstead, R.
158.085
Hunt, A. J.
033.021
Hunt, G. E.
011.065
099.012
Hunt, R.
142.052
161.004
Hunten, D. M.
082.014
099.077
100.006
Hunter, A.
008.043
010.022
Hunter, C.
151.021 .063
Hunter, J. H.
065.106
Hunter Jr., J. H.
099.002
Huntley, J.
158.084
Huntley, J. M.
151.003 .081
Huppert, H. E.
062.068
Huppler, D.
103.102
Hurford, G. J.
078.011
Hurkens, R.
119.007
Hurley, K. C.
099.015
Hurly, R. F.
047.029
Hurnik, H.
055.009
Hurst, G. M.
124.025 .103 .110
Hurst, R. B.
031.268
Hurta, L.
079.103
Hurukawa, K.
041.034
098.061
Husain, L.
094.432
Huss, G.
105.099
Husson, J. C.
032.531
Hutcheon, I. D.
094.441
106.055
Hutcheon, R. J.
074.007
076.006
Hutchings, J. B.
008.109
114.375
115.013
119.002 .013
142.112
Hutchinson, J. L.
064.054
122.091 .188
Hutchison, R.
105.013 .132
Huth, H.
124.103
Hutton, L.
121.039
Hutton, L. K.
031.218
141.143
Hyland, A. R.
114.392
141.623
Hynek, J. A.
005.007
Hynes, S. J.
153.012

Hysom, E. J.
010.012

Iacomo Jr., P.
077.053
Ianna, P. A.
113.016
153.011
Iannini, G. M.
079.108
Ibanoglu, C.
121.063
Ibragimov, N. B.
003.049
097.032
099.031
101.009
Ibrahim, M.
094.159
Ibrus, Ue.
034.097
Icenogle, H. W.
031.047
Ichimaru, S.
062.050
Icke, V.
155.041
Idel'son, N. I.
003.076
Ignat'ev, Yu. G.
062.065
Iguchi, T.
022.062
131.145
Ihrig, E.
162.083
Iijima, T.
032.543
084.257
103.114
Ikeda, M.
061.001
Ikee, R.
022.025
Ikeuchi, S.
151.094
160.048
Ikramuddin, M.
105.027 .138
Il'inov, A. S.
143.070
Ilk, K. H.
046.067
Ill, M.
082.056
Illarionova, N. V.
076.046
Illes-Almar, E.
054.005
Illing, R. M. E.
116.004
Illingworth, G.
125.106
154.010
Ilmas, M.
114.054
Ilovaisky, S. A.
142.228
Il'yasov, U. I.
080.071
Imamura, M.
105.157
Imbert, M.
121.080
159.001
Imbro, D. R.
082.028
Imhof, W. L.
084.404 .405
Imhoff, C. L.
064.029
115.012
Imnadze, M. P.
103.118
Imshennik, V. S.
003.077
065.067 .150
066.159
Ingel', L. Kh.
066.049 .098
Ingemann-Hilberg, C.
022.011
Ingersoll, A. P.
071.002
097.040 .050
Ingham, W.
141.112
Ingham, W. H.
126.021
Inglis, D. R.
062.030
Innanen, K. A.
151.013
Inoue, M.
033.042
Intriligator, D. S.
099.020 .023
Ip, W.-H.
098.043
103.102
Ipatov, A. P.
114.062
Ipatov, A. V.
079.101
141.116
157.003
Ipser, J. R.
066.001 .053
151.028
Iriarte Erro, B.
122.216
Iribarne, J. V.
003.078
Irkaev, B. N.
123.064
Irkaeva, Sh. N.
106.068
Iroshnikov, R. S.
080.017
Irshad, M.
022.087
Irvine, J. M.
003.079
Irvine, W. M.
100.018
Irwin, D. J. G.
022.046 .068
Isaeva, F. I.
098.094
Isakov, I. S.
032.056
Isaksen, I. S. A.
078.003
Isamutdinov, Sh. O.
104.078
Isern, J.
061.016
Isham, C. J.
012.048
Ishiguro, M.
033.030
Ishii, H.
045.011 .016
Ishii, K.
022.025
Ishii, T.
094.473
Iskudaryan, S. G.
158.109 .110 .240
Isles, J.
010.012
Isles, J. E.
123.078
Isles, J. S.
010.012
Isobe, S.
124.104
Israel, F. P.
132.017
Israel, G.
066.030
Ito, K.
032.543
Ito, N.
061.092
Itoh, N.
141.303 .320
Iuell, P.
072.075
Ivanchuk, V. I.
074.055
Ivanenko, D. D.
066.124
Ivanov, A. A.
143.072
Ivanov, A. I.
082.095 .096
Ivanov, E. V.
011.061
Ivanov, G. R.
122.030
123.051
Ivanov, K. G.
062.036
106.030 .032
Ivanov, L. N.
061.071
Ivanov, M. Yu.
066.140
Ivanov, V. M.
094.062
Ivanov, V. N.
034.076 .118
Ivanov, V. V.
063.027
064.052
Ivanov-Kholodnyj, G. S.
083.030 .075
Ivanova, E. Ya.
052.048
Ivanova, G. M.
003.080

Ivanova, L. N.
065.150
Ivanova, T. S.
062.057
072.082
Ivanova, T. V.
042.018
Ivanova, V.
052.020
Ivchenko, M. P.
084.242
Iversen, J. D.
097.037
Ives, J.
142.065
Ives, J. C.
142.016
Iwadate, K.
032.034
044.041
Iwanowska, W.
007.000
Iye, M.
125.026 .103
Izakov, M. N.
097.043
Izhakevich, E. M.
094.019
096.011
Izotov, A. A.
046.015
Izotov, Yu. I.
071.039

Jaakkola, T.
141.095
158.014 .052
Jaaniste, J.
155.047
Jacchia, L. G.
031.237
082.118
Jackson, A. R. G.
141.161
Jackson, B. W.
032.506
Jackson, J. C.
066.057
142.050 .052
160.009
Jackson, P. D.
155.026
Jacobs, J. A.
084.284
Jacobs, J. W.
094.438
Jacobs, K. C.
080.004
162.024
Jacobs, V. L.
022.019
Jacobs, W. W.
061.048
Jacoub, Sammani A.
074.020
Jacques, M.
124.103
Jaekel, C.
157.005
Jaffe, J.
066.014
Jaffe, W. J.
160.005
Jaggi, R. K.
084.901
Jagoda, N.
034.157
Jain, D. C.
022.004
Jakimiec, J.
076.054
Jamar, C.
131.126 .211
Jamar, J.
036.017
James, J. R.
033.039
James, W. A.
119.001
Jameson, R. F.
121.051
124.111
Janes, A. F.
032.525
Janes, K. A.
113.015
Janetzke, R. W.
106.076
Janin, G.
042.022
151.015
Janin, L.
035.018
Janson, M. M.
162.014
Janssen, M. A.
100.024
Jaques, A. L.
105.089
Jardim, J. O.
131.534
Jarosewich, E.
105.005 .199
Jarrett, A. H.
122.144 .145
Jarzebowski, T.
124.100
Jaschek, C.
002.008
117.005
Jasicek, H.
123.076
Jauho, P.
062.072
Jauncey, D.
141.135
142.222
Jauncey, D. L.
141.059 .082 .141
Jayanthi, U. B.
106.001
Jedwab, J.
105.032 .036
Jeffers, S.
124.104
Jefferts, K. B.
132.042
Jefferys, W. H.
100.205
Jeffries, A. J.
096.003
Jehanno, C.
032.541
Jekabsons, C.
103.136
Jelley, J. V.
142.119
Jenkins, E. B.
131.114
Jenkins, J.
103.100
Jennens, P. A.
113.011 .012
114.021
153.003
Jenner, D. C.
158.189
Jennings, R. E.
032.544
132.009
141.629
Jensen, E. B.
158.071 .072
Jernigan, G.
142.073 .075 .076 .078
.081 .084 .211
Jerome, D. Y.
105.119
Jerzykiewicz, M.
122.173
Jessberger, E.
094.466
Jessberger, E. K.
094.189 .451
Jetzer, F.
099.049
100.030
Jezewski, D. J.
042.047
Jimenez, J. J.
035.014
Jochum, K. P.
105.139
Jockers, K.
073.128
Joensen, P.
131.152
Joergensen, H. E.
121.042 .079 .083
Joergensen, N. T.
004.094
Joeveer, M.
155.045
Joeveer, M. M.
011.035
Johannin-Gilles, A.
022.010
Johanson, J. M.
162.048
Johansson, L. E. B.
131.136
Johler, J. R.
035.001
John, R. W.
066.145
John, T. L.
022.021
Johns, M. W.
065.056

Johns, O.
061.047
Johns, P. B.
033.033
Johnson, A. A.
105.039
Johnson, A. M.
031.078
Johnson, B.
141.135
Johnson, C.
123.022
Johnson, D. R.
022.056
Johnson, D. W.
097.051
Johnson, F. M.
131.067
Johnson, H. L.
113.069
Johnson, H. M.
114.321
142.163
Johnson, H. R.
064.055 .056 .060 .068
065.139
Johnson, M. B.
065.019
Johnson, T. V.
099.216 .242
Johnston, K. J.
131.003 .046 .057 .158
Joki, E. G.
076.007
Jokipii, J. R.
063.025 .037
080.002
091.004
099.047
106.074
Jolly, S. T.
033.068
Jones, A. F.
123.045 .046
Jones, A. V.
084.008
Jones, B.
103.140
114.304
Jones, B. J. T.
143.008 .009
Jones, B. W.
082.009
Jones, D. E.
093.007
Jones, D. H. P.
008.043
122.034
Jones, E. M.
125.012 .033
131.069
159.003
Jones, H. P.
034.039
Jones, J.
104.026 .060
Jones, J. E.
011.017
Jones, K. G.
003.081
Jones, P. B.
061.059
Jones, T.
158.309
Jones, T. B.
073.158
Jones, T. W.
073.020
131.073 .162
141.038
142.004
Jones, W. P.
094.151
Jordan, C.
074.056
076.011
114.318
Jordan, J. L.
094.405
Jordan, P.
081.021
Jordan, T. M.
014.004
Jorden, P. R.
124.104
Joshi, P.
131.047 .048
Joshi, U. C.
153.042
Joss, P. C.
142.189 .229
Josse, M.
031.295
Journet, A.
041.016
Jovanovic, M.
044.009 .043
Jovanovic, S.
094.422 .440
Joyce, G.
099.213
Joyce, R. M.
034.160
Joyce, R. R.
034.037
115.022
141.640
158.193
Joyce, R. S.
034.027 .124
Juillerat, R.
052.002
Julienne, P. S.
082.027
Juliusson, E.
143.017
Jull, A. J. T.
094.190 .191 .461
Jupp, A. H.
042.093
Jura, M.
131.044 .066 .903
Jurgitis, I.
032.052

Kabelac, I.
046.033
Kaburaki, O.
062.046
Kadla, Z. I.
154.006 .020
Kaehler, H.
065.114
Kaempfer, B.
061.042
Kaevitser, V. I.
093.023
094.198
Kafatos, M.
132.020
Kafka, P.
066.084
Kahl, S. B.
094.442
Kahler, S.
073.011 .053 .103 .104
074.073
Kahler, S. W.
073.001 .003 .009
074.008
Kahn, C.
004.074
Kahn, F.
004.074
Kahn, S. M.
142.037
Kai, K.
077.054
Kaifu, N.
131.086 .109
Kaiser, M. L.
077.052
082.001
Kaitchuck, R. H.
122.143
Kajdanovskij, N. L.
141.116
Kajmakov, E. A.
102.025
Kajsin, V. K.
151.006
Kakar, R. K.
131.055
Kakaras, G.
113.054
Kakati, B. N.
065.181
Kakhidze, G. P.
032.534
143.071
Kakkuri, J.
046.025
Kalata, K.
142.085
Kaledin, B. F.
031.061
Kalenichenko, V. V.
104.082
Kaler, J. B.
132.058
Kalikhevich, F. F.
041.062
098.008 .048
Kalinenkov, N. D.
034.062 .064 .071
Kalinin, A. P.
097.048
Kalinin, M. I.
022.091

Kalinina, I. M.
041.042
Kalinkin, L. F.
143.074
Kalinkov, M.
162.071
Kalinowski, J. K.
082.116
153.034
Kalinyak, A. A.
071.047
Kalitis, R.
034.078
Kalkofen, W.
080.014
Kallman, C.-G.
061.093
Kalloglyan, A. T.
158.237
Kalman, L. B.
022.032
Kalnajs, A. J.
131.541
Kalnay De Rivas, E.
093.027
Kaloshin, V. O.
033.016
077.014
Kaluzienski, L. J.
142.015 .025 .120 .153 .154 .202 .203 .227 .230 .239
Kalv, P.
021.018
031.238
121.084 .085
Kalybaev, A. A.
045.002 .015
Kalzhanov, B.
031.250
085.015
Kamath, N. S.
034.098
Kambu, F.
074.093
Kamel, A. A.
054.011
Kameswara Rao, N.
114.084
Kamide, Y.
084.019 .248
Kamijo, F.
022.062 .123
Kaminker, A. D.
062.023
Kamionko, L. A.
034.074
Kamperman, T. M.
032.537
114.117
Kanalina, O. M.
122.063
Kanda
103.130
Kanda, M.
033.073
Kandaurova, K. A.
072.041
Kane, R. P.
083.032
143.019
Kane, S. R.
012.003
Kangas, J.
084.012 .013
Kaniuth, K.
046.067
Kanno, M.
073.067
074.069
Kanno, S.
062.026
063.040
Kantz, M. L.
103.100 .130 .131 .133 .136 .140
Kap-herr, A. Von
See Von Kap-herr, A.
Kapahi, V. K.
162.006
Kapkov, L. A.
034.018
054.003
Kaplan, G. H.
047.039
Kaplan, J. I.
065.006 .189
Kaplan, S. A.
003.085
063.049
Karachentsev, I.
158.100
Karachentsev, I. D.
151.085
158.032
160.023 .042
Karachun, A. M.
033.014
Karaev, A. A.
073.150
Karamyan, L. M.
034.048
Karapetyan, B. O.
034.084
Karashtin, A. N.
062.058
Karimova, D. K.
112.004 .006 .008
142.242
Karipova, M.
066.048
Karipova, M. K.
042.074
066.105
Karitskaya, E. A.
142.169
Karle, J. H.
121.124
Karmadonov, I. P.
143.026
Karnitskij, P. N.
014.013
Karoji, H.
158.014
162.040
Karp, A. H.
061.027
063.004
064.036
065.009
114.379
122.082
Karpinskij, V. N.
031.074
Karpman, V. I.
141.306
Karr Jr., C.
003.082
Karsky, G.
046.021
Karszenbaum, H.
083.014
Kartalev, M. D.
106.033
Kartasheva, T. A.
113.064
121.008
Kartashov, V. F.
099.073 .074 .094 .096
Kartashova, L. G.
073.100
Karyagina, Z. V.
103.101
131.557 .558 .559
132.036 .059
Kasahara, I.
076.025
Kasahara, S.
103.100 .131
Kashcheev, B. L.
104.022 .068
Kasinskis, V.
074.124
Kasper, U.
066.146 .152
Kassal, T.
091.011
Kastel', G. R.
103.100 .111
Kastner, S. O.
022.017
073.069
082.002
Kato, S.
151.041
Kats, I. M.
062.041
Katsnel'son, S. S.
072.028
Katsufrakis, J. P.
084.300
Kattawar, G. W.
063.006
Katz, A.
065.003
Katz, J.
066.181
151.002
Katz, J. I.
124.102
142.200
Katz, L.
073.138
Kaufman, A. S.
074.062
Kaufman, L.
022.128
Kaufman, M.
131.553
Kaufman, Yu. G.
031.211
034.018
054.003

Kaufmann, J. P.
065.014
Kaufmann, P.
077.053
083.009
131.534
158.143
Kaula, W. M.
091.023
094.020 .125 .201
Kaunzner, W.
005.008
Kautzleben, H.
046.053
Kawabata, K.
093.041
Kawabata, S.
121.026
Kawaguchi, I.
071.056
Kawajiri, N.
142.213
Kawakami, K.
042.124
Kawakami, S.
061.092
Kawano, N.
033.042
Kawashima, N.
083.010
Kayser, S.
158.027
Kazanas, D.
134.009
Kazaryan, M. A.
132.019
Kazes, I.
132.015
Kazes, J.
131.163
Kazimirov, D. A.
097.027
Kearney, P. D.
074.072
Keefer, D. R.
079.100
Keenan, D. W.
053.006
151.013
Keene, G. T.
036.001
Kegel, W. H.
064.041
131.022 .040 .125
Kegeles, L. S.
141.334
Keil, K.
094.474
105.092 .099 .124
Keil, S. L.
071.072
Keinonen, J.
065.172
Keirle, P.
032.518
Kelker, D. H.
115.015
Kellen, P. F.
031.246
Kellermann, K. I.
033.007
Kellermann, K. I.
141.073 .074
Kelley, M. C.
083.072
Kellman, S. A.
155.023
Kellogg, E.
160.004
Kelly, H. P.
022.073
Kelly, W. R.
105.140
Kemic, S. B.
022.033
Kemp, J. C.
117.002
121.019
124.104
Kemper, E.
115.014
Kenderdine, S.
141.056
Kendziorra, E.
134.006 .012
Kennedy, G. C.
094.101
Kennedy, J. E.
004.097
Kennel, C. F.
084.276
099.009 .030
Kennel, F. C.
141.341
Keondzhyan, V. P.
107.007
Kepple, P.
073.037
Kepple, P. C.
073.036 .038 .039
Kernahan, J. A.
022.068
Kerr, A. R.
033.063 .064
Kerr, D. E.
032.514
Kerr, F. J.
155.069
Kerridge, J. F.
105.141
Kerstholt, J. H.
131.526
Keryk, R. E.
071.038
Kesson, S. E.
094.427
Kestenbaum, H. L.
073.137
141.346
142.136 .137
Ketelsen, D. A.
124.104
Kevanishvili, G. F.
155.057
Keyes, C. D.
071.015
Khabarov, A. M.
052.061
Khabibullin, Sh. T.
052.010
094.129
Khachaj, Yu. V.
094.139
Khachikyan, Eh. E.
132.019 .034
Khajkin, A. S.
031.079
082.024
Khakimova, M.
084.320
Khalatnikov, I. M.
066.066
162.008 .065
Khalilov, V. M.
114.087
Khalilov, V. R.
066.130
Khaliullin, Kh. F.
113.064
114.369
Khan, I.
099.006
158.025
Khanina, F. B.
098.007
Khantadze, A. G.
083.041
Kharadze, E. K.
011.039
Khare, B. N.
099.034
Kharichev, I. A.
065.073 .074
Kharitonov, A. V.
071.032
114.064 .088 .093 .107 .109
Kharitonova, V. Ya.
105.071 .072
Khasanova, M. Kh.
042.088
Khasyanov, A. F.
093.024
Khatisov, A. Sh.
034.109
074.110
097.203 .204
098.009
124.105
Khazanov, G. V.
083.037 .058
Khejfets, M. E.
081.027
Khentov, A. A.
042.067
091.028
Khersonskij, V. K.
131.187 .188 .189
Khetsuriani, Ts. S.
034.108
073.129
074.110 .111
Khilov, E. D.
004.113
Khlestkov, Yu. A.
066.045
Khlopov, M. Yu.
066.159
Khlystov, A. I.
071.024
Khocholava, G. M.
083.061

Khodyachikh, M. F.
141.120
Khokhlov, M. Z.
074.043 .093
084.309
Khokhlova, V. L.
034.082
064.044
114.056 .075 .086
Kholopov, P. N.
112.013
120.006
122.100 .176
123.012
124.104
152.006
Khommik, L. M.
041.043 .044
Khomyak, V. N.
084.031
Khotilovskaya, T. G.
078.014
Khotimskaya, E. Z.
042.007
Khotinok, R. L.
105.010 .029
Khozov, G. V.
034.086
113.019
122.087
Khramov, G. N.
033.070
Khrenov, L. S.
004.112
Khristiansen, G. B.
143.053 .072
Khromov, G. S.
133.009
Khromov, T. S.
003.133
Khrulev, V. V.
033.071 .072
Khrutskaya, E. V.
041.026
Khudyakova, T. N.
113.019
Kiasat, A.
121.099
Kibardina, M. I.
098.071
103.100
124.104
Kibblewhite, E. J.
034.146
Kiceniuk, T.
098.051
Kidiyarova, V. G.
084.294
Kieffaber, L. M.
082.042
Kieffer, H. H.
100.021
Kieffer, S. W.
094.119
105.008 .033 .162
Kiepenheuer, K. O.
008.036
051.015
082.007
Kiesl, W.
094.469
Kiforenko, B. N.
052.005
Kifune, T.
142.159
Kiko, J.
094.452 .470
Kikuchi, S.
124.108
125.103
Kiladze, R. I.
034.109
124.104
Kilambi, G. C.
122.147
153.039
Kilar, B.
012.024
046.039
Kilkenny, D.
112.002
113.046
153.002
Killen, R. M.
131.524
Kilmartin, P. M.
103.100 .119 .120 .125
.130 .133 .136
Kim, I. S.
074.066
Kinahan, B. F.
065.030
King, A. R.
066.183
King, C. H.
061.061
King, D. A.
004.103
King, D. S.
122.114 .190
King, E. A.
105.142
King, G. W.
065.182
King, I. R.
151.017 .058
King, J. H.
078.032
King, M. W.
117.067
119.012
121.114
124.104
King-Hele, D. G.
054.009
081.002 .016 .017
Kinman, T. D.
141.006
158.303 .320
Kinnucan, P. F.
031.421
Kintanar, R. L.
047.027 .028
074.044
Kiplinger, A. L.
141.066
Kippenhahn, R.
065.076
141.011 .092
Kipper, A.
162.043
Kipper, M.
114.365
Kipper, T.
114.080 .081
Kiraly, P.
143.086
Kirchgraber, U.
021.012
Kirichenko, A. G.
041.007 .008
055.013
Kirichuk, V. V.
082.068 .126
Kirkham, B.
131.143 .207
Kirnozov, F. F.
097.018
Kirova, O. A.
105.071
Kirsch, E.
078.035
084.412
Kirsch, K.
103.100
Kirschbaum, A. R.
094.442
Kirshner, R.
125.107
Kirshner, R. P.
125.018 .023 .067 .100
Kirsten, T.
094.189 .451 .452 .466
.470
105.129
Kir'yan, T. R.
032.018 .019 .020
Kir'yan, V. V.
032.019
Kisabeth, J. L.
084.253
Kisdi, D.
065.193
Kiselev, A. A.
041.004 .060
Kiselev, N.
098.072
Kiselev, N. N.
124.100 .104 .106
Kiseleva, T. K.
122.129
Kiseleva, T. P.
041.004 .061
100.202
Kishonkov, A. K.
074.016 .052
Kislyakov, A. G.
091.029
131.109
132.047
141.097
Kislyuk, V. S.
011.032
094.140
Kiss, E.
105.089
Kissel, J.
105.024
Kissick, W. A.
083.017
Kissin, S. A.
105.025

Kissling, M.
121.149
Kitago, H.
032.034
044.041
Kitai, R.
071.056
Kitamura, M.
121.148
Kivelson, M. G.
084.250
Kiyokawa, M.
121.148
Kizilirmak, A.
121.092
Klapisch, M.
151.002
Klebesadel, R. W.
142.021
Klein, C. A.
031.038
Klein, L.
094.117
Kleinhans, W. A.
031.071
Kleinmann, D. E.
032.547
131.503
141.605
158.087
Kleinmann, S. G.
141.610 .612 .624
Klement, G.
123.076
Klement'eva, A. Yu.
132.014
Klemola, A. R.
103.100 .140
Klemola, A, R.
111.001
Klenina, L. I.
052.067
Klepczynski, W. J.
032.045
Klevenskij, Yu. N.
014.032
Klimas, A. J.
073.160
Klimek, Z.
120.005
162.035 .074
Klimenko, V. V.
076.040 .048
Klimishin, I. A.
062.033
Klinglesmith, D. A.
031.286 .292 .294 .417
114.301
Klinkmann, W.
155.005
Kliore, A.
099.077
Klobuchar, J. A.
083.068
Klochek, N. V.
034.016
Klokkervoll, F.
082.122
Klokocnik, J.
052.019 .021
081.012
Klossa, J.
094.460
Klvana, M.
034.102
Knacke, R. F.
099.043
105.028
155.020
158.193
Knapp, G. R.
141.633
Knapp, S. L.
031.244
141.633
Knappenberger, P. H.
014.001
Kneer, F.
063.012
Kneubuehl, F. K.
071.037
Knezevic, Z.
032.028 .054
Kniffen, D.
061.011 .088
Kniffen; D. A.
141.318
155.003
Knigge, R.
123.059
Knight, C. A.
031.218
Knight, J. W.
085.005
Knoechel, R.
033.055
Knoepfel, K. V.
097.031
Knothe, M.
032.511
Knott, K.
084.279
Knowles, S.
131.148
Knowles, S. H.
131.003 .046
141.143
Kobayashi, K.
106.079
Kobayashi, S.
033.030
Kobzarev, I. Yu.
061.057
162.053
Kobzev, A. A.
082.024
Kobzev, V. A.
078.016
Koch, D.
160.004
Koch, J.
034.127
Koch, K. R.
021.019
081.043
Koch, R. H.
121.123
124.104
Koch-Miramond, L.
142.246
Kocharov, G. E.
012.023
Kocharov, G. E.
080.043 .046 .049
082.069
Kocheev, A. A.
082.089
Kockarts, G.
082.015
Kocourek
114.385
Kodaira, K.
125.026 .103
Koebke, K.
162.076
Koechlin, L.
031.283
Koehler, H. W.
053.003
097.021
099.061 .068
Koelbing, H. M.
004.051
Koen, M. A.
083.058
Koenig, B.
094.105 .186
Koennen, G. P.
082.053
Koeppen, J.
072.008
Koga, R.
076.019
Kogan, A. Yu.
052.063
Kogan, L. R.
033.070
Kogoshvili, N. G.
158.204
Kogut, J. A.
072.063
073.145
Kohl, J. L.
071.013
Kohl, J. W.
084.322
Kohler, P.
032.021
Kohno, T.
034.153
Kohoutek, L.
103.128 .132
113.005
132.008
Koishikawa, M.
103.130 .131 .140
124.103
Kojima, N.
103.100 .111 .130
124.109
Kojoian, G.
142.215
Kokarev, B. D.
031.298
Kokhan, E. K.
071.064
Kokubun, S.
084.257
Kokurin, Yu. L.
094.018
Kolaczek, B.
012.034
094.106

Kolchinskij, I. G.
082.030
Kolenkievicz, R.
046.043
Kolesnichenko, Ya. I.
073.116
Kolesnik, I. G.
065.024 .063 .064
Kolesnikov, S. M.
081.044
085.014
Kolesnikova, E. M.
081.044
085.014
Kollberg, E.
122.003
Kollovsky, B.
073.087
Kolomeets, E. V.
078.005 .016
Kolomejtseva, L. N.
105.070
Kolomenskij, V. D.
105.075
Kolomiets, G. I.
104.085
Kolosov, M. A.
051.007
094.199
097.058 .062
Kolosov, V. A.
143.072
Kolotilov, E. A.
114.058 .065 .069
122.018 .122 .205
Kolpakov, A. I.
031.405
Komar, N. M.
031.063
Komarek, B.
034.003
Komarov, N. S.
064.051
114.050 .059 .348 .381
Komberg, B. V.
011.028
141.098 .304
142.123 .144
158.041 .195
Komesaroff, M. M.
099.003
Kondo, Y.
114.003 .037 .375
117.035
121.049
Kondrashin, V. V.
055.025
Kondrashova, N. N.
071.025
Kondrat'ev, K. Ya.
003.083
097.028
Kondrat'ev, Yu. N.
031.049
Kondrat'eva, E. D.
103.106 .108
Kondrat'eva, L. N.
114.060
133.023 .033
Koning, P. A.
083.040
Konitz, H.
162.094
Konjevic, N.
022.020 .108
Konnikova, V. K.
141.121
Kononov, Eh. Ya.
073.115
076.026
Kononovich, Eh. V.
071.070
072.070
073.075
077.056
Konopleva, V. P.
102.015
Konradi, A.
084.303
Konstantinov, M. Yu.
066.129
Kontor, N. N.
078.014 .018
Konyukhov, A. S.
098.049
103.100
Koo, D. C.
112.016
Kopal, Z.
002.012
012.012
091.034
094.183
121.022 .078
Kopatskaya, E. N.
121.043
Kopecky, M.
072.052
Kopelyanskij, G. D.
033.070
Kopp, R. A.
073.113
Koppe, E. H.
077.053
Korchagin, V. I.
158.043
Korchak, A. A.
062.045
Korduba, B. M.
071.038
Kordylewski, K.
120.005
Korff, D.
082.082
Korff, S. A.
143.029
Korkina, M. P.
066.046 .071
Korneev, V. V.
076.054
Kornilova, T. A.
084.271
Korniski, R.
031.047
Korolev, O. S.
073.080
077.018 .020 .035 .043
Koroleva, L. S.
041.004
100.202
Korol'kov, D. V.
073.083
Korol'kov, D. V.
141.116
Korotev, R. L.
094.438
Korotkikh, T. N.
114.381
Korovkin, A. I.
034.067
Korsun', A. A.
041.003
Korsunova, L. P.
104.028 .057
Korzhavin, A. N.
079.102
Kosai, H.
103.008 .100 .130 .131
124.104 .109
Kose, R.
034.154
Koshelev, V. V.
083.046
Koski, A. T.
158.093 .094
Kostelecky, J.
046.021
Kostenko, V. I.
033.070
Kostik, R. I.
See Kostyk, R. I.
Kostina, L. D.
041.059
045.006
Kostyk, R. I.
073.143
Kostylev, K. K.
034.113
Kostyuk, I. P.
158.230
Kostyuk, N. D.
073.030
Kos'yanenko, A. B.
082.097
Kota, J.
143.042
Kotel'nikov, V. A.
091.026
Kothari, B. K.
094.412
105.143 .144
Kotlaric, S.
046.047
Kotlyarchuk, V. P.
045.025
Kotov, V. A.
080.054
Kottsov, V. A.
032.533
Koubsky, P.
114.385 .390
119.007
Koutchmy, S.
074.021
100.201
Kovac, L.
079.103
Kovacs, S. J.
066.051
Koval', I. K.
003.049
097.023 .026 .045

Koval', V. I.
014.006
Koval'chuk, G. U.
122.203
Kovalchuk, L. V.
082.079
Koval'chuk, M. A.
066.126
Kovalenko, O. N.
103.102 .114
121.044
Kovalenko, V. A.
074.123
Kovalenko, V. M.
098.073
103.102 .114
121.044
Kovalev, E. L.
034.076
Kovalev, V. A.
033.014
073.080
077.018
Kovalev, V. S.
082.024
Kovalevskij, I. V.
106.066
Kovalevsky, J.
009.003
041.045 .048
042.015
046.001
Kovesi-Domokos, S.
162.014
Kovetz, A.
065.078 .128
Kovner, A. Sh.
106.067
Kovner, M. S.
084.321
106.067
Kovrizhnykh, O. M.
076.047
Kovrygina, L. M.
084.410
Kovshov, V. I.
064.040
Kowal, C. T.
099.235 .237 .241
103.138
112.021
122.079
125.064 .107 .108
158.038
Kozai, Y.
054.015
Kozak, P. P.
071.045
Kozelov, V. P.
073.081 .082
Kozenko, A. V.
091.015
Kozhanchikov, V. I.
004.057
Kozhanov, T. S.
151.074 .075
Kozhevnikov, N. I.
072.024 .033 .068 .071
.072 .073
Koziel, K.
094.025
Koziel, K.
120.005
Koziowskie, M.
065.162
Kozlov, I. S.
042.059 .065
Kozlov, N. N.
151.080
Kozlov, V. I.
078.024
143.025
Kozlovsky, B.
076.055
Kozlowski, M.
065.149
Kozoderov, V. V.
063.044
Kracher, A.
105.145 .148
Kraehenbuehl, U.
094.467
Kraetschmer, W.
094.455
Krafft, F.
004.002 .031 .037
011.008
012.004
Kraft, R. P.
065.152
117.023
Krahn, D.
002.009
Krajcheva, Z.
047.022
Krajcheva, Z. T.
117.053
Krall, N. A.
077.001
Kramarovskij, Ya. M.
162.062
Kramynin, A. P.
072.020
Krashchin, M. D.
031.069
Krasikova, A. A.
041.018
Krasil'nikov, D. D.
143.072
Krasinski, A.
066.160 .161
080.075
162.029
Krasinskij, G. A.
041.039
Krasinsky, C. A.
046.062
Krasnenko, N. P.
041.025
Krasnobabtsev, V. I.
122.157
Krasnopol'skij, V. A.
097.034
Krasovskij, A. A.
082.128
Krass, M. S.
105.044
Krat, V. A.
071.064
080.042
Kraus, J. D.
141.091
Kraut, F.
105.146
Kravtsov, F. I.
103.100
Krawczyk, S.
031.420
124.104
Krechet, V. G.
066.125
Kreiner, J. M.
103.114
120.005
Krelowski, J.
122.023
Krempec, J.
114.077
Kremser, G.
078.035
084.012
Kren, G.
009.017
098.093
Kreplin, R. W.
003.122
Kresak, L.
104.051 .065
Kreuzer, J. L.
034.099
Kreye, W. C.
034.120
Kreysa, E.
141.604
Krieger, A.
073.058
074.104
076.035
Krieger, A. S.
073.001 .003 .009 .103
.104
074.008 .035 .073 .074
.075 .078 .081 .105
076.003 .037
Krienke Jr., O. K.
158.002
Krimigis, S. M.
074.035
078.008
084.322
092.010
099.085
Krinov, E. L.
005.021
011.049
105.063 .078 .147
Krisciunas, K.
112.011
Krishna Murthy, B. V.
083.033
Krishna Swamy, K. S.
103.102
114.329
Krishnaswamy, M. R.
061.092
Kriss, G.
141.353
142.080
Krivoshapkin, P. A.
143.014
Krivsky, L.
073.022 .023 .119 .130

Krol', V. A.
151.079
Kron, R. G.
112.016
Kronberg, P. P.
141.080
158.045
Kronrod, M. A.
097.046
Kropachev, Eh. P.
061.020 .065
Kropotkin, A. P.
084.241
Kropotkina, E. P.
082.077 .113
Krori, K. D.
061.085
Krpata, J.
119.007
Kruchinenko, V. G.
104.080
Kruegel, E.
131.530
Krueger, A.
073.118
Kruglov, E. M.
142.145 .171
Kruglov, V. I.
034.016
Krumenaker, L.
103.104 .111 .128
124.109
Krupenio, N. N.
097.009
Krupp, B. M.
064.055 .056
Krutov, V. V.
076.054
Krylov, A. L.
084.224 .225 .226
Krylova, S. N.
106.039
Krymskij, G. F.
084.287
143.014 .027
Krzeminski, W.
142.029
Ksanfomaliti, L. V.
097.008 .022
Ku, W.
141.346
142.136 .137
Kuan, P.
064.001 .017 .030 .047
124.104
Kubesov, A.
004.084
Kubicela, A.
034.093
122.218 .219
Kubierschky, K.
034.157
Kucheryavenkov, A. I.
093.023
094.198
Kuchowicz, B.
061.005 .046 .081
065.126
091.038
114.110
143.038
Kuchowicz, B.
162.038 .086
Kudak, K. A.
041.007
055.013
Kudela, K.
078.017
Kudin, G. I.
052.008
Kudritzki, R.-P.
064.007
Kudrya, Yu. N.
072.059
Kudryavtsev, M. I.
073.034
142.168 .180
Kudryavtseva, Z. A.
114.342
Kuehne, C.
031.004
032.022
Kuenzel, H.
072.004
073.118
Kugaenko, B. V.
082.055
Kuhi, L. V.
064.001
114.041 .336
Kuhn, P. M.
082.033
Kuhn, W.
005.018
Kuijpers, J.
077.040
Kuimov, K. V.
142.242
Kuiper, E. N. R.
131.055
141.618
Kuiper, G. P.
097.026
Kuiper, T. B. H.
131.055 .529
141.618 .633
Kukarkin, B. V.
120.006
121.069
122.047
124.104
154.018
Kukarkina, N. P.
120.006
Kukin, D. S.
097.062
Kuklina, N. Ya.
082.112
Kulakova, R. D.
052.062
Kul'chitskij, A. P.
034.066
Kuleshova, L. V.
143.071
Kuleshova, V. P.
084.296
Kuli-zade, D. M.
071.043
Kulibekov, O. A.
003.084
Kulidzhanishvili, V. I.
031.051
Kulidzhanishvili, V. I.
079.004
Kulikov, G. B.
074.094
Kulikov, G. V.
143.053
Kulikov, Yu. Yu.
082.121
Kulikovskij, P. G.
041.037
Kulinich, V. V.
063.049
Kulsrud, R. M.
074.096
143.064
155.024
Kumar, S.
093.004 .025
Kumer, J. B.
084.010
Kumsiashvili, M. I.
121.068
Kunashev, B. S.
082.110
097.022
Kunasz, P. B.
063.018 .039
064.042
Kundt, W.
011.021
Kundu, M. R.
072.018
073.063 .144
077.008 .042
125.010 .027
Kunitskij, R. V.
014.013
015.014
Kunitsyn, A. L.
042.099
Kunitzsch, P.
004.104 .106 .107
Kunkel, W. E.
122.007 .010
159.007
Kuntz, E.
046.069
Kuntz, P. J.
131.202
Kunz, A. B.
022.055
Kunz, L. W.
082.021
Kuo, K.
094.428
Kuo-Petravic, L. G.
141.351
Kupchishin, A. I.
078.016
Kupo, I.
142.238
Kuprevich, N. F.
034.080
Kupriyanova, I. E.
055.025
Kurat, G.
105.137 .145 .148 .166
Kurbasov, V. V.
031.410
Kurchakov, A. V.
132.050

Kuribayashi, A.
103.130
Kuril'chik, V. N.
142.111
Kuriliene, G.
113.056
Kurochka, E. V.
073.139 .140
Kurochka, L. N.
071.046
072.058
073.092 .139 .140
Kurochkin, N. E.
121.033
122.058
141.069 .070
158.124
Kuroczkin, D.
114.114
Kurokawa, H.
074.070
Kurosaki, T.
103.100 .130 .131
Kurpinska, M.
103.114
121.055 .107
141.099
Kurskov, A. A.
162.032
Kurt, V. G.
078.022
082.110
Kurth, R.
003.085
Kurth, W. S.
084.003
Kurtz, D.
122.090
Kurtz, D. W.
121.093
Kur'yanova, A. N.
041.003
Kustaanheimo, P. E.
042.010
Kuszell, A.
151.012
Kutepov, A. A.
091.027
Kuteva, Z. N.
034.014
Kutiev, I.
051.008
Kutner, M. L.
131.002
Kutsenko, A. V.
031.410
Kutuzov, A. L.
052.017
Kutuzov, I. A.
094.154
Kutuzov, S. A.
151.067
Kuvshinov, V. M.
116.010
142.031 .041
Kuwano, Y.
124.103 .109
Kuzakov, V. K.
004.065
Kuzhevskij, B. M.
073.156
Kuzhevskij, B. M.
078.026
080.076
084.274
Kuzmanoski, M.
098.034
Kuz'menko, G. I.
072.053
Kuz'menko, V. V.
052.005
Kuz'menkov, L. S.
066.123
Kuz'min, A. D.
099.069
Kuz'min, A. I.
106.028
143.014 .072
Kuz'min, R. O.
097.017 .030
Kuz'minykh, V. A.
052.012
Kuznetsov, B. G.
004.085
Kuznetsov, B. I.
093.024
Kuznetsov, D. A.
072.051
Kuznetsov, I. A.
055.025
Kuznetsov, I. V.
082.121
141.097
Kuznetsov, L. I.
052.068 .069
Kuznetsov, S. N.
078.025
083.029 .034
084.269
Kuznetsova, V. G.
084.244
Kvasha, L. G.
105.064 .081
Kvenvolden, K. A.
105.026
Kveton, J.
103.100
Kviz, Z.
031.269
Kwok, S.
131.201
141.632
Kyalyan, P. A.
034.049
Kynchev, P. Z.
122.056 .057

Laas, D. P.
041.019
Labeyrie, A.
031.295
032.035
Laborde, J. R.
123.075
154.015
Labs, D.
007.000
031.207
Lacis, A. A.
093.038
Lacroute, P.
032.529
041.065
Lacy, J. H.
131.522
141.617
Lacy, J. L.
143.043
Lada, C. J.
131.054 .159
Ladley, C. M.
075.021
Lafleur, H. T. J. A.
034.158
Lager', V. F.
052.049
Lagerkvist, C.-I.
098.050 .084
Lagerqvist, A.
022.069
Laget, M.
113.073
Lahaye, E.
005.014
Lahulla, J. F.
031.274
052.055
072.078
Laird, F. N.
125.020
Lakatos, S.
094.405
Lake, K.
066.157
Lal, D.
106.001
Lala, P.
052.021
Lam, C. S.
064.058
Lam, K. S.
022.120
Lamb, D. Q.
065.164
126.005
141.354
Lamb, F. K.
065.163 .164
141.354
142.128 .197
Lamb, S. A.
072.001
Lambeck, K.
081.003 .034
Lambert, D. L.
064.019
113.010
114.303 .346
122.206
131.033
Lambert, P.
105.149
Lamers, H. J. G. L. M.
064.002 .072
114.017 .018 .090 .117
Lammerer, M.
009.025
032.011
Lammlein, D.
094.109

Lampton, M.
032.515
125.021
131.064
142.134
Lamy, P. L.
100.201
Landau, R.
092.011
093.044
Landau, R. W.
062.073
Landecker, P. B.
071.011
073.048 .049 .137 .146
Landecker, T. L.
125.039
158.173
Landegren, G. F.
014.039
Landi Degl'Innocenti, E.
063.045 .046
Landi Degl'Innocenti, M.
063.046
Landini, M.
012.026
074.047 .075 .076 .104
076.052
Landis, H. J.
121.054
Landman, D. A.
022.061
Landsberg, P. T.
066.061
Landstreet, J.
122.194
Landstreet, J. D.
116.004 .014
126.018
Lang, B.
105.150
Lang, K. R.
162.048
Lange, E. F.
005.015
Lange, G.
122.211
Lange, G. A.
120.002
122.128
Langel, R. A.
084.313
Langer, S. H.
076.033
Langer, W. D.
131.177 .200
Langevin, Y.
094.187
143.073
Langford, W. R.
122.095
Langton, R. J. J.
075.007
Lanning, H. H.
119.001
142.066
Lantos, P.
077.050
Lantrewitz, W.
106.007
Lanzano, P.
042.034
Lanzerotti, J. L.
084.310
Lanzerotti, L. J.
084.258 .295 .302
092.010
143.075
Lapchinskij, V. G.
066.125
Lapushka, K. K.
031.408
Lapuska, K.
013.017
Larcher, S.
066.030
Lari, C.
141.058 .123
Larimer, J. W.
105.018 .140
Larin, V. N.
003.086
Larionov, M. G.
141.037 .157 .158
Larionova, L. V.
113.019
Laros, J. G.
142.086
Larson, E. E.
105.058
Larson, H. P.
034.029
098.044
Larson, R. B.
065.103
151.069
158.059
161.010
Larsson, L.
143.020
Lasher, G.
061.027
125.025 .056
Lasker, B. M.
126.024
Laslo, T. J.
055.013
Lass, H.
042.048
Latham, G.
094.109
Latimer, J.
046.018
Latour, J.
065.156
Lattimer, J. M.
065.167
Latyshev, I. N.
042.004
Lauberts, A.
113.044
Laucenieks, L.
011.068
015.040
Laufer, B.
142.187
Launay, F.
034.150
Laurie, P. S.
075.021
Lauter, E. A.
075.017
Lauterborn, D.
065.154
Laux, U.
082.109
Lavrov, A. P.
141.116
Lavrov, M. I.
121.007 .030 .031 .120 .132
Lavrova, E. V.
084.296
Lavrova, N. V.
121.007 .031 .132
158.108
Lavrovskij, Eh. K.
092.013
Lavrukhina, A. K.
061.051
105.069 .150 .151
Lawrence, J. K.
162.044
Lawton, A. T.
015.005
Layzer, D.
003.005
015.026
Lazareff, B.
065.145
131.039
Lazarev, A. I.
082.075
Lazutin, L. L.
084.269
Lazutin, V. I.
032.534
Le Borgne, J.-F.
121.149
Le Contel, J. M.
122.111 .117 .154
Le Denmat, G.
162.021 .051
Le Squeren, A. M.
132.015
Lea, S. M.
160.027
Leach, R. W.
142.001
Leacock, R. J.
141.142
Leaphart, D. K.
042.042
Leaton, B. R.
084.283
Leavitt, R. P.
082.082
Lebedev, D. S.
097.049
Lebedev, L. A.
053.014
Lebedev, L. O.
031.056 .060
Lebedev, N. N.
034.017
Lebedev, V. G.
031.235
034.088
Lebedev, V. V.
106.067

Lebedeva, L. N.
074.020
Lebedinets, V. N.
031.211
034.018
054.003
104.050
Leblanc, Y.
033.013
Lebofsky, M. J.
141.610 .612 .624
Lebovitz, N. R.
065.078
117.044
Lecacheux, A.
033.013
Lecacheux, J.
099.083
Lecar, M.
125.006
151.046
160.010
Lecoultre, P.
034.154
Ledden, J. E.
142.195
158.310 .317
Lederle, T.
047.018
Ledoux, P.
012.016
013.014
065.021 .100
Lee, L. C.
063.025 .037
Lee, L. P.
047.034
081.035
Lee, M.
066.076
Lee, P.
114.333
Lee, R. W.
066.111
Lee, T. J.
124.104
Lee, T.-S.
065.110
Lee, Y. C.
084.247
Lee, Y.-W.
074.013
Lees, R. M.
131.152
Lefevre, J.
131.032
Legg, A. J.
031.261
Legzdins, H.
009.027
Lehmann, H.-R.
083.070
Lehnert, B.
062.052
Leibert, J. W.
126.012
Leibowitz, H. W.
031.033
Leich, D. A.
094.442 .476
Leinert, C.
106.005 .056
Leitinger, R.
083.076
Lelievre, G.
141.029
Lemaitre, G.
031.031
Lemekhova, E. N.
099.239
Lemke, D.
034.009
106.008
122.070
131.505 .506
141.607 .634
Lena, P.
131.532
Lenderman, E. I.
122.203
Lengauer, G. G.
131.193
Lengel, R. K.
022.006
Lengyel-Frey, D.
131.134
Lennard, W. N.
022.125
Lenoble, J.
093.015 .046
Leon, G.
032.049
Leonov, A. N.
014.013
Leont'ev, L. V.
092.012
Leont'ev, V. V.
099.060
Lepping, R. P.
092.003
Lequeux, J.
061.031
Lerch, R. P.
004.102
Lerche, I.
061.009 .068
063.001 .011
161.001
Leroy, B.
162.018
Lesh, J. R.
009.004
122.096
Leshchenko, L. N.
076.049
083.075
Lessnoff, G. W.
066.080
Lester, D.
119.020
Lester, D. F.
100.013
121.019
Lester, J. B.
114.097 .331
Lester Jr., W. A.
022.044
Letfus, V.
076.020
Leubner, M.
084.324
Leung, C. M.
131.010 .068 .074 .081
Leung, K.-C.
121.050
Leushin, V. V.
074.020
082.048
Levashev, A. E.
066.047
Levato, H.
114.076
153.027 .032 .038
Levay, Z. G.
158.098
Leventhal, M.
125.058 .059
141.167
Levi, F. A.
105.152
Levi-Donati, G. R.
105.153
Levin, B. Yu.
094.197
105.009
Levine, A.
142.073 .075
Levine, A. M.
142.047
Levine, J. S.
099.067
Levine, R. H.
072.016
080.065
Levitskij, A. V.
052.013
Levkovskij, A. A.
080.049
Levy, A.
022.030
Levy, D. H.
022.018
Levy, J.
004.046
Levy, S. G.
034.041
Lewin, W.
141.326 .353
142.078 .081 .089
Lewin, W. H. G.
124.104
142.080 .186
Lewis, B. M.
008.113
047.040
158.145 .171 .182
160.046
Lewis, J. L.
093.056
Lewis, J. S.
099.056
Lewis, R. S.
105.083 .154 .189
Leydon, R.
121.149
Li, D.
031.023
Li, F.
141.326
142.081
Li, F. K.
142.009

Liberman, A. A.
142.110
Libin, I. Ya.
082.090
Lichnerowicz, A.
062.047
066.103
Lichtenstein, B. R.
094.180
Lichti, G.
082.049
134.005
Lichtman, J.
003.087
Liddy, J. P.
131.132
Lidov, M. G.
052.036
Lidov, M. L.
052.040 .048
Liebenberg, D. H.
032.540
074.041 .107 .108
Liebert, J.
126.007 .018
141.313
Liebert, J. W.
126.010 .019
158.055
Liebovitch, L.
143.056
Liebscher, D.-E.
066.120 .153
Liechti, C. A.
033.040
Lieske, J. H.
042.029
Liewer, K. M.
031.244
Lifshits, E. M.
066.066
Lightman, A. P.
066.039
142.038
Likin, O. B.
073.034
076.018 .021 .042
084.274
Liller, M. H.
123.022
141.015
142.193
154.024
Liller, W.
121.002
122.142
124.006 .104
141.015 .027
142.039 .138 .193 .240
143.135
158.307
Lilley, A. E.
131.159
Lillie, C.
103.100
Lillie, C. F.
053.015
064.054
103.101
122.188
Lim, J. C.
033.041
Lin, C. C.
151.061
Lin, R. P.
073.017 .020
077.047
094.157
Lin, Y.
073.076
Lincoln, J. V.
075.025
Lind, R. W.
066.004
Lindblad, B. A.
104.045
Lindegren, L.
124.104
Lindemann, E.
152.003
Lindgren, H.
113.040
124.104
Lindinger, W.
082.029
Lindsay, E. M.
159.006
Lindsay, J. F.
094.137
Lindsey, C. A.
071.010
Lindstam, S.
143.078
Lingenfelter, R. E.
073.016 .088
076.055
Link, H.
011.015
Linke, R. A.
131.170 .171
Linnell, A. P.
021.011
061.052 .074
065.061
Linnik, V. P.
004.051 .067
Linscott, I. R.
033.075
Linsky, J. L.
064.018 .027
071.031
073.027
074.098
114.320 .330 .350 .370
131.183
Liou, K.-N.
082.034
Lipa, B. J.
076.032
084.262
Lipa, J. A.
066.027
Lipatov, B. N.
041.046
044.006
Lipatov, S. V.
034.081 .082
Liperovskij, V. A.
084.273 .320
Lipin, N. A.
034.061 .074
Lipovetskij, V. A.
084.024
125.106
Lipovka, N. M.
156.003
157.003
Lippincott, S. L.
118.004 .015
Lipschutz, M. E.
105.027 .138
Lipskij, Yu. N.
097.027
Lisin, O. G.
097.043
Lisina, L. R.
094.142
List, M.
002.002
005.004 .006 .007
Liszt, H. S.
131.068 .109 .555 .562
141.625
Lites, B. W.
071.028
072.062
073.132 .133
Little, L. T.
034.162
141.628
Litvak, M. M.
131.119 .159
Litvinenko, O. A.
083.077
Litvinova, N. N.
032.057
Liu, J. J. F.
052.046
Liu, L.
052.015
Liu, S. C.
093.022
Liu, S.-Y.
077.042
Livesey, R. J.
085.006
Livingston, A. E.
022.068
Livingston, W.
080.027
Livingston, W. C.
034.131
074.017
Livio, M.
121.023
142.148
Livshits, G. Sh.
003.007
Livshits, M. A.
073.091
Lloyd, I. D.
046.054
Lloyd Evans, T.
See Evans, T. L.
Lo, K. Y.
114.328
131.058 .121 .158
132.029
141.110
Lobanov, A. I.
034.062

Lobanova, N. I.
042.109
Lobkova, N. I.
101.008
Locher, K.
021.025
121.149 .150 .152
Lochno, P.
103.100
Lockhart, J. M.
022.002
Lockman, F. J.
131.117 .515 .521
132.030
155.062
Lockwood, G. W.
080.050
124.104
Lodenquai, J.
065.194
Loenko, Yu. M.
052.026 .064
Loewe, M.
003.025
Loewenstein, R.
141.639
Loewenstein, R. F.
091.031
Logachev, Yu. I.
078.025 .026
106.065
LoGuirato, J.
003.088
098.082
Lohne, J. A.
004.004 .051
Lohsen, E.
141.348
Lokhin, B. I.
052.004
Lokhov, V. K.
074.110
Lomadze, R. D.
063.048
Lomb, N. R.
119.006 .015
Lomonosov, V. N.
074.027
Long, K. S.
142.072 .074
Long, S. A. T.
046.013
Longair, M. S.
010.022
141.087
Longhi, J.
094.404
Longman, I. M.
081.009
Longmore, A. J.
103.126
121.051
Lopert, P. B.
063.020
Lopez Arroyo, M.
072.078
075.009
LoPresto, J. C.
114.396
Lord, E. A.
162.085 .091
Lord, S. D.
162.048
Lorell, J.
097.006
Loren, R. B.
131.173
Lorin, J. C.
105.155
Lorrain, P.
066.060
Lorre, J. J.
132.049
160.014
Lortet-Zuckermann, M. C.
114.024
Loskutov, V. M.
063.030
Lotmar, W.
013.010
Lotova, N. A.
074.028 .044 .057
106.029
141.132
Lotrian, J.
022.010
Loucka, L.
124.109
Loughhead, R. E.
071.061
Loulergue, M.
074.100 .101
Loumos, G. L.
046.042
122.206
Lovas, F. J.
022.056
131.080
Lovas, M.
125.064
Lovell, L. P.
121.054
Lovering, J. F.
105.003
Low, B. C.
062.002
Low, F. J.
032.516
141.605 .609 .615 .626
158.008 .048
Lowenstein, P. L.
105.089
Lowenthal, D. D.
031.017
Lowne, C. M.
008.043
Loyola, P.
041.036
Lozei, G. P.
105.012
Lozej, G. P.
105.201
Lozinskaya, T. A.
132.014
Lozinskij, A. M.
103.102
124.104
Lucano, G.
066.087
Lucas, R.
131.124
142.215
Lucchin, F.
162.023
Luck, J. McK.
031.262
Luck, R. E.
114.303 .371
Lucke, R.
142.237
Lucke, R. L.
094.131
Lucy, L. B.
064.033
Ludwig, H.
046.067
Luest, R.
003.089
Lugmair, G. W.
105.057 .156
Lui, A. T. Y.
084.020 .312
106.901
Luinge, W.
032.521
Lukas, J.
124.103
Lukas, R.
103.100
123.007 .008
124.015 .103
Lukash, V. N.
162.063
Lukashevich, E. L.
042.063
Lukatskaya, F. I.
122.026 .124 .203
158.112
Lukes, M. A.
121.062
Lukin, D. S.
091.040
Lukin, I. V.
104.071
Lukkari, L.
084.012 .013
Lum, R. K. L.
094.195
Lundquist, C. A.
046.056
Lundstroem, I.
113.040
Lunev, V. V.
052.066
Lustig, G.
098.062 .078
123.076
Lutsenko, V. N.
078.019 .022
084.275
Lutz, B. L.
022.104
091.016
Lutz, J. H.
133.035
Lutz, T. E.
115.015 .021
Luud, L.
114.054 .082 .364
Luvsan, L.
046.034
Luyten, W. J.
112.017 .018 .019 .020

Luyten, W. J.
112.021
L'vov, Yu. A.
003.080
Lyamova, G. V.
072.037
Lyatskaya, A. M.
074.095
083.060
Lyatskij, V. B.
074.095
084.246
Lykov, A. A.
085.011
Lykova, V. V.
085.011
Lynas-Gray, A. E.
031.412
153.002
Lynch, D. K.
080.025
Lynden-Bell, D.
008.023
151.010 .030 .053
Lynds, C. R.
141.013
Lyne, A. G.
141.302 .327 .339
Lynn, K. J. W.
031.261 .276
Lyon, J.
131.065 .089
133.021
Lyon, J. G.
133.015
Lyons, D.
084.255
Lyons, L. R.
084.407
Lypkan', N. M.
031.410
Lyubimkov, L. S.
064.045
Lyubimov, G. P.
078.014 .018
Lyutenko, V. F.
034.091
Lyuty, V. M.
See Lyutyj, V. M.
Lyutyj, V. M.
121.121
123.049
142.027 .110 .157
158.015

Ma, C.
121.039
Ma Sung, L. S.
073.057
Mabuchi, H.
105.157
MacAlpine, G. M.
158.011 .216
Macau-Hercot, D.
114.116
MacDonald, G. J. F.
003.095
MacDonald, J. W.
121.062
MacDoran, P. F.
031.260
MacDougall, J. D.
105.141 .144 .158
MacFarlane, M.
158.218
Mach, J.
095.009
Machado, M. E.
073.027
Machovsky, Z.
079.103
Mack, H. M.
015.025
Mack, J. E.
014.036
Mackal, P. K.
099.063
Mackay, C. D.
034.137
Mackie, F. D.
065.167
MacLennan, C. G.
084.258
143.075
MacLeod, J. M.
131.511
158.306
MacPhie, R. H.
031.300
MacQueen, R. M.
094.005
Macquet, M.
113.042
Macy Jr., W.
099.214 .231
101.011
Macy Jr., W. W.
099.206
Mader, G. L.
131.046 .057
Madey, J. M. J.
022.002
Madore, B. F.
122.002
158.186
Madore, J.
066.006
Maeckle, R.
114.308
Maeder, A.
065.060 .101
116.002
Maeder, D. G.
066.090 .091
Maedler, H.
002.004
Maehl, R. C.
143.059 .066
Maeyama, Y.
004.048
Maezawa, K.
084.251
Maffei, P.
002.010
Magnan, C.
063.007 .020
Magun, A.
077.058
Mahaffy, J. H.
065.113
Mahaffy, J. H.
158.228
Maheswaran, M.
065.087
Mahoney, M.
131.152
Maier, E. J.
083.071
Maischberger, K.
066.083
Maitzen, H. M.
106.050
114.363
124.103
Majer, A. K.
074.110
Majstrov, L. E.
004.059
Makarenko, E. N.
123.004 .010 .052 .060
Makarov, B. P.
031.065 .068
Makarov, K. N.
143.072
Makarov, V. A.
031.411
Makarov, V. I.
073.112
Makarova, E. A.
071.032
074.121
Makarova, R. K.
034.014
Makarova, V. V.
073.112
Makeev, N. N.
052.027
Makino, F.
061.033
076.025
Maksimov, S. V.
143.072
Maksimova, I. I.
045.034
Maksumov, O. S.
034.063
Maksyukov, N. I.
066.124
Malacara, D.
031.076 .081 .083 .086
034.125 .126
Malacara H., D.
032.053
Malacara-Hernandez, D.
031.013
Malaise, D.
032.504
Malakpur, I.
124.100
Malarev, V. A.
032.005
034.015
Malaroda, S.
114.022 .076
153.027 .032
Maley, P.
103.100
Malezieux, J. M.
105.104
Malin, M. C.
097.065

Malin, S. R. C.
084.283
Malina, R.
125.021
142.134
155.058
Malinovsky, M.
022.057
Malkiehl', G. S.
083.029 .034
Mal'kov, M. V.
084.270
Malkus, W. V. R.
062.069
Mallama, A. D.
121.053 .073
124.103
Mallik, D. C. V.
131.902
Malone, R. C.
065.019 .184
Malov, I. F.
064.065
Maltby, P.
072.019
073.068
Maltsev, Yu. P.
084.246
Malville, K.
003.090
Malyavkin, L. P.
034.083
Malyshev, S. K.
033.071 .072
Malyshev, V. S.
122.061
Malysheva, T. V.
003.091
Malyshkin, V. N.
074.123
Malyuto, V.
114.094
155.046
Mamaev, S. G.
162.010
Mamakov, A. S.
031.223
034.047
Mamatkazina, A. Kh.
114.067 .068 .342
Mamedov, S. G.
073.110 .111 .150
Mammano, A.
114.012
122.042 .046
Mamrukova, V. P.
143.014
Manara, A.
031.215
Manchester, R. N.
141.340 .356
Mandel', O. E.
122.019 .043 .062
Mandel'shtam, S. D.
076.026
Mandel'shtam, S. L.
061.050
073.115
Mandeville, J. C.
094.200
Mandujano, J.
158.242
Manfroid, J.
113.043
131.533
Mango, S. A.
131.003 .046
Manhes, G.
105.159
Mansurov, V. N.
142.030
Mantegazza, L.
103.100
Manukin, A. B.
066.072
Maran, S. P.
034.034
103.102
114.330
132.020
Marchal, C.
042.035
052.042
Marchal, J.
034.152
Marechal, A.
012.013
031.008 .240
Marek, G.
055.015
Marek, J.
004.005 .051
022.075
Marek, K.-H.
046.022
Marette, G.
022.092
Margolis, J. S.
022.007
Margolis, S.
161.009
Margon, B.
032.515
061.006
141.313
142.134
158.309
Margrave, T. E.
121.062
123.036
124.104
Mariani, F.
106.052 .064
Marik, M.
073.033
Marino, B. F.
031.265
Marionni, P.
133.015
Marionni, P. A.
133.012 .021 .031
Maris, G.
073.151
075.010
Mariska, J. T.
073.060
074.103
Mark, J. W-K.
151.034 .051
Mark, N.
094.431
Markeev, A. K.
077.022 .043
Markeev, A. P.
052.036 .039
Markellos, V. V.
042.020 .041
Markelov, L. A.
103.110
Markert, T. H.
142.009 .047
Markey, P.
065.085
Markov, M. A.
066.070
Markova, L. G.
054.018
Markova, L. T.
103.102
141.068 .072
Markowitz, W.
045.017
Marlenskij, A. D.
014.033
Marochnik, L. S.
158.043
Marot, A.
121.149
Marov, M. Ya.
091.033
093.005 .006
097.015
Marquardt, C. L.
094.403
Marques Dos Santos, P.
077.053
Marscher, A. P.
142.060
Marsden, B. G.
098.020 .026 .053 .057
.059 .060
099.237
103.100 .108 .111 .119
.124 .126 .130 .131
.133 .134 .136 .140
Marsh, G. E.
066.061
Marsh, J. G.
046.011
052.033
081.024
Marsh, K. A.
141.061 .125
Marshall, R. F.
131.176
Marti, K.
105.057 .160
Martin, A.
031.018
Martin, D. H.
082.041
Martin, F.
031.305
Martin, L. J.
097.039
Martin, N.
159.001
Martin, P. G.
131.106 .139 .140
Martin, P. M.
105.014

Martin, R.
034.142
Martin, R. N.
131.141
Martin, S. F.
073.047 .061
Martin, T. W.
034.154
Martin, W.
066.041
Martin, W. L.
142.042
Martinet, L.
155.013 .042
Martinez Benjamin, J. J.
042.104
Martinic, N. J.
078.037
Martins, D. H.
113.072
121.155
Martinson, I.
022.037 .067
Martirosyan, M. A.
034.048
Martonchik, J. V.
093.043
Martres, M.-J.
003.122
010.028
075.008
077.005
Mart'yanov, A. S.
091.040
Martynenko, V. G.
066.071
Martynov, A. V.
041.007
Martynov, D. Ya.
123.063
124.104
Martynov, V. V.
082.110
Marvin, U. B.
094.436 .437
105.161
Marx, H.
124.104
Masajtis, V. L.
105.062
Masevich, A. G.
046.017
Mason, B. H.
105.006
Mason, D. P.
066.010
Mason, G. M.
143.022 .023 .034 .062
.063
Mason, K.
158.309
Mason, K. O.
141.313
142.016 .065 .158 .226
.233
Mass, C.
085.013
Massa, D.
116.012
Masse, D.
034.128
Massevitch, A.
117.052
Massey, P. L.
160.022
Masters, A. R.
065.165
Mastrantonio, G.
084.206 .211
Masuda, A.
105.001 .165
Masursky, H.
097.026
Materne, J.
122.069
Mather, R. S.
012.036
046.058
Mathews, G. J.
061.061
Mathewson, D. S.
158.183
Mathiesen, O.
143.020
Matilsky, T.
142.069 .078 .081 .084
.188 .212
Matilsky, T. A.
142.105
Matlock, L. T.
064.069
Matrosov, V. M.
052.028
Matsakis, D.
131.518
Matsakis, D. M.
131.062
Matson, D. L.
099.216 .242
Matsubayashi, T.
143.028
Matsumoto
103.100
Matsumoto, T.
032.543
103.114
Matsushita, S.
106.061
Mattei, J. A.
123.023 .079
124.022
Matteson, J. L.
142.077
Matthews, C. J.
132.021
Matthews, H. E.
131.133
Matthews, K.
158.056 .070
Mattig, W.
071.001 .050
072.005
Mattila, K.
131.085
Mattoo, S. K.
106.001
Matulajtite, S.
004.071
Matveenko, L. I.
006.000
033.070
134.003
Matveenko, S. M.
042.118
Matveev, G. D.
034.109
Matveev, M. I.
084.285 .308
Matvievskaya, G. P.
004.111
Matyagin, V. S.
155.052
Matyugov, S. S.
093.023
094.198
Matzner, R.
066.172
Matzner, R. A.
036.010
Maucherat, A. J.
131.551
Mauder, H.
010.010
113.004
123.034
142.034
Maugin, G. A.
066.003
Maupome, L.
105.033 .162
Maurer, P.
094.464 .467
Maurette, M.
094.187 .460
105.097
143.073
Maurin, L.
121.149
Mauron, N.
121.149
Mavraganis, A. G.
042.070
062.054
Mavridis, L. N.
122.146
Maxson, C. W.
034.167
Maxworthy, T.
099.005
May, J.
099.037
May, T. L.
162.001
Mayaud, P. N.
084.315
Mayer, B.
124.104
Mayer, E. H.
121.052
Mayer, F.
151.037
155.042
Mayer, P.
124.104
Mayer, W.
141.326
142.075 .076 .078 .081
.083 .084 .089
Mayer-Hasselwander,
H. A.
032.518
Maynard, N. C.
084.303

Mayor, M.
013.010
155.038
Mayrhofer, K.
047.012
Mazeh, T.
066.181
Mazo, V. L.
042.012
Mazurek, T. J.
125.029
Mazzitelli, I.
034.096
126.004 .008 .015 .016
Mazzoleni, F.
031.215
McAdam, W. B.
141.007
McAlary, C. W.
122.172
McAlister, H. A.
032.532
153.011
McAshan, M. S.
066.025
McBeath, K. B.
114.115
McCall, S. L.
125.058 .059
McCallister, R.
098.055
McCallister, R. A.
103.135
McCarroll, R.
022.059 .060 .084
McCarthy, C. C.
098.052 .055
099.233 .237 .238
103.100 .111 .123 .127
.135 .139
McCarthy, D. D.
098.011
McCarthy, D. W.
141.609 .626
McCarthy, M. F.
114.078
McClintock, J.
141.326
142.078 .081 .083 .084
.089 .220
McClintock, J. E.
142.098 .189
McClintock, W.
114.350 .370
131.060 .183
McClinton, A. T.
081.024
McCluskey Jr., G. E.
114.375
121.049
McClymont, A. N.
076.022
McColl, D. H.
105.035
McConnell, J. C.
093.002
McCord, T.
158.089
McCord, T. B.
031.421
034.134
McCormac, B. M.
012.007
McCoy, J. E.
094.157
McCracken, C.
103.100 .130
McCray, R.
064.015 .022
131.070
142.003 .070
McCrea, J. D.
066.005
McCrea, W. H.
107.008
131.197
McCrosky, R. E.
098.058
103.005 .100 .104 .112
.122 .128 .130 .131
.132 .134 .140
104.034
McCulloch, P. M.
099.003 .092
McCutcheon, W. H.
124.104
131.152
McDiarmid, I. B.
106.062
McDonald, F. B.
073.057
078.001
099.025
143.044 .058
McDowell, M. W.
031.073
McElhinny, M. W.
084.281
McEllin, M.
141.034
McElroy, J. H.
034.161
McElroy, M. B.
099.011
McEwan, M. J.
131.132
McEwan, N. J.
141.063
McFarland, J.
159.006
McGee, R. X.
131.120
McGehee, R.
042.080
McGimsey, B. Q.
141.142
McGovern, P. A.
033.043
McGrath, M.
124.103
McGraw, J. T.
126.006
McGruder III, C. H.
113.032
155.029
McGuire, J. P.
072.012 .015
074.030 .031
McGuire, R. E.
094.157
McHone, J. F.
105.120
McInnes, B.
010.012
McInnes, P. A.
033.034
McIntosh, B. A.
104.047
McIntosh, D. H.
084.028
McIntosh, P. S.
073.124
076.037
McKay, D. S.
094.118
McKee, C. F.
125.006 .019
141.109
158.134
McKeith, C. D.
114.397
131.143 .207
McKenna, S. J.
015.016
McKenzie, D. L.
073.048 .049 .146
McKibben, R. B.
099.022 .024
143.012
McLaren, R. A.
132.003
McLean, I. S.
031.226
034.013
114.341
142.067
McLernon, C. R.
047.034
McNabb, P. W.
084.282
McNally, D.
010.022
065.105
131.130
McNamara, B.
111.001
McNamara, D. H.
121.064
122.095
McNulty, G. R.
010.007
McPherron, R. L.
074.116
084.230
Meaburn, J.
034.020
132.010
Meadows, A. J.
085.002
091.036
Mebold, U.
131.510
141.024
158.167
Medvedev, A. B.
034.073
Medway, K. J.
073.077
Meegan, C. A.
142.118 .165
Meeks, M. L.
014.040

Meeus, J.
047.024
051.013
079.003 .105
091.039
096.002 .010
097.025
098.012 .079 .083 .087
102.010
Mefford, J. W.
123.036
Meffroy, J.
042.028
Meggers, W. F.
003.092
Mehltretter, J. P.
072.002
Mehrmann, R.
011.025
Meier, F.
094.469
Meier, W.
162.078
Meinhardt, R.
113.005
Meinschein, W. G.
094.418
Meinunger, L.
158.241
Meisel, D. D.
104.011
Meistas, E.
113.055
Mekler, Y. B.
099.065
Melchior, P.
044.035
081.023
Melik-Alaverdyan, Yu. K.
034.117
Melikhov, V. R.
105.077
Melikov, G. O.
071.040
Melioranskij, A. S.
076.018 .021 .042 .044 .046
142.168 .180
143.074
Melioranskij, V. M.
073.034
Meliorenskij, A. S.
076.021
Melkumov, V. S.
052.061
Mellema, J. P.
094.423
Melnick, J.
160.011 .043
Mel'nikov, O. A.
004.113
032.003
158.191 .231
Mel'nikov, V. N.
022.091
Melosh, H. J.
094.170
Melrose, D. B.
062.055
074.048
077.013
Melvin, P. J.
122.186
Menath, A.
005.005
Mende, S. B.
084.016
Mendell, W. W.
094.118
Mendes, G.
046.018
Mendez, R. H.
133.001
Mendis, D. A.
102.016
103.102
107.015
Mendoza V., E. E.
113.021 .023
114.042
Meng, C.-I.
084.298
Mennessier, M. O.
122.132
131.127
155.037
Menning, J.
022.070
Menon, M. G. K.
061.092
Menon, T. K.
141.016 .041
Menyavtseva, T. A.
105.031
Menzel, D. H.
003.093
004.052
066.060
Menzies, B.
123.001 .045 .046 .047
Mercado, R. I.
031.034 .080
Mercer, R. M.
031.294
Mercier, C.
077.005 .051
Merezhin, V. P.
063.015
121.029
Mergenthaler, J. L.
022.016
Meriwether, J. W.
082.006
Merman, G. A.
042.006
Merman, N. V.
032.023 .050
Mermilliod, M.
113.034
Merrill, K. M.
098.038
131.073 .075 .162 .514
133.005 .013
141.606 .608 .613 .630
Merrill, R. B.
105.130
Merrill, R. T.
084.281
Mersov, G. A.
052.058
Meshcheryakov, G. A.
094.013
Mestel, L.
010.022
065.084
141.092
Meszaros, P.
066.096
142.174
Metelev, M. I.
033.071
Methot, R. L.
105.005 .127
Metlov, V. G.
158.137 .178
Metlyaeva, Eh. A.
143.014
Metreveli, M. D.
131.165
Metz, K.
122.068
Metz, W. D.
141.001
142.100
Metzger, A. E.
051.010
Meurers, J.
008.114
009.019 .028
Mewaldt, R. A.
078.011
Mewe, R.
022.095
064.048
076.053
122.167
142.014 .045
Meyer, C.
041.051
096.001
Meyer, H. O. A.
094.457
Meyer, K.
011.008
012.004
Meyer, P.
143.017 .052
Mezger, P.
061.066
Mezger, P. G.
131.530
157.001
Mianes, P.
122.133
159.001
Michalodimitrakis, M.
151.040
Michalsky, J. J.
101.004
Michel, F. C.
091.010
Michel, G.
034.028
Michelson, E.
131.207
Middleditch, J.
141.312 .313
Mies, F. H.
102.024
Mietelski, J.
005.035
Mignani, R.
066.013

Mignard, F.
042.054
Migunov, V. M.
106.028
Mihalas, D.
063.009 .010 .018 .026 .039
064.028 .042 .061
Mihalov, J. D.
106.019
Mijatov, M.
See Miyatov, M.
Mikaelian, K. O.
065.031 .185
Mikami, T.
115.017
Mikerina, N. V.
106.030 .031 .032
Mikeshina, L. A.
004.125
Miketinac, M. J.
065.002
Mikhail, J. S.
094.142
Mikhailov, A. A.
See Mikhajlov, A. A.
Mikhajlov, A. A.
004.014 .053 .086
011.006
041.055
082.062
Mikhajlov, A. V.
083.030
Mikhajlov, Yu. V.
098.069
Mikhajlova, O. M.
036.007 .022 .023 .026
Mikhajlovskij, A. V.
047.019
Mikhajlutsa, V. P.
080.074
Mikheev, V. M.
081.044
Mikhel'son, N. N.
031.011
Mikhnevich, V. V.
084.294
Mikirtumova, G. G.
082.100
104.017
Miko, I.
079.100
Mikulasek, Z.
125.055
Milanovich, F. P.
022.045
Milbourn, S. W.
010.012
103.119 .130 .136
Mil'chenko, V. A.
033.071
Milcheva, S.
047.022
Miles, H.
010.012
Miles, P. A.
031.037
Milet, B.
103.100 .130 .131
Miley, G. K.
141.055
Miley, G. K.
158.018 .203
Militskij, Yu. A.
033.015
Milkey, R. W.
063.009 .010 .026
064.028
073.004
080.033 .052
Millar, T. J.
131.146
Millburn, J. R.
004.128
Miller, H. R.
099.052
121.155
158.311 .315
Miller, J. S.
158.078 .302
Miller, M. G.
031.246
Miller, M. J.
031.262
Miller, R. H.
117.027
151.016 .049
Miller, W.
103.100
Miller, W. C.
036.016
Miller, W. H.
022.044
Millikan, A. G.
036.009
Millington, R. E.
103.100
Millis, R. L.
099.232
124.104
Millman, P. M.
010.039
012.030
051.020
104.042 .043 .089
Mills, A. A.
105.014
Mills, J. S.
061.067
Milman, A. S.
131.160
Milne, R. G.
106.045
Milogradov-Turin, J.
092.017
141.096
Milon, D.
011.033
103.100
Milone, E. F.
121.101
Milovanov, V. N.
032.039
Milovanovic, V.
010.037
045.029
Milton, D. J.
105.163
Minasyants, G. S.
031.075
072.022 .057
Minbaeva, A.
014.008
Minchik, S.
003.012
Mineev, Yu. V.
084.410
Mingaleva, G. I.
073.081
Minkin, J.
094.421
Minster, J.-F.
105.102 .159
Mints, R. I.
094.429
Mira, H.
103.125
Mirabel, I. F.
155.054
Mirmovich, Eh. G.
083.047
Mironov, A. V.
031.234
114.109
121.044
154.003 .014
Mironov, N. T.
081.036
Miropol'skij, Yu. Z.
011.060
Miroshnichenko, L. I.
106.015
Mirtov, B. A.
083.045
Mirzoyan, L. V.
122.006
151.089
Misconi, N. Y.
106.022
Mishchenko, M. P.
044.003
Mishin, V. M.
084.285 .286 .289 .308
106.034
Mishin, V. V.
106.067
Mishurov, Yu. N.
151.001
Mis'kin, N. A.
103.100
Miskovic, V. V.
098.047
Misner, C. W.
066.081
Mison, K.
066.175
Missana, M.
063.028
Misyura, V. A.
083.077
Mitalas, R.
141.349
Mitchell, R. I.
113.069
Mitchell, R. J.
160.006
Mitic, L. A.
032.055
Mitrofanov, I. G.
102.021
106.044
141.622 .631

Mittelstrass, J.
004.048
Miyake, S.
061.092
Miyamoto, M.
151.066
155.030
Miyamoto, S.
097.026
Miyamoto, Y.
124.103
Miyatov, M.
031.029
032.025 .026
082.073
Mnatsakanyan, M. A.
063.022 .023
151.089
Mnatsakanyan, R. G.
125.060
Mo, J.
124.013
Mochizuki, H.
124.103
Modali, S. B.
082.002
Modena, I.
066.087
Modisette, J. L.
114.003 .037
Modrzejowski, T.
009.026
Moe, K.
082.028
Moe, M. M.
082.028
Moe, O. K.
065.034
076.038
Moehlmann, D.
081.026
083.069
Moellenhoff, C.
065.016
Moeller Pedersen, K.
004.095
Moeller-Pedersen, B.
033.013
Moergeli, M.
094.469
Moesgaard, K. P.
004.087
Moffat, A. F. J.
131.560
155.009
Moffat, J. W.
066.168
Moffet, A. T.
033.007
Mogilevskij, Eh. I.
074.118
077.033
080.053
Mogro-Campero, A.
099.024 .229
143.013
Mohanty, D. K.
141.901
Moiseev, I. G.
077.025
122.156
Moiseev, I. G.
132.047
141.067 .097
Moity, J.
022.109
Mojsya, R. I.
104.084 .085
Mokrov, V. S.
074.043
Molchanov, V. A.
082.095 .096
Molchanov, Yu. B.
004.124
Moles, M.
158.014
162.021 .051
Molnar, M. R.
113.062
114.016
Molodenskij, M. M.
062.039 .045
074.016
Monet, D. G.
151.043 .082
Monfils, A.
114.116
131.126 .211
Mongan, T. R.
022.128
Monin, A. S.
093.016
107.007
Monk, P. M.
103.010
Monnet, G.
031.009
114.024
151.057
158.157
Monsignori Fossi, B. C.
074.047 .075 .076 .104
076.052
Montag, H.
046.049
052.016
Montgomery, M. D.
074.072 .090 .113 .114
.115 .125
084.207
Moore, C. B.
105.117
Moore, C. E.
022.096
Moore, E.
103.130 .131 .136
Moore, E. G.
005.010
Moore, E. P.
032.013
Moore, P.
003.014
010.012
011.014
Moore, R. L.
073.044 .072
Moore, S. R.
033.052
Moore, W. E.
125.005
142.074 .178
Moore Jr., J. H.
032.522
Moorey, G. G.
033.021
Moorwood, A. F. M.
141.629
Moos, H. W.
114.332 .345 .350 .370
.398
131.183
Mora, D. J.
155.012
Moraal, H.
143.016
Morabito, D. D.
141.135
Morales, C.
153.018
Moran, J. M.
131.046 .057 .058 .158
141.145
Morando, B.
005.004
044.001
047.030
099.035 .215
Moranzino, C.
044.042
Morbey, C. L.
118.020
Mord, A. J.
082.021
Morel, P. J.
065.065
Moreno, G.
106.064
Moreno, H.
103.119
Morfill, G.
078.030
084.208 .406
Morgan, C. J.
094.110 .408
Morgan, J. W.
094.439 .468
105.084 .136
Morgan, P. J.
031.262
Morgan, T.
103.131
Morgan, T. H.
114.037
121.155
Morgan, W. W.
158.027
Morgenthaler, G. W.
012.042
Morguleff, N.
114.055
122.037 .181
Mori, H.
103.130 .131
Morison
142.215
Morison, I.
142.103
Moritz, H.
081.030 .039
Moriyama, F.
008.102

Moroz, V. I.
097.008 .016 .022 .026
.033 .044 .054 .055
.057
142.110
Morozhenko, A. V.
097.024
Morozhenko, N. N.
073.066 .084 .109
Morozov, A. M.
072.045
Morozov, V. I.
033.072
Morozova, T. I.
078.027
Morris, C. S.
103.100
Morris, G.
131.013
Morris, M.
131.008 .058 .086
158.009 .219
Morris, R. V.
094.475
Morris Jr., G. A.
100.023
Morrisby, A. G. F.
010.007
096.007
Morrison, D.
011.030
094.192
098.017
Morrison, L. V.
044.048
092.007
134.006
Morrison, N. D.
113.009
114.083
Morrison, P.
141.112
142.199
Morrison, R. H.
094.107
Morrison-Smith, D. J.
105.047
Morton, D. C.
114.120 .329
121.049
131.157 .206
158.075 .125
Morton, J. B.
094.169
Moseley, H.
031.256
Moseley, S. H.
091.031
141.639
Moses, R.
034.164
Mosher, J.
071.052
Moshkalev, V. G.
153.026
154.007
Moskalenko
124.111
Moskalenko, I. N.
032.534
Moskalenko, N. I.
093.020
097.028 .056
Moskaleva, G. V.
079.103
Moskaleva, L. P.
097.018
Mosley, D. R.
031.222
113.058
Moss, C.
160.040
Moss, D. L.
065.039 .086
Mostepanenko, V. M.
162.010
Moszkowski, S. A.
061.093
Motrunich, I. I.
055.013
Motta, S.
073.107
Mottmann, J.
094.179
Motz, L.
014.034
Moulding, M.
142.198
Mount, G. H.
071.031
114.320
Mouradian, Z.
073.026
Mouschovias, T. C.
131.122 .203
Moutsoulas, M.
002.012
Moyano, C.
134.005
Mozer, F. S.
083.072
084.025
Mrkos, A.
098.045
103.002 .100 .130 .131
.137
Mucke, H.
047.012
Mudroch, O.
105.183
Mueller, E. A.
071.005 .022 .037
Mueller, G.
015.028
Mueller, H.
053.007
Mueller, H. W.
094.470
Mueller, I. I.
046.003 .006 .020 .061
Mueller, O.
094.181 .453
105.164
Mueller, P.
003.094
Muench, G.
099.218
132.032
Muench, J. W.
084.412
Muffoletto, C. V.
031.046
Mufson, S. L.
131.138 .555
133.015 .021
141.625
Muhleman, D. O.
066.041
141.644
Mukai, S.
091.020
Mukhamednazarov, S.
104.025 .063
Mukhametnazarov, S.
104.018
Mukherjee, N. R.
094.164
Mukhin, L. M.
091.017
Mukins, E.
051.022
Mulholland, J. D.
012.032
031.255
041.017 .021
046.042
094.011 .012 .016 .021
Mullamaa, U.-A. R.
085.016
Mullan, D. J.
122.084 .135
Mullen, E. G.
082.103
Muller, A. B.
103.136
Muller, C. J.
010.007
Muller, P.
006.000
012.017
118.037
Muller, P. M.
043.001
Muller, R.
072.065
Mulligan, K.
119.018
Mumford, G. S.
121.025
Muminovic, M.
009.015 .016
Mundet
103.100
Mundet, J. M.
103.131
Munk, W. H.
003.095
Munro, R. H.
074.037
094.005
Muradyan, R. M.
162.026
Murakami, G.
032.034
044.041
Murcray, F. J.
099.209
Murdin, P.
082.078
099.210
142.016 .042 .043

Murdock, J. A.
042.122
Murnikova, V. P.
121.006
Murphy, G. L.
061.029
062.071
Murphy, R. E.
100.022
Murray, B. G.
092.006
Murray, C. A.
034.138
142.104
153.036
Murray, J. B.
099.203
Murray, J. D.
141.009 .146 .155
158.183
Murray, N. D.
121.062
Murray, S.
142.071
Murray, S. S.
142.001 .005
Murthy, A. G.
104.093
Murthy, B. S.
084.223
Murthy, B. V. Krishna
See Krishna Murthy, B. V.
Murthy, V. R.
094.406
Murty, M. V. R. K.
034.126
Mur'yanov, M. F.
004.109
Muscari, J. A.
034.021
Musen, P.
052.032
Mushotzky, R. F.
142.077 .087
Musij, V. I.
104.019
Mustel, E. R.
124.002
125.011 .031 .032 .051
Mutch, T. A.
097.010
Mutel, R. L.
106.059
141.136
Muzyka, D. F.
074.029
Muzylev, V.
065.071
Muzylev, V. V.
021.008 .009
153.005
Myachin, V. F.
042.003
Mychelkin, Eh. G.
085.015
Myers, K. A.
052.043
Myers, P. C.
131.123
Nadeau, L. B.
079.104
Nadezhin, D. K.
022.031 .058
065.067
066.159
125.041
Nadzhip, A. Eh.
097.033 .044
Naef, R. A.
047.011
Naftilan, S. A.
119.011
Nagai, R.
151.066
Nagamoto, C. T.
104.004
Nagasawa, H.
105.174
Nagata, T.
083.038
Nagayama, N.
084.201
Nagel, E.
046.067
Nagel, K.
094.181
Nagirner, D. I.
063.021
Nagle, J. S.
094.112
Nagnibeda, V. G.
079.101
Nagy, A. F.
093.022
Nagy, B.
094.480
Nagy, S.
054.007
Nagy, T. A.
121.110
Naito, I.
045.011
Nakada, M. P.
074.018 .065
Nakada, Y.
022.062
Nakagawa, Y.
062.002
074.067
Nakajima, K.
098.061
Nakamura, N.
105.001 .165
Nakamura, T.
151.094
Nakamura, Y.
094.109
105.157
Nakano, G. H.
084.404 .405
Namba, O.
079.100
Nance, W. B.
105.089
Nandy, K.
131.115 .126 .211
159.010
Nankivell, G. R.
031.025
Napier, W. McD.
121.003
141.004
Naraeva, M. K.
097.023 .049
Narain, U.
074.012 .126
Narasimham, V. S.
061.092
Narayan, S. Y.
033.068
Nariai, K.
042.053
Narihara, K.
066.109
Narizhnaya, N.
098.072
Narizhnaya, N. V.
124.106
Narlikar, J. V.
061.002
066.901
141.133
Nash, W. P.
105.055
Nasyrov, A. M.
031.411
034.113
Nasyrov, G. A.
082.114
Natali, G.
034.096
Nather, R. E.
124.111
Natsvlichvili
125.109
Naugol'naya, M. N.
034.053
Naumann, R. J.
104.035
Naumov, A. I.
141.097
Naumov, V. A.
041.057 .058
Naumova, A. A.
041.036
Navach, C.
158.036
Navara, P.
052.021
Nayak, V. K.
105.004
Nayar, P. S.
034.031
Nazarenko, A. I.
054.018
Nazarova, M. N.
085.009
Nazarova, N. I.
076.044 .046
Nebelitskij, V. B.
082.048
Nechitajlenko, V. A.
104.024 .068
Neckel, H.
113.006
115.003 .011
Nedyalkova, G. M.
097.042
Nees, W.
114.347

Nefed'eva, A. I.
082.031
Neff, J. S.
114.085 .372
122.193
124.104
Neighbours, J. E.
142.010
Nejman, Yu. M.
081.048
Nejshtadt, A. I.
042.061
Nelen, J.
105.166
Nelson, B.
004.009
Nelson, G. J.
077.059
Nelson, J.
141.312
Nelson, M. R.
124.102
Nelson, R.
093.049 .050
Nelyubin, N. F.
082.048
Nemeth, J.
065.193
Nemiro, A. A.
011.055
041.056
Nemo, A. A.
042.026
Nemtsova, Eh. I.
084.286
106.034
Neplokhov, E. M.
032.005
034.061
Nerney, S. F.
074.122
106.013 .053
Nerovnaya, V. S.
052.068
Nesci, R.
031.422
114.010
Neshpor, Yu. I.
142.150 .151 .177
Nesis, A.
072.005
Nesmyanovich, A. T.
074.021 .055 .087
Ness, N. F.
092.003 .015
Nesterov, N. S.
132.047
141.067
Nesterov, V. E.
143.069 .070
Nesterov, V. V.
003.101
Nestorov, G.
033.004
083.049
Nesvedov, S. A.
031.405
Netzer, H.
141.162
Neubert, R.
046.048
Neugebauer, G.
131.523
141.047 .630 .639
155.011
158.056 .070
Neugebauer, O.
003.096
004.048
Neukum, G.
094.105 .181 .186
Neupert, H. E.
126.024
Neupert, W. M.
072.081
073.069
074.018 .065 .067
Nevatia, J.
106.001
143.047
Nevo, I.
119.008
Nevskaya, N. I.
004.114
Nevskij, M. Yu.
074.020
Newberg, R. T.
031.037
Newbound, K. B.
034.101
Newburn, R. L.
091.032
Newcomb, J. S.
031.042 .415
Newell, H. E.
051.026
Newelski, L.
098.075
Newman, B. A.
033.026
Newman, C. E.
073.041
Newman, M. J.
065.046
080.003 .037 .067
Newman, W. I.
093.021
Newton, L. M.
131.120
Newton, R. R.
054.014
Ney, E. P.
103.100
124.104
141.619
Nezhinskij, E. M.
042.005
158.174
Ng, C. K.
078.021
Nguen Tuong Viet
035.014
Nguyen, T. D.
012.037
Nguyen-Hoe
062.018
Nicastro, A. J.
122.152
Nichik, V. M.
004.068
Nichiporuk, W.
105.167
Nicholls, R. W.
034.103
Nicholson, T. F.
104.015
Nicholson, W.
034.138
Nicolas, K. R.
076.038
Nicolet, B.
113.035
Nicolson, G. D.
009.018
Nidey, R. A.
034.156
Niederer, F.
094.471
Niekerke, J.
034.056
113.074
114.335
Niell, A.
121.039
Niell, A. E.
031.218
Nielsen, E.
078.002
Niemann, B.
031.003
Niemeyer, S.
094.442
Nieto, J. L.
162.021 .051
Nieto, M. M.
099.066
Nieuwenhuijzen, H.
012.041
031.418
Nightingale, H. C.
106.018
Niimi, Y.
098.061
Nikander, J.
093.014
Nikiforov, A. F.
061.062
Nikitin, L. V.
011.054
Nikitin, S. A.
051.017
053.009
Nikogosyan, A. G.
063.051
Nikolaev, A. G.
052.013
078.014
082.075
Nikolaeva, N. S.
084.275
Nikolov, N.
013.002
Nikolov, N. S.
013.003
122.030
Nikol'skaya, K. I.
074.086
Nikol'skij, G. M.
051.021
074.066
Nikol'skij, S. I.
143.072

Nikonov, V. B.
031.233
034.069
Nikonova, L. E.
042.025
Nikulin, A. M.
052.062
Nininger, H. H.
105.204
Nisbet, J. S.
083.011 .074
Nisenson, P.
031.242
Nishi, R.
124.103
Nishida, A.
084.201
Nishimura, A.
065.196
Nishimura, T.
032.543
Nissim-Sabat, C.
066.061
Nita, I. D.
073.151
075.010
Nitschke, A.
004.052
Nobis, H. M.
005.008
Noci, G.
074.079
Noel, F.
041.014
Noels, A.
065.021 .097 .100 .176
080.063
Noerdlinger, P. D.
065.044
141.130
Noergard, H.
061.012
Nolt, I. G.
117.002
121.019
Nolte, J. T.
074.035
076.037
Noma, M.
143.028
Noonan, A. F.
105.005
Noonan, T. W.
160.029
Norberg, R.
034.144
Nordin, J. A.
034.144
Nordlund, A.
064.026
Nordmann, J. C.
094.009
Nordstroem, B.
114.013
Nordtvedt Jr., K.
066.095
Norman, C. A.
062.006 .007 .067
141.306
Norman, E. B.
061.077
Norris, J.
122.165
126.003
154.022
North, J.
004.007
Northcutt, K. J.
051.010
Northover, K. J. E.
160.030
Norton, J. A.
141.322 .330 .333 .350
Noskov, B. N.
052.009 .018
Noskova, R. I.
158.144
Notarys, H. A.
034.036
Notni, P.
031.228
141.057
158.100
Nottale, L.
162.040
Novak, A. F.
062.027 .033
Novatskij, G.
003.050
November, L. J.
080.026 .033
Novick, R.
134.010
141.346
142.136 .137 .179
Novikov, I. D.
003.137
066.043 .068 .075
131.128 .167
158.041
162.054 .063
Novikov, L. S.
082.088
084.024
Novikov, S. B.
031.010
082.048
Novikov, V. D.
106.066
Novoselova, N. V.
104.022
Novotny, M.
051.001
Novotny, V.
123.066
Novozhilov, A. V.
031.068
Novozhilov, Y.
005.032
Nowak, T.
064.006
Nowakowski, L.
033.010
Noxon, J. F.
082.018 .039
Noyes, R. W.
074.005
076.002
141.605
Nozaki, M.
103.100
Nuccio, P. M.
105.110 .168
Nudzhenko, A. G.
122.103
Nugis, T.
114.009 .054
Nunes, P. D.
094.444
Nuraliev, B. T.
085.015
Nuritdinov, S. N.
151.086
Nusinov, M. D.
094.147 .148
Nussbaumer, H.
022.034 .081 .100
074.100 .101
076.058
Nussinov, M. D.
See Nusinov, M. D.
Nuzhdina, M. A.
071.059
072.064
Nylen, P.
034.001
Nys, O.
118.039

Obashev, S.
072.022
Oberbeck, V.
094.108
Oberbeck, V. R.
094.107
Oberg, D. L.
061.048
Obozov, V. I.
066.042
Obridko, V. N.
003.122
O'Brien, R. S.
033.044
O'Callaghan, F. G.
114.005
Occhionero, F.
162.042
Ochsenbein, F.
112.023
O'Connell, R. F.
066.012 .016 .032
141.301
O'Connell, R. W.
158.074
Odagiri, S.
103.100
Odell, D. M.
141.033
O'Dell, S. L.
141.038
Odintsova, I. N.
076.049
083.075
Oegelman, H. B.
141.318
Oemler Jr., A.
141.013
Oepik, E. J.
008.006
015.021
102.027

Oezisik, M. N.
022.070
Oganesyan, O. V.
133.019
Oganesyan, R. Kh.
114.046 .111
Ogir', M.
072.023
Ogir', M. B.
072.036
Ogorodnik, I. P.
045.033
Ogorodnikov, K. F.
004.110
Oguti, T.
084.026 .027
Ohki, K.
073.054 .055 .097
Ohnishi, T.
151.093
O'Hora, N. P. J.
044.034
Ohtani, H.
125.103
Oines, R. K.
014.004
Oishi, M.
103.114
Oja, H.
102.014
Okamoto, I.
044.041
064.039
Okazaki, K.
103.131
Oke, J. B.
124.111
125.100
158.232 .312 .314
O'Kelley, G. D.
051.010
094.426
Okhlopkov, V. P.
143.036
Okongwu, E. H.
031.300
Okuda, H.
103.114
Okuda, T.
008.070
044.041
Okun', L. B.
061.079
Olbert, S.
062.016
O'Leary, B.
093.036
098.026
Oleolenko, A. S.
041.007
Oleson, J. R.
031.214
Olevic, D.
098.035
118.028 .030
Olijnyk, P. A.
071.038
Oliver, J. P.
121.088 .154
Oliver, W. L.
083.007
Olivier, M.
035.016
Olivieri, G.
075.015
Olmr, J.
004.091
074.089
Olness, R. J.
062.003
099.001
Olsen, E. J.
105.169
Olsen, E. T.
141.137
Olsen, H. K.
141.357 .360
Ol'shevskij, V. A.
082.110
Olson, B. I.
115.004
Olson, E. C.
124.102
Olson, G. L.
065.158
Olson, R. H.
085.007
Olson, W. P.
011.074
Olthof, H.
132.041
Olvares, O.
079.104
Omarov, S. Z.
034.063
Omarov, T. B.
003.007 .097
012.033
151.074 .075
Omnes, R.
162.028
Onderlicka, B.
047.025
O'Neill, G. K.
015.023
O'Neill, L. H.
034.154
Oneto, J. L.
031.283
Onicescu, O.
066.133
Ono, T.
032.543
103.114
Onuma, N.
105.001 .174
Ooe, M.
045.010
Ookawa, H.
045.012
Oort, J.
158.018
Oort, J. H.
012.022
153.008 .025
158.050 .168
Opal, C. B.
032.522
103.102
104.011
Opfer, J. E.
066.025
Opher, R.
141.026 .078
Opolski, A.
122.029
Oppenheimer, M.
131.023 .045
Oproiu, T.
054.006
Oraevskaya, S. P.
034.019
Oran, E. S.
082.027
Orciuolo, M.
142.206
Ordway III, F. I.
003.098 .132
Orel'skaya, V. I.
041.020
098.065
Oriol, E.
082.045
Orlov, A. A.
042.119
Orlov, M. Ya.
122.024 .200 .202
Orlov, V. A.
143.072
Orlova, T. V.
073.143
Ormes, J. F.
143.059 .066
Orrall, F. Q.
073.135
Orth, C. D.
143.002 .050
Ortiz, R.
032.016
034.052
Ortolani, S.
099.090
Orton, G. S.
099.057 .058 .059
Orudzhev, Eh. Sh.
073.110 .150
Osada, K.
124.103
Osaki, Y.
080.032 .058
Osantowski, J. F.
031.036
Osawa, K.
044.049
Osborn, W.
115.006
Oshchepkov, V. A.
121.104
Osherov, R. S.
103.100 .101
Osherovich, V. A.
062.043 .044 .051 .056 .063
Osipenko, M. M.
041.008
Osipenko, V. P.
103.102
Osipkov, L. P.
151.005 .068
Osipov, A. K.
094.019
096.004 .011

Oskanian, V.
See Oskanyan, V.
Oskanyan, V.
122.037
Osorio, J. P.
052.054
Ostapenko, A. A.
084.268
Ostapenko, V. A.
071.046
073.139 .140
Oster, L.
141.309 .310
Osterbrock, D. E.
158.091 .092 .093 .094
.107 .213 .233
Ostriker, J. P.
061.038
066.019
125.019
155.032 .048
160.035
162.030
O'Sullivan, J. D.
131.133
Otgonsurehn, O.
105.043 .173
Otsuka, M.
022.025
Ott, E.
084.033
Ottelet, I.
081.045
Ounnas, C.
034.152
Ourassine, L.
See Urasin, L.
Ovenden, M. W.
042.075
091.003
Overbeek, M. D.
010.007
Ovezgel'dyev, O.
104.028 .057
Owen, F.
142.216
Owen, F. N.
141.028
142.017
158.224 .225
Owen, T.
022.104
091.016
099.042 .043
Owens, A. J.
106.049
Owens, D. A.
031.033
Oyaste, Ya.
034.087
Ozernoj, L. M.
141.093
155.051
158.116 .118 .123
160.024
162.032
Ozima, M.
082.004
Ozolina, V.
114.113
Ozolins, G.
011.068
072.077
077.062

Pacault, R.
066.030
Pacholczyk, A. G.
158.176 .226
Paciesas, W. S.
142.077
Pacini, F.
141.338
158.057
Paczynski, B.
065.068 .102 .137 .149
Padalia, T. D.
121.001 .076 .077 .094
Paddack, S. J.
106.073
Padevet, V.
104.044
Padrielli, L.
031.236
141.058
Page, C. G.
142.024 .046 .101 .210
Page, D. E.
080.060
Page, T. L.
103.102
113.061
132.055
Pagel, B. E. J.
114.006
Pagel, M.
105.170
Paige, E. G. S.
033.056
Paik, H.
066.033
Paik, H. J.
066.025
Paizis, C.
073.028
Pakvor, I.
032.030
Pakvor, L.
032.027
Pal, A.
082.058
Palagi, F.
077.027 .050
Pallavicini, R.
073.104
076.052
077.037
Palluconi, F. D.
012.028
100.019
Palmadesso, P.
084.299
Palme, H.
094.024
105.171
Palmer, M. D.
052.029
Palmer, P.
131.006 .008 .026 .086
158.009 .219
Palmieri, T. M.
134.004
142.164
Palowski, J.
009.026
Palter, R.
004.010
Pamjatnykh, A. A.
See Pamyatnykh, A. A.
Pamyatnykh, A.
064.032
Pamyatnykh, A. A.
064.014
122.035 .112
Panadzhyan, V. G.
106.021
Panagia, N.
131.041
132.035
141.147
Panasyuk, A. I.
052.044
Panasyuk, V. I.
052.044
Panchuk, V. E.
064.062
Panda, H. S.
162.092
Pandey, P.
153.042
Pandey, R. K.
022.117 .124
Pandharipande, V. R.
061.044
065.190
Panfilov, A. S.
093.011
Pang, K. D.
097.038
Panicucci, R.
073.154
Pankonin, V.
131.078
Pankov, V. M.
076.044 .046
Pankow, M.
009.026
Pankow, S.
009.026
Panova, G. V.
041.004
100.202
Pansecchi, L.
103.100
Papadopoulos, C.
036.004
Papagiannis, M. D.
072.063
073.145
Papaloizou, J. C. B.
117.014
Papoulis, A.
033.065
Paprotny, Z.
015.038
Paquet, P.
044.044
045.019
046.065
Parcelter, P.
035.017

Paresce, F.
032.515 .535
034.111
131.064
Parfait, C. M.
085.022
Parijskij, Yu. N.
141.116
Paritskaya, G. G.
034.061
Park, C. G.
084.317
Parker, E. A.
033.029
Parker, E. N.
062.049
076.039
080.022 .038
155.025
Parker, J. T.
085.003
Parker, R. L.
081.004
Parkinson, J. H.
076.005 .006
142.160
Parkinson, R. C.
053.008
Parkinson, T. D.
099.211
Parkinson, W. H.
071.013
Parra, G. T.
032.548
Parrish, A.
131.078 .512
Parsamyan, Eh. S.
122.104 .105
Parshina, Z. S.
005.016
Parsignault, D.
032.523
142.014 .088 .122
Parsignault, D. R.
034.157
142.001 .085 .121 .194
Parsons, S. B.
114.005 .103 .313 .343
115.020
117.035
Parthasarathy, M.
119.023
Partridge, R. B.
061.013
162.009
Pasachoff, J. M.
074.029
Pascal, M.
042.050
Paschke, J.
077.064 .065
Paschmann, G.
084.207
Pascu, D.
097.001
Paseka, A. M.
157.004
Pashchenko, M. I.
131.149 .196
Pasinetti, L. E.
114.340 .389
Pasinetti, L. E.
115.001
122.151
Patchett, B. E.
114.006
Paterno, L.
082.129
Paterson, R. A. H.
032.010
Patkos, L.
121.097
Patnaik, R. B.
042.052
Patriarchi, P.
077.027 .050
Patston, G. E.
122.156
Patterson, J. R.
142.074 .178
Patterson, R. S.
122.193
Paturel, G.
158.202
Paul, A. K.
083.076
Paul, J.
032.519
Paul, J. A.
032.518
Paulikas, G. A.
084.409
Pauliny-Toth, I. I. K.
033.007
141.010
Pauls, T.
157.001
Paupere, M.
021.022
077.062
Pavel'ev, A. G.
093.023
094.198
Pavlenko, P. P.
032.004
098.030
Pavlenko, V. N.
062.012
Pavlik, S. A.
041.010
Pavlov, A. F.
052.067
Pavlov, G. G.
062.023
116.008
117.029
Pavlov, N. V.
066.126
Pavlov, V. A.
082.085
Pavlova, L. A.
131.161
Pavlova, V. S.
142.030
Pavlovskaya, E. D.
112.004 .006 .008
142.242
153.019
Pavlovski, K.
104.030
Payzant, J. D.
022.063
Peach, G.
022.118
Peacock, A.
142.052 .202
Peale, S. J.
094.175
Pearce, J. A.
153.041
Pearson, P. K.
022.119
Pechinskaya, N. I.
071.064
Pecina, P.
104.009
Pecker, J.-C.
013.008 .012
162.039
Peckett, A.
094.467
Peckham, R. J.
141.339
Pedersen, K. M.
008.043
Pedersen, O.
003.099
Peebles, P.
003.100
Peebles, P. J. E.
151.036
158.001
160.012
Peery Jr., B. F.
122.004
Peimbert, M.
131.531 .561
132.044
133.010
155.066
Pelagatti, G.
115.001
Pell, C.
034.162
Pellas, P.
105.155 .172 .203
Pellat, R.
141.341
Pellerin Jr., C. J.
078.038
Pelling, M.
022.023
131.018
Pels, G.
153.008 .025
Pels-Kluyver, H. A.
153.008 .025
Pena, J. H.
065.158
Pendl, E.
034.008
Penegor, G.
121.091
Penfield, H.
131.054
Penfold, J. E.
142.053
Peng, P. W.
031.247
Pennell, W.
124.110
Penny, A.
004.073

Penny, A. J.
154.005
Penrose, R.
012.048
066.162
Penston, M. J.
142.042 .043
Penston, M. V.
113.002
141.018 .602
142.042 .104
158.141
Penzias, A. A.
131.104 .170 .171
Penzo, P. A.
100.020
Peralta, M.-T. C.
114.091
Peralta, R.
032.047
Percy, J. R.
047.014
113.031
122.097 .137 .172 .180
Perdang, J.
065.025 .171
Perelygin, V. P.
105.043 .173
Pereslegina, N. V.
078.014 .018
Pereverzentsev, A. F.
034.118
Pereyaslova, N. K.
085.009
Perez-Peraza, J.
073.064 .065
078.013 .031
Perissinotto, M.
123.038
Perkins, F. W.
083.024
Perko, L. M.
042.086
Perley, R. A.
125.013
Perola, G. C.
158.203
160.005
Perona, G. E.
076.060
Perov, V. A.
031.062 .063 .066
Perova, N. B.
121.121
123.003
Perrenoud, M. R.
077.031
Perrin, M. N.
114.072
Perry, C. L.
114.333
Perry, J. E.
008.043
Perry, J. J.
141.011 .092 .150 .163
Perry, M. A.
093.049
Persi, P.
071.023
Persson, S. E.
131.519 .527
Persson, S. E.
158.056 .070
Pesch, P.
122.130
124.110
141.614
158.047
Pesek, I.
041.002
Peskov, Yu. A.
046.038
Pesses, M. E.
106.072
Peter, H.
121.149
Peterova, N. G.
072.042
079.102
Peters, B.
143.048
Peters, C. F.
100.204
Peters, G. J.
065.049
114.387
Peters, W. L.
131.056
155.065
Petersen, J. O.
065.142
122.118
Peterson, A. W.
082.042
Peterson, B.
142.222
Peterson, B. A.
141.155 .156
Peterson, C. J.
158.006 .222
Peterson, F. W.
141.129
Peterson, L. E.
032.524
051.010
076.028
142.004 .082 .087
Peterson, R. C.
031.239
114.035
Peterson, S.
121.039
Peterson, S. D.
121.028
158.039
Peterson, W. K.
084.032
Pethick, C. J.
142.128
Petit, J.
032.048
Petit, J. P.
151.057
Petit, M.
123.041
Petit, P.
035.013
Petkov, A. P.
022.079 .102
071.051
Petkovic, V.
009.014
Petrasso, R.
073.053
074.073 .078
Petrasso, R. D.
073.003
Petravic, M.
141.351
Petrenko, I. E.
085.009
Petri, W.
005.006
Petrosian, V.
076.032 .033
Petrov, G. I.
091.018
105.030
Petrov, G. M.
093.024
094.198
Petrov, P. P.
031.234
122.122
Petrov, T. G.
061.008
Petrova, I. V.
078.015 .025
Petrova, N. S.
072.051
Petrovskaya, I. V.
151.039
Petrukhin, N. S.
080.023
Petrusevich, S. I.
062.022
Pettengill, G. H.
012.028
Pettersen, B. R.
103.102
122.075
Petterson, J. A.
142.114
Pettini, M.
131.143 .207
Petukhova, T. M.
094.429
Petviashvili, V. I.
077.020
Peyton, B. J.
034.161
Peytremann, E.
064.067
Pfau, W.
117.037
Pfeffermann, E.
032.518
Pfister, H.
072.080
Pfleiderer, J.
006.000
008.048
063.008
158.025
Pham Duc Wuong
022.071
Phelps III, F. M.
034.101
Philip, A. G. D.
036.021
113.047 .059
154.011

Phillips, J. D.
084.306
Phillips, J. G.
003.005
Phillips, K. J. H.
074.002
Phillips, M. M.
158.092 .093 .233
Phillips, R. B.
114.372
Phillips, R. J.
097.005
Phillips, T. G.
131.148
132.042
Philpotts, J. A.
105.174
Phinney, D.
094.442
Piana, A.
063.028
Piazza, L. R.
083.009
Picat, J. P.
074.061
Piccirillo, J.
113.057
Pick, M.
033.012
077.005
Pickup
124.104
Piddington, J. H.
080.001 .056
156.002
Pieczynski, L.
045.037
Pieper, G. F.
032.513
Pierce, D. A.
014.002
100.207
Pieri, D.
097.030
Pieri, P. E.
022.109
Pigeon, R.
022.087
Piirola, V.
034.122
121.095 .096
Pikel'ner, S. B.
073.029
080.072
Pilcher, F.
098.081 .085 .086
Pillinger, C. T.
094.190 .191 .461
Pilloud, F.
013.010
Pilnik, G. P.
044.019
Pines, D.
142.129 .197
Pinigin, G. I.
032.046
Pinnington, E. H.
022.068
Pinter, S.
073.022 .119
074.089
Pinto, G.
041.052
Piotrowski, S.
007.000
Piotrowski, S. L.
102.028
Pipes, B.
066.025
Pipher, J. L.
131.508 .552
Pippi, I.
082.122
Piran, T.
062.048
142.018
Pirraglia, J. A.
091.044
Pisacane, V. L.
052.070
Pisarenko, N. F.
073.034
076.018 .021 .042
084.274 .275
106.065
Piskunov, A. Eh.
153.005 .015 .016 .017
Pismis, P.
132.056
151.024
Pissarenko, N. F.
See Pisarenko, N. F.
Pit'ev, N. P.
042.084 .108
Pittich, E.
079.100
Pitzalis, C.
014.027
Pizarro, G.
103.133
Pizarro, O.
103.133
Pizzella, G.
099.028
Plagemann, S. H.
003.060
Plakhov, Yu. V.
042.017
Platais, I.
124.104
Platisa, M.
022.020 .108
Platov, Yu. V.
031.271
Platt, B. C.
031.047
Plaut, L.
012.001
122.169
123.039
Plavec, M.
114.326 .375
121.037
Plebanski, J. F.
066.180
Plocieniak, S.
076.054
Plotkin, H. H.
055.016
Plotnikov, V. S.
031.056 .065
Plyasova-Bakunina, T. A.
106.067
Plyugin, G. A.
117.071
Pochinok, B. D.
122.058
Pochukaev, V. N.
003.020
Podgornyj, I. M.
074.094
Podobed, V. V.
003.101
Podolak, M.
099.036
Podstrigach, T. S.
077.010
Poeldmaa, V. K.
085.016
Pogorzelski, R. J.
063.047
Pogulyaevskij, L. I.
083.054
Pohl, E.
121.092
Poland, A. I.
036.019
074.037
094.005
Poletto, G.
074.081 .105
Polidan, R. S.
114.032 .375 .387
Polkowski, G.
105.040
Pollack, J. B.
004.052
093.028 .042
097.014 .037
099.001 .033
100.026
Pollard, G.
142.233
Pollock, G. E.
105.026
Polnarev, A. G.
066.097
Polojentsev, D. D.
See Polozhentsev, D. D.
Polojentseva, T. A.
See Polozhentseva, T. A.
Polosukhina, N. S.
114.357
Polos'yants, B. A.
031.410
Polozhentsev, D. D.
032.047 .048
041.036
Polozhentseva, T. A.
032.047 .048
041.036
Polupan, P. N.
073.141
Polyakhova, E. N.
100.009
Polyakov, V. M.
083.058
Polyakov, V. S.
003.020
Polyakova, G. I.
113.037

Poma, A.
011.040
Pomerantz, M. A.
078.002
Pommereau, J.-P.
082.051
Pomphrey, R. B.
131.554
Poncet, J. L.
013.010
Ponnamperuma, C.
094.428
Ponomarev, E. A.
084.288
Ponomarev, V. N.
066.125
084.272
Ponomarev, Yu. N.
084.272
Ponomareva, V. I.
021.006
Poole, L. M. G.
104.015
Pooley, G. G.
141.118
Pop, P.
031.201
Pop, V.
031.201
Pope, T.
032.007
071.052
Popeko, A. G.
105.175
Popov, E. I.
066.072
Popov, G. M.
031.028
034.068
Popov, G. V.
083.037
106.034
Popov, M. V.
141.037
142.170
Popov, O. S.
074.021 .087
Popov, V. S.
032.003
158.191 .231
Popov, Yu. P.
125.040
Popova, L. V.
157.004
Popova, M.
122.031
Popova, R. I.
045.032
Popovic, B.
042.071 .072
Popovic, G. M.
118.018 .027 .029 .034
.036
Popovic, M. V.
022.020 .108
Popovici, C.
075.010
Poppoff, I. G.
097.063
Poroshin, N. D.
034.072
Porteous, R. L.
031.416
Portsevskij, K. A.
009.021
Portugal, E.
032.048
Posen, A.
104.034
Postan, M. Ya.
072.053
Postel, J.
034.152
Postl, A.
005.007
Potdar, A.
151.020
Pottasch, S. R.
132.041
Poulsen, O.
022.102
Poumeyrol, F.
010.038
Pounds, K.
142.234
Pounds, K. A.
076.006
134.006 .012
142.024 .048 .049 .052
.101 .102 .120 .153
.202
Poupeau, G.
094.110 .192
Poveda, A.
117.039
155.006
Poverennov, A. L.
031.272
Powell, A. L. T.
075.003
Powell, C.
051.003
Powell, N. R.
033.075
Pozdnyakov, I. G.
054.018
Praderie, F.
114.374
Prager, F.
004.052
Prakash, A.
084.222 .280
099.007
Pramesh Rao, A.
141.094
Prasad, S. S.
084.017 .023
099.018 .040
Prasanna, A. R.
162.079 .080
Pratt, N. M.
034.142
158.029
Pravdo, S.
142.196
Pravdo, S. H.
061.017
142.190 .202
Preite-Martinez, A.
132.035
Presnyakov, L. P.
022.022
Press, F.
045.001
Press, W. H.
031.203
066.073
117.062 .066
162.057
Preston, R. A.
141.135
Preuss, E.
033.007
141.010
Preussner, P.-R.
141.012
Prevot, L.
159.001
Prialnik, D.
065.128
Price, I.
103.100
Price, J. C.
074.011
Price, K. M.
141.022
Price, P. B.
143.051
Priest, E. R.
080.036
Priester, W.
008.017
Prihoda, P.
091.030
Prikhod'ko, A. N.
143.026
Prikhod'ko, V. A.
098.024
Prilutskij, O. F.
142.108 .181
Primini, F.
141.326
142.076 .078 .081 .089
.219
Pringle, J. E.
142.044 .051
Prinn, R. G.
093.053
099.056
Prinz, M.
094.474
105.177
Priser, J. B.
041.035
Proctor, M. R. E.
062.069
Proelss, G. W.
082.109
083.039
Projdakov, V. I.
033.071
Prokes, M.
052.021
082.023 .059
Prokhin, V. L.
143.069
Prokhorov, A. V.
103.100
Prokof'eva, V. V.
034.079
122.101
158.120

Prokudina, V. S.
072.030
091.024
099.081 .082
Pronik, I. I.
158.016
Pronik, V. I.
132.048
142.110
158.015 .032
Prorok, V. P.
031.211
034.018
054.003
Protic, M.
008.010
Protic-Benisek, V.
031.029
032.026
041.067
095.010
096.014 .015
Protitch, M. B.
096.014 .015
Proverbio, E.
012.029
014.028
035.005
045.021
Prutenskaya, E. I.
106.015
Pryakhin, E. A.
143.074
Prytkov, N. M.
033.016
077.014
Przybylski, A.
114.391
Ptacek, V.
044.050
Ptak, R.
158.053
Ptak, R. L.
151.525
Ptitsyn, D. A.
034.082
114.086
Ptuskin, V. S.
143.039
Puccilo, M.
121.046
Pucel, R. A.
034.128
Puch, A. A.
123.075
Pucillo, M.
121.005
Pugach, A. F.
121.011
122.021 .098 .204
Puget, J.-L.
131.072
Puigcerver, M.
082.045
Pulkkinen, K. F.
047.039
Puplett, E.
082.041
Pupyshev, Yu. A.
104.062
Purcell, G. H.
033.007
Purcell, I. M.
103.010
Purkins
124.104
Purton, C. R.
141.125
142.215
Pushkov, A. N.
084.242 .243 .291
Pustil'nik, L. A.
142.030
Pustylnik, I.
121.085 .086 .087
Puttkamer, J. Von
See Von Puttkamer, J.
Pyatigorskij, G. A.
142.171
Pyatnitskij, V. S.
073.149
Pyatunina, T. B.
131.556
Pye, J. P.
076.006
Pyle, K. R.
099.024
Pynzar', A. V.
106.047 .057
141.132
Pystina, N. S.
072.040

Qian, S.
073.076
Quaide, W.
094.108
Quenby, J. J.
061.067
142.019 .023
Querci, F.
114.334
Querci, M.
114.334
Quesada, V.
035.006
Quinones, J. A.
079.102
Quintana, H.
065.136
158.143
160.043

Raadu, M. A.
080.036
Rabache, P.
082.066 .083
Rachkovskaya, T. M.
122.159
Racine, R.
154.030
Radchenko, V. E.
066.131
Radford, G. A.
119.005 .014 .021
Radhakrishnan, V.
141.345
Radkov, R.
047.022
Radostitz, J. V.
117.002
Raedler, K.-H.
065.088
Raff, M. I.
124.103
Raghavan, N.
073.070
Raghuram, R.
084.300
Ragot, J.
032.508
Rahe, J.
102.004
Raimond, E.
066.108
Raine, W. L.
094.102
Rainov, G.
055.011
Raitt, W. J.
083.004 .005 .039 .064
Rajamohan, R.
119.023
Rajan, R. S.
105.059
Rajchl, J.
104.008
Rajen, V.
004.069
Rajkova, D.
047.022
Raju, P. K.
022.097
Rakhimov, A. G.
112.007
Rakos, K. D.
118.013
Rakosch, K.
118.002
Ralincourt, P.
121.149
Ramadurai, S.
011.044
066.901
Ramana, B. V.
074.132
Ramanujam, P. S.
022.102
Ramaty, R.
073.016 .087 .088 .108
.114
076.055
Rambaldi, E.
105.125 .176
Ramdohr, P.
094.454
105.177
Ramsayer, K.
046.071
Ramsey, H. E.
073.047
Ramsey, L.
114.302
Ramsey, L. W.
064.060
071.006
073.136
Randall, B. A.
099.026

Randich, E.
105.178
Randolph, J. E.
022.045
Rands, R.
031.277
Ranga Rao, M. P.
074.132
Rangarajan, R.
052.014
Ranieri, M.
142.206
Rank, D. M.
034.035
100.013
131.522
133.014
141.617
Ranzinger, P.
046.041
072.032
Rao, A. Pramesh
See Pramesh Rao, A.
Rao, D. R. K.
084.261
Rao, G. R.
105.004
Rao, J. R.
162.092
Rao, J. S.
033.023
Rao, M. N.
105.100 .131
Rao, M. Srirama
See Srirama Rao, M.
Rao, N. Kameswara
See Kameswara Rao, N.
Rao, U. R.
143.076
Rapavy, P.
095.005
Rapley, C. G.
076.007
125.016
Rapoport, I. D.
032.534
143.069 .071
Rapoport, V. O.
074.085
077.023
Rapp, R. H.
081.010 .015 .032 .033
Rappaport, S.
125.024 .034
141.326
142.073 .075 .081 .083
.089 .220 .231
Rappaport, S. A.
142.189
Rash, J. P. S.
062.009
Raskin, N. M.
002.001
Rasmussen, I. L.
143.048
Rasool, I.
012.028
Rassadovskij, V. A.
033.071
Rast, J.
071.037
Rastogi, R. G.
084.260
Rasulov, A. M.
072.049
Ratcliffe, J. A.
003.102
Ratier, G.
034.165
Ratkiewicz, R.
074.088
Ratnam, S. R.
104.093
Raudsaar, H. K.
098.002 .003
103.100
Rautureau, M.
105.109
Ravenhall, D. G.
065.167
Ray, S.
022.073
Raychaudhuri, A. K.
066.174
162.089
Raymond, J. C.
022.053
062.059
131.185
Rayo, J. F.
131.531 .561
Razin, V. A.
057.004
Raznik, R. M.
153.023
Reade, V.
008.077
Readhead, A. C. S.
131.042
Reagan, J. B.
084.404 .405
Reasenberg, R. D.
097.052
Reay, N. K.
051.006
131.036
Rebeirot, E.
159.001
Rebours, B.
082.066 .083
Rebrov, M. F.
003.103
Recami, E.
066.013
Reddish, V. C.
008.034
160.020
Redman, R. O.
119.013
Redyuk, T. I.
072.061
Reed, E. I.
082.026
Reed, M. A.
034.119
Reed, R. A.
099.033
Reed, S. J. B.
105.035
Reed Jr., G. W.
094.422 .440
Reedy, R. C.
051.010
Rees, M. H.
084.030
Rees, M. J.
010.022
066.085
142.044 .126
160.044
Reese, E. J.
099.088 .091
Reeves, E. M.
051.005
074.005
076.002
Reeves, H.
061.047
Reeves, R.
103.100
Reeves, R. R.
097.051
Refsdal, S.
065.033
117.008 .045
Regev, O.
065.062
Reginaldo
099.090
Reichel, A.
022.013
Reichhart, F.
045.014
Reichley, P. E.
097.035
Reichmann, E.
073.124
Reichmann, E. J.
072.012
Reid, A. M.
105.090 .091 .179
Reid, G. C.
078.003
Reid, M. J.
141.644
Reigber, C.
046.067
081.028
Reijmerink, F.
124.104
Reimers, D.
064.034
Reinhardt, M.
006.000
062.004
066.169
142.123 .149
158.019 .021
Reinhart, E.
021.017
046.070
Reitsema, H. J.
100.007
Rejnkhardt, M. I.
142.144
Remis, J.
121.149
Remo, J. L.
105.039
Remy, F.
022.111 .112 .113

Renschen, C. P.
099.045 .055
Renzini, A.
065.035
Reppin, C.
076.043
134.006 .012
Resch, G.
121.039
Resch, G. M.
141.049 .050 .143
Rettig, T. W.
154.033
Reutov, V. P.
062.058
Revah, I.
091.047
ReVelle, D. O.
104.003
Revenchuk, S. M.
062.012
Reynolds, J. M.
066.025
Reynolds, R. J.
103.102
125.057
Reynoso, R. R.
004.122
Reyss, J. L.
094.196
105.157
Reznikov, E. A.
103.104
Rhee, J. W.
106.073
Rhodes Jr., E. J.
074.034
Rhynsburger, R. W.
041.029
Ribes, J. C.
033.003
141.009
Rich, M. A.
033.046
Richard, A.-J.
045.027
Richards, G. R.
051.004
Richards, P. B.
012.051
Richards, W. G.
131.116
Richardson, J.
142.081 .188
Richardson, J. A.
141.326
142.089
Richer, H. B.
115.004
Richings, J. G.
031.275
Richou, J.
022.109
Richstone, D. O.
121.036
141.046
158.125
160.008
Richter, A. K.
074.022
Richter, G. A.
124.104
Richter, H.
094.451
Richter, N.
007.000
154.020
160.016
Rickard, J. J.
131.031
Rickard, L. J.
131.026
158.009 .219
Ricker, G.
141.353
142.080
Rickett, B. J.
074.050
141.324 .329
Ricketts, M. J.
142.024 .049 .101 .102
.120 .153
Ricort, G.
031.227
071.026 .027
Riddle, A. C.
073.127
077.046
099.044 .097
Ride, S. K.
131.063
Ridgway, S. T.
034.033
115.022
Ridley, E. C.
093.051
Ridpath, I.
003.104
015.004
Riedler, W.
078.035
084.012
Riegler, G. R.
061.023
113.051
Rieke, G. H.
100.011
141.610
158.008 .048 .320
Riepe, B. Y.
041.035
Riera, A.
022.059 .060
Ries, L. M.
100.205
Righini, A.
012.026
Righini, G.
012.025
Rigutti, M.
014.019 .025
073.154
Rihm, K.
011.025
Riihonen, M.
065.172
Riives, V.
103.001
Rikow, W. T.
066.050
Riley, J. M.
141.118
Rinehart, R.
141.073 .074
Ring, J.
031.285
131.036
Rinsland, C. P.
141.091
Rios, M.
061.004
Ritchings, R. T.
141.327
Ritter, H.
065.033
117.006 .007
Rizov, E. F.
082.115 .120 .121
Rizvanov, N.
092.002
Rizvanov, N. G.
041.012
098.071
124.104
Roach, G. F.
021.002
Roberti, G.
073.154
Roberts, J. A.
141.009
Roberts, K. V.
141.351
Roberts, M. S.
158.039 .061 .127 .220
Roberts, P. H.
062.053
Roberts, W. O.
085.007
Roberts Jr., C. E.
042.091
Roberts Jr., W. W.
151.054
Robertson
124.111
Robertson, D. S.
031.220 .261
Robertson, E. C.
003.105
Robertson, J. W.
143.065
Robertson, P. B.
105.180
Robinson, E. L.
114.316
124.001 .111
126.006
142.232
Robinson, J. H.
010.012
Robinson, K. L.
094.413
Robinson, L. B.
031.409
Robinson, R. D.
077.058 .059 .060
Roble, R. G.
083.001
Robley, R.
106.009
Robson, E. I.
162.012

Rocchia, R.
032.541
Rochester, M. G.
084.284
Rocznik, K.
005.005
Roddier, F.
031.305
071.007
Rodgers, A. W.
154.021
Rodionova, Zh. F.
097.027
Rodono, M.
122.011 .012 .207
Rodriguez, L. F.
132.024
Rodriguez, M. H.
122.024 .202
Rodriguez Kuiper, E. N.
See Kuiper, E. N. R.
Roeder, R. C.
066.157
Roederer, J. G.
003.106
Roelof, E. C.
074.035
078.001 .008
099.025
Roels, J.
042.049
Roemer, E.
098.052 .055 .058
099.233 .237 .238
103.100 .111 .121 .123
.124 .127 .134 .135
.138 .139
Roennaeng, B. O.
031.218
Roeser, H.-J.
141.011 .148
Roeser, M.
065.013 .173
Roesler, F. L.
103.102
Roessiger, S.
141.621
Roev, B. A.
031.058 .067 .068
Roger, R. S.
141.009 .146
Rogers, A.
121.039
Rogers, A. E. E.
031.218
Rogerson Jr., J. B.
064.002
Rogovich, M. D.
004.068
Rogovtsov, N. N.
063.050
Rogowski, J.
094.106
Rogowski, R. S.
099.067
Rogstad, D. H.
141.079
Roig, R. A.
022.038
Rojas, E.
119.022
Roland, G.
013.010
103.100
Rolland, R.
121.149
Romanchuk, P. R.
071.059
072.025 .060 .061 .064
Romani, L.
003.107
Romanishin, W.
142.235
Romano, G.
113.041
122.175
123.038
Romanov, S. A.
074.064
Romanov, Yu. S.
122.056 .126
Romanovskij, Yu. A.
083.044 .054
Romanteev, N. F.
052.013
Romanyuk, V. F.
104.056
Romashchenko, Yu. A.
084.287
Romney, J. D.
033.007
141.073 .074
Romoli, C.
121.149
Rompolt, B.
073.153
Ronan, C. A.
015.003
Ronchi, V.
004.051
Rood, H. J.
158.028 .040
160.015 .021 .901
Rood, R. T.
065.008
Roosen, R. G.
082.025 .116
106.037
Rose, J. R.
053.022
Rose, W. K.
061.030
141.355
155.026
Rosen, E.
004.030
005.002 .006
Rosen, J.
065.144
Rosen, N.
065.144
066.089 .118
Rosen, R. A.
131.213
Rosenbauer, H.
074.129
084.207
Rosenberg, F. D.
076.023
142.022 .093 .131
Rosenberg, G. D.
003.108
Rosenberg, T. J.
084.302
Rosenblum, A.
062.004
066.015 .169
141.334
Rosenkranz, P. W.
082.125
Rosenvinge, T. T. Von
See Von Rosenvinge, T. T.
Rosenzweig L., P.
122.225
Rosino, L.
004.028
007.000
114.323
122.042 .046 .220
124.101 .104
Rosinski, J.
104.004
Roskot, V.
079.103
Rosmasheva, L. N.
052.069
Rosolen, C.
033.013
Ross, C. L.
094.005
Ross, D. K.
066.115 .179
Ross, H. N.
141.031 .110 .126
Ross, R. R.
065.108
122.114
Rossada, V. M.
072.058
Rosser, W. G. V.
066.061
Rossi, B.
012.025
Rossi, F. S.
004.088
Rossner, H. H.
061.061
Rossow, W. B.
093.045
Rostoker, G.
084.253
Rosztoczy, F. E.
033.025 .027
Rotenberg, M.
003.011
Roth, E.
066.030
Roth, M. L.
117.008 .045
Rothenflug, R.
032.541
Rothschild, R. E.
142.097 .155 .190 .196
Rothwell, P.
082.043
Rothwell, P. L.
073.138
Rots, A. H.
158.149 .200 .201 .208
Rottman, G. J.
071.028
072.062
073.132 .133

Rouan, D.
131.532
Rouse, C. A.
080.024 .041
Rousell-Dupree, D.
071.028
Rousseau, J.
122.133
159.001
Roussel, K. M.
126.029
Rowan-Robinson, M.
010.022
011.003
031.210
131.019
Rowbotham, T. R.
033.033
Rowe, C.
010.024
Rowe, M. W.
105.058
Rowley, J. K.
080.045
Roxburgh, I. W.
065.077
Roxburgh, W.
080.007
Roy, D. W.
105.111 .181
Roy, J.-R.
076.024
Roy, N. L.
105.051
Roy, S. R.
066.007
Roychoudhuri, C.
031.077 .304
Royer, A.
121.149
Rozantsev, I. N.
076.044 .046
Roze, L.
003.050
011.069
031.273
Rozenberg, G. V.
082.100
Rozenbush, A. Eh.
122.124 .203
Rozenbush, V. K.
102.015
Rozenfel'd, B. A.
004.063
Rozental', I. L.
142.181
Rozhkovskij, D. A.
031.251
036.008
041.038
155.052
Rspaev, F. K.
031.251
Rubakha, N. R.
066.098
Ruban, E. V.
114.023
Rubashevskij, A. A.
064.037
114.063 .066
Rubenstein, D. M.
074.032
Rubin, V. C.
158.005 .006 .220 .222
Rublev, S. V.
114.008
Rubo, G. A.
074.055
Rudd, P. J.
075.021
Rudenko, V. N.
066.072
Rudenskaya, I. M.
084.244
Rudkjoebing, M.
022.011
131.129
Rudnick, L.
158.224 .225
Rudnicki, K.
011.070
158.236
161.005
Rudnikova, E. G.
073.085
Rudolph, D.
031.002
Rudolph, E.
066.104
Rudolph, L.
014.035
Rudy, R. J.
117.002
124.104
Ruffini, R.
012.014
066.056 .110
142.130
Rugge, H. R.
074.038 .128
082.036
Ruiz, M. T.
064.046
Rule, B. H.
036.014
Rummel, R.
044.032
Rumyantsev, A. A.
073.090
Rumyantsev, V. V.
042.066
Runcorn, S. K.
003.108
094.144
Ruppel, W.
003.046
Ruprecht, J.
047.025
Rusch, D. W.
084.001
Rusch, W. V. T.
033.074
Rusev, R. M.
114.058
122.053
Rusin, V.
074.026
Ruskol, E. L.
003.109
Russell, C. T.
074.116
Russell, C. T.
084.213 .230 .235 .319
094.180
Russell, J.
041.005
Russell, J. A.
104.041 .090
Russell, J. L.
004.048
112.014
124.104
Russell, R. W.
131.075
133.005 .013
141.606
Russev, R. M.
See Rusev, R. M.
Rust, B. W.
125.059
Rust, D.
076.035 .036
Rust, D. M.
073.013 .071
074.067
Rustad, B. M.
073.147
Rustamov, Yu. S.
114.087
Rutily, B.
122.037 .181
Rutten, R. J.
031.289
034.149
Rutter, G. H.
103.010 .100
Ruzdjak, V.
022.078
073.093
Ruzmaikina, T. V.
See Ruzmajkina, T. V.
Ruzmajkin, A. A.
022.093
061.078
072.082
141.336
Ruzmajkina, T. V.
141.603
Ryabchikova, T. A.
114.356
Ryabov, O. L.
093.005 .006
Ryabov, V. P.
041.007
Ryabov, Yu. A.
042.073 .119
101.006
Ryabtsev, A. N.
022.048
Ryan, J. M.
074.131
Ryan Jr., M. P.
036.010
Rybansky, M.
074.026 .091 .092
Rybka, E.
004.021 .054
Rybski, P. M.
114.005 .045
Ryckman, S.
141.353
142.080

Rydbeck, G.
066.076
Rydbeck, O. E. H.
031.218
Ryder, G.
094.437
Rydgren, A. E.
065.059
115.010
Ryle, M.
033.001
Ryle. M.
141.033
Ryl'kov, V. P.
114.112
Rylov, V. S.
034.060
Ryter, C.
131.072
Ryter, C. E.
142.058 .096
Ryzhova, T. A.
084.271
Rzhiga, O. N.
091.026

Saakyan, G. S.
003.110
066.067
Saakyan, K. A.
158.121
Saar, E.
155.047
Saari, D. G.
042.035 .081
Saba, J. L. R.
142.097
Sabbadin, F.
155.066
Sabitov, Sh. N.
032.037
132.037
Sadeh, D.
119.008
Sadler, D. H.
007.000
Sadokhin, V. P.
034.067
Sadzakov, S.
032.026
041.033 .067 .068 .069
Safko, J. L.
042.042
Safraj, A. S.
093.016 .018
Safronova, U. I.
061.071
076.041 .045
Sagan, C.
004.052
091.012
093.034 .045
097.002 .012 .026 .029
099.034
Sagar, R.
153.042
Saggion, A.
061.060
Saginashvili, M. G.
151.077
Sagnier, J. L.
042.030
Sahai, Y.
082.040
Saigusa, Y.
103.131
Saio, H.
065.026
Saito, M.
121.026
Saito, T.
143.028
Saito, T. T.
031.052
Saito, Y.
143.028
Sajfudinova, T. I.
084.308
Sajmon, T.
004.126
Sakai, S.
044.041
Sakashita, S.
064.049
Sakharov, V. I.
045.006
Sakhibullin, N. A.
133.008
Sakin, I. L.
031.049
Sakurai, K.
003.111
073.160 .161
106.077
Salem, M.
022.086
031.301 .302
Salemi, S.
031.423
Saletic, D.
032.026
041.033 .067 .068 .069
Salimzibarov, R. B.
076.040 .048
Salisbury, J. W.
002.012
Salisbury, W. W.
066.060
Salomonovich, A. E.
082.024
Salpeter, E. E.
061.027
133.026
Salstein, D. A.
061.084
Salukvadze, G. N.
074.110
097.203 .204
Salvati, M.
141.343
Samojlov, R. A.
033.071 .072
Samojlova-Yakhontova, N. S.
098.005 .064
Samokhin, A. A.
125.040
Samokhvalov, Yu. E.
157.004
Samovol, V. A.
094.199
Samoznaev, L. N.
094.199
Samson, A. M.
063.050
Samsonov, I. S.
143.027
Samsonova, Z. N.
143.027
Samus', N. N.
122.064 .169
154.003 .014
Sanadze, A. A.
082.069
Sanahuja, B.
061.016
Sanchez, G.
103.125
Sanchez Magro, C.
124.104
Sanchez-Sinencio, E.
031.303
Sancisi, R.
158.159 .160 .164
Sandage, A.
131.195
162.047
Sandakova, E. V.
104.081
Sandell, G.
131.085
Sanders, J.
098.051
Sanders, R. H.
151.081
158.084
Sanders, W.
113.017
Sanders, W. M.
032.540
074.041
Sanders, W. T.
155.053
Sandig, H.-U.
041.054
046.052
Sandler, D. G.
065.192
Sandlin, G.
076.034
Sandlin, G. D.
022.096
Sandmann, W.
114.038
Sandomirskij, A. B.
082.087
Sandqvist, A.
155.040
Sanduleak, N.
114.104
122.130
123.032
124.107
133.028 .032
158.047
Sanford, P.
142.065
158.309
Sanford, P. W.
141.313
142.016 .139 .226 .228 .233

Sanitt, N.
141.124
Sanjav, S.
055.012
Santaniello, M. A.
011.071
014.025
Santella, G.
065.107
Santhanan, K. S. V.
131.083
Santos, J.
079.102
Sapar, A.
022.085
Sapar, L.
121.087
Sapargalieva, L. M.
114.373
Sareyan, J. P.
122.111 .117 .154
Sargent, W. L. W.
158.038
Sari, J. W.
143.075
Sarkisyan, R. A.
031.224
158.111
Sarma, M. B. K.
112.010
Sarris, E. T.
099.085
106.072
143.011
Sarychev, V. A.
042.089
052.051 .052
Sasaki, S.
083.010
Sasao, T.
044.041
045.010
Saslaw, W. C.
151.003
158.065
Sasorov, P. V.
073.073
078.020
143.007
Sastri, H. J.
084.223
Sastri, N. S.
084.260
Sastry, K. S.
151.020
Sather, R.
103.134
Sato, K.
045.009 .022
Sato, T.
131.013
Sato, Y.
103.130 .140
Satterthwaite, G. E.
010.012
Satyvaldiev, V.
123.018
Sauerbrey, G.
062.042
Sauermost, R.
003.022
Saunders, R. S.
097.005
Savage, A.
141.018 .114
158.141
Savage, B. D.
131.004
Savage, H. F.
104.037
Savage, P. D.
162.048
Savchenko, A.Ya.
042.117
Savchenko, Yu. N.
084.292
Savedoff, M. P.
114.380
Savel'ev, V. N.
105.075
Savel'eva, M. V.
158.120
Savenko, I. A.
032.534
073.034 .156
076.042 .044 .046 .047
078.022 .025 .026
084.274
142.168 .180
143.069 .071 .074
Savich, N. A.
051.007
094.199
097.058
Sawa, T.
155.028
Sawyer, R. F.
065.010 .187
Saykally, R. J.
022.089
Sazhin, M. V.
066.059 .106
Scaddan, R. J.
131.036
Scalapino, D. J.
065.010
Scalo, J. M.
065.036
131.190
Scarf, F. L.
099.029
Scarfe, C. D.
122.178
Scargle, J. D.
022.088
132.025
158.099 .305
Scarsi, L.
032.518
Schaack, D.
099.033
Schaal, R. E.
077.053
Schaber, G. G.
094.414
Schaedlich, M.
046.050
Schaefer III, H. F.
022.119
Schaeffer, O. A.
094.432
Schaeffer, R. C.
082.039
Schalen, C.
131.051
Scharlemann, E. T.
063.013
Schatten, K. H.
010.024
074.042
080.077
Schatz, D. L.
009.005
Schawlow, A. L.
003.125
Schechter, P.
160.037
Scheepmaker, A.
141.353
142.080
Scheinin, N. B.
105.057 .156
Scherb, F.
103.102
Scherer, F. M.
031.259
055.014
Scherrer, P. H.
080.070
Scherrer, V.
076.034
Scherrer, V. E.
073.050
Schiff, H. I.
022.063
082.012
Schiffer, M.
003.009
Schiffers, N.
005.006
Schild, R.
121.002
158.089
Schilizzi, R. T.
033.007
141.007 .008 .073 .074
.090 .110
Schilling, G.
095.007
Schilling, R. F.
034.154
Schindler, K.
084.202
Schindler, S. M.
078.036
Schippers, J. W.
124.104
Schlegelmilch, R.
031.026
Schlereth, H.
003.112
Schlesinger, B. M.
065.004 .047
Schlichter, H. D.
034.168
Schlickeiser, R.
061.089
Schlosser, W.
103.100
106.050
122.015
124.103

Schlosser, W.
155.004
Schmadebeck, R.
032.520
Schmadebeck, R. L.
051.010
Schmadel, L. D.
002.009
031.020 .027 .401
Schmahl, E. J.
073.135
074.005
076.002
Schmahl, G.
031.001
Schmeidler, F.
004.043
Schmid-Burgk, J.
064.012
Schmidt, E. G.
114.095
122.089
131.034
Schmidt, G.
034.042
083.066 .076
Schmidt, G. D.
132.022 .031
142.224 .235
158.209
Schmidt, H. F.
046.067
Schmidt, K.-H.
013.004
158.026
Schmidt, M.
121.036
141.086 .630
142.005
155.015 .035
Schmidt, R. E.
153.010
Schmidt, R. R.
072.010
Schmidt, T.
159.002
Schmidt-Kaler, T.
008.016
113.001
155.004
Schmidtke, G.
032.511
Schmied, L.
075.004
Schmitt, H. H.
094.017 .447
Schmitt, R. A.
105.116
Schmoegner, M.
005.022
Schmutzer, E.
061.082
Schneck, L. J.
099.018
Schneeberger, T. J.
071.012
Schneider, D. P.
121.050
Schneider, E.
094.114 .116 .181
Schneider, H. E.
106.042
Schneider, M.
034.152
042.009
046.037 .067 .072
Schneider, S. H.
085.013
Schneider, W. C.
003.113
Schneiderman, A. M.
031.246
Schniewind, J.
066.030
Schnitzer, A.
034.044
Schnopper, H.
032.523
142.014 .075 .081 .121
.122 .194
Schnopper, H. W.
142.076 .083 .085 .088
Schnur, G.
103.100
Schnur, G. F. O.
034.007
Schober, H. J.
098.010 .028 .029
Schoembs, R.
034.012
Schoenberner, D.
064.009
114.362
Schoeneich, W.
116.010
Schoenfelder, V.
034.095
061.091
082.049
134.005
Schoenmaker, A. A.
103.100
Schoeps, D.
004.108
Scholer, M.
078.010 .030
084.208 .406 .412
Scholes, C. A.
032.058
Scholl, H.
002.009
042.008
098.019
Scholz, D.
075.017
077.064 .065
Scholz, G.
008.081
Schonfeld, E.
051.010
Schoolman, S.
032.007
Schorn, R. A.
071.021
Schott, P.
082.111
Schove, D. J.
102.008
Schramm, D. N.
061.037 .072 .076 .077
065.042 .162 .167
Schramm, D. N.
134.009
143.038
161.009
Schreier, E.
142.014 .026 .056 .122
.194 .230
Schreier, E. J.
142.001 .088 .121
Schrijver, J.
022.095
064.048
122.167
142.014 .045
Schroeder, L. W.
014.004
Schroeder, R.
122.065
Schroeter, E. H.
080.010
Schruefer, E.
141.362
Schubart, J.
098.041
Schubert, G.
003.116
084.297
094.178 .203
Schucking, E. L.
066.182
Schuenemann, K.
033.055
Schuiling, R. D.
081.047
Schukowski, M.
014.031
Schultz, G. V.
131.101
141.604
Schultz, L.
105.182
Schulz, M.
084.229
Schulz-Gulde, E.
022.082
Schumann, J. D.
114.011 .309
Schunk, R. W.
083.005
Schurmeier, H. M.
012.028
Schuster, H. E.
103.133 .136
113.044
133.030
Schutz, B. E.
031.263
053.021
Schutz, B. F.
021.021
066.038 .040
Schwarcz, H. P.
105.025 .183
Schwartz, D.
142.071
Schwartz, G.
098.058
103.005 .100 .104 .122
.123 .132 .134 .140
124.109

Schwartz, K.
084.297
094.203
Schwartz, P. R.
131.017 .046 .092 .518
Schwartz, R. A.
080.019
131.184
Schwartz, R. D.
113.022
132.054
Schwarz, J.
162.030
Schwarz, R.
033.057
Schwarz, U. J.
125.035
141.079
Schwarzschild, M.
065.029
Schwehm, G.
106.006
Schweitzer, E.
010.035 .036
120.004
Schweitzer, J. S.
061.004
Schweizer, F.
158.162
Schwob, J. L.
062.025
Sciama, D. W.
012.048
066.088
Sciambi, R.
078.029
Sciuto, V.
073.105 .106 .107
074.082
Sckopke, N.
084.207
Sclar, C. B.
094.472
Scott, E. R. D.
105.023 .184
Scott, J. S.
062.011
143.024 .065
158.226
Scott, P. F.
142.216
Scott, R. L.
141.142
Scott, S. D.
105.025
Scovil, C. E.
010.001
Scoville, N.
141.643
Scoville, N. Z.
131.016 .104
142.096
Scribner, B. F.
003.092
Scudder, J. D.
074.059
Scuflaire, R.
065.021 .035 .097 .100
.176
080.063
Seaquist, E. R.
142.007
Searle, L.
125.107
Sears, D. W.
105.034 .185
Sears, R. D.
084.006
Seaton, M. J.
061.083
Secco, L.
107.010
Sechrist Jr., C. F.
083.026
Seck, F.
004.052
Sedlmayr, E.
063.035
131.030 .040 .098 .107
Sedmak, G.
034.054
Sedwick, K. E.
071.014
Seeber, G.
008.046
046.068
Seeger, C. L.
015.007
Seeger, R. J.
005.008
Seehafer, N.
080.009
Segafimov, K.
051.008
Segal', V. M.
094.429
Segewitz, M. W.
094.102
Seggewiss, W.
064.004
114.011 .309
124.104
Sehnal, L.
052.021
082.022 .057
Seidel, B.
093.052
Seidelmann, P. K.
035.004
041.053
099.220
Seiler, M.
073.086
Seitter, W.
114.014
Seitter, W. C.
124.103
Sekanina, Z.
098.057
102.006 .017
103.100 .130 .131 .136
.140
Seki, T.
103.100 .111 .130 .131
.133 .140
Seldner, M.
160.012
Seling, T. V.
158.310
Selivanov, A. S.
097.023 .049
Selke, C.
046.048
Sellers, B.
073.138
Seltzer, S. M.
032.539
051.010
Selvelli, P. L.
114.307
Selyakov, K. I.
073.155
Semenenko, V. P.
105.074
Semeniuk, I.
124.104
Semenov, A. I.
082.065 .119
Semenov, V. P.
066.069
Semenova, E. V.
158.102
Semenyuk, T. P.
082.044
Sen, K. K.
063.033 .041
Sen Gupta, K.
083.033
Sen Gupta, S.
105.163
Senftle, F. E.
094.421
Sengbusch, K. v.
See Von Sengbusch, K.
Sengoer, A. M. C.
094.166
Sengupta, P. R.
073.008
Sentman, D. D.
099.084
Serafimov, K.
083.050
Sergeev, M. A.
046.073
Sergeev, V. A.
084.270
Sergeevich, V. N.
082.087
Serio, S.
142.247
Serkowski, K.
034.043
158.095 .313
Serlemitsos, P. J.
142.015 .025 .097 .120
.153 .154 .155 .190
.196 .202 .203 .217
.227 .230 .239
Serra, J. A.
085.022
Sersic, J. L.
158.143
Serverne, G.
151.012
Sevarlic, B.
008.010
010.037
014.012
Sevast'yanov, V. I.
052.013
082.075

Severinov, V. I.
084.403
Severny, A. B.
See Severnyj, A. B.
Severnyj, A. B.
071.035
077.025
080.054
094.152 .161
142.031 .041
Sevilla De Lerma, M. J.
046.014
Seward, F. D.
125.007 .008
134.004
142.059 .164
Sewell, M. H.
036.001 .009 .011
Sexl, R.
066.034
Sexl, R. U.
003.114
Shafer, Yu. G.
076.040 .048
Shaffer, D. B.
033.007
141.073 .074 .090
158.304
Shafranovskij, I. I.
004.051
Shaganyan, B. L.
123.009 .021
Shah, G. A.
132.062
Shaham, J.
142.018 .197
Shakhovskaya, N. I.
122.158
Shakhovskoj, N. M.
142.033
Shakht, N. A.
118.007
Shakirov, K. S.
094.007 .023
Shakirzade, A. A.
114.087
Shakura, N. I.
065.131
066.062
117.048
141.344
142.152 .169
Shakura, N. N.
077.056
Shalimov, V. P.
084.273
Shamakhov, B. F.
082.089
Shamolin, V. M.
073.034
076.018 .042
Shamolin, W. M.
076.021
Shandor, V. V.
094.172
Shane, W. W.
158.101 .151 .154 .165
.200
Shanin, G. I.
034.066 .089
114.305
Shanin, G. I.
122.122 .160
Shao, C. Y.
103.005 .100 .104 .123
.130 .131 .132 .134
.140
124.104 .109
Shapere, D.
004.013
Shapiro, A. E.
004.093
Shapiro, B. S.
083.047
Shapiro, Eh. Kh.
034.074
Shapiro, I.
121.039
Shapiro, I. I.
031.218
097.052
Shapiro, L. T.
009.023
Shapiro, M. M.
143.061
Shapiro, R. P.
131.088
Shapiro, S. L.
142.038
Shaporev, S.
103.143
Shaporev, S. D.
103.115
Shaposhnikov, D. P.
105.073
Shaposhnikova, T. M.
098.094
Shapovalova, A. I.
158.229
Shari, V. P.
093.019
Sharma, A.
097.003
Sharma, J. P.
126.027
Sharma, R. K.
042.039 .055
Sharov, A. S.
113.037
121.121
122.038
123.049 .050 .901
124.011 .012
Sharp, J. H.
009.001
Sharp, R. D.
084.253
Sharp, R. P.
097.065
Sharp, W. E.
084.001
Sharples, F. P.
031.054
Shatalov, V. A.
003.103
082.087
Shatsova, R. B.
112.015
118.011
Shaver, P. A.
131.112
Shaviv, G.
065.062 .078 .128 .141
121.023
142.148
Shavrin, P. I.
078.009 .023
Shavtvalishvili, I. A.
061.071
Shaw, D. M.
105.037 .133
Shaw, J. S.
121.111
122.074
Shaw, R. R.
084.301
Shawhan, S. D.
099.213
122.094
141.042
Shawl, S. J.
122.078
131.038
132.028
154.035
Shcharenskaya, Eh. G.
083.062
Shcheglov, P. V.
031.010
034.060
082.047
114.062
131.194
Shchekinov, Yu. A.
158.115
Shchekotov, B. V.
041.046
Shcherba, I. A.
084.294
Shcherbakov, A. G.
034.066 .089
114.305
122.032 .122 .160
Shcherbakova, Z. A.
073.101
Shcherbanovskij, A. L.
160.023 .042
Shcherbina, A. G.
034.074
Shcherbina-Samojlova,
I. S.
003.002
Shea, M. A.
003.122
Sheaffer, R.
015.017
Shealy, D. L.
031.043
Sheather, P. H.
083.004
Shectman, S. A.
158.011
Sheeley, N.
076.034
Sheeley Jr., N. R.
073.050
074.032
Sheffer
124.111
Shefov, N. N.
082.077

Sheftel', V. M.
083.062
Shelley, E. G.
084.253
Shelomentsev, V. V.
084.289 .290 .308
106.034
Shelton, D. H.
142.118 .165
Shelus, P. J.
031.254
041.017 .021
046.042
099.234
Shemilt, R. E.
122.097
Shenavrin, V. I.
132.014
142.110
Shepherd, G. G.
082.016
083.071
Sheptunov, G. S.
045.025
Sheptunova, A. N.
045.025
Sherbaum, L. M.
104.083
Sheridan, K. V.
141.153
Sherrod, C.
100.015
Sherwood, V. E.
012.001
122.022
Shestaka, I. S.
104.019
Shesterikov, V. F.
032.534
143.071
Shestopalov, D. I.
094.174
Shestopalov, I. P.
078.026
084.275
Shevchenko, V. S.
031.298
034.089
114.305
Shevnina, N. F.
083.060
Shi, Z.
073.076
Shibahashi, H.
080.032
Shibanov, Yu. A.
116.008
Shibasaki, H.
103.130 .131
Shidlovskaya, L. V.
106.043
Shields, G. A.
141.044
158.232
Shikin, I. S.
162.027
Shimamura, T.
106.079
Shimchuk, G. B.
118.011
Shimizu, M.
102.003
Shimizu, N.
105.186
Shimoyama, A.
094.428
Shine, R.
074.098
Shine, R. A.
063.009 .010 .026
064.028 .061
071.028
072.062
073.132 .133
080.057
Shiper, N. M.
103.100
Shipilova, V. I.
021.008 .009
Shirck, J.
105.187
Shirman, B. L.
084.031
Shirokov, M. F.
066.107
Shiryaev, A. V.
044.003
Shishov, V. I.
106.047 .057
Shishova, T. D.
106.047
Shivris, O. N.
141.116
Shkelev, E. I.
084.321
Shklovskij, I. S.
141.342
155.016
Shkodrov, V. G.
091.021 .022 .035
Shkol'nikov, V. A.
091.040
Shmilauehr, Ya.
083.042
Shmulevich, S. A.
033.071 .072
Shneer, V. S.
084.249
Shnol', Eh. Eh.
065.127
Shobbrook, R. R.
119.015
Shokin, Yu. A.
124.104
Sholin, G. V.
062.008
Sholomitskij, G. B.
034.116
Shor, V. A.
042.032
097.203
Shore, S. N.
121.040
Shornikov, O. E.
034.047
Short, N. M.
003.115
105.120
Shorthill, R. W.
099.204
Shortridge, K.
141.018
158.141
Shostak, G. S.
158.039
Shoub, E. C.
063.019
Shpilevsky, A. W.
See Szpilewski, A. W.
Shpital'naya, A. A.
072.040 .051
Shtaerman, V. L.
112.004
Shtejnshlejger, V. B.
132.047
141.067
Shtern, D. Ya.
094.199
Shu, F. H.
158.161
Shugarov, S. Yu.
121.013
122.123 .210 .212
123.014 .016 .017 .062
Shukla, P. N.
094.412
Shukstova, Z. N.
072.037
Shul'berg, A. M.
121.006
Shull, J. M.
151.008 .045
Shul'man, G. A.
065.140
Shul'man, L. M.
064.003 .063
102.015
Shulman, S.
142.011 .182 .183 .237
Shulov, O. S.
121.043
126.002
142.032
Shustarev, P. N.
034.063
Shustov, B. M.
021.006 .010
Shustova, Eh. I.
021.007
031.403
Shuter, W. L. H.
131.152
Shuvaeva, V. A.
097.027
Shvalagin, I. V.
055.013
Shvarev, V. V.
003.031
Shvartsman, V. F.
142.030
Shvetsov, A. A.
082.121
Sickels, B.
003.087
Siddesh, S. G.
125.035
Sidel'nikov, Yu. V.
073.115
076.026
Sidorenko, A. I.
094.199

Sidorenkov, N. S.
044.007
Sidorov, V. V.
104.072 .076
Sieber, W.
121.149
141.310
Siefert, P. T.
158.166
Siegel, H. P.
066.163
Siegman, B. C.
158.179
Sienkiewicz, R.
126.023
Siever, R.
081.011
Sigl, R.
003.117
046.067 .070
Sigler, R. D.
031.015
Signer, P.
105.182
Sikorska, I.
009.026
Silant'ev, N. A.
117.028 .046 .050
142.109
Silberberg, R.
143.061
Silich, S. A.
151.079
Silk, J.
131.144
142.061
Silk, J. K.
032.512
073.003 .009 .103 .104
074.073
076.003
Sil'kis, Eh. G.
034.083
Sill, G. T.
093.048
Silva, G.
032.047 .048
Silverberg, E. C.
031.253
046.042
115.019
Silverberg, R. F.
143.045
Silverman, S. M.
082.103
Sim, M. E.
103.126
Simkin, S. M.
141.138
158.054
161.002
Simmonds, P. G.
051.001
Simmons, G.
105.061
Simnett, G. M.
076.019
Simon, B.
094.176
Simon, G.
004.051
Simon, G.
005.007
073.026
Simon, G. W.
080.026 .033
Simon, J. L.
042.023 .062
Simon, K. P.
064.008
Simon, M.
158.193
Simon, N. R.
122.089
Simon, P.
003.122
010.008
Simon, R.
065.082
073.053
074.073
Simon, R. W.
073.009
Simon, T.
114.030 .070
122.072
158.035
Simoncini, A.
035.006
Simonenko, A. N.
003.118
105.009 .010
Simonneau, E.
114.374
Simonov, A. G.
083.028 .059
Simonov, R. A.
004.058
Simonson, G. E.
012.042
Simonson III, S. C.
155.033 .036
Simovljevic, J. L.
042.095 .096 .097
Simpson, A. B.
105.188
Simpson, J. A.
073.131
092.010
099.021 .022 .024
143.013 .022 .023 .034
.062 .063
Simpson, J. P.
132.025
Sincheskul, B. F.
096.012 .017
Sinclair, A. T.
042.033
047.042
Singer, S.
142.086
Singer, S. F.
003.119
101.001
Singh, M.
022.111 .113
Singh, P. D.
022.083
Singh, V. B.
097.003
Singleton, D. G.
083.056
Sinha, K. P.
162.085
Sinha, R. P.
155.026
Sinigaglia, G.
033.077
Sinitsyn, V. A.
052.038
Sinn, L. A.
158.181
Siomos, K.
022.064
Sion, E. M.
065.050
126.019
142.106
Sipler, D. P.
082.105
Siquig, R. A.
065.154
Siry, J. W.
046.055
Siscoe, G.
092.014
Siscoe, G. L.
084.239 .305 .314
092.015
Sistla, G.
141.105
Sitarski, G.
009.026
102.028
Sitnik, G. F.
073.092
Sitnikova, G. A.
094.130
Sivan, J. P.
113.073
Sivaram, C.
162.085
Sivertsen, S.
082.122
Sjogren, W. L.
097.006
Skadron, G.
143.003
Skarzhinskaya, A. V.
031.058 .059
Skarzynski, E.
162.036 .037
Skilling, J.
143.004 .030 .031 .057
Skimel', V. N.
042.111
Skinner, G. K.
142.022 .093 .131 .215
.225
Skjerve, L. J.
141.135
Sklar, E.
031.072
Sklyarov, Yu. A.
014.013
Skobel'tsyn, D. V.
012.044
Skomorovskij, V. I.
034.067
Skotele, S.
003.050
Skripin, G. V.
106.028

Skripin, G. V.
143.014 .026
Skul'skij, M. Yu.
117.019
121.048
Skumanich, A.
080.020
Skuratov, A. V.
033.069
Slade, M. A.
141.135
Slancikova, J.
104.064 .065
Slattery, W. L.
091.002
Slavatinskij, S. A.
143.068
Slavin, V. S.
072.028
Slaymaker, N. A.
033.047
Slee, O. B.
125.001
141.153
Sleptsov, I. E.
143.072
Sloanaker, R. M.
131.003
Slobodrian, R. J.
022.087
Slowey, J. W.
082.118
Slysh, V. I.
132.018
Smak, J.
114.358
117.064
121.106
Smarr, L. L.
117.062 .066
Smart, D.
003.122
Smartt, R. N.
031.044
Smerd, S. F.
077.060
Smeyers, P.
061.034
Smiley, C. H.
079.001
Smilowski, R.
009.026
Smirnov, A. I.
034.065
Smirnov, G. I.
131.150
Smirnov, G. T.
131.517
Smirnov, N. K.
034.090
Smirnov, R. V.
082.112
Smirnov, V. A.
055.025
104.018 .053 .054
Smirnov, V. N.
097.048
Smirnov, V. S.
084.268
Smirnova, T. M.
098.004 .066
Smirnova, T. V.
099.069
Smirnova, Ya. A.
104.054
Smith, A. G.
099.037 .052
141.142
Smith, A. M.
114.329
131.208
Smith, B.
097.026
Smith, B. A.
053.016
100.002 .007
Smith, B. F.
084.297
106.019
Smith, B. L.
065.001
Smith, B. W.
022.053
062.059
Smith, C. D.
014.001
Smith, D.
104.016
Smith, D. E.
046.043
Smith, D. F.
062.032
077.001 .046
Smith, D. W.
099.208
Smith, E. J.
074.034
099.099
Smith, E. W.
022.027
Smith, F. G.
141.314
Smith, G. R.
091.001
Smith, H. E.
131.507
141.075
158.031 .085
Smith, H. J.
003.032
158.319
Smith, H. M.
035.010
Smith, J. R.
010.012
075.007
094.416
114.031
132.027 .046
Smith, K. L.
033.031 .060
Smith, L. F.
132.017
Smith, M. A.
114.343
Smith, M. G.
133.027
154.035
158.197 .221
Smith, P. H.
084.303
Smith, P. J.
084.277
Smith, R. A.
099.078
Smith, R. C.
065.012 .023 .038
131.154
Smith, R. F.
103.131
Smith, S. P.
105.059 .060
Smith, V. V.
124.104
Smith, W. H.
022.105
099.080
Smith, Z. K.
106.019
Smith Jr., H.
125.009
Smith Jr., J. B.
072.012 .015
074.030 .031
Smithson, R. C.
034.032
Smolinski, J.
031.420
114.029
124.108
Smolkin, G. E.
034.084
Smoluchowski, R.
099.032
Smoot, G. F.
143.002 .050
Smorodinskij, Ya. A.
005.008
Smylie, D. E.
084.284
Smyth, J. R.
105.130
Smyth, M. J.
121.003
Smythe, C.
080.020
Snegerev, V. A.
063.044
Snell, R.
131.172
Snell, R. L.
113.010
Snezhko, L. I.
113.064
Snijders, M. A. J.
064.024
114.017 .018 .324
Snow, T. P.
114.380
Snow Jr., T. P.
114.374
131.090 .142
Snowden, M. S.
153.030 .035
154.013
Snyder, L. E.
022.056
033.017
131.052 .080 .174
Snyder, W.
142.182 .183

Snyder, W. A.
142.139
Soares, R. A.
033.061
Soberman, R. K.
053.020
Sobolev, S. V.
074.058
Sobolev, V. M.
071.041
Sobolev, V. V.
003.120
063.030
064.066
Soboleva, N. A.
134.011
Soboleva, N. S.
141.116
Sochilina, A. S.
054.017
055.026
Soederhjelm, S.
117.022
121.075
Soerensen, G.
022.079
Soerensen, O.
033.074
Sofia, S.
065.045
142.191
Soglasnova, V. A.
034.116
Sohval, A. R.
142.085
Soifer, B. T.
131.075 .079 .508 .514 .552
133.005 .013
141.606
Sokolov, D. D.
162.033
Sokolov, V. S.
072.028
Sokolova, V. S.
005.016
Sokol'skij, A. G.
052.039
Solaric, M.
046.026
052.021
Solc, M.
131.168
Solf, J.
034.005 .006
114.015
Solinger, A.
061.026
125.024 .034
Solomon, L. H.
122.156
Solomon, P. M.
131.016 .104
142.096
158.010
Solomonov, S. V.
082.024
Solovaya, N. A.
042.056
Solov'ev, A. A.
052.040
Solov'ev, A. A.
072.046
Solov'ev, V. Ya.
158.196
Somayajulu, B. L. K.
105.202
Somayajulu, V. V.
083.033
Somlo, P. I.
033.048
Sommer, R.
084.202
Somogyi, A.
156.004
Somogyi, A. J.
143.040
Son, A. T.
098.023
Sonett, C. P.
084.297
094.143
106.019
Sorgsepp, L.
121.086 .087
Sorochenko, R. L.
131.517
Sosna, F. M.
123.034
Sosnina, M. A.
032.023 .050
Sosonkin, M. G.
034.085
Soter, S.
097.201
Souchon, G.
082.051
Soulie, G.
041.013
South, R. H. H.
103.010
South, R. H. S.
103.123 .128
Southwick, R. G.
117.002
Southwood, D. J.
084.250
Southworth, R. B.
104.033 .034
Soutoul, A.
143.010
Souza, K. A.
015.025
Spada S., F.
051.023
Spangenberg, W. W.
034.050
Spangler, S. R.
122.093 .094 .196
141.136
Sparks, W. M.
065.032
124.007 .008 .009 .010 .104
126.009
142.006
Sparrow, J. G.
106.010
Specker, H. E.
005.005
Speer, R. J.
034.100
Speich, D. M.
072.012 .015
074.030 .031
Spencer
142.215
Spencer, J. H.
131.017 .518
141.014
Spencer, R. E.
141.115
142.103
Speraukas, Yu.
034.078
Speroni, N.
077.027
Spettel, B.
094.188
105.171
Spicer, D. S.
062.029
073.019 .035 .036
Spiegel, E. A.
063.038
065.156
Spight, L. D.
155.018
Spinrad, H.
141.003 .075 .313
158.007 .055 .085 .133
Spiridonov, Yu. G.
091.040
097.062
Spitzer Jr., L.
131.114
151.008 .009 .045
Spitzmesser, D. J.
141.135
Spjeldvik, W. N.
083.015
Splittgerber, E.
122.215
Spulgis, G. V.
034.075
Sramek, R.
066.034
158.113 .169
Sramek, R. A.
066.037
141.323
Sreekantan, B. V.
076.025
Sreenivasan, S. R.
062.070
065.083
Srinivasan, B.
105.083 .154 .189
Srirama Rao, M.
104.093
Srivastava, R. K.
121.001 .076 .077
Srnka, L. J.
074.019
094.137 .158
Stabell, R.
065.133
162.031
Stadsnes, J.
084.012
Staehle, V.
105.190

Stahlman, W. D.
004.020
Stair Jr., A. T.
084.009
Stakhanov, I. P.
106.017 .066
Stakheav, Y. I.
See Stakheev, Yu. I.
Stakheev, Yu. I.
105.150
Stalio, R.
114.307
Staller, R. F. A.
065.028
Standen, P. R.
103.126
Stannard, D.
099.013
141.115
Stanoev, I.
055.011
Stanyukovich, A. K.
105.076
Stanyukovich, K. P.
066.125 .128
Starbunov, Yu. N.
080.048
Starikov, K. Z.
066.044
Starikova, G. A.
118.014 .026
Stark, J. P.
142.064
Starkova, A. G.
083.045
Starobinskij, A. A.
162.033 .063
Starodubtsev, A. M.
034.091
Starrfield, S.
124.008
Starrfield, S. G.
124.007 .009 .010 .104
126.009
Startsev, P. A.
004.056
Starzhinskij, V. M.
042.112
Stassinopoulos, E. G.
078.032
Staubert, R.
134.006 .012
Staude, J.
072.004
Staveland, L.
072.075
Stavrev, K.
162.071
Stavroudis, O. N.
031.034
Stearns, L. P.
082.033
Stearns, S.
119.020
Stebbins, R. T.
031.214
080.018 .030
Stecher, R.
046.048
Stecher, T. P.
114.329
Stecher, T. P.
131.134
132.020 .053
133.012 .031
Stecker, F. W.
142.096
143.049
Steele-Perkins, E. M.
105.002
Stefanini, L.
118.040
Stefanovitch, D.
141.627
Steffen, P.
077.038
Stegena, L.
094.202
Stegmann, W.
105.007 .191
Steigman, G.
022.015
131.009
Steimer, P.
121.149
Stein, J. W.
121.113
124.104
Stein, R. F.
080.019
Stein, S.
094.177
Stein, W. A.
131.073 .162
141.100
Steinberg, J. L.
074.083
Steinert, K.-G.
013.004
014.030
Steinitz, R.
071.058
073.059 .078 .134
114.096
Steins, K.
013.016
Steklov, A. F.
151.085
Stella, G.
022.105
Stellingwerf, R. F.
065.153
122.071 .114 .185 .187
Stellmacher, G.
072.006 .055
Stenflo, J. O.
013.007
Stenflo, L.
062.005
Stenholm, B.
113.040
Stepanian, A. A.
See Stepanyan, A. A.
Stepanov, A. I.
010.033
Stepanov, A. V.
077.036
Stepanov, D. I.
104.012
Stepanyan, A. A.
142.150 .151 .177
Stepanyan, N. N.
073.101
085.009
Stephan, K.-H.
113.074
114.335
Stephens, S. A.
143.006
Stephens, T. C.
113.062
131.186
Stephenson, A.
094.423
Stephenson, C. B.
114.104
Stephenson, F. R.
125.028 .047
Stepien, K.
117.064
Steppel, S.
066.076
Sterken, C.
114.310
122.134
Sterling, K. J.
010.007
Stern, A.
010.008
Stern, D. P.
084.299
Stern, R.
032.515 .535
155.058
Sternberg, S.
162.093
Sternberk, B.
006.000
Sternlieb, A.
084.311
Stetsenko, S. G.
105.173
Stettler, A.
094.464 .467
Stettler, P.
071.037
Stevens, G.
073.012
Stevens, G. A.
076.059
Stevens, R. W.
032.041
Stevenson, D. J.
092.001
Stewart, J. A. C.
033.066
Stewart, J. M.
062.015
Stewart, P.
062.037
134.001
Stewart, R. T.
077.058
Stewartson, K.
062.053
Steyaert, H.
071.065 .067 .068 .069
.071
Sticker, B.
005.001 .003
011.008
012.004

Stickland, D. J.
114.353 .388
Stiening, R.
141.636
Stift, M. J.
031.291
065.011
122.182
Stiller, H.
107.016
Stilwell, D. E.
034.160
Stix, M.
156.001
Stobie, R. S.
124.104
Stocke, J. T.
014.015
062.011
Stockman Jr., H. S.
034.156
Stockton, A.
132.002
158.214
Stoddart, J.
103.100
Stodolkiewicz, J. S.
151.078
Stodolkiewicza, J.
014.009
Stoeckley, T. R.
065.054
114.049
Stoeffler, D.
105.040 .192
Stoeser, D. B.
094.437
Stohl, J.
162.055
Stoker, P. H.
143.016
Stokes, A. J.
010.001
Stol'berg, B. M.
082.087
Stolper, E.
105.041 .193
Stolper, E. M.
094.404
Stolpovskij, V. G.
083.029 .034
Stone, E. C.
078.011
143.021
Stone, G. E.
010.012
Stone, P. H.
093.026
Stone, R. C.
153.033
Stone, R. G.
082.001
Stone, R. P. S.
158.133
Stoner, R.
158.053
Stoner, R. E.
131.525
Storey, J. R.
084.262
Storey, P. J.
022.081
Storzer, D.
094.186
105.172 .196 .203
Stotskij, A. A.
031.299
141.116
Stoy, R. H.
007.000
Stoyanov, V. I.
066.141
Stoyko, A.
044.028 .031
Stoyko, N.
044.028 .031
Stozhkov, Yu. I.
143.037
Straizys, V.
113.055 .056
Straka, R. M.
072.063
073.145
Strand, K. A.
012.005
034.140
041.050 .070
111.004
123.027
Strassl, H.
008.074
Straton, P.
119.020
121.091
Straume, J.-I.
014.017
Straus, J. M.
065.042
082.010
Strecker, D. W.
093.042
114.031
132.025
Streeruwitz, E.
162.087
Streeter, G.
034.160
Strel'nitskij, V. S.
065.066
131.103 .131
Strickland, D.
073.037
Strid, G.
033.049
Stringer, D. V.
033.032
Strittmatter, P.
124.104
126.009
Strittmatter, P. A.
126.011 .013
141.107
158.082
Strobel, A.
124.104
Strobel, D. F.
082.027
091.005
Strohmeier, W.
008.007
123.059
Strom, K. M.
065.118
131.198
141.608
158.071 .072
Strom, R. G.
125.014
158.018
Strom, S. E.
065.118
131.198
141.608
151.035
158.071 .072
Strong, A. W.
155.072 .073
Strong, I. B.
061.014
142.021
Strugatskaya, A. A.
154.020
Strukov, I. A.
033.015
Stry, P. E.
133.036
Stubbe, P.
083.013
Stuchlik, J.
079.103
Stuewer, R. H.
003.121
Stull, M. A.
015.008
141.022
Stulov, V. P.
091.018
105.030
Sturrock, P. A.
073.062
074.060
080.029
085.005
141.325
Styazhkin, V. A.
084.274
Subba Rao, P. V.
042.039
Subbarao, P. V.
042.055
Subbotin, F. M.
034.090
Suchkov, A. A.
151.001
158.115
Suda, J.
072.011
Suda, K.
065.132
Suess, H. E.
105.134
Suess, S. T.
074.122
080.078
106.013 .053
Sugimoto, H.
143.028
Sukhanov, K. G.
052.034 .057
Sulejman, M. O.
124.104

Sulentic, J. W.
158.029
Sullivan III, W. T.
131.526
158.181
Sultanov, G. F.
042.016
098.094
Sumaruk, P. V.
083.060
106.046
Summers, A.
099.033
Summers, D.
074.102
Sumner, B.
103.100 .136
Sumzina, N. K.
097.203
Sung, C. C.
022.040
Suntzeff, N.
065.152
Sunyaev, R. A.
See Syunyaev, R. A.
Suri, A. N.
073.015
076.043
Surkov, Yu. A.
094.172
097.018
Surkova, L. P.
121.133
Surmelian, G. L.
022.126
Sutantyo, W.
117.063
142.140
Sutherland, P. G.
142.037
Sutter, B.
005.005
Sutton, G. H.
094.431
Sutton, J.
125.014
Suyarkova, O.
124.104
Suzuki, A.
065.191
Suzuki, I.
076.025
Suzuki, K.
103.100 .104 .130 .131
Suzuki, S.
103.131
Svalgaard, L.
076.056
080.070
106.011 .027
Svaton, E. M.
162.073
Svatos, J.
131.168
Svechnikov, M.
092.002
Svechnikov, M. A.
121.008
123.025
Sveshnikov, M. L.
041.064
Sveshnikova, E. S.
041.064
Svestka, Z.
003.122
073.021 .058
Svetchnikov, M.
See Svechnikov, M.
Svetlitchny, N. P.
083.077
Svoren, J.
103.102
Swamy, K. S. Krishna
See Krishna Swamy, K. S.
Swanenburg, B. N.
032.518
Swank, J. H.
142.190 .196
Swanson, P. N.
034.163
077.012
Swartz, M.
073.069
Swarup, G.
141.020
Sweeney, G. S. S.
062.037
134.001
Sweeney, M. A.
064.070
065.043
Sweet, P. A.
008.038
Sweetnam, D.
093.052
Swenson Jr., G. W.
033.007 .017
141.073 .074
Swensson, J. W.
022.035
Swerdlow, N. M.
004.078
Swider, A.
104.043
Swider, W.
083.012
Swift, C. D.
082.021
Swifte, R. H. D.
041.028
Swings, J.-P.
103.142
114.306
Swits, G.
094.445
Switzer, P.
084.307
Sy, A.
062.020
Sydow, J.
005.005
Sykes, M.
119.020
Sykora, J.
079.100
Sylwester, B.
076.054
Sylwester, J.
076.054
Synek, I.
021.013
Syrovatskij, S. I.
073.089
143.046
Sysoev, I. I.
014.013
Syunyaev, R. A.
065.131
066.062
117.004
142.094 .142
Szanto, P. G.
022.089
Szczepanski, M.
009.026
Sze, N. D.
093.003
Szebehely, V.
042.001
Szeidl, B.
122.049
Szekeres, P.
066.008
162.075
Szkody, P.
159.901
Szpilczynski, S.
004.051
Szpilewski, A. W.
004.117
015.037
Szpunar, W.
046.008
Szudy, J.
022.008
Szumiejko, E.
124.100
Szymanski, W.
009.026
072.074

Taam, R. E.
065.152
Tabarroni, G.
004.041
Taber, R. C.
066.025
Tacier, J.-D.
084.307
Tademaru, E.
141.331
142.113
Taff, L. G.
160.007
Tagi-zade, A. K.
004.064 .081 .099 .100
Tagirov, M. A.
042.110
Tagliaferri, G.
051.014
Taguchi, T.
103.100
Tai, M. H.
034.098
Tajbo, R.
032.047 .048 .049
Takada, M.
022.062
Takagi, S.
032.033 .034
044.021 .041

Takagi, S.
045.012 .024
082.076
Takahara, F.
151.094
160.048
Takahashi, H.
105.084 .136 .157
Takahashi, K.
042.124
Takahashi, Y.
065.191
Takakura, T.
073.014
076.014
Takashima, T.
091.007
Takeda, H.
094.473
105.179
Takeishi, M.
103.100
Takemori, M. T.
065.188
Takens, R. J.
142.223
Takhautdinova, S. S.
104.029
Takizawa, M.
083.038
Talbot Jr., R. J.
065.046
080.003 .037
151.084
Talon, R.
073.034
076.018 .021 .042
Talurau, B.
141.627
Tamao, T.
074.117
Tammann, G. A.
158.024 .049
Tamura, H.
132.033
Tanaka, H.
003.122
033.030
075.015
Tanaka, K.
073.043
Tanaka, R.
073.067
Tanaka, S.-I.
012.039
Tanaka, T.
105.001
Tanaka, Y.
076.025
142.166
Tananbaum, H. D.
142.001
Tandberg-Hanssen, E.
072.012 .015
073.127
074.030 .031
Tanenbaum, M. C.
054.016
Tang, F.
073.044
076.024
Tang, F.
080.028
Tanner, R. W.
094.182
Tanskanen, P.
084.012
Tantsova, N. N.
003.057
Tanzi, E. G.
124.104
Tapia, R.
032.047 .048
Tapia, S.
117.012
123.054
126.001
158.095 .313 .316 .318
Tapley, B. D.
031.263
052.043
053.021
Tarafdar, S. P.
080.039
114.002
131.047 .048
Tarasashvili, I. P.
042.101
Tarashchuk, V.
103.100
Tarashchuk, V. P.
103.102
Tarbell, T. D.
065.008
071.017
Tarcsai, G.
046.024
Tarenghi, M.
132.028
158.095 .176 .316 .318
160.002
Tarnstrom, G. L.
022.052
033.006
077.011 .016 .031
Tarraro, I.
161.005
Tarter, C. B.
065.020
141.109
Tartois, L.
007.000
010.028
Tarushkin, V. T.
052.025
Tarutin, V. A.
032.534
Tassoul, M.
065.143
117.060
Tatham, W. G.
004.127
Tatian, B.
031.041
Tatsumoto, M.
094.444
105.194
Tatum, J. B.
102.020 .023
Tauriainen, A.
083.066
Tavastsherna, K. N.
032.029
041.032 .036
Tavrizov, A. A.
032.534
Tayler, R. J.
065.085
Taylor, B. G.
032.518 .519
Taylor, F. W.
093.037 .057
Taylor, G. E.
010.012
035.012
047.006
097.053
Taylor, J. G.
003.123
Taylor, J. H.
141.332 .340
Taylor, J. W.
094.419
Taylor, K.
132.032
Taylor, L. A.
094.193
Taylor, M.
010.012
122.141
Taylor, R. S.
034.159
Taylor, S. R.
105.089
Taylor Jr., H. P.
105.020
Tchrakian, D. H.
066.009 .114
Tedesco, E. F.
103.130 .131 .136
Teegarden, B. J.
034.160
078.001
099.025
143.044 .058
Teerikorpi, P.
141.095
162.050
Teixeira, N. R.
082.040
Tejfel', V. G.
099.054
100.005
Teleki, D.
045.035
Teleki, G.
010.037
012.024
041.031
082.070 .072
Telesco, C. M.
091.031
155.064
158.086
Telfer, D. J.
094.430
Tel'nyuk-Adamchuk, V. V.
031.258
Temesvary, S. C.
065.159
Temkin, A.
022.019

Temnyj, V. V.
074.093
Tempesti, P.
121.127
122.149
123.035
124.104
Ten, A. P.
082.099
Tenn, J. S.
114.327
Tennakone, K.
066.178
Tentori, D.
031.076
Teplitskaya, R. B.
072.034
Teptin, G. M.
082.107
Ter Haar, D.
062.006 .067
141.306 .361
Tera, F.
094.194
Terent'eva, A. K.
104.021
Terez, Eh. I.
094.152 .161
103.102
Terez, G. A.
103.102
Ternovskaya, M. V.
084.403
Terrell, J.
141.053 .089
Terstiege, G.
141.012
Terzan, A.
122.037 .181
Terzi Belloni, A.
014.026
Terzian, Y.
132.063
142.216
Terzic, P.
009.013
Testard, O.
032.541
Testerman, L.
071.006 .033 .062
114.302
Tetruashvili, Eh. I.
074.110 .112
Teulet
041.013
Thacker, R.
105.176
Thackeray, A. D.
112.023
121.020
132.011
Thaddeus, P.
131.091
Thaenert, C.
123.065
The, P. S.
118.001
155.001
Theron, E.
031.073
Thielheim, K. O.
061.089
157.005
Thieman, J. R.
099.037
Thierry-Mieg, J.
066.079
Thirring, W.
042.125
Thomas, B. T.
084.214
Thomas, D. V.
008.043
Thomas, G. E.
092.008
106.024
Thomas, H.-C.
065.174
Thomas, J. B.
141.135
Thomas, J. C.
119.017
Thomas, L.
082.003
Thomas, R. J.
072.081
074.018 .065
076.004
Thomas, R. M.
142.141
Thomas, R. N.
080.055
Thomasson, P.
131.537
155.055
Thompson, D. C.
154.034
Thompson, D. J.
141.318 .335
155.003
Thompson, D. L.
062.070
Thompson, D. T.
099.232
Thompson, G. I.
131.115 .126 .211
Thompson, J. M. T.
061.094
Thompson, L. A.
125.106
158.080 .223
Thompson, R. I.
034.119
114.378
Thompson, R. J.
122.077
Thompson, R. T.
022.026
Thompson, T. W.
094.414
Thomsen, B.
131.538
162.024
Thonnard, N.
158.005 .220 .222 .227
Thoren, V. E.
004.044
Thorne, D. J.
141.302
Thorne, K. S.
065.007
Thorne, K. S.
066.051 .086
Thorne, R. M.
083.015 .022
Thornton, C. L.
066.041
Thorpe, A. N.
094.421
Thorstensen, J.
124.103
Thorstensen, J. R.
162.009
Thronson, H. A.
091.031
141.636 .639
Thuan, T. X.
155.032 .048
158.104 .129 .312 .314
Thum, C.
106.008
131.505 .506 .544
141.607 .634
Tifft, W. G.
160.002 .003 .036
Tifrea, E.
075.010
Tilford, S. G.
022.096 .116
Tilton, G. R.
105.113
Timashkova, G. M.
032.047 .048
Timofeev, A. N.
033.070
Timofeev, G. A.
078.027
Timofeeva, G. M.
079.101
141.116
Timothy, A. F.
073.009 .103
074.008 .073 .074 .078
.081 .105
076.003
Timothy, J. G.
034.023 .133
074.005
076.002 .029
Tinsley, B. A.
082.040
Tinsley, B. M.
061.038
125.062
158.033 .175
162.015
Title, A.
032.007
Title, A. M.
031.239
071.017
Titulaer, C.
053.011
Tkaczyk, W.
143.087
Todd, J. J.
114.118
Todoran, I.
122.153
Tofani, G.
077.003 .028 .037
141.147

Toft, A. R.
031.036
Tohmatsu, T.
083.038
Tokarev, O. D.
082.098 .099
Tokhtas'ev, S. S.
113.025 .026 .027
125.104
131.094
158.117
Tokis, J. N.
042.021
Tokovinin, A. A.
121.014
122.060
Toksoez, M. N.
094.126 .130 .177
Tokunaga, A. T.
099.043
Tolmacheva, T. A.
042.073
Tolomeo, L.
105.107 .112
Tolson, R. H.
094.204
Tomanov, V. P.
102.022
Tomasetti, G.
033.077
Tomasi, P.
141.147
Tombrello, T. A.
094.476
Tomić, A.
010.029
091.041
Tomilin, A. N.
003.124
Tomita, K.
103.008 .100 .130 .131
Tomkin, J.
064.019
114.303
122.206
124.104
Tomlinson, H. S.
032.545
Tondello, G.
022.038 .077 .080
Toomre, J.
065.156
Toor, A.
134.004
142.164
Top, Z.
105.048
Topaktas, L.
155.034
Toporov, Yu. P.
094.172
Topper, D. R.
014.038
Toraskar, J. R.
034.025
Torelli, M.
075.020
Torii, C.
033.030
Torochkov, V. Yu.
031.069
Toropova, T. P.
082.099
Toroshelidze, T. I.
082.101
Torr, D. G.
084.001 .032
Torr, M. R.
084.032
Torras, N.
103.100 .131
Torres, C.
098.059
103.102 .104 .119 .129
Torres-Peimbert, S.
131.531 .561
Torroja, J. M.
008.065
Tosa, M.
131.535 .545
Totochava, A. G.
122.025
Totomanov, I.
046.030
Tousey, R.
022.096
073.038 .050 .121
074.004
076.034 .050 .051
Tovmassian, H. M.
034.117
122.156
Towe, K. M.
162.013
Towlson, W. A.
032.545
Townes, C. H.
003.125
131.062 .522
Townsend, R. E.
153.031
Trafton, L.
099.214 .227 .228 .231
101.011
103.100
Trafton, L. M.
099.206
Trainor, H. J.
034.160
Trainor, J.
078.001
Trainor, J. B.
010.008
103.136
Trainor, J. H.
099.025 .064
143.044 .058
Traub, W. A.
093.030
Trauger, J.
103.102
Trautman, A.
066.082
Traving, G.
008.047
063.035
Travis, L. D.
093.047
Treder, H.-J.
004.045
066.147 .148 .149 .151
.154 .155 .156
Treder, H.-J.
162.004
Trefall, H.
084.012
Treffers, R. R.
022.094
Trefzger, C.
034.006
155.021
Tregaskis, T. B.
010.008
103.136
Trejo, J. G.
073.065
Tremaine, S. D.
160.035
Tremko, J.
103.102
121.125
Tret'yak, A. I.
042.113
Treumann, R.
083.070
Treutner, H.
032.011
Trifu, M.
054.006
Trigubov, Yu. V.
143.071
Trimble, V.
011.031
061.036
Trimble, W. L.
066.076
Trinks, H.
082.109
Tripathi, V. N.
066.007
Tripodi, P.
014.024
Tritton, D. J.
093.010
Troendle, H.
033.059
Trofimov, A. S.
031.267
Troispoux, G.
121.149
Troitskaya, V. A.
106.067
Trombka, J. I.
032.520
051.010
Tropp, Eh. A.
065.115
Troshenkov, M. K.
031.057 .065 .066
Troshichev, O. A.
084.265
Trubnikov, B. N.
084.294
True, G. A.
032.546
Truemper, J.
008.107
134.006 .012
Truemper, M.
066.002
Truran, J. W.
105.015
124.007 .009 .010 .104

Truran, J. W.
125.037
126.009
Trussoni, E.
062.066
Truttse, Yu. L.
084.018
Tsai, H.-M.
094.457
Tsai, W-C.
063.047
Tsap, T. T.
073.099
080.054
Tsaplin, V. S.
078.023
083.053
Tsarevskaya, R. L.
160.023 .042
Tsarevskij, G. S.
142.030
Tscharnuter, W.
065.016 .017
Tsedilina, E. E.
084.224 .226 .255
Tsel'man, F. Kh.
042.116
Tsesevich, V. P.
120.002
121.047 .134
122.119 .209
123.010 .060 .061
Tsikoudi, V.
122.163
Tsioumis, A.
155.022
Tsirs, V. E.
084.269
Tsitsin, F. A.
151.087
Tsubaki, T.
036.020
074.039 .049
Tsuchiya, K.
124.103
Tsukiji, Y.
033.030
Tsurutani, B. T.
083.072
Tsvetkov, Ts.
064.032
Tsvetkov, V.
105.022
Tsybko, Ya. G.
077.023
Tsygan, A. I.
065.125
141.305
Tsyganenko, N. A.
084.266
Tsyganov, A. N.
077.055
Tsytovich, V. N.
062.005 .006
073.029
141.306
Tubbs, D. L.
061.037 .076
Tubbs, E. F.
022.032
Tubbs, L. D.
031.039
Tubertini, C.
121.056 .057 .058 .153
Tucker, K. D.
131.002
Tulinov, G. F.
084.024
Tulinov, V. F.
082.088
084.024
Tull, R.
103.100
131.172
Tully, R. B.
158.081 .152 .156 .172
Tulyakov, V. V.
082.088
Tuman, V. S.
066.035
Tung, C. T.
063.041
Tuohy, I. R.
142.068
Tuominen, I.
065.071
Turchinovich, I. E.
097.042
Turcotte, D. L.
094.009
Turekian, K. K.
093.055
Turikov, V. G.
093.016
Turk, C. R. G.
010.007
Turkeeva, B. A.
083.027
Turland, B. D.
141.005
Turlo, Z.
031.413 .420
Turner, B. E.
131.008 .091 .109
158.009 .219
Turner, D. G.
113.049
Turner, E. L.
162.011
Turner, G.
094.458
Turner, J. A.
033.047 .061
Turner, K. C.
155.054
Turner, M. J.
142.049
Turner, M. J. L.
142.024 .048 .101 .102
.210 .211
Turnshek, D. A.
126.019
Turon, P.
131.127 .532
Turon Lacarrieu, C.
111.011
Tursky, H.
005.004 .005
Tutukov, A. V.
065.027
117.043
Tuzzolino, A. J.
099.024
Tverskaya, L. V.
084.410
Tverskoj, B. A.
074.027
078.014 .018
Tylenda, R.
124.100
Tyrnov, O. F.
083.077
Tyson, J. A.
066.111
161.003
Tyuflin, Yu. S.
094.155
Tyurin, V. S.
034.114
Tyuterev, G. S.
082.063

Uchida, J.
065.132
Uchida, Y.
073.032
Udal'tsov, V. A.
141.347
Uehera, M.
162.090
Uexkuell, M. v.
See Von Uexkuell, M.
Uhlmann, D. R.
094.117
Uhrhammer, R.
081.006
Uiterwaal, G. M.
124.025 .103
Ukhova, O. K.
098.025
Ulehla, I.
015.032
Ulich, B. L.
131.080 .174
Ulmer, M. P.
142.001
Ulmschneider, P.
064.006
080.014
Ulrich, M.-H.
141.123
158.234
Ulrich, R. K.
065.036
080.059
Ulrych, T. J.
125.053
141.076
Ulwick, J. C.
084.009
Ul'yanov, E. G.
083.054
Underhill, A. B.
114.119 .301 .311 .324
Underwood, J. H.
031.024
073.048 .049 .051 .146
Ungar, P.
009.007
Unno, W.
065.138

Unno, W.
080.032
Unruh, D. M.
105.194
Unsoeld, A.
162.020
Unti, T.
106.019
Upgren, A. R.
111.005 .007
155.050 .059
Upol'nikov, A. A.
034.091
Upson, W.
103.100
Uralov, A. M.
062.014
Uras, S.
045.021
Urasin, L.
092.002
Urasin, L. A.
113.028
124.104 .106
125.102
131.095 .096
158.108
Urasina, I. A.
081.013 .014
124.106
Urata, T.
103.100 .104 .130 .131
Urban, P.
004.032
Urbantke, H. K.
003.114
Ureche, V.
122.050
Urey, H. C.
012.012
094.022
Urnov, A. M.
076.041 .045
Uscinski, B. J.
131.020
Usher, P. D.
141.902
Usol'tseva, L. A.
032.038
Usov, V. S.
031.069
Usowicz, J.
033.011
Ustinova, G. K.
105.151
Utech, K.
105.050
Uus, U.
065.135 .170
Uvarov, V. B.
061.062
Uyama, K.
032.543

Vaaraniemi, P.
100.003
Vagner, Eh. A.
042.058
Vagnetti, F.
162.042
Vaiana, G.
073.009
076.035 .037
Vaiana, G. S.
032.507
073.001 .003 .103 .104
074.008 .073 .074 .075
.078 .081 .104 .105
076.003
Vaidya, P. C.
066.093
Vajner, Yu. G.
034.083
Vajsberg, O. L.
074.093
084.309
097.048
Vakhabov, S. A.
004.064 .100 .101
Vakulov, P. V.
084.410
Valbousquet, A.
041.040 .065
Valeev, S. G.
044.004
094.135
Valenza, M.
105.110 .168
Valetti, A.
003.126
Valiron, P.
022.084
Vallee, J. P.
141.080
Valnicek, B.
034.002 .003
Valtier, J. C.
122.111 .117
Valtier, J.-C.
122.154
Valtonen, M. J.
117.054 .055
158.066
Van Albada, G. D.
158.101
Van Albada, T. S.
151.022
158.160
Van Allen, J. A.
099.084
106.014
Van Altena, W. F.
031.219 .256 .296
032.532
036.013
111.002
Van Amerongen, H. J.
034.149
Van Apeldoorn, B. C. J.
009.011
Van Beek, H. F.
073.012
076.015 .022 .059
Van Bemmelen, R. W.
094.026
Van Breda, I. G.
031.412
131.135
Van De Kamp, P.
011.005
041.022
Van De Kamp, P.
042.013
117.016 .051 .070
Van Den Bergh, S.
015.020
065.048
122.079
125.054
154.023
155.043
158.013 .139
Van Den Heuvel, E. P. J.
085.023
142.127 .223
Van Der Borght, R.
065.037 .111
Van Der Hucht, K. A.
114.090 .117 .353 .361
.401
Van Der Waerden, B. L.
003.127
Van Der Wal, P. B.
118.024
Van Dessel, E. L.
071.053
117.069
Van Diest, H.
120.001
Van Duinen, R. J.
032.521
142.207
Van Flandern, T. C.
041.053
102.007
117.011
142.164
Van Gelder, G. P.
031.289
Van Genderen, A. M.
113.039 .067
122.150 .208
124.025 .103
125.100
Van Helden, A.
004.092
Van Hollebeke, M. A.
073.057
Van Hoosier, M. E.
074.004
Van Horn, H. M.
065.051
114.380
124.104
126.005
Van Hoven, G.
073.040
Van Kuilenburg, J.
034.147
Van Leeuwen, F.
121.070
Van Loo, F.
082.052
103.100
Van Maris, J.
103.010 .100
Van Paradijs, J.
114.079
Van Paradijs, J. A.
142.223
Van Riper, K. A.
065.195

Van Swearingen, F.
034.164
Van Swearingen, J.
094.102
Van Tend, W.
080.069
Van Wageningen, G.
035.003
Van Woerden, H.
158.159 .167
Vandakurov, Yu. V.
065.116
122.013 .155
Vanden Bout, P.
131.172
Vanden Bout, P. A.
131.056 .059 .077
Vandenberg, N. R.
141.049 .050 .143
Vandenberg, N. .
141.358
Vandervoort, P. O.
151.023 .038 .043 .048 .083
Vanicek, P.
046.010
Vanlandingham, F. G.
117.073
Van't Veer, F.
117.056
Vanysek, V.
011.059
131.168
Vapillon, L.
099.083
100.008
Vardanyan, R. A.
131.097
153.040
Vardavas, I. M.
022.013
Vardya, M. S.
080.039
Varenik, L. I.
054.017
Varfolomeev, D. I.
031.056 .061
Varin, M. P.
032.049
Varina, V. A.
032.049
Varshalovich, D. A.
131.151 .187 .188 .189
143.067
Varshavskij, V. I.
065.027 .069 .072
Varshni, Y. P.
064.058
141.065 .081 .127
Vartanyan, Yu. L.
126.026
Vashchenko, E. P.
104.081
Vashkevich, A. S.
098.031
Vashkov'yak, M. A.
042.057
052.036
Vasilenko, N. A.
082.046
Vasil'ev, K. N.
083.062
Vasil'ev, M. B.
094.199
Vasil'ev, N. V.
003.080
105.031
Vasil'ev, O. B.
080.071
114.023
Vasil'ev, S. N.
052.028
Vasil'ev, S. S.
080.049
Vasil'eva, G. Ya.
072.040 .051
Vass, G.
041.009
Vasudevan, R.
063.032
Vasyliunas, V. M.
074.099
084.218
099.076
Vaucher, C.
121.124
Vauclair, S.
003.018
065.147
Vaughn, L. M.
099.237
103.135
Vaulina, Eh. N.
010.014
Vazquez, M.
080.035
Vdovykin, G. P.
003.129
Vedeneev, Yu. B.
033.016
077.014
Vedren, Zh.
076.042
Vedrenne, G.
073.034
076.018 .021
Vehrenberg, H.
015.018
Veis, G.
046.060
Vejsmann, U. K.
034.077
Vekhov, A. A.
082.110
Velinov, P.
083.049
085.025
Velusamy, T.
125.010
Venis, T. E.
032.545
Venkatavaradan, V. S.
106.001
143.047
Venkatesan, D.
012.045
084.015 .312
106.901
143.075
Ventura, A.
022.076
Verbeiren, R.
041.071
046.065
Veres, F.
082.060
Vereshchagin, I. F.
052.023
Vereshkov, G. M.
011.047
Verheest, F.
062.038
Verigin, M. I.
106.016
Vermeesch, T.
003.130
032.043
Vermeesch, T. M.
009.010
Vermilion, J. R.
099.204
Vernazza, J. E.
074.005
076.002
Vernov, S. N.
078.018
106.065
143.037 .072
Veron, M. P.
141.023
158.301
Veron, P.
141.023
Veronese, G.
014.021
Verschure, P.-P. H.
082.054
Verschuur, G. L.
009.001
131.061 .539
Vesecky, J. F.
099.016
Veselova, G. V.
143.070
Veselovskij, I. N.
004.027
Vesely, C.
103.134
Vesic, D.
044.043
Vessot, R. F. C.
066.014 .021
Veverka, J.
004.052
097.011
100.206 .208
Victor, G. A.
022.001
Vidal, N. V.
141.025
142.040 .099 .238
160.045
Viehmann, W.
032.513
Vieira, R.
032.016
034.052
Viennet, J.
035.013
Viewing, D.
015.011

Vigier, J.-P.
158.014
162.021 .039 .040
Vigneau, J.
113.042
Vigroux, L.
061.031
Viik, T.
063.036
Vij, K. K.
084.015
Vila, S. C.
065.050
133.020
142.106
Vilcsek, E.
105.098
Vilkki, E. U.
111.002
Vil'koviskij, Eh. Ya.
158.068 .122
Villante, U.
106.052
Villard Jr., O. G.
084.262
Vin, A.
031.284
034.145
Vince, I.
031.278
034.130
122.218 .219
Vincent, S.
081.024
Viner, M.
121.039
Viner, M. R.
121.028
131.153 .546 .547
141.113 .144
Vingardt, P. M.
082.127
Vinh, N. X.
042.079
Vinogradov, A. L.
064.031
Vinogradov, A. P.
097.018
Vinogradov, A. V.
046.045
Vinogradov, V. A.
094.199
Vinokurov, S. K.
158.102
Viola Jr., V. E.
061.061
Viotti, R.
031.293
114.010
Virdefors, B.
071.049
Virgopia, N.
062.064
Virtamo, J.
062.072
Vishveshwara, C. V.
066.182
Visser, H. C. D.
158.150
Vitins, M.
021.012
Vitinskij, Yu. I.
072.041 .043 .047 .067
077.034
Vitkovskij, V. V.
079.101
Viton, M.
113.073
Vitrichenko, Eh. A.
034.089
Vitz, R. C.
114.332
Vivers, D.
032.049
Vives, T. J.
153.018
Vladimirov, Yu. S.
162.045
Vladimirova, I. A.
045.025
Vladimirskij, B. M.
142.150 .151 .177
Vlasov, M. N.
083.044
Vlasov, N. A.
065.151
Vleeming, G.
155.001
Vnuk, E.
055.009
Voelcker, K.
113.014
152.002
Voelk, H. J.
143.015
Voges, W.
032.518
Vogt, N.
122.040
131.560
155.009
Vogt, R. E.
078.011
143.021
Vogt, S.
131.172
Vogt, S. S.
096.005
122.005
Voigt, A.
003.131
Voigt, H. H.
008.039
010.010
Voigt, W.
077.064 .065
Voise, W.
004.033
Vojskovskij, M. I.
082.055
Vojtyuk, E. V.
098.069
Volkov, M. S.
042.002 .103
Vollendorf, F.
162.052
Vollmer, O.
078.004
Volodarskij, A. I.
004.060
Volodina, M. P.
119.009
Volonte, S.
022.098
Von Arx, B.
033.006
Von Borzeszowski, H.-H.
066.146
Von Braun, W.
003.132
Von Ehr, H.-L.
032.012
Von Gersdorff, N.
034.144
Von Hoerner, S.
015.009
033.002 .079
162.024
Von Kap-herr, A.
124.104
158.025
Von Puttkamer, J.
053.004
Von Rosenvinge, T. T.
143.044 .058
Von Sengbusch, K.
065.091
Von Uexkuell, M.
073.025
Von Zahn, U.
082.109
083.039
Vonbun, F. O.
046.055
Vondrak, R. R.
083.048
084.278
Vorob'ev, L. V.
034.116
Vorob'eva, E. Ya.
004.072
005.030
Voronenko, V. I.
098.048
Voronoj, E. M.
046.002
Voronov, N. A.
066.127
Voronov, V. N.
091.029
Vorontsov-Vel'yaminov,
B. A.
131.536
158.042 .119 .120 .144
.177
Voroshilov, Yu. V.
011.045
Vorpahl, J.
076.014
Vorpahl, J. A.
073.048 .049 .146
Voss, J.
011.069
Vozdvizhenskij, B. S.
093.012
Vranic-Kren, T.
009.017
Vrba, F. J.
132.022 .031
Vreux, J. M.
103.142
114.108
124.103

Vsekhsvyatskij, S. K.
003.003
074.021 .055
102.009
107.009
124.104
Vuillemin, A.
113.073
Vujnovic, V.
012.024
022.078
061.054
073.093
Vukicevic-Karabin, M.
061.055
073.094
Vyal'shin, G. F.
073.118
Vyater, Ya.
055.008
Vylev, G.
055.011
Vypirajlo, I. I.
123.009
Vyshlov, A. S.
094.199
Vytnov, V. A.
044.036

Waak, J. A.
131.518
Wachmann, A. A.
121.098
Wacker, W. K.
099.062
Waddington, C. J.
022.099
Waddington, E. D.
105.195
Wade, N.
031.030
141.308
Waelder, M.
121.131
Waelke, K.
011.072
Waenke, H.
094.024 .188
Waggett, P. C.
141.033
Wagner, C. A.
081.018 .019 .025
Wagner, C.-U.
075.017
Wagner, G. A.
105.196
Wagner, R.
155.007
Wagner, W. J.
074.033
Wagoner, R. V.
065.055
Wahlen, M.
094.471
Wahlgren, U.
022.119
Wakamatsu, K.
158.187
Wako, Y.
045.009 .022
Walborn, N. R.
114.339 .367 .368
131.015 .549
Walbridge, E.
094.160
Waldmeier, M.
074.120
075.001 .002 .014 .015 .016
Walker, A. R.
115.009
Walker, D.
094.404
Walker, D. M. C.
081.016 .017
082.081
Walker, E.
104.014
Walker, G.
094.430
Walker, G. A. H.
131.027
158.214
Walker, G. H.
064.016
Walker, G. S.
034.142
Walker, J. C. G.
003.057
082.013
084.032
093.054
Walker, R.
131.058
Walker, R. C.
131.158
Walker, R. L.
041.035
Walker, R. M.
094.115 .192
105.097
Walker, R. W.
105.187
Walker, W. S. G.
010.040
031.229 .265
Walker Jr., A. B.
074.128
Walker Jr., A. B. C.
074.038
076.009
131.063
Wall, J. V.
141.102
Wallace, L.
101.002
Wallerstein, G.
122.183
Walls, D. M.
046.011
Walraven, G. D.
142.118 .165
Walsh, D.
141.164
Walter, K.
121.015 .024
Walther, H.
022.064
Waltman, W. B.
131.046
Walton, J. R.
094.405
Wampler, E. J.
032.008 .014
Wamsteker, W.
114.384
Wang, H. T.
073.114
Wang, M. Y.
066.177
Wannier, P. G.
131.170 .171
Waranius, F. B.
002.012
Warburton, R. D. H.
066.052
Ward, B. D.
106.070
Ward, C. G.
092.007
Ward, D. B.
131.540
132.007
Ward, M. J.
142.104
Ward, W. R.
094.002
107.005
Warden, E. S.
114.332
Ware, N. G.
105.089
Warner, B.
008.025
032.032
116.005
121.020
122.085
124.024
Warner, D. J.
005.027
Warner, J.
094.410
Warner, J. W.
141.635
158.308
Warner, R. D.
094.474
Warren, P. R.
142.053
Warwick, J. W.
099.097
Wasilewski, P. J.
094.121 .417 .433
Wasserburg, G. J.
094.194 .407
105.059
Wasserman, I.
126.021
Wassmund, P.
034.039
Wasson, J. T.
003.106
094.111 .413
105.021 .023 .184 .197
Watanabe, T.
074.045
Waterfield, R. L.
103.010 .100 .123 .128 .130 .131

Watkins, J. R.
072.015
094.151
Watkins, S. N.
033.062
Watson, B.
074.108
Watson, D. E.
105.058
Watson, M.
142.052 .120 .153 .209
Watson, M. G.
142.202
Watson, W. D.
022.055
131.087 .176 .181
142.184
Wattenberg, D.
002.006
Wayman, P. A.
031.291
155.049
Wdowczyk, J.
143.035 .087
155.073
Weaver, H.
155.067 .071
Weaver, K. F.
093.039
Weaver, R.
064.015
131.070
Weaver, T. A.
125.022 .068
Webb, C. J.
114.038
Webb, D.
074.073
076.035 .036
Webber, J. C.
033.017
Webber, W. R.
078.001
099.025
143.044 .058
Webbink, R. F.
117.065
142.051
Weber, H.
105.098
Weber, H. W.
105.198
Weber, J.
066.023 .076 .099
Webrova, L.
044.050
Webster, B. L.
133.024
Webster, M. W. G.
046.009
Webster, W.
121.039
Webster Jr., W. J.
103.102
Wedekind, J.
105.040
Wedel, B.
098.018
Weedman, D. W.
158.004 .215
Wefer, F. L.
009.009
Wegener, A.
094.166
Wegner, W.
114.394
Wehinger, P. A.
103.100
124.103
125.103
142.221
158.212
Wehlau, A.
008.063
Wehmeyer, R.
132.008
Wehrse, R.
131.099
Weickmann, H. K.
082.033
Weidelt, R.
064.010
Weidemann, V.
008.051
011.012 .037
065.175
Weidenschilling, S. J.
097.041
Weidner, J. R.
094.185 .463
Weiffenbach, G.
012.034
Weigand, P. W.
105.093
Weigelt, G. P.
031.249
Weigert, A.
117.008 .045
Weihaupt, J. G.
091.048
Weiler, E. J.
121.061
Weiler, K. W.
066.108
Weill, G.
082.064
Weinberg, J. L.
106.010
Weinberger, R.
131.504
132.043
Weinstein, A.
114.332 .345
Weinstock, J.
062.048
083.067
Weir, A. D.
065.080
Weis, E. W.
111.007
Weisberg, J.
142.192
Weiser, H.
114.332
Weisheit, J. C.
158.210
Weiss, K.
074.128
Weiss, N.
011.034
Weisskopf, M. C.
073.137
142.037 .159
Weissman, P.
102.005
Weistrop, D.
154.028
155.017
Weitbrecht, R. H.
034.046
Welch, G. A.
158.217
160.901
Weliachew, L.
012.022
158.148 .159 .244
Welin, G.
123.042
Weller, W.
124.104
Wellington, K. J.
066.108
Wellmann, P.
008.073
Wells, D. C.
115.022
Wells, D. E.
046.010
Wells, E.
094.122
Wells, N.
082.043
Wempe, J.
008.081
031.228
101.005
Wendker, H. J.
125.050
132.017
142.020
158.024 .049
Wene, C.-O.
061.049
Wenig, O.
005.006
Wentzel, D. G.
074.040
155.026
Wenzel, K.-P.
143.021
Wenzel, W.
064.011
122.214
Wereb, D.
124.103
Wernecke, S. J.
099.087
141.022
Werner, A.
014.016
118.022 .023
121.118
Werner, M.
123.071
Werner, M. W.
131.025 .523
132.003
141.047 .637 .638
Werntz, C.
073.108

Wesemael, F.
114.380
Wesselius, P. R.
142.207
Wesson, P. S.
061.039 .040
131.050
162.017 .019
West, D. K.
131.134
West, R. M.
034.143
098.056 .057
103.006 .126 .133
113.044
West, R. W.
103.133
West Jr., H. I.
078.002
084.234
Westbrook, C. K.
065.020
Westbrook, W. E.
131.025 .523
141.047 .630
Westbrooke, W. J.
079.002
Westman, R. S.
004.048
011.009
Weston, E. B.
098.026
Wetherill, G. W.
105.053
Weymann, R.
034.115
Whang, Y. C.
092.003
Wheaton, W. A.
061.022
142.004 .082 .087
Wheeler, J. A.
066.074
Wheeler, J. C.
125.006
Whelan, J.
117.012
142.013 .162
Whelan, J. A. J.
142.161
Whippey, M. R.
010.012
Whipple, F. L.
098.040
102.001
107.013
Whitaker, R. W.
065.139
Whitcomb, S.
141.636
White, A. E.
141.054
142.035
White, B. R.
097.037
White, G. J.
141.628
White, J. D.
131.520
White, M.
143.086
White, M. K.
094.102
White, N. E.
142.226 .233
White, O. R.
071.028
072.062
073.113 .132 .133
White, R. A.
158.027
White, R. D.
097.052
White, R. E.
113.058
White, R. S.
034.095
076.019
White III, R. E.
031.222
White Jr., R. E.
031.222
Whitehurst, R. N.
158.127
Whitelock, P. A.
124.104
Whiteoak, J. B.
131.509 .516
141.036 .154
158.192
Whiteside, D. T.
004.036
Whitford-Stark, J. L.
094.415
Whitham, P. S.
033.019
Whitney, A.
121.039
Whitney, A. R.
031.218
Whitney, C.
003.133
Whitney, C. A.
122.140
Whitteker, J. H.
083.071
Whitten, R. C.
097.063
Whittet, D. C. B.
131.135
Whitworth, A.
011.016
Wiant, J. R.
115.019
Wibberenz, G.
106.075
Wickes, W. C.
117.015
118.003
Wickramasinghe, D. T.
126.013
142.013 .062 .161
Wickramasinghe, N. C.
066.901
102.016
107.015
114.002
124.020
131.035 .083
Widing, K. G.
073.006 .045 .046 .079
Wiedemann, E.
006.000
031.070
036.002
Wiehr, E.
072.006 .055
Wielebinski, R.
141.152
158.022
Wielen, R.
151.018 .056
Wiemer, W.
131.101
Wiesmeier, A.
034.010
Wiita, P. J.
117.062
Wijnbergen, J.
131.532
Wikierski, B.
031.414
Wilcox, J. M.
076.056
080.070
106.027
Wild, P.
047.011
125.106
Wildeman, K. J.
032.521
Wildey, R. L.
097.013
Wiley III, W.
084.255
Wilhelmsson, H.
062.005
Wilkening, L. L.
104.066
105.016 .160
Wilkins, G. A.
044.017
Wilkins Jr., J. E.
031.247
Wilkinson, A. W.
031.055
Wilkinson, P. N.
158.045
Will, C. M.
066.020
Willey, R. E.
052.070
Williamon, R. M.
121.115 .155
Williams, D. A.
131.146
Williams, D. J.
084.217 .303 .407
Williams, G. E.
015.024
Williams, G. P.
099.050
Williams, I. P.
003.134
107.003 .011
Williams, J. D.
009.024
Williams, J. G.
098.020
105.011
141.135

Williams, J. T.
104.034
Williams, P. M.
124.104
Williams, P. R.
034.142
Williams, R. E.
141.107
Williams, R. P.
054.019
Williams, T. B.
158.030 .073
Williams, V.
031.046
Williamson, F.
034.167
Willis, A. J.
114.118 .354
Willmore, A. P.
142.022 .093 .131 .215 .225 .236
Willner, S.
131.076
Willner, S. P.
141.620 .630
Wills, B. J.
141.040 .045 .131
158.190
Wills, D.
141.045
158.319
Wills, R. D.
032.518 .519
Willson, R. F.
132.005
141.149
Wilson, A. M.
142.225
Wilson, A. S.
143.018
Wilson, C. A.
004.040
Wilson, C. P.
158.058
Wilson, I.
003.047
Wilson, I. D. L.
131.116
Wilson, J. R.
065.183
066.165
Wilson, P. R.
072.007 .009
Wilson, R.
051.006
114.027 .118 .318 .354
131.126 .210 .211
158.036
159.010
Wilson, R. E.
117.042
162.049
Wilson, R. G.
031.014
Wilson, R. M.
072.012 .015
074.030 .031
Wilson, R. W.
131.170 .171
Wilson, S. J.
063.033 .041
Wilson, T. L.
131.169 .537
Wilson, W. E.
141.152
Wilson, W. J.
131.520 .554
Wimmer, H. K.
003.029
Winckler, J. R.
084.304
Wing, R. F.
114.325
Winiarski, M.
005.035
103.114
Wink, J. E.
124.104
Winkler, C. A.
005.020
Winkler, G. M. R.
044.026
Winkler, P. F.
125.020
Winkler Jr., P. F.
141.054
142.035
Winnberg, A.
131.133 .136
Winnewisser, G.
022.122
131.021 .029 .082 .164
Winnewisser, M.
131.021
Winningham, J. D.
084.245
Winston, R.
141.636
Winter, A. J. B.
158.012
Winzer, S. R.
094.195
Wirtanen, C. A.
111.001
Wisniewski, W. Z.
098.046
Wissinger, A. B.
032.549 .550
Withbroe, G. L.
072.016
073.060 .070
074.005 .103
076.002
080.031
Witkin, I.
097.019
Witt, A. N.
131.186
Witta, P. J.
117.066
Witteborn, F. C.
022.002
082.021
093.042
132.025
Wittels, J.
121.039
Wittels, J. J.
031.218
Wittmann, A.
080.064
082.008
Witzel, A.
033.007
141.010
Wlerick, G.
141.029
Wlotzka, F.
105.199
Woehl, H.
034.011
071.034
080.010 .064
082.008
Woerner, A.
035.019
Wofsy, S. C.
093.003
Wolf, B.
114.310 .317
Wolf, G. W.
114.003
Wolf, H.
081.043
Wolf, J. A.
162.093
Wolf, R. A.
084.901
Wolfe, A. M.
141.030 .043
Wolfe, E. W.
094.435
Wolfe, J. H.
099.019 .041
106.019
Wolfe, W. L.
031.047
Wolfendale, A. W.
143.035 .087
155.073
Wolff, C. L.
074.011
Wolff, R. J.
116.001
Wolff, R. S.
071.011
073.137
134.010
141.346
142.136 .137 .159
Wolff, S. C.
065.120
116.001 .007
Wolfson, R.
142.212
Wollenhaupt, W. R.
094.158 .176
Wollman, E. R.
131.522
141.617
Wolstencroft, R. D.
122.072
158.142
Woltjer, L.
141.083
154.008
Wong, A. S.
033.045
Wong, L.
097.006
Wong, W.
033.079

Woo, R.
093.035
106.020
Wood, D. B.
121.035
Wood, H. J.
114.347
116.015
Wood, J. A.
094.128 .434 .437 .448
Wood, J. E.
004.073
Wood, P. R.
122.195
Woodgate, B. E.
034.156
125.018
Woodman, D.
031.306
Woodman, J.
103.100
Woodman, J. H.
093.049
Woods, N. E.
121.019
Woods, R. C.
022.089
Woodsworth, A.
142.007 .175
Woodward, E. J.
121.123
Woolf, N.
122.194
Woolf, N. J.
141.616
Woolley, R.
008.025
Wootten, H. A.
131.059
Worden, S. P.
031.270
080.026 .033
115.018
122.088 .197
158.071
Worden Jr., P. W.
066.028
Wordsworth Hudson, J.
003.075
Worek, T. F.
117.068
Worswick, S. P.
031.285
Woszczyk, A.
103.100
124.104
Woyk (Chvojkova), E.
062.035
Wray, J. D.
114.005 .313
117.035
Wright, A. E.
065.012 .104
131.154
141.151 .155
142.222
151.062
155.039
158.179
Wright, A. N.
005.020
Wright, E. L.
032.516
122.142
131.503
141.605 .615
142.039 .135
158.087
Wright, F. W.
122.001
Wright, K. O.
008.109
013.013
031.280
034.148
121.126
Wright, M. C. H.
158.166
Wright, S. L.
021.026 .027
Wroblewski, H.
098.059
Wu, C. S.
074.901
Wu, C.-C.
117.021
142.207
Wu, S. T.
072.015
073.124
074.030 .031
Wu, Y.
094.121
Wuethrich, T.
121.149
Wussing, H.
003.135
005.019
Wyant, J. C.
011.046
031.045
Wyatt, C. L.
034.112
Wyatt, S. P.
133.006 .034
Wyckoff, S.
073.078
103.100
124.103
125.103
142.221
158.212
Wyllie, T. H. A.
031.226
Wynn-Williams, C. G.
141.630 .638 .639

Yabushita, S.
065.022
Yagovkin, A. R.
083.077
Yahil, A.
162.030
Yajima, N.
034.153
Yakhno, G. S.
014.033
Yakimov, S. P.
079.101
151.039 .065 .090
Yakimova, N. N.
122.030 .039 .138
Yakovkin, N. A.
071.039
Yakovlev, B. A.
143.071
Yakovlev, B. M.
032.534
Yakovlev, D. G.
117.017
Yakovlev, N. I.
042.089
052.052
Yakovlev, O. I.
051.007
093.023
094.198
097.062
Yakushev, A. K.
142.030
Yallop, B. D.
044.034
Yamaguchi, Y.
065.196
Yamanaka, T.
105.179
Yamasaki, A.
117.041
Yamashita, K.
142.166
Yamashita, Y.
113.030
Yang, K. S.
033.017
Yaniv, A.
094.405
Yanke, V. G.
082.090
Yao, B.
124.013
Yaroshevich, G. O.
042.100
Yaroslavskij, L. P.
097.046 .049
Yarygina, T.
114.082
Yasuhara, F.
084.245 .248
Yates, G. K.
073.138
Yatsenko, A. Yu.
042.025
Yatskiv, Ya. S.
041.003
045.007
Yavnel', A. A.
105.065 .068
Yeates, C. M.
092.015
Yen, J. L.
033.007
141.073 .074
Yentis, D.
142.237
Yentis, D. J.
142.011
Yeomans, D. K.
103.112
Yildizdogdu, S.
122.046

Yilmaz, H.
066.119
Yngvesson, S.
122.003
Yoder, C. F.
099.221
Yokosawa, M.
064.049
Yokoyama, K.
044.041
045.016
Yokoyama, Y.
094.196
105.157
York, D. G.
114.380
Yoshida
103.100
Yoshimi, N.
033.030
Yoshimura, H.
072.027
Yoshizawa, M.
158.187
Yoss, K. M.
115.021
Youmans, A. B.
141.143
Young, A.
093.008
117.033
121.089
Young, A. T.
071.021
093.040
Young, E.
158.193
Young, L.
093.008
Young, R.
093.028
Young, T. S. T.
084.259
Younger, F.
114.047
Yousef, S.
076.008
Yuan, S.
033.053
Yudin, I. A.
105.075
Yudina, I. V.
072.076
Yufarkin, V. Ya.
143.074
Yukhimchuk, A. K.
073.116
Yukhimuk, A. K.
084.256
085.019
106.071
Yumi, S.
041.034
045.016 .026
Yung, Y. L.
099.212
Yungelson, L.
117.052
Yungel'son, L. R.
117.043
Yurevich, V. A.
103.102
Yurkina, M. I.
046.066
Yurov, S. S.
010.014
Yurova, L. A.
031.404 .406 .407
Yurovskaya, L.
072.023
Yurovskaya, L. I.
077.026
079.104
Yushkov, B. Yu.
142.168 .180
Yuzefovich, I. A.
073.156

Zabolotskij, F. D.
082.126
Zagar, F.
005.004
012.029
044.022 .037
112.022
117.047
Zagatin, V. I.
141.067
Zaginayilo, Y. I.
082.079
Zagvodzkin, B. V.
083.077
Zaharia, R.
031.009
Zahn, J.-P.
065.079 .156
117.013
Zahn, U. Von
See Von Zahn, U.
Zajacz, G.
121.149
Zajdler, L.
004.116
008.111
009.026
010.020
Zajtsev, A. L.
094.198
Zajtsev, V. N.
031.410
Zajtsev, V. V.
074.085
077.019 .023 .036
Zajtsev, Yu. I.
003.136
Zajtseva, E. I.
131.166
Zajtseva, G. V.
121.121
122.018
Zajtseva, S. A.
084.408
Zakharchenko, V. F.
143.032
Zakirova, A. R.
093.020
Zambelli, P.
004.090
Zander, R.
082.050
Zanetti, V.
014.020
Zappala, R. A.
073.105 .106 .107
074.082
Zappala, V.
031.225
Zarnecki, J. C.
142.064
Zasova, L. V.
100.010
Zastenker, G. N.
074.043 .093
084.274 .309
Zatsepin, G. T.
143.005
Zavarzin, Yu. M.
106.036 .048 .063 .069
113.050 .068
Zavatti, F.
121.066
123.028
Zazulyak, P. M.
094.013
Zbinden, P. A.
083.063
Zdanavicius, K.
113.052 .053 .055
Zealey, W. J.
158.142
Zeau, Y.
100.008
Zech, G.
002.009
Zehntner, C.
077.016
Zeilik II, M.
131.503
141.605 .615
Zejnalov, R. A.
052.053
Zel'dovich, M. A.
073.156
078.022 .026
Zel'dovich, Ya. B.
003.137
061.079
141.344
162.054 .067
Zellner, B.
098.036 .046 .054
Zemanek, E. N.
072.039
Zemelman, M.
103.125
Zenker, A.
005.012
Zenkert, A.
014.010
Zentsova, A. S.
125.042
Zerefos, C. S.
085.001 .007 .018
Zertsalov, A. A.
084.309
Zerull, R.
063.016
106.006 .007
Zessewitsch, W.
120.005

Zevakina, R. A.
085.009
Zhagar, Yu. Kh.
052.047
Zhao, R.
073.076
Zharkov, V. N.
003.138
099.060
Zhegulev, V. S.
097.022 .044
Zheleznyakov, V. V.
074.071
077.044 .045
141.337
Zhilkin, A. M.
031.069
Zhilyaev, B. E.
122.201
Zhitetskij, A. A.
079.103
096.004 .011
Zhitnik, I. A.
073.115
076.026 .054
Zholud', T. P.
105.079
Zhongolovich, I.
031.266
Zhongolovich, I. D.
006.000
Zhuchenko, Yu. M.
084.024
Zhukov, G. V.
098.071
103.100
121.120 .135
132.014
141.068 .072
Zhukov, L. V.
041.004
Zhukov, V. I.
080.061
Zhukov, V. V.
104.023
Zhuravlev, L. T.
094.172
Zhuravlev, S. G.
042.083
Zhuravleva, L. V.
098.070
Zieba, A.
021.014
158.180
Zieba, S.
077.049
141.099
158.180
Zielinski, J. B.
046.012
Ziener, R.
031.228
Zietek, J.
009.026
Zigel', F.
099.053
Zigel', F. Yu.
014.013
Ziglin, S. L.
117.030
Zilitinkevich, S. S.
093.016
Zimmerman, B. A.
022.065
Zimmermann, H.
065.124
Zinchenko, I. I.
091.029
141.097
Zinchenko, Yu. F.
098.069
Zinichev, V. A.
077.023
Zinn, R.
122.165
Zinner, E.
094.192
Zinov'ev, V. A.
099.095
Zintchenko, G. N.
083.077
Ziolkowski, K.
021.020
Zipoy, D. M.
114.344
Zirin, H.
073.002
114.036
Zisk, S. H.
094.414
Zizka, F. R.
122.191
Ziznovsky, J.
133.003
Zlatoustova, V. A.
052.051
Zlobec, P.
077.024 .030
Zlobin, V. N.
141.347
Zlotnik, E. Ya.
074.071
077.044 .045
Zolondkovsky, N. D.
083.077
Zombeck, M.
074.073
Zombeck, M. V.
074.105
Zook, H. A.
106.025
Zosimova, A. G.
084.272
Zotkin, I. T.
105.067
Zuckerman, B.
131.006 .008 .013 .026
.055 .086 .548
141.618
158.009 .219
Zuev, V. E.
003.139
Zuiderwijk, E. J.
142.223
Zulevic, D. J.
118.033 .036
Zumwalt, R. W.
094.428
Zverev, M. S.
041.036
Zverev, Yu. K.
141.116
Zvereva, A. M.
094.152 .161
Zverko, J.
114.025
Zwaan, C.
012.026
072.066
Zweibel, E. G.
143.064
155.024
Zwicky, F.
004.050
158.038
Zyabrikova, L. V.
114.023
Zykov, I. A.
153.022
Zyrnicki, W.
022.005
Zytkow, A. N.
065.007

Subject Index

A Stars
Finding Lists
113.047
A Stars
Line Blocking
114.044
A Stars
Luminosities
115.016
A Stars
Peculiar
065.120
112.008
113.031 .036 .064
114.001 .075 .086
.087 .337 .343
.355 .363 .372
116.004 .007 .014
.015
117.010
A Stars
Photometry
113.031
A Stars
Proper Motions
155.057
A Stars
Space Distribution
155.057
A Stars
Spectra
114.044
A Stars
Spectrophotometry
114.071
Absolute Magnitudes
115.000
Absorption
Intergalactic Matter
161.005
Absorption
Interstellar Clouds
131.209
Absorption
Interstellar Matter
131.014 .043 .084
Absorption
Interstellar Matter
131.093 .094 .095
.096 .100 .127
.129 .135 .166
.193 .212 .548
Absorption
Stellar Associations
152.001
Absorption Coefficient
022.021
082.005
Absorption Lines
Eclipsing Variables
121.119
Absorption Lines
H I Clouds
131.039
Absorption Lines
Interstellar Matter
131.053 .155 .157
.206
Absorption Lines
Photosphere
071.038
Absorption Lines
Quasars
141.019 .030
Absorption Lines
Quasi-Stellar Objects
141.013 .017 .106
.107 .108 .109
.148
Absorption Lines
Solar Spectrum
071.045 .065 .069
.071
Accretion
Binaries
117.029
Accretion
Black Holes
066.039 .062 .073
.096 .137
Accretion
Clusters of Galaxies
160.048
Accretion
Neutron Stars
142.114 .128
Accretion
White Dwarfs
124.009
126.016 .023
142.106
Accretion
X-Ray Sources
062.015
142.003 .010 .038
.050 .052 .054
.058 .094 .113
.114 .116 .128
.132 .148 .174
Achondrites
105.027 .037
AI Velorum Stars
122.161
Airglow
082.000
Albedo
Jupiter
099.059
Albedo
Jupiter Satellites
099.204 .207 .223
Albedo
Mercury
092.005
Albedo
Minor Planets
098.074
099.202
Albedo
Moon
094.174 .446
Albedo
Planets
099.202

Albedo
Saturn
100.005
Alfven Waves
Chromosphere
073.126
Alfven Waves
Cosmic Rays
143.030 .031
Alfven Waves
Interplanetary Matter
074.022
106.049
Alfven Waves
Solar Wind
074.022 .053 .096
.113 .115 .130
Alfven Waves
Sunspots
072.009 .079
Algol
121.003
Algol Systems
117.014
121.051 .076 .090
Almanacs
047.000
Ammonia
Jupiter Atmosphere
099.031 .058
Andromeda Nebula
131.535 .541
141.014
158.055 .127 .139
.145 .153 .154
.155
Andromeda Nebula
Novae
124.011 .012
Andromeda Nebula
Variables
114.322 .323
Antimatter
143.050
162.049 .073
Artificial Satellites
054.000
Artificial Satellites
Moon
053.013
Artificial Satellites
Motion
052.010 .032 .033
.046 .047 .052
.070
082.022 .023
Artificial Satellites
Observations
055.000
Artificial Satellites
Orbits
052.011 .040 .043
.053 .054
054.013
Associations
010.000
Associations Stellar
152.000
Asteroids
See Minor Planets
Astrodynamics
052.000
Astrographs
032.031
Astrolabe Observations
Jupiter Satellites
041.016
Astrolabe Observations
Saturn
041.001
Astrolabe Observations
Star Catalogues
041.014
Astrometry
041.000
Astronomical Accessories
034.000
Astronomical Constants
043.000
Astronomical Geodesy
046.000
Astronomical Instruments
032.000
Astronomical Optics
031.000
Atlases
041.000
Atmospheres
Binaries
063.015
Atmospheres
Cool Stars
064.068
Atmospheres
Early-Type Stars
064.042 .045 .050
114.324
Atmospheres
F Stars
114.359
Atmospheres
Late-Type Stars
064.038
Atmospheres
M Giants
114.365
Atmospheres
M Stars
064.051
Atmospheres
Pulsating Stars
122.099
Atmospheres
White Dwarfs
126.008 .014 .015
.016
Atmospheres
Wolf-Rayet Stars
064.004 .065
Atmospheric Optics
031.016
034.030
Aurorae
084.000
Automation
031.400
B Stars
Finding Lists
113.047
B Stars
Luminosities
115.002
B Stars
Peculiar
114.043 .366
B Stars
Proper Motions
112.012
B Stars
Subdwarfs
126.003
B Stars
UV Spectra
114.307
B Stars
UV Spectrophotometry
114.111
B Supergiants
UV Spectra
114.119
Background Radiation
066.000
Balmer Lines
Solar Flares
073.092
Barium Stars
114.085
115.014
Be Stars
Photometry
113.018 .024 .065
.067
Be Stars
Polarization
114.338 .341
Be Stars
Pulsations
113.018
Be Stars
Rotation
116.012
Be Stars
Spectra
114.019 .067 .068
.115 .383
Be Stars
Spectrophotometry
114.373
Beta Cephei Stars
119.015
122.159
Beta Cephei Stars
Photometry
122.074 .097
Beta Cephei Stars
Pulsations
115.002
122.097
Beta Cephei Stars
Radial Velocities
122.097
Beta Cephei Stars
UV Spectra
122.096
Beta CMa Variables
115.002
122.134 .154
Beta Lyrae Stars
114.375
121.048 .119 .155

Bibliographical Publ
002.000
Binaries
117.000
Binaries
Accretion
117.029
Binaries
Atmospheres
063.015
Binaries
Black Holes
066.073 .096
114.375
Binaries
Close Binaries
042.053
113.029
114.047 .358 .375
117.001 .013 .014
.015 .017 .018
.019 .021 .025
.028 .038 .040
.041 .042 .045
.049 .052 .053
.056 .057 .060
.061 .062 .063
.073
119.013 .020
121.019 .046 .051
.073 .074 .078
.087 .106 .107
.122 .123
142.127
Binaries
Contact Binaries
117.012
121.086 .087
Binaries
Dwarf Novae
124.008 .017 .019
Binaries
Early-Type Stars
115.013
Binaries
Evolution
118.019
151.070
Binaries
Galactic Clusters
117.156
153.029
Binaries
Mass Loss
114.375
117.020
119.002
Binaries
Masses
115.013
117.048
Binaries
Models
117.044
Binaries
Neutron Stars
125.006
Binaries
Novae
114.358
117.064
Binaries
Novae
124.008
Binaries
Orbits
121.021
Binaries
Parallaxes
117.071
Binaries
Photometry
113.054
Binaries
Planetary Orbits
117.030
Binaries
Polarization
117.002
Binaries
Proper Motions
117.071
Binaries
Pulsars
031.203
066.095 .104
117.018
141.301 .305 .311
.312 .313 .345
Binaries
Radio Sources
121.028
141.028
Binaries
Rotation
117.057
Binaries
Spectra
117.072
Binaries
Star Clusters
151.064
Binaries
Stellar Systems
117.026 .027
151.014
Binaries
Stellar Winds
119.002 .003
Binaries
Supernovae
125.006
Binaries
Unseen Companions
117.051
Binaries
White Dwarfs
114.316
117.017
Binaries
Wolf-Rayet Stars
114.369
117.020
Binaries
X-Ray Sources
062.015
065.040 .134
114.315
117.001 .002 .004
.018 .028 .046
.048 .050 .062
.063
Binaries
X-Ray Sources
121.023 .044
125.006
142.002 .003 .006
.011 .013 .026
.027 .036 .039
.044 .046 .049
.051 .052 .054
.055 .062 .068
.107 .112 .113
.114 .116 .125
.126 .127 .130
.131 .132 .133
.135 .140 .148
.149 .152 .154
.159 .161 .169
.174 .176 .202
.203
Biography
005.000
BL Lacertae
158.301 .314
BL Lacertae Objects
141.006 .015 .027
.030 .066 .073
.075 .100
142.030
158.302 .315 .316
Black Dwarfs
065.028
Black Holes
066.074 .075 .138
.162 .166 .167
.168 .170 .171
.173 .177 .184
Black Holes
Accretion
066.039 .062 .073
.096 .137
Black Holes
Binaries
066.073 .096
114.375
Black Holes
Galaxies
151.003
158.065
162.016
Black Holes
Gamma-Ray Sources
142.018
Black Holes
Globular Clusters
066.019
154.028
Black Holes
Gravitation
066.073 .097
Black Holes
Motion
141.321
Black Holes
Neutron Stars
142.130
Black Holes
Solar Models
080.037
Black Holes
Spectroscopic Binaries
119.008

Black Holes
  Stellar Evolution
    080.037
Black Holes
  White Dwarfs
    066.134
Black Holes
  X-Ray Sources
    066.073 .096 .137
    142.038 .055 .061
      .116 .174
Blue Objects
    114.352
    160.020
Blue Stars
  Photometry
    113.046
Blue Stars
  Radial Velocities
    034.094
Blue Stars
  Variations
    122.101
Bolometric Corrections
  Late-Type Stars
    115.005
Books
    003.000
Bright Stars
  Photometry
    113.055
Bright Stars
  Spectrophotometry
    114.116

C-M Diagrams
  Galactic Clusters
    153.029
C-M Diagrams
  Globular Clusters
    154.002 .007 .012
      .015 .017
C-M Diagrams
  Star Clusters
    154.004
Calendars
    047.000
Calibrations
  Photometric
    031.228
Carbon Dioxide
  Mars Atmosphere
    093.002
    097.022 .040 .043
      .050 .051
Carbon Dioxide
  Venus Atmosphere
    093.002
    097.043
Carbon Monoxide
  H II Regions
    131.562
Carbon Monoxide
  Interstellar Matter
    131.001 .045 .056
      .117 .118 .148
      .160
    141.623
    158.009 .010
Carbon Monoxide
  Planetary Nebulae
    133.021
Carbon Stars
    065.128 .150
    113.070
    114.106 .113
    122.213
    123.015 .019
    141.625
Carbon Stars
  Chemical Composition
    114.077
    131.008
Carbon Stars
  Colors
    113.030
Carbon Stars
  Evolution
    065.143
Carbon Stars
  Galactic Clusters
    114.007
Carbon Stars
  Globular Clusters
    154.026
Carbon Stars
  Infrared Radiation
    122.087
Carbon Stars
  Infrared Spectra
    114.002 .004 .334
    141.623
Carbon Stars
  Luminosities
    115.004 .008 .017
Carbon Stars
  Nucleosynthesis
    065.143
Carbon Stars
  Spectra
    022.094
    114.077 .325
Carina Nebula
  Forbidden Lines
    131.549
Cataclysmic Variables
    117.049 .074
    121.025
Catalogues
  Clusters of Galaxies
    160.042
Catalogues
  Comparisons
    041.003
Catalogues
  Fundamental
    041.020
Catalogues
  Galaxies
    158.204
Catalogues
  OB Stars
    113.032
Catalogues
  Planetary Nebulae
    133.028
Catalogues
  Proper Motions
    041.000
Catalogues
  Pulsars
    141.340
Catalogues
  Solar Activity
    075.005 .006
Catalogues
  Star Positions
    041.000
Celestial Mechanics
    042.000
Cepheids
    064.064
    065.089
    122.030 .079 .115
      .118 .164 .165
      .206
    123.051
    153.021
Cepheids
  Colors
    122.077 .081 .116
      .133
Cepheids
  Diameters
    122.139
Cepheids
  Luminosities
    122.027 .029 .077
Cepheids
  Magellanic Clouds
    122.002 .138
    159.009
Cepheids
  Masses
    122.036
Cepheids
  Models
    063.004
    064.036
    065.009
    122.071 .082 .113
      .114 .116
Cepheids
  Periods
    122.027 .071 .081
      .100 .137 .161
Cepheids
  Photometry
    122.002 .027 .028
      .133 .136 .161
Cepheids
  Spectra
    122.032 .051 .082
      .160
Chemical Composition
  Carbon Stars
    114.077
    131.008
Chemical Composition
  Cosmic Rays
    143.013 .051 .066
Chemical Composition
  Earth
    081.000
Chemical Composition
  Globular Clusters
    154.014
Chemical Composition
  H I Regions
    131.114

Chemical Composition
H II Regions
131.114
Chemical Composition
Jupiter Atmosphere
099.012
Chemical Composition
Late-Type Stars
064.019 .038
114.072 .303 .320
.325
Chemical Composition
Meteorites
105.065 .068 .072
.083 .084
Chemical Composition
Moon
094.123 .438 .439
.440 .442
Chemical Composition
Planetary Atmospheres
091.005
Chemical Composition
Saturn Atmosphere
099.012
Chemical Composition
Stellar Evolution
065.030
Chemical Composition
White Dwarfs
126.028
Chemical Composition
Wolf-Rayet Stars
114.009
Chemical Evolution
Galaxy
061.038
Chemical Evolution
Solar Neighborhood
114.006
Chondrites
105.001 .008 .014
.015 .026 .028
.035 .037 .038
.045 .047 .048
.053 .058 .061
.074 .089 .092
.096 .097 .104
Chromosphere
Alfven Waves
073.126
Chromosphere
Emission Lines
074.005
Chromosphere
Fine Structure
073.071 .099
Chromosphere
Fluctuations
073.083
Chromosphere
Heating
061.034
071.027
Chromosphere
Helium Lines
073.109
Chromosphere
Line Formation
063.026
Chromosphere
Line Intensities
073.122
Chromosphere
Line Profiles
073.004 .078
Chromosphere
Magnetic Fields
071.036
Chromosphere
Models
073.033 .098 .123
Chromosphere
Network
073.095
Chromosphere
Oscillations
061.034
071.027
Chromosphere
Shock Waves
071.027
Chromosphere
Spectra
022.096
073.025 .112
Chromosphere
Structure
071.003
073.026 .100 .157
Chromosphere
Turbulence
073.031
Chromosphere
Velocities
073.068
Chromosphere
Wave Propagation
073.120 .126
Chromosphere-Corona
Transition Region
022.034 .081 .100
073.067
074.098
076.050 .051
Chromospheres
Stellar
064.018 .027 .028
113.010
114.020 .318 .350
.370
Chronology
004.000
Circumstellar Matter
064.063
122.023 .132
131.038
132.006 .029
141.061 .603 .625
153.004
Circumstellar Matter
Radiative Transfer
063.039
Circumstellar Shells
022.056
064.001 .003 .011
.046 .047
114.070 .302 .311
.344 .346
122.072 .083 .183
131.032 .198
Circumstellar Shells
H I Regions
141.626
Circumstellar Shells
Stellar Atmospheres
113.010
Clocks
035.000
Close Binaries
Apsidal Motion
119.020
Close Binaries
Emission Lines
117.021
Close Binaries
Evolution
117.018 .038 .052
.053 .063
121.123
142.127
Close Binaries
Gaseous Streams
117.049
121.051
Close Binaries
Interferometry
117.015
Close Binaries
Light Curves
113.029
117.025
121.046 .073 .074
.078 .106 .107
.122 .123
Close Binaries
Mass Exchange
119.013
121.073
Close Binaries
Mass Loss
042.053
117.001 .045 .052
121.051
Close Binaries
Masses
117.019
Close Binaries
MK Types
114.047
Close Binaries
Models
117.014 .041 .042
121.046 .051
Close Binaries
Neutron Stars
117.018
Close Binaries
Orbits
117.040
121.087
Close Binaries
Periods
117.062
Close Binaries
Photometry
114.358
121.107
Close Binaries
Polarization
117.028
121.019

Close Binaries
Reflection Effect
117.073
Close Binaries
Rotation
117.056 .060 .061
.062
Close Binaries
Stability
117.014 .060 .061
Close Binaries
Tidal Evolution
117.013 .017 .057
.063
Close Binaries
UV Spectrophotometry
114.375
Clouds
Jupiter Atmosphere
099.054 .074
Clouds
Mars Atmosphere
097.029 .045 .054
Clouds
Venus Atmosphere
093.003 .007 .014
.015 .031 .038
.040 .042 .043
.045 .047 .048
.050 .053
Clusters
Galactic
153.000
Clusters
Globular
154.000
Clusters
Moving Clusters
153.000
Clusters
Open Clusters
153.000
Clusters of Galaxies
160.000
Clusters of Galaxies
Accretion
160.048
Clusters of Galaxies
Catalogues
160.042
Clusters of Galaxies
Classification
158.027
160.026
Clusters of Galaxies
Coma
142.166
160.001 .002 .004
.010 .015 .016
.021 .027 .029
.032 .047
Clusters of Galaxies
Compact
160.022
Clusters of Galaxies
Distribution
160.023 .033 .034
Clusters of Galaxies
Dynamics
151.046
160.008 .009
Clusters of Galaxies
Evolution
160.035
Clusters of Galaxies
Formation
160.044
Clusters of Galaxies
Galaxy Distribution
160.007 .025 .028
.043 .046
Clusters of Galaxies
Intergalactic Matter
160.027 .030
161.003 .006 .007
.010 .011
Clusters of Galaxies
Local Group
158.194
160.028 .033 .034
Clusters of Galaxies
Luminosity Function
160.019
Clusters of Galaxies
Mass-Lumin Relation
160.029
Clusters of Galaxies
Missing Mass
160.024
Clusters of Galaxies
Orientation
158.001
Clusters of Galaxies
Perseus
160.004
Clusters of Galaxies
Photometry
160.022
Clusters of Galaxies
Quasars
141.095
Clusters of Galaxies
Radio Galaxies
141.003
160.002
Clusters of Galaxies
Radio Radiation
160.005
Clusters of Galaxies
Radio Sources
141.168
158.203
Clusters of Galaxies
Redshifts
160.001 .002 .003
.031 .032
162.040 .047
Clusters of Galaxies
Structure
160.016 .021
162.005
Clusters of Galaxies
Surface Brightnesses
160.031
Clusters of Galaxies
Virgo
160.004 .046
Clusters of Galaxies
X Rays
160.048
Clusters of Galaxies
X-Ray Sources
141.025
142.008 .099 .166
160.004 .006 .045
Collapse
Gravitation
061.037
065.105
066.068 .073 .164
Collapse
Interstellar Clouds
062.064
Collapsing Stars
Magnetic Fields
066.165
Collapsing Stars
Neutrinos
065.183
Colloquia Proceedings
012.000
Colloquia Reports
011.000
Color Excesses
122.133
Colors
113.000
Colors
Carbon Stars
113.030
Colors
Cepheids
122.077 .081 .116
.133
Colors
Delta Scuti Stars
122.081
Colors
Diffuse Nebulae
132.034
Colors
Early-Type Stars
113.021 .048 .066
Colors
Galaxies
158.055 .056 .196
Colors
OB Stars
114.046
131.101
Colors
Star Catalogues
113.025
Colors
Supergiants
159.004
Coma
Clusters of Galaxies
142.166
160.001 .002 .004
.010 .015 .016
.021 .027 .029
.032 .047
Comet 1866 I
Tempel-Tuttle
103.106
Comet 1884 III Wolf 1
103.112
Comet 1892 V Barnard 3
103.113

Comet 1913 VI Westphal
103.109
Comet 1916 II Neujmin 2
103.103
Comet 1919 III Brorsen-Metcalf
103.116
Comet 1925 II Schwassmann-Wachmann 1
103.137
Comet 1930 VI Schwassmann-Wachmann 3
103.115
Comet 1947 XIII Wirtanen
103.118
Comet 1954 XI Wirtanen
103.118
Comet 1954 XIII Harrington-Abell
103.124
Comet 1955 VII Perrine-Mrkos
103.138
Comet 1964 IX Everhart
103.110
Comet 1965 VII De Vico-Swift
103.143
Comet 1969 II Gunn
103.104
Comet 1969i Bennett
103.101
Comet 1970 II Bennett
103.101
Comet 1970 VIII Pons-Winnecke
103.108
Comet 1970 XI Kopff
103.141
Comet 1970 XII Kojima
103.105
Comet 1971 II Encke
103.120
Comet 1971c Kearns-Kwee
103.117
Comet 1972 VI Giacobini-Zinner
103.107
Comet 1972 XI Kearns-Kwee
103.117
Comet 1973 I Gehrels 1
103.139
Comet 1973 XII Kohoutek
103.102
Comet 1973b Tuttle-Giacobini-Kresak
103.142
Comet 1973f Kohoutek
103.102
Comet 1973l Schwassmann-Wachmann 2
103.122
Comet 1974b Bradfield
103.114
Comet 1974c Lovas
103.125
Comet 1974f Honda-Mrkos-Pajdusakova
103.129
Comet 1974g Van Den Bergh
103.135
Comet 1975a Boethin
103.123
Comet 1975b West-Kohoutek-Ikemura
103.128
Comet 1975c Kohoutek
103.132
Comet 1975d Bradfield
103.119
Comet 1975e Smirnova-Chernykh
103.111
Comet 1975f Wolf 1
103.112
Comet 1975g Longmore
103.126
Comet 1975h Kobayashi-Berger-Milon
103.100
Comet 1975i Churyumov-Gerasimenko
103.127
Comet 1975j Mori-Sato-Fujikawa
103.130
Comet 1975k Suzuki-Saigusa-Mori
103.131
Comet 1975l Harrington-Abell
103.124
Comet 1975m Arend
103.121
Comet 1975n West
103.133
Comet 1975o Gehrels 3
103.134
Comet 1975p Bradfield
103.136
Comet 1975q Sato
103.140
Cometary Nebulae
132.002 .019 .029
Comets
102.000
Comets
Atmospheres
102.003 .018
Comets
Dust
102.016
Comets
Infrared Radiation
102.016
Comets
Listed Objects
103.000
Comets
Nuclei
102.013 .025
Comets
Orbits
098.020
102.022 .028
Comets
Outbursts
102.026
Comets
Photometry
103.001
Comets
Short Period
102.017
Comets
Tails
102.015
Computing
021.000
Congress Proceedings
012.000
Congress Reports
011.000
Convection
Photosphere
071.059
Convection
Solar Atmosphere
080.002 .017 .038
Convection
Stellar Atmospheres
064.040
065.037
114.048
Convection
Stellar Envelopes
064.014 .070
Convection
Stellar Evolution
065.026 .060 .080
Convection
Stellar Interiors
065.004 .057 .062
.101 .111 .112
.132 .176 .177
Convection
Venus Atmosphere
093.010
Convection
White Dwarfs
126.004 .008 .014
.016
Cool Stars
Atmospheres
064.068
Cool Stars
Infrared Spectra
114.070
Cosmic Rays
143.000
Cosmic Rays
Acceleration
143.011
Cosmic Rays
Alfven Waves
143.030 .031
Cosmic Rays
Antiprotons
143.043
Cosmic Rays
Chemical Composition
143.013 .051 .066
Cosmic Rays
Diffusion
143.018
Cosmic Rays
Electrons
143.006 .018 .045
.046

Cosmic Rays
Element Abundances
061.036
143.002 .017 .022
.023 .058 .059
Cosmic Rays
Energy Spectra
143.006 .007 .008
.009 .020 .035
.043 .051 .053
.058
Cosmic Rays
Galaxies
143.005
Cosmic Rays
Galaxy
142.096 .118
155.003 .024 .026
156.002
Cosmic Rays
High Energy
143.005 .006 .086
Cosmic Rays
Interplanetary Space
143.019 .025
Cosmic Rays
Interstellar Matter
061.017
062.055
131.122
143.002 .043
Cosmic Rays
Jupiter
099.028
Cosmic Rays
Magnetic Fields
062.014
143.003 .007
Cosmic Rays
Moon
094.441
Cosmic Rays
Nuclear Reactions
143.059
Cosmic Rays
Nuclei
022.099
143.001 .010 .013
.017 .020 .021
.047 .051 .052
.087
Cosmic Rays
Origin
099.028
143.013 .024 .048
.049
156.002
Cosmic Rays
Positrons
143.043
Cosmic Rays
Propagation
106.075
143.003 .004 .011
.015 .022 .023
.079 .085
Cosmic Rays
Protons
143.005
Cosmic Rays
Pulsars
141.341
Cosmic Rays
Solar
078.000
Cosmic Rays
Solar Corona
078.004
Cosmic Rays
Solar Flares
073.156 .160 .161
078.011 .021 .025
Cosmic Rays
Solar Modulation
106.012
143.012 .014 .016
.023 .034
Cosmic Rays
Solar Wind
143.033 .054 .060
Cosmic Rays
Supernovae
143.081
Cosmic Rays
Variations
143.042 .075 .076
.080 .083 .084
Cosmogony
Planetary System
107.000
Cosmological Models
141.059 .077
160.031 .035 .044
162.001 .004 .009
.010 .013 .015
.016 .017 .018
.019 .022 .024
.026 .027 .029
.035 .036 .037
.038 .041 .042
.045 .046 .047
.048 .050 .052
.067 .068 .069
.070 .071 .074
.075 .078 .081
.082 .083 .084
.085 .086 .088
.089 .090 .092
.093 .094
Cosmology
162.000
Crab Nebula
134.000
Crab Nebula
Gamma Rays
134.005 .007 .013
142.165
Crab Nebula
Hydromagnetic Waves
022.088
Crab Nebula
Lunar Occultations
142.136 .137
Crab Nebula
Pulsars
022.093
061.009
062.037
141.143 .302 .305
.316 .328 .338
Crab Nebula
Pulsars
141.346 .348
142.063 .165 .241
Crab Nebula
Radio Radiation
134.003 .007 .011
Crab Nebula
Wave Propagation
134.001
Crab Nebula
X Rays
134.006 .007
142.063 .136 .137
.164 .165
143.018
Cygnus Loop
X Rays
142.139

Dark Clouds
113.002
131.051 .124 .135
Dark Nebulae
131.034 .085 .118
141.602
Data Processing
031.400
Delta Scuti Stars
122.033 .035 .036
.136 .181
Delta Scuti Stars
Colors
122.081
Delta Scuti Stars
Light Curves
122.117
Delta Scuti Stars
Periods
122.037 .172 .173
Delta Scuti Stars
Photometry
122.173
Detectors
034.131 .133 .135
.136
Diameter
Moon
094.141
Diameters
Cepheids
122.139
Diameters
Galaxies
158.202
Diameters
Saturn Satellites
100.208
Diameters
Stars
115.000
Diameters
Supergiants
115.018
Diameters
Supernova Remnants
125.009
Diffuse Nebulae
Colors
132.034

Diffuse Nebulae
Forbidden Lines
132.010
Diffuse Nebulae
H II Regions
132.010
Distance
Moon
094.003 .004 .010
Distances
Galactic Clusters
153.001 .002 .013
.030 .039
Distances
Galaxies
158.131 .202
162.051
Distances
Globular Clusters
154.001 .002 .015
.016 .017 .025
Distances
H II Regions
131.560
Distances
M Dwarfs
112.016
Distances
Quasars
141.084 .089
Distances
Radio Sources
141.146
Distances
Supernovae
125.048
Distances
X-Ray Sources
131.100
142.048 .162
Dust
Comets
102.016
Dust
H II Regions
131.041 .099 .508
.530
Dust
Interplanetary Matter
106.003 .009 .025
.037 .040 .073
.078 .079
Dust
Interstellar Matter
131.098 .099 .125
Dust
Mars Atmosphere
097.029 .038 .051
Dust
Planetary Nebulae
133.022 .026
Dust
Venus
093.033 .034
Dust Clouds
H I Regions
131.563
Dust Clouds
Interstellar Matter
131.011 .123
Dust Grains
Stellar Atmospheres
064.016
Dwarf Galaxies
122.165 .166
158.172
Dwarf Novae
Binaries
124.008 .017 .019
Dwarf Novae
Mass Loss
117.049
Dwarf Novae
Outbursts
124.008 .019
Dwarf Novae
Photometry
122.085
Dynamics
Clusters of Galaxies
151.046
160.008 .009
Dynamics
Earth-Moon System
042.011
Dynamics
Galactic Clusters
151.046
Dynamics
Galaxies
151.063
158.063 .064
Dynamics
Globular Clusters
151.008 .045 .046
Dynamics
Jupiter Atmosphere
099.005
Dynamics
Solar Atmosphere
080.055
Dynamics
Star Clusters
151.018
Dynamics
Stellar Atmospheres
064.005
Dynamics
Stellar Systems
151.000
Dynamics
Venus Atmosphere
093.026

Early-Type Stars
Atmospheres
064.042 .045 .050
114.324
Early-Type Stars
Binaries
115.013
Early-Type Stars
Colors
113.021 .048 .066
Early-Type Stars
H II Regions
114.024
Early-Type Stars
Infrared Radiation
114.349
Early-Type Stars
Line Profiles
114.016 .017 .018
.049
Early-Type Stars
Mass Loss
064.001 .002 .053
Early-Type Stars
MK Types
114.024
Early-Type Stars
Photometry
113.001 .047 .065
Early-Type Stars
Radial Velocities
112.002
Early-Type Stars
Radio Radiation
141.105
Early-Type Stars
Rotation
065.106
114.049 .071
116.002 .005 .006
Early-Type Stars
Spectra
113.047
114.017 .023
Early-Type Stars
Spectrophotometry
114.092 .345
Early-Type Stars
UV Spectra
114.003 .005 .018
.092 .324 .345
Earth
Chemical Composition
081.000
Earth
Figure
081.000
Earth
Gravity
081.000
Earth
Magnetic Field
084.200
Earth
Rotation
044.000
Earth Atmosphere
082.000
Earth Atmosphere
Density
082.000
Earth Atmosphere
Element Abundances
082.012
Earth Atmosphere
Extinction
082.000
Earth Atmosphere
Refraction
082.000
Earth Atmosphere
Scintillation
082.000
Earth-Moon System
Dynamics
042.011

Eclipses
Lunar
095.000
Eclipses
Solar
079.000
Eclipsing Variables
121.000
Eclipsing Variables
Absorption Lines
121.119
Eclipsing Variables
Evolution
117.012
Eclipsing Variables
Galactic Clusters
121.012 .050
Eclipsing Variables
Globular Clusters
121.004
Eclipsing Variables
Light Curves
121.002 .003 .006
.007 .008 .009
.010 .022 .030
.042 .050 .051
.052 .053 .073
.074 .076 .077
.078 .079 .083
.084 .085 .088
.090 .102 .106
.125 .148 .154
Eclipsing Variables
Mass Loss
121.049
Eclipsing Variables
Mass-Lumin Relation
111.009
Eclipsing Variables
Masses
121.071 .080 .081
.082 .102 .103
Eclipsing Variables
Models
121.088
Eclipsing Variables
Orbits
117.022
121.032 .042 .071
.076 .077 .081
.082 .089 .102
.132
Eclipsing Variables
Parallaxes
111.009
Eclipsing Variables
Periods
121.024 .031 .074
.091 .105 .106
.131
Eclipsing Variables
Photometry
121.001 .005 .020
.032 .042 .050
.054 .075 .079
.084 .091 .102
.106 .115 .123
.124 .133 .147
.148 .154
Eclipsing Variables
Polarization
121.019 .043
Eclipsing Variables
Radial Velocities
121.089
Eclipsing Variables
Radio Radiation
121.028
Eclipsing Variables
Spectra
121.026 .032 .045
.080 .081 .082
.102 .103 .122
Eclipsing Variables
Spectral Types
121.027
Eclipsing Variables
Spectrophotometry
114.316
Eclipsing Variables
UV Photometry
121.090
Einstein Equations
066.063
Element Abundances
061.000
Element Abundances
Cosmic Rays
061.036
143.002 .017 .022
.023 .058 .059
Element Abundances
Earth Atmosphere
082.012
Element Abundances
F Stars
114.359
Element Abundances
Gaseous Nebulae
132.044
Element Abundances
H II Regions
131.507
Element Abundances
Interstellar Matter
022.055
061.031 .036
131.044 .084 .087
.088 .141 .142
.143 .155 .156
.157 .159 .160
.515
132.041
Element Abundances
K Giants
114.340
Element Abundances
Late-Type Stars
114.057 .079 .081
.091
Element Abundances
Luminous Stars
114.371
Element Abundances
Mars Atmosphere
097.016
Element Abundances
Metallic-Line Stars
114.086
Element Abundances
Meteorites
105.027 .060 .069
.202
Element Abundances
Meteors
104.066
Element Abundances
Moon
094.120 .150
Element Abundances
Peculiar A Stars
114.086 .337
Element Abundances
Peculiar B Stars
114.043
Element Abundances
Photosphere
022.032
071.022 .031 .063
Element Abundances
Planetary Nebulae
133.004 .025 .031
.037
Element Abundances
Planetary System
061.036
Element Abundances
Red Giants
114.396
Element Abundances
Solar Atmosphere
071.005
Element Abundances
Solar Corona
074.018 .065
Element Abundances
Solar Cosmic Rays
078.001
Element Abundances
Solar Flares
073.018
Element Abundances
Solar Spectrum
071.021 .051
Element Abundances
Solar Wind
074.072
Element Abundances
Stellar Atmospheres
061.015
064.013 .019
114.362
122.206
Element Abundances
Stellar Interiors
065.003 .036
Element Abundances
Stellar Spectra
114.025
Elements
Origin
061.000
Emission Lines
Chromosphere
074.005
Emission Lines
Close Binaries
117.021

Emission Lines
Galaxies
158.047 .197
Emission Lines
Quasars
022.016
141.065
Emission Lines
Quasi-Stellar Objects
141.060
Emission Lines
Seyfert Galaxies
022.016
158.004 .011 .015
.016 .197
Emission Lines
Solar Atmosphere
080.020
Emission Lines
Solar Corona
022.097
074.007 .065 .066
.070 .091 .092
.107 .108
Emission Lines
Solar Flares
073.141
Emission Lines
Solar Spectrum
071.046
Emission Lines
Wolf-Rayet Stars
113.040
Emission Nebulae
132.000
Emission Nebulae
Infrared Radiation
132.031
Emission Nebulae
Radio Radiation
132.047
Emission Nebulae
Spectra
132.014
Emission-Line Stars
Infrared Spectra
114.360
Emission-Line Stars
Line Profiles
114.370
Emission-Line Stars
Peculiar
114.358
Emission-Line Stars
Radio Radiation
114.012
141.061 .151
Emission-Line Stars
Spectra
064.047
114.010 .012 .306
.312 .382
Ephemerides
047.000
Ephemerides
Mars
047.039
Ephemeris Time
044.004
Eta Carinae
064.046
Eta Carinae
Infrared Radiation
114.304
141.640
Evolution
Binaries
118.019
151.070
Evolution
Carbon Stars
065.143
Evolution
Close Binaries
117.018 .038 .052
.053 .063
121.123
142.127
Evolution
Clusters of Galaxies
160.035
Evolution
Eclipsing Variables
117.012
Evolution
Galactic Clusters
153.006
Evolution
Galaxies
061.036
155.048
158.129 .138 .139
.152
Evolution
Galaxy
155.000
Evolution
Helium-Rich Stars
065.173
114.362
Evolution
Jupiter
099.001
Evolution
Low-Mass Stars
065.028
Evolution
Massive Stars
064.042
065.004 .027 .072
.083 .096 .132
Evolution
Moon
094.017
Evolution
Planetary Nebulae
065.029 .128
133.006 .020
Evolution
Population I Stars
065.129
Evolution
Population II Stars
065.129 .133
Evolution
Stellar Interiors
065.020
Evolution
Stellar Models
065.071 .149 .173
Evolution
Stellar Systems
151.009 .010 .014
.019 .045 .089
Evolution
Supernova Remnants
125.017 .033
Evolution
White Dwarfs
125.038
126.005 .015
Evolution
Wolf-Rayet Stars
114.008
Evolution
X-Ray Sources
142.127
Evolution of Stars
065.000
Exhibitions
009.000
Expeditions Reports
011.000
Extinction
Earth Atmosphere
082.000
Extinction
Galactic Clusters
153.014
Extinction
Interstellar Matter
114.394
131.004 .024 .028
.037 .051 .088
.124 .126 .135
.162 .192 .195
.198 .211
Extraterrestrial
Research
051.000

F Stars
Atmospheres
114.359
F Stars
Element Abundances
114.359
F Stars
Metal Abundances
114.094
F Stars
MK Types
114.022
F Stars
Spectra
114.359
155.046
Faculae
072.000
Faculae
Models
072.065
Faculae
Photometry
072.076
Faculae
Photosphere
072.065

Faint Stars
  Parallaxes
    041.035
Faint Stars
  Photometry
    041.035
Figure
  Celestial Bodies
    042.000
Figure
  Earth
    081.000
Figure
  Moon
    094.173 .201
Figure
  Planets
    091.000
Figure
  Stars
    116.000
Figure
  Sun
    080.000
Filaments
    071.054
    073.143 .144 .145
      .148 .153
Filaments
  Galaxies
    158.142
Filters
    034.002 .027 .124
Flare Stars
    061.016
    122.005 .006 .007
      .008 .009 .010
      .011 .012 .013
      .014 .075 .084
      .135 .144 .145
      .146 .148 .152
      .158 .218 .219
    132.039
Flare Stars
  Photometry
    122.106
Flare Stars
  Pleiades
    122.104 .105 .157
      .216
Flare Stars
  Polarimetry
    122.106
Flare Stars
  Radio Radiation
    122.156
Flare Stars
  X Rays
    122.163 .167
Flocculi
    073.098
Forbidden Lines
  Carina Nebula
    131.549
Forbidden Lines
  Diffuse Nebulae
    132.010
Forbidden Lines
  Gaseous Nebulae
    132.035
Forbidden Lines
  H II Regions
    131.533 .540
Forbidden Lines
  Planetary Nebulae
    133.027 .033
Forbidden Lines
  Quasars
    141.167
Forbidden Lines
  Solar Spectrum
    071.021
Forbidden Lines
  Stellar Spectra
    114.351
Forbidden Lines
  Supernova Remnants
    125.018
Formaldehyde
  Interstellar Matter
    131.013
Four-Body Problem
    042.083
Fraunhofer Lines
  Photosphere
    071.004 .020 .039
Fraunhofer Lines
  Solar Spectrum
    071.029 .033 .045
      .064
Frequency Standards
    035.000
FU Orionis Stars
    122.205
    132.060 .061
Fundamental Catalogues
    041.020

G Dwarfs
  Kinematics
    155.044
G Giants
  Metal Abundances
    113.012
G Giants
  Photometry
    113.011 .012
G Stars
  Metal Abundances
    114.094
G Stars
  Spectra
    155.046
Galactic Center
  Gamma Rays
    142.118
Galactic Center
  H II Regions
    157.001
Galactic Center
  Infrared Radiation
    155.011
Galactic Center
  Molecules
    131.086
Galactic Center
  Neutral Hydrogen
    155.054
Galactic Center
  OH Sources
    131.133
Galactic Center
  Radio Lines
    131.029
Galactic Center
  Radio Sources
    141.110
Galactic Center
  Recombination Lines
    157.001
Galactic Clusters
    153.000
Galactic Clusters
  Ages
    153.024 .028 .030
      .031 .039
Galactic Clusters
  Binaries
    117.156
    153.029
Galactic Clusters
  C-M Diagrams
    153.029
Galactic Clusters
  Carbon Stars
    114.007
Galactic Clusters
  Distances
    153.001 .002 .013
      .030 .039
Galactic Clusters
  Distribution
    153.007
Galactic Clusters
  Dynamics
    151.046
Galactic Clusters
  Eclipsing Variables
    121.012 .050
Galactic Clusters
  Evolution
    153.006
Galactic Clusters
  Extinction
    153.014
Galactic Clusters
  H II Regions
    153.024
Galactic Clusters
  Kinematics
    153.036
Galactic Clusters
  Luminosity Function
    153.009 .025
Galactic Clusters
  Mass Distribution
    153.005
Galactic Clusters
  Masses
    153.017
Galactic Clusters
  Membership
    153.025 .026
Galactic Clusters
  Metal Abundances
    153.003 .028
Galactic Clusters
  Models
    153.024

Galactic Clusters
OB Stars
153.004
Galactic Clusters
Photometry
031.287
153.001 .002 .014
.028 .030 .031
.037 .039 .042
Galactic Clusters
Planetary Nebulae
133.003
Galactic Clusters
Proper Motions
041.024
153.021 .022 .025
.037
Galactic Clusters
Red Giants
115.006
153.006
Galactic Clusters
Spectral Types
153.032
Galactic Clusters
Star Encounters
151.039
Galactic Clusters
Stellar Evolution
065.060
Galactic Clusters
Stellar Spectra
153.027 .038
Galactic Clusters
Variables
122.031
Galactic Clusters
Wolf-Rayet Stars
153.013
Galactic Coronae
155.047
158.174
Galactic Disk
155.026
Galactic Disk
Gas Dynamics
151.054
Galactic Halo
154.008
155.035
Galactic Halo
Mass Density
155.035
Galactic Halo
Stellar Populations
155.007
Galactic Magnetic Field
141.080
155.024 .025
156.000
Galactic Nebulae
Electron Densities
132.050
Galactic Nebulae
Exciting Stars
114.024
Galactic Nebulae
Radio Radiation
132.015
Galactic Nuclei
Mass Loss
141.078
Galactic Nuclei
Molecules
158.009 .010
Galactic Nucleus
155.016 .023
Galactic Plane
Gamma Rays
142.096
Galactic Radio Radiation
157.000
Galactic Rotation
155.022 .068
Galactic Structure
064.013
114.013
131.016 .031 .100
.117 .539
142.096
151.001 .041
155.001 .003 .008
.009 .029 .033
.034 .035 .048
Galaxies
Barred Spirals
151.058 .059 .060
158.164 .165 .166
Galaxies
Black Holes
151.003
158.065
162.016
Galaxies
Catalogues
158.204
Galaxies
Chains
151.085
Galaxies
Classification
158.013 .033
Galaxies
Colors
158.055 .056 .196
Galaxies
Compact
141.100
158.013 .014 .032
.038 .065 .102
.182
Galaxies
Companions
158.028 .042
Galaxies
Cosmic Rays
143.005
Galaxies
Densities
158.178
Galaxies
Density Waves
151.001 .051 .056
158.043 .152 .155
.162 .163
Galaxies
Diameters
158.202
Galaxies
Distances
158.131 .202
162.051
Galaxies
Distribution
061.039 .040
158.119 .180
Galaxies
Dwarf Galaxies
122.165 .166
158.172
Galaxies
Dynamics
151.063
158.063 .064
Galaxies
Elliptical
151.063 .069
158.019 .037 .039
.049 .056 .058
.060 .126 .129
.181
Galaxies
Emission Lines
158.047 .197
Galaxies
Evolution
061.036
155.048
158.129 .138 .139
.152
Galaxies
Filaments
158.142
Galaxies
Formation
151.069
158.059 .060 .110
.126 .168
160.044
Galaxies
Gas Dynamics
158.150 .151
Galaxies
Groups of Galaxies
158.028 .119 .131
Galaxies
H I Regions
131.535
158.034 .039 .101
.235
Galaxies
H II Regions
131.507 .535 .541
.545
158.132 .157 .158
.198 .240
Galaxies
Infrared Photometry
158.008 .048 .056
.176
Galaxies
Infrared Radiation
158.057 .193
Galaxies
Interacting
151.055 .062 .080
155.033 .039
158.002 .012 .034
.042 .122 .131

Galaxies
Interacting
158.177 .188 .198
160.008
Galaxies
Internal Motions
151.043 .044
158.005 .006 .122
.125 .126
Galaxies
Interstellar Gas
158.128
Galaxies
Irregular
158.002 .167
Galaxies
Jets
158.066
Galaxies
Kinematics
151.060
Galaxies
Luminosity Function
158.243
Galaxies
Magnetic Fields
156.002
Galaxies
Markarian Galaxies
158.013 .019 .048
.100 .121 .237
Galaxies
Mass Distribution
151.066
Galaxies
Mass-Lumin Relation
158.032 .062 .127
.172 .201
Galaxies
Masses
158.030 .125 .140
.178 .194
Galaxies
Metal Abundances
158.133
Galaxies
Models
151.066
Galaxies
Nearby Galaxies
158.036
Galaxies
Neutral Hydrogen
131.545
141.032
158.049 .146 .148
.149 .151 .153
.154 .155 .157
.158 .159 .160
.165 .166 .167
.171 .172 .188
.200 .201
Galaxies
Nuclei
141.002 .100 .116
142.121
151.003
158.048 .055 .065
.066 .067 .068
.109 .112 .136
.182 .190 .197
Galaxies
Nuclei
158.240
Galaxies
Orientation
158.001
Galaxies
Pairs
141.024
151.020
158.032 .100 .199
Galaxies
Peculiar
158.186 .188
Galaxies
Photometry
131.195
158.002 .003 .029
.044 .051 .102
.142 .144 .176
.187 .229
160.031
Galaxies
Planetary Nebulae
158.189
Galaxies
Quasars
158.029
Galaxies
Quasi-Stellar Objects
141.024 .032
158.017
Galaxies
Radial Velocities
151.060
158.040 .100 .183
.184 .185
Galaxies
Radio Radiation
155.023
158.045 .113 .147
.170 .173 .179
.192 .235
Galaxies
Radio Sources
141.062 .116 .123
158.031 .134 .169
.181 .190
Galaxies
Redshifts
031.221
141.003
158.007 .014 .172
.182 .231
160.018
162.039 .040 .043
.048 .050
Galaxies
Rotation
151.060 .079
155.023
158.030 .043 .046
.051 .061 .062
.127 .137 .150
.178 .234 .235
161.001
162.026
Galaxies
RR Lyrae Stars
122.166
Galaxies
Seyfert Galaxies
022.016
141.062 .078
142.004
158.004 .005 .011
.013 .015 .016
.019 .035 .053
.054 .066 .125
.135 .141 .143
.193 .197 .232
.233 .239
Galaxies
Spectra
158.234 .238
Galaxies
Spectrophotometry
158.232 .233
Galaxies
Star Formation
131.535 .545
158.033 .057 .126
.161 .163
Galaxies
Stellar Associations
158.110
Galaxies
Stellar Orbits
151.023 .038 .040
.043 .044
Galaxies
Supernovae
125.026 .048
158.057
Galaxies
Surface Brightnesses
158.187
Galaxies
Surveys
113.044
Galaxies
Tidal Effects
155.068
Galaxies
UV Radiation
158.036
Galaxies
21 cm Radiation
141.024
158.012 .101 .127
.131 .146 .147
.150 .153 .154
.158 .159 .160
.165 .172 .200
.201
Galaxies Multiple
158.000
Galaxies Single
158.000
Galaxy
Chemical Evolution
061.038
Galaxy
Cosmic Rays
142.096 .118
155.003 .024 .026
156.002
Galaxy
Density Distribution
155.030

Galaxy
Density Waves
151.041
155.036 .038
Galaxy
Dust Distribution
131.127
Galaxy
Early History
162.023
Galaxy
Evolution
155.000
Galaxy
Gamma Rays
142.118
155.003 .072 .073
Galaxy
Gas-To-Dust Ratio
131.503
Galaxy
Globular Clusters
154.008
Galaxy
Interstellar Gas
158.128
Galaxy
Interstellar Matter
131.086
155.024 .025 .026
Galaxy
Kinematics
155.038
Galaxy
Magnetic Field
156.000
Galaxy
Mass Density
155.045
Galaxy
Mass Loss
155.006
Galaxy
Models
155.042
Galaxy
Neutral Hydrogen
155.069
Galaxy
Nucleosynthesis
061.031
Galaxy
Pulsars
141.332
Galaxy
Shock Waves
155.028
Galaxy
Spectral Surveys
114.013
Galaxy
Stellar Orbits
155.013 .042
Galaxy
Stellar Velocities
155.030
Galaxy
Structure
155.000
Galaxy
Supernova Remnants
125.033 .044 .046
Galaxy
Supernovae
125.054
Galaxy
21 cm Radiation
155.033 .069
Galilean Satellites
041.016
042.029 .030 .031
.033
099.203 .205 .207
.208 .219 .223
.225 .226 .232
.239
Gamma Radiation
Cosmic
032.519
061.045 .067 .091
142.181
Gamma Ray Background
034.095
Gamma Rays
Bursts
061.045 .067
142.018 .021 .119
162.049
Gamma Rays
Crab Nebula
134.005 .007 .013
142.165
Gamma Rays
Galactic Center
142.118
Gamma Rays
Galactic Plane
142.096
Gamma Rays
Galaxy
142.118
155.003 .072 .073
Gamma Rays
Interstellar Matter
131.122
Gamma Rays
Neutron Stars
065.125
141.335
Gamma Rays
Origin
142.021
Gamma Rays
Pulsars
141.318 .328 .335
142.095
Gamma Rays
Radio Sources
141.002
Gamma Rays
Solar
076.000
Gamma Rays
Solar Flares
073.015 .016 .114
076.018 .021 .040
.043 .048 .055
Gamma Rays
X-Ray Sources
142.095 .151 .177
Gamma Rays
X-Ray Sources
142.245 .247
Gamma-Ray Astronomy
061.000
Gamma-Ray Sources
142.000
Gamma-Ray Sources
Black Holes
142.018
Gamma-Ray Sources
Search
142.150
Gamma-Ray Sources
Surveys
142.095 .177
Gas
Intergalactic Matter
161.006
Gas Clouds
Intergalactic Matter
161.007
Gas Clouds
Molecules
132.009
Gas Dynamics
Galactic Disk
151.054
Gas Dynamics
Galaxies
158.150 .151
Gaseous Nebulae
Element Abundances
132.044
Gaseous Nebulae
Extragalactic
132.044
Gaseous Nebulae
Forbidden Lines
132.035
Gaseous Nebulae
Helium Abundance
132.048
Gaseous Nebulae
Ionization
132.063
Gaseous Nebulae
OH Absorption
132.018
Gaseous Nebulae
Oxygen Abundance
132.048
Gegenschein
106.010
Geminids
104.009 .042
Geomagnetic Field
084.200
Geomagnetic Field
Reversals
084.231 .306 .307
Geomagnetic Field
Solar Wind
084.210 .213 .221
Geopotential
081.010 .015 .016
.028
Giacobinids
104.065

Giants
Globular Clusters
154.022
Giants
Space Motions
155.050
Giants
Velocity Distribution
155.050
Globular Clusters
154.000
Globular Clusters
Black Holes
066.019
154.028
Globular Clusters
C-M Diagrams
154.002 .007 .012
.015 .017
Globular Clusters
Carbon Stars
154.026
Globular Clusters
Chemical Composition
154.014
Globular Clusters
Distances
154.001 .002 .015
.016 .017 .025
Globular Clusters
Dynamics
151.008 .045 .046
Globular Clusters
Eclipsing Variables
121.004
Globular Clusters
Galaxy
154.008
Globular Clusters
Giants
154.022
Globular Clusters
Horizontal Branches
122.162 .165
154.001 .021 .022
.025
Globular Clusters
Infrared Photometry
154.019
Globular Clusters
Luminosity Function
154.010 .023
Globular Clusters
M Stars
154.021
Globular Clusters
Magellanic Clouds
154.005
159.005
Globular Clusters
Masses
154.010
Globular Clusters
Metal Abundances
154.002 .009 .012
Globular Clusters
Models
154.027
Globular Clusters
Origin
154.014
Globular Clusters
Photometry
154.001 .002 .007
.016 .022 .025
Globular Clusters
Red Giants
154.006
Globular Clusters
RR Lyrae Stars
122.049 .073 .162
.217 .223
123.012
154.003 .021
Globular Clusters
Star Distribution
154.028 .038
Globular Clusters
Stellar Evolution
154.021 .022 .037
Globular Clusters
Variables
122.048 .053 .131
154.003 .024
Globular Clusters
X-Ray Sources
066.019
142.009 .010 .036
.040 .044 .047
.061 .100 .105
.117 .140 .163
154.024
Globules
131.034
Globules
H II Regions
131.130 .501
Gould's Belt
155.022
Grains
Interstellar Dust
131.050
Grains
Interstellar Matter
105.028
131.010 .014 .019
.028 .030 .032
.035 .041 .049
.088 .135 .139
.140 .145 .162
.190
132.046
Granulation
031.227
065.037
071.041 .050 .055
.061 .072
073.126
Gravitation
Black Holes
066.073 .097
Gravitation
Collapse
061.037
065.105
066.068 .073 .164
Gravitation
Jupiter
099.060
Gravitation Theory
066.000
Gravitational Deflection
066.037 .049 .108
.137
Gravitational
Instability
061.000
Gravitational Radiation
065.125
066.023 .024 .025
.029 .035 .051
.057 .072 .076
.079 .080 .084
.087 .092 .094
.109 .111 .121
.122 .179
155.006
Gravitational Waves
066.054 .059 .077
.078 .081 .083
.085 .088 .091
.098 .099 .103
162.072
Gravity
Earth
081.000
Gravity
Mars
097.006 .036 .052
Gravity
Moon
094.001 .171 .201
.204
Gravity
Planets
091.035
Groups of Galaxies
158.028 .119 .131
Gum Nebula
132.043

H I Absorption
131.510
H I Clouds
Absorption Lines
131.039
H I Regions
131.500
H I Regions
Chemical Composition
131.114
H I Regions
Dust Clouds
131.563
H I Regions
Galaxies
131.535
158.034 .039 .101
.235
H I Regions
Magellanic Clouds
131.535
H I Regions
Recombination Lines
131.512
H II Regions
131.500
132.012
H II Regions
Carbon Monoxide
131.562

H II Regions
Chemical Composition
131.114
H II Regions
Compact
131.136 .503 .508
.513 .514 .530
.548 .551 .556
.557 .558 .559
141.607 .634
H II Regions
Diffuse Nebulae
132.010
H II Regions
Distances
131.560
H II Regions
Dust
131.041 .099 .508
.530
H II Regions
Early-Type Stars
114.024
H II Regions
Element Abundances
131.507
H II Regions
Exciting Stars
131.560
H II Regions
Forbidden Lines
131.533 .540
H II Regions
Galactic
131.560 .561
H II Regions
Galactic Center
157.001
H II Regions
Galactic Clusters
153.024
H II Regions
Galaxies
131.507 .535 .541
.545
158.132 .157 .158
.198 .240
H II Regions
Globules
131.130 .501
H II Regions
Helium Abundance
131.538
H II Regions
Infrared Photometry
131.508
H II Regions
Infrared Radiation
131.503 .514 .530
.540
132.046
141.607 .629
H II Regions
Infrared Sources
131.548
141.605 .634
H II Regions
Internal Motions
131.551
H II Regions
Line Profiles
131.509
H II Regions
Magellanic Clouds
131.535
H II Regions
Photometry
131.538
H II Regions
Radial Velocities
132.043
H II Regions
Radio Lines
131.517 .529 .542
H II Regions
Radio Radiation
131.550
141.146
H II Regions
Radio Sources
141.104
H II Regions
Recombination Lines
131.511 .513 .515
.529 .537
155.055
H II Regions
Spectra
131.514
H II Regions
Structure
131.516
H II Regions
Water Vapor
131.003
Helium Abundance
Gaseous Nebulae
132.048
Helium Abundance
H II Regions
131.538
Helium Abundance
Seyfert Galaxies
022.016
Helium Abundance
Solar Wind
074.051
Helium Abundance
Stellar Atmospheres
064.045
Helium Burning
Stellar Evolution
065.033 .102 .137
.149 .173
Helium-Rich Stars
Evolution
065.173
114.362
Helium-Rich Stars
Fine Analyses
114.395
Helium-Rich Stars
Mass Loss
065.147
Helium-Rich Stars
Spectra
114.339
Helium-Rich Stars
Spectrophotometry
114.394
Helium-Rich Stars
Variations
065.147
Herbig-Haro Objects
065.118
131.092
132.004 .031 .060
.061
Hertzsprung Gap
065.030
High-Luminosity Stars
Photometry
113.042
High-Velocity Clouds
131.539
High-Velocity Stars
112.015
141.331
High-Velocity Stars
Luminosity Function
155.035
History of Astronomy
004.000
Horizontal Branches
Globular Clusters
122.162 .165
154.001 .021 .022
.025
Horizontal-Branch Stars
064.069
122.166
153.006
HR Diagrams
115.000
HR Diagrams
Peculiar B Stars
114.366
HR Diagrams
RV Tauri Stars
122.102
HR Diagrams
Supergiants
065.135
Hubble Constant
162.019 .021 .039
.040 .050 .051
Hyades
111.001
153.008 .025 .036
Hydrodynamics
062.000
Hydromagnetic Waves
Crab Nebula
022.088

Image Tubes
034.121
Infrared Astronomy
061.000
Infrared Excesses
M Stars
113.010
Infrared Photometry
Galaxies
158.008 .048 .056
.176
Infrared Photometry
Globular Clusters
154.019

Infrared Photometry
H II Regions
131.508
Infrared Photometry
Mars
097.044
Infrared Photometry
Minor Planets
098.038
Infrared Photometry
Planetary Nebulae
133.005
Infrared Photometry
Seyfert Galaxies
158.035
Infrared Photometry
Stellar Associations
113.002 .014
Infrared Photometry
Wolf-Rayet Stars
113.045
Infrared Radiation
Carbon Stars
122.087
Infrared Radiation
Comets
102.016
Infrared Radiation
Early-Type Stars
114.349
Infrared Radiation
Emission Nebulae
132.031
Infrared Radiation
Galactic Center
155.011
Infrared Radiation
Galaxies
158.057 .193
Infrared Radiation
H II Regions
131.503 .514 .530
.540
132.046
141.607 .629
Infrared Radiation
Jupiter
099.010 .058
Infrared Radiation
Moon
094.102
Infrared Radiation
Planetary Nebulae
133.009 .029
141.630
Infrared Radiation
Reflection Nebulae
132.009
Infrared Radiation
Solar Flares
073.142
Infrared Sources
131.140
141.600
Infrared Sources
Compact
141.628 .630
Infrared Sources
H II Regions
131.548
141.605 .634
Infrared Sources
Models
131.544
Infrared Sources
Orion Nebula
131.017 .019
132.046
Infrared Sources
Photometry
141.606 .627 .629
Infrared Sources
Polarization
122.072
141.631
Infrared Sources
Radio Radiation
141.630
Infrared Sources
Spectra
132.006
141.640
Infrared Sources
Structure
132.006
141.626 .629
Infrared Sources
Surveys
141.645
Infrared Spectra
Carbon Stars
114.002 .004 .334
141.623
Infrared Spectra
Cool Stars
114.070
Infrared Spectra
Emission-Line Stars
114.360
Infrared Spectra
Jupiter
099.033 .034 .039
Infrared Spectra
Minor Planets
098.044
Infrared Spectra
Orion Nebula
132.003 .007
Infrared Spectra
Photosphere
071.034
Infrared Spectra
Planetary Nebulae
133.022
Infrared Spectra
Saturn
100.011
Infrared Spectra
Saturn Satellites
100.203
Infrared Spectra
Solar Flares
073.097
Infrared Spectra
Venus Atmosphere
093.018
Infrared Stars
OH Emission
141.603
Infrared Stars
Photometry
113.019
Infrared Stars
Photometry
141.641
Infrared Stars
Polarimetry
113.019
Instability
Gravitational
061.000
Institutes
008.000
Instruments
Astronomical
032.000
Interferometers
034.030 .125 .126
.153
Interferometers
Intensity
064.050
Interferometers
Very-Long-Baseline
141.143
Interferometry
031.227 .283
Interferometry
Close Binaries
117.015
Interferometry
Pulsars
141.358
Interferometry
Solar Corona
074.121
Interferometry
Speckle Interferometry
031.249
Interferometry
Very-Long-Baseline
031.220 .260 .261
.266 .276
033.007
Intergalactic Matter
161.000
Intergalactic Matter
Absorption
161.005
Intergalactic Matter
Clusters of Galaxies
160.027 .030
161.003 .006 .007
.010 .011
Intergalactic Matter
Gas
161.006
Intergalactic Matter
Gas Clouds
161.007
Intergalactic Matter
Heating
161.006
Intergalactic Reddening
161.011
International
Cooperation
013.000
Interplanetary Magnetic
Field
106.000
143.025

Interplanetary Matter
106.000
Interplanetary Matter
Alfven Waves
074.022
106.049
Interplanetary Matter
Dust
106.003 .009 .025
.037 .040 .073
.078 .079
Interplanetary Matter
Electrons
143.044
Interplanetary Matter
Origin
106.025
Interplanetary Matter
Particles
097.041
106.001
Interplanetary Matter
Protons
106.072
143.044
Interplanetary Matter
Scintillations
106.020 .021 .045
.047 .057 .070
Interplanetary Matter
Shock Waves
073.119
106.066 .077
Interplanetary Matter
Solar Cosmic Rays
078.030
Interplanetary Matter
Solar Wind
074.001 .011 .015
.059 .088
078.004
106.013 .020 .070
Interplanetary Matter
Wave Propagation
106.067
Interplanetary Space
Cosmic Rays
143.019 .025
Interplanetary Space
Solar Wind
106.053
Interstellar Clouds
Absorption
131.209
Interstellar Clouds
Collapse
062.064
Interstellar Clouds
Electron Densities
131.200
Interstellar Clouds
Ionization
131.207
Interstellar Clouds
Molecules
131.048 .124
132.042
Interstellar Dust
Grains
131.050
Interstellar Gas
Densities
155.028
Interstellar Gas
Galaxies
158.128
Interstellar Gas
Galaxy
158.128
Interstellar Gas
Shock Waves
155.028
Interstellar Lines
131.012 .015 .033
.100 .108 .115
.134 .143 .157
.199 .204 .208
Interstellar Matter
131.000
Interstellar Matter
Absorption
131.014 .043 .084
.093 .094 .095
.096 .100 .127
.129 .135 .166
.193 .212 .548
Interstellar Matter
Absorption Lines
131.053 .155 .157
.206
Interstellar Matter
Carbon Monoxide
131.001 .045 .056
.117 .118 .148
.160
141.623
158.009 .010
Interstellar Matter
Cosmic Rays
061.017
062.055
131.122
143.002 .043
Interstellar Matter
Dust
131.098 .099 .125
Interstellar Matter
Dust Clouds
131.011 .123
Interstellar Matter
Element Abundances
022.055
061.031 .036
131.044 .084 .087
.088 .141 .142
.143 .155 .156
.157 .159 .160
.515
132.041
Interstellar Matter
Extinction
114.394
131.004 .024 .028
.037 .051 .088
.124 .126 .135
.162 .192 .195
.198 .211
Interstellar Matter
Formaldehyde
131.013
Interstellar Matter
Galactic Distribution
131.031
Interstellar Matter
Galaxy
131.086
155.024 .025 .026
Interstellar Matter
Gamma Rays
131.122
Interstellar Matter
Grains
105.028
131.010 .014 .019
.028 .030 .032
.035 .041 .049
.088 .135 .139
.140 .145 .162
.190
132.046
Interstellar Matter
Heating
131.098 .099
Interstellar Matter
Ionization
022.055
131.009 .089 .114
.129 .210
Interstellar Matter
Lyman Alpha
131.043
Interstellar Matter
Magellanic Clouds
131.148
Interstellar Matter
Magnetic Fields
131.138 .203
Interstellar Matter
Molecules
022.018 .054 .059
.060 .063 .084
.089
131.001 .002 .005
.006 .007 .008
.011 .012 .013
.016 .017 .018
.021 .023 .025
.026 .029 .033
.047 .052 .054
.055 .056 .083
.085 .086 .087
.090 .091 .104
.105 .109 .114
.116 .117 .119
.120 .122 .123
.132 .136 .141
.144 .145 .146
.150 .159 .163
.164 .187 .188
.189 .202 .516
132.016
158.009 .010
Interstellar Matter
Neutral Hydrogen
131.155
Interstellar Matter
OH
022.073
132.005

Interstellar Matter
Polarization
131.003 .027 .088
.106 .119 .139
.140
Interstellar Matter
Radiative Transfer
131.010 .014
Interstellar Matter
Radio Lines
131.087 .091 .120
Interstellar Matter
Radio Radiation
132.005
Interstellar Matter
Recombination Lines
131.112
Interstellar Matter
Scattering
131.042
Interstellar Matter
Scintillations
063.025 .037
131.020 .111
Interstellar Matter
Shock Waves
125.043
131.138
Interstellar Matter
Star Formation
065.118
131.056 .121 .130
.138
Interstellar Matter
Supernova Remnants
125.001 .033 .050
Interstellar Matter
Supernovae
131.190
Interstellar Matter
UV Radiation
131.114 .191
Interstellar Matter
Water
131.003 .046 .092
.121
141.628
Interstellar Matter
Wave Propagation
131.102
Interstellar Matter
X Rays
022.055
Interstellar Reddening
131.053 .101 .135
.195
154.013
155.034
Interstellar Reddening
Star Clusters
131.198
Interstellar Space
131.000
Intrinsic Variables
122.000
Io
099.052 .201 .206
.207 .209 .211
.212 .213 .214
.217 .227 .228
.231 .242
Ionosphere
083.000
Ionosphere
Jupiter
099.011
Ionosphere
Mars
097.058 .062 .063
Ionosphere
Venus
093.004 .022 .052

Jupiter
099.000
Jupiter
Albedo
099.059
Jupiter
Cosmic Rays
099.028
Jupiter
Electrons
099.021 .023 .026
Jupiter
Evolution
099.001
Jupiter
Gravitation
099.060
Jupiter
Infrared Radiation
099.010 .058
Jupiter
Infrared Spectra
099.033 .034 .039
Jupiter
Ionosphere
099.011
Jupiter
Limb Darkening
099.094
Jupiter
Magnetic Field
099.002 .022 .032
.066 .099
Jupiter
Magnetosphere
084.222
099.003 .007 .009
.019 .020 .023
.024 .025 .026
.029 .030 .064
.076 .078 .084
.085 .086 .087
.213
Jupiter
Models
099.036
Jupiter
Oblateness
099.077
Jupiter
Polarization
099.003
Jupiter
Radiation Belts
099.008 .027 .225
Jupiter
Radio Radiation
099.003 .013 .014
Jupiter
Radio Radiation
099.037 .038 .048
.052 .076 .087
.092 .093 .097
Jupiter
Red Spot
099.045 .055 .056
.082
Jupiter
Solar Wind
099.019 .020 .048
Jupiter
Stellar Occultations
099.083
Jupiter
X Rays
099.015 .016
Jupiter Atmosphere
099.011 .046 .056
.072
Jupiter Atmosphere
Ammonia
099.031 .058
Jupiter Atmosphere
Chemical Composition
099.012
Jupiter Atmosphere
Clouds
099.054 .074
Jupiter Atmosphere
Dynamics
099.005
Jupiter Atmosphere
Hydrocarbons
099.018 .075
Jupiter Atmosphere
Methane
099.058
Jupiter Atmosphere
Models
099.040 .075 .100
Jupiter Atmosphere
Molecules
099.010 .034 .039
.067 .070
Jupiter Atmosphere
Photometry
099.073 .074
Jupiter Atmosphere
Scattering
091.004
Jupiter Atmosphere
Temperatures
099.057 .058 .069
.083
Jupiter Atmosphere
Turbulence
099.050
Jupiter Satellites
099.200
Jupiter Satellites
Albedo
099.204 .207 .223
Jupiter Satellites
Astrolabe Observations
041.016
Jupiter Satellites
Atmospheres
099.214

Jupiter Satellites
Io
099.052 .201 .206
.207 .209 .211
.212 .213 .214
.217 .227 .228
.231 .242
Jupiter Satellites
Light Curves
099.207 .208
Jupiter Satellites
Light Variations
099.217
Jupiter Satellites
Motion
099.239
Jupiter Satellites
Occultations
099.204 .210 .216
.224 .226
Jupiter Satellites
Orbits
042.029 .030 .031
.033 .075
099.219
Jupiter Satellites
Photometry
099.210 .232
Jupiter Satellites
Polarimetry
099.205
Jupiter Satellites
Resonances
042.033
Jupiter Satellites
Rotation
099.223
Jupiter Satellites
Spectra
099.206 .209 .227
Jupiter Satellites
Surfaces
099.203 .224

K Dwarfs
Spectra
114.104
K Giants
Element Abundances
114.340
K Giants
Metal Abundances
113.012
114.021
K Giants
Photometry
113.011 .012
Kinematics
G Dwarfs
155.044
Kinematics
Galactic Clusters
153.036
Kinematics
Galaxies
151.060
Kinematics
Galaxy
155.038
Kinematics
Late-Type Stars
114.091
Kinematics
M Dwarfs
155.014 .015
Kinematics
Metal-Rich Stars
114.114
Kinematics
Planetary Nebulae
133.007 .027
Kinematics
Stellar Systems
151.000

Late-Type Stars
Atmospheres
064.038
Late-Type Stars
Bolometric Corrections
115.005
Late-Type Stars
Chemical Composition
064.019 .038
114.072 .303 .320
.325
Late-Type Stars
Cyanogen
114.091
Late-Type Stars
Element Abundances
114.057 .079 .081
.091
Late-Type Stars
Kinematics
114.091
Late-Type Stars
Luminosities
115.015
Late-Type Stars
Magellanic Clouds
114.078
Late-Type Stars
OH Emission
122.132
Late-Type Stars
Opacities
061.069
Late-Type Stars
Photometry
113.063
Late-Type Stars
Polarization
122.135
Late-Type Stars
Spectra
114.050 .072 .079
.080 .320 .350
.370
Latitude Determination
045.000
Leonids
104.047
Libration Clouds
Moon
094.005
Limb Brightening
077.012
Limb Darkening
Jupiter
099.094
Limb Darkening
Sun
031.214
Limb Darkening
Venus
093.021 .037
Line Blocking
A Stars
114.044
Line Blocking
Solar Spectrum
071.049
Line Broadening
Plasma
062.029
Line Broadening
Stark Effect
022.009 .020 .024
.027 .028 .029
.108 .109
062.029
Line Broadening
Stellar Atmospheres
063.007
Line Broadening
Turbulence
071.006
Line Formation
063.009 .035
Line Formation
Chromosphere
063.026
Line Formation
Magnetic Fields
063.045
Line Formation
Orion Nebula
132.001
Line Formation
Photosphere
071.022
080.033
Line Formation
Planetary Atmospheres
091.001
Line Formation
Solar Atmosphere
080.052 .057
Line Formation
Solar Corona
073.004
074.098 .100
Line Formation
Solar Spectrum
063.010
Line Formation
Stellar Atmospheres
063.039
064.027
114.048 .083
Line Intensities
Chromosphere
073.122
Line Intensities
Prominences
073.066 .084

Line Intensities
Solar Corona
074.107 .120
Line Intensities
Solar Flares
073.139
Line Intensities
Solar Spectrum
022.034
071.053
Line Profiles
Chromosphere
073.004 .078
Line Profiles
Early-Type Stars
114.016 .017 .018
.049
Line Profiles
Emission-Line Stars
114.370
Line Profiles
H II Regions
131.509
Line Profiles
Magnetic Stars
065.034
Line Profiles
O Stars
114.083
Line Profiles
Orion Nebula
132.016
Line Profiles
Peculiar A Stars
114.075 .343
Line Profiles
Photosphere
071.057
Line Profiles
Prominences
073.153
Line Profiles
Solar Corona
074.049 .108 .112
076.050
Line Profiles
Solar Flares
073.005
Line Profiles
Solar Spectrum
063.028
071.006 .046 .062
.064
073.125
Line Profiles
Spectroscopic Binaries
119.006
Line Profiles
Stellar Atmospheres
063.028
064.028
Line Profiles
Stellar Chromospheres
064.027
Line Profiles
Stellar Spectra
031.226
065.122
114.302 .374
Line Profiles
Supergiants
114.383
Local Group
Clusters of Galaxies
158.194
160.028 .033 .034
Low-Luminosity Stars
126.000
Low-Mass Stars
Evolution
065.028
Luminosities
A Stars
115.016
Luminosities
B Stars
115.002
Luminosities
Carbon Stars
115.004 .008 .017
Luminosities
Cepheids
122.027 .029 .077
Luminosities
Late-Type Stars
115.015
Luminosities
M Stars
155.001
Luminosities
Quasars
141.058 .064 .120
Luminosities
RV Tauri Stars
122.083
Luminosities
Stars
115.000
Luminosities
Wolf-Rayet Stars
114.008
Luminosity Function
Clusters of Galaxies
160.019
Luminosity Function
Galactic Clusters
153.009 .025
Luminosity Function
Galaxies
158.243
Luminosity Function
Globular Clusters
154.010 .023
Luminosity Function
High-Velocity Stars
155.035
Luminous Stars
Element Abundances
114.371
Luminous Stars
Photometry
113.013
Lunar Eclipses
095.000
Lunar Laser
031.253 .254 .262
094.018 .020 .021
Lunar Occultations
096.000
Lunar Occultations
Crab Nebula
142.136 .137
Lunar Occultations
Multiple Stars
117.011
Lunar Occultations
Radio Sources
141.122
Lunar Occultations
Saturn Satellites
100.208
Lunar Occultations
X-Ray Sources
142.246
Lunar Probes
053.000
Lunar Satellites
053.000
Lyman Alpha
Interstellar Matter
131.043

M Dwarfs
Distances
112.016
M Dwarfs
Kinematics
155.014 .015
M Dwarfs
Radial Velocities
112.016
M Dwarfs
Spectra
114.104
M Giants
Atmospheres
114.365
M Giants
Spectra
114.026
M Giants
Spectrophotometry
114.107
M Stars
Atmospheres
064.051
M Stars
Distribution
155.001
M Stars
Globular Clusters
154.021
M Stars
Infrared Excesses
113.010
M Stars
Luminosities
155.001
M Stars
Photometry
113.033
M Stars
Spectra
022.083
M Stars
Spectrophotometry
114.093

M Supergiants
OH Emission
141.601
M 82
158.045
Magellanic Clouds
158.139
159.000
Magellanic Clouds
Cepheids
122.002 .138
159.009
Magellanic Clouds
Globular Clusters
154.005
159.005
Magellanic Clouds
H I Regions
131.535
Magellanic Clouds
H II Regions
131.535
Magellanic Clouds
Interstellar Matter
131.148
Magellanic Clouds
Late-Type Stars
114.078
Magellanic Clouds
OB Stars
159.001
Magellanic Clouds
Open Clusters
159.005
Magellanic Clouds
Photometry
159.004
Magellanic Clouds
Planetary Nebulae
133.024
Magellanic Clouds
RR Lyrae Stars
122.052
Magellanic Clouds
Star Clusters
154.004
Magellanic Clouds
Supergiants
113.042
114.392
159.004
Magellanic Clouds
Supernova Remnants
159.003
Magellanic Clouds
UV Survey
159.010
Magellanic Clouds
Variables
114.317
122.001 .052 .171
Magellanic Clouds
VV Cephei Stars
122.130
Magellanic Clouds
X-Ray Sources
142.053 .172
Magnetic Field
Earth
084.200
Magnetic Field
Galaxy
156.000
Magnetic Field
Interplanetary
106.000
Magnetic Field
Jupiter
099.002 .022 .032
.066 .099
Magnetic Field
Mercury
092.001 .003 .010
Magnetic Field
Moon
094.101 .124 .134
.144 .157 .158
.168 .178 .180
Magnetic Fields
Active Regions
062.039
Magnetic Fields
Chromosphere
071.036
Magnetic Fields
Collapsing Stars
066.165
Magnetic Fields
Cosmic Rays
062.014
143.003 .007
Magnetic Fields
Galaxies
156.002
Magnetic Fields
Interstellar Matter
131.138 .203
Magnetic Fields
Line Formation
063.045
Magnetic Fields
Neutron Stars
061.068
065.006 .123 .125
142.094
Magnetic Fields
Orion Nebula
131.006 .007
Magnetic Fields
Peculiar A Stars
114.363
116.004 .014 .015
Magnetic Fields
Photosphere
071.036 .052
073.071
Magnetic Fields
Prominences
073.107
Magnetic Fields
Pulsars
022.093
062.001
141.320 .331 .336
Magnetic Fields
Pulsating Stars
122.182
Magnetic Fields
Radio Sources
141.088
Magnetic Fields
Solar Corona
022.061
074.003 .042 .060
.067 .077 .081
.083 .105 .109
Magnetic Fields
Solar Flares
062.046 .050
073.029
Magnetic Fields
Solar Radio Radiation
077.041
Magnetic Fields
Solar Wind
074.027 .060 .062
.095 .117 .119
.122
Magnetic Fields
Stability
062.039
Magnetic Fields
Stars
116.000
Magnetic Fields
Stellar Evolution
065.084 .085
Magnetic Fields
Stellar Interiors
065.140
Magnetic Fields
Stellar Models
065.002 .039 .086
.087 .088
Magnetic Fields
Sunspots
072.004 .026 .035
.036 .042 .045
.046 .049
073.076
Magnetic Fields
White Dwarfs
065.189
126.004 .018 .022
.029
Magnetic Stars
063.045
116.013 .015
Magnetic Stars
Circulation
065.086
Magnetic Stars
Line Profiles
065.034
Magnetic Stars
Models
065.011
116.003
Magnetic Stars
Polarization
116.008
Magnetic Stars
Rotation
062.031
116.004 .007
Magnetic Variables
114.377
Magnetohydrodynamics
062.000
Magnetometers
051.024

Magnetosphere
Jupiter
084.222
099.003 .007 .009
.019 .020 .023
.024 .025 .026
.029 .030 .064
.076 .078 .084
.085 .086 .087
.213
Magnetosphere
Mercury
092.008 .014 .015
Magnetosphere
Saturn
099.029
Magnetosphere
Solar Wind
074.019
084.202
Magnetospheres
Planets
091.010
Magnetospheres
Pulsars
061.068
062.037
063.011
099.009 .030
141.330 .333 .334
.337 .351
Magnitudes
Stars
113.000
Main-Sequence Stars
Envelopes
064.014 .032
Main-Sequence Stars
Pulsations
065.108
Main-Sequence Stars
Stability
065.108
Manganese Stars
113.036
114.043 .343 .361
Markarian Galaxies
158.013 .019 .048
.100 .121 .237
Mars
097.000
Mars
Channels
097.012
Mars
Climate
097.060
Mars
Craters
097.011 .017 .037
.041
Mars
Dust Storms
097.011 .039
Mars
Ephemerides
047.039
Mars
Gravity
097.006 .036 .052
Mars
Infrared Photometry
097.044
Mars
Internal Structure
097.036
Mars
Ionosphere
097.058 .062 .063
Mars
Mass
097.001
Mars
Motion
042.012
Mars
Orbit
097.004
Mars
Photometry
097.013
Mars
Pictures
097.030 .035 .049
Mars
Polar Caps
097.020
Mars
Polarimetry
097.024
Mars
Positions
041.004 .062
Mars
Soil
097.057
Mars
Stellar Occultations
097.053
Mars
Surface Structures
097.005 .007 .012
.026
Mars
Volcanism
097.036
Mars Atmosphere
097.028 .056
Mars Atmosphere
Carbon Dioxide
093.002
097.022 .040 .043
.050 .051
Mars Atmosphere
Clouds
097.029 .045 .054
Mars Atmosphere
Dust
097.029 .038 .051
Mars Atmosphere
Element Abundances
097.016
Mars Atmosphere
Models
097.003 .038 .055
Mars Atmosphere
Molecules
097.032
Mars Atmosphere
Solar Wind
097.048
Mars Atmosphere
Structure
097.034
Mars Atmosphere
Temperatures
091.044
097.003
Mars Atmosphere
Water
097.033 .040
Mars Satellites
097.200
Mars Satellites
Deimos
042.032
097.201 .202 .203
Mars Satellites
Orbits
042.032
097.001 .202 .203
Mars Satellites
Origin
097.201
Mars Satellites
Phobos
042.032
097.201 .202 .203
Mars Satellites
Positions
097.204
Mars Surface
097.013 .027 .030
.035 .037
Mars Surface
Photographs
097.023 .046
Mascons
Moon
094.171 .176
Masers
022.056
131.040
Masers
Interstellar
131.018 .119 .125
.136 .201 .213
Masers
OH Sources
131.025 .026 .046
.125 .136 .147
.149 .151 .158
.163
141.603 .643 .644
158.192
Mass
Mars
097.001
Mass Density
Galaxy
155.045
Mass Density
Solar Neighborhood
115.011
Mass Exchange
Close Binaries
119.013
121.073
Mass Loss
064.000

Mass Loss
Binaries
114.375
117.020
119.002
Mass Loss
Close Binaries
042.053
117.001 .045 .052
121.051
Mass Loss
Dwarf Novae
117.049
Mass Loss
Early-Type Stars
064.001 .002 .053
Mass Loss
Eclipsing Variables
121.049
Mass Loss
Galactic Nuclei
141.078
Mass Loss
Galaxy
155.006
Mass Loss
Helium-Rich Stars
065.147
Mass Loss
Quasars
141.078
Mass Loss
Red Giants
064.035
Mass Loss
Stellar Atmospheres
064.017
141.061 .625
Mass Loss
Stellar Evolution
061.031
064.033 .034
065.032
Mass Loss
Supergiants
064.035 .072
Mass Loss
Wolf-Rayet Stars
064.053
Mass Loss
X-Ray Sources
142.002 .003 .006
.055 .107 .132
Mass-Lumin Relation
Clusters of Galaxies
160.029
Mass-Lumin Relation
Eclipsing Variables
111.009
Mass-Lumin Relation
Galaxies
158.032 .062 .127
.172 .201
Masses
Binaries
115.013
117.048
Masses
Cepheids
122.036
Masses
Close Binaries
117.019
Masses
Eclipsing Variables
121.071 .080 .081
.082 .102 .103
Masses
Galactic Clusters
153.017
Masses
Galaxies
158.030 .125 .140
.178 .194
Masses
Globular Clusters
154.010
Masses
Neutron Stars
065.019 .110 .144
Masses
OH Sources
131.104
Masses
Prominences
073.124
Masses
Stars
115.000
Masses
Visual Binaries
118.005 .016
Masses
White Dwarfs
065.005
Masses
X-Ray Sources
142.176
Massive Stars
Evolution
064.042
065.004 .027 .072
.083 .096 .132
Massive Stars
Nuclear Reactions
065.181
Massive Stars
Pulsations
065.108
Mathematics
021.000
Measuring Machines
031.415
034.138 .139 .140
.141 .142 .143
.145 .146 .147
.149 .150
Meetings Proceedings
012.000
Meetings Reports
011.000
Mercury
092.000
Mercury
Albedo
092.005
Mercury
Atmosphere
092.008
Mercury
Craters
092.012
Mercury
Magnetic Field
092.001 .003 .010
Mercury
Magnetosphere
092.008 .014 .015
Mercury
Motion
099.082
Mercury
Pictures
092.018
Mercury
Polarization
092.011
Mercury
Rotation
092.013
Mercury
Solar Wind
092.008
Mercury
Surface
092.005 .011 .012
Mercury
Transit
077.025
092.002 .007
Meridian Circles
032.036 .047 .048
.055
Metal Abundances
F Stars
114.094
Metal Abundances
G Giants
113.012
Metal Abundances
G Stars
114.094
Metal Abundances
Galactic Clusters
153.003 .028
Metal Abundances
Galaxies
158.133
Metal Abundances
Globular Clusters
154.002 .009 .012
Metal Abundances
K Giants
113.012
114.021
Metal Abundances
Nearby Stars
114.006
Metal Abundances
RR Lyrae Stars
122.073 .223
Metal-Rich Stars
Galactic Distribution
114.114
Metal-Rich Stars
Kinematics
114.114
Metallic-Line Stars
Element Abundances
114.086

Metallic-Line Stars
Spectra
114.001
Meteor Streams
104.000
Meteor Trails
062.028
104.007 .015 .023
.024 .059 .084
.088
Meteorite Craters
105.000
Meteorites
105.000
Meteorites
Ages
105.059
Meteorites
Chemical Composition
105.065 .068 .072
.083 .084
Meteorites
Element Abundances
105.027 .060 .069
.202
Meteorites
Iron Meteorites
105.023 .025 .066
.072 .081 .105
.106
Meteorites
Mass Distribution
106.055
Meteorites
Micrometeorites
106.055
Meteorites
Stone Meteorites
105.013 .048 .056
.069
Meteoroids
Mass Distribution
094.177
Meteors
104.000
Meteors
Element Abundances
104.066
Meteors
Mass Distribution
104.032
Meteors
Orbits
094.177
104.001 .045 .061
105.011
Meteors
Origin
104.051
Meteors
Radar Echoes
104.033
Meteors
Spectra
104.039 .040 .043
.066
Methane
Jupiter Atmosphere
099.058
Methods of Observation
031.200
Methods of Reduction
031.200
Microturbulence
Solar Atmosphere
071.062
Minor Planets
098.000
Minor Planets
Albedo
098.074
099.202
Minor Planets
Infrared Photometry
098.038
Minor Planets
Infrared Spectra
098.044
Minor Planets
Kirkwood Gaps
098.019 .020 .039
.043
Minor Planets
Light Curves
098.028 .029 .033
.050
Minor Planets
Models
098.077
Minor Planets
Motion
098.024 .031
Minor Planets
Orbits
098.007 .039 .065
.068
Minor Planets
Photometry
098.046 .050
Minor Planets
Polarization
098.074
Minor Planets
Positions
098.002 .003 .004
.008 .009 .011
.076
Minor Planets
Resonances
098.016 .020 .043
Minor Planets
Rotation
098.028 .029 .050
Mira Variables
064.016
114.325
122.103 .132 .140
.168
Missing Mass
155.014 .015
158.026 .123 .139
.140
160.010
162.061
Missing Mass
Clusters of Galaxies
160.024
MK Types
Close Binaries
114.047
MK Types
Early-Type Stars
114.024
MK Types
F Stars
114.022
MK Types
Stellar Spectra
153.027
Molecules
Galactic Center
131.086
Molecules
Galactic Nuclei
158.009 .010
Molecules
Gas Clouds
132.009
Molecules
Interstellar Clouds
131.048 .124
132.042
Molecules
Interstellar Matter
022.018 .054 .059
.060 .063 .084
.089
131.001 .002 .005
.006 .007 .008
.011 .012 .013
.016 .017 .018
.021 .023 .025
.026 .029 .033
.047 .052 .054
.055 .056 .083
.085 .086 .087
.090 .091 .104
.105 .109 .114
.116 .117 .119
.120 .122 .123
.132 .136 .141
.144 .145 .146
.150 .159 .163
.164 .187 .188
.189 .202 .516
132.016
158.009 .010
Molecules
Jupiter Atmosphere
099.010 .034 .039
.067 .070
Molecules
Mars Atmosphere
097.032
Molecules
Planetary Atmospheres
091.046
Molecules
Planetary Nebulae
133.021
Molecules
Saturn Atmosphere
100.008
Molecules
Spectra
022.018 .083
131.052
Molecules
Stellar Atmospheres
114.325

Molecules
Stellar Spectra
114.085
Moon
Albedo
094.174 .446
Moon
Artificial Satellites
053.013
Moon
Atmosphere
094.139 .149 .162
.163 .164
Moon
Chemical Composition
094.123 .438 .439
.440 .442
Moon
Cosmic Rays
094.441
Moon
Craters
094.103 .104 .166
.181 .186
Moon
Diameter
094.141
Moon
Distance
094.003 .004 .010
Moon
Early History
094.432 .458
Moon
Electric Conductivity
094.143
Moon
Electric Fields
094.159
Moon
Element Abundances
094.120 .150
Moon
Evolution
094.017
Moon
Figure
094.173 .201
Moon
Gravity
094.001 .171 .201
.204
Moon
History
094.107 .123
Moon
Impacts
094.170
Moon
Infrared Radiation
094.102
Moon
Interior
094.101 .143
107.016
Moon
Landing Sites
094.435
Moon
Libration Clouds
094.005
Moon
Magnetic Field
094.101 .124 .134
.144 .157 .158
.168 .178 .180
Moon
Mare Origin
094.182 .415
Moon
Mascons
094.171 .176
Moon
Meteorite Impact
094.111 .114 .116
.156 .175 .177
.420 .437 .439
.456
Moon
Models
094.129 .143 .167
Moon
Origin
094.009 .026 .150
Moon
Polarimetry
094.174
Moon
Polarization
094.145
Moon
Positions
041.013
Moon
Potential
094.159
Moon
Regolith
094.107 .108 .109
.110 .111 .112
.113 .115 .119
.120 .122 .132
.147 .407 .413
.417 .420 .425
105.040
Moon
Rocks
094.156 .404 .412
.413 .432 .442
.446 .451 .474
.479
Moon
Rotation
094.002 .006 .008
.011 .023 .106
.170
Moon
Samples
094.121 .402 .403
.410 .411 .418
.419 .422 .423
.425 .426 .428
.429 .433 .436
.437 .439 .440
.444 .448 .449
.453 .456 .457
.476 .477 .480
Moon
Seismicity
094.126 .130
Moon
Soil
094.118 .132 .137
.172 .401 .405
.406 .409 .412
.424 .458 .475
Moon
Solar Wind
074.094
094.180
Moon
Structure
094.123 .150
Moon
Surface Brightness
094.152
Moon
Surface Structures
094.450
Moon
Temperatures
094.143 .151
Moon
Thermal History
094.169
Moon
Thermal Properties
094.151
Moon
Volcanism
094.103
Moon Dynamics
094.000
Moon Global Properties
094.100
Moon Local Properties
094.400
Moon Surface
094.148 .158 .197
.419
Moon Surface
Radar Echoes
094.198 .414
Moon Surface
Reflectivities
094.142 .145
Moon Surface
Spectrophotometry
094.138 .146
Moustaches
071.056
Multiple Galaxies
158.000
Multiple Stars
117.000
Multiple Stars
Lunar Occultations
117.011

N-Body Problem
042.003 .052 .057
.091 .103
151.014 .015 .016
.027
Navigation
046.000
Navigation
Space Vehicles
052.000

Nearby Dwarfs
Velocities
112.011
Nearby Galaxies
158.036
Nearby Stars
Metal Abundances
114.006
Neptune
101.000
Neptune
Atmosphere
101.011
Neptune
Satellites
042.054
Neutral Hydrogen
Galactic Center
155.054
Neutral Hydrogen
Galaxies
131.545
141.032
158.049 .146 .148
.149 .151 .153
.154 .155 .157
.158 .159 .160
.165 .166 .167
.171 .172 .188
.200 .201
Neutral Hydrogen
Galaxy
155.069
Neutral Hydrogen
Interstellar Matter
131.155
Neutrino Astronomy
061.000
Neutrinos
Collapsing Stars
065.183
Neutrinos
Neutron Stars
065.187
Neutrinos
Stellar Interiors
065.031 .117 .143
Neutron Stars
065.007 .117 .119
.184 .188 .189
.192 .193 .197
066.052 .112
126.029
Neutron Stars
Accretion
142.114 .128
Neutron Stars
Binaries
125.006
Neutron Stars
Black Holes
142.130
Neutron Stars
Close Binaries
117.018
Neutron Stars
Gamma Rays
065.125
141.335
Neutron Stars
Magnetic Fields
061.068
065.006 .123 .125
142.094
Neutron Stars
Masses
065.019 .110 .144
Neutron Stars
Models
065.010 .019 .022
.146
Neutron Stars
Neutrinos
065.187
Neutron Stars
Oscillations
065.125
Neutron Stars
Pulsars
065.019 .123 .136
.146
141.301 .303 .317
.331 .334 .337
.338 .343 .351
142.129
Neutron Stars
Rotation
065.040 .136
066.038
141.317
Neutron Stars
X Rays
141.346
Neutron Stars
X-Ray Sources
065.040
142.002 .010 .054
.094 .116 .128
.129 .140 .155
Noctilucent Clouds
082.080
Nova Aquilae 1975
124.110
Nova Aurigae 1964
124.107
Nova Cephei 1971
124.101
Nova Cygni 1975
124.104
Nova Delphini 1967
124.100
Nova Herculis 1934
124.102
Nova Monocerotis 1975
124.111
Nova Persei 1974
124.106
Nova Sagittarii 1974
124.105
Nova Sagittarii 1975
124.109
Nova Scuti 1970
124.108
Nova Scuti 1975
124.103
Novae
124.000
Novae
Andromeda Nebula
124.011 .012
Novae
Binaries
114.358
117.064
124.008
Novae
Dwarf Novae
117.049
122.085
124.008 .017 .019
Novae
Envelopes
124.003 .009 .010
Novae
Light Curves
124.001 .006
Novae
Models
124.003 .010
Novae
Outbursts
124.009 .010
Novae
Photometry
124.013
Novae
X Rays
124.005
Nuclear Reactions
Cosmic Rays
143.059
Nuclear Reactions
Massive Stars
065.181
Nuclear Reactions
Nucleosynthesis
061.047 .048
Nuclear Reactions
Solar Atmosphere
080.076
Nuclear Reactions
Solar Flares
061.016
Nuclear Reactions
Stellar Evolution
065.008
Nuclear Reactions
Stellar Interiors
061.036
065.010 .031 .058
.137 .151 .182
.186
Nucleosynthesis
061.012 .016
065.172
Nucleosynthesis
Carbon Stars
065.143
Nucleosynthesis
CNO Cycle
061.015 .030
Nucleosynthesis
Galaxy
061.031
Nucleosynthesis
Nuclear Reactions
061.047 .048
Nucleosynthesis
Primeval
162.018

Nucleosynthesis
  Stellar Evolution
    061.004
Nucleosynthesis
  Stellar Interiors
    065.018
Nucleosynthesis
  Supernovae
    124.010
    125.037

O Stars
  Line Profiles
    114.083
O Stars
  Photometry
    113.009 .020
O Stars
  Spectra
    114.083 .108
    119.002
OB Stars
  Catalogues
    113.032
OB Stars
  Classification
    114.013
OB Stars
  Colors
    114.046
    131.101
OB Stars
  Galactic Clusters
    153.004
OB Stars
  Magellanic Clouds
    159.001
OB Stars
  Photometry
    113.032
    155.029
    159.001
OB Stars
  Spectra
    114.016
OB Stars
  Velocities
    155.037
Obituaries
    007.000
Oblateness
  Jupiter
    099.077
Observatories
    008.000
Occultations
  Jupiter Satellites
    099.204 .210 .216
    .224 .226
Occultations
  Lunar
    096.000
Of Stars
    114.367
    119.002
Of Stars
  Spectra
    114.108 .315
OH
  Interstellar Matter
    022.073
    132.005
OH Absorption
  Gaseous Nebulae
    132.018
OH Emission
  Infrared Stars
    141.603
OH Emission
  Late-Type Stars
    122.132
OH Emission
  M Supergiants
    141.601
OH Emission
  Variables
    122.003
OH Sources
    022.056
    131.092 .196 .516
    141.036 .113 .642
OH Sources
  Galactic Center
    131.133
OH Sources
  Masers
    131.025 .026 .046
    .125 .136 .147
    .149 .151 .158
    .163
    141.603 .643 .644
    158.192
OH Sources
  Masses
    131.104
OH Sources
  Positions
    131.147
Opacities
  Late-Type Stars
    061.069
Opacities
  Solar Atmosphere
    080.039
Opacities
  Stellar Evolution
    065.026 .133 .139
Opacities
  Stellar Models
    061.037
    065.036 .133
Open Clusters
    153.000
Open Clusters
  Magellanic Clouds
    159.005
Optics
  Astronomical
    031.000
Orbit
  Mars
    097.004
Orbits
  Artificial Satellites
    052.011 .040 .043
    .053 .054
    054.013
Orbits
  Binaries
    121.021
Orbits
  Close Binaries
    117.040
    121.087
Orbits
  Comets
    098.020
    102.022 .028
Orbits
  Eclipsing Variables
    117.022
    121.032 .042 .071
    .076 .077 .081
    .082 .089 .102
    .132
Orbits
  Jupiter Satellites
    042.029 .030 .031
    .033 .075
    099.219
Orbits
  Mars Satellites
    042.032
    097.001 .202 .203
Orbits
  Meteors
    094.177
    104.001 .045 .061
    105.011
Orbits
  Minor Planets
    098.007 .039 .065
    .068
Orbits
  Periodic
    042.037 .038 .041
    .045 .046 .070
    .102
Orbits
  Spectroscopic Binaries
    118.020
    119.005 .009 .014
    .019 .021 .023
Orbits
  Visual Binaries
    118.005 .010 .015
    .016 .020 .030
    .031
Organizations
    010.000
Orion Nebula
    114.045
    131.005 .502 .538
    132.011 .012 .033
    .041 .042
Orion Nebula
  Infrared Sources
    131.017 .019
    132.046
Orion Nebula
  Infrared Spectra
    132.003 .007
Orion Nebula
  Line Formation
    132.001
Orion Nebula
  Line Profiles
    132.016

Orion Nebula
Magnetic Fields
131.006 .007
Orion Nebula
Polarization
132.040 .051
Orion Nebula
Radio Lines
131.013
132.013 .030
Orion Nebula
Radio Radiation
132.047
Oscillations
Chromosphere
061.034
071.027
Oscillations
Neutron Stars
065.125
Oscillations
Photosphere
071.007 .026 .027
.042
Oscillations
Solar Atmosphere
071.033 .042
080.021 .034 .058
.061
Oscillations
Solar Interior
080.032 .063
Oscillations
Stellar Evolution
065.021 .089 .095
.096 .097 .098
.099 .100 .138
117.013
Oscillations
Stellar Models
061.034
Oscillations
Stellar Systems
151.086
Oscillator Strengths
022.010 .103

P Cygni Stars
064.001
114.082 .364 .367
Parallaxes
Binaries
117.071
Parallaxes
Eclipsing Variables
111.009
Parallaxes
Faint Stars
041.035
Parallaxes
Star Catalogues
041.035
Parallaxes
Stars
111.000
Parallaxes
Trigonometric
111.006 .010 .011
Parallaxes
Visual Binaries
118.004 .005 .010
.015
Peculiar A Stars
065.120
113.031
114.001 .355
116.007
Peculiar A Stars
Element Abundances
114.086 .337
Peculiar A Stars
Light Curves
113.036
Peculiar A Stars
Line Profiles
114.075 .343
Peculiar A Stars
Magnetic Fields
114.363
116.004 .014 .015
Peculiar A Stars
Photometry
113.064
114.363
Peculiar A Stars
Proper Motions
112.008
Peculiar A Stars
Rotation
114.363
Peculiar A Stars
Spectra
114.363
117.010
Peculiar A Stars
Spectrophotometry
114.087 .372
Peculiar B Stars
Element Abundances
114.043
Peculiar B Stars
HR Diagrams
114.366
Peculiar Galaxies
158.186 .188
Peculiar Objects
158.300
Peculiar Stars
Spectral Types
114.076
Periodic Orbits
042.037 .038 .041
.045 .046 .070
Periodic Orbits
Stability
042.102
Periodicals
001.000
Perseids
104.041 .091
Perseus
Clusters of Galaxies
160.004
Personal Notes
006.000
Perturbation Theory
042.007 .036 .048
052.004 .009
Photography
036.000
Photometers
034.021 .048 .056
.096 .122 .123
.164 .166
Photometers
Electrophotometers
034.107
Photometers
Microphotometers
034.148
Photometric Standards
113.001
114.064 .381
Photometric Systems
031.238
113.011 .020 .021
.028 .037 .050
.052 .053 .056
.063 .069
Photometry
113.000 .071
Photometry
A Stars
113.031
Photometry
Be Stars
113.018 .024 .065
.067
Photometry
Beta Cephei Stars
122.074 .097
Photometry
Binaries
113.054
Photometry
Blue Stars
113.046
Photometry
Bright Stars
113.055
Photometry
Cepheids
122.002 .027 .028
.133 .136 .161
Photometry
Close Binaries
114.358
121.107
Photometry
Clusters of Galaxies
160.022
Photometry
Comets
103.001
Photometry
Delta Scuti Stars
122.173
Photometry
Dwarf Novae
122.085
Photometry
Early-Type Stars
113.001 .047 .065
Photometry
Eclipsing Variables
121.001 .005 .020
.032 .042 .050
.054 .075 .079
.084 .091 .102

Photometry
Eclipsing Variables
121.106 .115 .123
.124 .133 .147
.148 .154
Photometry
Extreme UV
113.051
Photometry
Faculae
072.076
Photometry
Faint Stars
041.035
Photometry
Flare Stars
122.106
Photometry
G Giants
113.011 .012
Photometry
Galactic Clusters
031.287
153.001 .002 .014
.028 .030 .031
.037 .039 .042
Photometry
Galaxies
131.195
158.002 .003 .029
.044 .051 .102
.142 .144 .176
.187 .229
160.031
Photometry
Globular Clusters
154.001 .002 .007
.016 .022 .025
Photometry
H II Regions
131.538
Photometry
High-Luminosity Stars
113.042
Photometry
Infrared Sources
141.606 .627 .629
Photometry
Infrared Stars
113.019
141.641
Photometry
Jupiter Atmosphere
099.073 .074
Photometry
Jupiter Satellites
099.210 .232
Photometry
K Giants
113.011 .012
Photometry
Late-Type Stars
113.063
Photometry
Luminous Stars
113.013
Photometry
M Stars
113.033
Photometry
Magellanic Clouds
159.004
Photometry
Mars
097.013
Photometry
Minor Planets
098.046 .050
Photometry
Multicolor
113.052 .069
122.133
Photometry
Novae
124.013
Photometry
O Stars
113.009 .020
Photometry
OB Stars
113.032
155.029
159.001
Photometry
Peculiar A Stars
113.064
114.363
Photometry
Planetary Nebulae
133.019
Photometry
Quasars
141.027 .066 .119
Photometry
Radio Sources
141.029 .142
Photometry
Red Dwarfs
113.072
Photometry
RR Lyrae Stars
122.057
Photometry
Saturn
100.005
Photometry
Saturn Satellites
100.201
Photometry
Seyfert Galaxies
158.015
Photometry
Solar Corona
074.021 .110
Photometry
Spectroscopic Binaries
119.001 .015 .020
Photometry
UBV
113.043
155.008
Photometry
Variables
122.085 .086
Photometry
Venus
093.046
Photometry
Wolf-Rayet Stars
113.040
Photometry
Wolf-Rayet Stars
114.314
121.002
Photometry
X-Ray Sources
121.044
142.041 .042 .043
.066 .110 .135
.172
Photometry
Zodiacal Light
106.050
Photomultipliers
034.118
Photosphere
Absorption Lines
071.038
Photosphere
Brightnesses
071.047
Photosphere
Convection
071.059
Photosphere
Element Abundances
022.032
071.022 .031 .063
Photosphere
Faculae
072.065
Photosphere
Fluctuations
071.055
Photosphere
Fraunhofer Lines
071.004 .020 .039
Photosphere
Infrared Spectra
071.034
Photosphere
Line Formation
071.022
080.033
Photosphere
Line Profiles
071.057
Photosphere
Magnetic Fields
071.036 .052
073.071
Photosphere
Models
071.030
Photosphere
Oscillations
071.007 .026 .027
.042
Photosphere
Spectra
071.023 .067 .068
Photosphere
Structure
071.003
Photosphere
Temperatures
071.002 .047
Photosphere
Turbulence
063.035
071.004

Photosphere
Velocities
071.025 .029 .059 .060
080.033
Photospheres
Stellar
064.020 .066
071.030
114.320
Physics
022.000
Plages
073.091 .095 .101
Planetaria
009.000
Planetary Atmospheres
082.094
091.019
Planetary Atmospheres
Chemical Composition
091.005
Planetary Atmospheres
Line Formation
091.001
Planetary Atmospheres
Molecules
091.046
Planetary Atmospheres
Radiative Transfer
091.001 .020 .047
Planetary Atmospheres
Scattering
091.004 .007 .011
Planetary Atmospheres
Stellar Occultations
091.025
Planetary Atmospheres
Turbulence
091.004
Planetary Companions
117.000
Planetary Nebulae
133.000
Planetary Nebulae
Carbon Monoxide
133.021
Planetary Nebulae
Catalogues
133.028
Planetary Nebulae
Central Stars
133.001 .002 .006
Planetary Nebulae
Distribution
061.040
133.024
Planetary Nebulae
Dust
133.022 .026
Planetary Nebulae
Element Abundances
133.004 .025 .031 .037
Planetary Nebulae
Evolution
065.029 .128
133.006 .020
Planetary Nebulae
Forbidden Lines
133.027 .033
Planetary Nebulae
Formation
133.036
Planetary Nebulae
Galactic Clusters
133.003
Planetary Nebulae
Galaxies
158.189
Planetary Nebulae
Infrared Photometry
133.005
Planetary Nebulae
Infrared Radiation
133.009 .029
141.630
Planetary Nebulae
Infrared Spectra
133.022
Planetary Nebulae
Kinematics
133.007 .027
Planetary Nebulae
Magellanic Clouds
133.024
Planetary Nebulae
Models
133.026
Planetary Nebulae
Molecules
133.021
Planetary Nebulae
Nuclei
064.042
133.008
Planetary Nebulae
Orientation
133.007
Planetary Nebulae
Photometry
133.019
Planetary Nebulae
Radial Velocities
034.092
Planetary Nebulae
Radio Radiation
141.103
Planetary Nebulae
Radio Spectra
133.022
Planetary Nebulae
Spectra
133.002 .006 .024 .025 .027 .030
Planetary Nebulae
Spectrophotometry
133.023
Planetary Nebulae
Surveys
113.044
133.028
Planetary Nebulae
UV Spectra
133.031
Planetary Probes
053.000
Planetary System
Cosmogony
107.000
Planetary System
Element Abundances
061.036
Planetary System
Formation
107.001
Planetary System
Physics
091.000
Planetary Theory
042.018 .023 .054 .062
066.050
Planets
Albedo
099.202
Planets
Figure
091.000
Planets
Formation
107.003 .011
Planets
Gravity
091.035
Planets
Interiors
091.002 .006
Planets
Magnetospheres
091.010
Planets
Minor
098.000
Planets
Organic Matter
091.017
Planets
Positions
041.013
Plasma
062.000
Plasma
Line Broadening
062.029
Plasma
Recombination Lines
131.113
Plasma
Scattering
062.026
063.040
Plasma
Turbulence
062.050
Pleiades
153.041
Pleiades
Flare Stars
122.104 .105 .157 .216
Pluto
101.000
Pluto
Motion
101.008
Pluto
Positions
041.052
Polar Motion
045.000

Polarimeters
034.013 .122 .130 .155 .165
Polarimetry
142.179
Polarimetry
Flare Stars
122.106
Polarimetry
Infrared Stars
113.019
Polarimetry
Jupiter Satellites
099.205
Polarimetry
Mars
097.024
Polarimetry
Moon
094.174
Polarization
Be Stars
114.338 .341
Polarization
Binaries
117.002
Polarization
Close Binaries
117.028
121.019
Polarization
Eclipsing Variables
121.019 .043
Polarization
Infrared Sources
122.072
141.631
Polarization
Interstellar Matter
131.003 .027 .088 .106 .119 .139 .140
Polarization
Jupiter
099.003
Polarization
Late-Type Stars
122.135
Polarization
Magnetic Stars
116.008
Polarization
Mercury
092.011
Polarization
Minor Planets
098.074
Polarization
Moon
094.145
Polarization
Orion Nebula
132.040 .051
Polarization
Pulsars
141.337 .356
Polarization
Radio Sources
125.039
141.009 .022 .033 .034 .080 .088
Polarization
Radio Sources
141.111
158.173
Polarization
Red Variables
122.078
Polarization
Reflection Nebulae
132.062
Polarization
Seyfert Galaxies
158.239
Polarization
Solar Corona
022.061
074.016 .052 .110
077.017
Polarization
Solar Radio Bursts
077.064 .065
Polarization
Solar X Rays
076.016
Polarization
Stellar Atmospheres
063.046
Polarization
Sunspots
063.046
Polarization
Supernova Remnants
125.010 .027
Polarization
Variables
131.097
Polarization
Venus
093.044
Polarization
Venus Atmosphere
093.041
Polarization
White Dwarfs
126.018 .022
Polarization
X-Ray Sources
117.046 .050
142.031 .032 .033 .041 .067 .109
Polarization
Zodiacal Light
106.010 .039 .044
Polarization of Starlight
131.000
Polytropes
061.052
065.002 .022
Polytropes
Models
065.148
Polytropes
Rotation
021.011
065.001 .061
Population I Stars
Evolution
065.129
Population II Stars
Evolution
065.129 .133
Positional Astronomy
041.000
Positions
Mars
041.004 .062
Positions
Mars Satellites
097.204
Positions
Minor Planets
098.002 .003 .004 .008 .009 .011 .076
Positions
Moon
041.013
Positions
OH Sources
131.147
Positions
Planets
041.013
Positions
Pluto
041.052
Positions
Radio Sources
141.008 .014 .023 .063 .071 .118 .141 .144
158.113
Positions
Saturn
041.004
Positions
Saturn Satellites
100.202
Positions
Star Catalogues
041.020 .040
112.013
Positions
Venus
041.004 .060 .061
Positions
X-Ray Sources
142.121 .147 .158
Praesepe
153.026
Pre-Main-Sequence Stars
065.118
Prominences
073.000
Prominences
Electron Densities
073.085
Prominences
Helium Lines
073.109
Prominences
Line Intensities
073.066 .084
Prominences
Line Profiles
073.153
Prominences
Magnetic Fields
073.107

Prominences
Masses
073.124
Prominences
Peculiar
074.017
Prominences
Spectra
073.121 .147 .149
.152 .155
Prominences
Structure
073.127 .128
Prominences
Temperatures
073.085
Proper Motion Surveys
112.017 .019 .020
.021
Proper Motions
112.000
Proper Motions
A Stars
155.057
Proper Motions
B Stars
112.012
Proper Motions
Binaries
117.071
Proper Motions
Catalogues
041.000
Proper Motions
Galactic Clusters
041.024
153.021 .022 .025
.037
Proper Motions
Peculiar A Stars
112.008
Proper Motions
Star Catalogues
041.024 .029 .069
112.013
Proper Motions
Variables
112.004
Proper Motions
Visual Binaries
112.003
Protogalaxies
141.148
158.138
Protoplanets
107.003 .010 .015
117.010
Protostars
065.020 .063 .073
.074
107.010
Pulsars
141.300
Pulsars
Binaries
031.203
066.095 .104
117.018
141.301 .305 .311
.312 .313 .345
Pulsars
Catalogues
141.340
Pulsars
Cosmic Rays
141.341
Pulsars
Crab Nebula
022.093
061.009
062.037
141.143 .302 .305
.316 .328 .338
.346 .348
142.063 .165 .241
Pulsars
Distribution
061.040
Pulsars
Formation
065.032
Pulsars
Galaxy
141.332
Pulsars
Gamma Rays
141.318 .328 .335
142.095
Pulsars
Interferometry
141.358
Pulsars
Magnetic Fields
022.093
062.001
141.320 .331 .336
Pulsars
Magnetospheres
061.068
062.037
063.011
099.009 .030
141.330 .333 .334
.337 .351
Pulsars
Mechanisms
141.306
Pulsars
Models
141.335 .337 .338
.343 .351
Pulsars
Motion
141.321
Pulsars
Neutron Stars
065.019 .123 .136
.146
141.301 .303 .317
.331 .334 .337
.338 .343 .351
142.129
Pulsars
Optical Pulses
141.312 .313 .328
Pulsars
Periods
141.348
Pulsars
Polarization
141.337 .356
Pulsars
Pulse Intensities
141.302
Pulsars
Pulse Structure
063.025
141.316 .327 .329
Pulsars
Radio Radiation
141.337 .347 .356
Pulsars
Radio Spectra
141.329
Pulsars
Rotation
141.333 .344
Pulsars
Scintillations
131.020 .111
141.352
Pulsars
Supernova Remnants
125.012
141.338
Pulsars
Surfaces
141.320
Pulsars
Surveys
141.332
Pulsars
Variations
141.329
Pulsars
Velocities
141.331 .339
Pulsars
X Rays
125.016
141.304 .313 .328
.342 .346
142.164 .241
Pulsating Stars
Atmospheres
122.099
Pulsating Stars
Magnetic Fields
122.182
Pulsating Stars
Periods
122.225
Pulsation Theory
122.000
Pulsations
Be Stars
113.018
Pulsations
Beta Cephei Stars
115.002
122.097
Pulsations
Main-Sequence Stars
065.108
Pulsations
Massive Stars
065.108
Pulsations
Rotating Stars
065.121

Pulsations
  Stellar Envelopes
    065.091
    122.114
Pulsations
  Stellar Evolution
    064.036
    065.009 .091 .093
    122.114 .117 .118
Pulsations
  Stellar Models
    065.109
Pulsations
  Variables
    065.089
Pulsations
  White Dwarfs
    126.002 .006
    142.106
Pulsations
  X-Ray Sources
    065.134
    142.011 .043 .094
      .098 .106 .113
      .241

Quarks
    022.128
    061.079
Quasars
    141.000
Quasars
  Absorption Lines
    141.019 .030
Quasars
  Clusters of Galaxies
    141.095
Quasars
  Distances
    141.084 .089
Quasars
  Distribution
    141.077 .086 .099
Quasars
  Emission Lines
    022.016
    141.065
Quasars
  Energy Distribution
    141.060
Quasars
  Envelopes
    141.092
Quasars
  Forbidden Lines
    141.167
Quasars
  Galaxies
    158.029
Quasars
  Identifications
    141.102
Quasars
  Luminosities
    141.058 .064 .120
Quasars
  Mass Loss
    141.078
Quasars
  Models
    141.078 .163
Quasars
  Nuclei
    158.066 .112
Quasars
  Photometry
    141.027 .066 .119
Quasars
  Radio Radiation
    141.089
Quasars
  Radio Spectra
    141.032 .058 .095
      .155
Quasars
  Redshifts
    141.015 .017 .026
      .030 .064 .065
      .081 .083 .084
      .086 .114 .120
      .133 .161
    160.018
    162.048
Quasars
  Spectra
    141.018 .159 .160
      .161 .162
Quasars
  Structure
    141.004
Quasars
  Variations
    141.001 .006 .124
Quasars
  X-Ray Sources
    142.005 .122
Quasi-Stellar Objects
  Absorption Lines
    141.013 .017 .106
      .107 .108 .109
      .148
Quasi-Stellar Objects
  Emission Lines
    141.060
Quasi-Stellar Objects
  Galaxies
    141.024 .032
    158.017
Quasi-Stellar Objects
  Optical Identification
    141.074
Quasi-Stellar Objects
  Radiative Transfer
    141.109
Quasi-Stellar Objects
  Redshifts
    141.013 .021 .085
      .106 .107 .108
      .130 .148 .150
Quasi-Stellar Objects
  Spectra
    141.107 .148
Quasi-Stellar Objects
  Spectrophotometry
    141.060 .169
Quasi-Stellar Objects
  Variations
    141.015 .027
R CrB Variables
    011.062
    064.003 .062 .063
    114.312 .362
    122.020 .021 .022
      .023 .024 .025
      .198 .200 .201
      .202 .203
Radar Echoes
  Meteors
    104.033
Radar Echoes
  Moon Surface
    094.198 .414
Radar Echoes
  Venus
    093.024
Radial Velocities
    112.000
Radial Velocities
  Beta Cephei Stars
    122.097
Radial Velocities
  Blue Stars
    034.094
Radial Velocities
  Early-Type Stars
    112.002
Radial Velocities
  Eclipsing Variables
    121.089
Radial Velocities
  Galaxies
    151.060
    158.040 .100 .183
      .184 .185
Radial Velocities
  H II Regions
    132.043
Radial Velocities
  M Dwarfs
    112.016
Radial Velocities
  Planetary Nebulae
    034.092
Radial Velocities
  Spectroscopic Binaries
    119.006 .007 .014
      .015 .021
Radial Velocities
  Variables
    122.183
Radiation Belts
    084.400
Radiation Belts
  Jupiter
    099.008 .027 .225
Radiative Transfer
    063.000
Radiative Transfer
  Circumstellar Matter
    063.039
Radiative Transfer
  Interstellar Matter
    131.010 .014
Radiative Transfer
  Planetary Atmospheres
    091.001 .020 .047
Radiative Transfer
  Quasi-Stellar Objects
    141.109

Radiative Transfer
Solar Atmosphere
063.026
Radiative Transfer
Stellar Atmospheres
063.014 .015
064.018 .027 .028 .052
122.115
Radiative Transfer
Venus Atmosphere
093.019
Radio Equipment
033.000
Radio Galaxies
141.055 .089
158.007 .018 .134 .173
Radio Galaxies
Clusters of Galaxies
141.003
160.002
Radio Galaxies
Structure
158.203
Radio Lines
Galactic Center
131.029
Radio Lines
H II Regions
131.517 .529 .542
Radio Lines
Interstellar Matter
131.087 .091 .120
Radio Lines
Orion Nebula
131.013
132.013 .030
Radio Radiation
Active Regions
077.006
Radio Radiation
Clusters of Galaxies
160.005
Radio Radiation
Crab Nebula
134.003 .007 .011
Radio Radiation
Early-Type Stars
141.105
Radio Radiation
Eclipsing Variables
121.028
Radio Radiation
Emission Nebulae
132.047
Radio Radiation
Emission-Line Stars
114.012
141.061 .151
Radio Radiation
Flare Stars
122.156
Radio Radiation
Galactic
157.000
Radio Radiation
Galactic Nebulae
132.015
Radio Radiation
Galaxies
155.023
158.045 .113 .147 .170 .173 .179 .192 .235
Radio Radiation
H II Regions
131.550
141.146
Radio Radiation
Infrared Sources
141.630
Radio Radiation
Interstellar Matter
132.005
Radio Radiation
Jupiter
099.003 .013 .014 .037 .038 .048 .052 .076 .087 .092 .093 .097
Radio Radiation
Orion Nebula
132.047
Radio Radiation
Planetary Nebulae
141.103
Radio Radiation
Pulsars
141.337 .347 .356
Radio Radiation
Quasars
141.089
Radio Radiation
Saturn
099.014
Radio Radiation
Seyfert Galaxies
158.143
Radio Radiation
Solar
077.000
Radio Radiation
Solar Corona
077.005 .017 .036
Radio Radiation
Solar Flares
073.017 .022 .104 .156
077.004 .060
Radio Radiation
Sunspots
077.026
Radio Radiation
Supernova Remnants
125.001 .013 .014 .027 .035 .036 .039 .044 .045 .046
141.056 .146
Radio Radiation
Variables
122.003
Radio Radiation
X-Ray Sources
142.001 .007 .017 .020 .060 .103 .170
Radio Sources
141.000
Radio Sources
Binaries
121.028
141.028
Radio Sources
Clusters of Galaxies
141.168
158.203
Radio Sources
Compact
141.033 .055 .090 .100 .110
158.190
Radio Sources
Counts
141.082
162.006
Radio Sources
Distances
141.146
Radio Sources
Distribution
141.037 .082
158.180
Radio Sources
Extragalactic
141.009 .020 .022 .111 .115 .118 .128 .129 .141 .142
162.006
Radio Sources
Fine Structure
125.035
141.056 .079 .113 .115 .118 .123 .143 .153
158.173
Radio Sources
Flux Densities
125.013 .039
131.542
141.014 .016 .073 .074 .097 .117 .121 .123 .131 .143 .144 .166
158.113 .170 .173
Radio Sources
Galactic
141.117
Radio Sources
Galactic Center
141.110
Radio Sources
Galactic Distribution
141.147
Radio Sources
Galaxies
141.062 .116 .123
158.031 .134 .169 .181 .190
Radio Sources
Gamma Rays
141.002
Radio Sources
H II Regions
141.104
Radio Sources
Identifications
141.101

Radio Sources
Light Curves
141.076
Radio Sources
Lunar Occultations
141.122
Radio Sources
Magnetic Fields
141.088
Radio Sources
Models
141.002 .035
Radio Sources
Optical Identification
031.236
141.008 .029 .057
.063 .087 .141
Radio Sources
Photometry
141.029 .142
Radio Sources
Polarization
125.039
141.009 .022 .033
.034 .080 .088
.111
158.173
Radio Sources
Positions
141.008 .014 .023
.063 .071 .118
.141 .144
158.113
Radio Sources
Scintillations
106.045 .047
Radio Sources
Spectra
131.542
134.007
141.059 .075 .104
.117 .128 .129
.131 .146 .147
.154 .156
158.170 .190
Radio Sources
Structure
141.005 .007 .016
.034 .055 .073
.090
Radio Sources
Surveys
141.082 .091 .096
.144 .147
Radio Sources
Synchrotron Radiation
062.011
141.128 .129
142.060
143.018
Radio Sources
Variations
141.031 .074 .075
.076 .094 .142
Radio Sources
X Rays
141.002
142.064
Radio Spectra
Planetary Nebulae
133.022
Radio Spectra
Pulsars
141.329
Radio Spectra
Quasars
141.032 .058 .095
.155
Radio Telescopes
033.000
Radioheliographs
033.012
Radiometers
034.161
Recombination Lines
Galactic Center
157.001
Recombination Lines
H I Regions
131.512
Recombination Lines
H II Regions
131.511 .513 .515
.529 .537
155.055
Recombination Lines
Interstellar Matter
131.112
Recombination Lines
Plasma
131.113
Recombination Lines
Radio Frequencies
022.110
131.512
132.030
Recombination Lines
Solar Spectrum
080.066
Red Dwarfs
Photometry
113.072
Red Giants
Element Abundances
114.396
Red Giants
Galactic Clusters
115.006
153.006
Red Giants
Globular Clusters
154.006
Red Giants
Mass Loss
064.035
Red Variables
Polarization
122.078
Redshifts
Clusters of Galaxies
160.001 .002 .003
.031 .032
162.040 .047
Redshifts
Galaxies
031.221
141.003
158.007 .014 .172
.182 .231
160.018
162.039 .040 .043
.048 .050
Redshifts
Quasars
141.015 .017 .026
.030 .064 .065
.081 .083 .084
.086 .114 .120
.133 .161
160.018
162.048
Redshifts
Quasi-Stellar Objects
141.013 .021 .085
.106 .107 .108
.130 .148 .150
Reflection Nebulae
131.002
132.000 .008
Reflection Nebulae
Exciting Stars
132.002
Reflection Nebulae
Infrared Radiation
132.009
Reflection Nebulae
Polarization
132.062
Refraction
Earth Atmosphere
082.000
Relativistic
Astrophysics
066.000
Relativistic Stars
Rotation
065.136
066.038 .053
Relativistic Stars
Stability
065.092 .109
066.001 .040
Relativity Theory
Tests
066.041
Resonances
Jupiter Satellites
042.033
Resonances
Minor Planets
098.016 .020 .043
Rotating Stars
Distortion
065.134
Rotating Stars
Instabilities
065.116
Rotating Stars
Models
116.002
Rotating Stars
Pulsations
065.121
Rotating Stars
Stability
065.023 .121
Rotating Stars
Turbulence
065.122
Rotation
Be Stars
116.012

Rotation
Binaries
117.057
Rotation
Close Binaries
117.056 .060 .061
.062
Rotation
Early-Type Stars
065.106
114.049 .071
116.002 .005 .006
Rotation
Earth
044.000
Rotation
Galaxies
151.060 .079
155.023
158.030 .043 .046
.051 .061 .062
.127 .137 .150
.178 .234 .235
161.001
162.026
Rotation
Jupiter Satellites
099.223
Rotation
Magnetic Stars
062.031
116.004 .007
Rotation
Mercury
092.013
Rotation
Minor Planets
098.028 .029 .050
Rotation
Moon
094.002 .006 .008
.011 .023 .106
.170
Rotation
Neutron Stars
065.040 .136
066.038
141.317
Rotation
Peculiar A Stars
114.363
Rotation
Polytropes
021.011
065.001 .061
Rotation
Pulsars
141.333 .344
Rotation
Relativistic Stars
065.136
066.038 .053
Rotation
Stars
116.000
Rotation
Stellar Evolution
065.077 .078 .079
.080 .082 .086
.120 .174
Rotation
Stellar Interiors
065.113
Rotation
Sun
080.000
Rotation
Universe
162.019
Rotation
Venus Atmosphere
093.029
Rotation
White Dwarfs
065.005
066.038
126.006 .026
RR Lyrae Stars
065.089 .091
120.002
122.047 .050 .054
.055 .056 .076
.080 .081 .141
.171
123.017
RR Lyrae Stars
Galaxies
122.166
RR Lyrae Stars
Globular Clusters
122.049 .073 .162
.217 .223
123.012
154.003 .021
RR Lyrae Stars
Magellanic Clouds
122.052
RR Lyrae Stars
Metal Abundances
122.073 .223
RR Lyrae Stars
Photometry
122.057
RV Tauri Stars
HR Diagrams
122.102
RV Tauri Stars
Luminosities
122.083

Satellite Geodesy
046.000
Saturn
100.000
Saturn
Albedo
100.005
Saturn
Astrolabe Observations
041.001
Saturn
Brightness Temperature
100.025
Saturn
Infrared Spectra
100.011
Saturn
Magnetosphere
099.029
Saturn
Photometry
100.005
Saturn
Positions
041.004
Saturn
Radio Radiation
099.014
Saturn
Rings
053.019
100.001 .002 .003
.009 .011 .012
.014 .017 .018
.019 .020 .021
.022 .023 .024
.025 .026 .027
.028 .029
Saturn Atmosphere
Chemical Composition
099.012
Saturn Atmosphere
Molecules
100.008
Saturn Atmosphere
Spectra
100.008 .013
Saturn Satellites
100.200
Saturn Satellites
Diameters
100.208
Saturn Satellites
Infrared Spectra
100.203
Saturn Satellites
Light Variations
099.217
Saturn Satellites
Lunar Occultations
100.208
Saturn Satellites
Motion
100.205
Saturn Satellites
Photometry
100.201
Saturn Satellites
Positions
100.202
Saturn Satellites
Titan
099.217
100.203
Scattering
063.000
Scattering
Compton Scattering
061.042
Scattering
Interstellar Matter
131.042
Scattering
Jupiter Atmosphere
091.004
Scattering
Planetary Atmospheres
091.004 .007 .011

Scattering
Plasma
062.026
063.040
Scattering
Stellar Atmospheres
063.028
Scattering
Stellar Photospheres
064.066
Scattering
Thomson Scattering
063.028 .040
Scattering
Venus Atmosphere
093.009
Scintillation
Earth Atmosphere
082.000
Scintillations
Interplanetary Matter
106.020 .021 .045
.047 .057 .070
Scintillations
Interstellar Matter
063.025 .037
131.020 .111
Scintillations
Pulsars
131.020 .111
141.352
Scintillations
Radio Sources
106.045 .047
Seyfert Galaxies
141.062
158.013 .019 .193
.233
Seyfert Galaxies
Emission Lines
022.016
158.004 .011 .015
.016 .197
Seyfert Galaxies
Helium Abundance
022.016
Seyfert Galaxies
Infrared Photometry
158.035
Seyfert Galaxies
Internal Motions
158.054
Seyfert Galaxies
Nuclei
141.078
158.004 .015 .016
.053 .066 .125
.135 .143 .232
.239
Seyfert Galaxies
Photometry
158.015
Seyfert Galaxies
Polarization
158.239
Seyfert Galaxies
Radio Radiation
158.143
Seyfert Galaxies
Spectra
158.005 .135 .141
Seyfert Galaxies
X Rays
142.004
158.143
Shell Stars
114.319 .338 .344
119.007
Shock Waves
Chromosphere
071.027
Shock Waves
Galaxy
155.028
Shock Waves
Interplanetary Matter
073.119
106.066 .077
Shock Waves
Interstellar Gas
155.028
Shock Waves
Interstellar Matter
125.043
131.138
Shock Waves
Solar Atmosphere
076.057
Shock Waves
Solar Flares
073.017
074.093 .102
Shock Waves
Solar Wind
073.119
074.019 .059 .084
.093 .102 .132
Shock Waves
Stellar Atmospheres
064.017
Shock Waves
Stellar Systems
155.028
Shock Waves
Supernova Remnants
125.002
Shock Waves
Supernovae
125.034 .037 .043
.068
Site Testing
082.000
Societies
010.000
Solar Activity
072.000
Solar Activity
Catalogues
075.005 .006
Solar Activity
Cycles
072.043 .050 .061
.067 .082
Solar Activity
Forecasts
072.040 .041 .053
.059
Solar Atmosphere
Acoustic Waves
080.023
Solar Atmosphere
Convection
080.002 .017 .038
Solar Atmosphere
Dynamics
080.055
Solar Atmosphere
Element Abundances
071.005
Solar Atmosphere
Emission Lines
080.020
Solar Atmosphere
Line Formation
080.052 .057
Solar Atmosphere
Microturbulence
071.062
Solar Atmosphere
Models
064.060 .061
080.021
Solar Atmosphere
Nuclear Reactions
080.076
Solar Atmosphere
Opacities
080.039
Solar Atmosphere
Oscillations
071.033 .042
080.021 .034 .058
.061
Solar Atmosphere
Radiative Transfer
063.026
Solar Atmosphere
Shock Waves
076.057
Solar Atmosphere
Structure
080.061
Solar Atmosphere
Wave Propagation
080.019
Solar Chromosphere
See Chromosphere
Solar Constant
071.032
080.050
Solar Corona
074.000
Solar Corona
Active Regions
073.008
074.002 .008 .009
.047 .070 .075
.076 .104
Solar Corona
Brightness
074.026
Solar Corona
Cosmic Rays
078.004
Solar Corona
Electron Densities
066.108
074.061
Solar Corona
Element Abundances
074.018 .065

Solar Corona
Emission Lines
022.097
074.007 .065 .066 .070 .091 .092 .107 .108
Solar Corona
Extreme UV
022.057
074.005 .045 .056 .067
076.050 .051
Solar Corona
Heating
074.003 .005 .106
Solar Corona
Holes
073.002
074.004 .006 .045 .060 .078 .079 .080 .098 .108
076.050 .051
077.027
Solar Corona
Instability
074.071
Solar Corona
Interferometry
074.121
Solar Corona
Ionization
074.126
Solar Corona
Line Formation
073.004
074.098 .100
Solar Corona
Line Intensities
074.107 .120
Solar Corona
Line Profiles
074.049 .108 .112
076.050
Solar Corona
Magnetic Fields
022.061
074.003 .042 .060 .067 .077 .081 .083 .105 .109
Solar Corona
Models
074.103 .104
Solar Corona
Photometry
074.021 .110
Solar Corona
Plasma Waves
074.071
Solar Corona
Polarization
022.061
074.016 .052 .110
077.017
Solar Corona
Radio Radiation
077.005 .017 .036
Solar Corona
Spectra
071.039
074.004 .068 .069 .100 .101
Solar Corona
Spectra
077.017
Solar Corona
Spectrophotometry
074.086
Solar Corona
Streamers
077.005
Solar Corona
Structure
073.070
074.020 .021 .048 .055 .061 .067 .068 .110 .111
Solar Corona
Temperatures
074.012 .046 .049
Solar Corona
Velocities
074.049
Solar Corona
X Rays
074.003 .047 .073 .074 .081 .105 .128
076.007 .009
Solar Cosmic Rays
078.000
Solar Cosmic Rays
Electrons
078.002 .022
Solar Cosmic Rays
Element Abundances
078.001
Solar Cosmic Rays
Interplanetary Matter
078.030
Solar Cosmic Rays
Nuclei
078.001 .038
Solar Cosmic Rays
Particles
078.034 .036
Solar Cosmic Rays
Propagation
078.021 .025 .030 .037
Solar Cosmic Rays
Protons
078.002 .003 .009 .010 .012 .013 .014 .031 .032 .033
143.034
Solar Cosmic Rays
Variations
078.035
106.031
Solar Eclipse
1965 May 30
079.106
Solar Eclipse
1968 September 22
079.107
Solar Eclipse
1970 March 7
079.102
Solar Eclipse
1972 July 10
079.101
Solar Eclipse
1973 June 30
079.100
Solar Eclipse
1973 December 24
079.104
Solar Eclipse
1975 May 11
079.103
Solar Eclipse
1975 November 3
079.108
Solar Eclipse
1976 April 29
079.105
Solar Eclipse
1976 October 23
079.109
Solar Eclipses
079.000
Solar Flares
073.000
122.008
Solar Flares
Balmer Lines
073.092
Solar Flares
Brightness Variations
073.117
Solar Flares
Cosmic Rays
073.156 .160 .161
078.011 .021 .025
Solar Flares
Electrons
073.028
Solar Flares
Element Abundances
073.018
Solar Flares
Emission Lines
073.141
Solar Flares
Extreme UV
073.006 .007 .022 .079 .158
076.002 .004
Solar Flares
Gamma Rays
073.015 .016 .114
076.018 .021 .040 .043 .048 .055
Solar Flares
Infrared Radiation
073.142
Solar Flares
Infrared Spectra
073.097
Solar Flares
Line Intensities
073.139
Solar Flares
Line Profiles
073.005
Solar Flares
Magnetic Fields
062.046 .050
073.029
Solar Flares
Models
073.019 .021 .027

Solar Flares
  Models
    073.030 .072
    080.036
Solar Flares
  Nuclear Reactions
    061.016
Solar Flares
  Origin
    073.032
Solar Flares
  Particles
    078.030
    106.055
Solar Flares
  Positrons
    073.108
Solar Flares
  Protons
    073.023 .028 .076
      .114 .119 .138
Solar Flares
  Radio Radiation
    073.017 .022 .104
      .156
    077.004 .060
Solar Flares
  Shock Waves
    073.017
    074.093 .102
Solar Flares
  Spectra
    073.009 .069
Solar Flares
  Structure
    073.140 .146
Solar Flares
  Temperatures
    073.010
Solar Flares
  White Light
    073.080
Solar Flares
  X Rays
    062.025
    073.001 .003 .007
      .008 .009 .011
      .012 .013 .014
      .017 .022 .072
      .097 .103 .104
      .146
    076.014 .016 .017
      .018 .024 .026
      .027 .046 .047
Solar Gamma Rays
    076.000
Solar Gamma Rays
  Bursts
    076.018
Solar Interior
  Oscillations
    080.032 .063
Solar Limb
    071.040
    073.157
    080.071
Solar Magnetic Fields
    062.002
    076.056
    080.001 .002 .009
      .020 .033 .036
Solar Magnetic Fields
    080.056 .065 .066
      .070 .074 .078
Solar Models
  Black Holes
    080.037
Solar Models
  Stability
    065.099 .100
    080.032
Solar Neighborhood
  Chemical Evolution
    114.006
Solar Neighborhood
  Mass Density
    115.011
Solar Neutrinos
    022.087
    065.021
    080.003 .004 .037
      .040 .041 .059
      .062 .063 .073
Solar Oblateness
    071.002
    080.005 .007 .018
Solar Patrol
    075.000
Solar Photosphere
  See Photosphere
Solar Radio Bursts
    074.078
    077.001 .004 .005
      .007 .009 .013
      .016 .017 .019
      .021 .022 .023
      .024 .030 .032
      .035 .036 .037
      .038 .039 .040
      .043 .046 .047
      .048 .049 .051
      .052 .053 .054
      .058 .059 .060
    085.001 .003
Solar Radio Bursts
  Polarization
    077.064 .065
Solar Radio Radiation
    077.000
Solar Radio Radiation
  Fine Structure
    077.044 .045
Solar Radio Radiation
  Fluctuations
    077.062
Solar Radio Radiation
  Magnetic Fields
    077.041
Solar Radio Radiation
  Variations
    077.008 .033 .034
Solar Rotation
    080.000
Solar Spectrum
    071.000
Solar Spectrum
  Absorption Lines
    071.045 .065 .069
      .071
Solar Spectrum
  Element Abundances
    071.021 .051
Solar Spectrum
  Emission Lines
    071.046
Solar Spectrum
  Extreme UV
    073.121
    076.001 .002 .058
Solar Spectrum
  Forbidden Lines
    071.021
Solar Spectrum
  Fraunhofer Lines
    071.029 .033 .045
      .064
Solar Spectrum
  Line Blocking
    071.049
Solar Spectrum
  Line Formation
    063.010
Solar Spectrum
  Line Intensities
    022.034
    071.053
Solar Spectrum
  Line Profiles
    063.028
    071.006 .046 .062
      .064
    073.125
Solar Spectrum
  Recombination Lines
    080.066
Solar Spectrum
  UV
    071.008
    074.009
Solar System
  See Planetary System
    107.000
Solar UV Radiation
    076.000
Solar Wind
    074.000
Solar Wind
  Alfven Waves
    074.022 .053 .096
      .113 .115 .130
Solar Wind
  Cosmic Rays
    143.033 .054 .060
Solar Wind
  Disturbances
    074.050
Solar Wind
  Electrons
    074.114
Solar Wind
  Element Abundances
    074.072
Solar Wind
  Energy Spectra
    074.043
Solar Wind
  Expansion
    074.088
Solar Wind
  Geomagnetic Field
    084.210 .213 .221

Solar Wind
Heating
074.125
Solar Wind
Helium Abundance
074.051
Solar Wind
Interplanetary Matter
074.001 .011 .015
.059 .088
078.004
106.013 .020 .070
Solar Wind
Interplanetary Space
106.053
Solar Wind
Jupiter
099.019 .020 .048
Solar Wind
Magnetic Fields
074.027 .060 .062
.095 .117 .119
.122
Solar Wind
Magnetosphere
074.019
084.202
Solar Wind
Mars Atmosphere
097.048
Solar Wind
Mercury
092.008
Solar Wind
Models
074.131
106.053
Solar Wind
Moon
074.094
094.180
Solar Wind
Protons
074.011 .014 .027
.062
Solar Wind
Shock Waves
073.119
074.019 .059 .084
.093 .102 .132
Solar Wind
Streams
074.085
Solar Wind
Structure
074.001 .044
Solar Wind
Turbulence
074.113
Solar Wind
Variations
074.099 .123
Solar Wind
Velocities
074.013 .028 .045
.057 .088
Solar X Rays
076.000
080.038
Solar X Rays
Active Regions
076.023
Solar X Rays
Bursts
062.032
073.003 .011
076.003 .004 .012
.013 .014 .016
.017 .021 .024
.025 .027 .039
.044 .052
077.037
Solar X Rays
Polarization
076.016
Solar X Rays
Spectra
076.005 .006 .007
.008 .009 .016
.053 .054
Solar X Rays
Variations
076.020
Solar-Terrestrial
Relations
085.000
Solar-Type Stars
Companions
117.072
Source Functions
Sunspots
072.034
Space Instrumentation
032.500
Space Motions
112.000
Space Motions
Giants
155.050
Space Motions
Subdwarfs
126.003
Space Probes
Observations
055.000
Space Vehicles
Navigation
052.000
Spaceflight
051.000
Spectral Classification
113.033
114.001 .026 .045
.047 .050
153.038
Spectral Types
Eclipsing Variables
121.027
Spectral Types
Galactic Clusters
153.032
Spectral Types
Peculiar Stars
114.076
Spectral Types
Wolf-Rayet Stars
153.013
Spectrographs
031.005
034.020 .092 .108
Spectroheliograms
Extreme UV
073.070
Spectrometers
032.537
034.028 .029 .119
.156 .162
Spectrophotometers
034.022
Spectrophotometry
031.293
114.088
Spectrophotometry
A Stars
114.071
Spectrophotometry
Be Stars
114.373
Spectrophotometry
Bright Stars
114.116
Spectrophotometry
Early-Type Stars
114.092 .345
Spectrophotometry
Eclipsing Variables
114.316
Spectrophotometry
Galaxies
158.232 .233
Spectrophotometry
Helium-Rich Stars
114.394
Spectrophotometry
M Giants
114.107
Spectrophotometry
M Stars
114.093
Spectrophotometry
Moon Surface
094.138 .146
Spectrophotometry
Peculiar A Stars
114.087 .372
Spectrophotometry
Planetary Nebulae
133.023
Spectrophotometry
Quasi-Stellar Objects
141.060 .169
Spectrophotometry
Solar Corona
074.086
Spectrophotometry
Supernova Remnants
125.067
Spectrophotometry
Supernovae
125.067
Spectrophotometry
Uranus
101.009
Spectrophotometry
UV
114.090
Spectrophotometry
Venus
093.049

Spectrophotometry
White Dwarfs
126.007
Spectroscopic Binaries
119.000
Spectroscopic Binaries
Black Holes
119.008
Spectroscopic Binaries
Line Identifications
119.004
Spectroscopic Binaries
Line Profiles
119.006
Spectroscopic Binaries
Orbits
118.020
119.005 .009 .014
.019 .021 .023
Spectroscopic Binaries
Periods
119.020
Spectroscopic Binaries
Photometry
119.001 .015 .020
Spectroscopic Binaries
Radial Velocities
119.006 .007 .014
.015 .021
Spectroscopic Binaries
Unseen Companions
119.008
Spectroscopic Binaries
X-Ray Sources
142.115
Spectroscopy
031.280
114.000
Spectrum Variables
113.036
114.301 .339 .355
.366
Spherules
105.031 .032
Spicules
073.071 .110 .111
.112 .113 .122
.150
074.103
076.051
Spiral Arms
155.008 .009
Spiral Arms
Stellar Associations
155.002
Spiral Structure
141.080
151.043 .049 .051
.052 .053 .054
.055 .056 .061
155.002 .034 .036
.037 .038
158.147 .152 .155
.156 .157 .161
.162 .163 .168
.201
Stability
Close Binaries
117.014 .060 .061
Stability
Magnetic Fields
062.039
Stability
Main-Sequence Stars
065.108
Stability
Periodic Orbits
042.102
Stability
Relativistic Stars
065.092 .109
066.001 .040
Stability
Rotating Stars
065.023 .121
Stability
Solar Models
065.099 .100
080.032
Stability
Stellar Evolution
065.029 .062 .085
.089 .094 .097
.098
Stability
Stellar Interiors
065.127
Stability
Stellar Models
065.005 .008 .025
.033 .078 .090
.093 .096 .114
.138
Stability
Stellar Systems
151.002 .012 .026
.089 .094
Stability
Supermassive Stars
065.035
Star Catalogues
Astrolabe Observations
041.014
Star Catalogues
Colors
113.025
Star Catalogues
Comparisons
041.040 .058
Star Catalogues
Corrections
041.043 .044
098.065
Star Catalogues
Declinations
041.025 .042 .058
.069
Star Catalogues
Parallaxes
041.035
Star Catalogues
Photometric
113.034 .035
Star Catalogues
Positions
041.020 .040
112.013
Star Catalogues
Proper Motions
041.024 .029 .069
Star Catalogues
Proper Motions
112.013
Star Catalogues
Right Ascensions
041.015 .036 .058
Star Clusters
Binaries
151.064
Star Clusters
C-M Diagrams
154.004
Star Clusters
Dynamics
151.018
Star Clusters
Interstellar Reddening
131.198
Star Clusters
Magellanic Clouds
154.004
Star Clusters
Relativistic
151.029
Star Clusters
Stellar Evolution
065.171
Star Clusters
Surveys
113.044
Star Formation
065.012 .020 .024
.103 .105 .115
113.002
122.165
131.005 .034 .501
Star Formation
Galaxies
131.535 .545
158.033 .057 .126
.161 .163
Star Formation
Interstellar Matter
065.118
131.056 .121 .130
.138
Stars
Diameters
115.000
Stars
Figure
116.000
Stars
Luminosities
115.000
Stars
Magnetic Fields
116.000
Stars
Magnitudes
113.000
Stars
Masses
115.000
Stars
Parallaxes
111.000
Stars
Rotation
116.000

Stars
Temperatures
114.000
Stellar Associations
152.000
Stellar Associations
Absorption
152.001
Stellar Associations
Ages
152.005
Stellar Associations
Galactic Distribution
155.002
Stellar Associations
Galaxies
158.110
Stellar Associations
Infrared Photometry
113.002 .014
Stellar Associations
Spiral Arms
155.002
Stellar Atmospheres
064.000
Stellar Atmospheres
Circumstellar Shells
113.010
Stellar Atmospheres
Convection
064.040
065.037
114.048
Stellar Atmospheres
Dust Grains
064.016
Stellar Atmospheres
Dynamics
064.005
Stellar Atmospheres
Element Abundances
061.015
064.013 .019
114.362
122.206
Stellar Atmospheres
Fine Analyses
064.064
Stellar Atmospheres
Helium Abundance
064.045
Stellar Atmospheres
Line Broadening
063.007
Stellar Atmospheres
Line Formation
063.039
064.027
114.048 .083
Stellar Atmospheres
Line Profiles
063.028
064.028
Stellar Atmospheres
Mass Loss
064.017
141.061 .625
Stellar Atmospheres
Models
064.018 .021 .026
.042 .060 .061
Stellar Atmospheres
Models
064.062 .064
113.009
114.353 .374
126.011
Stellar Atmospheres
Molecules
114.325
Stellar Atmospheres
Polarization
063.046
Stellar Atmospheres
Radiative Transfer
063.014 .015
064.018 .027 .028
.052
122.115
Stellar Atmospheres
Scattering
063.028
Stellar Atmospheres
Shock Waves
064.017
Stellar Atmospheres
Structure
064.060
Stellar Atmospheres
Temperatures
114.374
Stellar Atmospheres
Turbulence
122.199
Stellar Atmospheres
X Rays
064.048
073.002
Stellar Chains
152.003 .004
Stellar Chromospheres
064.018 .028
113.010
114.020 .318 .350
.370
Stellar Chromospheres
Line Profiles
064.027
Stellar Coronae
062.033
064.002 .048
113.051
114.318 .350 .351
142.134
Stellar Coronae
X Rays
142.045
Stellar Envelopes
064.000
Stellar Envelopes
Convection
064.014 .070
Stellar Envelopes
Pulsations
065.091
122.114
Stellar Envelopes
Spectra
114.367
Stellar Evolution
065.000
Stellar Evolution
Black Holes
080.037
Stellar Evolution
Carbon Burning
065.113 .128 .145
.172 .173
Stellar Evolution
Chemical Composition
065.030
Stellar Evolution
Circulation
065.081 .087
Stellar Evolution
Convection
065.026 .060 .080
Stellar Evolution
Crystallization
126.005
Stellar Evolution
Galactic Clusters
065.060
Stellar Evolution
Globular Clusters
154.021 .022 .037
Stellar Evolution
Helium Burning
065.033 .102 .137
.149 .173
Stellar Evolution
Magnetic Fields
065.084 .085
Stellar Evolution
Mass Loss
061.031
064.033 .034
065.032
Stellar Evolution
Nuclear Reactions
065.008
Stellar Evolution
Nucleosynthesis
061.004
Stellar Evolution
Opacities
065.026 .133 .139
Stellar Evolution
Oscillations
065.021 .089 .095
.096 .097 .098
.099 .100 .138
117.013
Stellar Evolution
Pre-Main Sequences
065.118
Stellar Evolution
Pulsations
064.036
065.009 .091 .093
122.114 .117 .118
Stellar Evolution
Rotation
065.077 .078 .079
.080 .082 .086
.120 .174
Stellar Evolution
Stability
065.029 .062 .085
.089 .094 .097
.098

Stellar Evolution
Star Clusters
065.171
Stellar Flares
122.163
Stellar Interiors
Convection
065.004 .057 .062
.101 .111 .112
.132 .176 .177
Stellar Interiors
Element Abundances
065.003 .036
Stellar Interiors
Evolution
065.020
Stellar Interiors
Magnetic Fields
065.140
Stellar Interiors
Models
065.007
Stellar Interiors
Neutrinos
065.031 .117 .143
Stellar Interiors
Nuclear Reactions
061.036
065.010 .031 .058
.137 .151 .182
.186
Stellar Interiors
Nucleosynthesis
065.018
Stellar Interiors
Rotation
065.113
Stellar Interiors
Stability
065.127
Stellar Interiors
Urca Processes
065.062 .117 .145
Stellar Models
Evolution
065.071 .149 .173
Stellar Models
Magnetic Fields
065.002 .039 .086
.087 .088
Stellar Models
Opacities
061.037
065.036 .133
Stellar Models
Oscillations
061.034
Stellar Models
Pulsations
065.109
Stellar Models
Stability
065.005 .008 .025
.033 .078 .090
.093 .096 .114
.138
Stellar Occultations
Jupiter
099.083
Stellar Occultations
Mars
097.053
Stellar Occultations
Planetary Atmospheres
091.025
Stellar Orbits
Galaxies
151.023 .038 .040
.043 .044
Stellar Orbits
Galaxy
155.013 .042
Stellar Photospheres
064.020
071.030
114.320
Stellar Photospheres
Scattering
064.066
Stellar Populations
122.165
158.139 .175
Stellar Populations
Galactic Halo
155.007
Stellar Rings
152.003
Stellar Spectra
114.000
Stellar Spectra
Element Abundances
114.025
Stellar Spectra
Forbidden Lines
114.351
Stellar Spectra
Galactic Clusters
153.027 .038
Stellar Spectra
Individual Stars
114.300
Stellar Spectra
Line Profiles
031.226
065.122
114.302 .374
Stellar Spectra
MK Types
153.027
Stellar Spectra
Molecules
114.085
Stellar Spectra
UV
114.335 .350 .353
119.003
Stellar Structure
065.000
Stellar Systems
Binaries
117.026 .027
151.014
Stellar Systems
Collisions
151.025
Stellar Systems
Dynamics
151.000
Stellar Systems
Evolution
151.009 .010 .014
.019 .045 .089
Stellar Systems
Formation
062.064
Stellar Systems
Kinematics
151.000
Stellar Systems
Models
151.032 .071 .086
Stellar Systems
Oscillations
151.086
Stellar Systems
Relativistic
151.025
Stellar Systems
Secular Stability
151.037
Stellar Systems
Shock Waves
155.028
Stellar Systems
Stability
151.002 .012 .026
.089 .094
Stellar Systems
Star Encounters
151.019
Stellar Systems
Structure
151.017
Stellar Velocities
Galaxy
155.030
Stellar Winds
062.016
064.015 .035 .039
.049 .053
065.040
116.012
131.110 .501
142.134
Stellar Winds
Binaries
119.002 .003
Subdwarfs
126.000
Subdwarfs
B Stars
126.003
Subdwarfs
Space Motions
126.003
Subdwarfs
Spectra
126.003 .011
Subgiants
114.109
Sun
Active Regions
072.067
073.063 .069 .070
.095 .096 .100
.106 .118
074.056
076.011
077.003 .006 .028

Sun
Active Regions
078.026
Sun
Brightness Distribution
077.012
Sun
Brightness Temperature
071.037
077.042 .050
Sun
Differential Rotation
080.016 .017 .035
Sun
Figure
080.000
Sun
Interior
080.000
Sun
Limb Darkening
031.214
Sun
Models
080.059
Sun
Rotation
080.000
Sun
Transition Regions
071.053
074.103 .106
Sundials
035.012 .018
Sunspot Groups
072.047 .052 .058
.064 .068 .069
.080
073.036 .095
Sunspots
072.000
Sunspots
Alfven Waves
072.009 .079
Sunspots
Brightness Variations
073.117
Sunspots
Distribution
072.060
Sunspots
Formation
062.053
Sunspots
Internal Motions
072.001
Sunspots
Magnetic Fields
072.004 .026 .035
.036 .042 .045
.046 .049
073.076
Sunspots
Models
072.007 .009 .057
Sunspots
Penumbrae
072.019 .044 .056
.075
Sunspots
Polarization
063.046
Sunspots
Radio Radiation
077.026
Sunspots
Source Functions
072.034
Sunspots
Spectra
072.039 .048
114.079
Sunspots
Statistics
072.052
Sunspots
Umbrae
072.004 .008 .011
.034 .035 .044
.055 .066 .075
Supergiants
Colors
159.004
Supergiants
Diameters
115.018
Supergiants
HR Diagrams
065.135
Supergiants
Line Profiles
114.383
Supergiants
Lists
114.105
Supergiants
Magellanic Clouds
113.042
114.392
159.004
Supergiants
Mass Loss
064.035 .072
Supergiants
Models
065.135 .170
Supergiants
Spectra
114.321 .322 .368
Supergiants
UV Spectra
064.071
Supermassive Stars
Stability
065.035
Supernova in NGC 1325
125.110
Supernova in NGC 2487
125.107
Supernova in NGC 3583
125.108
Supernova in NGC 4165
125.102
Supernova in NGC 4254
125.101
Supernova in NGC 4298
125.109
Supernova in NGC 4414
125.103
Supernova in NGC 5253
125.100
Supernova in NGC 5457
125.105
Supernova in NGC 7619
125.104
Supernova in NGC 7723
125.106
Supernova Remnants
125.000
Supernova Remnants
Diameters
125.009
Supernova Remnants
Distribution
125.050
Supernova Remnants
Evolution
125.017 .033
Supernova Remnants
Expansion
125.002
Supernova Remnants
Forbidden Lines
125.018
Supernova Remnants
Galaxy
125.033 .044 .046
Supernova Remnants
Interstellar Matter
125.001 .033 .050
Supernova Remnants
Magellanic Clouds
159.003
Supernova Remnants
Models
125.034
Supernova Remnants
Polarization
125.010 .027
Supernova Remnants
Pulsars
125.012
141.338
Supernova Remnants
Radio Radiation
125.001 .013 .014
.027 .035 .036
.039 .044 .045
.046
141.056 .146
Supernova Remnants
Shock Waves
125.002
Supernova Remnants
Spectrophotometry
125.067
Supernova Remnants
X Rays
125.005 .007 .008
.034
142.064 .173
Supernova Remnants
X-Ray Sources
125.016 .017 .018
142.059 .139
Supernovae
125.000
Supernovae
Binaries
125.006

Supernovae
Cosmic Rays
143.081
Supernovae
Distances
125.048
Supernovae
Envelopes
125.037 .049 .051
Supernovae
Explosions
125.029 .042
Supernovae
Formation
065.032
Supernovae
Galaxies
125.026 .048
158.057
Supernovae
Galaxy
125.054
Supernovae
Interstellar Matter
131.190
Supernovae
Light Curves
125.025 .066
141.342
Supernovae
Models
065.113
125.011 .025 .029
Supernovae
Nucleosynthesis
124.010
125.037
Supernovae
Presupernovae
125.029
Supernovae
Shells
125.032
Supernovae
Shock Waves
125.034 .037 .043
.068
Supernovae
Spectra
125.049
Supernovae
Spectrophotometry
125.067
Symbiotic Stars
114.065
119.013
122.041 .043 .044
.045 .046
Symposia Proceedings
012.000
Symposia Reports
011.000
Synchrotron Radiation
022.076
061.032
122.135
Synchrotron Radiation
Radio Sources
062.011
141.128 .129
142.060
Synchrotron Radiation
Radio Sources
143.018

T Tauri Stars
064.011 .017 .047
065.118
113.041
114.305 .349
122.016 .018 .019
.072 .122
123.034
131.092
132.004 .029 .031
.049 .061
Teaching In Astronomy
014.000
Telescopes
032.000
Telescopes
Infrared Telescopes
031.048
Telescopes
Schmidt Telescopes
031.031 .048
032.037
Television Systems
031.286
034.132 .137
Temperatures
Jupiter Atmosphere
099.057 .058 .069
.083
Temperatures
Mars Atmosphere
091.044
097.003
Temperatures
Moon
094.143 .151
Temperatures
Photosphere
071.002 .047
Temperatures
Prominences
073.085
Temperatures
Solar Corona
074.012 .046 .049
Temperatures
Solar Flares
073.010
Temperatures
Stars
114.000
Temperatures
Stellar Atmospheres
114.374
Temperatures
Uranus
101.002
Temperatures
Venus Atmosphere
093.005 .006
Temperatures
White Dwarfs
126.028
Three-Body Problem
042.024 .035 .037
.045 .046 .056
Three-Body Problem
042.073 .076 .077
.080 .081 .092
.099 .105 .106
.107
053.013
117.022 .030 .054
.055
Three-Body Problem
Restricted
042.020 .038 .039
.041 .042 .049
.053 .055 .061
.075 .086 .087
.094 .098 .102
Time
044.000
Titan
099.217
100.203
Transit Circles
032.047 .048 .055
Transition Probabilities
022.003 .005 .006
.032 .057 .059
.060 .061 .063
.064 .073 .074
.082 .100 .107
.125
074.101
076.058
Transplutonian Planets
101.000
Turbulence
Chromosphere
073.031
Turbulence
Jupiter Atmosphere
099.050
Turbulence
Line Broadening
071.006
Turbulence
Photosphere
063.035
071.004
Turbulence
Planetary Atmospheres
091.004
Turbulence
Plasma
062.050
Turbulence
Rotating Stars
065.122
Turbulence
Solar Wind
074.113
Turbulence
Stellar Atmospheres
122.199
Turbulence
Venus Atmosphere
093.035
Two-Body Problem
042.022 .093

U Geminorum Stars
122.039 .040
123.016

U Geminorum Stars
124.017
142.028
Universe
Density
162.005 .019
Universe
Early History
162.023 .030
Universe
Friedmann Universe
162.032 .033 .035
Universe
Lithium Abundance
061.061
Universe
Rotation
162.019
Universe Evolution
162.000
Universe Structure
162.000
Uranus
101.000
Uranus
Atmosphere
101.010
Uranus
Satellites
101.001 .003
Uranus
Spectra
101.010
Uranus
Spectrophotometry
101.009
Uranus
Temperatures
101.002
Uranus System
Formation
101.001
UV Objects
114.118
UV Photometry
Eclipsing Variables
121.090
UV Radiation
Galaxies
158.036
UV Radiation
Interstellar Matter
131.114 .191
UV Radiation
Solar
076.000
UV Spectra
B Stars
114.307
UV Spectra
B Supergiants
114.119
UV Spectra
Beta Cephei Stars
122.096
UV Spectra
Early-Type Stars
114.003 .005 .018
.092 .324 .345
UV Spectra
Planetary Nebulae
133.031
UV Spectra
Supergiants
064.071
UV Spectra
Variables
122.224
UV Spectra
Wolf-Rayet Stars
114.313
UV Spectrophotometry
B Stars
114.111
UV Spectrophotometry
Close Binaries
114.375
UV Stars
131.089
UV Survey
Magellanic Clouds
159.010

Variables
Andromeda Nebula
114.322 .323
Variables
Galactic Clusters
122.031
Variables
Globular Clusters
122.048 .053 .131
154.003 .024
Variables
Irregular
122.017 .018 .019
.021 .026 .038
142.067
Variables
Long Period
022.056
122.003 .086 .103
.183
Variables
Magellanic Clouds
114.317
122.001 .052 .171
Variables
OH Emission
122.003
Variables
Periods
117.064
Variables
Photometry
122.085 .086
Variables
Polarization
131.097
Variables
Proper Motions
112.004
Variables
Pulsations
065.089
Variables
Radial Velocities
122.183
Variables
Radio Radiation
122.003
Variables
Semiregular
122.025 .026 .079
Variables
Short Period
122.034
Variables
UV Spectra
122.224
Variables Catalogues
120.000
Variables Eclipsing
121.000
Variables Ephemerides
120.000
Variables Intrinsic
122.000
Variables Observations
123.000
Velocities
Chromosphere
073.068
Velocities
Nearby Dwarfs
112.011
Velocities
OB Stars
155.037
Velocities
Photosphere
071.025 .029 .059
.060
080.033
Velocities
Pulsars
141.331 .339
Velocities
Solar Corona
074.049
Velocities
Solar Wind
074.013 .028 .045
.057 .088
Venus
093.000
Venus
Brightness Distribution
093.014 .015
Venus
Dust
093.033 .034
Venus
Ionosphere
093.004 .022 .052
Venus
Limb Darkening
093.021 .037
Venus
Photometry
093.046
Venus
Polarization
093.044
Venus
Positions
041.004 .060 .061

Venus
  Radar Echoes
    093.024
Venus
  Spectrophotometry
    093.049
Venus
  Surface
    053.022
    093.011 .023 .024
Venus Atmosphere
    093.004 .020 .032
      .054 .055 .057
Venus Atmosphere
  Carbon Dioxide
    093.002
    097.043
Venus Atmosphere
  Circulation
    093.027 .029
Venus Atmosphere
  Clouds
    093.003 .007 .014
      .015 .031 .038
      .040 .042 .043
      .045 .047 .048
      .050 .053
Venus Atmosphere
  Convection
    093.010
Venus Atmosphere
  Dynamics
    093.026
Venus Atmosphere
  Helium Emission
    093.025
Venus Atmosphere
  Infrared Spectra
    093.018
Venus Atmosphere
  Models
    093.003 .015 .016
      .021 .028 .051
Venus Atmosphere
  Polarization
    093.041
Venus Atmosphere
  Radiative Transfer
    093.019
Venus Atmosphere
  Rotation
    093.029
Venus Atmosphere
  Scattering
    093.009
Venus Atmosphere
  Structure
    093.036
Venus Atmosphere
  Temperatures
    093.005 .006
Venus Atmosphere
  Turbulence
    093.035
Virgo
  Clusters of Galaxies
    160.004 .046
Visual Binaries
    118.000
Visual Binaries
  Masses
    118.005 .016
Visual Binaries
  Orbits
    118.005 .010 .015
      .016 .020 .030
      .031
Visual Binaries
  Parallaxes
    118.004 .005 .010
      .015
Visual Binaries
  Proper Motions
    112.003
Visual Multiple Stars
    118.000
VV Cephei Stars
  Magellanic Clouds
    122.130

W UMa Stars
    117.012 .041 .056
    121.021 .087 .104
      .107 .120 .122
W Virginis Stars
    122.047 .079 .115
Water
  Interstellar Matter
    131.003 .046 .092
      .121
    141.628
Water
  Mars Atmosphere
    097.033 .040
White Dwarfs
    126.000
White Dwarfs
  Accretion
    124.009
    126.016 .023
    142.106
White Dwarfs
  Atmospheres
    126.008 .014 .015
      .016
White Dwarfs
  Binaries
    114.316
    117.017
White Dwarfs
  Black Holes
    066.134
White Dwarfs
  Chemical Composition
    126.028
White Dwarfs
  Convection
    126.004 .008 .014
      .016
White Dwarfs
  Cooling
    126.004 .005 .011
White Dwarfs
  Envelopes
    126.023
White Dwarfs
  Evolution
    125.038
    126.005 .015
White Dwarfs
  Magnetic Fields
    065.189
    126.004 .018 .022
      .029
White Dwarfs
  Masses
    065.005
White Dwarfs
  Models
    126.015 .023
White Dwarfs
  Polarization
    126.018 .022
White Dwarfs
  Pulsations
    126.002 .006
    142.106
White Dwarfs
  Rotation
    065.005
    066.038
    126.006 .026
White Dwarfs
  Spectra
    126.001 .013
White Dwarfs
  Spectrophotometry
    126.007
White Dwarfs
  Temperatures
    126.028
White Dwarfs
  Variations
    126.006
White Dwarfs
  X-Ray Sources
    142.113
White Holes
    061.002
Wolf-Rayet Stars
  Atmospheres
    064.004 .065
Wolf-Rayet Stars
  Binaries
    114.369
    117.020
Wolf-Rayet Stars
  Chemical Composition
    114.009
Wolf-Rayet Stars
  Emission Lines
    113.040
Wolf-Rayet Stars
  Envelopes
    114.354
    132.017
Wolf-Rayet Stars
  Evolution
    114.008
Wolf-Rayet Stars
  Galactic Clusters
    153.013
Wolf-Rayet Stars
  Infrared Photometry
    113.045
Wolf-Rayet Stars
  Luminosities
    114.008

Wolf-Rayet Stars
Mass Loss
064.053
Wolf-Rayet Stars
Photometry
113.040
114.314
121.002
Wolf-Rayet Stars
Spectra
114.008 .011 .112
.314
Wolf-Rayet Stars
Spectral Types
153.013
Wolf-Rayet Stars
UV Spectra
114.313
Wolf-Rayet Stars
Variations
114.314

X Rays
Bursts
142.131 .180
X Rays
Clusters of Galaxies
160.048
X Rays
Crab Nebula
134.006 .007
142.063 .136 .137
.164 .165
143.018
X Rays
Cygnus Loop
142.139
X Rays
Energy Spectra
061.001 .033
X Rays
Flare Stars
122.163 .167
X Rays
Interstellar Matter
022.055
X Rays
Jupiter
099.015 .016
X Rays
Neutron Stars
141.346
X Rays
Novae
124.005
X Rays
Origin
061.001 .033
X Rays
Pulsars
125.016
141.304 .313 .328
.342 .346
142.164 .241
X Rays
Radio Sources
141.002
142.064
X Rays
Seyfert Galaxies
142.004
158.143
X Rays
Solar
076.000
X Rays
Solar Corona
074.003 .047 .073
.074 .081 .105
.128
076.007 .009
X Rays
Solar Flares
062.025
073.001 .003 .007
.008 .009 .011
.012 .013 .014
.017 .022 .072
.097 .103 .104
.146
076.014 .016 .017
.018 .024 .026
.027 .046 .047
X Rays
Stellar Atmospheres
064.048
073.002
X Rays
Stellar Coronae
142.045
X Rays
Supernova Remnants
125.005 .007 .008
.034
142.064 .173
X-Ray Astronomy
061.000
X-Ray Background
061.017
122.163 .167
142.012 .059 .134
161.001
162.030
X-Ray Sources
142.000
X-Ray Sources
Accretion
062.015
142.003 .010 .038
.050 .052 .054
.058 .094 .113
.114 .116 .128
.132 .148 .174
X-Ray Sources
Binaries
062.015
065.040 .134
114.315
117.001 .002 .004
.018 .028 .046
.048 .050 .062
.063
121.023 .044
125.006
142.002 .003 .006
.011 .013 .026
.027 .036 .039
.044 .046 .049
.051 .052 .054
X-Ray Sources
Binaries
142.055 .062 .068
.107 .112 .113
.114 .116 .125
.126 .127 .130
.131 .132 .133
.135 .140 .148
.149 .152 .154
.159 .161 .169
.174 .176 .202
.203
X-Ray Sources
Black Holes
066.073 .096 .137
142.038 .055 .061
.116 .174
X-Ray Sources
Bursts
142.023 .024 .025
X-Ray Sources
Clusters of Galaxies
141.025
142.008 .099 .166
160.004 .006 .045
X-Ray Sources
Compact
142.058 .124 .129
.148
X-Ray Sources
Counts
142.012
X-Ray Sources
Distances
131.100
142.048 .162
X-Ray Sources
Distribution
142.163 .164
X-Ray Sources
Energy Spectra
142.165
X-Ray Sources
Evolution
142.127
X-Ray Sources
Extragalactic
142.035
X-Ray Sources
Galactic
142.042 .093 .117
.124 .163
X-Ray Sources
Gamma Rays
142.095 .151 .177
.245 .247
X-Ray Sources
Globular Clusters
066.019
142.009 .010 .036
.040 .044 .047
.061 .100 .105
.117 .140 .163
154.024
X-Ray Sources
Identifications
142.008 .122 .168
X-Ray Sources
Light Curves
121.023
142.013 .039 .062

X-Ray Sources
Light Curves
142.112 .120
X-Ray Sources
Lunar Occultations
142.246
X-Ray Sources
Magellanic Clouds
142.053 .172
X-Ray Sources
Mass Loss
142.002 .003 .006
.055 .107 .132
X-Ray Sources
Masses
142.176
X-Ray Sources
Models
142.006 .055 .114
.128 .133
X-Ray Sources
Neutron Stars
065.040
142.002 .010 .054
.094 .116 .128
.129 .140 .155
X-Ray Sources
Optical Identification
031.236
113.067
141.025
142.005 .028 .042
.104 .158 .172
X-Ray Sources
Periods
142.202
X-Ray Sources
Photometry
121.044
142.041 .042 .043
.066 .110 .135
.172
X-Ray Sources
Polarization
117.046 .050
142.031 .032 .033
.041 .067 .109
X-Ray Sources
Positions
142.121 .147 .158
X-Ray Sources
Pulsations
065.134
142.011 .043 .094
.098 .106 .113
.241
X-Ray Sources
Pulse Structure
142.123
X-Ray Sources
Quasars
142.005 .122
X-Ray Sources
Radio Radiation
142.001 .007 .017
.020 .060 .103
.170
X-Ray Sources
Soft X Rays
142.178
X-Ray Sources
Spectra
125.007 .008
142.001 .004 .016
.065 .068 .097
.102 .131 .133
.158
160.004
X-Ray Sources
Spectroscopic Binaries
142.115
X-Ray Sources
Structure
142.136 .137
X-Ray Sources
Supernova Remnants
125.016 .017 .018
142.059 .139
X-Ray Sources
Surveys
142.059
X-Ray Sources
Transient
065.040
142.015 .022 .025
.051 .092 .093
.101 .102 .120
.147 .153 .160
.161 .166 .201
X-Ray Sources
Variations
142.001 .009 .010
.014 .016 .017
.022 .030 .035
.037 .063 .064
.097 .123 .125
.167
X-Ray Sources
White Dwarfs
142.113

Zodiacal Dust Cloud
106.009 .050
Zodiacal Light
106.000
Zodiacal Light
Brightnesses
106.003 .022
Zodiacal Light
Photometry
106.050
Zodiacal Light
Polarization
106.010 .039 .044

ASTRONOMY AND ASTROPHYSICS ABSTRACTS

A Publication of the
Astronomisches Rechen-Institut Heidelberg
Member of the Abstracting Board
of the International Council of Scientific Unions

Editors:
S. Böhme, U. Esser, W. Fricke, U. Güntzel-Lingner, I. Heinrich,
F. Henn, D. Krahn, L. Schmadel, H. Scholl, G. Zech

Vol. 1	Literature 1969, Part 1, X + 435 pp. (1969)
Vol. 2	Literature 1969, Part 2, X + 516 pp. (1970)
Vol. 3	Literature 1970, Part 1, X + 490 pp. (1970)
Vol. 4	Literature 1970, Part 2, X + 562 pp. (1971)
Vol. 5	Literature 1971, Part 1, X + 505 pp. (1971)
Vol. 6	Literature 1971, Part 2, X + 560 pp. (1972)
Vol. 7	Literature 1972, Part 1, X + 526 pp. (1972)
Vol. 8	Literature 1972, Part 2, X + 594 pp. (1973)
Vol. 9	Literature 1973, Part 1, X + 610 pp. (1973)
Vol. 10	Literature 1973, Part 2, X + 661 pp. (1974)
Vol. 11	Literature 1974, Part 1, X + 579 pp. (1974)
Vol. 12	Literature 1974, Part 2, X + 699 pp. (1975)
Vol. 13	Literature 1975, Part 1, X + 632 pp. (1975)
Vol. 14	Literature 1975, Part 2, X + 747 pp. (1976)